RECIPROCALS OF BASIC FUNCTIONS

18. $\displaystyle\int \frac{1}{1 \pm \sin u}\, du = \tan u \mp \sec u + C$

19. $\displaystyle\int \frac{1}{1 \pm \cos u}\, du = -\cot u \pm \csc u + C$

20. $\displaystyle\int \frac{1}{1 \pm \tan u}\, du = \tfrac{1}{2}(u \pm \ln|\cos u \pm \sin u|) + C$

21. $\displaystyle\int \frac{1}{\sin u \cos u}\, du = \ln|\tan u| + C$

22. $\displaystyle\int \frac{1}{1 \pm \cot u}\, du = \tfrac{1}{2}(u \mp \ln|$

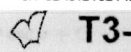 T3-BHD-771

23. $\displaystyle\int \frac{1}{1 \pm \sec u}\, du = u + \cot u$

24. $\displaystyle\int \frac{1}{1 \pm \csc u}\, du = u - \tan u \pm \sec u + C$

25. $\displaystyle\int \frac{1}{1 \pm e^u}\, du = u - \ln(1 \pm e^u) + C$

POWERS OF TRIGONOMETRIC FUNCTIONS

26. $\displaystyle\int \sin^2 u\, du = \tfrac{1}{2}u - \tfrac{1}{4}\sin 2u + C$

27. $\displaystyle\int \cos^2 u\, du = \tfrac{1}{2}u + \tfrac{1}{4}\sin 2u + C$

28. $\displaystyle\int \tan^2 u\, du = \tan u - u + C$

29. $\displaystyle\int \sin^n u\, du = -\frac{1}{n}\sin^{n-1} u \cos u + \frac{n-1}{n}\int \sin^{n-2} u\, du$

30. $\displaystyle\int \cos^n u\, du = \frac{1}{n}\cos^{n-1} u \sin u + \frac{n-1}{n}\int \cos^{n-2} u\, du$

31. $\displaystyle\int \tan^n u\, du = \frac{1}{n-1}\tan^{n-1} u - \int \tan^{n-2} u\, du$

32. $\displaystyle\int \cot^2 u\, du = -\cot u - u + C$

33. $\displaystyle\int \sec^2 u\, du = \tan u + C$

34. $\displaystyle\int \csc^2 u\, du = -\cot u + C$

35. $\displaystyle\int \cot^n u\, du = -\frac{1}{n-1}\cot^{n-1} u - \int \cot^{n-2} u\, du$

36. $\displaystyle\int \sec^n u\, du = \frac{1}{n-1}\sec^{n-2} u \tan u + \frac{n-2}{n-1}\int \sec^{n-2} u\, du$

37. $\displaystyle\int \csc^n u\, du = -\frac{1}{n-1}\csc^{n-2} u \cot u + \frac{n-2}{n-1}\int \csc^{n-2} u\, du$

PRODUCTS OF TRIGONOMETRIC FUNCTIONS

38. $\displaystyle\int \sin mu \sin nu\, du = -\frac{\sin(m+n)u}{2(m+n)} + \frac{\sin(m-n)u}{2(m-n)} + C$

39. $\displaystyle\int \cos mu \cos nu\, du = \frac{\sin(m+n)u}{2(m+n)} + \frac{\sin(m-n)u}{2(m-n)} + C$

40. $\displaystyle\int \sin mu \cos nu\, du = -\frac{\cos(m+n)u}{2(m+n)} - \frac{\cos(m-n)u}{2(m-n)} + C$

41. $\displaystyle\int \sin^m u \cos^n u\, du = -\frac{\sin^{m-1} u \cos^{n+1} u}{m+n} + \frac{m-1}{m+n}\int \sin^{m-2} u \cos^n u\, du$

$\displaystyle\qquad = \frac{\sin^{m+1} u \cos^{n-1} u}{m+n} + \frac{n-1}{m+n}\int \sin^m u \cos^{n-2} u\, du$

PRODUCTS OF TRIGONOMETRIC AND EXPONENTIAL FUNCTIONS

42. $\displaystyle\int e^{au} \sin bu\, du = \frac{e^{au}}{a^2 + b^2}(a \sin bu - b \cos bu) + C$

43. $\displaystyle\int e^{au} \cos bu\, du = \frac{e^{au}}{a^2 + b^2}(a \cos bu + b \sin bu) + C$

POWERS OF u MULTIPLYING OR DIVIDING BASIC FUNCTIONS

44. $\displaystyle\int u \sin u\, du = \sin u - u \cos u + C$

45. $\displaystyle\int u \cos u\, du = \cos u + u \sin u + C$

46. $\displaystyle\int u^2 \sin u\, du = 2u \sin u + (2 - u^2) \cos u + C$

47. $\displaystyle\int u^2 \cos u\, du = 2u \cos u + (u^2 - 2) \sin u + C$

48. $\displaystyle\int u^n \sin u\, du = -u^n \cos u + n\int u^{n-1} \cos u\, du$

49. $\displaystyle\int u^n \cos u\, du = u^n \sin u - n\int u^{n-1} \sin u\, du$

50. $\displaystyle\int u^n \ln u\, du = \frac{u^{n+1}}{(n+1)^2}[(n+1)\ln u - 1] + C$

51. $\displaystyle\int u e^u\, du = e^u(u - 1) + C$

52. $\displaystyle\int u^n e^u\, du = u^n e^u - n\int u^{n-1} e^u\, du$

53. $\displaystyle\int u^n a^u\, du = \frac{u^n a^u}{\ln a} - \frac{n}{\ln a}\int u^{n-1} a^u\, du + C$

54. $\displaystyle\int \frac{e^u\, du}{u^n} = -\frac{e^u}{(n-1)u^{n-1}} + \frac{1}{n-1}\int \frac{e^u\, du}{u^{n-1}}$

55. $\displaystyle\int \frac{a^u\, du}{u^n} = -\frac{a^u}{(n-1)u^{n-1}} + \frac{\ln a}{n-1}\int \frac{a^u\, du}{u^{n-1}}$

56. $\displaystyle\int \frac{du}{u \ln u} = \ln|\ln u| + C$

POLYNOMIALS MULTIPLYING BASIC FUNCTIONS

57. $\displaystyle\int p(u) e^{au}\, du = \frac{1}{a}p(u)e^{au} - \frac{1}{a^2}p'(u)e^{au} + \frac{1}{a^3}p''(u)e^{au} - \cdots$ [signs alternate: $+ - + - \cdots$]

58. $\displaystyle\int p(u) \sin au\, du = -\frac{1}{a}p(u)\cos au + \frac{1}{a^2}p'(u)\sin au + \frac{1}{a^3}p''(u)\cos au - \cdots$ [signs alternate in pairs after first term: $+ + - - + + - - \cdots$]

59. $\displaystyle\int p(u) \cos au\, du = \frac{1}{a}p(u)\sin au + \frac{1}{a^2}p'(u)\cos au - \frac{1}{a^3}p''(u)\sin au - \cdots$ [signs alternate in pairs: $+ + - - + + - - \cdots$]

Calculus

A NEW HORIZON

Calculus

A NEW HORIZON

SIXTH EDITION

HOWARD ANTON
Drexel University

JOHN WILEY & SONS, INC.
New York Chichester Brisbane Toronto Singapore

Mathematics Editor: Barbara Holland
Associate Editor: Sharon Smith
Freelance Production Manager: Jeanine Furino
Production and Text Design: HRS Electronic Text Management
Copy Editor: Lilian Brady
Photo Editor: Hilary Newman
Electronic Illustration: Techsetters, Inc.
Typesetting: Techsetters, Inc.
Cover Design: Madelyn Lesure
Cover Photo: © Dann Coffey/The Image Bank

This book was set in Times Roman by Techsetters, Inc., and printed and bound by Von Hoffmann Press, Inc. The cover was printed by The Phoenix Color Corp.

Recognizing the importance of preserving what has been written, it is a policy of John Wiley & Sons, Inc. to have books of enduring value published in the United States printed on acid-free paper, and we exert our best efforts to that end.

The paper on this book was manufactured by a mill whose forest management programs include sustained yield harvesting of its timberlands. Sustained yield harvesting principles ensure that the numbers of trees cut each year does not exceed the amount of new growth.

Copyright © 1999, John Wiley & Sons, Inc. All rights reserved.

No part of this publication may be reproduced, stored in a retrieval system, or transmitted in any form or by any means, electronic, mechanical, photocopying, recording, scanning, or otherwise, except as permitted under Sections 107 and 108 of the 1976 United States Copyright Act, without either the prior written permission of the Publisher, or authorization through payment of the appropriate per-copy fee to the Copyright Clearance Center, 222 Rosewood Drive, Danvers, MA 01923, (508) 750-8400, fax (508) 750-4470. Requests to the Publisher for permission should be addressed to the Permissions Department, John Wiley & Sons, Inc., 605 Third Avenue, New York, NY 10158-0012, (212) 850-6011, fax (212) 850-6008, E-mail: PERMREQ@WILEY.COM.

Derive is a registered trademark of Soft Warehouse, Inc.
Maple is a registered trademark of Waterloo Maple Software, Inc.
Mathematica is a registered trademark of Wolfram Research, Inc.

ISBN   0-471-15306-0

Printed in the United States of America

10 9 8 7 6 5 4 3 2

ABOUT
HOWARD ANTON

Howard Anton obtained his B.A. from Lehigh University, his M.A. from the University of Illinois, and his Ph.D. from the Polytechnic University of Brooklyn, all in mathematics. In the early 1960s he worked for Burroughs Corporation and Avco Corporation at Cape Canaveral, Florida, where he was involved with missile tracking problems for the manned space program. In 1968 he joined the Mathematics Department at Drexel University, where he taught full time until 1983. Since that time he has been an adjunct professor at Drexel and has devoted the majority of his time to textbook writing and activities for mathematical associations. Dr. Anton was President of the EPADEL Section of the Mathematical Association of America (MAA), served on the board of Governors of that organization, and guided the creation of the Student Chapters of the MAA. He has published numerous research papers in Functional Analysis, Approximation Theory, and Topology, as well as pedagogical papers on applications of mathematics. He is best known for his textbooks in mathematics, which are among the most widely used in the world. There are currently more than ninety versions of his books, including translations into Spanish, Arabic, Portuguese, Italian, Indonesian, French, Japanese, Chinese, Hebrew, and German. Dr. Anton has an avid interest in computer technology as it relates to mathematical education and publishing. He has developed pedagogical software for teaching calculus and linear algebra as well as various software programs for the publishing industry that automate the production of four-color mathematical text and art. For relaxation he enjoys traveling and photography.

To
My Wife Pat
My Children Brian, David, and Lauren

In Memory of
My Mother Shirley
Stephen Girard (1750–1831)—Benefactor
Albert Herr—Esteemed Colleague and Contributor

A NOTE FROM THE AUTHOR

When I began writing the first edition of this calculus text almost 25 years ago, the task, though daunting, was straightforward in that the content and organization of a standard calculus course was nearly universal—the challenge for me at that time was to present the material in a livelier style and with greater clarity than my predecessors. Since this calculus text is still among the most widely used in the world, I take comfort that the goals I set for myself as a young writer and mathematician have been achieved.

However, times are changing, and the era of a standard and universal calculus course seems destined for the repository of slide rules and three-cent stamps. We are witnessing a lot of experimentation with the content, organization, and goals of calculus—some of which has been successful and some of which has not. Thus, my challenge in writing the sixth edition has been to create a text that has all of the strengths of the earlier editions, yet incorporates those new ideas that are clearly important and have withstood the objective scrutiny of skilled and thoughtful teachers.

In preparing for this edition, I sought advice from outstanding teachers at a wide variety of institutions. Needless to say, I received a diversity of opinions—some reviewers advised against any major changes, arguing that the book was already clearly written and working well in the classroom, while others felt that major changes were required to incorporate technology and make the book more contemporary. I listened carefully, and the lively discussions that followed helped me formulate my philosophy for the new edition. Many of the specific changes are itemized in the preface, but here are some of the general goals:

- Add graphing calculator and CAS materials to the text in a way that will allow students who have those tools to use them but that will not prevent the text from being used by those students who do not have access to that technology.
- Place more emphasis on mathematical modeling and applications.
- Incorporate new examples and exercises that will be meaningful to today's students and will more accurately convey the role of calculus in the real world.
- Widen the variety of exercises to focus more on *conceptual understanding* through conjecture, multi-step analysis, expository writing, and what-if analysis.

In addition, I wanted to provide some optional innovative materials that would capture the student's interest and provide the kind of problem-solving experience that he or she might find in a research or industrial setting. This gave birth to an exciting set of modules that we have called *Expanding the Calculus Horizon*. These modules appear at the ends of selected chapters and each has an optional Internet component that we hope will grow dynamically over time with input from teachers and students.

In developing this edition I have stood firm on two principles that were adhered to in earlier editions:

- The text material is presented at a mathematical level that is suitable for students who will embark on careers in engineering and science.
- It remains a primary goal of the text to teach the student clear, logical, mathematical thinking. Informal discussions play an important motivational role in the exposition and are used extensively, but eventually I want the student to be able to read and understand the language of mathematics.

Although this edition has many changes and new features, they have been implemented in a *flexible* way that will accommodate a wide variety of teaching philosophies. Thus, I am confident that professors who have had positive experiences with earlier editions will be comfortable with this revision, and I am hopeful that those professors who are looking for a contemporary text with an established history of success in the classroom will be pleased with the innovations in this new edition.

Sincerely,

Howard Anton

At times the words of a complete stranger are difficult to accept. That is why I am about to take this first opportunity to introduce myself. Hopefully by revealing a bit about myself and how I relate to this textbook may help you find these words more compelling.

Hello, my name is Ajay Arora and I am an Electrical Engineering student at McMaster University in Hamilton, Canada. I too was in your place when I began my entry into the much dreaded field of *CALCULUS*. The vast amounts of rate of change and antiderivative problems were overwhelming. With a little struggle and hard work, I successfully completed that course only to be faced with three more advanced level calculus courses. What I am about to write is the unbiased truth on how you can be successful in calculus and how this textbook will assist you on your journey.

I have been a member of the Student Advisory Board for this textbook for over a year now. The committee came together as a venture from the authors and publisher to get more student input in the development stages instead of simply focusing on feedback when the book was published. After a chapter was completed by the author, each student committee member evaluated, commented, and in some cases, recommended alternative approaches. These tasks involved lots of special deliveries, E-mails, faxes, telephone calls, conference calls, and of course, a whole lot of calculus! But in the end it was a total rewarding experience.

How many times have you asked yourself, "Is math really useful?" Or how about, "Will I ever use calculus in the real world?" I know I have! This textbook will definitely help you answer some of these questions with true applications of the theories you learn. The modules entitled Expanding the Calculus Horizon have been included for precisely that purpose. Every module has been critiqued extensively by the Student Advisory Board, and I encourage you to try them. Not only will these applications of calculus surprise you, but they may actually help give you direction in a field that you might want to pursue after college.

I wish you success in this course, as well as the many others you will face during your college career. Good Luck!

Sincerely,

Ajay Arora
McMaster University

Best Wishes for Success from the Student Advisory Board

Dan Arndt, *University of Texas at Dallas*
Ajay Arora, *McMaster University*
Scott E. Barnett, *Wayne State University*

Fatenah Issa, *Loyola University of Chicago*
Laurie Haskell Messina, *University of Oklahoma*
Steven E. Pav, *Alfred University*

ABOUT THIS EDITION

This is a major revision. In keeping with current trends in calculus, the goal for this edition is to focus more on *conceptual understanding* and *applicability* of the subject matter. In designing this edition, we worked closely with a talented team of reviewers to ensure that the book is sufficiently *flexible* that it will continue to meet the needs of those using the last edition and at the same time provide a fresh approach for those instructors who are taking their calculus course in a new direction. Some of the more significant changes are as follows:

Technology   This edition provides extensive materials for instructors who want to use graphing calculators or computer algebra systems. However, these materials are implemented in a way that allows the text to be used in courses where technology is used heavily, moderately, or not at all. To provide a sound foundation for the technology material, I have added a new section entitled Graphing Functions on Calculators and Computers; Computer Algebra Systems (Section 1.3).

Horizon Modules   Selected chapters end with modules called Expanding the Calculus Horizon. As the name implies, these modules are intended to take the student a step beyond the traditional calculus text. The modules, all of which are optional, can be assigned either as individual or group projects and can be used by instructors to tailor the calculus course to meet their specific needs and teaching philosophies. For example, there are modules that touch on iteration and dynamical systems, modeling from experimental data by curve fitting, applications, expository report writing, and so forth.

Mathematical Modeling   Mathematical modeling plays a more prominent role in this edition. The concept of a mathematical model is introduced in Section 1.5 and is used extensively thereafter. The Horizon module for Chapter 5 discusses how to obtain mathematical models from experimental data. In Section 10.3 we discuss mathematical modeling with differential equations, and in Section 11.10 we discuss mathematical modeling with Taylor series. The Horizon module for Chapter 17 develops a mathematical model of a hurricane.

Applicability of Calculus   One of the goals in this edition is to link calculus more closely to the real world and to the student's own experience. This theme starts with the Introduction and is carried through in the modules, examples, and exercises. Applications appearing in exercises and examples are carefully chosen to be sufficiently simple that they do not divert time from learning important mathematical fundamentals. More extensive applications appear in various Horizon modules.

Earlier Differential Equations   Basic ideas about differential equations, initial-value problems, direction fields, and integral curves are introduced concurrently with integration and then revisited in more detail in Chapter 10.

Quicker Entry to Functions   Chapter 1 begins immediately with functions, and the precalculus material that formed the first chapter in earlier editions has been moved to the appendix.

For the Reader   This element is new. At various points in the exposition the student is assigned a brief task. Some tasks are appropriate for all readers, while others are appropriate only for readers who have a graphing calculator or a CAS. The tasks for all readers are designed to immerse the student more deeply into the text by asking them to think about an idea and reach some conclusion; the tasks for students using technology are designed to familiarize them with the procedures for using that technology by asking them to read their documentation and perform some text-related computation. Some instructors may want to make these tasks part of their assignments.

Earlier Logarithms and Exponentials   Logarithmic and exponential functions are introduced in Chapter 4 from the exponent point of view and then revisited in Section 7.9 from the integral point of view. This provides a richer variety of functions to work with earlier in the text, fits in better with the discussions of modeling, and makes for a less fragmented presentation of the analysis of graphs. However, for instructors who prefer a later presentation of logarithmic and exponential functions, there is a guide for doing this on pages xvi and xvii below.

Early Parametric Option   There is a new option for introducing parametric curves in Section 1.7 of Chapter 1 and revisiting the material in Chapter 12, where calculus-related issues are discussed. Instructors who prefer the traditional late discussion of parametric equations will have no problem teaching Section 1.7 as part of Chapter 12 or 13.

More Variety in Exercises   The exercise sets have been revised extensively to create a richer variety—there are many more exercises that include conjecture, exploration, multistep analysis, and expository writing. The goal has been to put more focus on *conceptual understanding*. There are also many new exercises that are intended to be solved using a graphing calculator or a CAS. These are marked with icons for easy identification.

Analysis of Functions   The old "curve-sketching" material has been replaced by Sections 5.1–5.3 on the Analysis of Functions. The name change reflects a more contemporary approach to the material—there is more emphasis on the interplay between technology and calculus and more focus on the problem of finding a *complete graph*, that is, a graph that contains all of the significant features of concern.

Principles of Integral Evaluation   The old "Techniques of Integration" has been renamed Principles of Integral Evaluation to reflect its more contemporary approach to the material. The chapter has been condensed and there is now more emphasis on general methods and less on tricks for evaluating complicated or obscure integrals. The section entitled Using Integral Tables and Computer Algebra Systems has been expanded and rewritten extensively.

Supplementary Exercises   Supplementary exercises have been added at the ends of chapters.

New Appendix on Solving Polynomial Equations   Appendix F, entitled Solving Polynomial Equations, is new. It reviews the Factor Theorem, the Remainder Theorem, and procedures for finding rational roots. Many students are weak on this material, yet it plays an important role in determining whether a polynomial graph generated on a calculator or computer is complete.

Rule of Four   The "rule of four" refers to the presentation of material from the verbal, algebraic, visual, and numerical points of view. It is used more extensively in this edition, where appropriate.

Internet An internet site http://www.wiley.com/college/anton has been established to complement the text. This site contains additional Horizon modules and technology materials. The site is experimental, but we expect it to grow dynamically over time.

OTHER FEATURES

Flexibility This edition has a built-in flexibility that is designed to serve a broad spectrum of calculus philosophies, ranging from traditional to reform. Graphing technology can be used heavily, moderately, or not at all; and the order of presentation of many sections can be permuted to accommodate specific course needs.

Trigonometry Review Deficiencies in trigonometry plague many students, so I have included a substantial trigonometry review in Appendix E.

Historical Notes The biographies and historical notes have been a hallmark of this text from its first edition and have been maintained in this edition. All of the biographical materials have been distilled from standard sources with the goal of capturing the personalities of the great mathematicians and bringing them to life for the student.

Graded Exercise Sets Section Exercise Sets are graded to begin with routine problems and progress gradually toward problems of greater difficulty. However, in the Supplementary Exercises I have opted not to grade the exercises by level of difficulty to avoid giving the student a predisposition about the level of effort required.

Rigor The challenge of writing a good calculus book is to strike the right balance between rigor and clarity. My goal is to present precise mathematics to the fullest extent possible for the freshman audience, but where clarity and rigor conflict I choose clarity. However, I believe it to be essential that the student understand the difference between a careful proof and an informal argument, so I try to make it clear to the reader when arguments are informal. Theory involving δ-ϵ arguments appear in separate sections, so they can be bypassed if desired.

Mathematical Level This book is written at a mathematical level that is suitable for students planning on careers in engineering or science.

Computer Graphics This edition makes extensive use of modern computer graphics to clarify concepts and to help develop the student's ability to visualize mathematical objects, particularly in 3-space. For those students who are working with graphing technology, there are exercises that are designed to develop the student's ability to generate mathematical graphics.

Student Review A Student Advisory Board was actively involved in the development process of this edition to provide information on pedagogical clarity and to advise on the development of examples, exercises, and modules that students would find interesting and relevant.

SOME ORGANIZATION CHANGES FROM FIFTH EDITION

▶ Much of the precalculus material has been moved to appendices to allow for an earlier presentation of functions. However, where appropriate, we have included quick summaries of review material in the body of the text.

▶ The material on logarithmic and exponential functions has been reorganized, so it can be covered in the first semester (an early transcendental presentation). There is a guide on the next page for implementing a late transcendental presentation.

▶ The first 13 chapters of the fifth edition are covered in the first 12 chapters of the sixth edition.

▶ The first 7 chapters of the fifth edition correspond to the first 9 chapters of the sixth edition. However, *the number of sections is about the same*, so there is *no increase in the number of lectures required to cover the material*. The new subdivision is more natural in that the chapter titles now reflect the chapter content more accurately.

▶ In the sixth edition, as in the fifth edition, instructors teaching on the semester system should have no trouble covering material on integration in the first semester.

▶ Chapter 11 on Infinite Series has been condensed from 12 sections to 10, and the material has been reorganized so that Taylor polynomials and Taylor series appear earlier. This makes it possible to cover these topics without covering the entire chapter.

▶ The material on analytic geometry and polar coordinates, which occupied Chapters 12 and 13 in the fifth edition, is covered in Chapter 12 of the sixth edition.

▶ L'Hôpital's rule was moved to an earlier position, so it can be used to analyze the end-behavior of logarithmic and exponential functions.

▶ The two parts to the Fundamental Theorem of Calculus, which appeared in separate sections of the fifth edition, now appear together in the same section (Section 7.6).

LATE TRANSCENDENTAL OPTION

In keeping with current trends, Chapters 1 to 8 of this text are organized so that the basic material on logarithmic and exponential functions is covered in the first semester (commonly called an "early transcendental" presentation). This is achieved by introducing logarithms informally from the exponent point of view (Section 4.2) and deferring the integral representation of the natural logarithm (Section 7.9). However, we have included the following guide for instructors who prefer to cover logarithmic and exponential functions in the second semester (as in the fifth edition). Depending on your preference, you can place the deferred material after Chapter 7 or after Chapter 8. The guide shows how to place it after Chapter 8. To place it after Chapter 7, ignore the exercise modifications listed for Chapter 8.

	Section	Text Modifications (bulleted)	Exercise Modifications
1	1.1	Functions and Analysis of Graphical Information	
2	1.2	Properties of Functions	
3	1.3	Graphing Functions on Calculators and Computers	
4	1.4	New Functions from Old	
5	1.5	Mathematical Models; Linear Models	
6	1.6	Families of Functions	
7	1.7	Parametric Equations	
8	2.1	Limits (Intuitive)	
9	2.2	Limits (Computational)	
10	2.3	Limits (Rigorous)	
11	2.4	Continuity	
12	2.5	Limits / Continuity of Trigonometric Functions	
13	3.1	Tangent Lines and Rates of Change	
14	3.2	The Derivative	
15	3.3	Techniques of Differentiation	
16	3.4	Derivatives of Trigonometric Functions	
17	3.5	The Chain Rule	
18	3.6	Local Linear Approximation; Differentials	
19	4.3	Implicit Differentiation • Defer the concluding subsection on derivatives of inverse functions (pp. 252–253).	Defer Exercises 10, 53–56.
20	4.6	Related Rates • Defer the alternative solution to Example 3 at the bottom of p. 272.	Defer Exercise 37. Defer Supplementary Exercises 1–6, 8–14, 16–24.
21	5.1	Analysis I: Increase, Decrease, Concavity • Defer Examples 6(a) and 6(c) on p. 295.	Defer Exercises 21–24, 38, 41, 53.
22	5.2	Analysis II: Relative Extrema	Defer Exercises 15, 31, 32, 39–42, 50, 51.
23	5.3	Analysis III: Applying Technology • Defer Example 8 and the discussion of logistic curves that follows it (pp. 316–319). • Defer the Horizon Module for Chapter 5.	Defer Exercises 39–48, 53–55, 69, 70. Defer Supplementary Exercises 17–24, 33, 37–39.
24	6.1	Absolute Maxima and Minima	Defer Exercises 31, 32, 44.
25	6.2	Applied Maximum and Minimum Problems	Defer Exercise 15.
26	6.3	Rectilinear Motion	Defer Exercise 16.
27	6.4	Newton's Method	Defer Exercises 14, 16
28	6.5	Rolle's Theorem; Mean-Value Theorem	Defer Exercise 36 Defer Supplementary Exercises 7(d), 8(d), 22.

29	7.1	An Overview of the Area Problem	Defer Exercise 9.
30	7.2	Indefinite Integral; Integral Curves; Direction Fields • Defer integration formulas (6), (10), (11) in Table 7.2.1 on p. 384. • Defer the last (fifth) integral in Example 2 on p. 385.	Defer Exercises l(b), 19, 20, 25, 34, 39(b).
31	7.3	Integration by Substitution • Defer Example 5, p. 393. • Defer Example 7, p. 394.	Defer Exercises 2(e), 3(c), (d), (e), 5, 6, 19–22, 27, 28, 35, 36, 45–48, 54.
32	7.4	Sigma Notation	Defer Exercises 2(c), (e), 39(b).
33	7.5	The Definite Integral	Defer Exercises 6, 13, 14, 33(b), 38(a), 45(a).
34	7.6	The Fundamental Theorem of Calculus • Defer the first three of the four integrals in Example 5, p. 419.	Defer Exercises 7, 8, 19, 20, 24, 28(b), 45(b), 46(b), 55(b), 59.
35	7.7	Rectilinear Motion Revisited	Defer Exercises 13(b), 14(a), 23, 24, 27, 28, 53, 54.
36	7.8	Evaluating Definite Integrals by Substitution • Defer Example 2(a), p. 422. • Defer Example 3, p. 443.	Defer Exercises 2(a, b), 11, 12, 16, 21, 37, 38, 45–48. Defer Supplementary Exercises 12, 13, 14(c), 37, 39–41, 49.
37	8.1	Area Between Two Curves	Defer Exercise 13.
38	8.2	Volumes by Slicing; Disks and Washers	Defer Exercises 11, 12, 38.
39	8.3	Volumes by Cylindrical Shells	Defer Exercise 11.
40	8.4	Length of a Plane Curve	Defer Exercises 7, 13, 14, 15, 16.
41	8.5	Surface Area	Defer Exercises 13, 15, 22.
42	8.6	Work	
43	8.7	Fluid Pressure and Force	
44	4.1	Inverse Functions • Pick up material deferred from Section 4.3.	**Pick up Exercises:** Section 4.3 (53–56) Chapter 4 Supplementary (1, 6).
45	4.2	Logarithmic and Exponential Functions	**Pick up Exercises:** Section 4.3 (10) Chapter 4 Supplementary (5, 9, 17, 20–22).
46	4.4	Derivatives of Log and Exponential Functions • Pick up material deferred from Section 7.2. • Pick up material deferred from Section 5.3.	**Pick up Exercises:** Section 4.6 (37) Chapter 4 Supplementary (10, 12, 14, 16, 19, 23, 24) Section 5.1 (21–24, 38, 41, 53) Section 5.2 (15, 31, 32, 39–42, 50, 51) Section 5.3 (44, 53–55) Chapter 5 Supplementary (20, 33, 39) Section 6.1 (31, 32, 44) Section 6.2 (15) Section 6.3 (16) Section 6.4 (14, 16) Section 6.5 (36) Chapter 6 Supplementary (7(d), 8(d), 22). Chapter 7 (all deferred) Section 5.3 (69, 70) Chapter 8 (all deferred).
47	4.5	Derivatives of Inverse Trig Functions • Pick up material deferred from Section 4.6.	**Pick up Exercises:** Chapter 4 Supplementary (2, 8, 13, 18).
48	7.9	Logarithmic Functions; Integral Point of View	
49	8.8	Hyperbolic Functions and Hanging Cables	
50	4.7	L'Hôpital's Rules	**Pick up Exercises:** Chapter 4 Supplementary (3, 4, 11) Section 5.3 (39–43, 45–48) Chapter 5 Supplementary (17–19, 21–24, 37, 38).
		• Pick up Horizon module from Chapter 5.	

RESOURCES FOR THE STUDENT

Student Resource and Survival CD　　　　　　　　　　0-471-24608-5

This CD for IBM compatibles or Macintosh platforms provides students with an electronic form of detailed solutions to odd-numbered exercises, multiple choice and true–false sample tests for each section and chapter of the text, precalculus review material, and a brief introduction to those aspects of linear algebra that are of immediate concern to the calculus student. Two demonstration modules from the Windows-based multimedia calculus program *Calculus Connections, A Multimedia Adventure* are also available on this CD.

Student Resource Manual　　　　　　　　　　　　　0-471-24616-6

This manual provides students with detailed solutions to odd-numbered exercises and multiple choice and true–false sample tests for each section and chapter of the text.

RESOURCES FOR THE INSTRUCTOR

Hard copy and electronic resources are available for the instructor. These can be obtained by sending a request on your institutional letterhead to Mathematics Marketing Manager, John Wiley & Sons, Inc., 605 Third Avenue, New York, NY 10158-0012, or by requesting them from your local Wiley representative.

OTHER RESOURCES

We are proud to offer special pricing of the student educational versions of *MAPLE*™ or *MATHEMATICA*™ packaged with the sixth edition of Howard Anton's *Calculus* textbook. For pricing information, you can contact your local Wiley representative, email us at math@wiley.com, or call us at (800) 225-5945.

ACKNOWLEDGMENTS

It has been my good fortune to have the advice and guidance of many talented people whose knowledge and skills have enhanced this book in many ways. For their valuable help I thank:

Reviewers and Contributors to Earlier Editions

Edith Ainsworth, *University of Alabama*

Loren Argabright, *Drexel University*

David Armacost, *Amherst College*

John Bailey, *Clark State Community College*

Robert C. Banash, *St. Ambrose University*

George R. Barnes, *University of Louisville*

Larry Bates, *University of Calgary*

John P. Beckwith, *Michigan Technological University*

Joan E. Bell, *Northeastern Oklahoma State University*

Irl C. Bivens, *Davidson College*

Harry N. Bixler, *Bernard M. Baruch College, CUNY*

Marilyn Blockus, *San Jose State University*

Ray Boersma, *Front Range Community College*

Barbara Bohannon, *Hofstra University*

David Bolen, *Virginia Military Institute*

Daniel Bonar, *Denison University*

George W. Booth, *Brooklyn College*

Phyllis Boutilier, *Michigan Technological University*

Mark Bridger, *Northeastern University*

John Brothers, *Indiana University*

Stephen L. Brown, *Olivet Nazarene University*

Virginia Buchanan, *Hiram College*

Robert C. Bueker, *Western Kentucky University*

Robert Bumcrot, *Hofstra University*

Christopher Butler, *Case Western Reserve University*

Carlos E. Caballero, *Winthrop University*

James Caristi, *Valparaiso University*

Stan R. Chadick, *Northwestern State University*

Hongwei Chen, *Christopher Newport University*

Chris Christensen, *Northern Kentucky University*

Robert D. Cismowski, *San Bernardino Valley College*

Patricia Clark, *Rochester Institute of Technology*

Hannah Clavner, *Drexel University*

David Clydesdale, *Sauk Valley Community College*

David Cohen, *University of California, Los Angeles*

Michael Cohen, *Hofstra University*

Pasquale Condo, *University of Lowell*

Robert Conley, *Precision Visuals*

Cecil J. Coone, *State Technical Institute at Memphis*

Norman Cornish, *University of Detroit*

Terrance Cremeans, *Oakland Community College*

Lawrence Cusick, *California State University–Fresno*

Michael Dagg, *Numerical Solutions, Inc.*

Stephen L. Davis, *Davidson College*

A. L. Deal, *Virginia Military Institute*

Charles Denlinger, *Millersville University*

William H. Dent, *Maryville College*

Blaise DeSesa, *Allentown College of St. Francis de Sales*

Dennis DeTurck, *University of Pennsylvania*

Jacqueline Dewar, *Loyola Marymount University*

Preston Dinkins, *Southern University*

Irving Drooyan, *Los Angeles Pierce College*

Tom Drouet, *East Los Angeles College*

Clyde Dubbs, *New Mexico Institute of Mining and Technology*

Della Duncan, *California State University–Fresno*

Ken Dunn, *Dalhousie University*

Sheldon Dyck, *Waterloo Maple Software*

Hugh B. Easler, *College of William and Mary*

Scott Eckert, *Cuyamaca College*

Joseph M. Egar, *Cleveland State University*

Judith Elkins, *Sweet Briar College*

Brett Elliott, *Southeastern Oklahoma State University*

Garret J. Etgen, *University of Houston*

Benny Evans, *Oklahoma State University*

James H. Fife, *Educational Testing Service*

Dorothy M. Fitzgerald, *Golden West College*

Barbara Flajnik, *Virginia Military Institute*

Daniel Flath, *University of South Alabama*

Ernesto Franco, *California State University–Fresno*

Nicholas E. Frangos, *Hofstra University*

Katherine Franklin, *Los Angeles Pierce College*

Marc Frantz, *Indiana University–Purdue University at Indianapolis*

Michael Frantz, *University of La Verne*

Susan L. Friedman, *Bernard M. Baruch College, CUNY*

William R. Fuller, *Purdue University*

Daniel B. Gallup, *Pasadena City College*

Mahmood Ghamsary, *Long Beach City College*

G. S. Gill, *Brigham Young University*

Michael Gilpin, *Michigan Technological University*

Kaplana Godbole, *Michigan Technological Institute*

S. B. Gokhale, *Western Illinois University*

Morton Goldberg, *Broome Community College*

Mardechai Goodman, *Rosary College*

Sid Graham, *Michigan Technological University*

Raymond Greenwell, *Hofstra University*

Gary Grimes, *Mt. Hood Community College*

Jane Grossman, *University of Lowell*

Michael Grossman, *University of Lowell*

Diane Hagglund, *Waterloo Maple Software*

Douglas W. Hall, *Michigan State University*

Nancy A. Harrington, *University of Lowell*

Kent Harris, *Western Illinois University*

Jim Hefferson, *St. Michael College*

Albert Herr, *Drexel University*

Peter Herron, *Suffolk County Community College*

Warland R. Hersey, *North Shore Community College*

Konrad J. Heuvers, *Michigan Technological University*

Robert Higgins, *Quantics Corporation*

Rebecca Hill, *Rochester Institute of Technology*

Edwin Hoefer, *Rochester Institute of Technology*

Louis F. Hoelzle, *Bucks County Community College*

Robert Homolka, *Kansas State University–Salina*

Jerry Johnson, *University of Nevada–Reno*

John M. Johnson, *George Fox College*

Wells R. Johnson, *Bowdoin College*

Herbert Kasube, *Bradley University*

Phil Kavanaugh, *Mesa State College*

Maureen Kelley, *Northern Essex Community College*

Harvey B. Keynes, *University of Minnesota*

Richard Krikorian, *Westchester Community College*

Paul Kumpel, *SUNY, Stony Brook*

Fat C. Lam, *Gallaudet University*

Leo Lampone, *Quantics Corporation*

James F. Lanahan, *University of Detroit–Mercy*

Bruce Landman, *University of North Carolina at Greensboro*

Kuen Hung Lee, *Los Angeles Trade–Technology College*

Marshall J. Leitman, *Case Western Reserve University*

Benjamin Levy, *Lexington H.S., Lexington, Mass.*

Darryl A. Linde, *Northeastern Oklahoma State University*

Phil Locke, *University of Maine, Orono*

Leland E. Long, *Muscatine Community College*

John Lucas, *University of Wisconsin–Oshkosh*

Stanley M. Lukawecki, *Clemson University*

Nicholas Macri, *Temple University*

Melvin J. Maron, *University of Louisville*

Mauricio Marroquin, *Los Angeles Valley College*

Majid Masso, *Brookdale Community College*

Larry Matthews, *Concordia College*

Thomas McElligott, *University of Lowell*

Phillip McGill, *Illinois Central College*

Judith McKinney, *California State Polytechnic University, Pomona*

Joseph Meier, *Millersville University*

Aileen Michaels, *Hofstra University*

Janet S. Milton, *Radford University*

Robert Mitchell, *Rowan College of New Jersey*

Marilyn Molloy, *Our Lady of the Lake University*

Ron Moore, *Ryerson Polytechnical Institute*

Barbara Moses, *Bowling Green State University*

David Nash, *VP Research, Autofacts, Inc.*

Kylene Norman, *Clark State Community College*

Roxie Novak, *Radford University*

Richard Nowakowski, *Dalhousie University*

Stanley Ocken, *City College–CUNY*

Donald Passman, *University of Wisconsin*

David Patterson, *West Texas A & M*

Walter M. Patterson, *Lander University*

Edward Peifer, *Ulster County Community College*

Robert Phillips, *University of South Carolina at Aiken*

Mark A. Pinsky, *Northeastern University*

Catherine H. Pirri, *Northern Essex Community College*

Father Bernard Portz, *Creighton University*

David Randall, *Oakland Community College*

Richard Remzowski, *Broome Community College*

Guanshen Ren, *College of Saint Scholastica*

William H. Richardson, *Wichita State University*

David Rollins, *University of Central Florida*

Naomi Rose, *Mercer County Community College*

Sharon Ross, *DeKalb College*

David Ryeburn, *Simon Fraser University*

David Sandell, *U.S. Coast Guard Academy*
Ned W. Schillow, *Lehigh County Community College*
Dennis Schneider, *Knox College*
Dan Seth, *Morehead State University*
George Shapiro, *Brooklyn College*
Parashu R. Sharma, *Grambling State University*
Donald R. Sherbert, *University of Illinois*
Howard Sherwood, *University of Central Florida*
Bhagat Singh, *University of Wisconsin Centers*
Martha Sklar, *Los Angeles City College*

John L. Smith, *Rancho Santiago Community College*
Wolfe Snow, *Brooklyn College*
Ian Spatz, *Brooklyn College*
Jean Springer, *Mount Royal College*
Norton Starr, *Amherst College*
Richard B. Thompson, *The University of Arizona*
William F. Trench, *Trinity University*
Walter W. Turner, *Western Michigan University*
Richard C. Vile, *Eastern Michigan University*
David Voss, *Western Illinois University*
Shirley Wakin, *University of New Haven*

James Warner, *Precision Visuals*
Peter Waterman, *Northern Illinois University*
Evelyn Weinstock, *Glassboro State College*
Candice A. Weston, *University of Lowell*
Bruce F. White, *Lander University*
Gary L. Wood, *Azusa Pacific University*
Yihren Wu, *Hofstra University*
Richard Yuskaitis, *Precision Visuals*
Michael Zeidler, *Milwaukee Area Technical College*
Michael L. Zwilling, *Mount Union College*

Development Team for the Sixth Edition

The following people critiqued and reviewed various parts of the manuscript and suggested many of the ideas that found their way into this new edition:

Judith Broadwin, *Jericho High School*
Christopher D. Butler, *Case Western Reserve University*
Larry Cusick, *California State University–Fresno*
Philip Farmer, *Diablo Valley College*
Sally E. Fischbeck, *Rochester Institute of Technology*
J. Derrick Head, *University of Minnesota–Morris*
Tommie Ann Hill-Natter, *Prairie View A&M University*
Holly Hirst, *Appalachian State University*
Dan Kemp, *South Dakota State University*
Holly A. Kresch, *Diablo Valley College*
Marshall Leitman, *Case Western Reserve University*
Thomas W. Mason, *Florida A&M University*

Gary L. Peterson, *James Madison University*
Douglas Quinney, *University of Keele*
William H. Richardson, *Wichita State University*
Lila F. Roberts, *Georgia Southern University*
Avinash Sathaye, *University of Kentucky*
Mary Margaret Shoaf-Grubbs, *College of New Rochelle*
Mark Stevenson, *Oakland Community College*
John A. Suvak, *Memorial University of Newfoundland*
Skip Thompson, *Radford University*
Bruce R. Wenner, *University of Missouri–Kansas City*

The following people participated in phone surveys that helped to answer important questions about organization, philosophy, technology, and content:

Linda Bridge, *Long Beach City College*
Ted Clinkenbeard, *Des Moines Area Community College*
Victor Feser, *University of Maryland*
David Gross, *University of Connecticut*
Dennis Hadah, *Saddleback Community College*
Henry Horton, *University of West Florida*
Emmett Johnson, *Grambling State University*

Bill Kavanagh, *Mesa College*
Phil Locke, *University of Maine*
Thomas W. Mason, *Florida A&M University*
Ralph Okojie, *Elizabeth City State University*
David Robbins, *Trinity College*
Skip Thompson, *Radford University*
Paul Vesce, *University of Missouri–Kansas City*
Ronald Wagoner, *California State University–Fresno*

The following people read the sixth edition at various stages for mathematical and pedagogical accuracy and/or assisted with the critically important job of preparing answers to exercises:

Larry Cusick, *California State University–Fresno*
Stephen Davis, *Davidson College*
Blaise DeSesa, *Allentown College of St. Francis de Sales*
William D. Emerson, *Metropolitan State College*
Beverly Fusfield
Susan Gerstein
Konrad Heuvers, *Michigan Technological University*
Dean Hickerson

Cecilia Knoll, *Florida Institute of Technology*
Majid Masso, *University of Delaware*
Kylene Norman, *Clark State Community College*
Irwin Pressman, *Carleton University*
David Ryeburn, *Simon Fraser University*
Thomas Vanden Eynden, *Thomas More College*
Shirley Wakin, *University of New Haven*
Neil Wigley, *University of Windsor*

The following students are members of the Student Advisory Board that critiqued the manuscript for clarity and provided valuable advice on making material interesting and relevant to today's students:

Dan Arndt, *University of Texas at Dallas*
Ajay Arora, *McMaster University*
Scott E. Barnett, *Wayne State University*

Fatenah Issa, *Loyola University of Chicago*
Laurie Haskell Messina, *University of Oklahoma*
Steven E. Pav, *Alfred University*

The following people created materials for tests and other supplements:

William H. Barker, *Bowdoin College*
Henry Smith, *Southeastern Louisiana University*

James E. Ward, *Bowdoin College*
Neil Wigley, *University of Windsor*

I would also like to thank Gary S. Stoudt of the *University of Indiana of Pennsylvania* for his assistance in locating various obscure historical materials.

Content Contributions

The following people assisted in the creation of modules and exercises or contributed valuable ideas that helped in the development of those materials:

Mary Ann Connors, *U.S. Military Academy at West Point*
Art Davis, *San Jose State University*
Gloria S. Dion, *Educational Testing Service*
Iris Brann Feta, *Clemson University*
Dixie Griffin, Jr., *Louisiana Tech University*
Dean Hickerson
Hugh E. Huntley, *University of Michigan*
Lynn Kiaer, *Rose-Hulman Institute of Technology*

Cecilia Knoll, *Florida Institute of Technology*
Michael Magill, *Purdue University*
Robert Meitz, *Arizona State University*
John Rickert, *Rose-Hulman Institute of Technology*
David Ryeburn, *Simon Fraser University*
Michael D. Shaw, *Florida Institute of Technology*
P. Narayana Swamy, *Southern Illinois University*
Josef S. Torok, *Rochester Institute of Technology*

I gratefully acknowledge the permission to adapt various exercises, examples, and text materials from the following publications:

Physics, 3rd ed., Cutnell and Johnson, John Wiley & Sons, Inc., 1995.
Fundamentals of Physics, 4th ed., Halliday, Resnick, and Walker, John Wiley & Sons, Inc., 1993.
Applications of Calculus, Philip Straffin, Ed., MAA Notes Number 29, The Mathematical Association of America, 1993.
Calculus Problems for a New Century, Robert Fraga, Ed., MAA Notes Number 28, Vol. 2, The Mathematical Association of America, 1993.
Engineering Mechanics, Meriam and Kraige, 3rd ed., Vol. 2, John Wiley & Sons, Inc., 1992.

Special Contributions

A special debt of gratitude to:

Barbara Holland, my editor, for sharing a vision and having unwavering faith in my work.

Ann Berlin, Pam Kennedy, and Jeanine Furino of the Wiley Production Department for a scheduling miracle.

Sharon Smith, for again successfully coordinating a complex set of supplements.

Maddy Lesure, for capturing the "horizon" theme so beautifully in the cover.

Hilary Newman, for unearthing my obscure photographic requests.

Lilian Brady, for her unerring eye for aesthetics and typography.

The group at HRS, for a beautiful design and their patience with an (occasionally) cranky author.

The group at Techsetters, for superb composition and illustration and their devotion to excellence.

Mary Ann Connors, Lynn Kiaer, John Rickert, and Josef S. Torok for their creative contributions to the modules.

Neil Wigley, for enjoyable E-mail humor and an outstanding job in writing the solutions manuals.

David Ryeburn, for his attention to detail and valued advice on applications of calculus.

Dean Hickerson, for his keen sense of mathematical precision and sound pedagogy.

My assistant, Dolores Morgan, whose superb organizational skills played a major role in keeping me on schedule.

CONTENTS

INTRODUCTION. **CALCULUS: A NEW HORIZON FROM ANCIENT ROOTS   1**

CHAPTER 1. **FUNCTIONS   15**

1.1   Functions and the Analysis of Graphical Information   16
1.2   Properties of Functions   24
1.3   Graphing Functions on Calculators and Computers;
        Computer Algebra Systems   35
1.4   New Functions from Old   47
1.5   Mathematical Models; Linear Models   61
1.6   Families of Functions   75
1.7   Parametric Equations   93
        Horizon Module: Iteration and Dynamical Systems   106

CHAPTER 2. **LIMITS AND CONTINUITY   111**

2.1   Limits (An Intuitive Introduction)   112
2.2   Limits (Computational Techniques)   127
2.3   Limits (Discussed More Rigorously)   138
2.4   Continuity   148
2.5   Limits and Continuity of Trigonometric Functions   159

CHAPTER 3. **THE DERIVATIVE   169**

3.1   Tangent Lines and Rates of Change   170
3.2   The Derivative   177
3.3   Techniques of Differentiation   189
3.4   Derivatives of Trigonometric Functions   200
3.5   The Chain Rule   204
3.6   Local Linear Approximation; Differentials   210
        Horizon Module: Robotics   221

CHAPTER 4. LOGARITHMIC AND EXPONENTIAL FUNCTIONS   225

4.1   Inverse Functions   226
4.2   Logarithmic and Exponential Functions   235
4.3   Implicit Differentiation   246
4.4   Derivatives of Logarithmic and Exponential Functions   255
4.5   Derivatives of Inverse Trigonometric Functions   261
4.6   Related Rates   270
4.7   L'Hôpital's Rule; Indeterminate Forms   277

CHAPTER 5. ANALYSIS OF FUNCTIONS AND THEIR GRAPHS   289

5.1   Analysis of Functions I: Increase, Decrease, and Concavity   290
5.2   Analysis of Functions II: Relative Extrema;
       First and Second Derivative Tests   299
5.3   Analysis of Functions III:
       Applying Technology and the Tools of Calculus   306
       Horizon Module: Functions from Data   324

CHAPTER 6. APPLICATIONS OF THE DERIVATIVE   329

6.1   Absolute Maxima and Minima   330
6.2   Applied Maximum and Minimum Problems   339
6.3   Rectilinear Motion (Motion Along a Line)   352
6.4   Newton's Method   363
6.5   Rolle's Theorem; Mean-Value Theorem   368

CHAPTER 7. INTEGRATION   377

7.1   An Overview of the Area Problem   378
7.2   The Indefinite Integral; Integral Curves and Direction Fields   382
7.3   Integration by Substitution   391
7.4   Sigma Notation   397
7.5   The Definite Integral   404
7.6   The Fundamental Theorem of Calculus   416
7.7   Rectilinear Motion Revisited; Average Value   427
7.8   Evaluating Definite Integrals by Substitution   441
7.9   Logarithmic Functions from the Integral Point of View   446
       Horizon Module: Blammo the Human Cannonball   457

CHAPTER 8. APPLICATIONS OF THE DEFINITE INTEGRAL IN GEOMETRY, SCIENCE, AND ENGINEERING 461

8.1   Area Between Two Curves   462
8.2   Volumes by Slicing; Disks and Washers   468
8.3   Volumes by Cylindrical Shells   475
8.4   Length of a Plane Curve   480
8.5   Area of a Surface of Revolution   485
8.6   Work   489
8.7   Fluid Pressure and Force   496
8.8   Hyperbolic Functions and Hanging Cables   500

CHAPTER 9. PRINCIPLES OF INTEGRAL EVALUATION 513

9.1   An Overview of Integration Methods   514
9.2   Integration by Parts   516
9.3   Trigonometric Integrals   522
9.4   Trigonometric Substitutions   530
9.5   Integrating Rational Functions by Partial Fractions   536
9.6   Using Tables of Integrals and Computer Algebra Systems   543
9.7   Numerical Integration; Simpson's Rule   553
9.8   Improper Integrals   564
       Horizon Module: Railroad Design   576

CHAPTER 10. MATHEMATICAL MODELING WITH DIFFERENTIAL EQUATIONS 579

10.1   First-Order Differential Equations and Applications   580
10.2   Direction Fields; Euler's Method   592
10.3   Modeling with Differential Equations   598

CHAPTER 11. INFINITE SERIES 615

11.1   Sequences   616
11.2   Monotone Sequences   626
11.3   Infinite Series   632
11.4   Convergence Tests   640
11.5   Taylor and Maclaurin Series   647
11.6   The Comparison, Ratio, and Root Tests   656
11.7   Alternating Series; Conditional Convergence   662
11.8   Power Series   670
11.9   Convergence of Taylor Series; Computational Methods   676
11.10 Differentiating and Integrating Power Series; Modeling with Taylor Series   686

CHAPTER 12. ANALYTIC GEOMETRY IN CALCULUS   699

12.1   Polar Coordinates   700
12.2   Tangent Lines and Arc Length for Parametric and Polar Curves   713
12.3   Area in Polar Coordinates   720
12.4   Conic Sections in Calculus   726
12.5   Conic Sections in Polar Coordinates   743
       Horizon Module: Comet Collision   755

CHAPTER 13. THREE-DIMENSIONAL SPACE; VECTORS   759

13.1   Rectangular Coordinates in 3-Space; Spheres; Cylindrical Surfaces   760
13.2   Vectors   765
13.3   Dot Product; Projections   776
13.4   Cross Product   785
13.5   Parametric Equations of Lines   795
13.6   Planes in 3-Space   801
13.7   Quadric Surfaces   808
13.8   Cylindrical and Spherical Coordinates   819

CHAPTER 14. VECTOR-VALUED FUNCTIONS   829

14.1   Introduction to Vector-Valued Functions   830
14.2   Calculus of Vector-Valued Functions   836
14.3   Change of Parameter; Arc Length   843
14.4   Unit Tangent, Normal, and Binormal Vectors   853
14.5   Curvature   858
14.6   Motion Along a Curve   867
14.7   Kepler's Laws of Planetary Motion   880

CHAPTER 15. PARTIAL DERIVATIVES   889

15.1   Functions of Two or More Variables   890
15.2   Limits and Continuity   901
15.3   Partial Derivatives   910
15.4   Differentiability and Chain Rules   920
15.5   Tangent Planes; Total Differentials for Functions of Two Variables   930
15.6   Directional Derivatives and Gradients for Functions of Two Variables   937
15.7   Differentiability, Directional Derivatives, and Gradients for Functions of Three or More Variables   945
15.8   Maxima and Minima of Functions of Two Variables   956
15.9   Lagrange Multipliers   967

CHAPTER 16. MULTIPLE INTEGRALS   977

16.1   Double Integrals   978
16.2   Double Integrals over Nonrectangular Regions   985
16.3   Double Integrals in Polar Coordinates   993
16.4   Parametric Surfaces; Surface Area   1000
16.5   Triple Integrals   1012
16.6   Centroid, Center of Gravity, Theorem of Pappus   1019
16.7   Triple Integrals in Cylindrical and Spherical Coordinates   1029
16.8   Change of Variables in Multiple Integrals; Jacobians   1039

CHAPTER 17. TOPICS IN VECTOR CALCULUS   1055

17.1   Vector Fields   1056
17.2   Line Integrals   1064
17.3   Independence of Path; Conservative Vector Fields   1078
17.4   Green's Theorem   1088
17.5   Surface Integrals   1094
17.6   Applications of Surface Integrals; Flux   1101
17.7   The Divergence Theorem   1109
17.8   Stokes' Theorem   1117
        Horizon Module: Hurricane Modeling   1126

APPENDIX A. REAL NUMBERS, INTERVALS, AND INEQUALITIES   A1

APPENDIX B. ABSOLUTE VALUE   A11

APPENDIX C. COORDINATE PLANES AND LINES   A16

APPENDIX D. DISTANCE, CIRCLES, AND QUADRATIC EQUATIONS   A29

APPENDIX E. TRIGONOMETRY REVIEW   A39

APPENDIX F. SOLVING POLYNOMIAL EQUATIONS   A52

APPENDIX G. SELECTED PROOFS   A57

ANSWERS   A65

INDEX   I1

PHOTO CREDITS   C1

Calculus is a compilation of ideas that provides a way of viewing and analyzing the physical world. As with all mathematics courses, calculus involves equations and formulas. However, if you successfully learn to use all of the formulas and solve all of the problems in this text but don't master the underlying ideas, you will have missed the most important part of calculus. Keep in mind that every single problem in this text has already been solved by somebody, so your ability to solve those problems gives you nothing unique. However, if you master the ideas of calculus, then you will have the tools to go beyond what other people have done, limited only by your own talents and creativity.

Before starting your studies, you may find it helpful to leaf through this text to get a general feeling for its different parts.

▶ At the beginning of each chapter you will find a page that gives an overview of the chapter, and at the beginning of each section you will find an introduction that gives an overview of that section. To help you locate specific information, sections are divided into topics described by headings in the margin.

▶ Each section ends with a set of exercises. The answers to most odd-numbered exercises appear in the back of the book. Worked-out solutions to the odd-numbered exercises are given in the *Student Resource Manual* and on a CD, which are available as supplements to the text.

▶ Some of the exercises are tagged with icons to indicate that some kind of technology is required for their solution. If your calculus course does not incorporate the use of technology, then your instructor will probably not assign these. Those exercises tagged with the icon ▨ require graphing technology, which might be either a graphing calculator or a computer program that produces graphs from equations. Those exercises tagged with the icon [c] require a computer algebra system (called a CAS), which is a program that can perform symbolic as well as numerical calculations. The most common CAS programs are *Mathematica*, *Maple*, and *Derive*. Some of the newer calculators incorporate CAS capabilities.

▶ Each chapter ends with a set of supplementary exercises, many of which involve a combination of ideas from various sections within the chapter.

▶ Near the end of the text you will find seven appendices. Appendices A–F review some precalculus material, including trigonometry, and Appendix G contains some proofs that may or may not be part of your course.

▶ There is also reference material on the endpapers that are inside the front and back covers of the text.

▶ Illustrations in the exposition are referenced using a triple-number system. For example, Figure 1.6.3 is the third figure in Section 1.6, and Figure 7.2.5 is the fifth figure in Section 7.2. The same numbering system is used for theorems and definitions. Illustrations in the exercises are identified by the exercise number with which they are associated. For example, in a particular exercise set, Figure Ex-7 would be associated with Exercise 7.

▶ The ideas in this text were created by real people with interesting personalities and backgrounds. Pictures and signatures of many of these people appear on the opening pages of the chapters, and biographical sketches of various mathematicians appear throughout the text as footnotes.

▶ At various places in the text you will see elements labeled "For the Reader," which are designed to reinforce ideas in the text. Some of these ask you to think about an idea, some ask you to perform a computation, and (for students using technology) some ask you to read your reference manual and then use the technology to perform a computation or to generate a graph.

As you read through this book, you will find some ideas that you understand immediately, others that you don't understand until you have read them several times, and others that you do not understand, even after numerous readings. Don't become discouraged—some calculus ideas take time to "percolate," and you may well find that the idea suddenly becomes clear later when you least expect it.

If you find that your answer to an exercise does not match that in the back of the book, do not presume immediately that your answer is incorrect—there may be more than one way to express the answer. For example, if your answer is $\sqrt{3}/3$ and the text answer is $1/\sqrt{3}$, then both are correct, since your answer can be obtained by rationalizing the text answer. In general, if your answer does not match that in the text, then your best first step is to look for an algebraic manipulation or a trigonometric identity that relates the two answers. In cases where the answer is a decimal approximation, your answer may differ from that in the text because of different choices in the number of decimal places used in the computations.

Some exercises require a verbal answer. Express those answers in complete, correctly punctuated, logical sentences—not fragmented phrases and formulas.

It is *not* essential to have graphing technology to read and use this text. Exercises requiring technology have been tagged with icons precisely so they can be omitted if necessary. Text elements requiring technology are relegated to the "For the Reader," so they can be omitted as well. If you have graphing technology, then you may want to use it as you read the text or to check your work in exercises that are not tagged with icons. However, it is not essential.

Introduction

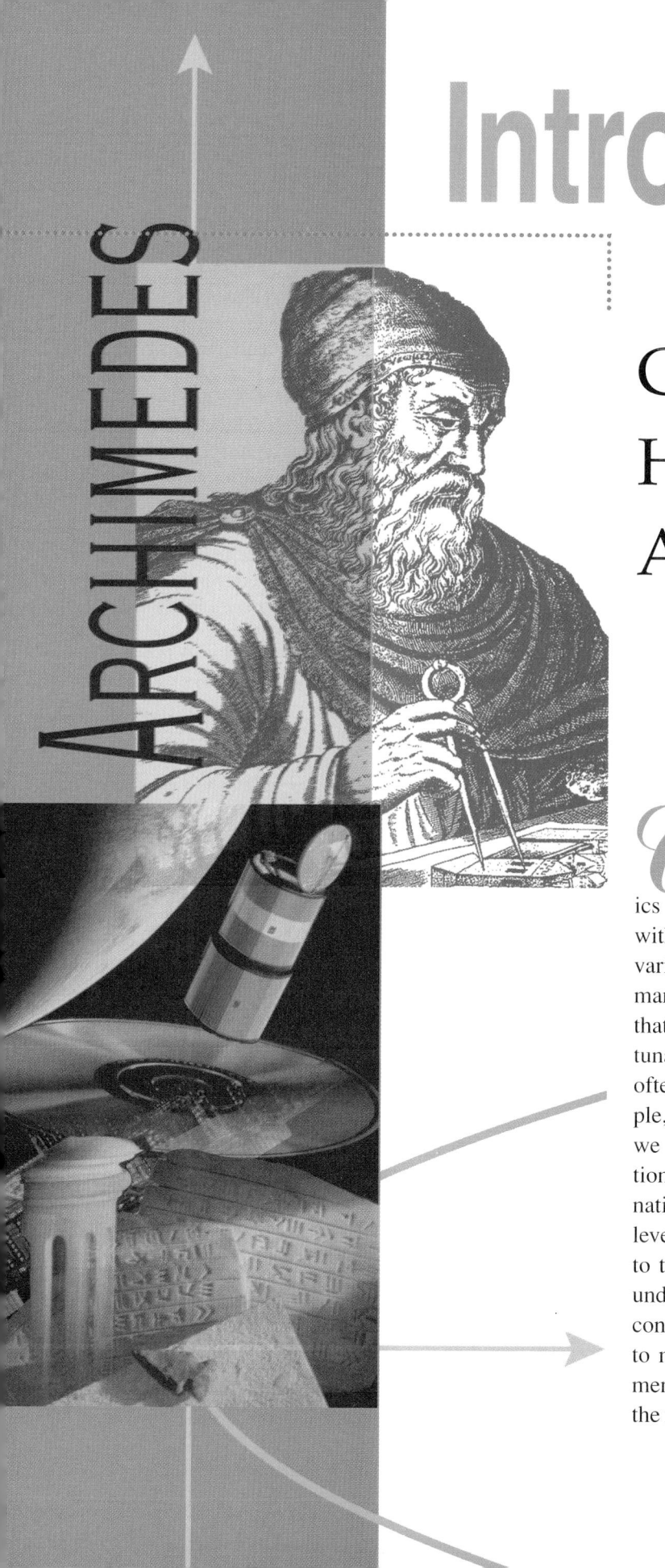

ARCHIMEDES

CALCULUS: A NEW HORIZON FROM ANCIENT ROOTS

Calculus, which is sometimes called the "mathematics of change," is the branch of mathematics concerned with describing the precise way in which changes in one variable relate to changes in another. In almost every human activity we encounter two types of variables: those that we can control directly and those that we cannot. Fortunately, those variables that we cannot control directly often respond in some way to those that we can. For example, the acceleration of a car responds to the way in which we control the flow of gasoline to the engine, the inflation rate of an economy responds to the way in which the national government controls the money supply, and the level of an antibiotic in a person's bloodstream responds to the dosage and timing of a doctor's prescription. By understanding quantitatively how the variables we cannot control directly respond to those that we can, we can hope to make predictions about the behavior of our environment and gain some mastery over it. Calculus is one of the fundamental mathematical tools used for this purpose.

Calculus has an enormous, but often unnoticed, impact on our daily lives. To provide some sense of how you and I are being affected by calculus, I have selected a few of its applications to fields of contemporary research. All of these applications involve other branches of science and mathematics, but they all use calculus in some essential way. The first three applications are based on a new and exciting area of mathematics called the theory of *wavelets*. Wavelets make it possible to capture and store mathematical representations of images and signals using much less data than previously possible. As a result, the current research literature is literally exploding with new applications of wavelets to such diverse fields as astronomy, acoustics, nuclear engineering, image processing, neurophysiology, music, medicine, speech synthesization, earthquake prediction, and pure mathematics, to name only a few.

FBI Fingerprint Compression — The U.S. Federal Bureau of Investigation began collecting fingerprints and handprints in 1924 and now has more than 30 million such prints in its files, all of which are being digitized for storage on computer. It takes about 0.6 megabyte of storage space to record a fingerprint and 6 megabytes to record a pair of handprints, so that digitizing the current FBI archive would result in about 200×10^{12} bytes of data to be stored, which is the capacity of roughly 138 million floppy disks. At today's prices for computer equipment, storage media, and labor, this would cost roughly 200 million dollars. To reduce this cost, the FBI's Criminal Justice Information Service Division began working in 1993 with the National Institute of Standards, the Los Alamos National Laboratory, and several other groups to devise compression methods for reducing the storage space. These methods, which are based on wavelets, are proving to be highly successful. Figure 1 is a good example—the image on the left is an original thumbprint and the one on the right is a mathematical reconstruction from a 26:1 data compression.

Music — Researchers with the Numerical Algorithms Research Group at Yale University have investigated the application of wavelets to sound synthesis (musical and voice). To approximate the sound of a musical instrument or voice, samples are taken and decomposed mathematically into numbers called *wavelet packet coefficients*. These coefficients can be stored on a computer and later the sound can be reconstructed (synthesized) from the computer data. This area of research makes it possible to reproduce complex sounds from a small amount of data and to transmit those data electronically in a highly compressed form. This research may eventually speed up the transmission of sound over the Internet, for example.

Removing Noise from Data — In fields ranging from planetary science to molecular spectroscopy, scientists are faced with the problem of recovering a true signal from incomplete or noisy data. For example, weak signals from deep space probes are often so overwhelmed with background noise that the signal itself is barely detectable, yet the signal must be used to produce a photograph or provide other information. Researchers at Stanford University and elsewhere have been working for several years on using wavelet methods to filter out such noise. For example, Figure 2 shows a signal from a medical imaging signal that has been cleaned up (de-noised) using wavelets.

Original Reconstruction

Figure 1

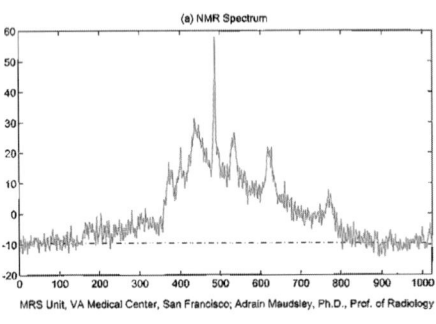 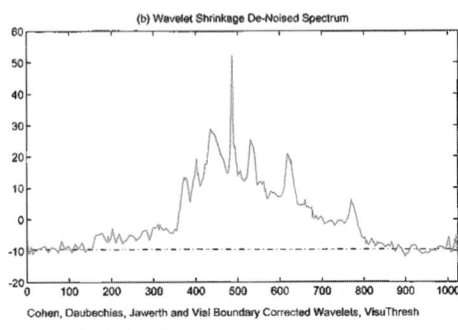

(a) NMR Spectrum

MRS Unit, VA Medical Center, San Francisco; Adrain Maudsley, Ph.D., Prof. of Radiology

Noisy signal

(b) Wavelet Shrinkage De-Noised Spectrum

Cohen, Daubechies, Jawerth and Vial Boundary Corrected Wavelets, VisuThresh

De-noised signal

Figure 2

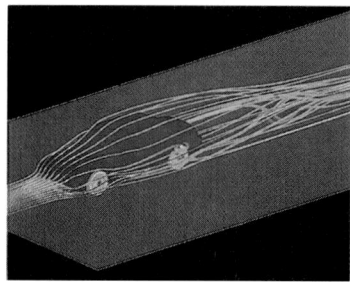

Airflow past a Saturn SL2

Figure 3

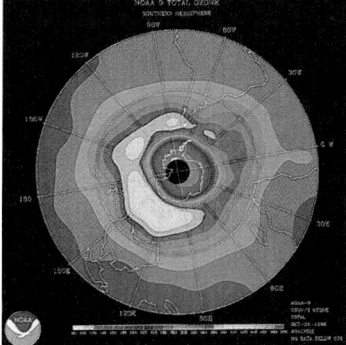

Ozone hole in the Southern Hemisphere

Figure 4

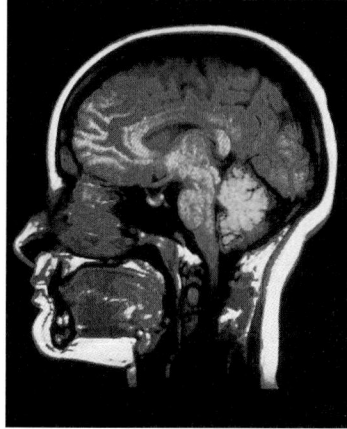

Magnetic resonance image

Figure 5

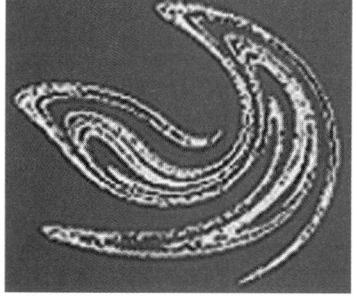

Chaotic ventricular fibrillation

Figure 6

Airflow Past an Automobile — Problems involving fluid flow (air, water, and blood, for example) are a major focus of scientific research. The Army High Performance Computing Research Center (AHPCRC) sponsors numerous unclassified research projects that involve teams of researchers from various science and engineering disciplines. One such project deals with airflow past an automobile (they use a General Motors Saturn SL2). The problem is quite complex since it takes into account the body contours, the wheels, the recessed headlights, and the spoiler. Figure 3 shows a simulation of airflow past an automobile that was produced using state-of-the-art mathematical methods and a Cray T3D supercomputer.

Weather Prediction — Modern meteorology is a marriage between mathematics and physics. Today's meteorologists are concerned with much more than predicting daily weather changes—their research delves into such areas as global warming, holes in the ozone layer (Figure 4), and weather patterns on other planets. In 1904 the Norwegian meteorologist Vilhelm Bjerknes (1862–1951) proposed that the state of the atmosphere at any future time can be determined by measuring appropriate variables at a single instant of time and then solving certain hydrodynamic equations. Although Bjerknes' idea is true in principle, it is difficult to apply because of uncertainties in measured variables, the enormous amounts of data to be processed, and technical complications involved with solving the equations. However, new mathematical discoveries have dramatically improved meteorological predictions and spawned enormous economic benefits. For example, it costs about 50 million dollars to prepare for a hurricane over 300 nautical miles of coastline, even if the hurricane does not hit the area. On the other hand, if the hurricane hits without adequate preparation, then the added costs can mount to billions of dollars (let alone the loss of life). Thus, each new mathematical breakthrough that produces more accurate hurricane prediction translates into enormous economic savings and preservation of human life.

Medical Imaging and DNA Structure — Advances in *nuclear magnetic resonance* (NMR) have made it possible to determine the structure of biological macromolecules, study DNA replication, and determine how proteins act as enzymes and antibodies. Related advances in *magnetic resonance imaging* (MRI) have made it possible to view internal human tissue without invasive surgery and to provide real-time images during surgical procedures (Figure 5). High-quality NMR and MRI would not be possible without mathematical discoveries that have occurred within the last decade.

Controlling Chaotic Behavior in the Human Heart — Chaos theory, which is one of the most exciting new branches of mathematics, is concerned with identifying regularities in phenomena that on the surface seem random and unpredictable (Figure 6). Today's research literature abounds with applications of chaos theory to almost every imaginable branch of science. Recently, researchers at the Applied Chaos Laboratory at Georgia Tech University collaborated with physicians at the Emory University Medical Center in applying chaos theory to control the chaotic behavior of heart tissue that is undergoing ventricular fibrillation (cardiac arrest). The research, though experimental, is already showing promising results.

The World Model of the Future — In anticipation of the 1992 United Nations Earth Summit, researchers at the Institute for Economic Analysis (IEA) at New York University were commissioned by a number of world leaders with the daunting task of creating a model that would predict the economic and environmental future of the world. They started with the World Model and World Database developed by Nobel laureate Wassily Leontief and his colleagues at Harvard in the 1970s, but they expanded on the model by incorporating such environmental factors as the cost of controlling pollutant emissions (from mining, energy creation, and automobiles, for example). They also accounted for the effect of population growth rates on the added demand for energy and other natural resources. Models such as this require a team effort by government, academic, and industrial experts in a variety of fields and play an important role in guiding the decisions of governmental agencies.

Deep Space Exploration — Alexander Wolszczan of Penn State University may go down in history as the first scientist to identify a planetary system beyond our own. While searching the radio sky, Professor Wolszczan discovered a new pulsar, PSR1257+12, that seemed to wobble as it traveled through space. As a result of an extensive mathematical analysis, many scientists are now convinced that the wobble is caused by two or three planets orbiting PSR1257+12. Although scientists have been able to detect pulsars for some time by searching for faint periodic radio signals from outer space, it is only recently that the mathematical techniques have been developed to analyze the data in a way that stands up to scientific scrutiny. Wolszczan predicts that the planets orbiting PSR1257+12 are barren and inhospitable because of stellar winds, but his methods open the possibility of discovering new planetary systems that may sustain intelligent life.

THE ROOTS OF CALCULUS

Today's exciting applications of calculus have roots that can be traced to the work of the Greek mathematician Archimedes, but the actual discovery of the fundamental principles of calculus was made independently by Isaac Newton and Gottfried Leibniz in the late seventeenth century. The work of Newton and Leibniz was motivated by four major classes of scientific and mathematical problems of the time:

- Find the tangent line to a general curve at a given point.
- Find the area of a general region, the length of a general curve, and the volume of a general solid.
- Find the maximum or minimum value of a quantity—for example, the maximum and minimum distances of a planet from the Sun, or the maximum range attainable for a projectile by varying its angle of fire.
- Given a formula for the distance traveled by a body in any specified amount of time, find the velocity and acceleration of the body at any instant. Conversely, given a formula that specifies the acceleration of velocity at any instant, find the distance traveled by the body in a specified period of time.

INFINITE PROCESSES

Even though these problems may seem diverse and unrelated, we will see later that they are all closely linked by the fundamental principles of calculus and that all of them involve *infinite processes* in some way. These same principles and processes underlie the contemporary applications that we discussed at the beginning of this section.

There is something very satisfying about starting a task and bringing it to a step-by-step conclusion. However, the real world is replete with processes that by their very nature cannot be completed in finitely many steps, and hence must be left unfinished in some sense. For example, whereas the complete decimal expansion of the fraction 1/8 can be obtained in three steps by long division,

$$1/8 = .125$$

the complete decimal expansion of $\sqrt{2}$ cannot be obtained in a finite number of steps by *any* procedure. Although there are numerous *algorithms* (i.e., step-by-step procedures) for approximating $\sqrt{2}$ to any desired degree of accuracy, none of them produces the exact value in finitely many steps. One such algorithm, called the *mechanic's rule*, is based on the formula

$$y_0 = 1, \quad y_{n+1} = \frac{1}{2}\left(y_n + \frac{2}{y_n}\right) \tag{1}$$

These equations can be used to generate an *infinite sequence* of approximations

$$y_0, y_1, y_2, y_3, \ldots \tag{2}$$

that get closer and closer to $\sqrt{2}$, achieving an arbitrary degree of accuracy in finitely many steps. This is done by first setting $y_0 = 1$ and then using the second part of Formula (1) to generate each new approximation y_{n+1} from the preceding approximation y_n. For example, Table 1 shows the first six approximations of $\sqrt{2}$ produced by the mechanic's rule. The fractions in the table were obtained using a computer program, called a *Computer Algebra*

System (CAS),[*] that is capable of performing algebraic operations exactly. We used the same program to convert the fractions to decimal approximations with 12 digits, but we could also have used a calculator. At $n = 4$ the decimal approximations began to repeat because we had reached the accuracy limit of a 12-digit display.

Table 1

n	$y_0 = 1, \quad y_{n+1} = \frac{1}{2}\left(y_n + \frac{2}{y_n}\right)$	DECIMAL APPROXIMATION
	$y_0 = 1$ \quad (Starting value)	1.00000000000
0	$y_1 = \frac{1}{2}\left[1 + \frac{2}{1}\right] = \frac{3}{2}$	1.50000000000
1	$y_2 = \frac{1}{2}\left[\frac{3}{2} + \frac{2}{3/2}\right] = \frac{17}{2}$	1.41666666667
2	$y_3 = \frac{1}{2}\left[\frac{17}{2} + \frac{2}{17/2}\right] = \frac{577}{408}$	1.41421568627
3	$y_4 = \frac{1}{2}\left[\frac{577}{408} + \frac{2}{577/408}\right] = \frac{665,857}{470,832}$	1.41421356237
4	$y_5 = \frac{1}{2}\left[\frac{665,857}{470,832} + \frac{2}{665,857/470,832}\right] = \frac{886,731,088,897}{627,013,566,048}$	1.41421356237

INFINITE SERIES

We learn in elementary arithmetic that the decimal expansion of $\frac{1}{3}$ is

$$\frac{1}{3} = .33333\ldots$$

(all decimal digits being 3). We can rewrite this equation as

$$\frac{1}{3} = .3 + .03 + .003 + .0003 + .00003 + \cdots$$

or alternatively, as

$$\frac{1}{3} = \frac{3}{10} + \frac{3}{10^2} + \frac{3}{10^3} + \frac{3}{10^4} + \frac{3}{10^5} + \cdots \tag{3}$$

This formula expresses the number $\frac{1}{3}$ as an unending sum with infinitely many terms; such sums are called *infinite series*. An infinite series denotes an addition process that cannot be completed in finitely many steps—one can add the first 10 terms, the first 100 terms, or even the first 10,000 terms (with the help of a computer), but one cannot add *all* of the terms in the usual sense because there are infinitely many of them. However, if we start at the beginning of the series and add terms one by one, then at each step the sum gets closer and closer to $\frac{1}{3}$. For example,

$$\frac{1}{3} \approx \frac{3}{10} = .3$$

$$\frac{1}{3} \approx \frac{3}{10} + \frac{3}{10^2} = .33$$

$$\frac{1}{3} \approx \frac{3}{10} + \frac{3}{10^2} + \frac{3}{10^3} = .333$$

$$\frac{1}{3} \approx \frac{3}{10} + \frac{3}{10^2} + \frac{3}{10^3} + \frac{3}{10^4} = .3333$$

---

[*]The most widely used CAS programs are *Mathematica*, by Wolfram Research, Inc.; *Maple*, by Waterloo Maple Software, Inc.; and *Derive*, by Soft Warehouse, Inc.

Thus, Formula (3) is interpreted to mean that $\frac{1}{3}$ can be approximated to any desired degree of accuracy by adding sufficiently many terms from the beginning of the series.

CALCULUS AND THE SEARCH FOR π

The need for an accurate approximation of π dates back to the surveyors of the early Babylonian and Egyptian civilizations. The Greek mathematician Archimedes calculated π to two decimal places by geometric means and later mathematicians obtained greater accuracy by improving his geometric methods. With the advent of calculus, various infinite series for π were discovered. The first such series was

$$\pi = 4 \left(1 - \tfrac{1}{3} + \tfrac{1}{5} - \tfrac{1}{7} + \tfrac{1}{9} - \cdots\right)$$

Although this series is quite beautiful in its simplicity, it has little practical value because enormous numbers of terms are required to achieve good approximations. For example, a computer computation of the sum of the first 5000 terms in this series yields the approximation

$$\pi \approx 3.14139$$

which is correct to only three decimal places in spite of the large amount of computation. (Compare this to the value of π that your calculator produces.) Later in this text you will encounter infinite series that are better suited for approximating π because they produce better accuracy using fewer terms. The activity of finding algorithms that produce better accuracy with less computation is an area of current mathematical research in a branch of mathematics called ***numerical analysis***.

AREA

Formulas for the areas of plane regions with straight-line boundaries (squares, rectangles, triangles, trapezoids, etc.) were well known in many early civilizations. However, obtaining formulas for regions with curvilinear boundaries (a circle being the simplest case) caused problems for early mathematicians. An idea for computing the area of a circle to an arbitrary degree of accuracy was suggested around 430 B.C. by the Greek scholar Antiphon and was later systematized by the Greek mathematician Eudoxus into an algorithm called the ***method of exhaustion***. That method, when applied to a circle of radius r, consists of inscribing a succession of regular polygons in the circle and allowing the number of sides n to increase indefinitely (Figure 7). As n increases, the polygons tend to "exhaust" the region inside the circle, and the areas of those polygons become better and better approximations to the exact area of the circle.

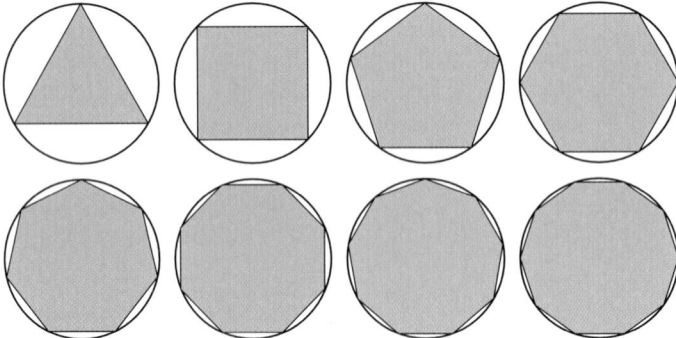

Figure 7

To see how this works numerically, let $p(n)$ denote the area of a regular n-sided polygon inscribed in a circle of radius r. We can find a formula for $p(n)$ by subdividing the polygon into n congruent triangles (Figure 8a) and adding the areas of those triangles to obtain the area of the entire polygon. Each triangle is isosceles, since two of its sides are radii of the circle; and the angle at the apex of each triangle is $2\pi/n$, since the triangles divide the

Table 2

n	$p(n)$
100	3.13952597647 r^2
200	3.14107590781 r^2
300	3.14136298250 r^2
400	3.14146346236 r^2
500	3.14150997084 r^2
600	3.14153523487 r^2
700	3.14155046835 r^2
800	3.14156035548 r^2
900	3.14156713408 r^2
1000	3.14157198278 r^2
2000	3.14158748588 r^2
3000	3.14159035683 r^2
4000	3.14159136166 r^2
5000	3.14159182676 r^2
6000	3.14159207940 r^2
7000	3.14159223174 r^2
8000	3.14159233061 r^2
9000	3.14159239839 r^2
10000	3.14159244688 r^2

central angle of the circle into n equal parts. Thus, with the help of some basic trigonometry (Figure 8b), we deduce that the area of each triangle is

$$\text{area} = \tfrac{1}{2} \cdot \text{base} \cdot \text{height}$$
$$= \tfrac{1}{2} \cdot 2(r \sin \pi/n)(r \cos \pi/n) = r^2 \sin(\pi/n) \cos(\pi/n)$$

from which it follows that the area of n triangles is

$$p(n) = nr^2 \sin(\pi/n) \cos(\pi/n) \tag{4}$$

As n increases, this formula should produce better and better approximations to the exact area of the circle. To see that this is so, we used a calculator set to the radian mode to generate Table 2. Later, using the tools of calculus, we will show definitively that $p(n)$ converges to the limit πr^2 as n increases.

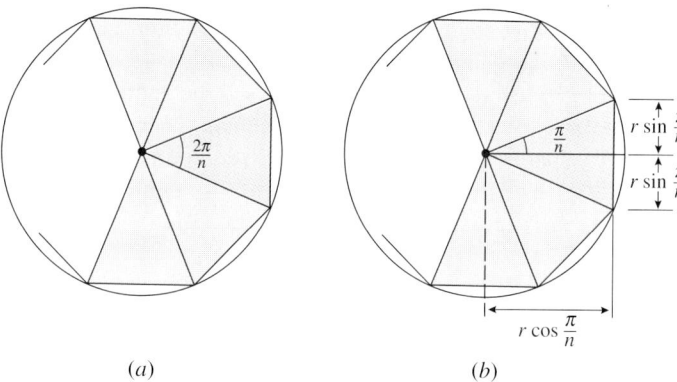

(a)   (b)

Figure 8

Figure 9 shows a variation of the method of exhaustion that appeared in a seventeenth century Japanese manuscript. In that manuscript the area of the circle is approximated by inscribed rectangles, rather than polygons.

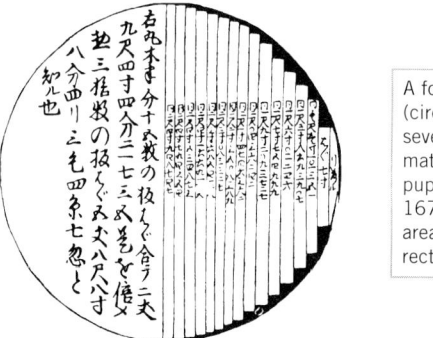

A form of calculus called *yenri* (circle principle) was developed in seventeenth century Japan by the mathematician Seki Kōwa and his pupils. This illustration, dating to 1670, shows an approximation of the area of a circle with inscribed rectangles.

Figure 9

TANGENT LINES

Tangent lines to general curves were of great interest to the mathematicians and scientists of the seventeenth century because of their application to the design of lenses. To determine how a ray of light passes through a lens using the laws of optics, one must know the angle at which the ray strikes the lens. This angle is measured between the ray and the normal line to the lens surface, the normal line being perpendicular to the tangent line (Figure 10). Thus, the study of various lens shapes led to the mathematical problem of finding the tangent line at a point on a general curve.

For circles, the concept of a tangent line is simple—a line is tangent to the circle if it meets the circle at precisely one point. However, this does not work for other kinds of curves (Figure 11).

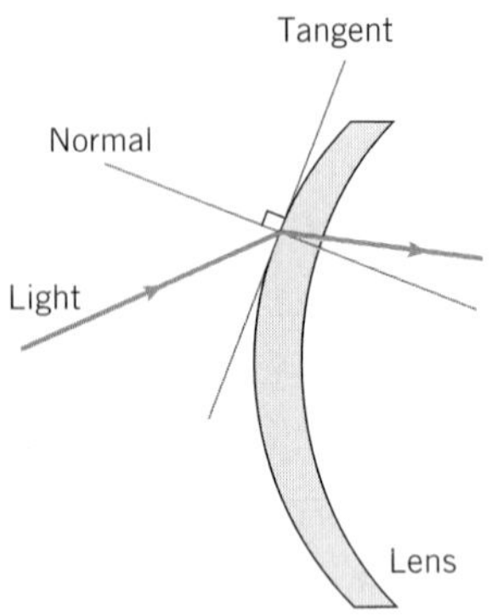

Tangent

Normal

Light

Lens

Figure 10

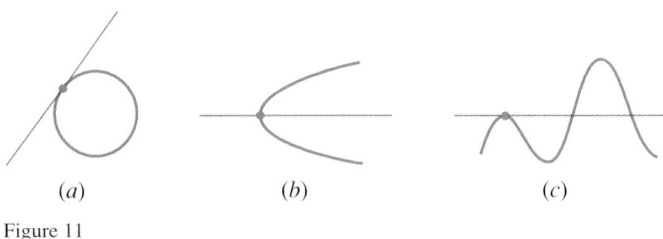

(a) (b) (c)

Figure 11

In order to apply the concept of a tangent line to curves other than circles, we must view tangent lines another way. For this purpose, suppose that we are interested in the tangent line at a point P on a circle, and let Q be any point on the circle different from P (Figure 12a). If we draw the secant line through P and Q, and then allow Q to move along the circle toward P, then intuition suggests that the secant line will rotate toward a "limiting position" that coincides with the tangent line at P. This viewpoint about tangent lines is important because it can be applied to more general curves (Figure 12b). Thus, the geometric problem of finding a tangent line leads to a problem involving an infinite process—finding the limiting position of secant lines.

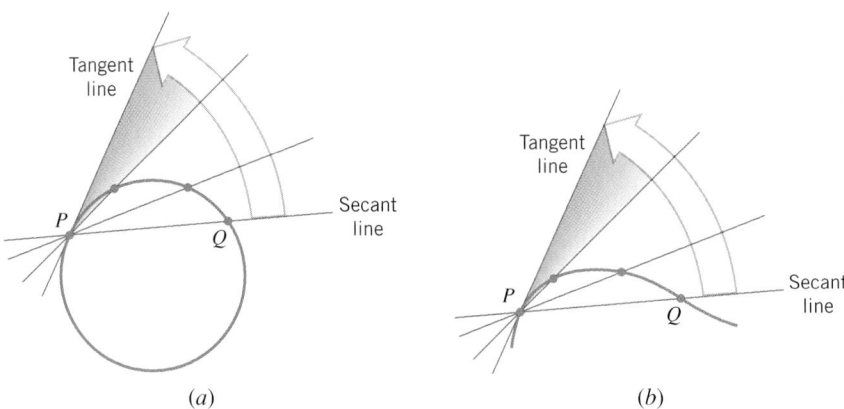

(a) (b)

Figure 12

CALCULUS AND THE MYSTERY OF CONTINUOUS MOTION

One of the early triumphs of calculus was its use in clarifying and quantifying continuous motion. The ancient Greeks had two schools of thought on the nature of space and time: the *discrete* and the *continuous*. From the discrete viewpoint, space and time are composed of small *indivisible* units (points and instants) and motion is a succession of small discrete jumps that gives the illusion of smoothness to the eye (like a movie). From the continuous viewpoint, every unit of space and time, no matter how small, can be further subdivided, and motion is a smooth continuous process.

The Greek philosopher Zeno (born c. 490 B.C.) raised perplexing questions about both theories of motion with some paradoxes. (A **paradox** is an argument that appears to be logically correct but that leads to a contradiction or reaches a conclusion that flies in the face of common sense.) Zeno questioned the discrete theory of motion with his *Arrow Paradox* and the continuous theory of motion with his *Paradox of Achilles and the Tortoise*:

Zeno's Arrow Paradox: *If time and space are discrete, then an arrow cannot move through the air. For at each instant of time the arrow is at a definite point and hence is at rest at that instant. Thus, the arrow is always at rest.*

Zeno's Paradox of Achilles and the Tortoise: *If time and space are continuous, and if a tortoise is given the slightest head start in a race with Achilles, then Achilles will never catch the tortoise. For when Achilles reaches the tortoise's starting point, the tortoise will have moved ahead to a point B. When Achilles reaches the point B the tortoise will have moved ahead to a point C—ad infinitum. Thus, the tortoise will always be ahead, even if by a hair* (Figure 13).

Figure 13

Even today the paradoxes of Zeno raise bothersome philosophical issues about the nature of motion. However, in the early fourteenth century the emphasis shifted away from the philosophical issues toward the quantitative study of speed and acceleration. The difficulty faced by mathematicians and scientists of that period was the lack of a precise *definition* of speed that could be used as a starting point for quantitative analysis—and that turned out to be a nontrivial matter.

To understand the difficulty, suppose that a car travels 75 miles in a 3-hour period. We say that the *average* speed of the car is 25 miles per hour ($75/3 = 25$ mi/h). More generally, the *average speed* of an object during a specified time interval is defined as

$$\text{average speed} = \frac{\text{distance traveled}}{\text{time elapsed}}$$

However, it is important to recognize that this is just an average—a car with an average speed of 25 mi/h on a trip need not travel at a constant speed of 25 mi/h—it may speed up and it may slow down. Moreover, average speed is not a very useful quantity in certain situations. For example, if the car happens to hit a tree, then the resulting damage will not be determined by the average speed up to the time of impact, but rather by the *instantaneous* speed at the precise moment of impact.

But exactly what do we mean by "instantaneous speed" and how do we compute it? We cannot simply carry over the process for computing average speed, since in any given instant the distance traveled is 0 and the time elapsed is 0, so the distance traveled divided by the time elapsed is $0/0$, which is meaningless. Thus, although instantaneous speed is a physical reality, there is difficulty computing it for lack of a precise definition.

Mathematicians and scientists ultimately resolved this difficulty by using the well-defined notion of average speed together with an infinite process to define the concept of instantaneous speed. The idea is as follows: Suppose that we are interested in the instantaneous speed of an object at some time t. Intuition suggests that over a *small* time interval the speed of the object should not change very much. Thus, if $t + h$ is a point in time slightly later than t, then the average speed over the time interval from t to $t + h$ should be very close to the instantaneous speed at time t. Moreover, the closer $t + h$ is to t, the better we should expect the approximation to be. This suggests that the instantaneous speed at time t be *defined* as the limiting value of the average speed computed over smaller and smaller time intervals starting at time t.

What is fascinating about this is the link between precise mathematical definition and real-world applications—once physicists were armed with the right definition of instantaneous velocity, they were ultimately able to find equations for the motion of the planets and develop fundamental theories about gravitational attraction.

THE ROLE OF RIGOR AND PROOF IN CALCULUS

Although the concept of deductive proof dates back to Euclid, most of the developments in mathematics from about 200 B.C. to 1870 were based on intuition and empirical discovery— the idea of proving new mathematical results rigorously was largely ignored. However, as calculus developed, concepts related to infinite processes began to challenge the reliability of intuition and eventually an emphasis on precise definitions and careful proof was reestablished.

To illustrate how intuition can fail when dealing with infinite processes, consider the infinite series

$$1 - 1 + 1 - 1 + 1 - 1 + \cdots$$

We might be tempted to conclude that the sum of this series is zero by grouping the terms as

$$(1 - 1) + (1 - 1) + (1 - 1) + \cdots = 0 + 0 + 0 + \cdots$$

However, we can also group the terms as

$$1 + (-1 + 1) + (-1 + 1) + (-1 + 1) + \cdots = 1 + 0 + 0 + 0 + \cdots$$

which suggests that the sum is 1. Something has to be wrong! (Later, we will see that neither conclusion is correct.) The difficulty is our lack of mathematical precision; we have not established a precise definition of what we mean by the sum of an infinite series, and we have assumed without justification that the rules of grouping for finitely many terms also apply to infinite series.

In this text many ideas will be introduced informally at first to develop our intuition, but eventually we will take great care to define terms precisely and state exact conditions under which results are valid. The preceding example should convince you that this is not an idle mathematical exercise but rather an essential part of avoiding serious mathematical errors.

THE DISCOVERY OF CALCULUS

The development of calculus was an evolutionary process that culminated in the discovery of a fundamental relationship between the problem of finding areas and the problem of finding tangent lines. The discovery of that result, which was made independently by Sir Isaac Newton (English) and Gottfried Wilhelm Leibniz (German), is considered to be the "discovery" of calculus. Newton made the discovery 10 years before Leibniz but did not publish his work until 20 years after Leibniz published his work. This situation led to a stormy debate over the rightful discoverer of calculus that engulfed Europe for half a century, with the scientists of the Continent supporting Leibniz and those from England supporting Newton. The conflict was extremely unfortunate because Newton's inferior notation badly hampered scientific development in England, and the Continent in turn lost the benefit of Newton's discoveries in astronomy and physics for nearly 50 years. In spite of it all, Newton and Leibniz were sincere admirers of each other's work.

ISAAC NEWTON (1642–1727)

Newton was born in the village of Woolsthorpe, England. His father died before he was born and his mother raised him on the family farm. As a youth he showed little evidence of his later brilliance, except for an unusual talent with mechanical devices—he apparently built a working water clock and a toy flour mill powered by a mouse. In 1661 he entered Trinity College in Cambridge with a deficiency in geometry. Fortunately, Newton caught the eye of Isaac Barrow, a gifted mathematician and teacher. Under Barrow's guidance Newton immersed himself in mathematics and science, but he graduated without any special distinction. Because the Plague was spreading rapidly through London, Newton returned to his home in Woolsthorpe and stayed there during the years of 1665 and 1666. In those two momentous years the entire framework of modern science was miraculously created in Newton's mind—he discovered calculus, recognized the underlying principles of planetary motion and gravity, and determined that "white" sunlight was composed of all colors, red to violet. For some reasons he kept his discoveries to himself. In 1667 he returned to Cambridge to obtain his Master's degree and upon graduation became a teacher at Trinity. Then in 1669

Newton succeeded his teacher, Isaac Barrow, to the Lucasian chair of mathematics at Trinity, one of the most honored chairs of mathematics in the world. Thereafter, brilliant discoveries flowed from Newton steadily. He formulated the law of gravitation and used it to explain the motion of the Moon, the planets, and the tides; he formulated basic theories of light, thermodynamics, and hydrodynamics; and he devised and constructed the first modern reflecting telescope.

Throughout his life Newton was hesitant to publish his major discoveries, revealing them only to a select circle of friends, perhaps because of a fear of criticism or controversy. In 1687, only after intense coaxing by the astronomer, Edmond Halley (Halley's comet), did Newton publish his masterpiece, *Philosophiae Naturalis Principia Mathematica* (The Mathematical Principles of Natural Philosophy). This work is generally considered to be the most important and influential scientific book ever written. In it Newton explained the workings of the solar system and formulated the basic laws of motion which to this day are fundamental in engineering and physics. However, not even the pleas of his friends could convince Newton to publish his discovery of calculus. Only after Leibniz published his results did Newton relent and publish his own work on calculus.

After 25 years as a professor, Newton suffered depression and a nervous breakdown. He gave up research in 1695 to accept a position as warden and later master of the London mint. During the 25 years that he worked at the mint, he did virtually no scientific or mathematical work. He was knighted in 1705 and on his death was buried in Westminster Abbey with all the honors his country could bestow. It is interesting to note that Newton was a learned theologian who viewed the primary value of his work to be its support of the existence of God. Throughout his life he worked passionately to date biblical events by relating them to astronomical phenomena. He was so consumed with this passion that he spent years searching the Book of Daniel for clues to the end of the world and the geography of hell.

Newton described his brilliant accomplishments as follows: "I seem to have been only like a boy playing on the seashore and diverting myself in now and then finding a smoother pebble or prettier shell than ordinary, whilst the great ocean of truth lay all undiscovered before me."

GOTTFRIED WILHELM LEIBNIZ (1646–1716)

This gifted genius was one of the last people to have mastered most major fields of knowledge—an impossible accomplishment in our own era of specialization. He was an expert in law, religion, philosophy, literature, politics, geology, metaphysics, alchemy, history, and mathematics.

Leibniz was born in Leipzig, Germany. His father, a professor of moral philosophy at the University of Leipzig, died when Leibniz was six years old. The precocious boy then gained access to his father's library and began reading voraciously on a wide range of subjects, a habit that he maintained throughout his life. At age 15 he entered the University of Leipzig as a law student and by the age of 20 received a doctorate from the University of Altdorf. Subsequently, Leibniz followed a career in law and international politics, serving as counsel to kings and princes.

During his numerous foreign missions, Leibniz came in contact with outstanding mathematicians and scientists who stimulated his interest in mathematics—most notably, the physicist Christian Huygens. In mathematics Leibniz was self-taught, learning the subject by reading papers and journals. As a result of this fragmented mathematical education, Leibniz often rediscovered the results of others, and this helped to fuel the debate over the discovery of calculus.

Leibniz never married. He was moderate in his habits, quick-tempered, but easily appeased, and charitable in his judgment of other people's work. In spite of his great achievements, Leibniz never received the honors showered on Newton, and he spent his final years as a lonely embittered man. At his funeral there was one mourner, his secretary. An eye-witness stated, "He was buried more like a robber than what he really was—an ornament of his country."

1. The repeating decimal $0.137137137\ldots$ can be expressed as a ratio of integers by writing

$$x = 0.137137137\ldots$$
$$1000x = 137.137137137\ldots$$

and subtracting to obtain $999x = 137$ or $x = \frac{137}{999}$. Use this idea, where needed, to express the following decimals as ratios of integers.

(a) $0.123123123\ldots$ (b) $12.7777\ldots$
(c) $38.07818181\ldots$ (d) $0.4296000\ldots(=0.4296)$

All decimals fall into two categories: ***repeating decimals*** and ***nonrepeating decimals***. In a repeating decimal there is some point after which a fixed block of integers repeats over and over. For example, all of the decimals in Exercise 1 are repeating. In a nonrepeating decimal there is no *fixed block* of digits that repeats over and over. For example, although the decimal $0.101001000100001\ldots$ has a definite pattern, it is nonrepeating. Those real numbers whose decimals are repeating are called ***rational numbers*** and those whose decimals are nonrepeating are called ***irrational numbers***. It can be proved that the rational numbers are precisely those real numbers that can be expressed as the ratio of two integers (as in Exercise 1). Some familiar irrational numbers are π, $\sqrt{2}$, $\sqrt{3}$, $\sqrt{5}$, ... Such numbers cannot be expressed as the ratio of two integers.

---

[c] **2.** (a) Use the preceding discussion to explain in one sentence why π cannot be equal to $\frac{22}{7}$.
   (b) The accompanying figure shows π to 500 decimal places. Use it to determine whether $\frac{22}{7}$ is greater than or less than π.
   (c) Use a CAS to duplicate the results in the accompanying figure.

3.14159265358979323846264338327950288419716939937510 5
82097494459230781640628620899862803482534211706798 2
14808651328230664709384460955058223172535940812848 1
11745028410270193852110555964462294895493038196442 8
81097566593344612847564823378678316527120190914564 8
56692346034861045432664821339360726024914127372458 7
00660631558817488152092096282925409171536436789259 0
36001133053054882046652138414695194151160943305727 0
36575959195309218611738193261179310511854807446237 9
96274956735188575272489122793818301194913

Figure Ex-2

3. The following are all famous approximations of π:

$$\frac{22}{7}, \quad \frac{223}{71}, \quad \frac{333}{106}, \quad \frac{355}{113}, \quad \frac{63}{25}\left(\frac{17+15\sqrt{5}}{7+15\sqrt{5}}\right)$$

(a) Use a calculating device to order these approximations according to size.
(b) Which of these approximations is closest to but larger than π?
(c) Which of these approximations is closest to but smaller than π?
(d) Which of these approximations is most accurate?
(e) The last approximation is due to a famous self-taught Indian mathematician, named Ramanujan (1887–1920). Do some reading about his fascinating but tragic life.

4. The Rhind Papyrus, which is a fragment of Egyptian mathematical writing from about 1650 B.C., is one of the oldest known examples of written mathematics. It is stated in the papyrus that the area A of a circle is related to its diameter D by

$$A = \left(\tfrac{8}{9}D\right)^2$$

(a) What approximation of π were the Egyptians using?
(b) Use a calculating device to determine if this approximation is better or worse than the approximation $\frac{22}{7}$.

5. In this section we stated that π can be expressed as the infinite series

$$\pi = 4\left(1 - \tfrac{1}{3} + \tfrac{1}{5} - \tfrac{1}{7} + \tfrac{1}{9} - \cdots\right)$$

However, this series is of little practical value because it *converges* too slowly; that is, too many terms are required to obtain a good approximation. A more practical approach is based on the following formula, discovered in 1706 by the English astronomer John Machin (1680–1752):

$$\pi = 16\left(\frac{1}{5} - \frac{1}{3\cdot5^3} + \frac{1}{5\cdot5^5} - \cdots\right)$$
$$-4\left(\frac{1}{239} - \frac{1}{3\cdot239^3} + \frac{1}{5\cdot239^5} - \cdots\right)$$

Machin's formula was used in 1949 on the ENIAC computer at the Ballistic Research Laboratories to produce the first computer calculation of π (2037 decimal places). Show that the terms shown in Machin's formula give a more accurate approximation of π than the sum of the first 10 terms of the first series. [*Suggestion:* Compare your calculated values to the approximation of π in Figure Ex-2.]

6. In each part, use a calculating device to find the decimal expansion of the fraction, and then use that expansion to express the fraction as an infinite series. (Show at least the first six terms of the series.)

    (a) $\frac{1}{9}$ (b) $\frac{5}{27}$ (c) $\frac{14}{45}$

7. Repeat the directions of Exercise 6 for

    (a) $\frac{7}{11}$ (b) $\frac{8}{33}$ (c) $\frac{5}{12}$.

In Formula (1) we gave an algorithm, called the *mechanic's rule*, for approximating $\sqrt{2}$ to any degree of accuracy. That algorithm is a special case of the following more general algorithm for approximating the square root of any positive number p to any degree of accuracy:

$$y_0 = 1, \quad y_{n+1} = \frac{1}{2}\left(y_n + \frac{p}{y_n}\right)$$

Use this result in Exercises 8 and 9.

---

8. In each part use the algorithm stated above to approximate the square root to four decimal places.

    (a) $\sqrt{3}$ (b) $\sqrt{5}$

9. Repeat the directions of Exercise 8 for

    (a) $\sqrt{7}$ (b) $\sqrt{50}$.

10. If a and b are distinct real numbers, say $a < b$, then it can be proved that there must be real numbers between a and b. One such number is the arithmetic average $\frac{1}{2}(a+b)$.

    (a) Explain why there must be infinitely many real numbers between any two distinct real numbers.

    (b) Do you think it is true that

    $$0.9999999\ldots < 1.0000000\ldots ?$$

    Explain your reasoning.

    (c) Find the decimal representation of the arithmetic average of $0.9999999\ldots$ and $1.0000000\ldots$. Is this result consistent with your answer in part (b)? Explain.

    (d) Use the method of Exercise 1 to express the decimal $0.9999999\ldots$ as a ratio of two integers. Is this result consistent with your answer in part (b)? Explain.

Rene' Descartes

FUNCTIONS

One of the important themes in calculus is the analysis of relationships between physical or mathematical quantities. Such relationships can be described in terms of graphs, formulas, numerical data, or words. In this chapter we will develop the concept of a *function*, which is the basic idea that underlies almost all mathematical and physical relationships, regardless of the form in which they are expressed. We will study properties of some of the most basic functions that occur in calculus, and we will examine some familiar ideas involving lines, polynomials, and trigonometric functions from viewpoints that may be new. We will also discuss ideas relating to the use of graphing utilities such as graphing calculators and graphing software for computers. Before you start reading, you may want to scan through the appendices, since they contain various kinds of precalculus material that may be helpful if you need to review some of those ideas.

1.1 FUNCTIONS AND THE ANALYSIS OF GRAPHICAL INFORMATION

In this section we will define and develop the concept of a function. Functions are used by mathematicians and scientists to describe the relationships between variable quantities and hence play a central role in calculus and its applications.

SCATTER PLOTS AND TABULAR DATA

Many scientific laws are discovered by collecting, organizing, and analyzing experimental data. Since graphs play a major role in studying data, we will begin by discussing the kinds of information that a graph can convey.

To start, we will focus on paired data. For example, Table 1.1.1 shows the top qualifying speed by year in the Indianapolis 500 auto race from 1975 to 1994. This table pairs up each year t between 1975 and 1994 with the top qualifying speed S for that year. This paired data can be represented graphically in a number of ways:

- One possibility is to plot the paired data points in a rectangular tS-coordinate system (t horizontal and S vertical), in which case we obtain a **scatter plot** of S versus t (Figure 1.1.1a).
- A second possibility is to enhance the scatter plot visually by joining successive points with straight-line segments, in which case we obtain a **line graph** (Figure 1.1.1b).
- A third possibility is to represent the paired data by a **bar graph** (Figure 1.1.1c).

All three graphical representations reveal an upward trend in the data, as one would expect with improvements in automotive technology.

Table 1.1.1

INDIANAPOLIS 500
QUALIFYING SPEEDS

YEAR t	SPEED S (mi/h)
1975	193.976
1976	188.957
1977	198.884
1978	202.156
1979	193.736
1980	192.256
1981	200.546
1982	207.004
1983	207.395
1984	210.029
1985	212.583
1986	216.828
1987	215.390
1988	219.198
1989	223.885
1990	225.301
1991	224.113
1992	232.482
1993	223.967
1994	228.011

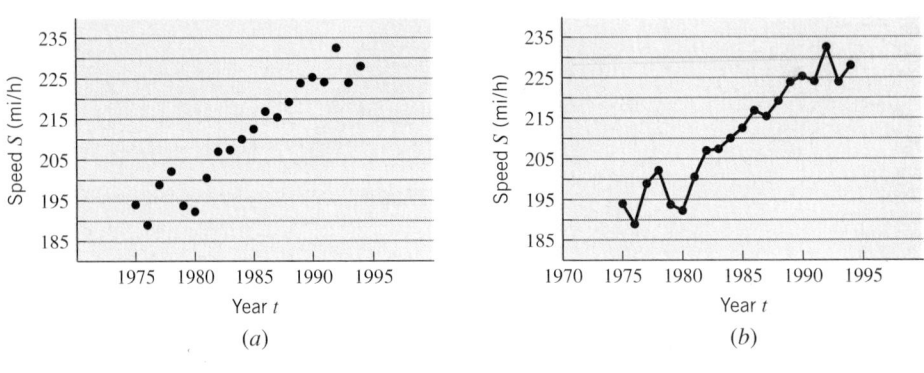

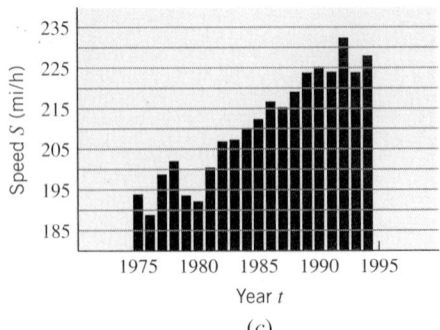

Figure 1.1.1

EXTRACTING INFORMATION FROM GRAPHS

One of the first books to use graphs for representing numerical data was *The Commercial and Political Atlas*, published in 1786 by the Scottish political economist William Playfair (1759–1823). Figure 1.1.2a shows an engraving from that work that compares exports and imports by England to Denmark and Norway (combined). In spite of its antiquity, the engraving is modern in spirit and provides a wealth of information. You should be able to extract the following information from Playfair's graphs:

- In the year 1700 imports were valued at about 70,000 pounds and exports at about 35,000 pounds.

- During the period from 1700 to about 1754 imports exceeded exports (a trade deficit for England).

- In the year 1754 the imports and exports were equal (a trade balance in today's economic terminology).

- From 1754 to 1780 exports exceeded imports (a trade surplus for England). The greatest surplus occurred in 1780, at which time exports exceeded imports by about 95,000 pounds.

- During the period from 1700 to 1725 imports were rising. They peaked in 1725, and then slowly fell until about 1760, at which time they bottomed out and began to rise again slowly until 1780.

- During the period from 1760 to 1780 exports and imports were both rising, but exports were rising more rapidly than imports, resulting in an ever-widening trade surplus for England.

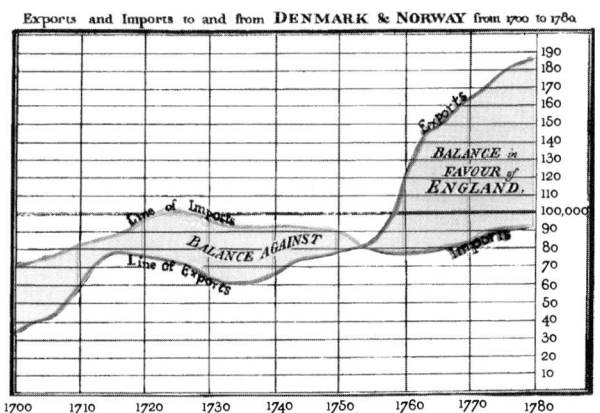

Playfair's Graph of 1786: The horizontal scale is in years from 1700 to 1780 and the vertical scale is in units of 1,000 pounds sterling from 0 to 200.

(*a*)

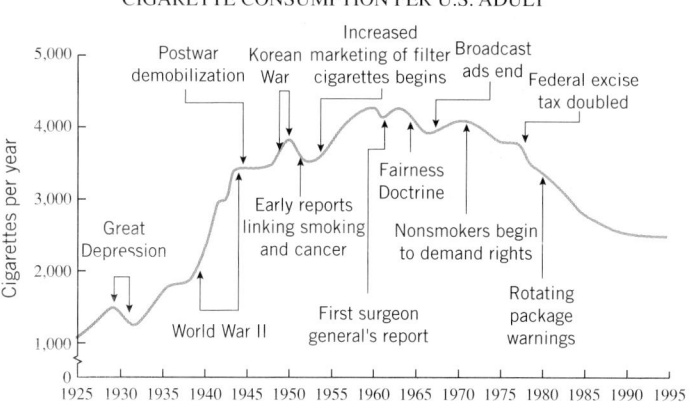

Source: U.S. Department of Health and Human Services.

(*b*)

Figure 1.1.2

Figure 1.1.2*b* is a more contemporary graph; it describes the per capita consumption of cigarettes in the United States between 1925 and 1995.

FOR THE READER.   Use the graph in Figure 1.1.2*b* to provide reasonable answers to the following questions:

- When did the maximum annual cigarette consumption per adult occur and how many were consumed?

- What factors are likely to cause sharp decreases in cigarette consumption?

- What factors are likely to cause sharp increases in cigarette consumption?

- What were the long- and short-term effects of the first surgeon general's report on the health risks of smoking?

Graphs can be used to describe mathematical equations as well as physical data. For example, consider the equation

$$y = x\sqrt{9 - x^2} \tag{1}$$

For each value of x in the interval $-3 \leq x \leq 3$, this equation produces a corresponding real value of y, which is obtained by substituting the value of x into the right side of the equation. Some typical values are shown in Table 1.1.2.

Table 1.1.2

x	-3	-2	-1	0	1	2	3
y	0	$-2\sqrt{5} \approx -4.47214$	$-2\sqrt{2} \approx -2.82843$	0	$2\sqrt{2} \approx 2.82843$	$2\sqrt{5} \approx 4.47214$	0

The set of *all* points in the xy-plane whose coordinates satisfy an equation in x and y is called the **graph** of that equation in the xy-plane. Figure 1.1.3 shows the graph of Equation (1) in the xy-plane. Notice that the graph extends only over the interval $[-3, 3]$. This is because values of x outside of this interval produce complex values of y, and in these cases the ordered pairs (x, y) do not correspond to points in the xy-plane. For example, if $x = 8$, then the corresponding value of y is $y = 8\sqrt{-55} = 8\sqrt{55}\,i$, and the ordered pair $(8, 8\sqrt{55}\,i)$ is not a point in the xy-plane.

Example 1

Figure 1.1.4 shows the graph of an unspecified equation that was used to obtain the values that appear in the shaded parts of the accompanying tables. Examine the graph and confirm that the values in the tables are reasonable approximations. ◄

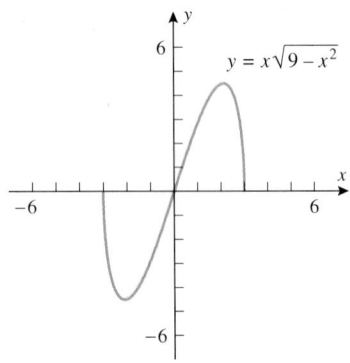

Figure 1.1.3

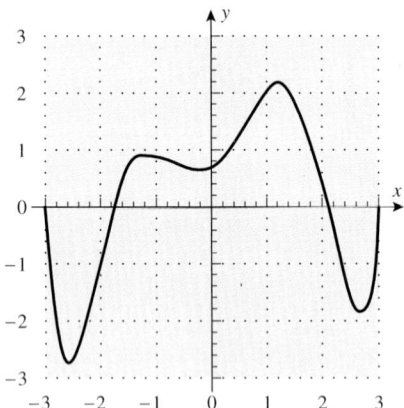

Figure 1.1.4

x	y
-3	0
-2	-1
-1	0.9
0	0.7
1	2
2	0.4
3	0

x	y
None	-3
$-2.3, -2.8$	-2
$-2, -2.9, 2.4, 2.9$	-1
$-3, -1.7, 2.1, 3$	0
$0.3, 1.8$	1
$1, 1.4$	2
None	3

Tables, graphs, and equations provide three methods for describing how one quantity depends on another—numerical, visual, and algebraic. The fundamental importance of this idea was recognized by Leibniz in 1673 when he coined the term *function* to describe the dependence of one quantity on another. The following examples illustrate how this term is used:

- The area A of a circle depends on its radius r by the equation $A = \pi r^2$, so we say that *A is a function of r*.

- The velocity v of a ball falling freely in the Earth's gravitational field increases with time t until it hits the ground, so we say that *v is a function of t.*

- In a bacteria culture, the number n of bacteria present after 1 hour of growth depends on the number n_0 of bacteria present initially, so we say that *n is a function of n_0.*

This idea is captured in the following definition.

1.1.1   DEFINITION.   If a variable y depends on a variable x in such a way that each value of x determines exactly one value of y, then we say that **y is a function of x.**

In the mid-eighteenth century the Swiss mathematician Leonhard Euler[*] (pronounced "oiler") conceived the idea of denoting functions by letters of the alphabet, thereby making it possible to describe functions without stating specific formulas, graphs, or tables. To understand Euler's idea, think of a function as a computer program that takes an *input x*, operates on it in some way, and produces exactly one *output y*. The computer program is an object in its own right, so we can give it a name, say f. Thus, the function f (the computer program) associates a unique output y with each input x (Figure 1.1.5). This suggests the following definition.

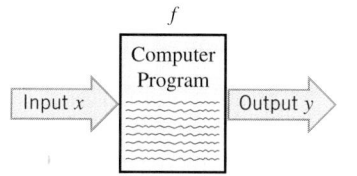

Figure 1.1.5

1.1.2   DEFINITION.   A **function** f is a rule that associates a unique output with each input. If the input is denoted by x, then the output is denoted by $f(x)$ (read "f of x").

REMARK.   In this definition the term *unique* means "exactly one." Thus, a function cannot assign two different outputs to the same input. For example, Figure 1.1.6 shows a scatter plot of weight versus age for a random sample of 100 college students. This scatter plot does not describe the weight W as a function of the age A because there are some values of A with more than one corresponding value of W. This is to be expected, since two people with the same age need not have the same weight. In contrast, Table 1.1.1 describes S as a function of t because there is only one top qualifying speed in a given year; similarly, Equation (1) describes y as a function of x because each input x in the interval $-3 \leq x \leq 3$ produces exactly one output $y = x\sqrt{9 - x^2}$.

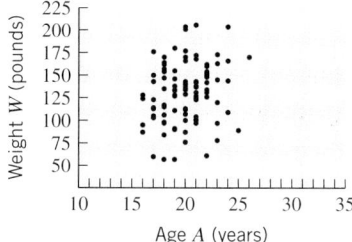

Figure 1.1.6

[*] LEONHARD EULER (1707–1783). Euler was probably the most prolific mathematician who ever lived. It has been said that "Euler wrote mathematics as effortlessly as most men breathe." He was born in Basel, Switzerland, and was the son of a Protestant minister who had himself studied mathematics. Euler's genius developed early. He attended the University of Basel, where by age 16 he obtained both a Bachelor of Arts degree and a Master's degree in philosophy. While at Basel, Euler had the good fortune to be tutored one day a week in mathematics by a distinguished mathematician, Johann Bernoulli. At the urging of his father, Euler then began to study theology. The lure of mathematics was too great, however, and by age 18 Euler had begun to do mathematical research. Nevertheless, the influence of his father and his theological studies remained, and throughout his life Euler was a deeply religious, unaffected person. At various times Euler taught at St. Petersburg Academy of Sciences (in Russia), the University of Basel, and the Berlin Academy of Sciences. Euler's energy and capacity for work were virtually boundless. His collected works form more than 100 quarto-sized volumes and it is believed that much of his work has been lost. What is particularly astonishing is that Euler was blind for the last 17 years of his life, and this was one of his most productive periods! Euler's flawless memory was phenomenal. Early in his life he memorized the entire *Aeneid* by Virgil and at age 70 could not only recite the entire work, but could also state the first and last sentence on each page of the book from which he memorized the work. His ability to solve problems in his head was beyond belief. He worked out in his head major problems of lunar motion that baffled Isaac Newton and once did a complicated calculation in his head to settle an argument between two students whose computations differed in the fiftieth decimal place.

Following the development of calculus by Leibniz and Newton, results in mathematics developed rapidly in a disorganized way. Euler's genius gave coherence to the mathematical landscape. He was the first mathematician to bring the full power of calculus to bear on problems from physics. He made major contributions to virtually every branch of mathematics as well as to the theory of optics, planetary motion, electricity, magnetism, and general mechanics.

Functions can be represented in four basic ways:

- Numerically by tables
- Geometrically by graphs
- Algebraically by formulas
- Verbally

The method of representation often depends on how the function arises. For example:

- Table 1.1.1 is a numerical representation of S as a function of t. This is the natural way in which data of this type are recorded.
- Figure 1.1.7 shows a seismic graph of an earthquake's intensity H as a function of the elapsed time t. In this case the function originates as a graph.
- Some of the most familiar examples of functions arise as formulas; for example, the formula $C = 2\pi r$ expresses the circumference C of a circle as a function of its radius r.
- Sometimes functions are described in words. For example, Isaac Newton's Universal Law of Gravitation is often stated as follows: The gravitational force of attraction between two bodies in the Universe is directly proportional to the product of their masses and inversely proportional to the square of the distance between them. This is the verbal description of the formula

$$F = G\frac{m_1 m_2}{r^2} \tag{2}$$

in which F is the force of attraction, m_1 and m_2 are the masses, r is the distance between them, and G is a constant.

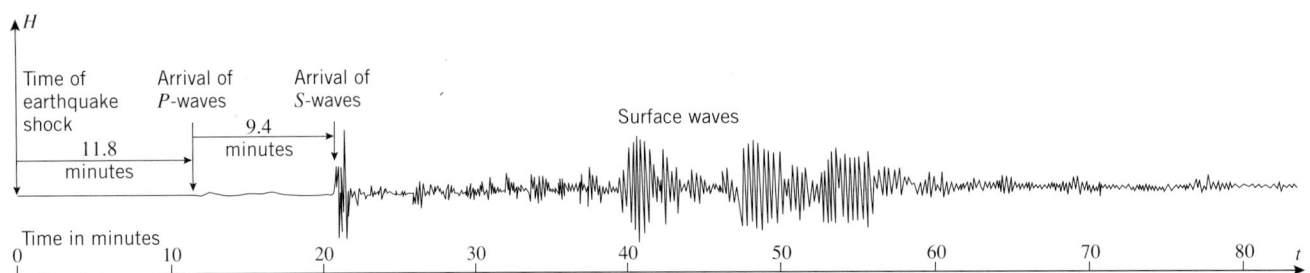

Figure 1.1.7

Table 1.1.3

U.S. POPULATION

YEAR t	POPULATION P (millions)
1790	3.9
1800	5.3
1810	7.2
1820	9.6
1830	12
1840	17
1850	23

Source: The World Almanac.

Sometimes it is desirable to convert one representation of a function into another. For example, in Figure 1.1.1 we converted the numerical relationship between S and t into a graphical relationship, and in writing Formula (2) we converted the verbal representation of the Universal Law of Gravitation into an algebraic relationship.

The problem of converting numerical representations of functions into algebraic formulas often requires special techniques known as **curve fitting**. For example, Table 1.1.3 gives the U.S. population at 10-year intervals from 1790 to 1850. This table is a numerical representation of the function $P = f(t)$ that relates the U.S. population P to the year t. If we plot P versus t, we obtain the scatter plot in Figure 1.1.8a, and if we use curve-fitting methods that will be discussed later, we can obtain the approximation

$$P \approx 3.94(1.03)^{t-1790}$$

Figure 1.1.8b shows the graph of this equation imposed on the scatter plot.

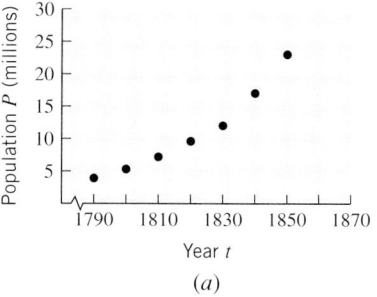
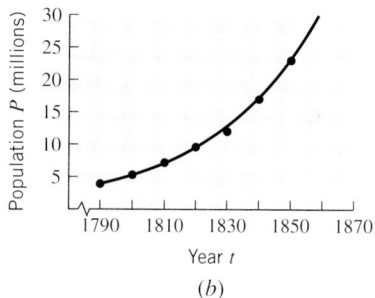

Figure 1.1.8

DISCRETE VERSUS CONTINUOUS DATA

Engineers and physicists distinguish between ***continuous data*** and ***discrete data***. Continuous data have values that vary *continuously* over an interval, whereas discrete data have values that make *discrete* jumps. For example, for the seismic data in Figure 1.1.7 both the time and intensity vary continuously, whereas in Table 1.1.3 and Figure 1.1.8*a* both the year and population make discrete jumps. As a rule, continuous data lead to graphs that are continuous, unbroken curves, whereas discrete data lead to scatter plots consisting of isolated points. Sometimes, as in Figure 1.1.8*b*, it is desirable to approximate a scatter plot by a continuous curve. This is useful for making conjectures about the values of the quantities between the recorded data points.

GRAPHS AS PROBLEM-SOLVING TOOLS

Sometimes a function is buried in the statement of a problem, and it is up to the problem solver to uncover it and use it in an appropriate way to solve the problem. Here is an example that illustrates the power of graphical representations of functions as a problem-solving tool.

Example 2

Figure 1.1.9*a* shows an offshore oil well located at a point W that is 5 km from the closest point A on a straight shoreline. Oil is to be piped from W to a shore point B that is 8 km from A. It costs $1,000,000/km to lay pipe under water and $500,000/km over land. In your role as project manager you receive three proposals for piping the oil from W to B. Proposal 1 claims that it is cheapest to pipe directly from W to B, since the shortest distance between two points is a straight line. Proposal 2 claims that it is cheapest to pipe directly to point A and then along the shoreline to B, thereby using the least amount of expensive underwater pipe. Proposal 3 claims that it is cheapest to compromise by piping under water to some well-chosen point between A and B, and then piping along the shoreline to B. Which proposal is correct?

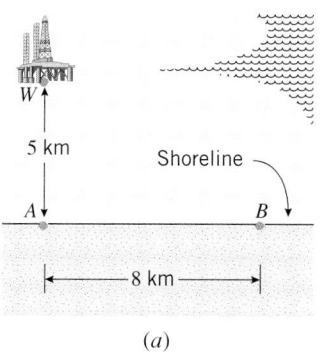
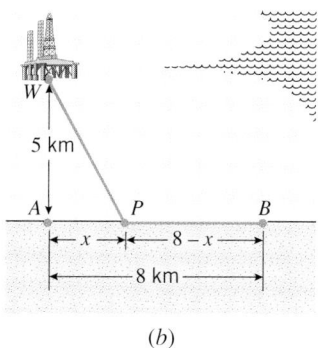
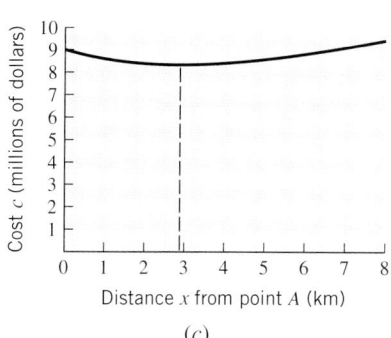

Figure 1.1.9

Solution. Let P be any point between A and B (Figure 1.1.9*b*), and let

$x =$ distance (in kilometers) between A and P

$c =$ cost (in millions of dollars) for the entire pipeline

Proposal 1 claims that $x = 8$ results in the least cost, Proposal 2 claims that it is $x = 0$, and Proposal 3 claims it is some value of x between 0 and 8. From Figure 1.1.9*b* the length of pipe along the shore is

$$8 - x \qquad\qquad (3)$$

and from the Theorem of Pythagoras, the length of pipe under water is

$$\sqrt{x^2 + 25} \qquad\qquad (4)$$

Thus, from (3) and (4) the total cost c (in millions of dollars) for the pipeline is

$$c = 1\left(\sqrt{x^2 + 25}\right) + 0.5(8 - x) = \sqrt{x^2 + 25} + 0.5(8 - x) \qquad\qquad (5)$$

where $0 \le x \le 8$. The graph of Equation (5), shown in Figure 1.1.9*c*, makes it clear that Proposal 3 is correct—the most cost-effective strategy is to pipe to a point a little less than 3 km from point A. ◀

EXERCISE SET 1.1 ⌇ Graphing Calculator

1. Use the cigarette consumption graph in Figure 1.1.2*b* to answer the following questions, making reasonable approximations where needed.
   (a) When did the annual cigarette consumption reach 3000 per adult for the first time?
   (b) When did the annual cigarette consumption per adult reach its peak, and what was the peak value?
   (c) Can you tell from the graph how many cigarettes were consumed in a given year? If not, what additional information would you need to make that determination?
   (d) What factors are likely to cause a sharp increase in annual cigarette consumption per adult?
   (e) What factors are likely to cause a sharp decline in annual cigarette consumption per adult?

2. The accompanying graph shows the median income in U.S. households (adjusted for inflation) between 1975 and 1995. Use the graph to answer the following questions, making reasonable approximations where needed.
   (a) When did the median income reach its maximum value, and what was the median income when that occurred?
   (b) When did the median income reach its minimum value, and what was the median income when that occurred?
   (c) The median income was declining during the 4-year period between 1989 and 1993. Was it declining more

rapidly during the first 2 years or the second 2 years of that period? Explain your reasoning.

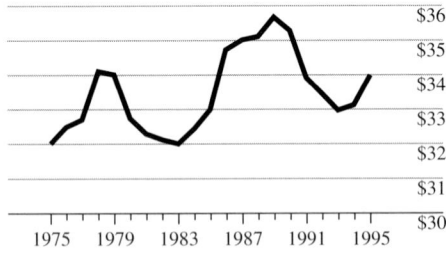

MEDIAN U.S. HOUSEHOLD INCOME IN
THOUSANDS OF CONSTANT 1995 DOLLARS

Source: Census Bureau, March 1996
[1996 measures 1995 income].

Figure Ex-2

3. Use the accompanying graph to answer the following questions, making reasonable approximations were needed.
   (a) For what values of x is $y = 1$?
   (b) For what values of x is $y = 3$?
   (c) For what values of y is $x = 3$?
   (d) For what values of x is $y \le 0$?
   (e) What are the maximum and minimum values of y and for what values of x do they occur?

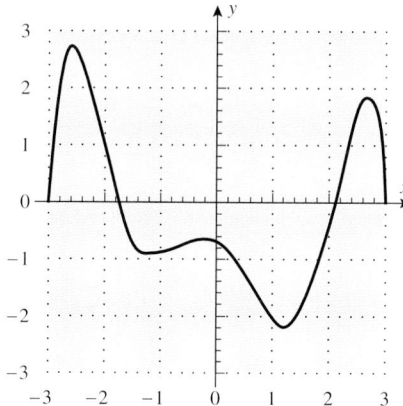

Figure Ex-3

4. Use the table in the accompanying figure to answer the questions posed in Exercise 3.

x	−2	−1	0	2	3	4	5	6
y	5	1	−2	7	−1	1	0	9

Figure Ex-4

5. Use the equation $y = x^2 - 6x + 8$ to answer the following questions.
(a) For what values of x is $y = 0$?
(b) For what values of x is $y = -10$?
(c) For what values of x is $y \geq 0$?
(d) Does y have a minimum value? A maximum value? If so, find them.

6. Use the equation $y = 1 + \sqrt{x}$ to answer the following questions.
(a) For what values of x is $y = 4$?
(b) For what values of x is $y = 0$?
(c) For what values of x is $y \geq 6$?
(d) Does y have a minimum value? A maximum value? If so, find them.

7. (a) If you had a device that could record the Earth's population continuously, would you expect the graph of population versus time to be a continuous (unbroken) curve? Explain what might cause breaks in the curve.
(b) Suppose that a hospital patient receives an injection of an antibiotic every 8 hours and that between injections the concentration C of the antibiotic in the bloodstream decreases as the antibiotic is absorbed by the tissues. What might the graph of C versus the elapsed time t look like?

8. (a) If you had a device that could record the temperature of a room continuously over a 24-hour period, would you expect the graph of temperature versus time to be a continuous (unbroken) curve? Explain your reasoning.
(b) If you had a computer that could track the number of boxes of cereal on the shelf of a market continuously over a 1-week period, would you expect the graph of the number of boxes on the shelf versus time to be a continuous (unbroken) curve? Explain your reasoning.

9. A construction company wants to build a rectangular enclosure with an area of 1000 square feet by fencing in three sides and using its office building as the fourth side. Your objective as supervising engineer is to design the enclosure so that it uses the least amount of fencing. Proceed as follows.
(a) Let x and y be the dimensions of the enclosure, and let L be the length of fencing required for those dimensions. Since the area must be 1000 square feet, we must have $xy = 1000$. Find a formula for L in terms of x and y, and then express L in terms of x alone by using the area equation.
(b) Are there any restrictions on the value of x? Explain.
(c) Make a graph of L versus x over a reasonable interval, and use the graph to estimate the value of x that results in the smallest value of L.
(d) Estimate the smallest value of L.

10. A manufacturer constructs open boxes from sheets of cardboard that are 6 inches square by cutting small squares from the corners and folding up the sides (as shown in the accompanying figure). The Research and Development Department asks you to determine the size of the square that produces a box of greatest volume. Proceed as follows.
(a) Let x be the length of a side of the square to be cut, and let V be the volume of the resulting box. Show that $V = x(6 - 2x)^2$.
(b) Are there any restrictions on the value of x? Explain.
(c) Make a graph of V versus x over an appropriate interval, and use the graph to estimate the value of x that results in the largest volume.
(d) Estimate the largest volume.

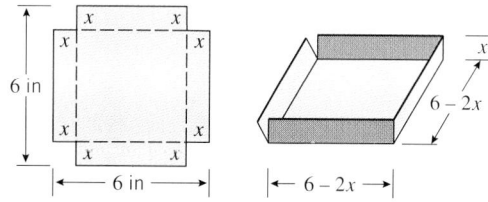

Figure Ex-10

11. A soup company wants to manufacture a can in the shape of a right circular cylinder that will hold 500 cm³ of liquid. The material for the top and bottom costs 0.02 cent/cm², and the material for the sides costs 0.01 cent/cm².
(a) Use the method of Exercises 9 and 10 to estimate the radius r and height h of the can that costs the least to manufacture. [*Suggestion:* Express the cost C in terms of r.]
(b) Suppose that the tops and bottoms of radius r are punched out from square sheets with sides of length $2r$ and the scraps are waste. If you allow for the cost of

the waste, would you expect the can of least cost to be taller or shorter than the one in part (a)? Explain.

(c) Estimate the radius, height, and cost of the can in part (b), and determine whether your conjecture was correct.

12. The designer of a sports facility wants to put a quarter-mile (1320 ft) running track around a football field, oriented as in the accompanying figure. The football field is 360 ft long (including the end zones) and 160 ft wide. The track consists of two straightaways and two semicircles.

(a) Show that it is possible to construct a quarter-mile track around the football field. [*Suggestion:* Find the shortest track that can be constructed around the field.]

(b) Let L be the length of a straightaway (in feet), and let x be the distance (in feet) between a sideline of the football field and a straightaway. Make a graph of L versus x.

(c) Use the graph to estimate the value of x that produces the shortest straightaways, and then find this value of x exactly.

(d) Use the graph to estimate the length of the longest possible straightaways, and then find that length exactly.

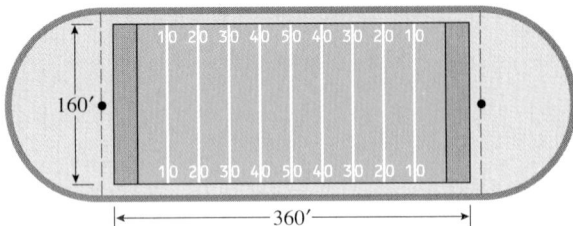

Figure Ex-12

1.2 PROPERTIES OF FUNCTIONS

In this section we will explore properties of functions in more detail. We will assume that you are familiar with the standard notation for intervals and the basic properties of absolute value. Reviews of these topics are provided in Appendices A and B.

INDEPENDENT AND DEPENDENT VARIABLES

Recall from the last section that a function f is a rule that associates a unique output $f(x)$ with each input x. This output is sometimes called the *value* of f at x or the *image* of x under f. Sometimes we will want to denote the output by a single letter, say y, and write

$$y = f(x)$$

This equation expresses y as a function of x; the variable x is called the *independent variable* (or *argument*) of f, and the variable y is called the *dependent variable* of f. This terminology is intended to suggest that x is free to vary, but that once x has a specific value a corresponding value of y is determined. For now we will only consider functions in which the independent and dependent variables are real numbers, in which case we say that f is a *real-valued function of a real variable*. Later, we will consider other kinds of functions as well.

Table 1.2.1 can be viewed as a numerical representation of a function of f. For this function we have

Table 1.2.1

x	0	1	2	3
y	3	4	-1	6

$f(0) = 3$ f associates $y = 3$ with $x = 0$.

$f(1) = 4$ f associates $y = 4$ with $x = 1$.

$f(2) = -1$ f associates $y = -1$ with $x = 2$.

$f(3) = 6$ f associates $y = 6$ with $x = 3$.

To illustrate how functions can be defined by equations, consider

$$y = 3x^2 - 4x + 2 \tag{1}$$

This equation has the form $y = f(x)$, where

$$f(x) = 3x^2 - 4x + 2 \tag{2}$$

The outputs of f (the y-values) are obtained by substituting numerical values for x in this formula. For example,

$$f(0) = 3(0)^2 - 4(0) + 2 = 2 \qquad \text{\small f associates $y = 2$ with $x = 0$.}$$

$$f(-1.7) = 3(-1.7)^2 - 4(-1.7) + 2 = 17.47 \qquad \text{\small f associates $y = 17.47$ with $x = -1.7$.}$$

$$f(\sqrt{2}) = 3(\sqrt{2})^2 - 4\sqrt{2} + 2 = 8 - 4\sqrt{2} \qquad \text{\small f associates $y = 8 - 4\sqrt{2}$ with $x = \sqrt{2}$.}$$

REMARK. Although f, x, and y are the most common notations for functions and variables, any letters can be used. For example, to indicate that the area A of a circle is a function of the radius r, it would be more natural to write $A = f(r)$ [where $f(r) = \pi r^2$]. Similarly, to indicate that the circumference C of a circle is a function of the radius r, we might write $C = g(r)$ [where $g(r) = 2\pi r$]. The area function and the circumference function are different, which is why we denoted them by different letters, f and g.

DOMAIN AND RANGE

If $y = f(x)$, then the set of all possible inputs (x-values) is called the ***domain*** of f, and the set of outputs (y-values) that result when x varies over the domain is called the ***range*** of f. For example, consider the equations

$$y = x^2 \quad \text{and} \quad y = x^2, \quad x \geq 2$$

In the first equation there is no restriction on x, so we may assume that any real value of x is an allowable input. Thus, the equation defines a function $f(x) = x^2$ with domain $-\infty < x < +\infty$. In the second equation, the inequality $x \geq 2$ restricts the allowable inputs to be greater than or equal to 2, so the equation defines a function $g(x) = x^2$, $x \geq 2$ with domain $2 \leq x < +\infty$.

As x varies over the domain of the function $f(x) = x^2$, the values of $y = x^2$ vary over the interval $0 \leq y < +\infty$, so this is the range of f. By comparison, as x varies over the domain of the function $g(x) = x^2$, $x \geq 2$, the values of $y = x^2$, $x \geq 2$ vary over the interval $4 \leq y < +\infty$, so this is the range of g.

It is important to understand here that even though $f(x) = x^2$ and $g(x) = x^2$, $x \geq 2$ involve the same formula, we regard them to be different functions because they have different domains. In short, *to fully describe a function you must not only specify the rule that relates the inputs and outputs, but you must also specify the domain, that is, the set of allowable inputs.*

GRAPHS OF FUNCTIONS

If f is a real-valued function of a real variable, then the ***graph*** of f in the xy-plane is defined to be the graph of the equation $y = f(x)$. For example, the graph of the function $f(x) = x$ is the graph of the equation $y = x$, shown in Figure 1.2.1. That figure also shows the graphs

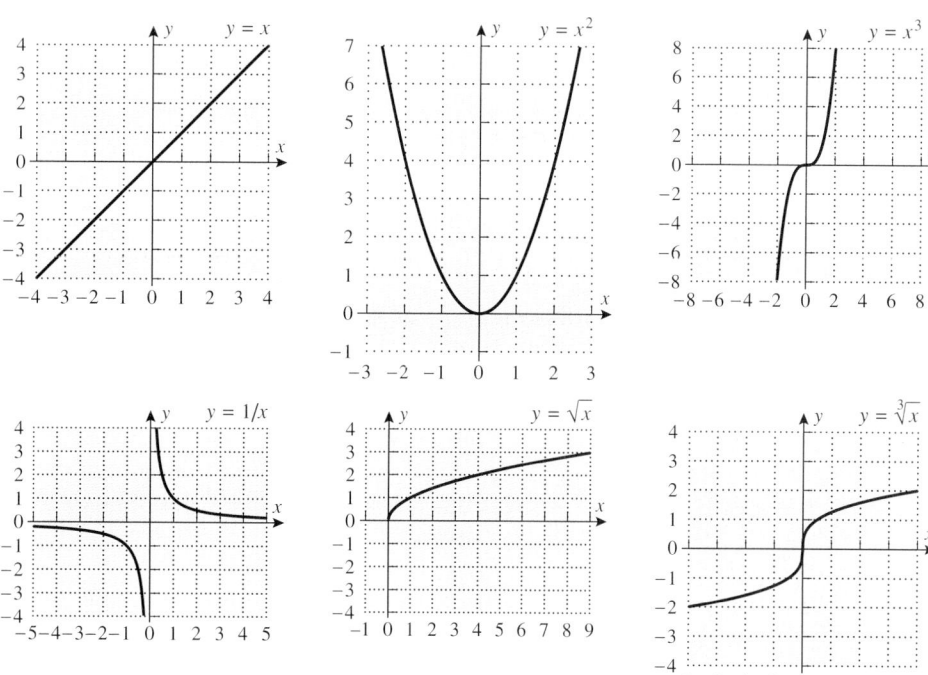

Figure 1.2.1

of some other basic functions that may already be familiar to you. Later in this chapter we will discuss techniques for graphing functions using graphing calculators and computers.

Graphs can provide useful visual information about a function. For example, because the graph of a function f in the xy-plane consists of all points whose coordinates satisfy the equation $y = f(x)$, the points on the graph of f are of the form $(x, f(x))$; hence each y-coordinate is the value of f at the x-coordinate (Figure 1.2.2a). Pictures of the domain and range of f can be obtained by projecting the graph of f onto the coordinate axes (Figure 1.2.2b). The values of x for which $f(x) = 0$ are the x-coordinates of the points where the graph of f intersects the x-axis (Figure 1.2.2c); these values of x are called the **zeros** of f, the **roots** of $f(x) = 0$, or the **x-intercepts** of $y = f(x)$.

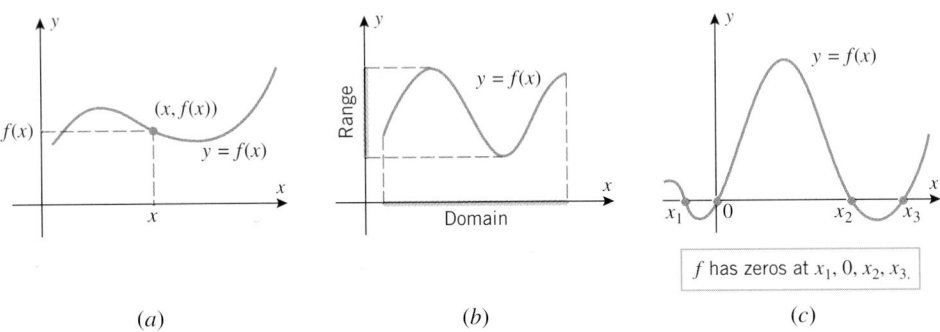

| (a) | (b) | (c) |

Figure 1.2.2

THE VERTICAL LINE TEST

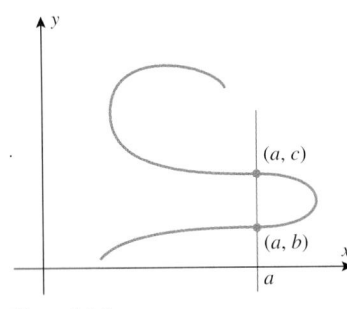

Figure 1.2.3

Not every curve in the xy-plane is the graph of a function. For example, consider the curve in Figure 1.2.3, which is cut at two distinct points, (a, b) and (a, c), by a vertical line. This curve cannot be the graph of $y = f(x)$ for any function f; otherwise, we would have

$$f(a) = b \quad \text{and} \quad f(a) = c$$

which is impossible, since f cannot assign two different values to a. Thus, there is no function f whose graph is the given curve. This illustrates the following general result, which we will call the **vertical line test**.

> **1.2.1**   THE VERTICAL LINE TEST.   *A curve in the xy-plane is the graph of some function f if and only if no vertical line intersects the curve more than once.*

Example 1

The graph of the equation

$$x^2 + y^2 = 25 \tag{3}$$

is a circle of radius 5, centered at the origin (see Appendix D for a review of circles), and hence there are vertical lines that cut the graph more than once. This can also be seen algebraically by solving (3) for y in terms of x:

$$y = \pm\sqrt{25 - x^2}$$

This equation does not define y as a function of x because the right side is "multiple valued" in the sense that values of x in the interval $(-5, 5)$ produce two corresponding values of y. For example, if $x = 4$, then $y = \pm 3$, and hence $(4, 3)$ and $(4, -3)$ are two points on the circle that lie on the same vertical line (Figure 1.2.4a). However, we can regard the circle as the union of two semicircles:

$$y = \sqrt{25 - x^2} \quad \text{and} \quad y = -\sqrt{25 - x^2}$$

(Figure 1.2.4b), each of which defines y as a function of x.   ◄

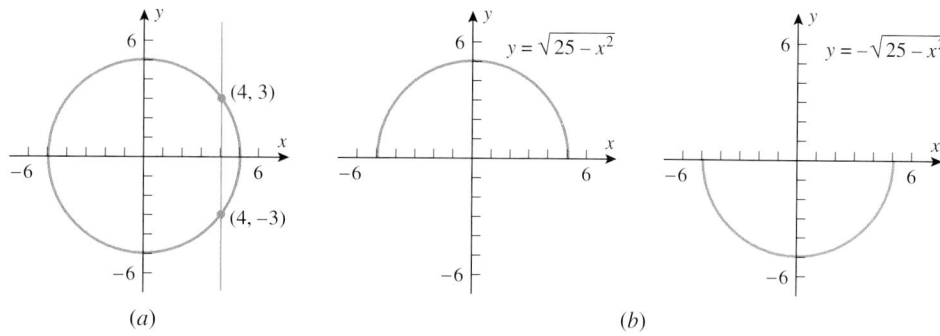

Figure 1.2.4

Recall that the ***absolute value*** or ***magnitude*** of a real number x is defined by

$$|x| = \begin{cases} x, & x \geq 0 \\ -x, & x < 0 \end{cases}$$

The effect of taking the absolute value of a number is to strip away the minus sign if the number is negative and to leave the number unchanged if it is nonnegative. Thus,

$$|5| = 5, \quad \left|-\tfrac{4}{7}\right| = \tfrac{4}{7}, \quad |0| = 0$$

A more detailed discussion of the properties of absolute value is given in Appendix B. However, for convenience we provide the following summary of its algebraic properties.

1.2.2 PROPERTIES OF ABSOLUTE VALUE. *If a and b are real numbers, then*

(a) $|-a| = |a|$ A number and its negative have the same absolute value.

(b) $|ab| = |a|\,|b|$ The absolute value of a product is the product of the absolute values.

(c) $|a/b| = |a|/|b|$ The absolute value of a ratio is the ratio of the absolute values.

(d) $|a + b| \leq |a| + |b|$ The ***triangle inequality***

REMARK. Symbols such as $+x$ and $-x$ are deceptive, since it is tempting to conclude that $+x$ is positive and $-x$ is negative. However, this need not be so, since x itself can be positive or negative. For example, if x is negative, say $x = -3$, then $-x = 3$ is positive and $+x = -3$ is negative.

The graph of the function $f(x) = |x|$ can be obtained by graphing the two parts of the equation

$$y = \begin{cases} x, & x \geq 0 \\ -x, & x < 0 \end{cases}$$

separately. For $x \geq 0$, the graph of $y = x$ is a ray of slope 1 with its endpoint at the origin, and for $x < 0$, the graph of $y = -x$ is a ray of slope -1 with its endpoint at the origin. Combining the two parts produces the V-shaped graph in Figure 1.2.5.

Absolute values have important relationships to square roots. To see why this is so, recall from algebra that every positive real number x has two square roots, one positive and one negative. By definition, the symbol $\sqrt{x}$ denotes the *positive* square root of x. To denote the negative square root you must write $-\sqrt{x}$. For example, the positive square root of 9 is $\sqrt{9} = 3$, and the negative square root is $-\sqrt{9} = -3$. (Do not make the mistake of writing $\sqrt{9} = \pm 3$.)

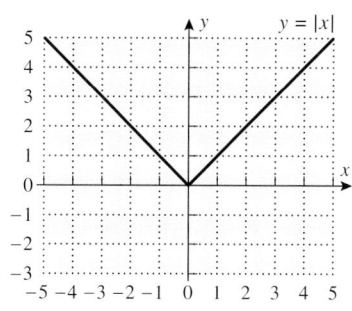

Figure 1.2.5

Care must be exercised in simplifying expressions of the form $\sqrt{x^2}$, since it is *not* always true that $\sqrt{x^2} = x$. This equation is correct if x is nonnegative, but it is false for negative x. For example, if $x = -4$, then

$$\sqrt{x^2} = \sqrt{(-4)^2} = \sqrt{16} = 4 \neq x$$

A statement that is correct for all real values of x is

$$\sqrt{x^2} = |x|$$

FOR THE READER.   Verify this relationship by using a graphing utility to show that the equations $y = \sqrt{x^2}$ and $y = |x|$ have the same graph.

....

FUNCTIONS DEFINED PIECEWISE

The absolute value function $f(x) = |x|$ is an example of a function that is defined *piecewise* in the sense that the formula for f changes, depending on the value of x.

Example 2

Sketch the graph of the function defined piecewise by the formula

$$f(x) = \begin{cases} 0, & x \leq -1 \\ \sqrt{1 - x^2}, & -1 < x < 1 \\ x, & x \geq 1 \end{cases}$$

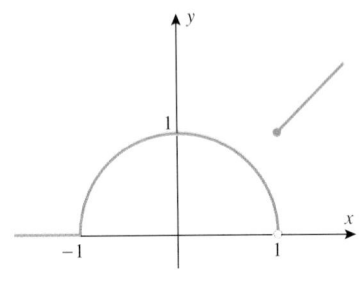

Figure 1.2.6

Solution. The formula for f changes at the points $x = -1$ and $x = 1$. (We call these the **breakpoints** for the formula.) A good procedure for graphing functions defined piecewise is to graph the function separately over the open intervals determined by the breakpoints, and then graph f at the breakpoints themselves. For the function f in this example the graph is the horizontal line segment $y = 0$ on the interval $(-\infty, -1)$, it is the semicircle $y = \sqrt{1 - x^2}$ on the interval $(-1, 1)$, and it is the line segment $y = x$ on the interval $(1, +\infty)$. The formula for f specifies that the equation $y = 0$ applies at the breakpoint -1 [so $y = f(-1) = 0$], and it specifies that the equation $y = x$ applies at the breakpoint 1 [so $y = f(1) = 1$]. The graph of f is shown in Figure 1.2.6. ◄

REMARK.   In Figure 1.2.6 the solid dot and open circle at the breakpoint $x = 1$ serve to emphasize that the point on the graph lies on the line segment and not the semicircle. There is no ambiguity at the breakpoint $x = -1$ because the two parts of the graph join together continuously there.

Example 3

Increasing the speed at which air moves over a person's skin increases the rate of moisture evaporation and makes the person feel cooler. (This is why we fan ourselves in hot weather.) The **windchill index** is the temperature at a wind speed of 4 mi/h that would produce the same sensation on exposed skin as the current temperature and wind speed combination. An empirical formula (i.e., a formula based on experimental data) for the windchill index W at $32°F$ for a wind speed of v mi/h is

$$W = \begin{cases} 32, & 0 \leq v \leq 4 \\ 91.4 + 59.4(0.0203v - 0.304\sqrt{v} - 0.474), & 4 < v < 45 \\ -3.6, & v \geq 45 \end{cases}$$

A computer-generated graph of $W(v)$ is shown in Figure 1.2.7.   ◄

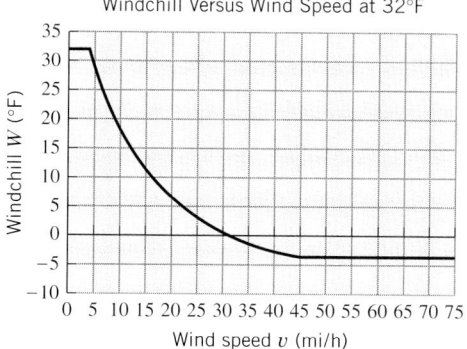

Figure 1.2.7

THE NATURAL DOMAIN

Sometimes, restrictions on the allowable values of an independent variable result from a mathematical formula that defines the function. For example, if $f(x) = 1/x$, then $x = 0$ must be excluded from the domain to avoid division by zero, and if $f(x) = \sqrt{x}$, then negative values of x must be excluded from the domain, since we are only considering real-valued functions of a real variable for now. We make the following definition.

> **1.2.3   DEFINITION.**   If a real-valued function of a real variable is defined by a formula, and if no domain is stated explicitly, then it is to be understood that the domain consists of all real numbers for which the formula yields a real value. This is called the ***natural domain*** of the function.

Example 4

Find the natural domain of

(a) $f(x) = x^3$ (b) $f(x) = 1/(x-1)(x-3)$

(c) $f(x) = \tan x$ (d) $f(x) = \sqrt{x^2 - 5x + 6}$

Solution (a). The function f has real values for all real x, so its natural domain is the interval $(-\infty, +\infty)$.

Solution (b). The function f has real values for all real x, except $x = 1$ and $x = 3$, where divisions by zero occur. Thus, the natural domain is

$$\{x : x \neq 1 \text{ and } x \neq 3\} = (-\infty, 1) \cup (1, 3) \cup (3, +\infty)$$

Solution (c). Since $f(x) = \tan x = \sin x / \cos x$, the function f has real values except where $\cos x = 0$, and this occurs when x is an odd integer multiple of $\pi/2$. Thus, the natural domain consists of all real numbers except

$$x = \pm\frac{\pi}{2}, \pm\frac{3\pi}{2}, \pm\frac{5\pi}{2}, \dots$$

Solution (d). The function f has real values, except when the expression inside the radical is negative. Thus the natural domain consists of all real numbers x such that

$$x^2 - 5x + 6 = (x-3)(x-2) \geq 0$$

This inequality is satisfied if $x \leq 2$ or $x \geq 3$ (verify), so the natural domain of f is

$$(-\infty, 2] \cup [3, +\infty)$$ ◀

REMARK.   In some problems we will want to limit the domain of a function by imposing specific restrictions. For example, by writing

$$f(x) = x^2, \quad x \geq 0$$

we can limit the domain of f to the positive x-axis (Figure 1.2.8).

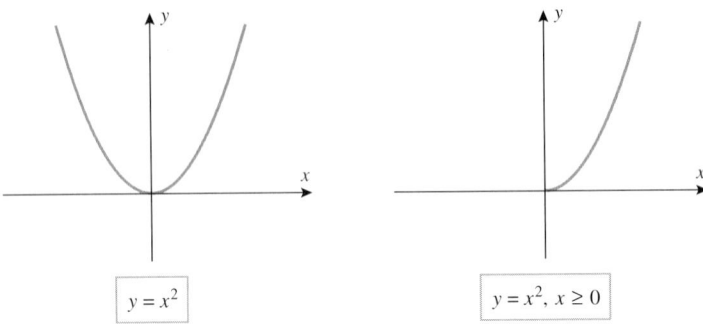

Figure 1.2.8

THE EFFECT OF ALGEBRAIC OPERATIONS ON THE DOMAIN

Algebraic expressions are frequently simplified by canceling common factors in the numerator and denominator. However, care must be exercised when simplifying formulas for functions in this way, since this process can alter the domain.

Example 5

The natural domain of the function

$$f(x) = \frac{x^2 - 4}{x - 2}$$

consists of all real x except $x = 2$. However, if we factor the numerator and then cancel the common factor in the numerator and denominator, we obtain

$$f(x) = \frac{(x - 2)(x + 2)}{x - 2} = x + 2$$

which *is* defined at $x = 2$ [since $f(2) = 4$ for the altered function f]. Thus, the algebraic simplification has altered the domain of the function. Geometrically, the graph of $y = x + 2$ is a line of slope 1 and y-intercept 2, whereas the graph of $y = (x^2 - 4)/(x - 2)$ is the same line, but with a hole in it at $x = 2$, since y is undefined there (Figure 1.2.9). Thus, the geometric effect of the algebraic cancellation is to eliminate the hole in the original graph. In some situations such minor alterations in the domain are irrelevant to the problem under consideration and can be ignored. However, if we wanted to preserve the domain in this example, then we would express the simplified form of the function as

$$f(x) = x + 2, \quad x \neq 2 \qquad \blacktriangleleft$$

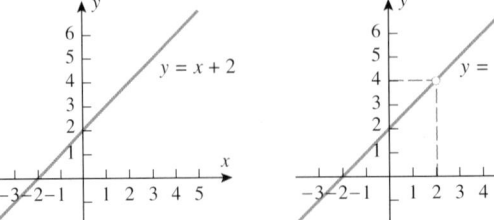

Figure 1.2.9

Example 6

Find the domain and range of

$\qquad$ (a) $f(x) = 2 + \sqrt{x-1}$ $\qquad$ (b) $f(x) = (x+1)/(x-1)$

Solution (a). Since no domain is stated explicitly, the domain of f is the natural domain $[1, +\infty)$. To determine the range, it will be convenient to introduce a dependent variable $y = 2 + \sqrt{x-1}$. As x varies over the interval $[1, +\infty)$, the value of $\sqrt{x-1}$ varies over the interval $[0, +\infty)$, so the value of $y = 2 + \sqrt{x-1}$ varies over the interval $[2, +\infty)$, which is the range of f. The domain and range are shown graphically in Figure 1.2.10a.

Solution (b). The given function f is defined for all real x, except $x = 1$, so the natural domain of f is

$\qquad$ $\{x : x \neq 1\} = (-\infty, 1) \cup (1, +\infty)$

As in the preceding part of this example, it will be convenient to introduce a dependent variable

$$y = \frac{x+1}{x-1} \qquad\qquad (4)$$

Although the set of possible y-values is not immediately evident from this equation, the graph of (4), which is shown in Figure 1.2.10b, suggests that the range of f consists of all y, except $y = 1$. To see that this is so, we solve (4) for x in terms of y:

$\qquad$ $(x-1)y = x + 1$

$\qquad\quad$ $xy - y = x + 1$

$\qquad\quad$ $xy - x = y + 1$

$\qquad$ $x(y-1) = y + 1$

$\qquad\qquad$ $x = \dfrac{y+1}{y-1}$

It is now evident from the right side of this equation that $y = 1$ is not in the range; otherwise we would have a division by zero. No other values of y are excluded by this equation, so the range of the function f is $\{y : y \neq 1\} = (-\infty, 1) \cup (1, +\infty)$, which agrees with the result obtained graphically. ◀

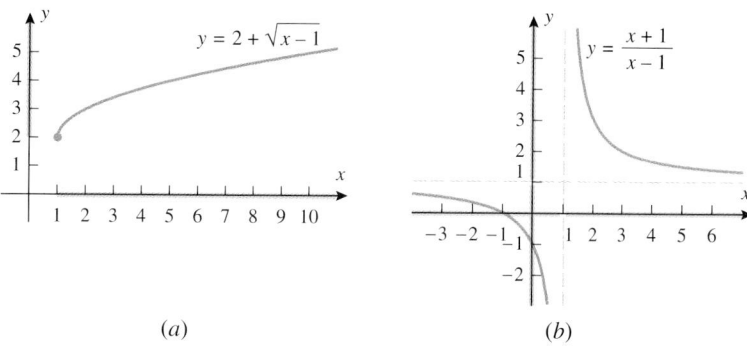

$\qquad\qquad\qquad$ (a) $\qquad\qquad\qquad\qquad\qquad\qquad\qquad\qquad$ (b)

Figure 1.2.10

In applications, physical considerations often impose restrictions on the domain and range of a function.

Example 7

An open box is to be made from a 16 in by 30 in piece of cardboard by cutting out squares of equal size from the four corners and bending up the sides (Figure 1.2.11a).

(a) Let V be the volume of the box that results when the squares have sides of length x. Find a formula for V as a function of x.

(b) Find the domain of V.

(c) Use the graph of V given in Figure 1.2.11c to estimate the range of V.

(d) Describe in words what the graph tells you about the volume.

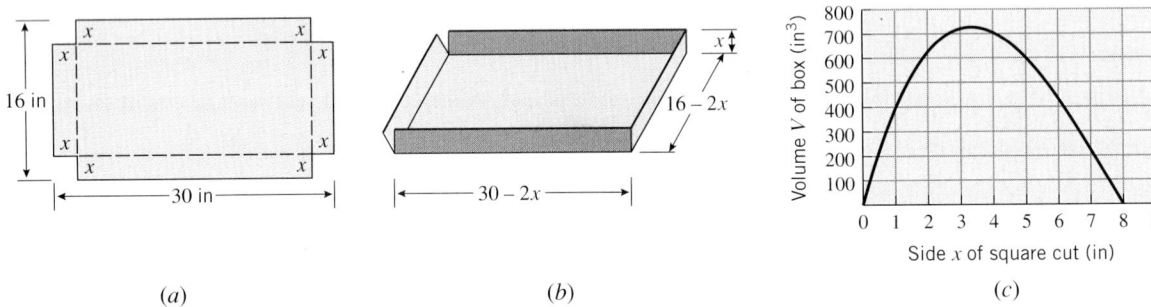

(a) (b) (c)

Figure 1.2.11

Solution (a). As shown in Figure 1.2.11b, the resulting box has dimensions $16 - 2x$ by $30 - 2x$ by x, so the volume $V(x)$ is given by

$$V(x) = (16 - 2x)(30 - 2x)x = 480x - 92x^2 + 4x^3$$

Solution (b). The domain is the set of x-values and the range is the set of V-values. Because x is a length, it must be nonnegative, and because we cannot cut out squares whose sides are more than 8 in long (why?), the x-values in the domain must satisfy

$$0 \le x \le 8$$

Solution (c). From the graph of V versus x in Figure 1.2.11c we estimate that the V-values in the range satisfy

$$0 \le V \le 725$$

Note that this is an approximation. Later we will show how to find the range exactly.

Solution (d). The graph tells us that the box of maximum volume occurs for a value of x that is between 3 and 4 and that the maximum volume is approximately 725 in^3. Moreover, the volume decreases toward zero as x gets closer to 0 or 8. ◄

In applications involving time, formulas for functions are often expressed in terms of a variable t whose starting value is taken to be $t = 0$.

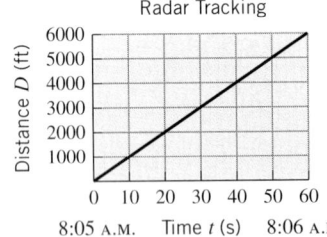

Radar Tracking

Figure 1.2.12

Example 8

At 8:05 A.M. a car is clocked at 100 ft/s by a radar detector that is positioned at the edge of a straight highway. Assuming that the car maintains a constant speed between 8:05 A.M. and 8:06 A.M., find a function $D(t)$ that expresses the distance traveled by the car during that time interval as a function of the time t.

Solution. It would be clumsy to use clock time for the variable t, so let us agree to measure the elapsed time in seconds, starting with $t = 0$ at 8:05 A.M. and ending with $t = 60$ at

8:06 A.M. At each instant, the distance traveled (in ft) is equal to the speed of the car (in ft/s) multiplied by the elapsed time (in s). Thus,

$$D(t) = 100t, \quad 0 \le t \le 60$$

The graph of D versus t is shown in Figure 1.2.12. ◀

ISSUES OF SCALE AND UNITS

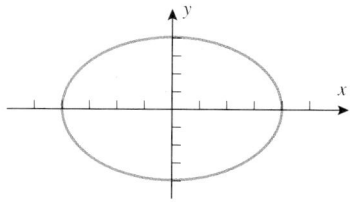

The circle is squashed because 1 unit on the y-axis has a smaller length than 1 unit on the x-axis.

Figure 1.2.13

In geometric problems where you want to preserve the "true" shape of a graph, you must use units of equal length on both axes. For example, if you graph a circle in a coordinate system in which 1 unit in the y-direction is smaller than 1 unit in the x-direction, then the circle will be squashed vertically into an elliptical shape (Figure 1.2.13). You must also use units of equal length when you want to apply the distance formula

$$d = \sqrt{(x_2 - x_1)^2 + (y_2 - y_1)^2}$$

to calculate the distance between two points (x_1, y_1) and (x_2, y_2) in the xy-plane.

However, sometimes it is inconvenient or impossible to display a graph using units of equal length. For example, consider the equation

$$y = x^2$$

If we want to show the portion of the graph over the interval $-3 \le x \le 3$, then there is no problem using units of equal length, since y only varies from 0 to 9 over that interval. However, if we want to show the portion of the graph over the interval $-10 \le x \le 10$, then there is a problem keeping the units equal in length, since the value of y varies between 0 and 100. In this case the only reasonable way to show all of the graph that occurs over the interval $-10 \le x \le 10$ is to compress the unit of length along the y-axis, as illustrated in Figure 1.2.14.

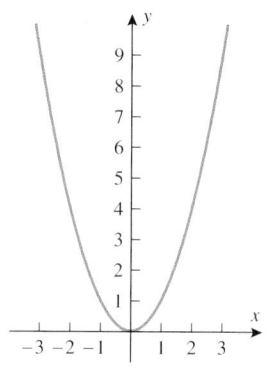
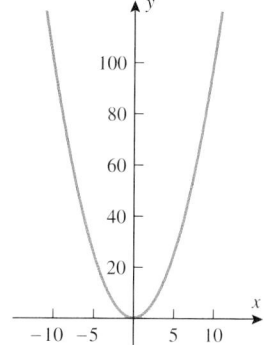

Figure 1.2.14

REMARK. In applications where the variables on the two axes have unrelated units (say, centimeters on the y-axis and seconds on the x-axis), then nothing is gained by requiring the units to have equal lengths; choose the lengths to make the graph as clear as possible.

EXERCISE SET 1.2 ☑ Graphing Calculator

1. Find $f(0)$, $f(2)$, $f(-2)$, $f(3)$, $f(\sqrt{2}\,)$, and $f(3t)$.

  (a) $f(x) = 3x^2 - 2$

  (b) $f(x) = \begin{cases} \dfrac{1}{x}, & x > 3 \\ 2x, & x \le 3 \end{cases}$

2. Find $g(3)$, $g(-1)$, $g(\pi)$, $g(-1.1)$, and $g(t^2 - 1)$.

  (a) $g(x) = \dfrac{x+1}{x-1}$

  (b) $g(x) = \begin{cases} \sqrt{x+1}, & x \ge 1 \\ 3, & x < 1 \end{cases}$

In Exercises 3–6, find the natural domain of the function algebraically, and confirm that your result is consistent with the graph produced by your graphing utility. [*Note:* Set your graphing utility to the radian mode when graphing trigonometric functions.]

3. (a) $f(x) = \dfrac{1}{x - 3}$ (b) $g(x) = \sqrt{x^2 - 3}$

(c) $G(x) = \sqrt{x^2 - 2x + 5}$ (d) $f(x) = \dfrac{x}{|x|}$

(e) $h(x) = \dfrac{1}{1 - \sin x}$

4. (a) $f(x) = \dfrac{1}{5x + 7}$ (b) $h(x) = \sqrt{x - 3x^2}$

(c) $G(x) = \sqrt{\dfrac{x^2 - 4}{x - 4}}$ (d) $f(x) = \dfrac{x^2 - 1}{x + 1}$

(e) $h(x) = \dfrac{3}{2 - \cos x}$

5. (a) $f(x) = \sqrt{3 - x}$ (b) $g(x) = \sqrt{4 - x^2}$
(c) $h(x) = 3 + \sqrt{x}$ (d) $G(x) = x^3 + 2$
(e) $H(x) = 3 \sin x$

6. (a) $f(x) = \sqrt{3x - 2}$ (b) $g(x) = \sqrt{9 - 4x^2}$

(c) $h(x) = \dfrac{1}{3 + \sqrt{x}}$ (d) $G(x) = \dfrac{3}{x}$

(e) $H(x) = \sin^2 \sqrt{x}$

7. In each part of the accompanying figure, determine whether the graph defines y as a function of x.

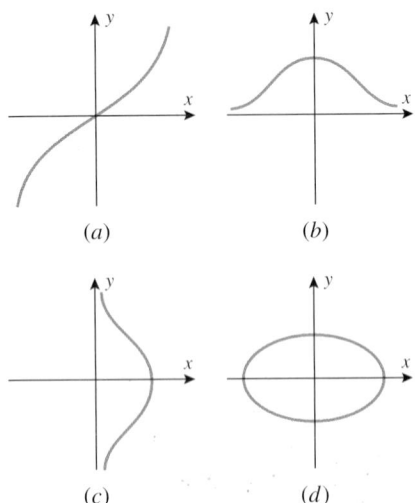

Figure Ex-7

8. Express the length L of a chord of a circle with radius 10 cm as a function of the central angle θ (see the accompanying figure).

9. As shown in the accompanying figure, a pendulum of constant length L makes an angle θ with its vertical position. Express the height h as a function of the angle θ.

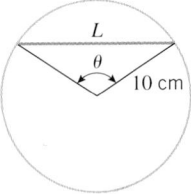

Figure Ex-8

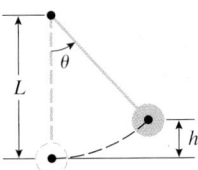

Figure Ex-9

10. A cup of hot coffee sits on a table. You pour in some cool milk and let it sit for an hour. Sketch a rough graph of the temperature of the coffee as a function of time.

11. A boat is bobbing up and down on some gentle waves. Suddenly it gets hit by a large wave and sinks. Sketch a rough graph of the height of the boat above the ocean floor as a function of time.

12. Make a rough sketch of your weight as a function of time from birth to the present.

In Exercises 13 and 14, express the function in piecewise form without using absolute values. [*Suggestion:* It may help to generate the graph of the function.]

13. (a) $f(x) = |x| + 3x + 1$ (b) $g(x) = |x| + |x - 1|$

14. (a) $f(x) = 3 + |2x - 5|$ (b) $g(x) = 3|x - 2| - |x + 1|$

15. As shown in the accompanying figure, an open box is to be constructed from a rectangular sheet of metal, 8 inches by 15 inches, by cutting out squares with sides of length x from each corner and bending up the sides.
  (a) Express the volume V as a function of x.
  (b) Find the natural domain and the range of the function, ignoring any physical restrictions on the values of the variables.
  (c) Modify the domain and range appropriately to account for the physical restrictions on the values of V and x.
  (d) In words, describe how the volume V of the box varies with x, and discuss how one might construct boxes of maximum volume and minimum volume.

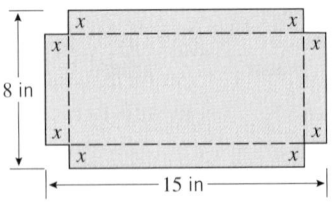

Figure Ex-15

16. As shown in the accompanying figure, a camera is mounted at a point 3000 ft from the base of a rocket launching pad.

The shuttle rises vertically when launched, and the camera's elevation angle is constantly adjusted to follow the bottom of the rocket.

(a) Choose letters to represent the height of the rocket and the elevation angle of the camera, and express the height as a function of the elevation angle.

(b) Find the natural domain and the range of the function, ignoring any physical restrictions on the values of the variables.

(c) Modify the domain and range appropriately to account for the physical restrictions on the values of the variables.

(d) Generate the graph of height versus the elevation on a graphing utility, and use it to estimate the height of the rocket when the elevation angle is $\pi/4 \approx 0.7854$ radian. Compare this estimate to the exact height. [*Suggestion:* If you are using a graphing calculator, the trace and zoom features will be helpful here.

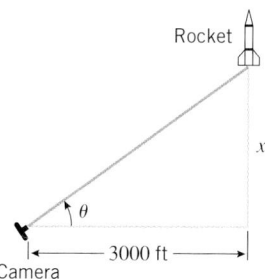

Camera

Figure Ex-16

In Exercises 17 and 18: (i) Explain why the function f has one or more holes in its graph, and state the x-values at which those holes occur. (ii) Find a function g whose graph is identical to that of f, but without the holes.

17. $f(x) = \dfrac{(x + 2)(x^2 - 1)}{(x + 2)(x - 1)}$ **18.** $f(x) = \dfrac{x + \sqrt{x}}{\sqrt{x}}$

19. For a given outside temperature T and wind speed v, the windchill index (WCI) is the equivalent temperature that exposed skin would feel with a wind speed of 4 mi/h. An empirical formula for the WCI (based on experience and observation) is

$$\text{WCI} = \begin{cases} T, & 0 \le v \le 4 \\ 91.4 + (91.4 - T)(0.0203v - 0.304\sqrt{v} - 0.474), & 4 < v < 45 \\ 1.6T - 55, & v \ge 45 \end{cases}$$

where T is the air temperature in $^\circ$F, v is the wind speed in mi/h, and WCI is the equivalent temperature in $^\circ$F. Find the WCI to the nearest degree if the air temperature is 25°F and

(a) $v = 3$ mi/h   (b) $v = 15$ mi/h
(c) $v = 46$ mi/h.

[Adapted from UMAP Module 658, *Windchill*, W. Bosch and L. Cobb, COMAP, Arlington, MA.]

In Exercises 20–22, use the formula for the windchill index described in Exercise 19.

20. Find the air temperature to the nearest degree if the WCI is reported as -60°F with a wind speed of 48 mi/h.

21. Find the air temperature to the nearest degree if the WCI is reported as -10°F with a wind speed of 8 mi/h.

22. Find the wind speed to the nearest mile per hour if the WCI is reported as -15°F with an air temperature of 20°F.

23. At 9:23 A.M. a lunar lander that is 1000 ft above the Moon's surface begins a vertical descent, touching down at 10:13 A.M. Assuming that the lander maintains a constant speed, find a function $D(t)$ that expresses the altitude of the lander above the Moon's surface as a function of t.

1.3 GRAPHING FUNCTIONS ON CALCULATORS AND COMPUTERS; COMPUTER ALGEBRA SYSTEMS

In this section we will discuss issues that relate to generating graphs of equations and functions with graphing utilities (graphing calculators and computers). Because graphing utilities vary widely, it is difficult to make general statements about them. Therefore, at various places in this section we will ask you to refer to the documentation for your own graphing utility for specific details about the way it operates.

GRAPHING CALCULATORS AND COMPUTER ALGEBRA SYSTEMS

The development of new technology has significantly changed how and where mathematicians, engineers, and scientists perform their work, as well as their approach to problem solving. Not only have portable computers and handheld calculators with graphing capabilities become standard tools in the scientific community, but there have been major new innovations in computer software. Among the most significant of these innovations are programs called **Computer Algebra Systems** (abbreviated CAS), the most common being *Mathematica*, *Maple*, and *Derive*.[*] Computer algebra systems not only have powerful graphing capabilities, but, as their name suggests, they can perform many of the symbolic computations that occur in algebra, calculus, and branches of higher mathematics. For example, it is a trivial task for a CAS to perform the factorization

$$x^6 + 23x^5 + 147x^4 - 139x^3 - 3464x^2 - 2112x + 23040 = (x + 5)(x - 3)^2(x + 8)^3$$

or the exact numerical computation

$$\left(\frac{63456}{3177295} - \frac{43907}{22854377} \right)^3 = \frac{22519124571642082912593202301228669 23}{382895955819369204449565945369203764688375}$$

Technology has also made it possible to generate graphs of equations and functions in seconds that in the past might have taken hours to produce. Graphing technology includes handheld graphing calculators, computer algebra systems, and software designed for that purpose. Figure 1.3.1 shows the graphs of the function $f(x) = x^4 - x^3 - 2x^2$ produced with various graphing utilities; the first two were generated with the CAS programs, *Mathematica* and *Maple*, and the third with a graphing calculator. Graphing calculators produce coarser graphs than most computer programs but have the advantage of being compact and portable.

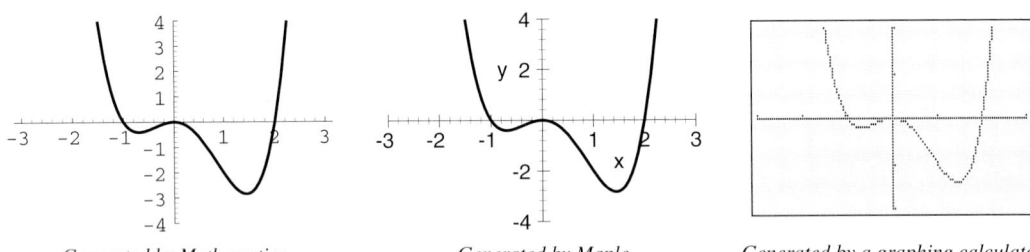

Generated by Mathematica   Generated by Maple   Generated by a graphing calculator

Figure 1.3.1

VIEWING WINDOWS

Graphing utilities can only show a portion of the *xy*-plane in the viewing screen, so the first step in graphing an equation is to determine which rectangular portion of the *xy*-plane you want to display. This region is called the *viewing window* (or *viewing rectangle*). For example, in Figure 1.3.1 the viewing window extends over the interval $[-3, 3]$ in the *x*-direction and over the interval $[-4, 4]$ in the *y*-direction, so we say that the viewing window is $[-3, 3] \times [-4, 4]$ (read "$[-3, 3]$ by $[-4, 4]$"). In general, if the viewing window is $[a, b] \times [c, d]$, then the window extends between $x = a$ and $x = b$ in the *x*-direction and between $y = c$ and $y = d$ in the *y*-direction. We will call $[a, b]$ the *x-interval* for the window and $[c, d]$ the *y-interval* for the window (Figure 1.3.2).

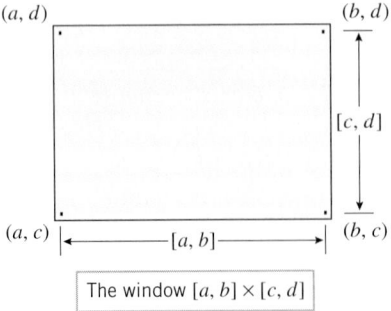

The window $[a, b] \times [c, d]$

Figure 1.3.2

[*]*Mathematica* is a product of Wolfram Research, Inc.; *Maple* is a product of Waterloo Maple Software, Inc.; and *Derive* is a product of Soft Warehouse, Inc.

Different graphing utilities designate viewing windows in different ways. For example, the first two graphs in Figure 1.3.1 were produced by the commands

```
Plot[x^4 - x^3 -2*x^2, {x, -3, 3}, PlotRange->{-4, 4}]
```
(*Mathematica*)

```
plot( x^4 - x^3 -2*x^2, x = -3..3, y = -4..4);
```
(*Maple*)

and the last graph was produced on a graphing calculator by pressing the GRAPH button after setting the following values for the variables that determine the x-interval and y-intervals:

$$x\text{Min} = -3, \quad x\text{Max} = 3, \quad y\text{Min} = -4, \quad y\text{Max} = 4$$

FOR THE READER.   Use your own graphing utility to generate the graph of the function $f(x) = x^4 - x^3 - 2x^2$ in the window $[-3, 3] \times [-4, 4]$.

TICK MARKS AND GRID LINES

To help locate points in a viewing window visually, graphing utilities provide methods for drawing ***tick marks*** (also called ***scale marks***) on the coordinate axes or at other locations in the viewing window. With computer programs such as *Mathematica* and *Maple*, there are specific commands for designating the spacing between tick marks, but if the user does not specify the spacing, then the programs make certain *default* choices. For example, in the first two parts of Figure 1.3.1, the tick marks shown were the default choices.

On graphing calculators the spacing between tick marks is determined by two ***scale variables*** (also called ***scale factors***), which we will denote by

$$x\text{Scl} \quad \text{and} \quad y\text{Scl}$$

(The notation varies among calculators.) These variables specify the spacing between the tick marks in the x- and y-directions, respectively. For example, in the third part of Figure 1.3.1 the window and tick marks were designated by the settings

$$x\text{Min} = -3 \qquad x\text{Max} = 3$$
$$y\text{Min} = -4 \qquad y\text{Max} = 4$$
$$x\text{Scl} = 1 \qquad y\text{Scl} = 1$$

Most graphing utilities allow for variations in the design and positioning of tick marks. For example, Figure 1.3.3 shows two variations of the graphs in Figure 1.3.1; the first was generated on a computer using an option for placing the ticks and numbers on the edges of a box, and the second was generated on a graphing calculator using an option for drawing grid lines to simulate graph paper.

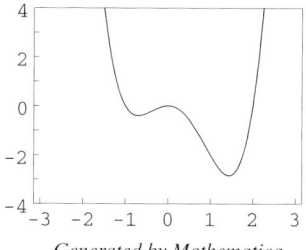
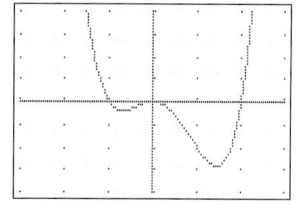

Generated by Mathematica   *Generated by a graphing calculator*

Figure 1.3.3

Example 1

Figure 1.3.4*a* shows the window $[-5, 5] \times [-5, 5]$ with the tick marks spaced .5 unit apart in the x-direction and 10 units apart in the y-direction. Note that no tick marks are actually visible in the y-direction because the tick mark at the origin is covered by the x-axis, and all other tick marks in the y-direction fall outside of the viewing window. ◄

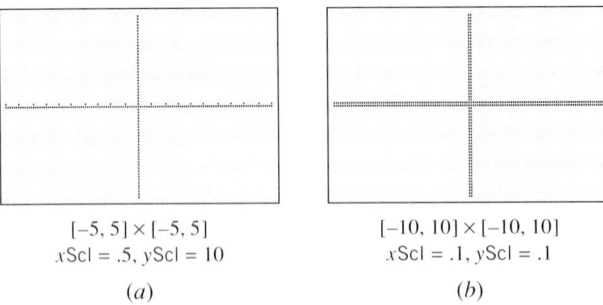

$[-5, 5] \times [-5, 5]$
$x\mathrm{Scl} = .5, y\mathrm{Scl} = 10$

(a)

$[-10, 10] \times [-10, 10]$
$x\mathrm{Scl} = .1, y\mathrm{Scl} = .1$

(b)

Figure 1.3.4

Example 2

Figure 1.3.4b shows the window $[-10, 10] \times [-10, 10]$ with the tick marks spaced .1 unit apart in the x- and y-directions. In this case the tick marks are so close together that they create the effect of thick lines on the coordinate axes. When this occurs you will usually want to increase the scale factors to reduce the number of tick marks and make them legible. ◄

FOR THE READER. Graphing calculators provide a way of clearing all settings and returning them to *default values*. For example, on the author's calculator the default window is $[-10, 10] \times [-10, 10]$ and the default scale factors are $x\mathrm{Scl} = 1$ and $y\mathrm{Scl} = 1$. Check your documentation to determine the default values for your calculator and how to reset the calculator to its default configuration. If you are using a computer program, check your documentation to determine the commands for specifying the spacing between tick marks.

CHOOSING A VIEWING WINDOW

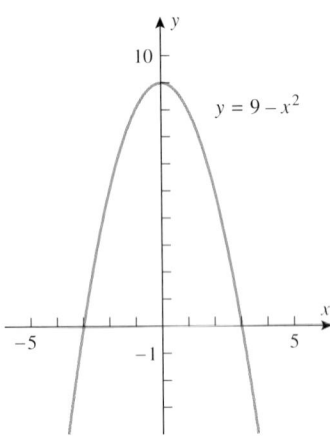

Figure 1.3.5

When the graph of a function extends indefinitely in some direction, no single viewing window can show the entire graph. In such cases the choice of the viewing window can drastically affect one's perception of how the graph looks. For example, Figure 1.3.5 shows a computer-generated graph of $y = 9 - x^2$, and Figure 1.3.6 shows four views of this graph generated on the author's calculator:

- In part (a) the graph falls completely outside of the window, so the window is blank (except for the ticks and axes).
- In part (b) the graph is broken into two pieces because it passes in and out of the window.
- In part (c) the graph appears to be a straight line because we have zoomed in on such a small segment of the curve.
- In part (d) we have a more complete picture of the graph shape because the window encompasses all of the important points, namely the high point on the graph and the intersections with the x-axis.

For a function whose graph does not extend indefinitely in either the x- or y-directions, the domain and range of the function can be used to obtain a viewing window that contains the entire graph.

Example 3

Use the domain and range of the function $f(x) = \sqrt{12 - 3x^2}$ to determine a viewing window that contains the entire graph.

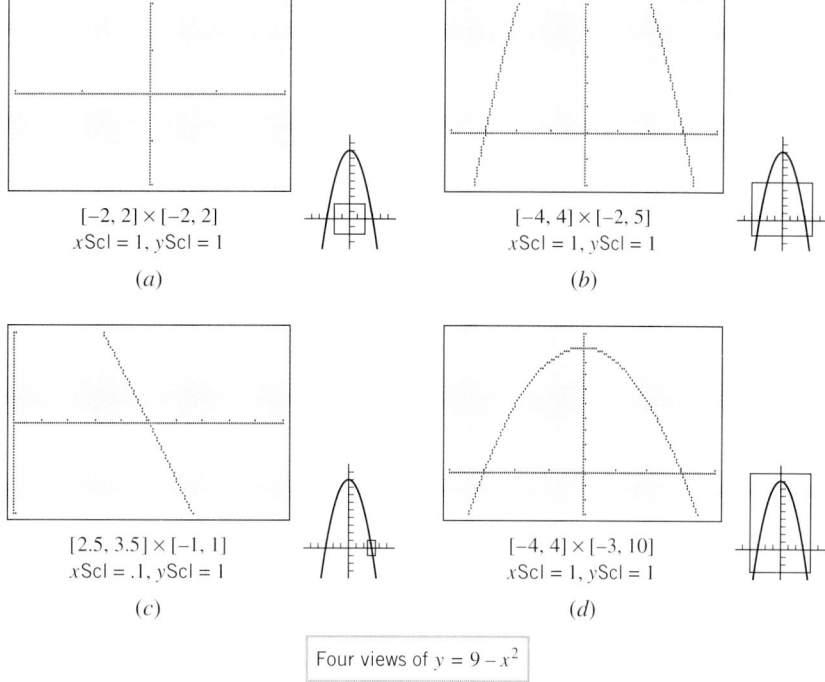

Four views of $y = 9 - x^2$

Figure 1.3.6

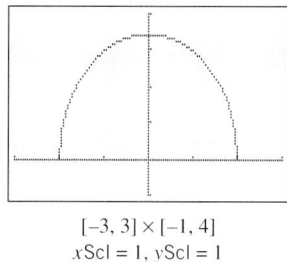

[−3, 3] × [−1, 4]
xScl = 1, yScl = 1

Figure 1.3.7

Solution. The natural domain of f is $[-2, 2]$ and the range is $[0, \sqrt{12}]$ (verify), so the entire graph will be contained in the viewing window $[-2, 2] \times [0, \sqrt{12}]$. For clarity, it is desirable to use a slightly larger window to avoid having the graph too close to the ends of the screen. For example, taking the viewing window to be $[-3, 3] \times [-1, 4]$ yields the graph in Figure 1.3.7. ◄

If the graph of f extends indefinitely in either the x- or y-direction, then it will not be possible to show the entire graph in any one viewing window. In such cases one tries to choose the window to show all of the important features for the problem at hand. (Of course, what is important in one problem may not be important in another, so the choice of the viewing window will often depend on the objectives in the problem.)

Example 4

Graph the equation $y = x^3 - 12x^2 + 18$ in the following windows and discuss the advantages and disadvantages of each window.

(a) $[-10, 10] \times [-10, 10]$ with xScl = 1, yScl = 1
(b) $[-20, 20] \times [-20, 20]$ with xScl = 1, yScl = 1
(c) $[-20, 20] \times [-300, 20]$ with xScl = 1, yScl = 20
(d) $[-5, 15] \times [-300, 20]$ with xScl = 1, yScl = 20
(e) $[1, 2] \times [-1, 1]$ with xScl = .1, yScl = .1

Solution (a). The window in Figure 1.3.8a has chopped off the portion of the graph that intersects the y-axis, and it shows only two of three possible real roots for the given cubic polynomial. To remedy these problems we need to widen the window in both the x- and y-directions.

Solution (b). The window in Figure 1.3.8b shows the intersection of the graph with the y-axis and the three real roots, but it has chopped off the portion of the graph between the two positive roots. Moreover, the ticks in the y-direction are nearly illegible because they are so close together. We need to extend the window in the negative y-direction and increase yScl. We do not know how far to extend the window, so some experimentation will be required to obtain what we want.

Solution (c). The window in Figure 1.3.8c shows all of the main features of the graph. However, we have some wasted space in the x-direction. We can improve the picture by shortening the window in the x-direction appropriately.

Solution (d). The window in Figure 1.3.8d shows all of the main features of the graph without a lot of wasted space. However, the window does not provide a clear view of the roots. To get a closer view of the roots we must forget about showing all of the main features of the graph and choose windows that zoom in on the roots themselves.

Solution (e). The window in Figure 1.3.8e displays very little of the graph, but it clearly shows that the root in the interval [1, 2] is slightly less than 1.3, say $x \approx 1.29$. ◀

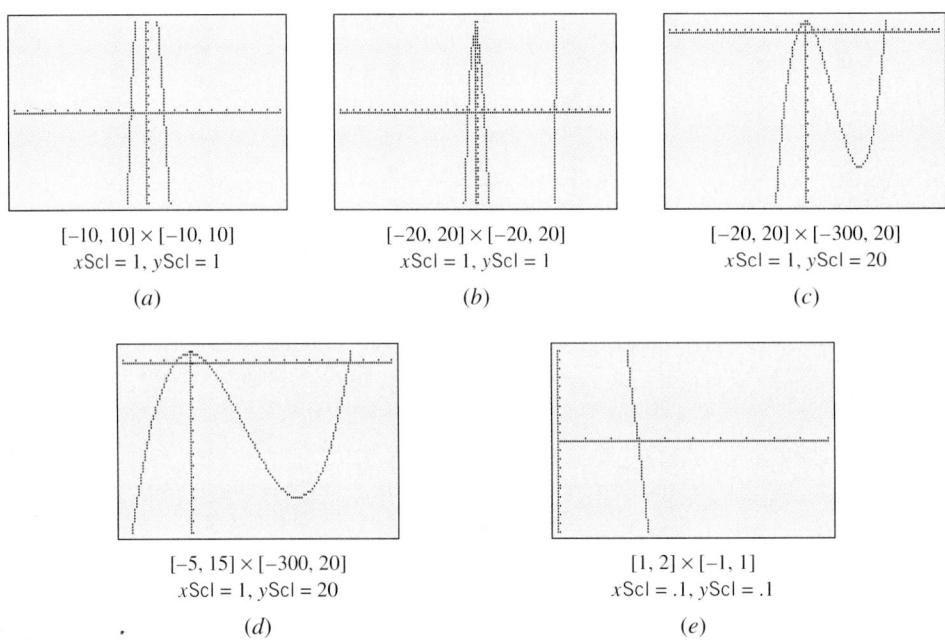

$[-10, 10] \times [-10, 10]$
xScl = 1, yScl = 1

(a)

$[-20, 20] \times [-20, 20]$
xScl = 1, yScl = 1

(b)

$[-20, 20] \times [-300, 20]$
xScl = 1, yScl = 20

(c)

$[-5, 15] \times [-300, 20]$
xScl = 1, yScl = 20

(d)

$[1, 2] \times [-1, 1]$
xScl = .1, yScl = .1

(e)

Figure 1.3.8

FOR THE READER.   Sometimes you will want to determine the viewing window by choosing the x-interval for the window and allowing the graphing utility to determine a y-interval that encompasses the maximum and minimum values of the function over the x-interval. Most graphing utilities provide some method for doing this, so check your documentation to determine how to use this feature. Allowing the graphing utility to determine the y-interval of the window takes some of the guesswork out of problems like that in part (b) of the preceding example.

ZOOMING

The process of enlarging or reducing the size of a viewing window is called *zooming*. If you reduce the size of the window, you see less of the graph as a whole, but more detail of the part shown; this is called *zooming in*. In contrast, if you enlarge the size of the window, you see more of the graph as a whole, but less detail of the part shown; this is called *zooming out*. Most graphing calculators provide menu items for zooming in or zooming out by fixed factors. For example, on the author's calculator the amount of enlargement or reduction is controlled by setting values for two *zoom factors*, xFact and yFact. If

$$x\text{Fact} = 10 \quad \text{and} \quad y\text{Fact} = 5$$

then each time a zoom command is executed the viewing window is enlarged or reduced by a factor of 10 in the x-direction and a factor of 5 in the y-direction. With computer programs such as *Mathematica* and *Maple*, zooming is controlled by adjusting the x-interval and y-interval directly; however, there are ways to automate this by programming.

FOR THE READER. If you are using a graphing calculator, read your documentation to determine how to use the zooming feature.

COMPRESSION

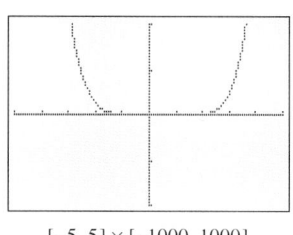

$[-5, 5] \times [-1000, 1000]$
$x\text{Scl} = 1, y\text{Scl} = 500$

(*a*)

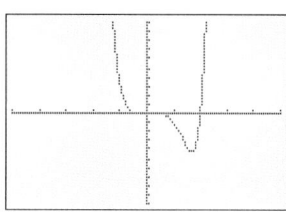

$[-5, 5] \times [-10, 10]$
$x\text{Scl} = 1, y\text{Scl} = 1$

(*b*)

Figure 1.3.9

Enlarging the viewing window for a graph has the geometric effect of compressing the graph, since more of the graph is packed into the calculator screen. If the compression is sufficiently great, then some of the detail in the graph may be lost. Thus, the choice of the viewing window frequently depends on whether you want to see more of the graph or more of the detail. Figure 1.3.9 shows two views of the equation

$$y = x^5(x - 2)$$

In part (*a*) of the figure the y-interval is very large, resulting in a vertical compression that obscures the detail in the vicinity of the x-axis. In part (*b*) the y-interval is smaller, and consequently we see more of the detail in the vicinity of the x-axis but less of the graph in the y-direction.

Example 5

Describe the graph of the function $f(x) = x + 0.01 \sin 50\pi x$; then graph the function in the following windows and explain why the graphs do or do not differ from your description.

(a) $[-10, 10] \times [-10, 10]$ (b) $[-1, 1] \times [-1, 1]$

(c) $[-.1, .1] \times [-.1, .1]$ (d) $[-.01, .01] \times [-.01, .01]$

Solution. The formula for f is the sum of the function x (whose graph is a straight line) and the function $0.01 \sin 50\pi x$ (whose graph is a sinusoidal curve with an amplitude of 0.01 and a period of $2\pi/50\pi = 0.04$). Intuitively, this suggests that the graph of f will follow the general path of the line $y = x$ but will have small bumps resulting from the contributions of the sinusoidal oscillations.

To generate the four graphs, we first set the calculator to the radian mode.* Because the windows in successive parts of this example are decreasing in size by a factor of 10, it will be convenient to use the zoom in feature of the calculator with the zoom factors set to 10 in the x- and y-directions. In Figure 1.3.10*a* the graph appears to be a straight line

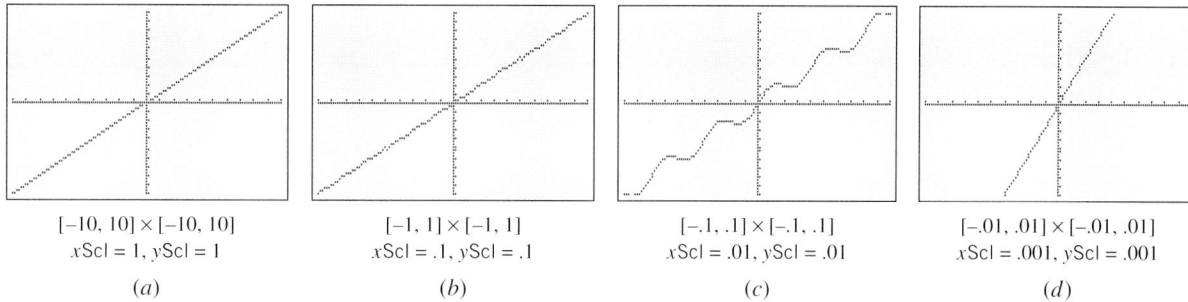

$[-10, 10] \times [-10, 10]$
$x\text{Scl} = 1, y\text{Scl} = 1$

(*a*)

$[-1, 1] \times [-1, 1]$
$x\text{Scl} = .1, y\text{Scl} = .1$

(*b*)

$[-.1, .1] \times [-.1, .1]$
$x\text{Scl} = .01, y\text{Scl} = .01$

(*c*)

$[-.01, .01] \times [-.01, .01]$
$x\text{Scl} = .001, y\text{Scl} = .001$

(*d*)

Figure 1.3.10

*In this text we follow the convention that angles are measured in radians unless degree measure is specified.

because compression has hidden the small sinusoidal oscillations. (Keep in mind that the amplitude of the sinusoidal portion of the function is only 0.01.) In part (*b*) the oscillations have begun to appear since the *y*-interval has been reduced, and in part (*c*) the oscillations have become very clear because the vertical scale is more in keeping with the amplitude of the oscillations. In part (*d*) the graph appears to be a line segment because we have zoomed in on such a small portion of the curve. ◀

ASPECT RATIO DISTORTION

Figure 1.3.11*a* shows a circle of radius 5 and two perpendicular lines graphed in the window $[-10, 10] \times [-10, 10]$ with $x\text{Scl} = 1$ and $y\text{Scl} = 1$. However, the circle is distorted and the lines do not appear perpendicular because the calculator has not used the same length for 1 unit on the *x*-axis and 1 unit on the *y*-axis. (Compare the spacing between the ticks on the axes.) This is called ***aspect ratio distortion***. Many calculators provide a menu item for automatically correcting the distortion by adjusting the viewing window appropriately. For example, the author's calculator makes this correction to the viewing window $[-10, 10] \times [-10, 10]$ by changing it to

$$[-16.9970674487, 16.9970674487] \times [-10, 10]$$

(Figure 1.3.11*b*). With computer programs such as *Mathematica* and *Maple*, aspect ratio distortion is controlled with adjustments to the physical dimensions of the viewing window on the computer screen, rather than altering the *x*- and *y*-intervals of the viewing window.

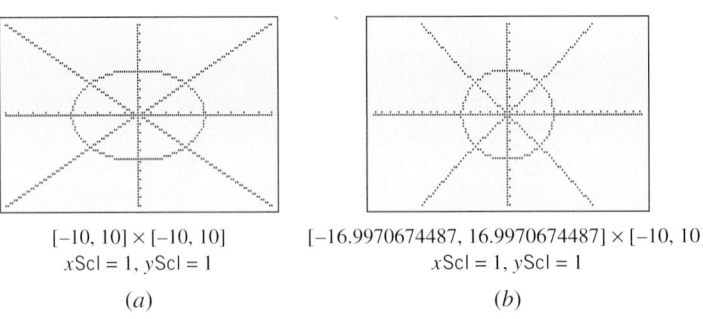

$[-10, 10] \times [-10, 10]$
$x\text{Scl} = 1, y\text{Scl} = 1$

(*a*)

$[-16.9970674487, 16.9970674487] \times [-10, 10]$
$x\text{Scl} = 1, y\text{Scl} = 1$

(*b*)

Figure 1.3.11

FOR THE READER.   Read the documentation for your graphing utility to determine how to control aspect ratio distortion.

PIXELS AND RESOLUTION

Sometimes graphing utilities produce unexpected results. For example, Figure 1.3.12 shows the graph of $y = \cos(10\pi x)$ generated on the author's graphing calculator in four different windows. (Your own calculator may produce different results.) The first graph has the correct shape, but the remaining three do not. To explain what is happening here we need to understand more precisely how graphing utilities generate graphs.

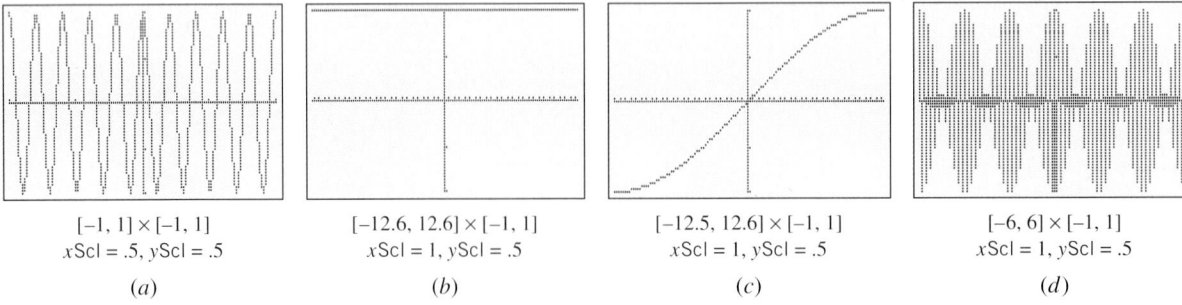

$[-1, 1] \times [-1, 1]$
$x\text{Scl} = .5, y\text{Scl} = .5$

(*a*)

$[-12.6, 12.6] \times [-1, 1]$
$x\text{Scl} = 1, y\text{Scl} = .5$

(*b*)

$[-12.5, 12.6] \times [-1, 1]$
$x\text{Scl} = 1, y\text{Scl} = .5$

(*c*)

$[-6, 6] \times [-1, 1]$
$x\text{Scl} = 1, y\text{Scl} = .5$

(*d*)

Figure 1.3.12

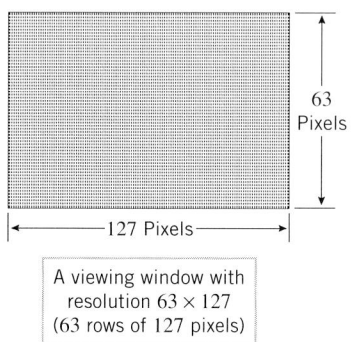

A viewing window with resolution 63 × 127 (63 rows of 127 pixels)

Figure 1.3.13

Screen displays for graphing utilities are divided into rows and columns of rectangular blocks, called *pixels*. For black-and-white displays each pixel has two possible states—an activated (or dark) state and a deactivated (or light) state. Since graphical elements are produced by activating pixels, the more pixels that a screen has to work with, the greater the amount of detail it can show. For example, the author's calculator has a *resolution* of 63 × 127, meaning that there are 63 rows with 127 pixels per row (Figure 1.3.13). In contrast, the author's computer screen has a resolution of 1024 × 1280 (1024 rows with 1280 pixels per row), so the computer screen is capable of displaying much smoother graphs than the calculator.

FOR THE READER. If you are using a graphing calculator, check the documentation to determine its resolution.

SAMPLING ERROR

The procedure that a graphing utility follows to generate a graph is similar to the procedure for plotting points by hand. When a viewing window is selected and an equation is entered, the graphing utility determines the x-coordinates of certain pixels on the x-axis and computes the corresponding points (x, y) on the graph. It then activates the pixels whose coordinates most closely match those of the calculated points and uses some built-in algorithm to activate additional intermediate pixels to create the curve shape. The point to keep in mind here is that *changing the window changes the points plotted by the graphing utility*. Thus, it is possible that a particular window will produce a false impression about the graph shape because significant characteristics of the graph occur *between* the plotted pixels. This is called *sampling error*. This is exactly what occurred in Figure 1.3.12 when we graphed $y = \cos(10\pi x)$. In part (*b*) of the figure the plotted pixels happened to fall at the peaks of the cosine curve, giving the false impression that the graph is a horizontal line at $y = 1$. In part (*c*) the plotted pixels fell at successively higher points along the graph, and in part (*d*) the plotted pixels fell in a strange way that created yet another misleading impression of the graph shape.

REMARK. Figure 1.3.12 suggests that for trigonometric graphs with rapid oscillations, restricting the x-interval to a few periods is likely to produce a more accurate representation about the graph shape.

FALSE GAPS

Sometimes graphs that are continuous appear to have gaps when they are generated on a calculator. These *false gaps* typically occur where the graph rises so rapidly that vertical space is opened up between successive pixels.

Example 6

Figure 1.3.14 shows the graph of the semicircle $y = \sqrt{9 - x^2}$ in two viewing windows. Although this semicircle has x-intercepts at the points $x = \pm 3$, part (*a*) of the figure shows false gaps at those points because there are no pixels with x-coordinates ± 3 in the window

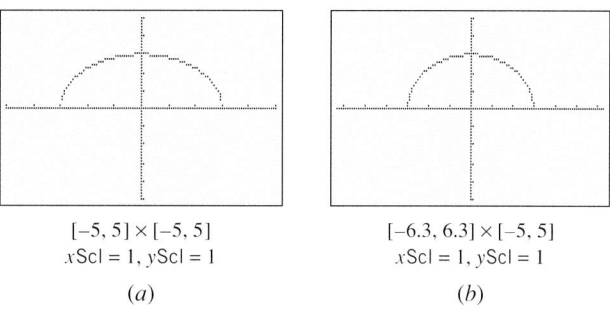

$[-5, 5] \times [-5, 5]$
xScl = 1, yScl = 1

(*a*)

$[-6.3, 6.3] \times [-5, 5]$
xScl = 1, yScl = 1

(*b*)

Figure 1.3.14

selected. In part (*b*) no gaps occur because there are pixels with *x*-coordinates $x = \pm 3$ in the window being used. ◀

In addition to creating false gaps in continuous graphs, calculators can err in the opposite direction by placing *false line segments* in the gaps of discontinuous curves.

Example 7

Figure 1.3.15*a* shows the graph of $y = 1/(x - 1)$ in the default window on the author's calculator. Although the graph appears to contain vertical line segments near $x = 1$, they should not be there. There is actually a gap in the curve at $x = 1$, since a division by zero occurs at that point (Figure 1.3.15*b*). ◀

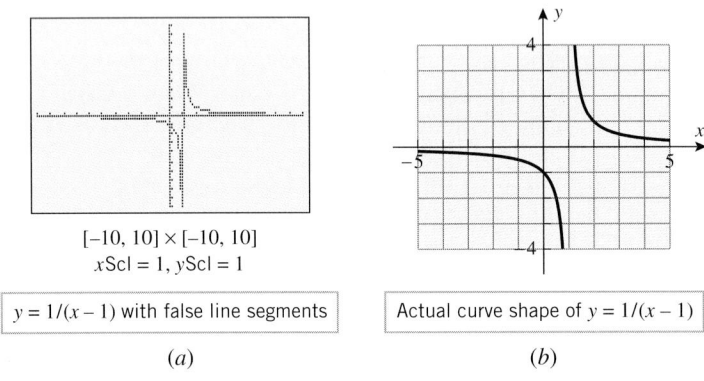

| $[-10, 10] \times [-10, 10]$ |
| xScl = 1, yScl = 1 |

| $y = 1/(x-1)$ with false line segments | Actual curve shape of $y = 1/(x-1)$ |

(*a*) (*b*)

Figure 1.3.15

Most graphing utilities use logarithms to evaluate functions with fractional exponents such as $f(x) = x^{2/3} = \sqrt[3]{x^2}$. However, because logarithms are only defined for positive numbers, many (but not all) graphing utilities will omit portions of the graphs of functions with fractional exponents. For example, the author's calculator graphs $y = x^{2/3}$ as in Figure 1.3.16*a*, whereas the actual graph is as in Figure 1.3.16*b*. (See the discussion preceding Exercise 29 for a way of circumventing this problem.)

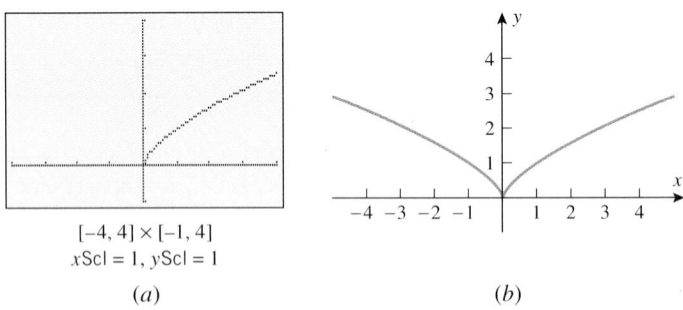

| $[-4, 4] \times [-1, 4]$ |
| xScl = 1, yScl = 1 |

(*a*) (*b*)

Figure 1.3.16

FOR THE READER.   Determine whether your graphing utility produces the complete graph of $y = x^{2/3}$.

Although graphing utilities are powerful tools for generating graphs quickly, they can produce misleading graphs as a result of compression, sampling error, false gaps, and false line segments. In short, *graphing utilities can suggest graph shapes*, *but they cannot establish them with certainty*. Thus, the more you know about the functions you are graphing, the easier it will be to choose good viewing windows, and the better you will be able to judge the reasonableness of the results produced by your graphing utility.

....
**MORE INFORMATION ON
GRAPHING AND CALCULATING
UTILITIES**

The main source of information about your graphing utility is its own documentation, and from time to time we will suggest that you refer to that documentation to learn some particular technique.

EXERCISE SET 1.3
....

1. Use a graphing utility to generate the graph of the function $f(x) = x^4 - x^2$ in the given viewing windows, and specify the window that you think gives the best view of the graph.
   (a) $-50 \le x \le 50$, $-50 \le y \le 50$
   (b) $-5 \le x \le 5$, $-5 \le y \le 5$
   (c) $-2 \le x \le 2$, $-2 \le y \le 2$
   (d) $-2 \le x \le 2$, $-1 \le y \le 1$
   (e) $-1.5 \le x \le 1.5$, $-0.5 \le y \le 0.5$

2. Use a graphing utility to generate the graph of the function $f(x) = x^5 - x^3$ in the given viewing windows, and specify the window that you think gives the best view of the graph.
   (a) $-50 \le x \le 50$, $-50 \le y \le 50$
   (b) $-5 \le x \le 5$, $-5 \le y \le 5$
   (c) $-2 \le x \le 2$, $-2 \le y \le 2$
   (d) $-2 \le x \le 2$, $-1 \le y \le 1$
   (e) $-1.5 \le x \le 1.5$, $-0.5 \le y \le 0.5$

3. Use a graphing utility to generate the graph of the function $f(x) = x^2 + 12$ in the given viewing windows, and specify the window that you think gives the best view of the graph.
   (a) $-1 \le x \le 1$, $13 \le y \le 15$
   (b) $-2 \le x \le 2$, $11 \le y \le 15$
   (c) $-4 \le x \le 4$, $10 \le y \le 28$
   (d) A window of your choice

4. Use a graphing utility to generate the graph of the function $f(x) = -12 - x^2$ in the given viewing windows, and specify the window that you think gives the best view of the graph.
   (a) $-1 \le x \le 1$, $-15 \le y \le -13$
   (b) $-2 \le x \le 2$, $-15 \le y \le -11$
   (c) $-4 \le x \le 4$, $-28 \le y \le -10$
   (d) A window of your choice

In Exercises 5 and 6, use the domain and range of f to determine a viewing window that contains the entire graph, and generate the graph in that window.

5. $f(x) = \sqrt{16 - 2x^2}$
6. $f(x) = \sqrt{3 - 2x - x^2}$

7. Graph the function $f(x) = x^3 - 15x^2 - 3x + 45$ using the stated windows and tick spacing, and discuss the advantages and disadvantages of each window.
   (a) $-10 \le x \le 10$, $-10 \le y \le 10$
   with xScl $= 1$ and yScl $= 1$

   (b) $-20 \le x \le 20$, $-20 \le y \le 20$
   with xScl $= 1$ and yScl $= 1$
   (c) $-5 \le x \le 20$, $-500 \le y \le 50$
   with xScl $= 5$ and yScl $= 50$
   (d) $-2 \le x \le -1$, $-1 \le y \le 1$
   with xScl $= 0.1$ and yScl $= 0.1$
   (e) $9 \le x \le 11$, $-486 \le y \le -484$
   with xScl $= 0.1$ and yScl $= 0.1$

8. Graph the function $f(x) = -x^3 - 12x^2 + 4x + 48$ using the stated windows and tick spacing, and discuss the advantages and disadvantages of each window.
   (a) $-10 \le x \le 10$, $-10 \le y \le 10$
   with xScl $= 1$ and yScl $= 1$
   (b) $-20 \le x \le 20$, $-20 \le y \le 20$
   with xScl $= 1$ and yScl $= 1$
   (c) $-16 \le x \le 4$, $-250 \le y \le 50$
   with xScl $= 2$ and yScl $= 25$
   (d) $-3 \le x \le -1$, $-1 \le y \le 1$
   with xScl $= 0.1$ and yScl $= 0.1$
   (e) $-9 \le x \le -7$, $-241 \le y \le -239$
   with xScl $= 0.1$ and yScl $= 0.1$

In Exercises 9–16, generate the graph of f in a viewing window that you think is appropriate.

9. $f(x) = x^2 - 9x - 36$
10. $f(x) = \dfrac{x + 7}{x - 9}$
11. $f(x) = 2\cos 80x$
12. $f(x) = 12\sin(x/80)$
13. $f(x) = 300 - 10x^2 + 0.01x^3$
14. $f(x) = x(30 - 2x)(25 - 2x)$
15. $f(x) = x^2 + \dfrac{1}{x}$
16. $f(x) = \sqrt{11x - 18}$

In Exercises 17 and 18, generate the graph of f and determine whether your graphs contain false line segments. Sketch the actual graph and see if you can make the false line segments disappear by changing the viewing window.

17. $f(x) = \dfrac{x}{x^2 - 1}$
18. $f(x) = \dfrac{x^2}{4 - x^2}$

19. The graph of the equation $x^2 + y^2 = 16$ is a circle of radius 4 centered at the origin.

(a) Find a function whose graph is the upper semicircle and graph it.

(b) Find a function whose graph is the lower semicircle and graph it.

(c) Graph the upper and lower semicircles together. If the combined graphs do not appear circular, see if you can adjust the viewing window to eliminate the aspect ratio distortion.

(d) Graph the portion of the circle in the first quadrant.

(e) Is there a function whose graph is the right half of the circle? Explain.

20. In each part, graph the equation by solving for y in terms of x and graphing the resulting functions together.

(a) $x^2/4 + y^2/9 = 1$ (b) $y^2 - x^2 = 1$

21. Read the documentation for your graphing utility to determine how to graph functions involving absolute values, and graph the given equation.

(a) $y = |x|$ (b) $y = |x - 1|$

(c) $y = |x| - 1$ (d) $y = |\sin x|$

(e) $y = \sin|x|$ (f) $y = |x| - |x + 1|$

22. Based on your knowledge of the absolute value function, sketch the graph of $f(x) = |x|/x$. Check your result using a graphing utility.

23. Make a conjecture about the relationship between the graph of $y = f(x)$ and the graph of $y = |f(x)|$; check your conjecture with some specific functions.

24. Make a conjecture about the relationship between the graph of $y = f(x)$ and the graph of $y = f(|x|)$; check your conjecture with some specific functions.

25. (a) Based on your knowledge of the absolute value function, sketch the graph of $y = |x - a|$, where a is a constant. Check your result using a graphing utility and some specific values of a.

(b) Sketch the graph of $y = |x - 1| + |x - 2|$; check your result with a graphing utility.

26. How are the graphs of $y = |x|$ and $y = \sqrt{x^2}$ related? Check your answer with a graphing utility.

Most graphing utilities provide some way of graphing functions that are defined piecewise; read the documentation for your graphing utility to find out how to do this. However, if your goal is just to find the general shape of the graph, you can graph each portion of the function separately and combine the pieces with a hand-drawn sketch. Use this method in Exercises 27 and 28.

27. Draw the graph of

$$f(x) = \begin{cases} \sqrt[3]{x - 2}, & x \le 2 \\ x^3 - 2x - 4, & x > 2 \end{cases}$$

28. Draw the graph of

$$f(x) = \begin{cases} x^3 - x^2, & x \le 1 \\ \dfrac{1}{1 - x}, & 1 < x < 4 \\ x^2 \cos\sqrt{x}, & 4 \le x \end{cases}$$

We noted in the text that for functions involving fractional exponents (or radicals), graphing utilities sometimes omit portions of the graph. If $f(x) = x^{p/q}$, where p/q is a positive fraction in *lowest terms*, then you can circumvent this problem as follows:

• If p is even and q is odd, then graph $g(x) = |x|^{p/q}$ instead of $f(x)$.

• If p is odd and q is odd, then graph $g(x) = (|x|/x)|x|^{p/q}$ instead of $f(x)$.

We will explain why this works in the exercises of the next section.

29. (a) Generate the graphs of $f(x) = x^{2/5}$ and $g(x) = |x|^{2/5}$, and determine whether your graphing utility missed part of the graph of f.

(b) Generate the graphs of the functions $f(x) = x^{1/5}$ and $g(x) = (|x|/x)|x|^{1/5}$, and then determine whether your graphing utility missed part of the graph of f.

(c) Generate a complete graph of the equation

$$y = (x - 1)^{4/5}$$

(d) Generate a complete graph of the equation

$$y = (x + 1)^{3/4}$$

30. The graphs of $y = (x^2 - 4)^{2/3}$ and $y = [(x^2 - 4)^2]^{1/3}$ should be the same. Does your graphing utility produce the same graph for both equations? If not, what do you think is happening?

31. In each part, graph the function for various values of c, and write a paragraph or two that describes how changes in c affect the graph in each case.

(a) $y = cx^2$ (b) $y = x^2 + cx$

(c) $y = x^2 + x + c$

32. The graph of an equation of the form $y^2 = x(x - a)(x - b)$ (where $0 < a < b$) is called a **bipartite cubic**. The accompanying figure shows a typical graph of this type.

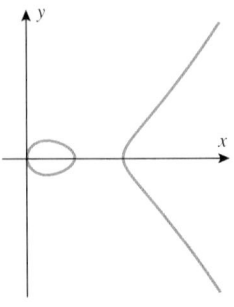

Bipartite cubic

Figure Ex-32

(a) Graph the bipartite cubic $y^2 = x(x-1)(x-2)$ by solving for y in terms of x and graphing the two resulting functions.

(b) Find the x-intercepts of the bipartite cubic

$$y^2 = x(x-a)(x-b)$$

and make a conjecture about how changes in the values of a and b would affect the graph. Test your conjecture by graphing the bipartite cubic for various values of a and b.

33. Based on your knowledge of the graphs of $y = x$ and $y = \sin x$, make a sketch of the graph of $y = x \sin x$. Check your conclusion using a graphing utility.

34. What do you think the graph of $y = \sin(1/x)$ looks like? Test your conclusion using a graphing utility. [*Suggestion:* Examine the graph on a succession of smaller and smaller intervals centered at $x = 0$.]

1.4 NEW FUNCTIONS FROM OLD

Just as numbers can be added, subtracted, multiplied, and divided to produce other numbers, so functions can be added, subtracted, multiplied, and divided to produce other functions. In this section we will discuss these operations and some others that have no analogs in ordinary arithmetic.

ARITHMETIC OPERATIONS ON FUNCTIONS

Two functions, f and g, can be added, subtracted, multiplied, and divided in a natural way to form new functions $f + g$, $f - g$, fg, and f/g. For example, $f + g$ is defined by the formula

$$(f + g)(x) = f(x) + g(x) \tag{1}$$

which states that for each input the value of $f + g$ is obtained by adding the values of f and g. For example, if

$$f(x) = x \quad \text{and} \quad g(x) = x^2$$

then

$$(f + g)(x) = f(x) + g(x) = x + x^2$$

Equation (1) provides a formula for $f + g$ but does not say anything about the domain of $f + g$. However, for the right side of this equation to be defined, x must lie in the domain of f *and* in the domain of g, so we define the domain of $f + g$ to be the intersection of those two domains. More generally, we make the following definition:

1.4.1 DEFINITION. Given functions f and g, we define

$$(f + g)(x) = f(x) + g(x)$$
$$(f - g)(x) = f(x) - g(x)$$
$$(fg)(x) \quad\;\; = f(x)g(x)$$
$$(f/g)(x) \;\; = f(x)/g(x)$$

For the functions $f + g$, $f - g$, and fg we define the domain to be the intersection of the domains of f and g, and for the function f/g we define the domain to be the intersection of the domains of f and g but with the points where $g(x) = 0$ excluded (to avoid division by zero).

REMARK. If f is a constant function, say $f(x) = c$ for all x, then the product of f and g is cg, so multiplying a function by a constant is a special case of multiplying two functions.

Example 1

Let

$$f(x) = 1 + \sqrt{x - 2} \quad \text{and} \quad g(x) = x - 3$$

Find $(f + g)(x)$, $(f - g)(x)$, $(fg)(x)$, $(f/g)(x)$, and $(7f)(x)$; state the domains of $f + g$, $f - g$, fg, f/g, and $7f$.

Solution. First, we will find formulas for the functions and then the domains. The formulas are

$$(f + g)(x) = f(x) + g(x) = (1 + \sqrt{x - 2}) + (x - 3) = x - 2 + \sqrt{x - 2} \qquad (2)$$

$$(f - g)(x) = f(x) - g(x) = (1 + \sqrt{x - 2}) - (x - 3) = 4 - x + \sqrt{x - 2} \qquad (3)$$

$$(fg)(x) = f(x)g(x) = (1 + \sqrt{x - 2})(x - 3) \qquad (4)$$

$$(f/g)(x) = f(x)/g(x) = \frac{1 + \sqrt{x - 2}}{x - 3} \qquad (5)$$

$$(7f)(x) = 7f(x) = 7 + 7\sqrt{x - 2} \qquad (6)$$

In all five cases the natural domain determined by the formula is the same as the domain specified in Definition 1.4.1, so there is no need to state the domain explicitly in any of these cases. For example, the domain of f is $[2, +\infty)$, the domain of g is $(-\infty, +\infty)$, and the natural domain for $f(x) + g(x)$ determined by Formula (2) is $[2, +\infty)$, which is precisely the intersection of the domains of f and g. ◀

REMARK. There are situations in which the natural domain associated with the formula resulting from an operation on two functions is not the correct domain for the new function. For example, if $f(x) = \sqrt{x}$ and $g(x) = \sqrt{x}$, then according to Definition 1.4.1 the domain of fg should be $[0, +\infty) \cap [0, +\infty) = [0, +\infty)$. However, $(fg)(x) = \sqrt{x}\sqrt{x} = x$, which has a natural domain of $(-\infty, +\infty)$. Thus, to be precise in describing the formula for fg, we must write $(fg)(x) = x$, $x \geq 0$.

STRETCHES AND COMPRESSIONS

Multiplying a function f by a *nonnegative* constant c has the geometric effect of stretching or compressing the graph of f vertically. For example, examine the graphs of $y = f(x)$, $y = 2f(x)$, and $y = \frac{1}{2}f(x)$ shown in Figure 1.4.1a. Multiplying by 2 doubles each y-coordinate, thereby stretching the graph, and multiplying by $\frac{1}{2}$ cuts each y-coordinate in

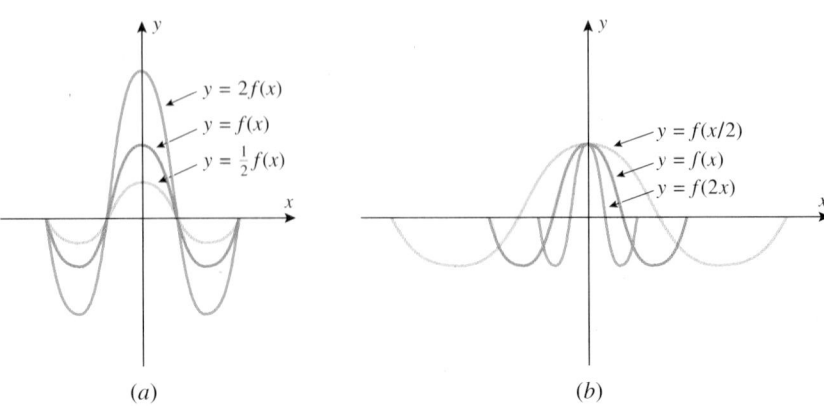

(a) $\qquad\qquad\qquad (b)$

Figure 1.4.1

half, thereby compressing the graph. In general, if $c > 0$, then the graph of $y = cf(x)$ can be obtained from the graph of $y = f(x)$ by compressing the graph of $y = f(x)$ vertically by a factor of $1/c$ if $0 < c < 1$, or stretching it by a factor of c if $c > 1$.

Analogously, multiplying the independent variable of a function f by a **nonnegative** constant c has the geometric effect of stretching or compressing the graph of f horizontally. For example, examine the graphs of $y = f(x)$, $y = f(2x)$, and $y = f(x/2)$ shown in Figure 1.4.1b. Multiplying x by 2 compresses the graph by a factor of 2 and multiplying x by $\frac{1}{2}$ stretches the graph by a factor of 2. [This is a little confusing, but think of it this way: The value of $2x$ changes twice as fast as the value of x, so a point moving along the x-axis will only have to move half as far from the origin for $y = f(2x)$ to have the same value as $y = f(x)$.] In general, if $c > 0$, then the graph of $y = f(cx)$ can be obtained from the graph of $y = f(x)$ by stretching the graph of $y = f(x)$ horizontally by a factor of c if $0 < c < 1$, or compressing it by a factor of c if $c > 1$.

SUMS OF FUNCTIONS

Adding two functions can be accomplished geometrically by adding the corresponding y-coordinates of their graphs. For example, Figure 1.4.2 shows line graphs of yearly new car sales $N(t)$ and used car sales $U(t)$ in the United States between 1985 and 1995. The sum of these functions, $T(t) = N(t) + U(t)$, represents the yearly total car sales for that period. As illustrated in the figure, the graph of $T(t)$ can be obtained by adding the values of $N(t)$ and $U(t)$ together at each time t and plotting the resulting value.

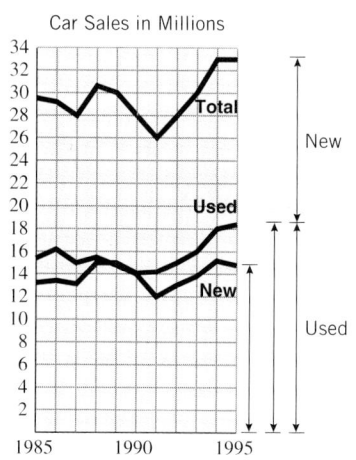

Car Sales in Millions

Source: NADA.

Figure 1.4.2

Example 2

Referring to Figure 1.2.2 for the graphs of $y = \sqrt{x}$ and $y = 1/x$, make a sketch that shows the general shape of the graph of $y = \sqrt{x} + 1/x$ for $x \geq 0$.

Solution.   To add the corresponding y-values of $y = \sqrt{x}$ and $y = 1/x$ graphically, just imagine them to be "stacked" on top of one another. This yields the sketch in Figure 1.4.3.   ◀

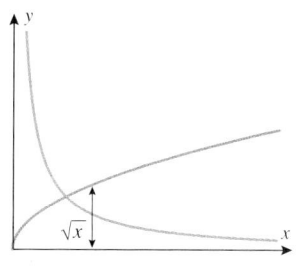 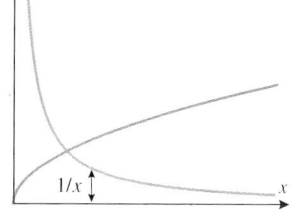 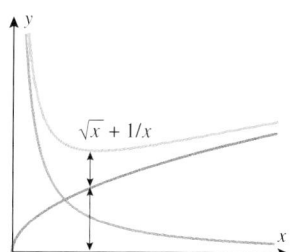

Figure 1.4.3

COMPOSITION OF FUNCTIONS

We now consider an operation on functions, called *composition*, which has no direct analog in ordinary arithmetic. Informally stated, the operation of composition is performed by substituting some function for the independent variable of another function. For example, suppose that

$$f(x) = x^2 \quad \text{and} \quad g(x) = x + 1$$

If we substitute $g(x)$ for x in the formula for f, we obtain a new function

$$f(g(x)) = (g(x))^2 = (x + 1)^2$$

which we denote by $f \circ g$. Thus,

$$(f \circ g)(x) = f(g(x)) = (g(x))^2 = (x + 1)^2$$

In general, we make the following definition.

1.4.2 DEFINITION. Given functions f and g, the **composition** of f with g, denoted by $f \circ g$, is the function defined by

$$(f \circ g)(x) = f(g(x))$$

The domain of $f \circ g$ is defined to consist of all x in the domain of g for which $g(x)$ is in the domain of f.

REMARK. Although the domain of $f \circ g$ may seem complicated at first glance, it makes sense intuitively: To compute $f(g(x))$ one needs x in the domain of g to compute $g(x)$, then one needs $g(x)$ in the domain of f to compute $f(g(x))$.

COMPOSITIONS VIEWED AS COMPUTER PROGRAMS

In Section 1.1 we noted that a function f can be viewed as a computer program that takes an input x, operates on it, and produces an output $f(x)$. From this viewpoint composition can be viewed as two programs, g and f, operating in succession: An input x is fed first to a program g, which produces the output $g(x)$; then this output is fed as input to a program f, which produces the output $f(g(x))$ (Figure 1.4.4). However, rather than have two separate programs operating in succession, we could create a **single** program that takes the input x and directly produces the output $f(g(x))$. This program is the composition $f \circ g$ since $(f \circ g)(x) = f(g(x))$.

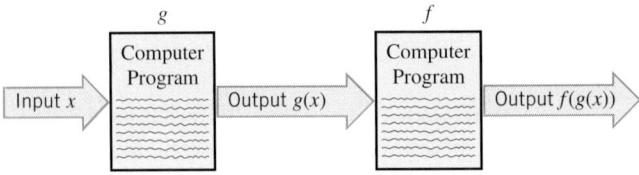

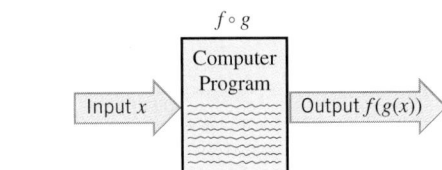

Figure 1.4.4

Example 3

Let $f(x) = x^2 + 3$ and $g(x) = \sqrt{x}$. Find

(a) $(f \circ g)(x)$ (b) $(g \circ f)(x)$

Solution (a). The formula for $f(g(x))$ is

$$f(g(x)) = [g(x)]^2 + 3 = (\sqrt{x}\,)^2 + 3 = x + 3$$

Since the domain of g is $[0, +\infty)$ and the domain of f is $(-\infty, +\infty)$, the domain of $f \circ g$ consists of all x in $[0, +\infty)$ such that $g(x) = \sqrt{x}$ lies in $(-\infty, +\infty)$; thus, the domain of $f \circ g$ is $[0, +\infty)$. Therefore,

$$(f \circ g)(x) = x + 3, \quad x \geq 0$$

Solution (b). The formula for $g(f(x))$ is

$$g(f(x)) = \sqrt{f(x)} = \sqrt{x^2 + 3}$$

Since the domain of f is $(-\infty, +\infty)$ and the domain of g is $[0, +\infty)$, the domain of $g \circ f$ consists of all x in $(-\infty, +\infty)$ such that $f(x) = x^2 + 3$ lies in $[0, +\infty)$. Thus, the domain of

$g \circ f$ is $(-\infty, +\infty)$. Therefore,

$$(g \circ f)(x) = \sqrt{x^2 + 3}$$

There is no need to indicate that the domain is $(-\infty, +\infty)$, since this is the natural domain of $\sqrt{x^2 + 3}$.   ◄

REMARK.   Note that the functions $f \circ g$ and $g \circ f$ in the preceding example are not the same. Thus, the order in which functions are composed can (and usually will) make a difference in the end result.

Compositions can also be defined for three or more functions; for example, $(f \circ g \circ h)(x)$ is computed as

$$(f \circ g \circ h)(x) = f(g(h(x)))$$

In other words, first find $h(x)$, then find $g(h(x))$, and then find $f(g(h(x)))$.

Example 4

Find $(f \circ g \circ h)(x)$ if

$$f(x) = \sqrt{x}, \quad g(x) = 1/x, \quad h(x) = x^3$$

Solution.

$$(f \circ g \circ h)(x) = f(g(h(x))) = f(g(x^3)) = f(1/x^3) = \sqrt{1/x^3} = 1/x^{3/2}$$ ◄

EXPRESSING A FUNCTION AS A COMPOSITION

Many problems in mathematics are attacked by "decomposing" functions into compositions of simpler functions. For example, consider the function h given by

$$h(x) = (x + 1)^2$$

To evaluate $h(x)$ for a given value of x, we would first compute $x + 1$ and then square the result. These two operations are performed by the functions

$$g(x) = x + 1 \quad \text{and} \quad f(x) = x^2$$

We can express h in terms of f and g by writing

$$h(x) = (x + 1)^2 = [g(x)]^2 = f(g(x))$$

so we have succeeded in expressing h as the composition $h = f \circ g$.

The thought process in this example suggests a general procedure for decomposing a function h into a composition $h = f \circ g$:

- Think about how you would evaluate $h(x)$ for a specific value of x, trying to break the evaluation into two steps performed in succession.
- The first operation in the evaluation will determine a function g and the second a function f.
- The formula for h can then be written as $h(x) = f(g(x))$.

For descriptive purposes, we will refer to g as the "inside function" and f as the "outside function" in the expression $f(g(x))$. The inside function performs the first operation and the outside function performs the second.

Example 5

Express $h(x) = (x - 4)^5$ as a composition of two functions.

Solution.   To evaluate $h(x)$ for a given value of x we would first compute $x - 4$ and then raise the result to the fifth power. Therefore, the inside function (first operation) is

$$g(x) = x - 4$$

and the outside function (second operation) is

$$f(x) = x^5$$

so $h(x) = f(g(x))$. As a check,

$$f(g(x)) = [g(x)]^5 = (x-4)^5 = h(x)$$ ◀

Example 6

Express $\sin(x^3)$ as a composition of two functions.

Solution. To evaluate $\sin(x^3)$, we would first compute x^3 and then take the sine, so $g(x) = x^3$ is the inside function and $f(x) = \sin x$ the outside function. Therefore,

$$\sin(x^3) = f(g(x)) \qquad \boxed{g(x) = x^3 \text{ and } f(x) = \sin x}$$ ◀

Example 7

Table 1.4.1 gives some more examples of decomposing functions into compositions.

Table 1.4.1

FUNCTION	$g(x)$ INSIDE	$f(x)$ OUTSIDE	COMPOSITION
$(x^2 + 1)^{10}$	$x^2 + 1$	x^{10}	$(x^2 + 1)^{10} = f(g(x))$
$\sin^3 x$	$\sin x$	x^3	$\sin^3 x = f(g(x))$
$\tan(x^5)$	x^5	$\tan x$	$\tan(x^5) = f(g(x))$
$\sqrt{4 - 3x}$	$4 - 3x$	$\sqrt{x}$	$\sqrt{4 - 3x} = f(g(x))$
$8 + \sqrt{x}$	$\sqrt{x}$	$8 + x$	$8 + \sqrt{x} = f(g(x))$
$\dfrac{1}{x + 1}$	$x + 1$	$\dfrac{1}{x}$	$\dfrac{1}{x + 1} = f(g(x))$

◀

REMARK. It should be noted that there is always more than one way to express a function as a composition. For example, here are two ways to express $(x^2 + 1)^{10}$ as a composition that differ from that in Table 1.4.1:

$$(x^2 + 1)^{10} = \left[(x^2 + 1)^2\right]^5 = f(g(x)) \qquad \boxed{g(x) = (x^2 + 1)^2 \text{ and } f(x) = x^5}$$

$$(x^2 + 1)^{10} = \left[(x^2 + 1)^3\right]^{10/3} = f(g(x)) \qquad \boxed{g(x) = (x^2 + 1)^3 \text{ and } f(x) = x^{10/3}}$$

SYMMETRY

Figure 1.4.5 shows the graphs of three curves that have certain obvious symmetries. The graph in part (*a*) is **symmetric about the x-axis** in the sense that for each point (x, y) on the graph the point $(x, -y)$ is also on the graph; the graph in part (*b*) is **symmetric about the y-axis** in the sense that for each point (x, y) on the graph the point $(-x, y)$ is also on the graph; and the graph in part (*c*) is **symmetric about the origin** in the sense that for each point (x, y) on the graph the point $(-x, -y)$ is also on the graph. Geometrically, symmetry about the origin occurs if rotating the graph $180°$ about the origin leaves the graph unchanged.

Symmetries can often be detected from the equation of a curve. For example, the graph of

$$y = x^3 \tag{7}$$

must be symmetric about the origin because for any point (x, y) whose coordinates satisfy (7), the coordinates of the point $(-x, -y)$ also satisfy (7), since substituting these

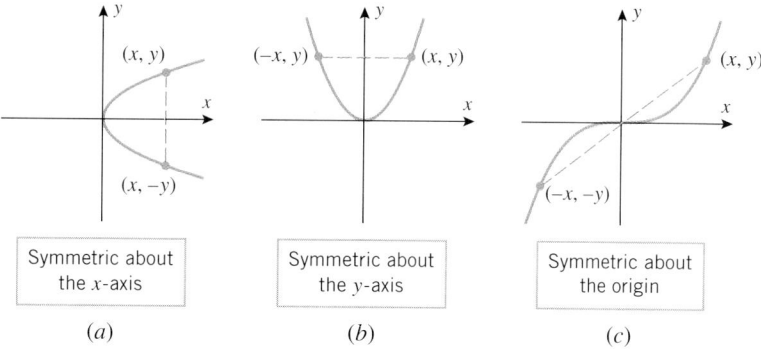

Figure 1.4.5

coordinates in (7) yields

$$-y = (-x)^3$$

which simplifies to (7). This suggests the following symmetry tests (Figure 1.4.6).

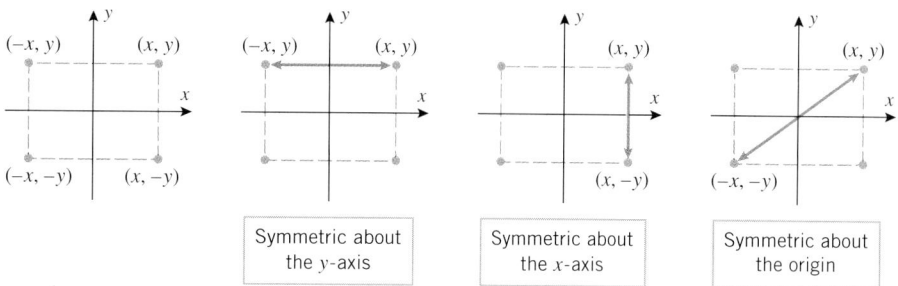

Figure 1.4.6

1.4.3 THEOREM (*Symmetry Tests*).

(a) *A plane curve is symmetric about the y-axis if and only if replacing x by −x in its equation produces an equivalent equation.*

(b) *A plane curve is symmetric about the x-axis if and only if replacing y by −y in its equation produces an equivalent equation.*

(c) *A plane curve is symmetric about the origin if and only if replacing both x by −x and y by −y in its equation produces an equivalent equation.*

EVEN AND ODD FUNCTIONS

For the graph of a function f to be symmetric about the y-axis, the equations $y = f(x)$ and $y = f(-x)$ must be equivalent; for this to happen we must have

$$f(x) = f(-x)$$

A function with this property is called an ***even function***. Some examples are x^2, x^4, x^6, and $\cos x$. Similarly, for the graph of a function f to be symmetric about the origin, the equations $y = f(x)$ and $-y = f(-x)$ must be equivalent; for this to happen we must have

$$f(x) = -f(-x)$$

A function with this property is called an ***odd function***. Some examples are x, x^3, x^5, and $\sin x$.

FOR THE READER. Explain why the graph of a nonzero function cannot by symmetric about the x-axis.

TRANSLATIONS

Once you know the graph of an equation $y = f(x)$, there are some techniques that can be used to help visualize the graphs of the equations

$$y = f(x) + c, \quad y = f(x) - c, \quad y = f(x + c), \quad y = f(x - c)$$

where c is any positive constant.

If a positive constant is added to or subtracted from $f(x)$, the geometric effect is to translate the graph of $y = f(x)$ parallel to the y-axis; addition translates the graph in the positive direction and subtraction translates it in the negative direction. This is illustrated in Table 1.4.2. Similarly, if a positive constant is added to or subtracted from the independent variable x, the geometric effect is to translate the graph of the function parallel to the x-axis; subtraction translates the graph in the positive direction, and addition translates it in the negative direction. This is also illustrated in Table 1.4.2.

Table 1.4.2

OPERATION ON $y = f(x)$	Add a positive constant c to $f(x)$	Subtract a positive constant c from $f(x)$	Add a positive constant c to x	Subtract a positive constant c from x
NEW EQUATION	$y = f(x) + c$	$y = f(x) - c$	$y = f(x + c)$	$y = f(x - c)$
GEOMETRIC EFFECT	Translates the graph of $y = f(x)$ up c units	Translates the graph of $y = f(x)$ down c units	Translates the graph of $y = f(x)$ left c units	Translates the graph of $y = f(x)$ right c units
EXAMPLE	$y = x^2 + 2$ $y = x^2$	$y = x^2$ $y = x^2 - 2$	$y = (x + 2)^2$ $y = x^2$	$y = (x - 2)^2$ $y = x^2$

Before proceeding to the following examples, it will be helpful to review the graphs in Figure 1.2.1.

Example 8

Sketch the graph of

(a) $y = \sqrt{x - 3}$ (b) $y = \sqrt{x + 3}$

Solution. The graph of the equation $y = \sqrt{x - 3}$ can be obtained by translating the graph of $y = \sqrt{x}$ right 3 units, and the graph of $y = \sqrt{x + 3}$ by translating the graph of $y = \sqrt{x}$ left 3 units (Figure 1.4.7). ◄

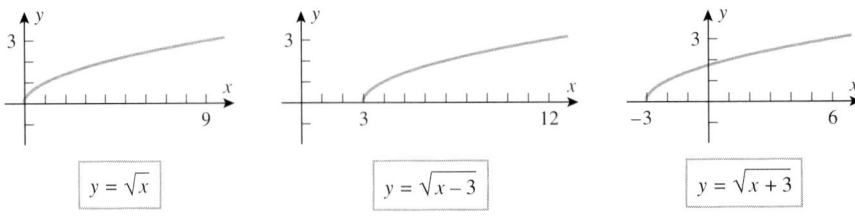

$y = \sqrt{x}$ $y = \sqrt{x - 3}$ $y = \sqrt{x + 3}$

Figure 1.4.7

Example 9

Sketch the graph of $y = |x - 3| + 2$.

Solution. The graph can be obtained by two translations: first translate the graph of $y = |x|$ right 3 units to obtain the graph of $y = |x - 3|$, then translate this graph up 2 units to obtain the graph of $y = |x - 3| + 2$ (Figure 1.4.8). ◄

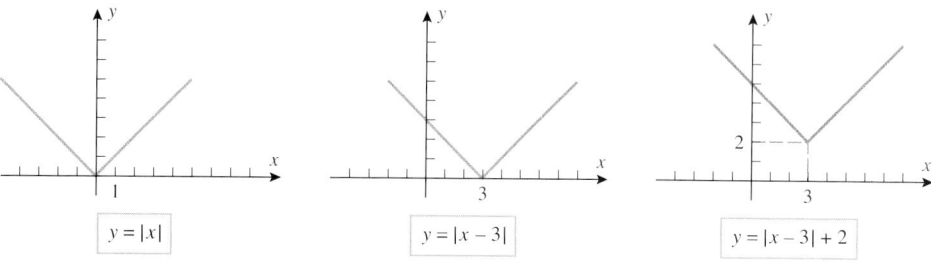

Figure 1.4.8

REMARK. The graph in the preceding example could also have been obtained by performing the translations in the opposite order: first translating the graph of $y = |x|$ up 2 units to obtain the graph of $y = |x| + 2$, then translating this graph right 3 units to obtain the graph of $y = |x - 3| + 2$.

Example 10

Sketch the graph of $y = x^2 - 4x + 5$.

Solution. Completing the square on the first two terms yields

$$y = (x^2 - 4x + 4) - 4 + 5 = (x - 2)^2 + 1$$

(see Appendix D for a review of this technique). In this form we see that the graph can be obtained by translating the graph of $y = x^2$ right 2 units because of the $x - 2$, and up 1 unit because of the $+1$ (Figure 1.4.9). ◄

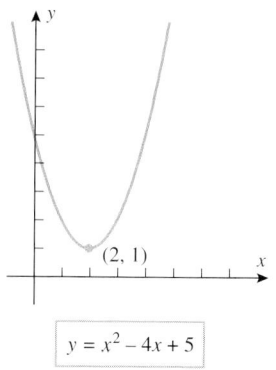

$y = x^2 - 4x + 5$

Figure 1.4.9

Example 11

By completing the square, an equation of the form $y = ax^2 + bx + c$ with $a \neq 0$ can be expressed as

$$y = a(x - h)^2 + k \tag{8}$$

Sketch the graph of this equation.

Solution. We can build up Equation (8) in three steps from the equation $y = x^2$. First, we can multiply by a to obtain $y = ax^2$. If $a > 0$, this operation has the geometric effect of stretching or compressing the graph of $y = x^2$; and if $a < 0$, it has the geometric effect of reflecting the graph about the x-axis, in addition to stretching or compressing it. Since stretching or compressing does not alter the general parabolic shape of the original curve, the graph of $y = ax^2$ looks roughly like one of those in Figure 1.4.10a. Next, we can subtract h from x to obtain the equation $y = a(x - h)^2$, and then we can add k to obtain $y = a(x - h)^2 + k$. Subtracting h causes a horizontal translation (right or left, depending on the sign of h), and adding k causes a vertical translation (up or down, depending on the sign of k). Thus, the graph of (8) looks roughly like one of those in Figure 1.4.10b, which are shown with $h > 0$ and $k > 0$ for simplicity. ◄

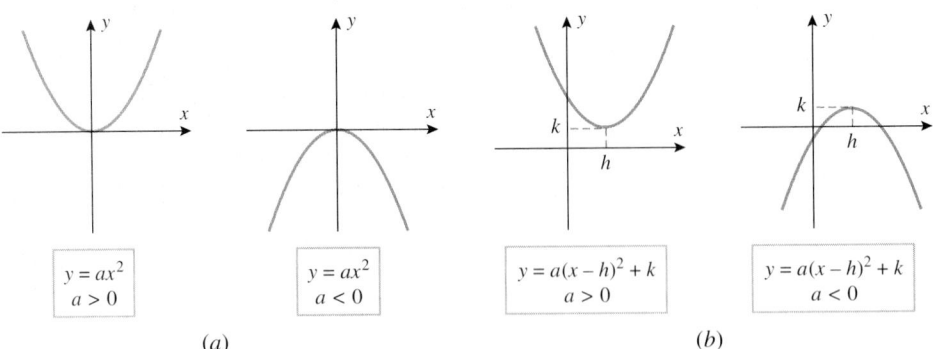

Figure 1.4.10

The graph of $y = f(-x)$ is the reflection of the graph of $y = f(x)$ about the y-axis, and the graph of $y = -f(x)$ [or equivalently, $-y = f(x)$] is the reflection of the graph of $y = f(x)$ about the x-axis. Thus, if you know what the graph of $y = f(x)$ looks like, you can obtain the graphs of $y = f(-x)$ and $y = -f(x)$ by making appropriate reflections. This is illustrated in Table 1.4.3.

Table 1.4.3

OPERATION ON $y = f(x)$	Replace x by $-x$	Multiply $f(x)$ by -1
NEW EQUATION	$y = f(-x)$	$y = -f(x)$
GEOMETRIC EFFECT	Reflects the graph of $y = f(x)$ about the y-axis	Reflects the graph of $y = f(x)$ about the x-axis
EXAMPLE	$y = \sqrt{-x}$ $\quad y = \sqrt{x}$	$y = \sqrt{x}$ $\quad y = -\sqrt{x}$

Example 12

Sketch the graph of $y = \sqrt[3]{2-x}$.

Solution. The graph can be obtained by a reflection and a translation: first reflect the graph of $y = \sqrt[3]{x}$ about the y-axis to obtain the graph of $y = \sqrt[3]{-x}$, then translate this graph right 2 units to obtain the graph of the equation $y = \sqrt[3]{-(x-2)} = \sqrt[3]{2-x}$ (Figure 1.4.11). ◀

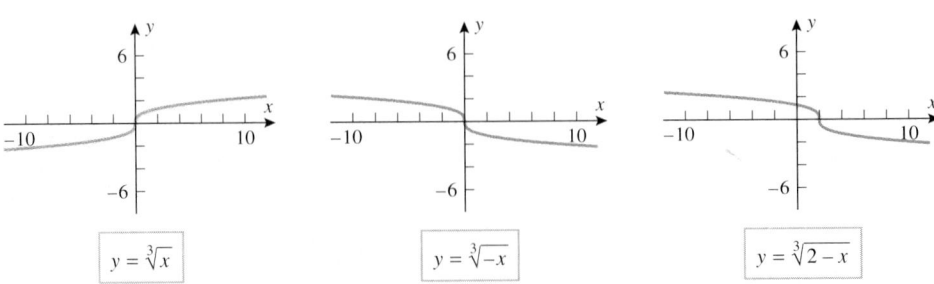

Figure 1.4.11

Example 13

Sketch the graph of $y = 4 - |x - 2|$.

Solution. The graph can be obtained by a reflection and two translations: first translate the graph of $y = |x|$ right 2 units to obtain the graph of $y = |x - 2|$; then reflect this graph about the x-axis to obtain the graph of $y = -|x - 2|$; and then translate this graph up 4 units to obtain the graph of the equation $y = -|x - 2| + 4 = 4 - |x - 2|$ (Figure 1.4.12).

◀

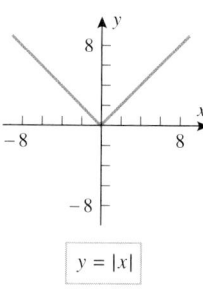

$y = |x|$

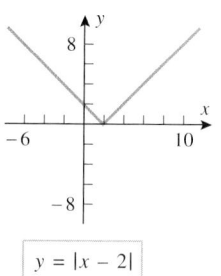

$y = |x - 2|$

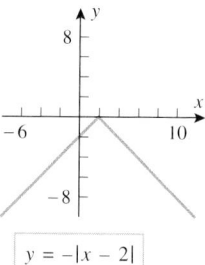

$y = -|x - 2|$

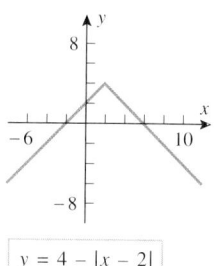

$y = 4 - |x - 2|$

Figure 1.4.12

EXERCISE SET 1.4 ⌇ Graphing Calculator

1. The graph of a function f is shown in the accompanying figure. Sketch the graphs of the following equations.
 (a) $y = f(x) - 1$ (b) $y = f(x - 1)$
 (c) $y = \frac{1}{2} f(x)$ (d) $y = f\left(-\frac{1}{2}x\right)$

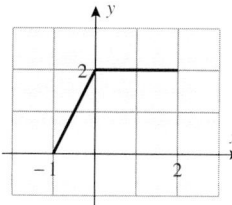

Figure Ex-1

2. Use the graph in Exercise 1.4.1 to sketch the graphs of the following equations.
 (a) $y = -f(-x)$ (b) $y = f(2 - x)$
 (c) $y = 1 - f(2 - x)$ (d) $y = \frac{1}{2} f(2x)$

3. The graph of a function f is shown in the accompanying figure. Sketch the graphs of the following equations.
 (a) $y = f(x + 1)$ (b) $y = f(2x)$
 (c) $y = |f(x)|$ (d) $y = 1 - |f(x)|$

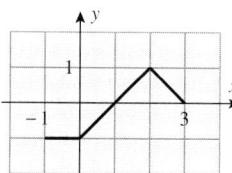

Figure Ex-3

4. Use the graph in Exercise 1.4.3 to sketch the graph of the equation $y = f(|x|)$.

In Exercises 5–12, sketch the graph of the equation by translating, reflecting, compressing, and stretching the graph of $y = x^2$ appropriately, and then use a graphing utility to confirm that your sketch is correct.

⌇ **5.** $y = 1 + (x - 2)^2$ ⌇ **6.** $y = 2 - (x + 1)^2$
⌇ **7.** $y = -2(x + 1)^2 - 3$ ⌇ **8.** $y = \frac{1}{2}(x - 3)^2 + 2$
⌇ **9.** $y = x^2 + 6x$ ⌇ **10.** $y = x^2 + 6x - 10$
⌇ **11.** $y = 1 + 2x - x^2$ ⌇ **12.** $y = \frac{1}{2}(x^2 - 2x + 3)$

In Exercises 13–16, sketch the graph of the equation by translating, reflecting, compressing, and stretching the graph of $y = \sqrt{x}$ appropriately, and then use a graphing utility to confirm that your sketch is correct.

⌇ **13.** $y = 3 - \sqrt{x + 1}$ ⌇ **14.** $y = 1 + \sqrt{x - 4}$
⌇ **15.** $y = \frac{1}{2}\sqrt{x} + 1$ ⌇ **16.** $y = -\sqrt{3x}$

In Exercises 17–20, sketch the graph of the equation by translating, reflecting, compressing, and stretching the graph of $y = 1/x$ appropriately, and then use a graphing utility to confirm that your sketch is correct.

⌇ **17.** $y = \dfrac{1}{x - 3}$ ⌇ **18.** $y = \dfrac{1}{1 - x}$

19. $y = 2 - \dfrac{1}{x+1}$

20. $y = \dfrac{x-1}{x}$

In Exercises 21–24, sketch the graph of the equation by translating, reflecting, compressing, and stretching the graph of $y = |x|$ appropriately, and then use a graphing utility to confirm that your sketch is correct.

21. $y = |x+2| - 2$

22. $y = 1 - |x-3|$

23. $y = |2x - 1| + 1$

24. $y = \sqrt{x^2 - 4x + 4}$

In Exercises 25–28, sketch the graph of the equation by translating, reflecting, compressing, and stretching the graph of $y = \sqrt[3]{x}$ appropriately, and then use a graphing utility to confirm that your sketch is correct.

25. $y = 1 - 2\sqrt[3]{x}$

26. $y = \sqrt[3]{x-2} - 3$

27. $y = 2 + \sqrt[3]{x+1}$

28. $y + \sqrt[3]{x-2} = 0$

29. (a) Sketch the graph of $y = x + |x|$ by adding the corresponding y-coordinates on the graphs of $y = x$ and $y = |x|$.

    (b) Express the equation $y = x + |x|$ in piecewise form with no absolute values, and confirm that the graph you obtained in part (a) is consistent with this equation.

30. Sketch the graph of $y = x + (1/x)$ by adding corresponding y-coordinates on the graphs of $y = x$ and $y = 1/x$. Use a graphing utility to confirm that your sketch is correct.

In Exercises 31–34, find formulas for $f + g$, $f - g$, fg, and f/g, and state the domains of the functions.

31. $f(x) = 2x$, $g(x) = x^2 + 1$

32. $f(x) = 3x - 2$, $g(x) = |x|$

33. $f(x) = 2\sqrt{x-1}$, $g(x) = \sqrt{x-1}$

34. $f(x) = \dfrac{x}{1+x^2}$, $g(x) = \dfrac{1}{x}$

35. Let $f(x) = \sqrt{x}$ and $g(x) = x^3 + 1$. Find

    (a) $f(g(2))$ (b) $g(f(4))$

    (c) $f(f(16))$ (d) $g(g(0))$.

36. Let $g(x) = \pi - x^2$ and $h(x) = \cos x$. Find

    (a) $g(h(0))$ (b) $h(g(\sqrt{\pi/2}\,))$

    (c) $g(g(1))$ (d) $h(h(\pi/2))$.

37. Let $f(x) = x^2 + 1$. Find

    (a) $f(t^2)$ (b) $f(t+2)$ (c) $f(x+2)$

    (d) $f\left(\dfrac{1}{x}\right)$ (e) $f(x+h)$ (f) $f(-x)$

    (g) $f(\sqrt{x}\,)$ (h) $f(3x)$.

38. Let $g(x) = \sqrt{x}$. Find

    (a) $g(5s+2)$ (b) $g(\sqrt{x}+2)$ (c) $3g(5x)$

    (d) $\dfrac{1}{g(x)}$ (e) $g(g(x))$ (f) $(g(x))^2 - g(x^2)$

    (g) $g(1/\sqrt{x}\,)$ (h) $g((x-1)^2)$.

In Exercises 39–44, find formulas for $f \circ g$ and $g \circ f$, and state the domains of the functions.

39. $f(x) = 2x + 1$, $g(x) = x^2 - x$

40. $f(x) = 2 - x^2$, $g(x) = x^3$

41. $f(x) = x^2$, $g(x) = \sqrt{1-x}$

42. $f(x) = \sqrt{x-3}$, $g(x) = \sqrt{x^2 + 3}$

43. $f(x) = \dfrac{1+x}{1-x}$, $g(x) = \dfrac{x}{1-x}$

44. $f(x) = \dfrac{x}{1+x^2}$, $g(x) = \dfrac{1}{x}$

In Exercises 45 and 46, find a formula for $f \circ g \circ h$.

45. $f(x) = x^2 + 1$, $g(x) = \dfrac{1}{x}$, $h(x) = x^3$

46. $f(x) = \dfrac{1}{1+x}$, $g(x) = \sqrt[3]{x}$, $h(x) = \dfrac{1}{x^3}$

In Exercises 47–50, express f as a composition of two functions; that is, find g and h such that $f = g \circ h$. [*Note:* Each exercise has more than one solution.]

47. (a) $f(x) = \sqrt{x+2}$ (b) $f(x) = |x^2 - 3x + 5|$

48. (a) $f(x) = x^2 + 1$ (b) $f(x) = \dfrac{1}{x-3}$

49. (a) $f(x) = \sin^2 x$ (b) $f(x) = \dfrac{3}{5 + \cos x}$

50. (a) $f(x) = 3\sin(x^2)$ (b) $f(x) = 3\sin^2 x + 4\sin x$

In Exercises 51 and 52, express F as a composition of three functions; that is, find f, g, and h such that $F = f \circ g \circ h$. [*Note:* Each exercise has more than one solution.]

51. (a) $F(x) = \left(1 + \sin(x^2)\right)^3$ (b) $F(x) = \sqrt{1 - \sqrt[3]{x}}$

52. (a) $F(x) = \dfrac{1}{1 - x^2}$ (b) $F(x) = |5 + 2x|$

53. Use the table in the accompanying figure to make a scatter plot of $y = f(g(x))$.

x	-3	-2	-1	0	1	2	3
$f(x)$	-4	-3	-2	-1	0	1	2
$g(x)$	-1	0	1	2	3	-2	-3

Figure Ex-53

54. Find the domain of $g \circ f$ for the functions f and g in Exercise 53.

55. Sketch the graph of $y = f(g(x))$ for the functions graphed in the accompanying figure.

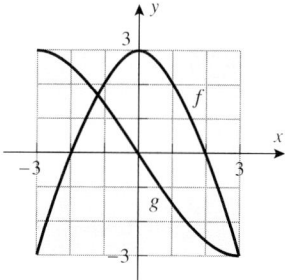

Figure Ex-55

56. Sketch the graph of $y = g(f(x))$ for the functions graphed in Exercise 55.

57. Use the graphs of f and g in Exercise 55 to estimate the solutions of the equations $f(g(x)) = 0$ and $g(f(x)) = 0$.

58. Use the table in Exercise 53 to solve the equations $f(g(x)) = 0$ and $g(f(x)) = 0$.

In Exercises 59–62, find
$$\frac{f(x+h) - f(x)}{h}$$
and simplify as much as possible.

59. $f(x) = 3x^2 - 5$ **60.** $f(x) = x^2 + 6x$

61. $f(x) = 1/x$ **62.** $f(x) = 1/x^2$

63. In each part of the accompanying figure determine whether the graph is symmetric about the x-axis, the y-axis, the origin, or none of the preceding.

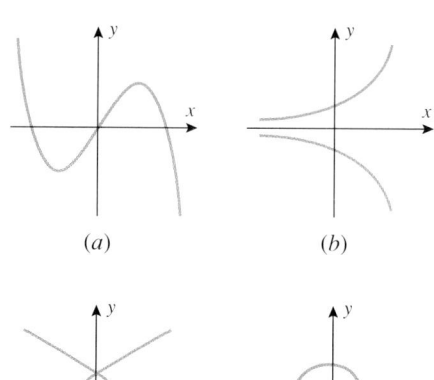

(a) (b)

(c) (d)

Figure Ex-63

64. The accompanying figure shows a portion of a graph. Complete the graph so that the entire graph is symmetric about
(a) the x-axis (b) the y-axis (c) the origin.

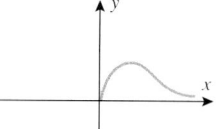

Figure Ex-64

65. Complete the table in the accompanying figure so that the graph of $y = f(x)$ (which is a scatter plot) is symmetric about
(a) the y-axis (b) the origin.

x	-3	-2	-1	0	1	2	3
$f(x)$	1		-1	0		-5	

Figure Ex-65

66. The accompanying figure shows a portion of the graph of a function f. Complete the graph assuming that
(a) f is an even function (b) f is an odd function.

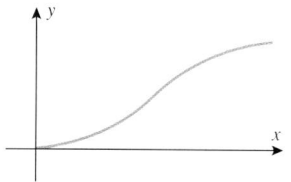

Figure Ex-66

67. Classify the functions graphed in the accompanying figure as even, odd, or neither.

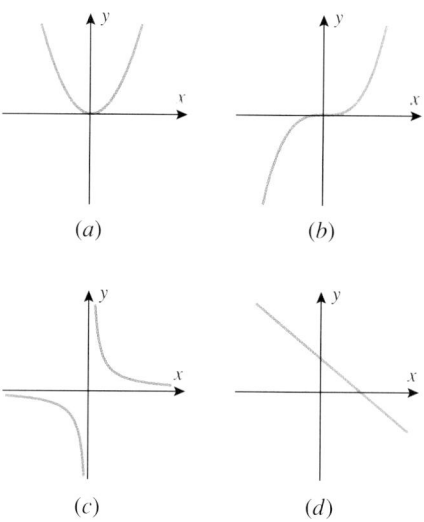

(a) (b)

(c) (d)

Figure Ex-67

68. Classify the functions whose values are given in the following table as even, odd, or neither.

x	-3	-2	-1	0	1	2	3
$f(x)$	5	3	2	3	1	-3	5
$g(x)$	4	1	-2	0	2	-1	-4
$h(x)$	2	-5	8	-2	8	-5	2

69. In each part, classify the function as even, odd, or neither.
   (a) $f(x) = x^2$ (b) $f(x) = x^3$
   (c) $f(x) = |x|$ (d) $f(x) = x + 1$
   (e) $f(x) = \dfrac{x^5 - x}{1 + x^2}$ (f) $f(x) = 2$

In Exercises 70 and 71, use Theorem 1.4.3 to determine whether the graph has symmetries about the x-axis, the y-axis, or the origin.

70. (a) $x = 5y^2 + 9$ (b) $x^2 - 2y^2 = 3$
   (c) $xy = 5$

71. (a) $x^4 = 2y^3 + y$ (b) $y = \dfrac{x}{3 + x^2}$
   (c) $y^2 = |x| - 5$

In Exercises 72 and 73: (i) Use a graphing utility to graph the equation in the first quadrant. [*Note:* To do this you will have to solve the equation for y in terms of x.] (ii) Use symmetry to make a hand-drawn sketch of the entire graph. (iii) Confirm your work by generating the graph of the equation in the remaining three quadrants.

72. $9x^2 + 4y^2 = 36$ **73.** $4x^2 + 16y^2 = 16$

74. The graph of the equation $x^{2/3} + y^{2/3} = 1$, which is shown in the accompanying figure, is called a *four-cusped hypocycloid*.
   (a) Use Theorem 1.4.3 to confirm that this graph is symmetric about the x-axis, the y-axis, and the origin.
   (b) Find a function f whose graph in the first quadrant coincides with the four-cusped hypocycloid, and use a graphing utility to confirm your work.
   (c) Repeat part (b) for the remaining three quadrants.

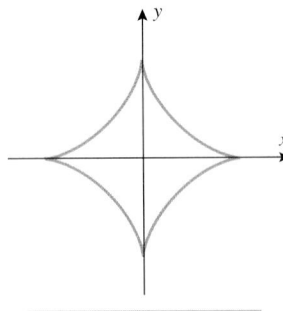

Four-cusped hypocycloid

Figure Ex-74

75. The equation $y = |f(x)|$ can be written as
$$y = \begin{cases} f(x), & f(x) \geq 0 \\ -f(x), & f(x) < 0 \end{cases}$$
which shows that the graph of $y = |f(x)|$ can be obtained from the graph of $y = f(x)$ by retaining the portion that lies on or above the x-axis and reflecting about the x-axis the portion that lies below the x-axis. Use this method to obtain the graph of $y = |2x - 3|$ from the graph of $y = 2x - 3$.

In Exercises 76 and 77, use the method described in Exercise 75.

76. Sketch the graph of $y = |1 - x^2|$.

77. Sketch the graph of
   (a) $f(x) = |\cos x|$ (b) $f(x) = \cos x + |\cos x|$.

78. The *greatest integer function*, $[x]$, is defined to be the greatest integer that is less than or equal to x. For example, $[2.7] = 2$, $[-2.3] = -3$, and $[4] = 4$. Sketch the graph of
   (a) $f(x) = [x]$ (b) $f(x) = [x^2]$
   (c) $f(x) = [x]^2$ (d) $f(x) = [\sin x]$.

79. Is it ever true that $f \circ g = g \circ f$ if f and g are nonconstant functions? If not, prove it; if so, give some examples for which it is true.

80. In the discussion preceding Exercise 29 of Section 1.3, we gave a procedure for generating a complete graph of $f(x) = x^{p/q}$ in which we suggested graphing the function $g(x) = |x|^{p/q}$ instead of $f(x)$ when p is even and q is odd and graphing $g(x) = (|x|/x)|x|^{p/q}$ if p is odd and q is odd. Show that in both cases $f(x) = g(x)$ if $x > 0$ or $x < 0$. [*Hint:* Show that $f(x)$ is an even function if p is even and q is odd and is an odd function if p is odd and q is odd.]

1.5 MATHEMATICAL MODELS; LINEAR MODELS

In this section we will discuss mathematical modeling, which is the process that is used to express scientific laws in mathematical form. We will also review some results about lines and apply those results to mathematical modeling.

This section includes a quick review of precalculus material on lines. Readers who want to review this material in more depth are referred to Appendix C.

MATHEMATICAL MODELS

A *mathematical model* of a physical law is a description of that law in the language of mathematics. The process of constructing a mathematical model is called *mathematical modeling*. For example, suppose that two variables, x and y, are related by some physical law that we would like to describe by a mathematical model. Models can be expressed in terms of graphs, tables, or equations, ranging from simple to extremely complicated. However, many important mathematical models are simply equations of the form

$$y = f(x)$$

that relate x and y. For such models the fundamental problem is to find a function f that accurately describes the physical relationship between the variables. Sometimes an appropriate function f might be suggested by experimental data, in which case we say that the model is obtained *inductively*, and sometimes it might be derived from some general theory proposed by a researcher, in which case we say that the model is obtained *deductively*.

The more factors one takes into account when creating a mathematical model, the more complicated the model tends to become, so there is always a balance to be struck between keeping a model mathematically simple and accounting for all of the physical factors that might affect the relationship between the variables. For example, if a meteorologist were trying to model the relationship between the speed of a raindrop when it hits the ground and the height of the cloud in which it was formed, then he or she would certainly have to take air resistance into account. However, the meteorologist would likely ignore the gravitational pull of the planet Pluto since its effect is so small.

Once a mathematical model of a physical law is obtained, it may be possible to use mathematical methods to deduce results about the physical world that are not self-evident or have never been observed. For example, the possibility of placing a satellite in orbit around the Earth was deduced mathematically from Isaac Newton's model of mechanics nearly 200 years before the launching of *Sputnik*, and Albert Einstein's relativistic model of mechanics in 1915 explained a precession (position shift) in the perihelion of the planet Mercury that was not confirmed by physical measurement until 1967.

A good mathematical model is one that produces results that are consistent with observations in the physical world. If a time comes when the mathematical results produced by the model do not agree with real-world observations, then the model must be abandoned in favor of a new model that does. This is the nature of the scientific method—old models constantly being replaced by new models that more accurately describe the real world.

A QUICK REVIEW OF LINES

An equation that is expressible in the form

$$Ax + By + C = 0 \tag{1}$$

where A and B are not both zero, is called a *first-degree equation* or a *linear equation* in x and y. It is shown in precalculus that every first-degree equation in x and y has a straight line as its graph and, conversely, every straight line can be represented by a first-degree equation in x and y. For this reason (1) is sometimes called the *general equation* of a line.

Recall that equations of lines can be written in several different forms:

$$y = mx + b \qquad \text{Slope-intercept form} \qquad (2)$$

$$y - y_1 = m(x - x_1) \qquad \text{Point-slope form} \qquad (3)$$

$$\frac{x}{a} + \frac{y}{b} = 1 \qquad \text{Double-intercept form} \qquad (4)$$

In these equations m is the slope of the line, a is the x-intercept, b is the y-intercept, and (x_1, y_1) is any point on the line (Figure 1.5.1). Keep in mind that these equations do not apply to vertical lines. For vertical lines the slope is *undefined*, or stated informally, a vertical line has **infinite slope**. Vertical and horizontal lines have particularly simple equations:

$$x = a \qquad \text{The vertical line with } x\text{-intercept } a \qquad (5)$$

$$y = b \qquad \text{The horizontal line with } y\text{-intercept } b \qquad (6)$$

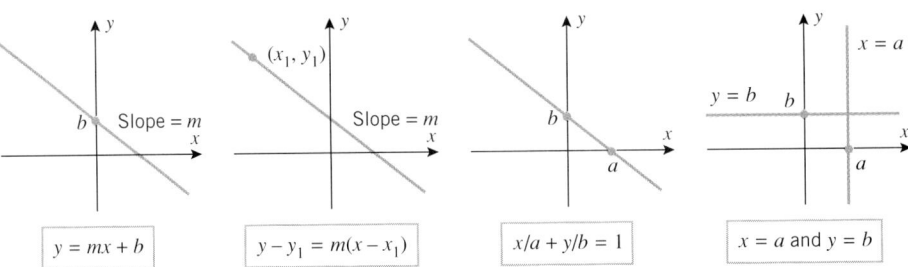

Figure 1.5.1

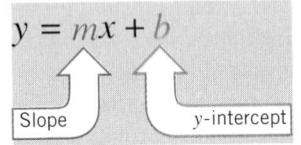

Figure 1.5.2

Equation (2) is especially useful because the slope and the y-intercept of the line can be determined by inspection: the slope is the coefficient of x, and the y-intercept is the constant term (Figure 1.5.2). This equation expresses y as a function of x, the function being $f(x) = mx + b$. A function of this form is called a **linear function** of x.

....

INTERPRETATIONS OF SLOPE

The slope m of a nonvertical line $y = mx + b$ has two important interpretations (which are related but different in viewpoint):

- m is a measure of the *steepness* of the line.
- m is the rate of change of y with respect to x.

The steepness interpretation has an analog in surveying. Surveyors measure the grade or slope of a hill as the ratio of its rise over its run (Figure 1.5.3a). The same idea applies to lines. Consider a particle that moves left to right along a nonvertical line from a point

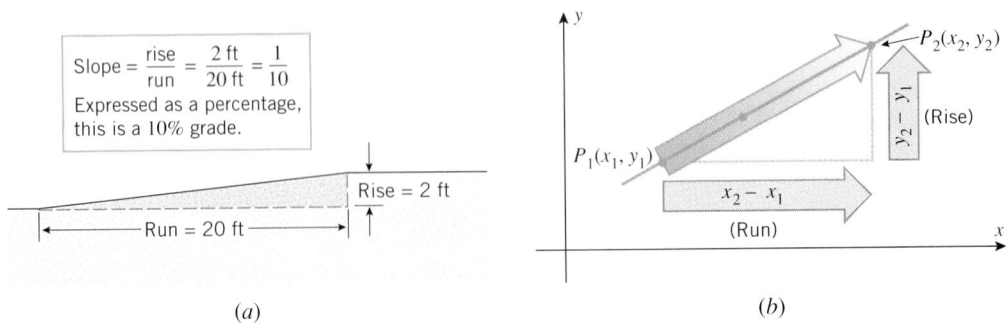

(a)

(b)

Figure 1.5.3

$P_1(x_1, y_1)$ to a point $P_2(x_2, y_2)$. In the course of its travel the point moves $y_2 - y_1$ units vertically as it travels $x_2 - x_1$ units horizontally (Figure 1.5.3b). The vertical change, which is denoted by $\Delta y = y_2 - y_1$, is called the **rise**, and the horizontal change, which is denoted by $\Delta x = x_2 - x_1$, is called the **run**. The ratio of the rise over the run is always equal to the slope, regardless of where the points P_1 and P_2 are located on the line; that is,

$$m = \frac{\Delta y}{\Delta x} = \frac{y_2 - y_1}{x_2 - x_1} \tag{7}$$

REMARK.   The symbols Δx and Δy should not be interpreted as products; rather, Δx should be viewed as a single entity representing the *change* in the value of x, and Δy as a single entity representing the *change* in the value of y. In general, if v is any variable whose value changes from an initial value of v_1 to a final value of v_2, then we call $\Delta v = v_2 - v_1$ (final value minus initial value) an **increment** in v. Increments can be positive or negative, depending on whether the final value is larger or smaller than the initial value.

ANGLE OF INCLINATION

The slope of a nonvertical line L is related to the angle that L makes with the positive x-axis. If ϕ is the smallest positive angle measured counterclockwise from the x-axis to L, then the slope of the line can be expressed as

$$m = \tan \phi \tag{8}$$

(Figure 1.5.4a). The angle ϕ, which is called the **angle of inclination** of the line, satisfies $0° \leq \phi < 180°$ in degree measure (or, equivalently, $0 \leq \phi < \pi$ in radian measure). If ϕ is an acute angle, then $m = \tan \phi$ is positive and the line slopes up to the right, and if ϕ is an obtuse angle, then $m = \tan \phi$ is negative and the line slopes down to the right. For example, a line whose angle of inclination is $45°$ has slope $m = \tan 45° = 1$, and a line whose angle of inclination is $135°$ has a slope of $m = \tan 135° = -1$ (Figure 1.5.4b). Figure 1.5.5 shows a convenient way of using the line $x = 1$ as a "ruler" for visualizing the relationship between lines of various slopes.

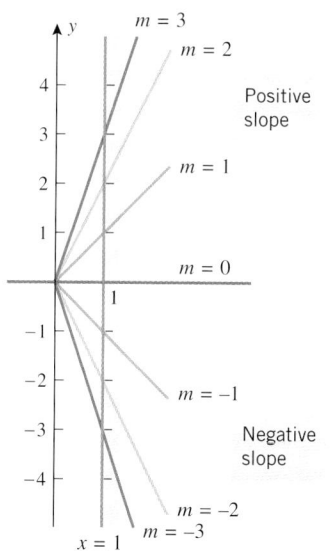

Figure 1.5.5

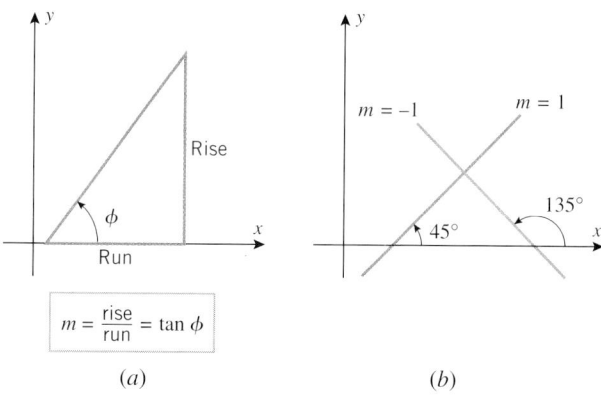

Figure 1.5.4

SLOPES OF LINES IN APPLIED PROBLEMS

In applied problems, changing the units of measurement can change the slope of a line, so it is essential to include the units when calculating the slope. The following example illustrates this.

Example 1

Suppose that a uniform rod of length 40 cm (= 0.4 m) is thermally insulated around the lateral surface and that the exposed ends of the rod are held at constant temperatures of $25°C$ and $5°C$, respectively (Figure 1.5.6a). It is shown in physics that under appropriate

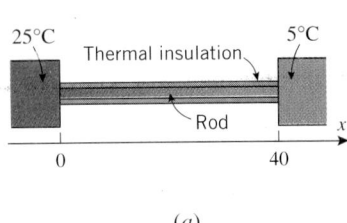

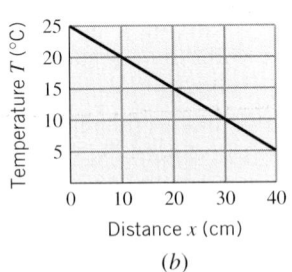

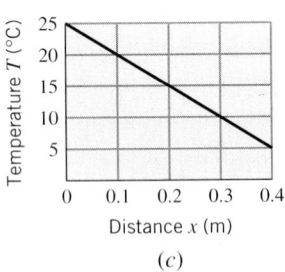

(a)   (b)   (c)

Figure 1.5.6

conditions the graph of the temperature T versus the distance x from the left-hand end of the rod will be a straight line. Parts (b) and (c) of Figure 1.5.6 show two such graphs: one in which x is measured in centimeters and one in which it is measured in meters. The slopes in the two cases are

$$m = \frac{5-25}{40-0} = \frac{-20}{40} = -0.5^\circ\text{C/cm} \tag{9}$$

$$m = \frac{5-25}{0.4-0} = \frac{-20}{0.4} = -50^\circ\text{C/m} \tag{10}$$

The slope in (9) implies that the temperature *decreases* at a rate of 0.5°C per centimeter of distance from the left end of the rod, and the slope in (10) implies that the temperature decreases at a rate of 50°C per meter of distance from the left end of the rod. The two statements are equivalent physically, even though the slopes differ.   ◄

Example 2

Find the slope-intercept form of the equation of the temperature distribution in the preceding example if the temperature T is measured in degrees Celsius ($^\circ$C) and the distance x is measured in (a) centimeters and (b) meters.

Solution (a). The slope is $m = -0.5^\circ$C/cm and the intercept on the T-axis is 25°, so

$$T = -0.5x + 25, \quad 0 \le x \le 40$$

where the restriction on x is required because the rod is 40 cm in length. The graph of this equation with the restriction is a line segment rather than a line.

Solution (b). The slope is $m = -50^\circ$C/m, the intercept on the T-axis is 25°, and the restriction on x is $0 \le x \le 0.4$. Thus, the equation is

$$T = -50x + 25, \quad 0 \le x \le 0.4 \qquad\qquad ◄$$

LINEAR MATHEMATICAL MODELS

If y is a linear function of x, say $y = mx + b$, then it follows from (7) that

$$\Delta y = m\,\Delta x$$

Thus, a 1-unit increase in x ($\Delta x = 1$) produces an m-unit change in y ($\Delta y = m$). Moreover, this is true at every point on the line (Figure 1.5.7), so we say that y changes at a *constant rate* with respect to x, and we call m the **rate of change of y with respect to x**. This idea can be summarized as follows.

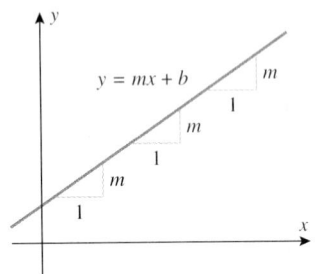

A 1-unit increase in x always produces an m-unit change in y.

Figure 1.5.7

1.5.1   LINEAR MATHEMATICAL MODELS.   *If a variable y is related to a variable x in such a way that the rate of change of y with respect to x is constant, say m, then y is a linear function of x of the form*

$$y = mx + b$$

*and we say that y is related to x by a **linear mathematical model**. Conversely, if y is a linear function of x whose graph has slope m, then the rate of change of y with respect to x is constant and equal to m.*

It follows from this that linear models are appropriate whenever experimentation or theory suggests that the rate of change of y with respect to x is constant.

UNIFORM RECTILINEAR MOTION

One of the important themes in calculus is the study of motion. To describe the motion of an object completely, one must specify its **speed** (how fast it is going) and the direction in which it is moving. The speed and the direction of motion together comprise what is called the **velocity** of the object. For example, knowing that the speed of an aircraft is 500 mi/h tells us how fast it is going, but not which way it is moving. In contrast, knowing that the velocity of the aircraft is 500 mi/h *due south* pins down the speed and the direction of motion.

Later, we will study the motion of particles that move along curves in two- or three-dimensional space, but for now we will focus on motion along a line; this is called **rectilinear motion**. In general rectilinear motion, a particle can move back and forth along the line (as with a piston moving up and down in a cylinder); however, for now we will only consider the simple case in which the particle moves in just *one direction* along a line (as with a car traveling on a straight road).

For simplicity, we will assume that the motion is along a coordinate line, such as an x-axis or y-axis, and that the particle is moving in the positive direction. In general discussions we will usually name the coordinate line the s-axis to avoid being specific. A graphical description of rectilinear motion along an s-axis can be obtained by making a plot of the s-coordinate of the particle versus the elapsed time t. This is called the **position versus time curve** for the particle. Figure 1.5.8a shows a typical position versus time curve for a car moving in the positive direction along an s-axis.

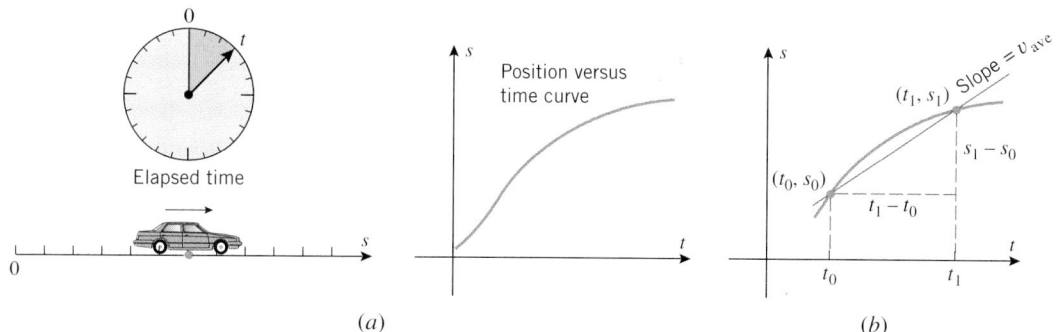

(a)   (b)

Figure 1.5.8

FOR THE READER. How can you tell from the position versus time curve in Figure 1.5.8a that the car does not reverse direction?

Because we are assuming that the particle is moving in the positive direction of the s-axis, there is no ambiguity about the direction of motion, and hence the terms "speed" and "velocity" can be used interchangeably. However, later, when we consider general rectilinear motion or motion along a curved path, it will be necessary to distinguish between these terms, since the direction of motion may vary.

For a particle in rectilinear motion along a coordinate axis, we define the **average velocity** v_{ave} of the particle during the time interval from t_0 to t_1 to be

$$v_{ave} = \frac{s_1 - s_0}{t_1 - t_0} = \frac{\Delta s}{\Delta t} \tag{11}$$

where s_0 and s_1 are the s-coordinates of the particle at times t_0 and t_1, respectively. Geometrically, this is the slope of the secant line connecting the points (t_0, s_0) and (t_1, s_1) on the position versus time curve (Figure 1.5.8b). The quantity $\Delta s = s_1 - s_0$ is called

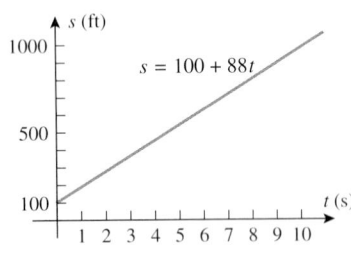

Figure 1.5.9

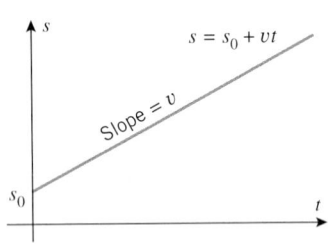

Position versus time curve for a particle with coordinate s_0 at time $t = 0$ and moving with constant velocity v

Figure 1.5.10

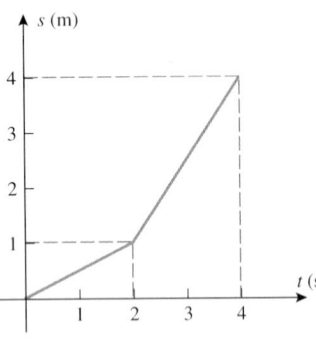

Figure 1.5.11

CONSTANT ACCELERATION

the *displacement* or *change in position* of the particle during the time interval from t_0 to t_1. With this terminology, Formula (11) states that for a particle in rectilinear motion *the average velocity over a time interval is the displacement during the time interval divided by the length of the time interval*. For example, if a car moving in one direction along a straight road travels 75 miles in 3 hours, then its average velocity is $75/3 = 25$ mi/h.

In the special case where the average velocity of a particle in rectilinear motion is the same over every time interval, the particle is said to have **constant velocity** and **uniform rectilinear motion**. If the average velocity over every time interval is v, then we will refer to v as the **velocity** of the particle (dropping the adjective "average").

For a particle with uniform rectilinear motion the displacement over *any* time interval is given by the formula

$$\text{displacement} = \text{velocity} \times \text{elapsed time} \tag{12}$$

Example 3

Suppose that a car moves with a constant velocity of 88 ft/s in the positive direction of an s-axis. Given that the s-coordinate of the car at time $t = 0$ is $s = 100$, find an equation for s as a function of t, and graph the position versus time curve.

Solution. It follows from (12) that in a period of t seconds, the car will move $88t$ feet from its starting point, so its coordinate s at time t will be

$$s = 100 + 88t$$

The graph of this equation is the line in Figure 1.5.9. ◀

It is not accidental that the position versus time curve turned out to be a line in the last example; this will always be the case for uniform rectilinear motion. To see why this is so, suppose that a particle moves with constant velocity v in the positive direction along an s-axis, starting at the point s_0 at time $t = 0$. It follows from (12) that in t units of time the particle will move vt units from its starting point s_0, so its coordinate s at time t will be

$$s = s_0 + vt$$

which is a line with s-intercept s_0 and slope v (Figure 1.5.10). It follows from this equation and 1.5.1 that we can view the velocity v as the rate of change of s with respect to t, that is, the rate of change of position with respect to time.

Example 4

Figure 1.5.11 shows the position versus time curve for a particle moving along an s-axis. Describe the motion of the particle in words.

Solution. At time $t = 0$ the particle is at the origin. From time $t = 0$ to $t = 2$ the slope of the line segment is $\frac{1}{2}$, so the particle is moving with a constant velocity of $\frac{1}{2} = 0.5$ m/s. At time $t = 2$ the particle is at the point $s = 1$ (i.e., 1 meter from the origin). From time $t = 2$ to $t = 4$ the slope of the line segment is $\frac{3}{2}$, so the particle is moving with a constant velocity of $\frac{3}{2} = 1.5$ m/s. At time $t = 4$ it is at the point $s = 4$. ◀

In everyday language we say that an object is "accelerating" if it is speeding up and "decelerating" if it is slowing down. Mathematically, the **acceleration** of a particle in rectilinear motion is defined to be the *rate of change of velocity with respect to time*, where the acceleration is positive if the velocity is increasing and negative if it is decreasing. Thus, for a particle that moves in the positive direction of an s-axis, negative acceleration means the particle is "decelerating" in everyday language. Acceleration, like velocity, can be variable or constant. For example, by pressing the gas pedal of a car toward the floor smoothly, the driver can make the car's velocity increase at a constant rate (a constant acceleration); however, if the driver suddenly slams the pedal to the floor, the car will lurch forward, reflecting a nonconstant acceleration. Later in the text we will study acceleration in more depth, but for now we will only consider the case in which acceleration is constant.

REMARK. The units of acceleration are units of velocity divided by units of time. For example, if the velocity of a particle is increasing at a rate of 3 feet per second each second, then its acceleration is 3 ft/s/s (velocity in ft/s divided by time in s); this is usually written as 3 ft/s^2 (read "3 feet per second per second" or "3 feet per second squared"). Similarly, if the velocity of a particle is decreasing at a rate of 3 feet per second each second, then it has an acceleration of -3 ft/s^2.

Graphical information about the acceleration of a particle can be obtained from the graph of velocity versus time; this is called the ***velocity versus time curve***. In the case where the particle has constant acceleration, the velocity versus time curve will be linear, and its slope, which is the rate of change of velocity with time, will be the acceleration (Figure 1.5.12).

Example 5

Suppose that a car moves in the positive direction of an s-axis in such a way that its velocity v increases at a constant rate of 2 ft/s^2.

(a) Assuming that the velocity of the car is 88 ft/s at time $t = 0$, find an equation for v as a function of t.

(b) Make a graph of velocity versus time, and mark the point on the graph at which the car attains a velocity of 100 ft/s.

Solution (a). Since the rate of change of v with respect to t is 2 ft/s^2, and since $v = 88$ ft/s if $t = 0$, the equation for velocity as a function of time is

$$v = 88 + 2t \tag{13}$$

Solution (b). To find the time it takes for the car to reach a velocity of 100 ft/s, we substitute $v = 100$ in (13) and solve for t. This yields $t = 6$. The graph of (13) and the point at which the velocity reaches 100 ft/s is shown in Figure 1.5.13. ◀

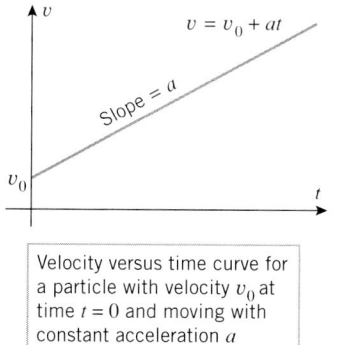

Velocity versus time curve for a particle with velocity v_0 at time $t = 0$ and moving with constant acceleration a

Figure 1.5.12

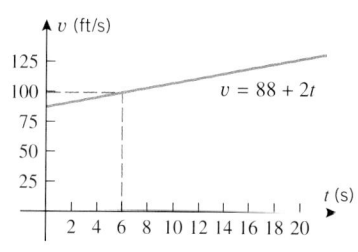

Velocity versus time curve for a particle with a velocity of 88 ft/s at time $t = 0$ and moving with a constant acceleration of 2 ft/s^2

Figure 1.5.13

LINEAR MODELS FROM DIRECT PROPORTION

Recall that a variable y is said to be ***directly proportional*** to a variable x if there is a positive constant k, called the ***constant of proportionality***, such that

$$y = kx \tag{14}$$

The graph of this equation is a line through the origin whose slope k is the constant of proportionality. Thus, linear models are appropriate in physical problems where one variable is directly proportional to another.

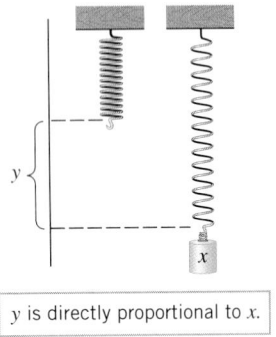

y is directly proportional to x.

Figure 1.5.14

Hooke's law[*] in physics provides a nice example of direct proportion. It follows from this law that if a weight of x units is suspended from a spring, then the spring will be stretched by an amount y that is directly proportional to x, that is, $y = kx$ (Figure 1.5.14). The constant k depends on the stiffness of the spring: the stiffer the spring, the smaller the value of k (why?).

Example 6

Figure 1.5.15 shows an old-fashioned spring scale that is calibrated in pounds.

(a) Given that the pound scale marks are 0.5 in apart, find an equation that expresses the length y that the spring is stretched (in inches) in terms of the suspended weight x in pounds).

(b) Graph the equation obtained in part (a).

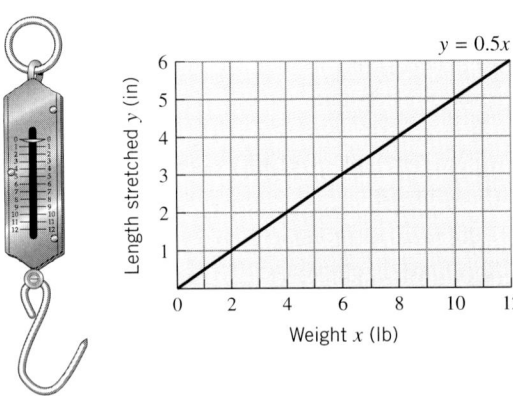

Figure 1.5.15

Solution (a). It follows from Hooke's law that y is related to x by an equation of the form $y = kx$. To find k we rewrite this equation as $k = y/x$, and use the fact that a weight of $x = 1$ lb stretches the spring $y = 0.5$ in. Thus,

$$k = \frac{y}{x} = \frac{0.5}{1} = 0.5 \qquad \text{and hence} \qquad y = 0.5x$$

Solution (b). The graph of the equation $y = 0.5x$ is shown in Figure 1.5.15. ◀

LINEAR MODELS FROM GRAPHICAL DATA

Sometimes linear models are suggested by graphical data. For example, Figure 1.5.16*a* shows a graph of temperature versus altitude that was transmitted by the *Magellan* spacecraft when it entered the atmosphere of Venus in October 1991. The graph strongly suggests that there is a linear relationship between temperature and altitude for altitudes between 35 km and 60 km.

Example 7

(a) Use the graph transmitted by the *Magellan* spacecraft to find a linear model of temperature versus altitude in the Venusian atmosphere that is valid for altitudes between 35 km and 60 km.

(b) Use the model to estimate the temperature at the surface of Venus, and discuss the assumptions you are making in obtaining the estimate.

---

[*] Hooke's law, named for the English physicist Robert Hooke (1635–1703), applies only for small displacements that do not stretch the spring to the point of permanently distorting it.

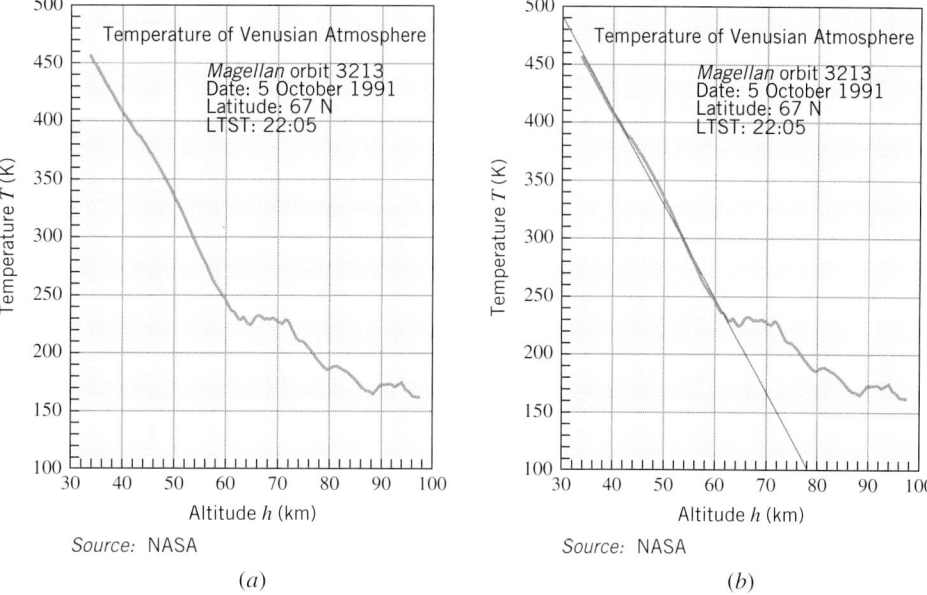

Figure 1.5.16

Solution (a). Let T be the temperature in kelvins and h the altitude in kilometers. We will first estimate the slope m of the linear portion of the graph, then estimate the coordinates of a data point (h_1, T_1) on that portion of the graph, and then use the point-slope form of a line

$$T - T_1 = m(h - h_1) \tag{15}$$

The graph nearly passes through the point $(60, 250)$, so we will take $h_1 \approx 60$ and $T_1 \approx 250$. In Figure 1.5.16*b* we have sketched a line that closely approximates the linear portion of the data. Using the intersections of that line with the edges of the grid box, we estimate the slope to be

$$m \approx \frac{100 - 490}{78 - 30} = -\frac{390}{48} \approx -8.125 \text{ K/km}$$

Substituting our estimates of h_1, T_1, and m into (15) yields the equation

$$T - 250 = -8.125(h - 60)$$

or equivalently,

$$T = -8.125h + 737.5 \tag{16}$$

Solution (b). The *Magellan* spacecraft stopped transmitting data at an altitude of approximately 35 km, so we cannot be certain that the linear model applies at lower altitudes. However, since we have no other data to work with, let us *assume* that the model is valid at all lower altitudes, in which case we can approximate the temperature at the surface of Venus by setting $h = 0$ in (16). We obtain $T \approx 737.5$ K. ◀

REMARK.   The method of the preceding example is crude, at best, since it relies on extracting rough estimates of numerical data from a graph. Nevertheless, the final result is quite good, since the most recent information from NASA places the surface temperature of Venus at about 740 K (hot enough to melt lead).

LINEAR MODELS FROM NUMERICAL DATA

One method for determining whether n points

$$(x_1, y_1), (x_2, y_2), \ldots, (x_n, y_n)$$

lie on a line is to compare the slopes of the line segments joining successive points. The points lie on a line if and only if those slopes are equal (Figure 1.5.17).

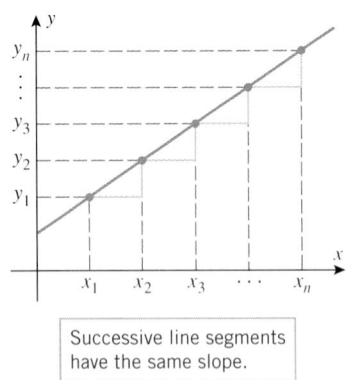

Successive line segments have the same slope.

Figure 1.5.17

Table 1.5.1

x	y
1.5	0.3
2.5	1.1
3.5	1.9
5.5	3.5
9.5	6.7

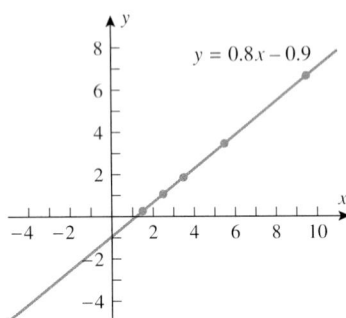

Figure 1.5.18

Example 8

Consider the data in Table 1.5.1.

(a) Explain why a linear model is appropriate for the data in the table.

(b) Find a linear equation that relates x and y, and graph the equation and the data together.

Solution (a). The five data points lie on a line, since each 1-unit increase in x produces a corresponding 0.8-unit increase in y. Thus, the slope of the line segment joining any two successive data points is

$$m = \frac{\Delta y}{\Delta x} = \frac{0.8}{1} = 0.8$$

Solution (b). A linear equation relating x and y can be obtained from the point-slope form of the line using the slope $m = 0.8$ calculated in part (a) and any one of the five data points. If we use the first data point, $(1.5, 0.3)$, we obtain

$$y - 0.3 = 0.8(x - 1.5)$$

or in slope-intercept form,

$$y = 0.8x - 0.9$$

The graph of this equation together with the given data are shown in Figure 1.5.18. ◀

REMARK.   Sometimes, data points that should theoretically lie on a line do not because of experimental error and other factors. In such cases curve-fitting techniques are used to find a line that most closely fits the data. Such techniques will be discussed later in the text.

Linear functions arise in a variety of practical problems. Here is a typical example.

OTHER APPLICATIONS OF LINEAR FUNCTIONS

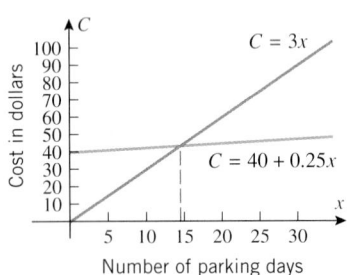

Number of parking days

Figure 1.5.19

Example 9

A university parking lot charges \$3.00 per day but offers a \$40.00 monthly sticker with which the student pays only \$0.25 per day.

(a) Find equations for the cost C of parking for x days per month under both payment methods, and graph the equations for $0 \le x \le 30$. (Treat C as a continuous function of x, even though x only assumes integer values.)

(b) Find the value of x for which the graphs intersect, and discuss the significance of this value.

Solution (a). The cost in dollars of parking for x days at \$3.00 per day is $C = 3x$, and the cost for the \$40.00 sticker plus x days at \$0.25 per day is $C = 40 + 0.25x$ (Figure 1.5.19).

Solution (b). The graphs intersect at the point where

$$3x = 40 + 0.25x$$

which is $x = 40/2.75 \approx 14.5$. This value of x is not an option for the student, since x must be an integer. However, it is the dividing point at which the monthly sticker method becomes less expensive than the daily payment method; that is, for $x \geq 15$ it is cheaper to buy the monthly sticker and for $x \leq 14$ it is cheaper to pay the daily rate. ◀

EXERCISE SET 1.5 ⌇ Graphing Calculator

· ·

Exercises 1–26 involve the basic properties of lines and slope. In some of these exercises you will need to use slopes to determine whether two lines are parallel or perpendicular. If you have forgotten how to do this, review Appendix C.

1. (a) Find the slopes of the sides of the triangle with vertices $(0, 3)$, $(2, 0)$, and $\left(6, \frac{8}{3}\right)$.
   (b) Is this a right triangle? Explain.

2. (a) Find the slopes of the sides of the quadrilateral with vertices $(-3, -1)$, $(5, -1)$, $(7, 3)$, and $(-1, 3)$.
   (b) Is this a parallelogram? Explain.

3. List the lines in the accompanying figure in the order of increasing slope.

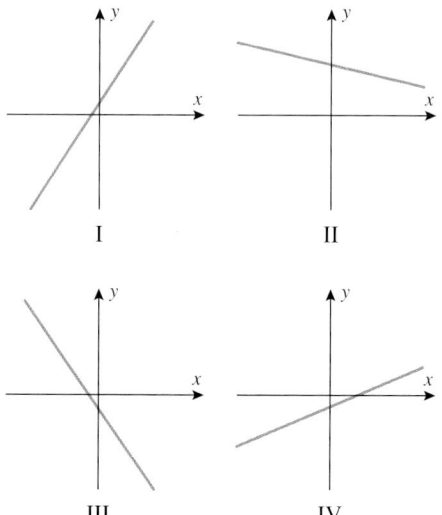

Figure Ex-3

4. List the lines in the accompanying figure in the order of increasing slope.

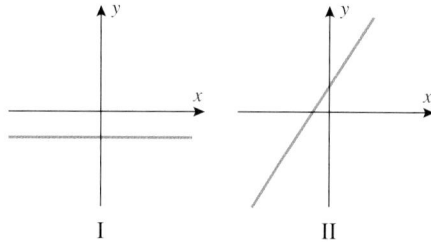

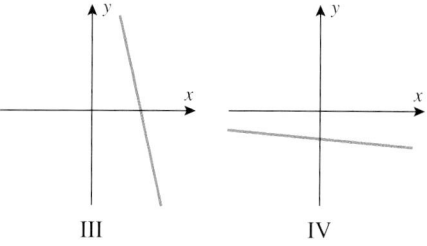

Figure Ex-4

5. Use slopes to determine whether the given points lie on the same line.
   (a) $(1, 1)$, $(-2, -5)$, and $(0, -1)$
   (b) $(-2, 4)$, $(0, 2)$, and $(1, 5)$

6. A particle, initially at $(7, 5)$, moves along a line of slope $m = -2$ to a new position (x, y).
   (a) Find y if $x = 9$.
   (b) Find x if $y = 12$.

7. A particle, initially at $(1, 2)$, moves along a line of slope $m = 3$ to a new position (x, y).
   (a) Find y if $x = 5$.
   (b) Find x if $y = -2$.

8. Find x and y if the line through $(0, 0)$ and (x, y) has slope $\frac{1}{2}$, and the line through (x, y) and $(7, 5)$ has slope 2.

9. Find x if the slope of the line through $(1, 2)$ and $(x, 0)$ is the negative of the slope of the line through $(4, 5)$ and $(x, 0)$.

In Exercises 10 and 11, find the angle of inclination of the line with slope m to the nearest degree. Use a calculating utility, where needed.

10. (a) $m = \frac{1}{2}$ (b) $m = -1$
(c) $m = 2$ (d) $m = -57$

11. (a) $m = -\frac{1}{2}$ (b) $m = 1$
(c) $m = -2$ (d) $m = 57$

In Exercises 12 and 13, find the angle of inclination of the line to the nearest degree. Use a calculating utility, where needed.

12. (a) $3y = 2 - \sqrt{3}x$ (b) $y - 4x + 7 = 0$

13. (a) $y = \sqrt{3}x + 2$ (b) $y + 2x + 5 = 0$

14. Find equations for the x- and y-axes.

In Exercises 15–22, find the slope-intercept form of the equation of the line satisfying the stated conditions, and check your answer using a graphing utility.

15. Slope $= -2$, y-intercept $= 4$

16. $m = 5$, $b = -3$

17. The line is parallel to $y = 4x - 2$ and its y-intercept is 7.

18. The line is parallel to $3x + 2y = 5$ and passes through $(-1, 2)$.

19. The line is perpendicular to the equation $y = 5x + 9$ and has y-intercept 6.

20. The line is perpendicular to $x - 4y = 7$ and passes through $(3, -4)$.

21. The line passes through $(2, 4)$ and $(1, -7)$.

22. The line passes through $(-3, 6)$ and $(-2, 1)$.

23. In each part, classify the lines as parallel, perpendicular, or neither.
(a) $y = 4x - 7$ and $y = 4x + 9$
(b) $y = 2x - 3$ and $y = 7 - \frac{1}{2}x$
(c) $5x - 3y + 6 = 0$ and $10x - 6y + 7 = 0$
(d) $Ax + By + C = 0$ and $Bx - Ay + D = 0$
(e) $y - 2 = 4(x - 3)$ and $y - 7 = \frac{1}{4}(x - 3)$

24. In each part, classify the lines as parallel, perpendicular, or neither.
(a) $y = -5x + 1$ and $y = 3 - 5x$
(b) $y - 1 = 2(x - 3)$ and $y - 4 = -\frac{1}{2}(x + 7)$
(c) $4x + 5y + 7 = 0$ and $5x - 4y + 9 = 0$
(d) $Ax + By + C = 0$ and $Ax + By + D = 0$
(e) $y = \frac{1}{2}x$ and $x = \frac{1}{2}y$

In Exercises 25 and 26, use the graph to find the equation of the line in slope-intercept form, and then check your result by using a graphing utility to graph the equation.

25.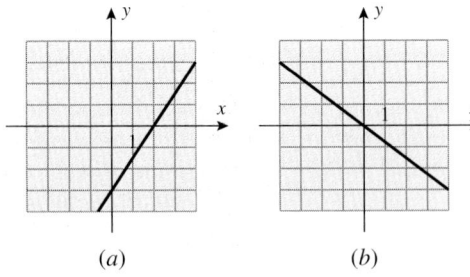

(a)   (b)

Figure Ex-25

26.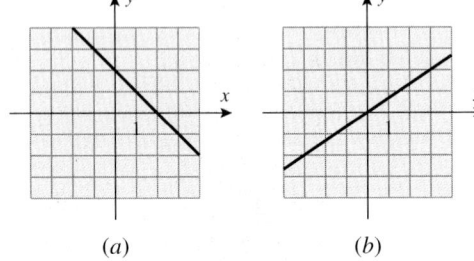

(a)   (b)

Figure Ex-26

27. The accompanying figure shows the position versus time curve for a particle moving along an x-axis.
(a) What is the velocity of the particle?
(b) What is the x-coordinate of the particle at time $t = 0$?
(c) What is the x-coordinate of the particle at time $t = 2$?
(d) At what time does the particle have an x-coordinate of $x = 4$?

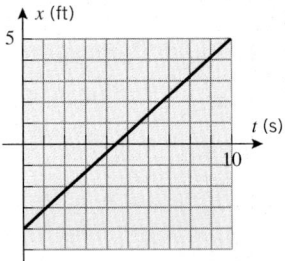

Figure Ex-27

28. A particle moving along an x-axis with constant velocity is at the point $x = 1$ when $t = 2$ and is at the point $x = 5$ when $t = 4$.
(a) Find the velocity of the particle if x is in meters and t is in seconds.
(b) Find an equation that expresses x as a function of t.
(c) What is the coordinate of the particle at time $t = 0$?

29. A particle moving along an x-axis with constant acceleration has velocity $v = 3$ ft/s at time $t = 1$ and velocity $v = -1$ ft/s at time $t = 4$.
(a) Find the acceleration of the particle.
(b) Find an equation that expresses v as a function of t.
(c) What is the velocity of the particle at time $t = 0$?

30. The accompanying figure shows the velocity versus time curve for a particle moving along the x-axis.
(a) What is the acceleration of the particle?
(b) What is the velocity of the particle at time $t = 0$?
(c) What is the velocity of the particle at time $t = 2$?
(d) At what time does the particle have a velocity of $v = 3$ ft/s?

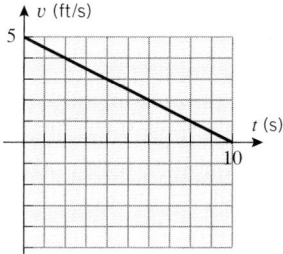

Figure Ex-30

31. The accompanying figure shows the position versus time curve for a particle moving along an x-axis.
(a) Describe the motion of the particle in words.
(b) Find the average velocity of the particle from $t = 0$ to $t = 10$.
(c) Find the average speed of the particle from $t = 0$ to $t = 10$.

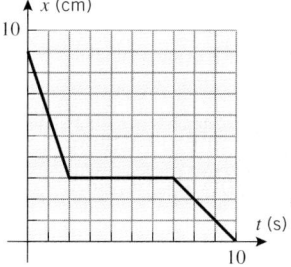

Figure Ex-31

32. The accompanying figure shows the velocity versus time curve for a particle moving along an x-axis. Describe the motion of the particle in words.

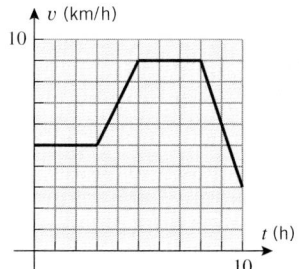

Figure Ex-32

33. A locomotive travels on a straight track at a constant speed of 40 mi/h, then reverses direction and returns to its starting point, traveling at a constant speed of 60 mi/h.
(a) What is the average velocity for the round-trip?
(b) What is the average speed for the round-trip?
(c) What is the total distance traveled by the train if the total trip took 5 h?

34. A ball is tossed straight up at time $t = 0$ with an initial velocity of 64 ft/s. We will show later using basic principles of physics that the velocity of the ball as a function of time is $v = 64 - 32t$.
(a) What direction is the ball traveling 3 s after it is released? Explain your reasoning.
(b) At what time does the ball reach its maximum height above the ground? Explain your reasoning.
(c) What can you say about the acceleration of the ball?

35. A car is stopped at a toll booth on a straight highway. Starting at time $t = 0$ it accelerates at a constant rate of 10 ft/s^2 for 10 s. It then travels at a constant speed of 100 ft/s for 90 s. At that time it begins to decelerate at a constant rate of 5 ft/s^2 for 20 s, at which point in time it reaches a full stop at a traffic light.
(a) Sketch the velocity versus time curve.
(b) Express v as a piecewise function of t.

36. Make a reasonable sketch of a position versus time curve for a particle that moves in the positive x-direction with positive constant acceleration.

37. A spring with a natural length of 15 in stretches to a length of 20 in when a 45-lb object is suspended from it.
(a) Use Hooke's law to find an equation that expresses the length y that the spring is stretched (in inches) in terms of the suspended weight x (in pounds).
(b) Graph the equation obtained in part (b).
(c) Find the length of the spring when a 100-lb object is suspended from it.
(d) What is the largest weight that can be suspended from the spring if the spring cannot be stretched to more than twice its natural length?

38. The spring in a heavy-duty shock absorber has a natural length of 3 ft and is compressed 0.2 ft by a load of 1 ton. An additional load of 5 tons compresses the spring an additional 1 ft.
(a) Assuming that Hooke's law applies to compression as well as extension, find an equation that expresses the length y that the spring is compressed from its natural length (in feet) in terms of the load x (in tons).
(b) Graph the equation obtained in part (a).
(c) Find the amount that the spring is compressed from its natural length by a load of 3 tons.
(d) Find the maximum load that can be applied if safety regulations prohibit compressing the spring to less than half its natural length.

In Exercises 39 and 40, confirm that a linear model is appropriate for the relationship between x and y. Find a linear equation relating x and y, and verify that the data points lie on the graph of your equation.

39.

x	0	1	2	4	6
y	2	3.2	4.4	6.8	9.2

40.

x	−1	0	2	5	8
y	12.6	10.5	6.3	0	−6.3

41. There are two common systems for measuring temperature, Celsius and Fahrenheit. Water freezes at $0°$Celsius ($0°$C) and $32°$Fahrenheit ($32°$F); it boils at $100°$C and $212°$F.
(a) Assuming that the Celsius temperature T_C and the Fahrenheit temperature T_F are related by a linear equation, find the equation.
(b) What is the slope of the line relating T_F and T_C if T_F is plotted on the horizontal axis?
(c) At what temperature is the Fahrenheit reading equal to the Celsius reading?
(d) Normal body temperature is $98.6°$F. What is it in $°$C?

42. Thermometers are calibrated using the so-called "triple point" of water, which is 273.16 K on the Kelvin scale and $0.01°$C on the Celsius scale. A one-degree difference on the Celsius scale is the same as a one-degree difference on the Kelvin scale, so there is a linear relationship between the temperature T_C in degrees Celsius and the temperature T_K in kelvins.
(a) Find an equation that relates T_C and T_K.
(b) Absolute zero (0 K on the Kelvin scale) is the temperature below which a body's temperature cannot be lowered. Express absolute zero in $°$C.

43. To the extent that water can be assumed to be incompressible, the pressure p in a body of water varies linearly with the distance h below the surface.
(a) Given that the pressure is 1 atmosphere (1 atm) at the surface and 5.9 atm at a depth of 50 m, find an equation that relates pressure to depth.
(b) At what depth is the pressure twice that at the surface?

44. A resistance thermometer is a device that determines temperature by measuring the resistance of a fine wire whose resistance varies with temperature. Suppose that the resistance R in ohms (Ω) varies linearly with the temperature T in $°$C and that $R = 123.4\ \Omega$ when $T = 20°$C and that $R = 133.9\ \Omega$ when $T = 45°$C.
(a) Find an equation for R in terms of T.
(b) If R is measured experimentally as $128.6\ \Omega$, what is the temperature?

45. Suppose that the mass of a spherical mothball decreases with time, due to evaporation, at a rate that is proportional to its surface area. Assuming that it always retains the shape of a sphere, it can be shown that the radius r of the sphere decreases linearly with the time t.
(a) If, at a certain instant, the radius is 0.80 mm and 4 days later it is 0.75 mm, find an equation for r (in millimeters) in terms of the elapsed time t (in days).
(b) How long will it take for the mothball to completely evaporate?

46. The accompanying figure shows three masses suspended from a spring: a mass of 11 g, a mass of 24 g, and an unknown mass of W g.
(a) What will the pointer indicate on the scale if no mass is suspended?
(b) Find W.

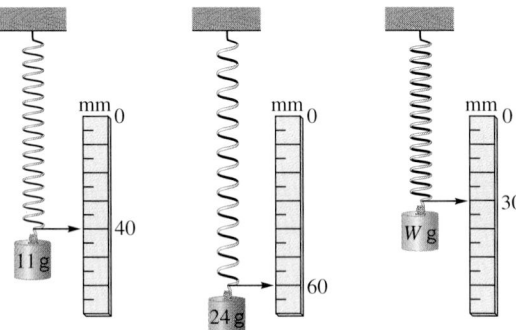

Figure Ex-46

47. The price for a round-trip bus ride from a university to center city is $2.00, but it is possible to purchase a monthly commuter pass for $25.00 with which each round-trip ride costs an additional $0.25.
(a) Find equations for the cost C of making x round-trips per month under both payment plans, and graph the equations for $0 \le x \le 30$ (treating C as a continuous function of x, even though x assumes only integer values).
(b) How many round-trips per month would a student have to make for the commuter pass to be worthwhile?

48. A student must decide between buying one of two used cars: car A for $4000 or car B for $5500. Car A gets 20 miles per gallon of gas, and car B gets 30 miles per gallon. The student estimates that gas will run $1.25 per gallon. Both cars are in excellent condition, so the student feels that repair costs should be negligible for the foreseeable future. How many miles would the student have to drive before car B becomes the better buy?

49. (**The Age of the Universe**) In the early 1900s the astronomer Edwin P. Hubble (1889–1953) noted an unexpected relationship between the radial velocity of a galaxy and its distance d from any reference point (Earth, for example). That relationship, now known as **Hubble's law**, states that the galaxies are receding with a velocity v that is directly proportional to the distance d. This is usually expressed as $v = Hd$, where H (the constant of proportionality) is called **Hubble's constant**. When applying this formula it is usual to express v in kilometers per second (km/s) and d in millions of light-years (Mly), in which case H has units of km/s/Mly. The accompanying figure shows an original plot and trend line of the velocity-distance relationship obtained by Hubble and a collaborator Milton L. Humason (1891–1972).

(a) Use the trend line in the figure to estimate Hubble's constant.

(b) An estimate of the age of the universe can be obtained by assuming that the galaxies move with constant velocity v, in which case v and d are related by $d = vt$. Assuming that the Universe began with a "big bang" that initiated its expansion, show that the Universe is roughly 1.5×10^{10} years old. [Take $H = 20$ km/s/Mly,

which is in keeping with current estimates that place H between 15 and 27 km/s/Mly. (Note that the current estimates are significantly less than that resulting from Hubble's data.)]

(c) In a more realistic model of the Universe, the velocity v would decrease with time. What effect would that have on your estimate in part (b)?

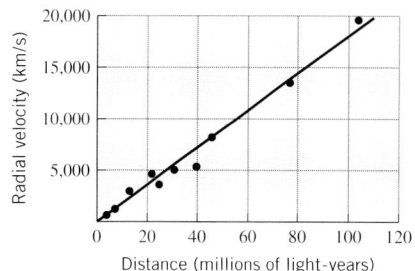

Figure Ex-49

1.6 FAMILIES OF FUNCTIONS

Functions are often grouped into families according to the form of their defining formulas or other common characteristics. In this section we will discuss some of the most basic families of functions.

This section includes quick reviews of precalculus material on polynomials and trigonometry. Readers who want to review this material in more depth are referred to Appendices E and F. Instructors who want to spend some additional time on precalculus review can divide this section into two parts, covering the trigonometry material in a second lecture.

FAMILIES OF LINES

A function f whose values are all the same is called a **constant function**. For example, the formula $f(x) = c$ defines the constant function whose value is c for all x. The graph of the constant function $f(x) = c$ is the horizontal line $y = c$ (Figure 1.6.1a). If we vary c, then we obtain a set or **family** of horizontal lines (Figure 1.6.1b).

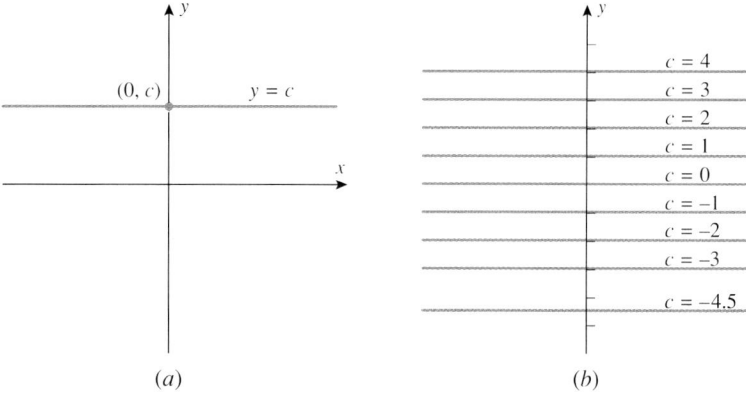

Figure 1.6.1

REMARK.   The expression $f(x) = c$ can be confusing because it can be interpreted either as an equation that is satisfied for certain x (as in $x^2 = c$) or as an identity that is satisfied for all x; it is the latter interpretation that defines a constant function. Thus, when you see an expression of the form $f(x) = c$, you will have to determine from its context whether it is intended as an equation or a constant function.

The quantities m and b in the equation $y = mx + b$ can be viewed as unspecified constants whose values may change from one application to another; such changeable constants are called *parameters*.

If we keep b fixed and vary the parameter m in the equation $y = mx + b$, then we obtain a family of lines whose members all have y-intercept b (Figure 1.6.2a); and if we keep m fixed and vary the parameter b, then we obtain a family of parallel lines whose members all have slope m (Figure 1.6.2b).

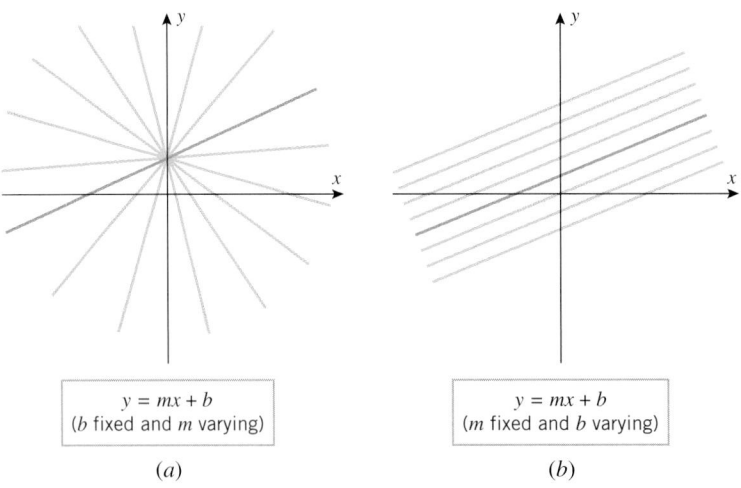

$$y = mx + b$$
(b fixed and m varying)

$$y = mx + b$$
(m fixed and b varying)

(a)

(b)

Figure 1.6.2

Example 1

(a)   Find an equation for the family of lines with slope $\frac{1}{2}$.

(b)   Find the member of the family in part (a) that passes through the point (4, 1).

(c)   Find an equation for the family of lines whose members are perpendicular to the lines in part (a).

Solution (a). The lines of slope $\frac{1}{2}$ are of the form

$$y = \tfrac{1}{2}x + b \tag{1}$$

where the parameter b can have any real value.

Solution (b). To find the line in the family that passes through the point (4, 1), we must find the value of b for which the coordinates $x = 4$ and $y = 1$ satisfy (1). Substituting these coordinates into (1) and solving for b yields $b = -1$, and hence the equation of the line is

$$y = \tfrac{1}{2}x - 1 \tag{2}$$

(Figure 1.6.3a).

Solution (c). Since the slopes of perpendicular lines are negative reciprocals, it follows that the lines perpendicular to those in part (a) have slope -2 and hence are of the form

$$y = -2x + b$$

Some typical lines in families (1) and (2) are graphed in Figure 1.6.3b.   ◀

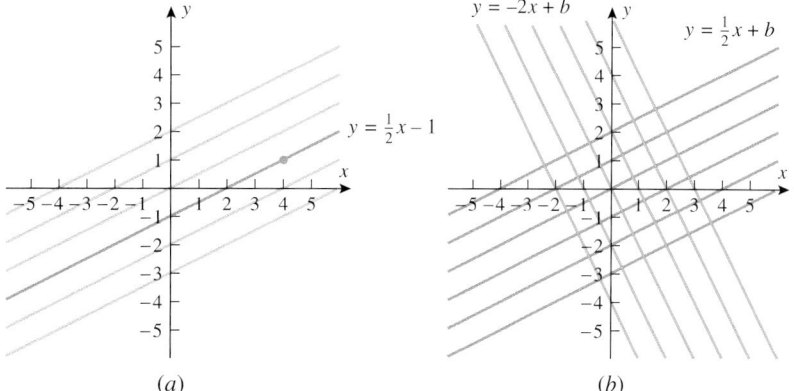

Figure 1.6.3

THE FAMILY **y = x^n**

A function of the form $f(x) = x^p$, where p is constant is called a ***power function***. If p is a positive integer, say $p = n$, then the power functions have the form $f(x) = x^n$. The graphs of the curves $y = x^n$ for $n = 1, 2, 3, 4$, and 5 are shown in Figure 1.6.4. The first graph is the line $y = x$ with slope 1 that passes through the origin, and the second is a parabola that opens up and has its vertex at the origin (see Appendix 2).

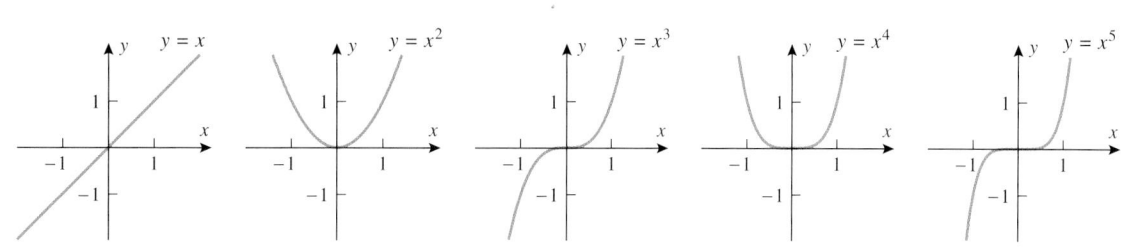

Figure 1.6.4

For $n > 2$ the shape of the graph of $y = x^n$ depends on whether n is even or odd (Figure 1.6.5). For even values of n the graphs have the same general shape as the parabola $y = x^2$ (though they are not actually parabolas if $n > 2$), and for odd values of n greater than 1 they have the same general shape as $y = x^3$. The graphs in the family $y = x^n$ share a number of important characteristics:

- For even values of n the functions $f(x) = x^n$ are even, and their graphs are symmetric about the y-axis; for odd values of n the functions $f(x) = x^n$ are odd, and their graphs are symmetric about the origin.
- For all values of n the graphs pass through the origin and the point $(1, 1)$. For even values of n the graphs pass through $(-1, 1)$, and for odd values of n they pass through $(-1, -1)$.
- Increasing n causes the graph to become flatter over the interval $-1 < x < 1$ and steeper over the intervals $x > 1$ and $x < -1$.

REMARK. The last characteristic can be explained numerically by considering the effect of raising a real number x to successively higher powers. If x is a fraction, that is, $-1 < x < 1$, then the absolute value of x^n *decreases* as n increases (try raising $\frac{1}{2}$ or $-\frac{1}{2}$ to higher and higher powers, for example). This explains why successive graphs in Figure 1.6.5 become flatter over the interval $-1 < x < 1$. On the other hand, if $x > 1$ or $x < -1$, then the

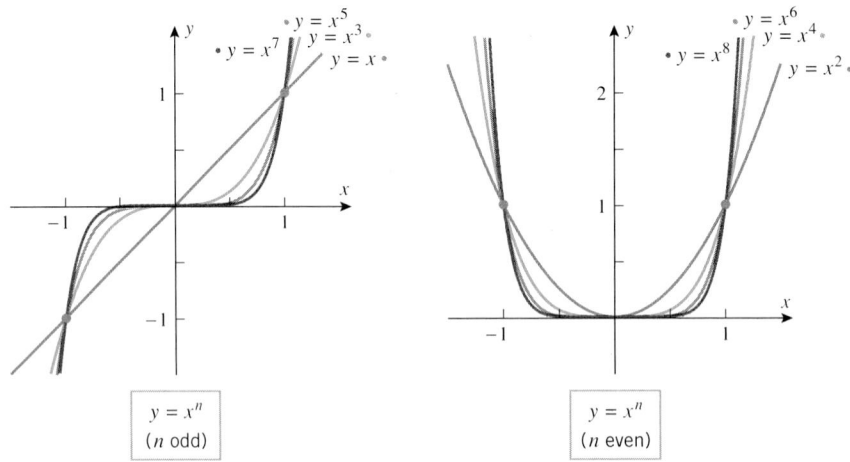

Figure 1.6.5

absolute value of x^n *increases* as n increases (try raising 2 or -2 to higher and higher powers). This explains why successive graphs become steeper if $x > 1$ or $x < -1$.

THE FAMILY $y = x^{-n}$

If p is a negative integer, say $p = -n$, then the power functions $f(x) = x^p$ have the form $f(x) = x^{-n} = 1/x^n$. Figure 1.6.6a shows the graphs of $y = 1/x$ and $y = 1/x^2$, and Figure 1.6.6b shows how these graphs relate to other members of the family. The graph of $y = 1/x$ is called an *equilateral hyperbola* (for reasons to be discussed later).

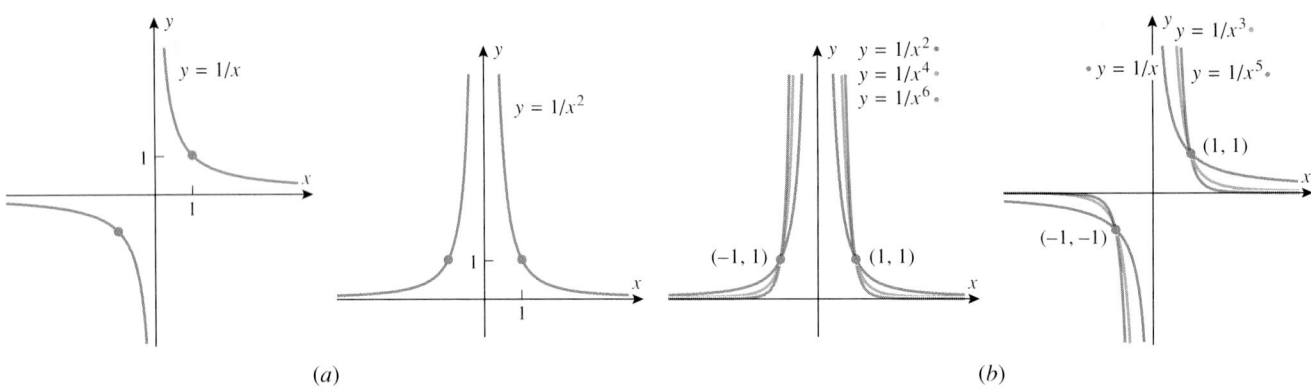

(a) (b)

Figure 1.6.6

For odd values of n the graphs have the same general shape as $y = 1/x$, and for even values of n they have the same general shape as $y = 1/x^2$. The graphs in the family $y = 1/x^n$ share a number of important characteristics:

- For even values of n the functions $f(x) = 1/x^n$ are even, and their graphs are symmetric about the y-axis; for odd values of n the functions $f(x) = x^n$ are odd, and their graphs are symmetric about the origin.
- For all values of n the graphs pass through the point $(1, 1)$ and have a break (called a *discontinuity*) at the origin. This is caused by the division by zero that occurs when $x = 0$. For even values of n the graphs pass through $(-1, 1)$, and for odd values of n they pass through $(-1, -1)$.
- Increasing n causes the graph to become steeper over the interval $-1 < x < 1$ and flatter over the intervals $x > 1$ and $x < -1$.

REMARK.   The last characteristic can be explained numerically by considering the effect of raising the reciprocal of a number x to successively higher powers. If x is a nonzero fraction, then it lies in the interval $-1 < x < 1$, and its reciprocal satisfies $1/x > 1$ or $1/x < -1$. Thus, as n increases the absolute value of $1/x^n$ also increases. This explains why successive graphs in Figure 1.6.6 become successively steeper over the interval $-1 < x < 1$. On the other hand, if $x > 1$ or $x < -1$, then $-1 < 1/x < 1$. Thus, as n increases the absolute value of $1/x^n$ *decreases*. This explains why successive graphs in Figure 1.6.6 get successively flatter if $x > 1$ or $x < -1$.

THE FAMILY $y = x^{1/n}$

If $p = 1/n$, where n is a positive integer, then the power functions $f(x) = x^p$ have the form $f(x) = x^{1/n} = \sqrt[n]{x}$. In particular, if $n = 2$, then $f(x) = \sqrt{x}$, and if $n = 3$, then $f(x) = \sqrt[3]{x}$. The graphs of these functions are shown in parts (*a*) and (*b*) of Figure 1.6.7. Observe that the graph of $y = \sqrt[3]{x}$ extends over the entire x-axis because $f(x) = \sqrt[3]{x}$ is defined for all real values of x (every real number has a cube root); in contrast, the graph of $y = \sqrt{x}$ only extends over the nonnegative x-axis (negative numbers have imaginary square roots). Observe also that the graph of $y = \sqrt{x}$ is the upper half of the parabola $x = y^2$ (Figure 1.6.7*c*).

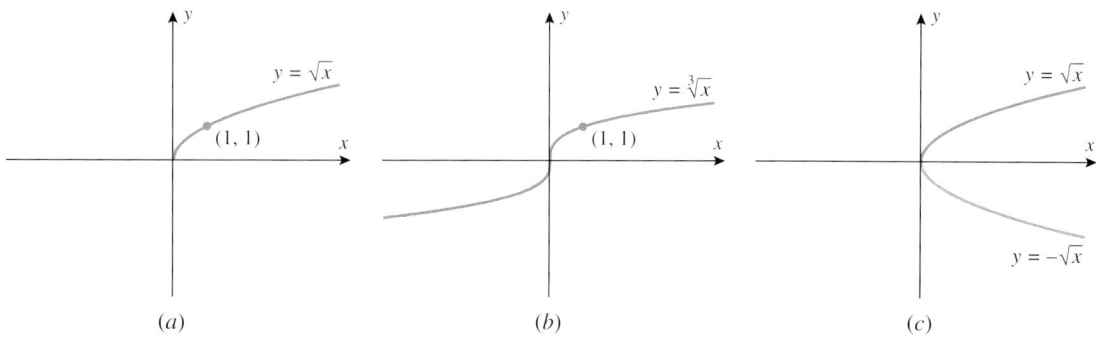

(*a*)   (*b*)   (*c*)

Figure 1.6.7

For even values of n the graphs of $y = \sqrt[n]{x}$ have the same general shape as $y = \sqrt{x}$, and for odd values of n they have the same general shape as $y = \sqrt[3]{x}$.

FOR THE READER.   Sketch the graphs of $y = \sqrt[n]{x}$ for $n = 2, 4, 6$ on one set of axes and for $n = 3, 5, 7$ on another set. Use a graphing device to check your work.

POWER FUNCTIONS WITH FRACTIONAL AND IRRATIONAL EXPONENTS

Power functions can also have fractional or irrational exponents. For example,

$$f(x) = x^{2/3}, \quad f(x) = \sqrt[5]{x^3}, \quad f(x) = x^{-7/8}, \quad \text{and} \quad f(x) = x^{\sqrt{2}} \tag{3}$$

are all power functions of this type; we will discuss power functions of these forms in later sections.

FOR THE READER.   Read the note preceding Exercise 29 of Section 1.3, and use a graphing utility to generate complete graphs of the functions in (3).

MODELS INVOLVING INVERSE PROPORTIONS

Recall that a variable y is said to be *inversely proportional to a variable* x if there is a positive constant k, called the *constant of proportionality*, such that

$$y = \frac{k}{x} \tag{4}$$

Since k is assumed to be positive, the graph of this equation has the same basic shape as $y = 1/x$ but is compressed or stretched in the x-direction.

Observe that in Formula (4) doubling x decreases y by a factor of $1/2$, tripling x decreases y by a factor of $1/3$, and, more generally, increasing x by a factor of r decreases y by a factor of $1/r$.

Models involving inverse proportion arise in various laws of physics. For example, **Boyle's law** in physics states that at a constant temperature the pressure P exerted by a fixed quantity of an ideal gas is inversely proportional to the volume V occupied by the gas, that is,

$$P = \frac{k}{V}$$

(Figure 1.6.8).

If y is inversely proportional to x, then it follows from (4) that the product of y and x is constant, since $yx = k$. This provides a useful way of identifying inverse proportion models in experimental data.

Example 2

Table 1.6.1 shows some experimental data.

Table 1.6.1

EXPERIMENTAL DATA

x	0.8	1	2.5	4	6.25	10
y	6.25	5	2	1.25	0.8	0.5

(a)   Explain why the data suggest that y is inversely proportional to x.

(b)   Express y as a function of x.

(c)   Graph your function and the data together for $x \geq 0$.

Solution.   For every data point we have $xy = 5$, so y is inversely proportional to x and $y = 5/x$. The graph of this equation with the data points is shown in Figure 1.6.9.   ◀

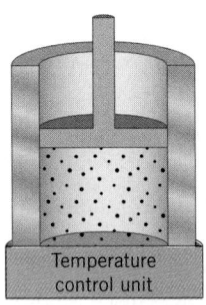

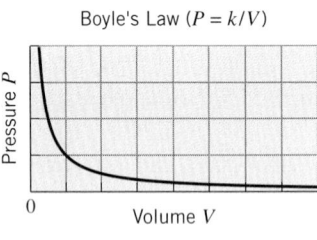

Boyle's Law ($P = k/V$)

Pressure P

Volume V

Temperature control unit

As the volume of the gas changes, the temperature control unit adds or removes heat to maintain a constant temperature.

Figure 1.6.8

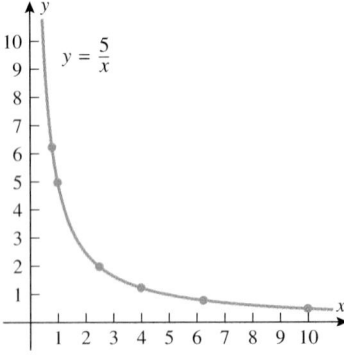

$y = \dfrac{5}{x}$

Figure 1.6.9

A QUICK REVIEW OF POLYNOMIALS

A detailed review of polynomials is given in Appendix F, but for convenience we will review some of the terminology here.

A **polynomial in x** is a function that is expressible as a sum of finitely many terms of the form cx^n, where c is a constant and n is a nonnegative integer. Some examples of polynomials are

$$2x + 1, \quad 3x^2 + 5x - \sqrt{2}, \quad x^3, \quad 4 \,(= 4x^0), \quad 5x^7 - x^4 + 3$$

The function $(x^2 - 4)^3$ is also a polynomial because it can be expanded by the binomial formula (see the inside front cover) and expressed as a sum of terms of the form cx^n:

$$(x^2 - 4)^3 = (x^2)^3 - 3(x^2)^2(4) + 3(x^2)(4^2) - (4^3) = x^6 - 12x^4 + 48x^2 - 64 \qquad (5)$$

A general polynomial can be written in either of the following forms, depending on whether one wants the powers of x in ascending or descending order:

$$c_0 + c_1 x + c_2 x^2 + \cdots + c_n x^n$$
$$c_n x^n + c_{n-1} x^{n-1} + \cdots + c_1 x + c_0$$

The constants $c_0, c_1, \ldots, c_n$ are called the ***coefficients*** of the polynomial. When a polynomial is expressed in one of these forms, the highest power of x that occurs with a nonzero coefficient is called the ***degree*** of the polynomial. Constants are considered to have degree 0, since we can write $c = cx^0$. Polynomials of degree 1, 2, 3, 4, and 5 are described as ***linear***, ***quadratic***, ***cubic***, ***quartic***, and ***quintic***, respectively. For example,

$$3 + 5x \qquad \text{Has degree 1 (linear)}$$

$$x^2 - 3x + 1 \qquad \text{Has degree 2 (quadratic)}$$

$$2x^3 - 7 \qquad \text{Has degree 3 (cubic)}$$

$$8x^4 - 9x^3 + 5x - 3 \qquad \text{Has degree 4 (quartic)}$$

$$\sqrt{3} + x^3 + x^5 \qquad \text{Has degree 5 (quintic)}$$

$$(x^2 - 4)^3 \qquad \text{Has degree 6 [see (5)]}$$

The natural domain of a polynomial in x is $(-\infty, +\infty)$, since the only operations involved are multiplication and addition; the range depends on the particular polynomial. We already know that the graphs of polynomials of degree 0 and 1 are lines and that the graphs of polynomials of degree 2 are parabolas. Figure 1.6.10 shows the graphs of some typical polynomials of higher degree. Later, we will discuss polynomial graphs in detail, but for now it suffices to observe that graphs of polynomials are very well behaved in the sense that they have no discontinuities or sharp corners. As illustrated in Figure 1.6.10, the graphs of polynomials wander up and down for awhile in a roller-coaster fashion, but eventually that behavior stops and the graphs steadily rise or fall indefinitely as one travels along the curve in either the positive or negative direction. We will see later that the number of peaks and valleys is determined by the degree of the polynomial.

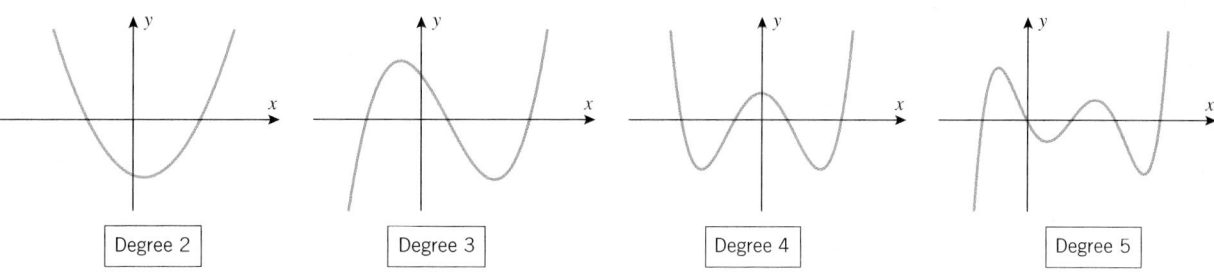

Degree 2   Degree 3   Degree 4   Degree 5

Figure 1.6.10

RATIONAL FUNCTIONS

A function that can be expressed as a ratio of two polynomials is called a ***rational function***. If $P(x)$ and $Q(x)$ are polynomials, then the domain of the rational function

$$f(x) = \frac{P(x)}{Q(x)}$$

consists of all values of x such that $Q(x) \neq 0$. For example, the domain of the rational function

$$f(x) = \frac{x^2 + 2x}{x^2 - 1}$$

consists of all values of x, except $x = 1$ and $x = -1$. Its graph is shown in Figure 1.6.11 along with the graphs of two other typical rational functions.

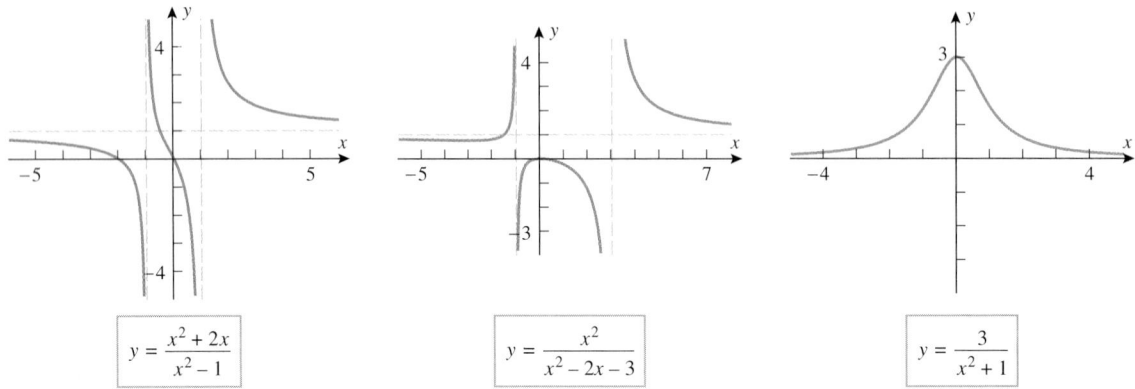

$$y = \frac{x^2 + 2x}{x^2 - 1} \qquad y = \frac{x^2}{x^2 - 2x - 3} \qquad y = \frac{3}{x^2 + 1}$$

Figure 1.6.11

The graphs of rational functions with nonconstant denominators differ from the graphs of polynomials in some essential ways:

- Unlike polynomials whose graphs are continuous (unbroken) curves, the graphs of rational functions have discontinuities at the points where the denominator is zero.

- As x gets closer and closer to a point of discontinuity, the graph rises or falls indefinitely, getting closer and closer to a vertical line, called a ***vertical asymptote***; these are represented by the dashed vertical lines in Figure 1.6.11.

- Unlike the graphs of polynomials, which eventually rise or fall indefinitely, the graphs of many (but not all) rational functions eventually get closer and closer to some horizontal line, called a ***horizontal asymptote***, as one travels along the curve in either the positive or negative direction; these are represented by the dashed horizontal lines in the first two parts of Figure 1.6.11. In the third part of the figure the x-axis is a horizontal asymptote.

ALGEBRAIC FUNCTIONS

Functions that can be constructed from polynomials by applying finitely many algebraic operations (addition, subtraction, division, and root extraction) are called ***algebraic functions***. Some examples are

$$f(x) = \sqrt{x^2 - 4}, \qquad f(x) = 3\sqrt[3]{x}(2 + x), \qquad f(x) = x^{2/3}(x + 2)^2$$

As illustrated in Figure 1.6.12, the graphs of algebraic functions vary widely, so it is difficult to make general statements about them. Later in this text we will develop general calculus methods for analyzing such functions.

A QUICK REVIEW OF TRIGONOMETRIC FUNCTIONS

A detailed review of trigonometric functions is given in Appendix E, but for convenience we will summarize some of the main ideas here.

It is often convenient to think of the trigonometric functions in terms of circles rather than triangles. For this purpose, consider a point that moves either clockwise or counter-

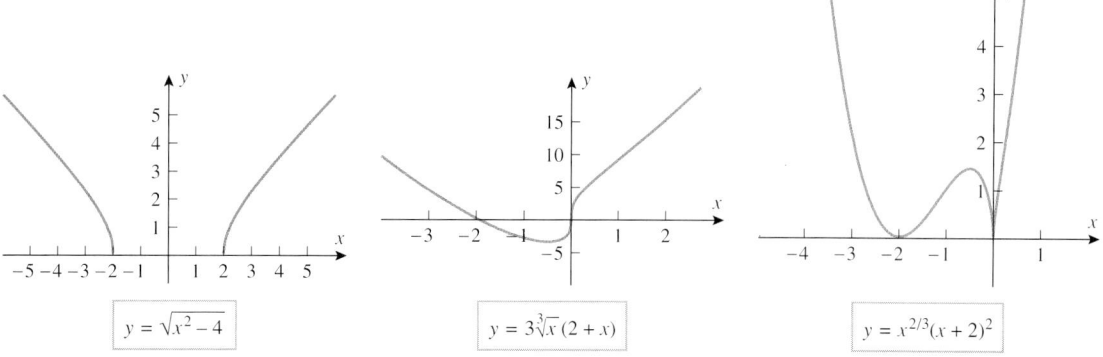

Figure 1.6.12

clockwise along the **unit circle** $u^2 + v^2 = 1$ in the uv-plane, starting at $(1, 0)$ and stopping at a point P (Figure 1.6.13a). Let x denote the **signed** arc length traveled by the moving point, taking x to be positive for counterclockwise motion and negative for clockwise motion. (We allow for the possibility that the point may traverse the circle more than once.) When convenient, the variable x can also be interpreted as the angle in radians that is swept out by the radial line from the origin to P, with the usual convention that angles are positive if generated by counterclockwise rotations and negative if generated by clockwise rotations. We can *define* $\cos x$ to be the u-coordinate of P and $\sin x$ to be the v-coordinate of P (Figure 1.6.13b).

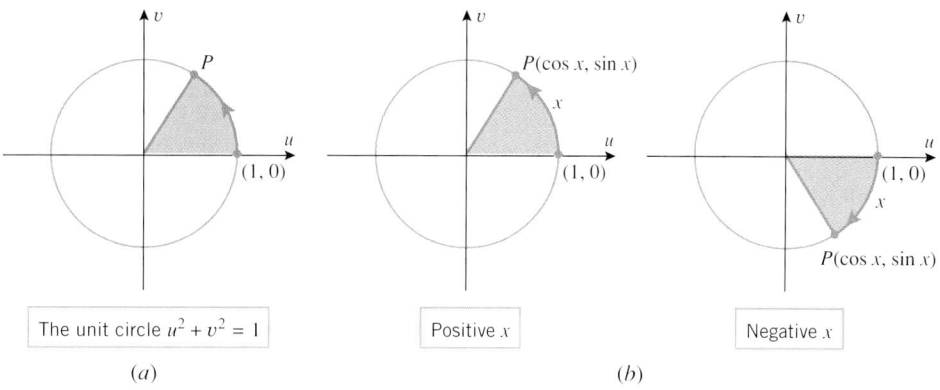

Figure 1.6.13

The remaining trigonometric functions can be defined in terms of the functions $\sin x$ and $\cos x$:

$$\tan x = \frac{\sin x}{\cos x} \qquad \cot x = \frac{\cos x}{\sin x}$$

$$\sec x = \frac{1}{\cos x} \qquad \csc x = \frac{1}{\sin x}$$

The graphs of the six trigonometric functions in Figure 1.6.14 should already be familiar to you, but try generating them using a graphing utility, making sure to use radian measure for x.

REMARK.   In this text we will always assume that the independent variable in a trigono-metric function is in radians unless specifically stated otherwise.

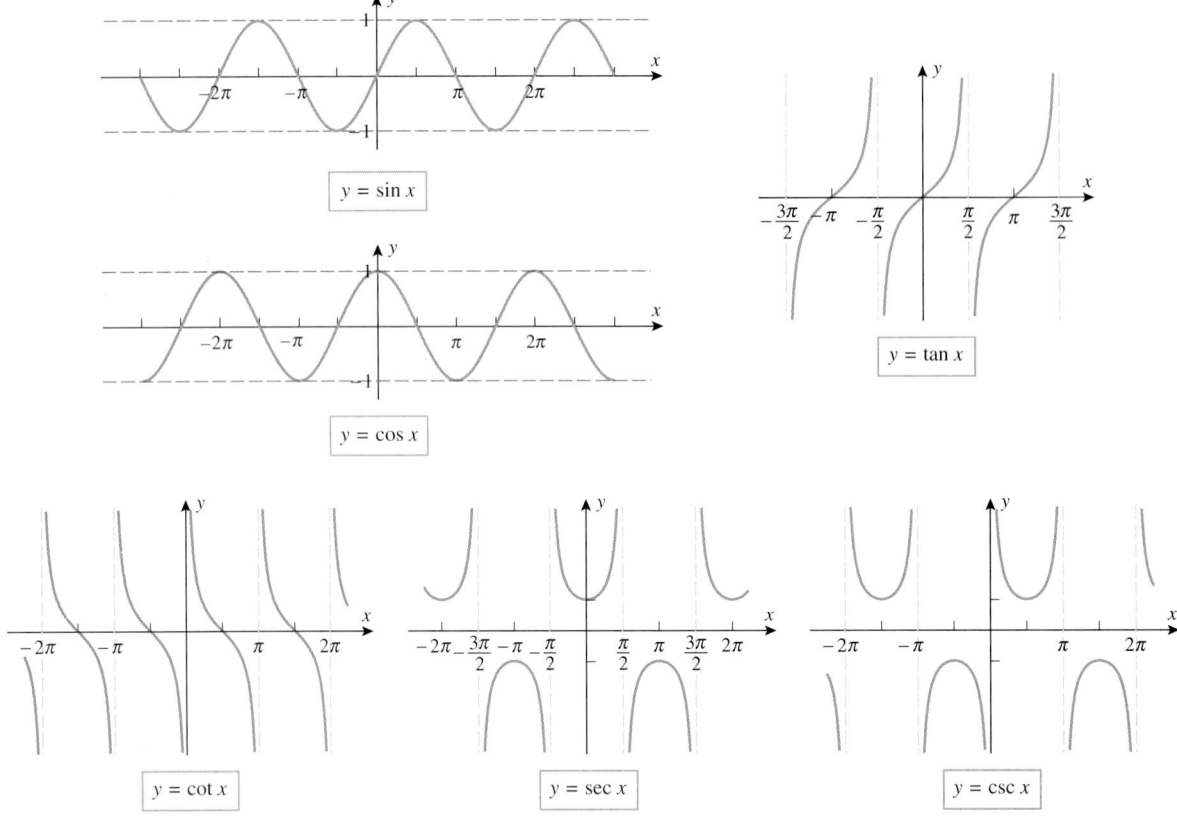

Figure 1.6.14

PROPERTIES OF sin x, cos x, AND tan x

Many of the basic properties of $\sin x$ and $\cos x$ can be deduced from the circle definitions of these functions. For example:

- As the point $P(\cos x, \sin x)$ moves around the unit circle, its coordinates vary between -1 and 1, and hence

$$-1 \le \sin x \le 1 \quad \text{and} \quad -1 \le \cos x \le 1$$

- If x increases or decreases by 2π radians, then the point $P(\cos x, \sin x)$ makes one complete revolution around the unit circle, and the coordinates return to their starting values. Thus, $\sin x$ and $\cos x$ have period 2π; that is,

$$\sin(x \pm 2\pi) = \sin x$$
$$\cos(x \pm 2\pi) = \cos x$$

- As $P(\cos x, \sin x)$ moves around the unit circle, $\sin x$ is zero when P is on the horizontal axis (which occurs when x is an integer multiple of π), and $\cos x$ is zero when P is on the vertical axis (which occurs when x is an odd multiple of $\pi/2$). Thus,

$$\sin x = 0 \quad \text{if and only if} \quad x = 0, \pm\pi, \pm2\pi, \pm3\pi, \ldots$$
$$\cos x = 0 \quad \text{if and only if} \quad x = \pm\pi/2, \pm3\pi/2, \pm5\pi/2, \ldots$$

- As $P(\cos x, \sin x)$ moves around the unit circle $u^2 + v^2 = 1$, its coordinates satisfy this equation for all x, which produces the fundamental trigonometric identity

$$\cos^2 x + \sin^2 x = 1$$

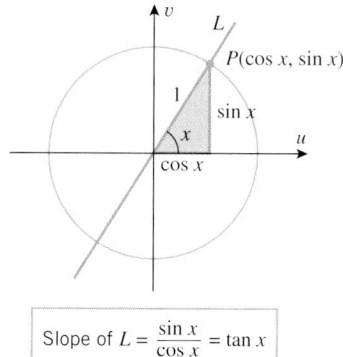

Slope of $L = \dfrac{\sin x}{\cos x} = \tan x$

Figure 1.6.15

Observe that the graph of $y = \tan x$ has vertical asymptotes at the points $x = \pm\pi/2$, $\pm 3\pi/2$, $\pm 5\pi/2$, ... This is to be expected since $\tan x = \sin x/\cos x$, and these are the values of x at which $\cos x$ is zero. What is less obvious, however, is the fact that $\tan x$ repeats every π radians (i.e., has period π), even though $\sin x$ and $\cos x$ have period 2π. This can be explained by interpreting

$$\tan x = \frac{\sin x}{\cos x}$$

as the slope of the line L that passes through the origin and the point $P(\cos x, \sin x)$ on the unit circle in the uv-plane (Figure 1.6.15). Each time x increases or decreases by π radians, the point P traverses half the circumference, and the line L rotates π radians, so its starting and ending slope are the same.

RADIANS AS A DIMENSIONLESS UNIT

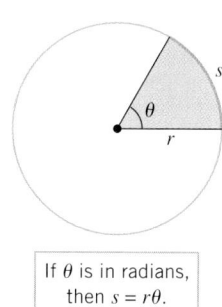

If θ is in radians, then $s = r\theta$.

Figure 1.6.16

The choice of radian measure as opposed to degree measure depends on the nature of the problem being considered; degree measure is usually chosen in engineering problems involving measurements of angles, and radian measure is usually chosen when the function properties of $\sin x$, $\cos x$, $\tan x$, ... are the primary focus. Radian measure is also usually chosen in problems involving arc lengths on circles because of the basic result in trigonometry which states that the arc length s of a sector with radius r and a central angle of θ (radians) is given by

$$s = r\theta \tag{6}$$

(Figure 1.6.16).

In applications involving angles, radians require special treatment to ensure that quantities are assigned proper units. To see why this is so, let us rewrite (6) as

$$\theta = \frac{s}{r}$$

The left side of this equation is in radians, and the right side is the ratio of two lengths, say meters/meters or feet/feet. However, because these units of length cancel, the right side of this equation is actually *dimensionless* (has no units). Thus, to ensure consistency between the two sides of the equation, we would have to omit the units of radians on the left side to make it dimensionless as well. In practical terms this means that units of radians can be used in intermediate computations, when convenient, but they need to be omitted in the end result to ensure consistency of units. This is confusing, to say the least, but the following example should clarify the idea.

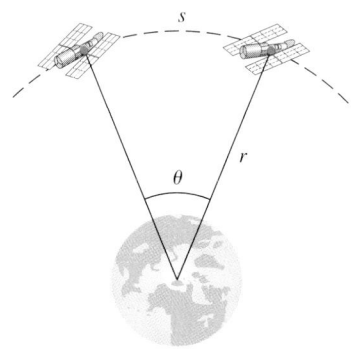

Figure 1.6.17

Example 3

Suppose that two satellites circle the equator in an orbit of radius $r = 4.23 \times 10^7$ m (Figure 1.6.17). Find the arc length s that separates the satellites if they have an angular separation of $\theta = 2.00°$.

Solution.   To apply Formula (6), we must convert the angular separation to radians:

$$2.00° = \frac{\pi}{180}(2.00) \approx 0.0349 \text{ rad}$$

Thus, from (6)

$$s = r\theta = (4.23 \times 10^7 \text{ m})(0.0349 \text{ rad}) = 1.48 \times 10^6 \text{ m}$$

In this computation the product $r\theta$ produces units of meters $\times$ radians, but if we treat radians as dimensionless, we have meters $\times$ radians = meters, which correctly produces units of meters (m) for the arc length s.   ◀

Many important applications lead to trigonometric functions of the form

$$f(x) = A \sin(Bx - C) \quad \text{and} \quad g(x) = A \cos(Bx - C) \tag{7}$$

where A, B, and C are nonzero constants. The graphs of such functions can be obtained by stretching, compressing, translating, and reflecting the graphs of $y = \sin x$ and $y = \cos x$ appropriately. To see why this is so, let us start with the case where $C = 0$ and consider how the graphs of the equations

$$y = A \sin Bx \quad \text{and} \quad y = A \cos Bx$$

relate to the graphs of $y = \sin x$ and $y = \cos x$. If A and B are positive, then the effect of the constant A is to stretch or compress the graphs of $y = \sin x$ and $y = \cos x$ vertically by a factor of A, and the effect of the constant B is to compress or stretch the graphs of $\sin x$ and $\cos x$ horizontally by a factor of B. For example, the graph of $y = 2 \sin 4x$ can be obtained by stretching the graph of $y = \sin x$ vertically by a factor of 2 and compressing it horizontally by a factor of 4. (Recall from Section 1.4 that the multiplier of x *stretches* when it is less than 1 and *compresses* when it is greater than 1.) Thus, as shown in Figure 1.6.18, the graph of $y = 2 \sin 4x$ varies between -2 and 2, and repeats every $2\pi/4 = \pi/2$ units.

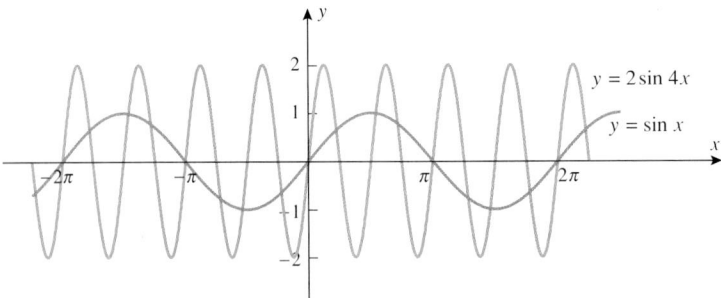

Figure 1.6.18

In general, if A and B are positive numbers, then the graphs of

$$y = A \sin Bx \quad \text{and} \quad y = A \cos Bx$$

oscillate between $-A$ and A and repeat every $2\pi/B$ units, so we say that these functions have **amplitude** A and **period** $2\pi/B$. In addition, we define the **frequency** of these functions to be the reciprocal of the period, that is, the frequency is $B/2\pi$. If A or B is negative, then these constants cause reflections of the graphs about the axes as well as compressing or stretching them; and in this case the amplitude, period, and frequency are given by $|A|$, $2\pi/|B|$, and $|B|/2\pi$, respectively.

Example 4

Make sketches of the following graphs that show the period and amplitude.

(a) $y = 3 \sin 2\pi x$ (b) $y = -3 \cos 0.5x$ (c) $y = 1 + \sin x$

Solution (a). The equation is of the form $y = A \sin Bx$ with $A = 3$ and $B = 2\pi$, so the graph has the shape of a sine function, but with amplitude $A = 3$ and period $2\pi/B = 2\pi/2\pi = 1$ (Figure 1.6.19a).

Solution (b). The equation is of the form $y = A \cos Bx$ with $A = -3$ and $B = 0.5$, so the graph has the shape of a cosine function that has been reflected about the x-axis (because $A = -3$ is negative), but with amplitude $|A| = 3$ and period $2\pi/B = 2\pi/0.5 = 4\pi$ (Figure 1.6.19b).

Solution (c). The graph has the shape of a sine function that has been translated up 1 unit (Figure 1.6.19c). ◄

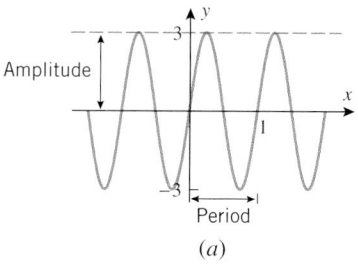 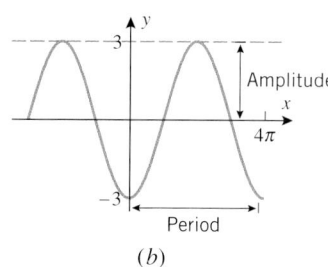 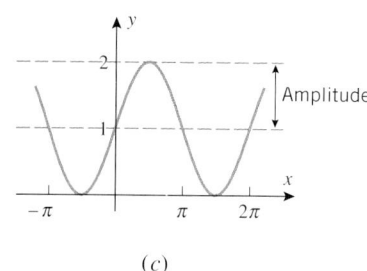

(a) (b) (c)

Figure 1.6.19

To investigate the graphs of the more general families

$$y = A \sin(Bx - C) \quad \text{and} \quad y = A \cos(Bx - C)$$

it will be helpful to rewrite these equations as

$$y = A \sin\left[B\left(x - \frac{C}{B} \right) \right] \quad \text{and} \quad y = A \cos\left[B\left(x - \frac{C}{B} \right) \right]$$

In this form we see that the graphs of these equations can be obtained by translating the graphs of $y = A \sin Bx$ and $y = A \cos Bx$ to the left or right, depending on the sign of C/B. For example, if $C/B > 0$, then the graph of

$$y = A \sin[B(x - C/B)] = A \sin(Bx - C)$$

can be obtained by translating the graph of $y = A \sin Bx$ to the right by C/B units (Figure 1.6.20). The quantity C/B is called the *phase shift* of the function; a positive phase shift corresponds to right translation, and a negative phase shift corresponds to a left translation.

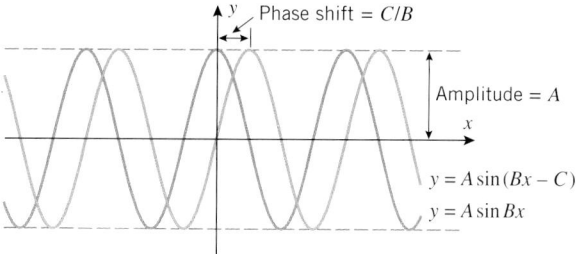

Figure 1.6.20

Example 5

Find the amplitude, period, and phase shift of

$$y = 3 \cos\left(2x + \frac{\pi}{2} \right)$$

and confirm your results by graphing the equation on a calculator or computer.

Solution. The equation can be rewritten as

$$y = 3 \cos\left[2x - \left(-\frac{\pi}{2} \right) \right] = 3 \cos\left[2\left(x - \left(-\frac{\pi}{4} \right) \right) \right]$$

which is of the form

$$y = A \cos\left[B\left(x - \frac{C}{B} \right) \right]$$

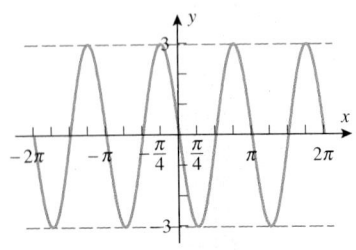

Figure 1.6.21

with $A = 3$, $B = 2$, and $C/B = -\pi/4$; thus, the graph has the shape of a cosine function, but with amplitude $A = 3$, period $2\pi/B = \pi$, and phase shift $C/B = -\pi/4$ (Figure 1.6.21). ◄

Example 6

Figure 1.6.22a shows a table and scatter plot of temperature data recorded over a 24-hour period in the city of Philadelphia.[*] Find a function that models the data, and graph your function and data together.

PHILADELPHIA TEMPERATURES
FROM 1:00 A.M. TO 12:00 MIDNIGHT ON 27 AUGUST 1993
(t = HOURS AFTER MIDNIGHT AND T = DEGREES FAHRENHEIT)

	A.M.		P.M.	
	t	T	t	T
1:00	1	78°	13	91°
2:00	2	77°	14	93°
3:00	3	77°	15	94°
4:00	4	76°	16	95°
5:00	5	76°	17	93°
6:00	6	75°	18	92°
7:00	7	75°	19	89°
8:00	8	77°	20	86°
9:00	9	79°	21	84°
10:00	10	83°	22	83°
11:00	11	87°	23	81°
12:00	12	90°	24	79°

Source: Philadelphia Inquirer, 28 August 1993.

Figure 1.6.22

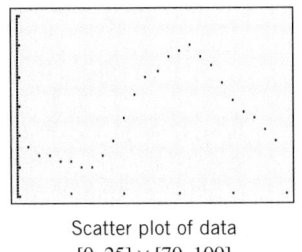

Scatter plot of data
$[0, 25] \times [70, 100]$
t T

(a)

Model for data
$T = 85 + 10 \sin[(\pi/12)(t - 10)]$
$[0, 25] \times [70, 100]$
t T

(b)

Solution. The pattern of the data suggests that the relationship between the temperature T and the time t can be modeled by a sinusoidal function that has been translated both horizontally and vertically, so we will look for an equation of the form

$$T = D + A\sin[Bt - C] = D + A\sin\left[B\left(t - \frac{C}{B}\right)\right] \tag{8}$$

Since the highest temperature is $95°\,$F and the lowest temperature is $75°\,$F, we take $2A = 20$ or $A = 10$. The midpoint between the high and low is $85°\,$F, so we have a vertical shift of $D = 85$. The period seems to be about 24, so $2\pi/B = 24$ or $B = \pi/12$. The phase shift appears to be about 10 (verify), so $C/B = 10$. Substituting these values in (8) yields the equation

$$T = 85 + 10\sin\left[\frac{\pi}{12}(t - 10)\right]$$

(Figure 1.6.22b). ◄

---

[*]This example is based on the article "Everybody Talks About It!—Weather Investigations," by Gloria S. Dion and Iris Brann Fetta, *The Mathematics Teacher*, Vol. 89, No. 2, February 1996, pp. 160–165.

OTHER FAMILIES

In addition to the functions mentioned in this section, there are exponential and logarithmic functions, which we will study later, and various special functions that arise in physics and engineering. There are also many kinds of functions that have no names; indeed, one of the important themes of calculus is to provide methods for analyzing new types of functions.

EXERCISE SET 1.6 ⌐∿ Graphing Calculator

1. (a) Find an equation for the family of lines whose members have slope $m = 3$.
   (b) Find an equation for the member of the family that passes through $(-1, 3)$.
   (c) Sketch some members of the family, and label them with their equations. Include the line in part (b).

2. Find an equation for the family of lines whose members are perpendicular to those in Exercise 1.

3. (a) Find an equation for the family of lines with y-intercept $b = 2$.
   (b) Find an equation for the member of the family whose angle of inclination is $135°$.
   (c) Sketch some members of the family, and label them with their equations. Include the line in part (b).

4. Find an equation for
   (a) the family of lines that pass through the origin
   (b) the family of lines with x-intercept $a = 1$
   (c) the family of lines that pass through the point $(1, -2)$
   (d) the family of lines parallel to $2x + 4y = 1$.

In Exercises 5 and 6, state a geometric property common to all lines in the family, and sketch five of the lines.

5. (a) The family $y = -x + b$
   (b) The family $y = mx - 1$
   (c) The family $y = m(x + 4) + 2$
   (d) The family $x - ky = 1$

6. (a) The family $y = b$
   (b) The family $Ax + 2y + 1 = 0$
   (c) The family $2x + By + 1 = 0$
   (d) The family $y - 1 = m(x + 1)$

7. Find an equation for the family of lines tangent to the circle with center at the origin and radius 3.

8. Find an equation for the family of lines that pass through the intersection of $5x - 3y + 11 = 0$ and $2x - 9y + 7 = 0$.

9. The U.S. Internal Revenue Service uses a 10-year linear depreciation schedule to determine the value of various business items. This means that an item is assumed to have a value of zero at the end of the tenth year and that at intermediate times the value is a linear function of the elapsed time. Sketch some typical depreciation lines, and explain the practical significance of the y-intercepts.

10. Find all lines through $(6, -1)$ for which the product of the x- and y-intercepts is 3.

11. In each part, match the equation with one of the accompanying graphs.
    (a) $y = \sqrt[5]{x}$
    (b) $y = 2x^5$
    (c) $y = -1/x^8$
    (d) $y = 8^x$
    (e) $y = \sqrt[4]{x - 2}$
    (f) $y = 1/8^x$

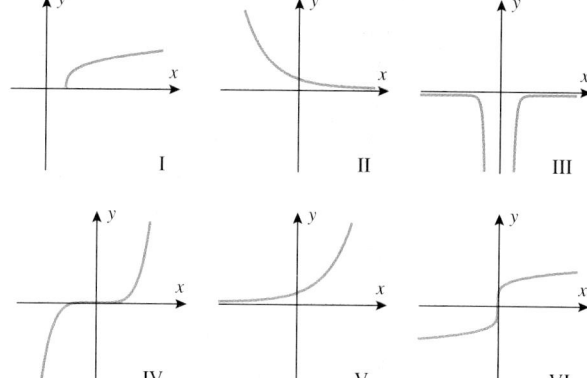

Figure Ex-11

12. The table in the accompanying figure gives approximate values of three functions: one of the form kx^2, one of the form kx^{-3}, and one of the form $kx^{3/2}$. Identify which is which, and estimate k in each case.

x	0.25	0.37	2.1	4.0	5.8	6.2	7.9	9.3
$f(x)$	640	197	1.08	0.156	0.0513	0.0420	0.0203	0.0124
$g(x)$	0.0312	0.0684	2.20	8.00	16.8	19.2	31.2	43.2
$h(x)$	0.250	0.450	6.09	16.0	27.9	30.9	44.4	56.7

Figure Ex-12

In Exercises 13 and 14, sketch the graph of the equation for $n = 1, 3$, and 5 in one coordinate system and for $n = 2, 4$, and 6 in another coordinate system. Check your work with a graphing utility.

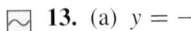 **13.** (a) $y = -x^n$ (b) $y = 2x^{-n}$ (c) $y = (x - 1)^{1/n}$

14. (a) $y = 2x^n$ (b) $y = -x^{-n}$

(c) $y = -3(x + 2)^{1/n}$

15. (a) Sketch the graph of $y = ax^2$ for $a = \pm 1, \pm 2$, and ± 3 in a single coordinate system.

(b) Sketch the graph of $y = x^2 + b$ for $b = \pm 1, \pm 2$, and ± 3 in a single coordinate system.

(c) Sketch some typical members of the family of curves $y = ax^2 + b$.

16. (a) Sketch the graph of $y = a\sqrt{x}$ for $a = \pm 1, \pm 2$, and ± 3 in a single coordinate system.

(b) Sketch the graph of $y = \sqrt{x} + b$ for $b = \pm 1, \pm 2$, and ± 3 in a single coordinate system.

(c) Sketch some typical members of the family of curves $y = a\sqrt{x} + b$.

In Exercises 17–20, sketch the graph of the equation by making appropriate transformations to the graph of a basic power function. Check your work with a graphing utility.

17. (a) $y = 2(x + 1)^2$ (b) $y = -3(x - 2)^3$

(c) $y = \dfrac{-3}{(x + 1)^2}$ (d) $y = \dfrac{1}{(x - 3)^5}$

18. (a) $y = 1 - \sqrt{x + 2}$ (b) $y = 1 - \sqrt[3]{x + 2}$

(c) $y = \dfrac{5}{(1 - x)^3}$ (d) $y = \dfrac{2}{(4 + x)^4}$

19. (a) $y = \sqrt[3]{x + 1}$ (b) $y = 1 - \sqrt{x - 2}$

(c) $y = (x - 1)^5 + 2$ (d) $y = \dfrac{x + 1}{x}$

20. (a) $y = 1 + \dfrac{1}{x - 2}$ (b) $y = \dfrac{1}{1 + 2x - x^2}$

(c) $y = -\dfrac{2}{x^7}$ (d) $y = x^2 + 2x$

21. Sketch the graph of $y = x^2 + 2x$ by completing the square and making appropriate transformations to the graph of $y = x^2$.

22. (a) Use the graph of $y = \sqrt{x}$ to help sketch the graph of $y = \sqrt{|x|}$.

(b) Use the graph of $y = \sqrt[3]{x}$ to help sketch the graph of $y = \sqrt[3]{|x|}$.

23. The table in the accompanying figure provides data about the relationship between distance d traveled in meters and elapsed time t in seconds for an object dropped near the Earth's surface. Plot time versus distance and make a guess at a "square-root function" that provides a reasonable model for t in terms of d. Use a graphing utility to confirm the reasonableness of your guess.

d (meters)	0	2.5	5	10	15	20	25
t (seconds)	0	0.7	1.0	1.4	1.7	2	2.3

Figure Ex-23

24. (a) The table below provides data on five moons of the planet Saturn. In this table r is the *orbital radius* (the average distance between the moon and Saturn) and t is the time in days required for the moon to complete one orbit around Saturn. For each data pair calculate $tr^{-3/2}$, and use your results to find a reasonable model for r as a function of t.

(b) Use the model from part (a) to estimate the orbital radius of the moon Enceladus, given that its orbit time is $t \approx 1.370$ days.

(c) Use the model from part (a) to estimate the orbit time of the moon Tethys, given that its orbital radius is $r \approx 2.9467 \times 10^5$ km.

Moon	Radius (100,000 km)	Orbit Time (days)
1980S28	1.3767	0.602
1980S27	1.3935	0.613
1980S26	1.4170	0.629
1980S3	1.5142	0.694
1980S1	1.5147	0.695

25. As discussed in this section, Boyle's law states that at a constant temperature the pressure P exerted by a gas is related to the volume V by the equation $P = k/V$.

(a) Find the appropriate units for the constant k if pressure (which is force per unit area) is in newtons per square meter (N/m^2) and volume is in cubic meters (m^3).

(b) Find k if the gas exerts a pressure of 20,000 N/m^2 when the volume is 1 liter (0.001 m^3).

(c) Make a table that shows the pressures for volumes of 0.25, 0.5, 1.0, 1.5, and 2.0 liters.

(d) Make a graph of P versus V.

26. A manufacturer of cardboard drink containers wants to construct a closed rectangular container that has a square base and will hold $\frac{1}{10}$ liter (100 cm^3). Estimate the dimension of the container that will require the least amount of material for its manufacture.

A variable y is said to be **inversely proportional to the square of a variable x** if y is related to x by an equation of the form $y = k/x^2$, where k is a nonzero constant, called the **constant of proportionality**. This terminology is used in Exercises 27 and 28.

27. According to Coulomb's law, the force F of attraction between positive and negative point charges is inversely proportional to the square of the distance x between them.

(a) Assuming that the force of attraction between two point charges is 0.0005 newton when the distance between them is 0.3 meter, find the constant of proportionality (with proper units).

(b) Find the force of attraction between the point charges when they are 3 meters apart.

(c) Make a graph of force versus distance for the two charges.

(d) What happens to the force as the particles get closer and closer together? What happens as they get farther and farther apart?

28. It follows from Newton's Universal Law of Gravitation that the weight W of an object (relative to the Earth) is inversely proportional to the square of the distance x between the object and the center of the Earth, that is, $W = C/x^2$.

(a) Assuming that a weather satellite weighs 2000 pounds on the surface of the Earth and that the Earth is a sphere of radius 4000 miles, find the constant C.

(b) Find the weight of the satellite when it is 1000 miles above the surface of the Earth.

(c) Make a graph of the satellite's weight versus its distance from the center of the Earth.

(d) Is there any distance from the center of the Earth at which the weight of the satellite is zero? Explain your reasoning.

29. In each part, match the equation with one of the accompanying graphs, and give the equations for the horizontal and vertical asymptotes.

(a) $y = \dfrac{x^2}{x^2 - x - 2}$

(b) $y = \dfrac{x - 1}{x^2 - x - 6}$

(c) $y = \dfrac{2x^4}{x^4 + 1}$

(d) $y = \dfrac{4}{(x + 2)^2}$

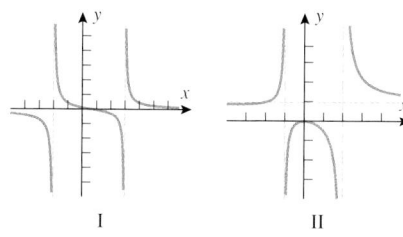

I   II

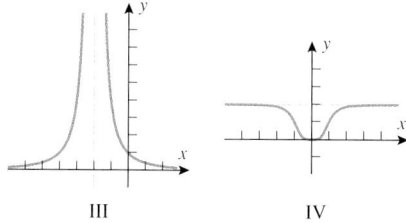

III   IV   Figure Ex-29

30. Find an equation of the form $y = k/(x^2 + bx + c)$ whose graph is a reasonable match to that in the accompanying figure. Check your work with a graphing utility.

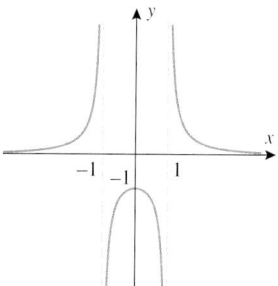

Figure Ex-30

In Exercises 31 and 32, draw a radial line from the origin with the given angle, and determine whether the six trigonometric functions are positive, negative, or undefined for that angle.

31. (a) $\dfrac{\pi}{3}$ (b) $-\dfrac{\pi}{2}$ (c) $\dfrac{2\pi}{3}$

(d) -1 (e) $\dfrac{5\pi}{4}$ (f) $\dfrac{11\pi}{6}$

32. (a) $\dfrac{3\pi}{2}$ (b) $-\dfrac{5\pi}{4}$ (c) π

(d) $\dfrac{5\pi}{2}$ (e) 4 (f) $-\dfrac{33\pi}{7}$

In Exercises 33 and 34, use a calculating utility set to the radian mode to confirm the approximations $\sin(\pi/5) \approx 0.588$ and $\cos(\pi/8) \approx 0.924$, and then use these values to approximate the given expressions by hand calculation. Check your answers using the trigonometric function operations of your calculating utility.

33. (a) $\sin\dfrac{4\pi}{5}$ (b) $\cos\left(-\dfrac{\pi}{8}\right)$ (c) $\sin\dfrac{11\pi}{5}$

(d) $\cos\dfrac{7\pi}{8}$ (e) $\sin\dfrac{2\pi}{5}$ (f) $\cos^2\dfrac{\pi}{5}$

34. (a) $\sin\dfrac{16\pi}{5}$ (b) $\cos\left(-\dfrac{17\pi}{8}\right)$ (c) $\sin\dfrac{41\pi}{5}$

(d) $\sin\left(-\dfrac{\pi}{16}\right)$ (e) $\cos\dfrac{27\pi}{8}$ (f) $\tan^2\dfrac{\pi}{8}$

35. Assuming that $\sin\alpha = a$, $\cos\beta = b$, and $\tan\gamma = c$, express the stated quantities in terms of a, b, and c.

(a) $\sin(-\alpha)$ (b) $\cos(-\beta)$ (c) $\tan(-\gamma)$

(d) $\sin\left(\dfrac{\pi}{2} - \alpha\right)$ (e) $\cos(\pi - \beta)$ (f) $\sin(\alpha + \pi)$

(g) $\sin(2\beta)$ (h) $\cos(2\beta)$ (i) $\sec(\beta + 2\pi)$

(j) $\csc(\alpha + \pi)$ (k) $\cot(\gamma + 5\pi)$ (l) $\sin^2\left(\dfrac{\beta}{2}\right)$

36. A ship travels from a point near Hawaii at 20° N latitude directly north to a point near Alaska at 56° N latitude.
  (a) Assuming the Earth to be a sphere of radius 4000 mi, find the actual distance traveled by the ship.
  (b) What fraction of the Earth's circumference did the ship travel?

37. The Moon completes one revolution around the Earth in approximately 29.5 days. Assuming that the Moon's orbit is a circle with a radius of 0.38×10^9 m from the center of the Earth, find the arc length traveled by the Moon in 1 day.

38. A spoked wheel with a diameter of 3 ft rolls along a flat road without slipping. How far along the road does the wheel roll if the spokes turn through 225°?

39. As illustrated in the accompanying figure, suppose that you hold one quarter flat against a table while you rotate a second quarter around it without slippage. Through what angle will the second quarter have turned about its own center when it returns to its original location?

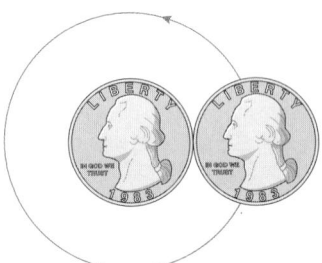

Figure Ex-39

40. Suppose that you begin cutting wedge-shaped pieces from a pie so that the arc length along the outer crust of each piece is equal to the radius. What fraction of the pie will remain after all pieces that can be cut in this way are eaten?

In Exercises 41 and 42, find an equation for the graph assuming that there is no phase shift.

41.

(a)

(b)

(c)

Figure Ex-41

42.

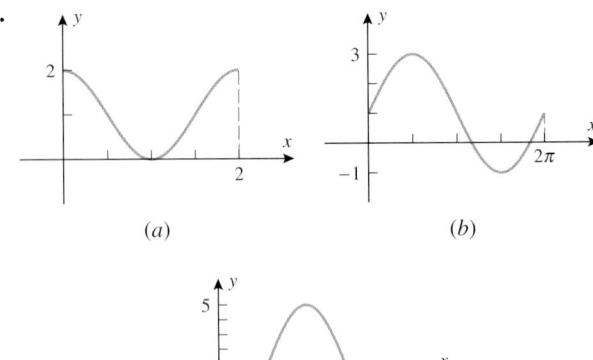

(a)   (b)

(c)

Figure Ex-42

43. In each part, find an equation for the graph that has the form $y = y_0 + A\sin(Bx - C)$.

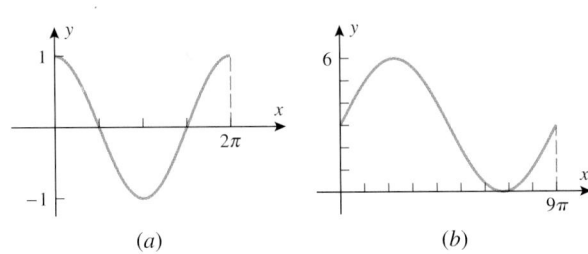

(a)   (b)

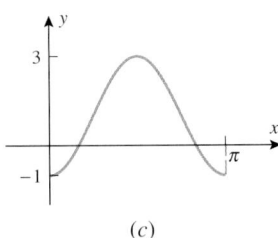

(c)

Figure Ex-43

44. In the United States, a standard electrical outlet supplies sinusoidal electrical current with a maximum voltage of $V = 120\sqrt{2}$ volts (V) at a frequency of 60 cycles per second. Write an equation that expresses V as a function of the time t, assuming that $V = 0$ if $t = 0$.

In Exercises 45 and 46, find the amplitude, period, and phase shift, and sketch at least two periods of the graph by hand. Check your work with a graphing utility.

45. (a) $y = 3\sin 4x$ (b) $y = -2\cos \pi x$
  (c) $y = 2 + \cos\left(\dfrac{x}{2}\right)$

46. (a) $y = -1 - 4\sin 2x$ (b) $y = \frac{1}{2}\cos(3x - \pi)$

   (c) $y = -4\sin\left(\dfrac{x}{3} + 2\pi\right)$

47. Equations of the form

$$x = A_1 \sin \omega t + A_2 \cos \omega t$$

arise in the study of vibrations and other periodic motion.

(a) Use the trigonometric identity for $\sin(\alpha + \beta)$ to show that this equation can be expressed in the form

$$x = A\sin(\omega t + \theta)$$

(b) State formulas that express A and θ in terms of the constants A_1, A_2, and ω.

(c) Express the equation

$$x = 5\sqrt{3}\sin 2\pi t + \frac{5}{2}\cos 2\pi t$$

in the form $x = A\sin(\omega t + \theta)$, and use a graphing utility to confirm that both equations have the same graph.

48. Determine the number of solutions of $x = 2\sin x$, and use a graphing or calculating utility to estimate them.

1.7 PARAMETRIC EQUATIONS

Thus far, our study of graphs has focused on graphs of functions. However, because such graphs must pass the vertical line test, this limitation precludes curves with self-intersections or even such basic curves as circles. In this section we will study an alternative method for describing curves algebraically that is not subject to the severe restriction of the vertical line test.

This material is placed here to provide an early parametric option. However, it can be deferred until Chapter 12, if preferred.

PARAMETRIC EQUATIONS

Suppose that a particle moves along a curve C in the xy-plane in such a way that its x- and y-coordinates, as functions of time, are

$$x = f(t), \quad y = g(t)$$

We call these the **parametric equations** of motion for the particle and refer to C as the **trajectory** of the particle or the **graph** of the equations (Figure 1.7.1). The variable t is called the **parameter** for the equations.

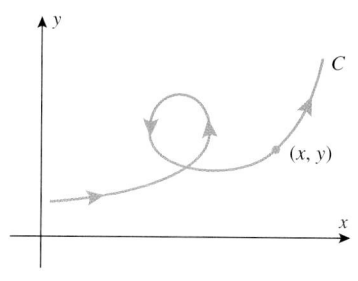

A moving particle with trajectory C

Figure 1.7.1

Example 1

Sketch the trajectory over the time interval $0 \le t \le 10$ of the particle whose parametric equations of motion are

$$x = t - 3\sin t, \quad y = 4 - 3\cos t \tag{1}$$

Solution. One way to sketch the trajectory is to choose a representative succession of times, plot the (x, y) coordinates of points on the trajectory at those times, and connect the points with a smooth curve. The trajectory in Figure 1.7.2 was obtained in this way from Table 1.7.1 in which the approximate coordinates of the particle are given at time increments of 1 unit. Observe that there is no t-axis in the picture; the values of t appear only as labels on the plotted points, and even these are usually omitted unless it is important to emphasize the location of the particle at specific times. ◄

FOR THE READER. Read the documentation for your graphing utility to learn how to graph parametric equations, and then generate the trajectory in Example 1. Explore the behavior of the particle beyond time $t = 10$.

Although parametric equations commonly arise in problems of motion with time as the parameter, they arise in other contexts as well. Thus, unless the problem dictates that the

Table 1.7.1

t	x	y
0	0.0	1.0
1	−1.5	2.4
2	−0.7	5.2
3	2.6	7.0
4	6.3	6.0
5	7.9	3.1
6	6.8	1.1
7	5.0	1.7
8	5.0	4.4
9	7.8	6.7
10	11.6	6.5

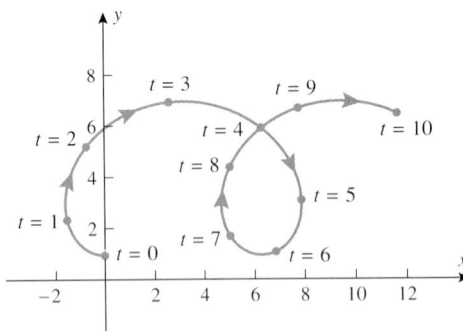

Figure 1.7.2

parameter t in the equations

$$x = f(t), \quad y = g(t)$$

represents time, it should be viewed simply as an independent variable that varies over some interval of real numbers. (In fact, there is no need to use the letter t for the parameter; any letter not reserved for another purpose can be used.) If no restrictions on the parameter are stated explicitly or implied by the equations, then it is understood that it varies from $-\infty$ to $+\infty$. To indicate that a parameter t is restricted to an interval $[a, b]$, we will write

$$x = f(t), \quad y = g(t) \qquad (a \le t \le b)$$

Example 2

Find the graph of the parametric equations

$$x = \cos t, \quad y = \sin t \qquad (0 \le t \le 2\pi) \tag{2}$$

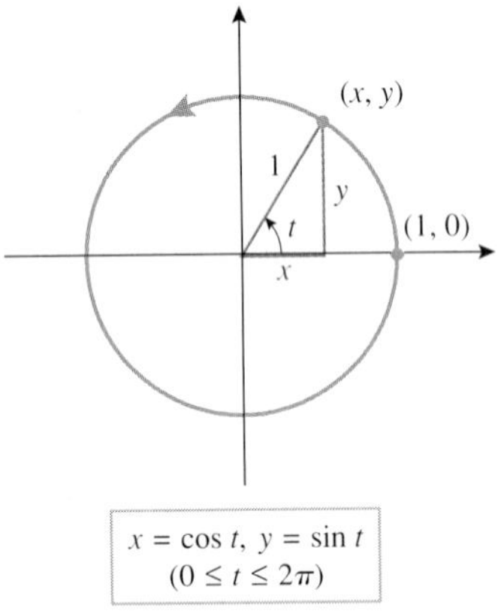

Solution.   One way to find the graph is to eliminate the parameter t by noting that

$$x^2 + y^2 = \sin^2 t + \cos^2 t = 1$$

Thus, the graph is the unit circle $x^2 + y^2 = 1$. This result can also be deduced geometrically by interpreting t as the angle swept out by the radial line from the origin to the point $(x, y) = (\cos t, \sin t)$ on the unit circle (Figure 1.7.3). As t increases from 0 to 2π, the point traces the circle counterclockwise, starting at $(1, 0)$ when $t = 0$ and completing one full revolution when $t = 2\pi$. One can obtain different portions of the circle by varying the interval over which the parameter varies. For example,

$$x = \cos t, \quad y = \sin t \qquad (0 \le t \le \pi) \tag{3}$$

represents just the upper semicircle in Figure 1.7.3.   ◄

$x = \cos t, \ y = \sin t$
$(0 \le t \le 2\pi)$

Figure 1.7.3

ORIENTATION

The direction in which the graph of a pair of parametric equations is traced as the parameter increases is called the ***direction of increasing parameter*** or sometimes the ***orientation*** imposed on the curve by the equations. Thus, we make a distinction between a ***curve***, which is a set of points, and a ***parametric curve***, which is a curve with an orientation imposed on it by a set of parametric equations. For example, we saw in Example 2 that the circle represented parametrically by (2) is traced counterclockwise as t increases and hence has *counterclockwise orientation*. As shown in Figures 1.7.2 and 1.7.3, the orientation of a parametric curve can be indicated by arrowheads.

To obtain parametric equations for the unit circle with *clockwise orientation*, we can replace t by $-t$ in (2), and use the identities $\cos(-t) = \cos t$ and $\sin(-t) = -\sin t$. This yields

$$x = \cos t, \quad y = -\sin t \qquad (0 \le t \le 2\pi)$$

Here, the circle is traced clockwise by a point that starts at $(1, 0)$ when $t = 0$ and completes one full revolution when $t = 2\pi$ (Figure 1.7.4).

FOR THE READER.   When parametric equations are graphed using a calculator, the orientation can often be determined by watching the direction in which the graph is traced on the screen. However, many computers graph so fast that it is often hard to discern the orientation. See if you can use your graphing utility to confirm that (3) has a counterclockwise orientation.

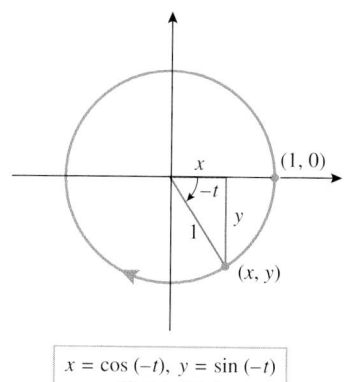

$x = \cos(-t), \ y = \sin(-t)$
$(0 \leq t \leq 2\pi)$

Figure 1.7.4

Example 3

Graph the parametric curve

$$x = 2t - 3, \quad y = 6t - 7$$

by eliminating the parameter, and indicate the orientation on the graph.

Solution.   To eliminate the parameter we will solve the first equation for t as a function of x, and then substitute this expression for t into the second equation:

$$t = \left(\tfrac{1}{2}\right)(x + 3)$$
$$y = 6\left(\tfrac{1}{2}\right)(x + 3) - 7$$
$$y = 3x + 2$$

Thus, the graph is a line of slope 3 and y-intercept 2. To find the orientation we must look to the original equations; the direction of increasing t can be deduced by observing that x increases as t increases *or* by observing that y increases as t increases. Either piece of information tells us that the line is traced left to right as shown in Figure 1.7.5.   ◀

REMARK.   Not all parametric equations produce curves with definite orientations; if the equations are badly behaved, then the point tracing the curve may leap around sporadically or move back and forth, failing to determine a definite direction. For example, if

$$x = \sin t, \quad y = \sin^2 t$$

then the point (x, y) moves along the parabola $y = x^2$. However, the value of x varies periodically between -1 and 1, so the point (x, y) moves periodically back and forth along the parabola between the points $(-1, 1)$ and $(1, 1)$ (as shown in Figure 1.7.6). Later in the text we will discuss restrictions that eliminate such erratic behavior, but for now we will just avoid such complications.

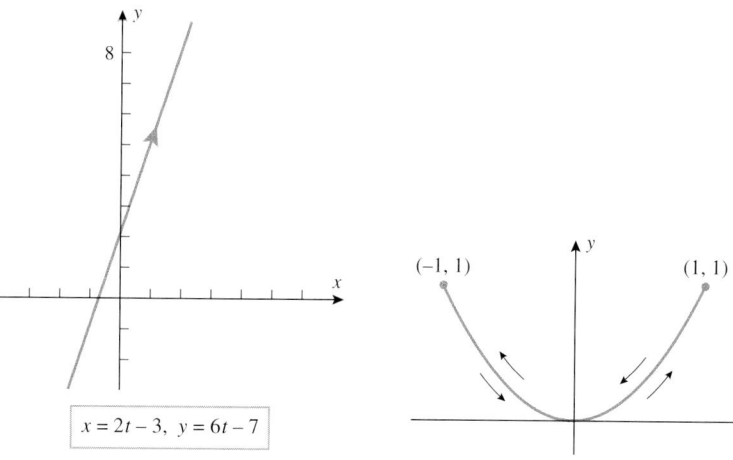

$x = 2t - 3, \ y = 6t - 7$

Figure 1.7.5

Figure 1.7.6

**EXPRESSING ORDINARY
FUNCTIONS PARAMETRICALLY**

An equation $y = f(x)$ can be expressed in parametric form by introducing the parameter $t = x$; this yields the parametric equations $x = t$, $y = f(t)$. For example, the portion of the curve $y = \cos x$ over the interval $[-2\pi, 2\pi]$ can be expressed parametrically as

$$x = t, \quad y = \cos t \qquad (-2\pi \le t \le 2\pi)$$

(Figure 1.7.7).

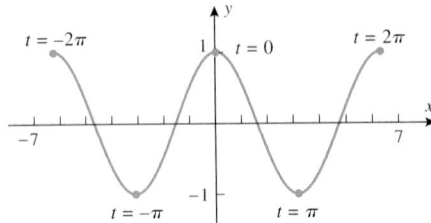

Figure 1.7.7

**GENERATING PARAMETRIC
CURVES WITH GRAPHING UTILITIES**

Many graphing utilities allow you to graph equations of the form $y = f(x)$ but not equations of the form $x = g(y)$. Sometimes you will be able to rewrite $x = g(y)$ in the form $y = f(x)$; however, if this is inconvenient or impossible, then you can graph $x = g(y)$ by introducing a parameter $t = y$ and expressing the equation in the parametric form $x = g(t)$, $y = t$. (You may have to experiment with various intervals for t to produce a complete graph.)

Example 4

Use a graphing utility to graph the equation $x = 3y^5 - 5y^3 + 1$.

Solution. If we let $t = y$ be the parameter, then the equation can be written in parametric form as

$$x = 3t^5 - 5t^3 + 1, \quad y = t$$

Figure 1.7.8 shows the graph of these equations for $-1.5 \le t \le 1.5$.   ◄

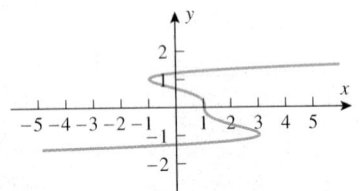

$x = 3t^5 - 5t^3 + 1, \ y = t$
$-1.5 \le t \le 1.5$

Figure 1.7.8

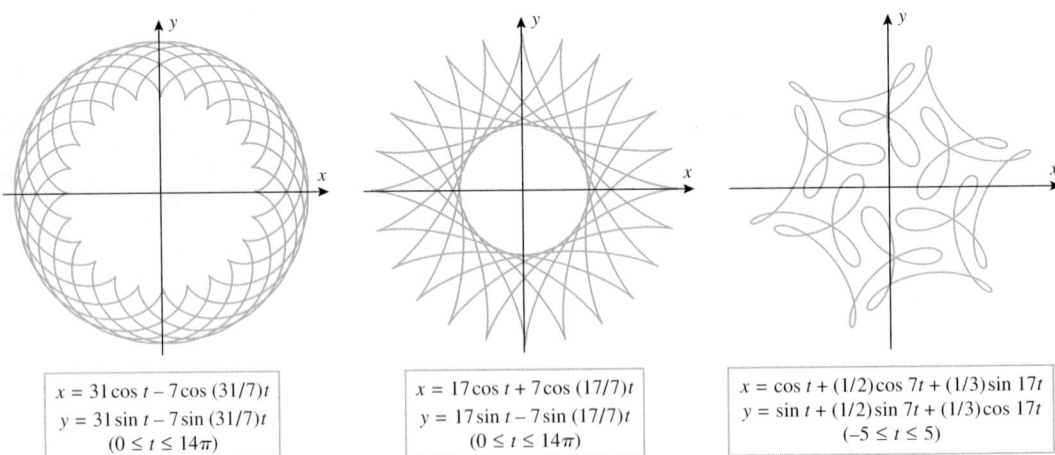

$x = 31\cos t - 7\cos(31/7)t$
$y = 31\sin t - 7\sin(31/7)t$
$(0 \le t \le 14\pi)$

$x = 17\cos t + 7\cos(17/7)t$
$y = 17\sin t - 7\sin(17/7)t$
$(0 \le t \le 14\pi)$

$x = \cos t + (1/2)\cos 7t + (1/3)\sin 17t$
$y = \sin t + (1/2)\sin 7t + (1/3)\cos 17t$
$(-5 \le t \le 5)$

Figure 1.7.9

Some parametric curves are so complex that it is virtually impossible to visualize them without using some kind of graphing utility. Figure 1.7.9 shows three such curves.

FOR THE READER.   Without spending too much time, try your hand at generating some parametric curves with a graphing utility that you think are interesting or beautiful.

TRANSLATION

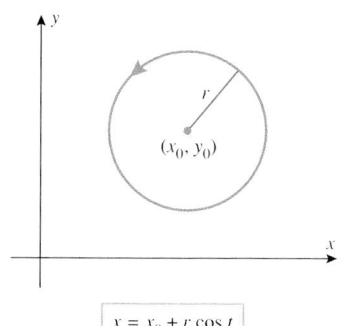

$$x = x_0 + r \cos t$$
$$y = y_0 + r \sin t$$
$$(0 \le t \le 2\pi)$$

Figure 1.7.10

If a parametric curve C is given by the equations $x = f(t)$, $y = g(t)$, then adding a constant to $f(t)$ translates the curve C in the x-direction, and adding a constant to $g(t)$ translates it in the y-direction. Thus, a circle of radius r, centered at (x_0, y_0) can be represented parametrically as

$$x = x_0 + r \cos t, \quad y = y_0 + r \sin t \quad (0 \le t \le 2\pi) \tag{4}$$

(Figure 1.7.10). If desired, we can eliminate the parameter from these equations by noting that

$$(x - x_0)^2 + (y - y_0)^2 = (r \cos t)^2 + (r \sin t)^2 = r^2$$

Thus, we have obtained the familiar equation in rectangular coordinates for a circle of radius r, centered at (x_0, y_0):

$$(x - x_0)^2 + (y - y_0)^2 = r^2 \tag{5}$$

FOR THE READER.   Use the parametric capability of your graphing utility to generate a circle of radius 5 that is centered at $(3, -2)$.

SCALING

If a parametric curve C is given by the equations $x = f(t)$, $y = g(t)$, then multiplying $f(t)$ by a constant stretches or compresses C in the x-direction, and multiplying $g(t)$ by a constant stretches or compresses C in the y-direction. For example, we would expect the parametric equations

$$x = 3 \cos t, \quad y = 2 \sin t \quad (0 \le t \le 2\pi)$$

to represent an ellipse, centered at the origin, since the graph of these equations results from stretching the unit circle

$$x = \cos t, \quad y = \sin t \quad (0 \le t \le 2\pi)$$

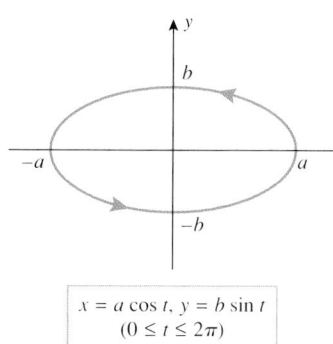

$$x = a \cos t, \, y = b \sin t$$
$$(0 \le t \le 2\pi)$$

Figure 1.7.11

by a factor of 3 in the x-direction and a factor of 2 in the y-direction. In general, if a and b are positive constants, then the parametric equations

$$x = a \cos t, \quad y = b \sin t \quad (0 \le t \le 2\pi) \tag{6}$$

represent an ellipse, centered at the origin, and extending between $-a$ and a on the x-axis and between $-b$ and b on the y-axis (Figure 1.7.11). The numbers a and b are called the *semiaxes* of the ellipse. If desired, we can eliminate the parameter t in (6) and rewrite the equations in rectangular coordinates as

$$\frac{x^2}{a^2} + \frac{y^2}{b^2} = 1 \tag{7}$$

FOR THE READER.   Use the parametric capability of your graphing utility to generate an ellipse that is centered at the origin and that extends between -4 and 4 in the x-direction and between -3 and 3 in the y-direction. Generate an ellipse with the same dimensions, but translated so that its center is at $(2, 3)$.

LISSAJOUS CURVES

In the mid-1850s the French physicist Jules Antoine Lissajous (1822–1880) became interested in parametric equations of the form

$$x = \sin at, \quad y = \sin bt \tag{8}$$

in the course of studying vibrations that combine two perpendicular sinusoidal motions. The first equation in (8) describes a sinusoidal oscillation in the x-direction with frequency $a/2\pi$, and the second describes a sinusoidal oscillation in the y-direction with frequency $b/2\pi$. If a/b is a rational number, then the combined effect of the oscillations is a periodic motion along a path called a *Lissajous curve*. Figure 1.7.12 shows some typical Lissajous curves.

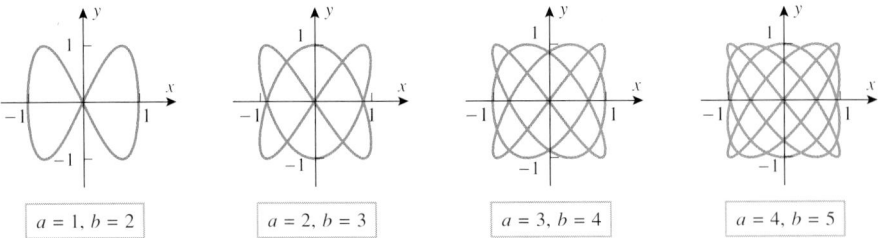

Figure 1.7.12

FOR THE READER.   Generate some Lissajous curves on your graphing utility, and also see
if you can figure out when each of the curves in Figure 1.7.12 begins to repeat.

CYCLOIDS

If a wheel rolls in a straight line along a flat road, then a point on the rim of the wheel will
trace a curve called a *cycloid* (Figure 1.7.13). This curve has a fascinating history, which
we will discuss shortly; but first we will show how to obtain parametric equations for it. For
this purpose, let us assume that the wheel has radius a and rolls along the positive x-axis of
a rectangular coordinate system. Let $P(x, y)$ be the point on the rim that traces the cycloid,
and assume that P is initially at the origin. We will take as our parameter the angle θ that
is swept out by the radial line to P as the wheel rolls (Figure 1.7.13). It is standard here to
regard θ to be positive, even though it is generated by a clockwise rotation.

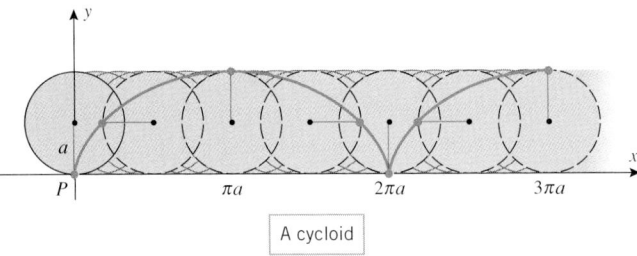

A cycloid

Figure 1.7.13

The motion of P is a combination of the movement of the wheel's center parallel to the
x-axis and the rotation of P around the center. As the radial line sweeps out an angle θ, the
point P traverses an arc of length $a\theta$, and the wheel moves a distance $a\theta$ along the x-axis
(why?). Thus, as suggested by Figure 1.7.14, the center moves to the point $(a\theta, a)$, and the
coordinates of $P(x, y)$ are

$$x = a\theta - a\sin\theta, \quad y = a - a\cos\theta \tag{9}$$

These are the equations of the cycloid in terms of the parameter θ.

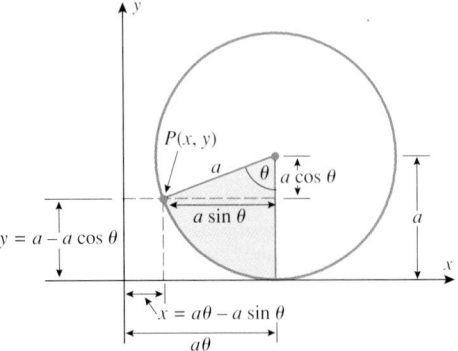

Figure 1.7.14

FOR THE READER.   Use your graphing utility to generate two "arches" of the cycloid produced by a point on the rim of a wheel of radius 1.

THE ROLE OF THE CYCLOID IN MATHEMATICS HISTORY

Figure 1.7.15

The cycloid is of interest because it provides the solution to two famous mathematical problems—the ***brachistochrone problem*** (from Greek words meaning "shortest time") and the ***tautochrone problem*** (from Greek words meaning "equal time"). The brachistochrone problem is to determine the shape of a wire along which a bead might slide from a point P to another point Q, not directly below, in the *shortest time*. The tautochrone problem is to find the shape of a wire from P to Q such that two beads started at any points on the wire between P and Q reach Q in the same amount of time (Figure 1.7.15). The solution to both problems turns out to be an inverted cycloid.

In June of 1696, Johann Bernoulli[*] posed the brachistochrone problem in the form of a challenge to other mathematicians. At first, one might conjecture that the wire should form a straight line, since that shape results in the shortest distance from P to Q. However, the inverted cycloid allows the bead to fall more rapidly at first, building up sufficient initial

---

[*] BERNOULLI. An amazing Swiss family that included several generations of outstanding mathematicians and scientists. Nikolaus Bernoulli (1623–1708), a druggist, fled from Antwerp to escape religious persecution and ultimately settled in Basel, Switzerland. There he had three sons, Jakob I (also called Jacques or James), Nikolaus, and Johann I (also called Jean or John). The Roman numerals are used to distinguish family members with identical names (see the family tree below). Following Newton and Leibniz, the Bernoulli brothers, Jakob I and Johann I, are considered by some to be the two most important founders of calculus. Jakob I was self-taught in mathematics. His father wanted him to study for the ministry, but he turned to mathematics and in 1686 became a professor at the University of Basel. When he started working in mathematics, he knew nothing of Newton's and Leibniz' work. He eventually became familiar with Newton's results, but because so little of Leibniz' work was published, Jakob duplicated many of Leibniz' results.

Jakob's younger brother Johann I was urged to enter into business by his father. Instead, he turned to medicine and studied mathematics under the guidance of his older brother. He eventually became a mathematics professor at Groningen in Holland, and then, when Jakob died in 1705, Johann succeeded him as mathematics professor at Basel. Throughout their lives, Jakob I and Johann I had a mutual passion for criticizing each other's work, which frequently erupted into ugly confrontations. Leibniz tried to mediate the disputes, but Jakob, who resented Leibniz' superior intellect, accused him of siding with Johann, and thus Leibniz became entangled in the arguments. The brothers often worked on common problems that they posed as challenges to one another. Johann, interested in gaining fame, often used unscrupulous means to make himself appear the originator of his brother's results; Jakob occasionally retaliated. Thus, it is often difficult to determine who deserves credit for many results. However, both men made major contributions to the development of calculus. In addition to his work on calculus, Jakob helped establish fundamental principles in probability, including the Law of Large Numbers, which is a cornerstone of modern probability theory.

Among the other members of the Bernoulli family, Daniel, son of Johann I, is the most famous. He was a professor of mathematics at St. Petersburg Academy in Russia and subsequently a professor of anatomy and then physics at Basel. He did work in calculus and probability, but is best known for his work in physics. A basic law of fluid flow, called Bernoulli's principle, is named in his honor. He won the annual prize of the French Academy 10 times for work on vibrating strings, tides of the sea, and kinetic theory of gases.

Johann II succeeded his father as professor of mathematics at Basel. His research was on the theory of heat and sound. Nikolaus I was a mathematician and law scholar who worked on probability and series. On the recommendation of Leibniz, he was appointed professor of mathematics at Padua and then went to Basel as a professor of logic and then law. Nikolaus II was professor of jurisprudence in Switzerland and then professor of mathematics at St. Petersburg Academy. Johann III was a professor of mathematics and astronomy in Berlin and Jakob II succeeded his uncle Daniel as professor of mathematics at St. Petersburg Academy in Russia. Truly an incredible family!

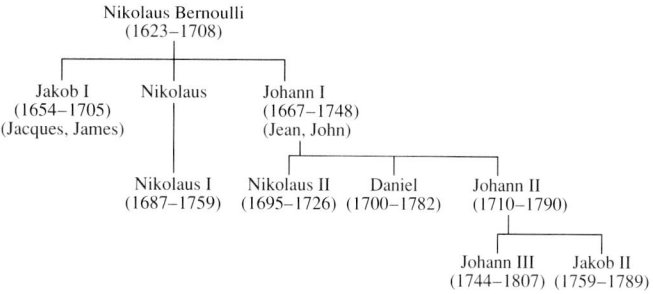

speed to reach Q in the shortest time, even though it travels a longer distance. The problem was solved by Newton and Leibniz as well as by Johann Bernoulli and his older brother Jakob; it was formulated and solved *incorrectly* years earlier by Galileo, who thought the answer was a circular arc.

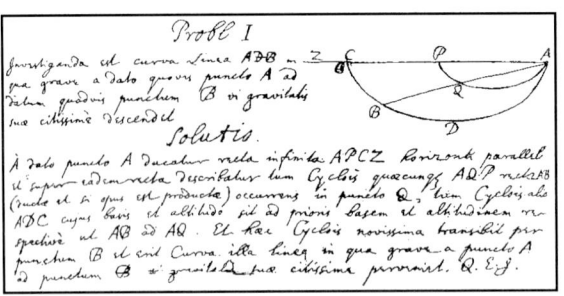

Newton's solution of the brachistochrone problem in his own handwriting

EXERCISE SET 1.7   ⌇ Graphing Calculator

1. (a) By eliminating the parameter, sketch the trajectory over the time interval $0 \leq t \leq 5$ of the particle whose parametric equations of motion are

    $$x = t - 1, \quad y = t + 1$$

   (b) Indicate the direction of motion on your sketch.

   (c) Make a table of x- and y-coordinates of the particle at times $t = 0, 1, 2, 3, 4, 5$.

   (d) Mark the position of the particle on the curve at the times in part (c), and label those positions with the values of t.

2. (a) By eliminating the parameter, sketch the trajectory over the time interval $0 \leq t \leq 1$ of the particle whose parametric equations of motion are

    $$x = \cos(\pi t), \quad y = \sin(\pi t)$$

   (b) Indicate the direction of motion on your sketch.

   (c) Make a table of x- and y-coordinates of the particle at times $t = 0, 0.25, 0.5, 0.75, 1$.

   (d) Mark the position of the particle on the curve at the times in part (c), and label those positions with the values of t.

In Exercises 3–12, sketch the curve by eliminating the parameter, and indicate the direction of increasing t.

3. $x = 3t - 4, \ y = 6t + 2$

4. $x = t - 3, \ y = 3t - 7 \quad (0 \leq t \leq 3)$

5. $x = 2\cos t, \ y = 5\sin t \quad (0 \leq t \leq 2\pi)$

6. $x = \sqrt{t}, \ y = 2t + 4$

7. $x = 3 + 2\cos t, \ y = 2 + 4\sin t \quad (0 \leq t \leq 2\pi)$

8. $x = \sec t, \ y = \tan t \quad (\pi \leq t < 3\pi/2)$

9. $x = \cos 2t, \ y = \sin t \quad (-\pi/2 \leq t \leq \pi/2)$

10. $x = 4t + 3, \ y = 16t^2 - 9$

11. $x = 2\sin^2 t, \ y = 3\cos^2 t$

12. $x = \sec^2 t, \ y = \tan^2 t$

In Exercises 13–18, find parametric equations for the curve, and check your work by generating the curve with a graphing utility.

⌇ **13.** A circle of radius 5, centered at the origin, oriented clockwise.

⌇ **14.** The portion of the circle $x^2 + y^2 = 1$ that lies in the third quadrant, oriented counterclockwise.

⌇ **15.** A vertical line intersecting the x-axis at $x = 2$, oriented upward.

⌇ **16.** The ellipse $\frac{x^2}{4} + \frac{y^2}{9} = 1$, oriented counterclockwise.

⌇ **17.** The portion of the parabola $x = y^2$ joining $(1, -1)$ and $(1, 1)$, oriented down to up.

⌇ **18.** The circle of radius 4, centered at $(1, -3)$, oriented counterclockwise.

19. In each part, match the parametric equation with one of the curves labeled (I)–(VI), and explain your reasoning.
   (a) $x = \sqrt{t}, \ y = \sin 3t$ (b) $x = 2\cos t, \ y = 3\sin t$

   (c) $x = t\cos t, \ y = t\sin t$ (d) $x = \dfrac{3t}{1 + t^3}, \ y = \dfrac{3t^2}{1 + t^3}$

   (e) $x = \dfrac{t^3}{1 + t^2}, \ y = \dfrac{2t^2}{1 + t^2}$ (f) $x = 2\cos t, \ y = \sin 2t$

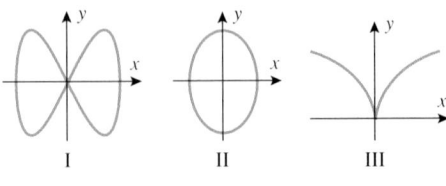

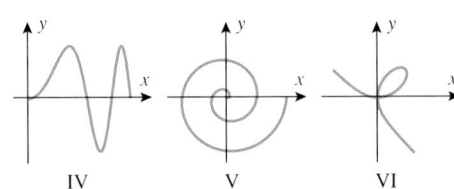

Figure Ex-19

20. Use a graphing utility to generate the curves in Exercise 19, and in each case identify the orientation.

21. (a) Use a graphing utility to generate the trajectory of a particle whose equations of motion over the time interval $0 \le t \le 5$ are

$$x = 6t - \tfrac{1}{2}t^3, \quad y = 1 + \tfrac{1}{2}t^2$$

(b) Make a table of x- and y-coordinates of the particle at times $t = 0, 1, 2, 3, 4, 5$.

(c) At what times is the particle on the y-axis?

(d) During what time interval is $y < 5$?

(e) At what time is the x-coordinate of the particle maximum?

22. (a) Use a graphing utility to generate the trajectory of a paper airplane whose equations of motion for $t \ge 0$ are

$$x = t - 2 \sin t, \quad y = 3 - 2 \cos t$$

(b) Assuming that the plane flies in a room in which the floor is at $y = 0$, explain why the plane will not crash into the floor. [For simplicity, ignore the physical size of the plane by treating it as a particle.]

(c) How high must the ceiling be to ensure that the plane does not touch or crash into it?

In Exercises 23 and 24, graph the equation using a graphing utility.

23. (a) $x = y^2 + 2y + 1$
(b) $x = \sin y, \ -2\pi \le y \le 2\pi$

24. (a) $x = y + 2y^3 - y^5$
(b) $x = \tan y, \ -\pi/2 < y < \pi/2$

25. (a) By eliminating the parameter, show that the equations

$$x = x_0 + (x_1 - x_0)t, \quad y = y_0 + (y_1 - y_0)t$$

represent the line passing through the points (x_0, y_0) and (x_1, y_1).

(b) Show that if $0 \le t \le 1$, then the equations in part (a) represent the line segment joining (x_0, y_0) and (x_1, y_1), oriented in the direction from (x_0, y_0) to (x_1, y_1).

(c) Use the result in part (b) to find parametric equations for the line segment joining the points $(1, -2)$ and $(2, 4)$, oriented in the direction from $(1, -2)$ to $(2, 4)$.

(d) Use the result in part (b) to find parametric equations for the line segment in part (c), but oriented in the direction from $(2, 4)$ to $(1, -2)$.

26. Use the result in Exercise 25 to find

(a) parametric equations for the line segment joining the points $(-3, -4)$ and $(-5, 1)$, oriented from $(-3, -4)$ to $(-5, 1)$

(b) parametric equations for the line segment traced from $(0, b)$ to $(a, 0)$, oriented from $(0, b)$ to $(a, 0)$.

27. (a) Suppose that the line segment from the point $P(x_0, y_0)$ to $Q(x_1, y_1)$ is represented parametrically by

$$x = x_0 + (x_1 - x_0)t,$$
$$\qquad\qquad\qquad (0 \le t \le 1)$$
$$y = y_0 + (y_1 - y_0)t$$

and that $R(x, y)$ is the point on the line segment corresponding to a specified value of t (see the accompanying figure). Show that $t = r/q$, where r is the distance from P to R and q is the distance from P to Q.

(b) What value of t produces the midpoint between points P and Q?

(c) What value of t produces the point that is three-fourths of the way from P to Q?

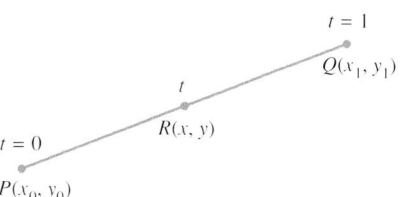

Figure Ex-27

28. Find parametric equations for the line segment joining $P(2, -1)$ and $Q(3, 1)$, and use the result in Exercise 27 to find

(a) the midpoint between P and Q

(b) the point that is one-fourth of the way from P to Q

(c) the point that is three-fourths of the way from P to Q.

29. Explain why the parametric curve

$$x = t^2, \quad y = t^4 \quad (-1 \le t \le 1)$$

does not have a definite orientation.

30. (a) In parts (a) and (b) of Exercise 25 we obtained parametric equations for a line segment in which the parameter varied from $t = 0$ to $t = 1$. Sometimes it is desirable to have parametric equations for a line segment in which the parameter varies over some other interval, say $t_0 \le t \le t_1$. Use the ideas in Exercise 25 to show that the line segment joining the points (x_0, y_0) and (x_1, y_1) can be represented parametrically as

$$x = x_0 + (x_1 - x_0)\frac{t - t_0}{t_1 - t_0},$$
$$\qquad\qquad\qquad (t_0 \le t \le t_1)$$
$$y = y_0 + (y_1 - y_0)\frac{t - t_0}{t_1 - t_0}$$

(b) Which way is the line segment oriented?

(c) Find parametric equations for the line segment traced from $(3, -1)$ to $(1, 4)$ as t varies from 1 to 2, and check your result with a graphing utility.

31. (a) By eliminating the parameter, show that if a and c are not both zero, then the graph of the parametric equations

$$x = at + b, \quad y = ct + d \quad (t_0 \le t \le t_1)$$

is a line segment.

(b) Sketch the parametric curve

$$x = 2t - 1, \quad y = t + 1 \quad (1 \le t \le 2)$$

and indicate its orientation.

32. (a) What can you say about the line in Exercise 31 if a or c (but not both) is zero?

(b) What do the equations represent if a and c are both zero?

33. Parametric curves can be defined piecewise by using different formulas for different values of the parameter. Sketch the curve that is represented piecewise by the parametric equations

$$x = 2t, \quad y = 4t^2 \quad \left(0 \leq t \leq \tfrac{1}{2}\right)$$

$$x = 2 - 2t, \quad y = 2t \quad \left(\tfrac{1}{2} \leq t \leq 1\right)$$

34. Find parametric equations for the rectangle in the accompanying figure, assuming that the rectangle is traced counterclockwise as t varies from 0 to 1, starting at $\left(\tfrac{1}{2}, \tfrac{1}{2}\right)$ when $t = 0$. [*Hint:* Represent the rectangle piecewise, letting t vary from 0 to $\tfrac{1}{4}$ for the first edge, from $\tfrac{1}{4}$ to $\tfrac{1}{2}$ for the second edge, and so forth.]

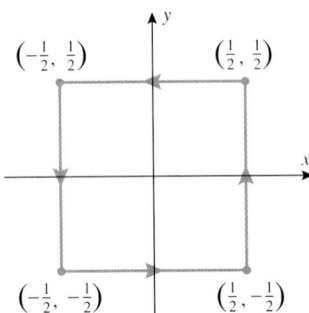

$\left(-\tfrac{1}{2}, \tfrac{1}{2}\right)$ $\left(\tfrac{1}{2}, \tfrac{1}{2}\right)$

$\left(-\tfrac{1}{2}, -\tfrac{1}{2}\right)$ $\left(\tfrac{1}{2}, -\tfrac{1}{2}\right)$

Figure Ex-34

35. (a) Find parametric equations for the ellipse that is centered at the origin and has intercepts $(4, 0)$, $(-4, 0)$, $(0, 3)$, and $(0, -3)$.

(b) Find parametric equations for the ellipse that results by translating the ellipse in part (a) so that its center is at $(-1, 2)$.

(c) Confirm your results in parts (a) and (b) using a graphing utility.

36. We will show later in the text that if a projectile is fired from ground level with an initial speed of v_0 meters per second at an angle α with the horizontal, and if air resistance is neglected, then its position after t seconds, relative to the coordinate system in the accompanying figure is

$$x = (v_0 \cos \alpha)t, \quad y = (v_0 \sin \alpha)t - \tfrac{1}{2}gt^2$$

where $g \approx 9.8 \text{ m/s}^2$.

(a) By eliminating the parameter, show that the trajectory is a parabola.

(b) Sketch the trajectory if $\alpha = 30°$ and $v_0 = 1000$ m/s.

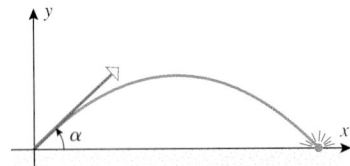

Figure Ex-36

37. A shell is fired from a cannon at an angle of $\alpha = 45°$ with an initial speed of $v_0 = 800$ m/s.

(a) Find parametric equations for the shell's trajectory relative to the coordinate system in Figure Ex-36.

(b) How high does the shell rise?

(c) How far does the shell travel horizontally?

38. A robot arm, designed to buff flat surfaces on an automobile, consists of two attached rods, one that moves back and forth horizontally, and a second, with the buffing pad at the end, that moves up and down (see the accompanying figure).

(a) Suppose that the horizontal arm of the robot moves so that the x-coordinate of the buffer's center at time t is $x = 25 \sin \pi t$ and the vertical arm moves so that the y-coordinate of the buffer's center at time t is $y = 12.5 \sin \pi t$. Graph the trajectory of the center of the buffing pad.

(b) Suppose that the x- and y-coordinates in part (a) are $x = 25 \sin \pi a t$ and $y = 12.5 \sin \pi b t$, where the constants a and b can be controlled by programming the robot arm. Graph the trajectory of the center of the pad if $a = 4$ and $b = 5$.

(c) Investigate the trajectories that result in part (b) for various choices of a and b.

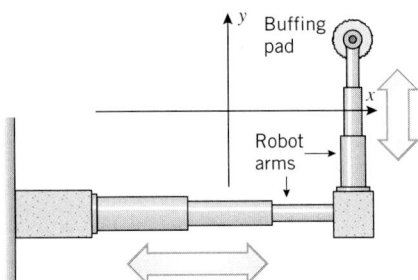

Buffing pad

Robot arms

Figure Ex-38

39. Describe the family of curves described by the parametric equations

$$x = a \cos t + h, \quad y = b \sin t + k \quad (0 \leq t \leq 2\pi)$$

if

(a) h and k are fixed but a and b can vary

(b) a and b are fixed but h and k can vary

(c) $a = 1$ and $b = 1$, but h and k vary so that $h = k + 1$.

40. A *hypocycloid* is a curve traced by a point P on the circumference of a circle that rolls inside a larger fixed circle. Suppose that the fixed circle has radius a, the rolling circle has radius b, and the fixed circle is centered at the origin. Let ϕ be the angle shown in the following figure, and assume that the point P is at $(a, 0)$ when $\phi = 0$. Show that the hypocycloid generated is given by the parametric equations

$$x = (a - b) \cos \phi + b \cos \left(\frac{a - b}{b}\phi\right)$$

$$y = (a - b) \sin \phi - b \sin \left(\frac{a - b}{b}\phi\right)$$

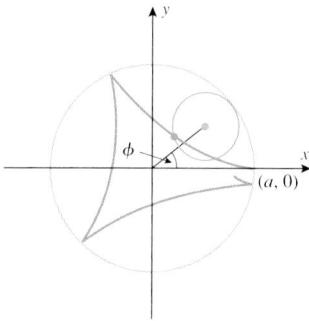

Figure Ex-40

41. If $b = \frac{1}{4}a$ in Exercise 40, then the resulting curve is called a four-cusped hypocycloid.

(a) Sketch this curve.

(b) Show that the curve is given by the parametric equations
$x = a \cos^3 \phi,\quad y = a \sin^3 \phi.$

(c) Show that the curve is given by the equation
$$x^{2/3} + y^{2/3} = a^{2/3}$$
in rectangular coordinates.

42. (a) Use a graphing utility to study how the curves in the family
$$x = 2a \cos^2 t,\quad y = 2a \cos t \sin t \quad (-2\pi < t < 2\pi)$$
change as a varies from 0 to 5.

(b) Confirm your conclusion algebraically.

(c) Write a brief paragraph that describes your findings.

SUPPLEMENTARY EXERCISES

1. Referring to the cigarette consumption graph in Figure 1.1.2b, during what 5-year period was the annual cigarette consumption per adult increasing most rapidly on average? Explain your reasoning.

2. Use the graphs of the functions f and g in the accompanying figure to solve the following problems.

(a) Find the values of $f(-1)$ and $g(3)$.

(b) For what values of x is $f(x) = g(x)$?

(c) For what values of x is $f(x) < 2$?

(d) What are the domain and range of f?

(e) What are the domain and range of g?

(f) Find the zeros of f and g.

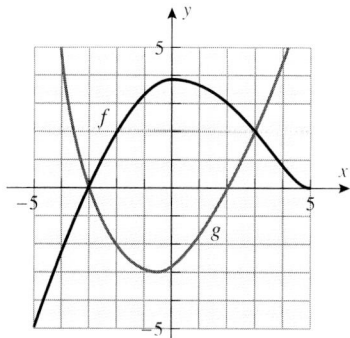

Figure Ex-2

3. A glass filled with water that has a temperature of $40°\,$F is placed in a room in which the temperature is a constant $70°\,$F. Sketch a rough graph that reasonably describes the temperature of the water in the glass as a function of the elapsed time.

4. A student begins driving toward school but 5 minutes into the trip remembers that he forgot his homework. He drives home hurriedly, retrieves his notes, and then drives at great speed toward school, hitting a tree 5 minutes after leaving home. Sketch a rough graph that reasonably describes the student's distance from home as a function of the elapsed time.

5. A rectangular storage container with an open top and a square base has a volume of 8 cubic meters. Material for the base costs $5 per square meter, and material for the sides $2 per square meter. Express the total cost of the materials as a function of the length of a side of the base.

6. You want to paint the top of a circular table. Find a formula that expresses the amount of paint required as a function of the radius, and discuss all of the assumptions you have made in finding the formula.

7. Sketch the graph of the function
$$f(x) = \begin{cases} -1, & x \leq -5 \\ \sqrt{25 - x^2}, & -5 < x < 5 \\ x - 5, & x \geq 5 \end{cases}$$

8. A ball of radius 3 inches is coated uniformly with plastic. Express the volume of the plastic as a function of its thickness.

9. A box with a closed top is to be made from a 6-ft by 10-ft piece of cardboard by cutting out four squares of equal size (see the accompanying figure), folding along the dashed lines, and tucking the two extra flaps inside.

(a) Find a formula that expresses the volume of the box as a function of the length of the sides of the cut-out squares.

(b) Find an inequality that specifies the domain of the function in part (a).

(c) Estimate the dimensions of the box of largest volume.

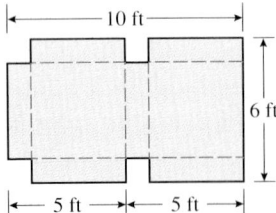

Figure Ex-9

10. Let $f(x) = -x^2$ and $g(x) = 1/\sqrt{x}$. Find the natural domains of $f \circ g$ and $g \circ f$.

11. Given that $f(x) = 2x - 5$ and $g(x) = 3x - 2$, find a value of x such that $f(g(x)) = g(f(x))$.

12. Let $f(x) = (2x - 1)/(x + 1)$ and $g(x) = 1/(x - 1)$.
(a) Find $f(g(x))$.
(b) Is the natural domain of the function $f(g(x))$ obtained in part (a) the same as the domain of $f \circ g$? Explain.

13. Find $f(g(h(x)))$, given that
$$f(x) = \frac{x}{x - 1}, \quad g(x) = \frac{1}{x}, \quad h(x) = x^2 - 1$$

14. Given that $f(x) = 2x + 1$ and $h(x) = 2x^2 + 4x + 1$, find a function g such that $f(g(x)) = h(x)$.

15. Complete the following table.

x	-4	-3	-2	-1	0	1	2	3	4
$f(x)$	0	-1	2	1	3	-2	-3	4	-4
$g(x)$	3	2	1	-3	-1	-4	4	-2	0
$(f \circ g)(x)$									
$(g \circ f)(x)$									

16. (a) Write an equation for the graph that is obtained by reflecting the graph of $y = |x - 1|$ about the y-axis, then stretching that graph vertically by a factor of 2, then translating that graph down 3 units, and then reflecting that graph about the x-axis.
(b) Sketch the original graph and the final graph.

17. In each part, classify the function as even, odd, or neither.
(a) $x^2 \sin x$ (b) $\sin^2 x$
(c) $x + x^2$ (d) $\sin x \tan x$

18. (a) Find exact values for all x-intercepts of
$$y = \cos x - \sin 2x$$
in the interval $-2\pi \le x \le 2\pi$.

(b) Find the coordinates of all intersections of the graphs of $y = \cos x$ and $y = \sin 2x$ if $-2\pi \le x \le 2\pi$, and use a graphing utility to check your answer.

19. (a) A surveyor measures the angle of elevation α of a tower from a point A due south of the tower and also measures the angle of elevation β from a point B that is d feet due east of the point A (see the accompanying figure). Show that the height h of the tower in feet is given by
$$h = \frac{d \sin \alpha \sin \beta}{\sqrt{\sin(\alpha + \beta) \sin(\alpha - \beta)}}$$
(b) Use a calculating utility to approximate the height of the tower to the nearest tenth of a foot if $\alpha = 17°$, $\beta = 12°$, and $d = 1000$ ft.

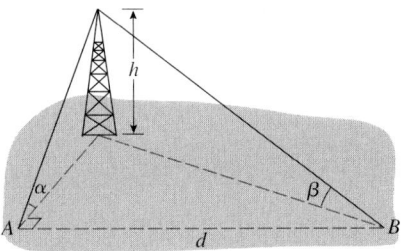

Figure Ex-19

20. Suppose that the expected low temperature in Anchorage, Alaska (in °F), is modeled by the equation
$$T = 50 \sin \frac{2\pi}{365}(t - 101) + 25$$
where t is in days and $t = 0$ corresponds to January 1.
(a) Sketch the graph of T versus t for $0 \le t \le 365$.
(b) Use the model to predict when the coldest day of the year will occur.
(c) Based on this model, how many days during the year would you expect the temperature to be below $0°$ F?

21. The accompanying figure shows the graph of the equation $y = \frac{1}{2}x + \sin x$ for $-2\pi \le x \le 2\pi$. Find the coordinates of the points A, B, C, and D. Explain your reasoning.

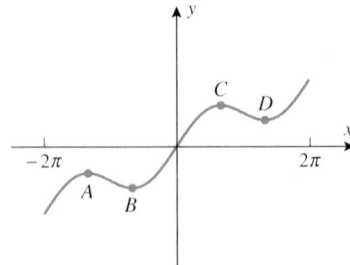

Figure Ex-21

22. The accompanying figure shows a model for the tide variation in an inlet to San Francisco Bay during a 24-hour period. Find an equation of the form $y = y_0 + y_1 \sin(at + b)$ for the model, assuming that $t = 0$ corresponds to midnight.

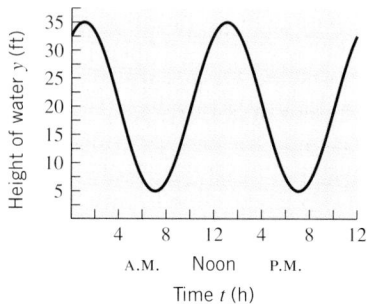

Figure Ex-22

23. In each part describe the family of curves.
(a) $(x - a)^2 + (y - a^2)^2 = 1$
(b) $y = a + (x - 2a)^2$

24. (a) Suppose that the equations $x = f(t)$, $y = g(t)$ describe a curve C as t increases from 0 to 1. Find parametric equations that describe the same curve C but traced in the opposite direction as t increases from 0 to 1.
(b) Check your work using the parametric graphing feature of a graphing utility by generating the line segment between $(1, 2)$ and $(4, 0)$ in both possible directions as t increases from 0 to 1.

25. Sketch the graph of the equation $x^2 - 4y^2 = 0$.

26. Find an equation for a parabola that passes through the points $(2, 0)$, $(8, 18)$, and $(-8, 18)$.

27. Sketch the curve described by the parametric equations
$$x = t \cos(2\pi t), \quad y = t \sin(2\pi t)$$
and check your result with a graphing utility.

28. The electrical resistance R in ohms (Ω) for a pure metal wire is related to its temperature T in $^\circ$C by the formula
$$R = R_0(1 + kT)$$
in which R_0 and k are positive constants.
(a) Make a hand-drawn sketch of the graph of R versus T, and explain the geometric significance of R_0 and k for your graph.
(b) In theory, the resistance R of a wire drops to zero when the temperature reaches absolute zero ($T = -273\,^\circ$C). What information does this give you about k?
(c) A tungsten bulb filament has a resistance of 1.1 Ω at a temperature of $20\,^\circ$C. What information does this give you about R_0 for the filament?
(d) At what temperature will a tungsten filament have a resistance of 1.5 Ω?

Most of the following exercises require access to graphing and calculating utilities. When you are asked to *find* an answer or to *solve* an equation, you may choose to find either an exact result or a numerical approximation, depending on the particular technology you are using and your own imagination.

29. Find the distance between the point $P(1, 2)$ and an arbitrary point $(x, \sqrt{x})$ on the curve $y = \sqrt{x}$. Graph this distance ver-sus x, and use the graph to find the x-coordinate of the point on the curve that is closest to the point P.

30. Find the distance between the point $P(1, 0)$ and an arbitrary point $(x, 1/x)$ on the curve $y = 1/x$, where $x > 0$. Graph this distance versus x, and use the graph to find the x-coordinate of the point on the curve that is closest to the point P.

In Exercises 31 and 32, use **Archimedes' principle**: *A body wholly or partially immersed in a fluid is buoyed up by a force equal to the weight of the fluid that it displaces.*

31. A hollow metal sphere of diameter 5 feet weighs 108 pounds and floats partially submerged in seawater. Assuming that seawater weighs 63.9 pounds per cubic foot, how far below the surface is the bottom of the sphere? [*Hint:* If a sphere of radius r is submerged to a depth h, then the volume V of the submerged portion is given by the formula $V = \pi h^2(r - h/3)$.]

32. Suppose that a hollow metal sphere of diameter 5 feet and weight w pounds floats in seawater. (See Exercise 31.)
(a) Graph w versus h for $0 \le h \le 5$.
(b) Find the weight of the sphere if exactly half of the sphere is submerged.

33. A breeding group of 20 bighorn sheep is released in a protected area in Colorado. It is expected that with careful management the number of sheep, N, after t years will be given by the formula
$$N = \frac{220}{1 + 10(0.83)^t}$$
and that the sheep population will be able to maintain itself without further supervision once the population reaches a size of 80.
(a) Graph N versus t.
(b) How many years must the state of Colorado maintain a program to care for the sheep?
(c) How many bighorn sheep can the environment in the protected area support? [*Hint:* Examine the graph of N versus t for large values of t.]

In Exercises 34 and 35, use the following empirical formula for the windchill index (WCI) [see Example 3 of Section 1.2]:
$$\text{WCI} = \begin{cases} T, & 0 \le v \le 4 \\ 91.4 + (91.4 - T)(0.0203v - 0.304\sqrt{v} - 0.474), & 4 < v < 45 \\ 1.6T - 55, & v \ge 45 \end{cases}$$
where T is the air temperature in $^\circ$F, v is the wind speed in mi/h, and WCI is the equivalent temperature in $^\circ$F.

34. (a) Graph T versus v over the interval $4 \le v \le 45$ for WCI $= 0$.
(b) Use your graph to estimate the values of T for WCI $= 0$ corresponding to $v = 10, 20, 30$, to the nearest degree.

35. (a) Graph WCI versus v over the interval $0 \le v \le 50$ for $T = 20$.

(b) Use your graph to estimate the values of the WCI corresponding to $v = 10, 20, 30, 40$, to the nearest degree.

(c) Use your graph to estimate the values of v corresponding to WCI $= -20, -10, 0, 10$, to the nearest mile per hour.

36. Find the domain and range of the function
$$f(x) = \frac{\sin x}{x^4 + x^3 + 5}$$

37. Find the domain and range of the function
$$f(x) = x^2 - \sqrt{1 + x - x^4}$$

38. An oven is preheated and then remains at a constant temperature. A potato is placed in the oven to bake. Suppose that the temperature T (in $°$F) of the potato t minutes later is given by $T = 400 - 325(0.97)^t$. The potato will be considered done when its temperature is anywhere between $260°$F and $280°$F.

(a) During what interval of time would the potato be considered done?

(b) How long does it take for the temperature of the potato to reach 95% of the oven temperature?

39. Suppose that a package of medical supplies is dropped from a helicopter straight down by parachute into a remote area. The velocity v (in feet per second) of the package t seconds after it is released is given by $v = 24.61(1 - (0.273^t))$.

(a) Graph v versus t.

(b) Show that the graph has a horizontal asymptote $v = c$.

(c) The constant c is called the **terminal velocity**. Explain what the terminal velocity means in practical terms.

(d) Can the package actually reach its terminal velocity? Explain.

(e) How long does it take for the package to reach 98% of its terminal velocity?

EXPANDING THE CALCULUS HORIZON

Iteration and Dynamical Systems

*What do the four figures below have in common? The answer is that all of them are of interest in contemporary research and all involve a mathematical process called **iteration**. In this module we will introduce this concept and touch on some of the fascinating ideas to which it leads.*

Barnsley's fern

The Sierpinski triangle

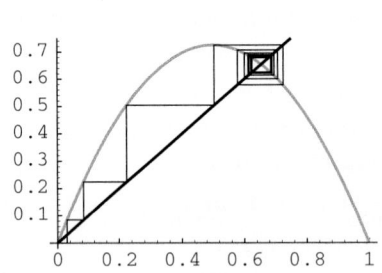
A cobweb diagram

A Julia set

Iterative Processes

Recall that in the notation $y = f(x)$, the variable x is called an **input** of the function f, and the variable y is called the corresponding **output**. Suppose that we start with some input, say $x = c$, and each time we compute an output we feed it back into f as an input. This generates the following sequence of numbers:

$$f(c), \quad f(f(c)), \quad f(f(f(c))), \quad f(f(f(f(c)))), \ldots$$

This is called an **iterated function sequence** for f (from the Latin word *iteratus*, meaning "to repeat"). The number c is called the **seed value** for the sequence, the terms in the sequence are called **iterates**, and each time f is applied we say that we have performed an **iteration**. Iterated function sequences arise in a wide variety of physical processes that are collectively called **dynamical systems**.

....
Exercise 1   Let $f(x) = x^2$.

(a) Calculate the first 10 iterates in the iterated function sequence for f, starting with seed values of $c = 0.5, 1,$ and 2. In each case make a conjecture about the ***long-term behavior*** of the iterates, that is, the behavior of the iterates as more and more iterations are performed.

(b) Try your own seed values, and make a conjecture about the effect of a seed value on the long-term behavior of the iterates.

Recursion Formulas

The proliferation of parentheses in an iterated function sequence can become confusing, so for simplicity let us introduce the following notation for the successive iterates

$$y_0 = c, \quad y_1 = f(c), \quad y_2 = f(f(c)), \quad y_3 = f(f(f(c))), \quad y_4 = f(f(f(f(c)))), \ldots$$

or expressed more simply,

$$y_0 = c, \quad y_1 = f(y_0), \quad y_2 = f(y_1), \quad y_3 = f(y_2), \quad y_4 = f(y_3), \ldots$$

Thus, successive terms in the sequence are related by the formulas

$$y_0 = c, \quad y_{n+1} = f(y_n) \quad (n = 0, 1, 2, 3, \ldots)$$

These two formulas, taken together, comprise what is called a *recursion formula* for the iterated function sequence. In general, a ***recursion formula*** is any formula or set of formulas that provides a method for generating the terms of a sequence from the preceding terms and a seed value. For example, the recursion formula for the iterated function sequence of $f(x) = x^2$ with seed value c is

$$y_0 = c, \quad y_{n+1} = y_n^2$$

As another example, review the formula in the discussion preceding Exercise 8 in the Introduction. As noted in that discussion, the recursion formula

$$y_0 = 1, \quad y_{n+1} = \frac{1}{2}\left(y_n + \frac{p}{y_n}\right) \qquad (1)$$

produces an iterated function sequence whose iterates can be used to approximate $\sqrt{p}$ to any degree of accuracy.

....
Exercise 2   Use (1) to approximate $\sqrt{5}$ by generating successive iterates on a calculator until you encounter two successive iterates that are the same. Compare this approximation of $\sqrt{5}$ to that produced directly by your calculator.

....
Exercise 3

(a) Find iterates y_1 up to y_6 of the sequence that is generated by the recursion formula

$$y_0 = 1, \quad y_{n+1} = \frac{1}{2}y_n$$

(b) By examining the terms generated in part (a), find a formula that expresses y_n as a function of n.

....
Exercise 4   Suppose that you deposit \$1000 in a bank at 5% interest per year and allow it to accumulate value without making withdrawals.

(a) If y_n denotes the value of the account at the end of the nth year, how could you find the value of y_{n+1} if you knew the value of y_n?

(b) Starting with $y_0 = 1000$ (dollars), use the result in part (a) to calculate $y_1, y_2, y_3, y_4,$ and y_5.

(c) Find a recursion formula for the sequence of yearly account values assuming that $y_0 = 1000$.

(d) Find a formula that expresses y_n as a function of n, and use that formula to calculate the value of the account at the end of the 15th year.

Exploring Iterated Function Sequences

Iterated function sequences for a function f can be explored in various ways. Here are three possibilities:

- Choose a specific seed value, and investigate the long-term behavior of the iterates (as in Exercise 1).
- Let the seed value be a variable x (in which case the iterates become functions of x), and investigate what happens to the graphs of the iterates as more and more iterations are performed.
- Choose a specific iterate, say the 10th, and investigate how the value of this iterate varies with different seed values.

Exercise 5   Let $f(x) = \sqrt{x}$.

(a) Find formulas for the first five iterates in the iterated function sequence for f, taking the seed value to be x.

(b) Graph the iterates in part (a) in the same coordinate system, and make a conjecture about the behavior of the graphs as more and more iterations are performed.

Continued Fractions and Fibonacci Sequences

If $f(x) = 1/x$, and the seed value is x, then the iterated function sequence for f flip-flops between x and $1/x$:

$$y_1 = \frac{1}{x}, \quad y_2 = \frac{1}{1/x} = x, \quad y_3 = \frac{1}{x}, \quad y_4 = \frac{1}{1/x} = x, \ldots$$

However, if $f(x) = 1/(x+1)$, then the iterated function sequence becomes a sequence of fractions that, if continued indefinitely, is an example of a **continued fraction**:

$$\frac{1}{1+x}, \quad \frac{1}{1+\dfrac{1}{1+x}}, \quad \frac{1}{1+\dfrac{1}{1+\dfrac{1}{1+x}}}, \quad \frac{1}{1+\dfrac{1}{1+\dfrac{1}{1+\dfrac{1}{1+x}}}}, \ldots$$

Exercise 6   Let $f(x) = 1/(x + 1)$ and $c = 1$.

(a) Find *exact values* for the first 10 terms in the iterated function sequence for f; that is, express each term as a fraction p/q with no common factors in the numerator and denominator.

(b) Write down the numerators from part (a) in sequence, and see if you can discover how each term after the first two is related to its predecessors. The sequence of numerators is called a **Fibonacci sequence** [in honor of its medieval discoverer Leonardo ("Fibonacci") da Pisa]. Do some research on Fibonacci and his sequence, and write a paper on the subject.

(c) Use the pattern you discovered in part (b) to write down the exact values of the second 10 terms in the iterated function sequence.

(d) Find a recursion formula that will generate all the terms in the Fibonacci sequence after the first two.

(e) It can be proved that the terms in the iterated function sequence for f get closer and closer to one of the two solutions of the equation $q = 1/(1 + q)$. Which solution is it? This solution is a number known as the **golden ratio**. Do some research on the golden ratio, and write a paper on the subject.

Applications to Ecology

There are numerous models for predicting the growth and decline of populations (flowers, plants, people, animals, etc.). One way to model populations is to give a recursion formula that describes how the number of individuals in each generation relates to the number of individuals in the preceding generation. One of the simplest such models, called the ***exponential model***, assumes that the number of individuals in each generation is a fixed percentage of the number of individuals in the preceding generation. Thus, if there are c individuals initially and if the number of individuals in any generation is r times the number of individuals in the preceding generation, then the growth through successive generations is given by the recursion formula

$$y_0 = c, \quad y_{n+1} = r y_n \quad (n = 0, 1, 2, 3, \ldots)$$

Exercise 7   Suppose that a population with an exponential growth model has c individuals initially.

(a) Express the iterates y_1, y_2, y_3, and y_4 in terms of c and r.

(b) Find a formula for y_{n+1} in terms of c and r.

(c) Describe the eventual fate of the population if $r = 1$, $r < 1$, and $r > 1$.

There is a more sophisticated model of population growth, called the ***logistic model***, that takes environmental constraints into account. In this model, it is assumed that there is some maximum population that can be supported by the environment, and the population is expressed as a fraction of the maximum. Thus, in each generation the population is represented as a number in the interval $0 \leq y_n \leq 1$. When y_n is near 0 the population has lots of room to grow, but when y_n is near 1 the population is close to the maximum and the environmental factors tend to inhibit further growth. Models of this type are given by recursion formulas of the form

$$y_0 = c, \quad y_{n+1} = k y_n (1 - y_n) \tag{2}$$

in which k is a positive constant that depends on the ecological conditions.

Figure 1 illustrates a graphical method for tracking the growth of a population described by (2). That figure, which is called a ***cobweb diagram***, shows graphs of the line $y = x$ and the curve $y = kx(1 - x)$.

Exercise 8   Explain why the values y_1, y_2, and y_3 are the populations for the first three generations of the logistic growth model given by (2).

Exercise 9   The cobweb diagram in Figure 2 tracks the growth of a population with a logistical growth model given by the recursion formula

$$y_0 = 0.1, \quad y_{n+1} = 2.9 y_n (1 - y_n)$$

(a) Find the populations y_1, y_2, ..., y_5 of the first five generations.

(b) What happens to the population over the long term?

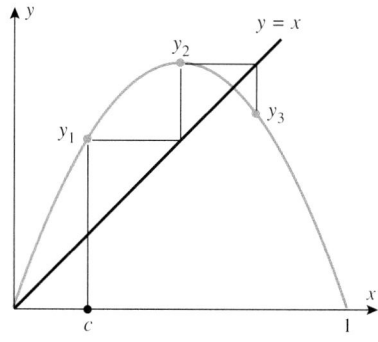

Figure 1

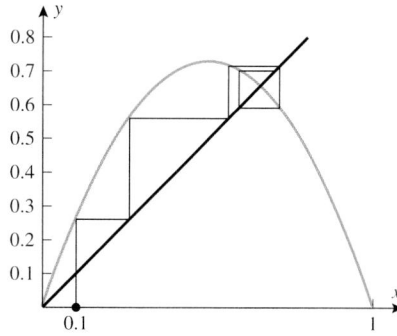

Figure 2

Chaos and Fractals

Observe that (2) is a recursion formula for the iterated function sequence of $f(x) = kx(1 - x)$. Iterated function sequences of this form are called ***iterated quadratic systems***. These are important not only in modeling populations but also in the study of ***chaos*** and ***fractals***—two important fields of contemporary research.

Module by: C. Lynn Kiaer, Rose-Hulman Institute of Technology
Howard Anton, Drexel University

Additional material for this module can be found on the World Wide Web at http://www.wiley.com/college/anton

Leonard Euler

2

LIMITS
AND
CONTINUITY

he development of calculus was stimulated by two geometric problems: finding areas of plane regions and finding tangent lines to curves. As discussed in the Introduction, both of these problems require a "limit process" for their general solution. However, limit processes occur in many other applications as well—so many, in fact that the concept of a "limit" is the fundamental building block on which all other calculus concepts are based.

In this chapter we will develop the concept of a limit in stages: In Section 2.1 we will develop the basic ideas informally, relying on our intuition; in Section 2.2 we will discuss methods for calculating limits; and in Section 2.3 we will give the precise mathematical definition of a limit. In Sections 2.4 and 2.5 we will apply limits to the study "continuous" curves. Such curves are important because they model the idea of a smooth flow without breaks or interruptions—the flow of time, the motion of an object in flight, or the gradual warming of a room on a sunny day, for example.

2.1 LIMITS (AN INTUITIVE INTRODUCTION)

As discussed in the introduction to this chapter, the concept of a limit is the fundamental building block on which all other calculus concepts are based. In this section we will study limits informally, with the goal of developing an "intuitive feel" for the basic ideas. In the next two sections we will focus on the computational methods and precise definitions.

THE TANGENT LINE, AREA, AND VELOCITY PROBLEMS

Many of the basic ideas in calculus can be motivated by the following three problems.

> THE TANGENT LINE PROBLEM.   Given a function f and a point $P(x_0, y_0)$ on its graph, find an equation of the line that is tangent to the graph at P (Figure 2.1.1).

> THE AREA PROBLEM.   Given a function f, find the area between the graph of f and an interval $[a, b]$ on the x-axis (Figure 2.1.2).

> THE INSTANTANEOUS VELOCITY PROBLEM.   Given the position versus time curve for a particle moving along a coordinate line, find the velocity of the particle at a specified instant of time.

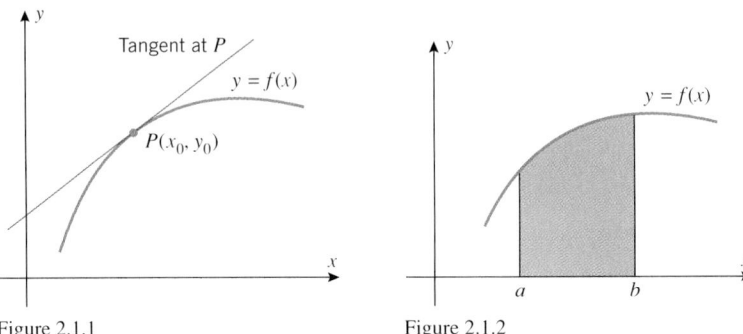

Figure 2.1.1   Figure 2.1.2

Traditionally, that portion of calculus arising from the tangent line problem is called ***differential calculus*** and that arising from the area problem is called ***integral calculus***. However, we will see later that the tangent line and area problems are so closely related that the distinction between differential and integral calculus is often hard to discern.

In order to solve the three problems posed above, it is necessary to have a more precise understanding of what the terms *tangent line*, *area*, and *velocity at an instant* actually mean. Let us begin with the notion of a tangent line.

TANGENT LINES AND LIMITS

In plane geometry, a line is called *tangent* to a circle if it meets the circle at precisely one point (Figure 2.1.3*a*). However, this definition is not appropriate for more general curves. For example, in Figure 2.1.3*b*, the line meets the curve exactly once but is obviously not what we would regard to be a tangent line; and in Figure 2.1.3*c*, the line appears to be tangent to the curve, yet it intersects the curve more than once.

To obtain a definition of a tangent line that applies to curves other than circles, we must view tangent lines another way. For this purpose, suppose that we are interested in the tangent line at a point P on a curve in the xy-plane and that Q is any point that lies on the curve and is different from P. The line through P and Q is called a ***secant line*** for the curve at P. Intuition suggests that if we move the point Q along the curve toward P, then the

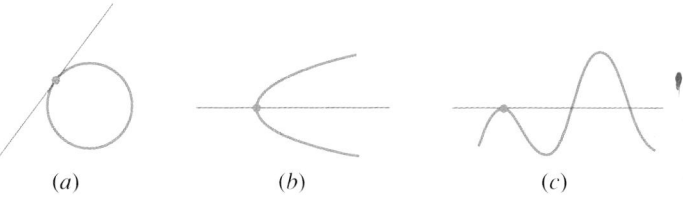

Figure 2.1.3

secant line will rotate toward a *limiting position*. The line in this limiting position is what we will consider to be the ***tangent line*** at P (Figure 2.1.4a). As suggested by Figure 2.1.4b, this new concept of a tangent line coincides with the traditional concept when applied to circles.

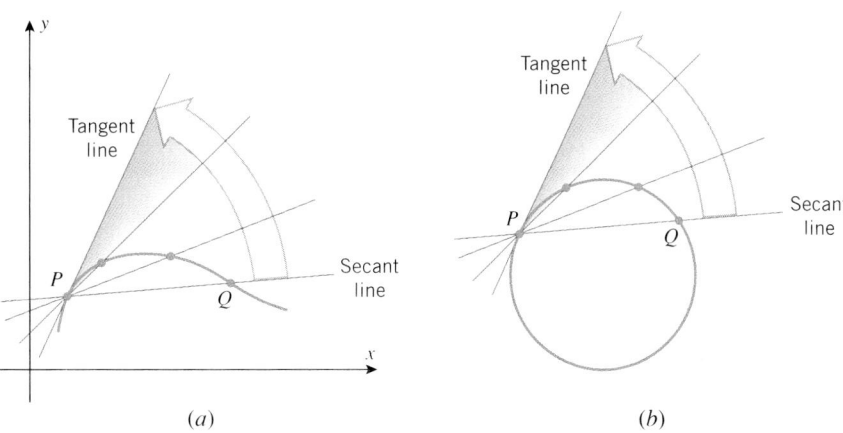

Figure 2.1.4

AREAS AND LIMITS

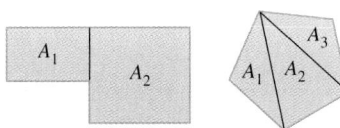

Figure 2.1.5

Just as the general notion of a tangent line leads to the concept of *limit*, so does the general notion of area. For many plane regions with straight-line boundaries, areas can be calculated by subdividing the region into rectangles or triangles and adding the areas of the constituent parts (Figure 2.1.5). However, for regions with curved boundaries, such as that in Figure 2.1.6a, a more general approach is needed. One such approach is to begin by approximating the area of the region by inscribing a number of rectangles of equal width under the curve and adding the areas of these rectangles (Figure 2.1.6b). Intuition suggests that if we repeat that approximation process using more and more rectangles, then the rectangles will tend to fill in the gaps under the curve, and the approximations will get closer and closer to the exact area under the curve (Figure 2.1.6c). This suggests that we can define the area under the curve to be the *limiting value* of these approximations.

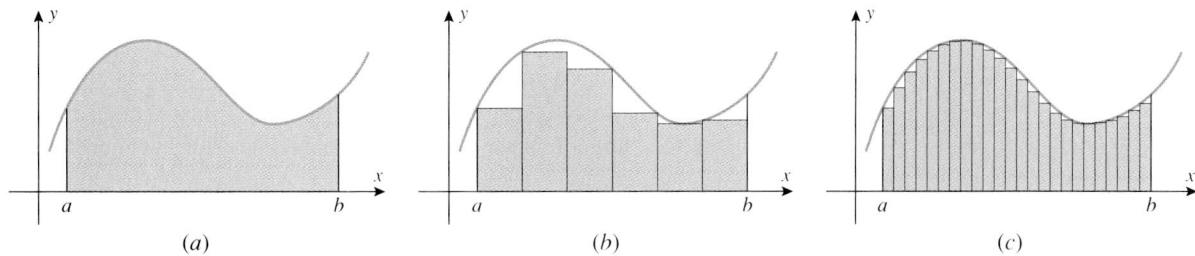

Figure 2.1.6

INSTANTANEOUS VELOCITY AND LIMITS

Recall from Formula (11) of Section 1.5 that if a particle moves along an s-axis, then its average velocity v_{ave} over the time interval from t_0 to t_1 is defined as

$$v_{\text{ave}} = \frac{\Delta s}{\Delta t} = \frac{s_1 - s_0}{t_1 - t_0} \tag{1}$$

where s_0 and s_1 are the coordinates of the particle at times t_0 and t_1, respectively. Geometrically, v_{ave} is the slope of the secant line joining the points (t_0, s_0) and (t_1, s_1) on the position versus time curve for the particle (Figure 2.1.7).

Suppose, however, that we are not interested in the average velocity over a time interval, but rather the velocity v_{inst} at a specific instant of time. It is not a simple matter of applying Formula (1), since the displacement and the elapsed time in an instant are both 0. However, intuition suggests that over a sufficiently small time interval, the velocity of the particle will not vary much; thus, there should not be much difference between the instantaneous velocity at an instant of time, say $t = t_0$, and the average velocity over a time interval from $t = t_0$ to $t = t_1$, provided that the time interval is small. This suggests that we can approximate v_{inst} as

$$v_{\text{inst}} \approx v_{\text{ave}} = \frac{s_1 - s_0}{t_1 - t_0} \tag{2}$$

Moreover, the closer t_1 is to t_0, the better the approximation. However, as t_1 gets closer and closer to t_0, the slope of the secant line in Figure 2.1.8 will approach the slope of the tangent line to the curve at time $t = t_0$; and this suggests that we can *define* the instantaneous velocity of the particle at time $t = t_0$ to be the slope of the tangent line to the position versus time curve at that point. Thus, once we know how to calculate slopes of tangent lines, we will have a method for calculating instantaneous velocities.

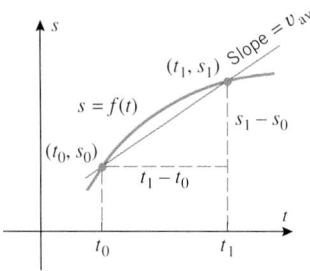

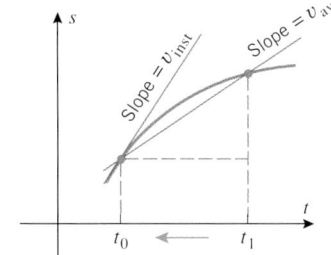

Figure 2.1.7   Figure 2.1.8

LIMITS

Now that we have seen how the concept of a limit enters into solving the tangent line, area, and instantaneous velocity problems, let us focus on the limit concept itself.

The most basic use of limits is to describe how a function behaves as the independent variable approaches a given value. For example, let us examine the behavior of the function

$$f(x) = x^2 - x + 1$$

as x gets closer and closer to 2. It is evident from the graph and table in Figure 2.1.9 that the values of $f(x)$ get closer and closer to 3 as x gets closer and closer to 2 from either the left side or the right side. Moreover, the graph and table both suggest that we can make the values of $f(x)$ as close as we like to 3 by making x sufficiently close to 2. We describe this by saying that the "limit of $x^2 - x + 1$ is 3 as x approaches 2 from either side," and we write

$$\lim_{x \to 2} (x^2 - x + 1) = 3 \tag{3}$$

Observe that in this limit analysis we are only concerned with the values of f *near* the point $x = 2$ and not the value of f *at* the point $x = 2$.

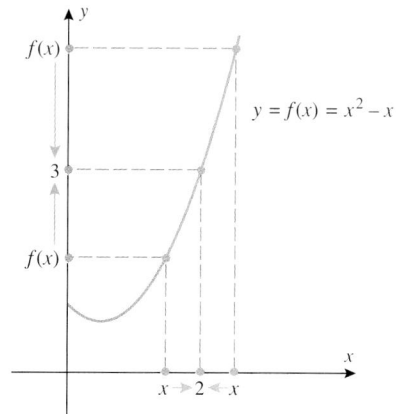

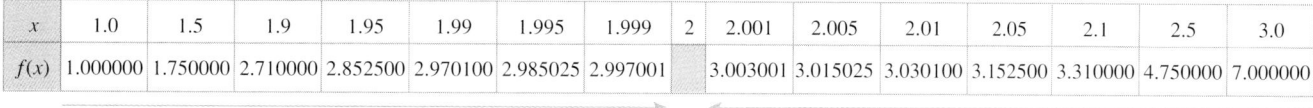

x	1.0	1.5	1.9	1.95	1.99	1.995	1.999	2	2.001	2.005	2.01	2.05	2.1	2.5	3.0
$f(x)$	1.000000	1.750000	2.710000	2.852500	2.970100	2.985025	2.997001		3.003001	3.015025	3.030100	3.152500	3.310000	4.750000	7.000000

Left side   Right side

Figure 2.1.9

This leads us to the following general idea.

> **2.1.1** LIMITS (AN INFORMAL VIEW).   If the values of $f(x)$ can be made as close as we like to L by making x sufficiently close to a (but not equal to a), then we write
> $$\lim_{x \to a} f(x) = L \qquad (4)$$
> which is read "the limit of $f(x)$ as x approaches a is L."

Expression (4) is also commonly written as

$$f(x) \to L \quad \text{as} \quad x \to a$$

With this notation we can express (3) as

$$x^2 - x + 1 \to 3 \quad \text{as} \quad x \to 2$$

Example 1

Make a conjecture about the value of the limit

$$\lim_{x \to 0} \frac{x}{\sqrt{x + 1} - 1} \qquad (5)$$

Solution. Observe that this function is undefined at $x = 0$. However, this has no bearing on the limit, since the limit is concerned with the behavior of f for x near, but not equal to, 0. Table 2.1.1 shows successions of x-values approaching 0 from the left side and the right side. In both cases the values of $f(x)$, calculated to six decimal places, appear to get closer

Table 2.1.1

x	−0.01	−0.001	−0.0001	−0.00001	0	0.00001	0.0001	0.001	0.01
$f(x)$	1.994987	1.999500	1.999950	1.999995		2.000005	2.000050	2.000500	2.004988

Left side   Right side

and closer to 2, and hence we conjecture that

$$\lim_{x \to 0} \frac{x}{\sqrt{x+1} - 1} = 2 \qquad (6)$$

However, it should be kept in mind that this conjecture is based on a limited amount of numerical evidence; we are *guessing* that if we were to extend the table and continue to let x get closer and closer to 0 from either side, then the values of $f(x)$ would continue to get closer and closer to 2. Fortunately, in this example we have other ways of confirming our conjecture. One possibility is to simplify Formula (5) algebraically by rationalizing the denominator. This yields

$$f(x) = \frac{x}{\sqrt{x+1} - 1} = \frac{x(\sqrt{x+1} + 1)}{(x+1) - 1} = \sqrt{x+1} + 1 \quad (x \neq 0) \qquad (7)$$

It is evident from this alternative formula for f that as x gets closer and closer to 0, the values of $f(x) = \sqrt{x+1} + 1$ get closer and closer to 2, confirming (6). Yet another confirmation of (6) can be obtained from the graph of f. It follows from (7) that the graph of f is identical to the graph of $y = \sqrt{x+1} + 1$, except for a hole at $x = 0$, where f is undefined (Figure 2.1.10). This figure suggests that as x moves along the x-axis toward 0 from either side, the values of $y = f(x)$ get closer and closer to 2, which again agrees with (6). ◀

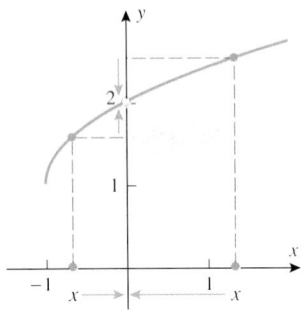

Figure 2.1.10

Example 2

Make a conjecture about the value of the limit

$$\lim_{x \to 0} \frac{\sin x}{x}$$

Solution. The function $f(x) = (\sin x)/x$ is undefined at $x = 0$, but, as discussed previously, this has no bearing on the limit. With the help of a calculating utility set to radian measure, we obtain Table 2.1.2, which suggests that

$$\lim_{x \to 0} \frac{\sin x}{x} = 1 \qquad (8)$$

This result is consistent with the graph of $f(x) = (\sin x)/x$ shown in Figure 2.1.11; but unlike the preceding example, where we were able to confirm the limit algebraically by simplifying the formula for the function, that is not possible here. However, later in this chapter we will give a geometric argument to prove that our conjecture is correct. ◀

Table 2.1.2

x (RADIANS)	$y = \dfrac{\sin x}{x}$
±1.0	0.84147
±0.9	0.87036
±0.8	0.89670
±0.7	0.92031
±0.6	0.94107
±0.5	0.95885
±0.4	0.97355
±0.3	0.98507
±0.2	0.99335
±0.1	0.99833
±0.01	0.99998

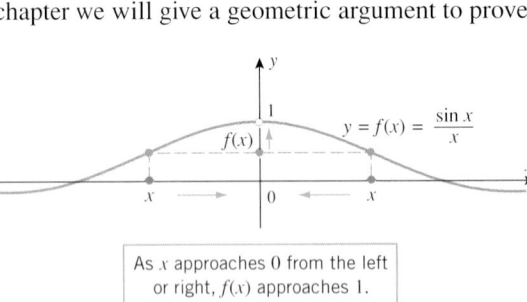

Figure 2.1.11

FOR THE READER.   Use a calculating utility to confirm limit (8). Does the limit change if x is in degrees?

NUMERICAL PITFALLS

Although numerical evidence is helpful for guessing at limits, it can lead to incorrect conclusions. For example, Table 2.1.3 shows values of $f(x) = \sin(\pi/x)$ at selected values of x on both sides of 0. The numerical data in that table suggest that

$$\lim_{x \to 0} \sin\left(\frac{\pi}{x}\right) = 0$$

However, this conclusion is incorrect, as evidenced by the graph of f shown in Figure 2.1.12. This graph shows that as $x \to 0$, the values of f oscillate between -1 and 1 with increasing rapidity, and hence do not approach a limit. The numerical data in Table 2.1.3 deceived us into believing the limit to be zero because we happened to choose values of x that were all x-intercepts.

Table 2.1.3

x (RADIANS)	$\dfrac{\pi}{x}$	$f(x) = \sin\left(\dfrac{\pi}{x}\right)$
$x = \pm 1$	$\pm \pi$	$\sin(\pm \pi) = 0$
$x = \pm 0.1$	$\pm 10\pi$	$\sin(\pm 10\pi) = 0$
$x = \pm 0.01$	$\pm 100\pi$	$\sin(\pm 100\pi) = 0$
$x = \pm 0.001$	$\pm 1000\pi$	$\sin(\pm 1000\pi) = 0$
$x = \pm 0.0001$	$\pm 10{,}000\pi$	$\sin(\pm 10{,}000\pi) = 0$
$\vdots$	$\vdots$	$\vdots$

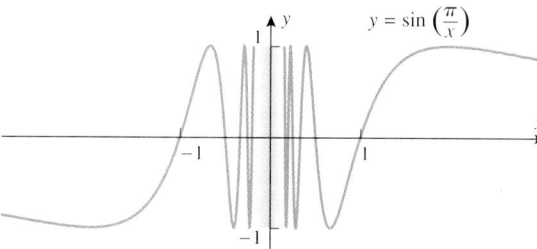

Figure 2.1.12

Numerical evidence can also lead to incorrect conclusions about limits because of round-off error or because the table of values used to find the limit is not extensive enough to reveal the behavior of the function completely. Thus, when a limit is conjectured from numerical data it is important to look for corroborating graphical or algebraic evidence to support the conjecture.

ONE-SIDED LIMITS

The limit in (4) is commonly called a ***two-sided limit*** because it requires the values of $f(x)$ to get closer and closer to L as x approaches a from *either* side. However, some functions exhibit different behaviors on the two sides of a point a, in which case it is necessary to distinguish whether x is near a on the left side or the right side for purposes of investigating the limiting behavior. For example, consider the function

$$f(x) = \frac{|x|}{x} = \begin{cases} 1, & x > 0 \\ -1, & x < 0 \end{cases}$$

(Figure 2.1.13). As x approaches 0 from the right side, the values of $f(x)$ approach 1 (in fact, they are exactly 1 for all such x), and as x approaches 0 from the left side, the values of $f(x)$ approach -1. We describe these two statements by saying that "the limit of $f(x) = |x|/x$ is 1 as x approaches 0 from the right" and that "the limit of $f(x) = |x|/x$ is -1 as x approaches 0 from the left"; we denote these limits by writing

$$\lim_{x \to 0^+} \frac{|x|}{x} = 1 \quad \text{and} \quad \lim_{x \to 0^-} \frac{|x|}{x} = -1 \tag{9--10}$$

With this notation, the superscript "+" indicates a limit from the right and the superscript "$-$" indicates a limit from the left.

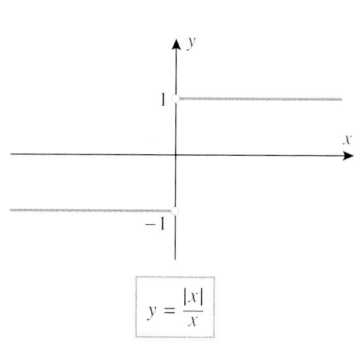

$$y = \frac{|x|}{x}$$

Figure 2.1.13

This leads us to the following general idea:

2.1.2 ONE-SIDED LIMITS (AN INFORMAL VIEW). If the values of $f(x)$ can be made as close as we like to L by making x sufficiently close to a (but greater than a), then we write

$$\lim_{x \to a^+} f(x) = L \tag{11}$$

which is read "the limit of $f(x)$ as x approaches a from the right is L." Similarly, if the values of $f(x)$ can be made as close as we like to L by making x sufficiently close to a (but less than a), then we write

$$\lim_{x \to a^-} f(x) = L \tag{12}$$

Expressions (11) and (12), which are called **one-sided limits**, are also commonly written as

$$f(x) \to L \text{ as } x \to a^+ \quad \text{and} \quad f(x) \to L \text{ as } x \to a^-$$

respectively. With this notation (9) and (10) can be expressed as

$$\frac{|x|}{x} \to 1 \text{ as } x \to 0^+ \quad \text{and} \quad \frac{|x|}{x} \to -1 \text{ as } x \to 0^-$$

THE RELATIONSHIP BETWEEN ONE-SIDED AND TWO-SIDED LIMITS

In general, there is no guarantee that a function will have a limit at a specified point, and there is some terminology to describe such situations. If the values of $f(x)$ do not get closer and closer to some *single* number L as $x \to a$, then we say that the limit of $f(x)$ as x approaches a **does not exist** (and similarly for one-sided limits). For example, the two-sided limit of $f(x) = |x|/x$ does not exist as $x \to 0$ because the values of $f(x)$ do not approach a single number—the values approach -1 from the left and 1 from the right.

In general, the following condition must be satisfied for the two-sided limit of a function to exist.

2.1.3 THE RELATIONSHIP BETWEEN ONE-SIDED AND TWO-SIDED LIMITS. The two-sided limit of a function f exists at a point a if and only if the one-sided limits exist at that point and have the same value; that is,

$$\lim_{x \to a} f(x) = L \quad \text{if and only if} \quad \lim_{x \to a^-} f(x) = L = \lim_{x \to a^+} f(x)$$

REMARK. Sometimes, one or both of the one-sided limits may fail to exist (which, in turn, implies that the two-sided limit does not exist). For example, we saw earlier that the one-sided limits of $f(x) = \sin(\pi/x)$ do not exist as x approaches 0 because the function keeps oscillating between -1 and 1, failing to settle in on a single value; and this implies that the two-sided limit does not exist as x approaches 0.

Example 3

For the functions in Figure 2.1.14, find the one-sided and two-sided limits at $x = a$ if they exist.

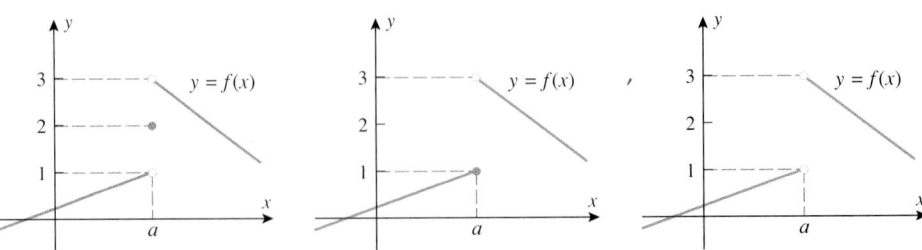

Figure 2.1.14

Solution. The functions in all three figures have the same one-sided limits as $x \to a$, since the functions are identical, except at $x = a$. These limits are

$$\lim_{x \to a^+} f(x) = 3 \quad \text{and} \quad \lim_{x \to a^-} f(x) = 1$$

In all three cases the two-sided limit does not exist as $x \to a$ because the one-sided limits are not equal.   ◄

Example 4

For the functions in Figure 2.1.15, find the one-sided and two-sided limits at $x = a$ if they exist.

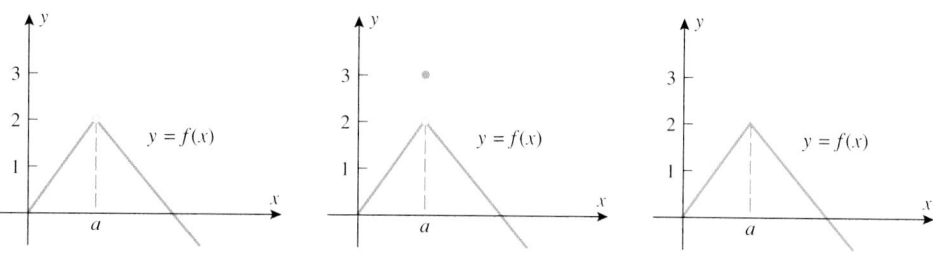

Figure 2.1.15

Solution. As in the preceding example, the value of f at $x = a$ has no bearing on the limits as $x \to a$, so that in all three cases we have

$$\lim_{x \to a^+} f(x) = 2 \quad \text{and} \quad \lim_{x \to a^-} f(x) = 2$$

Since the one-sided limits are equal, the two-sided limit exists and

$$\lim_{x \to a} f(x) = 2$$ ◄

A FIRST LOOK AT CONTINUITY

Plane curves can be divided into two categories—those that have breaks or holes and those that do not. Breaks or holes in a curve are called *discontinuities*; a curve with no discontinuities is called *continuous* (Figure 2.1.16).

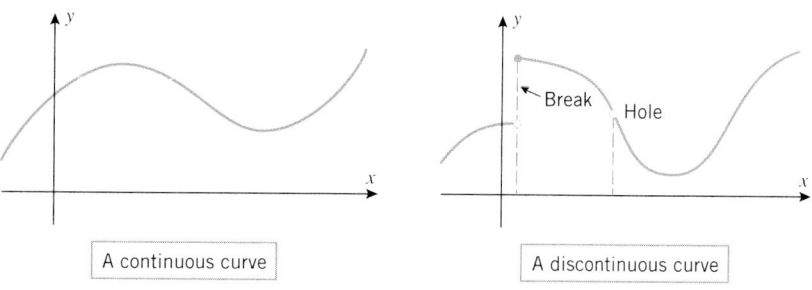

Figure 2.1.16

Examples 3 and 4 provide some useful insight into what it means for the graph of a function to be continuous. Of the six functions in those examples, only the last one does not have a break or hole in its graph at $x = a$. For the functions in Example 3, the break in the graph at $x = a$ results from the fact that the one-sided limits at that point have different values. A break of this type is called a *jump discontinuity* in the graph. For the first two functions in Example 4, the hole in the graph is caused by a mismatch between the value of the function at $x = a$ and the two-sided limit as x approaches a. In the first graph, the function is simply undefined at $x = a$, leaving a hole; and in the second graph, $f(a)$ is defined, but its value is different from the limit, resulting in a point that is displaced from the main part of the graph. A break due to a hole or a displaced point is called a ***removable discontinuity*** in the graph. The third graph is continuous at $x = a$, since the value of f at $x = a$ is the same as the two-sided limit of f as x approaches a, thereby ensuring that there is no jump or hole.

All of this suggests that three conditions must be satisfied to ensure that the graph of a function does not have a discontinuity at a given point:

- The function must be defined at the point.
- The two-sided limit must exist at the point.
- The value of the function and the value of the two-sided limit must be the same.

There will be more on this later.

INFINITE LIMITS AND VERTICAL ASYMPTOTES

Sometimes one-sided or two-sided limits will fail to exist because the values of the function increase or decrease indefinitely. For example, consider the behavior of the function $f(x) = 1/x$ as x gets closer and closer to 0. It is evident from the table and graph in Figure 2.1.17 that as x gets closer and closer to 0 from the right, the values of $f(x) = 1/x$ are positive and increase indefinitely; and as x gets closer and closer to 0 from the left, the values of $f(x)$ are negative and decrease indefinitely. We denote these limiting behaviors by writing

$$\lim_{x \to 0^+} \frac{1}{x} = +\infty \quad \text{and} \quad \lim_{x \to 0^-} \frac{1}{x} = -\infty$$

More generally:

2.1.4   INFINITE LIMITS (AN INFORMAL VIEW).   If the values of $f(x)$ increase indefinitely as x approaches a from the right or left, then we write

$$\lim_{x \to a^+} f(x) = +\infty \quad \text{or} \quad \lim_{x \to a^-} f(x) = +\infty$$

as appropriate, and we say that $f(x)$ ***increases without bound*** as $x \to a^+$ or $x \to a^-$. Similarly, if the values of $f(x)$ decrease indefinitely as x approaches a from the right or left, then we write

$$\lim_{x \to a^+} f(x) = -\infty \quad \text{or} \quad \lim_{x \to a^-} f(x) = -\infty$$

as appropriate, and say that $f(x)$ ***decreases without bound*** as $x \to a^+$ or $x \to a^-$. Moreover, if both one-sided limits are $+\infty$, then we write

$$\lim_{x \to a} f(x) = +\infty$$

and if both one-sided limits are $-\infty$, then we write

$$\lim_{x \to a} f(x) = -\infty$$

REMARK.   It should be emphasized that the symbols $+\infty$ and $-\infty$, as used here, describe the particular way in which the limits fail to exist; they are not numerical limits and consequently cannot be manipulated using rules of algebra. For example, it is *not* correct to write $(+\infty) - (+\infty) = 0$.

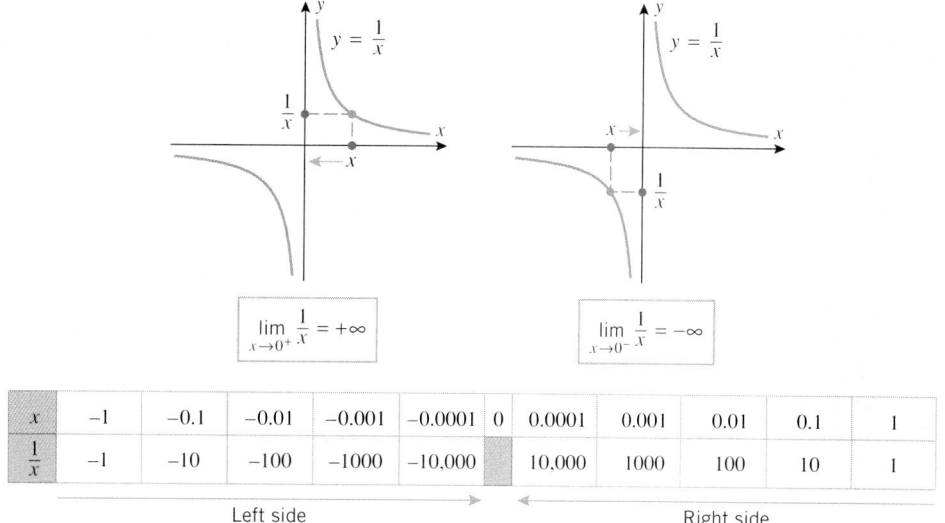

Figure 2.1.17

Example 5

For the functions in Figure 2.1.18, describe the limits at $x = a$ in appropriate limit notation.

Solution (a). In Figure 2.1.18a, the function increases indefinitely as x approaches a from the right and decreases indefinitely as x approaches a from the left. Thus,

$$\lim_{x \to a^+} \frac{1}{x - a} = +\infty \quad \text{and} \quad \lim_{x \to a^-} \frac{1}{x - a} = -\infty$$

Solution (b). In Figure 2.1.18b, the function increases indefinitely as x approaches a from both the left and right. Thus,

$$\lim_{x \to a} \frac{1}{(x - a)^2} = \lim_{x \to a^+} \frac{1}{(x - a)^2} = \lim_{x \to a^-} \frac{1}{(x - a)^2} = +\infty$$

Solution (c). In Figure 2.1.18c, the function decreases indefinitely as x approaches a from the right and increases indefinitely as x approaches a from the left. Thus,

$$\lim_{x \to a^+} \frac{-1}{x - a} = -\infty \quad \text{and} \quad \lim_{x \to a^-} \frac{-1}{x - a} = +\infty$$

Solution (d). In Figure 2.1.18d, the function decreases indefinitely as x approaches a from both the left and right. Thus,

$$\lim_{x \to a} \frac{-1}{(x - a)^2} = \lim_{x \to a^+} \frac{-1}{(x - a)^2} = \lim_{x \to a^-} \frac{-1}{(x - a)^2} = -\infty \qquad \blacktriangleleft$$

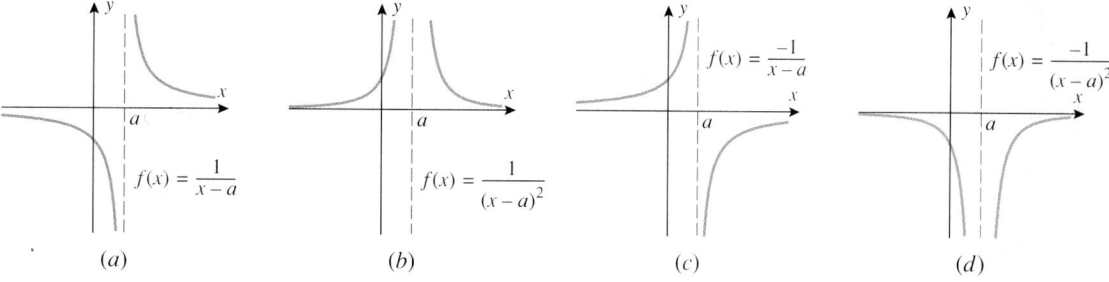

Figure 2.1.18

Geometrically, if $f(x) \to +\infty$ as x approaches a from the left or right, then the graph of $y = f(x)$ eventually gets closer and closer to the line $x = a$ as the graph is traversed in the positive y-direction; and if $f(x) \to -\infty$ as x approaches a from the left or right, then the graph of $y = f(x)$ eventually gets closer and closer to the line $x = a$ as the graph is traversed in the negative y-direction. We call this line a *vertical asymptote* (from the Greek *asymptotos*, meaning "nonintersecting").

2.1.5 DEFINITION.   A line $x = a$ is called a ***vertical asymptote*** of the graph of a function f if $f(x)$ approaches $+\infty$ or $-\infty$ as x approaches a from the left or right.

Example 6

The four functions graphed in Figure 2.1.18 all have a vertical asymptote at $x = a$, which is indicated by the dashed vertical lines in the figure.   ◀

LIMITS AT INFINITY AND HORIZONTAL ASYMPTOTES

Thus far, we have used limits to describe the behavior of $f(x)$ as x approaches a point $x = a$. However, sometimes we will not be concerned with the behavior of $f(x)$ near a specific point, but rather with how the values of $f(x)$ behave as x increases without bound or decreases without bound. This is sometimes called the ***end behavior*** of the function because it describes how the function behaves for values of x that are far from the origin. For example, it is evident from the table and graph in Figure 2.1.19 that as x increases without bound, the values of $f(x) = 1/x$ are positive, but get closer and closer to 0; and similarly, as x decreases without bound, the values of $f(x) = 1/x$ are negative, but also get closer and closer to 0. We denote these limiting behaviors by writing

$$\lim_{x \to +\infty} \frac{1}{x} = 0 \quad \text{and} \quad \lim_{x \to -\infty} \frac{1}{x} = 0$$

More generally:

2.1.6 LIMITS AT INFINITY (AN INFORMAL VIEW).   If the values of $f(x)$ eventually get closer and closer to a number L as x increases without bound, then we write

$$\lim_{x \to +\infty} f(x) = L \tag{13}$$

Similarly, if the values of $f(x)$ eventually get closer and closer to a number L as x decreases without bound, then we write

$$\lim_{x \to -\infty} f(x) = L \tag{14}$$

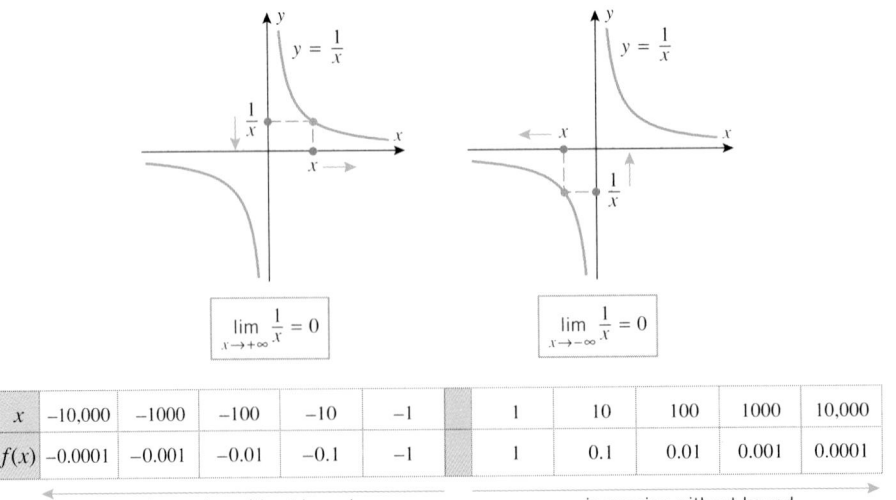

x	−10,000	−1000	−100	−10	−1		1	10	100	1000	10,000
$f(x)$	−0.0001	−0.001	−0.01	−0.1	−1		1	0.1	0.01	0.001	0.0001

x decreasing without bound   *x* increasing without bound

Figure 2.1.19

Geometrically, if $f(x) \to L$ as $x \to +\infty$, then the graph of $y = f(x)$ eventually gets closer and closer to the line $y = L$ as the graph is traversed in the positive direction (Figure 2.1.20a); and if $f(x) \to L$ as $x \to -\infty$, then the graph of $y = f(x)$ eventually gets closer and closer to the line $y = L$ as the graph is traversed in the negative x-direction (Figure 2.1.20b). In either case we call the line $y = L$ a *horizontal asymptote* of the graph of f. For example, the four functions in Figure 2.1.18 all have $y = 0$ as a horizontal asymptote.

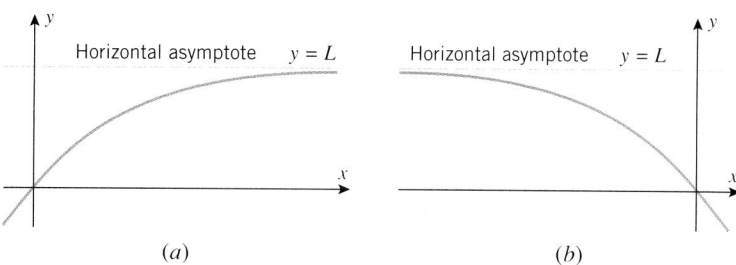

Figure 2.1.20

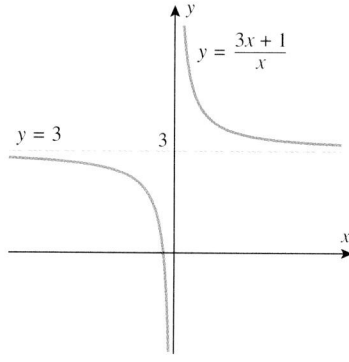

Figure 2.1.21

> **2.1.7 DEFINITION.** A line $y = L$ is called a *horizontal asymptote* of the graph of a function f if $f(x) \to L$ as $x \to +\infty$ or as $x \to -\infty$.

Sometimes the existence of a horizontal asymptote of a function f will be readily apparent from the formula for f. For example, it is evident that the function

$$f(x) = \frac{3x + 1}{x} = 3 + \frac{1}{x}$$

has a horizontal asymptote at $y = 3$ (Figure 2.1.21), since the value of $1/x$ approaches 0 as $x \to +\infty$ or $x \to -\infty$. For more complicated functions, algebraic manipulations or special techniques that we will study in the next section may have to be applied to confirm the existence of horizontal asymptotes.

HOW LIMITS AT INFINITY CAN FAIL TO EXIST

Limits at infinity can fail to exist for various reasons. One possibility is that the values of $f(x)$ may increase or decrease without bound as $x \to +\infty$ or as $x \to -\infty$. For example, the values of $f(x) = x^3$ increase without bound as $x \to +\infty$ and decrease without bound as $x \to -\infty$; and for $f(x) = -x^3$ the values decrease without bound as $x \to +\infty$ and increase without bound as $x \to -\infty$ (Figure 2.1.22). We denote this by writing

$$\lim_{x \to +\infty} x^3 = +\infty, \quad \lim_{x \to -\infty} x^3 = -\infty, \quad \lim_{x \to +\infty} (-x^3) = -\infty, \quad \lim_{x \to -\infty} (-x^3) = +\infty$$

More generally:

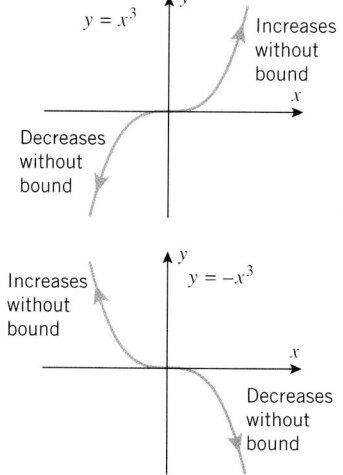

Figure 2.1.22

> **2.1.8 INFINITE LIMITS AT INFINITY (AN INFORMAL VIEW).** If the values of $f(x)$ increase without bound as $x \to +\infty$ or as $x \to -\infty$, then we write
>
> $$\lim_{x \to +\infty} f(x) = +\infty \quad \text{or} \quad \lim_{x \to -\infty} f(x) = +\infty$$
>
> as appropriate; and if the values of $f(x)$ decrease without bound as $x \to +\infty$ or as $x \to -\infty$, then we write
>
> $$\lim_{x \to +\infty} f(x) = -\infty \quad \text{or} \quad \lim_{x \to -\infty} f(x) = -\infty$$
>
> as appropriate.

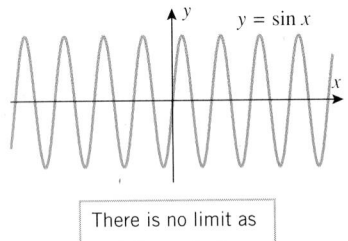

There is no limit as
$x \to +\infty$ or $x \to -\infty$.

Figure 2.1.23

Limits at infinity can also fail to exist because the graph of the function oscillates indefinitely in such a way that the values of the function do not approach a fixed number and do not increase or decrease without bound; the trigonometric functions $\sin x$ and $\cos x$ have this property, for example (Figure 2.1.23). In such cases we say that the limit *fails to exist because of oscillation*.

EXERCISE SET 2.1   ⌁ Graphing Calculator   [C] CAS

1. For the function f graphed in the accompanying figure, find
   (a) $\lim\limits_{x \to 3^-} f(x)$ (b) $\lim\limits_{x \to 3^+} f(x)$ (c) $\lim\limits_{x \to 3} f(x)$
   (d) $f(3)$ (e) $\lim\limits_{x \to -\infty} f(x)$ (f) $\lim\limits_{x \to +\infty} f(x)$.

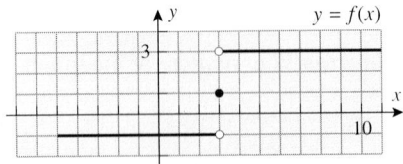

Figure Ex-1

2. For the function f graphed in the accompanying figure, find
   (a) $\lim\limits_{x \to 2^-} f(x)$ (b) $\lim\limits_{x \to 2^+} f(x)$ (c) $\lim\limits_{x \to 2} f(x)$
   (d) $f(2)$ (e) $\lim\limits_{x \to -\infty} f(x)$ (f) $\lim\limits_{x \to +\infty} f(x)$.

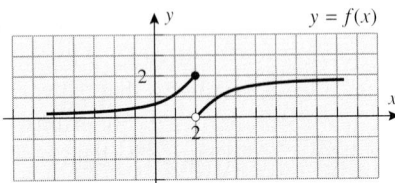

Figure Ex-2

3. For the function g graphed in the accompanying figure, find
   (a) $\lim\limits_{x \to 4^-} g(x)$ (b) $\lim\limits_{x \to 4^+} g(x)$ (c) $\lim\limits_{x \to 4} g(x)$
   (d) $g(4)$ (e) $\lim\limits_{x \to -\infty} g(x)$ (f) $\lim\limits_{x \to +\infty} g(x)$.

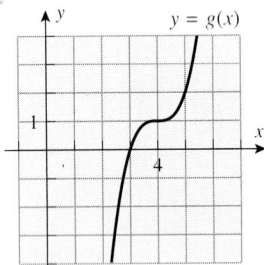

Figure Ex-3

4. For the function g graphed in the accompanying figure, find
   (a) $\lim\limits_{x \to 0^-} g(x)$ (b) $\lim\limits_{x \to 0^+} g(x)$ (c) $\lim\limits_{x \to 0} g(x)$
   (d) $g(0)$ (e) $\lim\limits_{x \to -\infty} g(x)$ (f) $\lim\limits_{x \to +\infty} g(x)$.

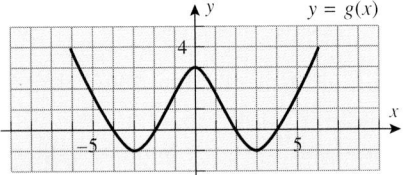

Figure Ex-4

5. For the function F graphed in the accompanying figure, find
   (a) $\lim\limits_{x \to -2^-} F(x)$ (b) $\lim\limits_{x \to -2^+} F(x)$ (c) $\lim\limits_{x \to -2} F(x)$
   (d) $F(-2)$ (e) $\lim\limits_{x \to -\infty} F(x)$ (f) $\lim\limits_{x \to +\infty} F(x)$.

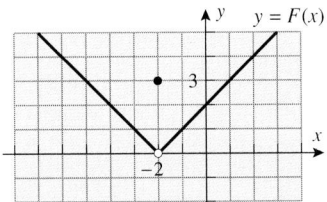

Figure Ex-5

6. For the function F graphed in the accompanying figure, find
   (a) $\lim\limits_{x \to 3^-} F(x)$ (b) $\lim\limits_{x \to 3^+} F(x)$ (c) $\lim\limits_{x \to 3} F(x)$
   (d) $F(3)$ (e) $\lim\limits_{x \to -\infty} F(x)$ (f) $\lim\limits_{x \to +\infty} F(x)$.

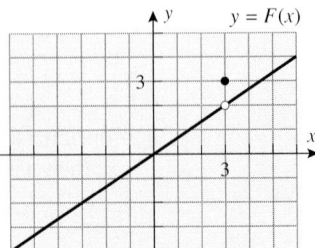

Figure Ex-6

7. For the function ϕ graphed in the accompanying figure, find
   (a) $\lim\limits_{x \to -2^-} \phi(x)$ (b) $\lim\limits_{x \to -2^+} \phi(x)$ (c) $\lim\limits_{x \to -2} \phi(x)$
   (d) $\phi(-2)$ (e) $\lim\limits_{x \to -\infty} \phi(x)$ (f) $\lim\limits_{x \to +\infty} \phi(x)$.

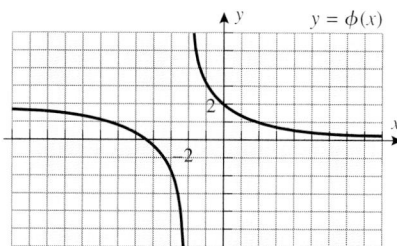

Figure Ex-7

8. For the function ϕ graphed in the accompanying figure, find
   (a) $\lim\limits_{x \to 4^-} \phi(x)$ (b) $\lim\limits_{x \to 4^+} \phi(x)$ (c) $\lim\limits_{x \to 4} \phi(x)$
   (d) $\phi(4)$ (e) $\lim\limits_{x \to -\infty} \phi(x)$ (f) $\lim\limits_{x \to +\infty} \phi(x)$.

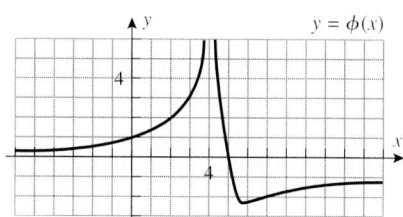

Figure Ex-8

9. For the function f graphed in the accompanying figure, find
   (a) $\lim\limits_{x \to 3^-} f(x)$ (b) $\lim\limits_{x \to 3^+} f(x)$ (c) $\lim\limits_{x \to 3} f(x)$
   (d) $f(3)$ (e) $\lim\limits_{x \to -\infty} f(x)$ (f) $\lim\limits_{x \to +\infty} f(x)$.

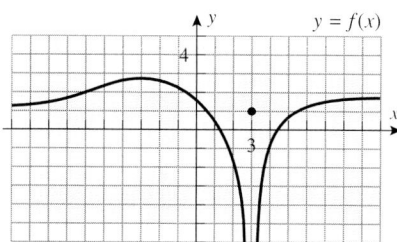

Figure Ex-9

10. For the function f graphed in the accompanying figure, find
    (a) $\lim\limits_{x \to 0^-} f(x)$ (b) $\lim\limits_{x \to 0^+} f(x)$ (c) $\lim\limits_{x \to 0} f(x)$
    (d) $f(0)$ (e) $\lim\limits_{x \to -\infty} f(x)$ (f) $\lim\limits_{x \to +\infty} f(x)$.

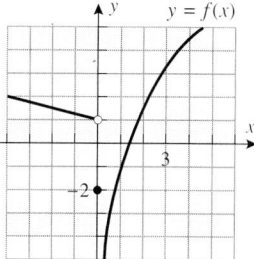

Figure Ex-10

11. For the function G graphed in the accompanying figure, find
    (a) $\lim\limits_{x \to 0^-} G(x)$ (b) $\lim\limits_{x \to 0^+} G(x)$ (c) $\lim\limits_{x \to 0} G(x)$
    (d) $G(0)$ (e) $\lim\limits_{x \to -\infty} G(x)$ (f) $\lim\limits_{x \to +\infty} G(x)$.

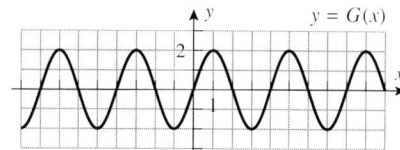

Figure Ex-11

12. For the function G graphed in the accompanying figure, find
    (a) $\lim\limits_{x \to 0^-} G(x)$ (b) $\lim\limits_{x \to 0^+} G(x)$ (c) $\lim\limits_{x \to 0} G(x)$
    (d) $G(0)$ (e) $\lim\limits_{x \to -\infty} G(x)$ (f) $\lim\limits_{x \to +\infty} G(x)$.

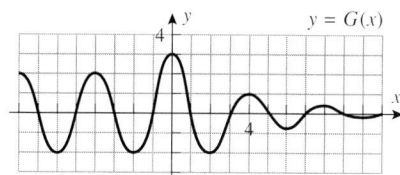

Figure Ex-12

13. Consider the function g graphed in the accompanying figure. For what values of x_0 does $\lim\limits_{x \to x_0} g(x)$ exist?

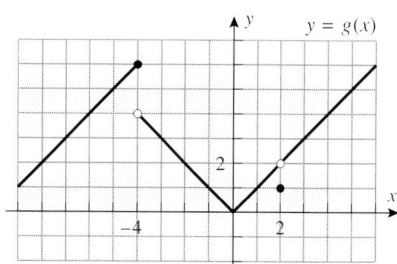

Figure Ex-13

14. Consider the function f graphed in the accompanying figure. For what values of x_0 does $\lim\limits_{x \to x_0} f(x)$ exist?

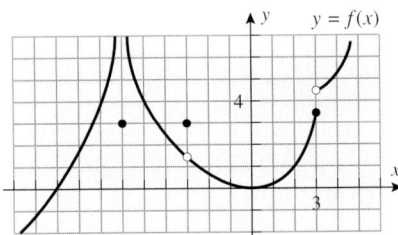

Figure Ex-14

In Exercises 15 and 16, find all points of discontinuity for the function, and for each such point state which of the three condition(s) for continuity fail to hold.

15. (a) The function f in Exercise 1
(b) The function F in Exercise 5
(c) The function f in Exercise 9

16. (a) The function f in Exercise 2
(b) The function F in Exercise 6
(c) The function f in Exercise 10

In Exercises 17–20: (i) Make a guess at the limit (if it exists) by evaluating the function at the specified points. (ii) Confirm your conclusions about the limit by graphing the function over an appropriate interval. (iii) If you have a CAS, then use it to find the limit. [*Note:* For the trigonometric functions, be sure to set your calculating and graphing utilities to the radian mode.]

[c] **17.** (a) $\lim\limits_{x \to 1} \dfrac{x-1}{x^3-1}$; $x = 2, 1.5, 1.1, 1.01, 1.001, 0, 0.5, 0.9,$
$0.99, 0.999$

(b) $\lim\limits_{x \to 1^+} \dfrac{x+1}{x^3-1}$; $x = 2, 1.5, 1.1, 1.01, 1.001, 1.0001$

(c) $\lim\limits_{x \to 1^-} \dfrac{x+1}{x^3-1}$; $x = 0, 0.5, 0.9, 0.99, 0.999, 0.9999$

[c] **18.** (a) $\lim\limits_{x \to 0} \dfrac{\sqrt{x+1}-1}{x}$; $x = \pm0.25, \pm0.1, \pm0.001,$
±0.0001

(b) $\lim\limits_{x \to 0^+} \dfrac{\sqrt{x+1}+1}{x}$; $x = 0.25, 0.1, 0.001, 0.0001$

(c) $\lim\limits_{x \to 0^-} \dfrac{\sqrt{x+1}+1}{x}$; $x = -0.25, -0.1, -0.001,$
-0.0001

[c] **19.** (a) $\lim\limits_{x \to 0} \dfrac{\sin 3x}{x}$; $x = \pm0.25, \pm0.1, \pm0.001, \pm0.0001$

(b) $\lim\limits_{x \to -1} \dfrac{\cos x}{x+1}$; $x = 0, -0.5, -0.9, -0.99, -0.999,$
$-1.5, -1.1, -1.01, -1.001$

[c] **20.** (a) $\lim\limits_{x \to -1} \dfrac{\tan(x+1)}{x+1}$; $x = 0, -0.5, -0.9, -0.99, -0.999,$
$-1.5, -1.1, -1.01, -1.001$

(b) $\lim\limits_{x \to 0} \dfrac{\sin(5x)}{\sin(2x)}$; $x = \pm0.25, \pm0.1, \pm0.001, \pm0.0001$

In Exercises 21 and 22: (i) Approximate the y-coordinates of all horizontal asymptotes of $y = f(x)$ by evaluating f at the points $\pm10, \pm100, \pm1000, \pm100,000,$ and $\pm100,000,000$. (ii) Confirm your conclusions by graphing $y = f(x)$ over an appropriate interval. (iii) If you have a CAS, then use it to find the horizontal asymptotes.

[c] **21.** (a) $f(x) = \dfrac{2x+3}{x+4}$ (b) $f(x) = \left(1 + \dfrac{3}{x}\right)^x$

(c) $f(x) = \dfrac{x^2+1}{x+1}$

[c] **22.** (a) $f(x) = \dfrac{x^2-1}{5x^2+1}$ (b) $f(x) = \left(2 + \dfrac{1}{x}\right)^x$

(c) $f(x) = \dfrac{\sin x}{x}$

In Exercises 23 and 24, express the limit as an equivalent limit in which $x \to 0^+$ or $x \to 0^-$, as appropriate. [You need not evaluate the limit.]

23. (a) $\lim\limits_{x \to +\infty} x \sin\left(\dfrac{1}{x}\right)$ (b) $\lim\limits_{x \to +\infty} \dfrac{1-x}{1+x}$

(c) $\lim\limits_{x \to -\infty} \left(1 + \dfrac{2}{x}\right)^x$

24. (a) $\lim\limits_{x \to +\infty} \dfrac{\cos(\pi/x)}{\pi/x}$ (b) $\lim\limits_{x \to +\infty} \dfrac{x}{1+x}$

(c) $\lim\limits_{x \to -\infty} (1+2x)^{1/x}$

25. (a) Sketch the graph of a function that has two horizontal asymptotes.

(b) Can the graph of a function intersect its horizontal asymptotes? If not, explain why. If so, sketch such a graph.

26. (a) Do any of the trigonometric functions, $\sin x$, $\cos x$, $\tan x$, $\cot x$, $\sec x$, $\csc x$, have horizontal asymptotes?

(b) Do any of them have vertical asymptotes? Where?

[c] **27.** (a) Let

$$f(x) = x^3 - \dfrac{3^x}{2000}$$

Make a conjecture about the limit of f as $x \to 0^+$ by evaluating f at the points $x = 1, 0.75, 0.5, 0.25,$ $0.1, 0.05.$

(b) Evaluate f at the points $x = 0.01, 0.001, 0.0001,$ 0.00001, 0.000001, and make another conjecture.

(c) What flaw does this reveal about using numerical evidence to make conjectures about limits?

(d) If you have a CAS, use it to show that the exact value of the limit is $-1/2000$.

> Roundoff error is one source of inaccuracy in calculator and computer computations. Another source of error, called *catastrophic subtraction*, occurs when two nearly equal numbers are subtracted, and the result is used as part of another calculation. For example, by hand calculation we have
>
> $$(0.123456789012345 - 0.123456789012344) \times 10^{15} = 1$$
>
> However, the author's calculator produces a value of 0 for this computation because it can only store 14 decimal digits, and the numbers being subtracted are identical in the first 14 decimal digits. Catastrophic subtraction can sometimes be avoided by rearranging formulas algebraically, but your best defense is to be aware that it can occur. Watch out for it in the next exercise.

[c] **28.** (a) Let

$$f(x) = \frac{x - \sin x}{x^3}$$

Make a conjecture about the limit of f as $x \to 0^+$ by evaluating f at the points $x = 0.1, 0.01, 0.001, 0.0001$.

(b) Evaluate f at the points $x = 0.00001, 0.0000001,$ 0.00000001, 0.000000001, 0.0000000001, and make another conjecture.

(c) What flaw does this reveal about using numerical evidence to make conjectures about limits?

(d) If you have a CAS, use it to show that the exact value of the limit is $\frac{1}{6}$.

29. (a) The accompanying figure shows graphs of the function from Exercise 28 over two different intervals. What is happening?

(b) Use your graphing utility to generate the graphs, and see whether the same problem occurs.

(c) Would you expect a similar problem to occur in the vicinity of $x = 0$ for the function

$$f(x) = \frac{1 - \cos x}{x}?$$

See if it does.

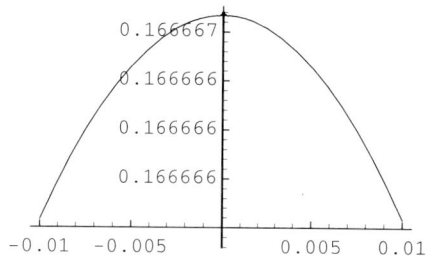

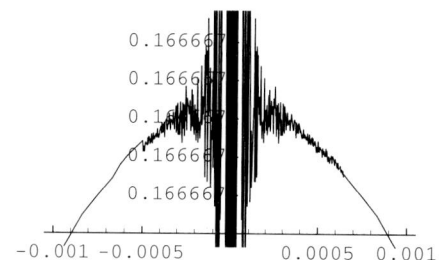

Erratic graph generated by Mathematica

Figure Ex-29

2.2 LIMITS (COMPUTATIONAL TECHNIQUES)

In the last section we discussed limits informally, focusing on the basic ideas. In this section, we will discuss algebraic methods for finding limits, reserving the discussion of the underlying theory behind these methods for the next section.

SOME BASIC LIMITS

Our strategy for finding limits algebraically has two parts:

- First we will establish the limits of some simple functions.

- Then we will develop a repertoire of theorems that will enable us to use the limits of those simple functions as building blocks for finding limits of more complicated functions.

The ten limits in the following theorem, all of which should be evident from Figure 2.2.1, will form our building blocks—three involve the constant function $f(x) = k$, three involve the linear function $f(x) = x$, and four involve the rational function $f(x) = 1/x$.

2.2.1 THEOREM.

$$\lim_{x \to a} k = k \qquad \lim_{x \to +\infty} k = k \qquad \lim_{x \to -\infty} k = k$$

$$\lim_{x \to a} x = a \qquad \lim_{x \to +\infty} x = +\infty \qquad \lim_{x \to -\infty} x = -\infty$$

$$\lim_{x \to 0^+} \frac{1}{x} = +\infty \qquad \lim_{x \to 0^-} \frac{1}{x} = -\infty \qquad \lim_{x \to +\infty} \frac{1}{x} = 0 \qquad \lim_{x \to -\infty} \frac{1}{x} = 0$$

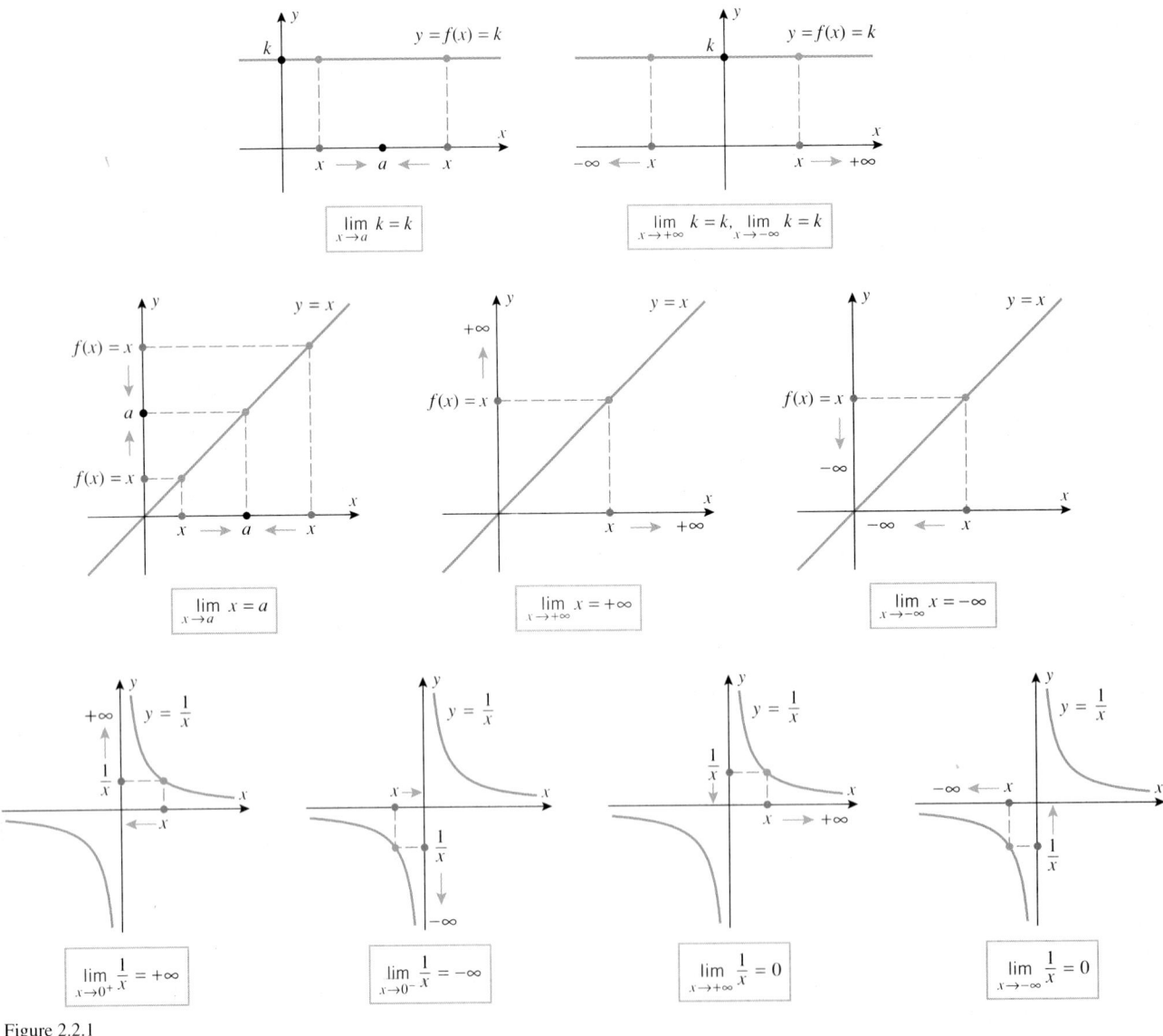

Figure 2.2.1

In the case of the constant function $f(x) = k$, the values of $f(x)$ do not change as x varies, which explains why the limit of $f(x)$ is k, regardless of whether the limit is computed at a point a or as x approaches $+\infty$ or $-\infty$. For example,

$$\lim_{x \to 2} 3 = 3, \quad \lim_{x \to -2} 3 = 3, \quad \lim_{x \to +\infty} 3 = 3, \quad \lim_{x \to +\infty} 0 = 0, \quad \lim_{x \to -\infty} 3 = 3, \quad \lim_{x \to -\infty} 0 = 0$$

The limits of the function $f(x) = 1/x$ should make sense to you intuitively, based on your experience with fractions: making the denominator closer to zero increases the numerical size of the fraction (i.e., increases its absolute value), and increasing the numerical size of the denominator makes the numerical size of the fraction closer to zero. This is illustrated in Table 2.2.1.

Table 2.2.1

	VALUES						CONCLUSION
x	1	.1	.01	.001	.0001	$\cdots$	As $x \to 0^+$ the value of $1/x$
$1/x$	1	10	100	1000	10,000	$\cdots$	increases without bound.
x	-1	$-.1$	$-.01$	$-.001$	$-.0001$	$\cdots$	As $x \to 0^-$ the value of $1/x$
$1/x$	-1	-10	-100	-1000	$-10,000$	$\cdots$	decreases without bound.
x	1	10	100	1000	10,000	$\cdots$	As $x \to +\infty$ the value of $1/x$
$1/x$	1	.1	.01	.001	.0001	$\cdots$	decreases toward zero.
x	-1	-10	-100	-1000	$-10,000$	$\cdots$	As $x \to -\infty$ the value of $1/x$
$1/x$	-1	$-.1$	$-.01$	$-.001$	$-.0001$	$\cdots$	increases toward zero.

The following theorem, parts of which are proved in Appendix G, will be our basic tool for finding limits algebraically.

2.2.2 THEOREM. *Let* $\lim$ *stand for one of the limits* $\lim\limits_{x \to a}$, $\lim\limits_{x \to a^-}$, $\lim\limits_{x \to a^+}$, $\lim\limits_{x \to +\infty}$, *or* $\lim\limits_{x \to -\infty}$. *If* $L_1 = \lim f(x)$ *and* $L_2 = \lim g(x)$ *both exist, then*

(a) $\lim [f(x) + g(x)] = \lim f(x) + \lim g(x) = L_1 + L_2$

(b) $\lim [f(x) - g(x)] = \lim f(x) - \lim g(x) = L_1 - L_2$

(c) $\lim [f(x)g(x)] = \lim f(x) \lim g(x) = L_1 L_2$

(d) $\lim \dfrac{f(x)}{g(x)} = \dfrac{\lim f(x)}{\lim g(x)} = \dfrac{L_1}{L_2}$ *if* $L_2 \neq 0$

(e) $\lim \sqrt[n]{f(x)} = \sqrt[n]{\lim f(x)} = \sqrt[n]{L_1}$ *provided* $L_1 \geq 0$ *if* n *is even.*

In words, this theorem states:

(a) *The limit of a sum is the sum of the limits.*

(b) *The limit of a difference is the difference of the limits.*

(c) *The limit of a product is the product of the limits.*

(d) *The limit of a quotient is the quotient of the limits provided the limit of the denominator is not zero.*

(e) *The limit of an nth root is the nth root of the limits.*

REMARK. Although results (a) and (c) are stated for two functions f and g, these results hold as well for any finite number of functions; that is, if the limits $\lim f_1(x)$,

$\lim f_2(x), \ldots, \lim f_n(x)$ all exist, then

$$\lim [f_1(x) + f_2(x) + \cdots + f_n(x)] = \lim f_1(x) + \lim f_2(x) + \cdots + \lim f_n(x) \qquad (1)$$

$$\lim [f_1(x) f_2(x) \cdots f_n(x)] = \lim f_1(x) \lim f_2(x) \cdots \lim f_n(x) \qquad (2)$$

In particular, if $f_1, f_2, \ldots, f_n$ are all the same function f, then (2) reduces to

$$\lim [f(x)]^n = [\lim f(x)]^n \qquad (3)$$

It follows from this result that

$$\lim_{x \to a} x^n = [\lim_{x \to a} x]^n = a^n \qquad (4)$$

and

$$\lim_{x \to +\infty} \frac{1}{x^n} = \left(\lim_{x \to +\infty} \frac{1}{x} \right)^n = 0 \qquad \lim_{x \to -\infty} \frac{1}{x^n} = \left(\lim_{x \to -\infty} \frac{1}{x} \right)^n = 0 \qquad (5)$$

For example,

$$\lim_{x \to 3} x^4 = 3^4 = 81, \qquad \lim_{x \to +\infty} \frac{1}{x^4} = 0, \qquad \lim_{x \to -\infty} \frac{1}{x^4} = 0$$

Another useful result follows from part (c) of Theorem 2.2.2 in the special case where one of the factors is a constant k:

$$\lim k f(x) = \lim k \lim f(x) = k \lim f(x) \qquad (6)$$

In words, the first and last expressions in (6) state:

> *A constant factor can be moved through a limit sign.*

LIMITS OF POLYNOMIALS
AS $x \to a$

Example 1

Find $\lim_{x \to 5} (x^2 - 4x + 3)$ and justify each step.

Solution.

$$
\begin{aligned}
\lim_{x \to 5} (x^2 - 4x + 3) &= \lim_{x \to 5} x^2 - \lim_{x \to 5} 4x + \lim_{x \to 5} 3 & \text{Theorem 2.2.2}(a),(b) \\
&= \lim_{x \to 5} x^2 - 4 \lim_{x \to 5} x + \lim_{x \to 5} 3 & \text{Equation (6)} \\
&= 5^2 - 4(5) + 3 & \text{Equation (4)} \\
&= 8 &
\end{aligned}
$$
◄

Our next result will show that the limit of a polynomial $p(x)$ at a point $x = a$ is the same as the value of the polynomial at that point. This greatly simplifies the computation of limits of polynomials by allowing us to evaluate the polynomial instead. Moreover, as discussed in the last section, this result also establishes that graphs of polynomials are continuous curves (see the discussion in the subsection of Section 2.1 entitled *A First Look at Continuity*).

2.2.3   THEOREM.   *For any polynomial*

$$p(x) = c_0 + c_1 x + \cdots + c_n x^n$$

and any real number a,

$$\lim_{x \to a} p(x) = c_0 + c_1 a + \cdots + c_n a^n = p(a)$$

Proof.

$$\lim_{x \to a} p(x) = \lim_{x \to a} \left(c_0 + c_1 x + \cdots + c_n x^n \right)$$

$$= \lim_{x \to a} c_0 + \lim_{x \to a} c_1 x + \cdots + \lim_{x \to a} c_n x^n$$

$$= \lim_{x \to a} c_0 + c_1 \lim_{x \to a} x + \cdots + c_n \lim_{x \to a} x^n$$

$$= c_0 + c_1 a + \cdots + c_n a^n = p(a) \qquad \blacksquare$$

Example 2

If we apply Theorem 2.2.3 to the problem in Example 1, we can bypass the intermediate steps and write immediately

$$\lim_{x \to 5} (x^2 - 4x + 3) = 5^2 - 4(5) + 3 = 8 \qquad \blacktriangleleft$$

LIMITS OF x^n AS $x \to +\infty$ OR $x \to -\infty$

In Figure 2.2.2 we have graphed the polynomials of the form x^n for $n = 1, 2, 3,$ and 4; and below each figure we have indicated the limits as $x \to +\infty$ and $x \to -\infty$. The results in the figure are special cases of the following general results:

$$\lim_{x \to +\infty} x^n = +\infty, \quad n = 1, 2, 3, \ldots \tag{7}$$

$$\lim_{x \to -\infty} x^n = \begin{cases} +\infty, & n = 2, 4, 6, \ldots \\ -\infty, & n = 1, 3, 5, \ldots \end{cases} \tag{8}$$

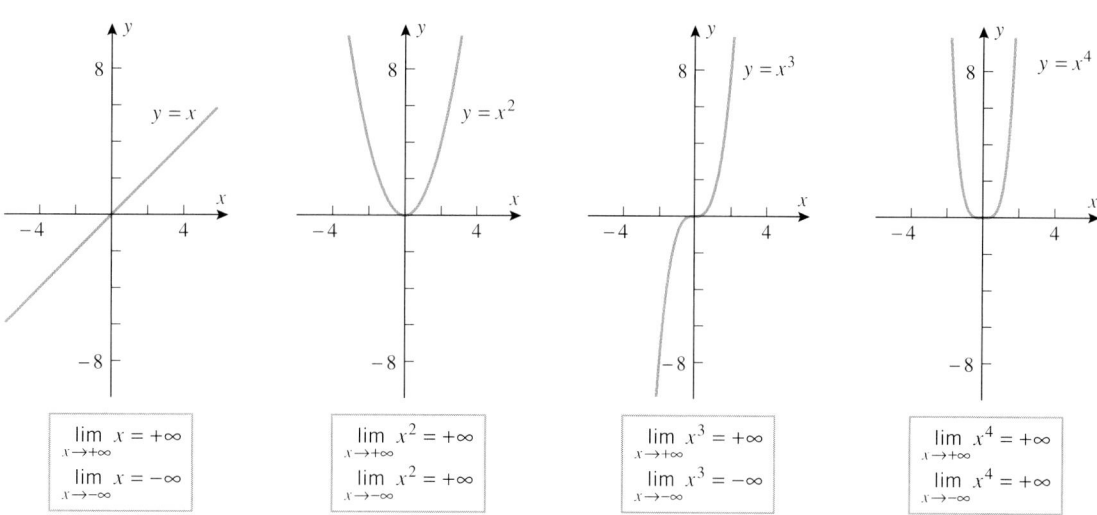

Figure 2.2.2

Multiplying x^n by a positive real number does not affect limits (7) and (8), but multiplying by a negative real number reverses the signs.

Example 3

$$\lim_{x \to +\infty} 2x^5 = +\infty, \qquad \lim_{x \to -\infty} 2x^5 = -\infty$$

$$\lim_{x \to +\infty} -7x^6 = -\infty, \qquad \lim_{x \to -\infty} -7x^6 = -\infty \qquad \blacktriangleleft$$

There is a useful principle about polynomials which, expressed informally, states that:

> *A polynomial behaves like its term of highest degree as $x \to +\infty$ or $x \to -\infty$.*

Stated more precisely, if $c_n \neq 0$, then

$$\lim_{x \to +\infty} \left(c_0 + c_1 x + \cdots + c_n x^n\right) = \lim_{x \to +\infty} c_n x^n \tag{9}$$

$$\lim_{x \to -\infty} \left(c_0 + c_1 x + \cdots + c_n x^n\right) = \lim_{x \to -\infty} c_n x^n \tag{10}$$

We can motivate these results by factoring out the highest power of x from the polynomial and examining the limit of the factored expression. Thus,

$$c_0 + c_1 x + \cdots + c_n x^n = x^n \left(\frac{c_0}{x^n} + \frac{c_1}{x^{n-1}} + \cdots + c_n \right)$$

As $x \to +\infty$ or $x \to -\infty$, it follows from (5) that all of the terms with positive powers of x in the denominator approach 0, so (9) and (10) are certainly plausible.

Example 4

$$\lim_{x \to -\infty} (7x^5 - 4x^3 + 2x - 9) = \lim_{x \to -\infty} 7x^5 = -\infty$$

$$\lim_{x \to -\infty} (-4x^8 + 17x^3 - 5x + 1) = \lim_{x \to -\infty} -4x^8 = -\infty \qquad \blacktriangleleft$$

Recall that a rational function is the ratio of two polynomials. Theorem 2.2.3 and Theorem 2.2.2(d) can often be used in combination to compute limits of rational functions.

Example 5

Find $\displaystyle\lim_{x \to 2} \frac{5x^3 + 4}{x - 3}$.

Solution.

$$\lim_{x \to 2} \frac{5x^3 + 4}{x - 3} = \frac{\displaystyle\lim_{x \to 2} (5x^3 + 4)}{\displaystyle\lim_{x \to 2} (x - 3)} = \frac{5 \cdot 2^3 + 4}{2 - 3} = -44 \qquad \blacktriangleleft$$

The method of the preceding example will not work if the limit of the denominator is zero, since Theorem 2.2.2(d) is not applicable in this situation. However, if the numerator and denominator *both* approach zero as x approaches a, then the numerator and denominator will have a common factor of $x - a$ and the limit can often be obtained by first canceling the common factors. The following example illustrates this technique.

Example 6

Find $\displaystyle\lim_{x \to 2} \frac{x^2 - 4}{x - 2}$.

Solution. The numerator and denominator both have a limit of zero as x approaches 2, so they share a common factor of $x - 2$. The limit can be obtained as follows:

$$\lim_{x \to 2} \frac{x^2 - 4}{x - 2} = \lim_{x \to 2} \frac{(x - 2)(x + 2)}{x - 2} = \lim_{x \to 2} (x + 2) = 4 \qquad \blacktriangleleft$$

REMARK. Although correct, the second equality in the preceding computation needs some justification, since canceling the factor $x - 2$ alters the function. However, as discussed in Example 5 of Section 1.2, the two functions are identical, except at $x = 2$ (Figure 1.2.9); and we know from our discussions in the last section that this difference has no effect on the limit as x approaches 2.

Example 7

Find

$$\text{(a)} \quad \lim_{x \to 3} \frac{x^2 - 6x + 9}{x - 3} \qquad \text{(b)} \quad \lim_{x \to -4} \frac{2x + 8}{x^2 + x - 12}$$

Solution (a). The numerator and denominator both have a limit of zero as x approaches 3, so there is a common factor of $x - 3$. We proceed as follows:

$$\lim_{x \to 3} \frac{x^2 - 6x + 9}{x - 3} = \lim_{x \to 3} \frac{(x - 3)^2}{x - 3} = \lim_{x \to 3} (x - 3) = 0$$

Solution (b). The numerator and denominator both have a limit of zero as x approaches -4, so there is a common factor of $x - (-4) = x + 4$. We proceed as follows:

$$\lim_{x \to -4} \frac{2x + 8}{x^2 + x - 12} = \lim_{x \to -4} \frac{2(x + 4)}{(x + 4)(x - 3)} = \lim_{x \to -4} \frac{2}{x - 3} = -\frac{2}{7} \qquad \blacktriangleleft$$

If the limit of the denominator is zero, but the limit of the numerator is not, then there are three possibilities for the limit of the rational function as $x \to a$:

- The limit may be $+\infty$.
- The limit may be $-\infty$.
- The limit may be $+\infty$ from one side and $-\infty$ from the other.

Figure 2.2.3 illustrates this graphically for functions of the form $1/(x - a)$, $1/(x - a)^2$, and $-1/(x - a)^2$.

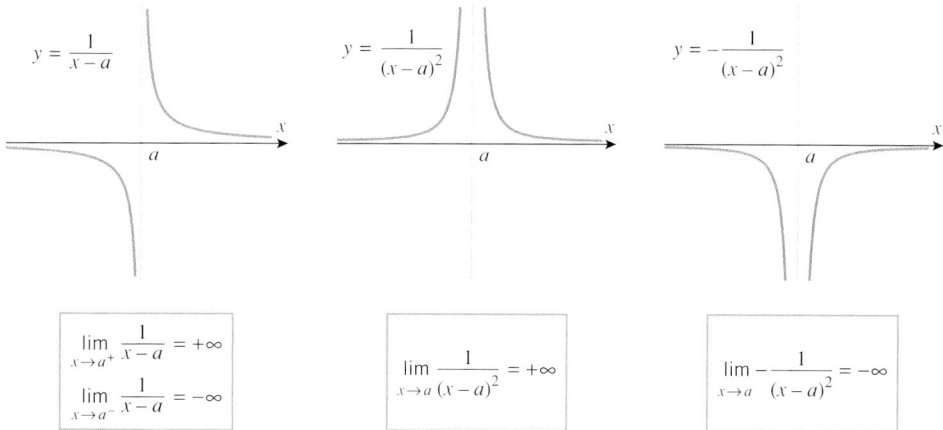

Figure 2.2.3

Example 8

Find

$$\text{(a)} \quad \lim_{x \to 4^+} \frac{2 - x}{(x - 4)(x + 2)} \qquad \text{(b)} \quad \lim_{x \to 4^-} \frac{2 - x}{(x - 4)(x + 2)} \qquad \text{(c)} \quad \lim_{x \to 4} \frac{2 - x}{(x - 4)(x + 2)}$$

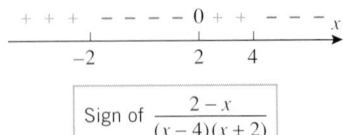

Figure 2.2.4

Solution. In all three parts the limit of the numerator is -2, and the limit of the denominator is 0, so the limit of the ratio does not exist. To be more specific than this, we need to analyze the sign of the ratio. The sign of the ratio, which is given in Figure 2.2.4, is determined by the signs of $2 - x$, $x - 4$, and $x + 2$. (The method of test points, discussed in Appendix A, provides a simple way of finding the sign of the ratio here.) It follows from this figure that as x approaches 4 from the right, the ratio is always negative; and as x approaches 4 from the left, the ratio is eventually positive (after x exceeds 2). Thus,

$$\lim_{x \to 4^+} \frac{2 - x}{(x - 4)(x + 2)} = -\infty \quad \text{and} \quad \lim_{x \to 4^-} \frac{2 - x}{(x - 4)(x + 2)} = +\infty$$

Because the one-sided limits have opposite signs, all we can say about the two-sided limit is that it does not exist. ◄

LIMITS OF RATIONAL FUNCTIONS AS $x \to +\infty$ OR $x \to -\infty$

If we divide the numerator and denominator of a rational function by the highest power of x that occurs in the denominator, then all the powers of x in the denominator become constants or powers of $1/x$. The following examples show how this observation together with (5), (9), and (10) can be used to find limits of rational functions as $x \to +\infty$ or $x \to -\infty$.

Example 9

Find $\displaystyle\lim_{x \to +\infty} \frac{3x + 5}{6x - 8}$.

Solution. Divide the numerator and denominator by the highest power of x that occurs in the denominator; this is $x^1 = x$. We obtain

$$\lim_{x \to +\infty} \frac{3x + 5}{6x - 8} = \lim_{x \to +\infty} \frac{3 + 5/x}{6 - 8/x} = \frac{\lim\limits_{x \to +\infty} (3 + 5/x)}{\lim\limits_{x \to +\infty} (6 - 8/x)}$$

$$= \frac{\lim\limits_{x \to +\infty} 3 + \lim\limits_{x \to +\infty} 5/x}{\lim\limits_{x \to +\infty} 6 - \lim\limits_{x \to +\infty} 8/x} = \frac{3 + 5 \lim\limits_{x \to +\infty} 1/x}{6 - 8 \lim\limits_{x \to +\infty} 1/x}$$

$$= \frac{3 + (5 \cdot 0)}{6 - (8 \cdot 0)} = \frac{1}{2} \qquad \blacktriangleleft$$

Example 10

Find

(a) $\displaystyle\lim_{x \to -\infty} \frac{4x^2 - x}{2x^3 - 5}$ (b) $\displaystyle\lim_{x \to -\infty} \frac{5x^3 - 2x^2 + 1}{3x + 5}$

Solution (a). Divide the numerator and denominator by the highest power of x that occurs in the denominator, namely x^3. We obtain

$$\lim_{x \to -\infty} \frac{4x^2 - x}{2x^3 - 5} = \lim_{x \to -\infty} \frac{4/x - 1/x^2}{2 - 5/x^3} = \frac{\lim\limits_{x \to -\infty} (4/x - 1/x^2)}{\lim\limits_{x \to -\infty} (2 - 5/x^3)}$$

$$= \frac{(4 \cdot 0) - 0}{2 - (5 \cdot 0)} = \frac{0}{2} = 0$$

Solution (b). Divide the numerator and denominator by x to obtain

$$\lim_{x \to -\infty} \frac{5x^3 - 2x^2 + 1}{3x + 5} = \lim_{x \to -\infty} \frac{5x^2 - 2x + 1/x}{3 + 5/x} = +\infty$$

where the final step is justified by the fact that

$$5x^2 - 2x \to +\infty, \quad 1/x \to 0, \quad \text{and} \quad 3 + 5/x \to 3$$

as $x \to -\infty$.   ◀

A QUICK METHOD FOR FINDING LIMITS OF RATIONAL FUNCTIONS AS $x \to +\infty$ OR $x \to -\infty$

Since a polynomial behaves like its term of highest degree as $x \to +\infty$ or $x \to -\infty$, it follows that a rational function behaves like the ratio of the terms of highest degree in the numerator and denominator as $x \to +\infty$ or $x \to -\infty$; that is, if $c_n \neq 0$ and $d_n \neq 0$, then

$$\lim_{x \to +\infty} \frac{c_0 + c_1 x + \cdots + c_n x^n}{d_0 + d_1 x + \cdots + d_m x^m} = \lim_{x \to +\infty} \frac{c_n x^n}{d_m x^m} \tag{11}$$

and

$$\lim_{x \to -\infty} \frac{c_0 + c_1 x + \cdots + c_n x^n}{d_0 + d_1 x + \cdots + d_m x^m} = \lim_{x \to -\infty} \frac{c_n x^n}{d_m x^m} \tag{12}$$

Example 11

Use Formulas (11) and (12) to find

    (a) $\displaystyle\lim_{x \to +\infty} \frac{3x + 5}{6x - 8}$ (b) $\displaystyle\lim_{x \to -\infty} \frac{4x^2 - x}{2x^3 - 5}$ (c) $\displaystyle\lim_{x \to +\infty} \frac{3 - 2x^4}{x + 1}$

Solution (a).

$$\lim_{x \to +\infty} \frac{3x + 5}{6x - 8} = \lim_{x \to +\infty} \frac{3x}{6x} = \lim_{x \to +\infty} \frac{1}{2} = \frac{1}{2}$$

which agrees with the result obtained in Example 9.

Solution (b).

$$\lim_{x \to -\infty} \frac{4x^2 - x}{2x^3 - 5} = \lim_{x \to -\infty} \frac{4x^2}{2x^3} = \lim_{x \to -\infty} \frac{2}{x} = 0$$

which agrees with the result obtained in Example 10.

Solution (c).

$$\lim_{x \to +\infty} \frac{3 - 2x^4}{x + 1} = \lim_{x \to +\infty} \frac{-2x^4}{x} = \lim_{x \to +\infty} -2x^3 = -\infty \qquad\qquad ◀$$

REMARK.   We emphasize that Formulas (11) and (12) are only applicable if $x \to +\infty$ or $x \to -\infty$; they do not apply to limits in which x approaches a *finite* number a.

LIMITS INVOLVING RADICALS

Example 12

Find $\displaystyle\lim_{x \to +\infty} \sqrt[3]{\frac{3x + 5}{6x - 8}}$.

Solution.

$$\lim_{x \to +\infty} \sqrt[3]{\frac{3x + 5}{6x - 8}} = \sqrt[3]{\lim_{x \to +\infty} \frac{3x + 5}{6x - 8}} = \sqrt[3]{\frac{1}{2}}$$

                       Theorem 2.2.2(*e*)   Example 9   ◀

Example 13

Find

    (a) $\displaystyle\lim_{x \to +\infty} \frac{\sqrt{x^2 + 2}}{3x - 6}$ (b) $\displaystyle\lim_{x \to -\infty} \frac{\sqrt{x^2 + 2}}{3x - 6}$

In both parts it would be helpful to manipulate the function so that the powers of x become powers of $1/x$. This can be achieved in both cases by dividing the numerator and denominator by $|x|$ and using the fact that $\sqrt{x^2} = |x|$.

Solution (a). As $x \to +\infty$, the values of x are eventually positive, so we can replace $|x|$ by x where helpful. We obtain

$$\lim_{x \to +\infty} \frac{\sqrt{x^2 + 2}}{3x - 6} = \lim_{x \to +\infty} \frac{\sqrt{x^2 + 2}/|x|}{(3x - 6)/|x|} = \lim_{x \to +\infty} \frac{\sqrt{x^2 + 2}/\sqrt{x^2}}{(3x - 6)/x}$$

$$= \lim_{x \to +\infty} \frac{\sqrt{1 + 2/x^2}}{3 - 6/x} = \frac{\lim_{x \to +\infty} \sqrt{1 + 2/x^2}}{\lim_{x \to +\infty} (3 - 6/x)}$$

$$= \frac{\sqrt{\lim_{x \to +\infty} (1 + 2/x^2)}}{\lim_{x \to +\infty} (3 - 6/x)} = \frac{\sqrt{\lim_{x \to +\infty} 1 + 2 \lim_{x \to +\infty} 1/x^2}}{\lim_{x \to +\infty} 3 - 6 \lim_{x \to +\infty} 1/x}$$

$$= \frac{\sqrt{1 + (2 \cdot 0)}}{3 - (6 \cdot 0)} = \frac{1}{3}$$

Solution (b). As $x \to -\infty$, the values of x are eventually negative, so we can replace $|x|$ by $-x$ where helpful. We obtain

$$\lim_{x \to -\infty} \frac{\sqrt{x^2 + 2}}{3x - 6} = \lim_{x \to -\infty} \frac{\sqrt{x^2 + 2}/|x|}{(3x - 6)/|x|} = \lim_{x \to -\infty} \frac{\sqrt{x^2 + 2}/\sqrt{x^2}}{(3x - 6)/(-x)}$$

$$= \lim_{x \to -\infty} \frac{\sqrt{1 + 2/x^2}}{(6/x) - 3} = -\frac{1}{3} \quad \blacktriangleleft$$

LIMITS OF FUNCTIONS DEFINED PIECEWISE

For functions that are defined piecewise, a two-sided limit at a point where the formula changes is best obtained by first finding the one-sided limits at the point.

Example 14

Find $\lim_{x \to 3} f(x)$ for $f(x) = \begin{cases} x^2 - 5, & x \le 3 \\ \sqrt{x + 13}, & x > 3. \end{cases}$

Solution. As x approaches 3 from the left, the formula for f is

$$f(x) = x^2 - 5$$

so that

$$\lim_{x \to 3^-} f(x) = \lim_{x \to 3^-} (x^2 - 5) = 3^2 - 5 = 4$$

As x approaches 3 from the right, the formula for f is

$$f(x) = \sqrt{x + 13}$$

so that

$$\lim_{x \to 3^+} f(x) = \lim_{x \to 3^+} \sqrt{x + 13} = \sqrt{\lim_{x \to 3^+} (x + 13)} = \sqrt{16} = 4$$

Since the one-sided limits are equal, we have

$$\lim_{x \to 3} f(x) = 4 \quad \blacktriangleleft$$

EXERCISE SET 2.2

1. Given that

$$\lim_{x \to a} f(x) = 2, \quad \lim_{x \to a} g(x) = -4, \quad \lim_{x \to a} h(x) = 0$$

   find the limits that exist. If the limit does not exist, explain why.

   (a) $\lim_{x \to a} [f(x) + 2g(x)]$ (b) $\lim_{x \to a} [h(x) - 3g(x) + 1]$

   (c) $\lim_{x \to a} [f(x)g(x)]$ (d) $\lim_{x \to a} [g(x)]^2$

   (e) $\lim_{x \to a} \sqrt[3]{6 + f(x)}$ (f) $\lim_{x \to a} \dfrac{2}{g(x)}$

   (g) $\lim_{x \to a} \dfrac{3f(x) - 8g(x)}{h(x)}$ (h) $\lim_{x \to a} \dfrac{7g(x)}{2f(x) + g(x)}$

2. Use the graphs of f and g in the accompanying figure to find the limits that exist. If the limit does not exist, explain why.

   (a) $\lim_{x \to 2} [f(x) + g(x)]$ (b) $\lim_{x \to 0} [f(x) + g(x)]$

   (c) $\lim_{x \to 0^+} [f(x) + g(x)]$ (d) $\lim_{x \to 0^-} [f(x) + g(x)]$

   (e) $\lim_{x \to 2} \dfrac{f(x)}{1 + g(x)}$ (f) $\lim_{x \to 2} \dfrac{1 + g(x)}{f(x)}$

   (g) $\lim_{x \to 0^+} \sqrt{f(x)}$ (h) $\lim_{x \to 0^-} \sqrt{f(x)}$

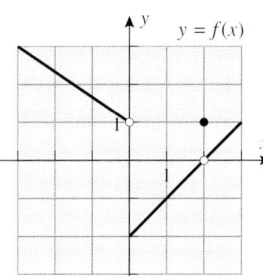 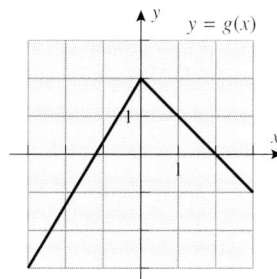

Figure Ex-2

3. In each part, find the limit by inspection.

   (a) $\lim_{x \to 8} 7$ (b) $\lim_{x \to -\infty} (-3)$ (c) $\lim_{x \to 0^+} \pi$

   (d) $\lim_{x \to -2} 3x$ (e) $\lim_{y \to 3^+} 12y$ (f) $\lim_{h \to +\infty} (-2h)$

4. In each part, find the stated limit of $f(x) = x/|x|$ by inspection.

   (a) $\lim_{x \to 5} f(x)$ (b) $\lim_{x \to -5} f(x)$ (c) $\lim_{x \to +\infty} f(x)$

   (d) $\lim_{x \to -\infty} f(x)$ (e) $\lim_{x \to 0^+} f(x)$ (f) $\lim_{x \to 0^-} f(x)$

   Find the limits in Exercises 5–48.

5. $\lim_{y \to 2^-} \dfrac{(y - 1)(y - 2)}{y + 1}$

6. $\lim_{x \to 3} \dfrac{x^2 - 2x}{x + 1}$

7. $\lim_{x \to 4} \dfrac{x^2 - 16}{x - 4}$

8. $\lim_{x \to 0} \dfrac{6x - 9}{x^3 - 12x + 3}$

9. $\lim_{x \to 1^+} \dfrac{x^4 - 1}{x - 1}$

10. $\lim_{t \to -2} \dfrac{t^3 + 8}{t + 2}$

11. $\lim_{x \to -1} \dfrac{x^2 + 6x + 5}{x^2 - 3x - 4}$

12. $\lim_{x \to 2} \dfrac{x^2 - 4x + 4}{x^2 + x - 6}$

13. $\lim_{x \to +\infty} \dfrac{3x + 1}{2x - 5}$

14. $\lim_{t \to 1} \dfrac{t^3 + t^2 - 5t + 3}{t^3 - 3t + 2}$

15. $\lim_{y \to -\infty} \dfrac{3}{y + 4}$

16. $\lim_{x \to +\infty} \dfrac{1}{x - 12}$

17. $\lim_{x \to -\infty} \dfrac{x - 2}{x^2 + 2x + 1}$

18. $\lim_{x \to +\infty} \dfrac{5x^2 + 7}{3x^2 - x}$

19. $\lim_{x \to -\infty} \dfrac{\sqrt{5x^2 - 2}}{x + 3}$

20. $\lim_{s \to +\infty} \sqrt[3]{\dfrac{3s^7 - 4s^5}{2s^7 + 1}}$

21. $\lim_{y \to -\infty} \dfrac{2 - y}{\sqrt{7 + 6y^2}}$

22. $\lim_{x \to +\infty} \dfrac{\sqrt{5x^2 - 2}}{x + 3}$

23. $\lim_{x \to -\infty} \dfrac{\sqrt{3x^4 + x}}{x^2 - 8}$

24. $\lim_{y \to +\infty} \dfrac{2 - y}{\sqrt{7 + 6y^2}}$

25. $\lim_{x \to 3^+} \dfrac{x}{x - 3}$

26. $\lim_{x \to +\infty} \dfrac{\sqrt{3x^4 + x}}{x^2 - 8}$

27. $\lim_{x \to 3} \dfrac{x}{x - 3}$

28. $\lim_{x \to 3^-} \dfrac{x}{x - 3}$

29. $\lim_{x \to 2^-} \dfrac{x}{x^2 - 4}$

30. $\lim_{x \to 2^+} \dfrac{x}{x^2 - 4}$

31. $\lim_{y \to 6^+} \dfrac{y + 6}{y^2 - 36}$

32. $\lim_{x \to 2} \dfrac{x}{x^2 - 4}$

33. $\lim_{y \to 6} \dfrac{y + 6}{y^2 - 36}$

34. $\lim_{y \to 6^-} \dfrac{y + 6}{y^2 - 36}$

35. $\lim_{x \to 4^-} \dfrac{3 - x}{x^2 - 2x - 8}$

36. $\lim_{x \to 4^+} \dfrac{3 - x}{x^2 - 2x - 8}$

37. $\lim_{x \to +\infty} \dfrac{7 - 6x^5}{x + 3}$

38. $\lim_{x \to 4} \dfrac{3 - x}{x^2 - 2x - 8}$

39. $\lim_{t \to +\infty} \dfrac{6 - t^3}{7t^3 + 3}$

40. $\lim_{t \to -\infty} \dfrac{5 - 2t^3}{t^2 + 1}$

41. $\lim_{x \to 9} \dfrac{x - 9}{\sqrt{x} - 3}$

42. $\lim_{x \to 3^-} \dfrac{1}{|x - 3|}$

43. $\lim_{x \to +\infty} \sqrt{x}$

44. $\lim_{y \to 4} \dfrac{4 - y}{2 - \sqrt{y}}$

45. $\lim_{x \to -\infty} (3 - x)$

46. $\lim_{x \to -\infty} \sqrt{5 - x}$

47. $\lim_{x \to +\infty} (1 + 2x - 3x^5)$

48. $\lim_{x \to +\infty} (2x^3 - 100x + 5)$

49. Let

$$f(x) = \begin{cases} x - 1, & x \le 3 \\ 3x - 7, & x > 3 \end{cases}$$

   Find

   (a) $\lim_{x \to 3^-} f(x)$ (b) $\lim_{x \to 3^+} f(x)$ (c) $\lim_{x \to 3} f(x)$.

50. Let
$$g(t) = \begin{cases} t^2, & t \geq 0 \\ t - 2, & t < 0 \end{cases}$$
Find
(a) $\lim\limits_{t \to 0^-} g(t)$ (b) $\lim\limits_{t \to 0^+} g(t)$ (c) $\lim\limits_{t \to 0} g(t)$.

51. Let $f(x) = \dfrac{x^3 - 1}{x - 1}$.
(a) Find $\lim\limits_{x \to 1} f(x)$.
(b) Sketch the graph of $y = f(x)$.

52. Let
$$f(x) = \begin{cases} \dfrac{x^2 - 9}{x + 3}, & x \neq -3 \\ k, & x = -3 \end{cases}$$
(a) Find k so that $F(-3) = \lim\limits_{x \to -3} F(x)$.
(b) With k assigned the value $\lim\limits_{x \to -3} F(x)$, show that $F(x)$ can be expressed as a polynomial.

53. (a) Explain why the following calculation is incorrect.
$$\lim_{x \to 0^+} \left(\frac{1}{x} - \frac{1}{x^2} \right) = \lim_{x \to 0^+} \frac{1}{x} - \lim_{x \to 0^+} \frac{1}{x^2}$$
$$= +\infty - (+\infty) = 0$$
(b) Show that $\lim\limits_{x \to 0^+} \left(\dfrac{1}{x} - \dfrac{1}{x^2} \right) = -\infty$.

54. Find $\lim\limits_{x \to 0^-} \left(\dfrac{1}{x} + \dfrac{1}{x^2} \right)$.

In Exercises 55 and 56, first rationalize the numerator, then find the limit.

55. $\lim\limits_{x \to 0} \dfrac{\sqrt{x + 4} - 2}{x}$ **56.** $\lim\limits_{x \to 0} \dfrac{\sqrt{x^2 + 4} - 2}{x}$

Find the limits in Exercises 57–60.

57. $\lim\limits_{x \to +\infty} (\sqrt{x^2 + 3} - x)$

58. $\lim\limits_{x \to +\infty} (\sqrt{x^2 - 3x} - x)$

59. $\lim\limits_{x \to +\infty} (\sqrt{x^2 + ax} - x)$

60. $\lim\limits_{x \to +\infty} (\sqrt{x^2 + ax} - \sqrt{x^2 + bx})$

61. Discuss the limits of $p(x) = (1 - x)^n$ as $x \to +\infty$ and $x \to -\infty$ for positive integer values of n.

62. Let $p(x) = (1 - x)^n$ and $q(x) = (1 - x)^m$. Discuss the limits of $p(x)/q(x)$ as $x \to +\infty$ and $x \to -\infty$ for positive integer values of m and n.

63. Let $p(x)$ be a polynomial of degree n. Discuss the limits of $p(x)/x^m$ as $x \to +\infty$ and $x \to -\infty$ for positive integer values of m.

64. In each part, find examples of polynomials $p(x)$ and $q(x)$ that satisfy the stated condition and such that $p(x) \to +\infty$ and $q(x) \to +\infty$ as $x \to +\infty$.
(a) $\lim\limits_{x \to +\infty} \dfrac{p(x)}{q(x)} = 1$ (b) $\lim\limits_{x \to +\infty} \dfrac{p(x)}{q(x)} = 0$
(c) $\lim\limits_{x \to +\infty} \dfrac{p(x)}{q(x)} = +\infty$ (d) $\lim\limits_{x \to +\infty} [p(x) - q(x)] = 3$

65. Let $p(x)$ and $q(x)$ be polynomials, and suppose $q(x_0) = 0$. Discuss the behavior of the graph of $y = p(x)/q(x)$ in the vicinity of the point $x = x_0$. Give examples to support your conclusions.

66. Find
$$\lim_{x \to +\infty} \frac{c_0 + c_1 x + \cdots + c_n x^n}{d_0 + d_1 x + \cdots + d_m x^m}$$
where $c_n \neq 0$ and $d_m \neq 0$. [*Hint:* Your answer will depend on whether $m < n$, $m = n$, or $m > n$.]

2.3 LIMITS (DISCUSSED MORE RIGOROUSLY)

Thus far, our discussion of limits has been based on our intuitive feeling of what it means for the values of a function to get closer and closer to a limiting value. However, this level of informality can only take us so far, so our goal in this section is to define limits precisely. From a purely mathematical point of view these definitions are needed to establish limits with certainty and to prove theorems about them. However, they will also provide us with a deeper understanding of the limit concept, making it possible for us to visualize some of the more subtle properties of functions.

DEFINITION OF A LIMIT

In earlier sections we interpreted the limit
$$\lim_{x \to a} f(x) = L$$
to mean that we can force the values of $f(x)$ closer and closer to L by making x closer and closer (but not equal) to a. Our goal here is to try to make the notion of a limit more precise by giving the informal phrase "closer and closer to" a precise mathematical meaning. However,

the concept is subtle, so we will build up to it by giving two preliminary definitions that capture the essential ideas, and then giving the final definition as it is commonly stated.

To start, consider the function f graphed in Figure 2.3.1a for which $f(x) \to L$ as $x \to a$. We have intentionally placed a hole in the graph at $x = a$ to emphasize that the function f need not be defined at $x = a$ to have a limit there.

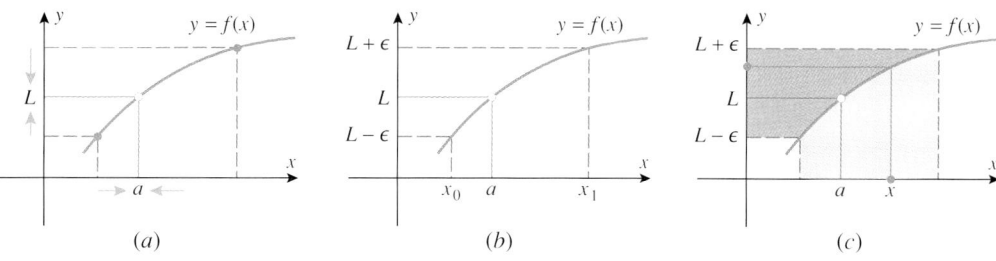

(a)   (b)   (c)

Figure 2.3.1

To motivate an appropriate definition for a two-sided limit, suppose that we choose *any* positive number, say ϵ, and draw horizontal lines from the points $L + \epsilon$ and $L - \epsilon$ on the y-axis to the curve $y = f(x)$ and then draw vertical lines from those points on the curve to the x-axis. As shown in Figure 2.3.1b, let x_0 and x_1 be points where the vertical lines intersect the x-axis.

Next, imagine that x gets closer and closer to a (from either side). Eventually, x will lie inside the interval (x_0, x_1), which is marked by the green band in Figure 2.3.1c; and when this happens, the value of $f(x)$ will fall between $L - \epsilon$ and $L + \epsilon$, marked by the red band in the figure. Thus, we conclude:

> *If $f(x) \to L$ as $x \to a$, then for any positive number ϵ, we can find an open interval on the x-axis that contains the point $x = a$ and has the property that for each x in that interval (except possibly for $x = a$), the value of $f(x)$ is between $L - \epsilon$ and $L + \epsilon$.*

What is important about this result is that it holds no matter how small we make ϵ. However, making ϵ smaller and smaller forces $f(x)$ *closer and closer* to L—which is precisely the concept we were trying to capture mathematically. This suggests the following definition of a two-sided limit.

2.3.1 LIMIT (FIRST PRELIMINARY DEFINITION). Let $f(x)$ be defined for all x in some open interval containing the number a, with the possible exception that $f(x)$ need not be defined at a. We will write

$$\lim_{x \to a} f(x) = L$$

if given any number $\epsilon > 0$ we can find an open interval (x_0, x_1) containing the point a such that $f(x)$ satisfies

$$L - \epsilon < f(x) < L + \epsilon$$

for each x in the interval (x_0, x_1), except possibly $x = a$.

Observe that in Figure 2.3.1c the interval (x_0, x_1) extends farther on the right side of a than on the left side. However, for many purposes it is preferable to have an interval that extends the same distance on both sides of a. For this purpose, let us choose any positive number δ that is smaller than both $x_1 - a$ and $a - x_0$, and consider the interval $(a - \delta, a + \delta)$. This interval extends the same distance δ on both sides of a and lies inside of the interval (x_0, x_1) (Figure 2.3.2). Moreover, the condition $L - \epsilon < f(x) < L + \epsilon$ holds for every x in this interval (except possibly $x = a$), since this condition holds on the larger interval (x_0, x_1). This suggests the following reformulation of Definition 2.3.1.

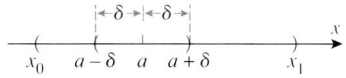

Figure 2.3.2

2.3.2 LIMIT (SECOND PRELIMINARY DEFINITION). Let $f(x)$ be defined for all x in some open interval containing the number a, with the possible exception that $f(x)$ need not be defined at a. We will write

$$\lim_{x \to a} f(x) = L$$

if given any number $\epsilon > 0$ we can find a number $\delta > 0$ such that $f(x)$ satisfies

$$L - \epsilon < f(x) < L + \epsilon$$

for each x in the interval $(a - \delta, a + \delta)$, except possibly $x = a$.

For our final version of the limit definition, we note that in Definition 2.3.2 the condition $L - \epsilon < f(x) < L + \epsilon$ can be expressed as

$$|f(x) - L| < \epsilon$$

and the condition that x lies in the interval $(a - \delta, a + \delta)$, but $x \neq a$, can be expressed as

$$0 < |x - a| < \delta$$

Thus, we can rewrite Definition 2.3.2 as follows.

2.3.3 LIMIT DEFINITION (FINAL FORM). Let $f(x)$ be defined for all x in some open interval containing the number a, with the possible exception that $f(x)$ need not be defined at a. We will write

$$\lim_{x \to a} f(x) = L$$

if given any number $\epsilon > 0$ we can find a number $\delta > 0$ such that

$$|f(x) - L| < \epsilon \quad \text{if} \quad 0 < |x - a| < \delta$$

REMARK. This defines a two-sided limit. The definitions for one-sided limits are similar, the difference being that the condition $|f(x)-L| < \epsilon$ is only required to hold on the interval $a < x < a + \delta$ for right-sided limits and on the interval $a - \delta < x < a$ for left-sided limits.

In the preceding sections we illustrated various numerical and graphical methods for *guessing* at limits. Now that we have a precise definition to work with, we can actually confirm the validity of those guesses with mathematical proof. Here is a typical example of such a proof.

Example 1

Use Definition 2.3.3 to prove that $\lim_{x \to 2} (3x - 5) = 1$.

Solution. We must show that given any positive number ϵ, we can find a positive number δ such that

$$|\underbrace{(3x - 5)}_{f(x)} - \underbrace{1}_{L}| < \epsilon \quad \text{if} \quad 0 < |x - \underbrace{2}_{a}| < \delta \tag{1}$$

There are two things to do. First, we must *discover* a value of δ for which this statement holds, and then we must *prove* that the statement holds for that δ. For the discovery part we begin by simplifying (1) and writing it as

$$|3x - 6| < \epsilon \quad \text{if} \quad 0 < |x - 2| < \delta$$

Next, we will rewrite this statement in a form that will facilitate the discovery of an appropriate δ:

$$
\begin{aligned}
3|x-2| &< \epsilon && \text{if} \quad 0 < |x-2| < \delta \\
|x-2| &< \epsilon/3 && \text{if} \quad 0 < |x-2| < \delta
\end{aligned}
\tag{2}
$$

It should be self-evident that this last statement holds if $\delta = \epsilon/3$, which completes the discovery portion of our work. Now we need to prove that (1) holds for this choice of δ. However, statement (1) is equivalent to (2), and (2) holds with $\delta = \epsilon/3$, so (1) also holds with $\delta = \epsilon/3$. This proves that $\lim_{x \to 2} (3x - 5) = 1$. ◀

REMARK.   This example illustrates the general form of a limit proof: We *assume* that we are given a positive number ϵ, and we try to *prove* that we can find a positive number δ such that

$$
|f(x) - L| < \epsilon \quad \text{if} \quad 0 < |x - a| < \delta
\tag{3}
$$

This is done by first discovering δ, and then proving that the discovered δ works. Since the argument has to be general enough to work for all positive values of ϵ, the quantity δ has to be expressed as a function of ϵ. In Example 1 we found the function $\delta = \epsilon/3$ by some simple algebra; however, most limit proofs require a little more algebraic and logical ingenuity. Thus, if you find our ensuing discussion of "δ-ϵ" proofs challenging, do not become discouraged; the concepts and techniques are intrinsically difficult. In fact, a precise understanding of limits evaded the finest mathematical minds for more than 150 years after the basic concepts of calculus were discovered.

Example 2

Prove that $\lim_{x \to 0^+} \sqrt{x} = 0$.

Solution.   We must show that given $\epsilon > 0$, there exists a $\delta > 0$ such that

$$
|\sqrt{x} - 0| < \epsilon \quad \text{if} \quad 0 < x < 0 + \delta
$$

or more simply,

$$
\sqrt{x} < \epsilon \quad \text{if} \quad 0 < x < \delta
\tag{4}
$$

But, by squaring both sides of the inequality $\sqrt{x} < \epsilon$, we can rewrite (4) as

$$
x < \epsilon^2 \quad \text{if} \quad 0 < x < \delta
\tag{5}
$$

It should be self-evident that (5) is true if $\delta = \epsilon^2$; and since (5) is a reformulation of (4), we have shown that (4) holds with $\delta = \epsilon^2$. This proves that $\lim_{x \to 0^+} \sqrt{x} = 0$. ◀

REMARK.   In this example we were only concerned with the limit from the right because $f(x) = \sqrt{x}$ has imaginary values for $x < 0$. Thus, the limit from the left and the two-sided limit are not applicable at $x = 0$.

THE VALUE OF δ IS NOT UNIQUE

In preparation for our next example, we note that the value of δ in Definition 2.3.3 is not unique; once we have found a value of δ that fulfills the requirements of the definition, then any *smaller* positive number δ_1 will also fulfill those requirements. That is, if it is true that

$$
|f(x) - L| < \epsilon \quad \text{if} \quad 0 < |x - a| < \delta
$$

then it will also be true that

$$
|f(x) - L| < \epsilon \quad \text{if} \quad 0 < |x - a| < \delta_1
$$

This is because $\{x : 0 < |x - a| < \delta_1\}$ is a subset of $\{x : 0 < |x - a| < \delta\}$ (Figure 2.3.3), and hence if $|f(x) - L| < \epsilon$ is satisfied for all x in the larger set, then it will automatically be satisfied for all x in the subset. Thus, in Example 1, where we used $\delta = \epsilon/3$, we could have used any smaller value of δ such as $\delta = \epsilon/4$, $\delta = \epsilon/5$, or $\delta = \epsilon/6$.

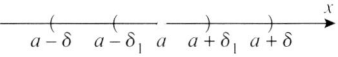

Figure 2.3.3

Example 3

Prove that $\lim\limits_{x \to 3} x^2 = 9$.

Solution. We must show that given any positive number ϵ, we can find a positive number δ such that

$$|x^2 - 9| < \epsilon \quad \text{if} \quad 0 < |x - 3| < \delta \tag{6}$$

Because $|x - 3|$ occurs on the right side of this "if statement," it will be helpful to factor the left side to introduce a factor of $|x - 3|$. This yields the following alternative form of (6):

$$|x + 3|\,|x - 3| < \epsilon \quad \text{if} \quad 0 < |x - 3| < \delta \tag{7}$$

To make this statement hold we need to find a δ that "controls" the size of both factors on the left side. However, the condition on the right side gives us direct control on the size of $|x - 3|$ but not of $|x + 3|$. To circumvent this difficulty, let us *temporarily* replace the factor $|x + 3|$ by a positive constant k and look for a δ such that

$$k|x - 3| < \epsilon \quad \text{if} \quad 0 < |x - 3| < \delta \tag{8}$$

This statement can be rewritten as

$$|x - 3| < \epsilon/k \quad \text{if} \quad 0 < |x - 3| < \delta$$

which can be satisfied by taking

$$\delta = \epsilon/k \tag{9}$$

Now let us assume that k can be chosen so that

$$|x + 3| < k \tag{10}$$

in which case

$$|x + 3|\,|x - 3| < \epsilon \quad \text{if} \quad k|x - 3| < \epsilon$$

Thus, if we can find k so that (10) holds, then choosing δ as in (9) will make (8) hold, and this in turn will make (7) hold.

To find k, let us *arbitrarily* agree that we will choose δ so that $\delta \leq 1$. This is justified because of our earlier observation that once a value of δ is found, then any smaller positive value of δ can be used. Thus, if it so happens that $\delta > 1$ in (9), we can use $\delta = 1$ instead, thereby guaranteeing that $\delta \leq 1$. If we impose this restriction on δ, then it will follow from the right side of (8) that

$$|x - 3| < 1 \quad \text{or} \quad 2 < x < 4 \quad \text{or} \quad 5 < x + 3 < 7$$

from which we can conclude that

$$|x + 3| < 7$$

Thus, given $\epsilon > 0$, we can take $k = 7$ in (8), and hence from (9) we can take $\delta = \epsilon/7$ (or smaller), subject to the restriction that $\delta \leq 1$. We can achieve this by taking δ to be the minimum of the numbers $\epsilon/7$ and 1, which is sometimes written as $\delta = \min(\epsilon/7, 1)$. This proves that $\lim\limits_{x \to 3} x^2 = 9$. ◄

REMARK. You may have wondered how we knew to make the restriction $\delta \leq 1$ (as opposed to $\delta \leq \frac{1}{2}$ or $\delta \leq 5$, for example). Actually, it does not matter; any restriction of the form $\delta \leq c$ would work equally well.

....

LIMITS AS $x \to +\infty$ OR $x \to -\infty$

In Section 2.1 we discussed the limits

$$\lim\limits_{x \to +\infty} f(x) = L \quad \text{and} \quad \lim\limits_{x \to -\infty} f(x) = L$$

from an intuitive viewpoint. We interpreted the first statement to mean that the values of $f(x)$ eventually get closer and closer to L as x increases indefinitely, and we interpreted the second statement to mean that the values of $f(x)$ eventually get closer and closer to L as x decreases indefinitely. These ideas are captured more precisely in the following definitions and are illustrated in Figure 2.3.4.

2.3.4 DEFINITION. Let $f(x)$ be defined for all x in some infinite open interval extending in the positive x-direction. We will write

$$\lim_{x \to +\infty} f(x) = L$$

if given any number $\epsilon > 0$, there corresponds a positive number N such that

$$|f(x) - L| < \epsilon \quad \text{if} \quad x > N$$

2.3.5 DEFINITION. Let $f(x)$ be defined for all x in some infinite open interval extending in the negative x-direction. We will write

$$\lim_{x \to -\infty} f(x) = L$$

if given any number $\epsilon > 0$, there corresponds a negative number N such that

$$|f(x) - L| < \epsilon \quad \text{if} \quad x < N$$

To see how these definitions relate to our informal concepts of these limits, suppose that $f(x) \to L$ as $x \to +\infty$, and for a given ϵ let N be the positive number described in Definition 2.3.4. If x is allowed to increase indefinitely, then eventually x will lie in the interval $(N, +\infty)$, which is marked by the green band in Figure 2.3.4a; when this happens, the value of $f(x)$ will fall between $L - \epsilon$ and $L + \epsilon$, marked by the red band in the figure. Since this is true for all positive values of ϵ (no matter how small), we can force the values of $f(x)$ as close as we like to L by making N sufficiently large. This agrees with our informal concept of this limit. Similarly, Figure 2.3.4b illustrates Definition 2.3.5.

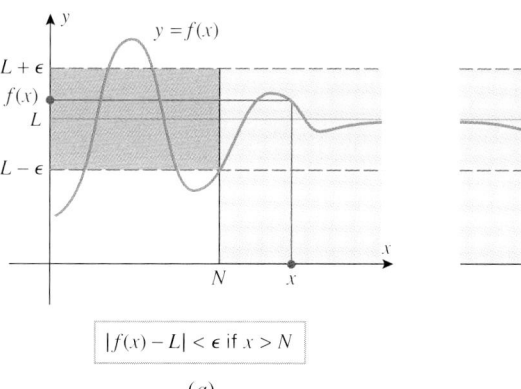

$$|f(x) - L| < \epsilon \text{ if } x > N \qquad\qquad |f(x) - L| < \epsilon \text{ if } x < N$$

$$(a) \qquad\qquad\qquad (b)$$

Figure 2.3.4

Example 4

Prove that $\displaystyle\lim_{x \to +\infty} \frac{1}{x} = 0$.

Solution. Applying Definition 2.3.4 with $f(x) = 1/x$ and $L = 0$, we must show that given $\epsilon > 0$, we must find a number $N > 0$ such that

$$\left| \frac{1}{x} - 0 \right| < \epsilon \quad \text{if} \quad x > N \tag{11}$$

Because $x \to +\infty$ we can assume that $x > 0$. Thus, we can eliminate the absolute values in this statement and rewrite it as

$$\frac{1}{x} < \epsilon \quad \text{if} \quad x > N$$

or, on taking reciprocals,

$$x > \frac{1}{\epsilon} \quad \text{if} \quad x > N \tag{12}$$

It is self-evident that $N = 1/\epsilon$ satisfies this requirement, and since (12) is equivalent to (11) for $x > 0$, the proof is complete. ◄

INFINITE LIMITS

In Section 2.1 we discussed limits of the following type from an intuitive viewpoint:

$$\lim_{x \to a} f(x) = +\infty, \qquad \lim_{x \to a} f(x) = -\infty \tag{13}$$

$$\lim_{x \to a^+} f(x) = +\infty, \qquad \lim_{x \to a^+} f(x) = -\infty \tag{14}$$

$$\lim_{x \to a^-} f(x) = +\infty, \qquad \lim_{x \to a^-} f(x) = -\infty \tag{15}$$

Recall that each of these expressions describes a particular way in which the limit fails to exist. The $+\infty$ indicates that the limit fails to exist because $f(x)$ increases without bound, and the $-\infty$ indicates that the limit fails to exist because $f(x)$ decreases without bound. These ideas are captured more precisely in the following definitions and are illustrated in Figure 2.3.5.

2.3.6 DEFINITION. Let $f(x)$ be defined for all x in some open interval containing a, except that $f(x)$ need not be defined at a. We will write

$$\lim_{x \to a} f(x) = +\infty$$

if given any positive number M, we can find a number $\delta > 0$ such that $f(x)$ satisfies

$$f(x) > M \quad \text{if} \quad 0 < |x - a| < \delta$$

2.3.7 DEFINITION. Let $f(x)$ be defined for all x in some open interval containing a, except that $f(x)$ need not be defined at a. We will write

$$\lim_{x \to a} f(x) = -\infty$$

if given any negative number M, we can find a number $\delta > 0$ such that $f(x)$ satisfies

$$f(x) < M \quad \text{if} \quad 0 < |x - a| < \delta$$

To see how these definitions relate to our informal concepts of these limits, suppose that $f(x) \to +\infty$ as $x \to a$, and for a given M let δ be the corresponding positive number described in Definition 2.3.6. Next, imagine that x gets closer and closer to a (from either side). Eventually, x will lie in the interval $(a - \delta, a + \delta)$, which is marked by the green band in Figure 2.3.5a; when this happens the value of $f(x)$ will be greater than M, marked by the red band in the figure. Since this is true for any positive value of M (no matter how large), we can force the values of $f(x)$ to be as large as we like by making x sufficiently close to a. This agrees with our informal concept of this limit. Similarly, Figure 2.3.5b illustrates Definition 2.3.7.

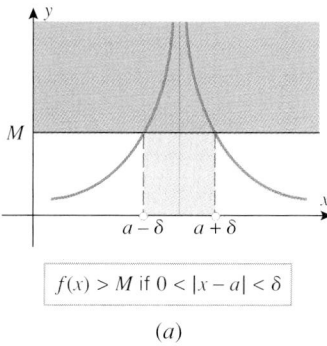

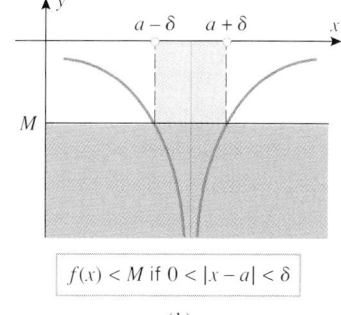

$$\boxed{f(x) > M \text{ if } 0 < |x - a| < \delta}$$

$$\boxed{f(x) < M \text{ if } 0 < |x - a| < \delta}$$

(a)   (b)

Figure 2.3.5

REMARK.   The definitions for the one-sided limits in (14) and (15) are similar, the difference being that the conditions $f(x) > M$ and $f(x) < M$ are only required to hold on the interval $a < x < a + \delta$ for the right-sided limits and on the interval $a - \delta < x < a$ for the left-sided limits.

Example 5

Prove that $\lim\limits_{x \to 0} \dfrac{1}{x^2} = +\infty$.

Solution.   Applying Definition 2.3.6 with $f(x) = 1/x^2$ and $a = 0$, we must show that given a number $M > 0$, we can find a number $\delta > 0$ such that

$$\frac{1}{x^2} > M \quad \text{if} \quad 0 < |x - 0| < \delta \tag{16}$$

or, on taking reciprocals and simplifying,

$$x^2 < \frac{1}{M} \quad \text{if} \quad 0 < |x| < \delta \tag{17}$$

But $x^2 < 1/M$ if $|x| < 1/\sqrt{M}$, so that $\delta = 1/\sqrt{M}$ satisfies (17). Since (16) is equivalent to (17), the proof is complete.   ◀

FOR THE READER.   How would you define

$$\lim_{x \to +\infty} f(x) = +\infty, \qquad \lim_{x \to +\infty} f(x) = -\infty$$

$$\lim_{x \to -\infty} f(x) = +\infty, \qquad \lim_{x \to -\infty} f(x) = -\infty? \tag{18}$$

EXERCISE SET 2.3 ⊠ Graphing Calculator
· ·

1. (a) Find the largest open interval, centered at the origin on the x-axis, such that for each point x in the interval the value of the function $f(x) = x + 2$ is within 0.1 unit of the number $f(0) = 2$.

(b) Find the largest open interval, centered at the point $x = 3$, such that for each point x in the interval the value of the function $f(x) = 4x - 5$ is within 0.01 unit of the number $f(3) = 7$.

(c) Find the largest open interval, centered at the point $x = 4$, such that for each point x in the interval the value of the function $f(x) = x^2$ is within 0.001 unit of the number $f(4) = 16$.

2. In each part, find the largest open interval, centered at the point $x = 0$, such that for each point x in the interval the value of $f(x) = 2x + 3$ is within ϵ units of the number $f(0) = 3$.

(a) $\epsilon = 0.1$ (b) $\epsilon = 0.01$

(c) $\epsilon = 0.0012$

3. (a) Find the values of x_1 and x_2 in the accompanying figure.
   (b) Find a positive number δ such that $|\sqrt{x} - 2| < 0.05$ if $0 < |x - 4| < \delta$.

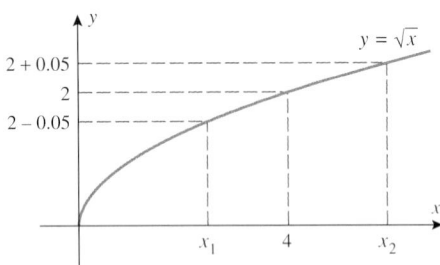

Not drawn to scale

Figure Ex-3

4. (a) Find the values of x_1 and x_2 in the accompanying figure.
   (b) Find a positive number δ such that $|(1/x) - 1| < 0.1$ if $0 < |x - 1| < \delta$.

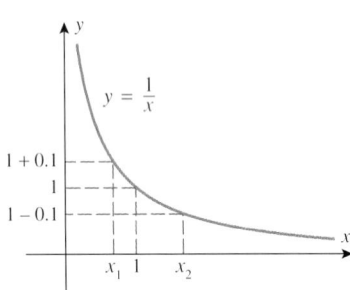

Not drawn to scale

Figure Ex-4

In Exercises 5–14, a positive number ϵ and the limit L of a function f at a point a are given. Find a number δ such that $|f(x) - L| < \epsilon$ if $0 < |x - a| < \delta$.

5. $\lim\limits_{x \to 4} 2x = 8$; $\epsilon = 0.1$

6. $\lim\limits_{x \to -2} \dfrac{1}{2}x = -1$; $\epsilon = 0.1$

7. $\lim\limits_{x \to -1} (7x + 5) = -2$; $\epsilon = 0.01$

8. $\lim\limits_{x \to 3} (5x - 2) = 13$; $\epsilon = 0.01$

9. $\lim\limits_{x \to 2} \dfrac{x^2 - 4}{x - 2} = 4$; $\epsilon = 0.05$

10. $\lim\limits_{x \to -1} \dfrac{x^2 - 1}{x + 1} = -2$; $\epsilon = 0.05$

11. $\lim\limits_{x \to 4} x^2 = 16$; $\epsilon = 0.001$

12. $\lim\limits_{x \to 9} \sqrt{x} = 3$; $\epsilon = 0.001$

13. $\lim\limits_{x \to 5} \dfrac{1}{x} = \dfrac{1}{5}$; $\epsilon = 0.05$

14. $\lim\limits_{x \to 0} |x| = 0$; $\epsilon = 0.05$

In Exercises 15–28, use Definition 2.3.3 to prove that the stated limit is correct.

15. $\lim\limits_{x \to 5} 3x = 15$

16. $\lim\limits_{x \to 3} (4x - 5) = 7$

17. $\lim\limits_{x \to 2} (2x - 7) = -3$

18. $\lim\limits_{x \to -1} (2 - 3x) = 5$

19. $\lim\limits_{x \to 0} \dfrac{x^2 + x}{x} = 1$

20. $\lim\limits_{x \to -3} \dfrac{x^2 - 9}{x + 3} = -6$

21. $\lim\limits_{x \to 1} 2x^2 = 2$

22. $\lim\limits_{x \to 3} (x^2 - 5) = 4$

23. $\lim\limits_{x \to 1/3} \dfrac{1}{x} = 3$

24. $\lim\limits_{x \to -2} \dfrac{1}{x + 1} = -1$

25. $\lim\limits_{x \to 4} \sqrt{x} = 2$

26. $\lim\limits_{x \to 6} \sqrt{x + 3} = 3$

27. $\lim\limits_{x \to 1} f(x) = 3$, where $f(x) = \begin{cases} x + 2, & x \neq 1 \\ 10, & x = 1 \end{cases}$

28. $\lim\limits_{x \to 2} (x^2 + 3x - 1) = 9$

29. (a) Find the smallest positive number N such that for each point x in the interval $(N, +\infty)$, the value of the function $f(x) = 1/x^2$ is within 0.1 unit of $L = 0$.
   (b) Find the smallest positive number N such that for each point x in the interval $(N, +\infty)$, the value of $f(x) = x/(x + 1)$ is within 0.01 unit of $L = 1$.
   (c) Find the largest negative number N such that for each point x in the interval $(-\infty, N)$, the value of the function $f(x) = 1/x^3$ is within 0.001 unit of $L = 0$.
   (d) Find the largest negative number N such that for each point x in the interval $(-\infty, N)$, the value of the function $f(x) = x/(x + 1)$ is within 0.01 unit of $L = 1$.

30. In each part, find the smallest positive value of N such that for each point x in the interval $(N, +\infty)$, the function $f(x) = 1/x^3$ is within ϵ units of the number $L = 0$.
   (a) $\epsilon = 0.1$ (b) $\epsilon = 0.01$ (c) $\epsilon = 0.001$

31. (a) Find the values of x_1 and x_2 in the accompanying figure.
   (b) Find a positive number N such that
$$\left| \dfrac{x^2}{1 + x^2} - 1 \right| < \epsilon$$
   for $x > N$.
   (c) Find a negative number N such that
$$\left| \dfrac{x^2}{1 + x^2} - 1 \right| < \epsilon$$
   for $x < N$.

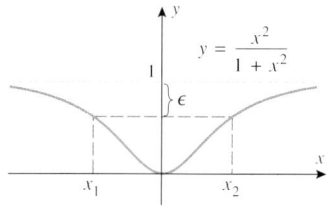

$$y = \frac{x^2}{1 + x^2}$$

Not drawn to scale　　　　Figure Ex-31

32. (a) Find the values of x_1 and x_2 in the accompanying figure.
(b) Find a positive number N such that
$$\left| \frac{1}{\sqrt[3]{x}} - 0 \right| = \left| \frac{1}{\sqrt[3]{x}} \right| < \epsilon$$
for $x > N$.
(c) Find a negative number N such that
$$\left| \frac{1}{\sqrt[3]{x}} - 0 \right| = \left| \frac{1}{\sqrt[3]{x}} \right| < \epsilon$$
for $x < N$.

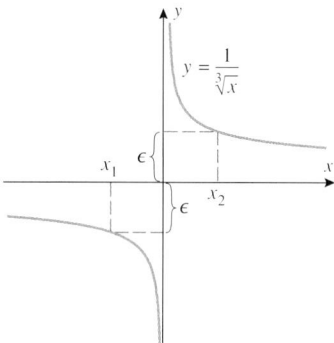

$$y = \frac{1}{\sqrt[3]{x}}$$

Figure Ex-32

In Exercises 33–36, a positive number ϵ and the limit L of a function f at $+\infty$ are given. Find a positive number N such that $|f(x) - L| < \epsilon$ if $x > N$.

33. $\lim\limits_{x \to +\infty} \dfrac{1}{x^2} = 0$; $\epsilon = 0.01$

34. $\lim\limits_{x \to +\infty} \dfrac{1}{x + 2} = 0$; $\epsilon = 0.005$

35. $\lim\limits_{x \to +\infty} \dfrac{x}{x + 1} = 1$; $\epsilon = 0.001$

36. $\lim\limits_{x \to +\infty} \dfrac{4x - 1}{2x + 5} = 2$; $\epsilon = 0.1$

In Exercises 37–40, a positive number ϵ and the limit L of a function f at $-\infty$ are given. Find a negative number N such that $|f(x) - L| < \epsilon$ if $x < N$.

37. $\lim\limits_{x \to -\infty} \dfrac{1}{x + 2} = 0$; $\epsilon = 0.005$

38. $\lim\limits_{x \to -\infty} \dfrac{1}{x^2} = 0$; $\epsilon = 0.01$

39. $\lim\limits_{x \to -\infty} \dfrac{4x - 1}{2x + 5} = 2$; $\epsilon = 0.1$

40. $\lim\limits_{x \to -\infty} \dfrac{x}{x + 1} = 1$; $\epsilon = 0.001$

In Exercises 41–48, use Definitions 2.3.4 and 2.3.5 to prove that the stated limit is correct.

41. $\lim\limits_{x \to +\infty} \dfrac{1}{x^2} = 0$

42. $\lim\limits_{x \to -\infty} \dfrac{1}{x} = 0$

43. $\lim\limits_{x \to -\infty} \dfrac{1}{x + 2} = 0$

44. $\lim\limits_{x \to +\infty} \dfrac{1}{x + 2} = 0$

45. $\lim\limits_{x \to +\infty} \dfrac{x}{x + 1} = 1$

46. $\lim\limits_{x \to -\infty} \dfrac{x}{x + 1} = 1$

47. $\lim\limits_{x \to -\infty} \dfrac{4x - 1}{2x + 5} = 2$

48. $\lim\limits_{x \to +\infty} \dfrac{4x - 1}{2x + 5} = 2$

49. (a) Find the largest open interval, centered at the origin on the x-axis, such that for each point x in the interval, other than the center, the values of $f(x) = 1/x^2$ are greater than 100.
(b) Find the largest open interval, centered at the point $x = 1$, such that for each point x in the interval, other than the center, the values of the function
$$f(x) = 1/|x - 1|$$
are greater than 1000.
(c) Find the largest open interval, centered at the point $x = 3$, such that for each point x in the interval, other than the center, the values of the function
$$f(x) = -1/(x - 3)^2$$
are less than -1000.
(d) Find the largest open interval, centered at the origin on the x-axis, such that for each point x in the interval, other than the center, the values of $f(x) = -1/x^4$ are less than $-10,000$.

50. In each part, find the largest open interval, centered at the point $x = 1$, such that for each point x in the interval the value of $f(x) = 1/(x - 1)^2$ is greater than M.
(a) $M = 10$　　(b) $M = 1000$　　(c) $M = 100,000$

In Exercises 51–56, use Definitions 2.3.6 and 2.3.7 to prove that the stated limit is correct.

51. $\lim\limits_{x \to 3} \dfrac{1}{(x - 3)^2} = +\infty$

52. $\lim\limits_{x \to 3} \dfrac{-1}{(x - 3)^2} = -\infty$

53. $\lim\limits_{x \to 0} \dfrac{1}{|x|} = +\infty$

54. $\lim\limits_{x \to 1} \dfrac{1}{|x - 1|} = +\infty$

55. $\lim\limits_{x \to 0} \left(-\dfrac{1}{x^4} \right) = -\infty$

56. $\lim\limits_{x \to 0} \dfrac{1}{x^4} = +\infty$

In Exercises 57–62, use the remark following Definition 2.3.3 to prove that the stated limit is correct.

57. $\lim\limits_{x \to 2^+} (x + 1) = 3$

58. $\lim\limits_{x \to 1^-} (3x + 2) = 5$

59. $\lim\limits_{x \to 4^+} \sqrt{x - 4} = 0$

60. $\lim\limits_{x \to 0^-} \sqrt{-x} = 0$

61. $\lim\limits_{x \to 2^+} f(x) = 2$, where $f(x) = \begin{cases} x, & x > 2 \\ 3x, & x \le 2 \end{cases}$

62. $\lim\limits_{x \to 2^-} f(x) = 6$, where $f(x) = \begin{cases} x, & x > 2 \\ 3x, & x \le 2 \end{cases}$

> In Exercises 63 and 64, use the remark following Definitions 2.3.6 and 2.3.7 to prove that the stated limit is correct.

63. (a) $\lim\limits_{x \to 1^+} \dfrac{1}{1 - x} = -\infty$ (b) $\lim\limits_{x \to 1^-} \dfrac{1}{1 - x} = +\infty$

64. (a) $\lim\limits_{x \to 0^+} \dfrac{1}{x} = +\infty$ (b) $\lim\limits_{x \to 0^-} \dfrac{1}{x} = -\infty$

> For Exercises 65 and 66, write out definitions of the four limits in (18), and use your definitions to prove that the stated limit is correct.

65. (a) $\lim\limits_{x \to +\infty} (x + 1) = +\infty$ (b) $\lim\limits_{x \to -\infty} (x + 1) = -\infty$

66. (a) $\lim\limits_{x \to +\infty} (x^2 - 3) = +\infty$ (b) $\lim\limits_{x \to -\infty} (x^3 + 5) = -\infty$

67. Prove the result in Example 3 under the assumption that $\delta \le 2$ rather than $\delta \le 1$.

68. (a) In Definition 2.3.3 there is a condition requiring that $f(x)$ be defined for all x in some open interval containing a, except possibly at a itself. What is the purpose of this requirement?

(b) Why is $\lim\limits_{x \to 0} \sqrt{x} = 0$ an incorrect statement?

(c) Is $\lim\limits_{x \to 0.01} \sqrt{x} = 0.1$ a correct statement?

69. Generate the graph of $f(x) = x^3 - 4x + 5$ with a graphing utility, and use the graph to find a number δ such that $|f(x) - 2| < 0.05$ if $0 < |x - 1| < \delta$. [*Hint:* Show that the inequality $|f(x) - 2| < 0.05$ can be rewritten as $1.95 < x^3 - 4x + 5 < 2.05$, and estimate the values of x for which $x^3 - 4x + 5 = 1.95$ and $x^3 - 4x + 5 = 2.05$.]

70. Use the method of Exercise 69 to find a number δ such that $|\sqrt{5x + 1} - 4| < 0.5$ if $0 < |x - 3| < \delta$.

2.4 CONTINUITY

A moving object cannot vanish at some point and reappear someplace else to continue its motion. Thus, we perceive the path of a moving object as a continuous curve, that is, a curve without gaps, breaks, or holes. Earlier, we discussed continuity from an intuitive viewpoint; in this section we will define this concept precisely and develop some fundamental properties of continuous curves.

DEFINITION OF CONTINUITY

Recall from Section 2.1 that the graph of a function f will have a hole or a break in it at a point c if any of the following situations occur:

- The function f is undefined at c (Figure 2.4.1a).
- The limit of $f(x)$ does not exist as x approaches c (Figures 2.4.1b, 2.4.1c).
- The value of the function and the value of the limit at c are different (Figure 2.4.1d).

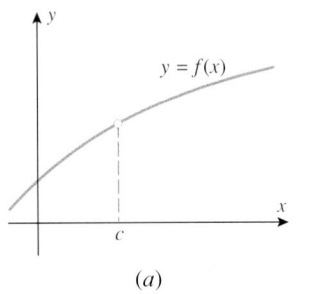
(a)

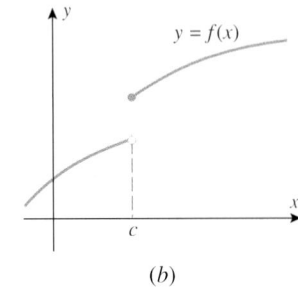
(b)

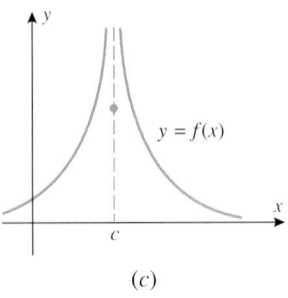
(c)

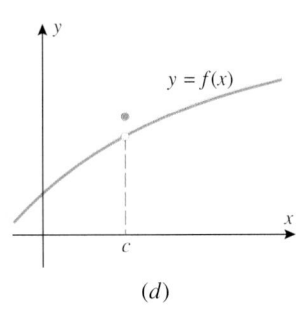
(d)

Figure 2.4.1

This suggests the following definition.

2.4.1 DEFINITION. A function f is said to be ***continuous at a point c*** if the following conditions are satisfied:

1. $f(c)$ is defined.
2. $\lim_{x \to c} f(x)$ exists.
3. $\lim_{x \to c} f(x) = f(c)$.

If one or more of the conditions in this definition fails to hold, then we will say that f has a ***discontinuity*** at the point $x = c$. If f is continuous at each point of an open interval (a, b), then we will say that f is ***continuous on*** $(\boldsymbol{a}, \boldsymbol{b})$. This definition also applies to infinite open intervals of the form $(a, +\infty)$, $(-\infty, b)$, and $(-\infty, +\infty)$. In the case where f is continuous on $(-\infty, +\infty)$, we will say that f is ***continuous everywhere***. If f is continuous on an open interval, but the particular interval is not important for the discussion, we will say that f is ***continuous*** (without referencing the interval).

REMARK. The first two conditions in Definition 2.4.1 are actually superfluous, since it is implicit in the third condition that $f(c)$ is defined and the limit exists (otherwise the equality would make no sense). We have included the first two conditions for emphasis and clarity, but, as a practical matter, you need only confirm that the third condition holds when you want to show that a function f is continuous at a point c.

Example 1

Determine whether the following functions are continuous at the point $x = 2$.

$$f(x) = \frac{x^2 - 4}{x - 2}, \qquad g(x) = \begin{cases} \dfrac{x^2 - 4}{x - 2}, & x \neq 2 \\ 3, & x = 2, \end{cases} \qquad h(x) = \begin{cases} \dfrac{x^2 - 4}{x - 2}, & x \neq 2 \\ 4, & x = 2 \end{cases}$$

Solution. In each case we must determine whether the limit of the function as $x \to 2$ is the same as the value of the function at $x = 2$. In all three cases the functions are identical, except at the point $x = 2$, and hence all three have the same limit at $x = 2$, namely

$$\lim_{x \to 2} f(x) = \lim_{x \to 2} g(x) = \lim_{x \to 2} h(x) = \lim_{x \to 2} \frac{x^2 - 4}{x - 2} = \lim_{x \to 2} (x + 2) = 4$$

The function f is undefined at $x = 2$, and hence is not continuous at that point (Figure 2.4.2a). The function g is defined at $x = 2$, but its value there is $g(2) = 3$, which is not the same as the limit at that point; hence, g is also not continuous at $x = 2$ (Figure 2.4.2b). The value of the function h at $x = 2$ is $h(2) = 4$, which is the same as the limit at that

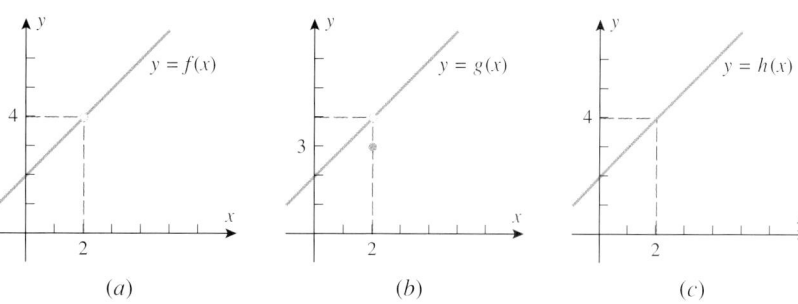

Figure 2.4.2

point; hence, h is continuous at $x = 2$ (Figure 2.4.2c). (Note that the function h could have been written more simply as $h(x) = x + 2$, but we wrote it in piecewise form to emphasize its relationship to f and g.) ◀

CONTINUITY IN APPLICATIONS

In applications, discontinuities often signal the occurrence of important physical phenomena. For example, Figure 2.4.3a is a graph of voltage versus time for an underground cable that is accidentally cut by a work crew at time $t = t_0$ (the voltage drops to zero when the line is cut). Figure 2.4.3b shows the graph of inventory versus time for a company that restocks its warehouse to y_1 units when the inventory falls to y_0 units. The discontinuities occur at those times when restocking occurs.

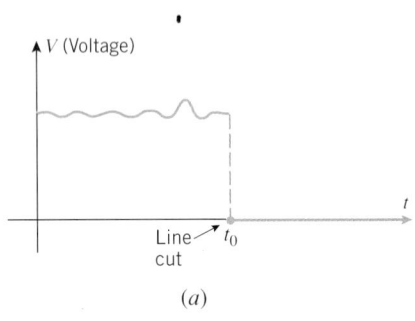

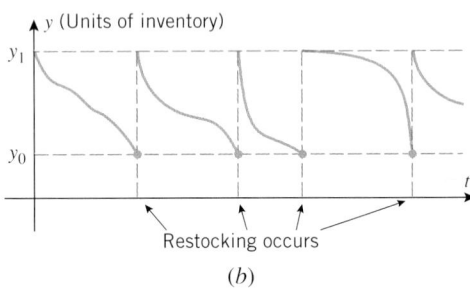

(a)   (b)

Figure 2.4.3

Given the possible physical significance of discontinuities, it is important to be able to identify points of discontinuity for specific functions, and to be able to make general statements about the continuity properties of entire families of functions. This is our next goal.

CONTINUITY OF POLYNOMIALS

The general procedure for showing that a function is continuous everywhere is to show that it is continuous at an *arbitrary* point. For example, we showed in Theorem 2.2.3 that if $p(x)$ is a polynomial and a is any real number, then

$$\lim_{x \to a} p(x) = p(a)$$

Thus, we have the following result.

2.4.2 THEOREM. *Polynomials are continuous everywhere.*

Example 2

Show that $|x|$ is continuous everywhere (Figure 1.2.5).

Solution. We can write $|x|$ as

$$|x| = \begin{cases} x & \text{if } x > 0 \\ 0 & \text{if } x = 0 \\ -x & \text{if } x < 0 \end{cases}$$

so $|x|$ is the same as the polynomial x on the interval $(0, +\infty)$ and is the same as the polynomial $-x$ on the interval $(-\infty, 0)$. But polynomials are continuous functions, so $x = 0$ is the only possible point of discontinuity for $|x|$. At this point we have $|0| = 0$, so to prove the continuity at $x = 0$ we must show that

$$\lim_{x \to 0} |x| = 0 \tag{1}$$

Because the formula for $|x|$ changes at 0, it will be helpful to consider the one-sided limits at 0 rather than the two-sided limit. We obtain

$$\lim_{x \to 0^+} |x| = \lim_{x \to 0^+} x = 0 \quad \text{and} \quad \lim_{x \to 0^-} |x| = \lim_{x \to 0^-} (-x) = 0$$

Thus, (1) holds and $|x|$ is continuous at $x = 0$.   ◀

SOME PROPERTIES OF CONTINUOUS FUNCTIONS

The following theorem, which is a consequence of Theorem 2.2.2, will enable us to reach conclusions about the continuity of functions that are obtained by adding, subtracting, multiplying, and dividing continuous functions.

2.4.3   THEOREM.   *If the functions f and g are continuous at c, then*

(a)   *f + g is continuous at c.*

(b)   *f − g is continuous at c.*

(c)   *fg is continuous at c.*

(d)   *f/g is continuous at c if $g(c) \neq 0$ and has a discontinuity at c if $g(c) = 0$.*

We will prove part (d). The remaining proofs are similar and will be omitted.

Proof.   First, consider the case where $g(c) = 0$. In this case $f(c)/g(c)$ is undefined, so the function f/g has a discontinuity at c.

Next, consider the case where $g(c) \neq 0$. To prove that f/g is continuous at c, we must show that

$$\lim_{x \to c} \frac{f(x)}{g(x)} = \frac{f(c)}{g(c)} \tag{2}$$

Since f and g are continuous at c,

$$\lim_{x \to c} f(x) = f(c) \quad \text{and} \quad \lim_{x \to c} g(x) = g(c)$$

Thus, by Theorem 2.2.2(d)

$$\lim_{x \to c} \frac{f(x)}{g(x)} = \frac{\displaystyle\lim_{x \to c} f(x)}{\displaystyle\lim_{x \to c} g(x)} = \frac{f(c)}{g(c)}$$

which proves (2).   ■

CONTINUITY OF RATIONAL FUNCTIONS

Since polynomials are continuous functions, and since rational functions are ratios of polynomials, part (d) of Theorem 2.4.3 yields the following result.

2.4.4   THEOREM.   *A rational function is continuous everywhere except at the points where the denominator is zero.*

Example 3

For what values of x is there a hole or a gap in the graph of

$$y = \frac{x^2 - 9}{x^2 - 5x + 6}?$$

Solution.   The function being graphed is a rational function, and hence is continuous everywhere except at the points where the denominator is zero. Solving the equation

$$x^2 - 5x + 6 = 0$$

yields two points of discontinuity, $x = 2$ and $x = 3$.   ◀

FOR THE READER.   If you use a graphing utility to generate the graph of the equation in this example, then there is a good chance that you will see the discontinuity at $x = 2$ but not at $x = 3$. Try it, and explain what you think is happening.

CONTINUITY OF COMPOSITIONS

The following theorem, whose proof is given in Appendix G, will be useful for calculating limits of compositions of functions.

2.4.5   THEOREM.   *Let* lim *stand for one of the limits* $\lim_{x \to c}$, $\lim_{x \to c^-}$, $\lim_{x \to c^+}$, $\lim_{x \to +\infty}$, *or* $\lim_{x \to -\infty}$. *If* $\lim g(x) = L$ *and if the function f is continuous at L, then* $\lim f(g(x)) = f(L)$. *That is,* $\lim f(g(x)) = f(\lim g(x))$.

In words, this theorem states:

A limit symbol can be moved through a function sign provided the limit of the expression inside the function sign exists and the function is continuous at this limit.

Example 4

Suppose that $\lim g(x)$ exists, where lim stands for any of the limits in Theorem 2.4.5. We know from Example 2 that the function $|x|$ is continuous everywhere; thus, it follows that

$$\lim |g(x)| = |\lim g(x)| \tag{3}$$

that is, a limit symbol can be moved through an absolute value sign, provided the limit of the expression inside the absolute value signs exists. For example,

$$\lim_{x \to 3} |5 - x^2| = |\lim_{x \to 3} (5 - x^2)| = |-4| = 4 \qquad \blacktriangleleft$$

The following theorem is concerned with the continuity of compositions of functions; the first part deals with continuity at a specific point, and the second part with continuity everywhere.

2.4.6   THEOREM.

(a)   *If the function g is continuous at the point c, and the function f is continuous at the point $g(c)$, then the composition $f \circ g$ is continuous at c.*

(b)   *If the function g is continuous everywhere and the function f is continuous everywhere, then the composition $f \circ g$ is continuous everywhere.*

Proof.   We will prove part (a) only; the proof of part (b) can be obtained by applying part (a) at an arbitrary point c. To prove that $f \circ g$ is continuous at c, we must show that the value of $f \circ g$ and the value of its limit are the same at $x = c$. But this is so, since we can write

$$\lim_{x \to c} (f \circ g)(x) = \lim_{x \to c} f(g(x)) = f(\lim_{x \to c} g(x)) = f(g(c)) = (f \circ g)(c)$$

| Theorem 2.4.5 | g is continuous at c. |

We know from Example 2 that the function $|x|$ is continuous everywhere. Thus, if $g(x)$ is continuous at the point c, then by part (a) of Theorem 2.4.6, the function $|g(x)|$ must also be continuous at the point c; and, more generally, if $g(x)$ is continuous everywhere, then so is $|g(x)|$. Stated informally:

The absolute value of a continuous function is continuous.

For example, the polynomial $g(x) = 4 - x^2$ is continuous everywhere, so we can conclude that the function $|4 - x^2|$ is also continuous everywhere (Figure 2.4.4).

FOR THE READER.   Can the absolute value of a function that is not continuous be continuous? Justify your answer.

CONTINUITY FROM THE LEFT AND RIGHT

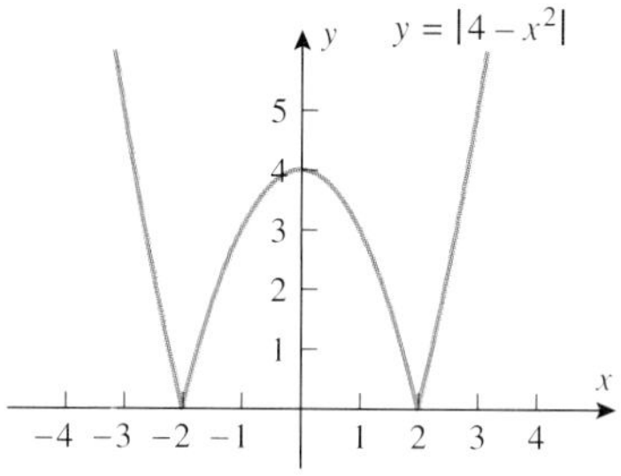

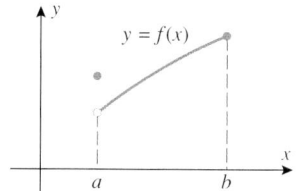

$y = |4 - x^2|$

Figure 2.4.4

Because Definition 2.4.1 involves a two-sided limit, that definition does not generally apply at the endpoints of a closed interval $[a, b]$ or at the endpoint of an interval of the form $[a, b)$, $(a, b]$, $(-\infty, b]$, or $[a, +\infty)$. To remedy this problem, we will agree that a function is continuous at an endpoint of an interval if its value at the endpoint is equal to the appropriate one-sided limit at that point. For example, the function graphed in Figure 2.4.5 is continuous at the right endpoint of the interval $[a, b]$ because

$$\lim_{x \to b^-} f(x) = f(b)$$

but it is not continuous at the left endpoint because

$$\lim_{x \to a^+} f(x) \neq f(a)$$

In general, we will say a function f is ***continuous from the left*** at a point c if

$$\lim_{x \to c^-} f(x) = f(c)$$

and is ***continuous from the right*** at a point c if

$$\lim_{x \to c^+} f(x) = f(c)$$

Using this terminology we define continuity on a closed interval as follows.

$y = f(x)$

Figure 2.4.5

2.4.7 DEFINITION.   A function f is said to be ***continuous on a closed interval*** $[a, b]$ if the following conditions are satisfied:

1. f is continuous on (a, b).
2. f is continuous from the right at a.
3. f is continuous from the left at b.

FOR THE READER.   We leave it for you to modify this definition appropriately so that it applies to intervals of the form $[a, +\infty)$, $(-\infty, b]$, $(a, b]$, and $[a, b)$.

Example 5

What can you say about the continuity of the function $f(x) = \sqrt{9 - x^2}$?

Solution. Because the natural domain of this function is the closed interval $[-3, 3]$, we will need to investigate the continuity of f on the open interval $(-3, 3)$ and at the two endpoints. If c is any point in the interval $(-3, 3)$, then it follows from Theorem 2.2.2(e) that

$$\lim_{x \to c} f(x) = \lim_{x \to c} \sqrt{9 - x^2} = \sqrt{\lim_{x \to c}(9 - x^2)} = \sqrt{9 - c^2} = f(c)$$

which proves f is continuous at each point of the interval $(-3, 3)$. The function f is also continuous at the endpoints since

$$\lim_{x \to 3^-} f(x) = \lim_{x \to 3^-} \sqrt{9 - x^2} = \sqrt{\lim_{x \to 3^-}(9 - x^2)} = 0 = f(3)$$

$$\lim_{x \to -3^+} f(x) = \lim_{x \to -3^+} \sqrt{9 - x^2} = \sqrt{\lim_{x \to -3^+}(9 - x^2)} = 0 = f(-3)$$

Thus, f is continuous on the closed interval $[-3, 3]$. ◀

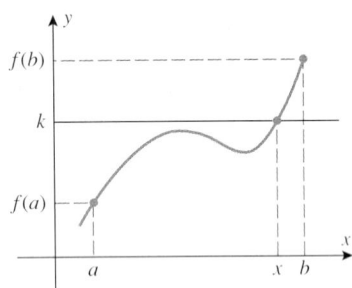

Figure 2.4.6

Figure 2.4.6 shows the graph of a function that is continuous on the closed interval $[a, b]$. The figure suggests that if we draw any horizontal line $y = k$, where k is between $f(a)$ and $f(b)$, then that line will cross the curve $y = f(x)$ line at least once over the $[a, b]$. Stated in numerical terms, if f is continuous on $[a, b]$, then the function f must take on every value k between $f(a)$ and $f(b)$ at least once as x varies from a to b. For example, the polynomial $p(x) = x^5 - x + 3$ has a value of 3 at $x = 1$ and a value of 33 at $x = 2$. Thus, it follows from the continuity of p that the equation $x^5 - x + 3 = k$ has at least one solution in the interval $[1, 2]$ for every value of k between 3 and 33. This idea is stated more precisely in the following theorem.

2.4.8   THEOREM (*Intermediate-Value Theorem*).   *If f is continuous on a closed interval $[a, b]$ and k is any number between $f(a)$ and $f(b)$, inclusive, then there is at least one number x in the interval $[a, b]$ such that $f(x) = k$.*

Although this theorem is intuitively obvious, its proof depends on a mathematically precise development of the real number system, which is beyond the scope of this text.

A variety of problems can be reduced to solving an equation $f(x) = 0$ for its roots. Sometimes it is possible to solve for the roots exactly using algebra, but often this is not possible and one must settle for decimal approximations of the roots. One procedure for approximating roots is based on the following consequence of the Intermediate-Value Theorem.

2.4.9   THEOREM.   *If f is continuous on $[a, b]$, and if $f(a)$ and $f(b)$ are nonzero and have opposite signs, then there is at least one solution of the equation $f(x) = 0$ in the interval (a, b).*

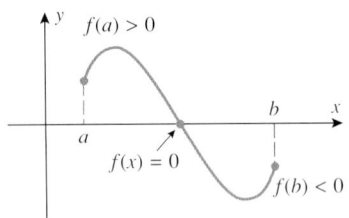

Figure 2.4.7

This result, which is illustrated in Figure 2.4.7, can be proved as follows.

Proof. Since $f(a)$ and $f(b)$ have opposite signs, 0 is between $f(a)$ and $f(b)$. Thus, by the Intermediate-Value Theorem there is at least one number x in the interval $[a, b]$ such that $f(x) = 0$. However, $f(a)$ and $f(b)$ are nonzero, so x must lie in the interval (a, b), which completes the proof. ■

Before we illustrate how this theorem can be used to approximate roots, it will be helpful to discuss some standard terminology for describing errors in approximations. If x is an approximation to a quantity x_0, then we call

$$\epsilon = |x - x_0|$$

the ***absolute error*** or (less precisely) the ***error*** in the approximation. The following terminology is used to describe the size of such errors:

Table 2.4.1

ERROR	DESCRIPTION
$\|x - x_0\| \leq 0.1$	x approximates x_0 with an error of at most 0.1.
$\|x - x_0\| \leq 0.01$	x approximates x_0 with an error of at most 0.01.
$\|x - x_0\| \leq 0.001$	x approximates x_0 with an error of at most 0.001.
$\|x - x_0\| \leq 0.0001$	x approximates x_0 with an error of at most 0.0001.
$\|x - x_0\| \leq 0.5$	x approximates x_0 to the nearest integer.
$\|x - x_0\| \leq 0.05$	x approximates x_0 to 1 decimal place (i.e., to the nearest tenth).
$\|x - x_0\| \leq 0.005$	x approximates x_0 to 2 decimal places (i.e., to the nearest hundredth).
$\|x - x_0\| \leq 0.0005$	x approximates x_0 to 3 decimal places (i.e., to the nearest thousandth).

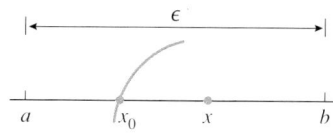

Every number x in the interval $[a, b]$ differs from x_0 by at most ϵ, and the midpoint of the interval differs from x_0 by at most $\epsilon/2$.

Figure 2.4.8

We will also need the following result, which should be evident geometrically from Figure 2.4.8.

2.4.10   APPROXIMATION PRINCIPLE.   Suppose that the equation $f(x) = 0$ has a root x_0 in the interval $[a, b]$ and that this interval has length $\epsilon = b - a$. Then any number x in the interval $[a, b]$ approximates x_0 with an error of at most ϵ, and the midpoint of the interval approximates x_0 with an error of at most $\epsilon/2$.

Example 6

The equation

$$x^3 - x - 1 = 0$$

cannot be solved algebraically very easily because the left side has no simple factors. However, if we graph $p(x) = x^3 - x - 1$ with a graphing utility (Figure 2.4.9), then we are led to conjecture that there is one real root and that this root lies inside the interval $[1, 2]$. The existence of a root in this interval is also confirmed by Theorem 2.4.8, since $p(1) = -1$ and $p(2) = 5$ have opposite signs. Approximate this root to two decimal-place accuracy.

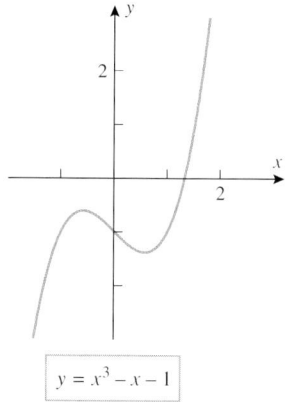

$y = x^3 - x - 1$

Figure 2.4.9

Solution.   Our objective is to approximate the unknown root x_0 with an error of at most 0.005. It follows from the Approximation Principle (2.4.10) that if we can find an interval of length 0.01 that contains the root, then the midpoint of that interval will approximate the root with an error of at most $0.01/2 = 0.005$, which will achieve the desired accuracy.

We know that the root x_0 lies in the interval $[1, 2]$. However, this interval has length 1, which is too large. We can pinpoint the location of the root more precisely by dividing the interval $[1, 2]$ into 10 equal parts and evaluating p at the points of subdivision using a calculating utility (Table 2.4.2). In this table $p(1.3)$ and $p(1.4)$ have opposite signs, so we know that the root lies in the interval $[1.3, 1.4]$. This interval has length 0.1, which is still too large, so we repeat the process by dividing the interval $[1.3, 1.4]$ into 10 parts and evaluating p at the points of subdivision; this yields Table 2.4.3, which tells us that the root is inside the interval $[1.32, 1.33]$. Since this interval has length 0.01, its midpoint 1.325 will approximate the root with an error of at most 0.005. Thus, $x_0 \approx 1.325$ to two decimal-place accuracy.   ◀

Table 2.4.2

x	1	1.1	1.2	1.3	1.4	1.5	1.6	1.7	1.8	1.9	2
$f(x)$	−1	−0.77	−0.47	−0.10	0.34	0.88	1.50	2.21	3.03	3.96	5

Table 2.4.3

x	1.3	1.31	1.32	1.33	1.34	1.35	1.36	1.37	1.38	1.39	1.4
$f(x)$	−0.103	−0.062	−0.020	0.023	0.066	0.110	0.155	0.201	0.248	0.296	0.344

APPROXIMATING ROOTS BY ZOOMING WITH A GRAPHING UTILITY

The method illustrated in Example 6 can also be implemented with a graphing utility as follows.

Step 1.   Figure 2.4.10*a* shows the graph of f in the window $[-5, 5] \times [-5, 5]$ with $x\mathrm{Scl} = 1$ and $y\mathrm{Scl} = 1$. That graph places the root between $x = 1$ and $x = 2$.

Step 2.   Since we know that the root lies between $x = 1$ and $x = 2$, we will zoom in by regraphing f over an x-interval that extends between these points and in which $x\text{Scl} = .1$. The y-interval and $y\text{Scl}$ are not critical, as long as the y-interval extends above and below the x-axis. Figure 2.4.10b shows the graph of f in the window $[1, 2] \times [-1, 1]$ with $x\text{Scl} = .1$ and $y\text{Scl} = .1$. That graph places the root between $x = 1.3$ and $x = 1.4$.

Step 4.   Since we know that the root lies between $x = 1.3$ and $x = 1.4$, we will zoom in again by regraphing f over an x-interval that extends between these points and in which $x\text{Scl} = .01$. Figure 2.4.10c shows the graph of f in the window $[1.3, 1.4] \times [-.1, .1]$ with $x\text{Scl} = .01$ and $y\text{Scl} = .01$. That graph places the root between $x = 1.32$ and $x = 1.33$.

Step 5.   Since the interval in Step 3 has length $.01$, its midpoint 1.325 approximates the root with an error of at most 0.005, so $x_0 \approx 1.325$ to two decimal-place accuracy.

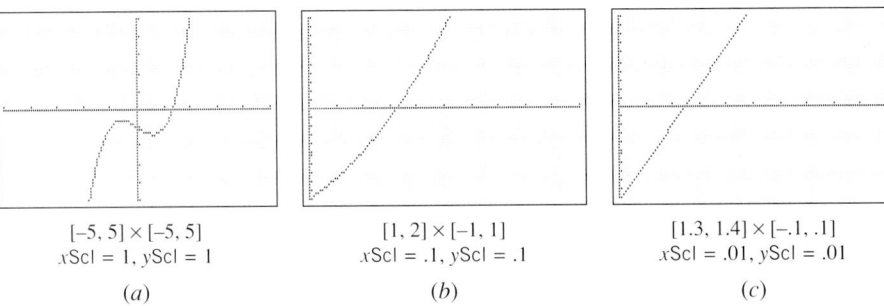

$[-5, 5] \times [-5, 5]$
$x\text{Scl} = 1, y\text{Scl} = 1$

(a)

$[1, 2] \times [-1, 1]$
$x\text{Scl} = .1, y\text{Scl} = .1$

(b)

$[1.3, 1.4] \times [-.1, .1]$
$x\text{Scl} = .01, y\text{Scl} = .01$

(c)

Figure 2.4.10

REMARK.   To say that x approximates x_0 to n decimal places does *not* mean that the first n decimal places of x and x_0 will be the same when the numbers are rounded to n decimal places. For example, $x = 1.084$ approximates $x_0 = 1.087$ to two decimal places because $|x - x_0| = 0.003 (<0.005)$. However, if we round these values to two decimal places, then we obtain $x \approx 1.08$ and $x_0 \approx 1.09$. Thus, if you approximate a number to n decimal places, then you should display that approximation to at least $n + 1$ decimal places to preserve the accuracy.

FOR THE READER.   Use a graphing or calculating utility to show that the root x_0 in Example 6 can be approximated as $x_0 \approx 1.3245$ to three decimal-place accuracy.

EXERCISE SET 2.4   ~ Graphing Calculator

In Exercises 1–4, let f be the function whose graph is shown. On which of the following intervals, if any, is f continuous?
(a) $[1, 3]$ (b) $(1, 3)$ (c) $[1, 2]$
(d) $(1, 2)$ (e) $[2, 3]$ (f) $(2, 3)$
On those intervals where f is not continuous, state where the discontinuities occur.

1.
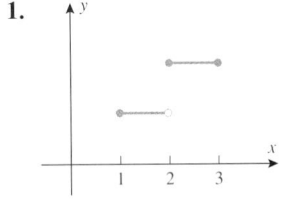

2.
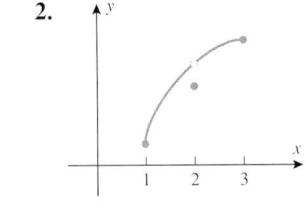

3.
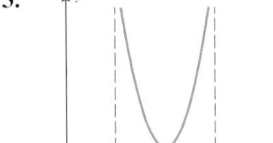

4.
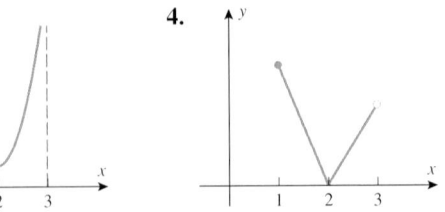

5. Suppose that f and g are continuous functions such that $f(2) = 1$ and $\lim_{x \to 2} [f(x) + 4g(x)] = 13$. Find
(a) $g(2)$ (b) $\lim_{x \to 2} g(x)$.

6. Suppose that f and g are continuous functions such that $\lim_{x \to 3} g(x) = 5$ and $f(3) = -2$. Find $\lim_{x \to 3} [f(x)/g(x)]$.

7. In each part sketch the graph of a function f that satisfies the stated conditions.
  (a) f is continuous everywhere except at $x = 3$, at which point it is continuous from the right.
  (b) f has a two-sided limit at $x = 3$, but it is not continuous at that point.
  (c) f is not continuous at $x = 3$, but if its value at $x = 3$ is changed from $f(3) = 1$ to $f(3) = 0$, it becomes continuous at $x = 3$.
  (d) f is continuous on the interval $[0, 3)$ and is defined on the closed interval $[0, 3]$; but f is not continuous on the interval $[0, 3]$.

8. Find formulas for some functions that are continuous on the intervals $(-\infty, 0)$ and $(0, +\infty)$, but are not continuous on the interval $(-\infty, +\infty)$.

9. A student parking lot at a university charges \$2.00 for the first half hour (or any part) and \$1.00 for each subsequent half hour (or any part) up to a daily maximum of \$10.00.
  (a) Sketch a graph of cost as a function of the time parked.
  (b) Discuss the significance of the discontinuities in the graph to a student who parks there.

10. In each part determine whether the function is continuous or not, and explain your reasoning.
  (a) The Earth's population as a function of time
  (b) Your exact height as a function of time
  (c) The cost of a taxi ride in your city as a function of the distance traveled
  (d) The volume of a melting ice cube as a function of time

In Exercises 11–22, find the points of discontinuity, if any.

11. $f(x) = x^3 - 2x + 3$ **12.** $f(x) = (x - 5)^{17}$

13. $f(x) = \dfrac{x}{x^2 + 1}$ **14.** $f(x) = \dfrac{x}{x^2 - 1}$

15. $f(x) = \dfrac{x - 4}{x^2 - 16}$ **16.** $f(x) = \dfrac{3x + 1}{x^2 + 7x - 2}$

17. $f(x) = \dfrac{x}{|x| - 3}$ **18.** $f(x) = \dfrac{5}{x} + \dfrac{2x}{x + 4}$

19. $f(x) = |x^3 - 2x^2|$ **20.** $f(x) = \dfrac{x + 3}{|x^2 + 3x|}$

21. $f(x) = \begin{cases} 2x + 3, & x \le 4 \\ 7 + \dfrac{16}{x}, & x > 4 \end{cases}$

22. $f(x) = \begin{cases} \dfrac{3}{x - 1}, & x \ne 1 \\ 3, & x = 1 \end{cases}$

23. Find a value for the constant k, if possible, that will make the function continuous.
  (a) $f(x) = \begin{cases} 7x - 2, & x \le 1 \\ kx^2, & x > 1 \end{cases}$
  (b) $f(x) = \begin{cases} kx^2, & x \le 2 \\ 2x + k, & x > 2 \end{cases}$

24. On which of the following intervals is
$$f(x) = \dfrac{1}{\sqrt{x - 2}}$$
continuous?
  (a) $[2, +\infty)$ (b) $(-\infty, +\infty)$
  (c) $(2, +\infty)$ (d) $[1, 2)$

A function f is said to have a ***removable discontinuity*** at $x = c$ if $\lim_{x \to c} f(x)$ exists, but
$$f(c) \ne \lim_{x \to c} f(x)$$
either because $f(c)$ is undefined or the value of $f(c)$ differs from the value of the limit. This terminology will be needed in Exercises 25–28.

25. (a) Sketch the graph of a function with a removable discontinuity at $x = c$ for which $f(c)$ is undefined.
  (b) Sketch the graph of a function with a removable discontinuity at $x = c$ for which $f(c)$ is defined.

26. (a) The terminology *removable discontinuity* is appropriate because a removable discontinuity of a function f at a point $x = c$ can be "removed" by redefining the value of f appropriately at $x = c$. What value for $f(c)$ removes the discontinuity?
  (b) Show that the following functions have removable discontinuities at $x = 1$, and sketch their graphs.
$$f(x) = \dfrac{x^2 - 1}{x - 1} \quad \text{and} \quad g(x) = \begin{cases} 1, & x > 1 \\ 0, & x = 1 \\ 1, & x < 1 \end{cases}$$
  (c) What values should be assigned to $f(1)$ and $g(1)$ to remove the discontinuities?

In Exercises 27 and 28, find the points of discontinuity, and determine whether the discontinuities are removable.

27. (a) $f(x) = \dfrac{|x|}{x}$ (b) $f(x) = \dfrac{x^2 + 3x}{x + 3}$
  (c) $f(x) = \dfrac{x - 2}{|x| - 2}$

28. (a) $f(x) = \dfrac{x^2 - 4}{x^3 - 8}$
  (b) $f(x) = \begin{cases} 2x - 3, & x \le 2 \\ x^2, & x > 2 \end{cases}$
  (c) $f(x) = \begin{cases} 3x^2 + 5, & x \ne 1 \\ 6, & x = 1 \end{cases}$

29. (a) Use a graphing utility to generate the graph of the function $f(x) = (x + 3)/(2x^2 + 5x - 3)$, and then use the graph to make a conjecture about the number and location of all discontinuities.
  (b) Check your conjecture by factoring the denominator.

30. (a) Use a graphing utility to generate the graph of the function $f(x) = x/(x^3 - x + 2)$, and then use the graph to make a conjecture about the number and location of all discontinuities.

(b) Use the Intermediate-Value Theorem to approximate the location of all points of discontinuity to two decimal places.

31. Prove that $f(x) = x^{3/5}$ is continuous everywhere, carefully justifying each step.

32. Prove that $f(x) = 1/\sqrt{x^4 + 7x^2 + 1}$ is continuous everywhere, carefully justifying each step.

33. Let f and g be discontinuous at c. Give examples to show that
(a) $f + g$ can be continuous or discontinuous at c
(b) fg can be continuous or discontinuous at c.

34. Prove Theorem 2.4.4.

35. Prove:
(a) part (a) of Theorem 2.4.3
(b) part (b) of Theorem 2.4.3
(c) part (c) of Theorem 2.4.3

36. Prove: If f and g are continuous on $[a, b]$, and $f(a) > g(a)$, $f(b) < g(b)$, then there is at least one solution of the equation $f(x) = g(x)$ in (a, b). [*Hint:* Consider $f(x) - g(x)$.]

37. Give an example of a function f that is defined at every point in a closed interval, and whose values at the endpoints have opposite signs, but for which the equation $f(x) = 0$ has no solution in the interval.

38. Use the Intermediate-Value Theorem to show that there is a square with a diagonal length that is between r and $2r$ and an area that is half the area of a circle of radius r.

39. Use the Intermediate-Value Theorem to show that there is a right circular cylinder of height h and radius less than r whose volume is equal to that of a right circular cone of height h and radius r.

In Exercises 40 and 41, show that the equation has at least one solution in the given interval.

40. $x^3 - 4x + 1 = 0$; $[1, 2]$

41. $x^3 + x^2 - 2x = 1$; $[-1, 1]$

42. Prove: If $p(x)$ is a polynomial of odd degree, then the equation $p(x) = 0$ has at least one real solution.

43. The accompanying figure shows the graph of $y = x^4 + x - 1$. Use the method of Example 6 to approximate the x-intercepts with an error of at most 0.05.

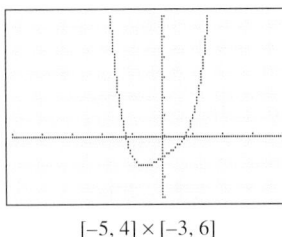

$[-5, 4] \times [-3, 6]$
$x\text{Scl} = 1, y\text{Scl} = 1$

Figure Ex-43

44. Use a graphing utility to solve the problem in Exercise 43 by zooming.

45. The accompanying figure shows the graph of $y = 5 - x - x^4$. Use the method of Example 6 to approximate the roots of the equation $5 - x - x^4 = 0$ to two decimal-place accuracy.

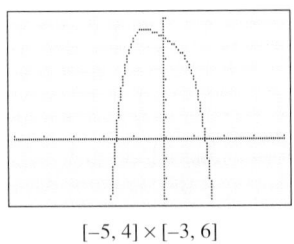

$[-5, 4] \times [-3, 6]$
$x\text{Scl} = 1, y\text{Scl} = 1$

Figure Ex-45

46. Use a graphing utility to solve the problem in Exercise 45 by zooming.

47. Use the fact that $\sqrt{5}$ is a solution of $x^2 - 5 = 0$ to approximate $\sqrt{5}$ with an error of at most 0.005.

48. Prove that if a and b are positive, then the equation
$$\frac{a}{x - 1} + \frac{b}{x - 3} = 0$$
has at least one solution in the interval $(1, 3)$.

49. A sphere of unknown radius x consists of a spherical core and a coating that is 1 cm thick (see the accompanying figure). Given that the volume of the coating and the volume of the core are the same, approximate the radius of the sphere to three decimal-place accuracy.

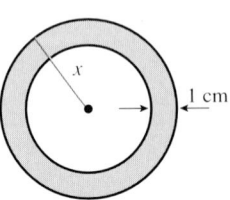

Figure Ex-49

50. A monk begins walking up a mountain road at 12:00 noon and reaches the top at 12:00 midnight. He meditates and rests until 12:00 noon the next day, at which time he begins walking down the same road, reaching the bottom at 12:00 midnight. Show that there is at least one point on the road that he reaches at the same time of day on the way up as on the way down.

51. Let f be defined at c. Prove that f is continuous at c if, given $\epsilon > 0$, there exists a $\delta > 0$ such that $|f(x) - f(c)| < \epsilon$ if $|x - c| < \delta$.

2.5 LIMITS AND CONTINUITY OF TRIGONOMETRIC FUNCTIONS

In this section we will investigate the continuity properties of the trigonometric functions, and we will discuss some important limits involving these functions.

CONTINUITY OF TRIGONOMETRIC FUNCTIONS

Before we begin, recall that in the expressions $\sin x$, $\cos x$, $\tan x$, $\cot x$, $\sec x$, and $\csc x$ it is understood that x is in radian measure.

In trigonometry, the graphs of $\sin x$ and $\cos x$ are drawn as continuous curves (Figure 2.5.1). To actually prove that these functions are continuous everywhere, we must show that the following equalities hold for every real number c:

$$\lim_{x \to c} \sin x = \sin c \quad \text{and} \quad \lim_{x \to c} \cos x = \cos c \tag{1-2}$$

Although we will not formally prove these results, we can make them plausible by considering the behavior of the point $P(\cos x, \sin x)$ as it moves around the unit circle. For this purpose, view c as a fixed angle in radian measure, and let $Q(\cos c, \sin c)$ be the corresponding point on the unit circle. As $x \to c$ (i.e., as the angle x approaches the angle c), the point P moves along the circle toward Q, and this implies that the coordinates of P approach the corresponding coordinates of Q; that is, $\cos x \to \cos c$, and $\sin x \to \sin c$ (Figure 2.5.2).

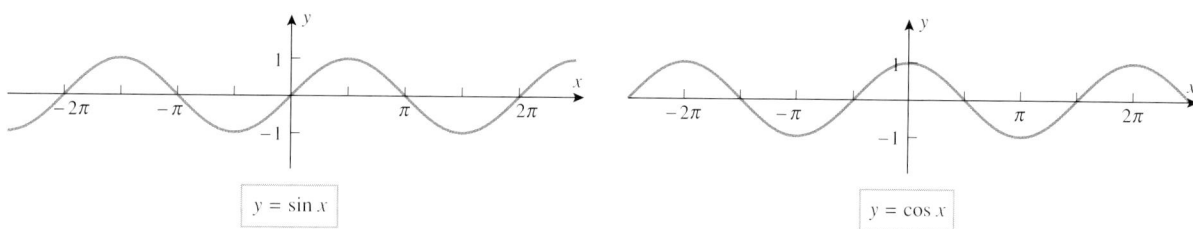

$$y = \sin x \qquad\qquad y = \cos x$$

Figure 2.5.1

Formulas (1) and (2) can be used to find limits of the remaining trigonometric functions by expressing them in terms of $\sin x$ and $\cos x$; for example, if $\cos c \neq 0$, then

$$\lim_{x \to c} \tan x = \lim_{x \to c} \frac{\sin x}{\cos x} = \frac{\sin c}{\cos c} = \tan c$$

Thus, we are led to the following theorem.

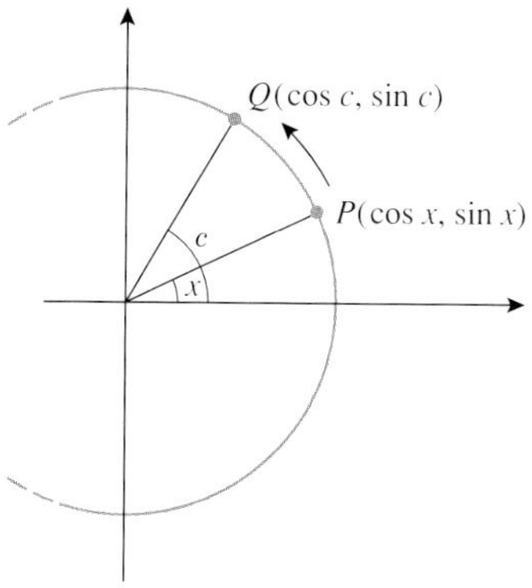

2.5.1 THEOREM. *If c is any number in the natural domain of the stated trigonometric function, then*

$$\lim_{x \to c} \sin x = \sin c \qquad \lim_{x \to c} \cos x = \cos c \qquad \lim_{x \to c} \tan x = \tan c$$

$$\lim_{x \to c} \csc x = \csc c \qquad \lim_{x \to c} \sec x = \sec c \qquad \lim_{x \to c} \cot x = \cot c$$

It follows from this theorem, for example, that $\sin x$ and $\cos x$ are continuous everywhere and that $\tan x$ is continuous, except at the points where it is undefined.

Q(cos c, sin c)

P(cos x, sin x)

c

x

Figure 2.5.2

Example 1

Find the limit

$$\lim_{x \to 1} \cos\left(\frac{x^2 - 1}{x - 1}\right)$$

Solution. Recall from the last section that a limit symbol can be moved through a function sign if the function is continuous and the limit of the expression inside the function sign exists. Thus,

$$\lim_{x \to 1} \cos\left(\frac{x^2 - 1}{x - 1}\right) = \lim_{x \to 1} \cos(x + 1) = \cos(\lim_{x \to 1} (x + 1)) = \cos 2 \qquad \blacktriangleleft$$

OBTAINING LIMITS BY SQUEEZING

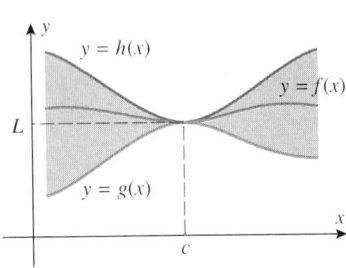

Figure 2.5.3

In Section 2.1 we used the numerical evidence in Table 2.1.2 to *conjecture* that

$$\lim_{x \to 0} \frac{\sin x}{x} = 1 \qquad (3)$$

However, it is not a simple matter to establish this limit with certainty. The difficulty is that the numerator and denominator both approach zero as $x \to 0$; such limits are called *indeterminate forms of type **0/0***. Sometimes indeterminate forms of this type can be established by manipulating the ratio algebraically (as in Example 7 of Section 2.2); but in this case no simple algebraic manipulation will work, so we must look for other methods.

The problem with indeterminate forms of type 0/0 is that there are two conflicting influences at work: as the numerator approaches 0 it drives the magnitude of the ratio toward 0, and as the denominator approaches 0 it drives the magnitude of the ratio toward $\pm\infty$ (depending on the sign of the expression). The limiting behavior of the ratio is determined by the precise way in which these influences offset each other. Later in this text we will discuss general methods for attacking indeterminate forms, but for the limit in (3) we can use a method called *squeezing*.

In the method of squeezing one proves that a function f has a limit L at a point c by trapping the function between two other functions, g and h, whose limits at c are known to be L (Figure 2.5.3). This is the idea behind the following theorem, which we state without proof.

2.5.2 THEOREM (*The Squeezing Theorem*). *Let f, g, and h be functions satisfying*

$$g(x) \le f(x) \le h(x)$$

for all x in some open interval containing the point c, with the possible exception that the inequalities need not hold at c. If g and h have the same limit as x approaches c, say

$$\lim_{x \to c} g(x) = \lim_{x \to c} h(x) = L$$

then f also has this limit as x approaches c, that is,

$$\lim_{x \to c} f(x) = L$$

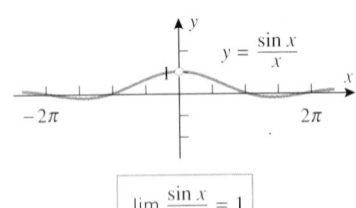

FOR THE READER. The Squeezing Theorem also holds for one-sided limits and limits at $+\infty$ and $-\infty$. How do you think the hypotheses of the theorem would change in those cases?

The usefulness of the Squeezing Theorem will be evident in our proof of the following theorem (Figure 2.5.4).

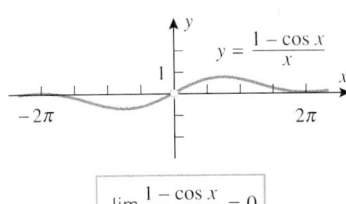

Figure 2.5.4

2.5.3 THEOREM.

(a) $\displaystyle\lim_{x \to 0} \frac{\sin x}{x} = 1$ $\qquad(b)$ $\displaystyle\lim_{x \to 0} \frac{1 - \cos x}{x} = 0$

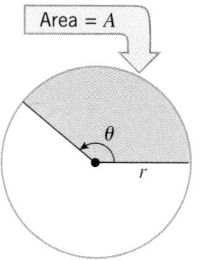

Figure 2.5.5

However, before giving the proof, it will be helpful to review the formula for the area A of a sector with radius r and a central angle of θ radians (Figure 2.5.5). The area of the sector can be derived by setting up the following proportion to the area of the entire circle:

$$\frac{A}{\pi r^2} = \frac{\theta}{2\pi} \qquad \left[\frac{\text{area of the sector}}{\text{area of the circle}} = \frac{\text{central angle of the sector}}{\text{central angle of the circle}} \right]$$

From this we obtain the formula

$$A = \tfrac{1}{2}r^2\theta \tag{4}$$

Now we are ready for the proof of Theorem 2.5.3.

Proof (a). In this proof we will interpret x as an angle in radian measure, and we will assume to start that $0 < x < \pi/2$. It follows from Formula (4) that the area of a sector of radius 1 and central angle x is $x/2$. Moreover, it is suggested by Figure 2.5.6 that the area of this sector lies between the areas of two triangles, one with area $(\tan x)/2$ and one with area $(\sin x)/2$. Thus,

$$\frac{\tan x}{2} \geq \frac{x}{2} \geq \frac{\sin x}{2}$$

Multiplying through by $2/(\sin x)$ yields

$$\frac{1}{\cos x} \geq \frac{x}{\sin x} \geq 1$$

and then taking reciprocals and reversing the inequalities yields

$$\cos x \leq \frac{\sin x}{x} \leq 1 \tag{5}$$

Moreover, these inequalities also hold for $-\pi/2 < x < 0$, since replacing x by $-x$ in (5) and using the identities $\sin(-x) = -\sin x$ and $\cos(-x) = \cos x$ leaves the inequalities unchanged (verify). Finally, since the functions $\cos x$ and 1 both have limits of 1 as $x \to 0$, it follows from the Squeezing Theorem that $(\sin x)/x$ also has a limit of 1 as $x \to 0$.

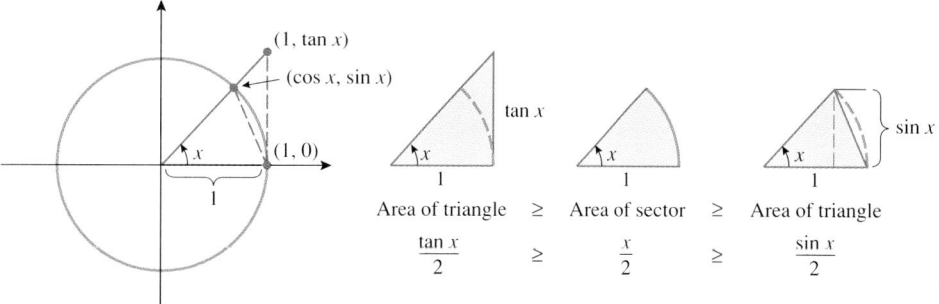

Figure 2.5.6

Proof (b). For this proof we will use the limit in part (a), the continuity of the sine function, and the trigonometric identity $\sin^2 x = 1 - \cos^2 x$. We obtain

$$\lim_{x \to 0} \frac{1 - \cos x}{x} = \lim_{x \to 0} \left[\frac{1 - \cos x}{x} \cdot \frac{1 + \cos x}{1 + \cos x} \right] = \lim_{x \to 0} \frac{\sin^2 x}{(1 + \cos x)x}$$

$$= \left(\lim_{x \to 0} \frac{\sin x}{x} \right) \left(\lim_{x \to 0} \frac{\sin x}{1 + \cos x} \right) = (1)\left(\frac{0}{1 + 1} \right) = 0 \qquad \blacksquare$$

Example 2

Find

(a) $\lim\limits_{x \to 0} \dfrac{\tan x}{x}$ (b) $\lim\limits_{\theta \to 0} \dfrac{\sin 2\theta}{\theta}$ (c) $\lim\limits_{x \to 0} \dfrac{\sin 3x}{\sin 5x}$

Solution (*a*).

$$\lim_{x \to 0} \frac{\tan x}{x} = \lim_{x \to 0} \left(\frac{\sin x}{x} \cdot \frac{1}{\cos x} \right) = (1)(1) = 1$$

Solution (*b*). The trick is to multiply and divide by 2, which will make the denominator the same as the argument of the sine function [just as in Theorem 2.5.3(*a*)]:

$$\lim_{\theta \to 0} \frac{\sin 2\theta}{\theta} = \lim_{\theta \to 0} 2 \cdot \frac{\sin 2\theta}{2\theta} = 2 \lim_{\theta \to 0} \frac{\sin 2\theta}{2\theta}$$

Now make the substitution $x = 2\theta$, and use the fact that $x \to 0$ as $\theta \to 0$. This yields

$$\lim_{\theta \to 0} \frac{\sin 2\theta}{\theta} = 2 \lim_{\theta \to 0} \frac{\sin 2\theta}{2\theta} = 2 \lim_{x \to 0} \frac{\sin x}{x} = 2(1) = 2$$

Solution (*c*).

$$\lim_{x \to 0} \frac{\sin 3x}{\sin 5x} = \lim_{x \to 0} \frac{\dfrac{\sin 3x}{x}}{\dfrac{\sin 5x}{x}} = \lim_{x \to 0} \frac{3 \cdot \dfrac{\sin 3x}{3x}}{5 \cdot \dfrac{\sin 5x}{5x}} = \frac{3 \cdot 1}{5 \cdot 1} = \frac{3}{5}$$ ◄

FOR THE READER.   Use a graphing utility to confirm the limits in the last example graphically, and if you have a CAS, then use it to obtain the limits.

Example 3

Make conjectures about the limits

(a) $\lim\limits_{x \to 0} \sin\left(\dfrac{1}{x} \right)$ (b) $\lim\limits_{x \to 0} x \sin\left(\dfrac{1}{x} \right)$

and confirm your conclusions by generating the graphs of the functions near $x = 0$ using a graphing utility.

Solution (*a*). Since $1/x \to +\infty$ as $x \to 0^+$, we can view $\sin(1/x)$ as the sine of an angle that increases indefinitely as $x \to 0^+$. As this angle increases, the function $\sin(1/x)$ keeps oscillating between -1 and 1 without approaching a limit. Similarly, there is no limit from the left since $1/x \to -\infty$ as $x \to 0^-$. These conclusions are consistent with the graph of $y = \sin(1/x)$ shown in Figure 2.5.7*a*. Observe that the oscillations become more and more rapid as x approaches 0 because $1/x$ increases (or decreases) more and more rapidly as x approaches 0.

Solution (*b*). The values of $x \sin(1/x)$ oscillate between x and $-x$, both of which approach 0 as x approaches 0. Thus, the Squeezing Theorem suggests that $x \sin(1/x) \to 0$ as $x \to 0$. This is consistent with Figure 2.5.7*b*. ◄

REMARK.   It follows from part (b) of this example that the function

$$f(x) = \begin{cases} x \sin(1/x), & x \neq 0 \\ 0, & x = 0 \end{cases}$$

is continuous at $x = 0$, since the value of the function and the value of the limit are the same at that point. This shows that the behavior of a function can be very complex at a point of continuity.

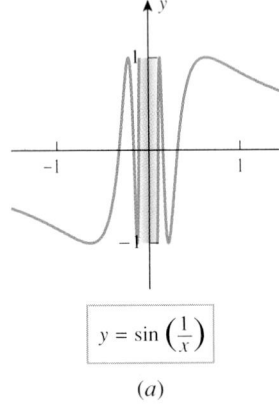

$y = \sin\left(\frac{1}{x}\right)$

(*a*)

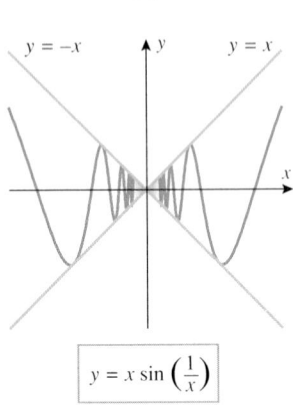

$y = x \sin\left(\frac{1}{x}\right)$

(*b*)

Figure 2.5.7

In Exercises 1–10, find the points of discontinuity, if any.

1. $f(x) = \sin(x^2 - 2)$

2. $f(x) = \cos\left(\dfrac{x}{x - \pi}\right)$

3. $f(x) = \cot x$

4. $f(x) = \sec x$

5. $f(x) = \csc x$

6. $f(x) = \dfrac{1}{1 + \sin^2 x}$

7. $f(x) = |\cos x|$

8. $f(x) = \sqrt{2 + \tan^2 x}$

9. $f(x) = \dfrac{1}{1 - 2\sin x}$

10. $f(x) = \dfrac{3}{5 + 2\cos x}$

11. Use Theorem 2.4.6 to show that the following functions are continuous everywhere by expressing them as compositions of simpler functions that are known to be continuous.
(a) $\sin(x^3 + 7x + 1)$ (b) $|\sin x|$
(c) $\cos^3(x + 1)$ (d) $\sqrt{3 + \sin 2x}$
(e) $\sin(\sin x)$ (f) $\cos^5 x - 2\cos^3 x + 1$

12. (a) Prove that if $g(x)$ is continuous everywhere, then so are $\sin(g(x))$, $\cos(g(x))$, $g(\sin(x))$, and $g(\cos(x))$.
(b) Illustrate the result in part (a) with some of your own choices for g.

Find the limits in Exercises 13–35.

13. $\displaystyle\lim_{x \to +\infty} \cos\left(\dfrac{1}{x}\right)$

14. $\displaystyle\lim_{x \to +\infty} \sin\left(\dfrac{2}{x}\right)$

15. $\displaystyle\lim_{x \to +\infty} \sin\left(\dfrac{\pi x}{2 - 3x}\right)$

16. $\displaystyle\lim_{h \to 0} \dfrac{\sin h}{2h}$

17. $\displaystyle\lim_{\theta \to 0} \dfrac{\sin 3\theta}{\theta}$

18. $\displaystyle\lim_{\theta \to 0^+} \dfrac{\sin \theta}{\theta^2}$

19. $\displaystyle\lim_{x \to 0^-} \dfrac{\sin x}{|x|}$

20. $\displaystyle\lim_{x \to 0} \dfrac{\sin^2 x}{3x^2}$

21. $\displaystyle\lim_{x \to 0^+} \dfrac{\sin x}{5\sqrt{x}}$

22. $\displaystyle\lim_{x \to 0} \dfrac{\sin 6x}{\sin 8x}$

23. $\displaystyle\lim_{x \to 0} \dfrac{\tan 7x}{\sin 3x}$

24. $\displaystyle\lim_{\theta \to 0} \dfrac{\sin^2 \theta}{\theta}$

25. $\displaystyle\lim_{h \to 0} \dfrac{h}{\tan h}$

26. $\displaystyle\lim_{h \to 0} \dfrac{\sin h}{1 - \cos h}$

27. $\displaystyle\lim_{\theta \to 0} \dfrac{\theta^2}{1 - \cos \theta}$

28. $\displaystyle\lim_{x \to 0} \dfrac{x}{\cos\left(\frac{1}{2}\pi - x\right)}$

29. $\displaystyle\lim_{\theta \to 0} \dfrac{\theta}{\cos \theta}$

30. $\displaystyle\lim_{t \to 0} \dfrac{t^2}{1 - \cos^2 t}$

31. $\displaystyle\lim_{h \to 0} \dfrac{1 - \cos 5h}{\cos 7h - 1}$

32. $\displaystyle\lim_{x \to 0^+} \sin\left(\dfrac{1}{x}\right)$

33. $\displaystyle\lim_{x \to 0^+} \cos\left(\dfrac{1}{x}\right)$

34. $\displaystyle\lim_{x \to 0} \dfrac{x^2 - 3\sin x}{x}$

35. $\displaystyle\lim_{x \to 0} \dfrac{2x + \sin x}{x}$

36. Find a value for the constant k that makes
$$f(x) = \begin{cases} \dfrac{\sin 3x}{x}, & x \neq 0 \\ k, & x = 0 \end{cases}$$
continuous at $x = 0$.

37. Find a nonzero value for the constant k that makes
$$f(x) = \begin{cases} \dfrac{\tan kx}{x}, & x < 0 \\ 3x + 2k^2, & x \geq 0 \end{cases}$$
continuous at $x = 0$.

38. Is
$$f(x) = \begin{cases} \dfrac{\sin x}{|x|}, & x \neq 0 \\ 1, & x = 0 \end{cases}$$
continuous at $x = 0$?

39. In each part, find the limit by making the indicated substitution.
(a) $\displaystyle\lim_{x \to +\infty} x \sin \dfrac{1}{x}$. $\left[\textit{Hint: Let } t = \dfrac{1}{x}.\right]$

(b) $\displaystyle\lim_{x \to -\infty} x\left(1 - \cos \dfrac{1}{x}\right)$. $\left[\textit{Hint: Let } t = \dfrac{1}{x}.\right]$

(c) $\displaystyle\lim_{x \to \pi} \dfrac{\pi - x}{\sin x}$.   [*Hint: Let* $t = \pi - x$.]

40. Find $\displaystyle\lim_{x \to 2} \dfrac{\cos(\pi/x)}{x - 2}$. $\left[\textit{Hint: Let } t = \dfrac{\pi}{2} - \dfrac{\pi}{x}.\right]$

41. Find $\displaystyle\lim_{x \to 1} \dfrac{\sin(\pi x)}{x - 1}$.   **42.** Find $\displaystyle\lim_{x \to \pi/4} \dfrac{\tan x - 1}{x - \pi/4}$.

43. Use the Squeezing Theorem to show that
$$\lim_{x \to 0} x \cos \dfrac{50\pi}{x} = 0$$
and illustrate the principle involved by using a graphing utility to graph $y = x$, $y = -x$, and $y = x\cos(50\pi/x)$ on the same screen over the x-interval from -1 to 1.

44. Use the Squeezing Theorem to show that
$$\lim_{x \to 0} x^2 \sin\left(\dfrac{50\pi}{\sqrt[3]{x}}\right) = 0$$
and illustrate the principle involved by using a graphing utility to graph $y = x^2$, $y = -x^2$, and $y = x^2 \sin(50\pi/\sqrt[3]{x})$ on the same screen over the x-interval from -0.5 to 0.5.

45. Sketch the graphs of $y = 1 - x^2$, $y = \cos x$, and $y = f(x)$, where f is any continuous function that satisfies the inequalities
$$1 - x^2 \leq f(x) \leq \cos x$$
for all x in the interval $(-\pi/2, \pi/2)$. What can you say about the limit of $f(x)$ as $x \to 0$? Explain your reasoning.

46. Sketch the graphs of $y = 1/x$, $y = -1/x$, and $y = f(x)$ in one coordinate system, where f is any continuous function that satisfies the inequalities

$$-\frac{1}{x} \le f(x) \le \frac{1}{x}$$

for all x in the interval $[1, +\infty)$. What can you say about the limit of $f(x)$ as $x \to +\infty$? Explain your reasoning.

47. Find formulas for functions g and h such that $g(x) \to 0$ and $h(x) \to 0$ as $x \to +\infty$ and such that

$$g(x) \le \frac{\sin x}{x} \le h(x)$$

for positive values of x. What can you say about the limit

$$\lim_{x \to +\infty} \frac{\sin x}{x}?$$

Explain your reasoning.

48. Draw pictures analogous to Figure 2.5.3 that illustrate the Squeezing Theorem for limits of the form $\lim_{x \to +\infty} f(x)$ and $\lim_{x \to -\infty} f(x)$.

Recall that unless stated otherwise the variable x in trigonometric functions such as $\sin x$ and $\cos x$ are assumed to be in radian measure. The limits in Theorem 2.5.3 are based on that assumption. Exercises 49 and 50 explore what happens to those limits if degree measure is used for x.

49. (a) Show that if x is in degrees, then

$$\lim_{x \to 0} \frac{\sin x}{x} = \frac{\pi}{180}$$

(b) Confirm that the limit in part (a) is consistent with the results produced by your calculating utility by setting the utility to degree measure and calculating $(\sin x)/x$ for some values of x that get closer and closer to 0.

50. What is the limit of $(1 - \cos x)/x$ as $x \to 0$ if x is in degrees?

51. It follows from part (*a*) of Theorem 2.5.3 that if θ is small (near zero) and measured in radians, then one should expect the approximation

$$\sin \theta \approx \theta$$

to be good.
(a) Find $\sin 10°$ using a calculating utility.
(b) Find $\sin 10°$ using the approximation above.

52. (a) Use the approximation of $\sin \theta$ that is given in Exercise 51 together with the identity $\cos 2\alpha = 1 - 2\sin^2 \alpha$ with $\alpha = \theta/2$ to show that if θ is small (near zero) and measured in radians, then one should expect the approximation

$$\cos \theta \approx 1 - \tfrac{1}{2}\theta^2$$

to be good.
(b) Find $\cos 10°$ using a calculating utility.
(c) Find $\cos 10°$ using the approximation above.

53. It follows from part (a) of Example 2 that if θ is small (near zero) and measured in radians, then one should expect the

approximation

$$\tan \theta \approx \theta$$

to be good.
(a) Find $\tan 5°$ using a calculating utility.
(b) Find $\tan 5°$ using the approximation above.

54. Referring to the accompanying figure, suppose that the angle of elevation of the top of a building, as measured from a point L feet from its base, is found to be α degrees.
(a) Use the relationship $h = L \tan \alpha$ to calculate the height of a building for which $L = 500$ ft and $\alpha = 6°$.
(b) Show that if L is large compared to the building height h, then one should expect good results in approximating h by $h \approx \pi L \alpha / 180$.
(c) Use the result in part (b) to approximate the building height h in part (a).

Figure Ex-54

55. (a) Use the Intermediate-Value Theorem to show that the equation $x = \cos x$ has at least one solution in the interval $[0, \pi/2]$.
(b) Show graphically that there is exactly one solution in the interval.
(c) Approximate the solution to three decimal places.

56. (a) Use the Intermediate-Value Theorem to show that the equation $x + \sin x = 1$ has at least one solution in the interval $[0, \pi/6]$.
(b) Show graphically that there is exactly one solution in the interval.
(c) Approximate the solution to three decimal places.

57. In the study of falling objects near the surface of the Earth, the **acceleration g due to gravity** is commonly taken to be 9.8 m/s² or 32 ft/s². However, the elliptical shape of the Earth and other factors cause variations in this constant that are latitude dependent. The following formula, known as the Geodetic Reference Formula of 1967, is commonly used to predict the value of g at a latitude of ϕ degrees (either north or south of the equator):

$$g = 9.7803185(1.0 + 0.005278895 \sin^2 \phi \\ - 0.000023462 \sin^4 \phi)\ \text{m/s}^2$$

(a) Observe that g is an even function of ϕ. What does this suggest about the shape of the Earth, as modeled by the Geodetic Reference Formula?
(b) Show that $g = 9.8$ m/s² somewhere between latitudes of $38°$ and $39°$.

58. Let

$$f(x) = \begin{cases} 1 & \text{if } x \text{ is a rational number} \\ 0 & \text{if } x \text{ is an irrational number} \end{cases}$$

(a) Make a conjecture about the limit of $f(x)$ as $x \to 0$.
(b) Make a conjecture about the limit of $xf(x)$ as $x \to 0$.
(c) Prove your conjectures.

SUPPLEMENTARY EXERCISES

1. For the function f graphed in the accompanying figure, find the limit if it exists.
   (a) $\lim\limits_{x \to 1} f(x)$ (b) $\lim\limits_{x \to 2} f(x)$ (c) $\lim\limits_{x \to 3} f(x)$
   (d) $\lim\limits_{x \to 4} f(x)$ (e) $\lim\limits_{x \to +\infty} f(x)$ (f) $\lim\limits_{x \to -\infty} f(x)$
   (g) $\lim\limits_{x \to 3^+} f(x)$ (h) $\lim\limits_{x \to 3^-} f(x)$ (i) $\lim\limits_{x \to 0} f(x)$

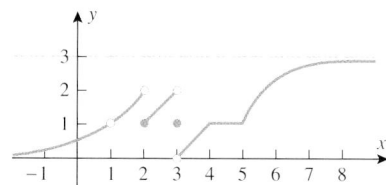

Figure Ex-1

2. (a) Find a formula for a rational function that has a vertical asymptote at $x = 1$ and a horizontal asymptote at $y = 2$.
   (b) Check your work by using a graphing utility to graph the function.

3. (a) Write a paragraph or two that describes how the limit of a function can fail to exist at a point $x = a$. Accompany your description with some specific examples.
   (b) Write a paragraph or two that describes how the limit of a function can fail to exist as $x \to +\infty$ or $x \to -\infty$. Also, accompany your description with some specific examples.
   (c) Write a paragraph or two that describes how a function can fail to be continuous at a point $x = a$. Accompany your description with some specific examples.

4. Show that the Intermediate-Value Theorem is false if f is not continuous on the interval $[a, b]$.

5. In each part, evaluate the function for the stated values of x, and make a conjecture about the value of the limit. Confirm your conjecture by finding the limit algebraically.
   (a) $f(x) = \dfrac{x-2}{x^2-4}$; $\lim\limits_{x \to 2^+} f(x)$; $x = 2.5, 2.1, 2.01,$
   $ 2.001, 2.0001, 2.00001$
   (b) $f(x) = \dfrac{\tan 4x}{x}$; $\lim\limits_{x \to 0} f(x)$; $x = \pm1.0, \pm0.1, \pm0.01,$
   $ \pm0.001, \pm0.0001, \pm0.00001$

6. In each part, find the horizontal asymptotes, if any.
   (a) $y = \dfrac{2x-7}{x^2-4x}$ (b) $y = \dfrac{x^3-x^2+10}{3x^2-4x}$
   (c) $y = \dfrac{2x^2-6}{x^2+5x}$

7. (a) Approximate the value for the limit
   $$\lim\limits_{x \to 0} \frac{3^x - 2^x}{x}$$
   to three decimal places by constructing an appropriate table of values.

(b) Confirm your approximation using graphical evidence.

8. According to Ohm's law, when a voltage of V volts is applied across a resistor with a resistance of R ohms, a current of $I = V/R$ amperes flows through the resistor.
   (a) How much current flows if a voltage of 3.0 volts is applied across a resistance of 7.5 ohms?
   (b) If the resistance varies by ± 0.1 ohm, and the voltage remains constant at 3.0 volts, what is the resulting range of values for the current?
   (c) If temperature variations cause the resistance to vary by $\pm\delta$ from its value of 7.5 ohms, and the voltage remains constant at 3.0 volts, what is the resulting range of values for the current?
   (d) If the current is not allowed to vary by more than $\epsilon = \pm 0.001$ ampere at a voltage of 3.0 volts, what variation of $\pm\delta$ from the value of 7.5 ohms is allowable?
   (e) Certain alloys become **superconductors** as their temperature approaches absolute zero ($-273°$ C), meaning that their resistance approaches zero. If the voltage remains constant, what happens to the current in a superconductor as $R \to 0^+$?

9. Suppose that f is continuous on the interval $[0, 1]$ and that $0 \le f(x) \le 1$ for all x in this interval.
   (a) Sketch the graph of $y = x$ together with a possible graph for f over the interval $[0, 1]$.
   (b) Use the Intermediate-Value Theorem to help prove that there is at least one number c in the interval $[0, 1]$ such that $f(c) = c$.

10. Use algebraic methods to find
    (a) $\lim\limits_{\theta \to 0} \tan\left(\dfrac{1-\cos\theta}{\theta}\right)$ (b) $\lim\limits_{t \to 1} \dfrac{t-1}{\sqrt{t}-1}$
    (c) $\lim\limits_{x \to +\infty} \dfrac{(2x-1)^5}{(3x^2+2x-7)(x^3-9x)}$
    (d) $\lim\limits_{\theta \to 0} \cos\left(\dfrac{\sin(\theta+\pi)}{2\theta}\right)$.

11. Suppose that f is continuous on the interval $[0, 1]$, that $f(0) = 2$, and that f has no zeros in the interval. Prove that $f(x) > 0$ for all x in $[0, 1]$.

12. Suppose that
    $$f(x) = \begin{cases} -x^4 + 3, & x \le 2 \\ x^2 + 9, & x > 2 \end{cases}$$
    Is f continuous everywhere? Justify your conclusion.

13. Show that the equation $x^4 + 5x^3 + 5x - 1 = 0$ has at least two real solutions in the interval $[-6, 2]$.

14. Use the Intermediate-Value Theorem to approximate $\sqrt{11}$ to three decimal places, and check your answer by finding the root directly with a calculating utility.

15. Suppose that f is continuous and that $f(x_0) > 0$. Give either a δ-ϵ proof or a convincing verbal argument to show

that there must be an open interval containing x_0 on which $f(x) > 0$.

16. Sketch the graph of $f(x) = |x^2 - 4|/(x^2 - 4)$.

17. In each part, approximate the points of discontinuity of f to three decimal places.

(a) $f(x) = \dfrac{\sqrt{x+1}}{x^2 + 2x - 5}$

(b) $f(x) = \dfrac{x+3}{|2\sin x - x|}$

18. In Example 3 of Section 2.5 we used the Squeezing Theorem to prove that

$$\lim_{x \to 0} x \sin\left(\frac{1}{x}\right) = 0$$

Why couldn't we have obtained the same result by writing

$$\lim_{x \to 0} x \sin\left(\frac{1}{x}\right) = \lim_{x \to 0} x \cdot \lim_{x \to 0} \sin\left(\frac{1}{x}\right)$$

$$= 0 \cdot \lim_{x \to 0} \sin\left(\frac{1}{x}\right) = 0?$$

In Exercises 19 and 20, find $\lim\limits_{x \to a} f(x)$, if it exists, for
$$a = 0,\ 5^+,\ -5^-,\ -5,\ 5,\ -\infty,\ +\infty$$

19. (a) $f(x) = \sqrt{5-x}$ (b) $f(x) = (x^2 - 25)/(x-5)$

20. (a) $f(x) = (x+5)/(x^2 - 25)$

(b) $f(x) = \begin{cases} (x-5)/|x-5|, & x \neq 5 \\ 0, & x = 5 \end{cases}$

In Exercises 21–28, find the indicated limit, if it exists.

21. $\lim\limits_{x \to 0} \dfrac{\tan ax}{\sin bx}$ $(a \neq 0, b \neq 0)$

22. $\lim\limits_{x \to 0} \dfrac{\sin 3x}{\tan 3x}$

23. $\lim\limits_{\theta \to 0} \dfrac{\sin 2\theta}{\theta^2}$

24. $\lim\limits_{x \to 0} \dfrac{x \sin x}{1 - \cos x}$

25. $\lim\limits_{x \to 0^+} \dfrac{\sin x}{\sqrt{x}}$

26. $\lim\limits_{x \to 0} \dfrac{\sin^2(kx)}{x^2}$, $k \neq 0$

27. $\lim\limits_{x \to 0} \dfrac{3x - \sin(kx)}{x}$, $k \neq 0$

28. $\lim\limits_{x \to +\infty} \dfrac{2x + x\sin 3x}{5x^2 - 2x + 1}$

29. The author's dictionary describes a continuous function as "one whose value at each point is closely approached by its values at neighboring points."

(a) How would you explain the meaning of the terms "neighboring points" and "closely approached" to a nonmathematician?

(b) Write a paragraph that explains why the dictionary definition is consistent with the definition given in the text.

30. (a) Show by rationalizing the numerator that

$$\lim_{x \to 0} \frac{\sqrt{x^2 + 4} - 2}{x^2} = \frac{1}{4}$$

(b) Evaluate $f(x)$ for

$$x = \pm 1.0,\ \pm 0.1,\ \pm 0.01,\ \pm 0.001,\ \pm 0.0001,\ \pm 0.00001$$

and explain why the values are not getting closer and closer to the limit.

(c) The accompanying figure shows the graph of f generated with a graphing utility and zooming in on the origin. Explain what is happening.

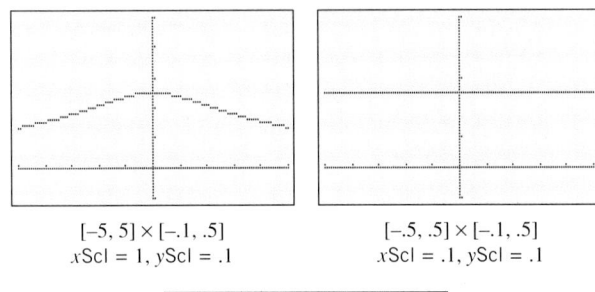

$[-5, 5] \times [-.1, .5]$
$x\text{Scl} = 1, y\text{Scl} = .1$

$[-.5, .5] \times [-.1, .5]$
$x\text{Scl} = .1, y\text{Scl} = .1$

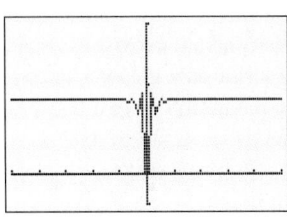

$[-5 \times 10^{-6}, 5 \times 10^{-6}] \times [-.1, .5]$
$x\text{Scl} = 10^{-6}, y\text{Scl} = .1$

Figure Ex-30

In Exercises 31–36, approximate the limit of the function by looking at its graph and calculating values for some appropriate choices of x. Compare your answer with the value produced by a CAS.

C 31. $\lim\limits_{x \to 0} (1+x)^{1/x}$

C 32. $\lim\limits_{x \to 3} \dfrac{2^x - 8}{x - 3}$

C 33. $\lim\limits_{x \to 1} \dfrac{\sin x - \sin 1}{x - 1}$

C 34. $\lim\limits_{x \to 0^+} x^{-2}(1.001)^{-1/x}$

C 35. $\lim\limits_{x \to +\infty} \left(\sqrt{x + \sqrt{x}} - \sqrt{x}\right)$

C 36. $\lim\limits_{x \to +\infty} \left(3^x + 5^x\right)^{1/x}$

37. The limit

$$\lim_{x \to 0} \frac{\sin x}{x} = 1$$

ensures that there is a number δ such that

$$\left| \frac{\sin x}{x} - 1 \right| < 0.001$$

if $0 < |x| < \delta$. Estimate the largest such δ.

38. If \$1000 is invested in an account that pays 7% interest compounded n times each year, then in 10 years there will be $1000(1 + 0.07/n)^{10n}$ dollars in the account. How much money will be in the account in 10 years if the interest is compounded quarterly ($n = 4$)? Monthly ($n = 12$)? Daily ($n = 365$)? How much money will be in the account in 10 years if the interest is compounded *continuously*, that is, as $n \to +\infty$?

39. There are various numerical methods other than the method discussed in Section 2.4 to obtain approximate solutions of equations of the form $f(x) = 0$. One such method requires that the equation be expressed in the form $x = g(x)$, so that a solution $x = c$ can be interpreted as the value of x where the line $y = x$ intersects the curve $y = g(x)$, as shown in the accompanying figure. If x_1 is an initial estimate of c and the graph of $y = g(x)$ is not too steep in the vicinity of c, then a better approximation can be obtained from $x_2 = g(x_1)$ (see the figure). An even better approximation is obtained from $x_3 = g(x_2)$, and so forth. The formula $x_{n+1} = g(x_n)$ for $n = 1, 2, 3, \ldots$ generates successive approximations $x_2, x_3, x_4, \ldots$ that get closer and closer to c.
(a) The equation $x^3 - x - 1 = 0$ has only one real solution. Show that this equation can be written as

$$x = g(x) = \sqrt[3]{x + 1}$$

(b) Graph $y = x$ and $y = g(x)$ in the same coordinate system for $-1 \le x \le 3$.

(c) Starting with an arbitrary point x_1, make a sketch that shows the location of the successive iterates

$$x_2 = g(x_1), \quad x_3 = g(x_2), \ldots$$

(d) Use $x_1 = 1$ and calculate $x_2, x_3, \ldots$, continuing until you obtain two consecutive values that differ by less than 10^{-4}. Experiment with other starting values such as $x_1 = 2$ or $x_1 = 1.5$.

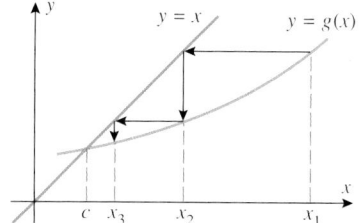

Figure Ex-39

40. The method described in Exercise 39 will not always work.
(a) The equation $x^3 - x - 1 = 0$ can be expressed as $x = g(x) = x^3 - 1$. Graph $y = x$ and $y = g(x)$ in the same coordinate system. Starting with an arbitrary point x_1, make a sketch illustrating the location of the successive iterates $x_2 = g(x_1), x_3 = g(x_2), \ldots$.
(b) Use $x_1 = 1$ and calculate the successive iterates x_n for $n = 2, 3, 4, 5, 6$.

In Exercises 41 and 42, use the method of Exercise 39 to approximate the roots of the equation.

41. $x^5 - x - 2 = 0$ **42.** $x - \cos x = 0$

EXPANDING THE CALCULUS HORIZON

For additional material relating to this chapter, visit the Anton Website at http://www.wiley.com/college/anton

3

THE DERIVATIVE

Sir Isaac Newton

*M*any physical phenomena involve changing quantities—the speed of a rocket, the inflation of currency, the number of bacteria in a culture, the shock intensity of an earthquake, the voltage of an electrical signal, and so forth. In this chapter we will develop the concept of a *derivative*, which is the mathematical tool that is used to study rates at which physical quantities change. In Section 3.1 we will show that there is a close relationship between rates of change and tangent lines to graphs, and we will show how the familiar idea of velocity can be viewed as a rate of change. In Sections 3.2 to 3.5 we will define the concept of a derivative precisely and develop the mathematical tools for calculating them.

One of the important themes in applied science is developing methods for approximating quantities that are difficult to calculate exactly. In Section 3.6 we will show how derivatives can be applied to certain kinds of approximation problems.

3.1 TANGENT LINES AND RATES OF CHANGE

In this section we will establish a basic relationship between tangent lines and rates of change. Our work here is intended to be informal and introductory, and all of the ideas that we develop will be revisited in more detail in later sections.

SLOPE OF A TANGENT LINE

In Section 2.1 we observed informally that if a secant line is drawn between two distinct points P and Q on a curve $y = f(x)$, and if Q is allowed to move along the curve toward P, then we can expect the secant line to rotate toward a *limiting position*, which can be regarded as the tangent line to the curve at the point P (Figure 3.1.1). In the next section we will give a precise mathematical definition of a tangent line, but for now this intuitive idea will suffice.

In many problems we will be more concerned with the *slope* of the tangent line than with the tangent line itself, so it will be helpful to understand the relationship between the slope $m_{\tan}$ of the tangent line at P and the slope $m_{\sec}$ of the secant line between P and Q as the point Q moves along the curve $y = f(x)$ toward P. For this purpose, suppose that the secant line passes through the distinct points $P(x_0, f(x_0))$ and $Q(x_1, f(x_1))$, in which case its slope is

$$m_{\sec} = \frac{f(x_1) - f(x_0)}{x_1 - x_0} \tag{1}$$

(Figure 3.1.2). As this figure suggests, the point Q moves along the curve toward P if and only if x_1 approaches x_0. Thus, from (1) the slope of the tangent line at P is

$$m_{\tan} = \lim_{x_1 \to x_0} \frac{f(x_1) - f(x_0)}{x_1 - x_0}$$

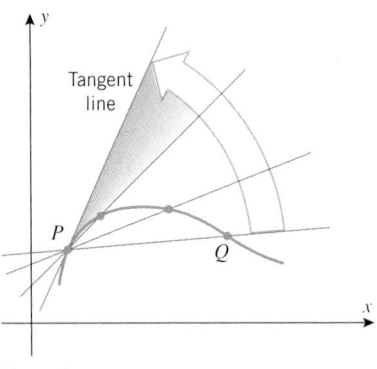

Figure 3.1.1

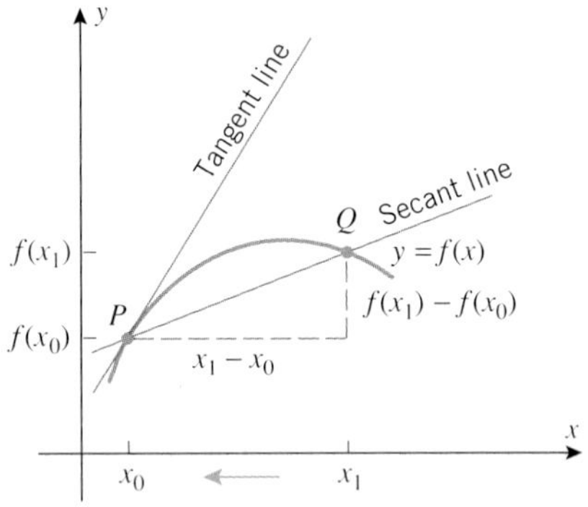

Figure 3.1.2

AVERAGE VERSUS INSTANTANEOUS VELOCITY

Although tangent lines are of interest as a matter of pure geometry, much of the impetus for studying them arose in the seventeenth century when scientists recognized their importance in studying the motion of objects that move with nonconstant velocity. Some of the relevant ideas were discussed in Section 1.5, but it will be helpful to review them here.

Recall that a particle moving along a line, say an s-axis, is said to have ***rectilinear motion***. In the most general kind of rectilinear motion the particle may move back and forth on the line; however, here, as in Section 1.5, we will assume that the particle moves in one direction only—the positive direction of the s-axis. As discussed in Section 1.5, this allows us to use the terms *speed* and *velocity* interchangeably, since there is only one possible direction of motion. General rectilinear motion will be discussed later.

We showed in Section 1.5 that if a particle has uniform rectilinear motion, that is, it moves with constant velocity v along a line, then its position versus time curve is a line of slope v; conversely, if the position versus time curve for a particle in rectilinear motion is a line of slope v, then the particle has constant velocity v. Here, we will consider the more general case of a particle moving in the positive s-direction with *variable* velocity, in which case the position versus time curve need not be linear. For this purpose we will need to examine the meaning of the term *velocity* more critically.

If a car travels 75 miles over a straight road in a 3-hour period, then its average velocity during the trip is $75/3 = 25$ mi/h. However, this does not mean that the car travels at 25 mi/h for the entire trip; it may speed up and slow down at various times. Thus, the average velocity provides information about the velocity of the car over the entire trip but no information about its velocity at specific times during the trip.

Although average velocity is useful for many purposes, there are many situations in which it is of no help. For example, if a car strikes a tree during a trip, the damage sustained is not determined by the average velocity up to the time of impact, but rather by the *instantaneous velocity* at the precise moment of impact. However, the concept of instantaneous velocity is subtle, and a clear understanding of its meaning evaded scientists until the advent of calculus in the seventeenth century.

A nice explanation of the difficulty in defining and calculating instantaneous velocity was given by Morris Kline[*] who wrote:

> *In contrasting average velocity with instantaneous velocity we implicitly utilize a distinction between interval and instant.... An average velocity is one that concerns what happens over an interval of time—3 hours, 5 seconds, one-half second, and so forth. The interval may be small or large, but it does represent the passage of a definite amount of time. We use the word instant, however, to state the fact that something happens so fast that no time elapses. The event is momentary. When we say, for example, that it is 3 o'clock, we refer to an instant, a precise moment. If the lapse of time is pictured by length along a line, then an interval (of time) is represented by a line segment, whereas an instant corresponds to a point. The notion of an instant, although it is used in everyday life, is strictly a mathematical idealization.*
>
> *Our ways of thinking about real events cause us to speak in terms of instants and velocity at an instant, but closer examination shows that the concept of velocity at an instant presents difficulties. Average velocity, which is simply the distance traveled during some interval of time divided by that amount of time, is easily calculated. Suppose, however, that we try to carry over this process to instantaneous velocity. The distance an automobile travels in one instant is 0 and the time that elapses during one instant is also 0. Hence the distance divided by the time is 0/0, which is meaningless. Thus, although instantaneous velocity is a physical reality, there seems to be a difficulty in calculating it, and unless we can calculate it, we cannot work with it mathematically.*

Our goal, then, is to define the concept of instantaneous velocity in a way that it can be calculated and worked with mathematically. For this purpose, consider a car that moves in a single direction along a straight road, and assume that an s-axis has been introduced with its positive direction in the direction of motion. As shown in Figure 3.1.3, suppose that a clock tracks the elapsed time t, starting at $t = 0$, and that the coordinate of the car as a function of t is $s = f(t)$. The function f is called the **position function** of the car, and the graph of $s = f(t)$ is what we have been calling the position versus time curve. The third part of Figure 3.1.3 shows a typical position versus time curve for a car whose coordinate

---

[*] MORRIS KLINE (1908–1992). American mathematician, scholar, and educator. Kline made numerous contributions to mathematical thought, wrote extensively on education, especially mathematics education, and taught, lectured, and served as a consultant throughout his very active career. He was the author of many popular books including *Mathematical Thought from Ancient to Modern Times* and *Why Johnny Can't Add: The Failure of the New Mathematics*.

at time $t = 0$ is s_0. Observe that we have drawn the curve so that s increases with t. This is because we have assumed the car to be traveling in the positive direction, and decreasing values of s would imply a motion in the negative direction.

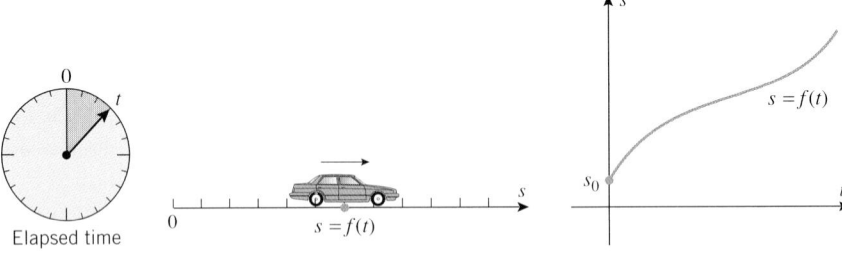

Elapsed time

Figure 3.1.3

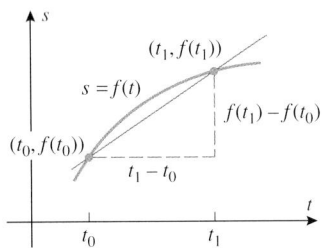

Figure 3.1.4

The position versus time curve provides a simple geometric interpretation of the average velocity of the car over a time interval, say from t_0 to t_1. If the car has a coordinate $s_0 = f(t_0)$ at time t_0 and coordinate $s_1 = f(t_1)$ at time t_1, where $t_1 > t_0$, then the distance traveled during the time interval is $s_1 - s_0$ and the time elapsed is $t_1 - t_0$. Thus, the average velocity during the time interval, denoted by v_{ave}, is

$$v_{\text{ave}} = \frac{s_1 - s_0}{t_1 - t_0} = \frac{f(t_1) - f(t_0)}{t_1 - t_0} \tag{2}$$

which is just the slope of the secant line connecting the points (t_0, s_0) and (t_1, s_1) on the position versus time curve (Figure 3.1.4).

Now suppose that we are interested in the instantaneous velocity of the car at time t_0. Intuition suggests that over a small time interval the velocity of the car cannot vary much, so if t_1 is close to t_0, then the average velocity of the car over the time interval from t_0 to t_1 should closely approximate the instantaneous velocity of the car at time t_0. Moreover, the smaller the time interval between t_0 and t_1, the better the approximation. This suggests that if we let t_1 get closer and closer to t_0, then the average velocity of the car over the time interval from t_0 to t_1 should get closer and closer to the instantaneous velocity at time t_0. Thus, if we denote the instantaneous velocity of the car at time t_0 by v_{inst}, we have

$$v_{\text{inst}} = \lim_{t_1 \to t_0} v_{\text{ave}} = \lim_{t_1 \to t_0} \frac{f(t_1) - f(t_0)}{t_1 - t_0} \tag{3}$$

Since v_{ave} is the slope of the secant line joining the points $(t_0, f(t_0))$ and $(t_1, f(t_1))$ on the position versus time curve $s = f(t)$, and since the point $(t_1, f(t_1))$ moves along this curve toward $(t_0, f(t_0))$ as $t_1 \to t_0$ (Figure 3.1.5), it follows from (3) that v_{inst} can be interpreted as the slope of the tangent line to the position versus time curve at the point $(t_0, f(t_0))$.

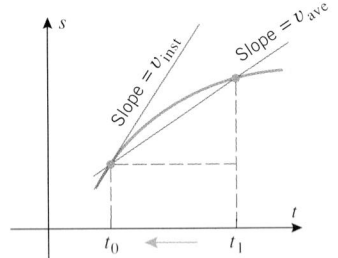

Figure 3.1.5

These ideas are illustrated numerically in Table 3.1.1. The first part of the table shows the coordinates of a particle moving along an s-axis over the time interval from $t = 4.00$ to

Table 3.1.1

t (s)	4.00	4.50	5.00	5.50	5.80	5.90	5.95	5.98	6.00
s (ft)	1.00	1.25	2.00	3.25	4.24	4.61	4.80	4.92	5.00

TIME INTERVAL	[4.00, 6.00]	[4.50, 6.00]	[5.00, 6.00]	[5.50, 6.00]	[5.80, 6.00]	[5.90, 6.00]	[5.95, 6.00]	[5.98, 6.00]
AVERAGE VELOCITY (ft/s)	1.00	2.50	3.00	3.50	3.80	3.90	4.00	4.00

$t = 6.00$. From these values we can calculate the average velocity of the particle over a succession of shrinking time intervals ending at time $t = 6.00$ s. For example, the calculations for the average velocity over the time interval $[4.50, 6.00]$ are

$$v_{\text{ave}} = \frac{5.00 - 1.25}{6.00 - 4.50} = \frac{3.75}{1.50} = 2.50 \text{ ft/s}$$

The resulting average velocities in the second part of the table suggest that to two decimal places the instantaneous velocity at time $t = 6.00$ s is 4.00 ft/s.

The main ideas in the preceding discussion can be summarized as follows.

3.1.1 GEOMETRIC INTERPRETATION OF AVERAGE VELOCITY. *If a particle moves in the positive direction along an s-axis, and if the position versus time curve is $s = f(t)$, then the average velocity of the particle between times t_0 and t_1 is represented geometrically by the slope of the secant line joining the points $(t_0, f(t_0))$ and $(t_1, f(t_1))$.*

3.1.2 GEOMETRIC INTERPRETATION OF INSTANTANEOUS VELOCITY. *If a particle moves in the positive direction along an s-axis, and if the position versus time curve is $s = f(t)$, then the instantaneous velocity of the particle at time t_0 is represented geometrically by the slope of the tangent line to the curve at the point $(t_0, f(t_0))$.*

AVERAGE AND INSTANTANEOUS RATES OF CHANGE

Velocity can be viewed as a *rate of change*—the rate of change of position with time, or in algebraic terms, the rate of change of s with t. Rates of change occur in many applications. For example:

- A microbiologist might be interested in the rate at which the number of bacteria in a colony changes with time.
- An engineer might be interested in the rate at which the length of a metal rod changes with temperature.
- An economist might be interested in the rate at which production cost changes with the quantity of a product that is manufactured.
- A medical researcher might be interested in the rate at which the radius of an artery changes with the concentration of alcohol in the bloodstream.

In general, if x and y are any quantities related by an equation $y = f(x)$, we can consider the rate at which y changes with x. As with velocity, we distinguish between an average rate of change represented by the slope of a secant line and an instantaneous rate of change represented by the slope of the tangent line. More precisely, we make the following definitions.

3.1.3 DEFINITION. If $y = f(x)$, then the ***average rate of change of y with respect to x over the interval $[x_0, x_1]$*** is the slope m_{sec} of the secant line joining the points $(x_0, f(x_0))$ and $(x_1, f(x_1))$ on the graph of f (Figure 3.1.6a); that is,

$$m_{\text{sec}} = \frac{f(x_1) - f(x_0)}{x_1 - x_0} \tag{4}$$

3.1.4 DEFINITION. If $y = f(x)$, then the ***instantaneous rate of change of y with respect to x at the point x_0*** is the slope m_{tan} of the tangent line to the graph of f at the point x_0 (Figure 3.1.6*b*); that is,

$$m_{tan} = \lim_{x_1 \to x_0} \frac{f(x_1) - f(x_0)}{x_1 - x_0} \tag{5}$$

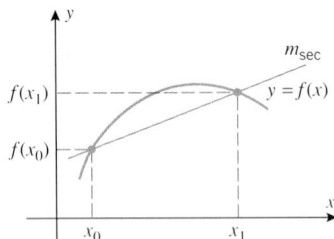

m_{sec} is the average rate of change of y with respect to x over the interval $[x_0, x_1]$.

(*a*)

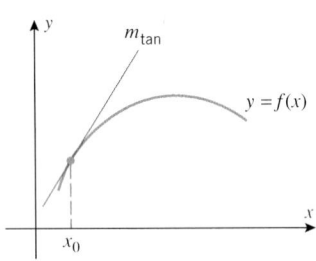

m_{tan} is the instantaneous rate of change of y with respect to x at the point x_0.

(*b*)

Figure 3.1.6

Example 1

Let $y = x^2 + 1$.

(a) Find the average rate of change of y with respect to x over the interval $[3, 5]$.

(b) Find the instantaneous rate of change of y with respect to x at the point $x = -4$.

(c) Find the instantaneous rate of change of y with respect to x at a general point $x = x_0$.

Solution (a). We will apply Formula (4) with $f(x) = x^2 + 1$, $x_0 = 3$, and $x_1 = 5$. This yields

$$m_{sec} = \frac{f(x_1) - f(x_0)}{x_1 - x_0} = \frac{f(5) - f(3)}{5 - 3} = \frac{26 - 10}{5 - 3} = 8$$

Thus, on the average, y increases 8 units per unit increase in x over the interval $[3, 5]$.

Solution (b). We will apply Formula (5) with $f(x) = x^2 + 1$ and $x_0 = -4$. This yields

$$m_{tan} = \lim_{x_1 \to x_0} \frac{f(x_1) - f(x_0)}{x_1 - x_0} = \lim_{x_1 \to -4} \frac{(x_1^2 + 1) - 17}{x_1 + 4}$$

$$= \lim_{x_1 \to -4} \frac{x_1^2 - 16}{x_1 + 4} = \lim_{x_1 \to -4} (x_1 - 4) = -8$$

Because the instantaneous rate of change is negative, y is *decreasing* at the point $x = -4$; it is decreasing at a rate of 8 units per unit increase in x.

Solution (c). We proceed as in part (b).

$$m_{tan} = \lim_{x_1 \to x_0} \frac{f(x_1) - f(x_0)}{x_1 - x_0} = \lim_{x_1 \to x_0} \frac{(x_1^2 + 1) - (x_0^2 + 1)}{x_1 - x_0}$$

$$= \lim_{x_1 \to x_0} \frac{x_1^2 - x_0^2}{x_1 - x_0} = \lim_{x_1 \to x_0} (x_1 + x_0) = 2x_0$$

Thus, the instantaneous rate of change of y with respect to x at $x = x_0$ is $2x_0$. Observe that the result in part (b) can be obtained from this more general result by letting $x_0 = -4$. ◀

RATES OF CHANGE IN APPLICATIONS

In applied problems, average and instantaneous rates of change must be accompanied by appropriate units. In general, the units for a rate of change of y with respect to x are obtained by "dividing" the units of y by the units of x and then simplifying according to the standard rules of algebra. Here are some examples:

• If y is in degrees Fahrenheit (°F) and x is in inches (in), then a rate of change of y with respect to x has units of degrees Fahrenheit per inch (°F/in).

• If y is in feet per second (ft/s) and x is in seconds (s), then a rate of change of y with respect to x has units of feet per second per second (ft/s/s), which would usually be written as ft/s².

• If y is in newton-meters (N·m) and x is in meters (m), then a rate of change of y with respect to x has units of newtons (N), since N·m/m = N.

• If y is in foot-pounds (ft·lb) and x is in hours (h), then a rate of change of y with respect to x has units of foot-pounds per hour (ft·lb/h).

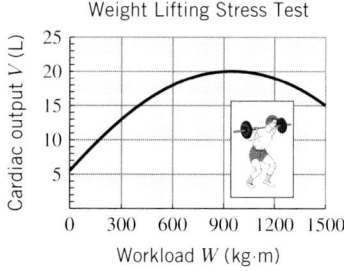

Figure 3.1.7

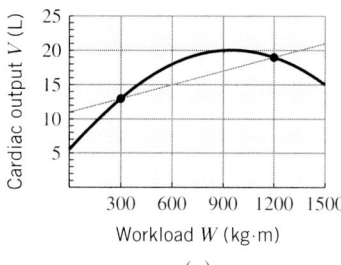

(a)

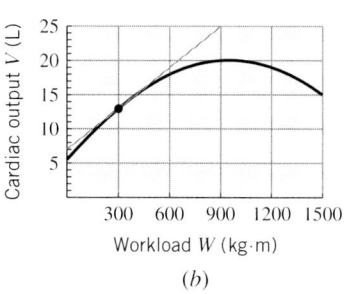

(b)

Figure 3.1.8

Example 2

The limiting factor in athletic endurance is cardiac output, that is, the volume of blood that the heart can pump per unit of time during an athletic competition. Figure 3.1.7 shows a stress-test graph of cardiac output V in liters (L) of blood versus workload W in kilogram-meters (kg·m) for 1 minute of weight lifting. This graph illustrates the known medical fact that cardiac output increases with the workload, but after reaching a peak value begins to decrease.

(a) Use the secant line shown in Figure 3.1.8a to estimate the average rate of change of cardiac output with respect to workload as the workload increases from 300 to 1200 kg·m.

(b) Use the tangent line shown in Figure 3.1.8b to estimate the instantaneous rate of change of cardiac output with respect to workload at the point where the workload is 300 kg·m.

Solution (a). Using the estimated points $(300, 13)$ and $(1200, 19)$, the slope of the secant line indicated in Figure 3.1.8a is

$$m_{\text{sec}} \approx \frac{19 - 13}{1200 - 300} \approx 0.0067 \frac{\text{L}}{\text{kg·m}}$$

Thus, the average rate of change of cardiac output with respect to workload over the interval is approximately 0.0067 L/kg·m. This means that on the average a 1-unit increase in workload produced a 0.0067-L increase in cardiac output over the interval.

Solution (b). Using the estimated tangent line in Figure 3.1.8b and the estimated points $(0, 7)$ and $(900, 25)$ on this tangent line, we obtain

$$m_{\text{tan}} \approx \frac{25 - 7}{900 - 0} \approx 0.02 \frac{\text{L}}{\text{kg·m}}$$

Thus, the instantaneous rate of change of cardiac output with respect to workload is approximately 0.02 L/kg·m. ◀

EXERCISE SET 3.1

· ·

In Exercises 1–4, a function $y = f(x)$ and values of x_0 and x_1 are given.

(a) Find the average rate of change of y with respect to x over the interval $[x_0, x_1]$.

(b) Find the instantaneous rate of change of y with respect to x at the given value of x_0.

(c) Find the instantaneous rate of change of y with respect to x at a general point x_0.

(d) Sketch the graph of $y = f(x)$ together with the secant and tangent lines whose slopes are given by the results in parts (a) and (b).

1. $y = \frac{1}{2}x^2$; $x_0 = 3$, $x_1 = 4$

2. $y = x^3$; $x_0 = 1$, $x_1 = 2$

3. $y = 1/x$; $x_0 = 2$, $x_1 = 3$

4. $y = 1/x^2$; $x_0 = 1$, $x_1 = 2$

In Exercises 5–8, a function f and a value of x_0 are given.

(a) Find the slope of the tangent to the graph of f at a general point x_0.

(b) Use the result in part (a) to find the slope of the tangent line at the given value of x_0.

5. $f(x) = x^2 + 1$; $x_0 = 2$

6. $f(x) = x^2 + 3x + 2$; $x_0 = 2$

7. $f(x) = \sqrt{x}$; $x_0 = 1$

8. $f(x) = 1/\sqrt{x}$; $x_0 = 4$

9. The accompanying figure shows the position versus time curve for an elevator that moves upward a distance of 60 m and then discharges its passengers.
   (a) Estimate the instantaneous velocity of the elevator at $t = 10$ s.
   (b) Sketch a velocity versus time curve for the motion of the elevator for $0 \le t \le 20$.

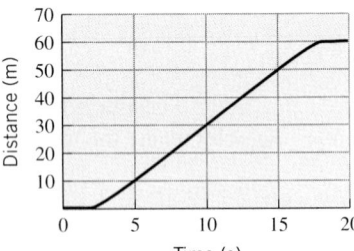

Figure Ex-9

10. The accompanying figure shows the position versus time curve for a certain particle moving along a straight line. Estimate each of the following from the graph:
    (a) the average velocity over the interval $0 \le t \le 3$
    (b) the values of t at which the instantaneous velocity is zero
    (c) the values of t at which the instantaneous velocity is either a maximum or a minimum
    (d) the instantaneous velocity when $t = 3$ s.

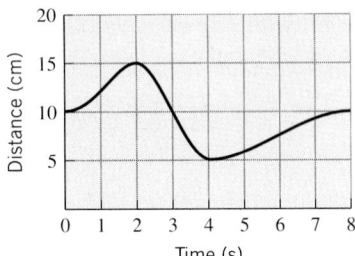

Figure Ex-10

11. The accompanying figure shows the position versus time curve for a certain particle moving on a straight line.
    (a) Is the particle moving faster at time t_0 or time t_2? Explain.
    (b) At the origin, the tangent is horizontal. What does this tell us about the initial velocity of the particle?
    (c) Is the particle speeding up or slowing down in the interval $[t_0, t_1]$? Explain.
    (d) Is the particle speeding up or slowing down in the interval $[t_1, t_2]$? Explain.

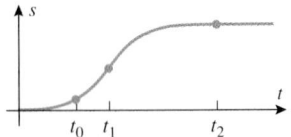

Figure Ex-11

12. An automobile, initially at rest, begins to move along a straight track. The velocity increases steadily until suddenly the driver sees a concrete barrier in the road and applies the brakes sharply at time t_0. The car decelerates rapidly, but it is too late—the car crashes into the barrier at time t_1 and instantaneously comes to rest. Sketch a position versus time curve that might represent the motion of the car.

13. If a particle moves at constant velocity, what can you say about its position versus time curve?

14. The accompanying figure shows the position versus time curves of four different particles moving on a straight line. For each particle, determine whether its instantaneous velocity is increasing or decreasing with time.

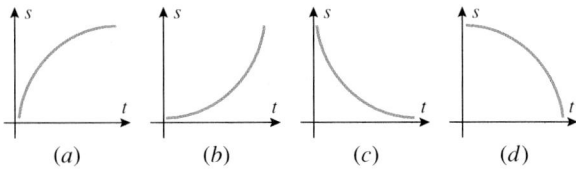

Figure Ex-14

15. Suppose that the outside temperature versus time curve over a 24-hour period is as shown in the accompanying figure.
    (a) Estimate the maximum temperature and the time at which it occurs.
    (b) The temperature rise is fairly linear from 8 A.M. to 2 P.M. Estimate the rate at which the temperature is increasing during this time period.
    (c) Estimate the time at which the temperature is decreasing most rapidly. Estimate the instantaneous rate of change of temperature with respect to time at this instant.

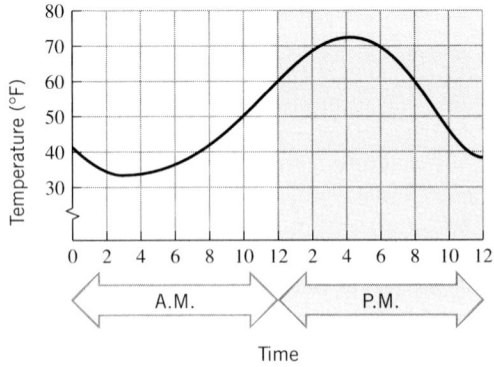

Figure Ex-15

16. The accompanying figure shows the graph of the pressure p in atmospheres (atm) versus the volume V in liters (L) of 1 mole of an ideal gas at a constant temperature of 300 K (kelvins). Use the tangent lines shown in the figure to estimate the rate of change of pressure with respect to volume at the points where $V = 10$ L and $V = 25$ L.

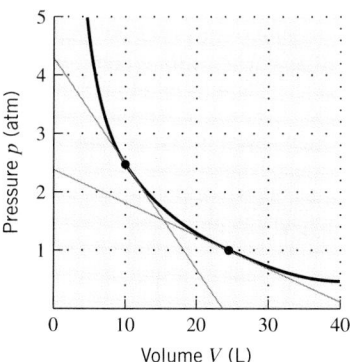

Figure Ex-16

17. The accompanying figure shows the graph of the height h in centimeters versus the age t in years of an individual from birth to age 20.

(a) When is the growth rate greatest?

(b) Estimate the growth rate at age 5.

(c) At approximately what age between 10 and 20 is the growth rate greatest? Estimate the growth rate at this age.

(d) Draw a rough graph of the growth rate versus age.

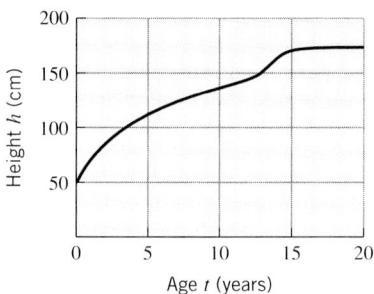

Figure Ex-17

In Exercises 18–21, use Formulas (2) and (3) to find the average and instantaneous velocity.

18. A rock is dropped from a height of 576 ft and falls toward Earth in a straight line. In t seconds the rock drops a distance of $s = 16t^2$ ft.

(a) How many seconds after release does the rock hit the ground?

(b) What is the average velocity of the rock during the time it is falling?

(c) What is the average velocity of the rock for the first 3 s?

(d) What is the instantaneous velocity of the rock when it hits the ground?

19. During the first 40 s of a rocket flight, the rocket is propelled straight up so that in t seconds it reaches a height of $s = 5t^3$ ft.

(a) How high does the rocket travel in 40 s?

(b) What is the average velocity of the rocket during the first 40 s?

(c) What is the average velocity of the rocket during the first 135 ft of its flight?

(d) What is the instantaneous velocity of the rocket at the end of 40 s?

20. A particle moves on a line away from its initial position so that after t hours it is $s = 3t^2 + t$ miles from its initial position.

(a) Find the average velocity of the particle over the interval $[1, 3]$.

(b) Find the instantaneous velocity at $t = 1$.

21. A particle moves in the positive direction along a straight line so that after t minutes its distance is $s = 6t^4$ feet from the origin.

(a) Find the average velocity of the particle over the interval $[2, 4]$.

(b) Find the instantaneous velocity at $t = 2$.

3.2 THE DERIVATIVE

In this section we will introduce the concept of a derivative, which is the primary mathematical tool that is used to calculate rates of change and slopes of tangent lines.

TANGENT LINES DEFINED PRECISELY

In the preceding section we showed informally that the slope of the tangent line to the graph of $y = f(x)$ at the point x_0 is given by

$$m_{\tan} = \lim_{x_1 \to x_0} \frac{f(x_1) - f(x_0)}{x_1 - x_0} \qquad (1)$$

However, for computational purposes it will be more convenient to express this formula in a different form by introducing a new variable $h = x_1 - x_0$. It follows that $x_1 = x_0 + h$, and consequently $x_1 \to x_0$ as $h \to 0$. Thus, (1) can be expressed as

$$m_{\tan} = \lim_{h \to 0} \frac{f(x_0 + h) - f(x_0)}{h}$$

(Figure 3.2.1). This suggests the following formal definition of a tangent line.

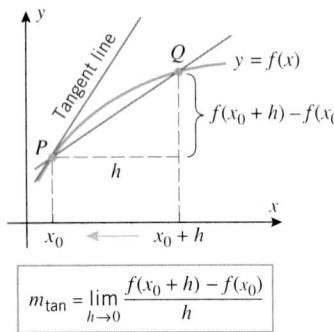

$$m_{\tan} = \lim_{h \to 0} \frac{f(x_0 + h) - f(x_0)}{h}$$

Figure 3.2.1

3.2.1 DEFINITION. If $P(x_0, y_0)$ is a point on the graph of a function f, then the *tangent line to the graph of f at P*, also called the *tangent line to the graph of f at x_0*, is defined to be the line through P with slope

$$m_{\tan} = \lim_{h \to 0} \frac{f(x_0 + h) - f(x_0)}{h} \tag{2}$$

provided this limit exists. If the limit does not exist, then by agreement the graph has no tangent line at P.

It follows from this definition that the point-slope form of the equation of the tangent line at x_0 is

$$y - y_0 = m_{\tan}(x - x_0) \tag{3}$$

Example 1

Find the equation of the tangent line to the graph of $y = x^2 + 1$ at the point $(2, 5)$ (Figure 3.2.2).

Solution. First, we will find the slope of the tangent line using (2) with $f(x) = x^2 + 1$ and $x_0 = 2$, and then we will find the equation by using (3). We obtain

$$m_{\tan} = \lim_{h \to 0} \frac{f(2 + h) - f(2)}{h} = \lim_{h \to 0} \frac{[(2 + h)^2 + 1] - 5}{h}$$

$$= \lim_{h \to 0} \frac{(5 + 4h + h^2) - 5}{h} = \lim_{h \to 0} \frac{4h + h^2}{h}$$

$$= \lim_{h \to 0} (4 + h) = 4$$

Thus, from (3) with $x_0 = 2$, $y_0 = 5$, and $m_{\tan} = 4$, the point-slope form of the equation of the tangent line is

$$y - 5 = 4(x - 2)$$

which we can write in slope-intercept form as $y = 4x - 3$. ◄

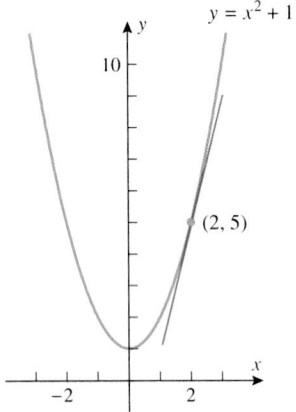

Figure 3.2.2

SLOPES OF TANGENT LINES BY ZOOMING

Slopes of tangent lines can be estimated by zooming with graphing utilities. The idea is to zoom in on the point of tangency until the surrounding curve segment appears to be a straight line that nearly coincides with the tangent line (Figure 3.2.3). The utility's trace operation can then be used to estimate the slope. Figure 3.2.4 illustrates this procedure for the tangent line in Example 1. The first part of the figure shows the graph of $y = x^2 + 1$ in the window[*] $[-6.3, 6.3] \times [0, 6.2]$, and the second part shows the graph after we have zoomed in on the point $(2, 5)$ by a factor of 10. The trace operation produces the points $(2.05, 5.2025)$ and $(1.95, 4.8025)$ on the line, so the slope of the tangent line can be approximated as

$$m \approx \frac{5.2025 - 4.8025}{2.05 - 1.95} = \frac{0.4}{0.1} = 4.0$$

which happens to agree exactly with the result in Example 1. It is important to understand, however, that the exact agreement in this case is accidental; in general, this method will not produce exact results because of roundoff errors in the computations, and also because the magnified curve segment may have a slight curvature, even though it appears to be a straight line.

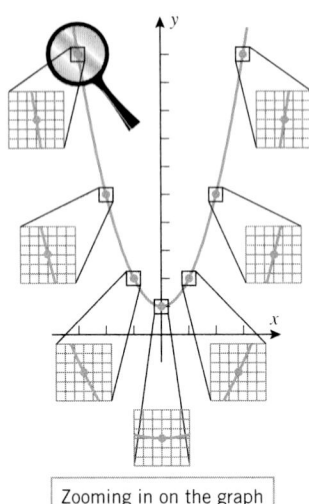

Zooming in on the graph of $y = x^2 + 1$

Figure 3.2.3

---

[*]The window $[-6.3, 6.3] \times [0, 6.2]$ was chosen because it contains the point of tangency $(2, 5)$ and produces convenient steps on the author's calculator when the trace operation is applied. Books on graphing calculators sometimes call these "friendly windows."

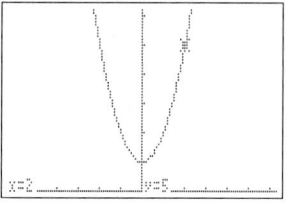 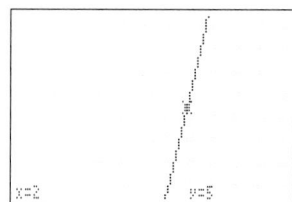 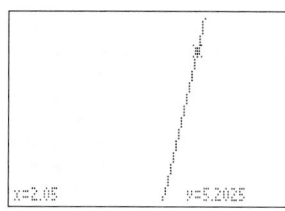 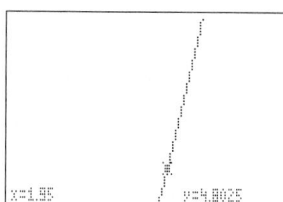

Figure 3.2.4

THE DERIVATIVE

In general, the slope of a tangent line to a curve $y = f(x)$ will depend on the point x at which the slope is being computed; thus, the slope is itself a function of x. To illustrate this, let us use (2) to compute $m_{\tan}$ at a general point x for the curve $y = x^2 + 1$. The computations are similar to those in Example 1, except that now we let x_0 have an arbitrary value $x_0 = x$, whereas in Example 1 we had $x_0 = 2$. We obtain

$$m_{\tan} = \lim_{h \to 0} \frac{f(x+h) - f(x)}{h} = \lim_{h \to 0} \frac{[(x+h)^2 + 1] - [x^2 + 1]}{h}$$

$$= \lim_{h \to 0} \frac{x^2 + 2xh + h^2 + 1 - x^2 - 1}{h} = \lim_{h \to 0} \frac{2xh + h^2}{h}$$

$$= \lim_{h \to 0} (2x + h) = 2x \tag{4}$$

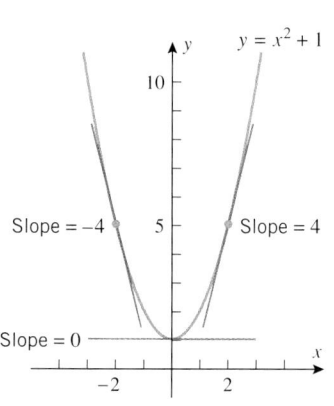

Figure 3.2.5

Now we can use the general formula $m_{\tan} = 2x$ to compute the slope of the tangent line at any point along the curve $y = x^2 + 1$ simply by substituting the appropriate value for x. For example, if $x = 2$, then we obtain $m_{\tan} = 2x = 4$, which agrees with the result in Example 1. Similarly, if $x = 0$, then $m_{\tan} = 0$; and if $x = -2$, then $m_{\tan} = -4$ (Figure 3.2.5).

To generalize this idea, the slope of the tangent line to the graph of $y = f(x)$ at a general point x can be obtained by setting $x_0 = x$ in (2), which yields the formula

$$m_{\tan} = \lim_{h \to 0} \frac{f(x+h) - f(x)}{h}$$

This "slope-producing function" is so important that it has some notation and terminology associated with it.

3.2.2 DEFINITION. The function f' defined by the formula

$$f'(x) = \lim_{h \to 0} \frac{f(x+h) - f(x)}{h} \tag{5}$$

is called the ***derivative of f with respect to x***. The domain of f' consists of all x for which the limit exists.

Recalling from the last section that the slope of a tangent line to the graph of $y = f(x)$ can be interpreted as the instantaneous rate of change of y with respect to x, it follows that the derivative of a function f can be interpreted in two ways:

Two interpretations of the Derivative. The derivative f' of a function f can be interpreted either as a function whose value at x is the slope of the tangent line to the graph of $y = f(x)$ at x, or, alternatively, it can be interpreted as a function whose value at x is the instantaneous rate of change of y with respect to x at the point x.

Example 2

(a) Find the derivative with respect to x of $f(x) = x^3 - x$.

(b) Graph f and f' together, and discuss the relationship between the two graphs.

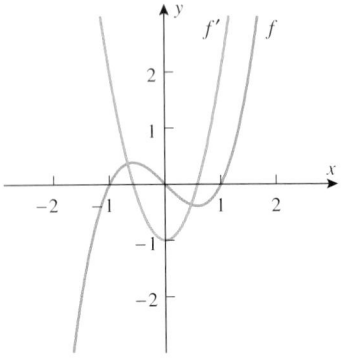

Figure 3.2.6

Solution (a). Later in this chapter we will develop efficient methods for finding derivatives, but for now we will find the derivative directly from Formula (5) in the definition of f'. The computations are as follows:

$$f'(x) = \lim_{h \to 0} \frac{f(x + h) - f(x)}{h} = \lim_{h \to 0} \frac{[(x + h)^3 - (x + h)] - (x^3 - x)}{h}$$

$$= \lim_{h \to 0} \frac{x^3 + 3x^2h + 3xh^2 + h^3 - x - h - x^3 + x}{h}$$

$$= \lim_{h \to 0} \frac{3x^2h + 3xh^2 + h^3 - h}{h}$$

$$= \lim_{h \to 0} (3x^2 + 3xh + h^2 - 1) = 3x^2 - 1$$

Solution (b). Since $f'(x)$ can be interpreted as the slope of the tangent line to the graph of $y = f(x)$ at the point x, the derivative $f'(x)$ is positive where the tangent line $y = f(x)$ has positive slope, it is negative where the tangent line has negative slope, and it is zero where the tangent line is horizontal. We leave it for the reader to verify that this is consistent with the graphs of $f(x) = x^3 - x$ and $f'(x) = 3x^2 - 1$ shown in Figure 3.2.6. ◀

Example 3

At each point x, the tangent line to a line $y = mx + b$ coincides with the line itself (Figure 3.2.7), and hence all tangent lines have slope m. This suggests geometrically that if $f(x) = mx + b$, then $f'(x) = m$ for all x. This is confirmed by the following computations:

$$f'(x) = \lim_{h \to 0} \frac{f(x + h) - f(x)}{h} = \lim_{h \to 0} \frac{[m(x + h) + b] - (mx + b)}{h}$$

$$= \lim_{h \to 0} \frac{mx + mh + b - mx - b}{h} = \lim_{h \to 0} \frac{mh}{h} = \lim_{h \to 0} m = m$$ ◀

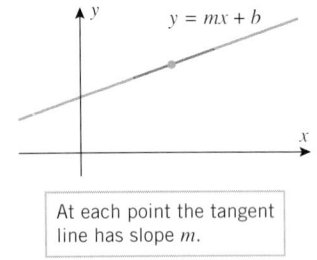

At each point the tangent line has slope m.

Figure 3.2.7

Example 4

(a) Find the derivative with respect to x of $f(x) = \sqrt{x}$.

(b) Find the slope of the tangent line to $y = \sqrt{x}$ at $x = 9$.

(c) Find the limits of $f'(x)$ as $x \to 0^+$ and as $x \to +\infty$, and explain what those limits say about the graph of f.

Solution (a). From Definition 3.2.2,

$$f'(x) = \lim_{h \to 0} \frac{f(x + h) - f(x)}{h} = \lim_{h \to 0} \frac{\sqrt{x + h} - \sqrt{x}}{h}$$

$$= \lim_{h \to 0} \frac{(\sqrt{x + h} - \sqrt{x})(\sqrt{x + h} + \sqrt{x})}{h(\sqrt{x + h} + \sqrt{x})} = \lim_{h \to 0} \frac{(x + h) - x}{h(\sqrt{x + h} + \sqrt{x})}$$

$$= \lim_{h \to 0} \frac{h}{h(\sqrt{x + h} + \sqrt{x})} = \lim_{h \to 0} \frac{1}{\sqrt{x + h} + \sqrt{x}}$$

$$= \frac{1}{\sqrt{x} + \sqrt{x}} = \frac{1}{2\sqrt{x}}$$

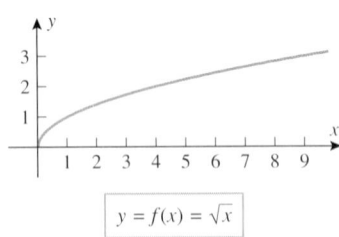

$y = f(x) = \sqrt{x}$

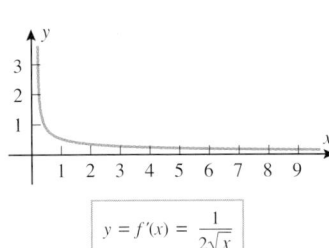

$y = f'(x) = \dfrac{1}{2\sqrt{x}}$

Figure 3.2.8

Solution (b). The slope of the tangent line at $x = 9$ is $f'(9)$, and thus from part (a) this slope is $f'(9) = 1/(2\sqrt{9}) = \frac{1}{6}$.

Solution (c). The graphs of $f(x) = \sqrt{x}$ and $f'(x) = 1/(2\sqrt{x})$ are shown in Figure 3.2.8. Observe that $f'(x) > 0$ if $x > 0$, which means that all tangent lines to the graph of $y = \sqrt{x}$

have positive slope over this interval. Since

$$\lim_{x \to 0^+} \frac{1}{2\sqrt{x}} = +\infty \quad \text{and} \quad \lim_{x \to +\infty} \frac{1}{2\sqrt{x}} = 0$$

the tangent lines become more and more vertical as $x \to 0^+$, and they become more and more horizontal as $x \to +\infty$. ◄

FOR THE READER. Use a graphing utility to estimate the slope of the tangent line to $y = \sqrt{x}$ at $x = 9$ by zooming, and compare your result to the exact value obtained in the last example. If you have a CAS, read the documentation to determine how it can be used to find derivatives, and then use it to confirm the derivatives obtained in Examples 2, 3, and 4.

DIFFERENTIABILITY

Recall from Definition 3.2.2 that the derivative of a function f is defined at those points where the limit (5) exists. Points where this limit exists are called ***points of differentiability*** for f, and points where this limit does not exist are called ***points of nondifferentiability*** for f.

If x_0 is a point of differentiability for f, then we say that f is ***differentiable at x_0*** or that ***the derivative of f exists at x_0***; and if x_0 is a point of nondifferentiability for f, then we say that ***the derivative of f does not exist at x_0***. If f is differentiable at every point in an open interval (a, b), then we will say that f is ***differentiable on (a, b)***. This definition also applies to infinite open intervals of the form $(a, +\infty)$, $(-\infty, b)$, and $(-\infty, +\infty)$. In the case where f is differentiable on $(-\infty, +\infty)$ we will say that f is ***differentiable everywhere***. If f is differentiable on an open interval but the particular interval is not important for the discussion, then we will say that f is ***differentiable*** (without referencing the interval).

Geometrically, the points of differentiability of f are the points where the curve $y = f(x)$ has a tangent line, and the points of nondifferentiability are the points where the curve does not have a tangent line. Informally stated, the most commonly encountered points of nondifferentiability can be classified as

- Corners
- Points of vertical tangency
- Points of discontinuity

Figure 3.2.9 illustrates each of these situations.

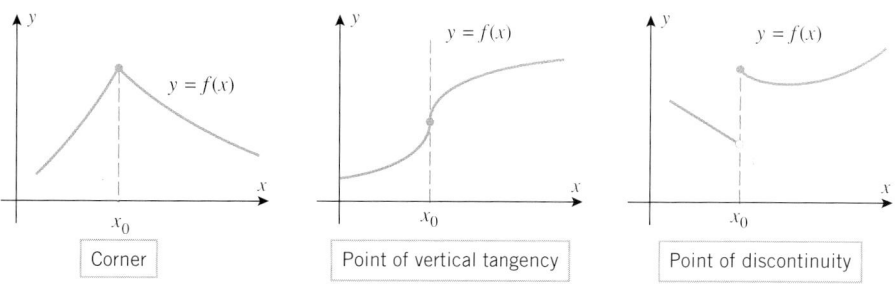

Figure 3.2.9

It makes sense intuitively that corners are points of nondifferentiability, since there is no reasonable way to draw a unique tangent line at such points. For example, Figure 3.2.10a shows a typical corner point $P(x_0, f(x_0))$ on the graph of a function f. At this point the secant lines joining P and Q have different limiting positions, depending on whether Q approaches P from the left or right; hence the slopes of the secant lines do not have a two-sided limit.

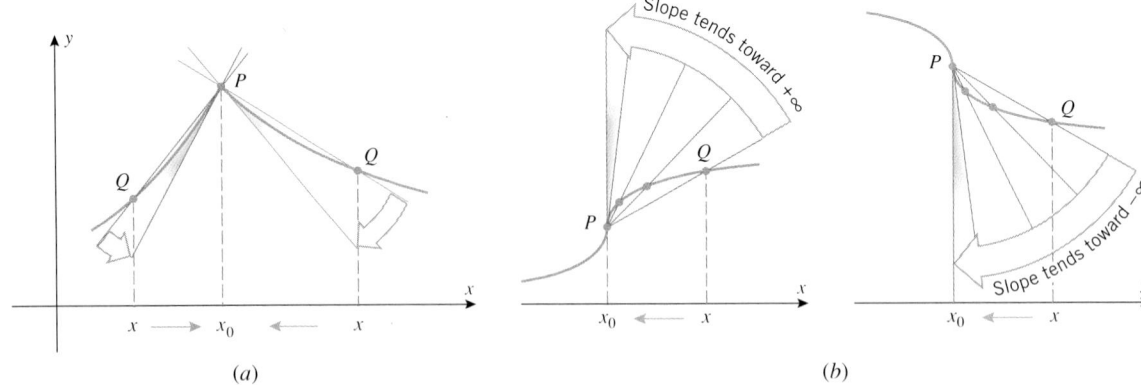

Figure 3.2.10

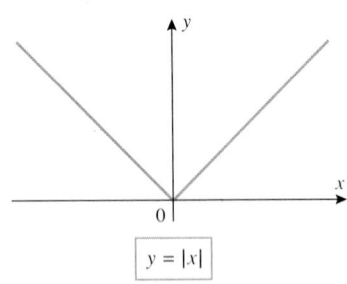

$y = |x|$

Figure 3.2.11

By a point of *vertical tangency* we mean a place on the curve where the secant lines approach a vertical limiting position. At such points, the only reasonable candidate for the tangent line is the vertical line at the point. But vertical lines have infinite slope, so the derivative (were it to exist) would not have a finite real value there, which explains intuitively why the derivative does not exist at points of vertical tangency (Figure 3.2.10*b*).

Example 5

The graph of $y = |x|$ in Figure 3.2.11 suggests that there is a corner at $x = 0$, and this implies that $f(x) = |x|$ is not differentiable at that point.

(a) Prove that $f(x) = |x|$ is not differentiable at $x = 0$ by showing that the limit in Definition 3.2.2 does not exist at that point.

(b) Find a formula for $f'(x)$.

Solution (*a*). From Formula (5) with $x = 0$, the value of $f'(0)$, if it were to exist, would be given by

$$f'(0) = \lim_{h \to 0} \frac{f(0 + h) - f(0)}{h} = \lim_{h \to 0} \frac{f(h) - f(0)}{h} = \lim_{h \to 0} \frac{|h| - |0|}{h} = \lim_{h \to 0} \frac{|h|}{h}$$

But

$$\frac{|h|}{h} = \begin{cases} 1, & h > 0 \\ -1, & h < 0 \end{cases}$$

so that

$$\lim_{h \to 0^-} \frac{|h|}{h} = -1 \quad \text{and} \quad \lim_{h \to 0^+} \frac{|h|}{h} = 1$$

Thus,

$$f'(0) = \lim_{h \to 0} \frac{|h|}{h}$$

does not exist because the one-sided limits are not equal. Consequently, $f(x) = |x|$ is not differentiable at $x = 0$.

Solution (*b*). A formula for the derivative of $f(x) = |x|$ can be obtained by writing $|x|$ in piecewise form and treating the cases $x > 0$ and $x < 0$ separately. If $x > 0$, then $f(x) = x$ and $f'(x) = 1$; and if $x < 0$, then $f(x) = -x$ and $f'(x) = -1$. Thus,

$$f'(x) = \begin{cases} 1, & x > 0 \\ -1, & x < 0 \end{cases}$$

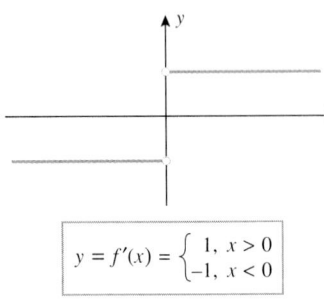

$y = f'(x) = \begin{cases} 1, & x > 0 \\ -1, & x < 0 \end{cases}$

Figure 3.2.12

The graph of f' is shown in Figure 3.2.12. Observe that f' is not a continuous function, so this example shows that the derivative of a continuous function need not be continuous. ◀

**RELATIONSHIP BETWEEN
DIFFERENTIABILITY AND
CONTINUITY**

It makes sense intuitively that a function f cannot be differentiable at a point of discontinuity, since there is no reasonable way to draw a unique tangent line at such points. The following theorem shows that a function f must be continuous at each point where it is differentiable (or stated another way, a function f cannot be differentiable at a point of discontinuity).

3.2.3 THEOREM. *If f is differentiable at a point x_0, then f is also continuous at x_0.*

Proof. We are given that f is differentiable at x_0, so it follows from (5) that $f'(x_0)$ exists and is given by

$$f'(x_0) = \lim_{h \to 0} \left[\frac{f(x_0 + h) - f(x_0)}{h} \right] \tag{6}$$

To show that f is continuous at x_0, we must show that $\lim_{x \to x_0} f(x) = f(x_0)$, or equivalently,

$$\lim_{x \to x_0} [f(x) - f(x_0)] = 0$$

Expressing this in terms of the variable $h = x - x_0$, we must prove that

$$\lim_{h \to 0} [f(x_0 + h) - f(x_0)] = 0$$

However, this can be proved using (6) as follows:

$$\lim_{h \to 0} [f(x_0 + h) - f(x_0)] = \lim_{h \to 0} \left[\frac{f(x_0 + h) - f(x_0)}{h} \cdot h \right]$$
$$= \lim_{h \to 0} \left[\frac{f(x_0 + h) - f(x_0)}{h} \right] \cdot \lim_{h \to 0} h$$
$$= f'(x_0) \cdot 0 = 0 \quad \blacksquare$$

REMARK. Theorem 3.2.3 shows that differentiability at a point implies continuity at that point. However, the converse is false; that is, *a function may be continuous at a point but not differentiable there*. In fact, this occurs at any point where the function is continuous and has a corner. For example, we saw in Example 5 that the function $f(x) = |x|$ is continuous at $x = 0$, yet not differentiable there.

The relationship between continuity and differentiability was of great historical significance in the development of calculus. In the early nineteenth century mathematicians believed that the graph of a continuous function could not have too many points of nondifferentiability bunched up. They felt that if a continuous function had many points of nondifferentiability, these points, like the tips of a sawblade, would have to be separated from each other and joined by smooth curve segments (Figure 3.2.13). This misconception was shattered by a series of discoveries beginning in 1834. In that year a Bohemian priest, philosopher, and mathematician named Bernhard Bolzano[*] discovered a procedure for constructing a continuous function that is not differentiable at any point. Later, in 1860, the great

[*] BERNHARD BOLZANO (1781–1848). Bolzano, the son of an art dealer, was born in Prague, Bohemia (Czech Republic). He was educated at the University of Prague, and eventually won enough mathematical fame to be recommended for a mathematics chair there. However, Bolzano became an ordained Roman Catholic priest, and in 1805 he was appointed to a chair of Philosophy at the University of Prague. Bolzano was a man of great human compassion; he spoke out for educational reform, he voiced the right of individual conscience over government demands, and he lectured on the absurdity of war and militarism. His views so disenchanted Emperor Franz I of Austria that the emperor pressed the Archbishop of Prague to have Bolzano recant his statements. Bolzano refused and was then forced to retire in 1824 on a small pension. Bolzano's main contribution to mathematics was philosophical. His work helped convince mathematicians that sound mathematics must ultimately rest on rigorous proof rather than intuition. In addition to his work in mathematics, Bolzano investigated problems concerning space, force, and wave propagation.

German mathematician, Karl Weierstrass[*] produced the first formula for such a function. The graphs of such functions are impossible to draw; it is as if the corners are so numerous that any segment of the curve, when suitably enlarged, reveals more corners. The discovery of these pathological functions was important in that it made mathematicians distrustful of their geometric intuition and more reliant on precise mathematical proof. However, they remained only mathematical curiosities until the early 1980s, when applications of them began to emerge. During the past 10 years they have started to play a fundamental role in the study of geometric objects called **_fractals_**. Fractals have revealed an order to natural phenomena that were previously dismissed as random and chaotic.

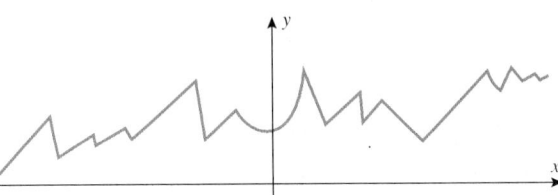

Figure 3.2.13

DERIVATIVE NOTATION

The process of finding a derivative is called **_differentiation_**. You can think of differentiation as an operation on functions that associates a function f' with a function f. When the independent variable is x, the differentiation operation is often denoted by

$$\frac{d}{dx}[f(x)]$$

which is read "**_the derivative of f (x) with respect to x_.**" Thus,

$$\frac{d}{dx}[f(x)] = f'(x) \tag{7}$$

For example, with this notation the derivatives obtained in Examples 2, 3, and 4 can be expressed as

$$\frac{d}{dx}[x^3 - x] = 3x^2 - 1, \quad \frac{d}{dx}[mx + b] = m, \quad \frac{d}{dx}[\sqrt{x}] = \frac{1}{2\sqrt{x}} \tag{8}$$

To denote the value of the derivative at a specific point x_0 with the notation in (7), we would

---

[*] KARL WEIERSTRASS (1815–1897). Weierstrass, the son of a customs officer, was born in Ostenfelde, Germany. As a youth Weierstrass showed outstanding skills in languages and mathematics. However, at the urging of his dominant father, Weierstrass entered the law and commerce program at the University of Bonn. To the chagrin of his family, the rugged and congenial young man concentrated instead on fencing and beer drinking. Four years later he returned home without a degree. In 1839 Weierstrass entered the Academy of Münster to study for a career in secondary education, and he met and studied under an excellent mathematician named Christof Gudermann. Gudermann's ideas greatly influenced the work of Weierstrass. After receiving his teaching certificate, Weierstrass spent the next 15 years in secondary education teaching German, geography, and mathematics. In addition, he taught handwriting to small children. During this period much of Weierstrass's mathematical work was ignored because he was a secondary schoolteacher and not a college professor. Then, in 1854, he published a paper of major importance that created a sensation in the mathematics world and catapulted him to international fame overnight. He was immediately given an honorary Doctorate at the University of Königsberg and began a new career in college teaching at the University of Berlin in 1856. In 1859 the strain of his mathematical research caused a temporary nervous breakdown and led to spells of dizziness that plagued him for the rest of his life. Weierstrass was a brilliant teacher and his classes overflowed with multitudes of auditors. In spite of his fame, he never lost his early beer-drinking congeniality and was always in the company of students, both ordinary and brilliant. Weierstrass was acknowledged as the leading mathematical analyst in the world. He and his students opened the door to the modern school of mathematical analysis.

write

$$\frac{d}{dx}[f(x)]\bigg|_{x=x_0} = f'(x_0) \tag{9}$$

For example, from (8)

$$\frac{d}{dx}[x^3 - x]\bigg|_{x=1} = 3(1^2) - 1 = 2, \quad \frac{d}{dx}[mx+b]\bigg|_{x=5} = m, \quad \frac{d}{dx}[\sqrt{x}]\bigg|_{x=9} = \frac{1}{2\sqrt{9}} = \frac{1}{6}$$

Notations (7) and (9) are convenient when no dependent variable is involved. However, if there is a dependent variable, say $y = f(x)$, then (7) and (9) can be written as

$$\frac{d}{dx}[y] = f'(x) \quad \text{and} \quad \frac{d}{dx}[y]\bigg|_{x=x_0} = f'(x_0)$$

It is common to omit the brackets on the left side and write these expressions as

$$\frac{dy}{dx} = f'(x) \qquad \text{and} \qquad \frac{dy}{dx}\bigg|_{x=x_0} = f'(x_0)$$

where dy/dx is read as "the derivative of y with respect to x." For example, if $y = \sqrt{x}$, then

$$\frac{dy}{dx} = \frac{1}{2\sqrt{x}}, \quad \frac{dy}{dx}\bigg|_{x=x_0} = \frac{1}{2\sqrt{x_0}}, \quad \frac{dy}{dx}\bigg|_{x=9} = \frac{1}{2\sqrt{9}} = \frac{1}{6}$$

REMARK.   Later, the symbols dy and dx will be defined separately. However, for the time being, dy/dx should not be regarded as a ratio; rather, it should be considered as a single symbol denoting the derivative.

When letters other than x and y are used for the independent and dependent variables, then the various notations for the derivative must be adjusted accordingly. For example, if $y = f(u)$, then the derivative with respect to u would be written as

$$\frac{d}{du}[f(u)] = f'(u) \quad \text{and} \quad \frac{dy}{du} = f'(u)$$

In particular, if $y = \sqrt{u}$, then

$$\frac{dy}{du} = \frac{1}{2\sqrt{u}}, \quad \frac{dy}{du}\bigg|_{u=u_0} = \frac{1}{2\sqrt{u_0}}, \quad \frac{dy}{du}\bigg|_{u=9} = \frac{1}{2\sqrt{9}} = \frac{1}{6}$$

OTHER NOTATIONS

Some writers denote the derivative as $D_x[f(x)] = f'(x)$, but we will not use this notation in this text. In problems where the name of the independent variable is clear from the context, there are some other possible notations for the derivative. For example, if $y = f(x)$, but it is clear from the problem that the independent variable is x, then the derivative with respect to x might be denoted by y' or f'.

Often, you will see Definition 3.2.2 expressed using Δx (delta x) rather than h for the varying quantity, in which case (5) has the form

$$f'(x) = \lim_{\Delta x \to 0} \frac{f(x + \Delta x) - f(x)}{\Delta x} \tag{10}$$

If $y = f(x)$, then it is also common to let

$$\Delta y = f(x + \Delta x) - f(x)$$

in which case

$$\frac{dy}{dx} = \lim_{\Delta x \to 0} \frac{\Delta y}{\Delta x} = \lim_{\Delta x \to 0} \frac{f(x + \Delta x) - f(x)}{\Delta x} \tag{11}$$

The geometric interpretations of Δx and Δy are shown in Figure 3.2.14.

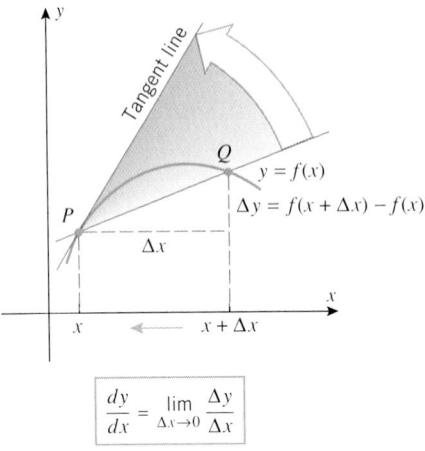

$$\boxed{\frac{dy}{dx} = \lim_{\Delta x \to 0} \frac{\Delta y}{\Delta x}}$$

Figure 3.2.14

DERIVATIVES AT THE ENDPOINTS OF AN INTERVAL

If a function f is defined on a closed interval $[a, b]$ and is not defined outside of that interval, then the derivative $f'(x)$ is not defined at the endpoints a and b because

$$f'(x) = \lim_{h \to 0} \frac{f(x + h) - f(x)}{h}$$

is a two-sided limit and only one-sided limits make sense at the endpoints. To deal with this situation, we define *derivatives from the left and right*. These are denoted by f'_- and f'_+, respectively, and are defined by

$$f'_-(x) = \lim_{h \to 0^-} \frac{f(x + h) - f(x)}{h} \quad \text{and} \quad f'_+(x) = \lim_{h \to 0^+} \frac{f(x + h) - f(x)}{h}$$

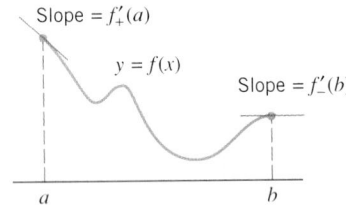

Figure 3.2.15

At points where $f'_+(x)$ exists we say that the function f is *differentiable from the right*, and at points where $f'_-(x)$ exists we say that f is *differentiable from the left*. Geometrically, $f'_+(x)$ is the limit of the slopes of the secant lines approaching x from the right, and $f'_-(x)$ is the limit of the slopes of the secant lines approaching x from the left (Figure 3.2.15).

It can be proved that a function f is continuous from the left at those points where it is differentiable from the left and is continuous from the right at those points where it is differentiable from the right.

We will call a function f *differentiable* on an interval of the form $[a, b], [a, +\infty), (-\infty, b]$, $[a, b)$, or $(a, b]$ if it is differentiable at all points inside the interval, and it is differentiable at the endpoint(s) from the left or right, as appropriate.

EXERCISE SET 3.2 ☒ Graphing Calculator

1. Use the graph of $y = f(x)$ in the accompanying figure to estimate the value of $f'(1)$, $f'(3)$, $f'(5)$, and $f'(6)$.

2. For the function graphed in the accompanying figure, arrange the numbers 0, $f'(-3)$, $f'(0)$, $f'(2)$, and $f'(4)$ in increasing order.

3. (a) If you are given an equation for the tangent line at the point $(a, f(a))$ on a curve $y = f(x)$, how would you go about finding $f'(a)$?

(b) Given that the tangent line to the graph of $y = f(x)$ at the point $(2, 5)$ has the equation $y = 3x + 1$, find $f'(2)$.

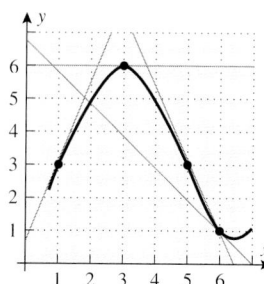

Figure Ex-1

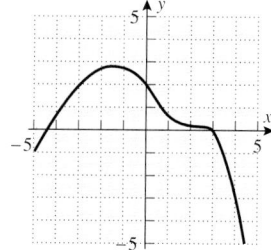

Figure Ex-2

(c) For the equation in part (b), what is the instantaneous rate of change of y with respect to x at $x = 2$?

4. Given that the tangent line to $y = f(x)$ at the point $(-1, 3)$ passes through the point $(0, 4)$, find $f'(-1)$.

5. Sketch the graph of a function f for which $f(0) = 1$, $f'(0) = 0$, $f'(x) > 0$ if $x < 0$, and $f'(x) < 0$ if $x > 0$.

6. Sketch the graph of a function f for which $f(0) = 0$, $f'(0) = 0$, and $f'(x) > 0$ if $x < 0$ or $x > 0$.

7. Given that $f(3) = -1$ and $f'(3) = 5$, find an equation for the tangent line to the graph of $y = f(x)$ at the point where $x = 3$.

8. Given that $f(-2) = 3$ and $f'(-2) = -4$, find an equation for the tangent line to the graph of $y = f(x)$ at the point where $x = -2$.

In Exercises 9–14, use Definition 3.2.2 to find $f'(x)$, and then find the equation of the tangent line to $y = f(x)$ at the point $x = a$.

9. $f(x) = 3x^2$; $a = 3$

10. $f(x) = x^2 - x$; $a = 2$

11. $f(x) = x^3$; $a = 0$

12. $f(x) = 2x^3 + 1$; $a = -1$

13. $f(x) = \sqrt{x + 1}$; $a = 8$ **14.** $f(x) = x^4$; $a = -2$

In Exercises 15–20, use Formula (11) to find dy/dx.

15. $y = \dfrac{1}{x}$ **16.** $y = \dfrac{1}{x^2}$

17. $y = ax^2 + b$ (a, b constants) **18.** $y = \dfrac{1}{x + 1}$

19. $y = \dfrac{1}{\sqrt{x}}$ **20.** $y = x^{1/3}$

In Exercises 21 and 22, use Definition 3.2.2 (with the appropriate change in notation) to obtain the derivative requested.

21. Find $f'(t)$ if $f(t) = 4t^2 + t$.

22. Find dV/dr if $V = \frac{4}{3}\pi r^3$.

23. Match the graphs of the functions shown in (a)–(f) with the graphs of their derivatives in (A)–(F).

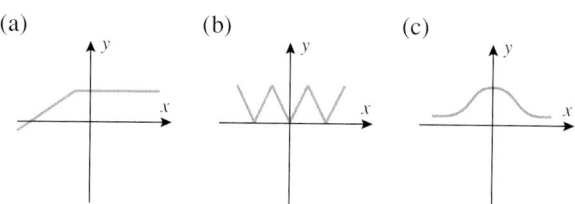

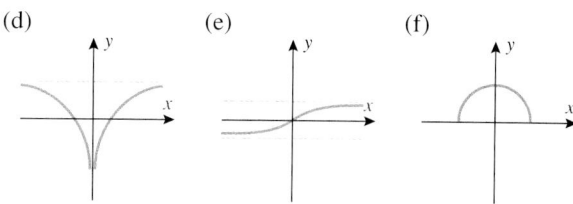

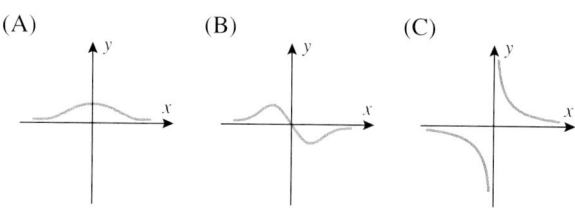

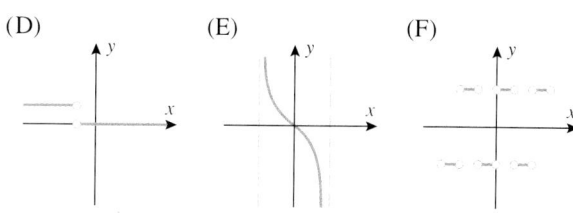

24. Find a function f such that $f'(x) = 1$ for all x, and give an informal argument to justify your answer.

In Exercises 25 and 26, sketch the graph of the derivative of the function whose graph is shown.

25. (a)   (b)   (c)

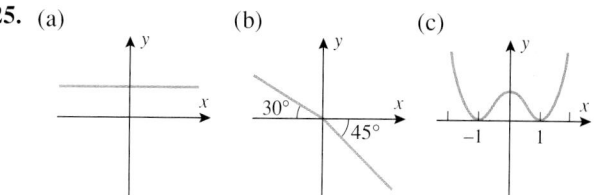

26. (a)   (b)   (c)

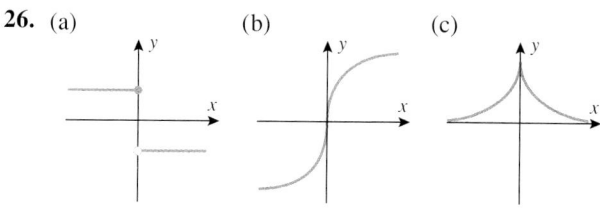

In Exercises 27 and 28, the limit represents $f'(a)$ for some function f and some number a. Find $f(x)$ and a in each case.

27. (a) $\displaystyle\lim_{h\to 0} \frac{(3+h)^2 - 9}{h}$ (b) $\displaystyle\lim_{\Delta x\to 0} \frac{\sqrt{1+\Delta x} - 1}{\Delta x}$

28. (a) $\displaystyle\lim_{h\to 0} \frac{\cos(\pi + h) + 1}{h}$ (b) $\displaystyle\lim_{x\to 1} \frac{x^7 - 1}{x - 1}$

29. Find $dy/dx|_{x=1}$, given that $y = 4x^2 + 1$.

30. Find $dy/dx|_{x=-2}$, given that $y = (5/x) + 1$.

31. Find an equation for the line that is tangent to the curve $y = x^3 - 2x + 1$ at the point $(0, 1)$, and use a graphing utility to graph the curve and its tangent line on the same screen.

32. Use a graphing utility to graph the following on the same screen: the curve $y = x^2/4$, the tangent line to this curve at $x = 1$, and the secant line joining the points $(0, 0)$ and $(2, 1)$ on this curve.

33. Let $f(x) = 2^x$. Estimate $f'(1)$ by
(a) using a graphing utility to zoom in at an appropriate point until the graph looks like a straight line, and then estimating the slope
(b) using a calculating utility to estimate the limit in Definition 3.2.2 by making a table of values for a succession of smaller and smaller values of h.

34. Let $f(x) = \sin x$. Estimate $f'(\pi/4)$ by
(a) using a graphing utility to zoom in at an appropriate point until the graph looks like a straight line, and then estimating the slope
(b) using a calculating utility to estimate the limit in Definition 3.2.2 by making a table of values for a succession of smaller and smaller values of h.

35. Suppose that the cost of drilling x feet for an oil well is $C = f(x)$ dollars.
(a) What are the units of $f'(x)$?
(b) In practical terms, what does $f'(x)$ mean in this case?
(c) What can you say about the sign of $f'(x)$?
(d) Estimate the cost of drilling an additional foot, starting at a depth of 300 ft, given that $f'(300) = 1000$.

36. A paint manufacturing company estimates that it can sell $g = f(p)$ gallons of paint at a price of p dollars.
(a) What are the units of dg/dp?
(b) In practical terms, what does dg/dp mean in this case?
(c) What can you say about the sign of dg/dp?
(d) Given that $dg/dp|_{p=10} = -100$, what can you say about the effect of increasing the price from \$10 per gallon to \$11 per gallon?

37. It is a fact that when a flexible rope is wrapped around a rough cylinder, a small force of magnitude F_0 at one end can resist a large force of magnitude F at the other end. The size of F depends on the angle θ through which the rope is wrapped around the cylinder (see the accompanying figure). That figure shows the graph of F (in pounds) versus θ (in radians), where F is the magnitude of the force that can be resisted by a force with magnitude $F_0 = 10$ lb for a certain rope and cylinder.
(a) Estimate the values of F and $dF/d\theta$ when the angle $\theta = 10$ radians.
(b) It can be shown that the force F satisfies the equation $dF/d\theta = \mu F$, where the constant μ is called the *coefficient of friction*. Use the results in part (a) to estimate the value of μ.

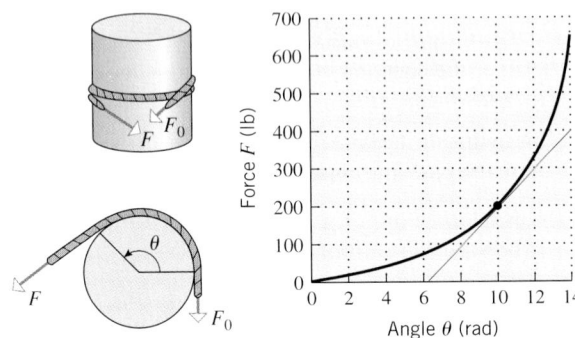

Figure Ex-37

38. According to *The World Almanac and the Book of Facts* (1987), the estimated world population, N, in millions for the years 1850, 1900, 1950, and 1985 was 1175, 1600, 2490, and 4843, respectively. Although the increase in population is not a continuous function of the time t, we can apply the ideas in this section if we are willing to approximate the graph of N versus t by a continuous curve, as shown in the accompanying figure.
(a) Use the estimated tangent line shown in the figure at the point where $t = 1950$ to approximate the value of dN/dt there. Describe your result as a rate of change.
(b) At any instant, the *growth rate* is defined as

$$\frac{dN/dt}{N}$$

Use your answer to part (a) to approximate the growth rate in 1950. Express the result as a percentage and include the proper units.

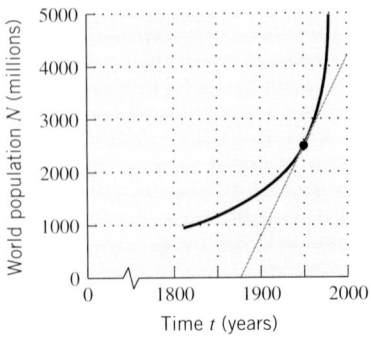

Figure Ex-38

39. According to *Newton's Law of Cooling*, the rate of change of an object's temperature is proportional to the difference between the temperature of the object and that of the surrounding medium. The accompanying figure shows the graph of the temperature T (in degrees Fahrenheit) versus time t (in minutes) for a cup of coffee, initially with a temperature of $200°$ F, that is allowed to cool in a room with a constant temperature of $75°$ F.
(a) Estimate T and dT/dt when $t = 10$ min.
(b) Newton's Law of Cooling can be expressed as

$$\frac{dT}{dt} = k(T - T_0)$$

where k is the constant of proportionality and T_0 is the temperature (assumed constant) of the surrounding medium. Use the results in part (a) to estimate the value of k.

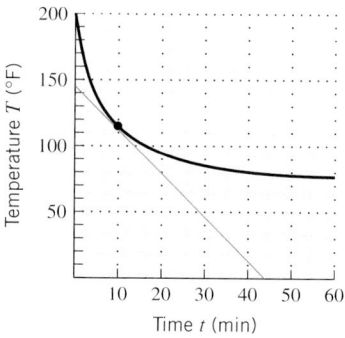

Figure Ex-39

40. Write a paragraph that explains what it means for a function to be differentiable. Include some examples of functions that are not differentiable, and explain the relationship between differentiability and continuity.

41. Show that $f(x) = \sqrt[3]{x}$ is continuous at $x = 0$ but not differentiable at $x = 0$. Sketch the graph of f.

42. Show that $f(x) = \sqrt[3]{(x-2)^2}$ is continuous at $x = 2$ but not differentiable at $x = 2$. Sketch the graph of f.

43. Show that

$$f(x) = \begin{cases} x^2 + 1, & x \le 1 \\ 2x, & x > 1 \end{cases}$$

is continuous and differentiable at $x = 1$. Sketch the graph of f.

44. Show that

$$f(x) = \begin{cases} x^2 + 2, & x \le 1 \\ x + 2, & x > 1 \end{cases}$$

is continuous but not differentiable at $x = 1$. Sketch the graph of f.

45. Suppose that a function f is differentiable at $x = 1$ and $\lim_{h \to 0} \frac{f(1+h)}{h} = 5$. Find $f(1)$ and $f'(1)$.

46. Suppose that f is a differentiable function with the property that $f(x + y) = f(x) + f(y) + 5xy$ and $\lim_{h \to 0} \frac{f(h)}{h} = 3$. Find $f(0)$ and $f'(x)$.

47. Suppose that f has the property $f(x + y) = f(x)f(y)$ for all values of x and y and that $f(0) = f'(0) = 1$. Show that f is differentiable and $f'(x) = f(x)$. [*Hint:* Start by expressing $f'(x)$ as a limit.]

3.3 TECHNIQUES OF DIFFERENTIATION

In the last section we defined the derivative of a function f as a limit, and we used that limit to calculate a few simple derivatives. In this section we will develop some important theorems that will enable us to calculate derivatives more efficiently.

DERIVATIVE OF A CONSTANT

The graph of a constant function $f(x) = c$ is the horizontal line $y = c$, and hence the tangent line to this graph has slope 0 at every point x (Figure 3.3.1). Thus, we should expect the derivative of a constant function to be 0 for all x.

3.3.1 THEOREM. *The derivative of a constant function is 0; that is, if c is any real number, then*

$$\frac{d}{dx}[c] = 0$$

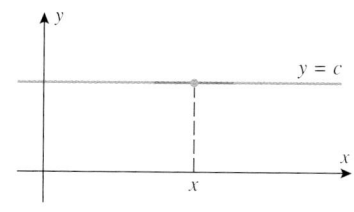

The tangent line to the graph of $f(x) = c$ has slope 0 for all x.

Figure 3.3.1

Proof. Let $f(x) = c$. Then from the definition of a derivative,

$$\frac{d}{dx}[c] = f'(x) = \lim_{h \to 0} \frac{f(x+h) - f(x)}{h} = \lim_{h \to 0} \frac{c - c}{h} = \lim_{h \to 0} 0 = 0 \quad ■$$

Example 1

If $f(x) = 5$ for all x, then $f'(x) = 0$ for all x; that is,

$$\frac{d}{dx}[5] = 0 \qquad \blacktriangleleft$$

DERIVATIVE OF x TO A POWER

3.3.2   THEOREM (*The Power Rule*).   *If n is a positive integer, then*

$$\frac{d}{dx}[x^n] = nx^{n-1}$$

Proof. Let $f(x) = x^n$. Thus, from the definition of a derivative and the binomial theorem for expanding the expression $(x + h)^n$, we obtain

$$\frac{d}{dx}[x^n] = f'(x) = \lim_{h \to 0} \frac{f(x+h) - f(x)}{h} = \lim_{h \to 0} \frac{(x+h)^n - x^n}{h}$$

$$= \lim_{h \to 0} \frac{\left[x^n + nx^{n-1}h + \frac{n(n-1)}{2!}x^{n-2}h^2 + \cdots + nxh^{n-1} + h^n\right] - x^n}{h}$$

$$= \lim_{h \to 0} \frac{nx^{n-1}h + \frac{n(n-1)}{2!}x^{n-2}h^2 + \cdots + nxh^{n-1} + h^n}{h}$$

$$= \lim_{h \to 0} \left[nx^{n-1} + \frac{n(n-1)}{2!}x^{n-2}h + \cdots + nxh^{n-2} + h^{n-1}\right]$$

$$= nx^{n-1} + 0 + \cdots + 0 + 0$$

$$= nx^{n-1} \qquad \blacksquare$$

REMARK.   In words, *to differentiate x to a positive integer power, multiply that power by x raised to the next lower integer power.*

Example 2

$$\frac{d}{dx}[x^5] = 5x^4, \qquad \frac{d}{dx}[x] = 1 \cdot x^0 = 1, \qquad \frac{d}{dx}[x^{12}] = 12x^{11} \qquad \blacktriangleleft$$

DERIVATIVE OF A CONSTANT TIMES A FUNCTION

3.3.3   THEOREM.   *If f is differentiable at x and c is any real number, then cf is also differentiable at x and*

$$\frac{d}{dx}[cf(x)] = c\frac{d}{dx}[f(x)]$$

Proof.

$$\frac{d}{dx}[cf(x)] = \lim_{h \to 0} \frac{cf(x+h) - cf(x)}{h} = \lim_{h \to 0} c\left[\frac{f(x+h) - f(x)}{h}\right]$$

$$= c\lim_{h \to 0} \frac{f(x+h) - f(x)}{h} = c\frac{d}{dx}[f(x)] \qquad \blacksquare$$

A constant factor can be moved through a limit sign.

In function notation, Theorem 3.3.3 states

$$(cf)' = cf'$$

REMARK. In words, *a constant factor can be moved through a derivative sign.*

Example 3

$$\frac{d}{dx}[4x^8] = 4\frac{d}{dx}[x^8] = 4[8x^7] = 32x^7$$

$$\frac{d}{dx}[-x^{12}] = (-1)\frac{d}{dx}[x^{12}] = -12x^{11}$$

$$\frac{d}{dx}\left[\frac{x}{\pi}\right] = \frac{1}{\pi}\frac{d}{dx}[x] = \frac{1}{\pi}$$

◀

DERIVATIVES OF SUMS AND DIFFERENCES

3.3.4 THEOREM. *If f and g are differentiable at x, then so are f + g and f − g and*

$$\frac{d}{dx}[f(x) + g(x)] = \frac{d}{dx}[f(x)] + \frac{d}{dx}[g(x)]$$

$$\frac{d}{dx}[f(x) - g(x)] = \frac{d}{dx}[f(x)] - \frac{d}{dx}[g(x)]$$

Proof.

$$\frac{d}{dx}[f(x) + g(x)] = \lim_{h \to 0} \frac{[f(x+h) + g(x+h)] - [f(x) + g(x)]}{h}$$

$$= \lim_{h \to 0} \frac{[f(x+h) - f(x)] + [g(x+h) - g(x)]}{h}$$

$$= \lim_{h \to 0} \frac{f(x+h) - f(x)}{h} + \lim_{h \to 0} \frac{g(x+h) - g(x)}{h} \qquad \text{The limit of a sum is the sum of the limits.}$$

$$= \frac{d}{dx}[f(x)] + \frac{d}{dx}[g(x)]$$

The proof for $f - g$ is similar. ∎

In function notation, Theorem 3.3.4 states

$$(f + g)' = f' + g' \qquad (f - g)' = f' - g'$$

REMARK. In words, *the derivative of a sum equals the sum of the derivatives,* and *the derivative of a difference equals the difference of the derivatives.*

Example 4

$$\frac{d}{dx}[x^4 + x^2] = \frac{d}{dx}[x^4] + \frac{d}{dx}[x^2] = 4x^3 + 2x$$

$$\frac{d}{dx}[6x^{11} - 9] = \frac{d}{dx}[6x^{11}] - \frac{d}{dx}[9] = 66x^{10} - 0 = 66x^{10}$$

Although Theorem 3.3.4 was stated for sums and differences of two terms, it can be extended to any mixture of finitely many sums and differences of differentiable functions. For example,

$$\frac{d}{dx}[3x^8 - 2x^5 + 6x + 1] = \frac{d}{dx}[3x^8] + \frac{d}{dx}[-2x^5] + \frac{d}{dx}[6x] + \frac{d}{dx}[1]$$

$$= 24x^7 - 10x^4 + 6 \qquad \blacktriangleleft$$

DERIVATIVE OF A PRODUCT

> **3.3.5** THEOREM (*The Product Rule*). *If f and g are differentiable at x, then so is the product $f \cdot g$, and*
>
> $$\frac{d}{dx}[f(x)g(x)] = f(x)\frac{d}{dx}[g(x)] + g(x)\frac{d}{dx}[f(x)]$$

Proof. The earlier proofs in this section were straightforward applications of the definition of the derivative. However, this proof requires a trick—adding and subtracting the quantity $f(x + h)g(x)$ to the numerator in the derivative definition as follows:

$$\frac{d}{dx}[f(x)g(x)] = \lim_{h \to 0} \frac{f(x + h) \cdot g(x + h) - f(x) \cdot g(x)}{h}$$

$$= \lim_{h \to 0} \frac{f(x + h)g(x + h) - f(x + h)g(x) + f(x + h)g(x) - f(x)g(x)}{h}$$

$$= \lim_{h \to 0} \left[f(x + h) \cdot \frac{g(x + h) - g(x)}{h} + g(x) \cdot \frac{f(x + h) - f(x)}{h} \right]$$

$$= \lim_{h \to 0} f(x + h) \cdot \lim_{h \to 0} \frac{g(x + h) - g(x)}{h} + \lim_{h \to 0} g(x) \cdot \lim_{h \to 0} \frac{f(x + h) - f(x)}{h}$$

$$= [\lim_{h \to 0} f(x + h)]\frac{d}{dx}[g(x)] + [\lim_{h \to 0} g(x)]\frac{d}{dx}[f(x)]$$

$$= f(x)\frac{d}{dx}[g(x)] + g(x)\frac{d}{dx}[f(x)]$$

[*Note:* In the last step $f(x + h) \to f(x)$ as $h \to 0$ because f is continuous at x by Theorem 3.2.3, and $g(x) \to g(x)$ as $h \to 0$ because $g(x)$ does not involve h and hence remains constant.] ∎

The product rule can be written in function notation as

$$(f \cdot g)' = f \cdot g' + g \cdot f'$$

REMARK. In words, *the derivative of a product of two functions is the first function times the derivative of the second plus the second function times the derivative of the first.*

WARNING. Note that it is *not* true in general that $(f \cdot g)' = f' \cdot g'$; that is, the derivative of a product is *not* generally the product of the derivatives!

Example 5

Find dy/dx if $y = (4x^2 - 1)(7x^3 + x)$.

Solution. There are two methods that can be used to find dy/dx. We can either use the product rule or we can multiply out the factors in y and then differentiate. We will give both methods.

Method I. (*Using the Product Rule*)

$$\frac{dy}{dx} = \frac{d}{dx}[(4x^2 - 1)(7x^3 + x)]$$

$$= (4x^2 - 1)\frac{d}{dx}[7x^3 + x] + (7x^3 + x)\frac{d}{dx}[4x^2 - 1]$$

$$= (4x^2 - 1)(21x^2 + 1) + (7x^3 + x)(8x) = 140x^4 - 9x^2 - 1$$

Method II. (*Multiplying First*)

$$y = (4x^2 - 1)(7x^3 + x) = 28x^5 - 3x^3 - x$$

Thus,

$$\frac{dy}{dx} = \frac{d}{dx}[28x^5 - 3x^3 - x] = 140x^4 - 9x^2 - 1$$

which agrees with the result obtained using the product rule.   ◀

DERIVATIVE OF A QUOTIENT

3.3.6 THEOREM (*The Quotient Rule*).   *If f and g are differentiable at x and $g(x) \neq 0$, then f/g is differentiable at x and*

$$\frac{d}{dx}\left[\frac{f(x)}{g(x)}\right] = \frac{g(x)\dfrac{d}{dx}[f(x)] - f(x)\dfrac{d}{dx}[g(x)]}{[g(x)]^2}$$

Proof.

$$\frac{d}{dx}\left[\frac{f(x)}{g(x)}\right] = \lim_{h \to 0} \frac{\dfrac{f(x+h)}{g(x+h)} - \dfrac{f(x)}{g(x)}}{h} = \lim_{h \to 0} \frac{f(x+h) \cdot g(x) - f(x) \cdot g(x+h)}{h \cdot g(x) \cdot g(x+h)}$$

Adding and subtracting $f(x) \cdot g(x)$ in the numerator yields

$$\frac{d}{dx}\left[\frac{f(x)}{g(x)}\right] = \lim_{h \to 0} \frac{f(x+h) \cdot g(x) - f(x) \cdot g(x) - f(x) \cdot g(x+h) + f(x) \cdot g(x)}{h \cdot g(x) \cdot g(x+h)}$$

$$= \lim_{h \to 0} \frac{\left[g(x) \cdot \dfrac{f(x+h) - f(x)}{h}\right] - \left[f(x) \cdot \dfrac{g(x+h) - g(x)}{h}\right]}{g(x) \cdot g(x+h)}$$

$$= \frac{\displaystyle\lim_{h \to 0} g(x) \cdot \lim_{h \to 0} \frac{f(x+h) - f(x)}{h} - \lim_{h \to 0} f(x) \cdot \lim_{h \to 0} \frac{g(x+h) - g(x)}{h}}{\displaystyle\lim_{h \to 0} g(x) \cdot \lim_{h \to 0} g(x+h)}$$

$$= \frac{[\displaystyle\lim_{h \to 0} g(x)] \cdot \dfrac{d}{dx}[f(x)] - [\displaystyle\lim_{h \to 0} f(x)] \cdot \dfrac{d}{dx}[g(x)]}{\displaystyle\lim_{h \to 0} g(x) \cdot \lim_{h \to 0} g(x+h)}$$

$$= \frac{g(x)\dfrac{d}{dx}[f(x)] - f(x)\dfrac{d}{dx}[g(x)]}{[g(x)]^2}$$

[See the note at the end of the proof of Theorem 3.3.5 for an explanation of the last step.] ∎

The quotient rule can be written in function notation as

$$\left(\frac{f}{g}\right)' = \frac{g \cdot f' - f \cdot g'}{g^2}$$

REMARK. In words, *the derivative of a quotient of two functions is the denominator times the derivative of the numerator minus the numerator times the derivative of the denominator, all divided by the denominator squared.*

WARNING. Note that it is *not* generally true that $(f/g)' = f'/g'$; that is, the derivative of a quotient is *not* generally the quotient of the derivatives.

Example 6

Let $f(x) = \dfrac{x^2 - 1}{x^4 + 1}$.

(a) Graph $y = f(x)$, and use your graph to make rough estimates of the locations of all horizontal tangent lines.

(b) By differentiating, find the exact locations of the horizontal tangent lines.

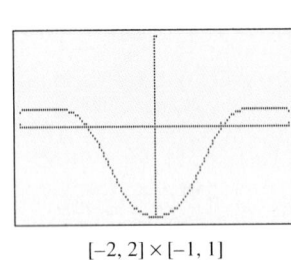

$[-2, 2] \times [-1, 1]$
xScl $= 1$, yScl $= 1$

$$y = \frac{x^2 - 1}{x^4 + 1}$$

Figure 3.3.2

Solution (a). Figure 3.3.2 shows the graph of $y = f(x)$ in the window $[-2, 2] \times [-1, 1]$. This graph suggests that horizontal tangent lines occur at $x = 0$, $x \approx 1.5$, and $x \approx -1.5$.

Solution (b). To find the exact location of the horizontal tangent lines, we must find the points where $dy/dx = 0$. We start by finding dy/dx:

$$\frac{dy}{dx} = \frac{d}{dx}\left[\frac{x^2 - 1}{x^4 + 1}\right] = \frac{(x^4 + 1)\dfrac{d}{dx}[x^2 - 1] - (x^2 - 1)\dfrac{d}{dx}[x^4 + 1]}{\left(x^4 + 1\right)^2}$$

$$= \frac{(x^4 + 1)(2x) - (x^2 - 1)(4x^3)}{\left(x^4 + 1\right)^2}$$

> The differentiation is complete. The rest is simplification.

$$= \frac{-2x^5 + 4x^3 + 2x}{\left(x^4 + 1\right)^2} = -\frac{2x(x^4 - 2x^2 - 1)}{\left(x^4 + 1\right)^2}$$

Now we will set $dy/dx = 0$ and solve for x. We obtain

$$-\frac{2x(x^4 - 2x^2 - 1)}{\left(x^4 + 1\right)^2} = 0$$

The solutions of this equation are the values of x for which the numerator is 0:

$$2x(x^4 - 2x^2 - 1) = 0$$

The first factor yields the solution $x = 0$. Other solutions can be found by solving the equation

$$x^4 - 2x^2 - 1 = 0$$

This can be treated as a quadratic equation in x^2 and solved by the quadratic formula. This yields

$$x^2 = \frac{2 \pm \sqrt{8}}{2} = 1 \pm \sqrt{2}$$

The minus sign yields imaginary values of x, which we ignore since they are not relevant to the problem. The plus sign yields the solutions

$$x = \pm\sqrt{1 + \sqrt{2}}$$

In summary, horizontal tangent lines occur at

$$x = 0, \quad x = \sqrt{1 + \sqrt{2}} \approx 1.55, \quad \text{and} \quad x = -\sqrt{1 + \sqrt{2}} \approx -1.55$$

which is consistent with the rough estimates that we obtained graphically in part (a). ◀

DERIVATIVE OF A RECIPROCAL

The special case of Theorem 3.3.6 in which f is the constant function 1 is of interest in its own right. We leave it for the reader to deduce the following result from Theorem 3.3.6.

3.3.7   THEOREM (*The Reciprocal Rule*).   *If g is differentiable at x and $g(x) \neq 0$, then $1/g$ is differentiable at x and*

$$\frac{d}{dx}\left[\frac{1}{g(x)}\right] = -\frac{\dfrac{d}{dx}[g(x)]}{[g(x)]^2}$$

The reciprocal rule can be written in function notation as

$$\left(\frac{1}{g}\right)' = -\frac{g'}{g^2}$$

REMARK.   In words, *the derivative of the reciprocal of a function is the negative of the derivative of the function divided by the function squared.*

Example 7

$$\frac{d}{dx}\left[\frac{1}{x}\right] = -\frac{\dfrac{d}{dx}[x]}{x^2} = -\frac{1}{x^2}$$

$$\frac{d}{dx}\left[\frac{1}{x^3 + 2x - 3}\right] = -\frac{\dfrac{d}{dx}[x^3 + 2x - 3]}{\left(x^3 + 2x - 3\right)^2} = -\frac{3x^2 + 2}{\left(x^3 + 2x - 3\right)^2} \qquad \blacktriangleleft$$

REMARK.   The computations in the preceding example could have been done using the quotient rule, but this would have been more work. Where it applies, the reciprocal rule is preferable to the quotient rule.

THE POWER RULE FOR INTEGER EXPONENTS

In Theorem 3.3.2 we established the formula

$$\frac{d}{dx}[x^n] = nx^{n-1}$$

for *positive* integer values of n. Eventually, we will show that this formula applies if n is any real number. As our first step in this direction we will show that it applies for *all integer* values of n.

3.3.8   THEOREM.   *If n is any integer, then*

$$\frac{d}{dx}[x^n] = nx^{n-1} \tag{1}$$

Proof.   The result has already been established in the case where $n > 0$. If $n < 0$, then let $m = -n$ so that

$$f(x) = x^{-m} = \frac{1}{x^m}$$

From Theorem 3.3.7,

$$f'(x) = \frac{d}{dx}\left[\frac{1}{x^m}\right] = -\frac{\frac{d}{dx}[x^m]}{(x^m)^2}$$

Since $n < 0$, it follows that $m > 0$, so x^m can be differentiated using Theorem 3.3.2. Thus,

$$f'(x) = -\frac{mx^{m-1}}{x^{2m}} = -mx^{m-1-2m} = -mx^{-m-1} = nx^{n-1}$$

which proves (1). In the case $n = 0$ Formula (1) reduces to

$$\frac{d}{dx}[1] = 0 \cdot x^{-1} = 0$$

which is correct by Theorem 3.3.1. ∎

Example 8

$$\frac{d}{dx}[x^{-9}] = -9x^{-9-1} = -9x^{-10}$$

$$\frac{d}{dx}\left[\frac{1}{x}\right] = \frac{d}{dx}[x^{-1}] = (-1)x^{-1-1} = -x^{-2} = -\frac{1}{x^2}$$

Note that the last result agrees with that obtained in Example 7. ◀

In Example 4 of Section 3.2 we showed that

$$\frac{d}{dx}[\sqrt{x}] = \frac{1}{2\sqrt{x}} \tag{2}$$

which shows that Formula (1) also works with $n = \frac{1}{2}$, since

$$\frac{d}{dx}[x^{1/2}] = \frac{1}{2x^{1/2}} = \frac{1}{2}x^{-1/2}$$

HIGHER DERIVATIVES

If the derivative f' of a function f is itself differentiable, then the derivative of f' is denoted by f'' and is called the ***second derivative*** of f. As long as we have differentiability, we can continue the process of differentiating derivatives to obtain third, fourth, fifth, and even higher derivatives of f. The successive derivatives of f are denoted by

$$f', \quad f'' = (f')', \quad f''' = (f'')', \quad f^{(4)} = (f''')', \quad f^{(5)} = (f^{(4)})', \dots$$

These are called the first derivative, the second derivative, the third derivative, and so forth. Beyond the third derivative, it is too clumsy to continue using primes, so we switch from primes to integers in parentheses to denote the ***order*** of the derivative. In this notation it is easy to denote a derivative of arbitrary order by writing

$$f^{(n)} \qquad \boxed{\text{The } n\text{th derivative of } f}$$

The significance of the derivatives of order 2 and higher will be discussed later.

Example 9

If $f(x) = 3x^4 - 2x^3 + x^2 - 4x + 2$, then

$$f'(x) = 12x^3 - 6x^2 + 2x - 4$$
$$f''(x) = 36x^2 - 12x + 2$$
$$f'''(x) = 72x - 12$$
$$f^{(4)}(x) = 72$$
$$f^{(5)}(x) = 0$$
$$\vdots$$
$$f^{(n)}(x) = 0 \quad (n \geq 5)$$

◀

Successive derivatives can also be denoted as follows:

$$f'(x) = \frac{d}{dx}[f(x)]$$

$$f''(x) = \frac{d}{dx}\left[\frac{d}{dx}[f(x)]\right] = \frac{d^2}{dx^2}[f(x)]$$

$$f'''(x) = \frac{d}{dx}\left[\frac{d^2}{dx^2}[f(x)]\right] = \frac{d^3}{dx^3}[f(x)]$$

$$\vdots \qquad\qquad \vdots$$

In general, we write

$$f^{(n)}(x) = \frac{d^n}{dx^n}[f(x)]$$

which is read "the nth derivative of f with respect to x."

When a dependent variable is involved, say $y = f(x)$, then successive derivatives can be denoted by writing

$$\frac{dy}{dx}, \quad \frac{d^2y}{dx^2}, \quad \frac{d^3y}{dx^3}, \quad \frac{d^4y}{dx^4}, \ldots, \frac{d^ny}{dx^n}, \ldots$$

or more briefly,

$$y', \quad y'', \quad y''', \quad y^{(4)}, \ldots, y^{(n)}, \ldots$$

EXERCISE SET 3.3 ⟋ Graphing Calculator [c] CAS

In Exercises 1–12, find dy/dx.

1. $y = 4x^7$ **2.** $y = -3x^{12}$

3. $y = 3x^8 + 2x + 1$ **4.** $y = \frac{1}{2}(x^4 + 7)$

5. $y = \pi^3$ **6.** $y = \sqrt{2}x + (1/\sqrt{2})$

7. $y = -\frac{1}{3}(x^7 + 2x - 9)$ **8.** $y = \dfrac{x^2 + 1}{5}$

9. $y = ax^3 + bx^2 + cx + d$ (a, b, c, d constant)

10. $y = \dfrac{1}{a}\left(x^2 + \dfrac{1}{b}x + c\right)$ (a, b, c constant)

11. $y = -3x^{-8} + 2\sqrt{x}$ **12.** $y = 7x^{-6} - 5\sqrt{x}$

In Exercises 13–20, find $f'(x)$.

13. $f(x) = x^{-3} + \dfrac{1}{x^7}$ **14.** $f(x) = \sqrt{x} + \dfrac{1}{x}$

15. $f(x) = (3x^2 + 6)\left(2x - \frac{1}{4}\right)$

16. $f(x) = (2 - x - 3x^3)(7 + x^5)$

17. $f(x) = (x^3 + 7x^2 - 8)(2x^{-3} + x^{-4})$

18. $f(x) = \left(\dfrac{1}{x} + \dfrac{1}{x^2}\right)(3x^3 + 27)$

19. $f(x) = (3x^2 + 1)^2$ **20.** $f(x) = (x^5 + 2x)^2$

In Exercises 21 and 22, find $y'(1)$.

21. $y = \dfrac{1}{5x - 3}$ **22.** $y = \dfrac{3}{\sqrt{x} + 2}$

In Exercises 23 and 24, find dx/dt.

23. $x = \dfrac{3t}{2t + 1}$ **24.** $x = \dfrac{t^2 + 1}{3t}$

In Exercises 25–28, find $dy/dx|_{x=1}$.

25. $y = \dfrac{2x - 1}{x + 3}$ **26.** $y = \dfrac{4x + 1}{x^2 - 5}$

27. $y = \left(\dfrac{3x + 2}{x}\right)(x^{-5} + 1)$

28. $y = (2x^7 - x^2)\left(\dfrac{x - 1}{x + 1}\right)$

In Exercises 29–32, find the indicated derivative.

29. $\dfrac{d}{dt}[16t^2]$ **30.** $\dfrac{dC}{dr}$, where $C = 2\pi r$

31. $V'(r)$, where $V = \pi r^3$ **32.** $\dfrac{d}{d\alpha}[2\alpha^{-1} + \alpha]$

33. A spherical balloon is being inflated.
    (a) Find a general formula for the instantaneous rate of change of the volume V with respect to the radius r.
    (b) Find the rate of change of V with respect to r at the instant when the radius is $r = 5$.

34. Use a CAS to check the answers to the problems you solved in Exercises 1–32.

35. Find $g'(4)$ given that $f(4) = 3$ and $f'(4) = -5$.
    (a) $g(x) = \sqrt{x} f(x)$ (b) $g(x) = \dfrac{f(x)}{x}$

36. Find $g'(3)$ given that $f(3) = -2$ and $f'(3) = 4$.
    (a) $g(x) = 3x^2 - 5f(x)$ (b) $g(x) = \dfrac{2x + 1}{f(x)}$

37. Find $F'(2)$ given that $f(2) = -1$, $f'(2) = 4$, $g(2) = 1$, and $g'(2) = -5$.
    (a) $F(x) = 5f(x) + 2g(x)$ (b) $F(x) = f(x) - 3g(x)$
    (c) $F(x) = f(x)g(x)$ (d) $F(x) = f(x)/g(x)$

38. Find an equation for the line that is tangent to the curve $y = (1 - x)/(1 + x)$ at the point where $x = 2$.

39. Find an equation of the tangent line to the graph of $y = f(x)$ at the point where $x = -3$ if $f(-3) = 2$ and $f'(-3) = 5$.

40. Find $\dfrac{d}{d\lambda}\left[\dfrac{\lambda\lambda_0 + \lambda^6}{2 - \lambda_0}\right]$ (λ_0 is constant).

In Exercises 41 and 42, find $d^2 y/dx^2$.

41. (a) $y = 7x^3 - 5x^2 + x$ (b) $y = 12x^2 - 2x + 3$
    (c) $y = \dfrac{x + 1}{x}$ (d) $y = (5x^2 - 3)(7x^3 + x)$

42. (a) $y = 4x^7 - 5x^3 + 2x$ (b) $y = 3x + 2$
    (c) $y = \dfrac{3x - 2}{5x}$ (d) $y = (x^3 - 5)(2x + 3)$

In Exercises 43 and 44, find y'''.

43. (a) $y = x^{-5} + x^5$ (b) $y = 1/x$
    (c) $y = ax^3 + bx + c$ (a, b, c constant)

44. (a) $y = 5x^2 - 4x + 7$ (b) $y = 3x^{-2} + 4x^{-1} + x$
    (c) $y = ax^4 + bx^2 + c$ (a, b, c constant)

45. Find
    (a) $f'''(2)$, where $f(x) = 3x^2 - 2$
    (b) $\dfrac{d^2 y}{dx^2}\bigg|_{x=1}$, where $y = 6x^5 - 4x^2$
    (c) $\dfrac{d^4}{dx^4}[x^{-3}]\bigg|_{x=1}$

46. Find
    (a) $y'''(0)$, where $y = 4x^4 + 2x^3 + 3$
    (b) $\dfrac{d^4 y}{dx^4}\bigg|_{x=1}$, where $y = \dfrac{6}{x^4}$.

47. Show that $y = x^3 + 3x + 1$ satisfies $y''' + xy'' - 2y' = 0$.

48. Show that if $x \neq 0$, then $y = 1/x$ satisfies the equation $x^3 y'' + x^2 y' - xy = 0$.

49. Find a general formula for $F''(x)$ if $F(x) = xf(x)$ and f and f' are differentiable at x.

50. Use a CAS to check the answers to the problems you solved in Exercises 41–46.

In Exercises 51 and 52, use a graphing utility to make rough estimates of the locations of all horizontal tangent lines, and then find their exact locations by differentiating.

51. $y = \frac{1}{3}x^3 - \frac{3}{2}x^2 + 2x$ **52.** $y = \dfrac{x}{x^2 + 9}$

In Exercises 53 and 54, use Definition 3.2.2 to approximate $f'(1)$ by choosing a small value of h to approximate the limit, and then find the exact value of $f'(1)$ by differentiating.

53. $f(x) = x^3 - 3x + 1$ **54.** $f(x) = x\sqrt{x}$

In Exercises 55 and 56, estimate the value of $f'(1)$ by zooming in on the graph of f, and then compare your estimate to the exact value obtained by differentiating.

55. $f(x) = \dfrac{x}{x^2 + 1}$ **56.** $f(x) = \dfrac{x^2 - 1}{x^2 + 1}$

57. Find a function $y = ax^2 + bx + c$ whose graph has an x-intercept of 1, a y-intercept of -2, and a tangent line with a slope of -1 at the y-intercept.

58. Find k if the curve $y = x^2 + k$ is tangent to the line $y = 2x$.

59. Find the x-coordinate of the point on the graph of $y = x^2$ where the tangent line is parallel to the secant line that cuts the curve at $x = -1$ and $x = 2$.

60. Find the x-coordinate of the point on the graph of $y = \sqrt{x}$ where the tangent line is parallel to the secant line that cuts the curve at $x = 1$ and $x = 4$.

61. Find the coordinate of all points on the graph of $y = 1 - x^2$ at which the tangent line passes through the point $(2, 0)$.

62. Show that any two tangent lines to the parabola $y = ax^2$, $a \neq 0$, intersect at a point that is on the vertical line halfway between the points of tangency.

63. Suppose that L is the tangent line at $x = x_0$ to the graph of the cubic equation $y = ax^3 + bx$. Find the x-coordinate of the point where L intersects the graph a second time.

64. Show that the segment of the tangent line to the graph of $y = 1/x$ that is cut off by the coordinate axes is bisected by the point of tangency.

65. Show that the triangle that is formed by any tangent line to the graph of $y = 1/x$, $x > 0$, and the coordinate axes has an area of 2 square units.

66. Find conditions on a, b, c, and d so that the graph of the polynomial $f(x) = ax^3 + bx^2 + cx + d$ has
    (a) exactly two horizontal tangents
    (b) exactly one horizontal tangent
    (c) no horizontal tangents.

67. Newton's Law of Gravitation states that the magnitude F of the force exerted by a point with mass M on a point with mass m is

$$F = \frac{GmM}{r^2}$$

where G is a constant and r is the distance between the bodies. Assuming that the points are moving, find a formula for the instantaneous rate of change of F with respect to r.

68. In the temperature range between $0°C$ and $700°C$ the resistance R [in ohms (Ω)] of a certain platinum resistance thermometer is given by

$$R = 10 + 0.04124T - 1.779 \times 10^{-5}T^2$$

where T is the temperature in degrees Celsius. Where in the interval from $0°C$ to $700°C$ is the resistance of the thermometer most sensitive and least sensitive to temperature changes? [*Hint:* Consider the size of dR/dT in the interval $0 \leq T \leq 700$.]

In Exercises 69 and 70, use a graphing utility to make rough estimates of the intervals on which $f'(x) > 0$, and then find those intervals exactly by differentiating.

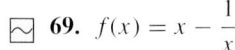 **69.** $f(x) = x - \dfrac{1}{x}$ **70.** $f(x) = \dfrac{5x}{x^2 + 4}$

71. Apply the product rule (3.3.5) twice to show that if f, g, and h are differentiable functions, then $f \cdot g \cdot h$ is differentiable, and

$$(f \cdot g \cdot h)' = f' \cdot g \cdot h + f \cdot g' \cdot h + f \cdot g \cdot h'$$

72. Based on the result in Exercise 71, make a conjecture about a formula for differentiating a product of n functions.

73. Use the formula in Exercise 71 to find

(a) $\dfrac{d}{dx}\left[(2x + 1)\left(1 + \dfrac{1}{x}\right)(x^{-3} + 7)\right]$

(b) $\dfrac{d}{dx}\left[(x^7 + 2x - 3)^3\right]$.

74. Use the formula you obtained in Exercise 72 to find

(a) $\dfrac{d}{dx}\left[x^{-5}(x^2 + 2x)(4 - 3x)(2x^9 + 1)\right]$

(b) $\dfrac{d}{dx}\left[(x^2 + 1)^{50}\right]$.

In Exercises 75–79, you will have to determine whether a function f is differentiable at a point x_0 where the formula for f changes. Use the following result:

Theorem. *Let f be continuous at x_0 and suppose that*

$$\lim_{x \to x_0^+} f'(x) \quad and \quad \lim_{x \to x_0^-} f'(x)$$

exist. Then f is differentiable at x_0 if and only if these limits are equal. Moreover, in the case of equality

$$f'(x_0) = \lim_{x \to x_0^+} f'(x) = \lim_{x \to x_0^-} f'(x)$$

75. Let

$$f(x) = \begin{cases} x^2, & x \leq 1 \\ \sqrt{x}, & x > 1 \end{cases}$$

Determine whether f is differentiable at $x = 1$. If so, find the value of the derivative there.

76. Let

$$f(x) = \begin{cases} x^3 + \frac{1}{16}, & x < \frac{1}{2} \\ \frac{3}{4}x^2, & x \geq \frac{1}{2} \end{cases}$$

Determine whether f is differentiable at $x = \frac{1}{2}$. If so, find the value of the derivative there.

77. Let

$$f(x) = \begin{cases} 3x^2, & x \leq 1 \\ ax + b, & x > 1 \end{cases}$$

Find the values of a and b so that f will be differentiable at $x = 1$.

78. (a) Let

$$f(x) = \begin{cases} x^2, & x \leq 0 \\ x^2 + 1, & x > 0 \end{cases}$$

Show that

$$\lim_{x \to 0^-} f'(x) = \lim_{x \to 0^+} f'(x)$$

but that $f'(0)$ does not exist.

(b) Let

$$f(x) = \begin{cases} x^2, & x \leq 0 \\ x^3, & x > 0 \end{cases}$$

Show that $f'(0)$ exists but $f''(0)$ does not.

79. Find all points where f fails to be differentiable. Justify your answer.

(a) $f(x) = |3x - 2|$ (b) $f(x) = |x^2 - 4|$

80. In each part compute f', f'', f''' and then state the formula for $f^{(n)}$.

(a) $f(x) = 1/x$ (b) $f(x) = 1/x^2$

[*Hint:* The expression $(-1)^n$ has a value of 1 if n is even and -1 if n is odd. Use this expression in your answer.]

81. (a) Prove:

$$\frac{d^2}{dx^2}[cf(x)] = c\frac{d^2}{dx^2}[f(x)]$$

$$\frac{d^2}{dx^2}[f(x) + g(x)] = \frac{d^2}{dx^2}[f(x)] + \frac{d^2}{dx^2}[g(x)]$$

(b) Do the results in part (a) generalize to nth derivatives? Justify your answer.

82. Prove:

$$(f \cdot g)'' = f'' \cdot g + 2f' \cdot g' + f \cdot g''$$

83. (a) Find $f^{(n)}(x)$ if $f(x) = x^n$.
   (b) Find $f^{(n)}(x)$ if $f(x) = x^k$ and $n > k$, where k is a positive integer.
   (c) Find $f^{(n)}(x)$ if

$$f(x) = a_0 + a_1 x + a_2 x^2 + \cdots + a_n x^n$$

84. Let $f(x) = x^8 - 2x + 3$; find

$$\lim_{h \to 0} \frac{f'(2 + h) - f'(2)}{h}$$

85. (a) Prove: If $f''(x)$ exists for each x in (a, b), then both f and f' are continuous on (a, b).
   (b) What can be said about the continuity of f and its derivatives if $f^{(n)}(x)$ exists for each x in (a, b)?

3.4 DERIVATIVES OF TRIGONOMETRIC FUNCTIONS

The main objective of this section is to obtain formulas for the derivatives of trigonometric functions.

DERIVATIVES OF THE TRIGONOMETRIC FUNCTIONS

For the purpose of finding derivatives of the trigonometric functions $\sin x$, $\cos x$, $\tan x$, $\cot x$, $\sec x$, and $\csc x$, we will assume that x is measured in radians. We will also need the following limits, which were stated in Theorem 2.5.3 (with x rather than h as the variable):

$$\lim_{h \to 0} \frac{\sin h}{h} = 1 \quad \text{and} \quad \lim_{h \to 0} \frac{1 - \cos h}{h} = 0$$

We begin with the problem of differentiating $\sin x$. From the definition of a derivative we have

$$\frac{d}{dx}[\sin x] = \lim_{h \to 0} \frac{\sin(x + h) - \sin x}{h}$$

$$= \lim_{h \to 0} \frac{\sin x \cos h + \cos x \sin h - \sin x}{h}$$

$$= \lim_{h \to 0} \left[\sin x \left(\frac{\cos h - 1}{h} \right) + \cos x \left(\frac{\sin h}{h} \right) \right]$$

$$= \lim_{h \to 0} \left[\cos x \left(\frac{\sin h}{h} \right) - \sin x \left(\frac{1 - \cos h}{h} \right) \right]$$

Since $\sin x$ and $\cos x$ do not involve h, they remain constant as $h \to 0$; thus,

$$\lim_{h \to 0} (\sin x) = \sin x \quad \text{and} \quad \lim_{h \to 0} (\cos x) = \cos x$$

Consequently,

$$\frac{d}{dx}[\sin x] = \cos x \cdot \lim_{h \to 0} \left(\frac{\sin h}{h} \right) - \sin x \cdot \lim_{h \to 0} \left(\frac{1 - \cos h}{h} \right)$$

$$= \cos x \cdot (1) - \sin x \cdot (0) = \cos x$$

Thus, we have shown that

$$\frac{d}{dx}[\sin x] = \cos x \tag{1}$$

The derivative of $\cos x$ can be obtained similarly, resulting in the formula

$$\frac{d}{dx}[\cos x] = -\sin x \tag{2}$$

The derivatives of the remaining trigonometric functions are

$$\frac{d}{dx}[\tan x] = \sec^2 x \qquad \frac{d}{dx}[\sec x] = \sec x \tan x \tag{3-4}$$

$$\frac{d}{dx}[\cot x] = -\csc^2 x \qquad \frac{d}{dx}[\csc x] = -\csc x \cot x \qquad (5\text{-}6)$$

These can all be obtained from (1) and (2) using the relationships

$$\tan x = \frac{\sin x}{\cos x}, \quad \cot x = \frac{\cos x}{\sin x}, \quad \sec x = \frac{1}{\cos x}, \quad \csc x = \frac{1}{\sin x}$$

For example,

$$\frac{d}{dx}[\tan x] = \frac{d}{dx}\left[\frac{\sin x}{\cos x}\right] = \frac{\cos x \cdot \frac{d}{dx}[\sin x] - \sin x \cdot \frac{d}{dx}[\cos x]}{\cos^2 x}$$

$$= \frac{\cos x \cdot \cos x - \sin x \cdot (-\sin x)}{\cos^2 x} = \frac{\cos^2 x + \sin^2 x}{\cos^2 x} = \frac{1}{\cos^2 x} = \sec^2 x$$

REMARK.   The derivative formulas for the trigonometric functions should be memorized. An easy way of doing this is discussed in Exercise 42. Moreover, we emphasize again that in all of the derivative formulas for the trigonometric functions, x is measured in radians.

Example 1

Find $f'(x)$ if $f(x) = x^2 \tan x$.

Solution.   Using the product rule and Formula (3), we obtain

$$f'(x) = x^2 \cdot \frac{d}{dx}[\tan x] + \tan x \cdot \frac{d}{dx}[x^2] = x^2 \sec^2 x + 2x \tan x \qquad \blacktriangleleft$$

Example 2

Find dy/dx if $y = \dfrac{\sin x}{1 + \cos x}$.

Solution.   Using the quotient rule together with Formulas (1) and (2) we obtain

$$\frac{dy}{dx} = \frac{(1 + \cos x) \cdot \frac{d}{dx}[\sin x] - \sin x \cdot \frac{d}{dx}[1 + \cos x]}{(1 + \cos x)^2}$$

$$= \frac{(1 + \cos x)(\cos x) - (\sin x)(-\sin x)}{(1 + \cos x)^2}$$

$$= \frac{\cos x + \cos^2 x + \sin^2 x}{(1 + \cos x)^2} = \frac{\cos x + 1}{(1 + \cos x)^2} = \frac{1}{1 + \cos x} \qquad \blacktriangleleft$$

Example 3

Find $y''(\pi/4)$ if $y(x) = \sec x$.

Solution.

$$y'(x) = \sec x \tan x$$

$$y''(x) = \sec x \cdot \frac{d}{dx}[\tan x] + \tan x \cdot \frac{d}{dx}[\sec x]$$

$$= \sec x \cdot \sec^2 x + \tan x \cdot \sec x \tan x$$

$$= \sec^3 x + \sec x \tan^2 x$$

Thus,

$$y''(\pi/4) = \sec^3(\pi/4) + \sec(\pi/4) \tan^2(\pi/4)$$

$$= (\sqrt{2})^3 + (\sqrt{2})(1)^2 = 3\sqrt{2} \qquad \blacktriangleleft$$

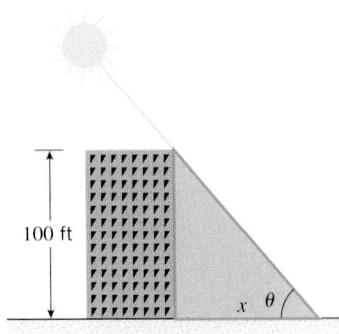

100 ft

x θ

Figure 3.4.1

Example 4

Suppose that the rising Sun passes directly over a building that is 100 feet high, and let θ be the Sun's angle of elevation (Figure 3.4.1). Find the rate at which the length x of the building's shadow is changing with respect to θ when $\theta = 45°$. Express the answer in units of feet/degree.

Solution. The variables x and θ are related by $\tan\theta = 100/x$, or equivalently,

$$x = 100\cot\theta \tag{7}$$

If θ is measured in radians, then Formula (5) is applicable, which yields

$$\frac{dx}{d\theta} = -100\csc^2\theta$$

which is the rate of change of shadow length with respect to the elevation angle θ in units of feet/radian. When $\theta = 45°$ (or equivalently, $\theta = \pi/4$ radians), we obtain

$$\left.\frac{dx}{d\theta}\right|_{\theta=\pi/4} = -100\csc^2(\pi/4) = -200 \text{ feet/radian}$$

Converting radians (rad) to degrees (deg) yields

$$-200\frac{\text{ft}}{\text{rad}} \cdot \frac{\pi}{180}\frac{\text{rad}}{\text{deg}} = -\frac{10}{9}\pi \approx -3.49 \text{ ft/deg}$$

Thus, when $\theta = 45°$, the shadow length is decreasing (because of the minus sign) at an approximate rate of 3.49 ft/deg increase in the angle of elevation. ◄

EXERCISE SET 3.4   ⌇ Graphing Calculator   [C] CAS

In Exercises 1–18, find $f'(x)$.

1. $f(x) = 2\cos x - 3\sin x$

2. $f(x) = \sin x \cos x$

3. $f(x) = \dfrac{\sin x}{x}$

4. $f(x) = x^2 \cos x$

5. $f(x) = x^3 \sin x - 5\cos x$

6. $f(x) = \dfrac{\cos x}{x \sin x}$

7. $f(x) = \sec x - \sqrt{2}\tan x$

8. $f(x) = (x^2 + 1)\sec x$

9. $f(x) = \sec x \tan x$

10. $f(x) = \dfrac{\sec x}{1 + \tan x}$

11. $f(x) = \csc x \cot x$

12. $f(x) = x - 4\csc x + 2\cot x$

13. $f(x) = \dfrac{\cot x}{1 + \csc x}$

14. $f(x) = \dfrac{\csc x}{\tan x}$

15. $f(x) = \sin^2 x + \cos^2 x$

16. $f(x) = \dfrac{1}{\cot x}$

17. $f(x) = \dfrac{\sin x \sec x}{1 + x \tan x}$

18. $f(x) = \dfrac{(x^2 + 1)\cot x}{3 - \cos x \csc x}$

In Exercises 19–24, find d^2y/dx^2.

19. $y = x\cos x$

20. $y = \csc x$

21. $y = x\sin x - 3\cos x$

22. $y = x^2\cos x + 4\sin x$

23. $y = \sin x \cos x$

24. $y = \tan x$

[C] 25. Use a CAS to check the answers to the problems you solved in Exercises 1–24.

26. Find the equation of the line tangent to the graph of $\sin x$ at the point where
   (a) $x = 0$ (b) $x = \pi$ (c) $x = \pi/4$.

27. Find the equation of the line tangent to the graph of $\tan x$ at the point where
   (a) $x = 0$ (b) $x = \pi/4$ (c) $x = -\pi/4$.

28. (a) Show that $y = \cos x$ and $y = \sin x$ are solutions of the equation $y'' + y = 0$.
   (b) Show that $y = A\sin x + B\cos x$ is a solution for all constants A and B.

29. Find all points in the interval $[-2\pi, 2\pi]$ at which the graph of f has a horizontal tangent line.
   (a) $f(x) = \sin x$ (b) $f(x) = x + \cos x$
   (c) $f(x) = \tan x$ (d) $f(x) = \sec x$

⌇ 30. (a) Use a graphing utility to make rough estimates of the points in the interval $[0, 2\pi]$ at which the graph of $y = \sin x \cos x$ has a horizontal tangent line.

(b) Find the exact locations of the points where the graph has a horizontal tangent line.

31. A 10-ft ladder leans against a wall at an angle θ with the horizontal, as shown in the accompanying figure. The top of the ladder is x feet above the ground. If the bottom of the ladder is pushed toward the wall, find the rate at which x changes with respect to θ when $\theta = 60°$. Express the answer in units of feet/degree.

32. An airplane is flying on a horizontal path at a height of 3800 ft, as shown in the accompanying figure. At what rate is the distance s between the airplane and the fixed point P changing with respect to θ when $\theta = 30°$? Express the answer in units of feet/degree.

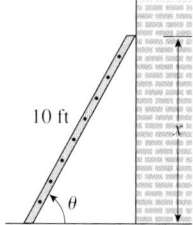

10 ft

Figure Ex-31

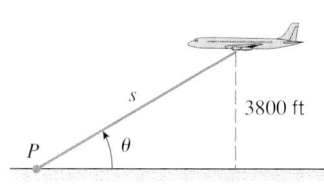

s

3800 ft

P

Figure Ex-32

33. A searchlight is located 50 m from a straight wall, as shown in the accompanying figure. Find the rate at which the distance D is changing with θ when $\theta = 45°$. Express the answer in units of meters/degree.

34. An Earth-observing satellite can see only a portion of the Earth's surface. The satellite has horizon sensors that can detect the angle θ shown in the accompanying figure. Let r be the radius of the Earth (assumed spherical) and h the distance of the satellite from the Earth's surface.
  (a) Show that $h = r(\csc\theta - 1)$.
  (b) Using $r = 6378$ km, and assuming that the satellite is getting closer to the Earth, find the rate at which h is changing with respect to θ when $\theta = 30°$. Express the answer in units of kilometers/degree. [Adapted from *Space Mathematics*, NASA, 1985.]

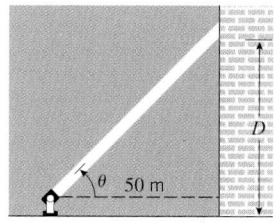

θ 50 m

Figure Ex-33

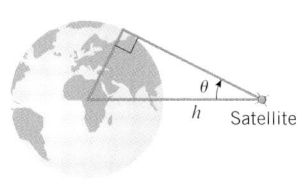

θ

h Satellite

Earth

Figure Ex-34

In Exercises 35 and 36, make a conjecture about the derivative by calculating the first few derivatives and observing the resulting pattern.

35. (a) $\dfrac{d^{87}}{dx^{87}}[\sin x]$ (b) $\dfrac{d^{100}}{dx^{100}}[\cos x]$

36. $\dfrac{d^{17}}{dx^{17}}[x \sin x]$

37. In each part, determine where f is differentiable.
  (a) $f(x) = \sin x$ (b) $f(x) = \cos x$
  (c) $f(x) = \tan x$ (d) $f(x) = \cot x$
  (e) $f(x) = \sec x$ (f) $f(x) = \csc x$
  (g) $f(x) = \dfrac{1}{1 + \cos x}$ (h) $f(x) = \dfrac{1}{\sin x \cos x}$
  (i) $f(x) = \dfrac{\cos x}{2 - \sin x}$

38. (a) Derive Formula (2) using the definition of a derivative.
  (b) Use Formulas (1) and (2) to obtain (5).
  (c) Use Formula (2) to obtain (4).
  (d) Use Formula (1) to obtain (6).

39. Let $f(x) = \cos x$. Find all positive integers n for which $f^{(n)}(x) = \sin x$.

40. (a) Show that $\displaystyle\lim_{h\to 0}\frac{\tan h}{h} = 1$.
  (b) Use the result in part (a) to help derive the formula for the derivative of $\tan x$ directly from the definition of a derivative.

41. Without using any trigonometric identities, find
$$\lim_{x\to 0}\frac{\tan(x + y) - \tan y}{x}$$
[*Hint:* Relate the given limit to the definition of the derivative of an appropriate function of y.]

42. Let us agree to call the functions $\cos x$, $\cot x$, and $\csc x$ the **cofunctions** of $\sin x$, $\tan x$, and $\sec x$, respectively. Convince yourself that the derivative of any cofunction can be obtained from the derivative of the corresponding function by introducing a minus sign and replacing each function in the derivative by its cofunction. Memorize the derivatives of $\sin x$, $\tan x$, and $\sec x$ and then use the above observation to deduce the derivatives of the cofunctions.

43. The derivative formulas for $\sin x$, $\cos x$, $\tan x$, $\cot x$, $\sec x$, and $\csc x$ were obtained under the assumption that x is measured in radians. This exercise shows that different (more complicated) formulas result if x is measured in degrees. Prove that if h and x are degree measures, then
  (a) $\displaystyle\lim_{h\to 0}\frac{\cos h - 1}{h} = 0$ (b) $\displaystyle\lim_{h\to 0}\frac{\sin h}{h} = \frac{\pi}{180}$
  (c) $\dfrac{d}{dx}[\sin x] = \dfrac{\pi}{180}\cos x$.

3.5 THE CHAIN RULE

In this section we will derive a formula that expresses the derivative of a composition $f \circ g$ in terms of the derivatives of f and g. This formula will enable us to differentiate complicated functions using known derivatives of simpler functions.

DERIVATIVES OF COMPOSITIONS

3.5.1   PROBLEM. *If we know the derivatives of f and g, how can we use this information to find the derivative of the composition $f \circ g$?*

The key to solving this problem is to introduce dependent variables

$$y = (f \circ g)(x) = f(g(x)) \quad \text{and} \quad u = g(x)$$

so that $y = f(u)$. We are interested in using the known derivatives

$$\frac{dy}{du} = f'(u) \quad \text{and} \quad \frac{du}{dx} = g'(x)$$

to find the unknown derivative

$$\frac{dy}{dx} = \frac{d}{dx}[f(g(x))]$$

Stated another way, we are interested in using the known rates of change dy/du and du/dx to find the unknown rate of change dy/dx. But intuition suggests that rates of change multiply. For example, if y changes at 4 times the rate of change of u and u changes at 2 times the rate of change of x, then y changes at $4 \times 2 = 8$ times the rate of change of x. Thus, Figure 3.5.1 suggests that

$$\frac{dy}{dx} = \frac{dy}{du} \cdot \frac{du}{dx}$$

These ideas are formalized in the following theorem.

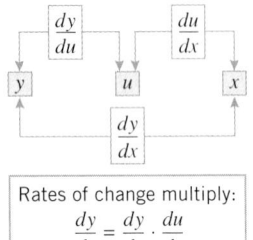

Rates of change multiply:
$$\frac{dy}{dx} = \frac{dy}{du} \cdot \frac{du}{dx}$$

Figure 3.5.1

3.5.2   THEOREM (*The Chain Rule*).   *If g is differentiable at the point x and f is differentiable at the point $g(x)$, then the composition $f \circ g$ is differentiable at the point x. Moreover, if*

$$y = f(g(x)) \quad \text{and} \quad u = g(x)$$

then $y = f(u)$ and

$$\frac{dy}{dx} = \frac{dy}{du} \cdot \frac{du}{dx} \tag{1}$$

The proof of this result is given in Appendix G.

Example 1
Find dy/dx if $y = 4\cos(x^3)$.

Solution.   Let $u = x^3$ so that

$$y = 4\cos u$$

By the chain rule,

$$\frac{dy}{dx} = \frac{dy}{du} \cdot \frac{du}{dx} = \frac{d}{du}[4\cos u] \cdot \frac{d}{dx}[x^3]$$

$$= (-4\sin u) \cdot (3x^2) = (-4\sin(x^3)) \cdot (3x^2) = -12x^2\sin(x^3) \quad \blacktriangleleft$$

REMARK. Formula (1) is easy to remember because the left side is exactly what results if we "cancel" the du's on the right side. This "canceling" device provides a good way to remember the chain rule when variables other than x, y, and u are used.

Example 2

Find dw/dt if $w = \tan x$ and $x = 4t^3 + t$.

Solution. In this case the chain rule takes the form

$$\frac{dw}{dt} = \frac{dw}{dx} \cdot \frac{dx}{dt} = \frac{d}{dx}[\tan x] \cdot \frac{d}{dt}[4t^3 + t]$$

$$= (\sec^2 x)(12t^2 + 1) = (12t^2 + 1)\sec^2(4t^3 + t) \qquad ◀$$

GENERALIZED DERIVATIVE FORMULAS

Although Formula (1) is useful, it is sometimes unwieldy because it involves so many dependent variables. A simpler version of the chain rule can be obtained by noting that $y = f(u)$ in (1), so

$$\frac{dy}{dx} = \frac{d}{dx}[f(u)] \quad \text{and} \quad \frac{dy}{du} = f'(u)$$

Substituting these expressions in (1) yields the following alternative form of the chain rule:

$$\frac{d}{dx}[f(u)] = f'(u)\frac{du}{dx} \qquad (2)$$

This very powerful formula vastly extends our differentiation capabilities. For example, to differentiate the function

$$f(x) = \left(x^2 - x + 1\right)^{23} \qquad (3)$$

we can let $u = x^2 - x + 1$, so (3) becomes $f(u) = u^{23}$, then apply (2) to obtain

$$\frac{d}{dx}\left[(x^2 - x + 1)^{23}\right] = \frac{d}{dx}[u^{23}] = \underbrace{23u^{22}}_{f'(u)}\frac{du}{dx}$$

$$= 23\left(x^2 - x + 1\right)^{22}\frac{d}{dx}[x^2 - x + 1]$$

$$= 23\left(x^2 - x + 1\right)^{22} \cdot (2x - 1)$$

More generally, if u were any other differentiable function of x, the pattern of computations would be virtually the same. For example, if $u = \cos x$, then

$$\frac{d}{dx}[\cos^{23} x] = \frac{d}{dx}[u^{23}] = 23u^{22}\frac{du}{dx} = 23\cos^{22} x\frac{d}{dx}[\cos x]$$

$$= 23\cos^{22} x \cdot (-\sin x) = -23\sin x \cos^{22} x$$

In both of the preceding computations, the chain rule took the form

$$\frac{d}{dx}[u^{23}] = 23u^{22}\frac{du}{dx} \qquad (4)$$

This formula is a generalization of the more basic formula

$$\frac{d}{dx}[x^{23}] = 23x^{22} \qquad (5)$$

In fact, in the special case where $u = x$, Formula (4) reduces to (5) since

$$\frac{d}{dx}[u^{23}] = 23u^{22}\frac{du}{dx} = 23x^{22}\frac{d[x]}{dx} = 23x^{22}$$

Table 3.5.1 contains a list of *generalized derivative formulas* that are consequences of (2).

Table 3.5.1

GENERALIZED DERIVATIVE FORMULAS

$$\frac{d}{dx}[u^n] = nu^{n-1}\frac{du}{dx} \quad (n \text{ an integer})$$

$$\frac{d}{dx}[\sqrt{u}] = \frac{1}{2\sqrt{u}}\frac{du}{dx}$$

$$\frac{d}{dx}[\sin u] = \cos u\frac{du}{dx}$$

$$\frac{d}{dx}[\cos u] = -\sin u\frac{du}{dx}$$

$$\frac{d}{dx}[\tan u] = \sec^2 u\frac{du}{dx}$$

$$\frac{d}{dx}[\cot u] = -\csc^2 u\frac{du}{dx}$$

$$\frac{d}{dx}[\sec u] = \sec u \tan u\frac{du}{dx}$$

$$\frac{d}{dx}[\csc u] = -\csc u \cot u\frac{du}{dx}$$

Example 3

Find

(a) $\frac{d}{dx}[\sin(2x)]$ (b) $\frac{d}{dx}[\tan(x^2+1)]$

(c) $\frac{d}{dx}[\sqrt{x^3 + \csc x}]$ (d) $\frac{d}{dx}\left[(1 + x^5 \cot x)^{-8}\right]$

Solution (a). Taking $u = 2x$ in the generalized derivative formula for $\sin u$ yields

$$\frac{d}{dx}[\sin(2x)] = \frac{d}{dx}[\sin u] = \cos u\frac{du}{dx} = \cos 2x \cdot \frac{d}{dx}[2x] = \cos 2x \cdot 2 = 2\cos 2x$$

Solution (b). Taking $u = x^2 + 1$ in the generalized derivative formula for $\tan u$ yields

$$\frac{d}{dx}[\tan(x^2+1)] = \frac{d}{dx}[\tan u] = \sec^2 u\frac{du}{dx}$$

$$= \sec^2(x^2+1) \cdot \frac{d}{dx}[x^2+1] = \sec^2(x^2+1) \cdot 2x$$

$$= 2x\sec^2(x^2+1)$$

Solution (c). Taking $u = x^3 + \csc x$ in the generalized derivative formula for $\sqrt{u}$ yields

$$\frac{d}{dx}[\sqrt{x^3+\csc x}] = \frac{d}{dx}[\sqrt{u}] = \frac{1}{2\sqrt{u}}\frac{du}{dx} = \frac{1}{2\sqrt{x^3+\csc x}} \cdot \frac{d}{dx}[x^3+\csc x]$$

$$= \frac{1}{2\sqrt{x^3+\csc x}} \cdot (3x^2 - \csc x \cot x) = \frac{3x^2 - \csc x \cot x}{2\sqrt{x^3+\csc x}}$$

Solution (d). Taking $u = 1 + x^5 \cot x$ in the generalized derivative formula for u^{-8} yields

$$\frac{d}{dx}\left[(1+x^5\cot x)^{-8}\right] = \frac{d}{dx}[u^{-8}] = -8u^{-9}\frac{du}{dx}$$

$$= -8\left(1+x^5\cot x\right)^{-9} \cdot \frac{d}{dx}[1+x^5\cot x]$$

$$= -8\left(1+x^5\cot x\right)^{-9} \cdot (x^5(-\csc^2 x) + 5x^4\cot x)$$

$$= (8x^5\csc^2 x - 40x^4\cot x)\left(1+x^5\cot x\right)^{-9} \quad \blacktriangleleft$$

Sometimes you will have to make adjustments in notation or apply the chain rule more than once to calculate a derivative.

Example 4

Find

(a) $\frac{d}{dx}[\sin(\sqrt{1+\cos x})]$ (b) $\frac{d\mu}{dt}$ if $u = \sec\sqrt{\omega t}$ (ω constant)

Solution (a). Taking $u = \sqrt{1 + \cos x}$ in the generalized derivative formula for $\sin u$ yields

$$\frac{d}{dx}[\sin(\sqrt{1 + \cos x})] = \frac{d}{dx}[\sin u] = \cos u \frac{du}{dx}$$

$$= \cos(\sqrt{1 + \cos x}) \cdot \frac{d}{dx}[\sqrt{1 + \cos x}]$$

We use the generalized derivative formula for $\sqrt{u}$ with $u = 1 + \cos x$.

$$= \cos(\sqrt{1 + \cos x}) \cdot \frac{-\sin x}{2\sqrt{1 + \cos x}}$$

$$= -\frac{\sin x \cos(\sqrt{1 + \cos x})}{2\sqrt{1 + \cos x}}$$

Solution (b).

$$\frac{d\mu}{dt} = \frac{d}{dt}[\sec \sqrt{\omega t}] = \sec \sqrt{\omega t} \tan \sqrt{\omega t} \frac{d}{dt}[\sqrt{\omega t}]$$

We used the generalized derivative formula for $\sec u$ with $u = \sqrt{\omega t}$.

$$= \sec \sqrt{\omega t} \tan \sqrt{\omega t} \frac{\omega}{2\sqrt{\omega t}}$$

We used the generalized derivative formula for $\sqrt{u}$ with $u = \omega t$.   ◀

AN ALTERNATIVE APPROACH TO USING THE CHAIN RULE

As you become more comfortable with the chain rule, you may want to dispense with actually writing out the expression for u in your computations. To accomplish this, it is helpful to express Formula (2) in words. If we call u the "inside function" and f the "outside function" in the composition $f(u)$, then (2) states:

The derivative of $f(u)$ is the derivative of the outside function evaluated at the inside function times the derivative of the inside function.

For example,

$$\frac{d}{dx}[\cos(x^2 + 9)] = \underbrace{-\sin(x^2 + 9)}_{\substack{\text{Derivative of the} \\ \text{outside evaluated} \\ \text{at the inside}}} \cdot \underbrace{2x}_{\substack{\text{Derivative} \\ \text{of the inside}}}$$

$$\frac{d}{dx}[\tan^2 x] = \frac{d}{dx}\left[(\tan x)^2\right] = \underbrace{(2 \tan x)}_{\substack{\text{Derivative of} \\ \text{the outside} \\ \text{evaluated at} \\ \text{the inside}}} \cdot \underbrace{(\sec^2 x)}_{\substack{\text{Derivative} \\ \text{of the inside}}} = 2 \tan x \sec^2 x$$

In general, if $f(g(x))$ is a composition of functions in which the inside function g and the outside function f are differentiable, then

$$\frac{d}{dx}[f(g(x))] = \underbrace{f'(g(x))}_{\substack{\text{Derivative of} \\ \text{the outside} \\ \text{evaluated at} \\ \text{the inside}}} \cdot \underbrace{g'(x)}_{\substack{\text{Derivative} \\ \text{of the inside}}} \tag{6}$$

DIFFERENTIATING USING COMPUTER ALGEBRA SYSTEMS

Although the chain rule makes it possible to differentiate extremely complicated functions, the computations can be time-consuming to execute by hand. For complicated derivatives engineers and scientists often use computer algebra systems such as *Mathematica*, *Maple*, and *Derive*. For example, although we have all of the mathematical tools to perform the differentiation

$$\frac{d}{dx}\left[\frac{(x^2+1)^{10}\sin^3(\sqrt{x})}{\sqrt{1+\csc x}}\right]$$ (7)

by hand, the computations are sufficiently tedious that it would be more efficient to use a computer algebra system.

FOR THE READER.   If you have a CAS, use it to obtain the derivatives in Examples 2, 3, and 4, and also to perform the differentiation in (7).

EXERCISE SET 3.5   ~ Graphing Calculator   [c] CAS

In Exercises 1–24, find $f'(x)$.

1. $f(x) = (x^3 + 2x)^{37}$

2. $f(x) = (3x^2 + 2x - 1)^6$

3. $f(x) = \left(x^3 - \dfrac{7}{x}\right)^{-2}$

4. $f(x) = \dfrac{1}{(x^5 - x + 1)^9}$

5. $f(x) = \dfrac{4}{(3x^2 - 2x + 1)^3}$

6. $f(x) = \sqrt{x^3 - 2x + 5}$

7. $f(x) = \sqrt{4 + \sqrt{3x}}$

8. $f(x) = \sin^3 x$

9. $f(x) = \sin(x^3)$

10. $f(x) = \cos^2(3\sqrt{x})$

11. $f(x) = \tan(4x^2)$

12. $f(x) = 3\cot^4 x$

13. $f(x) = 4\cos^5 x$

14. $f(x) = \csc(x^3)$

15. $f(x) = \sin\left(\dfrac{1}{x^2}\right)$

16. $f(x) = \tan^4(x^3)$

17. $f(x) = 2\sec^2(x^7)$

18. $f(x) = \cos^3\left(\dfrac{x}{x+1}\right)$

19. $f(x) = \sqrt{\cos(5x)}$

20. $f(x) = \sqrt{3x - \sin^2(4x)}$

21. $f(x) = [x + \csc(x^3 + 3)]^{-3}$

22. $f(x) = [x^4 - \sec(4x^2 - 2)]^{-4}$

23. $f(x) = x^2\sqrt{5 - x^2}$

24. $f(x) = \dfrac{x}{\sqrt{1 - x^2}}$

In Exercises 25–39, find dy/dx.

25. $y = x^3 \sin^2(5x)$

26. $y = \sqrt{x}\,\tan^3(\sqrt{x})$

27. $y = x^5 \sec(1/x)$

28. $y = \dfrac{\sin x}{\sec(3x + 1)}$

29. $y = \cos(\cos x)$

30. $y = \sin(\tan 3x)$

31. $y = \cos^3(\sin 2x)$

32. $y = \dfrac{1 + \csc(x^2)}{1 - \cot(x^2)}$

33. $y = (5x + 8)^{13}(x^3 + 7x)^{12}$

34. $y = (2x - 5)^2(x^2 + 4)^3$

35. $y = \left(\dfrac{x - 5}{2x + 1}\right)^3$

36. $y = \left(\dfrac{1 + x^2}{1 - x^2}\right)^{17}$

37. $y = \dfrac{(2x + 3)^3}{(4x^2 - 1)^8}$

38. $y = [1 + \sin^3(x^5)]^{12}$

39. $y = [x\sin 2x + \tan^4(x^7)]^5$

In Exercises 40–43, find d^2y/dx^2.

40. $y = \sin(3x^2)$

41. $y = x\cos(5x) - \sin^2 x$

42. $y = x\tan\left(\dfrac{1}{x}\right)$

43. $y = \dfrac{1 + x}{1 - x}$

[c] **44.** Use a CAS to check the answers to the problems you solved in Exercises 1–43.

In Exercises 45–48, find an equation for the tangent line to the graph at the specified point.

45. $y = x\cos 3x$, $x = \pi$

46. $y = \sin(1 + x^3)$, $x = -3$

47. $y = \sec^3\left(\dfrac{\pi}{2} - x\right)$, $x = -\dfrac{\pi}{2}$

48. $y = \left(x - \dfrac{1}{x}\right)^3$, $x = 2$

In Exercises 49–52, find the indicated derivative.

49. $y = \cot^3(\pi - \theta)$; find $\dfrac{dy}{d\theta}$.

50. $\lambda = \left(\dfrac{au + b}{cu + d}\right)^6$; find $\dfrac{d\lambda}{du}$ (a, b, c, d constants).

51. $\dfrac{d}{d\omega}[a\cos^2 \pi\omega + b\sin^2 \pi\omega]$ (a, b constants).

52. $x = \csc^2\left(\dfrac{\pi}{3} - y\right)$; find $\dfrac{dx}{dy}$.

53. (a) Use a graphing utility to obtain the graph of the function $f(x) = x\sqrt{4 - x^2}$.
   (b) Use the graph in part (a) to make a rough sketch of the graph of f'.
   (c) Find $f'(x)$, and then check your work in part (b) by using the graphing utility to obtain the graph of f'.
   (d) Find the equation of the tangent line to the graph of f at $x = 1$, and graph f and the tangent line together.

54. (a) Use a graphing utility to obtain the graph of the function $f(x) = \sin x^2 \cos x$ over the interval $[-\pi/2, \pi/2]$.
   (b) Use the graph in part (a) to make a rough sketch of the graph of f' over the interval.
   (c) Find $f'(x)$, and then check your work in part (b) by using the graphing utility to obtain the graph of f' over the interval.
   (d) Find the equation of the tangent line to the graph of f at $x = 1$, and graph f and the tangent line together over the interval.

55. If an object suspended from a spring is displaced vertically from its equilibrium position by a small amount and released, and if the air resistance and the mass of the spring are ignored, then the resulting oscillation of the object is called **simple harmonic motion**. Under appropriate conditions the displacement y from equilibrium in terms of time t is given by

$$y = A \cos \omega t$$

where A is the initial displacement at time $t = 0$, and ω is a constant that depends on the mass of the object and the stiffness of the spring (see the accompanying figure). The constant $|A|$ is called the **amplitude** of the motion and ω the **angular frequency**.
   (a) Show that
   $$\dfrac{d^2 y}{dt^2} = -\omega^2 y$$
   (b) The **period** T is the time required to make one complete oscillation. Show that $T = 2\pi/\omega$.
   (c) The **frequency** f of the vibration is the number of oscillations per unit time. Find f in terms of the period T.
   (d) Find the amplitude, period, and frequency of an object that is executing simple harmonic motion given by $y = 0.6 \cos 15t$, where t is in seconds and y is in centimeters.

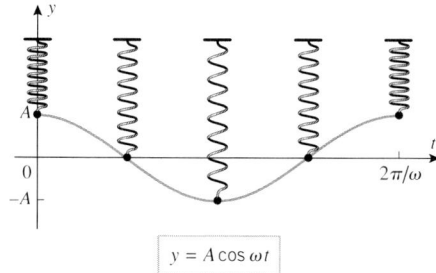

$y = A \cos \omega t$

Figure Ex-55

56. Find the value of the constant A so that $y = A \sin 3t$ satisfies the equation

$$\dfrac{d^2 y}{dt^2} + 2y = 4 \sin 3t$$

57. The accompanying figure shows the graph of atmospheric pressure p (lb/in^2) versus the altitude h (mi) above sea level.
   (a) From the graph and the tangent line at $h = 2$ shown on the graph, estimate the values of p and dp/dh at an altitude of 2 mi.
   (b) If the altitude of a space vehicle is increasing at the rate of 0.3 mi/s at the instant when it is 2 mi above sea level, how fast is the pressure changing with time at this instant?

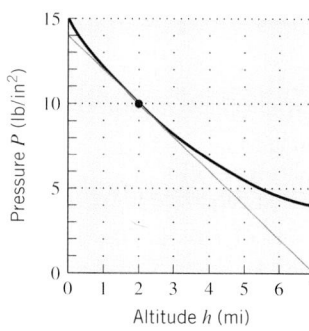

Figure Ex-57

58. The force F (in pounds) acting at an angle θ with the horizontal that is needed to drag a crate weighing W pounds along a horizontal surface at a constant velocity is given by

$$F = \dfrac{\mu W}{\cos \theta + \mu \sin \theta}$$

where μ is a constant called the **coefficient of sliding friction** between the crate and the surface (see the accompanying figure). Suppose that the crate weighs 150 lb and that $\mu = 0.3$.
   (a) Find $dF/d\theta$ when $\theta = 30°$. Express the answer in units of pounds/degree.
   (b) Find dF/dt when $\theta = 30°$ if θ is decreasing at the rate of $0.5°/s$ at this instant.

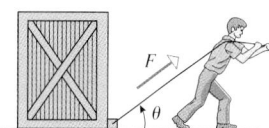

Figure Ex-58

59. Recall that

$$\dfrac{d}{dx}(|x|) = \begin{cases} 1, & x > 0 \\ -1, & x < 0 \end{cases}$$

Use this result and the chain rule to find

$$\dfrac{d}{dx}(|\sin x|)$$

for nonzero x in the interval $(-\pi, \pi)$.

60. Use the derivative formula for $\sin x$ and the identity

$$\cos x = \sin\left(\frac{\pi}{2} - x\right)$$

to obtain the derivative formula for $\cos x$.

61. Let

$$f(x) = \begin{cases} x \sin\dfrac{1}{x}, & x \neq 0 \\ 0, & x = 0 \end{cases}$$

(a) Find $f'(x)$ for $x \neq 0$.
(b) Show that f is continuous at $x = 0$.
(c) Use Definition 3.2.2 to show that $f'(0)$ does not exist.

62. Let

$$f(x) = \begin{cases} x^2 \sin\dfrac{1}{x}, & x \neq 0 \\ 0, & x = 0 \end{cases}$$

(a) Find $f'(x)$ for $x \neq 0$.
(b) Show that f is continuous at $x = 0$.
(c) Use Definition 3.2.2 to find $f'(0)$.
(d) Show that f' is not continuous at $x = 0$.

63. Given the following table of values, find the indicated derivatives in parts (a) and (b).

x	$f(x)$	$f'(x)$
2	1	7
8	5	-3

(a) $g'(2)$, where $g(x) = [f(x)]^3$
(b) $h'(2)$, where $h(x) = f(x^3)$

64. Given the following table of values, find the indicated derivatives in parts (a) and (b).

x	$f(x)$	$f'(x)$	$g(x)$	$g'(x)$
-1	2	3	2	-3
2	0	4	1	-5

(a) $F'(-1)$, where $F(x) = f(g(x))$
(b) $G'(-1)$, where $G(x) = g(f(x))$

65. Given that $f'(0) = 2$, $g(0) = 0$, and $g'(0) = 3$, find $(f \circ g)'(0)$.

66. Given that $f'(x) = \sqrt{3x + 4}$ and $g(x) = x^2 - 1$, find $F'(x)$ if $F(x) = f(g(x))$.

67. Given that $f'(x) = \dfrac{x}{x^2 + 1}$ and $g(x) = \sqrt{3x - 1}$, find $F'(x)$ if $F(x) = f(g(x))$.

68. Find $f'(x^2)$ if $\dfrac{d}{dx}[f(x^2)] = x^2$.

69. Find $\dfrac{d}{dx}[f(x)]$ if $\dfrac{d}{dx}[f(3x)] = 6x$.

70. Recall that a function f is **even** if $f(-x) = f(x)$ and **odd** if $f(-x) = -f(x)$, for all x in the domain of f. Assuming that f is differentiable, prove:
(a) f' is odd if f is even
(b) f' is even if f is odd.

71. Draw some pictures to illustrate the results in Exercise 70, and write a paragraph that gives an informal explanation of why the results are true.

72. Let $y = f_1(u)$, $u = f_2(v)$, $v = f_3(w)$, and $w = f_4(x)$. Express dy/dx in terms of dy/du, dw/dx, du/dv, and dv/dw.

73. Find a formula for

$$\frac{d}{dx}[f(g(h(x)))]$$

3.6 LOCAL LINEAR APPROXIMATION; DIFFERENTIALS

Up to now we have been interpreting dy/dx as a single entity representing the derivative of y with respect to x. In this section we will give the quantities dy and dx separate meanings that will allow us to treat dy/dx as a ratio. We will also show how derivatives can be used to approximate functions by simpler linear functions.

INCREMENTS

If the value of a variable changes from one number to another, then the final value minus the initial value is called an ***increment*** in the variable. It is traditional in calculus to denote an increment in a variable x by Δx (read "delta x"). Thus, if the initial value of x is x_0 and the final value is x_1, then

$$\Delta x = x_1 - x_0$$

In this notation the expression Δx is not the product of Δ and x; rather, it is a single entity representing the *change* in the value of x. This notation can be used with any variable; for example, increments in y, t, and θ would be denoted as Δy, Δt, and $\Delta\theta$.

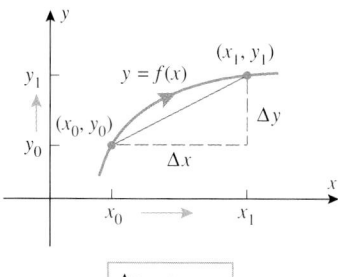

Figure 3.6.1

If $y = f(x)$, and if x changes from an initial value x_0 to a final value x_1, then there is a corresponding change in the value of y from $y_0 = f(x_0)$ to $y_1 = f(x_1)$. Stated another way, the increment $\Delta x = x_1 - x_0$ in x produces a corresponding increment $\Delta y = y_1 - y_0$ in y, where

$$\Delta y = y_1 - y_0 = f(x_1) - f(x_0) \tag{1}$$

(Figure 3.6.1).

Increments can be positive, negative, or zero, depending on the relative positions of the initial and final points—an increment in x is positive if the final point is to the right of the initial point, negative if the final point is to the left of the initial point, and zero if the initial and final points coincide. In Figure 3.6.1, both Δx and Δy are positive.

Observe that the expressions $\Delta x = x_1 - x_0$ and $\Delta y = y_1 - y_0$ can be rewritten as

$$x_1 = x_0 + \Delta x \quad \text{and} \quad y_1 = y_0 + \Delta y$$

which simply states that the *final value of a variable is equal to its initial value plus its increment*. With this notation we can express (1) as

$$\Delta y = f(x_0 + \Delta x) - f(x_0) \tag{2}$$

Sometimes, it is convenient to dispense with subscripts on the initial and final values of a variable, in which case the initial and final values of x would be denoted as x and $x + \Delta x$, and the initial and final values of the variable y would be denoted as y and $y + \Delta y$ (Figure 3.6.2). With this notation the symbols x and y play dual roles—they serve as the names as well as the initial values of the variables. However, this rarely causes any confusion.

With the subscripts omitted, Formula (2) becomes

$$\Delta y = f(x + \Delta x) - f(x) \tag{3}$$

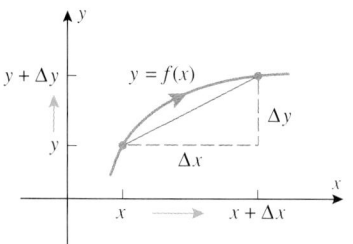

Figure 3.6.2

The ratio $\Delta y / \Delta x$ can be interpreted as the slope of the secant line joining the points (x, y) and $(x + \Delta x, y + \Delta y)$, and hence the derivative of y with respect to x can be expressed as

$$\frac{dy}{dx} = \lim_{\Delta x \to 0} \frac{\Delta y}{\Delta x} = \lim_{\Delta x \to 0} \frac{f(x + \Delta x) - f(x)}{\Delta x} \tag{4}$$

(Figure 3.6.3). This is consistent with (11) in Section 3.2.

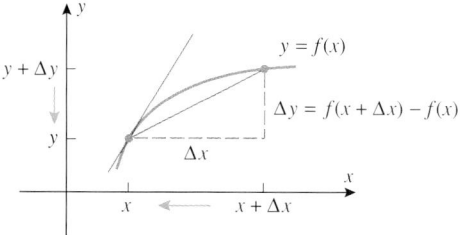

Figure 3.6.3

DIFFERENTIALS

When Newton and Leibniz independently published their discoveries of calculus, they each used different notations for the derivative, and battles raged for more than 50 years over which notation was better. In the end the **Leibniz notation** dy/dx won out because it produced correct formulas in a natural way; the chain rule

$$\frac{dy}{dx} = \frac{dy}{du} \cdot \frac{du}{dx}$$

is a good example.

The symbols "dy" and "dx" that appear in the derivative dy/dx are called **differentials**, and our next objective is to define these symbols so that dy/dx can actually be treated as a

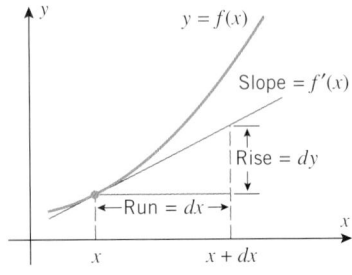

Figure 3.6.4

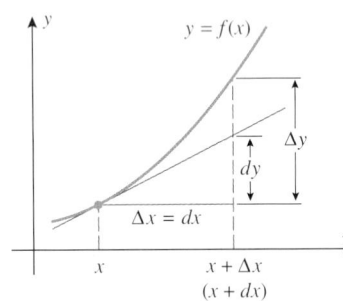

Figure 3.6.5

ratio. For this purpose, regard x as fixed and *define dx* to be an independent variable that can be assigned an arbitrary value. If f is differentiable at x, then we *define dy* by the formula

$$dy = f'(x)\,dx \qquad (5)$$

If $dx \neq 0$, then we can divide both sides of (5) by dx to obtain

$$\frac{dy}{dx} = f'(x)$$

Thus, we have achieved our goal of defining dy and dx so that their ratio is $f'(x)$.

Because

$$\frac{dy}{dx} = f'(x) = m_{\text{tan}}$$

where m_{tan} is the slope of the tangent to $y = f(x)$ at x, the differentials dy and dx can be viewed as a corresponding rise and run of this tangent line (Figure 3.6.4).

It is important to understand the distinction between the increment Δy and the differential dy. To see the difference, let us assign the independent variables dx and Δx the same value, so $dx = \Delta x$. Then Δy represents the change in y that occurs when we start at x and travel *along the curve* $y = f(x)$ until we have moved $\Delta x \,(= dx)$ units in the x-direction, while dy represents the change in y that occurs if we start at x and travel *along the tangent line* until we have moved $dx \,(= \Delta x)$ units in the x-direction (Figure 3.6.5).

Example 1

If $y = x^2$, then the relation $dy/dx = 2x$ can be written in the *differential form*

$$dy = 2x\,dx$$

When $x = 3$, this becomes

$$dy = 6\,dx$$

This tells us that if we travel along the tangent to the curve $y = x^2$ at $x = 3$, then a change of dx units in x produces a change of $6\,dx$ units in y. For example, if the change in x is $dx = 4$, then the change in y along the tangent is

$$dy = 6(4) = 24 \quad \text{units} \qquad \blacktriangleleft$$

Example 2

Let $y = \sqrt{x}$. Find dy and Δy at $x = 4$ with $dx = \Delta x = 3$. Then make a sketch of $y = \sqrt{x}$, showing dy and Δy in the picture.

Solution. From (3) with $f(x) = \sqrt{x}$ we obtain

$$\Delta y = \sqrt{x + \Delta x} - \sqrt{x} = \sqrt{7} - \sqrt{4} \approx 0.65$$

If $y = \sqrt{x}$, then

$$\frac{dy}{dx} = \frac{1}{2\sqrt{x}}, \quad \text{so} \quad dy = \frac{1}{2\sqrt{x}}\,dx = \frac{1}{2\sqrt{4}}(3) = \frac{3}{4} = 0.75$$

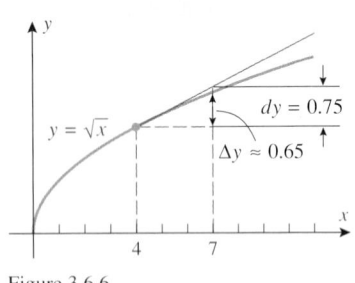

Figure 3.6.6

Figure 3.6.6 shows the curve $y = \sqrt{x}$ together with dy and Δy. $\qquad \blacktriangleleft$

LOCAL LINEAR APPROXIMATION

Points of differentiability for a function f can be described informally in terms of the behavior of the graph of f under magnification: If P is a point of differentiability for a function f, then stronger and stronger magnifications at P eventually make the curve segment containing P look more and more like a nonvertical line, the line being the tangent line at P. For this reason, a function that is differentiable at a point P is said to be *locally linear* at P (Figure 3.6.7).

It follows from the preceding observations that if f is differentiable at x_0, then the tangent line through $(x_0, f(x_0))$ closely approximates the graph of f for values of x near

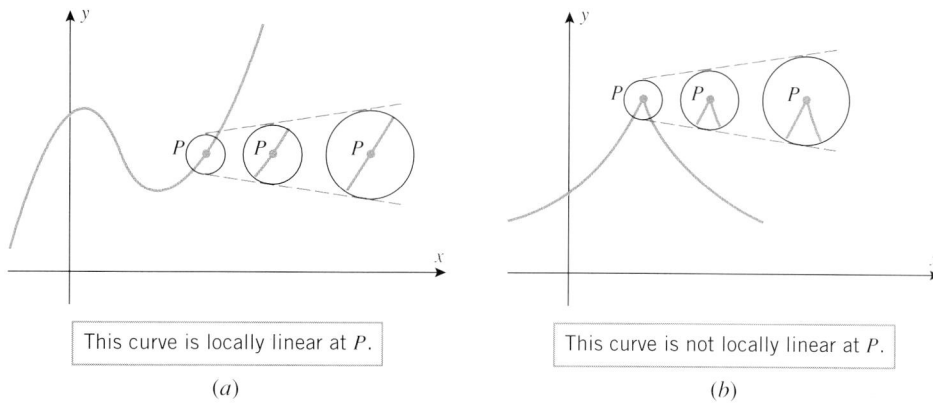

This curve is locally linear at P.	This curve is not locally linear at P.
(a)	(b)

Figure 3.6.7

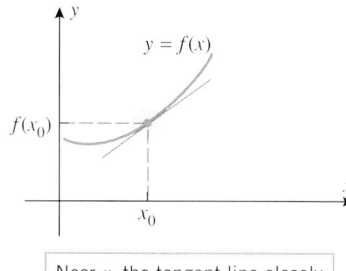

Near x_0 the tangent line closely approximates the curve.

Figure 3.6.8

x_0 (Figure 3.6.8). To capture this intuitive idea analytically, observe that the tangent line through the point $(x_0, f(x_0))$ has slope $f'(x_0)$, so the point-slope form of its equation is

$$y - f(x_0) = f'(x_0)(x - x_0)$$

which we can rewrite as

$$y = f(x_0) + f'(x_0)(x - x_0)$$

To say that this line closely approximates the curve $y = f(x)$ for values of x near x_0, we mean that the approximation

$$f(x) \approx f(x_0) + f'(x_0)(x - x_0) \tag{6}$$

gets better and better as $x \to x_0$. We call (6) the **local linear approximation of f at x_0**. An alternative version of this formula can be obtained by letting $\Delta x = x - x_0$, in which case (6) can be expressed as

$$f(x_0 + \Delta x) \approx f(x_0) + f'(x_0) \, \Delta x \tag{7}$$

Example 3

(a) Find the local linear approximation of $f(x) = \sin x$ at $x_0 = 0$.

(b) Use the local linear approximation obtained in part (a) to approximate $\sin 2°$, and compare your approximation to the result produced directly by your calculating device.

Solution (a). Since $f'(x) = \cos x$, it follows from (6) that the local linear approximation of $\sin x$ at a point x_0 is

$$\sin x \approx \sin x_0 + (\cos x_0)(x - x_0)$$

Thus, the local linear approximation at $x_0 = 0$ is

$$\sin x \approx \sin 0 + (\cos 0)(x - 0)$$

which simplifies to

$$\sin x \approx x \tag{8}$$

Solution (b). In (8), the variable x is in radian measure, so we must first convert $2°$ to radians before we can apply this formula. Since

$$2° = 2(\pi/180) = \pi/90 \text{ radians}$$

it follows from (8) with $x = \pi/90$ that

$$\sin 2° = \sin(\pi/90) \approx \pi/90 \approx 0.0349066$$

This is quite close to the value

$$\sin 2° \approx 0.0348995$$

produced directly on the author's calculator. ◀

Example 4

(a) Find the local linear approximation of $f(x) = \sqrt{x}$ at $x_0 = 1$.

(b) Use the local linear approximation obtained in part (a) to approximate $\sqrt{1.1}$, and compare your approximation to the result produced directly by your calculating device.

Solution (a). Since $f'(x) = 1/(2\sqrt{x})$, it follows from (6) that the local linear approximation of $\sqrt{x}$ at a point x_0 is

$$\sqrt{x} \approx \sqrt{x_0} + \frac{1}{2\sqrt{x_0}}(x - x_0)$$

Thus, the local linear approximation at $x_0 = 1$ is

$$\sqrt{x} \approx 1 + \tfrac{1}{2}(x - 1) \tag{9}$$

Solution (b). Applying Formula (9) with $x = 1.1$ yields

$$\sqrt{1.1} \approx 1 + \tfrac{1}{2}(0.1) = 1.05$$

which compares favorably with the approximation

$$\sqrt{1.1} \approx 1.04881$$

produced directly on the author's calculator. ◀

REMARK.   In the last two examples we used Formula (6) for the local linear approximation. We could just as well have used Formula (7). For example, with this formula the local linear approximation of $f(x) = \sqrt{x}$ at x_0 is

$$\sqrt{x_0 + \Delta x} \approx \sqrt{x_0} + \frac{1}{2\sqrt{x_0}}\Delta x$$

Thus, to approximate $\sqrt{1.1}$ with this formula, we take $x_0 = 1$ and $\Delta x = 0.1$, which yields

$$\sqrt{1.1} \approx 1 + \tfrac{1}{2}(0.1) = 1.05$$

This agrees with the result in Example 4.

ERROR IN LOCAL LINEAR APPROXIMATIONS

As a general rule, the accuracy of the local linear approximation to $f(x)$ at a point x_0 will deteriorate as x gets progressively farther from x_0. To illustrate this for approximation (8) in Example 3, let us graph the function

$$E(x) = |\sin x - x|$$

which is the absolute value of the error in the approximation (Figure 3.6.9).

In Figure 3.6.9, the graph shows how the absolute error in the local linear approximation of $\sin x$ at 0 increases as x moves progressively farther from 0 in either the positive or negative direction. The graph also tells us that for values of x between the two vertical lines the absolute error does not exceed 0.01. Thus, for example, we could use the local linear approximation $\sin x \approx x$ for all values of x in the interval $-0.35 < x < 0.35$ (radians) with confidence that the approximation is within ± 0.01 of the exact value.

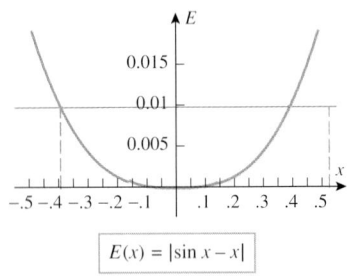

$$E(x) = |\sin x - x|$$

Figure 3.6.9

In applications, small errors invariably occur in measured quantities. When these quantities are used in computations, those errors are propagated in turn to the computed quantities; this is called ***error propagation***. We will now show how to use a local linear approximation to estimate the error in a computed quantity from estimates of the error in the measured quantity. For this purpose, suppose that

> x is the quantity being measured
>
> $y = f(x)$ is the quantity being computed
>
> x_0 is the true value of x
>
> y_0 is the true value of y
>
> Δx is the measurement error in x
>
> Δy is the propagated error in y

Thus, the measured value of x is $x_0 + \Delta x$, and the computed value of y is $y_0 + \Delta y$; and we are interested in using an estimate of Δx to find an estimate of Δy. To do this, we will start with version (7) of the local linear approximation of f at x_0:

$$f(x_0 + \Delta x) \approx f(x_0) + f'(x_0)\,\Delta x \tag{10}$$

In this formula, $f(x_0) = y_0$ is the true value of y, and $f(x_0 + \Delta x) = y_0 + \Delta y$ is the computed value of y, so we can rewrite (10) as

$$y_0 + \Delta y \approx y_0 + f'(x_0)\,\Delta x$$

or

$$\Delta y \approx f'(x_0)\,\Delta x$$

Moreover, if we agree to let $dx = \Delta x$, then we can rewrite this as

$$\Delta y \approx f'(x_0)\,dx = dy \tag{11}$$

which tells us that *the propagated error in y can be estimated by the differential of y at x_0 with dx interpreted as the measurement error in x.*

Although Formula (11) looks nice on the surface, it is useless in applied problems because the true value x_0 is unknown! (Keep in mind that the only value of x that is available to the researcher is the measured value $x_0 + dx = x_0 + \Delta x$.) To get around this roadblock researchers use the observed value $x_0 + dx$ rather than the true value x_0 in computing the differential. This is usually satisfactory if dx is small, since x_0 and $x_0 + dx$ are close in value. We will illustrate how this works in the next example, but it will simplify our computations if we drop the subscript in (11) and write the formula as

$$\Delta y \approx dy - f'(x)\,dx \tag{12}$$

Example 5

The radius of a sphere is measured to be 50 cm with a measurement error of ± 0.02 cm. Estimate the error in the computed volume of the sphere.

Solution. The volume of a sphere is

$$V = \tfrac{4}{3}\pi r^3$$

We are given that the error in the radius is $\Delta r = \pm 0.02$, and we want to find the error ΔV in V. If we consider Δr to be small and if we let $dr = \Delta r$, then ΔV can be approximated by dV. Thus, from (12),

$$\Delta V \approx dV = 4\pi r^2\,dr \tag{13}$$

Substituting $r = 50$ and $dr = \pm 0.02$ in (13), we obtain

$$\Delta V \approx 4\pi(2500)(\pm 0.02) \approx \pm 628.32$$

Therefore, the error in the calculated volume is approximately ± 628.32 cubic centimeters (cm³).   ◀

If the true value of a quantity is q and a measurement or calculation produces an error Δq, then $\Delta q / q$ is called the ***relative error*** in the measurement or calculation; when expressed as a percentage, $\Delta q / q$ is called the ***percentage error***. As a practical matter, the true value q is usually unknown, so that the measured or calculated value of q is used instead; and the relative error is approximated by dq / q.

Example 6

The side of a square is measured with a percentage error of $\pm 5\%$. Estimate the percentage error in the calculated area of the square.

Solution.   The area A of a square with side x is

$$A = x^2$$

so

$$dA = 2x\, dx$$

We are given that $dx/x = \pm 0.05$, and we want to find dA/A. But it follows from the two preceding formulas that

$$\frac{dA}{A} = \frac{2x\, dx}{A} = \frac{2x\, dx}{x^2} = 2\frac{dx}{x} = 2(\pm 0.05) = \pm 0.1 \tag{14}$$

Thus, the percentage error in the calculated area of the square is $\pm 10\%$.   ◀

FOR THE READER.   In (14) we saw that $dA/A = 2(dx/x)$, which tells us that as a rule of thumb the percentage error in the calculated area of a square is twice the percentage error in the measured side. What rule of thumb relates the percentage error in the computed volume of a cube to the percentage error in the measured side? Why?

DIFFERENTIAL FORMULAS

Now that we have defined differentials, every derivative formula has a corresponding differential formula. For example, if $y = \sin x$, then the derivative formula $dy/dx = \cos x$ can also be expressed as

$$dy = \cos x\, dx$$

Moreover, all of the general rules of differentiation have corresponding differential versions:

DERIVATIVE FORMULA	DIFFERENTIAL FORMULA
$\dfrac{d}{dx}[c] = 0$	$d[c] = 0$
$\dfrac{d}{dx}[cf] = c\dfrac{df}{dx}$	$d[cf] = c\, df$
$\dfrac{d}{dx}[f+g] = \dfrac{df}{dx} + \dfrac{dg}{dx}$	$d[f+g] = df + dg$
$\dfrac{d}{dx}[fg] = f\dfrac{dg}{dx} + g\dfrac{df}{dx}$	$d[fg] = f\, dg + g\, df$
$\dfrac{d}{dx}\left[\dfrac{f}{g}\right] = \dfrac{g\dfrac{df}{dx} - f\dfrac{dg}{dx}}{g^2}$	$d\left[\dfrac{f}{g}\right] = \dfrac{g\, df - f\, dg}{g^2}$

EXERCISE SET 3.6 ⊠ Graphing Calculator [C] CAS

1. (a) Let $y = x^2$. Find dy and Δy at $x = 2$ with $dx = \Delta x = 1$.
   (b) Sketch the graph of $y = x^2$, showing dy and Δy in the picture.

2. (a) Let $y = x^3$. Find dy and Δy at $x = 1$ with $dx = \Delta x = 1$.
   (b) Sketch the graph of $y = x^3$, showing dy and Δy in the picture.

3. (a) Let $y = 1/x$. Find dy and Δy at $x = 1$ with $dx = \Delta x = -0.5$.
   (b) Sketch the graph of $y = 1/x$, showing dy and Δy in the picture.

4. (a) Let $y = \sqrt{x}$. Find dy and Δy at $x = 9$ with $dx = \Delta x = -1$.
   (b) Sketch the graph of $y = \sqrt{x}$, showing dy and Δy in the picture.

In Exercises 5–8, find formulas for dy and Δy at a general point x.

5. $y = x^3$
6. $y = 8x - 4$
7. $y = x^2 - 2x + 1$
8. $y = \sin x$

In Exercises 9–12, find the differential dy.

9. (a) $y = 4x^3 - 7x^2$ (b) $y = x \cos x$
10. (a) $y = 1/x$ (b) $y = 5 \tan x$
11. (a) $y = x\sqrt{1-x}$ (b) $y = (1+x)^{-17}$
12. (a) $y = \dfrac{1}{x^3 - 1}$ (b) $y = \dfrac{1-x^3}{2-x}$

13. (a) Use Formula (6) to obtain the local linear approximation of x^3 at $x_0 = 1$.
    (b) Use Formula (7) to rewrite the approximation obtained in part (a) in terms of Δx.
    (c) Use the result obtained in part (a) to approximate $(1.02)^3$, and confirm that the formula obtained in part (b) produces the same result.

14. (a) Use Formula (6) to obtain the local linear approximation of $1/x$ at $x_0 = 2$.
    (b) Use Formula (7) to rewrite the approximation obtained in part (a) in terms of Δx.
    (c) Use the result obtained in part (a) to approximate $1/2.05$, and confirm that the formula obtained in part (b) produces the same result.

15. (a) Find the local linear approximation of $f(x) = \sqrt{1+x}$ at $x_0 = 0$, and use it to approximate $\sqrt{0.9}$ and $\sqrt{1.1}$.
    (b) Graph f and its tangent line at x_0 together, and use the graphs to illustrate the relationship between the exact values and the approximations of $\sqrt{0.9}$ and $\sqrt{1.1}$.

16. (a) Find the local linear approximation of $f(x) = 1/\sqrt{x}$ at $x_0 = 4$, and use it to approximate $1/\sqrt{3.9}$ and $1/\sqrt{4.1}$.

(b) Graph f and its tangent line at x_0 together, and use the graphs to illustrate the relationship between the exact values and the approximations of $1/\sqrt{3.9}$ and $1/\sqrt{4.1}$.

In Exercises 17–20, confirm that the stated formula is the local linear approximation at $x_0 = 0$.

17. $(1+x)^{15} \approx 1 + 15x$
18. $\dfrac{1}{\sqrt{1-x}} \approx 1 + \tfrac{1}{2}x$
19. $\tan x \approx x$
20. $\dfrac{1}{1+x} \approx 1 - x$

In Exercises 21–24, confirm that the stated formula is the local linear approximation of f at $x_0 = 1$, where $\Delta x = x - 1$.

21. $f(x) = x^4$; $(1 + \Delta x)^4 \approx 1 + 4x^3 \Delta x$
22. $f(x) = \sqrt{x}$; $\sqrt{1 + \Delta x} \approx 1 + \tfrac{1}{2}\Delta x$
23. $f(x) = \dfrac{1}{2+x}$; $\dfrac{1}{3 + \Delta x} \approx \dfrac{1}{3} - \dfrac{1}{9}\Delta x$
24. $f(x) = (4+x)^3$; $(5 + \Delta x)^3 \approx 125 + 75\Delta x$

25. (a) Use the local linear approximation of $\sin x$ at $x_0 = 0$ obtained in Example 3 to approximate $\sin 1°$, and compare the approximation to the result produced directly by your calculating device.
    (b) How would you choose x_0 to approximate $\sin 44°$?
    (c) Approximate $\sin 44°$; compare the approximation to the result produced directly by your calculating device.

26. (a) Use the local linear approximation of $\tan x$ at $x_0 = 0$ to approximate $\tan 2°$, and compare the approximation to the result produced directly by your calculating device.
    (b) How would you choose x_0 to approximate $\tan 61°$?
    (c) Approximate $\tan 61°$; compare the approximation to the result produced directly by your calculating device.

In Exercises 27–35, use an appropriate local linear approximation to estimate the value of the given quantity.

27. $(3.02)^4$ 28. $(1.97)^3$ 29. $\sqrt{65}$
30. $\sqrt{24}$ 31. $\sqrt{80.9}$ 32. $\sqrt{36.03}$
33. $\sin 0.1$ 34. $\tan 0.2$ 35. $\cos 31°$

36. The approximation $(1 + x)^k \approx 1 + kx$ is commonly used by engineers for quick calculations.
    (a) Derive this result, and use it to make a rough estimate of $(1.001)^{37}$.
    (b) Compare your estimate to that produced directly by your calculating device.
    (c) Show that this formula produces a very bad estimate of $(1.1)^{37}$, and explain why.

In Exercises 37–40, confirm that the formula is a local linear approximation at $x_0 = 0$, and use a graphing utility to estimate an interval of x-values on which the error in the approximation is at most ± 0.1.

37. $\sqrt{x+3} \approx \sqrt{3} + \dfrac{1}{2\sqrt{3}}x$

38. $\dfrac{1}{\sqrt{9-x}} \approx \dfrac{1}{3} + \dfrac{1}{54}x$

39. $\tan x \approx x$

40. $\dfrac{1}{(1+2x)^5} \approx 1 - 10x$

In Exercises 41–44, use dy to approximate Δy when x changes as indicated.

41. $y = \sqrt{3x-2}$; from $x = 2$ to $x = 2.03$

42. $y = \sqrt{x^2+8}$; from $x = 1$ to $x = 0.97$

43. $y = \dfrac{x}{x^2+1}$; from $x = 2$ to $x = 1.96$

44. $y = x\sqrt{8x+1}$; from $x = 3$ to $x = 3.05$

45. The side of a square is measured to be 10 ft, with a possible error of ± 0.1 ft.
  (a) Use differentials to estimate the error in the calculated area.
  (b) Estimate the percentage errors in the side and the area.

46. The side of a cube is measured to be 25 cm, with a possible error of ± 1 cm.
  (a) Use differentials to estimate the error in the calculated volume.
  (b) Estimate the percentage errors in the side and volume.

47. The hypotenuse of a right triangle is known to be 10 in exactly, and one of the acute angles is measured to be $30°$, with a possible error of $\pm 1°$.
  (a) Use differentials to estimate the errors in the sides opposite and adjacent to the measured angle.
  (b) Estimate the percentage errors in the sides.

48. One side of a right triangle is known to be 25 cm exactly. The angle opposite to this side is measured to be $60°$, with a possible error of $\pm 0.5°$.
  (a) Use differentials to estimate the errors in the adjacent side and the hypotenuse.
  (b) Estimate the percentage errors in the adjacent side and hypotenuse.

49. The electrical resistance R of a certain wire is given by $R = k/r^2$, where k is a constant and r is the radius of the wire. Assuming that the radius r has a possible error of $\pm 5\%$, use differentials to estimate the percentage error in R. (Assume k is exact.)

50. A 12-foot ladder leaning against a wall makes an angle θ with the floor. If the top of the ladder is h feet up the wall, express h in terms of θ and then use dh to estimate the change in h if θ changes from $60°$ to $59°$.

51. The area of a right triangle with a hypotenuse of H is calculated using the formula $A = \frac{1}{4}H^2 \sin 2\theta$, where θ is one of the acute angles. Use differentials to approximate the error in calculating A if $H = 4$ cm (exactly) and $\theta = 30° \pm 15'$.

52. The side of a square is measured with a possible percentage error of $\pm 1\%$. Use differentials to estimate the percentage error in the area.

53. The side of a cube is measured with a possible percentage error of $\pm 2\%$. Use differentials to estimate the percentage error in the volume.

54. The volume of a sphere is to be computed from a measured value of its radius. Estimate the maximum permissible percentage error in the measurement if the percentage error in the volume must be kept within $\pm 3\%$. ($V = \frac{4}{3}\pi r^3$ is the volume of a sphere of radius r.)

55. The area of a circle is to be computed from a measured value of its diameter. Estimate the maximum permissible percentage error in the measurement if the percentage error in the area must be kept within $\pm 1\%$.

56. A steel cube with 1-in sides is coated with 0.01 in of copper. Use differentials to estimate the volume of copper in the coating. [*Hint:* Let ΔV be the change in the volume of the cube.]

57. A metal rod 15 cm long and 5 cm in diameter is to be covered (except for the ends) with insulation that is 0.001 cm thick. Use differentials to estimate the volume of insulation. [*Hint:* Let ΔV be the change in volume of the rod.]

58. The time required for one complete oscillation of a pendulum is called its *period*. If the length L of the pendulum is measured in feet and the period P in seconds, then the period is given by $P = 2\pi\sqrt{L/g}$, where g is a constant called *the acceleration due to gravity*. Use differentials to show that the percentage error in P is approximately half the percentage error in L.

59. If the temperature T of a metal rod of length L is changed by an amount ΔT, then the length will change by the amount $\Delta L = \alpha L \,\Delta T$, where α is called the *coefficient of linear expansion*. For moderate changes in temperature α is taken as constant.
  (a) Suppose that a rod 40 cm long at $20°$C is found to be 40.006 cm long when the temperature is raised to $30°$C. Find α.
  (b) If an aluminum pole is 180 cm long at $15°$C, how long is the pole if the temperature is raised to $40°$C? [Take $\alpha = 2.3 \times 10^{-5}/°$C.]

60. If the temperature T of a solid or liquid of volume V is changed by an amount ΔT, then the volume will change by the amount $\Delta V = \beta V \,\Delta T$, where β is called the *coefficient of volume expansion*. For moderate changes in temperature β is taken as constant. Suppose that a tank truck loads 4000 gallons of ethyl alcohol at a temperature of $35°$C and delivers its load sometime later at a temperature of $15°$C. Using $\beta = 7.5 \times 10^{-4}/°$C for ethyl alcohol, find the number of gallons delivered.

SUPPLEMENTARY EXERCISES

1. State the definition of a derivative, and give two interpretations of it.

2. Explain the difference between average and instantaneous rate of change, and discuss how they are calculated.

3. Given that $y = f(x)$, explain the difference between dy and Δy. Draw a picture that illustrates the relationship between these quantities.

4. Use the definition of a derivative to find dy/dx, and check your answer by calculating the derivative using appropriate derivative formulas.
(a) $y = \sqrt{9 - 4x}$
(b) $y = \dfrac{x}{x + 1}$

In Exercises 5–8, find the values of x at which the curve $y = f(x)$ has a horizontal tangent line.

5. $f(x) = (2x + 7)^6(x - 2)^5$
6. $f(x) = \dfrac{(x - 3)^4}{x^2 + 2x}$

7. $f(x) = \sqrt{3x + 1}(x - 1)^2$
8. $f(x) = \left(\dfrac{3x + 1}{x^2}\right)^3$

9. The accompanying figure shows the graph of $y = f'(x)$ for an unspecified function f.
(a) For what values of x does the curve $y = f(x)$ have a horizontal tangent line?
(b) Over what intervals does the curve $y = f(x)$ have tangent lines with positive slope?
(c) Over what intervals does the curve $y = f(x)$ have tangent lines with negative slope?
(d) Given that $g(x) = f(x) \sin x$, and $f(0) = -1$, find $g''(0)$.

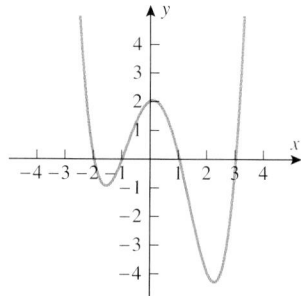

Figure Ex-9

10. In each part, evaluate the expression given that $f(1) = 1$, $g(1) = -2$, $f'(1) = 3$, and $g'(1) = -1$.
(a) $\dfrac{d}{dx}[f(x)g(x)]\Big|_{x=1}$
(b) $\dfrac{d}{dx}\left[\dfrac{f(x)}{g(x)}\right]\Big|_{x=1}$
(c) $\dfrac{d}{dx}[\sqrt{f(x)}]\Big|_{x=1}$
(d) $\dfrac{d}{dx}[f(1)g'(1)]$

11. Find the equations of all lines through the origin that are tangent to the curve $y = x^3 - 9x^2 - 16x$.

12. Find all values of x for which the tangent line to $y = 2x^3 - x^2$ is perpendicular to the line $x + 4y = 10$.

13. Find all values of x for which the line that is tangent to $y = 3x - \tan x$ is parallel to the line $y - x = 2$.

14. Suppose that $f(x) = \begin{cases} x^2 - 1, & x \le 1 \\ k(x - 1), & x > 1. \end{cases}$

For what values of k is f
(a) continuous
(b) differentiable?

15. Let $f(x) = x^2$. Show that for any distinct values of a and b, the slope of the tangent line to $y = f(x)$ at $x = \frac{1}{2}(a + b)$ is equal to the slope of the secant line through the points (a, a^2) and (b, b^2). Draw a picture to illustrate this result.

16. A car is traveling on a straight road that is 120 mi long. For the first 100 mi the car travels at an average velocity of 50 mi/h. Show that no matter how fast the car travels for the final 20 mi it cannot bring the average velocity up to 60 mi/h for the entire trip.

17. In each part, use the given information to find Δx, Δy, and dy.
(a) $y = 1/(x - 1)$; x decreases from 2 to 1.5.
(b) $y = \tan x$; x increases from $-\pi/4$ to 0.
(c) $y = \sqrt{25 - x^2}$; x increases from 0 to 3.

18. Use the formula $V = l^3$ for the volume of a cube of side l to find
(a) the average rate at which the volume of a cube changes with l as l increases from $l = 2$ to $l = 4$
(b) the instantaneous rate at which the volume of a cube changes with l when $l = 5$.

19. The amount of water in a tank t minutes after it has started to drain is given by $W = 100(t - 15)^2$ gal.
(a) At what rate is the water running out at the end of 5 min?
(b) What is the average rate at which the water flows out during the first 5 min?

20. Use an appropriate local linear approximation to estimate the value of $\cot 46°$, and compare your answer to the value obtained with a calculating device.

21. The base of the Great Pyramid at Giza is a square that is 230 m on each side.
(a) As illustrated in the accompanying figure, suppose that an archaeologist standing at the center of a side measures the angle of elevation of the apex to be $\phi = 51°$ with an error of $\pm 0.5°$. What can the archaeologist reasonably say about the height of the pyramid?
(b) Use differentials to estimate the allowable error in the elevation angle that will ensure an error in the height is at most ± 5 m.

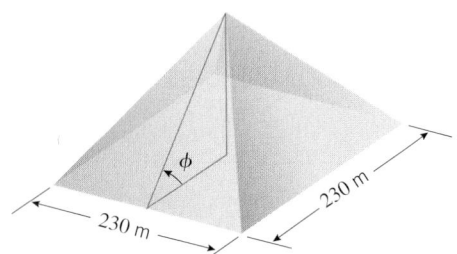

Figure Ex-21

22. The period T of a clock pendulum (i.e., the time required for one back-and-forth movement) is given in terms of its length L by $T = 2\pi\sqrt{L/g}$, where g is the gravitational constant.

(a) Assuming that the length of a clock pendulum can vary (say, due to temperature changes), find the rate of change of the period T with respect to the length L.

(b) If L is in meters (m) and T is in seconds (s), what are the units for the rate of change in part (a)?

(c) If a pendulum clock is running slow, should the length of the pendulum be increased or decreased to correct the problem?

(d) The constant g generally decreases with altitude. If you move a pendulum clock from sea level to a higher elevation, will it run faster or slower?

(e) Assuming the length of the pendulum to be constant, find the rate of change of the period T with respect to g.

(f) Assuming that T is in seconds (s) and g is in meters per second per second (m/s²), find the units for the rate of change in part (e).

In Exercises 23 and 24, zoom in on the graph of f on an interval containing $x = x_0$ until the graph looks like a straight line. Estimate the slope of this line and then check your answer by finding the exact value of $f'(x_0)$.

23. (a) $f(x) = x^2 - 1$, $x_0 = 1.8$

(b) $f(x) = \dfrac{x^2}{x - 2}$, $x_0 = 3.5$

24. (a) $f(x) = x^3 - x^2 + 1$, $x_0 = 2.3$

(b) $f(x) = \dfrac{x}{x^2 + 1}$, $x_0 = -0.5$

In Exercises 25 and 26, approximate $f'(2)$ by using the limit in Definition 3.2.2 with small values of h. If you have a CAS, see if it can find the exact value of the limit.

25. $f(x) = 2^x$

26. $f(x) = x^{\sin x}$

27. At time $t = 0$ a car moves into the passing lane to pass a slow-moving truck. The average velocity of the car from $t = 1$ to $t = 1 + h$ is

$$v_{\text{ave}} = \frac{3(h + 1)^{2.5} + 580h - 3}{10h}$$

Estimate the instantaneous velocity of the car at $t = 1$, where time is in seconds and distance is in feet.

28. A sky diver jumps from an airplane. Suppose that the distance she falls during the period before her parachute opens is $s(t) = 986((0.835)^t - 1) + 176t$, where s is in feet, t is in seconds, and $t \geq 1$. Graph s versus t for $1 \leq t \leq 20$, and use your graph to estimate the instantaneous velocity at $t = 15$.

29. Approximate the values of x at which the tangent line to the graph of $y = x^3 - \sin x$ is horizontal.

30. Use a graphing utility to graph the function

$$f(x) = |x^4 - x - 1| - x$$

and find the values of x where the derivative of this function does not exist.

31. Use a CAS to find the derivative of f from the definition

$$f'(x) = \lim_{h \to 0} \frac{f(x + h) - f(x)}{h}$$

and check the result by finding the derivative by hand.

(a) $f(x) = x^5$

(b) $f(x) = 1/x$

(c) $f(x) = 1/\sqrt{x}$

(d) $f(x) = \dfrac{2x + 1}{x - 1}$

(e) $f(x) = \sqrt{3x^2 + 5}$

(f) $f(x) = \sin 3x$

In Exercises 32–37: (a) use a CAS to find $f'(x)$, and check the result by hand; (b) use the CAS to find $f''(x)$.

32. $f(x) = x^2 \sin x$

33. $f(x) = \sqrt{x} + \cos^2 x$

34. $f(x) = \dfrac{2x^2 - x + 5}{3x + 2}$

35. $f(x) = \dfrac{\tan x}{1 + x^2}$

36. $f(x) = \dfrac{1}{x} - \sin \sqrt{x}$

37. $f(x) = \dfrac{\sqrt{x^4 - 3x + 2}}{x(2 - \cos x)}$

EXPANDING THE CALCULUS HORIZON

Robotics

*R*obin *designs and sells room dividers to defray college expenses. She is soon overwhelmed with orders and decides to build a robot to spray paint her dividers. As in most engineering projects, Robin begins with a simplified model that she will eventually refine to be more realistic. However, Robin quickly discovers that robotics (the design and control of robots) involves a considerable amount of mathematics, some of which we will discuss in this module.*

The Design Plan

Robin's plan is to develop a two-dimensional version of the robot arm in Figure 1. As shown in Figure 2, Robin's robot arm will consist of two links of fixed length, each of which will rotate independently about a pivot point. A paint sprayer will be attached to the end of the second link, and a computer will vary the angles θ_1 and θ_2, thereby allowing the robot to paint a region of the xy-plane.

The Mathematical Analysis

To analyze the motion of the robot arm, Robin denotes the coordinates of the paint sprayer by (x, y), as in Figure 3, and she derives the following equations that express x and y in terms of the angles θ_1 and θ_2 and the lengths l_1 and l_2 of the links:

$$x = l_1 \cos \theta_1 + l_2 \cos(\theta_1 + \theta_2)$$
$$y = l_1 \sin \theta_1 + l_2 \sin(\theta_1 + \theta_2)$$

$$(1)$$

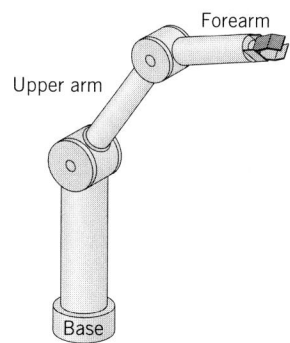

Figure 1

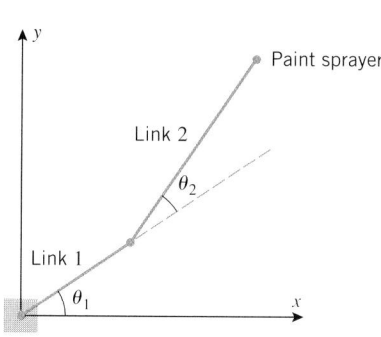

Figure 2

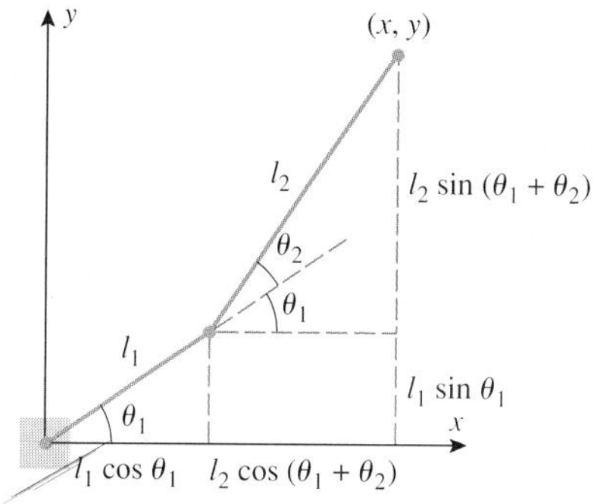

Figure 3

....
Exercise 1   Use Figure 3 to confirm the equations in (1).

In the language of robotics, θ_1 and θ_2 are called the **control angles**, the point (x, y) is called the **end effector**, and the equations in (1) are called the **forward kinematic equations** (from the Greek word *kinema*, meaning "motion").

....
Exercise 2   What is the region of the plane that can be reached by the end effector if:
(a) $l_1 = l_2$, (b) $l_1 > l_2$, and (c) $l_1 < l_2$?

....
Exercise 3   What are the coordinates of the end effector if $l_1 = 2$, $l_2 = 3$, $\theta_1 = \pi/4$, and $\theta_2 = \pi/6$?

Simulating Paint Patterns

Robin recognizes that if θ_1 and θ_2 are regarded as functions of time, then the forward kinematic equations can be expressed as

$$x = l_1 \cos\theta_1(t) + l_2 \cos(\theta_1(t) + \theta_2(t))$$
$$y = l_1 \sin\theta_1(t) + l_2 \sin(\theta_1(t) + \theta_2(t))$$

which are parametric equations for the curve traced by the end effector. For example, if the arms extend horizontally along the positive x-axis at time $t = 0$, and if links 1 and 2 rotate at the constant rates of ω_1 and ω_2 radians per second (rad/s), respectively, then

$$\theta_1(t) = \omega_1 t \quad \text{and} \quad \theta_2(t) = \omega_2 t$$

and the parametric equations of motion for the end effector become

$$x = l_1 \cos\omega_1 t + l_2 \cos(\omega_1 t + \omega_2 t)$$
$$y = l_1 \sin\omega_1 t + l_2 \sin(\omega_1 t + \omega_2 t)$$

Exercise 4   Show that if $l_1 = l_2 = 1$, and if $\omega_1 = 2$ rad/s and $\omega_2 = 3$ rad/s, then the parametric equations of motion are

$$x = \cos 2t + \cos 5t$$
$$y = \sin 2t + \sin 5t$$

Use a graphing utility to show that the curve traced by the end effector over the time interval $0 \leq t \leq 12\pi$ is as shown in Figure 4. This would be the painting pattern of Robin's paint sprayer.

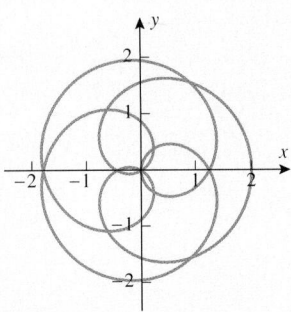

Figure 4

Exercise 5   Use a graphing utility to explore how the rotation rates of the links affect the spray patterns of a robot arm for which $l_1 = l_2 = 1$.

Exercise 6   Suppose that $l_1 = l_2 = 1$, and a malfunction in the robot arm causes the second link to lock at $\theta_2 = 0$, while the first link rotates at a constant rate of 1 rad/s. Make a conjecture about the path of the end effector, and confirm your conjecture by finding parametric equations for its motion.

Controlling the Position of the End Effector

Robin's plan is to make the robot paint the dividers in vertical strips, sweeping from the bottom up. After a strip is painted, she will have the arm return to the bottom of the divider and then move horizontally to position itself for the next upward sweep. Since the sections of her dividers will be 3 ft wide by 5 ft high, Robin decides on a robot with two 3-ft links whose base is positioned near the lower left corner of a divider section, as in Figure 5a. Since the fully extended links span a radius of 6 ft, she feels that this arrangement will work.

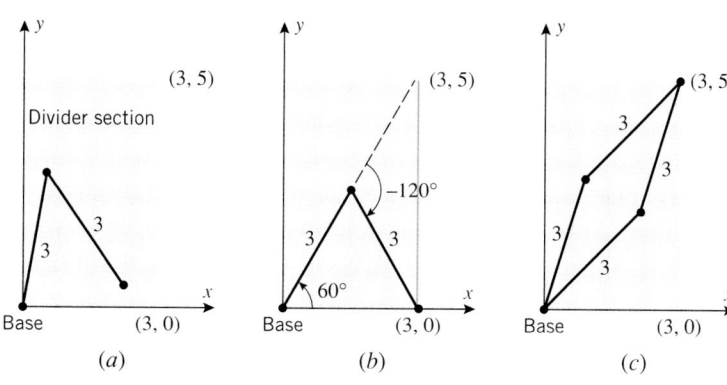

Figure 5

Robin starts with the problem of painting the far right edge from $(3, 0)$ to $(3, 5)$. With the help of some basic geometry (Figure 5b), she determines that the end effector can be placed at the point $(3, 0)$ by taking the control angles to be $\theta_1 = \pi/3$ $(= 60°)$ and $\theta_2 = -2\pi/3$ $(= -120°)$ (verify). However, the problem of finding the control angles that correspond to the point $(3, 5)$ is more complicated, so she starts by substituting the link lengths $l_1 = l_2 = 3$ into the forward kinematic equations in (1) to obtain

$$
\begin{aligned}
x &= 3 \cos \theta_1 + 3 \cos(\theta_1 + \theta_2) \\
y &= 3 \sin \theta_1 + 3 \sin(\theta_1 + \theta_2)
\end{aligned}
\tag{2}
$$

Thus, to put the end effector at the point $(3, 5)$, the control angles must satisfy the equations

$$
\begin{aligned}
\cos \theta_1 + \cos(\theta_1 + \theta_2) &= 1 \\
3 \sin \theta_1 + 3 \sin(\theta_1 + \theta_2) &= 5
\end{aligned}
\tag{3}
$$

Solving these equations for θ_1 and θ_2 challenges Robin's algebra and trigonometry skills, but she manages to do it using the procedure in the following exercise.

....

Exercise 7

(a) Use the equations in (3) and the identity

$$
\sin^2(\theta_1 + \theta_2) + \cos^2(\theta_1 + \theta_2) = 1
$$

to show that

$$
15 \sin \theta_1 + 9 \cos \theta_1 = 17
$$

(b) Solve the last equation for $\sin \theta_1$ in terms of $\cos \theta_1$ and substitute in the identity

$$
\sin^2 \theta_1 + \cos^2 \theta_1 = 1
$$

to obtain

$$
153 \cos^2 \theta_1 - 153 \cos \theta_1 + 32 = 0
$$

(c) Treat this as a quadratic equation in $\cos \theta_1$, and use the quadratic formula to obtain

$$
\cos \theta_1 = \frac{1}{2} \pm \frac{5\sqrt{17}}{102}
$$

(d) Use the arccosine (inverse cosine) operation of a calculating utility to solve the equations in part (c) to obtain

$$
\theta_1 \approx 0.792436 \text{ rad} \approx 45.4032° \quad \text{and} \quad \theta_1 \approx 1.26832 \text{ rad} \approx 72.6694°
$$

(e) Substitute each of these angles into the first equation in (3), and solve for the corresponding values of θ_2.

At first, Robin was surprised that the solutions for θ_1 and θ_2 were not unique, but her sketch in Figure 5c quickly made it clear that there will always be two ways of positioning the links to put the end effector at a specified point.

■■■ **Controlling the Motion of the End Effector**

Now that Robin has figured out how to place the end effector at the points $(3, 0)$ and $(3, 5)$, she turns to the problem of making the robot paint the vertical line segment between those points. She recognizes that not only must she make the end effector move on a vertical line, but she must control its velocity—if the end effector moves too quickly, the paint will be too thin, and if it moves too slowly, the paint will be too thick.

After some experimentation, she decides that the end effector should have a constant velocity of 1 ft/s. Thus, Robin's mathematical problem is to determine the rotation rates $d\theta_1/dt$ and $d\theta_2/dt$ (in rad/s) that will make $dx/dt = 0$ and $dy/dt = 1$. The first condition will ensure that the end effector moves vertically (no horizontal velocity), and the second condition will ensure that it moves upward at 1 ft/s.

To find formulas for dx/dt and dy/dt, Robin uses the chain rule to differentiate the forward kinematic equations in (2) and obtains

$$\frac{dx}{dt} = -3\sin\theta_1 \frac{d\theta_1}{dt} - [3\sin(\theta_1 + \theta_2)]\left(\frac{d\theta_1}{dt} + \frac{d\theta_2}{dt}\right)$$

$$\frac{dy}{dt} = 3\cos\theta_1 \frac{d\theta_1}{dt} + [3\cos(\theta_1 + \theta_2)]\left(\frac{d\theta_1}{dt} + \frac{d\theta_2}{dt}\right)$$

She uses the forward kinematic equations again to simplify these formulas and she then substitutes $dx/dt = 0$ and $dy/dt = 1$ to obtain

$$-y\frac{d\theta_1}{dt} - 3\sin(\theta_1 + \theta_2)\frac{d\theta_2}{dt} = 0$$

$$x\frac{d\theta_1}{dt} + 3\cos(\theta_1 + \theta_2)\frac{d\theta_2}{dt} = 1$$

(4)

· · · · · · · · · ·
Exercise 8   Confirm Robin's computations.

The equations in (4) will be used in the following way: At a given time t, the robot will report the control angles θ_1 and θ_2 of its links to the computer, the computer will use the forward kinematic equations in (2) to calculate the x- and y-coordinates of the end effector, and then the values of $\theta_1, \theta_2, x,$ and y will be substituted into (4) to produce two equations in the two unknowns $d\theta_1/dt$ and $d\theta_2/dt$. The computer will solve these equations to determine the required rotation rates for the links.

· · · · · · · · · ·
Exercise 9   In each part, use the given information to sketch the position of the links, and then calculate the rotation rates for the links in rad/s that will make the end effector of Robin's robot move upward with a velocity of 1 ft/s from that position.

   (a) $\theta_1 = \pi/3$, $\theta_2 = -2\pi/3$ (b) $\theta_1 = \pi/2$, $\theta_2 = -\pi/2$

· ·
Module by Mary Ann Connors, USMA, West Point, and Howard Anton, Drexel University, and based on the article "Moving a Planar Robot Arm" by Walter Meyer, MAA Notes Number 29, The Mathematical Association of America, 1993.

Pierre de Fermat

4

LOGARITHMIC AND EXPONENTIAL FUNCTIONS

*I*n this chapter we will study logarithms and exponents from the function point of view. These functions have applications in the study of population growth, sound, heating and cooling, earthquakes, and carbon dating, to name a few. We will review the algebraic aspects of logarithms and exponents, but we will focus mainly on those aspects of logarithmic and exponential functions that relate to calculus. The heart of this chapter is Section 4.1 on inverse functions, in which we develop fundamental ideas that link logarithmic and exponential functions together numerically, algebraically, and graphically. We also apply inverse functions to the study of inverse trigonometric functions (Section 4.5) and to the problem of differentiating functions whose formulas cannot be expressed in the form $y = f(x)$ (Section 4.3). We show how these methods of differentiation can be applied to problems involving rates of change (Section 4.6); and finally, we develop a powerful tool for evaluating limits, especially limits involving logarithmic and exponential functions.

4.1 INVERSE FUNCTIONS

*In everyday language the term "inversion" conveys the idea of a reversal. For example, in meteorology a temperature inversion is a reversal in the usual temperature properties of air layers; in music an inversion is a recurring theme that uses the same notes in reverse order; and in grammar an inversion is a reversal of the normal order of words. In mathematics the term **inverse** is used to describe functions that are reverses of one another in the sense that each undoes the effect of the other. The purpose of this section is to discuss this fundamental mathematical idea.*

INVERSE FUNCTIONS

The idea of solving an equation $y = f(x)$ for x as a function of y, say $x = g(y)$, is one of the most important ideas in mathematics. Sometimes, solving an equation is a simple process; for example, using basic algebra the equation

$$y = x^3 + 1 \qquad \boxed{y = f(x)}$$

can be solved for x as a function of y:

$$x = \sqrt[3]{y - 1} \qquad \boxed{x = g(y)}$$

The first equation is better for computing y if x is known, and the second is better for computing x if y is known (Figure 4.1.1).

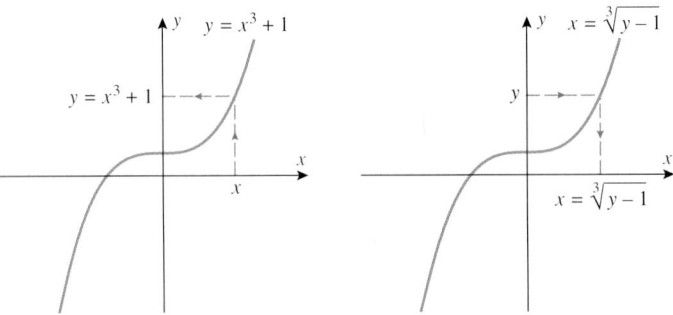

Figure 4.1.1

Our primary interest in this section is to identify relationships that may exist between the functions f and g when an equation $y = f(x)$ is expressed as $x = g(y)$, or conversely. For example, consider the functions $f(x) = x^3 + 1$ and $g(y) = \sqrt[3]{y - 1}$ discussed above. When these functions are composed in either order they cancel out the effect of one another in the sense that

$$g(f(x)) = \sqrt[3]{f(x) - 1} = \sqrt[3]{(x^3 + 1) - 1} = x$$
$$f(g(y)) = [g(y)]^3 + 1 = (\sqrt[3]{y - 1})^3 + 1 = y$$

(1)

The first of these equations states that each output of the composition $g(f(x))$ is the same as the input, and the second states that each output of the composition $f(g(y))$ is the same as the input. Pairs of functions with these two properties are so important that there is some terminology for them.

4.1.1   DEFINITION.   If the functions f and g satisfy the two conditions

$g(f(x)) = x$ for every x in the domain of f

$f(g(y)) = y$ for every y in the domain of g

then we say that f and g are **inverse functions**. Moreover, we call **f an inverse of g** and **g an inverse of f**.

Example 1

It follows from (1) that $f(x) = x^3 + 1$ and $g(y) = \sqrt[3]{y - 1}$ are inverse functions. ◀

It can be shown that a function cannot have two different inverses. Thus, if a function f has an inverse, then the inverse is unique, and we are entitled to talk about *the* inverse of f. The inverse of a function f is commonly denoted by f^{-1} (read "f inverse"). Thus, instead of using g in Example 1, the inverse of $f(x) = x^3$ could have been expressed as $f^{-1}(y) = \sqrt[3]{y - 1}$.

WARNING. The symbol f^{-1} should always be interpreted as the inverse of f and *never* as the reciprocal $1/f$.

It is important to understand that a function is determined by the relationship that it establishes between its inputs and outputs and not by the letter used for the independent variable. Thus, even though the formulas $f(x) = 3x$ and $f(y) = 3y$ use different independent variables, they define the *same* function f, since the two formulas have the same "form" and hence assign the same value to each input; for example, in either notation $f(2) = 6$. As we progress through this text, there will be certain occasions on which we will want the independent variables for f and f^{-1} to be the same, and other occasions on which we will want them to be different. Thus, in Example 1 we could have expressed the inverse of $f(x) = x^3 + 1$ as $f^{-1}(x) = \sqrt[3]{x - 1}$ had we wanted f and f^{-1} to have the same independent variable.

If we use the notation f^{-1} (rather than g) in Definition 4.1.1, and if we use x as the independent variable in the formulas for both f and f^{-1}, then the defining equations relating these functions are

$$f^{-1}(f(x)) = x \quad \text{for every } x \text{ in the domain of } f$$
$$f(f^{-1}(x)) = x \quad \text{for every } x \text{ in the domain of } f^{-1} \tag{2}$$

Example 2

Confirm each of the following.

(a)   The inverse of $f(x) = 2x$ is $f^{-1}(x) = \frac{1}{2}x$.

(b)   The inverse of $f(x) = x^3$ is $f^{-1}(x) = x^{1/3}$.

Solution (a).

$$f^{-1}(f(x)) = f^{-1}(2x) = \tfrac{1}{2}(2x) = x$$

$$f(f^{-1}(x)) = f\left(\tfrac{1}{2}x\right) = 2\left(\tfrac{1}{2}x\right) = x$$

Solution (b).

$$f^{-1}(f(x)) = f^{-1}(x^3) = \left(x^3\right)^{1/3} = x$$

$$f(f^{-1}(x)) = f(x^{1/3}) = \left(x^{1/3}\right)^3 = x$$ ◀

REMARK. The results in Example 2 should make sense to you intuitively, since the operations of multiplying by 2 and multiplying by $\frac{1}{2}$ in either order cancel the effect of one another, as do the operations of cubing and taking a cube root.

DOMAIN AND RANGE OF INVERSE FUNCTIONS

The equations in (2) imply certain relationships between the domains and ranges of f and f^{-1}. For example, in the first equation the quantity $f(x)$ is an input of f^{-1}, so points in the range of f lie in the domain of f^{-1}; and in the second equation the quantity $f^{-1}(x)$ is an input of f, so points in the range of f^{-1} lie in the domain of f. All of this suggests the following relationships, which we state without formal proof:

$$\text{domain of } f^{-1} = \text{range of } f$$
$$\text{range of } f^{-1} = \text{domain of } f$$

(3)

At the beginning of this section we solved the equation $y = f(x) = x^3 + 1$ for x as a function of y to obtain $x = g(y) = \sqrt[3]{y - 1}$, and we observed in Example 1 that g is the inverse of f. This was not accidental—whenever an equation $y = f(x)$ is solved for x as a function of y, say $x = g(y)$, then f and g will be inverses. We can see why this is so by making two substitutions:

- Substitute $y = f(x)$ into $x = g(y)$. This yields $x = g(f(x))$, which is the first equation in Definition 4.1.1.

- Substitute $x = g(y)$ into $y = f(x)$. This yields $y = f(g(y))$, which is the second equation in Definition 4.1.1.

Since f and g satisfy the two conditions in Definition 4.1.1, we conclude that they are inverses. Thus, we have the following result.

4.1.2   THEOREM.   *If an equation $y = f(x)$ can be solved for x as a function of y, then f has an inverse and the resulting equation is $x = f^{-1}(y)$.*

A METHOD FOR FINDING INVERSES

Example 3

Find the inverse of $f(x) = \sqrt{3x - 2}$.

Solution.   From Theorem 4.1.2 we can find a formula for $f^{-1}(y)$ by solving the equation

$$y = \sqrt{3x - 2}$$

for x as a function of y. The computations are

$$y^2 = 3x - 2$$
$$x = \tfrac{1}{3}(y^2 + 2)$$

from which it follows that

$$f^{-1}(y) = \tfrac{1}{3}(y^2 + 2)$$

At this point we have successfully produced a formula for f^{-1}; however, we are not quite done, since there is no guarantee that the natural domain associated with this formula is the correct domain for f^{-1}. To determine whether this is so, we will examine the range of $y = f(x) = \sqrt{3x - 2}$. The range consists of all y in the interval $[0, +\infty)$, so from (3) this interval is also the domain of $f^{-1}(y)$; thus, the inverse of f is given by the formula

$$f^{-1}(y) = \tfrac{1}{3}(y^2 + 2), \quad y \geq 0$$ ◀

REMARK.   When a formula for f^{-1} is obtained by solving the equation $y = f(x)$ for x as a function of y, the resulting formula has y as the independent variable. If it is preferable to have x as the independent variable for f^{-1}, then there are two ways to proceed: you can solve $y = f(x)$ for x as a function of y, and then replace y by x in the *final* formula for f^{-1}, or you can interchange x and y in the *original* equation and solve the equation $x = f(y)$ for y in terms of x, in which case the final equation will be $y = f^{-1}(x)$. In Example 3, either of these procedures will produce $f^{-1}(x) = \tfrac{1}{3}(x^2 + 2), x \geq 0$.

Theorem 4.1.2 not only provides a method for finding the inverse of a function f, but it also provides an interpretation of what the values of f^{-1} represent. The theorem tells us that for a given y, the quantity $f^{-1}(y)$ is that number x with the property that $f(x) = y$. For example, if $f^{-1}(1) = 4$, then you know that $f(4) = 1$; and similarly, if $f(3) = 7$, then you know that $f^{-1}(7) = 3$.

....
EXISTENCE OF INVERSE FUNCTIONS

Not every function has an inverse. In general, in order for a function f to have an inverse it must assign distinct outputs to distinct inputs. To see why this is so, consider the function $f(x) = x^2$. Since $f(2) = f(-2) = 4$, the function f assigns the same output to two distinct inputs. If f were to have an inverse, then the equation $f(2) = 4$ would imply that $f^{-1}(4) = 2$, and the equation $f(-2) = 4$ would imply that $f^{-1}(4) = -2$. This is obviously impossible, since we cannot have two different values for $f^{-1}(4)$. Thus, $f(x) = x^2$ has no inverse. Another way to see that $f(x) = x^2$ has no inverse is to attempt to find the inverse by solving the equation $y = x^2$ for x in terms of y. We run into trouble immediately because the resulting equation, $x = \pm\sqrt{y}$, does not express x as a *single* function of y.

Functions that assign distinct outputs to distinct inputs are sufficiently important that there is a name for them—they are said to be **one-to-one** or **invertible**. Stated algebraically, a function f is one-to-one if $f(x_1) \neq f(x_2)$ whenever $x_1 \neq x_2$; and stated geometrically, a function f is one-to-one if the graph of $y = f(x)$ is cut at most once by any horizontal line (Figure 4.1.2).

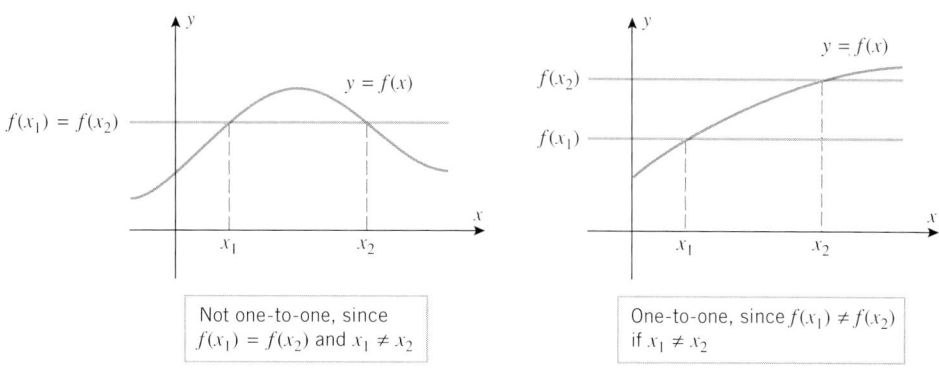

Not one-to-one, since $f(x_1) = f(x_2)$ and $x_1 \neq x_2$

One-to-one, since $f(x_1) \neq f(x_2)$ if $x_1 \neq x_2$

Figure 4.1.2

One can prove that a function f has an inverse if and only if it is one-to-one, and this provides us with the following geometric test for determining whether a function has an inverse.

4.1.3 THEOREM (*The Horizontal Line Test*). *A function f has an inverse if and only if its graph is cut at most once by any horizontal line.*

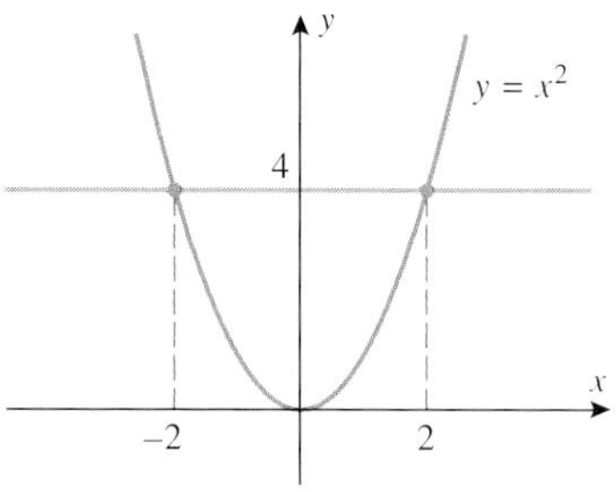

Example 4

We observed above that the function $f(x) = x^2$ does not have an inverse. This is confirmed by the horizontal line test, since the graph of $y = x^2$ is cut more than once by certain horizontal lines (Figure 4.1.3). ◄

figure 4.1.3

Example 5

We saw in Example 2(b) that the function $f(x) = x^3$ has an inverse [namely, $f^{-1}(x) = x^{1/3}$]. The existence of an inverse is confirmed by the horizontal line test, since the graph of $y = x^3$ is cut at most once by any horizontal line (Figure 4.1.4). ◄

Figure 4.1.3

Example 6

Explain why the function f that is graphed in Figure 4.1.5 has an inverse, and find $f^{-1}(3)$.

Solution.   The function f has an inverse since its graph passes the horizontal line test. To evaluate $f^{-1}(3)$, we view $f^{-1}(3)$ as that number x for which $f(x) = 3$. From the graph we see that $f(2) = 3$, so $f^{-1}(3) = 2$.   ◄

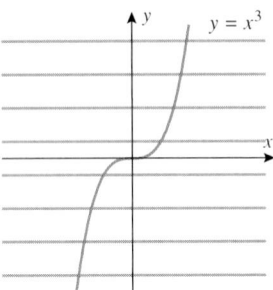

Figure 4.1.4

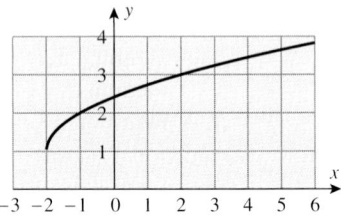

Figure 4.1.5

GRAPHS OF INVERSE FUNCTIONS

Our next objective is to explore the relationship between the graphs of f and f^{-1}. For this purpose, it will be desirable to use x as the independent variable for both functions, which means that we will be comparing the graphs of $y = f(x)$ and $y = f^{-1}(x)$.

If (a, b) is a point on the graph $y = f(x)$, then $b = f(a)$. This is equivalent to the statement that $a = f^{-1}(b)$, which means that (b, a) is a point on the graph of $y = f^{-1}(x)$. In short, reversing the coordinates of a point on the graph of f produces a point on the graph of f^{-1}. Similarly, reversing the coordinates of a point on the graph of f^{-1} produces a point on the graph of f (verify). However, the geometric effect of reversing the coordinates of a point is to reflect that point about the line $y = x$ (Figure 4.1.6), and hence the graphs of $y = f(x)$ and $y = f^{-1}(x)$ are reflections of one another about this line (Figure 4.1.7). In summary, we have the following result.

4.1.4   THEOREM.   *If f has an inverse, then the graphs of $y = f(x)$ and $y = f^{-1}(x)$ are reflections of one another about the line $y = x$; that is, each is the mirror image of the other with respect to that line.*

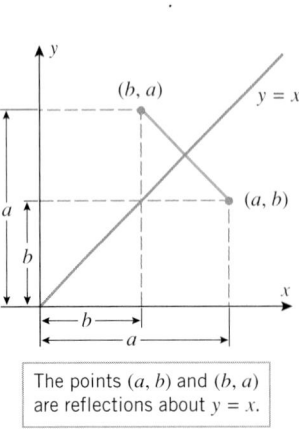

The points (a, b) and (b, a) are reflections about $y = x$.

Figure 4.1.6

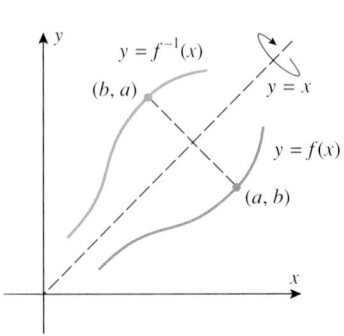

Figure 4.1.7

Example 7

Figure 4.1.8 shows the graphs of the inverse functions discussed in Examples 2 and 3.

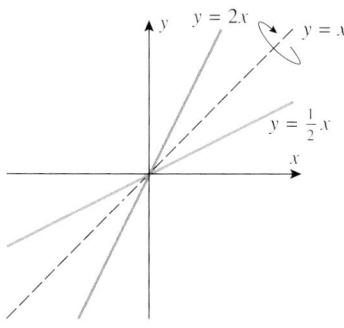
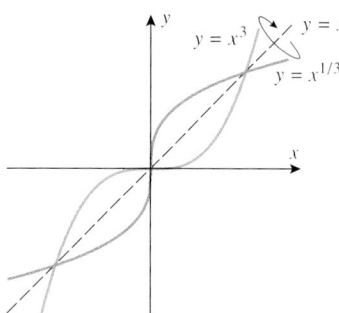
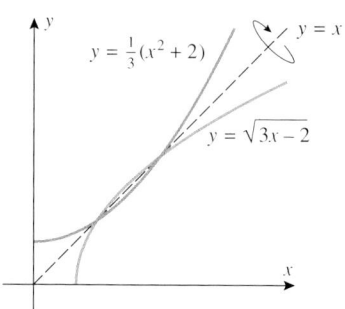

Figure 4.1.8

INCREASING OR DECREASING FUNCTIONS HAVE INVERSES

If the graph of a function f is always increasing or always decreasing over the domain of f, then the graph of f can be cut at most once by any horizontal line and consequently the function f must have an inverse. One way to tell whether the graph of a function is increasing or decreasing over an interval is by examining the slopes of its tangent lines. We will prove in the next chapter that the graph of f must be increasing on any interval where $f'(x) > 0$ (since the tangent lines have positive slope) and must be decreasing on any interval where $f'(x) < 0$ (since the tangent lines have negative slope) (Figure 4.1.9). These intuitive observations suggest the following theorem, which we state without formal proof.

> **4.1.5**   THEOREM.   *If the domain of f is an interval on which $f'(x) > 0$ or on which $f'(x) < 0$, then the function f has an inverse.*

Figure 4.1.9

Example 8

The graph of $f(x) = x^5 + x + 1$ is always increasing on $(-\infty, +\infty)$, since

$$f'(x) = 5x^4 + 1 > 0$$

for all x. However, there is no easy way to solve the equation $y = x^5 + x + 1$ for x in terms of y (try it), so even though we know that f has an inverse, we cannot produce a formula for it.   ◀

REMARK.   What is important to understand here is that our inability to find a formula for the inverse does not negate the existence of the inverse; indeed, one of our goals in later sections will be to develop ways of finding properties of functions in which there are no explicit formulas for the functions to work with.

RESTRICTING DOMAINS TO MAKE FUNCTIONS INVERTIBLE

Sometimes a function that is not one-to-one can be made one-to-one by restricting its domain. For example, although the function $f(x) = x^2$ is not one-to-one, the functions

$$g(x) = x^2, \quad x \geq 0$$
$$h(x) = x^2, \quad x \leq 0$$

which result from restricting the domain of f, are one-to-one since their graphs pass the horizontal line test [the graph of g is the right half of the parabola $y = x^2$ and the graph of h is the left half (Figure 4.1.10)]. The inverses of g and h can be found by solving each of the equations $y = g(x)$ and $y = h(x)$ for x as a function of y. For example, to find the inverse of g we solve

$$y = x^2, \quad x \geq 0$$

for x, which yields $x = \sqrt{y}$; hence, $g^{-1}(y) = \sqrt{y}$. Similarly, $h^{-1}(y) = -\sqrt{y}$. Geometrically, the graphs of $g(x) = x^2$, $x \geq 0$ and $g^{-1}(x) = \sqrt{x}$ are reflections of one another about the line $y = x$ (Figure 4.1.11), which reveals that the graph of $y = \sqrt{x}$ is a portion of a reflected parabola.

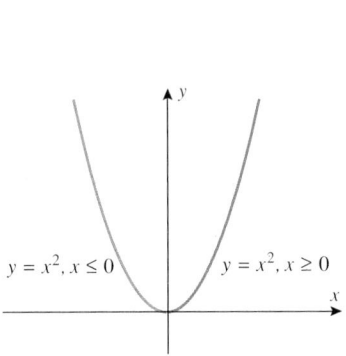

Figure 4.1.10

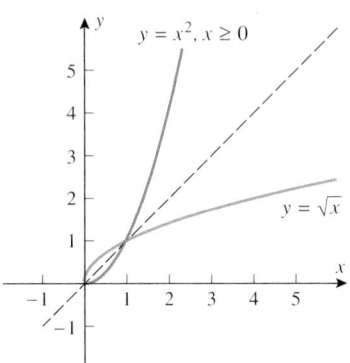

Figure 4.1.11

CONTINUITY OF INVERSE FUNCTIONS

Because the graphs of f and f^{-1} are reflections of one another about the line $y = x$, it is intuitively obvious that if the graph of f has no breaks, then neither will the graph of f^{-1}. This suggests the following result, which we state without proof.

> **4.1.6** THEOREM. *If a function f is continuous and has an inverse, then f^{-1} is also continuous.*

For example, even though we cannot find a formula for f^{-1} in Example 8, the continuity of the polynomial f guarantees that f^{-1} is a continuous function.

DIFFERENTIABILITY OF INVERSE FUNCTIONS

Suppose that f is a continuous one-to-one function. Speaking informally, the points of nondifferentiability of f^{-1} occur most commonly at corners or points of vertical tangency in the graph of $y = f^{-1}(x)$. However, the graph of $y = f^{-1}(x)$ is the reflection about $y = x$ of the graph of $y = f(x)$; hence, corners in the graph of f^{-1} are reflections of corners in the graph of f, and points of vertical tangency in the graph of f^{-1} are reflections of points of horizontal tangency in the graph of f. This suggests that if f is a differentiable function whose derivative is nonzero, then f^{-1} will be a differentiable function. The following theorem, which we state without proof, makes this idea precise.

> **4.1.7** THEOREM (***Differentiability of Inverse Functions***). *Suppose that the function f is invertible and differentiable on an interval I. Then f^{-1} is differentiable at any point x where $f'(f^{-1}(x)) \neq 0$.*

Example 9

We showed in Example 8 that the function $f(x) = x^5 + x + 1$ has an inverse. Use Theorem 4.1.7 to show that f^{-1} is differentiable on the interval $(-\infty, +\infty)$.

Solution. Let I denote the interval $(-\infty, +\infty)$. We must show that for each x in I, the function f has a nonzero derivative at the point $f^{-1}(x)$. But this is so because the derivative of f is

$$f'(x) = 5x^4 + 1$$

which is nonzero for all x. ◄

GRAPHING INVERSE FUNCTIONS WITH GRAPHING UTILITIES

Most graphing utilities cannot graph inverse functions directly. However, there is a way of graphing inverse functions by expressing the graph parametrically. To see how this can be done, suppose that we are interested in graphing the inverse of a one-to-one function f. We observed in Section 1.7 that the equation $y = f(x)$ can be expressed parametrically as

$$x = t, \quad y = f(t) \tag{4}$$

Moreover, we know that the graph of f^{-1} can be obtained by interchanging x and y, since this reflects the graph of f about the line $y = x$. Thus, from (4) the graph of f^{-1} can be represented parametrically as

$$x = f(t), \quad y = t \tag{5}$$

For example, Figure 4.1.12 shows the graph of $f(x) = x^5 + x + 1$ and its inverse generated with a graphing utility. The graph of f was generated from the parametric equations

$$x = t, \quad y = t^5 + t + 1$$

and the graph of f^{-1} was generated from the parametric equations

$$x = t^5 + t + 1, \quad y = t$$

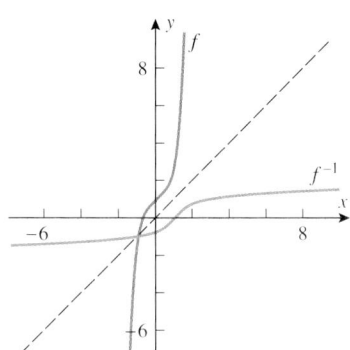

Figure 4.1.12

EXERCISE SET 4.1   ∿ Graphing Calculator

1. In (a)–(d), determine whether f and g are inverse functions.
   (a) $f(x) = 4x, \ g(x) = \frac{1}{4}x$
   (b) $f(x) = 3x + 1, \ g(x) = 3x - 1$
   (c) $f(x) = \sqrt[3]{x - 2}, \ g(x) = x^3 + 2$
   (d) $f(x) = x^4, \ g(x) = \sqrt[4]{x}$

∿ 2. Check your answers to Exercise 1 with a graphing utility by determining whether the graphs of f and g are reflections of one another about the line $y = x$.

3. In each part, determine whether the function f defined by the table is one-to-one.

(a)

x	1	2	3	4	5	6
$f(x)$	−2	−1	0	1	2	3

(b)

x	1	2	3	4	5	6
$f(x)$	4	−7	6	−3	1	4

4. In each part, determine whether the function f is one-to-one, and justify your answer.
   (a) $f(t)$ is the number of people in line at a movie theater at time t.
   (b) $f(x)$ is your weight on your xth birthday.
   (c) $f(v)$ is the weight of v cubic inches of lead.

5. In each part, use the horizontal line test to determine whether the function f is one-to-one.
   (a) $f(x) = 3x + 2$ (b) $f(x) = \sqrt{x - 1}$
   (c) $f(x) = |x|$ (d) $f(x) = x^3$
   (e) $f(x) = x^2 - 2x + 2$ (f) $f(x) = \sin x$

∿ 6. In each part, generate the graph of the function f with a graphing utility, and determine whether f is one-to-one.
   (a) $f(x) = x^3 - 3x + 2$ (b) $f(x) = x^3 - 3x^2 + 3x - 1$

7. In each part, determine whether f is one-to-one.
   (a) $f(x) = \tan x$
   (b) $f(x) = \tan x, \ -\pi < x < \pi$
   (c) $f(x) = \tan x, \ -\pi/2 < x < \pi/2$

8. In each part, determine whether f is one-to-one.
  (a) $f(x) = \cos x$
  (b) $f(x) = \cos x$, $-\pi/2 \leq x \leq \pi/2$
  (c) $f(x) = \cos x$, $0 \leq x \leq \pi$

9. (a) The accompanying figure shows the graph of a function f over its domain $-8 \leq x \leq 8$. Explain why f has an inverse, and use the graph to find $f^{-1}(2)$, $f^{-1}(-1)$, and $f^{-1}(0)$.
  (b) Find the domain and range of f^{-1}.
  (c) Sketch the graph of f^{-1}.

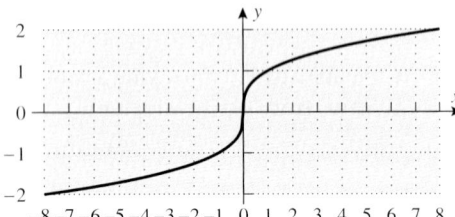

Figure Ex-9

10. (a) Explain why the function f graphed in the accompanying figure has no inverse on its domain $-3 \leq x \leq 4$.
  (b) Subdivide the domain into three adjacent intervals on each of which the function f has an inverse.

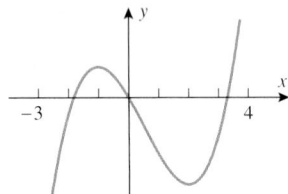

Figure Ex-10

In Exercises 11 and 12, determine whether the function f is one-to-one by examining the sign of $f'(x)$.

11. (a) $f(x) = x^2 + 8x + 1$
  (b) $f(x) = 2x^5 + x^3 + 3x + 2$
  (c) $f(x) = 2x + \sin x$

12. (a) $f(x) = x^3 + 3x^2 - 8$
  (b) $f(x) = x^5 + 8x^3 + 2x - 1$
  (c) $f(x) = \dfrac{x}{x + 1}$

In Exercises 13–23, find a formula for $f^{-1}(x)$.

13. $f(x) = x^5$ **14.** $f(x) = 6x$

15. $f(x) = 7x - 6$ **16.** $f(x) = \dfrac{x + 1}{x - 1}$

17. $f(x) = 3x^3 - 5$ **18.** $f(x) = \sqrt[5]{4x + 2}$

19. $f(x) = \sqrt[3]{2x - 1}$

20. $f(x) = 5/(x^2 + 1)$, $x \geq 0$

21. $f(x) = 3/x^2$, $x < 0$

22. $f(x) = \begin{cases} 2x, & x \leq 0 \\ x^2, & x > 0 \end{cases}$

23. $f(x) = \begin{cases} 5/2 - x, & x < 2 \\ 1/x, & x \geq 2 \end{cases}$

24. Find a formula for $p^{-1}(x)$, given that
$$p(x) = x^3 - 3x^2 + 3x - 1$$

In Exercises 25–29, find a formula for $f^{-1}(x)$, and state the domain of f^{-1}.

25. $f(x) = (x + 2)^4$, $x \geq 0$

26. $f(x) = \sqrt{x + 3}$ **27.** $f(x) = -\sqrt{3 - 2x}$

28. $f(x) = 3x^2 + 5x - 2$, $x \geq 0$

29. $f(x) = x - 5x^2$, $x \geq 1$

30. The formula $F = \frac{9}{5}C + 32$, where $C \geq -273.15°$C expresses the Fahrenheit temperature F as a function of the Celsius temperature C.
  (a) Find a formula for the inverse function.
  (b) In words, what does the inverse function tell you?
  (c) Find the domain and range of the inverse function.

31. (a) One meter is about 6.214×10^{-4} miles. Find a formula $y = f(x)$ that expresses a length x in meters as a function of the same length y in miles.
  (b) Find a formula for the inverse of f.
  (c) In practical terms, what does the formula $x = f^{-1}(y)$ tell you?

32. Suppose that f is a one-to-one, continuous function such that $\lim_{x \to 3} f(x) = 7$. Find $\lim_{x \to 7} f^{-1}(x)$, and justify your reasoning.

33. Let $f(x) = x^2$, $x > 1$, and $g(x) = \sqrt{x}$.
  (a) Show that $f(g(x)) = x$, $x > 1$, and $g(f(x)) = x$, $x > 1$.
  (b) Show that f and g are *not* inverses of one another by showing that the graphs of $y = f(x)$ and $y = g(x)$ are not reflections of one another about $y = x$.
  (c) Do parts (a) and (b) contradict one another? Explain.

34. Let $f(x) = ax^2 + bx + c$, $a > 0$. Find f^{-1} if the domain of f is restricted to
  (a) $x \geq -b/(2a)$ (b) $x \leq -b/(2a)$.

35. (a) Show that $f(x) = (3 - x)/(1 - x)$ is its own inverse.
  (b) What does the result in part (a) tell you about the graph of f?

36. Suppose that a line of nonzero slope m intersects the x-axis at $(x_0, 0)$. Find an equation for the reflection of this line about $y = x$.

37. (a) Show that $f(x) = x^3 - 3x^2 + 2x$ is not one-to-one on $(-\infty, +\infty)$.
  (b) Find the largest value of k such that f is one-to-one on the interval $(-k, k)$.

38. (a) Show that the function $f(x) = x^4 - 2x^3$ is not one-to-one on $(-\infty, +\infty)$.
   (b) Find the smallest value of k such that f is one-to-one on the interval $[k, +\infty)$.

39. Let $f(x) = 2x^3 + 5x + 3$. Find x if $f^{-1}(x) = 1$.

40. Let $f(x) = \dfrac{x^3}{x^2 + 1}$. Find x if $f^{-1}(x) = 2$.

> In Exercises 41–44, use a graphing utility and parametric equations to display the graphs of f and f^{-1} on the same screen.

41. $f(x) = x^3 + 0.2x - 1, \quad -1 \le x \le 2$

42. $f(x) = \sqrt{x^2 + 2} + x, \quad -5 \le x \le 5$

43. $f(x) = \cos(\cos 0.5x), \quad 0 \le x \le 3$

44. $f(x) = x + \sin x, \quad 0 \le x \le 6$

45. Prove that if $a^2 + bc \ne 0$, then the graph of
$$f(x) = \frac{ax + b}{cx - a}$$
is symmetric about the line $y = x$.

46. (a) Prove: If f and g are one-to-one, then so is the composition $f \circ g$.
   (b) Prove: If f and g are one-to-one, then
   $$(f \circ g)^{-1} = g^{-1} \circ f^{-1}$$

47. Sketch the graph of a function that is one-to-one on $(-\infty, +\infty)$, yet not increasing on $(-\infty, +\infty)$ and not decreasing on $(-\infty, +\infty)$.

48. Prove: A one-to-one function f cannot have two different inverses.

49. Let $F(x) = f(2g(x))$ where $f(x) = x^4 + x^3 + 1$ for $0 \le x \le 2$, and $g(x) = f^{-1}(x)$. Find $F(3)$.

4.2 LOGARITHMIC AND EXPONENTIAL FUNCTIONS

When logarithms were introduced in the seventeenth century as a computational tool, they provided scientists of that period computing power that was previously unimaginable. Although computers and calculators have largely replaced logarithms for numerical calculations, the logarithmic functions and their relatives have wide-ranging applications in mathematics and science. Some of these will be introduced in this section.

IRRATIONAL EXPONENTS

In algebra, integer and rational powers of a number b are defined by

$$b^n = b \times b \times \cdots \times b \quad (n \text{ factors}), \qquad b^{-n} = \frac{1}{b^n}, \qquad b^0 = 1,$$

$$b^{p/q} = \sqrt[q]{b^p} = (\sqrt[q]{b})^p, \qquad b^{-p/q} = \frac{1}{b^{p/q}}.$$

If b is negative, then some of the fractional powers of b will have imaginary values; for example, $(-2)^{1/2} = \sqrt{-2}$. To avoid this complication we will assume throughout this section that $b \ge 0$, even if it is not stated explicitly.

Observe that the preceding definitions do not include *irrational* powers of b such as

$$2^\pi, \quad 3^{\sqrt{2}}, \quad \text{and} \quad \pi^{-\sqrt{7}}$$

There are various methods for defining irrational powers. One approach is to define irrational powers of b as limits of rational powers of b. For example, to define 2^π we can start with the decimal representation of π, namely,

$$3.1415926\ldots$$

From this decimal we can form a sequence of rational numbers that gets closer and closer to π, namely,

$$3.1, \quad 3.14, \quad 3.141, \quad 3.1415, \quad 3.14159$$

and from these we can form a sequence of *rational* powers of 2:

$$2^{3.1}, \quad 2^{3.14}, \quad 2^{3.141}, \quad 2^{3.1415}, \quad 2^{3.14159}$$

Table 4.2.1

x	2^x
3	8.000000
3.1	8.574188
3.14	8.815241
3.141	8.821353
3.1415	8.824411
3.14159	8.824962
3.141592	8.824974

Since the exponents of the terms in this sequence approach a limit of π, it seems plausible that the terms themselves approach a limit, and it would seem reasonable to *define* 2^π to be this limit. Table 4.2.1 provides numerical evidence that the sequence does, in fact, have a limit and that to four decimal places the value of this limit is $2^\pi \approx 8.8250$. More generally, for any irrational exponent p and positive number b, we can define b^p as the limit of the rational powers of b created from the decimal expansion of p.

FOR THE READER.   Confirm the approximation $2^\pi \approx 8.8250$ by computing 2^π directly using your calculating utility.

Although our definition of b^p for irrational p certainly seems reasonable, there is a lot of tedious mathematical detail required to make the definition precise. We will not be concerned with such matters here and will accept without proof that the following familiar laws hold for all real exponents:

$$b^p b^q = b^{p+q}, \quad \frac{b^p}{b^q} = b^{p-q}, \quad \left(b^p\right)^q = b^{pq}$$

THE FAMILY OF EXPONENTIAL FUNCTIONS

A function of the form $f(x) = b^x$, where $b > 0$ and $b \neq 1$, is called an *exponential function with base b*. Some examples are

$$f(x) = 2^x, \quad f(x) = \left(\tfrac{1}{2}\right)^x, \quad f(x) = \pi^x$$

Note that an exponential function has a constant base and variable exponent. Thus, functions such as $f(x) = x^2$ and $f(x) = x^\pi$ would not be classified as exponential functions, since they have a variable base and a constant exponent. Functions of this type, which are called *power functions*, will be studied later.

It can be shown that exponential functions are continuous and have one of the basic two shapes shown in Figure 4.2.1a, depending on whether $0 < b < 1$ or $b > 1$. Figure 4.2.1b shows the graphs of some specific exponential functions.

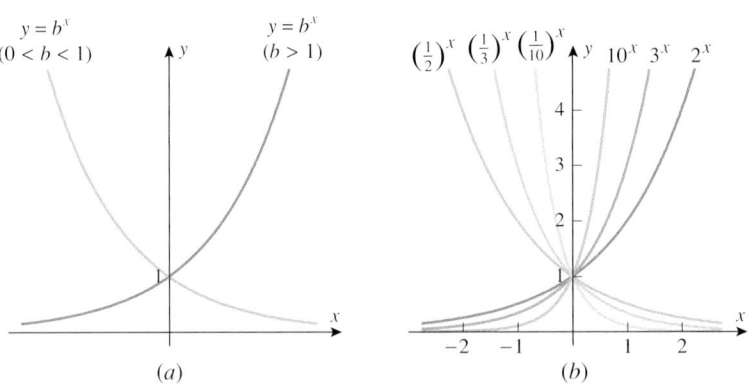

(a)   (b)

Figure 4.2.1

REMARK.   If $b = 1$, then the function b^x is constant, since $b^x = 1^x = 1$. This case is of no interest to us here, so we have excluded it from the family of exponential functions.

FOR THE READER.   Use your graphing utility to confirm that the graphs $y = \left(\tfrac{1}{2}\right)^x$ and $y = 2^x$ agree with Figure 4.2.1b, and explain why the two graphs are reflections of one another about the y-axis.

Since it is not our objective in this section to develop the properties of exponential functions in rigorous mathematical detail, we will simply observe without proof that the following properties of exponential functions are consistent with the graphs shown in Figure 4.2.1.

> **4.2.1** THEOREM. *If $b > 0$ and $b \neq 1$, then:*
>
> (*a*) *The function $f(x) = b^x$ is defined for all real values of x, so its natural domain is $(-\infty, +\infty)$.*
>
> (*b*) *The function $f(x) = b^x$ is continuous on the interval $(-\infty, +\infty)$, and its range is $(0, +\infty)$.*

LOGARITHMS

Recall from algebra that a logarithm is an exponent. More precisely, if $b > 0$ and $b \neq 1$, then for positive values of x the ***logarithm to the base b of x*** is denoted by

$$\log_b x$$

and is defined to be that exponent to which b must be raised to produce x. For example,

$$\log_{10} 100 = 2, \quad \log_{10}(1/1000) = -3, \quad \log_2 16 = 4, \quad \log_b 1 = 0, \quad \log_b b = 1$$

$$\underbrace{}_{10^2 = 100} \qquad \underbrace{}_{10^{-3} = 1/1000} \qquad \underbrace{}_{2^4 = 16} \qquad \underbrace{}_{b^0 = 1} \qquad \underbrace{}_{b^1 = b}$$

Historically, the first logarithms ever studied were the logarithms with base 10, called ***common logarithms***. For such logarithms it is usual to suppress explicit reference to the base and write $\log x$ rather than $\log_{10} x$. More recently, logarithms with base 2 have played a role in computer science, since they arise naturally in the binary number system. However, the most widely used logarithms in applications are the ***natural logarithms***, which have an irrational base denoted by the letter e in honor of the Swiss mathematician Leonard Euler (p. 19), who first suggested its application to logarithms in an unpublished paper written in 1728. This constant, whose value to six decimal places is

$$e \approx 2.718282 \tag{1}$$

arises as the horizontal asymptote of the graph of the equation

$$y = \left(1 + \frac{1}{x}\right)^x \tag{2}$$

(Figure 4.2.2).

THE VALUES OF $(1 + 1/x)^x$ APPROACH e

x	$1 + \frac{1}{x}$	$\left(1 + \frac{1}{x}\right)^x$
1	2	≈ 2.000000
10	1.1	2.593742
100	1.01	2.704814
1000	1.001	2.716924
10,000	1.0001	2.718146
100,000	1.00001	2.718268
1,000,000	1.000001	2.718280

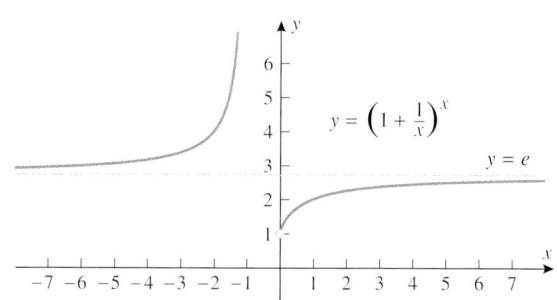

Figure 4.2.2

The fact that $y = e$ is a horizontal asymptote of (2) as $x \to +\infty$ and as $x \to -\infty$ is expressed by the limits

$$e = \lim_{x \to +\infty} \left(1 + \frac{1}{x}\right)^x \qquad \text{and} \qquad e = \lim_{x \to -\infty} \left(1 + \frac{1}{x}\right)^x \tag{3–4}$$

Later, we will show that these limits can be derived from the limit

$$e = \lim_{x \to 0} (1 + x)^{1/x} \tag{5}$$

which is sometimes taken as the definition of the number e.

It is standard to denote the natural logarithm of x by $\ln x$ (read "ell en of x"), rather than $\log_e x$. Thus, $\ln x$ can be viewed as that power to which e must be raised to produce x. For example,

$$\ln 1 = 0, \qquad \ln e = 1, \qquad \ln 1/e = -1, \qquad \ln(e^2) = 2$$

| Since $e^0 = 1$ | Since $e^1 = e$ | Since $e^{-1} = 1/e$ | Since $e^2 = e^2$ |

In general, the statements

$$y = \ln x \quad \text{and} \quad x = e^y$$

are equivalent.

The exponential function $f(x) = e^x$ is called the **natural exponential function**. To simplify typography, this function is sometimes written as $\exp x$. Thus, for example, you might see the relationship $e^{x_1 + x_2} = e^{x_1} e^{x_2}$ expressed as

$$\exp(x_1 + x_2) = \exp(x_1) \exp(x_2)$$

This notation is also used by graphing and calculating utilities, and it is typical to access the function e^x with some variation of the command EXP.

FOR THE READER.   Most scientific calculating utilities provide some way of evaluating common logarithms, natural logarithms, and powers of e. Check your documentation to see how this is done, and then confirm the approximation $e \approx 2.718282$ and the values that appear in the table in Figure 4.2.2.

LOGARITHMIC FUNCTIONS

Figure 4.2.1a suggests that if $b > 0$ and $b \neq 1$, then the graph of $y = b^x$ passes the horizontal line test, and this implies that the function $f(x) = b^x$ has an inverse. To find a formula for this inverse (with x as the independent variable), we can solve the equation $x = b^y$ for y as a function of x. This can be done by taking the logarithm to the base b of both sides of this equation. This yields

$$\log_b x = \log_b(b^y) \tag{6}$$

However, if we think of $\log_b(b^y)$ as that exponent to which b must be raised to produce b^y, then it becomes evident that $\log_b(b^y) = y$. Thus, (6) can be rewritten as

$$y = \log_b x$$

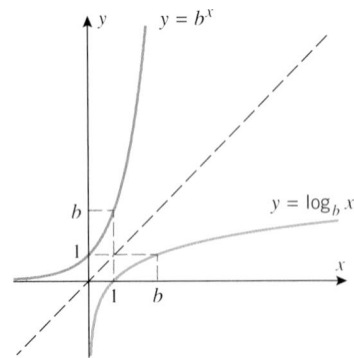

from which we conclude that the inverse of $f(x) = b^x$ is $f^{-1}(x) = \log_b x$. This implies that the graphs of $y = b^x$ and $y = \log_b x$ are reflections of one another about the line $y = x$ (Figure 4.2.3). We call $\log_b x$ the **logarithmic function with base b**.

Recall from Section 4.1 that a one-to-one function f and its inverse satisfy the equations

$$f^{-1}(f(x)) = x \quad \text{for every } x \text{ in the domain of } f$$
$$f(f^{-1}(x)) = x \quad \text{for every } x \text{ in the domain of } f^{-1}$$

Figure 4.2.3

In particular, if we take $f(x) = b^x$ and $f^{-1}(x) = \log_b x$, and if we keep in mind that the domain of f^{-1} is the same as the range of f, then we obtain

$$\log_b(b^x) = x \quad \text{for all real values of } x$$
$$b^{\log x} = x \quad \text{for } x > 0 \tag{7}$$

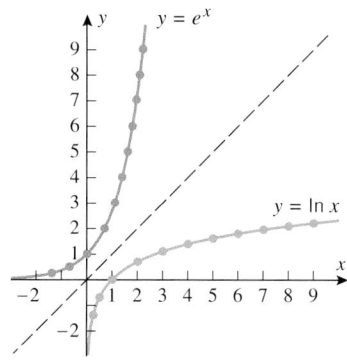

In the special case where $b = e$, these equations become

$$\ln(e^x) = x \quad \text{for all real values of } x$$
$$e^{\ln x} = x \quad \text{for } x > 0$$

(8)

In words, the equations in (7) tell us that the functions b^x and $\log_b x$ cancel out the effect of one another when composed in either order; for example,

$$\log 10^x = x, \quad 10^{\log x} = x, \quad \ln e^x = x, \quad e^{\ln x} = x, \quad \ln e^5 = 5, \quad e^{\ln \pi} = \pi$$

FOR THE READER.   Figure 4.2.4 shows computer-generated tables and graphs of $y = e^x$ and $y = \ln x$. Use your calculating and graphing utilities to generate the graphs and table values.

The inverse relationship between b^x and $\log_b x$ allows us to translate properties of exponential functions into properties of logarithmic functions, and vice versa.

x	$y = \ln x$	x	$y = e^x$
0.25	−1.39	−1.39	0.25
0.50	−0.69	−0.69	0.50
1	0	0	1
2	0.69	0.69	2
3	1.10	1.10	3
4	1.39	1.39	4
5	1.61	1.61	5
6	1.79	1.79	6
7	1.95	1.95	7
8	2.08	2.08	8
9	2.20	2.20	9

Figure 4.2.4

4.2.2   THEOREM (*Comparison of Exponential and Logarithmic Functions for $b > 1$*).

$b^0 = 1$	$\log_b 1 = 0$
$b^1 = b$	$\log_b b = 1$
range $b^x = (0, +\infty)$	domain $\log_b x = (0, +\infty)$
domain $b^x = (-\infty, +\infty)$	range $\log_b x = (-\infty, +\infty)$
$0 < b^x < 1 \quad \text{if } x < 0$	$\log_b x < 0 \quad \text{if } 0 < x < 1$

It follows from Theorem 4.1.2 that the equation $y = e^x$ can be solved for x in terms of y as $x = \ln y$, provided (of course) that y is in the domain of the natural logarithm function and x is in the domain of the natural exponential function; that is, $y > 0$ and x is any real number. Thus,

$$y = e^x \text{ is equivalent to } x = \ln y \text{ if } y > 0 \text{ and } x \text{ is any real number}$$

More generally, if $b > 0$ and $b \neq 1$, then

$$y = b^x \text{ is equivalent to } x = \log_b y \text{ if } y > 0 \text{ and } x \text{ is any real number}$$

You should recall the following algebraic properties of logarithms from your earlier studies.

4.2.3   THEOREM (*Algebraic Properties of Logarithms*).

$\log_b(ac) = \log_b a + \log_b c$	Product property
$\log_b(a/c) = \log_b a - \log_b c$	Quotient property
$\log_b(a^r) = r \log_b a$	Power property
$\log_b(1/c) = -\log_b c$	Reciprocal property

These properties are often used to expand a single logarithm into sums, differences, and multiples of other logarithms and, conversely, to condense sums, differences, and multiples of logarithms into a single logarithm. For example,

$$\log \frac{xy^5}{\sqrt{z}} = \log xy^5 - \log \sqrt{z} = \log x + \log y^5 - \log z^{1/2} = \log x + 5\log y - \tfrac{1}{2}\log z$$

$$5\log 2 + \log 3 - \log 8 = \log 32 + \log 3 - \log 8 = \log \frac{32 \cdot 3}{8} = \log 12$$

$$\tfrac{1}{3}\ln x - \ln(x^2 - 1) + 2\ln(x + 3) = \ln x^{1/3} - \ln(x^2 - 1) + \ln(x + 3)^2 = \ln \frac{\sqrt[3]{x}(x+3)^2}{x^2 - 1}$$

REMARK.   Expressions of the form $\log_b(u + v)$ and $\log_b(u - v)$ have no useful simplifications in terms of $\log_b u$ and $\log_b v$. In particular,

$$\log_b(u + v) \neq \log_b u + \log_b v$$

$$\log_b(u - v) \neq \log_b u - \log_b v$$

SOLVING EQUATIONS INVOLVING EXPONENTIALS AND LOGARITHMS

Equations of the form $\log_b x = k$ can be solved by converting them to the exponential form $x = b^k$, and equations of the form $b^x = k$ can be solved by taking a logarithm of both sides (usually log or ln).

Example 1

Find x such that

  (a) $\log x = \sqrt{2}$ (b) $\ln(x + 1) = 5$ (c) $5^x = 7$

Solution (a). Converting the equation to exponential form yields

$$x = 10^{\sqrt{2}} \approx 25.95$$

Solution (b). Converting the equation to exponential form yields

$$x + 1 = e^5 \quad \text{or} \quad x = e^5 - 1 \approx 147.41$$

Solution (c). Taking the natural logarithm of both sides and using the power property of logarithms yields

$$x\ln 5 = \ln 7 \quad \text{or} \quad x = \frac{\ln 7}{\ln 5} \approx 1.21 \qquad \blacktriangleleft$$

Example 2

A satellite that requires 7 watts of power to operate at full capacity is equipped with a radioisotope power supply whose power output in watts is given by the equation

$$P = 75e^{-t/125}$$

where t is the time in days that the supply is used. How long can the satellite operate at full capacity?

Solution. The power P will fall to 7 watts when

$$7 = 75e^{-t/125}$$

The solution for t is as follows:

$$7/75 = e^{-t/125}$$
$$\ln(7/75) = \ln(e^{-t/125})$$
$$\ln(7/75) = -t/125$$
$$t = -125\ln(7/75) \approx 296.4$$

so the satellite can operate at full capacity for about 296 days. $\qquad \blacktriangleleft$

Here is a more complicated example.

Example 3

Solve $\dfrac{e^x - e^{-x}}{2} = 1$ for x.

Solution. Multiplying both sides of the given equation by 2 yields

$$e^x - e^{-x} = 2$$

or equivalently,

$$e^x - \frac{1}{e^x} = 2$$

Multiplying through by e^x yields

$$e^{2x} - 1 = 2e^x \quad \text{or} \quad e^{2x} - 2e^x - 1 = 0$$

This is really a quadratic equation in disguise, as can be seen by rewriting it in the form

$$\left(e^x\right)^2 - 2e^x - 1 = 0$$

and letting $u = e^x$ to obtain

$$u^2 - 2u - 1 = 0$$

Solving for u by the quadratic formula yields

$$u = \frac{2 \pm \sqrt{4+4}}{2} = \frac{2 \pm \sqrt{8}}{2} = 1 \pm \sqrt{2}$$

or, since $u = e^x$,

$$e^x = 1 \pm \sqrt{2}$$

But e^x cannot be negative, so we discard the negative value $1 - \sqrt{2}$; thus,

$$e^x = 1 + \sqrt{2}$$
$$\ln e^x = \ln(1 + \sqrt{2})$$
$$x = \ln(1 + \sqrt{2}) \approx 0.881 \qquad \blacktriangleleft$$

CHANGE OF BASE FORMULA FOR LOGARITHMS

Scientific calculators generally provide keys for evaluating common logarithms and natural logarithms but have no keys for evaluating logarithms with other bases. However, this is not a serious deficiency because it is possible to express a logarithm with any base in terms of logarithms with any other base (see Exercise 40). For example, the following formula expresses a logarithm with base b in terms of natural logarithms:

$$\log_b x = \frac{\ln x}{\ln b} \tag{9}$$

We can derive this result by letting $y = \log_b x$, from which it follows that $b^y = x$. Taking the natural logarithm of both sides of this equation we obtain $y \ln b = \ln x$, from which (9) follows.

Example 4

Use a calculating utility to evaluate $\log_2 5$ by expressing this logarithm in terms of natural logarithms.

Solution. From (9) we obtain

$$\log_2 5 = \frac{\ln 5}{\ln 2} \approx 2.321928 \qquad \blacktriangleleft$$

LOGARITHMIC SCALES IN SCIENCE AND ENGINEERING

Logarithms are used in science and engineering to deal with quantities whose units vary over an excessively wide range of values. For example, the "loudness" of a sound can be measured by its ***intensity*** I (in watts per square meter), which is related to the energy transmitted by the sound wave—the greater the intensity, the greater the transmitted energy, and the louder the sound is perceived by the human ear. However, intensity units are unwieldy because they vary over an enormous range. For example, a sound at the threshold of human hearing has an intensity of about 10^{-12} W/m^2, a close whisper has an intensity that is about 100 times the hearing threshold, and a jet engine at 50 meters has an intensity that is about $1,000,000,000,000 = 10^{12}$ times the hearing threshold. To see how logarithms can be used to reduce this wide spread, observe that if

$$y = \log x$$

then increasing x by a *factor* of 10 *adds* 1 unit to y since

$$\log 10x = \log 10 + \log x = 1 + y$$

Physicists and engineers take advantage of this property by measuring loudness in terms of the ***sound level*** β, which is defined by

$$\beta = 10\log(I/I_0)$$

where $I_0 = 10^{-12}$ W/m^2 is a reference intensity close to the threshold of human hearing. The units of β are ***decibels*** (dB), named in honor of the telephone inventor Alexander Graham Bell. With this scale of measurement, *multiplying* the intensity I by a factor of 10 *adds* 10 dB to the sound level β (verify). This results in a more tractable scale than intensity for measuring sound loudness (Table 4.2.2). Some other familiar logarithmic scales are the ***Richter scale*** used to measure earthquake intensity and the **pH** *scale* used to measure acidity in chemistry, both of which are discussed in the exercises.

Table 4.2.2

β (dB)	I/I_0
0	$10^0 = 1$
10	$10^1 = 10$
20	$10^2 = 100$
30	$10^3 = 1,000$
40	$10^4 = 10,000$
50	$10^5 = 100,000$
$\vdots$	$\vdots$
120	$10^{12} = 1,000,000,000,000$

Example 5

In 1976 the rock group The Who set the record for the loudest concert: 120 dB. By comparison, a jackhammer positioned at the same spot as The Who would have produced a sound level of 92 dB. What is the ratio of the sound intensity of The Who to the sound intensity of a jackhammer?

Solution. Let I_1 and $\beta_1 (= 120$ dB) denote the intensity and sound level of The Who, and let I_2 and β_2 $(= 92$ dB) denote the intensity and sound level of the jackhammer. Then

$$I_1/I_2 = (I_1/I_0)/(I_2/I_0)$$
$$\log(I_1/I_2) = \log(I_1/I_0) - \log(I_2/I_0)$$
$$10\log(I_1/I_2) = 10\log(I_1/I_0) - 10\log(I_2/I_0)$$
$$10\log(I_1/I_2) = \beta_1 - \beta_2 = 120 - 92 = 28$$
$$\log(I_1/I_2) = 2.8$$

Thus, $I_1/I_2 = 10^{2.8} \approx 631$, which tells us that the sound intensity of The Who was 631 times greater than a jackhammer!   ◄

Peter Townsend of the Who sustained permanent hearing reduction due to the high decibel level of his band's music.

EXPONENTIAL AND LOGARITHMIC GROWTH

The growth patterns of e^x and $\ln x$ illustrated by Table 4.2.3 are worth noting. Both functions increase as x increases, but they increase in dramatically different ways—e^x increases extremely rapidly and $\ln x$ increases extremely slowly. For example, at $x = 10$ the value of e^x is over 22,000, but at $x = 1000$ the value of $\ln x$ has not even reached 7.

The table strongly suggests that $e^x \to +\infty$ as $x \to +\infty$. However, the growth of $\ln x$ is so slow that its limiting behavior as $x \to +\infty$ is not clear from the table. However, in spite of its slow growth, it is still true that $\ln x \to +\infty$ as $x \to +\infty$. To see that this is so, choose any positive number M (as large as you like). The value of $\ln x$ will reach M when $x = e^M$, since

$$\ln x = \ln(e^M) = M$$

Since $\ln x$ increases as x increases, we can conclude that $\ln x > M$ for $x > e^M$; hence, $\ln x \to +\infty$ as $x \to +\infty$ since the values of $\ln x$ eventually exceed any positive number M (Figure 4.2.5).

In summary,

$$\lim_{x \to +\infty} e^x = +\infty \qquad \lim_{x \to +\infty} \ln x = +\infty \qquad (10\text{--}11)$$

The following limits, which are consistent with Figure 4.2.5, can be deduced numerically by constructing appropriate tables of values (verify):

$$\lim_{x \to -\infty} e^x = 0 \qquad \lim_{x \to 0^+} \ln x = -\infty \qquad (12\text{--}13)$$

The following limits can be deduced numerically, but they can be seen more readily by noting that the graph of $y = e^{-x}$ is the reflection about the y-axis of the graph of $y = e^x$ (Figure 4.2.6):

$$\lim_{x \to +\infty} e^{-x} = 0 \qquad \lim_{x \to -\infty} e^{-x} = +\infty \qquad (14\text{--}15)$$

Table 4.2.3

x	e^x	$\ln x$
1	2.72	0.00
2	7.39	0.69
3	20.09	1.10
4	54.60	1.39
5	148.41	1.61
6	403.43	1.79
7	1096.63	1.95
8	2980.96	2.08
9	8103.08	2.20
10	22026.47	2.30
100	2.69×10^{43}	4.61
1000	1.97×10^{434}	6.91

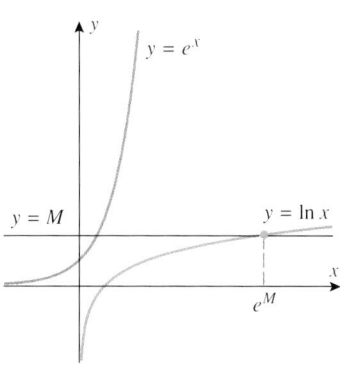

Figure 4.2.5

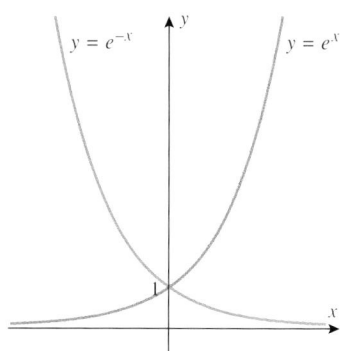

Figure 4.2.6

EXERCISE SET 4.2 ⌇ Graphing Calculator

In Exercises 1 and 2, simplify the expression without using a calculating utility.

1. (a) $-8^{2/3}$ (b) $(-8)^{2/3}$ (c) $8^{-2/3}$

2. (a) 2^{-4} (b) $4^{1.5}$ (c) $9^{-0.5}$

In Exercises 3 and 4, use a calculating utility to approximate the expression. Round your answer to four decimal places.

3. (a) $2^{1.57}$ (b) $5^{-2.1}$

4. (a) $\sqrt[5]{24}$ (b) $\sqrt[8]{0.6}$

In Exercises 5 and 6, find the exact value of the expression without using a calculating utility.

5. (a) $\log_2 16$ (b) $\log_2 \left(\frac{1}{32}\right)$
      (c) $\log_4 4$ (d) $\log_9 3$

6. (a) $\log_{10}(0.001)$ (b) $\log_{10}(10^4)$
      (c) $\ln(e^3)$ (d) $\ln(\sqrt{e})$

In Exercises 7 and 8, use a calculating utility to approximate the expression. Round your answer to four decimal places.

7. (a) $\log 23.2$ (b) $\ln 0.74$

8. (a) $\log 0.3$ (b) $\ln \pi$

In Exercises 9 and 10 use the logarithm properties in Theorem 4.2.3 to rewrite the expression in terms of r, s, and t, where $r = \ln a$, $s = \ln b$, and $t = \ln c$.

9. (a) $\ln a^2 \sqrt{bc}$ (b) $\ln \dfrac{b}{a^3 c}$

10. (a) $\ln \dfrac{\sqrt[3]{c}}{ab}$ (b) $\ln \sqrt{\dfrac{ab^3}{c^2}}$

In Exercises 11 and 12, expand the logarithm in terms of sums, differences, and multiples of simpler logarithms.

11. (a) $\log(10x\sqrt{x-3})$ (b) $\ln \dfrac{x^2 \sin^3 x}{\sqrt{x^2+1}}$

12. (a) $\log \dfrac{\sqrt[3]{x+2}}{\cos 5x}$ (b) $\ln \sqrt{\dfrac{x^2+1}{x^3+5}}$

In Exercises 13–15, rewrite the expression as a single logarithm.

13. $4\log 2 - \log 3 + \log 16$

14. $\frac{1}{2}\log x - 3\log(\sin 2x) + 2$

15. $2\ln(x+1) + \frac{1}{3}\ln x - \ln(\cos x)$

In Exercises 16–25, solve for x without using a calculating utility.

16. $\log_{10}(1+x) = 3$ **17.** $\log_{10}(\sqrt{x}) = -1$

18. $\ln(x^2) = 4$ **19.** $\ln(1/x) = -2$

20. $\log_3(3^x) = 7$ **21.** $\log_5(5^{2x}) = 8$

22. $\log_{10} x^2 + \log_{10} x = 30$

23. $\log_{10} x^{3/2} - \log_{10} \sqrt{x} = 5$

24. $\ln 4x - 3\ln(x^2) = \ln 2$

25. $\ln(1/x) + \ln(2x^3) = \ln 3$

In Exercises 26–31, solve for x without using a calculating utility. Use the natural logarithm anywhere that logarithms are needed.

26. $3^x = 2$ **27.** $5^{-2x} = 3$

28. $3e^{-2x} = 5$ **29.** $2e^{3x} = 7$

30. $e^x - 2xe^x = 0$ **31.** $xe^{-x} + 2e^{-x} = 0$

In Exercises 32 and 33, rewrite the given equation as a quadratic equation in u, where $u = e^x$; then solve for x.

32. $e^{2x} - e^x = 6$ **33.** $e^{-2x} - 3e^{-x} = -2$

In Exercises 34–36, sketch the graph of the equation without using a graphing utility.

34. (a) $y = 1 + \ln(x-2)$ (b) $y = 3 + e^{x-2}$

35. (a) $y = \left(\frac{1}{2}\right)^{x-1} - 1$ (b) $y = \ln|x|$

36. (a) $y = 1 - e^{-x+1}$ (b) $y = 3\ln\sqrt[3]{x-1}$

37. Use a calculating utility and the change of base formula (9) to find the values of $\log_2 7.35$ and $\log_5 0.6$, rounded to four decimal places.

In Exercises 38 and 39, graph the functions on the same screen of a graphing utility. [Use the change of base formula (9), where needed].

38. $y = \ln x$, $y = e^x$, $\log x$, 10^x

39. $y = \log_2 x$, $\ln x$, $\log_5 x$, $\log x$

40. (a) Derive the general change of base formula

$$\log_b x = \frac{\log_a x}{\log_a b}$$

(b) Use the result in part (a) to find the exact value of $(\log_2 81)(\log_3 32)$ without using a calculating utility. [*Hint:* Take $x = a$.]

41. Use a graphing utility to estimate where the graphs of $y = x^{0.2}$ and $y = \ln x$ intersect.

42. The United States public debt D, in billions of dollars, has been modeled as $D = 0.051517(1.1306727)^x$, where x is the number of years since 1900. Based on this model, when did the debt first reach one trillion dollars?

43. (a) Is the curve in the accompanying figure the graph of an exponential function? Explain your reasoning.
(b) Find the equation of an exponential function that passes through the point $(4, 2)$.
(c) Find the equation of an exponential function that passes through the point $\left(2, \frac{1}{4}\right)$.
(d) Use a graphing utility to generate the graph of an exponential function that passes through the point $(2, 5)$.

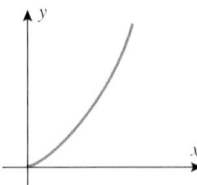

Figure Ex-43

44. (a) Make a conjecture about the general shape of the graph of $y = \log(\log x)$, and sketch the graph of this equation and $y = \log x$ in the same coordinate system.
(b) Check your work in part (a) with a graphing utility.

45. Find the fallacy in the following "proof" that $\frac{1}{8} > \frac{1}{4}$. Multiply both sides of the inequality $3 > 2$ by $\log \frac{1}{2}$ to get

$$3 \log \tfrac{1}{2} > 2 \log \tfrac{1}{2}$$

$$\log \left(\tfrac{1}{2}\right)^3 > \log \left(\tfrac{1}{2}\right)^2$$

$$\log \tfrac{1}{8} > \log \tfrac{1}{4}$$

$$\tfrac{1}{8} > \tfrac{1}{4}$$

46. Prove the four algebraic properties of logarithms in Theorem 4.2.3.

47. If equipment in the satellite of Example 2 requires 15 watts to operate correctly, what is the operational lifetime of the power supply?

48. The equation $Q = 12e^{-0.055t}$ gives the mass Q in grams of radioactive potassium-42 that will remain from some initial quantity after t hours of radioactive decay.
  (a) How many grams were there initially?
  (b) How many grams remain after 4 hours?
  (c) How long will it take to reduce the amount of radioactive potassium-42 to half of the initial amount?

49. The acidity of a substance is measured by its pH value, which is defined by the formula

$$\text{pH} = -\log[H^+]$$

where the symbol $[H^+]$ denotes the concentration of hydrogen ions measured in moles per liter. Distilled water has a pH of 7; a substance is called *acidic* if it has pH < 7 and *basic* if it has pH > 7. Find the pH of each of the following substances and state whether it is acidic or basic.

SUBSTANCE	$[H^+]$
(a) Arterial blood	3.9×10^{-8} mol/L
(b) Tomatoes	6.3×10^{-5} mol/L
(c) Milk	4.0×10^{-7} mol/L
(d) Coffee	1.2×10^{-6} mol/L

50. Use the definition of pH in Exercise 49 to find $[H^+]$ in a solution having a pH equal to
  (a) 2.44   (b) 8.06

51. The perceived loudness β of a sound in decibels (dB) is related to its intensity I in watts/square meter (W/m^2) by the equation

$$\beta = 10 \log(I/I_0)$$

where $I_0 = 10^{-12}$ W/m^2. Damage to the average ear occurs at 90 dB or greater. Find the decibel level of each of the following sounds and state whether it will cause ear damage.

SOUND	I
(a) Jet aircraft (from 500 ft)	1.0×10^2 W/m^2
(b) Amplified rock music	1.0 W/m^2
(c) Garbage disposal	1.0×10^{-4} W/m^2
(d) TV (mid volume from 10 ft)	3.2×10^{-5} W/m^2

In Exercises 52–54, use the definition of the decibel level of a sound (see Exercise 51).

52. If one sound is three times as intense as another, how much greater is its decibel level?

53. According to one source, the noise inside a moving automobile is about 70 dB, while an electric blender generates 93 dB. Find the ratio of the intensity of the noise of the blender to that of the automobile.

54. Suppose that the decibel level of an echo is $\frac{2}{3}$ the decibel level of the original sound. If each echo results in another echo, how many echoes will be heard from a 120-dB sound given that the average human ear can hear a sound as low as 10 dB?

55. On the ***Richter scale***, the magnitude M of an earthquake is related to the released energy E in joules (J) by the equation

$$\log E = 4.4 + 1.5M$$

  (a) Find the energy E of the 1906 San Francisco earthquake that registered $M = 8.2$ on the Richter scale.
  (b) If the released energy of one earthquake is 10 times that of another, how much greater is its magnitude on the Richter scale?

56. Suppose that the magnitudes of two earthquakes differ by 1 on the Richter scale. Find the ratio of the released energy of the larger earthquake to that of the smaller earthquake. [*Note:* See Exercise 55 for terminology.]

In Exercises 57 and 58, use Formula (3) or (5), as appropriate, to find the limit.

57. Find $\lim\limits_{x \to 0} (1 - 2x)^{1/x}$. [*Hint:* Let $t = -2x$.]

58. Find $\lim\limits_{x \to +\infty} (1 + 3/x)^x$. [*Hint:* Let $t = 3/x$.]

4.3 IMPLICIT DIFFERENTIATION

In earlier sections we were concerned with differentiating functions that were given by equations of the form $y = f(x)$. In this section we will consider methods for differentiating functions for which it is inconvenient or impossible to express them in this form.

FUNCTIONS DEFINED EXPLICITLY AND IMPLICITLY

Up to now, we have been concerned with differentiating functions that are expressed in the form $y = f(x)$. An equation of this form is said to define y **explicitly** as a function of x, because the variable y appears alone on one side of the equation. However, sometimes functions are defined by equations in which y is not alone on one side; for example, the equation

$$yx + y + 1 = x \tag{1}$$

is not of the form $y = f(x)$. However, this equation still defines y as a function of x since it can be rewritten as

$$y = \frac{x - 1}{x + 1}$$

Thus, we say that (1) defines y **implicitly** as a function of x, the function being

$$f(x) = \frac{x - 1}{x + 1}$$

An equation in x and y can implicitly define more than one function of x; for example, if we solve the equation

$$x^2 + y^2 = 1 \tag{2}$$

for y in terms of x, we obtain $y = \pm\sqrt{1 - x^2}$, so we have found two functions that are defined implicitly by (2), namely

$$f_1(x) = \sqrt{1 - x^2} \quad \text{and} \quad f_2(x) = -\sqrt{1 - x^2} \tag{3}$$

The graphs of these functions are the upper and lower semicircles of the circle $x^2 + y^2 = 1$ (Figure 4.3.1).

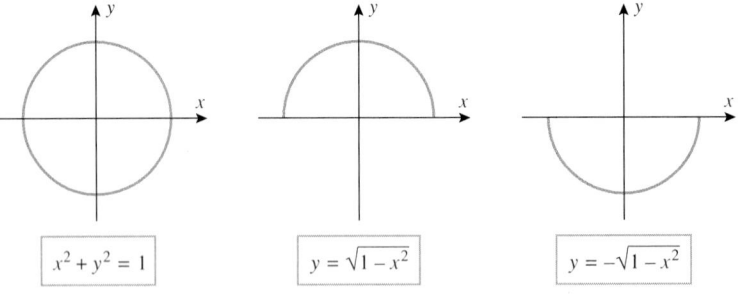

Figure 4.3.1

Observe that the complete circle $x^2 + y^2 = 1$ does not pass the vertical line test, and hence is not itself the graph of a function of x. However, the upper and lower semicircles (which are only portions of the entire circle) do pass the vertical line test, and hence are graphs of functions. In general, if we have an equation in x and y, then any segment of its graph that passes the vertical line test can be viewed as the graph of a function defined by the equation. Thus, we make the following definition.

4.3.1 DEFINITION. We will say that a given equation in x and y defines the function f **implicitly** if the graph of $y = f(x)$ coincides with some segment of the graph of the equation.

Thus, for example, the equation $x^2 + y^2 = 1$ defines the functions $f_1(x) = \sqrt{1 - x^2}$ and $f_2(x) = -\sqrt{1 - x^2}$ implicitly, since the graphs of these functions are segments of the circle $x^2 + y^2 = 1$.

Sometimes it may be difficult or impossible to solve an equation in x and y for y in terms of x. For example, with persistence the equation

$$x^3 + y^3 = 3xy \tag{4}$$

can be solved for y in terms of x, but the algebra is tedious and the resulting formulas are complicated. On the other hand, the equation

$$\sin(xy) = y$$

cannot be solved for y in terms of x by any elementary method. Thus, even though an equation in x and y may define one or more functions of x, it may not be practical or possible to find explicit formulas for those functions.

GRAPHS OF EQUATIONS IN x AND y

When an equation in x and y cannot be solved for y in terms of x (or x in terms of y), it may be difficult or time-consuming to obtain even a rough sketch of the graph, so the graphing of such equations is usually best left for graphing utilities. In particular, the CAS programs *Mathematica* and *Maple* both have "implicit plot" capabilities for graphing such equations. For example, Figure 4.3.2 shows the graph of Equation (4), which is called the *Folium of Descartes*.

FOR THE READER. Figure 4.3.3 shows the graphs of two functions (in solid color) that are defined implicitly by (4). Sketch some more.

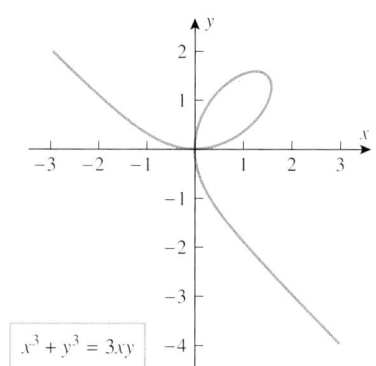

$x^3 + y^3 = 3xy$

Figure 4.3.2

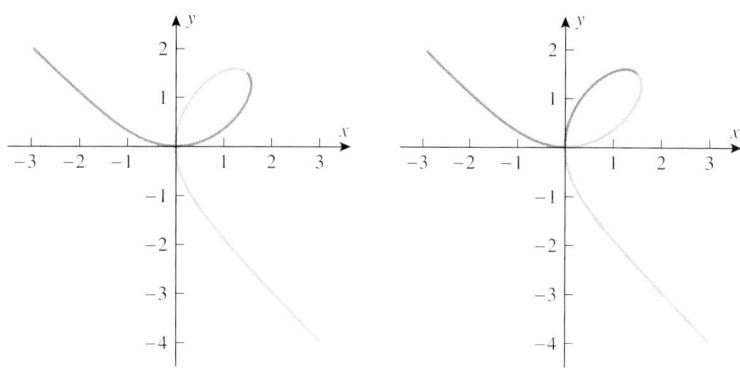

Figure 4.3.3

IMPLICIT DIFFERENTIATION

In general, it is not necessary to solve an equation for y in terms of x in order to differentiate the functions defined implicitly by the equation. To illustrate this, let us consider the simple equation

$$xy = 1 \tag{5}$$

One way to find dy/dx is to rewrite this equation as

$$y = \frac{1}{x} \tag{6}$$

from which it follows that

$$\frac{dy}{dx} = -\frac{1}{x^2} \tag{7}$$

However, there is another way to obtain this derivative. We can differentiate both sides of

(5) *before* solving for y in terms of x, treating y as a (temporarily unspecified) differentiable function of x. With this approach we obtain

$$\frac{d}{dx}[xy] = \frac{d}{dx}[1]$$

$$x\frac{d}{dx}[y] + y\frac{d}{dx}[x] = 0$$

$$x\frac{dy}{dx} + y = 0$$

$$\frac{dy}{dx} = -\frac{y}{x}$$

If we now substitute (6) into the last expression, we obtain

$$\frac{dy}{dx} = -\frac{1}{x^2}$$

which agrees with (7). This method of obtaining derivatives is called ***implicit differentiation***.

Example 1

Use implicit differentiation to find dy/dx if $5y^2 + \sin y = x^2$.

$$\frac{d}{dx}[5y^2 + \sin y] = \frac{d}{dx}[x^2]$$

$$5\frac{d}{dx}[y^2] + \frac{d}{dx}[\sin y] = 2x$$

$$5\left(2y\frac{dy}{dx}\right) + (\cos y)\frac{dy}{dx} = 2x \qquad \boxed{\begin{array}{l}\text{The chain rule was}\\ \text{used here because}\\ y \text{ is a function of } x.\end{array}}$$

$$10y\frac{dy}{dx} + (\cos y)\frac{dy}{dx} = 2x$$

Solving for dy/dx we obtain

$$\frac{dy}{dx} = \frac{2x}{10y + \cos y} \tag{8}$$

Note that this formula involves both x and y. In order to obtain a formula for dy/dx that involves x alone, we would have to solve the original equation for y in terms of x and then substitute in (8). However, it is impossible to do this, so we are forced to leave the formula for dy/dx in terms of x and y. ◀

Example 2

Use implicit differentiation to find d^2y/dx^2 if $4x^2 - 2y^2 = 9$.

Solution. Differentiating both sides of $4x^2 - 2y^2 = 9$ implicitly yields

$$8x - 4y\frac{dy}{dx} = 0$$

from which we obtain

$$\frac{dy}{dx} = \frac{2x}{y} \tag{9}$$

Differentiating both sides of (9) implicitly yields

$$\frac{d^2y}{dx^2} = \frac{(y)(2) - (2x)(dy/dx)}{y^2} \tag{10}$$

Substituting (9) into (10) and simplifying using the original equation, we obtain

$$\frac{d^2y}{dx^2} = \frac{2y - 2x(2x/y)}{y^2} = \frac{2y^2 - 4x^2}{y^3} = -\frac{9}{y^3} \qquad ◀$$

In Examples 1 and 2, the resulting formulas for dy/dx involved both x and y. Although it is usually more desirable to have the formula for dy/dx expressed in terms of x alone, having the formula in terms of x and y is not an impediment to finding slopes and equations of tangent lines provided the x- and y-coordinates of the point of tangency are known. This is illustrated in the following example.

Example 3

Find the slopes of the tangent lines at $(2, -1)$ and $(2, 1)$ to $y^2 - x + 1 = 0$.

Solution. We could proceed by solving the equation for y in terms of x, and then evaluating the derivative of $y = \sqrt{x-1}$ at $(2, 1)$ and the derivative of $y = -\sqrt{x-1}$ at $(2, -1)$ (Figure 4.3.4). However, implicit differentiation is more efficient since it gives the slopes of *both* functions. Differentiating implicitly yields

$$\frac{d}{dx}[y^2 - x + 1] = \frac{d}{dx}[0]$$

$$\frac{d}{dx}[y^2] - \frac{d}{dx}[x] + \frac{d}{dx}[1] = \frac{d}{dx}[0]$$

$$2y\frac{dy}{dx} - 1 = 0$$

$$\frac{dy}{dx} = \frac{1}{2y}$$

At $(2, -1)$ we have $y = -1$, and at $(2, 1)$ we have $y = 1$, so the slopes of the tangent lines at those points are

$$m_{\tan} = \left.\frac{dy}{dx}\right|_{\substack{x=2\\y=-1}} = -\frac{1}{2} \quad \text{and} \quad m_{\tan} = \left.\frac{dy}{dx}\right|_{\substack{x=2\\y=1}} = \frac{1}{2} \qquad \blacktriangleleft$$

Example 4

(a) Use implicit differentiation to find dy/dx for the Folium of Descartes $x^3 + y^3 = 3xy$.

(b) Find an equation for the tangent line to the Folium of Descartes at the point $\left(\frac{3}{2}, \frac{3}{2}\right)$.

(c) At what points is the tangent line to the Folium of Descartes horizontal?

Solution (a). Differentiating both sides of the given equation implicitly yields

$$\frac{d}{dx}[x^3 + y^3] = \frac{d}{dx}[3xy]$$

$$3x^2 + 3y^2\frac{dy}{dx} = 3x\frac{dy}{dx} + 3y$$

$$x^2 + y^2\frac{dy}{dx} = x\frac{dy}{dx} + y$$

$$(y^2 - x)\frac{dy}{dx} = y - x^2$$

$$\frac{dy}{dx} = \frac{y - x^2}{y^2 - x} \tag{11}$$

Solution (b). At the point $\left(\frac{3}{2}, \frac{3}{2}\right)$, we have $x = \frac{3}{2}$ and $y = \frac{3}{2}$, so from (11) the slope $m_{\tan}$ of the tangent line at this point is

$$m_{\tan} = \left.\frac{dy}{dx}\right|_{\substack{x=3/2\\y=3/2}} = \frac{(3/2) - (3/2)^2}{(3/2)^2 - (3/2)} = -1$$

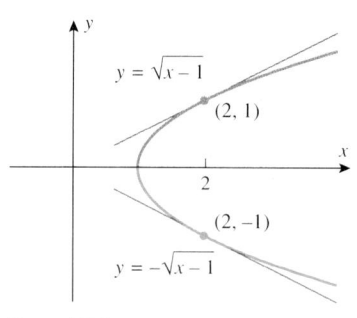

$y = \sqrt{x-1}$

$(2, 1)$

$(2, -1)$

$y = -\sqrt{x-1}$

Figure 4.3.4

Thus, the equation of the tangent line at the point $\left(\frac{3}{2}, \frac{3}{2}\right)$ is

$$y - \tfrac{3}{2} = -1\left(x - \tfrac{3}{2}\right) \quad \text{or} \quad x + y = 3$$

which is consistent with Figure 4.3.5.

Solution (*c*). The tangent line is horizontal at the points where $dy/dx = 0$, and from (11) this occurs where $y - x^2 = 0$ or

$$y = x^2 \tag{12}$$

Substituting this expression for y in the equation $x^3 + y^3 = 3xy$ for the curve yields

$$x^3 + \left(x^2\right)^3 = 3x^3$$
$$x^6 - 2x^3 = 0$$
$$x^3(x^3 - 2) = 0$$

whose solutions are $x = 0$ and $x = 2^{1/3}$. Thus, from (12), the tangent line is horizontal at the points $(0, 0)$ and $(2^{1/3}, 2^{2/3}) \approx (1.26, 1.59)$, which is consistent with Figure 4.3.6. ◀

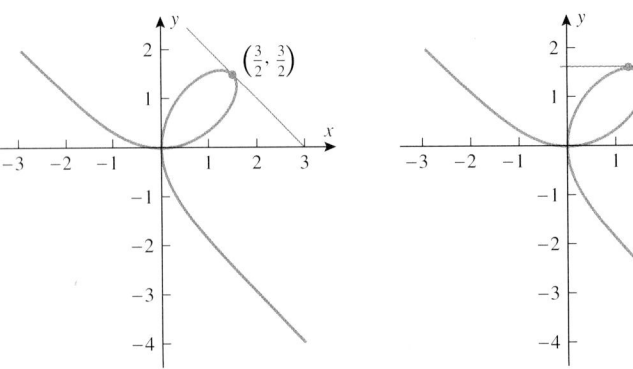

Figure 4.3.5   Figure 4.3.6

DIFFERENTIABILITY OF FUNCTIONS DEFINED IMPLICITLY

When differentiating implicitly, it is assumed that y represents a differentiable function of x. If this is not so, then the resulting calculations may be nonsense. For example, if we differentiate the equation

$$x^2 + y^2 + 1 = 0 \tag{13}$$

we obtain

$$2x + 2y\frac{dy}{dx} = 0 \quad \text{or} \quad \frac{dy}{dx} = -\frac{x}{y}$$

However, this derivative is meaningless because (13) does not define a function of x. (The left side of the equation is greater than zero.)

Sometimes it is possible to identify points of nondifferentiability graphically. For example, the first function in Figure 4.3.3 is differentiable at each point of its domain because there are no corners, discontinuities, or points of vertical tangency; however, the second function is not differentiable at the origin.

In general, it can be difficult to determine analytically whether functions defined implicitly are differentiable, so we will leave such matters for more advanced courses.

DERIVATIVES OF RATIONAL POWERS OF x

In Theorem 3.3.8 and the discussion immediately following it, we showed that the formula

$$\frac{d}{dx}[x^n] = nx^{n-1} \tag{14}$$

holds for integer values of n and for $n = \frac{1}{2}$. We will now use implicit differentiation to show that this formula holds for any rational exponent. More precisely, we will show that if r is a rational number, then

$$\frac{d}{dx}[x^r] = rx^{r-1} \tag{15}$$

wherever x^r and x^{r-1} are defined. For now, we will assume without proof that x^r is differentiable; the justification for this will be considered later.

Let $y = x^r$. Since r is a rational number, it can be expressed as a ratio of integers $r = m/n$. Thus, $y = x^r = x^{m/n}$ can be written as

$$y^n = x^m \quad \text{so that} \quad \frac{d}{dx}[y^n] = \frac{d}{dx}[x^m]$$

By differentiating implicitly with respect to x and using (14), we obtain

$$ny^{n-1}\frac{dy}{dx} = mx^{m-1} \tag{16}$$

But

$$y^{n-1} = \left[x^{m/n}\right]^{n-1} = x^{m-(m/n)}$$

Thus, (16) can be written as

$$nx^{m-(m/n)}\frac{dy}{dx} = mx^{m-1}$$

so that

$$\frac{dy}{dx} = \frac{m}{n}x^{(m/n)-1} = rx^{r-1}$$

which establishes (15).

Example 5

From (15)

$$\frac{d}{dx}[x^{4/5}] = \frac{4}{5}x^{(4/5)-1} = \frac{4}{5}x^{-1/5}$$

$$\frac{d}{dx}[x^{-7/8}] = -\frac{7}{8}x^{(-7/8)-1} = -\frac{7}{8}x^{-15/8}$$

$$\frac{d}{dx}[\sqrt[3]{x}] = \frac{d}{dx}[x^{1/3}] = \frac{1}{3}x^{-2/3} = \frac{1}{3\sqrt[3]{x^2}} \quad \blacktriangleleft$$

If u is a differentiable function of x, and r is a rational number, then the chain rule yields the following generalization of (15):

$$\frac{d}{dx}[u^r] = ru^{r-1} \cdot \frac{du}{dx} \tag{17}$$

Example 6

$$\frac{d}{dx}\left[x^2 - x + 2\right]^{3/4} = \frac{3}{4}\left(x^2 - x + 2\right)^{-1/4} \cdot \frac{d}{dx}[x^2 - x + 2]$$

$$= \frac{3}{4}\left(x^2 - x + 2\right)^{-1/4}(2x - 1)$$

$$\frac{d}{dx}[(\sec \pi x)^{-4/5}] = -\frac{4}{5}(\sec \pi x)^{-9/5} \cdot \frac{d}{dx}[\sec \pi x]$$

$$= -\frac{4}{5}(\sec \pi x)^{-9/5} \cdot \sec \pi x \tan \pi x \cdot \pi$$

$$= -\frac{4\pi}{5}(\sec \pi x)^{-4/5} \tan \pi x \quad \blacktriangleleft$$

We conclude this section with a brief discussion of the general relationship between the derivatives of f and f^{-1}. For this purpose, suppose that both functions are differentiable, and let

$$y = f^{-1}(x) \tag{18}$$

Rewriting this equation as

$$x = f(y) \tag{19}$$

and differentiating implicitly with respect to x yields

$$\frac{d}{dx}[x] = \frac{d}{dx}[f(y)]$$

$$1 = f'(y) \cdot \frac{dy}{dx}$$

$$\frac{dy}{dx} = \frac{1}{f'(y)} \tag{20}$$

Thus, from (18) we obtain the following formula that relates the derivative of f^{-1} to the derivative of f.

$$\frac{d}{dx}[f^{-1}(x)] = \frac{1}{f'(f^{-1}(x))} \tag{21}$$

For example, if $f^{-1}(x) = \sqrt{x}$, then $f(x) = x^2$, so $f'(x) = 2x$; this formula implies that

$$\frac{d}{dx}[\sqrt{x}] = \frac{1}{2(f^{-1}(x))} = \frac{1}{2\sqrt{x}}$$

which is consistent with the known derivative formula for $\sqrt{x}$.

An alternative version of Formula (21) that uses dependent variables can be obtained by using (19) to rewrite $f'(y)$ as dx/dy, in which case (21) becomes

$$\frac{dy}{dx} = \frac{1}{dx/dy} \tag{22}$$

For example, if $y = \sqrt{x}$, then $x = y^2$. Thus, $dx/dy = 2y$, and (22) implies that

$$\frac{dy}{dx} = \frac{1}{2y} = \frac{1}{2\sqrt{x}}$$

which again is consistent with the known derivative formula for $y = \sqrt{x}$.

If an explicit formula can be obtained for the inverse of a function, then the differentiability and the derivative of the inverse can usually be deduced from that formula. However, if no explicit formula for the inverse can be obtained, then Theorem 4.1.7 is the primary mathematical tool for establishing differentiability of the inverse. Once differentiability has been established, a formula for the derivative of the inverse can be obtained either by differentiating implicitly or by using Formulas (21) or (22). The following example illustrates this.

Example 7

We showed in Example 9 of Section 4.1 that the inverse of the function $f(x) = x^5 + x + 1$ is differentiable on the interval $(-\infty, +\infty)$. However, there is no way to obtain an explicit formula for f^{-1}, so we must resort to indirect methods to differentiate this function.

(a)　Find the derivative of f^{-1} by using Formula (22).

(b)　Find the derivative of f^{-1} by differentiating implicitly.

Solution (*a*). If we let $y = f^{-1}(x)$, then

$$x = f(y) = y^5 + y + 1 \tag{23}$$

from which it follows that

$$\frac{dx}{dy} = 5y^4 + 1$$

$$\frac{dy}{dx} = \frac{1}{dx/dy} = \frac{1}{5y^4 + 1} \qquad (24)$$

Although it would be preferable to have dy/dx expressed as a function of x, we are forced to leave it in terms of y, since we cannot solve (23) for y in terms of x.

Solution (b). Differentiating (23) implicitly with respect to x yields

$$\frac{d}{dx}[x] = \frac{d}{dx}[y^5 + y + 1]$$

$$1 = 5y^4 \frac{dy}{dx} + \frac{dy}{dx}$$

$$1 = (5y^4 + 1)\frac{dy}{dx}$$

$$\frac{dy}{dx} = \frac{1}{5y^4 + 1}$$

which agrees with (24). ◀

EXERCISE SET 4.3 C CAS

In Exercises 1–8, find dy/dx.

1. $y = \sqrt[3]{2x - 5}$ 　　　　**2.** $y = \sqrt[3]{2 + \tan(x^2)}$

3. $y = \left(\dfrac{x-1}{x+2}\right)^{3/2}$ 　　**4.** $y = \sqrt{\dfrac{x^2+1}{x^2-5}}$

5. $y = x^3\left(5x^2 + 1\right)^{-2/3}$ 　　**6.** $y = \dfrac{(3-2x)^{4/3}}{x^2}$

7. $y = [\sin(3/x)]^{5/2}$ 　　**8.** $y = \left[\cos(x^3)\right]^{-1/2}$

In Exercises 9 and 10: (a) Find dy/dx by differentiating implicitly. (b) Solve the equation for y as a function of x, and find dy/dx from that equation. (c) Confirm that the two results are consistent by expressing the derivative in part (a) as a function of x alone.

9. $x^3 + xy - 2x = 1$ 　　**10.** $\sqrt{y} - e^x = 2$

In Exercises 11–20, find dy/dx by implicit differentiation.

11. $x^2 + y^2 = 100$ 　　　**12.** $x^3 - y^3 = 6xy$

13. $x^2y + 3xy^3 - x = 3$ 　**14.** $x^3y^2 - 5x^2y + x = 1$

15. $\dfrac{1}{y} + \dfrac{1}{x} = 1$ 　　　**16.** $x^2 = \dfrac{x+y}{x-y}$

17. $\sin(x^2y^2) = x$ 　　**18.** $x^2 = \dfrac{\cot y}{1 + \csc y}$

19. $\tan^3(xy^2 + y) = x$ 　**20.** $\dfrac{xy^3}{1 + \sec y} = 1 + y^4$

In Exercises 21–26, find d^2y/dx^2 by implicit differentiation.

21. $3x^2 - 4y^2 = 7$ 　　　**22.** $x^3 + y^3 = 1$

23. $x^3y^3 - 4 = 0$ 　　　**24.** $2xy - y^2 = 3$

25. $y + \sin y = x$ 　　　**26.** $x\cos y = y$

In Exercises 27 and 28, find the slope of the tangent line to the curve at the given points in two ways: first by solving for y in terms of x and differentiating and then by implicit differentiation.

27. $x^2 + y^2 = 1$; $(1/\sqrt{2}, 1/\sqrt{2})$, $(1/\sqrt{2}, -1/\sqrt{2})$

28. $y^2 - x + 1 = 0$; $(10, 3)$, $(10, -3)$

In Exercises 29–32, use implicit differentiation to find the slope of the tangent line to the curve at the specified point, and check that your answer is consistent with the accompanying graph.

29. $x^4 + y^4 = 16$; $(1, \sqrt[4]{15})$ [*Lamé's special quartic*]

30. $y^3 + yx^2 + x^2 - 3y^2 = 0$; $(0, 3)$ [*trisectrix*]

31. $2(x^2 + y^2)^2 = 25(x^2 - y^2)$; $(3, 1)$ [*lemniscate*]

32. $x^{2/3} + y^{2/3} = 4$; $(-1, 3\sqrt{3})$ [*four-cusped hypocycloid*]

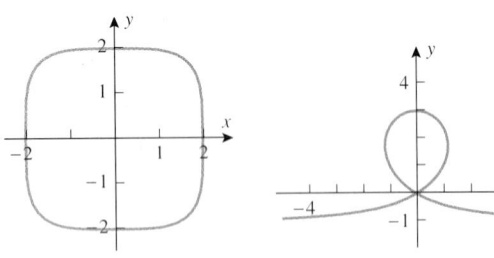

Figure Ex-29   Figure Ex-30

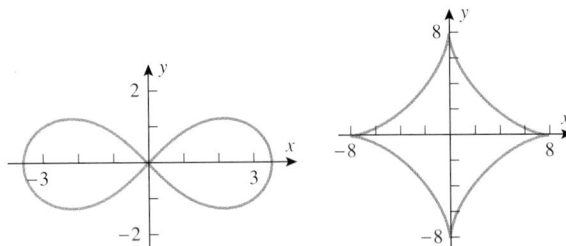

Figure Ex-31   Figure Ex-32

c **33.** If you have a CAS, read the documentation on "implicit plotting," and then generate the four curves in Exercises 29–32.

c **34.** Curves with equations of the form $y^2 = x(x - a)(x - b)$, where $a < b$ are called **bipartite cubics**.
   (a) Use the implicit plotting capability of a CAS to graph the bipartite cubic $y^2 = x(x - 1)(x - 2)$.
   (b) At what points does the curve in part (a) have a horizontal tangent line?
   (c) Solve the equation in part (a) for y in terms of x, and use the result to explain why the graph consists of two separate parts (i.e., is *bipartite*).
   (d) Graph the equation in part (a) without using the implicit plotting capability of the CAS.

c **35.** (a) Use the implicit plotting capability of a CAS to graph the rotated ellipse $x^2 - xy + y^2 = 4$.
   (b) Use the graph to estimate the x-coordinates of all horizontal tangent lines.
   (c) Find the exact values for the x-coordinates in part (b).

In Exercises 36–39, use implicit differentiation to find the specified derivative.

36. $\sqrt{u} + \sqrt{v} = 5$; du/dv **37.** $a^4 - t^4 = 6a^2 r$; da/dt

38. $y = \sin x$; dx/dy.

39. $a^2 \omega^2 + b^2 \lambda^2 = 1$ $(a, b$ constants$)$; $d\omega/d\lambda$

40. At what point(s) is the tangent line to the curve $y^2 = 2x^3$ perpendicular to the line $4x - 3y + 1 = 0$?

41. Find the values of a and b for the curve $x^2 y + ay^2 = b$ if the point $(1, 1)$ is on its graph and the tangent line at $(1, 1)$ has the equation $4x + 3y = 7$.

42. Find the coordinates of the point in the first quadrant at which the tangent line to the curve $x^3 - xy + y^3 = 0$ is parallel to the x-axis.

43. Find equations for two lines through the origin that are tangent to the curve $x^2 - 4x + y^2 + 3 = 0$.

44. Use implicit differentiation to show that the equation of the tangent line to the curve $y^2 = kx$ at (x_0, y_0) is

$$y_0 y = \tfrac{1}{2} k(x + x_0)$$

45. Find dy/dx if

$$2y^3 t + t^3 y = 1 \quad \text{and} \quad \frac{dt}{dx} = \frac{1}{\cos t}$$

In Exercises 46 and 47, find dy/dt in terms of x, y, and dx/dt, assuming that x and y are differentiable functions of the variable t. [*Hint:* Differentiate both sides of the given equation with respect to t.]

46. $x^3 y^2 + y = 3$ **47.** $xy^2 = \sin 3x$

48. (a) Show that $f(x) = x^{4/3}$ is differentiable at 0, but not twice differentiable at 0.
   (b) Show that $f(x) = x^{7/3}$ is twice differentiable at 0, but not three times differentiable at 0.
   (c) Find an exponent k such that $f(x) = x^k$ is $(n-1)$ times differentiable at 0, but not n times differentiable at 0.

In Exercises 49 and 50, find all rational values of r such that $y = x^r$ satisfies the given equation.

49. $3x^2 y'' + 4xy' - 2y = 0$ **50.** $16x^2 y'' + 24xy' + y = 0$

Two curves are said to be **orthogonal** if their tangent lines are perpendicular at each point of intersection, and two families of curves are said to be **orthogonal trajectories** of one another if each member of one family is orthogonal to each member of the other family. This terminology is used in Exercises 51 and 52.

51. The accompanying figure shows some typical members of the families of circles $x^2 + (y - c)^2 = c^2$ (black curves) and $(x - k)^2 + y^2 = k^2$ (gray curves). Show that these families are orthogonal trajectories of one another. [*Hint:* For the tangent lines to be perpendicular at a point of intersection, the slopes of those tangent lines must be negative reciprocals of one another.]

52. The accompanying figure shows some typical members of the families of hyperbolas $xy = c$ (black curves) and $x^2 - y^2 = k$ (gray curves), where $c \neq 0$ and $k \neq 0$. Use the hint in Exercise 51 to show that these families are orthogonal trajectories of one another.

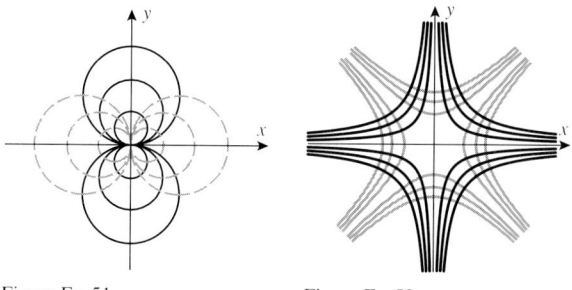

Figure Ex-51   Figure Ex-52

In Exercises 53–56, find the derivative of f^{-1} by using Formula (22), and check your result by differentiating implicitly.

53. $f(x) = 5x^3 + x - 7$ **54.** $f(x) = 1/x^2, \ x > 0$

55. $f(x) = 2x^5 + x^3 + 1$ **56.** $f(x) = 5x - \sin 2x$

4.4 DERIVATIVES OF LOGARITHMIC AND EXPONENTIAL FUNCTIONS

In this section we will obtain derivative formulas for logarithmic and exponential functions, and we will discuss the general relationship between the derivative of a one-to-one function and its inverse.

DERIVATIVES OF LOGARITHMIC FUNCTIONS

The natural logarithm plays a special role in calculus that can be motivated by differentiating $\log_b x$, where b is an arbitrary base. For this purpose, we will *assume* that $\log_b x$ is differentiable, and hence continuous, for $x > 0$. We will also need the limit

$$\lim_{v \to 0} (1 + v)^{1/v} = e$$

that was given in Formula (5) of Section 4.2 (with x rather than v as the variable).

Using the definition of a derivative, we obtain

$$\frac{d}{dx}[\log_b x] = \lim_{h \to 0} \frac{\log_b(x + h) - \log_b x}{h}$$

$$= \lim_{h \to 0} \frac{1}{h} \log_b \left(\frac{x + h}{x} \right) \qquad \text{\small The quotient property of logarithms in Theorem 4.2.3}$$

$$= \lim_{h \to 0} \frac{1}{h} \log_b \left(1 + \frac{h}{x} \right)$$

$$= \lim_{v \to 0} \frac{1}{vx} \log_b(1 + v) \qquad \text{\small Let } v = h/x \text{ and note that } v \to 0 \text{ as } h \to 0.$$

$$= \frac{1}{x} \lim_{v \to 0} \frac{1}{v} \log_b(1 + v) \qquad \text{\small } 1/x \text{ does not vary with } v, \text{ so it can be moved through the limit sign.}$$

$$= \frac{1}{x} \lim_{v \to 0} \log_b(1 + v)^{1/v} \qquad \text{\small The power property of logarithms in Theorem 4.2.3}$$

$$= \frac{1}{x} \log_b \left[\lim_{v \to 0} (1 + v)^{1/v} \right] \qquad \text{\small } \log_b x \text{ is continuous, so we can move the limit through the function symbol.}$$

$$= \frac{1}{x} \log_b e$$

Thus,

$$\frac{d}{dx}[\log_b x] = \frac{1}{x} \log_b e, \quad x > 0$$

But from Formula (9) of Section 4.2 we have $\log_b e = 1/\ln b$, so we can rewrite this derivative formula as

$$\frac{d}{dx}[\log_b x] = \frac{1}{x \ln b}, \quad x > 0 \tag{1}$$

In the special case where $b = e$, we have $\log_b e = \ln e = 1$, so this formula becomes

$$\frac{d}{dx}[\ln x] = \frac{1}{x}, \quad x > 0 \tag{2}$$

Thus, among all possible bases, the base $b = e$ produces the simplest derivative formula for $\log_b x$. This is one of the reasons why the natural logarithm function is preferred over other logarithms in calculus.

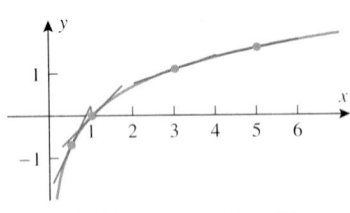

$y = \ln x$ with tangent lines

Figure 4.4.1

Example 1

(a) Figure 4.4.1 shows the graph of $y = \ln x$ and its tangent lines at the points $x = \frac{1}{2}$, 1, 3, and 5. Find the slopes of those tangent lines.

(b) Do you think that the graph of $y = \ln x$ has any horizontal tangent lines? Use the derivative of $\ln x$ to justify your answer.

Solution (*a*). From (2), the slopes of the tangent lines at the points $x = \frac{1}{2}$, 1, 3, and 5 are $1/x = 2$, 1, $\frac{1}{3}$, and $\frac{1}{5}$, which is consistent with Figure 4.4.1.

Solution (*b*). From the graph of $y = \ln x$, it does not appear that there are any horizontal tangent lines. This is confirmed by the fact that $dy/dx = 1/x$ is not equal to zero for any real value of x. ◀

If u is a differentiable function of x, and if $u(x) > 0$, then applying the chain rule to (1) and (2) produces the following generalized derivative formulas:

$$\frac{d}{dx}[\log_b u] = \frac{1}{u \ln b} \cdot \frac{du}{dx} \quad \text{and} \quad \frac{d}{dx}[\ln u] = \frac{1}{u} \cdot \frac{du}{dx} \tag{3-4}$$

Example 2

Find $\dfrac{d}{dx}[\ln(x^2 + 1)]$.

Solution. From (4) with $u = x^2 + 1$,

$$\frac{d}{dx}[\ln(x^2 + 1)] = \frac{1}{x^2 + 1} \cdot \frac{d}{dx}[x^2 + 1] = \frac{1}{x^2 + 1} \cdot 2x = \frac{2x}{x^2 + 1} \quad ◀$$

When possible, the properties of logarithms in Theorem 4.2.3 should be used to convert products, quotients, and exponents into sums, differences, and constant multiples *before* differentiating a function involving logarithms.

Example 3

$$\frac{d}{dx}\left[\ln\left(\frac{x^2 \sin x}{\sqrt{1 + x}}\right)\right] = \frac{d}{dx}\left[2\ln x + \ln(\sin x) - \frac{1}{2}\ln(1 + x)\right]$$

$$= \frac{2}{x} + \frac{\cos x}{\sin x} - \frac{1}{2(1 + x)}$$

$$= \frac{2}{x} + \cot x - \frac{1}{2 + 2x} \quad ◀$$

Example 4

Find $\dfrac{d}{dx}[\ln |x|]$.

Solution. The function $\ln |x|$ is defined for all x, except $x = 0$; we will consider the cases $x > 0$ and $x < 0$ separately.

If $x > 0$, then $|x| = x$, so

$$\frac{d}{dx}[\ln |x|] = \frac{d}{dx}[\ln x] = \frac{1}{x}$$

If $x < 0$, then $|x| = -x$, so from (4) we have

$$\frac{d}{dx}[\ln |x|] = \frac{d}{dx}[\ln(-x)] = \frac{1}{(-x)} \cdot \frac{d}{dx}[-x] = \frac{1}{x}$$

Since the same formula results in both cases, we have shown that

$$\frac{d}{dx}[\ln |x|] = \frac{1}{x} \quad \text{if } x \neq 0 \tag{5}$$

Example 5

From (5) and the chain rule,

$$\frac{d}{dx}[\ln |\sin x|] = \frac{1}{\sin x} \cdot \frac{d}{dx}[\sin x] = \frac{\cos x}{\sin x} = \cot x \qquad \blacktriangleleft$$

LOGARITHMIC DIFFERENTIATION

We now consider a technique called *logarithmic differentiation* that is useful for differentiating functions that are composed of products, quotients, and powers.

Example 6

The derivative of

$$y = \frac{x^2 \sqrt[3]{7x - 14}}{\left(1 + x^2\right)^4} \tag{6}$$

is messy to calculate directly. However, if we first take the natural logarithm of both sides and then use its properties, we can write

$$\ln y = 2 \ln x + \tfrac{1}{3} \ln(7x - 14) - 4 \ln(1 + x^2)$$

Differentiating both sides with respect to x yields

$$\frac{1}{y}\frac{dy}{dx} = \frac{2}{x} + \frac{7/3}{7x - 14} - \frac{8x}{1 + x^2} \tag{7}$$

Thus, on solving for dy/dx and using (6) we obtain

$$\frac{dy}{dx} = \frac{x^2 \sqrt[3]{7x - 14}}{\left(1 + x^2\right)^4} \left[\frac{2}{x} + \frac{1}{3x - 6} - \frac{8x}{1 + x^2} \right] \tag{8}$$

REMARK. Since $\ln y$ is defined only for $y > 0$, logarithmic differentiation of $y = f(x)$ is valid only on intervals where $f(x)$ is positive. Thus, the derivative obtained in the preceding example is valid on the interval $(2, +\infty)$, since the given function is positive for $x > 2$. However, the formula is actually valid on the interval $(-\infty, 2)$ as well. This can be seen by taking absolute values before proceeding with the logarithmic differentiation and noting that $\ln |y|$ is defined for all y except $y = 0$. If we do this and simplify using properties of logarithms and absolute values, we obtain

$$\ln |y| = 2 \ln |x| + \tfrac{1}{3} \ln |7x - 14| - 4 \ln |1 + x^2|$$

Differentiating both sides with respect to x yields (7), and hence results in (8).

In general, if the derivative of $y = f(x)$ is to be obtained by logarithmic differentiation, then the same formula for dy/dx will result regardless of whether one first takes absolute values or not. Thus, a derivative formula obtained by logarithmic differentiation will be valid except perhaps at points where $f(x)$ is zero. The formula may, in fact, be valid at those points as well, but it is not guaranteed.

DERIVATIVES OF IRRATIONAL POWERS OF x

We know from Formula (15) of Section 4.3 that the differentiation formula

$$\frac{d}{dx}[x^r] = rx^{r-1} \tag{9}$$

holds for rational values of r. We will now use logarithmic differentiation to show that this formula holds if r is *any* real number (rational or irrational). In our computations we will assume that x^r is a differentiable function and that the familiar laws of exponents hold for real exponents.

Let $y = x^r$, where r is a real number. The derivative dy/dx can be obtained by logarithmic differentiation as follows:

$$\ln y = \ln x^r = r \ln x$$

$$\frac{d}{dx}[\ln y] = \frac{d}{dx}[r \ln x]$$

$$\frac{1}{y}\frac{dy}{dx} = \frac{r}{x}$$

$$\frac{dy}{dx} = \frac{r}{x}y = \frac{r}{x}x^r = rx^{r-1}$$

which establishes (9) for real values of r. Thus, for example,

$$\frac{d}{dx}[x^\pi] = \pi x^{\pi-1} \quad \text{and} \quad \frac{d}{dx}\left[x^{\sqrt{2}}\right] = \sqrt{2}x^{\sqrt{2}-1} \tag{10}$$

DERIVATIVES OF EXPONENTIAL FUNCTIONS

To obtain a derivative formula for the exponential function

$$y = b^x \tag{11}$$

we rewrite this equation as

$$x = \log_b y$$

and differentiate implicitly using (3) to obtain

$$1 = \frac{1}{y \ln b} \cdot \frac{dy}{dx}$$

which we can rewrite using (11) as

$$\frac{dy}{dx} = y \ln b = b^x \ln b$$

Thus, we have shown that if b^x is a differentiable function, then its derivative with respect to x is

$$\frac{d}{dx}[b^x] = b^x \ln b \tag{12}$$

In the special case where $b = e$ we have $\ln e = 1$, so that (12) becomes

$$\frac{d}{dx}[e^x] = e^x \tag{13}$$

Moreover, if u is a differentiable function of x, then it follows from (12) and (13) that

$$\frac{d}{dx}[b^u] = b^u \ln b \cdot \frac{du}{dx} \quad \text{and} \quad \frac{d}{dx}[e^u] = e^u \cdot \frac{du}{dx} \tag{14--15}$$

REMARK.   It is important to distinguish between differentiating b^x (variable exponent and constant base) and x^b (variable base and constant exponent). For example, compare the derivative of x^π in (10) to the following derivative of π^x, which is obtained from (12):

$$\frac{d}{dx}[\pi^x] = \pi^x \ln \pi$$

Example 7

The following computations use (14) and (15).

$$\frac{d}{dx}[2^{\sin x}] = (2^{\sin x})(\ln 2) \cdot \frac{d}{dx}[\sin x] = (2^{\sin x})(\ln 2)(\cos x)$$

$$\frac{d}{dx}[e^{-2x}] = e^{-2x} \cdot \frac{d}{dx}[-2x] = -2e^{-2x}$$

$$\frac{d}{dx}\left[e^{x^3}\right] = e^{x^3} \cdot \frac{d}{dx}[x^3] = 3x^2 e^{x^3}$$

$$\frac{d}{dx}[e^{\cos x}] = e^{\cos x} \cdot \frac{d}{dx}[\cos x] = -(\sin x)e^{\cos x} \qquad \blacktriangleleft$$

Example 8

A glass of lemonade with a temperature of $40°$ F sits in a room whose temperature is a constant $70°$ F. Using a principle of physics, called ***Newton's Law of Cooling***, one can show that if the temperature of the lemonade reaches $52°$ F in 1 hour, then the temperature T of the lemonade as a function of the elapsed time t is modeled approximately by the equation

$$T = 70 - 30e^{-0.5t}$$

where T is in $°$F and t is in hours. The graph of this equation, shown in Figure 4.4.2, confirms our everyday experience that the temperature of the lemonade gradually approaches the temperature of the room.

(a)   In words, what happens to the *rate* of temperature rise over time?

(b)   Use a derivative to confirm your conclusion.

Solution (a). The rate of change of temperature with respect to time is the slope of the tangent line to the graph of T versus t. As t increases, these slopes decrease, so the temperature rises at an ever-decreasing rate.

Solution (b). The rate of change of temperature with respect to time is

$$\frac{dT}{dt} = \frac{d}{dt}[70 - 30e^{-0.5t}] = -30(-0.5)e^{-0.5t} = 15e^{-0.5t}$$

As t increases, this derivative decreases, which confirms the conclusion in part (a). $\blacktriangleleft$

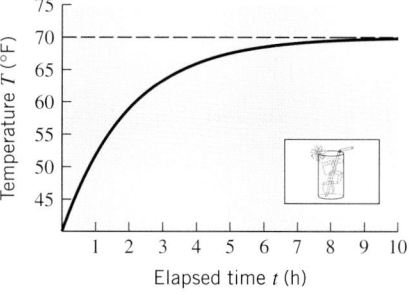

Figure 4.4.2

EXERCISE SET 4.4 ⌇ Graphing Calculator

In Exercises 1–30, find dy/dx.

1. $y = \ln 2x$

2. $y = \ln(x^3)$

3. $y = (\ln x)^2$

4. $y = \ln(\sin x)$

5. $y = \ln|\tan x|$

6. $y = \ln(2 + \sqrt{x})$

7. $y = \ln\left(\dfrac{x}{1+x^2}\right)$

8. $y = \ln(\ln x)$

9. $y = \ln|x^3 - 7x^2 - 3|$

10. $y = x^3 \ln x$

11. $y = \sqrt{\ln x}$

12. $y = \sqrt{1 + \ln^2 x}$

13. $y = \cos(\ln x)$

14. $y = \sin^2(\ln x)$

15. $y = x^3 \log_2(3 - 2x)$

16. $y = x\left[\log_2(x^2 - 2x)\right]^3$

17. $y = \dfrac{x^2}{1 + \log x}$

18. $y = \dfrac{\log x}{1 + \log x}$

19. $y = e^{7x}$

20. $y = e^{-5x^2}$

21. $y = x^3 e^x$

22. $y = e^{1/x}$

23. $y = \dfrac{e^x - e^{-x}}{e^x + e^{-x}}$

24. $y = \sin(e^x)$

25. $y = e^{x \tan x}$

26. $y = \dfrac{e^x}{\ln x}$

27. $y = e^{(x - e^{3x})}$

28. $y = \exp(\sqrt{1 + 5x^3})$

29. $y = \ln(1 - xe^{-x})$

30. $y = \ln(\cos e^x)$

In Exercises 31 and 32, find dy/dx by implicit differentiation.

31. $y + \ln xy = 1$

32. $y = \ln(x \tan y)$

In Exercises 33 and 34, use the method of Example 3 to help perform the indicated differentiation.

33. $\dfrac{d}{dx}\left[\ln \dfrac{\cos x}{\sqrt{4 - 3x^2}}\right]$

34. $\dfrac{d}{dx}\left[\ln \sqrt{\dfrac{x-1}{x+1}}\right]$

In Exercises 35–38, find dy/dx using the method of logarithmic differentiation.

35. $y = x\sqrt[3]{1 + x^2}$

36. $y = \sqrt[5]{\dfrac{x-1}{x+1}}$

37. $y = \dfrac{(x^2 - 8)^{1/3}\sqrt{x^3 + 1}}{x^6 - 7x + 5}$

38. $y = \dfrac{\sin x \cos x \tan^3 x}{\sqrt{x}}$

In Exercises 39–42, find $f'(x)$ by Formula (14) and then by logarithmic differentiation.

39. $f(x) = 2^x$

40. $f(x) = 3^{-x}$

41. $f(x) = \pi^{\sin x}$

42. $f(x) = \pi^{x \tan x}$

In Exercises 43–46, find dy/dx using the method of logarithmic differentiation.

43. $y = (x^3 - 2x)^{\ln x}$

44. $y = x^{\sin x}$

45. $y = (\ln x)^{\tan x}$

46. $y = (x^2 + 3)^{\ln x}$

47. Show that for any constants A and B, the function

$$y = Ae^{2x} + Be^{-4x}$$

satisfies the equation

$$y'' + 2y' - 8y = 0$$

48. Show that for any constants A and k, the function $y = Ae^{kt}$ satisfies the equation $dy/dt = ky$.

49. Let $f(x) = e^{kx}$ and $g(x) = e^{-kx}$. Find
 (a) $f^{(n)}(x)$
 (b) $g^{(n)}(x)$.

50. Find dy/dt if $y = e^{-\lambda t}(A \sin \omega t + B \cos \omega t)$, where A, B, λ, and ω are constants.

51. Find $f'(x)$ if

$$f(x) = \dfrac{1}{\sqrt{2\pi}\sigma}\exp\left[-\dfrac{1}{2}\left(\dfrac{x - \mu}{\sigma}\right)^2\right]$$

where μ and σ are constants and $\sigma \neq 0$.

52. Show that
 (a) $y = xe^{-x}$ satisfies the equation $xy' = (1 - x)y$
 (b) $y = xe^{-x^2/2}$ satisfies the equation $xy' = (1 - x^2)y$.

53. Find
 (a) $\dfrac{d}{dx}[\log_x e]$
 (b) $\dfrac{d}{dx}[\log_x 2]$.

54. Recall from Section 4.2 that the loudness β of a sound in decibels (db) is given by $\beta = 10\log(I/I_0)$, where I is the intensity of the sound in watts per square meter (W/m²) and I_0 is a constant that is approximately the intensity of a sound at the threshold of human hearing. Find the rate of change of β with respect to I at the point where
 (a) $I/I_0 = 10$ (b) $I/I_0 = 100$ (c) $I/I_0 = 1000$

55. The equilibrium constant k of a balanced chemical reaction changes with the absolute temperature T according to the law

$$k = k_0 \exp\left(-\dfrac{q(T - T_0)}{2T_0 T}\right)$$

where k_0, q, and T_0 are constants. Find the rate of change of k with respect to T.

56. (a) Explain why Formula (12) cannot be used to find $(d/dx)[x^x]$.
 (b) Find this derivative by logarithmic differentiation.

57. Find $f'(x)$ if $f(x) = x^e$.

58. Find a point on the graph of $y = e^{3x}$ at which the tangent line passes through the origin.

In Exercises 59 and 60, find the limit by interpreting the expression as an appropriate derivative.

59. (a) $\displaystyle\lim_{h \to 0} \frac{\ln(1 + h)}{h}$ (b) $\displaystyle\lim_{h \to 0} \frac{10^h - 1}{h}$

60. (a) $\displaystyle\lim_{h \to 0} \frac{\ln(e^2 + h) - 2}{h}$ (b) $\displaystyle\lim_{x \to 1} \frac{2^x - 2}{x - 1}$

 61. (a) Make a conjecture about the shape of the graph of $y = \frac{1}{2}x - \ln x$, and draw a rough sketch.
   (b) Check your conjecture by graphing the equation over the interval $0 < x < 5$ with a graphing utility.
   (c) Show that the slopes of the tangent lines to the curve at $x = 1$ and $x = e$ have opposite signs.
   (d) What does part (c) imply about the existence of a horizontal tangent line to the curve? Explain your reasoning.
   (e) Find the exact x-coordinates of all horizontal tangent lines to the curve.

 62. (a) Use a graphing utility to graph the function
$$f(x) = 2x^3 + x^2 - 20x + 4$$
   over the interval $-5 < x < 5$.
   (b) Working with the graph in part (a), make a rough sketch of the graph of $f'(x)$ over the interval $-5 < x < 5$.
   (c) Check your work in part (b) by generating the graph of $f'(x)$ with a graphing utility.
   (d) Find the exact locations of the horizontal tangent lines to the graph of f over the interval $-5 < x < 5$.
   (e) Confirm that the result in part (d) is consistent with the graph of $f'(x)$ in part (c).

63. (a) Sketch the curves $y = e^x$ and $y = -e^x$ in the same coordinate system; then make a conjecture about the general shape of the equation $y = e^x \cos \pi x$ for $x \geq 0$, and sketch its graph in the same coordinate system as the two exponential functions.
   (b) Check your conjecture in part (a) by using a graphing utility to generate the graphs of $y = e^x$, $y = -e^x$, and $y = e^x \cos \pi x$ in the same window for $0 \leq x \leq 3$.

64. Suppose that the population of oxygen-dependent bacteria in a pond is modeled by the equation
$$P(t) = \frac{60}{5 + 7e^{-t}}$$
where $P(t)$ is the population (in billions) t days after an initial observation at time $t = 0$.
   (a) Use a graphing utility to graph the function $P(t)$.
   (b) In words, explain what happens to the population over time? Check your conclusion by finding $\displaystyle\lim_{t \to +\infty} P(t)$.
   (c) In words, what happens to the *rate* of population growth over time? Check your conclusion by graphing $P'(t)$.

65. Suppose that the population of deer on an island is modeled by the equation
$$P(t) = \frac{95}{5 - 4e^{-t/4}}$$
where $P(t)$ is the number of deer t weeks after an initial observation at time $t = 0$.
   (a) Use a graphing utility to graph the function $P(t)$.
   (b) In words, explain what happens to the population over time. Check your conclusion by finding $\displaystyle\lim_{t \to +\infty} P(t)$.
   (c) In words, what happens to the *rate* of population growth over time? Check your conclusion by graphing $P'(t)$.

4.5 DERIVATIVES OF INVERSE TRIGONOMETRIC FUNCTIONS

A common problem in trigonometry is to find an angle whose trigonometric functions are known. As you may recall, problems of this type involve the computation of "arc functions" such as arcsin x, arccos x, arctan x, *and so forth. In this section we will consider this idea from the viewpoint of inverse functions, with the goal of developing derivative formulas for the inverse trigonometric functions.*

INVERSE TRIGONOMETRIC FUNCTIONS

None of the six basic trigonometric functions is one-to-one because they all repeat periodically and hence do not pass the horizontal line test. Thus, to define inverse trigonometric functions we must first restrict the domains of the trigonometric functions to make them one-to-one. The top part of Figure 4.5.1 shows how these restrictions are made for $\sin x$, $\cos x$, $\tan x$, and $\sec x$. (Inverses of $\cot x$ and $\csc x$ are of lesser importance and will be left for the exercises.) The inverses of these restricted functions are denoted by

$$\sin^{-1} x, \quad \cos^{-1} x, \quad \tan^{-1} x, \quad \sec^{-1} x$$

(or alternatively by arcsin x, arccos x, arctan x, arcsec x) and are defined as follows:

4.5.1 DEFINITION. The *inverse sine function*, denoted by $\sin^{-1}$, is defined to be the inverse of the restricted sine function

$$\sin x, \quad -\pi/2 \leq x \leq \pi/2$$

4.5.2 DEFINITION. The *inverse cosine function*, denoted by $\cos^{-1}$, is defined to be the inverse of the restricted cosine function

$$\cos x, \quad 0 \leq x \leq \pi$$

4.5.3 DEFINITION. The *inverse tangent function*, denoted by $\tan^{-1}$, is defined to be the inverse of the restricted tangent function

$$\tan x, \quad -\pi/2 < x < \pi/2$$

4.5.4 DEFINITION.* The *inverse secant function*, denoted by $\sec^{-1}$, is defined to be the inverse of the restricted secant function

$$\sec x, \quad 0 \leq x \leq \pi \text{ with } x \neq \pi/2$$

REMARK. The notations $\sin^{-1} x$, $\cos^{-1} x$, ... are reserved exclusively for the inverse trigonometric functions and are not used for reciprocals of the trigonometric functions. For example, to denote the reciprocal $1/\sin x$ in exponent form, we would write $(\sin x)^{-1}$ and *never* $\sin^{-1} x$.

The graphs of the inverse trigonometric functions, which are shown in the bottom part of Figure 4.5.1, are obtained by reflecting the graphs in the top part of the figure about the line $y = x$. If you have trouble visualizing these relationships, then look at Figure 4.5.2

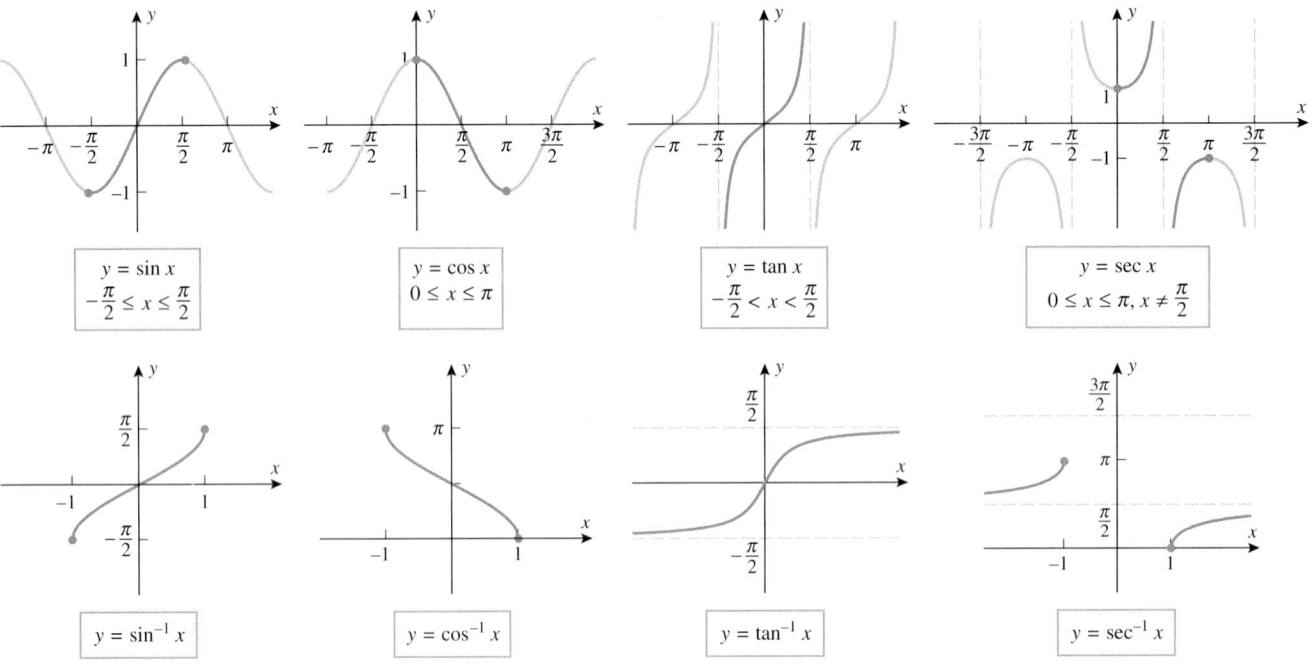

Figure 4.5.1

*There is no universal agreement on the definition of $\sec^{-1} x$, and some mathematicians prefer to restrict the domain of $\sec x$ so that $0 \leq x < \pi/2$ or $\pi \leq x < 3\pi/2$, which was the definition used in earlier editions of this text. Each definition has advantages and disadvantages, but we have changed to the current definition to conform with the conventions used by the CAS programs *Mathematica*, *Maple*, and *Derive*.

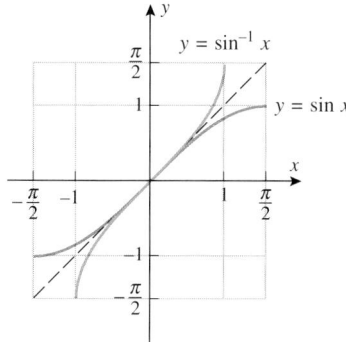

Figure 4.5.2

for a more detailed illustration for the inverse sine. It may also help to keep in mind that reflection about $y = x$ converts vertical lines to horizontal lines, and vice versa, and that x-intercepts reflect into y-intercepts, and vice versa.

Table 4.5.1 summarizes the basic properties of the inverse sine, cosine, tangent, and secant functions. You should confirm that the domains and ranges listed in this table are consistent with the graphs in the bottom part of Figure 4.5.1.

Table 4.5.1

FUNCTION	DOMAIN	RANGE	BASIC RELATIONSHIPS
$\sin^{-1}$	$[-1, 1]$	$[-\pi/2, \pi/2]$	$\sin^{-1}(\sin x) = x$ if $-\pi/2 \le x \le \pi/2$ $\sin(\sin^{-1} x) = x$ if $-1 \le x \le 1$
$\cos^{-1}$	$[-1, 1]$	$[0, \pi]$	$\cos^{-1}(\cos x) = x$ if $0 \le x \le \pi$ $\cos(\cos^{-1} x) = x$ if $-1 \le x \le 1$
$\tan^{-1}$	$(-\infty, +\infty)$	$(-\pi/2, \pi/2)$	$\tan^{-1}(\tan x) = x$ if $-\pi/2 < x < \pi/2$ $\tan(\tan^{-1} x) = x$ if $-\infty < x < +\infty$
$\sec^{-1}$	$(-\infty, -1] \cup [1, +\infty)$	$[0, \pi/2) \cup (\pi/2, \pi]$	$\sec^{-1}(\sec x) = x$ if $0 \le x \le \pi, x \ne \pi/2$ $\sec(\sec^{-1} x) = x$ if $|x| \ge 1$

EVALUATING INVERSE TRIGONOMETRIC FUNCTIONS

It follows from Theorem 4.1.2 that the equations $y = \sin^{-1} x$ and $x = \sin y$ are equivalent provided (of course) that y is in the domain of the restricted sine function and x is in the domain of the inverse sine function; that is, $-\pi/2 \le y \le \pi/2$ and $-1 \le x \le 1$. Thus,

$$y = \sin^{-1} x \text{ is equivalent to } \sin y = x \text{ if } \begin{cases} -1 \le x \le 1 \\ -\pi/2 \le y \le \pi/2 \end{cases}$$

Similarly,

$$y = \cos^{-1} x \text{ is equivalent to } \cos y = x \text{ if } \begin{cases} -1 \le x \le 1 \\ 0 \le y \le \pi \end{cases}$$

$$y = \tan^{-1} x \text{ is equivalent to } \tan y = x \text{ if } \begin{cases} -\infty < x < +\infty \\ -\pi/2 < y < \pi/2 \end{cases}$$

$$y = \sec^{-1} x \text{ is equivalent to } \sec y = x \text{ if } \begin{cases} x \ge 1 \\ 0 \le y < \pi/2 \end{cases} \text{ or } \begin{cases} x \le -1 \\ \pi/2 < y \le \pi \end{cases}$$

A common problem in trigonometry is to find an angle whose sine is known. For example, you might want to find an angle θ in radian measure such that

$$\sin \theta = \tfrac{1}{2} \tag{1}$$

and, more generally, for a given value of y in the interval $-1 \le y \le 1$ you might want to solve the equation

$$\sin \theta = y \tag{2}$$

Because $\sin \theta$ repeats periodically, such equations have infinitely many solutions for θ; however, if we solve this equation as

$$\theta = \sin^{-1} y$$

then we isolate the specific solution that lies in the interval $[-\pi/2, \pi/2]$, since this is the range of the inverse sine. For example, Figure 4.5.3 shows four solutions of Equation (1), namely, $-11\pi/6$, $-7\pi/6$, $\pi/6$, and $5\pi/6$. Of these, $\pi/6$ is the solution in the interval $[-\pi/2, \pi/2]$, so

$$\sin^{-1} \left(\tfrac{1}{2} \right) = \pi/6 \tag{3}$$

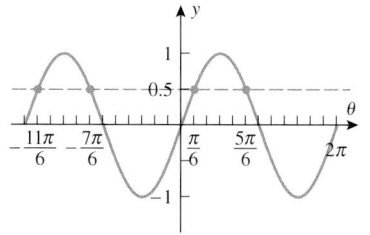

Figure 4.5.3

FOR THE READER.   Refer to the documentation for your calculating utility to determine how to calculate inverse sines, inverse cosines, and inverse tangents; and then confirm Equation (3) numerically by showing that

$$\sin^{-1}(0.5) \approx 0.523598775598\ldots \approx \pi/6$$

In general, if we view $\theta = \sin^{-1} y$ as an angle in radian measure whose sine is y, then the restriction $-\pi/2 \le \theta \le \pi/2$ imposes the geometric requirement that the angle θ terminate in either the first or fourth quadrant or on an axis adjacent to those quadrants.

Example 1

Find exact values of

    (a) $\sin^{-1}(1/\sqrt{2})$ (b) $\sin^{-1}(-1)$

by inspection, and confirm your results numerically using a calculating utility.

Solution (*a*). Because $\sin^{-1}(1/\sqrt{2}) > 0$, we can view $\theta = \sin^{-1}(1/\sqrt{2})$ as that angle in the first quadrant such that $\sin \theta = 1/\sqrt{2}$. Thus, $\sin^{-1}(1/\sqrt{2}) = \pi/4$. You can confirm this with your calculating utility by showing that $\sin^{-1}(1/\sqrt{2}) \approx 0.785 \approx \pi/4$.

Solution (*b*). Because $\sin^{-1}(-1) < 0$, we can view $\theta = \sin^{-1}(-1)$ as an angle in the fourth quadrant (or an adjacent axis) such that $\sin \theta = -1$. Thus, $\sin^{-1}(-1) = -\pi/2$. You can confirm this with your calculating utility by showing that $\sin^{-1}(-1) \approx -1.57 \approx -\pi/2$. ◄

FOR THE READER.   If $\theta = \cos^{-1} y$ is viewed as an angle in radian measure whose cosine is y, in what possible quadrants can θ lie? Answer the same question for $\theta = \tan^{-1} y$ and $\theta = \sec^{-1} y$.

FOR THE READER.   Most calculators do not provide a direct method for calculating inverse secants. In such situations the identity

$$\sec^{-1} x = \cos^{-1}(1/x) \tag{4}$$

is useful (Exercise 16). Use this formula to show that

$$\sec^{-1}(2.25) \approx 1.11 \quad \text{and} \quad \sec^{-1}(-2.25) \approx 2.03$$

If you have a calculating utility (such as a CAS) that can find $\sec^{-1} x$ directly, use it to check these values.

IDENTITIES FOR INVERSE TRIGONOMETRIC FUNCTIONS

If we interpret $\sin^{-1} x$ as an angle in radian measure whose sine is x, and if that angle is *nonnegative*, then we can represent $\sin^{-1} x$ geometrically as an angle in a right triangle in which the hypotenuse has length 1 and the side opposite to the angle $\sin^{-1} x$ has length x (Figure 4.5.4*a*). By the Theorem of Pythagoras the side adjacent to the angle $\sin^{-1} x$ has length $\sqrt{1 - x^2}$. Moreover, the angle opposite to $\sin^{-1} x$ is $\cos^{-1} x$, since the cosine of that angle is x (Figure 4.5.4*b*). This triangle motivates a number of useful identities involving inverse trigonometric functions that are valid for $-1 \le x \le 1$; for example,

$$\sin^{-1} x + \cos^{-1} x = \frac{\pi}{2} \tag{5}$$

$$\cos(\sin^{-1} x) = \sqrt{1 - x^2} \tag{6}$$

$$\sin(\cos^{-1} x) = \sqrt{1 - x^2} \tag{7}$$

$$\tan(\sin^{-1} x) = \frac{x}{\sqrt{1 - x^2}} \tag{8}$$

In a similar manner, $\tan^{-1} x$ and $\sec^{-1} x$ can be represented as angles in the right triangles shown in Figures 4.5.4*c* and 4.5.4*d* (verify). Those triangles reveal more useful identities; for example,

$$\sec(\tan^{-1} x) = \sqrt{1 + x^2} \tag{9}$$

$$\sin(\sec^{-1} x) = \frac{\sqrt{x^2 - 1}}{x} \quad (x \ge 1) \tag{10a}$$

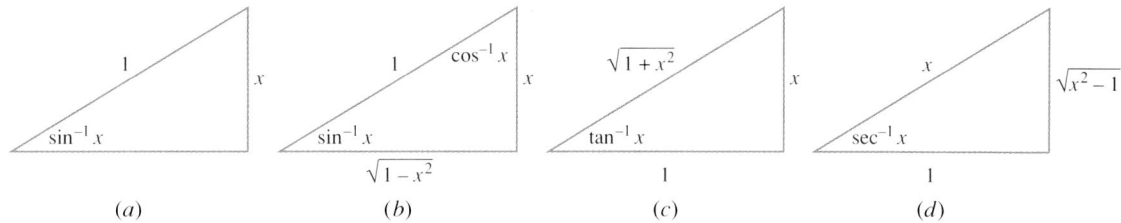

Figure 4.5.4

REMARK.   We leave it as an exercise to use (4) and (7) to obtain the following identity that is valid for $x \geq 1$ and $x \leq 1$ (Exercise 48):

$$\sin(\sec^{-1} x) = \frac{\sqrt{x^2 - 1}}{|x|} \qquad (|x| \geq 1) \tag{10b}$$

REMARK.   There is nothing to be gained by memorizing these identities; what is important to understand is the *method* that was used to obtain them.

Referring to Figure 4.5.1, observe that the inverse sine and inverse tangent are odd functions; that is,

$$\sin^{-1}(-x) = -\sin^{-1}(x) \quad \text{and} \quad \tan^{-1}(-x) = -\tan^{-1}(x) \tag{11–12}$$

Example 2

Figure 4.5.5 shows a computer-generated graph of $y = \sin^{-1}(\sin x)$. One might think that this graph should be the line $y = x$, since $\sin^{-1}(\sin x) = x$. Why isn't it?

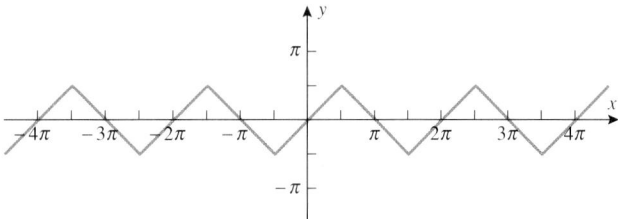

Figure 4.5.5

Solution.   The relationship $\sin^{-1}(\sin x) = x$ is valid on the interval $-\pi/2 \leq x \leq \pi/2$, so we can say with certainty that the graphs of $y = \sin^{-1}(\sin x)$ and $y = x$ coincide on this interval (which is confirmed by Figure 4.5.5). However, outside of this interval the relationship $\sin^{-1}(\sin x) = x$ need not hold. For example, if x lies in the interval $\pi/2 \leq x \leq 3\pi/2$, then the quantity $x - \pi$ lies in the interval $-\pi/2 \leq x \leq \pi/2$, so

$$\sin^{-1}[\sin(x - \pi)] = x - \pi$$

Thus, by using the identity $\sin(x - \pi) = -\sin x$ and the fact that $\sin^{-1}$ is an odd function, we can express $\sin^{-1}(\sin x)$ as

$$\sin^{-1}(\sin x) = \sin^{-1}[-\sin(x - \pi)] = -\sin^{-1}[\sin(x - \pi)] = -(x - \pi)$$

This shows that on the interval $\pi/2 \leq x \leq 3\pi/2$ the graph of $y = \sin^{-1}(\sin x)$ coincides with the line $y = -(x - \pi)$, which has slope -1 and an x-intercept at $x = \pi$. This agrees with Figure 4.5.5.   ◀

DERIVATIVES OF THE INVERSE TRIGONOMETRIC FUNCTIONS

Recall that if f is a one-to-one function whose derivative is known, then there are two basic ways to obtain a derivative formula for $f^{-1}(x)$—we can rewrite the equation $y = f^{-1}(x)$ as $x = f(y)$, and differentiate implicitly, or we can apply Formula (21) or (22) of Section 4.3. Here we will use implicit differentiation to obtain the derivative formula for $y = \sin^{-1} x$. Rewriting this equation as $x = \sin y$ and differentiating implicitly, we obtain

$$\frac{d}{dx}[x] = \frac{d}{dx}[\sin y]$$

$$1 = \cos y \cdot \frac{dy}{dx}$$

$$\frac{dy}{dx} = \frac{1}{\cos y} = \frac{1}{\cos(\sin^{-1} x)}$$

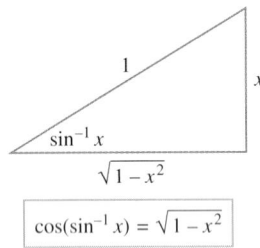

$$\cos(\sin^{-1} x) = \sqrt{1 - x^2}$$

Figure 4.5.6

At this point we have succeeded in obtaining the derivative; however, this derivative formula can be simplified by applying Formula (6), which is derived from the triangle in Figure 4.5.6. This yields

$$\frac{dy}{dx} = \frac{1}{\sqrt{1 - x^2}}$$

Thus, we have shown that

$$\frac{d}{dx}[\sin^{-1} x] = \frac{1}{\sqrt{1 - x^2}} \tag{13}$$

If u is a differentiable function of x, then (13) and the chain rule produce the following generalized derivative formula:

$$\frac{d}{dx}[\sin^{-1} u] = \frac{1}{\sqrt{1 - u^2}} \frac{du}{dx} \tag{14}$$

The method used to obtain this formula can also be used to obtain generalized derivative formulas for the other inverse trigonometric functions. These formulas, which are valid for $-1 < u < 1$, are

$$\frac{d}{dx}[\sin^{-1} u] = \frac{1}{\sqrt{1 - u^2}} \frac{du}{dx}, \qquad \frac{d}{dx}[\cos^{-1} u] = -\frac{1}{\sqrt{1 - u^2}} \frac{du}{dx} \tag{15--16}$$

$$\frac{d}{dx}[\tan^{-1} u] = \frac{1}{1 + u^2} \frac{du}{dx}, \qquad \frac{d}{dx}[\cot^{-1} u] = -\frac{1}{1 + u^2} \frac{du}{dx} \tag{17--18}$$

$$\frac{d}{dx}[\sec^{-1} u] = \frac{1}{|u|\sqrt{u^2 - 1}} \frac{du}{dx}, \qquad \frac{d}{dx}[\csc^{-1} u] = -\frac{1}{|u|\sqrt{u^2 - 1}} \frac{du}{dx} \tag{19--20}$$

DIFFERENTIABILITY OF THE INVERSE TRIGONOMETRIC FUNCTIONS

In the derivation of (13) we *assumed* that $\sin^{-1} x$ is differentiable. However, we can establish the differentiability with the help of Theorem 4.1.7. Since $f(x) = \sin x$ and $f'(x) = \cos x$, it follows from that theorem that the function $f^{-1}(x) = \sin^{-1} x$ will be differentiable at any point x where $\cos(\sin^{-1} x) \neq 0$ or from (6) where $\sqrt{1 - x^2} \neq 0$. Thus, $\sin^{-1} x$ is differentiable on the interval $(-1, 1)$. The differentiability of the remaining inverse trigonometric functions can be deduced similarly.

REMARK.   Observe that $\sin^{-1} x$ is only differentiable on the interval $(-1, 1)$, even though its domain is $[-1, 1]$. However, it can be seen geometrically that $\sin^{-1}$ cannot be differentiable at $x = \pm 1$. Just observe that the graph of $y = \sin x$ has horizontal tangent lines at $(\pi/2, 1)$ and $(-\pi/2, -1)$ and that these become points of vertical tangency for $y = \sin^{-1} x$ when reflected around the line $y = x$.

Example 3

Find dy/dx if

(a) $y = \sin^{-1}(x^3)$ (b) $y = \sec^{-1}(e^x)$

Solution (a). From (14)

$$\frac{dy}{dx} = \frac{1}{\sqrt{1 - (x^3)^2}}(3x^2) = \frac{3x^2}{\sqrt{1 - x^6}}$$

Solution (b). From (19)

$$\frac{dy}{dx} = \frac{1}{e^x\sqrt{(e^x)^2 - 1}}(e^x) = \frac{1}{\sqrt{e^{2x} - 1}}$$

◀

EXERCISE SET 4.5   ⬚ Graphing Calculator

1. Find the exact value of
   (a) $\sin^{-1}(-1)$
   (b) $\cos^{-1}(-1)$
   (c) $\tan^{-1}(-1)$
   (d) $\sec^{-1}(1)$.

2. Find the exact value of
   (a) $\sin^{-1}\left(\frac{1}{2}\sqrt{3}\right)$
   (b) $\cos^{-1}\left(\frac{1}{2}\right)$
   (c) $\tan^{-1}(1)$
   (d) $\sec^{-1}(-2)$.

3. Given that $\theta = \sin^{-1}\left(-\frac{1}{2}\sqrt{3}\right)$, find the exact values of $\cos\theta$, $\tan\theta$, $\cot\theta$, $\sec\theta$, and $\csc\theta$.

4. Given that $\theta = \cos^{-1}\left(\frac{1}{2}\right)$, find the exact values of $\sin\theta$, $\tan\theta$, $\cot\theta$, $\sec\theta$, and $\csc\theta$.

5. Given that $\theta = \tan^{-1}\left(\frac{4}{3}\right)$, find the exact values of $\sin\theta$, $\cos\theta$, $\cot\theta$, $\sec\theta$, and $\csc\theta$.

6. Make a table that lists the six inverse trigonometric functions together with their domains and ranges.

7. Find the exact value of
   (a) $\sin^{-1}(\sin\pi/7)$
   (b) $\sin^{-1}(\sin\pi)$
   (c) $\sin^{-1}(\sin 5\pi/7)$
   (d) $\sin^{-1}(\sin 630)$.

8. Find the exact value of
   (a) $\cos^{-1}(\cos\pi/7)$
   (b) $\cos^{-1}(\cos\pi)$
   (c) $\cos^{-1}(\cos 12\pi/7)$
   (d) $\cos^{-1}(\cos 200)$.

9. For which values of x is it true that
   (a) $\cos^{-1}(\cos x) = x$
   (b) $\cos(\cos^{-1} x) = x$
   (c) $\tan^{-1}(\tan x) = x$
   (d) $\tan(\tan^{-1} x) = x$

In Exercises 10 and 11, find the exact value of the given quantity.

10. $\sec\left[\sin^{-1}\left(-\frac{3}{4}\right)\right]$ 11. $\sin\left[2\cos^{-1}\left(\frac{3}{5}\right)\right]$

In Exercises 12 and 13, complete the identities using the triangle method (Figure 4.5.4).

12. (a) $\sin(\cos^{-1} x) =\,?$ (b) $\tan(\cos^{-1} x) =\,?$
    (c) $\csc(\tan^{-1} x) =\,?$ (d) $\sin(\tan^{-1} x) =\,?$

13. (a) $\cos(\tan^{-1} x) =\,?$ (b) $\tan(\cot^{-1} x) =\,?$
    (c) $\sin(\sec^{-1} x) =\,?$ (d) $\cot(\csc^{-1} x) =\,?$

⬚ 14. (a) Use a calculating utility set to radian measure to make tables of values of $y = \sin^{-1} x$ and $y = \cos^{-1} x$ for $x = -1, -0.8, -0.6, \ldots, 0, 0.2, \ldots, 1$. Round your answers to two decimal places.
    (b) Plot the points obtained in part (a), and use the points to sketch the graphs of $y = \sin^{-1} x$ and $y = \cos^{-1} x$. Confirm that your sketches agree with those in Figure 4.5.1.
    (c) Use your graphing utility to graph $y = \sin^{-1} x$ and $y = \cos^{-1} x$; confirm that the graphs agree with those in Figure 4.5.1.

The function $\cot^{-1} x$ is defined to be the inverse of the restricted cotangent function

$$\cot x, \quad 0 < x < \pi$$

and the function $\csc^{-1} x$ is defined to be the inverse of the restricted cosecant function

$$\csc x, \quad -\pi/2 < x < \pi/2, \quad x \neq 0$$

Use these definitions in Exercises 15 and 16 and in all subsequent exercises that involve these functions.

15. (a) Sketch the graphs of $\cot^{-1} x$ and $\csc^{-1} x$.
    (b) Find the domain and range of $\cot^{-1} x$ and $\csc^{-1} x$.

16. Show that

(a) $\cot^{-1} x = \tan^{-1} \dfrac{1}{x}$, if $x > 0$

(b) $\sec^{-1} x = \cos^{-1} \dfrac{1}{x}$, if $|x| \geq 1$

(c) $\csc^{-1} x = \sin^{-1} \dfrac{1}{x}$, if $|x| \geq 1$.

17. Most scientific calculators have keys for the values of only $\sin^{-1} x$, $\cos^{-1} x$, and $\tan^{-1} x$. The formulas in Exercise 16 show how a calculator can be used to obtain values of $\cot^{-1} x$, $\sec^{-1} x$, and $\csc^{-1} x$ for positive values of x. Use these formulas and a calculator to find numerical values for each of the following inverse trigonometric functions. Express your answers in degrees, rounded to the nearest tenth of a degree.

(a) $\cot^{-1} 0.7$ (b) $\sec^{-1} 1.2$ (c) $\csc^{-1} 2.3$

In Exercises 18–20, use a calculating utility to approximate the solution of the equation. Where radians are used, express your answer to four decimal places, and where degrees are used, express it to the nearest tenth of a degree. [*Note:* In each part, the solution is not in the range of the relevant inverse trigonometric function.]

18. (a) $\sin x = 0.37$, $\pi/2 < x < \pi$
(b) $\sin \theta = -0.61$, $180° < \theta < 270°$

19. (a) $\cos x = -0.85$, $\pi < x < 3\pi/2$
(b) $\cos \theta = 0.23$, $-90° < \theta < 0°$

20. (a) $\tan x = 3.16$, $-\pi < x < -\pi/2$
(b) $\tan \theta = -0.45$, $90° < \theta < 180°$

In Exercises 21–28, find dy/dx.

21. (a) $y = \sin^{-1}\left(\frac{1}{3}x\right)$ (b) $y = \cos^{-1}(2x + 1)$

22. (a) $y = \tan^{-1}(x^2)$ (b) $y = \cot^{-1}(\sqrt{x})$

23. (a) $y = \sec^{-1}(x^7)$ (b) $y = \csc^{-1}(e^x)$

24. (a) $y = (\tan x)^{-1}$ (b) $y = \dfrac{1}{\tan^{-1} x}$

25. (a) $y = \sin^{-1}(1/x)$ (b) $y = \cos^{-1}(\cos x)$

26. (a) $y = \ln(\cos^{-1} x)$ (b) $y = \sqrt{\cot^{-1} x}$

27. (a) $y = e^x \sec^{-1} x$ (b) $y = x^2 \left(\sin^{-1} x\right)^3$

28. (a) $y = \sin^{-1} x + \cos^{-1} x$ (b) $y = \sec^{-1} x + \csc^{-1} x$

In Exercises 29 and 30, find dy/dx by implicit differentiation.

29. $x^3 + x \tan^{-1} y = e^y$

30. $\sin^{-1}(xy) = \cos^{-1}(x - y)$

31. (a) Referring to the graph of $y = \sin^{-1} x$ in Figure 4.5.1, make a rough sketch of the graph of dy/dx.
(b) Check your work in part (a) using a graphing utility to generate the graph of dy/dx.

32. (a) Referring to the graph of $y = \tan^{-1} x$ in Figure 4.5.1, make a rough sketch of the graph of dy/dx.
(b) Check your work in part (a) using a graphing utility to generate the graph of dy/dx.

33. (a) Make a conjecture about the shape of the graph of

$$y = \cos^{-1}(\cos x)$$

and sketch the graph for $-4\pi \leq x \leq 4\pi$.
(b) Check your work in part (a) using a graphing utility to generate the graph.

34. (a) Use a calculating utility to evaluate $\sin^{-1}(\sin^{-1} 0.25)$ and $\sin^{-1}(\sin^{-1} 0.9)$, and explain what you think is happening in the second calculation.
(b) For what values of x in the interval $-1 \leq x \leq 1$ will your calculating utility produce a real value for the function $\sin^{-1}(\sin^{-1} x)$?

35. In each part, sketch the graph and check your work with a graphing utility.
(a) $y = \sin^{-1} 2x$ (b) $y = \tan^{-1} \frac{1}{2}x$

36. In each part, express x in terms of k and an appropriate inverse trigonometric function. [*Note:* x may not be in the range of the inverse trigonometric function.]
(a) $\cos x = k$, if $0 < k < 1$ and $3\pi/2 < x < 2\pi$
(b) $\tan x = k$, if $k < 0$ and $\pi/2 < x < \pi$
(c) $\sin 2x = k$, if $0 < k < 1$ and $0 < x < \pi/2$.
[*Hint:* Consider the following cases: $0 < 2x < \pi/2$ and $\pi/2 < 2x < \pi$.]

37. An Earth-observing satellite has horizon sensors that can measure the angle θ shown in the accompanying figure. Let R be the radius of the Earth (assumed spherical) and h the distance between the satellite and the Earth's surface.
(a) Show that $\sin \theta = \dfrac{R}{R + h}$.
(b) Find θ, to the nearest degree, for a satellite that is 10,000 km from the Earth's surface (use $R = 6378$ km).

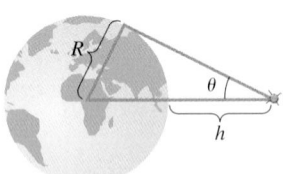

Earth

Figure Ex-37

38. The number of hours of daylight on a given day at a given point on the Earth's surface depends on the latitude λ of the point, the angle γ through which the Earth has moved in its orbital plane during the time period from the vernal equinox (March 21), and the angle of inclination ϕ of the Earth's axis of rotation measured from ecliptic north ($\phi \approx 23.55°$). The number of hours of daylight h can be approximated by the formula

$$h = \begin{cases} 24, & D \geq 1 \\ 12 + \frac{2}{15}\sin^{-1}D, & |D| < 1 \\ 0, & D \leq -1 \end{cases}$$

where

$$D = \frac{\sin\phi\,\sin\gamma\,\tan\lambda}{\sqrt{1 - \sin^2\phi\,\sin^2\gamma}}$$

and $\sin^{-1}D$ is in degree measure. Given that Fairbanks, Alaska, is located at a latitude of $\lambda = 65^\circ$ N and also that $\gamma = 90^\circ$ on June 20 and $\gamma = 270^\circ$ on December 20, approximate

(a) the maximum number of daylight hours at Fairbanks to one decimal place
(b) the minimum number of daylight hours at Fairbanks to one decimal place.

[*Note:* This problem was adapted from *TEAM, A Path to Applied Mathematics*, The Mathematical Association of America, Washington, D.C., 1985.]

39. A soccer player kicks a ball with an initial speed of 14 m/s at an angle θ with the horizontal (see the accompanying figure). The ball lands 18 m down the field. If air resistance is neglected, then the ball will have a parabolic trajectory and the horizontal range R will be given by

$$R = \frac{v^2}{g}\sin 2\theta$$

where v is the initial speed of the ball and g is the acceleration due to gravity. Using $g = 9.8$ m/s^2, approximate two values of θ, to the nearest degree, at which the ball could have been kicked. Which angle results in the shorter time of flight? Why?

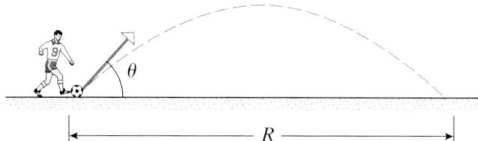

Figure Ex-39

40. The *law of cosines* states that

$$c^2 = a^2 + b^2 - 2ab\cos\theta$$

where a, b, and c are the lengths of the sides of a triangle and θ is the angle formed by sides a and b. Find θ, to the nearest degree, for the triangle with $a = 2$, $b = 3$, and $c = 4$.

41. An airplane is flying at a constant height of 3000 ft above water at a speed of 400 ft/s. The pilot is to release a survival package so that it lands in the water at a sighted point P. If air resistance is neglected, then the package will follow a parabolic trajectory whose equation relative to the coordinate system in the accompanying figure is

$$y = 3000 - \frac{g}{2v^2}x^2$$

where g is the acceleration due to gravity and v is the speed

of the airplane. Using $g = 32$ ft/s^2, find the "line of sight" angle θ, to the nearest degree, that will result in the package hitting the target point.

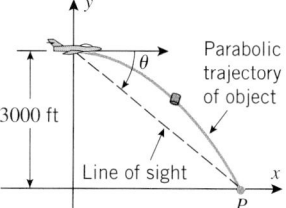

Figure Ex-41

42. A camera is positioned x feet from the base of a missile launching pad (see the accompanying figure). If a missile of length a feet is launched vertically, show that when the base of the missile is b feet above the camera lens, the angle θ subtended at the lens by the missile is

$$\theta = \cot^{-1}\frac{x}{a+b} - \cot^{-1}\frac{x}{b}$$

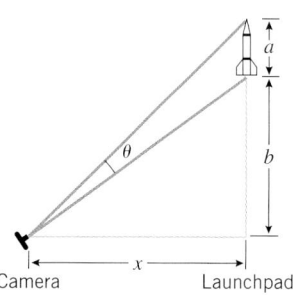

Figure Ex-42

43. Prove:
(a) $\sin^{-1}(-x) = -\sin^{-1}x$
(b) $\tan^{-1}(-x) = -\tan^{-1}x$.

44. Prove:
(a) $\cos^{-1}(-x) = \pi - \cos^{-1}x$
(b) $\sec^{-1}(-x) = \pi - \sec^{-1}x$, if $|x| \geq 1$.

45. Prove:
(a) $\sin^{-1}x = \tan^{-1}\dfrac{x}{\sqrt{1-x^2}}$
(b) $\cos^{-1}x = \dfrac{\pi}{2} - \tan^{-1}\dfrac{x}{\sqrt{1-x^2}}$.

46. Prove:

$$\tan^{-1}x + \tan^{-1}y = \tan^{-1}\left(\frac{x+y}{1-xy}\right)$$

provided $-\pi/2 < \tan^{-1}x + \tan^{-1}y < \pi/2$. [*Hint:* Use an identity for $\tan(\alpha + \beta)$.]

47. Use the result in Exercise 46 to show that
(a) $\tan^{-1}\frac{1}{2} + \tan^{-1}\frac{1}{3} = \pi/4$
(b) $2\tan^{-1}\frac{1}{3} + \tan^{-1}\frac{1}{7} = \pi/4$.

48. Use identities (4) and (7) to obtain identity (10b).

4.6 RELATED RATES

In this section we will study related rates problems. In such problems one tries to find the rate at which some quantity is changing by relating it to other quantities whose rates of change are known.

RATES OF CHANGE USING THE CHAIN RULE

Figure 4.6.1 shows a liquid draining through a conical filter. As the liquid drains, its volume V, height h, and radius r are functions of the elapsed time t, and at each instant these variables are related by the equation

$$V = \frac{\pi}{3} r^2 h$$

If we differentiate both sides of this equation implicitly with respect to t, then we obtain

$$\frac{dV}{dt} = \frac{\pi}{3}\left[r^2 \frac{dh}{dt} + h\left(2r \frac{dr}{dt}\right)\right] = \frac{\pi}{3}\left(r^2 \frac{dh}{dt} + 2rh \frac{dr}{dt}\right)$$

Thus, if the values of $r, h, dh/dt$, and dr/dt are known, then this equation can be used to find dV/dt. Here are some specific examples that use this basic idea.

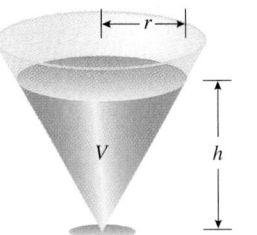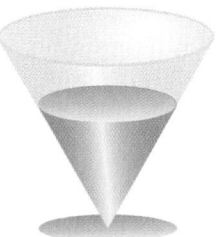

Figure 4.6.1

Example 1

Assume that oil spilled from a ruptured tanker spreads in a circular pattern whose radius increases at a constant rate of 2 ft/s. How fast is the area of the spill increasing when the radius of the spill is 60 ft?

Solution. Let

t = number of seconds elapsed from the time of the spill
r = radius of the spill in feet after t seconds
A = area of the spill in square feet after t seconds

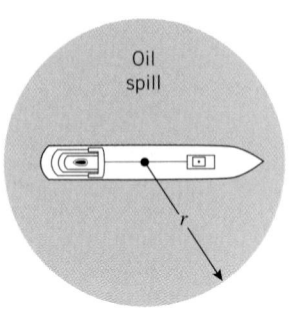

Figure 4.6.2

(Figure 4.6.2). We know the rate at which the radius is increasing, and we want to find the rate at which the area is increasing at the instant when $r = 60$; that is, we want to find

$$\left.\frac{dA}{dt}\right|_{r=60} \qquad \text{given that} \qquad \frac{dr}{dt} = 2 \text{ ft/s}$$

From the formula for the area of a circle we obtain

$$A = \pi r^2 \qquad\qquad (1)$$

Because A and r are functions of t, we can differentiate both sides of (1) implicitly with respect to t to obtain

$$\frac{dA}{dt} = 2\pi r \frac{dr}{dt}$$

Thus, when $r = 60$ the area of the spill is increasing at the rate of

$$\frac{dA}{dt}\bigg|_{r=60} = 2\pi(60)(2) = 240\pi \text{ ft}^2/\text{s}$$

or approximately 754 ft²/s. ◀

With only minor variations, the method used in Example 1 can be used to solve a variety of related rates problems. The method consists of five steps:

A Strategy for Solving Related Rates Problems

Step 1. Draw a figure and label the quantities that vary.

Step 2. Identify the rates of change that are known and the rate of change that is to be found.

Step 3. Find an equation that relates the quantity whose rate of change is to be found to the quantities whose rates of change are known.

Step 4. Differentiate both sides of this equation with respect to time and solve for the derivative that will give the unknown rate of change.

Step 5. Evaluate this derivative at the appropriate point.

Example 2

A baseball diamond is a square whose sides are 90 ft long (Figure 4.6.3). Suppose that a player running from second base to third base has a speed of 30 ft/s at the instant when he is 20 ft from third base. At what rate is the player's distance from home plate changing at that instant?

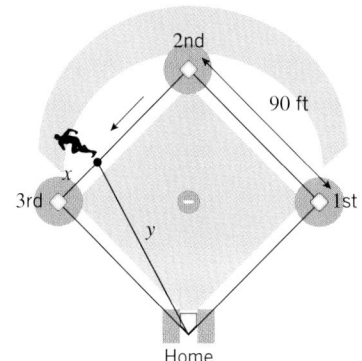

2nd
90 ft
3rd
1st
Home

Figure 4.6.3

Solution. Let

t = number of seconds after the player leaves second base
x = distance in feet from third base
y = distance in feet from home plate

(Figure 4.6.3). The rate at which the distance from third base is changing is dx/dt, and the rate at which the distance from home plate is changing is dy/dt. We want to find

$$\frac{dy}{dt}\bigg|_{x=20} \quad \text{given that} \quad \frac{dx}{dt}\bigg|_{x=20} = -30 \text{ ft/s}$$

(Note that dx/dt is negative because x is decreasing with respect to t.) From the Theorem of Pythagoras we have

$$x^2 + 90^2 = y^2 \tag{2}$$

Differentiating both sides of this equation with respect to t using the chain rule yields

$$2x\frac{dx}{dt} = 2y\frac{dy}{dt} \quad \text{or} \quad \frac{dy}{dt} = \frac{x}{y}\frac{dx}{dt} \tag{3}$$

When $x = 20$, it follows from (2) that

$$y = \sqrt{20^2 + 90^2} = \sqrt{8500} = 10\sqrt{85}$$

so that (3) yields

$$\frac{dy}{dt}\bigg|_{x=20} = \frac{20}{10\sqrt{85}}(-30) = -\frac{60}{\sqrt{85}} \approx -6.51 \text{ ft/s}$$

The negative sign in the answer tells us that y is decreasing, which makes sense physically from Figure 4.6.3. ◀

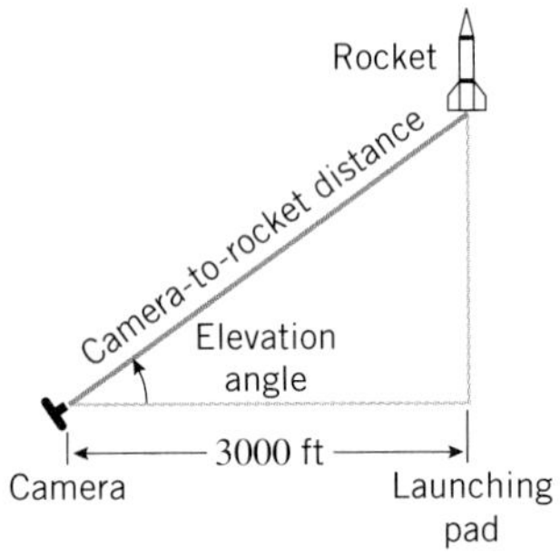

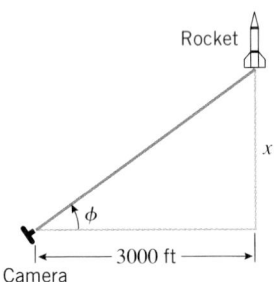

Rocket

Camera-to-rocket distance

Elevation angle

3000 ft

Camera   Launching pad

Figure 4.6.4

In Figure 4.6.4 we have shown a camera mounted at a point 3000 ft from the base of a rocket launching pad. Let us assume that the rocket rises vertically and the camera is to take a series of photographs of the rocket. Because the rocket will be rising, the elevation angle of the camera will have to vary at just the right rate to keep the rocket in sight. Moreover, because the camera-to-rocket distance will be changing constantly, the camera focusing mechanism will also have to vary at just the right rate to keep the picture sharp. The focusing problem is considered in the exercises, and the elevation problem is addressed in the following example:

Example 3

If the rocket shown in Figure 4.6.4 is rising vertically at 880 ft/s when it is 4000 ft up, how fast must the camera elevation angle change at that instant to keep the rocket in sight?

Solution. Let

t = number of seconds elapsed from the time of launch
ϕ = camera elevation angle in radians after t seconds
x = height of the rocket in feet after t seconds

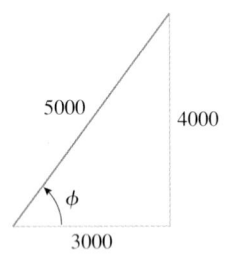

Rocket

x

ϕ

3000 ft

Camera

Figure 4.6.5

(Figure 4.6.5). At each instant the rate at which the camera elevation angle must change is $d\phi/dt$, and the rate at which the rocket is rising is dx/dt. We want to find

$$\left.\frac{d\phi}{dt}\right|_{x=4000} \quad \text{given that} \quad \left.\frac{dx}{dt}\right|_{x=4000} = 880 \text{ ft/s}$$

From Figure 4.6.5 we see that

$$\tan\phi = \frac{x}{3000} \tag{4}$$

Because ϕ and x are functions of t, we can differentiate both sides of (4) with respect to t to obtain

$$(\sec^2\phi)\frac{d\phi}{dt} = \frac{1}{3000}\frac{dx}{dt} \quad \text{or} \quad \frac{d\phi}{dt} = \frac{1}{3000\sec^2\phi}\frac{dx}{dt} \tag{5}$$

When $x = 4000$, it follows that

$$\sec\phi = \frac{5000}{3000} = \frac{5}{3}$$

5000   4000

ϕ

3000

Figure 4.6.6

(Figure 4.6.6), so that from (5)

$$\left.\frac{d\phi}{dt}\right|_{x=4000} = \frac{1}{3000\left(\frac{5}{3}\right)^2}\cdot 880 = \frac{66}{625} \approx 0.11 \text{ radian/s} \approx 6.05 \text{ degrees/s}$$

Alternative Solution. Instead of differentiating both sides of (4), we could have first solved the equation for ϕ and then differentiated:

$$\phi = \tan^{-1}\left(\frac{x}{3000}\right)$$

so

$$\frac{d\phi}{dt} = \frac{1}{1+\left(\frac{x}{3000}\right)^2}\cdot\frac{1}{3000}\frac{dx}{dt}$$

Thus,

$$\left.\frac{d\phi}{dt}\right|_{x=4000} = \frac{1}{1+\left(\frac{4000}{3000}\right)^2}\cdot\frac{880}{3000} = \frac{66}{625} \approx 0.11 \text{ radian/s} \approx 6.05 \text{ degrees/s}$$

which agrees with our previous result. ◀

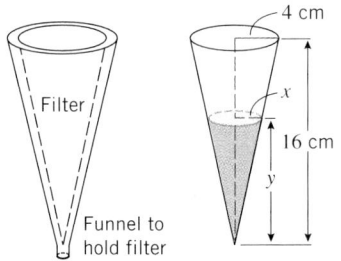

Figure 4.6.7

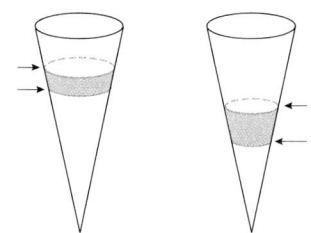

The same volume has drained, but the change in height is greater near the bottom than near the top.

Figure 4.6.8

Example 4

Suppose that liquid is to be cleared of sediment by pouring it through a conical filter that is 16 cm high and has a radius of 4 cm at the top (Figure 4.6.7). Suppose also that the liquid flows out of the cone at a constant rate of 2 cm^3/min.

(a)   Do you think that the depth of the liquid will decrease at a constant rate? Give a verbal argument that justifies your conclusion.

(b)   Find a formula that expresses the rate of change to the depth of the liquid in terms of the depth, and use that formula to determine whether your conclusion in part (a) is correct.

(c)   At what rate is the depth of the liquid changing at the instant when the level is 8 cm deep?

Solution (a). For the volume of liquid to decrease by a *fixed amount*, it requires a greater decrease in depth when the cone is near empty than when it is near full (Figure 4.6.8). This suggests that for the volume to decrease at a constant rate, the depth must decrease at an increasing rate.

Solution (b). Let

$$t = \text{time elapsed from the initial observation (min)}$$
$$V = \text{volume of liquid in the cone at time } t \text{ (cm}^3)$$
$$y = \text{depth of the liquid in the cone at time } t \text{ (cm)}$$
$$x = \text{radius of the liquid surface at time } t \text{ (cm)}$$

(Figure 4.6.7). At each instant the rate at which the volume of liquid is changing is dV/dt, and the rate at which the depth is changing is dy/dt. We want to express dy/dt in terms of y given that dV/dt has a constant value of $dV/dt = -2$. (We must use a minus sign here because V *decreases* as t increases.)

From the formula for the volume of a cone, the volume V, the radius x, and the depth y are related by

$$V = \tfrac{1}{3}\pi x^2 y \tag{6}$$

If we differentiate both sides of (6) with respect to t, the right side will involve the quantity dx/dt. Since we have no direct information about dx/dt, it is desirable to eliminate x from (6) before differentiating. This can be done using similar triangles. From Figure 4.6.7 we see that

$$\frac{x}{y} = \frac{4}{16} \quad \text{or} \quad x = \frac{1}{4}y.$$

Substituting this expression in (6) gives

$$V = \frac{\pi}{48}y^3 \tag{7}$$

Differentiating both sides of (7) with respect to t we obtain

$$\frac{dV}{dt} = \frac{\pi}{48}\left(3y^2\frac{dy}{dt}\right)$$

or

$$\frac{dy}{dt} = \frac{16}{\pi y^2}\frac{dV}{dt} = \frac{16}{\pi y^2}(-2) = -\frac{32}{\pi y^2} \tag{8}$$

which expresses dy/dt in terms of y. The minus sign tells us that y is decreasing with time, and

$$\left|\frac{dy}{dt}\right| = \frac{32}{\pi y^2}$$

tells us how fast y is decreasing. From this formula we see that $|dy/dt|$ increases as y decreases, which confirms our conjecture in part (a) that the depth of the liquid decreases at an increasing rate as the liquid drains through the filter.

Solution (c). The rate at which the depth is changing when the depth is 8 cm can be obtained from (8) with $y = 8$:

$$\left.\frac{dy}{dt}\right|_{y=8} = -\frac{32}{\pi(8^2)} = -\frac{1}{2\pi} \approx -0.16 \text{ cm/min} \qquad \blacktriangleleft$$

EXERCISE SET 4.6

1. Let A be the area of a square whose sides have length x, and assume that x varies with the time t.
   (a) Draw a picture of the square with the labels A and x placed appropriately.
   (b) Write an equation that relates A and x.
   (c) Use the equation in part (b) to find an equation that relates dA/dt and dx/dt.
   (d) At a certain instant the sides are 3 ft long and increasing at a rate of 2 ft/min. How fast is the area increasing at that instant?

2. Let A be the area of a circle of radius r, and assume that r increases with the time t.
   (a) Draw a picture of the circle with the labels A and r placed appropriately.
   (b) Write an equation that relates A and r.
   (c) Use the equation in part (b) to find an equation that relates dA/dt and dr/dt.
   (d) At a certain instant the radius is 5 cm and increasing at the rate of 2 cm/s. How fast is the area increasing at that instant?

3. Let V be the volume of a cylinder having height h and radius r, and assume that h and r vary with time.
   (a) How are dV/dt, dh/dt, and dr/dt related?
   (b) At a certain instant, the height is 6 in and increasing at 1 in/s, while the radius is 10 in and decreasing at 1 in/s. How fast is the volume changing at that instant? Is the volume increasing or decreasing at that instant?

4. Let l be the length of a diagonal of a rectangle whose sides have lengths x and y, and assume that x and y vary with time.
   (a) How are dl/dt, dx/dt, and dy/dt related?
   (b) If x increases at a constant rate of $\frac{1}{2}$ ft/s and y decreases at a constant rate of $\frac{1}{4}$ ft/s, how fast is the size of the diagonal changing when $x = 3$ ft and $y = 4$ ft? Is the diagonal increasing or decreasing at that instant?

5. Let θ (in radians) be an acute angle in a right triangle, and let x and y, respectively, be the lengths of the sides adjacent and opposite θ. Suppose also that x and y vary with time.
   (a) How are $d\theta/dt$, dx/dt, and dy/dt related?

(b) At a certain instant, $x = 2$ units and is increasing at 1 unit/s, while $y = 2$ units and is decreasing at $\frac{1}{4}$ unit/s. How fast is θ changing at that instant? Is θ increasing or decreasing at that instant?

6. Suppose that $z = x^3 y^2$, where both x and y are changing with time. At a certain instant when $x = 1$ and $y = 2$, x is decreasing at the rate of 2 units/s, and y is increasing at the rate of 3 units/s. How fast is z changing at this instant? Is z increasing or decreasing?

7. The minute hand of a certain clock is 4 in long. Starting from the moment when the hand is pointing straight up, how fast is the area of the sector that is swept out by the hand increasing at any instant during the next revolution of the hand?

8. A stone dropped into a still pond sends out a circular ripple whose radius increases at a constant rate of 3 ft/s. How rapidly is the area enclosed by the ripple increasing at the end of 10 s?

9. Oil spilled from a ruptured tanker spreads in a circle whose area increases at a constant rate of 6 mi²/h. How fast is the radius of the spill increasing when the area is 9 mi²?

10. A spherical balloon is inflated so that its volume is increasing at the rate of 3 ft³/min. How fast is the diameter of the balloon increasing when the radius is 1 ft?

11. A spherical balloon is to be deflated so that its radius decreases at a constant rate of 15 cm/min. At what rate must air be removed when the radius is 9 cm?

12. A 17-ft ladder is leaning against a wall. If the bottom of the ladder is pulled along the ground away from the wall at a constant rate of 5 ft/s, how fast will the top of the ladder be moving down the wall when it is 8 ft above the ground?

13. A 13-ft ladder is leaning against a wall. If the top of the ladder slips down the wall at a rate of 2 ft/s, how fast will the foot be moving away from the wall when the top is 5 ft above the ground?

14. A 10-ft plank is leaning against a wall. If at a certain instant the bottom of the plank is 2 ft from the wall and is being

pushed toward the wall at the rate of 6 in/s, how fast is the acute angle that the plank makes with the ground increasing?

15. A softball diamond is a square whose sides are 60 ft long. Suppose that a player running from first to second base has a speed of 25 ft/s at the instant when she is 10 ft from second base. At what rate is the player's distance from home plate changing at that instant?

16. A rocket, rising vertically, is tracked by a radar station that is on the ground 5 mi from the launchpad. How fast is the rocket rising when it is 4 mi high and its distance from the radar station is increasing at a rate of 2000 mi/h?

17. For the camera and rocket shown in Figure 4.6.4, at what rate is the camera-to-rocket distance changing when the rocket is 4000 ft up and rising vertically at 880 ft/s?

18. For the camera and rocket shown in Figure 4.6.4, at what rate is the rocket rising when the elevation angle is $\pi/4$ radians and increasing at a rate of 0.2 radian/s?

19. A satellite is in an elliptical orbit around the Earth. Its distance r (in miles) from the center of the Earth is given by

$$r = \frac{4995}{1 + 0.12\cos\theta}$$

where θ is the angle measured from the point on the orbit nearest the Earth's surface (see the accompanying figure).
    (a) Find the altitude of the satellite at **perigee** (the point nearest the surface of the Earth) and at **apogee** (the point farthest from the surface of the Earth). Use 3960 mi as the radius of the Earth.
    (b) At the instant when θ is $120°$, the angle θ is increasing at the rate of $2.7°$/min. Find the altitude of the satellite and the rate at which the altitude is changing at this instant. Express the rate in units of mi/min.

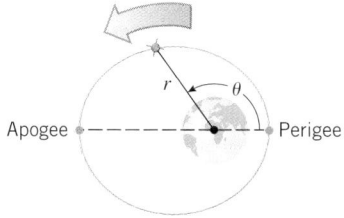

Figure Ex-19

20. An aircraft is flying horizontally at a constant height of 4000 ft above a fixed observation point (see the accompanying figure). At a certain instant the angle of elevation θ is $30°$ and decreasing, and the speed of the aircraft is 300 mi/h.
    (a) How fast is θ decreasing at this instant? Express the result in units of degrees/s.
    (b) How fast is the distance between the aircraft and the observation point changing at this instant? Express the result in units of ft/s. Use 1 mi = 5280 ft.

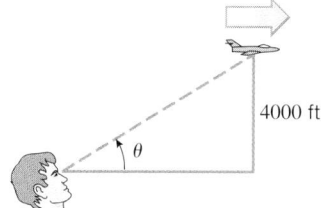

4000 ft

θ

Figure Ex-20

21. A conical water tank with vertex down has a radius of 10 ft at the top and is 24 ft high. If water flows into the tank at a rate of 20 ft^3/min, how fast is the depth of the water increasing when the water is 16 ft deep?

22. Grain pouring from a chute at the rate of 8 ft^3/min forms a conical pile whose altitude is always twice its radius. How fast is the altitude of the pile increasing at the instant when the pile is 6 ft high?

23. Sand pouring from a chute forms a conical pile whose height is always equal to the diameter. If the height increases at a constant rate of 5 ft/min, at what rate is sand pouring from the chute when the pile is 10 ft high?

24. Wheat is poured through a chute at the rate of 10 ft^3/min, and falls in a conical pile whose bottom radius is always half the altitude. How fast will the circumference of the base be increasing when the pile is 8 ft high?

25. An aircraft is climbing at a $30°$ angle to the horizontal. How fast is the aircraft gaining altitude if its speed is 500 mi/h?

26. A boat is pulled into a dock by means of a rope attached to a pulley on the dock (see the accompanying figure). The rope is attached to the bow of the boat at a point 10 ft below the pulley. If the rope is pulled through the pulley at a rate of 20 ft/min, at what rate will the boat be approaching the dock when 125 ft of rope is out?

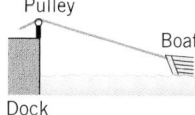

Pulley

Boat

Dock

Figure Ex-26

27. For the boat in Exercise 26, how fast must the rope be pulled if we want the boat to approach the dock at a rate of 12 ft/min at the instant when 125 ft of rope is out?

28. A man 6 ft tall is walking at the rate of 3 ft/s toward a streetlight 18 ft high (see the accompanying figure).
    (a) At what rate is his shadow length changing?
    (b) How fast is the tip of his shadow moving?

Figure Ex-28

29. A beacon that makes one revolution every 10 s is located on a ship anchored 4 kilometers from a straight shoreline. How fast is the beam moving along the shoreline when it makes an angle of 45° with the shore?

30. An aircraft is flying at a constant altitude with a constant speed of 600 mi/h. An antiaircraft missile is fired on a straight line perpendicular to the flight path of the aircraft so that it will hit the aircraft at a point P (see the accompanying figure). At the instant the aircraft is 2 mi from the impact point P the missile is 4 mi from P and flying at 1200 mi/h. At that instant, how rapidly is the distance between missile and aircraft decreasing?

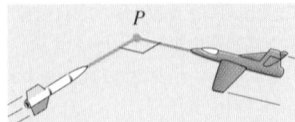

Figure Ex-30

31. Solve Exercise 30 under the assumption that the angle between the flight paths is 120° instead of the assumption that the paths are perpendicular. [*Hint:* Use the law of cosines.]

32. A police helicopter is flying due north at 100 mi/h and at a constant altitude of $\frac{1}{2}$ mi. Below, a car is traveling west on a highway at 75 mi/h. At the moment the helicopter crosses over the highway the car is 2 mi east of the helicopter.
   (a) How fast is the distance between the car and helicopter changing at the moment the helicopter crosses the highway?
   (b) Is the distance between the car and helicopter increasing or decreasing at that moment?

33. A particle is moving along the curve whose equation is

$$\frac{xy^3}{1+y^2} = \frac{8}{5}$$

Assume that the x-coordinate is increasing at the rate of 6 units/s when the particle is at the point $(1, 2)$.
   (a) At what rate is the y-coordinate of the point changing at that instant?
   (b) Is the particle rising or falling at that instant?

34. A point P is moving along the curve whose equation is $y = \sqrt{x^3 + 17}$. When P is at $(2, 5)$, y is increasing at the rate of 2 units/s. How fast is x changing?

35. A point P is moving along the line whose equation is $y = 2x$. How fast is the distance between P and the point $(3, 0)$ changing at the instant when P is at $(3, 6)$ if x is decreasing at the rate of 2 units/s at that instant?

36. A point P is moving along the curve whose equation is $y = \sqrt{x}$. Suppose that x is increasing at the rate of 4 units/s when $x = 3$.
   (a) How fast is the distance between P and the point $(2, 0)$ changing at this instant?
   (b) How fast is the angle of inclination of the line segment from P to $(2, 0)$ changing at this instant?

37. A particle is moving along the curve $y = x \ln x$. Find all values of x at which the rate of change of y with respect to time is three times that of x. [Assume that dx/dt is never zero.]

38. A particle is moving along the curve $16x^2 + 9y^2 = 144$. Find all points (x, y) at which the rates of change of x and y with respect to time are equal. [Assume that dx/dt and dy/dt are never both zero at the same point.]

39. The **thin lens equation** in physics is

$$\frac{1}{s} + \frac{1}{S} = \frac{1}{f}$$

where s is the object distance from the lens, S is the image distance from the lens, and f is the focal length of the lens. Suppose that a certain lens has a focal length of 6 cm and that an object is moving toward the lens at the rate of 2 cm/s. How fast is the image distance changing at the instant when the object is 10 cm from the lens? Is the image moving away from the lens or toward the lens?

40. Water is stored in a cone-shaped reservoir (vertex down). Assuming the water evaporates at a rate proportional to the surface area exposed to the air, show that the depth of the water will decrease at a constant rate that does not depend on the dimensions of the reservoir.

41. A meteorite enters the Earth's atmosphere and burns up at a rate that, at each instant, is proportional to its surface area. Assuming that the meteorite is always spherical, show that the radius decreases at a constant rate.

42. On a certain clock the minute hand is 4 in long and the hour hand is 3 in long. How fast is the distance between the tips of the hands changing at 9 o'clock?

43. Coffee is poured at a uniform rate of 20 cm³/s into a cup whose inside is shaped like a truncated cone (see the accompanying figure). If the upper and lower radii of the cup are 4 cm and 2 cm and the height of the cup is 6 cm, how fast will the coffee level be rising when the coffee is halfway up? [*Hint:* Extend the cup downward to form a cone.]

Figure Ex-43

4.7 L'HÔPITAL'S RULE; INDETERMINATE FORMS

In this section we will discuss a general method for using derivatives to find limits. This method will enable us to establish limits with certainty that earlier in the text we were only able to conjecture using numerical or graphical evidence. The method that we will discuss in this section is an extremely powerful tool that is used internally by many computer programs to calculate limits of various types.

INDETERMINATE FORMS OF TYPE 0/0

In earlier sections we discussed limits that can be determined by inspection or by some appropriate algebraic manipulation. In this section we will be concerned with limits that cannot be obtained by such methods. For example, in Theorem 2.5.3 we were able to show that

$$\lim_{x \to 0} \frac{\sin x}{x} = 1 \tag{1}$$

but it required the Squeezing Theorem (2.5.2) and some tricky manipulation of inequalities. Our goal here is to develop a more straightforward method.

What makes the limit in (1) bothersome is the fact that the numerator and denominator both approach 0 as $x \to 0$. Such limits are called *indeterminate forms of type* **0/0**. In limits of this type there are two tendencies working against each other: as the numerator approaches 0 it tends to drive the ratio toward 0, and as the denominator approaches 0 it tends to drive the ratio toward $+\infty$ or $-\infty$. What happens in (1) is that these conflicting tendencies offset each other in such a way that the limit is 1.

Although the limit in (1) is not self-evident, it can be conjectured from numerical evidence, as in Table 2.1.2. However, it can also be conjectured from the local linear approximation of $\sin x$ at 0. To see this, recall from Formula (5) of Section 3.6 that if a function f is differentiable at a point x_0, then for values of x near x_0, the values of $f(x)$ can be approximated as

$$f(x) \approx f(x_0) + f'(x_0)(x - x_0)$$

where the approximation tends to get better and better as $x \to x_0$. In particular, we showed in Example 3 of Section 3.6 that the local linear approximation of $\sin x$ at $x_0 = 0$ is

$$\sin x \approx x$$

This suggests that the value of $(\sin x)/x$ gets closer and closer to 1 as $x \to 0$, and hence we can reasonably conclude that

$$\lim_{x \to 0} \frac{\sin x}{x} = \lim_{x \to 0} \frac{x}{x} = 1$$

L'HÔPITAL'S RULE

The idea of using local linear approximations to evaluate indeterminate forms of type 0/0 can be used to motivate a more general procedure for finding such limits. For this purpose, suppose that

$$\lim_{x \to x_0} \frac{f(x)}{g(x)}$$

is an indeterminate form of type 0/0, that is,

$$\lim_{x \to x_0} f(x) = 0 \quad \text{and} \quad \lim_{x \to x_0} g(x) = 0 \tag{2}$$

For simplicity, let us also assume that f and g are differentiable at $x = x_0$ and that f' and g' are continuous at $x = x_0$. The differentiability of f and g at $x = x_0$ implies that f and g are continuous at $x = x_0$, and hence from (2)

$$f(x_0) = \lim_{x \to x_0} f(x) = 0 \quad \text{and} \quad g(x_0) = \lim_{x \to x_0} g(x) = 0 \tag{3}$$

Moreover, the continuity of f' and g' at $x = x_0$ implies that

$$\lim_{x \to x_0} f'(x) = f'(x_0) \quad \text{and} \quad \lim_{x \to x_0} g'(x) = g'(x_0) \tag{4}$$

Thus, from (3) and (4) and the local linear approximations of f and g at $x = x_0$, we have

$$\lim_{x \to x_0} \frac{f(x)}{g(x)} = \lim_{x \to x_0} \frac{f(x_0) + f'(x_0)(x - x_0)}{g(x_0) + g'(x_0)(x - x_0)}$$

$$= \lim_{x \to x_0} \frac{f'(x_0)(x - x_0)}{g'(x_0)(x - x_0)} = \frac{f'(x_0)}{g'(x_0)}$$

which from (4) can be expressed as

$$\lim_{x \to x_0} \frac{f(x)}{g(x)} = \lim_{x \to x_0} \frac{f'(x)}{g'(x)} \tag{5}$$

This result, called **L'Hôpital's** [*] **rule**, converts an indeterminate form of type $0/0$ into a new limit involving derivatives that in many situations can be evaluated by inspection or by algebraic methods. For example,

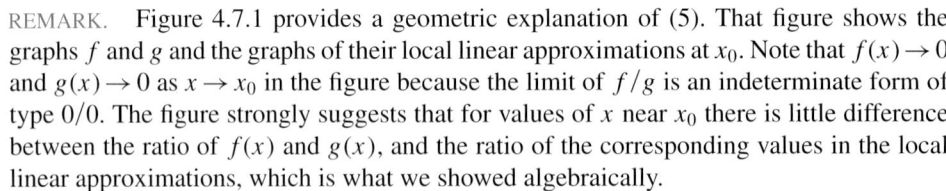

$$\lim_{x \to 0} \frac{1 - \cos x}{x} = \lim_{x \to 0} \frac{\dfrac{d}{dx}[1 - \cos x]}{\dfrac{d}{dx}[x]} = \lim_{x \to 0} \frac{\sin x}{1} = \sin 0 = 0$$

which agrees with the result in Theorem 2.5.3.

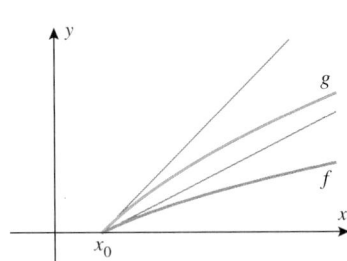

The graphs of f and g together with their local linear approximations at the point x_0

Figure 4.7.1

REMARK.   Figure 4.7.1 provides a geometric explanation of (5). That figure shows the graphs f and g and the graphs of their local linear approximations at x_0. Note that $f(x) \to 0$ and $g(x) \to 0$ as $x \to x_0$ in the figure because the limit of f/g is an indeterminate form of type $0/0$. The figure strongly suggests that for values of x near x_0 there is little difference between the ratio of $f(x)$ and $g(x)$, and the ratio of the corresponding values in the local linear approximations, which is what we showed algebraically.

Although we motivated Formula (5) by assuming that f and g have continuous derivatives at $x = x_0$, the result is true without this assumption. Moreover, the result is also valid for one-sided limits and limits at $+\infty$ and $-\infty$. We omit the formal proof.

4.7.1   THEOREM (*L'Hôpital's Rule for Form* **0/0**).   *Let* lim *stand for one of the limits* $\lim\limits_{x \to a}$, $\lim\limits_{x \to a^+}$, $\lim\limits_{x \to a^-}$, $\lim\limits_{x \to +\infty}$, *or* $\lim\limits_{x \to -\infty}$, *and suppose that* $\lim f(x) = 0$ *and* $\lim g(x) = 0$. *If* $\lim [f'(x)/g'(x)]$ *has a finite value* L, *or if this limit is* $+\infty$ *or* $-\infty$, *then*

$$\lim \frac{f(x)}{g(x)} = \lim \frac{f'(x)}{g'(x)}$$

REMARK.   Note that in L'Hôpital's rule the numerator and denominator are differentiated separately, which is not the same as differentiating $f(x)/g(x)$.

---

[*] GUILLAUME FRANCOIS ANTOINE DE L'HÔPITAL (1661–1704). French mathematician. L'Hôpital, born to parents of the French high nobility, held the title of Marquis de Sainte-Mesme Comte d'Autrement. He showed mathematical talent quite early and at age 15 solved a difficult problem about cycloids posed by Pascal. As a young man he served briefly as a cavalry officer, but resigned because of nearsightedness. In his own time he gained fame as the author of the first textbook ever published on differential calculus, *L'Analyse des Infiniment Petits pour l'Intelligence des Lignes Courbes* (1696). L'Hôpital's rule appeared for the first time in that book. Actually, L'Hôpital's rule and most of the material in the calculus text were due to John Bernoulli, who was L'Hôpital's teacher. L'Hôpital dropped his plans for a book on integral calculus when Leibniz informed him that he intended to write such a text. L'Hôpital was apparently generous and personable, and his many contacts with major mathematicians provided the vehicle for disseminating major discoveries in calculus throughout Europe.

In the following examples we will apply L'Hôpital's rule using the following three-step process:

Step 1.   Check that $\lim f(x)/g(x)$ is an indeterminate form. If it is not, then L'Hôpital's rule cannot be used.

Step 2.   Differentiate f and g separately.

Step 3.   Find $\lim f'(x)/g'(x)$. If this limit is finite, $+\infty$, or $-\infty$, then it is equal to $\lim f(x)/g(x)$.

Example 1

In each part confirm that the limit is an indeterminate form of type 0/0, and evaluate it using L'Hôpital's rule.

(a) $\displaystyle\lim_{x \to 2} \frac{x^2 - 4}{x - 2}$ (b) $\displaystyle\lim_{x \to 0} \frac{\sin 2x}{x}$ (c) $\displaystyle\lim_{x \to \pi/2} \frac{1 - \sin x}{\cos x}$ (d) $\displaystyle\lim_{x \to 0} \frac{e^x - 1}{x^3}$

(e) $\displaystyle\lim_{x \to 0^-} \frac{\tan x}{x^2}$ (f) $\displaystyle\lim_{x \to 0} \frac{1 - \cos x}{x^2}$ (g) $\displaystyle\lim_{x \to +\infty} \frac{x^{-4/3}}{\sin(1/x)}$

Solution (a). The numerator and denominator have a limit of 0, so L'Hôpital's rule applies and yields

$$\lim_{x \to 2} \frac{x^2 - 4}{x - 2} = \lim_{x \to 2} \frac{\dfrac{d}{dx}[x^2 - 4]}{\dfrac{d}{dx}[x - 2]} = \lim_{x \to 2} \frac{2x}{1} = 4$$

Observe that this particular limit could also have been obtained by factoring

$$\lim_{x \to 2} \frac{x^2 - 4}{x - 2} = \lim_{x \to 2} \frac{(x - 2)(x + 2)}{x - 2} = \lim_{x \to 2} (x + 2) = 4$$

Solution (b). The numerator and denominator have a limit of 0, so L'Hôpital's rule applies and yields

$$\lim_{x \to 0} \frac{\sin 2x}{x} = \lim_{x \to 0} \frac{\dfrac{d}{dx}[\sin 2x]}{\dfrac{d}{dx}[x]} = \lim_{x \to 0} \frac{2 \cos 2x}{1} = 2$$

Observe that this result agrees with that obtained by substitution in Example 2(b) of Section 2.5.

Solution (c). The numerator and denominator have a limit of 0, so L'Hôpital's rule applies and yields

$$\lim_{x \to \pi/2} \frac{1 - \sin x}{\cos x} = \lim_{x \to \pi/2} \frac{\dfrac{d}{dx}[1 - \sin x]}{\dfrac{d}{dx}[\cos x]} = \lim_{x \to \pi/2} \frac{-\cos x}{-\sin x} = \frac{0}{-1} = 0$$

Solution (d). The numerator and denominator have a limit of 0, so L'Hôpital's rule applies and yields

$$\lim_{x \to 0} \frac{e^x - 1}{x^3} = \lim_{x \to 0} \frac{\dfrac{d}{dx}[e^x - 1]}{\dfrac{d}{dx}[x^3]} = \lim_{x \to 0} \frac{e^x}{3x^2} = +\infty$$

Solution (e). The numerator and denominator have a limit of 0, so L'Hôpital's rule applies and yields

$$\lim_{x \to 0^-} \frac{\tan x}{x^2} = \lim_{x \to 0^-} \frac{\sec^2 x}{2x} = -\infty$$

Solution (f). The numerator and denominator have a limit of 0, so L'Hôpital's rule applies and yields

$$\lim_{x \to 0} \frac{1 - \cos x}{x^2} = \lim_{x \to 0} \frac{\sin x}{2x}$$

Since the new limit is another indeterminate form of type $0/0$, we apply L'Hôpital's rule again:

$$\lim_{x \to 0} \frac{1 - \cos x}{x^2} = \lim_{x \to 0} \frac{\sin x}{2x} = \lim_{x \to 0} \frac{\cos x}{2} = \frac{1}{2}$$

Solution (g). The numerator and denominator have a limit of 0, so L'Hôpital's rule applies and yields

$$\lim_{x \to +\infty} \frac{x^{-4/3}}{\sin(1/x)} = \lim_{x \to +\infty} \frac{-\frac{4}{3}x^{-7/3}}{(-1/x^2)\cos(1/x)} = \lim_{x \to +\infty} \frac{\frac{4}{3}x^{-1/3}}{\cos(1/x)} = \frac{0}{1} = 0 \qquad \blacktriangleleft$$

WARNING. Applying L'Hôpital's rule to limits that are not indeterminate forms can lead to incorrect results. For example, in the limit

$$\lim_{x \to 0} \frac{x + 6}{x + 2} = \frac{6}{2} = 3$$

the numerator approaches 6 and the denominator approaches 2, so the limit is not an indeterminate form of type $0/0$. However, if we ignore this and blindly apply L'Hôpital's rule, we reach the following *erroneous* conclusion:

$$\lim_{x \to 0} \frac{\dfrac{d}{dx}[x + 6]}{\dfrac{d}{dx}[x + 2]} = \lim_{x \to 0} \frac{1}{1} = 1$$

INDETERMINATE FORMS OF TYPE ∞/∞

When we want to indicate that the limit (or the one-sided limits) of a function are $+\infty$ or $-\infty$ without being specific about the sign, we will say that the limit is ∞. For example,

$$\lim_{x \to a^+} f(x) = \infty \quad \text{means} \quad \lim_{x \to a^+} f(x) = +\infty \quad \text{or} \quad \lim_{x \to a^+} f(x) = -\infty$$

$$\lim_{x \to +\infty} f(x) = \infty \quad \text{means} \quad \lim_{x \to +\infty} f(x) = +\infty \quad \text{or} \quad \lim_{x \to +\infty} f(x) = -\infty$$

$$\lim_{x \to a} f(x) = \infty \quad \text{means} \quad \lim_{x \to a^+} f(x) = \pm\infty \quad \text{and} \quad \lim_{x \to a^-} f(x) = \pm\infty$$

The limit of a ratio, $f(x)/g(x)$, in which the numerator has limit ∞ and the denominator has limit ∞ is called an *indeterminate form of type ∞/∞*. The following version of L'Hôpital's rule, which we state without proof, can often be used to evaluate limits of this type.

4.7.2 THEOREM (*L'Hôpital's Rule for Form ∞/∞*). *Let* lim *stand for one of the limits* $\lim_{x \to a}$, $\lim_{x \to a^+}$, $\lim_{x \to a^-}$, $\lim_{x \to +\infty}$, *or* $\lim_{x \to -\infty}$, *and suppose that* $\lim f(x) = \infty$ *and* $\lim g(x) = \infty$. *If* $\lim[f'(x)/g'(x)]$ *has a finite value* L, *or if this limit is* $+\infty$ *or* $-\infty$, *then*

$$\lim \frac{f(x)}{g(x)} = \lim \frac{f'(x)}{g'(x)}$$

Example 2

In each part confirm that the limit is an indeterminate form of type ∞/∞ and apply L'Hôpital's rule.

(a) $\displaystyle\lim_{x \to +\infty} \frac{x}{e^x}$ (b) $\displaystyle\lim_{x \to 0^+} \frac{\ln x}{\csc x}$

Solution (a). The numerator and denominator both have a limit of $+\infty$, so we have an indeterminate form of type ∞/∞. Applying L'Hôpital's rule yields

$$\lim_{x \to +\infty} \frac{x}{e^x} = \lim_{x \to +\infty} \frac{1}{e^x} = 0$$

Solution (b). The numerator has a limit of $-\infty$ and the denominator has a limit of $+\infty$, so we have an indeterminate form of type ∞/∞. Applying L'Hôpital's rule yields

$$\lim_{x \to 0^+} \frac{\ln x}{\csc x} = \lim_{x \to 0^+} \frac{1/x}{-\csc x \cot x} \qquad (6)$$

This last limit is again an indeterminate form of type ∞/∞. Moreover, any additional applications of L'Hôpital's rule will yield powers of $1/x$ in the numerator and expressions involving $\csc x$ and $\cot x$ in the denominator; thus, repeated application of L'Hôpital's rule simply produces new indeterminate forms. We must try something else. The last limit in (6) can be rewritten as

$$\lim_{x \to 0^+} \left(-\frac{\sin x}{x} \tan x \right) = -\lim_{x \to 0^+} \frac{\sin x}{x} \cdot \lim_{x \to 0^+} \tan x = -(1)(0) = 0$$

Thus,

$$\lim_{x \to 0^+} \frac{\ln x}{\csc x} = 0 \qquad \blacktriangleleft$$

ANALYZING THE GROWTH OF EXPONENTIAL FUNCTIONS USING L'HÔPITAL'S RULE

If n is any positive integer, then $x^n \to +\infty$ as $x \to +\infty$. Such integer powers of x are sometimes used as "measuring sticks" to describe how rapidly other functions grow. For example, we know that $e^x \to +\infty$ as $x \to +\infty$ and that the growth of e^x is very rapid (Table 4.2.3); however, the growth of x^n is also rapid when n is a high power, so it is reasonable to ask whether high powers of x grow more or less rapidly than e^x. One way to investigate this is to examine the behavior of the ratio x^n/e^x as $x \to +\infty$. For example, Figure 4.7.2a shows the graph of $y = x^5/e^x$. This graph suggests that $x^5/e^x \to 0$ as $x \to +\infty$, and this implies that the growth of the function e^x is sufficiently rapid that its values eventually overtake those of x^5 and force the ratio toward zero. Stated informally, "e^x eventually grows more rapidly than x^5." The same conclusion could have been reached by putting e^x on top and examining the behavior of e^x/x^5 as $x \to +\infty$ (Figure 4.7.2b). In this case the values of e^x eventually overtake those of x^5 and force the ratio toward $+\infty$. More generally, we can use L'Hôpital's rule to show that e^x *eventually grows more rapidly than any positive integer power of x*, that is,

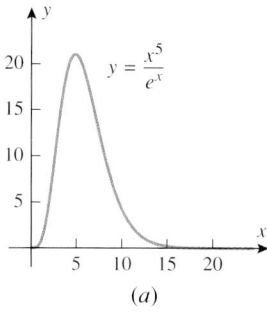

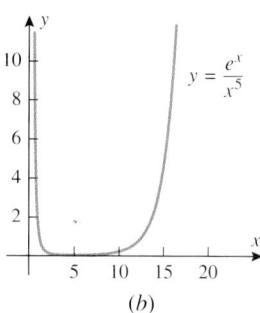

(a)   (b)

Figure 4.7.2

$$\lim_{x \to +\infty} \frac{x^n}{e^x} = 0 \qquad \text{and} \qquad \lim_{x \to +\infty} \frac{e^x}{x^n} = +\infty \qquad (7\text{--}8)$$

Both limits are indeterminate forms of type ∞/∞ that can be evaluated using L'Hôpital's rule. For example, to establish (7), we will need to apply L'Hôpital's rule n times. For this purpose, observe that successive differentiations of x^n reduce the exponent by 1 each time, thus producing a constant for the nth derivative. For example, the successive derivatives of x^3 are $3x^2, 6x$, and 6. In general, the nth derivative of x^n is the constant $n(n-1)(n-2)\cdots 1 = n!$ (verify).[*] Thus, applying L'Hôpital's rule n times to (7) yields

$$\lim_{x \to +\infty} \frac{x^n}{e^x} = \lim_{x \to +\infty} \frac{n!}{e^x} = 0$$

Limit (8) can be established similarly.

INDETERMINATE FORMS OF TYPE $0 \cdot \infty$

Thus far we have discussed indeterminate forms of type $0/0$ and ∞/∞. However, these are not the only possibilities; in general, the limit of an expression that has one of the forms

$$\frac{f(x)}{g(x)}, \quad f(x) \cdot g(x), \quad f(x)^{g(x)}, \quad f(x) - g(x), \quad f(x) + g(x)$$

is called an *indeterminate form* if the limits of $f(x)$ and $g(x)$ individually exert conflicting influences on the limit of the entire expression. For example, the limit

$$\lim_{x \to 0^+} x \ln x$$

is an ***indeterminate form of type $0 \cdot \infty$*** because the limit of the first factor is 0, the limit of the second factor is $-\infty$, and these two limits exert conflicting influences on the product. On the other hand, the limit

$$\lim_{x \to +\infty} [\sqrt{x}(1 - x^2)]$$

is not an indeterminate form because the first factor has a limit of $+\infty$, the second factor has a limit of $-\infty$, and these influences work together to produce a limit of $-\infty$ for the product.

WARNING.   It is tempting to argue that an indeterminate form of type $0 \cdot \infty$ has value 0 since "zero times anything is zero." However, this is fallacious since $0 \cdot \infty$ is not a product of numbers, but rather a statement about limits. For example, the following limits are of the form $0 \cdot \infty$:

$$\lim_{x \to 0^+} x \cdot \frac{1}{x} = 1, \quad \lim_{x \to 0^+} x^2 \cdot \frac{1}{x} = 0, \quad \lim_{x \to 0^+} \sqrt{x} \cdot \frac{1}{x} = +\infty$$

Indeterminate forms of type $0 \cdot \infty$ can sometimes be evaluated by rewriting the product as a ratio, and then applying L'Hôpital's rule for indeterminate forms of type $0/0$ or ∞/∞.

Example 3

Evaluate

(a) $\lim_{x \to 0^+} x \ln x$ (b) $\lim_{x \to \pi/4} (1 - \tan x) \sec 2x$

Solution (a). The factor x has a limit of 0 and the factor $\ln x$ has a limit of $-\infty$, so the stated problem is an indeterminate form of type $0 \cdot \infty$. There are two possible approaches: we can rewrite the limit as

$$\lim_{x \to 0^+} \frac{\ln x}{1/x} \quad \text{or} \quad \lim_{x \to 0^+} \frac{x}{1/\ln x}$$

the first being an indeterminate form of type ∞/∞ and the second an indeterminate form of

---
[*] Recall that for $n \geq 1$ the expression $n!$ is read ***n-factorial*** and denotes the product of the first n integers.

type $0/0$. However, the first form is the preferred initial choice because the derivative of $1/x$ is less complicated than the derivative of $1/\ln x$. That choice yields

$$\lim_{x \to 0^+} x \ln x = \lim_{x \to 0^+} \frac{\ln x}{1/x} = \lim_{x \to 0^+} \frac{1/x}{-1/x^2} = \lim_{x \to 0^+} (-x) = 0$$

Solution (b). The stated problem is an indeterminate form of type $0 \cdot \infty$. We will convert it to an indeterminate form of type ∞/∞:

$$\lim_{x \to \pi/4} (1 - \tan x) \sec 2x = \lim_{x \to \pi/4} \frac{1 - \tan x}{1/\sec 2x} = \lim_{x \to \pi/4} \frac{1 - \tan x}{\cos 2x}$$

$$= \lim_{x \to \pi/4} \frac{-\sec^2 x}{-2 \sin 2x} = \frac{-2}{-2} = 1 \qquad \blacktriangleleft$$

INDETERMINATE FORMS OF TYPE $\infty - \infty$

A limit problem that leads to one of the expressions

$$(+\infty) - (+\infty), \quad (-\infty) - (-\infty),$$

$$(+\infty) + (-\infty), \quad (-\infty) + (+\infty)$$

is called an *indeterminate form of type $\infty - \infty$*. Such limits are indeterminate because the two terms exert conflicting influences on the expression: one pushes it in the positive direction and the other pushes it in the negative direction. However, limit problems that lead to one of the expressions

$$(+\infty) + (+\infty), \qquad (+\infty) - (-\infty),$$

$$(-\infty) + (-\infty), \qquad (-\infty) - (+\infty)$$

are not indeterminate, since the two terms work together (those on the top produce a limit of $+\infty$ and those on the bottom produce a limit of $-\infty$).

Indeterminate forms of type $\infty - \infty$ can sometimes be evaluated by combining the terms and manipulating the result to produce an indeterminate form of type $0/0$ or ∞/∞.

Example 4

Evaluate $\displaystyle\lim_{x \to 0^+} \left(\frac{1}{x} - \frac{1}{\sin x} \right)$.

Solution. Both terms have a limit of $+\infty$, so the stated problem is an indeterminate form of type $\infty - \infty$. Combining the two terms yields

$$\lim_{x \to 0^+} \left(\frac{1}{x} - \frac{1}{\sin x} \right) = \lim_{x \to 0^+} \left(\frac{\sin x - x}{x \sin x} \right)$$

which is an indeterminate form of type $0/0$. Applying L'Hôpital's rule twice yields

$$\lim_{x \to 0^+} \left(\frac{\sin x - x}{x \sin x} \right) = \lim_{x \to 0^+} \frac{\cos x - 1}{\sin x + x \cos x}$$

$$= \lim_{x \to 0^+} \frac{-\sin x}{\cos x + \cos x - x \sin x} = \frac{0}{2} = 0 \qquad \blacktriangleleft$$

INDETERMINATE FORMS OF TYPE 0^0, ∞^0, 1^∞

Limits of the form

$$\lim f(x)^{g(x)}$$

give rise to *indeterminate forms of the types 0^0, ∞^0, and 1^∞*. (The meaning of these symbols should be clear.) For example, the limit

$$\lim_{x \to 0^+} (1 + x)^{1/x}$$

whose value we know to be e [see Formula (5) of Section 4.2] is an indeterminate form of type 1^∞. It is indeterminate because the expressions $1 + x$ and $1/x$ exert two conflicting influences: the first approaches 1, which drives the expression toward 1, and the second approaches $+\infty$, which drives the expression toward $+\infty$.

Indeterminate forms of types 0^0, ∞^0, and 1^∞ can sometimes be evaluated by first introducing a dependent variable

$$y = f(x)^{g(x)}$$

and then calculating the limit of $\ln y$ by expressing it as

$$\lim \ln y = \lim [\ln(f(x)^{g(x)})] = \lim [g(x) \ln f(x)]$$

Once the limit of $\ln y$ is known, the limit of $y = f(x)^{g(x)}$ itself can generally be obtained by a method that we will illustrate in the next example.

Example 5

Show that $\lim\limits_{x \to 0} (1 + x)^{1/x} = e$.

Solution. As discussed above, we begin by introducing a dependent variable

$$y = (1 + x)^{1/x}$$

and taking the natural logarithm of both sides:

$$\ln y = \ln(1 + x)^{1/x} = \frac{1}{x} \ln(1 + x) = \frac{\ln(1 + x)}{x}$$

Thus,

$$\lim_{x \to 0} \ln y = \lim_{x \to 0} \frac{\ln(1 + x)}{x}$$

which is an indeterminate form of type $0/0$, so by L'Hôpital's rule

$$\lim_{x \to 0} \ln y = \lim_{x \to 0} \frac{\ln(1 + x)}{x} = \lim_{x \to 0} \frac{1/(1 + x)}{1} = 1$$

Since we have shown that $\ln y \to 1$ as $x \to 0$, the continuity of the exponential function implies that $e^{\ln y} \to e^1$ as $x \to 0$, and this implies that $y \to e$ as $x \to 0$. Thus,

$$\lim_{x \to 0} (1 + x)^{1/x} = e$$ ◀

EXERCISE SET 4.7   ◪ Graphing Calculator   ☐c CAS

In Exercises 1 and 2, evaluate the given limit without using L'Hôpital's rule, and then check that your answer is correct using L'Hôpital's rule.

1. (a) $\lim\limits_{x \to 2} \dfrac{x^2 - 4}{x^2 + 2x - 8}$ (b) $\lim\limits_{x \to +\infty} \dfrac{2x - 5}{3x + 7}$

2. (a) $\lim\limits_{x \to 0} \dfrac{\sin x}{\tan x}$ (b) $\lim\limits_{x \to 1} \dfrac{x^2 - 1}{x^3 - 1}$

In Exercises 3–36, find the limit.

3. $\lim\limits_{x \to 1} \dfrac{\ln x}{x - 1}$ **4.** $\lim\limits_{x \to 0} \dfrac{\sin 2x}{\sin 5x}$

5. $\lim\limits_{x \to 0} \dfrac{e^x - 1}{\sin x}$

6. $\lim\limits_{x \to 3} \dfrac{x - 3}{3x^2 - 13x + 12}$

7. $\lim\limits_{\theta \to 0} \dfrac{\tan \theta}{\theta}$

8. $\lim\limits_{t \to 0} \dfrac{te^t}{1 - e^t}$

9. $\lim\limits_{x \to \pi^+} \dfrac{\sin x}{x - \pi}$

10. $\lim\limits_{x \to 0^+} \dfrac{\sin x}{x^2}$

11. $\lim\limits_{x \to +\infty} \dfrac{\ln x}{x}$

12. $\lim\limits_{x \to +\infty} \dfrac{e^{3x}}{x^2}$

13. $\lim\limits_{x \to 0^+} \dfrac{\cot x}{\ln x}$

14. $\lim\limits_{x \to 0^+} \dfrac{1 - \ln x}{e^{1/x}}$

15. $\lim\limits_{x \to +\infty} \dfrac{x^{100}}{e^x}$

16. $\lim\limits_{x \to 0^+} \dfrac{\ln(\sin x)}{\ln(\tan x)}$

17. $\lim\limits_{x \to 0} \dfrac{\sin^{-1} 2x}{x}$

18. $\lim\limits_{x \to 0} \dfrac{x - \tan^{-1} x}{x^3}$

19. $\lim\limits_{x \to +\infty} x e^{-x}$

20. $\lim\limits_{x \to \pi^-} (x - \pi) \tan \tfrac{1}{2}x$

21. $\lim\limits_{x \to +\infty} x \sin \dfrac{\pi}{x}$

22. $\lim\limits_{x \to 0^+} \tan x \ln x$

23. $\lim\limits_{x \to \pi/2^-} \sec 3x \cos 5x$

24. $\lim\limits_{x \to \pi} (x - \pi) \cot x$

25. $\lim\limits_{x \to +\infty} (1 - 3/x)^x$

26. $\lim\limits_{x \to 0} (1 + 2x)^{-3/x}$

27. $\lim\limits_{x \to 0} (e^x + x)^{1/x}$

28. $\lim\limits_{x \to +\infty} (1 + a/x)^{bx}$

29. $\lim\limits_{x \to 1} (2 - x)^{\tan(\pi/2)x}$

30. $\lim\limits_{x \to +\infty} [\cos(2/x)]^{x^2}$

31. $\lim\limits_{x \to 0} (\csc x - 1/x)$

32. $\lim\limits_{x \to 0} \left(\dfrac{1}{x^2} - \dfrac{\cos 3x}{x^2} \right)$

33. $\lim\limits_{x \to +\infty} (\sqrt{x^2 + x} - x)$

34. $\lim\limits_{x \to 0} \left(\dfrac{1}{x} - \dfrac{1}{e^x - 1} \right)$

35. $\lim\limits_{x \to +\infty} [x - \ln(x^2 + 1)]$

36. $\lim\limits_{x \to +\infty} [\ln x - \ln(1 + x)]$

c 37. Use a CAS to check the answers you obtained in Exercises 31–36.

38. Show that for any positive integer n

(a) $\lim\limits_{x \to +\infty} \dfrac{\ln x}{x^n} = 0$

(b) $\lim\limits_{x \to +\infty} \dfrac{x^n}{\ln x} = +\infty$

39. (a) Find the error in the following calculation:

$$\lim\limits_{x \to 1} \dfrac{x^3 - x^2 + x - 1}{x^3 - x^2} = \lim\limits_{x \to 1} \dfrac{3x^2 - 2x + 1}{3x^2 - 2x}$$

$$= \lim\limits_{x \to 1} \dfrac{6x - 2}{6x - 2} = 1$$

(b) Find the correct answer.

40. Find $\lim\limits_{x \to 1} \dfrac{x^4 - 4x^3 + 6x^2 - 4x + 1}{x^4 - 3x^3 + 3x^2 - x}$.

In Exercises 41–44, make a conjecture about the limit by graphing the function involved with a graphing utility; then check your conjecture using L'Hôpital's rule.

⊠ 41. $\lim\limits_{x \to +\infty} \dfrac{\ln(\ln x)}{\sqrt{x}}$

⊠ 42. $\lim\limits_{x \to 0^+} x^x$

⊠ 43. $\lim\limits_{x \to 0^+} (\sin x)^{3/\ln x}$

⊠ 44. $\lim\limits_{x \to (1/2)\pi^-} \dfrac{4 \tan x}{1 + \sec x}$

In Exercises 45–48, make a conjecture about the equations of horizontal asymptotes, if any, by graphing the equation with a graphing utility; then check your answer using L'Hôpital's rule.

⊠ 45. $y = \ln x - e^x$

⊠ 46. $y = x - \ln(1 + 2e^x)$

⊠ 47. $y = (\ln x)^{1/x}$

⊠ 48. $y = \left(\dfrac{x + 1}{x + 2} \right)^x$

49. Limits of the type

$$0/\infty, \quad \infty/0, \quad 0^\infty, \quad \infty \cdot \infty, \quad +\infty + (+\infty),$$

$$+\infty - (-\infty), \quad -\infty + (-\infty), \quad -\infty - (+\infty)$$

are *not* indeterminate forms. Find the following limits by inspection.

(a) $\lim\limits_{x \to 0^+} \dfrac{x}{\ln x}$

(b) $\lim\limits_{x \to +\infty} \dfrac{x^3}{e^{-x}}$

(c) $\lim\limits_{x \to (1/2)\pi^-} (\cos x)^{\tan x}$

(d) $\lim\limits_{x \to 0^+} (\ln x) \cot x$

(e) $\lim\limits_{x \to 0^+} \left(\dfrac{1}{x} - \ln x \right)$

(f) $\lim\limits_{x \to -\infty} (x + x^3)$

50. There is a myth that circulates among beginning calculus students which states that all indeterminate forms of types 0^0, ∞^0, and 1^∞ have value 1 because "anything to the zero power is 1" and "1 to any power is 1." The fallacy is that 0^0, ∞^0, and 1^∞ are not powers of numbers, but rather descriptions of limits. The following examples, which were transmitted to me by Prof. Jack Staib of Drexel University, show that such indeterminate forms can have any positive real value:

(a) $\lim\limits_{x \to 0^+} \left[x^{(\ln a)/(1 + \ln x)} \right] = 0^0 = a$

(b) $\lim\limits_{x \to +\infty} \left[x^{(\ln a)/(1 + \ln x)} \right] = \infty^0 = a$

(c) $\lim\limits_{x \to 0} \left[(x + 1)^{(\ln a)/x} \right] = 1^\infty = a$.

Prove these results.

In Exercises 51–54, verify that L'Hôpital's rule is of no help in finding the limit, then find the limit, if it exists, by some other method.

51. $\lim\limits_{x \to +\infty} \dfrac{x + \sin 2x}{x}$

52. $\lim\limits_{x \to +\infty} \dfrac{2x - \sin x}{3x + \sin x}$

53. $\lim\limits_{x \to +\infty} \dfrac{x(2 + \sin 2x)}{x + 1}$

54. $\lim\limits_{x \to +\infty} \dfrac{x(2 + \sin x)}{x^2 + 1}$

55. The accompanying schematic diagram represents an electrical circuit consisting of an electromotive force that produces a voltage V, a resistor with resistance R, and an inductor with inductance L. It is shown in electrical circuit theory that if the voltage is first applied at time $t = 0$, then the current I flowing through the circuit at time t is given by

$$I = \dfrac{V}{R}(1 - e^{-Rt/L})$$

What is the effect on the current at a fixed time t if the resistance approaches 0 (i.e., $R \to 0^+$)?

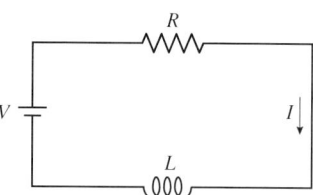

Figure Ex-55

56. (a) Show that $\lim\limits_{x \to \pi/2} (\pi/2 - x) \tan x = 1$.

(b) Show that

$$\lim_{x \to \pi/2} \left(\frac{1}{\pi/2 - x} - \tan x \right) = 0$$

(c) It follows from part (b) that the approximation

$$\tan x \approx \frac{1}{\pi/2 - x}$$

should be good for values of x near $\pi/2$. Use a calculator to find $\tan x$ and $1/(\pi/2 - x)$ for $x = 1.57$; compare the results.

[c] 57. (a) Use a CAS to show that if k is a positive constant, then

$$\lim_{x \to +\infty} x(k^{1/x} - 1) = \ln k$$

(b) Confirm this result using L'Hôpital's rule. [*Hint:* Express the limit in terms of $t = 1/x$.]

(c) If n is a positive integer, then it follows from part (a) with $x = n$ that the approximation

$$n(\sqrt[n]{k} - 1) \approx \ln k$$

should be good when n is large. Use this result and the square root key on a calculator to approximate the values of $\ln 0.3$ and $\ln 2$ with $n = 1024$, then compare

the values obtained with values of the logarithms generated directly from the calculator. [*Hint:* The nth roots for which n is a power of 2 can be obtained as successive square roots.]

[~] 58. Let $f(x) = x^2 \sin(1/x)$.
(a) Are the limits $\lim_{x \to 0^+} f(x)$ and $\lim_{x \to 0^-} f(x)$ indeterminate forms?
(b) Use a graphing utility to generate the graph of f, and use the graph to make conjectures about the limits in part (a).
(c) Use the Squeezing Theorem (2.5.2) to confirm that your conjectures in part (b) are correct.

59. Find all values of k and l such that

$$\lim_{x \to 0} \frac{k + \cos lx}{x^2} = -4$$

60. (a) Explain why L'Hôpital's rule does not apply to the problem

$$\lim_{x \to 0} \frac{x^2 \sin(1/x)}{\sin x}$$

(b) Find the limit.

61. Find $\lim_{x \to 0^+} \dfrac{x \sin(1/x)}{\sin x}$ if it exists.

SUPPLEMENTARY EXERCISES

1. (a) State conditions under which two functions, f and g, will be inverses, and give several examples of such functions.
   (b) In words, what is the relationship between the graphs of $y = f(x)$ and $y = g(x)$ when f and g are inverse functions?
   (c) What is the relationship between the domains and ranges of inverse functions f and g?
   (d) What condition must be satisfied for a function f to have an inverse? Give some examples of functions that do not have inverses.
   (e) If f and g are inverse functions and f is continuous, must g be continuous? Give a reasonable informal argument to support your answer.
   (f) If f and g are inverse functions and f is differentiable, must g be differentiable? Give a reasonable informal argument to support your answer.

2. (a) State the restrictions on the domains of $\sin x$, $\cos x$, $\tan x$, and $\sec x$ that are imposed to make those functions one-to-one in the definitions of $\sin^{-1} x$, $\cos^{-1} x$, $\tan^{-1} x$, and $\sec^{-1} x$.
   (b) Sketch the graphs of the restricted trigonometric functions in part (a) and their inverses.

3. (a) Under what conditions will a limit of the form

$$\lim [f(x)/g(x)]$$

   be an indeterminate form?
   (b) If $\lim g(x) = 0$, must $\lim [f(x)/g(x)]$ be an indeterminate form? Give some examples to support your answer.

4. Suppose that $\lim f(x) = \pm\infty$ and $\lim g(x) = \pm\infty$. In each of the four possible cases, state whether $\lim [f(x) - g(x)]$ is an indeterminate form, and give a reasonable informal argument to support your answer.

5. In each part, find $f^{-1}(x)$ if the inverse exists.
   (a) $f(x) = 8x^3 - 1$ (b) $f(x) = x^2 - 2x + 1$
   (c) $f(x) = (e^x)^2 + 1$ (d) $f(x) = (x + 2)/(x - 1)$

6. Let $f(x) = (ax + b)/(cx + d)$. What conditions on a, b, c, d guarantee that f^{-1} exists? Find $f^{-1}(x)$.

7. In each part, find the equation of the tangent line at the specified point.
   (a) $x^{2/3} - y^{2/3} - y = 1$; $(1, -1)$
   (b) $\sin xy = y$; $(\pi/2, 1)$

8. In each part, find the exact numerical value of the given expression.
(a) $\cos[\cos^{-1}(4/5) + \sin^{-1}(5/13)]$
(b) $\sin[\sin^{-1}(4/5) + \cos^{-1}(5/13)]$

9. Express the following function as a rational function of x:

$$3 \ln \left(e^{2x}(e^x)^3\right) + 2 \exp(\ln 1)$$

10. Suppose that $y = Ce^{kt}$, where C and k are constants, and let $Y = \ln y$. Show that the graph of Y versus t is a line, and state its slope and Y-intercept.

11. In each part, find the limit.
(a) $\lim\limits_{x \to +\infty} (e^x - x^2)$
(b) $\lim\limits_{x \to 1} \sqrt{\dfrac{\ln x}{x^4 - 1}}$
(c) $\lim\limits_{x \to 0} \dfrac{a^x - 1}{x}, \quad a > 0$

12. Show that the function $y = e^{ax} \sin bx$ satisfies

$$y'' - 2ay' + (a^2 + b^2)y = 0$$

for any real constants a and b.

13. Show that the function $y = \tan^{-1} x$ satisfies

$$y'' = -2 \sin y \cos^3 y$$

14. Show that the rate of change of $y = 3^{2x}5^{7x}$ is proportional to y.

15. The hypotenuse of a right triangle is growing at a rate of a cm/s and one leg is decreasing at a rate of b cm/s. How fast is the acute angle between the hypotenuse and the other leg changing at the instant when both legs are 1 cm?

16. In each part, find $(f^{-1})'(x)$ using Formula (21) of Section 4.3, and check your answer by differentiating f^{-1} directly.
(a) $f(x) = 3/(x + 1)$
(b) $f(x) = \sqrt{e^x}$

17. (a) Sketch the curves $y = \pm e^{-x/2}$ and $y = e^{-x/2} \sin 2x$ for $-\pi/2 \le x \le 3\pi/2$ in the same coordinate system, and check your work using a graphing utility.
(b) Find all x-intercepts of the curve $y = e^{-x/2} \sin 2x$ in the stated interval, and find the x-coordinates of all points where this curve intersects the curves $y = \pm e^{-x/2}$.

18. In each part, sketch the graph, and check your work with a graphing utility.
(a) $f(x) = 3 \sin^{-1}(x/2)$

(b) $f(x) = \cos^{-1} x - \pi/2$
(c) $f(x) = 2 \tan^{-1}(-3x)$
(d) $f(x) = \cos^{-1} x + \sin^{-1} x$

19. In each part, use any appropriate method to find dy/dx.
(a) $y = (1 + x)^{1/x}$
(b) $y = x^{(e^x)}$
(c) $y = e^{\ln(x^3 + 1)}$
(d) $y = \dfrac{a}{1 + be^{-x}}$
(e) $xy^{2/3} + yx^{2/3} = x^2$
(f) $y = \ln \left(\dfrac{\sqrt{x} \sqrt[3]{x + 1}}{\sin x \sec x} \right)$

20. (a) Suppose that the graph of $y = \log x$ is drawn with equal scales of 1 inch per unit in both the x- and y-directions. If a bug wants to walk along the graph until it reaches a height of 5 ft above the x-axis, how many miles to the right of the origin will it have to travel?
(b) Suppose that the graph of $y = 10^x$ is drawn with equal scales of 1 inch per unit in both the x- and y-directions. If a bug wants to walk along the graph until it reaches a height of 100 mi above the x-axis, how many feet to the right of the origin will it have to travel?

21. (a) Show that the graphs of $y = \ln x$ and $y = x^{0.2}$ intersect.
(b) Approximate the solution(s) of the equation $\ln x = x^{0.2}$ to three decimal places.

22. (a) Show that for $x > 0$ and $k \ne 0$ the equations

$$x^k = e^x \quad \text{and} \quad \dfrac{\ln x}{x} = \dfrac{1}{k}$$

have the same solutions.
(b) Use the graph of $y = (\ln x)/x$ to determine the values of k for which the equation $x^k = e^x$ has two distinct positive solutions.
(c) Find the positive solution(s) of $x^8 = e^x$.

23. Find the value of b so that the line $y = x$ is tangent to the graph of $y = \log_b x$. Confirm your result by graphing both $y = x$ and $y = \log_b x$ in the same coordinate system.

24. In each part, find the value of k for which the graphs of $y = f(x)$ and $y = \ln x$ share a common tangent line at their point of intersection. Confirm your result by graphing $y = f(x)$ and $y = \ln x$ in the same coordinate system.
(a) $f(x) = \sqrt{x} + k$
(b) $f(x) = k\sqrt{x}$

EXPANDING THE CALCULUS HORIZON

For additional material relating to this chapter, visit the Anton Website at http://www.wiley.com/college/anton

5

ANALYSIS

OF FUNCTIONS

AND THEIR GRAPHS

Maria Agnesi

$\mathcal{I}$n this chapter we will use methods of calculus to analyze functions and their graphs. We will be concerned here with such matters as identifying where the graph of a function is increasing or decreasing, where its high and low points occur, which way it bends, and what its limiting behavior is at important points.

One of the major goals of this chapter is to show how calculus and graphing utilities, working together, can provide most of the important information about the behavior of functions. Although graphing utilities can give us general information about the shape of a graph, such graphs lack perfect precision, since they are based on numerical approximations that can be affected by compression, distortion, and sampling error—it requires calculus to pin down the *exact* location of the key features and to reveal the nature of the fine detail. On the other hand, graphs produced by graphing utilities often provide information that is useful in pointing the calculus analysis in the right direction.

5.1 ANALYSIS OF FUNCTIONS I: INCREASE, DECREASE, AND CONCAVITY

Although graphing utilities are useful for determining the general shape of a graph, many problems require more precision than graphing utilities are capable of producing. The purpose of this section is to develop mathematical tools that can be used to determine the exact shape of a graph and the precise location of its key features.

INCREASING AND DECREASING FUNCTIONS

The terms *increasing*, *decreasing*, and *constant* are used to describe the behavior of a function over an interval as we travel left to right along its graph. For example, the function graphed in Figure 5.1.1 can be described as increasing on the interval $(-\infty, 0]$, decreasing on the interval $[0, 2]$, increasing again on the interval $[2, 4]$, and constant on the interval $[4, +\infty)$.

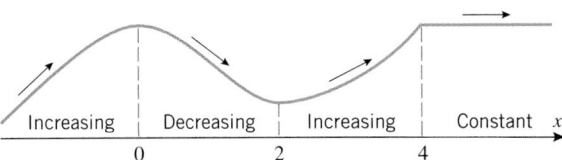

Figure 5.1.1

The following definition, which is illustrated in Figure 5.1.2, expresses these intuitive ideas precisely.

5.1.1 DEFINITION. Let f be defined on an interval, and let x_1 and x_2 denote points in that interval.

(a) f is **increasing** on the interval if $f(x_1) < f(x_2)$ whenever $x_1 < x_2$.

(b) f is **decreasing** on the interval if $f(x_1) > f(x_2)$ whenever $x_1 < x_2$.

(c) f is **constant** on the interval if $f(x_1) = f(x_2)$ for all points x_1 and x_2.

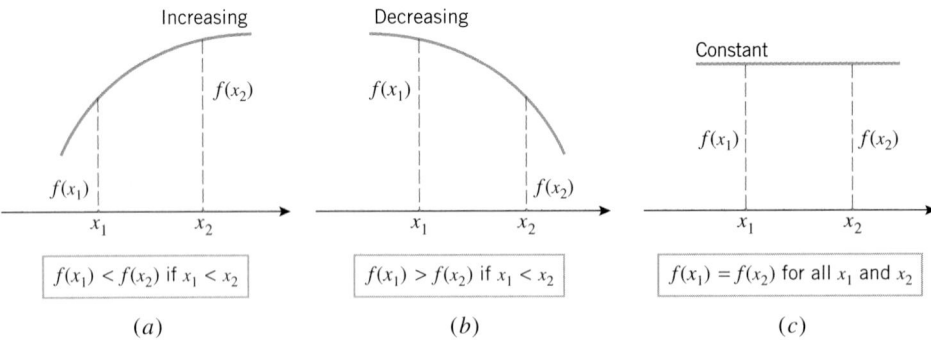

Figure 5.1.2

Figure 5.1.3 suggests that a differentiable function f is increasing on any interval where its graph has tangent lines with positive slope, is decreasing on any interval where its graph has tangent lines with negative slope, and is constant on any interval where its graph has tangent lines with zero slope. This intuitive observation suggests the following important theorem that will be proved in Section 6.5.

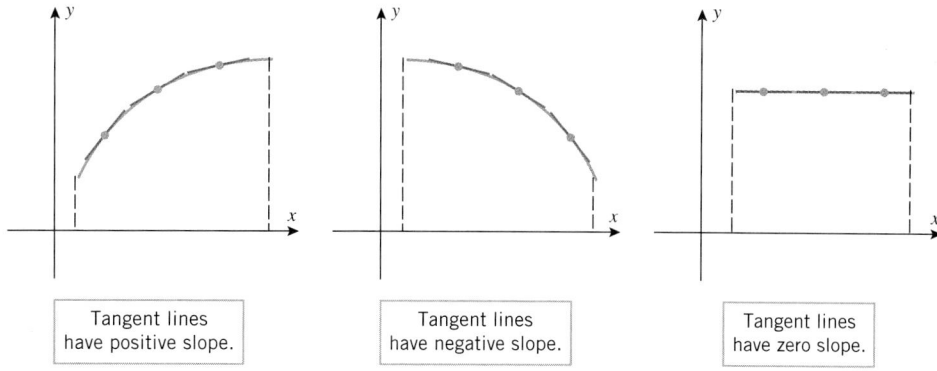

Figure 5.1.3

5.1.2 THEOREM. *Let f be a function that is continuous on a closed interval $[a, b]$ and differentiable on the open interval (a, b).*

(a) If $f'(x) > 0$ for every value of x in (a, b), then f is increasing on $[a, b]$.

(b) If $f'(x) < 0$ for every value of x in (a, b), then f is decreasing on $[a, b]$.

(c) If $f'(x) = 0$ for every value of x in (a, b), then f is constant on $[a, b]$.

REMARK.   Observe that in Theorem 5.1.2 it is only necessary to examine the derivative of f on the open interval (a, b) to determine whether f is increasing, decreasing, or constant on the closed interval $[a, b]$. Moreover, although this theorem was stated for a closed interval $[a, b]$, it is applicable to any interval I on which f is continuous and inside of which f is differentiable. For example, if f is continuous on $(a, +\infty)$ and $f'(x) > 0$ for each x in the interval $(a, +\infty)$, then f is increasing on $[a, +\infty)$; and if $f'(x) < 0$ on $(-\infty, +\infty)$, then f is decreasing on $(-\infty, +\infty)$ [the continuity on $(-\infty, +\infty)$ follows from the differentiability].

Example 1

Find the intervals on which the following functions are increasing and the intervals on which they are decreasing.

(a) $f(x) = x^2 - 4x + 3$ (b) $f(x) = x^3$

Solution (a). The graph of f in Figure 5.1.4 suggests that f is decreasing for $x \le 2$ and increasing for $x \ge 2$. To confirm this, we differentiate f to obtain

$$f'(x) = 2x - 4 = 2(x - 2)$$

It follows that

$$f'(x) < 0 \quad \text{if} \quad -\infty < x < 2$$
$$f'(x) > 0 \quad \text{if} \quad 2 < x < +\infty$$

Since f is continuous at $x = 2$, it follows from Theorem 5.1.2 and the subsequent remark that

f is decreasing on $(-\infty, 2]$

f is increasing on $[2, +\infty)$

These conclusions are consistent with the graph of f in Figure 5.1.4.

Solution (b). The graph of f in Figure 5.1.5 suggests that f is increasing over the entire x-axis. To confirm this, we differentiate f to obtain $f'(x) = 3x^2$. Thus,

$$f'(x) > 0 \quad \text{if} \quad -\infty < x < 0$$
$$f'(x) > 0 \quad \text{if} \quad 0 < x < +\infty$$

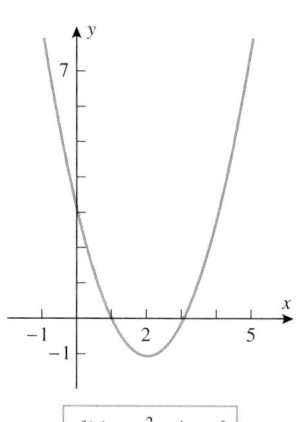

$f(x) = x^2 - 4x + 3$

Figure 5.1.4

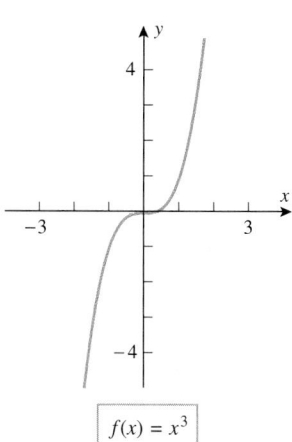

$f(x) = x^3$

Figure 5.1.5

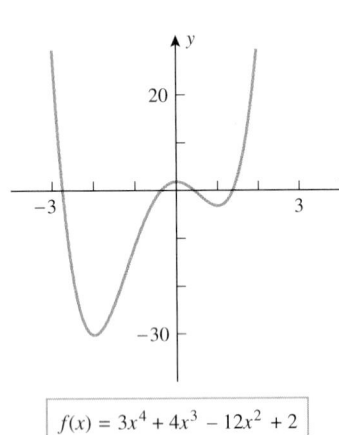

$f(x) = 3x^4 + 4x^3 - 12x^2 + 2$

Figure 5.1.6

Since f is continuous at $x = 0$,

f is increasing on $(-\infty, 0]$

f is increasing on $[0, +\infty)$

Hence f is increasing over the entire interval $(-\infty, +\infty)$, which is consistent with the graph in Figure 5.1.5 (see Exercise 51). ◄

Example 2

(a) Use the graph of $f(x) = 3x^4 + 4x^3 - 12x^2 + 2$ in Figure 5.1.6 to make a conjecture about the intervals on which f is increasing or decreasing.

(b) Use Theorem 5.1.2 to determine whether your conjecture is correct.

Solution (*a*). The graph suggests that f is decreasing if $x \le -2$, increasing if $-2 \le x \le 0$, decreasing if $0 \le x \le 1$, and increasing if $x \ge 1$.

Solution (*b*). Differentiating f we obtain

$$f'(x) = 12x^3 + 12x^2 - 24x = 12x(x^2 + x - 2) = 12x(x + 2)(x - 1)$$

The sign analysis of f' in Table 5.1.1 can be obtained using the method of test points discussed in Appendix A. The conclusions in that table confirm the conjecture in part (a). ◄

Table 5.1.1

INTERVAL	$12x$	$x + 2$	$x - 1$	f'	CONCLUSION
$x < -2$	−	−	−	−	f is decreasing on $(-\infty, -2]$
$-2 < x < 0$	−	+	−	+	f is increasing on $[-2, 0]$
$0 < x < 1$	+	+	−	−	f is decreasing on $[0, 1]$
$1 < x$	+	+	+	+	f is increasing on $[1, +\infty)$

CONCAVITY

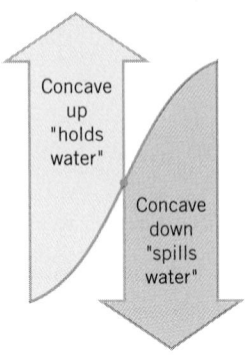

Figure 5.1.7

Although the sign of the derivative of f reveals where the graph of f is increasing or decreasing, it does not reveal the direction of *curvature*. For example, on both sides of the point in Figure 5.1.7 the graph is increasing, but on the left side it has an upward curvature ("holds water") and on the right side it has a downward curvature ("spills water"). On intervals where the graph of f has upward curvature we say that f is *concave up*, and on intervals where the graph has downward curvature we say that f is *concave down*.

For differentiable functions, the direction of curvature can be characterized in terms of the tangent lines in two ways: As suggested by Figure 5.1.8, the graph of a function f has upward curvature on intervals where the graph lies above its tangent lines, and it has downward curvature on intervals where it lies below its tangent lines. Alternatively, the graph has upward curvature on intervals where the tangent lines have increasing slopes and downward curvature on intervals where they have decreasing slopes. We will use this latter characterization as our formal definition.

> **5.1.3** DEFINITION. If f is differentiable on an open interval I, then f is said to be *concave up* on I if f' is increasing on I, and f is said to be *concave down* on I if f' is decreasing on I.

To apply this definition we need some way to determine the intervals on which f' is increasing or decreasing. One way to do this is to apply Theorem 5.1.2 (and the remark that follows it) to the function f'. It follows from that theorem and remark that f' will be

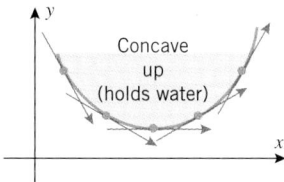

Concave up (holds water)

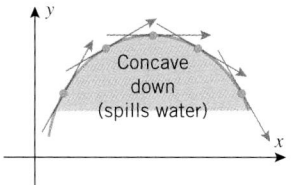

Concave down (spills water)

Figure 5.1.8

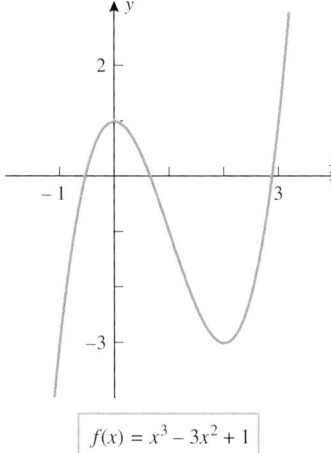

$f(x) = x^3 - 3x^2 + 1$

Figure 5.1.9

INFLECTION POINTS

increasing where its derivative f'' is positive and will be decreasing where its derivative f'' is negative. This is the idea behind the following theorem.

> **5.1.4** THEOREM. *Let f be twice differentiable on an open interval I.*
> (a) *If $f''(x) > 0$ on I, then f is concave up on I.*
> (b) *If $f''(x) < 0$ on I, then f is concave down on I.*

Example 3

Find open intervals on which the following functions are concave up and open intervals on which they are concave down.

    (a) $f(x) = x^2 - 4x + 3$ (b) $f(x) = x^3$ (c) $f(x) = x^3 - 3x^2 + 1$

Solution (a). Calculating the first two derivatives we obtain

$$f'(x) = 2x - 4 \quad \text{and} \quad f''(x) = 2$$

Since $f''(x) > 0$ for all x, the function f is concave up on $(-\infty, +\infty)$. This is consistent with Figure 5.1.4.

Solution (b). Calculating the first two derivatives we obtain

$$f'(x) = 3x^2 \quad \text{and} \quad f''(x) = 6x$$

Since $f''(x) < 0$ if $x < 0$ and $f''(x) > 0$ if $x > 0$, the function f is concave down on $(-\infty, 0)$ and concave up on $(0, +\infty)$. This is consistent with Figure 5.1.5.

Solution (c). Calculating the first two derivatives we obtain

$$f'(x) = 3x^2 - 6x \quad \text{and} \quad f''(x) = 6x - 6 = 6(x - 1)$$

Since $f''(x) > 0$ if $x > 1$ and $f''(x) < 0$ if $x < 1$, we conclude that

    f is concave up on $(1, +\infty)$

    f is concave down on $(-\infty, 1)$

which is consistent with the graph in Figure 5.1.9.   ◄

Points where a graph changes from concave up to concave down, or vice versa, are of special interest, so there is some terminology associated with them.

> **5.1.5** DEFINITION. If f is continuous on an open interval containing the point x_0, and if f changes the direction of its concavity at that point, then we say that f has an ***inflection point at x_0***, and we call the point $(x_0, f(x_0))$ on the graph of f an ***inflection point*** of f (Figure 5.1.10).

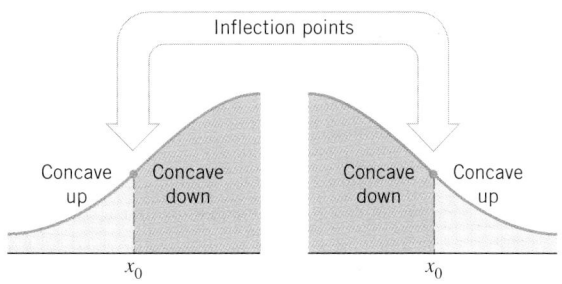

Figure 5.1.10

For example, the function $f(x) = x^3$ has an inflection point at $x = 0$ (Figure 5.1.5), the function $f(x) = x^3 - 3x^2 + 1$ has an inflection point at $x = 1$ (Figure 5.1.9), and the function $f(x) = x^2 - 4x + 3$ has no inflection points (Figure 5.1.4).

Example 4

Use the graph in Figure 5.1.6 to make rough estimates of the locations of the inflection points of $f(x) = 3x^4 + 4x^3 - 12x^2 + 2$, and check your estimates by finding the exact location of the inflection points.

Solution. The graph changes from concave up to concave down somewhere between -2 and -1, say roughly at $x = -1.25$; and the graph changes from concave down to concave up somewhere between 0 and 1, say roughly at $x = 0.5$. To find the exact location of the inflection points, we start by calculating the second derivative of f:

$$f'(x) = 12x^3 + 12x^2 - 24x$$
$$f''(x) = 36x^2 + 24x - 24 = 12(3x^2 + 2x - 2)$$

We could analyze the sign of f'' by factoring this function and applying the method of test points (as in Table 5.1.1). However, here is another approach. The graph of f'' is a parabola that opens up, and the quadratic formula shows that the equation $f'' = 0$ has the roots

$$x = \frac{-1 - \sqrt{7}}{3} \approx -1.22 \quad \text{and} \quad x = \frac{-1 + \sqrt{7}}{3} \approx 0.55 \tag{1}$$

(verify). Thus, from the rough graph of f'' in Figure 5.1.11 we obtain the sign analysis of f'' in Table 5.1.2; this implies that f has inflection points at the points in (1).   ◀

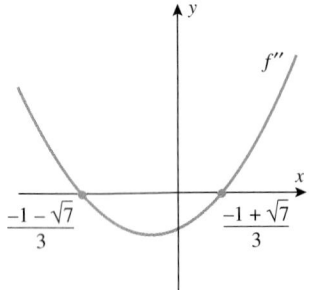

Figure 5.1.11

Table 5.1.2

INTERVAL	SIGN OF f''	CONCLUSION
$x < \dfrac{-1 - \sqrt{7}}{3}$	$+$	f is concave up
$\dfrac{-1 - \sqrt{7}}{3} < x < \dfrac{-1 + \sqrt{7}}{3}$	$-$	f is concave down
$x > \dfrac{-1 + \sqrt{7}}{3}$	$+$	f is concave up

In the preceding example the inflection points of f occurred at points where $f''(x) = 0$. However, inflection points do not always occur at points where $f''(x) = 0$. For example, if the graph of f'' happens to touch the x-axis at a point without crossing over it, then f'' will not change sign at that point, and hence no change in the concavity of f will occur at that point. Here is a specific example.

Example 5

Find the inflection points of $f(x) = x^4$.

Solution. Calculating the first two derivatives of f we obtain

$$f'(x) = 4x^3, \quad f''(x) = 12x^2$$

Here $f''(x) > 0$ for $x < 0$ and for $x > 0$, which implies that f is concave up for $x < 0$ and for $x > 0$. Thus, there are no inflection points; and in particular, there is no inflection point at $x = 0$, even though $f''(0) = 0$ (Figure 5.1.12).   ◀

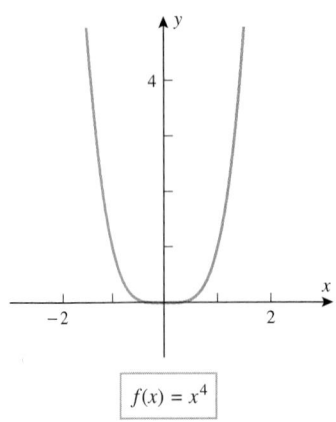

$f(x) = x^4$

Figure 5.1.12

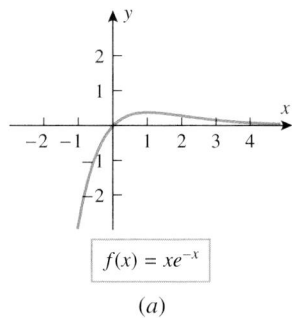

$$f(x) = xe^{-x}$$

(a)

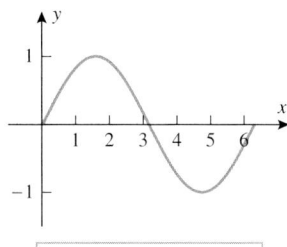

$$f(x) = \sin x, \ 0 \le x \le 2\pi$$

(b)

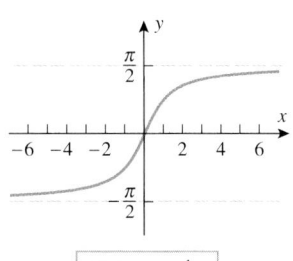

$$f(x) = \tan^{-1} x$$

(c)

Figure 5.1.13

Example 6

Find the inflection points of the following functions, and confirm that your results are consistent with the graphs of the functions.

(a) $f(x) = xe^{-x}$ (b) $f(x) = \sin x, \ 0 \le x \le 2\pi$ (c) $f(x) = \tan^{-1} x$

Solution (a). Calculating the first two derivatives of f we obtain

$$f'(x) = (1 - x)e^{-x}, \quad f''(x) = (x - 2)e^{-x}$$

(verify). Keeping in mind that e^{-x} is always positive, it follows that the sign of f'' is determined by the factor $x - 2$. Thus, $f''(x) < 0$ if $x < 2$, and $f''(x) > 0$ if $x > 2$, which implies that the graph is concave down for $x < 2$ and concave up for $x > 2$. Thus, there is an inflection point at $x = 2$ (Figure 5.1.13a).

Solution (b). Calculating the first two derivatives of f we obtain

$$f'(x) = \cos x, \quad f''(x) = -\sin x$$

Thus, $f''(x) < 0$ if $0 < x < \pi$, and $f''(x) > 0$ if $\pi < x < 2\pi$, which implies that the graph is concave down for $0 < x < \pi$ and concave up for $\pi < x < 2\pi$. Thus, there is an inflection point at $x = \pi \approx 3.14$ (Figure 5.1.13b).

Solution (c). Calculating the first two derivatives of f we obtain

$$f'(x) = \frac{1}{1 + x^2}, \quad f''(x) = -\frac{2x}{\left(1 + x^2\right)^2}$$

(verify). Thus, $f''(x) > 0$ if $x < 0$, and $f''(x) < 0$ if $x > 0$, which implies that the graph is concave up for $x < 0$ and concave down for $x > 0$. Thus, there is an inflection point at $x = 0$ (Figure 5.1.13c). ◄

FOR THE READER. If you have a CAS, devise a method for using it to find exact values for the inflection points of a function f, and use your method to find the inflection points of $f(x) = x/(x^2 + 1)$. Verify that your results are consistent with the graph of f.

INFLECTION POINTS IN APPLICATIONS

Up to now we have viewed the inflection points of a curve $y = f(x)$ as those points where the curve changes the direction of its concavity. However, inflection points also mark the points on the curve where the slopes of the tangent lines change from increasing to decreasing, or vice versa (Figure 5.1.14); stated another way:

Inflection points mark the places on the curve $y = f(x)$ where the rate of change of y with respect to x changes from increasing to decreasing, or vice versa.

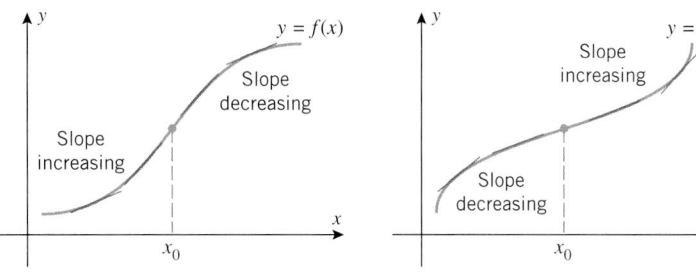

Figure 5.1.14

Note that we are dealing with a rather subtle concept here—a change of a rate of change. However, the following physical example should help to clarify the idea: Suppose that water is added to the flask in Figure 5.1.15 in such a way that the volume increases at a constant rate, and let us examine the rate at which the water level y rises with the time t. Initially, the level y will rise at a slow rate because of the wide base. However, as the diameter of the flask narrows, the rate at which the level y rises will increase until the level is at the narrow point in the neck. From that point on the rate at which the level rises will decrease as the diameter gets wider and wider. Thus, the narrow point in the neck is the point at which the rate of change of y with respect to t changes from increasing to decreasing.

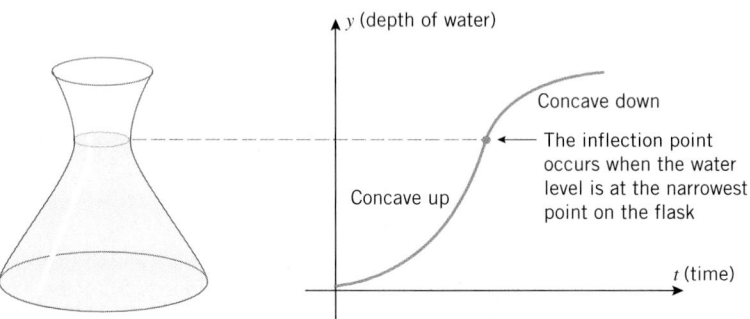

Figure 5.1.15

EXERCISE SET 5.1   ◠ Graphing Calculator   [C] CAS

1. In each part, sketch the graph of a function f with the stated properties, and discuss the signs of f' and f''.
   (a) The function f is concave up and increasing on the interval $(-\infty, +\infty)$.
   (b) The function f is concave down and increasing on the interval $(-\infty, +\infty)$.
   (c) The function f is concave up and decreasing on the interval $(-\infty, +\infty)$.
   (d) The function f is concave down and decreasing on the interval $(-\infty, +\infty)$.

2. In each part, sketch the graph of a function f with the stated properties.
   (a) f is increasing on $(-\infty, +\infty)$, has an inflection point at the origin, and is concave up on $(0, +\infty)$.
   (b) f is increasing on $(-\infty, +\infty)$, has an inflection point at the origin, and is concave down on $(0, +\infty)$.
   (c) f is decreasing on $(-\infty, +\infty)$, has an inflection point at the origin, and is concave up on $(0, +\infty)$.
   (d) f is decreasing on $(-\infty, +\infty)$, has an inflection point at the origin, and is concave down on $(0, +\infty)$.

3. Use the graph of the equation $y = f(x)$ in the accompanying figure to find the signs of dy/dx and d^2y/dx^2 at the points A, B, and C.

4. Use the graph of the equation $y = f'(x)$ in the accompanying figure to find the signs of dy/dx and d^2y/dx^2 at the points A, B, and C.

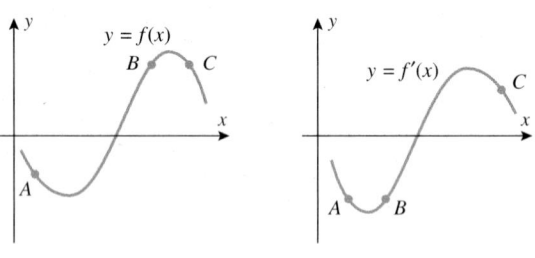

Figure Ex-3   Figure Ex-4

5. Use the graph of $y = f''(x)$ in the accompanying figure to determine the x-coordinates of all inflection points of f. Explain your reasoning.

6. Use the graph of $y = f'(x)$ in the accompanying figure to replace the question mark with $<$, $=$, or $>$, as appropriate. Explain your reasoning.
   (a) $f(0)$? $f(1)$ (b) $f(1)$? $f(2)$ (c) $f'(0)$? 0
   (d) $f'(1)$? 0   (e) $f''(0)$? 0   (f) $f''(2)$? 0

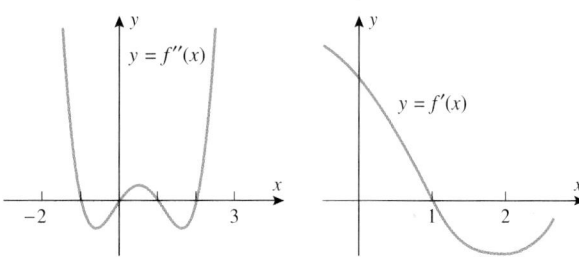

Figure Ex-5   Figure Ex-6

7. In each part, use the graph of $y = f(x)$ in the accompanying figure to find the requested information.
   (a) Find the intervals on which f is increasing.
   (b) Find the intervals on which f is decreasing.
   (c) Find the open intervals on which f is concave up.
   (d) Find the open intervals on which f is concave down.
   (e) Find all values of x at which f has an inflection point.

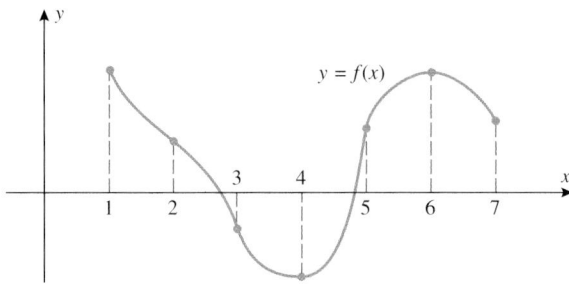

Figure Ex-7

8. Use the graph in Exercise 7 to make a table that shows the signs of f' and f'' over the intervals $(1, 2)$, $(2, 3)$, $(3, 4)$, $(4, 5)$, $(5, 6)$, and $(6, 7)$.

In Exercises 9–24, find: (a) the intervals on which f is increasing, (b) the intervals on which f is decreasing, (c) the open intervals on which f is concave up, (d) the open intervals on which f is concave down, and (e) the x-coordinates of all inflection points.

9. $f(x) = x^2 - 5x + 6$ 10. $f(x) = 4 - 3x - x^2$

11. $f(x) = (x + 2)^3$ 12. $f(x) = 5 + 12x - x^3$

13. $f(x) = 3x^4 - 4x^3$ 14. $f(x) = x^4 - 8x^2 + 16$

15. $f(x) = \dfrac{x^2}{x^2 + 2}$ 16. $f(x) = \dfrac{x}{x^2 + 2}$

17. $f(x) = \sqrt[3]{x + 2}$ 18. $f(x) = x^{2/3}$

19. $f(x) = x^{1/3}(x + 4)$ 20. $f(x) = x^{4/3} - x^{1/3}$

21. $f(x) = e^{-x^2/2}$ 22. $f(x) = xe^{x^2}$

23. $f(x) = \ln(1 + x^2)$ 24. $f(x) = x^2 \ln x$

In Exercises 25–30, analyze the trigonometric function f over the specified interval, stating where f is increasing, decreasing, concave up, and concave down, and stating the x-coordinates of all inflection points. Confirm that your results are consistent with the graph of f generated with a graphing utility.

25. $f(x) = \cos x$; $[0, 2\pi]$

26. $f(x) = \sin^2 2x$; $[0, \pi]$

27. $f(x) = \tan x$; $(-\pi/2, \pi/2)$

28. $f(x) = 2x + \cot x$; $(0, \pi)$

29. $f(x) = \sin x \cos x$; $[0, \pi]$

30. $f(x) = \cos^2 x - 2\sin x$; $[0, 2\pi]$

31. In each part sketch a continuous curve $y = f(x)$ with the stated properties.
   (a) $f(2) = 4$, $f'(2) = 0$, $f''(x) > 0$ for all x
   (b) $f(2) = 4$, $f'(2) = 0$, $f''(x) < 0$ for $x < 2$, $f''(x) > 0$ for $x > 2$
   (c) $f(2) = 4$, $f''(x) < 0$ for $x \neq 2$ and $\lim\limits_{x \to 2^+} f'(x) = +\infty$, $\lim\limits_{x \to 2^-} f'(x) = -\infty$

32. In each part sketch a continuous curve $y = f(x)$ with the stated properties.
   (a) $f(2) = 4$, $f'(2) = 0$, $f''(x) < 0$ for all x
   (b) $f(2) = 4$, $f'(2) = 0$, $f''(x) > 0$ for $x < 2$, $f''(x) < 0$ for $x > 2$
   (c) $f(2) = 4$, $f''(x) > 0$ for $x \neq 2$ and $\lim\limits_{x \to 2^+} f'(x) = -\infty$, $\lim\limits_{x \to 2^-} f'(x) = +\infty$

33. In each part, assume that a is a constant and find the inflection points, if any.
   (a) $f(x) = (x - a)^3$ (b) $f(x) = (x - a)^4$

34. Given that a is a constant and n is a positive integer, what can you say about the existence of inflection points of the function $f(x) = (x - a)^n$? Justify your answer.

If f is increasing on an interval $[0, b)$, then it follows from Definition 5.1.1 that $f(0) < f(x)$ for each x in the interval. Use this result in Exercises 35–38.

35. Show that $\sqrt[3]{1 + x} < 1 + \frac{1}{3}x$ if $x > 0$, and confirm the inequality with a graphing utility. [*Hint:* Show that the function $f(x) = 1 + \frac{1}{3}x - \sqrt[3]{1 + x}$ is increasing on $[0, +\infty)$.]

36. Show that $x < \tan x$ if $0 < x < \pi/2$, and confirm the inequality with a graphing utility. [*Hint:* Show that the function $f(x) = \tan x - x$ is increasing on $[0, \pi/2)$.]

37. Use a graphing utility to make a conjecture about the relative sizes of x and $\sin x$ for $x \geq 0$, and prove your conjecture.

38. (a) Show that $e^x \geq 1 + x$ if $x \geq 0$.
   (b) Show that $e^x \geq 1 + x + \frac{1}{2}x^2$ if $x \geq 0$.
   (c) Confirm the inequalities in parts (a) and (b) with a graphing utility.

In Exercises 39 and 40, use a graphing utility to generate the graphs of f' and f'' over the stated interval; then use those graphs to estimate the x-coordinates of the inflection points of f, the intervals on which f is concave up or down, and the intervals on which f is increasing or decreasing. Check your estimates by graphing f.

39. $f(x) = x^4 - 24x^2 + 12x$, $-5 \le x \le 5$

40. $f(x) = \dfrac{1}{1 + x^2}$, $-5 \le x \le 5$

41. For the function $f(x) = e^x/(1 + x^2)$, use the method of Example 6 in Section 2.4 to approximate the x-coordinates of the inflection points to two decimal places.

42. For the function f in Exercise 40, use the method of Example 6 in Section 2.4 to approximate the x-coordinates of the inflection points to two decimal places.

In Exercises 43 and 44, use a CAS to find f'', and then use the method of Example 6 in Section 2.4 to approximate the x-coordinates of the inflection points to one decimal place. Confirm that your answer is consistent with the graph of f.

43. $f(x) = \dfrac{10x - 3}{3x^2 - 5x + 8}$ **44.** $f(x) = \dfrac{x^3 - 8x + 7}{\sqrt{x^2 + 1}}$

45. Use Definition 5.1.1 to prove that $f(x) = x^2$ is increasing on $[0, +\infty)$.

46. Use Definition 5.1.1 to prove that $f(x) = 1/x$ is decreasing on $(0, +\infty)$.

47. In each part, determine whether the statement is true or false. If it is false, find functions for which the statement fails to hold.
(a) If f and g are increasing on an interval, then so is $f + g$.
(b) If f and g are increasing on an interval, then so is $f \cdot g$.

48. In each part, find functions f and g that are increasing on $(-\infty, +\infty)$ and for which $f - g$ has the stated property.
(a) $f - g$ is decreasing on $(-\infty, +\infty)$.
(b) $f - g$ is constant on $(-\infty, +\infty)$.
(c) $f - g$ is increasing on $(-\infty, +\infty)$.

49. (a) Prove that a general cubic polynomial
$$f(x) = ax^3 + bx^2 + cx + d \quad (a \ne 0)$$
has exactly one inflection point.
(b) Prove that if a cubic polynomial has three x-intercepts, then the inflection point occurs at the average value of the intercepts.

(c) Use the result in part (b) to find the inflection point of the cubic polynomial $f(x) = x^3 - 3x^2 + 2x$, and check your result by using f'' to determine where f is concave up and concave down.

50. From Exercise 49, the polynomial $f(x) = x^3 + bx^2 + 1$ has one inflection point. Use a graphing utility to reach a conclusion about the effect of the constant b on the location of the inflection point. Use f'' to explain what you have observed graphically.

51. Use Definition 5.1.1 to prove:
(a) If f is increasing on the intervals $(a, c]$ and $[c, b)$, then f is increasing on (a, b).
(b) If f is decreasing on the intervals $(a, c]$ and $[c, b)$, then f is decreasing on (a, b).

52. Use part (a) of Exercise 51 to show that $f(x) = x + \sin x$ is increasing on the interval $(-\infty, +\infty)$.

53. Suppose that the spread of a flu virus on a college campus is modeled by the function
$$y(t) = \dfrac{1000}{1 + 999e^{-0.9t}}$$
where $y(t)$ is the number of infected students at time t (in days, starting with $t = 0$). Use a graphing utility to estimate the day on which the virus is spreading most rapidly.

54. Let $y = 1/(1 + x^2)$. Find the values of x for which y is increasing and decreasing most rapidly.

In Exercises 55 and 56, suppose that water is flowing at a constant rate into the container shown. Make a rough sketch of the graph of the water level y versus the time t. Make sure that your sketch conveys where the graph is concave up and concave down, and label the y-coordinates of the inflection points.

55.

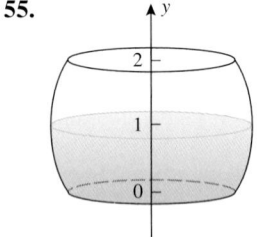

56.

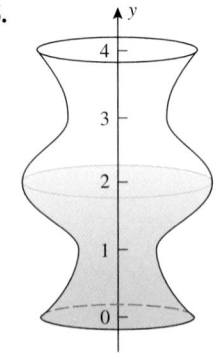

5.2 ANALYSIS OF FUNCTIONS II: RELATIVE EXTREMA; FIRST AND SECOND DERIVATIVE TESTS

In this section we will discuss methods for finding the high and low points on the graph of a function. The ideas we develop here will have important applications.

RELATIVE MAXIMA AND MINIMA

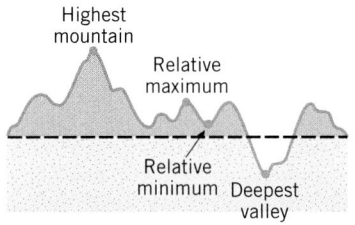

Highest mountain
Relative maximum
Relative minimum Deepest valley

Figure 5.2.1

If we imagine the graph of a function f to be a two-dimensional mountain range with hills and valleys, then the tops of the hills are called *relative maxima*, and the bottoms of the valleys are called *relative minima* (Figure 5.2.1).

The relative maxima are the high points in their *immediate vicinity*, and the relative minima are the low points. Note that a relative maximum need not be the highest point in the entire mountain range, and a relative minimum need not be the lowest point—they are just high and low points *relative* to the nearby terrain. These ideas are captured in the following definition.

> **5.2.1** DEFINITION. A function f is said to have a ***relative maximum*** at x_0 if there is an open interval containing x_0 on which $f(x_0)$ is the largest value, that is, $f(x_0) \geq f(x)$ for all x in the interval. Similarly, f is said to have a ***relative minimum*** at x_0 if there is an open interval containing x_0 on which $f(x_0)$ is the smallest value, that is, $f(x_0) \leq f(x)$ for all x in the interval. If f has either a relative maximum or a relative minimum at x_0, then f is said to have a ***relative extremum*** at x_0.

Example 1

Locate the relative extrema of the four functions graphed in Figure 5.2.2.

Solution.

(a) The function $f(x) = x^2$ has a relative minimum at $x = 0$ but no relative maxima.

(b) The function $f(x) = x^3$ has no relative extrema.

(c) The function $f(x) = x^3 - 3x + 3$ has a relative maximum at $x = -1$ and a relative minimum at $x = 1$.

(d) The function $f(x) = \cos x$ has relative maxima at all even multiples of π and relative minima at all odd multiples of π. ◀

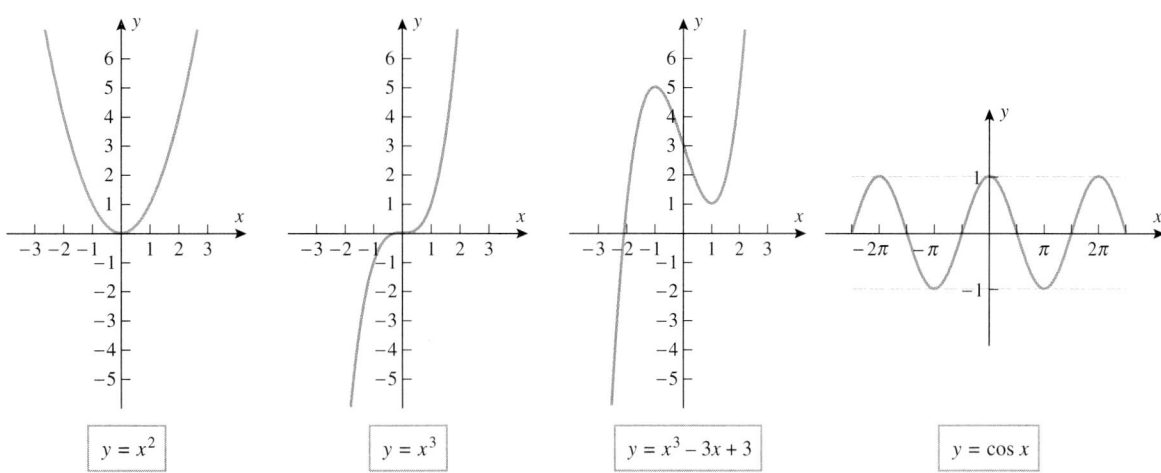

$$y = x^2$$ $$y = x^3$$ $$y = x^3 - 3x + 3$$ $$y = \cos x$$

Figure 5.2.2

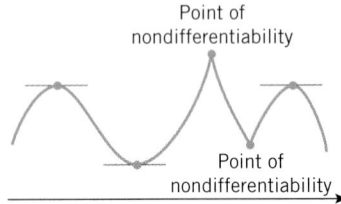

Figure 5.2.3

Relative extrema can be viewed as the transition points that separate the regions where a graph is increasing from those where it is decreasing. As suggested by Figure 5.2.3, the relative extrema of a continuous function f occur either at corners or at points where the graph of f has a horizontal tangent line. This is the content of the following theorem, whose proof is given in Appendix G.

5.2.2 THEOREM. *If a function f has any relative extrema, then they occur either at points where $f'(x) = 0$ or at points where f is not differentiable.*

The points at which either $f'(x) = 0$ or f is not differentiable are called the ***critical points*** of f, so that Theorem 5.2.2 can be rephrased as follows:

> *The relative extrema of a function, if any, occur at critical points.*

CRITICAL POINTS

Sometimes we will want to distinguish the critical points at which $f'(x) = 0$ from those points where f is not differentiable, in which case we will call the critical points at which $f'(x) = 0$ the ***stationary points*** of f.

It is important not to read too much into Theorem 5.2.2—the theorem asserts that the relative extrema must occur at critical points, but it does not say that a relative extremum occurs at *every* critical point; that is, there may be critical points at which a relative extremum does not occur. For example, for the eight critical points shown in Figure 5.2.4, relative extrema occur at all of the points in the top row, but not at any of the points in the bottom row.

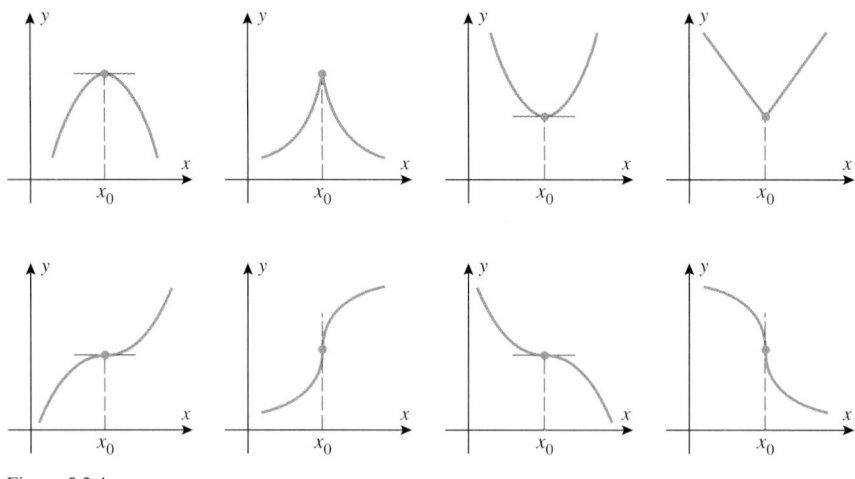

Figure 5.2.4

FIRST DERIVATIVE TEST

To develop an effective method for finding critical points of a function f, we need some criteria that will enable us to distinguish between the critical points where relative extrema occur and those where they do not. One such criterion can be motivated by examining the sign of the first derivative of f on each side of the eight critical points in Figure 5.2.4:

- At the two relative maxima in the top row, f' is positive to the left of x_0 and negative to the right.
- At the two relative minima in the top row, f' is negative to the left of x_0 and positive to the right.
- At the first two critical points in the bottom row, f' is positive on both sides of x_0.
- At the last two critical points in the bottom row, f' is negative on both sides of x_0.

These observations suggest that relative extrema of a function f occur at those critical points, and only those critical points, where f' changes sign. Moreover, if the sign changes from positive to negative, then a relative maximum occurs; and if the sign changes from negative to positive, then a relative minimum occurs. This is the content of the following theorem, whose proof is given at the end of this section.

5.2.3 THEOREM (*First Derivative Test*). *Suppose f is continuous at a critical point x_0.*

(a) *If $f'(x) > 0$ on an open interval extending left from x_0 and $f'(x) < 0$ on an open interval extending right from x_0, then f has a relative maximum at x_0.*

(b) *If $f'(x) < 0$ on an open interval extending left from x_0 and $f'(x) > 0$ on an open interval extending right from x_0, then f has a relative minimum at x_0.*

(c) *If $f'(x)$ has the same sign [either $f'(x) > 0$ or $f'(x) < 0$] on an open interval extending left from x_0 and on an open interval extending right from x_0, then f does not have a relative extremum at x_0.*

Example 2

(a) Locate the relative maxima and minima of $f(x) = 3x^{5/3} - 15x^{2/3}$.

(b) Confirm that the results in part (a) agree with the graph of f.

Solution (a). The function f is defined and continuous for all real values of x, and its derivative is

$$f'(x) = 5x^{2/3} - 10x^{-1/3} = 5x^{-1/3}(x - 2) = \frac{5(x - 2)}{x^{1/3}}$$

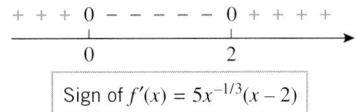

Sign of $f'(x) = 5x^{-1/3}(x - 2)$

Figure 5.2.5

Since $f'(x)$ does not exist if $x = 0$, and since $f'(x) = 0$ if $x = 2$, there are critical points at $x = 0$ and $x = 2$. To apply the first derivative test, we examine the sign of $f'(x)$ on intervals extending to the left and right of the critical points (Figure 5.2.5). Since the sign of the derivative changes from positive to negative at $x = 0$, there is a relative maximum there, and since it changes from negative to positive at $x = 2$, there is a relative minimum there.

Solution (b). The result in part (a) agrees with the graph of f shown in Figure 5.2.6. ◄

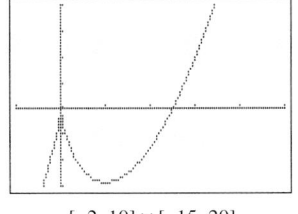

$[-2, 10] \times [-15, 20]$
xScl = 2, yScl = 5

Figure 5.2.6

FOR THE READER. As discussed in the subsection of Section 1.3 entitled Errors of Omission, many graphing utilities omit portions of the graphs of functions with fractional exponents and must be "tricked" into producing complete graphs; and indeed, for the function in the last example the author's calculator and CAS both failed to produce the portion of the graph over the negative x-axis. To generate the graph in Figure 5.2.6, the author had to apply the techniques discussed in Exercise 29 of Section 1.3 to each term in the formula for f. Use a graphing utility to generate this graph.

Example 3

Locate the relative extrema of $f(x) = x^3 - 3x^2 + 3x - 1$, if any.

Solution. Since f is differentiable everywhere, the only possible critical points are stationary points. Differentiating f yields

$$f'(x) = 3x^2 - 6x + 3 = 3(x - 1)^2$$

Solving $f'(x) = 0$ yields $x = 1$ as the only stationary point. However, $3(x - 1)^2 \geq 0$ for all x, so $f'(x)$ does not change sign at $x = 1$; consequently, f does not have a relative extremum at $x = 1$. Thus, f has no relative extrema (Figure 5.2.7). ◄

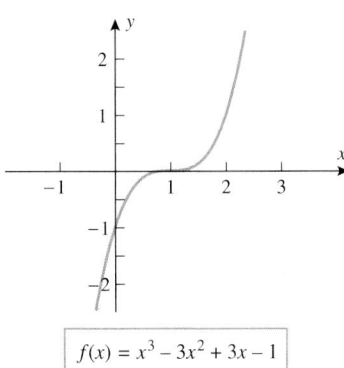

$f(x) = x^3 - 3x^2 + 3x - 1$

Figure 5.2.7

FOR THE READER. How many relative extrema can a polynomial of degree n have? Explain your reasoning.

SECOND DERIVATIVE TEST

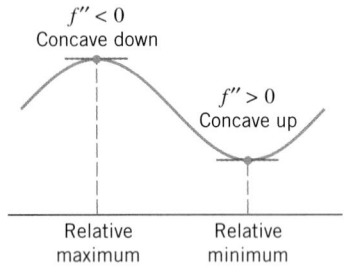

$f'' < 0$
Concave down

$f'' > 0$
Concave up

Relative maximum   Relative minimum

Figure 5.2.8

There is another test for relative extrema that is often easier to apply than the first derivative test. It is based on the geometric observation that a function f has a relative maximum at a stationary point if the graph of f is concave down on an open interval containing the point, and it has a relative minimum if it is concave up (Figure 5.2.8).

5.2.4   THEOREM (*Second Derivative Test*).   *Suppose that f is twice differentiable at the point x_0.*

(a)   *If $f'(x_0) = 0$ and $f''(x_0) > 0$, then f has a relative minimum at x_0.*

(b)   *If $f'(x_0) = 0$ and $f''(x_0) < 0$, then f has a relative maximum at x_0.*

(c)   *If $f'(x_0) = 0$ and $f''(x_0) = 0$, then the test is inconclusive; that is, f may have a relative maximum, a relative minimum, or neither at x_0.*

REMARK.   The proof of parts (a) and (b) is given at the end of this section. For part (c), consider the functions $f(x) = x^3$, $f(x) = x^4$, and $f(x) = -x^4$. In all three cases we have $f'(0) = 0$ and $f''(0) = 0$ (verify); but from Figure 1.6.4, $f(x) = x^4$ has a relative minimum at $x = 0$, $f(x) = -x^4$ has a relative maximum at $x = 0$ (why?), and $f(x) = x^3$ has neither a relative maximum nor a relative minimum at $x = 0$.

Example 4

Locate the relative maxima and minima of $f(x) = x^4 - 2x^2$, and confirm that your results are consistent with the graph of f.

Solution.

$$f'(x) = 4x^3 - 4x = 4x(x - 1)(x + 1)$$
$$f''(x) = 12x^2 - 4$$

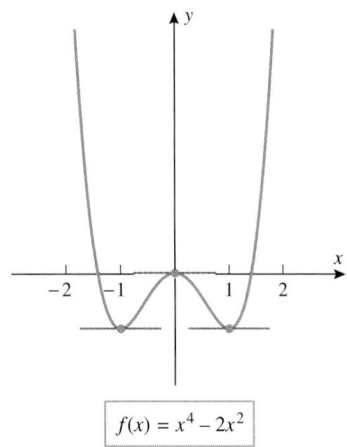

$f(x) = x^4 - 2x^2$

Figure 5.2.9

Solving $f'(x) = 0$ yields the stationary points $x = 0$, $x = 1$, and $x = -1$. Evaluating f'' at these points yields

$$f''(0) = -4 < 0$$
$$f''(1) = 8 > 0$$
$$f''(-1) = 8 > 0$$

so there is a relative maximum at $x = 0$ and relative minima at $x = 1$ and $x = -1$ (Figure 5.2.9).   ◀

MORE ON THE SIGNIFICANCE OF INFLECTION POINTS

In Section 5.1 we observed that the inflection points of a curve $y = f(x)$ mark the points where the slopes of the tangent lines change from increasing to decreasing, or vice versa. Thus, in the case where f is twice differentiable, the inflection points mark the places on the curve $y = f(x)$ where $f'(x)$ has a relative maximum or minimum (Figure 5.2.10); stated another way:

Inflection points mark the places on the curve $y = f(x)$ at which the rate of change of y with respect to x has a relative maximum or minimum; that is, they are the places where y is increasing or decreasing most rapidly in the immediate vicinity.

As an illustration of this principle, consider the flask shown in Figure 5.1.15. We observed in Section 5.1 that if water is poured into the flask so that the volume increases at a constant rate, then the graph of y versus t has an inflection point when y is at the narrow point in the neck. However, this is also the place where the water level is rising most rapidly.

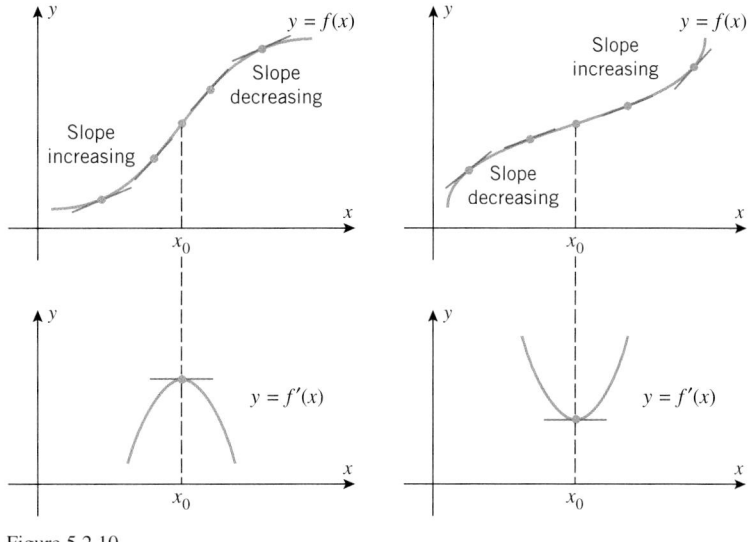

Figure 5.2.10

▪ PROOF OF THE FIRST DERIVATIVE TEST (*Theorem 5.2.3*)

Proof. We will prove part (*a*) and leave parts (*b*) and (*c*) as exercises. We are assuming that $f'(x) > 0$ on the interval (a, x_0) and that $f'(x) < 0$ on the interval (x_0, b), and we want to show that

$$f(x_0) \geq f(x)$$

for all x in the interval (a, b). However, the two hypotheses, together with Theorem 5.1.2 (and its following remark) imply that f is increasing on the interval $(a, x_0]$ and decreasing on the interval $[x_0, b)$. Thus, $f(x_0) \geq f(x)$ for all x in (a, b) with equality only at x_0. ▪

▪ PROOF OF THE SECOND DERIVATIVE TEST (*Theorem 5.2.4*)

Proof. We will prove part (*a*) and leave part (*b*) as an exercise. We want to show that if $f'(x_0) = 0$ and $f''(x_0) > 0$, then f has a relative minimum at x_0; that is, there is an open interval (a, b) containing x_0 on which

$$f(x) \geq f(x_0)$$

For simplicity, we will assume that f'' is continuous at x_0. The proof for the case where f is twice differentiable at x_0 is left for more advanced courses. Observe first that the tangent line at x_0 is horizontal [since $f'(x_0) = 0$], and hence its equation is $y = f(x_0)$. Moreover, since $f''(x_0) > 0$, and since f'' is continuous at x_0, there is an open interval (a, b) containing x_0 on which $f''(x) > 0$. This implies that f is concave up on (a, b), and hence its graph lies above the tangent line $y = f(x_0)$ over the interval (a, b). This shows that $f(x) \geq f(x_0)$ on the interval (a, b). ▪

EXERCISE SET 5.2 ⌇ Graphing Calculator [c] CAS

1. In each part, sketch the graph of a continuous function f with the stated properties.
   (a) f is concave up on the interval $(-\infty, +\infty)$ and has exactly one relative extremum.
   (b) f is concave up on the interval $(-\infty, +\infty)$ and has no relative extrema.
   (c) The function f has exactly two relative extrema on the interval $(-\infty, +\infty)$, and $f(x) \to +\infty$ as $x \to +\infty$.
   (d) The function f has exactly two relative extrema on the interval $(-\infty, +\infty)$, and $f(x) \to -\infty$ as $x \to +\infty$.

2. In each part, sketch the graph of a continuous function f with the stated properties.
   (a) f has exactly one relative extremum on $(-\infty, +\infty)$, and $f(x) \to 0$ as $x \to +\infty$ and as $x \to -\infty$.
   (b) f has exactly two relative extrema on $(-\infty, +\infty)$, and $f(x) \to 0$ as $x \to +\infty$ and as $x \to -\infty$.
   (c) f has exactly one inflection point and one relative extremum on $(-\infty, +\infty)$.
   (d) f has infinitely many relative extrema, and $f(x) \to 0$ as $x \to +\infty$ and as $x \to -\infty$.

3. (a) Use both the first and second derivative tests to show that $f(x) = 3x^2 - 6x + 1$ has a relative minimum at $x = 1$.
   (b) Use both the first and second derivative tests to show that $f(x) = x^3 - 3x + 3$ has a relative minimum at $x = 1$ and a relative maximum at $x = -1$.

4. (a) Use both the first and second derivative tests to show that $f(x) = \sin^2 x$ has a relative minimum at $x = 0$.
   (b) Use both the first and second derivative tests to show that $g(x) = \tan^2 x$ has a relative minimum at $x = 0$.
   (c) Give an informal verbal argument to explain without calculus why the functions in parts (a) and (b) have relative minima at $x = 0$.

5. (a) Show that both of the functions $f(x) = (x - 1)^4$ and $g(x) = x^3 - 3x^2 + 3x - 2$ have stationary points at $x = 1$.
   (b) What does the second derivative test tell you about the nature of these stationary points?
   (c) What does the first derivative test tell you about the nature of these stationary points?

6. (a) Show that $f(x) = 1 - x^5$ and $g(x) = 3x^4 - 8x^3$ both have stationary points at $x = 0$.
   (b) What does the second derivative test tell you about the nature of these stationary points?
   (c) What does the first derivative test tell you about the nature of these stationary points?

In Exercises 7–12, locate the critical points, and classify them as stationary points or points of nondifferentiability.

7. (a) $f(x) = x^3 + 3x^2 - 9x + 1$
   (b) $f(x) = x^4 - 6x^2 - 3$

8. (a) $f(x) = 2x^3 - 6x + 7$ (b) $f(x) = 3x^4 - 4x^3$

9. (a) $f(x) = \dfrac{x}{x^2 + 2}$ (b) $f(x) = x^{2/3}$

10. (a) $f(x) = \dfrac{x^2 - 3}{x^2 + 1}$ (b) $f(x) = \sqrt[3]{x + 2}$

11. (a) $f(x) = x^{1/3}(x + 4)$ (b) $f(x) = \cos 3x$

12. (a) $f(x) = x^{4/3} - 6x^{1/3}$ (b) $f(x) = |\sin x|$

In Exercises 13 and 14, use the graph of f' shown in the figure to estimate all values of x at which f has (a) relative minima, (b) relative maxima, and (c) inflection points.

13.
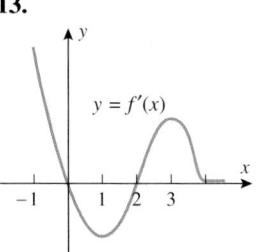

14.
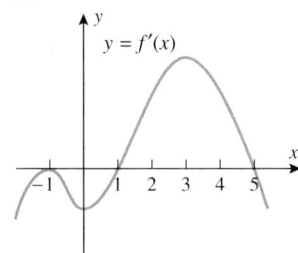

In Exercises 15 and 16, use the given derivative to find the x-coordinates of all critical points of f, and determine whether a relative maximum, relative minimum, or neither occurs there.

15. (a) $f'(x) = x^3(x^2 - 5)$ (b) $f'(x) = xe^{-x}$

16. (a) $f'(x) = x^2(2x + 1)(x - 1)$
    (b) $f'(x) = \dfrac{9 - 4x^2}{\sqrt[3]{x + 1}}$

In Exercises 17–20, find the relative extrema using both the first and second derivative tests.

17. $f(x) = 1 - 4x - x^2$ 18. $f(x) = 2x^3 - 9x^2 + 12x$

19. $f(x) = \sin^2 x$, $0 < x < 2\pi$

20. $f(x) = \frac{1}{2}x - \sin x$, $0 < x < 2\pi$

In Exercises 21–34, use any method to find the relative extrema of the function f.

21. $f(x) = x^3 + 5x - 2$ 22. $f(x) = x^4 - 2x^2 + 7$

23. $f(x) = x(x - 1)^2$ 24. $f(x) = x^4 + 2x^3$

25. $f(x) = 2x^2 - x^4$ 26. $f(x) = (2x - 1)^5$

27. $f(x) = x^{4/5}$ 28. $f(x) = 2x + x^{2/3}$

29. $f(x) = \dfrac{x^2}{x^2 + 1}$ 30. $f(x) = \dfrac{x}{x + 2}$

31. $f(x) = \ln(1 + x^2)$

32. $f(x) = x^2 e^x$

33. $f(x) = |x^2 - 4|$

34. $f(x) = \begin{cases} 9 - x, & x \le 3 \\ x^2 - 3, & x > 3 \end{cases}$

In Exercises 35–38, find the relative extrema in the interval $0 < x < 2\pi$, and confirm that your results are consistent with the graph of f generated with a graphing utility.

35. $f(x) = |\sin 2x|$

36. $f(x) = \sqrt{3}x + 2\sin x$

37. $f(x) = \cos^2 x$

38. $f(x) = \dfrac{\sin x}{2 - \cos x}$

In Exercises 39–42, use a graphing utility to make a conjecture about the relative extrema of f, and then check your conjecture using either the first or second derivative test.

39. $f(x) = x \ln x$

40. $f(x) = \dfrac{2}{e^x + e^{-x}}$

41. $f(x) = x^2 e^{-2x}$

42. $f(x) = 10 \ln x - x$

In Exercises 43 and 44, use a graphing utility to generate the graphs of f' and f'' over the stated interval, and then use those graphs to estimate the x-coordinates of the relative extrema of f. Check that your estimates are consistent with the graph of f.

43. $f(x) = x^4 - 24x^2 + 12x, \quad -5 \le x \le 5$

44. $f(x) = \sin \frac{1}{2}x \cos x, \quad -\pi/2 \le x \le \pi/2$

45. For the function f in Exercise 43, use the method of Example 6 in Section 2.4 to approximate the x-coordinates of the relative maxima to two decimal places.

46. For the function f in Exercise 44, use the method of Example 6 in Section 2.4 to approximate the x-coordinates of the relative maxima to two decimal places.

In Exercises 47 and 48, use a CAS to graph f' and f'' over the stated interval. Use those graphs to make a conjecture about the locations and nature of the relative extrema of f, and check your conjecture by graphing f.

47. $f(x) = \dfrac{10x - 3}{3x^2 - 5x + 8}$

48. $f(x) = \dfrac{x^3 - 8x + 7}{\sqrt{x^2 + 1}}$

49. In each part, find k so that f has a relative extremum at the point $x = 3$.

(a) $f(x) = x^2 + \dfrac{k}{x}$

(b) $f(x) = \dfrac{x}{x^2 + k}$

50. Functions of the form

$$f(x) = cx^n e^{-x}, \quad x > 0$$

where n is a positive integer and $c = 1/n!$, arise in the statistical study of traffic flow.

(a) Use a graphing utility to generate the graph of f for $n = 2, 3, 4,$ and 5, and make a conjecture about the number and locations of the relative extrema of f.

(b) Confirm your conjecture using the first derivative test.

51. Functions of the form

$$f(x) = \dfrac{1}{\sqrt{2\pi}} e^{-x^2/2}$$

arise in a wide variety of statistical problems.

(a) Use the first derivative test to show that f has a relative maximum at $x = 0$, and confirm this by using a graphing utility to graph f.

(b) Sketch the graph of

$$f(x) = \dfrac{1}{\sqrt{2\pi}} e^{-(x-\mu)^2/2}$$

where μ is a constant, and label the coordinates of the relative extrema.

52. (a) Use a CAS to graph the function

$$f(x) = \dfrac{x^4 + 1}{x^2 + 1}$$

and use the graph to estimate the x-coordinates of the relative extrema.

(b) Find the exact x-coordinates by using the CAS to solve the equation $f'(x) = 0$.

53. Find values of $a, b, c,$ and d so that the function

$$f(x) = ax^3 + bx^2 + cx + d$$

has a relative minimum at $(0, 0)$ and a relative maximum at $(1, 1)$.

54. Let h and g have relative maxima at x_0. Prove or disprove:

(a) $h + g$ has a relative maximum at x_0

(b) $h - g$ has a relative maximum at x_0.

55. Sketch some curves that show that the three parts of the first derivative test (Theorem 5.2.3) can be false without the assumption that f is continuous at x_0.

5.3 ANALYSIS OF FUNCTIONS III: APPLYING TECHNOLOGY AND THE TOOLS OF CALCULUS

In this section we will discuss how to use technology and the tools of calculus that we developed in the last two sections to analyze various types of graphs that occur in applications.

This section contains a brief review of material on polynomials. Readers who want to review this material in more depth are referred to Appendix F. Instructors who want to spend more time on this section can divide the section into two parts, treating the analysis of polynomials and rational functions in one lecture and the remaining topics in a second lecture.

PROPERTIES OF GRAPHS

In many problems, the properties of interest in the graph of a function are:

- symmetries
- x-intercepts
- relative extrema
- intervals of increase and decrease
- asymptotes

- periodicity
- y-intercepts
- inflection points
- concavity
- behavior as $x \to +\infty$ or $x \to -\infty$

Some of these properties may not be relevant in certain cases; for example, asymptotes are characteristic of rational functions but not of polynomials, and periodicity is characteristic of trigonometric functions but not of logarithmic or exponential functions. Thus, when analyzing the graph of a function f, it helps to know something about the general properties of the family to which it belongs.

In a given problem you will usually have a definite objective for your analysis. For example, you may be interested in finding a ***complete graph*** of $y = f(x)$, that is, a graph that shows all of the important characteristics of f; or you may be interested in something specific, say the exact location of the relative extrema or the behavior of the graph as $x \to +\infty$ or $x \to -\infty$. However, regardless of your objectives, you will usually find it helpful to begin your analysis by generating the graph with a graphing utility. As discussed in Section 1.3, this graph may or may not be complete, and some of the important characteristics may be obscured by compression or resolution problems. However, with this graph as a starting point, you can often use calculus to complete the analysis and resolve any ambiguities.

A PROCEDURE FOR ANALYZING GRAPHS

There are no hard and fast rules that are guaranteed to produce all of the information you may need about the graph of a function f, but here is one possible way of organizing the analysis of a function (the order of the steps can be varied).

Step 1. Use a graphing utility to generate the graph of f in some reasonable window, taking advantage of any general knowledge you have about the function to help in choosing the window.

Step 2. See if the graph suggests the existence of symmetries, periodicity, or domain restrictions. If so, try to confirm those properties analytically.

Step 3. Find the intercepts, if needed.

Step 4. Investigate the behavior of the graph as $x \to +\infty$ and as $x \to -\infty$, and identify all horizontal and vertical asymptotes, if any.

Step 5. Calculate $f'(x)$ and $f''(x)$, and use these derivatives to determine the critical points, the intervals on which f is increasing or decreasing, the intervals on which f is concave up and concave down, and the inflection points.

Step 6. If you have discovered that some of the significant features did not fall within the graphing window in Step 1, then try adjusting the window to include them. However, it is possible that compression or resolution problems may prevent you from showing all of the features of interest in a single window, in which case you may need to use different windows to focus on different features. In some cases you may even find that a hand-drawn sketch labeled with the location of the significant features is clearer or more informative than a graph generated with a graphing utility.

A BRIEF REVIEW OF POLYNOMIALS

Recall that if n is a nonnegative integer, then a ***polynomial of degree n*** is a function that can be written in the following forms, depending on whether you want the powers of x in ascending or descending order:

$$c_0 + c_1 x + c_2 x^2 + \cdots + c_n x^n \quad (c_n \neq 0)$$
$$c_n x^n + c_{n-1} x^{n-1} + \cdots + c_1 x + c_0 \quad (c_n \neq 0)$$

The numbers $c_0, c_1, \ldots, c_n$ are called the ***coefficients*** of the polynomial. The coefficient c_n (which multiplies the highest power of x) is called the ***leading coefficient***, the term $c_n x^n$ is called the ***leading term***, and the coefficient c_0 is called the ***constant term***. Polynomials of degree 1, 2, 3, 4, and 5 are called ***linear, quadratic, cubic, quartic***, and ***quintic***, respectively. For simplicity, general polynomials of low degree are often written without subscripts on the coefficients:

$p(x) = a$	Constant polynomial
$p(x) = ax + b \quad (a \neq 0)$	Linear polynomial
$p(x) = ax^2 + bx + c \quad (a \neq 0)$	Quadratic polynomial
$p(x) = ax^3 + bx^2 + cx + d \quad (a \neq 0)$	Cubic polynomial

When you attempt to factor a polynomial completely, one of three things can happen:

- You may be able to decompose the polynomial into distinct linear factors using only real numbers; for example,

$$x^3 + x^2 - 2x = x(x^2 + x - 2) = x(x - 1)(x + 2)$$

- You may be able to decompose the polynomial into linear factors using only real numbers, but some of the factors may be repeated; for example,

$$x^6 - 3x^4 + 2x^3 = x^3(x^3 - 3x + 2) = x^3(x - 1)^2(x + 2) \tag{1}$$

- You may be able to decompose the polynomial into linear and quadratic factors using only real numbers, but you may not be able to decompose the quadratic factors into linear factors without using imaginary numbers (such quadratic factors are said to be ***irreducible*** over the real numbers); for example,

$$x^4 - 1 = (x^2 - 1)(x^2 + 1) = (x - 1)(x + 1)(x^2 + 1)$$
$$= (x - 1)(x + 1)(x - i)(x + i)$$

Here, the factor $x^2 + 1$ is irreducible over the real numbers.

In general, if $p(x)$ is a polynomial of degree n with leading coefficient a, and if imaginary numbers are allowed, then $p(x)$ can be factored as

$$p(x) = a(x - r_1)(x - r_2) \cdots (x - r_n) \tag{2}$$

where $r_1, r_2, \ldots, r_n$ are called the **zeros** of $p(x)$ or the **roots** of the equation $p(x) = 0$, and (2) is called the **complete linear factorization** of $p(x)$. If some of the factors in (2) are repeated, then they can be combined; for example, if the first k factors are distinct and the rest are repetitions of the first k, then (2) can be expressed in the form

$$p(x) = a(x - r_1)^{m_1}(x - r_2)^{m_2} \cdots (x - r_k)^{m_k} \tag{3}$$

where $r_1, r_2, \ldots, r_k$ are the *distinct* roots of $p(x) = 0$. The exponents $m_1, m_2, \ldots, m_k$ tell us how many times the various factors occur in the complete linear factorization; for example, in (3) the factor $(x - r_1)$ occurs m_1 times, the factor $(x - r_2)$ occurs m_2 times, and so forth. Some techniques for factoring polynomials are discussed in Appendix F. In general, if a factor $(x - r)$ occurs m times in the complete linear factorization of a polynomial, then we say that r is a root or zero of **multiplicity m**, and if $(x - r)$ has no repetitions (i.e., r has multiplicity 1), then we say that r is a **simple** root or zero. For example, it follows from (1) that the equation $x^6 - 3x^4 + 2x^3 = 0$ can be expressed as

$$x^3(x - 1)^2(x + 2) = 0 \tag{4}$$

so this equation has three distinct roots—a root $x = 0$ of multiplicity 3, a root $x = 1$ of multiplicity 2, and a simple root $x = -2$.

Note that in (3) the multiplicities of the roots must add up to n, since $p(x)$ has degree n; that is,

$$m_1 + m_2 + \cdots + m_k = n$$

For example, in (4) the multiplicities add up to 6, which is the same as the degree of the polynomial.

It follows from (2) that a polynomial of degree n can have at most n distinct roots; if all of the roots are simple, then there will be *exactly* n, but if some are repeated, then there will be fewer than n. However, when counting the roots of a polynomial, it is standard practice to count multiplicities, since that convention allows us to say that a polynomial of degree n has n roots. For example, from (1) the six roots of the polynomial $p(x) = x^6 - 3x^4 + 2x^3$ are

$$r = 0, \quad 0, \quad 0, \quad 1, \quad 1, \quad -2$$

In summary, we have the following important theorem.

5.3.1 THEOREM. *If imaginary roots are allowed, and if roots are counted according to their multiplicity, then a polynomial of degree n has exactly n roots.*

ANALYSIS OF POLYNOMIALS

Polynomials are among the simplest functions to graph and analyze, since their only significant features are intercepts, relative extrema, inflection points, and the behavior as $x \to +\infty$ and $x \to -\infty$. Figure 5.3.1 shows the graphs of four typical polynomials in x.

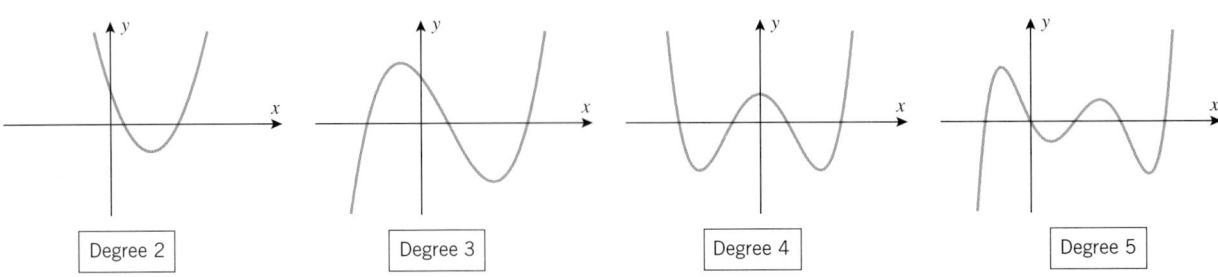

Degree 2   Degree 3   Degree 4   Degree 5

Figure 5.3.1

The graphs in Figure 5.3.1 have properties that are common to all polynomials:

- The natural domain of a polynomial in x is the entire x-axis, since the only operations involved in its formula are additions, subtractions, and multiplications; the range depends on the particular polynomial.
- Graphs of polynomials are continuous since polynomials are continuous functions.
- Graphs of polynomials have no sharp corners or points of vertical tangency, since polynomials are differentiable functions.
- The graph of a polynomial eventually increases or decreases without bound as $x \to +\infty$ or $x \to -\infty$, since the limit of a polynomial as $x \to +\infty$ or $x \to -\infty$ is $\pm\infty$ (see the subsection in Section 2.2 entitled Limits of Polynomials as $x \to +\infty$ or $x \to -\infty$).
- The graph of a polynomial of degree n has at most n x-intercepts, at most $n - 1$ relative extrema, and at most $n - 2$ inflection points.

The last property is a consequence of the fact that the x-intercepts, relative extrema, and inflection points occur at real roots of $p(x) = 0$, $p'(x) = 0$, and $p''(x) = 0$, respectively, so if $p(x)$ has degree n greater than 1, then $p'(x)$ has degree $n - 1$ and $p''(x)$ has degree $n - 2$. Thus, for example, the graph of a quadratic polynomial has at most two x-intercepts, one relative extremum, and no inflection points; and the graph of a cubic polynomial has at most three x-intercepts, two relative extrema, and one inflection point.

FOR THE READER.   For each of the graphs in Figure 5.3.1, count the number of x-intercepts, relative extrema, and inflection points, and confirm that your count is consistent with the degree of the polynomial.

Example 1

Figure 5.3.2 shows the graph of

$$y = x^3 - x^2 - 2x$$

produced on a graphing calculator. Confirm that the graph is complete, that is, it is not missing any significant features.

Solution.   We can be confident that the graph is complete because the polynomial has degree 3, and three roots, two relative extrema, and one inflection point are accounted for. Moreover, the graph exhibits the correct behavior as $x \to +\infty$ and $x \to -\infty$, since

$$\lim_{x \to +\infty} (x^3 - x^2 - 2x) = \lim_{x \to +\infty} x^3 = +\infty$$

$$\lim_{x \to -\infty} (x^3 - x^2 - 2x) = \lim_{x \to -\infty} x^3 = -\infty \qquad \blacktriangleleft$$

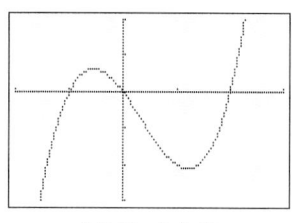

[–2, 3] × [–3, 2]
xScl = 1, yScl = 1

Figure 5.3.2

GEOMETRIC IMPLICATIONS OF MULTIPLICITY

For polynomials, there is a close relationship between the multiplicity of a root and the behavior of the graph in the vicinity of the root. For example, observe that the polynomial $p(x) = x^n$ has a root of multiplicity n at $x = 0$, and observe that the graphs in Figure 1.6.4 have the following geometric properties:

- When n is even, the graph of $y = p(x)$ is tangent to the x-axis at the origin but does not cross the x-axis there.
- When n is odd and greater than 1, the graph is tangent to the x-axis at the origin, has an inflection point at the origin, and crosses the x-axis there.
- When $n = 1$, the graph crosses the x-axis at the origin but is not tangent to the x-axis there.

These properties of $p(x) = x^n$ at $x = 0$ are special cases of the following more general result, which we state without formal proof (Figure 5.3.3).

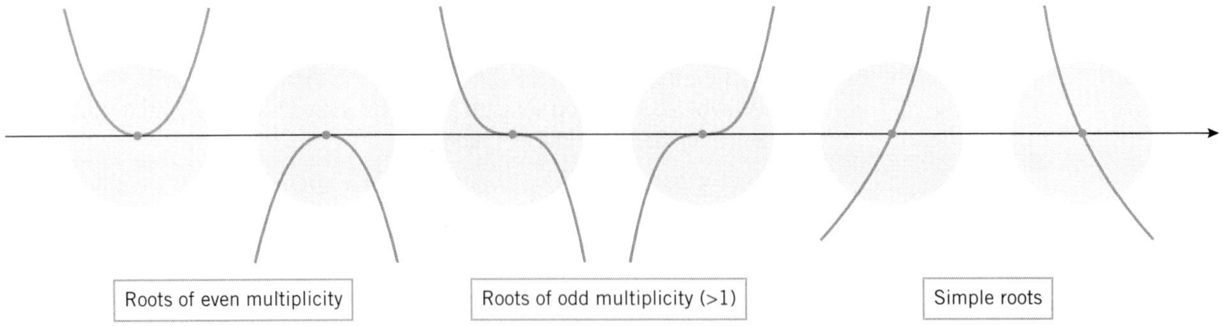

| Roots of even multiplicity | Roots of odd multiplicity (>1) | Simple roots |

Figure 5.3.3

5.3.2 THE GEOMETRIC IMPLICATIONS OF MULTIPLICITY. *Suppose that $p(x)$ is a polynomial with a root of multiplicity m at $x = r$.*

(a) *If m is even, then the graph of $y = p(x)$ is tangent to the x-axis at $x = r$ and does not cross the x-axis there.*

(b) *If m is odd and greater than 1, then the graph is tangent to the x-axis at $x = r$, has an inflection point there, and also crosses the x-axis there.*

(c) *If $m = 1$ (so that the root is simple), then the graph crosses the x-axis at $x = r$ but is not tangent to the x-axis there.*

Example 2

Make a conjecture about the behavior of the graph of

$$y = x^3(3x - 4)(x + 2)^2$$

in the vicinity of its x-intercepts, and test your conjecture by generating the graph.

Solution. The x-intercepts occur at $x = 0$, $x = \frac{4}{3}$, and $x = -2$. The root $x = 0$ has multiplicity 3, which is odd, so at that point the graph should be tangent to the x-axis, cross the x-axis, and have an inflection point. The root $x = -2$ has multiplicity 2, which is even, so the graph should be tangent to but not cross the x-axis there. The root $x = \frac{4}{3}$ is simple, so at that point the curve should cross the x-axis without being tangent to it. All of this is consistent with the graph in Figure 5.3.4. ◄

Example 3

Generate or sketch a complete graph of the equation

$$y = x^3 - 3x + 2$$

and identify the exact location of the intercepts, relative extrema, and inflection points.

Solution. Figure 5.3.5 shows a graph of the given equation produced with a graphing utility. We can be reasonably confident that the graph is complete since the polynomial has degree 3, and all roots, relative extrema, and inflection points are accounted for in the graph: There are three roots (a simple negative root and a positive root of multiplicity 2), and there are two relative extrema and one inflection point. The following analysis will confirm that the graph is complete and identify the exact location of the intercepts, relative extrema, and inflection points.

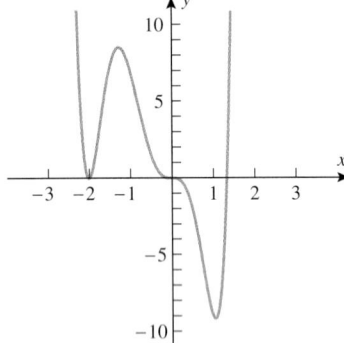

Figure 5.3.4

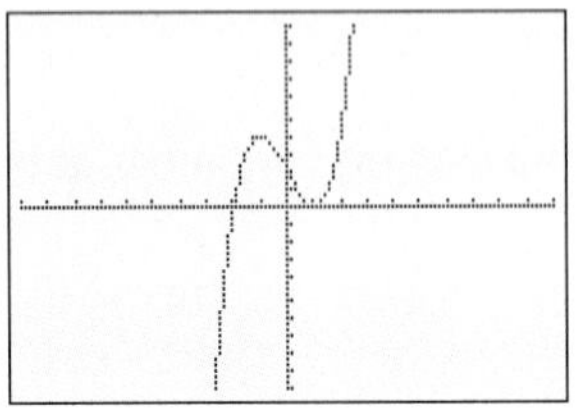

$[-10, 10] \times [-10, 10]$
$x\text{Scl} = 1, y\text{Scl} = 1$

Figure 5.3.5

- *x-intercepts:* Setting $y = 0$ yields

$$x^3 - 3x + 2 = (x + 2)(x^2 - 2x + 1) = (x + 2)(x - 1)^2 = 0$$

so there is a simple root at $x = -2$ and a root of multiplicity 2 at $x = 1$.

- *y-intercept:* Setting $x = 0$ yields $y = 2$.
- *Behavior as $x \to +\infty$ and $x \to -\infty$:* The graph in Figure 5.3.5 suggests that the graph increases without bound as $x \to +\infty$ and decreases without bound as $x \to -\infty$. This is confirmed by the limits

$$\lim_{x \to +\infty} (x^3 - 3x + 2) = \lim_{x \to +\infty} x^3 = +\infty$$

$$\lim_{x \to -\infty} (x^3 - 3x + 2) = \lim_{x \to -\infty} x^3 = -\infty$$

- *Derivatives:*

$$\frac{dy}{dx} = 3x^2 - 3 = 3(x - 1)(x + 1)$$

$$\frac{d^2y}{dx^2} = 6x$$

- *Intervals of increase and decrease; relative extrema:* Figure 5.3.6 shows the sign pattern of the first and second derivatives and what they imply about the graph shape. In the first part of the figure the upward arrows indicate intervals where the graph is increasing, the downward arrows indicate intervals where the graph is decreasing, and the horizontal arrows indicate the stationary points. The second part of the figure shows what the sign pattern of the second derivative implies about the concavity. The third part of the figure shows what the first and second derivatives together imply about the graph shape.

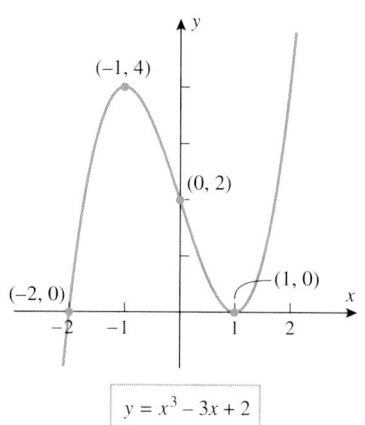

$$y = x^3 - 3x + 2$$

Figure 5.3.7

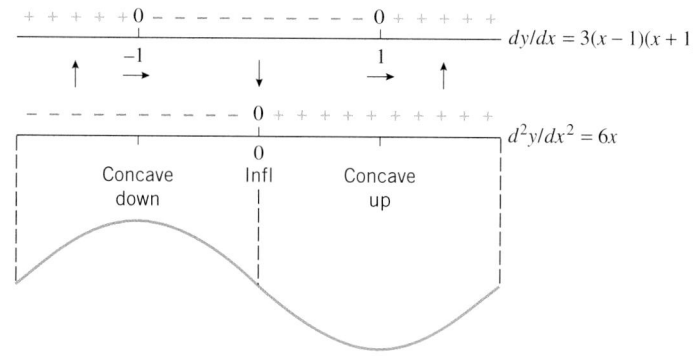

Figure 5.3.6

Figure 5.3.7 shows the complete graph labeled with the coordinates of the intercepts, relative extrema, and inflection point.   ◄

GRAPHING RATIONAL FUNCTIONS

Rational functions (ratios of polynomials) are more complicated to graph than polynomials because they have discontinuities and asymptotes.

Example 4

Generate or sketch a complete graph of the equation

$$y = \frac{2x^2 - 8}{x^2 - 16}$$

and identify the exact location of the intercepts, relative extrema, inflection points, and asymptotes.

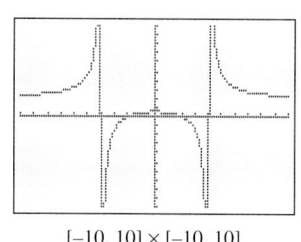

$[-10, 10] \times [-10, 10]$
$x\text{Scl} = 1, y\text{Scl} = 1$

Figure 5.3.8

Solution.   Figure 5.3.8 shows a calculator-generated graph of the equation in the window $[-10, 10] \times [-10, 10]$. The figure suggests that the graph is symmetric about the y-axis and has two vertical asymptotes and a horizontal asymptote. The figure also suggests that there is a relative maximum at $x = 0$ and two x-intercepts. There do not seem to be any inflection points. The following analysis will identify the exact location of the key features and help us determine whether the graph in Figure 5.3.8 is complete.

- *Symmetries:* Replacing x by $-x$ does not change the equation, so the graph is symmetric about the y-axis.
- *x-intercepts:* Setting $y = 0$ yields the x-intercepts $x = -2$ and $x = 2$.
- *y-intercept:* Setting $x = 0$ yields the y-intercept $y = 1/2$.
- *Vertical asymptotes:* Setting $x^2 - 16 = 0$ yields the vertical asymptotes $x = -4$ and $x = 4$.
- *Horizontal asymptotes:* The limits

$$\lim_{x \to +\infty} \frac{2x^2 - 8}{x^2 - 16} = \lim_{x \to +\infty} \frac{2x^2}{x^2} = 2$$

$$\lim_{x \to -\infty} \frac{2x^2 - 8}{x^2 - 16} = \lim_{x \to -\infty} \frac{2x^2}{x^2} = 2$$

yield the horizontal asymptote $y = 2$.

The set of points where x-intercepts or vertical asymptotes occur is $\{-4, -2, 2, 4\}$. These points divide the x-axis into the open intervals

$$(-\infty, -4), \quad (-4, -2), \quad (-2, 2), \quad (2, 4), \quad (4, +\infty)$$

Over each of these intervals, y cannot change sign (why?). We can find the sign of y on each interval by choosing an arbitrary test point in the interval and evaluating $y = f(x)$ at the test points (Table 5.3.1).

Table 5.3.1

INTERVAL	TEST POINT	$y = \dfrac{2x^2 - 8}{x^2 - 16}$	SIGN OF y
$(-\infty, -4)$	$x = -5$	$y = 14/3$	$+$
$(-4, -2)$	$x = -3$	$y = -10/7$	$-$
$(-2, 2)$	$x = 0$	$y = 1/2$	$+$
$(2, 4)$	$x = 3$	$y = -10/7$	$-$
$(4, +\infty)$	$x = 5$	$y = 14/3$	$+$

The information in Table 5.3.1 is consistent with Figure 5.3.8, so we can be certain that the calculator graph has not missed any sign changes. The next step is to use the first and second derivatives to determine whether the calculator graph has missed any relative extrema or changes in concavity.

- *Derivatives:*

$$\frac{dy}{dx} = \frac{(x^2 - 16)(4x) - (2x^2 - 8)(2x)}{\left(x^2 - 16\right)^2} = -\frac{48x}{\left(x^2 - 16\right)^2}$$

$$\frac{d^2y}{dx^2} = \frac{48(16 + 3x^2)}{\left(x^2 - 16\right)^3} \quad \text{(verify)}$$

- *Intervals of increase and decrease; relative extrema:* A sign analysis of dy/dx yields

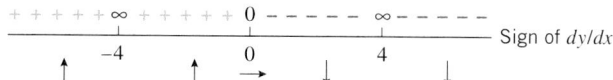

Thus, the graph is increasing on the intervals $(-\infty, -4)$ and $(-4, 0]$; and it is decreasing on the intervals $[0, 4)$ and $(4, +\infty)$. There is a relative maximum at $x = 0$.

- *Concavity:* A sign analysis of d^2y/dx^2 yields

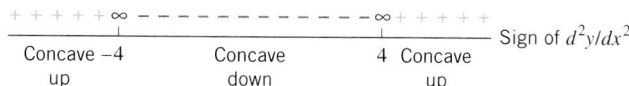

There are changes in concavity at the vertical asymptotes, $x = -4$ and $x = 4$, but there are no inflection points.

This analysis confirms that our calculator-generated graph was, in fact, complete. Figure 5.3.9 shows a complete graph of the equation with the asymptotes, intercepts, and relative maximum identified. ◀

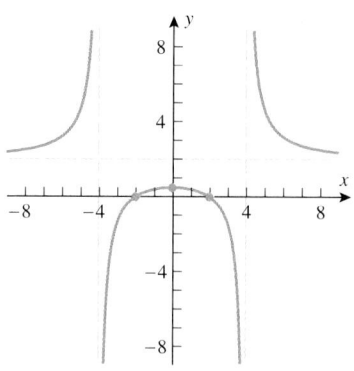

Figure 5.3.9

Example 5

Generate or sketch a complete graph of

$$y = \frac{x^2 - 1}{x^3}$$

and identify the exact location of all asymptotes, intercepts, relative extrema, and inflection points.

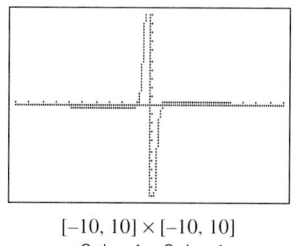

[−10, 10] × [−10, 10]
xScl = 1, yScl = 1

(a)

[−4, 4] × [−2, 2]
xScl = 1, yScl = 1

(b)

Figure 5.3.10

Solution. Figure 5.3.10*a* shows a calculator-generated graph of the given equation in the window $[-10, 10] \times [-10, 10]$, and Figure 5.3.10*b* shows a second version of the graph that gives more detail in the vicinity of the x-axis. These figures suggest that the graph is symmetric about the origin. They also suggest that there are two relative extrema, two inflection points, two x-intercepts, a vertical asymptote at $x = 0$, and a horizontal asymptote at $y = 0$. The following analysis will identify the exact location of all the key features and will determine whether the calculator-generated graphs in Figure 5.3.10 have missed any of these features.

- *Symmetries:* Replacing x by $-x$ and y by $-y$ yields an equation that simplifies back to the original equation, so the graph is symmetric about the origin.
- *x-intercepts:* Setting $y = 0$ yields the x-intercepts $x = -1$ and $x = 1$.
- *y-intercept:* Setting $x = 0$ leads to a division by zero, so that there is no y-intercept.
- *Vertical asymptotes:* Setting $x^3 = 0$ yields the vertical asymptote $x = 0$.
- *Horizontal asymptotes:* The limits

$$\lim_{x \to +\infty} \frac{x^2 - 1}{x^3} = \lim_{x \to +\infty} \frac{x^2}{x^3} = \lim_{x \to +\infty} \frac{1}{x} = 0$$

$$\lim_{x \to -\infty} \frac{x^2 - 1}{x^3} = \lim_{x \to -\infty} \frac{x^2}{x^3} = \lim_{x \to -\infty} \frac{1}{x} = 0$$

yield the horizontal asymptote $y = 0$.

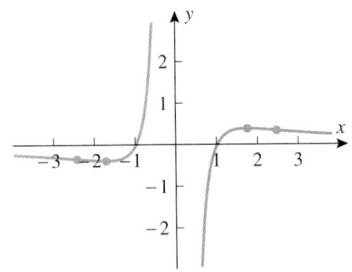

x	$y = \dfrac{x^2 - 1}{x^3}$
$-\sqrt{6} \approx -2.45$	$-\dfrac{5\sqrt{6}}{36} \approx -0.34$
$-\sqrt{3} \approx -1.73$	$-\dfrac{2\sqrt{3}}{9} \approx -0.38$
$\sqrt{3} \approx 1.73$	$\dfrac{2\sqrt{3}}{9} \approx 0.38$
$\sqrt{6} \approx 2.45$	$\dfrac{5\sqrt{6}}{36} \approx 0.34$

Figure 5.3.11

- *Derivatives:*

$$\frac{dy}{dx} = \frac{x^3(2x) - (x^2 - 1)(3x^2)}{\left(x^3\right)^2} = \frac{3 - x^2}{x^4}$$

$$\frac{d^2y}{dx^2} = \frac{x^4(-2x) - (3 - x^2)(4x^3)}{\left(x^4\right)^2} = \frac{2(x^2 - 6)}{x^5}$$

- *Intervals of increase and decrease; relative extrema:*

$$- - - - - \; 0 \; + + + + + \; \infty \; + + + + + \; 0 \; - - - - - \quad \text{Sign of } dy/dx$$
$$\downarrow \quad -\sqrt{3} \quad \uparrow \quad 0 \quad \uparrow \quad \sqrt{3} \quad \downarrow$$

This analysis reveals a relative minimum at $x = -\sqrt{3}$ and a relative maximum at $x = \sqrt{3}$.

- *Concavity:*

$$- - - - - \; 0 \; + + + + + \; \infty \; - - - - - \; 0 \; + + + + + \quad \text{Sign of } d^2y/dx^2$$
$$-\sqrt{6} \qquad 0 \qquad \sqrt{6}$$

| Concave down | Infl | Concave up | Concave down | Infl | Concave up |

This analysis reveals that changes in concavity occur at the vertical asymptote $x = 0$ and at the inflection points $x = -\sqrt{6}$ and $x = \sqrt{6}$.

Figure 5.3.11 shows a table of coordinate values at the relative extrema and inflection points together with a complete graph of the equation on which we have emphasized these points. ◀

GRAPHS WITH VERTICAL TANGENTS AND CUSPS

Figure 5.3.12 shows four curve elements that are commonly found in the graphs of functions that involve radicals or fractional exponents. In all four cases x_0 is a point of nondifferentiability, and in all four cases the tangent line at a point x approaches a vertical limiting position as x approaches x_0 from either side. Thus, we will call x_0 a *point of vertical tangency* for the function. In parts (c) and (d) of the figure the curve segments form what is called a *cusp*.

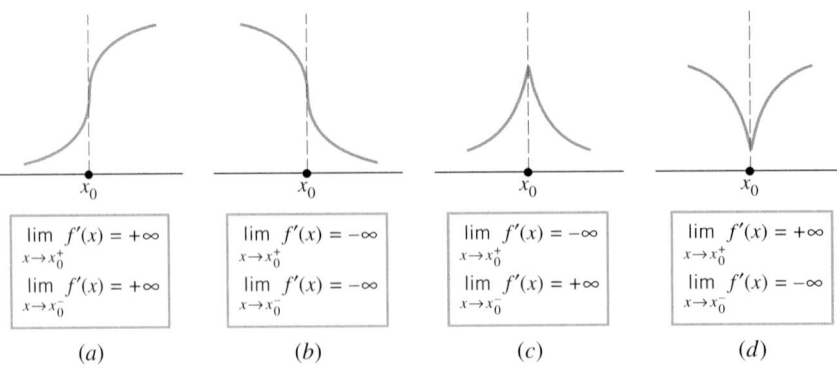

Figure 5.3.12

The following definition makes these ideas precise.

5.3.3 DEFINITION. The graph of a function f is said to have a **vertical tangent line** at x_0, and x_0 is called a **point of vertical tangency** for f if f is continuous at x_0 and $f'(x)$ approaches either $+\infty$ or $-\infty$ as $x \to x_0^+$ and as $x \to x_0^-$. In the case where $f'(x)$ approaches $+\infty$ from one side and $-\infty$ from the other side, the function f is said to have a **cusp** at x_0.

REMARK. It is important to observe that vertical tangent lines occur at points of nondifferentiability, whereas nonvertical tangent lines occur at points of differentiability.

Example 6

Generate or sketch a complete graph of $y = (x - 4)^{2/3}$.

Solution. Figure 5.3.13 shows a computer-generated graph of the equation $y = (x-4)^{2/3}$. (As suggested in the discussion preceding Exercise 29 of Section 1.3, we had to trick the computer into producing the left branch by graphing the equation $y = |x - 4|^{2/3}$.) To determine whether this graph is complete, we let $f(x) = (x-4)^{2/3}$ and proceed as follows.

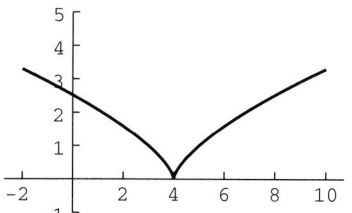

Generated by Mathematica

Figure 5.3.13

- *Symmetries:* There are no symmetries about the coordinate axes or the origin (verify). However, the graph of $y = (x - 4)^{2/3}$ is symmetric about the line $x = 4$, since it is a translation (four units to the right) of the graph of $y = x^{2/3}$, which is symmetric about the y-axis.

- *x-intercepts:* Setting $y = 0$ yields the x-intercept $x = 4$.

- *y-intercepts:* Setting $x = 0$ yields the y-intercept $y = \sqrt[3]{16}$.

- *Vertical asymptotes:* None, since $f(x) = (x - 4)^{2/3}$ is a continuous function.

- *Horizontal asymptotes:* None, since

$$\lim_{x \to +\infty} (x - 4)^{2/3} = +\infty \quad \text{and} \quad \lim_{x \to -\infty} (x - 4)^{2/3} = +\infty$$

- *Derivatives:*

$$\frac{dy}{dx} = f'(x) = \frac{2}{3}(x - 4)^{-1/3} = \frac{2}{3(x - 4)^{1/3}}$$

$$\frac{d^2y}{dx^2} = f''(x) = -\frac{2}{9}(x - 4)^{-4/3} = -\frac{2}{9(x - 4)^{4/3}}$$

- *Relative extrema; concavity:* There is a critical point at $x = 4$, since f is not differentiable there; and by the first derivative test there is a relative minimum at that critical point, since $f'(x) < 0$ if $x < 4$ and $f'(x) > 0$ if $x > 4$. Since $f''(x) < 0$ if $x \neq 4$, the graph is concave down for $x < 4$ and for $x > 4$.

- *Vertical tangent lines:* There is a vertical tangent line and cusp at $x = 4$ of the type in Figure 5.3.12*d* since $f(x) = (x - 4)^{2/3}$ is continuous at $x = 4$ and

$$\lim_{x \to 4^+} f'(x) = \lim_{x \to 4^+} \frac{2}{3(x - 4)^{1/3}} = +\infty$$

$$\lim_{x \to 4^-} f'(x) = \lim_{x \to 4^-} \frac{2}{3(x - 4)^{1/3}} = -\infty$$

Combining the preceding information with a sign analysis of the first and second derivatives yields Figure 5.3.14. This confirms that the computer-generated graph in Figure 5.3.13 is complete. ◄

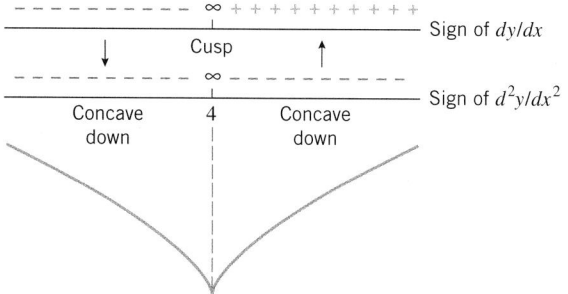

Figure 5.3.14

Example 7

Generate or sketch a complete graph of $y = 6x^{1/3} + 3x^{4/3}$.

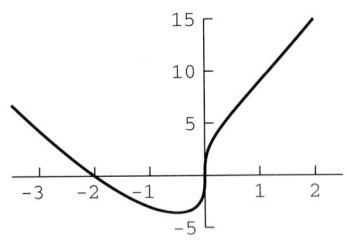

Generated by Mathematica

Figure 5.3.15

Solution. Figure 5.3.15 shows a computer-generated graph of the equation. Once again, we had to call on the discussion preceding Exercise 29 of Section 1.3 to trick the computer into graphing a portion of the graph over the negative x-axis. (See if you can figure out how to do this.) To determine whether the graph in Figure 5.3.15 is complete, we let

$$f(x) = 6x^{1/3} + 3x^{4/3} = 3x^{1/3}(2 + x)$$

and proceed as follows.

- *Symmetries:* There are no symmetries about the coordinate axes or the origin (verify).
- *x-intercepts:* Setting $y = 3x^{1/3}(2 + x) = 0$ yields the x-intercepts $x = 0$ and $x = -2$.
- *y-intercept:* Setting $x = 0$ yields the y-intercept $y = 0$.
- *Vertical asymptotes:* None, since $f(x) = 6x^{1/3} + 3x^{4/3}$ is continuous.
- *Horizontal asymptotes:* None, since

$$\lim_{x \to +\infty} (6x^{1/3} + 3x^{4/3}) = \lim_{x \to +\infty} 3x^{1/3}(2 + x) = +\infty$$

$$\lim_{x \to -\infty} (6x^{1/3} + 3x^{4/3}) = \lim_{x \to -\infty} 3x^{1/3}(2 + x) = +\infty$$

- *Derivatives:*

$$\frac{dy}{dx} = f'(x) = 2x^{-2/3} + 4x^{1/3} = 2x^{-2/3}(1 + 2x) = \frac{2(2x + 1)}{x^{2/3}}$$

$$\frac{d^2y}{dx^2} = f''(x) = -\frac{4}{3}x^{-5/3} + \frac{4}{3}x^{-2/3} = \frac{4}{3}x^{-5/3}(-1 + x) = \frac{4(x - 1)}{3x^{5/3}}$$

There are critical points at $x = 0$ and $x = -\frac{1}{2}$. From the first derivative test and the sign analysis of dy/dx in Figure 5.3.16, there is a relative minimum at $x = -\frac{1}{2}$. There is a point of vertical tangency at $x = 0$, since

$$\lim_{x \to 0^+} f'(x) = \lim_{x \to 0^+} \frac{2(2x + 1)}{x^{2/3}} = +\infty$$

$$\lim_{x \to 0^-} f'(x) = \lim_{x \to 0^-} \frac{2(2x + 1)}{x^{2/3}} = +\infty$$

From the sign analysis of d^2y/dx^2 in Figure 5.3.16, the graph is concave up for $x < 0$, concave down for $0 < x < 1$, and concave up again for $x > 1$.

- *Intervals of increase and decrease; concavity:* Combining the preceding information with a sign analysis of the first and second derivatives yields the graph shape shown in Figure 5.3.16.

This confirms that the computer-generated graph in Figure 5.3.15 is complete, except for the fact that it did not reveal the very subtle inflection point at $x = 1$. In this case the artistic rendering of the curve in Figure 5.3.16 describes the subtleties of the graph shape more effectively than the computer-generated graph. ◀

Example 8

Generate or sketch a complete graph of $y = e^{-x^2/2}$ and identify the exact location of all relative extrema and inflection points.

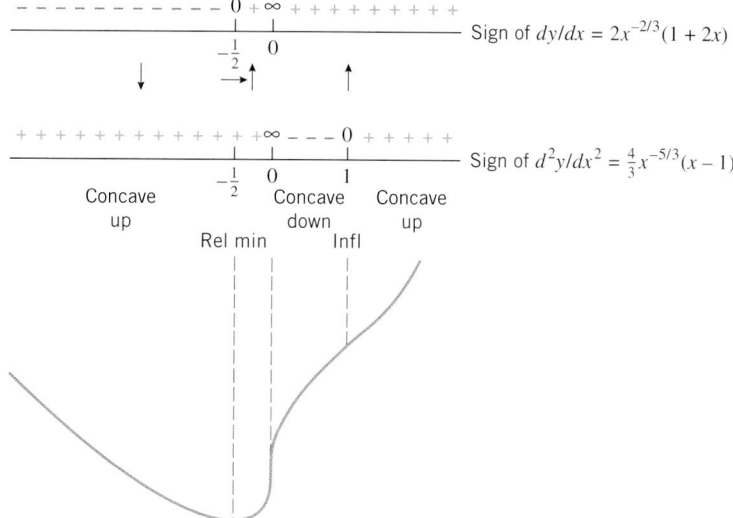

Figure 5.3.16

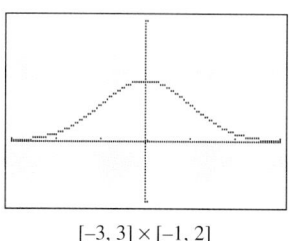

$[-3, 3] \times [-1, 2]$
xScl = 1, yScl = 1

Figure 5.3.17

Solution. Figure 5.3.17 shows a calculator-generated graph of the equation in the window $[-3, 3] \times [-1, 2]$. This figure suggests that the graph is symmetric about the y-axis and has a relative maximum at $x = 0$, a horizontal asymptote at $y = 0$, and two inflection points. The following analysis will identify the exact location of these features and determine whether the graph is complete.

- *Symmetries:* Replacing x by $-x$ does not change the equation, so the graph is symmetric about the y-axis.

- *x-intercepts:* Setting $y = 0$ yields the equation $e^{-x^2/2} = 0$, which has no solutions since all powers of e have positive values; thus, there are no x-intercepts.

- *y-intercepts:* Setting $x = 0$ yields the y-intercept $y = 1$.

- *Vertical asymptotes:* None, since $e^{-x^2/2}$ is a continuous function.

- *Horizontal asymptotes:* Since $x^2/2 \to +\infty$ as $x \to +\infty$ or $x \to -\infty$, it follows from Formula (14) of Section 4.2 that

$$\lim_{x \to +\infty} e^{-x^2/2} = \lim_{x \to -\infty} e^{-x^2/2} = 0$$

Thus, $y = 0$ is a horizontal asymptote.

- *Derivatives:*

$$\frac{dy}{dx} = e^{-x^2/2} \frac{d}{dx}\left[-\frac{x^2}{2}\right] = -xe^{-x^2/2}$$

$$\frac{d^2y}{dx^2} = -x\frac{d}{dx}\left[e^{-x^2/2}\right] + e^{-x^2/2}\frac{d}{dx}[-x]$$

$$= x^2 e^{-x^2/2} - e^{-x^2/2}$$

$$= (x^2 - 1)e^{-x^2/2}$$

- *Intervals of increase and decrease:* Since $e^{-x^2/2} > 0$ for all x, the sign of dy/dx is the same as that of $-x$.

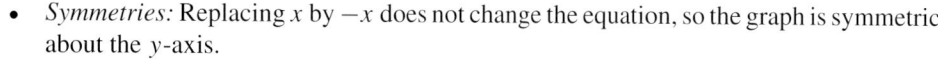

This analysis reveals a relative maximum at $x = 0$.

- *Concavity:* Since $e^{-x^2/2} > 0$ for all x, the sign of d^2y/dx^2 is the same as that of $x^2 - 1$.

$$
\begin{array}{ccccc}
+ + + + & 0 & - - - - & 0 & + + + + \\
& | & & | & \\
\hline
& -1 & & 1 & \\
\text{Concave} & \text{Infl} & \text{Concave} & \text{Infl} & \text{Concave} \\
\text{up} & & \text{down} & & \text{up}
\end{array}
\qquad \text{Sign of } x^2 - 1 \text{ and } d^2y/dx^2
$$

Thus, the inflection points occur at $x = -1$ and $x = 1$. At these points the corresponding y-values are $y = e^{-1/2} \approx 0.61$, which seems consistent with Figure 5.3.17.   ◄

....

LOGISTIC CURVES

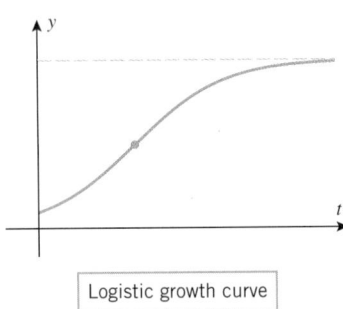

Logistic growth curve

Figure 5.3.18

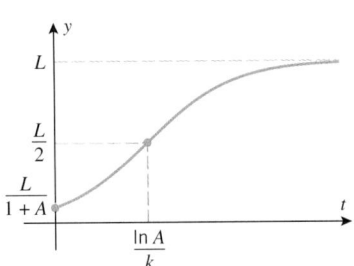

Figure 5.3.19

When a population grows in an environment in which space or food is limited, the graph of population versus time is typically an S-shaped curve of the form shown in Figure 5.3.18. The scenario described by this curve is a population that grows slowly at first and then more and more rapidly as the number of individuals producing offspring increases. However, at a certain point in time (where the inflection point occurs) the environmental factors begin to show their effect, and the growth rate begins a steady decline. Over an extended period of time the population approaches a limiting value that represents the upper limit on the number of individuals that the available space or food can sustain. Population growth curves of this type are called *logistic growth curves*.

Example 9

We will show in a later chapter that logistic growth curves arise from equations of the form

$$
y = \frac{L}{1 + Ae^{-kt}} \tag{5}
$$

where y is the population at time t $(t \geq 0)$ and $A, k,$ and L are positive constants. Show that Figure 5.3.19 correctly describes the graph of this equation.

Solution. We leave it for you to confirm that at time $t = 0$ the value of y is

$$
y = \frac{L}{1 + A}
$$

and that for $t \geq 0$ the population y satisfies

$$
\frac{L}{1 + A} \leq y < L
$$

This is consistent with the graph in Figure 5.3.19. The horizontal asymptote at $y = L$ is confirmed by the limit

$$
\lim_{t \to +\infty} \frac{L}{1 + Ae^{-kt}} = \frac{L}{1 + 0} = L
$$

Physically, L represents the upper limit on the size of the population.

To investigate intervals of increase or decrease, concavity, and inflection points, we need the first and second derivatives of y with respect to t. We leave it for you to confirm that

$$
\frac{dy}{dt} = \frac{k}{L} y (L - y) \tag{6}
$$

$$
\frac{d^2y}{dt^2} = \frac{k^2}{L^2} (L - y)(L - 2y) \tag{7}
$$

Since $k > 0$, $y > 0$, and $L - y > 0$, it follows from (6) that $dy/dt > 0$ for all t. Thus, y is always increasing and there are no stationary points, which is consistent with Figure 5.3.19.

Since $y > 0$ and $L - y > 0$, it follows from (7) that

$$\frac{d^2y}{dt^2} > 0 \quad \text{if} \quad L - 2y > 0$$

$$\frac{d^2y}{dt^2} < 0 \quad \text{if} \quad L - 2y < 0$$

Thus, the graph of y versus t is concave up if $y < L/2$, concave down if $y > L/2$, and has an inflection point where $y = L/2$, all of which is consistent with Figure 5.3.19.

Finally, we leave it as an exercise for you to confirm that the inflection point occurs at time

$$t = \frac{1}{k} \ln A = \frac{\ln A}{k} \tag{8}$$

by solving the equation

$$\frac{L}{2} = \frac{L}{1 + Ae^{-kt}}$$

for t. ◀

EXERCISE SET 5.3 ◪ Graphing Calculator Ⓒ CAS

· ·

In Exercises 1–10, give a complete graph of the polynomial, and label the coordinates of the stationary points and inflection points. Check your work with a graphing utility.

1. $x^2 - 2x - 3$ **2.** $1 + x - x^2$
3. $x^3 - 3x + 1$ **4.** $2x^3 - 3x^2 + 12x + 9$
5. $x^4 + 2x^3 - 1$ **6.** $x^4 - 2x^2 - 12$
7. $3x^5 - 5x^3$ **8.** $3x^4 + 4x^3$
9. $x(x - 1)^3$ **10.** $x^5 + 5x^4$

In Exercises 11–19, give a complete graph of the rational function, and label the coordinates of the stationary points and inflection points. Show the horizontal and vertical asymptotes, and label them with their equations. Check your work with a graphing utility.

11. $\dfrac{2x}{x - 3}$ **12.** $\dfrac{x}{x^2 - 1}$ **13.** $\dfrac{x^2}{x^2 - 1}$
14. $\dfrac{x^2 - 1}{x^2 + 1}$ **15.** $x^2 - \dfrac{1}{x}$ **16.** $\dfrac{2x^2 - 1}{x^2}$
17. $\dfrac{x^3 - 1}{x^3 + 1}$ **18.** $\dfrac{8}{4 - x^2}$ **19.** $\dfrac{x - 1}{x^2 - 4}$

In Exercises 20–22, the graph of the rational function crosses a horizontal asymptote. Give a complete graph of the function, and label the coordinates of the stationary points and inflection points. Show the horizontal and vertical asymptotes, and label the point(s) where the graph crosses a horizontal asymptote. Check your work with a graphing utility.

20. $\dfrac{3x^2 - 4x - 4}{x^2}$ **21.** $\dfrac{(x - 1)^2}{x^2}$ **22.** $2 + \dfrac{3}{x} - \dfrac{1}{x^3}$

23. In each part, match the functions with graphs I–VI without using a graphing utility, and then use a graphing utility to generate the graphs.

(a) $x^{1/3}$ (b) $x^{1/4}$ (c) $x^{1/5}$
(d) $x^{2/5}$ (e) $x^{4/3}$ (f) $x^{-1/3}$

I
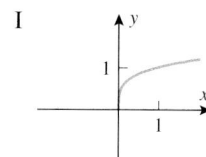

II
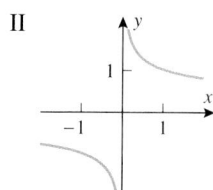

III
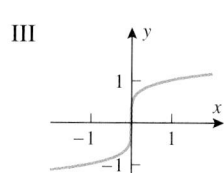

IV
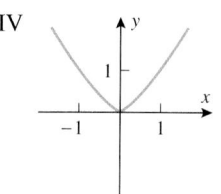

V
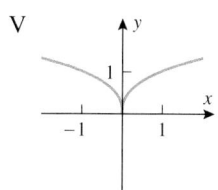

VI
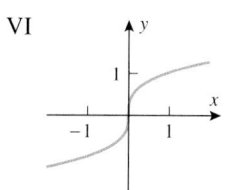

24. Sketch the general shape of the graph of $y = x^{1/n}$, and then explain in words what happens to the shape of the graph as n increases if
(a) n is a positive even integer
(b) n is a positive odd integer.

In Exercises 25–32, give a complete graph of the function, and identify the location of all critical points and inflection points. Check your work with a graphing utility.

25. $\sqrt{x^2 - 1}$ **26.** $\sqrt[3]{x^2 - 4}$

27. $2x + 3x^{2/3}$ **28.** $4x - 3x^{4/3}$

29. $x\sqrt{3 - x}$ **30.** $4x^{1/3} - x^{4/3}$

31. $\dfrac{8(\sqrt{x} - 1)}{x}$ **32.** $\dfrac{1 + \sqrt{x}}{1 - \sqrt{x}}$

In Exercises 33–38, give a complete graph of the function, and identify the location of all relative extrema and inflection points. Check your work with a graphing utility.

33. $x + \sin x$ **34.** $x - \cos x$

35. $\sin x + \cos x$ **36.** $\sqrt{3}\cos x + \sin x$

37. $\sin^2 x$, $0 \le x \le 2\pi$

38. $x \tan x$, $-\pi/2 < x < \pi/2$

In Exercises 39–44: (a) Find the limits of the function as $x \to +\infty$ and $x \to -\infty$. (b) Give a complete graph of the function, and identify the location of all relative extrema and inflection points. Check your work with a graphing utility.

39. xe^x **40.** xe^{-2x} **41.** $x^2 e^{-2x}$

42. $x^2 e^{2x}$ **43.** xe^{x^2} **44.** e^{-1/x^2}

In Exercises 45–48: (a) Find the limits of the function as $x \to 0^+$ and $x \to +\infty$. (b) Give a complete graph of the function, and identify the location of all relative extrema and inflection points. Check your work with a graphing utility.

45. $x \ln x$ **46.** $x^2 \ln x$ **47.** $\dfrac{\ln x}{x^2}$ **48.** $\dfrac{\ln x}{\sqrt{x}}$

49. In each part: (i) Make a conjecture about the behavior of the graph in the vicinity of its x-intercepts. (ii) Make a rough sketch of the graph based on your conjecture and the limits of the polynomials as $x \to +\infty$ and $x \to -\infty$. (iii) Compare your sketch to the graph generated with a graphing utility.
(a) $y = x(x - 1)(x + 1)$ (b) $y = x^2(x - 1)^2(x + 1)^2$
(c) $y = x^2(x - 1)^2(x + 1)^3$ (d) $y = x(x - 1)^5(x + 1)^4$

50. Sketch the graph of $y = (x - a)^m(x - b)^n$ for the stated values of m and n, assuming that $a \ne b$ (six graphs in total).
(a) $m = 1$, $n = 1, 2, 3$ (b) $m = 2$, $n = 2, 3$
(c) $m = 3$, $n = 3$

51. In each part, make a rough sketch of the graph using asymptotes and appropriate limits but no derivatives. Compare

your sketch to that generated with a graphing utility.

(a) $y = \dfrac{3x^2 - 8}{x^2 - 4}$ (b) $y = \dfrac{x^2 + 2x}{x^2 - 1}$

(c) $y = \dfrac{2x - x^2}{x^2 + x - 2}$ (d) $y = \dfrac{x^2}{x^2 - x - 2}$

52. Sketch the graph of
$$y = \frac{1}{(x - a)(x - b)}$$
assuming that $a \ne b$.

53. Consider the family of curves $y = xe^{-bx}$ $(b > 0)$.
(a) Use a graphing utility to generate some members of this family.
(b) Discuss the effect of varying b on the shape of the graph, and discuss the locations of the relative extrema and inflection points.

54. Consider the family of curves $y = e^{-bx^2}$ $(b > 0)$.
(a) Use a graphing utility to generate some members of this family.
(b) Discuss the effect of varying b on the shape of the graph, and discuss the locations of the relative extrema and inflection points.

55. (a) Determine whether the following limits exist, and if so, find them:
$$\lim_{x \to +\infty} e^x \cos x, \qquad \lim_{x \to -\infty} e^x \cos x$$
(b) Sketch the graphs of $y = e^x$, $y = e^{-x}$, and $y = e^x \cos x$ in the same coordinate system, and label any points of intersection.
(c) Use a graphing utility to generate some members of the family $y = e^{ax} \cos bx$ $(a > 0$ and $b > 0)$, and discuss the effect of varying a and b on the shape of the curve.

56. (**Oblique Asymptotes**) If a rational function $P(x)/Q(x)$ is such that the degree of the numerator exceeds the degree of the denominator by *one*, then the graph of $P(x)/Q(x)$ will have an **oblique asymptote**, that is, an asymptote that is neither vertical nor horizontal. To see why, we perform the division of $P(x)$ by $Q(x)$ to obtain
$$\frac{P(x)}{Q(x)} = (ax + b) + \frac{R(x)}{Q(x)}$$
where $ax + b$ is the quotient and $R(x)$ is the remainder. Use the fact that the degree of the remainder $R(x)$ is less than the degree of the divisor $Q(x)$ to help prove
$$\lim_{x \to +\infty} \left[\frac{P(x)}{Q(x)} - (ax + b) \right] = 0$$
$$\lim_{x \to -\infty} \left[\frac{P(x)}{Q(x)} - (ax + b) \right] = 0$$

As illustrated in the accompanying figure, these results tell us that the graph of the equation $y = P(x)/Q(x)$ "approaches" the line (an oblique asymptote) $y = ax + b$ as $x \to +\infty$ or $x \to -\infty$.

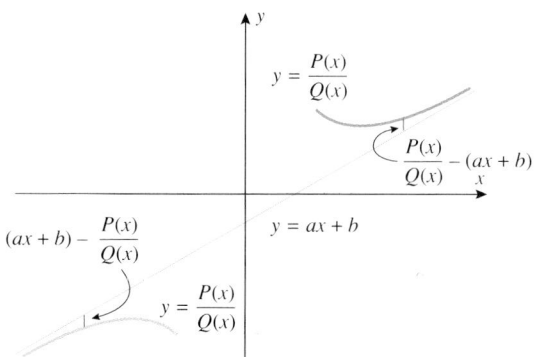

Figure Ex-56

In Exercises 57–61, sketch the graph of the rational function. Show all vertical, horizontal, and oblique asymptotes (see Exercise 56).

57. $\dfrac{x^2 - 2}{x}$ **58.** $\dfrac{x^2 - 2x - 3}{x + 2}$ **59.** $\dfrac{(x - 2)^3}{x^2}$

60. $\dfrac{4 - x^3}{x^2}$ **61.** $x + 1 - \dfrac{1}{x} - \dfrac{1}{x^2}$

62. Find all values of x where the graph of

$$y = \frac{2x^3 - 3x + 4}{x^2}$$

crosses its oblique asymptote. [See Exercise 56.]

63. Let $f(x) = (x^3 + 1)/x$. Show that the graph of $y = f(x)$ approaches the curve $y = x^2$ "asymptotically" in the sense that

$$\lim_{x \to +\infty} [f(x) - x^2] = 0 \quad \text{and} \quad \lim_{x \to -\infty} [f(x) - x^2] = 0$$

Sketch the graph of $y = f(x)$ showing this asymptotic behavior.

64. Let $f(x) = (2 + 3x - x^3)/x$. Show that $y = f(x)$ approaches the curve $y = 3 - x^2$ asymptotically in the sense described in Exercise 63. Sketch the graph of $y = f(x)$ showing this asymptotic behavior.

65. A rectangular plot of land is to be fenced off so that the area enclosed will be 400 ft^2. Let L be the length of fencing needed and x the length of one side of the rectangle. Show

that $L = 2x + 800/x$ for $x > 0$, and sketch the graph of L versus x for $x > 0$.

66. A box with a square base and open top is to be made from sheet metal so that its volume is 500 in^3. Let S be the area of the surface of the box and x the length of a side of the square base. Show that $S = x^2 + 2000/x$ for $x > 0$, and sketch the graph of S versus x for $x > 0$.

67. The accompanying figure shows a computer-generated graph of the polynomial $y = 0.1x^5(x - 1)$ using a viewing window of $[-2, 2.5] \times [-1, 5]$. Show that the choice of the vertical scale caused the computer to miss important features of the graph. Find the features that were missed and make your own sketch of the graph that shows the missing features.

68. The accompanying figure shows a computer-generated graph of the polynomial $y = 0.1x^5(x + 1)^2$ using a viewing window of $[-2, 1.5] \times [-0.2, 0.2]$. Show that the choice of the vertical scale caused the computer to miss important features of the graph. Find the features that were missed and make your own sketch of the graph that shows the missing features.

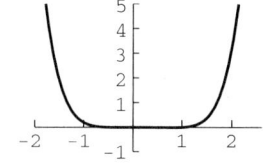

Generated by Mathematica

Figure Ex-67

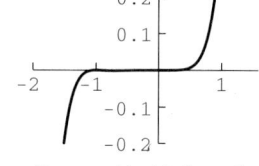

Generated by Mathematica

Figure Ex-68

69. Suppose that a population y grows according to the logistic model given by Formula (5).
   (a) At what rate is y increasing at time $t = 0$?
   (b) In words, describe how the rate of growth of P varies with time.
   (c) At what time is the population growing most rapidly?

70. Show that the inflection point of the logistic growth curve in Example 9 occurs at the time t given by Formula (8).

SUPPLEMENTARY EXERCISES

1. (a) If $x_1 < x_2$, what relationship must hold between $f(x_1)$ and $f(x_2)$ for f to be increasing on an interval containing x_1 and x? Decreasing? Constant?
   (b) What condition on f' ensures that f is increasing on an interval $[a, b]$? Decreasing? Constant?

2. (a) What condition on f' ensures that f is concave up on an open interval I? Concave down?
   (b) What condition on f'' ensures that f is concave up on an open interval I? Concave down?
   (c) In words, what is an inflection point of f?

3. (a) Where on the graph of $y = f(x)$ would you expect y to be increasing or decreasing most rapidly with respect to x?

(b) In words, what is a relative extremum?

(c) State a procedure for determining where the relative extrema of f occur.

4. Determine whether the statement is true or false. If it is false, give an example that illustrates why.

(a) If f has a relative maximum at x_0, then $f(x_0)$ is the largest value that $f(x)$ can have.

(b) If $f(x_0)$ is the largest value for f on the interval (a, b), then f has a relative maximum at x_0.

(c) A function f has a relative extremum at each of its critical points.

5. (a) According to the first derivative test, what conditions ensure that f has a relative maximum at x_0? A relative minimum?

(b) According to the second derivative test, what conditions ensure that f has a relative maximum at x_0? A relative minimum?

6. In each part, sketch a continuous curve $y = f(x)$ with the stated properties.

(a) $f(2) = 4$, $f'(2) = 1$, $f''(x) < 0$ for $x < 2$, $f''(x) > 0$ for $x > 2$

(b) $f(2) = 4$, $f''(x) > 0$ for $x < 2$, $f''(x) < 0$ for $x > 2$, and $\lim_{x \to 2^-} f'(x) = +\infty$, $\lim_{x \to 2^+} f'(x) = +\infty$

(c) $f(2) = 4$, $f''(x) < 0$ for $x \neq 2$, and $\lim_{x \to 2^-} f'(x) = 1$, $\lim_{x \to 2^+} f'(x) = -1$

7. In each part, find the location of all critical points, and use the first derivative test to classify them as relative maxima, relative minima, or neither.

(a) $f(x) = x^{1/3}(x - 7)^2$

(b) $f(x) = 2 \sin x - \cos 2x$, $0 \leq x \leq 2\pi$

(c) $f(x) = 3x - (x - 1)^{3/2}$

8. In each part, find the location of all critical points, and use the second derivative test (where possible) to classify them as relative maxima, relative minima, or neither.

(a) $f(x) = x^{-1/2} + \frac{1}{9}x^{1/2}$

(b) $f(x) = x^2 + 8/x$

(c) $f(x) = \sin^2 x - \cos x$, $0 \leq x \leq 2\pi$

In Exercises 9–24, give a complete graph of f, and identify the limits as $x \to \pm\infty$, as well as locations of all relative extrema, inflection points, and asymptotes (as appropriate).

9. $f(x) = x^4 - 3x^3 + 3x^2 + 1$

10. $f(x) = x^5 - 4x^4 + 4x^3$

11. $f(x) = \tan(x^2 + 1)$

12. $f(x) = x - \cos x$

13. $f(x) = \dfrac{x^2}{x^2 + 2x + 5}$

14. $f(x) = \dfrac{25 - 9x^2}{x^3}$

15. $f(x) = \begin{cases} \frac{1}{2}x^2, & x \leq 0 \\ -x^2, & x > 0 \end{cases}$

16. $f(x) = (1 + x)^{2/3}(3 - x)^{1/3}$

17. $f(x) = x \ln x$

18. $f(x) = x^2 \ln x$

19. $f(x) = \dfrac{\ln x}{x^2}$

20. $f(x) = \ln(x^2 + 1)$

21. $f(x) = \dfrac{e^x}{x}$

22. $f(x) = xe^{-x}$

23. $f(x) = x^2 e^{1-x}$

24. $f(x) = x^3 e^{x-1}$

> When using a graphing utility, important features of a graph may be missed if the viewing window is not chosen appropriately. This is illustrated in Exercises 25 and 26.

25. (a) Generate the graph of $f(x) = \frac{1}{3}x^3 - \frac{1}{400}x$ over the interval $[-5, 5]$, and make a conjecture about the location and nature of all critical points.

(b) Find the exact location of all critical points, and classify them as relative maxima, relative minima, or neither.

(c) Confirm the results in part (b) by graphing f over an appropriate interval.

26. (a) Generate the graph of

$$f(x) = \frac{1}{5}x^5 - \frac{7}{8}x^4 + \frac{1}{3}x^3 + \frac{7}{2}x^2 - 6x$$

over the interval $[-5, 5]$, and make a conjecture about the location and nature of all critical points.

(b) Find the exact location of all critical points, and classify them as relative maxima, relative minima, or neither.

(c) Confirm the results in part (b) by graphing portions of f over appropriate intervals. [*Note:* It will not be possible to find a single window in which all of the critical points are clearly visible.]

27. (a) Use a graphing utility to generate the graphs of $y = x$ and $y = (x^3 - 8)/(x^2 + 1)$ together over the interval $[-5, 5]$, and make a conjecture about the relationship between the two graphs.

(b) Use Exercise 56 of Section 5.3 to confirm your conjecture in part (a).

28. In parts (a)–(d), the complete graph of a polynomial with degree at most 6 is given. Find equations for polynomials that produce graphs with these shapes, and check your answers with a graphing utility.

(a)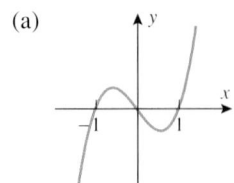

(b)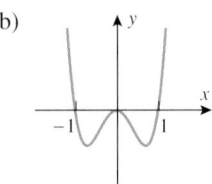

(c) 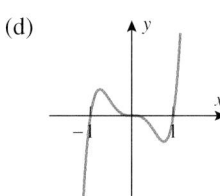   (d)

29. Find the equations of the tangent lines at all inflection points of the graph of

$$f(x) = x^4 - 6x^3 + 12x^2 - 8x + 3$$

30. Use implicit differentiation to show that a function defined implicitly by $\sin x + \cos y = 2y$ has a critical point whenever $\cos x = 0$. Then use either the first or second derivative test to classify these critical points as relative maxima or minima.

31. Let

$$f(x) = \frac{2x^3 + x^2 - 15x + 7}{(2x-1)(3x^2 + x - 1)}$$

Graph $y = f(x)$, and find the equations of all horizontal and vertical asymptotes. Explain why there is no vertical asymptote at $x = \frac{1}{2}$, even though the denominator of f is zero at that point.

C 32. Let

$$f(x) = \frac{x^5 - x^4 - 3x^3 + 2x + 4}{x^7 - 2x^6 - 3x^5 + 6x^4 + 4x - 8}$$

(a) Use a CAS to factor the numerator and denominator of f, and use the results to determine the locations of all vertical asymptotes.

(b) Confirm that your answer is consistent with the graph of f.

33. (a) By inspection, find the largest and smallest possible values for $f(x) = e^{\sin x}$, and then confirm that your answers are consistent with the graph of f.

(b) Find the exact locations of the relative extrema.

(c) Estimate the locations of the inflection points in the interval $0 < x < 2\pi$ from the graph of f''.

34. For the general cubic polynomial

$$f(x) = ax^3 + bx^2 + cx + d \quad (a \neq 0)$$

find conditions on a, b, c, and d to ensure that f is always increasing or always decreasing on $(-\infty, +\infty)$.

35. In each part, approximate the coordinates (x, y) of the relative extrema, and confirm that your answers are consistent with the graph of f.

(a) $f(x) = x^2 - \sin x$

(b) $f(x) = \sqrt{x^4 + 1} - \sqrt{x^2 + 1}$

(c) $f(x) = \dfrac{x}{x^2 - \sin x + 1}$

36. (a) Approximate to two decimal places the largest value of k such that the function $f(x) = 1 + 2x + x^3 - x^4$ is one-to-one for $x \leq k$.

(b) For the value of k found in part (a), find the domain and range of f^{-1} and the value of $f^{-1}(-1)$ for the function $f(x) = 1 + 2x + x^3 - x^4$, $x \leq k$.

37. Consider the family of curves $y = xe^{-ax}$, $x \geq 0$, where a is a positive constant.

(a) Use a graphing utility to graph some members of this family.

(b) Find the value of y at $x = 0$ and the limiting value of y as $x \to +\infty$; confirm that these values are consistent with your graphs.

(c) Find formulas for the coordinates of the relative extrema and inflection points, and confirm that these formulas are consistent with the graphs.

(d) How does increasing a affect the graph?

38. Consider the family of curves $y = e^{-(x-a)/2b}$, where a and b are constants and $b > 0$.

(a) Use a graphing utility to graph some members of this family, first keeping a fixed and varying b, and then keeping b fixed and varying a.

(b) Find the value of y at $x = a$ and the limiting values of y as $x \to \pm\infty$; confirm that these values are consistent with your graphs.

(c) Find formulas for the coordinates of the relative extrema and inflection points, and confirm that these formulas are consistent with the graphs.

(d) If a is kept fixed, how does increasing b affect the shape of the graph?

(e) If b is kept fixed, how does varying a affect the graph?

39. Show that for successive positive integer values of n, the number $(1 + 1/n)^{n+1}$ is smaller than its predecessor. [*Hint:* Consider the function $f(x) = (x + 1)\ln(1 + 1/x)$.]

EXPANDING THE CALCULUS HORIZON

Functions from Data

One of the most important procedures in applied science is using experimental data to discover relationships between variables. In this module we will discuss some mathematical techniques for doing this, and we will use these ideas to investigate principles of planetary motion and the cooling of liquids.

Fitting Curves to Data

Suppose that a scientist is looking for a relationship between two variables x and y and that measurements of corresponding values of these variables have produced a set of n data points

$$(x_1, y_1), \quad (x_2, y_2), \quad (x_3, y_3), \ldots, \quad (x_n, y_n)$$

If the scientist uses the data in some way to obtain a relationship $y = f(x)$ between x and y, then this equation is called a ***mathematical model*** for the data.

One way to obtain a mathematical model for a set of data is to look for a function f whose graph passes through all of the data points; this is called an ***interpolating function***. Although interpolating functions are appropriate in certain situations, they do not adequately account for measurement errors in the data. For example, suppose that the relationship between x and y is known to be linear but that accuracy limitations in the measuring devices and random variations in experimental conditions produce a scatter plot such as that shown in Figure 1a. With the help of a computer, one can find a polynomial of degree 10 whose graph passes through all of the data points (Figure 1b). However, this polynomial model does not successfully convey the underlying linear relationship; a better approach is to look for a linear equation $y = mx + b$ whose graph more accurately describes the linear relationship, even if it does not pass through all (or any) of the data points (Figure 1c).

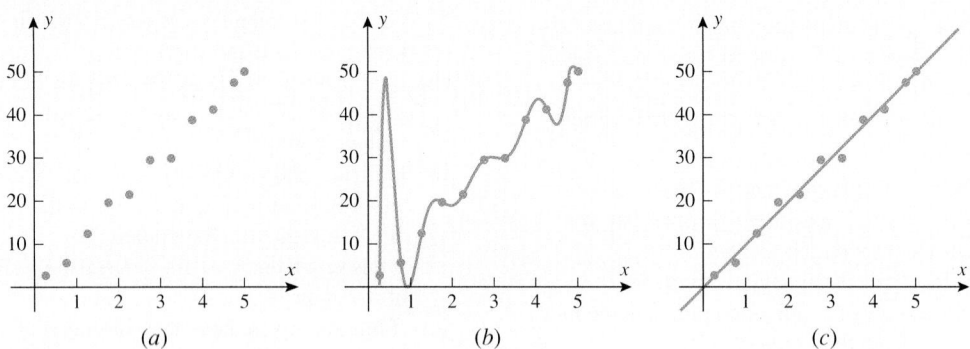

Figure 1

Finding Mathematical Models

The most challenging part of finding a model $y = f(x)$ for experimental data is coming up with an appropriate form for the function f. Sometimes the form of the function will be suggested by the visual appearance of the scatter plot, and sometimes it will be dictated by a known physical law that relates x and y. For example, Figure 1a strongly suggests that the relationship between x and y is linear, so in absence of additional information it would be natural to look for a linear model $y = mx + b$. In contrast, the scatter plot of U.S. population growth in Figure 2 strongly suggests some nonlinear relationship, so we must look for a nonlinear function for the model. The possibilities for nonlinear models are endless; however, there are theories in the study of population growth which suggest that in absence of environmental constraints, populations of people can be modeled over time by equations of the form $P = P_0 e^{kt}$, so in this case we might look for an equation of this form to model the data.

Linear Models

The most important methods for finding linear models are based on the following idea: For any proposed linear model $y = mx + b$, draw a vertical connector from each data point (x_i, y_i) to the line, and consider the differences $y_i - y$ (Figure 3). These differences, which are called **residuals**, may be viewed as "errors" that result when the line is used to model the data. Points above the line have positive errors, points below the line have negative errors, and points on the line have no error.

One way to choose a linear model is to look for a line $y = mx + b$ in which the sum of the residuals is zero, the logic being that this makes the positive and negative errors balance out. However, one can find examples where this procedure produces unacceptably poor models, so for reasons that we cannot discuss here the most common method for finding a linear model is to look for a line $y = mx + b$ in which the *sum of the squares* of the residuals is as small as possible. This is called the **least-squares line of best fit** or the **regression line**.

· · · · · · · · · · ·

Exercise 1   One of the lines in Figure 4 is the regression line. Which one is it?

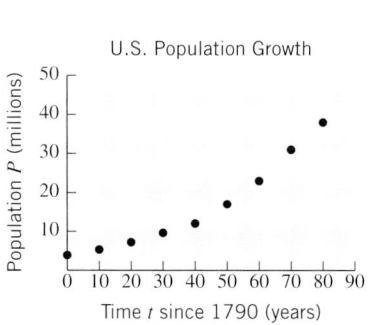

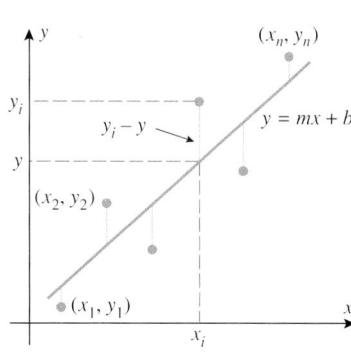

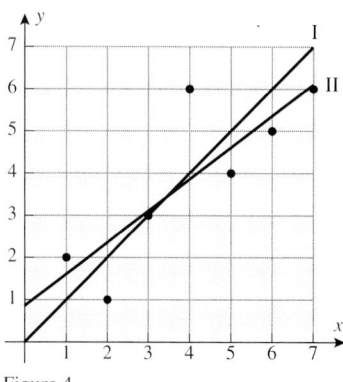

Figure 2   Figure 3   Figure 4

· · · · · · · · · · ·

Exercise 2

(a) Most scientific calculators and CAS programs provide a method for finding regression lines. Read the documentation for your calculator or CAS to determine how to do this, and then find the regression line for the following (x, y) data:

x	1.0	1.5	2.0	2.5	3.0	3.5	4.0
y	1.0	2.5	6.0	9.0	10.5	14.5	15.0

(b) Make a scatter plot of the data together with the regression line.

How Good Is the Linear Model?

It is possible to compute a regression line, even in cases where the data have no apparent linear pattern. Thus, it is important to have some quantitative method of determining whether a linear model is appropriate for the data. The most common measure of linearity in data is called the **correlation coefficient**, which is usually denoted by the letter r. A detailed explanation of correlation coefficients and the formula used to compute them is outside the scope of this text. However, here are some of the basic ideas:

- The values of r are in the interval $-1 \leq r \leq 1$, where r has the same sign as the slope of the regression line.
- If $r = \pm 1$, then the data points all lie on a line, so a linear model is a perfect fit for the data.
- If $r = 0$, then the data points exhibit no linear tendency, so a linear model is inappropriate for the data.

The closer r is to ± 1, the more tightly the data points hug the regression line and the more appropriate the regression line is as a model; the closer r is to 0, the more scattered the points and the less appropriate the regression line is as a model (Figure 5). Roughly stated, the value of r^2 is a measure of the percentage of data points that fall in a "tight linear band." Thus, $r = 0.5$ means that 25% of the points fall in a tight linear band, and $r = 0.9$ means that 81% of the points fall in a tight linear band. A precise explanation of what is meant by a "tight linear band" requires ideas from statistics.

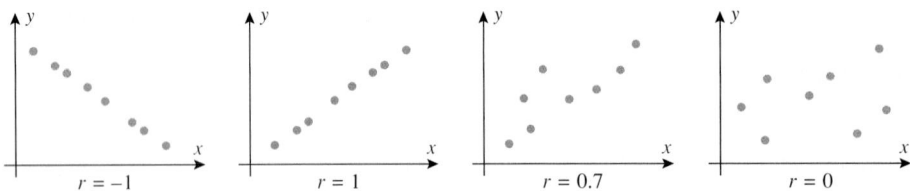

Figure 5

Exercise 3   If you have a scientific calculator, read the documentation to determine whether it produces the correlation coefficient when it computes a regression line. If you have a CAS, then the chances are that it will not automatically produce the correlation coefficient. However, on our website we have provided some CAS "miniprograms" that can be used to find regression lines with their associated correlation coefficients.

Exercise 4   Find the correlation coefficient for the data in Exercise 2.

Exercise 5

(a) Table 1.1.1 of Chapter 1 gives the Indianapolis 500 qualifying speeds S from 1975 to 1994. Take 1975 to be $t = 0$, and find the regression line and correlation coefficient for S versus t.

(b) Do you think that a linear model is reasonable for the data? Explain your reasoning.

(c) Predict the qualifying speed for the year 2000.

(d) What assumptions did you make in part (c)?

Nonlinear Models

Three of the most important nonlinear models are

- **Exponential models** ($y = ae^{bx}$)
- **Logarithmic models** ($y = a + b \ln x$)
- **Power function models** ($y = ax^b$)

Many scientific calculators and computer programs can fit models of these types to data by the method of least squares. However, a useful alternative approach is to use logarithms to transform the original data into a form where linear models can be applied. This procedure, called **linearizing** the data, is based on the following idea:

- A set of (x_i, y_i) data will have an exponential model if the transformed data $(x_i, \log y_i)$ have a linear model.
- A set of (x_i, y_i) data will have a logarithmic model if the transformed data $(\log x_i, y_i)$ have a linear model.
- A set of (x_i, y_i) data will have a power function model if the transformed data $(\log x_i, \log y_i)$ have a linear model.

The following exercise explains the reason for this.

. . . .
Exercise 6

(a) Suppose that $y = ae^{bx}$, and let $Y = \ln y$. Show that the graph of Y versus x is a line of slope b and Y-intercept $\ln a$.

(b) Suppose that $y = a + b \ln x$, and let $X = \ln x$. Show that the graph of y versus X is a line of slope b and y-intercept a.

(c) Suppose that $y = ax^b$, and let $Y = \ln y$ and $X = \ln x$. Show that the graph of Y versus X is a line of slope b and Y-intercept $\ln a$.

(d) Show that in parts (a), (b), and (c) the statements remain true if the natural logarithm "ln" is replaced by the common logarithm "log".

. . . .
Exercise 7

(a) Find an exponential model $y = ae^{bx}$ for the following data by linearizing the data, finding the regression line for the linearized data, and then applying part (a) of Exercise 6 to find a and b.

x	0	1	2	3	4	5	6	7
y	3.9	5.3	7.2	9.6	12	17	23	31

(b) Make a scatter plot of the data together with the exponential model.

. . . .
Exercise 8   The table in Figure 6 shows the relationship between the time T that it takes for each planet in our solar system to make one revolution around the Sun and the mean distance d between the planet and the Sun during one revolution. The graph in Figure 6 is a plot of $\log T$ versus $\log d$.

(a) What type of model for T as a function of d is suggested by the graph?

(b) Find the regression line for the $(\log d, \log T)$ data.

(c) Use the appropriate part of Exercise 6 to express T as a function of d.

(d) In part (c) you discovered Kepler's Third Law of Planetary Motion. Find some information about this law, and state the law in words.

PLANET	MEAN DISTANCE d FROM THE SUN	TIME T FOR ONE REVOLUTION
Mercury	0.387	0.241
Venus	0.723	0.615
Earth	1.000	1.000
Mars	1.523	1.881
Jupiter	5.203	11.861
Saturn	9.541	29.457
Uranus	19.190	84.008
Neptune	30.086	164.784
Pluto	39.507	248.350

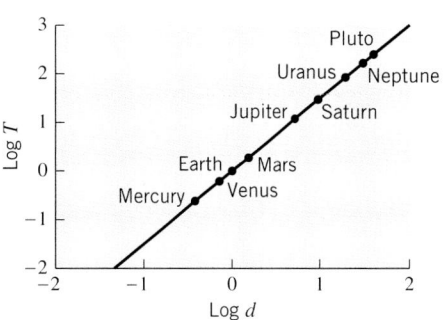

Note: Distances are measured in astronomical units (AU); 1 AU ≈ 93,000,000 mi. Time is measured in Earth years.

Figure 6

Modeling Cooling

If a cup of hot coffee is left on a table to cool, then the graph of its temperature T versus the elapsed time t will have the general shape shown in Figure 7. The graph suggests that the coffee will cool quickly at first and then more and more slowly as its temperature approaches that of the room. To be more precise, it is shown in Physics that if the temperature of a liquid at time $t = 0$ is T_0 and if the room has a constant temperature of T_a, where $T_a < T_0$ (the room is cooler than the liquid), then the temperature T of the liquid at time t is given by

$$T = T_a + (T_0 - T_a)e^{-kt}$$

where k is a *negative* constant whose value depends on the physical characteristics of the liquid. This equation, called ***Newton's Law of Cooling***, can also be written in the form

$$T - T_a = (T_0 - T_a)e^{-kt}$$

which states that the difference between the temperature of the liquid and the temperature of the room has an exponential model.

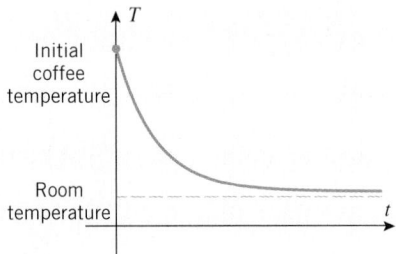

Figure 7

. . . .
Exercise 9   Table 1 shows temperature measurements of a cup of coffee at 1-minute intervals after it was placed in a room with a constant temperature of $27°$C.

(a) Find a model for the temperature T as a function of the elapsed time t.

(b) Estimate the temperature of the coffee at the time it was placed in the room.

(c) Approximately how long will it take until the coffee temperature is within $5°$C of the room temperature?

Table 1

t (min)	1	2	3	4	5	6	7	8	9	10
T (°C)	82.2	79.6	77.3	75.0	73.1	70.7	69.2	66.9	65.3	63.3

Module by Mary Ann Connors, USMA, West Point, and Howard Anton, Drexel University

Additional material for this module can be found on the World Wide Web at http://www.wiley.com/college/anton

6

Isaac Barrow

APPLICATIONS OF THE DERIVATIVE

*I*n this chapter we will study various applications of the derivative. For example, we will investigate problems that are concerned with finding the "best" way to perform a task—these are called *optimization problems*. Many optimization problems are concerned with time and cost. For example, if time is the main consideration in a problem, we might be interested in finding the quickest way to perform a task, and if cost is the main consideration, we might be interested in finding the most profitable way to perform the task. Mathematically, optimization problems can be reduced to finding the largest or smallest value of a function on some interval and determining where the largest or smallest value occurs; thus, part of our work in this chapter will focus on developing the mathematical tools for solving such problems. We will also use the derivative to study the motion of a particle moving along a line, and we will show how the derivative can be used to approximate solutions of equations.

6.1 ABSOLUTE MAXIMA AND MINIMA

At the beginning of Section 5.2 we observed that if the graph of a function f is viewed as a two-dimensional mountain range (Figure 5.2.1), then the relative maxima and minima correspond to the tops of the hills and the bottoms of the valleys; that is, they are the high and low points in their immediate vicinity. In this section we will be concerned with the more encompassing problem of finding the highest and lowest points over the entire mountain range, that is, we will be looking for the top of the highest hill and the bottom of the deepest valley. In mathematical terms, we will be looking for the largest and smallest values of a function over an interval.

ABSOLUTE EXTREMA

We will be concerned here with finding the largest and smallest values of a function over a finite or infinite interval I. We begin with some terminology.

6.1.1 DEFINITION. A function f is said to have an **absolute maximum** on an interval I at the point x_0 if $f(x_0)$ is the largest value of f on I; that is, $f(x_0) \geq f(x)$ for all x in I. Similarly, f is said to have an **absolute minimum** on I at the point x_0 if $f(x_0)$ is the smallest value of f on I; that is, $f(x_0) \leq f(x)$ for all x in I. If f has either an absolute maximum or absolute minimum on I at x_0, then f is said to have an **absolute extremum** on I at x_0.

As illustrated in Figure 6.1.1, there is no guarantee that a function f will have absolute extrema on a given interval.

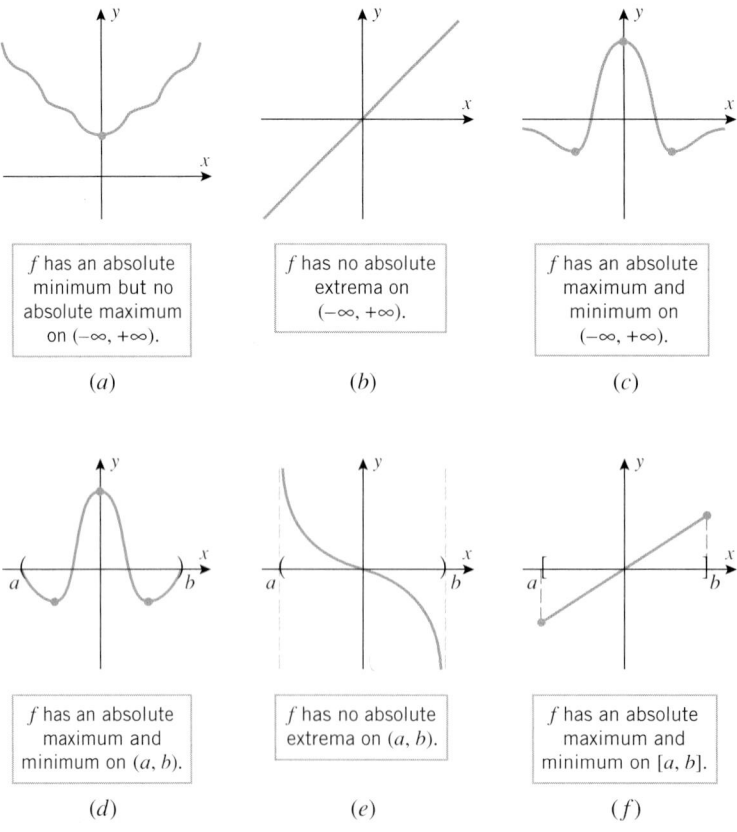

f has an absolute minimum but no absolute maximum on $(-\infty, +\infty)$.

(a)

f has no absolute extrema on $(-\infty, +\infty)$.

(b)

f has an absolute maximum and minimum on $(-\infty, +\infty)$.

(c)

f has an absolute maximum and minimum on (a, b).

(d)

f has no absolute extrema on (a, b).

(e)

f has an absolute maximum and minimum on $[a, b]$.

(f)

Figure 6.1.1

EXISTENCE OF ABSOLUTE
EXTREMA

The remainder of this section will focus on the following problem.

> **6.1.2** PROBLEM.
> (a)   Determine whether a function f has any absolute extrema on a given interval I.
> (b)   If there are absolute extrema, determine where they occur and what the absolute maximum and minimum values are.

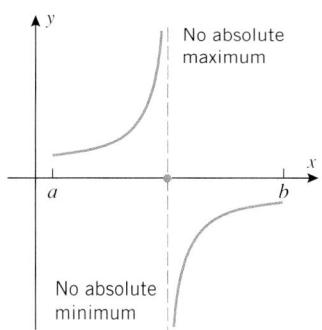

Figure 6.1.2

Parts (a)–(e) of Figure 6.1.1 show that a continuous function may or may not have relative maxima or minima on an infinite interval or on a finite open interval. However, the following theorem shows that a continuous function must have both an absolute maximum and an absolute minimum on every *finite closed* interval [see part (f) of Figure 6.1.1].

> **6.1.3** THEOREM (*Extreme-Value Theorem*).   *If a function f is continuous on a finite closed interval $[a, b]$, then f has both an absolute maximum and an absolute minimum on $[a, b]$.*

FOR THE READER.   Although the proof of this theorem is too difficult to include here, you should be able to convince yourself of its validity with a little experimentation—try graphing various continuous functions over the interval $[0, 1]$, and convince yourself that there is no way to avoid having a highest and lowest point on the graph. As a physical analogy, if you imagine the graph to be a roller coaster track starting at $x = 0$ and ending at $x = 1$, the roller coaster will have to pass through a highest point and a lowest point during the trip.

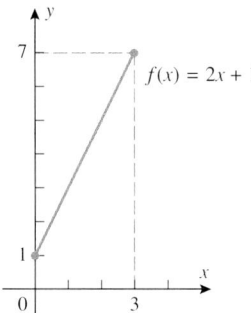

Figure 6.1.3

The hypotheses in the Extreme-Value Theorem are essential; for example, if f is not continuous, then we could encounter a situation such as that in Figure 6.1.2, and if f is continuous but the interval is not closed and finite, then we could encounter situations such as those in Figure 6.1.1. This is illustrated further in the following example.

Example 1

The function $f(x) = 2x + 1$ is continuous, and hence is guaranteed to have both an absolute maximum and an absolute minimum on every finite closed interval and, in particular, on the interval $[0, 3]$. For this interval an absolute minimum occurs at $x = 0$ and an absolute maximum occurs at $x = 3$, at which points the absolute minimum and maximum values are $f(0) = 1$ and $f(3) = 7$ (Figure 6.1.3).

However, if we consider this same function on the half-open interval $[0, 3)$, then there is no longer an absolute maximum. To see why this so, observe that $f(3) = 7$ is no longer the absolute maximum because we have removed the point $x = 3$ from the interval. However, f cannot have an absolute maximum in the interval at a point x_0 that is *less* than 3, because f will have a larger value at any point in the interval to the right of x_0 (Figure 6.1.4). Thus, f has no absolute maximum on the interval $[0, 3)$.   ◄

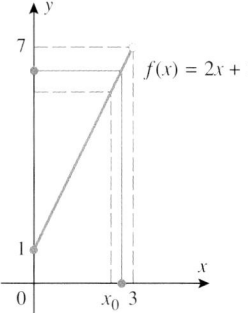

Figure 6.1.4

FINDING ABSOLUTE EXTREMA ON
FINITE CLOSED INTERVALS

The Extreme-Value Theorem is an example of what mathematicians call an ***existence theorem***. Such theorems state conditions under which something exists, in this case absolute extrema. However, knowing that something exists and finding it are two separate things, so we will now address the problem of finding the absolute extrema.

If f is continuous on the finite closed interval $[a, b]$, then the absolute extrema of f can occur either at the endpoints of the interval or inside on the open interval (a, b). If the absolute extrema happen to fall inside, then the following theorem tells us that they must occur at critical points of f.

> **6.1.4** THEOREM. *If f has an absolute extremum on an open interval (a, b), then it must occur at a critical point of f.*

Proof. If f has an absolute maximum on (a, b) at x_0, then $f(x_0)$ is also a relative maximum for f; for if $f(x_0)$ is the largest value of f on all of (a, b), then $f(x_0)$ is certainly the largest value for f in the immediate vicinity of x_0. Thus, x_0 is a critical point of f by Theorem 5.2.2. The proof for absolute minima is similar. ∎

It follows from this theorem, that if f is continuous on the finite closed interval $[a, b]$, then the absolute extrema occur either at the endpoints of the interval or at critical points inside the interval (Figure 6.1.5). Thus, we can use the following procedure to find the absolute extrema of a continuous function on a finite closed interval $[a, b]$.

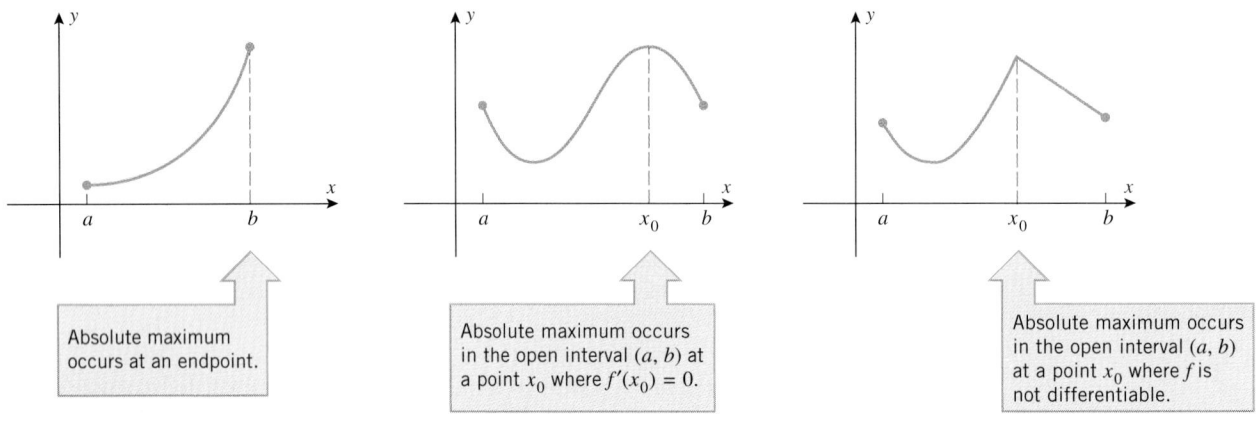

Absolute maximum occurs at an endpoint.

Absolute maximum occurs in the open interval (a, b) at a point x_0 where $f'(x_0) = 0$.

Absolute maximum occurs in the open interval (a, b) at a point x_0 where f is not differentiable.

Figure 6.1.5

A Procedure for Finding the Absolute Extrema of a Continuous Function f on a Finite Closed Interval $[a, b]$.

Step 1. Find the critical points of f in (a, b).

Step 2. Evaluate f at all the critical points and at the endpoints a and b.

Step 3. The largest of the values in Step 2 is the absolute maximum value of f on $[a, b]$ and the smallest value is the absolute minimum.

Example 2

Find the absolute maximum and minimum values of $f(x) = 2x^3 - 15x^2 + 36x$ on the interval $[1, 5]$, and determine where these values occur.

Solution. Since f is differentiable, the absolute extrema must occur either at the endpoints of the interval $[1, 5]$ or at stationary points in the open interval $(1, 5)$. To find the stationary points, we must solve the equation $f'(x) = 0$, which can be written as

$$6x^2 - 30x + 36 = 6(x^2 - 5x + 6) = 6(x - 3)(x - 2) = 0$$

Thus, there are stationary points at $x = 2$ and $x = 3$. Evaluating f at the endpoints and the

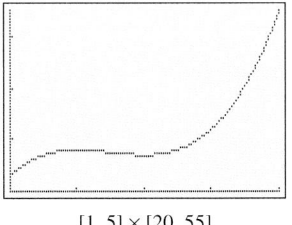

$[1, 5] \times [20, 55]$
xScl = 1, yScl = 10

Figure 6.1.6

stationary points yields

$$f(1) = 2(1)^3 - 15(1)^2 + 36(1) = 23$$
$$f(2) = 2(2)^3 - 15(2)^2 + 36(2) = 28$$
$$f(3) = 2(3)^3 - 15(3)^2 + 36(3) = 27$$
$$f(5) = 2(5)^3 - 15(5)^2 + 36(5) = 55$$

from which we conclude that an absolute minimum of $f(x) = 23$ occurs at $x = 1$ and an absolute maximum of $f(x) = 55$ occurs at $x = 5$. This is consistent with the graph of f in Figure 6.1.6. ◀

Example 3

Find the absolute extrema of $f(x) = 6x^{4/3} - 3x^{1/3}$ on the interval $[-1, 1]$, and determine where these values occur.

Solution. Differentiating we obtain

$$f'(x) = 8x^{1/3} - x^{-2/3} = x^{-2/3}(8x - 1) = \frac{8x - 1}{x^{2/3}}$$

Table 6.1.1

x	-1	0	$\frac{1}{8}$	1
$f(x)$	9	0	$-\frac{9}{8}$	3

Thus, $f'(x) = 0$ at $x = \frac{1}{8}$, and there is a point of nondifferentiability at $x = 0$. Evaluating f at these critical points and the endpoints yields Table 6.1.1, from which we conclude that an absolute minimum of $f(x) = -\frac{9}{8}$ occurs at $x = \frac{1}{8}$, and an absolute maximum of $f(x) = 9$ occurs at $x = -1$. ◀

ABSOLUTE EXTREMA ON INFINITE INTERVALS

We observed earlier that a continuous function may or may not have absolute extrema on an infinite interval (see Figure 6.1.1). However, certain conclusions about the existence of absolute extrema of a continuous function f on $(-\infty, +\infty)$ can be drawn from the behavior of $f(x)$ as $x \to -\infty$ and $x \to +\infty$ (Table 6.1.2).

Table 6.1.2

LIMITS	$\lim_{x \to -\infty} f(x) = +\infty$ $\lim_{x \to +\infty} f(x) = +\infty$	$\lim_{x \to -\infty} f(x) = -\infty$ $\lim_{x \to +\infty} f(x) = -\infty$	$\lim_{x \to -\infty} f(x) = -\infty$ $\lim_{x \to +\infty} f(x) = +\infty$	$\lim_{x \to -\infty} f(x) = +\infty$ $\lim_{x \to +\infty} f(x) = -\infty$
CONCLUSION IF f IS CONTINUOUS	f has an absolute minimum but no absolute maximum on $(-\infty, +\infty)$.	f has an absolute maximum but no absolute minimum on $(-\infty, +\infty)$.	f has neither an absolute maximum nor an absolute minimum on $(-\infty, +\infty)$.	f has neither an absolute maximum nor an absolute minimum on $(-\infty, +\infty)$.
GRAPH				

Example 4

What can you say about the existence of absolute extrema on $(-\infty, +\infty)$ for polynomials?

Solution. If $p(x)$ is a polynomial of odd degree, then

$$\lim_{x \to +\infty} p(x) \quad \text{and} \quad \lim_{x \to -\infty} p(x) \tag{1}$$

have opposite signs (one is $+\infty$ and the other is $-\infty$), so there are no absolute extrema. On the other hand, if $p(x)$ has even degree, then the limits in (1) have the same sign (both $+\infty$ or both $-\infty$). If the leading coefficient is positive, then both limits are $+\infty$, and there is an absolute minimum; if the leading coefficient is negative, then both limits are $-\infty$, and there is an absolute maximum.   ◄

Example 5

Determine by inspection whether $p(x) = 3x^4 + 4x^3$ has any absolute extrema. If so, find them, and state where they occur.

Solution. Since $p(x)$ has even degree and the leading coefficient is positive, $p(x) \to +\infty$ as $x \to \pm\infty$. Thus, there is an absolute minimum but no absolute maximum. From Theorem 6.1.4 [applied to the interval $(-\infty, +\infty)$], the absolute minimum must occur at a critical point of p. Since p is differentiable, all critical points are stationary points, so we can find them by solving the equation $p'(x) = 0$. This equation is

$$12x^3 + 12x^2 = 12x^2(x + 1) = 0$$

from which we conclude that stationary points occur at $x = 0$ and $x = -1$. Evaluating p at the stationary points yields

$$p(0) = 0 \quad \text{and} \quad p(-1) = -1$$

from which we conclude that p has an absolute minimum of $p(x) = -1$, and this occurs at $x = -1$ (Figure 6.1.7).   ◄

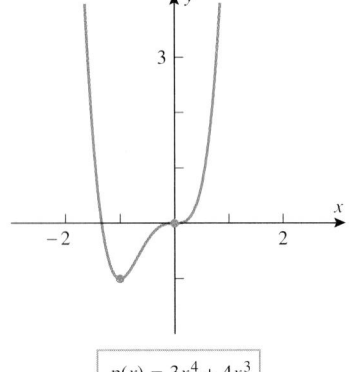

$p(x) = 3x^4 + 4x^3$

Figure 6.1.7

ABSOLUTE EXTREMA ON OPEN INTERVALS

We know that a continuous function may or may not have absolute extrema on an open interval. However, certain conclusions about the existence of absolute extrema of a continuous function f on a finite open interval (a, b) can be drawn from the behavior of $f(x)$ as $x \to a^+$ and as $x \to b^-$ (Table 6.1.3).

Table 6.1.3

LIMITS	$\lim\limits_{x \to a^+} f(x) = +\infty$ $\lim\limits_{x \to b^-} f(x) = +\infty$	$\lim\limits_{x \to a^+} f(x) = -\infty$ $\lim\limits_{x \to b^-} f(x) = -\infty$	$\lim\limits_{x \to a^+} f(x) = -\infty$ $\lim\limits_{x \to b^-} f(x) = +\infty$	$\lim\limits_{x \to a^+} f(x) = +\infty$ $\lim\limits_{x \to b^-} f(x) = -\infty$
CONCLUSION IF f IS CONTINUOUS ON (a, b)	f has an absolute minimum but no absolute maximum on (a, b).	f has an absolute maximum but no absolute minimum on (a, b).	f has neither an absolute maximum nor an absolute minimum on (a, b).	f has neither an absolute maximum nor an absolute minimum on (a, b).
GRAPH				

Example 6

Determine whether the function

$$f(x) = \frac{1}{x^2 - x}$$

has any absolute extrema on the interval $(0, 1)$. If so, find them and state where they occur.

Solution. Since f is continuous on the interval $(0, 1)$ and

$$\lim_{x \to 0^+} f(x) = \lim_{x \to 0^+} \frac{1}{x^2 - x} = \lim_{x \to 0^+} \frac{1}{x(x-1)} = -\infty$$

$$\lim_{x \to 1^-} f(x) = \lim_{x \to 1^-} \frac{1}{x^2 - x} = \lim_{x \to 1^-} \frac{1}{x(x-1)} = -\infty$$

the function f has an absolute maximum but no absolute minimum on the interval $(0, 1)$. By Theorem 6.1.4 the absolute maximum must occur at a critical point of f, so we need to look for stationary points or points of nondifferentiability in the interval $(0, 1)$. We have

$$f'(x) = -\frac{2x - 1}{\left(x^2 - x\right)^2}$$

so the only solution of the equation $f'(x) = 0$ is $x = \frac{1}{2}$. The denominator of f' is zero if $x = 0$ or $x = 1$, but these critical points are of no concern here because they fall outside of the open interval $(0, 1)$. Thus, the absolute maximum occurs at $x = \frac{1}{2}$, and this absolute maximum is

$$f\left(\tfrac{1}{2}\right) = \frac{1}{\left(\tfrac{1}{2}\right)^2 - \tfrac{1}{2}} = -4$$

(Figure 6.1.8). ◀

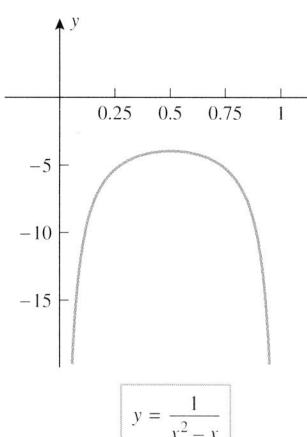

$$y = \frac{1}{x^2 - x}$$

Figure 6.1.8

ABSOLUTE EXTREMA OF FUNCTIONS WITH ONE RELATIVE EXTREMUM

If a continuous function has only one relative extremum on a finite or infinite interval I, then that relative extremum must of necessity also be an absolute extremum. To understand why this is so, suppose that f has a relative maximum at a point x_0 on an interval I, and there are no other relative extrema of f on I. If $f(x_0)$ is *not* the absolute maximum of f on I, then the graph of f has to make an upward turn somewhere on I to rise above $f(x_0)$. However, this cannot happen because in the process of making an upward turn it would produce a second relative extremum on I (Figure 6.1.9). Thus, $f(x_0)$ must be the absolute maximum as well as a relative maximum. This idea is captured in the following theorem, which we state without proof.

6.1.5 THEOREM. *Suppose that f is continuous and has exactly one relative extremum on an interval I, say at x_0.*
(a) If f has a relative minimum at x_0, then $f(x_0)$ is the absolute minimum of f on I.
(b) If f has a relative maximum at x_0, then $f(x_0)$ is the absolute maximum of f on I.

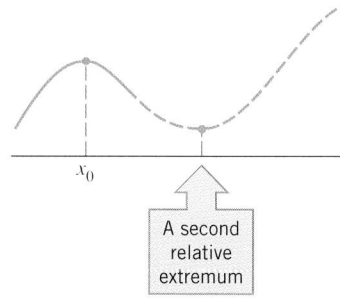

Figure 6.1.9

This theorem is often helpful in situations where other methods are difficult or tedious to apply.

Example 7

Find all absolute extrema of the function $f(x) = x^3 - 3x^2 + 4$ on the interval

(a) $(-\infty, +\infty)$ (b) $(0, +\infty)$

Solution (a). Because f is a polynomial of odd degree, it follows from the discussion in Example 4 that there are no absolute extrema on the interval $(-\infty, +\infty)$.

Solution (b). Since

$$\lim_{x \to +\infty} (x^3 - 3x^2 + 4) = +\infty$$

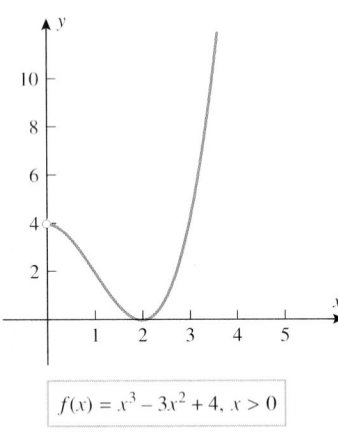

$$f(x) = x^3 - 3x^2 + 4, \ x > 0$$

Figure 6.1.10

we know that f cannot have an absolute maximum on the interval $(0, +\infty)$. However, the limit

$$\lim_{x \to 0^+} (x^3 - 3x^2 + 4) = 4$$

is not infinite, so there is a possibility that f may have an absolute minimum on this interval. In this case it would have to occur at a stationary point, which suggests that we look for solutions of the equation $f'(x) = 0$. But,

$$f'(x) = 3x^2 - 6x = 3x(x - 2)$$

so f has stationary points at $x = 0$ and $x = 2$. However, $x = 0$ falls outside of the interval $(0, +\infty)$, so only the stationary point at $x = 2$ lies in the interval $(0, +\infty)$. Thus, Theorem 6.1.5 is applicable here. Since

$$f''(x) = 6x - 6$$

we have $f''(2) = 6 > 0$, so a relative minimum occurs at $x = 2$ by the second derivative test. Thus, $f(x)$ has an absolute minimum at $x = 2$, and this absolute minimum is $f(2) = 0$ (Figure 6.1.10). ◀

ABSOLUTE EXTREMA AND PARAMETRIC CURVES

Suppose that a curve C is given parametrically by the equations

$$x = f(t), \quad y = g(t) \qquad (a \le t \le b)$$

where f and g are *continuous* on the finite closed interval $[a, b]$. It follows from the Extreme-Value Theorem that $f(t)$ and $g(t)$ have absolute maxima and absolute minima for $a \le t \le b$; this means that a particle moving along the curve cannot move away from the origin indefinitely—there must be a smallest and largest x-coordinate and a smallest and largest y-coordinate. Geometrically, the entire curve is contained within a box determined by these smallest and largest coordinates.

Example 8

Suppose that the equations of motion for a paper airplane during its first 10 seconds of flight are

$$x = t - 3\sin t, \quad y = 4 - 3\cos t \qquad (0 \le t \le 10)$$

What are the highest and lowest points in the trajectory, and when is the airplane at those points?

Solution. The trajectory, pictured in Figure 6.1.11, is shown in more detail in Figure 1.7.2. We want to find the absolute maximum and minimum values of y over the time interval $[0, 10]$ and the values of t for which these absolute extrema occur. The absolute extrema

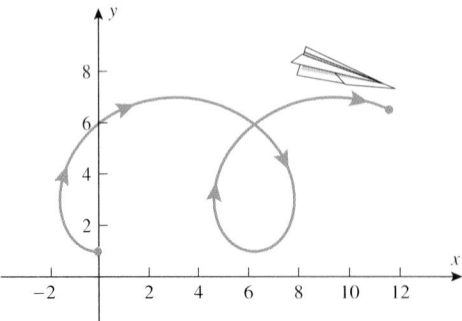

Figure 6.1.11

must occur either at the endpoints of the closed interval [0, 10] or at stationary points in the open interval (0, 10). To find the stationary points, we must solve the equation $dy/dt = 0$, which is

$$3\sin t = 0$$

Thus, there are stationary points in the interval (0, 10) at $t = \pi, 2\pi$, and 3π. Evaluating $y = 4 - 3\cos t$ at the endpoints and the stationary points yields

$$y = 4 - 3\cos 0 = 4 - 3 = 1$$
$$y = 4 - 3\cos \pi = 4 - (-3) = 7$$
$$y = 4 - 3\cos 2\pi = 4 - 3 = 1$$
$$y = 4 - 3\cos 3\pi = 4 - (-3) = 7$$
$$y = 4 - 3\cos 10 \approx 6.517$$

Thus, a high point of $y = 7$ is reached at times $t = \pi$ and $t = 3\pi$, and a low point of $y = 1$ is reached at times $t = 0$ and $t = 2\pi$. This is consistent with Figure 1.7.2. ◀

EXERCISE SET 6.1 Graphing Calculator C CAS

In Exercises 1–2, use the graph to find x-coordinates of the relative extrema and absolute extrema of f.

1.

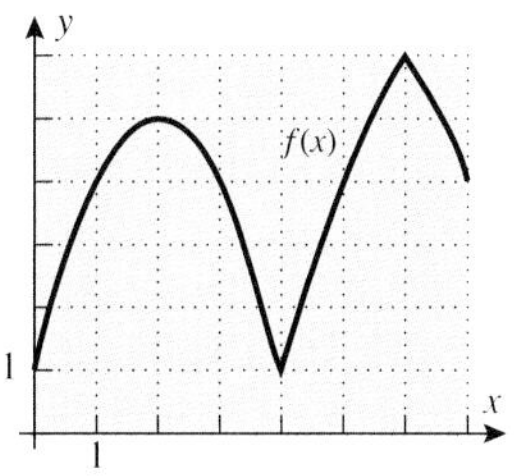

2.

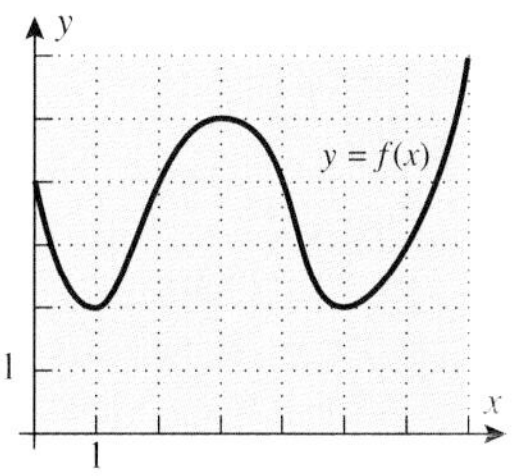

3. In each part, sketch the graph of a continuous function f with the stated properties on the interval [0, 10].
(a) f has an absolute minimum at $x = 0$ and an absolute maximum at $x = 10$.
(b) f has an absolute minimum at $x = 2$ and an absolute maximum at $x = 7$.
(c) f has a relative minima at $x = 1$ and $x = 8$, has relative maxima at $x = 3$ and $x = 7$, has an absolute minimum at $x = 5$, and has an absolute maximum at $x = 10$.

4. In each part, sketch the graph of a continuous function f with the stated properties on the interval $(-\infty, +\infty)$.
(a) f has no relative extrema or absolute extrema.
(b) f has an absolute minimum at $x = 0$ but no absolute maximum.
(c) f has an absolute maximum at $x = -5$ and an absolute minimum at $x = 5$.

In Exercises 5–14, find the absolute maximum and minimum values of f on the given closed interval, and state where those values occur.

5. $f(x) = 4x^2 - 4x + 1$; [0, 1]

6. $f(x) = 8x - x^2$; [0, 6]

7. $f(x) = (x - 1)^3$; [0, 4]

8. $f(x) = 2x^3 - 3x^2 - 12x$; [−2, 3]

9. $f(x) = \dfrac{3x}{\sqrt{4x^2 + 1}}$; [−1, 1]

10. $f(x) = (x^2 + x)^{2/3}$; [−2, 3]

11. $f(x) = x - \tan x$; $[-\pi/4, \pi/4]$

12. $f(x) = \sin x - \cos x$; $[0, \pi]$

13. $f(x) = 1 + |9 - x^2|$; [−5, 1]

14. $f(x) = |6 - 4x|$; [−3, 3]

In Exercises 15–22, find the absolute maximum and minimum values of f, if any, on the given interval, and state where those values occur.

15. $f(x) = x^2 - 3x - 1$; $(-\infty, +\infty)$

16. $f(x) = 3 - 4x - 2x^2$; $(-\infty, +\infty)$

17. $f(x) = 4x^3 - 3x^4$; $(-\infty, +\infty)$

18. $f(x) = x^4 + 4x$; $(-\infty, +\infty)$

19. $f(x) = x^3 - 3x - 2$; $(-\infty, +\infty)$

20. $f(x) = x^3 - 9x + 1$; $(-\infty, +\infty)$

21. $f(x) = \dfrac{x^2}{x+1}$; $(-5, -1)$ **22.** $f(x) = \dfrac{x+3}{x-3}$; $[-5, 5]$

In Exercises 23–34, use a graphing utility to estimate the absolute maximum and minimum values of f, if any, on the stated interval, and then use calculus methods to find the exact values.

23. $f(x) = (x^2 - 1)^2$; $(-\infty, +\infty)$

24. $f(x) = (x-1)^2(x+2)^2$; $(-\infty, +\infty)$

25. $f(x) = x^{2/3}(20 - x)$; $[-1, 20]$

26. $f(x) = \dfrac{x}{x^2 + 2}$; $[-1, 4]$

27. $f(x) = 1 + \dfrac{1}{x}$; $(0, +\infty)$

28. $f(x) = \dfrac{x}{x^2 + 1}$; $[0, +\infty)$

29. $f(x) = 2 \sec x - \tan x$; $[0, \pi/4]$

30. $f(x) = \sin^2 x + \cos x$; $[-\pi, \pi]$

31. $f(x) = x^3 e^{-2x}$; $[1, 4]$

32. $f(x) = \dfrac{\ln x}{x}$; $[1, e]$

33. $f(x) = \sin(\cos x)$; $[0, 2\pi]$

34. $f(x) = \cos(\sin x)$; $[0, \pi]$

35. Find the absolute maximum and minimum values of

$$f(x) = \begin{cases} 4x - 2, & x < 1 \\ (x-2)(x-3), & x \geq 1 \end{cases}$$

on $\left[\frac{1}{2}, \frac{7}{2}\right]$.

36. Let $f(x) = x^2 + px + q$. Find the values of p and q such that $f(1) = 3$ is an extreme value of f on $[0, 2]$. Is this value a maximum or minimum?

If f is a periodic function, then the locations of all absolute extrema on the interval $(-\infty, +\infty)$ can be obtained by finding the locations of the absolute extrema for one period and using the periodicity to locate the rest. Use this idea in Exercise 37 and 38 to find the absolute maximum and minimum values of the function, and state the x-values at which they occur.

37. $f(x) = 2 \sin 2x + \sin 4x$ **38.** $f(x) = 3 \cos \dfrac{x}{3} + 2 \cos \dfrac{x}{2}$

One way of proving that $f(x) \leq g(x)$ for all x in a given interval is to show that $0 \leq g(x) - f(x)$ for all x in the interval; and one way of proving the latter inequality is to show that the absolute minimum value of $g(x) - f(x)$ on the interval is nonnegative. Use this idea to prove the inequalities in Exercises 39 and 40.

39. Prove that $\sin x \leq x$ for all x in the interval $[0, 2\pi]$.

40. Prove that $\ln x \leq x - 1$ on the interval $(0, +\infty)$.

41. What is the smallest possible slope for a tangent to the equation $y = x^3 - 3x^2 + 5x$?

42. (a) Show that

$$f(x) = \frac{64}{\sin x} + \frac{27}{\cos x}$$

has a minimum value, but no maximum value on the interval $(0, \pi/2)$.

(b) Find the minimum value.

[c] 43. Use a CAS to show that the absolute minimum value of

$$f(x) = x^2 + \frac{16x^2}{(8-x)^2}, \quad x > 8$$

occurs at $x = 4(2 + \sqrt[3]{2})$ by using it to find $f'(x)$ and to solve the equation $f'(x) = 0$.

44. The concentration $C(t)$ of a drug in the bloodstream t hours after it has been injected is commonly modeled by an equation of the form

$$C(t) = \frac{K(e^{-bt} - e^{-at})}{a - b}$$

where $K > 0$ and $a > b > 0$.

(a) At what time does the maximum concentration occur?

(b) Let $K = 1$ for simplicity, and use a graphing utility to check your result in part (a) by graphing $C(t)$ for various values of a and b.

45. It can be proved that if f is differentiable on (a, b) and L is a line that does not intersect the curve $y = f(x)$ over an interval (a, b), then the points at which the curve is closest to or farthest from the line L, if any, occur at points where the tangent line to the curve is parallel to L (see the accompanying figure). Use this result to find the points on the graph of $y = -x^2$ that are closest to and farthest from the line $y = 2 - x$ for $-1 \leq x \leq \frac{3}{2}$.

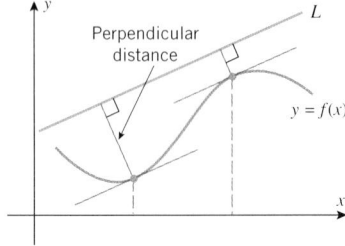

Figure Ex-45

46. Use the idea discussed in Exercise 45 to find the coordinates of all points on the graph of $y = x^3$ closest to and farthest from the line $y = \frac{4}{3}x - 1$ for $-1 \leq x \leq 1$.

47. Suppose that the equations of motion of a paper airplane during the first 12 seconds of flight are

$$x = t - 2\sin t, \quad y = 2 - 2\cos t \quad (0 \leq t \leq 12)$$

What are the highest and lowest points in the trajectory, and when is the airplane at those points?

48. The accompanying figure shows the path of a fly whose equations of motion are

$$x = \frac{\cos t}{2 + \sin t}, \quad y = 3 + \sin(2t) - 2\sin^2 t \quad (0 \leq t \leq 2\pi)$$

(a) How high and low does it fly?
(b) How far left and right of the origin does it fly?

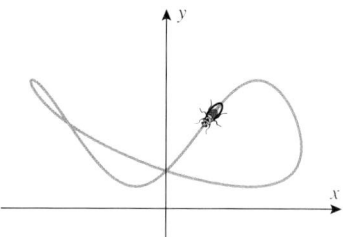

Figure Ex-48

49. Let $f(x) = ax^2 + bx + c$, where $a > 0$. Prove that $f(x) \geq 0$ for all x if and only if $b^2 - 4ac \leq 0$. [*Hint:* Find the minimum of $f(x)$.]

50. Prove Theorem 6.1.4 in the case where the extreme value is a minimum.

6.2 APPLIED MAXIMUM AND MINIMUM PROBLEMS

In this section we will show how the methods discussed in the last section can be used to solve various applied optimization problems.

CLASSIFICATION OF OPTIMIZATION PROBLEMS

The applied optimization problems that we will consider in this section fall into the following two categories:

- Problems that reduce to maximizing or minimizing a continuous function over a finite closed interval.

- Problems that reduce to maximizing or minimizing a continuous function over an infinite interval or a finite interval that is not closed.

For problems of the first type the Extreme-Value Theorem (6.1.3) guarantees that the problem has a solution, and we know that the solution can be obtained by examining the values of the function at the critical points and the endpoints. However, for problems of the second type there may or may not be a solution. Thus, part of the attack on such problems is to determine whether there actually is a solution. If the function is continuous and has exactly one relative extremum on the interval, then Theorem 6.1.5 guarantees the existence of a solution and provides a method for finding it. In cases where this theorem is not applicable some ingenuity may be required to solve the problem.

PROBLEMS INVOLVING FINITE CLOSED INTERVALS

Example 1

Find the dimensions of a rectangle with perimeter 100 ft whose area is as large as possible.

Solution. Let

x = length of the rectangle (ft)
y = width of the rectangle (ft)
A = area of the rectangle (ft^2)

Then

$$A = xy \tag{1}$$

Since the perimeter of the rectangle is 100 ft, the variables x and y are related by the equation

$$2x + 2y = 100 \quad \text{or} \quad y = 50 - x \tag{2}$$

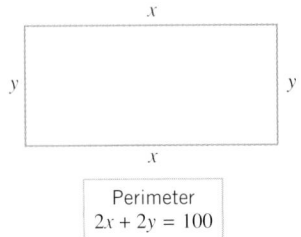

Perimeter
$2x + 2y = 100$

Figure 6.2.1

(See Figure 6.2.1.) Substituting (2) in (1) yields

$$A = x(50 - x) = 50x - x^2 \qquad (3)$$

Because x represents a length it cannot be negative, and because the two sides of length x cannot have a combined length exceeding the total perimeter of 100 ft, the variable x must satisfy

$$0 \le x \le 50 \qquad (4)$$

Thus, we have reduced the problem to that of finding the value (or values) of x in $[0, 50]$, for which A is maximum. Since A is a polynomial in x, it is continuous on $[0, 50]$, and so the maximum must occur at an endpoint of this interval or at a stationary point.

From (3) we obtain

$$\frac{dA}{dx} = 50 - 2x$$

Setting $dA/dx = 0$ we obtain

$$50 - 2x = 0$$

or $x = 25$. Thus, the maximum occurs at one of the points

$$x = 0, \quad x = 25, \quad x = 50$$

Substituting these values in (3) yields Table 6.2.1, which tells us that the maximum area of 625 ft^2 occurs at $x = 25$, which is consistent with the graph of (3) in Figure 6.2.2. From (2) the corresponding value of $y = 25$, so the rectangle of perimeter 100 ft with greatest area is a square with sides of length 25 ft. ◀

Table 6.2.1

x	0	25	50
A	0	625	0

REMARK.   In this example we included $x = 0$ and $x = 50$ as possible values for x, even though both values lead to rectangles with two sides of length zero. Whether or not these values should be allowed will depend on our objective in the problem. If we view this purely as a mathematical problem, then there is nothing wrong with allowing sides of length zero. However, if we view this as an applied problem in which the rectangle will be formed from physical material, then these values should be excluded.

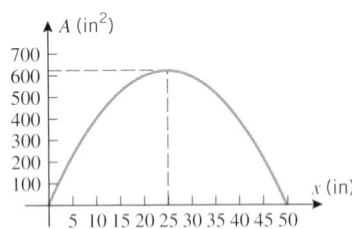

Figure 6.2.2

Example 1 illustrates the following five-step procedure that can be used for solving many applied maximum and minimum problems.

Step 1. Draw an appropriate figure and label the quantities relevant to the problem.

Step 2. Find a formula for the quantity to be maximized or minimized.

Step 3. Using the conditions stated in the problem to eliminate variables, express the quantity to be maximized or minimized as a function of one variable.

Step 4. Find the interval of possible values for this variable from the physical restrictions in the problem.

Step 5. If applicable, use the techniques of the preceding section to obtain the maximum or minimum.

Example 2

An open box is to be made from a 16-inch by 30-inch piece of cardboard by cutting out squares of equal size from the four corners and bending up the sides (Figure 6.2.3). What size should the squares be to obtain a box with largest possible volume?

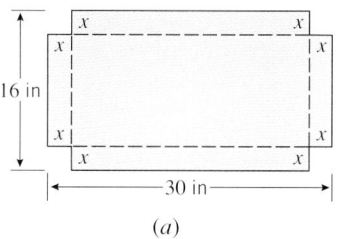
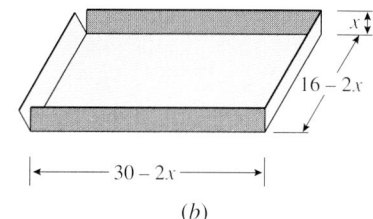

(a)   (b)

Figure 6.2.3

Solution. Let

$$x = \text{length (in inches) of the sides of the squares to be cut out}$$

$$V = \text{volume (in cubic inches) of the resulting box}$$

Because we are removing a square of side x from each corner, the resulting box will have dimensions $16 - 2x$ by $30 - 2x$ by x (Figure 6.2.3*b*). Since the volume of a box is the product of its dimensions, we have

$$V = (16 - 2x)(30 - 2x)x = 480x - 92x^2 + 4x^3 \tag{5}$$

The variable x in this expression is subject to certain restrictions. Because x represents a length it cannot be negative, and because the width of the cardboard is 16 inches we cannot cut out squares whose sides are more than 8 inches long. Thus, the variable x in (5) must satisfy

$$0 \le x \le 8$$

Table 6.2.2

x	0	$\frac{10}{3}$	8
V	0	$\frac{19600}{27} \approx 726$	0

and hence we have reduced our problem to finding the value (or values) of x in the interval $[0, 8]$ for which (5) is maximum. From (5) we obtain

$$\frac{dV}{dx} = 480 - 184x + 12x^2 = 4(120 - 46x + 3x^2)$$

Setting $dV/dx = 0$ yields

$$120 - 46x + 3x^2 = 0$$

which can be solved by the quadratic formula to obtain the critical points

$$x = \tfrac{10}{3} \quad \text{and} \quad x = 12$$

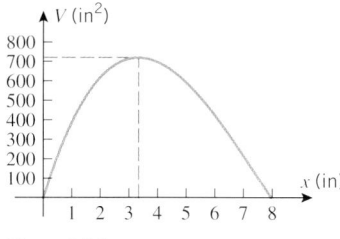

Figure 6.2.4

Since $x = 12$ falls outside the interval $[0, 8]$, the maximum value of V occurs either at the critical point $x = \frac{10}{3}$ or at one of the endpoints $x = 0$, $x = 8$. Substituting these values in (5) yields Table 6.2.2, which tells us that the greatest possible volume $V = \frac{19600}{27}$ in³ ≈ 726 in³ occurs when we cut out squares whose sides have length $\frac{10}{3}$ inches. This is consistent with the graph of (5) shown in Figure 6.2.4. ◄

In Example 2 of Section 1.1 we used approximate graphical methods to solve a problem of piping oil from an offshore well to a point on the shore with minimal cost. We will now show how to solve that problem exactly using calculus.

Example 3

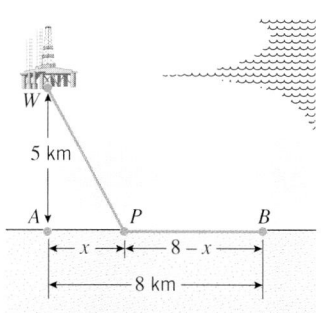

Figure 6.2.5

Figure 6.2.5 shows an offshore oil well located at a point W that is 5 km from the closest point A on a straight shoreline. Oil is to be piped from W to a shore point B that is 8 km from A by piping it on a straight line under water from W to some shore point P between A and B and then on to B via pipe along the shoreline. If the cost of laying pipe is \$1,000,000/km under water and \$500,000/km over land, where should the point P be located to minimize the cost of laying the pipe?

Solution. Let

$\qquad x = $ distance (in kilometers) between A and P

$\qquad c = $ cost (in millions of dollars) for the entire pipeline

From Figure 6.2.5 the length of pipe under water is the distance between W and P. By the Theorem of Pythagoras, that length is

$$\sqrt{x^2 + 25} \qquad\qquad (6)$$

Also from Figure 6.2.5, the length of pipe over land is the distance between P and B, which is

$$8 - x \qquad\qquad (7)$$

From (6) and (7) it follows that the total cost c (in millions of dollars) for the pipeline is

$$c = 1(\sqrt{x^2 + 25}) + \tfrac{1}{2}(8 - x) = \sqrt{x^2 + 25} + \tfrac{1}{2}(8 - x) \qquad\qquad (8)$$

Because the distance between A and B is 8 km, the distance x between A and P must satisfy

$$0 \le x \le 8$$

We have thus reduced our problem to finding the value (or values) of x in the interval $[0, 8]$ for which (8) is a minimum. Since c is a continuous function of x on the closed interval $[0, 8]$, we can use the methods developed in the preceding section to find the minimum.

From (8) we obtain

$$\frac{dc}{dx} = \frac{x}{\sqrt{x^2 + 25}} - \frac{1}{2}$$

Setting $dc/dx = 0$ and solving for x yields

$$\frac{x}{\sqrt{x^2 + 25}} = \frac{1}{2} \qquad\qquad (9)$$

$$x^2 = \frac{1}{4}(x^2 + 25)$$

$$x = \pm\frac{5}{\sqrt{3}}$$

The number $-5/\sqrt{3}$ is not a solution of (9) and must be discarded, leaving $x = 5/\sqrt{3}$ as the only critical point. Since this point lies in the interval $[0, 8]$, the minimum must occur at one of the points

$$x = 0, \quad x = 5/\sqrt{3}, \quad x = 8$$

Substituting these values in (8) yields Table 6.2.3, which tells us that the least possible cost of the pipeline (to the nearest dollar) is $c = \$8,330,127$, and this occurs when the point P is located at a distance of $5/\sqrt{3} \approx 2.89$ km from A. This is consistent with the graph in Figure 1.1.9c. ◀

Table 6.2.3

x	0	$\frac{5}{\sqrt{3}}$	8
c	9	$\frac{10}{\sqrt{3}} + \left(4 - \frac{5}{2\sqrt{3}}\right) \approx 8.330127$	$\sqrt{89} \approx 9.433981$

FOR THE READER.   If you have a CAS, use it to check all of the computations in this example. Specifically, differentiate c with respect to x, solve the equation $dc/dx = 0$, and perform all of the numerical calculations.

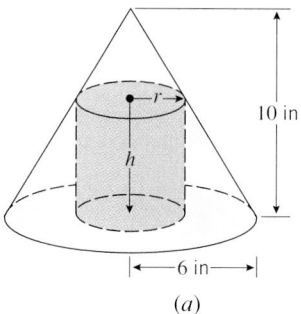

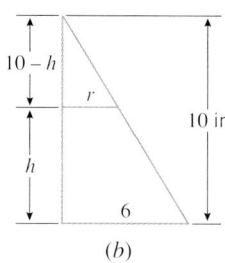

Figure 6.2.6

Table 6.2.4

r	0	4	6
V	0	$\frac{160}{3}\pi$	0

....

PROBLEMS INVOLVING INTERVALS THAT ARE NOT FINITE AND CLOSED

Example 4

Find the radius and height of the right circular cylinder of largest volume that can be inscribed in a right circular cone with radius 6 inches and height 10 inches (Figure 6.2.6a).

Solution. Let

r = radius (in inches) of the cylinder

h = height (in inches) of the cylinder

V = volume (in cubic inches) of the cylinder

The formula for the volume of the inscribed cylinder is

$$V = \pi r^2 h \tag{10}$$

To eliminate one of the variables in (10) we need a relationship between r and h. Using similar triangles (Figure 6.2.6b) we obtain

$$\frac{10-h}{r} = \frac{10}{6} \quad \text{or} \quad h = 10 - \tfrac{5}{3}r \tag{11}$$

Substituting (11) into (10) we obtain

$$V = \pi r^2 \left(10 - \tfrac{5}{3}r\right) = 10\pi r^2 - \tfrac{5}{3}\pi r^3 \tag{12}$$

which expresses V in terms of r alone. Because r represents a radius it cannot be negative, and because the radius of the inscribed cylinder cannot exceed the radius of the cone, the variable r must satisfy

$$0 \le r \le 6$$

Thus, we have reduced the problem to that of finding the value (or values) of r in [0, 6] for which (12) is a maximum. Since V is a continuous function of r on [0, 6], the methods developed in the preceding section apply.

From (12) we obtain

$$\frac{dV}{dr} = 20\pi r - 5\pi r^2 = 5\pi r(4 - r)$$

Setting $dV/dr = 0$ gives

$$5\pi r(4 - r) = 0$$

so $r = 0$ and $r = 4$ are critical points. Since these lie in the interval [0, 6], the maximum must occur at one of the points

$$r = 0, \quad r = 4, \quad r = 6$$

Substituting these values in (12) yields Table 6.2.4, which tells us that the maximum volume $V = \frac{160}{3}\pi \approx 168 \text{ in}^3$ occurs when the inscribed cylinder has radius 4 in. When $r = 4$ it follows from (11) that $h = \frac{10}{3}$. Thus, the inscribed cylinder of largest volume has radius $r = 4$ in and height $h = \frac{10}{3}$ in. ◀

Example 5

A closed cylindrical can is to hold 1 liter (1000 cm³) of liquid. How should we choose the height and radius to minimize the amount of material needed to manufacture the can?

Solution. Let

h = height (in cm) of the can

r = radius (in cm) of the can

S = surface area (in cm²) of the can

Assuming there is no waste or overlap, the amount of material needed for manufacture will be the same as the surface area of the can. Since the can consists of two circular disks of radius r and a rectangular sheet with dimensions h by $2\pi r$ (Figure 6.2.7), the surface area will be

$$S = 2\pi r^2 + 2\pi rh \tag{13}$$

Since S depends on two variables, r and h, we will look for some condition in the problem that will allow us to express one of these variables in terms of the other. For this purpose, observe that the volume of the can is 1000 cm³, so it follows from the formula $V = \pi r^2 h$ for the volume of a cylinder that

$$1000 = \pi r^2 h \quad \text{or} \quad h = \frac{1000}{\pi r^2} \tag{14–15}$$

Substituting (15) in (13) yields

$$S = 2\pi r^2 + \frac{2000}{r} \tag{16}$$

Thus, we have reduced the problem to finding a value of r in the interval $(0, +\infty)$ for which S is minimum, provided there actually is a minimum.[*] However, S is a continuous function of r on the interval $(0, +\infty)$ and

$$\lim_{r \to 0^+} \left(2\pi r^2 + \frac{2000}{r}\right) = +\infty \quad \text{and} \quad \lim_{r \to +\infty} \left(2\pi r^2 + \frac{2000}{r}\right) = +\infty$$

so the analysis in Table 6.1.3 implies that S does have a minimum on the interval $(0, +\infty)$. Since this minimum must occur at a critical point, we calculate

$$\frac{dS}{dr} = 4\pi r - \frac{2000}{r^2} \tag{17}$$

Setting $dS/dr = 0$ gives

$$4\pi r - \frac{2000}{r^2} = 0 \quad \text{or} \quad r = \frac{10}{\sqrt[3]{2\pi}} \tag{18}$$

Since (18) is the only critical point in the interval $(0, +\infty)$, this value of r yields the minimum value of S. From (15) the value of h corresponding to this r is

$$h = \frac{1000}{\pi(10/\sqrt[3]{2\pi})^2} = \frac{20}{\sqrt[3]{2\pi}} = 2r$$

It is not accidental here that the minimum occurs when the height of the can is equal to the diameter of its base (Exercise 27).

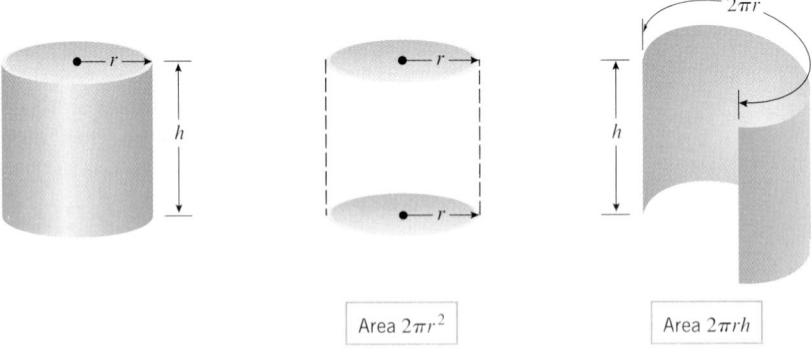

Area $2\pi r^2$ Area $2\pi rh$

Figure 6.2.7

Second Solution. The conclusion that a minimum occurs at the value of r in (18) can be deduced from Theorem 6.1.5 and the second derivative test by noting that

---

[*] The value $r = 0$ must be excluded because a cylindrical can of radius 0 cm cannot have a volume of 1000 cm³ [see (14)].

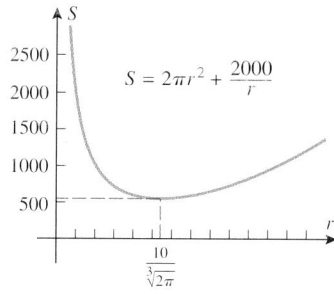

Figure 6.2.8

$$\frac{d^2S}{dr^2} = 4\pi + \frac{4000}{r^3}$$

is positive if $r > 0$ and hence is positive if $r = 10/\sqrt[3]{2\pi}$. This implies that a relative minimum, and therefore a minimum, occurs at the critical point $r = 10/\sqrt[3]{2\pi} \approx 5.4$.

Third Solution. The existence of a minimum is implied by the graph of S versus r in Figure 6.2.8. As shown in (18), this minimum occurs at $r = 10/\sqrt[3]{2\pi}$. ◀

REMARK.   Note that S has no maximum on $(0, +\infty)$. Thus, had we asked for the dimensions of the can requiring the maximum amount of material for its manufacture, there would have been no solution to the problem. Optimization problems with no solution are sometimes called *ill posed*.

Example 6

Find a point on the curve $y = x^2$ that is closest to the point $(18, 0)$.

Solution.   The distance L between $(18, 0)$ and an arbitrary point (x, y) on the curve $y = x^2$ (Figure 6.2.9) is given by

$$L = \sqrt{(x - 18)^2 + (y - 0)^2}$$

Since (x, y) lies on the curve, x and y satisfy $y = x^2$; thus,

$$L = \sqrt{(x - 18)^2 + x^4} \tag{19}$$

Because there are no restrictions on x, the problem reduces to finding a value of x in $(-\infty, +\infty)$ for which (19) is minimum, provided such a value exists.

In problems of minimizing or maximizing a distance, there is a trick that is helpful for simplifying the computations. It is based on the observation that the distance and the square of the distance have their maximum or minimum at the same point (see Exercise 60). Thus, the minimum value of L in (19) and the minimum value of

$$S = L^2 = (x - 18)^2 + x^4 \tag{20}$$

occur at the same x-value.

From (20),

$$\frac{dS}{dx} = 2(x - 18) + 4x^3 = 4x^3 + 2x - 36 \tag{21}$$

so that the critical points satisfy $4x^3 + 2x - 36 = 0$ or, equivalently,

$$2x^3 + x - 18 = 0 \tag{22}$$

To solve for x we will begin by checking the divisors of -18 to see whether the polynomial on the left side has any integer roots (see Appendix F). These divisors are $\pm 1, \pm 2, \pm 3, \pm 6, \pm 9$, and ± 18. A check of these values shows that $x = 2$ is a root, so that $x - 2$ is a factor of the polynomial. After dividing the polynomial by this factor we can rewrite (22) as

$$(x - 2)(2x^2 + 4x + 9) = 0$$

Thus, the remaining solutions of (22) satisfy the quadratic equation

$$2x^2 + 4x + 9 = 0$$

But these solutions are imaginary numbers (use the quadratic formula), so that $x = 2$ is the only real solution of (22) and consequently the only critical point of S. To determine the nature of this critical point we will use the second derivative test. From (21),

$$\frac{d^2S}{dx^2} = 12x^2 + 2, \quad \text{so} \quad \left.\frac{d^2S}{dx^2}\right|_{x=2} = 50 > 0$$

which shows that a relative minimum occurs at $x = 2$. Since $x = 2$ is the only relative

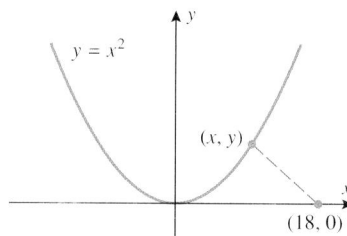

Figure 6.2.9

extremum for L, it follows from Theorem 6.1.5 that an absolute minimum value of L also occurs at $x = 2$. Thus, the point on the curve $y = x^2$ closest to $(18, 0)$ is

$$(x, y) = (x, x^2) = (2, 4) \qquad \blacktriangleleft$$

....

AN APPLICATION TO ECONOMICS

Three functions of importance to an economist or a manufacturer are

$C(x) =$ total cost of producing x units of a product during some time period

$R(x) =$ total revenue from selling x units of the product during the time period

$P(x) =$ total profit obtained by selling x units of the product during the time period

These are called, respectively, the *cost function*, *revenue function*, and *profit function*. If all units produced are sold, then these are related by

$$P(x) = R(x) - C(x) \tag{23}$$

[profit] = [revenue] − [cost]

The total cost $C(x)$ of producing x units can be expressed as a sum

$$C(x) = a + M(x) \tag{24}$$

where a is a constant, called *overhead*, and $M(x)$ is a function representing *manufacturing cost*. The overhead, which includes such fixed costs as rent and insurance, does not depend on x; it must be paid even if nothing is produced. On the other hand, the manufacturing cost $M(x)$, which includes such items as cost of materials and labor, depends on the number of items manufactured. It is shown in economics that with suitable simplifying assumptions, $M(x)$ can be expressed in the form

$$M(x) = bx + cx^2$$

where b and c are constants. Substituting this in (24) yields

$$C(x) = a + bx + cx^2 \tag{25}$$

If a manufacturing firm can sell all the items it produces for p dollars apiece, then its total revenue $R(x)$ (in dollars) will be

$$R(x) = px \tag{26}$$

and its total profit $P(x)$ (in dollars) will be

$$P(x) = [\text{total revenue}] - [\text{total cost}] = R(x) - C(x) = px - C(x)$$

Thus, if the cost function is given by (25),

$$P(x) = px - (a + bx + cx^2) \tag{27}$$

Depending on such factors as number of employees, amount of machinery available, economic conditions, and competition, there will be some upper limit l on the number of items a manufacturer is capable of producing and selling. Thus, during a fixed time period the variable x in (27) will satisfy

$$0 \leq x \leq l$$

By determining the value or values of x in $[0, l]$ that maximize (27), the firm can determine how many units of its product must be manufactured and sold to yield the greatest profit. This is illustrated in the following numerical example.

Example 7

A liquid form of penicillin manufactured by a pharmaceutical firm is sold in bulk at a price of $200 per unit. If the total production cost (in dollars) for x units is

$$C(x) = 500{,}000 + 80x + 0.003x^2$$

and if the production capacity of the firm is at most 30,000 units in a specified time, how many units of penicillin must be manufactured and sold in that time to maximize the profit?

Solution.   Since the total revenue for selling x units is $R(x) = 200x$, the profit $P(x)$ on x units will be

$$P(x) = R(x) - C(x) = 200x - (500{,}000 + 80x + 0.003x^2) \qquad (28)$$

Since the production capacity is at most 30,000 units, x must lie in the interval $[0, 30{,}000]$. From (28)

$$\frac{dP}{dx} = 200 - (80 + 0.006x) = 120 - 0.006x$$

Setting $dP/dx = 0$ gives

$$120 - 0.006x = 0 \quad \text{or} \quad x = 20{,}000$$

Since this critical point lies in the interval $[0, 30{,}000]$, the maximum profit must occur at one of the points

$$x = 0, \quad x = 20{,}000, \quad \text{or} \quad x = 30{,}000$$

Substituting these values in (28) yields Table 6.2.5, which tells us that the maximum profit $P = \$700{,}000$ occurs when $x = 20{,}000$ units are manufactured and sold in the specified time.   ◄

Table 6.2.5

x	0	20,000	30,000
$P(x)$	−500,000	700,000	400,000

MARGINAL ANALYSIS

Economists call $P'(x)$, $R'(x)$, and $C'(x)$ the ***marginal profit***, ***marginal revenue***, and ***marginal cost***, respectively; and they interpret these quantities as the *additional* profit, revenue, and cost that result from producing and selling one additional unit of the product when the production and sales levels are at x units. These interpretations follow from the local linear approximations of the profit, revenue, and cost functions. For example, it follows from Formula (7) of Section 3.6 that when the production and sales levels are at x units the local linear approximation of the profit function is

$$P(x + \Delta x) \approx P(x) + P'(x)\Delta x$$

Thus, if $\Delta x = 1$ (one additional unit produced and sold), this formula implies

$$P(x + 1) \approx P(x) + P'(x)$$

and hence the *additional* profit that results from producing and selling one additional unit can be approximated as

$$P(x + 1) - P(x) \approx P'(x)$$

A BASIC PRINCIPLE OF ECONOMICS

It follows from (23) that $P'(x) = 0$ at those points where $C'(x) = R'(x)$, and this implies that the maximum profit must occur at a point where the marginal revenue is equal to the marginal cost; that is:

> *The maximum profit occurs at a point where the cost of manufacturing and selling an additional unit of a product is exactly equal to the revenue generated by the additional unit.*

This is one of the basic principles of economics.

EXERCISE SET 6.2

1. Express the number 10 as a sum of two nonnegative numbers whose product is as large as possible.

2. How should two nonnegative numbers be chosen so that their sum is 1 and the sum of their squares is
   (a) as large as possible
   (b) as small as possible?

3. Find a number in the closed interval $\left[\frac{1}{2}, \frac{3}{2}\right]$ such that the sum of the number and its reciprocal is
   (a) as small as possible
   (b) as large as possible.

4. A rectangular field is to be bounded by a fence on three sides and by a straight stream on the fourth side. Find the dimensions of the field with maximum area that can be enclosed with 1000 feet of fence.

5. A rectangular plot of land is to be fenced in using two kinds of fencing. Two opposite sides will use heavy-duty fencing selling for $3 a foot, while the remaining two sides will use standard fencing selling for $2 a foot. What are the dimensions of the rectangular plot of greatest area that can be fenced in at a cost of $6000?

6. A rectangle is to be inscribed in a right triangle having sides of length 6 in, 8 in, and 10 in. Find the dimensions of the rectangle with greatest area assuming the rectangle is positioned as in the accompanying figure.

7. Solve the problem in Exercise 6 assuming the rectangle is positioned as in the accompanying figure.

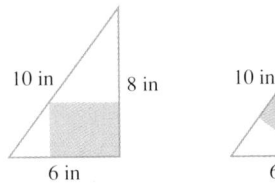
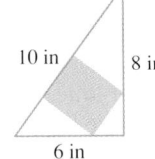

Figure Ex-6   Figure Ex-7

8. A rectangle has its two lower corners on the x-axis and its two upper corners on the curve $y = 16 - x^2$. For all such rectangles, what are the dimensions of the one with largest area?

9. Find the dimensions of the rectangle with maximum area that can be inscribed in a circle of radius 10.

10. Find the dimensions of the rectangle of greatest area that can be inscribed in a semicircle of radius R as shown in the accompanying figure.

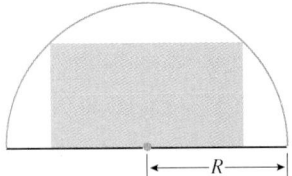

$\leftarrow\!\!-R\!\!-\!\!\rightarrow$ Figure Ex-10

11. A rectangular area of 3200 ft² is to be fenced off. Two opposite sides will use fencing costing $1 per foot and the remaining sides will use fencing costing $2 per foot. Find the dimensions of the rectangle of least cost.

12. Show that among all rectangles with perimeter p, the square has the maximum area.

13. Show that among all rectangles with area A, the square has the minimum perimeter.

14. A wire of length 12 in can be bent into a circle, bent into a square, or cut into two pieces to make both a circle and a square. How much wire should be used for the circle if the total area enclosed by the figure(s) is to be
    (a) a maximum   (b) a minimum?

15. Suppose that the number of bacteria in a culture at time t is given by $N = 5000(25 + te^{-t/20})$.
    (a) Find the largest and smallest number of bacteria in the culture during the time interval $0 \le t \le 100$.
    (b) At what time during the time interval in part (a) is the number of bacteria decreasing most rapidly?

16. A church window consisting of a rectangle topped by a semicircle is to have a perimeter p. Find the radius of the semicircle if the area of the window is to be maximum.

17. A sheet of cardboard 12 in square is used to make an open box by cutting squares of equal size from the four corners and folding up the sides. What size squares should be cut to obtain a box with largest possible volume?

18. A square sheet of cardboard of side k is used to make an open box by cutting squares of equal size from the four corners and folding up the sides. What size squares should be cut from the corners to obtain a box with largest possible volume?

19. An open box is to be made from a 3-ft by 8-ft rectangular piece of sheet metal by cutting out squares of equal size from the four corners and bending up the sides. Find the maximum volume that the box can have.

20. A closed rectangular container with a square base is to have a volume of 2250 in³. The material for the top and bottom of the container will cost $2 per in², and the material for the sides will cost $3 per in². Find the dimensions of the container of least cost.

21. A closed rectangular container with a square base is to have a volume of 2000 cm³. It costs twice as much per square centimeter for the top and bottom as it does for the sides. Find the dimensions of the container of least cost.

22. A container with square base, vertical sides, and open top is to be made from 1000 ft² of material. Find the dimensions of the container with greatest volume.

23. A rectangular container with two square sides and an open top is to have a volume of V cubic units. Find the dimensions of the container with minimum surface area.

24. Find the dimensions of the right circular cylinder of largest volume that can be inscribed in a sphere of radius R.

25. Find the dimensions of the right circular cylinder of greatest surface area that can be inscribed in a sphere of radius R.

26. Show that the right circular cylinder of greatest volume that can be inscribed in a right circular cone has volume that is $\frac{4}{9}$ the volume of the cone (Figure Ex-26).

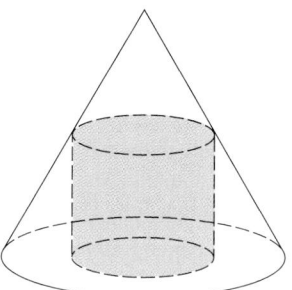

Figure Ex-26

27. A closed, cylindrical can is to have a volume of V cubic units. Show that the can of minimum surface area is achieved when the height is equal to the diameter of the base.

28. A closed cylindrical can is to have a surface area of S square units. Show that the can of maximum volume is achieved when the height is equal to the diameter of the base.

29. A cylindrical can, open at the top, is to hold 500 cm³ of liquid. Find the height and radius that minimize the amount of material needed to manufacture the can.

30. A soup can in the shape of a right circular cylinder of radius r and height h is to have a prescribed volume V. The top and bottom are cut from squares as shown in the accompanying figure. If the shaded corners are wasted, but there is no other waste, find the ratio r/h for the can requiring the least material (including waste).

31. A box-shaped wire frame consists of two identical wire squares whose vertices are connected by four straight wires of equal length (Figure Ex-31). If the frame is to be made from a wire of length L, what should the dimensions be to obtain a box of greatest volume?

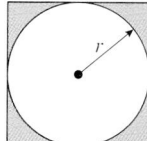 

Figure Ex-30   Figure Ex-31

32. Suppose that the sum of the surface areas of a sphere and a cube is a constant.
  (a) Show that the sum of their volumes is smallest when the diameter of the sphere is equal to the length of an edge of the cube.
  (b) When will the sum of their volumes be greatest?

33. Find the height and radius of the cone of slant height L whose volume is as large as possible.

34. A cone is made from a circular sheet of radius R by cutting out a sector and gluing the cut edges of the remaining piece together (Figure Ex-34). What is the maximum volume attainable for the cone?

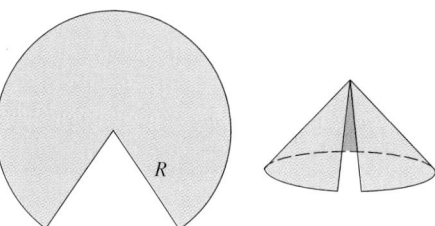

Figure Ex-34

35. A cone-shaped paper drinking cup is to hold 10 cm³ of water. Find the height and radius of the cup that will require the least amount of paper.

36. Find the dimensions of the isosceles triangle of least area that can be circumscribed about a circle of radius R.

37. Find the height and radius of the right circular cone with least volume that can be circumscribed about a sphere of radius R.

38. A trapezoid is inscribed in a semicircle of radius 2 so that one side is along the diameter (Figure Ex-38). Find the maximum possible area for the trapezoid. [*Hint:* Express the area of the trapezoid in terms of θ.]

39. A drainage channel is to be made so that its cross section is a trapezoid with equally sloping sides (Figure Ex-39). If the sides and bottom all have a length of 5 ft, how should the angle θ ($0 \le \theta \le \pi/2$) be chosen to yield the greatest cross-sectional area?

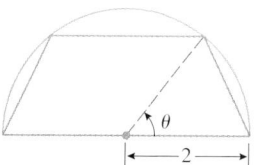   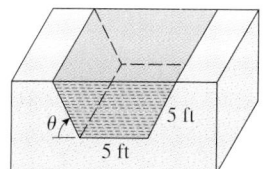

Figure Ex-38   Figure Ex-39

40. A lamp is suspended above the center of a round table of radius r. How high above the table should the lamp be placed to achieve maximum illumination at the edge of the table? [Assume that the illumination I is directly proportional to the cosine of the angle of incidence ϕ of the light rays and inversely proportional to the square of the distance l from the light source (Figure Ex-40).]

41. A plank is used to reach over a fence 8 ft high to support a wall that is 1 ft behind the fence (Figure Ex-41). What is the length of the shortest plank that can be used? [*Hint:* Express the length of the plank in terms of the angle θ shown in the figure.]

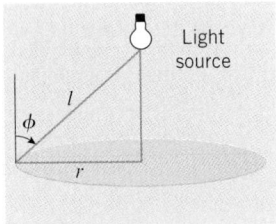

Figure Ex-40   Figure Ex-41

42. A commercial cattle ranch currently allows 20 steers per acre of grazing land; on the average its steers weigh 2000 lb at market. Estimates by the Agriculture Department indicate that the average market weight per steer will be reduced by 50 lb for each additional steer added per acre of grazing land. How many steers per acre should be allowed in order for the ranch to get the largest possible total market weight for its cattle?

43. (a) A chemical manufacturer sells sulfuric acid in bulk at a price of $100 per unit. If the daily total production cost in dollars for x units is

$$C(x) = 100{,}000 + 50x + 0.0025x^2$$

and if the daily production capacity is at most 7000 units, how many units of sulfuric acid must be manufactured and sold daily to maximize the profit?

(b) Would it benefit the manufacturer to expand the daily production capacity?

44. A firm determines that x units of its product can be sold daily at p dollars per unit, where

$$x = 1000 - p$$

The cost of producing x units per day is

$$C(x) = 3000 + 20x$$

(a) Find the revenue function $R(x)$.
(b) Find the profit function $P(x)$.
(c) Assuming that the production capacity is at most 500 units per day, determine how many units the company must produce and sell each day to maximize the profit.
(d) Find the maximum profit.
(e) What price per unit must be charged to obtain the maximum profit?

45. In a certain chemical manufacturing process, the daily weight y of defective chemical output depends on the total weight x of all output according to the empirical formula

$$y = 0.01x + 0.00003x^2$$

where x and y are in pounds. If the profit is $100 per pound

of nondefective chemical produced and the loss is $20 per pound of defective chemical produced, how many pounds of chemical should be produced daily to maximize the total daily profit?

46. The cost c (in dollars per hour) to run an ocean liner at a constant speed v (in miles per hour) is given by $c = a + bv^n$, where a, b, and n are positive constants with $n > 1$. Find the speed needed to make the cheapest 3000-mi run.

47. Two particles, A and B, are in motion in the xy-plane. Their coordinates at each instant of time t ($t \geq 0$) are given by $x_A = t$, $y_A = 2t$, $x_B = 1 - t$, and $y_B = t$. Find the minimum distance between A and B.

48. Follow the directions of Exercise 47, with $x_A = t$, $y_A = t^2$, $x_B = 2t$, and $y_B = 2$.

49. Prove that $(1, 0)$ is the closest point on the curve $x^2 + y^2 = 1$ to $(2, 0)$.

50. Find all points on the curve $y = \sqrt{x}$ for $0 \leq x \leq 3$ that are closest to, and at the greatest distance from, the point $(2, 0)$.

51. Find all points on the curve $x^2 - y^2 = 1$ closest to $(0, 2)$.

52. Find a point on the curve $x = 2y^2$ closest to $(0, 9)$.

53. Find the coordinates of the point P on the curve

$$y = \frac{1}{x^2} \quad (x > 0)$$

where the segment of the tangent line at P that is cut off by the coordinate axes has its shortest length.

54. Find the x-coordinate of the point P on the parabola

$$y = 1 - x^2 \quad (0 < x \leq 1)$$

where the triangle that is enclosed by the tangent line at P and the coordinate axes has the smallest area.

55. Where on the curve $y = (1 + x^2)^{-1}$ does the tangent line have the greatest slope?

56. A man is on the bank of a river that is 1 mile wide. He wants to travel to a town on the opposite bank, but 1 mile upstream. He intends to row on a straight line to some point P on the opposite bank and then walk the remaining distance along the bank (Figure Ex-56). To what point should he row in order to reach his destination in the least time if
(a) he can walk 5 mi/h and row 3 mi/h
(b) he can walk 5 mi/h and row 4 mi/h?

57. A pipe of negligible diameter is to be carried horizontally around a corner from a hallway 8 ft wide into a hallway 4 ft wide (Figure Ex-57). What is the maximum length that the pipe can have? [An interesting discussion of this problem in the case where the diameter of the pipe is not neglected is given by Norman Miller in the *American Mathematical Monthly*, Vol. 56, 1949, pp. 177–179.]

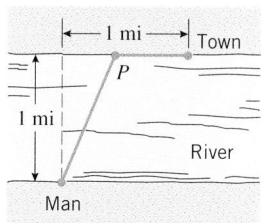

Figure Ex-56

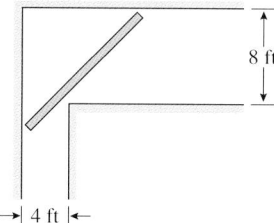

Figure Ex-57

58. In an unknown physical quantity x is measured n times, the measurements $x_1, x_2, \ldots, x_n$ often vary because of uncontrollable factors such as temperature, atmospheric pressure, and so forth. Thus, a scientist is often faced with the problem of using n different observed measurements to obtain an estimate $\bar{x}$ of an unknown quantity x. One method for making such an estimate is based on the ***least squares principle***, which states that the estimate $\bar{x}$ should be chosen to minimize

$$s = (x_1 - \bar{x})^2 + (x_2 - \bar{x})^2 + \cdots + (x_n - \bar{x})^2$$

which is the sum of the squares of the deviations between the estimate $\bar{x}$ and the measured values. Show that the estimate resulting from the least squares principle is

$$\bar{x} = \frac{1}{n}(x_1 + x_2 + \cdots + x_n)$$

that is, $\bar{x}$ is the arithmetic average of the observed values.

59. Suppose that the intensity of a point light source is directly proportional to the strength of the source and inversely proportional to the square of the distance from the source. Two point light sources with strengths of S and $8S$ are separated by a distance of 90 cm. Where on the line segment between the two sources is the intensity a minimum?

60. Prove: If $f(x) \geq 0$ on an interval I and if $f(x)$ has a maximum value on I at x_0, then $\sqrt{f(x)}$ also has a maximum value at x_0. Similarly for minimum values. [*Hint:* Use the fact that $\sqrt{x}$ is an increasing function on the interval $[0, +\infty)$.]

61. Fermat's* (biography on pp. 352–353) principle in optics states that light traveling from one point to another follows that path for which the total travel time is minimum. In a uniform medium, the paths of "minimum time" and "shortest distance" turn out to be the same, so that light, if unobstructed, travels along a straight line. Assume that we have a light source, a flat mirror, and an observer in a uniform medium. If a light ray leaves the source, bounces off the mirror, and travels on to the observer, then its path will consist of two line segments, as shown in Figure Ex-61. According to Fermat's principle, the path will be such that the total travel time t is minimum or, since the medium is uniform, the path will be such that the total distance traveled from A to P to B is as small as possible. Assuming the minimum

occurs when $dt/dx = 0$, show that the light ray will strike the mirror at the point P where the "angle of incidence" θ_1 equals the "angle of reflection" θ_2.

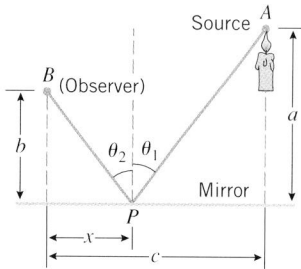

Figure Ex-61

62. Fermat's principle (Exercise 61) also explains why light rays traveling between air and water undergo bending (refraction). Imagine that we have two uniform media (such as air and water) and a light ray traveling from a source A in one medium to an observer B in the other medium (Figure Ex-62). It is known that light travels at a constant speed in a uniform medium, but more slowly in a dense medium (such as water) than in a thin medium (such as air). Consequently, the path of shortest time from A to B is not necessarily a straight line, but rather some broken line path A to P to B allowing the light to take greatest advantage of its higher speed through the thin medium. Snell's† (biography on p. 353) law of refraction states that the path of the light ray will be such that

$$\frac{\sin \theta_1}{v_1} = \frac{\sin \theta_2}{v_2}$$

where v_1 is the speed of light in the first medium, v_2 is the speed of light in the second medium, and θ_1 and θ_2 are the angles shown in Figure Ex-62. Show that this follows from the assumption that the path of minimum time occurs when $dt/dx = 0$.

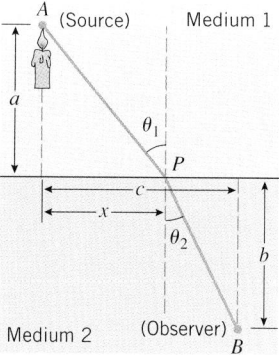

Figure Ex-62

63. A farmer wants to walk at a constant rate from her barn to a straight river, fill her pail, and carry it to her house in the least time.

(a) Explain how this problem relates to Fermat's principle and the light-reflection problem in Exercise 61.

(b) Use the result of Exercise 61 to describe geometrically the best path for the farmer to take.

(c) Use part (b) to determine where the farmer should fill her pail if her house and barn are located as in Figure Ex-63.

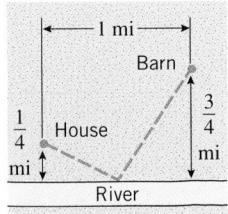

Figure Ex-63

6.3 RECTILINEAR MOTION (MOTION ALONG A LINE)

In Section 1.5 we discussed the motion of a particle moving with constant velocity in one direction along a line, and in Section 3.1 we discussed the motion of a particle moving with variable velocity in one direction along a line. In this section we will investigate the more general situation in which a particle may move back and forth with variable velocity along a line. Some examples are a piston moving up and down in a cylinder, a buoy bobbing up and down in the waves, or an object attached to a vibrating spring.

TERMINOLOGY

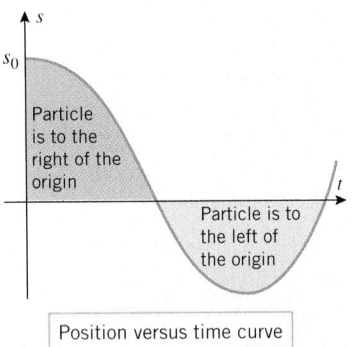

Position versus time curve

Figure 6.3.1

In this section we will assume that a particle representing some object is allowed to move in either direction along a coordinate line. This is called ***rectilinear motion***. The coordinate line might be an x-axis, a y-axis, or an axis that is inclined at some angle. To avoid being specific, we will denote the coordinate line as the s-axis. We will assume that units are chosen for measuring distance and time and that we begin observing the particle at time $t = 0$. As the particle moves along the s-axis, its coordinate is some function of the elapsed time t, say $s = s(t)$. We call $s(t)$ the ***position function*** of the particle, and we call the graph of s versus t the ***position versus time curve***.

Figure 6.3.1 shows a typical position versus time curve for a particle in rectilinear motion. We can tell from that graph that the coordinate of the particle at time $t = 0$ is s_0, and we can tell from the sign of s when the particle is to the left or right of the origin as it moves along the coordinate line.

---

*PIERRE DE FERMAT (1601–1665). Fermat, the son of a successful French leather merchant, was a lawyer who practiced mathematics as a hobby. He received a Bachelor of Civil Laws degree from the University of Orleans in 1631 and subsequently held various government positions, including a post as councillor to the Toulouse parliament. Although he was apparently financially successful, confidential documents of that time suggest that his performance in office and as a lawyer was poor, perhaps because he devoted so much time to mathematics. Throughout his life, Fermat fought all his efforts to have his mathematical results published. He had the unfortunate habit of scribbling his work in the margins of books and often sent his results to friends without keeping copies for himself. As a result, he never received credit for many major achievements until his name was raised from obscurity in the mid-nineteenth century. It is now known that Fermat, simultaneously and independently of Descartes, developed analytic geometry. Unfortunately, Descartes and Fermat argued bitterly over various problems so that there was never any real cooperation between these two great geniuses.

Fermat solved many fundamental calculus problems. He obtained the first procedure to differentiating polynomials, and solved many important maximization, minimization, area, and tangent problems. His work served to inspire Isaac Newton. Fermat is best known for his work in number theory, the study of properties and relationships between whole numbers. He was the first mathematician to make substantial contributions to this field after the ancient Greek mathematician Diophantus. Unfortunately, none of Fermat's contemporaries appreciated his work in this area, a fact that eventually pushed Fermat into isolation and obscurity in later life. In addition to his work*

Example 1

Figure 6.3.2 shows the position versus time curve for a jackrabbit moving along an s-axis. In words, describe how the position of the rabbit changes with time.

Solution. The rabbit is at the point $s = -3$ at time $t = 0$. It moves in the positive direction until time $t = 4$, since s is increasing. At time $t = 4$ the rabbit is at the point $s = 3$. At that time it turns around and travels in the negative direction until time $t = 7$, since s is decreasing. At time $t = 7$ the rabbit is at the point $s = -1$, and it remains stationary at that point thereafter, since s is constant for $t > 7$. ◀

INSTANTANEOUS VELOCITY

In rectilinear motion, the rate at which the coordinate of a particle changes with time is called the *velocity* of the particle. More precisely, we make the following definition.

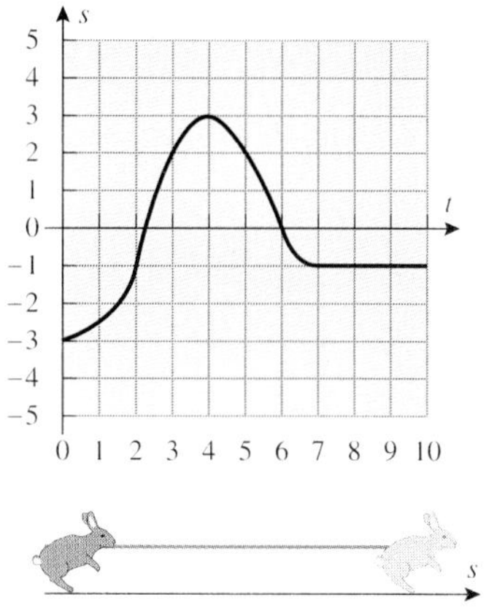

6.3.1 DEFINITION. If $s(t)$ is the position function of a particle moving on a coordinate line, then the ***instantaneous velocity*** of the particle at time t is defined by

$$v(t) = s'(t) = \frac{ds}{dt} \tag{1}$$

Geometrically, the instantaneous velocity at a given time is the slope of the tangent line to the position versus time curve at that time, and hence the sign of the velocity tells which way the particle is moving—a positive velocity means that s is increasing with time, so the particle is moving in the positive direction; a negative velocity means that s is decreasing with time, so the particle is moving in the negative direction (Figure 6.3.3). For example, in Figure 6.3.2 the rabbit is moving in the positive direction between times $t = 0$ and $t = 4$ and is moving in the negative direction between times $t = 4$ and $t = 7$.

Figure 6.3.2

SPEED VERSUS VELOCITY

Recall from our discussion of uniform rectilinear motion in Section 1.5 that there is a distinction between the terms *speed* and *velocity*—speed describes how fast an object is moving without regard to direction, whereas velocity describes how fast it is moving and in what direction. Mathematically, we define the ***instantaneous speed*** of a particle to be the absolute value of its instantaneous velocity; that is,

$$\begin{bmatrix} \text{instantaneous} \\ \text{speed} \end{bmatrix} = |v(t)| = \left| \frac{ds}{dt} \right| \tag{2}$$

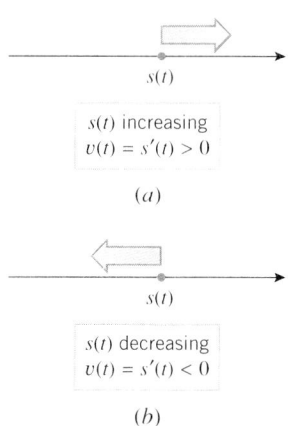

Figure 6.3.3

in calculus and number theory, Fermat was one of the founders of probability theory and made major contributions to the theory of optics. Outside mathematics, Fermat was a classical scholar of some note, was fluent in French, Italian, Spanish, Latin, and Greek, and he composed a considerable amount of Latin poetry.

One of the great mysteries of mathematics is shrouded in Fermat's work in number theory. In the margin of a book by Diophantus, Fermat scribbled that for integer values of n greater than 2, the equation $x^n + y^n = z^n$ has no nonzero integer solutions for x, y, and z. He stated, "I have discovered a truly marvelous proof of this, which however the margin is not large enough to contain." This result, which became known as "Fermat's last theorem," appeared to be true, but its proof evaded the greatest mathematical geniuses for 300 years until Professor Andrew Wiles of Princeton University presented a proof in June 1993 in a dramatic series of three lectures that drew international media attention (see *New York Times*, June 27, 1993). A prize of 100,000 German marks was offered in 1908 for the solution, but it is worthless today because of inflation.

† WILLEBRORD VAN ROIJEN SNELL (1591–1626). Dutch mathematician. Snell, who succeeded his father to the post of Professor of Mathematics at the University of Leiden in 1613, is most famous for the result of light refraction that bears his name. Although this phenomenon was studied as far back as the ancient Greek astronomer Ptolemy, until Snell's work the relationship was incorrectly thought to be $\theta_1/v_1 = \theta_2/v_2$. Snell's law was published by Descartes in 1638 without giving proper credit to Snell. Snell also discovered a method for determining distances by triangulation that founded the modern technique of mapmaking.

For example, if two particles on the same coordinate line are moving with velocities $v = 5$ m/s and $v = -5$ m/s, respectively, then the particles are moving in opposite directions, but they both have a speed of $|v| = 5$ m/s.

Example 2

Let $s(t) = t^3 - 6t^2$ be the position function of a particle moving along an s-axis, where s is in meters and t is in seconds. Find the instantaneous velocity and speed, and show the graphs of position, velocity, and speed versus time.

Solution. From (1) and (2), the instantaneous velocity and speed are given by

$$v(t) = \frac{ds}{dt} = 3t^2 - 12t \quad \text{and} \quad |v(t)| = |3t^2 - 12t|$$

The graphs of position, velocity, and speed versus time are shown in Figure 6.3.4. Observe that velocity and speed both have units of meters per second (m/s), since s is in meters (m) and time is in seconds (s). ◄

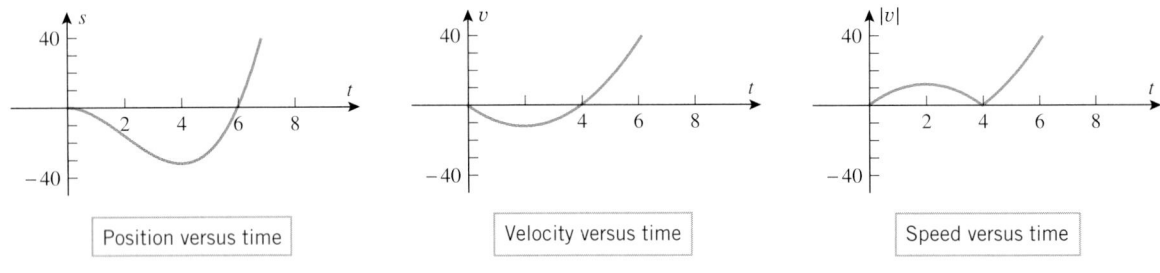

Figure 6.3.4

The graphs in Figure 6.3.4 provide a wealth of visual information about the motion of the particle. For example, the position versus time curve tells us that the particle is to the left of the origin for $0 < t < 6$, is to the right of the origin for $t > 6$, and is at the origin at times $t = 0$ and $t = 6$. The velocity versus time curve tells us that the particle is moving in the negative direction if $0 < t < 4$, is moving in the positive direction if $t > 4$, and is momentarily stopped at times $t = 0$ and $t = 4$ (the velocity is zero at those times). The speed versus time curve tells us that the speed of the particle is increasing for $0 < t < 2$, decreasing for $2 < t < 4$, and increasing again for $t > 4$.

ACCELERATION

In rectilinear motion, the rate at which the velocity of a particle changes with time is called its *acceleration*. More precisely, we make the following definition.

6.3.2 DEFINITION. If $s(t)$ is the position function of a particle moving on a coordinate line, then the instantaneous acceleration of the particle at time t is defined by

$$a(t) = v'(t) = \frac{dv}{dt} \tag{3}$$

or alternatively, since $v(t) = s'(t)$,

$$a(t) = s''(t) = \frac{d^2s}{dt^2} \tag{4}$$

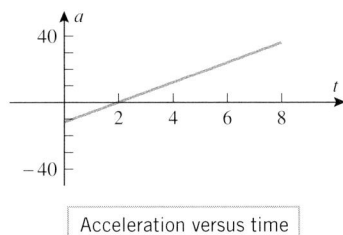

Acceleration versus time

Figure 6.3.5

Example 3

Let $s(t) = t^3 - 6t^2$ be the position function of a particle moving along an s-axis, where s is in meters and t is in seconds. Find the instantaneous acceleration $a(t)$, and show the graph of acceleration versus time.

Solution.   From Example 2, the instantaneous velocity of the particle is $v(t) = 3t^2 - 12t$, so the instantaneous acceleration is

$$a(t) = \frac{dv}{dt} = 6t - 12$$

and the acceleration versus time curve is the line shown in Figure 6.3.5. Note that in this example the acceleration has units of m/s², since v is in meters per second (m/s) and time is in seconds (s).   ◄

INTERPRETING THE SIGN OF ACCELERATION

We will say that a particle in rectilinear motion is **speeding up** when its instantaneous speed is increasing and is **slowing down** when its instantaneous speed is decreasing. In everyday language an object that is speeding up is said to be "accelerating" and an object that is slowing down is said to be "decelerating"; thus, one might expect that a particle in rectilinear motion will be speeding up when its instantaneous acceleration is positive and slowing down when it is negative. Although this is true for a particle moving in the positive direction, it is *not* true for a particle moving in the negative direction—a particle with negative velocity is speeding up when its acceleration is negative and slowing down when its acceleration is positive. This is because a positive acceleration implies an increasing velocity, and increasing a negative velocity decreases its absolute value; similarly, a negative acceleration implies a decreasing velocity, and decreasing a negative velocity increases its absolute value. In summary:

> **6.3.3**   INTERPRETING THE SIGN OF ACCELERATION.   *A particle in rectilinear motion is speeding up when its velocity and acceleration have the same sign and slowing down when they have opposite signs.*

FOR THE READER.   For a particle in rectilinear motion, what is happening when $v(t) = 0$? When $a(t) = 0$?

Example 4

In Examples 2 and 3 we found the velocity versus time curve and the acceleration versus time curve for a particle with position function $s(t) = t^3 - 6t^2$. Use those curves to determine when the particle is speeding up and slowing down, and confirm that your results are consistent with the speed versus time curve obtained in Example 2.

Solution.   Over the time interval $0 < t < 2$ the velocity and acceleration are negative, so the particle is speeding up. This is consistent with the speed versus time curve, since the speed is increasing over this time interval. Over the time interval $2 < t < 4$ the velocity is negative and the acceleration is positive, so the particle is slowing down. This is also consistent with the speed versus time curve, since the speed is decreasing over this time interval. Finally, on the time interval $t > 4$ the velocity and acceleration are positive, so the particle is speeding up, which again is consistent with the speed versus time curve.   ◄

ANALYZING THE POSITION VERSUS TIME CURVE

The position versus time curve contains all of the significant information about the position and velocity of a particle in rectilinear motion:

- Where $s(t) > 0$, the particle is on the positive side of the s-axis.
- Where $s(t) < 0$, the particle is on the negative side of the s-axis.

- The slope of the tangent line at a point in time is the instantaneous velocity at that time.
- Where the tangent line has positive slope, the velocity is positive and the particle is moving in the positive direction.
- Where the tangent line has negative slope, the velocity is negative, and the particle is moving in the negative direction.
- Where the tangent line is horizontal, the velocity is zero, and the particle is momentarily stopped.

Information about the acceleration of a particle in rectilinear motion can also be deduced from the position versus time curve by examining its concavity. To see why this is so, observe that the position versus time curve will be concave up on intervals where $s''(t) > 0$, and it will be concave down on intervals where $s''(t) < 0$. But we know from (4) that $s''(t)$ is the instantaneous acceleration, so that on intervals where the position versus time curve is concave up the particle has a positive acceleration, and on intervals where it is concave down the particle has a negative acceleration.

Table 6.3.1 summarizes our observations about the position versus time curve.

Table 6.3.1

POSITION VERSUS TIME CURVE	CHARACTERISTICS OF THE CURVE AT $t = t_0$	BEHAVIOR OF THE PARTICLE AT TIME $t = t_0$
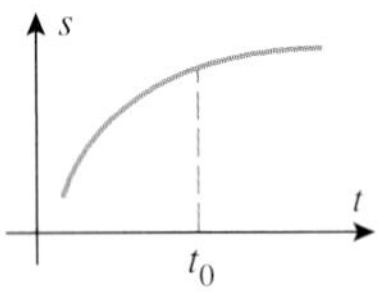	• $s(t_0) > 0$ • Tangent line has positive slope. • Curve is concave down.	• Particle is on the positive side of the origin. • Particle is moving in the positive direction. • Velocity is decreasing. • Particle is slowing down.
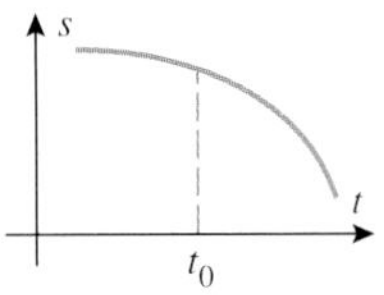	• $s(t_0) > 0$ • Tangent line has negative slope. • Curve is concave down.	• Particle is on the positive side of the origin. • Particle is moving in the negative direction. • Velocity is decreasing. • Particle is speeding up.
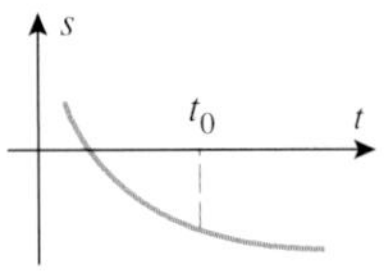	• $s(t_0) < 0$ • Tangent line has negative slope. • Curve is concave up.	• Particle is on the negative side of the origin. • Particle is moving in the negative direction. • Velocity is increasing. • Particle is slowing down.
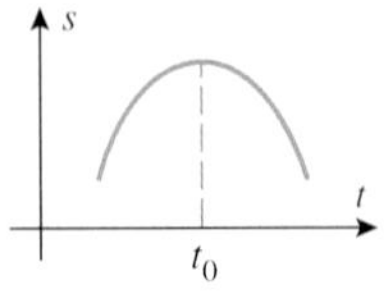	• $s(t_0) > 0$ • Tangent line has zero slope. • Curve is concave down.	• Particle is on the positive side of the origin. • Particle is momentarily stopped. • Velocity is decreasing.

Example 5

Use the position versus time curve in Figure 6.3.2 to determine when the jackrabbit in Example 1 is speeding up and slowing down.

Solution. From $t = 0$ to $t = 2$, the acceleration and velocity are positive, so the rabbit is speeding up. From $t = 2$ to $t = 4$, the acceleration is negative and the velocity is positive, so the rabbit is slowing down. At $t = 4$, the velocity is zero, so the rabbit has momentarily stopped. From $t = 4$ to $t = 6$, the acceleration is negative and the velocity is negative, so the rabbit is speeding up. From $t = 6$ to $t = 7$, the acceleration is positive and the velocity is negative, so the rabbit is slowing down. Thereafter, the velocity is zero, so the rabbit has stopped.   ◄

Example 6

Suppose that the position function of a particle moving on a coordinate line is given by $s(t) = 2t^3 - 21t^2 + 60t + 3$. Analyze the motion of the particle for $t \geq 0$.

Solution. The velocity and acceleration at time t are

$$v(t) = s'(t) = 6t^2 - 42t + 60 = 6(t - 2)(t - 5)$$
$$a(t) = v'(t) = 12t - 42 = 12\left(t - \tfrac{7}{2}\right)$$

At each instant we can determine the direction of motion from the sign of $v(t)$ and whether the particle is speeding up or slowing down from the signs of $v(t)$ and $a(t)$ together (Figures 6.3.6a and 6.3.6b). The motion of the particle is described schematically by the curved line in Figure 6.3.6c. At time $t = 0$ the particle is at the point $s(0) = 3$ moving right with velocity $v(0) = 60$ ft/s, but slowing down with acceleration $a(0) = -42$ ft/s^2. The particle continues moving right until time $t = 2$, when it stops at the point $s(2) = 55$, reverses

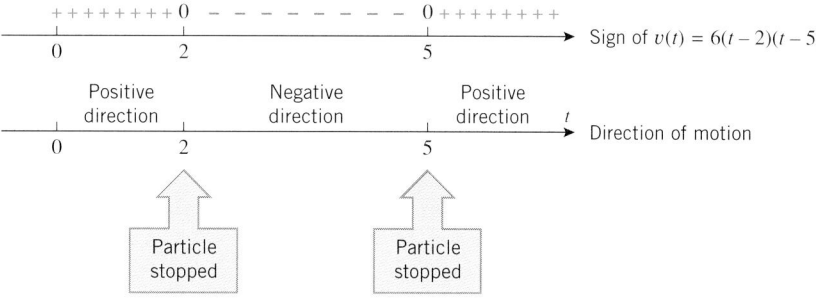

Figure 6.3.6a

Analysis of the particle's direction

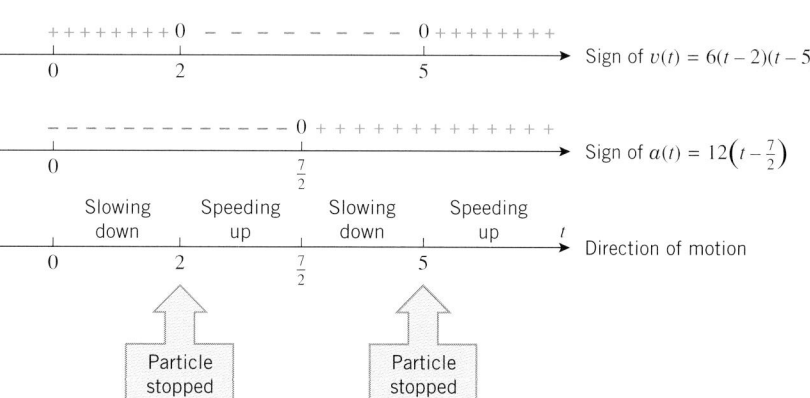

Figure 6.3.6b

Analysis of the particle's speed

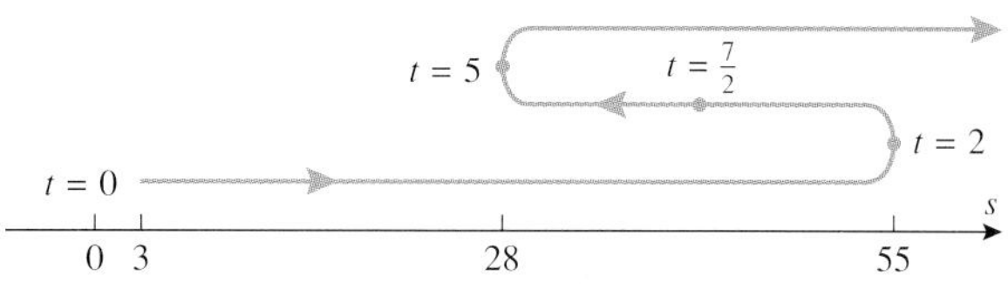

Figure 6.3.6c

direction, and begins to speed up with an acceleration of $a(2) = -18$ ft/s². At time $t = \frac{7}{2}$ the particle begins to slow down, but continues moving left until time $t = 5$, when it stops at the point $s(5) = 28$, reverses direction again, and begins to speed up with acceleration $a(5) = 18$ ft/s². The particle then continues moving right thereafter with increasing speed.

◄

REMARK. The curved line in Figure 6.3.6c is descriptive only. The actual path of the particle is back and forth on the coordinate line.

FREE-FALL MOTION

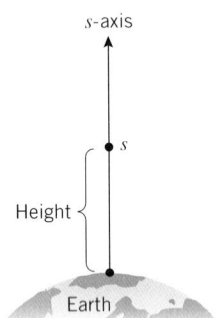

s-axis

Height

Earth

Figure 6.3.7

We will now discuss how some of the ideas in this section can be applied to the study of *free-fall motion*, which is the motion that occurs when an object near the Earth is imparted some initial vertical velocity (up or down), and thereafter moves on a vertical line. In modeling free-fall motion it is assumed that the only force acting on the object is the Earth's gravity and that the object stays sufficiently close to the Earth's surface so that the gravitational force is constant. In particular, air resistance and the gravitational pull of other celestial bodies are neglected.

In our study of free-fall motion, we will ignore the physical size of the object by treating it as a particle, and we will assume that the object moves along an s-axis whose origin is at the surface of the Earth and whose positive direction is up. With this convention, the s-coordinate of the particle is the height of the particle above the Earth's surface (Figure 6.3.7). The following result will be derived later using calculus and some basic principles of physics.

6.3.4 THE FREE-FALL MODEL. Suppose that at time $t = 0$ an object at a height of s_0 above the Earth's surface is imparted an upward or downward velocity of v_0 and thereafter moves vertically subject only to the force of the Earth's gravity. If the positive direction of the s-axis is up, and if the origin is at the surface of the Earth, then at any time t the height $s = s(t)$ of the object is given by the formula

$$s = s_0 + v_0 t - \tfrac{1}{2}gt^2 \tag{5}$$

where g is a constant, called the *acceleration due to gravity*. In this text we will use the following approximations for g, depending on the units of measurement:

$g = 9.8$ m/s² [distance in meters and time in seconds]

$g = 32$ ft/s² [distance in feet and time in seconds]

It follows from (5) that the instantaneous velocity and acceleration of an object in free-fall motion are

$$v = \frac{ds}{dt} = v_0 - gt \tag{6}$$

$$a = \frac{dv}{dt} = -g \tag{7}$$

REMARK. Because we have chosen the positive direction of the s-axis to be up, a positive velocity implies an upward motion and a negative velocity a downward motion. Thus, it makes sense that instantaneous acceleration $-g$ is negative, since an upward-moving object has positive velocity and negative acceleration, which implies that it is slowing down; and a downward-moving object has negative velocity and negative acceleration, which implies that it is speeding up. (It is a little confusing that the positive constant g is called the *acceleration due to gravity* in 6.3.4, given that the instantaneous acceleration is actually the negative constant $-g$. This mismatch in terminology is caused by the upward orientation of the s-axis in Figure 6.3.7; had we chosen the positive direction to be down, then the instantaneous acceleration would have turned out to be g. However, our orientation has the advantage of allowing us to interpret s as the height of the object.)

Nolan Ryan's rookie baseball card

Example 7

Nolan Ryan, one of the fastest baseball pitchers of all time, was capable of throwing a baseball 150 ft/s (over 102 mi/h). Could Nolan Ryan have hit the 208-ft ceiling of the Houston Astrodome if he were capable of giving a baseball an upward velocity of 100 ft/s from a height of 7 ft?

Solution.   Taking $g = 32$ ft/s^2, $v_0 = 100$ ft/s, and $s_0 = 7$ ft in (5) and (6) yields the equations

$$s = 7 + 100t - 16t^2 \quad \text{and} \quad v = 100 - 32t \qquad (8\text{--}9)$$

whose graphs are shown in Figure 6.3.8. It is evident from the graph of s versus t that the maximum height of the baseball is less than 208 ft, so Ryan could not have hit the ceiling. However, let us go a step further and determine exactly how high the ball will go. The maximum height s occurs at the stationary point obtained by solving the equation $ds/dt = 0$. However, $ds/dt = v$, which means that the maximum height occurs when $v = 0$, which from (9) can be expressed as

$$100 - 32t = 0 \qquad (10)$$

Solving this equation yields $t = 25/8$. To find the height s at this time we substitute this value of t in (8), from which we obtain

$$s = 7 + 100(25/8) - 16(25/8)^2 = 163.25 \text{ ft}$$

which is roughly 45 ft short of hitting the ceiling.   ◀

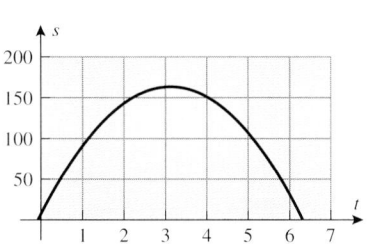

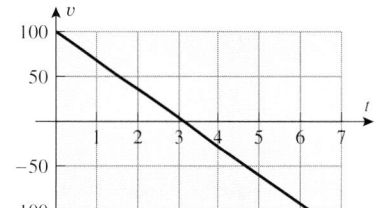

Figure 6.3.8

REMARK.   Equation (10) can also be deduced by physical reasoning: The ball is moving up when the velocity is positive and moving down when the velocity is negative, so it makes sense that the velocity is zero when the ball reaches its peak.

EXERCISE SET 6.3   ☒ Graphing Calculator

1. The graphs of three position functions are shown in the accompanying figure. In each case determine the sign of the velocity and acceleration, then determine whether the particle is speeding up or slowing down.

2. The graphs of three velocity functions are shown in the accompanying figure. In each case determine the sign of the acceleration, then determine whether the particle is speeding up or slowing down.

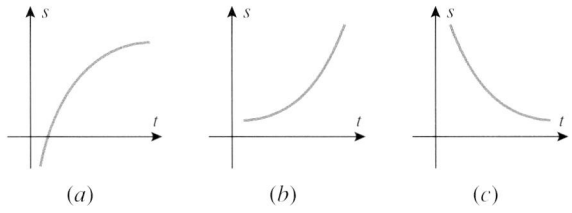

Figure Ex-1

Figure Ex-2

3. The position function of a particle moving on a horizontal x-axis is shown in the accompanying figure.
 (a) Is the particle moving left or right at time t_0?
 (b) Is the acceleration positive or negative at time t_0?
 (c) Is the particle speeding up or slowing down at time t_0?
 (d) Is the particle speeding up or slowing down at time t_1?

Figure Ex-3

4. For the graphs in the accompanying figure, match the position functions with their corresponding velocity functions.

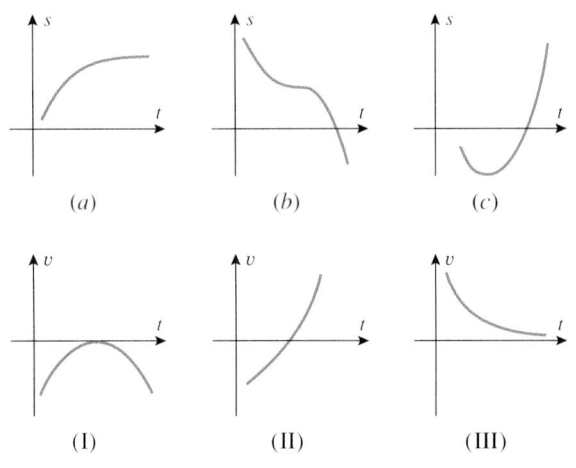

Figure Ex-4

5. Sketch a reasonable graph of s versus t for a mouse that is trapped in a narrow corridor (an s-axis with the positive direction to the right) and scurries back and forth as follows. It runs right with a constant speed of 1.2 m/s for awhile, then gradually slows down to 0.6 m/s, then quickly speeds up to 2.0 m/s, then gradually slows to a stop but immediately reverses direction and quickly speeds up to 1.2 m/s.

6. The accompanying figure shows the graph of s versus t for an ant that moves along a narrow vertical pipe (an s-axis with the positive direction up).
 (a) When, if ever, is the ant above the origin?
 (b) When, if ever, does the ant have velocity zero?
 (c) When, if ever, is the ant moving down the pipe?

7. The accompanying figure shows the graph of velocity versus time for a particle moving along a coordinate line. Make a rough sketch of the graphs of speed versus time and acceleration versus time.

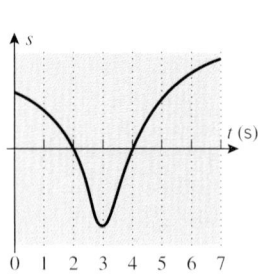 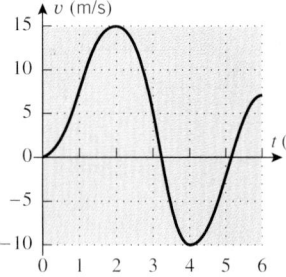

Figure Ex-6   Figure Ex-7

8. The accompanying figure shows the position versus time graph for an elevator that ascends 40 m from one stop to the next.
 (a) Estimate the velocity when the elevator is halfway up.
 (b) Sketch rough graphs of the velocity versus time curve and the acceleration versus time curve.

9. The accompanying figure shows the velocity versus time graph for a test run on the Grand Prix GTP. Using this graph, estimate
 (a) the acceleration at 60 mi/h (in units of ft/s^2)
 (b) the time at which the maximum acceleration occurs.
 [Data from *Car and Driver Magazine*, October 1990.]

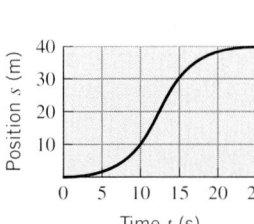 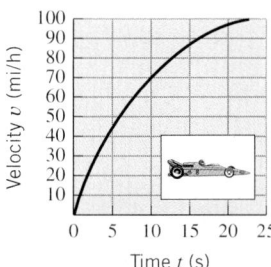

Figure Ex-8   Figure Ex-9

10. Let $s(t) = \sin(\pi t/4)$ be the position function of a particle moving along a coordinate line, where s is in meters and t is in seconds.
 (a) Make a table showing the position, velocity, and acceleration to two decimal places at times $t = 1, 2, 3, 4,$ and 5.
 (b) At each of the times in part (a), determine whether the particle is stopped; if it is not, state its direction of motion.
 (c) At each of the times in part (a), determine whether the particle is speeding up, slowing down, or neither.

In Exercises 11–14, the position function of a particle moving along a coordinate line is given, where s is in feet and t is in seconds.

(a) Find the velocity and acceleration functions.
(b) Find the position, velocity, speed, and acceleration at time $t = 1$.
(c) At what times is the particle stopped?
(d) When is the particle speeding up? Slowing down?
(e) Find the total distance traveled by the particle from time $t = 0$ to time $t = 5$.

11. $s(t) = t^3 - 6t^2$, $\quad t \geq 0$

12. $s(t) = t^4 - 4t + 2$, $\quad t \geq 0$

13. $s(t) = 3\cos(\pi t/2)$, $\quad 0 \leq t \leq 5$

14. $s(t) = \dfrac{t}{t^2 + 4}$, $\quad t \geq 0$

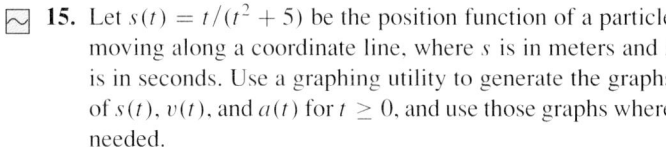 **15.** Let $s(t) = t/(t^2 + 5)$ be the position function of a particle moving along a coordinate line, where s is in meters and t is in seconds. Use a graphing utility to generate the graphs of $s(t)$, $v(t)$, and $a(t)$ for $t \geq 0$, and use those graphs where needed.

(a) Use the appropriate graph to make a rough estimate of the time at which the particle first reverses the direction of its motion; and then find the time exactly.
(b) Find the exact position of the particle when it first reverses the direction of its motion.
(c) Use the appropriate graphs to make a rough estimate of the time intervals on which the particle is speeding up and on which it is slowing down; and then find those time intervals exactly.

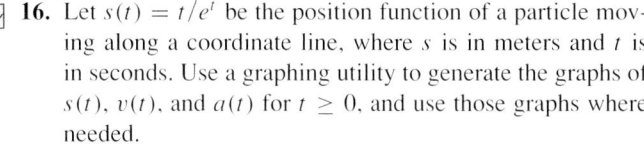 **16.** Let $s(t) = t/e^t$ be the position function of a particle moving along a coordinate line, where s is in meters and t is in seconds. Use a graphing utility to generate the graphs of $s(t)$, $v(t)$, and $a(t)$ for $t \geq 0$, and use those graphs where needed.

(a) Use the appropriate graph to make a rough estimate of the time at which the particle first reverses the direction of its motion; and then find the time exactly.
(b) Find the exact position of the particle when it first reverses the direction of its motion.
(c) Use the appropriate graphs to make a rough estimate of the time intervals on which the particle is speeding up and on which it is slowing down; and then find those time intervals exactly.

In Exercises 17–22, the position function of a particle moving along a coordinate line is given. Use the method of Example 6 to analyze the motion of the particle for $t \geq 0$, and give a schematic picture of the motion (as in Figure 6.3.6).

17. $s = -3t + 2$

18. $s = t^3 - 6t^2 + 9t + 1$

19. $s = t^3 - 9t^2 + 24t$

20. $s = t + \dfrac{9}{t + 1}$

21. $s = \begin{cases} \cos t, & 0 \leq t \leq 2\pi \\ 1, & t > 2\pi \end{cases}$

22. $s = \sqrt{t}(4 - 4t + 2t^2)$

23. Let $s(t) = 5t^2 - 22t$ be the position function of a particle moving along a coordinate line, where s is in feet and t is in seconds.

(a) Find the maximum speed of the particle during the time interval $1 \leq t \leq 3$.
(b) When, during the time interval $1 \leq t \leq 3$, is the particle farthest from the origin? What is its position at that instant?

24. Let $s = 100/(t^2 + 12)$ be the position function of a particle moving along a coordinate line, where s is in feet and t is in seconds. Find the maximum speed of the particle for $t \geq 0$, and find the direction of motion of the particle when it has its maximum speed.

In Exercises 25–29, assume that the free-fall model applies and that the positive direction is up, so that Formulas (5), (6), and (7) can be used. In those problems stating that an object is "dropped" or "released from rest," you should interpret that to mean that the initial velocity of the object is zero. Take $g = 32$ ft/s^2 or $g = 9.8$ m/s^2, depending on the units.

25. A wrench is accidentally dropped at the top of an elevator shaft in a tall building.

(a) How many meters does the wrench fall in 1.5 s?
(b) What is the velocity of the wrench at that time?
(c) How long does it take for the wrench to reach a speed of 12 m/s?
(d) How long does it take for the wrench to fall 100 m?

26. In 1939, Joe Sprinz of the San Francisco Seals Baseball Club attempted to catch a ball dropped from a blimp at a height of 800 ft (for the purpose of breaking the record for catching a ball dropped from the greatest height set the preceding year by members of the Cleveland Indians).

(a) How long does it take for a ball to drop 800 ft?
(b) What is the velocity of a ball in miles per hour after an 800-ft drop (88 ft/s = 60 mi/h)?

[*Note:* As a practical matter, it is unrealistic to ignore wind resistance in this problem; however, even with the slowing effect of wind resistance, the impact of the ball slammed Sprinz's glove hand into his face, fractured his upper jaw in 12 places, broke five teeth, and knocked him unconscious. He dropped the ball!]

27. A projectile is launched upward from ground level with an initial speed of 60 m/s.

(a) How long does it take for the projectile to reach its highest point?
(b) How high does the projectile go?
(c) How long does it take for the projectile to drop back to the ground from its highest point?
(d) What is the speed of the projectile when it hits the ground?

28. (a) Use the results in Exercise 27 to make a conjecture about the relationship between the initial and final speeds of a projectile that is launched upward from ground level and returns to ground level.
    (b) Prove your conjecture.

29. In Example 7, how fast would Nolan Ryan have to throw a ball upward from a height of 7 feet in order to hit the ceiling of the Astrodome?

30. The free-fall formulas (5) and (6) can be combined and rearranged in various useful ways. Derive the following variations of those formulas.
    (a) $v^2 = v_0^2 - 2g(s - s_0)$ (b) $s = s_0 + \frac{1}{2}(v_0 + v)t$
    (c) $s = s_0 + vt + \frac{1}{2}gt^2$

31. A rock, dropped from an unknown height, strikes the ground with a speed of 24 m/s. Use the formula in part (a) of Exercise 30 to find the unknown height.

32. A rock thrown downward with an unknown initial velocity from a height of 1000 ft reaches the ground in 5 s. Use the formula in part (c) of Exercise 30 to find the velocity of the rock when it hits the ground.

33. (a) A ball is thrown upward from a height s_0 with an initial velocity of v_0. Use the formula in part (a) of Exercise 30 to show that the maximum height of the ball is $s_{\max} = s_0 + v_0^2/2g$.
    (b) Use this result to solve Exercise 29.

34. Let $s = t^3 - 6t^2 + 1$.
    (a) Find s and v when $a = 0$.
    (b) Find s and a when $v = 0$.

35. Let $s = \sqrt{2t^2 + 1}$ be the position function of a particle moving along a coordinate line.
    (a) Use a graphing utility to generate the graph of v versus t, and make a conjecture about the velocity of the particle as $t \to +\infty$.
    (b) Check your conjecture by finding $\lim\limits_{t \to +\infty} v$.

36. (a) Use the chain rule to show that for a particle in rectilinear motion $a = v(dv/ds)$.
    (b) Let $s = \sqrt{3t + 7}, t \geq 0$. Find a formula for v in terms of s and use the equation in part (a) to find the acceleration when $s = 5$.

37. Suppose that the position function of two particles, P_1 and P_2, in motion along the same line are

$$s_1 = \frac{1}{2}t^2 - t + 3 \quad \text{and} \quad s_2 = -\frac{1}{4}t^2 + t + 1$$

respectively, for $t \geq 0$.
(a) Prove that P_1 and P_2 do not collide.
(b) How close can P_1 and P_2 get to one another?
(c) During what intervals of time are they moving in opposite directions?

38. Let $s_A = 15t^2 + 10t + 20$ and $s_B = 5t^2 + 40t, t \geq 0$, be the position functions of cars A and B that are moving along parallel straight lanes of a highway.
    (a) How far is car A ahead of car B when $t = 0$?
    (b) At what instants of time are the cars next to one another?
    (c) At what instant of time do they have the same velocity? Which car is ahead at this instant?

39. The accompanying figure shows the velocity versus distance graph for a 222 Remington Magnum 55 grain pointed soft point bullet.
    (a) Use the graph to estimate the value of dv/ds when the velocity is 2000 ft/s.
    (b) Use the result in part (a) and the chain rule to approximate the acceleration when the velocity is 2000 ft/s. [*Hint:* See Exercise 36.]
    [Data from the *Shooter's Bible*, No. 82, Stoeger Publishing Co., 1991.]

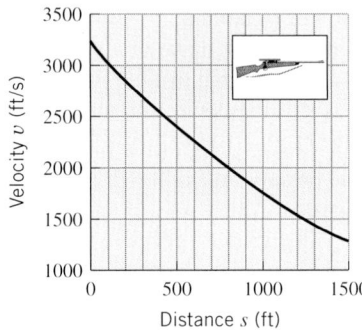

Figure Ex-39

40. Prove that a particle is speeding up if the velocity and acceleration have the same sign, and slowing down if they have opposite signs. [*Hint:* Let $r(t) = |v(t)| = \sqrt{v^2(t)}$, and find $r'(t)$.]

6.4 NEWTON'S METHOD

In Section 2.4 we showed how to approximate the roots of an equation $f(x) = 0$ by using the Intermediate-Value Theorem and also by zooming in on the x-intercepts of $y = f(x)$ with a graphing utility. In this section we will study a technique, called Newton's Method, that is usually more efficient than either of those methods. Newton's Method is the technique used by many commercial and scientific computer programs for finding roots.

NEWTON'S METHOD

In beginning algebra one learns that the solution of a first-degree equation $ax + b = 0$ is given by the formula $x = -b/a$, and the solutions of a second-degree equation

$$ax^2 + bx + c = 0$$

are given by the quadratic formula. Formulas also exist for the solutions of all third- and fourth-degree equations, although they are too complicated to be of practical use. In 1826 it was shown by the Norwegian mathematician Niels Henrik Abel[*] that it is impossible to construct a similar formula for the solutions of a *general* fifth-degree equation or higher. Thus, for a *specific* fifth-degree polynomial equation such as

$$x^5 - 9x^4 + 2x^3 - 5x^2 + 17x - 8 = 0$$

it may be difficult or impossible to find exact values for all of the solutions. Similar difficulties occur for trigonometric equations such as

$$x - \cos x = 0$$

as well as equations of other types. For such equations the solutions are generally approximated in some way, often by the method we will now discuss.

Suppose that we are trying to find a root r of the equation $f(x) = 0$, and suppose that by some method we are able to obtain a rough initial estimate of r, say by generating the graph of $y = f(x)$ with a graphing utility and examining the x-intercepts. If we let x_1 denote our

---

[*] NIELS HENRIK ABEL (1802–1829). Norwegian mathematician. Abel was the son of a poor Lutheran minister and a remarkably beautiful mother from whom he inherited strikingly good looks. In his brief life of 26 years Abel lived in virtual poverty and suffered a succession of adversities; yet he managed to prove major results that altered the mathematical landscape forever. At the age of thirteen he was sent away from home to a school whose better days had long passed. By a stroke of luck the school had just hired a teacher named Bernt Michael Holmboe, who quickly discovered that Abel had extraordinary mathematical ability. Together, they studied the calculus texts of Euler and works of Newton and the later French mathematicians. By the time he graduated, Abel was familiar with most of the great mathematical literature. In 1820 his father died, leaving the family in dire financial straits. Abel was able to enter the University of Christiania in Oslo only because he was granted a free room and several professors supported him directly from their salaries. The University had no advanced courses in mathematics, so Abel took a preliminary degree in 1822 and then continued to study mathematics on his own. In 1824 he published at his own expense the proof that it is impossible to solve the general fifth-degree polynomial equation algebraically. With the hope that this landmark paper would lead to his recognition and acceptance by the European mathematical community, Abel sent the paper to the great German mathematician Gauss, who casually declared it to be a "monstrosity" and tossed it aside. However, in 1826 Abel's paper on the fifth-degree equation and other work was published in the first issue of a new journal, founded by his friend, Leopold Crelle. In the summer of 1826 he completed a landmark work on transcendental functions, which he submitted to the French Academy of Sciences in the hope of establishing himself as a major mathematician, for many young mathematicians had gained quick distinction by having their work accepted by the Academy. However, Abel waited in vain because the paper was either ignored or misplaced by one of the referees, and it did not surface again until two years after his death. That paper was later described by one major mathematician as "... the most important mathematical discovery that has been made in our century. ..." After submitting his paper, Abel returned to Norway, ill with tuberculosis and in heavy debt. While eking out a meager living as a tutor, he continued to produce great work and his fame spread. Soon great efforts were being made to secure a suitable mathematical position for him. Fearing that his great work had been lost by the Academy, he mailed a proof of the main results to Crelle in January of 1829. In April he suffered a violent hemorrhage and died. Two days later Crelle wrote to inform him that an appointment had been secured for him in Berlin and his days of poverty were over! Abel's great paper was finally published by the Academy twelve years after his death.

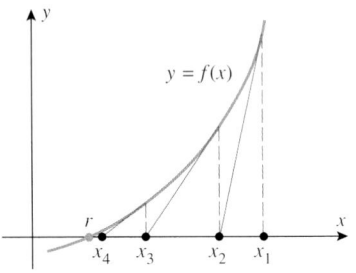

Figure 6.4.1

initial approximation to r, then we can generally improve on this approximation by moving along the tangent line to $y = f(x)$ at x_1 until we meet the x-axis at a point x_2 (Figure 6.4.1). Usually, x_2 will be closer to r than x_1. To improve the approximation further, we can repeat the process by moving along the tangent line to $y = f(x)$ at x_2 until we meet the x-axis at a point x_3. Continuing in this way we can generate a succession of values $x_1, x_2, x_3, x_4, \ldots$ that will usually get closer and closer to r. This procedure for approximating r is called ***Newton's Method***.

To implement Newton's Method analytically, we must derive a formula that will tell us how to calculate each improved approximation from the preceding approximation. For this purpose, we note that the point-slope form of the tangent line to $y = f(x)$ at the initial approximation x_1 is

$$y - f(x_1) = f'(x_1)(x - x_1) \tag{1}$$

If $f'(x_1) \neq 0$, then this line is not parallel to the x-axis and consequently it crosses the x-axis at some point $(x_2, 0)$. Substituting the coordinates of this point in (1) yields

$$-f(x_1) = f'(x_1)(x_2 - x_1)$$

Solving for x_2 we obtain

$$x_2 = x_1 - \frac{f(x_1)}{f'(x_1)} \tag{2}$$

The next approximation can be obtained more easily. If we view x_2 as the starting approximation and x_3 the new approximation, we can simply apply (2) with x_2 in place of x_1 and x_3 in place of x_2. This yields

$$x_3 = x_2 - \frac{f(x_2)}{f'(x_2)} \tag{3}$$

provided $f'(x_2) \neq 0$. In general, if x_n is the nth approximation, then it is evident from the pattern in (2) and (3) that the improved approximation x_{n+1} is given by

> **Newton's Method**
> $$x_{n+1} = x_n - \frac{f(x_n)}{f'(x_n)}, \quad n = 1, 2, 3, \ldots \tag{4}$$

Example 1

Use Newton's Method to approximate the real solutions of

$$x^3 - x - 1 = 0$$

Solution. Let $f(x) = x^3 - x - 1$, so $f'(x) = 3x^2 - 1$ and (4) becomes

$$x_{n+1} = x_n - \frac{x_n^3 - x_n - 1}{3x_n^2 - 1} \tag{5}$$

From the graph of f in Figure 6.4.2, we see that the given equation has only one real solution. This solution lies between 1 and 2 because $f(1) = -1 < 0$ and $f(2) = 5 > 0$. We will use $x_1 = 1.5$ as our first approximation ($x_1 = 1$ or $x_1 = 2$ would also be reasonable choices).

Letting $n = 1$ in (5) and substituting $x_1 = 1.5$ yields

$$x_2 = 1.5 - \frac{(1.5)^3 - 1.5 - 1}{3(1.5)^2 - 1} = 1.34782609$$

(We used a calculator that displays nine digits.) Next, we let $n = 2$ in (5) and substitute $x_2 = 1.34782609$ to obtain

$$x_3 = 1.34782609 - \frac{(1.34782609)^3 - (1.34782609) - 1}{3(1.34782609)^2 - 1} = 1.32520040$$

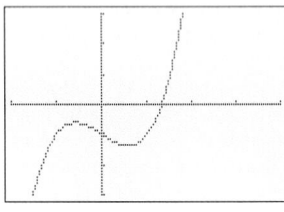

$[-2, 4] \times [-3, 3]$
xScl $= 1$, yScl $= 1$

$y = x^3 - x - 1$

Figure 6.4.2

If we continue this process until two identical approximations are generated in succession, we obtain

$$x_1 = 1.5$$
$$x_2 = 1.34782609$$
$$x_3 = 1.32520040$$
$$x_4 = 1.32471817$$
$$x_5 = 1.32471796$$
$$x_6 = 1.32471796$$

At this stage there is no need to continue further because we have reached the accuracy limit of our calculator, and all subsequent approximations that the calculator generates will be the same. Thus, the solution is approximately $x \approx 1.32471796$. ◄

Example 2

It is evident from Figure 6.4.3 that if x is in radians, then the equation

$$\cos x = x$$

has a solution between 0 and 1. Use Newton's Method to approximate it.

Solution. Rewrite the equation as

$$x - \cos x = 0$$

and apply (4) with $f(x) = x - \cos x$. Since $f'(x) = 1 + \sin x$, (4) becomes

$$x_{n+1} = x_n - \frac{x_n - \cos x_n}{1 + \sin x_n} \qquad (6)$$

From Figure 6.4.3, the solution seems closer to $x = 1$ than $x = 0$, so we will use $x_1 = 1$ (radian) as our initial approximation. Letting $n = 1$ in (6) and substituting $x_1 = 1$ yields

$$x_2 = 1 - \frac{1 - \cos 1}{1 + \sin 1} = .750363868$$

Next, letting $n = 2$ in (6) and substituting this value of x_2 yields

$$x_3 = .750363868 - \frac{.750363868 - \cos(.750363868)}{1 + \sin(.750363868)} = .739112891$$

If we continue this process until two identical approximations are generated in succession, we obtain

$$x_1 = 1$$
$$x_2 = .750363868$$
$$x_3 = .739112891$$
$$x_4 = .739085133$$
$$x_5 = .739085133$$

Thus, to the accuracy limit of our calculator, the solution of the equation $\cos x = x$ is $x \approx .739085133$. ◄

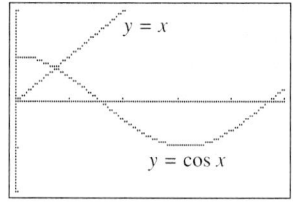

$[0, 5] \times [-2, 2]$
xScl = 1, yScl = 1

Figure 6.4.3

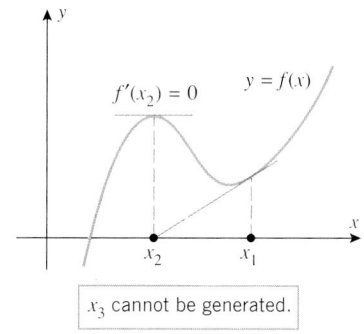

Figure 6.4.4

SOME DIFFICULTIES WITH NEWTON'S METHOD

When Newton's Method works, the approximations usually converge toward the solution with dramatic speed. However, there are situations in which the method fails. For example, if $f'(x_n) = 0$ for some n, then (4) involves a division by zero, making it impossible to generate x_{n+1}. However, this is to be expected because the tangent line to $y = f(x)$ is parallel to the x-axis where $f'(x_n) = 0$, and hence this tangent line does not cross the x-axis to generate the next approximation (Figure 6.4.4).

Newton's Method can fail for other reasons as well; sometimes it may overlook the root you are trying to find and converge to a different root, and sometimes it may fail to converge altogether. For example, consider the equation

$$x^{1/3} = 0$$

which has $x = 0$ as its only solution, and try to approximate this solution by Newton's Method with a starting value of $x_0 = 1$. Letting $f(x) = x^{1/3}$, Formula (4) becomes

$$x_{n+1} = x_n - \frac{(x_n)^{1/3}}{\frac{1}{3}(x_n)^{-2/3}} = x_n - 3x_n = -2x_n$$

Beginning with $x_1 = 1$, the successive values generated by this formula are

$$x_1 = 1, \quad x_2 = -2, \quad x_3 = 4, \quad x_4 = -8, \dots$$

which obviously do not converge to $x = 0$. Figure 6.4.5 illustrates what is happening geometrically in this situation.

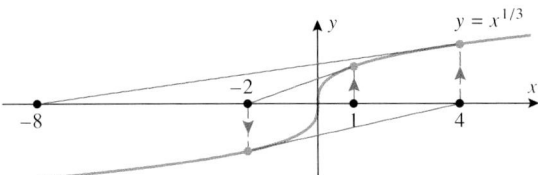

Figure 6.4.5

To learn more about the conditions under which Newton's Method converges and for a discussion of error questions, you should consult a book on numerical analysis. For a more in-depth discussion of Newton's Method and its relationship to contemporary studies of chaos and fractals, you may want to read the article, "Newton's Method and Fractal Patterns," by Phillip Straffin, which appears in *Applications of Calculus*, MAA Notes, Vol. 3, No. 29, 1993, published by the Mathematical Association of America.

EXERCISE SET 6.4 ⊠ Graphing Calculator

In this exercise set, use a calculator, and keep as many decimal places as it can display.

1. Approximate $\sqrt{2}$ by applying Newton's Method to the equation $x^2 - 2 = 0$.

2. Approximate $\sqrt{7}$ by applying Newton's Method to the equation $x^2 - 7 = 0$.

3. Approximate $\sqrt[3]{6}$ by applying Newton's Method to the equation $x^3 - 6 = 0$.

4. To what equation would you apply Newton's Method to approximate the nth root of a?

In Exercises 5–8, the equation has one real solution. Approximate it by Newton's Method.

5. $x^3 - x + 3 = 0$

6. $x^3 + x - 1 = 0$

7. $x^5 + x^4 - 5 = 0$

8. $x^5 - x + 1 = 0$

In Exercises 9–14, use a graphing utility to determine how many solutions the equation has, and then use Newton's Method to approximate the solution that satisfies the stated condition.

⊠ **9.** $x^4 + x - 3 = 0$; $x < 0$

⊠ **10.** $x^5 - 5x^3 - 2 = 0$; $x > 0$

⊠ **11.** $2 \sin x = x$; $x > 0$ ⊠ **12.** $\sin x = x^2$; $x > 0$

⊠ **13.** $x - \tan x = 0$; $\pi/2 < x < 3\pi/2$

⊠ **14.** $1 - e^x \cos x = 0$; $0 < x < \pi$

In Exercises 15–18, use a graphing utility to determine the number of times the curves intersect; and then apply Newton's Method, where needed, to approximate the x-coordinates of all intersections.

15. $y = x^3$ and $y = \frac{1}{2}x - 1$

16. $y = e^{-x}$ and $y = \ln x$

17. $y = x^2$ and $y = \sqrt{2x + 1}$

18. $y = \frac{1}{8}x^3 + 1$ and $y = \cos 2x$

19. The *mechanic's rule* for approximating square roots states that $\sqrt{a} \approx x_{n+1}$, where

$$x_{n+1} = \frac{1}{2}\left(x_n + \frac{a}{x_n}\right), \quad n = 1, 2, 3, \ldots$$

and x_1 is any positive approximation to $\sqrt{a}$.
  (a) Apply Newton's Method to

$$f(x) = x^2 - a$$

  to derive the mechanic's rule.
  (b) Use the mechanic's rule to approximate $\sqrt{10}$.

20. Many calculators compute reciprocals using the approximation $1/a \approx x_{n+1}$, where

$$x_{n+1} = x_n(2 - ax_n), \quad n = 1, 2, 3, \ldots$$

and x_1 is an initial approximation to $1/a$. This formula makes it possible to perform divisions using multiplications and subtractions, which is a faster procedure than dividing directly.
  (a) Apply Newton's Method to

$$f(x) = (1/x) - a$$

  to derive this approximation.
  (b) Use the formula to approximate $\frac{1}{17}$.

21. Use Newton's Method to find the absolute minimum of

$$f(x) = \frac{1}{4}x^4 + x^2 + 5x$$

22. Use Newton's Method to find the absolute maximum of $f(x) = x \sin x$ on the interval $[0, \pi]$.

23. Use Newton's Method to find the coordinates of the point on the parabola $y = x^2$ that is closest to the point $(1, 0)$.

24. Use Newton's Method to find the dimensions of the rectangle of largest area that can be inscribed under the curve $y = \cos x$ for $0 \le x \le \pi/2$, as shown in the accompanying figure.

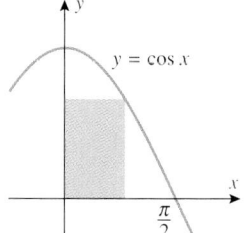

Figure Ex-24

25. (a) Show that on a circle of radius r, the central angle θ that subtends an arc whose length is 1.5 times the length L

of its chord satisfies the equation $\theta = 3 \sin(\theta/2)$ (see the accompanying figure).
  (b) Use Newton's Method to approximate θ.

26. A *segment* of a circle is the region enclosed by an arc and its chord (see the accompanying figure). If r is the radius of the circle and θ the angle subtended at the center of the circle, then it can be shown that the area A of the segment is $A = \frac{1}{2}r^2(\theta - \sin\theta)$, where θ is in radians. Find the value of θ for which the area of the segment is one-fourth the area of the circle. Give θ to the nearest degree.

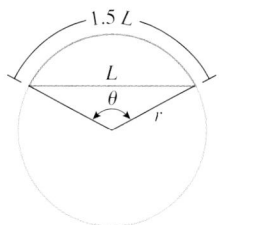

Figure Ex-25   Figure Ex-26

In Exercises 27 and 28, use Newton's Method to approximate all real values of y satisfying the given equation for the indicated value of x.

27. $xy^4 + x^3y = 1$; $x = 1$

28. $xy - \cos\left(\frac{1}{2}xy\right) = 0$; $x = 2$

29. An *annuity* is a sequence of equal payments that are paid or received at regular time intervals. For example, you may want to deposit equal amounts at the end of each year into an interest-bearing account for the purpose of accumulating a lump sum at some future time. If, at the end of each year, interest of $i \times 100\%$ on the account balance for that year is added to the account, then the account is said to pay $i \times 100\%$ interest, *compounded annually*. It can be shown that if payments of Q dollars are deposited at the end of each year into an account that pays $i \times 100\%$ compounded annually, then at the time when the nth payment and the accrued interest for the past year are deposited, the amount $S(n)$ in the account is given by the formula

$$S(n) = \frac{Q}{i}[(1 + i)^n - 1]$$

Suppose that you can invest \$5000 in an interest-bearing account at the end of each year, and your objective is to have \$250,000 on the 25th payment. What annual compound interest rate must the account pay for you to achieve your goal? [*Hint:* Show that the interest rate i satisfies the equation $50i = (1 + i)^{25} - 1$, and solve it using Newton's Method.]

30. (a) Use a graphing utility to generate the graph of

$$f(x) = \frac{x}{x^2 + 1}$$

  and use it to explain what happens if you apply Newton's Method with a starting value of $x_1 = 2$. Check your conclusion by computing x_2, x_3, x_4, and x_5.

(b) Use the graph generated in part (a) to explain what happens if you apply Newton's Method with a starting value of $x_1 = 0.5$. Check your conclusion by computing x_2, x_3, x_4, and x_5.

31. (a) Apply Newton's Method to the function $f(x) = x^2 + 1$ with a starting value of $x_1 = 0.5$, and determine if the values of $x_2, \ldots, x_{10}$ appear to converge.

(b) Explain what is happening.

6.5 ROLLE'S THEOREM; MEAN-VALUE THEOREM

In this section we will discuss a result called the Mean-Value Theorem. This theorem has so many important consequences that it is regarded as one of the major principles in calculus.

ROLLE'S THEOREM

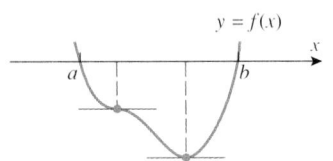

Figure 6.5.1

We will begin with a special case of the Mean-Value Theorem, called Rolle's Theorem, in honor of the mathematician Michel Rolle.[*] This theorem states the geometrically obvious fact that if the graph of a differentiable function crosses the x-axis at two points, a and b, then somewhere between those points there must be at least one place where the tangent line is horizontal (Figure 6.5.1). The precise statement of the theorem is as follows:

6.5.1 THEOREM (*Rolle's Theorem*). *Let f be differentiable on (a, b) and continuous on $[a, b]$. If $f(a) = f(b) = 0$, then there is at least one point c in (a, b) where $f'(c) = 0$.*

Proof. Either $f(x)$ is equal to zero for all x in $[a, b]$ or it is not. If it is, then $f'(x) = 0$ for all x in (a, b), since f is constant on (a, b). Thus, for any c in (a, b)

$$f'(c) = 0$$

If $f(x)$ is not equal to zero for all x in $[a, b]$, then there must be a point x in (a, b) where $f(x) > 0$ or $f(x) < 0$. We will consider the first case and leave the second as an exercise.

Since f is continuous on $[a, b]$, it follows from the Extreme-Value Theorem (6.1.3) that f has a maximum value at some point c in $[a, b]$. Since $f(a) = f(b) = 0$ and $f(x) > 0$ at some point in (a, b), the point c cannot be an endpoint; it must lie in (a, b). By hypothesis, f is differentiable everywhere on (a, b). In particular, it is differentiable at c so that $f'(c) = 0$ by Theorem 6.1.4. ∎

[*] MICHEL ROLLE (1652-1719), French mathematician. Rolle, the son of a shopkeeper, received only an elementary education. He married early and as a young man struggled hard to support his family on the meager wages of a transcriber for notaries and attorneys. In spite of his financial problems and minimal education, Rolle studied algebra and Diophantine analysis (a branch of number theory) on his own. Rolle's fortune changed dramatically in 1682 when he published an elegant solution of a difficult, unsolved problem in Diophantine analysis. The public recognition of his achievement led to a patronage under minister Louvois, a job as an elementary mathematics teacher, and eventually to a short-term administrative post in the Ministry of War. In 1685 he joined the Académie des Sciences in a low-level position for which he received no regular salary until 1699. He stayed there until he died of apoplexy in 1719.

While Rolle's forté was always Diophantine analysis, his most important work was a book on the algebra of equations, called *Traité d'algèbre*, published in 1690. In that book Rolle firmly established the notation $\sqrt[n]{a}$ [earlier written as $\sqrt{n}\ a$] for the nth root of a, and proved a polynomial version of the theorem that today bears his name. (Rolle's Theorem was named by Giusto Bellavitis in 1846.) Ironically, Rolle was one of the most vocal early antagonists of calculus. He strove intently to demonstrate that it gave erroneous results and was based on unsound reasoning. He quarreled so vigorously on the subject that the Académie des Sciences was forced to intervene on several occasions. Among his several achievements, Rolle helped advance the currently accepted size order for negative numbers. Descartes, for example, viewed -2 as smaller than -5. Rolle preceded most of his contemporaries by adopting the current convention in 1691.

Example 1

The function $f(x) = \sin x$ has roots at $x = 0$ and $x = 2\pi$. Moreover, f is continuous and differentiable everywhere, so it is differentiable on $(0, 2\pi)$ and continuous on $[0, 2\pi]$. Thus, Rolle's Theorem guarantees that there is at least one point c in the interval $(0, 2\pi)$ where the tangent line to the graph of $y = \sin x$ is horizontal. Since $dy/dx = \cos x$, we can find c by solving the equation $\cos c = 0$ on the interval $(0, 2\pi)$. This yields two values for c, namely $c_1 = \pi/2$ and $c_2 = 3\pi/2$ (Figure 6.5.2). ◀

REMARK. In the preceding example, we were able to find the values of c because the equation $f'(c) = 0$ was easy to solve. However, if this equation cannot be solved, then you will not be able to find values of c, even though you know they exist. This will rarely cause problems because usually one is more interested in knowing that the values of c exist than in finding them.

The hypotheses in Rolle's Theorem are critical—if f fails to be differentiable at even one point in the interval, then the theorem may fail. For example, the function $f(x) = |x| - 1$ has roots at $x = \pm 1$, yet there is no horizontal tangent line to the graph of f over the interval $(-1, 1)$ (Figure 6.5.3).

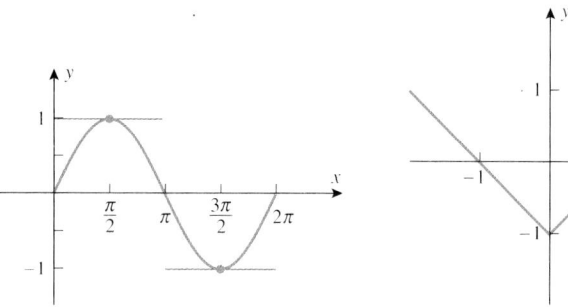

Figure 6.5.2   Figure 6.5.3

THE MEAN-VALUE THEOREM

Rolle's Theorem is a special case of the **Mean-Value Theorem**, which states that between any two points A and B on the graph of a differentiable function, there must be at least one place where the tangent line to the curve is parallel to the secant line joining A and B (Figure 6.5.4).

Noting that the slope of the secant line joining $A(a, f(a))$ and $B(b, f(b))$ is

$$\frac{f(b) - f(a)}{b - a}$$

and the slope of the tangent at c is $f'(c)$, the Mean-Value Theorem can be stated precisely as follows.

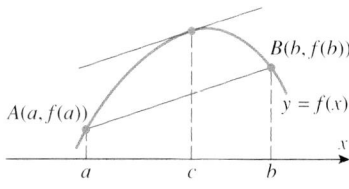

Figure 6.5.4

6.5.2 THEOREM (*Mean-Value Theorem*). *Let f be differentiable on (a, b) and continuous on $[a, b]$. Then there is at least one point c in (a, b) where*

$$f'(c) = \frac{f(b) - f(a)}{b - a} \qquad (1)$$

VELOCITY INTERPRETATION OF THE MEAN-VALUE THEOREM

There is a nice interpretation of the Mean-Value Theorem in the situation where $x = f(t)$ is the position versus time curve for a car moving along a straight road. In this case, the right side of (1) is the average velocity of the car over the time interval from $a \le t \le b$, and the left side is the instantaneous velocity at time $t = c$. Thus, the Mean-Value Theorem implies that at least once during the time interval the instantaneous velocity must equal the average velocity. This agrees with our real-world experience—if the average velocity for a trip is 40 mi/h, then sometime during the trip the speedometer has to read 40 mi/h.

Example 2

You are driving on a straight highway on which the speed limit is 55 mi/h. At 8:05 A.M. a police car clocks your velocity at 50 mi/h and at 8:10 A.M. a second police car posted 5 mi down the road clocks your velocity at 55 mi/h. Explain why the police have a right to charge you with a speeding violation.

Solution. You traveled 5 mi in 5 min $(= \frac{1}{12}$ h), so your average velocity was 60 mi/h. However, the Mean-Value Theorem guarantees the police that your instantaneous velocity was 60 mi/h at least once over the 5-mi section of highway. ◄

PROOF OF THE MEAN-VALUE THEOREM

Motivation for the Proof of Theorem 6.5.2. Figure 6.5.4 suggests that (1) will hold (i.e., the tangent line will be parallel to the secant line) at a point c where the vertical distance between the curve and the secant line is maximum. Thus, to prove the Mean-Value Theorem it is natural to begin by looking for a formula for the vertical distance $v(x)$ between the curve $y = f(x)$ and the secant line joining $(a, f(a))$ and $(b, f(b))$.

Proof of Theorem 6.5.2. Since the two-point form of the equation of the secant line joining $(a, f(a))$ and $(b, f(b))$ is

$$y - f(a) = \frac{f(b) - f(a)}{b - a}(x - a)$$

or equivalently,

$$y = \frac{f(b) - f(a)}{b - a}(x - a) + f(a)$$

the difference $v(x)$ between the height of the graph of f and the height of the secant line is

$$v(x) = f(x) - \left[\frac{f(b) - f(a)}{b - a}(x - a) + f(a) \right] \tag{2}$$

Since $f(x)$ is continuous on $[a, b]$ and differentiable on (a, b), so is $v(x)$. Moreover,

$$v(a) = 0 \quad \text{and} \quad v(b) = 0$$

so that $v(x)$ satisfies the hypotheses of Rolle's Theorem on the interval $[a, b]$. Thus, there is a point c in (a, b) such that $v'(c) = 0$. But from Equation (2)

$$v'(x) = f'(x) - \frac{f(b) - f(a)}{b - a}$$

so

$$v'(c) = f'(c) - \frac{f(b) - f(a)}{b - a}$$

Thus, at the point c in (a, b), where $v'(c) = 0$, we have

$$f'(c) = \frac{f(b) - f(a)}{b - a}$$
■

Example 3

(a) Generate the graph of $f(x) = (x^3/4) + 1$ over the interval $[0, 2]$, and use it to determine the number of tangent lines to the graph of f over the interval $(0, 2)$ that are parallel to the secant line joining the endpoints of the graph.

(b) Show that f satisfies the hypotheses of the Mean-Value Theorem on the interval $[0, 2]$, and find all values of c in the interval $(0, 2)$ whose existence is guaranteed by the Mean-Value Theorem. Confirm that these values of c are consistent with your graph in part (a).

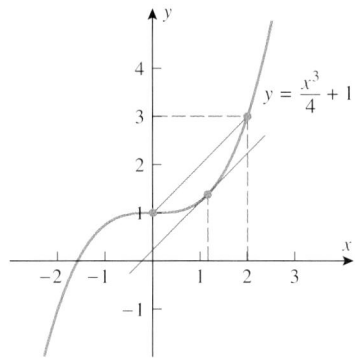

Figure 6.5.5

Solution (*a*). The graph of f in Figure 6.5.5 suggests that there is only one tangent line over the interval $(0, 2)$ that is parallel to the secant line joining the endpoints.

Solution (*b*). The function f is continuous and differentiable everywhere because it is a polynomial. In particular, f is continuous on $[0, 2]$ and differentiable on $(0, 2)$, so the hypotheses of the Mean-Value Theorem are satisfied with $a = 0$ and $b = 2$. But

$$f(a) = f(0) = 1, \quad f(b) = f(2) = 3$$

$$f'(x) = \frac{3x^2}{4}, \qquad f'(c) = \frac{3c^2}{4}$$

so in this case Formula (1) becomes

$$\frac{3c^2}{4} = \frac{3-1}{2-0} \quad \text{or} \quad 3c^2 = 4$$

which has the two solutions $c = \pm 2/\sqrt{3} \approx \pm 1.15$. However, only the positive solution lies in the interval $[0, 2]$; this value of c is consistent with Figure 6.5.5. ◀

CONSEQUENCES OF THE MEAN-VALUE THEOREM

We stated at the beginning of this section that the Mean-Value Theorem is the starting point for many important results in calculus. As an example of this, we will use it to prove Theorem 5.1.2, which was one of our fundamental tools for analyzing graphs of functions.

5.1.2 THEOREM (*Revisited*). *Let f be a function that is continuous on a closed interval $[a, b]$ and differentiable on the open interval (a, b).*

(*a*) *If $f'(x) > 0$ for every value of x in (a, b), then f is increasing on $[a, b]$.*

(*b*) *If $f'(x) < 0$ for every value of x in (a, b), then f is decreasing on $[a, b]$.*

(*c*) *If $f'(x) = 0$ for every value of x in (a, b), then f is constant on $[a, b]$.*

Proof (*a*). Suppose that x_1 and x_2 are points in $[a, b]$ such that $x_1 < x_2$. We must show that $f(x_1) < f(x_2)$. Because the hypotheses of the Mean-Value Theorem are satisfied on the entire interval $[a, b]$, they are satisfied on the subinterval $[x_1, x_2]$. Thus, there is some point c in the open interval (x_1, x_2) such that

$$f'(c) = \frac{f(x_2) - f(x_1)}{x_2 - x_1}$$

or equivalently,

$$f(x_2) - f(x_1) = f'(c)(x_2 - x_1) \tag{3}$$

Since c is in the open interval (x_1, x_2), it follows that $a < c < b$; thus, $f'(c) > 0$. However, $x_2 - x_1 > 0$ since we assumed that $x_1 < x_2$. It follows from (3) that $f(x_2) - f(x_1) > 0$ or, equivalently, $f(x_1) < f(x_2)$, which is what we were to prove. The proofs of parts (*b*) and (*c*) are similar and are left as exercises. ∎

THE CONSTANT DIFFERENCE THEOREM

We know from our earliest study of derivatives that the derivative of a constant is zero. Part (*c*) of Theorem 5.1.2 is the converse of that result; that is, a function whose derivative is zero on an interval must be constant on that interval. If we apply this to the difference of two functions, we obtain the following useful theorem.

6.5.3 THEOREM (*The Constant Difference Theorem*). *If f and g are continuous on a closed interval $[a, b]$, and if $f'(x) = g'(x)$ for all x in the open interval (a, b), then f and g differ by a constant on $[a, b]$; that is, there is a constant k such that $f(x) - g(x) = k$ for all x in $[a, b]$.*

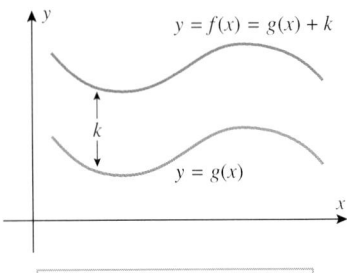

$y = f(x) = g(x) + k$

$y = g(x)$

If $f'(x) = g'(x)$ on an interval, then the graphs of f and g are vertical translations of one another.

Figure 6.5.6

Proof. Let $h(x) = f(x) - g(x)$. Then for every x in (a, b)

$$h'(x) = f'(x) - g'(x) = 0$$

Thus, $h(x) = f(x) - g(x)$ is constant on $[a, b]$ by Theorem 5.1.2(c). ∎

REMARK.   This theorem remains true if the closed interval $[a, b]$ is replaced by a finite or infinite interval (a, b), $[a, b)$, or $(a, b]$, provided f and g are differentiable on (a, b) and continuous on the entire interval.

The Constant Difference Theorem has a simple geometric interpretation—it tells us that if f and g have the same derivative on an interval, then there is a constant k such that $f(x) = g(x) + k$ for each x in the interval; that is, the graphs of f and g can be obtained from one another by a vertical translation (Figure 6.5.6).

EXERCISE SET 6.5   ⌢ Graphing Calculator   [C] CAS

In Exercises 1 and 2, use the graph of f to find an interval $[a, b]$ on which Rolle's Theorem applies, and find all values of c in that interval that satisfy the conclusion of the theorem.

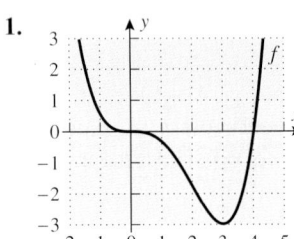

1.

2.

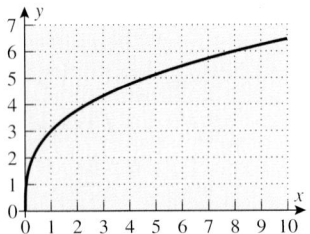

Figure Ex-9

10. Use the graph of f in Exercise 9 to estimate all values of c that satisfy the conclusion of the Mean-Value Theorem on the interval $[0, 4]$.

In Exercises 3–8, verify that the hypotheses of Rolle's Theorem are satisfied on the given interval, and find all values of c in that interval that satisfy the conclusion of the theorem.

In Exercises 11–16, verify that the hypotheses of the Mean-Value Theorem are satisfied on the given interval, and find all values of c in that interval that satisfy the conclusion of the theorem.

3. $f(x) = x^2 - 6x + 8$; $[2, 4]$

4. $f(x) = x^3 - 3x^2 + 2x$; $[0, 2]$

5. $f(x) = \cos x$; $[\pi/2, 3\pi/2]$

6. $f(x) = \dfrac{x^2 - 1}{x - 2}$; $[-1, 1]$

7. $f(x) = \frac{1}{2}x - \sqrt{x}$; $[0, 4]$

8. $f(x) = \dfrac{1}{x^2} - \dfrac{4}{3x} + \dfrac{1}{3}$; $[1, 3]$

9. Use the graph of f in the accompanying figure to estimate all values of c that satisfy the conclusion of the Mean-Value Theorem on the interval $[0, 8]$.

11. $f(x) = x^2 + x$; $[-4, 6]$

12. $f(x) = x^3 + x - 4$; $[-1, 2]$

13. $f(x) = \sqrt{x + 1}$; $[0, 3]$ **14.** $f(x) = x + \dfrac{1}{x}$; $[3, 4]$

15. $f(x) = \sqrt{25 - x^2}$; $[-5, 3]$

16. $f(x) = \dfrac{1}{x - 1}$; $[2, 5]$

⌢ **17.** (a) Find an interval $[a, b]$ on which

$$f(x) = x^4 + x^3 - x^2 + x - 2$$

satisfies the hypotheses of Rolle's Theorem.

(b) Generate the graph of $f'(x)$, and use it to make rough estimates of all values of c in the interval obtained in part (a) that satisfy the conclusion of Rolle's Theorem.

(c) Use Newton's Method to improve on the rough estimates obtained in part (b).

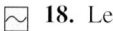 **18.** Let $f(x) = x^3 + 4x$.

(a) Find the equation of the secant line through the points $(-2, f(-2))$ and $(1, f(1))$.

(b) Show that there is only one number c in the interval $(-2, 1)$ that satisfies the conclusion of the Mean-Value Theorem for the secant line in part (a).

(c) Find the equation of the tangent line to the graph of f at the point $(c, f(c))$.

(d) Use a graphing utility to generate the secant line in part (a) and the tangent line in part (c) in the same coordinate system, and confirm visually that the two lines seem parallel.

19. Let $f(x) = \tan x$.

(a) Show that there is no point c in the interval $(0, \pi)$ such that $f'(c) = 0$, even though $f(0) = f(\pi) = 0$.

(b) Explain why the result in part (a) does not violate Rolle's Theorem.

20. Let $f(x) = x^{2/3}$, $a = -1$, and $b = 8$.

(a) Show that there is no point c in (a, b) such that

$$f'(c) = \frac{f(b) - f(a)}{b - a}$$

(b) Explain why the result in part (a) does not violate the Mean-Value Theorem.

21. (a) Show that if f is differentiable on $(-\infty, +\infty)$, and if $y = f(x)$ and $y = f'(x)$ are graphed in the same coordinate system, then between any two x-intercepts of f there is at least one x-intercept of f'.

(b) Give some examples that illustrate this.

22. Review Definitions 3.1.3 and 3.1.4 of average and instantaneous rate of change of y with respect to x, and use the Mean-Value Theorem to show that if f is differentiable on $(-\infty, +\infty)$, then in any interval $[x_0, x_1]$ there is at least one point where the instantaneous rate of change of y with respect to x is equal to the average rate of change over the interval.

In Exercises 23–25, use the result of Exercise 22.

23. An automobile travels 4 mi along a straight road in 5 min. Show that the speedometer read exactly 48 mi/h at least once during the trip.

24. At 11 A.M. on a certain morning the outside temperature was 76°F. At 11 P.M. that evening it had dropped to 52°F.

(a) Show that at some instant during this period the temperature was decreasing at the rate of 2°F/h.

(b) Suppose that you know that the temperature reached a high of 88°F sometime between 11 A.M. and 11 P.M.

Show that at some instant during this period the temperature was decreasing at a rate greater than 3°F/h.

25. Suppose that two runners in a 100-m dash finish in a tie. Show that they had the same velocity at least once during the race.

26. Use the fact that

$$\frac{d}{dx}(x^6 - 2x^2 + x) = 6x^5 - 4x + 1$$

to show that the equation $6x^5 - 4x + 1 = 0$ has at least one solution in the interval $(0, 1)$.

27. (a) Use the Constant Difference Theorem (6.5.3) to show that if $f'(x) = g'(x)$ for all x in the interval $(-\infty, +\infty)$, and if f and g have the same value at any point x_0, then $f(x) = g(x)$ for all x in $(-\infty, +\infty)$.

(b) Use the result in part (a) to prove the trigonometric identity $\sin^2 x + \cos^2 x = 1$.

28. (a) Use the Constant Difference Theorem (6.5.3) to show that if $f'(x) = g'(x)$ for all x in $(-\infty, +\infty)$, and if $f(x_0) - g(x_0) = c$ at some point x_0, then

$$f(x) - g(x) = c$$

for all x in $(-\infty, +\infty)$.

(b) Use the result in part (a) to show that the function

$$h(x) = (x - 1)^3 - (x^2 + 3)(x - 3)$$

is constant for all x in $(-\infty, +\infty)$, and find the constant.

(c) Check the result in part (b) by multiplying out and simplifying the formula for $h(x)$.

29. (a) Use the Mean-Value Theorem to show that if f is differentiable on an interval I, and if $|f'(x)| \le M$ for all values of x in I, then

$$|f(x) - f(y)| \le M|x - y|$$

for all values of x and y in I.

(b) Use the result in part (a) to show that

$$|\sin x - \sin y| \le |x - y|$$

for all real values of x and y.

30. (a) Use the Mean-Value Theorem to show that if f is differentiable on an open interval I, and if $|f'(x)| \ge M$ for all values of x in I, then

$$|f(x) - f(y)| \ge M|x - y|$$

for all values of x and y in I.

(b) Use the result in part (a) to show that

$$|\tan x - \tan y| \ge |x - y|$$

for all values of x and y in the interval $(-\pi/2, \pi/2)$.

(c) Use the result in part (b) to show that

$$|\tan x + \tan y| \ge |x + y|$$

for all values of x and y in the interval $(-\pi/2, \pi/2)$.

31. (a) Use the Mean-Value Theorem to show that

$$\sqrt{y} - \sqrt{x} < \frac{y - x}{2\sqrt{x}}$$

if $0 < x < y$.

(b) Use the result in part (a) to show that if x and y are positive, then $\sqrt{xy} < \frac{1}{2}(x + y)$.

32. Show that if f is differentiable on an open interval I and $f'(x) \neq 0$ on I, the equation $f(x) = 0$ can have at most one real root in I.

33. Use the result in Exercise 32 to show the following:
(a) The equation $x^3 + 4x - 1 = 0$ has exactly one real root.
(b) If $b^2 - 3ac < 0$ and if $a \neq 0$, then the equation

$$ax^3 + bx^2 + cx + d = 0$$

has exactly one real root (possibly repeated).

34. Use the Mean-Value Theorem to prove that

$$1.71 < \sqrt{3} < 1.75$$

[*Hint:* Let $f(x) = \sqrt{x}, a = 3$, and $b = 4$ in the Mean-Value Theorem.]

35. (a) Show that if f and g are functions for which

$$f'(x) = g(x) \quad \text{and} \quad g'(x) = -f(x)$$

for all x, then $f^2(x) + g^2(x)$ is a constant.

(b) Give an example of functions f and g with this property.

36. (a) Show that if f and g are functions for which

$$f'(x) = g(x) \quad \text{and} \quad g'(x) = f(x)$$

for all x, then $f^2(x) - g^2(x)$ is a constant.

(b) Show that the function $f(x) = \frac{1}{2}(e^x + e^{-x})$ and the function $g(x) = \frac{1}{2}(e^x - e^{-x})$ have this property.

37. Let $g(x) = x^3 - 4x + 6$. Find $f(x)$ so that $f'(x) = g'(x)$ and $f(1) = 2$.

38. Let f and g be continuous on $[a, b]$ and differentiable on (a, b). Prove: If $f(a) = g(a)$ and $f(b) = g(b)$, then there is a point c in (a, b) where $f'(c) = g'(c)$.

39. Illustrate the result in Exercise 38 by drawing an appropriate picture.

40. (a) Prove: If $f''(x) > 0$ for all x in (a, b), then $f'(x) = 0$ at most once in (a, b).

(b) Give a geometric interpretation of the result in (a).

41. Prove part (*b*) of Theorem 5.1.2.

42. Prove part (*c*) of Theorem 5.1.2.

SUPPLEMENTARY EXERCISES

1. (a) What inequality must $f(x)$ satisfy for the function f to have an absolute maximum on an interval I at x_0?

(b) What inequality must $f(x)$ satisfy for f to have an absolute minimum on I at x_0?

(c) What is the difference between an absolute extremum and a relative extremum?

2. According to the Extreme-Value Theorem, what conditions on a function f and an interval I guarantee that f will have both an absolute maximum and an absolute minimum on I?

3. In each part, determine whether the statement is true or false, and justify your answer.
(a) If f is differentiable on the open interval (a, b), and if f has an absolute extremum on that interval, then it must occur at a stationary point of f.
(b) If f is continuous on the open interval (a, b), and if f is an absolute extremum on that interval, then it must occur at a stationary point of f.

4. Is it true or false that a particle in rectilinear motion is speeding up when its velocity is increasing and slowing down when its velocity is decreasing? Justify your answer.

5. Suppose that f is continuous on the closed interval $[a, b]$ and differentiable on the open interval (a, b), and suppose that $f(a) = f(b)$. Is it true or false that f must have at least one stationary point in (a, b)? Justify your answer.

6. Draw an appropriate picture, and describe the basic idea of Newton's Method without using any formulas.

7. In each part, find the absolute minimum m and the absolute maximum M of f on the given interval (if they exist), and state where the absolute extrema occur.
(a) $f(x) = 1/x$; $[-2, -1]$
(b) $f(x) = x^3 - x^4$; $\left[-1, \frac{3}{2}\right]$
(c) $f(x) = x^2(x - 2)^{1/3}$; $(0, 3]$
(d) $f(x) = e^x/x^2$; $(0, +\infty)$

8. In each part, find the absolute minimum m and the absolute maximum M of f on the given interval (if they exist), and state where the absolute extrema occur.
(a) $f(x) = 2x/(x^2 + 3)$; $(0, 2]$
(b) $f(x) = 2x^5 - 5x^4 + 7$; $(-1, 3)$
(c) $f(x) = -|x^2 - 2x|$; $[1, 3]$
(d) $f(x) = x^x$; $[0, +\infty)$

9. Use Newton's Method to approximate the smallest positive solution of $\sin x + \cos x = 0$.

10. Use Newton's Method to approximate all three solutions of $x^3 - 4x + 1 = 0$.

11. In each part, determine whether all of the hypotheses of Rolle's Theorem are satisfied on the stated interval. If not, state which hypotheses fail; if so, find all values of c guaranteed in the conclusion of the theorem.
(a) $f(x) = \sqrt{4 - x^2}$ on $[-2, 2]$

(b) $f(x) = x^{2/3} - 1$ on $[-1, 1]$

(c) $f(x) = \sin(x^2)$ on $[0, \sqrt{\pi}]$

12. In each part, determine whether all of the hypotheses of the Mean-Value Theorem are satisfied on the stated interval. If not, state which hypotheses fail; if so, find all values of c guaranteed in the conclusion of the theorem.

(a) $f(x) = |x - 1|$ on $[-2, 2]$

(b) $f(x) = \dfrac{x + 1}{x - 1}$ on $[2, 3]$

(c) $f(x) = \begin{cases} 3 - x^2 & \text{if } x \le 1 \\ 2/x & \text{if } x > 1 \end{cases}$ on $[0, 2]$

13. A church window consists of a blue semicircular section surmounting a clear rectangular section as shown in the accompanying figure. The blue glass lets through half as much light per unit area as the clear glass. Find the radius r of the window that admits the most light if the perimeter of the entire window is to be P feet.

14. Find the dimensions of the rectangle of maximum area that can be inscribed inside the ellipse $(x/4)^2 + (y/3)^2 = 1$ (see the accompanying figure).

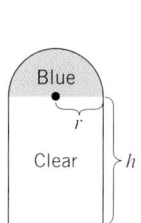

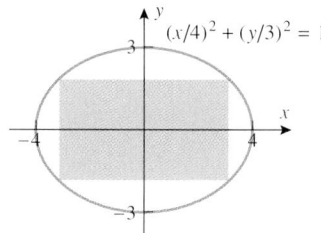

Figure Ex-13   Figure Ex-14

15. (a) Can an object in rectilinear motion reverse direction if its acceleration is constant? Justify your answer using a velocity versus time curve.

(b) Can an object in rectilinear motion have increasing speed and decreasing acceleration? Justify your answer using a velocity versus time curve.

 16. Suppose that the position function of a particle in rectilinear motion is given by the formula $s(t) = t/(t^2 + 5)$ for $t \ge 0$.

(a) Use a graphing utility to generate the position, velocity, and acceleration versus time curves.

(b) Use the appropriate graph to make a rough estimate of the time when the particle reverses direction, and then find that time exactly.

(c) Find the position, velocity, and acceleration at the instant when the particle reverses direction.

(d) Use the appropriate graphs to make rough estimates of the time intervals on which the particle is speeding up and the time intervals on which it is slowing down, and then find those time intervals exactly.

(e) When does the particle have its maximum and minimum velocities?

17. A basketball player, standing near the basket to grab a rebound, jumps 76.0 cm vertically.

(a) How much time does the player spend in the top 15.0 cm of the jump and how much time in the bottom 15.0 cm?

(b) In words, explain why basketball players seem to be suspended in air when they jump.

18. (a) Suppose that an object is released from rest from the top of a high building. Assuming that a free-fall model applies and that time is in seconds and distance is in meters, make a table that shows the distance traveled by the object and its speed to one decimal place at 1-second increments from $t = 0$ to $t = 4$.

(b) Confirm that doubling the elapsed time doubles the velocity, and explain why this happens.

(c) Confirm that doubling the elapsed time increases the distance traveled by a factor of 4, and explain why this happens.

[c] **19.** Let

$$f(x) = \frac{x^3 + 2}{x^4 + 1}$$

(a) Generate the graph of $y = f(x)$, and use the graph to make rough estimates of the coordinates of the absolute extrema.

(b) Use a CAS to solve the equation $f'(x) = 0$ and then use it to make more accurate approximations of the coordinates in part (a).

[c] **20.** As shown in the accompanying figure, suppose that a boat enters the river at the point $(1, 0)$ and maintains a heading toward the origin. As a result of the strong current, the boat follows the path

$$y = \frac{x^{10/3} - 1}{2x^{2/3}}$$

where x and y are in miles.

(a) Graph the path taken by the boat.

(b) Can the boat reach the origin? If not, discuss its fate and find how close it comes to the origin.

(c) What is the velocity of the boat in the x-direction at the instant when it is closest to the origin if the velocity in the y-direction is -4 mi/h at this instant?

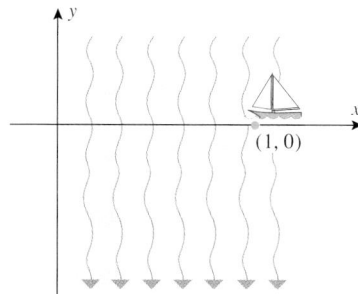

Figure Ex-20

21. Suppose that the position function of a particle in rectilinear motion is given by the formula

$$s(t) = \frac{t^2 + 1}{t^4 + 1}, \quad t \geq 0$$

(a) Use a CAS to find simplified formulas for the velocity $v(t)$ and the acceleration $a(t)$.

(b) Graph the position, velocity, and acceleration versus time curves.

(c) Use the appropriate graph to make a rough estimate of the time at which the particle is farthest from the origin and its distance from the origin at that time.

(d) Use the appropriate graph to make a rough estimate of the time interval during which the particle is moving in the positive direction.

(e) Use the appropriate graphs to make rough estimates of the time intervals during which the particle is speeding up and the time intervals during which it is slowing down.

(f) Use the appropriate graph to make a rough estimate of the maximum speed of the particle and the time at which the maximum speed occurs.

22. Suppose that the number of individuals at time t in a certain wildlife population is given by

$$N(t) = \frac{340}{1 + 9(0.77)^t}, \quad t \geq 0$$

where t is in years. At approximately what instant of time is the size of the population increasing most rapidly?

23. According to **Kepler's law**, the planets in our solar system move in elliptical orbits around the Sun. If a planet's closest approach to the Sun occurs at time $t = 0$, then the distance r from the center of the planet to the center of the Sun at some later time t can be determined from the equation

$$r = a(1 - e\cos\phi)$$

where a is the average distance between centers, e is a positive constant that measures the "flatness" of the elliptical orbit, and ϕ is the solution of *Kepler's equation*

$$\frac{2\pi t}{T} = \phi - e\sin\phi$$

in which T is the time it takes for one complete orbit of the planet. Estimate the distance from the Earth to the Sun when $t = 90$ days. [First find ϕ from Kepler's equation, and then use this value of ϕ to find the distance. Use $a = 150 \times 10^6$ km, $e = 0.0167$, and $T = 365$ days.]

24. Using the formulas in Exercise 23, find the distance from the planet Mars to the Sun when $t = 1$ year. For Mars use $a = 228 \times 10^6$ km, $e = 0.934$, and $T = 1.88$ years.

EXPANDING THE CALCULUS HORIZON

For additional material relating to this chapter, visit the Anton Website at http://www.wiley.com/college/anton

7

INTEGRATION

Gottfried Leibniz

*T*raditionally, that portion of calculus concerned with finding tangent lines and rates of change is called ***differential calculus*** and that portion concerned with finding areas is called ***integral calculus***. However, we will see in this chapter that the two problems are so closely related that the distinction between differential and integral calculus is often hard to discern.

In this chapter we will begin with an overview of the problem of finding areas—we will discuss what the term "area" means, and we will outline two approaches to defining and calculating areas. Following this overview, we will discuss the "Fundamental Theorem of Calculus", which is the theorem that relates the problems of finding tangent lines and areas, and we will discuss techniques for calculating areas. Finally, we will use the ideas in this chapter to continue our study of rectilinear motion and to reexamine the concept of a natural logarithm.

7.1 AN OVERVIEW OF THE AREA PROBLEM

In this introductory section we will give an overview of the problem of defining and calculating areas of plane regions with curvilinear boundaries. All of the results in this section will be reexamined in more detail later in this chapter, so our purpose here is to introduce the fundamental concepts.

DEFINING AREA

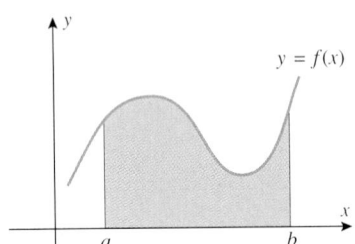

Figure 7.1.1

The main goal of this chapter is to study the following major problem of calculus:

> **7.1.1 THE AREA PROBLEM.** Given a function f that is continuous and nonnegative on an interval $[a, b]$, find the area between the graph of f and the interval $[a, b]$ on the x-axis (Figure 7.1.1).

Area formulas for basic geometric figures, such as rectangles, polygons, and circles, date back to the earliest written records of mathematics. The first real advance beyond the elementary level of area computation was made by the Greek mathematician, Archimedes,[*] who devised an ingenious but cumbersome technique, called the *method of exhaustion*, for finding areas of regions bounded by parabolas, spirals, and various other curves.

---

[*] ARCHIMEDES (287 B.C.–212 B.C.). Greek mathematician and scientist. Born in Syracuse, Sicily, Archimedes was the son of the astronomer Pheidias and possibly related to Heiron II, king of Syracuse. Most of the facts about his life come from the Roman biographer, Plutarch, who inserted a few tantalizing pages about him in the massive biography of the Roman soldier, Marcellus. In the words of one writer, "the account of Archimedes is slipped like a tissue-thin shaving of ham in a bull-choking sandwich."

Archimedes ranks with Newton and Gauss as one of the three greatest mathematicians who ever lived, and he is certainly the greatest mathematician of antiquity. His mathematical work is so modern in spirit and technique that it is barely distinguishable from that of a seventeenth-century mathematician, yet it was all done without benefit of algebra or a convenient number system. Among his mathematical achievements, Archimedes developed a general method (exhaustion) for finding areas and volumes, and he used the method to find areas bounded by parabolas and spirals and to find volumes of cylinders, paraboloids, and segments of spheres. He gave a procedure for approximating π and bounded its value between $3\frac{10}{71}$ and $3\frac{1}{7}$. In spite of the limitations of the Greek numbering system, he devised methods for finding square roots and invented a method based on the Greek myriad (10,000) for representing numbers as large as 1 followed by 80 million billion zeros.

Of all his mathematical work, Archimedes was most proud of his discovery of the method for finding the volume of a sphere—he showed that the volume of a sphere is two-thirds the volume of the smallest cylinder that can contain it. At his request, the figure of a sphere and cylinder was engraved on his tombstone.

In addition to mathematics, Archimedes worked extensively in mechanics and hydrostatics. Nearly every schoolchild knows Archimedes as the absent-minded scientist who, on realizing that a floating object displaces its weight of liquid, leaped from his bath and ran naked through the streets of Syracuse shouting, "Eureka, Eureka!"—(meaning, "I have found it!"). Archimedes actually created the discipline of hydrostatics and used it to find equilibrium positions for various floating bodies. He laid down the fundamental postulates of mechanics, discovered the laws of levers, and calculated centers of gravity for various flat surfaces and solids. In the excitement of discovering the mathematical laws of the lever, he is said to have declared, "Give me a place to stand and I will move the earth."

Although Archimedes was apparently more interested in pure mathematics than its applications, he was an engineering genius. During the second Punic war, when Syracuse was attacked by the Roman fleet under the command of Marcellus, it was reported by Plutarch that Archimedes' military inventions held the fleet at bay for three years. He invented super catapults that showered the Romans with rocks weighing a quarter ton or more, and fearsome mechanical devices with iron "beaks and claws" that reached over the city walls, grasped the ships, and spun them against the rocks. After the first repulse, Marcellus called Archimedes a "geometrical Briareus (a hundred-armed mythological monster) who uses our ships like cups to ladle water from the sea."

Eventually the Roman army was victorious and contrary to Marcellus' specific orders the 75-year-old Archimedes was killed by a Roman soldier. According to one report of the incident, the soldier cast a shadow across the sand in which Archimedes was working on a mathematical problem. When the annoyed Archimedes yelled, "Don't disturb my circles," the soldier flew into a rage and cut the old man down.

With his death the Greek gift of mathematics passed into oblivion, not to be fully resurrected again until the sixteenth century. Unfortunately, there is no known accurate likeness or statue of this great man.

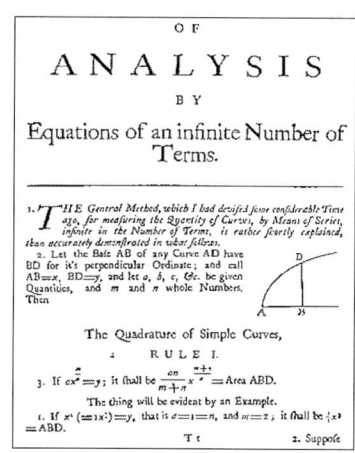

First page of the 1745 English translation of Newton's *De Analysi*

By the seventeenth century, several mathematicians had discovered how to obtain such areas more simply by calculating limits. However, the method of exhaustion and its successors lacked generality—for each different problem one had to devise special procedures. The major breakthrough in obtaining a general method for calculating areas was made independently by Newton and Leibniz, both of whom discovered that areas could be obtained by reversing the process of differentiation. This discovery, which is regarded as the beginning of calculus, was circulated by Newton in 1669 and published in 1711 in a paper entitled, *De Analysi per Aequationes Numero Terminorum Infinitas* (*On the Analysis by Means of Equations with Infinitely Many Terms*); and it was discovered by Leibniz around 1673 and stated in an unpublished manuscript dated November 11, 1675.

Before one can talk logically about methods for calculating areas, it is necessary to have a precise definition of what the term *area* means. To avoid a lot of mathematical formality, let us assume that the areas of geometric figures with straight boundaries, such as rectangles, triangles, and polygons, are defined and computed using the standard formulas for such figures. However, the problem of defining and computing areas of figures with *curvilinear* boundaries is more complicated and will require various limiting processes. For example, in the introductory section of this text we showed that the area of a circle could be viewed as a limit of areas of inscribed polygons (Figure 7 in the Introduction). Thus, once a definition is established for the area of a polygon, the area of a circle can be *defined* as a limit of areas of polygons.

THE RECTANGLE METHOD FOR FINDING AREAS

There are two basic methods for finding the area of the region having the form shown in Figure 7.1.1—the *rectangle method* and the *antiderivative method*. The idea behind the rectangle method is as follows:

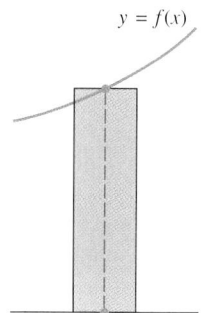

Figure 7.1.2

- Divide the interval $[a, b]$ into n equal subintervals, and over each subinterval construct a rectangle that extends from the x-axis to any point on the curve $y = f(x)$ that is above the subinterval; the particular point does not matter—it can be above the center, above an endpoint, or above any other point in the subinterval. In Figure 7.1.2 it is above the center.
- For each n, the total area of the rectangles can be viewed as an *approximation* to the exact area under the curve over the interval $[a, b]$. Moreover, it is evident intuitively that as n increases these approximations will get better and better and will approach the exact area as a limit (Figure 7.1.3).

This procedure serves both as a mathematical definition and a method of computation—we can *define* the area under $y = f(x)$ over the interval $[a, b]$ as the limit of the areas of the approximating rectangles, and we can use the method itself to approximate this area.

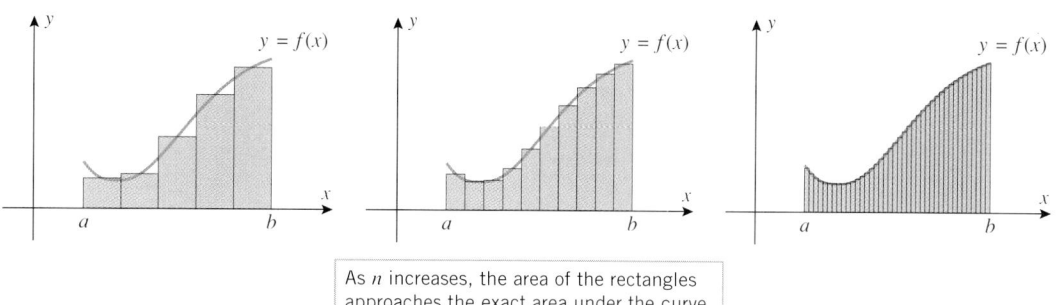

As n increases, the area of the rectangles approaches the exact area under the curve.

Figure 7.1.3

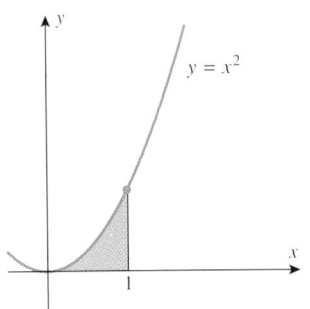

Figure 7.1.4

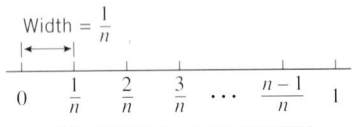

Figure 7.1.5

To illustrate this idea, we will use the rectangle method to approximate the area under the curve $y = x^2$ over the interval $[0, 1]$ (Figure 7.1.4). We will begin by dividing the interval $[0, 1]$ into n equal subintervals, from which it follows that each subinterval has length $1/n$; the endpoints of the subintervals occur at

$$0, \quad \frac{1}{n}, \quad \frac{2}{n}, \quad \frac{3}{n}, \ldots, \quad \frac{n-1}{n}, \quad 1$$

(Figure 7.1.5). We want to construct a rectangle over each of these intervals whose height is the value of the function $f(x) = x^2$ at any point in the interval. To be specific, let us use the right endpoints, in which case the heights of our rectangles will be

$$\left(\frac{1}{n}\right)^2, \quad \left(\frac{2}{n}\right)^2, \quad \left(\frac{3}{n}\right)^2, \ldots, \quad 1$$

and since each rectangle has a base of width $1/n$, the total area A_n of the n rectangles will be

$$A_n = \left[\left(\frac{1}{n}\right)^2 + \left(\frac{2}{n}\right)^2 + \left(\frac{3}{n}\right)^2 + \cdots + 1^2\right]\left(\frac{1}{n}\right) \tag{1}$$

For example, if $n = 4$, then the total area of the four approximating rectangles would be

$$A_4 = \left[\left(\tfrac{1}{4}\right)^2 + \left(\tfrac{2}{4}\right)^2 + \left(\tfrac{3}{4}\right)^2 + 1^2\right]\left(\tfrac{1}{4}\right) = \tfrac{15}{32} = 0.46875$$

Table 7.1.1 shows the result of evaluating (1) on a computer for some increasingly large values of n. These computations suggest that the exact area is close to $\frac{1}{3}$.

Table 7.1.1

n	4	10	100	1000	10,000	100,000
A_n	0.468750	0.385000	0.338350	0.333834	0.333383	0.333338

FOR THE READER. Use your calculating utility to confirm the value of A_{10} given in Table 7.1.1.

THE ANTIDERIVATIVE METHOD FOR FINDING AREAS

The antiderivative method for finding areas reflects the genius of Newton and Leibniz—they suggested that to find the area under the curve in Figure 7.1.1, one should first consider the more general problem of finding the area $A(x)$ under the curve from the point a to an arbitrary point x in the interval $[a, b]$ (Figure 7.1.6). Newton and Leibniz discovered independently that the *derivative* of the function $A(x)$ is easy to find, so that if one can figure out how to find $A(x)$ from $A'(x)$, then the area under the curve from a to b can be obtained by substituting $x = b$ in the area formula $A(x)$.

To illustrate how all of this works, let us begin with the problem of finding

$$A'(x) = \lim_{h \to 0} \frac{A(x+h) - A(x)}{h} \tag{2}$$

For simplicity, consider the case where $h > 0$. The numerator on the right side of (2) is the difference of two areas: the area between a and $x + h$ minus the area between a and x (Figure 7.1.7a). If we let c be the midpoint between x and $x + h$, then this difference of areas can be approximated by the area of a rectangle with base h and height $f(c)$ (Figure 7.1.7b). Thus,

$$\frac{A(x+h) - A(x)}{h} \approx \frac{f(c) \cdot h}{h} = f(c) \tag{3}$$

It seems plausible from Figure 7.1.7b that the error in approximation (3) will approach

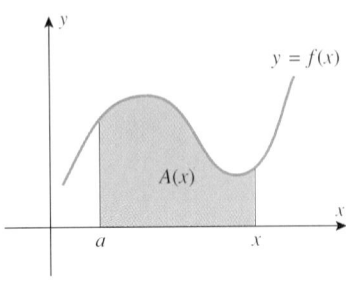

Figure 7.1.6

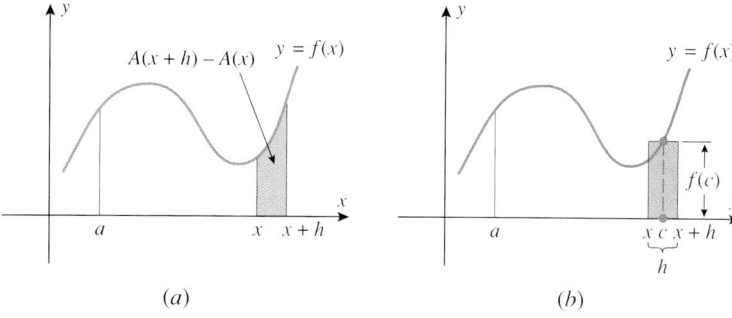

Figure 7.1.7

approach zero as $h \to 0$. If we accept this to be so, then it follows from (2) and (3) that

$$A'(x) = \lim_{h \to 0} \frac{A(x+h) - A(x)}{h} = \lim_{h \to 0} f(c) \tag{4}$$

Since c is the midpoint between x and $x + h$, it follows that $c \to x$ as $h \to 0$. But we have assumed f to be a continuous function, so $f(c) \to f(x)$ as $c \to x$. Therefore,

$$\lim_{h \to 0} f(c) = f(x)$$

Thus, it follows from (4) that

$$A'(x) = f(x) \tag{5}$$

This is the result we were looking for; it tells us that *the derivative of the area function $A(x)$ is the function whose graph forms the upper boundary of the region.*

To illustrate how the antiderivative method works, let us apply it to the same problem we investigated with the rectangle method—finding the area under $y = x^2$ over the interval $[0, 1]$. The upper boundary of the region is the graph of $f(x) = x^2$, so it follows from (5) that the derivative of the area function is

$$A'(x) = x^2 \tag{6}$$

Thus, to find $A(x)$ we must look for a function whose derivative is x^2. This is called an ***antidifferentiation*** problem because we are trying to find $A(x)$ by "undoing" a differentiation. By simply guessing we see that

$$A(x) = \tfrac{1}{3}x^3$$

is one solution to (6). But this is not the only solution, since it follows from Theorem 6.5.3 that

$$A(x) = \tfrac{1}{3}x^3 + C \tag{7}$$

also satisfies (6) for any real value of C. We still have some work to do since this formula involves an unknown constant C that must be determined. This is where the decision to solve the area problem for a general right-hand endpoint helps. If we consider the case where $x = 0$, then the interval $[0, x]$ reduces to a single point. If we agree that the area above a single point should be taken as zero, then it follows on substituting $x = 0$ in (7) that

$$A(0) = 0 + C = 0 \quad \text{or} \quad C = 0$$

so (7) simplifies to

$$A(x) = \tfrac{1}{3}x^3 \tag{8}$$

which is the formula for the area under $y = x^2$ over the interval $[0, x]$. For the area over the interval $[0, 1]$ we set $x = 1$ in (8), which yields $A(1) = \frac{1}{3}$ for the exact area under the curve. This confirms definitely what was suggested numerically in Table 7.1.1.

REMARK.   Our success in finding the exact area under the curve $y = x^2$ hinged on our ability to guess at a function $A(x)$ whose derivative is x^2. Had we not been able to find such a function, then the antiderivative method would have failed and we would have been forced to rely on the rectangle method. Thus, whereas earlier in this text we were concerned with the process of differentiation, we will now also be concerned with the process of antidifferentiation.

EXERCISE SET 7.1

In Exercises 1–4, use an appropriate formula from plane geometry to find the exact area between the graph of f and the given interval; and then use the rectangle method to make a table of approximations $A_1, A_2, \ldots, A_{10}$ to the exact area, where A_n is the approximation that results by dividing the interval into n subintervals and constructing a rectangle over each subinterval whose height is the y-coordinate of the curve $y = f(x)$ at the right endpoint.

1. $f(x) = x$; $[0, 1]$

2. $f(x) = 4 - 2x$; $[0, 2]$

3. $f(x) = 6x + 2$; $[0, 2]$

4. $f(x) = \sqrt{1 - x^2}$; $[0, 1]$

5. Let $A(x) = x^2/2$. Confirm that $A'(x) = x$, and use the antiderivative method to find the exact area in Exercise 1.

6. Let $A(x) = 4x - x^2$. Confirm that $A'(x) = 4 - 2x$, and use the antiderivative method to find the exact area in Exercise 2.

7. Let $A(x) = 3x^2 + 2x$. Confirm that $A'(x) = 6x + 2$, and use the antiderivative method to find the exact area in Exercise 3.

8. Let $A(x) = \frac{1}{2}x\sqrt{1 - x^2} + \frac{1}{2}\sin^{-1} x$. Then confirm that $A'(x) = \sqrt{1 - x^2}$, and use the antiderivative method to find the exact area in Exercise 4.

9. Use the antiderivative method to find the exact area between the curve $y = e^x$ and the interval $[0, 1]$.

10. Use the antiderivative method to find the exact area between the curve $y = \sin x$ and the interval $[0, \pi]$.

7.2 THE INDEFINITE INTEGRAL; INTEGRAL CURVES AND DIRECTION FIELDS

In the last section we saw that antidifferentiation plays an important role in finding exact areas. In this section we will develop some fundamental results about antidifferentiation that will ultimately lead us to systematic procedures for finding a function from its derivative.

7.2.1 DEFINITION.   A function F is called an ***antiderivative*** of a function f on a given interval I if $F'(x) = f(x)$ for all x in the interval.

For example, the function $F(x) = \frac{1}{3}x^3$ is an antiderivative of $f(x) = x^2$ on the interval $(-\infty, +\infty)$ because for each x in this interval

$$F'(x) = \frac{d}{dx}\left[\frac{1}{3}x^3\right] = x^2 = f(x)$$

However, this is not the only antiderivative of F on this interval. If we add any constant C to $\frac{1}{3}x^3$, then the function $F(x) = \frac{1}{3}x^3 + C$ is also an antiderivative of f on $(-\infty, +\infty)$, since

$$F'(x) = \frac{d}{dx}\left[\frac{1}{3}x^3 + C\right] = x^2 + 0 = f(x)$$

In general, once any single antiderivative of a function is known, other antiderivatives can be obtained by adding constants to the known antiderivative. Thus,

$$\tfrac{1}{3}x^3, \quad \tfrac{1}{3}x^3 + 2, \quad \tfrac{1}{3}x^3 - 5, \quad \tfrac{1}{3}x^3 + \sqrt{2}$$

are all antiderivatives of $f(x) = x^2$.

WARNING. Do not confuse derivatives and antiderivatives—the *derivative* of the function $f(x) = x^2$ is $f'(x) = 2x$, but the functions $F(x) = \tfrac{1}{3}x^3 + C$ are *antiderivatives* of f.

It is reasonable to ask if there are antiderivatives of a function f that cannot be obtained by adding some constant to a known antiderivative F. The answer is *no*—once a single antiderivative of f on an interval I is known, all other antiderivatives on that interval are obtainable by adding constants to that antiderivative. This is so because Theorem 6.5.3 tells us that if two functions have the same derivative on an interval, then they differ by a constant on that interval. The following theorem summarizes these observations.

7.2.2 THEOREM. *If $F(x)$ is any antiderivative of $f(x)$ on an interval I, then for any constant C the function $F(x) + C$ is also an antiderivative of $f(x)$ on that interval. Moreover, each antiderivative of $f(x)$ on the interval I can be expressed in the form $F(x) + C$ by choosing the constant C appropriately.*

THE INDEFINITE INTEGRAL

The process of finding antiderivatives is called ***antidifferentiation*** or ***integration***. Thus, if

$$\frac{d}{dx}[F(x)] = f(x)$$

then integrating (or antidifferentiating) $f(x)$ produces the antiderivatives $F(x) + C$. We denote this by writing

$$\int f(x)\,dx = F(x) + C \tag{1}$$

For example, the antiderivatives of $f(x) = x^2$ are the functions $F(x) = \tfrac{1}{3}x^3 + C$, so

$$\int x^2\,dx = \tfrac{1}{3}x^3 + C$$

The "elongated s" that appears on the left side of (1) is called an ***integral sign***[*] or an ***indefinite integral***, the function $f(x)$ is called the ***integrand***, and the constant C is called the ***constant of integration***. You should read Equation (1) as "the integral of $f(x)$ with respect to x is equal to $F(x) + C$." The adjective "indefinite" emphasizes that the integration process does not produce a *definite* function, but rather a whole set of functions.

The dx symbols in the differentiation and antidifferentiation operations

$$\frac{d}{dx}[\ \ ] \quad \text{and} \quad \int [\ \ ]\,dx$$

serve to identify the independent variable. If an independent variable other than x is used, say t, then the notation must be adjusted appropriately. Thus,

$$\frac{d}{dt}[F(t)] = f(t) \quad \text{and} \quad \int f(t)\,dt = F(t) + C$$

are equivalent statements.

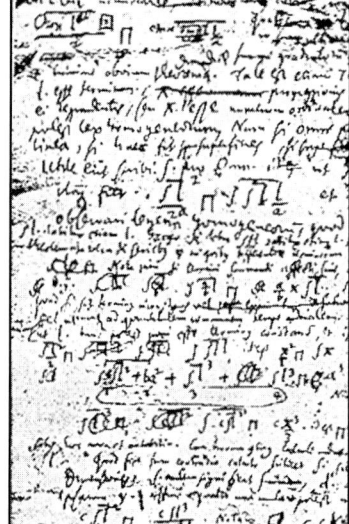

Extract from the manuscript of Leibniz dated October 29, 1675 in which the integral sign first appeared.

---

[*]This notation was devised by Leibniz. In his early papers Leibniz used the notation "omn." (an abbreviation for the Latin word "omnes") to denote integration. Then on October 29, 1675 he wrote, "It will be useful to write $\int$ for omn., thus $\int \ell$ for omn. ℓ ...." Two or three weeks later he refined the notation further and wrote $\int[\ \ ]\,dx$ rather than $\int$ alone. This notation is so useful and so powerful that its development by Leibniz must be regarded as a major milestone in the history of mathematics and science.

Example 1

DERIVATIVE FORMULA	EQUIVALENT INTEGRATION FORMULA
$\frac{d}{dx}[x^3] = 3x^2$	$\int 3x^2\,dx = x^3 + C$
$\frac{d}{dx}[\sqrt{x}] = \frac{1}{2\sqrt{x}}$	$\int \frac{1}{2\sqrt{x}}\,dx = \sqrt{x} + C$
$\frac{d}{dt}[\tan t] = \sec^2 t$	$\int \sec^2 t\,dt = \tan t + C$
$\frac{d}{du}[u^{3/2}] = \frac{3}{2}u^{1/2}$	$\int \frac{3}{2}u^{1/2}\,du = u^{3/2} + C$

For simplicity, the dx is sometimes absorbed into the integrand. For example,

$$\int 1\,dx \quad \text{can be written as} \quad \int dx$$

$$\int \frac{1}{x^2}\,dx \quad \text{can be written as} \quad \int \frac{dx}{x^2}$$

INTEGRATION FORMULAS

Integration is essentially educated guesswork—given the derivative of a function f, one tries to guess what the function f is. However, many basic integration formulas can be obtained directly from their companion differentiation formulas. Some of the most important ones are given in Table 7.2.1.

Table 7.2.1

DIFFERENTIATION FORMULA	INTEGRATION FORMULA
1. $\frac{d}{dx}[x] = 1$	$\int dx = x + C$
2. $\frac{d}{dx}\left[\frac{x^{r+1}}{r+1}\right] = x^r \ (r \neq -1)$	$\int x^r\,dx = \left[\frac{x^{r+1}}{r+1}\right] + C \ (r \neq -1)$
3. $\frac{d}{dx}[\sin x] = \cos x$	$\int \cos x\,dx = \sin x + C$
4. $\frac{d}{dx}[-\cos x] = \sin x$	$\int \sin x\,dx = -\cos x + C$
5. $\frac{d}{dx}[\tan x] = \sec^2 x$	$\int \sec^2 x\,dx = \tan x + C$
6. $\frac{d}{dx}[-\cot x] = \csc^2 x$	$\int \csc^2 x\,dx = -\cot x + C$
7. $\frac{d}{dx}[\sec x] = \sec x \tan x$	$\int \sec x \tan x\,dx = \sec x + C$
8. $\frac{d}{dx}[-\csc x] = \csc x \cot x$	$\int \csc x \cot x\,dx = -\csc x + C$
9. $\frac{d}{dx}[e^x] = e^x$	$\int e^x\,dx = e^x + C$
10. $\frac{d}{dx}\left[\frac{b^x}{\ln b}\right] = b^x$	$\int b^x\,dx = \frac{b^x}{\ln b} + C$
11. $\frac{d}{dx}[\ln|x|] = \frac{1}{x}$	$\int \frac{dx}{x} = \ln|x| + C$

Example 2

The second integration formula in this table will be easy to remember if you express it in words: *to integrate a power of x (other than -1), add 1 to the power and divide by the new power.* Here are some examples:

$$\int x^2\, dx = \frac{x^3}{3} + C \qquad \boxed{r = 2}$$

$$\int x^3\, dx = \frac{x^4}{4} + C \qquad \boxed{r = 3}$$

$$\int \frac{1}{x^5}\, dx = \int x^{-5}\, dx = \frac{x^{-5+1}}{-5+1} + C = -\frac{1}{4x^4} + C \qquad \boxed{r = -5}$$

$$\int \sqrt{x}\, dx = \int x^{\frac{1}{2}}\, dx = \frac{x^{\frac{1}{2}+1}}{\frac{1}{2}+1} + C = \frac{2}{3}x^{\frac{3}{2}} + C = \frac{2}{3}(\sqrt{x})^3 + C \qquad \boxed{r = \frac{1}{2}}$$

$$\int x^{-1}\, dx = \int \frac{dx}{x} = \ln|x| + C \qquad \blacktriangleleft$$

PROPERTIES OF THE INDEFINITE INTEGRAL

If we differentiate an antiderivative of $f(x)$, we obtain $f(x)$ back again. Thus,

$$\frac{d}{dx}\left[\int f(x)\, dx\right] = f(x) \tag{2}$$

This result is helpful for proving the following basic properties of antiderivatives.

7.2.3 THEOREM.

(a) *A constant factor can be moved through an integral sign; that is,*

$$\int cf(x)\, dx = c\int f(x)\, dx$$

(b) *An antiderivative of a sum is the sum of the antiderivatives; that is,*

$$\int [f(x) + g(x)]\, dx = \int f(x)\, dx + \int g(x)\, dx$$

(c) *An antiderivative of a difference is the difference of the antiderivatives; that is,*

$$\int [f(x) - g(x)]\, dx = \int f(x)\, dx - \int g(x)\, dx$$

Proof. In each part we must show that the expression on the right side of the equation is an antiderivative of the integrand on the left side of the equation. This can be done using (2) as follows:

$$\frac{d}{dx}\left[c\int f(x)\, dx\right] = c\frac{d}{dx}\left[\int f(x)\, dx\right] = cf(x)$$

$$\frac{d}{dx}\left[\int f(x)\, dx + \int g(x)\, dx\right] = \frac{d}{dx}\left[\int f(x)\, dx\right] + \frac{d}{dx}\left[\int g(x)\, dx\right]$$
$$= f(x) + g(x)$$

$$\frac{d}{dx}\left[\int f(x)\, dx - \int g(x)\, dx\right] = \frac{d}{dx}\left[\int f(x)\, dx\right] - \frac{d}{dx}\left[\int g(x)\, dx\right]$$
$$= f(x) - g(x) \qquad \blacksquare$$

When applying Theorem 7.2.3, it is best to put in the constant of integration at the *very end* of the computations to obtain the simplest form of the answer. This is illustrated in the following example.

Example 3

Evaluate

(a) $\int 4\cos x\,dx$ (b) $\int (x + x^2)\,dx$

Solution (a).

$$\int 4\cos x\,dx = 4\int \cos x\,dx = 4(\sin x + C) = 4\sin x + 4C$$

| Theorem 7.2.3(a) | Table 7.2.1 |

Since C is an arbitrary constant, so is $4C$. However, this latter form is unnecessarily complicated and can be avoided by deferring the insertion of the constant until the end of the computations; this procedure yields

$$\int 4\cos x\,dx = 4\int \cos x\,dx = 4\sin x + C$$

Solution (b).

$$\int (x + x^2)\,dx = \int x\,dx + \int x^2\,dx = \frac{x^2}{2} + \frac{x^3}{3} + C$$ ◄

| Theorem 7.2.3(b) | Table 7.2.1 |

Parts (b) and (c) of Theorem 7.2.3 can be extended to more than two functions, which in combination with part (a) results in the following general formula:

$$\int [c_1 f_1(x) + c_2 f_2(x) + \cdots + c_n f_n(x)]\,dx$$

$$= c_1 \int f_1(x)\,dx + c_2 \int f_2(x)\,dx + \cdots + c_n \int f_n(x)\,dx \qquad (3)$$

Example 4

$$\int (3x^6 - 2x^2 + 7x + 1)\,dx = 3\int x^6\,dx - 2\int x^2\,dx + 7\int x\,dx + \int 1\,dx$$

$$= \frac{3x^7}{7} - \frac{2x^3}{3} + \frac{7x^2}{2} + x + C$$ ◄

Sometimes it is useful to rewrite an integrand in a different form before performing the integration.

Example 5

Evaluate

(a) $\int \dfrac{\cos x}{\sin^2 x}\,dx$ (b) $\int \dfrac{t^2 - 2t^4}{t^4}\,dt$

Solution (a).

$$\int \frac{\cos x}{\sin^2 x}\,dx = \int \frac{1}{\sin x}\frac{\cos x}{\sin x}\,dx = \int \csc x \cot x\,dx = -\csc x + C$$

| Formula 8 in Table 7.2.1 |

Solution (b).

$$\int \frac{t^2 - 2t^4}{t^4} \, dt = \int \left(\frac{1}{t^2} - 2 \right) dt = \int (t^{-2} - 2) \, dt$$

$$= \frac{t^{-1}}{-1} - 2t + C = -\frac{1}{t} - 2t + C \qquad \blacktriangleleft$$

INTEGRAL CURVES

Graphs of antiderivatives of a function f are called *integral curves* of f. We know from Theorem 7.2.2 that if $y = F(x)$ is any integral curve of $f(x)$, then all other integral curves are vertical translations of this curve, since they have equations of the form $y = F(x) + C$. For example, $y = \frac{1}{3}x^3$ is one integral curve for $f(x) = x^2$, so all the other integral curves have equations of the form $y = \frac{1}{3}x^3 + C$; conversely, the graph of any equation of this form is an integral curve (Figure 7.2.1).

In many problems one is interested in finding a function whose derivative satisfies specified conditions. The following example illustrates a geometric problem of this type.

Example 6

Suppose that a point moves along some unknown curve $y = f(x)$ in the xy-plane in such a way that at each point (x, y) on the curve, the tangent line has slope x^2. Find an equation for the curve given that it passes through the point $(2, 1)$.

Solution. We know that $dy/dx = x^2$, so

$$y = \int x^2 \, dx = \frac{1}{3}x^3 + C$$

Since the curve passes through $(2, 1)$, a specific value for C can be found by using the fact that $y = 1$ if $x = 2$. Substituting these values in the above equation yields

$$1 = \frac{1}{3}(2^3) + C \quad \text{or} \quad C = -\frac{5}{3}$$

so the curve is $y = \frac{1}{3}x^3 - \frac{5}{3}$. $\quad \blacktriangleleft$

Observe that in this example the requirement that the unknown curve pass through the point $(2, 1)$ enabled us to determine a specific value for the constant of integration, thereby isolating the single integral curve $y = \frac{1}{3}x^3 - \frac{5}{3}$ from the family $y = \frac{1}{3}x^3 + C$ (Figure 7.2.2).

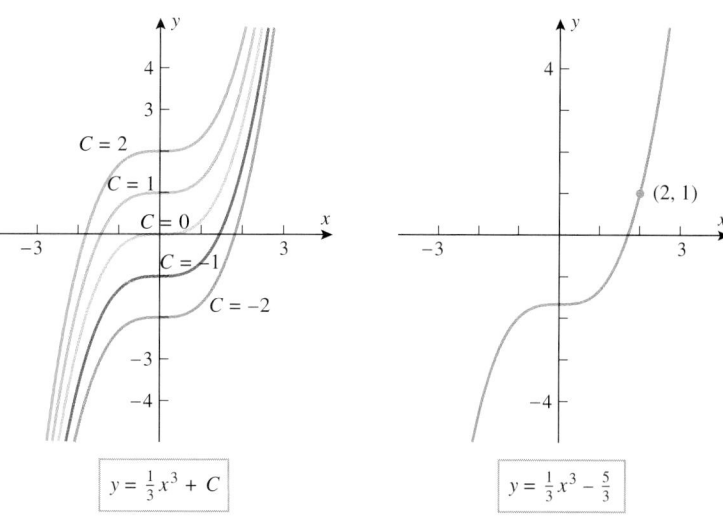

Figure 7.2.1   Figure 7.2.2

We will now consider another way of looking at integration that will be useful in our later work. Suppose that $f(x)$ is a known function and we are interested in finding a function $F(x)$ such that $y = F(x)$ satisfies the equation

$$\frac{dy}{dx} = f(x) \tag{4}$$

The solutions of this equation are the antiderivatives of $f(x)$, and we know that these can be obtained by integrating $f(x)$. For example, the solutions of the equation

$$\frac{dy}{dx} = x^2 \tag{5}$$

are

$$y = \int x^2 \, dx = \frac{x^3}{3} + C$$

Equation (4) is called a ***differential equation*** because it involves a derivative of an unknown function. Differential equations are different from the kinds of equations we have encountered so far in that the unknown is a *function* and not a *number* as in an equation such as $x^2 + 5x - 6 = 0$.

Sometimes we will not be interested in finding all of the solutions of (4), but rather we will want only the solution whose integral curve passes through a specified point (x_0, y_0). For example, in Example 6 we solved (5) for the integral curve that passed through the point $(2, 1)$.

For simplicity, it is common in the study of differential equations to denote a solution of $dy/dx = f(x)$ as $y(x)$ rather than $F(x)$, as earlier. With this notation, the problem of finding a function $y(x)$ whose derivative is $f(x)$ and whose integral curve passes through the point (x_0, y_0) is expressed as

$$\frac{dy}{dx} = f(x), \quad y(x_0) = y_0 \tag{6}$$

For reasons that will be explained later, this is called an ***initial-value problem***, and the requirement that $y(x_0) = y_0$ is called the ***initial condition*** for the problem.

Example 7

Solve the initial-value problem

$$\frac{dy}{dx} = \cos x, \quad y(0) = 1$$

Solution. The solution of the differential equation is

$$y = \int \cos x \, dx = \sin x + C \tag{7}$$

The initial condition $y(0) = 1$ implies that $y = 1$ if $x = 0$; substituting these values in (7) yields

$$1 = \sin(0) + C \quad \text{or} \quad C = 1$$

Thus, the solution of the initial-value problem is $y = \sin x + 1$. ◄

If we interpret dy/dx as the slope of a tangent line, then at a point (x, y) on an integral curve of the equation $dy/dx = f(x)$, the slope of the tangent line is $f(x)$. What is interesting about this is that the slopes of the tangent lines to the integral curves can be obtained without actually solving the differential equation. For example, if

$$\frac{dy}{dx} = \sqrt{x^2 + 1}$$

then we know without solving the equation that at the point where $x = 1$ the tangent line to an integral curve has slope $\sqrt{1^2 + 1} = \sqrt{2}$; and more generally, at a point where $x = a$, the tangent line to an integral curve has slope $\sqrt{a^2 + 1}$.

A geometric description of the integral curves of a differential equation $dy/dx = f(x)$ can be obtained by choosing a rectangular grid of points in the xy-plane, calculating the slopes of the tangent lines to the integral curves at the gridpoints, and drawing small portions of the tangent lines at those points. The resulting picture, which is called a ***direction field*** or ***slope field*** for the equation, shows the "direction" of the integral curves at the gridpoints. With sufficiently many gridpoints it is often possible to visualize the integral curves themselves; for example, Figure 7.2.3a shows a direction field for the differential equation $dy/dx = x^2$, and Figure 7.2.3b shows that same field with the integral curves imposed on it—the more gridpoints that are used, the more completely the direction field reveals the shape of the integral curves. However, the amount of computation can be considerable, so computers are usually used when direction fields with many gridpoints are needed.

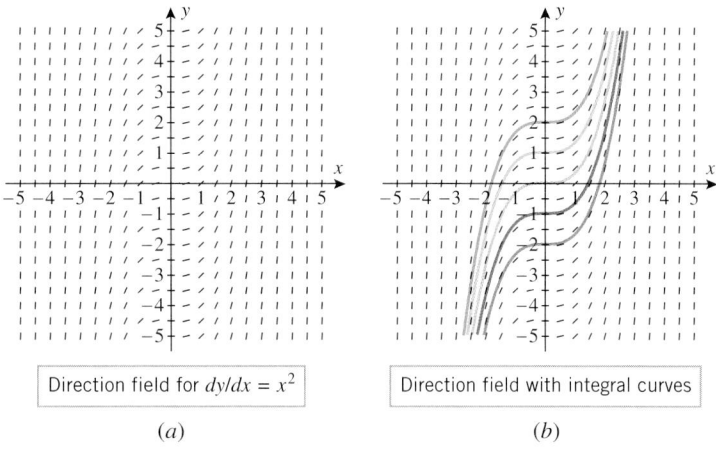

Direction field for $dy/dx = x^2$

(a)

Direction field with integral curves

(b)

Figure 7.2.3

EXERCISE SET 7.2   ⊠ Graphing Calculator   [c] CAS

1. In each part, confirm that the formula is correct, and state a corresponding integration formula.

   (a) $\dfrac{d}{dx}[\sqrt{1 + x^2}] = \dfrac{x}{\sqrt{1 + x^2}}$

   (b) $\dfrac{d}{dx}[xe^x] = (x + 1)e^x$

2. In each part, confirm that the stated formula is correct by differentiating.

   (a) $\displaystyle\int x \sin x \, dx = \sin x - x \cos x + C$

   (b) $\displaystyle\int \dfrac{dx}{(1 - x^2)^{3/2}} = \dfrac{x}{\sqrt{1 - x^2}} + C$

In Exercises 3–6, find the derivative and state a corresponding integration formula.

3. $\dfrac{d}{dx}[\sqrt{x^3 + 5}]$

4. $\dfrac{d}{dx}\left[\dfrac{x}{x^2 + 3}\right]$

5. $\dfrac{d}{dx}[\sin(2\sqrt{x})]$

6. $\dfrac{d}{dx}[\sin x - x \cos x]$

In Exercises 7 and 8, evaluate the integral by rewriting the integrand appropriately, if required, and then applying Formula 2 in Table 7.2.1.

7. (a) $\displaystyle\int x^8 \, dx$ (b) $\displaystyle\int x^{5/7} \, dx$ (c) $\displaystyle\int x^3 \sqrt{x} \, dx$

8. (a) $\displaystyle\int \sqrt[3]{x^2} \, dx$ (b) $\displaystyle\int \dfrac{1}{x^6} \, dx$ (c) $\displaystyle\int x^{-7/8} \, dx$

In Exercises 9–12, evaluate the integral by applying Theorem 7.2.3 and Formula 2 in Table 7.2.1 appropriately.

9. (a) $\displaystyle\int \dfrac{1}{2x^3} \, dx$ (b) $\displaystyle\int (u^3 - 2u + 7) \, du$

10. $\displaystyle\int (x^{2/3} - 4x^{-1/5} + 4) \, dx$

11. $\int (x^{-3} + \sqrt{x} - 3x^{1/4} + x^2)\, dx$

12. $\int \left(\dfrac{7}{y^{3/4}} - \sqrt[3]{y} + 4\sqrt{y} \right) dy$

> In Exercises 13–30, evaluate the integral, and check your answer by differentiating.

13. $\int x(1 + x^3)\, dx$

14. $\int (2 + y^2)^2\, dy$

15. $\int x^{1/3}(2 - x)^2\, dx$

16. $\int (1 + x^2)(2 - x)\, dx$

17. $\int \dfrac{x^5 + 2x^2 - 1}{x^4}\, dx$

18. $\int \dfrac{1 - 2t^3}{t^3}\, dt$

19. $\int \left[\dfrac{2}{x} + 3e^x \right] dx$

20. $\int \left[\dfrac{1}{2t} - \sqrt{2}e^t \right] dt$

21. $\int [4\sin x + 2\cos x]\, dx$

22. $\int [4\sec^2 x + \csc x \cot x]\, dx$

23. $\int \sec x(\sec x + \tan x)\, dx$ **24.** $\int \sec x(\tan x + \cos x)\, dx$

25. $\int \left[\dfrac{1}{\theta} - 2e^{\theta} - \csc^2 \theta \right] d\theta$ **26.** $\int \dfrac{dy}{\csc y}$

27. $\int \dfrac{\sin x}{\cos^2 x}\, dx$

28. $\int \left[\phi + \dfrac{2}{\sin^2 \phi} \right] d\phi$

29. $\int [1 + \sin^2 \theta \csc \theta]\, d\theta$ **30.** $\int \dfrac{\sin 2x}{\cos x}\, dx$

31. Evaluate the integral

$$\int \dfrac{1}{1 + \sin x}\, dx$$

by multiplying the numerator and denominator by an appropriate expression.

c **32.** For each of the integrals you evaluated in Exercises 13–31, use a CAS to check your answer. If the answer produced by the CAS does not match yours, show that the two answers are equivalent.

33. (a) Graph some representative integral curves of $f(x) = x$.
(b) Find an equation for the integral curve that passes through the point $(4, 7)$.

34. (a) Graph some representative integral curves of the function $f(x) = e^x/2$.
(b) Find an equation for the integral curve that passes through the point $(0, 1)$.

35. Use a graphing utility to generate some representative integral curves of the function $f(x) = 5x^4 - \sec^2 x$ over the interval $(-\pi/2, \pi/2)$.

36. Use a graphing utility to generate some representative integral curves of $f(x) = (x - 1)/x$ over the interval $(0, 5)$.

37. Suppose that a point moves along a curve $y = f(x)$ in the xy-plane in such a way that at each point (x, y) on the curve

the tangent line has slope $-\sin x$. Find an equation for the curve, given that it passes through the point $(0, 2)$.

38. Suppose that a point moves along a curve $y = f(x)$ in the xy-plane in such a way that at each point (x, y) on the curve the tangent line has slope $(x + 1)^2$. Find an equation for the curve, given that it passes through the point $(-2, 8)$.

> In Exercises 39 and 40, solve the initial-value problems.

39. (a) $\dfrac{dy}{dx} = \sqrt[3]{x}, \; y(1) = 2$ (b) $\dfrac{dy}{dt} = \dfrac{1}{t}, \; y(-1) = 5$

(c) $\dfrac{dy}{dx} = \dfrac{x + 1}{\sqrt{x}}, \; y(1) = 0$

40. (a) $\dfrac{dy}{dx} = \dfrac{1}{(2x)^3}, \; y(1) = 0$

(b) $\dfrac{dy}{dt} = \sec^2 t - \sin t, \; y\left(\dfrac{\pi}{4}\right) = 1$

(c) $\dfrac{dy}{dx} = x^2\sqrt{x^3}, \; y(0) = 0$

41. Find the general form of a function whose second derivative is $\sqrt{x}$. [*Hint:* Solve the equation $f''(x) = \sqrt{x}$ for $f(x)$ by integrating both sides twice.]

42. Find a function f such that $f''(x) = x + \cos x$ and such that $f(0) = 1$ and $f'(0) = 2$. [*Hint:* Integrate both sides of the equation twice.]

> In Exercises 43–45, find an equation of the curve that satisfies the given conditions.

43. At each point (x, y) on the curve the slope is $2x + 1$; the curve passes through the point $(-3, 0)$.

44. At each point (x, y) on the curve the slope equals the square of the distance between the point and the y-axis; the point $(-1, 2)$ is on the curve.

45. At each point (x, y) on the curve, y satisfies the condition $d^2y/dx^2 = 6x$; the line $y = 5 - 3x$ is tangent to the curve at the point where $x = 1$.

46. Suppose that a uniform metal rod 50 cm long is insulated laterally, and the temperatures at the exposed ends are maintained at $25°C$ and $85°C$, respectively. Assume that an x-axis is chosen as in the accompanying figure and that the temperature $T(x)$ at each point x satisfies the equation

$$\dfrac{d^2T}{dx^2} = 0$$

Find $T(x)$ for $0 \le x \le 50$.

Figure Ex-46

47. (a) Show that

$$F(x) = \tfrac{1}{6}(3x+4)^2 \quad \text{and} \quad G(x) = \tfrac{3}{2}x^2 + 4x$$

differ by a constant by showing that they are antiderivatives of the same function.

(b) Find the constant C such that $F(x) - G(x) = C$ by evaluating $F(x)$ and $G(x)$ at some point x_0.

(c) Check your answer in part (b) by simplifying the expression $F(x) - G(x)$ algebraically.

48. Follow the directions of Exercise 47 with

$$F(x) = \frac{x^2}{x^2+5} \quad \text{and} \quad G(x) = -\frac{5}{x^2+5}$$

In Exercises 49 and 50, use a trigonometric identity to help evaluate the integral.

49. $\displaystyle\int \tan^2 x \, dx$ **50.** $\displaystyle\int \cot^2 x \, dx$

51. Use the identities $\cos 2\theta = 1 - 2\sin^2\theta = 2\cos^2\theta - 1$ to help evaluate the integrals

(a) $\displaystyle\int \sin^2(x/2)\, dx$ (b) $\displaystyle\int \cos^2(x/2)\, dx$

52. Let F and G be the functions defined piecewise by

$$F(x) = \begin{cases} x, & x > 0 \\ -x, & x < 0 \end{cases} \quad \text{and} \quad G(x) = \begin{cases} x+2, & x > 0 \\ -x+3, & x < 0 \end{cases}$$

(a) Show that F and G have the same derivative.

(b) Show that $G(x) \neq F(x) + C$ for any constant C.

(c) Do parts (a) and (b) violate Theorem 7.2.2? Explain.

53. The speed of sound in air at $0°C$ (or 273 K on the Kelvin scale) is 1087 ft/s, but the speed v increases as the temperature T rises. Experimentation has shown that the rate of change of v with respect to T is

$$\frac{dv}{dT} = \frac{1087}{2\sqrt{273}} T^{-1/2}$$

where v is in feet per second and T is in kelvins (K). Find a formula that expresses v as a function of T.

7.3 INTEGRATION BY SUBSTITUTION

*In this section we will study a technique, called **substitution**, that can often be used to transform complicated integration problems into simpler ones.*

u-SUBSTITUTION

The method of substitution can be motivated by examining the chain rule from the viewpoint of antidifferentiation. For this purpose, suppose that F is an antiderivative of f and that g is a differentiable function. The chain rule implies that the derivative of $F(g(x))$ can be expressed as

$$\frac{d}{dx}[F(g(x))] = F'(g(x))g'(x)$$

which we can write in integral form as

$$\int F'(g(x))g'(x)\, dx = F(g(x)) + C \tag{1}$$

or since F is an antiderivative of f,

$$\int f(g(x))g'(x)\, dx = F(g(x)) + C \tag{2}$$

For our purposes it will be useful to let $u = g(x)$ and to write $du/dx = g'(x)$ in the differential form $du = g'(x)\, dx$. With this notation (1) can be expressed as

$$\int f(u)\, du = F(u) + C \tag{3}$$

The process of evaluating an integral of form (2) by converting it into form (3) with the substitution

$$u = g(x) \quad \text{and} \quad du = g'(x)\, dx$$

is called the **method of u-substitution**. The following example illustrates how the method works.

Example 1

Evaluate $\int (x^2 + 1)^{50} \cdot 2x\, dx$.

Solution. If we let $u = x^2 + 1$, then $du/dx = 2x$, which implies that $du = 2x\, dx$. Thus, the given integral can be written as

$$\int (x^2 + 1)^{50} \cdot 2x\, dx = \int u^{50}\, du = \frac{u^{51}}{51} + C = \frac{(x^2 + 1)^{51}}{51} + C \qquad \blacktriangleleft$$

It is important to realize that in the method of u-substitution you have control over the choice of u, but once you make that choice you have no control over the resulting expression for du. Thus, in the last example we *chose* $u = x^2 + 1$ but $du = 2x\, dx$ was *computed*. Fortunately, our choice of u, combined with the computed du, worked out perfectly to produce an integral involving u that was easy to evaluate. However, in general, the method of u-substitution will fail if the chosen u and the computed du do not produce an integrand in which no expressions involving x remain, or if you cannot evaluate the resulting integral. Thus, for example, the substitution $u = x^2 + 1$, $du = 2x\, dx$ will not work for the integral

$$\int (x^2 + 1)^{50} \cdot 2x \cos x\, dx$$

because this substitution results in the integral

$$\int u^{50} \cos x\, du$$

which still contains an expression involving x.

In general, there are no hard and fast rules for choosing u, and in some problems no choice of u will work. In such cases other methods need to be used, some of which will be discussed later. Making appropriate choices for u will come with experience, but you may find the following guidelines, combined with a mastery of the basic integrals in Table 7.2.1, helpful.

Integration by Substitution

Step 1. Make a choice for u, say $u = g(x)$.

Step 2. Compute $du/dx = g'(x)$.

Step 3. Make the substitution $u = g(x)$, $du = g'(x)\, dx$.

At this stage, the *entire* integral must be in terms of u; no x's should remain. If this is not the case, try a different choice of u.

Step 4. Evaluate the resulting integral, if possible.

Step 5. Replace u by $g(x)$, so that the final answer is in terms of x.

Example 2

The easiest substitutions occur when the integrand is the derivative of a known function, except for a constant added to or subtracted from the independent variable. For example,

$$\int \sin(x + 9)\, dx = \int \sin u\, du = -\cos u + C = -\cos(x + 9) + C$$

$$\boxed{\begin{array}{l} u = x + 9 \\ du = 1 \cdot dx = dx \end{array}}$$

$$\int (x - 8)^{23}\, dx = \int u^{23}\, du = \frac{u^{24}}{24} + C = \frac{(x - 8)^{24}}{24} + C \qquad \blacktriangleleft$$

$$\boxed{\begin{array}{l} u = x - 8 \\ du = 1 \cdot dx = dx \end{array}}$$

Another easy *u*-substitution occurs when the integrand is the derivative of a known function, except for a constant that multiplies or divides the independent variable. The following example illustrates two ways to evaluate such integrals.

Example 3

Evaluate $\int \cos 5x\, dx$.

Solution.

$$\int \cos 5x\, dx = \int (\cos u)\cdot\frac{1}{5}du = \frac{1}{5}\int \cos u\, du = \frac{1}{5}\sin u + C = \frac{1}{5}\sin 5x + C$$

$u = 5x$
$du = 5\,dx \text{ or } dx = \frac{1}{5}du$

Alternative Solution. There is a variation of the preceding method that some people prefer. The substitution $u = 5x$ requires $du = 5\,dx$. If there were a factor of 5 in the integrand, then we could group the 5 and dx together to form the du required by the substitution. Since there is no factor of 5, we will insert one and compensate by putting a factor of $\frac{1}{5}$ in front of the integral. The computations are as follows:

$$\int \cos 5x\, dx = \frac{1}{5}\int \cos 5x \cdot 5\, dx = \frac{1}{5}\int \cos u\, du = \frac{1}{5}\sin u + C = \frac{1}{5}\sin 5x + C \quad \blacktriangleleft$$

$u = 5x$
$du = 5\,dx$

Example 4

Evaluate $\int \sin^2 x \cos x\, dx$.

Solution. If we let $u = \sin x$, then

$$\frac{du}{dx} = \cos x, \quad \text{so} \quad du = \cos x\, dx$$

Thus,

$$\int \sin^2 x \cos x\, dx = \int u^2\, du = \frac{u^3}{3} + C = \frac{\sin^3 x}{3} + C \quad \blacktriangleleft$$

Example 5

Evaluate $\int \frac{e^{\sqrt{x}}}{\sqrt{x}}\, dx$.

Solution. If we let $u = \sqrt{x}$, then

$$\frac{du}{dx} = \frac{1}{2\sqrt{x}}, \quad \text{so} \quad du = \frac{1}{2\sqrt{x}}\, dx \quad \text{or} \quad 2\,du = \frac{1}{\sqrt{x}}\, dx$$

Thus,

$$\int \frac{e^{\sqrt{x}}}{\sqrt{x}}\, dx = \int 2e^u\, du = 2\int e^u\, du = 2e^u + C = 2e^{\sqrt{x}} + C \quad \blacktriangleleft$$

Example 6

$$\int \frac{dx}{\left(\frac{1}{3}x - 8\right)^5} = \int \frac{3\,du}{u^5} = 3\int u^{-5}\, du = -\frac{3}{4}u^{-4} + C = -\frac{3}{4}\left(\frac{1}{3}x - 8\right)^{-4} + C \quad \blacktriangleleft$$

$u = \frac{1}{3}x - 8$
$du = \frac{1}{3}dx \text{ or } dx = 3\,du$

Example 7

With the help of Theorem 7.2.3, a complicated integral can sometimes be computed by expressing it as a sum of simpler integrals. For example,

$$\int \left(\frac{1}{x} + \sec^2 \pi x \right) dx = \int \frac{dx}{x} + \int \sec^2 \pi x \, dx = \ln |x| + \int \sec^2 \pi x \, dx$$

$$= \ln |x| + \frac{1}{\pi} \int \sec^2 u \, du$$

$$\boxed{\begin{array}{c} u = \pi x \\ du = \pi \, dx \text{ or } dx = \frac{1}{\pi} \, du \end{array}}$$

$$= \ln |x| + \frac{1}{\pi} \tan u + C = \ln |x| + \frac{1}{\pi} \tan \pi x + C \qquad \blacktriangleleft$$

Example 8

Evaluate $\displaystyle\int t^4 \sqrt[3]{3 - 5t^5} \, dt$.

Solution. After some possible false starts most readers would eventually hit on the following substitution:

$$\int t^4 \sqrt[3]{3 - 5t^5} \, dt = -\frac{1}{25} \int \sqrt[3]{u} \, du = -\frac{1}{25} \int u^{1/3} \, du$$

$$\boxed{\begin{array}{c} u = 3 - 5t^5 \\ du = -25t^4 \, dt \text{ or } -\frac{1}{25} \, du = t^4 \, dt \end{array}}$$

$$= -\frac{1}{25} \frac{u^{4/3}}{4/3} + C = -\frac{3}{100} \left(3 - 5t^5 \right)^{4/3} + C \qquad \blacktriangleleft$$

Example 9

Evaluate $\displaystyle\int x^2 \sqrt{x - 1} \, dx$.

Solution. Let

$$u = x - 1 \quad \text{so that} \quad du = dx \qquad (4)$$

From the first equality in (4)

$$x^2 = (u + 1)^2 = u^2 + 2u + 1$$

so that

$$\int x^2 \sqrt{x - 1} \, dx = \int (u^2 + 2u + 1) \sqrt{u} \, du = \int (u^{5/2} + 2u^{3/2} + u^{1/2}) \, du$$

$$= \tfrac{2}{7} u^{7/2} + \tfrac{4}{5} u^{5/2} + \tfrac{2}{3} u^{3/2} + C$$

$$= \tfrac{2}{7} (x - 1)^{7/2} + \tfrac{4}{5} (x - 1)^{5/2} + \tfrac{2}{3} (x - 1)^{3/2} + C \qquad \blacktriangleleft$$

REMARK. Not every function can be integrated in terms of familiar functions using *u*-substitutions. For example, you will not find any *u*-substitution that will integrate

$$\int \sin(x^2) \, dx$$

in terms of functions encountered thus far in this text (try).

The advent of computer algebra systems has made it possible to evaluate many kinds of integrals that would be laborious to evaluate by hand. For example, *Mathematica*, *Maple*, and *Derive* all produce the following result in a matter of seconds:

$$\int \sqrt{2x - x^2}\, dx = \tfrac{1}{2}(x-1)\sqrt{2x-x^2} - \tfrac{1}{2}\sin^{-1}(1-x) + C$$

However, just as one would not want to rely on a calculator to compute $2 + 2$, so one would not want to use a CAS to integrate a simple function such as $f(x) = x^2$. Thus, even if you have a CAS, you will want to develop a reasonable level of competence in evaluating basic integrals. Moreover, the mathematical techniques that we will introduce for evaluating basic integrals are precisely the techniques that computer algebra systems use to evaluate more complicated integrals.

FOR THE READER. If you have a CAS, use it to calculate the integrals in the examples of this section. If your CAS produces a form of the answer that is different from the one in the text, then confirm algebraically that the two answers agree. Your CAS has various commands for simplifying answers. Explore the effect of using the CAS to simplify the expressions it produces for the integrals.

EXERCISE SET 7.3 ⌇ Graphing Calculator   ⓒ CAS

In Exercises 1–4, evaluate the integrals by making the indicated substitutions.

1. (a) $\displaystyle\int 2x\left(x^2+1\right)^{23} dx;\ u = x^2 + 1$

(b) $\displaystyle\int \cos^3 x \sin x\, dx;\ u = \cos x$

(c) $\displaystyle\int \frac{1}{\sqrt{x}} \sin\sqrt{x}\, dx;\ u = \sqrt{x}$

(d) $\displaystyle\int \frac{3x\, dx}{\sqrt{4x^2+5}};\ u = 4x^2 + 5$

(e) $\displaystyle\int \frac{x^2}{x^3-4}\, dx;\ u = x^3 - 4$

2. (a) $\displaystyle\int \sec^2(4x+1)\, dx;\ u = 4x + 1$

(b) $\displaystyle\int y\sqrt{1+2y^2}\, dy;\ u = 1 + 2y^2$

(c) $\displaystyle\int \sqrt{\sin \pi\theta}\, \cos \pi\theta\, d\theta;\ u = \sin \pi\theta$

(d) $\displaystyle\int (2x+7)(x^2+7x+3)^{4/5}\, dx;\ u = x^2 + 7x + 3$

(e) $\displaystyle\int \frac{e^x}{1+e^x}\, dx;\ u = 1 + e^x$

3. (a) $\displaystyle\int \cot x\, \csc^2 x\, dx;\ u = \cot x$

(b) $\displaystyle\int (1+\sin t)^9 \cos t\, dt;\ u = 1 + \sin t$

(c) $\displaystyle\int \frac{dx}{x \ln x};\ u = \ln x$

(d) $\displaystyle\int e^{-5x}\, dx;\ u = -5x$

(e) $\displaystyle\int \frac{\sin 3\theta}{1 + \cos 3\theta}\, d\theta;\ u = 1 + \cos 3\theta$

4. (a) $\displaystyle\int x^2\sqrt{1+x}\, dx;\ u = 1 + x$

(b) $\displaystyle\int [\csc(\sin x)]^2 \cos x\, dx;\ u = \sin x$

(c) $\displaystyle\int e^{\tan x} \sec^2 x\, dx;\ u = \tan x$

(d) $\displaystyle\int e^{2t}\sqrt{1+e^{2t}}\, dt;\ u = 1 + e^{2t}$

(e) $\displaystyle\int \frac{5x^4}{x^5+1}\, dx;\ u = x^5 + 1$

In Exercises 5–36, evaluate the integrals by making appropriate substitutions.

5. $\displaystyle\int e^{2x}\, dx$

6. $\displaystyle\int \frac{dx}{2x}$

7. $\displaystyle\int x\left(2 - x^2\right)^3 dx$

8. $\displaystyle\int (3x-1)^5\, dx$

9. $\displaystyle\int \cos 8x\, dx$

10. $\displaystyle\int \sin 3x\, dx$

11. $\int \sec 4x \tan 4x \, dx$

12. $\int \sec^2 5x \, dx$

13. $\int t\sqrt{7t^2 + 12} \, dt$

14. $\int \dfrac{x}{\sqrt{4 - 5x^2}} \, dx$

15. $\int \dfrac{x^2}{\sqrt{x^3 + 1}} \, dx$

16. $\int \dfrac{1}{(1 - 3x)^2} \, dx$

17. $\int \dfrac{x}{(4x^2 + 1)^3} \, dx$

18. $\int x \cos(3x^2) \, dx$

19. $\int e^{\sin x} \cos x \, dx$

20. $\int x^3 e^{x^4} \, dx$

21. $\int x^2 e^{-2x^3} \, dx$

22. $\int \dfrac{e^x + e^{-x}}{e^x - e^{-x}} \, dx$

23. $\int \dfrac{\sin(5/x)}{x^2} \, dx$

24. $\int \dfrac{\sec^2(\sqrt{x})}{\sqrt{x}} \, dx$

25. $\int x^2 \sec^2(x^3) \, dx$

26. $\int \cos^3 2t \sin 2t \, dt$

27. $\int \dfrac{dx}{e^x}$

28. $\int \sqrt{e^x} \, dx$

29. $\int \sin^5 3t \cos 3t \, dt$

30. $\int \dfrac{\sin 2\theta}{(5 + \cos 2\theta)^3} \, d\theta$

31. $\int \cos 4\theta \sqrt{2 - \sin 4\theta} \, d\theta$

32. $\int \tan^3 5x \sec^2 5x \, dx$

33. $\int \sec^3 2x \tan 2x \, dx$

34. $\int [\sin(\sin \theta)] \cos \theta \, d\theta$

35. $\int \dfrac{e^{\sqrt{y}}}{\sqrt{y}} \, dy$

36. $\int \dfrac{dy}{\sqrt{y} e^{\sqrt{y}}}$

c **37.** For each of the integrals you evaluated in Exercises 5–36, use a CAS to check your answer. If the answer produced by the CAS does not match your own, show that the two answers are equivalent. [*Suggestion:* You may be able to obtain a match by applying the CAS "simplify" commands to the answer.]

In Exercises 38 and 39, evaluate the integrals assuming that n is a positive integer and $b \neq 0$.

38. $\int \sqrt[n]{a + bx} \, dx \quad (b \neq 0)$

39. $\int \sin^n(a + bx) \cos(a + bx) \, dx$

c **40.** Use a CAS to check the answers you obtained in Exercises 38 and 39. If the answer produced by the CAS does not match yours, show that the two answers are equivalent. [*Suggestion: Mathematica* users may find it helpful to apply the Simplify command to the answer.]

In Exercises 41 and 42, evaluate the integrals by making the indicated substitutions.

41. $\int x\sqrt{x - 3} \, dx; \ u = x - 3$

42. $\int \dfrac{y \, dy}{\sqrt{y + 1}}; \ u = y + 1$

The integrals in Exercises 43–48 are a little trickier than those you have encountered thus far. To evaluate these integrals you will have to apply a trigonometric identity or modify the form of the integrand algebraically before making a substitution.

43. $\int \tan^2 3\theta \, d\theta$

44. $\int \sin^3 2\theta \, d\theta$

45. $\int \dfrac{t + 1}{t} \, dt$

46. $\int e^{2 \ln x} \, dx$

47. $\int [\ln(e^x) + \ln(e^{-x})] \, dx$

48. $\int \cot x \, dx$

49. (a) Evaluate the integral $\int \sin x \cos x \, dx$ by two methods: first by letting $u = \sin x$, then by letting $u = \cos x$.

(b) Explain why the two apparently different answers obtained in part (a) are really equivalent.

50. (a) Evaluate $\int (5x - 1)^2 \, dx$ by two methods: first square and integrate, then let $u = 5x - 1$.

(b) Explain why the two apparently different answers obtained in part (a) are really equivalent.

In Exercises 51 and 52, solve the initial-value problems.

51. $\dfrac{dy}{dx} = \sqrt{3x + 1}; \ y(1) = 5$

52. $\dfrac{dy}{dx} = 6 - 5 \sin 2x; \ y(0) = 3$

53. Find a function f such that the slope of the tangent line at a point (x, y) on the curve $y = f(x)$ is $\sqrt{3x + 1}$, and the curve passes through the point $(0, 1)$.

54. Use a graphing utility to generate some typical integral curves of $f(x) = x/(x^2 + 1)$ over the interval $(-5, 5)$.

55. Suppose that a population p of frogs is estimated at the start of 1995 to be 100,000, and the growth model for the population assumes that the rate of growth (in thousands) after t years will be $p'(t) = (4 + 0.15t)^{3/2}$. Estimate the projected population at the start of the year 2000.

7.4 SIGMA NOTATION

In this section we will digress briefly from the main theme of this chapter to introduce a notation that can be used to write lengthy sums in a compact form. This material will be needed in many of the later chapters.

SIGMA NOTATION

The notation we will discuss in this section is called **sigma notation** or **summation notation** because it uses the uppercase Greek letter Σ (sigma) to denote various kinds of sums. To illustrate how this notation works, consider the sum

$$1^2 + 2^2 + 3^2 + 4^2 + 5^2$$

in which each term is of the form k^2, where k is one of the integers from 1 to 5. In sigma notation this sum can be written as

$$\sum_{k=1}^{5} k^2$$

which is read "the summation of k^2, where k runs from 1 to 5." The notation tells us to form the sum of the terms that result when we substitute successive integers for k in the expression k^2, starting with $k = 1$ and ending with $k = 5$.

More generally, if $f(k)$ is a function of k, and if m and n are integers such that $m \leq n$, then

$$\sum_{k=m}^{n} f(k) \tag{1}$$

denotes the sum of the terms that result when we substitute successive integers for k, starting with $k = m$ and ending with $k = n$ (Figure 7.4.1).

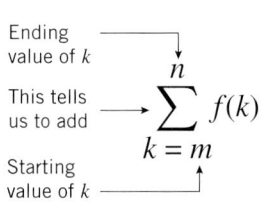

Ending value of k — n

This tells us to add — $\sum f(k)$ — $k = m$

Starting value of k

Figure 7.4.1

Example 1

$$\sum_{k=4}^{8} k^3 = 4^3 + 5^3 + 6^3 + 7^3 + 8^3$$

$$\sum_{k=1}^{5} 2k = 2 \cdot 1 + 2 \cdot 2 + 2 \cdot 3 + 2 \cdot 4 + 2 \cdot 5 = 2 + 4 + 6 + 8 + 10$$

$$\sum_{k=0}^{5} (2k + 1) = 1 + 3 + 5 + 7 + 9 + 11$$

$$\sum_{k=0}^{5} (-1)^k (2k + 1) = 1 - 3 + 5 - 7 + 9 - 11$$

$$\sum_{k=-3}^{1} k^3 = (-3)^3 + (-2)^3 + (-1)^3 + 0^3 + 1^3 = -27 - 8 - 1 + 0 + 1$$

$$\sum_{k=1}^{3} k \sin\left(\frac{k\pi}{5}\right) = \sin\frac{\pi}{5} + 2\sin\frac{2\pi}{5} + 3\sin\frac{3\pi}{5} \quad \blacktriangleleft$$

The numbers m and n in (1) are called, respectively, the **lower** and **upper limits of summation**; and the letter k is called the **index of summation**. It is not essential to use k as the index of summation; any letter not reserved for another purpose will do. For example,

$$\sum_{i=1}^{6} \frac{1}{i}, \quad \sum_{j=1}^{6} \frac{1}{j}, \quad \text{and} \quad \sum_{n=1}^{6} \frac{1}{n}$$

all denote the sum

$$1 + \frac{1}{2} + \frac{1}{3} + \frac{1}{4} + \frac{1}{5} + \frac{1}{6}$$

If the upper and lower limits of summation are the same, then the "sum" in (1) reduces to one term. For example,

$$\sum_{k=2}^{2} k^3 = 2^3 \quad \text{and} \quad \sum_{i=1}^{1} \frac{1}{i+2} = \frac{1}{1+2} = \frac{1}{3}$$

In the sums

$$\sum_{i=1}^{5} 2, \quad \sum_{k=3}^{6} 7, \quad \text{and} \quad \sum_{j=0}^{2} x^3$$

the expression to the right of the Σ sign does not involve the index of summation. In such cases, we take all the terms in the sum to be the same, with one term for each allowable value of the summation index. Thus,

$$\sum_{i=1}^{5} 2 = 2 + 2 + 2 + 2 + 2$$

$$\sum_{k=3}^{6} 7 = 7 + 7 + 7 + 7$$

$$\sum_{j=0}^{2} x^3 = x^3 + x^3 + x^3$$

A sum can be written in more than one way with sigma notation by changing the limits of summation. For example, the sum of the first five positive even integers can be written in the following ways:

$$\sum_{k=1}^{5} 2k = 2 + 4 + 6 + 8 + 10$$

$$\sum_{k=0}^{4} (2k+2) = 2 + 4 + 6 + 8 + 10$$

$$\sum_{k=2}^{6} (2k-2) = 2 + 4 + 6 + 8 + 10$$

CHANGING THE INDEX OF SUMMATION

On occasion we will want to change the sigma notation for a given sum to a sigma notation with different limits of summation. The following example illustrates a method for doing this.

Example 2

Express

$$\sum_{k=3}^{7} 5^{k-2}$$

in sigma notation so that the lower limit of summation is 0 rather than 3.

Solution. If we define a new summation index j by means of the formula

$$j = k - 3 \tag{2}$$

then j runs from 0 up to 4 as k runs from 3 up to 7. From (2), $k = j + 3$, so

$$\sum_{k=3}^{7} 5^{k-2} = \sum_{j=0}^{4} 5^{(j+3)-2} = \sum_{j=0}^{4} 5^{j+1}$$

As a check, the reader can verify that

$$\sum_{j=0}^{4} 5^{j+1} \quad \text{and} \quad \sum_{k=3}^{7} 5^{k-2}$$

both denote the sum $5 + 5^2 + 5^3 + 5^4 + 5^5$. ◄

REMARK. In the solution of Example 2 the summation index was changed from k to j. If it is desirable to keep the same symbol for the summation index, we can change the j back to k *at the very end* and express the final result as

$$\sum_{k=0}^{4} 5^{k+1} \quad \text{instead of} \quad \sum_{j=0}^{4} 5^{j+1}$$

When we want to represent a general sum we will use letters with subscripts. For example, a general sum with five terms might be written as

$$a_1 + a_2 + a_3 + a_4 + a_5$$

or in sigma notation as

$$\sum_{k=1}^{5} a_k, \quad \sum_{j=1}^{5} a_j, \quad \text{or} \quad \sum_{m=1}^{5} a_m$$

A general sum with n terms might be written as

$$b_1 + b_2 + \cdots + b_n$$

or in sigma notation as

$$\sum_{k=1}^{n} b_k, \quad \sum_{j=1}^{n} b_j, \quad \text{or} \quad \sum_{m=1}^{n} b_m$$

PROPERTIES OF SIGMA NOTATION

The following properties of sigma notation will help to manipulate sums:

7.4.1 THEOREM.

(a) $\displaystyle\sum_{k=1}^{n} ca_k = c \sum_{k=1}^{n} a_k$

(b) $\displaystyle\sum_{k=1}^{n} (a_k + b_k) = \sum_{k=1}^{n} a_k + \sum_{k=1}^{n} b_k$

(c) $\displaystyle\sum_{k=1}^{n} (a_k - b_k) = \sum_{k=1}^{n} a_k - \sum_{k=1}^{n} b_k$

We will prove parts (a) and (b) and leave part (c) as an exercise.

Proof (a).

$$\sum_{k=1}^{n} ca_k = ca_1 + ca_2 + \cdots + ca_n = c(a_1 + a_2 + \cdots + a_n) = c\sum_{k=1}^{n} a_k$$

Proof (b).

$$\sum_{k=1}^{n}(a_k + b_k) = (a_1 + b_1) + (a_2 + b_2) + \cdots + (a_n + b_n)$$

$$= (a_1 + a_2 + \cdots + a_n) + (b_1 + b_2 + \cdots + b_n)$$

$$= \sum_{k=1}^{n} a_k + \sum_{k=1}^{n} b_k \qquad \blacksquare$$

REMARK. Loosely phrased, this theorem states: *A constant factor can be moved through a sigma sign; sigma of a sum equals the sum of the sigmas; and sigma of a difference equals the difference of the sigmas.*

SUMMATION FORMULAS

The following formulas will be used in our later work.

7.4.2 THEOREM.

(a) $\displaystyle\sum_{k=1}^{n} k = 1 + 2 + 3 + \cdots + n = \frac{n(n+1)}{2}$

(b) $\displaystyle\sum_{k=1}^{n} k^2 = 1^2 + 2^2 + 3^2 + \cdots + n^2 = \frac{n(n+1)(2n+1)}{6}$

(c) $\displaystyle\sum_{k=1}^{n} k^3 = 1^3 + 2^3 + 3^3 + \cdots + n^3 = \left[\frac{n(n+1)}{2}\right]^2$

We will prove parts (*a*) and (*b*) and leave part (*c*) as an exercise.

Proof (a). If we write the terms of

$$\sum_{k=1}^{n} k = 1 + 2 + 3 + \cdots + (n-2) + (n-1) + n \tag{3}$$

in the opposite order, we obtain

$$\sum_{k=1}^{n} k = n + (n-1) + (n-2) + \cdots + 3 + 2 + 1 \tag{4}$$

Adding (3) and (4) term by term yields

$$2\sum_{k=1}^{n} k = \underbrace{(n+1) + (n+1) + (n+1) + \cdots + (n+1)}_{n\text{ terms}} = n(n+1)$$

Thus,

$$\sum_{k=1}^{n} k = \frac{n(n+1)}{2}$$

Proof (b). This proof begins with a trick. Since

$$(k+1)^3 - k^3 = k^3 + 3k^2 + 3k + 1 - k^3 = 3k^2 + 3k + 1$$

we obtain

$$\sum_{k=1}^{n}[(k+1)^3 - k^3] = \sum_{k=1}^{n}(3k^2 + 3k + 1) \tag{5}$$

Writing out the left side of (5) yields

$$[2^3 - 1^3] + [3^3 - 2^3] + [4^3 - 3^3] + \cdots + [(n+1)^3 - n^3] \tag{6}$$

Observe that in (6) the 2^3 in the first term cancels out the -2^3 in the second term, the 3^3 in the second term cancels out the -3^3 in the third term, and so forth, so that the entire sum collapses like a folding telescope (hence, is called a ***telescoping sum***), leaving only $-1^3 + (n + 1)^3$. Thus, (5) can be rewritten as

$$-1 + (n + 1)^3 = \sum_{k=1}^{n} (3k^2 + 3k + 1) \tag{7}$$

or, from Theorem 7.4.1,

$$-1 + (n + 1)^3 = 3 \sum_{k=1}^{n} k^2 + 3 \sum_{k=1}^{n} k + \sum_{k=1}^{n} 1 \tag{8}$$

But

$$\sum_{k=1}^{n} 1 = \underbrace{1 + 1 + \cdots + 1}_{n \text{ terms}} = n$$

and by part (a) of this theorem

$$\sum_{k=1}^{n} k = \frac{n(n + 1)}{2}$$

Thus, (8) can be written as

$$-1 + (n + 1)^3 = 3 \sum_{k=1}^{n} k^2 + 3 \frac{n(n + 1)}{2} + n$$

Therefore,

$$\sum_{k=1}^{n} k^2 = \frac{1}{3} \left[(n + 1)^3 - 3 \frac{n(n + 1)}{2} - (n + 1) \right]$$

$$= \frac{n + 1}{6} [2(n + 1)^2 - 3n - 2]$$

$$= \frac{n + 1}{6} (2n^2 + n) = \frac{n(n + 1)(2n + 1)}{6}$$

Example 3

Evaluate $\displaystyle\sum_{k=1}^{30} k(k + 1)$.

Solution.

$$\sum_{k=1}^{30} k(k + 1) = \sum_{k=1}^{30} (k^2 + k) = \sum_{k=1}^{30} k^2 + \sum_{k=1}^{30} k$$

$$= \frac{30(31)(61)}{6} + \frac{30(31)}{2} = 9920 \qquad \text{Theorem 7.4.2}(a),(b) \qquad \blacktriangleleft$$

REMARK. In formulas such as

$$\sum_{k=1}^{n} k^2 = \frac{n(n + 1)(2n + 1)}{6}$$

or

$$1^2 + 2^2 + \cdots + n^2 = \frac{n(n + 1)(2n + 1)}{6}$$

the left side of the equality is said to express the sum in ***open form*** and the right side is said to express it in ***closed form***; the open form just indicates the terms to be added, while the closed form is an explicit formula for their sum.

Example 4

Express $\displaystyle\sum_{k=1}^{n}(3+k)^2$ in closed form.

Solution.

$$\sum_{k=1}^{n}(3+k)^2 = \sum_{k=1}^{n}(9+6k+k^2) = \sum_{k=1}^{n}9 + 6\sum_{k=1}^{n}k + \sum_{k=1}^{n}k^2$$

$$= 9n + 6\frac{n(n+1)}{2} + \frac{n(n+1)(2n+1)}{6}$$

$$= \frac{1}{3}n^3 + \frac{7}{2}n^2 + \frac{73}{6}n \qquad \blacktriangleleft$$

FOR THE READER.   Your numerical calculating utility probably provides some way of evaluating sums that can be expressed in sigma notation. Check your documentation to find out how to do this, and then use your utility to confirm that the numerical result obtained in Example 3 is correct. If you have access to a CAS, then it provides some method for finding closed forms for sums such as those in Theorem 7.4.2. Use your CAS to confirm the formulas in that theorem, and then find closed forms for

$$\sum_{k=1}^{n}k^4 \quad \text{and} \quad \sum_{k=1}^{n}k^5$$

EXERCISE SET 7.4   [C] CAS

1. Evaluate

(a) $\displaystyle\sum_{k=1}^{3}k^3$ (b) $\displaystyle\sum_{j=2}^{6}(3j-1)$ (c) $\displaystyle\sum_{i=-4}^{1}(i^2-i)$

(d) $\displaystyle\sum_{n=0}^{5}1$ (e) $\displaystyle\sum_{k=0}^{4}(-2)^k$ (f) $\displaystyle\sum_{n=1}^{6}\sin n\pi.$

2. Evaluate

(a) $\displaystyle\sum_{k=1}^{4}k\sin\frac{k\pi}{2}$ (b) $\displaystyle\sum_{j=0}^{5}(-1)^j$ (c) $\displaystyle\sum_{i=7}^{20}e^2$

(d) $\displaystyle\sum_{m=3}^{5}2^{m+1}$ (e) $\displaystyle\sum_{n=1}^{6}\ln n$ (f) $\displaystyle\sum_{k=0}^{10}\cos k\pi.$

In Exercises 3–12, write each expression in sigma notation, but do not evaluate.

3. $1+2+3+\cdots+10$

4. $3\cdot1+3\cdot2+3\cdot3+\cdots+3\cdot20$

5. $1\cdot2+2\cdot3+3\cdot4+\cdots+49\cdot50$

6. $1+2+2^2+2^3+2^4$

7. $2+4+6+8+\cdots+20$

8. $1+3+5+7+\cdots+15$

9. $1-3+5-7+9-11$ **10.** $1-\dfrac{1}{2}+\dfrac{1}{3}-\dfrac{1}{4}+\dfrac{1}{5}$

11. $-1+\dfrac{1}{2}-\dfrac{1}{3}+\dfrac{1}{4}-\dfrac{1}{5}$

12. $1+\cos\dfrac{\pi}{7}+\cos\dfrac{2\pi}{7}+\cos\dfrac{3\pi}{7}$

13. (a) Express the sum of the even integers from 2 to 100 in sigma notation.

   (b) Express the sum of the odd integers from 1 to 99 in sigma notation.

14. Express in sigma notation.

   (a) $a_1-a_2+a_3-a_4+a_5$

   (b) $-b_0+b_1-b_2+b_3-b_4+b_5$

   (c) $a_0+a_1x+a_2x^2+\cdots+a_nx^n$

   (d) $a^5+a^4b+a^3b^2+a^2b^3+ab^4+b^5$

In Exercises 15–22, use Theorem 7.4.2 to evaluate the sums, and check your answers using the summation feature of a calculating utility.

15. $\displaystyle\sum_{k=1}^{100}k$ **16.** $\displaystyle\sum_{k=3}^{100}k$ **17.** $\displaystyle\sum_{k=1}^{20}k^2$

18. $\displaystyle\sum_{k=1}^{100}(7k+1)$ **19.** $\displaystyle\sum_{k=1}^{6}(4k^3-2k+1)$ **20.** $\displaystyle\sum_{k=4}^{20}k^2$

21. $\displaystyle\sum_{k=1}^{30} k(k-2)(k+2)$ **22.** $\displaystyle\sum_{k=1}^{6}(k-k^3)$

In Exercises 23–28, express the sums in closed form.

23. $\displaystyle\sum_{k=1}^{n}(4k-3)$ **24.** $\displaystyle\sum_{k=1}^{n-1} k^2$ **25.** $\displaystyle\sum_{k=1}^{n} \frac{3k}{n}$

26. $\displaystyle\sum_{k=1}^{n-1} \frac{k^2}{n}$ **27.** $\displaystyle\sum_{k=1}^{n-1} \frac{k^3}{n^2}$ **28.** $\displaystyle\sum_{k=1}^{n}\left(\frac{5}{n}-\frac{2k}{n}\right)$

[c] **29.** For each of the sums that you obtained in Exercises 23–28, use a CAS to check your answer. If the answer produced by the CAS does not match your own, show that the two answers are equivalent.

30. Let
$$S = \sum_{k=0}^{n} ar^k$$
Show that $S - rS = a - ar^{n+1}$ and hence that
$$\sum_{k=0}^{n} ar^k = \frac{a - ar^{n+1}}{1-r} \quad (r \neq 1)$$
(A sum of this form is called a ***geometric sum***.)

31. In each part, rewrite the sum, if necessary, so that the lower limit is 0, and then use the formula derived in Exercise 30 to evaluate the sum. Check your answers using the summation feature of a calculating utility.

(a) $\displaystyle\sum_{k=1}^{20} 3^k$ (b) $\displaystyle\sum_{k=5}^{30} 2^k$ (c) $\displaystyle\sum_{k=0}^{100}(-1)^{k+1}\frac{1}{2^k}$

[c] **32.** In each part, make a conjecture about the limit by using a CAS to evaluate the sum for $n = 10$, 20, and 50; and then check your conjecture by using the formula in Exercise 30 to express the sum in closed form, and then finding the limit exactly.

(a) $\displaystyle\lim_{n \to +\infty}\sum_{k=0}^{n} \frac{1}{2^k}$ (b) $\displaystyle\lim_{n \to +\infty}\sum_{k=1}^{n}\left(\frac{3}{4}\right)^k$

In Exercises 33–36, express the function of n in closed form, and then use L'Hôpital's rule to find the limit. [*Note:* L'Hôpital's rule was derived for functions of a real-valued variable x, whereas here the variable n assumes only integer values. Thus, strictly speaking, L'Hôpital's rule cannot be used without justifying that it applies to functions of integer-valued variables. We will do this later in the text.]

33. $\displaystyle\lim_{n \to +\infty}\frac{1+2+3+\cdots+n}{n^2}$

34. $\displaystyle\lim_{n \to +\infty}\frac{1^2+2^2+3^2+\cdots+n^2}{n^3}$

35. $\displaystyle\lim_{n \to +\infty}\sum_{k=1}^{n} \frac{5k}{n^2}$ **36.** $\displaystyle\lim_{n \to +\infty}\sum_{k=1}^{n-1} \frac{2k^2}{n^3}$

37. Express $1 + 2 + 2^2 + 2^3 + 2^4 + 2^5$ in sigma notation with
(a) $j = 0$ as the lower limit of summation

(b) $j = 1$ as the lower limit of summation
(c) $j = 2$ as the lower limit of summation.

38. Express
$$\sum_{k=5}^{9} k2^{k+4}$$
in sigma notation with
(a) $k = 1$ as the lower limit of summation
(b) $k = 13$ as the upper limit of summation.

39. Change the limits of summation appropriately to simplify
(a) $\displaystyle\sum_{k=11}^{28} \sin\left(\frac{\pi}{k-10}\right)$ (b) $\displaystyle\sum_{k=6}^{12} e^{k-6}$

40. Show that the sum of the first n consecutive positive odd integers is n^2.

41. The accompanying figure shows a square that is n units by n units that has been subdivided into a one-unit square and $n-1$ "*L*-shaped" regions. Use this figure to derive the result in Exercise 40.

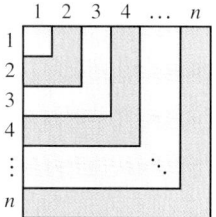

Figure Ex-41

42. Solve the equation $\displaystyle\sum_{k=1}^{n} k = 465$.

When part of each term of a sum cancels part of the next term, leaving only portions of the first and last terms at the end, the sum is said to ***telescope***. In Exercises 43–46, evaluate the telescoping sum.

43. $\displaystyle\sum_{k=5}^{17}(3^k - 3^{k-1})$ **44.** $\displaystyle\sum_{k=1}^{50}\left(\frac{1}{k}-\frac{1}{k+1}\right)$

45. $\displaystyle\sum_{k=2}^{20}\left(\frac{1}{k^2}-\frac{1}{(k-1)^2}\right)$ **46.** $\displaystyle\sum_{k=1}^{100}(2^{k+1}-2^k)$

47. (a) Show that
$$\frac{1}{1\cdot3}+\frac{1}{3\cdot5}+\cdots+\frac{1}{(2n-1)(2n+1)}=\frac{n}{2n+1}$$
$$\left[Hint: \frac{1}{(2n-1)(2n+1)}=\frac{1}{2}\left(\frac{1}{2n-1}-\frac{1}{2n+1}\right).\right]$$

(b) Use the result in part (a) to find
$$\lim_{n \to +\infty}\sum_{k=1}^{n} \frac{1}{(2k-1)(2k+1)}$$

48. (a) Show that

$$\frac{1}{1\cdot 2} + \frac{1}{2\cdot 3} + \frac{1}{3\cdot 4} + \cdots + \frac{1}{n(n+1)} = \frac{n}{n+1}$$

$$\left[\textit{Hint:}\ \frac{1}{n(n+1)} = \frac{1}{n} - \frac{1}{n+1}. \right]$$

(b) Use the result in part (a) to find

$$\lim_{n \to +\infty} \sum_{k=1}^{n} \frac{1}{k(k+1)}$$

49. By writing out the sums, determine whether the following are valid identities.

(a) $\displaystyle\int \left[\sum_{i=1}^{n} f_i(x) \right] dx = \sum_{i=1}^{n} \left[\int f_i(x)\,dx \right]$

(b) $\displaystyle\frac{d}{dx}\left[\sum_{i=1}^{n} f_i(x) \right] = \sum_{i=1}^{n}\left[\frac{d}{dx}[f_i(x)] \right]$

50. Which of the following are valid identities?

(a) $\displaystyle\sum_{i=1}^{n} a_i b_i = \sum_{i=1}^{n} a_i \sum_{i=1}^{n} b_i$

(b) $\displaystyle\sum_{i=1}^{n} \frac{a_i}{b_i} = \sum_{i=1}^{n} a_i \Big/ \sum_{i=1}^{n} b_i$

(c) $\displaystyle\sum_{i=1}^{n} a_i^2 = \left(\sum_{i=1}^{n} a_i \right)^2$

51. Let $\bar{x}$ denote the arithmetic average of the n numbers $x_1, x_2, \ldots, x_n$. Use Theorem 7.4.1 to prove that

$$\sum_{i=1}^{n} (x_i - \bar{x}) = 0$$

52. Prove part (c) of Theorem 7.4.1.

53. Prove part (c) of Theorem 7.4.2. [*Hint:* Begin with the difference $(k+1)^4 - k^4$ and follow the steps used to prove part (b) of the theorem.]

54. An artist wants to create a rough triangular design using uniform square tiles glued edge to edge. She places n tiles in a row to form the base of the triangle and then makes each successive row two tiles shorter than the preceding row. Find a formula for the number of tiles used in the design. [*Hint:* Your answer will depend on whether n is even or odd.]

55. An artist wants to create a sculpture by gluing together uniform spheres. She creates a rough rectangular base that has 50 spheres along one edge and 30 spheres along the other. She then creates successive layers by gluing spheres in the grooves of the preceding layer. How many spheres will there be in the sculpture?

7.5 THE DEFINITE INTEGRAL

Recall from the informal discussion in Section 7.1 that if a function f is continuous and nonnegative on an interval $[a, b]$, then the area under the graph of f over the interval $[a, b]$ can be obtained by either the "rectangle method" or "the antiderivative method." In this section we will discuss the rectangle method in more detail, and we will introduce the concept of a "definite integral," which will link the concept of area to other important concepts such as length, volume, density, probability, and work.

A DEFINITION OF AREA

Our first goal in this section is to define formally what we mean by the area of a region R that is bounded below by the x-axis, bounded on the sides by the vertical lines $x = a$ and $x = b$, and bounded above by the curve $y = f(x)$, where f is continuous and nonnegative on the interval $[a, b]$ (Figure 7.5.1). We will start by defining the area of a rectangle to be the product of its length and width and defining the area of a region composed of finitely many rectangles to be the sum of the areas of those rectangles. To define the area of the region R, we will use these definitions and the rectangle method of Section 7.1. The basic idea is as follows (Figure 7.5.2):

- Divide the interval $[a, b]$ into n equal subintervals.

- Over each subinterval construct a rectangle whose height is the value of f at any point in the subinterval.

- The union of these rectangles forms a region R_n whose area can be regarded as an approximation to the "area" A of the region R.

- Repeat the process using more and more subdivisions.

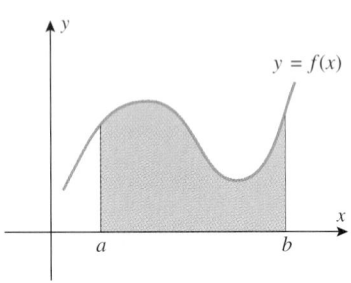

$y = f(x)$

Figure 7.5.1

- Define the area of R to be the limit of the areas of the approximating regions, R_n; that is,

$$A = \text{area}(R) = \lim_{n \to +\infty} [\text{area}(R_n)] \tag{1}$$

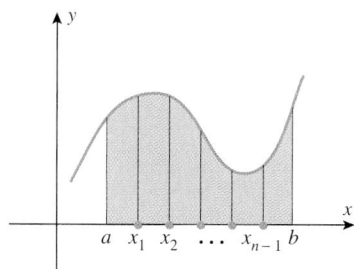

Figure 7.5.2

To make all of this more precise, it will be helpful to capture this procedure in mathematical notation. For this purpose, suppose that we divide the interval $[a, b]$ into n subintervals by inserting $n - 1$ equally spaced points between a and b, say

$$x_1, x_2, \ldots, x_{n-1}$$

(Figure 7.5.2). Each of these intervals has width $(b - a)/n$, which it is customary to denote by

$$\Delta x = \frac{b - a}{n}$$

In each subinterval we need to choose a point at which to evaluate the function f to determine the height of a rectangle over that interval. If we denote those points by

$$x_1^*, x_2^*, \ldots, x_n^*$$

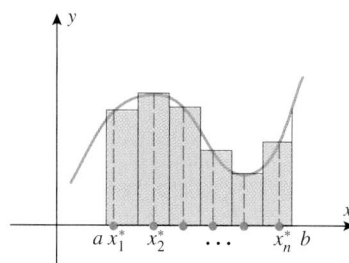

Figure 7.5.3

(Figure 7.5.3), then the areas of the rectangles constructed over these intervals will be

$$f(x_1^*)\Delta x, \quad f(x_2^*)\Delta x, \ldots, \quad f(x_n^*)\Delta x$$

(Figure 7.5.4), and the total area of the region R_n will be

$$\text{area}(R_n) = f(x_1^*)\Delta x + f(x_2^*)\Delta x + \cdots + f(x_n^*)\Delta x$$

or in sigma notation,

$$\text{area}(R_n) = \sum_{k=1}^{n} f(x_k^*)\Delta x$$

With this notation (1) can be expressed as

$$A = \lim_{n \to +\infty} \sum_{k=1}^{n} f(x_k^*)\Delta x$$

which suggests the following definition of the area of the region R.

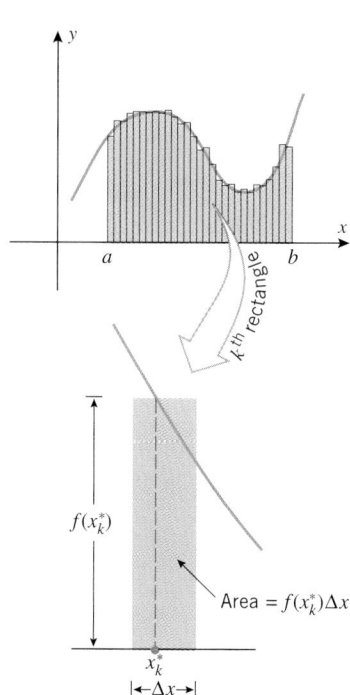

Figure 7.5.4

7.5.1 DEFINITION (*Area Under a Curve*).   If the function f is continuous on $[a, b]$ and if $f(x) \geq 0$ for all x in $[a, b]$, then the *area* under the curve $y = f(x)$ over the interval $[a, b]$ is defined by

$$A = \lim_{n \to +\infty} \sum_{k=1}^{n} f(x_k^*)\Delta x \tag{2}$$

REMARK.   Although this definition is satisfactory for our present purposes, there are some issues that would have to be resolved before it could be regarded as a rigorous mathematical definition. For example, we would have to prove that the limit actually exists and that its value does not depend on how the points $x_1^*, x_2^*, \ldots, x_n^*$ are chosen. It can be proved that this is true if f is continuous on $[a, b]$, but the details are beyond the scope of this text.

The limit in Formula (2) is often difficult or impossible to find, so that when an *exact* area is needed the antiderivative method, which we will discuss in the next section, is the method of choice. However, if an *approximation* to the area will suffice, then instead of taking the limit we can approximate the area as

$$A \approx \sum_{k=1}^{n} f(x_k^*)\Delta x$$

where n is sufficiently large to produce the required accuracy. For this purpose it is convenient to rewrite this sum as

$$\sum_{k=1}^{n} f(x_k^*)\Delta x = \Delta x \sum_{k=1}^{n} f(x_k^*) = \Delta x[f(x_1^*) + f(x_2^*) + \cdots + f(x_n^*)] \tag{3}$$

where $\Delta x = (b - a)/n$. The calculation here involves only the sum of the values of the function at n points, followed by a multiplication by Δx. The points $x_1^*, x_2^*, \ldots, x_n^*$ can be chosen arbitrarily in successive subintervals; however, the most common choices are at the left endpoints, the right endpoints, or the centers of the subintervals, in which cases Formula (3) is called the *left endpoint approximation*, the *right endpoint approximation*, or the *midpoint approximation* of the exact area (Figure 7.5.5).

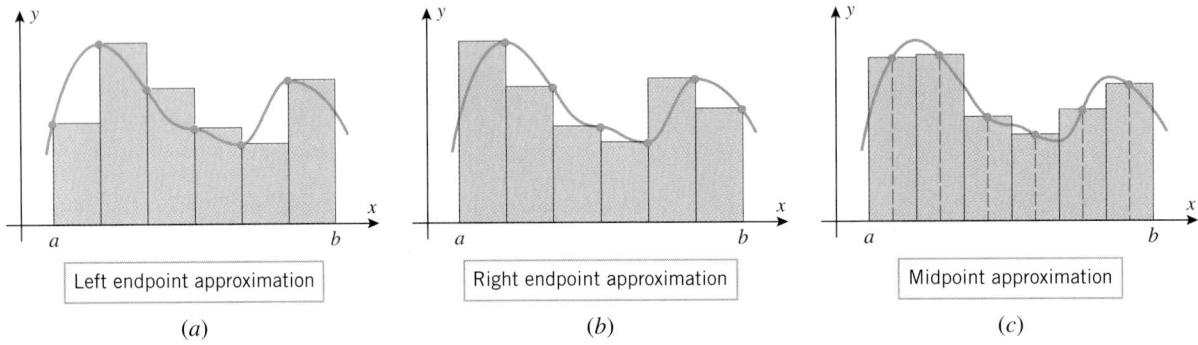

Left endpoint approximation	Right endpoint approximation	Midpoint approximation
(a)	(b)	(c)

Figure 7.5.5

Example 1

Find the left endpoint, right endpoint, and midpoint approximations of the area under the curve $y = 9 - x^2$ over the interval $[0, 3]$ with $n = 10$, $n = 20$, and $n = 50$ (Figure 7.5.6).

Solution. Details of the computations for the case $n = 10$ are shown to six decimal places in Table 7.5.1 and the results of all computations are given in Table 7.5.2. ◀

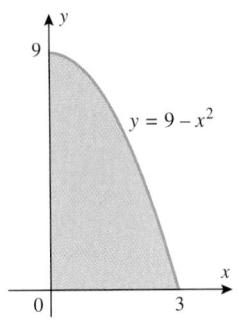

Figure 7.5.6

Table 7.5.1

$n = 10$, $\Delta x = (b - a)/n = (3 - 0)/10 = 0.3$

k	LEFT ENDPOINT APPROXIMATION x_k^*	$9 - (x_k^*)^2$	RIGHT ENDPOINT APPROXIMATION x_k^*	$9 - (x_k^*)^2$	MIDPOINT APPROXIMATION x_k^*	$9 - (x_k^*)^2$
1	0.0	9.000000	0.3	8.910000	0.15	8.977500
2	0.3	8.910000	0.6	8.640000	0.45	8.797500
3	0.6	8.640000	0.9	8.190000	0.75	8.437500
4	0.9	8.190000	1.2	7.560000	1.05	7.897500
5	1.2	7.560000	1.5	6.750000	1.35	7.177500
6	1.5	6.750000	1.8	5.760000	1.65	6.277500
7	1.8	5.760000	2.1	4.590000	1.95	5.197500
8	2.1	4.590000	2.4	3.240000	2.25	3.937500
9	2.4	3.240000	2.7	1.710000	2.55	2.497500
10	2.7	1.710000	3.0	0.000000	2.85	0.877500
		64.350000		55.350000		60.075000
$\Delta x \sum_{k=1}^{n} f(x_k^*)$		(.3)(64.350000) = 19.305000		(.3)(55.350000) = 16.605000		(.3)(60.075000) = 18.022500

Table 7.5.2

n	LEFT ENDPOINT APPROXIMATION	RIGHT ENDPOINT APPROXIMATION	MIDPOINT APPROXIMATION
10	19.305000	16.605000	18.022500
20	18.663750	17.313750	18.005625
50	18.268200	17.728200	18.000900

REMARK.   We will show in the next section that the exact area under $y = 9 - x^2$ over the interval [0, 3] is 18 (i.e., 18 square units), so that in the preceding example the midpoint approximation is more accurate than either of the endpoint approximations. This can also be seen geometrically from the approximating rectangles: Since the graph of $y = 9 - x^2$ is decreasing over the interval [0, 3], each left endpoint approximation overestimates the area, each right endpoint approximation underestimates the area, and each midpoint approximation falls between the overestimate and the underestimate (Figure 7.5.7). This is consistent with the values in Table 7.5.2. Later in the text we will investigate the error that results when an area is approximated by the midpoint rule.

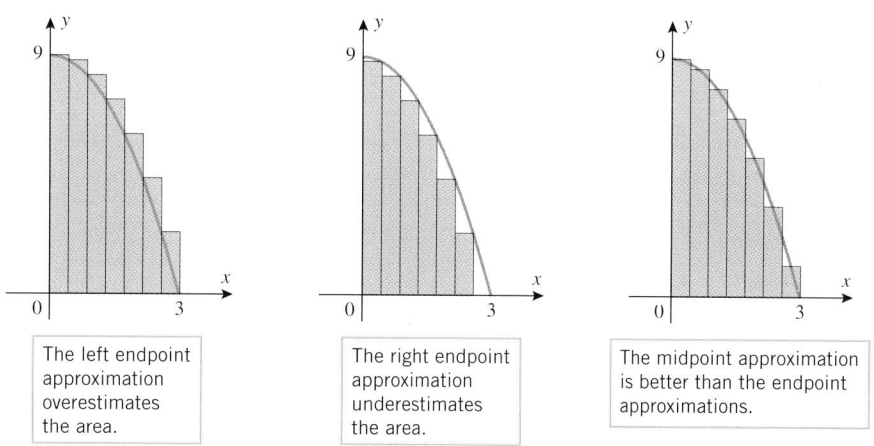

The left endpoint approximation overestimates the area.

The right endpoint approximation underestimates the area.

The midpoint approximation is better than the endpoint approximations.

Figure 7.5.7

THE DEFINITE INTEGRAL OF A CONTINUOUS FUNCTION

In Definition 7.5.1 we assumed that f is continuous and nonnegative on the interval $[a, b]$. If f is continuous and assumes both positive and negative values on $[a, b]$, then the limit

$$\lim_{n \to +\infty} \sum_{k=1}^{n} f(x_k^*)\Delta x \tag{4}$$

no longer represents the area between the curve $y = f(x)$ and the interval $[a, b]$; rather it represents a difference of areas—the area of the region that is above the interval $[a, b]$ and below the curve $y = f(x)$ minus the area of the region that is below the interval $[a, b]$ and above the curve $y = f(x)$. We call this the ***net signed area*** between the graph of $y = f(x)$ and the interval $[a, b]$. For example, in Figure 7.5.8a, the net signed area between the curve $y = f(x)$ and the interval $[a, b]$ is

$$(A_I + A_{III}) - A_{II} = \big[\text{area above } [a, b]\big] - \big[\text{area below } [a, b]\big]$$

To explain why the limit in (4) represents this net signed area, let us subdivide the interval $[a, b]$ in Figure 7.5.8a into n equal subintervals and examine the terms in the sum

$$\sum_{k=1}^{n} f(x_k^*)\Delta x \tag{5}$$

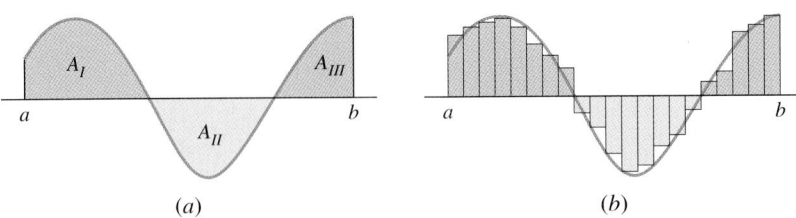

Figure 7.5.8   (a)   (b)

If $f(x_k^*)$ is positive, then the product $f(x_k^*)\Delta x$ represents the area of the rectangle with height $f(x_k^*)$ and base Δx (the beige rectangles in Figure 7.5.8b). However, if $f(x_k^*)$ is negative, then the product $f(x_k^*)\Delta x$ is the *negative* of the area of the rectangle with height $|f(x_k^*)|$ and base Δx (the green rectangles in Figure 7.5.8b). Thus, (5) represents the total area of the beige rectangles minus the total area of the green rectangles. As n increases, the beige rectangles fill out the regions with areas A_I and A_{III} and the green rectangles fill out the region with area A_{II}, which explains why the limit in (4) represents the signed area between $y = f(x)$ and the interval $[a, b]$.

The limit in (4) is so important that there is some terminology and notation associated with it. We will denote this limit by the symbol

$$\int_a^b f(x)\, dx = \lim_{n \to +\infty} \sum_{k=1}^{n} f(x_k^*)\Delta x \tag{6}$$

which is called the **definite integral** of f from a to b. Geometrically, the definite integral represents the signed area between $y = f(x)$ and the interval $[a, b]$, and in the case where $f(x)$ is nonnegative on the interval $[a, b]$, the definite integral represents the area under the curve over the interval $[a, b]$. The numbers a and b are called the **lower limit of integration** and **upper limit of integration**, respectively, and $f(x)$ is called the **integrand**. The reason for the integral sign will become clear in the next section, where we will establish a link between the definite integral and the indefinite integral studied earlier.

In the simplest cases, definite integrals can be calculated using formulas from plane geometry to compute the signed areas.

Example 2

Sketch the region whose area is represented by the definite integral, and evaluate the integral using an appropriate formula from geometry.

(a) $\displaystyle\int_1^4 2\, dx$ (b) $\displaystyle\int_{-1}^2 (x+2)\, dx$ (c) $\displaystyle\int_0^1 \sqrt{1-x^2}\, dx$

Solution (a). The graph of the integrand is the horizontal line $y = 2$, so the region is a rectangle of height 2 extending over the interval from 1 to 4 (Figure 7.5.9a). Thus,

$$\int_1^4 2\, dx = (\text{area of rectangle}) = 2(3) = 6$$

Solution (b). The graph of the integrand is the line $y = x+2$, so the region is a trapezoid whose base extends from $x = -1$ to $x = 2$ (Figure 7.5.9b). Thus,

$$\int_{-1}^2 (x+2)\, dx = (\text{area of trapezoid}) = \tfrac{1}{2}(3)(1+4) = \tfrac{15}{2}$$

Solution (c). The graph of $y = \sqrt{1-x^2}$ is the upper semicircle of radius 1, centered at the origin, so the region is the right quarter-circle extending from $x = 0$ to $x = 1$ (Figure 7.5.9c).

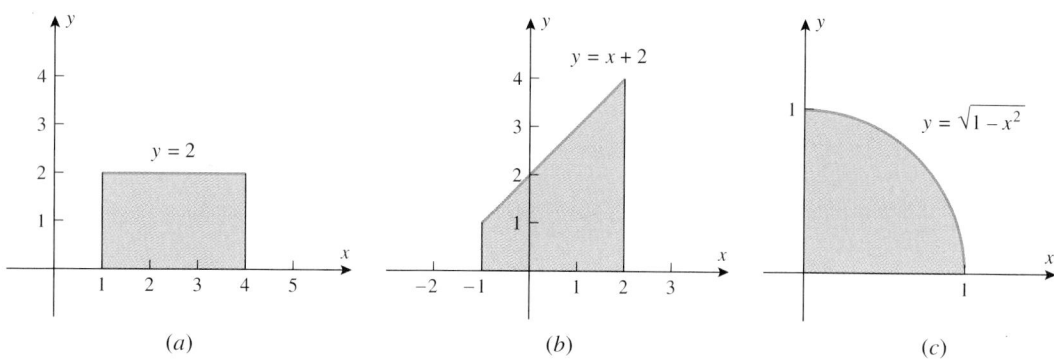

Figure 7.5.9

Thus,

$$\int_0^1 \sqrt{1 - x^2}\, dx = \text{(area of quarter-circle)} = \tfrac{1}{4}\pi(1^2) = \frac{\pi}{4} \qquad \blacktriangleleft$$

Example 3

Evaluate

(a) $\displaystyle\int_0^2 (x - 1)\, dx$ (b) $\displaystyle\int_0^1 (x - 1)\, dx$

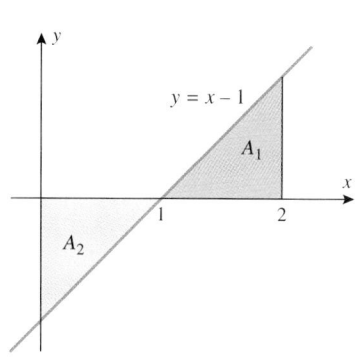

Figure 7.5.10

Solution. The graph of $y = x - 1$ is shown in Figure 7.5.10, and we leave it for you to verify that the shaded triangular regions both have area $\tfrac{1}{2}$. Over the interval $[0, 2]$ the net signed area is $A_1 - A_2 = \tfrac{1}{2} - \tfrac{1}{2} = 0$, and over the interval $[0, 1]$ the net signed area is $-A_2 = -\tfrac{1}{2}$. Thus,

$$\int_0^2 (x - 1)\, dx = 0 \quad \text{and} \quad \int_0^1 (x - 1)\, dx = -\tfrac{1}{2} \qquad \blacktriangleleft$$

THE RIEMANN INTEGRAL

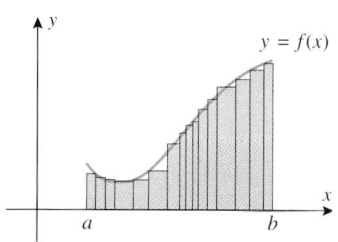

Figure 7.5.11

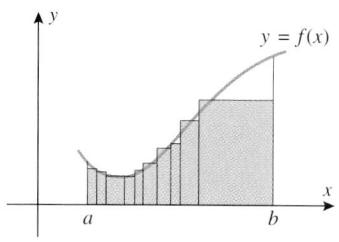

Figure 7.5.12

It is assumed in (6) that the function f is continuous on the interval $[a, b]$ and that for each n this interval is subdivided into n subintervals of equal length to create bases for the approximating rectangles. Although equal lengths are useful for computations, this restriction is not essential. That is, the signed area between $y = f(x)$ and $[a, b]$ can be obtained using rectangles with different widths provided that successive subdivisions are constructed in such a way that the widths of the rectangles approach zero as n increases (Figure 7.5.11). Thus, we must preclude the kind of situation that occurs in Figure 7.5.12 in which the right half of the interval is never subdivided. If this kind of subdivision were allowed, the error in the approximation would not approach zero as n increased.

To provide for the added generality of unequal intervals, suppose that the interval $[a, b]$ is subdivided into n subintervals whose widths are

$$\Delta x_1, \Delta x_2, \ldots, \Delta x_n$$

and let max Δx_k denote the largest of the subinterval widths, which is read "the maximum of the Δx_k's." The subintervals are said to form a *partition* of the interval $[a, b]$, and max Δx_k is called the *mesh size* of the partition. For example, Figure 7.5.13 shows a partition of the interval $[0, 6]$ into four subintervals with a mesh size of 2.

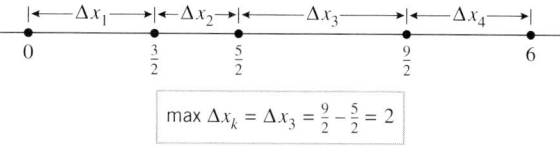

$$\max \Delta x_k = \Delta x_3 = \tfrac{9}{2} - \tfrac{5}{2} = 2$$

Figure 7.5.13

To generalize (6) so that it allows for unequal subinterval widths, we must replace the constant interval length Δx by the variable interval length Δx_k, and we must replace $n \to +\infty$ by an expression to specify that the lengths of all the subintervals approach zero. We will use the expression $\max \Delta x_k \to 0$ for this purpose. With these modifications in notation (6) becomes

$$\int_a^b f(x)\,dx = \lim_{\max \Delta x_k \to 0} \sum_{k=1}^n f(x_k^*)\Delta x_k \tag{7}$$

The sum that appears in this expression is called a ***Riemann*** [*] ***sum***, and the limit is sometimes called the ***Riemann integral*** in honor of the German mathematician Bernhard Riemann who formulated many of the basic concepts of integration.

REMARK. Some writers use the symbol $\|\Delta\|$ rather than $\max \Delta x_k$ for the mesh size of the partition, in which case (7) would be written as

$$\int_a^b f(x)\,dx = \lim_{\|\Delta\| \to 0} \sum_{k=1}^n f(x_k^*)\Delta x_k$$

INTEGRABILITY

Because the definite integral is defined as a limit, it is possible that the limit may not exist, in which case the definite integral would not exist. Thus, we make the following definition:

7.5.2 DEFINITION. A function f is said to be ***Riemann integrable*** or more simply ***integrable*** on a finite closed interval $[a, b]$ if the limit

$$\int_a^b f(x)\,dx = \lim_{\max \Delta x_k \to 0} \sum_{k=1}^n f(x_k^*)\Delta x_k$$

exists and does not depend on choice of the partitions or on the points x_k^* in the subintervals.

At the end of this section we will discuss various conditions that ensure integrability, but for now suffice it to say that a function that is continuous on a finite closed interval $[a, b]$ is integrable on that interval.

---

[*] GEORG FRIEDRICH BERNHARD RIEMANN (1826–1866). German mathematician. Bernhard Riemann, as he is commonly known, was the son of a Protestant minister. He received his elementary education from his father and showed brilliance in arithmetic at an early age. In 1846 he enrolled at Göttingen University to study theology and philology, but he soon transferred to mathematics. He studied physics under W. E. Weber and mathematics under Karl Friedrich Gauss, whom some people consider to be the greatest mathematician who ever lived. In 1851 Riemann received his Ph.D. under Gauss, after which he remained at Göttingen to teach. In 1862, one month after his marriage, Riemann suffered an attack of pleuritis, and for the remainder of his life was an extremely sick man. He finally succumbed to tuberculosis in 1866 at age 39.

An interesting story surrounds Riemann's work in geometry. For his introductory lecture prior to becoming an associate professor, Riemann submitted three possible topics to Gauss. Gauss surprised Riemann by choosing the topic Riemann liked the least, the foundations of geometry. The lecture was like a scene from a movie. The old and failing Gauss, a giant in his day, watching intently as his brilliant and youthful protégé skillfully pieced together portions of the old man's own work into a complete and beautiful system. Gauss is said to have gasped with delight as the lecture neared its end, and on the way home he marveled at his student's brilliance. Gauss died shortly thereafter. The results presented by Riemann that day eventually evolved into a fundamental tool that Einstein used some 50 years later to develop relativity theory.

In addition to his work in geometry, Riemann made major contributions to the theory of complex functions and mathematical physics. The notion of the definite integral, as it is presented in most basic calculus courses, is due to him. Riemann's early death was a great loss to mathematics, for his mathematical work was brilliant and of fundamental importance.

PROPERTIES OF THE DEFINITE INTEGRAL

It is assumed in Definition 7.5.2 that $[a, b]$ is a finite closed interval with $a < b$, and hence the upper limit of integration in the definite integral is greater than the lower limit of integration. However, it will be convenient to extend this definition to allow for cases in which the upper and lower limits of integration are equal or the lower limit of integration is greater than the upper limit of integration. For this purpose we make the following special definitions.

7.5.3 DEFINITION.

(a) If a is in the domain of f, we define

$$\int_a^a f(x)\, dx = 0$$

(b) If f is integrable on $[a, b]$, then we define

$$\int_b^a f(x)\, dx = -\int_a^b f(x)\, dx$$

REMARK. Part (a) of this definition is consistent with the intuitive idea that the area between a point on the x-axis and a curve $y = f(x)$ should be zero (Figure 7.5.14). Part (b) of the definition is simply a useful convention; it states that interchanging the limits of integration reverses the sign of the integral.

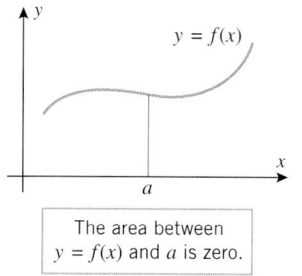

The area between $y = f(x)$ and a is zero.

Figure 7.5.14

Example 4

(a) $\displaystyle\int_1^1 x^2\, dx = 0$

(b) $\displaystyle\int_1^0 \sqrt{1 - x^2}\, dx = -\int_0^1 \sqrt{1 - x^2}\, dx = -\frac{\pi}{4}$ ◀

Example 2(c)

Because definite integrals are defined as limits, they inherit many of the properties of limits. For example, we know that constants can be moved through limit signs and that the limit of a sum or difference is the sum or difference of the limits. Thus, you should not be surprised by the following theorem, which we state without formal proof.

7.5.4 THEOREM. *If f and g are integrable on $[a, b]$ and if c is a constant, then cf, $f + g$, and $f - g$ are integrable on $[a, b]$ and*

(a) $\displaystyle\int_a^b cf(x)\, dx = c\int_a^b f(x)\, dx$

(b) $\displaystyle\int_a^b [f(x) + g(x)]\, dx = \int_a^b f(x)\, dx + \int_a^b g(x)\, dx$

(c) $\displaystyle\int_a^b [f(x) - g(x)]\, dx = \int_a^b f(x)\, dx - \int_a^b g(x)\, dx$

Part (b) of this theorem can be extended to more than two functions. More precisely,

$$\int_a^b [f_1(x) + f_2(x) + \cdots + f_n(x)]\,dx$$
$$= \int_a^b f_1(x)\,dx + \int_a^b f_2(x)\,dx + \cdots + \int_a^b f_n(x)\,dx \tag{8}$$

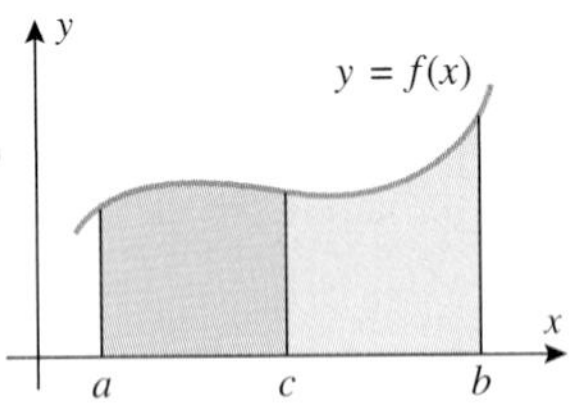

Some properties of definite integrals can be motivated by interpreting the integral as an area. For example, if f is continuous and nonnegative on the interval $[a, b]$, and if c is a point between a and b, then the area under $y = f(x)$ over the interval $[a, b]$ can be split into two parts and expressed as the area under the graph from a to c plus the area under the graph from c to b (Figure 7.5.15), that is,

$$\int_a^b f(x)\,dx = \int_a^c f(x)\,dx + \int_c^b f(x)\,dx$$

This is a special case of the following theorem about definite integrals, which we state without proof.

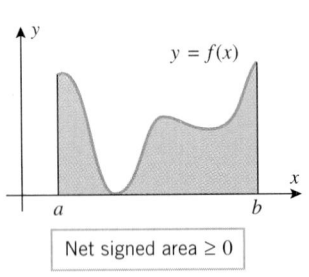

Figure 7.5.15

7.5.5 THEOREM. *If f is integrable on a closed interval containing the three points a, b, and c, then*

$$\int_a^b f(x)\,dx = \int_a^c f(x)\,dx + \int_c^b f(x)\,dx \tag{9}$$

no matter how the points are ordered.

The following theorem, which we state without formal proof, can also be motivated by interpreting definite integrals as areas.

7.5.6 THEOREM.

(a) *If f is integrable on $[a, b]$ and $f(x) \geq 0$ for all x in $[a, b]$, then*

$$\int_a^b f(x)\,dx \geq 0$$

(b) *If f and g are integrable on $[a, b]$ and $f(x) \geq g(x)$ for all x in $[a, b]$, then*

$$\int_a^b f(x)\,dx \geq \int_a^b g(x)\,dx$$

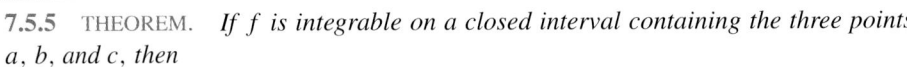

Net signed area ≥ 0

Figure 7.5.16

Geometrically, part (a) of this theorem states the obvious fact that if f is nonnegative on $[a, b]$, then the net signed area between the graph of f and the interval $[a, b]$ is also nonnegative (Figure 7.5.16). Part (b) has its simplest interpretation when f and g are nonnegative on $[a, b]$, in which case the theorem states that if the graph of f does not go below the graph of g, then the area under the graph of f is at least as large as the area under the graph of g (Figure 7.5.17).

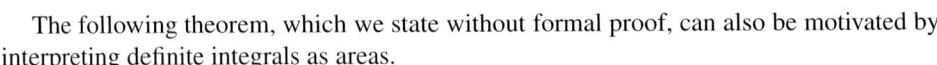

Area under f ≥ area under g

Figure 7.5.17

REMARK. In words, part (b) of this theorem states that one can integrate both sides of the inequality $f(x) \geq g(x)$ without altering the sense of the inequality. We also note that in the case where $b > a$, both parts of the theorem remain true if $\geq$ is replaced by $\leq$, $>$, or $<$ throughout.

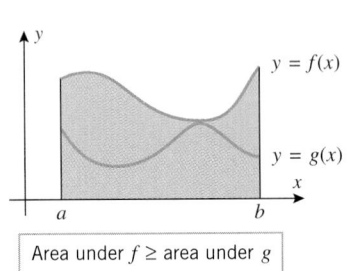

Example 5

Evaluate

$$\int_0^1 (5 - 3\sqrt{1 - x^2})\, dx$$

Solution. From parts (*a*) and (*c*) of Theorem 7.5.4 we can write

$$\int_0^1 (5 - 3\sqrt{1 - x^2})\, dx = \int_0^1 5\, dx - \int_0^1 3\sqrt{1 - x^2}\, dx = \int_0^1 5\, dx - 3\int_0^1 \sqrt{1 - x^2}\, dx$$

The first integral can be interpreted as the area of a rectangle of height 5 and base 1, so its value is 5, and from Example 2 the value of the second integral is $\pi/4$. Thus,

$$\int_0^1 (5 - 3\sqrt{1 - x^2})\, dx = 5 - 3\left(\frac{\pi}{4}\right) = 5 - \frac{3\pi}{4}$$ ◀

CONDITIONS FOR INTEGRABILITY

The problem of determining precisely which functions are integrable is quite complex and beyond the scope of this text. However, there are a few basic results about integrability that are important to know; we begin with a definition.

> **7.5.7** DEFINITION. A function f is said to be ***bounded*** on an interval I if there is a positive number M such that
>
> $$-M \le f(x) \le M$$
>
> for all x in the interval I. Geometrically, this means that the graph of f over the interval I lies between the lines $y = -M$ and $y = M$.

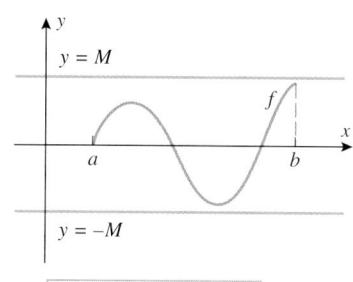

f is bounded on $[a, b]$.

Figure 7.5.18

For example, a continuous function f is bounded on *every* finite closed interval because the Extreme-Value Theorem (6.1.3) implies that f has an absolute maximum and an absolute minimum on the interval; hence, its graph will lie between the line $y = -M$ and $y = M$, provided we make M large enough (Figure 7.5.18). In contrast, a function that has a vertical asymptote inside of an interval is not bounded on that interval because its graph over the interval cannot be made to lie between the lines $y = -M$ and $y = M$, no matter how large we make the value of M (Figure 7.5.19).

The following theorem, which we state without proof, lists three of the most important facts about integrability.

> **7.5.8** THEOREM. *Let f be a function that is defined at all points in the finite closed interval $[a, b]$.*
> (a) *If f is continuous on $[a, b]$, then f is integrable on $[a, b]$.*
> (b) *If f has finitely many points of discontinuity on $[a, b]$ but is bounded on $[a, b]$, then f is integrable on $[a, b]$.*
> (c) *If f is not bounded on $[a, b]$, then f is not integrable on $[a, b]$.*

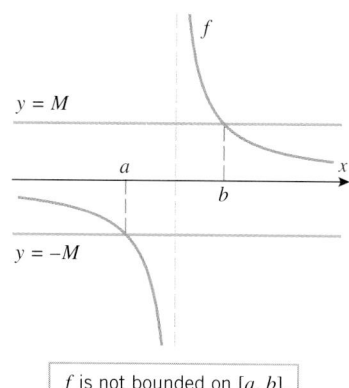

f is not bounded on $[a, b]$.

Figure 7.5.19

FOR THE READER.   Sketch the graph of a function over the interval $[0, 1]$ that has the properties stated in part (*b*) of this theorem.

EXERCISE SET 7.5 ☐ C CAS

1. (a) Use an appropriate geometric formula to find the exact area A under the line $x + y = 4$ over the interval $[0, 4]$.

   (b) Sketch the rectangles for the left endpoint approximation to the area A using $n = 4$ subintervals. Is that approximation greater than, less than, or equal to A? Explain your reasoning, and check your conclusion by calculating the left endpoint approximation.

   (c) Sketch the rectangles for the right endpoint approximation to the area A using $n = 4$ subintervals. Is that approximation greater than, less than, or equal to A? Explain your reasoning, and check your conclusion by calculating the right endpoint approximation.

   (d) Sketch the rectangles for the midpoint approximation to the area A using $n = 4$ subintervals. Is that approximation greater than, less than, or equal to A? Explain your reasoning, and check your conclusion by calculating the midpoint approximation.

2. Follow the directions of Exercise 1 for the area A under the line $y = 3x$ over the interval $[2, 6]$.

3. Find the left endpoint, right endpoint, and midpoint approximations of the area under the curve $y = x^2 + 1$ over the interval $[0, 5]$ using $n = 5$ subintervals.

4. Find the left endpoint, right endpoint, and midpoint approximations of the area under the curve $y = x^3$ over the interval $[1, 6]$ using $n = 5$ subintervals.

5. Find the left endpoint, right endpoint, and midpoint approximations of the area under the curve $y = \cos x$ over the interval $[-\pi/2, \pi/2]$ using $n = 4$ subintervals.

6. Find the left endpoint, right endpoint, and midpoint approximations of the area under the curve $y = e^x$ over the interval $[0, 5]$ using $n = 5$ subintervals.

7. The accompanying figure shows five points on the graph of an unknown function f. Devise a strategy for using the known points to approximate the area A under the graph of $y = f(x)$ over the interval $[1, 5]$. Describe your strategy, and use it to approximate A.

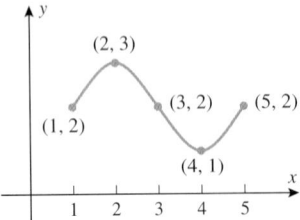

Figure Ex-7

8. (a) Use an appropriate geometric formula to find the exact area A under the line $y = 3x + 1$ over the interval $[1, 5]$.

   (b) Show that the exact area is equal to the average value of the left endpoint and right endpoint approximations of A obtained using $n = 4$ subintervals.

   (c) What is the explanation of the result in part (b)?

In Exercises 9–14, use a calculating utility to find the left endpoint, right endpoint, and midpoint approximations to the area under the curve $y = f(x)$ over the stated interval using $n = 10$ subintervals.

9. $y = 1/x$; $[1, 2]$

10. $y = 1/x^2$; $[1, 3]$

11. $y = \sin x$; $[0, \pi/2]$

12. $y = \sqrt{x}$; $[0, 4]$

13. $y = \ln x$; $[1, 2]$

14. $y = e^x$; $[0, 1]$

15. If you have a programmable calculator, create a program for calculating the midpoint approximation of the area under a curve $y = f(x)$ over an interval $[a, b]$ using n subintervals, and use the program to find midpoint approximations in Exercises 9–14 with

   (a) $n = 25$ (b) $n = 50$ (c) $n = 100$.

C 16. If you have a CAS, devise a procedure for using it to calculate the midpoint approximation of the area under a curve $y = f(x)$ over an interval $[a, b]$ using n subintervals, and use the procedure to find the midpoint approximations in Exercises 9–14 with

   (a) $n = 25$ (b) $n = 50$ (c) $n = 100$.

In Exercises 17–20, sketch the region whose signed area is represented by the definite integral, and evaluate the integral using an appropriate formula from geometry, where needed.

17. (a) $\displaystyle\int_0^3 x\,dx$ (b) $\displaystyle\int_{-2}^{-1} x\,dx$

   (c) $\displaystyle\int_{-1}^4 x\,dx$ (d) $\displaystyle\int_{-5}^5 x\,dx$

18. (a) $\displaystyle\int_0^2 \left(1 - \tfrac{1}{2}x\right) dx$ (b) $\displaystyle\int_{-1}^1 \left(1 - \tfrac{1}{2}x\right) dx$

   (c) $\displaystyle\int_2^3 \left(1 - \tfrac{1}{2}x\right) dx$ (d) $\displaystyle\int_0^3 \left(1 - \tfrac{1}{2}x\right) dx$

19. (a) $\displaystyle\int_0^5 2\,dx$ (b) $\displaystyle\int_0^\pi \cos x\,dx$

   (c) $\displaystyle\int_{-1}^2 |2x - 3|\,dx$ (d) $\displaystyle\int_{-1}^1 \sqrt{1 - x^2}\,dx$

20. (a) $\displaystyle\int_{-10}^{-5} 6\,dx$ (b) $\displaystyle\int_{-\pi/3}^{\pi/3} \sin x\,dx$

   (c) $\displaystyle\int_0^3 |x - 2|\,dx$ (d) $\displaystyle\int_0^2 \sqrt{4 - x^2}\,dx$

21. Use the areas shown in the accompanying figure to find

   (a) $\displaystyle\int_a^b f(x)\,dx$ (b) $\displaystyle\int_b^c f(x)\,dx$

   (c) $\displaystyle\int_a^c f(x)\,dx$ (d) $\displaystyle\int_a^d f(x)\,dx$.

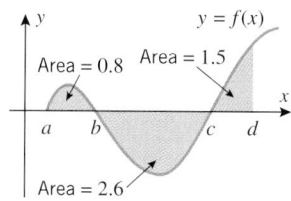

Figure Ex-21

22. In each part, evaluate the integral, given that

$$f(x) = \begin{cases} 2x, & x \le 1 \\ 2, & x > 1 \end{cases}$$

(a) $\displaystyle\int_0^1 f(x)\,dx$

(b) $\displaystyle\int_{-1}^1 f(x)\,dx$

(c) $\displaystyle\int_1^{10} f(x)\,dx$

(d) $\displaystyle\int_{1/2}^5 f(x)\,dx$

23. Find $\displaystyle\int_{-1}^2 [f(x) + 2g(x)]\,dx$ if

$$\int_{-1}^2 f(x)\,dx = 5 \quad \text{and} \quad \int_{-1}^2 g(x)\,dx = -3$$

24. Find $\displaystyle\int_1^4 [3f(x) - g(x)]\,dx$ if

$$\int_1^4 f(x)\,dx = 2 \quad \text{and} \quad \int_1^4 g(x)\,dx = 10$$

25. Find $\displaystyle\int_1^5 f(x)\,dx$ if

$$\int_0^1 f(x)\,dx = -2 \quad \text{and} \quad \int_0^5 f(x)\,dx = 1$$

26. Find $\displaystyle\int_3^{-2} f(x)\,dx$ if

$$\int_{-2}^1 f(x)\,dx = 2 \quad \text{and} \quad \int_1^3 f(x)\,dx = -6$$

In Exercises 27 and 28, use Theorem 7.5.4 and appropriate formulas from geometry to evaluate the integrals.

27. (a) $\displaystyle\int_0^1 (x + 2\sqrt{1-x^2})\,dx$ (b) $\displaystyle\int_{-1}^3 (4-5x)\,dx$

28. (a) $\displaystyle\int_{-3}^0 (2 + \sqrt{9-x^2})\,dx$ (b) $\displaystyle\int_{-2}^2 (1 - 3|x|)\,dx$

In Exercises 29 and 30, use Theorem 7.5.6 to determine whether the value of the integral is positive or negative.

29. (a) $\displaystyle\int_2^3 \frac{\sqrt{x}}{1-x}\,dx$ (b) $\displaystyle\int_0^4 \frac{x^2}{3 - \cos x}\,dx$

30. (a) $\displaystyle\int_{-3}^{-1} \frac{x^4}{\sqrt{3-x}}\,dx$ (b) $\displaystyle\int_{-2}^2 \frac{x^3 - 9}{|x|+1}\,dx$

In Exercises 31 and 32, evaluate the integrals by completing the square and applying appropriate formulas from geometry.

31. $\displaystyle\int_0^{10} \sqrt{10x - x^2}\,dx$ **32.** $\displaystyle\int_0^3 \sqrt{6x - x^2}\,dx$

In Exercises 33 and 34, express the limits as definite integrals over the interval $[a, b]$. Do not try to evaluate the integrals.

33. (a) $\displaystyle\lim_{\max \Delta x_k \to 0} \sum_{k=1}^n 4x_k^*(1 - 3x_k^*)\Delta x_k;\ a = -3, b = 3$

(b) $\displaystyle\lim_{\max \Delta x_k \to 0} \sum_{k=1}^n e^{x_k^*}\,\Delta x_k;\ a = 0, b = 1$

34. (a) $\displaystyle\lim_{\max \Delta x_k \to 0} \sum_{k=1}^n (x_k^*)^3 \Delta x_k;\ a = 1, b = 2$

(b) $\displaystyle\lim_{\max \Delta x_k \to 0} \sum_{k=1}^n (\sin^2 x_k^*)\Delta x_k;\ a = 0, b = \pi/2$

In Exercises 35 and 36, evaluate the limit over the interval $[a, b]$ by expressing it as a definite integral and applying an appropriate formula from geometry.

35. $\displaystyle\lim_{\max \Delta x_k \to 0} \sum_{k=1}^n (3x_k^* + 1)\Delta x_k;\ a = 0, b = 1$

36. $\displaystyle\lim_{\max \Delta x_k \to 0} \sum_{k=1}^n \sqrt{4 - (x_k^*)^2}\,\Delta x_k;\ a = -2, b = 2$

In Exercises 37 and 38, use Formula (7) to express the integrals as limits of Riemann sums. Do not try to evaluate the integrals.

37. (a) $\displaystyle\int_1^2 2x\,dx$ (b) $\displaystyle\int_0^1 \frac{x}{x+1}\,dx$

38. (a) $\displaystyle\int_1^2 \ln x\,dx$ (b) $\displaystyle\int_{-\pi/2}^{\pi/2} (1 + \cos x)\,dx$

39. In this exercise you will find the area A under the graph of $y = x$ over the interval $[1, 2]$ by calculating the limit of right endpoint approximations. For this particular problem, the area can be found much more easily using a formula from geometry, so our purpose here is not to provide a practical method for calculating the area, but rather to illustrate the idea that underlies the concept of a definite integral.

(a) Suppose that the interval $[1, 2]$ is subdivided into n equal subintervals of length $\Delta x = 1/n$ and that the points $x_1^*, x_2^*, \ldots, x_n^*$ are the right endpoints of the subintervals. Show that the right endpoint of the kth subinterval is

$$x_k^* = 1 + \frac{k}{n}$$

[*Suggestion:* Find x_1^*, x_2^*, and x_3^*, and then look for the pattern.]

(b) Show that with n subintervals the right endpoint approximation of the area A is

$$\sum_{k=1}^{n} f(x_k^*)\Delta x = \sum_{k=1}^{n}\left[\left(1+\frac{k}{n}\right)\frac{1}{n}\right]$$

(c) Use Theorem 7.4.2 to show that the right endpoint approximation can be expressed as

$$\sum_{k=1}^{n} f(x_k^*)\Delta x = \frac{3}{2}+\frac{1}{2n}$$

(d) From (2), the area A is

$$A = \lim_{n\to+\infty}\sum_{k=1}^{n} f(x_k^*)\Delta x$$

Find this limit, and check your answer by using a formula from geometry to calculate A.

40. Find the area A in Exercise 39 as a limit of left endpoint approximations.

In Exercises 41–44, use the method of Exercise 39 to find the area under the curve $y = f(x)$ over the interval $[a, b]$ as a limit of right and left endpoint approximations.

41. $y = x^2$; $a = 0, b = 1$
42. $y = 4 - \frac{1}{4}x^2$; $a = 0, b = 3$
43. $y = x^3$; $a = 2, b = 6$
44. $y = 1 - x^3$; $a = -3, b = -1$
45. In each part, use Theorem 7.5.8 to determine whether the function f is integrable on the interval $[-1, 1]$.
(a) $f(x) = e^x \cos x$

(b) $f(x) = \begin{cases} x/|x|, & x \neq 0 \\ 0, & x = 0 \end{cases}$

(c) $f(x) = \begin{cases} 1/x^2, & x \neq 0 \\ 0, & x = 0 \end{cases}$

(d) $f(x) = \begin{cases} \sin 1/x, & x \neq 0 \\ 0, & x = 0 \end{cases}$

46. It can be shown that every interval contains both rational and irrational numbers. Accepting this to be so, do you believe that the function

$$f(x) = \begin{cases} 1 & \text{if } x \text{ is rational} \\ 0 & \text{if } x \text{ is irrational} \end{cases}$$

is integrable on a closed interval $[a, b]$? Explain your reasoning.

47. It can be shown that the limit in Formula (7) has all of the limit properties stated in Theorem 2.2.2. Accepting this to be so, show that

(a) $\int_a^b cf(x)\,dx = c\int_a^b f(x)\,dx$

(b) $\int_a^b [f(x)+g(x)]\,dx = \int_a^b f(x)\,dx + \int_a^b g(x)\,dx$

48. Find the smallest and largest values that the Riemann sum

$$\sum_{k=1}^{3} f(x_k^*)\Delta x_k$$

can have on the interval $[0, 4]$ if $f(x) = x^2 - 3x + 4$ and $\Delta x_1 = 1, \Delta x_2 = 2, \Delta x_3 = 1$.

7.6 THE FUNDAMENTAL THEOREM OF CALCULUS

In this section we will establish two basic relationships between definite and indefinite integrals that together constitute a result called the Fundamental Theorem of Calculus. One part of this theorem will relate the rectangle and antiderivative methods for calculating areas, and the second part will provide a powerful method for evaluating definite integrals using antiderivatives.

THE FUNDAMENTAL THEOREM OF CALCULUS

To motivate the results we are looking for, let us begin by assuming that f is nonnegative and continuous on the interval $[a, b]$, in which case the area A under the graph of f over the interval $[a, b]$ is represented by the definite integral

$$A = \int_a^b f(x)\,dx \tag{1}$$

(Figure 7.6.1).

Recall from our discussion of the antiderivative method in Section 7.1 that if $A(x)$ is the area under the graph of f from a to x (Figure 7.6.2), then:

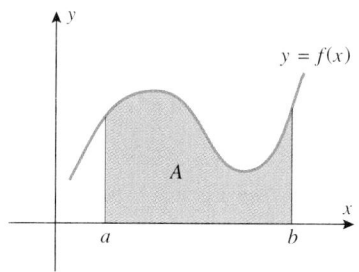

Figure 7.6.1

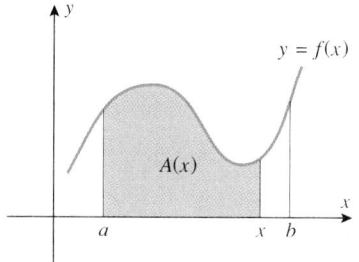

Figure 7.6.2

- $A'(x) = f(x)$
- $A(a) = 0$ The area under the curve from a to a is the area above the single point a, and hence is zero.
- $A(b) = A$ The area under the curve from a to b is A.

The formula $A'(x) = f(x)$ states that $A(x)$ is an antiderivative of $f(x)$, which implies that every other antiderivative of $f(x)$ can be obtained by adding a constant to $A(x)$. Accordingly, let

$$F(x) = A(x) + C$$

be any antiderivative of $f(x)$, and consider what happens when we subtract $F(a)$ from $F(b)$. We obtain

$$F(b) - F(a) = [A(b) + C] - [A(a) + C] = A(b) - A(a) = A - 0 = A$$

and hence (1) can be expressed as

$$\int_a^b f(x)\,dx = F(b) - F(a)$$

In words, this equation states that the definite integral can be evaluated by finding any antiderivative of the integrand and then subtracting the value of this antiderivative at the lower limit of integration from its value at the upper limit of integration. Although we derived this result subject to the assumption that f is nonnegative on $[a, b]$, this assumption is not essential, as we will prove in the following theorem, which is the main tool used to evaluate definite integrals.

7.6.1 THEOREM (*The Fundamental Theorem of Calculus, Part 1*). *If f is continuous on $[a, b]$, and if F is any antiderivative of f on $[a, b]$, then*

$$\int_a^b f(x)\,dx = F(b) - F(a) \tag{2}$$

Proof. Let $x_1, x_2, \ldots, x_{n-1}$ be any points in $[a, b]$ such that

$$a < x_1 < x_2 < \cdots < x_{n-1} < b$$

These points divide $[a, b]$ into n subintervals

$$[a, x_1], [x_1, x_2], \ldots, [x_{n-1}, b] \tag{3}$$

whose lengths, as usual, we denote by

$$\Delta x_1, \Delta x_2, \ldots, \Delta x_n$$

By hypothesis, $F'(x) = f(x)$ for all x in $[a, b]$, so F satisfies the hypotheses of the Mean-Value Theorem (6.5.2) on each subinterval in (3). Hence, we can find points $x_1^*, x_2^*, \ldots, x_n^*$ in the respective subintervals in (3) such that

$$F(x_1) - F(a) = F'(x_1^*)(x_1 - a) = f(x_1^*)\Delta x_1$$
$$F(x_2) - F(x_1) = F'(x_2^*)(x_2 - x_1) = f(x_2^*)\Delta x_2$$
$$F(x_3) - F(x_2) = F'(x_3^*)(x_3 - x_2) = f(x_3^*)\Delta x_3$$
$$\vdots \qquad\qquad \vdots \qquad\qquad \vdots$$
$$F(b) - F(x_{n-1}) = F'(x_n^*)(b - x_{n-1}) = f(x_n^*)\Delta x_n$$

Adding the preceding equations yields

$$F(b) - F(a) = \sum_{k=1}^{n} f(x_k^*)\Delta x_k \tag{4}$$

Let us now increase n in such a way that max $\Delta x_k \to 0$. Since f is assumed to be continuous, the right side of (4) approaches $\int_a^b f(x)\,dx$, by Theorem 7.5.8(a) and Formula (7) of Section 7.5. However, the left side of (4) is a constant that is independent of n; thus,

$$F(b) - F(a) = \lim_{\max \Delta x_k \to 0} \sum_{k=1}^{n} f(x_k^*)\Delta x_k = \int_a^b f(x)\,dx$$

It is standard to denote the difference $F(b) - F(a)$ as

$$F(x)\Big]_a^b = F(b) - F(a) \quad \text{or} \quad \left[F(x)\right]_a^b = F(b) - F(a)$$

For example, using the first of these notations we can express (2) as

$$\int_a^b f(x)\,dx = F(x)\Big]_a^b \tag{5}$$

Example 1

Evaluate $\displaystyle\int_1^2 x\,dx$.

Solution. The function $F(x) = \frac{1}{2}x^2$ is an antiderivative of $f(x) = x$; thus, from (2)

$$\int_1^2 x\,dx = \frac{1}{2}x^2\Big]_1^2 = \frac{1}{2}(2)^2 - \frac{1}{2}(1)^2 = 2 - \frac{1}{2} = \frac{3}{2} \qquad \blacktriangleleft$$

Example 2

In Example 1 of the last section we approximated the area under the graph of $y = 9 - x^2$ over the interval $[0, 3]$ using left endpoint, right endpoint, and midpoint approximations, all of which produced an approximation of roughly 18 (square units); and in the remark following that example we stated without proof that the exact area A is 18 (square units). We can now confirm this using the Fundamental Theorem of Calculus as follows:

$$A = \int_0^3 (9 - x^2)\,dx = 9x - \frac{x^3}{3}\Big]_0^3 = \left(27 - \frac{27}{3}\right) - 0 = 18 \qquad \blacktriangleleft$$

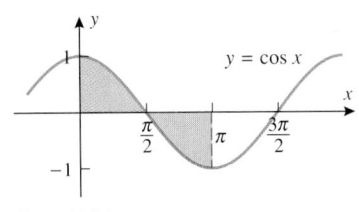

Figure 7.6.3

Example 3

(a)   Find the area under the curve $y = \cos x$ over the interval $[0, \pi/2]$ (Figure 7.6.3).

(b)   Make a conjecture about the value of the integral

$$\int_0^{\pi} \cos x\,dx$$

and confirm your conjecture using the Fundamental Theorem of Calculus.

Solution (a). Since $\cos x \geq 0$ over the interval $[0, \pi/2]$, the area A under the curve is

$$A = \int_0^{\pi/2} \cos x\,dx = \sin x\Big]_0^{\pi/2} = \sin\frac{\pi}{2} - \sin 0 = 1$$

Solution (b). The given integral can be interpreted as the signed area between the graph of $y = \cos x$ and the interval $[0, \pi]$. The graph in Figure 7.6.3 suggests that over the interval $[0, \pi]$ the portion of area above the x-axis is the same as the portion of area below the x-axis, so we conjecture that the signed area is zero; this implies that the value of the integral is zero. This is confirmed by the computations

$$\int_0^{\pi} \cos x\,dx = \sin x\Big]_0^{\pi} = \sin\pi - \sin 0 = 0 \qquad \blacktriangleleft$$

Observe that in the preceding examples we did not include a constant of integration in the antiderivatives. In general, when applying the Fundamental Theorem of Calculus there is no need to include a constant of integration because it will drop out anyhow. To see that this is so, let F be any antiderivative of the integrand on $[a, b]$, and let C be any constant; then

$$\int_a^b f(x)\,dx = F(x) + C \Big]_a^b = [F(b) + C] - [F(a) + C] = F(b) - F(a)$$

Thus, for purposes of evaluating a definite integral we can omit the constant of integration in

$$\int f(x)\,dx = F(x) + C$$

and express (5) as

$$\int_a^b f(x)\,dx = \left[\int f(x)\,dx\right]_a^b \tag{6}$$

which relates the definite and indefinite integrals.

Example 4

$$\int_1^9 \sqrt{x}\,dx = \int \sqrt{x}\,dx \Big]_1^9 = \int x^{1/2}\,dx \Big]_1^9 = \frac{2}{3}x^{3/2}\Big]_1^9 = \frac{2}{3}(27 - 1) = \frac{52}{3} \qquad \blacktriangleleft$$

REMARK. Usually, we will dispense with the step of displaying the indefinite integral explicitly and write the antiderivative immediately, as in our first three examples.

Example 5

Table 7.2.1 will be helpful for the following computations.

$$\int_0^{\ln 3} 5e^x\,dx = 5\int_0^{\ln 3} e^x\,dx = 5e^x\Big]_0^{\ln 3} = 5(e^{\ln 3} - e^0) = 5(3 - 1) = 10$$

$$\int_1^2 \frac{1}{x}\,dx = \ln|x|\Big]_1^2 = \ln|2| - \ln|1| = \ln 2 - \ln 1 = \ln 2$$

$$\int_{-2}^{-1} \frac{1}{x}\,dx = \ln|x|\Big]_{-2}^{-1} = \ln|-1| - \ln|-2| = \ln 1 - \ln 2 = -\ln 2$$

$$\int_{-\pi/4}^{\pi/4} \sec x \tan x\,dx = \sec x\Big]_{-\pi/4}^{\pi/4} = \sec\left(\frac{\pi}{4}\right) - \sec\left(-\frac{\pi}{4}\right) = \frac{2}{\sqrt{2}} - \frac{2}{\sqrt{2}} = 0 \qquad \blacktriangleleft$$

WARNING. The requirement in the Fundamental Theorem of Calculus that f be continuous on $[a, b]$ is important to keep in mind, for if you attempt to apply this theorem in cases where the integrand is not continuous on the interval of integration, then you may obtain erroneous results. For example, the function $f(x) = 1/x^2$ has a discontinuity at $x = 0$, so the Fundamental Theorem of Calculus cannot be used to integrate f on any interval that contains $x = 0$. However, if we ignore this and blindly apply the theorem over the interval $[-1, 1]$, we obtain

$$\int_{-1}^1 \frac{1}{x^2}\,dx = -\frac{1}{x}\Big]_{-1}^1 = -[1 - (-1)] = -2$$

which is clearly erroneous because $f(x) = 1/x^2$ is a nonnegative function and hence cannot possibly produce a negative definite integral.

FOR THE READER.   If you have a CAS, read the documentation on evaluating definite integrals, and then check the results in the preceding examples.

The Fundamental Theorem of Calculus can be applied without modification to definite integrals in which the lower limit of integration is greater than or equal to the upper limit of integration.

Example 6

$$\int_1^1 x^2\,dx = \left.\frac{x^3}{3}\right]_1^1 = \frac{1}{3}-\frac{1}{3}=0$$

$$\int_4^0 x\,dx = \left.\frac{x^2}{2}\right]_4^0 = \left[\frac{0}{2}-\frac{16}{2}\right]=-8$$

The latter result is consistent with the result that would be obtained by first reversing the limits of integration in accordance with Definition 7.5.3(b):

$$\int_4^0 x\,dx = -\int_0^4 x\,dx = -\left.\frac{x^2}{2}\right]_0^4 = -\left[\frac{16}{2}-\frac{0}{2}\right]=-8 \qquad \blacktriangleleft$$

To integrate a continuous function that is defined piecewise on an interval $[a,b]$, split this interval into subintervals at the breakpoints of the function, and integrate separately over each subinterval in accordance with Theorem 7.5.5.

Example 7

Evaluate $\int_0^6 f(x)\,dx$ if

$$f(x) = \begin{cases} x^2, & x<2 \\ 3x-2, & x\ge 2 \end{cases}$$

Solution. From Theorem 7.5.5

$$\int_0^6 f(x)\,dx = \int_0^2 f(x)\,dx + \int_2^6 f(x)\,dx = \int_0^2 x^2\,dx + \int_2^6 (3x-2)\,dx$$

$$= \left.\frac{x^3}{3}\right]_0^2 + \left[\frac{3x^2}{2}-2x\right]_2^6 = \left(\frac{8}{3}-0\right)+(42-2)=\frac{128}{3} \qquad \blacktriangleleft$$

Example 8

Evaluate $\int_{-1}^2 |x|\,dx$.

Solution. Since $|x|=x$ when $x\ge 0$ and $|x|=-x$ when $x\le 0$,

$$\int_{-1}^2 |x|\,dx = \int_{-1}^0 |x|\,dx + \int_0^2 |x|\,dx$$

$$= \int_{-1}^0 (-x)\,dx + \int_0^2 x\,dx$$

$$= \left.-\frac{x^2}{2}\right]_{-1}^0 + \left.\frac{x^2}{2}\right]_0^2 = \frac{1}{2}+2=\frac{5}{2} \qquad \blacktriangleleft$$

DUMMY VARIABLES

To evaluate a definite integral using the Fundamental Theorem of Calculus, one needs to be able to find an antiderivative of the integrand; thus, it is important to know what kinds of functions have antiderivatives. It is our next objective to show that all continuous functions have antiderivatives, but to do this we will need some preliminary results.

Formula (6) shows that there is a close relationship between the integrals

$$\int_a^b f(x)\, dx \quad \text{and} \quad \int f(x)\, dx$$

However, the definite and indefinite integrals differ in some important ways. For one thing, the two integrals are different kinds of objects—the definite integral is a *number* (the signed area between the graph of $y = f(x)$ and the interval $[a, b]$), whereas the indefinite integral is a *function*, or more accurately a set of functions [the antiderivatives of $f(x)$]. However, the two types of integrals also differ in the role played by the variable of integration. In an indefinite integral, the variable of integration is "passed through" to the antiderivative in the sense that integrating a function of x produces a function of x, integrating a function of t produces a function of t, and so forth. For example,

$$\int x^2\, dx = \frac{x^3}{3} + C \quad \text{and} \quad \int t^2\, dt = \frac{t^3}{3} + C$$

In contrast, the variable of integration in a definite integral is not passed through to the end result, since the end result is a number. Thus, integrating a function of x over an interval and integrating the same function of t over the same interval of integration produces the same value for the integral. For example,

$$\int_1^3 x^2\, dx = \frac{x^3}{3}\bigg]_{x=1}^3 = \frac{27}{3} - \frac{1}{3} = \frac{26}{3} \quad \text{and} \quad \int_1^3 t^2\, dt = \frac{t^3}{3}\bigg]_{t=1}^3 = \frac{27}{3} - \frac{1}{3} = \frac{26}{3}$$

However, this latter result should not be surprising, since the area under the graph of the curve $y = f(x)$ over an interval $[a, b]$ on the x-axis is the same as the area under the graph of the curve $y = f(t)$ over the interval $[a, b]$ on the t-axis (Figure 7.6.4).

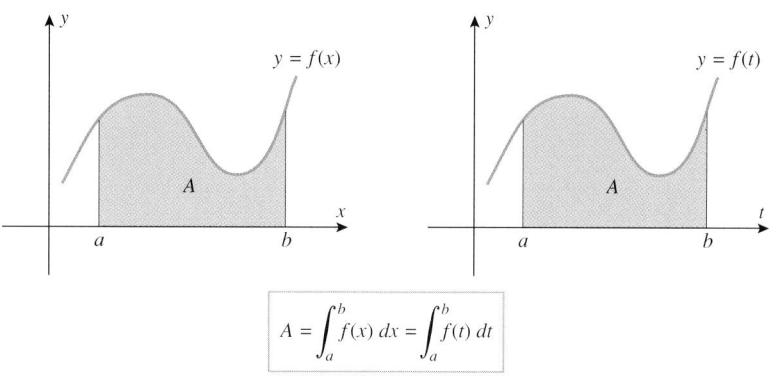

Figure 7.6.4

Because the variable of integration in a definite integral plays no role in the end result, it is often referred to as a ***dummy variable***. In summary:

> *Whenever you find it convenient to change the letter used for the variable of integration in a definite integral, you can do so without changing the value of the integral.*

THE MEAN-VALUE THEOREM FOR INTEGRALS

To reach our goal of showing that continuous functions have antiderivatives, we will need to develop a basic property of definite integrals, known as the *Mean-Value Theorem for Integrals*. In the next section we will use this theorem to extend the familiar idea of "average value" so that it applies to continuous functions, but here we will need it as a tool for developing other results.

Let f be a continuous nonnegative function on $[a, b]$, and let m and M be the minimum and maximum values of $f(x)$ on this interval. Consider the rectangle of heights m and M

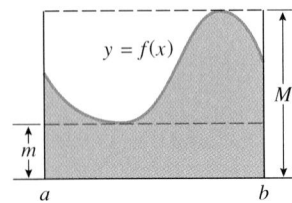

Figure 7.6.5

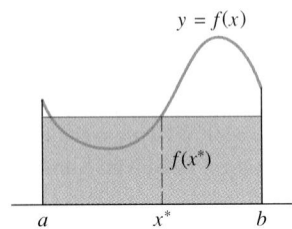

Figure 7.6.6

over the interval $[a, b]$ (Figure 7.6.5). It is clear geometrically from this figure that the area

$$A = \int_a^b f(x)\, dx$$

under $y = f(x)$ is at least as large as the area of the rectangle of height m and no larger than the area of the rectangle of height M. It seems reasonable, therefore, that there is a rectangle over the interval $[a, b]$ of some appropriate height $f(x^*)$ between m and M whose area is precisely A; that is,

$$\int_a^b f(x)\, dx = f(x^*)(b - a)$$

(Figure 7.6.6). This is a special case of the following result.

7.6.2   THEOREM (*The Mean-Value Theorem for Integrals*).   *If f is continuous on a closed interval $[a, b]$, then there is at least one number x^* in $[a, b]$ such that*

$$\int_a^b f(x)\, dx = f(x^*)(b - a) \tag{7}$$

Proof. By the Extreme-Value Theorem (6.1.3), f assumes a maximum value M and a minimum value m on $[a, b]$. Thus, for all x in $[a, b]$,

$$m \le f(x) \le M$$

and from Theorem 7.5.6(*b*)

$$\int_a^b m\, dx \le \int_a^b f(x)\, dx \le \int_a^b M\, dx$$

or

$$m(b - a) \le \int_a^b f(x)\, dx \le M(b - a) \tag{8}$$

or

$$m \le \frac{1}{b - a} \int_a^b f(x)\, dx \le M$$

This implies that

$$\frac{1}{b - a} \int_a^b f(x)\, dx \tag{9}$$

is a number between m and M, and since $f(x)$ assumes the values m and M on $[a, b]$, it follows from the Intermediate-Value Theorem (2.4.8) that $f(x)$ must assume the value (9) at some point x^* in $[a, b]$; that is,

$$\frac{1}{b - a} \int_a^b f(x)\, dx = f(x^*) \quad \text{or} \quad \int_a^b f(x)\, dx = f(x^*)(b - a) \qquad \blacksquare$$

Example 9

Since $f(x) = x^2$ is continuous on the interval $[1, 4]$, the Mean-Value Theorem for Integrals guarantees that there is a number x^* in $[1, 4]$ such that

$$\int_1^4 x^2\, dx = f(x^*)(4 - 1) = (x^*)^2(4 - 1) = 3(x^*)^2$$

But

$$\int_1^4 x^2\, dx = \frac{x^3}{3} \Big]_1^4 = 21$$

so that

$$3(x^*)^2 = 21 \quad \text{or} \quad (x^*)^2 = 7 \quad \text{or} \quad x^* = \pm\sqrt{7}$$

Thus, $x^* = \sqrt{7} \approx 2.65$ is the number in the interval $[1, 4]$ whose existence is guaranteed by the Mean-Value Theorem for Integrals. ◀

PART 2 OF THE FUNDAMENTAL THEOREM OF CALCULUS

In Section 7.1 we gave an informal argument to show that if f is continuous and nonnegative on $[a, b]$, and if $A(x)$ is the area under the graph of $y = f(x)$ over the interval $[a, x]$ (Figure 7.6.2), then $A'(x) = f(x)$. But $A(x)$ can be expressed as the definite integral

$$A(x) = \int_a^x f(t)\,dt$$

(where we have used t rather than x as the variable of integration to avoid a conflict with the x that appears as the upper limit of integration). Thus, the relationship $A'(x) = f(x)$ can be expressed as

$$\frac{d}{dx}\left[\int_a^x f(t)\,dt\right] = f(x)$$

This is a special case of the following more general result, which applies even if f has negative values.

7.6.3 THEOREM (*The Fundamental Theorem of Calculus, Part 2*). *If f is continuous on an interval I, then f has an antiderivative on I. In particular, if a is any point in I, then the function F defined by*

$$F(x) = \int_a^x f(t)\,dt$$

is an antiderivative of f on I; that is, $F'(x) = f(x)$ for each x in I, or in an alternative notation

$$\frac{d}{dx}\left[\int_a^x f(t)\,dt\right] = f(x) \tag{10}$$

Proof. We will show first that $F(x)$ is defined at each point x in the interval I. If $x > a$ and x is in the interval I, then Theorem 7.5.8(a) applied to the interval $[a, x]$ and the continuity of f on I ensures that $F(x)$ is defined; and if x is in the interval I and $x \le a$, then Definition 7.5.3(b) combined with Theorem 7.5.8(a) ensures that $F(x)$ is defined. Thus, $F(x)$ is defined for all x in I.

Next we will show that $F'(x) = f(x)$ for each x in the interval I. If x is not an endpoint of I, then it follows from the definition of a derivative that

$$F'(x) = \lim_{h \to 0} \frac{F(x+h) - F(x)}{h}$$

$$= \lim_{h \to 0} \frac{1}{h}\left[\int_a^{x+h} f(t)\,dt - \int_a^x f(t)\,dt\right]$$

$$= \lim_{h \to 0} \frac{1}{h}\left[\int_a^{x+h} f(t)\,dt + \int_x^a f(t)\,dt\right]$$

$$= \lim_{h \to 0} \frac{1}{h}\int_x^{x+h} f(t)\,dt \qquad \boxed{\text{Theorem 7.5.5}}$$

Applying the Mean-Value Theorem for Integrals (7.6.2) to the last expression, we obtain

$$F'(x) = \lim_{h \to 0} \frac{1}{h}[f(t^*) \cdot h] = \lim_{h \to 0} f(t^*) \tag{11}$$

where t^* is some number between x and $x + h$. Because t^* is between x and $x + h$, it follows that $t^* \to x$ as $h \to 0$. Thus, $f(t^*) \to f(x)$ as $h \to 0$, since f is assumed continuous at x. Therefore, it follows from (11) that $F'(x) = f(x)$. If x is an endpoint of the interval I, then the two-sided limits in the proof must be replaced by the appropriate one-sided limits, but otherwise the arguments are identical. ∎

In words, Formula (10) states:

If a definite integral has a variable upper limit of integration and a continuous integrand, then the derivative of the integral with respect to its upper limit is equal to the integrand evaluated at the upper limit.

Example 10

Find

$$\frac{d}{dx}\left[\int_1^x t^3\, dt\right]$$

by applying Part 2 of the Fundamental Theorem of Calculus, and then confirm the result by performing the integration and then differentiating.

Solution. The integrand is a continuous function, so from (10)

$$\frac{d}{dx}\left[\int_1^x t^3\, dt\right] = x^3$$

Alternatively, evaluating the integral and then differentiating yields

$$\int_1^x t^3\, dt = \frac{t^4}{4}\Big]_{t=1}^x = \frac{x^4}{4} - \frac{1}{4}, \qquad \frac{d}{dx}\left[\frac{x^4}{4} - \frac{1}{4}\right] = x^3$$

so the two methods for differentiating the integral agree. ◄

Example 11

Since

$$f(x) = \frac{\sin x}{x}$$

is continuous on any interval that does not contain the origin, it follows from (10) that on the interval $(0, +\infty)$ we have

$$\frac{d}{dx}\left[\int_1^x \frac{\sin t}{t}\, dt\right] = \frac{\sin x}{x}$$

Unlike the preceding example, there is no way to evaluate the integral in terms of familiar functions, so Formula (10) provides the only simple method for finding the derivative. ◄

....

DIFFERENTIATION AND INTEGRATION ARE INVERSE PROCESSES

The two parts of the Fundamental Theorem of Calculus, when taken together, tell us that differentiation and integration are inverse processes in the sense that each undoes the effect of the other. To see why this is so, note that Part 1 of the Fundamental Theorem of Calculus (7.6.1) implies that

$$\int_a^x f'(t)\, dt = f(x) - f(a)$$

which tells us that if the value of $f(a)$ is known, then function f can be recovered from its derivative f' by integrating. Conversely, Part 2 of the Fundamental Theorem of Calculus

(7.6.3) states that

$$\frac{d}{dx}\left[\int_a^x f(t)\,dt\right] = f(x)$$

which tells us that the function f can be recovered from its integral by differentiating. Thus, differentiation and integration can be viewed as inverse processes.

It is common to treat parts 1 and 2 of the Fundamental Theorem of Calculus as a single theorem, and refer to it simply as the *Fundamental Theorem of Calculus*. This theorem ranks as one of the greatest discoveries in the history of science, and its formulation by Newton and Leibniz is generally regarded to be the "discovery of calculus."

EXERCISE SET 7.6 ⋀ Graphing Calculator [c] CAS

1. In each part, use a definite integral to find the area of the region, and check your answer using an appropriate formula from geometry.

(a)   (b)   (c)

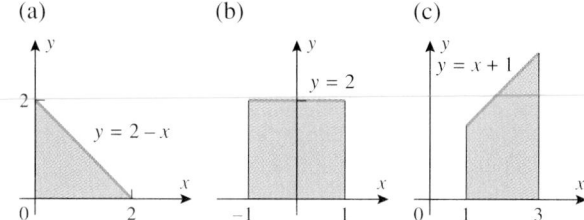

2. In each part, use a definite integral to find the area under the curve $y = f(x)$ over the stated interval, and check your answer using an appropriate formula from geometry.
  (a) $f(x) = x$; $[0, 5]$
  (b) $f(x) = 5$; $[3, 9]$
  (c) $f(x) = x + 3$; $[-1, 2]$

In Exercises 3–8, find the area under the curve $y = f(x)$ over the stated interval.

3. $f(x) = x^3$; $[2, 3]$ **4.** $f(x) = x^4$; $[-1, 1]$

5. $f(x) = \sqrt{x}$; $[1, 9]$ **6.** $f(x) = x^{-3/5}$; $[1, 4]$

7. $f(x) = e^x$; $[1, 3]$ **8.** $f(x) = \dfrac{1}{x}$; $[1, 5]$

In Exercises 9–24, evaluate the integrals using Part 1 of the Fundamental Theorem of Calculus.

9. $\displaystyle\int_{-3}^0 (x^2 - 4x + 7)\,dx$ **10.** $\displaystyle\int_{-1}^2 x(1 + x^3)\,dx$

11. $\displaystyle\int_1^3 \frac{1}{x^2}\,dx$ **12.** $\displaystyle\int_1^2 \frac{1}{x^6}\,dx$

13. $\displaystyle\int_4^9 2x\sqrt{x}\,dx$ **14.** $\displaystyle\int_1^8 (5x^{2/3} - 4x^{-2})\,dx$

15. $\displaystyle\int_{-\pi/2}^{\pi/2} \sin\theta\,d\theta$ **16.** $\displaystyle\int_0^{\pi/4} \sec^2\theta\,d\theta$

17. $\displaystyle\int_{-\pi/4}^{\pi/4} \cos x\,dx$ **18.** $\displaystyle\int_0^1 (x - \sec x\tan x)\,dx$

19. $\displaystyle\int_{\ln 2}^3 5e^x\,dx$ **20.** $\displaystyle\int_{1/2}^1 \frac{1}{2x}\,dx$

21. $\displaystyle\int_1^4 \left(\frac{3}{\sqrt{t}} - 5\sqrt{t} - t^{-3/2}\right)dt$

22. $\displaystyle\int_4^9 (4y^{-1/2} + 2y^{1/2} + y^{-5/2})\,dy$

23. $\displaystyle\int_{\pi/6}^{\pi/2} \left(x + \frac{2}{\sin^2 x}\right)dx$

24. $\displaystyle\int_1^2 (x^{-1} + \sqrt{2}e^x - \csc x\cot x)\,dx$

[c] **25.** For each of the integrals you evaluated in Exercises 9–24, use a CAS to check your answer. [*Note:* CAS programs have commands for evaluating definite integrals exactly or approximately. Use the exact evaluation here.]

[c] **26.** Use a CAS to evaluate the integral

$$\int_a^{4a} (a^{1/2} - x^{1/2})\,dx$$

and check the answer by hand.

In Exercises 27–29, use Theorem 7.5.5 to evaluate the given integrals.

27. (a) $\displaystyle\int_0^2 |2x - 3|\,dx$ (b) $\displaystyle\int_0^{3\pi/4} |\cos x|\,dx$

28. (a) $\displaystyle\int_{-1}^2 \sqrt{2 + |x|}\,dx$ (b) $\displaystyle\int_{-1}^1 |e^x - 1|\,dx$

29. $\int_{-2}^{3} f(x)\,dx$, where $f(x) = \begin{cases} -x, & x \geq 0 \\ x^2, & x < 0 \end{cases}$

C **30.** CAS programs provide methods for entering functions that are defined piecewise. Check your documentation to see how this is done, and then use the CAS to evaluate

$$\int_{0}^{4} f(x)\,dx, \quad \text{where} \quad f(x) = \begin{cases} \sqrt{x}, & 0 \leq x < 1 \\ 1/x^2, & x \geq 1 \end{cases}$$

Check the answer by hand.

In Exercises 31–33, use a calculating utility to find the mid-point approximation of the integral using $n = 20$ subinter-vals, and then find the exact value of the integral using Part 1 of the Fundamental Theorem of Calculus.

---

31. $\int_{1}^{3} \frac{1}{x^2}\,dx$ **32.** $\int_{0}^{\pi/2} \sin x\,dx$ **33.** $\int_{1}^{3} \frac{1}{x}\,dx$

C **34.** Compare the answers obtained by the midpoint rule in Exercises 31–33 to those obtained using the numerical (approx-imate) integration command of a CAS.

35. Find the area under the curve $y = x^2 + 1$ over the interval $[0, 3]$. Make a sketch of the region.

36. Find the area that is above the x-axis, but below the curve $y = (1 - x)(x - 2)$. Make a sketch of the region.

37. Find the area under the curve $y = 3 \sin x$ over the interval $[0, 2\pi/3]$. Sketch the region.

38. Find the area below the interval $[-2, -1]$, but above the curve $y = x^3$. Make a sketch of the region.

39. Find the total area between the curve $y = x^2 - 3x - 10$ and the interval $[-3, 8]$. Make a sketch of the region. [*Hint:* Find the portion of area above the interval and the portion of area below the interval separately.]

40. (a) Use a graphing utility to generate the graph of

$$f(x) = \frac{1}{100}(x + 2)(x + 1)(x - 3)(x - 5)$$

and use the graph to make a conjecture about the sign of the integral

$$\int_{-2}^{5} f(x)\,dx$$

(b) Check your conjecture by evaluating the integral.

41. (a) Let f be an odd function; that is, $f(-x) = -f(x)$. In-vent a theorem that makes a statement about the value of an integral of the form

$$\int_{-a}^{a} f(x)\,dx$$

(b) Confirm that your theorem works for the integrals

$$\int_{-1}^{1} x^3\,dx \quad \text{and} \quad \int_{-\pi/2}^{\pi/2} \sin x\,dx$$

(c) Let f be an even function; that is, $f(-x) = f(x)$. In-vent a theorem that makes a statement about the rela-

tionship between the integrals

$$\int_{-a}^{a} f(x)\,dx \quad \text{and} \quad \int_{0}^{a} f(x)\,dx$$

(d) Confirm that your theorem works for the integrals

$$\int_{-1}^{1} x^2\,dx \quad \text{and} \quad \int_{-\pi/2}^{\pi/2} \cos x\,dx$$

C **42.** Use the theorem you invented in Exercise 41(a) to evaluate the integral

$$\int_{-5}^{5} \frac{x^7 - x^5 + x}{x^4 + x^2 + 7}\,dx$$

and check your answer with a CAS.

43. Define $F(x)$ by

$$F(x) = \int_{1}^{x} (t^3 + 1)\,dt$$

(a) Use Part 2 of the Fundamental Theorem of Calculus to find $F'(x)$.

(b) Check the result in part (a) by first integrating and then differentiating.

44. Define $F(x)$ by

$$F(x) = \int_{\pi/4}^{x} \cos 2t\,dt$$

(a) Use Part 2 of the Fundamental Theorem of Calculus to find $F'(x)$.

(b) Check the result in part (a) by first integrating and then differentiating.

In Exercises 45–48, use Part 2 of the Fundamental Theorem of Calculus to find the derivative.

---

45. (a) $\dfrac{d}{dx} \int_{1}^{x} \sin(\sqrt{t})\,dt$ (b) $\dfrac{d}{dx} \int_{0}^{x} e^{t^2}\,dt$

46. (a) $\dfrac{d}{dx} \int_{0}^{x} \dfrac{dt}{1 + \sqrt{t}}$ (b) $\dfrac{d}{dx} \int_{1}^{x} \ln t\,dt$

47. $\dfrac{d}{dx} \int_{x}^{0} \dfrac{t}{\cos t}\,dt$ [*Hint:* Use Definition 7.5.3(b).]

48. $\dfrac{d}{du} \int_{0}^{u} |x|\,dx$

49. Let $F(x) = \int_{2}^{x} \sqrt{3t^2 + 1}\,dt$. Find

(a) $F(2)$ (b) $F'(2)$ (c) $F''(2)$

50. Let $F(x) = \int_{0}^{x} \dfrac{\cos t}{t^2 + 3}\,dt$. Find

(a) $F(0)$ (b) $F'(0)$ (c) $F''(0)$

51. Let $F(x) = \int_{0}^{x} \dfrac{t - 3}{t^2 + 7}\,dt$ for $-\infty < x < +\infty$.

(a) Find the value of x where F attains its minimum value.

(b) Find intervals over which F is only increasing or only decreasing.

(c) Find open intervals over which F is only concave up or only concave down.

$\boxed{\text{c}}$ **52.** Use the plotting and numerical integration commands of a CAS to generate the graph of the function F in Exercise 51 over the interval $-20 \le x \le 20$, and confirm that the graph is consistent with the results obtained in that exercise.

53. (a) Over what open interval does the formula

$$F(x) = \int_1^x \frac{dt}{t}$$

represent an antiderivative of $f(x) = 1/x$?

(b) Find a point where the graph of F crosses the x-axis.

54. (a) Over what open interval does the formula

$$F(x) = \int_1^x \frac{1}{t^2 - 9}\, dt$$

represent an antiderivative of

$$f(x) = \frac{1}{x^2 - 9}?$$

(b) Find a point where the graph of F crosses the x-axis.

In Exercises 55 and 56, find all values of x^* in the stated interval that satisfy Equation (7) in the Mean-Value Theorem for Integrals (7.6.2), and explain what these numbers represent.

---

55. (a) $f(x) = \sqrt{x}$; $[0, 9]$ (b) $f(x) = 1/x$; $[1, e]$

56. (a) $f(x) = \sin x$; $[-\pi, \pi]$ (b) $f(x) = 1/x^2$; $[1, 3]$

It was shown in the proof of the Mean-Value Theorem for Integrals that if f is continuous on $[a, b]$, and if $m \le f(x) \le M$ on $[a, b]$, then

$$m(b - a) \le \int_a^b f(x)\, dx \le M(b - a)$$

[see (8)]. These inequalities make it possible to obtain bounds on the size of a definite integral from bounds on the size of its integrand. This is illustrated in Exercises 57–59.

---

57. Find the maximum and minimum values of $\sqrt{x^3 + 2}$ for $0 \le x \le 3$, and use these values to find bounds on the value of the integral

$$\int_0^3 \sqrt{x^3 + 2}\, dx$$

58. Find values of m and M such that $m \le x \sin x \le M$ for $0 \le x \le \pi$, and use these values to find bounds on the value of the integral

$$\int_0^\pi x \sin x\, dx$$

59. Show that

$$0 \le \int_1^5 \ln x\, dx \le 4 \ln 5$$

60. Prove:

(a) $[c F(x)]_a^b = c[F(x)]_a^b$

(b) $[F(x) + G(x)]_a^b = F(x)]_a^b + G(x)]_a^b$

(c) $[F(x) - G(x)]_a^b = F(x)]_a^b - G(x)]_a^b$.

7.7 RECTILINEAR MOTION REVISITED; AVERAGE VALUE

In Section 6.3 we used the derivative to define the notions of instantaneous velocity and acceleration for a particle moving along a line. In this section we will resume the study of such motion using the tools of integration. We will also investigate the general problem of integrating a rate of change, and we will show how the definite integral can be used to define the average value of a continuous function. More applications of integration will be given in Chapter 8.

FINDING POSITION AND VELOCITY BY INTEGRATION

Recall from Definitions 6.3.1 and 6.3.2 that if $s(t)$ is the position function of a particle moving on a coordinate line, then the instantaneous velocity and acceleration of the particle are given by the formulas

$$v(t) = s'(t) = \frac{ds}{dt} \quad \text{and} \quad a(t) = v'(t) = \frac{dv}{dt} = \frac{d^2 s}{dt^2}$$

It follows from these formulas that $s(t)$ is an antiderivative of $v(t)$ and $v(t)$ is an antideriva-

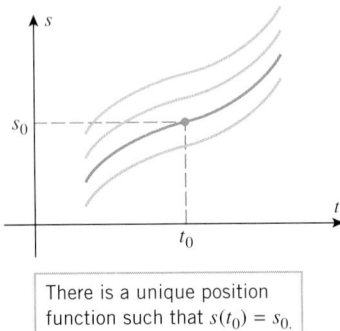

There is a unique position function such that $s(t_0) = s_0$.

Figure 7.7.1

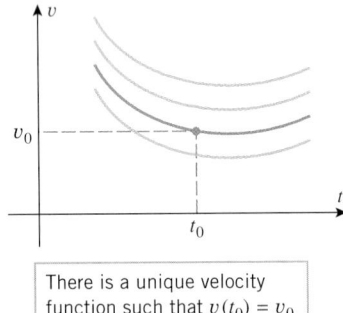

There is a unique velocity function such that $v(t_0) = v_0$.

Figure 7.7.2

tive of $a(t)$; that is,

$$s(t) = \int v(t)\,dt \qquad \text{and} \qquad v(t) = \int a(t)\,dt \qquad (1\text{–}2)$$

Thus, if the velocity of a particle is known, then its position function can be obtained from (1) by integration, provided there is sufficient additional information to determine the constant of integration. In particular, we can determine the constant of integration if we know the position s_0 of the particle at some time t_0, since this information determines a unique antiderivative $s(t)$ (Figure 7.7.1). Similarly, if the acceleration function of the particle is known, then its velocity function can be obtained from (2) by integration if we know the velocity v_0 of the particle at some time t_0 (Figure 7.7.2).

Example 1

Find the position function of a particle that is moving with velocity $v(t) = \cos \pi t$ along a coordinate line, assuming that the particle has coordinates $s = 4$ at time $t = 0$.

Solution. The position function is

$$s(t) = \int v(t)\,dt = \int \cos \pi t\,dt = \frac{1}{\pi} \sin \pi t + C$$

Since $s = 4$ when $t = 0$, it follows that

$$4 = s(0) = \frac{1}{\pi} \sin 0 + C = C$$

Thus,

$$s(t) = \frac{1}{\pi} \sin \pi t + 4 \qquad \blacktriangleleft$$

UNIFORMLY ACCELERATED MOTION

One of the most important cases of rectilinear motion occurs when a particle has constant acceleration. We call this *uniformly accelerated motion*.

We will show that if a particle moves with constant acceleration along an s-axis, and if the position and velocity of the particle are known at some point in time, say when $t = 0$, then it is possible to derive formulas for the position $s(t)$ and the velocity $v(t)$ at any time t. To see how this can be done, suppose that the particle has constant acceleration

$$a(t) = a \qquad (3)$$

and

$$s = s_0 \quad \text{when} \quad t = 0 \qquad (4)$$

$$v = v_0 \quad \text{when} \quad t = 0 \qquad (5)$$

where s_0 and v_0 are known. We call (4) and (5) the *initial conditions* for the motion.

With (3) as a starting point, we can integrate $a(t)$ to obtain $v(t)$, and we can integrate $v(t)$ to obtain $s(t)$, using an initial condition in each case to determine the constant of integration. The computations are as follows:

$$v(t) = \int a(t)\,dt = \int a\,dt = at + C_1 \qquad (6)$$

To determine the constant of integration C_1 we apply initial condition (5) to this equation to obtain

$$v_0 = v(0) = a \cdot 0 + C_1 = C_1$$

Substituting this in (6) and putting the constant term first yields

$$v(t) = v_0 + at$$

Since v_0 is constant, it follows that

$$s(t) = \int v(t)\,dt = \int (v_0 + at)\,dt = v_0 t + \tfrac{1}{2}at^2 + C_2 \qquad (7)$$

To determine the constant C_2 we apply initial condition (4) to this equation to obtain

$$s_0 = s(0) = v_0 \cdot 0 + \tfrac{1}{2}a \cdot 0 + C_2 = C_2$$

Substituting this in (7) and putting the constant term first yields

$$s(t) = s_0 + v_0t + \tfrac{1}{2}at^2$$

In summary, we have the following result.

7.7.1   UNIFORMLY ACCELERATED MOTION.   *If a particle moves with constant acceleration a along an s-axis, and if the position and velocity at time $t = 0$ are s_0 and v_0, respectively, then the position and velocity functions of the particle are*

$$s(t) = s_0 + v_0t + \tfrac{1}{2}at^2 \tag{8}$$

$$v(t) = v_0 + at \tag{9}$$

FOR THE READER.   How can you tell from the velocity versus time curve whether a particle moving along a line has uniformly accelerated motion?

Example 2

Suppose that an intergalactic spacecraft uses a sail and the "solar wind" to produce a constant acceleration of 0.032 m/s^2. Assuming that the spacecraft has a velocity of 10,000 m/s when the sail is first raised, how far will the spacecraft travel in 1 hour, and what will its velocity be at that time?

Solution.   In this problem the choice of a coordinate axis is at our discretion, so we will choose it to make the computations as simple as possible. Accordingly, let us introduce an s-axis whose positive direction is in the direction of motion, and let us take the origin to coincide with the position of the spacecraft at the time $t = 0$ when the sail is raised. Thus, the Formulas (8) and (9) for uniformly accelerated motion apply with

$$s_0 = s(0) = 0, \quad v_0 = v(0) = 10{,}000, \quad \text{and} \quad a = 0.032$$

Since 1 hour corresponds to $t = 3600$ s, it follows from (8) that in 1 hour the spacecraft travels a distance of

$$s(3600) = 10{,}000(3600) + \tfrac{1}{2}(0.032)(3600)^2 \approx 36{,}207{,}400 \text{ m}$$

and it follows from (9) that after 1 hour its velocity is

$$v(3600) = 10{,}000 + (0.032)(3600) \approx 10{,}115 \text{ m/s} \qquad \blacktriangleleft$$

Example 3

A bus has stopped to pick up riders, and a woman is running at a constant velocity of 5 m/s to catch it. When she is 11 m behind the front door the bus pulls away with a constant acceleration of 1 m/s^2. From that point in time, how long will it take for the woman to reach the front door of the bus if she keeps running with a velocity of 5 m/s?

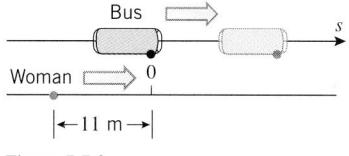

Figure 7.7.3

Solution.   As shown in Figure 7.7.3, choose the s-axis so that the bus and the woman are moving in the positive direction, and the front door of the bus is at the origin at the time $t = 0$ when the bus begins to pull away. To catch the bus at some later time t, the woman will have to cover a distance $s_w(t)$ that is equal to 11 m plus the distance $s_b(t)$ traveled by the bus; that is, the woman will catch the bus when

$$s_w(t) = s_b(t) + 11 \tag{10}$$

Since the woman has a constant velocity of 5 m/s, the distance she travels in t seconds is $s_w(t) = 5t$. Thus, (10) can be written as

$$s_b(t) = 5t - 11 \tag{11}$$

Since the bus has a constant acceleration of $a = 1$ m/s^2, and since $s_0 = v_0 = 0$ at time $t = 0$ (why?), it follows from (8) that

$$s_b(t) = \tfrac{1}{2}t^2$$

Substituting this equation into (11) and reorganizing the terms yields the quadratic equation

$$\tfrac{1}{2}t^2 - 5t + 11 = 0 \quad \text{or} \quad t^2 - 10t + 22 = 0$$

Solving this equation for t using the quadratic formula yields two solutions:

$$t = 5 - \sqrt{3} \approx 3.3 \quad \text{and} \quad t = 5 + \sqrt{3} \approx 6.7$$

(verify). Thus, the woman can reach the door at two different times, $t = 3.3$ s and $t = 6.7$ s. The reason that there are two solutions can be explained as follows: When the woman first reaches the door, she is running faster than the bus and can run past it if the driver does not see her. However, as the bus speeds up, it eventually catches up to her, and she has another chance to flag it down. ◄

THE FREE-FALL MODEL

In Section 6.3 we discussed the free-fall model of motion near the surface of the Earth with the promise that we would derive Formula (5) of that section later in the text; we will now show how to do this. As stated in 6.3.4 and illustrated in Figure 6.3.7, we will assume that the object moves on an s-axis whose origin is at the surface of the Earth and whose positive direction is up; and we will assume that the position and velocity of the object at time $t = 0$ are s_0 and v_0, respectively.

It is a fact of physics that a particle moving on a vertical line near the Earth's surface and subject only to the force of the Earth's gravity moves with constant acceleration. The magnitude of this constant, denoted by the letter g, is approximately 9.8 m/s^2 or 32 ft/s^2, depending on whether distance is measured in meters or feet.*

Recall that a particle is speeding up when its velocity and acceleration have the same sign and is slowing down when they have opposite signs. Thus, because we have chosen the positive direction to be up, it follows that the acceleration $a(t)$ of a particle in free fall is negative for all values of t. To see that this is so, observe that an upward-moving particle (positive velocity) is slowing down, so its acceleration must be negative; and a downward-moving particle (negative velocity) is speeding up, so its acceleration must also be negative. Thus, we conclude that

$$a(t) = -g$$

and hence it follows from (8) and (9) that the position and velocity functions of an object in free fall are

$$s(t) = s_0 + v_0 t - \tfrac{1}{2}g t^2 \tag{12}$$

$$v(t) = v_0 - g t \tag{13}$$

FOR THE READER. Had we chosen the positive direction of the s-axis to be down, then the acceleration would have been $a(t) = g$ (why?). How would this have affected Formulas (12) and (13)?

Example 4

A ball is thrown directly upward with an initial velocity of 49 m/s and is released from a point that is 8 m above the ground. Assuming that the free-fall model applies, how high will the ball travel?

---

*Strictly speaking, the constant g varies with the latitude and the distance from the Earth's center. However, for motion at a fixed latitude and near the surface of the Earth, the assumption of a constant g is satisfactory for many applications.

Solution. Since distance is in meters, we take $g = 9.8$ m/s^2. Initially, we have $s_0 = 8$ and $v_0 = 49$, so from (12) and (13)

$$v(t) = -9.8t + 49$$
$$s(t) = -4.9t^2 + 49t + 8$$

The ball will rise until $v(t) = 0$, that is, until $-9.8t + 49 = 0$ or $t = 5$. At this instant the height above the ground will be

$$s(5) = -4.9(5)^2 + 49(5) + 8 = 130.5 \text{ m}$$ ◄

Example 5

A penny is released from rest near the top of the Empire State Building at a point that is 1250 ft above the ground (Figure 7.7.4). Assuming that the free-fall model applies, how long does it take for the penny to hit the ground, and what is its speed at the time of impact?

Solution. Since distance is in feet, we take $g = 32$ ft/s^2. Initially, we have $s_0 = 1250$ and $v_0 = 0$, so from (12)

$$s(t) = -16t^2 + 1250 \tag{14}$$

Impact occurs when $s(t) = 0$. Solving this equation for t, we obtain

$$-16t^2 + 1250 = 0$$
$$t^2 = \frac{1250}{16} = \frac{625}{8}$$
$$t = \pm\frac{25}{\sqrt{8}} \approx \pm 8.8 \text{ s}$$

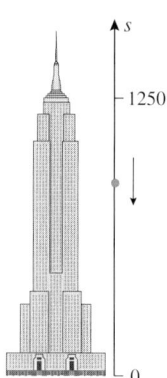

Figure 7.7.4

Since $t \geq 0$, we can discard the negative solution and conclude that it takes $25/\sqrt{8} \approx 8.8$ s for the penny to hit the ground. To obtain the velocity at the time of impact, we substitute $t = 25/\sqrt{8}$, $v_0 = 0$, and $g = 32$ in (13) to obtain

$$v\left(\frac{25}{\sqrt{8}}\right) = 0 - 32\left(\frac{25}{\sqrt{8}}\right) = -200\sqrt{2} \approx -282.8 \text{ ft/s}$$

Thus, the speed at the time of impact is

$$\left| v\left(\frac{25}{\sqrt{8}}\right) \right| = 200\sqrt{2} \approx 282.8 \text{ ft/s}$$

which is more than 192 mi/h.   ◄

INTEGRATING RATES OF CHANGE

The Fundamental Theorem of Calculus

$$\int_a^b f(x)\,dx = F(b) - F(a) \tag{15}$$

has a useful interpretation that can be seen by rewriting it in a slightly different form. Since F is an antiderivative of f on the interval $[a, b]$, we can use the relationship $F'(x) = f(x)$ to rewrite (15) as

$$\int_a^b F'(x)\,dx = F(b) - F(a) \tag{16}$$

In this formula we can view $F'(x)$ as the rate of change of $F(x)$ with respect to x, and we can view $F(b) - F(a)$ as the *change* in the value of $F(x)$ as x increases from a to b (Figure 7.7.5). Thus, we have the following useful principle.

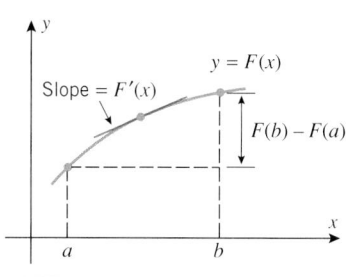

Integrating the slope of $y = F(x)$ over the interval $[a, b]$ produces the change $F(b) - F(a)$ in the value of $F(x)$.

Figure 7.7.5

7.7.2   INTEGRATING A RATE OF CHANGE.   Integrating the rate of change of $F(x)$ with respect to x over an interval $[a, b]$ produces the change in the value of $F(x)$ that occurs as x increases from a to b.

Here are some examples of this idea:

- If $P(t)$ is a population (e.g., plants, animals, or people) at time t, then $P'(t)$ is the rate at which the population is changing at time t, and

$$\int_{t_1}^{t_2} P'(t)\,dt = P(t_2) - P(t_1)$$

is the change in the population between times t_1 and t_2.

- If $A(t)$ is the area of an oil spill at time t, then $A'(t)$ is the rate at which the area of the spill is changing at time t, and

$$\int_{t_1}^{t_2} A'(t)\,dt = A(t_2) - A(t_1)$$

is the change in the area of the spill between times t_1 and t_2.

- If $P'(x)$ is the marginal profit that results from producing and selling x units of a product (see Section 6.2), then

$$\int_{x_1}^{x_2} P'(x)\,dx = P(x_2) - P(x_1)$$

is the change in the profit that results when the production level increases from x_1 units to x_2 units.

DISPLACEMENT IN RECTILINEAR MOTION

As another application of (16), suppose that $s(t)$ and $v(t)$ are the position and velocity functions of a particle moving on a coordinate line. Since $v(t)$ is the rate of change of $s(t)$ with respect to t, it follows from the principle in 7.7.2 that integrating $v(t)$ over an interval $[t_0, t_1]$ will produce the change in the value of $s(t)$ as t increases from t_0 to t_1; that is,

$$\int_{t_0}^{t_1} v(t)\,dt = \int_{t_0}^{t_1} s'(t)\,dt = s(t_1) - s(t_0) \tag{17}$$

The expression $s(t_1) - s(t_0)$ in this formula is called the ***displacement*** or ***change in position*** of the particle over the time interval $[t_0, t_1]$. For a particle moving horizontally, the displacement is positive if the final position of the particle is to the right of its initial position, negative if it is to the left of its initial position, and zero if it coincides with the initial position (Figure 7.7.6).

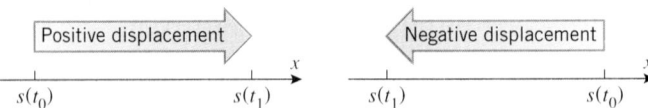

Figure 7.7.6

REMARK.   In physical problems it is important to associate the correct units with definite integrals. In general, the units for the definite integral

$$\int_a^b f(x)\,dx$$

will be units of $f(x)$ times units of x. This is because the definite integral is a limit of Riemann sums each of whose terms is a product of the form $f(x) \cdot \Delta x$. For example, if time is measured in seconds (s) and velocity is measured in meters per second (m/s), then integrating velocity over a time interval will produce a result whose units are in meters, since m/s × s = m. Note that this is consistent with Formula (17), since displacement has units of length.

In general, the displacement of a particle is not the same as the distance traveled by the particle. For example, a particle that travels 100 units in the positive direction and then 100 units in the negative direction travels a distance of 200 units but has a displacement of zero, since it returns to its starting point. The only case in which the displacement and the distance traveled are the same occurs when the particle moves in the positive direction without reversing the direction of its motion.

FOR THE READER.   What is the relationship between the displacement of a particle and the distance it travels if the particle moves in the negative direction without reversing the direction of motion?

From (17), integrating the velocity function of a particle over a time interval yields the displacement of a particle over that time interval. In contrast, to find the *total distance* traveled by the particle over the time interval (the distance traveled in the positive direction plus the distance traveled in the negative direction), we must integrate the *absolute value* of the velocity function; that is, we must integrate the speed:

$$\begin{bmatrix} \text{total distance} \\ \text{traveled during} \\ \text{time interval} \\ [t_0, t_1] \end{bmatrix} = \int_{t_0}^{t_1} |v(t)|\, dt \tag{18}$$

Example 6

A particle moves on a coordinate line so that its velocity at time t is $v(t) = t^2 - 2t$ m/s.

(a)   Find the displacement of the particle during the time interval $0 \le t \le 3$.

(b)   Find the distance traveled by the particle during the time interval $0 \le t \le 3$.

Solution (a).   From (17) the displacement is

$$\int_0^3 v(t)\, dt = \int_0^3 (t^2 - 2t)\, dt = \left[\frac{t^3}{3} - t^2 \right]_0^3 = 0$$

Thus, the particle is at the same position at time $t = 3$ as at $t = 0$.

Solution (b).   The velocity can be written as $v(t) = t^2 - 2t = t(t - 2)$, from which we see that $v(t) \le 0$ for $0 \le t \le 2$ and $v(t) \ge 0$ for $2 \le t \le 3$. Thus, it follows from (18) that the distance traveled is

$$\int_0^3 |v(t)|\, dt = \int_0^2 -v(t)\, dt + \int_2^3 v(t)\, dt$$

$$= \int_0^2 -(t^2 - 2t)\, dt + \int_2^3 (t^2 - 2t)\, dt$$

$$= -\left[\frac{t^3}{3} - t^2 \right]_0^2 + \left[\frac{t^3}{3} - t^2 \right]_2^3 = \frac{4}{3} + \frac{4}{3} = \frac{8}{3} \text{ m} \qquad \blacktriangleleft$$

In Section 6.3 we showed how to use the position versus time curve to obtain information about the behavior of a particle moving on a coordinate line (Table 6.3.1). Similarly, there is valuable information that can be obtained from the *velocity versus time curve*. For example, the integral in (17) can be interpreted geometrically as the net signed area between the graph of $v(t)$ and the interval $[t_0, t_1]$, and it can be interpreted physically as the displacement of the particle over this interval. Thus, we have the following result.

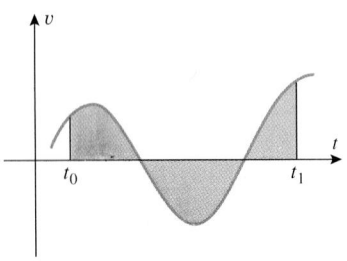

The net signed area is the displacement of the particle during the interval $[t_0, t_1]$.

Figure 7.7.7

7.7.3 FINDING DISPLACEMENT FROM THE VELOCITY VERSUS TIME CURVE.   For a particle in rectilinear motion, the net signed area between the velocity versus time curve and an interval $[t_0, t_1]$ on the t-axis represents the displacement of the particle over that time interval (Figure 7.7.7).

Example 7

Figure 7.7.8 shows three velocity versus time curves for a particle in rectilinear motion along a horizontal line. In each case, find the displacement of the particle over the time interval $0 \leq t \leq 4$, and explain what it tells you about the motion of the particle.

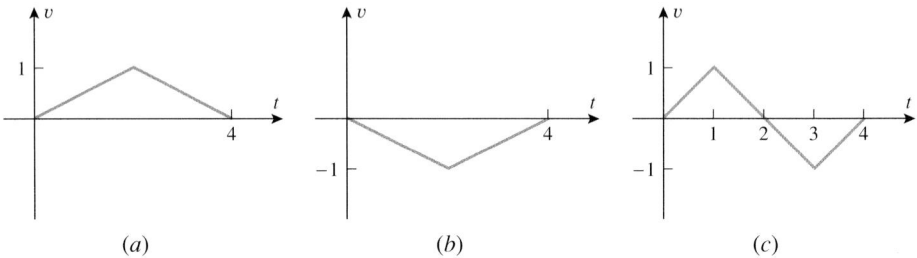

Figure 7.7.8

Solution.   In part (*a*) of Figure 7.7.8 the net signed area under the curve is 2, so the particle is 2 units to the right of its starting point at the end of the time period. In part (*b*) the net signed area under the curve is -2, so the particle is 2 units to the left of its starting point at the end of the time period. In part (*c*) the net signed area under the curve is 0, so the particle is back at its starting point at the end of the time period.   ◀

Sometimes we will not want the net signed area between a curve $y = f(x)$ and an interval $[a, b]$, but rather the total area between the curve and the interval. This can be found by integrating $|f(x)|$ rather than $f(x)$ over the interval $[a, b]$.

Example 8

Find the total area between the curve $y = 1 - x^2$ and the x-axis over the interval $[0, 2]$ (Figure 7.7.9).

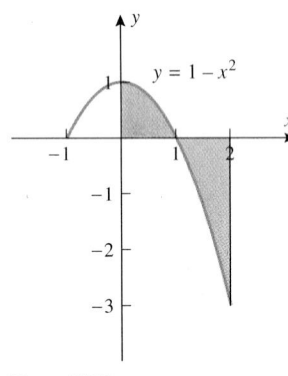

Figure 7.7.9

Solution.   The area A is given by

$$A = \int_0^2 |1 - x^2|\, dx = \int_0^1 (1 - x^2)\, dx + \int_1^2 -(1 - x^2)\, dx$$

$$= \left[x - \frac{x^3}{3}\right]_0^1 - \left[x - \frac{x^3}{3}\right]_1^2$$

$$= \frac{2}{3} - \left(-\frac{4}{3}\right) = 2 \qquad ◀$$

From (18), integrating the speed $|v(t)|$ over a time interval $[t_0, t_1]$ produces the distance traveled by the particle during the time interval. However, we can also interpret the integral in (18) as the total area between the velocity versus time curve and the interval $[t_0, t_1]$ on the t-axis. Thus, we have the following result.

7.7.4 FINDING DISTANCE TRAVELED FROM THE VELOCITY VERSUS TIME CURVE..   For a particle in rectilinear motion, the total area between the velocity versus time curve and an interval $[t_0, t_1]$ on the t-axis represents the distance traveled by the particle over that time interval.

Example 9

For each of the velocity versus time curves in Figure 7.7.8 find the total distance traveled by the particle over the time interval $0 \leq t \leq 4$.

Solution.   In all three parts of Figure 7.7.8 the total area between the curve and the interval [0, 4] is 2, so the particle travels a distance of 2 units during the time period in all three cases, even though the displacement is different in each case, as discussed in Example 7.   ◄

**AVERAGE VALUE OF A
CONTINUOUS FUNCTION**

In scientific work, numerical information is often summarized by computing some sort of *average* or *mean* value of the observed data. There are various kinds of averages, but the most common is the ***arithmetic mean*** or ***arithmetic average***, which is formed by adding the data and dividing by the number of data points. Thus, the arithmetic average $\bar{a}$ of n numbers $a_1, a_2, \ldots, a_n$ is

$$\bar{a} = \frac{1}{n}(a_1 + a_2 + \cdots + a_n) = \frac{1}{n}\sum_{k=1}^{n} a_k$$

In the case where the a_k's are values of a function f, say,

$$a_1 = f(x_1), a_2 = f(x_2), \ldots, a_n = f(x_n)$$

then the arithmetic average $\bar{a}$ of these function values is

$$\bar{a} = \frac{1}{n}\sum_{k=1}^{n} f(x_k)$$

We will now show how to extend this concept so that we can compute not only the arithmetic average of finitely many function values but an average of *all* values of $f(x)$ as x varies over a closed interval $[a, b]$. For this purpose recall the Mean-Value Theorem for Integrals (7.6.2), which states that if f is continuous on the interval $[a, b]$, then there is at least one point x^* in this interval such that

$$\int_a^b f(x)\,dx = f(x^*)(b - a)$$

The quantity

$$f(x^*) = \frac{1}{b-a}\int_a^b f(x)\,dx \tag{19}$$

will be our candidate for the average value of f over the interval $[a, b]$. To explain what motivates this, divide the interval $[a, b]$ into n subintervals of equal length

$$\Delta x = \frac{b-a}{n} \tag{20}$$

and choose arbitrary points $x_1^*, x_2^*, \ldots, x_n^*$ in successive subintervals. Then the arithmetic average of the numbers $f(x_1^*), f(x_2^*), \ldots, f(x_n^*)$ is

$$\text{ave} = \frac{1}{n}[f(x_1^*) + f(x_2^*) + \cdots + f(x_n^*)]$$

or from (20)

$$\text{ave} = \frac{1}{b-a}[f(x_1^*)\Delta x + f(x_2^*)\Delta x + \cdots + f(x_n^*)\Delta x] = \frac{1}{b-a}\sum_{k=1}^{n} f(x_k^*)\Delta x$$

Taking the limit as $n \to +\infty$ yields

$$\lim_{n \to +\infty} \frac{1}{b-a}\sum_{k=1}^{n} f(x_k^*)\Delta x = \frac{1}{b-a}\int_a^b f(x)\,dx$$

Since this equation describes what happens when we compute the average of "more and more" values of $f(x)$, we are led to the following definition.

7.7.5 DEFINITION. If f is continuous on $[a, b]$, then the **average value** (or **mean value**) of f on $[a, b]$ is defined to be

$$f_{\text{ave}} = \frac{1}{b-a} \int_a^b f(x)\, dx \qquad (21)$$

REMARK. When f is nonnegative on $[a, b]$, the quantity f_{ave} has a simple geometric interpretation, which can be seen by writing (21) as

$$f_{\text{ave}} \cdot (b - a) = \int_a^b f(x)\, dx$$

The left side of this equation is the area of a rectangle with a height of f_{ave} and base of length $b - a$, and the right side is the area under $y = f(x)$ over $[a, b]$. Thus, f_{ave} is the height of a rectangle constructed over the interval $[a, b]$, whose area is the same as the area under the graph of f over that interval (Figure 7.7.10). Note also that the Mean-Value Theorem, when expressed in form (21), ensures that there is always at least one point x^* in $[a, b]$ at which the value of f is equal to the average value of f over the interval.

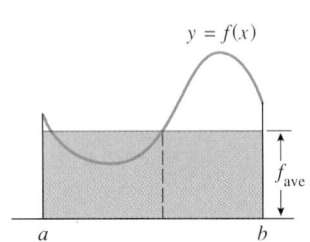

$y = f(x)$

f_{ave}

a b

Figure 7.7.10

Example 10

Find the average value of the function $f(x) = \sqrt{x}$ over the interval $[1, 4]$, and find all points in the interval at which the value of f is the same as the average.

Solution.

$$f_{\text{ave}} = \frac{1}{b-a} \int_a^b f(x)\, dx = \frac{1}{4-1} \int_1^4 \sqrt{x}\, dx = \frac{1}{3} \left[\frac{2x^{3/2}}{3} \right]_1^4$$

$$= \frac{1}{3} \left[\frac{16}{3} - \frac{2}{3} \right] = \frac{14}{9} \approx 1.6$$

The x-values at which $f(x) = \sqrt{x}$ is the same as the average satisfy $\sqrt{x} = 14/9$, from which we obtain $x = 196/81 \approx 2.4$ (Figure 7.7.11). ◄

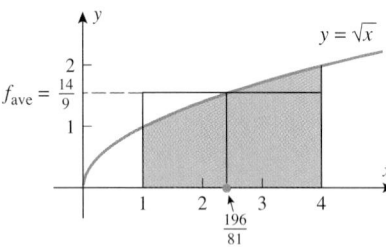

Figure 7.7.11

AVERAGE VELOCITY REVISITED

In Section 3.1 we considered the motion of a particle moving in the *positive direction* along a coordinate line, and we motivated the concept of instantaneous velocity in that special case by viewing it as the limit of average velocities over smaller and smaller time intervals. That discussion led us to conclude that the average velocity of the particle over a time interval could be interpreted as the slope of a secant line and the instantaneous velocity as the slope of a tangent line to the position versus time curve (Figure 3.1.5). We will now show that the same results are true in the more general case where the particle can move in either direction along the coordinate line.

For this purpose, suppose that $s(t)$ and $v(t)$ are the position and velocity functions of such a particle, and let us use Formula (21) to calculate the average velocity of the particle over a time interval $[t_0, t_1]$. This yields

$$v_{\text{ave}} = \frac{1}{t_1 - t_0} \int_{t_0}^{t_1} v(t)\, dt = \frac{1}{t_1 - t_0} \int_{t_0}^{t_1} s'(t)\, dt = \frac{s(t_1) - s(t_0)}{t_1 - t_0}$$

Thus, *the average velocity over a time interval is the displacement divided by the elapsed time*. Geometrically, this is the slope of the secant line shown in Figure 7.7.12. Moreover, if we allow t_1 to approach t_0, then the slopes of the secant lines approach the slope of the tangent line at t_0, which is the instantaneous velocity at that instant. Thus, the relationship between average and instantaneous velocity developed in Section 3.1 also applies to general rectilinear motion.

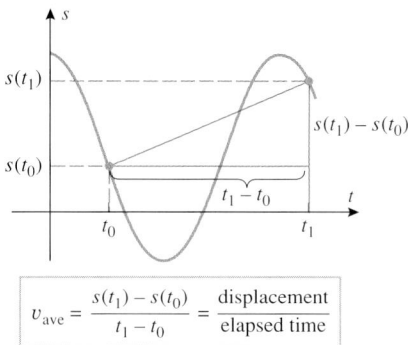

Figure 7.7.12

$$v_{\text{ave}} = \frac{s(t_1) - s(t_0)}{t_1 - t_0} = \frac{\text{displacement}}{\text{elapsed time}}$$

EXERCISE SET 7.7 ⊠ Graphing Calculator ⃞c CAS

1. (a) If $h'(t)$ is the rate of change of a child's height measured in inches per year, what does the integral $\int_0^{10} h'(t)\, dt$ represent, and what are its units?

   (b) If $r'(t)$ is the rate of change of the radius of a spherical balloon measured in centimeters per second, what does the integral $\int_1^2 r'(t)\, dt$ represent, and what are its units?

   (c) If $H(t)$ is the rate of change of the speed of sound with respect to temperature measured in ft/s per °F, what does the integral $\int_{32}^{100} H(t)\, dt$ represent, and what are its units?

   (d) If $v(t)$ is the velocity of a particle in rectilinear motion, measured in cm/h, what does the integral $\int_{t_1}^{t_2} v(t)\, dt$ represent, and what are its units?

2. (a) Suppose that sludge is emptied into a river at the rate of $V(t)$ gallons per minute, starting at time $t = 0$. Write an integral that represents the total volume of sludge that is emptied into the river during the first hour.

   (b) Suppose that the tangent line to a curve $y = f(x)$ has slope $m(x)$ at the point x. What does the integral $\int_{x_1}^{x_2} m(x)\, dx$ represent?

3. In each part, the velocity versus time curve is given for a particle moving along a line. Use the curve to find the displacement and the distance traveled by the particle over the time interval $0 \le t \le 3$.

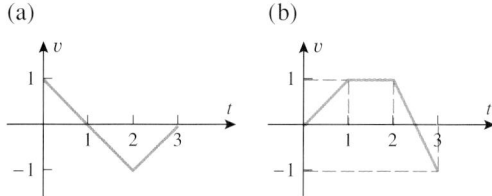

(a)   (b)

4. Sketch a velocity versus time curve for a particle that travels a distance of 5 units along a coordinate line during the time interval $0 \le t \le 10$ and has a displacement of 0 units.

5. The accompanying figure shows the acceleration versus time curve for a particle moving along a coordinate line. If the initial velocity of the particle is 20 m/s, estimate
   (a) the velocity at time $t = 4$ s
   (b) the velocity at time $t = 6$ s.

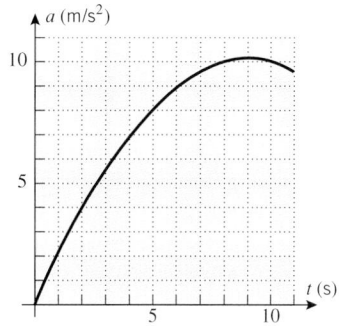

Figure Ex-5

6. Determine whether the particle in Exercise 5 is speeding up or slowing down at times $t = 4$ s and $t = 6$ s.

In Exercises 7–10, a particle moves along an s-axis. Use the given information to find the position function of the particle.

---

7. (a) $v(t) = t^3 - 2t^2 + 1$; $s(0) = 1$
  (b) $a(t) = 4\cos 2t$; $v(0) = -1$; $s(0) = -3$

8. (a) $v(t) = 1 + \sin t$; $s(0) = -3$
  (b) $a(t) = t^2 - 3t + 1$; $v(0) = 0$; $s(0) = 0$

9. (a) $v(t) = 2t - 3$; $s(1) = 5$
  (b) $a(t) = \cos t$; $v(\pi/2) = 2$; $s(\pi/2) = 0$

10. (a) $v(t) = t^{2/3}$; $s(8) = 0$
  (b) $a(t) = \sqrt{t}$; $v(4) = 1$; $s(4) = -5$

In Exercises 11–14, a particle moves with a velocity of $v(t)$ m/s along an s-axis. Find the displacement and the distance traveled by the particle during the given time interval.

---

11. (a) $v(t) = \sin t$; $0 \leq t \leq \pi/2$
  (b) $v(t) = \cos t$; $\pi/2 \leq t \leq 2\pi$

12. (a) $v(t) = 2t - 4$; $0 \leq t \leq 6$
  (b) $v(t) = |t - 3|$; $0 \leq t \leq 5$

13. (a) $v(t) = t^3 - 3t^2 + 2t$; $0 \leq t \leq 3$
  (b) $v(t) = e^t - 2$; $0 \leq t \leq 3$

14. (a) $v(t) = \frac{1}{2} - 1/t$; $1 \leq t \leq 3$
  (b) $v(t) = 3/\sqrt{t}$; $4 \leq t \leq 9$

In Exercises 15–18, a particle moves with acceleration $a(t)$ m/s^2 along an s-axis and has velocity v_0 m/s at time $t = 0$. Find the displacement and the distance traveled by the particle during the given time interval.

---

15. $a(t) = -2$; $v_0 = 3$; $1 \leq t \leq 4$

16. $a(t) = t - 2$; $v_0 = 0$; $1 \leq t \leq 5$

17. $a(t) = 1/\sqrt{5t + 1}$; $v_0 = 2$; $0 \leq t \leq 3$

18. $a(t) = \sin t$; $v_0 = 1$; $\pi/4 \leq t \leq \pi/2$

19. In each part use the given information to find the position, velocity, speed, and acceleration at time $t = 1$.
  (a) $v = \sin \frac{1}{2}\pi t$; $s = 0$ when $t = 0$
  (b) $a = -3t$; $s = 1$ and $v = 0$ when $t = 0$

20. The accompanying figure shows the velocity versus time curve over the time interval $1 \leq t \leq 5$ for a particle moving along a horizontal coordinate line.
  (a) What can you say about the sign of the acceleration over the time interval?
  (b) When is the particle speeding up? Slowing down?
  (c) What can you say about the location of the particle at time $t = 5$ relative to its location at time $t = 1$? Explain your reasoning.

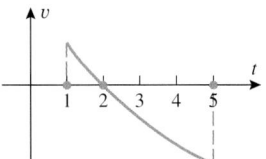

Figure Ex-20

In Exercises 21–24, sketch the curve and find the total area between the curve and the given interval on the x-axis.

---

21. $y = x^2 - 1$; $[0, 3]$

22. $y = \sin x$; $[0, 3\pi/2]$

23. $y = e^x - 1$; $[-1, 1]$

24. $y = \dfrac{x - 1}{x}$; $\left[\frac{1}{2}, 2\right]$

25. Suppose that the velocity function of a particle moving along an s-axis is $v(t) = 20t^2 - 100t + 50$ ft/s and that the particle is at the origin at time $t = 0$. Use a graphing utility to generate the graphs of $s(t)$, $v(t)$, and $a(t)$ for the first 6 s of motion.

26. Suppose that the acceleration function of a particle moving along an s-axis is $a(t) = 4t - 30$ m/s and that the position and velocity at time $t = 0$ are $s_0 = -5$ m and $v_0 = 3$ m/s. Use a graphing utility to generate the graphs of $s(t)$, $v(t)$, and $a(t)$ for the first 25 s of motion.

27. Let the velocity function for a particle that is at the origin initially and moves along an s-axis be $v(t) = 0.5 - te^{-t}$.
  (a) Generate the velocity versus time curve, and use it to make a conjecture about the sign of the displacement over the time interval $0 \leq t \leq 5$.
  (b) Use a CAS to find the displacement.

28. Let the velocity function for a particle that is at the origin initially and moves along an s-axis be $v(t) = t \ln(t + 0.1)$.
  (a) Generate the velocity versus time curve, and use it to make a conjecture about the sign of the displacement over the time interval $0 \leq t \leq 1$.
  (b) Use a CAS to find the displacement.

29. Suppose that at time $t = 0$ a particle is at the origin of an x-axis and has a velocity of $v_0 = 25$ cm/s. For the first 4 s thereafter it has no acceleration, and then it is acted on by a retarding force that produces a constant negative acceleration of $a = -10$ cm/s^2.
  (a) Sketch the acceleration versus time curve over the interval $0 \leq t \leq 12$.
  (b) Sketch the velocity versus time curve over the time interval $0 \leq t \leq 12$.
  (c) Find the x-coordinate of the particle at times $t = 8$ s and $t = 12$ s.
  (d) What is the maximum x-coordinate of the particle over the time interval $0 \leq t \leq 12$?

30. Formulas (8) and (9) for uniformly accelerated motion can be rearranged in various useful ways. For simplicity, let $s = s(t)$ and $v = v(t)$, and derive the following variations of those formulas.

(a) $a = \dfrac{v^2 - v_0^2}{2(s - s_0)}$ (b) $t = \dfrac{2(s - s_0)}{v_0 + v}$

(c) $s = s_0 + vt - \frac{1}{2}at^2$ [Note how this differs from (8).]

Exercises 31–38 involve uniformly accelerated motion. In these exercises assume that the object is moving in the positive direction of a coordinate line, and apply Formulas (8) and (9) or those from Exercise 30, as appropriate. In some of these problems you will need the fact that 88 ft/s = 60 mi/h.

31. (a) An automobile traveling on a straight road decelerates uniformly from 55 mi/h to 25 mi/h in 30 s. Find its acceleration in ft/s².

    (b) A bicycle rider traveling on a straight path accelerates uniformly from rest to 30 km/h in 1 min. Find his acceleration in km/s².

32. A car traveling 60 mi/h along a straight road decelerates at a constant rate of 10 ft/s².

    (a) How long will it take until the speed is 45 mi/h?

    (b) How far will the car travel before coming to a stop?

33. Spotting a police car, you hit the brakes on your new Porsche to reduce your speed from 90 mi/h to 60 mi/h at a constant rate over a distance of 200 ft.

    (a) Find the acceleration in ft/s².

    (b) How long does it take for you to reduce your speed to 55 mi/h?

    (c) At the acceleration obtained in part (a), how long would it take for you to bring your Porsche to a complete stop from 90 mi/h?

34. A particle moving along a straight line is accelerating at a constant rate of 3 m/s². Find the initial velocity if the particle moves 40 m in the first 4 s.

35. A motorcycle, starting from rest, speeds up with a constant acceleration of 2.6 m/s². After it has traveled 120 m, it slows down with a constant acceleration of −1.5 m/s² until it attains a speed of 12 m/s. What is the distance traveled by the motorcycle at that point?

36. A sprinter in a 100-m race explodes out of the starting block with an acceleration of 4.0 m/s², which she sustains for 2.0 s. Her acceleration then drops to zero for the rest of race.

    (a) What is her time for the race?

    (b) Make a graph of her distance from the starting block versus time.

37. A car that has stopped at a toll booth leaves the booth with a constant acceleration of 2 ft/s². At the time the car leaves the booth it is 5000 ft behind a truck traveling with a constant velocity of 50 ft/s. How long will it take for the car to catch the truck, and how far will the car be from the toll booth at that time?

38. In the final sprint of a rowing race the challenger is rowing at a constant speed of 12 m/s. At the point where the leader is 100 m from the finish line and the challenger is 15 m behind, the leader is rowing at 8 m/s but starts accelerating at a constant 0.5 m/s². Who wins?

In Exercises 39–48, assume that a free-fall model applies. Solve these exercises by applying Formulas (12) and (13) or, if appropriate, use those from Exercise 30 with $a = -g$. In these exercises take $g = 32$ ft/s² or $g = 9.8$ m/s², depending on the units.

39. A projectile is launched vertically upward from ground level with an initial velocity of 112 ft/s.

    (a) Find the velocity at $t = 3$ s and $t = 5$ s.

    (b) How high will the projectile rise?

    (c) Find the speed of the projectile when it hits the ground.

40. A projectile fired downward from a height of 112 ft reaches the ground in 2 s. What is its initial velocity?

41. A projectile is fired vertically upward from ground level with an initial velocity of 16 ft/s.

    (a) How long will it take for the projectile to hit the ground?

    (b) How long will the projectile be moving upward?

42. A rock is dropped from the top of the Washington Monument, which is 555 ft high.

    (a) How long will it take for the rock to hit the ground?

    (b) What is the speed of the rock at impact?

43. A helicopter pilot drops a package when the helicopter is 200 ft above the ground and rising at a speed of 20 ft/s.

    (a) How long will it take for the package to hit the ground?

    (b) What will be its speed at impact?

44. A stone is thrown downward with an initial speed of 96 ft/s from a height of 112 ft.

    (a) How long will it take for the stone to hit the ground?

    (b) What will be its speed at impact?

45. A projectile is fired vertically upward with an initial velocity of 49 m/s from a tower 150 m high.

    (a) How long will it take for the projectile to reach its maximum height?

    (b) What is the maximum height?

    (c) How long will it take for the projectile to pass its starting point on the way down?

    (d) What is the velocity when it passes the starting point on the way down?

    (e) How long will it take for the projectile to hit the ground?

    (f) What will be its speed at impact?

46. A man drops a stone from a bridge. What is the height of the bridge if

    (a) the stone hits the water 4 s later

    (b) the sound of the splash reaches the man 4 s later? [Take 1080 ft/s as the speed of sound.]

47. In the final stages of a Moon landing, a lunar module fires its retrorockets and descends to a height of $h = 5$ m above the lunar surface (Figure Ex-47). At that point the retrorockets are cut off, and the module goes into free fall. Given that the Moon's gravity is 1/6 of the Earth's, find the speed of the module when it touches the lunar surface.

Figure Ex-47

48. Given that the Moon's gravity is 1/6 of the Earth's, how much faster would a projectile have to be launched upward from the surface of the Earth than from the surface of the Moon to reach a height of 1000 ft?

In Exercises 49–54, find the average value of the function over the given interval.

49. $f(x) = 3x$; $[1, 3]$ **50.** $f(x) = x^2$; $[-1, 2]$

51. $f(x) = \sin x$; $[0, \pi]$ **52.** $f(x) = \cos x$; $[0, \pi]$

53. $f(x) = 1/x$; $[1, e]$ **54.** $f(x) = e^x$; $[-1, \ln 5]$

55. (a) Find f_{ave} of $f(x) = x^2$ over $[0, 2]$.
   (b) Find a point x^* in $[0, 2]$ such that $f(x^*) = f_{\text{ave}}$.
   (c) Sketch the graph of $f(x) = x^2$ over $[0, 2]$ and construct a rectangle over the interval whose area is the same as the area under the graph of f over the interval.

56. (a) Find f_{ave} of $f(x) = 2x$ over $[0, 4]$.
   (b) Find a point x^* in $[0, 4]$ such that $f(x^*) = f_{\text{ave}}$.
   (c) Sketch the graph of $f(x) = 2x$ over $[0, 4]$ and construct a rectangle over the interval whose area is the same as the area under the graph of f over the interval.

57. (a) Suppose that the velocity function of a particle moving along a coordinate line is $v(t) = 3t^3 + 2$. Find the average velocity of the particle over the time interval $1 \le t \le 4$ by integrating.
   (b) Suppose that the position function of a particle moving along a coordinate line is $s(t) = 6t^2 + t$. Find the average velocity of the particle over the time interval $1 \le t \le 4$ algebraically.

58. (a) Suppose that the acceleration function of a particle moving along a coordinate line is $a(t) = t + 1$. Find the average acceleration of the particle over the time interval $0 \le t \le 5$ by integrating.
   (b) Suppose that the velocity function of a particle moving along a coordinate line is $v(t) = \cos t$. Find the average acceleration of the particle over the time interval $0 \le t \le \pi/4$ algebraically.

59. Water is run at a constant rate of 1 ft³/min to fill a cylindrical tank of radius 3 ft and height 5 ft. Assuming that the tank is empty initially, make a conjecture about the average weight of the water in the tank over the time period required to fill

it, and then check your conjecture by integrating. [Take the weight density of water to be 62.4 lb/ft³.]

60. (a) The temperature of a 10-m-long metal bar is 15°C at one end and 30°C at the other end. Assuming that the temperature increases linearly from the cooler end to the hotter end, what is the average temperature of the bar?
   (b) Explain why there must be a point on the bar where the temperature is the same as the average, and find it.

61. (a) Suppose that a reservoir supplies water to an industrial park at a constant rate of $r = 4$ gallons per minute (gal/min) between 8:30 A.M. and 9:00 A.M. How much water does the reservoir supply during that time period?
   (b) Suppose that one of the industrial plants increases its water consumption between 9:00 A.M. and 10:00 A.M. and that the rate at which the reservoir supplies water increases linearly, as shown in the accompanying figure. How much water does the reservoir supply during that 1-hour time period?
   (c) Suppose that from 10:00 A.M. to 12 noon the rate at which the reservoir supplies water is given by the formula $r(t) = 10 + \sqrt{t}$ gal/min, where $t = 0$ corresponds to 10:00 A.M. How much water does the reservoir supply during that 2-hour time period?

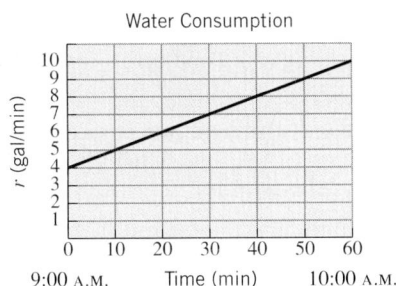

Water Consumption

Figure Ex-61

62. A traffic engineer monitors the rate at which cars enter the main highway during the afternoon rush hour. From her data she estimates that between 4:30 P.M. and 5:30 P.M. the rate $R(t)$ at which cars enter the highway is given by the formula $R(t) = 100(1 - 0.0001t^2)$ cars per minute, where $t = 0$ corresponds to 4:30 P.M.
   (a) When does the peak traffic flow into the highway occur?
   (b) Find the number of cars that enter the highway during the rush hour.

63. (a) Prove: If f is continuous on $[a, b]$, then
$$\int_a^b [f(x) - f_{\text{ave}}] \, dx = 0$$
   (b) Does there exist a constant $c \ne f_{\text{ave}}$ such that
$$\int_a^b [f(x) - c] \, dx = 0?$$

7.8 EVALUATING DEFINITE INTEGRALS BY SUBSTITUTION

In this section we will discuss two methods for evaluating definite integrals in which a substitution is required.

TWO METHODS FOR MAKING SUBSTITUTIONS IN DEFINITE INTEGRALS

Recall from Section 7.3 that indefinite integrals of the form

$$\int f(g(x))g'(x)\,dx$$

can sometimes be evaluated by making the u-substitution

$$u = g(x), \quad du = g'(x)\,dx \tag{1}$$

which converts the integral to the form

$$\int f(u)\,du$$

To apply this method to a definite integral of the form

$$\int_a^b f(g(x))g'(x)\,dx$$

we need to account for the effect that the substitution has on the x-limits of integration. There are two ways of doing this.

Method 1

First evaluate the indefinite integral

$$\int f(g(x))g'(x)\,dx$$

by substitution, and then use the relationship

$$\int_a^b f(g(x))g'(x)\,dx = \left[\int f(g(x))g'(x)\,dx\right]_a^b$$

to evaluate the definite integral. This procedure does not require any modification of the x-limits of integration.

Method 2

Make the substitution (1) directly in the definite integral, and then use the relationship $u = g(x)$ to replace the x-limits, $x = a$ and $x = b$, by corresponding u-limits, $u = g(a)$ and $u = g(b)$. This produces a new definite integral

$$\int_{g(a)}^{g(b)} f(u)\,du$$

that is expressed entirely in terms of u.

Example 1

Use the two methods above to evaluate $\displaystyle\int_0^2 x(x^2+1)^3\,dx$.

Solution by Method 1.   If we let

$$u = x^2 + 1 \quad \text{so that} \quad du = 2x\,dx \tag{2}$$

then we obtain

$$\int x(x^2+1)^3\,dx = \frac{1}{2}\int u^3\,du = \frac{u^4}{8} + C = \frac{(x^2+1)^4}{8} + C$$

Thus,

$$\int_0^2 x(x^2+1)^3\,dx = \left[\int x(x^2+1)^3\,dx\right]_{x=0}^{2} = \frac{(x^2+1)^4}{8}\Bigg]_{x=0}^{2}$$

$$= \frac{625}{8} - \frac{1}{8} = 78$$

Solution by Method 2. If we make the substitution $u = x^2+1$ in (2), then

$u = 1$ if $x = 0$

$u = 5$ if $x = 2$

Thus,

$$\int_0^2 x(x^2+1)^3\,dx = \frac{1}{2}\int_1^5 u^3\,du = \frac{u^4}{8}\Bigg]_{u=1}^{5} = \frac{625}{8} - \frac{1}{8} = 78$$

which agrees with the result obtained by Method 1. ◀

The following theorem states precise conditions under which Method 2 can be used. The proof is a straightforward application of the chain rule and the Fundamental Theorem of Calculus, but we will omit the details.

7.8.1 THEOREM. *If g' is continuous on $[a, b]$ and f is continuous and has an anti-derivative on an interval containing the values of $g(x)$ for $a \le x \le b$, then*

$$\int_a^b f(g(x))g'(x)\,dx = \int_{g(a)}^{g(b)} f(u)\,du$$

The choice of methods for evaluating definite integrals by substitution is generally a matter of taste, but in the following examples we will use the second method, since the idea is new.

Example 2

Evaluate

$$\text{(a)} \int_0^{3/4} \frac{dx}{1-x} \qquad \text{(b)} \int_0^{\pi/8} \sin^5 2x \cos 2x\,dx$$

Solution (a). Let

$u = 1 - x$ so that $du = -dx$

With this substitution we have

$u = 1$ if $x = 0$

$u = \frac{1}{4}$ if $x = \frac{3}{4}$

Thus,

$$\int_0^{3/4} \frac{dx}{1-x} = -\int_1^{1/4} \frac{du}{u} = -\ln|u|\Bigg]_{u=1}^{1/4}$$

$$= -\left[\ln\left(\frac{1}{4}\right) - \ln(1)\right] = \ln 4$$

Solution (b). Let

$$u = \sin 2x \quad \text{so that} \quad du = 2\cos 2x\, dx \quad (\text{or } \tfrac{1}{2} du = \cos 2x\, dx)$$

With this substitution we have

$$u = \sin(0) = 0 \quad \text{if} \quad x = 0$$
$$u = \sin(\pi/4) = 1/\sqrt{2} \quad \text{if} \quad x = \pi/8$$

so

$$\int_0^{\pi/8} \sin^5 2x \cos 2x\, dx = \frac{1}{2} \int_0^{1/\sqrt{2}} u^5\, du = \frac{1}{2} \cdot \frac{u^6}{6}\Bigg]_0^{1/\sqrt{2}}$$

$$= \frac{1}{2}\left[\frac{1}{6(\sqrt{2})^6} - 0\right] = \frac{1}{96} \quad\blacktriangleleft$$

Example 3

In Example 8 of Section 4.4 we stated the following model for the temperature T in degrees Fahrenheit ($^\circ$F) of a glass of lemonade t hours after being placed in a room with a constant temperature of 70°F, given that the initial temperature of the lemonade was 40°F:

$$T = 70 - 30e^{-0.5t}$$

Find the average temperature T_{ave} of the lemonade over the first 5 hours.

Solution. From Definition 7.7.5 the average value of T over the time interval $[0, 5]$ is

$$T_{\text{ave}} = \frac{1}{5} \int_0^5 (70 - 30e^{-0.5t})\, dt \tag{3}$$

To evaluate this integral, we make the substitution

$$u = -0.5t \quad \text{so that} \quad du = -0.5\, dt \quad [\text{or} \ dt = -(1/0.5)\, du]$$

With this substitution we have

$$u = 0 \quad \text{if} \quad t = 0$$
$$u = -(0.5)5 = -2.5 \quad \text{if} \quad t = 5$$

Thus, (3) can be expressed as

$$T_{\text{ave}} = \frac{1}{5}\int_0^{-2.5}(70 - 30e^u)\left(-\frac{1}{0.5}\right)du = -\frac{1}{2.5}\int_0^{-2.5}(70 - 30e^u)\,du$$

$$= -\frac{1}{2.5}\big[70u - 30e^u\big]_{u=0}^{-2.5} = -\frac{1}{2.5}\big[(-175 - 30e^{-2.5}) - (-30)\big]$$

$$= 58 + 12e^{-2.5} \approx 58.99^\circ\text{F} \quad\blacktriangleleft$$

REMARK. Observe that the u-substitution in this example produced an integral in which the upper u-limit of integration was smaller than the lower u-limit of integration. In our computations we left the limits of integration in that order, but had we wanted to we could have reversed the order to put the larger limit on top and compensated by reversing the sign of the integral in accordance with Definition 7.5.3(b). The choice of procedures is a matter of taste; both produce the same result (verify).

FOR THE READER. If you have a CAS, use it to evaluate the integral in the last example. See whether it makes any difference in the form of the answer if you express the exponent as $-t/2$ rather than $-0.5t$.

In Exercises 1 and 2, express the integral in terms of the variable u, but do not evaluate it.

1. (a) $\displaystyle\int_0^2 (x+1)^7\,dx; \quad u = x+1$

    (b) $\displaystyle\int_{-1}^2 x\sqrt{8-x^2}\,dx; \quad u = 8-x^2$

    (c) $\displaystyle\int_{-1}^1 \sin(\pi\theta)\,d\theta; \quad u = \pi\theta$

    (d) $\displaystyle\int_0^3 (x+2)(x-3)^{20}\,dx; \quad u = x-3$

2. (a) $\displaystyle\int_0^1 e^{2x-1}\,dx; \quad u = 2x-1$

    (b) $\displaystyle\int_e^{e^2} \frac{\ln x}{x}\,dx; \quad u = \ln x$

    (c) $\displaystyle\int_0^{\pi/4} \tan^2 x \sec^2 x\,dx; \quad u = \tan x$

    (d) $\displaystyle\int_0^1 x^3\sqrt{x^2+3}\,dx; \quad u = x^2+3$

In Exercises 3–12, evaluate the definite integral two ways: first by a u-substitution in the definite integral and then by a u-substitution in the corresponding indefinite integral.

3. $\displaystyle\int_0^1 (2x+1)^4\,dx$

4. $\displaystyle\int_1^2 (4x-2)^3\,dx$

5. $\displaystyle\int_{-1}^0 (1-2x)^3\,dx$

6. $\displaystyle\int_1^2 (4-3x)^8\,dx$

7. $\displaystyle\int_0^8 x\sqrt{1+x}\,dx$

8. $\displaystyle\int_{-5}^0 x\sqrt{4-x}\,dx$

9. $\displaystyle\int_0^{\pi/2} 4\sin(x/2)\,dx$

10. $\displaystyle\int_0^{\pi/6} 2\cos 3x\,dx$

11. $\displaystyle\int_{-\ln 3}^{\ln 3} \frac{e^x}{e^x+4}\,dx$

12. $\displaystyle\int_0^{\ln 5} e^x(3-4e^x)\,dx$

In Exercises 13–16, evaluate the definite integral by expressing it in terms of u and evaluating the resulting integral using a formula from geometry.

13. $\displaystyle\int_0^{5/3} \sqrt{25-9x^2}\,dx; \quad u = 3x$

14. $\displaystyle\int_0^2 x\sqrt{16-x^4}\,dx; \quad u = x^2$

15. $\displaystyle\int_{\pi/3}^{\pi/2} \sin\theta\sqrt{1-4\cos^2\theta}\,d\theta; \quad u = 2\cos\theta$

16. $\displaystyle\int_{e^{-6}}^{e^6} \frac{\sqrt{36-(\ln x)^2}}{x}\,dx; \quad u = \ln x$

17. Find the area under the curve $y = \sin\pi x$ over the interval $[0, 1]$.

18. Find the area under the curve $y = 3\cos 2x$ over the interval $[0, \pi/8]$.

19. Find the area under the curve $y = 1/(x+5)^2$ over the interval $[3, 7]$.

20. Find the area under the curve $y = 1/(3x+1)^2$ over the interval $[0, 1]$.

21. Find the average value of $f(x) = e^{-2x}$ over the interval $[0, 4]$.

22. Find the average value of $f(x) = \sec^2 \pi x$ over the interval $\left[-\frac{1}{4}, \frac{1}{4}\right]$.

In Exercises 23–38, evaluate the integrals by any method.

23. $\displaystyle\int_0^1 \frac{dx}{\sqrt{3x+1}}$

24. $\displaystyle\int_1^2 \sqrt{5x-1}\,dx$

25. $\displaystyle\int_{-1}^1 \frac{x^2\,dx}{\sqrt{x^3+9}}$

26. $\displaystyle\int_{-1}^0 6t^2(t^3+1)^{19}\,dt$

27. $\displaystyle\int_1^3 \frac{x+2}{\sqrt{x^2+4x+7}}\,dx$

28. $\displaystyle\int_1^2 \frac{dx}{x^2-6x+9}$

29. $\displaystyle\int_{-3\pi/4}^{\pi/4} \sin x \cos x\,dx$

30. $\displaystyle\int_0^{\pi/4} \sqrt{\tan x}\,\sec^2 x\,dx$

31. $\displaystyle\int_0^{\sqrt{\pi}} 5x\cos(x^2)\,dx$

32. $\displaystyle\int_{\pi^2}^{4\pi^2} \frac{1}{\sqrt{x}}\sin\sqrt{x}\,dx$

33. $\displaystyle\int_{\pi/12}^{\pi/9} \sec^2 3\theta\,d\theta$

34. $\displaystyle\int_0^{\pi/2} \sin^2 3\theta \cos 3\theta\,d\theta$

35. $\displaystyle\int_0^1 \frac{y^2\,dy}{\sqrt{4-3y}}$

36. $\displaystyle\int_{-1}^4 \frac{x\,dx}{\sqrt{5+x}}$

37. $\displaystyle\int_0^e \frac{dx}{x+e}$

38. $\displaystyle\int_1^{\sqrt{2}} xe^{-x^2}\,dx$

C **39.** For each of the integrals you evaluated in Exercises 23–38, check your answer using a CAS.

C **40.** Use a CAS to find the exact value of the integral

$$\int_{-3}^1 \sqrt{3-2x-x^2}\,dx$$

and then confirm the result by hand calculation. [*Hint:* Complete the square.]

41. (a) Find $\displaystyle\int_0^1 f(3x+1)\,dx$ if $\displaystyle\int_1^4 f(x)\,dx = 5$.

    (b) Find $\displaystyle\int_0^3 f(3x)\,dx$ if $\displaystyle\int_0^9 f(x)\,dx = 5$.

    (c) Find $\displaystyle\int_{-2}^0 xf(x^2)\,dx$ if $\displaystyle\int_0^4 f(x)\,dx = 1$.

42. Given that m and n are positive integers, show that

$$\int_0^1 x^m (1-x)^n \, dx = \int_0^1 x^n (1-x)^m \, dx$$

by making a substitution. Do not attempt to evaluate the integrals.

43. Given that n is a positive integer, show that

$$\int_0^{\pi/2} \sin^n x \, dx = \int_0^{\pi/2} \cos^n x \, dx$$

by using a trigonometric identity and making a substitution. Do not attempt to evaluate the integrals.

44. Given that n is a positive integer, evaluate the integral

$$\int_0^1 x(1-x)^n \, dx$$

45. Suppose that at time $t = 0$ there are 750 bacteria in a growth medium and the bacteria population $y(t)$ grows at the rate $y'(t) = 802.137 e^{1.528t}$ bacteria per hour. How many bacteria will there be in 12 hours?

46. Suppose that the value of a yacht in dollars after t years of use is $V(t) = 275{,}000 e^{-0.17t}$. What is the average value of the yacht over its first 10 years of use?

47. Suppose that a particle moving along a coordinate line has velocity $v(t) = 25 + 10 e^{-0.05t}$ ft/s.
  (a) What is the distance traveled by the particle from time $t = 0$ to time $t = 10$?
  (b) Does the term $10 e^{-0.05t}$ have much effect on the distance traveled by the particle over that time interval? Explain your reasoning.

48. Find a positive value of k such that the area under the graph of $y = e^{2x}$ over the interval $[0, k]$ is 3 square units.

49. Electricity is supplied to homes in the form of *alternating current*, which means that the voltage has a sinusoidal waveform described by an equation of the form

$$V = V_p \sin(2\pi ft)$$

(see the accompanying figure). In this equation, V_p is called the *peak voltage* or *amplitude* of the current, f is called its *frequency*, and $1/f$ is called its *period*. The voltages V and V_p are measured in volts (V), the time t is measured in seconds (s), and the frequency is measured in hertz (Hz) or sometimes in cycles per second. (A *cycle* is the electrical term for one period of the waveform.) Alternating current voltmeters read what is called the *rms* or *root-mean-square* value of V. By definition, this is the square root of the average value of V^2 over one period.
  (a) Show that

$$V_{\text{rms}} = \frac{V_p}{\sqrt{2}}$$

  [*Hint:* Compute the average over the cycle from $t = 0$ to $t = 1/f$, and use the identity $\sin^2 \theta = \frac{1}{2}(1 - \cos 2\theta)$ to help evaluate the integral.]

  (b) In the United States, electrical outlets supply alternating current with an rms voltage of 120 V at a frequency of 60 Hz. What is the peak voltage at such an outlet?

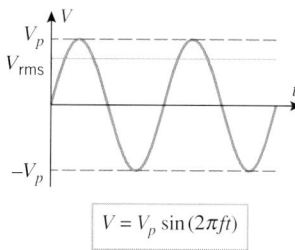

$$V = V_p \sin(2\pi ft)$$

Figure Ex-49

50. Show that if f and g are continuous functions, then

$$\int_0^t f(t-x)g(x)\,dx = \int_0^t f(x)g(t-x)\,dx$$

51. (a) Let $I = \displaystyle\int_0^a \frac{f(x)}{f(x) + f(a - x)}\,dx$. Show that $I = a/2$.
      [*Hint:* Let $u = a - x$, and then express the integrand as the sum of two fractions.]
  (b) Use the result of part (a) to find

$$\int_0^3 \frac{\sqrt{x}}{\sqrt{x} + \sqrt{3 - x}}\,dx$$

  (c) Use the result of part (a) to find

$$\int_0^{\pi/2} \frac{\sin x}{\sin x + \cos x}\,dx$$

52. Let $I = \displaystyle\int_{-1}^1 \frac{1}{1 + x^2}\,dx$. Show that the substitution $x = 1/u$ results in

$$I = -\int_{-1}^1 \frac{1}{1 + u^2}\,du = -I$$

so $2I = 0$, which implies that $I = 0$. However, this is impossible since the integrand of the given integral is positive over the interval of integration. Where is the error?

53. Find the limit

$$\lim_{n \to +\infty} \sum_{k=1}^n \frac{\sin(k\pi/n)}{n}$$

by evaluating an appropriate definite integral over the interval $[0, 1]$.

[C] **54.** Check your answer to Exercise 53 by evaluating the limit directly with a CAS.

55. (a) Prove that if f is an odd function, then

$$\int_{-a}^a f(x)\,dx = 0$$

and give a geometric explanation of this result. [*Hint:* One way to prove that a quantity q is zero is to show that $q = -q$.]

(b) Prove that if f is an even function, then

$$\int_{-a}^{a} f(x)\, dx = 2 \int_{0}^{a} f(x)\, dx$$

and give a geometric explanation of this result. [*Hint:* Split the interval of integration from $-a$ to a into two parts at 0.]

7.9 LOGARITHMIC FUNCTIONS FROM THE INTEGRAL POINT OF VIEW

In Section 4.2 we discussed natural logarithms from the viewpoint of exponents; that is, we regarded $y = \ln x$ to mean that $e^y = x$. In this section we will show that $\ln x$ can also be expressed as an integral with a variable upper limit. This integral representation of $\ln x$ is important mathematically because it provides a convenient way of establishing properties such as differentiability and continuity. However, it is also important in applications because it provides a way of recognizing when integral solutions of problems can be expressed as natural logarithms.

THE LINK BETWEEN NATURAL LOGARITHMS AND INTEGRALS

The connection between natural logarithms and integrals was made in the middle of the seventeenth century in the course of investigating areas under the curve $y = 1/t$. The problem being considered was to find values of $t_1, t_2, t_3, \ldots, t_n, \ldots$ for which the areas $A_1, A_2, A_3, \ldots, A_n, \ldots$ in Figure 7.9.1a would be equal. Through the combined work of Isaac Newton, the Belgian Jesuit priest, Gregory of St. Vincent (1584–1667), and Gregory's student, Alfons A. de Sarasa (1618–1667), it was shown that by taking the points to be

$$t_1 = e, \quad t_2 = e^2, \quad t_3 = e^3, \ldots, \quad t_n = e^n, \ldots$$

each of the areas would be 1 (Figure 7.9.1b). Thus, in modern integral notation

$$\int_{1}^{e^n} \frac{1}{t}\, dt = n$$

which can be expressed as

$$\int_{1}^{e^n} \frac{1}{t}\, dt = \ln(e^n)$$

By comparing the upper limit of the integral and the expression inside the logarithm, it is a natural leap to the more general result

$$\int_{1}^{x} \frac{1}{t}\, dt = \ln x$$

which today we take as the formal definition of the natural logarithm.

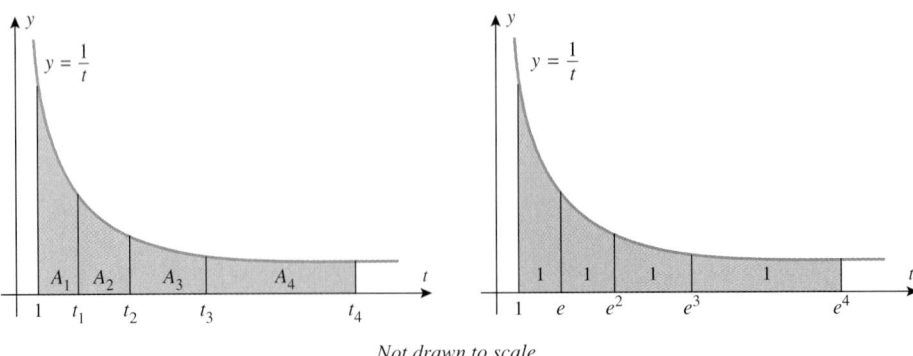

Not drawn to scale

Figure 7.9.1   (*a*)   (*b*)

7.9.1   DEFINITION.   The *natural logarithm* of x is denoted by $\ln x$ and is defined by the integral

$$\ln x = \int_1^x \frac{1}{t}\, dt, \quad x > 0 \tag{1}$$

Geometrically, $\ln x$ is the area under the curve $y = 1/t$ from $t = 1$ to $t = x$ when $x > 1$, and $\ln x$ is the negative of the area under the curve $y = 1/t$ from $t = x$ to $t = 1$ when $0 < x < 1$ (Figure 7.9.2). If $x = 1$, then $\ln x = 0$, since the upper and lower limits in (1) are the same. All of this is consistent with the computer-generated graph of $y = \ln x$ in Figure 4.2.4.

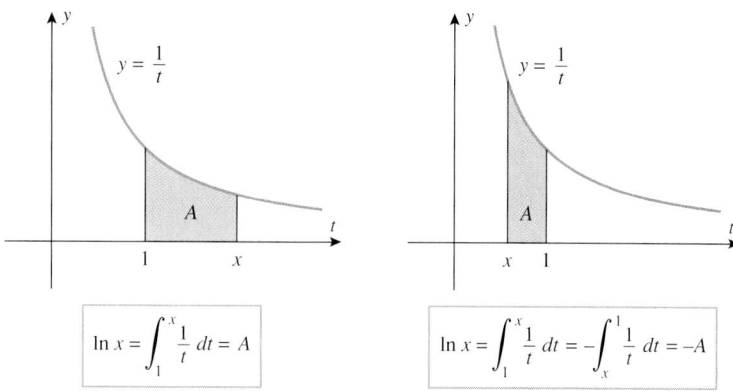

Figure 7.9.2

FOR THE READER.   Review Theorem 7.5.8, and then explain why x is required to be positive in Definition 7.9.1.

APPROXIMATING ln x NUMERICALLY

Table 7.9.1

$$n = 10$$
$$\Delta t = (b - a)/n = (2 - 1)/10 = 0.1$$

k	t_k^*	$1/t_k^*$
1	1.05	0.952381
2	1.15	0.869565
3	1.25	0.800000
4	1.35	0.740741
5	1.45	0.689655
6	1.55	0.645161
7	1.65	0.606061
8	1.75	0.571429
9	1.85	0.540541
10	1.95	0.512821
		6.928355

$$\Delta t \sum_{k=1}^{n} f(t_k^*) = (0.1)(6.928355)$$
$$= 0.692836$$

For specific values of x, the value of $\ln x$ can be approximated numerically by approximating the definite integral in (1), say by using the midpoint approximation that was discussed in Section 7.5.

Example 1

Approximate $\ln 2$ using the midpoint approximation with $n = 10$.

Solution.   From (1), the exact value of $\ln 2$ is represented by the integral

$$\ln 2 = \int_1^2 \frac{1}{t}\, dt$$

The midpoint rule is given in Formula (3) of Section 7.5. Expressed in terms of t, that formula is

$$\int_a^b f(t)\, dt \approx \Delta t \sum_{k=1}^{n} f(t_k^*)$$

where Δt is the width of each subinterval and $t_1^*, t_2^*, \ldots, t_n^*$ are the midpoints. In this case we have 10 subintervals, so $\Delta t = (2 - 1)/10 = 0.1$. The computations to six decimal places are shown in Table 7.9.1. By comparison, a calculator set to display six decimal places gives $\ln 2 \approx 0.693147$, so the magnitude of the error in the midpoint approximation is about 0.000311. Greater accuracy in the midpoint approximation can be obtained by increasing n. For example, the midpoint approximation with $n = 100$ yields $\ln 2 \approx 0.693144$, which is correct to five decimal places.   ◄

Definition 7.9.1 is not only useful for approximating values of $\ln x$, but it is the key to establishing many of the fundamental properties of the natural logarithm. For example, in Section 4.4 we obtained the derivative

$$\frac{d}{dx}[\ln x] = \frac{1}{x} \quad (x > 0) \tag{2}$$

by *assuming* that $f(x) = \ln x$ is differentiable for $x > 0$. However, now that we have Definition 7.9.1 to work with, both the differentiability of $\ln x$ and Formula (2) follow immediately from Part 2 of the Fundamental Theorem of Calculus (7.6.3). Moreover, since differentiable functions are continuous, this also shows that $\ln x$ is continuous for $x > 0$.

Although it is not our objective to prove all of the properties of the functions we encounter, it is worthwhile to understand in principle how the differentiability and continuity of $\ln x$ can be used to establish differentiability and continuity of other important functions. For example, since the exponential function e^x is the inverse of $\ln x$, it follows from Theorem 4.1.7, with $f(x) = \ln x$ and $f^{-1}(x) = e^x$, that e^x is differentiable at any point x where $f'(f^{-1}(x)) = 1/e^x \neq 0$. Since this holds for all x, it follows that e^x is differentiable and hence continuous everywhere.

The differentiability $\ln x$ for $x > 0$ can be used to prove the differentiability of $\log_b x$ for $x > 0$ by using Formula (9) of Section 4.2 to express $\log_b x$ in terms of $\ln x$, and the differentiability of e^x can be used to prove the differentiability of b^x by expressing b^x in terms of e^x as $b^x = e^{x \ln b}$. We omit the details.

In Formulas (3), (4), and (5) of Section 4.2 we gave three limits for e, but at that time we did not have the mathematical tools to prove the existence of those limits; the following theorem does this.

7.9.2 THEOREM.

(a) $\displaystyle\lim_{x \to 0} (1 + x)^{1/x} = e$ $\qquad (b)$ $\displaystyle\lim_{x \to +\infty} \left(1 + \frac{1}{x}\right)^x = e$ $\qquad (c)$ $\displaystyle\lim_{x \to -\infty} \left(1 + \frac{1}{x}\right)^x = e$

Proof. We will prove part (a), and leave the proofs of the other parts for the exercises. Our proof will build on the differentiability of $\ln x$, and more specifically on the derivative of $\ln x$ at the point $x = 1$, namely

$$\frac{d}{dx}[\ln x]\Big|_{x=1} = \frac{1}{x}\Big|_{x=1} = 1$$

If we express this relationship using the definition of a derivative, we obtain

$$1 = \lim_{h \to 0} \frac{\ln(1 + h) - \ln 1}{h} = \lim_{h \to 0} \frac{\ln(1 + h)}{h} = \lim_{h \to 0} \ln(1 + h)^{1/h}$$

Thus, it follows that

$$e = e^{\lim_{h \to 0} \ln(1+h)^{1/h}}$$

which from the continuity of e^x can be written as

$$e = \lim_{h \to 0} e^{\ln(1+h)^{1/h}} = \lim_{h \to 0} (1 + h)^{1/h}$$

Except for a difference in notation, this is what we wanted to prove. ∎

The functions that we have dealt with thus far in this text are called ***elementary functions***; they include polynomials, rational functions, power functions, exponential functions, logarithmic functions, trigonometric functions, and all other functions that can be obtained from these by addition, subtraction, multiplication, division, root extraction, composition, and by taking inverses.

However, there are many important functions that do not fall into this category. Such functions occur in many ways, but they commonly arise in the course of solving initial-value problems of the form

$$\frac{dy}{dx} = f(x), \quad y(x_0) = y_0 \tag{3}$$

Recall from Example 7 of Section 7.2 and the discussion preceding it that the basic method for solving (3) is to integrate $f(x)$, and then use the initial condition to determine the constant of integration. It can be proved that if f is continuous, then (3) has a unique solution and that this procedure produces it. However, there is another approach: Instead of solving each initial-value problem individually, we can find a general formula for the solution of (3), and then apply that formula to solve specific problems. We will now show that

$$y(x) = y_0 + \int_{x_0}^{x} f(t)\, dt \tag{4}$$

is a formula for the solution of (3). To confirm that this is so we must show that $dy/dx = f(x)$ and that $y(x_0) = y_0$. The computations are as follows:

$$\frac{dy}{dx} = \frac{d}{dx}\left[y_0 + \int_{x_0}^{x} f(t)\, dt \right] = 0 + f(x) = f(x)$$

$$y(x_0) = y_0 + \int_{x_0}^{x_0} f(t)\, dt = y_0 + 0 = y_0$$

Example 2

In Example 7 of Section 7.2 we showed that the solution of the initial-value problem

$$\frac{dy}{dx} = \cos x, \quad y(0) = 1$$

is $y(x) = 1 + \sin x$. This initial-value problem can also be solved by applying Formula (4) with $f(x) = \cos x$, $x_0 = 0$, and $y_0 = 1$. This yields

$$y(x) = 1 + \int_{0}^{x} \cos t\, dt = 1 + \left[\sin t \right]_{t=0}^{x} = 1 + \sin x \qquad \blacktriangleleft$$

In the last example we were able to perform the integration in Formula (4) and express the solution of the initial-value problem as an elementary function. However, sometimes this will not be possible, in which case the solution of the initial-value problem must be left in terms of an "unevaluated" integral. For example, from (4), the solution of the initial-value problem

$$\frac{dy}{dx} = e^{-x^2}, \quad y(0) = 1$$

is

$$y(x) = 1 + \int_{0}^{x} e^{-t^2}\, dt$$

However, it can be shown that there is no way to express the integral in this solution as an elementary function. Thus, we have encountered a *new* function, which we regard to be *defined* by the integral. A close relative of this function, known as the **error function**, plays an important role in probability and statistics; it is denoted by $\operatorname{erf}(x)$ and is defined as

$$\operatorname{erf}(x) = \frac{2}{\sqrt{\pi}} \int_{0}^{x} e^{-t^2}\, dt \tag{5}$$

Indeed, many of the most important functions in science and engineering are defined as integrals that have special names and notations associated with them. For example, the functions defined by

$$S(x) = \int_0^x \sin\left(\frac{\pi t^2}{2}\right) dt \quad \text{and} \quad C(x) = \int_0^x \cos\left(\frac{\pi t^2}{2}\right) dt \tag{6–7}$$

are called the ***Fresnel sine and cosine functions***, respectively, in honor of the French physicist Augustin Fresnel (1788–1827), who first encountered them in his study of diffraction of light waves.

EVALUATING AND GRAPHING FUNCTIONS DEFINED BY INTEGRALS

The following values of $S(1)$ and $C(1)$ were produced by a CAS that has a built-in algorithm for approximating definite integrals:

$$S(1) = \int_0^1 \sin\left(\frac{\pi t^2}{2}\right) dt \approx 0.438259, \qquad C(1) = \int_0^1 \cos\left(\frac{\pi t^2}{2}\right) dt \approx 0.779893$$

To generate graphs of functions defined by integrals, computer programs choose a set of x-values in the domain, approximate the integral for each of those values, and then plot the resulting points. Thus, there is a lot of computation involved in generating such graphs, since each plotted point requires the approximation of an integral. The graphs of the Fresnel functions in Figure 7.9.3 were generated in this way using a CAS.

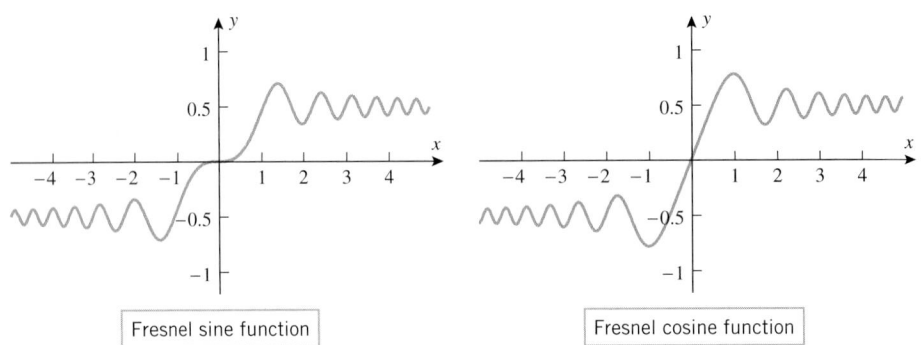

Fresnel sine function

Fresnel cosine function

Figure 7.9.3

REMARK. Although it required a considerable amount of computation to generate the graphs of the Fresnel functions, the derivatives of $S(x)$ and $C(x)$ are easy to obtain using Part 2 of the Fundamental Theorem of Calculus (7.6.3); they are

$$S'(x) = \sin\left(\frac{\pi x^2}{2}\right) \quad \text{and} \quad C'(x) = \cos\left(\frac{\pi x^2}{2}\right) \tag{8–9}$$

These derivatives can be used to determine the locations of the relative extrema and inflection points and to investigate other properties of $S(x)$ and $C(x)$.

INTEGRALS WITH FUNCTIONS AS LIMITS OF INTEGRATION

Various applications can lead to integrals in which one or both of the limits of integration is a function of x. Some examples are

$$\int_x^1 \sqrt{\sin t}\, dt, \qquad \int_{x^2}^{\sin x} \sqrt{t^3 + 1}\, dt, \qquad \int_{\ln x}^{\pi} \frac{dt}{t^7 - 8}$$

We will complete this section by showing how to differentiate integrals of the form

$$\int_a^{g(x)} f(t)\, dt \tag{10}$$

where a is constant. Derivatives of other kinds of integrals with functions as limits of integration will be discussed in the exercises.

To differentiate (10) we can view the integral as a composition $F(g(x))$, where

$$F(x) = \int_a^x f(t)\, dt$$

If we now apply the chain rule, we obtain

$$\frac{d}{dx}\left[\int_a^{g(x)} f(t)\, dt\right] = \frac{d}{dx}[F(g(x))] = F'(g(x))g'(x) = f(g(x))g'(x)$$

Theorem 7.6.3

Thus,

$$\frac{d}{dx}\left[\int_a^{g(x)} f(t)\, dt\right] = f(g(x))g'(x) \tag{11}$$

REMARK.   In words, *to differentiate an integral with a constant lower limit and a function as the upper limit, substitute the upper limit into the integrand, and multiply by the derivative of the upper limit.*

Example 3

$$\frac{d}{dx}\left[\int_1^{\sin x} (1 - t^2)\, dt\right] = (1 - \sin^2 x)\cos x = \cos^3 x \qquad \blacktriangleleft$$

EXERCISE SET 7.9   ⬚ Graphing Calculator   Ⓒ CAS
. . . .

1. Sketch the curve $y = 1/t$, and shade a region under the curve whose area is
   (a) $\ln 2$ (b) $-\ln 0.5$ (c) 2.

2. Sketch the curve $y = 1/t$, and shade two different regions under the curve whose area is $\ln 1.5$.

3. Given that $\ln a = 2$ and $\ln c = 5$, find
   (a) $\displaystyle\int_1^{ac} \frac{1}{t}\, dt$ (b) $\displaystyle\int_1^{1/c} \frac{1}{t}\, dt$
   (c) $\displaystyle\int_1^{a/c} \frac{1}{t}\, dt$ (d) $\displaystyle\int_1^{a^3} \frac{1}{t}\, dt$.

4. Given that $\ln a = 4$, find
   (a) $\displaystyle\int_1^{\sqrt{a}} \frac{1}{t}\, dt$ (b) $\displaystyle\int_1^{2a} \frac{1}{t}\, dt$
   (c) $\displaystyle\int_1^{2/a} \frac{1}{t}\, dt$ (d) $\displaystyle\int_2^{a} \frac{1}{t}\, dt$.

5. Approximate $\ln 5$ using the midpoint rule with $n = 10$, and estimate the magnitude of the error by comparing your answer to that produced directly by a calculating utility.

6. Approximate $\ln 3$ using the midpoint rule with $n = 20$, and estimate the magnitude of the error by comparing your answer to that produced directly by a calculating utility.

7. Simplify the expression and state the values of x for which your simplification is valid.
   (a) $e^{-\ln x}$ (b) $e^{\ln x^2}$
   (c) $\ln\left(e^{-x^2}\right)$ (d) $\ln(1/e^x)$
   (e) $\exp(3\ln x)$ (f) $\ln(xe^x)$
   (g) $\ln\left(e^{x-\sqrt[3]{x}}\right)$ (h) $e^{x-\ln x}$

8. (a) Let $f(x) = e^{-2x}$. Find the simplest exact value of the function $f(\ln 3)$.
   (b) Let $f(x) = e^x + 3e^{-x}$. Find the simplest exact value of the function $f(\ln 2)$.

In Exercises 9 and 10, express the given quantity as a power of e.

9. (a) 3^π (b) $2^{\sqrt{2}}$

10. (a) π^{-x} (b) $x^{2x}, \quad x > 0$

In Exercises 11 and 12, find the limits by making appropriate substitutions in the limits given in Theorem 7.9.2.

11. (a) $\displaystyle\lim_{x \to +\infty} \left(1 + \frac{1}{x}\right)^{2x}$ (b) $\displaystyle\lim_{x \to 0} (1 + 2x)^{1/x}$

12. (a) $\displaystyle\lim_{x \to +\infty} \left(1 + \frac{1}{3x}\right)^x$ **(b)** $\displaystyle\lim_{x \to 0} (1 + x)^{1/3x}$

In Exercises 13 and 14, find $g'(x)$ using Part 2 of the Fundamental Theorem of Calculus, and check your answer by evaluating the integral and then differentiating.

13. $g(x) = \displaystyle\int_1^x (t^2 - t)\,dt$ **14.** $g(x) = \displaystyle\int_\pi^x (1 - \cos t)\,dt$

In Exercises 15 and 16, find the derivative using Formula (11), and check your answer by evaluating the integral and then differentiating.

15. (a) $\displaystyle\frac{d}{dx} \int_1^{x^3} \frac{1}{t}\,dt$ **(b)** $\displaystyle\frac{d}{dx} \int_1^{\ln x} e^t\,dt$

16. (a) $\displaystyle\frac{d}{dx} \int_{-1}^{x^2} \sqrt{t+1}\,dt$ **(b)** $\displaystyle\frac{d}{dx} \int_\pi^{1/x} \sin t\,dt$

17. Let $F(x) = \displaystyle\int_0^x \frac{\cos t}{t^2 + 3}\,dt$. Find

(a) $F(0)$ (b) $F'(0)$ (c) $F''(0)$.

18. Let $F(x) = \displaystyle\int_2^x \sqrt{3t^2 + 1}\,dt$. Find

(a) $F(2)$ (b) $F'(2)$ (c) $F''(2)$.

c **19. (a)** Use Formula (11) to find

$$\frac{d}{dx} \int_1^{x^2} t\sqrt{1+t}\,dt$$

(b) Use a CAS to evaluate the integral and differentiate the resulting function.

(c) Use the simplification command of the CAS, if necessary, to confirm that answers in parts (a) and (b) are the same.

20. Show that

(a) $\displaystyle\frac{d}{dx}\left[\int_x^a f(t)\,dt\right] = -f(x)$

(b) $\displaystyle\frac{d}{dx}\left[\int_{g(x)}^a f(t)\,dt\right] = -f(g(x))g'(x).$

In Exercises 21 and 22, use the results in Exercise 20 to find the derivative.

21. (a) $\displaystyle\frac{d}{dx} \int_x^1 \sin(t^2)\,dt$ **(b)** $\displaystyle\frac{d}{dx} \int_{\tan x}^3 \frac{t^2}{1+t^2}\,dt$

22. (a) $\displaystyle\frac{d}{dx} \int_x^0 (t^2 + 1)^{40}\,dt$ **(b)** $\displaystyle\frac{d}{dx} \int_{1/x}^\pi \cos^3 t\,dt$

23. Find

$$\frac{d}{dx}\left[\int_{3x}^{x^2} \frac{t-1}{t^2+1}\,dt\right]$$

by writing

$$\int_{3x}^{x^2} \frac{t-1}{t^2+1}\,dt = \int_{3x}^0 \frac{t-1}{t^2+1}\,dt + \int_0^{x^2} \frac{t-1}{t^2+1}\,dt$$

24. Use Exercise 20(b) and the idea in Exercise 23 to show that

$$\frac{d}{dx} \int_{h(x)}^{g(x)} f(t)\,dt = f(g(x))g'(x) - f(h(x))h'(x)$$

25. Use the result obtained in Exercise 24 to perform the following differentiations:

(a) $\displaystyle\frac{d}{dx} \int_{x^2}^{x^3} \sin^2 t\,dt$ **(b)** $\displaystyle\frac{d}{dx} \int_{-x}^x \frac{1}{1+t}\,dt.$

26. Prove that the function

$$F(x) = \int_x^{3x} \frac{1}{t}\,dt$$

is constant on the interval $(0, +\infty)$ by using Exercise 24 to find $F'(x)$. What is that constant?

27. Let $F(x) = \int_0^x f(t)\,dt$, where f is the function whose graph is shown in the accompanying figure.

(a) Find $F(0)$, $F(3)$, $F(5)$, $F(7)$, and $F(10)$.

(b) On what subintervals of the interval $[0, 10]$ is F increasing? Decreasing?

(c) Where does F have its maximum value? Its minimum value?

(d) Sketch the graph of F.

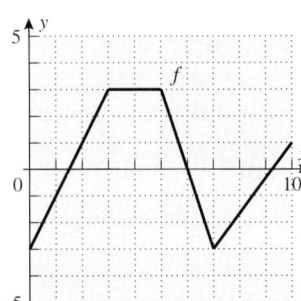

Figure Ex-27

28. Use the appropriate values found in part (a) of Exercise 27 to find the average value of f over the interval $[0, 10]$.

In Exercises 29 and 30, express $F(x)$ in a piecewise form that does not involve an integral.

29. $F(x) = \displaystyle\int_{-1}^x |t|\,dt$

30. $F(x) = \displaystyle\int_0^x f(t)\,dt$, where $f(x) = \begin{cases} x, & 0 \le x \le 2 \\ 2, & x > 2 \end{cases}$

In Exercises 31–34, use Formula (4) to solve the initial-value problem.

31. $\dfrac{dy}{dx} = \sqrt[3]{x};\ y(1) = 2$

32. $\dfrac{dy}{dx} = \dfrac{x+1}{\sqrt{x}};\ y(1) = 0$

33. $\dfrac{dy}{dx} = \sec^2 x - \sin x;\ y(\pi/4) = 1$

34. $\dfrac{dy}{dx} = xe^{x^2};\ y(0) = 0$

35. Suppose that at time $t = 0$ there are P_0 individuals who have disease X, and suppose that a certain model for the spread of the disease predicts that the disease will spread at the rate of $r(t)$ individuals per day. Write a formula for the number of individuals who will have disease X after x days.

36. Suppose that $v(t)$ is the velocity function of a particle moving along an s-axis. Write a formula for the coordinate of the particle at time T if the particle is at the point s_1 at time $t = 1$.

37. The accompanying figure shows the graphs of $y = f(x)$ and $y = \int_0^x f(t)\,dt$. Determine which graph is which, and explain your reasoning.

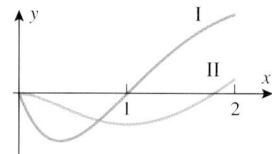

Figure Ex-37

38. (a) Make a conjecture about the value of the limit

$$\lim_{k \to 0} \int_1^x t^{k-1}\,dt \quad (x > 0)$$

(b) Check your conjecture by evaluating the integral, and then using L'Hôpital's rule to find the limit.

39. Let $F(x) = \int_0^x f(t)\,dt$, where f is the function graphed in the accompanying figure.
(a) Where do the relative minima of F occur?
(b) Where do the relative maxima of F occur?
(c) Where does the absolute maximum of F on the interval $[0, 5]$ occur?
(d) Where does the absolute minimum of F on the interval $[0, 5]$ occur?
(e) Where is F concave up? Concave down?
(f) Sketch the graph of F.

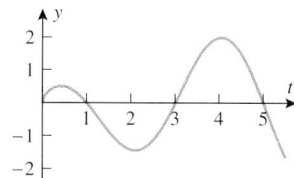

Figure Ex-39

c 40. CAS programs have commands for working with most of the important nonelementary functions. Check your CAS

documentation for information about the error function $\mathrm{erf}(x)$ [see Formula (5)], and then complete the following.
(a) Generate the graph of $\mathrm{erf}(x)$.
(b) Use the graph to make a conjecture about the existence and location of any relative maxima and minima of $\mathrm{erf}(x)$.
(c) Check your conjecture in part (b) using the derivative of $\mathrm{erf}(x)$.
(d) Use the graph to make a conjecture about the existence and location of any inflection points of $\mathrm{erf}(x)$.
(e) Check your conjecture in part (d) using the second derivative of $\mathrm{erf}(x)$.
(f) Use the graph to make a conjecture about the existence of horizontal asymptotes of $\mathrm{erf}(x)$.
(g) Check your conjecture in part (f) by using the CAS to find the limits of $\mathrm{erf}(x)$ as $x \to \pm\infty$.

41. The Fresnel sine and cosine functions $S(x)$ and $C(x)$ were defined in Formulas (6) and (7) and graphed in Figure 7.9.3. Their derivatives were given in Formulas (8) and (9).
(a) At what points does $C(x)$ have relative minima? Relative maxima?
(b) Where do the inflection points of $C(x)$ occur?
(c) Confirm that your answers in parts (a) and (b) are consistent with the graph of $C(x)$.

42. Find the limit

$$\lim_{h \to 0} \frac{1}{h} \int_x^{x+h} \ln t\,dt$$

43. Find a function f and a number a such that

$$2 + \int_a^x f(t)\,dt = e^{3x}$$

44. (a) Give a geometric argument to show that

$$\frac{1}{x+1} < \int_x^{x+1} \frac{1}{t}\,dt < \frac{1}{x}, \quad x > 0$$

(b) Use the result in part (a) to prove that

$$\frac{1}{x+1} < \ln\left(1 + \frac{1}{x}\right) < \frac{1}{x}, \quad x > 0$$

(c) Use the result in part (b) to prove that

$$e^{\frac{x}{x+1}} < \left(1 + \frac{1}{x}\right)^x < e, \quad x > 0$$

and hence that

$$\lim_{x \to +\infty} \left(1 + \frac{1}{x}\right)^x = e$$

(d) Use the inequality in part (c) to prove that

$$\left(1 + \frac{1}{x}\right)^x < e < \left(1 + \frac{1}{x}\right)^{x+1}, \quad x > 0$$

45. Use a graphing utility to generate the graph of

$$y = \left(1 + \frac{1}{x}\right)^{x+1} - \left(1 + \frac{1}{x}\right)^x$$

in the window $[0, 100] \times [0, 0.2]$, and use that graph and part (d) of Exercise 44 to make a rough estimate of the error in the approximation

$$e \approx \left(1 + \frac{1}{50}\right)^{50}$$

46. Prove: If f is continuous on an open interval I and a is any point in I, then

$$F(x) = \int_a^x f(t)\, dt$$

is continuous on I.

SUPPLEMENTARY EXERCISES

1. Write a paragraph that describes the *rectangle method* for defining the area under a curve $y = f(x)$ over an interval $[a, b]$.

2. What is an *integral curve* of a function f? How are two integral curves of a function f related?

3. The *definite integral* of f over the interval $[a, b]$ is defined as the limit

$$\int_a^b f(x)\, dx = \lim_{\max \Delta x_k \to 0} \sum_{k=1}^{n} f(x_k^*)\Delta x_k$$

Explain what the various symbols on the right side of this equation mean.

4. State the two parts of the Fundamental Theorem of Calculus, and explain what is meant by the phrase "differentiation and integration are inverse processes."

5. Derive the formulas for the position and velocity functions of a particle that moves with uniformly accelerated motion along a coordinate line.

6. (a) Devise a procedure for finding upper and lower estimates of the area of the region in the accompanying figure (in cm²).
(b) Use your procedure to find upper and lower estimates of the area.
(c) Improve on the estimates you obtained in part (b).

Figure Ex-6

7. Suppose that

$$\int_0^1 f(x)\, dx = \frac{1}{2}, \quad \int_1^2 f(x)\, dx = \frac{1}{4},$$

$$\int_0^3 f(x)\, dx = -1, \quad \int_0^1 g(x)\, dx = 2$$

In each part, use this information to evaluate the given integral, if possible. If there is not enough information to evaluate the integral, then say so.

(a) $\int_0^2 f(x)\, dx$ (b) $\int_1^3 f(x)\, dx$ (c) $\int_2^3 5f(x)\, dx$

(d) $\int_1^0 g(x)\, dx$ (e) $\int_0^1 g(2x)\, dx$ (f) $\int_0^1 [g(x)]^2\, dx$

8. In each part, use the information in Exercise 7 to evaluate the given integral. If there is not enough information to evaluate the integral, then say so.

(a) $\int_0^1 [f(x) + g(x)]\, dx$ (b) $\int_0^1 f(x)g(x)\, dx$

(c) $\int_0^1 \frac{f(x)}{g(x)}\, dx$ (d) $\int_0^1 [4g(x) - 3f(x)]\, dx$

9. In each part, evaluate the integral. Where appropriate, you may use a geometric formula.

(a) $\int_{-1}^1 1 + \sqrt{1 - x^2}\, dx$

(b) $\int_0^3 (x\sqrt{x^2 + 1} - \sqrt{9 - x^2})\, dx$

(c) $\int_0^1 x\sqrt{1 - x^4}\, dx$

10. Evaluate the integral $\int_0^1 |2x - 1|\, dx$, and sketch the region whose area it represents.

11. One of the numbers π, $\pi/2$, $35\pi/128$, $1 - \pi$ is the correct value of the integral

$$\int_0^\pi \sin^8 x\, dx$$

Use the accompanying graph of $y = \sin^8 x$ and a logical process of elimination to find the correct value. [Do not attempt to evaluate the integral.]

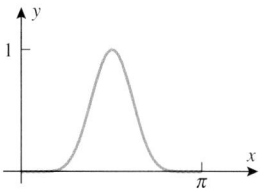

Figure Ex-11

12. Evaluate

$$\int \frac{e^{2x}}{e^x + 3} \, dx$$

[*Hint:* Divide $e^x + 3$ into e^{2x}.]

13. Give a convincing geometric argument to show that

$$\int_1^e \ln x \, dx + \int_0^1 e^x \, dx = e$$

14. In each part, find the limit by interpreting it as a limit of Riemann sums in which the interval $[0, 1]$ is divided into n subintervals of equal length.

(a) $\displaystyle\lim_{n \to +\infty} \frac{\sqrt{1} + \sqrt{2} + \sqrt{3} + \cdots + \sqrt{n}}{n^{3/2}}$

(b) $\displaystyle\lim_{n \to +\infty} \frac{1^4 + 2^4 + 3^4 + \cdots + n^4}{n^5}$

(c) $\displaystyle\lim_{n \to +\infty} \frac{e^{1/n} + e^{2/n} + e^{3/n} + \cdots + e^{n/n}}{n}$

15. (a) Divide the interval $[1, 2]$ into 5 subintervals of equal length, and use appropriate Riemann sums to show that

$$0.2 \left[\tfrac{1}{1.2} + \tfrac{1}{1.4} + \tfrac{1}{1.6} + \tfrac{1}{1.8} + \tfrac{1}{2.0} \right] < \ln 2$$

$$< 0.2 \left[\tfrac{1}{1.0} + \tfrac{1}{1.2} + \tfrac{1}{1.4} + \tfrac{1}{1.6} + \tfrac{1}{1.8} \right]$$

(b) Show that if the interval $[1, 2]$ is divided into n subintervals of equal length, then

$$\sum_{k=1}^{n} \frac{1}{n+k} < \ln 2 < \sum_{k=0}^{n-1} \frac{1}{n+k}$$

(c) Show that the difference between the two sums in part (b) is $1/2n$, and use this result to show that the sums in part (a) approximate $\ln 2$ with an error of at most 0.1.

(d) How large must n be to ensure that the sums in part (b) approximate $\ln 2$ to three decimal places?

16. The accompanying figure shows the direction field for a differential equation $dy/dx = f(x)$. Which of the following functions is most likely to be $f(x)$?

$$\sqrt{x}, \quad \sin x, \quad x^4, \quad x$$

Explain your reasoning.

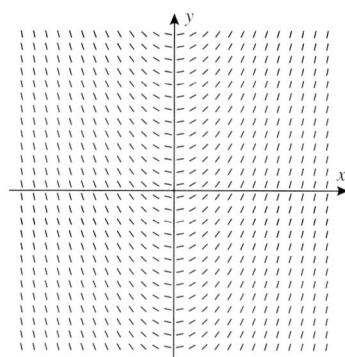

Figure Ex-16

17. In each part, confirm the stated equality.

(a) $1 \cdot 2 + 2 \cdot 3 + \cdots + n(n+1) = \tfrac{1}{3}n(n+1)(n+2)$

(b) $\displaystyle\lim_{n \to +\infty} \sum_{k=1}^{n-1} \left(\frac{9}{n} - \frac{k}{n^2} \right) = \frac{17}{2}$

(c) $\displaystyle\sum_{i=1}^{3} \left(\sum_{j=1}^{2} (i+j) \right) = 21$

18. Express

$$\sum_{k=4}^{18} k(k-3)$$

in sigma notation with

(a) $k = 0$ as the lower limit of summation

(b) $k = 5$ as the lower limit of summation.

19. (a) Show that the substitutions $u = \sec x$ and $u = \tan x$ produce different values for the integral

$$\int \sec^2 x \tan x \, dx$$

(b) Explain why both are correct.

20. Use the two substitutions in Exercise 19 to evaluate the definite integral

$$\int_0^{\pi/4} \sec^2 x \tan x \, dx$$

and confirm that they produce the same result.

21. Evaluate the integral

$$\int \sqrt{1 + x^{-2/3}} \, dx$$

by making the substitution $u = 1 + x^{2/3}$.

22. (a) Express Formula 8 of Section 7.5 in sigma notation.

(b) If $c_1, c_2, \ldots, c_n$ are constants and $f_1, f_2, \ldots, f_n$ are integrable functions on $[a, b]$, do you think it is always true that

$$\int_a^b \left(\sum_{k=1}^{n} c_k f_k(x) \right) dx = \sum_{k=1}^{n} \left[c_k \int_a^b f_k(x) \, dx \right]?$$

Explain your reasoning.

23. Find an integral formula for the antiderivative of $1/(1+x^2)$ on the interval $(-\infty, +\infty)$ whose value at $x = 1$ is (a) 0 and (b) 2.

C **24.** Let $F(x) = \displaystyle\int_0^x \frac{t-3}{t^2+7}\, dt$.

(a) Find the intervals on which F is increasing. Decreasing.

(b) Find the open intervals on which F is concave up. Concave down.

(c) Find the x-values, if any, at which the function F has absolute extrema.

(d) Use a CAS to graph F, and confirm that the results in parts (a), (b), and (c) are consistent with the graph.

25. Prove that the function

$$F(x) = \int_0^x \frac{1}{1+t^2}\, dt + \int_0^{1/x} \frac{1}{1+t^2}\, dt$$

is constant on the interval $(0, +\infty)$.

26. What is the natural domain of the function

$$F(x) = \int_1^x \frac{1}{t^2-9}\, dt?$$

Explain your reasoning.

27. In each part, determine the values of x for which $F(x)$ is positive, negative, or zero without performing the integration; explain your reasoning.

(a) $F(x) = \displaystyle\int_1^x \frac{t^4}{t^2+3}\, dt$ (b) $F(x) = \displaystyle\int_{-1}^x \sqrt{4-t^2}\, dt$

28. Find a formula (defined piecewise) for the upper boundary of the trapezoid shown in the accompanying figure, and then integrate that function to derive the formula for the area of the trapezoid given on the inside front cover of this text.

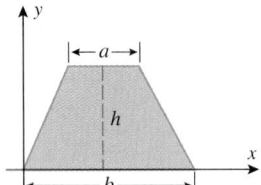

Figure Ex-28

29. An engineer studying the power consumption of a manufacturing plant determines that the plant's daily rate of electricity usage in kilowatts per hour (kW/h) can be reasonably modeled by the formula

$$R(t) = 2000e^{-t/48} + 500\sin\left(\tfrac{\pi}{12}t\right) \quad (0 \le t \le 24)$$

(a) How many kilowatts of electricity does the plant use in a 24-hour period?

(b) Find the average rate of electricity usage over the first 8 hours of operation.

(c) Generate the graph of $R(t)$ over the first 8-hour period, and use it to make a rough estimate of the maximum

rate of electricity usage during that period and when it occurs.

(d) Determine the maximum rate of electricity usage during the first 8-hour period to two decimal places.

30. Suppose that a tumor grows at the rate of $r(t) = t/7$ grams (g) per week. When, during the second 26 weeks of growth, is the weight of the tumor the same as its average weight during that period?

31. The velocity of a particle moving along an s-axis is measured at 5-s intervals for 40 s, and the velocity function is modeled by a smooth curve drawn through the data points, as shown in the accompanying figure.

(a) Does the particle have constant acceleration? Explain your reasoning.

(b) Is there any 15-s time interval during which the acceleration is constant? Explain your reasoning.

(c) Estimate the average velocity of the particle over the 40-s time period.

(d) Estimate the distance traveled by the particle from time $t = 0$ to time $t = 40$.

(e) Is the particle ever slowing down during the 40-s time period? Explain your reasoning.

(f) Is there sufficient information for you to determine the s-coordinate of the particle at time $t = 10$? If so, find it. If not, explain what additional information you need.

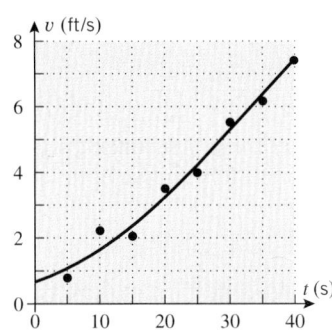

Figure Ex-31

32. Suppose that a particle moves along the x-axis so that its x-coordinate at time t is given by $x = ae^{kt} + be^{-kt}$.

(a) Show that the acceleration is proportional to x.

(b) Assuming that the velocity of the particle at time $t = 0$ is v_0, find a formula for the acceleration function in terms of a, b, x, and v_0.

In Exercises 33–42, evaluate the integrals by hand, and check your answers with a CAS if you have one.

33. $\displaystyle\int \frac{\cos 3x}{\sqrt{5+2\sin 3x}}\, dx$

34. $\displaystyle\int \frac{\sqrt{3+\sqrt{x}}}{\sqrt{x}}\, dx$

35. $\displaystyle\int \frac{x^2}{(ax^3+b)^2}\, dx$

36. $\displaystyle\int x \sec^2(ax^2)\, dx$

37. $\int [\ln(e^x) + \ln(e^{-x})]\, dx$

38. $\int_{-2}^{-1} \left(u^{-4} + 3u^{-2} - \dfrac{1}{u^5} \right) du$

39. $\int_{e}^{e^2} \dfrac{dx}{x \ln x}$

40. $\int_{0}^{1} \dfrac{dx}{\sqrt{e^x}}$

41. $\int_{0}^{\ln \sqrt{2}} \dfrac{1 + \cos(e^{-2x})}{e^{2x}}\, dx$

42. $\int_{0}^{1} \sin^2(\pi x) \cos(\pi x)\, dx$

c **43.** Use a CAS to approximate the area of the region in the first quadrant that lies below the curve $y = x + x^2 - x^3$ and above the x-axis.

c **44.** In each part, use a CAS to solve the initial-value problem.

(a) $\dfrac{dy}{dx} = x^2 \cos 3x; \ \ y(\pi/2) = -1$

(b) $\dfrac{dy}{dx} = \dfrac{x^3}{(4 + x^2)^{3/2}}; \ \ y(0) = -2$

c **45.** In each part, use a CAS, where needed, to solve for k.

(a) $\int_{1}^{k} (x^3 - 2x - 1)\, dx = 0, \ \ k > 1$

(b) $\int_{0}^{k} (x^2 + \sin 2x)\, dx = 3, \ \ k \geq 0$

c **46.** Use a CAS to approximate the largest and smallest values of the integral

$$\int_{-1}^{x} \dfrac{t}{\sqrt{2 + t^3}}\, dt$$

for $1 \leq x \leq 3$.

c **47.** The function J_0 defined by

$$J_0(x) = \frac{1}{\pi} \int_{0}^{\pi} \cos(x \sin t)\, dt$$

is called the **Bessel function of order zero**.
(a) Use a CAS to graph the equation $y = J_0(x)$ over the interval $0 \leq x \leq 8$.
(b) Find $J_0(1)$.
(c) Find the smallest positive zero of $J_0(x)$.

c **48.** Let A be the area under the curve $y = x^2$ over the interval $[0, 1]$.
(a) Find A by using Part 1 of the Fundamental Theorem of Calculus.
(b) Find A by computing the limit of the left endpoint approximations by hand, and then find the limit using a CAS.
(c) Find A by computing the limit of the right endpoint approximations by hand, and then find the limit using a CAS.

c **49.** In number theory, $\pi(n)$ denotes the number of prime numbers that are less than or equal to the positive integer n. For example, it can be shown with the help of a computer that $\pi(100,000) = 9592$; that is, there are 9592 prime numbers that are less than or equal to 100,000. There are two useful approximations to $\pi(n)$ that are appropriate for large values of n:

$$\pi(n) \approx \frac{n}{\ln n} \quad \text{and} \quad \pi(n) \approx \int_{2}^{n} \frac{1}{\ln t}\, dt$$

Use a CAS to determine which of these approximations produces the better estimate of $\pi(100,000)$.

EXPANDING THE CALCULUS HORIZON

Blammo the Human Cannonball

*Blammo the Human Cannonball will be fired from a cannon and hopes to land in a small net at the opposite end of the circus arena. Your job as Blammo's manager is to do the mathematical calculations that will allow Blammo to perform his death-defying act safely. The methods that you will use are from the field of **ballistics** (the study of projectile motion).*

■■■■■ **The Problem**

Blammo's cannon has a **muzzle velocity** of 35 m/s, which means that Blammo will leave the muzzle with that velocity. The muzzle opening will be 5 m above the ground, and Blammo's objective is to land in a net that is also 5 m above the ground and that extends a distance of 10 m between 90 m and 100 m from the cannon opening (Figure 1). Your mathematical problem is to determine the **elevation angle** α of the cannon (the angle from the horizontal to the cannon barrel) that will make Blammo land in the net.

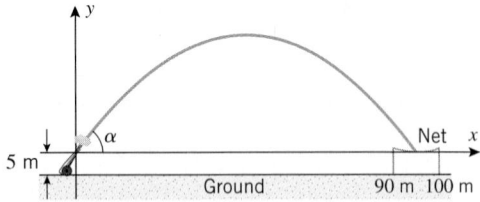

Figure 1

Modeling Assumptions

Blammo's trajectory will be determined by his initial velocity, the elevation angle of the cannon, and the forces that act on him after he leaves the muzzle. We will assume that the only force acting on Blammo after he leaves the muzzle is the downward force of the Earth's gravity. In particular, we will ignore the effect of air resistance. It will be convenient to introduce the xy-coordinate system shown in Figure 1 and to assume that Blammo is at the origin at time $t = 0$. We will also assume that Blammo's motion can be decomposed into two independent components, a horizontal component parallel to the x-axis and a vertical component parallel to the y-axis. We will analyze the horizontal and vertical components of Blammo's motion separately, and then we will combine the information to obtain a complete picture of his trajectory.

Blammo's Equations of Motion

We will denote the position and velocity functions for Blammo's horizontal component of motion by $x(t)$ and $v_x(t)$, and we will denote the position and velocity functions for his vertical component of motion by $y(t)$ and $v_y(t)$.

Since the only force acting on Blammo after he leaves the muzzle is the downward force of the Earth's gravity, there are no horizontal forces to alter his initial horizontal velocity $v_x(0)$. Thus, Blammo will have a constant velocity of $v_x(0)$ in the x-direction; this implies that

$$x(t) = v_x(0)t \tag{1}$$

In the y-direction Blammo is acted on only by the downward force of the Earth's gravity. Thus, his motion in this direction is governed by the free-fall model; hence, from (12) in Section 7.7 his vertical position function is

$$y(t) = y(0) + v_y(0)t - \tfrac{1}{2}gt^2$$

Taking $g = 9.8$ m/s^2, and using the fact that $y(0) = 0$, this equation can be written as

$$y(t) = v_y(0)t - 4.9t^2 \tag{2}$$

....

Exercise 1   At time $t = 0$ Blammo's velocity is 35 m/s, and this velocity is directed at an angle α with the horizontal. It is a fact of physics that the initial velocity components $v_x(0)$ and $v_y(0)$ can be obtained geometrically from the muzzle velocity and the angle of elevation using the triangle shown in Figure 2. We will justify this later in the text, but for now use this fact to show that Equations (1) and (2) can be expressed as

$$x(t) = (35\cos\alpha)t$$
$$y(t) = (35\sin\alpha)t - 4.9t^2$$

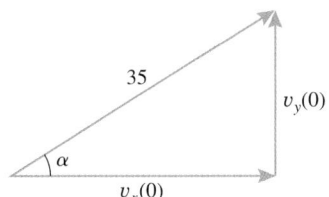

Figure 2

• • • • • • • • • • •

Exercise 2

(a) Use the result in Exercise 1 to find the velocity functions $v_x(t)$ and $v_y(t)$ in terms of the elevation angle α.

(b) Find the time t at which Blammo is at his maximum height above the x-axis, and show that this maximum height (in meters) is

$$y_{\max} = 62.5 \sin^2 \alpha$$

• • • • • • • • • • •

Exercise 3   The equations obtained in Exercise 1 can be viewed as parametric equations for Blammo's trajectory. Show, by eliminating the parameter t, that if $0 < \alpha < \pi/2$, then Blammo's trajectory is given by the equation

$$y = (\tan \alpha)x - \frac{0.004}{\cos^2 \alpha} x^2$$

Explain why Blammo's trajectory is a parabola.

Finding the Elevation Angle

Define Blammo's **horizontal range** R to be the horizontal distance he travels until he returns to the height of the muzzle opening ($y = 0$). Your objective is to find elevation angles that will make the horizontal range fall between 90 m and 100 m, thereby ensuring that Blammo lands in the net (Figure 3).

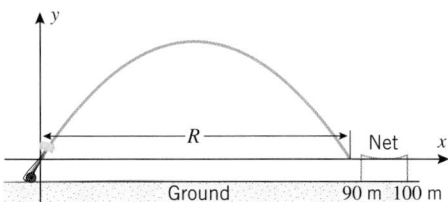

Figure 3

• • • • • • • • • • •

Exercise 4   Use a graphing utility and either the parametric equations obtained in Exercise 1 or the single equation obtained in Exercise 3 to generate Blammo's trajectories, taking elevation angles at increments of $10°$ from $15°$ to $85°$. In each case, determine visually whether Blammo lands in the net.

• • • • • • • • • • •

Exercise 5   Find the time required for Blammo to return to his starting height ($y = 0$), and use that result to show that Blammo's range R is given by the formula

$$R = 125 \sin 2\alpha$$

···········
Exercise 6

(a) Use the result in Exercise 5 to find two elevation angles that will allow Blammo to hit the midpoint of the net 95 m away.

(b) The tent is 55 m high. Explain why the larger elevation angle cannot be used.

···········
Exercise 7   How much can the smaller elevation angle in Exercise 6 vary and still have Blammo hit the net between 90 m and 100 m?

▬ Blammo's Shark Trick

Blammo is to be fired from 5 m above ground level with a muzzle velocity of 35 m/s over a flaming wall that is 20 m high and past a 5-m-high shark pool (Figure 4). To make the feat impressive, the pool will be made as long as possible. Your job as Blammo's manager is to determine the length of the pool, how far to place the cannon from the wall, and what elevation angle to use to ensure that Blammo clears the pool.

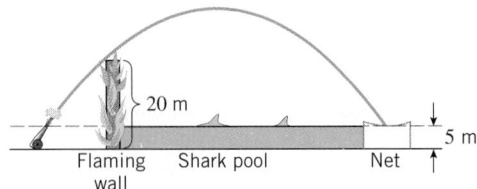

Flaming wall   Shark pool   Net   5 m

Figure 4

···········
Exercise 8   Prepare a written presentation of the problem and your solution of it that is at an appropriate level for an engineer, physicist, or mathematician to read. Your presentation should contain the following elements: an explanation of all notation, a list and description of all formulas that will be used, a diagram that shows the orientation of any coordinate systems that will be used, a description of any assumptions you make to solve the problem, graphs that you think will enhance the presentation, and a clear step-by-step explanation of your solution.

···
Module by: John Rickert, Rose-Hulman Institute of Technology
           Howard Anton, Drexel University

Additional material for this module can be found on the World Wide Web at http://www.wiley.com/college/anton

G.F.B. Riemann

8

APPLICATIONS OF THE DEFINITE INTEGRAL IN GEOMETRY, SCIENCE, AND ENGINEERING

*I*n the last chapter we introduced the definite integral as the limit of Riemann sums in the context of finding areas. However, Riemann sums and definite integrals have applications that extend far beyond the area problem. In this chapter we will show how Riemann sums and definite integrals arise in such problems as finding the volume and surface area of a solid, finding the length of a plane curve, calculating the work done by a force, finding the pressure and force exerted by a fluid on a submerged object, and finding properties of suspended cables.

Although these problems are diverse, the required calculations can all be approached by the same procedure that we used to find areas—breaking the required calculation into "small parts," making an approximation that is good because the part is small, adding the approximations from the parts to produce a Riemann sum that approximates the entire quantity to be calculated, and then taking the limit of the Riemann sums to produce an exact result.

8.1 AREA BETWEEN TWO CURVES

In the last chapter we showed how to find the area between a curve $y = f(x)$ and an interval on the x-axis. Here we will show how to find the area between two curves.

Before we consider the problem of finding the area between two curves it will be helpful to review the basic principle that underlies the calculation of area as a definite integral. Recall that if f is continuous and nonnegative on $[a, b]$, then the definite integral for the area A under $y = f(x)$ over the interval $[a, b]$ is obtained in four steps (Figure 8.1.1):

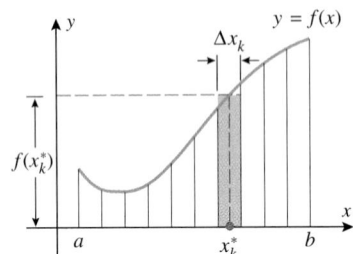

Figure 8.1.1

- Divide the interval $[a, b]$ into n subintervals, and use those subintervals to divide the area under the curve $y = f(x)$ into n strips.

- Assuming that the width of the kth strip is Δx_k, approximate the area of that strip by the area of a rectangle of width Δx_k and height $f(x_k^*)$, where x_k^* is any point in the kth subinterval.

- Add the approximate areas of the strips to approximate the entire area A by the Riemann sum:

$$A \approx \sum_{k=1}^{n} f(x_k^*)\Delta x_k$$

- Take the limit of the Riemann sums as the number of subintervals increases and their widths approach zero. This causes the error in the approximations to approach zero and produces the following definite integral for the exact area A:

$$A = \lim_{\max \Delta x_k \to 0} \sum_{k=1}^{n} f(x_k^*)\Delta x_k = \int_a^b f(x)\,dx$$

Observe the effect that the limit process has on the various parts of the Riemann sum:

- The quantity x_k^* in the Riemann sum becomes the variable x in the definite integral.
- The interval width Δx_k in the Riemann sum becomes the dx in the definite integral.
- The endpoints of the interval $[a, b]$ do not appear in the Riemann sum, but they become the limits of integration in the definite integral.

We will now consider the following extension of the area problem.

8.1.1 FIRST AREA PROBLEM. Suppose that f and g are continuous functions on an interval $[a, b]$ and

$$f(x) \geq g(x) \quad \text{for} \quad a \leq x \leq b$$

[This means that the curve $y = f(x)$ lies above the curve $y = g(x)$ and that the two can touch but not cross.] Find the area A of the region bounded above by $y = f(x)$, below by $y = g(x)$, and on the sides by the lines $x = a$ and $x = b$ (Figure 8.1.2a).

To solve this problem we divide the interval $[a, b]$ into n subintervals, which has the effect of subdividing the region into n strips (Figure 8.1.2b). If we assume that the width of the kth strip is Δx_k, then the area of the strip can be approximated by the area of a rectangle of width Δx_k and height $f(x_k^*) - g(x_k^*)$, where x_k^* is any point in the kth subinterval. Adding these approximations yields the following Riemann sum that approximates the area A:

$$A \approx \sum_{k=1}^{n} [f(x_k^*) - g(x_k^*)]\Delta x_k$$

Taking the limit as n increases and the widths of the subintervals approach zero yields the

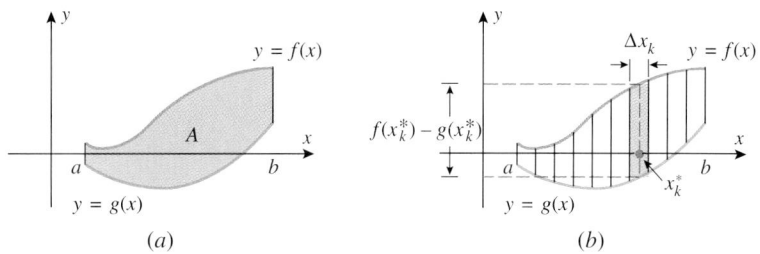

Figure 8.1.2   (a)   (b)

following definite integral for the area A between the curves:

$$A = \lim_{\max \Delta x_k \to 0} \sum_{k=1}^{n} [f(x_k^*) - g(x_k^*)] \Delta x_k = \int_a^b [f(x) - g(x)]\, dx$$

In summary, we have the following result:

8.1.2   AREA FORMULA.   If f and g are continuous functions on the interval $[a, b]$, and if $f(x) \geq g(x)$ for all x in $[a, b]$, then the area of the region bounded above by $y = f(x)$, below by $y = g(x)$, on the left by the line $x = a$, and on the right by the line $x = b$ is

$$A = \int_a^b [f(x) - g(x)]\, dx \tag{1}$$

In the case where f and g are *nonnegative* on the interval $[a, b]$, the formula

$$A = \int_a^b [f(x) - g(x)]\, dx = \int_a^b f(x)\, dx - \int_a^b g(x)\, dx$$

states that the area A between the curves can be obtained by subtracting the area under $y = g(x)$ from the area under $y = f(x)$ (Figure 8.1.3).

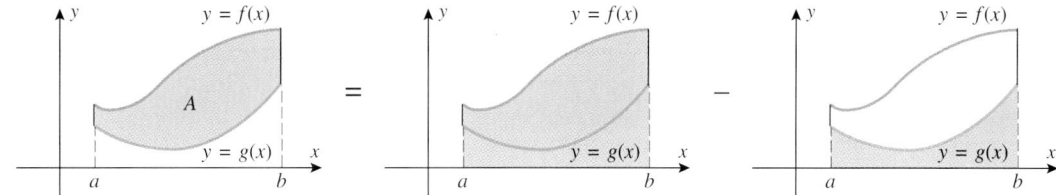

Figure 8.1.3

When the region is complicated, it may require some careful thought to determine the integrand and limits of integration in (1). Here is a systematic procedure that you can follow to set up this formula.

Step 1.   Sketch the region and then draw a vertical line segment through the region at an arbitrary point x, connecting the top and bottom boundaries (Figure 8.1.4a).

Step 2.   The top endpoint of the line segment sketched in Step 1 will be $f(x)$, the bottom one $g(x)$, and the length of the line segment will be $f(x) - g(x)$. This is the integrand in (1).

Step 3.   To determine the limits of integration, imagine moving the line segment left and then right. The leftmost position at which the line segment intersects the region is $x = a$ and the rightmost is $x = b$ (Figures 8.1.4b and 8.1.4c).

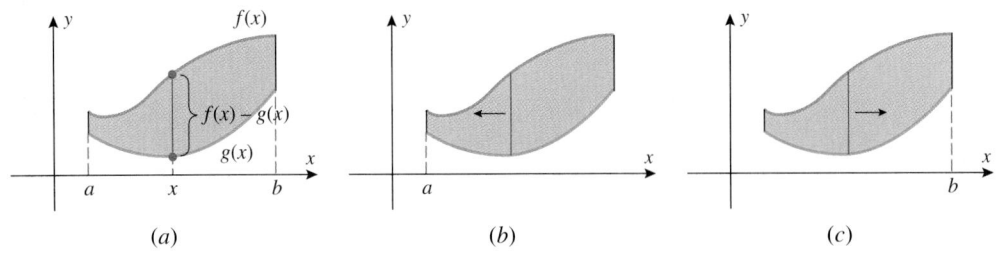

Figure 8.1.4

(a)   (b)   (c)

REMARK. It is not necessary to make an extremely accurate sketch in Step 1; the only purpose of the sketch is to determine which curve is the upper boundary and which is the lower boundary.

REMARK. There is a useful way of thinking about this procedure: If you view the vertical line segment as the "cross section" of the region at the point x, then Formula (1) states that the area between the curves is obtained by integrating the length of the cross section over the interval from a to b.

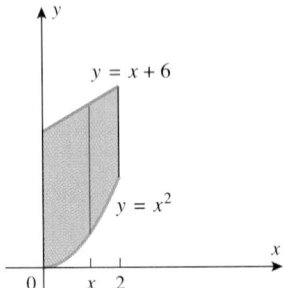

Figure 8.1.5

Example 1

Find the area of the region bounded above by $y = x + 6$, bounded below by $y = x^2$, and bounded on the sides by the lines $x = 0$ and $x = 2$.

Solution. The region and a cross section are shown in Figure 8.1.5. The cross section extends from $g(x) = x^2$ on the bottom to $f(x) = x + 6$ on the top. If the cross section is moved through the region, then its leftmost position will be $x = 0$ and its rightmost position will be $x = 2$. Thus, from (1)

$$A = \int_0^2 [(x + 6) - x^2]\,dx = \left[\frac{x^2}{2} + 6x - \frac{x^3}{3}\right]_0^2 = \frac{34}{3} - 0 = \frac{34}{3} \qquad \blacktriangleleft$$

It is possible that the upper and lower boundaries of a region may intersect at one or both endpoints, in which case the sides of the region will be points, rather than vertical line segments (Figure 8.1.6). When that occurs you will have to determine the points of intersection to obtain the limits of integration.

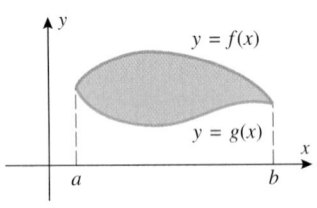

Both side boundaries reduce to points.

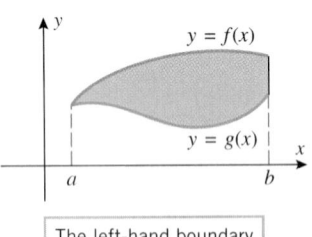

The left-hand boundary reduces to a point.

Figure 8.1.6

Example 2

Find the area of the region that is enclosed between the curves $y = x^2$ and $y = x + 6$.

Solution. A sketch of the region (Figure 8.1.7) shows that the lower boundary is $y = x^2$ and the upper boundary is $y = x + 6$. At the endpoints of the region, the upper and lower boundaries have the same y-coordinates; thus, to find the endpoints we equate

$$y = x^2 \quad \text{and} \quad y = x + 6 \tag{2}$$

This yields

$$x^2 = x + 6 \quad \text{or} \quad x^2 - x - 6 = 0 \quad \text{or} \quad (x + 2)(x - 3) = 0$$

from which we obtain

$$x = -2 \quad \text{and} \quad x = 3$$

Although the y-coordinates of the endpoints are not essential to our solution, they may be obtained from (2) by substituting $x = -2$ and $x = 3$ in either equation. This yields $y = 4$ and $y = 9$, so the upper and lower boundaries intersect at $(-2, 4)$ and $(3, 9)$.

From (1) with $f(x) = x + 6$, $g(x) = x^2$, $a = -2$, and $b = 3$, we obtain the area

$$A = \int_{-2}^3 [(x + 6) - x^2]\,dx = \left[\frac{x^2}{2} + 6x - \frac{x^3}{3}\right]_{-2}^3 = \frac{27}{2} - \left(-\frac{22}{3}\right) = \frac{125}{6} \qquad \blacktriangleleft$$

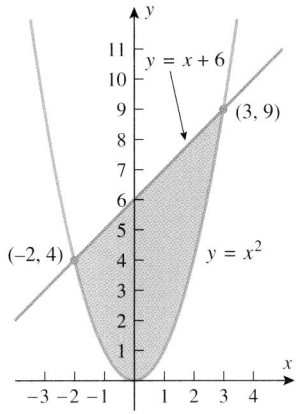

Figure 8.1.7

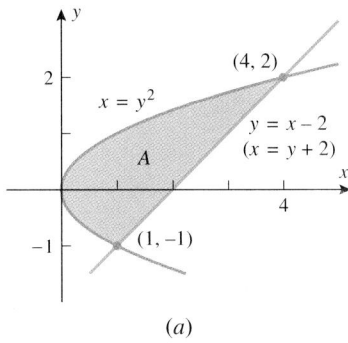

(a)

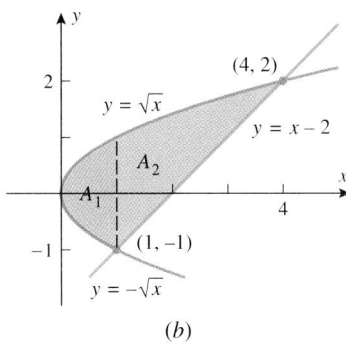

(b)

Figure 8.1.8

It is possible for the upper or lower boundary of a region to consist of two or more different curves, in which case it will be necessary to subdivide the region into smaller pieces in order to apply Formula (1). This is illustrated in the next example.

Example 3

Find the area of the region enclosed by $x = y^2$ and $y = x - 2$.

Solution. To make an accurate sketch of the region, we need to know where the curves $x = y^2$ and $y = x - 2$ intersect. In Example 2 we found intersections by equating the expressions for y. Here it is easier to rewrite the latter equation as $x = y + 2$ and equate the expressions for x, namely

$$x = y^2 \quad \text{and} \quad x = y + 2 \tag{3}$$

This yields

$$y^2 = y + 2 \quad \text{or} \quad y^2 - y - 2 = 0 \quad \text{or} \quad (y+1)(y-2) = 0$$

from which we obtain $y = -1$, $y = 2$. Substituting these values in either equation in (3) we see that the corresponding x-values are $x = 1$ and $x = 4$, respectively, so the points of intersection are $(1, -1)$ and $(4, 2)$ (Figure 8.1.8*a*).

To apply Formula (1), the equations of the boundaries must be written so that y is expressed explicitly as a function of x. The upper boundary can be written as $y = \sqrt{x}$ (rewrite $x = y^2$ as $y = \pm\sqrt{x}$ and choose the $+$ for the upper portion of the curve). The lower portion of the boundary consists of two parts: $y = -\sqrt{x}$ for $0 \le x \le 1$ and $y = x - 2$ for $1 \le x \le 4$ (Figure 8.1.8*b*). Because of this change in the formula for the lower boundary, it is necessary to divide the region into two parts and find the area of each part separately.

From (1) with $f(x) = \sqrt{x}$, $g(x) = -\sqrt{x}$, $a = 0$, and $b = 1$, we obtain

$$A_1 = \int_0^1 [\sqrt{x} - (-\sqrt{x})]\,dx = 2\int_0^1 \sqrt{x}\,dx = 2\left[\frac{2}{3}x^{3/2}\right]_0^1 = \frac{4}{3} - 0 = \frac{4}{3}$$

From (1) with $f(x) = \sqrt{x}$, $g(x) = x - 2$, $a = 1$, and $b = 4$, we obtain

$$A_2 = \int_1^4 [\sqrt{x} - (x - 2)]\,dx = \int_1^4 (\sqrt{x} - x + 2)\,dx$$

$$= \left[\frac{2}{3}x^{3/2} - \frac{1}{2}x^2 + 2x\right]_1^4 = \left(\frac{16}{3} - 8 + 8\right) - \left(\frac{2}{3} - \frac{1}{2} + 2\right) = \frac{19}{6}$$

Thus, the area of the entire region is

$$A = A_1 + A_2 = \frac{4}{3} + \frac{19}{6} = \frac{9}{2} \qquad \blacktriangleleft$$

FOR THE READER. It is assumed in Formula (1) that $f(x) \ge g(x)$ for all x in the interval $[a, b]$. What do you think that the integral represents if this condition is not satisfied, that is, the graphs of f and g cross one another over the interval? Explain your reasoning, and give an example to support your conclusion.

Example 4

Figure 8.1.9 shows velocity versus time curves for two race cars that move along a straight track, starting from rest at the same line. What does the area A between the curves over the interval $0 \le t \le T$ represent?

Solution. From (1)

$$A = \int_0^T [v_2(t) - v_1(t)]\,dt = \int_0^T v_2(t)\,dt - \int_0^T v_1(t)\,dt$$

But from 7.7.4, the first integral is the distance traveled by car 2 during the time interval, and the second integral is the distance traveled by car 1. Thus, A is the distance by which car 2 is ahead of car 1 at time T. $\blacktriangleleft$

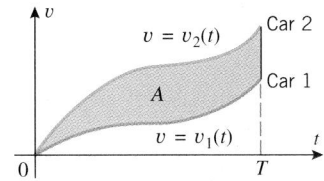

Figure 8.1.9

REVERSING THE ROLES OF x AND y

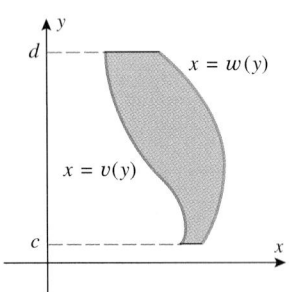

Figure 8.1.10

Sometimes it is possible to avoid splitting a region into parts by integrating with respect to y rather than x. We will now show how this can be done.

8.1.3 SECOND AREA PROBLEM. Suppose that w and v are continuous functions of y on an interval $[c, d]$ and that

$$w(y) \geq v(y) \quad \text{for} \quad c \leq y \leq d$$

[This means that the curve $x = w(y)$ lies to the right of the curve $x = v(y)$ and that the two can touch but not cross.] Find the area A of the region bounded on the left by $x = v(y)$, on the right by $x = w(y)$, and above and below by the lines $y = d$ and $y = c$ (Figure 8.1.10).

Proceeding as in the derivation of (1), but with the roles of x and y reversed, leads to the following analog of 8.1.2.

8.1.4 AREA FORMULA. If w and v are continuous functions and if $w(y) \geq v(y)$ for all y in $[c, d]$, then the area of the region bounded on the left by $x = v(y)$, on the right by $x = w(y)$, below by $y = c$, and above by $y = d$ is

$$A = \int_c^d [w(y) - v(y)]\, dy \tag{4}$$

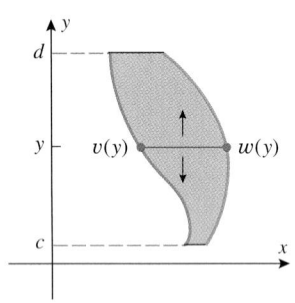

Figure 8.1.11

The guiding principle in applying this formula is the same as with (1): The integrand in (4) can be viewed as the length of the horizontal cross section at the point y, in which case Formula (4) states that the area can be obtained by integrating the length of the horizontal cross section over the interval $[c, d]$ on the y-axis (Figure 8.1.11).

In Example 3, where we integrated with respect to x to find the area of the region enclosed by $x = y^2$ and $y = x - 2$, we had to split the region into parts and evaluate two integrals. In the next example we will see that by integrating with respect to y no splitting of the region is necessary.

Example 5

Find the area of the region enclosed by $x = y^2$ and $y = x - 2$, integrating with respect to y.

Solution. From Figure 8.1.8 the left boundary is $x = y^2$, the right boundary is $y = x - 2$, and the region extends over the interval $-1 \leq y \leq 2$. However, to apply (4) the equations for the boundaries must be written so that x is expressed explicitly as a function of y. Thus, we rewrite $y = x - 2$ as $x = y + 2$. It now follows from (4) that

$$A = \int_{-1}^{2} [(y + 2) - y^2]\, dy = \left[\frac{y^2}{2} + 2y - \frac{y^3}{3} \right]_{-1}^{2} = \frac{9}{2}$$

which agrees with the result obtained in Example 3. ◀

REMARK. The choice between Formulas (1) and (4) is generally dictated by the shape of the region, and one would usually choose the formula that requires the least amount of splitting. However, if the integral(s) resulting by one method are difficult to evaluate, then the other method might be preferable, even if it requires more splitting.

EXERCISE SET 8.1 ☑ Graphing Calculator ⊡ CAS

In Exercises 1–4, find the area of the shaded region.

1.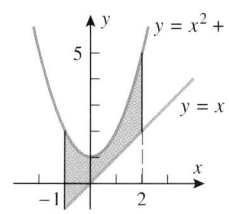

2.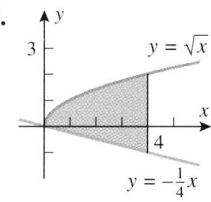

3.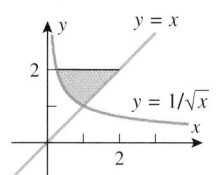

4.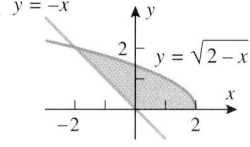

5. Find the area of the region enclosed by the curves $y = x^2$ and $y = 4x$ by integrating
   (a) with respect to x (b) with respect to y.

6. Find the area of the region enclosed by the curves $y^2 = 4x$ and $y = 2x - 4$ by integrating
   (a) with respect to x (b) with respect to y.

In Exercises 7–16, sketch the region enclosed by the curves, and find its area.

7. $y = x^2$, $y = \sqrt{x}$, $x = 1/4$, $x = 1$

8. $y = x^3 - 4x$, $y = 0$, $x = 0$, $x = 2$

9. $y = \cos 2x$, $y = 0$, $x = \pi/4$, $x = \pi/2$

10. $y = \sec^2 x$, $y = 2$, $x = -\pi/4$, $x = \pi/4$

11. $x = \sin y$, $x = 0$, $y = \pi/4$, $y = 3\pi/4$

12. $x^2 = y$, $x = y - 2$

13. $y = e^x$, $y = e^{2x}$, $x = 0$, $x = \ln 2$

14. $x = 1/y$, $x = 0$, $y = 1$, $y = e$

15. $y = 2 + |x - 1|$, $y = -\frac{1}{5}x + 7$

16. $y = x$, $y = 4x$, $y = -x + 2$

In Exercises 17–22, use a graphing utility, where helpful, to find the area of the region enclosed by the curves.

☑ **17.** $y = x^3 - 4x^2 + 3x$, $y = 0$, $x = 0$, $x = 3$

☑ **18.** $y = x^3 - 2x^2$, $y = 2x^2 - 3x$, $x = 0$, $x = 3$

☑ **19.** $y = \sin x$, $y = \cos x$, $x = 0$, $x = 2\pi$

☑ **20.** $y = x^3 - 4x$, $y = 0$, $x = -2$, $x = 2$

☑ **21.** $x = y^3 - y$, $x = 0$

☑ **22.** $x = y^3 - 4y^2 + 3y$, $x = y^2 - y$

⊡ **23.** Use a CAS to find the area enclosed by $y = 3 - 2x$ and $y = x^6 + 2x^5 - 3x^4 + x^2$.

⊡ **24.** Use a CAS to find the exact area enclosed by the curves $y = x^5 - 2x^3 - 3x$ and $y = x^3$.

25. Find a horizontal line $y = k$ that divides the area between $y = x^2$ and $y = 9$ into two equal parts.

26. Find a vertical line $x = k$ that divides the area enclosed by $x = \sqrt{y}$, $x = 2$, and $y = 0$ into two equal parts.

27. (a) Find the area of the region enclosed by the parabola $y = 2x - x^2$ and the x-axis.
   (b) Find the value of m so that the line $y = mx$ divides the region in part (a) into two regions of equal area.

28. Find the area between the curve $y = \sin x$ and the line segment joining the points $(0, 0)$ and $(5\pi/6, 1/2)$ on the curve.

29. Suppose that f and g are integrable on $[a, b]$, but neither $f(x) \geq g(x)$ nor $g(x) \geq f(x)$ holds for all x in $[a, b]$ [i.e., the curves $y = f(x)$ and $y = g(x)$ are intertwined].
   (a) What is the geometric significance of the integral
   $$\int_a^b [f(x) - g(x)]\, dx?$$
   (b) What is the geometric significance of the integral
   $$\int_a^b |f(x) - g(x)|\, dx?$$

30. Let $A(n)$ be the area in the first quadrant enclosed by the curves $y = \sqrt[n]{x}$ and $y = x$.
   (a) By considering how the graph of $y = \sqrt[n]{x}$ changes as n increases, make a conjecture about the limit of $A(n)$ as $n \to +\infty$.
   (b) Confirm your conjecture by calculating the limit.

In Exercises 31 and 32, use Newton's Method (Section 6.4), where needed, to approximate the x-coordinates of the intersections of the curves to at least four decimal places; and then use those approximations to approximate the area of the region.

31. The region that lies below the curve $y = \sin x$ and above the line $y = 0.2x$, where $x \geq 0$.

32. The region enclosed by the graphs of $y = x^2$ and $y = \cos x$.

33. The accompanying figure shows velocity versus time curves for two cars that move along a straight track, accelerating from rest at a common starting line.
   (a) How far apart are the cars after 60 seconds?
   (b) How far apart are the cars after T seconds, where $0 \leq T \leq 60$?

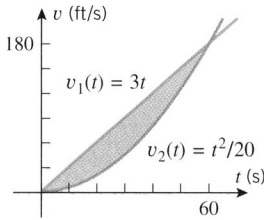

Figure Ex-33

34. The accompanying figure shows acceleration time curves for two cars that move along a straight track, accelerating from rest at the starting line. What does the area A between the curves over the interval $0 \le t \le T$ represent? Justify your answer.

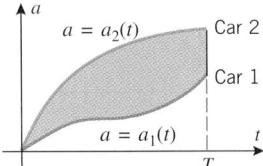

Figure Ex-34

35. Find the area of the region enclosed between the curve $x^{1/2} + y^{1/2} = a^{1/2}$ and the coordinate axes.

36. Show that the area of the ellipse in the accompanying figure is πab. [*Hint:* Use a formula from geometry.]

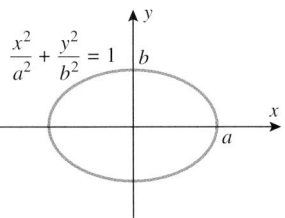

Figure Ex-36

37. A rectangle with edges parallel to the coordinate axes has one vertex at the origin and the diagonally opposite vertex on the curve $y = kx^m$ at the point where $x = b$ ($b > 0, k > 0$, and $m \ge 0$). Show that the fraction of the area of the rectangle that lies between the curve and the x-axis depends on m but not on k or b.

8.2 VOLUMES BY SLICING; DISKS AND WASHERS

In the last section we showed that the area of a plane region bounded by two curves can be obtained by integrating the length of a general cross section over an appropriate interval. In this section we will see that the same basic principle can be used to find volumes of certain three-dimensional solids.

VOLUMES BY SLICING

Recall that the underlying principle for finding the area of a plane region is to divide the region into thin strips, approximate the area of each strip by the area of a rectangle, add the approximations to form a Riemann sum, and take the limit of the Riemann sums to produce an integral for the area. Under appropriate conditions, the same strategy can be used to find the volume of a solid. The idea is to divide the solid into thin slabs, approximate the volume of each slab, add the approximations to form a Riemann sum, and take the limit of the Riemann sums to produce an integral for the volume (Figure 8.2.1).

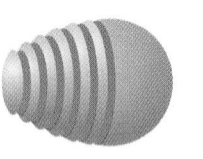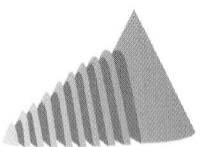

Figure 8.2.1

What makes this method work is the fact that a *thin* slab has cross sections that do not vary much in size or shape, which, as we will see, makes its volume easy to approximate (Figure 8.2.2). Moreover, the thinner the slab, the less variation in its cross sections and the better the approximation. Thus, once we approximate the volumes of the slabs, we can set up a Riemann sum whose limit is the volume of the entire solid. We will give the details shortly, but first we need to discuss how to find the volume of a solid whose cross sections do not vary in size and shape (i.e., are congruent).

One of the simplest examples of a solid with congruent cross sections is a right circular cylinder of radius r, since all cross sections taken perpendicular to the central axis are circular regions of radius r. The volume V of a right circular cylinder of radius r and height

In a thin slab, the cross sections do not vary much in size and shape.

Figure 8.2.2

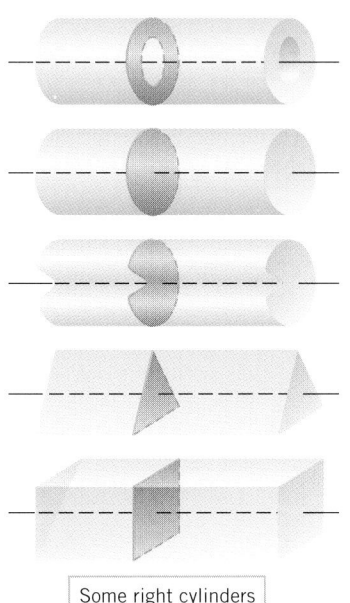

Some right cylinders

Figure 8.2.3

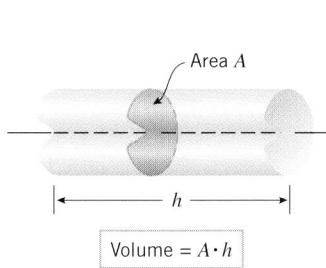

Area A

h

Volume $= A \cdot h$

Figure 8.2.4

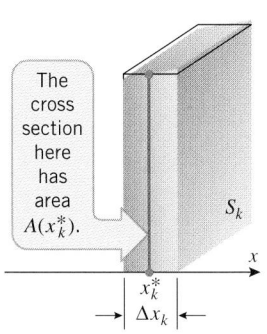

The cross section here has area $A(x_k^*)$.

S_k

x_k^*

Δx_k

Figure 8.2.6

h can be expressed in terms of the height and the area of a cross section as

$$V = \pi r^2 h = [\text{area of a cross section}] \times [\text{height}] \qquad (1)$$

This is a special case of a more general volume formula that applies to solids called *right cylinders*. A **right cylinder** is a solid that is generated when a plane region is translated along a line or **axis** that is perpendicular to the region (Figure 8.2.3). The distance h that the region is translated is called the **height** or sometimes the **width** of the cylinder, and each cross section is a duplicate of the translated region. We will assume that the volume V of a right cylinder with cross-sectional area A and height h is given by

$$V = A \cdot h = [\text{area of a cross section}] \times [\text{height}] \qquad (2)$$

(Figure 8.2.4). Note that this is consistent with Formula (1) for the volume of a right circular cylinder. We now have all of the tools required to solve the following problem.

8.2.1 PROBLEM. Let S be a solid that extends along the x-axis and is bounded on the left and right, respectively, by the planes that are perpendicular to the x-axis at $x = a$ and $x = b$ (Figure 8.2.5a). Find the volume V of the solid, assuming that its cross-sectional area $A(x)$ is known at each point x in the interval $[a, b]$.

To solve this problem we divide the interval $[a, b]$ into n subintervals, which has the effect of dividing the solid into n slabs (Figure 8.2.5b).

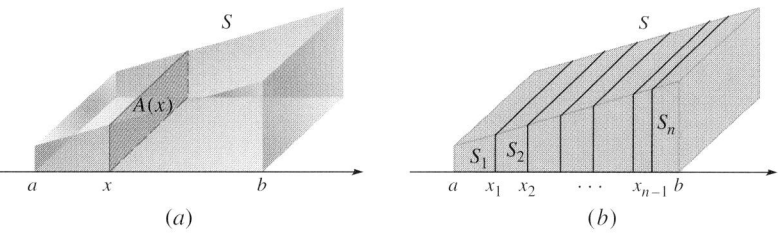

(a) $\qquad\qquad$ (b)

Figure 8.2.5

If we assume that the width of the kth slab is Δx_k, then the volume of the slab can be approximated by the volume of a right cylinder of width (height) Δx_k and cross-sectional area $A(x_k^*)$, where x_k^* is any point in the kth subinterval (Figure 8.2.6). Adding these approximations yields the following Riemann sum that approximates the volume V:

$$V \approx \sum_{k=1}^{n} A(x_k^*) \Delta x_k$$

Taking the limit as n increases and the widths of the subintervals approach zero yields the definite integral

$$V = \lim_{\max \Delta x_k \to 0} \sum_{k=1}^{n} A(x_k^*) \Delta x_k = \int_a^b A(x)\, dx$$

In summary, we have the following result:

8.2.2 VOLUME FORMULA. Let S be a solid bounded by two parallel planes perpendicular to the x-axis at $x = a$ and $x = b$. If, for each x in $[a, b]$, the cross-sectional area of S perpendicular to the x-axis is $A(x)$, then the volume of the solid is

$$V = \int_a^b A(x)\, dx \qquad (3)$$

provided $A(x)$ is integrable.

There is a similar result for cross sections perpendicular to the y-axis.

8.2.3 VOLUME FORMULA. Let S be a solid bounded by two parallel planes perpendicular to the y-axis at $y = c$ and $y = d$. If, for each y in $[c, d]$, the cross-sectional area of S perpendicular to the y-axis is $A(y)$, then the volume of the solid is

$$V = \int_c^d A(y)\, dy \tag{4}$$

provided $A(y)$ is integrable.

REMARK. In words, these formulas state that the volume of the solid can be obtained by integrating the cross-sectional area from one end of the solid to the other.

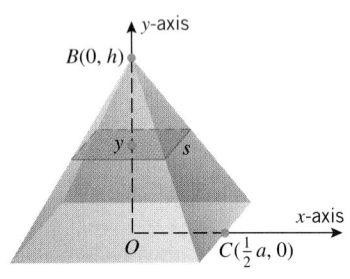

(a)

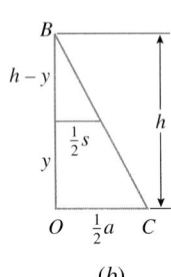

(b)

Figure 8.2.7

Example 1

Derive the formula for the volume of a right pyramid whose altitude is h and whose base is a square with sides of length a.

Solution. As illustrated in Figure 8.2.7a, we introduce a rectangular coordinate system in which the y-axis passes through the apex and is perpendicular to the base, and the x-axis passes through the base and is parallel to a side of the base.

At any point y in the interval $[0, h]$ on the y-axis, the cross section perpendicular to the y-axis is a square. If s denotes the length of a side of this square, then by similar triangles (Figure 8.2.7b)

$$\frac{\frac{1}{2}s}{\frac{1}{2}a} = \frac{h - y}{h} \quad \text{or} \quad s = \frac{a}{h}(h - y)$$

Thus, the area $A(y)$ of the cross section at y is

$$A(y) = s^2 = \frac{a^2}{h^2}(h - y)^2$$

and by (4) the volume is

$$V = \int_0^h A(y)\, dy = \int_0^h \frac{a^2}{h^2}(h - y)^2\, dy = \frac{a^2}{h^2} \int_0^h (h - y)^2\, dy$$

$$= \frac{a^2}{h^2}\left[-\frac{1}{3}(h - y)^3 \right]_{y=0}^{h} = \frac{a^2}{h^2}\left[0 + \frac{1}{3}h^3 \right] = \frac{1}{3}a^2 h$$

That is, the volume is $\frac{1}{3}$ of the area of the base times the altitude. ◄

SOLIDS OF REVOLUTION

A *solid of revolution* is a solid that is generated by revolving a plane region about a line that lies in the same plane as the region; the line is called the *axis of revolution*. Many familiar solids are of this type (Figure 8.2.8).

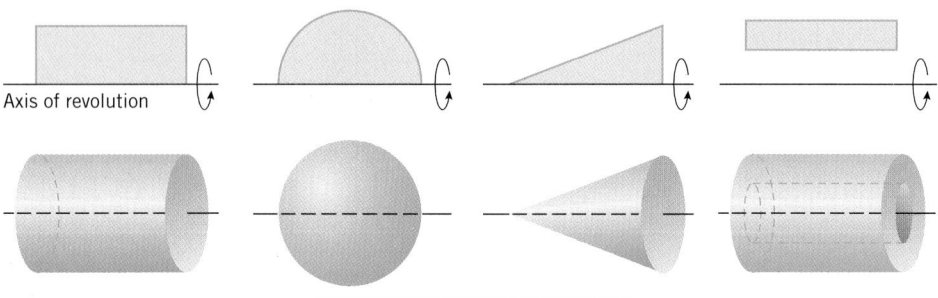

Some familiar solids of revolution

Figure 8.2.8

We will be interested in the following general problem:

> **8.2.4** PROBLEM. Let f be continuous and nonnegative on $[a, b]$, and let R be the region that is bounded above by $y = f(x)$, below by the x-axis, and on the sides by the lines $x = a$ and $x = b$ (Figure 8.2.9a). Find the volume of the solid of revolution that is generated by revolving the region R about the x-axis.

We can solve this problem by slicing. For this purpose, observe that the cross section of the solid taken perpendicular to the x-axis at the point x is a circular disk of radius $f(x)$ (Figure 8.2.9b). The area of this region is

$$A(x) = \pi[f(x)]^2$$

Thus, from (3) the volume of the solid is

$$V = \int_a^b \pi[f(x)]^2\, dx \tag{5}$$

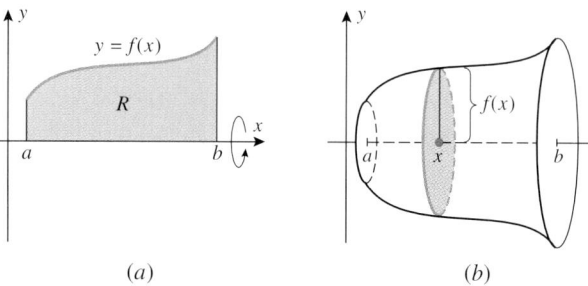

Figure 8.2.9   (a)   (b)

VOLUMES BY DISKS PERPENDICULAR TO THE x-AXIS

Because the cross sections are disk shaped, the application of this formula is called the *method of disks*.

Example 2

Find the volume of the solid that is obtained when the region under the curve $y = \sqrt{x}$ over the interval $[1, 4]$ is revolved about the x-axis (Figure 8.2.10).

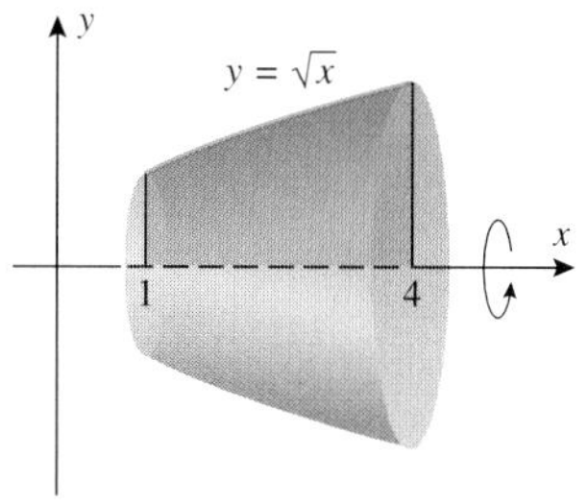

Solution. From (5), the volume is

$$V = \int_a^b \pi[f(x)]^2\, dx = \int_1^4 \pi x\, dx = \left. \frac{\pi x^2}{2} \right]_1^4 = 8\pi - \frac{\pi}{2} = \frac{15\pi}{2} \qquad \blacktriangleleft$$

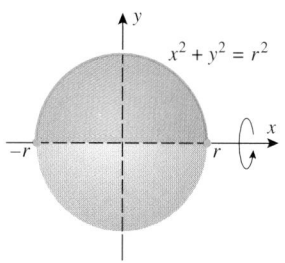

Figure 8.2.10

Example 3

Derive the formula for the volume of a sphere of radius r.

Solution. As indicated in Figure 8.2.11, a sphere of radius r can be generated by revolving the upper semicircular disk enclosed between the x-axis and

$$x^2 + y^2 = r^2$$

about the x-axis. Since the upper half of this circle is the graph of $y = f(x) = \sqrt{r^2 - x^2}$, it follows from (5) that the volume of the sphere is

$$V = \int_a^b \pi[f(x)]^2\, dx = \int_{-r}^r \pi(r^2 - x^2)\, dx = \pi \left[r^2 x - \frac{x^3}{3} \right]_{-r}^r = \frac{4}{3}\pi r^3 \qquad \blacktriangleleft$$

Figure 8.2.11

VOLUMES BY WASHERS PERPENDICULAR TO THE x-AXIS

Not all solids of revolution have solid interiors; some have holes or channels that create interior surfaces, as in the last part of Figure 8.2.8. Thus, we will be interested in problems of the following type.

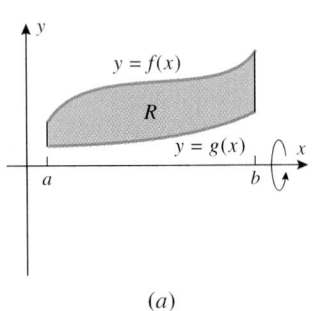

(a)

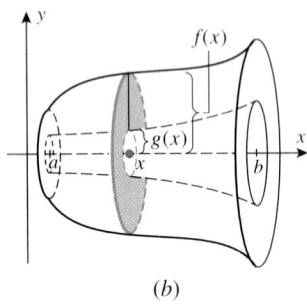

(b)

Figure 8.2.12

8.2.5 PROBLEM. Let f and g be continuous and nonnegative on $[a, b]$, and suppose that $f(x) \geq g(x)$ for all x in the interval $[a, b]$. Let R be the region that is bounded above by $y = f(x)$, below by $y = g(x)$, and on the sides by the lines $x = a$ and $x = b$ (Figure 8.2.12a). Find the volume of the solid of revolution that is generated by revolving the region R about the x-axis.

We can solve this problem by slicing. For this purpose, observe that the cross section of the solid taken perpendicular to the x-axis at the point x is the annular or "washer-shaped" region with inner radius $g(x)$ and outer radius $f(x)$ (Figure 8.2.12b); hence its area is

$$A(x) = \pi[f(x)]^2 - \pi[g(x)]^2 = \pi([f(x)]^2 - [g(x)]^2)$$

Thus, from (3) the volume of the solid is

$$V = \int_a^b \pi([f(x)]^2 - [g(x)]^2)\,dx \qquad (6)$$

Because the cross sections are washer shaped, the application of this formula is called the *method of washers*.

Example 4

Find the volume of the solid generated when the region between the graphs of the equations $f(x) = \frac{1}{2} + x^2$ and $g(x) = x$ over the interval $[0, 2]$ is revolved about the x-axis (Figure 8.2.13).

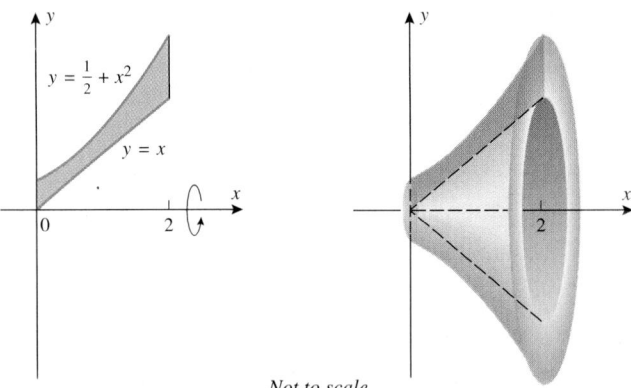

Figure 8.2.13   *Not to scale*

Solution. From (6) the volume is

$$V = \int_a^b \pi([f(x)]^2 - [g(x)]^2)\,dx = \int_0^2 \pi\left(\left[\tfrac{1}{2} + x^2\right]^2 - x^2\right)dx$$

$$= \int_0^2 \pi\left(\frac{1}{4} + x^4\right)dx = \pi\left[\frac{x}{4} + \frac{x^5}{5}\right]_0^2 = \frac{69\pi}{10} \qquad \blacktriangleleft$$

··

VOLUMES BY DISKS AND WASHERS PERPENDICULAR TO THE y-AXIS

The methods of disks and washers have analogs for regions that are revolved about the y-axis (Figures 8.2.14 and 8.2.15). Using the method of slicing and Formula (4), you should have no trouble deducing the following formulas for the volumes of the solids in the figures.

$$V = \int_c^d \pi[u(y)]^2\,dy \qquad\qquad V = \int_c^d \pi([w(y)]^2 - [v(y)]^2)\,dy \qquad (7\text{–}8)$$

             Disks   Washers

Example 5

Find the volume of the solid generated when the region enclosed by $y = \sqrt{x}$, $y = 2$, and $x = 0$ is revolved about the y-axis (Figure 8.2.16).

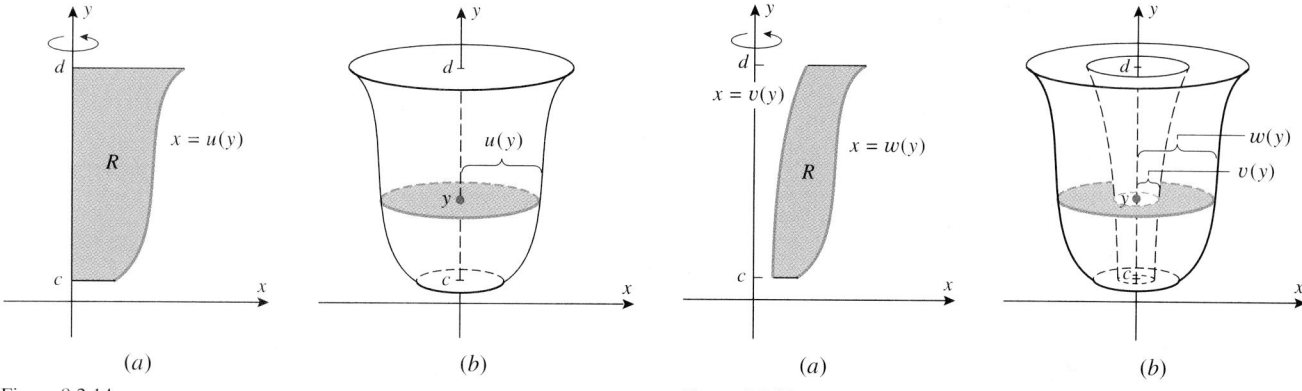

Figure 8.2.14   Figure 8.2.15

Solution. The cross sections taken perpendicular to the y-axis are disks, so we will apply (7). But first we must rewrite $y = \sqrt{x}$ as $x = y^2$. Thus, from (7) with $u(y) = y^2$, the volume is

$$V = \int_c^d \pi [u(y)]^2 \, dy = \int_0^2 \pi y^4 \, dy = \frac{\pi y^5}{5} \Big]_0^2 = \frac{32\pi}{5} \qquad \blacktriangleleft$$

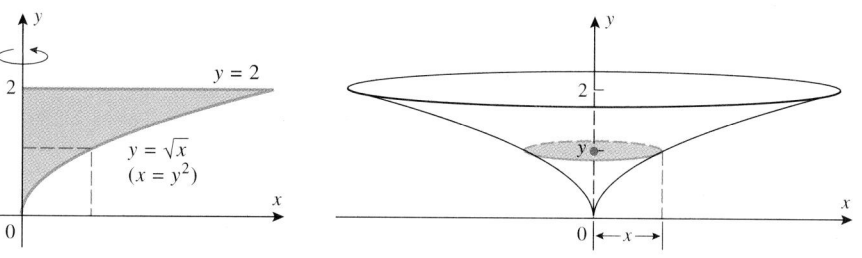

Figure 8.2.16

EXERCISE SET 8.2 □c CAS

In Exercises 1–4, find the volume of the solid that results when the shaded region is revolved about the indicated axis.

1.
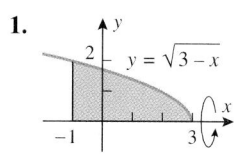
$y = \sqrt{3-x}$

2.
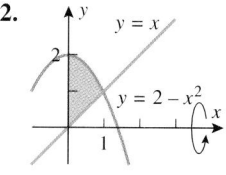
$y = x$, $y = 2 - x^2$

3.
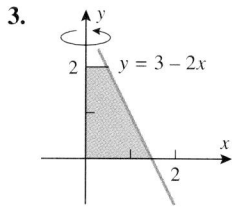
$y = 3 - 2x$

4.
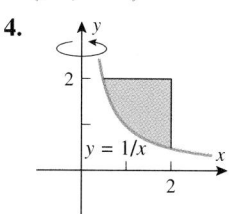
$y = 1/x$

In Exercises 5–14, find the volume of the solid that results when the region enclosed by the given curves is revolved about the x-axis.

5. $y = x^2$, $x = 0$, $x = 2$, $y = 0$

6. $y = \sec x$, $x = \pi/4$, $x = \pi/3$, $y = 0$

7. $y = \sqrt{\cos x}$, $x = \pi/4$, $x = \pi/2$, $y = 0$

8. $y = x^2$, $y = x^3$

9. $y = \sqrt{25 - x^2}$, $y = 3$

10. $y = 9 - x^2$, $y = 0$

11. $y = e^x$, $y = 0$, $x = 0$, $x = \ln 3$

12. $y = e^{-2x}$, $y = 0$, $x = 0$, $x = 1$

13. $x = \sqrt{y}$, $x = y/4$

14. $y = \sin x$, $y = \cos x$, $x = 0$, $x = \pi/4$. [*Hint:* Use the identity $\cos 2x = \cos^2 x - \sin^2 x$.]

In Exercises 15–22, find the volume of the solid that results when the region enclosed by the given curves is revolved about the y-axis.

15. $y = x^3$, $x = 0$, $y = 1$

16. $x = 1 - y^2$, $x = 0$

17. $x = \sqrt{1 + y}$, $x = 0$, $y = 3$

18. $y = x^2 - 1$, $x = 2$, $y = 0$

19. $x = \csc y$, $y = \pi/4$, $y = 3\pi/4$, $x = 0$

20. $y = x^2$, $x = y^2$

21. $x = y^2$, $x = y + 2$

22. $x = 1 - y^2$, $x = 2 + y^2$, $y = -1$, $y = 1$

23. Find the volume of the solid that results when the region above the x-axis and below the ellipse

$$\frac{x^2}{a^2} + \frac{y^2}{b^2} = 1 \quad (a > 0, b > 0)$$

is revolved about the x-axis.

24. Let V be the volume of the solid that results when the region enclosed by $y = 1/x$, $y = 0$, $x = 2$, and $x = b$ $(0 < b < 2)$ is revolved about the x-axis. Find the value of b for which $V = 3$.

25. Find the volume of the solid generated when the region enclosed by $y = \sqrt{x + 1}$, $y = \sqrt{2x}$, and $y = 0$ is revolved about the x-axis. [*Hint:* Split the solid into two parts.]

26. Find the volume of the solid generated when the region enclosed by $y = \sqrt{x}$, $y = 6 - x$, and $y = 0$ is revolved about the x-axis. [*Hint:* Split the solid into two parts.]

27. Find the volume of the solid that results when the region enclosed by $y = \sqrt{x}$, $y = 0$, and $x = 9$ is revolved about the line $x = 9$.

28. Find the volume of the solid that results when the region in Exercise 27 is revolved about the line $y = 3$.

29. Find the volume of the solid that results when the region enclosed by $x = y^2$ and $x = y$ is revolved about the line $y = -1$.

30. Find the volume of the solid that results when the region in Exercise 29 is revolved about the line $x = -1$.

31. A nose cone for a space reentry vehicle is designed so that a cross section, taken x ft from the tip and perpendicular to the axis of symmetry, is a circle of radius $\frac{1}{4}x^2$ ft. Find the volume of the nose cone given that its length is 20 ft.

32. A certain solid is 1 ft high, and a horizontal cross section taken x ft above the bottom of the solid is an annulus of inner radius x^2 and outer radius $\sqrt{x}$. Find the volume of the solid.

33. Find the volume of the solid whose base is the region bounded between the curves $y = x$ and $y = x^2$, and whose cross sections perpendicular to the x-axis are squares.

34. The base of a certain solid is the region enclosed by $y = \sqrt{x}$, $y = 0$, and $x = 4$. Every cross section perpendicular to the x-axis is a semicircle with its diameter across the base. Find the volume of the solid.

35. Find the volume of the solid whose base is enclosed by the circle $x^2 + y^2 = 1$ and whose cross sections taken perpendicular to the base are

(a) semicircles   (b) squares

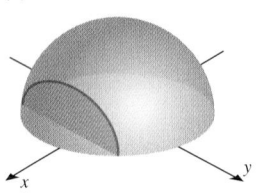
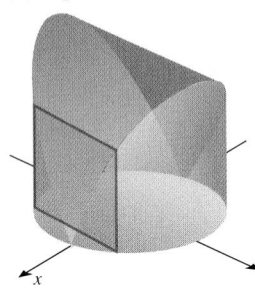

(c) equilateral triangles.

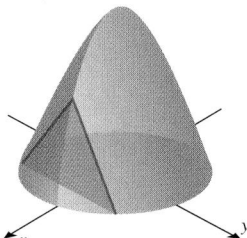

36. Derive the formula for the volume of a right circular cone with radius r and height h.

In Exercises 37 and 38, use a CAS to find the volume of the solid that results when the region enclosed by the curves is revolved about the stated axis.

c 37. $y = \sin^8 x$, $y = 2x/\pi$, $x = 0$, $x = \pi/2$; x-axis

c 38. $y = e^x$, $x = 1$, $y = 1$; y-axis

39. The accompanying figure shows a **spherical cap** of radius ρ and height h cut from a sphere of radius r. Show that the volume V of the spherical cap can be expressed as
(a) $V = \frac{1}{3}\pi h^2(3r - h)$ (b) $V = \frac{1}{6}\pi h(3\rho^2 + h^2)$

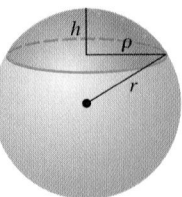

Figure Ex-39

40. If fluid enters a hemispherical vat with a radius of 10 ft at a rate of $\frac{1}{2}$ ft^3/min, how fast will the fluid be rising when the depth is 5 ft? [*Hint:* See Exercise 39.]

41. The accompanying figure shows the dimensions of a small lightbulb at 10 equally spaced points.
(a) Use formulas from geometry to make a rough estimate of the volume enclosed by the glass portion of the bulb.

(b) Use the average of left and right endpoint approximations to approximate the volume.

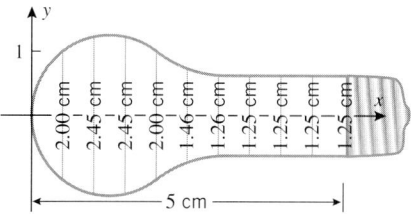

Figure Ex-41

Figure Ex-43

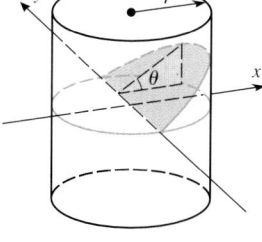

Figure Ex-45

42. Use the result in Exercise 39 to find the volume of the solid that remains when a hole of radius $r/2$ is drilled through the center of a sphere of radius r, and then check your answer by integrating.

43. As shown in the accompanying figure, a cocktail glass with a bowl shaped like a hemisphere of diameter 8 cm contains a cherry with a diameter of 2 cm. If the glass is filled to a depth of h cm, what is the volume of liquid it contains? [*Hint:* First consider the case where the cherry is partially submerged, then the case where it is totally submerged.]

44. Find the volume of the torus that results when the region enclosed by the circle of radius r with center at $(h, 0)$, $h > r$, is revolved about the y-axis. [*Hint:* Use an appropriate formula from plane geometry to help evaluate the definite integral.]

45. A wedge is cut from a right circular cylinder of radius r by two planes, one perpendicular to the axis of the cylinder and the other making an angle θ with the first. Find the volume of the wedge by slicing perpendicular to the y-axis as shown in the accompanying figure.

46. Find the volume of the wedge described in Exercise 45 by slicing perpendicular to the x-axis.

47. Two right circular cylinders of radius r have axes that intersect at right angles. Find the volume of the solid common to the two cylinders. [*Hint:* One-eighth of the solid is sketched in the accompanying figure.]

48. In 1635 Bonaventura Cavalieri, a student of Galileo, stated the following result, called ***Cavalieri's principle***: *If two solids have the same height, and if the areas of their cross sections taken parallel to and at equal distances from their bases are always equal, then the solids have the same volume.* Use this result to find the volume of the oblique cylinder in the accompanying figure.

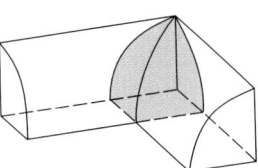

Figure Ex-47

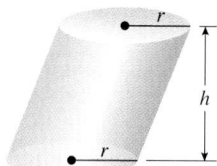

Figure Ex-48

8.3 VOLUMES BY CYLINDRICAL SHELLS

The methods for computing volumes that have been discussed so far depend on our ability to compute the cross-sectional area of the solid and to integrate that area across the solid. In this section we will develop another method for finding volumes that may be applicable when the cross-sectional area cannot be found or the integration is too difficult.

CYLINDRICAL SHELLS

In this section we will be interested in the following problem:

8.3.1 PROBLEM. Let f be continuous and nonnegative on $[a, b]$, and let R be the region that is bounded above by $y = f(x)$, below by the x-axis, and on the sides by the lines $x = a$ and $x = b$. Find the volume V of the solid of revolution S that is generated by revolving the region R about the y-axis (Figure 8.3.1).

Sometimes problems of this type can be solved by the method of disks or washers perpendicular to the y-axis, but when that method is not applicable or the resulting integral is difficult, the *method of cylindrical shells,* which we will discuss here, will often work.

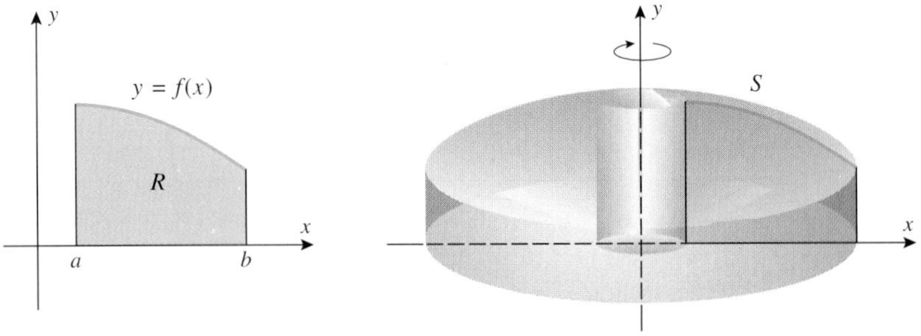

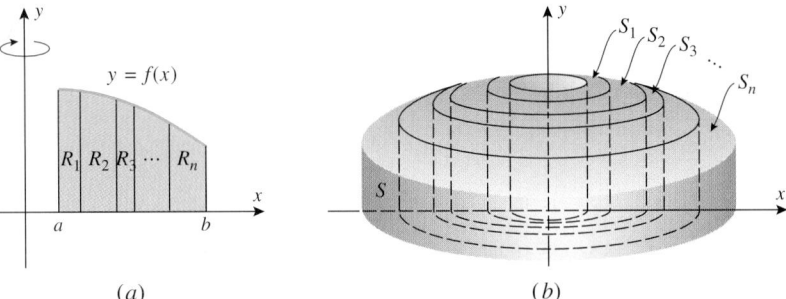

Figure 8.3.1

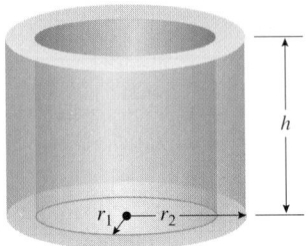

Figure 8.3.2

A **cylindrical shell** is a solid enclosed by two concentric right circular cylinders (Figure 8.3.2). The volume V of a cylindrical shell with inner radius r_1, outer radius r_2, and height h can be written as

$$V = [\text{area of cross section}] \cdot [\text{height}] = (\pi r_2^2 - \pi r_1^2)h$$
$$= \pi(r_2 + r_1)(r_2 - r_1)h = 2\pi \cdot \left[\tfrac{1}{2}(r_1 + r_2)\right] \cdot h \cdot (r_2 - r_1)$$

But $\tfrac{1}{2}(r_1 + r_2)$ is the average radius of the shell and $r_2 - r_1$ is its thickness, so

$$V = 2\pi \cdot [\text{average radius}] \cdot [\text{height}] \cdot [\text{thickness}] \tag{1}$$

We will now show how this formula can be used to solve the problem posed above. The underlying idea is to divide the interval $[a, b]$ into n subintervals, thereby subdividing the region R into n strips, $R_1, R_2, \ldots, R_n$ (Figure 8.3.3a). When the region R is revolved about the y-axis, these strips generate "tube-like" solids $S_1, S_2, \ldots, S_n$ that are nested one inside the other and together comprise the entire solid S (Figure 8.3.3b). Thus, the volume V of the solid can be obtained by adding together the volumes of the tubes; that is,

$$V = V(S_1) + V(S_2) + \cdots + V(S_n)$$

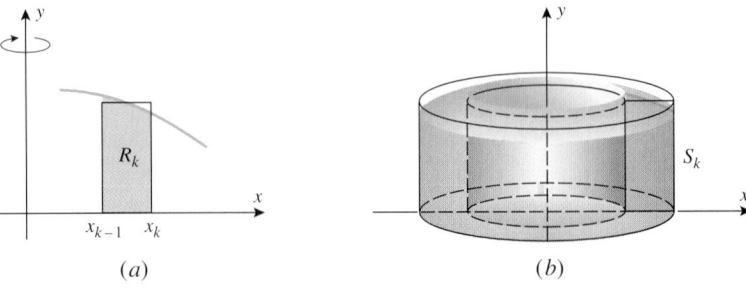

Figure 8.3.3 (a) (b)

As a rule, the tubes will have curved upper surfaces, so there will be no simple formulas for their volumes. However, if the strips are thin, then we can approximate each strip by a rectangle (Figure 8.3.4a). These rectangles, when revolved about the y-axis, will produce cylindrical shells whose volumes closely approximate the volumes generated by the original strips (Figure 8.3.4b). We will show that by adding the volumes of the cylindrical shells we

Figure 8.3.4 (a) (b)

can obtain a Riemann sum that approximates the volume V, and by taking the limit of the Riemann sums we can obtain an integral for the exact volume V.

To implement this idea, suppose that the kth strip extends from the point x_{k-1} to the point x_k and that the width of this strip is

$$\Delta x_k = x_k - x_{k-1}$$

If we let x_k^* be the *midpoint* of the interval $[x_{k-1}, x_k]$, and if we construct a rectangle of height $f(x_k^*)$ over the interval, then revolving this rectangle about the y-axis produces a cylindrical shell of height $f(x_k^*)$, average radius x_k^*, and thickness Δx_k (Figure 8.3.5). From (1), the volume V_k of this cylindrical shell is

$$V_k = 2\pi x_k^* f(x_k^*) \Delta x_k$$

Adding the volumes of the n cylindrical shells yields the following Riemann sum that approximates the volume V:

$$V \approx \sum_{k=1}^{n} 2\pi x_k^* f(x_k^*) \Delta x_k$$

Taking the limit as n increases and the widths of the subintervals approach zero yields the definite integral

$$V = \lim_{\max \Delta x_k \to 0} \sum_{k=1}^{n} 2\pi x_k^* f(x_k^*) \Delta x_k = \int_a^b 2\pi x f(x)\, dx$$

In summary, we have the following result.

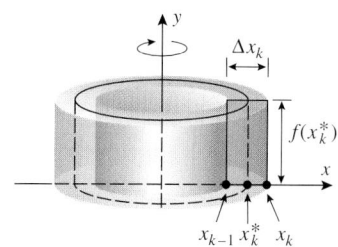

Figure 8.3.5

8.3.2 VOLUME BY CYLINDRICAL SHELLS ABOUT THE y-AXIS. Let f be continuous and nonnegative on $[a, b]$, and let R be the region that is bounded above by $y = f(x)$, below by the x-axis, and on the sides by the lines $x = a$ and $x = b$. Then the volume V of the solid of revolution that is generated by revolving the region R about the y-axis is given by

$$V = \int_a^b 2\pi x f(x)\, dx \qquad (2)$$

Example 1

Use cylindrical shells to find the volume of the solid generated when the region enclosed between $y = \sqrt{x}$, $x = 1$, $x = 4$, and the x-axis is revolved about the y-axis (Figure 8.3.6).

Solution. Since $f(x) = \sqrt{x}$, $a = 1$, and $b = 4$, Formula (2) yields

$$V = \int_1^4 2\pi x \sqrt{x}\, dx = 2\pi \int_1^4 x^{3/2}\, dx = \left[2\pi \cdot \frac{2}{5} x^{5/2} \right]_1^4 = \frac{4\pi}{5}[32 - 1] = \frac{124\pi}{5} \quad \blacktriangleleft$$

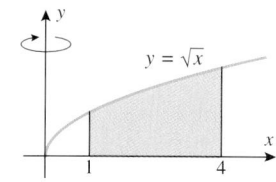

Cutaway view of the solid

Figure 8.3.6

VARIATIONS OF THE METHOD OF CYLINDRICAL SHELLS

The method of cylindrical shells is applicable in a variety of situations that do not fit the conditions required by Formula (2). For example, the region may be enclosed between two curves, or the axis of revolution may be some line other than the y-axis. However, rather than develop a separate formula for every possible situation, we will give a general way of thinking about the method of cylindrical shells that can be adapted to each new situation as it arises.

For this purpose, we will need to reexamine the integrand in Formula (2): At each point x in the interval $[a, b]$, the vertical line segment from the x-axis to the curve $y = f(x)$ can be viewed as the cross section of the region R at x (Figure 8.3.7a). When the region R is revolved about the y-axis, the cross section at x sweeps out the *surface* of a right circular cylinder of height $f(x)$ and radius x (Figure 8.3.7b). The area of this surface is

$$2\pi x f(x)$$

(Figure 8.3.7c), which is the integrand in (2). Thus, Formula (2) can be viewed informally in the following way.

> **8.3.3**   AN INFORMAL VIEWPOINT ABOUT CYLINDRICAL SHELLS.   The volume V of a solid of revolution that is generated by revolving a region R about an axis can be obtained by integrating the area of the surface generated by an arbitrary cross section of R taken parallel to the axis of revolution.

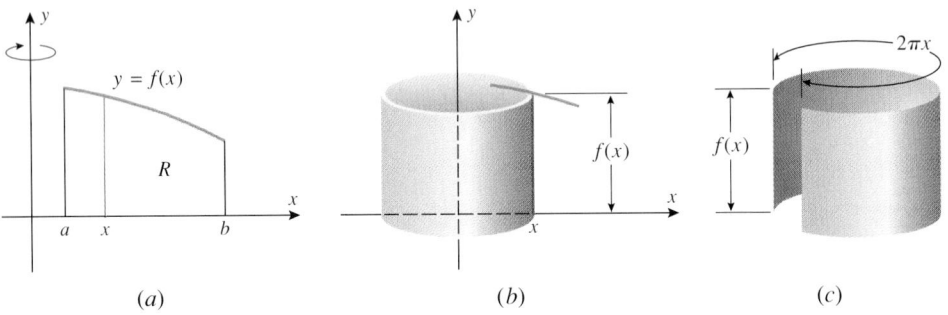

(a)   (b)   (c)

Figure 8.3.7

The following examples illustrate how to apply this result in situations where Formula (2) is not applicable:

Example 2

Use cylindrical shells to find the volume of the solid generated when the region R in the first quadrant enclosed between $y = x$ and $y = x^2$ is revolved about the y-axis (Figure 8.3.8).

Solution.   At each x in $[0, 1]$ the cross section of R parallel to the y-axis generates a cylindrical surface of height $x - x^2$ and radius x. Since the area of this surface is

$$2\pi x(x - x^2)$$

the volume of the solid is

$$V = \int_0^1 2\pi x(x - x^2)\,dx = 2\pi \int_0^1 (x^2 - x^3)\,dx = 2\pi \left[\frac{x^3}{3} - \frac{x^4}{4} \right]_0^1 = 2\pi \left[\frac{1}{3} - \frac{1}{4} \right] = \frac{\pi}{6}$$

◀

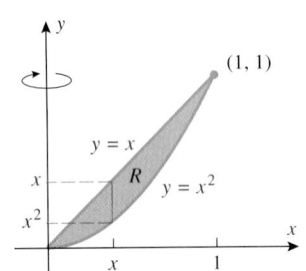

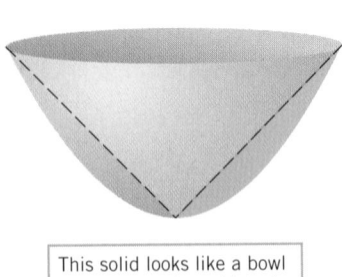

This solid looks like a bowl with a cone-shaped interior.

Figure 8.3.8

FOR THE READER.   The volume in this example can also be obtained by the method of washers. Confirm that the volume produced by that method agrees with the volume obtained by cylindrical shells.

Example 3

Use cylindrical shells to find the volume of the solid generated when the region R under $y = x^2$ over the interval $[0, 2]$ is revolved about the x-axis (Figure 8.3.9).

Solution.   At each y in the interval $0 \le y \le 4$, the cross section of R parallel to the x-axis generates a cylindrical surface of height $2 - \sqrt{y}$ and radius y. Since the area of this surface is $2\pi y(2 - \sqrt{y})$, the volume of the solid is

$$V = \int_0^4 2\pi y(2 - \sqrt{y})\,dy = 2\pi \int_0^4 (2y - y^{3/2})\,dy = 2\pi \left[y^2 - \frac{2}{5}y^{5/2} \right]_0^4 = \frac{32\pi}{5}$$

◀

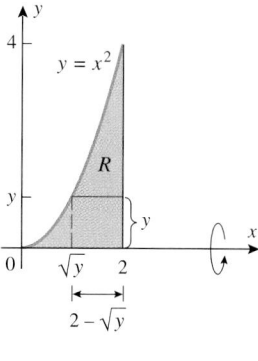

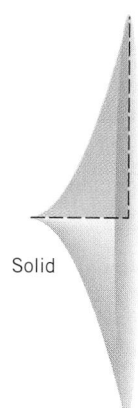

Solid

Figure 8.3.9

FOR THE READER.   The volume in this example can also be obtained by the method of disks. Confirm that the volume produced by that method agrees with the volume obtained by cylindrical shells.

EXERCISE SET 8.3   C CAS

In Exercises 1–4, use cylindrical shells to find the volume of the solid generated when the shaded region is revolved about the indicated axis.

1.

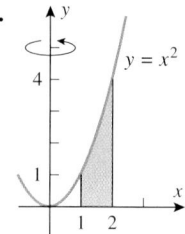

$y = x^2$

2.

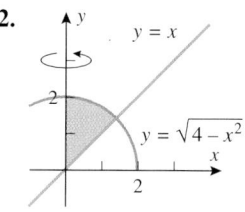

$y = x$

$y = \sqrt{4 - x^2}$

3.

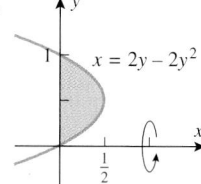

$x = 2y - 2y^2$

4.

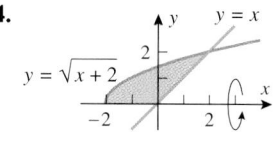

$y = \sqrt{x + 2}$ $y = x$

In Exercises 5–12, use cylindrical shells to find the volume of the solid generated when the region enclosed by the given curves is revolved about the y-axis.

5. $y = x^3$, $x = 1$, $y = 0$

6. $y = \sqrt{x}$, $x = 4$, $x = 9$, $y = 0$

7. $y = 1/x$, $y = 0$, $x = 1$, $x = 3$

8. $y = \cos(x^2)$, $x = 0$, $x = \frac{1}{2}\sqrt{\pi}$, $y = 0$

9. $y = 2x - 1$, $y = -2x + 3$, $x = 2$

10. $y = \dfrac{1}{x^2 + 1}$, $x = 0$, $x = 1$, $y = 0$

11. $y = e^{x^2}$, $x = 1$, $x = \sqrt{3}$, $y = 0$

12. $y = 2x - x^2$, $y = 0$

In Exercises 13–16, use cylindrical shells to find the volume of the solid generated when the region enclosed by the given curves is revolved about the x-axis.

13. $y^2 = x$, $y = 1$, $x = 0$

14. $x = 2y$, $y = 2$, $y = 3$, $x = 0$

15. $y = x^2$, $x = 1$, $y = 0$

16. $xy = 4$, $x + y = 5$

C **17.** Use a CAS to find the volume of the solid generated when the region enclosed by $y = \sin x$ and $y = 0$ for $0 \le x \le \pi$ is revolved about the y-axis.

C **18.** Use a CAS to find the volume of the solid generated when the region enclosed by $y = \cos x$, $y = 0$, and $x = 0$ for $0 \le x \le \pi/2$ is revolved about the y-axis.

19. (a) Use cylindrical shells to find the volume of the solid that is generated when the region under the curve

$$y = x^3 - 3x^2 + 2x$$

over [0, 1] is revolved about the y-axis.
(b) For this problem, is the method of cylindrical shells easier or harder than the method of slicing discussed in the last section? Explain.

20. Use cylindrical shells to find the volume of the solid that is generated when the region that is enclosed by $y = 1/x^3$, $x = 1$, $x = 2$, $y = 0$ is revolved about the line $x = -1$.

21. Use cylindrical shells to find the volume of the solid that is generated when the region that is enclosed by $y = x^3$, $y = 1$, $x = 0$ is revolved about the line $y = 1$.

22. Let R_1 and R_2 be regions of the form shown in the accompanying figure. Use cylindrical shells to find a formula for the volume of the solid that results when
(a) region R_1 is revolved about the y-axis
(b) region R_2 is revolved about the x-axis.

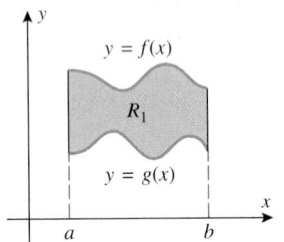

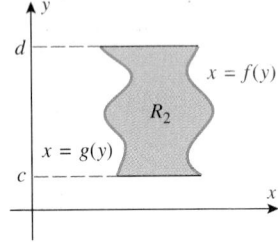

Figure Ex-22

23. Use cylindrical shells to find the volume of the cone generated when the triangle with vertices $(0, 0)$, $(0, r)$, $(h, 0)$, where $r > 0$ and $h > 0$, is revolved about the x-axis.

24. The region enclosed between the curve $y^2 = kx$ and the line $x = \frac{1}{4}k$ is revolved about the line $x = \frac{1}{2}k$. Use cylindrical shells to find the volume of the resulting solid. (Assume $k > 0$.)

25. A round hole of radius a is drilled through the center of a solid sphere of radius r. Use cylindrical shells to find the volume of the portion removed. (Assume $r > a$.)

26. Use cylindrical shells to find the volume of the torus obtained by revolving the circle $x^2 + y^2 = a^2$ about the line $x = b$, where $b > a > 0$. [*Hint:* It may help in the integration to think of an integral as an area.]

27. Let V_x and V_y be the volumes of the solids that result when the region enclosed by $y = 1/x$, $y = 0$, $x = \frac{1}{2}$, and $x = b$ $\left(b > \frac{1}{2}\right)$ is revolved about the x-axis and y-axis, respectively. Is there a value of b for which $V_x = V_y$?

8.4 LENGTH OF A PLANE CURVE

In this section we will consider the problem of finding the length of a plane curve.

ARC LENGTH

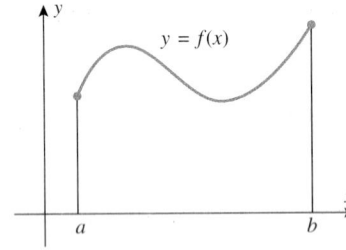

Figure 8.4.1

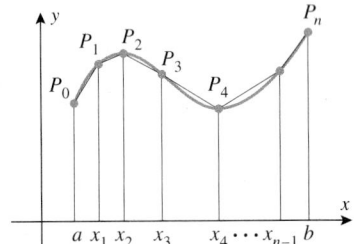

Figure 8.4.2

Although formulas for lengths of circular arcs appear in early historical records, very little was known about the lengths of more general curves until the mid-seventeenth century. About that time formulas were discovered for a few specific curves such as the length of an arch of a cycloid. However, such basic problems as finding the length of an ellipse defied the mathematicians of that period, and almost no progress was made on the general problem of finding lengths of curves until the advent of calculus in the next century.

Our first objective in this section is to *define* what we mean by the length (also called the *arc length*) of a plane curve $y = f(x)$ over an interval $[a, b]$ (Figure 8.4.1). Once that is done we will be able to focus on computational matters. To avoid some complications that would otherwise occur, we will impose the requirement that f' be continuous on $[a, b]$, in which case we will say that $y = f(x)$ is a *smooth curve* on $[a, b]$ (or that f is a *smooth function* on $[a, b]$).

We will be concerned with the following problem:

> **8.4.1** ARC LENGTH PROBLEM. Suppose that $y = f(x)$ is a smooth curve on the interval $[a, b]$. Define and find a formula for the arc length L of the curve $y = f(x)$ over the interval $[a, b]$.

The basic idea for defining arc length is to break up the curve into small segments, approximate the curve segments by line segments, add the lengths of the line segments to form a Riemann sum that approximates the arc length L, and take the limit of the Riemann sums to obtain an integral for L.

To implement this idea, divide the interval $[a, b]$ into n subintervals by inserting points $x_1, x_2, \ldots, x_{n-1}$ between a and b. As shown in Figure 8.4.2, let $P_0, P_1, \ldots, P_n$ be the points

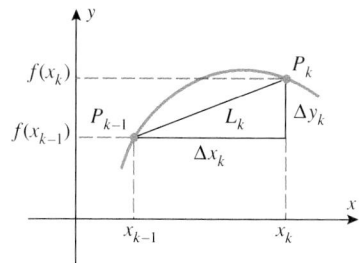

Figure 8.4.3

on the curve with x-coordinates $a, x_1, x_2, \ldots, x_{n-1}, b$ and join these points with straight line segments. These line segments form a ***polygonal path*** that we can regard as an approximation to the curve $y = f(x)$. As suggested by Figure 8.4.3, the length L_k of the kth line segment in the polygonal path is

$$L_k = \sqrt{(\Delta x_k)^2 + (\Delta y_k)^2} = \sqrt{(\Delta x_k)^2 + [f(x_k) - f(x_{k-1})]^2} \tag{1}$$

If we now add the lengths of these line segments, we obtain the following approximation to the length L of the curve

$$L \approx \sum_{k=1}^{n} L_k = \sum_{k=1}^{n} \sqrt{(\Delta x_k)^2 + [f(x_k) - f(x_{k-1})]^2} \tag{2}$$

To put this in the form of a Riemann sum we will apply the Mean-Value Theorem (6.5.2). This theorem implies that there is a point x_k^* between x_{k-1} and x_k such that

$$\frac{f(x_k) - f(x_{k-1})}{x_k - x_{k-1}} = f'(x_k^*) \quad \text{or} \quad f(x_k) - f(x_{k-1}) = f'(x_k^*)\Delta x_k$$

and hence we can rewrite (2) as

$$L \approx \sum_{k=1}^{n} \sqrt{1 + [f'(x_k^*)]^2}\, \Delta x_k$$

Thus, taking the limit as n increases and the widths of the subintervals approach zero yields the following integral that defines the arc length L:

$$L = \lim_{\max \Delta x_k \to 0} \sum_{k=1}^{n} \sqrt{1 + [f'(x_k^*)]^2}\, \Delta x_k = \int_a^b \sqrt{1 + [f'(x)]^2}\, dx$$

In summary, we have the following definition:

8.4.2 DEFINITION. If $y = f(x)$ is a smooth curve on the interval $[a, b]$, then the arc length L of this curve over $[a, b]$ is defined as

$$L = \int_a^b \sqrt{1 + [f'(x)]^2}\, dx \tag{3}$$

This result provides both a definition and a formula for computing arc lengths. Where convenient, (3) can also be expressed as

$$L = \int_a^b \sqrt{1 + [f'(x)]^2}\, dx = \int_a^b \sqrt{1 + \left(\frac{dy}{dx}\right)^2}\, dx \tag{4}$$

Moreover, for a curve expressed in the form $x = g(y)$, where g' is continuous on $[c, d]$, the arc length L from $y = c$ to $y = d$ can be expressed as

$$L = \int_c^d \sqrt{1 + [g'(y)]^2}\, dy = \int_c^d \sqrt{1 + \left(\frac{dx}{dy}\right)^2}\, dy \tag{5}$$

Example 1

Find the arc length of the curve $y = x^{3/2}$ from $(1, 1)$ to $(2, 2\sqrt{2})$ (Figure 8.4.4) in two ways: (a) using Formula (4) and (b) using Formula (5).

Solution (a). Since

$$\frac{dy}{dx} = \tfrac{3}{2}x^{1/2}$$

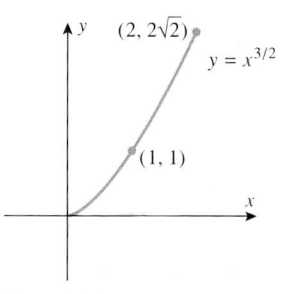

Figure 8.4.4

and since the curve extends from $x = 1$ to $x = 2$, it follows from (4) that

$$L = \int_1^2 \sqrt{1 + \frac{9}{4}x}\, dx$$

To evaluate this integral we make the u-substitution

$$u = 1 + \tfrac{9}{4}x, \quad du = \tfrac{9}{4}\, dx$$

and then change the x-limits of integration $(x = 1, x = 2)$ to the corresponding u-limits $\left(u = \frac{13}{4}, u = \frac{22}{4}\right)$:

$$L = \frac{4}{9}\int_{13/4}^{22/4} u^{1/2}\, du = \frac{8}{27}u^{3/2}\Big]_{13/4}^{22/4} = \frac{8}{27}\left[\left(\frac{22}{4}\right)^{3/2} - \left(\frac{13}{4}\right)^{3/2}\right]$$

$$= \frac{22\sqrt{22} - 13\sqrt{13}}{27} \approx 2.09$$

Solution (b). To apply Formula (5) we must first rewrite the equation $y = x^{3/2}$ so that x is expressed as a function of y. This yields $x = y^{2/3}$ and

$$\frac{dx}{dy} = \tfrac{2}{3}y^{-1/3}$$

Since the curve extends from $y = 1$ to $y = 2\sqrt{2}$, it follows from (5) that

$$L = \int_1^{2\sqrt{2}} \sqrt{1 + \frac{4}{9}y^{-2/3}}\, dy = \frac{1}{3}\int_1^{2\sqrt{2}} y^{-1/3}\sqrt{9y^{2/3} + 4}\, dy$$

To evaluate this integral we make the u-substitution

$$u = 9y^{2/3} + 4, \quad du = 6y^{-1/3}\, dy$$

and change the y-limits of integration $(y = 1, y = 2\sqrt{2})$ to the corresponding u-limits $(u = 13, u = 22)$. This gives

$$L = \frac{1}{18}\int_{13}^{22} u^{1/2}\, du = \frac{1}{27}u^{3/2}\Big]_{13}^{22} = \frac{1}{27}[(22)^{3/2} - (13)^{3/2}] = \frac{22\sqrt{22} - 13\sqrt{13}}{27}$$

This result agrees with that in part (a); however, the integration here is more tedious. In problems where there is a choice between using (4) or (5), it is often the case that one of the formulas leads to a simpler integral than the other. ◄

ARC LENGTH OF CURVES DEFINED PARAMETRICALLY

The following result provides a formula for finding the arc length of a curve from parametric equations for the curve. Its derivation is similar to that of Formula (3) and will be omitted.

8.4.3 ARC LENGTH FORMULA FOR PARAMETRIC CURVES. If no segment of the curve represented by the parametric equations

$$x = x(t), \quad y = y(t) \quad (a \le t \le b)$$

is traced more than once as t increases from a to b, and if dx/dt and dy/dt are continuous functions for $a \le t \le b$, then the arc length L of the curve is given by

$$L = \int_a^b \sqrt{\left(\frac{dx}{dt}\right)^2 + \left(\frac{dy}{dt}\right)^2}\, dt \tag{6}$$

REMARK. Note that Formulas (4) and (5) are special cases of (6). For example, Formula (4) can be obtained from (6) by writing $y = f(x)$ parametrically as $x = t$, $y = f(t)$; similarly, Formula (5) can be obtained from (6) by writing $x = g(y)$ parametrically as $x = g(t)$, $y = t$. We leave the details as exercises.

Example 2

Use (6) to find the circumference of a circle of radius a from the parametric equations

$$x = a \cos t, \quad y = a \sin t \qquad (0 \le t \le 2\pi)$$

Solution.

$$L = \int_0^{2\pi} \sqrt{\left(\frac{dx}{dt}\right)^2 + \left(\frac{dy}{dt}\right)^2} \, dt = \int_0^{2\pi} \sqrt{(-a \sin t)^2 + (a \cos t)^2} \, dt$$

$$= \int_0^{2\pi} a \, dt = at \Big]_0^{2\pi} = 2\pi a \qquad \blacktriangleleft$$

FINDING ARC LENGTH BY NUMERICAL METHODS

As a rule, the integrals that arise in calculating arc length tend to be impossible to evaluate in terms of elementary functions, so it will often be necessary to approximate the integral using a numerical method such as the midpoint approximation (discussed in Section 7.5) or some other comparable method. Examples 1 and 2 are rare exceptions.

Example 3

From (4), the arc length of $y = \sin x$ from $x = 0$ to $x = \pi$ is given by the integral

$$L = \int_0^{\pi} \sqrt{1 + (\cos x)^2} \, dx$$

This integral cannot be evaluated in terms of elementary functions; however, using a calculating utility with a numerical integration capability yields the approximation $L \approx 3.8202$.

$\blacktriangleleft$

FOR THE READER. In Figure 8.4.5, the scale on both axes is 2 centimeters per unit. Confirm that the result in Example 3 is reasonable by laying a piece of string as closely as possible along the curve in the figure and measuring its length in centimeters.

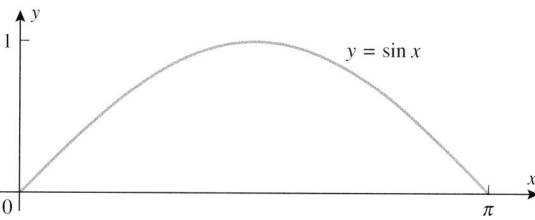

Figure 8.4.5

FOR THE READER. Computer algebra systems and some scientific calculators have commands for evaluating integrals numerically, and some scientific calculators have built-in commands for approximating arc lengths. If you have a scientific calculator with one of these capabilities or a CAS, read the documentation, and then use your calculator or CAS to check the result in Example 3.

EXERCISE SET 8.4 ⌁ Graphing Calculator [c] CAS

1. Use the Theorem of Pythagoras to find the length of the line segment $y = 2x$ from $(1, 2)$ to $(2, 4)$, and confirm that the value is consistent with the length computed using
   (a) Formula (4) (b) Formula (5).

2. Use the Theorem of Pythagoras to find the length of the line segment $x = t, y = 5t$ $(0 \le t \le 1)$, and confirm that the value is consistent with the length computed using Formula (6).

In Exercises 3–8, find the exact arc length of the curve over the stated interval.

3. $y = 3x^{3/2} - 1$ from $x = 0$ to $x = 1$

4. $x = \frac{1}{3}(y^2 + 2)^{3/2}$ from $y = 0$ to $y = 1$

5. $y = x^{2/3}$ from $x = 1$ to $x = 8$

6. $y = (x^6 + 8)/16x^2$ from $x = 2$ to $x = 3$

7. $y = \frac{1}{2}(e^x + e^{-x})$ from $x = 0$ to $x = 3$

8. $x = \frac{1}{8}y^4 + \frac{1}{4}y^{-2}$ from $y = 1$ to $y = 4$

In Exercises 9–14, find the exact arc length of the parametric curve without eliminating the parameter.

9. $x = \frac{1}{3}t^3$, $y = \frac{1}{2}t^2$ $(0 \le t \le 1)$

10. $x = (1+t)^2$, $y = (1+t)^3$ $(0 \le t \le 1)$

11. $x = \cos 2t$, $y = \sin 2t$ $(0 \le t \le \pi/2)$

12. $x = \cos t + t \sin t$, $y = \sin t - t \cos t$ $(0 \le t \le \pi)$

13. $x = e^t \cos t$, $y = e^t \sin t$ $(0 \le t \le \pi/2)$

14. $x = e^t(\sin t + \cos t)$, $y = e^t(\cos t - \sin t)$ $(1 \le t \le 4)$

In Exercises 15 and 16, express the exact arc length of the curve over the given interval as an integral that has been simplified to eliminate the radical, and then evaluate the integral using a CAS.

[c] **15.** $y = \ln(\sec x)$ from $x = 0$ to $x = \pi/4$

[c] **16.** $y = \ln(\sin x)$ from $x = \pi/4$ to $\pi/2$

[c] **17.** (a) Recall from Section 1.7 that a cycloid is the path traced by a point on the rim of a wheel that rolls along a line (Figure 1.7.13). Use the parametric equations in Formula (9) of that section to show that the length L of one arch of a cycloid is given by the integral

$$L = a \int_0^{2\pi} \sqrt{2(1 - \cos \theta)}\, d\theta$$

(b) Use a CAS to show that L is eight times the radius of the wheel (see the accompanying figure).

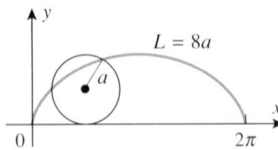

Figure Ex-17

[~] **18.** It was stated in Exercise 41 of Section 1.7 that the curve given parametrically by the equations

$$x = a \cos^3 \phi, \quad y = a \sin^3 \phi$$

is called a *four-cusped hypocycloid* (also called an **astroid**).
(a) Use a graphing utility to generate the graph in the case where $a = 1$, so that it is traced exactly once.
(b) Find the exact arc length of the curve in part (a).

19. Consider the curve $y = x^{2/3}$.
(a) Sketch the portion of the curve between $x = -1$ and $x = 8$.
(b) Explain why Formula (4) cannot be used to find the arc length of the curve sketched in part (a).
(c) Find the arc length of the curve sketched in part (a).

20. Derive Formulas (4) and (5) from Formula (6) by choosing appropriate parametrizations of the curves.

In Exercises 21 and 22, use the midpoint approximation with $n = 20$ subintervals to approximate the arc length of the curve over the given interval.

21. $y = x^2$ from $x = 0$ to $x = 2$

22. $x = \sin y$ from $y = 0$ to $y = \pi$

[c] **23.** Use a CAS or a scientific calculator with numerical integration capabilities to approximate the arc lengths in Exercises 21 and 22.

24. Let $y = f(x)$ be a smooth curve on the closed interval $[a, b]$. Prove that if there are nonnegative numbers m and M such that $m \le f'(x) \le M$ for all x in $[a, b]$, then the arc length L of $y = f(x)$ over the interval $[a, b]$ satisfies the inequalities

$$(b - a)\sqrt{1 + m^2} \le L \le (b - a)\sqrt{1 + M^2}$$

25. Use the result of Exercise 24 to show that the arc length L of $y = \sin x$ over the interval $0 \le x \le \pi/4$ satisfies

$$\frac{\pi}{4}\sqrt{\frac{3}{2}} \le L \le \frac{\pi}{4}\sqrt{2}$$

26. Show that the total arc length of the ellipse $x = a \cos t$, $y = b \sin t$, $0 \le t \le 2\pi$ for $a > b > 0$ is given by

$$4a \int_0^{\pi/2} \sqrt{1 - k^2 \cos^2 t}\, dt$$

where $k = \sqrt{a^2 - b^2}/a$.

[c] **27.** (a) Show that the total arc length of the ellipse

$$x = 2\cos t, \quad y = \sin t \quad (0 \le t \le 2\pi)$$

is given by

$$4 \int_0^{\pi/2} \sqrt{1 + 3\sin^2 t}\, dt$$

(b) Use a CAS or a scientific calculator with numerical integration capabilities to approximate the arc length in part (a). Round your answer to two decimal places.
(c) Suppose that the parametric equations in part (a) describe the path of a particle moving in the xy-plane, where t is time in seconds and x and y are in centimeters. Use a CAS or a scientific calculator with numerical integration capabilities to approximate the distance traveled by the particle from $t = 1.5$ s to $t = 4.8$ s. Round your answer to two decimal places.

[c] **28.** A basketball player makes a successful shot from the free throw line. Suppose that the path of the ball from the mo-

ment of release to the moment it enters the hoop is described by

$$y = 2.15 + 2.09x - 0.41x^2, \quad 0 \le x \le 4.6$$

where x is the horizontal distance (in meters) from the point of release, and y is the vertical distance (in meters) above the floor. Use a CAS or a scientific calculator with numerical integration capabilities to approximate the distance the ball travels from the moment it is released to the moment it enters the hoop. Round your answer to two decimal places.

[c] **29.** Find a positive value of k (to two decimal places) such that the curve $y = k \sin x$ has an arc length of $L = 5$ units over the interval from $x = 0$ to $x = \pi$. [*Hint:* Find an integral for the arc length L in terms of k, and then use a CAS or a scientific calculator with a numeric integration capability to find integer values of k at which the values of $L - 5$ have opposite signs. Complete the solution by using the Intermediate-Value Theorem (2.4.8) to approximate the value of k to two decimal places.]

8.5 AREA OF A SURFACE OF REVOLUTION

In this section we will consider the problem of finding the area of a surface that is generated by revolving a plane curve about a line.

SURFACE AREA

A *surface of revolution* is a surface that is generated by revolving a plane curve about an axis that lies in the same plane as the curve. For example, the surface of a sphere can be generated by revolving a semicircle about its diameter, and the lateral surface of a right circular cylinder can be generated by revolving a line segment about an axis that is parallel to it (Figure 8.5.1).

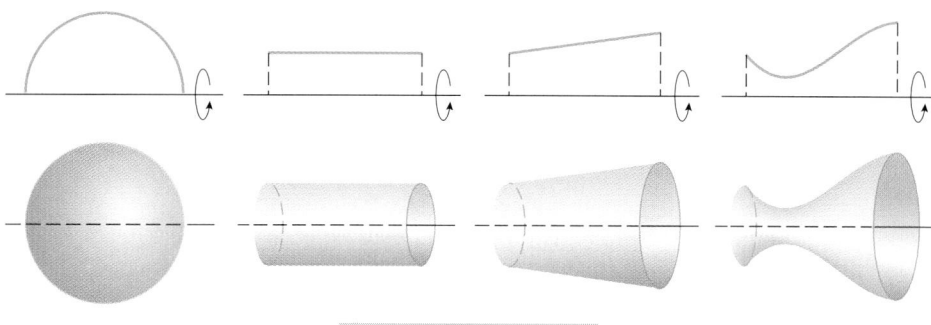

Some surfaces of revolution

Figure 8.5.1

In this section we will be concerned with the following problem:

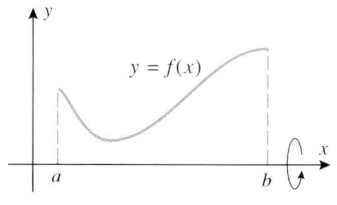

8.5.1 SURFACE AREA PROBLEM. Suppose that f is a smooth, nonnegative function on $[a, b]$ and that a surface of revolution is generated by revolving the portion of the curve $y = f(x)$ between $x = a$ and $x = b$ about the x-axis (Figure 8.5.2). Define what is meant by the *area* S of the surface, and find a formula for computing it.

To motivate an appropriate definition for the area S of a surface of revolution, we will decompose the surface into small sections whose areas can be approximated by elementary formulas, add the approximations of the areas of the sections to form a Riemann sum that approximates S, and then take the limit of the Riemann sums to obtain an integral for the exact value of S.

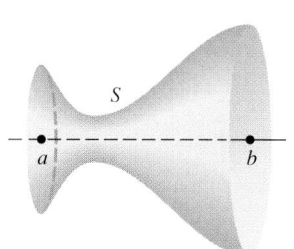

Figure 8.5.2

To implement this idea, divide the interval $[a, b]$ into n subintervals by inserting points $x_1, x_2, \ldots, x_{n-1}$ between a and b. As illustrated in Figure 8.5.3a, these points define a polygonal path that approximates the curve $y = f(x)$ over the interval $[a, b]$. When this

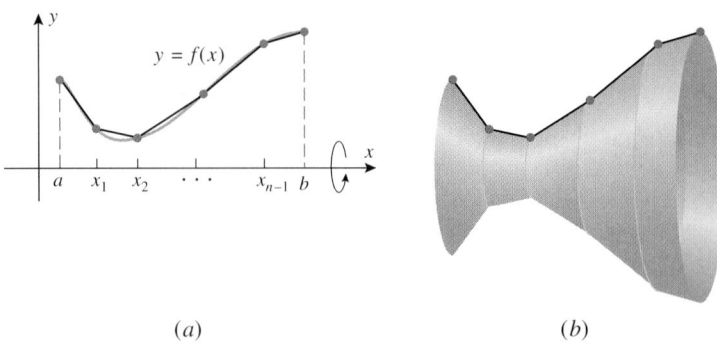

(a)   (b)

Figure 8.5.3

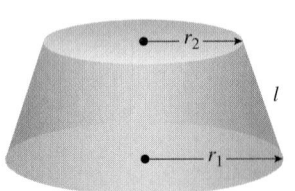

Figure 8.5.4

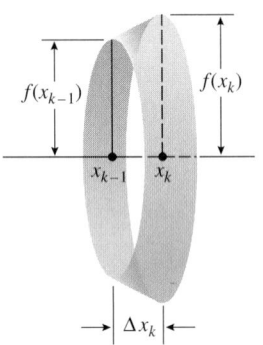

Figure 8.5.5

polygonal path is revolved about the x-axis, it generates a surface consisting of n parts, each of which is a frustum of a right circular cone (Figure 8.5.3b). Thus, the area of each part of the approximating surface can be obtained from the formula

$$S = \pi(r_1 + r_2)l \tag{1}$$

for the lateral area S of a frustum of slant height l and base radii r_1 and r_2 (Figure 8.5.4). As suggested by Figure 8.5.5, the kth frustum has radii $f(x_{k-1})$ and $f(x_k)$ and height Δx_k. Its slant height is the length L_k of the kth line segment in the polygonal path, which from Formula (1) of Section 8.4 is

$$L_k = \sqrt{(\Delta x_k)^2 + [f(x_k) - f(x_{k-1})]^2}$$

Thus, the lateral area S_k of the kth frustum is

$$S_k = \pi[f(x_{k-1}) + f(x_k)]\sqrt{(\Delta x_k)^2 + [f(x_k) - f(x_{k-1})]^2}$$

If we add these areas, we obtain the following approximation to the area S of the entire surface:

$$S \approx \sum_{k=1}^{n} \pi[f(x_{k-1}) + f(x_k)]\sqrt{(\Delta x_k)^2 + [f(x_k) - f(x_{k-1})]^2} \tag{2}$$

To put this in the form of a Riemann sum we will apply the Mean-Value Theorem (6.5.2). This theorem implies that there is a point x_k^* between x_{k-1} and x_k such that

$$\frac{f(x_k) - f(x_{k-1})}{x_k - x_{k-1}} = f'(x_k^*) \quad \text{or} \quad f(x_k) - f(x_{k-1}) = f'(x_k^*)\Delta x_k$$

and hence we can rewrite (2) as

$$S \approx \sum_{k=1}^{n} \pi[f(x_{k-1}) + f(x_k)]\sqrt{1 + [f'(x_k^*)]^2}\,\Delta x_k \tag{3}$$

However, this is not yet a Riemann sum because it involves the variables x_{k-1} and x_k. To eliminate these variables from the expression, observe that the average value of the numbers $f(x_{k-1})$ and $f(x_k)$ lies between these numbers, so the continuity of f and the Intermediate-Value Theorem (2.4.8) imply that there is a point x_k^{**} between x_{k-1} and x_k such that

$$\tfrac{1}{2}[f(x_{k-1}) + f(x_k)] = f(x_k^{**})$$

Thus, (2) can be expressed as

$$S \approx \sum_{k=1}^{n} 2\pi f(x_k^{**})\sqrt{1 + [f'(x_k^*)]^2}\,\Delta x_k$$

Although this expression is close to a Riemann sum in form, it is not a true Riemann sum because it involves two variables x_k^* and x_k^{**}, rather than x_k^* alone. However, it is proved in advanced calculus courses that this has no effect on the limit because of the continuity of f. Thus, we can assume that $x_k^{**} = x_k^*$ when taking the limit, and this suggests that S can be defined as

$$S = \lim_{\max \Delta x_k \to 0} \sum_{k=1}^{n} 2\pi f(x_k^{**})\sqrt{1 + [f'(x_k^*)]^2}\,\Delta x_k = \int_a^b 2\pi f(x)\sqrt{1 + [f'(x)]^2}\,dx$$

In summary, we have the following definition:

> **8.5.2** DEFINITION. If f is a smooth, nonnegative function on $[a, b]$, then the surface area S of the surface of revolution that is generated by revolving the portion of the curve $y = f(x)$ between $x = a$ and $x = b$ about the x-axis is defined as
>
> $$S = \int_a^b 2\pi f(x)\sqrt{1 + [f'(x)]^2}\,dx$$

This result provides both a definition and formula for computing surface areas. Where convenient, this formula can also be expressed as

$$S = \int_a^b 2\pi f(x)\sqrt{1 + [f'(x)]^2}\,dx = \int_a^b 2\pi y \sqrt{1 + \left(\frac{dy}{dx}\right)^2}\,dx \qquad (4)$$

Moreover, if g is nonnegative and $x = g(y)$ is a smooth curve on the interval $[c, d]$, then the area of the surface that is generated by revolving the portion of a curve $x = g(y)$ between $y = c$ and $y = d$ about the y-axis can be expressed as

$$S = \int_c^d 2\pi g(y)\sqrt{1 + [g'(y)]^2}\,dy = \int_c^d 2\pi x \sqrt{1 + \left(\frac{dx}{dy}\right)^2}\,dy \qquad (5)$$

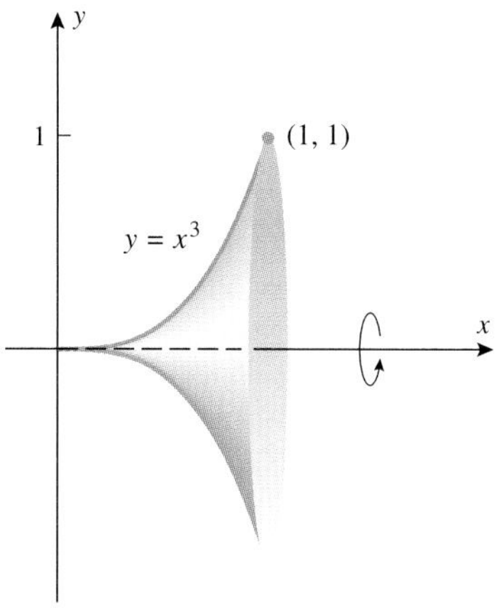

Example 1

Find the area of the surface that is generated by revolving the portion of the curve $y = x^3$ between $x = 0$ and $x = 1$ about the x-axis (Figure 8.5.6).

Solution. Since $y = x^3$, we have $dy/dx = 3x^2$, and hence from (4) the surface area S is

$$S = \int_0^1 2\pi y \sqrt{1 + \left(\frac{dy}{dx}\right)^2}\,dx$$

$$= \int_0^1 2\pi x^3 \sqrt{1 + (3x^2)^2}\,dx$$

$$= 2\pi \int_0^1 x^3 (1 + 9x^4)^{1/2}\,dx$$

$$= \frac{2\pi}{36} \int_1^{10} u^{1/2}\,du \qquad \boxed{\begin{array}{l} u = 1 + 9x^4 \\ du = 36x^3\,dx \end{array}}$$

$$= \frac{2\pi}{36} \cdot \frac{2}{3} u^{3/2} \Big]_{u=1}^{10} = \frac{\pi}{27}(10^{3/2} - 1) \approx 3.56 \qquad \blacktriangleleft$$

Example 2

Find the area of the surface that is generated by revolving the portion of the curve $y = x^2$ between $x = 1$ and $x = 2$ about the y-axis (Figure 8.5.7).

Solution. Because the curve is revolved about the y-axis we will apply Formula (5). Toward this end, we rewrite $y = x^2$ as $x = \sqrt{y}$ and observe that the y-values corresponding to $x = 1$ and $x = 2$ are $y = 1$ and $y = 4$. Since $x = \sqrt{y}$, we have $dx/dy = 1/(2\sqrt{y})$, and hence from (5) the surface area S is

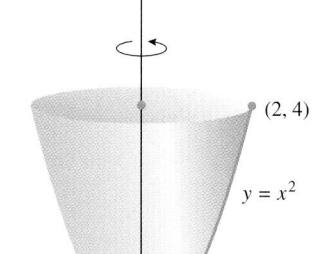

Figure 8.5.6

Figure 8.5.7

$$S = \int_1^4 2\pi x \sqrt{1 + \left(\frac{dx}{dy}\right)^2}\, dy$$

$$= \int_1^4 2\pi \sqrt{y} \sqrt{1 + \left(\frac{1}{2\sqrt{y}}\right)^2}\, dy$$

$$= \pi \int_1^4 \sqrt{4y + 1}\, dy$$

$$= \frac{\pi}{4} \int_5^{17} u^{1/2}\, du \qquad \boxed{\begin{array}{l} u = 4y + 1 \\ du = 4\, dy \end{array}}$$

$$= \frac{\pi}{4} \cdot \frac{2}{3} u^{3/2} \Big]_{u=5}^{17} = \frac{\pi}{6}(17^{3/2} - 5^{3/2}) \approx 30.85 \qquad \blacktriangleleft$$

EXERCISE SET 8.5   C CAS

In Exercises 1–4, find the area of the surface generated by revolving the given curve about the x-axis.

1. $y = 7x,\ 0 \le x \le 1$

2. $y = \sqrt{x},\ 1 \le x \le 4$

3. $y = \sqrt{4 - x^2},\ -1 \le x \le 1$

4. $x = \sqrt[3]{y},\ 1 \le y \le 8$

In Exercises 5–8, find the area of the surface generated by revolving the given curve about the y-axis.

5. $x = 9y + 1,\ 0 \le y \le 2$

6. $x = y^3,\ 0 \le y \le 1$

7. $x = \sqrt{9 - y^2},\ -2 \le y \le 2$

8. $x = 2\sqrt{1 - y},\ -1 \le y \le 0$

In Exercises 9–12, use a CAS to find the exact area of the surface generated by revolving the curve about the stated axis.

C **9.** $y = \sqrt{x} - \frac{1}{3}x^{3/2},\ 1 \le x \le 3;\ x$-axis

C **10.** $y = \frac{1}{3}x^3 + \frac{1}{4}x^{-1},\ 1 \le x \le 2;\ x$-axis

C **11.** $8xy^2 = 2y^6 + 1,\ 1 \le y \le 2;\ y$-axis

C **12.** $x = \sqrt{16 - y},\ 0 \le y \le 15;\ y$-axis

In Exercises 13–16, use a CAS or a scientific calculator with numerical integration capabilities to approximate the area of the surface generated by revolving the curve about the stated axis. Round your answer to two decimal places.

C **13.** $y = e^x,\ 0 \le x \le 1;\ x$-axis

C **14.** $y = \sin x,\ 0 \le x \le \pi;\ x$-axis

C **15.** $y = e^x,\ 1 \le y \le e;\ y$-axis

C **16.** $x = \tan y,\ 0 \le y \le \pi/4;\ y$-axis

17. Use Formula (4) to show that the lateral area S of a right circular cone with height h and base radius r is
$$S = \pi r \sqrt{r^2 + h^2}$$

18. Show that the area of the surface of a sphere of radius r is $4\pi r^2$. [*Hint:* Revolve the semicircle $y = \sqrt{r^2 - x^2}$ about the x-axis.]

19. (a) The figure in Exercise 39 of Section 8.2 shows a spherical cap of height h cut from a sphere of radius r. Show that the surface area S of the cap is $S = 2\pi r h$. [*Hint:* Revolve an appropriate portion of the circle $x^2 + y^2 = r^2$ about the y-axis.]

(b) The portion of a sphere that is cut by two parallel planes is called a **zone**. Use the result in part (a) to show that the surface area of a zone depends on the radius of the sphere and the distance between the planes, but not on the location of the zone.

Exercises 20–26 require the formulas developed in the following discussion: If $x'(t)$ and $y'(t)$ are continuous functions and if no segment of the curve
$$x = x(t), \quad y = y(t) \qquad (a \le t \le b)$$
is traced more than once, then it can be shown that the area of the surface generated by revolving this curve about the x-axis is
$$S = \int_a^b 2\pi y(t)\sqrt{[x'(t)]^2 + [y'(t)]^2}\, dt \qquad (A)$$
and the area of the surface generated by revolving the curve about the y-axis is
$$S = \int_a^b 2\pi x(t)\sqrt{[x'(t)]^2 + [y'(t)]^2}\, dt \qquad (B)$$

20. Derive Formulas (4) and (5) from Formulas (A) and (B) above by choosing appropriate parametrizations for the curves $y = f(x)$ and $x = g(y)$.

21. Find the area of the surface generated by revolving the parametric curve $x = t^2$, $y = 2t$, $0 \le t \le 4$ about the x-axis.

C **22.** Use a CAS to find the area of the surface generated by revolving the parametric curve $x = e^t \cos t$, $y = e^t \sin t$, $0 \le t \le \pi/2$ about the x-axis.

23. Find the area of the surface generated by revolving the parametric curve $x = t$, $y = 2t^2$, $0 \le t \le 1$ about the y-axis.

24. Find the area of the surface generated by revolving the equations $x = \cos^2 t$, $y = \sin^2 t$, $0 \le t \le \pi/2$ about the y-axis.

25. By revolving the semicircle

$$x = r \cos t, \quad y = r \sin t \quad (0 \le t \le \pi)$$

about the x-axis, show that the surface area of a sphere of radius r is $4\pi r^2$.

26. The equations

$$x = a\phi - a \sin\phi, \quad y = a - a\cos\phi \quad (0 \le \phi \le 2\pi)$$

represent one arch of a cycloid. Show that the surface area generated by revolving this curve about the x-axis is $S = 64\pi a^2/3$. [*Hint:* Use the identities $\sin^2 \dfrac{\phi}{2} = \dfrac{1 - \cos\phi}{2}$ and $\sin^3 \phi = (1 - \cos^2 \phi) \sin\phi$ to help with the integration.]

27. (a) If a cone of slant height l and base radius r is cut along a lateral edge and laid flat, then as shown in the accompanying figure it becomes a sector of a circle of radius l. Use the formula $A = \tfrac{1}{2}l^2\theta$ for the area of a sector with radius l and central angle θ (in radians) to show that the lateral surface area of the cone is $\pi r l$.

(b) Use the result in part (a) to obtain Formula (1) for the lateral surface area of a frustum.

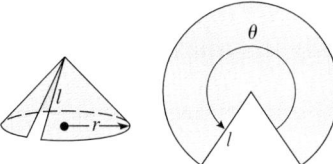

Figure Ex-27

28. Let $y = f(x)$ be a smooth curve on the interval $[a, b]$ and assume that $f(x) \ge 0$ for $a \le x \le b$. By the Extreme-Value Theorem 6.1.3, the function f has a maximum value K and a minimum value k on $[a, b]$. Prove: If L is the arc length of the curve $y = f(x)$ between $x = a$ and $x = b$ and if S is the area of the surface that is generated by revolving this curve about the x-axis, then

$$2\pi k L \le S \le 2\pi K L$$

29. Let $y = f(x)$ be a smooth curve on $[a, b]$ and assume that $f(x) \ge 0$ for $a \le x \le b$. Let A be the area under the curve $y = f(x)$ between $x = a$ and $x = b$ and let S be the area of the surface obtained when this section of curve is revolved about the x-axis.
(a) Prove that $2\pi A \le S$.
(b) For what functions f is $2\pi A = S$?

8.6 WORK

In this section we will use the integration tools developed in the preceding chapter to study some of the basic principles of "work," which is one of the fundamental concepts in physics and engineering.

THE ROLE OF WORK IN PHYSICS AND ENGINEERING

In this section we will be concerned with two related concepts, *work* and *energy*. To put these ideas in a familiar setting, when you push a stalled car for a certain distance you are performing work, and the effect of your work is to make the car move. The energy of motion caused by the work is called the *kinetic energy* of the car. The exact relationship between work and kinetic energy is governed by a principle of physics, called the *work–energy theorem*. Although we will touch on this idea in this section, a detailed study of the relationship between work and energy will be left for courses in physics and engineering. Our primary goal here will be to explain the role of integration in the study of work.

WORK DONE BY A CONSTANT FORCE APPLIED IN THE DIRECTION OF MOTION

When a stalled car is pushed, the speed that the car attains depends on the force F with which it is pushed and the distance d over which that force is applied (Figure 8.6.1). Thus, force and distance are the ingredients of work in the following definition.

> **8.6.1** DEFINITION. If a constant force of magnitude F is applied in the direction of motion of an object, and if that object moves a distance d, then we define the **work** W performed by the force on the object to be
>
> $$W = F \cdot d \tag{1}$$

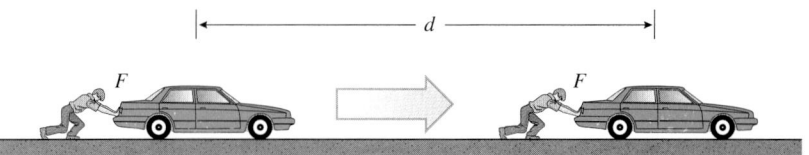

Figure 8.6.1

FOR THE READER. If you push against an immovable object, such as a brick wall, you may tire yourself out, but you will perform no work. Why?

Common units for measuring force are newtons (N) in the International System of Units (SI), dynes (dyn) in the CGS system, and pounds (lb) in the British Engineering system. One newton is the force required to give a mass of 1 kg an acceleration of 1 m/s^2, one dyne is the force required to give a mass of 1 g an acceleration of 1 cm/s^2, and one pound of force is the force required to give a mass of 1 slug an acceleration of 1 ft/s^2.

It follows from Definition 8.6.1 that work has units of force times distance. The most common units of work are newton-meters (N·m), dyne-centimeters (dyn·cm), and foot-pounds (ft·lb). As indicated in Table 8.6.1, one newton-meter is also called a ***joule*** (J), and one dyne-centimeter is also called an ***erg***. One foot-pound is approximately 1.36 J.

Table 8.6.1

SYSTEM	FORCE	×	DISTANCE	=	WORK
SI	newton (N)		meter (m)		joule (J)
CGS	dyne (dyn)		centimeter (cm)		erg
BE	pound (lb)		foot (ft)		foot-pound (ft·lb)

CONVERSION FACTORS:
$1 \text{ N} = 10^5 \text{ dyn} \approx 0.225 \text{ lb}$ $1 \text{ lb} \approx 4.45 \text{ N}$
$1 \text{ J} = 10^7 \text{ erg} \approx 0.738 \text{ ft·lb}$ $1 \text{ ft·lb} \approx 1.36 \text{ J} = 1.36 \times 10^7 \text{ erg}$

Example 1

An object moves 5 ft along a line while subjected to a constant force of 100 lb in its direction of motion. The work done is

$$W = F \cdot d = 100 \cdot 5 = 500 \text{ ft·lb}$$

An object moves 25 m along a line while subjected to a constant force of 4 N in its direction of motion. The work done is

$$W = F \cdot d = 4 \cdot 25 = 100 \text{ N·m} = 100 \text{ J} \qquad \blacktriangleleft$$

Example 2

In the 1976 Olympics, Vasili Alexeev astounded the world by lifting a record-breaking 562 lb from the floor to above his head (about 2 m). Equally astounding was the feat of strongman Paul Anderson, who in 1957 braced himself on the floor and used his back to lift 6270 lb of lead and automobile parts a distance of 1 cm. Who did more work?

Solution. To lift an object one must apply sufficient force to overcome the gravitational force that the Earth exerts on that object. The force that the Earth exerts on an object is that object's weight; thus, in performing their feats, Alexeev applied a force of 562 lb over a distance of 2 m and Anderson applied a force of 6270 lb over a distance of 1 cm. Since pounds are units in the BE system, meters are units in SI, and centimeters are units in the CGS system, we will need to decide on the measurement system we want to use and be consistent. Let us agree to use SI and express the work of the two men in joules. Using the conversion factor in Table 8.6.1 we obtain

Vasili Alexeev lifting a record-breaking 562 lb in the 1976 Olympics

$$562 \text{ lb} \approx 562 \text{ lb} \times 4.45 \text{ N/lb} = 2500.9 \text{ N}$$

$$6270 \text{ lb} \approx 6270 \text{ lb} \times 4.45 \text{ N/lb} = 27{,}901.5 \text{ N}$$

Using these values and the fact that $1 \text{ cm} = 0.01 \text{ m}$ we obtain

$$\text{Alexeev's work} = (2500.9 \text{ N}) \times (2 \text{ m}) \approx 5002 \text{ J}$$

$$\text{Anderson's work} = (27{,}901.5 \text{ N}) \times (0.01 \text{ m}) \approx 279 \text{ J}$$

Therefore, even though Anderson's lift required a tremendous upward force, it was applied over such a short distance that Alexeev did more work. ◀

WORK DONE BY A VARIABLE FORCE APPLIED IN THE DIRECTION OF MOTION

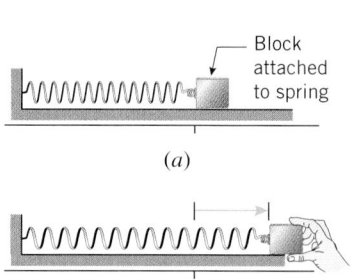

Block attached to spring

(a)

(b)

Figure 8.6.2

Many important problems are concerned with finding the work done by a *variable* force that is applied in the direction of motion. For example, Figure 8.6.2a shows a spring in its natural state (neither compressed nor stretched). If we want to pull the block horizontally so that it moves with a uniform speed (Figure 8.6.2b), then we would have to apply more and more force to the block to overcome the increasing force of the stretching spring. Thus, our next objective is to define what is meant by the work performed by a variable force and to find a formula for computing it. This will require calculus.

8.6.2 PROBLEM. Suppose that an object moves in the positive direction along a coordinate line while subjected to a variable force $F(x)$ that is applied in the direction of motion. Define what is meant by the *work* W performed by the force on the object as the object moves from $x = a$ to $x = b$, and find a formula for computing the work.

The basic idea for solving this problem is to break up the interval $[a, b]$ into subintervals that are sufficiently small that the force does not vary much on each subinterval. This will allow us to treat the force as constant on each subinterval and to approximate the work on each subinterval using Formula (1). By adding the approximations to the work on the subintervals, we will obtain a Riemann sum that approximates the work W over the entire interval, and by taking the limit of the Riemann sums we will obtain an integral for W.

To implement this idea, divide the interval $[a, b]$ into n subintervals by inserting points $x_1, x_2, \ldots, x_{n-1}$ between a and b. We can use Formula (1) to approximate the work W_k done in the kth subinterval by choosing any point x_k^* in this interval and regarding the force to have a constant value $F(x_k^*)$ throughout the interval. Since the width of the kth subinterval is $x_k - x_{k-1} = \Delta x_k$, this yields the approximation

$$W_k \approx F(x_k^*)\Delta x_k$$

Adding these approximations yields the following Riemann sum that approximates the work W done over the entire interval:

$$W \approx \sum_{k=1}^{n} F(x_k^*)\Delta x_k$$

Taking the limit as n increases and the widths of the subintervals approach zero yields the definite integral

$$W = \lim_{\max \Delta x_k \to 0} \sum_{k=1}^{n} F(x_k^*)\Delta x_k = \int_a^b F(x)\,dx$$

In summary, we have the following result:

8.6.3 DEFINITION. Suppose that an object moves in the positive direction along a coordinate line over the interval $[a, b]$ while subjected to a variable force $F(x)$ that is applied in the direction of motion. Then we define the **work** W performed by the force on the object to be

$$W = \int_a^b F(x)\,dx \qquad (2)$$

Hooke's law [Robert Hooke (1635–1703), English physicist] states that under appropriate conditions a spring that is stretched x units beyond its natural length pulls back with a force

$$F(x) = kx$$

where k is a constant (called the ***spring constant*** or ***spring stiffness***). The value of k depends on such factors as the thickness of the spring and the material used in its composition. Since $k = F(x)/x$, the constant k has units of force per unit length.

Example 3

A spring exerts a force of 5 N when stretched 1 m beyond its natural length.

(a)   Find the spring constant k.

(b)   How much work is required to stretch the spring 1.8 m beyond its natural length?

Solution (a). From Hooke's law,

$$F(x) = kx$$

From the data, $F(x) = 5$ N when $x = 1$ m, so $5 = k \cdot 1$. Thus, the spring constant is $k = 5$ newtons per meter (N/m). This means that the force $F(x)$ required to stretch the spring x meters is

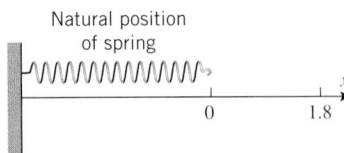

Natural position
of spring

Figure 8.6.3

$$F(x) = 5x \tag{3}$$

Solution (b). Place the spring along a coordinate line as shown in Figure 8.6.3. We want to find the work W required to stretch the spring over the interval from $x = 0$ to $x = 1.8$. From (2) and (3) the work W required is

$$W = \int_a^b F(x)\,dx = \int_0^{1.8} 5x\,dx = \left. \frac{5x^2}{2} \right]_0^{1.8} = 8.1 \text{ J} \qquad \blacktriangleleft$$

Example 4

An astronaut's *weight* (or more precisely, *Earth weight*) is the force exerted on the astronaut by the Earth's gravity. As the astronaut moves upward into space, the gravitational pull of the Earth decreases, and hence so does his or her weight. We will show later in the text that if the Earth is assumed to be a sphere of radius 4000 mi, then an astronaut who weighs 150 lb on Earth will have a weight of

$$w(x) = \frac{2{,}400{,}000{,}000}{x^2} \text{ lb}$$

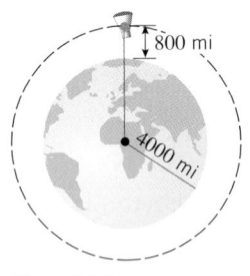

Figure 8.6.4

at a distance of x mi from the Earth's center. Use this formula to determine the work in foot-pounds required to lift the astronaut to a point that is 800 mi above the surface of the Earth (Figure 8.6.4).

Solution. Since the Earth has a radius of 4000 mi, the astronaut is lifted from a point that is 4000 mi from the Earth's center to a point that is 4800 mi from the Earth's center. Thus, from (2), the work W required to lift the astronaut is

$$W = \int_{4000}^{4800} \frac{2{,}400{,}000{,}000}{x^2}\,dx$$

$$= -\left. \frac{2{,}400{,}000{,}000}{x} \right]_{4000}^{4800}$$

$$= -500{,}000 + 600{,}000$$

$$= 100{,}000 \text{ mile-pounds}$$

$$= (100{,}000 \text{ mi·lb}) \times (5280 \text{ ft/mi})$$

$$= 5.28 \times 10^8 \text{ ft·lb} \qquad \blacktriangleleft$$

Some problems cannot be solved by mechanically substituting into formulas, and one must return to basic principles to obtain solutions. This is illustrated in the next example.

Example 5

A cylindrical water tank of radius 10 ft and height 30 ft is half filled with water. How much work is required to pump all of the water out through a hole in the top of the tank?

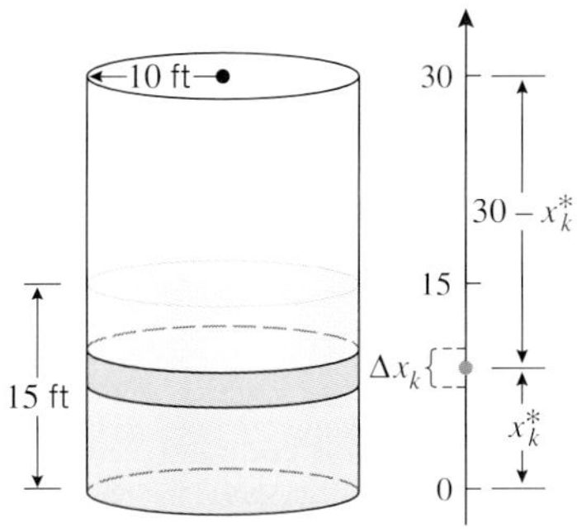

Figure 8.6.5

Solution. Our strategy will be to divide the water into thin layers, approximate the work required to move each layer to the top of the tank, add the approximations for the layers to obtain a Riemann sum that approximates the total work, and then take the limit of the Riemann sums to produce an integral for the total work.

To implement this idea, introduce an x-axis as shown in Figure 8.6.5, and divide the water into n layers with Δx_k denoting the thickness of the kth layer. Although the upper and lower surfaces of the kth layer are at different distances from the top, the difference will be small if the layer is thin, and we can reasonably assume that the entire layer is concentrated at a single point x_k^* (Figure 8.6.5). Thus, the work W_k required to move the kth layer to the top of the tank is approximately

$$W_k \approx F_k(30 - x_k^*) \tag{4}$$

where F_k is the force required to lift the kth layer. But the force required to lift the kth layer is the force needed to overcome gravity, and this is the same as the weight of the layer. To find the weight F_k of the kth layer we will multiply its volume by the weight density of water (62.4 lb/ft^3). This yields

$$F_k = (\pi(10)^2 \Delta x_k)(62.4) = 6240\pi \Delta x_k$$

Thus, from (4)

$$W_k \approx \underbrace{(30 - x_k^*)}_{\text{Distance}} \cdot \underbrace{6240\pi \Delta x_k}_{\text{Force}}$$

and hence the work W required to move all n layers is approximately

$$W = \sum_{k=1}^{n} W_k \approx \sum_{k=1}^{n} (30 - x_k^*)(6240\pi) \Delta x_k$$

To find the *exact* value of the work we take the limit as max $\Delta x_k \to 0$. This yields

$$W = \lim_{\max \Delta x_k \to 0} \sum_{k=1}^{n} (30 - x_k^*)(6240\pi) \Delta x_k = \int_0^{15} (30 - x)(6240\pi) \, dx$$

$$= 6240\pi \left(30x - \frac{x^2}{2} \right)\Bigg]_0^{15} = 2{,}106{,}000\pi \text{ ft·lb} \approx 6{,}616{,}194 \text{ ft·lb} \quad \blacktriangleleft$$

When you see an object in motion, you can be certain that somehow work has been expended to create that motion. For example, when you drop a stone from a building the stone gathers speed because the force of the Earth's gravity is performing work on it, and when a hockey player strikes a puck with a hockey stick, the work performed on the puck during the brief period of contact with the stick creates the enormous speed of the puck across the ice.

The linkage between work and motion is based on the concept of *kinetic energy*, which we will now define. If a particle of mass m is moving with speed v at a certain instant, then its **kinetic energy** K at that instant is defined as

$$K = \tfrac{1}{2} m v^2 \tag{5}$$

It is the following fundamental principle of physics that relates work and kinetic energy.

> **8.6.4** WORK–ENERGY THEOREM. *When a force does work on an object, it causes a change in the kinetic energy of the object that is equal to the work performed; that is, if W is the work performed on an object of mass m, and if the initial and final speeds of the object are v_i and v_f, respectively, then*
>
> $$W = \tfrac{1}{2}mv_f^2 - \tfrac{1}{2}mv_i^2 \qquad (6)$$

The units of kinetic energy are the same as the units of work. For example, in the Standard International system kinetic energy is measured in joules (J).

Example 6

A space probe of mass $m = 5.00 \times 10^4$ kg travels in deep space subjected only to the force of its own engine. Starting at a time when the speed of the probe is $v = 1.10 \times 10^4$ m/s, the engine is fired continuously over a distance of 2.50×10^6 m with a constant force of 4.00×10^5 N in the direction of motion. What is the final speed of the probe?

Solution. Since the force applied by the engine is constant and in the direction of motion, the work W expended by the engine on the probe is

$$W = \text{force} \times \text{distance} = (4.00 \times 10^5 \text{ N}) \times (2.50 \times 10^6 \text{ m}) = 1.00 \times 10^{12} \text{ J}$$

From (6), the final kinetic energy $K_f = \tfrac{1}{2}mv_f^2$ of the probe can be expressed in terms of the work W and the initial kinetic energy $K_i = \tfrac{1}{2}mv_i^2$ as

$$K_f = W + K_i$$

Thus, from the known mass and initial speed we have

$$K_f = (1.00 \times 10^{12} \text{ J}) + \tfrac{1}{2}(5.00 \times 10^4 \text{ kg})(1.10 \times 10^4 \text{ m/s})^2 \approx 4.03 \times 10^{12} \text{ J}$$

The final kinetic energy is $K_f = \tfrac{1}{2}mv_f^2$, so the final speed of the probe is

$$v_f = \sqrt{\frac{2K_f}{m}} = \sqrt{\frac{2(4.03 \times 10^{12})}{5.00 \times 10^4}} \approx 1.27 \times 10^4 \text{ m/s} \qquad \blacktriangleleft$$

EXERCISE SET 8.6

1. Find the work done when
   (a) a constant force of 30 lb in the positive x-direction moves an object from $x = -2$ to $x = 5$ ft
   (b) a variable force of $F(x) = 1/x^2$ lb in the positive x-direction moves an object from $x = 1$ to $x = 6$ ft.

2. A variable force $F(x)$ in the positive x-direction is graphed in the accompanying figure. Find the work done by the force on a particle that moves from $x = 0$ to $x = 5$.

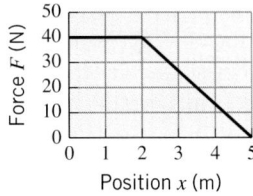

Position x (m)   Figure Ex-2

3. A constant force of 10 lb in the positive x-direction is applied to a particle whose velocity versus time curve is shown in the accompanying figure. Find the work done by the force on the particle from time $t = 0$ to $t = 5$.

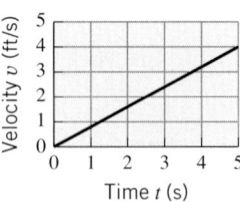

Time t (s)   Figure Ex-3

4. A spring whose natural length is 15 cm exerts a force of 45 N when stretched to a length of 20 cm.
   (a) Find the spring constant (in newtons/meter).
   (b) Find the work that is done in stretching the spring 3 cm beyond its natural length.
   (c) Find the work done in stretching the spring from a length of 20 cm to a length of 25 cm.

5. A spring exerts a force of 100 N when it is stretched 0.2 m beyond its natural length. How much work is required to stretch the spring 0.8 m beyond its natural length?

6. Assume that a force of 6 N is required to compress a spring from a natural length of 4 m to a length of $3\frac{1}{2}$ m. Find the work required to compress the spring from its natural length to a length of 2 m. (Hooke's law applies to compression as well as extension.)

7. Assume that 10 ft·lb of work is required to stretch a spring 1 ft beyond its natural length. What is the spring constant?

8. A cylindrical tank of radius 5 ft and height 9 ft is two-thirds filled with water. Find the work required to pump all the water over the upper rim.

9. Solve Exercise 8 assuming that the tank is two-thirds filled with a liquid that weighs ρ lb/ft^3.

10. A cone-shaped water reservoir is 20 ft in diameter across the top and 15 ft deep. If the reservoir is filled to a depth of 10 ft, how much work is required to pump all the water to the top of the reservoir?

11. The vat shown in the accompanying figure contains water to a depth of 2 m. Find the work required to pump all the water to the top of the vat. [Use 9810 N/m^3 as the weight density of water.]

12. The cylindrical tank shown in the accompanying figure is filled with a liquid weighing 50 lb/ft^3. Find the work required to pump all the liquid to a level 1 ft above the top of the tank.

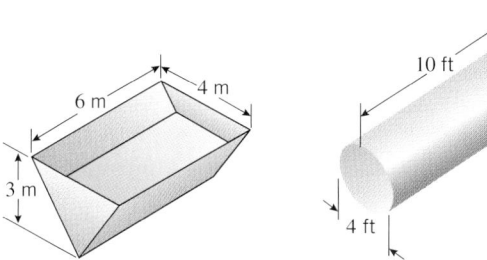

Figure Ex-11   Figure Ex-12

13. A swimming pool is built in the shape of a rectangular parallelepiped 10 ft deep, 15 ft wide, and 20 ft long.
   (a) If the pool is filled to 1 ft below the top, how much work is required to pump all the water into a drain at the top edge of the pool?
   (b) A one-horsepower motor can do 550 ft·lb of work per second. What size motor is required to empty the pool in 1 hour?

14. A rocket weighing 3 tons is filled with 40 tons of liquid fuel. In the initial part of the flight, fuel is burned off at a constant rate of 2 tons per 1000 ft of vertical height. How much work is done in lifting the rocket to 3000 ft?

15. A 100-ft length of steel chain weighing 15 lb/ft is dangling from a pulley. How much work is required to wind the chain onto the pulley?

16. It follows from Coulomb's law in physics that two like electrostatic charges repel each other with a force inversely proportional to the square of the distance between them. Suppose that two charges A and B repel with a force of k newtons when they are positioned at points $A(-a, 0)$ and

$B(a, 0)$, where a is measured in meters. Find the work W required to move charge A along the x-axis to the origin if charge B remains stationary.

17. It is a law of physics that the gravitational force exerted by the Earth on an object varies inversely as the square of its distance from the Earth's center. Thus, an object's weight $w(x)$ is related to its distance x from the Earth's center by a formula of the form

$$w(x) = \frac{k}{x^2}$$

where k is a constant of proportionality that depends on the mass of the object.
   (a) Use this fact and the assumption that the Earth is a sphere of radius 4000 mi to obtain the formula for $w(x)$ in Example 4.
   (b) Find a formula for the weight $w(x)$ of a satellite that is x mi from the Earth's surface if its weight is 6000 lb.
   (c) How much work is required to lift the satellite from the surface of the Earth to an orbital position that is 1000 mi high?

18. (a) The formula $w(x) = k/x^2$ in Exercise 17 is applicable to all celestial bodies. Assuming that the Moon is a sphere of radius 1080 mi, find the force that the Moon exerts on an astronaut who is x mi from the surface of the Moon if her weight on the Moon's surface is 20 lb.
   (b) How much work is required to lift the astronaut to a point that is 10.8 mi above the Moon's surface?

19. The Yamanashi Maglev Test Line in Japan that runs between Sakaigawa and Akiyama is currently testing magnetic levitation (MAGLEV) trains that are designed to levitate inches above powerful magnetic fields. Suppose that a MAGLEV train has a mass of $m = 4.00 \times 10^5$ kg and that starting at a time when the train has a speed of 20 m/s the engine applies a force of 6.40×10^5 N in the direction of motion over a distance of 3.00×10^3 m. Use the Work–Energy Theorem (8.6.4) to find the final speed of the train.

20. Assume that a Mars probe of mass $m = 2.00 \times 10^3$ kg is subjected only to the force of its own engine. Starting at a time when the speed of the probe is $v = 1.00 \times 10^4$ m/s, the engine is fired continuously over a distance of 2.00×10^5 m with a constant force of 2.00×10^5 N in the direction of motion. Use the Work–Energy Theorem (8.6.4) to find the final speed of the probe.

21. On August 10, 1972 a meteorite with an estimated mass of 4×10^6 kg and an estimated speed of 15 km/s skipped across the atmosphere above the western United States and Canada but fortunately did not hit the Earth.
   (a) Assuming that the meteorite had hit the Earth with a speed of 15 km/s, what would have been its change in kinetic energy in joules (J)?
   (b) Express the energy as a multiple of the explosive energy of 1 megaton of TNT, which is 4.2×10^{15} J.
   (c) The energy associated with the Hiroshima atomic bomb was 13 kilotons of TNT. To how many such bombs would the meteorite impact have been equivalent?

8.7 FLUID PRESSURE AND FORCE

In this section we will use the integration tools developed in the preceding chapter to study the pressures and forces exerted by fluids on submerged objects.

WHAT IS A FLUID?

A *fluid* is a substance that flows to conform to the boundaries of any container in which it is placed. Fluids include *liquids*, such as water, oil, and mercury, as well as *gases*, such as helium, oxygen, and air. The study of fluids falls into two categories: *fluid statics* (the study of fluids at rest) and *fluid dynamics* (the study of fluids in motion). In this section we will be concerned only with fluid statics; toward the end of this text we will investigate problems in fluid dynamics.

THE CONCEPT OF PRESSURE

The effect that a force has on an object depends on how that force is spread over the surface of the object. For example, when you walk on soft snow with boots, the weight of your body crushes the snow and you sink into it. However, if you put on a pair of skis to spread the weight of your body over a greater surface area, then the weight of your body has less of a crushing effect on the snow, and you are able to glide across the surface. The concept that accounts for both the magnitude of a force and the area over which it is applied is called *pressure*.

> **8.7.1** DEFINITION. If a force of magnitude F is applied to a surface of area A, then we define the *pressure* P exerted by the force on the surface to be
>
> $$P = \frac{F}{A} \tag{1}$$

It follows from this definition that pressure has units of force per unit area. The most common units of pressure are newtons per square meter (N/m^2) in SI and pounds per square inch (lb/in^2) or pounds per square foot (lb/ft^2) in the BE system. As indicated in Table 8.7.1, one newton per square meter is called a *pascal*[*] (Pa). A pressure of 1 Pa is quite small ($1\ Pa = 1.45 \times 10^{-4}\ lb/in^2$), so in countries using SI, tire pressure gauges are usually calibrated in kilopascals (kPa), which is 1000 pascals.

In this section we will be interested in pressures and forces on objects submerged in fluids. Pressures themselves have no directional characteristics, but the forces that they create always act perpendicular to the face of the submerged object. Thus, in Figure 8.7.1 the water pressure creates horizontal forces on the sides of the tank, vertical forces on the bottom of the tank, and forces that vary in direction, so as to be perpendicular to the different parts of the swimmer's body.

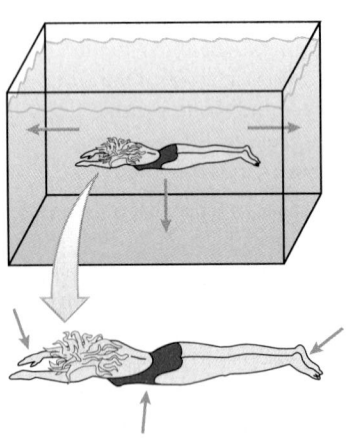

Fluid forces always act perpendicular to the surface of a submerged object.

Figure 8.7.1

---

[*] BLAISE PASCAL (1623–1662). French mathematician and scientist. Pascal's mother died when he was three years old and his father, a highly educated magistrate, personally provided the boy's early education. Although Pascal showed an inclination for science and mathematics, his father refused to tutor him in those subjects until he mastered Latin and Greek. Pascal's sister and primary biographer claimed that he independently discovered the first thirty-two propositions of Euclid without ever reading a book on geometry. (However, it is generally agreed that the story is apocryphal.) Nevertheless, the precocious Pascal published a highly respected essay on conic sections by the time he was sixteen years old. Descartes, who read the essay, thought it so brilliant that he could not believe that it was written by such a young man. By age 18 his health began to fail and until his death he was in frequent pain. However, his creativity was unimpaired.

Pascal's contributions to physics include the discovery that air pressure decreases with altitude and the principle of fluid pressure that bears his name. However, the originality of his work is questioned by some historians. Pascal made major contributions to a branch of mathematics called "projective geometry," and he helped to develop probability theory through a series of letters with Fermat.

In 1646, Pascal's health problems resulted in a deep emotional crisis that led him to become increasingly concerned with religious matters. Although born a Catholic, he converted to a religious doctrine called Jansenism and spent most of his final years writing on religion and philosophy.

Table 8.7.1

SYSTEM	FORCE	÷	AREA	=	PRESSURE
SI	newton (N)		square meter (m^2)		pascal (Pa)
BE	pound (lb)		square foot (ft^2)		lb/ft^2
BE	pound (lb)		square inch (in^2)		lb/in^2 (psi)

CONVERSION FACTORS:
$1 \text{ Pa} \approx 1.45 \times 10^{-4} \text{ lb/in}^2 \approx 2.09 \times 10^{-2} \text{ lb/ft}^2$
$1 \text{ lb/in}^2 \approx 6.90 \times 10^3 \text{ Pa}$ $1 \text{ lb/ft}^2 \approx 47.9 \text{ Pa}$

Example 1

Referring to Figure 8.7.1, suppose that the back of the swimmer's hand has a surface area of 8.4×10^{-3} m^2 and that the pressure acting on it is 1.2×10^5 Pa (a realistic value near the bottom of a deep diving pool). Find the force that acts on the swimmer's hand.

Solution. From (1), the force F is

$$F = PA = (1.2 \times 10^5 \text{ N/m}^2)(8.4 \times 10^{-3} \text{ m}^2) \approx 1.0 \times 10^3 \text{ N}$$

This is quite a large force (about 230 lb in the BE system). ◀

FLUID DENSITY

Table 8.7.2

WEIGHT DENSITIES

SI	N/m^3
Machine oil	4,708
Gasoline	6,602
Fresh water	9,810
Seawater	10,045
Mercury	133,416

BE SYSTEM	lb/ft^3
Machine oil	30.0
Gasoline	42.0
Fresh water	62.4
Seawater	64.0
Mercury	849.0

All densities are affected by variations in temperature and pressure. Weight densities are affected by variations in g.

FLUID PRESSURE

Scuba divers know that the deeper they dive, the greater the pressure and the forces that they feel on their bodies. This sense of pressure and force is caused by the weight of the water and air above—the deeper the diver goes, the greater the weight above and hence the greater the pressure and force that he or she feels.

To calculate pressures and forces on submerged objects, we need to know something about the characteristics of the fluids in which they are submerged. For simplicity, we will assume that the fluids under consideration are *homogeneous*, by which we mean that any two samples of the fluid with the same volume have the same mass. It follows from this assumption that the mass per unit volume is a constant δ that depends on the physical characteristics of the fluid but not on the size or location of the sample; we call

$$\delta = \frac{m}{V} \tag{2}$$

the ***mass density*** of the fluid. Sometimes it is more convenient to work with weight per unit volume than with mass per unit volume. Thus, we define the ***weight density*** ρ of a fluid to be

$$\rho = \frac{w}{V} \tag{3}$$

where w is the weight of a fluid sample of volume V. Thus, if the weight density of a fluid is known, then the weight w of a fluid sample of volume V can be computed from the formula $w = \rho V$. Table 8.7.2 shows some typical weight densities.

To calculate fluid pressures and forces we will need ***Pascal's principle***, which states that *fluid pressure at a given depth is the same in all directions* (Figure 8.7.2). This implies, for example, that at the bottom corner of a swimming pool the pressure on the two side walls is the same as the pressure on the bottom.

It is a straightforward matter to calculate fluid force and pressure on a flat surface that is submerged *horizontally* because each point on the surface is at the same depth. If a flat surface of area A is submerged horizontally at a depth h in a container of fluid with weight density ρ, then the fluid exerts a force F that is perpendicular to the surface and is

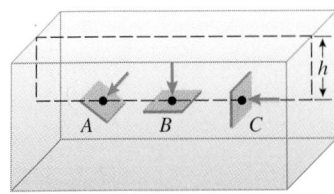

By Pascal's principle the pressure at points A, B, and C is the same.

Figure 8.7.2

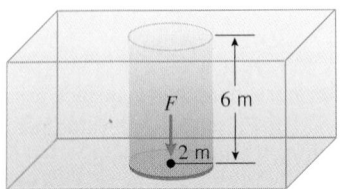

The fluid force is the fluid pressure times the area.

Figure 8.7.3

FLUID FORCE ON A VERTICAL SURFACE

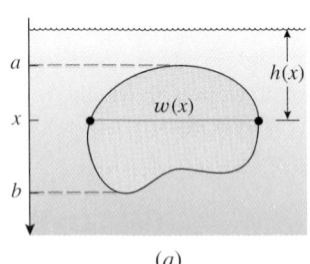

(a)

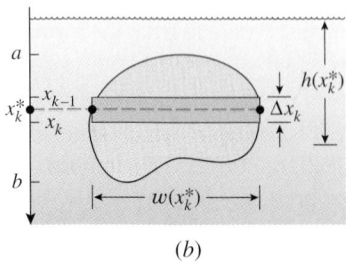

(b)

Figure 8.7.4

given by

$$F = \rho h A \tag{4}$$

Thus, it follows from (1) that the pressure P on the surface created by this force is

$$P = \frac{F}{A} = \rho h \tag{5}$$

Example 2

Find the fluid pressure and force on the top of a flat circular plate of radius 2 m that is submerged horizontally in water at a depth of 6 m (Figure 8.7.3).

Solution. Since the weight density of water is $\rho = 9810 \text{ N/m}^3$, it follows from (5) that the fluid pressure is

$$P = \rho h = (9810)(6) = 58,860 \text{ Pa}$$

and it follows from (4) that the fluid force is

$$F = \rho h A = \rho h(\pi r^2) = (9810)(6)(4\pi) = 235,440\pi \approx 739,657 \text{ N} \quad \blacktriangleleft$$

It was easy to calculate the fluid force on the horizontal plate in Example 2 because each point on the plate was at the same depth. The problem of finding the fluid force on a vertical surface is more complicated because the depth, and hence the pressure, is not constant over the surface. To find the fluid force on a vertical surface we will need calculus.

> **8.7.2** PROBLEM. Suppose that a flat surface is immersed vertically in a fluid of weight density ρ and that the submerged portion of the surface extends from $x = a$ to $x = b$ along an x-axis whose positive direction is down (Figure 8.7.4a). For $a \le x \le b$, suppose that $w(x)$ is the width of the surface and that $h(x)$ is the depth of the point x. Define what is meant by the *fluid force* F on the surface, and find a formula for computing it.

The basic idea for solving this problem is to break up the interval $[a, b]$ into subintervals that are sufficiently small that the depth does not vary much on each subinterval. This has the effect of dividing the plate into strips over each of which the depth can be treated as constant. This assumption will allow us to use Formula (4) to approximate the fluid force on each strip. By adding the approximations to the forces on the strips we will obtain a Riemann sum that approximates the total force F on the entire surface, and by taking the limit of the Riemann sums we will obtain an integral for F.

To implement this idea, divide the interval $[a, b]$ into n subintervals by inserting points $x_1, x_2, \ldots, x_{n-1}$ between a and b. To approximate the force on the kth strip we choose any point x_k^* in the kth interval and approximate the strip by a rectangle of length $w(x_k^*)$ and width $\Delta x_k = x_k - x_{k-1}$ (Figure 8.7.4b).

Although the top and bottom of the rectangle are at different depths, the difference will be small if the strip is thin and we can reasonably assume that the entire strip is at depth $h(x_k^*)$. Thus, from (4) we can approximate the force F_k on the kth strip as

$$F_k \approx \rho \underbrace{h(x_k^*)}_{\text{Depth}} \cdot \underbrace{w(x_k^*)\Delta x_k}_{\text{Area of rectangle}}$$

Adding these approximations yields the following Riemann sum that approximates the total force F on the surface:

$$F = \sum_{k=1}^{n} F_k \approx \sum_{k=1}^{n} \rho h(x_k^*) w(x_k^*) \Delta x_k$$

Taking the limit as n increases and the widths of the subintervals approach zero yields the definite integral

$$F = \lim_{\max \Delta x_k \to 0} \sum_{k=1}^{n} \rho h(x_k^*) w(x_k^*) \Delta x_k = \int_a^b \rho h(x) w(x)\, dx$$

In summary, we have the following result:

8.7.3   DEFINITION.   Suppose that a flat surface is immersed vertically in a fluid of weight density ρ and that the submerged portion of the surface extends from $x = a$ to $x = b$ along an x-axis whose positive direction is down (Figure 8.7.4b). For $a \le x \le b$, suppose that $w(x)$ is the width of the surface and that $h(x)$ is the depth of the point x. Then we define the **fluid force** F on the surface to be

$$F = \int_a^b \rho h(x) w(x)\, dx \tag{6}$$

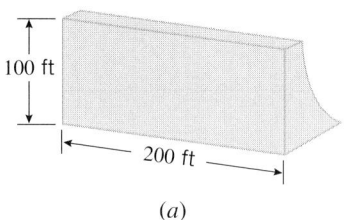

(a)

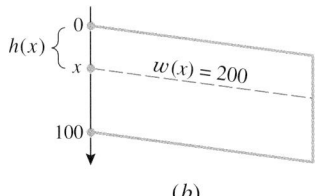

(b)

Figure 8.7.5

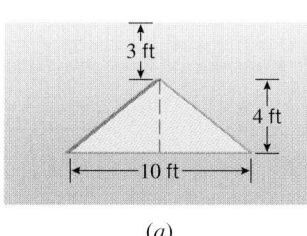

Example 3

The face of a dam is a vertical rectangle of height 100 ft and width 200 ft (Figure 8.7.5a). Find the total fluid force exerted on the face when the water surface is level with the top of the dam.

Solution.   Introduce an x-axis with its origin at the water surface as shown in Figure 8.7.5b. At a point x on this axis, the width of the dam in feet is $w(x) = 200$ and the depth in feet is $h(x) = x$. Thus, from (6) with $\rho = 62.4\ \text{lb/ft}^3$ (the weight density of water) we obtain as the total force on the face

$$F = \int_0^{100} (62.4)(x)(200)\, dx = 12{,}480 \int_0^{100} x\, dx = 12{,}480\, \frac{x^2}{2} \Big]_0^{100} = 62{,}400{,}000\ \text{lb}$$

◀

Example 4

A plate in the form of an isosceles triangle with base 10 ft and altitude 4 ft is submerged vertically in machine oil as shown in Figure 8.7.6a. Find the fluid force F against the plate surface if the oil has weight density $\rho = 30\ \text{lb/ft}^3$.

Solution.   Introduce an x-axis as shown in Figure 8.7.6b. By similar triangles, the width of the plate, in feet, at a depth of $h(x) = (3 + x)$ ft satisfies

$$\frac{w(x)}{10} = \frac{x}{4}, \quad \text{so} \quad w(x) = \frac{5}{2}x$$

Thus, it follows from (6) that the force on the plate is

$$F = \int_a^b \rho h(x) w(x)\, dx = \int_0^4 (30)(3 + x) \left(\frac{5}{2} x \right) dx$$

$$= 75 \int_0^4 (3x + x^2)\, dx = 75 \left[\frac{3x^2}{2} + \frac{x^3}{3} \right]_0^4 = 3400\ \text{lb}$$

◀

Figure 8.7.6

In this exercise set, refer to Table 8.7.2 for weight densities of fluids, when needed.

1.   A flat rectangular plate is submerged horizontally in a liquid.
(a)   Find the force (in lb) and the pressure (in lb/ft^2) on the top surface of the plate if its area is 100 ft^2, the liquid is water, and the surface is at a depth of 5 ft.

(b) Find the force (in N) and the pressure (in Pa) on the top surface of the plate if its area is 25 m², the liquid is water, and the surface is at a depth of 10 m.

2. (a) Find the force (in N) on the deck of a sunken ship if its area is 160 m² and the pressure acting on it is 6.0×10^5 Pa.

(b) Find the force (in lb) on a diver's face mask if its area is 60 in² and the pressure acting on it is 100 lb/in².

In Exercises 3–8, the flat surfaces shown are submerged vertically in water. Find the fluid force against the surface.

3.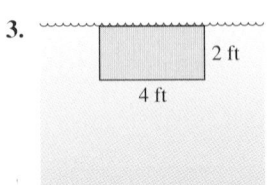
2 ft
4 ft

4.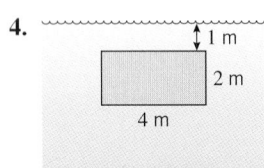
1 m
2 m
4 m

5.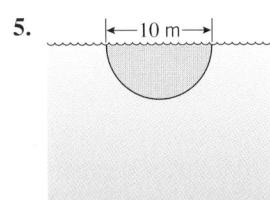
—10 m—

6.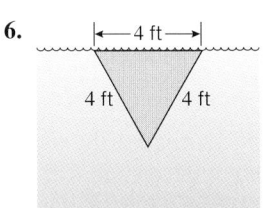
—4 ft—
4 ft   4 ft

7.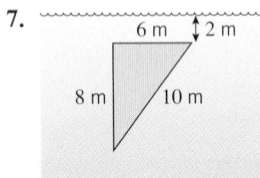
6 m   2 m
8 m   10 m

8.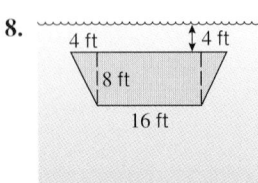
4 ft   4 ft
8 ft
16 ft

9. Suppose that a flat surface is immersed vertically in a fluid of weight density ρ. If ρ is doubled, is the force on the plate also doubled? Explain your reasoning.

10. An oil tank is shaped like a right circular cylinder of diameter 4 ft. Find the total fluid force against one end when the axis is horizontal and the tank is half filled with oil of weight density 50 lb/ft³.

11. A square plate of side a feet is dipped in a liquid of weight density ρ lb/ft³. Find the fluid force on the plate if a vertex is at the surface and a diagonal is perpendicular to the surface.

12. The accompanying figure shows a dam whose face is an inclined rectangle. Find the fluid force on the face when the water is level with the top of this dam.

13. The accompanying figure shows a rectangular swimming pool whose bottom is an inclined plane. Find the fluid force on the bottom when the pool is filled to the top.

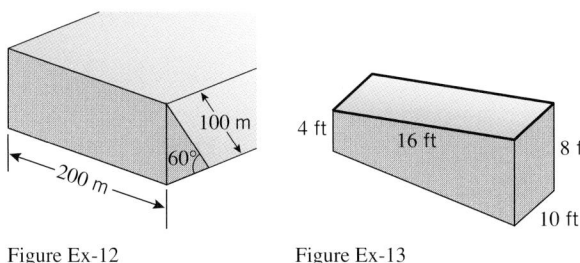

Figure Ex-12   Figure Ex-13

14. An observation window on a submarine is a square with 2-ft sides. Using ρ_0 for the weight density of seawater, find the fluid force on the window when the submarine has descended so that the window is vertical and its top is at a depth of h feet.

15. (a) Show: If the submarine in Exercise 14 descends vertically at a constant rate, then the fluid force on the window increases at a constant rate.

(b) At what rate is the force on the window increasing if the submarine is descending vertically at 20 ft/min?

8.8 HYPERBOLIC FUNCTIONS AND HANGING CABLES

In this section we will study certain combinations of e^x and e^{-x}, called "hyperbolic functions." These functions, which arise in various engineering applications, have many properties in common with the trigonometric functions. This similarity is somewhat surprising, since there is little on the surface to suggest that there should be any relationship between exponential and trigonometric functions. This is because the relationship occurs within the context of complex numbers, a topic which we will leave for more advanced courses.

DEFINITIONS OF HYPERBOLIC FUNCTIONS

To introduce the hyperbolic functions, observe that the function e^x can be expressed in the following way as the sum of an even function and an odd function:

$$e^x = \underbrace{\frac{e^x + e^{-x}}{2}}_{\text{Even}} + \underbrace{\frac{e^x - e^{-x}}{2}}_{\text{Odd}}$$

These functions are sufficiently important that there are names and notation associated with them: the odd function is called the *hyperbolic sine* of x and the even function is called the *hyperbolic cosine* of x. They are denoted by

$$\sinh x = \frac{e^x - e^{-x}}{2} \quad \text{and} \quad \cosh x = \frac{e^x + e^{-x}}{2}$$

where sinh is pronounced "cinch" and cosh rhymes with "gosh." From these two building blocks we can create four more functions to produce the following set of six **hyperbolic functions**.

8.8.1 DEFINITION.

Hyperbolic sine	$\sinh x = \dfrac{e^x - e^{-x}}{2}$
Hyperbolic cosine	$\cosh x = \dfrac{e^x + e^{-x}}{2}$
Hyperbolic tangent	$\tanh x = \dfrac{\sinh x}{\cosh x} = \dfrac{e^x - e^{-x}}{e^x + e^{-x}}$
Hyperbolic cotangent	$\coth x = \dfrac{\cosh x}{\sinh x} = \dfrac{e^x + e^{-x}}{e^x - e^{-x}}$
Hyperbolic secant	$\operatorname{sech} x = \dfrac{1}{\cosh x} = \dfrac{2}{e^x + e^{-x}}$
Hyperbolic cosecant	$\operatorname{csch} x = \dfrac{1}{\sinh x} = \dfrac{2}{e^x - e^{-x}}$

REMARK.   The terms "tanh," "sech," and "csch" are pronounced "tanch," "seech," and "coseech," respectively.

Example 1

$$\sinh 0 = \frac{e^0 - e^{-0}}{2} = \frac{1 - 1}{2} = 0$$

$$\cosh 0 = \frac{e^0 + e^{-0}}{2} = \frac{1 + 1}{2} = 1$$

$$\sinh 2 = \frac{e^2 - e^{-2}}{2} \approx 3.6269 \qquad\qquad \blacktriangleleft$$

GRAPHS OF THE HYPERBOLIC FUNCTIONS

The graphs of the hyperbolic functions, which are shown in Figure 8.8.1, can be generated with a graphing utility, but it is worthwhile to observe that the general shape of the graph of $y = \cosh x$ can be obtained by sketching the graphs of $y = \frac{1}{2}e^x$ and $y = \frac{1}{2}e^{-x}$ separately and adding the corresponding y-coordinates [see part (a) of the figure]. Similarly, the general shape of the graph of $y = \sinh x$ can be obtained by sketching the graphs of $y = \frac{1}{2}e^x$ and $y = -\frac{1}{2}e^{-x}$ separately and adding corresponding y-coordinates [see part (b) of the figure].

Observe that $\sinh x$ has a domain of $(-\infty, +\infty)$ and a range of $(-\infty, +\infty)$, whereas $\cosh x$ has a domain of $(-\infty, +\infty)$ and a range of $[1, +\infty)$. Observe also that $y = \frac{1}{2}e^x$ and $y = \frac{1}{2}e^{-x}$ are **curvilinear asymptotes** for $y = \cosh x$ in the sense that the graph of $y = \cosh x$ gets closer and closer to the graph of $y = \frac{1}{2}e^x$ as $x \to +\infty$ and gets closer and closer to the graph of $y = \frac{1}{2}e^{-x}$ as $x \to -\infty$. Similarly, $y = \frac{1}{2}e^x$ is a curvilinear asymptote for $y = \sinh x$ as $x \to +\infty$ and $y = -\frac{1}{2}e^{-x}$ is a curvilinear asymptote as $x \to -\infty$. Other properties of the hyperbolic functions are explored in the exercises.

The design of the Gateway Arch near St. Louis is based on an inverted hyperbolic cosine curve.

Figure 8.8.1

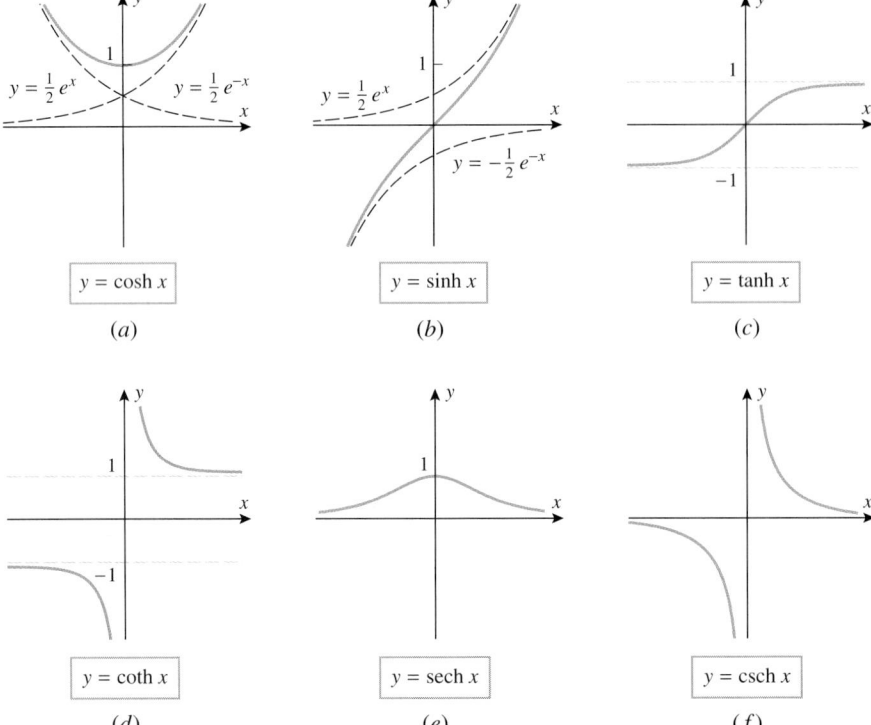

$y = \cosh x$

(a)

$y = \sinh x$

(b)

$y = \tanh x$

(c)

$y = \coth x$

(d)

$y = \operatorname{sech} x$

(e)

$y = \operatorname{csch} x$

(f)

HANGING CABLES AND OTHER APPLICATIONS

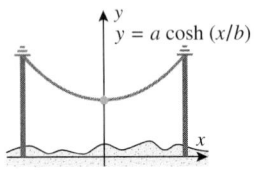

Figure 8.8.2

Hyperbolic functions arise in vibratory motions inside elastic solids and more generally in many problems where mechanical energy is gradually absorbed by a surrounding medium. They also occur when a homogeneous, flexible cable is suspended between two points, as with a telephone line hanging between two poles. Such a cable forms a curve, called a **catenary** (from the Latin *catena*, meaning "chain"). If, as in Figure 8.8.2, a coordinate system is introduced so that the low point of the cable lies on the y-axis, then it can be shown using principles of physics that the cable has an equation of the form

$$y = a \cosh \left(\frac{x}{b} \right)$$

HYPERBOLIC IDENTITIES

The hyperbolic functions satisfy various identities that are similar to identities for trigonometric functions. The most fundamental of these is

$$\cosh^2 x - \sinh^2 x = 1 \tag{1}$$

which can be proved by writing

$$\cosh^2 x - \sinh^2 x = \left(\frac{e^x + e^{-x}}{2} \right)^2 - \left(\frac{e^x - e^{-x}}{2} \right)^2$$

$$= \tfrac{1}{4}(e^{2x} + 2e^0 + e^{-2x}) - \tfrac{1}{4}(e^{2x} - 2e^0 + e^{-2x})$$

$$= 1$$

Other hyperbolic identities can be derived in a similar manner or, alternatively, by performing algebraic operations on known identities. For example, if we divide (1) by $\cosh^2 x$, we obtain

$$1 - \tanh^2 x = \operatorname{sech}^2 x$$

and if we divide (1) by $\sinh^2 x$, we obtain

$$\coth^2 x - 1 = \operatorname{csch}^2 x$$

The following theorem summarizes some of the more useful hyperbolic identities. The proofs of those not already obtained are left as exercises.

8.8.2   THEOREM.

$$\cosh x + \sinh x = e^x$$

$$\cosh x - \sinh x = e^{-x}$$

$$\cosh^2 x - \sinh^2 x = 1$$

$$1 - \tanh^2 x = \operatorname{sech}^2 x$$

$$\coth^2 x - 1 = \operatorname{csch}^2 x$$

$$\cosh(-x) = \cosh x$$

$$\sinh(-x) = -\sinh x$$

$$\sinh(x + y) = \sinh x \cosh y + \cosh x \sinh y$$

$$\cosh(x + y) = \cosh x \cosh y + \sinh x \sinh y$$

$$\sinh(x - y) = \sinh x \cosh y - \cosh x \sinh y$$

$$\cosh(x - y) = \cosh x \cosh y - \sinh x \sinh y$$

$$\sinh 2x = 2 \sinh x \cosh x$$

$$\cosh 2x = \cosh^2 x + \sinh^2 x$$

$$\cosh 2x = 2 \sinh^2 x + 1$$

$$\cosh 2x = 2 \cosh^2 x - 1$$

WHY THEY ARE CALLED HYPERBOLIC FUNCTIONS

Recall that the parametric equations

$$x = \cos t, \quad y = \sin t \qquad (0 \le t \le 2\pi)$$

represent the unit circle $x^2 + y^2 = 1$ (Figure 8.8.3a), as may be seen by writing

$$x^2 + y^2 = \cos^2 t + \sin^2 t = 1$$

If $0 \le t \le 2\pi$, then the parameter t can be interpreted as the angle in radians from the positive x-axis to the point $(\cos t, \sin t)$ or, alternatively, as twice the shaded area of the sector in Figure 8.8.3a (verify). Analogously, the parametric equations

$$x = \cosh t, \quad y = \sinh t \qquad (-\infty < t < +\infty)$$

represent a portion of the curve $x^2 - y^2 = 1$, as may be seen by writing

$$x^2 - y^2 = \cosh^2 t - \sinh^2 t = 1$$

and observing that $x = \cosh t > 0$. This curve, which is shown in Figure 8.8.3b, is the right half of a larger curve called the **unit hyperbola**; this is the reason why the functions in this section are called *hyperbolic* functions. It can be shown that if $t \ge 0$, then the parameter t can be interpreted as twice the shaded area in Figure 8.8.3b (We omit the details.)

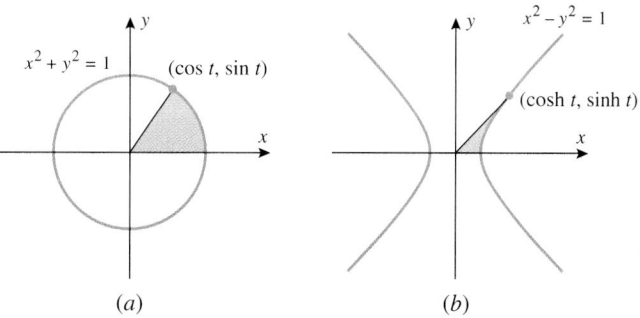

(a)   *(b)*

Figure 8.8.3

DERIVATIVE AND INTEGRAL FORMULAS

Derivative formulas for $\sinh x$ and $\cosh x$ can be obtained by expressing these functions in terms of e^x and e^{-x}:

$$\frac{d}{dx}[\sinh x] = \frac{d}{dx}\left[\frac{e^x - e^{-x}}{2}\right] = \frac{e^x + e^{-x}}{2} = \cosh x$$

$$\frac{d}{dx}[\cosh x] = \frac{d}{dx}\left[\frac{e^x + e^{-x}}{2}\right] = \frac{e^x - e^{-x}}{2} = \sinh x$$

Derivatives of the remaining hyperbolic functions can be obtained by expressing them in terms of sinh and cosh and applying appropriate identities. For example,

$$\frac{d}{dx}[\tanh x] = \frac{d}{dx}\left[\frac{\sinh x}{\cosh x}\right] = \frac{\cosh x \dfrac{d}{dx}[\sinh x] - \sinh x \dfrac{d}{dx}[\cosh x]}{\cosh^2 x}$$

$$= \frac{\cosh^2 x - \sinh^2 x}{\cosh^2 x} = \frac{1}{\cosh^2 x} = \mathrm{sech}^2\, x$$

The following theorem provides a complete list of the generalized derivative formulas and corresponding integration formulas for the hyperbolic functions.

8.8.3   THEOREM.

$$\frac{d}{dx}[\sinh u] = \cosh u \frac{du}{dx} \qquad\qquad \int \sinh u\, du = \cosh u + C$$

$$\frac{d}{dx}[\cosh u] = \sinh u \frac{du}{dx} \qquad\qquad \int \cosh u\, du = \sinh u + C$$

$$\frac{d}{dx}[\tanh u] = \mathrm{sech}^2\, u \frac{du}{dx} \qquad\qquad \int \mathrm{sech}^2\, u\, du = \tanh u + C$$

$$\frac{d}{dx}[\coth u] = -\mathrm{csch}^2\, u \frac{du}{dx} \qquad\qquad \int \mathrm{csch}^2\, u\, du = -\coth u + C$$

$$\frac{d}{dx}[\mathrm{sech}\, u] = -\mathrm{sech}\, u \tanh u \frac{du}{dx} \qquad \int \mathrm{sech}\, u \tanh u\, du = -\mathrm{sech}\, u + C$$

$$\frac{d}{dx}[\mathrm{csch}\, u] = -\mathrm{csch}\, u \coth u \frac{du}{dx} \qquad \int \mathrm{csch}\, u \coth u\, du = -\mathrm{csch}\, u + C$$

Example 2

$$\frac{d}{dx}[\cosh(x^3)] = \sinh(x^3) \cdot \frac{d}{dx}[x^3] = 3x^2 \sinh(x^3)$$

$$\frac{d}{dx}[\ln(\tanh x)] = \frac{1}{\tanh x} \cdot \frac{d}{dx}[\tanh x] = \frac{\mathrm{sech}^2\, x}{\tanh x} \qquad\qquad \blacktriangleleft$$

Example 3

$$\int \sinh^5 x \cosh x\, dx = \tfrac{1}{6} \sinh^6 x + C \qquad \boxed{\begin{array}{l} u = \sinh x \\ du = \cosh x\, dx \end{array}}$$

$$\int \tanh x\, dx = \int \frac{\sinh x}{\cosh x} dx$$

$$= \ln|\cosh x| + C \qquad \boxed{\begin{array}{l} u = \cosh x \\ du = \sinh x\, dx \end{array}}$$

$$= \ln(\cosh x) + C$$

We were justified in dropping the absolute value signs since $\cosh x > 0$ for all x. $\qquad \blacktriangleleft$

Example 4

Find the length of the catenary $y = 10\cosh(x/10)$ from $x = -10$ to $x = 10$ (Figure 8.8.4).

Solution.   From Formula (4) of Section 8.4, the length L of the catenary is

$$L = \int_{-10}^{10} \sqrt{1 + \left(\frac{dy}{dx}\right)^2}\, dx$$

$$= 2 \int_{0}^{10} \sqrt{1 + \left(\frac{dy}{dx}\right)^2}\, dx \qquad \boxed{\begin{array}{l}\text{By symmetry} \\ \text{about the } y\text{-axis}\end{array}}$$

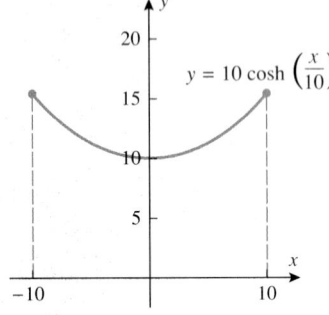

$y = 10\cosh\left(\dfrac{x}{10}\right)$

Figure 8.8.4

$$= 2 \int_0^{10} \sqrt{1 + \sinh^2 \left(\frac{x}{10}\right)} \, dx$$

$$= 2 \int_0^{10} \cosh \left(\frac{x}{10}\right) dx \qquad \boxed{\begin{array}{l}\text{By (1) and the fact}\\ \text{that } \cosh x > 0 \end{array}}$$

$$= 20 \sinh \left(\frac{x}{10}\right) \Big]_0^{10}$$

$$= 20[\sinh 1 - \sinh 0] = 20 \sinh 1 = 20 \left(\frac{e - e^{-1}}{2}\right) \approx 23.50 \qquad \blacktriangleleft$$

REMARK.   Computer algebra systems, such as *Mathematica*, *Maple*, and *Derive* have built-in capabilities for evaluating hyperbolic functions directly, but some calculators do not. However, if you need to evaluate a hyperbolic function on a calculator, you can do so by expressing it in terms of exponential functions, as in this example.

INVERSES OF HYPERBOLIC FUNCTIONS

Referring to Figure 8.8.1, it is evident that the graphs of $\sinh x$, $\tanh x$, $\coth x$, and $\operatorname{csch} x$ pass the horizontal line test, but the graphs of $\cosh x$ and $\operatorname{sech} x$ do not. In the latter case restricting x to be nonnegative makes the functions invertible (Figure 8.8.5). The graphs of the six inverse hyperbolic functions in Figure 8.8.6 were obtained by reflecting the graphs of the hyperbolic functions (with the appropriate restrictions) about the line $y = x$.

   Table 8.8.1 summarizes the basic properties of the inverse hyperbolic functions. You should confirm that the domains and ranges listed in this table agree with the graphs in Figure 8.8.6.

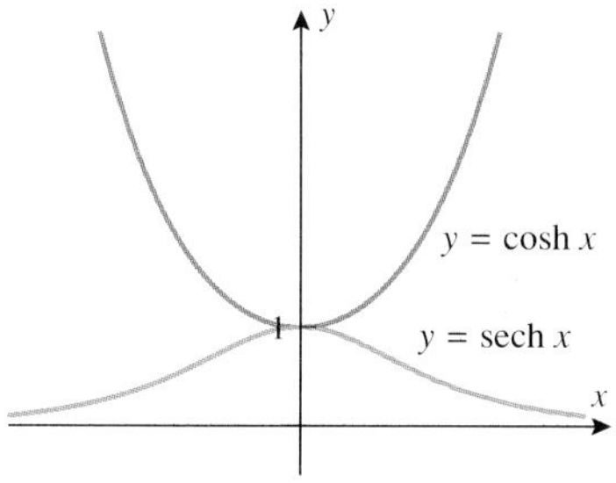

$y = \cosh x$

$y = \operatorname{sech} x$

With the restriction that $x \geq 0$, the curves $y = \cosh x$ and $y = \operatorname{sech} x$ pass the horizontal line test.

Figure 8.8.5

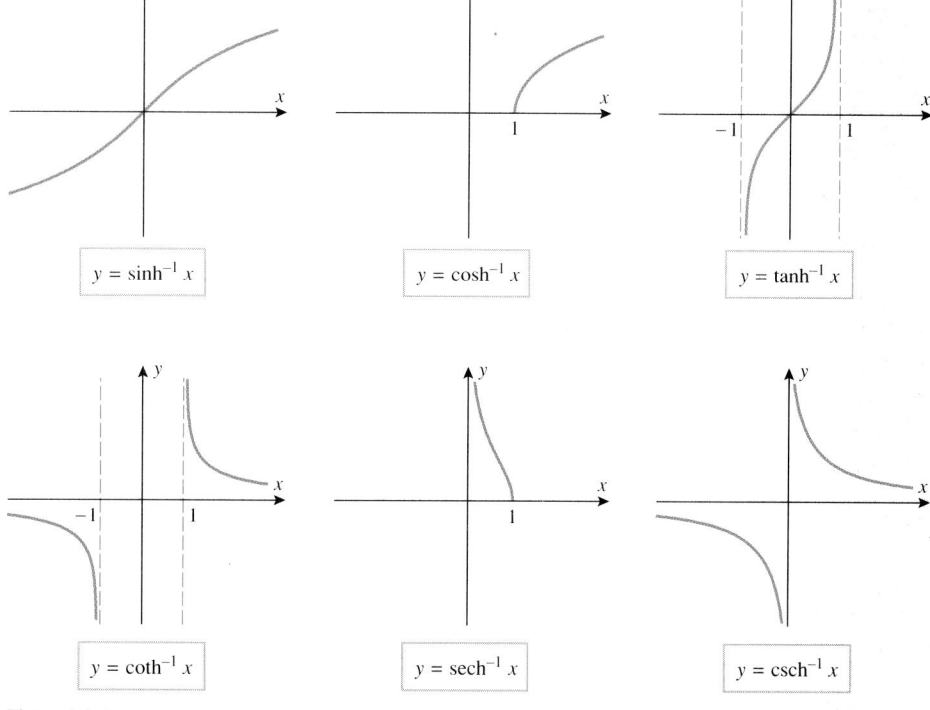

$y = \sinh^{-1} x$

$y = \cosh^{-1} x$

$y = \tanh^{-1} x$

$y = \coth^{-1} x$

$y = \operatorname{sech}^{-1} x$

$y = \operatorname{csch}^{-1} x$

Figure 8.8.6

LOGARITHMIC FORMS OF INVERSE HYPERBOLIC FUNCTIONS

Because the hyperbolic functions are expressible in terms of e^x, it should not be surprising that the inverse hyperbolic functions are expressible in terms of natural logarithms; the next theorem shows that this is so.

Table 8.8.1

FUNCTION	DOMAIN	RANGE	BASIC RELATIONSHIPS
$\sinh^{-1} x$	$(-\infty, +\infty)$	$(-\infty, +\infty)$	$\sinh^{-1}(\sinh x) = x$ if $-\infty < x < +\infty$ $\sinh(\sinh^{-1} x) = x$ if $-\infty < x < +\infty$
$\cosh^{-1} x$	$[1, +\infty)$	$[0, +\infty)$	$\cosh^{-1}(\cosh x) = x$ if $x \geq 0$ $\cosh(\cosh^{-1} x) = x$ if $x \geq 1$
$\tanh^{-1} x$	$(-1, 1)$	$(-\infty, +\infty)$	$\tanh^{-1}(\tanh x) = x$ if $-\infty < x < +\infty$ $\tanh(\tanh^{-1} x) = x$ if $-1 < x < 1$
$\coth^{-1} x$	$(-\infty, -1) \cup (1, +\infty)$	$(-\infty, 0) \cup (0, +\infty)$	$\coth^{-1}(\coth x) = x$ if $x < 0$ or $x > 0$ $\coth(\coth^{-1} x) = x$ if $x < -1$ or $x > 1$
$\text{sech}^{-1} x$	$(0, 1]$	$[0, +\infty)$	$\text{sech}^{-1}(\text{sech}\, x) = x$ if $x \geq 0$ $\text{sech}(\text{sech}^{-1} x) = x$ if $0 < x \leq 1$
$\text{csch}^{-1} x$	$(-\infty, 0) \cup (0, +\infty)$	$(-\infty, 0) \cup (0, +\infty)$	$\text{csch}^{-1}(\text{csch}\, x) = x$ if $x < 0$ or $x > 0$ $\text{csch}(\text{csch}^{-1} x) = x$ if $x < 0$ or $x > 0$

**8.8.4   THEOREM.   *The following relationships hold for all x in the domain of the stated inverse hyperbolic function:*

$$\sinh^{-1} x = \ln(x + \sqrt{x^2 + 1}) \qquad \cosh^{-1} x = \ln(x + \sqrt{x^2 - 1})$$

$$\tanh^{-1} x = \frac{1}{2}\ln\left(\frac{1+x}{1-x}\right) \qquad \coth^{-1} x = \frac{1}{2}\ln\left(\frac{x+1}{x-1}\right)$$

$$\text{sech}^{-1} x = \ln\left(\frac{1 + \sqrt{1 - x^2}}{x}\right) \qquad \text{csch}^{-1} x = \ln\left(\frac{1}{x} + \frac{\sqrt{1 + x^2}}{|x|}\right)$$

We will show how to derive the first formula in this theorem, and leave the rest as exercises. The basic idea is to write the equation $x = \sinh y$ in terms of exponential functions and solve this equation for y as a function of x. This will produce the equation $y = \sinh^{-1} x$ with $\sinh^{-1} x$ expressed in terms of natural logarithms. Expressing $x = \sinh y$ in terms of exponentials yields

$$x = \sinh y = \frac{e^y - e^{-y}}{2}$$

which can be rewritten as

$$e^y - 2x - e^{-y} = 0$$

Multiplying this equation through by e^y we obtain

$$e^{2y} - 2xe^y - 1 = 0$$

and applying the quadratic formula yields

$$e^y = \frac{2x \pm \sqrt{4x^2 + 4}}{2} = x \pm \sqrt{x^2 + 1}$$

Since $e^y > 0$, the solution involving the minus sign is extraneous and must be discarded. Thus,

$$e^y = x + \sqrt{x^2 + 1}$$

Taking natural logarithms yields

$$y = \ln(x + \sqrt{x^2 + 1}) \quad \text{or} \quad \sinh^{-1} x = \ln(x + \sqrt{x^2 + 1})$$

Example 5

$$\sinh^{-1} 1 = \ln(1 + \sqrt{1^2 + 1}) = \ln(1 + \sqrt{2}) \approx 0.8814$$

$$\tanh^{-1}\left(\frac{1}{2}\right) = \frac{1}{2}\ln\left(\frac{1 + \frac{1}{2}}{1 - \frac{1}{2}}\right) = \frac{1}{2}\ln 3 \approx 0.5493 \qquad \blacktriangleleft$$

DERIVATIVES AND INTEGRALS INVOLVING INVERSE HYPERBOLIC FUNCTIONS

Theorem 4.1.7 can be used to establish the differentiability of the inverse hyperbolic functions (we omit the details), and formulas for the derivatives can be obtained from Theorem 8.8.4. For example,

$$\frac{d}{dx}[\sinh^{-1} x] = \frac{d}{dx}[\ln(x + \sqrt{x^2 + 1})] = \frac{1}{x + \sqrt{x^2 + 1}}\left(1 + \frac{x}{\sqrt{x^2 + 1}}\right)$$

$$= \frac{\sqrt{x^2 + 1} + x}{(x + \sqrt{x^2 + 1})(\sqrt{x^2 + 1})} = \frac{1}{\sqrt{x^2 + 1}}$$

This computation leads to two integral formulas, a formula that involves $\sinh^{-1} x$ and an equivalent formula that involves logarithms:

$$\int \frac{dx}{\sqrt{x^2 + 1}} = \sinh^{-1} x + C = \ln(x + \sqrt{x^2 + 1}) + C$$

FOR THE READER. The derivative of $\sinh^{-1} x$ can also be obtained by letting $y = \sinh^{-1} x$ and differentiating the equation $x = \sinh y$ implicitly. Try it.

The following two theorems list the generalized derivative formulas and corresponding integration formulas for the inverse hyperbolic functions. Some of the proofs appear as exercises.

8.8.5 THEOREM.

$$\frac{d}{dx}(\sinh^{-1} u) = \frac{1}{\sqrt{1 + u^2}}\frac{du}{dx} \qquad\qquad \frac{d}{dx}(\coth^{-1} u) = \frac{1}{1 - u^2}\frac{du}{dx}, \quad |u| > 1$$

$$\frac{d}{dx}(\cosh^{-1} u) = \frac{1}{\sqrt{u^2 - 1}}\frac{du}{dx}, \quad u > 1 \qquad \frac{d}{dx}(\operatorname{sech}^{-1} u) = -\frac{1}{u\sqrt{1 - u^2}}\frac{du}{dx}, \quad 0 < u < 1$$

$$\frac{d}{dx}(\tanh^{-1} u) = \frac{1}{1 - u^2}\frac{du}{dx}, \quad |u| < 1 \qquad \frac{d}{dx}(\operatorname{csch}^{-1} u) = -\frac{1}{|u|\sqrt{1 + u^2}}\frac{du}{dx}, \quad u \neq 0$$

8.8.6 THEOREM.

$$\int \frac{du}{\sqrt{1 + u^2}} = \sinh^{-1} u + C = \ln(u + \sqrt{u^2 + 1}) + C$$

$$\int \frac{du}{\sqrt{u^2 - 1}} = \cosh^{-1} u + C = \ln(u + \sqrt{u^2 - 1}) + C, \quad u > 1$$

$$\int \frac{du}{1 - u^2} = \begin{cases} \tanh^{-1} u + C, & |u| < 1 \\ \coth^{-1} u + C, & |u| > 1 \end{cases} = \frac{1}{2}\ln\left|\frac{1 + u}{1 - u}\right| + C$$

$$\int \frac{du}{u\sqrt{1 - u^2}} = -\operatorname{sech}^{-1}|u| + C = -\ln\left(\frac{1 + \sqrt{1 - u^2}}{|u|}\right) + C, \quad 0 < |u| < 1$$

$$\int \frac{du}{u\sqrt{1 + u^2}} = -\operatorname{csch}^{-1}|u| + C = -\ln\left(\frac{1 + \sqrt{1 + u^2}}{|u|}\right) + C, \quad u \neq 0$$

Example 6

Evaluate $\int \dfrac{dx}{\sqrt{4x^2 - 1}}, \; x > \dfrac{1}{2}.$

Solution. Let $u = 2x$. Thus, $du = 2\,dx$ and

$$\int \frac{dx}{\sqrt{4x^2 - 1}} = \frac{1}{2} \int \frac{2\,dx}{\sqrt{4x^2 - 1}} = \frac{1}{2} \int \frac{du}{\sqrt{u^2 - 1}}$$

$$= \frac{1}{2} \cosh^{-1} u + C = \frac{1}{2} \cosh^{-1}(2x) + C$$

Alternatively, we can use the logarithmic equivalent of $\cosh^{-1}(2x)$ and express the answer as

$$\int \frac{dx}{\sqrt{4x^2 - 1}} = \frac{1}{2} \ln(2x + \sqrt{4x^2 - 1}) + C \qquad \blacktriangleleft$$

EXERCISE SET 8.8   ⊠ Graphing Calculator   [C] CAS

In Exercises 1 and 2, approximate the expression to four decimal places.

1. (a) $\sinh 3$ (b) $\cosh(-2)$ (c) $\tanh(\ln 4)$
   (d) $\sinh^{-1}(-2)$ (e) $\cosh^{-1} 3$ (f) $\tanh^{-1} \frac{3}{4}$

2. (a) $\operatorname{csch}(-1)$ (b) $\operatorname{sech}(\ln 2)$ (c) $\coth 1$
   (d) $\operatorname{sech}^{-1} \frac{1}{2}$ (e) $\coth^{-1} 3$ (f) $\operatorname{csch}^{-1}(-\sqrt{3})$

3. In each part, find the exact numerical value of the expression.
   (a) $\sinh(\ln 3)$ (b) $\cosh(-\ln 2)$
   (c) $\tanh(2 \ln 5)$ (d) $\sinh(-3 \ln 2)$

4. In each part, rewrite the expression as a ratio of polynomials.
   (a) $\cosh(\ln x)$ (b) $\sinh(\ln x)$
   (c) $\tanh(2 \ln x)$ (d) $\cosh(-\ln x)$

5. In each part, a value for one of the hyperbolic functions is given at an unspecified positive number x_0. Use appropriate identities to find the exact values of the remaining five hyperbolic functions at x_0.
   (a) $\sinh x_0 = 2$ (b) $\cosh x_0 = \frac{5}{4}$ (c) $\tanh x_0 = \frac{4}{5}$

6. Obtain the derivative formulas for $\operatorname{csch} x$, $\operatorname{sech} x$, and $\coth x$ from the derivative formulas for $\sinh x$, $\cosh x$, and $\tanh x$.

7. Find the derivatives of $\sinh^{-1} x$, $\cosh^{-1} x$, and $\tanh^{-1} x$ by differentiating the equations $x = \sinh y$, $x = \cosh y$, and $x = \tanh y$ implicitly.

[C] **8.** Use a CAS to find the derivatives of $\sinh^{-1} x$, $\cosh^{-1} x$, $\tanh^{-1} x$, $\coth^{-1} x$, $\operatorname{sech}^{-1} x$, and $\operatorname{csch}^{-1} x$, and confirm that your answers are consistent with those in Theorem 8.8.5.

In Exercises 9–28, find dy/dx.

9. $y = \sinh(4x - 8)$ **10.** $y = \cosh(x^4)$
11. $y = \coth(\ln x)$ **12.** $y = \ln(\tanh 2x)$

13. $y = \operatorname{csch}(1/x)$ **14.** $y = \operatorname{sech}(e^{2x})$

15. $y = \sqrt{4x + \cosh^2(5x)}$ **16.** $y = \sinh^3(2x)$

17. $y = x^3 \tanh^2(\sqrt{x})$ **18.** $y = \sinh(\cos 3x)$

19. $y = \sinh^{-1}\left(\frac{1}{3}x\right)$ **20.** $y = \sinh^{-1}(1/x)$

21. $y = \ln(\cosh^{-1} x)$ **22.** $y = \cosh^{-1}(\sinh^{-1} x)$

23. $y = \dfrac{1}{\tanh^{-1} x}$ **24.** $y = (\coth^{-1} x)^2$

25. $y = \cosh^{-1}(\cosh x)$ **26.** $y = \sinh^{-1}(\tanh x)$

27. $y = e^x \operatorname{sech}^{-1} \sqrt{x}$ **28.** $y = (1 + x \operatorname{csch}^{-1} x)^{10}$

[C] **29.** Use a CAS to find the derivatives in Example 2. If the answers produced by the CAS do not match those in the text, then use appropriate identities to show that the answers are equivalent.

[C] **30.** For each of the derivatives you obtained in Exercises 9–28, use a CAS to check your answer. If the answer produced by the CAS does not match your own, show that the two answers are equivalent.

In Exercises 31–46, evaluate the integrals.

31. $\displaystyle\int \sinh^6 x \cosh x \, dx$ **32.** $\displaystyle\int \cosh(2x - 3) \, dx$

33. $\displaystyle\int \sqrt{\tanh x} \, \operatorname{sech}^2 x \, dx$ **34.** $\displaystyle\int \operatorname{csch}^2(3x) \, dx$

35. $\displaystyle\int \tanh x \, dx$ **36.** $\displaystyle\int \coth^2 x \, \operatorname{csch}^2 x \, dx$

37. $\displaystyle\int_{\ln 2}^{\ln 3} \tanh x \, \operatorname{sech}^3 x \, dx$ **38.** $\displaystyle\int_0^{\ln 3} \frac{e^x - e^{-x}}{e^x + e^{-x}} \, dx$

39. $\displaystyle\int \frac{dx}{\sqrt{1 + 9x^2}}$ **40.** $\displaystyle\int \frac{dx}{\sqrt{x^2 - 2}} \quad (x > \sqrt{2})$

41. $\int \dfrac{dx}{\sqrt{1 - e^{2x}}}$ $(x < 0)$ **42.** $\int \dfrac{\sin \theta \, d\theta}{\sqrt{1 + \cos^2 \theta}}$

43. $\int \dfrac{dx}{x\sqrt{1 + 4x^2}}$ **44.** $\int \dfrac{dx}{\sqrt{9x^2 - 25}}$ $(x > 5/3)$

45. $\int_0^{1/2} \dfrac{dx}{1 - x^2}$ **46.** $\int_0^{\sqrt{3}} \dfrac{dt}{\sqrt{t^2 + 1}}$

[c] **47.** For each of the integrals you evaluated in Exercises 31–46, use a CAS to check your answer. If the answer produced by the CAS does not match your own, show that the two answers are equivalent.

48. Use a graphing utility to generate the graphs of $\sinh x$, $\cosh x$, and $\tanh x$ by expressing these functions in terms of e^x and e^{-x}. If your graphing utility can graph the hyperbolic functions directly, then generate the graphs that way as well.

49. Find the area enclosed by $y = \sinh 2x$, $y = 0$, and $x = \ln 3$.

50. Find the volume of the solid that is generated when the region enclosed by $y = \operatorname{sech} x$, $y = 0$, $x = 0$, and $x = \ln 2$ is revolved about the x-axis.

51. Find the volume of the solid that is generated when the region enclosed by $y = \cosh 2x$, $y = \sinh 2x$, $x = 0$, and $x = 5$ is revolved about the x-axis.

52. Use Newton's Method to approximate the positive value of the constant a such that the area enclosed by $y = \cosh ax$, $y = 0$, $x = 0$, and $x = 1$ is 2 square units. Express your answer to at least five decimal places.

53. Find the arc length of $y = \cosh x$ between $x = 0$ and $x = \ln 2$.

54. Find the arc length of the catenary $y = a \cosh(x/a)$ between $x = 0$ and $x = x_1$ $(x_1 > 0)$.

55. Prove that $\sinh x$ is an odd function of x and that $\cosh x$ is an even function of x, and check that this is consistent with the graphs in Figure 8.8.1.

In Exercises 56 and 57, prove the identities.

56. (a) $\cosh x + \sinh x = e^x$
   (b) $\cosh x - \sinh x = e^{-x}$
   (c) $\sinh(x + y) = \sinh x \cosh y + \cosh x \sinh y$
   (d) $\sinh 2x = 2 \sinh x \cosh x$
   (e) $\cosh(x + y) = \cosh x \cosh y + \sinh x \sinh y$
   (f) $\cosh 2x = \cosh^2 x + \sinh^2 x$
   (g) $\cosh 2x = 2 \sinh^2 x + 1$
   (h) $\cosh 2x = 2 \cosh^2 x - 1$

57. (a) $1 - \tanh^2 x = \operatorname{sech}^2 x$
   (b) $\tanh(x + y) = \dfrac{\tanh x + \tanh y}{1 + \tanh x \tanh y}$
   (c) $\tanh 2x = \dfrac{2 \tanh x}{1 + \tanh^2 x}$

58. Prove:
   (a) $\cosh^{-1} x = \ln(x + \sqrt{x^2 - 1})$, $x \geq 1$
   (b) $\tanh^{-1} x = \dfrac{1}{2} \ln\left(\dfrac{1 + x}{1 - x}\right)$, $-1 < x < 1$.

59. Use Exercise 58 to obtain the derivative formulas for $\cosh^{-1} x$ and $\tanh^{-1} x$.

60. Prove:
$$\operatorname{sech}^{-1} x = \cosh^{-1}(1/x), \quad 0 < x \leq 1$$
$$\coth^{-1} x = \tanh^{-1}(1/x), \quad |x| > 1$$
$$\operatorname{csch}^{-1} x = \sinh^{-1}(1/x), \quad x \neq 0$$

61. Use Exercise 60 to express the integral
$$\int \frac{du}{1 - u^2}$$
entirely in terms of $\tanh^{-1}$.

62. Show that
   (a) $\dfrac{d}{dx}[\operatorname{sech}^{-1}|x|] = -\dfrac{1}{x\sqrt{1 - x^2}}$
   (b) $\dfrac{d}{dx}[\operatorname{csch}^{-1}|x|] = -\dfrac{1}{x\sqrt{1 + x^2}}$.

63. Find the limits, and confirm that they are consistent with the graphs in Figures 8.8.1 and 8.8.6.
   (a) $\lim\limits_{x \to +\infty} \sinh x$ (b) $\lim\limits_{x \to -\infty} \sinh x$
   (c) $\lim\limits_{x \to +\infty} \tanh x$ (d) $\lim\limits_{x \to -\infty} \tanh x$
   (e) $\lim\limits_{x \to +\infty} \sinh^{-1} x$ (f) $\lim\limits_{x \to 1^-} \tanh^{-1} x$

64. In each part, find the limit.
   (a) $\lim\limits_{x \to +\infty} (\cosh^{-1} x - \ln x)$ (b) $\lim\limits_{x \to +\infty} \dfrac{\cosh x}{e^x}$

65. Use the first and second derivatives to show that the graph of $y = \tanh^{-1} x$ is always increasing and has an inflection point at the origin.

66. The integration formulas for $1/\sqrt{u^2 - 1}$ in Theorem 8.8.6 are valid for $u > 1$. Show that the following formula is valid for $u < -1$:
$$\int \frac{du}{\sqrt{u^2 - 1}} = -\cosh^{-1}(-u) + C = \ln|u + \sqrt{u^2 - 1}| + C$$

67. Show that $(\sinh x + \cosh x)^n = \sinh nx + \cosh nx$.

68. Show that
$$\int_{-a}^{a} e^{tx} \, dx = \frac{2 \sinh at}{t}$$

69. A cable is suspended between two poles as shown in Figure 8.8.2. The equation of the curve formed by the cable is $y = a \cosh(x/a)$, where a is a positive constant. Suppose that the x-coordinates of the points of support are $x = -b$ and $x = b$, where $b > 0$.
   (a) Show that the length L of the cable is given by
$$L = 2a \sinh \frac{b}{a}$$
   (b) Show that the sag S (the vertical distance between the highest and lowest points on the cable) is given by
$$S = a \cosh \frac{b}{a} - a$$

Exercises 70 and 71 refer to the hanging cable described in Exercise 69.

70. Assuming that the cable is 120 ft long and the poles are 100 ft apart, approximate the sag in the cable by using Newton's Method to approximate a. Express your final answer to the nearest tenth of a foot. [*Hint:* First let $u = 50/a$.]

71. Assuming that the poles are 400 ft apart and the sag in the cable is 30 ft, approximate the length of the cable by using Newton's Method to approximate a. Express your final answer to the nearest tenth of a foot. [*Hint:* First let $u = 200/a$.]

72. The accompanying figure shows a person pulling a boat by holding a rope of length a attached to the bow and walking along the edge of a dock. If we assume that the rope is always tangent to the curve traced by the bow of the boat, then this curve, which is called a **tractrix**, has the property that the segment of the tangent line between the curve and the y-axis has a constant length a. It can be proved that the equation of this tractrix is

$$y = a \operatorname{sech}^{-1} \frac{x}{a} - \sqrt{a^2 - x^2}$$

(a) Show that to move the bow of the boat to a point (x, y), the person must walk a distance

$$D = a \operatorname{sech}^{-1} \frac{x}{a}$$

from the origin.

(b) If the rope has a length of 15 m, how far must the person walk from the origin to bring the boat 10 m from the dock? Round your answer to two decimal places.

(c) Find the distance traveled by the bow along the tractrix as it moves from its initial position to the point where it is 5 m from the dock.

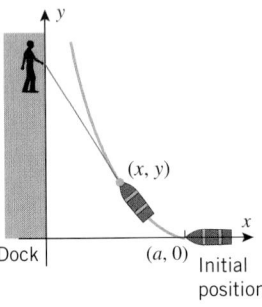

Figure Ex-72

SUPPLEMENTARY EXERCISES

1. State an integral formula for finding the arc length of a smooth curve $y = f(x)$ over an interval $[a, b]$, and use Riemann sums to derive the formula.

2. Describe the method of slicing for finding volumes, and use that method to derive an integral formula for finding volumes by the method of disks.

3. State an integral formula for finding a volume by the method of cylindrical shells, and use Riemann sums to derive the formula.

4. State an integral formula for the work W done by a variable force $F(x)$ applied in the direction of motion to an object moving from $x = a$ to $x = b$, and use Riemann sums to derive the formula.

5. State an integral formula for the fluid force F exerted on a vertical flat surface immersed in a fluid of weight density ρ, and use Riemann sums to derive the formula.

6. Let R be the region in the first quadrant enclosed by $y = x^2$, $y = 2 + x$, and $x = 0$. In each part, set up, but *do not evaluate*, an integral or a sum of integrals that will solve the problem.

(a) Find the area of R by integrating with respect to x.

(b) Find the area of R by integrating with respect to y.

(c) Find the volume of the solid generated by revolving R about the x-axis by integrating with respect to x.

(d) Find the volume of the solid generated by revolving R about the x-axis by integrating with respect to y.

(e) Find the volume of the solid generated by revolving R about the y-axis by integrating with respect to x.

(f) Find the volume of the solid generated by revolving R about the y-axis by integrating with respect to y.

7. (a) Set up a sum of definite integrals that represents the total shaded area between the curves $y = f(x)$ and $y = g(x)$ in the accompanying figure.

(b) Find the total area enclosed between $y = x^3$ and $y = x$ over the interval $[-1, 2]$.

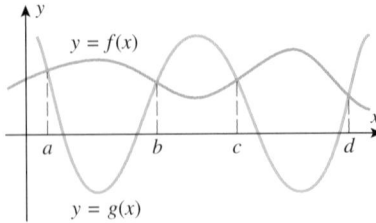

Figure Ex-7

8. Let C be the curve $27x - y^3 = 0$ between $y = 0$ and $y = 2$. In each part, set up, but *do not evaluate*, an integral or a sum of integrals that solves the problem.

(a) Find the area of the surface generated by revolving C about the y-axis by integrating with respect to x.

(b) Find the area of the surface generated by revolving C about the y-axis by integrating with respect to y.

(c) Find the area of the surface generated by revolving C about the line $y = -2$ by integrating with respect to y.

9. Find the arc length in the second quadrant of the curve $x^{2/3} + y^{2/3} = a^{2/3}$ from the point $x = -a$ to $x = -\frac{1}{8}a$ ($a > 0$).

10. As shown in the accompanying figure, a cathedral dome is designed with three semicircular supports of radius r so that each horizontal cross section is a regular hexagon. Show that the volume of the dome is $r^3\sqrt{3}$.

11. As shown in the accompanying figure, a cylindrical hole is drilled all the way through the center of a sphere. Show that the volume of the remaining solid depends only on the length L of the hole, not on the size of the sphere.

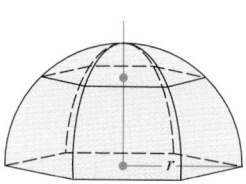
Figure Ex-10

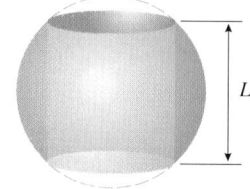
Figure Ex-11

12. A football has the shape of the solid generated by revolving the region bounded between the x-axis and the parabola $y = 4R(x^2 - \frac{1}{4}L^2)/L^2$ about the x-axis. Find its volume.

13. The design of the Gateway Arch in St. Louis, Missouri, by architect Eero Saarinen was implemented using equations provided by Dr. Hannskarl Badel. The equation used for the centerline of the arch was

$$y = 693.8597 - 68.7672\cosh(0.0100333x)\ \text{ft}$$

for x between -299.2239 and 299.2239.

(a) Use a graphing utility to graph the centerline of the arch.

(b) Find the length of the centerline to four decimal places.

(c) For what values of x is the height of the arch 100 ft? Round your answers to four decimal places.

(d) Approximate, to the nearest degree, the acute angle that the tangent line to the centerline makes with the ground at the ends of the arch.

14. A golfer makes a successful chip shot to the green. Suppose that the path of the ball from the moment it is struck to the moment it hits the green is described by

$$y = 12.54x - 0.41x^2$$

where x is the horizontal distance (in yards) from the point where the ball is struck, and y is the vertical distance (in yards) above the fairway. Find the distance the ball travels from the moment it is struck to the moment it hits the green. Assume that the fairway and green are at the same level and round your answer to two decimal places.

15. Derive integration formulas for

$$\int \frac{du}{\sqrt{a^2 + u^2}},\quad \int \frac{du}{\sqrt{u^2 - a^2}},\quad \int \frac{du}{a^2 - u^2}$$

and use those formulas to evaluate

(a) $\displaystyle\int \frac{dx}{\sqrt{4 + x^2}}$

(b) $\displaystyle\int \frac{dx}{\sqrt{x^2 - 9}}$

(c) $\displaystyle\int \frac{dx}{2 - x^2}$

(d) $\displaystyle\int \frac{dx}{\sqrt{16 + 5x^2}}.$

16. In each part, prove the identity.

(a) $\cosh 3x = 4\cosh^3 x - 3\cosh x$

(b) $\cosh \frac{1}{2}x = \sqrt{\frac{1}{2}(\cosh x + 1)}$

(c) $\sinh \frac{1}{2}x = \pm\sqrt{\frac{1}{2}(\cosh x - 1)}$

17. (a) A spring exerts a force of 0.5 N when stretched 0.25 m beyond its natural length. Assuming that Hooke's law applies, how much work was performed in stretching the spring to this length?

(b) How far beyond its natural length can the spring be stretched with 25 J of work?

18. A boat is anchored so that the anchor is 150 ft below the surface of the water. In the water, the anchor weighs 2000 lb and the chain weighs 30 lb/ft. How much work is required to raise the anchor to the surface?

19. In each part, set up, but *do not evaluate*, an integral that solves the problem.

(a) Find the fluid force exerted on a side of a box that has a 3-m-square base and is filled to a depth of 1 m with a liquid of weight density ρ N/m^3.

(b) Find the fluid force exerted by a liquid of weight density ρ lb/ft^3 on a face of the vertical plate shown in part (a) of the accompanying figure.

(c) Find the fluid force exerted on the parabolic dam in part (b) of the accompanying figure by water that extends to the top of the dam.

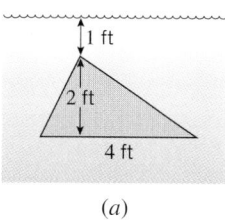
(a)

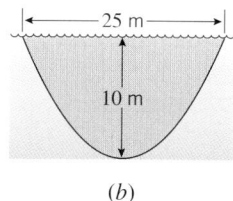
(b)

Figure Ex-19

20. Suppose that a hollow tube rotates with a constant angular velocity of ω rad/s about a horizontal axis at one end of the tube, as shown in the accompanying figure (next page). Assume that an object is free to slide without friction in the tube while the tube is rotating. Let r be the distance from the object to the pivot point at time $t \geq 0$, and assume that the object is at rest and $r = 0$ when $t = 0$. It can be shown that if the tube is horizontal at time $t = 0$, then

$$r = \frac{g}{2\omega^2}[\sinh(\omega t) - \sin(\omega t)]$$

during the period that the object is in the tube. Assume that t is in seconds and r is in meters, and use $g = 9.8$ m/s^2 and $\omega = 2$ rad/s.

(a) Graph r versus t for $0 \leq t \leq 1$.

(b) Assuming that the tube has a length of 1 m, approximately how long does it take for the object to reach the end of the tube?

(c) Use the result of part (b) to approximate dr/dt at the instant that the object reaches the end of the tube.

21. As shown in the accompanying figure, a horizontal beam with dimensions 2 in × 6 in × 16 ft is fixed at both ends and is subjected to a uniformly distributed load of 120 lb/ft. As a result of the load, the centerline of the beam undergoes a deflection that is described by

$$y = -1.67 \times 10^{-8}(x^4 - 2Lx^3 + L^2x^2)$$

$(0 \leq x \leq 192)$, where $L = 192$ inches is the length of the unloaded beam, x is the horizontal distance along the beam measured in inches from the left end, and y is the deflection of the centerline in inches.

(a) Graph y versus x for $0 \leq x \leq 192$.

(b) Find the maximum deflection of the centerline.

(c) Use a CAS or a calculator with a numerical integration capability to find the length of the centerline of the loaded beam. Round your answer to two decimal places.

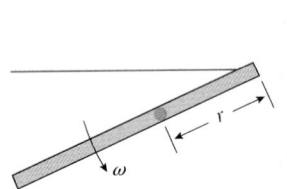

Figure Ex-20

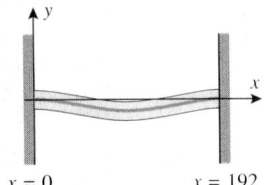

Figure Ex-21

Exercises 22–25 lead to equations that cannot be solved exactly. Use any method you want to approximate the solutions of those equations, and round your answers to two decimal places.

22. Find the area of the region enclosed by the curves $y = x^2 - 1$ and $y = 2 \sin x$.

23. Referring to the accompanying figure, find the value of k so that the areas of the shaded regions are equal. [*Note:* This exercise is based on Problem A1 of the Fifty-Fourth Annual William Lowell Putnam Mathematical Competition.]

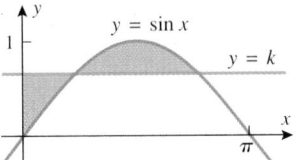

Figure Ex-23

24. Consider the region to the left of the vertical line $x = k$ $(0 < k < \pi)$ and between the curve $y = \sin x$ and the x-axis. Use a CAS to find the value of k so that the solid generated by revolving the region about the y-axis has a volume of 8 cubic units.

25. Suppose that an object moves in the positive direction on an x-axis while subject to the force

$$F(x) = \frac{x}{\sqrt{1 + x^3}}, \quad x \geq 0$$

where x is in meters and F is in newtons. The object moves 2 m from an unspecified starting point $x = a$ $(a \geq 0)$.

(a) Find a definite integral that gives the work done by F as a function of a.

(b) Find the value of a for which the work done by F is maximum. What is that maximum work? [*Hint:* See Exercise 24, Section 7.9.]

EXPANDING THE CALCULUS HORIZON

For additional material relating to this chapter, visit the Anton Website at http://www.wiley.com/college/anton

9

PRINCIPLES OF INTEGRAL EVALUATION

Augustin Cauchy

$\mathscr{I}$n earlier chapters we obtained many basic integration formulas from the corresponding differentiation formulas. For example, knowing that the derivative of $\sin x$ is $\cos x$ enabled us to deduce that the integral of $\cos x$ is $\sin x$. Subsequently, we expanded our integration repertoire by introducing the method of u-substitution. That method enabled us to integrate many functions by transforming the integrand of an unfamiliar integral into a familiar form. However, u-substitution alone is not adequate to handle the wide variety of integrals that arise in applications, so additional integration techniques are still needed. In this chapter we will discuss some of those techniques, and we will provide a more systematic procedure for attacking unfamiliar integrals. We will talk more about numerical approximations of definite integrals, and we will explore the idea of integrating over infinite intervals.

9.1 AN OVERVIEW OF INTEGRATION METHODS

In this section we will give a brief overview of methods for evaluating integrals, and we will review the integration formulas that were discussed in earlier sections.

METHODS FOR APPROACHING INTEGRATION PROBLEMS

There are three basic approaches for evaluating unfamiliar integrals:

- **Technology**—CAS programs such as *Mathematica*, *Maple*, and *Derive* are capable of evaluating extremely complicated integrals, and more and more modern research facilities are being equipped with such programs.

- **Tables**—Prior to the development of CAS programs, scientists relied heavily on tables to evaluate difficult integrals arising in applications. Such tables were compiled over many years, incorporating the skills and experience of many people. One such table appears in the endpapers of this text, but more comprehensive tables appear in various reference books such as the *CRC Standard Mathematical Tables and Formulae*, CRC Press, Inc., 1991.

- **Transformation Methods**—Transformation methods are methods for converting unfamiliar integrals into familiar integrals. These include *u*-substitution, algebraic manipulation of the integrand, and other methods that we will discuss in this chapter.

None of the three methods is perfect; for example, CAS programs often encounter integrals that they cannot evaluate and they sometimes produce answers that are excessively complicated, tables are not exhaustive and hence may not include a particular integral of interest, and transformation methods rely on human ingenuity that may prove to be inadequate in difficult problems.

In this chapter we will focus on transformation methods and tables, so it will *not be necessary* to have a CAS such as *Mathematica*, *Maple*, or *Derive*. However, if you have a CAS, then you can use it to confirm the results in the examples, and there are exercises that are designed to be solved with a CAS. If you have a CAS, keep in mind that many of the algorithms that it uses are based on the methods we will discuss here, so an understanding of these methods will help you to use your technology in a more informed way.

A REVIEW OF FAMILIAR INTEGRATION FORMULAS

The following is a list of basic integrals that we have encountered thus far:

CONSTANTS, POWERS, EXPONENTIALS

$$1. \int du = u + C \qquad\qquad 2. \int a\,du = a\int du = au + C$$

$$3. \int u^r\,du = \frac{u^{r+1}}{r+1} + C,\ r \neq -1 \qquad 4. \int \frac{du}{u} = \ln|u| + C$$

$$5. \int e^u\,du = e^u + C \qquad\qquad 6. \int b^u\,du = \frac{b^u}{\ln b} + C,\ b > 0, b \neq 1$$

TRIGONOMETRIC FUNCTIONS

$$7. \int \sin u\,du = -\cos u + C \qquad 8. \int \cos u\,du = \sin u + C$$

$$9. \int \sec^2 u\,du = \tan u + C \qquad 10. \int \csc^2 u\,du = -\cot u + C$$

$$11. \int \sec u \tan u\,du = \sec u + C \qquad 12. \int \csc u \cot u\,du = -\csc u + C$$

$$13. \int \tan u\,du = -\ln|\cos u| + C \qquad 14. \int \cot u\,du = \ln|\sin u| + C$$

HYPERBOLIC FUNCTIONS

15. $\displaystyle\int \sinh u \, du = \cosh u + C$ **16.** $\displaystyle\int \cosh u \, du = \sinh u + C$

17. $\displaystyle\int \operatorname{sech}^2 u \, du = \tanh u + C$ **18.** $\displaystyle\int \operatorname{csch}^2 u \, du = -\coth u + C$

19. $\displaystyle\int \operatorname{sech} u \tanh u \, du = -\operatorname{sech} u + C$ **20.** $\displaystyle\int \operatorname{csch} u \coth u \, du = -\operatorname{csch} u + C$

ALGEBRAIC FUNCTIONS ($a > 0$)

21. $\displaystyle\int \frac{du}{\sqrt{a^2 - u^2}} = \sin^{-1}\frac{u}{a} + C \qquad (|u| < a)$

22. $\displaystyle\int \frac{du}{a^2 + u^2} = \frac{1}{a}\tan^{-1}\frac{u}{a} + C$

23. $\displaystyle\int \frac{du}{u\sqrt{u^2 - a^2}} = \frac{1}{a}\sec^{-1}\left|\frac{u}{a}\right| + C \qquad (0 < a < |u|)$

24. $\displaystyle\int \frac{du}{\sqrt{a^2 + u^2}} = \ln(u + \sqrt{u^2 + a^2}) + C$

25. $\displaystyle\int \frac{du}{\sqrt{u^2 - a^2}} = \ln\left|u + \sqrt{u^2 - a^2}\right| + C \qquad (0 < a < |u|)$

26. $\displaystyle\int \frac{du}{a^2 - u^2} = \frac{1}{2a}\ln\left|\frac{a+u}{a-u}\right| + C$

27. $\displaystyle\int \frac{du}{u\sqrt{a^2 - u^2}} = -\frac{1}{a}\ln\left|\frac{a + \sqrt{a^2 - u^2}}{u}\right| + C \qquad (0 < |u| < a)$

28. $\displaystyle\int \frac{du}{u\sqrt{a^2 + u^2}} = -\frac{1}{a}\ln\left|\frac{a + \sqrt{a^2 + u^2}}{u}\right| + C$

REMARK.   Formulas 24–28 are generalizations of those in Theorem 8.8.6. Readers who did not cover that section can ignore those formulas for now, since we will develop other methods for obtaining them in this chapter.

EXERCISE SET 9.1
· ·

Review: Without looking at the text, complete the following integration formulas and then check your results by referring to the list of formulas at the beginning of this section.

Constants, Powers, Exponentials

$\displaystyle\int du =$ $\displaystyle\int a \, du =$

$\displaystyle\int u^r \, du =$ $\displaystyle\int \frac{du}{u} =$

$\displaystyle\int e^u \, du =$ $\displaystyle\int b^u \, du =$

Trigonometric Functions

$\displaystyle\int \sin u \, du =$ $\displaystyle\int \cos u \, du =$

$\displaystyle\int \sec^2 u \, du =$ $\displaystyle\int \csc^2 u \, du =$

$\displaystyle\int \sec u \tan u \, du =$ $\displaystyle\int \csc u \cot u \, du =$

$\displaystyle\int \tan u \, du =$ $\displaystyle\int \cot u \, du =$

Algebraic Functions

$$\int \frac{du}{\sqrt{1-u^2}} = \qquad \int \frac{du}{1+u^2} =$$

$$\int \frac{du}{u\sqrt{u^2-1}} = \qquad \int \frac{du}{\sqrt{1+u^2}} =$$

$$\int \frac{du}{\sqrt{u^2-1}} = \qquad \int \frac{du}{1-u^2} =$$

$$\int \frac{du}{u\sqrt{1-u^2}} = \qquad \int \frac{du}{u\sqrt{1+u^2}} =$$

Hyperbolic Functions

$$\int \sinh u \, du = \qquad \int \cosh u \, du =$$

$$\int \operatorname{sech}^2 u \, du = \qquad \int \operatorname{csch}^2 u \, du =$$

$$\int \operatorname{sech} u \tanh u \, du =$$

$$\int \operatorname{csch} u \coth u \, du =$$

In Exercises 1–30, evaluate the integrals by making appropriate u-substitutions and applying the formulas reviewed in this section.

1. $\displaystyle\int (3-2x)^3 \, dx$

2. $\displaystyle\int \sqrt{4+9x} \, dx$

3. $\displaystyle\int x \sec^2(x^2) \, dx$

4. $\displaystyle\int 4x \tan(x^2) \, dx$

5. $\displaystyle\int \frac{\sin 3x}{2+\cos 3x} \, dx$

6. $\displaystyle\int \frac{1}{4+9x^2} \, dx$

7. $\displaystyle\int e^x \sinh(e^x) \, dx$

8. $\displaystyle\int \frac{\sec(\ln x)\tan(\ln x)}{x} \, dx$

9. $\displaystyle\int e^{\cot x} \csc^2 x \, dx$

10. $\displaystyle\int \frac{x}{\sqrt{1-x^4}} \, dx$

11. $\displaystyle\int \cos^5 7x \sin 7x \, dx$

12. $\displaystyle\int \frac{\cos x}{\sin x \sqrt{\sin^2 x + 1}} \, dx$

13. $\displaystyle\int \frac{e^x}{\sqrt{4+e^{2x}}} \, dx$

14. $\displaystyle\int \frac{e^{\tan^{-1} x}}{1+x^2} \, dx$

15. $\displaystyle\int \frac{e^{\sqrt{x-2}}}{\sqrt{x-2}} \, dx$

16. $\displaystyle\int (3x+1)\cot(3x^2+2x) \, dx$

17. $\displaystyle\int \frac{\cosh\sqrt{x}}{\sqrt{x}} \, dx$

18. $\displaystyle\int \frac{dx}{x\ln x}$

19. $\displaystyle\int \frac{dx}{\sqrt{x}\,3^{\sqrt{x}}}$

20. $\displaystyle\int \sec(\sin\theta)\tan(\sin\theta)\cos\theta \, d\theta$

21. $\displaystyle\int \frac{\operatorname{csch}^2(2/x)}{x^2} \, dx$

22. $\displaystyle\int \frac{dx}{\sqrt{x^2-3}}$

23. $\displaystyle\int \frac{e^{-x}}{4-e^{-2x}} \, dx$

24. $\displaystyle\int \frac{\cos(\ln x)}{x} \, dx$

25. $\displaystyle\int \frac{e^x}{\sqrt{1-e^{2x}}} \, dx$

26. $\displaystyle\int \frac{\sinh(x^{-1/2})}{x^{3/2}} \, dx$

27. $\displaystyle\int \frac{x}{\sec(x^2)} \, dx$

28. $\displaystyle\int \frac{e^x}{\sqrt{4-e^{2x}}} \, dx$

29. $\displaystyle\int x4^{-x^2} \, dx$

30. $\displaystyle\int 2^{\pi x} \, dx$

31. (a) Use Formulas (15), (17), and (19) of Section 4.5 to derive integration formulas for

$$\int \frac{dx}{\sqrt{1-x^2}}, \quad \int \frac{dx}{1+x^2}, \quad \int \frac{dx}{x\sqrt{x^2-1}}$$

(b) Use the integration formulas you obtained in part (a) to derive Formulas (21), (22), and (23) in this section.

9.2 INTEGRATION BY PARTS

In this section we will discuss an integration technique that is essentially the antiderivative formulation of the formula for differentiating a product of two functions.

DERIVATION OF THE FORMULA FOR INTEGRATION BY PARTS

If f and g are differentiable functions, then by the rule for differentiating products

$$\frac{d}{dx}[f(x)g(x)] = f(x)g'(x) + g(x)f'(x)$$

Integrating both sides we obtain

$$\int \frac{d}{dx}[f(x)g(x)] \, dx = \int f(x)g'(x) \, dx + \int g(x)f'(x) \, dx$$

or

$$f(x)g(x) + C = \int f(x)g'(x)\,dx + \int g(x)f'(x)\,dx$$

or

$$\int f(x)g'(x)\,dx = f(x)g(x) - \int g(x)f'(x)\,dx + C$$

Since the integral on the right will produce another constant of integration, there is no need to keep the C in this last equation; thus, we obtain

$$\int f(x)g'(x)\,dx = f(x)g(x) - \int g(x)f'(x)\,dx \qquad (1)$$

which is called the formula for ***integration by parts***. By using this formula we can sometimes reduce a hard integration problem to an easier one.

In practice, it is usual to rewrite (1) by letting

$$u = f(x), \quad du = f'(x)\,dx$$
$$v = g(x), \quad dv = g'(x)\,dx$$

This yields the following alternative form for (1):

$$\int u\,dv = uv - \int v\,du \qquad (2)$$

Example 1

Evaluate $\displaystyle\int xe^x\,dx$.

Solution. To apply (2) we must write the integral in the form

$$\int u\,dv$$

One way to do this is to let

$$u = x \quad \text{and} \quad dv = e^x\,dx$$

so that

$$du = dx \quad \text{and} \quad v = \int e^x\,dx = e^x$$

Thus, from (2)

$$\int xe^x\,dx = \int \underbrace{x}_{u}\,\underbrace{e^x\,dx}_{dv} = \underbrace{x}_{u}\,\underbrace{e^x}_{v} - \int \underbrace{e^x}_{v}\,\underbrace{dx}_{du} = xe^x - e^x + C \qquad \blacktriangleleft$$

REMARK.   In the calculation of v from dv above, we omitted the constant of integration and wrote $v = \int e^x\,dx = e^x$. Had we included a constant of integration and written $v = \int e^x\,dx = e^x + C_1$, the constant C_1 would have eventually canceled out [Exercise 58(a)]. This is always the case in integration by parts [Exercise 58(b)], so we will usually omit the constant when calculating v from dv.

To use integration by parts successfully, the choice of u and dv must be made so that the new integral is easier than the original. For example, had we decided above to let

$$u = e^x, \quad dv = x\,dx, \quad du = e^x\,dx, \quad v = \int x\,dx = \frac{x^2}{2}$$

then we would have obtained

$$\int xe^x\,dx = \int u\,dv = uv - \int v\,du = \frac{x^2}{2}e^x - \frac{1}{2}\int x^2 e^x\,dx$$

For this choice of u and dv the new integral is actually more complicated than the original.

It is difficult to give hard and fast rules for choosing u and dv. It is a matter of experience that comes with lots of practice.

The next example shows that it is sometimes necessary to use integration by parts more than once in the same problem.

Example 2

Evaluate $\displaystyle\int x^2 e^{-x}\, dx$.

Solution. Let

$$u = x^2, \quad dv = e^{-x}\, dx, \quad du = 2x\, dx, \quad v = \int e^{-x}\, dx = -e^{-x}$$

so that

$$\int x^2 e^{-x}\, dx = \int u\, dv = uv - \int v\, du = -x^2 e^{-x} + 2\int x e^{-x}\, dx \qquad (3)$$

The last integral is similar to the original except that we have replaced x^2 by x. Another integration by parts applied to $\int x e^{-x}\, dx$ will complete the problem. We let

$$u = x, \quad dv = e^{-x}\, dx, \quad du = dx, \quad v = \int e^{-x}\, dx = -e^{-x}$$

so that

$$\int x e^{-x}\, dx = \int u\, dv = uv - \int v\, du$$

$$= -x e^{-x} + \int e^{-x}\, dx$$

$$= -x e^{-x} - e^{-x} + C_1$$

Substituting in (3) we obtain

$$\int x^2 e^{-x}\, dx = -x^2 e^{-x} + 2(-x e^{-x} - e^{-x} + C_1)$$

$$= -x^2 e^{-x} - 2x e^{-x} - 2e^{-x} + 2C_1$$

$$= -(x^2 + 2x + 2)e^{-x} + C$$

where $C = 2C_1$. ◀

Example 3

Evaluate $\displaystyle\int \ln x\, dx$.

Solution. Let

$$u = \ln x, \quad dv = dx, \quad du = \frac{1}{x}\, dx, \quad v = \int dx = x$$

so that

$$\int \ln x\, dx = \int u\, dv = uv - \int v\, du = x \ln x - \int x\left(\frac{1}{x}\right) dx$$

$$= x \ln x - \int dx = x \ln x - x + C \qquad\qquad ◀$$

Example 4

Evaluate $\displaystyle\int e^x \cos x\, dx$.

Solution. Let

$$u = e^x, \quad dv = \cos x\, dx, \quad du = e^x\, dx, \quad v = \int \cos x\, dx = \sin x$$

Thus,

$$\int e^x \cos x \, dx = \int u \, dv = uv - \int v \, du = e^x \sin x - \int e^x \sin x \, dx \qquad (4)$$

Since the integral $\int e^x \sin x \, dx$ is similar in form to the original integral $\int e^x \cos x \, dx$, it seems that nothing has been accomplished. However, let us integrate this new integral by parts. We let

$$u = e^x, \quad dv = \sin x \, dx, \quad du = e^x \, dx, \quad v = \int \sin x \, dx = -\cos x$$

Thus,

$$\int e^x \sin x \, dx = \int u \, dv = uv - \int v \, du = -e^x \cos x + \int e^x \cos x \, dx$$

Substituting in (4) yields

$$\int e^x \cos x \, dx = e^x \sin x - \left[-e^x \cos x + \int e^x \cos x \, dx \right]$$

or

$$\int e^x \cos x \, dx = e^x \sin x + e^x \cos x - \int e^x \cos x \, dx$$

which is an equation we can solve for the unknown integral. We obtain

$$2 \int e^x \cos x \, dx = e^x \sin x + e^x \cos x$$

and hence

$$\int e^x \cos x \, dx = \tfrac{1}{2} e^x \sin x + \tfrac{1}{2} e^x \cos x + C \qquad \blacktriangleleft$$

INTEGRATION BY PARTS FOR DEFINITE INTEGRALS

For definite integrals the formula corresponding to (2) is

$$\int_a^b u \, dv = uv \Big]_a^b - \int_a^b v \, du \qquad (5)$$

REMARK.   It is important to keep in mind that the variables u and v in this formula are functions of x and that the limits of integration in (5) are limits on the variable x. Sometimes it is helpful to emphasize this by writing (5) as

$$\int_{x=a}^{x=b} u \, dv = uv \Big]_{x=a}^{x=b} - \int_{x=a}^{x=b} v \, du \qquad (6)$$

The next example illustrates how integration by parts can be used to integrate the inverse trigonometric functions.

Example 5

Evaluate $\displaystyle\int_0^1 \tan^{-1} x \, dx$.

Solution.   Let

$$u = \tan^{-1} x, \quad dv = dx, \quad du = \frac{1}{1+x^2} \, dx, \quad v = \int dx = x$$

Thus,

$$\int_0^1 \tan^{-1} x \, dx = \int_0^1 u \, dv = uv \Big]_0^1 - \int_0^1 v \, du$$

> The limits of integration refer to x; that is, $x = 0$ and $x = 1$.

$$= x \tan^{-1} x \Big]_0^1 - \int_0^1 \frac{x}{1+x^2} \, dx$$

But

$$\int_0^1 \frac{x}{1+x^2}\, dx = \frac{1}{2} \int_0^1 \frac{2x}{1+x^2}\, dx = \frac{1}{2} \ln(1+x^2) \Big]_0^1 = \frac{1}{2} \ln 2$$

so

$$\int_0^1 \tan^{-1} x\, dx = x \tan^{-1} x \Big]_0^1 - \frac{1}{2} \ln 2 = \left(\frac{\pi}{4} - 0\right) - \frac{1}{2} \ln 2 = \frac{\pi}{4} - \ln\sqrt{2} \qquad \blacktriangleleft$$

REDUCTION FORMULAS

Integration by parts can be used to derive *reduction formulas* for integrals. These are formulas that express an integral involving a power of a function in terms of an integral that involves a *lower* power of that function. For example, if n is a positive integer and $n \geq 2$, then integration by parts can be used to obtain the reduction formulas

$$\int \sin^n x\, dx = -\frac{1}{n} \sin^{n-1} x \cos x + \frac{n-1}{n} \int \sin^{n-2} x\, dx \qquad (7)$$

$$\int \cos^n x\, dx = \frac{1}{n} \cos^{n-1} x \sin x + \frac{n-1}{n} \int \cos^{n-2} x\, dx \qquad (8)$$

To illustrate how such formulas can be obtained, let us derive (8). We begin by writing $\cos^n x$ as $\cos^{n-1} x \cdot \cos x$ and letting

$$u = \cos^{n-1} x \qquad\qquad\qquad dv = \cos x\, dx$$

$$du = (n-1) \cos^{n-2} x(-\sin x)\, dx \qquad v = \int \cos x\, dx = \sin x$$

$$= -(n-1) \cos^{n-2} x \sin x\, dx$$

so that

$$\int \cos^n x\, dx = \int \cos^{n-1} x \cos x\, dx = \int u\, dv = uv - \int v\, du$$

$$= \cos^{n-1} x \sin x + (n-1) \int \sin^2 x \cos^{n-2} x\, dx$$

$$= \cos^{n-1} x \sin x + (n-1) \int (1 - \cos^2 x) \cos^{n-2} x\, dx$$

$$= \cos^{n-1} x \sin x + (n-1) \int \cos^{n-2} x\, dx - (n-1) \int \cos^n x\, dx$$

Transposing the last term on the right to the left side yields

$$n \int \cos^n x\, dx = \cos^{n-1} x \sin x + (n-1) \int \cos^{n-2} x\, dx$$

from which (8) follows.

Reduction formulas (7) and (8) reduce the exponent of sine (or cosine) by 2. Thus, if the formulas are applied repeatedly, the exponent can eventually be reduced to 0 if n is even or 1 if n is odd, at which point the integration can be completed. We will discuss this method in more detail in the next section, but for now, here is an example that illustrates how reduction formulas work.

Example 6

Evaluate $\int \cos^4 x\, dx.$

Solution. From (8) with $n = 4$

$$\int \cos^4 x \, dx = \frac{1}{4} \cos^3 x \sin x + \frac{3}{4} \int \cos^2 x \, dx \qquad \boxed{\begin{array}{l}\text{Now apply (8)}\\ \text{with } n = 2.\end{array}}$$

$$= \frac{1}{4} \cos^3 x \sin x + \frac{3}{4} \left(\frac{1}{2} \cos x \sin x + \frac{1}{2} \int dx \right)$$

$$= \frac{1}{4} \cos^3 x \sin x + \frac{3}{8} \cos x \sin x + \frac{3}{8} x + C \qquad \blacktriangleleft$$

EXERCISE SET 9.2 ☐c CAS

In Exercises 1–40, evaluate the integral.

1. $\displaystyle\int x e^{-x} \, dx$

2. $\displaystyle\int x e^{3x} \, dx$

3. $\displaystyle\int x^2 e^x \, dx$

4. $\displaystyle\int x^2 e^{-2x} \, dx$

5. $\displaystyle\int x \sin 2x \, dx$

6. $\displaystyle\int x \cos 3x \, dx$

7. $\displaystyle\int x^2 \cos x \, dx$

8. $\displaystyle\int x^2 \sin x \, dx$

9. $\displaystyle\int \sqrt{x} \ln x \, dx$

10. $\displaystyle\int x \ln x \, dx$

11. $\displaystyle\int (\ln x)^2 \, dx$

12. $\displaystyle\int \frac{\ln x}{\sqrt{x}} \, dx$

13. $\displaystyle\int \ln(2x + 3) \, dx$

14. $\displaystyle\int \ln(x^2 + 4) \, dx$

15. $\displaystyle\int \sin^{-1} x \, dx$

16. $\displaystyle\int \cos^{-1}(2x) \, dx$

17. $\displaystyle\int \tan^{-1}(2x) \, dx$

18. $\displaystyle\int x \tan^{-1} x \, dx$

19. $\displaystyle\int e^x \sin x \, dx$

20. $\displaystyle\int e^{2x} \cos 3x \, dx$

21. $\displaystyle\int e^{ax} \sin bx \, dx$

22. $\displaystyle\int e^{-3\theta} \sin 5\theta \, d\theta$

23. $\displaystyle\int \sin(\ln x) \, dx$

24. $\displaystyle\int \cos(\ln x) \, dx$

25. $\displaystyle\int x \sec^2 x \, dx$

26. $\displaystyle\int x \tan^2 x \, dx$

27. $\displaystyle\int x^3 e^{x^2} \, dx$

28. $\displaystyle\int \frac{x e^x}{(x+1)^2} \, dx$

29. $\displaystyle\int_0^1 x e^{-5x} \, dx$

30. $\displaystyle\int_0^2 x e^{2x} \, dx$

31. $\displaystyle\int_1^e x^2 \ln x \, dx$

32. $\displaystyle\int_{\sqrt{e}}^e \frac{\ln x}{x^2} \, dx$

33. $\displaystyle\int_{-2}^2 \ln(x + 3) \, dx$

34. $\displaystyle\int_0^{1/2} \sin^{-1} x \, dx$

35. $\displaystyle\int_2^4 \sec^{-1} \sqrt{\theta} \, d\theta$

36. $\displaystyle\int_1^2 x \sec^{-1} x \, dx$

37. $\displaystyle\int_0^{\pi/2} x \sin 4x \, dx$

38. $\displaystyle\int_0^\pi (x + x \cos x) \, dx$

39. $\displaystyle\int_1^3 \sqrt{x} \tan^{-1} \sqrt{x} \, dx$

40. $\displaystyle\int_0^2 \ln(x^2 + 1) \, dx$

41. In each part, evaluate the integral by making a u-substitution and then integrating by parts.

(a) $\displaystyle\int e^{\sqrt{x}} \, dx$

(b) $\displaystyle\int \cos \sqrt{x} \, dx$

☐c **42.** For each of the integrals you evaluated in Exercises 1–41, use a CAS to check your answer. If the answer produced by the CAS does not match your own, show that the two answers are equivalent.

43. (a) Find the area of the region enclosed by $y = \ln x$, the line $x = e$, and the x-axis.

(b) Find the volume of the solid generated when the region in part (a) is revolved about the x-axis.

44. Find the area of the region between $y = x \sin x$ and $y = x$ for $0 \le x \le \pi/2$.

45. Find the volume of the solid generated when the region between $y = \sin x$ and $y = 0$ for $0 \le x \le \pi$ is revolved about the y-axis.

46. Find the volume of the solid generated when the region enclosed between $y = \cos x$ and $y = 0$ for $0 \le x \le \pi/2$ is revolved about the y-axis.

47. A particle moving along the x-axis has velocity function $v(t) = t^2 e^{-t}$. How far does the particle travel from time $t = 0$ to $t = 5$?

48. The study of sawtooth waves in electrical engineering leads to integrals of the form

$$\int_{-\pi/\omega}^{\pi/\omega} t \sin(k\omega t) \, dt$$

where k is an integer and ω is a nonzero constant. Evaluate the integral.

49. Use reduction formula (7) to evaluate

(a) $\displaystyle\int \sin^3 x \, dx$

(b) $\displaystyle\int_0^{\pi/4} \sin^4 x \, dx$.

50. Use reduction formula (8) to evaluate

(a) $\displaystyle\int \cos^5 x \, dx$ (b) $\displaystyle\int_0^{\pi/2} \cos^6 x \, dx$.

51. Derive reduction formula (7).

52. In each part, use integration by parts or other methods to derive the reduction formula.

(a) $\displaystyle\int \sec^n x \, dx = \frac{\sec^{n-2} x \tan x}{n-1} + \frac{n-2}{n-1} \int \sec^{n-2} x \, dx$

(b) $\displaystyle\int \tan^n x \, dx = \frac{\tan^{n-1} x}{n-1} - \int \tan^{n-2} x \, dx$

(c) $\displaystyle\int x^n e^x \, dx = x^n e^x - n \int x^{n-1} e^x \, dx$

In Exercises 53 and 54, use the reduction formulas in Exercise 52 to evaluate the integrals.

53. (a) $\displaystyle\int \tan^4 x \, dx$ (b) $\displaystyle\int \sec^4 x \, dx$ (c) $\displaystyle\int x^3 e^x \, dx$

54. (a) $\displaystyle\int x^2 e^{3x} \, dx$ (b) $\displaystyle\int_0^1 x e^{-\sqrt{x}} \, dx$

[*Hint:* First make a substitution.]

55. Let f be a function whose second derivative is continuous on $[-1, 1]$. Show that

$$\int_{-1}^1 x f''(x) \, dx = f'(1) + f'(-1) - f(1) + f(-1)$$

56. Recall from Theorem 4.1.5 and the discussion preceding it that if $f'(x) > 0$, then the function f is increasing and has an inverse. The purpose of this problem is to show that if this condition is satisfied and if f' is continuous, then a definite integral of f^{-1} can be expressed in terms of a definite integral of f.

(a) Use integration by parts to show that

$$\int_a^b f(x) \, dx = bf(b) - af(a) - \int_a^b x f'(x) \, dx$$

(b) Use the result in part (a) to show that if $y = f(x)$, then

$$\int_a^b f(x) \, dx = bf(b) - af(a) - \int_{f(a)}^{f(b)} f^{-1}(y) \, dy$$

(c) Show that if we let $\alpha = f(a)$ and $\beta = f(b)$, then the result in part (b) can be written as

$$\int_\alpha^\beta f^{-1}(x) \, dx = \beta f^{-1}(\beta) - \alpha f^{-1}(\alpha) - \int_{f^{-1}(\alpha)}^{f^{-1}(\beta)} f(x) \, dx$$

57. In each part, use the result in Exercise 56 to obtain the equation, and then confirm that the equation is correct by performing the integrations.

(a) $\displaystyle\int_0^{1/2} \sin^{-1} x \, dx = \frac{1}{2} \sin^{-1}\left(\frac{1}{2}\right) - \int_0^{\pi/6} \sin x \, dx$

(b) $\displaystyle\int_e^{e^2} \ln x \, dx = (2e^2 - e) - \int_1^2 e^x \, dx$

58. (a) In Example 1, let

$$u = x, \quad dv = e^x \, dx,$$

$$du = dx, \quad v = \int e^x \, dx = e^x + C_1$$

and show that the constant C_1 cancels out, thus giving the same solution obtained by omitting C_1.

(b) Show that in general

$$uv - \int v \, du = u(v + C_1) - \int (v + C_1) \, du$$

thereby justifying the omission of the constant of integration when calculating v in integration by parts.

9.3 TRIGONOMETRIC INTEGRALS

In the last section we derived reduction formulas for integrating positive integer powers of sine, cosine, tangent, and secant. In this section we will show how to work with those reduction formulas, and we will discuss methods for integrating other kinds of integrals that involve trigonometric functions.

INTEGRATING POWERS OF SINE AND COSINE

In the preceding section we derived the reduction formulas

$$\int \sin^n x \, dx = -\frac{1}{n} \sin^{n-1} x \cos x + \frac{n-1}{n} \int \sin^{n-2} x \, dx \qquad (1)$$

$$\int \cos^n x \, dx = \frac{1}{n} \cos^{n-1} x \sin x + \frac{n-1}{n} \int \cos^{n-2} x \, dx \qquad (2)$$

In the case where $n = 2$, these formulas yield

$$\int \sin^2 x \, dx = -\frac{1}{2} \sin x \cos x + \frac{1}{2} \int dx = \frac{1}{2} x - \frac{1}{2} \sin x \cos x + C \qquad (3)$$

$$\int \cos^2 x \, dx = \frac{1}{2} \cos x \sin x + \frac{1}{2} \int dx = \frac{1}{2}x + \frac{1}{2} \sin x \cos x + C \qquad (4)$$

Alternative forms of these integration formulas can be derived from the trigonometric identities

$$\sin^2 x = \tfrac{1}{2}(1 - \cos 2x) \qquad \text{and} \qquad \cos^2 x = \tfrac{1}{2}(1 + \cos 2x) \qquad (5\text{–}6)$$

which follow from the double-angle formulas

$$\cos 2x = 1 - 2\sin^2 x \quad \text{and} \quad \cos 2x = 2\cos^2 x - 1$$

These identities yield

$$\int \sin^2 x \, dx = \frac{1}{2} \int (1 - \cos 2x) \, dx = \frac{1}{2}x - \frac{1}{4} \sin 2x + C \qquad (7)$$

$$\int \cos^2 x \, dx = \frac{1}{2} \int (1 + \cos 2x) \, dx = \frac{1}{2}x + \frac{1}{4} \sin 2x + C \qquad (8)$$

Observe that the antiderivatives in Formulas (3) and (4) involve both sines and cosines, whereas those in (7) and (8) involve sines alone. However, the apparent discrepancy is easy to resolve by using the identity

$$\sin 2x = 2 \sin x \cos x$$

to rewrite (7) and (8) in forms (3) and (4), or conversely.

In the case where $n = 3$, the reduction formulas for integrating $\sin^3 x$ and $\cos^3 x$ yield

$$\int \sin^3 x \, dx = -\frac{1}{3} \sin^2 x \cos x + \frac{2}{3} \int \sin x \, dx = -\frac{1}{3} \sin^2 x \cos x - \frac{2}{3} \cos x + C \quad (9)$$

$$\int \cos^3 x \, dx = \frac{1}{3} \cos^2 x \sin x + \frac{2}{3} \int \cos x \, dx = \frac{1}{3} \cos^2 x \sin x + \frac{2}{3} \sin x + C \quad (10)$$

If desired, Formula (9) can be expressed in terms of cosines alone by using the identity $\sin^2 x = 1 - \cos^2 x$, and Formula (10) can be expressed in terms of sines alone by using the identity $\cos^2 x = 1 - \sin^2 x$. We leave it for you to do this and confirm that

$$\int \sin^3 x \, dx = \tfrac{1}{3} \cos^3 x - \cos x + C \qquad (11)$$

$$\int \cos^3 x \, dx = \sin x - \tfrac{1}{3} \sin^3 x + C \qquad (12)$$

FOR THE READER. When asked to integrate $\sin^3 x$ and $\cos^3 x$, the *Maple* CAS produces forms (11) and (12). However, the *Mathematica* CAS produces

$$\int \sin^3 x \, dx = -\tfrac{3}{4} \cos x + \tfrac{1}{12} \cos 3x + C$$

$$\int \cos^3 x \, dx = \tfrac{3}{4} \sin x + \tfrac{1}{12} \sin 3x + C$$

See if you can reconcile *Mathematica*'s results with (11) and (12).

We leave it as an exercise to obtain the following formulas by first applying the reduction formulas, and then using appropriate trigonometric identities.

$$\int \sin^4 x \, dx = \tfrac{3}{8}x - \tfrac{1}{4} \sin 2x + \tfrac{1}{32} \sin 4x + C \qquad (13)$$

$$\int \cos^4 x \, dx = \tfrac{3}{8}x + \tfrac{1}{4} \sin 2x + \tfrac{1}{32} \sin 4x + C \qquad (14)$$

Example 1

Find the volume V of the solid that is obtained when the region under the curve $y = \sin^2 x$ over the interval $[0, \pi]$ is revolved about the x-axis (Figure 9.3.1).

Solution. Using the method of disks, Formula (5) of Section 8.2 yields

$$V = \int_0^\pi \pi \sin^4 x \, dx = \pi \left[\tfrac{3}{8}x - \tfrac{1}{4}\sin 2x + \tfrac{1}{32}\sin 4x \right]_0^\pi = \tfrac{3}{8}\pi^2 \qquad \blacktriangleleft$$

INTEGRATING PRODUCTS OF SINES AND COSINES

If m and n are positive integers, then the integral

$$\int \sin^m x \cos^n x \, dx$$

can be evaluated by one of the three procedures stated in Table 9.3.1, depending on whether m and n are odd or even.

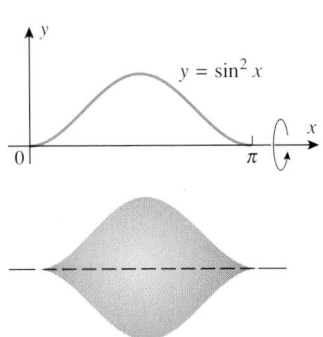

Figure 9.3.1

Example 2

Evaluate

(a) $\displaystyle\int \sin^4 x \cos^5 x \, dx$ (b) $\displaystyle\int \sin^4 x \cos^4 x \, dx$

Solution (a). Since $n = 5$ is odd, we will follow the first procedure in Table 9.3.1:

$$\int \sin^4 x \cos^5 x \, dx = \int \sin^4 x \cos^4 x \cos x \, dx$$

$$= \int \sin^4 x (1 - \sin^2 x)^2 \cos x \, dx$$

$$= \int u^4 (1 - u^2)^2 \, du$$

$$= \int (u^4 - 2u^6 + u^8) \, du$$

$$= \tfrac{1}{5}u^5 - \tfrac{2}{7}u^7 + \tfrac{1}{9}u^9 + C$$

$$= \tfrac{1}{5}\sin^5 x - \tfrac{2}{7}\sin^7 x + \tfrac{1}{9}\sin^9 x + C$$

Solution (b). Since $m = n = 4$, both exponents are even, so we will follow the third procedure in Table 9.3.1:

$$\int \sin^4 x \cos^4 x \, dx = \int (\sin^2 x)^2 (\cos^2 x)^2 \, dx$$

$$= \int \left(\frac{1}{2}[1 - \cos 2x] \right)^2 \left(\frac{1}{2}[1 + \cos 2x] \right)^2 dx$$

$$= \frac{1}{16} \int (1 - \cos^2 2x)^2 \, dx$$

$$= \frac{1}{16} \int \sin^4 2x \, dx \qquad \boxed{\begin{array}{l}\text{Note that this can be obtained more} \\ \text{directly from the original integral using} \\ \text{the identity } \sin x \cos x = \tfrac{1}{2}\sin 2x.\end{array}}$$

$$= \frac{1}{32} \int \sin^4 u \, du \qquad \boxed{\begin{array}{l} u = 2x \\ du = 2\,dx \text{ or } dx = \tfrac{1}{2}\,du \end{array}}$$

$$= \frac{1}{32} \left(\frac{3}{8}u - \frac{1}{4}\sin 2u + \frac{1}{32}\sin 4u \right) + C \qquad \boxed{\text{Formula (13)}}$$

$$= \frac{3}{128}x - \frac{1}{128}\sin 4x + \frac{1}{1024}\sin 8x + C \qquad \blacktriangleleft$$

Table 9.3.1

$\int \sin^m x \cos^n x\,dx$	PROCEDURE	RELEVANT IDENTITIES
n odd	• Split off a factor of $\cos x$. • Apply the relevant identity. • Make the substitution $u = \sin x$.	$\cos^2 x = 1 - \sin^2 x$
m odd	• Split off a factor of $\sin x$. • Apply the relevant identity. • Make the substitution $u = \cos x$.	$\sin^2 x = 1 - \cos^2 x$
$\begin{cases} m \text{ even} \\ n \text{ even} \end{cases}$	• Use the relevant identities to reduce the powers on $\sin x$ and $\cos x$.	$\begin{cases} \sin^2 x = \frac{1}{2}(1 - \cos 2x) \\ \cos^2 x = \frac{1}{2}(1 + \cos 2x) \end{cases}$

Integrals of the form

$$\int \sin mx \cos nx\,dx, \quad \int \sin mx \sin nx\,dx, \quad \int \cos mx \cos nx\,dx \tag{15}$$

can be found by using the trigonometric identities

$$\sin\alpha\cos\beta = \tfrac{1}{2}[\sin(\alpha - \beta) + \sin(\alpha + \beta)] \tag{16}$$

$$\sin\alpha\sin\beta = \tfrac{1}{2}[\cos(\alpha - \beta) - \cos(\alpha + \beta)] \tag{17}$$

$$\cos\alpha\cos\beta = \tfrac{1}{2}[\cos(\alpha - \beta) + \cos(\alpha + \beta)] \tag{18}$$

to express the integrand as a sum or difference of sines and cosines.

Example 3

Evaluate $\int \sin 7x \cos 3x\,dx$.

Solution. Using (16) yields

$$\int \sin 7x \cos 3x\,dx = \frac{1}{2}\int (\sin 4x + \sin 10x)\,dx = -\frac{1}{8}\cos 4x - \frac{1}{20}\cos 10x + C \quad \blacktriangleleft$$

INTEGRATING POWERS OF TANGENT AND SECANT

The procedures for integrating powers of tangent and secant closely parallel those for sine and cosine. The idea is to use the following reduction formulas (which were derived in Exercise 52 of Section 9.2) to reduce the exponent in the integrand until the resulting integral can be evaluated:

$$\int \tan^n x\,dx = \frac{\tan^{n-1} x}{n-1} - \int \tan^{n-2} x\,dx \tag{19}$$

$$\int \sec^n x\,dx = \frac{\sec^{n-2} x \tan x}{n-1} + \frac{n-2}{n-1}\int \sec^{n-2} x\,dx \tag{20}$$

In the case where n is odd, the exponent can be reduced to 1, leaving us with the problem of integrating $\tan x$ or $\sec x$. These integrals are given by

$$\int \tan x\,dx = \ln|\sec x| + C \tag{21}$$

$$\int \sec x\,dx = \ln|\sec x + \tan x| + C \tag{22}$$

Formula (21) can be obtained by writing

$$\int \tan x \, dx = \int \frac{\sin x}{\cos x} \, dx$$

$$= -\ln |\cos x| + C \qquad \boxed{\begin{array}{l} u = \cos x \\ du = -\sin x \, dx \end{array}}$$

$$= \ln |\sec x| + C \qquad \boxed{\ln |\cos x| = -\ln \frac{1}{|\cos x|}}$$

Formula (22) requires a trick. We write

$$\int \sec x \, dx = \int \sec x \left(\frac{\sec x + \tan x}{\sec x + \tan x} \right) dx = \int \frac{\sec^2 x + \sec x \tan x}{\sec x + \tan x} \, dx$$

$$= \ln |\sec x + \tan x| + C \qquad \boxed{\begin{array}{l} u = \sec x + \tan x \\ du = (\sec^2 x + \sec x \tan x) \, dx \end{array}}$$

The following basic integrals occur frequently and are worth noting:

$$\int \tan^2 x \, dx = \tan x - x + C \tag{23}$$

$$\int \sec^2 x \, dx = \tan x + C \tag{24}$$

Formula (24) is already known to us, since the derivative of $\tan x$ is $\sec^2 x$. Formula (23) can be obtained by applying reduction formula (19) with $n = 2$ (verify) or, alternatively, by using the identity

$$1 + \tan^2 x = \sec^2 x$$

to write

$$\int \tan^2 x \, dx = \int (\sec^2 x - 1) \, dx = \tan x - x + C$$

The formulas

$$\int \tan^3 x \, dx = \tfrac{1}{2} \tan^2 x - \ln |\sec x| + C \tag{25}$$

$$\int \sec^3 x \, dx = \tfrac{1}{2} \sec x \tan x + \tfrac{1}{2} \ln |\sec x + \tan x| + C \tag{26}$$

can be deduced from (21), (22), and reduction formulas (19) and (20) as follows:

$$\int \tan^3 x \, dx = \frac{1}{2} \tan^2 x - \int \tan x \, dx = \frac{1}{2} \tan^2 x - \ln |\sec x| + C$$

$$\int \sec^3 x \, dx = \frac{1}{2} \sec x \tan x + \frac{1}{2} \int \sec x \, dx = \frac{1}{2} \sec x \tan x + \frac{1}{2} \ln |\sec x + \tan x| + C$$

INTEGRATING PRODUCTS OF TANGENTS AND SECANTS

If m and n are positive integers, then the integral

$$\int \tan^m x \sec^n x \, dx$$

can be evaluated by one of the three procedures stated in Table 9.3.2, depending on whether m and n are odd or even.

Example 4

Evaluate

(a) $\displaystyle\int \tan^2 x \sec^4 x \, dx$ (b) $\displaystyle\int \tan^3 x \sec^3 x \, dx$ (c) $\displaystyle\int \tan^2 x \sec x \, dx$

Table 9.3.2

$\int \tan^m x \sec^n x\, dx$	PROCEDURE	RELEVANT IDENTITIES
n even	• Split off a factor of $\sec^2 x$. • Apply the relevant identity. • Make the substitution $u = \tan x$.	$\sec^2 x = \tan^2 x + 1$
m odd	• Split off a factor of $\sec x \tan x$. • Apply the relevant identity. • Make the substitution $u = \sec x$.	$\tan^2 x = \sec^2 x - 1$
$\begin{cases} m \text{ even} \\ n \text{ odd} \end{cases}$	• Use the relevant identities to reduce the integrand to powers of $\sec x$ alone. • Then use the reduction formula for powers of $\sec x$.	$\tan^2 x = \sec^2 x - 1$

Solution (a). Since $n = 4$ is even, we will follow the first procedure in Table 9.3.2:

$$\int \tan^2 x \sec^4 x\, dx = \int \tan^2 x \sec^2 x \sec^2 x\, dx$$

$$= \int \tan^2 x (\tan^2 x + 1) \sec^2 x\, dx$$

$$= \int u^2 (u^2 + 1)\, du$$

$$= \tfrac{1}{5} u^5 + \tfrac{1}{3} u^3 + C = \tfrac{1}{5} \tan^5 x + \tfrac{1}{3} \tan^3 x + C$$

Solution (b). Since $m = 3$ is odd, we will follow the second procedure in Table 9.3.2:

$$\int \tan^3 x \sec^3 x\, dx = \int \tan^2 x \sec^2 x (\sec x \tan x)\, dx$$

$$= \int (\sec^2 x - 1) \sec^2 x (\sec x \tan x)\, dx$$

$$= \int (u^2 - 1) u^2\, du$$

$$= \tfrac{1}{5} u^5 - \tfrac{1}{3} u^3 + C = \tfrac{1}{5} \sec^5 x - \tfrac{1}{3} \sec^3 x + C$$

Solution (c). Since $m = 2$ is even and $n = 1$ is odd, we will follow the third procedure in Table 9.3.2:

$$\int \tan^2 x \sec x\, dx = \int (\sec^2 x - 1) \sec x\, dx$$

$$= \int \sec^3 x\, dx - \int \sec x\, dx \qquad \boxed{\text{See (26) and (22).}}$$

$$= \tfrac{1}{2} \sec x \tan x + \tfrac{1}{2} \ln |\sec x + \tan x| - \ln |\sec x + \tan x| + C$$

$$= \tfrac{1}{2} \sec x \tan x - \tfrac{1}{2} \ln |\sec x + \tan x| + C \qquad \blacktriangleleft$$

AN ALTERNATIVE METHOD FOR INTEGRATING POWERS OF SINE, COSINE, TANGENT, AND SECANT

The methods in Tables 9.3.1 and 9.3.2 can sometimes be applied if $m = 0$ or $n = 0$ to integrate positive integer powers of sine, cosine, tangent, and secant without reduction formulas. For example, instead of using the reduction formula to integrate $\sin^3 x$, we can apply the second procedure in Table 9.3.1.

$$\int \sin^3 x\, dx = \int (\sin^2 x)\sin x\, dx$$

$$= \int (1 - \cos^2 x)\sin x\, dx \qquad \boxed{\begin{array}{l} u = \cos x \\ du = -\sin x\, dx \end{array}}$$

$$= -\int (1 - u^2)\, du$$

$$= \tfrac{1}{3}u^3 - u + C = \tfrac{1}{3}\cos^3 x - \cos x + C$$

which agrees with (11).

REMARK.   With the aid of the identity $1 + \cot^2 x = \csc^2 x$ the techniques in Table 9.3.2 can be adapted to treat integrals of the form

$$\int \cot^m x\ \csc^n x\, dx$$

Also, there are reduction formulas for powers of cosecant and cotangent that are analogous to Formulas (19) and (20).

MERCATOR'S MAP OF THE WORLD

The integral of $\sec x$ plays an important role in the design of navigational maps for charting nautical and aeronautical courses. Sailors and pilots usually chart their courses along paths with constant compass headings; for example, the course might be $30°$ northeast or $135°$ southwest. Except for courses that are parallel to the equator or run due north or south, a course with constant compass heading spirals around the Earth toward one of the poles (as in Figure 9.3.2a). However, in 1569 the Flemish mathematician and geographer Gerhard Kramer (1512–1594) (better known by the Latin name Mercator) devised a world map, called the *Mercator projection*, in which spirals of constant compass headings appear as straight lines. This was extremely important because it enabled sailors to determine compass headings between two points by connecting them with a straight line on a map (Figure 9.3.2b).

A flight with constant compass heading from New York City to Moscow as it appears on a globe

(a)

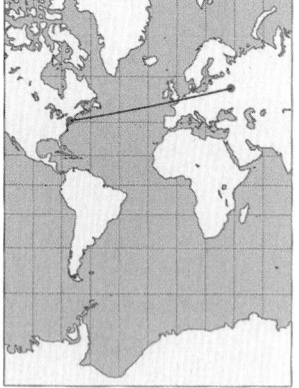

A flight with constant compass heading from New York City to Moscow as it appears on a Mercator projection

(b)

Figure 9.3.2

If the Earth is assumed to be a sphere of radius 4000 mi, then the lines of latitude at $1°$ increments are equally spaced about 70 mi apart (why?). However, in the Mercator projection, the lines of latitude become wider apart toward the poles, so that two widely spaced latitude lines near the poles may be actually the same distance apart on the Earth as two closely spaced latitude lines near the equator. It can be proved that on a Mercator map in which the equatorial line has length L, the vertical distance D_β on the map between the equator (latitude $0°$) and the line of latitude $\beta°$ is

$$D_\beta = \frac{L}{2\pi} \int_0^\beta \sec x \, dx \qquad (27)$$

(see Exercises 59 and 60).

EXERCISE SET 9.3   ☐c CAS

In Exercises 1–52, evaluate the integral.

1. $\displaystyle\int \cos^5 x \sin x \, dx$ **2.** $\displaystyle\int \sin^4 3x \cos 3x \, dx$

3. $\displaystyle\int \sin ax \cos ax \, dx$ **4.** $\displaystyle\int \cos^2 3x \, dx$

5. $\displaystyle\int \sin^2 5\theta \, d\theta$ **6.** $\displaystyle\int \cos^3 at \, dt$

7. $\displaystyle\int \cos^5 \theta \, d\theta$ **8.** $\displaystyle\int \sin^3 x \cos^3 x \, dx$

9. $\displaystyle\int \sin^2 2t \cos^3 2t \, dt$ **10.** $\displaystyle\int \sin^3 2x \cos^2 2x \, dx$

11. $\displaystyle\int \sin^2 x \cos^2 x \, dx$ **12.** $\displaystyle\int \sin^2 x \cos^4 x \, dx$

13. $\displaystyle\int \sin x \cos 2x \, dx$ **14.** $\displaystyle\int \sin 3\theta \cos 2\theta \, d\theta$

15. $\displaystyle\int \sin x \cos(x/2) \, dx$ **16.** $\displaystyle\int \cos^{1/5} x \sin x \, dx$

17. $\displaystyle\int_0^{\pi/4} \cos^3 x \, dx$ **18.** $\displaystyle\int_0^{\pi/2} \sin^2 \frac{x}{2} \cos^2 \frac{x}{2} \, dx$

19. $\displaystyle\int_0^{\pi/3} \sin^4 3x \cos^3 3x \, dx$ **20.** $\displaystyle\int_{-\pi}^{\pi} \cos^2 5\theta \, d\theta$

21. $\displaystyle\int_0^{\pi/6} \sin 2x \cos 4x \, dx$ **22.** $\displaystyle\int_0^{2\pi} \sin^2 kx \, dx$

23. $\displaystyle\int \sec^2(3x+1) \, dx$ **24.** $\displaystyle\int \tan 5x \, dx$

25. $\displaystyle\int e^{-2x} \tan(e^{-2x}) \, dx$ **26.** $\displaystyle\int \cot 3x \, dx$

27. $\displaystyle\int \sec 2x \, dx$ **28.** $\displaystyle\int \frac{\sec(\sqrt{x})}{\sqrt{x}} \, dx$

29. $\displaystyle\int \tan^2 x \sec^2 x \, dx$ **30.** $\displaystyle\int \tan^5 x \sec^4 x \, dx$

31. $\displaystyle\int \tan^3 4x \sec^4 4x \, dx$ **32.** $\displaystyle\int \tan^4 \theta \sec^4 \theta \, d\theta$

33. $\displaystyle\int \sec^5 x \tan^3 x \, dx$ **34.** $\displaystyle\int \tan^5 \theta \sec \theta \, d\theta$

35. $\displaystyle\int \tan^4 x \sec x \, dx$ **36.** $\displaystyle\int \tan^2 \frac{x}{2} \sec^3 \frac{x}{2} \, dx$

37. $\displaystyle\int \tan 2t \sec^3 2t \, dt$ **38.** $\displaystyle\int \tan x \sec^5 x \, dx$

39. $\displaystyle\int \sec^4 x \, dx$ **40.** $\displaystyle\int \sec^5 x \, dx$

41. $\displaystyle\int \tan^4 x \, dx$ **42.** $\displaystyle\int \tan^3 4x \, dx$

43. $\displaystyle\int \sqrt{\tan x} \sec^4 x \, dx$ **44.** $\displaystyle\int \tan x \sec^{3/2} x \, dx$

45. $\displaystyle\int_0^{\pi/6} \tan^2 2x \, dx$ **46.** $\displaystyle\int_0^{\pi/6} \sec^3 \theta \tan \theta \, d\theta$

47. $\displaystyle\int_0^{\pi/2} \tan^5 \frac{x}{2} \, dx$ **48.** $\displaystyle\int_0^{1/4} \sec \pi x \tan \pi x \, dx$

49. $\displaystyle\int \cot^3 x \csc^3 x \, dx$ **50.** $\displaystyle\int \cot^2 3t \sec 3t \, dt$

51. $\displaystyle\int \cot^3 x \, dx$ **52.** $\displaystyle\int \csc^4 x \, dx$

53. Let m, n be distinct nonnegative integers. Use Formulas (16)–(18) to prove:

    (a) $\displaystyle\int_0^{2\pi} \sin mx \cos nx \, dx = 0$

    (b) $\displaystyle\int_0^{2\pi} \cos mx \cos nx \, dx = 0$

    (c) $\displaystyle\int_0^{2\pi} \sin mx \sin nx \, dx = 0.$

☐c **54.** For each of the integrals you evaluated in Exercises 1–52, use a CAS to check your answer. If the answer produced by the CAS does not match your own, show that the two answers are equivalent.

55. Find the arc length of the curve $y = \ln(\cos x)$ over the interval $[0, \pi/4]$.

56. Find the volume of the solid generated when the region enclosed by $y = \tan x$, $y = 1$, and $x = 0$ is revolved about the x-axis.

57. Find the volume of the solid that results when the region enclosed by $y = \cos x$, $y = \sin x$, $x = 0$, and $x = \pi/4$ is revolved about the x-axis.

58. The region bounded below by the x-axis and above by the portion of $y = \sin x$ from $x = 0$ to $x = \pi$ is revolved about the x-axis. Find the volume of the resulting solid.

59. Use Formula (27) to show that if the length of the equatorial line on a Mercator projection is L, then the vertical distance D between the latitude lines at $\alpha°$ and $\beta°$ on the same side of the equator (where $\alpha < \beta$) is

$$D = \frac{L}{2\pi} \ln \left| \frac{\sec \beta + \tan \beta}{\sec \alpha + \tan \alpha} \right|$$

60. Suppose that the equator has a length of 100 cm on a Mercator projection. In each part, use the result in Exercise 59 to answer the question.
  (a) What is the vertical distance on the map between the equator and the line at $25°$ north latitude?
  (b) What is the vertical distance on the map between New Orleans, Louisiana, at $30°$ north latitude and Winnepeg, Canada, at $50°$ north latitude?

61. (a) Show that

$$\int \csc x \, dx = -\ln|\csc x + \cot x| + C$$

  (b) Show that the result in part (a) can also be written as

$$\int \csc x \, dx = \ln|\csc x - \cot x| + C$$

  and

$$\int \csc x \, dx = \ln\left|\tan \tfrac{1}{2}x\right| + C$$

62. Rewrite $\sin x + \cos x$ in the form

$$A \sin(x + \phi)$$

  and use your result together with Exercise 61 to evaluate

$$\int \frac{dx}{\sin x + \cos x}$$

63. Use the method of Exercise 62 to evaluate

$$\int \frac{dx}{a \sin x + b \cos x} \qquad (a, b \text{ not both zero})$$

64. (a) Use Formula (7) in Section 9.2 to show that

$$\int_0^{\pi/2} \sin^n x \, dx = \frac{n-1}{n} \int_0^{\pi/2} \sin^{n-2} x \, dx$$

  (b) Use this result to derive the **Wallis sine formulas**:

$$\int_0^{\pi/2} \sin^n x \, dx = \frac{\pi}{2} \cdot \frac{1 \cdot 3 \cdot 5 \cdots (n-1)}{2 \cdot 4 \cdot 6 \cdots n} \quad \left(\begin{array}{c} n \text{ even} \\ \text{and} \geq 2 \end{array}\right)$$

$$\int_0^{\pi/2} \sin^n x \, dx = \frac{2 \cdot 4 \cdot 6 \cdots (n-1)}{3 \cdot 5 \cdot 7 \cdots n} \quad \left(\begin{array}{c} n \text{ odd} \\ \text{and} \geq 3 \end{array}\right)$$

65. Use the Wallis formulas in Exercise 64 to evaluate

  (a) $\displaystyle\int_0^{\pi/2} \sin^3 x \, dx$ (b) $\displaystyle\int_0^{\pi/2} \sin^4 x \, dx$

  (c) $\displaystyle\int_0^{\pi/2} \sin^5 x \, dx$ (d) $\displaystyle\int_0^{\pi/2} \sin^6 x \, dx.$

66. Use Formula (8) in Section 9.2 and the method of Exercise 64 to derive the **Wallis cosine formulas**:

$$\int_0^{\pi/2} \cos^n x \, dx = \frac{\pi}{2} \cdot \frac{1 \cdot 3 \cdot 5 \cdots (n-1)}{2 \cdot 4 \cdot 6 \cdots n} \quad \left(\begin{array}{c} n \text{ even} \\ \text{and} \geq 2 \end{array}\right)$$

$$\int_0^{\pi/2} \cos^n x \, dx = \frac{2 \cdot 4 \cdot 6 \cdots (n-1)}{3 \cdot 5 \cdot 7 \cdots n} \quad \left(\begin{array}{c} n \text{ odd} \\ \text{and} \geq 3 \end{array}\right)$$

9.4 TRIGONOMETRIC SUBSTITUTIONS

In this section we will discuss a method for evaluating integrals containing radicals by making substitutions involving trigonometric functions. We will also show how integrals containing quadratic polynomials can sometimes be evaluated by completing the square.

THE METHOD OF TRIGONOMETRIC SUBSTITUTION

To start, we will be concerned with integrals that contain expressions of the form

$$\sqrt{a^2 - x^2}, \quad \sqrt{x^2 + a^2}, \quad \sqrt{x^2 - a^2}$$

in which a is a positive constant. The basic idea for evaluating such integrals is to make a substitution for x that will eliminate the radical. For example, to eliminate the radical in the expression $\sqrt{a^2 - x^2}$, we can make the substitution

$$x = a \sin\theta, \quad -\pi/2 \leq \theta \leq \pi/2 \qquad (1)$$

which yields

$$\sqrt{a^2 - x^2} = \sqrt{a^2 - a^2 \sin^2\theta} = \sqrt{a^2(1 - \sin^2\theta)}$$

$$= a\sqrt{\cos^2\theta} = a|\cos\theta| = a\cos\theta \qquad \boxed{\cos\theta \geq 0 \text{ since } -\pi/2 \leq \theta \leq \pi/2}$$

The restriction on θ in (1) serves two purposes—it enables us to replace $|\cos\theta|$ by $\cos\theta$ to simplify the calculations, and it also ensures that the substitutions can be rewritten as $\theta = \sin^{-1}(x/a)$, if needed.

Example 1

Evaluate $\displaystyle\int \frac{dx}{x^2\sqrt{4-x^2}}$.

Solution. To eliminate the radical we make the substitution

$$x = 2\sin\theta, \quad dx = 2\cos\theta \, d\theta$$

This yields

$$\int \frac{dx}{x^2\sqrt{4-x^2}} = \int \frac{2\cos\theta \, d\theta}{(2\sin\theta)^2\sqrt{4-4\sin^2\theta}}$$

$$= \int \frac{2\cos\theta \, d\theta}{(2\sin\theta)^2(2\cos\theta)} = \frac{1}{4}\int \frac{d\theta}{\sin^2\theta}$$

$$= \frac{1}{4}\int \csc^2\theta \, d\theta = -\frac{1}{4}\cot\theta + C \qquad (2)$$

At this point we have completed the integration; however, because the original integral was expressed in terms of x, it is desirable to express $\cot\theta$ in terms of x as well. This can be done using trigonometric identities, but the expression can also be obtained by writing the substitution $x = 2\sin\theta$ as $\sin\theta = x/2$ and representing it geometrically as in Figure 9.4.1. From that figure we obtain

$$\cot\theta = \frac{\sqrt{4-x^2}}{x}$$

Substituting this in (2) yields

$$\int \frac{dx}{x^2\sqrt{4-x^2}} = -\frac{1}{4}\frac{\sqrt{4-x^2}}{x} + C \qquad \blacktriangleleft$$

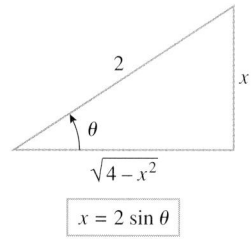

$$x = 2\sin\theta$$

Figure 9.4.1

Example 2

Evaluate $\displaystyle\int_1^{\sqrt{2}} \frac{dx}{x^2\sqrt{4-x^2}}$.

Solution. There are two possible approaches: we can make the substitution in the indefinite integral (as in Example 1) and then evaluate the definite integral using the x-limits of integration, or we can make the substitution in the definite integral and convert the x-limits to the corresponding θ-limits.

Method 1. Using the result from Example 1 with the x-limits of integration yields

$$\int_1^{\sqrt{2}} \frac{dx}{x^2\sqrt{4-x^2}} = -\frac{1}{4}\left[\frac{\sqrt{4-x^2}}{x}\right]_1^{\sqrt{2}} = -\frac{1}{4}[1-\sqrt{3}] = \frac{\sqrt{3}-1}{4}$$

Method 2. The substitution $x = 2\sin\theta$ can be expressed as $x/2 = \sin\theta$ or $\theta = \sin^{-1}(x/2)$, so the θ-limits that correspond to $x = 1$ and $x = \sqrt{2}$ are

$$x = 1: \quad \theta = \sin^{-1}(1/2) = \pi/6$$

$$x = \sqrt{2}: \quad \theta = \sin^{-1}(\sqrt{2}/2) = \pi/4$$

Thus, from (2) in Example 1 we obtain

$$\int_1^{\sqrt{2}} \frac{dx}{x^2\sqrt{4-x^2}} = \int_{\pi/6}^{\pi/4} \frac{2\cos\theta\,d\theta}{(2\sin\theta)^2\sqrt{4-4\sin^2\theta}} = \frac{1}{4}\int_{\pi/6}^{\pi/4} \frac{d\theta}{\sin^2\theta}$$

$$= \frac{1}{4}\int_{\pi/6}^{\pi/4} \csc^2\theta\,d\theta = -\frac{1}{4}\Big[\cot\theta\Big]_{\pi/6}^{\pi/4}$$

$$= -\frac{1}{4}[1-\sqrt{3}\,] = \frac{\sqrt{3}-1}{4} \qquad \blacktriangleleft$$

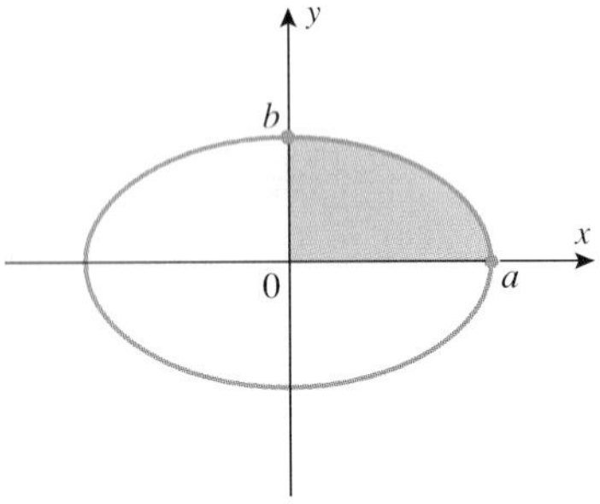

Example 3

Find the area of the ellipse

$$\frac{x^2}{a^2} + \frac{y^2}{b^2} = 1$$

Figure 9.4.2

Solution.   Because the ellipse is symmetric about both axes, its area A is four times the area in the first quadrant (Figure 9.4.2). If we solve the equation of the ellipse for y in terms of x, we obtain

$$y = \pm\frac{b}{a}\sqrt{a^2-x^2}$$

where the positive square root gives the equation of the upper half. Thus, the area A is given by

$$A = 4\int_0^a \frac{b}{a}\sqrt{a^2-x^2}\,dx = \frac{4b}{a}\int_0^a \sqrt{a^2-x^2}\,dx$$

To evaluate this integral, we will make the substitution $x = a\sin\theta$ $(dx = a\cos\theta\,d\theta)$ and convert the x-limits of integration to θ-limits. Since the substitution can be expressed as $\theta = \sin^{-1}(x/a)$, the θ-limits of integration are

$$x = 0: \quad \theta = \sin^{-1}(0) = 0$$

$$x = a: \quad \theta = \sin^{-1}(1) = \pi/2$$

Thus, we obtain

$$A = \frac{4b}{a}\int_0^a \sqrt{a^2-x^2}\,dx = \frac{4b}{a}\int_0^{\pi/2} a\cos\theta \cdot a\cos\theta\,d\theta$$

$$= 4ab\int_0^{\pi/2} \cos^2\theta\,d\theta = 4ab\int_0^{\pi/2} \frac{1}{2}(1+\cos2\theta)\,d\theta$$

$$= 2ab\Big[\theta + \frac{1}{2}\sin2\theta\Big]_0^{\pi/2} = 2ab\Big[\frac{\pi}{2} - 0\Big] = \pi ab \qquad \blacktriangleleft$$

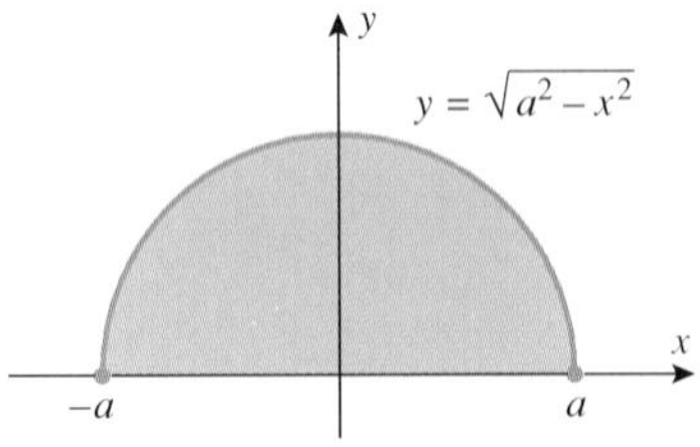

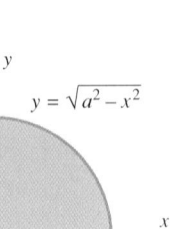

Figure 9.4.3

REMARK.   In the special case where $a = b$, the ellipse becomes a circle of radius a, and the area formula becomes $A = \pi a^2$, as expected. It is worth noting that

$$\int_{-a}^a \sqrt{a^2-x^2}\,dx = \tfrac{1}{2}\pi a^2 \tag{3}$$

since this integral represents the area of the upper semicircle (Figure 9.4.3).

FOR THE READER.   If you have a calculating utility with a numerical integration capability, use it and Formula (3) to approximate π to three decimal places.

Thus far, we have focused on using the substitution $x = a\sin\theta$ to evaluate integrals involving radicals of the form $\sqrt{a^2-x^2}$. Table 9.4.1 summarizes this method and describes some other substitutions of this type.

Table 9.4.1

EXPRESSION IN THE INTEGRAND	SUBSTITUTION	RESTRICTION ON θ	SIMPLIFICATION
$\sqrt{a^2 - x^2}$	$x = a\sin\theta$	$-\pi/2 \le \theta \le \pi/2$	$a^2 - x^2 = a^2 - a^2\sin^2\theta = a^2\cos^2\theta$
$\sqrt{a^2 + x^2}$	$x = a\tan\theta$	$-\pi/2 < \theta < \pi/2$	$a^2 + x^2 = a^2 + a^2\tan^2\theta = a^2\sec^2\theta$
$\sqrt{x^2 - a^2}$	$x = a\sec\theta$	$\begin{cases} 0 \le \theta < \pi/2 & (\text{if } x \ge a) \\ \pi/2 < \theta \le \pi & (\text{if } x \le -a) \end{cases}$	$x^2 - a^2 = a^2\sec^2\theta - a^2 = a^2\tan^2\theta$

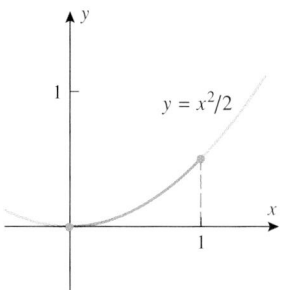

Figure 9.4.4

Example 4

Find the arc length of the curve $y = x^2/2$ from $x = 0$ to $x = 1$ (Figure 9.4.4).

Solution. From Formula (4) of Section 8.4 the arc length L of the curve is

$$L = \int_0^1 \sqrt{1 + \left(\frac{dy}{dx}\right)^2}\, dx = \int_0^1 \sqrt{1 + x^2}\, dx$$

The integrand involves a radical of the form $\sqrt{a^2 + x^2}$ with $a = 1$, so from Table 9.4.1 we make the substitution

$$x = \tan\theta, \quad -\pi/2 < \theta < \pi/2$$
$$\frac{dx}{d\theta} = \sec^2\theta \quad \text{or} \quad dx = \sec^2\theta\, d\theta$$

Since this substitution can be expressed as $\theta = \tan^{-1} x$, the θ-limits of integration that correspond to the x-limits, $x = 0$ and $x = 1$, are

$$x = 0: \quad \theta = \tan^{-1} 0 = 0$$
$$x = 1: \quad \theta = \tan^{-1} 1 = \pi/4$$

Thus,

$$L = \int_0^1 \sqrt{1 + x^2}\, dx = \int_0^{\pi/4} \sqrt{1 + \tan^2\theta}\, \sec^2\theta\, d\theta$$

$$= \int_0^{\pi/4} \sqrt{\sec^2\theta}\, \sec^2\theta\, d\theta$$

$$= \int_0^{\pi/4} |\sec\theta| \sec^2\theta\, d\theta$$

$$= \int_0^{\pi/4} \sec^3\theta\, d\theta \qquad \text{sec}\,\theta > 0 \text{ since } -\pi/2 < \theta < \pi/2$$

$$= \left[\tfrac{1}{2}\sec\theta\tan\theta + \tfrac{1}{2}\ln|\sec\theta + \tan\theta|\right]_0^{\pi/4} \qquad \begin{array}{l}\text{Formula (26)}\\\text{of Section 9.3}\end{array}$$

$$= \tfrac{1}{2}[\sqrt{2} + \ln(\sqrt{2} + 1)] \approx 1.148 \qquad \blacktriangleleft$$

Example 5

Evaluate $\displaystyle\int \frac{\sqrt{x^2 - 25}}{x}\, dx$, assuming that $x \ge 5$.

Solution. The integrand involves a radical of the form $\sqrt{x^2 - a^2}$ with $a = 5$, so from Table 9.4.1 we make the substitution

$$x = 5\sec\theta, \quad 0 \le \theta < \pi/2$$
$$\frac{dx}{d\theta} = 5\sec\theta\tan\theta \quad \text{or} \quad dx = 5\sec\theta\tan\theta\, d\theta$$

Thus,

$$\int \frac{\sqrt{x^2 - 25}}{x} \, dx = \int \frac{\sqrt{25 \sec^2 \theta - 25}}{5 \sec \theta} (5 \sec \theta \tan \theta) \, d\theta$$

$$= \int \frac{5|\tan \theta|}{5 \sec \theta} (5 \sec \theta \tan \theta) \, d\theta$$

$$= 5 \int \tan^2 \theta \, d\theta \qquad \boxed{\begin{array}{l} \tan \theta \geq 0 \text{ since} \\ 0 \leq \theta < \pi/2 \end{array}}$$

$$= 5 \int (\sec^2 \theta - 1) \, d\theta = 5 \tan \theta - 5\theta + C$$

To express the solution in terms of x, we will represent the substitution $x = 5 \sec \theta$ geometrically by the triangle in Figure 9.4.5, from which we obtain

$$\tan \theta = \frac{\sqrt{x^2 - 25}}{5}$$

From this and the fact that the substitution can be expressed as $\theta = \sec^{-1}(x/5)$, we obtain

$$\int \frac{\sqrt{x^2 - 25}}{x} \, dx = \sqrt{x^2 - 25} - 5 \sec^{-1}\left(\frac{x}{5}\right) + C \qquad \blacktriangleleft$$

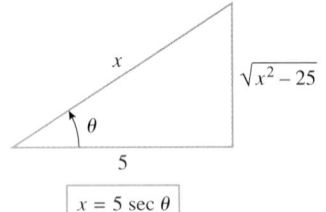

$x = 5 \sec \theta$

Figure 9.4.5

INTEGRALS INVOLVING
$ax^2 + bx + c$

Integrals that involve a quadratic expression $ax^2 + bx + c$, where $a \neq 0$ and $b \neq 0$, can often be evaluated by first completing the square, then making an appropriate substitution. The following examples illustrate this idea:

Example 6

Evaluate $\displaystyle\int \frac{x}{x^2 - 4x + 8} \, dx$.

Solution. Completing the square yields

$$x^2 - 4x + 8 = (x^2 - 4x + 4) + 8 - 4 = (x - 2)^2 + 4$$

Thus, the substitution

$$u = x - 2, \quad du = dx$$

yields

$$\int \frac{x}{x^2 - 4x + 8} \, dx = \int \frac{x}{(x-2)^2 + 4} \, dx = \int \frac{u + 2}{u^2 + 4} \, du$$

$$= \int \frac{u}{u^2 + 4} \, du + 2 \int \frac{du}{u^2 + 4}$$

$$= \frac{1}{2} \int \frac{2u}{u^2 + 4} \, du + 2 \int \frac{du}{u^2 + 4}$$

$$= \frac{1}{2} \ln(u^2 + 4) + 2 \left(\frac{1}{2}\right) \tan^{-1} \frac{u}{2} + C$$

$$= \frac{1}{2} \ln[(x-2)^2 + 4] + \tan^{-1}\left(\frac{x-2}{2}\right) + C \qquad \blacktriangleleft$$

Example 7

Evaluate $\displaystyle\int \frac{dx}{\sqrt{5 - 4x - 2x^2}}$.

Solution. Completing the square yields

$$5 - 4x - 2x^2 = 5 - 2(x^2 + 2x) = 5 - 2(x^2 + 2x + 1) + 2$$

$$= 5 - 2(x + 1)^2 + 2 = 7 - 2(x + 1)^2$$

Thus,

$$\int \frac{dx}{\sqrt{5-4x-2x^2}} = \int \frac{dx}{\sqrt{7-2(x+1)^2}}$$

$$= \int \frac{du}{\sqrt{7-2u^2}} \quad \begin{matrix} u = x+1 \\ du = dx \end{matrix}$$

$$= \frac{1}{\sqrt{2}} \int \frac{du}{\sqrt{(7/2)-u^2}}$$

$$= \frac{1}{\sqrt{2}} \sin^{-1}\left(\frac{u}{\sqrt{7/2}}\right) + C \quad \begin{matrix} \text{Formula (21), Section 9.1} \\ \text{with } a = \sqrt{7/2} \end{matrix}$$

$$= \frac{1}{\sqrt{2}} \sin^{-1}(\sqrt{2/7}(x+1)) + C \quad \blacktriangleleft$$

EXERCISE SET 9.4 © CAS

In Exercises 1–26, evaluate the integral.

1. $\int \sqrt{4-x^2}\,dx$

2. $\int \sqrt{1-4x^2}\,dx$

3. $\int \frac{x^2}{\sqrt{9-x^2}}\,dx$

4. $\int \frac{dx}{x^2\sqrt{16-x^2}}$

5. $\int \frac{dx}{(4+x^2)^2}$

6. $\int \frac{x^2}{\sqrt{5+x^2}}\,dx$

7. $\int \frac{\sqrt{x^2-9}}{x}\,dx$

8. $\int \frac{dx}{x^2\sqrt{x^2-16}}$

9. $\int \frac{x^3}{\sqrt{2-x^2}}\,dx$

10. $\int x^3\sqrt{5-x^2}\,dx$

11. $\int \frac{dx}{x^2\sqrt{4x^2-9}}$

12. $\int \frac{\sqrt{1+t^2}}{t}\,dt$

13. $\int \frac{dx}{(1-x^2)^{3/2}}$

14. $\int \frac{dx}{x^2\sqrt{x^2+25}}$

15. $\int \frac{dx}{\sqrt{x^2-1}}$

16. $\int \frac{dx}{1+2x^2+x^4}$

17. $\int \frac{dx}{(9x^2-1)^{3/2}}$

18. $\int \frac{x^2}{\sqrt{x^2-25}}\,dx$

19. $\int e^x\sqrt{1-e^{2x}}\,dx$

20. $\int \frac{\cos\theta}{\sqrt{2-\sin^2\theta}}\,d\theta$

21. $\int_0^4 x^3\sqrt{16-x^2}\,dx$

22. $\int_0^{1/3} \frac{dx}{(4-9x^2)^2}$

23. $\int_{\sqrt{2}}^2 \frac{dx}{x^2\sqrt{x^2-1}}$

24. $\int_{\sqrt{2}}^2 \frac{\sqrt{2x^2-4}}{x}\,dx$

25. $\int_1^3 \frac{dx}{x^4\sqrt{x^2+3}}$

26. $\int_0^3 \frac{x^3}{(3+x^2)^{5/2}}\,dx$

27. The integral

$$\int \frac{x}{x^2+4}\,dx$$

can be evaluated either by a trigonometric substitution or by the substitution $u = x^2+4$. Do it both ways and show that the results are equivalent.

© **28.** For each of the integrals you evaluated in Exercises 1–27, use a CAS to check your answer. If the answer produced by the CAS does not match your own, show that the two answers are equivalent.

29. Find the arc length of the curve $y = \ln x$ from $x = 1$ to $x = 2$.

30. Find the arc length of the curve $y = x^2$ from $x = 0$ to $x = 1$.

31. Find the area of the surface generated when the curve in Exercise 30 is revolved about the x-axis.

32. Find the volume of the solid generated when the region enclosed by $x = y(1-y^2)^{1/4}$, $y = 0$, $y = 1$, and $x = 0$ is revolved about the y-axis.

In Exercise 33, the trigonometric substitutions $x = a\sec\theta$ and $x = a\tan\theta$ lead to difficult integrals; for such integrals it is sometimes possible to use the **hyperbolic substitutions**

$x = a\sinh u$ for integrals involving $\sqrt{x^2+a^2}$

$x = a\cosh u$ for integrals involving $\sqrt{x^2-a^2}$

These substitutions are useful because in each case the hyperbolic identity

$a^2\cosh^2 u - a^2\sinh^2 u = a^2$

removes the radical.

33. (a) Evaluate

$$\int \frac{dx}{\sqrt{x^2+9}}$$

using the hyperbolic substitution that is suggested above.

(b) Evaluate the integral in part (a) by a trigonometric substitution and show that the results in parts (a) and (b) agree.

(c) Use a hyperbolic substitution to evaluate

$$\int \sqrt{x^2 - 1}\, dx, \quad x \geq 1$$

34. In Example 3 we found the area of an ellipse by making the substitution $x = a \sin\theta$ in the required integral. Find the area by making the substitution $x = a \cos\theta$, and discuss any restrictions on θ that are needed.

In Exercises 35–46, evaluate the integral.

35. $\displaystyle\int \frac{dx}{x^2 - 4x + 13}$

36. $\displaystyle\int \frac{dx}{\sqrt{2x - x^2}}$

37. $\displaystyle\int \frac{dx}{\sqrt{8 + 2x - x^2}}$

38. $\displaystyle\int \frac{dx}{16x^2 + 16x + 5}$

39. $\displaystyle\int \frac{dx}{\sqrt{x^2 - 6x + 10}}$

40. $\displaystyle\int \frac{x}{x^2 + 6x + 10}\, dx$

41. $\displaystyle\int \sqrt{3 - 2x - x^2}\, dx$

42. $\displaystyle\int \frac{e^x}{\sqrt{1 + e^x + e^{2x}}}\, dx$

43. $\displaystyle\int \frac{dx}{2x^2 + 4x + 7}$

44. $\displaystyle\int \frac{2x + 3}{4x^2 + 4x + 5}\, dx$

45. $\displaystyle\int_1^2 \frac{dx}{\sqrt{4x - x^2}}$

46. $\displaystyle\int_0^1 \sqrt{x(4 - x)}\, dx$

C **47.** For each of the integrals you evaluated in Exercises 35–46, use a CAS to check your answer. If the answer produced by the CAS does not match your own, show that the two answers are equivalent.

In Exercises 48–50, there is a good chance that your CAS will not be able to evaluate the integral as stated. If this is so, make a substitution that converts the integral into one that your CAS can evaluate.

C **48.** $\displaystyle\int (x \cos x + \sin x)\sqrt{1 + x^2 \sin^2 x}\, dx$

C **49.** $\displaystyle\int \cos x \sin x \sqrt{1 - \sin^4 x}\, dx$

C **50.** $\displaystyle\int_0^1 3^x \sqrt{9^x - 1}\, dx$

9.5 INTEGRATING RATIONAL FUNCTIONS BY PARTIAL FRACTIONS

Recall that a rational function is a ratio of two polynomials. In this section we will give a general method for integrating rational functions that is based on the idea of decomposing a rational function into a sum of simple rational functions that can be integrated by the methods studied in earlier sections.

PARTIAL FRACTIONS

In algebra one learns to combine two or more fractions into a single fraction by finding a common denominator. For example,

$$\frac{2}{x - 4} + \frac{3}{x + 1} = \frac{2(x + 1) + 3(x - 4)}{(x - 4)(x + 1)} = \frac{5x - 10}{x^2 - 3x - 4} \tag{1}$$

However, for purposes of integration, the left side of (1) is preferable to the right side since each of the terms is easy to integrate:

$$\int \frac{5x - 10}{x^2 - 3x - 4}\, dx = \int \frac{2}{x - 4}\, dx + \int \frac{3}{x + 1}\, dx = 2\ln|x - 4| + 3\ln|x + 1| + C$$

Thus, it is desirable to have some method that will enable us to obtain the left side of (1), starting with the right side. To illustrate how this can be done, we begin by noting that on the left side the numerators are constants and the denominators are the factors of the denominator on the right side. Thus, to find the left side of (1), starting from the right side, we could factor the denominator of the right side and look for constants A and B such that

$$\frac{5x - 10}{(x - 4)(x + 1)} = \frac{A}{x - 4} + \frac{B}{x + 1} \tag{2}$$

One way to find the constants A and B is to multiply (2) through by $(x - 4)(x + 1)$ to clear fractions. This yields

$$5x - 10 = A(x + 1) + B(x - 4) \tag{3}$$

This relationship holds for all x, so it holds in particular if $x = 4$ or $x = -1$. Substituting

$x = 4$ in (3) makes the second term on the right drop out and yields the equation $10 = 5A$ or $A = 2$; and substituting $x = -1$ in (3) makes the first term on the right drop out and yields the equation $-15 = -5B$ or $B = 3$. Substituting these values in (2) we obtain

$$\frac{5x - 10}{(x - 4)(x + 1)} = \frac{2}{x - 4} + \frac{3}{x + 1} \tag{4}$$

which agrees with (1).

A second method for finding the constants A and B is to multiply out the right side of (3) and collect like powers of x to obtain

$$5x - 10 = (A + B)x + (A - 4B)$$

Since the polynomials on the two sides are identical, their corresponding coefficients must be the same. Equating the corresponding coefficients on the two sides yields the following system of equations in the unknowns A and B:

$$\begin{aligned} A + B &= 5 \\ A - 4B &= -10 \end{aligned}$$

Solving this system yields $A = 2$ and $B = 3$ as before (verify).

The terms on the right side of (4) are called ***partial fractions*** of the expression on the left side because they each constitute *part* of that expression. To find those partial fractions we first had to make a guess about their form, and then we had to find the unknown constants. Our next objective is to extend this idea to general rational functions. For this purpose, suppose that $P(x)/Q(x)$ is a ***proper rational function***, by which we mean that the degree of the numerator is less than the degree of the denominator. There is a theorem in advanced algebra which states that every proper rational function can be expressed as a sum

$$\frac{P(x)}{Q(x)} = F_1(x) + F_2(x) + \cdots + F_n(x)$$

where $F_1(x), F_2(x), \ldots, F_n(x)$ are rational functions of the form

$$\frac{A}{(ax + b)^k} \quad \text{or} \quad \frac{Ax + B}{(ax^2 + bx + c)^k}$$

in which the denominators are factors of $Q(x)$. The sum is called the ***partial fraction decomposition*** of $P(x)/Q(x)$, and the terms are called ***partial fractions***. As in our opening example, there are two parts to finding a partial fraction decomposition: determining the exact form of the decomposition and finding the unknown constants.

FINDING THE FORM OF A PARTIAL FRACTION DECOMPOSITION

The first step in finding the form of the partial fraction decomposition of a proper rational function $P(x)/Q(x)$ is to factor $Q(x)$ completely into linear and irreducible quadratic factors, and then collect all repeated factors so that $Q(x)$ is expressed as a product of *distinct* factors of the form

$$(ax + b)^m \quad \text{and} \quad (ax^2 + bx + c)^m$$

From these factors we can determine the form of the partial fraction decomposition using two rules that we will now discuss.

LINEAR FACTORS

If all of the factors of $Q(x)$ are linear, then the partial fraction decomposition of $P(x)/Q(x)$ can be determined by using the following rule:

LINEAR FACTOR RULE. For each factor of the form $(ax + b)^m$, the partial fraction decomposition contains the following sum of m partial fractions:

$$\frac{A_1}{ax + b} + \frac{A_2}{(ax + b)^2} + \cdots + \frac{A_m}{(ax + b)^m}$$

where $A_1, A_2, \ldots, A_m$ are constants to be determined. In the case where $m = 1$, only the first term in the sum appears.

Example 1

Evaluate $\displaystyle\int \frac{dx}{x^2 + x - 2}$.

Solution. The integrand is a proper rational function that can be written as

$$\frac{1}{x^2 + x - 2} = \frac{1}{(x - 1)(x + 2)}$$

The factors $x - 1$ and $x + 2$ are both linear and appear to the first power, so each contributes one term to the partial fraction decomposition by the linear factor rule. Thus, the decomposition has the form

$$\frac{1}{(x - 1)(x + 2)} = \frac{A}{x - 1} + \frac{B}{x + 2} \tag{5}$$

where A and B are constants to be determined. Multiplying this expression through by $(x - 1)(x + 2)$ yields

$$1 = A(x + 2) + B(x - 1) \tag{6}$$

As discussed earlier, there are two methods for finding A and B: we can substitute values of x that are chosen to make terms on the right drop out, or we can multiply out on the right and equate corresponding coefficients on the two sides to obtain a system of equations that can be solved for A and B. We will use the first approach.

Setting $x = 1$ makes the second term in (6) drop out and yields $1 = 3A$ or $A = \frac{1}{3}$; and setting $x = -2$ makes the first term in (6) drop out and yields $1 = -3B$ or $B = -\frac{1}{3}$. Substituting these values in (5) yields the partial fraction decomposition

$$\frac{1}{(x - 1)(x + 2)} = \frac{\frac{1}{3}}{x - 1} + \frac{-\frac{1}{3}}{x + 2}$$

The integration can now be completed as follows:

$$\int \frac{dx}{(x - 1)(x + 2)} = \frac{1}{3} \int \frac{dx}{x - 1} - \frac{1}{3} \int \frac{dx}{x + 2}$$

$$= \frac{1}{3} \ln|x - 1| - \frac{1}{3} \ln|x + 2| + C = \frac{1}{3} \ln\left|\frac{x - 1}{x + 2}\right| + C \quad \blacktriangleleft$$

If the factors of $Q(x)$ are linear and none are repeated, as in the last example, then the recommended method for finding the constants in the partial fraction decomposition is to substitute appropriate values of x to make terms drop out. However, if some of the linear factors are repeated, then it will not be possible to find all of the constants in this way. In this case the recommended procedure is to find as many constants as possible by substitution and then find the rest by equating coefficients. This is illustrated in the next example.

Example 2

Evaluate $\displaystyle\int \frac{2x + 4}{x^3 - 2x^2}\, dx$.

Solution. The integrand can be rewritten as

$$\frac{2x + 4}{x^3 - 2x^2} = \frac{2x + 4}{x^2(x - 2)}$$

Although x^2 is a quadratic factor, it is *not* irreducible since $x^2 = xx$. Thus, by the linear factor rule, x^2 introduces two terms (since $m = 2$) of the form

$$\frac{A}{x} + \frac{B}{x^2}$$

and the factor $x - 2$ introduces one term (since $m = 1$) of the form

$$\frac{C}{x - 2}$$

so the partial fraction decomposition is

$$\frac{2x + 4}{x^2(x - 2)} = \frac{A}{x} + \frac{B}{x^2} + \frac{C}{x - 2} \tag{7}$$

Multiplying by $x^2(x - 2)$ yields

$$2x + 4 = Ax(x - 2) + B(x - 2) + Cx^2 \tag{8}$$

which, after multiplying out and collecting like powers of x, becomes

$$2x + 4 = (A + C)x^2 + (-2A + B)x - 2B \tag{9}$$

Setting $x = 0$ in (8) makes the first and third terms drop out and yields $B = -2$, and setting $x = 2$ in (8) makes the first and second terms drop out and yields $C = 2$ (verify). However, there is no substitution in (8) that produces A directly, so we look to Equation (9) to find this value. This can be done by equating the coefficients of x^2 on the two sides to obtain

$$A + C = 0 \quad \text{or} \quad A = -C = -2$$

Substituting the values $A = -2$, $B = -2$, and $C = 2$ in (7) yields the partial fraction decomposition

$$\frac{2x + 4}{x^2(x - 2)} = \frac{-2}{x} + \frac{-2}{x^2} + \frac{2}{x - 2}$$

Thus,

$$\int \frac{2x + 4}{x^2(x - 2)} \, dx = -2 \int \frac{dx}{x} - 2 \int \frac{dx}{x^2} + 2 \int \frac{dx}{x - 2}$$

$$= -2 \ln |x| + \frac{2}{x} + 2 \ln |x - 2| + C = 2 \ln \left| \frac{x - 2}{x} \right| + \frac{2}{x} + C \quad \blacktriangleleft$$

QUADRATIC FACTORS

If some of the factors of $Q(x)$ are irreducible quadratics, then the contribution of those factors to the partial fraction decomposition of $P(x)/Q(x)$ can be determined from the following rule:

QUADRATIC FACTOR RULE. For each factor of the form $(ax^2 + bx + c)^m$, the partial fraction decomposition contains the following sum of m partial fractions:

$$\frac{A_1x + B_1}{ax^2 + bx + c} + \frac{A_2x + B_2}{(ax^2 + bx + c)^2} + \cdots + \frac{A_mx + B_m}{(ax^2 + bx + c)^m}$$

where $A_1, A_2, \ldots, A_m, B_1, B_2, \ldots, B_m$ are constants to be determined. In the case where $m = 1$, only the first term in the sum appears.

Example 3

Evaluate $\displaystyle\int \frac{x^2 + x - 2}{3x^3 - x^2 + 3x - 1} \, dx.$

Solution. The denominator in the integrand can be factored by grouping:

$$\frac{x^2 + x - 2}{3x^3 - x^2 + 3x - 1} = \frac{x^2 + x - 2}{x^2(3x - 1) + (3x - 1)} = \frac{x^2 + x - 2}{(3x - 1)(x^2 + 1)}$$

By the linear factor rule, the factor $3x - 1$ introduces one term; namely

$$\frac{A}{3x - 1}$$

and by the quadratic factor rule, the factor $x^2 + 1$ introduces one term; namely

$$\frac{Bx + C}{x^2 + 1}$$

Thus, the partial fraction decomposition is

$$\frac{x^2 + x - 2}{(3x - 1)(x^2 + 1)} = \frac{A}{3x - 1} + \frac{Bx + C}{x^2 + 1} \tag{10}$$

Multiplying by $(3x - 1)(x^2 + 1)$ yields

$$x^2 + x - 2 = A(x^2 + 1) + (Bx + C)(3x - 1) \tag{11}$$

We could find A by substituting $x = \frac{1}{3}$ to make the last term drop out, and then find the rest of the constants by equating corresponding coefficients. However, in this case it is just as easy to find *all* of the constants by equating coefficients and solving the resulting system. For this purpose we multiply out the right side of (11) and collect like terms:

$$x^2 + x - 2 = (A + 3B)x^2 + (-B + 3C)x + (A - C)$$

Equating corresponding coefficients gives

$$
\begin{aligned}
A + 3B \quad\;\; &= \;\; 1 \\
-\,B + 3C &= \;\; 1 \\
A \quad\quad\; -\,C &= -2
\end{aligned}
$$

To solve this system, subtract the third equation from the first to eliminate A. Then use the resulting equation together with the second equation to solve for B and C. Finally, determine A from the first or third equation. This yields (verify)

$$A = -\tfrac{7}{5}, \quad B = \tfrac{4}{5}, \quad C = \tfrac{3}{5}$$

Thus, (10) becomes

$$\frac{x^2 + x - 2}{(3x - 1)(x^2 + 1)} = \frac{-\frac{7}{5}}{3x - 1} + \frac{\frac{4}{5}x + \frac{3}{5}}{x^2 + 1}$$

and

$$\int \frac{x^2 + x - 2}{(3x - 1)(x^2 + 1)}\,dx = -\frac{7}{5}\int \frac{dx}{3x - 1} + \frac{4}{5}\int \frac{x}{x^2 + 1}\,dx + \frac{3}{5}\int \frac{dx}{x^2 + 1}$$

$$= -\frac{7}{15}\ln|3x - 1| + \frac{2}{5}\ln(x^2 + 1) + \frac{3}{5}\tan^{-1} x + C \quad\blacktriangleleft$$

FOR THE READER.   Computer algebra systems have built-in capabilities for finding partial fraction decompositions. If you have a CAS, read the documentation on partial fraction decompositions, and use your CAS to find the decompositions in Examples 1, 2, and 3.

Example 4

Evaluate $\displaystyle\int \frac{3x^4 + 4x^3 + 16x^2 + 20x + 9}{(x + 2)(x^2 + 3)^2}\,dx.$

Solution.   Observe that the integrand is a proper rational function since the numerator has degree 4 and the denominator has degree 5. Thus, the method of partial fractions is applicable. By the linear factor rule, the factor $x + 2$ introduces the single term

$$\frac{A}{x + 2}$$

and by the quadratic factor rule, the factor $(x^2 + 3)^2$ introduces two terms (since $m = 2$):

$$\frac{Bx + C}{x^2 + 3} + \frac{Dx + E}{(x^2 + 3)^2}$$

Thus, the partial fraction decomposition of the integrand is

$$\frac{3x^4 + 4x^3 + 16x^2 + 20x + 9}{(x + 2)(x^2 + 3)^2} = \frac{A}{x + 2} + \frac{Bx + C}{x^2 + 3} + \frac{Dx + E}{(x^2 + 3)^2} \qquad (12)$$

Multiplying by $(x + 2)(x^2 + 3)^2$ yields

$$3x^4 + 4x^3 + 16x^2 + 20x + 9$$
$$= A(x^2 + 3)^2 + (Bx + C)(x^2 + 3)(x + 2) + (Dx + E)(x + 2) \qquad (13)$$

which, after multiplying out and collecting like powers of x, becomes

$$3x^4 + 4x^3 + 16x^2 + 20x + 9$$
$$= (A + B)x^4 + (2B + C)x^3 + (6A + 3B + 2C + D)x^2$$
$$+ (6B + 3C + 2D + E)x + (9A + 6C + 2E) \qquad (14)$$

Equating corresponding coefficients in (14) yields the following system of five linear equations in five unknowns:

$$\begin{array}{rcl}
A + B &=& 3 \\
2B + C &=& 4 \\
6A + 3B + 2C + D &=& 16 \\
6B + 3C + 2D + E &=& 20 \\
9A + 6C + 2E &=& 9
\end{array} \qquad (15)$$

Efficient methods for solving systems of linear equations such as this are studied in a branch of mathematics called *linear algebra*; those methods are outside the scope of this text. However, as a practical matter most linear systems of any size are solved by computer, and most computer algebra systems have commands that in many cases can solve linear systems exactly. In this particular case we can simplify the work by first substituting $x = -2$ in (13), which yields $A = 1$. Substituting this known value of A in (15) yields the simpler system

$$\begin{array}{rcl}
B &=& 2 \\
2B + C &=& 4 \\
3B + 2C + D &=& 10 \\
6B + 3C + 2D + E &=& 20 \\
6C + 2E &=& 0
\end{array} \qquad (16)$$

This system can be solved by starting at the top and working down, first substituting $B = 2$ in the second equation to get $C = 0$, then substituting the known values of B and C in the third equation to get $D = 4$, and so forth. This yields

$$A = 1, \quad B = 2, \quad C = 0, \quad D = 4, \quad E = 0$$

Thus, (12) becomes

$$\frac{3x^4 + 4x^3 + 16x^2 + 20x + 9}{(x + 2)(x^2 + 3)^2} = \frac{1}{x + 2} + \frac{2x}{x^2 + 3} + \frac{4x}{(x^2 + 3)^2}$$

and so

$$\int \frac{3x^4 + 4x^3 + 16x^2 + 20x + 9}{(x + 2)(x^2 + 3)^2}\, dx$$
$$= \int \frac{dx}{x + 2} + \int \frac{2x}{x^2 + 3}\, dx + 4\int \frac{x}{(x^2 + 3)^2}\, dx$$
$$= \ln|x + 2| + \ln(x^2 + 3) - \frac{2}{x^2 + 3} + C \qquad \blacktriangleleft$$

Although the method of partial fractions only applies to proper rational functions, an improper rational function can be integrated by performing a long division and expressing the function as the quotient plus the remainder over the divisor. The remainder over the divisor will be a proper rational function, which can then be decomposed into partial fractions. This idea is illustrated in the following example:

Example 5

Evaluate $\displaystyle\int \frac{3x^4 + 3x^3 - 5x^2 + x - 1}{x^2 + x - 2}\, dx$.

Solution. The integrand is an improper rational function since the numerator has degree 4 and the denominator has degree 2. Thus, we first perform the long division

$$
\begin{array}{r}
3x^2 + 1 \\
x^2 + x - 2 \overline{\left)\; 3x^4 + 3x^3 - 5x^2 + x - 1\right.} \\
\underline{3x^4 + 3x^3 - 6x^2} \\
x^2 + x - 1 \\
\underline{x^2 + x - 2} \\
1
\end{array}
$$

It follows that the integrand can be expressed as

$$\frac{3x^4 + 3x^3 - 5x^2 + x - 1}{x^2 + x - 2} = (3x^2 + 1) + \frac{1}{x^2 + x - 2}$$

and hence

$$\int \frac{3x^4 + 3x^3 - 5x^2 + x - 1}{x^2 + x - 2}\, dx = \int (3x^2 + 1)\, dx + \int \frac{dx}{x^2 + x - 2}$$

The second integral on the right now involves a proper rational function and can thus be evaluated by a partial fraction decomposition. Using the result of Example 1 we obtain

$$\int \frac{3x^4 + 3x^3 - 5x^2 + x - 1}{x^2 + x - 2}\, dx = x^3 + x + \frac{1}{3}\ln\left|\frac{x - 1}{x + 2}\right| + C \qquad \blacktriangleleft$$

There are some cases in which the method of partial fractions is inappropriate. For example, it would be illogical to use partial fractions to perform the integration

$$\int \frac{3x^2 + 2}{x^3 + 2x - 8}\, dx = \ln|x^3 + 2x - 8| + C$$

since the substitution $u = x^3 + 2x - 8$ is more direct. Similarly, the integration

$$\int \frac{2x - 1}{x^2 + 1}\, dx = \int \frac{2x}{x^2 + 1}\, dx - \int \frac{dx}{x^2 + 1} = \ln(x^2 + 1) - \tan^{-1} x + C$$

requires only a little algebra since the integrand is already in partial-fraction form.

EXERCISE SET 9.5 ☐c CAS

In Exercises 1–8, write out the form of the partial fraction decomposition. (Do not find the numerical values of the coefficients.)

1. $\dfrac{3x - 1}{(x - 2)(x + 5)}$

2. $\dfrac{5}{x(x^2 - 9)}$

3. $\dfrac{2x - 3}{x^3 - x^2}$

4. $\dfrac{x^2}{(x + 2)^3}$

5. $\dfrac{1 - 5x^2}{x^3(x^2 + 1)}$

6. $\dfrac{2x}{(x - 1)(x^2 + 5)}$

7. $\dfrac{4x^3 - x}{(x^2 + 5)^2}$

8. $\dfrac{1 - 3x^4}{(x - 2)(x^2 + 1)^2}$

In Exercises 9–32, evaluate the integral.

9. $\displaystyle\int \frac{dx}{x^2 + 3x - 4}$

10. $\displaystyle\int \frac{dx}{x^2 + 8x + 7}$

11. $\displaystyle \int \frac{11x + 17}{2x^2 + 7x - 4} \, dx$

12. $\displaystyle \int \frac{5x - 5}{3x^2 - 8x - 3} \, dx$

13. $\displaystyle \int \frac{2x^2 - 9x - 9}{x^3 - 9x} \, dx$

14. $\displaystyle \int \frac{dx}{x(x^2 - 1)}$

15. $\displaystyle \int \frac{x^2 + 2}{x + 2} \, dx$

16. $\displaystyle \int \frac{x^2 - 4}{x - 1} \, dx$

17. $\displaystyle \int \frac{3x^2 - 10}{x^2 - 4x + 4} \, dx$

18. $\displaystyle \int \frac{x^2}{x^2 - 3x + 2} \, dx$

19. $\displaystyle \int \frac{x^5 + 2x^2 + 1}{x^3 - x} \, dx$

20. $\displaystyle \int \frac{2x^5 - x^3 - 1}{x^3 - 4x} \, dx$

21. $\displaystyle \int \frac{2x^2 + 3}{x(x - 1)^2} \, dx$

22. $\displaystyle \int \frac{3x^2 - x + 1}{x^3 - x^2} \, dx$

23. $\displaystyle \int \frac{x^2 + x - 16}{(x + 1)(x - 3)^2} \, dx$

24. $\displaystyle \int \frac{2x^2 - 2x - 1}{x^3 - x^2} \, dx$

25. $\displaystyle \int \frac{x^2}{(x + 2)^3} \, dx$

26. $\displaystyle \int \frac{2x^2 + 3x + 3}{(x + 1)^3} \, dx$

27. $\displaystyle \int \frac{2x^2 - 1}{(4x - 1)(x^2 + 1)} \, dx$

28. $\displaystyle \int \frac{dx}{x^3 + x}$

29. $\displaystyle \int \frac{x^3 + 3x^2 + x + 9}{(x^2 + 1)(x^2 + 3)} \, dx$

30. $\displaystyle \int \frac{x^3 + x^2 + x + 2}{(x^2 + 1)(x^2 + 2)} \, dx$

31. $\displaystyle \int \frac{x^3 - 3x^2 + 2x - 3}{x^2 + 1} \, dx$

32. $\displaystyle \int \frac{x^4 + 6x^3 + 10x^2 + x}{x^2 + 6x + 10} \, dx$

In Exercises 33 and 34, evaluate the integral by making a substitution that converts the integrand to a rational function.

33. $\displaystyle \int \frac{\cos\theta}{\sin^2\theta + 4\sin\theta - 5} \, d\theta$

34. $\displaystyle \int \frac{e^t}{e^{2t} - 4} \, dt$

35. Find the volume of the solid generated when the region enclosed by $y = x^2/(9 - x^2)$, $y = 0$, $x = 0$, and $x = 2$ is revolved about the x-axis.

36. Find the area of the region under the curve $y = 1/(1 + e^x)$, over the interval $[-\ln 5, \ln 5]$. [*Hint:* Make a substitution that converts the integrand to a rational function.]

In Exercises 37 and 38, use a CAS to evaluate the integral in two ways: (i) integrate directly; (ii) use the CAS to find the partial fraction decomposition and integrate the decomposition. Integrate by hand to check the results.

C **37.** $\displaystyle \int \frac{x^2 + 1}{(x^2 + 2x + 3)^2} \, dx$

C **38.** $\displaystyle \int \frac{x^5 + x^4 + 4x^3 + 4x^2 + 4x + 4}{(x^2 + 2)^3} \, dx$

In Exercises 39 and 40, integrate by hand and check your answers using a CAS.

C **39.** $\displaystyle \int \frac{dx}{x^4 - 3x^3 - 7x^2 + 27x - 18}$

C **40.** $\displaystyle \int \frac{dx}{16x^3 - 4x^2 + 4x - 1}$

41. Show that

$$\int_0^1 \frac{x}{x^4 + 1} \, dx = \frac{\pi}{8}$$

42. Use partial fractions to derive the integration formula

$$\int \frac{1}{a^2 - x^2} \, dx = \frac{1}{2a} \ln\left|\frac{a + x}{a - x}\right| + C$$

9.6 USING TABLES OF INTEGRALS AND COMPUTER ALGEBRA SYSTEMS

In this section we will discuss how to integrate using tables, and we will address some of the issues that relate to using computer algebra systems for integration. Readers who are not using computer algebra systems can skip that material with no problem.

INTEGRAL TABLES

Tables of integrals are useful for eliminating tedious hand computation. The endpapers of this text contain a relatively brief table of integrals that we will refer to as the ***Endpaper Integral Table***; more comprehensive tables are published in standard reference books such as the *CRC Standard Mathematical Tables and Formulae*, CRC Press, Inc., 1991.

All integral tables have their own scheme for classifying integrals according to the form of the integrand. For example, the Endpaper Integral Table classifies the integrals into 15 categories; *Basic Functions, Reciprocals of Basic Functions, Powers of Trigonometric Functions, Products of Trigonometric Functions,* and so forth. The first step in working with tables is to read through the classifications so that you understand the classification scheme and know where to look in the table for integrals of different types.

If you are lucky, the integral you are attempting to evaluate will match up perfectly with one of the forms in the table. However, when looking for matches you may have to make an adjustment for the variable of integration. For example, the integral

$$\int x^2 \sin x \, dx$$

is a perfect match with Formula (46) in the Endpaper Integral Table, except for the letter used for the variable of integration. Thus, to apply Formula (46) to the given integral we need to change the variable of integration in the formula from u to x. With that minor modification we obtain

$$\int x^2 \sin x \, dx = 2x \sin x + (2 - x^2) \cos x + C$$

Here are some more examples of perfect matches:

Example 1

Use the Endpaper Integral Table to evaluate

(a) $\displaystyle\int \sin 7x \cos 2x \, dx$ (b) $\displaystyle\int x^2 \sqrt{7 + 3x} \, dx$

(c) $\displaystyle\int \frac{\sqrt{2 - x^2}}{x} \, dx$ (d) $\displaystyle\int (x^3 + 7x + 1) \sin \pi x \, dx$

Solution (a). The integrand can be classified as a product of trigonometric functions. Thus, from Formula (40) with $m = 7$ and $n = 2$ we obtain

$$\int \sin 7x \cos 2x \, dx = -\frac{\cos 9x}{18} - \frac{\cos 5x}{10} + C$$

Solution (b). The integrand can be classified as a power of x multiplying $\sqrt{a + bx}$. Thus, from Formula (103) with $a = 7$ and $b = 3$ we obtain

$$\int x^2 \sqrt{7 + 3x} \, dx = \frac{2}{2835}(135x^2 - 252x + 392)(7 + 3x)^{3/2} + C$$

Solution (c). The integrand can be classified as a power of x dividing $\sqrt{a^2 - x^2}$. Thus, from Formula (79) with $a = \sqrt{2}$ we obtain

$$\int \frac{\sqrt{2 - x^2}}{x} \, dx = \sqrt{2 - x^2} - \sqrt{2} \ln \left| \frac{\sqrt{2} + \sqrt{2 - x^2}}{x} \right| + C$$

Solution (d). The integrand can be classified as a polynomial multiplying a trigonometric function. Thus, we apply Formula (58) with $p(x) = x^3 + 7x + 1$ and $a = \pi$. The successive nonzero derivatives of $p(x)$ are

$$p'(x) = 3x^2 + 7, \quad p''(x) = 6x, \quad p'''(x) = 6$$

and hence

$$\int (x^3 + 7x + 1) \sin \pi x \, dx$$

$$= -\frac{x^3 + 7x + 1}{\pi} \cos \pi x + \frac{3x^2 + 7}{\pi^2} \sin \pi x + \frac{6x}{\pi^3} \cos \pi x - \frac{6}{\pi^4} \sin \pi x + C \quad \blacktriangleleft$$

Sometimes an integral that does not match any table entry can be made to match by making an appropriate substitution. Here are some examples.

Example 2

Use the Endpaper Integral Table to evaluate $\displaystyle\int \sqrt{x - 4x^2} \, dx$.

Solution. The integrand does not match any of the forms in the table precisely. It comes closest to matching Formula (112), but it misses because of the factor of 4 multiplying x^2 inside the radical. However, if we make the substitution

$$u = 2x, \quad du = 2\,dx$$

then the $4x^2$ will become a u^2, and the transformed integral will be

$$\int \sqrt{x - 4x^2}\,dx = \frac{1}{2}\int \sqrt{\tfrac{1}{2}u - u^2}\,du$$

which matches Formula (112) with $a = \frac{1}{4}$. Thus, we obtain

$$\int \sqrt{x - 4x^2}\,dx = \frac{1}{2}\left[\frac{u - \frac{1}{4}}{2}\sqrt{\tfrac{1}{2}u - u^2} + \frac{1}{32}\sin^{-1}\left(\frac{u - \frac{1}{4}}{\frac{1}{4}}\right)\right] + C$$

$$= \frac{1}{2}\left[\frac{2x - \frac{1}{4}}{2}\sqrt{x - 4x^2} + \frac{1}{32}\sin^{-1}\left(\frac{2x - \frac{1}{4}}{\frac{1}{4}}\right)\right] + C$$

$$= \frac{8x - 1}{16}\sqrt{x - 4x^2} + \frac{1}{64}\sin^{-1}(8x - 1) + C \qquad \blacktriangleleft$$

Example 3

Use the Endpaper Integral Table to evaluate

(a) $\displaystyle\int e^{\pi x}\sin^{-1}(e^{\pi x})\,dx$ (b) $\displaystyle\int x\sqrt{x^2 - 4x + 5}\,dx$

Solution (a). The integrand does not even come close to matching any of the forms in the table. However, a little thought suggests the substitution

$$u = e^{\pi x}, \quad du = \pi e^{\pi x}\,dx$$

from which we obtain

$$\int e^{\pi x}\sin^{-1}(e^{\pi x})\,dx = \frac{1}{\pi}\int \sin^{-1}u\,du$$

The integrand is now a basic function, and Formula (7) yields

$$\int e^{\pi x}\sin^{-1}(e^{\pi x})\,dx = \frac{1}{\pi}[u\sin^{-1}u + \sqrt{1 - u^2}] + C$$

$$= \frac{1}{\pi}[e^{\pi x}\sin^{-1}(e^{\pi x}) + \sqrt{1 - e^{2\pi x}}] + C$$

Solution (b). Again, the integrand does not closely match any of the forms in the table. However, a little thought suggests that it may be possible to bring the integrand closer to the form $x\sqrt{x^2 + a^2}$ by completing the square to eliminate the term involving x inside the radical. Doing this yields

$$\int x\sqrt{x^2 - 4x + 5}\,dx = \int x\sqrt{(x^2 - 4x + 4) + 1}\,dx = \int x\sqrt{(x - 2)^2 + 1}\,dx \qquad (1)$$

At this point we are closer to the form $x\sqrt{x^2 + a^2}$, but we are not quite there because of the $(x - 2)^2$ rather than x^2 inside the radical. However, we can resolve that problem with the substitution

$$u = x - 2, \quad du = dx$$

With this substitution we have $x = u + 2$, so (1) can be expressed in terms of u as

$$\int x\sqrt{x^2 - 4x + 5}\,dx = \int (u + 2)\sqrt{u^2 + 1}\,du = \int u\sqrt{u^2 + 1}\,du + 2\int \sqrt{u^2 + 1}\,du$$

The first integral on the right is now a perfect match with Formula (84) with $a = 1$, and the second is a perfect match with Formula (72) with $a = 1$. Thus, applying these formulas and dropping the unnecessary absolute value signs we obtain

$$\int x\sqrt{x^2 - 4x + 5}\, dx = \left[\frac{1}{3}(u^2 + 1)^{3/2}\right] + 2\left[\frac{u}{2}\sqrt{u^2 + 1} + \frac{1}{2}\ln(u + \sqrt{u^2 + 1})\right] + C$$

If we now replace u by $x - 2$ (in which case $u^2 + 1 = x^2 - 4x + 5$), we obtain

$$\int x\sqrt{x^2 - 4x + 5}\, dx = \tfrac{1}{3}(x^2 - 4x + 5)^{3/2} + (x - 2)\sqrt{x^2 - 4x + 5}$$
$$+ \ln(x - 2 + \sqrt{x^2 - 4x + 5}) + C$$

Although correct, this form of the answer has an unnecessary mixture of radicals and fractional exponents. If desired, we can "clean up" the answer by writing

$$(x^2 - 4x + 5)^{3/2} = (x^2 - 4x + 5)\sqrt{x^2 - 4x + 5}$$

from which it follows that (verify)

$$\int x\sqrt{x^2 - 4x + 5}\, dx = \tfrac{1}{3}(x^2 - x - 1)\sqrt{x^2 - 4x + 1}$$
$$+ \ln(x - 2 + \sqrt{x^2 - 4x + 5}) + C \qquad \blacktriangleleft$$

MATCHES REQUIRING REDUCTION FORMULAS

In cases where the entry in an integral table is a reduction formula, that formula will have to be applied first to reduce the given integral to a form in which it can be evaluated.

Example 4

Use the Endpaper Integral Table to evaluate $\displaystyle\int \frac{x^3}{\sqrt{1 + x}}\, dx$.

Solution. The integrand can be classified as a power of x multiplying the reciprocal of $\sqrt{a + bx}$. Thus, from reduction formula (107) with $a = 1$, $b = 1$, and $n = 3$, followed by Formula (106), we obtain

$$\int \frac{x^3}{\sqrt{1 + x}}\, dx = \frac{2x^3\sqrt{1 + x}}{7} - \frac{6}{7}\int \frac{x^2}{\sqrt{1 + x}}\, dx$$

$$= \frac{2x^3\sqrt{1 + x}}{7} - \frac{6}{7}\left[\frac{2}{15}(3x^2 - 4x + 8)\sqrt{1 + x}\right] + C$$

$$= \left(\frac{2x^3}{7} - \frac{12x^2}{35} + \frac{16x}{35} - \frac{32}{35}\right)\sqrt{1 + x} + C \qquad \blacktriangleleft$$

MATCHES REQUIRING SPECIAL SUBSTITUTIONS

The Endpaper Integral Table has two entries involving an exponent of $3/2$ and numerous entries involving square roots (exponent $1/2$), but it has no entries with other fractional exponents. However, integrals involving fractional powers of x can often be simplified by making the substitution $u = x^{1/n}$ in which n is the least common multiple of the denominators of the exponents. Here are some examples.

Example 5

Evaluate

(a) $\displaystyle\int \frac{\sqrt{x}}{1 + \sqrt[3]{x}}\, dx$ (b) $\displaystyle\int \frac{dx}{2 + 2\sqrt{x}}$ (c) $\displaystyle\int \sqrt{1 + e^x}\, dx$

Solution (a). The integrand contains $x^{1/2}$ and $x^{1/3}$, so we make the substitution $u = x^{1/6}$, from which we obtain

$$x = u^6, \quad dx = 6u^5\, du$$

Thus,

$$\int \frac{\sqrt{x}}{1 + \sqrt[3]{x}}\, dx = \int \frac{(u^6)^{1/2}}{1 + (u^6)^{1/3}} (6u^5)\, du = 6 \int \frac{u^8}{1 + u^2}\, du$$

By long division

$$\frac{u^8}{1 + u^2} = u^6 - u^4 + u^2 - 1 + \frac{1}{1 + u^2}$$

from which it follows that

$$\int \frac{\sqrt{x}}{1 + \sqrt[3]{x}}\, dx = 6 \int \left(u^6 - u^4 + u^2 - 1 + \frac{1}{1 + u^2} \right) du$$

$$= \tfrac{6}{7} u^7 - \tfrac{6}{5} u^5 + 2u^3 - 6u + 6\tan^{-1} u + C$$

$$= \tfrac{6}{7} x^{7/6} - \tfrac{6}{5} x^{5/6} + 2x^{1/2} - 6x^{1/6} + 6\tan^{-1}(x^{1/6}) + C$$

Solution (b). The integrand contains $x^{1/2}$ but does not match any of the forms in the Endpaper Integral Table. Thus, we make the substitution $u = x^{1/2}$, from which we obtain

$$x = u^2, \quad dx = 2u\, du$$

Making this substitution yields

$$\int \frac{dx}{2 + 2\sqrt{x}} = \int \frac{2u}{2 + 2u}\, du$$

$$= \int \left(1 - \frac{1}{1 + u} \right) du \qquad \boxed{\text{Long division}}$$

$$= u - \ln|1 + u| + C$$

$$= \sqrt{x} - \ln(1 + \sqrt{x}) + C \qquad \boxed{\text{Absolute value not needed}}$$

Solution (c). Again, the integral does not match any of the forms in the Endpaper Integral Table. However, the integrand contains $(1 + e^x)^{1/2}$, which is analogous to the situation in part (b), except that here it is $1 + e^x$ rather than x that is raised to the $1/2$ power. This suggests the substitution $u = (1 + e^x)^{1/2}$, from which we obtain (verify)

$$x = \ln(u^2 - 1), \quad dx = \frac{2u}{u^2 - 1}\, du$$

Thus,

$$\int \sqrt{1 + e^x}\, dx = \int u \left(\frac{2u}{u^2 - 1} \right) du$$

$$= \int \frac{2u^2}{u^2 - 1}\, du$$

$$= \int \left(2 + \frac{2}{u^2 - 1} \right) du \qquad \boxed{\text{Long division}}$$

$$= 2u + \int \left(\frac{1}{u - 1} - \frac{1}{u + 1} \right) du \qquad \boxed{\text{Partial fractions}}$$

$$= 2u + \ln|u - 1| - \ln|u + 1| + C$$

$$= 2u + \ln \left| \frac{u - 1}{u + 1} \right| + C$$

$$= 2\sqrt{1 + e^x} + \ln \left[\frac{\sqrt{1 + e^x} - 1}{\sqrt{1 + e^x} + 1} \right] + C \qquad \boxed{\begin{array}{c}\text{Absolute value}\\\text{not needed}\end{array}}$$

◀

Functions that consist of finitely many sums, differences, quotients, and products of $\sin x$ and $\cos x$ are called *rational functions of sin x and cos x*. Some examples are

$$\frac{\sin x + 3 \cos^2 x}{\cos x + 4 \sin x}, \quad \frac{\sin x}{1 + \cos x - \cos^2 x}, \quad \frac{3 \sin^5 x}{1 + 4 \sin x}$$

The Endpaper Integral Table gives a few formulas for integrating rational functions of $\sin x$ and $\cos x$ under the heading *Reciprocals of Basic Functions*. For example, it follows from Formula (18) that

$$\int \frac{1}{1 + \sin x} \, dx = \tan x - \sec x + C \tag{2}$$

However, since the integrand is a rational function of $\sin x$, it may be desirable in a particular application to express the value of the integral in terms of $\sin x$ and $\cos x$ and rewrite (2) as

$$\int \frac{1}{1 + \sin x} \, dx = \frac{\sin x - 1}{\cos x} + C$$

Many rational functions of $\sin x$ and $\cos x$ can be evaluated by an ingenious method that was discovered by the mathematician Karl Weierstrass (see p. 184). The idea is to make the substitution

$$u = \tan(x/2), \quad -\pi/2 < x/2 < \pi/2$$

from which it follows that

$$x = 2 \tan^{-1} u, \quad dx = \frac{2}{1 + u^2} \, du$$

To implement this substitution we need to express $\sin x$ and $\cos x$ in terms of u. For this purpose we will use the identities

$$\sin x = 2 \sin(x/2) \cos(x/2) \tag{3}$$
$$\cos x = \cos^2(x/2) - \sin^2(x/2) \tag{4}$$

and the following relationships suggested by Figure 9.6.1:

$$\sin(x/2) = \frac{u}{\sqrt{1 + u^2}} \quad \text{and} \quad \cos(x/2) = \frac{1}{\sqrt{1 + u^2}}$$

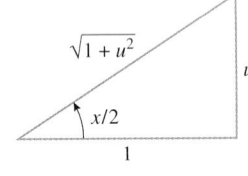

Figure 9.6.1

Substituting these expressions in (3) and (4) yields

$$\sin x = 2 \left(\frac{u}{\sqrt{1 + u^2}} \right) \left(\frac{1}{\sqrt{1 + u^2}} \right) = \frac{2u}{1 + u^2}$$

$$\cos x = \left(\frac{1}{\sqrt{1 + u^2}} \right)^2 - \left(\frac{u}{\sqrt{1 + u^2}} \right)^2 = \frac{1 - u^2}{1 + u^2}$$

In summary, we have shown that the substitution $u = \tan(x/2)$ can be implemented in a rational function of $\sin x$ and $\cos x$ by letting

$$\sin x = \frac{2u}{1 + u^2}, \quad \cos x = \frac{1 - u^2}{1 + u^2}, \quad dx = \frac{2}{1 + u^2} \, du \tag{5}$$

Example 6

Evaluate $\displaystyle\int \frac{dx}{1 - \sin x + \cos x}$.

Solution. The integrand is a rational function of $\sin x$ and $\cos x$ that does not match any of the formulas in the Endpaper Integral Table, so we make the substitution $u = \tan(x/2)$.

Thus, from (5) we obtain

$$\int \frac{dx}{1 - \sin x + \cos x} = \int \frac{\dfrac{2\,du}{1 + u^2}}{1 - \left(\dfrac{2u}{1 + u^2}\right) + \left(\dfrac{1 - u^2}{1 + u^2}\right)}$$

$$= \int \frac{2\,du}{(1 + u^2) - 2u + (1 - u^2)}$$

$$= \int \frac{du}{1 - u} = -\ln|1 - u| + C = -\ln|1 - \tan(x/2)| + C \quad \blacktriangleleft$$

REMARK. The substitution $u = \tan(x/2)$ will convert any rational function of $\sin x$ and $\cos x$ to an ordinary rational function of u. However, the method can lead to cumbersome partial fraction decompositions, so it may be worthwhile to explore the existence of simpler methods when hand computation is to be used.

INTEGRATING WITH COMPUTER ALGEBRA SYSTEMS

Integration tables are rapidly giving way to computerized integration using computer algebra systems. However, computerized integration is very much like computerized chess—the computer can sort through myriads of possibilities quickly, but the approach is sometimes mechanical and lacking in the imagination and judgment of human thought. As a result, answers produced by computer integration are sometimes less satisfactory than those in integral tables that have been refined over many years by many excellent mathematical minds.

Sometimes computer algebra systems do not produce the most general form of an indefinite integral. For example, the integral formula

$$\int \frac{dx}{x - 1} = \ln|x - 1| + C$$

which can be obtained by inspection or by using the substitution $u = x - 1$ is valid for $x \neq 1$. However, *Mathematica, Maple*, and *Derive* evaluate this integral as[*]

$$\ln(1 - x), \qquad \ln(x - 1), \qquad \ln(x - 1)$$
$$\quad\textit{Mathematica} \qquad\quad \textit{Maple} \qquad\quad \textit{Derive}$$

Observe that none of the systems put in the constant of integration—it is just assumed to be there. Observe also that none of the systems put in the absolute value signs; consequently, for *Maple* and *Derive* the resulting antiderivative is only valid if $x > 1$, and for *Mathematica* it is only valid if $x < 1$. Thus, although the computer algebra systems all produced a correct antiderivative, none of them produced the most general antiderivative.

Now let us examine how *Mathematica, Maple*, and *Derive* handle the integral

$$\int x\sqrt{x^2 - 4x + 5}\,dx = \tfrac{1}{3}(x^2 - x - 1)\sqrt{x^2 - 4x + 5}$$
$$+ \ln(x - 2 + \sqrt{x^2 - 4x + 5}) \qquad (6)$$

which we obtained in Example 3(b) (with the constant of integration included). *Derive* produces this result in a slightly different algebraic form, and *Maple* produces the result

$$\int x\sqrt{x^2 - 4x + 5}\,dx = \tfrac{1}{3}(x^2 - 4x + 5)^{3/2} + \tfrac{1}{2}(2x - 4)\sqrt{x^2 - 4x + 5} + \sinh^{-1}(x - 2)$$

This can be rewritten as (6) by expressing the fractional exponent in radical form and expressing $\sinh^{-1}(x - 2)$ in logarithmic form using Theorem 8.8.4 (verify). *Mathematica*

---

[*]Results produced by *Mathematica, Maple*, and *Derive* may vary depending on the version of the software that is used.

produces the result

$$\int x\sqrt{x^2 - 4x + 5}\, dx = \tfrac{1}{3}(x^2 - x - 1)\sqrt{x^2 - 4x + 5} - \sinh^{-1}(2 - x)$$

which can be rewritten in form (6) by using Theorem 8.8.4 together with the identity $\sinh^{-1}(-x) = -\sinh^{-1} x$ (verify).

Computer algebra systems can sometimes produce inconvenient or unnatural answers to integration problems. For example, when *Mathematica*, *Maple*, and *Derive* are asked to integrate $(x + 1)^7$, they produce the following results:

$$\underset{\textit{Maple}}{\frac{(x + 1)^8}{8}}, \quad \underset{\textit{Derive}}{\frac{(x + 1)^8}{8}}, \quad \underset{\textit{Mathematica}}{x + \tfrac{7}{2}x^2 + 7x^3 + \tfrac{35}{4}x^4 + 7x^5 + \tfrac{7}{2}x^6 + x^7 + \tfrac{1}{8}x^8}$$

The answers produced by *Maple* and *Derive* are in keeping with the hand computation

$$\int (x + 1)^7\, dx = \frac{(x + 1)^8}{8} + C$$

that uses the substitution $u = x + 1, du = dx$, whereas the answer produced by *Mathematica* is based on expanding $(x + 1)^7$ and integrating term by term.

FOR THE READER.   If you expand the answers produced by *Maple* and *Derive*, you will discover that they contain the term $\tfrac{1}{8}$ that did not appear in the *Mathematica* result. What is the explanation?

In Example 2(a) of Section 9.3 we showed that

$$\int \sin^4 x \cos^5 x\, dx = \tfrac{1}{5}\sin^5 x - \tfrac{2}{7}\sin^7 x + \tfrac{1}{9}\sin^9 x + C$$

In contrast, *Mathematica* integrates this as

$$\tfrac{1}{80640}(1890 \sin x - 420 \sin 3x - 252 \sin 5x + 45 \sin 7x + 35 \sin 9x)$$

and *Maple* and *Derive* essentially integrate it as

$$-\tfrac{1}{9}\sin^3 x \cos^6 x - \tfrac{1}{21}\sin x \cos^6 x + \tfrac{1}{105}\cos^4 x \sin x + \tfrac{4}{315}\cos^2 x \sin x + \tfrac{8}{315}\sin x$$

Although the three results look quite different, they can be obtained from one another using appropriate trigonometric identities.

COMPUTER ALGEBRA SYSTEMS CAN FAIL

Every computer algebra system has a library of functions that it can use to construct antiderivatives. Such libraries contain elementary functions, such as polynomials, rational functions, trigonometric functions, as well as various nonelementary functions that arise in engineering, physics, and other applied fields. If the result of an integration cannot be expressed in terms of the functions in the program's library, then the program will give some indication that it cannot evaluate the integral. For example, when asked to evaluate the integral

$$\int (1 + \ln x)\sqrt{1 + (x \ln x)^2}\, dx \tag{7}$$

Mathematica, *Maple*, and *Derive* all respond by displaying some form of the unevaluated integral as an answer to indicate that they could not perform the integration.

FOR THE READER.   Sometimes integrals that cannot be evaluated by a CAS in their given form can be evaluated by first rewriting them in a different form or by making a substitution. Make a u-substitution in (7) that will enable you to evaluate the integral with your CAS.

Sometimes computer algebra systems respond by expressing an integral in terms of another integral. For example, if you try to integrate e^{x^2} using *Mathematica*, *Maple*, or

Derive, you will obtain an expression involving erf (which stands for ***error function***). This function is defined as

$$\text{erf}(x) = \frac{2}{\sqrt{\pi}} \int_0^x e^{-t^2}\, dt$$

so that the three programs essentially do nothing but express the given integral in terms of a closely related integral.

Example 7

A particle moves along an x-axis in such a way that its velocity $v(t)$ at time t is

$$v(t) = 30 \cos^7 t \sin^4 t \quad (t \geq 0)$$

Graph the position versus time curve for the particle, given that the particle is at the point $x = 1$ when $t = 0$.

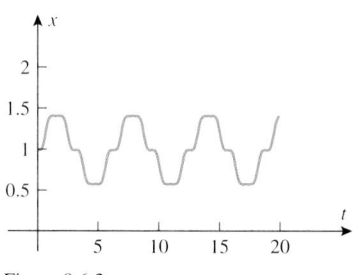

Figure 9.6.2

Solution. Since $dx/dt = v(t)$, the position function $x(t)$ is obtained by integrating the velocity function. We will perform the integration using *Mathematica*, but the procedure would be the same for *Maple* or *Derive*. Integrating $v(t)$ and adding the missing constant of integration yields

$$x = \int 30 \cos^7 t \sin^4 t\, dt$$

$$= \tfrac{1}{39424}(16170 \sin t - 2310 \sin 3t - 2541 \sin 5t - 165 \sin 7t + 385 \sin 9t$$
$$+ 105 \sin 11t) + C$$

Since $x = 1$ if $t = 0$, it follows on substituting these values in this equation that $C = 1$. Thus, with this value for C the graph of x versus t is as shown in Figure 9.6.2. ◀

EXERCISE SET 9.6 [C] CAS

In Exercises 1–24:
(a) Use the Endpaper Integral Table to evaluate the integral.
(b) If you have a CAS, use it to evaluate the integral, and then confirm that the result is equivalent to the one that you found in part (a).

1. $\displaystyle\int \frac{3x}{4x-1}\, dx$

2. $\displaystyle\int \frac{x}{(2-3x)^2}\, dx$

3. $\displaystyle\int \frac{1}{x(2x+5)}\, dx$

4. $\displaystyle\int \frac{1}{x^2(1-5x)}\, dx$

5. $\displaystyle\int x\sqrt{2x-3}\, dx$

6. $\displaystyle\int \frac{x}{\sqrt{2-x}}\, dx$

7. $\displaystyle\int \frac{1}{x\sqrt{4-3x}}\, dx$

8. $\displaystyle\int \frac{1}{x\sqrt{3x-4}}\, dx$

9. $\displaystyle\int \frac{1}{5-x^2}\, dx$

10. $\displaystyle\int \frac{1}{x^2-9}\, dx$

11. $\displaystyle\int \sqrt{x^2-3}\, dx$

12. $\displaystyle\int \frac{\sqrt{x^2+5}}{x^2}\, dx$

13. $\displaystyle\int \frac{x^2}{\sqrt{x^2+4}}\, dx$

14. $\displaystyle\int \frac{1}{x^2\sqrt{x^2-2}}\, dx$

15. $\displaystyle\int \sqrt{9-x^2}\, dx$

16. $\displaystyle\int \frac{\sqrt{4-x^2}}{x^2}\, dx$

17. $\displaystyle\int \frac{\sqrt{3-x^2}}{x}\, dx$

18. $\displaystyle\int \frac{1}{x\sqrt{6x-x^2}}\, dx$

19. $\displaystyle\int \sin 3x \sin 2x\, dx$

20. $\displaystyle\int \sin 2x \cos 5x\, dx$

21. $\displaystyle\int x^3 \ln x\, dx$

22. $\displaystyle\int \frac{\ln x}{\sqrt{x}}\, dx$

23. $\displaystyle\int e^{-2x} \sin 3x\, dx$

24. $\displaystyle\int e^x \cos 2x\, dx$

In Exercises 25–36:
(a) Make the indicated u-substitution, and then use the Endpaper Integral Table to evaluate the integral.
(b) If you have a CAS, use it to evaluate the integral, and then confirm that the result is equivalent to the one that you found in part (a).

25. $\displaystyle\int \frac{e^{4x}}{(4-3e^{2x})^2}\, dx, \ u = e^{2x}$

26. $\displaystyle\int \frac{\cos 2x}{(\sin 2x)(3-\sin 2x)}\, dx, \ u = \sin 2x$

27. $\displaystyle\int \frac{1}{\sqrt{x}(9x+4)}\,dx,\ u=3\sqrt{x}$

28. $\displaystyle\int \frac{\cos 4x}{9+\sin^2 4x}\,dx,\ u=\sin 4x$

29. $\displaystyle\int \frac{1}{\sqrt{9x^2-4}}\,dx,\ u=3x$

30. $\displaystyle\int x\sqrt{2x^4+3}\,dx,\ u=\sqrt{2}x^2$

31. $\displaystyle\int \frac{x^5}{\sqrt{5-9x^4}}\,dx,\ u=3x^2$

32. $\displaystyle\int \frac{1}{x^2\sqrt{3-4x^2}}\,dx,\ u=2x$

33. $\displaystyle\int \frac{\sin^2(\ln x)}{x}\,dx,\ u=\ln x$

34. $\displaystyle\int e^{-2x}\cos^2(e^{-2x})\,dx,\ u=e^{-2x}$

35. $\displaystyle\int xe^{-2x}\,dx,\ u=-2x$

36. $\displaystyle\int \ln(5x-1)\,dx,\ u=5x-1$

In Exercises 37–48:
(a) Make an appropriate u-substitution, and then use the Endpaper Integral Table to evaluate the integral.
(b) If you have a CAS, use it to evaluate the integral (no substitution), and then confirm that the result is equivalent to that in part (a).

37. $\displaystyle\int \frac{\sin 3x}{(\cos 3x)(\cos 3x+1)^2}\,dx$

38. $\displaystyle\int \frac{\ln x}{x\sqrt{4\ln x-1}}\,dx$

39. $\displaystyle\int \frac{x}{16x^4-1}\,dx$

40. $\displaystyle\int \frac{e^x}{3-4e^{2x}}\,dx$

41. $\displaystyle\int e^x\sqrt{3-4e^{2x}}\,dx$

42. $\displaystyle\int \frac{\sqrt{4-9x^2}}{x^2}\,dx$

43. $\displaystyle\int \sqrt{5x-9x^2}\,dx$

44. $\displaystyle\int \frac{1}{x\sqrt{x-5x^2}}\,dx$

45. $\displaystyle\int x\sin 3x\,dx$

46. $\displaystyle\int \cos\sqrt{x}\,dx$

47. $\displaystyle\int e^{-\sqrt{x}}\,dx$

48. $\displaystyle\int x\ln(2-3x^2)\,dx$

In Exercises 49–52:
(a) Complete the square, make an appropriate u-substitution, and then use the Endpaper Integral Table to evaluate the integral.
(b) If you have a CAS, use it to evaluate the integral (no substitution or square completion), and then confirm that the result is equivalent to that in part (a).

49. $\displaystyle\int \frac{1}{x^2+4x-5}\,dx$

50. $\displaystyle\int \sqrt{3-2x-x^2}\,dx$

51. $\displaystyle\int \frac{x}{\sqrt{5+4x-x^2}}\,dx$

52. $\displaystyle\int \frac{x}{x^2+6x+13}\,dx$

In Exercises 53–66:
(a) Make an appropriate u-substitution of the form $u=x^{1/n}$, $u=(x+a)^{1/n}$, or $u=x^n$, and then use the Endpaper Integral Table to evaluate the integral.
(b) If you have a CAS, use it to evaluate the integral, and then confirm that the result is equivalent to the one that you found in part (a).

53. $\displaystyle\int x\sqrt{x-2}\,dx$

54. $\displaystyle\int \frac{x}{\sqrt{x+1}}\,dx$

55. $\displaystyle\int x^5\sqrt{x^3+1}\,dx$

56. $\displaystyle\int \frac{1}{x\sqrt{x^3-1}}\,dx$

57. $\displaystyle\int \frac{dx}{\sqrt{x}+\sqrt[3]{x}}$

58. $\displaystyle\int \frac{dx}{x-x^{3/5}}$

59. $\displaystyle\int \frac{dx}{x(1-x^{1/4})}$

60. $\displaystyle\int \frac{x^{2/3}}{x+1}\,dx$

61. $\displaystyle\int \frac{dx}{x^{1/2}-x^{1/3}}$

62. $\displaystyle\int \frac{1+\sqrt{x}}{1-\sqrt{x}}\,dx$

63. $\displaystyle\int \frac{x^3}{\sqrt{1+x^2}}\,dx$

64. $\displaystyle\int \frac{x}{(x+3)^{1/5}}\,dx$

65. $\displaystyle\int \sin\sqrt{x}\,dx$

66. $\displaystyle\int e^{\sqrt{x}}\,dx$

In Exercises 67–72:
(a) Make u-substitution (5) to convert the integrand to a rational function of u, and then use the Endpaper Integral Table to evaluate the integral.
(b) If you have a CAS, use it to evaluate the integral (no substitution), and then confirm that the result is equivalent to that in part (a).

67. $\displaystyle\int \frac{dx}{1+\sin x+\cos x}$

68. $\displaystyle\int \frac{dx}{2+\sin x}$

69. $\displaystyle\int \frac{d\theta}{1-\cos\theta}$

70. $\displaystyle\int \frac{dx}{4\sin x-3\cos x}$

71. $\displaystyle\int \frac{\cos x}{2-\cos x}\,dx$

72. $\displaystyle\int \frac{dx}{\sin x+\tan x}$

In Exercises 73 and 74, use any method to solve for x.

73. $\displaystyle\int_2^x \frac{1}{t(4-t)}\,dt=0.5,\ 2<x<4$

74. $\displaystyle\int_1^x \frac{1}{t\sqrt{2t-1}}\,dt=1,\ x>\frac{1}{2}$

In Exercises 75–78, use any method to find the area of the region enclosed by the curves.

75. $y = \sqrt{25 - x^2}$, $y = 0$, $x = 0$, $x = 4$

76. $y = \sqrt{9x^2 - 4}$, $y = 0$, $x = 2$

77. $y = \dfrac{1}{25 - 16x^2}$, $y = 0$, $x = 0$, $x = 1$

78. $y = \sqrt{x} \ln x$, $y = 0$, $x = 4$

In Exercises 79–82, use any method to find the volume of the solid generated when the region enclosed by the curves is revolved about the y-axis.

79. $y = \cos x$, $y = 0$, $x = 0$, $x = \pi/2$

80. $y = \sqrt{x - 4}$, $y = 0$, $x = 8$

81. $y = e^{-x}$, $y = 0$, $x = 0$, $x = 3$

82. $y = \ln x$, $y = 0$, $x = 5$

In Exercises 83 and 84, use any method to find the arc length of the curve.

83. $y = 2x^2$, $0 \le x \le 2$

84. $y = 3 \ln x$, $1 \le x \le 3$

In Exercises 85 and 86, use any method to find the area of the surface generated by revolving the curve about the x-axis.

85. $y = \sin x$, $0 \le x \le \pi$

86. $y = 1/x$, $1 \le x \le 4$

In Exercises 87 and 88, information is given about the motion of a particle moving along a coordinate line.
(a) Use a CAS to find the position function of the particle for $t \ge 0$. You may approximate the constants of integration, where necessary.
(b) Graph the position versus time curve.

87. $v(t) = 20 \cos^6 t \sin^3 t$, $s(0) = 2$

88. $a(t) = e^{-t} \sin 2t \sin 4t$, $v(0) = 0$, $s(0) = 10$

89. (a) Use the substitution $u = \tan(x/2)$ to show that

$$\int \sec x \, dx = \ln \left| \frac{1 + \tan(x/2)}{1 - \tan(x/2)} \right| + C$$

and confirm that this is consistent with Formula (22) of Section 9.3.
(b) Use the result in part (a) to show that

$$\int \sec x \, dx = \ln \left| \tan \left(\frac{\pi}{4} + \frac{x}{2} \right) \right| + C$$

90. Use the substitution $u = \tan(x/2)$ to show that

$$\int \csc x \, dx = \frac{1}{2} \ln \left[\frac{1 - \cos x}{1 + \cos x} \right] + C$$

and confirm that this is consistent with the result in Exercise 61(a) of Section 9.3.

91. Find a substitution that can be used to integrate rational functions of $\sinh x$ and $\cosh x$ and use your substitution to evaluate

$$\int \frac{dx}{2 \cosh x + \sinh x}$$

without expressing the integrand in terms of e^x and e^{-x}.

9.7 NUMERICAL INTEGRATION; SIMPSON'S RULE

The usual procedure for evaluating a definite integral is to find an antiderivative of the integrand and apply the Fundamental Theorem of Calculus. However, if an antiderivative of the integrand cannot be found, then we must settle for a numerical approximation of the integral. In earlier sections we discussed three procedures for approximating areas using Riemann sums—left endpoint approximation, right endpoint approximation, and midpoint approximation. In this section we will adapt those ideas to approximating general definite integrals, and we will discuss some new approximation methods that often provide more accuracy with less computation.

A REVIEW OF RIEMANN SUM APPROXIMATIONS

Recall from Formula (6) of Section 7.5 that the definite integral of a function f over an interval $[a, b]$ is defined as

$$\int_a^b f(x) \, dx = \lim_{n \to +\infty} \sum_{k=1}^{n} f(x_k^*) \Delta x$$

where the sum that appears on the right side is called a Riemann sum. In this formula, the interval $[a, b]$ is divided into n subintervals of width $\Delta x = (b - a)/n$, and x_k^* denotes an arbitrary point in the kth subinterval. It follows that as n increases the Riemann sum will eventually be a good approximation to the integral, which we denote by writing

$$\int_a^b f(x)\,dx \approx \sum_{k=1}^n f(x_k^*)\Delta x$$

or, equivalently,

$$\int_a^b f(x)\,dx \approx \Delta x \left[f(x_1^*) + f(x_2^*) + \cdots + f(x_n^*) \right]$$

In this section we will denote the values of f at the endpoints of the subintervals by

$$y_0 = f(a), \quad y_1 = f(x_1), \quad y_2 = f(x_2), \ldots, \quad y_{n-1} = f(x_{n-1}), \quad y_n = f(b)$$

and we will denote the values of f at the midpoints of the subintervals by

$$y_{m_1}, y_{m_2}, \ldots, y_{m_n}$$

(Figure 9.7.1).

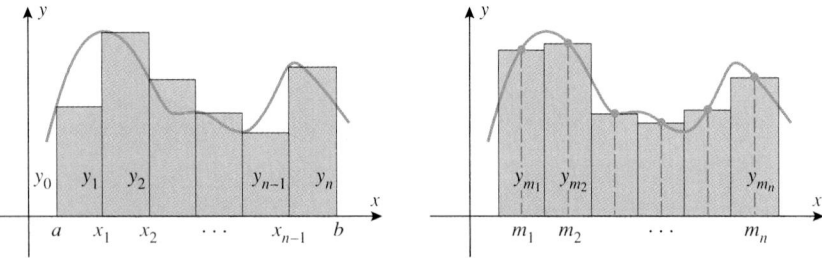

Figure 9.7.1

With this notation the left endpoint, right endpoint, and midpoint approximations discussed in Section 7.5 can be expressed as shown in Table 9.7.1.

Table 9.7.1

LEFT ENDPOINT APPROXIMATION	RIGHT ENDPOINT APPROXIMATION	MIDPOINT APPROXIMATION
$\int_a^b f(x)\,dx \approx \left(\dfrac{b-a}{n}\right)[y_0 + y_1 + \cdots + y_{n-1}]$	$\int_a^b f(x)\,dx \approx \left(\dfrac{b-a}{n}\right)[y_1 + y_2 + \cdots + y_n]$	$\int_a^b f(x)\,dx \approx \left(\dfrac{b-a}{n}\right)[y_{m_1} + y_{m_2} + \cdots + y_{m_n}]$

TRAPEZOIDAL APPROXIMATION

The left-hand and right-hand endpoint approximations are rarely used in applications; however, if we take the average of the left-hand and right-hand endpoint approximations, we obtain a result, called the ***trapezoidal approximation***, which is commonly used:

Trapezoidal Approximation

$$\int_a^b f(x)\,dx \approx \left(\frac{b-a}{2n}\right)[y_0 + 2y_1 + \cdots + 2y_{n-1} + y_n] \tag{1}$$

The name *trapezoidal approximation* can be explained by considering the case in which $f(x) \geq 0$ on $[a, b]$, so that $\int_a^b f(x)\, dx$ represents the area under $f(x)$ over $[a, b]$. Geometrically, the trapezoidal approximation formula results if we approximate this area by the sum of the trapezoidal areas shown in Figure 9.7.2 (Exercise 41).

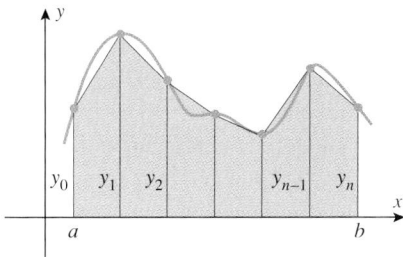

Figure 9.7.2

Trapezoidal approximation

Example 1

In Table 9.7.2 we have approximated

$$\ln 2 = \int_1^2 \frac{1}{x}\, dx$$

using the midpoint approximation and the trapezoidal approximation. In each case we used $n = 10$ subdivisions of the interval $[1, 2]$, so that

$$\underbrace{\frac{b - a}{n} = \frac{2 - 1}{10} = 0.1}_{\text{Midpoint}} \quad \text{and} \quad \underbrace{\frac{b - a}{2n} = \frac{2 - 1}{20} = 0.05}_{\text{Trapezoidal}} \qquad \blacktriangleleft$$

REMARK. In Example 1 we rounded the numerical values to nine places to the right of the decimal point; we will follow this procedure throughout this section. If your calculator cannot produce this many places, then you will have to make the appropriate adjustments. What is important here is that you understand the principles involved.

Table 9.7.2

Midpoint Approximation			Trapezoidal Approximation				
i	MIDPOINT m_i	$y_{m_i} = f(m_i) = 1/m_i$	i	ENDPOINT x_i	$y_i = f(x_i) = 1/x_i$	MULTIPLIER w_i	$w_i y_i$
1	1.05	0.952380952	0	1.0	1.000000000	1	1.000000000
2	1.15	0.869565217	1	1.1	0.909090909	2	1.818181818
3	1.25	0.800000000	2	1.2	0.833333333	2	1.666666667
4	1.35	0.740740741	3	1.3	0.769230769	2	1.538461538
5	1.45	0.689655172	4	1.4	0.714285714	2	1.428571429
6	1.55	0.645161290	5	1.5	0.666666667	2	1.333333333
7	1.65	0.606060606	6	1.6	0.625000000	2	1.250000000
8	1.75	0.571428571	7	1.7	0.588235294	2	1.176470588
9	1.85	0.540540541	8	1.8	0.555555556	2	1.111111111
10	1.95	0.512820513	9	1.9	0.526315789	2	1.052631579
		6.928353603	10	2.0	0.500000000	1	0.500000000
							13.875428063

$$\int_1^2 \frac{1}{x}\, dx \approx (0.1)(6.928353603) = 0.692835360$$

$$\int_1^2 \frac{1}{x}\, dx \approx (0.05)(13.875428063) = 0.693771403$$

The value of ln 2 rounded to nine decimal places is

$$\ln 2 = \int_1^2 \frac{1}{x}\, dx \approx 0.693147181 \tag{2}$$

so that the midpoint approximation in Example 1 produced a more accurate result than the trapezoidal approximation (verify). To see why this should be so, we need to look at the midpoint approximation from another viewpoint. [For simplicity in the explanations, we will assume that $f(x) \geq 0$, but the conclusions will be true without this assumption.] For differentiable functions, the midpoint approximation is sometimes called the *tangent line approximation* because over each subinterval the area of the rectangle used in the midpoint approximation is equal to the area of the trapezoid whose upper boundary is the tangent line to $y = f(x)$ at the midpoint of the interval (Figure 9.7.3). The equality of these areas follows from the fact that the shaded triangles in Figure 9.7.3 are congruent.

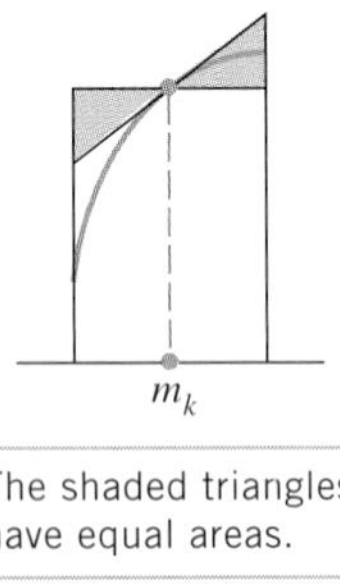

In this section we will denote the midpoint and trapezoidal approximations of $\int_a^b f(x)\, dx$ with n subintervals by M_n and T_n, respectively, and we will denote the errors in these approximations by

$$|E_M| = \left| \int_a^b f(x)\, dx - M_n \right| \quad \text{and} \quad |E_T| = \left| \int_a^b f(x)\, dx - T_n \right|$$

In Figure 9.7.4a we have isolated a subinterval of $[a, b]$ on which the graph of a function f is concave down, and we have shaded the areas that represent the errors in the midpoint and trapezoidal approximations over the subinterval. In Figure 9.7.4b we show a succession of four illustrations which make it evident that the error from the midpoint approximation is less than that from the trapezoidal approximation. If the graph of f were concave up, analogous figures would lead to the same conclusion. (This argument, due to Frank Buck, appeared in *The College Mathematics Journal*, Vol. 16, No. 1, 1985.)

m_k

The shaded triangles have equal areas.

Figure 9.7.3

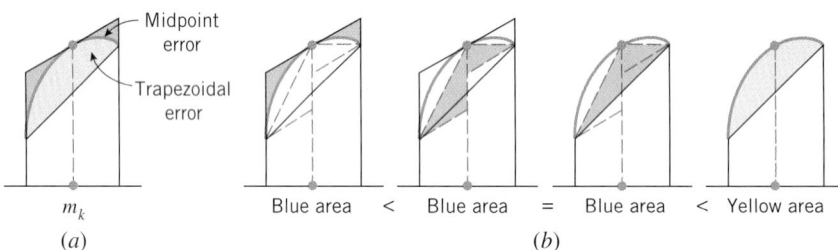

Figure 9.7.4

Blue area < Blue area = Blue area < Yellow area

(a)   (b)

Figure 9.7.4a also suggests that on a subinterval where the graph is concave down, the midpoint approximation is larger than the value of the integral and the trapezoidal approximation is smaller. On an interval where the graph is concave up it is the other way around. In summary, we have the following result, which we state without formal proof:

9.7.1   THEOREM.   *Let f be continuous on $[a, b]$, and let $|E_M|$ and $|E_T|$ be the absolute errors that result from the midpoint and trapezoidal approximations of $\int_a^b f(x)\, dx$ using n subintervals.*

(a)   *If the graph of f is either concave up or concave down on (a, b), then $|E_M| < |E_T|$, that is, the error from the midpoint approximation is less than that from the trapezoidal approximation.*

(b)   *If the graph of f is concave down on (a, b), then*

$$T_n < \int_a^b f(x)\, dx < M_n$$

(c)   *If the graph of f is concave up on (a, b), then*

$$M_n < \int_a^b f(x)\, dx < T_n$$

Example 2

We observed earlier that the midpoint approximation of ln 2 obtained in Example 1 was more accurate than the trapezoidal approximation. This is consistent with part (a) of Theorem 9.7.1, since $f(x) = 1/x$ is continuous on $[1, 2]$ and concave up on $(1, 2)$. Moreover, a comparison of the two approximations to (2) shows that the midpoint approximation is smaller than ln 2 and the trapezoidal approximation is larger. This is consistent with part (c) of Theorem 9.7.1. ◄

REMARK. Do not erroneously conclude that the midpoint approximation is always better than the trapezoidal approximation; for functions with inflection points, the trapezoidal approximation can be more accurate.

SIMPSON'S RULE

Intuition suggests that we might improve on the midpoint and trapezoidal approximations by replacing the linear upper boundaries of the approximating strips in Figure 9.7.2 by curved upper boundaries chosen to fit the shape of the curve $y = f(x)$ more closely. This is the idea behind ***Simpson's*** * ***rule***, which uses parabolic curves of the form

$$y = ax^2 + bx + c \tag{3}$$

to approximate sections of the curve $y = f(x)$. [Recall from (7) in Appendix D that (3) is the equation of a parabola with axis of symmetry parallel to the y-axis.]

To simplify the description of Simpson's rule we will assume that $f(x) \geq 0$ on $[a, b]$ so that we can interpret $\int_a^b f(x)\,dx$ as an area. However, the method is valid without this assumption. The heart of Simpson's rule is the formula

$$A = \frac{h}{3}[Y_0 + 4Y_1 + Y_2] \tag{4}$$

which gives the area under the curve

$$y = ax^2 + bx + c$$

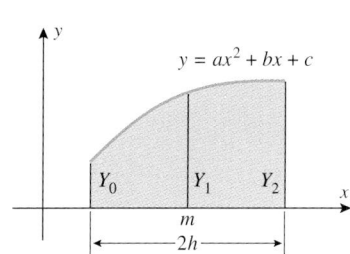

Figure 9.7.5

over an arbitrary interval of width $2h$. In this formula Y_0, Y_1, and Y_2 represent the y-values at the left-hand endpoint, the midpoint m, and the right-hand endpoint of the interval (Figure 9.7.5).

To derive (4), observe that the left-hand endpoint of the interval is $m - h$ and the right-hand endpoint is $m + h$, so the area A under $y = ax^2 + bx + c$ over this interval is

$$A = \int_{m-h}^{m+h} (ax^2 + bx + c)\,dx = \frac{a}{3}x^3 + \frac{b}{2}x^2 + cx \Big]_{m-h}^{m+h}$$

$$= \frac{a}{3}[(m+h)^3 - (m-h)^3] + \frac{b}{2}[(m+h)^2 - (m-h)^2] + c[(m+h) - (m-h)]$$

---

* THOMAS SIMPSON (1710–1761). English mathematician. Simpson was the son of a weaver. He was trained to follow in his father's footsteps and had little formal education in his early life. His interest in science and mathematics was aroused in 1724, when he witnessed an eclipse of the Sun and received two books from a peddler, one on astrology and the other on arithmetic. Simpson quickly absorbed their contents and soon became a successful local fortune teller. His improved financial situation enabled him to give up weaving and marry his landlady, an older woman. Then in 1733 some mysterious "unfortunate incident" forced him to move. He settled in Derby, where he taught in an evening school and worked at weaving during the day. In 1736 he moved to London and published his first mathematical work in a periodical called the *Ladies' Diary* (of which he later became the editor). In 1737 he published a successful calculus textbook that enabled him to give up weaving completely and concentrate on textbook writing and teaching. His fortunes improved further in 1740 when one Robert Heath accused him of plagiarism. The publicity was marvelous, and Simpson proceeded to dash off a succession of best-selling textbooks: *Algebra* (ten editions plus translations), *Geometry* (twelve editions plus translations), *Trigonometry* (five editions plus translations), and numerous others.

It is interesting to note that Simpson did not discover the rule that bears his name. It was a well-known result by Simpson's time.

or on simplifying,

$$A = \frac{h}{3}[a(6m^2 + 2h^2) + b(6m) + 6c] \tag{5}$$

But the values of $y = ax^2 + bx + c$ at the left-hand endpoint, the midpoint, and the right-hand endpoint are, respectively,

$$Y_0 = a(m - h)^2 + b(m - h) + c$$
$$Y_1 = am^2 + bm + c$$
$$Y_2 = a(m + h)^2 + b(m + h) + c$$

from which it follows that

$$Y_0 + 4Y_1 + Y_2 = a(6m^2 + 2h^2) + b(6m) + 6c \tag{6}$$

Thus, (4) follows from (5) and (6).

Simpson's rule is obtained by dividing the interval $[a, b]$ into an *even* number of subintervals of equal width h and applying Formula (4) to approximate the area under $y = f(x)$ over successive pairs of subintervals. The sum of these approximations then serves as an estimate of $\int_a^b f(x)\, dx$. More precisely, let $[a, b]$ be divided into n subintervals of width $h = (b - a)/n$ (n even) and let

$$y_0, y_1, \ldots, y_n$$

be the values of $y = f(x)$ at the subinterval endpoints

$$a = x_0, x_1, \ldots, x_n = b$$

By (4) the area under $y = f(x)$ over the first two subintervals is approximately

$$\frac{h}{3}[y_0 + 4y_1 + y_2]$$

and the area over the second pair of subintervals is approximately

$$\frac{h}{3}[y_2 + 4y_3 + y_4]$$

and the area over the last pair of subintervals is approximately

$$\frac{h}{3}[y_{n-2} + 4y_{n-1} + y_n]$$

Adding all the approximations, collecting terms, and replacing h by $(b - a)/n$ yields

Simpson's Rule

$$\int_a^b f(x)\, dx \approx \left(\frac{b - a}{3n}\right)[y_0 + 4y_1 + 2y_2 + 4y_3 + 2y_4 + \cdots + 2y_{n-2} + 4y_{n-1} + y_n]$$

We will denote the Simpson's rule approximation with n subintervals by S_n and the error in this approximation by

$$|E_S| = \left| \int_a^b f(x)\, dx - S_n \right|$$

Example 3

In Table 9.7.3 we have approximated

$$\ln 2 = \int_1^2 \frac{1}{x}\, dx$$

by Simpson's rule using $n = 10$ subdivisions so that

$$\frac{b - a}{3n} = \frac{2 - 1}{3(10)} = \frac{1}{30}$$

Observe, by comparing this result to (2), that Simpson's rule produced a more accurate approximation of ln 2 than either of the methods in Example 1. ◀

Table 9.7.3 Simpson's Rule

	ENDPOINT		MULTIPLIER	
i	x_i	$y_i = f(x_i) = 1/x_i$	w_i	$w_i y_i$
0	1.0	1.000000000	1	1.000000000
1	1.1	0.909090909	4	3.636363636
2	1.2	0.833333333	2	1.666666667
3	1.3	0.769230769	4	3.076923077
4	1.4	0.714285714	2	1.428571429
5	1.5	0.666666667	4	2.666666667
6	1.6	0.625000000	2	1.250000000
7	1.7	0.588235294	4	2.352941176
8	1.8	0.555555556	2	1.111111111
9	1.9	0.526315789	4	2.105263158
10	2.0	0.500000000	1	0.500000000
				20.794506921

$$\int_1^2 \frac{1}{x}\, dx \approx \left(\tfrac{1}{30}\right)(20.794506921) = 0.693150231$$

ERROR ESTIMATES

With all the methods studied in this section, there are two sources of error: the *intrinsic* or *truncation error* due to the approximation formula and the *roundoff* error introduced in the calculations. In general, increasing n reduces the truncation error but increases the roundoff error, since more computations are required for larger n. In practical applications, it is important to know how large n must be taken to ensure that a specified degree of accuracy is obtained. The analysis of roundoff error is complicated and will not be considered here. However, the following theorems, which are proved in books on **numerical analysis**, provide upper bounds on the truncation errors in the midpoint, trapezoidal, and Simpson's rule approximations.

9.7.2 THEOREM (*Midpoint and Trapezoidal Error Estimates*). *If f'' is continuous on $[a, b]$ and if K_2 is the maximum value of $|f''(x)|$ on $[a, b]$, then for n subdivisions of $[a, b]$*

$$(a)\ |E_M| \leq \frac{(b-a)^3 K_2}{24n^2} \qquad (b)\ |E_T| \leq \frac{(b-a)^3 K_2}{12n^2} \tag{7–8}$$

9.7.3 THEOREM (*Simpson Error Estimate*). *If $f^{(4)}$ is continuous on $[a, b]$ and if K_4 is the maximum value of $|f^{(4)}(x)|$ on $[a, b]$, then for n subdivisions of $[a, b]$*

$$|E_S| \leq \frac{(b-a)^5 K_4}{180n^4} \tag{9}$$

Example 4

Find an upper bound on the absolute error that results from approximating

$$\ln 2 = \int_1^2 \frac{1}{x}\, dx$$

using $n = 10$ subintervals by: (a) the trapezoidal approximation, (b) the midpoint approximation, and (c) Simpson's rule.

Solution. We will apply Formulas (7), (8), and (9) with

$$f(x) = \frac{1}{x}, \quad a = 1, \quad b = 2, \quad \text{and} \quad n = 10$$

We have

$$f'(x) = -\frac{1}{x^2}, \quad f''(x) = \frac{2}{x^3}, \quad f'''(x) = -\frac{6}{x^4}, \quad f^{(4)}(x) = \frac{24}{x^5}$$

Thus,

$$|f''(x)| = \left|\frac{2}{x^3}\right| = \frac{2}{x^3}, \quad |f^{(4)}(x)| = \left|\frac{24}{x^5}\right| = \frac{24}{x^5} \tag{10-11}$$

where we have dropped the absolute values because $f''(x)$ and $f^{(4)}(x)$ have positive values for $1 \leq x \leq 2$. Since (10) and (11) are continuous and decreasing on $[1, 2]$, both functions have their maximum values at $x = 1$; for (10) this maximum value is 2 and for (11) it is 24, so we can take $K_2 = 2$ in (7) and (8) and $K_4 = 24$ in (9). This yields

$$|E_T| \leq \frac{(b-a)^3 K_2}{12n^2} = \frac{1^3 \cdot 2}{12 \cdot 10^2} \approx 0.001666667$$

$$|E_M| \leq \frac{(b-a)^3 K_2}{24n^2} = \frac{1^3 \cdot 2}{24 \cdot 10^2} \approx 0.000833333$$

$$|E_S| \leq \frac{(b-a)^5 K_4}{180n^4} = \frac{1^5 \cdot 24}{180 \cdot 10^4} \approx 0.000013333 \qquad \blacktriangleleft$$

Table 9.7.4 shows that the estimates in the preceding example are consistent with the computations in Examples 1 and 3. In the table we have obtained approximate values of $|E_T|, |E_M|,$ and $|E_S|$ by computing the absolute value of the difference between the value of $\ln 2$ (to nine decimal places) and the approximations obtained in Examples 1 and 3. Observe that these values of $|E_T|, |E_M|,$ and $|E_S|$ satisfy the upper bounds obtained in Example 4; in fact, they are considerably smaller than the upper bounds. It is quite common that the actual errors in the approximations are substantially smaller than the upper bounds.

Table 9.7.4

ln 2 (NINE DECIMAL PLACES)	APPROXIMATION	ABSOLUTE VALUE OF THE DIFFERENCE
0.693147181	$T_{10} = 0.693771403$	$|E_T| \approx 0.000624222$
0.693147181	$M_{10} = 0.692835360$	$|E_M| \approx 0.000311821$
0.693147181	$S_{10} = 0.693150231$	$|E_S| \approx 0.000003050$

Example 5

How many subintervals should be used in approximating

$$\ln 2 = \int_1^2 \frac{1}{x} \, dx$$

by Simpson's rule for five decimal-place accuracy?

Solution. To obtain five decimal-place accuracy, we must choose the number of subintervals so that

$$|E_S| \leq 0.000005 = 5 \times 10^{-6}$$

From (9), this can be achieved by taking n in Simpson's rule to satisfy

$$\frac{(b-a)^5 K_4}{180n^4} \leq 5 \times 10^{-6}$$

Taking $a = 1$, $b = 2$, and $K_4 = 24$ (found in Example 4) in this inequality yields

$$\frac{24}{180n^4} \leq 5 \times 10^{-6}$$

which, on taking reciprocals, can be rewritten as

$$n^4 \geq \frac{2 \times 10^6}{75}$$

With the help of a calculating utility, and keeping in mind that n must be an even integer, you can verify that the smallest value of n that satisfies this requirement is $n = 14$. Thus, 14 subintervals will produce five decimal-place accuracy. ◀

REMARK.   In cases where it is difficult to find the values of K_2 and K_4 required in Formulas (7), (8), and (9), these constants may be replaced by any larger constants if such constants are easier to find. For example, if $K_2 < K$, then

$$|E_T| \leq \frac{(b-a)^3 K_2}{12n^2} < \frac{(b-a)^3 K}{12n^2} \tag{12}$$

so the right side of (12) is also an upper bound on the value of $|E_T|$ (although it is larger and therefore less desirable than the upper bound using K_2).

Example 6

How many subintervals should be used in approximating

$$\int_0^1 \cos(x^2)\, dx$$

by the midpoint approximation for three decimal-place accuracy?

Solution.   To obtain three decimal-place accuracy, we must choose n so that

$$|E_M| \leq 0.0005 = 5 \times 10^{-4} \tag{13}$$

However, from (7) with $f(x) = \cos(x^2)$, $a = 0$, and $b = 1$, an upper bound on the error $|E_M|$ is given by

$$|E_M| \leq \frac{K_2}{24n^2} \tag{14}$$

where K_2 is the maximum value of $|f''(x)|$ on the interval $[0, 1]$. But,

$$f'(x) = -2x \sin(x^2)$$
$$f''(x) = -4x^2 \cos(x^2) - 2\sin(x^2) = -(4x^2 \cos(x^2) + 2\sin(x^2))$$

so that

$$|f''(x)| = |4x^2 \cos(x^2) + 2\sin(x^2)| \tag{15}$$

It would be tedious to look for the maximum value of this function on the interval $[0, 1]$ analytically. However, it is evident from the graph of $|f''(x)|$ shown in Figure 9.7.6 that

$$|f''(x)| < 4 \qquad \text{for} \qquad 0 \leq x \leq 1$$

Thus, it follows from (14) that

$$|E_M| \leq \frac{K_2}{24n^2} < \frac{4}{24n^2} = \frac{1}{4n^2}$$

and hence we can satisfy (13) by choosing n so that

$$\frac{1}{6n^2} < 5 \times 10^{-4}$$

which, on taking reciprocals, can be written as

$$n^2 > \frac{10^4}{30}$$

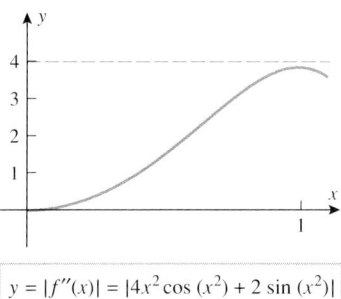

$y = |f''(x)| = |4x^2 \cos(x^2) + 2\sin(x^2)|$

Figure 9.7.6

With the help of a calculating utility, you can show that the smallest value of n that satisfies this requirement is $n = 19$. Thus, 19 subintervals will produce three decimal-place accuracy. ◀

A COMPARISON OF THE THREE METHODS

Of the three methods studied in this section, Simpson's rule generally produces more accurate results than the midpoint or trapezoidal approximations for the same amount of work. To make this plausible, let us express (7), (8), and (9) in terms of the subinterval width

$$\Delta x = \frac{b - a}{n}$$

We obtain

$$|E_M| \leq \frac{1}{24} K_2 (b - a)(\Delta x)^2 \tag{16}$$

$$|E_T| \leq \frac{1}{12} K_2 (b - a)(\Delta x)^2 \tag{17}$$

$$|E_S| \leq \frac{1}{180} K_4 (b - a)(\Delta x)^4 \tag{18}$$

(verify). Thus, for Simpson's rule the upper bound on the absolute error is proportional to $(\Delta x)^4$, whereas it is proportional to $(\Delta x)^2$ for the midpoint and trapezoidal approximations. Thus, reducing the interval width by a factor of 10, for example, reduces the error bound by a factor of 100 for the midpoint and trapezoidal approximations but reduces it by a factor of 10,000 for Simpson's rule. This suggests that the accuracy of Simpson's rule improves much more rapidly than that of the other approximations as n increases.

As a final note, observe that if $f(x)$ is a polynomial of degree 3 or less, then we have $f^{(4)}(x) = 0$ for all x, so $K_4 = 0$ in (9) and consequently $|E_S| = 0$. Thus, Simpson's rule gives exact results for polynomials of degree 3 or less. Similarly, the midpoint and trapezoidal approximations give exact results for polynomials of degree 1 or less. (You should also be able to see that this is so geometrically.)

EXERCISE SET 9.7 [C] CAS

In Exercises 1–6, use $n = 10$ subdivisions to approximate the integral by (a) the midpoint rule, (b) the trapezoidal rule, and (c) Simpson's rule. In each case find the exact value of the integral and approximate the absolute error. Express your answers to at least four decimal places.

1. $\int_0^3 \sqrt{x + 1}\, dx$ **2.** $\int_1^4 \frac{1}{\sqrt{x}}\, dx$

3. $\int_0^\pi \sin x\, dx$ **4.** $\int_0^1 \cos x\, dx$

5. $\int_1^3 e^{-x}\, dx$ **6.** $\int_{-1}^1 \frac{1}{2x + 3}\, dx$

In Exercises 7–12, use inequalities (7), (8), and (9) to find upper bounds on the errors in parts (a), (b), and (c) of the indicated exercise.

7. Exercise 1   **8.** Exercise 2

9. Exercise 3   **10.** Exercise 4

11. Exercise 5   **12.** Exercise 6

In Exercises 13–18, use inequalities (7), (8), and (9) to find a value for n to ensure that the absolute error will be less than the given value if n subdivisions are used to approximate the integral by (a) the midpoint rule, (b) the trapezoidal rule, and (c) Simpson's rule.

13. Exercise 1; 5×10^{-4} **14.** Exercise 2; 5×10^{-4}

15. Exercise 3; 10^{-3} **16.** Exercise 4; 10^{-3}

17. Exercise 5; 10^{-6} **18.** Exercise 6; 10^{-6}

In Exercises 19–24, approximate the integral using Simpson's rule with $n = 10$ subdivisions, and compare the answer to that produced by a calculating utility with a numerical integration capability. Express your answers to at least four decimal places.

19. $\int_0^1 e^{-x^2}\, dx$ **20.** $\int_0^2 \frac{x}{\sqrt{1 + x^3}}\, dx$

21. $\int_1^2 \sqrt{1 + x^3}\, dx$ **22.** $\int_0^\pi \frac{1}{2 - \sin x}\, dx$

23. $\displaystyle\int_0^2 \sin(x^2)\,dx$ **24.** $\displaystyle\int_1^3 \sqrt{\ln x}\,dx$

In Exercises 25 and 26, the exact value of the integral is π (verify). Use $n = 10$ subdivisions to approximate the integral by (a) the midpoint rule, (b) the trapezoidal rule, and (c) Simpson's rule. Find an upper bound on the absolute error, and express your answers to at least four decimal places.

25. $\displaystyle\int_0^1 \frac{4}{1+x^2}\,dx$ **26.** $\displaystyle\int_0^2 \sqrt{4-x^2}\,dx$

27. In Example 5 we showed that taking $n = 14$ subdivisions ensures that the approximation of

$$\ln 2 = \int_1^2 \frac{1}{x}\,dx$$

by Simpson's rule is accurate to five decimal places. Confirm this by comparing the approximation of ln 2 produced by Simpson's rule with $n = 14$ to the value produced directly by your calculating utility.

28. In parts (a) and (b), determine whether an approximation of the integral by the trapezoidal rule would be less than or would be greater than the exact value of the integral.

(a) $\displaystyle\int_1^2 e^{-x^2}\,dx$ (b) $\displaystyle\int_0^{0.5} e^{-x^2}\,dx$

In Exercises 29 and 30, find a value for n to ensure that the absolute error in approximating the integral by the midpoint rule will be less than 10^{-4}.

29. $\displaystyle\int_0^2 x\sin x\,dx$ **30.** $\displaystyle\int_0^1 e^{\cos x}\,dx$

In Exercises 31 and 32, show that inequalities (7) and (8) are of no value in finding an upper bound on the absolute error that results from approximating the integral by either the midpoint rule or the trapezoidal rule.

31. $\displaystyle\int_0^1 \sqrt{x}\,dx$ **32.** $\displaystyle\int_0^1 \sin\sqrt{x}\,dx$

In Exercises 33 and 34, use Simpson's rule with $n = 10$ subdivisions to approximate the length of the curve. Express your answers to at least four decimal places.

33. $y = \sin x$, $0 \le x \le \pi$ **34.** $y = 1/x$, $1 \le x \le 3$

Numerical integration methods can be used in problems where only measured or experimentally determined values of the integrand are available. In Exercises 35–40, use Simpson's rule to estimate the value of the integral.

35. A graph of the speed v versus time t for a test run of an Infiniti G20 automobile is shown in the accompanying figure. Estimate the speeds at $t = 0, 5, 10, 15,$ and 20 s from the graph, convert to ft/s using 1 mi/h = 22/15 ft/s, and use these speeds to approximate the number of feet traveled

during the first 20 s. Round your answer to the nearest foot. [*Hint:* Distance traveled $= \int_0^{20} v(t)\,dt$.] [Data from *Road and Track*, October 1990.]

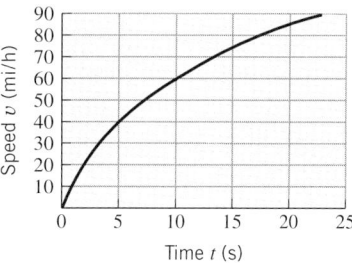

Figure Ex-35

36. A graph of the acceleration a versus time t for an object moving on a straight line is shown in the accompanying figure. Estimate the accelerations at $t = 0, 1, 2, \ldots, 8$ s from the graph and use them to approximate the change in velocity from $t = 0$ to $t = 8$ s. Round your answer to the nearest tenth cm/s. [*Hint:* Change in velocity $= \int_0^8 a(t)\,dt$.]

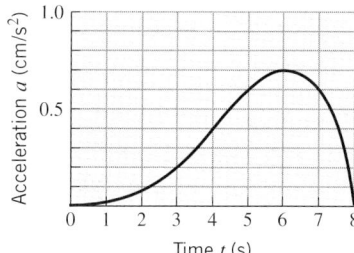

Figure Ex-36

37. The table in the accompanying figure gives the speeds, in miles per second, at various times for a test rocket that was fired upward from the surface of the Earth. Use these values to approximate the number of miles traveled during the first 180 s. Round your answer to the nearest tenth of a mile. [*Hint:* Distance traveled $= \int_0^{180} v(t)\,dt$.]

38. The table in the accompanying figure gives the speeds of a bullet at various distances from the muzzle of a rifle. Use these values to approximate the number of seconds for the bullet to travel 1800 ft. Express your answer to the nearest hundredth of a second. [*Hint:* If v is the speed of the bullet and x is the distance traveled, then $v = dx/dt$ so that $dt/dx = 1/v$ and $t = \int_0^{1800} (1/v)\,dx$.]

TIME t (S)	SPEED v (mi/s)	DISTANCE x (ft)	SPEED v (ft/s)
0	0.00	0	3100
30	0.03	300	2908
60	0.08	600	2725
90	0.16	900	2549
120	0.27	1200	2379
150	0.42	1500	2216
180	0.65	1800	2059

Figure Ex-37   Figure Ex-38

39. Measurements of a pottery shard recovered from an archaeological dig reveal that the shard came from a pot with a flat bottom and circular cross sections (see the accompanying figure). The figure shows interior radius measurements of the shard made every 4 cm from the bottom of the pot to the top. Use those values to approximate the interior volume of the pot to the nearest tenth of a liter (1 L = 1000 cm^3). [*Hint:* Use 8.2.3 (volume by cross sections) to set up an appropriate integral for the volume.]

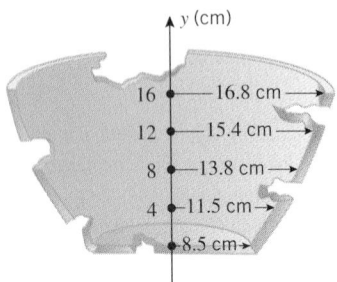

Figure Ex-39

40. Engineers want to construct a straight and level road 600 ft long and 75 ft wide by making a vertical cut through an intervening hill (see the accompanying figure). Heights of the hill above the centerline of the proposed road, as obtained at various points from a contour map of the region, are shown in the accompanying table. To estimate the construction costs, the engineers need to know the volume of earth that must be removed. Approximate this volume, rounded to the nearest cubic foot. [*Hint:* First, set up an integral for the cross-sectional area of the cut along the centerline of the road, then assume that the height of the hill does not vary between the centerline and edges of the road.]

HORIZONTAL DISTANCE x (ft)	HEIGHT h (ft)
0	0
100	7
200	16
300	24
400	25
500	16
600	0

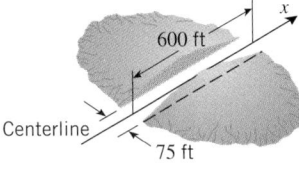

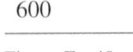

Figure Ex-40

41. Derive the trapezoidal rule by summing the areas of the trapezoids in Figure 9.7.2.

42. Let f be a function that is positive, continuous, decreasing, and concave down on the interval $[a, b]$. Assuming that $[a, b]$ is subdivided into n equal subintervals, arrange the following approximations of $\int_a^b f(x)\,dx$ in order of increasing value: left endpoint, right endpoint, midpoint, and trapezoidal.

C **43.** Let $f(x) = \cos(x^2)$.
  (a) Use a CAS to approximate the maximum value of $|f''(x)|$ on the interval $[0, 1]$.
  (b) How large must n be in the midpoint approximation of $\int_0^1 f(x)\,dx$ to ensure that the absolute error is less than 5×10^{-4}? Compare your result with that obtained in Example 6.
  (c) Evaluate the integral using the midpoint approximation with the value of n obtained in part (b).

C **44.** Let $f(x) = \sqrt{1 + x^3}$.
  (a) Use a CAS to approximate the maximum value of $|f''(x)|$ on the interval $[0, 1]$.
  (b) How large must n be in the trapezoidal approximation of $\int_0^1 f(x)\,dx$ to ensure that the absolute error is less than 10^{-3}?
  (c) Evaluate the integral using the trapezoidal approximation with the value of n obtained in part (b).

C **45.** Let $f(x) = \cos(x^2)$.
  (a) Use a CAS to approximate the maximum value of $|f^{(4)}(x)|$ on the interval $[0, 1]$.
  (b) How large must the value of n be in the approximation of $\int_0^1 f(x)\,dx$ by Simpson's rule to ensure that the absolute error is less than 10^{-4}?
  (c) Evaluate the integral using Simpson's rule with the value of n obtained in part (b).

C **46.** Let $f(x) = \sqrt{1 + x^3}$.
  (a) Use a CAS to approximate the maximum value of $|f^{(4)}(x)|$ on the interval $[0, 1]$.
  (b) How large must the value of n be in the approximation of $\int_0^1 f(x)\,dx$ by Simpson's rule to ensure that the absolute error is less than 10^{-5}?
  (c) Evaluate the integral using Simpson's rule with the value of n obtained in part (b).

9.8 IMPROPER INTEGRALS

Up to now we have focused on definite integrals with continuous integrands and finite intervals of integration. In this section we will extend the concept of a definite integral to include infinite intervals of integration and integrands that become infinite within the interval of integration.

It is assumed in the definition of the definite integral

$$\int_a^b f(x)\, dx$$

that $[a, b]$ is a finite interval and that the limit that defines the integral exists; that is, the function f is integrable. We observed in Theorem 7.5.8 that continuous functions are integrable, as are bounded functions with finitely many points of discontinuity. We also observed in that theorem that functions that are not bounded on the interval of integration are not integrable. Thus, for example, a function with a vertical asymptote within the interval of integration would not be integrable.

Our main objective in this section is to extend the concept of a definite integral to allow for infinite intervals of integration and integrands with vertical asymptotes within the interval of integration. We will call the vertical asymptotes *infinite discontinuities*, and we will call integrals with infinite intervals of integration or infinite discontinuities within the interval of integration *improper integrals*. Here are some examples:

- Improper integrals with infinite intervals of integration:

$$\int_1^{+\infty} \frac{dx}{x^2}, \quad \int_{-\infty}^0 e^x\, dx, \quad \int_{-\infty}^{+\infty} \frac{dx}{1+x^2}$$

- Improper integrals with infinite discontinuities in the interval of integration:

$$\int_{-3}^3 \frac{dx}{x^2}, \quad \int_1^2 \frac{dx}{x-1}, \quad \int_0^\pi \tan x\, dx$$

- Improper integrals with infinite discontinuities and infinite intervals of integration:

$$\int_0^{+\infty} \frac{dx}{\sqrt{x}}, \quad \int_{-\infty}^{+\infty} \frac{dx}{x^2-9}, \quad \int_1^{+\infty} \sec x\, dx$$

To motivate a reasonable definition for improper integrals of the form

$$\int_a^{+\infty} f(x)\, dx$$

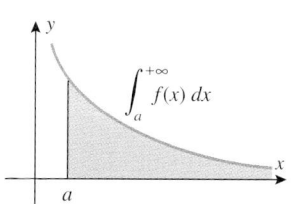

Figure 9.8.1

let us begin with the case where f is continuous and nonnegative on $[a, +\infty)$, so we can think of the integral as the area under the curve $y = f(x)$ over the interval $[a, +\infty)$ (Figure 9.8.1). At first, you might be inclined to argue that this area is infinite because the region has infinite extent. However, such an argument would be based on vague intuition rather than precise mathematical logic, since the concept of area has only been defined over intervals of *finite extent*. Thus, before we can make any reasonable statements about the area of the region in Figure 9.8.1, we need to begin by defining what we mean by the area of this region. For that purpose, it will help to focus on a specific example.

Suppose we are interested in the area A of the region that lies below the curve $y = 1/x^2$ and above the interval $[1, +\infty)$ on the x-axis. Instead of trying to find the entire area at once, let us begin by calculating the portion of the area that lies above a finite interval $[1, l]$, where $l > 1$ is arbitrary. That area is

$$\int_1^l \frac{dx}{x^2} = -\frac{1}{x}\Big]_1^l = 1 - \frac{1}{l}$$

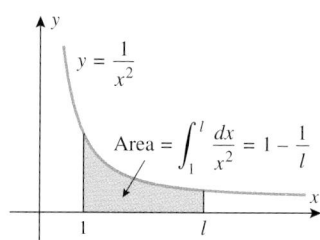

Figure 9.8.2

(Figure 9.8.2). If we now allow l to increase so that $l \to +\infty$, then the portion of the area over the interval $[1, l]$ will begin to fill out the area over the entire interval $[1, +\infty)$ (Figure 9.8.3), and hence we can reasonably define the area A under $y = 1/x^2$ over the interval $[1, +\infty)$ to be

$$A = \int_1^{+\infty} \frac{dx}{x^2} = \lim_{l \to +\infty} \int_1^l \frac{dx}{x^2} = \lim_{l \to +\infty}\left(1 - \frac{1}{l}\right) = 1 \tag{1}$$

Thus, the area has a finite value of 1 and is not infinite as we first conjectured.

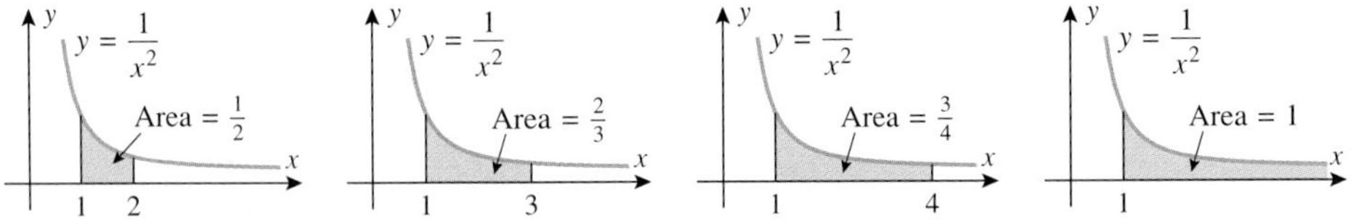

Figure 9.8.3

With the preceding discussion as our guide, we make the following definition (which is applicable to functions with both positive and negative values):

9.8.1 DEFINITION.   The **improper integral of f over the interval $[a, +\infty)$** is defined as

$$\int_{a}^{+\infty} f(x)\, dx = \lim_{l \to +\infty} \int_{a}^{l} f(x)\, dx$$

In the case where the limit exists, the improper integral is said to **converge**, and the limit is defined to be the value of the integral. In the case where the limit does not exist, the improper integral is said to **diverge**, and it is not assigned a value.

If f is nonnegative on $[a, +\infty)$ and the improper integral converges, then the value of the integral is regarded to be the area under the graph of f over the interval $[a, +\infty)$; and if the integral diverges, then the area under the graph of f over the interval $[a, +\infty)$ is regarded to be infinite.

Example 1

Evaluate

(a) $\displaystyle\int_{1}^{+\infty} \frac{dx}{x^3}$ (b) $\displaystyle\int_{1}^{+\infty} \frac{dx}{x}$

Solution (a). Following the definition, we replace the infinite upper limit by a finite upper limit l, and then take the limit of the resulting integral. This yields

$$\int_{1}^{+\infty} \frac{dx}{x^3} = \lim_{l \to +\infty} \int_{1}^{l} \frac{dx}{x^3} = \lim_{l \to +\infty} \left[-\frac{1}{2x^2} \right]_{1}^{l} = \lim_{l \to +\infty} \left(\frac{1}{2} - \frac{1}{2l^2} \right) = \frac{1}{2}$$

Solution (b).

$$\int_{1}^{+\infty} \frac{dx}{x} = \lim_{l \to +\infty} \int_{1}^{l} \frac{dx}{x} = \lim_{l \to +\infty} \left[\ln x \right]_{1}^{l} = \lim_{l \to +\infty} \ln l = +\infty$$

In this case the integral diverges and hence has no value.   ◀

Because the functions $1/x^3$, $1/x^2$, and $1/x$ are nonnegative over the interval $[1, +\infty)$, it follows from (1) and the last example that over this interval the area under $y = 1/x^3$ is $\frac{1}{2}$, the area under $y = 1/x^2$ is 1, and the area under $y = 1/x$ is infinite. However, on the surface the graphs of the three functions seem very much alike (Figure 9.8.4), and there is nothing to suggest why one of the areas should be infinite and the other two finite. One explanation is that $1/x^3$ and $1/x^2$ approach zero more rapidly than $1/x$ as $x \to +\infty$, so that the area over the interval $[1, l]$ accumulates less rapidly under the curves $y = 1/x^3$ and $y = 1/x^2$ than under $y = 1/x$ as $l \to +\infty$, and the difference is just enough that the first two areas are finite and the third is infinite.

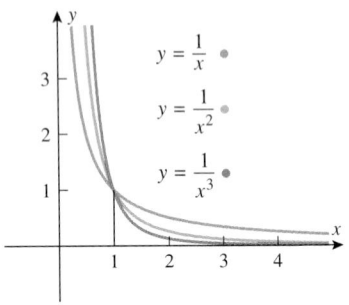

Figure 9.8.4

Example 2

For what values of p does the integral $\displaystyle\int_{1}^{+\infty} \frac{dx}{x^p}$ converge?

Solution. We know from the preceding example that the integral diverges if $p = 1$, so let us assume that $p \neq 1$. In this case we have

$$\int_{1}^{+\infty} \frac{dx}{x^p} = \lim_{l \to +\infty} \int_{1}^{l} x^{-p}\, dx = \lim_{l \to +\infty} \frac{x^{1-p}}{1-p}\bigg]_{1}^{l} = \lim_{l \to +\infty}\left[\frac{l^{1-p}}{1-p} - \frac{1}{1-p}\right]$$

If $p > 1$, then the exponent $1 - p$ is negative and $l^{1-p} \to 0$ as $l \to +\infty$; and if $p < 1$, then the exponent $1 - p$ is positive and $l^{1-p} \to +\infty$ as $l \to +\infty$. Thus, the integral converges if $p > 1$ and diverges otherwise. In the convergent case the value of the integral is

$$\int_{1}^{+\infty} \frac{dx}{x^p} = \left[0 - \frac{1}{1-p}\right] = \frac{1}{p-1} \quad (p > 1) \qquad \blacktriangleleft$$

The following theorem summarizes this result:

9.8.2   THEOREM.

$$\int_{1}^{+\infty} \frac{dx}{x^p} = \begin{cases} \dfrac{1}{p-1} & \text{if} \quad p > 1 \\ \text{diverges} & \text{if} \quad p \leq 1 \end{cases}$$

Example 3

Evaluate $\displaystyle\int_{0}^{+\infty} (1-x)e^{-x}\, dx$.

Solution.   Integrating by parts with $u = 1 - x$ and $dv = e^{-x}\, dx$ yields

$$\int (1-x)e^{-x}\, dx = -e^{-x}(1-x) - \int e^{-x}\, dx = -e^{-x} + xe^{-x} + e^{-x} + C = xe^{-x} + C$$

Thus,

$$\int_{0}^{+\infty} (1-x)e^{-x}\, dx = \lim_{l \to +\infty}\left[xe^{-x}\right]_{0}^{l} = \lim_{l \to +\infty} \frac{l}{e^l}$$

The limit is an indeterminate form of type ∞/∞, so we will apply L'Hôpital's rule by differentiating the numerator and denominator with respect to l. This yields

$$\int_{0}^{+\infty} (1-x)e^{-x}\, dx = \lim_{l \to +\infty} \frac{1}{e^l} = 0$$

An explanation of why this integral is zero can be obtained by interpreting the integral as the net signed area between the graph of $y = (1-x)e^{-x}$ and the interval $[0, +\infty)$ (Figure 9.8.5). $\blacktriangleleft$

We also make the following definition:

9.8.3   DEFINITION.   The *improper integral of f over the interval* $(-\infty, b]$ is defined as

$$\int_{-\infty}^{b} f(x)\, dx = \lim_{l \to -\infty} \int_{l}^{b} f(x)\, dx \qquad\qquad (2)$$

The integral is said to *converge* if the limit exists and *diverge* if it does not. The *improper integral of f over the interval* $(-\infty, +\infty)$ is defined as

$$\int_{-\infty}^{+\infty} f(x)\, dx = \int_{-\infty}^{c} f(x)\, dx + \int_{c}^{+\infty} f(x)\, dx \qquad\qquad (3)$$

where c is any real number. The improper integral is said to *converge* if both terms converge and *diverge* if either term diverges.

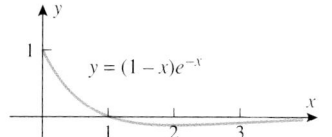

$y = (1-x)e^{-x}$

The net signed area between the graph and the interval $[0, +\infty)$ is zero.

Figure 9.8.5

REMARK.   In this definition, if f is nonnegative on the interval of integration, then the improper integral is regarded to be the area under the graph of f over that interval; the area has a finite value if the integral converges and is infinite if it diverges. We also note that in (3) it is usual to choose $c = 0$, but the choice does not matter; it can be proved that neither the convergence nor the value of the integral depends on the choice of c.

Example 4

Evaluate $\displaystyle\int_{-\infty}^{+\infty} \frac{dx}{1 + x^2}$.

Solution. We will evaluate the integral by choosing $c = 0$ in (3). With this value for c we obtain

$$\int_{0}^{+\infty} \frac{dx}{1 + x^2} = \lim_{l \to +\infty} \int_{0}^{l} \frac{dx}{1 + x^2} = \lim_{l \to +\infty} \left[\tan^{-1} x\right]_{0}^{l} = \lim_{l \to +\infty} (\tan^{-1} l) = \frac{\pi}{2}$$

$$\int_{-\infty}^{0} \frac{dx}{1 + x^2} = \lim_{l \to -\infty} \int_{l}^{0} \frac{dx}{1 + x^2} = \lim_{l \to -\infty} \left[\tan^{-1} x\right]_{l}^{0} = \lim_{l \to -\infty} (-\tan^{-1} l) = \frac{\pi}{2}$$

Thus, the integral converges and its value is

$$\int_{-\infty}^{+\infty} \frac{dx}{1 + x^2} = \int_{-\infty}^{0} \frac{dx}{1 + x^2} + \int_{0}^{+\infty} \frac{dx}{1 + x^2} = \frac{\pi}{2} + \frac{\pi}{2} = \pi$$

Since the integrand is nonnegative on the interval $(-\infty, +\infty)$, the integral represents the area of the region shown in Figure 9.8.6.   ◄

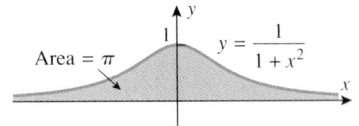

Area $= \pi$ $y = \dfrac{1}{1 + x^2}$

Figure 9.8.6

INTEGRALS WHOSE INTEGRANDS HAVE INFINITE DISCONTINUITIES

Next we will consider improper integrals whose integrands have infinite discontinuities. We will start with the case where the interval of integration is a finite interval $[a, b]$ and the infinite discontinuity occurs at the right-hand endpoint.

To motivate an appropriate definition for such an integral let us consider the case where f is nonnegative on $[a, b]$, so we can interpret the improper integral $\int_{a}^{b} f(x)\, dx$ as the area of the region in Figure 9.8.7a. The problem of finding the area of this region is complicated by the fact that it extends indefinitely in the positive y-direction. However, instead of trying to find the entire area at once, we can proceed indirectly by calculating the portion of the area over the interval $[a, l]$ and then letting l approach b to fill out the area of the entire region (Figure 9.8.7b). Motivated by this idea, we make the following definition:

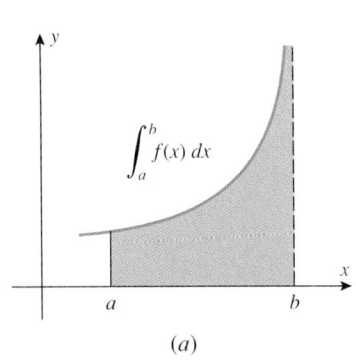

$\displaystyle\int_{a}^{b} f(x)\, dx$

(a)

9.8.4 DEFINITION.   If f is continuous on the interval $[a, b]$, except for an infinite discontinuity at b, then the ***improper integral of f over the interval $[a, b]$*** is defined as

$$\int_{a}^{b} f(x)\, dx = \lim_{l \to b^-} \int_{a}^{l} f(x)\, dx \tag{4}$$

In the case where the limit exists, the improper integral is said to ***converge***, and the limit is defined to be the value of the integral. In the case where the limit does not exist, the improper integral is said to ***diverge***, and it is not assigned a value.

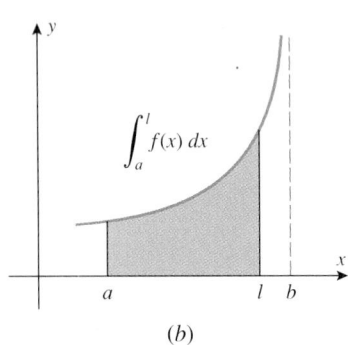

$\displaystyle\int_{a}^{l} f(x)\, dx$

(b)

Figure 9.8.7

Example 5

Evaluate $\displaystyle\int_{0}^{1} \frac{dx}{\sqrt{1 - x}}$.

Solution. The integral is improper because the integrand approaches $+\infty$ as x approaches the upper limit 1 from the left. From (4),

$$\int_0^1 \frac{dx}{\sqrt{1-x}} = \lim_{l \to 1^-} \int_0^l \frac{dx}{\sqrt{1-x}} = \lim_{l \to 1^-} \left[-2\sqrt{1-x} \right]_0^l$$

$$= \lim_{l \to 1^-} [-2\sqrt{1-l} + 2] = 2 \qquad \blacktriangleleft$$

Improper integrals with an infinite discontinuity at the left-hand endpoint or inside the interval of integration are defined as follows.

> **9.8.5** DEFINITION. If f is continuous on the interval $[a, b]$, except for an infinite discontinuity at a, then the ***improper integral of f over the interval $[a, b]$*** is defined as
>
> $$\int_a^b f(x)\,dx = \lim_{l \to a^+} \int_l^b f(x)\,dx \qquad (5)$$
>
> The integral is said to ***converge*** if the limit exists and ***diverge*** if it does not. If f is continuous on the interval $[a, b]$, except for an infinite discontinuity at a point c in (a, b), then the ***improper integral of f over the interval $[a, b]$*** is defined as
>
> $$\int_a^b f(x)\,dx = \int_a^c f(x)\,dx + \int_c^b f(x)\,dx \qquad (6)$$
>
> The improper integral is said to ***converge*** if both terms converge and ***diverge*** if either term diverges (Figure 9.8.8).

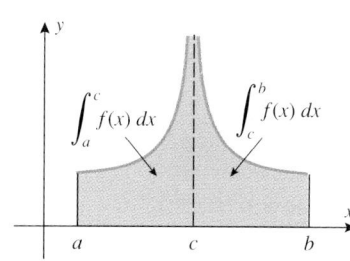

Figure 9.8.8

Example 6

Evaluate

(a) $\displaystyle\int_1^2 \frac{dx}{1-x}$ (b) $\displaystyle\int_1^4 \frac{dx}{(x-2)^{2/3}}$ (c) $\displaystyle\int_0^{+\infty} \frac{dx}{\sqrt{x}(x+1)}$

Solution (a). The integral is improper because the integrand approaches $-\infty$ as x approaches the lower limit 1 from the right (Figure 9.8.9). From Definition 9.8.5 we obtain

$$\int_1^2 \frac{dx}{1-x} = \lim_{l \to 1^+} \int_l^2 \frac{dx}{1-x} = \lim_{l \to 1^+} \left[-\ln|1-x| \right]_l^2$$

$$= \lim_{l \to 1^+} \left[-\ln|-1| + \ln|1-l| \right] = \lim_{l \to 1^+} \ln|1-l| = -\infty$$

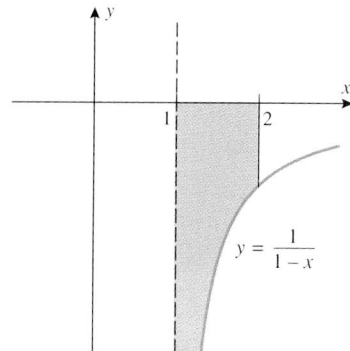

Figure 9.8.9

so the integral diverges.

Solution (b). The integral is improper because the integrand approaches $+\infty$ at the point $x = 2$, which is inside the interval of integration. From Definition 9.8.5 we obtain

$$\int_1^4 \frac{dx}{(x-2)^{2/3}} = \int_1^2 \frac{dx}{(x-2)^{2/3}} + \int_2^4 \frac{dx}{(x-2)^{2/3}} \qquad (7)$$

But

$$\int_1^2 \frac{dx}{(x-2)^{2/3}} = \lim_{l \to 2^-} \int_1^l \frac{dx}{(x-2)^{2/3}} = \lim_{l \to 2^-} [3(l-2)^{1/3} - 3(1-2)^{1/3}] = 3$$

$$\int_2^4 \frac{dx}{(x-2)^{2/3}} = \lim_{l \to 2^+} \int_l^4 \frac{dx}{(x-2)^{2/3}} = \lim_{l \to 2^+} [3(4-2)^{1/3} - 3(l-2)^{1/3}] = 3\sqrt[3]{2}$$

Thus, from (7)

$$\int_1^4 \frac{dx}{(x-2)^{2/3}} = 3 + 3\sqrt[3]{2}$$

Solution (c). This integral is improper for two reasons—the interval of integration is infinite, and there is an infinite discontinuity at $x = 0$. To evaluate this integral we will split the interval of integration at a convenient point, say $x = 1$, and write

$$\int_0^{+\infty} \frac{dx}{\sqrt{x}(x+1)} = \int_0^1 \frac{dx}{\sqrt{x}(x+1)} + \int_1^{+\infty} \frac{dx}{\sqrt{x}(x+1)}$$

The integrand in these two improper integrals does not match any of the forms in the Endpaper Integral Table, but the radical suggests the substitution $x = u^2, \, dx = 2u \, du$, from which we obtain

$$\int \frac{dx}{\sqrt{x}(x+1)} = \int \frac{2u \, du}{u(u^2+1)} = 2 \int \frac{du}{u^2+1}$$

$$= 2 \tan^{-1} u + C = 2 \tan^{-1} \sqrt{x} + C$$

Thus,

$$\int_0^{+\infty} \frac{dx}{\sqrt{x}(x+1)} = 2 \lim_{l \to 0^+} \left[\tan^{-1} \sqrt{x}\right]_l^1 + 2 \lim_{l \to +\infty} \left[\tan^{-1} \sqrt{x}\right]_1^l$$

$$= 2 \left[\frac{\pi}{4} - 0\right] + 2 \left[\frac{\pi}{2} - \frac{\pi}{4}\right] = \pi \qquad \blacktriangleleft$$

WARNING. It is sometimes tempting to apply the Fundamental Theorem of Calculus directly to an improper integral without taking the appropriate limits. To illustrate what can go wrong with this procedure, suppose we ignore the fact that the integral

$$\int_0^2 \frac{dx}{(x-1)^2} \tag{8}$$

is improper and write

$$\int_0^2 \frac{dx}{(x-1)^2} = -\frac{1}{x-1}\Bigg]_0^2 = -1 - (1) = -2$$

This result is clearly nonsense because the integrand is never negative and consequently the integral cannot be negative! To evaluate (8) correctly we should write

$$\int_0^2 \frac{dx}{(x-1)^2} = \int_0^1 \frac{dx}{(x-1)^2} + \int_1^2 \frac{dx}{(x-1)^2}$$

But

$$\int_0^1 \frac{dx}{(x-1)^2} = \lim_{l \to 1^-} \int_0^l \frac{dx}{(x-1)^2} = \lim_{l \to 1^-} \left[-\frac{1}{l-1} - 1\right] = +\infty$$

so that (8) diverges.

THE APPLICATION OF IMPROPER INTEGRALS TO ARC LENGTH AND SURFACE AREA

In Definitions 8.4.2 and 8.5.2 for arc length and surface area we required the function f to be smooth (continuous first derivative) to ensure the integrability in the resulting formula. However, smoothness is overly restrictive since some of the most basic formulas in geometry involve functions that are not smooth but lead to convergent improper integrals. Accordingly, let us agree to extend the definitions of arc length and surface area to allow functions that are not smooth, but for which the resulting integral in the formula converges.

Example 7

Derive the formula for the circumference of a circle of radius r.

Solution. For convenience, let us assume that the circle is centered at the origin, in which case its equation is $x^2 + y^2 = r^2$. We will find the arc length of the portion of the circle that lies in the first quadrant and then multiply by 4 to obtain the total circumference (Figure 9.8.10).

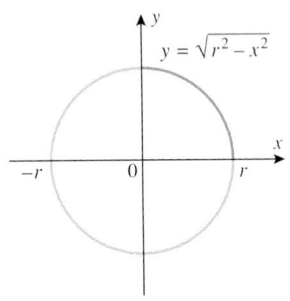

Figure 9.8.10

Since the equation of the upper semicircle is $y = \sqrt{r^2 - x^2}$, it follows from Formula (4) of Section 8.4 that the circumference C is

$$C = 4 \int_0^r \sqrt{1 + (dy/dx)^2}\, dx = 4 \int_0^r \sqrt{1 + \left(-\frac{x}{\sqrt{r^2 - x^2}} \right)^2}\, dx$$

$$= 4r \int_0^r \frac{dx}{\sqrt{r^2 - x^2}}$$

This integral is improper because of the infinite discontinuity at $x = r$, and hence we evaluate it by writing

$$C = 4r \lim_{l \to r^-} \int_0^l \frac{dx}{\sqrt{r^2 - x^2}}$$

$$= 4r \lim_{l \to r^-} \left[\sin^{-1}\left(\frac{x}{r}\right) \right]_0^l \qquad \text{Formula (77) in the Endpaper Integral Table}$$

$$= 4r \lim_{l \to r^-} \left[\sin^{-1}\left(\frac{l}{r}\right) - \sin^{-1} 0 \right]$$

$$= 4r[\sin^{-1} 1 - \sin^{-1} 0] = 4r \left(\frac{\pi}{2} - 0 \right) = 2\pi r$$

◀

EXERCISE SET 9.8   ☐ Graphing Calculator   ☐C CAS

1. In each part, determine whether the integral is improper, and if so, explain why.

(a) $\displaystyle\int_1^5 \frac{dx}{x-3}$ (b) $\displaystyle\int_1^5 \frac{dx}{x+3}$ (c) $\displaystyle\int_0^1 \ln x\, dx$

(d) $\displaystyle\int_1^{+\infty} e^{-x}\, dx$ (e) $\displaystyle\int_{-\infty}^{+\infty} \frac{dx}{\sqrt[3]{x-1}}$ (f) $\displaystyle\int_0^{\pi/4} \tan x\, dx$

2. In each part, determine all values of p for which the integral is improper.

(a) $\displaystyle\int_0^1 \frac{dx}{x^p}$ (b) $\displaystyle\int_1^2 \frac{dx}{x-p}$ (c) $\displaystyle\int_0^1 e^{-px}\, dx$

In Exercises 3–30, evaluate the integrals that converge.

3. $\displaystyle\int_0^{+\infty} e^{-x}\, dx$

4. $\displaystyle\int_{-1}^{+\infty} \frac{x}{1+x^2}\, dx$

5. $\displaystyle\int_4^{+\infty} \frac{2}{x^2-1}\, dx$

6. $\displaystyle\int_0^{+\infty} xe^{-x^2}\, dx$

7. $\displaystyle\int_c^{+\infty} \frac{1}{x \ln^3 x}\, dx$

8. $\displaystyle\int_2^{+\infty} \frac{1}{x\sqrt{\ln x}}\, dx$

9. $\displaystyle\int_{-\infty}^0 \frac{dx}{(2x-1)^3}$

10. $\displaystyle\int_{-\infty}^2 \frac{dx}{x^2+4}$

11. $\displaystyle\int_{-\infty}^0 e^{3x}\, dx$

12. $\displaystyle\int_{-\infty}^0 \frac{e^x\, dx}{3 - 2e^x}$

13. $\displaystyle\int_{-\infty}^{+\infty} x^3\, dx$

14. $\displaystyle\int_{-\infty}^{+\infty} \frac{x}{\sqrt{x^2+2}}\, dx$

15. $\displaystyle\int_{-\infty}^{+\infty} \frac{x}{(x^2+3)^2}\, dx$

16. $\displaystyle\int_{-\infty}^{+\infty} \frac{e^{-t}}{1+e^{-2t}}\, dt$

17. $\displaystyle\int_3^4 \frac{dx}{(x-3)^2}$

18. $\displaystyle\int_0^8 \frac{dx}{\sqrt[3]{x}}$

19. $\displaystyle\int_0^{\pi/2} \tan x\, dx$

20. $\displaystyle\int_0^9 \frac{dx}{\sqrt{9-x}}$

21. $\displaystyle\int_0^1 \frac{dx}{\sqrt{1-x^2}}$

22. $\displaystyle\int_{-3}^1 \frac{x\, dx}{\sqrt{9-x^2}}$

23. $\displaystyle\int_0^{\pi/6} \frac{\cos x}{\sqrt{1-2\sin x}}\, dx$

24. $\displaystyle\int_0^{\pi/4} \frac{\sec^2 x}{1 - \tan x}\, dx$

25. $\displaystyle\int_0^3 \frac{dx}{x-2}$

26. $\displaystyle\int_{-2}^2 \frac{dx}{x^2}$

27. $\displaystyle\int_{-1}^8 x^{-1/3}\, dx$

28. $\displaystyle\int_0^4 \frac{dx}{(x-2)^{2/3}}$

29. $\displaystyle\int_0^{+\infty} \frac{1}{x^2}\, dx$

30. $\displaystyle\int_1^{+\infty} \frac{dx}{x\sqrt{x^2-1}}$

In Exercises 31–34, make the u-substitution and evaluate the resulting definite integral.

31. $\displaystyle\int_0^{+\infty} \frac{e^{-\sqrt{x}}}{\sqrt{x}}\, dx; \ u = \sqrt{x}$ [*Note:* $u \to +\infty$ as $x \to +\infty$.]

32. $\displaystyle\int_0^{+\infty} \frac{dx}{\sqrt{x}(x+4)}; \ u = \sqrt{x}$

33. $\displaystyle\int_0^{+\infty} \frac{e^{-x}}{\sqrt{1-e^{-x}}}\, dx; \ u = 1 - e^{-x}$
[*Note:* $u \to 1$ as $x \to +\infty$.]

34. $\displaystyle\int_0^{+\infty} \frac{e^{-x}}{\sqrt{1-e^{-2x}}}\, dx; \ u = e^{-x}$

C **35.** Read your CAS documentation to determine how to evaluate definite integrals with infinite limits of integration, and then for each of the integrals you evaluated in Exercises 1–34, check your answer with your CAS.

In Exercises 36 and 37, express the improper integral as a limit, and then evaluate that limit with a CAS. Confirm the answer by evaluating the integral directly with the CAS.

C **36.** $\displaystyle\int_0^{+\infty} xe^{-3x}\,dx$ **C** **37.** $\displaystyle\int_0^{+\infty} e^{-x}\cos x\,dx$

C **38.** In each part, confirm the result with a CAS.

(a) $\displaystyle\int_0^{+\infty} \frac{\sin x}{\sqrt{x}}\,dx = \sqrt{\frac{\pi}{2}}$ (b) $\displaystyle\int_{-\infty}^{+\infty} e^{-x^2}\,dx = \sqrt{\pi}$

(c) $\displaystyle\int_0^1 \frac{\ln x}{1+x}\,dx = -\frac{\pi^2}{12}$

C **39.** In each part, try to evaluate the integral exactly with a CAS. If your result is not a simple numerical answer, then use the CAS to find a numerical approximation of the integral.

(a) $\displaystyle\int_{-\infty}^{+\infty} \frac{1}{x^8+x+1}\,dx$ (b) $\displaystyle\int_0^{+\infty} \frac{1}{\sqrt{1+x^3}}\,dx$

(c) $\displaystyle\int_1^{+\infty} \frac{\ln x}{e^x}\,dx$ (d) $\displaystyle\int_1^{+\infty} \frac{\sin x}{x^2}\,dx$

40. Find the length of the curve $y = \sqrt{9-x^2}$ over the interval $[0, 3]$.

In Exercises 41 and 42, use L'Hôpital's rule to help evaluate the improper integral.

41. $\displaystyle\int_0^1 \ln x\,dx$ **42.** $\displaystyle\int_1^{+\infty} \frac{\ln x}{x^2}\,dx$

43. Find the area of the region between the x-axis and the curve $y = e^{-3x}$ for $x \geq 0$.

44. Find the area of the region between the x-axis and the curve $y = 8/(x^2 - 4)$ for $x \geq 3$.

45. Suppose that the region between the x-axis and the curve $y = e^{-x}$ for $x \geq 0$ is revolved about the x-axis.
(a) Find the volume of the solid that is generated.
(b) Find the surface area of the solid.

46. Suppose that f and g are continuous functions and that

$$0 \leq f(x) \leq g(x)$$

if $x \geq a$. Give a reasonable informal argument using areas to explain why the following results are true.
(a) If $\int_a^{+\infty} f(x)\,dx$ diverges, then $\int_a^{+\infty} g(x)\,dx$ diverges.

(b) If $\int_a^{+\infty} g(x)\,dx$ converges, then $\int_a^{+\infty} f(x)\,dx$ converges and $\int_a^{+\infty} f(x)\,dx \leq \int_a^{+\infty} g(x)\,dx$.
[*Note:* The results in this exercise are sometimes called *comparison tests* for improper integrals.]

In Exercises 47–51, use the results in Exercise 46.

47. (a) Confirm graphically and algebraically that $e^{-x^2} \leq e^{-x}$ if $x \geq 1$.
(b) Evaluate the integral

$$\int_1^{+\infty} e^{-x}\,dx$$

(c) What does the result obtained in part (b) tell you about the integral

$$\int_1^{+\infty} e^{-x^2}\,dx?$$

48. (a) Confirm graphically and algebraically that

$$\frac{1}{2x+1} \leq \frac{e^x}{2x+1} \quad (x \geq 0)$$

(b) Evaluate the integral

$$\int_0^{+\infty} \frac{dx}{2x+1}$$

(c) What does the result obtained in part (b) tell you about the integral

$$\int_0^{+\infty} \frac{e^x}{2x+1}\,dx?$$

49. Let R be the region to the right of $x = 1$ that is bounded by the x-axis and the curve $y = 1/x$. When this region is revolved about the x-axis it generates a solid whose surface is known as **Gabriel's Horn** (for reasons that should be clear from the accompanying figure). Show that the solid has a finite volume but its surface has an infinite area. [*Note:* It has been suggested that if one could saturate the interior of the solid with paint and allow it to seep through to the surface, then one could paint an infinite surface with a finite amount of paint! What do you think?]

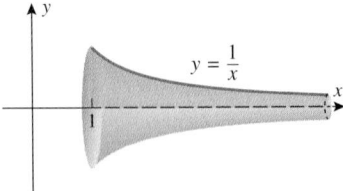

Figure Ex-49

50. In each part, use Exercise 46 to determine whether the integral converges or diverges. If it converges, then use part (b) of that exercise to find an upper bound on the value of the integral.

(a) $\displaystyle\int_2^{+\infty} \frac{\sqrt{x^3+1}}{x}\,dx$ (b) $\displaystyle\int_2^{+\infty} \frac{x}{x^5+1}\,dx$

(c) $\displaystyle\int_0^{+\infty} \frac{e^x}{2x+1}\,dx$

51. Show that

$$\lim_{x \to +\infty} \frac{\int_0^{2x} \sqrt{1 + t^3}\, dt}{x^{5/2}}$$

is an indeterminate form of type ∞/∞, and then use L'Hôpital's rule to find the limit.

52. (a) Give a reasonable informal argument, based on areas, that explains why the integrals

$$\int_0^{+\infty} \sin x\, dx \quad \text{and} \quad \int_0^{+\infty} \cos x\, dx$$

diverge.

(b) Show that $\displaystyle\int_0^{+\infty} \frac{\cos\sqrt{x}}{\sqrt{x}}\, dx$ diverges.

53. In electromagnetic theory, the magnetic potential at a point on the axis of a circular coil is given by

$$u = \frac{2\pi NIr}{k} \int_a^{+\infty} \frac{dx}{(r^2 + x^2)^{3/2}}$$

where N, I, r, k, and a are constants. Find u.

c **54.** The *average speed*, $\bar{v}$, of the molecules of an ideal gas is given by

$$\bar{v} = \frac{4}{\sqrt{\pi}} \left(\frac{M}{2RT}\right)^{3/2} \int_0^{+\infty} v^3 e^{-Mv^2/(2RT)}\, dv$$

and the *root-mean-square speed*, v_{rms}, by

$$v_{\text{rms}}^2 = \frac{4}{\sqrt{\pi}} \left(\frac{M}{2RT}\right)^{3/2} \int_0^{+\infty} v^4 e^{-Mv^2/(2RT)}\, dv$$

where v is the molecular speed, T is the gas temperature, M is the molecular weight of the gas, and R is the gas constant.

(a) Use a CAS to show that

$$\int_0^{+\infty} x^3 e^{-a^2 x^2}\, dx = \frac{1}{2a^4}, \quad a > 0$$

and use this result to show that $\bar{v} = \sqrt{8RT/\pi M}$.

(b) Use a CAS to show that

$$\int_0^{+\infty} x^4 e^{-a^2 x^2}\, dx = \frac{3\sqrt{\pi}}{8a^5}, \quad a > 0$$

and use this result to show that $v_{\text{rms}} = \sqrt{3RT/M}$.

55. In Exercise 17 of Section 8.6, we determined the work required to lift a 6000-lb satellite to an orbital position that is 1000 mi above the Earth's surface. The ideas discussed in that exercise will be needed here.

(a) Find a definite integral that represents the work required to lift a 6000-lb satellite to a position l miles above the Earth's surface.

(b) Find a definite integral that represents the work required to lift a 6000-lb satellite an "infinite distance" above the Earth's surface. Evaluate the integral. [*Note:* The result obtained here is sometimes called the work required to "escape" the Earth's gravity.]

A *transform* is a formula that converts or "transforms" one function into another. Transforms are used in applications to convert a difficult problem into an easier problem whose solution can then be used to solve the original difficult problem. The *Laplace transform* of a function $f(t)$, which plays an important role in the study of differential equations, is denoted by $\mathcal{L}\{f(t)\}$ and is defined by

$$\mathcal{L}\{f(t)\} = \int_0^{+\infty} e^{-st} f(t)\, dt$$

In this formula s is treated as a constant in the integration process; thus, the Laplace transform has the effect of transforming $f(t)$ into a function of s. Use this formula in Exercises 56 and 57.

56. Show that

(a) $\mathcal{L}\{1\} = \dfrac{1}{s}$, $s > 0$ (b) $\mathcal{L}\{e^{2t}\} = \dfrac{1}{s-2}$, $s > 2$

(c) $\mathcal{L}\{\sin t\} = \dfrac{1}{s^2 + 1}$, $s > 0$

(d) $\mathcal{L}\{\cos t\} = \dfrac{s}{s^2 + 1}$, $s > 0$.

57. In each part, find the Laplace transform.

(a) $f(t) = t$, $s > 0$ (b) $f(t) = t^2$, $s > 0$

(c) $f(t) = \begin{cases} 0, & t < 3 \\ 1, & t \geq 3 \end{cases}$, $s > 0$

58. Later in the text, we will show that

$$\int_0^{+\infty} e^{-x^2}\, dx = \tfrac{1}{2}\sqrt{\pi}$$

Confirm that this is reasonable by using a CAS or a calculator with a numerical integration capability.

59. Use the result in Exercise 58 to show that

(a) $\displaystyle\int_{-\infty}^{+\infty} e^{-ax^2}\, dx = \sqrt{\frac{\pi}{a}}$, $a > 0$

(b) $\displaystyle\frac{1}{\sqrt{2\pi}\sigma} \int_{-\infty}^{+\infty} e^{-x^2/2\sigma^2}\, dx = 1$, $\sigma > 0$.

A convergent improper integral over an infinite interval can be approximated by first replacing the infinite limit(s) of integration by finite limit(s), then using a numerical integration technique, such as Simpson's rule, to approximate the integral with finite limit(s). This technique is illustrated in Exercises 60 and 61.

60. Suppose that the integral in Exercise 58 is approximated by first writing it as

$$\int_0^{+\infty} e^{-x^2}\, dx = \int_0^K e^{-x^2}\, dx + \int_K^{+\infty} e^{-x^2}\, dx$$

then dropping the second term, and then applying Simpson's rule to the integral

$$\int_0^K e^{-x^2}\, dx$$

The resulting approximation has two sources of error: the error from Simpson's rule and the error

$$E = \int_K^{+\infty} e^{-x^2} dx$$

that results from discarding the second term. We call E the ***truncation error***.

(a) Approximate the integral in Exercise 58 by applying Simpson's rule with $n = 10$ subdivisions to the integral

$$\int_0^3 e^{-x^2} dx$$

Round your answer to four decimal places and compare it to $\frac{1}{2}\sqrt{\pi}$ rounded to four decimal places.

(b) Use the result that you obtained in Exercise 46 and the fact that $e^{-x^2} \leq \frac{1}{3}xe^{-x^2}$ for $x \geq 3$ to show that the truncation error for the approximation in part (a) satisfies $0 < E < 2.1 \times 10^{-5}$.

61. (a) It can be shown that

$$\int_0^{+\infty} \frac{1}{x^6 + 1} dx = \frac{\pi}{3}$$

Approximate this integral by applying Simpson's rule with $n = 20$ subdivisions to the integral

$$\int_0^4 \frac{1}{x^6 + 1} dx$$

Round your answer to three decimal places and compare it to $\pi/3$ rounded to three decimal places.

(b) Use the result that you obtained in Exercise 46 and the fact that $1/(x^6 + 1) < 1/x^6$ for $x \geq 4$ to show that the truncation error for the approximation in part (a) satisfies $0 < E < 2 \times 10^{-4}$.

62. For what values of p does $\int_0^{+\infty} e^{px} dx$ converge?

63. Show that $\int_0^1 \frac{dx}{x^p}$ converges if $p < 1$ and diverges if $p \geq 1$.

c **64.** It is sometimes possible to convert an improper integral into a "proper" integral having the same value by making an appropriate substitution. Evaluate the following integral by making the indicated substitution, and investigate what happens if you evaluate the integral directly using a CAS.

$$\int_0^1 \sqrt{\frac{1 + x}{1 - x}} \, dx; \quad u = \sqrt{1 - x}$$

In Exercises 65 and 66, transform the given improper integral into a proper integral by making the stated u-substitution, then approximate the proper integral by Simpson's rule with $n = 10$ subdivisions. Round your answer to three decimal places.

65. $\int_0^1 \frac{\cos x}{\sqrt{x}} \, dx; \quad u = \sqrt{x}$

66. $\int_0^1 \frac{\sin x}{\sqrt{1 - x}} \, dx; \quad u = \sqrt{1 - x}$

SUPPLEMENTARY EXERCISES

1. Consider the following methods for evaluating integrals: u-substitution, integration by parts, partial fractions, reduction formulas, and trigonometric substitutions. In each part, state the approach that you would try first to evaluate the integral. If none of them seems appropriate, then say so. You need not evaluate the integral.

(a) $\int x \sin x \, dx$

(b) $\int \cos x \sin x \, dx$

(c) $\int \tan^7 x \, dx$

(d) $\int \tan^7 x \sec^2 x \, dx$

(e) $\int \frac{3x^2}{x^3 + 1} \, dx$

(f) $\int \frac{3x^2}{(x + 1)^3} \, dx$

(g) $\int \tan^{-1} x \, dx$

(h) $\int \sqrt{4 - x^2} \, dx$

(i) $\int x\sqrt{4 - x^2} \, dx$

2. Consider the following trigonometric substitutions:

$$x = 3\sin\theta, \quad x = 3\tan\theta, \quad x = 3\sec\theta$$

In each part, state the substitution that you would try first to evaluate the integral. If none seems appropriate, then state a trigonometric substitution that you would use. You need

not evaluate the integral.

(a) $\int \sqrt{9 + x^2} \, dx$

(b) $\int \sqrt{9 - x^2} \, dx$

(c) $\int \sqrt{1 - 9x^2} \, dx$

(d) $\int \sqrt{x^2 - 9} \, dx$

(e) $\int \sqrt{9 + 3x^2} \, dx$

(f) $\int \sqrt{1 + (9x)^2} \, dx$

3. (a) What condition must a rational function satisfy for the method of partial fractions to be applicable directly?

(b) If the condition in part (a) is not satisfied, what must you do if you want to use partial fractions?

4. What is an improper integral?

5. In each part, find the number of the formula in the Endpaper Integral Table that you would apply to evaluate the integral. You need not evaluate the integral.

(a) $\int \sin 7x \cos 9x \, dx$

(b) $\int (x^7 - x^5)e^{9x} \, dx$

(c) $\int x\sqrt{x - x^2} \, dx$

(d) $\int \frac{dx}{x\sqrt{4x + 3}}$

(e) $\int x^9 \pi^x \, dx$

(f) $\int \frac{3x - 1}{2 + x^2} \, dx$

6. Evaluate the integral $\displaystyle\int_0^1 \frac{x^3}{\sqrt{x^2+1}}\,dx$ using

(a) integration by parts

(b) the substitution $u = \sqrt{x^2+1}$.

7. In each part, evaluate the integral by making an appropriate substitution and applying a reduction formula.

(a) $\displaystyle\int \sin^4 2x\,dx$ (b) $\displaystyle\int x\cos^5(x^2)\,dx$

8. Consider the integral $\displaystyle\int \frac{1}{x^3-x}\,dx$.

(a) Evaluate the integral using the substitution $x = \sec\theta$. For what values of x is your result valid?

(b) Evaluate the integral using the substitution $x = \sin\theta$. For what values of x is your result valid?

(c) Evaluate the integral using the method of partial fractions. For what values of x is your result valid?

9. (a) Evaluate the integral

$$\int \frac{1}{\sqrt{2x-x^2}}\,dx$$

three ways: using the substitution $u = \sqrt{x}$, using the substitution $u = \sqrt{2-x}$, and completing the square.

(b) Show that the answers in part (a) are equivalent.

10. Find the area of the region that is enclosed by the curves $y = (x-3)/(x^3+x^2)$, $y = 0$, $x = 1$, and $x = 2$.

11. Sketch the region whose area is $\displaystyle\int_0^{+\infty} \frac{dx}{1+x^2}$, and use your sketch to show that

$$\int_0^{+\infty} \frac{dx}{1+x^2} = \int_0^1 \sqrt{\frac{1-y}{y}}\,dy$$

12. Find the area that is enclosed between the x-axis and the curve $y = (\ln x - 1)/x^2$ for $x \geq e$.

13. Find the volume of the solid that is generated when the region between the x-axis and the curve $y = e^{-x}$ for $x \geq 0$ is revolved about the y-axis.

14. Find a positive value of a that satisfies the equation

$$\int_0^{+\infty} \frac{1}{x^2+a^2}\,dx = 1$$

In Exercises 15–30, evaluate the integral.

15. $\displaystyle\int \sqrt{\cos\theta}\,\sin\theta\,d\theta$ **16.** $\displaystyle\int_0^{\pi/4} \tan^7\theta\,d\theta$

17. $\displaystyle\int x\tan^2(x^2)\sec^2(x^2)\,dx$ **18.** $\displaystyle\int_{-1/\sqrt{2}}^{1/\sqrt{2}} (1-2x^2)^{3/2}\,dx$

19. $\displaystyle\int \frac{dx}{(3+x^2)^{3/2}}$

20. $\displaystyle\int \frac{\cos\theta}{\sin^2\theta - 6\sin\theta + 12}\,d\theta$

21. $\displaystyle\int \frac{x+3}{\sqrt{x^2+2x+2}}\,dx$ **22.** $\displaystyle\int \frac{\sec^2\theta}{\tan^3\theta - \tan^2\theta}\,d\theta$

23. $\displaystyle\int \frac{dx}{(x-1)(x+2)(x-3)}$ **24.** $\displaystyle\int \frac{dx}{x(x^2+x+1)}$

25. $\displaystyle\int_4^8 \frac{\sqrt{x-4}}{x}\,dx$ **26.** $\displaystyle\int_0^9 \frac{\sqrt{x}}{x+9}\,dx$

27. $\displaystyle\int \frac{1}{\sqrt{e^x+1}}\,dx$ **28.** $\displaystyle\int_0^{\ln 2} \sqrt{e^x-1}\,dx$

29. $\displaystyle\int_a^{+\infty} \frac{x\,dx}{(x^2+1)^2}$

30. $\displaystyle\int_0^{+\infty} \frac{dx}{a^2+b^2x^2}, \quad a,b > 0$

Some integrals that can be evaluated by hand cannot be evaluated by all computer algebra systems. In Exercises 31–34, evaluate the integral by hand, and determine if it can be evaluated on your CAS.

C **31.** $\displaystyle\int \frac{x^3}{\sqrt{1-x^8}}\,dx$

C **32.** $\displaystyle\int (\cos^{32}x\sin^{30}x - \cos^{30}x\sin^{32}x)\,dx$

C **33.** $\displaystyle\int \sqrt{x - \sqrt{x^2-4}}\,dx$. [*Hint:* $\frac{1}{2}(\sqrt{x+2} - \sqrt{x-2})^2 = ?$]

C **34.** $\displaystyle\int \frac{1}{x^{10}+x}\,dx$. [*Hint:* Rewrite the denominator as $x^{10}(1+x^{-9})$.]

C **35.** Let

$$f(x) = \frac{-2x^5 + 26x^4 + 15x^3 + 6x^2 + 20x + 43}{x^6 - x^5 - 18x^4 - 2x^3 - 39x^2 - x - 20}$$

(a) Use a CAS to factor the denominator, and then write down the form of the partial fraction decomposition. You need not find the values of the constants.

(b) Check your answer in part (a) by using the CAS to find the partial fraction decomposition of f.

(c) Integrate f by hand, and then check your answer by integrating with the CAS.

36. The *Gamma function*, $\Gamma(x)$, is defined as

$$\Gamma(x) = \int_0^{+\infty} t^{x-1}e^{-t}\,dt$$

It can be shown that this improper integral converges if and only if $x > 0$.

(a) Find $\Gamma(1)$.

(b) Prove: $\Gamma(x+1) = x\Gamma(x)$ for all $x > 0$. [*Hint:* Use integration by parts.]

(c) Use the results in parts (a) and (b) to find $\Gamma(2)$, $\Gamma(3)$, and $\Gamma(4)$; and then make a conjecture about $\Gamma(n)$ for positive integer values of n.

(d) Show that $\Gamma\left(\frac{1}{2}\right) = \sqrt{\pi}$. [*Hint:* See Exercise 58 of Section 9.8.]

(e) Use the results obtained in parts (b) and (d) to show that $\Gamma\left(\frac{3}{2}\right) = \frac{1}{2}\sqrt{\pi}$ and $\Gamma\left(\frac{5}{2}\right) = \frac{3}{4}\sqrt{\pi}$.

37. Refer to the Gamma function defined in Exercise 36 to show that

    (a) $\int_0^1 (\ln x)^n \, dx = (-1)^n \Gamma(n+1), \quad n > 0.$

    [*Hint:* Let $t = -\ln x$.]

    (b) $\int_0^{+\infty} e^{-x^n} \, dx = \Gamma\left(\dfrac{n+1}{n}\right), \quad n > 0.$

    [*Hint:* Let $t = x^n$. Use the result in Exercise 36(b).]

c 38. A *simple pendulum* consists of a mass that swings in a vertical plane at the end of a massless rod of length L, as shown in the accompanying figure. Suppose that a simple pendulum is displaced through an angle θ_0 and released from rest. It can be shown that in the absence of friction, the time T required for the pendulum to make one complete back-and-forth swing, called the *period*, is given by

$$T = \sqrt{\frac{8L}{g}} \int_0^{\theta_0} \frac{1}{\sqrt{\cos\theta - \cos\theta_0}} \, d\theta \tag{1}$$

where $\theta = \theta(t)$ is the angle the pendulum makes with the vertical at time t. The improper integral in (1) is difficult to evaluate numerically. By a substitution outlined below it can be shown that the period can be expressed as

$$T = 4\sqrt{\frac{L}{g}} \int_0^{\pi/2} \frac{1}{\sqrt{1 - k^2 \sin^2\phi}} \, d\phi \tag{2}$$

where $k = \sin(\theta_0/2)$. The integral in (2) is called a *complete elliptic integral of the first kind* and is more easily evaluated by numerical methods.

(a) Obtain (2) from (1) by substituting

$$\cos\theta = 1 - 2\sin^2(\theta/2)$$
$$\cos\theta_0 = 1 - 2\sin^2(\theta_0/2)$$
$$k = \sin(\theta_0/2)$$

and then making the change of variable

$$\sin\phi = \sin(\theta/2)/\sin(\theta_0/2) = \sin(\theta/2)/k$$

(b) Use (2) and the numerical integration capability of your CAS to find the period of a simple pendulum for which $L = 1.5$ ft, $\theta_0 = 20°$, and $g = 32$ ft/s^2.

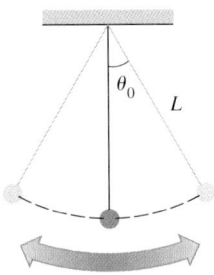

Figure Ex-38

EXPANDING THE CALCULUS HORIZON

Railroad Design

Your company has a contract to construct a track bed for a railroad line between towns A and B shown on the contour map in Figure 1. The bed can be created by cutting trenches through the surface or by using some combination of trenches and tunnels. As chief engineer, your assignment is to analyze the costs of trenches and tunnels and to propose a design strategy for minimizing the total construction cost.

Engineering Requirements

The Transportation Board submits the following engineering requirements to your company:

- The track bed is to be straight and 10 m wide. The grade is to increase at a constant rate from the existing elevation of 100 m at town A to an elevation of 110 m at point M and then decrease at a constant rate to the existing elevation of 88 m at town B.

- From town A to point M and from point N to town B the track bed is to be created by excavating a trench whose vertical cross sections are trapezoids with the dimensions shown in Figure 2.

- Between points M and N your company must decide whether to excavate a trench of the type in Figure 2 or to excavate a tunnel whose vertical cross sections have the dimensions shown in Figure 3.

CONTOUR MAP
ELEVATIONS IN METERS

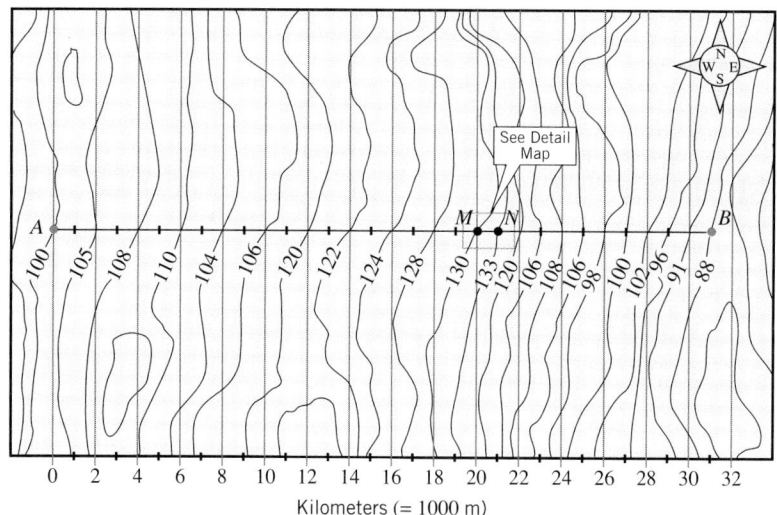

DETAIL MAP
PROPOSED TUNNEL CONSTRUCTION

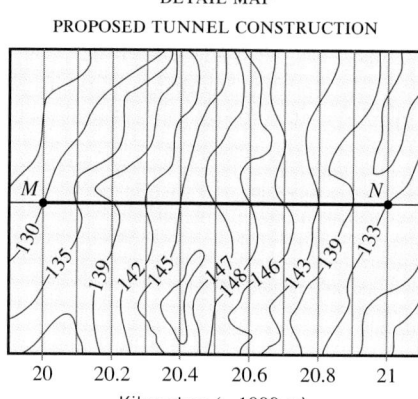

Figure 1

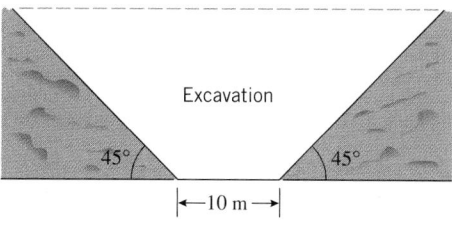

Figure 2

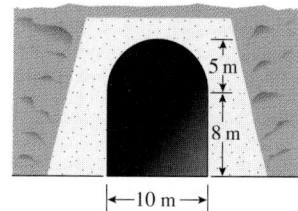

Figure 3

Cost Factors

Surface excavation of railbeds is performed using bulldozers, hydraulic excavators (backhoes), loading tractors, and other specialized equipment. Typically, the excavated dirt is piled at the side of the tracks to form sloped embankments, and the excavation cost is estimated from the volume of dirt to be removed and piled.

Tunnels in rock are often excavated by drilling shafts and inserting boring machines (called *moles*) to loosen and remove rock and dirt. Tunnels in soft ground are often excavated by starting at the tunnel face and using bucket or rotary excavators housed inside of shields. As the excavator progresses, tunnel liners are inserted behind it to support the earth and prevent cave-ins. Dirt removal is performed using conveyors or sometimes using railcars (called *muck cars*) that run on specially constructed tracks. Ventilation and air compression are other factors that add to the excavation cost of tunnels. In general, the excavation cost for a tunnel can be estimated from two components, the total volume of dirt to be removed and a cost that increases with the distance to the tunnel opening.

Make the following cost assumptions:

- The excavation and dirt-piling cost for a trench is $4.00 per cubic meter.

- The drilling and dirt-piling cost for a tunnel is $8.00 per cubic meter, and the costs involved in moving a load of dirt inside the tunnel a distance of 1 m toward the entrance along the track line is $0.06 per cubic meter.

■ Cost Analysis of Trenches

Assume that variations in elevation are negligible for short distances at right angles to the track, so that the cross sections of the dirt to be excavated always have the trapezoidal shape shown in Figure 2 (straight horizontal edges at the surface).

....

Exercise 1   Complete Table 1, and then use the table and Simpson's rule with $n = 10$ to approximate the cost of a trench from town A to point M.

Table 1

DISTANCE x FROM TOWN A (m)	TERRAIN ELEVATION (m)	TRACK ELEVATION (m)	DEPTH OF CUT (m)	CROSS-SECTIONAL AREA $f(x)$ OF CUT (m^2)
0	100	100	0	0
2,000	105	101	4	56
4,000				
6,000				
8,000				
10,000				
12,000				
14,000				
16,000				
18,000				
20,000				

....

Exercise 2   As in Exercise 1, use Simpson's rule with $n = 10$ to approximate the cost of constructing a trench from (a) point M to point N, and (b) point N to town B.

....

Exercise 3   Find the total cost of the project if a trench is used along the entire line from town A to town B.

■ Cost Analysis of a Tunnel

....

Exercise 4

(a) Find the volume of dirt that must be removed from the tunnel, and calculate the drilling and dirt-piling cost.

(b) Find an integral for the cost of moving all of the dirt inside the tunnel to the tunnel entrance. [*Suggestion:* Use Riemann sums.]

(c) Find the total cost of excavating the tunnel.

....

Exercise 5   Find the total cost of the project using a trench from town A to point M, a tunnel from point M to point N, and a trench from point N to town B. Compare the cost to that obtained in Exercise 3 and state which method is cheaper.

Module by: *C. Lynn Kiaer, Rose-Hulman Institute of Technology*
             David Ryeburn, Simon Fraser University
             Howard Anton, Drexel University
             Peter Dunn, Railroad Construction Company, Inc., Paterson, NJ

10

MATHEMATICAL MODELING

WITH DIFFERENTIAL

EQUATIONS

Colin Maclaurin

*M*any of the principles in science and engineering concern relationships between changing quantities. Since rates of change are represented mathematically by derivatives, it should not be surprising that such principles are often expressed in terms of differential equations. We introduced the concept of a differential equation in Section 7.2, but in this chapter we will go into more detail. We will discuss some important mathematical models that involve differential equations, and we will discuss some methods for solving and approximating solutions of some of the basic types of differential equations. However, we will only be able to touch the surface of this topic, leaving many important topics in differential equations to courses that are devoted completely to the subject.

10.1 FIRST-ORDER DIFFERENTIAL EQUATIONS AND APPLICATIONS

In this section we will introduce some basic terminology and concepts concerning differential equations. We will also discuss methods for solving certain basic types of differential equations, and we will give some applications of our work.

TERMINOLOGY

Recall from Section 7.2 that a ***differential equation*** is an equation involving one or more derivatives of an unknown function. In this section we will denote the unknown function by $y = y(x)$ unless the differential equation arises from an applied problem involving time, in which case we will denote it by $y = y(t)$. The ***order*** of a differential equation is the order of the highest derivative that it contains. Here are some examples:

DIFFERENTIAL EQUATION	ORDER
$\dfrac{dy}{dx} = 3y$	1
$\dfrac{d^2y}{dx^2} - 6\dfrac{dy}{dx} + 8y = 0$	2
$\dfrac{d^3y}{dx^3} - t\dfrac{dy}{dt} + (t^2 - 1)y = e^t$	3
$y' - y = e^{2x}$	1
$y'' + y' = \cos t$	2

In the last two equations the derivatives of y are expressed in "prime" notation. You will usually be able to tell from the equation itself or the context in which it arises whether to interpret y' as dy/dx or as dy/dt.

SOLUTIONS OF DIFFERENTIAL EQUATIONS

A function $y = y(x)$ is a ***solution*** of a differential equation on a given interval if the equation is satisfied for every x in that interval when y and its derivatives are substituted in the equation. For example, $y = e^{2x}$ is a solution of the differential equation

$$\frac{dy}{dx} - y = e^{2x} \qquad (1)$$

on the interval $(-\infty, +\infty)$, since substituting y and its derivative into the left side of this equation yields

$$\frac{dy}{dx} - y = \frac{d}{dx}[e^{2x}] - e^{2x} = 2e^{2x} - e^{2x} = e^{2x}$$

for all real values of x. However, this is not the only solution on the interval $(-\infty, +\infty)$; for example, the function

$$y = Ce^x + e^{2x} \qquad (2)$$

is also a solution for every real value of the constant C, since

$$\frac{dy}{dx} - y = \frac{d}{dx}[Ce^x + e^{2x}] - (Ce^x + e^{2x}) = (Ce^x + 2e^{2x}) - (Ce^x + e^{2x}) = e^{2x}$$

One can prove that *all* solutions of (1) on $(-\infty, +\infty)$ can be obtained by substituting values for the constant C in (2). On a given interval, a solution of a differential equation from which all solutions on that interval can be derived by substituting values for arbitrary constants is called the ***general solution*** of the equation on the interval. Thus, (2) is the general solution of (1) on the interval $(-\infty, +\infty)$.

REMARK.   Usually, the general solution of an nth-order differential equation on an interval will contain n arbitrary constants. Although we will not prove this, it makes sense intuitively because n integrations are needed to recover a function from its nth derivative, and each integration introduces an arbitrary constant. For example, (2) has one arbitrary constant, which is consistent with the fact that it is the general solution of the *first-order* equation (1).

The graph of a solution of a differential equation is called an ***integral curve*** for the equation, so the general solution of a differential equation produces a family of integral curves corresponding to the different possible choices for the arbitrary constants. For example, Figure 10.1.1 shows some integral curves for (1), which were obtained by assigning values to the arbitrary constant in (2).

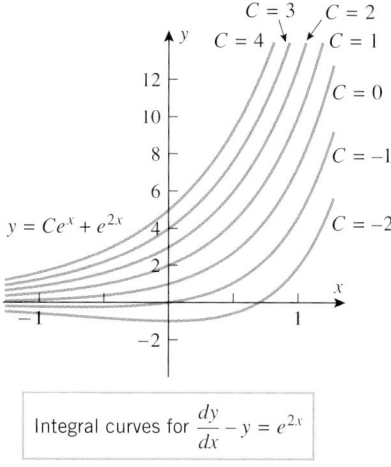

Integral curves for $\dfrac{dy}{dx} - y = e^{2x}$

Figure 10.1.1

INITIAL-VALUE PROBLEMS

When an applied problem leads to a differential equation, there are usually conditions in the problem that determine specific values for the arbitrary constants. As a rule of thumb, it requires n conditions to determine values for all n arbitrary constants in the general solution of an nth-order differential equation (one condition for each constant). For a first-order equation, the single arbitrary constant can be determined by specifying the value of the unknown function $y(x)$ at an arbitrary point x_0, say $y(x_0) = y_0$. This is called an ***initial condition***, and the problem of solving a first-order equation subject to an initial condition is called a ***first-order initial-value problem***. Geometrically, the initial condition $y(x_0) = y_0$ has the effect of isolating the integral curve that passes through the point (x_0, y_0) from the complete family of integral curves.

Example 1

The solution of the initial-value problem

$$\frac{dy}{dx} - y = e^{2x}, \quad y(0) = 3$$

can be obtained by substituting the initial condition $x = 0$, $y = 3$ in the general solution (2) to find C. We obtain

$$3 = Ce^0 + e^0 = C + 1$$

Thus, $C = 2$, and the solution of the initial-value problem, which is obtained by substituting this value of C in (2), is

$$y = 2e^x + e^{2x}$$

Geometrically, the graph of this solution is the integral curve in Figure 10.1.1 that passes through the point $(0, 3)$.   ◄

FIRST-ORDER EQUATIONS

The simplest first-order equations are those that can be written in the form

$$\frac{dy}{dx} = f(x) \tag{3}$$

Such equations can often be solved by integration. For example, if

$$\frac{dy}{dx} = x^3 \tag{4}$$

then

$$y = \int x^3 \, dx = \frac{x^4}{4} + C$$

is the general solution of (4) on the interval $(-\infty, +\infty)$.

FIRST-ORDER SEPARABLE EQUATIONS

Equation (4) can be solved by integrating because the right side is a function of x. However, if the right side involves both x and y, as with

$$\frac{dy}{dx} = \sin(xy)$$

then direct integration is not possible and other methods must be used. In general, such equations can be complicated to solve exactly, and often one must settle for numerical approximations of solutions, as we will discuss in the next section. However, if the equation can be expressed in the form

$$h(y)\frac{dy}{dx} = g(x) \tag{5}$$

then we say that the equation is **separable**, and we can often find the general solution by first rewriting the equation in the differential form

$$h(y) \, dy = g(x) \, dx \tag{6}$$

(all y's on one side and all x's on the other), and then integrating both sides to obtain

$$\int h(y) \, dy = \int g(x) \, dx \tag{7}$$

If the equation that results when these integrations are performed can be solved for y as a function of x, then this function provides an explicit formula for the general solution of (5). However, if the equation that results when these integrations are performed cannot be solved for y as a function of x, then the equation still defines solutions of (5), but it defines them implicitly.

The process of obtaining (6) from (5) is called **separating variables**, and the method we have just discussed for solving (5) is called **separation of variables**. A more detailed explanation of why this method works is given in the exercises.

Example 2

Solve the differential equation

$$\frac{dy}{dx} = -4xy^2$$

and then solve the initial-value problem

$$\frac{dy}{dx} = -4xy^2, \quad y(0) = 1$$

Solution. Separating variables and integrating yields

$$\frac{1}{y^2}\, dy = -4x\, dx$$

$$\int \frac{1}{y^2}\, dy = \int -4x\, dx$$

$$-\frac{1}{y} = -2x^2 + C$$

> The integration on the left produces a constant c_1, and the integration on the right produces a constant c_2. We have combined these constants into the constant $C = c_2 - c_1$.

Solving for y as a function of x, we obtain

$$y = \frac{1}{2x^2 - C} \tag{8}$$

The initial condition $y(0) = 1$ requires that $y = 1$ when $x = 0$. Substituting these values in (8) yields $C = -1$ (verify). Thus, the solution of the initial-value problem is

$$y = \frac{1}{2x^2 + 1}$$

Some typical integral curves and the solution of the initial-value problem are graphed in Figure 10.1.2.   ◀

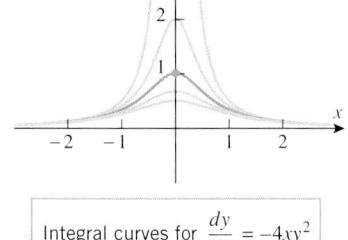

Integral curves for $\dfrac{dy}{dx} = -4xy^2$

Figure 10.1.2

Example 3

Solve the initial-value problem

$$(4y - \cos y)\frac{dy}{dx} - 3x^2 = 0, \quad y(0) = 0$$

Solution. First, we solve the differential equation. Separating variables and integrating yields

$$(4y - \cos y)\frac{dy}{dx} = 3x^2$$

$$(4y - \cos y)\, dy = 3x^2\, dx$$

$$\int (4y - \cos y)\, dy = \int 3x^2\, dx$$

$$2y^2 - \sin y = x^3 + C \tag{9}$$

Equation (9) defines the solutions of the differential equation implicitly; it cannot be solved explicitly for y as a function of x.

For the initial-value problem, the initial condition $y(0) = 0$ requires that $y = 0$ if $x = 0$. Substituting these values in (9) to determine the constant of integration yields $C = 0$ (verify). Thus, the solution of the initial-value problem is

$$2y^2 - \sin y = x^3 \qquad\qquad ◀$$

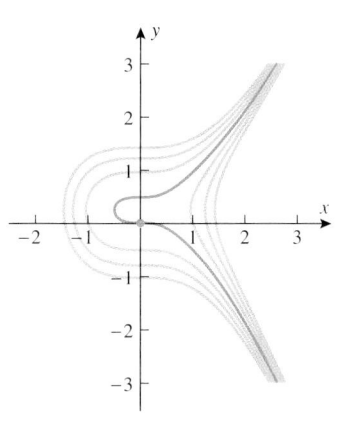

Integral curves for
$(4y - \cos y)\dfrac{dy}{dx} - 3x^2 = 0$

Figure 10.1.3

FOR THE READER.   Some computer algebra systems can graph implicit equations. For example, Figure 10.1.3 shows the graphs of (9) for $C = 0, \pm 1, \pm 2$, and ± 3, with emphasis on the solution of the initial-value problem. If you have a CAS that can graph implicit equations, read the documentation on graphing them and try to duplicate this figure. Also, try to determine which values of C produce which curves.

FIRST-ORDER LINEAR EQUATIONS

Not every first-order differential equation is separable. For example, it is impossible to separate the variables in the equation

$$\frac{dy}{dx} + 2xy = xe^{-x^2}$$

However, this equation can be solved by a different method that we will now consider.

A first-order differential equation is called *linear* if it is expressible in the form

$$\frac{dy}{dx} + p(x)y = q(x) \tag{10}$$

where the functions $p(x)$ and $q(x)$ are *continuous* and may or may not be constant. Some examples are

$$\frac{dy}{dx} + x^2 y = e^x, \qquad \frac{dy}{dx} + (\sin x)y + x^3 = 0, \qquad \frac{dy}{dx} + 5y = 2$$

$$\boxed{p(x) = x^2, q(x) = e^x} \qquad \boxed{p(x) = \sin x, q(x) = -x^3} \qquad \boxed{p(x) = 5, q(x) = 2}$$

One procedure for solving (10) is based on the observation that if we define $\mu = \mu(x)$ by

$$\mu = e^{\int p(x)\,dx}$$

then

$$\frac{d\mu}{dx} = e^{\int p(x)\,dx} \cdot \frac{d}{dx} \int p(x)\,dx = \mu p(x)$$

Thus,

$$\frac{d}{dx}(\mu y) = \mu \frac{dy}{dx} + \frac{d\mu}{dx}y = \mu \frac{dy}{dx} + \mu p(x)y \tag{11}$$

If (10) is multiplied through by μ, and then simplified using (11), it becomes

$$\mu \frac{dy}{dx} + \mu p(x)y = \mu q(x)$$

$$\frac{d}{dx}(\mu y) = \mu q(x)$$

This equation can be solved by integrating both sides to obtain

$$\mu y = \int \mu q(x)\,dx + C \quad \text{or} \quad y = \frac{1}{\mu}\left[\int \mu q(x)\,dx + C\right]$$

To summarize, (10) can be solved in three steps, called *the method of integrating factors*:

The Method of Integrating Factors

Step 1.   Calculate

$$\mu = e^{\int p(x)\,dx}$$

This is called the *integrating factor*. Since any μ will suffice, we can take the constant of integration to be zero in this step.

Step 2.   Multiply both sides of (10) by μ and express the result as

$$\frac{d}{dx}[\mu y] = \mu q(x)$$

Step 3.   Integrate both sides of the equation obtained in Step 2 and then solve for y. Be sure to include a constant of integration in this step.

Example 4

Solve the differential equation $\dfrac{dy}{dx} - 2y = 0$.

Solution.   This equation is separable, but it is also linear, since it is of form (10) with $p(x) = -2$ and $q(x) = 0$; thus, we can solve it by separation of variables or by the method of integrating factors. We will solve it both ways, using the method of integrating factors first. The integrating factor is

$$\mu = e^{\int -2\,dx} = e^{-2x}$$

If we multiply the differential equation through by μ and follow Step 2 of the method of integrating factors, then we obtain

$$\frac{d}{dx}[e^{-2x}y] = 0$$

Integrating both sides of this equation yields

$$e^{-2x}y = C$$

which can be rewritten as

$$y = Ce^{2x}$$

Alternative Solution. Separating variables and integrating yields

$$\int \frac{dy}{y} = \int 2\,dx$$

$\ln|y| = 2x + c$ We have used c as the constant of integration here to reserve C for the constant in the final result.

$$|y| = e^{2x+c}$$

$$|y| = e^c e^{2x}$$

$$y = \pm e^c e^{2x}$$

$y = Ce^{2x}$ Letting $C = \pm e^c$

which agrees with the answer obtained above.   ◀

REMARK.   For first-order equations that are both linear and separable, the method of integrating factors is usually simpler than separation of variables, provided the integrating factor can be found easily. Moreover, the careful reader may have observed in the alternative solution of Example 4 that the constant $C = \pm e^c$ is not truly arbitrary, since $C = 0$ is not an allowable value. Thus, separation of variables missed the solution $y = 0$, which the method of integrating factors did not. This problem occurred because we had to divide by y to separate the variables.

Example 5

Solve the initial-value problem

$$x\frac{dy}{dx} - y = x, \quad y(1) = 2$$

Solution. The differential equation can be rewritten in form (10) by dividing through by x. This yields

$$\frac{dy}{dx} - \frac{1}{x}y = 1 \tag{12}$$

Comparing this to (10), we have $p(x) = -1/x$ and $q(x) = 1$. However, there is a difficulty here because the method of integrating factors requires that $p(x)$ and $q(x)$ be continuous, and $p(x)$ has a discontinuity at $x = 0$. Thus, the method of integrating factors can be applied if $x > 0$ or if $x < 0$, but not on an interval containing $x = 0$. However, the initial condition $y(1) = 2$ is imposed at $x = 1$, so we will assume that $x > 0$. With this assumption, the integrating factor is

$$\mu = e^{\int -(1/x)\,dx} = e^{-\ln|x|} = \frac{1}{|x|} = \frac{1}{x}$$

If we multiply (12) through by μ, then from Step 2 of the method of integrating factors we obtain

$$\frac{d}{dx}\left(\frac{1}{x}y\right) = \frac{1}{x}$$

Integrating both sides of this equation yields

$$\frac{1}{x}y = \ln x + C$$

or

$$y = x\ln x + Cx \qquad (13)$$

The initial condition $y(1) = 2$ requires that $y = 2$ if $x = 1$. Substituting these values in (13) and solving for C yields $C = 2$ (verify), so the solution of the initial-value problem is

$$y = x\ln x + 2x \qquad \blacktriangleleft$$

APPLICATIONS IN GEOMETRY

We conclude this section with some applications of first-order differential equations.

Example 6

Find a curve in the xy-plane that passes through $(0, 3)$ and whose tangent line at a point (x, y) has slope $2x/y^2$.

Solution. Since the slope of the tangent line is dy/dx, we have

$$\frac{dy}{dx} = \frac{2x}{y^2} \qquad (14)$$

and, since the curve passes through $(0, 3)$, we have the initial condition

$$y(0) = 3 \qquad (15)$$

Equation (14) is separable and can be written as

$$y^2\,dy = 2x\,dx$$

so

$$\int y^2\,dy = \int 2x\,dx \quad \text{or} \quad \tfrac{1}{3}y^3 = x^2 + C$$

It follows from the initial condition (15) that $y = 3$ if $x = 0$. Substituting these values in the last equation yields $C = 9$ (verify), so the equation of the desired curve is

$$\tfrac{1}{3}y^3 = x^2 + 9 \quad \text{or} \quad y = (3x^2 + 27)^{1/3} \qquad \blacktriangleleft$$

MIXING PROBLEMS

In a typical mixing problem, a tank is filled to a specified level with a solution that contains a known amount of some soluble substance (say salt). The thoroughly stirred solution is allowed to drain from the tank at a known rate, and at the same time a solution with a known concentration of the soluble substance is added to the tank at a known rate that may or may not differ from the draining rate. As time progresses, the amount of the soluble substance in the tank will generally change, and the usual mixing problem seeks to determine the amount of the substance in the tank at a specified time. This type of problem serves as a model for many kinds of problems: discharge and filtration of pollutants in a river, injection and absorption of medication in the bloodstream, and migrations of species into and out of an ecological system, for example.

Example 7

At time $t = 0$, a tank contains 4 lb of salt dissolved in 100 gal of water. Suppose that brine containing 2 lb of salt per gallon of brine is allowed to enter the tank at a rate of 5 gal/min and that the mixed solution is drained from the tank at the same rate (Figure 10.1.4). Find the amount of salt in the tank after 10 minutes.

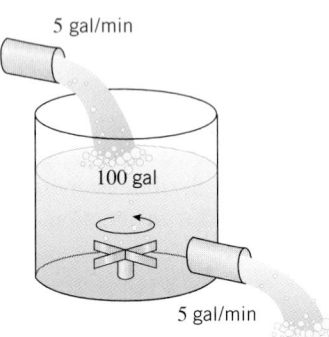

5 gal/min

100 gal

5 gal/min

Figure 10.1.4

Solution. Let $y(t)$ be the amount of salt (in pounds) after t minutes. We are given that $y(0) = 4$, and we want to find $y(10)$. We will begin by finding a differential equation that is satisfied by $y(t)$. To do this, observe that dy/dt, which is the rate at which the amount of salt in the tank changes with time, can be expressed as

$$\frac{dy}{dt} = \text{rate in} - \text{rate out} \tag{16}$$

where *rate in* is the rate at which salt enters the tank and *rate out* is the rate at which salt leaves the tank. But the rate at which salt enters the tank is

$$\text{rate in} = (2 \text{ lb/gal}) \cdot (5 \text{ gal/min}) = 10 \text{ lb/min}$$

Since brine enters and drains from the tank at the same rate, the volume of brine in the tank stays constant at 100 gal. Thus, after t minutes have elapsed, the tank contains $y(t)$ lb of salt per 100 gal of brine, and hence the rate at which salt leaves the tank at that instant is

$$\text{rate out} = \left(\frac{y(t)}{100} \text{ lb/gal} \right) \cdot (5 \text{ gal/min}) = \frac{y(t)}{20} \text{ lb/min}$$

Therefore, (16) can be written as

$$\frac{dy}{dt} = 10 - \frac{y}{20} \quad \text{or} \quad \frac{dy}{dt} + \frac{y}{20} = 10$$

which is a first-order linear differential equation satisfied by $y(t)$. Since we are given that $y(0) = 4$, the function $y(t)$ can be obtained by solving the initial-value problem

$$\frac{dy}{dt} + \frac{y}{20} = 10, \quad y(0) = 4$$

The integrating factor for the differential equation is

$$\mu = e^{\int (1/20)\,dt} = e^{t/20}$$

If we multiply the differential equation through by μ, then from Step 2 of the method of integrating factors we obtain

$$\frac{d}{dt}(e^{t/20}y) = 10e^{t/20}$$

$$e^{t/20}y = \int 10e^{t/20}dt = 200e^{t/20} + C$$

$$y(t) = 200 + Ce^{-t/20} \tag{17}$$

The initial condition states that $y = 4$ when $t = 0$. Substituting these values in (17) and solving for C yields $C = -196$ (verify), so

$$y(t) = 200 - 196e^{-t/20} \tag{18}$$

Thus, at time $t = 10$ the amount of salt in the tank is

$$y(10) = 200 - 196e^{-0.5} \approx 81.1 \text{ lb} \qquad \blacktriangleleft$$

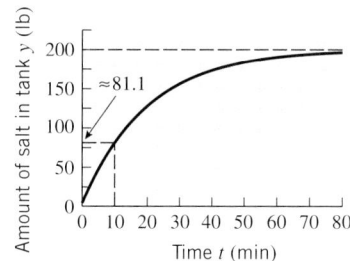

Figure 10.1.5

FOR THE READER.   Figure 10.1.5 shows the graph of (18). Observe that $y(t) \to 200$ as $t \to +\infty$, which means that over an extended period of time the amount of salt in the tank tends toward 200 lb. Give an informal physical argument to explain why this result is to be expected.

A MODEL OF FREE-FALL MOTION RETARDED BY AIR RESISTANCE

In Section 6.3 we considered the free-fall model of an object moving along a vertical axis near the surface of the Earth. It was assumed in that model that there is no air resistance and that the only force acting on the object is the Earth's gravity. Our goal here is to find a model that takes air resistance into account. For this purpose we make the following assumptions:

- The object moves along a vertical s-axis whose origin is at the surface of the Earth and whose positive direction is up (Figure 6.3.7).

- At time $t = 0$ the height of the object is s_0 and the velocity is v_0.
- The only forces on the object are the force $F_G = -mg$ of the Earth's gravity acting down and the force F_R of air resistance acting opposite to the direction of motion. The force F_R is called the **drag force**.

We will also need the following result from physics:

10.1.1 NEWTON'S SECOND LAW OF MOTION.   If an object with mass m is subjected to a force F, then the object undergoes an acceleration a that satisfies the equation

$$F = ma \tag{19}$$

In the case of free-fall motion retarded by air resistance, the net force acting on the object is

$$F_G + F_R = -mg + F_R$$

and the acceleration is d^2s/dt^2, so Newton's second law implies that

$$-mg + F_R = m\frac{d^2s}{dt^2} \tag{20}$$

Experimentation has shown that the force F_R of air resistance depends on the shape of the object and its speed—the greater the speed, the greater the drag force. There are many possible models for air resistance, but one of the most basic assumes that the drag force F_R is proportional to the velocity of the object, that is,

$$F_R = -cv$$

where c is a positive constant that depends on the object's shape and properties of the air.[*] (The minus sign ensures that the drag force is opposite to the direction of motion.) Substituting this in (20) and writing d^2s/dt^2 as dv/dt, we obtain

$$-mg - cv = m\frac{dv}{dt}$$

or on dividing by m and rearranging we obtain

$$\frac{dv}{dt} + \frac{c}{m}v = -g$$

which is a first-order linear differential equation in the unknown function $v = v(t)$ with $p(t) = c/m$ and $q(t) = -g$ [see (10)]. For a specific object, the coefficient c can be determined experimentally, so we can assume that m, g, and c are known constants. Thus, the velocity function $v = v(t)$ can be obtained by solving the initial-value problem

$$\frac{dv}{dt} + \frac{c}{m}v = -g, \quad v(0) = v_0 \tag{21}$$

Once the velocity function is found, the position function $s = s(t)$ can be obtained by solving the initial-value problem

$$\frac{ds}{dt} = v(t), \quad s(0) = s_0 \tag{22}$$

In Exercise 47 we will ask you to solve (21) and show that

$$v(t) = e^{-ct/m}\left(v_0 + \frac{mg}{c}\right) - \frac{mg}{c} \tag{23}$$

---

[*]Other common models assume that $F_R = -cv^2$ or, more generally, $F_R = -cv^p$ for some value of p.

Note that

$$\lim_{t \to +\infty} v(t) = -\frac{mg}{c} \tag{24}$$

(verify). Thus, the speed $|v(t)|$ does not increase indefinitely, as in free fall; rather, because of the air resistance, it approaches a finite limiting speed v_τ given by

$$v_\tau = \left| -\frac{mg}{c} \right| = \frac{mg}{c} \tag{25}$$

This is called the **terminal speed** of the object, and (24) is called its **terminal velocity**.

REMARK.   Intuition suggests that near the limiting velocity, the velocity $v(t)$ changes very slowly; that is, $dv/dt \approx 0$. Thus, it should not be surprising that the limiting velocity can be obtained informally from (21) by setting $dv/dt = 0$ in the differential equation and solving for v. This yields

$$v = -\frac{mg}{c}$$

which agrees with (24).

EXERCISE SET 10.1   ☑ Graphing Calculator   [c] CAS

1. Confirm that $y = 2e^{x^3/3}$ is a solution of the initial-value problem $y' = x^2 y$, $y(0) = 2$.

2. Confirm that $y = \frac{1}{4}x^4 + 2\cos x + 1$ is a solution of the initial-value problem $y' = x^3 - 2\sin x$, $y(0) = 3$.

In Exercises 3 and 4, state the order of the differential equation, and confirm that the functions in the given family are solutions.

3. (a) $(1 + x)\dfrac{dy}{dx} = y$; $y = c(1 + x)$

   (b) $y'' + y = 0$; $y = c_1 \sin t + c_2 \cos t$

4. (a) $2\dfrac{dy}{dx} + y = x - 1$; $y = ce^{-x/2} + x - 3$

   (b) $y'' - y = 0$; $y = c_1 e^t + c_2 e^{-t}$

In Exercises 5 and 6, use implicit differentiation to confirm that the equation defines implicit solutions of the differential equation.

5. $\ln y = xy + C$; $\dfrac{dy}{dx} = \dfrac{y^2}{1 - xy}$

6. $x^2 + xy^2 = C$; $2x + y^2 + 2xy\dfrac{dy}{dx} = 0$

In Exercises 7 and 8, solve the differential equation by the method of integrating factors and by separation of variables, and confirm that the two solutions are the same.

7. (a) $\dfrac{dy}{dx} + 3y = 0$ (b) $\dfrac{dy}{dt} - 2y = 0$

8. (a) $\dfrac{dy}{dx} - 4xy = 0$ (b) $\dfrac{dy}{dt} + y = 0$

In Exercises 9–18, solve the differential equation by separation of variables. Where reasonable, express the family of solutions as explicit functions of x.

9. $\dfrac{dy}{dx} = \dfrac{y}{x}$

10. $\dfrac{dy}{dx} = (1 + y^2)x^2$

11. $\dfrac{\sqrt{1 + x^2}}{1 + y}\dfrac{dy}{dx} = -x$

12. $(1 + x^4)\dfrac{dy}{dx} = \dfrac{x^3}{y}$

13. $(1 + y^2)y' = e^x y$

14. $y' = -xy$

15. $e^{-y}\sin x - y'\cos^2 x = 0$

16. $y' - (1 + x)(1 + y^2) = 0$

17. $\dfrac{dy}{dx} - \dfrac{y^2 - y}{\sin x} = 0$

18. $3\tan y - \dfrac{dy}{dx}\sec x = 0$

In Exercises 19–24, solve the differential equation by the method of integrating factors.

19. $\dfrac{dy}{dx} + 3y = e^{-2x}$

20. $\dfrac{dy}{dx} + 2xy = x$

21. $y' + y = \cos(e^x)$

22. $2\dfrac{dy}{dx} + 4y = 1$

23. $(x^2 + 1)\dfrac{dy}{dx} + xy = 0$

24. $\dfrac{dy}{dx} + y - \dfrac{1}{1 + e^x} = 0$

25. In each part, find the solution of the differential equation

$$x\frac{dy}{dx} + y = x$$

that satisfies the initial condition.

   (a) $y(1) = 2$ (b) $y(-1) = 2$

26. In each part, find the solution of the differential equation

$$\frac{dy}{dx} = xy$$

that satisfies the initial condition.
(a) $y(0) = 1$ (b) $y(0) = \frac{1}{2}$

In Exercises 27–32, solve the initial-value problem by any method.

---

27. $\dfrac{dy}{dx} - xy = x$, $y(0) = 3$

28. $\dfrac{dy}{dt} + y = 2$, $y(0) = 1$

29. $y' = \dfrac{4x^2}{y + \cos y}$, $y(1) = \pi$

30. $y' - xe^y = 2e^y$, $y(0) = 0$

31. $\dfrac{dy}{dt} = \dfrac{2t + 1}{2y - 2}$, $y(0) = -1$

32. $y' \cosh x + y \sinh x = \cosh^2 x$, $y(0) = \frac{1}{4}$

33. (a) Sketch some typical integral curves of the differential equation $y' = y/2x$.
   (b) Find an equation for the integral curve that passes through the point $(2, 1)$.

34. (a) Sketch some typical integral curves of the differential equation $y' = -x/y$.
   (b) Find an equation for the integral curve that passes through the point $(3, 4)$.

In Exercises 35 and 36, solve the differential equation, and then use a graphing utility to generate the integral curves for $C = -2, -1, 0, 1, 2$.

---

35. $(x^2 + 4)\dfrac{dy}{dx} + xy = 0$ 36. $y' + 2y - 3e^t = 0$

If you have a CAS that can graph implicit equations, solve the differential equations in Exercises 37 and 38, and then use the CAS to generate the integral curves for $C = -2, -1, 0, 1, 2$.

---

37. $y' = \dfrac{x^2}{1 - y^2}$ 38. $y' = \dfrac{y}{1 + y^2}$

If you have a CAS, read the documentation on solving differential equations and initial-value problems, and then use the CAS in Exercises 39 and 40.

---

39. Use a CAS to solve the differential equations in the odd-numbered Exercises 9–23, and confirm that the answers are consistent with those in the answer section of the text.

40. Use a CAS to solve the initial-value problems in the odd-numbered Exercises 25–31, and confirm that the answers are consistent with those in the answer section of the text.

41. Find an equation of a curve with x-intercept 2 whose tangent line at any point (x, y) has slope xe^y.

42. Use a graphing utility to generate a curve that passes through the point $(1, 1)$ and whose tangent line at (x, y) is perpendicular to the line through (x, y) with slope $-2y/(3x^2)$.

43. At time $t = 0$, a tank contains 25 ounces of salt dissolved in 50 gal of water. Then brine containing 4 ounces of salt per gallon of brine is allowed to enter the tank at a rate of 2 gal/min and the mixed solution is drained from the tank at the same rate.
   (a) How much salt is in the tank at an arbitrary time t?
   (b) How much salt is in the tank after 25 min?

44. A tank initially contains 200 gal of pure water. Then at time $t = 0$ brine containing 5 lb of salt per gallon of brine is allowed to enter the tank at a rate of 10 gal/min and the mixed solution is drained from the tank at the same rate.
   (a) How much salt is in the tank at an arbitrary time t?
   (b) How much salt is in the tank after 30 min?

45. A tank with a 1000-gal capacity initially contains 500 gal of water that is polluted with 50 lb of particulate matter. At time $t = 0$, pure water is added at a rate of 20 gal/min and the mixed solution is drained off at a rate of 10 gal/min. How much particulate matter is in the tank when it reaches the point of overflowing?

46. The water in a polluted lake initially contains 1 lb of mercury salts per 100,000 gal of water. The lake is circular with diameter 30 m and uniform depth 3 m. Polluted water is pumped from the lake at a rate of 1000 gal/h and is replaced with fresh water at the same rate. Construct a table that shows the amount of mercury in the lake (in lbs) at the end of each hour over a 12-hour period. Discuss any assumptions you made. [Use 264 gal/m^3.]

47. (a) Use the method of integrating factors to confirm that (23) is the solution of initial-value problem (21). [*Note:* Keep in mind that c, m, and g are constants.]
   (b) Show that (23) can be expressed in terms of the terminal speed (25) as
   $$v(t) = e^{-gt/v_\tau}(v_0 + v_\tau) - v_\tau$$
   (c) Show that if $s(0) = s_0$, then the position function of the object can be expressed as
   $$s(t) = s_0 - v_\tau t + \frac{v_\tau}{g}(v_0 + v_\tau)(1 - e^{-gt/v_\tau})$$

48. Based on the air resistance model discussed in this section, a fully equipped sky diver weighing 240 lb would have a terminal speed of approximately 120 ft/s with a closed parachute and approximately 24 ft/s with an open parachute. Suppose that such a sky diver is dropped from an airplane at an altitude of 10,000 ft, falls for 25 s with a closed parachute, and then falls the rest of the way with an open parachute.
   (a) Assuming that the sky diver's initial vertical velocity is zero, use Exercise 47 to find the sky diver's vertical velocity and height at the time the parachute opens. [Take $g = 32$ ft/s^2.]
   (b) Approximate the total time, to the nearest second, that the sky diver is in the air. [*Hint:* You will not be able to solve for the time exactly, so consider using Newton's Method or the method of Example 6 of Section 2.4.]

49. The accompanying figure is a schematic diagram of a basic RL series electrical circuit that contains a power source with a time-dependent voltage of $V(t)$ volts (V), a resistor with a constant resistance of R ohms (Ω), and an inductor with a constant inductance of L henrys (H). If you don't know anything about electrical circuits, don't worry; all you need to know is that electrical theory states that a current of $I(t)$ amperes (A) flows through the circuit where $I(t)$ satisfies the differential equation

$$L\frac{dI}{dt} + RI = V(t)$$

(a) Find $I(t)$ if $R = 10\,\Omega$, $L = 4$ H, V is a constant 12 V, and $I(0) = 0$ A.

(b) What happens to the current over a long period of time?

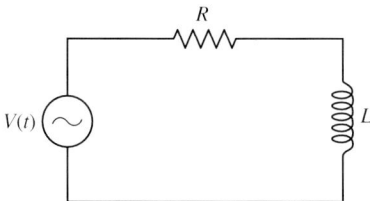

Figure Ex-49

50. Find $I(t)$ for the electrical circuit in Exercise 49 if $R = 6\,\Omega$, $L = 3$ H, $V(t) = 3\sin t$ V, and $I(0) = 15$ A.

51. A rocket, fired upward from rest at time $t = 0$, has an initial mass of m_0 (including its fuel). Assuming that the fuel is consumed at a constant rate k, the mass m of the rocket, while fuel is being burned, will be given by $m = m_0 - kt$. It can be shown that if air resistance is neglected and the fuel gases are expelled at a constant speed c relative to the rocket, then the velocity v of the rocket will satisfy the equation

$$m\frac{dv}{dt} = ck - mg$$

where g is the acceleration due to gravity.

(a) Find $v(t)$ keeping in mind that the mass m is a function of t.

(b) Suppose that the fuel accounts for 80% of the initial mass of the rocket and that all of the fuel is consumed in 100 s. Find the velocity of the rocket in meters per second at the instant the fuel is exhausted. [Take $g = 9.8$ m/s^2 and $c = 2500$ m/s.]

52. A bullet of mass m, fired straight up with an initial velocity of v_0, is slowed by the force of gravity and a drag force of air resistance kv^2, where g is the constant acceleration due to gravity and k is a positive constant. As the bullet moves upward, its velocity v satisfies the equation

$$m\frac{dv}{dt} = -(kv^2 + mg)$$

(a) Show that if $x = x(t)$ is the height of the bullet above

the barrel opening at time t, then

$$mv\frac{dv}{dx} = -(kv^2 + mg)$$

(b) Express x in terms of v given that $x = 0$ when $v = v_0$.

(c) Assuming that

$$v_0 = 988 \text{ m/s}, \quad g = 9.8 \text{ m/s}^2$$
$$m = 3.56 \times 10^{-3} \text{ kg}, \quad k = 7.3 \times 10^{-6} \text{ kg/m}$$

use the result in part (b) to find out how high the bullet rises. [*Hint:* Find the velocity of the bullet at its highest point.]

The following discussion is needed for Exercises 53 and 54. Suppose that a tank containing a liquid is vented to the air at the top and has an outlet at the bottom through which the liquid can drain. It follows from **Torricelli's law** in physics that if the outlet is opened at time $t = 0$, then at each instant the depth of the liquid $h(t)$ and the area $A(h)$ of the liquid's surface are related by

$$A(h)\frac{dh}{dt} = -k\sqrt{h}$$

where k is a positive constant that depends on such factors as the viscosity of the liquid and the cross-sectional area of the outlet. Use this result in Exercises 53 and 54, assuming that h is in feet, $A(h)$ is in square feet, and t is in seconds. A calculator will be useful.

---

53. Suppose that the cylindrical tank in the accompanying figure is filled to a depth of 4 feet at time $t = 0$ and that the constant in Torricelli's law is $k = 0.025$.

(a) Find $h(t)$.

(b) How many minutes will it take for the tank to drain completely?

54. Follow the directions of Exercise 53 for the cylindrical tank in the accompanying figure, assuming that the tank is filled to a depth of 4 feet at time $t = 0$ and that the constant in Torricelli's law is $k = 0.025$.

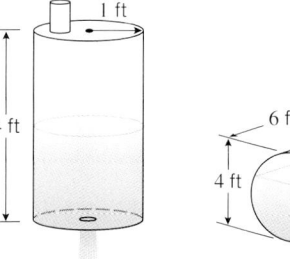

Figure Ex-53   Figure Ex-54

55. Suppose that a particle moving along the x-axis encounters a resisting force that results in an acceleration of $a = dv/dt = -0.04v^2$. Given that $x = 0$ cm and $v = 50$ cm/s at time $t = 0$, find the velocity v and position x as a function of t for $t \geq 0$.

56. Suppose that a particle moving along the x-axis encounters a resisting force that results in an acceleration of $a = dv/dt = -0.02\sqrt{v}$. Given that $x = 0$ cm and $v = 9$ cm/s at time $t = 0$, find the velocity v and position x as a function of t for $t \geq 0$.

57. Find an initial-value problem whose solution is

$$y = \cos x + \int_0^x e^{-t^2}\, dt$$

58. Derive Formula (7) for solving the separable differential equation

$$h(y)\frac{dy}{dx} = g(x)$$

by making the substitution $y = y(x), dy = y'(x)\, dx$ in the integral

$$\int h(y)\, dy$$

10.2 DIRECTION FIELDS; EULER'S METHOD

In this section we will reexamine the concept of a direction field, and we will discuss a method for approximating solutions of first-order equations numerically. Numerical approximations are important in cases where the differential equation cannot be solved exactly.

FUNCTIONS OF TWO VARIABLES

We will be concerned here with first-order equations that are expressed with the derivative by itself on one side of the equation. For example,

$$y' = x^3 \quad \text{and} \quad y' = \sin(xy)$$

The first of these equations involves only x on the right side, so it has the form $y' = f(x)$. However, the second equation involves both x and y on the right side, so it has the form $y' = f(x, y)$, where the symbol $f(x, y)$ stands for a function of the two variables x and y. Later in the text we will study functions of two variables in more depth, but for now it will suffice to think of $f(x, y)$ as a formula that produces a unique output when values of x and y are given as inputs. For example, if

$$f(x, y) = x^2 + 3y$$

and if the inputs are $x = 2$ and $y = -4$, then the output is

$$f(2, -4) = 2^2 + 3(-4) = 4 - 12 = -8$$

REMARK. In applied problems involving time, it is usual to use t as the independent variable, in which case we would be concerned with equations of the form $y' = f(t, y)$, where $y' = dy/dt$.

DIRECTION FIELDS

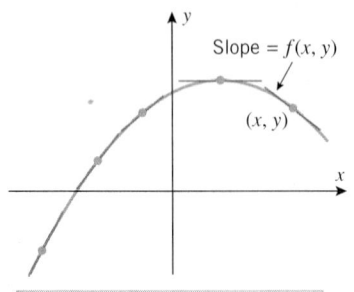

At each point (x, y) on an integral curve of $y' = f(x, y)$, the tangent line has slope $f(x, y)$.

Figure 10.2.1

In Section 7.2 we introduced the concept of a direction field in the context of differential equations of the form $y' = f(x)$; the same principles apply to differential equations of the form

$$y' = f(x, y)$$

To see why this is so, let us review the basic idea. If we interpret y' as the slope of a tangent line, then the differential equation states that at each point (x, y) on an integral curve, the slope of the tangent line is equal to the value of f at that point (Figure 10.2.1). For example, suppose that $f(x, y) = y - x$, in which case we have the differential equation

$$y' = y - x \tag{1}$$

A geometric description of the set of integral curves can be obtained by choosing a rectangular grid of points in the xy-plane, calculating the slopes of the tangent lines to the integral curves at the gridpoints, and drawing small segments of the tangent lines at those points. The resulting picture is called a ***direction field*** or a ***slope field*** for the differential equation because it shows the "direction" or "slope" of the integral curves at the gridpoints. The more gridpoints that are used, the better the description of the integral curves. For example, Figure 10.2.2 shows two direction fields for (1)—the first was obtained by hand calculation using the 49 gridpoints shown in the accompanying table, and the second, which gives a clearer picture of the integral curves, was obtained using 625 gridpoints and a CAS.

VALUES OF $f(x, y) = y - x$

	$y = -3$	$y = -2$	$y = -1$	$y = 0$	$y = 1$	$y = 2$	$y = 3$
$x = -3$	0	1	2	3	4	5	6
$x = -2$	-1	0	1	2	3	4	5
$x = -1$	-2	-1	0	1	2	3	4
$x = 0$	-3	-2	-1	0	1	2	3
$x = 1$	-4	-3	-2	-1	0	1	2
$x = 2$	-5	-4	-3	-2	-1	0	1
$x = 3$	-6	-5	-4	-3	-2	-1	0

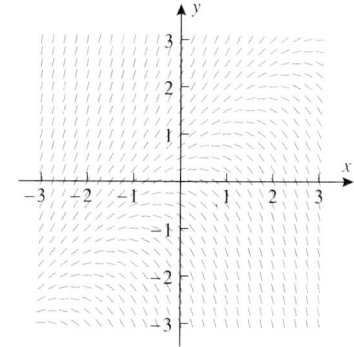

Figure 10.2.2

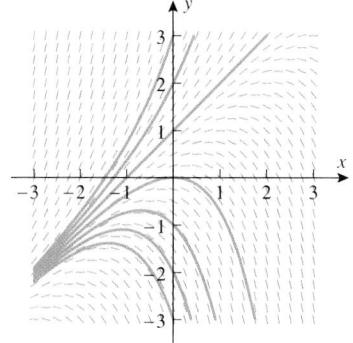

Figure 10.2.3

It so happens that Equation (1) can be solved exactly, since it can be written as

$$y' - y = -x$$

which, by comparison with Equation (10) in Section 10.1, is a first-order linear equation with $p(x) = -1$ and $q(x) = -x$. We leave it for you to use the method of integrating factors to show that the general solution of this equation is

$$y = x + 1 + Ce^x \tag{2}$$

Figure 10.2.3 shows some of the integral curves superimposed on the direction field. Observe, however, that it was not necessary to have the general solution to construct the direction field. Indeed, direction fields are important precisely because they can be constructed in cases where the differential equation cannot be solved exactly.

FOR THE READER.   Confirm that the first direction field in Figure 10.2.2 is consistent with the values in the accompanying table.

Example 1

In Example 7 of Section 10.1 we considered a mixing problem in which the amount of salt $y(t)$ in a tank at time t was shown to satisfy the differential equation

$$\frac{dy}{dt} + \frac{y}{20} = 10$$

which can be rewritten as

$$y' = 10 - \frac{y}{20} \tag{3}$$

We subsequently found the general solution of this equation to be

$$y(t) = 200 + Ce^{-t/20} \tag{4}$$

and then we found the value of the arbitrary constant C from the initial condition in the problem [the known amount of salt $y(0)$ at time $t = 0$]. However, it follows from (4) that

$$\lim_{t \to +\infty} y(t) = 200$$

for all values of C, so regardless of the amount of salt that is present in the tank initially, the amount of salt in the tank will eventually begin to stabilize at 200 lb. This can also be seen geometrically from the direction field for (3) shown in Figure 10.2.4. This direction field suggests that if the amount of salt present in the tank is greater than 200 lb initially, then the amount of salt will decrease steadily over time toward a limiting value of 200 lb; and if it is less than 200 lb initially, then it will increase steadily toward a limiting value of 200 lb. The direction field also suggests that if the amount present initially is exactly 200 lb, then the amount of salt in the tank will stay constant at 200 lb. This can also be seen from (4), since $C = 0$ in this case (verify).

Observe that for the direction field shown in Figure 10.2.4 the tangent segments along any horizontal line are parallel. This occurs because the differential equation has the form $y' = f(y)$ with t absent from the right side [see (3)]. Thus, for a fixed y the slope y' does not change as time varies. Because of this time independence of slope, differential equations of the form $y' = f(y)$ are said to be **autonomous** (from the Greek word *autonomous*, meaning "independent"). ◀

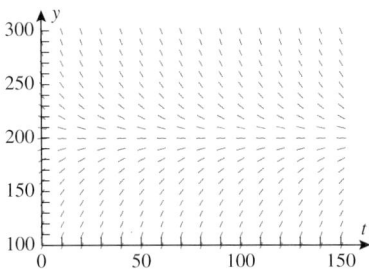

Figure 10.2.4

EULER'S METHOD

Our next objective is to develop a method for approximating the solution of an initial-value problem of the form

$$y' = f(x, y), \quad y(x_0) = y_0$$

We will not attempt to approximate $y(x)$ for all values of x; rather, we will choose some small increment h and focus on approximating the values of $y(x)$ at a succession of x-values spaced h units apart, starting from x_0. We will denote these x-values by

$$x_1 = x_0 + h, \quad x_2 = x_1 + h, \quad x_3 = x_2 + h, \quad x_4 = x_3 + h, \ldots$$

and we will denote the approximations of $y(x)$ at these points by

$$y_1 \approx y(x_1), \quad y_2 \approx y(x_2), \quad y_3 \approx y(x_3), \quad y_4 \approx y(x_4), \ldots$$

The technique that we will describe for obtaining these approximations is called **Euler's Method**. Although there are better approximation methods available, many of them use Euler's Method as a starting point, so the underlying concepts are important to understand.

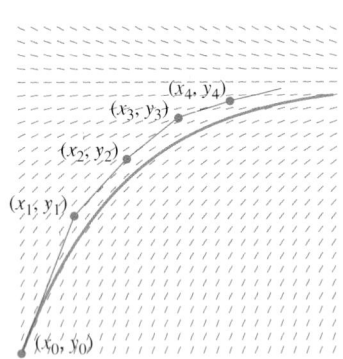

Figure 10.2.5

The basic idea behind Euler's Method is to start at the known initial point (x_0, y_0) and draw a line segment in the direction determined by the direction field until we reach the point (x_1, y_1) with x-coordinate $x_1 = x_0 + h$ (Figure 10.2.5). If h is small, then it is reasonable to expect that this line segment will not deviate much from the integral curve $y = y(x)$, and thus y_1 should closely approximate $y(x_1)$. To obtain the subsequent approximations, we repeat the process using the direction field as a guide at each step. Starting at the endpoint (x_1, y_1), we draw a line segment determined by the direction field until we reach the point (x_2, y_2) with x-coordinate $x_2 = x_1 + h$, and from that point we draw a line segment determined by the direction field to the point (x_3, y_3) with x-coordinate $x_3 = x_2 + h$, and so forth. As indicated in Figure 10.2.5, this procedure produces a polygonal path that tends to follow the integral curve closely, so it is reasonable to expect that the y-values $y_2, y_3, y_4, \ldots$ will closely approximate $y(x_2), y(x_3), y(x_4), \ldots$.

To explain how the approximations $y_1, y_2, y_3, \ldots$ can be computed, let us focus on a typical line segment. As indicated in Figure 10.2.6, assume that we have found the point (x_n, y_n), and we are trying to determine the next point (x_{n+1}, y_{n+1}), where $x_{n+1} = x_n + h$.

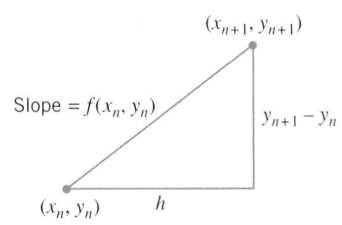

Figure 10.2.6

Since the slope of the line segment joining the points is determined by the direction field at the starting point, the slope is $f(x_n, y_n)$, and hence

$$\frac{y_{n+1} - y_n}{h} = f(x_n, y_n)$$

which we can rewrite as

$$y_{n+1} = y_n + f(x_n, y_n)h$$

This formula, which is the heart of Euler's Method, tells us how to use each approximation to compute the next approximation.

Euler's Method

To approximate the solution of the initial-value problem

$$y' = f(x, y), \quad y(x_0) = y_0$$

proceed as follows:

Step 1. Choose a nonzero number h to serve as an *increment* or *step size* along the x-axis, and let

$$x_1 = x_0 + h, \quad x_2 = x_1 + h, \quad x_3 = x_2 + h, \ldots$$

Step 2. Compute successively

$$
\begin{aligned}
y_1 &= y_0 + f(x_0, y_0)h \\
y_2 &= y_1 + f(x_1, y_1)h \\
y_3 &= y_2 + f(x_2, y_2)h \\
&\vdots \\
y_{n+1} &= y_n + f(x_n, y_n)h
\end{aligned}
$$

The numbers $y_1, y_2, y_3, \ldots$ in these equations are the approximations of $y(x_1), y(x_2), y(x_3), \ldots$.

Example 2

Use Euler's Method with a step size of 0.1 to make a table of approximate values of the solution of the initial-value problem

$$y' = y - x, \quad y(0) = 2 \tag{5}$$

over the interval $0 \leq x \leq 1$.

Solution. In this problem we have $f(x, y) = y - x$, $x_0 = 0$, and $y_0 = 2$. Moreover, since the step size is 0.1, the x-values at which the approximate values will be obtained are

$$x_1 = 0.1, \quad x_2 = 0.2, \quad x_3 = 0.3, \ldots, \quad x_9 = 0.9, \quad x_{10} = 1$$

The first three approximations are

$$y_1 = y_0 + f(x_0, y_0)h = 2 + (2 - 0)(0.1) = 2.2$$

$$y_2 = y_1 + f(x_1, y_1)h = 2.2 + (2.2 - 0.1)(0.1) = 2.41$$

$$y_3 = y_2 + f(x_2, y_2)h = 2.41 + (2.41 - 0.2)(0.1) = 2.631$$

Here is a way of organizing all 10 approximations rounded to five decimal places:

EULER'S METHOD FOR $y' = y - x$, $y(0) = 2$ WITH $h = 0.1$

n	x_n	y_n	$f(x_n, y_n)h$	$y_{n+1} = y_n + f(x_n, y_n)h$
0	0	2.00000	0.20000	2.20000
1	0.1	2.20000	0.21000	2.41000
2	0.2	2.41000	0.22100	2.63100
3	0.3	2.63100	0.23310	2.86410
4	0.4	2.86410	0.24641	3.11051
5	0.5	3.11051	0.26105	3.37156
6	0.6	3.37156	0.27716	3.64872
7	0.7	3.64872	0.29487	3.94359
8	0.8	3.94359	0.31436	4.25795
9	0.9	4.25795	0.33579	4.59374
10	1.0	4.59374	—	—

Observe that each entry in the last column becomes the next entry in the third column. ◄

ACCURACY OF EULER'S METHOD

It follows from (5) and the initial condition $y(0) = 2$ that the exact solution of the initial-value problem in Example 2 is

$$y = x + 1 + e^x$$

Thus, in this case we can compare the approximate values of $y(x)$ produced by Euler's Method with decimal approximations of the exact values (Table 10.2.1). In Table 10.2.1 the *absolute error* is calculated as

|exact value − approximation|

and the *percentage error* as

$$\frac{|\text{exact value} - \text{approximation}|}{|\text{exact value}|} \times 100\%$$

REMARK.   As a rough rule of thumb, the absolute error in an approximation produced by Euler's Method is proportional to the step size; thus, reducing the step size by half reduces the absolute error (and hence the percentage error) by roughly half. However, reducing the step size also increases the amount of computation, thereby increasing the potential for round-off error. We will leave a detailed study of error issues for courses in differential equations or numerical analysis.

Table 10.2.1

x	EXACT SOLUTION	EULER APPROXIMATION	ABSOLUTE ERROR	PERCENTAGE ERROR
0	2.00000	2.00000	0.00000	0.00
0.1	2.20517	2.20000	0.00517	0.23
0.2	2.42140	2.41000	0.01140	0.47
0.3	2.64986	2.63100	0.01886	0.71
0.4	2.89182	2.86410	0.02772	0.96
0.5	3.14872	3.11051	0.03821	1.21
0.6	3.42212	3.37156	0.05056	1.48
0.7	3.71375	3.64872	0.06503	1.75
0.8	4.02554	3.94359	0.08195	2.04
0.9	4.35960	4.25795	0.10165	2.33
1.0	4.71828	4.59374	0.12454	2.64

1. Sketch the direction field for $y' = xy/8$ at the gridpoints (x, y), where $x = 0, 1, \ldots, 4$ and $y = 0, 1, \ldots, 4$.

2. Sketch the direction field for $y' + y = 2$ at the gridpoints (x, y), where $x = 0, 1, \ldots, 4$ and $y = 0, 1, \ldots, 4$.

3. A direction field for the differential equation $y' = 1 - y$ is shown in the accompanying figure. In each part, sketch the graph of the solution that satisfies the initial condition.
   (a) $y(0) = -1$ (b) $y(0) = 1$ (c) $y(0) = 2$

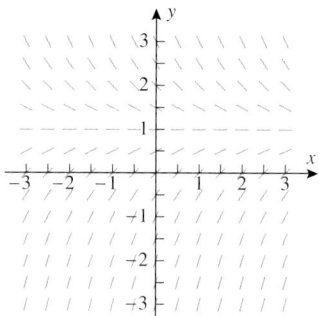

Figure Ex-3

⌐ 4. Solve the initial-value problems in Exercise 3, and use a graphing utility to confirm that the integral curves for these solutions are consistent with the sketches you obtained from the direction field.

5. A direction field for the differential equation $y' = 2y - x$ is shown in the accompanying figure. In each part, sketch the graph of the solution that satisfies the initial condition.
   (a) $y(1) = 1$ (b) $y(0) = -1$ (c) $y(-1) = 0$

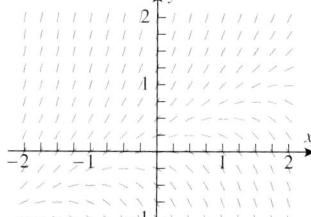

Figure Ex-5

⌐ 6. Solve the initial-value problems in Exercise 5, and use a graphing utility to confirm that the integral curves for these solutions are consistent with the sketches you obtained from the direction field.

7. Use the direction field in Exercise 3 to make a conjecture about the behavior of the solutions of $y' = 1 - y$ as $x \to +\infty$, and confirm your conjecture by examining the general solution of the equation.

8. Use the direction field in Exercise 5 to make a conjecture about the effect of y_0 on the behavior of the solution of the initial-value problem $y' = 2y - x$, $y(0) = y_0$ as $x \to +\infty$, and check your conjecture by examining the solution of the initial-value problem.

9. In each part, match the differential equation with the direction field (see next page), and explain your reasoning.
   (a) $y' = 1/x$ (b) $y' = 1/y$ (c) $y' = e^{-x^2}$
   (d) $y' = y^2 - 1$ (e) $y' = \dfrac{x + y}{x - y}$
   (f) $y' = (\sin x)(\sin y)$

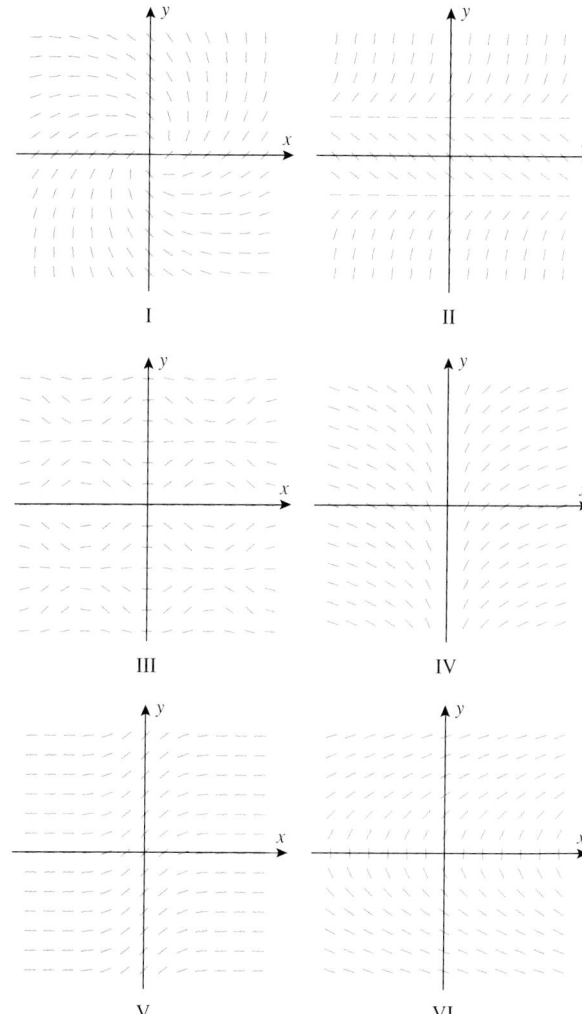

Figure Ex-9

[c] 10. If you have a CAS or a graphing utility that can generate direction fields, read the documentation on how to do it and check your answers in Exercise 9 by generating the direction fields for the differential equations.

11. (a) Use Euler's Method with a step size of $h = 0.2$ to approximate the solution of the initial-value problem
   $$y' = x + y, \quad y(0) = 1$$
   over the interval $0 \le x \le 1$.

(b) Solve the initial-value problem exactly, and calculate the error and the percentage error in each of the approximations in part (a).

(c) Sketch the exact solution and the approximate solution together.

12. It was stated at the end of this section that reducing the step size in Euler's Method by half reduces the error in each approximation by about half. Confirm that the error in $y(1)$ is reduced by about half if a step size of $h = 0.1$ is used in Exercise 11.

In Exercises 13–16, use Euler's Method with the given step size h to approximate the solution of the initial-value problem over the stated interval. Present your answer as a table and as a graph.

13. $dy/dx = \sqrt{y}$, $y(0) = 1$, $0 \le x \le 4$, $h = 0.5$

14. $dy/dx = x - y^2$, $y(0) = 1$, $0 \le x \le 2$, $h = 0.25$

15. $dy/dt = \sin y$, $y(0) = 1$, $0 \le t \le 2$, $h = 0.5$

16. $dy/dt = e^{-y}$, $y(0) = 0$, $0 \le t \le 1$, $h = 0.1$

17. Consider the initial-value problem

$$y' = \cos 2\pi t, \quad y(0) = 1$$

Use Euler's Method with five steps to approximate $y(1)$.

18. (a) Show that the solution of the initial-value problem
$y' = e^{-x^2}$, $y(0) = 0$ is

$$y(x) = \int_0^x e^{-t^2}\, dt$$

(b) Use Euler's Method with $h = 0.05$ to approximate the value of

$$y(1) = \int_0^1 e^{-t^2}\, dt$$

and compare the answer to that produced by a calculating utility with a numerical integration capability.

19. The accompanying figure shows a direction field for the differential equation $y' = -x/y$.

(a) Use the direction field to estimate $y(\frac{1}{2})$ for the solution that satisfies the given initial condition $y(0) = 1$.

(b) Compare your estimate to the exact value of $y(\frac{1}{2})$.

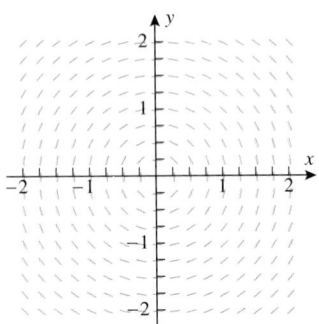

Figure Ex-19

20. Consider the initial-value problem

$$\frac{dy}{dx} = \frac{\sqrt{y}}{2}, \quad y(0) = 1$$

(a) Use Euler's Method with step sizes of $h = 0.2, 0.1$, and 0.05 to obtain three approximations of $y(1)$.

(b) Plot the three approximations versus h, and make a conjecture about the exact value of $y(1)$. Explain your reasoning.

(c) Check your conjecture by finding $y(1)$ exactly.

10.3 MODELING WITH DIFFERENTIAL EQUATIONS

Since many of the fundamental laws of the physical and social sciences involve rates of change, it should not be surprising that such laws are modeled by differential equations. In this section we will discuss the general idea of modeling with differential equations, and we will investigate some important models that can be applied to population growth, carbon dating, medicine, ecology, and the vibration of springs.

POPULATION GROWTH

One of the simplest models of population growth is based on the observation that when populations (people, plants, bacteria, and fruit flies, for example) are not constrained by environmental limitations, they tend to grow at a rate that is proportional to the size of the population—the larger the population, the more rapidly it grows.

To translate this principle into a mathematical model, suppose that $y = y(t)$ denotes the population at time t. At each point in time, the rate of increase of the population with respect to time is dy/dt, so the assumption that the rate of growth is proportional to the population is described by the differential equation

$$\frac{dy}{dt} = ky \tag{1}$$

where k is a positive constant of proportionality that can usually be determined experimentally. Thus, if the population is known at some point in time, say $y = y_0$ at time $t = 0$, then a general formula for the population $y(t)$ can be obtained by solving the initial-value problem

$$\frac{dy}{dt} = ky, \quad y(0) = y_0$$

PHARMACOLOGY

When a drug (say, penicillin or aspirin) is administered to an individual, it enters the bloodstream and then is absorbed by the body over time. Medical research has shown that the amount of a drug that is present in the bloodstream tends to decrease at a rate that is proportional to the amount of the drug present—the more of the drug that is present in the bloodstream, the more rapidly it is absorbed by the body.

To translate this principle into a mathematical model, suppose that $y = y(t)$ is the amount of the drug present in the bloodstream at time t. At each point in time, the rate of change in y with respect to t is dy/dt, so the assumption that the rate of decrease is proportional to the amount y in the bloodstream translates into the differential equation

$$\frac{dy}{dt} = -ky \tag{2}$$

where k is a positive constant of proportionality that depends on the drug and can be determined experimentally. The negative sign is required because y decreases with time. Thus, if the initial dosage of the drug is known, say $y = y_0$ at time $t = 0$, then a general formula for $y(t)$ can be obtained by solving the initial-value problem

$$\frac{dy}{dt} = -ky, \quad y(0) = y_0$$

SPREAD OF DISEASE

Suppose that a disease begins to spread in a population of L individuals. Logic suggests that at each point in time the rate at which the disease spreads will depend on how many individuals are already affected and how many are not—as more individuals are affected, the opportunity to spread the disease tends to increase, but at the same time there are fewer individuals who are not affected, so the opportunity to spread the disease tends to decrease. Thus, there are two conflicting influences on the rate at which the disease spreads.

To translate this into a mathematical model, suppose that $y = y(t)$ is the number of individuals who have the disease at time t, so of necessity the number of individuals who do not have the disease at time t is $L - y$. As the value of y increases, the value of $L - y$ decreases, so the conflicting influences of the two factors on the rate of spread dy/dt are taken into account by the differential equation

$$\frac{dy}{dt} = ky(L - y)$$

where k is a positive constant of proportionality that depends on the nature of the disease and the behavior patterns of the individuals and can be determined experimentally. Thus, if the number of affected individuals is known at some point in time, say $y = y_0$ at time $t = 0$, then a general formula for $y(t)$ can be obtained by solving the initial-value problem

$$\frac{dy}{dt} = ky(L - y), \quad y(0) = y_0 \tag{3}$$

INHIBITED POPULATION GROWTH

The population growth model that we discussed at the beginning of this section was predicated on the assumption that the population $y = y(t)$ is not constrained by the environment. For this reason, it is sometimes called the ***uninhibited growth model***. However, in the real world this assumption is usually not valid—populations generally grow within ecological systems that can only support a certain number of individuals; the number L of such individuals is called the ***carrying capacity*** of the system. Thus, when $y > L$, the population exceeds the capacity of the ecological system and tends to decrease toward L; when $y < L$, the population is below the capacity of the ecological system and tends to increase toward L; and when $y = L$, the population is in balance with the capacity of the ecological system and tends to remain stable.

To translate this into a mathematical model, we must look for a differential equation in which

$$\frac{dy}{dt} < 0 \quad \text{if} \quad \frac{y}{L} > 1$$

$$\frac{dy}{dt} > 0 \quad \text{if} \quad \frac{y}{L} < 1$$

$$\frac{dy}{dt} = 0 \quad \text{if} \quad \frac{y}{L} = 1$$

Moreover, logic suggests that when the population is far below the carrying capacity (i.e., $y/L \approx 0$), then the environmental constraints should have little effect, and the growth rate should behave very much like the uninhibited model. Thus, we want

$$\frac{dy}{dt} \approx ky \quad \text{if} \quad \frac{y}{L} \approx 0$$

A simple differential equation that meets all of these requirements is

$$\frac{dy}{dt} = k\left(1 - \frac{y}{L}\right)y$$

where k is a positive constant of proportionality. Thus, if k and L can be determined experimentally, and if the population is known at some point in time, say $y(0) = y_0$, then a general formula for the population $y(t)$ can be determined by solving the initial-value problem

$$\frac{dy}{dt} = k\left(1 - \frac{y}{L}\right)y, \quad y(0) = y_0 \tag{4}$$

This theory of population growth is due to the Belgian mathematician, P. F. Verhulst (1804–1849), who introduced it in 1838 and described it as "logistic growth."[*] Thus, the differential equation in (4) is called the ***logistic differential equation***, and the growth model described by (4) is called the ***logistic model*** or the ***inhibited growth model***.

REMARK.   Observe that the differential equation in (3) can be expressed as

$$\frac{dy}{dt} = kL\left(1 - \frac{y}{L}\right)y$$

which is a logistic equation with kL rather than k as the constant of proportionality. Thus, this model for the spread of disease is also a logistic or inhibited growth model.

EXPONENTIAL GROWTH AND DECAY MODELS

Equations (1) and (2) are examples of a general class of models called *exponential models*. In general, exponential models arise in situations where a quantity increases or decreases at a rate that is proportional to the amount of the quantity present. More precisely, we make the following definition:

---

[*]Verhulst's model fell into obscurity for nearly a hundred years because he did not have sufficient census data to test its validity. However, interest in the model was revived in the 1930s when biologists used it successfully to describe the growth of fruit fly and flour beetle populations. Verhulst himself used the model to predict that an upper limit on Belgium's population would be approximately 9,400,000. In 1994 the population was about 10,118,000.

10.3.1 DEFINITION. A quantity $y = y(t)$ is said to have an *exponential growth model* if it increases at a rate that is proportional to the amount of the quantity present, and it is said to have an *exponential decay model* if it decreases at a rate that is proportional to the amount of the quantity present. Thus, for an exponential growth model, the quantity $y(t)$ satisfies an equation of the form

$$\frac{dy}{dt} = ky \quad (k > 0) \tag{5}$$

and for an exponential decay model, the quantity $y(t)$ satisfies an equation of the form

$$\frac{dy}{dt} = -ky \quad (k > 0) \tag{6}$$

The constant k is called the *growth constant* or the *decay constant*, as appropriate.

Equations (5) and (6) are first-order linear equations, since they can be rewritten as

$$\frac{dy}{dt} - ky = 0 \quad \text{and} \quad \frac{dy}{dt} + ky = 0$$

both of which have the form of Equation (10) in Section 10.1 (but with t rather than x as the independent variable); in the first equation we have $p(t) = -k$ and $q(t) = 0$, and in the second we have $p(t) = k$ and $q(t) = 0$.

To illustrate how these equations can be solved, suppose that a quantity $y = y(t)$ has an exponential growth model and we know the amount of the quantity at some point in time, say $y = y_0$ when $t = 0$. Thus, a general formula for $y(t)$ can be obtained by solving the initial-value problem

$$\frac{dy}{dt} - ky = 0, \quad y(0) = y_0$$

Multiplying the differential equation through by the integrating factor

$$\mu = e^{\int -k\, dt} = e^{-kt}$$

yields

$$\frac{d}{dt}(e^{-kt} y) = 0$$

and then integrating with respect to t yields

$$e^{-kt} y = C \quad \text{or} \quad y = Ce^{kt}$$

The initial condition implies that $y = y_0$ when $t = 0$, from which it follows that $C = y_0$ (verify). Thus, the solution of the initial-value problem is

$$y = y_0 e^{kt} \tag{7}$$

We leave it for you to show that if $y = y(t)$ has an exponential decay model, and if $y(0) = y_0$, then

$$y = y_0 e^{-kt} \tag{8}$$

INTERPRETING THE GROWTH AND DECAY CONSTANTS

The significance of the constant k in Formulas (7) and (8) can be understood by reexamining the differential equations that gave rise to these formulas. For example, in the case of the exponential growth model, Equation (5) can be rewritten as

$$k = \frac{dy/dt}{y}$$

which states that the growth rate as a fraction of the entire population remains constant over time, and this constant is k. For this reason, k is called the *relative growth rate* of the population. It is usual to express the relative growth rate as a percentage. Thus, a relative growth rate of 3% per unit of time in an exponential growth model means that $k = 0.03$. Similarly, the constant k in an exponential decay model is called the **relative decay rate**.

REMARK.   It is standard practice in applications to call the relative growth rate the *growth rate*, even though it is not really correct (the growth rate is dy/dt). However, the practice is so common that we will follow it here.

Example 1

According to United Nations data, the world population at the beginning of 1990 was approximately 5.3 billion and growing at a rate of about 2% per year. Assuming an exponential growth model, estimate the world population at the beginning of the year 2015.

Solution. Let

$t =$ time elapsed from the beginning of 1990 (in years)

$y =$ world population (in billions)

Since the beginning of 1990 corresponds to $t = 0$, it follows from the given data that

$$y_0 = y(0) = 5.3 \text{ (billion)}$$

Since the growth rate is 2% ($k = 0.02$), it follows from (7) that the world population at time t will be

$$y(t) = y_0 e^{kt} = 5.3 e^{0.02t} \qquad (9)$$

Since the beginning of the year 2015 corresponds to an elapsed time of $t = 25$ years ($2015 - 1990 = 25$), it follows from (9) that the world population by the year 2015 will be

$$y(25) = 5.3 e^{0.02(25)} = 5.3 e^{0.5} \approx 8.7$$

which is a population of approximately 8.7 billion. ◀

REMARK.   In this example, the growth rate was given, so there was no need to calculate it. If the growth rate or decay rate in an exponential model is unknown, then it can be calculated using the initial condition and the value of y at one other point in time (Exercise 42).

DOUBLING TIME AND HALF-LIFE

If a quantity y has an exponential growth model, then the time required for the original size to double is called the **doubling time**, and if y has an exponential decay model, then the time required for the original size to reduce by half is called the **half-life**. As it turns out, doubling time and half-life depend only on the growth or decay rate and not on the amount present initially. To see why this is so, suppose that $y = y(t)$ has an exponential growth model

$$y = y_0 e^{kt} \qquad (10)$$

and let T denote the amount of time required for y to double in size. Thus, at time $t = T$ the value of y will be $2y_0$, and hence from (10)

$$2y_0 = y_0 e^{kT} \quad \text{or} \quad e^{kT} = 2$$

Taking the natural logarithm of both sides yields $kT = \ln 2$, which implies that the doubling time is

$$T = \frac{1}{k} \ln 2 \qquad (11)$$

We leave it as an exercise to show that Formula (11) also gives the half-life of an exponential decay model. Observe that this formula does not involve the initial amount y_0, so that in an exponential growth or decay model, the quantity y doubles (or reduces by half) every T units (Figure 10.3.1).

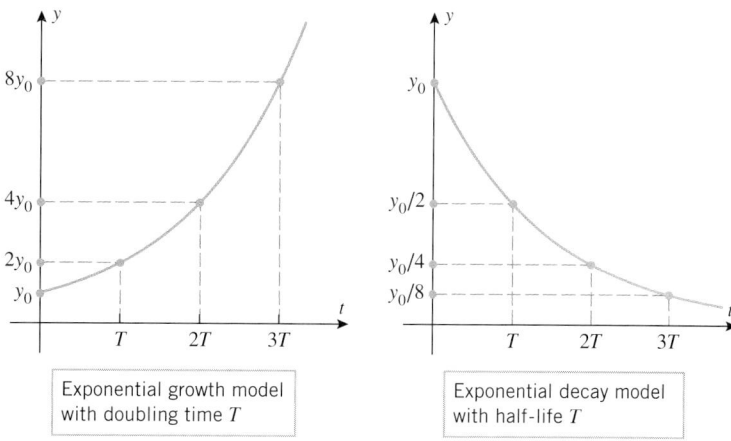

Figure 10.3.1

Example 2

It follows from (11) that with a continued growth rate of 2% per year, the doubling time for the world population will be

$$T = \frac{1}{0.02} \ln 2 \approx 34.657$$

or approximately 35 years. Thus, with a continued 2% annual growth rate the population of 5.3 billion in 1990 will double to 10.6 billion by the year 2025 and will double again to 21.2 billion by 2060.   ◄

RADIOACTIVE DECAY

It is a fact of physics that radioactive elements disintegrate spontaneously in a process called *radioactive decay*. Experimentation has shown that the rate of disintegration is proportional to the amount of the element present, which implies that the amount $y = y(t)$ of a radioactive element present as a function of time has an exponential decay model.

Every radioactive element has a specific half-life; for example, the half-life of radioactive carbon-14 is about 5730 years. Thus, from (11), the decay constant for this element is

$$k = \frac{1}{T} \ln 2 = \frac{\ln 2}{5730} \approx 0.000121$$

and this implies that if there are y_0 units of carbon-14 present at time $t = 0$, then the number of units present at a time t will be approximately

$$y(t) = y_0 e^{-0.000121t} \tag{12}$$

Example 3

If 100 grams of radioactive carbon-14 are stored in a cave for 1000 years, how many grams will be left at that time?

Solution.   From (12) with $y_0 = 100$ and $t = 1000$, we obtain

$$y(1000) = 100 e^{-0.000121(1000)} = 100 e^{-0.121} \approx 88.6$$

Thus, about 88.6 grams will be left.   ◄

CARBON DATING

When the nitrogen in the Earth's upper atmosphere is bombarded by cosmic radiation, the radioactive element carbon-14 is produced. This carbon-14 combines with oxygen to form carbon dioxide, which is ingested by plants, which in turn are eaten by animals. In this way all living plants and animals absorb quantities of radioactive carbon-14. In 1947 the

American nuclear scientist W. F. Libby[*] proposed the theory that the percentage of carbon-14 in the atmosphere and in living tissues of plants is the same. When a plant or animal dies, the carbon-14 in the tissue begins to decay. Thus, the age of an artifact that contains plant or animal material can be estimated by determining what percentage of its original carbon-14 content remains. Various procedures, called **carbon dating** or **carbon-14 dating**, have been developed for measuring this percentage.

Example 4

In 1988 the Vatican authorized the British Museum to date a cloth relic known as the Shroud of Turin, possibly the burial shroud of Jesus of Nazareth. This cloth, which first surfaced in 1356, contains the negative image of a human body that was widely believed to be that of Jesus. The report of the British Museum showed that the fibers in the cloth contained between 92% and 93% of their original carbon-14. Use this information to estimate the age of the shroud.

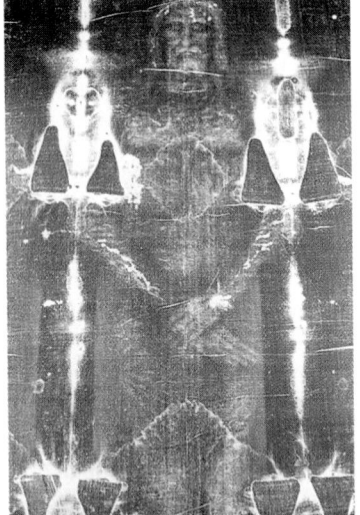

The Shroud of Turin

Solution. From (12), the fraction of the original carbon-14 that remains after t years is

$$\frac{y(t)}{y_0} = e^{-0.000121t}$$

Taking the natural logarithm of both sides and solving for t, we obtain

$$t = -\frac{1}{0.000121} \ln\left(\frac{y(t)}{y_0}\right)$$

Thus, taking $y(t)/y_0$ to be 0.93 and 0.92, we obtain

$$t = -\frac{1}{0.000121} \ln(0.93) \approx 600$$

$$t = -\frac{1}{0.000121} \ln(0.92) \approx 689$$

This means that when the test was done in 1988, the shroud was between 600 and 689 years old, thereby placing its origin between 1299 A.D. and 1388 A.D. Thus, if one accepts the validity of carbon-14 dating, the Shroud of Turin cannot be the burial shroud of Jesus of Nazareth. ◄

LOGISTIC MODELS

Recall that the logistic model of population growth in an ecological system with carrying capacity L is determined by initial-value problem (4). To illustrate how this initial-value problem can be solved for $y(t)$, let us focus on the differential equation

$$\frac{dy}{dt} = k\left(1 - \frac{y}{L}\right)y \tag{13}$$

It will be convenient to rewrite Equation (13) as

$$\frac{dy}{dt} = \frac{k}{L}(L - y)y = \frac{k}{L}y(L - y)$$

This equation is separable, since it can be rewritten in differential form as

$$\frac{L}{y(L - y)}\, dy = k\, dt$$

Integrating both sides yields the equation

$$\int \frac{L}{y(L - y)}\, dy = \int k\, dt$$

---

[*] W. F. Libby, "Radiocarbon Dating," *American Scientist*, Vol. 44, 1956, pp. 98–112.

Using partial fractions on the left side, we can rewrite this equation as (verify)

$$\int \left(\frac{1}{y} + \frac{1}{L-y} \right) dy = \int k\, dt$$

Integrating and rearranging the form of the result, we obtain

$$\ln |y| - \ln |L - y| = kt + C$$

$$\ln \left| \frac{y}{L-y} \right| = kt + C$$

$$\left| \frac{y}{L-y} \right| = e^{kt+C}$$

$$\left| \frac{L-y}{y} \right| = e^{-kt-C} = e^{-C} e^{-kt}$$

$$\frac{L-y}{y} = \pm e^{-C} e^{-kt}$$

$$\frac{L}{y} - 1 = Ae^{-kt} \quad \text{(where } A = \pm e^{-C}\text{)}$$

Solving this equation for y yields (verify)

$$y = \frac{L}{1 + Ae^{-kt}} \tag{14}$$

As the final step, we want to use the initial condition in (4) to determine the constant A. But the initial condition implies that $y = y_0$ if $t = 0$, so from (14)

$$y_0 = \frac{L}{1 + A}$$

from which we obtain

$$A = \frac{L - y_0}{y_0}$$

Thus, the solution of the initial-value problem (4) is

$$y = \frac{L}{1 + \left(\dfrac{L - y_0}{y_0} \right) e^{-kt}}$$

which can be rewritten more simply as

$$y = \frac{y_0 L}{y_0 + (L - y_0)e^{-kt}} \tag{15}$$

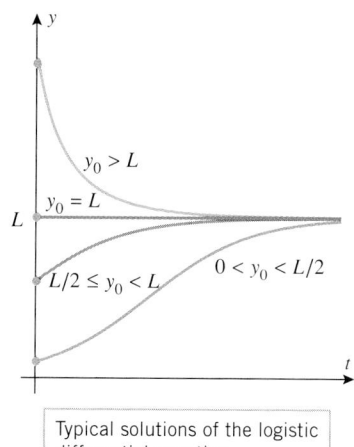

$y_0 > L$

$y_0 = L$

$L/2 \le y_0 < L$

$0 < y_0 < L/2$

Typical solutions of the logistic differential equation

Figure 10.3.2

The graph of (15) has one of four general shapes, depending on the relationship between the initial population y_0 and the carrying capacity L (Figure 10.3.2).

Example 5

Figure 10.3.3 shows the graph of a population $y = y(t)$ with a logistic growth model. Estimate the values of y_0, L, and k, and use the estimates to deduce a formula for y as a function of t.

Solution. The graph suggests that the carrying capacity is $L = 5$, and the population at time $t = 0$ is $y_0 = 1$. Thus, from (15), the equation has the form

$$y = \frac{5}{1 + 4e^{-kt}} \tag{16}$$

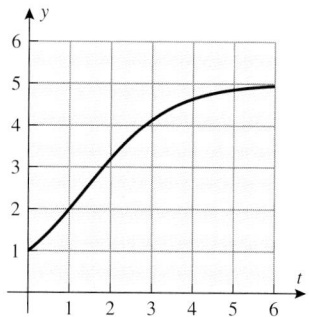

Figure 10.3.3

where k must still be determined. However, the graph passes through the point $(1, 2)$, which tells us that $y = 2$ if $t = 1$. Substituting these values in (16) yields

$$2 = \frac{5}{1 + 4e^{-k}}$$

Solving for k we obtain (verify)

$$k = \log \tfrac{8}{3} \approx 0.98$$

and substituting this in (16) yields

$$y = \frac{5}{1 + 4e^{-0.98t}}$$

◀

VIBRATIONS OF SPRINGS

We conclude this section with an engineering model that leads to a second-order differential equation.

As shown in Figure 10.3.4, consider a block of mass m that is suspended from a vertical spring and allowed to settle into an *equilibrium position*. Assume that the block is then set into vertical vibratory motion by pulling or pushing on it and releasing it at time $t = 0$. We will be interested in finding a mathematical model that describes the vibratory motion of the block over time.

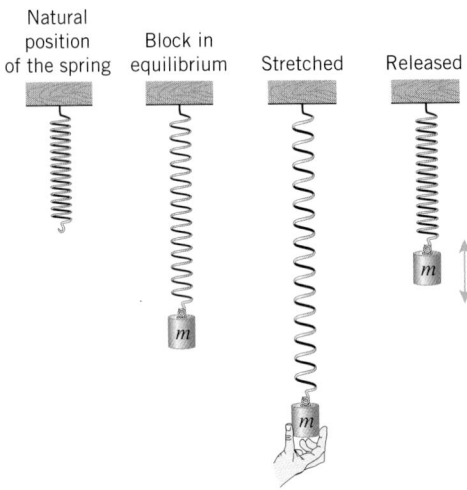

Figure 10.3.4

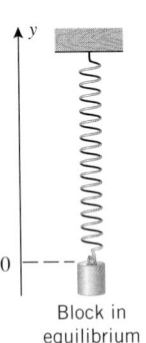

Figure 10.3.5

To translate this problem into mathematical form, we introduce a vertical y-axis whose positive direction is up and whose origin is at the connection of the spring to the block when the block is in equilibrium (Figure 10.3.5). Our goal is to find the coordinate $y = y(t)$ of the top of the block as a function of time. For this purpose we will need Newton's Second Law of Motion,

$$F = ma$$

[see (19) in Section 10.1], as well as the following two results from physics:

10.3.2 HOOKE'S LAW. If a spring is stretched (or compressed) l units beyond its natural position, then it pulls (or pushes) with a force of magnitude

$$F = kl$$

where k is a positive constant, called the *spring constant*. This constant, which is measured in units of force per unit length, depends on such factors as the thickness of the spring and its composition. The force exerted by the spring is called the *restoring force*.

> **10.3.3**   WEIGHT.   The gravitational force exerted by the Earth on an object is called the object's **weight** (or more precisely, its **Earth weight**). It follows from Newton's Second Law of Motion that an object with mass m has a weight w of magnitude mg, where g is the acceleration due to gravity. However, if the positive direction is up, as we are assuming here, then the force of the Earth's gravity is in the negative direction, so
>
> $$w = -mg$$
>
> The weight of an object is measured in units of force.

The motion of the block in Figure 10.3.4 will depend on how far it is stretched or compressed initially and the forces that act on it while it moves. In our model we will assume that there are only two such forces: its weight w and the restoring force F_s of the spring. In particular, we will ignore such forces as air resistance, internal frictional forces in the spring, forces due to movement of the spring support, and so forth. With these assumptions, the model is called the **simple harmonic model** and the motion of the block is called **simple harmonic motion**.

Our goal is to produce a differential equation whose solution gives the position function $y(t)$ of the block as a function of time. We will do this by determining the net force $F(t)$ acting on the block at a general time t and then applying Newton's Second Law of Motion. Since the only forces acting on the block are its weight $w = -mg$ and the restoring force F_s of the spring, and since the acceleration of the block at time t is $y''(t)$, it follows from Newton's second law that

$$F_s(t) - mg = my''(t) \tag{17}$$

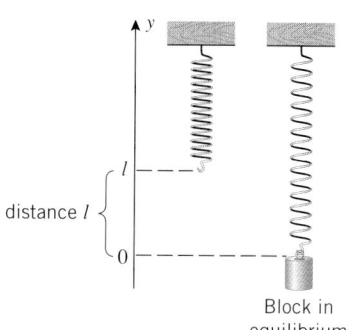

distance l

Figure 10.3.6

Block in equilibrium

To express $F_s(t)$ in terms of $y(t)$, we will begin by examining the forces on the block when it is in its equilibrium position. In this position the downward force of the weight is perfectly balanced by the upward restoring force of the spring, so that the sum of these two forces must be zero. Thus, if we assume that the spring constant is k and that the spring is stretched a distance of l units beyond its natural length when the block is in equilibrium (Figure 10.3.6), then

$$kl - mg = 0 \tag{18}$$

Now let us examine the restoring force acting on the block when the connection point has coordinate $y(t)$. At this point the end of the spring is displaced $l - y(t)$ units from its natural position (Figure 10.3.7), so Hooke's law implies that the restoring force is

$$F_s(t) = k(l - y(t)) = kl - ky(t)$$

which from (18) can be rewritten as

$$F_s(t) = mg - ky(t)$$

Substituting this in (17) and canceling the mg terms yields

$$-ky(t) = my''(t)$$

which we can rewrite as

$$y''(t) + \left(\frac{k}{m}\right) y(t) = 0 \tag{19}$$

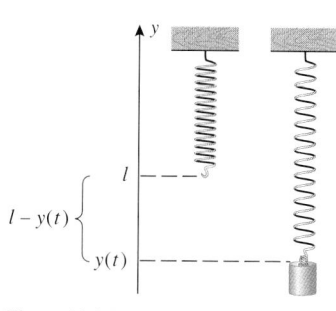

$l - y(t)$

$y(t)$

Figure 10.3.7

This is a second-order differential equation whose solution is the position function of the block, and it is shown in courses on differential equations that the general solution of this equation is

$$y(t) = c_1 \cos\left(\sqrt{\frac{k}{m}}\, t\right) + c_2 \sin\left(\sqrt{\frac{k}{m}}\, t\right) \tag{20}$$

where c_1 and c_2 are arbitrary constants. Note that there are two arbitrary constants because (19) is a second-order differential equation.

| FOR THE READER.   Confirm that the functions in family (20) are solutions of (19).

Since (20) has two arbitrary constants, it requires two initial conditions to determine $y(t)$ uniquely. These can be obtained from the initial position and velocity of the block. Specifically, we will ask you to show in Exercise 40 that if the position of the block at time $t = 0$ is y_0, and if the initial velocity of the block is zero (i.e., it is *released* from rest), then

$$y(t) = y_0 \cos\left(\sqrt{\frac{k}{m}}\,t\right) \tag{21}$$

This formula describes a periodic vibration with an amplitude of $|y_0|$, a period T given by

$$T = \frac{2\pi}{\sqrt{k/m}} = 2\pi\sqrt{m/k} \tag{22}$$

and a frequency f given by

$$f = \frac{1}{T} = \frac{\sqrt{k/m}}{2\pi} \tag{23}$$

(Figure 10.3.8).

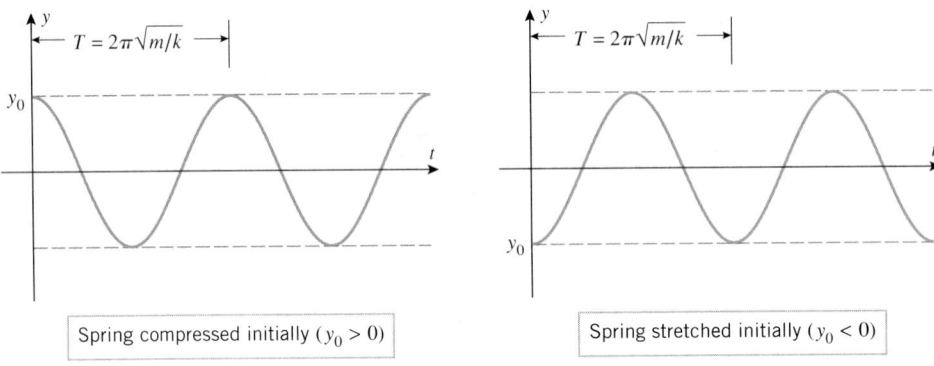

Figure 10.3.8

Example 6

Suppose that the block in Figure 10.3.4 stretches the spring 0.2 m in equilibrium. Suppose also that the block is pulled 0.5 m below its equilibrium position and released at time $t = 0$.
(a)   Find the position function $y(t)$ of the block.
(b)   Find the amplitude, period, and frequency of the vibration.

Solution (a). The appropriate formula is (21). Although we are not given the mass m of the block or the spring constant k, it does not matter because we can use the equilibrium condition (18) to find the ratio k/m without having values for k and m. Specifically, we are given that in equilibrium the block stretches the spring $l = 0.2$ m, and we know that $g = 9.8$ m/s^2. Thus, (18) implies that

$$\frac{k}{m} = \frac{g}{l} = \frac{9.8}{0.2} = 49 \text{ s}^{-2} \tag{24}$$

Substituting this in (21) yields

$$y(t) = y_0 \cos 7t$$

where y_0 is the coordinate of the block at time $t = 0$. However, we are given that the block is initially 0.5 m *below* the equilibrium position, so $y_0 = -0.5$ and hence the position function of the block is $y(t) = -0.5 \cos 7t$

Solution (*b*). The amplitude of the vibration is

$$\text{amplitude} = |y_0| = |-0.5| = 0.5 \text{ m}$$

and from (22), (23), and (24) the period and frequency are

$$\text{period} = T = 2\pi\sqrt{\frac{m}{k}} = 2\pi\sqrt{\frac{1}{49}} = \frac{2\pi}{7} \text{ s} \qquad \text{frequency} = f = \frac{1}{T} = \frac{7}{2\pi} \text{ Hz} \quad \blacktriangleleft$$

EXERCISE SET 10.3 ⬚ Graphing Calculator Ⓒ CAS

1. (a) Suppose that a quantity $y = y(t)$ increases at a rate that is proportional to the square of the amount present, and suppose that at time $t = 0$, the amount present is y_0. Find an initial-value problem whose solution is $y(t)$.
  (b) Suppose that a quantity $y = y(t)$ decreases at a rate that is proportional to the square of the amount present, and suppose that at a time $t = 0$, the amount present is y_0. Find an initial-value problem whose solution is $y(t)$.

2. (a) Suppose that a quantity $y = y(t)$ changes in such a way that $dy/dt = k\sqrt{y}$, where $k > 0$. Describe how y changes in words.
  (b) Suppose that a quantity $y = y(t)$ changes in such a way that $dy/dt = -ky^3$, where $k > 0$. Describe how y changes in words.

3. (a) Suppose that a particle moves along an s-axis in such a way that its velocity $v(t)$ is always half of $s(t)$. Find a differential equation whose solution is $s(t)$.
  (b) Suppose that an object moves along an s-axis in such a way that its acceleration $a(t)$ is always twice the velocity. Find a differential equation whose solution is $s(t)$.

4. Suppose that a body moves along an s-axis through a resistive medium in such a way that the velocity $v = v(t)$ decreases at a rate that is twice the square of the velocity.
  (a) Find a differential equation whose solution is the velocity $v(t)$.
  (b) Find a differential equation whose solution is the position $s(t)$.

5. Suppose that an initial population of 10,000 bacteria grows exponentially at a rate of 1% per hour and that $y = y(t)$ is the number of bacteria present t hours later.
  (a) Find an initial-value problem whose solution is $y(t)$.
  (b) Find a formula for $y(t)$.
  (c) How long does it take for the initial population of bacteria to double?
  (d) How long does it take for the population of bacteria to reach 45,000?

6. A cell of the bacterium *E. coli* divides into two cells every 20 minutes when placed in a nutrient culture. Let $y = y(t)$ be the number of cells that are present t minutes after a single cell is placed in the culture. Assume that the growth of

the bacteria is approximated by a continuous exponential growth model.
  (a) Find an initial-value problem whose solution is $y(t)$.
  (b) Find a formula for $y(t)$.
  (c) How many cells are present after 2 hours?
  (d) How long does it take for the number of cells to reach 1,000,000?

7. Radon-222 is a radioactive gas with a half-life of 3.83 days. This gas is a health hazard because it tends to get trapped in the basements of houses, and many health officials suggest that homeowners seal their basements to prevent entry of the gas. Assume that 5.0×10^7 radon atoms are trapped in a basement at the time it is sealed and that $y(t)$ is the number of atoms present t days later.
  (a) Find an initial-value problem whose solution is $y(t)$.
  (b) Find a formula for $y(t)$.
  (c) How many atoms will be present after 30 days?
  (d) How long will it take for 90% of the original quantity of gas to decay?

8. Polonium-210 is a radioactive element with a half-life of 140 days. Assume that 10 milligrams of the element are placed in a lead container and that $y(t)$ is the number of milligrams present t days later.
  (a) Find an initial-value problem whose solution is $y(t)$.
  (b) Find a formula for $y(t)$.
  (c) How many milligrams will be present after 10 weeks?
  (d) How long will it take for 70% of the original sample to decay?

9. Suppose that 100 fruit flies are placed in a breeding container that can support at most 5000 flies. Assuming that the population grows exponentially at a rate of 2% per day, how long will it take for the container to reach capacity?

10. Suppose that the town of Grayrock had a population of 10,000 in 1987 and a population of 12,000 in 1997. Assuming an exponential growth model, in what year will the population reach 20,000?

11. A scientist wants to determine the half-life of a certain radioactive substance. She determines that in exactly 5 days a 10.0-milligram sample of the substance decays to 3.5 milligrams. Based on these data, what is the half-life?

12. Suppose that 40% of a certain radioactive substance decays in 5 years.
   (a) What is the half-life of the substance in years?
   (b) Suppose that a certain quantity of this substance is stored in a cave. What percentage of it will remain after t years?

13. In each part, find an exponential growth model $y = y_0 e^{kt}$ that satisfies the stated conditions.
   (a) $y_0 = 2$; doubling time $T = 5$
   (b) $y(0) = 5$; growth rate 1.5%
   (c) $y(1) = 1$; $y(10) = 100$
   (d) $y(1) = 1$; doubling time $T = 5$

14. In each part, find an exponential decay model $y = y_0 e^{-kt}$ that satisfies the stated conditions.
   (a) $y_0 = 10$; half-life $T = 5$
   (b) $y(0) = 10$; decay rate 1.5%
   (c) $y(1) = 100$; $y(10) = 1$
   (d) $y(1) = 10$; half-life $T = 5$

 15. (a) Make a conjecture about the effect on the graphs of $y = y_0 e^{kt}$ and $y = y_0 e^{-kt}$ of varying k and keeping y_0 fixed. Confirm your conjecture with a graphing utility.
   (b) Make a conjecture about the effect on the graphs of $y = y_0 e^{kt}$ and $y = y_0 e^{-kt}$ of varying y_0 and keeping k fixed. Confirm your conjecture with a graphing utility.

16. (a) What effect does increasing y_0 and keeping k fixed have on the doubling time or half-life of an exponential model? Justify your answer.
   (b) What effect does increasing k and keeping y_0 fixed have on the doubling time and half-life of an exponential model? Justify your answer.

17. (a) There is a trick, called the ***Rule of 70***, that can be used to get a quick estimate of the doubling time or half-life of an exponential model. According to this rule, the doubling time or half-life is roughly 70 divided by the percentage growth or decay rate. For example, we showed in Example 2 that with a continued growth rate of 2% per year the world population would double every 35 years. This result agrees with the Rule of 70, since $70/2 = 35$. Explain why this rule works.
   (b) Use the Rule of 70 to estimate the doubling time of a population that grows exponentially at a rate of 1% per year.
   (c) Use the Rule of 70 to estimate the half-life of a population that decreases exponentially at a rate of 3.5% per hour.
   (d) Use the Rule of 70 to estimate the growth rate that would be required for a population growing exponentially to double every 10 years.

18. Find a formula for the tripling time of an exponential growth model.

19. In 1950, a research team digging near Folsom, New Mexico, found charred bison bones along with some leaf-shaped

projectile points (called the "Folsom points") that had been made by a Paleo-Indian hunting culture. It was clear from the evidence that the bison had been cooked and eaten by the makers of the points, so that carbon-14 dating of the bones made it possible for the researchers to determine when the hunters roamed North America. Tests showed that the bones contained between 27% and 30% of their original carbon-14. Use this information to show that the hunters lived roughly between 9000 B.C. and 8000 B.C.

 20. (a) Use a graphing utility to make a graph of p_{rem} versus t, where p_{rem} is the percentage of carbon-14 that remains in an artifact after t years.
   (b) Use the graph to estimate the percentage of carbon-14 that would have to have been present in the 1988 test of the Shroud of Turin for it to have been the burial shroud of Jesus. [See Example 4.]

In Exercises 21 and 22, the graph of a logistic model
$$y = \frac{y_0 L}{y_0 + (L - y_0)e^{-kt}}$$
is shown. Estimate y_0, L, and k.

21.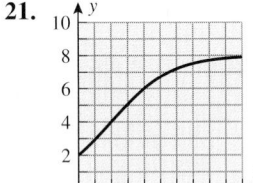

22.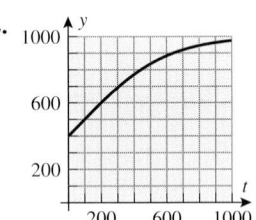

23. Suppose that the growth of a population $y = y(t)$ is given by the logistic equation
$$y = \frac{60}{5 + 7e^{-t}}$$
   (a) What is the population at time $t = 0$?
   (b) What is the carrying capacity L?
   (c) What is the constant k?
   (d) When does the population reach half of the carrying capacity?
   (e) Find an initial-value problem whose solution is $y(t)$.

24. Suppose that the growth of a population $y = y(t)$ is given by the logistic equation
$$y = \frac{1000}{1 + 999e^{-0.9t}}$$
   (a) What is the population at time $t = 0$?
   (b) What is the carrying capacity L?
   (c) What is the constant k?
   (d) When does the population reach 75% of the carrying capacity?
   (e) Find an initial-value problem whose solution is $y(t)$.

25. Suppose that a population $y(t)$ grows in accordance with the logistic model

$$\frac{dy}{dt} = 10(1 - 0.1y)y$$

(a) What is the carrying capacity?
(b) What is the value of k?
(c) For what value of y is the population growing most rapidly?

26. Suppose that a population $y(t)$ grows in accordance with the logistic model

$$\frac{dy}{dt} = 50y - 0.001y^2$$

(a) What is the carrying capacity?
(b) What is the value of k?
(c) For what value of y is the population growing most rapidly?

27. Suppose that a college residence hall houses 1000 students. Following the semester break, 20 students in the hall return with the flu, and 5 days later 35 students have the flu.
(a) Use model (4) to set up an initial-value problem whose solution is the number of students who will have had the flu t days after the return from the break. [*Note:* The differential equation in this case will involve a constant of proportionality.]
(b) Solve the initial-value problem, and use the given data to find the constant of proportionality.
(c) Make a table that illustrates how the flu spreads day to day over a 2-week period.
(d) Use a graphing utility to generate a graph that illustrates how the flu spreads over a 2-week period.

28. It has been observed experimentally that at a constant temperature the rate of change of the atmospheric pressure p with respect to the altitude h above sea level is proportional to the pressure.
(a) Assuming that the pressure at sea level is p_0, find an initial-value problem whose solution is $p(h)$. [*Note:* The differential equation in this case will involve a constant of proportionality.]
(b) Find a formula for $p(h)$ in atmospheres (atm) if the pressure at sea level is 1 atm and the pressure at 5000 ft above sea level is 0.83 atm.

Newton's Law of Cooling states that the rate at which the temperature of a cooling object decreases and the rate at which a warming object increases are proportional to the difference between the temperature of the object and the temperature of the surrounding medium. Use this result in Exercises 29–32.

29. A cup of water with a temperature of $95°C$ is placed in a room with a constant temperature $21°C$.
(a) Assuming that Newton's Law of Cooling applies, set up and solve an initial-value problem whose solution is the

temperature of the water t minutes after it is placed in the room. [*Note:* The differential equation will involve a constant of proportionality.]
(b) How many minutes will it take for the water to reach a temperature of $51°C$ if it cools to $85°C$ in 1 minute?

30. A glass of lemonade with a temperature of $40°F$ is placed in a room with a constant temperature of $70°F$, and 1 hour later its temperature is $52°F$. We stated in Example 8 of Section 4.4 that t hours after the lemonade is placed in the room its temperature is given by $T = 70 - 30e^{-0.5t}$. Confirm this using Newton's Law of Cooling and the method used in Exercise 29.

31. The great detective Sherlock Holmes and his assistant Dr. Watson are discussing the murder of actor Cornelius McHam. McHam was shot in the head, and his understudy, Barry Moore, was found standing over the body with the murder weapon in hand. Let's listen in.

Watson: Open-and-shut case Holmes—Moore is the murderer.
Holmes: Not so fast Watson—you are forgetting Newton's Law of Cooling!
Watson: Huh?
Holmes: Elementary my dear Watson—Moore was found standing over McHam at 10:06 P.M., at which time the coroner recorded a body temperature of $77.9°F$ and noted that the room thermostat was set to $72°F$. At 11:06 P.M. the coroner took another reading and recorded a body temperature of $75.6°F$. Since McHam's normal temperature is $98.6°F$, and since Moore was on stage between 6:00 P.M. and 8:00 P.M., Moore is obviously innocent.
Watson: Huh?
Holmes: Sometimes you are so dull Watson. Ask any calculus student to figure it out for you.
Watson: Hrrumph. . . .

32. Suppose that at time $t = 0$ an object with temperature T_0 is placed in a room with constant temperature T_a. If $T_0 < T_a$, then the temperature of the object will increase, and if $T_0 > T_a$, then the temperature will decrease. Assuming that Newton's Law of Cooling applies, show that in both cases the temperature $T(t)$ at time t is given by

$$T(t) = T_a + (T_0 - T_a)e^{-kt}$$

where k is a positive constant.

Exercises 33–38 involve vibrations of the block pictured in Figure 10.3.4. Assume that the y-axis is as shown in Figure 10.3.5 and that the simple harmonic model applies.

33. Suppose that the block has a mass of 1 kg, the spring constant is $k = 0.25$ N/m, and the block is pushed 0.3 m above its equilibrium position and released at time $t = 0$.
(a) Find the position function $y(t)$ of the block.

(b) Find the period and frequency of the vibration.

(c) Sketch the graph of $y(t)$.

(d) At what time does the block first pass through the equilibrium position?

(e) At what time does the block first reach its maximum distance below the equilibrium position?

34. Suppose that the block has a weight of 64 lb, the spring constant is $k = 0.25$ lb/ft, and the block is pushed 1 ft above its equilibrium position and released at time $t = 0$.

(a) Find the position function $y(t)$ of the block.

(b) Find the period and frequency of the vibration.

(c) Sketch the graph of $y(t)$.

(d) At what time does the block first pass through the equilibrium position?

(e) At what time does the block first reach its maximum distance below the equilibrium position?

35. Suppose that the block stretches the spring 0.05 m in equilibrium, and the block is pulled 0.12 m below the equilibrium position and released at time $t = 0$.

(a) Find the position function $y(t)$ of the block.

(b) Find the period and frequency of the vibration.

(c) Sketch the graph of $y(t)$.

(d) At what time does the block first pass through the equilibrium position?

(e) At what time does the block first reach its maximum distance above the equilibrium position?

36. Suppose that the block stretches the spring 0.5 ft in equilibrium, and is pulled 1.5 ft below the equilibrium position and released at time $t = 0$.

(a) Find the position function $y(t)$ of the block.

(b) Find the period and frequency of the vibration.

(c) Sketch the graph of $y(t)$.

(d) At what time does the block first pass through the equilibrium position?

(e) At what time does the block first reach its maximum distance above the equilibrium position?

37. (a) For what values of y would you expect the block in Exercise 36 to have its maximum speed? Confirm your answer to this question mathematically.

(b) For what values of y would you expect the block to have its minimum speed? Confirm your answer to this question mathematically.

38. Suppose that the block weighs w pounds and vibrates with a period of 3 s when it is pulled below the equilibrium position and released. Suppose also that if the process is repeated with an additional 4 lb of weight, then the period is 5 s.

(a) Find the spring constant.

(b) Find w.

39. As shown in the accompanying figure, suppose that a toy cart of mass m is attached to a wall by a spring with spring constant k, and let a horizontal x-axis be introduced with its origin at the connection point of the spring and cart when the cart is in equilibrium. Suppose that the cart is pulled or pushed horizontally to a point x_0 and then released at time $t = 0$. Find an initial-value problem whose solution is the position function of the cart, and state any assumptions you have made.

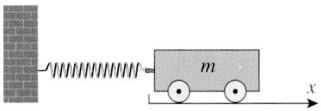

Figure Ex-39

40. Use the initial position $y(0) = y_0$ and the initial velocity $v(0) = 0$ to find the constants c_1 and c_2 in (20).

41. (a) Show that if $b > 1$, then the equation $y = y_0 b^t$ can be expressed as $y = y_0 e^{kt}$ for some positive constant k. [*Note:* This shows that if $b > 1$, and if y grows in accordance with the equation $y = y_0 b^t$, then y has an exponential growth model.]

(b) Show that if $0 < b < 1$, then the equation $y = y_0 b^t$ can be expressed as $y = y_0 e^{-kt}$ for some positive constant k. [*Note:* This shows that if $0 < b < 1$, and if y decays in accordance with the equation $y = y_0 b^t$, then y has an exponential decay model.]

(c) Express $y = 4(2^t)$ in the form $y = y_0 e^{kt}$.

(d) Express $y = 4(0.5^t)$ in the form $y = y_0 e^{-kt}$.

42. Suppose that a quantity y has an exponential growth model $y = y_0 e^{kt}$ or an exponential decay model $y = y_0 e^{-kt}$, and it is known that $y = y_1$ if $t = t_1$. In each case find a formula for k in terms of y_0, y_1, and t_1, assuming that $t_1 \neq 0$.

SUPPLEMENTARY EXERCISES

1. What is the relationship between the order of a differential equation and the number of arbitrary constants in its general solution? Give an informal explanation of why one would expect such a relationship.

2. Write a paragraph that describes Euler's Method.

3. (a) List the steps in the method of integrating factors for solving first-order linear differential equations.

(b) What would you do if you had to solve an important initial-value problem involving a first-order linear differential equation whose integrating factor could not be obtained because of the complexity of the integration?

4. Which of the following differential equations are separable?

(a) $\dfrac{dy}{dx} = f(x)g(y)$ (b) $\dfrac{dy}{dx} = \dfrac{f(x)}{g(y)}$

(c) $\dfrac{dy}{dx} = f(x) + g(y)$ (d) $\dfrac{dy}{dx} = \sqrt{f(x)g(y)}$

5. Classify the following first-order differential equations as separable, linear, both, or neither.

(a) $\dfrac{dy}{dx} - 3y = \sin x$ (b) $\dfrac{dy}{dx} + xy = x$

(c) $y\dfrac{dy}{dx} - x = 1$ (d) $\dfrac{dy}{dx} + xy^2 = \sin(xy)$

6. Confirm that the methods of integrating factors and separation of variables produce the same solution of the differential equation

$$\frac{dy}{dx} - 4xy = x$$

7. Consider the model $dy/dt = ky(L - y)$ for the spread of a disease, where $k > 0$ and $0 < y \le L$. For what value of y is the disease spreading most rapidly, and at what rate is it spreading?

8. (a) Show that if a quantity $y = y(t)$ has an exponential model, and if $y(t_1) = y_1$ and $y(t_2) = y_2$, then the doubling time or the half-life T is

$$T = \left| \frac{(t_2 - t_1)\ln 2}{\ln(y_2/y_1)} \right|$$

(b) In a certain 1-hour period the number of bacteria in a colony increases by 25%. Assuming an exponential growth model, what is the doubling time for the colony?

9. Assume that a spherical meteoroid burns up at a rate that is proportional to its surface area. Given that the radius is originally 4 m and 1 min later its radius is 3 m, find a formula for the radius as a function of time.

10. A tank contains 1000 gal of fresh water. At time $t = 0$ brine containing 5 ounces of salt per gallon of brine is poured into the tank at a rate of 10 gal/min, and the mixed solution is drained from the tank at the same rate. After 15 min that process is stopped and fresh water is poured into the tank at the rate of 5 gal/min, and the mixed solution is drained from the tank at the same rate. Find the amount of salt in the tank at time $t = 30$.

11. Suppose that a room containing $1200\ \text{ft}^3$ of air is free of carbon monoxide. At time $t = 0$ cigarette smoke containing 4% carbon monoxide is introduced at the rate of $0.1\ \text{ft}^3/\text{min}$, and the well-circulated mixture is vented from the room at the same rate.

(a) Find a formula for the percentage of carbon monoxide in the room at time t.

(b) Extended exposure to air containing 0.012% carbon monoxide is considered dangerous. How long will it take to reach this level? [This is based on a problem from William E. Boyce and Richard C. DiPrima, *Ele-*

mentary Differential Equations, 6th ed., John Wiley & Sons, 1997.]

In Exercises 12–16, solve the initial-value problem.

12. $y' = 1 + y^2$, $y(0) = 1$

13. $y' = \dfrac{y^5}{x(1 + y^4)}$, $y(1) = 1$

14. $xy' + 2y = 4x^2$, $y(1) = 2$

15. $y' = 4y^2 \sec^2 2x$, $y(\pi/8) = 1$

16. $y' = 6 - 5y + y^2$, $y(0) = \ln 2$

c **17.** (a) Solve the initial-value problem

$$y' - y = x \sin 3x, \quad y(0) = 1$$

by the method of integrating factors, using a CAS to perform any difficult integrations.

(b) Use the CAS to solve the initial-value problem directly, and confirm that the answer is consistent with that obtained in part (a).

(c) Graph the solution.

c **18.** Use a CAS to derive Formula (23) of Section 10.1 by solving initial-value problem (21).

19. (a) It is currently accepted that the half-life of carbon-14 might vary ± 40 years from its nominal value of 5730 years. Does this variation make it possible that the Shroud of Turin dates to the time of Jesus of Nazareth? [See Example 4 of Section 10.3.]

(b) Review the subsection of Section 3.6 entitled Error Propagation in Applications, and then estimate the percentage error that results in the computed age of an artifact from an $r\%$ error in the half-life of carbon-14.

20. (a) Use Euler's Method with a step-size of $h = 0.1$ to approximate the solution of the initial-value problem

$$y' = 1 + 5t - y, \quad y(1) = 5$$

over the interval $[1, 2]$.

(b) Find the percentage error in the values computed.

21. (a) Confirm that e^x and e^{-x} are solutions of the second-order differential equation $y'' - y = 0$.

(b) Find some more solutions.

(c) Find a solution $y(x)$ such that $y(0) = 1$ and $y'(0) = 1$.

22. (a) Sketch the integral curve of $2yy' = 1$ that passes through the point $(0, 1)$ and the integral curve that passes through the point $(0, -1)$.

(b) Sketch the integral curve of $y' = -2xy^2$ that passes through the point $(0, 1)$.

23. Suppose that a herd of 19 deer is moved to a small island whose estimated carrying capacity is 95 deer, and assume that the population has a logistic growth model.

(a) Given that 1 year later the population is 25, how long will it take for the deer population to reach 80% of the island's carrying capacity?

(b) Find an initial-value problem whose solution gives the deer population as a function of time.

 **24.** If the block in Figure 10.3.4 is displaced y_0 units from its equilibrium position and given an initial velocity of v_0, rather than being released with an initial velocity of 0, then its position function $y(t)$ given in Equation (20) of Section 10.3 must satisfy the initial conditions $y(0) = y_0$ and $y'(0) = v_0$.

(a) Show that

$$y(t) = y_0 \cos\left(\sqrt{\frac{k}{m}}\,t\right) + v_0\sqrt{\frac{m}{k}}\sin\left(\sqrt{\frac{k}{m}}\,t\right)$$

(b) Suppose that a block with a mass of 1 kg stretches the spring 0.5 m in equilibrium. Use a graphing utility to graph the position function of the block if it is set in motion by pulling it down 1 m and imparting it an initial upward velocity of 0.25 m/s.

(c) What is the maximum displacement of the block from the equilibrium position?

25. A block attached to a vertical spring is displaced from its equilibrium position and released, thereby causing it to vibrate with amplitude $|y_0|$ and period T.

(a) Show that the velocity of the block has maximum magnitude $2\pi|y_0|/T$ and that the maximum occurs when the block is at its equilibrium position.

(b) Show that the acceleration of the block has maximum magnitude $4\pi^2|y_0|/T^2$ and that the maximum occurs when the block is at a top or bottom point of its motion.

26. Suppose that P dollars is invested at an annual interest rate of $r \times 100\%$. If the accumulated interest is credited to the account at the end of the year, then the interest is said to be *compounded annually*; if it is credited at the end of each 6-month period, then it is said to be *compounded semiannu-*

ally; and if it is credited at the end of each 3-month period, then it is said to be *compounded quarterly*. The more frequently the interest is compounded, the better it is for the investor since more of the interest is itself earning interest.

(a) Show that if interest is compounded n times a year at equally spaced intervals, then the value A of the investment after t years is

$$A = P\left(1 + \frac{r}{n}\right)^{nt}$$

(b) One can imagine interest to be compounded each day, each hour, each minute, and so forth. Carried to the limit one can conceive of interest compounded at each instant of time; this is called **continuous compounding**. Thus, from part (a), the value A of P dollars after t years when invested at an annual rate of $r \times 100\%$, compounded continuously, is

$$A = \lim_{n \to +\infty} P\left(1 + \frac{r}{n}\right)^{nt}$$

Use the fact that $\lim_{x \to 0}(1 + x)^{1/x} = e$ to prove that $A = Pe^{rt}$.

(c) Use the result in part (b) to show that money invested at continuous compound interest increases at a rate proportional to the amount present.

27. (a) If $1000 is invested at 8% per year compounded continuously (Exercise 26), what will the investment be worth after 5 years?

(b) If it is desired that an investment at 8% per year compounded continuously should have a value of $10,000 after 10 years, how much should be invested now?

(c) How long does it take for an investment at 8% per year compounded continuously to double in value?

EXPANDING THE CALCULUS HORIZON

For additional material relating to this chapter, visit the Anton Website at http://www.wiley.com/college/anton

11

INFINITE SERIES

Brook Taylor

*I*n this chapter we will be concerned with *infinite series*, which are sums that involve infinitely many terms. Infinite series play a fundamental role in both mathematics and science—they are used, for example, to approximate trigonometric functions and logarithms, to solve differential equations, to evaluate difficult integrals, to create new functions, and to construct mathematical models of physical laws. Since it is impossible to add up infinitely many numbers directly, our first goal will be to define exactly what we mean by the sum of an infinite series. However, unlike finite sums, it turns out that not all infinite series actually have a sum, so we will need to develop tools for determining which infinite series have sums and which do not. Once the basic ideas have been developed we will begin to apply our work; we will show how infinite series are used to evaluate such quantities as $\sin 17°$ and $\ln 5$, how they are used to create functions, and finally, how they are used to model physical laws.

11.1 SEQUENCES

In everyday language, the term "sequence" means a succession of things in a definite order—chronological order, size order, or logical order, for example. In mathematics, the term "sequence" is commonly used to denote a succession of numbers whose order is determined by a rule or a function. In this section, we will develop some of the basic ideas concerning sequences of numbers.

DEFINITION OF A SEQUENCE

Stated informally, an ***infinite sequence***, or more simply a ***sequence***, is an unending succession of numbers, called ***terms***. It is understood that the terms have a definite order; that is, there is a first term a_1, a second term a_2, a third term a_3, a fourth term a_4, and so forth. Such a sequence would typically be written as

$$a_1, a_2, a_3, a_4, \ldots$$

where the dots are used to indicate that the sequence continues indefinitely. Some specific examples are

$$1, 2, 3, 4, \ldots, \qquad 1, \tfrac{1}{2}, \tfrac{1}{3}, \tfrac{1}{4}, \ldots,$$
$$2, 4, 6, 8, \ldots, \qquad 1, -1, 1, -1, \ldots$$

Each of these sequences has a definite pattern that makes it easy to generate additional terms if we assume that those terms follow the same pattern as the displayed terms. However, such patterns can be deceiving, so it is better to have a rule or formula for generating the terms. One way of doing this is to look for a function that relates each term in the sequence to its term number. For example, in the sequence

$$2, 4, 6, 8, \ldots$$

each term is twice the term number; that is, the nth term in the sequence is given by the formula $2n$. We denote this by writing the sequence as

$$2, 4, 6, 8, \ldots, 2n, \ldots$$

We call the function $f(n) = 2n$ the *general term* of this sequence. Now, if we want to know a specific term in the sequence, we need only substitute its term number in the formula for the general term. For example, the 37th term in the sequence is $2 \cdot 37 = 74$.

Example 1

In each part, find the general term of the sequence.

(a) $\tfrac{1}{2}, \tfrac{2}{3}, \tfrac{3}{4}, \tfrac{4}{5}, \ldots$ 　　　　(b) $\tfrac{1}{2}, \tfrac{1}{4}, \tfrac{1}{8}, \tfrac{1}{16}, \ldots$

(c) $\tfrac{1}{2}, -\tfrac{2}{3}, \tfrac{3}{4}, -\tfrac{4}{5}, \ldots$ 　　　(d) $1, 3, 5, 7, \ldots$

Solution (a). In Table 11.1.1, the four known terms have been placed below their term numbers, from which we see that the numerator is the same as the term number and the denominator is one greater than the term number. This suggests that the nth term has numerator n and denominator $n + 1$, as indicated in the table. Thus, the sequence can be expressed as

$$\frac{1}{2}, \frac{2}{3}, \frac{3}{4}, \frac{4}{5}, \ldots, \frac{n}{n+1}, \ldots$$

Solution (b). In Table 11.1.2, the denominators of the four known terms have been expressed as powers of 2 and placed below their term numbers, from which we see that the exponent in the denominator is the same as the term number. This suggests that the denominator of the nth term is 2^n, as indicated in the table. Thus, the sequence can be expressed as

$$\frac{1}{2}, \frac{1}{4}, \frac{1}{8}, \frac{1}{16}, \ldots, \frac{1}{2^n}, \ldots$$

Table 11.1.1

TERM NUMBER	1	2	3	4	$\cdots$	n	$\cdots$
TERM	$\dfrac{1}{2}$	$\dfrac{2}{3}$	$\dfrac{3}{4}$	$\dfrac{4}{5}$	$\cdots$	$\dfrac{n}{n+1}$	$\cdots$

Table 11.1.2

TERM NUMBER	1	2	3	4	$\cdots$	n	$\cdots$
TERM	$\dfrac{1}{2}$	$\dfrac{1}{2^2}$	$\dfrac{1}{2^3}$	$\dfrac{1}{2^4}$	$\cdots$	$\dfrac{1}{2^n}$	$\cdots$

Solution (c). This sequence is identical to that in part (a), except for the alternating signs. Thus, the nth term in the sequence can be obtained by multiplying the nth term in part (a) by $(-1)^{n+1}$. This factor produces the correct alternating signs, since its successive values, starting with $n = 1$, are $1, -1, 1, -1, \ldots$. Thus, the sequence can be written as

$$\frac{1}{2}, -\frac{2}{3}, \frac{3}{4}, -\frac{4}{5}, \ldots, (-1)^{n+1}\frac{n}{n+1}, \ldots$$

Solution (d). In Table 11.1.3, the denominators of the four known terms have been placed below their term numbers, from which we see that each term is one less than twice its term number. This suggests that the nth term in the sequence is $2n - 1$, as indicated in the table. Thus, the sequence can be expressed as

$$1, 3, 5, 7, \ldots, 2n - 1, \ldots \qquad \blacktriangleleft$$

Table 11.1.3

TERM NUMBER	1	2	3	4	$\cdots$	n	$\cdots$
TERM	1	3	5	7	$\cdots$	$2n - 1$	$\cdots$

FOR THE READER. Consider the sequence whose general term is

$$f(n) = \tfrac{1}{3}(3 - 5n + 6n^2 - n^3)$$

Calculate the first three terms, and make a conjecture about the fourth term. Check your conjecture by calculating the fourth term. What message does this convey?

When the general term of a sequence

$$a_1, a_2, a_3, \ldots, a_n, \ldots \qquad (1)$$

is known, there is no need to write out the initial terms, and it is common to write only the general term enclosed in braces. Thus, (1) might be written as

$$\{a_n\}_{n=1}^{+\infty}$$

For example, here are the four sequences in Example 1 expressed in brace notation.

SEQUENCE	BRACE NOTATION
$\dfrac{1}{2}, \dfrac{2}{3}, \dfrac{3}{4}, \dfrac{4}{5}, \ldots, \dfrac{n}{n+1}, \ldots$	$\left\{\dfrac{n}{n+1}\right\}_{n=1}^{+\infty}$
$\dfrac{1}{2}, \dfrac{1}{4}, \dfrac{1}{8}, \dfrac{1}{16}, \ldots, \dfrac{1}{2^n}, \ldots$	$\left\{\dfrac{1}{2^n}\right\}_{n=1}^{+\infty}$
$\dfrac{1}{2}, -\dfrac{2}{3}, \dfrac{3}{4}, -\dfrac{4}{5}, \ldots, (-1)^{n+1}\dfrac{n}{n+1}, \ldots$	$\left\{(-1)^{n+1}\dfrac{n}{n+1}\right\}_{n=1}^{+\infty}$
$1, 3, 5, 7, \ldots, 2n - 1, \ldots$	$\{2n - 1\}_{n=1}^{+\infty}$

The letter n in (1) is called the ***index*** for the sequence. It is not essential to use n for the index; any letter not reserved for another purpose can be used. For example, we might view the general term of the sequence $a_1, a_2, a_3, \ldots$ to be the kth term, in which case we would denote this sequence as $\{a_k\}_{k=1}^{+\infty}$. Moreover, it is not essential to start the index at

1; sometimes it is more convenient to start it at 0 (or some other integer). For example, consider the sequence

$$1, \frac{1}{2}, \frac{1}{2^2}, \frac{1}{2^3}, \ldots$$

One way to write this sequence is

$$\left\{\frac{1}{2^{n-1}}\right\}_{n=1}^{+\infty}$$

However, the general term will be simpler if we think of the initial term in the sequence as the zeroth term, in which case we can write the sequence as

$$\left\{\frac{1}{2^n}\right\}_{n=0}^{+\infty}$$

REMARK.   In general discussions that involve sequences in which the specific terms and the starting point for the index are not important, it is common to write $\{a_n\}$ rather than $\{a_n\}_{n=1}^{+\infty}$ or $\{a_n\}_{n=0}^{+\infty}$. Moreover, we can distinguish between different sequences by using different letters for their general terms; thus, $\{a_n\}$, $\{b_n\}$, and $\{c_n\}$ denote three different sequences.

We began this section by describing a sequence as an unending succession of numbers. Although this conveys the general idea, it is not a satisfactory mathematical definition because it relies on the term "succession," which is itself an undefined term. To motivate a precise definition, consider the sequence

$$2, 4, 6, 8, \ldots, 2n, \ldots$$

If we denote the general term by $f(n) = 2n$, then we can write this sequence as

$$f(1), f(2), f(3), \ldots, f(n), \ldots$$

which is a "list" of values of the function

$$f(n) = 2n, \quad n = 1, 2, 3, \ldots$$

whose domain is the set of positive integers. This suggests the following definition.

11.1.1   DEFINITION.   A ***sequence*** is a function whose domain is a set of integers. Specifically, we will regard the expression $\{a_n\}_{n=1}^{+\infty}$ to be an alternative notation for the function $f(n) = a_n, n = 1, 2, 3, \ldots$.

GRAPHS OF SEQUENCES

Since sequences are functions, it makes sense to talk about the graph of a sequence. For example, the graph of the sequence $\{1/n\}_{n=1}^{+\infty}$ is the graph of the equation

$$y = \frac{1}{n}, \quad n = 1, 2, 3, \ldots$$

Because the right side of this equation is defined only for positive integer values of n, the graph consists of a succession of isolated points (Figure 11.1.1a). This is in distinction to

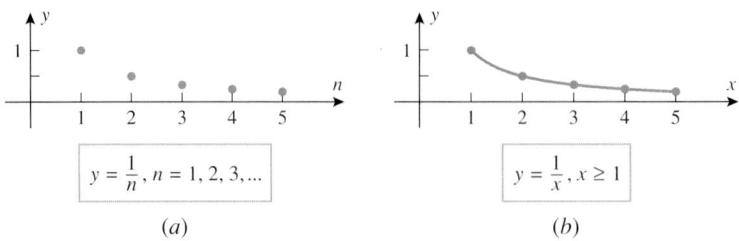

$$y = \frac{1}{n}, n = 1, 2, 3, \ldots$$

(a)

$$y = \frac{1}{x}, x \geq 1$$

(b)

Figure 11.1.1

the graph of

$$y = \frac{1}{x}, \quad x \geq 1$$

which is a continuous curve (Figure 11.1.1*b*).

LIMIT OF A SEQUENCE

Since sequences are functions, we can inquire about their limits. However, because a sequence $\{a_n\}$ is only defined for integer values of n, the only limit that makes sense is the limit of a_n as $n \to +\infty$. In Figure 11.1.2 we have shown the graphs of four sequences, each of which behaves differently as $n \to +\infty$:

- The terms in the sequence $\{n + 1\}$ increase without bound.
- The terms in the sequence $\{(-1)^{n+1}\}$ oscillate between -1 and 1.
- The terms in the sequence $\{n/(n + 1)\}$ increase toward a "limiting value" of 1.
- The terms in the sequence $\left\{1 + \left(-\frac{1}{2}\right)^n\right\}$ also tend toward a "limiting value" of 1, but do so in an oscillatory fashion.

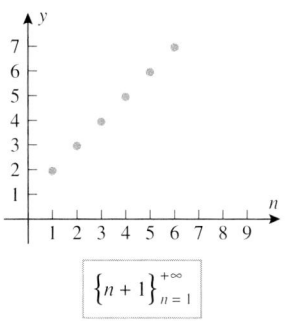
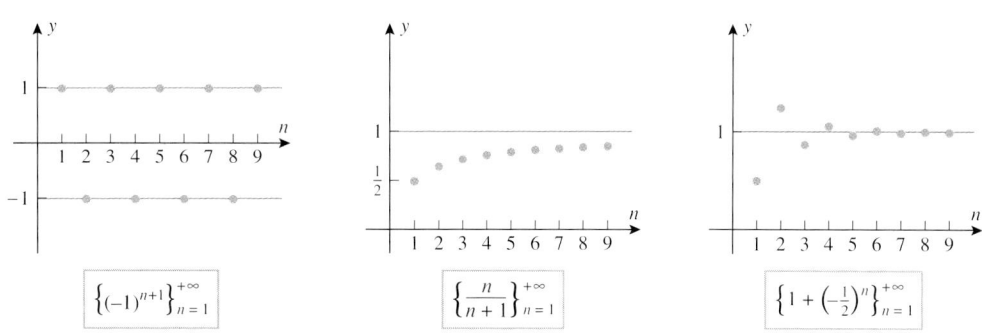

$$\left\{n + 1\right\}_{n=1}^{+\infty} \qquad \left\{(-1)^{n+1}\right\}_{n=1}^{+\infty} \qquad \left\{\frac{n}{n+1}\right\}_{n=1}^{+\infty} \qquad \left\{1 + \left(-\frac{1}{2}\right)^n\right\}_{n=1}^{+\infty}$$

Figure 11.1.2

Informally speaking, the limit of a sequence $\{a_n\}$ is intended to describe how a_n behaves as $n \to +\infty$. To be more specific, we will say that *a sequence $\{a_n\}$ approaches a limit L if the terms in the sequence eventually become arbitrarily close to L.* Geometrically, this means that for any positive number ϵ there is a point in the sequence after which all terms lie between the lines $y = L - \epsilon$ and $y = L + \epsilon$ (Figure 11.1.3).

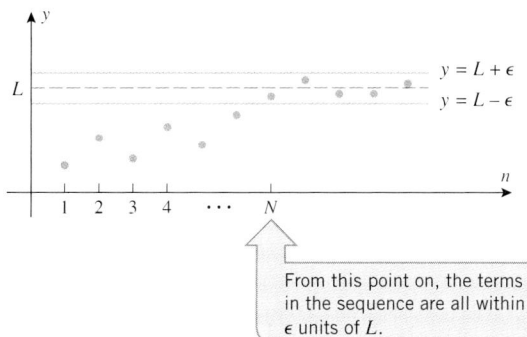

From this point on, the terms in the sequence are all within ϵ units of L.

Figure 11.1.3

The following definition makes these ideas precise.

11.1.2 DEFINITION. A sequence $\{a_n\}$ is said to **converge** to the **limit** L if given any $\epsilon > 0$, there is a positive integer N such that $|a_n - L| < \epsilon$ for $n \geq N$. In this case we write

$$\lim_{n \to +\infty} a_n = L$$

A sequence that does not converge to some finite limit is said to **diverge**.

Example 2

The first two sequences in Figure 11.1.2 diverge, and the second two converge to 1; that is,

$$\lim_{n \to +\infty} \frac{n}{n+1} = 1 \quad \text{and} \quad \lim_{n \to +\infty} \left(1 + \left(-\tfrac{1}{2}\right)^n\right) = 1 \quad \blacktriangleleft$$

The following theorem, which we state without proof, shows that the familiar properties of limits apply to sequences. This theorem ensures that the algebraic techniques used to find limits of the form $\lim_{x \to +\infty}$ can also be used for limits of the form $\lim_{n \to +\infty}$.

11.1.3 THEOREM. *Suppose that the sequences $\{a_n\}$ and $\{b_n\}$ converge to limits L_1 and L_2, respectively, and c is a constant. Then*

(a) $\quad \lim_{n \to +\infty} c = c$

(b) $\quad \lim_{n \to +\infty} ca_n = c \lim_{n \to +\infty} a_n = cL_1$

(c) $\quad \lim_{n \to +\infty} (a_n + b_n) = \lim_{n \to +\infty} a_n + \lim_{n \to +\infty} b_n = L_1 + L_2$

(d) $\quad \lim_{n \to +\infty} (a_n - b_n) = \lim_{n \to +\infty} a_n - \lim_{n \to +\infty} b_n = L_1 - L_2$

(e) $\quad \lim_{n \to +\infty} (a_n b_n) = \lim_{n \to +\infty} a_n \cdot \lim_{n \to +\infty} b_n = L_1 L_2$

(f) $\quad \lim_{n \to +\infty} \left(\frac{a_n}{b_n}\right) = \dfrac{\lim_{n \to +\infty} a_n}{\lim_{n \to +\infty} b_n} = \dfrac{L_1}{L_2} \quad (if\ L_2 \neq 0)$

Example 3

In each part, determine whether the sequence converges or diverges. If it converges, find the limit.

(a) $\left\{ \dfrac{n}{2n+1} \right\}_{n=1}^{+\infty}$ (b) $\left\{ (-1)^{n+1} \dfrac{n}{2n+1} \right\}_{n=1}^{+\infty}$

(c) $\left\{ (-1)^{n+1} \dfrac{1}{n} \right\}_{n=1}^{+\infty}$ (d) $\{8 - 2n\}_{n=1}^{+\infty}$

Solution (a). Dividing numerator and denominator by n yields

$$\lim_{n \to +\infty} \frac{n}{2n+1} = \lim_{n \to +\infty} \frac{1}{2 + 1/n} = \frac{\lim_{n \to +\infty} 1}{\lim_{n \to +\infty}(2 + 1/n)} = \frac{\lim_{n \to +\infty} 1}{\lim_{n \to +\infty} 2 + \lim_{n \to +\infty} 1/n}$$

$$= \frac{1}{2+0} = \frac{1}{2}$$

Thus, the sequence converges to $\frac{1}{2}$.

Solution (b). This sequence is the same as that in part (a), except for the factor of $(-1)^{n+1}$, which oscillates between $+1$ and -1. Thus, the terms in this sequence oscillate between positive and negative values, with the odd-numbered terms being identical to those in part (a) and the even-numbered terms being the negatives of those in part (a). Since the sequence in part (a) has a limit of $\frac{1}{2}$, it follows that the odd-numbered terms in this sequence approach $\frac{1}{2}$, while the even-numbered terms approach $-\frac{1}{2}$. Therefore, this sequence has no limit—it diverges.

Solution (c). Since $\lim_{n \to +\infty} 1/n = 0$, the product $(-1)^{n+1}(1/n)$ oscillates between positive and negative values, with the odd-numbered terms approaching 0 through positive values and the even-numbered terms approaching 0 through negative values. Thus,

$$\lim_{n \to +\infty} (-1)^{n+1} \frac{1}{n} = 0$$

so the sequence converges to 0.

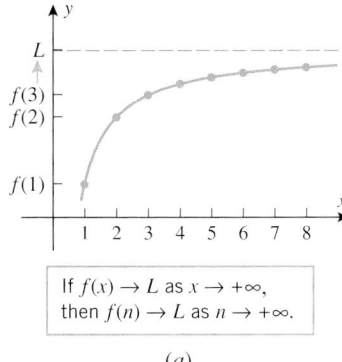

If $f(x) \to L$ as $x \to +\infty$, then $f(n) \to L$ as $n \to +\infty$.

(a)

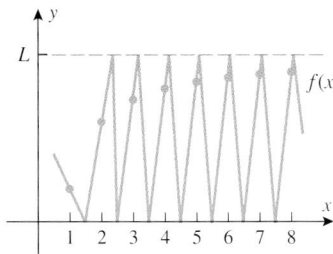

$f(n) \to L$ as $n \to +\infty$, but $f(x)$ diverges by oscillation as $x \to +\infty$.

(b)

Figure 11.1.4

Solution (d). $\lim\limits_{n \to +\infty} (8 - 2n) = -\infty$, so the sequence $\{8 - 2n\}_{n=1}^{+\infty}$ diverges. ◀

If the general term of a sequence is $f(n)$, and if we replace n by x, where x can vary over the entire interval $[1, +\infty)$, then the values of $f(n)$ can be viewed as "sample values" of $f(x)$ taken at the positive integers. Thus, if $f(x) \to L$ as $x \to +\infty$, then it must also be true that $f(n) \to L$ as $n \to +\infty$ (Figure 11.1.4a). However, the converse is not true; that is, one cannot infer that $f(x) \to L$ as $x \to +\infty$ from the fact that $f(n) \to L$ as $n \to +\infty$ (Figure 11.1.4b).

Example 4

In each part, determine whether the sequence converges, and if so, find its limit.

(a) $1, \dfrac{1}{2}, \dfrac{1}{2^2}, \dfrac{1}{2^3}, \ldots, \dfrac{1}{2^n}, \ldots$　　(b) $1, 2, 2^2, 2^3, \ldots, 2^n, \ldots$

Solution. Replacing n by x in the first sequence produces the power function $(1/2)^x$, and replacing n by x in the second sequence produces the power function 2^x. Now recall that if $0 < b < 1$, then $b^x \to 0$ as $x \to +\infty$, and if $b > 1$, then $b^x \to +\infty$ as $x \to +\infty$ (Figure 4.2.1). Thus,

$$\lim_{n \to +\infty} \frac{1}{2^n} = 0 \quad \text{and} \quad \lim_{n \to +\infty} 2^n = +\infty$$ ◀

Example 5

Find the limit of the sequence $\left\{\dfrac{n}{e^n}\right\}_{n=1}^{+\infty}$.

Solution. The expression n/e^n is an indeterminate form of type ∞/∞ as $n \to +\infty$, so L'Hôpital's rule is called for. However, we cannot apply this rule directly to n/e^n because the functions n and e^n are only defined at the positive integers, and hence are not differentiable functions. To circumvent this problem, we will replace n by x, and apply L'Hôpital's rule to the function x/e^x. This yields

$$\lim_{x \to +\infty} \frac{x}{e^x} = \lim_{x \to +\infty} \frac{1}{e^x} = 0$$

from which we can conclude that

$$\lim_{n \to +\infty} \frac{n}{e^n} = 0$$ ◀

Example 6

Show that $\lim\limits_{n \to +\infty} \sqrt[n]{n} = 1$.

Solution.

$$\lim_{n \to +\infty} \sqrt[n]{n} = \lim_{n \to +\infty} n^{1/n} = \lim_{n \to +\infty} e^{(1/n)\ln n} = e^0 = 1 \qquad \boxed{\begin{array}{l}\text{By L'Hôpital's rule applied} \\ \text{to } (1/x)\ln x\end{array}}$$ ◀

Sometimes the even-numbered and odd-numbered terms of a sequence behave sufficiently differently that it is desirable to investigate their convergence separately. The following theorem, whose proof is omitted, is helpful for that purpose.

11.1.4 THEOREM. *A sequence converges to a limit L if and only if the sequences of even-numbered terms and odd-numbered terms both converge to L.*

Example 7

The sequence

$$\frac{1}{2}, \frac{1}{3}, \frac{1}{2^2}, \frac{1}{3^2}, \frac{1}{2^3}, \frac{1}{3^3}, \ldots$$

converges to 0, since the even-numbered terms and the odd-numbered terms both converge to 0, and the sequence

$$1, \tfrac{1}{2}, 1, \tfrac{1}{3}, 1, \tfrac{1}{4}, \ldots$$

diverges, since the odd-numbered terms converge to 1 and the even-numbered terms converge to 0. ◀

THE SQUEEZING THEOREM FOR SEQUENCES

The following theorem, which we state without proof, is an adaptation of the Squeezing Theorem (2.5.2) to sequences. This theorem will be useful for finding limits of sequences that cannot be obtained directly.

11.1.5   THEOREM (*The Squeezing Theorem for Sequences*).   *Let $\{a_n\}$, $\{b_n\}$, and $\{c_n\}$ be sequences such that*

$$a_n \le b_n \le c_n \quad (\text{for all values of } n \text{ beyond some index } N)$$

If the sequences $\{a_n\}$ and $\{c_n\}$ have a common limit L as $n \to +\infty$, then $\{b_n\}$ also has the limit L as $n \to +\infty$.

Table 11.1.4

n	$\dfrac{n!}{n^n}$
1	1.0000000000
2	0.5000000000
3	0.2222222222
4	0.0937500000
5	0.0384000000
6	0.0154320988
7	0.0061198990
8	0.0024032593
9	0.0009366567
10	0.0003628800
11	0.0001399059
12	0.0000537232

Example 8

Use numerical evidence to make a conjecture about the limit of the sequence[*]

$$\left\{ \frac{n!}{n^n} \right\}_{n=1}^{+\infty}$$

and then confirm that your conjecture is correct.

Solution.   Table 11.1.4, which was obtained with a calculating utility, suggests that the limit of the sequence may be 0. To confirm this we need to examine the limit of

$$a_n = \frac{n!}{n^n}$$

as $n \to +\infty$. Although this is an indeterminate form of type ∞/∞, L'Hôpital's rule is not helpful because we have no definition of $x!$ for values of x that are not integers. However, let us write out some of the initial terms and the general term in the sequence:

$$a_1 = 1, \quad a_2 = \frac{1 \cdot 2}{2 \cdot 2}, \quad a_3 = \frac{1 \cdot 2 \cdot 3}{3 \cdot 3 \cdot 3}, \ldots, \quad a_n = \frac{1 \cdot 2 \cdot 3 \cdots n}{n \cdot n \cdot n \cdots n}, \ldots$$

We can rewrite the general term as

$$a_n = \frac{1}{n} \left(\frac{2 \cdot 3 \cdots n}{n \cdot n \cdots n} \right)$$

from which it is evident that

$$0 \le a_n \le \frac{1}{n}$$

However, the two outside expressions have a limit of 0 as $n \to +\infty$; thus, the Squeezing Theorem for Sequences implies that $a_n \to 0$ as $n \to +\infty$, which confirms our conjecture. ◀

The following theorem is often useful for finding the limit of a sequence with both positive and negative terms—it states that if the sequence $\{|a_n|\}$ that is obtained by taking

---

[*]Recall that if n is a positive integer, then the symbol $n!$ (read "n factorial") denotes the product of the first n positive integers; that is,

$$n! = 1 \cdot 2 \cdot 3 \cdots n \quad \text{or equivalently,} \quad n! = n(n-1)(n-2)\cdots 1$$

Moreover, it is agreed by convention that $0! = 1$.

the absolute value of each term in the sequence $\{a_n\}$ converges to 0, then $\{a_n\}$ also converges to 0.

11.1.6 THEOREM. *If* $\lim\limits_{n \to +\infty} |a_n| = 0,$ *then* $\lim\limits_{n \to +\infty} a_n = 0.$

Proof. Depending on the sign of a_n, either $a_n = |a_n|$ or $a_n = -|a_n|$. Thus, in all cases we have

$$-|a_n| \le a_n \le |a_n|$$

However, the limit of the two outside terms is 0, and hence the limit of a_n is 0 by the Squeezing Theorem for Sequences. ∎

Example 9

Consider the sequence

$$1, \ -\frac{1}{2}, \ \frac{1}{2^2}, \ -\frac{1}{2^3}, \ \dots, \ (-1)^n \frac{1}{2^n}, \dots$$

If we take the absolute value of each term, we obtain the sequence

$$1, \ \frac{1}{2}, \ \frac{1}{2^2}, \ \frac{1}{2^3}, \ \dots, \ \frac{1}{2^n}, \dots$$

which, as shown in Example 4, converges to 0. Thus, from Theorem 11.1.6 we have

$$\lim_{n \to +\infty} \left[(-1)^n \frac{1}{2^n} \right] = 0 \qquad \blacktriangleleft$$

SEQUENCES DEFINED RECURSIVELY

Some sequences do not arise from a formula for the general term, but rather from a formula or set of formulas that specify how to generate each term in the sequence from terms that precede it; such sequences are said to be defined *recursively*, and the defining formulas are called *recursion formulas*. A good example is the mechanic's rule for approximating square roots. In Formula (1) of the Introduction, we stated that the recursion formulas

$$y_0 = 1, \quad y_{n+1} = \frac{1}{2}\left(y_n + \frac{2}{y_n} \right), \qquad n = 0, 1, 2, \dots \tag{2}$$

generate a sequence $\{y_n\}$ that converges to $\sqrt{2}$, and in Table 1 of that section we used these recursion formulas to generate some of the terms in the sequence.

It would take us too far afield to investigate the convergence of sequences defined recursively, but we will conclude this section with a useful technique that can sometimes be used to compute limits of such sequences.

Example 10

Assuming that the sequence generated by (2) converges, show that the limit is $\sqrt{2}$.

Solution. Assume that $y_n \to L$, where L is to be determined. Since $n + 1 \to +\infty$ as $n \to +\infty$, it is also true that $y_{n+1} \to L$ as $n \to +\infty$; thus, if we take the limit of the expression

$$y_{n+1} = \frac{1}{2}\left(y_n + \frac{2}{y_n} \right)$$

as $n \to +\infty$, we obtain

$$L = \frac{1}{2}\left(L + \frac{2}{L} \right)$$

which can be rewritten as $L^2 = 2$. The negative solution of this equation is extraneous because $y_n > 0$ for all n, so $L = \sqrt{2}$. $\blacktriangleleft$

EXERCISE SET 11.1 ~ Graphing Calculator   [c] CAS

1. In each part, find a formula for the general term of the sequence, starting with $n = 1$.

   (a) $1, \dfrac{1}{3}, \dfrac{1}{9}, \dfrac{1}{27}, \ldots$ (b) $1, -\dfrac{1}{3}, \dfrac{1}{9}, -\dfrac{1}{27}, \ldots$

   (c) $\dfrac{1}{2}, \dfrac{3}{4}, \dfrac{5}{6}, \dfrac{7}{8}, \ldots$ (d) $\dfrac{1}{\sqrt{\pi}}, \dfrac{4}{\sqrt[3]{\pi}}, \dfrac{9}{\sqrt[4]{\pi}}, \dfrac{16}{\sqrt[5]{\pi}}, \ldots$

2. In each part, find two formulas for the general term of the sequence, one starting with $n = 1$ and the other with $n = 0$.

   (a) $1, -r, r^2, -r^3, \ldots$ (b) $r, -r^2, r^3, -r^4, \ldots$

3. (a) Write out the first four terms of the sequence $\{1 + (-1)^n\}$, starting with $n = 0$.

   (b) Write out the first four terms of the sequence $\{\cos n\pi\}$, starting with $n = 0$.

   (c) Use the results in parts (a) and (b) to express the general term of the sequence $4, 0, 4, 0, \ldots$ in two different ways, starting with $n = 0$.

4. In each part, find a formula for the general term using factorials and starting with $n = 1$.

   (a) $1 \cdot 2, \; 1 \cdot 2 \cdot 3 \cdot 4, \; 1 \cdot 2 \cdot 3 \cdot 4 \cdot 5 \cdot 6,$
   $1 \cdot 2 \cdot 3 \cdot 4 \cdot 5 \cdot 6 \cdot 7 \cdot 8, \ldots$

   (b) $1, \; 1 \cdot 2 \cdot 3, \; 1 \cdot 2 \cdot 3 \cdot 4 \cdot 5, \; 1 \cdot 2 \cdot 3 \cdot 4 \cdot 5 \cdot 6 \cdot 7, \ldots$

> In Exercises 5–22, write out the first five terms of the sequence, determine whether the sequence converges, and if so find its limit.

5. $\left\{\dfrac{n}{n+2}\right\}_{n=1}^{+\infty}$ 6. $\left\{\dfrac{n^2}{2n+1}\right\}_{n=1}^{+\infty}$ 7. $\{2\}_{n=1}^{+\infty}$

8. $\left\{\ln\left(\dfrac{1}{n}\right)\right\}_{n=1}^{+\infty}$ 9. $\left\{\dfrac{\ln n}{n}\right\}_{n=1}^{+\infty}$ 10. $\left\{n \sin \dfrac{\pi}{n}\right\}_{n=1}^{+\infty}$

11. $\{1 + (-1)^n\}_{n=1}^{+\infty}$ 12. $\left\{\dfrac{(-1)^{n+1}}{n^2}\right\}_{n=1}^{+\infty}$

13. $\left\{(-1)^n \dfrac{2n^3}{n^3+1}\right\}_{n=1}^{+\infty}$ 14. $\left\{\dfrac{n}{2^n}\right\}_{n=1}^{+\infty}$

15. $\left\{\dfrac{(n+1)(n+2)}{2n^2}\right\}_{n=1}^{+\infty}$ 16. $\left\{\dfrac{\pi^n}{4^n}\right\}_{n=1}^{+\infty}$

17. $\left\{\cos \dfrac{3}{n}\right\}_{n=1}^{+\infty}$ 18. $\left\{\cos \dfrac{\pi n}{2}\right\}_{n=1}^{+\infty}$

19. $\{n^2 e^{-n}\}_{n=1}^{+\infty}$ 20. $\{\sqrt{n^2+3n}-n\}_{n=1}^{+\infty}$

21. $\left\{\left(\dfrac{n+3}{n+1}\right)^n\right\}_{n=1}^{+\infty}$ 22. $\left\{\left(1 - \dfrac{2}{n}\right)^n\right\}_{n=1}^{+\infty}$

> In Exercises 23–30, find the general term of the sequence, starting with $n = 1$, determine whether the sequence converges, and if so find its limit.

23. $\dfrac{1}{2}, \dfrac{3}{4}, \dfrac{5}{6}, \dfrac{7}{8}, \ldots$ 24. $0, \dfrac{1}{2^2}, \dfrac{2}{3^2}, \dfrac{3}{4^2}, \ldots$

25. $\dfrac{1}{3}, \dfrac{1}{9}, \dfrac{1}{27}, \dfrac{1}{81}, \ldots$ 26. $-1, 2, -3, 4, -5, \ldots$

27. $\left(1 - \dfrac{1}{2}\right), \left(\dfrac{1}{2} - \dfrac{1}{3}\right), \left(\dfrac{1}{3} - \dfrac{1}{4}\right), \left(\dfrac{1}{4} - \dfrac{1}{5}\right), \ldots$

28. $3, \dfrac{3}{2}, \dfrac{3}{2^2}, \dfrac{3}{2^3}, \ldots$

29. $(\sqrt{2} - \sqrt{3}), (\sqrt{3} - \sqrt{4}), (\sqrt{4} - \sqrt{5}), \ldots$

30. $\dfrac{1}{3^5}, -\dfrac{1}{3^6}, \dfrac{1}{3^7}, -\dfrac{1}{3^8}, \ldots$

[c] 31. Read your CAS documentation to determine how to find limits approaching $+\infty$, and use the CAS to check the limits you calculated in Exercises 5–30.

[c] 32. (a) Use numerical evidence to make a conjecture about the limit of the sequence $\{\sqrt[n]{n^3}\}_{n=2}^{+\infty}$.

   (b) Use a CAS to confirm your conjecture.

33. (a) Starting with $n = 1$, write out the first six terms of the sequence $\{a_n\}$, where

$$a_n = \begin{cases} 1, & \text{if } n \text{ is odd} \\ n, & \text{if } n \text{ is even} \end{cases}$$

   (b) Starting with $n = 1$, and considering the even and odd terms separately, find a formula for the general term of the sequence

$$1, \dfrac{1}{2^2}, 3, \dfrac{1}{2^4}, 5, \dfrac{1}{2^6}, \ldots$$

   (c) Starting with $n = 1$, and considering the even and odd terms separately, find a formula for the general term of the sequence

$$1, \dfrac{1}{3}, \dfrac{1}{3}, \dfrac{1}{5}, \dfrac{1}{5}, \dfrac{1}{7}, \dfrac{1}{7}, \dfrac{1}{9}, \dfrac{1}{9}, \ldots$$

   (d) Determine whether the sequences in parts (a), (b), and (c) converge. For those that do, find the limit.

34. For what positive values of b does the sequence $b, 0, b^2, 0, b^3, 0, b^4, 0, \ldots$ converge? Justify your answer.

35. In the discussion preceding Exercise 8 of the Introduction, we implied that the sequence defined recursively by

$$y_0 = 1, \quad y_{n+1} = \dfrac{1}{2}\left(y_n + \dfrac{p}{y_n}\right)$$

converges to $\sqrt{p}$. Assuming that this sequence converges, use the method of Example 10 to confirm that this is so.

36. Consider the sequence

$$a_1 = \sqrt{6}$$

$$a_2 = \sqrt{6 + \sqrt{6}}$$

$$a_3 = \sqrt{6 + \sqrt{6 + \sqrt{6}}}$$

$$a_4 = \sqrt{6 + \sqrt{6 + \sqrt{6 + \sqrt{6}}}}$$

$$\vdots$$

(a) Find a recursion formula for a_{n+1}.

(b) Assuming that the sequence converges, use the method of Example 10 to find the limit.

37. Consider the sequence $\{a_n\}_{n=1}^{+\infty}$, where

$$a_n = \frac{1}{n^2} + \frac{2}{n^2} + \cdots + \frac{n}{n^2}$$

(a) Find a_1, a_2, a_3, and a_4.

(b) Use numerical evidence to make a conjecture about the limit of the sequence.

(c) Confirm your conjecture by expressing a_n in closed form and calculating the limit.

38. Follow the directions in Exercise 37 with

$$a_n = \frac{1^2}{n^3} + \frac{2^2}{n^3} + \cdots + \frac{n^2}{n^3}$$

In Exercises 39 and 40, use numerical evidence to make a conjecture about the limit of the sequence, and then use the Squeezing Theorem for Sequences (Theorem 11.1.5) to confirm that your conjecture is correct.

39. $\lim\limits_{n \to +\infty} \dfrac{\sin^2 n}{n}$ **40.** $\lim\limits_{n \to +\infty} \left(\dfrac{1+n}{2n}\right)^n$

41. (a) A bored student enters the number 0.5 in a calculator display and then repeatedly computes the square of the number in the display. Taking $a_0 = 0.5$, find a formula for the general term of the sequence $\{a_n\}$ of numbers that appear in the display.

(b) Try this with a calculator and make a conjecture about the limit of a_n.

(c) Confirm your conjecture by finding the limit of a_n.

(d) For what values of a_0 will this procedure produce a convergent sequence?

42. Let

$$f(x) = \begin{cases} 2x, & 0 \le x < 0.5 \\ 2x - 1, & 0.5 \le x < 1 \end{cases}$$

Does the sequence $f(0.2), f(f(0.2)), f(f(f(0.2))), \ldots$ converge? Justify your reasoning.

43. (a) Use a graphing utility to generate the graph of the equation $y = (2^x + 3^x)^{1/x}$, and then use the graph to make

a conjecture about the limit of the sequence

$$\{(2^n + 3^n)^{1/n}\}_{n=1}^{+\infty}$$

(b) Confirm your conjecture by calculating the limit.

44. Consider the sequence $\{a_n\}_{n=1}^{+\infty}$ whose nth term is

$$a_n = \frac{1}{n} \sum_{k=1}^{n} \frac{1}{1 + (k/n)}$$

Show that $\lim\limits_{n \to +\infty} a_n = \ln 2$ by interpreting a_n as the Riemann sum of a definite integral.

45. Let a_n be the average value of $f(x) = 1/x$ over the interval $[1, n]$. Determine whether the sequence $\{a_n\}$ converges, and if so find its limit.

46. The sequence whose terms are $1, 1, 2, 3, 5, 8, 13, 21, \ldots$ is called the **Fibonacci sequence** in honor of Leonardo ("Fibonacci") da Pisa (c. 1170–1250). This sequence has the property that after starting with two 1's, each term is the sum of the preceding two.

(a) Denoting the sequence by $\{a_n\}$ and starting with $a_1 = 1$ and $a_2 = 1$, show that

$$\frac{a_{n+2}}{a_{n+1}} = 1 + \frac{a_n}{a_{n+1}} \quad \text{if } n \ge 1$$

(b) Give a reasonable informal argument to show that if the sequence $\{a_{n+1}/a_n\}$ converges to some limit L, then the sequence $\{a_{n+2}/a_{n+1}\}$ must also converge to L.

(c) Assuming that the sequence $\{a_{n+1}/a_n\}$ converges, show that its limit is $(1 + \sqrt{5})/2$.

47. If we accept the fact that the sequence $\{1/n\}_{n=1}^{+\infty}$ converges to the limit $L = 0$, then according to Definition 11.1.2, for every $\epsilon > 0$, there exists an integer N such that $|a_n - L| = |(1/n) - 0| < \epsilon$ when $n \ge N$. In each part, find the smallest possible value of N for the given value of ϵ.

(a) $\epsilon = 0.5$ (b) $\epsilon = 0.1$ (c) $\epsilon = 0.001$

48. If we accept the fact that the sequence

$$\left\{ \frac{n}{n+1} \right\}_{n=1}^{+\infty}$$

converges to the limit $L = 1$, then according to Definition 11.1.2, for every $\epsilon > 0$ there exists an integer N such that

$$|a_n - L| = \left| \frac{n}{n+1} - 1 \right| < \epsilon$$

when $n \ge N$. In each part, find the smallest value of N for the given value of ϵ.

(a) $\epsilon = 0.25$ (b) $\epsilon = 0.1$ (c) $\epsilon = 0.001$

49. Use Definition 11.1.2 to prove that

(a) the sequence $\{1/n\}_{n=1}^{+\infty}$ converges to 0

(b) the sequence $\left\{ \dfrac{n}{n+1} \right\}_{n=1}^{+\infty}$ converges to 1.

50. Find $\lim\limits_{n \to +\infty} r^n$, where r is a real number. [*Hint:* Consider the cases $|r| < 1$, $|r| > 1$, $r = 1$, and $r = -1$ separately.]

11.2 MONOTONE SEQUENCES

There are many situations in which it is important to know whether a sequence converges, but the limit itself is not relevant to the problem at hand. In this section we will study several techniques that can be used to determine whether a sequence converges.

TERMINOLOGY

We begin with some terminology.

> **11.2.1   DEFINITION.**   A sequence $\{a_n\}_{n=1}^{+\infty}$ is called
> **strictly increasing** if $a_1 < a_2 < a_3 < \cdots < a_n < \cdots$
> **increasing** if $a_1 \leq a_2 \leq a_3 \leq \cdots \leq a_n \leq \cdots$
> **strictly decreasing** if $a_1 > a_2 > a_3 > \cdots > a_n > \cdots$
> **decreasing** if $a_1 \geq a_2 \geq a_3 \geq \cdots \geq a_n \geq \cdots$

In words, a sequence is strictly increasing if each term is larger than its predecessor, increasing if each term is the same as or larger than its predecessor, strictly decreasing if each term is smaller than its predecessor, and decreasing if each term is the same as or smaller than its predecessor. It follows that every strictly increasing sequence is increasing (but not conversely), and every strictly decreasing sequence is decreasing (but not conversely). A sequence that is either strictly increasing or strictly decreasing is called **strictly monotone**, and a sequence that is either increasing or decreasing is called **monotone**.

Example 1

SEQUENCE	DESCRIPTION
$\dfrac{1}{2}, \dfrac{2}{3}, \dfrac{3}{4}, \ldots, \dfrac{n}{n+1}, \ldots$	Strictly increasing
$1, \dfrac{1}{2}, \dfrac{1}{3}, \ldots, \dfrac{1}{n}, \ldots$	Strictly decreasing
$1, 1, 2, 2, 3, 3, \ldots$	Increasing; not strictly increasing
$1, 1, \dfrac{1}{2}, \dfrac{1}{2}, \dfrac{1}{3}, \dfrac{1}{3}, \ldots$	Decreasing; not strictly decreasing
$1, -\dfrac{1}{2}, \dfrac{1}{3}, -\dfrac{1}{4}, \ldots, (-1)^{n+1}\dfrac{1}{n}, \ldots$	Neither increasing nor decreasing

The first and second sequences are strictly monotone, and the third and fourth sequences are monotone but not strictly monotone. The fifth sequence is not monotone.   ◀

FOR THE READER.   Can a sequence be both increasing and decreasing? Explain.

TESTING FOR MONOTONICITY

In order for a sequence to be strictly increasing, *all* pairs of successive terms, a_n and a_{n+1}, must satisfy $a_n < a_{n+1}$ or, equivalently, $a_{n+1} - a_n > 0$. More generally, monotone sequences can be classified as follows:

DIFFERENCE BETWEEN SUCCESSIVE TERMS	CLASSIFICATION
$a_{n+1} - a_n > 0$	Strictly increasing
$a_{n+1} - a_n < 0$	Strictly decreasing
$a_{n+1} - a_n \geq 0$	Increasing
$a_{n+1} - a_n \leq 0$	Decreasing

Frequently, one can *guess* whether a sequence is monotone or strictly monotone by writing out some of the initial terms. However, to be certain that the guess is correct, one must give a precise mathematical argument. The following example illustrates one method for doing this.

Example 2

Show that
$$\frac{1}{2}, \frac{2}{3}, \frac{3}{4}, \dots, \frac{n}{n+1}, \dots$$
is a strictly increasing sequence.

Solution. The pattern of the initial terms suggests that the sequence is strictly increasing. To prove that this is so, let
$$a_n = \frac{n}{n+1}$$
We can obtain a_{n+1} by replacing n by $n+1$ in this formula. This yields
$$a_{n+1} = \frac{n+1}{(n+1)+1} = \frac{n+1}{n+2}$$
Thus, for $n \geq 1$
$$a_{n+1} - a_n = \frac{n+1}{n+2} - \frac{n}{n+1} = \frac{n^2 + 2n + 1 - n^2 - 2n}{(n+1)(n+2)} = \frac{1}{(n+1)(n+2)} > 0$$
which proves that the sequence is strictly increasing. ◀

If a_n and a_{n+1} are any successive terms in a strictly increasing sequence, then $a_n < a_{n+1}$. If the terms in the sequence are all positive, then we can divide both sides of this inequality by a_n to obtain $1 < a_{n+1}/a_n$ or, equivalently, $a_{n+1}/a_n > 1$. More generally, monotone sequences with *positive* terms can be classified as follows:

RATIO OF SUCCESSIVE TERMS	CONCLUSION
$a_{n+1}/a_n > 1$	Strictly increasing
$a_{n+1}/a_n < 1$	Strictly decreasing
$a_{n+1}/a_n \geq 1$	Increasing
$a_{n+1}/a_n \leq 1$	Decreasing

Example 3

Show that the sequence in Example 2 is strictly increasing by examining the ratio of successive terms.

Solution. As shown in the solution of Example 2,
$$a_n = \frac{n}{n+1} \quad \text{and} \quad a_{n+1} = \frac{n+1}{n+2}$$
Thus,
$$\frac{a_{n+1}}{a_n} = \frac{(n+1)/(n+2)}{n/(n+1)} = \frac{n+1}{n+2} \cdot \frac{n+1}{n} = \frac{n^2 + 2n + 1}{n^2 + 2n} \tag{1}$$
Since the numerator in (1) exceeds the denominator, it follows that $a_{n+1}/a_n > 1$ for $n \geq 1$. This proves that the sequence is strictly increasing. ◀

The following example illustrates still a third technique for determining whether a sequence is strictly monotone.

Example 4

In Examples 2 and 3 we proved that the sequence

$$\frac{1}{2}, \frac{2}{3}, \frac{3}{4}, \ldots, \frac{n}{n+1}, \ldots$$

is strictly increasing by considering the difference and ratio of successive terms. Alternatively, we can proceed as follows. Let

$$f(x) = \frac{x}{x+1}$$

so that the nth term in the given sequence is $a_n = f(n)$. The function f is increasing for $x \geq 1$ since

$$f'(x) = \frac{(x+1)(1) - x(1)}{(x+1)^2} = \frac{1}{(x+1)^2} > 0$$

Thus,

$$a_n = f(n) < f(n+1) = a_{n+1}$$

which proves that the given sequence is strictly increasing. ◀

In general, if $f(n) = a_n$ is the nth term of a sequence, and if f is differentiable for $x \geq 1$, then we have the following results:

DERIVATIVE OF f FOR $x \geq 1$	CONCLUSION FOR THE SEQUENCE WITH $a_n = f(n)$
$f'(x) > 0$	Strictly increasing
$f'(x) < 0$	Strictly decreasing
$f'(x) \geq 0$	Increasing
$f'(x) \leq 0$	Decreasing

PROPERTIES THAT HOLD EVENTUALLY

Sometimes a sequence will behave erratically at first and then settle down into a definite pattern. For example, the sequence

$$9, -8, -17, 12, 1, 2, 3, 4, \ldots \tag{2}$$

is strictly increasing from the fifth term on, but the sequence as a whole cannot be classified as strictly increasing because of the erratic behavior of the first four terms. To describe such sequences, we introduce the following terminology.

> **11.2.2 DEFINITION.** If discarding finitely many terms from the beginning of a sequence produces a sequence with a certain property, then the original sequence is said to have that property **eventually**.

For example, although we cannot say that sequence (2) is strictly increasing, we can say that it is eventually strictly increasing.

Example 5

Show that the sequence $\left\{ \dfrac{10^n}{n!} \right\}_{n=1}^{+\infty}$ is eventually strictly decreasing.

Solution. We have

$$a_n = \frac{10^n}{n!} \quad \text{and} \quad a_{n+1} = \frac{10^{n+1}}{(n+1)!}$$

so

$$\frac{a_{n+1}}{a_n} = \frac{10^{n+1}/(n+1)!}{10^n/n!} = \frac{10^{n+1}n!}{10^n(n+1)!} = 10\frac{n!}{(n+1)n!} = \frac{10}{n+1} \tag{3}$$

From (3), $a_{n+1}/a_n < 1$ for all $n \geq 10$, so the sequence is eventually strictly decreasing. ◀

AN INTUITIVE VIEW OF CONVERGENCE

Informally stated, the convergence or divergence of a sequence does not depend on the behavior of its *initial terms*, but rather on how the terms behave *eventually*. For example, the sequence

$$3,\ -9,\ -13,\ 17,\ 1,\ \frac{1}{2},\ \frac{1}{3},\ \frac{1}{4}, \ldots$$

eventually behaves like the sequence

$$1,\ \frac{1}{2},\ \frac{1}{3}, \ldots, \frac{1}{n}, \ldots$$

and hence has a limit of 0.

CONVERGENCE OF MONOTONE SEQUENCES

The following two theorems, whose proofs are discussed at the end of this section, show that a monotone sequence either converges or becomes infinite—divergence by oscillation cannot occur.

11.2.3 THEOREM. *If a sequence $\{a_n\}$ is eventually increasing, then there are two possibilities*:

(a) *There is a constant M, called an **upper bound** for the sequence, such that $a_n \leq M$ for all n, in which case the sequence converges to a limit L satisfying $L \leq M$.*

(b) *No upper bound exists, in which case $\displaystyle\lim_{n \to +\infty} a_n = +\infty$.*

11.2.4 THEOREM. *If a sequence $\{a_n\}$ is eventually decreasing, then there are two possibilities*:

(a) *There is a constant M, called a **lower bound** for the sequence, such that $a_n \geq M$ for all n, in which case the sequence converges to a limit L satisfying $L \geq M$.*

(b) *No lower bound exists, in which case $\displaystyle\lim_{n \to +\infty} a_n = -\infty$.*

Note that these results do not give a method for obtaining limits; they tell us only whether a limit exists.

Example 6

Show that the sequence $\left\{ \dfrac{10^n}{n!} \right\}_{n=1}^{+\infty}$ converges and find its limit.

Solution. We showed in Example 5 that the sequence is eventually strictly decreasing. Since all terms in the sequence are positive, it is bounded below by $M = 0$, and hence Theorem 11.2.4 guarantees that it converges to a nonnegative limit L. However, the limit is not evident directly from the formula $10^n/n!$ for the nth term, so we will need some ingenuity to obtain it.

Recall from Formula (3) of Example 5 that successive terms in the given sequence are related by the recursion formula

$$a_{n+1} = \frac{10}{n+1}a_n \tag{4}$$

where $a_n = 10^n/n!$. We will take the limit as $n \to +\infty$ of both sides of (4) and use the fact that

$$\lim_{n \to +\infty} a_{n+1} = \lim_{n \to +\infty} a_n = L$$

We obtain

$$L = \lim_{n \to +\infty} a_{n+1} = \lim_{n \to +\infty} \left(\frac{10}{n+1} a_n \right) = \lim_{n \to +\infty} \frac{10}{n+1} \lim_{n \to +\infty} a_n = 0 \cdot L = 0$$

so that

$$L = \lim_{n \to +\infty} \frac{10^n}{n!} = 0 \qquad \blacktriangleleft$$

REMARK.   In the exercises we will show that the technique illustrated in this example can be adapted to obtain the limit

$$\lim_{n \to +\infty} \frac{x^n}{n!} = 0 \tag{5}$$

for any real value of x (Exercise 26). This result, which shows that $n!$ eventually increases more rapidly than any positive integer power of x, will be useful in our later work.

THE COMPLETENESS AXIOM

In this text we have accepted the familiar properties of real numbers without proof, and indeed, we have not even attempted to define the term *real number*. Although this is sufficient for many purposes, it was recognized by the late nineteenth century that the study of limits and functions in calculus requires a precise axiomatic formulation of the real numbers analogous to the axiomatic development of Euclidean geometry. Although we will not attempt to pursue this development, we will need to discuss one of the axioms about real numbers in order to prove Theorems 11.2.3 and 11.2.4. But first we will introduce some terminology.

If S is a nonempty set of real numbers, then we call u an ***upper bound*** for S if u is greater than or equal to every number in S, and we call l a ***lower bound*** for S if l is smaller than or equal to every number in S. For example, if S is the set of numbers in the interval $(1, 3)$, then $u = 4, 10,$ and 100 are upper bounds for S and $l = -10, 0,$ and $\frac{1}{2}$ are lower bounds for S. Observe also that $u = 3$ is the smallest of all upper bounds and $l = 1$ is the largest of all lower bounds. The existence of a smallest upper bound and a greatest lower bound for S is not accidental; it is a consequence of the following axiom.

11.2.5   AXIOM (***The Completeness Axiom***).   *If a nonempty set S of real numbers has an upper bound, then it has a smallest upper bound (called the **least upper bound**), and if a nonempty set S of real numbers has a lower bound, then it has a largest lower bound (called the **greatest lower bound**).*

Proof of Theorem 11.2.3.

(*a*)   Assume there exists a number M such that $a_n \le M$ for $n = 1, 2, \ldots$. Then M is an upper bound for the set of terms in the sequence. By the Completeness Axiom there is a least upper bound for the terms, call it L. Now let ϵ be any positive number. Since L is the least upper bound for the terms, $L - \epsilon$ is not an upper bound for the terms, which means that there is at least one term a_N such that

$$a_N > L - \epsilon$$

Moreover, since $\{a_n\}$ is an increasing sequence, we must have

$$a_n \ge a_N > L - \epsilon \tag{6}$$

when $n \geq N$. But a_n cannot exceed L since L is an upper bound for the terms. This observation together with (6) tells us that $L \geq a_n > L - \epsilon$ for $n \geq N$, so all terms from the Nth on are within ϵ units of L. This is exactly the requirement to have

$$\lim_{n \to +\infty} a_n = L$$

Finally, $L \leq M$ since M is an upper bound for the terms and L is the least upper bound. This proves part (a).

(b)   If there is no number M such that $a_n \leq M$ for $n = 1, 2, \ldots$, then no matter how large we choose M, there is a term a_n such that

$$a_N > M$$

and, since the sequence is increasing,

$$a_n \geq a_N > M$$

when $n \geq N$. Thus, the terms in the sequence become arbitrarily large as n increases. That is,

$$\lim_{n \to +\infty} a_n = +\infty$$

∎

The proof of Theorem 11.2.4 will be omitted since it is similar to that of 11.2.3.

EXERCISE SET 11.2

In Exercises 1–6, use $a_{n+1} - a_n$ to show that the given sequence $\{a_n\}$ is strictly increasing or strictly decreasing.

1. $\left\{ \dfrac{1}{n} \right\}_{n=1}^{+\infty}$ **2.** $\left\{ 1 - \dfrac{1}{n} \right\}_{n=1}^{+\infty}$ **3.** $\left\{ \dfrac{n}{2n+1} \right\}_{n=1}^{+\infty}$

4. $\left\{ \dfrac{n}{4n-1} \right\}_{n=1}^{+\infty}$ **5.** $\{n - 2^n\}_{n=1}^{+\infty}$ **6.** $\{n - n^2\}_{n=1}^{+\infty}$

In Exercises 7–12, use a_{n+1}/a_n to show that the given sequence $\{a_n\}$ is strictly increasing or strictly decreasing.

7. $\left\{ \dfrac{n}{2n+1} \right\}_{n=1}^{+\infty}$ **8.** $\left\{ \dfrac{2^n}{1+2^n} \right\}_{n=1}^{+\infty}$ **9.** $\{ne^{-n}\}_{n=1}^{+\infty}$

10. $\left\{ \dfrac{10^n}{(2n)!} \right\}_{n=1}^{+\infty}$ **11.** $\left\{ \dfrac{n^n}{n!} \right\}_{n=1}^{+\infty}$ **12.** $\left\{ \dfrac{5^n}{2^{(n^2)}} \right\}_{n=1}^{+\infty}$

In Exercises 13–18, use differentiation to show that the sequence is strictly increasing or strictly decreasing.

13. $\left\{ \dfrac{n}{2n+1} \right\}_{n=1}^{+\infty}$ **14.** $\left\{ 3 - \dfrac{1}{n} \right\}_{n=1}^{+\infty}$

15. $\left\{ \dfrac{1}{n + \ln n} \right\}_{n=1}^{+\infty}$ **16.** $\{ne^{-2n}\}_{n=1}^{+\infty}$

17. $\left\{ \dfrac{\ln(n+2)}{n+2} \right\}_{n=1}^{+\infty}$ **18.** $\{\tan^{-1} n\}_{n=1}^{+\infty}$

In Exercises 19–24, use any method to show that the given sequence is eventually strictly increasing or eventually strictly decreasing.

19. $\{2n^2 - 7n\}_{n=1}^{+\infty}$ **20.** $\{n^3 - 4n^2\}_{n=1}^{+\infty}$

21. $\left\{ \dfrac{n}{n^2 + 10} \right\}_{n=1}^{+\infty}$ **22.** $\left\{ n + \dfrac{17}{n} \right\}_{n=1}^{+\infty}$

23. $\left\{ \dfrac{n!}{3^n} \right\}_{n=1}^{+\infty}$ **24.** $\{n^5 e^{-n}\}_{n=1}^{+\infty}$

25. (a) Suppose that $\{a_n\}$ is a monotone sequence such that $1 \leq a_n \leq 2$. Must the sequence converge? If so, what can you say about the limit?

(b) Suppose that $\{a_n\}$ is a monotone sequence such that $a_n \leq 2$. Must the sequence converge? If so, what can you say about the limit?

26. The goal in this exercise is to prove Formula (5) in this section. The case where $x = 0$ is obvious, so we will focus on the case where $x \neq 0$.

(a) Let $a_n = |x|^n/n!$. Show that

$$a_{n+1} = \dfrac{|x|}{n+1} a_n$$

(b) Show that the sequence $\{a_n\}$ is eventually strictly decreasing.

(c) Show that the sequence $\{a_n\}$ converges.

(d) Use the results in parts (a) and (c) to show that $a_n \to 0$ as $n \to +\infty$.

(e) Obtain Formula (5) from the result in part (d).

27. Let $\{a_n\}$ be the sequence defined recursively by $a_1 = \sqrt{2}$ and $a_{n+1} = \sqrt{2 + a_n}$ for $n \geq 1$.
  (a) List the first three terms of the sequence.
  (b) Show that $a_n < 2$ for $n \geq 1$.
  (c) Show that $a_{n+1}^2 - a_n^2 = (2 - a_n)(1 + a_n)$ for $n \geq 1$.
  (d) Use the results in parts (b) and (c) to show that $\{a_n\}$ is a strictly increasing sequence. [*Hint:* If x and y are positive real numbers such that $x^2 - y^2 > 0$, then it follows by factoring that $x - y > 0$.]
  (e) Show that $\{a_n\}$ converges and find its limit L.

28. Let $\{a_n\}$ be the sequence defined recursively by $a_1 = 1$ and $a_{n+1} = \frac{1}{2}[a_n + (3/a_n)]$ for $n \geq 1$.
  (a) Show that $a_n \geq \sqrt{3}$ for $n \geq 2$. [*Hint:* What is the minimum value of $\frac{1}{2}[x + (3/x)]$ for $x > 0$?]
  (b) Show that $\{a_n\}$ is eventually decreasing. [*Hint:* Examine $a_{n+1} - a_n$ or a_{n+1}/a_n and use the result in part (a).]
  (c) Show that $\{a_n\}$ converges and find its limit L.

29. (a) Compare appropriate areas in the accompanying figure to deduce the following inequalities for $n \geq 2$:
$$\int_1^n \ln x \, dx < \ln n! < \int_1^{n+1} \ln x \, dx$$

(b) Use the result in part (a) to show that
$$\frac{n^n}{e^{n-1}} < n! < \frac{(n+1)^{n+1}}{e^n}, \quad n > 1$$

(c) Use the Squeezing Theorem for Sequences (Theorem 11.1.5) and the result in part (b) to show that
$$\lim_{n \to +\infty} \frac{\sqrt[n]{n!}}{n} = \frac{1}{e}$$

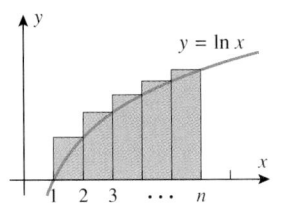 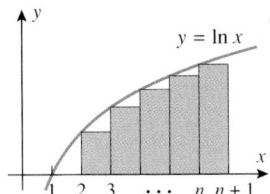

Figure Ex-29

30. Use the left inequality in Exercise 29(b) to show that
$$\lim_{n \to +\infty} \sqrt[n]{n!} = +\infty$$

11.3 INFINITE SERIES

The purpose of this section is to discuss sums that contain infinitely many terms. The most familiar examples of such sums occur in the decimal representation of real numbers. For example, when we write $\frac{1}{3}$ in the decimal form $\frac{1}{3} = 0.3333\ldots$, we mean

$$\frac{1}{3} = 0.3 + 0.03 + 0.003 + 0.0003 + \cdots$$

which suggests that the decimal representation of $\frac{1}{3}$ can be viewed as a sum of infinitely many real numbers.

SUMS OF INFINITE SERIES

Our first objective is to define what is meant by the "sum" of infinitely many real numbers. We begin with some terminology.

> **11.3.1** DEFINITION. An ***infinite series*** is an expression that can be written in the form
> $$\sum_{k=1}^{\infty} u_k = u_1 + u_2 + u_3 + \cdots + u_k + \cdots$$
> The numbers $u_1, u_2, u_3, \ldots$ are called the ***terms*** of the series.

Since it is impossible to add infinitely many numbers together directly, sums of infinite series are defined and computed by an indirect limiting process. To motivate the basic idea, consider the decimal

$$0.3333\ldots \tag{1}$$

This can be viewed as the infinite series

$$0.3 + 0.03 + 0.003 + 0.0003 + \cdots$$

or, equivalently,

$$\frac{3}{10} + \frac{3}{10^2} + \frac{3}{10^3} + \frac{3}{10^4} + \cdots \tag{2}$$

Since (1) is the decimal expansion of $\frac{1}{3}$, any reasonable definition for the sum of an infinite series should yield $\frac{1}{3}$ for the sum of (2). To obtain such a definition, consider the following sequence of (finite) sums:

$$s_1 = \frac{3}{10} = 0.3$$

$$s_2 = \frac{3}{10} + \frac{3}{10^2} = 0.33$$

$$s_3 = \frac{3}{10} + \frac{3}{10^2} + \frac{3}{10^3} = 0.333$$

$$s_4 = \frac{3}{10} + \frac{3}{10^2} + \frac{3}{10^3} + \frac{3}{10^4} = 0.3333$$

$$\vdots$$

The sequence of numbers $s_1, s_2, s_3, s_4, \ldots$ can be viewed as a succession of approximations to the "sum" of the infinite series, which we want to be $\frac{1}{3}$. As we progress through the sequence, more and more terms of the infinite series are used, and the approximations get better and better, suggesting that the desired sum of $\frac{1}{3}$ might be the *limit* of this sequence of approximations. To see that this is so, we must calculate the limit of the general term in the sequence of approximations, namely

$$s_n = \frac{3}{10} + \frac{3}{10^2} + \cdots + \frac{3}{10^n} \tag{3}$$

The problem of calculating

$$\lim_{n \to +\infty} s_n = \lim_{n \to +\infty} \left(\frac{3}{10} + \frac{3}{10^2} + \cdots + \frac{3}{10^n} \right)$$

is complicated by the fact that both the last term and the number of terms in the sum change with n. It is best to rewrite such limits in a closed form in which the number of terms does not vary, if possible. (See the remark following Example 3 in Section 7.4.) To do this, we multiply both sides of (3) by $\frac{1}{10}$ to obtain

$$\frac{1}{10} s_n = \frac{3}{10^2} + \frac{3}{10^3} + \cdots + \frac{3}{10^n} + \frac{3}{10^{n+1}} \tag{4}$$

and then subtract (4) from (3) to obtain

$$s_n - \frac{1}{10} s_n = \frac{3}{10} - \frac{3}{10^{n+1}}$$

$$\frac{9}{10} s_n = \frac{3}{10} \left(1 - \frac{1}{10^n} \right)$$

$$s_n = \frac{1}{3} \left(1 - \frac{1}{10^n} \right)$$

Since $1/10^n \to 0$ as $n \to +\infty$, it follows that

$$\lim_{n \to +\infty} s_n = \lim_{n \to +\infty} \frac{1}{3} \left(1 - \frac{1}{10^n} \right) = \frac{1}{3}$$

which we denote by writing

$$\frac{1}{3} = \frac{3}{10} + \frac{3}{10^2} + \frac{3}{10^3} + \cdots + \frac{3}{10^n} + \cdots$$

Motivated by the preceding example, we are now ready to define the general concept of the "sum" of an infinite series

$$u_1 + u_2 + u_3 + \cdots + u_k + \cdots$$

We begin with some terminology: Let s_n denote the sum of the first n terms of the series. Thus,

$$s_1 = u_1$$
$$s_2 = u_1 + u_2$$
$$s_3 = u_1 + u_2 + u_3$$
$$\vdots$$
$$s_n = u_1 + u_2 + u_3 + \cdots + u_n = \sum_{k=1}^{n} u_k$$

The number s_n is called the **nth partial sum** of the series and the sequence $\{s_n\}_{n=1}^{+\infty}$ is called the **sequence of partial sums**.

WARNING.   In everyday language the words "sequence" and "series" are often used interchangeably. However, this is not so in mathematics—mathematically, a sequence is a *succession* and a series is a *sum*. It is essential that you keep this distinction in mind.

As n increases, the partial sum $s_n = u_1 + u_2 + \cdots + u_n$ includes more and more terms of the series. Thus, if s_n tends toward a limit as $n \to +\infty$, it is reasonable to view this limit as the sum of *all* the terms in the series. This suggests the following definition.

11.3.2   DEFINITION.   Let $\{s_n\}$ be the sequence of partial sums of the series

$$u_1 + u_2 + u_3 + \cdots + u_k + \cdots$$

If the sequence $\{s_n\}$ converges to a limit S, then the series is said to **converge** to S, and S is called the **sum** of the series. We denote this by writing

$$S = \sum_{k=1}^{\infty} u_k$$

If the sequence of partial sums diverges, then the series is said to **diverge**. A divergent series has no sum.

REMARK.   Sometimes it will be desirable to start the summation index in an infinite series at $k = 0$ rather than $k = 1$, in which case we will view u_0 as the zeroth term and $s_0 = u_0$ as the zeroth partial sum. It can be proved that changing the starting value for the index has no effect on the convergence or divergence of an infinite series.

Example 1

Determine whether the series

$$1 - 1 + 1 - 1 + 1 - 1 + \cdots$$

converges or diverges. If it converges, find the sum.

Solution.   It is tempting to conclude that the sum of the series is zero by arguing that the positive and negative terms cancel one another. However, this is *not correct*; the problem is that algebraic operations that hold for finite sums do not carry over to infinite series in all cases. Later, we will discuss conditions under which familiar algebraic operations can be applied to infinite series, but for this example we turn directly to Definition 11.3.2. The partial sums are

$$s_1 = 1$$
$$s_2 = 1 - 1 = 0$$
$$s_3 = 1 - 1 + 1 = 1$$
$$s_4 = 1 - 1 + 1 - 1 = 0$$

and so forth. Thus, the sequence of partial sums is

$$1, 0, 1, 0, 1, 0, \ldots$$

Since this is a divergent sequence, the given series diverges and consequently has no sum.   ◀

GEOMETRIC SERIES

In many important geometric series, each term is obtained by multiplying the preceding term by some fixed constant. Thus, if the initial term of the series is a and each term is obtained by multiplying the preceding term by r, then the series has the form

$$\sum_{k=0}^{\infty} ar^k = a + ar + ar^2 + ar^3 + \cdots + ar^k + \cdots \quad (a \neq 0)$$

Such series are called *geometric series*, and the number r is called the *ratio* for the series. Here are some examples:

$$1 + 2 + 4 + 8 + \cdots + 2^k + \cdots \qquad \boxed{a = 1, r = 2}$$

$$\frac{3}{10} + \frac{3}{10^2} + \frac{3}{10^3} + \cdots + \frac{3}{10^k} + \cdots \qquad \boxed{a = \tfrac{3}{10}, r = \tfrac{1}{10}}$$

$$\frac{1}{2} - \frac{1}{4} + \frac{1}{8} - \frac{1}{16} + \cdots + (-1)^{k+1}\frac{1}{2^k} + \cdots \qquad \boxed{a = \tfrac{1}{2}, r = -\tfrac{1}{2}}$$

$$1 + 1 + 1 + \cdots + 1 + \cdots \qquad \boxed{a = 1, r = 1}$$

$$1 - 1 + 1 - 1 + \cdots + (-1)^{k+1} + \cdots \qquad \boxed{a = 1, r = -1}$$

$$1 + x + x^2 + x^3 + \cdots + x^k + \cdots \qquad \boxed{a = 1, r = x}$$

REMARK.   In some of these series we started the index of summation at $k = 0$ and in others at $k = 1$, depending on which choice produced the simpler general term.

The following theorem is the fundamental result on convergence of geometric series.

11.3.3   THEOREM.   *A geometric series*

$$\sum_{k=0}^{\infty} ar^k = a + ar + ar^2 + \cdots + ar^k + \cdots \quad (a \neq 0)$$

converges if $|r| < 1$ and diverges if $|r| \geq 1$. If the series converges, then the sum is

$$\sum_{k=0}^{\infty} ar^k = \frac{a}{1 - r}$$

Proof.   Let us treat the case $|r| = 1$ first. If $r = 1$, then the series is

$$a + a + a + a + \cdots$$

so the nth partial sum is $s_n = (n + 1)a$ and $\displaystyle\lim_{n \to +\infty} s_n = \lim_{n \to +\infty} (n + 1)a = \pm\infty$ (the sign depending on whether a is positive or negative). This proves divergence. If $r = -1$, the series is

$$a - a + a - a + \cdots$$

so the sequence of partial sums is

$$a, 0, a, 0, a, 0, \ldots$$

which diverges.

Now let us consider the case where $|r| \neq 1$. The nth partial sum of the series is

$$s_n = a + ar + ar^2 + \cdots + ar^n \tag{5}$$

Multiplying both sides of (5) by r yields

$$rs_n = ar + ar^2 + \cdots + ar^n + ar^{n+1} \tag{6}$$

and subtracting (6) from (5) gives

$$s_n - rs_n = a - ar^{n+1}$$

or

$$(1 - r)s_n = a - ar^{n+1} \tag{7}$$

Since $r \neq 1$ in the case we are considering, this can be rewritten as

$$s_n = \frac{a - ar^{n+1}}{1 - r} = \frac{a}{1 - r} - \frac{ar^{n+1}}{1 - r} \tag{8}$$

If $|r| < 1$, then $\lim\limits_{n \to +\infty} r^{n+1} = 0$ (can you see why?), so $\{s_n\}$ converges. From (8)

$$\lim_{n \to +\infty} s_n = \frac{a}{1 - r}$$

If $|r| > 1$, then either $r > 1$ or $r < -1$. In the case $r > 1$, $\lim\limits_{n \to +\infty} r^{n+1} = +\infty$, and in the case $r < -1$, r^{n+1} oscillates between positive and negative values that grow in magnitude, so $\{s_n\}$ diverges in both cases. ■

Example 2

The series

$$\sum_{k=0}^{\infty} \frac{5}{4^k} = 5 + \frac{5}{4} + \frac{5}{4^2} + \cdots + \frac{5}{4^k} + \cdots$$

is a geometric series with $a = 5$ and $r = \frac{1}{4}$. Since $|r| = \frac{1}{4} < 1$, the series converges and the sum is

$$\frac{a}{1 - r} = \frac{5}{1 - \frac{1}{4}} = \frac{20}{3} \qquad \blacktriangleleft$$

Example 3

Find the rational number represented by the repeating decimal

$$0.784784784\ldots$$

Solution. We can write

$$0.784784784\ldots = 0.784 + 0.000784 + 0.000000784 + \cdots$$

so the given decimal is the sum of a geometric series with $a = 0.784$ and $r = 0.001$. Thus,

$$0.784784784\ldots = \frac{a}{1 - r} = \frac{0.784}{1 - 0.001} = \frac{0.784}{0.999} = \frac{784}{999} \qquad \blacktriangleleft$$

Example 4

In each part, determine whether the series converges, and if so find its sum.

(a) $\displaystyle\sum_{k=1}^{\infty} 3^{2k} 5^{1-k}$ (b) $\displaystyle\sum_{k=0}^{\infty} x^{k}$

Solution (a). This is a geometric series in a concealed form, since we can rewrite it as

$$\sum_{k=1}^{\infty} 3^{2k} 5^{1-k} = \sum_{k=1}^{\infty} \frac{9^{k}}{5^{k-1}} = \sum_{k=1}^{\infty} 9\left(\frac{9}{5}\right)^{k-1}$$

Since $r = \frac{9}{5} > 1$, the series diverges.

Solution (b). The expanded form of the series is

$$\sum_{k=0}^{\infty} x^{k} = 1 + x + x^{2} + \cdots + x^{k} + \cdots$$

The series is a geometric series with $a = 1$ and $r = x$, so it converges if $|x| < 1$ and diverges otherwise. When the series converges its sum is

$$\sum_{k=0}^{\infty} x^{k} = \frac{1}{1-x}$$ ◀

Example 5

Determine whether the series

$$\sum_{k=1}^{\infty} \frac{1}{k(k+1)} = \frac{1}{1\cdot 2} + \frac{1}{2\cdot 3} + \frac{1}{3\cdot 4} + \frac{1}{4\cdot 5} + \cdots$$

converges or diverges. If it converges, find the sum.

Solution. The nth partial sum of the series is

$$s_{n} = \sum_{k=1}^{n} \frac{1}{k(k+1)} = \frac{1}{1\cdot 2} + \frac{1}{2\cdot 3} + \frac{1}{3\cdot 4} + \cdots + \frac{1}{n(n+1)}$$

To calculate $\displaystyle\lim_{n\to+\infty} s_{n}$ we will rewrite s_{n} in closed form. This can be accomplished by using the method of partial fractions to obtain (verify)

$$\frac{1}{k(k+1)} = \frac{1}{k} - \frac{1}{k+1}$$

from which we obtain the telescoping sum

$$\begin{aligned}
s_{n} &= \sum_{k=1}^{n} \left(\frac{1}{k} - \frac{1}{k+1}\right) \\
&= \left(1 - \frac{1}{2}\right) + \left(\frac{1}{2} - \frac{1}{3}\right) + \left(\frac{1}{3} - \frac{1}{4}\right) + \cdots + \left(\frac{1}{n} - \frac{1}{n+1}\right) \\
&= 1 + \left(-\frac{1}{2} + \frac{1}{2}\right) + \left(-\frac{1}{3} + \frac{1}{3}\right) + \cdots + \left(-\frac{1}{n} + \frac{1}{n}\right) - \frac{1}{n+1} \\
&= 1 - \frac{1}{n+1}
\end{aligned}$$

so

$$\sum_{k=1}^{\infty} \frac{1}{k(k+1)} = \lim_{n\to+\infty} s_{n} = \lim_{n\to+\infty} \left(1 - \frac{1}{n+1}\right) = 1$$ ◀

FOR THE READER. If you have a CAS, read the documentation to determine how to find sums of infinite series; then use the CAS to check the results in Example 5.

One of the most important of all diverging series is the ***harmonic series***,

$$\sum_{k=1}^{\infty} \frac{1}{k} = 1 + \frac{1}{2} + \frac{1}{3} + \frac{1}{4} + \frac{1}{5} + \cdots$$

which arises in connection with the overtones produced by a vibrating musical string. It is not immediately evident that this series diverges. However, the divergence will become apparent when we examine the partial sums in detail. Because the terms in the series are all positive, the partial sums

$$s_1 = 1, \quad s_2 = 1 + \frac{1}{2}, \quad s_3 = 1 + \frac{1}{2} + \frac{1}{3}, \quad s_4 = 1 + \frac{1}{2} + \frac{1}{3} + \frac{1}{4}, \ldots$$

form a strictly increasing sequence

$$s_1 < s_2 < s_3 < \cdots < s_n < \cdots$$

Thus, by Theorem 11.2.3 we can prove divergence by demonstrating that there is no constant M that is greater than or equal to *every* partial sum. To this end, we will consider some selected partial sums, namely $s_2, s_4, s_8, s_{16}, s_{32}, \ldots$. Note that the subscripts are successive powers of 2, so that these are the partial sums of the form s_{2^n}. These partial sums satisfy the inequalities

$$s_2 = 1 + \frac{1}{2} > \frac{1}{2} + \frac{1}{2} = \frac{2}{2}$$

$$s_4 = s_2 + \frac{1}{3} + \frac{1}{4} > s_2 + \left(\frac{1}{4} + \frac{1}{4}\right) = s_2 + \frac{1}{2} > \frac{3}{2}$$

$$s_8 = s_4 + \frac{1}{5} + \frac{1}{6} + \frac{1}{7} + \frac{1}{8} > s_4 + \left(\frac{1}{8} + \frac{1}{8} + \frac{1}{8} + \frac{1}{8}\right) = s_4 + \frac{1}{2} > \frac{4}{2}$$

$$s_{16} = s_8 + \frac{1}{9} + \frac{1}{10} + \frac{1}{11} + \frac{1}{12} + \frac{1}{13} + \frac{1}{14} + \frac{1}{15} + \frac{1}{16}$$

$$> s_8 + \left(\frac{1}{16} + \frac{1}{16} + \frac{1}{16} + \frac{1}{16} + \frac{1}{16} + \frac{1}{16} + \frac{1}{16} + \frac{1}{16}\right) = s_8 + \frac{1}{2} > \frac{5}{2}$$

$$\vdots$$

$$s_{2^n} > \frac{n+1}{2}$$

If M is any constant, we can find a positive integer n such that $(n+1)/2 > M$. But for this n

$$s_{2^n} > \frac{n+1}{2} > M$$

so that no constant M is greater than or equal to *every* partial sum of the harmonic series. This proves divergence.

This divergence proof, which predates the discovery of calculus, is due to a French bishop and teacher, Nicole Oresme (1323–1382). This series eventually attracted the interest of Johann and Jakob Bernoulli (p. 99) and led them to begin thinking about the general concept of convergence, which was a new idea at that time.

This is a proof of the divergence of the harmonic series, as it appeared in an appendix of Jakob Bernoulli's posthumous publication, *Ars Conjectandi*, which appeared in 1713.

1. In each part, find exact values for the first four partial sums, find a closed form for the nth partial sum, and determine whether the series converges by calculating the limit of the nth partial sum. If the series converges, then state its sum.

   (a) $2 + \frac{2}{5} + \frac{2}{5^2} + \cdots + \frac{2}{5^{k-1}} + \cdots$

   (b) $\frac{1}{4} + \frac{2}{4} + \frac{2^2}{4} + \cdots + \frac{2^{k-1}}{4} + \cdots$

   (c) $\frac{1}{2 \cdot 3} + \frac{1}{3 \cdot 4} + \frac{1}{4 \cdot 5} + \cdots + \frac{1}{(k+1)(k+2)} + \cdots$

2. In each part, find exact values for the first four partial sums, find a closed form for the nth partial sum, and determine whether the series converges by calculating the limit of the nth partial sum. If the series converges, then state its sum.

   (a) $\sum_{k=1}^{\infty} \left(\frac{1}{4}\right)^k$ (b) $\sum_{k=1}^{\infty} 4^{k-1}$ (c) $\sum_{k=1}^{\infty} \left(\frac{1}{k+3} - \frac{1}{k+4}\right)$

   In Exercises 3–14, determine whether the series converges, and if so, find its sum.

3. $\displaystyle\sum_{k=1}^{\infty}\left(-\frac{3}{4}\right)^{k-1}$

4. $\displaystyle\sum_{k=1}^{\infty}\left(\frac{2}{3}\right)^{k+2}$

5. $\displaystyle\sum_{k=1}^{\infty}(-1)^{k-1}\frac{7}{6^{k-1}}$

6. $\displaystyle\sum_{k=1}^{\infty}\left(-\frac{3}{2}\right)^{k+1}$

7. $\displaystyle\sum_{k=1}^{\infty}\frac{1}{(k+2)(k+3)}$

8. $\displaystyle\sum_{k=1}^{\infty}\left(\frac{1}{2^{k}}-\frac{1}{2^{k+1}}\right)$

9. $\displaystyle\sum_{k=1}^{\infty}\frac{1}{9k^{2}+3k-2}$

10. $\displaystyle\sum_{k=2}^{\infty}\frac{1}{k^{2}-1}$

11. $\displaystyle\sum_{k=3}^{\infty}\frac{1}{k-2}$

12. $\displaystyle\sum_{k=5}^{\infty}\left(\frac{e}{\pi}\right)^{k-1}$

13. $\displaystyle\sum_{k=1}^{\infty}\frac{4^{k+2}}{7^{k-1}}$

14. $\displaystyle\sum_{k=1}^{\infty}5^{3k}7^{1-k}$

In Exercises 15–20, express the given repeating decimal as a fraction.

15. $0.4444\ldots$

16. $0.9999\ldots$

17. $5.373737\ldots$

18. $0.159159159\ldots$

19. $0.782178217821\ldots$

20. $0.451141414\ldots$

[c] 21. Use a CAS to check your answers to Exercises 1–14.

[c] 22. In each part, use a CAS to find the sum of the series if it converges, and then confirm the result by hand calculation.

(a) $\displaystyle\sum_{k=1}^{\infty}(-1)^{k+1}2^{k}3^{2-k}$ (b) $\displaystyle\sum_{k=1}^{\infty}\frac{3^{3k}}{5^{k-1}}$ (c) $\displaystyle\sum_{k=1}^{\infty}\frac{1}{4k^{2}-1}$

23. A ball is dropped from a height of 10 m. Each time it strikes the ground it bounces vertically to a height that is $\frac{3}{4}$ of the preceding height. Find the total distance the ball will travel if it is assumed to bounce infinitely often.

24. The accompanying figure shows an "infinite staircase" constructed from cubes. Find the total volume of the staircase, given that the largest cube has a side of length 1 and each successive cube has a side whose length is half that of the preceding cube.

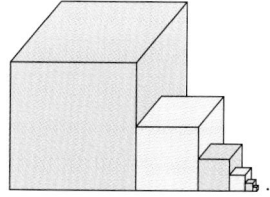

Figure Ex-24

25. In each part, find a closed form for the nth partial sum of the series, and determine whether the series converges. If so, find its sum.

(a) $\ln\dfrac{1}{2}+\ln\dfrac{2}{3}+\ln\dfrac{3}{4}+\cdots+\ln\dfrac{n}{n+1}+\cdots$

(b) $\ln\left(1-\dfrac{1}{4}\right)+\ln\left(1-\dfrac{1}{9}\right)+\ln\left(1-\dfrac{1}{16}\right)+\cdots$

$\qquad +\ln\left(1-\dfrac{1}{(k+1)^{2}}\right)+\cdots$

26. Use geometric series to show that

(a) $\displaystyle\sum_{k=0}^{\infty}(-1)^{k}x^{k}=\frac{1}{1+x}$ if $-1<x<1$

(b) $\displaystyle\sum_{k=0}^{\infty}(x-3)^{k}=\frac{1}{4-x}$ if $2<x<4$

(c) $\displaystyle\sum_{k=0}^{\infty}(-1)^{k}x^{2k}=\frac{1}{1+x^{2}}$ if $-1<x<1$.

27. In each part, find all values of x for which the series converges, and find the sum of the series for those values of x.

(a) $x-x^{3}+x^{5}-x^{7}+x^{9}-\cdots$

(b) $\dfrac{1}{x^{2}}+\dfrac{2}{x^{3}}+\dfrac{4}{x^{4}}+\dfrac{8}{x^{5}}+\dfrac{16}{x^{6}}+\cdots$

(c) $e^{-x}+e^{-2x}+e^{-3x}+e^{-4x}+e^{-5x}+\cdots$

28. Show: $\displaystyle\sum_{k=1}^{\infty}\frac{\sqrt{k+1}-\sqrt{k}}{\sqrt{k^{2}+k}}=1.$

29. Show: $\displaystyle\sum_{k=1}^{\infty}\left(\frac{1}{k}-\frac{1}{k+2}\right)=\frac{3}{2}.$

30. Show: $\dfrac{1}{1\cdot3}+\dfrac{1}{2\cdot4}+\dfrac{1}{3\cdot5}+\cdots=\dfrac{3}{4}.$

31. Show: $\dfrac{1}{1\cdot3}+\dfrac{1}{3\cdot5}+\dfrac{1}{5\cdot7}+\cdots=\dfrac{1}{2}.$

32. Show that for all real values of x

$\sin x-\dfrac{1}{2}\sin^{2}x+\dfrac{1}{4}\sin^{3}x-\dfrac{1}{8}\sin^{4}x+\cdots=\dfrac{2\sin x}{2+\sin x}$

33. Let a_{1} be any real number, and let $\{a_{n}\}$ be the sequence defined recursively by

$$a_{n+1}=\tfrac{1}{2}(a_{n}+1)$$

Make a conjecture about the limit of the sequence, and confirm your conjecture by expressing a_{n} in terms of a_{1} and taking the limit.

34. Recall that a *terminating decimal* is a decimal whose digits are all 0 from some point on ($0.5=0.50000\ldots$, for example). Show that a decimal of the form $0.a_{1}a_{2}\ldots a_{n}9999\ldots$, where $a_{n}\neq9$, can be expressed as a terminating decimal.

35. The great Swiss mathematician Leonhard Euler (biography on p. 19) sometimes reached incorrect conclusions in his pioneering work on infinite series. For example, Euler deduced that

$$\tfrac{1}{2}=1-1+1-1+\cdots$$

and

$$-1=1+2+4+8+\cdots$$

by substituting $x=-1$ and $x=2$ in the formula

$$\frac{1}{1-x}=1+x+x^{2}+x^{3}+\cdots$$

What was the problem with his reasoning?

36. As shown in the accompanying figure, suppose that lines L_1 and L_2 form an angle θ, $0 < \theta < \pi/2$, at their point of intersection P. A point P_0 is chosen that is on L_1 and a units from P. Starting from P_0 a zig-zag path is constructed by successively going back and forth between L_1 and L_2 along a perpendicular from one line to the other. Find the following sums in terms of θ.

(a) $P_0P_1 + P_1P_2 + P_2P_3 + \cdots$

(b) $P_0P_1 + P_2P_3 + P_4P_5 + \cdots$

(c) $P_1P_2 + P_3P_4 + P_5P_6 + \cdots$

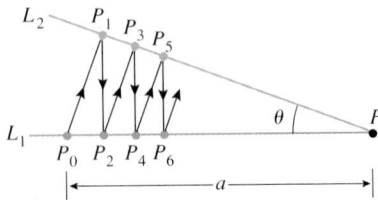

Figure Ex-36

37. As shown in the accompanying figure, suppose that an angle θ is bisected using a straightedge and compass to produce ray R_1, then the angle between R_1 and the initial side is bisected to produce ray R_2. Thereafter, rays R_3, R_4, R_5, ... are constructed in succession by bisecting the angle between the preceding two rays. Show that the sequence of angles that these rays make with the initial side has a limit of $\theta/3$. [This problem is based on *Trisection of an Angle in an Infinite Number of Steps* by Eric Kincannon, which appeared in *The College Mathematics Journal*, Vol. 21, No. 5, November 1990.]

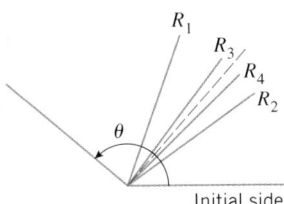

Figure Ex-37

38. In his *Treatise on the Configurations of Qualities and Motions* (written in the 1350s), the French Bishop of Lisieux, Nicole Oresme, used a geometric method to find the sum of the series

$$\sum_{k=1}^{\infty} \frac{k}{2^k} = \frac{1}{2} + \frac{2}{4} + \frac{3}{8} + \frac{4}{16} + \cdots$$

In part (a) of the accompanying figure, each term in the series is represented by the area of a rectangle, and in part (b) the configuration in part (a) has been divided into rectangles with areas A_1, A_2, A_3, ... Find the sum $A_1 + A_2 + A_3 + \cdots$.

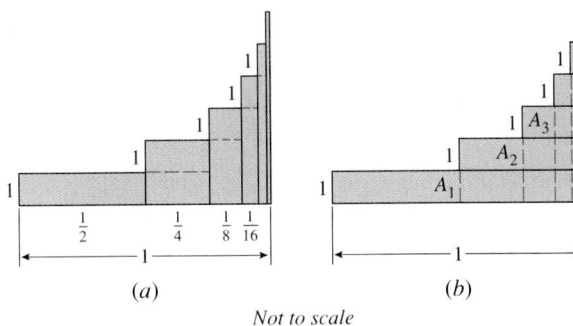

(a)   (b)

Not to scale

Figure Ex-38

39. (a) See if your CAS can find the sum of the series

$$\sum_{k=1}^{\infty} \frac{6^k}{(3^{k+1} - 2^{k+1})(3^k - 2^k)}$$

(b) Find A and B such that

$$\frac{6^k}{(3^{k+1} - 2^{k+1})(3^k - 2^k)} = \frac{2^k A}{3^k - 2^k} + \frac{2^k B}{3^{k+1} - 2^{k+1}}$$

(c) Use the result in part (b) to find a closed form for the nth partial sum, and then find the sum of the series. [This exercise is adapted from a problem that appeared in the Forty-Fifth Annual William Lowell Putnam Competition.]

11.4 CONVERGENCE TESTS

In the last section we showed how to find the sum of a series by finding a closed form for the nth partial sum and taking its limit. However, it is relatively rare that one can find a closed form for the nth partial sum of a series, so alternative methods are needed for finding sums of series. One possibility is to prove that the series converges, and then approximate the sum by a partial sum with sufficiently many terms to achieve the desired degree of accuracy. In this section we will develop various tests that can be used to determine whether a given series converges or diverges.

The kth term in an infinite series $\sum u_k$ is called the ***general term*** of the series. The following theorem establishes a relationship between the limit of the general term and the convergence properties of a series.

11.4.1 THEOREM (*The Divergence Test*).

(a) *If* $\lim\limits_{k \to +\infty} u_k \neq 0$, *then the series* $\sum u_k$ *diverges*.

(b) *If* $\lim\limits_{k \to +\infty} u_k = 0$, *then the series* $\sum u_k$ *may either converge or diverge*.

Proof (a). To prove this result, it suffices to show that if the series converges, then $\lim\limits_{k \to +\infty} u_k = 0$ (why?). We will prove this alternative form of (a).

Let us assume that the series converges. The general term u_k can be written as

$$u_k = s_k - s_{k-1} \tag{1}$$

where s_k is the sum of the first k terms and s_{k-1} is the sum of the first $k - 1$ terms. If S denotes the sum of the series, then $\lim\limits_{k \to +\infty} s_k = S$, and since $(k - 1) \to +\infty$ as $k \to +\infty$, we also have $\lim\limits_{k \to +\infty} s_{k-1} = S$. Thus, from (1)

$$\lim_{k \to +\infty} u_k = \lim_{k \to +\infty} (s_k - s_{k-1}) = S - S = 0$$

Proof (b). To prove this result, it suffices to produce both a convergent series and a divergent series for which $\lim\limits_{k \to +\infty} u_k = 0$. The following series both have this property:

$$\frac{1}{2} + \frac{1}{2^2} + \cdots + \frac{1}{2^k} + \cdots \quad \text{and} \quad 1 + \frac{1}{2} + \frac{1}{3} + \cdots + \frac{1}{k} + \cdots$$

The first is a convergent geometric series and the second is the divergent harmonic series. ∎

The alternative form of part (a) given in the preceding proof is sufficiently important that we state it separately for future reference.

11.4.2 THEOREM. *If the series* $\sum u_k$ *converges, then* $\lim\limits_{k \to +\infty} u_k = 0$.

Example 1

The series

$$\sum_{k=1}^{\infty} \frac{k}{k+1} = \frac{1}{2} + \frac{2}{3} + \frac{3}{4} + \cdots + \frac{k}{k+1} + \cdots$$

diverges since

$$\lim_{k \to +\infty} \frac{k}{k+1} = \lim_{k \to +\infty} \frac{1}{1 + 1/k} = 1 \neq 0 \qquad \blacktriangleleft$$

WARNING. The converse of Theorem 11.4.2 is false. To prove that a series converges it does not suffice to show that $\lim\limits_{k \to +\infty} u_k = 0$, since this property may hold for divergent as well as convergent series, as we saw in the proof of part (b) of Theorem 11.4.1.

For brevity, the proof of the following result is omitted.

11.4.3   THEOREM.

(a) If $\sum u_k$ and $\sum v_k$ are convergent series, then $\sum(u_k + v_k)$ and $\sum(u_k - v_k)$ are convergent series and the sums of these series are related by

$$\sum_{k=1}^{\infty}(u_k + v_k) = \sum_{k=1}^{\infty} u_k + \sum_{k=1}^{\infty} v_k$$

$$\sum_{k=1}^{\infty}(u_k - v_k) = \sum_{k=1}^{\infty} u_k - \sum_{k=1}^{\infty} v_k$$

(b) If c is a nonzero constant, then the series $\sum u_k$ and $\sum cu_k$ both converge or both diverge. In the case of convergence, the sums are related by

$$\sum_{k=1}^{\infty} cu_k = c \sum_{k=1}^{\infty} u_k$$

(c) Convergence or divergence is unaffected by deleting a finite number of terms from a series; in particular, for any positive integer K, the series

$$\sum_{k=1}^{\infty} u_k = u_1 + u_2 + u_3 + \cdots$$

$$\sum_{k=K}^{\infty} u_k = u_K + u_{K+1} + u_{K+2} + \cdots$$

both converge or both diverge.

REMARK.   Do not read too much into part (c) of this theorem. Although the convergence is not affected when a finite number of terms is deleted from the beginning of a convergent series, the *sum* of a convergent series is changed by the removal of these terms.

Example 2

Find the sum of the series

$$\sum_{k=1}^{\infty}\left(\frac{3}{4^k} - \frac{2}{5^{k-1}}\right)$$

Solution.   The series

$$\sum_{k=1}^{\infty}\frac{3}{4^k} = \frac{3}{4} + \frac{3}{4^2} + \frac{3}{4^3} + \cdots$$

is a convergent geometric series $\left(a = \frac{3}{4}, r = \frac{1}{4}\right)$, and the series

$$\sum_{k=1}^{\infty}\frac{2}{5^{k-1}} = 2 + \frac{2}{5} + \frac{2}{5^2} + \frac{2}{5^3} + \cdots$$

is also a convergent geometric series $\left(a = 2, r = \frac{1}{5}\right)$. Thus, from Theorems 11.4.3(a) and 11.3.3 the given series converges and

$$\sum_{k=1}^{\infty}\left(\frac{3}{4^k} - \frac{2}{5^{k-1}}\right) = \sum_{k=1}^{\infty}\frac{3}{4^k} - \sum_{k=1}^{\infty}\frac{2}{5^{k-1}} = \frac{\frac{3}{4}}{1 - \frac{1}{4}} - \frac{2}{1 - \frac{1}{5}} = -\frac{3}{2} \quad \blacktriangleleft$$

Example 3

Determine whether the following series converge or diverge.

(a) $\displaystyle\sum_{k=1}^{\infty}\frac{5}{k} = 5 + \frac{5}{2} + \frac{5}{3} + \cdots + \frac{5}{k} + \cdots$ (b) $\displaystyle\sum_{k=10}^{\infty}\frac{1}{k} = \frac{1}{10} + \frac{1}{11} + \frac{1}{12} + \cdots$

Solution. The first series is a constant times the divergent harmonic series, and hence diverges by part (*b*) of Theorem 11.4.3. The second series results by deleting the first nine terms from the divergent harmonic series, and hence diverges by part (*c*) of Theorem 11.4.3. ◄

THE INTEGRAL TEST

The expressions

$$\sum_{k=1}^{\infty} \frac{1}{k^2} \quad \text{and} \quad \int_{1}^{+\infty} \frac{1}{x^2}\, dx$$

are related in that the integrand in the improper integral results when the index k in the general term of the series is replaced by x and the limits of summation in the series are replaced by the corresponding limits of integration. The following theorem shows that there is a relationship between the convergence of the series and the integral.

> **11.4.4** THEOREM (*The Integral Test*). *Let $\sum u_k$ be a series with positive terms, and let $f(x)$ be the function that results when k is replaced by x in the general term of the series. If f is decreasing and continuous on the interval $[a, +\infty)$, then*
>
> $$\sum_{k=1}^{\infty} u_k \quad \text{and} \quad \int_{a}^{+\infty} f(x)\, dx$$
>
> *both converge or both diverge.*

Example 4

Use the integral test to determine whether the following series converge or diverge.

(a) $\displaystyle\sum_{k=1}^{\infty} \frac{1}{k}$ (b) $\displaystyle\sum_{k=1}^{\infty} \frac{1}{k^2}$

Solution (a). We already know that this is the divergent harmonic series, so the integral test will simply provide another way of establishing the divergence. If we replace k by x in the general term $1/k$, we obtain the function $f(x) = 1/x$, which is decreasing and continuous for $x \geq 1$ (as required to apply the integral test with $a = 1$). Since

$$\int_{1}^{+\infty} \frac{1}{x}\, dx = \lim_{l \to +\infty} \int_{1}^{l} \frac{1}{x}\, dx = \lim_{l \to +\infty} [\ln l - \ln 1] = +\infty$$

the integral diverges and consequently so does the series.

Solution (b). If we replace k by x in the general term $1/k^2$, we obtain the function $f(x) = 1/x^2$, which is decreasing and continuous for $x \geq 1$. Since

$$\int_{1}^{+\infty} \frac{1}{x^2}\, dx = \lim_{l \to +\infty} \int_{1}^{l} \frac{dx}{x^2} = \lim_{l \to +\infty} \left[-\frac{1}{x}\right]_{1}^{l} = \lim_{l \to +\infty} \left[1 - \frac{1}{l}\right] = 1$$

the integral converges and consequently the series converges by the integral test with $a = 1$. ◄

REMARK. In part (b) of the last example, do *not* erroneously conclude that the sum of the series is 1 because the value of the corresponding integral is 1. It can be proved that the sum of the series is actually $\pi^2/6$ and, indeed, the sum of the first two terms alone exceeds 1.

p-SERIES

The series in Example 4 are special cases of a class of series called ***p-series*** or ***hyperharmonic series***. A *p*-series is an infinite series of the form

$$\sum_{k=1}^{\infty} \frac{1}{k^p} = 1 + \frac{1}{2^p} + \frac{1}{3^p} + \cdots + \frac{1}{k^p} + \cdots$$

where $p > 0$. Examples of p-series are

$$\sum_{k=1}^{\infty} \frac{1}{k} = 1 + \frac{1}{2} + \frac{1}{3} + \cdots + \frac{1}{k} + \cdots \qquad \boxed{p = 1}$$

$$\sum_{k=1}^{\infty} \frac{1}{k^2} = 1 + \frac{1}{2^2} + \frac{1}{3^2} + \cdots + \frac{1}{k^2} + \cdots \qquad \boxed{p = 2}$$

$$\sum_{k=1}^{\infty} \frac{1}{\sqrt{k}} = 1 + \frac{1}{\sqrt{2}} + \frac{1}{\sqrt{3}} + \cdots + \frac{1}{\sqrt{k}} + \cdots \qquad \boxed{p = \tfrac{1}{2}}$$

The following theorem tells when a p-series converges.

11.4.5 THEOREM (*Convergence of p-Series*).

$$\sum_{k=1}^{\infty} \frac{1}{k^p} = 1 + \frac{1}{2^p} + \frac{1}{3^p} + \cdots + \frac{1}{k^p} + \cdots$$

converges if $p > 1$ and diverges if $0 < p \leq 1$.

Proof. To establish this result when $p \neq 1$, we will use the integral test.

$$\int_1^{+\infty} \frac{1}{x^p}\,dx = \lim_{l \to +\infty} \int_1^l x^{-p}\,dx = \lim_{l \to +\infty} \frac{x^{1-p}}{1-p}\Big]_1^l = \lim_{l \to +\infty} \left[\frac{l^{1-p}}{1-p} - \frac{1}{1-p} \right]$$

If $p > 1$, then $1 - p < 0$, so $l^{1-p} \to 0$ as $l \to +\infty$. Thus, the integral converges [its value is $-1/(1-p)$] and consequently the series also converges. For $0 < p < 1$, it follows that $1 - p > 0$ and $l^{1-p} \to +\infty$ as $l \to +\infty$, so the integral and the series diverge. The case $p = 1$ is the harmonic series, which was previously shown to diverge. ■

Example 5

$$1 + \frac{1}{\sqrt[3]{2}} + \frac{1}{\sqrt[3]{3}} + \cdots + \frac{1}{\sqrt[3]{k}} + \cdots$$

diverges since it is a p-series with $p = \frac{1}{3} < 1$. ◀

PROOF OF THE INTEGRAL TEST

Before we can prove the integral test, we need a basic result about convergence of series with *nonnegative* terms. If $u_1 + u_2 + u_3 + \cdots + u_k + \cdots$ is such a series, then its sequence of partial sums is increasing, that is,

$$s_1 \leq s_2 \leq s_3 \leq \cdots \leq s_n \leq \cdots$$

Thus, from Theorem 11.2.3 the sequence of partial sums converges to a limit S if and only if it has some upper bound M, in which case $S \leq M$. If no upper bound exists, then the sequence of partial sums diverges. Since convergence of the sequence of partial sums corresponds to convergence of the series, we have the following theorem.

11.4.6 THEOREM. *If $\sum u_k$ is a series with nonnegative terms, and if there is a constant M such that*

$$s_n = u_1 + u_2 + \cdots + u_n \leq M$$

for every n, then the series converges and the sum S satisfies $S \leq M$. If no such M exists, then the series diverges.

In words, this theorem implies that *a series with nonnegative terms converges if and only if its sequence of partial sums is bounded above.*

Proof of Theorem 11.4.4. We need only show that the series converges when the integral converges and that the series diverges when the integral diverges. For simplicity, we will limit the proof to the case where $a = 1$. Assume that $f(x)$ satisfies the hypotheses of the theorem for $x \geq 1$. Since

$$f(1) = u_1, \ f(2) = u_2, \dots, \ f(n) = u_n, \dots$$

the values of $u_1, u_2, \dots, u_n, \dots$ can be interpreted as the areas of the rectangles shown in Figure 11.4.1.

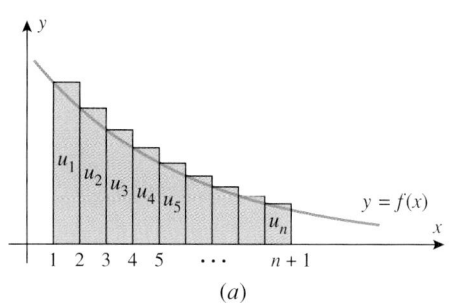 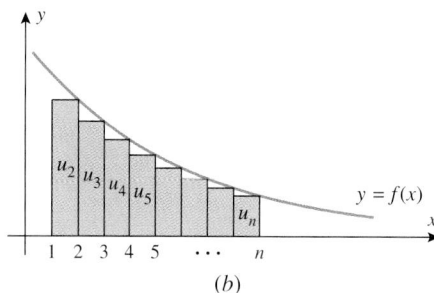

(a)	(b)

Figure 11.4.1

The following inequalities result by comparing the areas under the curve $y = f(x)$ to the areas of the rectangles in Figure 11.4.1 for $n > 1$:

$$\int_1^{n+1} f(x)\,dx < u_1 + u_2 + \cdots + u_n = s_n \qquad \text{Figure 11.4.1}a$$

$$s_n - u_1 = u_2 + u_3 + \cdots + u_n < \int_1^n f(x)\,dx \qquad \text{Figure 11.4.1}b$$

These inequalities can be combined as

$$\int_1^{n+1} f(x)\,dx < s_n < u_1 + \int_1^n f(x)\,dx \qquad (2)$$

If the integral $\int_1^\infty f(x)\,dx$ converges to a finite value L, then from the right-hand inequality in (2)

$$s_n < u_1 + \int_1^n f(x)\,dx < u_1 + \int_1^\infty f(x)\,dx = u_1 + L$$

Thus, each partial sum is less than the finite constant $u_1 + L$, and the series converges by Theorem 11.4.6. On the other hand, if the integral $\int_1^\infty f(x)\,dx$ diverges, then

$$\lim_{n \to +\infty} \int_1^{n+1} f(x)\,dx = +\infty$$

so that from the left-hand inequality in (2), $\displaystyle\lim_{n \to +\infty} s_n = +\infty$. This implies that the series also diverges. ∎

EXERCISE SET 11.4 ⌇ Graphing Calculator   C CAS
. . . .

1. In each part, use Theorem 11.4.3 to find the sum of the series.

(a) $\left(\dfrac{1}{2} + \dfrac{1}{4}\right) + \left(\dfrac{1}{2^2} + \dfrac{1}{4^2}\right) + \cdots + \left(\dfrac{1}{2^k} + \dfrac{1}{4^k}\right) + \cdots$

(b) $\displaystyle\sum_{k=1}^{\infty} \left(\dfrac{1}{5^k} - \dfrac{1}{k(k+1)}\right)$

2. In each part, use Theorem 11.4.3 to find the sum of the series.

(a) $\displaystyle\sum_{k=2}^{\infty} \left[\dfrac{1}{k^2 - 1} - \dfrac{7}{10^{k-1}}\right]$ (b) $\displaystyle\sum_{k=1}^{\infty} \left[7^{-k}3^{k+1} - \dfrac{2^{k+1}}{5^k}\right]$

> In Exercises 3 and 4, various *p*-series are given. In each case, find *p* and determine whether the series converges.

3. (a) $\displaystyle\sum_{k=1}^{\infty} \frac{1}{k^3}$ **(b)** $\displaystyle\sum_{k=1}^{\infty} \frac{1}{\sqrt{k}}$ **(c)** $\displaystyle\sum_{k=1}^{\infty} k^{-1}$ **(d)** $\displaystyle\sum_{k=1}^{\infty} k^{-2/3}$

4. (a) $\displaystyle\sum_{k=1}^{\infty} k^{-4/3}$ **(b)** $\displaystyle\sum_{k=1}^{\infty} \frac{1}{\sqrt[4]{k}}$ **(c)** $\displaystyle\sum_{k=1}^{\infty} \frac{1}{\sqrt[3]{k^5}}$ **(d)** $\displaystyle\sum_{k=1}^{\infty} \frac{1}{k^{\pi}}$

In Exercises 5 and 6, apply the divergence test, and state what it tells you about the series.

5. (a) $\displaystyle\sum_{k=1}^{\infty} \frac{k^2 + k + 3}{2k^2 + 1}$ **(b)** $\displaystyle\sum_{k=1}^{\infty} \left(1 + \frac{1}{k}\right)^k$

(c) $\displaystyle\sum_{k=1}^{\infty} \cos k\pi$ **(d)** $\displaystyle\sum_{k=1}^{\infty} \frac{1}{k!}$

6. (a) $\displaystyle\sum_{k=1}^{\infty} \frac{k}{e^k}$ **(b)** $\displaystyle\sum_{k=1}^{\infty} \ln k$ **(c)** $\displaystyle\sum_{k=1}^{\infty} \frac{1}{\sqrt{k}}$ **(d)** $\displaystyle\sum_{k=1}^{\infty} \frac{\sqrt{k}}{\sqrt{k} + 3}$

In Exercises 7 and 8, confirm that the integral test is applicable, and use it to determine whether the series converges.

7. (a) $\displaystyle\sum_{k=1}^{\infty} \frac{1}{5k + 2}$ **(b)** $\displaystyle\sum_{k=1}^{\infty} \frac{1}{1 + 9k^2}$

8. (a) $\displaystyle\sum_{k=1}^{\infty} \frac{k}{1 + k^2}$ **(b)** $\displaystyle\sum_{k=1}^{\infty} \frac{1}{(4 + 2k)^{3/2}}$

In Exercises 9–24, use any method to determine whether the series converges.

9. $\displaystyle\sum_{k=1}^{\infty} \frac{1}{k + 6}$ **10.** $\displaystyle\sum_{k=1}^{\infty} \frac{3}{5k}$ **11.** $\displaystyle\sum_{k=1}^{\infty} \frac{1}{\sqrt{k + 5}}$

12. $\displaystyle\sum_{k=1}^{\infty} \frac{1}{\sqrt[k]{e}}$ **13.** $\displaystyle\sum_{k=1}^{\infty} \frac{1}{\sqrt[3]{2k - 1}}$ **14.** $\displaystyle\sum_{k=3}^{\infty} \frac{\ln k}{k}$

15. $\displaystyle\sum_{k=1}^{\infty} \frac{k}{\ln(k + 1)}$ **16.** $\displaystyle\sum_{k=1}^{\infty} ke^{-k^2}$ **17.** $\displaystyle\sum_{k=1}^{\infty} \left(1 + \frac{1}{k}\right)^{-k}$

18. $\displaystyle\sum_{k=1}^{\infty} \frac{k^2 + 1}{k^2 + 3}$ **19.** $\displaystyle\sum_{k=1}^{\infty} \frac{\tan^{-1} k}{1 + k^2}$ **20.** $\displaystyle\sum_{k=1}^{\infty} \frac{1}{\sqrt{k^2 + 1}}$

21. $\displaystyle\sum_{k=1}^{\infty} k^2 \sin^2\left(\frac{1}{k}\right)$ **22.** $\displaystyle\sum_{k=1}^{\infty} k^2 e^{-k^3}$

23. $\displaystyle\sum_{k=5}^{\infty} 7k^{-1.01}$ **24.** $\displaystyle\sum_{k=1}^{\infty} \operatorname{sech}^2 k$

In Exercises 25 and 26, use the integral test to investigate the relationship between the value of p and the convergence of the series.

25. $\displaystyle\sum_{k=2}^{\infty} \frac{1}{k(\ln k)^p}$ **26.** $\displaystyle\sum_{k=3}^{\infty} \frac{1}{k(\ln k)[\ln(\ln k)]^p}$

$\boxed{c}$ **27.** Use a CAS to confirm that
$$\sum_{k=1}^{\infty} \frac{1}{k^2} = \frac{\pi^2}{6} \quad \text{and} \quad \sum_{k=1}^{\infty} \frac{1}{k^4} = \frac{\pi^4}{90}$$
and then use these results in each part to find the sum of the

series.

(a) $\displaystyle\sum_{k=1}^{\infty} \frac{3k^2 - 1}{k^4}$ **(b)** $\displaystyle\sum_{k=3}^{\infty} \frac{1}{k^2}$ **(c)** $\displaystyle\sum_{k=2}^{\infty} \frac{1}{(k - 1)^4}$

28. Suppose that the series $\sum u_k$ converges and the series $\sum v_k$ diverges.

(a) Show that the series $\sum(u_k + v_k)$ and $\sum(u_k - v_k)$ both diverge. [*Hint:* Assume that each series converges and use Theorem 11.4.3 to obtain a contradiction.]

(b) Find examples to show that if $\sum u_k$ and $\sum v_k$ both diverge, then the series $\sum(u_k + v_k)$ and $\sum(u_k - v_k)$ may either converge or diverge.

29. In each part, use the results in Exercise 28, if needed, to determine whether the series diverges.

(a) $\displaystyle\sum_{k=1}^{\infty} \left[\left(\frac{2}{3}\right)^{k-1} + \frac{1}{k}\right]$ **(b)** $\displaystyle\sum_{k=1}^{\infty} \left[\frac{1}{3k + 2} - \frac{1}{k^{3/2}}\right]$

(c) $\displaystyle\sum_{k=2}^{\infty} \left[\frac{1}{k(\ln k)^2} - \frac{1}{k^2}\right]$

Exercise 30 will show how a partial sum can be used to obtain upper and lower bounds on the sum of the series when the hypotheses of the integral test are satisfied. This result will be needed in Exercises 31–35.

30. (a) Let $\sum_{k=1}^{\infty} u_k$ be a convergent series with positive terms, let $f(x)$ be the function that results when k is replaced by x in the general term of the series, and suppose that f satisfies the hypotheses of the integral test for $x \geq n$ (Theorem 11.4.4). Use an area argument and the accompanying figure (following Exercise 35) to show that
$$\int_{n+1}^{+\infty} f(x)\, dx < \sum_{k=n+1}^{\infty} u_k < \int_{n}^{+\infty} f(x)\, dx$$

(b) Show that if S is the sum of the series $\sum_{k=1}^{\infty} u_k$ and s_n is the nth partial sum, then
$$s_n + \int_{n+1}^{+\infty} f(x)\, dx < S < s_n + \int_{n}^{+\infty} f(x)\, dx$$

31. (a) It was stated in Exercise 27 that
$$\sum_{k=1}^{\infty} \frac{1}{k^2} = \frac{\pi^2}{6}$$
Show that if s_n is the nth partial sum of this series, then
$$s_n + \frac{1}{n + 1} < \frac{\pi^2}{6} < s_n + \frac{1}{n}$$

(b) Calculate s_3 exactly, and then use the result in part (a) to show that
$$\frac{29}{18} < \frac{\pi^2}{6} < \frac{61}{36}$$

(c) Use a calculating utility to confirm that the inequalities in part (b) are correct.

(d) Find upper and lower bounds on the error that results if the sum of the series is approximated by the 10th partial sum.

32. In each part, find upper and lower bounds on the error that results if the sum of the series is approximated by the 10th partial sum.

(a) $\displaystyle\sum_{k=1}^{\infty} \frac{1}{(2k+1)^2}$ (b) $\displaystyle\sum_{k=1}^{\infty} \frac{1}{k^2+1}$ (c) $\displaystyle\sum_{k=1}^{\infty} \frac{k}{e^k}$

33. Our objective in this problem is to approximate the sum of the series $\sum_{k=1}^{\infty} 1/k^3$ to two decimal-place accuracy.

(a) Show that if S is the sum of the series and s_n is the nth partial sum, then
$$s_n + \frac{1}{2(n+1)^2} < S < s_n + \frac{1}{2n^2}$$

(b) For two decimal-place accuracy, the error must be less than 0.005 (see Table 2.4.1 on p. 154). According to the Approximation Principle (2.4.10), we can achieve this by finding an interval of length 0.01 (or less) that contains S and approximating S by the midpoint of that interval. Find the smallest value of n such that the interval containing S in part (a) has a length of 0.01 or less.

(c) Approximate S to two decimal-place accuracy.

34. (a) Use the method of Exercise 33 to approximate the sum of the series $\sum_{k=1}^{\infty} 1/k^4$ to two decimal-place accuracy.

(b) It was stated in Exercise 27 that the sum of this series is $\pi^4/90$. Use a calculating utility to confirm that your answer in part (a) is accurate to two decimal places.

35. We showed in Section 11.3 that the harmonic series $\sum_{k=1}^{\infty} 1/k$ diverges. Our objective in this problem is to demonstrate that although the partial sums of this series approach $+\infty$, they increase extremely slowly.

(a) Use inequality (2) to show that for $n \geq 2$
$$\ln(n+1) < s_n < 1 + \ln n$$

(b) Use the inequalities in part (a) to find upper and lower bounds on the sum of the first million terms in the series.

(c) Show that the sum of the first billion terms in the series is less than 22.

(d) Find a value of n so that the sum of the first n terms is greater than 100.

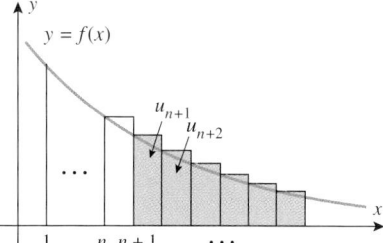

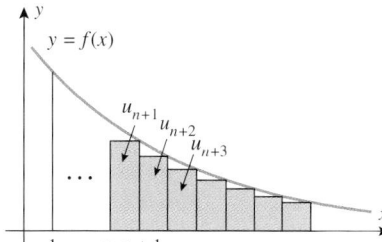

Figure Ex-30

36. Investigate the relationship between the value of a and the convergence of the series $\sum_{k=1}^{\infty} k^{-\ln a}$.

37. Use a graphing utility to confirm that the integral test applies to the series $\sum_{k=1}^{\infty} k^2 e^{-k}$, and then determine whether the series converges.

38. (a) Show that the integral test applies to the series $\sum_{k=1}^{\infty} 1/(k^3+1)$.

(b) Use a CAS and the integral test to confirm that the series converges.

(c) Construct a table of partial sums for $n = 10, 20, 30, \ldots, 100$, showing at least six decimal places.

(d) Based on your table, make a conjecture about the sum of the series to three decimal-place accuracy.

(e) Use part (b) of Exercise 30 to check your conjecture.

11.5 TAYLOR AND MACLAURIN SERIES

In this section we will discuss methods for approximating values of trigonometric and logarithmic functions. This will lead us to the more general problem of approximating functions by polynomials and then to the problem of finding infinite series that converge to specific functions. Issues of convergence and error estimation will be discussed in later sections. Moreover, if desired, this section can be covered after Section 11.7 or 11.8. We have placed it here for those who want an earlier discussion of Maclaurin and Taylor polynomials.

LOCAL QUADRATIC APPROXIMATIONS

Recall from Formula (6) in Section 3.6 that the local linear approximation of a function f at a point x_0 is
$$f(x) \approx f(x_0) + f'(x_0)(x - x_0) \tag{1}$$

In this formula, the approximating function

$$p(x) = f(x_0) + f'(x_0)(x - x_0)$$

is a first-degree polynomial whose value at x_0 is $f(x_0)$ and whose derivative at x_0 is $f'(x_0)$ (verify). Thus, the local linear approximation of f at x_0 has the property that its value and that of its first derivative match those of f at x_0.

If the graph of a function f has a pronounced "bend" at a point x_0, then we can expect that the accuracy of the local linear approximation of f at x_0 will decrease rapidly as we progress away from x_0 (Figure 11.5.1). One way to deal with this problem is to approximate the function f at x_0 by a polynomial p of degree 2 with the property that the value of p and the value of its first two derivatives match those of f at x_0. This ensures that the graphs of f and p not only have the same tangent line at x_0, but they also bend in the same direction at that point (both concave up or concave down). As a result, we can expect that the graph of p will remain close to the graph of f over a larger interval around x_0 than the graph of the local linear approximation. The polynomial p is called the ***local quadratic approximation of f at the point x = x_0***.

To illustrate this idea, let us try to find a formula for the local quadratic approximation of a function f at the point $x = 0$. This approximation has the form

$$f(x) \approx c_0 + c_1 x + c_2 x^2 \qquad (2)$$

where c_0, c_1, and c_2 must be chosen so that the values of

$$p(x) = c_0 + c_1 x + c_2 x^2$$

and its first two derivatives match those of f at 0. Thus, we want

$$p(0) = f(0), \quad p'(0) = f'(0), \quad p''(0) = f''(0) \qquad (3)$$

But the values of $p(0)$, $p'(0)$, and $p''(0)$ are as follows:

$$p(x) = c_0 + c_1 x + c_2 x^2 \qquad p(0) = c_0$$
$$p'(x) = c_1 + 2c_2 x \qquad p'(0) = c_1$$
$$p''(x) = 2c_2 \qquad p''(0) = 2c_2$$

Thus, it follows from (3) that

$$c_0 = f(0), \quad c_1 = f'(0), \quad c_2 = \frac{f''(0)}{2}$$

and substituting these in (2) yields the following formula for the local quadratic approximation of f at $x = 0$:

$$f(x) \approx f(0) + f'(0)x + \frac{f''(0)}{2}x^2 \qquad (4)$$

REMARK.   Observe that with $x_0 = 0$, Formula (1) becomes

$$f(x) \approx f(0) + f'(0)x \qquad (5)$$

and hence the linear part of the local quadratic approximation of f at 0 is the local linear approximation of f at 0.

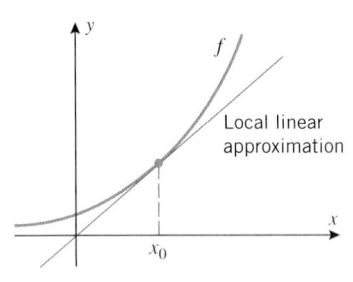

Figure 11.5.1

Local linear approximation

Example 1

Find the local linear and quadratic approximations of e^x at $x = 0$, and graph e^x and the two approximations together.

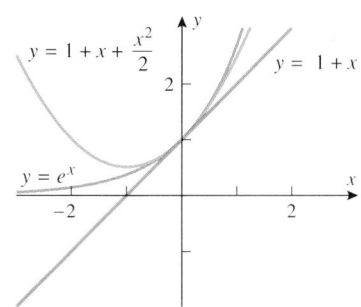

Figure 11.5.2

Solution.   If we let $f(x) = e^x$, then $f'(x) = f''(x) = e^x$; and hence

$$f(0) = f'(0) = f''(0) = e^0 = 1$$

Thus, from (4) the local quadratic approximation of e^x at $x = 0$ is

$$e^x \approx 1 + x + \frac{x^2}{2}$$

and the local linear approximation (which is the linear part of the local quadratic approximation) is

$$e^x \approx 1 + x$$

The graphs of e^x and the two approximations are shown in Figure 11.5.2. As expected, the local quadratic approximation is more accurate than the local linear approximation near $x = 0$.   ◀

MACLAURIN POLYNOMIALS

It is natural to ask whether one can improve on the accuracy of a local quadratic approximation by using a polynomial of degree 3. Specifically, one might look for a polynomial of degree 3 with the property that its value and the values of its first three derivatives match those of f at a point; and if this provides an improvement in accuracy, why not go on to polynomials of even higher degree? Thus, we are led to consider the following general problem.

> **11.5.1   PROBLEM.**   Given a function f that can be differentiated n times at a point x_0, find a polynomial p of degree n with the property that the value of p and the values of its first n derivatives match those of f at the point x_0.

We will begin by solving this problem in the case where $x_0 = 0$. Thus, we want a polynomial

$$p(x) = c_0 + c_1 x + c_2 x^2 + c_3 x^3 + \cdots + c_n x^n \tag{6}$$

such that

$$f(0) = p(0), \quad f'(0) = p'(0), \quad f''(0) = p''(0), \quad \ldots, \quad f^{(n)}(0) = p^{(n)}(0) \tag{7}$$

But

$$p(x) \ = c_0 + c_1 x + c_2 x^2 + c_3 x^3 + \cdots + c_n x^n$$
$$p'(x) \ = c_1 + 2c_2 x + 3c_3 x^2 + \cdots + nc_n x^{n-1}$$
$$p''(x) \ = 2c_2 + 3 \cdot 2c_3 x + \cdots + n(n-1)c_n x^{n-2}$$
$$p'''(x) \ = 3 \cdot 2c_3 + \cdots + n(n-1)(n-2)c_n x^{n-3}$$
$$\vdots$$
$$p^{(n)}(x) = n(n-1)(n-2) \cdots (1)c_n$$

Thus, to satisfy (7) we must have

$$f(0) \ = p(0) \ = c_0$$
$$f'(0) \ = p'(0) \ = c_1$$
$$f''(0) \ = p''(0) \ = 2c_2 = 2!c_2$$
$$f'''(0) \ = p'''(0) \ = 3 \cdot 2c_3 = 3!c_3$$
$$\vdots$$
$$f^{(n)}(0) = p^{(n)}(0) = n(n-1)(n-2) \cdots (1)c_n = n!c_n$$

which yields the following values for the coefficients of $p(x)$:

$$c_0 = f(0), \quad c_1 = f'(0), \quad c_2 = \frac{f''(0)}{2!}, \quad c_3 = \frac{f'''(0)}{3!}, \quad \ldots, \quad c_n = \frac{f^{(n)}(0)}{n!}$$

The polynomial that results by using these coefficients in (6) is called the *nth Maclaurin** *polynomial for f*.

11.5.2 DEFINITION. If f can be differentiated n times at 0, then we define the **nth Maclaurin polynomial for f** to be

$$p_n(x) = f(0) + f'(0)x + \frac{f''(0)}{2!}x^2 + \frac{f'''(0)}{3!}x^3 + \cdots + \frac{f^{(n)}(0)}{n!}x^n \qquad (8)$$

This polynomial has the property that its value and the values of its first n derivatives match the values of f and its first n derivatives at $x = 0$.

REMARK. Observe that $p_1(x)$ is the local linear approximation of f at 0 and $p_2(x)$ is the local quadratic approximation of f at $x = 0$.

Example 2

Find the Maclaurin polynomials p_0, p_1, p_2, p_3, and p_n for e^x.

Solution. Let $f(x) = e^x$. Thus,

$$f'(x) = f''(x) = f'''(x) = \cdots = f^{(n)}(x) = e^x$$

and

$$f(0) = f'(0) = f''(0) = f'''(0) = \cdots = f^{(n)}(0) = e^0 = 1$$

Therefore,

$$p_0(x) = f(0) = 1$$

$$p_1(x) = f(0) + f'(0)x = 1 + x$$

$$p_2(x) = f(0) + f'(0)x + \frac{f''(0)}{2!}x^2 = 1 + x + \frac{x^2}{2!} = 1 + x + \frac{1}{2}x^2$$

$$p_3(x) = f(0) + f'(0)x + \frac{f''(0)}{2!}x^2 + \frac{f'''(0)}{3!}x^3$$

$$= 1 + x + \frac{x^2}{2!} + \frac{x^3}{3!} = 1 + x + \frac{1}{2}x^2 + \frac{1}{6}x^3$$

$$p_n(x) = f(0) + f'(0)x + \frac{f''(0)}{2!}x^2 + \cdots + \frac{f^{(n)}(0)}{n!}x^n$$

$$= 1 + x + \frac{x^2}{2!} + \cdots + \frac{x^n}{n!} \qquad \blacktriangleleft$$

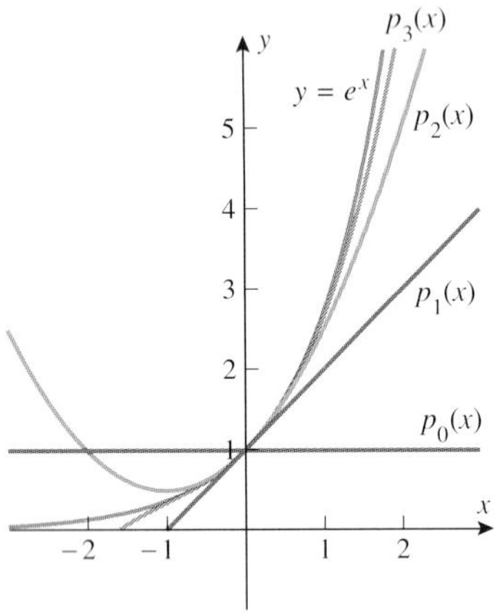

Figure 11.5.3

Figure 11.5.3 shows the graphs of e^x (in blue) and the graphs of the first four Maclaurin polynomials. Note that the graphs of $p_1(x)$, $p_2(x)$, and $p_3(x)$ are virtually indistinguishable from the graph of e^x near the origin, so that these polynomials are good approximations of e^x for x near 0. However, the farther x is from 0, the poorer these approximations become. This is typical of the Maclaurin polynomials for a function $f(x)$; they provide good

---

*COLIN MACLAURIN (1698–1746). Scottish mathematician. Maclaurin's father, a minister, died when the boy was only six months old, and his mother when he was nine years old. He was then raised by an uncle who was also a minister. Maclaurin entered Glasgow University as a divinity student, but transferred to mathematics after one year. He received his Master's degree at age 17 and, in spite of his youth, began teaching at Marischal College in Aberdeen, Scotland. He met Isaac Newton during a visit to London in 1719 and from that time on became Newton's disciple. During that era, some of Newton's analytic methods were bitterly attacked by major mathematicians and much of Maclaurin's important mathematical work resulted from his efforts to defend Newton's ideas geometrically. Maclaurin's work, *A Treatise of Fluxions* (1742), was the first systematic formulation of Newton's methods. The treatise was so carefully done that it was a standard of mathematical rigor in calculus until the work of Cauchy in 1821.

Maclaurin was an outstanding experimentalist. He devised numerous ingenious mechanical devices, made important astronomical observations, performed actuarial computations for insurance societies, and helped to improve maps of the islands around Scotland.

approximations of $f(x)$ near 0, but the accuracy diminishes as x progresses away from 0. However, it is usually the case that the higher the degree of the polynomial, the larger the interval on which it provides a specified accuracy. Accuracy issues will be investigated later.

TAYLOR POLYNOMIALS

Up to now we have focused on approximating a function f in the vicinity of the origin. Now we will consider the more general case of approximating f in the vicinity of an arbitrary point x_0. The basic idea is the same as before; we want to find an nth-degree polynomial p with the property that its value and the values of its first n derivatives match those of f at x_0. However, rather than expressing $p(x)$ in powers of x, it will simplify the computations if we express it in powers of $x - x_0$; that is,

$$p(x) = c_0 + c_1(x - x_0) + c_2(x - x_0)^2 + \cdots + c_n(x - x_0)^n \tag{9}$$

We will leave it as an exercise for you to imitate the computations used in the case where $x_0 = 0$ to show that

$$c_0 = f(x_0), \quad c_1 = f'(x_0), \quad c_2 = \frac{f''(x_0)}{2!}, \quad c_3 = \frac{f'''(x_0)}{3!}, \quad \dots, \quad c_n = \frac{f^{(n)}(x_0)}{n!}$$

Substituting these values in (9) we obtain a polynomial called the *nth Taylor* * *polynomial about $x = x_0$ for f*.

11.5.3 DEFINITION.   If f can be differentiated n times at x_0, then we define the ***nth Taylor polynomial for f about x = x_0*** to be

$$p_n(x) = f(x_0) + f'(x_0)(x - x_0) + \frac{f''(x_0)}{2!}(x - x_0)^2$$

$$+ \frac{f'''(x_0)}{3!}(x - x_0)^3 + \cdots + \frac{f^{(n)}(x_0)}{n!}(x - x_0)^n \tag{10}$$

REMARK.   Observe that the Maclaurin polynomials are special cases of the Taylor polynomials; that is, the nth-order Maclaurin polynomial is the nth-order Taylor polynomial about $x = 0$. Observe also that $p_1(x)$ is the local linear approximation of f at $x = x_0$ and $p_2(x)$ is the local quadratic approximation of f at $x = x_0$.

Example 3

Find the first four Taylor polynomials for $\ln x$ about $x = 2$.

Solution. Let $f(x) = \ln x$. Thus,

$$
\begin{aligned}
f(x) &= \ln x & f(2) &= \ln 2 \\
f'(x) &= 1/x & f'(2) &= 1/2 \\
f''(x) &= -1/x^2 & f''(2) &= -1/4 \\
f'''(x) &= 2/x^3 & f'''(2) &= 1/4
\end{aligned}
$$

---

* BROOK TAYLOR (1685–1731). English mathematician. Taylor was born of well-to-do parents. Musicians and artists were entertained frequently in the Taylor home, which undoubtedly had a lasting influence on young Brook. In later years, Taylor published a definitive work on the mathematical theory of perspective and obtained major mathematical results about the vibrations of strings. There also exists an unpublished work, *On Musick*, that was intended to be part of a joint paper with Isaac Newton. Taylor's life was scarred with unhappiness, illness, and tragedy. Because his first wife was not rich enough to suit his father, the two men argued bitterly and parted ways. Subsequently, his wife died in childbirth. Then, after he remarried, his second wife also died in childbirth, though his daughter survived. Taylor's most productive period was from 1714 to 1719, during which time he wrote on a wide range of subjects—magnetism, capillary action, thermometers, perspective, and calculus. In his final years, Taylor devoted his writing efforts to religion and philosophy. According to Taylor, the results that bear his name were motivated by coffeehouse conversations about works of Newton on planetary motion and works of Halley ("Halley's comet") on roots of polynomials.

Taylor's writing style was so terse and hard to understand that he never received credit for many of his innovations.

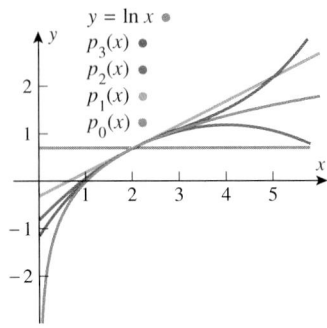

$y = \ln x$ •
$p_3(x)$ •
$p_2(x)$ •
$p_1(x)$ •
$p_0(x)$ •

Figure 11.5.4

Substituting in (10) with $x_0 = 2$ yields

$$p_0(x) = f(2) = \ln 2$$

$$p_1(x) = f(2) + f'(2)(x - 2) = \ln 2 + \tfrac{1}{2}(x - 2)$$

$$p_2(x) = f(2) + f'(2)(x - 2) + \frac{f''(2)}{2!}(x - 2)^2 = \ln 2 + \tfrac{1}{2}(x - 2) - \tfrac{1}{8}(x - 2)^2$$

$$p_3(x) = f(2) + f'(2)(x - 2) + \frac{f''(2)}{2!}(x - 2)^2 + \frac{f'''(2)}{3!}(x - 2)^3$$

$$= \ln 2 + \tfrac{1}{2}(x - 2) - \tfrac{1}{8}(x - 2)^2 + \tfrac{1}{24}(x - 2)^3$$

The graph of $\ln x$ (in blue) and its first four Taylor polynomials about $x = 2$ are shown in Figure 11.5.4. As expected, these polynomials produce their best approximations of $\ln x$ near 2.   ◀

Frequently, we will want to express Formula (10) in sigma notation. To do this, we use the notation $f^{(k)}(x_0)$ to denote the kth derivative of f at $x = x_0$, and we make the convention that $f^{(0)}(x_0)$ denotes $f(x_0)$. This enables us to write

$$\sum_{k=0}^{n} \frac{f^{(k)}(x_0)}{k!}(x - x_0)^k = f(x_0) + f'(x_0)(x - x_0)$$

$$+ \frac{f''(x_0)}{2!}(x - x_0)^2 + \cdots + \frac{f^{(n)}(x_0)}{n!}(x - x_0)^n \quad (11)$$

In particular, we can write the nth-order Maclaurin polynomial for $f(x)$ as

$$\sum_{k=0}^{n} \frac{f^{(k)}(0)}{k!}x^k = f(0) + f'(0)x + \frac{f''(0)}{2!}x^2 + \cdots + \frac{f^{(n)}(0)}{n!}x^n \quad (12)$$

For a fixed value of x near x_0, one would expect the approximation of $f(x)$ by its Taylor polynomial $p_n(x)$ about $x = x_0$ to improve as n increases, since increasing n has the effect of matching higher and higher derivatives of $f(x)$ with those of $p_n(x)$ at $x = x_0$. Indeed, it seems plausible that one might be able to achieve any desired degree of accuracy at a point x by choosing n sufficiently large; that is, the values of $p_n(x)$ might actually converge to $f(x)$ as $n \to +\infty$. Should this happen, we would have

$$f(x) = \lim_{n \to +\infty} \sum_{k=0}^{n} \frac{f^{(k)}(x_0)}{k!}(x - x_0)^k = \sum_{k=0}^{\infty} \frac{f^{(k)}(x_0)}{k!}(x - x_0)^k$$

Later we will study conditions under which the series on the right actually converges to $f(x)$. For the remainder of this section, we will focus on the computational aspects of finding these series. We make the following definition.

11.5.4   DEFINITION.   If f has derivatives of all orders at x_0, then we call the series

$$\sum_{k=0}^{\infty} \frac{f^{(k)}(x_0)}{k!}(x - x_0)^k = f(x_0) + f'(x_0)(x - x_0) + \frac{f''(x_0)}{2!}(x - x_0)^2$$

$$+ \cdots + \frac{f^{(k)}(x_0)}{k!}(x - x_0)^k + \cdots \quad (13)$$

the **Taylor series for f about $x = x_0$**. In the special case where $x_0 = 0$ this series becomes

$$\sum_{k=0}^{\infty} \frac{f^{(k)}(0)}{k!}x^k = f(0) + f'(0)x + \frac{f''(0)}{2!}x^2 + \cdots + \frac{f^{(k)}(0)}{k!}x^k + \cdots \quad (14)$$

in which case we call it the **Maclaurin series for f**.

REMARK.   Because the summation index in (13) starts at $k = 0$, it is convenient to call the initial term in this series the zeroth term. Thus, a Taylor series has a zeroth partial sum, a first partial sum, a second partial sum, and so forth. With this convention the nth partial sum of a Taylor series is the nth Taylor polynomial and the nth partial sum of a Maclaurin series is the nth Maclaurin polynomial [see (11) and (12)].

Example 4

Find the Maclaurin series for

$\quad$ (a) e^x $\quad$ (b) $\sin x$ $\quad$ (c) $\cos x$ $\quad$ (d) $\dfrac{1}{1-x}$

Solution (a). In Example 2 we found the nth Maclaurin polynomial for the function e^x to be

$$\sum_{k=0}^{n} \frac{x^k}{k!} = 1 + x + \frac{x^2}{2!} + \cdots + \frac{x^n}{n!}$$

Thus, the Maclaurin series for e^x is

$$\sum_{k=0}^{\infty} \frac{x^k}{k!} = 1 + x + \frac{x^2}{2!} + \frac{x^3}{3!} + \cdots + \frac{x^k}{k!} + \cdots$$

Solution (b). In the Maclaurin polynomials for $\sin x$, only the odd powers of x appear explicitly. To see this, let $f(x) = \sin x$; thus,

$$
\begin{aligned}
f(x) &= \sin x & f(0) &= 0 \\
f'(x) &= \cos x & f'(0) &= 1 \\
f''(x) &= -\sin x & f''(0) &= 0 \\
f'''(x) &= -\cos x & f'''(0) &= -1
\end{aligned}
$$

Since $f^{(4)}(x) = \sin x = f(x)$, the pattern $0, 1, 0, -1$ will repeat as we evaluate successive derivatives at 0. Therefore, the successive Maclaurin polynomials for $\sin x$ are

$$p_0(x) = 0$$
$$p_1(x) = 0 + x$$
$$p_2(x) = 0 + x + 0$$
$$p_3(x) = 0 + x + 0 - \frac{x^3}{3!}$$
$$p_4(x) = 0 + x + 0 - \frac{x^3}{3!} + 0$$
$$p_5(x) = 0 + x + 0 - \frac{x^3}{3!} + 0 + \frac{x^5}{5!}$$
$$p_6(x) = 0 + x + 0 - \frac{x^3}{3!} + 0 + \frac{x^5}{5!} + 0$$
$$p_7(x) = 0 + x + 0 - \frac{x^3}{3!} + 0 + \frac{x^5}{5!} + 0 - \frac{x^7}{7!}$$
$$\vdots$$

Because of the zero terms, each even-order Maclaurin polynomial [after $p_0(x)$] is the same as the preceding odd-order Maclaurin polynomial; that is,

$$p_{2k+1}(x) = p_{2k+2}(x) = x - \frac{x^3}{3!} + \frac{x^5}{5!} - \frac{x^7}{7!} + \cdots + (-1)^k \frac{x^{2k+1}}{(2k+1)!} \quad (k = 0, 1, 2, \ldots)$$

Thus, the Maclaurin series for $\sin x$ is

$$\sum_{k=0}^{\infty}(-1)^k\frac{x^{2k+1}}{(2k+1)!} = x - \frac{x^3}{3!} + \frac{x^5}{5!} - \frac{x^7}{7!} + \cdots + (-1)^k\frac{x^{2k+1}}{(2k+1)!} + \cdots$$

The graphs of $\sin x$, $p_1(x)$, $p_3(x)$, $p_5(x)$, and $p_7(x)$ are shown in Figure 11.5.5.

Solution (c). In the Maclaurin polynomials for $\cos x$, only the even powers of x appear explicitly; the computations are similar to those in part (b). The reader should be able to show that

$$p_0(x) = p_1(x) = 1$$

$$p_2(x) = p_3(x) = 1 - \frac{x^2}{2!}$$

$$p_4(x) = p_5(x) = 1 - \frac{x^2}{2!} + \frac{x^4}{4!}$$

$$p_6(x) = p_7(x) = 1 - \frac{x^2}{2!} + \frac{x^4}{4!} - \frac{x^6}{6!}$$

In general, the Maclaurin polynomials for $\cos x$ are

$$p_{2k}(x) = p_{2k+1}(x) = 1 - \frac{x^2}{2!} + \frac{x^4}{4!} - \cdots + (-1)^k\frac{x^{2k}}{(2k)!} \quad (k = 0, 1, 2, \ldots)$$

from which it follows that the Maclaurin series for $\cos x$ is

$$\sum_{k=0}^{\infty}(-1)^k\frac{x^{2k}}{(2k)!} = 1 - \frac{x^2}{2!} + \frac{x^4}{4!} - \frac{x^6}{6!} + \cdots + (-1)^k\frac{x^{2k}}{(2k)!} + \cdots$$

The graphs of $\cos x$, $p_0(x)$, $p_2(x)$, $p_4(x)$, and $p_6(x)$ are shown in Figure 11.5.6.

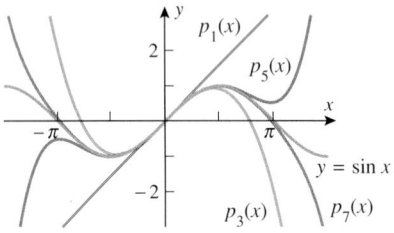

Figure 11.5.5

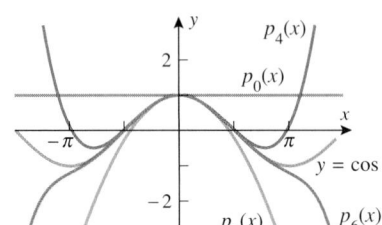

Figure 11.5.6

Solution (d). Let $f(x) = 1/(1 - x)$. Thus, the values of f and its first k derivatives at $x = 0$ are as follows:

$$
\begin{array}{ll}
f(x) = \dfrac{1}{1-x} & f(0) = 1 = 0! \\[2mm]
f'(x) = \dfrac{1}{(1-x)^2} & f'(0) = 1 = 1! \\[2mm]
f''(x) = \dfrac{2}{(1-x)^3} & f''(0) = 2 = 2! \\[2mm]
f'''(x) = \dfrac{3 \cdot 2}{(1-x)^4} & f'''(0) = 3! \\[2mm]
f^{(4)}(x) = \dfrac{4 \cdot 3 \cdot 2}{(1-x)^5} & f^{(4)}(0) = 4! \\[2mm]
\vdots & \vdots \\[2mm]
f^{(k)}(x) = \dfrac{k!}{(1-x)^{k+1}} & f^{(k)}(0) = k! \\[2mm]
\vdots & \vdots
\end{array}
$$

Substituting $f^{(k)}(0) = k!$ in Formula (14) yields the Maclaurin series

$$\sum_{k=0}^{\infty} x^k = 1 + x + x^2 + \cdots + x^k + \cdots$$

Thus, the Maclaurin series for $1/(1-x)$ happens to be the geometric series with initial term 1 and ratio x. ◀

Example 5

Find the Taylor series about $x = 1$ for $1/x$.

Solution. Let $f(x) = 1/x$. The computations are similar to those in part (d) of Example 4. We leave it for you to show that

$$f(1) = 1, \quad f'(1) = -1, \quad f''(1) = 2!, \quad f'''(1) = -3!,$$
$$f^{(4)}(1) = 4!, \quad \ldots, \quad f^{(k)}(1) = (-1)^k k!$$

Thus, substituting $f^{(k)}(1) = (-1)^k k!$ into Formula (13) with $x_0 = 1$ yields the Taylor series

$$\sum_{k=0}^{\infty} (-1)^k (x-1)^k = 1 - (x-1) + (x-1)^2 - (x-1)^3 + \cdots$$ ◀

FOR THE READER. CAS programs have commands for generating Taylor polynomials of any specified degree. If you have a CAS, read the documentation to determine how this is done, and then use the CAS to confirm the computations in the examples in this section.

EXERCISE SET 11.5 ☒ Graphing Calculator [c] CAS

1. In each part, find the local quadratic approximation of f at $x = x_0$, and use that approximation to find the local linear approximation of f at x_0.
   (a) $f(x) = e^{-x}$; $x_0 = 0$
   (b) $f(x) = \cos x$; $x_0 = 0$
   (c) $f(x) = \sin x$; $x_0 = \pi/2$
   (d) $f(x) = \sqrt{x}$; $x_0 = 1$

[c] 2. In each part, use a CAS to find the local quadratic approximation of f at $x = x_0$, and use that approximation to find the local linear approximation of f at $x = x_0$.
   (a) $f(x) = e^{\sin x}$; $x_0 = 0$
   (b) $f(x) = \sqrt{x}$; $x_0 = 9$
   (c) $f(x) = \sec^{-1} x$; $x_0 = 2$
   (d) $f(x) = \sin^{-1} x$; $x_0 = 0$

3. (a) Find the local quadratic approximation of $\sqrt{x}$ at $x_0 = 1$.
   (b) Use the result obtained in part (a) to approximate $\sqrt{1.1}$, and compare your approximation to that produced directly by your calculating utility. [See Example 4 of Section 3.6.]

4. (a) Find the local quadratic approximation of $\cos x$ at $x_0 = 0$.
   (b) Use the result obtained in part (a) to approximate $\cos 2°$, and compare the approximation to that produced directly by your calculating utility.

5. Use an appropriate local quadratic approximation to approximate $\tan 61°$, and compare the result to that produced directly by your calculating utility.

6. Use an appropriate local quadratic approximation to approximate $\sqrt{36.03}$, and compare the result to that produced directly by your calculating utility.

In Exercises 7–16, find the Maclaurin polynomials of orders $n = 0, 1, 2, 3,$ and 4, and then find the Maclaurin series for the function in sigma notation.

7. e^{-x}

8. e^{ax}

9. $\cos \pi x$

10. $\sin \pi x$

11. $\ln(1 + x)$

12. $\dfrac{1}{1 + x}$

13. $\cosh x$

14. $\sinh x$

15. $x \sin x$

16. xe^x

17. (a) Find the Maclaurin series for the polynomial
   $$f(x) = 1 + 2x - x^2 + x^3.$$
   (b) Find the Maclaurin series for the polynomial
   $$f(x) = c_0 + c_1 x + c_2 x^2 + \cdots + c_n x^n.$$

[c] 18. For each of the Exercises 7–16 that you worked on, use a CAS to check the Maclaurin polynomial of order $n = 4$.

In Exercises 19–26, find the Taylor polynomials of orders $n = 0, 1, 2, 3,$ and 4 about $x = x_0$, and then find the Taylor series for the function in sigma notation.

19. e^x; $x_0 = 1$

20. e^{-x}; $x_0 = \ln 2$

21. $\dfrac{1}{x}$; $x_0 = -1$

22. $\dfrac{1}{x + 2}$; $x_0 = 3$

23. $\sin \pi x;\ x_0 = \dfrac{1}{2}$ **24.** $\cos x;\ x_0 = \dfrac{\pi}{2}$

25. $\ln x;\ x_0 = 1$ **26.** $\ln x;\ x_0 = e$

27. (a) Find the Taylor series about $x = 1$ for the polynomial
$$f(x) = 1 + 2(x - 1) - (x - 1)^2 + (x - 1)^3$$

(b) Find the Taylor series about $x = x_0$ for the polynomial
$$f(x) = c_0 + c_1(x - x_0) + c_2(x - x_0)^2 + \cdots + c_n(x - x_0)^n$$

C 28. For each of the Exercises 19–26 that you worked on, use a CAS to check the Taylor polynomial of order $n = 4$.

> In Exercises 29–32, find the first four distinct Taylor polynomials about $x = x_0$, and use a graphing utility to graph the given function and the Taylor polynomials on the same screen.

29. $f(x) = e^{-2x};\ x_0 = 0$ **30.** $f(x) = \sin x;\ x_0 = \pi/2$

31. $f(x) = \cos x;\ x_0 = \pi$ **32.** $\ln(x + 1);\ x_0 = 0$

33. Which of the functions graphed in the following figure is most likely to have $p(x) = 1 - x + 2x^2$ as its second-order Maclaurin polynomial? Explain your reasoning.

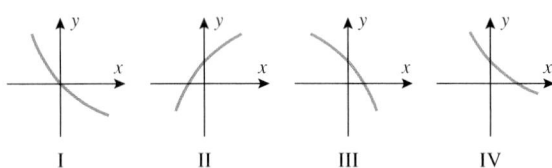

I   II   III   IV

34. Suppose that the values of a function f and its first three derivatives at $x = 1$ are
$$f(1) = 2, \quad f'(1) = -3, \quad f''(1) = 0, \quad f'''(1) = 6$$
Find as many Taylor polynomials for f as you can about $x = 1$.

35. Show that the Taylor series for $\sinh x$ about $x = \ln 4$ is
$$\sum_{k=0}^{\infty} \frac{16 - (-1)^k}{8k!}(x - \ln 4)^k$$

36. (a) The accompanying figure shows a sector of radius r and central angle 2α. Assuming that the angle α is small, use the local quadratic approximation of $\cos \alpha$ at $\alpha = 0$ to show that $x \approx r\alpha^2/2$.

(b) Assuming that the Earth is a sphere of radius 4000 mi, use the result in part (a) to approximate the maximum amount by which a 100-mi arc along the equator will diverge from its chord.

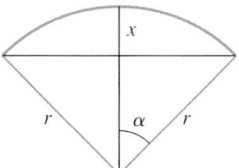

Figure Ex-36

37. Let $p_1(x)$ and $p_2(x)$ be the local linear and local quadratic approximations of $f(x) = e^{\sin x}$ at $x = 0$.

(a) Use a graphing utility to generate the graphs of $f(x)$, $p_1(x)$, and $p_2(x)$ on the same screen for $-1 \le x \le 1$.

(b) Construct a table of values of $f(x)$, $p_1(x)$, and $p_2(x)$ for $x = -1.00, -0.75, -0.50, -0.25, 0, 0.25, 0.50, 0.75, 1.00$. Round the values to three decimal places.

(c) Generate the graph of $|f(x) - p_1(x)|$, and use the graph to determine an interval on which $p_1(x)$ approximates $f(x)$ with an error of at most ± 0.01. [*Suggestion:* Review the discussion relating to Figure 3.6.9.]

(d) Generate the graph of $|f(x) - p_2(x)|$, and use the graph to determine an interval on which $p_2(x)$ approximates $f(x)$ with an error of at most ± 0.01.

11.6 THE COMPARISON, RATIO, AND ROOT TESTS

In this section we will develop some more basic convergence tests for series with non-negative terms. Later, we will use some of these tests to study the convergence of Taylor series.

THE COMPARISON TEST

We will begin with a test that is useful in its own right and is also the building block for other important convergence tests. The underlying idea of this test is to use the known convergence or divergence of a series to deduce the convergence or divergence of another series.

> **11.6.1 THEOREM (*The Comparison Test*).** Let $\sum_{k=1}^{\infty} a_k$ and $\sum_{k=1}^{\infty} b_k$ be series with nonnegative terms and suppose that
> $$a_1 \le b_1,\ a_2 \le b_2,\ a_3 \le b_3, \ldots, a_k \le b_k, \ldots$$
> (*a*)   If the "bigger series" $\sum b_k$ converges, then the "smaller series" $\sum a_k$ also converges.
> (*b*)   If the "smaller series" $\sum a_k$ diverges, then the "bigger series" $\sum b_k$ also diverges.

We have left the proof of this theorem for the exercises; however, it is easy to visualize why the theorem is true by interpreting the terms in the series as areas of rectangles (Figure 11.6.1). The comparison test states that if the total area $\sum b_k$ is finite, then the total area $\sum a_k$ must also be finite; and if the total area $\sum a_k$ is infinite, then the total area $\sum b_k$ must also be infinite.

REMARK.   As one would expect, it is not essential in Theorem 11.6.1 that the condition $a_k \le b_k$ hold for all k, as stated; the conclusions of the theorem remain true if this condition is eventually true.

USING THE COMPARISON TEST

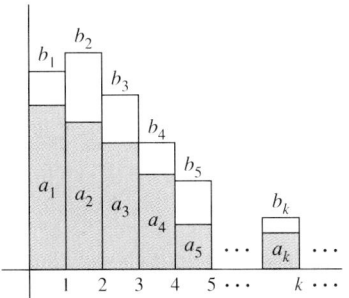

For each rectangle, b_k is the entire area and a_k is the area of the blue portion.

Figure 11.6.1

There are two steps required for using the comparison test to determine whether a series $\sum u_k$ with positive terms converges:

- Guess at whether the series $\sum u_k$ converges or diverges.
- Find a series that proves the guess to be correct. That is, if the guess is divergence, we must find a divergent series whose terms are "smaller" than the corresponding terms of $\sum u_k$, and if the guess is convergence, we must find a convergent series whose terms are "bigger" than the corresponding terms of $\sum u_k$.

To help with the guessing process in the first step, we have formulated two principles that sometimes *suggest* whether a series is likely to converge or diverge. We have called these "informal principles" because they are not intended as formal theorems. In fact, we will not guarantee that they *always* work. However, they work often enough to be useful.

11.6.2   INFORMAL PRINCIPLE.   *Constant terms in the denominator of u_k can usually be deleted without affecting the convergence or divergence of the series.*

11.6.3   INFORMAL PRINCIPLE.   *If a polynomial in k appears as a factor in the numerator or denominator of u_k, all but the leading term in the polynomial can usually be discarded without affecting the convergence or divergence of the series.*

Example 1

Use the comparison test to determine whether the following series converge or diverge.

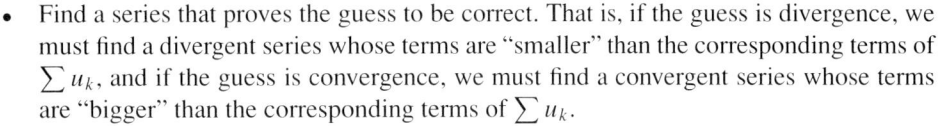

(a) $\sum_{k=1}^{\infty} \dfrac{1}{\sqrt{k} - \frac{1}{2}}$ (b) $\sum_{k=1}^{\infty} \dfrac{1}{2k^2 + k}$

Solution (a). According to Principle 11.6.2, we should be able to drop the constant in the denominator without affecting the convergence or divergence. Thus, the given series is likely to behave like

$$\sum_{k=1}^{\infty} \frac{1}{\sqrt{k}} \tag{1}$$

which is a divergent p-series $\left(p = \frac{1}{2}\right)$. Thus, we will guess that the given series diverges and try to prove this by finding a divergent series that is "smaller" than the given series. However, series (1) does the trick since

$$\frac{1}{\sqrt{k} - \frac{1}{2}} > \frac{1}{\sqrt{k}} \quad \text{for } k = 1, 2, \ldots$$

Thus, we have proved that the given series diverges.

Solution (b). According to Principle 11.6.3, we should be able to discard all but the leading term in the polynomial without affecting the convergence or divergence. Thus, the given series is likely to behave like

$$\sum_{k=1}^{\infty} \frac{1}{2k^2} = \frac{1}{2} \sum_{k=1}^{\infty} \frac{1}{k^2} \tag{2}$$

which converges since it is a constant times a convergent p-series ($p = 2$). Thus, we will guess that the given series converges and try to prove this by finding a convergent series that is "bigger" than the given series. However, series (2) does the trick since

$$\frac{1}{2k^2 + k} < \frac{1}{2k^2} \quad \text{for } k = 1, 2, \ldots$$

Thus, we have proved that the given series converges.   ◀

THE LIMIT COMPARISON TEST

In the last example, Principles 11.6.2 and 11.6.3 provided the guess about convergence or divergence as well as the series needed to apply the comparison test. Unfortunately, it is not always so straightforward to find the series required for comparison, so we will now consider an alternative to the comparison test that is usually easier to apply. The proof is given in Appendix G.

11.6.4   THEOREM (*The Limit Comparison Test*).   *Let $\sum a_k$ and $\sum b_k$ be series with positive terms and suppose that*

$$\rho = \lim_{k \to +\infty} \frac{a_k}{b_k}$$

If ρ is finite and $\rho > 0$, then the series both converge or both diverge.

The cases where $\rho = 0$ or $\rho = +\infty$ are discussed in the exercises (Exercise 54).

Example 2

Use the limit comparison test to determine whether the following series converge or diverge.

(a) $\displaystyle\sum_{k=2}^{\infty} \frac{1}{\sqrt{k}-1}$ (b) $\displaystyle\sum_{k=1}^{\infty} \frac{1}{2k^2+k}$ (c) $\displaystyle\sum_{k=1}^{\infty} \frac{3k^3 - 2k^2 + 4}{k^7 - k^3 + 2}$

Solution (a). As in Example 1, Principle 11.6.2 suggests that the series is likely to behave like the divergent p-series (1). To prove that the given series diverges, we will apply the limit comparison test with

$$a_k = \frac{1}{\sqrt{k}-1} \quad \text{and} \quad b_k = \frac{1}{\sqrt{k}}$$

We obtain

$$\rho = \lim_{k \to +\infty} \frac{a_k}{b_k} = \lim_{k \to +\infty} \frac{\sqrt{k}}{\sqrt{k}-1} = \lim_{k \to +\infty} \frac{1}{1 - \dfrac{1}{\sqrt{k}}} = 1$$

Since ρ is finite and positive, it follows from Theorem 11.6.4 that the given series diverges, which agrees with the conclusion reached in Example 1 using the comparison test.

Solution (b). As in Example 1, Principle 11.6.3 suggests that the series is likely to behave like the convergent series (2). To prove that the given series converges, we will apply the limit comparison test with

$$a_k = \frac{1}{2k^2+k} \quad \text{and} \quad b_k = \frac{1}{2k^2}$$

We obtain

$$\rho = \lim_{k \to +\infty} \frac{a_k}{b_k} = \lim_{k \to +\infty} \frac{2k^2}{2k^2 + k} = \lim_{k \to +\infty} \frac{2}{2 + \dfrac{1}{k}} = 1$$

Since ρ is finite and positive, it follows from Theorem 11.6.4 that the given series converges, which agrees with the conclusion reached in Example 1 using the comparison test.

Solution (*c*). From Principle 11.6.3, the series is likely to behave like

$$\sum_{k=1}^{\infty} \frac{3k^3}{k^7} = \sum_{k=1}^{\infty} \frac{3}{k^4} \tag{3}$$

which converges since it is a constant times a convergent *p*-series. Thus, the given series is likely to converge. To prove this, we will apply the limit comparison test to series (3) and the given series. We obtain

$$\rho = \lim_{k \to +\infty} \frac{\dfrac{3k^3 - 2k^2 + 4}{k^7 - k^3 + 2}}{\dfrac{3}{k^4}} = \lim_{k \to +\infty} \frac{3k^7 - 2k^6 + 4k^4}{3k^7 - 3k^3 + 6} = 1$$

Since ρ is finite and nonzero, it follows from Theorem 11.6.4 that the given series converges, since (3) converges. ◄

THE RATIO TEST

The comparison test and the limit comparison test hinge on first making a guess about convergence and then finding an appropriate series for comparison, both of which can be difficult tasks in cases where Principles 11.6.2 and 11.6.3 cannot be applied. In such cases the next test can often be used, since it works exclusively with the terms of the given series—it requires neither an initial guess about convergence nor the discovery of a series for comparison. Its proof is given in Appendix G.

11.6.5 THEOREM (*The Ratio Test*). *Let $\sum u_k$ be a series with positive terms and suppose that*

$$\rho = \lim_{k \to +\infty} \frac{u_{k+1}}{u_k}$$

(*a*) *If $\rho < 1$, the series converges.*

(*b*) *If $\rho > 1$ or $\rho = +\infty$, the series diverges.*

(*c*) *If $\rho = 1$, the series may converge or diverge, so that another test must be tried.*

Example 3

Use the ratio test to determine whether the following series converge or diverge.

(a) $\displaystyle\sum_{k=1}^{\infty} \frac{1}{k!}$ (b) $\displaystyle\sum_{k=1}^{\infty} \frac{k}{2^k}$ (c) $\displaystyle\sum_{k=1}^{\infty} \frac{k^k}{k!}$ (d) $\displaystyle\sum_{k=3}^{\infty} \frac{(2k)!}{4^k}$ (e) $\displaystyle\sum_{k=1}^{\infty} \frac{1}{2k-1}$

Solution (*a*). The series converges, since

$$\rho = \lim_{k \to +\infty} \frac{u_{k+1}}{u_k} = \lim_{k \to +\infty} \frac{1/(k+1)!}{1/k!} = \lim_{k \to +\infty} \frac{k!}{(k+1)!} = \lim_{k \to +\infty} \frac{1}{k+1} = 0 < 1$$

Solution (*b*). The series converges, since

$$\rho = \lim_{k \to +\infty} \frac{u_{k+1}}{u_k} = \lim_{k \to +\infty} \frac{k+1}{2^{k+1}} \cdot \frac{2^k}{k} = \frac{1}{2} \lim_{k \to +\infty} \frac{k+1}{k} = \frac{1}{2} < 1$$

Solution (*c*). The series diverges, since

$$\rho = \lim_{k \to +\infty} \frac{u_{k+1}}{u_k} = \lim_{k \to +\infty} \frac{(k+1)^{k+1}}{(k+1)!} \cdot \frac{k!}{k^k} = \lim_{k \to +\infty} \frac{(k+1)^k}{k^k} = \lim_{k \to +\infty} \left(1 + \frac{1}{k}\right)^k = e > 1$$

See Theorem 7.9.2(*b*)

Solution (*d*). In the preceding parts, the index of summation started at 1. Although we could rewrite this series to make the index start at 1, it is not necessary to do so, since the requirements for the ratio test need only hold eventually. The series diverges, since

$$\rho = \lim_{k \to +\infty} \frac{u_{k+1}}{u_k} = \lim_{k \to +\infty} \frac{[2(k+1)]!}{4^{k+1}} \cdot \frac{4^k}{(2k)!} = \lim_{k \to +\infty} \left(\frac{(2k+2)!}{(2k)!} \cdot \frac{1}{4}\right)$$

$$= \frac{1}{4} \lim_{k \to +\infty} (2k+2)(2k+1) = +\infty$$

Solution (*e*). The ratio test is of no help since

$$\rho = \lim_{k \to +\infty} \frac{u_{k+1}}{u_k} = \lim_{k \to +\infty} \frac{1}{2(k+1)-1} \cdot \frac{2k-1}{1} = \lim_{k \to +\infty} \frac{2k-1}{2k+1} = 1$$

However, the integral test proves that the series diverges since

$$\int_1^{+\infty} \frac{dx}{2x-1} = \lim_{l \to +\infty} \int_1^l \frac{dx}{2x-1} = \lim_{l \to +\infty} \left. \frac{1}{2} \ln(2x-1) \right]_1^l = +\infty$$

Both the comparison test and the limit comparison test would also have worked here (verify). ◀

THE ROOT TEST

In cases where it is difficult or inconvenient to find the limit required for the ratio test, the next test is sometimes useful. Since its proof is similar to the proof of the ratio test, we will omit it.

11.6.6 THEOREM (*The Root Test*).   *Let* $\sum u_k$ *be a series with positive terms and suppose that*

$$\rho = \lim_{k \to +\infty} \sqrt[k]{u_k} = \lim_{k \to +\infty} (u_k)^{1/k}$$

(*a*) *If* $\rho < 1$, *the series converges.*

(*b*) *If* $\rho > 1$, *or* $\rho = +\infty$, *the series diverges.*

(*c*) *If* $\rho = 1$, *the series may converge or diverge, so that another test must be tried.*

Example 4

Use the root test to determine whether the following series converge or diverge.

(a) $\displaystyle\sum_{k=2}^{\infty} \left(\frac{4k-5}{2k+1}\right)^k$ (b) $\displaystyle\sum_{k=1}^{\infty} \frac{1}{(\ln(k+1))^k}$

Solution (*a*). The series diverges, since

$$\rho = \lim_{k \to +\infty} (u_k)^{1/k} = \lim_{k \to +\infty} \frac{4k-5}{2k+1} = 2 > 1$$

Solution (*b*). The series converges, since

$$\rho = \lim_{k \to +\infty} (u_k)^{1/k} = \lim_{k \to +\infty} \frac{1}{\ln(k+1)} = 0 < 1 \qquad ◀$$

EXERCISE SET 11.6 ☐ CAS

In Exercises 1 and 2, make a guess about the convergence or divergence of the series, and confirm your guess using the comparison test.

1. (a) $\displaystyle\sum_{k=1}^{\infty} \frac{1}{5k^2 - k}$ (b) $\displaystyle\sum_{k=1}^{\infty} \frac{3}{k - \frac{1}{4}}$

2. (a) $\displaystyle\sum_{k=2}^{\infty} \frac{k+1}{k^2 - k}$ (b) $\displaystyle\sum_{k=1}^{\infty} \frac{2}{k^4 + k}$

3. In each part, use the comparison test to show that the series converges.

(a) $\displaystyle\sum_{k=1}^{\infty} \frac{1}{3^k + 5}$ (b) $\displaystyle\sum_{k=1}^{\infty} \frac{5\sin^2 k}{k!}$

4. In each part, use the comparison test to show that the series diverges.

(a) $\displaystyle\sum_{k=1}^{\infty} \frac{\ln k}{k}$ (b) $\displaystyle\sum_{k=1}^{\infty} \frac{k}{k^{3/2} - \frac{1}{2}}$

In Exercises 5–10, use the limit comparison test to determine whether the series converges.

5. $\displaystyle\sum_{k=1}^{\infty} \frac{4k^2 - 2k + 6}{8k^7 + k - 8}$ **6.** $\displaystyle\sum_{k=1}^{\infty} \frac{1}{9k + 6}$

7. $\displaystyle\sum_{k=1}^{\infty} \frac{5}{3^k + 1}$ **8.** $\displaystyle\sum_{k=1}^{\infty} \frac{k(k+3)}{(k+1)(k+2)(k+5)}$

9. $\displaystyle\sum_{k=1}^{\infty} \frac{1}{\sqrt[3]{8k^2 - 3k}}$ **10.** $\displaystyle\sum_{k=1}^{\infty} \frac{1}{(2k+3)^{17}}$

In Exercises 11–16, use the ratio test to determine whether the series converges. If the test is inconclusive, then say so.

11. $\displaystyle\sum_{k=1}^{\infty} \frac{3^k}{k!}$ **12.** $\displaystyle\sum_{k=1}^{\infty} \frac{4^k}{k^2}$ **13.** $\displaystyle\sum_{k=1}^{\infty} \frac{1}{5k}$

14. $\displaystyle\sum_{k=1}^{\infty} k\left(\frac{1}{2}\right)^k$ **15.** $\displaystyle\sum_{k=1}^{\infty} \frac{k!}{k^3}$ **16.** $\displaystyle\sum_{k=1}^{\infty} \frac{k}{k^2 + 1}$

In Exercises 17–20, use the root test to determine whether the series converges. If the test is inconclusive, then say so.

17. $\displaystyle\sum_{k=1}^{\infty} \left(\frac{3k+2}{2k-1}\right)^k$ **18.** $\displaystyle\sum_{k=1}^{\infty} \left(\frac{k}{100}\right)^k$

19. $\displaystyle\sum_{k=1}^{\infty} \frac{k}{5^k}$ **20.** $\displaystyle\sum_{k=1}^{\infty} (1 - e^{-k})^k$

In Exercises 21–44, use any method to determine whether the series converges.

21. $\displaystyle\sum_{k=0}^{\infty} \frac{7^k}{k!}$ **22.** $\displaystyle\sum_{k=1}^{\infty} \frac{1}{2k + 1}$ **23.** $\displaystyle\sum_{k=1}^{\infty} \frac{k^2}{5^k}$

24. $\displaystyle\sum_{k=1}^{\infty} \frac{k!10^k}{3^k}$ **25.** $\displaystyle\sum_{k=1}^{\infty} k^{50}e^{-k}$ **26.** $\displaystyle\sum_{k=1}^{\infty} \frac{k^2}{k^3 + 1}$

27. $\displaystyle\sum_{k=1}^{\infty} \frac{\sqrt{k}}{k^3 + 1}$ **28.** $\displaystyle\sum_{k=1}^{\infty} \frac{4}{2 + 3^k k}$

29. $\displaystyle\sum_{k=1}^{\infty} \frac{1}{\sqrt{k(k+1)}}$ **30.** $\displaystyle\sum_{k=1}^{\infty} \frac{2 + (-1)^k}{5^k}$

31. $\displaystyle\sum_{k=1}^{\infty} \frac{2 + \sqrt{k}}{(k+1)^3 - 1}$ **32.** $\displaystyle\sum_{k=1}^{\infty} \frac{4 + |\cos k|}{k^3}$

33. $\displaystyle\sum_{k=1}^{\infty} \frac{1}{1 + \sqrt{k}}$ **34.** $\displaystyle\sum_{k=1}^{\infty} \frac{k!}{k^k}$ **35.** $\displaystyle\sum_{k=1}^{\infty} \frac{\ln k}{e^k}$

36. $\displaystyle\sum_{k=1}^{\infty} \frac{k!}{e^{k^2}}$ **37.** $\displaystyle\sum_{k=0}^{\infty} \frac{(k+4)!}{4!k!4^k}$ **38.** $\displaystyle\sum_{k=1}^{\infty} \left(\frac{k}{k+1}\right)^{k^2}$

39. $\displaystyle\sum_{k=1}^{\infty} \frac{1}{4 + 2^{-k}}$ **40.** $\displaystyle\sum_{k=1}^{\infty} \frac{\sqrt{k}\ln k}{k^3 + 1}$ **41.** $\displaystyle\sum_{k=1}^{\infty} \frac{\tan^{-1} k}{k^2}$

42. $\displaystyle\sum_{k=1}^{\infty} \frac{5^k + k}{k! + 3}$ **43.** $\displaystyle\sum_{k=0}^{\infty} \frac{(k!)^2}{(2k)!}$ **44.** $\displaystyle\sum_{k=1}^{\infty} \frac{(k!)^2 2^k}{(2k+2)!}$

In Exercises 45 and 46, find the general term of the series, and use the ratio test to show that the series converges.

45. $1 + \dfrac{1 \cdot 2}{1 \cdot 3} + \dfrac{1 \cdot 2 \cdot 3}{1 \cdot 3 \cdot 5} + \dfrac{1 \cdot 2 \cdot 3 \cdot 4}{1 \cdot 3 \cdot 5 \cdot 7} + \cdots$

46. $1 + \dfrac{1 \cdot 3}{3!} + \dfrac{1 \cdot 3 \cdot 5}{5!} + \dfrac{1 \cdot 3 \cdot 5 \cdot 7}{7!} + \cdots$

In Exercises 47 and 48, use a CAS to investigate the convergence of the series.

☐ **47.** $\displaystyle\sum_{k=1}^{\infty} \frac{\ln k}{3^k}$ ☐ **48.** $\displaystyle\sum_{k=1}^{\infty} \frac{[\pi(k+1)]^k}{k^{k+1}}$

49. (a) Make a conjecture about the convergence of the series $\sum_{k=1}^{\infty} \sin(\pi/k)$ by considering the local linear approximation of $\sin x$ near $x = 0$.

(b) Try to confirm your conjecture using the limit comparison test.

50. (a) Make a conjecture about the convergence of the series

$$\sum_{k=1}^{\infty} \left[1 - \cos\left(\frac{1}{k}\right)\right]$$

by considering the local quadratic approximation of $\cos x$ near $x = 0$.

(b) Try to confirm your conjecture using the limit comparison test.

51. Show that $\ln x < \sqrt{x}$ if $x > 0$, and use this result to investigate the convergence of

(a) $\displaystyle\sum_{k=1}^{\infty} \frac{\ln k}{k^2}$ (b) $\displaystyle\sum_{k=2}^{\infty} \frac{1}{(\ln k)^2}$

52. For which positive values of α does the series $\sum_{k=1}^{\infty}(\alpha^k/k^\alpha)$ converge?

53. Use Theorem 11.4.6 to prove the comparison test (Theorem 11.6.1).

54. Let $\sum a_k$ and $\sum b_k$ be series with positive terms. Prove:

(a) If $\lim_{k \to +\infty}(a_k/b_k) = 0$ and $\sum b_k$ converges, then $\sum a_k$ converges.

(b) If $\lim_{k \to +\infty}(a_k/b_k) = +\infty$ and $\sum b_k$ diverges, then $\sum a_k$ diverges.

11.7 ALTERNATING SERIES; CONDITIONAL CONVERGENCE

Up to now we have focused exclusively on series with nonnegative terms. In this section we will discuss series that contain both positive and negative terms.

ALTERNATING SERIES

Series whose terms alternate between positive and negative, called **alternating series**, are of special importance. Some examples are

$$\sum_{k=1}^{\infty}(-1)^{k+1}\frac{1}{k} = 1 - \frac{1}{2} + \frac{1}{3} - \frac{1}{4} + \frac{1}{5} - \cdots$$

$$\sum_{k=1}^{\infty}(-1)^{k}\frac{1}{k} = -1 + \frac{1}{2} - \frac{1}{3} + \frac{1}{4} - \frac{1}{5} + \cdots$$

In general, an alternating series has one of the following two forms:

$$\sum_{k=1}^{\infty}(-1)^{k+1}a_k = a_1 - a_2 + a_3 - a_4 + \cdots \tag{1}$$

$$\sum_{k=1}^{\infty}(-1)^{k}a_k = -a_1 + a_2 - a_3 + a_4 - \cdots \tag{2}$$

where the a_k's are assumed to be positive in both cases.

The following theorem is the key result on convergence of alternating series.

11.7.1 THEOREM (*Alternating Series Test*). *An alternating series of either form* (1) *or form* (2) *converges if the following two conditions are satisfied:*

(a) $a_1 > a_2 > a_3 > \cdots > a_k > \cdots$

(b) $\lim_{k \to +\infty} a_k = 0$

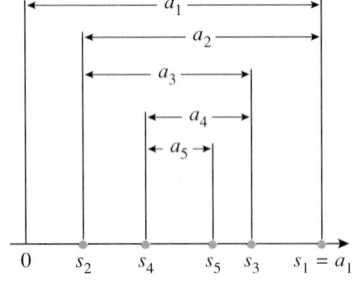

Figure 11.7.1

Proof. We will consider only alternating series of form (1). The idea of the proof is to show that if conditions (*a*) and (*b*) hold, then the sequences of even-numbered and odd-numbered partial sums converge to a common limit S. It will then follow from Theorem 11.1.4 that the entire sequence of partial sums converges to S.

Figure 11.7.1 shows how successive partial sums satisfying conditions (*a*) and (*b*) appear when plotted on a horizontal axis. The even-numbered partial sums

$$s_2, s_4, s_6, s_8, \ldots, s_{2n}, \ldots$$

form an increasing sequence bounded above by a_1, and the odd-numbered partial sums

$$s_1, s_3, s_5, \ldots, s_{2n-1}, \ldots$$

form a decreasing sequence bounded below by 0. Thus, by Theorems 11.2.3 and 11.2.4, the even-numbered partial sums converge to some limit S_E and the odd-numbered partial sums converge to some limit S_O. To complete the proof we must show that $S_E = S_O$. But the $(2n)$-th term in the series is $-a_{2n}$, so that $s_{2n} - s_{2n-1} = -a_{2n}$, which can be written as

$$s_{2n-1} = s_{2n} + a_{2n}$$

However, $2n \to +\infty$ and $2n - 1 \to +\infty$ as $n \to +\infty$, so that

$$S_O = \lim_{n \to +\infty} s_{2n-1} = \lim_{n \to +\infty} (s_{2n} + a_{2n}) = S_E + 0 = S_E$$

which completes the proof. ∎

REMARK. As might be expected, it is not essential for condition (*a*) in the alternating series test to hold for all terms; an alternating series will converge if condition (*b*) is true and condition (*a*) holds eventually.

Example 1

Use the alternating series test to show that the following series converge.

(a) $\displaystyle\sum_{k=1}^{\infty} (-1)^{k+1} \frac{1}{k}$ (b) $\displaystyle\sum_{k=1}^{\infty} (-1)^{k+1} \frac{k+3}{k(k+1)}$

Solution (*a*). The two conditions in the alternating series test are satisfied since

$$a_k = \frac{1}{k} > \frac{1}{k+1} = a_{k+1} \quad \text{and} \quad \lim_{k \to +\infty} a_k = \lim_{k \to +\infty} \frac{1}{k} = 0$$

Solution (*b*). The two conditions in the alternating series test are satisfied since

$$\frac{a_{k+1}}{a_k} = \frac{k+4}{(k+1)(k+2)} \cdot \frac{k(k+1)}{k+3} = \frac{k^2 + 4k}{k^2 + 5k + 6} = \frac{k^2 + 4k}{(k^2 + 4k) + (k + 6)} < 1$$

so

$$a_k > a_{k+1}$$

and

$$\lim_{k \to +\infty} a_k = \lim_{k \to +\infty} \frac{k+3}{k(k+1)} = \lim_{k \to +\infty} \frac{\dfrac{1}{k} + \dfrac{3}{k^2}}{1 + \dfrac{1}{k}} = 0 \qquad ◄$$

REMARK. The series in part (a) of the last example is called the ***alternating harmonic series***. Observe that this series converges, whereas the harmonic series diverges.

REMARK. If an alternating series violates condition (*b*) of the alternating series test, then the series must diverge by the divergence test (Theorem 11.4.1). However, if condition (*b*) is satisfied, but condition (*a*) is not, the series can either converge or diverge.[*]

....

APPROXIMATING SUMS OF ALTERNATING SERIES

The following theorem is concerned with the error that results when the sum of an alternating series is approximated by a partial sum.

11.7.2 THEOREM. *If an alternating series satisfies the hypotheses of the alternating series test, and if S is the sum of the series, then:*

(*a*) *S lies between any two successive partial sums; that is, either*

$$s_n < S < s_{n+1} \quad or \quad s_{n+1} < S < s_n \qquad (3)$$

depending on which partial sum is larger.

(*b*) *If S is approximated by s_n, then the absolute error $|S - s_n|$ satisfies*

$$|S - s_n| < a_{n+1} \qquad (4)$$

Moreover, the sign of the error $S - s_n$ is the same as that of the coefficient of a_{n+1}.

---

[*]The interested reader will find some nice examples in an article by R. Lariviere, "On a Convergence Test for Alternating Series," *Mathematics Magazine*, Vol. 29, 1956, p. 88.

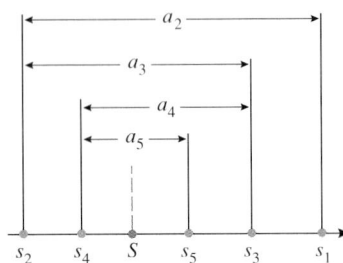

Figure 11.7.2

Proof. We will prove the theorem for series of form (1). Referring to Figure 11.7.2 and keeping in mind our observation in the proof of Theorem 11.7.1 that the odd-numbered partial sums form a decreasing sequence converging to S and the even-numbered partial sums form an increasing sequence converging to S, we see that successive partial sums oscillate from one side of S to the other in smaller and smaller steps with the odd-numbered partial sums being larger than S and the even-numbered partial sums being smaller than S. Thus, depending on whether n is even or odd, we have

$$s_n < S < s_{n+1} \quad \text{or} \quad s_{n+1} < S < s_n$$

which proves (3). Moreover, in either case we have

$$|S - s_n| < |s_{n+1} - s_n| \tag{5}$$

But $s_{n+1} - s_n = \pm a_{n+1}$ (the sign depending on whether n is even or odd). Thus, it follows from (5) that $|S - s_n| < a_{n+1}$, which proves (4). Finally, since the odd-numbered partial sums are larger than S and the even-numbered partial sums are smaller than S, it follows that $S - s_n$ has the same sign as the coefficient of a_{n+1} (verify). ■

REMARK. In words, inequality (4) states that for a series satisfying the hypotheses of the alternating series test, the magnitude of the error that results from approximating S by s_n is less than that of the first term that is *not* included in the partial sum.

Example 2

Later in this chapter we will show that the sum of the alternating harmonic series is

$$\ln 2 = 1 - \frac{1}{2} + \frac{1}{3} - \frac{1}{4} + \cdots + (-1)^{k+1}\frac{1}{k} + \cdots$$

(a) Accepting this to be so, find an upper bound on the magnitude of the error that results if $\ln 2$ is approximated by the sum of the first eight terms in the series.

(b) Find a partial sum that approximates $\ln 2$ to one decimal-place accuracy (the nearest tenth).

Solution (a). It follows from (4) that

$$|\ln 2 - s_8| < a_9 = \frac{1}{9} < 0.12 \tag{6}$$

As a check, let us compute s_8 exactly. We obtain

$$s_8 = 1 - \frac{1}{2} + \frac{1}{3} - \frac{1}{4} + \frac{1}{5} - \frac{1}{6} + \frac{1}{7} - \frac{1}{8} = \frac{533}{840}$$

Thus, with the help of a calculator

$$|\ln 2 - s_8| = \left|\ln 2 - \frac{533}{840}\right| \approx 0.059$$

This shows that the error is well under the estimate provided by upper bound (6).

Solution (b). For one decimal-place accuracy, we must choose n so that $|\ln 2 - s_n| \le 0.05$. However, it follows from (4) that

$$|\ln 2 - s_n| < a_{n+1}$$

so it suffices to choose n so that $a_{n+1} \le 0.05$.

One way to find n is to use a calculating utility to obtain numerical values for $a_1, a_2, a_3, \ldots$ until you encounter the first value that is less than or equal to 0.05. If you do this, you will find that it is $a_{20} = 0.05$; this tells us that partial sum s_{19} will provide the desired accuracy. Another way to find n is to solve the inequality

$$\frac{1}{n+1} \le 0.05$$

algebraically. We can do this by taking reciprocals, reversing the sense of the inequality, and then simplifying to obtain $n \ge 19$. Thus, s_{19} will provide the required accuracy, which is consistent with the previous result.

With the help of a calculating utility, the value of s_{19} is approximately $s_{19} \approx 0.7$ and the value of $\ln 2$ obtained directly is approximately $\ln 2 \approx 0.69$, which agrees with s_{19} when rounded to one decimal place. ◄

REMARK.   As this example illustrates, the alternating harmonic series does not provide an efficient way to approximate $\ln 2$, since too much computation is required to achieve reasonable accuracy. Later, we will develop better ways to approximate logarithms.

ABSOLUTE CONVERGENCE

The series

$$1 - \frac{1}{2} - \frac{1}{2^2} + \frac{1}{2^3} + \frac{1}{2^4} - \frac{1}{2^5} - \frac{1}{2^6} + \cdots$$

does not fit in any of the categories studied so far—it has mixed signs, but is not alternating. We will now develop some convergence tests that can be applied to such series.

11.7.3 DEFINITION.   A series

$$\sum_{k=1}^{\infty} u_k = u_1 + u_2 + \cdots + u_k + \cdots$$

is said to **converge absolutely** if the series of absolute values

$$\sum_{k=1}^{\infty} |u_k| = |u_1| + |u_2| + \cdots + |u_k| + \cdots$$

converges and is said to **diverge absolutely** if the series of absolute values diverges.

Example 3

Determine whether the following series converge absolutely.

(a) $1 - \dfrac{1}{2} - \dfrac{1}{2^2} + \dfrac{1}{2^3} + \dfrac{1}{2^4} - \dfrac{1}{2^5} - \cdots$ (b) $1 - \dfrac{1}{2} + \dfrac{1}{3} - \dfrac{1}{4} + \dfrac{1}{5} - \cdots$

Solution (a). The series of absolute values is the convergent geometric series

$$1 + \frac{1}{2} + \frac{1}{2^2} + \frac{1}{2^3} + \frac{1}{2^4} + \frac{1}{2^5} + \cdots$$

so the given series converges.

Solution (b). The series of absolute values is the divergent harmonic series

$$1 + \frac{1}{2} + \frac{1}{3} + \frac{1}{4} + \frac{1}{5} + \cdots$$

so the given series diverges absolutely.   ◄

It is important to distinguish between the notions of convergence and absolute convergence. For example, the series in part (b) of Example 3 converges, since it is the alternating harmonic series, yet we demonstrated that it does not converge absolutely. However, the following theorem shows that *if a series converges absolutely, then it converges.*

11.7.4 THEOREM.   *If the series*

$$\sum_{k=1}^{\infty} |u_k| = |u_1| + |u_2| + \cdots + |u_k| + \cdots$$

converges, then so does the series

$$\sum_{k=1}^{\infty} u_k = u_1 + u_2 + \cdots + u_k + \cdots$$

Proof. Our proof is based on a trick. We will write the series $\sum u_k$ as

$$\sum_{k=1}^{\infty} u_k = \sum_{k=1}^{\infty} [(u_k + |u_k|) - |u_k|] \tag{7}$$

We are assuming that $\sum |u_k|$ converges, so that if we can show that $\sum (u_k + |u_k|)$ converges, then it will follow from (7) and Theorem 11.4.3(a) that $\sum u_k$ converges. However, the value of $u_k + |u_k|$ is either 0 or $2|u_k|$, depending on the sign of u_k. Thus, in all cases it is true that

$$0 \leq u_k + |u_k| \leq 2|u_k|$$

But $\sum 2|u_k|$ converges, since it is a constant times the convergent series $\sum |u_k|$; hence $\sum (u_k + |u_k|)$ converges by the comparison test.   ▮

Theorem 11.7.4 is important because it provides a way of inferring convergence of a series with positive and negative terms from the convergence of a series with nonnegative terms (the series of absolute values). This is important because most of the convergence tests we have developed apply only to series with nonnegative terms.

Example 4
Show that the following series converge.

$$\text{(a)} \quad 1 - \frac{1}{2} - \frac{1}{2^2} + \frac{1}{2^3} + \frac{1}{2^4} - \frac{1}{2^5} - \frac{1}{2^6} + \cdots \qquad \text{(b)} \quad \sum_{k=1}^{\infty} \frac{\cos k}{k^2}$$

Solution (a). Observe that this is not an alternating series because the signs alternate in pairs after the first term. Thus, we have no convergence test that can be applied directly. However, we showed in Example 3(a) that the series converges absolutely, so Theorem 11.7.4 implies that it converges.

Solution (b). With the help of a calculating utility, you will be able to verify that the signs of the terms in this series vary irregularly. Thus, we will test for absolute convergence. The series of absolute values is

$$\sum_{k=1}^{\infty} \left| \frac{\cos k}{k^2} \right|$$

However,

$$\left| \frac{\cos k}{k^2} \right| \leq \frac{1}{k^2}$$

But $\sum 1/k^2$ is a convergent p-series $(p = 2)$, so the series of absolute values converges by the comparison test. Thus, the given series converges absolutely and hence converges.   ◄

....

CONDITIONAL CONVERGENCE

Although Theorem 11.7.4 is a useful tool for series that converge absolutely, it provides no information about the convergence or divergence of a series that diverges absolutely. For example, consider the two series

$$1 - \frac{1}{2} + \frac{1}{3} - \frac{1}{4} + \cdots + (-1)^{k+1}\frac{1}{k} + \cdots \tag{8}$$

$$-1 - \frac{1}{2} - \frac{1}{3} - \frac{1}{4} - \cdots - \frac{1}{k} - \cdots \tag{9}$$

Both of these series diverge absolutely, since in each case the series of absolute values is the divergent harmonic series

$$1 + \frac{1}{2} + \frac{1}{3} + \cdots + \frac{1}{k} + \cdots$$

However, series (8) converges, since it is the alternating harmonic series, and series (9) diverges, since it is a constant times the divergent harmonic series. As a matter of terminology, a series that converges but diverges absolutely is said to **converge conditionally** (or to be **conditionally convergent**). Thus, (9) is a conditionally convergent series.

THE RATIO TEST FOR ABSOLUTE CONVERGENCE

Although one cannot generally infer convergence or divergence of a series from absolute divergence, the following variation of the ratio test provides a way of deducing divergence from absolute divergence in certain situations. We omit the proof.

11.7.5 THEOREM (*Ratio Test for Absolute Convergence*). *Let $\sum u_k$ be a series with nonzero terms and suppose that*

$$\rho = \lim_{k \to +\infty} \frac{|u_{k+1}|}{|u_k|}$$

(*a*) *If $\rho < 1$, then the series $\sum u_k$ converges absolutely and therefore converges.*

(*b*) *If $\rho > 1$ or if $\rho = +\infty$, then the series $\sum u_k$ diverges.*

(*c*) *If $\rho = 1$, no conclusion about convergence or absolute convergence can be drawn from this test.*

Example 5

Use the ratio test for absolute convergence to determine whether the series converges.

(a) $\sum_{k=1}^{\infty} (-1)^k \frac{2^k}{k!}$ (b) $\sum_{k=1}^{\infty} (-1)^k \frac{(2k-1)!}{3^k}$

Solution (a). Taking the absolute value of the general term u_k we obtain

$$|u_k| = \left| (-1)^k \frac{2^k}{k!} \right| = \frac{2^k}{k!}$$

Thus,

$$\rho = \lim_{k \to +\infty} \frac{|u_{k+1}|}{|u_k|} = \lim_{k \to +\infty} \frac{2^{k+1}}{(k+1)!} \cdot \frac{k!}{2^k} = \lim_{k \to +\infty} \frac{2}{k+1} = 0 < 1$$

which implies that the series converges absolutely and therefore converges.

Solution (b). Taking the absolute value of the general term u_k we obtain

$$|u_k| = \left| (-1)^k \frac{(2k-1)!}{3^k} \right| = \frac{(2k-1)!}{3^k}$$

Thus,

$$\rho = \lim_{k \to +\infty} \frac{|u_{k+1}|}{|u_k|} = \lim_{k \to +\infty} \frac{[2(k+1)-1]!}{3^{k+1}} \cdot \frac{3^k}{(2k-1)!}$$

$$= \lim_{k \to +\infty} \frac{1}{3} \cdot \frac{(2k+1)!}{(2k-1)!} = \frac{1}{3} \lim_{k \to +\infty} (2k)(2k+1) = +\infty$$

which implies that the series diverges. ◀

SUMMARY OF CONVERGENCE TESTS

We conclude this section with a summary of convergence tests that can be used for reference.

Summary of Convergence Tests

NAME	STATEMENT	COMMENTS
Divergence Test (11.4.1)	If $\lim\limits_{k \to +\infty} u_k \neq 0$, then $\sum u_k$ diverges.	If $\lim\limits_{k \to +\infty} u_k = 0$, then $\sum u_k$ may or may not converge.
Integral Test (11.4.4)	Let $\sum u_k$ be a series with positive terms, and let $f(x)$ be the function that results when k is replaced by x in the general term of the series. If f is decreasing and continuous for $x \geq 1$, then $$\sum_{k=1}^{\infty} u_k \quad \text{and} \quad \int_1^{+\infty} f(x)\, dx$$ both converge or both diverge.	This test only applies to series that have positive terms. Try this test when $f(x)$ is easy to integrate.
Comparison Test (11.6.1)	Let $\sum_{k=1}^{\infty} a_k$ and $\sum_{k=1}^{\infty} b_k$ be series with nonnegative terms such that $$a_1 \leq b_1,\ a_2 \leq b_2,\ \ldots,\ a_k \leq b_k,\ \ldots$$ If $\sum b_k$ converges, then $\sum a_k$ converges, and if $\sum a_k$ diverges, then $\sum b_k$ diverges.	This test only applies to series with nonnegative terms. Try this test as a last resort; other tests are often easier to apply.
Ratio Test (11.6.5)	Let $\sum u_k$ be a series with positive terms and suppose that $$\rho = \lim_{k \to +\infty} \frac{u_{k+1}}{u_k}$$ (a) Series converges if $\rho < 1$. (b) Series diverges if $\rho > 1$ or $\rho = +\infty$. (c) The test is inconclusive if $\rho = 1$.	Try this test when u_k involves factorials or kth powers.
Root Test (11.6.6)	Let $\sum u_k$ be a series with positive terms such that $$\rho = \lim_{k \to +\infty} \sqrt[k]{u_k}$$ (a) The series converges if $\rho < 1$. (b) The series diverges if $\rho > 1$ or $\rho = +\infty$. (c) The test is inconclusive if $\rho = 1$.	Try this test when u_k involves kth powers.
Limit Comparison Test (11.6.4)	Let $\sum a_k$ and $\sum b_k$ be series with positive terms such that $$\rho = \lim_{k \to +\infty} \frac{a_k}{b_k}$$ If $0 < \rho < +\infty$, then both series converge or both diverge.	This is easier to apply than the comparison test, but still requires some skill in choosing the series $\sum b_k$ for comparison.
Alternating Series Test (11.7.1)	If $a_k > 0$ for $k = 1, 2, 3, \ldots$, then the series $$a_1 - a_2 + a_3 - a_4 + \cdots$$ $$-a_1 + a_2 - a_3 + a_4 - \cdots$$ converge if the following conditions hold: (a) $a_1 > a_2 > a_3 > \cdots$ (b) $\lim\limits_{k \to +\infty} a_k = 0$	This test applies only to alternating series.
Ratio Test for Absolute Convergence (11.7.5)	Let $\sum u_k$ be a series with nonzero terms such that $$\rho = \lim_{k \to +\infty} \frac{|u_{k+1}|}{|u_k|}$$ (a) The series converges absolutely if $\rho < 1$. (b) The series diverges absolutely if $\rho > 1$ or $\rho = +\infty$. (c) The test is inconclusive if $\rho = 1$.	The series need not have positive terms and need not be alternating to use this test.

EXERCISE SET 11.7 ∿ Graphing Calculator [c] CAS

In Exercises 1 and 2 show that the series converges by confirming that it satisfies the hypotheses of the alternating series test (Theorem 11.7.1).

1. $\displaystyle\sum_{k=1}^{\infty} \frac{(-1)^{k+1}}{2k+1}$

2. $\displaystyle\sum_{k=1}^{\infty} (-1)^{k+1} \frac{k}{3^k}$

In Exercises 3–6, determine whether the alternating series converges, and justify your answer.

3. $\displaystyle\sum_{k=1}^{\infty} (-1)^{k+1} \frac{k+1}{3k+1}$

4. $\displaystyle\sum_{k=1}^{\infty} (-1)^{k+1} \frac{k+1}{\sqrt{k}+1}$

5. $\displaystyle\sum_{k=1}^{\infty} (-1)^{k+1} e^{-k}$

6. $\displaystyle\sum_{k=3}^{\infty} (-1)^k \frac{\ln k}{k}$

In Exercises 7–12, use the ratio test for absolute convergence (Theorem 11.7.5) to determine whether the series converges or diverges. If the test is inconclusive, then say so.

7. $\displaystyle\sum_{k=1}^{\infty} \left(-\frac{3}{5}\right)^k$

8. $\displaystyle\sum_{k=1}^{\infty} (-1)^{k+1} \frac{2^k}{k!}$

9. $\displaystyle\sum_{k=1}^{\infty} (-1)^{k+1} \frac{3^k}{k^2}$

10. $\displaystyle\sum_{k=1}^{\infty} (-1)^k \frac{k}{5^k}$

11. $\displaystyle\sum_{k=1}^{\infty} (-1)^k \frac{k^3}{e^k}$

12. $\displaystyle\sum_{k=1}^{\infty} (-1)^{k+1} \frac{k^k}{k!}$

In Exercises 13–30, classify the series as absolutely convergent, conditionally convergent, or divergent.

13. $\displaystyle\sum_{k=1}^{\infty} \frac{(-1)^{k+1}}{3k}$

14. $\displaystyle\sum_{k=1}^{\infty} \frac{(-1)^{k+1}}{k^{4/3}}$

15. $\displaystyle\sum_{k=1}^{\infty} \frac{(-4)^k}{k^2}$

16. $\displaystyle\sum_{k=1}^{\infty} \frac{(-1)^{k+1}}{k!}$

17. $\displaystyle\sum_{k=1}^{\infty} \frac{\cos k\pi}{k}$

18. $\displaystyle\sum_{k=3}^{\infty} \frac{(-1)^k \ln k}{k}$

19. $\displaystyle\sum_{k=1}^{\infty} (-1)^{k+1} \frac{k+2}{k(k+3)}$

20. $\displaystyle\sum_{k=1}^{\infty} \frac{(-1)^{k+1} k^2}{k^3+1}$

21. $\displaystyle\sum_{k=1}^{\infty} \sin \frac{k\pi}{2}$

22. $\displaystyle\sum_{k=1}^{\infty} \frac{\sin k}{k^3}$

23. $\displaystyle\sum_{k=2}^{\infty} \frac{(-1)^k}{k \ln k}$

24. $\displaystyle\sum_{k=1}^{\infty} \frac{(-1)^k}{\sqrt{k(k+1)}}$

25. $\displaystyle\sum_{k=2}^{\infty} \left(-\frac{1}{\ln k}\right)^k$

26. $\displaystyle\sum_{k=1}^{\infty} \frac{(-1)^{k+1}}{\sqrt{k+1}+\sqrt{k}}$

27. $\displaystyle\sum_{k=2}^{\infty} \frac{(-1)^k (k^2+1)}{k^3+2}$

28. $\displaystyle\sum_{k=1}^{\infty} \frac{k \cos k\pi}{k^2+1}$

29. $\displaystyle\sum_{k=1}^{\infty} \frac{(-1)^{k+1} k!}{(2k-1)!}$

30. $\displaystyle\sum_{k=1}^{\infty} (-1)^{k+1} \frac{3^{2k-1}}{k^2+1}$

In Exercises 31–34, the series satisfies the hypotheses of the alternating series test. For the stated value of n, find an upper bound on the absolute error that results if the sum of the series is approximated by the nth partial sum.

31. $\displaystyle\sum_{k=1}^{\infty} \frac{(-1)^{k+1}}{k}; \; n=7$

32. $\displaystyle\sum_{k=1}^{\infty} \frac{(-1)^{k+1}}{k!}; \; n=5$

33. $\displaystyle\sum_{k=1}^{\infty} \frac{(-1)^{k+1}}{\sqrt{k}}; \; n=99$

34. $\displaystyle\sum_{k=1}^{\infty} \frac{(-1)^{k+1}}{(k+1)\ln(k+1)}; \; n=3$

In Exercises 35–38, the series satisfies the hypotheses of the alternating series test. Find a value of n for which the nth partial sum is ensured to approximate the sum of the series to the stated accuracy.

35. $\displaystyle\sum_{k=1}^{\infty} \frac{(-1)^{k+1}}{k}; \; |\text{error}| < 0.0001$

36. $\displaystyle\sum_{k=1}^{\infty} \frac{(-1)^{k+1}}{k!}; \; |\text{error}| < 0.00001$

37. $\displaystyle\sum_{k=1}^{\infty} \frac{(-1)^{k+1}}{\sqrt{k}}; \; \text{two decimal places}$

38. $\displaystyle\sum_{k=1}^{\infty} \frac{(-1)^{k+1}}{(k+1)\ln(k+1)}; \; \text{one decimal place}$

In Exercises 39 and 40, find an upper bound on the absolute error that results if s_{10} is used to approximate the sum of the given *geometric* series. Compute s_{10} rounded to four decimal places and compare this value with the exact sum of the series.

39. $\dfrac{3}{4} - \dfrac{3}{8} + \dfrac{3}{16} - \dfrac{3}{32} + \cdots$

40. $1 - \dfrac{2}{3} + \dfrac{4}{9} - \dfrac{8}{27} + \cdots$

In Exercises 41–44, the series satisfies the hypotheses of the alternating series test. Approximate the sum of the series to two decimal-place accuracy.

41. $1 - \dfrac{1}{3!} + \dfrac{1}{5!} - \dfrac{1}{7!} + \cdots$

42. $1 - \dfrac{1}{2!} + \dfrac{1}{4!} - \dfrac{1}{6!} + \cdots$

43. $\dfrac{1}{1 \cdot 2} - \dfrac{1}{2 \cdot 2^2} + \dfrac{1}{3 \cdot 2^3} - \dfrac{1}{4 \cdot 2^4} + \cdots$

44. $\dfrac{1}{1^5 + 4 \cdot 1} - \dfrac{1}{3^5 + 4 \cdot 3} + \dfrac{1}{5^5 + 4 \cdot 5} - \dfrac{1}{7^5 + 4 \cdot 7} + \cdots$

[c] **45.** The purpose of this exercise is to show that the error bound in part (b) of Theorem 11.7.2 can be overly conservative in certain cases.

(a) Use a CAS to confirm that

$$\frac{\pi}{4} = 1 - \frac{1}{3} + \frac{1}{5} - \frac{1}{7} + \cdots$$

(b) Use the CAS to show that $|(\pi/4) - s_{26}| < 10^{-2}$.

(c) According to the error bound in part (b) of Theorem 11.7.2, what value of n is required to ensure that $|(\pi/4) - s_n| < 10^{-2}$?

46. Show that the alternating p-series

$$1 - \frac{1}{2^p} + \frac{1}{3^p} - \frac{1}{4^p} + \cdots + (-1)^{k+1}\frac{1}{k^p} + \cdots$$

converges absolutely if $p > 1$, converges conditionally if $0 < p \leq 1$, and diverges if $p \leq 0$.

It can be proved that any series that is constructed from an absolutely convergent series by rearranging the terms is absolutely convergent and has the same sum as the original series. Use this fact together with parts (a) and (b) of Theorem 11.4.3 in Exercises 47 and 48.

---

47. It was stated in Exercise 27 of Section 11.4 that

$$\frac{\pi^2}{6} = 1 + \frac{1}{2^2} + \frac{1}{3^2} + \frac{1}{4^2} + \cdots$$

Use this to show that

$$\frac{\pi^2}{8} = 1 + \frac{1}{3^2} + \frac{1}{5^2} + \frac{1}{7^2} + \cdots$$

48. It was stated in Exercise 27 of Section 11.4 that

$$\frac{\pi^4}{90} = 1 + \frac{1}{2^4} + \frac{1}{3^4} + \frac{1}{4^4} + \cdots$$

Use this to show that

$$\frac{\pi^4}{96} = 1 + \frac{1}{3^4} + \frac{1}{5^4} + \frac{1}{7^4} + \cdots$$

49. It can be proved that the terms of any conditionally convergent series can be rearranged to give either a divergent series or a conditionally convergent series whose sum is any given number S. For example, we stated in Example 2 that

$$\ln 2 = 1 - \frac{1}{2} + \frac{1}{3} - \frac{1}{4} + \frac{1}{5} - \frac{1}{6} + \cdots$$

Show that we can rearrange this series so that its sum is $\frac{1}{2} \ln 2$ by rewriting it as

$$\left(1 - \frac{1}{2} - \frac{1}{4}\right) + \left(\frac{1}{3} - \frac{1}{6} - \frac{1}{8}\right) + \left(\frac{1}{5} - \frac{1}{10} - \frac{1}{12}\right) + \cdots$$

[*Hint:* Add the first two terms in each set of parentheses.]

50. (a) Use a graphing utility to graph

$$f(x) = \frac{4x - 1}{4x^2 - 2x}, \quad x \geq 1$$

(b) Based on your graph, do think that the series

$$\sum_{k=1}^{\infty} (-1)^{k+1} \frac{4k - 1}{4k^2 - 2k}$$

converges? Explain your reasoning.

51. As illustrated in the accompanying figure, a bug, starting at point A on a 180-cm wire, walks the length of the wire, stops and walks in the opposite direction for half the length of the wire, stops again and walks in the opposite direction for one-third the length of the wire, stops again and walks in the opposite direction for one-fourth the length of the wire, and so forth until it stops for the 1000th time.

(a) Give upper and lower bounds on the distance between the bug and point A when it finally stops. [*Hint:* As stated in Example 2, assume that the sum of the alternating harmonic series is $\ln 2$.]

(b) Give upper and lower bounds on the total distance that the bug has traveled when it finally stops. [*Hint:* Use inequality (2) of Section 11.4.]

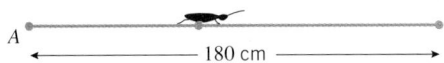

Figure Ex-51

52. (a) Prove that if $\sum a_k$ converges absolutely, then $\sum a_k^2$ converges.

(b) Show that the converse of part (a) is false by giving a counterexample.

11.8 POWER SERIES

In the last two sections we focused exclusively on series whose terms are numbers. In this section we will consider series whose terms are functions with the objective of developing the mathematical tools needed to investigate the convergence of Taylor and Maclaurin series.

POWER SERIES IN x

If $c_0, c_1, c_2, \ldots$ are constants and x is a variable, then a series of the form

$$\sum_{k=0}^{\infty} c_k x^k = c_0 + c_1 x + c_2 x^2 + \cdots + c_k x^k + \cdots \tag{1}$$

is called a *power series in x*. Some examples are

$$\sum_{k=0}^{\infty} x^k = 1 + x + x^2 + x^3 + \cdots$$

$$\sum_{k=0}^{\infty} \frac{x^k}{k!} = 1 + x + \frac{x^2}{2!} + \frac{x^3}{3!} + \cdots$$

$$\sum_{k=0}^{\infty} (-1)^k \frac{x^{2k}}{(2k)!} = 1 - \frac{x^2}{2!} + \frac{x^4}{4!} - \frac{x^6}{6!} + \cdots$$

More generally, every Maclaurin series

$$\sum_{k=0}^{\infty} \frac{f^{(k)}(0)}{k!} x^k = f(0) + f'(0)x + \frac{f''(0)}{2!} x^2 + \cdots + \frac{f^{(k)}(0)}{k!} x^k + \cdots$$

is a power series in x.

RADIUS AND INTERVAL OF CONVERGENCE

If a numerical value is substituted for x in a power series $\sum c_k x^k$, then the resulting series of numbers may either converge or diverge. This leads to the problem of determining the set of x-values for which a given power series converges; this is called its ***convergence set***.

Observe that every power series in x converges at $x = 0$, since substituting this value in (1) produces the series

$$c_0 + 0 + 0 + 0 + \cdots + 0 + \cdots$$

whose sum is c_0. In rare cases $x = 0$ may be the only point in the convergence set, but more usually the convergence set is some finite or infinite interval containing $x = 0$. This is the content of the following theorem, whose proof will be omitted.

11.8.1   THEOREM.   *For any power series in x, exactly one of the following is true:*

(a)   The series converges only for $x = 0$.

(b)   The series converges absolutely (and hence converges) for all real values of x.

(c)   The series converges absolutely (and hence converges) for all x in some finite open interval $(-R, R)$, and diverges if $x < -R$ or $x > R$. At either of the points $x = R$ or $x = -R$, the series may converge absolutely, converge conditionally, or diverge, depending on the particular series.

This theorem states that the convergence set for a power series in x is always an interval centered at $x = 0$ (possibly just the point $x = 0$ itself or possibly infinite). For this reason, the convergence set of a power series in x is called the ***interval of convergence***. In the case where the convergence set is the single point $x = 0$ we say that the series has ***radius of convergence 0***, in the case where the convergence set is $(-\infty, +\infty)$ we say that the series has ***radius of convergence $+\infty$***, and in the case where the convergence set extends between $-R$ and R we say that the series has ***radius of convergence R*** (Figure 11.8.1).

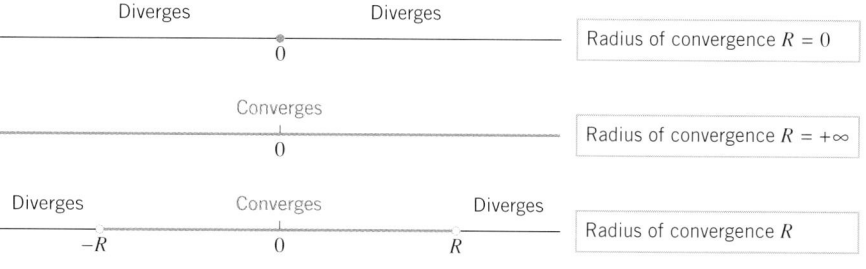

Figure 11.8.1

The usual procedure for finding the interval of convergence of a power series is to apply the ratio test for absolute convergence (Theorem 11.7.5). The following example illustrates how this works.

Example 1

Find the interval of convergence and radius of convergence of the following power series.

(a) $\displaystyle\sum_{k=0}^{\infty} x^k$ 　(b) $\displaystyle\sum_{k=0}^{\infty} \frac{x^k}{k!}$ 　(c) $\displaystyle\sum_{k=0}^{\infty} k!\,x^k$ 　(d) $\displaystyle\sum_{k=0}^{\infty} \frac{(-1)^k x^k}{3^k(k+1)}$

Solution (a). We apply the ratio test for absolute convergence. We have

$$\rho = \lim_{k \to +\infty} \left|\frac{u_{k+1}}{u_k}\right| = \lim_{k \to +\infty} \left|\frac{x^{k+1}}{x^k}\right| = \lim_{k \to +\infty} |x| = |x|$$

so the series converges absolutely if $\rho = |x| < 1$ and diverges if $\rho = |x| > 1$. The test is inconclusive if $|x| = 1$ (i.e., if $x = 1$ or $x = -1$), which means that we will have to investigate convergence at these points separately. At these points the series becomes

$$\sum_{k=0}^{\infty} 1^k = 1 + 1 + 1 + 1 + \cdots \qquad \boxed{x = 1}$$

$$\sum_{k=0}^{\infty} (-1)^k = 1 - 1 + 1 - 1 + \cdots \qquad \boxed{x = -1}$$

both of which diverge; thus, the interval of convergence for the given power series is $(-1, 1)$, and the radius of convergence is $R = 1$.

Solution (b). Applying the ratio test for absolute convergence, we obtain

$$\rho = \lim_{k \to +\infty} \left|\frac{u_{k+1}}{u_k}\right| = \lim_{k \to +\infty} \left|\frac{x^{k+1}}{(k+1)!} \cdot \frac{k!}{x^k}\right| = \lim_{k \to +\infty} \left|\frac{x}{k+1}\right| = 0$$

Since $\rho < 1$ for all x, the series converges absolutely for all x. Thus, the interval of convergence is $(-\infty, +\infty)$ and the radius of convergence is $R = +\infty$.

Solution (c). If $x \neq 0$, then the ratio test for absolute convergence yields

$$\rho = \lim_{k \to +\infty} \left|\frac{u_{k+1}}{u_k}\right| = \lim_{k \to +\infty} \left|\frac{(k+1)!\,x^{k+1}}{k!\,x^k}\right| = \lim_{k \to +\infty} |(k+1)x| = +\infty$$

Therefore, the series diverges for all nonzero values of x. Thus, the interval of convergence is the single point $x = 0$ and the radius of convergence is $R = 0$.

Solution (d). Since $|(-1)^k| = |(-1)^{k+1}| = 1$, we obtain

$$\rho = \lim_{k \to +\infty} \left|\frac{u_{k+1}}{u_k}\right| = \lim_{k \to +\infty} \left|\frac{x^{k+1}}{3^{k+1}(k+2)} \cdot \frac{3^k(k+1)}{x^k}\right|$$

$$= \lim_{k \to +\infty} \left[\frac{|x|}{3} \cdot \left(\frac{k+1}{k+2}\right)\right]$$

$$= \frac{|x|}{3} \lim_{k \to +\infty} \left(\frac{1 + (1/k)}{1 + (2/k)}\right) = \frac{|x|}{3}$$

The ratio test for absolute convergence implies that the series converges absolutely if $|x| < 3$ and diverges if $|x| > 3$. The ratio test fails to provide any information when $|x| = 3$, so the cases $x = -3$ and $x = 3$ need separate analyses. Substituting $x = -3$ in the given series yields

$$\sum_{k=0}^{\infty} \frac{(-1)^k(-3)^k}{3^k(k+1)} = \sum_{k=0}^{\infty} \frac{(-1)^k(-1)^k 3^k}{3^k(k+1)} = \sum_{k=0}^{\infty} \frac{1}{k+1}$$

which is the divergent harmonic series $1 + \frac{1}{2} + \frac{1}{3} + \frac{1}{4} + \cdots$. Substituting $x = 3$ in the given series yields

$$\sum_{k=0}^{\infty} \frac{(-1)^k 3^k}{3^k (k+1)} = \sum_{k=0}^{\infty} \frac{(-1)^k}{k+1} = 1 - \frac{1}{2} + \frac{1}{3} - \frac{1}{4} + \cdots$$

which is the conditionally convergent alternating harmonic series. Thus, the interval of convergence for the given series is $(-3, 3]$ and the radius of convergence is $R = 3$. ◀

POWER SERIES IN $x - x_0$

If x_0 is a constant, and if x is replaced by $x - x_0$ in (1), then the resulting series has the form

$$\sum_{k=0}^{\infty} c_k (x - x_0)^k = c_0 + c_1(x - x_0) + c_2(x - x_0)^2 + \cdots + c_k(x - x_0)^k + \cdots$$

This is called a **power series in $x - x_0$**. Some examples are

$$\sum_{k=0}^{\infty} \frac{(x-1)^k}{k+1} = 1 + \frac{(x-1)}{2} + \frac{(x-1)^2}{3} + \frac{(x-1)^3}{4} + \cdots \qquad \boxed{x_0 = 1}$$

$$\sum_{k=0}^{\infty} \frac{(-1)^k (x+3)^k}{k!} = 1 - (x+3) + \frac{(x+3)^2}{2!} - \frac{(x+3)^3}{3!} + \cdots \qquad \boxed{x_0 = -3}$$

The first of these is a power series in $x - 1$ and the second is a power series in $x + 3$. Note that a power series in x is a power series in $x - x_0$ in which $x_0 = 0$. More generally, the Taylor series

$$\sum_{k=0}^{\infty} \frac{f^{(k)}(x_0)}{k!} (x - x_0)^k$$

is a power series in $x - x_0$.

The main result on convergence of a power series in $x - x_0$ can be obtained by substituting $x - x_0$ for x in Theorem 11.8.1. This leads to the following theorem.

11.8.2 THEOREM. *For a power series $\sum c_k (x - x_0)^k$, exactly one of the following statements is true:*

(a) *The series converges only for $x = x_0$.*

(b) *The series converges absolutely (and hence converges) for all real values of x.*

(c) *The series converges absolutely (and hence converges) for all x in some finite open interval $(x_0 - R, x_0 + R)$ and diverges if $x < x_0 - R$ or $x > x_0 + R$. At either of the points $x = x_0 - R$ or $x = x_0 + R$, the series may converge absolutely, converge conditionally, or diverge, depending on the particular series.*

It follows from this theorem that the set of values for which a power series in $x - x_0$ converges is always an interval centered at $x = x_0$; we call this the **interval of convergence** (Figure 11.8.2). In part (a) of Theorem 11.8.2 the interval of convergence reduces to the single point $x = x_0$, in which case we say that the series has **radius of convergence $R = 0$**; in part (b) the interval of convergence is infinite (the entire real line), in which case we say that

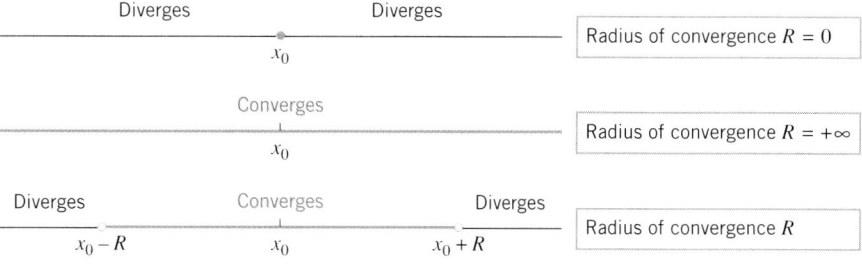

Figure 11.8.2

the series has ***radius of convergence R = +∞***; and in part (*c*) the interval extends between $x_0 - R$ and $x_0 + R$, in which case we say that the series has ***radius of convergence R***.

Example 2

Find the interval of convergence and radius of convergence of the series

$$\sum_{k=1}^{\infty} \frac{(x-5)^k}{k^2}$$

Solution.   We apply the ratio test for absolute convergence.

$$\rho = \lim_{k \to +\infty} \left| \frac{u_{k+1}}{u_k} \right| = \lim_{k \to +\infty} \left| \frac{(x-5)^{k+1}}{(k+1)^2} \cdot \frac{k^2}{(x-5)^k} \right|$$

$$= \lim_{k \to +\infty} \left[|x-5| \left(\frac{k}{k+1} \right)^2 \right]$$

$$= |x-5| \lim_{k \to +\infty} \left(\frac{1}{1+(1/k)} \right)^2 = |x-5|$$

Thus, the series converges absolutely if $|x - 5| < 1$, or $-1 < x - 5 < 1$, or $4 < x < 6$. The series diverges if $x < 4$ or $x > 6$.

To determine the convergence behavior at the endpoints $x = 4$ and $x = 6$, we substitute these values in the given series. If $x = 6$, the series becomes

$$\sum_{k=1}^{\infty} \frac{1^k}{k^2} = \sum_{k=1}^{\infty} \frac{1}{k^2} = 1 + \frac{1}{2^2} + \frac{1}{3^2} + \frac{1}{4^2} + \cdots$$

which is a convergent *p*-series ($p = 2$). If $x = 4$, the series becomes

$$\sum_{k=1}^{\infty} \frac{(-1)^k}{k^2} = -1 + \frac{1}{2^2} - \frac{1}{3^2} + \frac{1}{4^2} - \cdots$$

Since this series converges absolutely, the interval of convergence for the given series is [4, 6]. The radius of convergence is $R = 1$ (Figure 11.8.3).   ◄

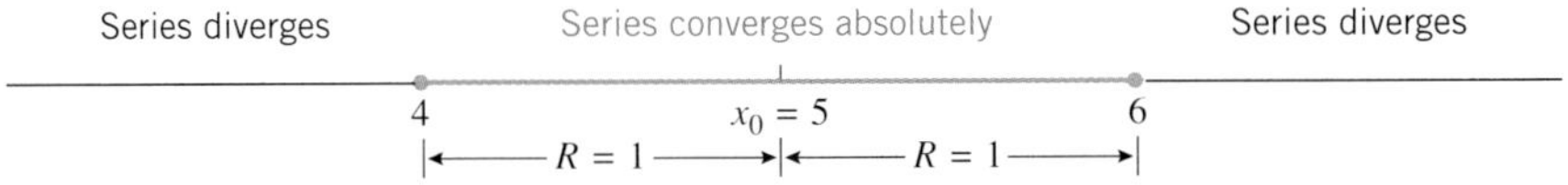

Series diverges	Series converges absolutely	Series diverges

$$\begin{array}{ccc} 4 & x_0 = 5 & 6 \\ |\!\!\longleftarrow R = 1 \longrightarrow\!|\!\longleftarrow R = 1 \longrightarrow\!| \end{array}$$

Figure 11.8.3

FOR THE READER.   It will always be a waste of time to test for convergence at the endpoints of the interval of convergence using the ratio test, since ρ will always be 1 at those points if $\rho = \lim_{n \to +\infty} |a_{n+1}/a_n|$ exists. Explain why this must be so.

FUNCTIONS DEFINED BY POWER SERIES

If a function f is expressed as a power series on some interval, then we say that f is ***represented*** by the power series on that interval. For example, we saw in Example 4 of Section 11.3 that

$$\frac{1}{1-x} = 1 + x + x^2 + \cdots + x^k + \cdots$$

so that this power series represents the function $1/(1 - x)$ on the interval $-1 < x < 1$.

Sometimes new functions actually originate as power series, and the properties of the functions are developed by working with their power series representations. For example, the functions

$$J_0(x) = \sum_{k=0}^{\infty} \frac{(-1)^k x^{2k}}{2^{2k}(k!)^2} = 1 - \frac{x^2}{2^2(1!)^2} + \frac{x^4}{2^4(2!)^2} - \frac{x^6}{2^6(3!)^2} + \cdots \qquad (2)$$

and

$$J_1(x) = \sum_{k=0}^{\infty} \frac{(-1)^k x^{2k+1}}{2^{2k+1}(k!)(k+1)!} = \frac{x}{2} - \frac{x^3}{2^3(1!)(2!)} + \frac{x^5}{2^5(2!)(3!)} - \cdots \qquad (3)$$

which are called **Bessel functions** in honor of the German mathematician and astronomer Friedrich Wilhelm Bessel (1784–1846), arise naturally in the study of planetary motion and in various problems that involve heat flow.

To find the domains of these functions, we must determine where their defining power series converge. For example, in the case $J_0(x)$ we have

$$\rho = \lim_{k \to +\infty} \left| \frac{u_{k+1}}{u_k} \right| = \lim_{k \to +\infty} \left| \frac{x^{2(k+1)}}{2^{2(k+1)}[(k+1)!]^2} \cdot \frac{2^{2k}(k!)^2}{x^{2k}} \right|$$

$$= \lim_{k \to +\infty} \left| \frac{x^2}{4(k+1)^2} \right| = 0 < 1$$

so that the series converges for all x; that is, the domain of $J_0(x)$ is $(-\infty, +\infty)$. We leave it as an exercise to show that the power series for $J_1(x)$ also converges for all x.

FOR THE READER.   Many CAS programs have the Bessel functions as part of their libraries. If you have a CAS, read the documentation to determine whether it can graph $J_0(x)$ and $J_1(x)$; if so, generate the graphs shown in Figure 11.8.4.

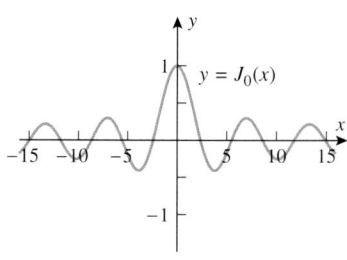

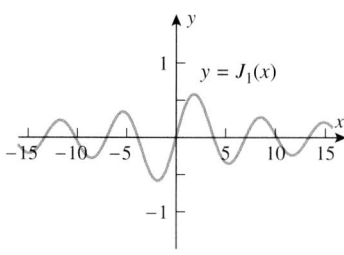

Figure 11.8.4

EXERCISE SET 11.8   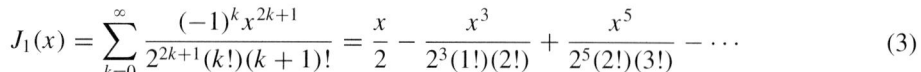 Graphing Calculator

In Exercises 1–4, find the interval of convergence of the power series, and find a familiar function that is represented by the power series on that interval.

1. $1 - x + x^2 - x^3 + \cdots + (-1)^k x^k + \cdots$

2. $1 + x^2 + x^4 + \cdots + x^{2k} + \cdots$

3. $1 + (x-2) + (x-2)^2 + \cdots + (x-2)^k + \cdots$

4. $1 - (x+3) + (x+3)^2 - (x+3)^3 + \cdots + (-1)^k (x+3)^k + \cdots$

5. Suppose that the function f is represented by the power series

$$f(x) = 1 - \frac{x}{2} + \frac{x^2}{4} - \frac{x^3}{8} + \cdots + (-1)^k \frac{x^k}{2^k} + \cdots$$

(a) Find the domain of f.
(b) Find $f(0)$ and $f(1)$.

6. Suppose that the function f is represented by the power series

$$f(x) = 1 - \frac{x-5}{3} + \frac{(x-5)^2}{3^2} - \frac{(x-5)^3}{3^3} + \cdots$$

(a) Find the domain of f.
(b) Find $f(3)$ and $f(6)$.

In Exercises 7–30, find the radius of convergence and the interval of convergence.

7. $\displaystyle\sum_{k=0}^{\infty} \frac{x^k}{k+1}$

8. $\displaystyle\sum_{k=0}^{\infty} 3^k x^k$

9. $\displaystyle\sum_{k=0}^{\infty} \frac{(-1)^k x^k}{k!}$

10. $\displaystyle\sum_{k=0}^{\infty} \frac{k!}{2^k} x^k$

11. $\displaystyle\sum_{k=1}^{\infty} \frac{5^k}{k^2} x^k$

12. $\displaystyle\sum_{k=2}^{\infty} \frac{x^k}{\ln k}$

13. $\displaystyle\sum_{k=1}^{\infty} \frac{x^k}{k(k+1)}$

14. $\displaystyle\sum_{k=0}^{\infty} \frac{(-2)^k x^{k+1}}{k+1}$

15. $\displaystyle\sum_{k=1}^{\infty} (-1)^{k-1} \frac{x^k}{\sqrt{k}}$

16. $\displaystyle\sum_{k=0}^{\infty} \frac{(-1)^k x^{2k}}{(2k)!}$

17. $\displaystyle\sum_{k=0}^{\infty} (-1)^k \frac{x^{2k+1}}{(2k+1)!}$

18. $\displaystyle\sum_{k=1}^{\infty} (-1)^k \frac{x^{3k}}{k^{3/2}}$

19. $\displaystyle\sum_{k=0}^{\infty} \frac{3^k}{k!} x^k$

20. $\displaystyle\sum_{k=2}^{\infty} (-1)^{k+1} \frac{x^k}{k(\ln k)^2}$

21. $\displaystyle\sum_{k=0}^{\infty} \frac{x^k}{1+k^2}$

22. $\displaystyle\sum_{k=0}^{\infty} \frac{(x-3)^k}{2^k}$

23. $\displaystyle\sum_{k=1}^{\infty} (-1)^{k+1} \frac{(x+1)^k}{k}$

24. $\displaystyle\sum_{k=0}^{\infty} (-1)^k \frac{(x-4)^k}{(k+1)^2}$

25. $\displaystyle\sum_{k=0}^{\infty} \left(\frac{3}{4} \right)^k (x+5)^k$

26. $\displaystyle\sum_{k=1}^{\infty} \frac{(2k+1)!}{k^3} (x-2)^k$

27. $\displaystyle\sum_{k=1}^{\infty} (-1)^k \frac{(x+1)^{2k+1}}{k^2+4}$

28. $\displaystyle\sum_{k=1}^{\infty} \frac{(\ln k)(x-3)^k}{k}$

29. $\displaystyle\sum_{k=0}^{\infty} \frac{\pi^k (x-1)^{2k}}{(2k+1)!}$

30. $\displaystyle\sum_{k=0}^{\infty} \frac{(2x-3)^k}{4^{2k}}$

31. Use the root test to find the interval of convergence of

$$\sum_{k=2}^{\infty} \frac{x^k}{(\ln k)^k}$$

32. Find the domain of the function

$$f(x) = \sum_{k=1}^{\infty} \frac{1 \cdot 3 \cdot 5 \cdots (2k-1)}{(2k-2)!} x^k$$

33. If a function f is represented by a power series on an interval, then the graphs of the partial sums can be used as approximations to the graph of f.

    (a) Use a graphing utility to generate the graph of $1/(1-x)$ together with the graphs of the first four partial sums of its Maclaurin series over the interval $(-1, 1)$.

    (b) In general terms, where are the graphs of the partial sums the most accurate?

34. Show that the power series representation of the Bessel function $J_1(x)$ converges for all x [Formula (3)].

35. Show that if p is a positive integer, then the power series

$$\sum_{k=0}^{\infty} \frac{(pk)!}{(k!)^p} x^k$$

has a radius of convergence of $1/p^p$.

36. Show that if p and q are positive integers, then the power series

$$\sum_{k=0}^{\infty} \frac{(k+p)!}{k!(k+q)!} x^k$$

has a radius of convergence of $+\infty$.

37. (a) Suppose that the power series $\sum c_k(x - x_0)^k$ has radius of convergence R and p is a nonzero constant. What can you say about the radius of convergence of the power series $\sum pc_k(x - x_0)^k$? Explain your reasoning. [*Hint:* See Theorem 11.4.3.]

    (b) Suppose that the power series $\sum c_k(x - x_0)^k$ has a finite radius of convergence R, and the power series $\sum d_k(x - x_0)^k$ has a radius of convergence of $+\infty$. What can you say about the radius of convergence of $\sum(c_k + d_k)(x - x_0)^k$? Explain your reasoning.

    (c) Suppose that the power series $\sum c_k(x - x_0)^k$ has a finite radius of convergence R_1 and the power series $\sum d_k(x - x_0)^k$ has a finite radius of convergence R_2. What can you say about the radius of convergence of $\sum(c_k + d_k)(x - x_0)^k$? Explain your reasoning.

38. Prove: If $\lim_{k \to +\infty} |c_k|^{1/k} = L$, where $L \neq 0$, then $1/L$ is the radius of convergence of the power series $\sum_{k=0}^{\infty} c_k x^k$.

39. Prove: If the power series $\sum_{k=0}^{\infty} c_k x^k$ has radius of convergence R, then the series $\sum_{k=0}^{\infty} c_k x^{2k}$ has radius of convergence $\sqrt{R}$.

40. Prove: If the interval of convergence of the series $\sum_{k=0}^{\infty} c_k(x - x_0)^k$ is $(x_0 - R, x_0 + R]$, then the series converges conditionally at $x_0 + R$.

11.9 CONVERGENCE OF TAYLOR SERIES; COMPUTATIONAL METHODS

In Section 11.5 we anticipated the possibility that a Taylor series for a function might actually converge to the function on some interval. In this section we will study the convergence of Taylor series, and we will show how they can be used to approximate trigonometric, exponential, and logarithmic functions.

THE nTH REMAINDER

Recall that the nth Taylor polynomial for a function f about $x = x_0$ has the property that its value and the values of its first n derivatives match those of f at x_0. As n increases, more and more derivatives match up, so it is reasonable to hope that for values of x near x_0 the values of the Taylor polynomials might converge to the value of $f(x)$; that is,

$$\sum_{k=0}^{n} \frac{f^{(k)}(x_0)}{k!}(x - x_0)^k \to f(x) \quad \text{as } n \to +\infty \qquad (1)$$

However, the nth Taylor polynomial for f is the nth partial sum of the Taylor series for f, so (1) is equivalent to stating that the Taylor series for f converges at the point x, and its sum is $f(x)$. Thus, we are led to consider the following problem.

11.9.1 PROBLEM. Given a function f that has derivatives of all orders at a point x_0, determine whether there is an open interval containing x_0 such that $f(x)$ is the sum of its Taylor series about $x = x_0$ at each point in the interval; that is,

$$f(x) = \sum_{k=0}^{\infty} \frac{f^{(k)}(x_0)}{k!}(x - x_0)^k \qquad (2)$$

FOR THE READER. Show that (2) holds at $x = x_0$, regardless of the function f.

To determine whether (2) holds on some open interval containing x_0, it will be convenient to consider the difference between $f(x)$ and its nth Taylor polynomial about $x = x_0$. This difference is called the ***n*th remainder for *f* about *x* = *x*$_0$**, and is denoted by

$$R_n(x) = f(x) - \sum_{k=0}^{n} \frac{f^{(k)}(x_0)}{k!}(x - x_0)^k \tag{3}$$

This can also be written as

$$f(x) = \sum_{k=0}^{n} \frac{f^{(k)}(x_0)}{k!}(x - x_0)^k + R_n(x) \tag{4}$$

which is called ***Taylor's formula with remainder***.

One can think of $R_n(x)$ as the error that results at the point x when f is approximated by its nth Taylor polynomial. Thus, for the Taylor polynomials about x_0 to converge to f at a point x as $n \to +\infty$, the error $R_n(x)$ must approach 0; conversely, if $R_n(x) \to 0$ as $n \to +\infty$, then the Taylor polynomials converge to f at the point x. More precisely, we have the following theorem.

11.9.2 THEOREM. *The equality*

$$f(x) = \sum_{k=0}^{\infty} \frac{f^{(k)}(x_0)}{k!}(x - x_0)^k$$

holds at a point x if and only if $\lim\limits_{n \to +\infty} R_n(x) = 0$.

ESTIMATING THE *n*TH REMAINDER

It is relatively rare that one can prove directly that $R_n(x) \to 0$ as $n \to +\infty$. Usually, this is proved indirectly by finding appropriate bounds on $|R_n(x)|$ and applying the Squeezing Theorem for Sequences. The following theorem, which is proved in Appendix G, provides a bound that can be used for this purpose.

11.9.3 THEOREM (*The Remainder Estimation Theorem*). *If the function f can be differentiated $n + 1$ times on an interval I containing the point x_0, and if $|f^{(n+1)}(x)| \leq M$ for all x in I, then*

$$|R_n(x)| \leq \frac{M}{(n+1)!}|x - x_0|^{n+1} \tag{5}$$

for all x in I.

The following example illustrates how this theorem is applied.

Example 1

Show that the Maclaurin series for $\cos x$ converges to $\cos x$ for all x; that is,

$$\cos x = \sum_{k=0}^{\infty} (-1)^k \frac{x^{2k}}{(2k)!} = 1 - \frac{x^2}{2!} + \frac{x^4}{4!} - \frac{x^6}{6!} + \cdots \qquad (-\infty < x < +\infty)$$

Solution. From Theorem 11.9.2 we must show that $R_n(x) \to 0$ for all x as $n \to +\infty$. For this purpose let $f(x) = \cos x$, so that for all x we have

$$f^{(n+1)}(x) = \pm \cos x \quad \text{or} \quad f^{(n+1)}(x) = \pm \sin x$$

In all cases we have

$$|f^{(n+1)}(x)| \leq 1$$

so we can apply Theorem 11.9.3 with $M = 1$ and $x_0 = 0$ to conclude that

$$0 \leq |R_n(x)| \leq \frac{|x|^{n+1}}{(n+1)!} \tag{6}$$

However, it follows from Formula (5) of Section 11.2 with $n + 1$ in place of n and $|x|$ in place of x that

$$\lim_{n \to +\infty} \frac{|x|^{n+1}}{(n+1)!} = 0 \tag{7}$$

Thus, it follows from (6) and the Squeezing Theorem for Sequences (Theorem 11.1.5) that $|R_n(x)| \to 0$ as $n \to +\infty$; this implies that $R_n(x) \to 0$ as $n \to +\infty$ by Theorem 11.1.6. Since this is true for all x, we have proved that the Maclaurin series for $\cos x$ converges to $\cos x$ for all x. This is illustrated in Figure 11.9.1, where we can see how successive partial sums approximate the cosine curve more and more closely. ◄

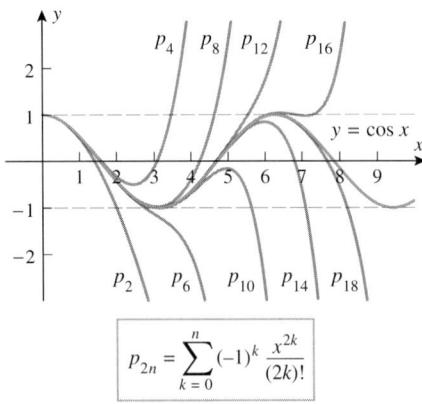

Figure 11.9.1

REMARK.   The method used in Example 1 can be easily modified to prove that the Taylor series for $\cos x$ about any point $x = x_0$ converges to $\cos x$ for all x, and similarly that the Taylor series for $\sin x$ about any point $x = x_0$ converges to $\sin x$ for all x (Exercises 25 and 26). For reference, there is a list of some of the most important Maclaurin series in Table 11.9.1 at the end of this section.

APPROXIMATING TRIGONOMETRIC FUNCTIONS

In general, to approximate the value of a function f at a point x using a Taylor series, there are two basic questions that must be answered:

- About what point x_0 should the Taylor series be expanded?
- How many terms in the series should be used to achieve the desired accuracy?

In response to the first question, x_0 needs to be a point where the derivatives of f can be evaluated easily, since these values are needed for the coefficients in the Taylor series. Furthermore, if the function f is being evaluated at the point x, then x_0 should be chosen as close as possible to x, since Taylor series tend to converge more rapidly near x_0. For example, to approximate $\sin 3°$ $(= \pi/60$ radians), it would be reasonable to take $x_0 = 0$, since $\pi/60$ is close to 0 and the derivatives of $\sin x$ are easy to evaluate at 0. On the other hand, to approximate $\sin 85°$ $(= 17\pi/36$ radians), it would be more natural to take $x_0 = \pi/2$, since $17\pi/36$ is close to $\pi/2$ and the derivatives of $\sin x$ are easy to evaluate at $\pi/2$.

In response to the second question posed above, the number of terms required to achieve a specific accuracy needs to be determined on a problem-by-problem basis. The next example gives two methods for doing this.

Example 2

Use the Maclaurin series for $\sin x$ to approximate $\sin 3°$ to five decimal-place accuracy.

Solution. In the Maclaurin series

$$\sin x = \sum_{k=0}^{\infty} (-1)^k \frac{x^{2k+1}}{(2k+1)!} = x - \frac{x^3}{3!} + \frac{x^5}{5!} - \frac{x^7}{7!} + \cdots \tag{8}$$

the angle x is assumed to be in radians (because the differentiation formulas for the trigonometric functions were derived with this assumption). Since $3° = \pi/60$ radians, it follows from (8) that

$$\sin 3° = \sin \frac{\pi}{60} = \left(\frac{\pi}{60}\right) - \frac{(\pi/60)^3}{3!} + \frac{(\pi/60)^5}{5!} - \frac{(\pi/60)^7}{7!} + \cdots \tag{9}$$

We must now determine how many terms in the series are required to achieve five decimal-place accuracy. We will consider two possible approaches, one using the Remainder Estimation Theorem (Theorem 11.9.3) and the other using the fact that (9) satisfies the hypotheses of the alternating series test (Theorem 11.7.1).

Method 1 (The Remainder Estimation Theorem). Since we want to achieve five decimal-place accuracy, our goal is to choose n so that the absolute value of the nth remainder at $x = \pi/60$ does not exceed $0.000005 = 5 \times 10^{-5}$; that is,

$$\left| R_n\left(\frac{\pi}{60}\right) \right| \le 0.000005 \tag{10}$$

However, if we let $f(x) = \sin x$, then $f^{(n+1)}(x)$ is either $\pm\sin x$ or $\pm\cos x$, and in either case $|f^{(n+1)}(x)| \le 1$ for all x. Thus, it follows from the Remainder Estimation Theorem with $M = 1$, $x_0 = 0$, and $x = \pi/60$ that

$$\left| R_n\left(\frac{\pi}{60}\right) \right| \le \frac{|\pi/60|^{n+1}}{(n+1)!}$$

Thus, we can satisfy (10) by choosing n so that

$$\frac{|\pi/60|^{n+1}}{(n+1)!} \le 0.000005$$

With the help of a calculating utility you can verify that the smallest value of n that meets this criterion is $n = 3$. Thus, to achieve five decimal-place accuracy we need only keep terms up to the third power in (9). This yields

$$\sin 3° \approx \left(\frac{\pi}{60}\right) - \frac{(\pi/60)^3}{3!} \approx 0.05234 \tag{11}$$

(verify). As a check, the author's calculator gives $\sin 3° \approx 0.05233595624$, which agrees with (11) when rounded to five decimal places.

Method 2 (The Alternating Series Test). We leave it for you to check that (9) satisfies the hypotheses of the alternating series test (Theorem 11.7.1).

Let s_n denote the sum of the terms in (9) up to and including the nth power of $\pi/60$. Since the exponents in the series are odd integers, the integer n must be odd, and the exponent of the first term *not* included in the sum s_n must be $n + 2$. Thus, it follows from part (b) of Theorem 11.7.2 that

$$|\sin 3° - s_n| < \frac{(\pi/60)^{n+2}}{(n+2)!}$$

This means that for five decimal-place accuracy we must look for the first positive odd integer n such that

$$\frac{(\pi/60)^{n+2}}{(n+2)!} \le 0.000005$$

With the help of a calculating utility you can verify that the smallest value of n that meets this criterion is $n = 3$. This agrees with the result obtained above using the Remainder Estimation Theorem and hence leads to approximation (11) as before. ◄

ROUNDOFF AND TRUNCATION ERROR

There are two types of errors that occur when computing with series. The first, called *truncation error*, is the error that results when a series is approximated by a partial sum; and the second, called *roundoff error*, is the error that arises from approximations in numerical computations. For example, in our derivation of (11) we took $n = 3$ to keep the truncation error below 0.000005. However, to evaluate the partial sum we had to approximate π, thereby introducing roundoff error. Had we not exercised some care in choosing this approximation, the roundoff error could easily have degraded the final result.

Methods for estimating and controlling roundoff error are studied in a branch of mathematics called *numerical analysis*. However, as a rule of thumb, to achieve n decimal-place accuracy in a final result, all intermediate calculations must be accurate to at least $n + 1$ decimal places. Thus, in (11) at least six decimal-place accuracy in π is required to achieve the five decimal-place accuracy in the final numerical result. As a practical matter, a good working procedure is to perform all intermediate computations with the maximum number of digits that your calculating utility can handle and then round at the end.

APPROXIMATING EXPONENTIAL FUNCTIONS

Example 3

Show that the Maclaurin series for e^x converges to e^x for all x; that is,

$$e^x = \sum_{k=0}^{\infty} \frac{x^k}{k!} = 1 + x + \frac{x^2}{2!} + \frac{x^3}{3!} + \cdots + \frac{x^k}{k!} + \cdots \qquad (-\infty < x < +\infty)$$

Solution. Let $f(x) = e^x$, so that

$$f^{(n+1)}(x) = e^x$$

We want to show that $R_n(x) \to 0$ as $n \to +\infty$ for all x in the interval $-\infty < x < +\infty$. However, it will be helpful here to consider the cases $x \le 0$ and $x > 0$ separately. If $x \le 0$, then we will take the interval I in Theorem 11.9.3 to be $[x, 0]$, and if $x > 0$, then we will take it to be $[0, x]$. Since $f^{(n+1)}(x) = e^x$ is an increasing function, it follows that if c is in the interval $[x, 0]$, then

$$|f^{(n+1)}(c)| \le |f^{(n+1)}(0)| = e^0 = 1$$

and if c is in the interval $[0, x]$, then

$$|f^{(n+1)}(c)| \le |f^{(n+1)}(x)| = e^x$$

Thus, we can apply Theorem 11.9.3 with $M = 1$ in the case where $x \le 0$ and with $M = e^x$ in the case where $x > 0$. This yields

$$0 \le |R_n(x)| \le \frac{|x|^{n+1}}{(n+1)!} \qquad \text{if } x \le 0$$

$$0 \le |R_n(x)| \le e^x \frac{|x|^{n+1}}{(n+1)!} \qquad \text{if } x > 0$$

Thus, in both cases it follows from (7) and the Squeezing Theorem for Sequences that $|R_n(x)| \to 0$ as $n \to +\infty$, which in turn implies that $R_n(x) \to 0$ as $n \to +\infty$. Since this is true for all x, we have proved that the Maclaurin series for e^x converges to e^x for all x. ◄

Example 4

Use the Maclaurin series for e^x to approximate e to five decimal-place accuracy.

Solution. If we substitute $x = 1$ in the Maclaurin series

$$e^x = 1 + x + \frac{x^2}{2!} + \frac{x^3}{3!} + \cdots + \frac{x^k}{k!} + \cdots$$

we obtain

$$e = 1 + 1 + \frac{1}{2!} + \frac{1}{3!} + \cdots + \frac{1}{k!} + \cdots \tag{12}$$

and hence we can approximate e to any degree of accuracy using an appropriate partial sum

$$e \approx 1 + 1 + \frac{1}{2!} + \frac{1}{3!} + \cdots + \frac{1}{n!}$$

Thus, our problem is to determine how many terms in this partial sum are required to achieve five decimal-place accuracy; that is, we want to choose n so that the absolute value of the nth remainder at $x = 1$ in the Maclaurin series satisfies

$$|R_n(1)| \leq 0.000005$$

To determine n we will apply the Remainder Estimation Theorem with $f(x) = e^x$, $x = 1$, $x_0 = 0$, and I being the interval $[0, 1]$. In this case it follows from Formula (5) that

$$|R_n(1)| \leq \frac{M}{(n+1)!} \tag{13}$$

where M is an upper bound on the value of $f^{(n+1)}(x) = e^x$ for x in the interval $[0, 1]$. However, e^x is an increasing function, so its maximum value on the interval $[0, 1]$ occurs at $x = 1$; that is, $e^x \leq e$ on this interval. Thus, we can take $M = e$ in (13) to obtain

$$|R_n(1)| \leq \frac{e}{(n+1)!} \tag{14}$$

Unfortunately, this inequality is not very useful because it involves e, which is the very quantity we are trying to approximate. However, if we accept that $e < 3$, then we can replace (14) with the following less precise, but more useful, inequality:

$$|R_n(1)| < \frac{3}{(n+1)!}$$

Thus, we can achieve five decimal-place accuracy by choosing n so that

$$\frac{3}{(n+1)!} \leq 0.000005$$

With the help of a calculating utility you can verify that the smallest value of n that meets this criterion is $n = 9$. Thus, to five decimal-place accuracy

$$e \approx 1 + 1 + \frac{1}{2!} + \frac{1}{3!} + \frac{1}{4!} + \frac{1}{5!} + \frac{1}{6!} + \frac{1}{7!} + \frac{1}{8!} + \frac{1}{9!} \approx 2.71828$$

(verify). As a check, the author's calculator gives $e \approx 2.71828182846$, which agrees with the preceding approximation when rounded to five decimal places. ◀

APPROXIMATING LOGARITHMS

The Maclaurin series

$$\ln(1 + x) = x - \frac{x^2}{2} + \frac{x^3}{3} - \frac{x^4}{4} + \cdots \qquad (-1 < x \leq 1) \tag{15}$$

is the starting point for the approximation of natural logarithms. Unfortunately, the usefulness of this series is limited because of its slow convergence and the restriction $-1 < x \leq 1$. However, if we replace x by $-x$ in this series, we obtain

$$\ln(1 - x) = -x - \frac{x^2}{2} - \frac{x^3}{3} - \frac{x^4}{4} - \cdots \qquad (-1 \leq x < 1) \tag{16}$$

and on subtracting (16) from (15) we obtain

$$\ln\left(\frac{1+x}{1-x}\right) = 2\left(x + \frac{x^3}{3} + \frac{x^5}{5} + \frac{x^7}{7} + \cdots\right) \qquad (-1 < x < 1) \qquad (17)$$

Series (17), first obtained by James Gregory[*] in 1668, can be used to compute the natural logarithm of any positive number y by letting

$$y = \frac{1+x}{1-x}$$

or, equivalently,

$$x = \frac{y-1}{y+1} \qquad (18)$$

and noting that $-1 < x < 1$. For example, to compute $\ln 2$ we let $y = 2$ in (18), which yields $x = \frac{1}{3}$. Substituting this value in (17) gives

$$\ln 2 = 2\left[\frac{1}{3} + \frac{\left(\frac{1}{3}\right)^3}{3} + \frac{\left(\frac{1}{3}\right)^5}{5} + \frac{\left(\frac{1}{3}\right)^7}{7} + \cdots\right] \qquad (19)$$

In Exercise 23 we will ask you to show that five decimal-place accuracy can be achieved using the partial sum with terms up to and including the 13th power of $\frac{1}{3}$. Thus, to five decimal-place accuracy

$$\ln 2 \approx 2\left[\frac{1}{3} + \frac{\left(\frac{1}{3}\right)^3}{3} + \frac{\left(\frac{1}{3}\right)^5}{5} + \frac{\left(\frac{1}{3}\right)^7}{7} + \cdots + \frac{\left(\frac{1}{3}\right)^{13}}{13}\right] \approx 0.69315$$

(verify). As a check, the author's calculator gives $\ln 2 \approx 0.69314718056$, which agrees with the preceding approximation when rounded to five decimal places.

REMARK.   In Example 2 of Section 11.7, we stated without proof that

$$\ln 2 = 1 - \frac{1}{2} + \frac{1}{3} - \frac{1}{4} + \frac{1}{5} - \cdots$$

This result can be obtained letting $x = 1$ in (15). However, this series converges too slowly to be of practical value.

APPROXIMATING π

In the next section we will show that

$$\tan^{-1} x = x - \frac{x^3}{3} + \frac{x^5}{5} - \frac{x^7}{7} + \cdots \qquad (-1 \le x \le 1) \qquad (20)$$

Letting $x = 1$, we obtain

$$\frac{\pi}{4} = \tan^{-1} 1 = 1 - \frac{1}{3} + \frac{1}{5} - \frac{1}{7} + \cdots$$

or

$$\pi = 4\left[1 - \frac{1}{3} + \frac{1}{5} - \frac{1}{7} + \cdots\right]$$

This famous series, obtained by Leibniz in 1674, converges too slowly to be of computational value. A more practical procedure for approximating π uses the identity

$$\frac{\pi}{4} = \tan^{-1} \frac{1}{2} + \tan^{-1} \frac{1}{3} \qquad (21)$$

which was derived in Exercise 47 of Section 4.5. By using this identity and series (20)

_____

[*] JAMES GREGORY (1638–1675). Scottish mathematician and astronomer. Gregory, the son of a minister, was famous in his time as the inventor of the Gregorian reflecting telescope, so named in his honor. Although he is not generally ranked with the great mathematicians, much of his work relating to calculus was studied by Leibniz and Newton and undoubtedly influenced some of their discoveries. There is a manuscript, discovered posthumously, which shows that Gregory had anticipated Taylor series well before Taylor.

to approximate $\tan^{-1}\frac{1}{2}$ and $\tan^{-1}\frac{1}{3}$, the value of π can be approximated efficiently to any degree of accuracy.

If m is a real number, then the Maclaurin series for $(1+x)^m$ is called the **binomial series**; it is given by (verify)

$$1 + mx + \frac{m(m-1)}{2!}x^2 + \frac{m(m-1)(m-2)}{3!}x^3 + \cdots + \frac{m(m-1)\cdots(m-k+1)}{k!} + \cdots$$

In the case where m is a nonnegative integer, the function $f(x) = (1+x)^m$ is a polynomial of degree m, so

$$f^{(m+1)}(0) = f^{(m+2)}(0) = f^{(m+3)}(0) = \cdots = 0$$

and the binomial series reduces to the familiar binomial expansion

$$(1+x)^m = 1 + mx + \frac{m(m-1)}{2!}x^2 + \frac{m(m-1)(m-2)}{3!}x^3 + \cdots + x^m$$

which is valid for $-\infty < x < +\infty$.

It can be proved that if m is not a nonnegative integer, then the binomial series converges to $(1+x)^m$ if $|x| < 1$. Thus, for such values of x

$$(1+x)^m = 1 + mx + \frac{m(m-1)}{2!}x^2 + \cdots + \frac{m(m-1)\cdots(m-k+1)}{k!}x^k + \cdots \qquad (22)$$

or in sigma notation,

$$(1+x)^m = 1 + \sum_{k=1}^{\infty} \frac{m(m-1)\cdots(m-k+1)}{k!}x^k \quad \text{if } |x| < 1 \qquad (23)$$

Example 5

Find binomial series for

(a) $\dfrac{1}{(1+x)^2}$ (b) $\dfrac{1}{\sqrt{1+x}}$

Solution (a). Since the general term of the binomial series is complicated, you may find it helpful to write out some of the beginning terms of the series, as in Formula (22), to see developing patterns. Substituting $m = -2$ in this formula yields

$$\frac{1}{(1+x)^2} = (1+x)^{-2} = 1 + (-2)x + \frac{(-2)(-3)}{2!}x^2$$

$$+ \frac{(-2)(-3)(-4)}{3!}x^3 + \frac{(-2)(-3)(-4)(-5)}{4!}x^4 + \cdots$$

$$= 1 - 2x + \frac{3!}{2!}x^2 - \frac{4!}{3!}x^3 + \frac{5!}{4!}x^4 + \cdots$$

$$= 1 - 2x + 3x^2 - 4x^3 + 5x^4 + \cdots$$

$$= \sum_{k=0}^{\infty}(-1)^k(k+1)x^k$$

Solution (b). Substituting $m = -\frac{1}{2}$ in (22) yields

$$\frac{1}{\sqrt{1+x}} = 1 - \frac{1}{2}x + \frac{\left(-\frac{1}{2}\right)\left(-\frac{1}{2}-1\right)}{2!}x^2 + \frac{\left(-\frac{1}{2}\right)\left(-\frac{1}{2}-1\right)\left(-\frac{1}{2}-2\right)}{3!}x^3 - \cdots$$

$$= 1 - \frac{1}{2}x + \frac{1\cdot 3}{2^2\cdot 2!}x^2 - \frac{1\cdot 3\cdot 5}{2^3\cdot 3!}x^3 + \cdots$$

$$= 1 + \sum_{k=1}^{\infty}(-1)^k\frac{1\cdot 3\cdot 5\cdots(2k-1)}{2^k k!}x^k \qquad \blacktriangleleft$$

For reference, Table 11.9.1 lists the Maclaurin series for some of the most important functions, together with a specification of the intervals over which the Maclaurin series converge to those functions. Some of these results are derived in the exercises and others will be derived in the next section using some special techniques that we will develop.

Table 11.9.1

MACLAURIN SERIES	INTERVAL OF CONVERGENCE
$\dfrac{1}{1-x} = \displaystyle\sum_{k=0}^{\infty} x^k = 1 + x + x^2 + x^3 + \cdots$	$-1 < x < 1$
$\dfrac{1}{1+x^2} = \displaystyle\sum_{k=0}^{\infty} (-1)^k x^{2k} = 1 - x^2 + x^4 - x^6 + \cdots$	$-1 < x < 1$
$e^x = \displaystyle\sum_{k=0}^{\infty} \dfrac{x^k}{k!} = 1 + x + \dfrac{x^2}{2!} + \dfrac{x^3}{3!} + \dfrac{x^4}{4!} + \cdots$	$-\infty < x < +\infty$
$\sin x = \displaystyle\sum_{k=0}^{\infty} (-1)^k \dfrac{x^{2k+1}}{(2k+1)!} = x - \dfrac{x^3}{3!} + \dfrac{x^5}{5!} - \dfrac{x^7}{7!} + \cdots$	$-\infty < x < +\infty$
$\cos x = \displaystyle\sum_{k=0}^{\infty} (-1)^k \dfrac{x^{2k}}{(2k)!} = 1 - \dfrac{x^2}{2!} + \dfrac{x^4}{4!} - \dfrac{x^6}{6!} + \cdots$	$-\infty < x < +\infty$
$\ln(1+x) = \displaystyle\sum_{k=1}^{\infty} (-1)^{k+1} \dfrac{x^k}{k} = x - \dfrac{x^2}{2} + \dfrac{x^3}{3} - \dfrac{x^4}{4} + \cdots$	$-1 < x \le 1$
$\tan^{-1} x = \displaystyle\sum_{k=0}^{\infty} (-1)^k \dfrac{x^{2k+1}}{2k+1} = x - \dfrac{x^3}{3} + \dfrac{x^5}{5} - \dfrac{x^7}{7} + \cdots$	$-1 \le x \le 1$
$\sinh x = \displaystyle\sum_{k=0}^{\infty} \dfrac{x^{2k+1}}{(2k+1)!} = x + \dfrac{x^3}{3!} + \dfrac{x^5}{5!} + \dfrac{x^7}{7!} + \cdots$	$-\infty < x < +\infty$
$\cosh x = \displaystyle\sum_{k=0}^{\infty} \dfrac{x^{2k}}{(2k)!} = 1 + \dfrac{x^2}{2!} + \dfrac{x^4}{4!} + \dfrac{x^6}{6!} + \cdots$	$-\infty < x < +\infty$
$(1+x)^m = 1 + \displaystyle\sum_{k=1}^{\infty} \dfrac{m(m-1)\cdots(m-k+1)}{k!} x^k$	$-1 < x < 1^*$ ($m \ne 0, 1, 2, \ldots$)

*The behavior at the endpoints depends on m: For $m > 0$ the series converges absolutely at both endpoints; for $m \le -1$ the series diverges at both endpoints; and for $-1 < m < 0$ the series converges conditionally at $x = 1$ and diverges at $x = -1$.

EXERCISE SET 11.9   ⌇ Graphing Calculator   ⊡ CAS

1. Use both of the methods given in Example 2 to approximate $\sin 4°$ to five decimal-place accuracy, and check your work by comparing your answer to that produced directly by your calculating utility.

2. Use both of the methods given in Example 2 to approximate $\cos 3°$ to three decimal-place accuracy, and check your work by comparing your answer to that produced directly by your calculating utility.

3. Use the method of Example 4 to approximate $\sqrt{e}$ to four decimal-place accuracy, and check your work by comparing your answer to that produced directly by your calculating utility. [*Suggestion:* Write $\sqrt{e}$ as $e^{0.5}$.]

4. Use the method of Example 4 to approximate $1/e$ to three decimal-place accuracy, and check your work by comparing your answer to that produced directly by your calculating utility.

5. Use the Maclaurin series for $\cos x$ to approximate $\cos 0.1$ to five decimal-place accuracy, and check your work by comparing your answer to that produced directly by your calculating utility.

6. Use the Maclaurin series for $\tan^{-1} x$ to approximate $\tan^{-1} 0.1$ to three decimal-place accuracy, and check your work by comparing your answer to that produced directly by your calculating utility.

7. Use an appropriate Taylor series to approximate $\sin 85°$ to four decimal-place accuracy, and check your work by comparing your answer to that produced directly by your calculating utility.

8. Use a Taylor series to approximate $\cos(-175°)$ to four decimal-place accuracy, and check your work by comparing your answer to that produced directly by your calculating utility.

9. Use the Maclaurin series for $\sinh x$ to approximate $\sinh 0.5$ to three decimal-place accuracy. Check your work by computing $\sinh 0.5$ with a calculating utility.

10. Use the Maclaurin series for $\cosh x$ to approximate $\cosh 0.1$ to three decimal-place accuracy. Check your work by computing $\cosh 0.1$ with a calculating utility.

11. Use the Remainder Estimation Theorem and the method of Example 1 to prove that the Taylor series for $\sin x$ about $x = \pi/4$ converges to $\sin x$ for all x.

12. Use the Remainder Estimation Theorem and the method of Example 3 to prove that the Taylor series for e^x about $x = 1$ converges to e^x for all x.

13. (a) Use Formula (17) in the text to find a series that converges to $\ln 1.25$.
 (b) Approximate $\ln 1.25$ using the first two terms of the series. Round your answer to three decimal places, and compare the result to that produced directly by your calculating utility.

14. (a) Use Formula (17) to find a series that converges to $\ln 3$.
 (b) Approximate $\ln 3$ using the first two terms of the series. Round your answer to three decimal places, and compare the result to that produced directly by your calculating utility.

15. (a) Use the Maclaurin series for $\tan^{-1} x$ to approximate $\tan^{-1} \frac{1}{2}$ and $\tan^{-1} \frac{1}{3}$ to three decimal-place accuracy.
 (b) Use the results in part (a) and Formula (21) to approximate π.
 (c) Would you be willing to guarantee that your answer in part (b) is accurate to three decimal places? Explain your reasoning.
 (d) Compare your answer in part (b) to that produced by your calculating utility.

16. Use an appropriate Taylor series for $\sqrt[3]{x}$ to approximate $\sqrt[3]{28}$ to three decimal-place accuracy, and check your answer by comparing it to that produced directly by your calculating utility.

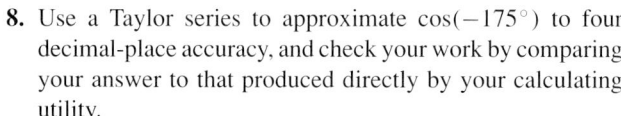 **17.** (a) Use the Remainder Estimation Theorem to find an interval containing $x = 0$ over which $\sin x$ can be approximated by $x - (x^3/3!)$ to three decimal-place accuracy throughout the interval.
 (b) Check your answer in part (a) by graphing
$$\left| \sin x - \left(x - \frac{x^3}{3!} \right) \right|$$
over the interval you obtained.

18. (a) Find an interval $[0, b]$ over which e^x can be approximated by $1 + x + (x^2/2!)$ to three decimal-place accuracy throughout the interval.
 (b) Check your answer in part (a) by graphing
$$\left| e^x - \left(1 + x + \frac{x^2}{2!} \right) \right|$$
over the interval you obtained.

19. (a) Find an upper bound on the error that can result if $\cos x$ is approximated by $1 - (x^2/2!) + (x^4/4!)$ over the interval $[-0.2, 0.2]$.
 (b) Check your answer in part (a) by graphing
$$\left| \cos x - \left(1 - \frac{x^2}{2!} + \frac{x^4}{4!} \right) \right|$$
over the interval.

20. (a) Find an upper bound on the error that can result if $\ln(1 + x)$ is approximated by x over the interval $[-0.01, 0.01]$.
 (b) Check your answer in part (a) by graphing
$$| \ln(1 + x) - x |$$
over the interval.

21. Use Formula (22) for the binomial series to obtain the Maclaurin series for
 (a) $\dfrac{1}{1 + x}$ (b) $\sqrt[3]{1 + x}$ (c) $\dfrac{1}{(1 + x)^3}$.

22. If m is any real number, and k is a nonnegative integer, then we define the ***binomial coefficient***
$$\binom{m}{k}$$ by the formulas $\binom{m}{0} = 1$ and
$$\binom{m}{k} = \frac{m(m - 1)(m - 2) \cdots (m - k + 1)}{k!}$$
for $k \geq 1$. Express Formula (22) in the text in terms of binomial coefficients.

23. In this exercise we will use the Remainder Estimation Theorem to determine the number of terms that are required in Formula (19) to approximate $\ln 2$ to five decimal-place accuracy. For this purpose let
$$f(x) = \ln \frac{1 + x}{1 - x} = \ln(1 + x) - \ln(1 - x) \quad (-1 < x < 1)$$
 (a) Show that
$$f^{(n+1)}(x) = n! \left[\frac{(-1)^n}{(1 + x)^{n+1}} + \frac{1}{(1 - x)^{n+1}} \right]$$
 (b) Use the triangle inequality [Theorem 1.2.2(d)] to show that
$$|f^{(n+1)}(x)| \leq n! \left[\frac{1}{(1 + x)^{n+1}} + \frac{1}{(1 - x)^{n+1}} \right]$$
 (c) Since we want to achieve five decimal-place accuracy, our goal is to choose n so that the absolute value of the nth remainder at $x = \frac{1}{3}$ does not exceed the value

$0.000005 = 0.5 \times 10^{-5}$; that is, $\left| R_n\left(\frac{1}{3}\right) \right| \le 0.000005$. Use the Remainder Estimation Theorem to show that this condition will be satisfied if n is chosen so that

$$\frac{M}{(n+1)!}\left(\frac{1}{3}\right)^{n+1} \le 0.000005$$

where $|f^{(n+1)}(x)| \le M$ on the interval $\left[0, \frac{1}{3}\right]$.

(d) Use the result in part (b) to show that M can be taken as

$$M = n!\left[1 + \frac{1}{\left(\frac{2}{3}\right)^{n+1}}\right]$$

(e) Use the results in parts (c) and (d) to show that five decimal-place accuracy will be achieved if n satisfies

$$\frac{1}{n+1}\left[\left(\frac{1}{3}\right)^{n+1} + \left(\frac{1}{2}\right)^{n+1}\right] \le 0.000005$$

and then show that the smallest value of n that satisfies this condition is $n = 13$.

24. Use Formula (17) and the method of Exercise 23 to approximate $\ln\left(\frac{5}{3}\right)$ to five decimal-place accuracy. Then check your work by comparing your answer to that produced directly by your calculating utility.

25. Prove: The Taylor series for $\cos x$ about any point $x = x_0$ converges to $\cos x$ for all x.

26. Prove: The Taylor series for $\sin x$ about any point $x = x_0$ converges to $\sin x$ for all x.

C **27.** (a) In 1706 the British astronomer and mathematician John Machin discovered the following formula for $\pi/4$, called **Machin's formula**:

$$\frac{\pi}{4} = 4\tan^{-1}\frac{1}{5} - \tan^{-1}\frac{1}{239}$$

Use a CAS to approximate $\pi/4$ using Machin's formula to 25 decimal places.

(b) In 1914 the brilliant Indian mathematician Srinivasa Ramanujan (1887–1920) showed that

$$\frac{1}{\pi} = \frac{\sqrt{8}}{9801}\sum_{k=0}^{\infty}\frac{(4k)!(1103 + 26{,}390k)}{(k!)^4 396^{4k}}$$

Use a CAS to compute the first four partial sums in **Ramanujan's formula**.

28. The purpose of this exercise is to show that the Taylor series of a function f may possibly converge to a value different from $f(x)$ for certain x. Let

$$f(x) = \begin{cases} e^{-1/x^2}, & x \ne 0 \\ 0, & x = 0 \end{cases}$$

(a) Use the definition of a derivative to show that $f'(0) = 0$.

(b) With some difficulty it can be shown that $f^{(n)}(0) = 0$ for $n \ge 2$. Accepting this fact, show that the Maclaurin series of f converges for all x, but converges to $f(x)$ only at the point $x = 0$.

11.10 DIFFERENTIATING AND INTEGRATING POWER SERIES; MODELING WITH TAYLOR SERIES

In this section we will discuss methods for finding power series for derivatives and integrals of functions, and we will discuss some practical methods for finding Taylor series that can be used in situations where it is difficult or impossible to find the series directly.

DIFFERENTIATING POWER SERIES

We begin by considering the following problem:

> **11.10.1** PROBLEM. Suppose that a function f is represented by a power series on an open interval. How can we use the power series to find the derivative of f on that interval?

The solution to this problem can be motivated by considering the Maclaurin series for $\sin x$:

$$\sin x = x - \frac{x^3}{3!} + \frac{x^5}{5!} - \frac{x^7}{7!} + \cdots \qquad (-\infty < x < +\infty)$$

Of course, we already know that the derivative of $\sin x$ is $\cos x$; however, we are concerned here with using the Maclaurin series to deduce this. The solution is easy—all we need to do is differentiate the Maclaurin series term by term and observe that the resulting series is the Maclaurin series for $\cos x$:

$$\frac{d}{dx}\left[x - \frac{x^3}{3!} + \frac{x^5}{5!} - \frac{x^7}{7!} + \cdots\right] = 1 - 3\frac{x^2}{3!} + 5\frac{x^4}{5!} - 7\frac{x^6}{7!} + \cdots$$

$$= 1 - \frac{x^2}{2!} + \frac{x^4}{4!} - \frac{x^6}{6!} + \cdots = \cos x$$

Here is another example.

$$\frac{d}{dx}[e^x] = \frac{d}{dx}\left[1 + x + \frac{x^2}{2!} + \frac{x^3}{3!} + \frac{x^4}{4!} + \cdots\right]$$

$$= 1 + 2\frac{x}{2!} + 3\frac{x^2}{3!} + 4\frac{x^3}{4!} + \cdots = 1 + x + \frac{x^2}{2!} + \frac{x^3}{3!} + \cdots = e^x$$

FOR THE READER.   See whether you can use this method to find the derivative of $\cos x$.

The preceding computations suggest that if a function f is represented by a power series on an open interval, then a power series representation of f' on that interval can be obtained by differentiating the power series for f term by term. This is stated more precisely in the following theorem, which we give without proof.

11.10.2   THEOREM (*Differentiation of Power Series*).   *Suppose that a function f is represented by a power series in $x - x_0$ that has a nonzero radius of convergence R; that is,*

$$f(x) = \sum_{k=0}^{\infty} c_k(x - x_0)^k \qquad (x_0 - R < x < x_0 + R)$$

Then:

(a)   *The function f is differentiable on the interval $(x_0 - R, x_0 + R)$.*

(b)   *If the power series representation for f is differentiated term by term, then the resulting series has radius of convergence R and converges to f' on the interval $(x_0 - R, x_0 + R)$; that is,*

$$f'(x) = \sum_{k=0}^{\infty} \frac{d}{dx}[c_k(x - x_0)^k] \qquad (x_0 - R < x < x_0 + R)$$

This theorem has an important implication about the differentiability of functions that are represented by power series. According to the theorem, the power series for f' has the same radius of convergence as the power series for f, and this means that the theorem can be applied to f' as well as f. However, if we do this, then we conclude that f' is differentiable on the interval $(x_0 - R, x_0 + R)$, and the power series for f'' has the same radius of convergence as the power series for f and f'. We can now repeat this process ad infinitum, applying the theorem successively to f'', f''', ..., $f^{(n)}$, ... to conclude that f has derivatives of all orders on the interval $(x_0 - R, x_0 + R)$. Thus, we have established the following result.

11.10.3   THEOREM.   *If a function f can be represented by a power series in $x - x_0$ with a nonzero radius of convergence R, then f has derivatives of all orders on the interval $(x_0 - R, x_0 + R)$.*

In short, it is only the most "well-behaved" functions that can be represented by power series; that is, if a function f does not possess derivatives of all orders on an interval $(x_0 - R, x_0 + R)$, then it cannot be represented by a power series in $x - x_0$ on that interval.

Example 1

In Section 11.8, we showed that the Bessel function $J_0(x)$ is represented by the power series

$$J_0(x) = \sum_{k=0}^{\infty} \frac{(-1)^k x^{2k}}{2^{2k}(k!)^2} \tag{1}$$

with radius of convergence $+\infty$ [see Formula (2) of that section and the related discussion]. Thus, $J_0(x)$ has derivatives of all orders on the interval $(-\infty, +\infty)$, and these can be obtained by differentiating the series term by term. For example, if we write (1) as

$$J_0(x) = 1 + \sum_{k=1}^{\infty} \frac{(-1)^k x^{2k}}{2^{2k}(k!)^2}$$

and differentiate term by term, we obtain

$$J_0'(x) = \sum_{k=1}^{\infty} \frac{(-1)^k (2k) x^{2k-1}}{2^{2k}(k!)^2} = \sum_{k=1}^{\infty} \frac{(-1)^k x^{2k-1}}{2^{2k-1} k!(k-1)!} \qquad \blacktriangleleft$$

REMARK. The computations in this example use some techniques that are worth noting. First, when a power series is expressed in sigma notation, the formula for the general term of the series will often not be of a form that can be used for differentiating the constant term. Thus, if the series has a nonzero constant term, as here, it is usually a good idea to split it off from the summation before differentiating. Second, observe how we simplified the final formula by canceling the factor k from one of the factorials in the denominator. This is a standard simplification technique.

INTEGRATING POWER SERIES

Since the derivative of a function that is represented by a power series can be obtained by differentiating the series term by term, it should not be surprising that an antiderivative of a function represented by a power series can be obtained by integrating the series term by term. For example, we know that $\sin x$ is an antiderivative of $\cos x$. Here is how this result can be obtained by integrating the Maclaurin series for $\cos x$ term by term:

$$\int \cos x \, dx = \int \left[1 - \frac{x^2}{2!} + \frac{x^4}{4!} - \frac{x^6}{6!} + \cdots \right] dx$$

$$= \left[x - \frac{x^3}{3(2!)} + \frac{x^5}{5(4!)} - \frac{x^7}{(6!)} + \cdots \right] + C$$

$$= \left[x - \frac{x^3}{3!} + \frac{x^5}{5!} - \frac{x^7}{7!} + \cdots \right] + C = \sin x + C$$

The same idea applies to definite integrals. For example, by direct integration we have

$$\int_0^1 \frac{dx}{1+x^2} = \tan^{-1} x \bigg]_0^1 = \tan^{-1} 1 - \tan 0 = \frac{\pi}{4} - 0 = \frac{\pi}{4}$$

and we will show later in this section that

$$\frac{\pi}{4} = 1 - \frac{1}{3} + \frac{1}{5} - \frac{1}{7} + \cdots \tag{2}$$

Thus,

$$\int_0^1 \frac{dx}{1+x^2} = 1 - \frac{1}{3} + \frac{1}{5} - \frac{1}{7} + \cdots$$

Here is how this result can be obtained by integrating the Maclaurin series for $1/(1+x^2)$ term by term (see Table 11.9.1):

$$\int_0^1 \frac{dx}{1+x^2} = \int_0^1 [1 - x^2 + x^4 - x^6 + \cdots] \, dx$$

$$= x - \frac{x^3}{3} + \frac{x^5}{5} - \frac{x^7}{7} + \cdots \bigg]_0^1 = 1 - \frac{1}{3} + \frac{1}{5} - \frac{1}{7} + \cdots$$

The preceding computations are justified by the following theorem, which we give without proof.

11.10.4   THEOREM (*Integration of Power Series*).   *Suppose that a function f is represented by a power series in $x - x_0$ that has a nonzero radius of convergence R; that is,*

$$f(x) = \sum_{k=0}^{\infty} c_k (x - x_0)^k \qquad (x_0 - R < x < x_0 + R)$$

(a)   *If the power series representation of f is integrated term by term using an indefinite integral, then the resulting series has radius of convergence R and converges to $\int f(x)\, dx$ on the interval $(x_0 - R, x_0 + R)$; that is,*

$$\int f(x)\, dx = \sum_{k=0}^{\infty} \left[\int c_k (x - x_0)^k\, dx \right] + C \qquad (x_0 - R < x < x_0 + R)$$

(b)   *If α and β are points in the interval $(x_0 - R, x_0 + R)$, and if the power series representation of f is integrated term by term from α to β, then the resulting series of numbers converges absolutely on the interval $(x_0 - R, x_0 + R)$ and*

$$\int_{\alpha}^{\beta} f(x)\, dx = \sum_{k=0}^{\infty} \left[\int_{\alpha}^{\beta} c_k (x - x_0)^k\, dx \right]$$

POWER SERIES REPRESENTATIONS MUST BE TAYLOR SERIES

For many functions it is difficult or impossible to find the derivatives that are required to obtain a Taylor series. For example, to find the Maclaurin series for $1/(1 + x^2)$ directly would require some tedious derivative computations (try it). A more practical approach is to substitute $-x^2$ for x in the geometric series

$$\frac{1}{1 - x} = 1 + x + x^2 + x^3 + x^4 + \cdots \qquad (-1 < x < 1)$$

to obtain

$$\frac{1}{1 + x^2} = 1 - x^2 + x^4 - x^6 + x^8 - \cdots$$

However, there are two questions of concern with this procedure:

- Where does the power series that we obtained for $1/(1 + x^2)$ actually converge to $1/(1 + x^2)$?

- How do we know that the power series we have obtained is actually the Maclaurin series for $1/(1 + x^2)$?

The first question is easy to resolve. Since the geometric series converges to $1/(1 - x)$ if $|x| < 1$, the second series will converge to $1/(1 + x^2)$ if $|-x^2| < 1$ or $|x^2| < 1$. However, this is true if and only if $|x| < 1$, so the power series we obtained for the function $1/(1 + x^2)$ converges to this function if $-1 < x < 1$.

The second question is more difficult to answer and leads us to the following general problem.

11.10.5   PROBLEM.   Suppose that a function f is represented by a power series in $x - x_0$ that has a nonzero radius of convergence. What relationship exists between the given power series and the Taylor series for f about $x = x_0$?

The answer is that they are the same; and here is the theorem that proves it.

11.10.6 THEOREM. *If a function f is represented by a power series in $x - x_0$ on some open interval containing x_0, then that power series is the Taylor series for f about $x = x_0$.*

Proof. Suppose that

$$f(x) = c_0 + c_1(x - x_0) + c_2(x - x_0)^2 + \cdots + c_k(x - x_0)^k + \cdots$$

for all x in some open interval containing x_0. To prove that this is the Taylor series for f about $x = x_0$, we must show that

$$c_k = \frac{f^{(k)}(x_0)}{k!} \quad \text{for} \quad k = 0, 1, 2, 3, \ldots$$

However, the assumption that the series converges to $f(x)$ on an open interval containing x_0 ensures that it has a nonzero radius of convergence R; hence we can differentiate term by term in accordance with Theorem 11.10.2. Thus,

$$f(x) \; = c_0 + c_1(x - x_0) + c_2(x - x_0)^2 + c_3(x - x_0)^3 + c_4(x - x_0)^4 + \cdots$$
$$f'(x) \; = c_1 + 2c_2(x - x_0) + 3c_3(x - x_0)^2 + 4c_4(x - x_0)^3 + \cdots$$
$$f''(x) = 2!c_2 + (3 \cdot 2)c_3(x - x_0) + (4 \cdot 3)c_4(x - x_0)^2 + \cdots$$
$$f'''(x) = 3!c_3 + (4 \cdot 3 \cdot 2)c_4(x - x_0) + \cdots$$
$$\vdots$$

On substituting $x = x_0$, all the powers of $x - x_0$ drop out, leaving

$$f(x_0) = c_0, \quad f'(x_0) = c_1, \quad f''(x_0) = 2!c_2, \quad f'''(x_0) = 3!c_3, \quad \ldots$$

from which we obtain

$$c_0 = f(x_0), \quad c_1 = f'(x_0), \quad c_2 = \frac{f''(x_0)}{2!}, \quad c_3 = \frac{f'''(x_0)}{3!}, \quad \ldots$$

which shows that the coefficients $c_0, c_1, c_2, c_3, \ldots$ are precisely the coefficients in the Taylor series about x_0 for $f(x)$. ∎

REMARK. This theorem tells us that no matter how we arrive at a power series representation of a function f, be it by substitution, by differentiation, by integration, or by some sort of algebraic manipulation, that series will be the Taylor series for f about $x = x_0$, provided that it converges to f on some open interval containing x_0.

SOME PRACTICAL WAYS TO FIND TAYLOR SERIES

Example 2

Find the Maclaurin series for $\tan^{-1} x$.

Solution. It would be tedious to find the Maclaurin series directly. A better approach is to start with the formula

$$\int \frac{1}{1 + x^2} \, dx = \tan^{-1} x + C$$

and integrate the Maclaurin series

$$\frac{1}{1 + x^2} = 1 - x^2 + x^4 - x^6 + x^8 - \cdots \qquad (-1 < x < 1)$$

term by term. This yields

$$\tan^{-1} x + C = \int \frac{1}{1 + x^2} \, dx = \int [1 - x^2 + x^4 - x^6 + x^8 - \cdots] \, dx$$

or

$$\tan^{-1} x = \left[x - \frac{x^3}{3} + \frac{x^5}{5} - \frac{x^7}{7} + \frac{x^9}{9} - \cdots \right] - C$$

The constant of integration can be evaluated by substituting $x = 0$ and using the condition $\tan^{-1} 0 = 0$. This gives $C = 0$, so that

$$\tan^{-1} x = x - \frac{x^3}{3} + \frac{x^5}{5} - \frac{x^7}{7} + \frac{x^9}{9} - \cdots \qquad (-1 < x < 1) \tag{3}$$

◄

REMARK.   Observe that neither Theorem 11.10.2 nor Theorem 11.10.3 addresses what happens at the endpoints of the interval of convergence. However, it can be proved that if the Taylor series for f about $x = x_0$ converges to $f(x)$ for all x in the interval $(x_0 - R, x_0 + R)$, and if the Taylor series converges at the right endpoint $x_0 + R$, then the value that it converges to at that point is the limit of $f(x)$ as $x \to x_0 + R$ from the left; and if the Taylor series converges at the left endpoint $x_0 - R$, then the value that it converges to at that point is the limit of $f(x)$ as $x \to x_0 - R$ from the right.

For example, the Maclaurin series for $\tan^{-1} x$ given in (3) converges at both $x = -1$ and $x = 1$, since the hypotheses of the alternating series test (Theorem 11.7.1) are satisfied at those points. Thus, the continuity of $\tan^{-1} x$ on the interval $[-1, 1]$ implies that at $x = 1$ the Maclaurin series converges to

$$\lim_{x \to 1^-} \tan^{-1} x = \tan^{-1} 1 = \frac{\pi}{4}$$

and at $x = -1$ it converges to

$$\lim_{x \to -1^+} \tan^{-1} x = \tan^{-1}(-1) = -\frac{\pi}{4}$$

This shows that the Maclaurin series for $\tan^{-1} x$ actually converges to $\tan^{-1} x$ on the interval $-1 \leq x \leq 1$. Moreover, the convergence at $x = 1$ establishes Formula (2).

Taylor series provide an alternative to Simpson's rule and other numerical methods for approximating definite integrals.

Example 3

Approximate the integral

$$\int_0^1 e^{-x^2} \, dx$$

to three decimal-place accuracy by expanding the integrand in a Maclaurin series and integrating term by term.

Solution.   The simplest way to obtain the Maclaurin series for e^{-x^2} is to replace x by $-x^2$ in the Maclaurin series

$$e^x = 1 + x + \frac{x^2}{2!} + \frac{x^3}{3!} + \frac{x^4}{4!} + \cdots$$

to obtain

$$e^{-x^2} = 1 - x^2 + \frac{x^4}{2!} - \frac{x^6}{3!} + \frac{x^8}{4!} - \cdots$$

Therefore,

$$\int_0^1 e^{-x^2} \, dx = \int_0^1 \left[1 - x^2 + \frac{x^4}{2!} - \frac{x^6}{3!} + \frac{x^8}{4!} - \cdots \right] dx$$

$$= \left[x - \frac{x^3}{3} + \frac{x^5}{5(2!)} - \frac{x^7}{7(3!)} + \frac{x^9}{9(4!)} - \cdots \right]_0^1$$

$$= 1 - \frac{1}{3} + \frac{1}{5 \cdot 2!} - \frac{1}{7 \cdot 3!} + \frac{1}{9 \cdot 4!} - \cdots$$

$$= \sum_{k=0}^{\infty} \frac{(-1)^k}{(2k + 1)k!}$$

Since this series clearly satisfies the hypotheses of the alternating series test (Theorem 11.7.1), it follows from Theorem 11.7.2 that if we approximate the integral by s_n (the nth partial sum of the series), then

$$\left| \int_0^1 e^{-x^2} \, dx - s_n \right| < \frac{1}{[2(n+1)+1](n+1)!} = \frac{1}{(2n+3)(n+1)!}$$

Thus, for three decimal-place accuracy we must choose n such that

$$\frac{1}{(2n+3)(n+1)!} \leq 0.0005 = 5 \times 10^{-4}$$

With the help of a calculating utility you can show that the smallest value of n that satisfies this condition is $n = 5$. Thus, the value of the integral to three decimal-place accuracy is

$$\int_0^1 e^{-x^2} \, dx \approx 1 - \frac{1}{3} + \frac{1}{5 \cdot 2!} - \frac{1}{7 \cdot 3!} + \frac{1}{9 \cdot 4!} \approx 0.747$$

As a check, the author's CAS produced the approximation 0.746824, which agrees with our result when rounded to three decimal places. ◄

FOR THE READER.   What advantages does the method in this example have over Simpson's rule? What are its disadvantages?

FINDING MACLAURIN SERIES BY MULTIPLICATION AND DIVISION

The following examples illustrate some algebraic techniques that are sometimes useful for finding Taylor series.

Example 4

Find the first three nonzero terms in the Maclaurin series for the function $f(x) = e^{-x^2} \tan^{-1} x$.

$$
\begin{array}{r}
1 - x^2 + \dfrac{x^4}{2} - \cdots \\[4pt]
\times \quad x - \dfrac{x^3}{3} + \dfrac{x^5}{5} - \cdots \\[4pt]
\hline
x - x^3 + \dfrac{x^5}{2} - \cdots \\[4pt]
-\dfrac{x^3}{3} + \dfrac{x^5}{3} - \dfrac{x^7}{6} + \cdots \\[4pt]
\dfrac{x^5}{5} - \dfrac{x^7}{5} + \cdots \\[4pt]
\hline
x - \dfrac{4}{3}x^3 + \dfrac{31}{30}x^5 - \cdots
\end{array}
$$

Solution.   Using the series for e^{-x^2} and $\tan^{-1} x$ obtained in Examples 2 and 3 gives

$$e^{-x^2} \tan^{-1} x = \left(1 - x^2 + \frac{x^4}{2} - \cdots\right)\left(x - \frac{x^3}{3} + \frac{x^5}{5} - \cdots\right)$$

Multiplying, as shown in the margin, we obtain

$$e^{-x^2} \tan^{-1} x = x - \frac{4}{3}x^3 + \frac{31}{30}x^5 - \cdots$$

More terms in the series can be obtained by including more terms in the factors. Moreover, one can prove that a series obtained by this method converges at each point in the intersection of the intervals of convergence of the factors (and possibly on a larger interval). Thus, we can be certain that the series we have obtained converges for all x in the interval $-1 \leq x \leq 1$ (why?). ◄

$$
\begin{array}{r}
x + \dfrac{x^3}{3} + \dfrac{2x^5}{15} + \cdots \\[6pt]
\hline
1 - \dfrac{x^2}{2} + \dfrac{x^4}{24} - \cdots \,\bigg)\; x - \dfrac{x^3}{6} + \dfrac{x^5}{120} - \cdots \\[4pt]
x - \dfrac{x^3}{2} + \dfrac{x^5}{24} - \cdots \\[4pt]
\hline
\dfrac{x^3}{3} - \dfrac{x^5}{30} + \cdots \\[4pt]
\dfrac{x^3}{3} - \dfrac{x^5}{6} + \cdots \\[4pt]
\hline
\dfrac{2x^5}{15} + \cdots
\end{array}
$$

FOR THE READER.   If you have a CAS, read the documentation about multiplying polynomials, and then use the CAS to duplicate the result in the last example.

Example 5

Find the first three nonzero terms in the Maclaurin series for $\tan x$.

Solution.   Using the first three terms in the Maclaurin series for $\sin x$ and $\cos x$, we can express $\tan x$ as

$$\tan x = \frac{\sin x}{\cos x} = \frac{x - \dfrac{x^3}{3!} + \dfrac{x^5}{5!} - \cdots}{1 - \dfrac{x^2}{2!} + \dfrac{x^4}{4!} - \cdots}$$

Dividing, as shown in the margin, we obtain

$$\tan x = x + \frac{x^3}{3} + \frac{2x^5}{15} + \cdots$$ ◀

MODELING PHYSICAL LAWS WITH TAYLOR SERIES

Taylor series provide an important way of modeling physical laws. To illustrate the idea we will consider the problem of modeling the period of a simple pendulum (Figure 11.10.1). As explained in Exercise 38 of the supplementary exercises to Chapter 9, the period T of such a pendulum is given by

$$T = 4\sqrt{\frac{L}{g}} \int_0^{\pi/2} \frac{1}{\sqrt{1 - k^2 \sin^2 \phi}} \, d\phi \tag{4}$$

where

L = length of the supporting rod

g = acceleration due to gravity

$k = \sin(\theta_0/2)$, where θ_0 is the initial angle of displacement from the vertical

$k \sin \phi = \sin(\theta/2)$, where θ is the displacement from the vertical at time t

The integral, which is called a ***complete elliptic integral of the first kind***, cannot be expressed in terms of elementary functions and is often approximated by numerical methods. Unfortunately, numerical values are so specific that they often give little insight into general physical principles. However, if we expand the integrand of (4) in a Maclaurin series and integrate term by term, then we can generate an infinite series that can be used to construct various mathematical models for the period T that give a deeper understanding of the behavior of the pendulum.

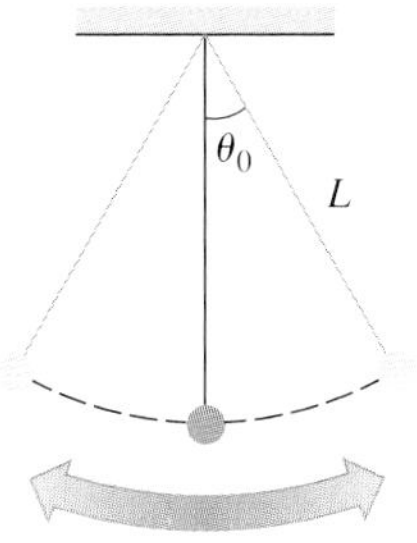

To obtain the Maclaurin series for the integrand, we will substitute $-k^2 \sin^2 \phi$ for x in the binomial series for $1/\sqrt{1 + x}$ that we derived in Example 5 of Section 11.9. If we do this, then we can rewrite (4) as

$$T = 4\sqrt{\frac{L}{g}} \int_0^{\pi/2} \left[1 + \frac{1}{2}k^2 \sin^2 \phi + \frac{1 \cdot 3}{2^2 2!}k^4 \sin^4 \phi + \frac{1 \cdot 3 \cdot 5}{2^3 3!}k^6 \sin^6 \phi + \cdots \right] d\phi \tag{5}$$

If we integrate term by term, then we can produce a Maclaurin series that converges to the period T. However, one of the most important cases of pendulum motion occurs when the initial displacement is small, in which case all subsequent displacements are small, and we can assume that $\phi \approx 0$. In this case we expect the convergence of the Maclaurin series for T to be rapid, and we can approximate the sum of the series by dropping all but the constant term in (5). This yields

$$T = 2\pi \sqrt{\frac{L}{g}} \tag{6}$$

Figure 11.10.1

which is called the ***first-order model*** of T or the model for ***small vibrations***. This model can be improved on by using more terms in the series. For example, if we use the first two terms in the Maclaurin series, we obtain the ***second-order model***

$$T = 2\pi \sqrt{\frac{L}{g}} \left(1 + \frac{k^2}{4} \right) \tag{7}$$

(verify).

1. In each part, obtain the Maclaurin series for the function by making an appropriate substitution in the Maclaurin series for $1/(1 - x)$. Include the general term in your answer, and state the radius of convergence of the series.

(a) $\dfrac{1}{1 + x}$ (b) $\dfrac{1}{1 - x^2}$ (c) $\dfrac{1}{1 - 2x}$ (d) $\dfrac{1}{2 - x}$

2. In each part, obtain the Maclaurin series for the function by making an appropriate substitution in the Maclaurin series for $\ln(1 + x)$. Include the general term in your answer, and state the radius of convergence of the series.
 (a) $\ln(1 - x)$
 (b) $\ln(1 + x^2)$
 (c) $\ln(1 + 2x)$
 (d) $\ln(2 + x)$

3. In each part, obtain the first four nonzero terms of the Maclaurin series for the function by making an appropriate substitution in one of the binomial series obtained in Example 5 of Section 11.9.
 (a) $(2 + x)^{-1/2}$
 (b) $(1 - x^2)^{-2}$

4. (a) Use the Maclaurin series for $1/(1 - x)$ to find the Maclaurin series for $1/(a - x)$, where $a \neq 0$, and state the radius of convergence of the series.
 (b) Use the binomial series for $1/(1 + x)^2$ obtained in Example 5 of Section 11.9 to find the first four nonzero terms in the Maclaurin series for $1/(a + x)^2$, where $a \neq 0$, and state the radius of convergence of the series.

In Exercises 5–8, obtain the first four nonzero terms of the Maclaurin series for the function by making an appropriate substitution in a known Maclaurin series and performing any algebraic operations that are required. State the radius of convergence of the series.

5. (a) $\sin 2x$ (b) e^{-2x} (c) e^{x^2} (d) $x^2 \cos \pi x$

6. (a) $\cos 2x$ (b) $x^2 e^x$ (c) xe^{-x} (d) $\sin(x^2)$

7. (a) $\dfrac{x^2}{1 + 3x}$ (b) $x \sinh 2x$ (c) $x(1 - x^2)^{3/2}$

8. (a) $\dfrac{x}{x - 1}$ (b) $3 \cosh(x^2)$ (c) $\dfrac{x}{(1 + 2x)^3}$

In Exercises 9 and 10, find the first four nonzero terms of the Maclaurin series for the function by using an appropriate trigonometric identity or property of logarithms and then substituting in a known Maclaurin series.

9. (a) $\sin^2 x$ (b) $\ln[(1 + x^3)^{12}]$

10. (a) $\cos^2 x$ (b) $\ln\left(\dfrac{1 - x}{1 + x}\right)$

11. (a) Use a known Maclaurin series to find the Taylor series of $1/x$ about $x = 1$ by expressing this function as
$$\frac{1}{x} = \frac{1}{1 - (1 - x)}$$
 (b) Find the interval of convergence of the Taylor series.

12. Use the method of Exercise 11 to find the Taylor series of $1/x$ about $x = x_0$, and state the interval of convergence of the Taylor series.

In Exercises 13 and 14, find the first four nonzero terms of the Maclaurin series for the function by multiplying the Maclaurin series of the factors.

13. (a) $e^x \sin x$ (b) $\sqrt{1 + x} \ln(1 + x)$

14. (a) $e^{-x^2} \cos x$ (b) $(1 + x^2)^{4/3}(1 + x)^{1/3}$

In Exercises 15 and 16, find the first four nonzero terms of the Maclaurin series for the function by dividing appropriate Maclaurin series.

15. (a) $\sec x$ $\left(= \dfrac{1}{\cos x}\right)$ (b) $\dfrac{\sin x}{e^x}$

16. (a) $\dfrac{\tan^{-1} x}{1 + x}$ (b) $\dfrac{\ln(1 + x)}{1 - x}$

17. Use the Maclaurin series for e^x and e^{-x} to derive the Maclaurin series for $\sinh x$ and $\cosh x$. Include the general terms in your answers and state the radius of convergence of each series.

18. Use the Maclaurin series for $\sinh x$ and $\cosh x$ to obtain the first four nonzero terms in the Maclaurin series for $\tanh x$.

In Exercises 19 and 20, find the first five nonzero terms of the Maclaurin series for the function by using partial fractions and a known Maclaurin series.

19. $\dfrac{4x - 2}{x^2 - 1}$ **20.** $\dfrac{x^3 + x^2 + 2x - 2}{x^2 - 1}$

In Exercises 21 and 22, confirm the derivative formula by differentiating the appropriate Maclaurin series term by term.

21. (a) $\dfrac{d}{dx}[\cos x] = -\sin x$ (b) $\dfrac{d}{dx}[\ln(1 + x)] = \dfrac{1}{1 + x}$

22. (a) $\dfrac{d}{dx}[\sinh x] = \cosh x$ (b) $\dfrac{d}{dx}[\tan^{-1} x] = \dfrac{1}{1 + x^2}$

In Exercises 23 and 24, confirm the integration formula by integrating the appropriate Maclaurin series term by term.

23. (a) $\displaystyle\int e^x \, dx = e^x + C$ (b) $\displaystyle\int \sinh x \, dx = \cosh x + C$

24. (a) $\displaystyle\int \sin x \, dx = -\cos x + C$
 (b) $\displaystyle\int \dfrac{1}{1 + x} \, dx = \ln(1 + x) + C$

25. (a) Use the Maclaurin series for $1/(1 - x)$ to find the Maclaurin series for
$$f(x) = \frac{x}{1 - x^2}$$
 (b) Use the Maclaurin series obtained in part (a) to find $f^{(5)}(0)$ and $f^{(6)}(0)$.
 (c) What can you say about the value of $f^{(n)}(0)$?

26. Let $f(x) = x^2 \cos 2x$. Use the method of Exercise 25 to find $f^{(99)}(0)$.

The limit of an indeterminate form as $x \to x_0$ can sometimes be found without using L'Hôpital's rule by expanding the functions involved in Taylor series about $x = x_0$ and taking the limit of the series term by term. Use this method to find the limits in Exercises 27 and 28.

27. (a) $\displaystyle\lim_{x \to 0} \frac{\sin x}{x}$ (b) $\displaystyle\lim_{x \to 0} \frac{\tan^{-1} x - x}{x^3}$

28. (a) $\displaystyle\lim_{x \to 0} \frac{1 - \cos x}{\sin x}$ (b) $\displaystyle\lim_{x \to 0} \frac{\ln\sqrt{1+x} - \sin 2x}{x}$

In Exercises 29–32, use Maclaurin series to approximate the integral to three decimal-place accuracy.

29. $\displaystyle\int_0^1 \sin(x^2)\, dx$ **30.** $\displaystyle\int_0^{1/2} \tan^{-1}(2x^2)\, dx$

31. $\displaystyle\int_0^{0.2} \sqrt[3]{1+x^4}\, dx$ **32.** $\displaystyle\int_0^{1/2} \frac{dx}{\sqrt[4]{x^2+1}}$

33. (a) Differentiate the Maclaurin series for $1/(1-x)$, and use the result to show that
$$\sum_{k=1}^{\infty} k x^k = \frac{x}{(1-x)^2} \quad \text{for } -1 < x < 1$$

(b) Integrate the Maclaurin series for $1/(1-x)$, and use the result to show that
$$\sum_{k=1}^{\infty} \frac{x^k}{k} = -\ln(1-x) \quad \text{for } -1 < x < 1$$

(c) Use the result in part (b) to show that
$$\sum_{k=1}^{\infty} (-1)^{k+1} \frac{x^k}{k} = \ln(1+x) \quad \text{for } -1 < x < 1$$

(d) Show that the series in part (c) converges if $x = 1$.

(e) Use the remark following Example 2 to show that
$$\sum_{k=1}^{\infty} (-1)^{k+1} \frac{x^k}{k} = \ln(1+x) \quad \text{for } -1 < x \le 1$$

34. In each part, use the results in Exercise 33 to find the sum of the series.

(a) $\displaystyle\sum_{k=1}^{\infty} \frac{k}{3^k} = \frac{1}{3} + \frac{2}{3^2} + \frac{3}{3^3} + \frac{4}{3^4} + \cdots$

(b) $\displaystyle\sum_{k=1}^{\infty} \frac{1}{k(4^k)} = \frac{1}{4} + \frac{1}{2(4^2)} + \frac{1}{3(4^3)} + \frac{1}{4(4^4)} + \cdots$

(c) $\displaystyle\sum_{k=1}^{\infty} (-1)^{k+1} \frac{1}{k} = 1 - \frac{1}{2} + \frac{1}{3} - \frac{1}{4} + \cdots$

35. (a) Use the relationship
$$\int \frac{1}{\sqrt{1+x^2}}\, dx = \sinh^{-1} x + C$$
to find the first four nonzero terms in the Maclaurin series for $\sinh^{-1} x$.

(b) Express the series in sigma notation.

(c) What is the radius of convergence?

36. (a) Use the relationship
$$\int \frac{1}{\sqrt{1-x^2}}\, dx = \sin^{-1} x + C$$
to find the first four nonzero terms in the Maclaurin series for $\sin^{-1} x$.

(b) Express the series in sigma notation.

(c) What is the radius of convergence?

37. We showed by Formula (12) of Section 10.3 that if there are y_0 units of radioactive carbon-14 present at time $t = 0$, then the number of units present t years later is
$$y(t) = y_0 e^{-0.000121t}$$

(a) Express $y(t)$ as a Maclaurin series.

(b) Use the first two terms in the series to show that the number of units present after 1 year is approximately $(0.999879)y_0$.

(c) Compare this to the value produced by the formula for $y(t)$.

38. In Section 10.1 we studied the motion of a falling object that has mass m and is retarded by air resistance. We showed that if the initial velocity is v_0 and the drag force F_R is proportional to the velocity, that is, $F_R = -cv$, then the velocity of the object at time t is
$$v(t) = e^{-ct/m}\left(v_0 + \frac{mg}{c}\right) - \frac{mg}{c}$$
where g is the acceleration due to gravity [see Formula (23) of Section 10.1].

(a) Use a Maclaurin series to show that if $ct/m \approx 0$, then the velocity can be approximated as
$$v(t) \approx v_0 - \left(\frac{cv_0}{m} + g\right)t$$

(b) Improve on the approximation in part (a).

39. Suppose that a simple pendulum with a length of $L = 1$ meter is given an initial displacement of $\theta_0 = 5°$ from the vertical.

(a) Approximate the period of the pendulum using Formula (6) for the first-order model. [Take $g = 9.8$ m/s².]

(b) Approximate the period of the pendulum using Formula (7) for the second-order model.

(c) Use the numerical integration capability of a CAS to approximate the period of the pendulum from Formula (4), and compare it to the values obtained in parts (a) and (b).

40. Use the first three nonzero terms in Formula (5) and the Wallis sine formula in the Endpaper Integral Table (Formula 122) to obtain a model for the period of a simple pendulum.

41. Recall that the gravitational force exerted by the Earth on an object is called the object's *weight* (or more precisely, its *Earth weight*). We noted in statement 10.3.3 that if an object has mass m, then the magnitude of its weight is mg. However, this result presumes that the object is on the surface of the Earth (mean sea level). A more general formula for the magnitude of the gravitational force that the Earth exerts on an object of mass m is

$$F = \frac{mgR^2}{(R+h)^2}$$

where R is the radius of the Earth and h is the height of the object above the Earth's surface.

(a) Use the binomial series for $1/(1+x)^2$ obtained in Example 5 of Section 11.9 to express F as a Maclaurin series in powers of h/R.

(b) Show that if $h = 0$, then $F = mg$.

(c) Show that if $h/R \approx 0$, then $F \approx mg - (2mgh/R)$. [*Note:* The quantity $2mgh/R$ can be thought of as a "correction term" for the weight that takes the object's height above the Earth's surface into account.]

(d) If we assume that the Earth is a sphere of radius $R = 4000$ mi at mean sea level, by approximately what percentage does a person's weight change in going from mean sea level to the top of Mt. Everest (29,028 ft)?

42. (a) Show that the Bessel function $J_0(x)$ given by Formula (2) of Section 11.8 satisfies the differential equation $xy'' + y' + xy = 0$. (This is called the ***Bessel equation of order zero.***)

(b) Show that the Bessel function $J_1(x)$ given by Formula (3) of Section 11.8 satisfies the differential equation $x^2y'' + xy' + (x^2 - 1)y = 0$. (This is called the ***Bessel equation of order one.***)

(c) Show that $J_0'(x) = -J_1(x)$.

43. Prove: If the power series $\sum_{k=0}^{\infty} a_k x^k$ and $\sum_{k=0}^{\infty} b_k x^k$ have the same sum on an interval $(-r, r)$, then $a_k = b_k$ for all values of k.

SUPPLEMENTARY EXERCISES

1. What is the difference between an infinite sequence and an infinite series?

2. What is meant by the sum of an infinite series?

3. (a) What is a geometric series? Give some examples of convergent and divergent geometric series.

(b) What is a p-series? Give some examples of convergent and divergent p-series.

4. (a) Write down the formula for the Maclaurin series for f in sigma notation.

(b) Write down the formula for the Taylor series for f about $x = x_0$ in sigma notation.

5. State conditions under which an alternating series is guaranteed to converge.

6. (a) What does it mean to say that an infinite series converges absolutely?

(b) What relationship exists between convergence and absolute convergence of an infinite series?

7. If a power series in $x - x_0$ has radius of convergence R, what can you say about the set of x-values at which it converges?

8. State the Remainder Estimation Theorem, and describe some of its uses.

9. Are the following statements true or false? If true, state a theorem to justify your conclusion; if false, then give a counterexample.

(a) If $\sum u_k$ converges, then $u_k \to 0$ as $k \to +\infty$.

(b) If $u_k \to 0$ as $k \to +\infty$, then $\sum u_k$ converges.

(c) If $f(n) = a_n$ for $n = 1, 2, 3, \ldots$, and if $a_n \to L$ as $n \to +\infty$, then $f(x) \to L$ as $x \to +\infty$.

(d) If $f(n) = a_n$ for $n = 1, 2, 3, \ldots$, and if $f(x) \to L$ as $x \to +\infty$, then $a_n \to L$ as $n \to +\infty$.

(e) If $0 < a_n < 1$, then $\{a_n\}$ converges.

(f) If $0 < u_k < 1$, then $\sum u_k$ converges.

(g) If $\sum u_k$ and $\sum v_k$ converge, then $\sum (u_k + v_k)$ diverges.

(h) If $\sum u_k$ and $\sum v_k$ diverge, then $\sum (u_k - v_k)$ converges.

(i) If $0 \le u_k \le v_k$ and $\sum v_k$ converges, then $\sum u_k$ converges.

(j) If $0 \le u_k \le v_k$ and $\sum u_k$ diverges, then $\sum v_k$ diverges.

(k) If an infinite series converges, then it converges absolutely.

(l) If an infinite series diverges absolutely, then it diverges.

10. State whether each of the following is true or false. Justify your answers.

(a) The function $f(x) = x^{1/3}$ has a Maclaurin series.

(b) $1 + \frac{1}{2} - \frac{1}{2} + \frac{1}{3} - \frac{1}{3} + \frac{1}{4} - \frac{1}{4} + \cdots = 1$

(c) $1 + \frac{1}{2} - \frac{1}{2} + \frac{1}{2} - \frac{1}{2} + \frac{1}{2} - \frac{1}{2} + \cdots = 1$

In Exercises 11–14, use any method to determine whether the series converge.

11. (a) $\displaystyle\sum_{k=1}^{\infty} \frac{1}{5^k}$ (b) $\displaystyle\sum_{k=1}^{\infty} \frac{1}{5^k + 1}$ (c) $\displaystyle\sum_{k=1}^{\infty} \frac{9}{\sqrt{k} + 1}$

12. (a) $\displaystyle\sum_{k=1}^{\infty} (-1)^{k+1} \frac{k+4}{k^2 + k}$ (b) $\displaystyle\sum_{k=1}^{\infty} (-1)^{k+1} \left(\frac{k+2}{3k-1}\right)^k$

(c) $\displaystyle\sum_{k=1}^{\infty} \frac{k^{-1/2}}{2 + \sin^2 k}$

13. (a) $\displaystyle\sum_{k=1}^{\infty} \frac{1}{k^3 + 2k + 1}$ (b) $\displaystyle\sum_{k=1}^{\infty} \frac{1}{(3+k)^{2/5}}$

(c) $\displaystyle\sum_{k=1}^{\infty} \frac{\cos(1/k)}{k^2}$

14. (a) $\displaystyle\sum_{k=1}^{\infty} \frac{\ln k}{k\sqrt{k}}$ (b) $\displaystyle\sum_{k=1}^{\infty} \frac{k^{4/3}}{8k^2 + 5k + 1}$ (c) $\displaystyle\sum_{k=1}^{\infty} \frac{(-1)^{k+1}}{k^2 + 1}$

15. Find a formula for the exact error that results when the sum of the geometric series $\sum_{k=0}^{\infty} (1/5)^k$ is approximated by the sum of the first 100 terms in the series.

16. Does the series $1 - \frac{2}{3} + \frac{3}{5} - \frac{4}{7} + \frac{5}{9} + \cdots$ converge? Justify your answer.

17. (a) Find the first five Maclaurin polynomials of the function $p(x) = 1 - 7x + 5x^2 + 4x^3$.
    (b) Make a general statement about the Maclaurin polynomials of a polynomial of degree n.

18. Use a Maclaurin series and properties of alternating series to show that $|\ln(1 + x) - x| \le x^2/2$ if $0 < x < 1$.

19. Show that the approximation
$$\sin x \approx x - \frac{x^3}{3!} + \frac{x^5}{5!}$$
is accurate to four decimal places if $0 \le x \le \pi/4$.

20. Use Maclaurin series to approximate the integral
$$\int_0^1 \frac{1 - \cos x}{x}\, dx$$
to three decimal-place accuracy.

21. It can be proved that
$$\lim_{n \to +\infty} \sqrt[n]{n!} = +\infty \quad \text{and} \quad \lim_{n \to +\infty} \frac{\sqrt[n]{n!}}{n} = \frac{1}{e}$$
In each part, use these limits and the root test to determine whether the series converges.
    (a) $\displaystyle\sum_{k=0}^{\infty} \frac{2^k}{k!}$ (b) $\displaystyle\sum_{k=0}^{\infty} \frac{k^k}{k!}$

22. (a) Show that $k^k \ge k!$.
    (b) Use the comparison test to show that $\sum_{k=1}^{\infty} k^{-k}$ converges.
    (c) Use the root test to show that the series converges.

23. Suppose that $\displaystyle\sum_{k=1}^{n} u_k = 2 - \frac{1}{n}$. Find
    (a) u_{100} (b) $\displaystyle\lim_{k \to +\infty} u_k$ (c) $\displaystyle\sum_{k=1}^{\infty} u_k$.

24. In each part, determine whether the series converges; if so, find its sum.
    (a) $\displaystyle\sum_{k=1}^{\infty} \left(\frac{3}{2^k} - \frac{2}{3^k} \right)$ (b) $\displaystyle\sum_{k=1}^{\infty} [\ln(k + 1) - \ln k]$
    (c) $\displaystyle\sum_{k=1}^{\infty} \frac{1}{k(k + 2)}$ (d) $\displaystyle\sum_{k=1}^{\infty} [\tan^{-1}(k+1) - \tan^{-1} k]$

25. In each part, find the sum of the series by associating it with some Maclaurin series.
    (a) $2 + \dfrac{4}{2!} + \dfrac{8}{3!} + \dfrac{16}{4!} + \cdots$
    (b) $\pi - \dfrac{\pi^3}{3!} + \dfrac{\pi^5}{5!} - \dfrac{\pi^7}{7!} + \cdots$

(c) $1 - \dfrac{e^2}{2!} + \dfrac{e^4}{4!} - \dfrac{e^6}{6!} + \cdots$

(d) $1 - \ln 3 + \dfrac{(\ln 3)^2}{2!} - \dfrac{(\ln 3)^3}{3!} + \cdots$

26. Suppose that the sequence $\{a_k\}$ is defined recursively by
$$a_0 = c, \quad a_{k+1} = \sqrt{a_k}$$
Assuming that the sequence converges, find its limit if
    (a) $c = \frac{1}{2}$ (b) $c = \frac{3}{2}$.

27. Research has shown that the proportion p of the population with IQs (intelligence quotients) between α and β is approximately
$$p = \frac{1}{16\sqrt{2\pi}} \int_\alpha^\beta e^{-\frac{1}{2}\left(\frac{x-100}{16}\right)^2}\, dx$$
Use the first three terms of an appropriate Maclaurin series to estimate the proportion of the population that has IQs between 100 and 110.

28. Differentiate the Maclaurin series for xe^x and use the result to show that
$$\sum_{k=0}^{\infty} \frac{k + 1}{k!} = 2e$$

29. Given: $\dfrac{\pi^2}{6} = 1 + \dfrac{1}{2^2} + \dfrac{1}{3^2} + \dfrac{1}{4^2} + \cdots$.

Show: $\dfrac{\pi^2}{12} = 1 - \dfrac{1}{2^2} + \dfrac{1}{3^2} - \dfrac{1}{4^2} + \cdots$.

30. Let a, b, and p be positive constants. For which values of p does the series $\displaystyle\sum_{k=1}^{\infty} \frac{1}{(a + bk)^p}$ converge?

31. In each part, write out the first four terms of the series, and then find the radius of convergence.
    (a) $\displaystyle\sum_{k=1}^{\infty} \frac{1 \cdot 2 \cdot 3 \cdots k}{1 \cdot 4 \cdot 7 \cdots (3k - 2)} x^k$
    (b) $\displaystyle\sum_{k=1}^{\infty} (-1)^k \frac{1 \cdot 2 \cdot 3 \cdots k}{1 \cdot 3 \cdot 5 \cdots (2k - 1)} x^{2k+1}$

32. Find the interval of convergence of
$$\sum_{k=0}^{\infty} \frac{(x - x_0)^k}{b^k} \quad (b > 0)$$

33. Show that the series
$$1 - \frac{x}{2!} + \frac{x^2}{4!} - \frac{x^3}{6!} + \cdots$$
converges to the function
$$f(x) = \begin{cases} \cos \sqrt{x}, & x \ge 0 \\ \cosh \sqrt{-x}, & x < 0 \end{cases}$$
[*Hint:* Use the Maclaurin series for $\cos x$ and $\cosh x$ to obtain series for $\cos \sqrt{x}$, where $x \ge 0$, and $\cosh \sqrt{-x}$, where $x \le 0$.]

34. Prove:

(a) If f is an even function, then all odd powers of x in its Maclaurin series have coefficient 0.

(b) If f is an odd function, then all even powers of x in its Maclaurin series have coefficient 0.

35. In Section 8.6 we defined the kinetic energy K of a particle with mass m and velocity v to be $K = \frac{1}{2}mv^2$ [see Formula (5) of that section]. In this formula the mass m is assumed to be constant, and K is called the **Newtonian Kinetic Energy**. However, in Albert Einstein's relativity theory the mass m increases with the velocity and the kinetic energy K is given by the formula

$$ K = m_0 c^2 \left[\frac{1}{\sqrt{1 - (v/c)^2}} - 1 \right] $$

in which m_0 is the mass of the particle when its velocity is zero, and c is the speed of light. This is called the **relativistic kinetic energy**. Use an appropriate binomial series to show that if the velocity is small compared to the speed of light (i.e., $v/c \approx 0$), then the Newtonian and relativistic kinetic energies are in close agreement.

 36. If the constant p in the general p-series is replaced by a variable x for $x > 1$, then the resulting function is called the **Riemann zeta function** and is denoted by

$$ \zeta(x) = \sum_{k=1}^{\infty} \frac{1}{k^x} $$

(a) Let s_n be the nth partial sum of the series for $\zeta(3.7)$. Find n such that s_n approximates $\zeta(3.7)$ to two decimal-place accuracy, and calculate s_n using this value of n. [*Hint:* Use the right inequality in Exercise 30(b) of Section 11.4 with $f(x) = 1/x^{3.7}$.]

(b) Determine whether your CAS can evaluate the Riemann zeta function directly. If so, compare the value produced by the CAS to the value of s_n obtained in part (a).

EXPANDING THE CALCULUS HORIZON

For additional material relating to this chapter, visit the Anton Website at http://www.wiley.com/college/anton

12

ANALYTIC GEOMETRY IN CALCULUS

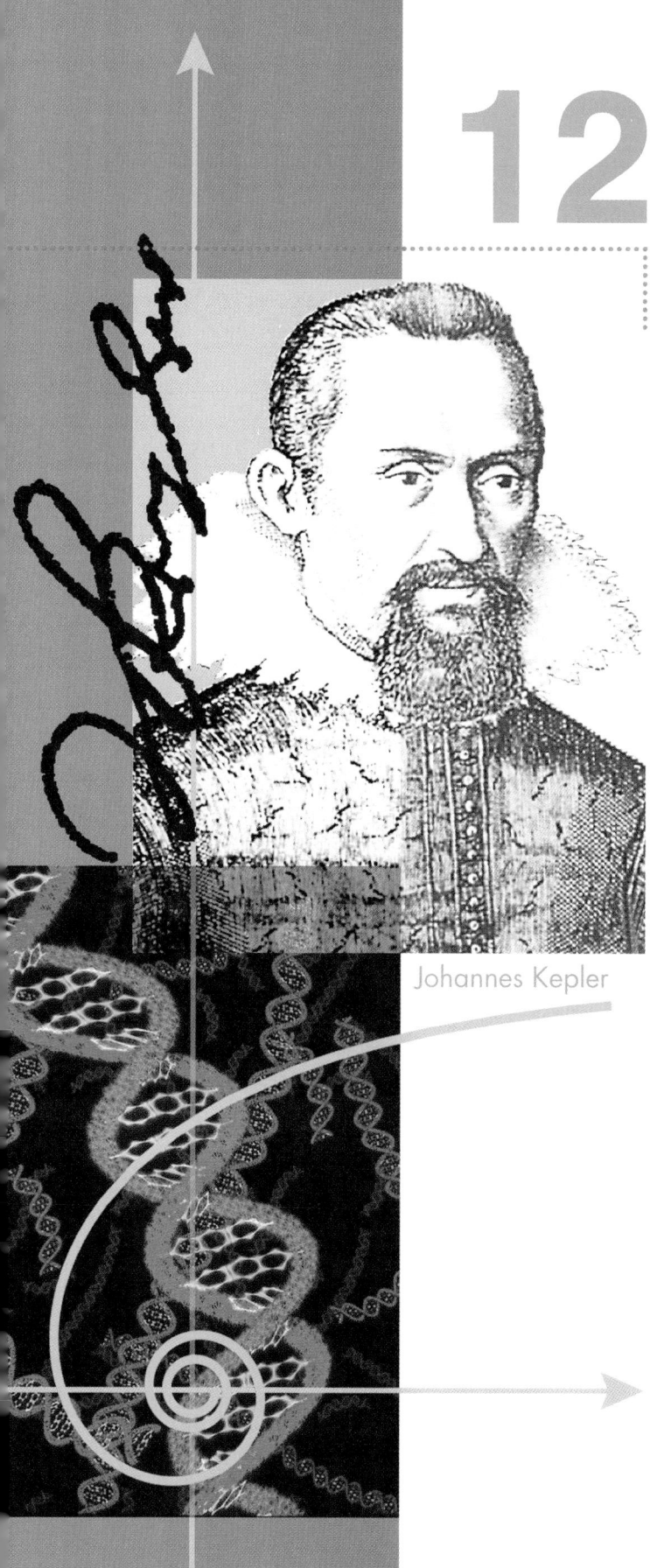

Johannes Kepler

*I*n this chapter we will study aspects of analytic geometry that are important in applications of calculus. We will begin by introducing *polar coordinate systems*, which are used, for example, in tracking the motion of planets and satellites, in identifying the location of objects from information on radar screens, and in the design of antennas. We will then discuss relationships between curves in polar coordinates and parametric curves in rectangular coordinates, and we will discuss methods for finding areas in polar coordinates and tangent lines to curves given in polar coordinates or parametrically in rectangular coordinates. We will then review the basic properties of parabolas, ellipses, and hyperbolas and discuss these curves in the context of polar coordinates. Finally, we will give some basic applications of our work in astronomy.

12.1 POLAR COORDINATES

Up to now we have specified the location of a point in the plane by means of coordinates relative to two perpendicular coordinate axes. However, sometimes a moving point has a special affinity for some fixed point, such as a planet moving in an orbit under the central attraction of the Sun. In such cases, the path of the particle is best described by its angular direction and its distance from the fixed point. In this section we will discuss a new kind of coordinate system that is based on this idea.

POLAR COORDINATE SYSTEMS

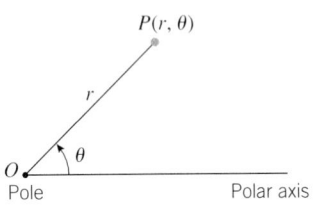

Figure 12.1.1

A **polar coordinate system** in a plane consists of a fixed point O, called the **pole** (or **origin**), and a ray emanating from the pole, called the **polar axis**. In such a coordinate system we can associate with each point P in the plane a pair of **polar coordinates** (r, θ), where r is the distance from P to the pole and θ is an angle from the polar axis to the ray OP (Figure 12.1.1). The number r is called the **radial coordinate** of P and the number θ the **angular coordinate** (or **polar angle**) of P. In Figure 12.1.2, the points $(6, 45°)$, $(5, 120°)$, $(3, 225°)$, and $(4, 330°)$ are plotted in polar coordinate systems. If P is the pole, then $r = 0$, but there is no clearly defined polar angle. We will agree that an arbitrary angle can be used in this case; that is, $(0, \theta)$ are polar coordinates of the pole for all choices of θ.

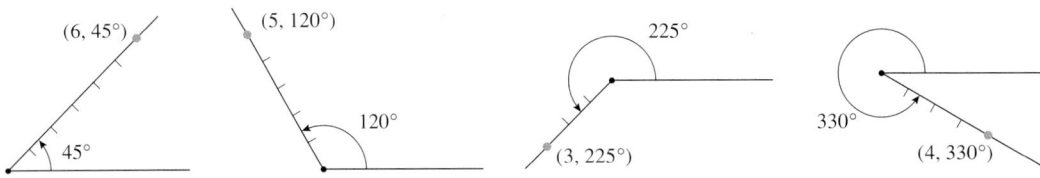

Figure 12.1.2

The polar coordinates of a point are not unique. For example, the polar coordinates

$$(1, 315°), \quad (1, -45°), \quad \text{and} \quad (1, 675°)$$

all represent the same point (Figure 12.1.3). In general, if a point P has polar coordinates (r, θ), then

$$(r, \theta + n \cdot 360°) \quad \text{and} \quad (r, \theta - n \cdot 360°)$$

are also polar coordinates of P for any nonnegative integer n. Thus, every point has infinitely many pairs of polar coordinates.

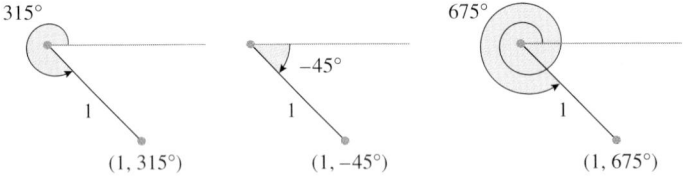

Figure 12.1.3

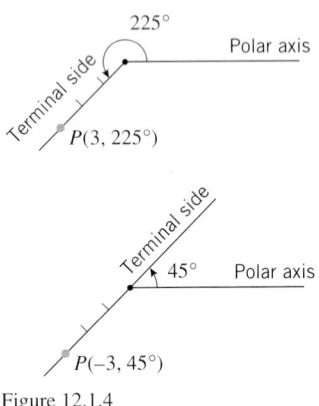

Figure 12.1.4

As defined above, the radial coordinate r of a point P is nonnegative, since it represents the distance from P to the pole. However, it will be convenient to allow for negative values of r as well. To motivate an appropriate definition, consider the point P with polar coordinates $(3, 225°)$. As shown in Figure 12.1.4, we can reach this point by rotating the polar axis through an angle of $225°$ and then moving 3 units from the pole along the terminal side of the angle, or we can reach the point P by rotating the polar axis through an angle of $45°$ and then moving 3 units from the pole along the extension of the terminal side. This suggests that the point $(3, 225°)$ might also be denoted by $(-3, 45°)$, with the minus sign serving to indicate that the point is on the *extension* of the angle's terminal side rather than on the terminal side itself.

In general, the terminal side of the angle $\theta + 180°$ is the extension of the terminal side of θ, so we define negative radial coordinates by agreeing that

$$(-r, \theta) \quad \text{and} \quad (r, \theta + 180°)$$

are polar coordinates of the same point.

FOR THE READER.   For many purposes it does not matter whether polar angles are measured in degrees or radians. However, in problems that involve derivatives or integrals they must be measured in radians, since the derivatives of the trigonometric functions were derived under this assumption. Henceforth, we will use radian measure for polar angles, except in certain applications where it is not required and degree measure is more convenient.

RELATIONSHIP BETWEEN POLAR AND RECTANGULAR COORDINATES

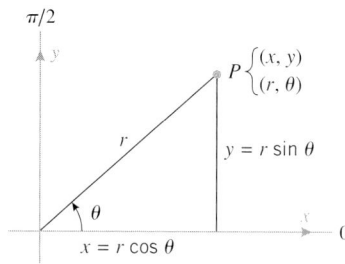

Figure 12.1.5

Frequently, it will be useful to superimpose a rectangular xy-coordinate system on top of a polar coordinate system, making the positive x-axis coincide with the polar axis. If this is done, then every point P will have both rectangular coordinates (x, y) and polar coordinates (r, θ). As suggested by Figure 12.1.5, these coordinates are related by the equations

$$x = r \cos \theta, \quad y = r \sin \theta \tag{1}$$

These equations are well suited for finding x and y when r and θ are known. However, to find r and θ when x and y are known, it is preferable to use the identities $\sin^2 \theta + \cos^2 \theta = 1$ and $\tan \theta = \sin \theta / \cos \theta$ to rewrite (1) as

$$r^2 = x^2 + y^2, \quad \tan \theta = \frac{y}{x} \tag{2}$$

Example 1

Find the rectangular coordinates of the point P whose polar coordinates are $(6, 2\pi/3)$.

Solution.   Substituting the polar coordinates $r = 6$ and $\theta = 2\pi/3$ in (1) yields

$$x = 6 \cos \frac{2\pi}{3} = 6 \left(-\frac{1}{2} \right) = -3$$

$$y = 6 \sin \frac{2\pi}{3} = 6 \left(\frac{\sqrt{3}}{2} \right) = 3\sqrt{3}$$

Thus, the rectangular coordinates of P are $(-3, 3\sqrt{3})$ (Figure 12.1.6).   ◀

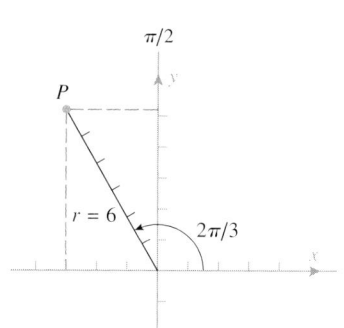

Figure 12.1.6

Example 2

Find polar coordinates of the point P whose rectangular coordinates are $(-2, 2\sqrt{3})$.

Solution.   We will find the polar coordinates (r, θ) of P that satisfy the conditions $r > 0$ and $0 \leq \theta < 2\pi$. From the first equation in (2),

$$r^2 = x^2 + y^2 = (-2)^2 + (2\sqrt{3})^2 = 4 + 12 = 16$$

so $r = 4$. From the second equation in (2),

$$\tan \theta = \frac{y}{x} = \frac{2\sqrt{3}}{-2} = -\sqrt{3}$$

From this and the fact that $(-2, 2\sqrt{3})$ lies in the second quadrant, it follows that the angle satisfying the requirement $0 \leq \theta < 2\pi$ is $\theta = 2\pi/3$. Thus, $(4, 2\pi/3)$ are polar coordinates of P. All other polar coordinates of P are expressible in the form

$$\left(4, \frac{2\pi}{3} + 2n\pi \right) \quad \text{or} \quad \left(-4, \frac{5\pi}{3} + 2n\pi \right)$$

where n is an integer.   ◀

GRAPHS IN POLAR COORDINATES

We will now consider the problem of graphing equations of the form $r = f(\theta)$ in polar coordinates, where θ is assumed to be measured in radians. Some examples of such equations are

$$r = 2\cos\theta, \quad r = \frac{4}{1 - 3\sin\theta}, \quad r = \theta$$

In a rectangular coordinate system the graph of an equation $y = f(x)$ consists of all points whose coordinates (x, y) satisfy the equation. However, in a polar coordinate system, points have infinitely many different pairs of polar coordinates, so that a given point may have some polar coordinates that satisfy the equation $r = f(\theta)$ and others that do not. Taking this into account, we define the **graph of $r = f(\theta)$ in polar coordinates** to consist of all points with *at least one* pair of coordinates (r, θ) that satisfy the equation.

The most elementary way to graph an equation $r = f(\theta)$ in polar coordinates is to plot points. The idea is to choose some typical values of θ, calculate the corresponding values of r, and then plot the resulting pairs (r, θ) in a polar coordinate system. Here are some examples.

Example 3

Sketch the graph of the equation $r = \sin\theta$ in polar coordinates by plotting points.

Solution. Table 12.1.1 shows the coordinates of points on the graph at increments of $\pi/6 \,(= 30°)$.

Table 12.1.1

θ (RADIANS)	0	$\frac{\pi}{6}$	$\frac{\pi}{3}$	$\frac{\pi}{2}$	$\frac{2\pi}{3}$	$\frac{5\pi}{6}$	π	$\frac{7\pi}{6}$	$\frac{4\pi}{3}$	$\frac{3\pi}{2}$	$\frac{5\pi}{3}$	$\frac{11\pi}{6}$	2π
$r = \sin\theta$	0	$\frac{1}{2}$	$\frac{\sqrt{3}}{2}$	1	$\frac{\sqrt{3}}{2}$	$\frac{1}{2}$	0	$-\frac{1}{2}$	$-\frac{\sqrt{3}}{2}$	-1	$-\frac{\sqrt{3}}{2}$	$-\frac{1}{2}$	0
(r, θ)	$(0,0)$	$\left(\frac{1}{2},\frac{\pi}{6}\right)$	$\left(\frac{\sqrt{3}}{2},\frac{\pi}{3}\right)$	$\left(1,\frac{\pi}{2}\right)$	$\left(\frac{\sqrt{3}}{2},\frac{2\pi}{3}\right)$	$\left(\frac{1}{2},\frac{5\pi}{6}\right)$	$(0,\pi)$	$\left(-\frac{1}{2},\frac{7\pi}{6}\right)$	$\left(-\frac{\sqrt{3}}{2},\frac{4\pi}{3}\right)$	$\left(-1,\frac{3\pi}{2}\right)$	$\left(-\frac{\sqrt{3}}{2},\frac{5\pi}{3}\right)$	$\left(-\frac{1}{2},\frac{11\pi}{6}\right)$	$(0,2\pi)$

These points are plotted in Figure 12.1.7. Note, however, that there are 13 points listed in the table but only 6 distinct plotted points. This is because the pairs from $\theta = \pi$ on yield duplicates of the preceding points. For example, $(-1/2, 7\pi/6)$ and $(1/2, \pi/6)$ represent the same point. ◀

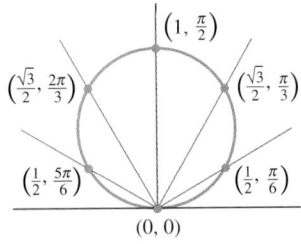

$\left(1, \frac{\pi}{2}\right)$ $\left(\frac{\sqrt{3}}{2}, \frac{2\pi}{3}\right)$ $\left(\frac{\sqrt{3}}{2}, \frac{\pi}{3}\right)$ $\left(\frac{1}{2}, \frac{5\pi}{6}\right)$ $\left(\frac{1}{2}, \frac{\pi}{6}\right)$ $(0, 0)$

Figure 12.1.7

Observe that the points in Figure 12.1.7 appear to lie on a circle. We can confirm that this is so by expressing the polar equation $r = \sin\theta$ in terms of x and y. To do this, we multiply the equation through by r to obtain

$$r^2 = r\sin\theta$$

which now allows us to apply Formulas (1) and (2) to rewrite the equation as

$$x^2 + y^2 = y$$

Rewriting this equation as $x^2 + y^2 - y = 0$ and then completing the square yields

$$x^2 + \left(y - \tfrac{1}{2}\right)^2 = \tfrac{1}{4}$$

which is a circle of radius $\frac{1}{2}$ centered at the point $\left(0, \frac{1}{2}\right)$ in the xy-plane.

Just because an equation $r = f(\theta)$ involves the variables r and θ does not mean that it has to be graphed in a polar coordinate system. When useful, this equation can also be graphed in a rectangular coordinate system. For example, Figure 12.1.8 shows the graph of $r = \sin\theta$ in a rectangular θr-coordinate system. This graph can actually help to visualize how the polar graph in Figure 12.1.7 is generated:

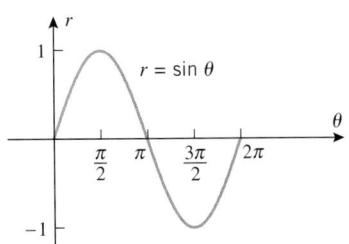

Figure 12.1.8

- At $\theta = 0$ we have $r = 0$, which corresponds to the pole $(0, 0)$ on the polar graph.
- As θ varies from 0 to $\pi/2$, the value of r increases from 0 to 1, so the point (r, θ) moves along the circle from the pole to the high point at $(1, \pi/2)$.
- As θ varies from $\pi/2$ to π, the value of r decreases from 1 back to 0, so the point (r, θ) moves along the circle from the high point back to the pole.
- As θ varies from π to $3\pi/2$, the values of r are negative, varying from 0 to -1. Thus, the point (r, θ) moves along the circle from the pole to the high point at $(1, \pi/2)$, which is the same as the point $(-1, 3\pi/2)$. This duplicates the motion that occurred for $0 \le \theta \le \pi/2$.
- As θ varies from $3\pi/2$ to 2π, the value of r varies from -1 to 0. Thus, the point (r, θ) moves along the circle from the high point back to the pole, duplicating the motion that occurred for $\pi/2 \le \theta \le \pi$.

Example 4

Sketch the graph of $r = \cos 2\theta$ in polar coordinates.

Solution. Instead of plotting points, we will use the graph of $r = \cos 2\theta$ in rectangular coordinates (Figure 12.1.9) to visualize how the polar graph of this equation is generated. The analysis and the resulting polar graph are shown in Figure 12.1.10. This curve is called a *four-petal rose*. ◀

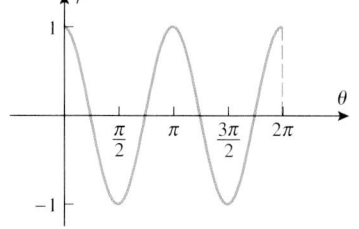

Figure 12.1.9

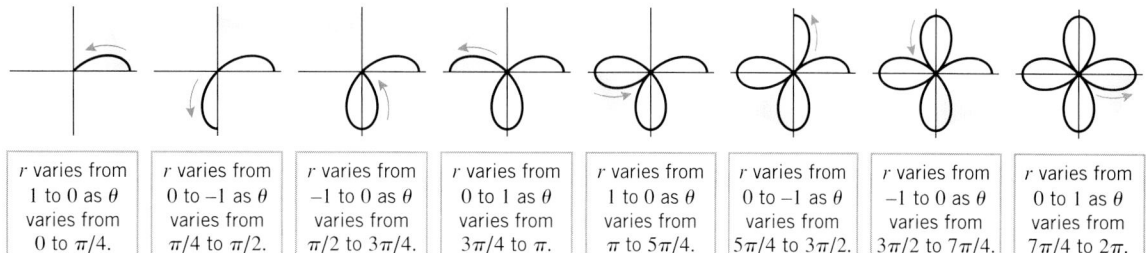

| r varies from 1 to 0 as θ varies from 0 to $\pi/4$. | r varies from 0 to -1 as θ varies from $\pi/4$ to $\pi/2$. | r varies from -1 to 0 as θ varies from $\pi/2$ to $3\pi/4$. | r varies from 0 to 1 as θ varies from $3\pi/4$ to π. | r varies from 1 to 0 as θ varies from π to $5\pi/4$. | r varies from 0 to -1 as θ varies from $5\pi/4$ to $3\pi/2$. | r varies from -1 to 0 as θ varies from $3\pi/2$ to $7\pi/4$. | r varies from 0 to 1 as θ varies from $7\pi/4$ to 2π. |

Figure 12.1.10

SYMMETRY TESTS

Observe that the polar graph of $r = \cos 2\theta$ in Figure 12.1.10 is symmetric about the x-axis and the y-axis. This symmetry could have been predicted from the following theorem, which is suggested by Figure 12.1.11 (we omit the proof).

> **12.1.1** THEOREM (*Symmetry Tests*).
> (*a*) *A curve in polar coordinates is symmetric about the x-axis if replacing θ by $-\theta$ in its equation produces an equivalent equation* (Figure 12.1.11*a*).
> (*b*) *A curve in polar coordinates is symmetric about the y-axis if replacing θ by $\pi - \theta$ in its equation produces an equivalent equation* (Figure 12.1.11*b*).
> (*c*) *A curve in polar coordinates is symmetric about the origin if replacing r by $-r$ in its equation produces an equivalent equation* (Figure 12.1.11*c*).

Example 5

Use Theorem 12.1.1 to confirm that the graph of $r = \cos 2\theta$ in Figure 12.1.10 is symmetric about the x-axis and y-axis.

Solution. To test for symmetry about the x-axis, we replace θ by $-\theta$. This yields

$$r = \cos(-2\theta) = \cos 2\theta$$

Thus, replacing θ by $-\theta$ does not alter the equation.

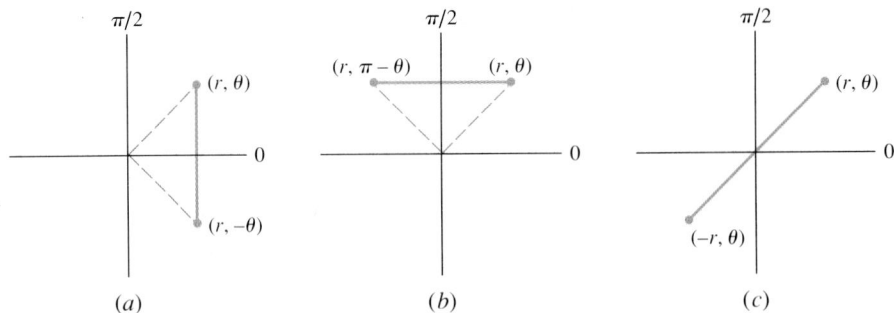

Figure 12.1.11

To test for symmetry about the y-axis, we replace θ by $\pi - \theta$. This yields

$$r = \cos 2(\pi - \theta) = \cos(2\pi - 2\theta) = \cos(-2\theta) = \cos 2\theta$$

Thus, replacing θ by $\pi - \theta$ does not alter the equation. ◄

Example 6

Sketch the graph of $r = a(1 - \cos\theta)$ in polar coordinates, assuming a to be a positive constant.

Solution. Observe first that replacing θ by $-\theta$ does not alter the equation, so we know in advance that the graph is symmetric about the polar axis. Thus, if we graph the upper half of the curve, then we can obtain the lower half by reflection about the polar axis.

As in our previous examples, we will first graph the equation in rectangular coordinates. This graph, which is shown in Figure 12.1.12*a*, can be obtained by rewriting the given equation as $r = a - a\cos\theta$, from which we see that the graph in rectangular coordinates can be obtained by first reflecting the graph of $r = a\cos\theta$ about the x-axis to obtain the graph of $r = -a\cos\theta$, and then translating that graph up a units to obtain the graph of $r = a - a\cos\theta$. Now we can see that:

- As θ varies from 0 to $\pi/3$, r increases from 0 to $a/2$.
- As θ varies from $\pi/3$ to $\pi/2$, r increases from $a/2$ to a.
- As θ varies from $\pi/2$ to $2\pi/3$, r increases from a to $3a/2$.
- As θ varies from $2\pi/3$ to π, r increases from $3a/2$ to $2a$.

This produces the polar curve shown in Figure 12.1.12*b*. The rest of the curve can be obtained by continuing the preceding analysis from π to 2π or, as noted above, by reflecting the portion already graphed about the x-axis (Figure 12.1.12*c*). This heart-shaped curve is called a ***cardioid*** (from the Greek word "kardia" for heart). ◄

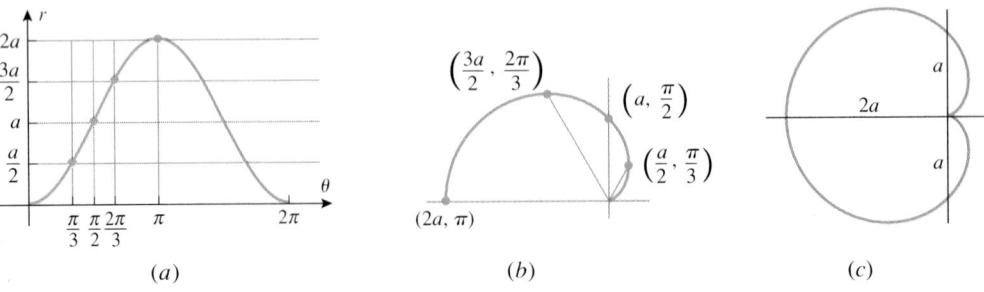

Figure 12.1.12

Example 7

Sketch the curves

(a) $r = 1$ (b) $\theta = \dfrac{\pi}{4}$ (c) $r = \theta$ $(\theta \geq 0)$

in polar coordinates.

Solution (*a*). For all values of θ, the point $(1, \theta)$ is 1 unit away from the pole. Thus, the graph is the circle of radius 1 centered at the pole (Figure 12.1.13*a*).

Solution (*b*). For all values of r, the point $(r, \pi/4)$ lies on a line that makes an angle of $\pi/4$ with the polar axis (Figure 12.1.13*b*). Positive values of r correspond to points on the line in the first quadrant and negative values of r to points on the line in the third quadrant. Thus, in absence of any restriction on r, the graph is the entire line. Observe, however, that had we imposed the restriction $r \geq 0$, the graph would have been just the ray in the first quadrant.

Solution (*c*). Observe that as θ increases, so does r; thus, the graph is a curve that spirals out from the pole as θ increases. A reasonably accurate sketch of the spiral can be obtained by plotting the intersections with the x- and y-axes for values of θ that are multiples of $\pi/2$, keeping in mind that the value of r is always equal to the value of θ (Figure 12.1.13*c*). ◄

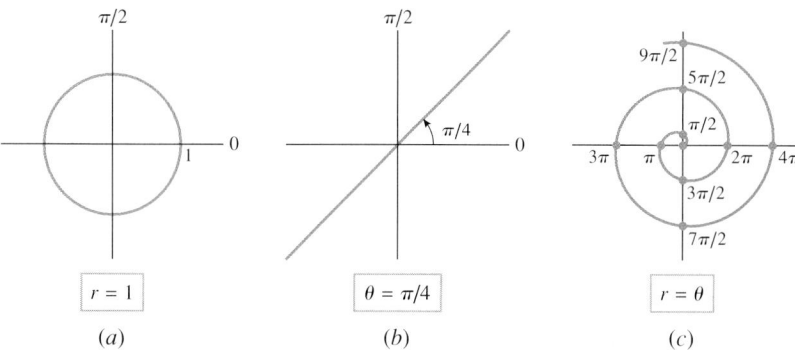

Figure 12.1.13

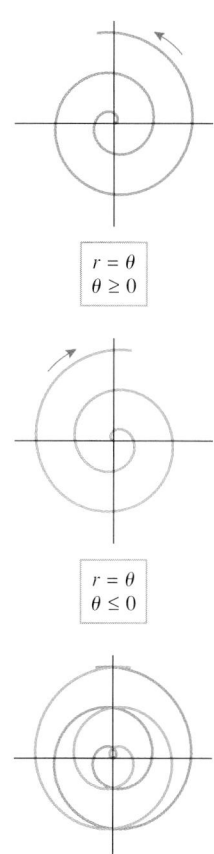

$r = \theta$
$\theta \geq 0$

$r = \theta$
$\theta \leq 0$

$r = \theta$
$-\infty < \theta < +\infty$

Figure 12.1.14

REMARK.   The spiral in Figure 12.1.13*c*, which belongs to the family of ***Archimedean spirals*** $r = a\theta$, coils counterclockwise around the pole because of the restriction $\theta \geq 0$. Had we made the restriction $\theta \leq 0$, the spiral would have coiled clockwise, and had we allowed both positive and negative values of θ, the clockwise and counterclockwise spirals would have been superimposed to form a double Archimedean spiral (Figure 12.1.14).

Example 8

Sketch the graph of $r^2 = 4 \cos 2\theta$ in polar coordinates.

Solution. This equation does not express r as a function of θ, since solving for r in terms of θ yields two functions:

$$r = 2\sqrt{\cos 2\theta} \quad \text{and} \quad r = -2\sqrt{\cos 2\theta}$$

Thus, to graph the equation $r^2 = 4 \cos 2\theta$ we will have to graph the two functions separately and then combine those graphs.

We will start with the graph of $r = 2\sqrt{\cos 2\theta}$. Observe first that this equation is not changed if we replace θ by $-\theta$ or if we replace θ by $\pi - \theta$. Thus, the graph is symmetric about the x-axis and the y-axis. This means that the entire graph can be obtained by graphing the portion in the first quadrant, reflecting that portion about the y-axis to obtain the portion in the second quadrant and then reflecting those two portions about the x-axis to obtain the portions in the third and fourth quadrants.

To begin the analysis, we will graph the equation $r = 2\sqrt{\cos 2\theta}$ in rectangular coordinates (see Figure 12.1.15a). Note that there are gaps in that graph over the intervals $\pi/4 < \theta < 3\pi/4$ and $5\pi/4 < \theta < 7\pi/4$ because $\cos 2\theta$ is negative for those values of θ. From this graph we can see that:

- As θ varies from 0 to $\pi/4$, r decreases from 2 to 0.

- As θ varies from $\pi/4$ to $\pi/2$, no points are generated on the polar graph.

This produces the portion of the graph shown in Figure 12.1.15b. As noted above, we can complete the graph by a reflection about the y-axis followed by a reflection about the x-axis (12.1.15c). The resulting propeller-shaped graph is called a **lemniscate** (from the Greek word "lemniscos" for a looped ribbon resembling the number 8). We leave it for you to verify that the equation $r = 2\sqrt{\cos 2\theta}$ has the same graph as $r = -2\sqrt{\cos 2\theta}$, but traced in a diagonally opposite manner. Thus, the graph of the equation $r^2 = 4\cos 2\theta$ consists of two identical superimposed lemniscates. ◀

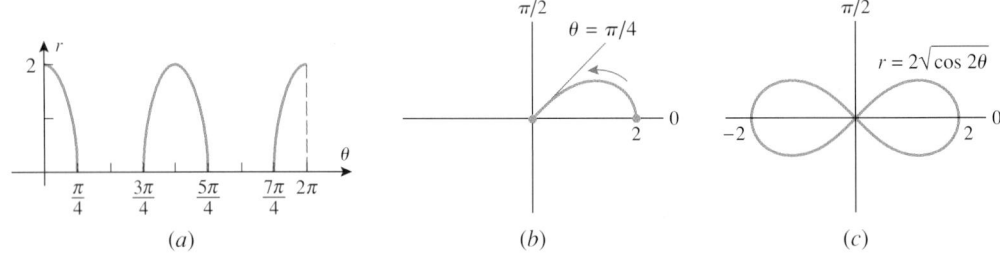

Figure 12.1.15

FAMILIES OF LINES AND RAYS THROUGH THE POLE

If θ_0 is a fixed angle, then for all values of r the point (r, θ_0) lies on the line that makes an angle of $\theta = \theta_0$ with the polar axis; and, conversely, every point on this line has a pair of polar coordinates of the form (r, θ_0). Thus, the equation $\theta = \theta_0$ represents the line that passes through the pole and makes an angle of θ_0 with the polar axis (Figure 12.1.16a). If r is restricted to be nonnegative, then the graph of the equation $\theta = \theta_0$ is the ray that emanates from the pole and makes an angle of θ_0 with the polar axis (Figure 12.1.16b). Thus, as θ_0 varies, the equation $\theta = \theta_0$ produces either a family of lines through the pole or a family of rays through the pole, depending on the restrictions on r.

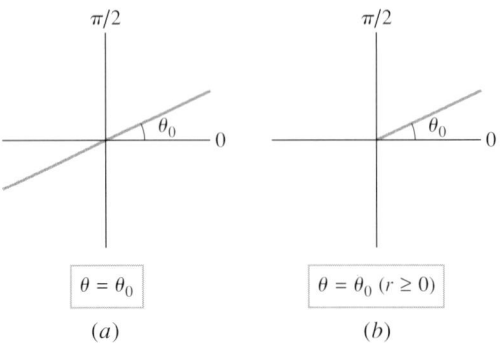

Figure 12.1.16

FAMILIES OF CIRCLES

We will consider three families of circles in which a is assumed to be a positive constant:

$$r = a \qquad r = 2a \cos \theta \qquad r = 2a \sin \theta \qquad\qquad (3\text{--}5)$$

The equation $r = a$ represents a circle of radius a centered at the pole (Figure 12.1.17a). Thus, as a varies, this equation produces a family of circles centered at the pole. For families (4) and (5), recall from plane geometry that a triangle that is inscribed in a circle with a diameter of the circle for a side must be a right triangle. Thus, as indicated in Figures 12.1.17b and 12.1.17c, the equation $r = 2a \cos \theta$ represents a circle of radius a, centered on the x-axis and tangent to the y-axis at the origin; similarly, the equation $r = 2a \sin \theta$ represents a circle of radius a, centered on the y-axis and tangent to the x-axis at the origin. Thus, as a varies, Equations (4) and (5) produce the families illustrated in Figures 12.1.17d and 12.1.17e.

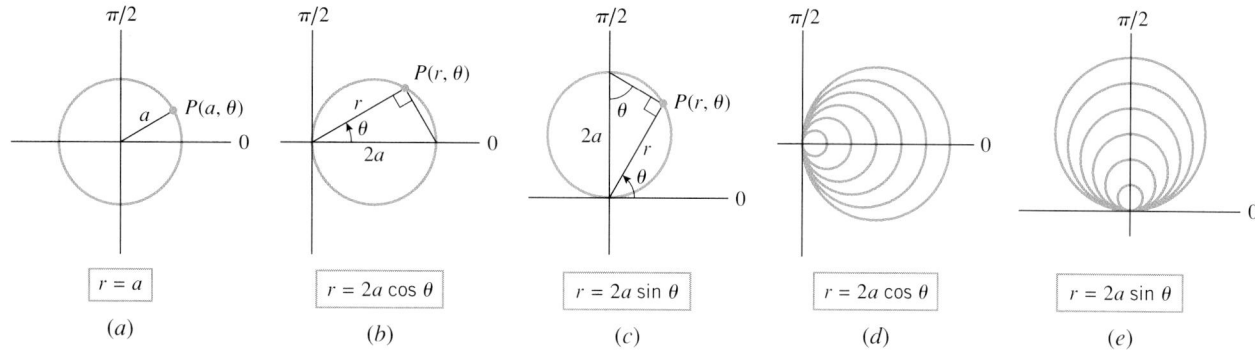

Figure 12.1.17

REMARK. Observe that replacing θ by $-\theta$ does not change the equation $r = 2a \cos \theta$, and replacing θ by $\pi - \theta$ does not change the equation $r = 2a \sin \theta$. This explains why the circles in Figure 12.1.17d are symmetric about the x-axis and those in Figure 12.1.17e are symmetric about the y-axis.

FAMILIES OF ROSE CURVES

In polar coordinates, equations of the form

$$r = a \sin n\theta \qquad r = a \cos n\theta \qquad\qquad (6\text{--}7)$$

in which $a > 0$ and n is a positive integer represent families of flower-shaped curves called **roses** (Figure 12.1.18). The rose consists of n equally spaced petals of radius a if n is odd

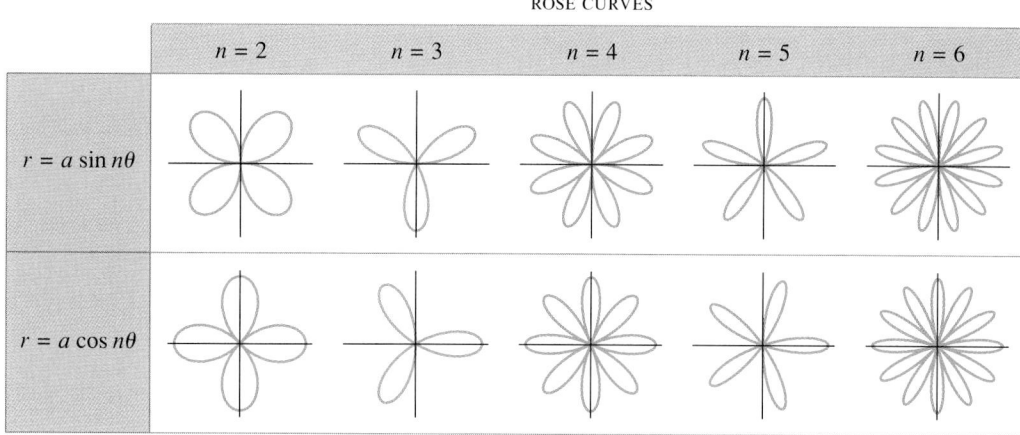

Figure 12.1.18

and $2n$ equally spaced petals of radius a if n is positive and even. It can be shown that a rose with an even number of petals is traced out exactly once as θ varies over the interval $0 \le \theta < 2\pi$ and a rose with an odd number of petals is traced out exactly once as θ varies over the interval $0 \le \theta < \pi$ (Exercise 73). A four-petal rose of radius 1 was graphed in Example 4.

⋮ FOR THE READER.   What do the graphs of the one-petal roses look like?

FAMILIES OF CARDIOIDS AND LIMAÇONS

Equations with any of the four forms

$$r = a \pm b \sin\theta \qquad r = a \pm b \cos\theta \tag{8–9}$$

in which $a > 0$ and $b > 0$ represent polar curves called *limaçons* (from the Latin word "limax" for a snail-like creature that is commonly called a slug). There are four possible shapes for a limaçon that are determined by the ratio a/b (Figure 12.1.19). If $a = b$ (the case $a/b = 1$), then the limaçon is called a *cardioid* because of its heart-shaped appearance, as noted in Example 6.

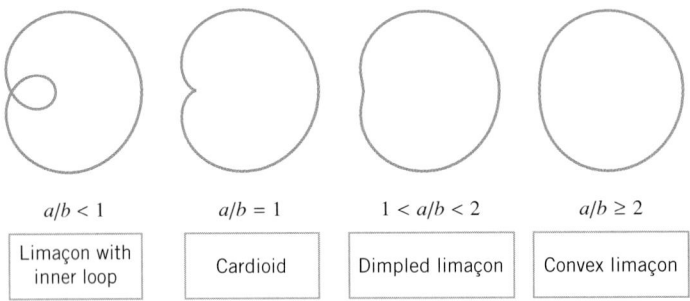

$a/b < 1$	$a/b = 1$	$1 < a/b < 2$	$a/b \ge 2$
Limaçon with inner loop	Cardioid	Dimpled limaçon	Convex limaçon

Figure 12.1.19

Example 9

Figure 12.1.20 shows the family of limaçons $r = a + \cos\theta$ with the constant a varying from 0.25 to 2.50 in steps of 0.25. In keeping with Figure 12.1.19, the limaçons evolve from the loop type to the convex type. As a increases from the starting value of 0.25, the loops get smaller and smaller until the cardioid is reached at $a = 1$. As a increases further, the limaçons evolve through the dimpled type into the convex type.   ◄

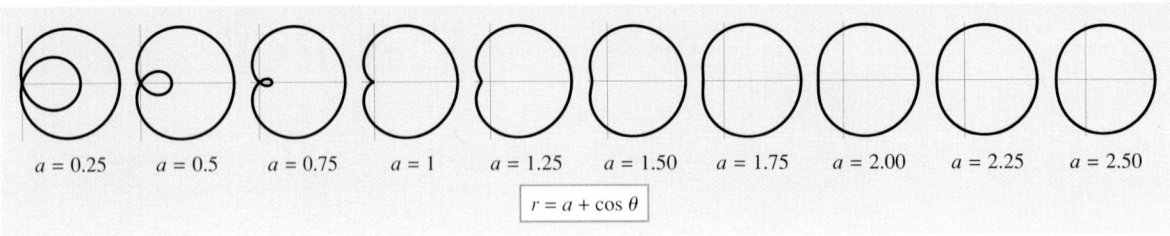

$a = 0.25 \qquad a = 0.5 \qquad a = 0.75 \qquad a = 1 \qquad a = 1.25 \qquad a = 1.50 \qquad a = 1.75 \qquad a = 2.00 \qquad a = 2.25 \qquad a = 2.50$

$$r = a + \cos\theta$$

Figure 12.1.20

FAMILIES OF SPIRALS

A *spiral* is a curve that coils around a central point. As illustrated in Figure 12.1.14, spirals generally have "left-hand" and "right-hand" versions that coil in opposite directions, depending on the restrictions on the polar angle and the signs of constants that appear in their equations. Some of the more common types of spirals are shown in Figure 12.1.21 for nonnegative values of θ, a, and b.

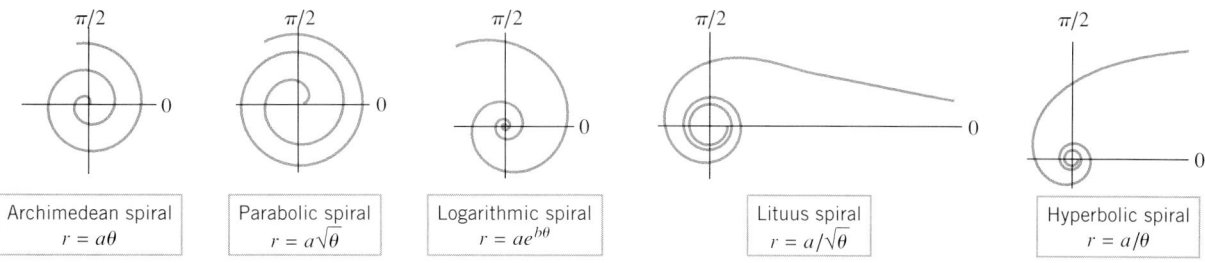

| Archimedean spiral $r = a\theta$ | Parabolic spiral $r = a\sqrt{\theta}$ | Logarithmic spiral $r = ae^{b\theta}$ | Lituus spiral $r = a/\sqrt{\theta}$ | Hyperbolic spiral $r = a/\theta$ |

Figure 12.1.21

SPIRALS IN NATURE

Spirals of many kinds occur in nature. For example, the shell of the chambered nautilus (*below*) forms a logarithmic spiral, and a coiled sailor's rope forms an Archimedean spiral. Spirals also occur in flowers, the tusks of certain animals, and in the shapes of galaxies.

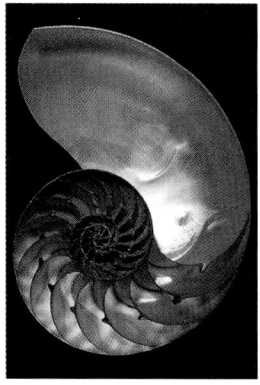

The shell of the chambered nautilus reveals a logarithmic spiral. The animal lives in the outermost chamber.

A sailor's coiled rope forms an Archimedean spiral.

GENERATING POLAR CURVES WITH GRAPHING UTILITIES

For polar curves that are too complicated for hand computation, graphing utilities must be used. Although many graphing utilities are capable of graphing polar curves directly, some are not. However, if a graphing utility is capable of graphing parametric equations, then it can be used to graph a polar curve $r = f(\theta)$ by converting this equation to parametric form. This can be done by substituting $f(\theta)$ for r in (1). This yields

$$x = f(\theta)\cos\theta, \quad y = f(\theta)\sin\theta \tag{10}$$

which is a pair of parametric equations for the polar curve in terms of the parameter θ.

Example 10

Express the polar equation

$$r = 2 + \cos\frac{5\theta}{2}$$

parametrically, and generate the polar graph from the parametric equations using a graphing utility.

Solution. Substituting the given expression for r in $x = r\cos\theta$ and $y = r\sin\theta$ yields the parametric equations

$$x = \left[2 + \cos\frac{5\theta}{2}\right]\cos\theta, \quad y = \left[2 + \cos\frac{5\theta}{2}\right]\sin\theta$$

Next, we need to find an interval over which to vary θ to produce the entire graph. To find such an interval, we will look for the smallest number of complete revolutions that must occur until the value of r begins to repeat. Algebraically, this amounts to finding the smallest positive integer n such that

$$2 + \cos\left(\frac{5(\theta + 2n\pi)}{2}\right) = 2 + \cos\frac{5\theta}{2}$$

or

$$\cos\left(\frac{5\theta}{2} + 5n\pi\right) = \cos\frac{5\theta}{2}$$

For this equality to hold, the quantity $5n\pi$ must be an even multiple of π; the smallest n for which this occurs is $n = 2$. Thus, the entire graph will be traced in two revolutions, which means it can be generated from the parametric equations

$$x = \left[2 + \cos\frac{5\theta}{2}\right]\cos\theta, \quad y = \left[2 + \cos\frac{5\theta}{2}\right]\sin\theta \quad (0 \le \theta \le 4\pi)$$

This yields the graph in Figure 12.1.22. ◀

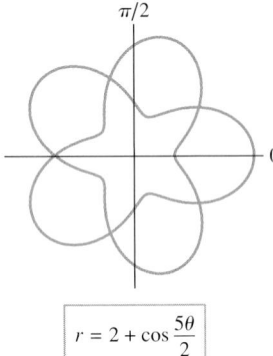

$r = 2 + \cos\dfrac{5\theta}{2}$

Figure 12.1.22

FOR THE READER.   Some graphing utilities require that t be used for the parameter. If this is true of your graphing utility, then you will have to replace θ by t in (10) to generate graphs in polar coordinates. Use a graphing utility to duplicate the curve in Figure 12.1.22.

EXERCISE SET 12.1   ~ Graphing Calculator

In Exercises 1 and 2, plot the points in polar coordinates.

1. (a) $(3, \pi/4)$ (b) $(5, 2\pi/3)$ (c) $(1, \pi/2)$
   (d) $(4, 7\pi/6)$ (e) $(-6, -\pi)$ (f) $(-1, 9\pi/4)$

2. (a) $(2, -\pi/3)$ (b) $(3/2, -7\pi/4)$ (c) $(-3, 3\pi/2)$
   (d) $(-5, -\pi/6)$ (e) $(2, 4\pi/3)$ (f) $(0, \pi)$

In Exercises 3 and 4, find the rectangular coordinates of the points whose polar coordinates are given.

3. (a) $(6, \pi/6)$ (b) $(7, 2\pi/3)$ (c) $(-6, -5\pi/6)$
   (d) $(0, -\pi)$ (e) $(7, 17\pi/6)$ (f) $(-5, 0)$

4. (a) $(-8, \pi/4)$ (b) $(7, -\pi/4)$ (c) $(8, 9\pi/4)$
   (d) $(5, 0)$ (e) $(-2, -3\pi/2)$ (f) $(0, \pi)$

5. In each part, a point is given in rectangular coordinates. Find two pairs of polar coordinates for the point, one pair satisfying $r \ge 0$ and $0 \le \theta < 2\pi$, and the second pair satisfying $r \ge 0$ and $-\pi < \theta \le \pi$.
   (a) $(-5, 0)$ (b) $(2\sqrt{3}, -2)$ (c) $(0, -2)$
   (d) $(-8, -8)$ (e) $(-3, 3\sqrt{3})$ (f) $(1, 1)$

6. In each part find polar coordinates satisfying the stated conditions for the point whose rectangular coordinates are $(-\sqrt{3}, 1)$.
   (a) $r \ge 0$ and $0 \le \theta < 2\pi$

(b) $r \le 0$ and $0 \le \theta < 2\pi$
(c) $r \ge 0$ and $-2\pi < \theta \le 0$
(d) $r \le 0$ and $-\pi < \theta \le \pi$

In Exercises 7 and 8, use a calculating utility, where needed, to approximate the polar coordinates of the points whose rectangular coordinates are given.

7. (a) $(4, 3)$ (b) $(2, -5)$ (c) $(1, \tan^{-1}1)$

8. (a) $(-3, 4)$ (b) $(-3, 1.7)$ (c) $\left(2, \sin^{-1}\frac{1}{2}\right)$

In Exercises 9 and 10, identify the curve by transforming the given polar equation to rectangular coordinates.

9. (a) $r = 2$ (b) $r\sin\theta = 4$
   (c) $r = 3\cos\theta$ (d) $r = \dfrac{6}{3\cos\theta + 2\sin\theta}$

10. (a) $r = 5\sec\theta$ (b) $r = 2\sin\theta$
    (c) $r = 4\cos\theta + 4\sin\theta$ (d) $r = \sec\theta\tan\theta$

In Exercises 11 and 12, express the given equations in polar coordinates.

11. (a) $x = 7$ (b) $x^2 + y^2 = 9$
    (c) $x^2 + y^2 - 6y = 0$ (d) $4xy = 9$

12. (a) $y = -3$ (b) $x^2 + y^2 = 5$
    (c) $x^2 + y^2 + 4x = 0$ (d) $x^2(x^2 + y^2) = y^2$

In Exercises 13–16, a graph is given in a rectangular θr-coordinate system. Sketch the corresponding graph in polar coordinates.

13.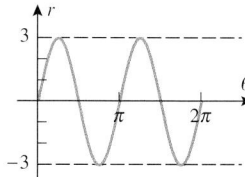

14.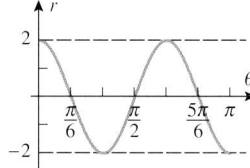

15.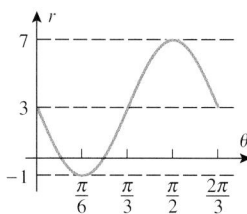

16.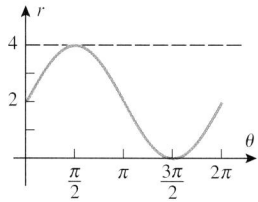

In Exercises 17–20, find an equation for the given polar graph.

17. (a) 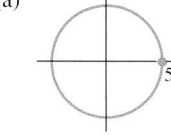 (b) (c)

Circle   Circle   Cardioid

18. (a) 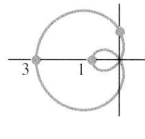 (b) 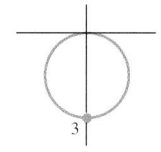 (c)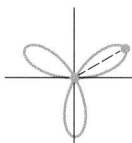

Limaçon   Circle   Three-petal rose

19. (a) 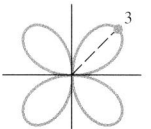 (b) 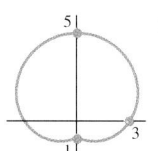 (c)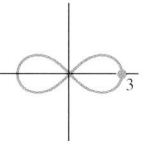

Four-petal rose   Limaçon   Lemniscate

20. (a) 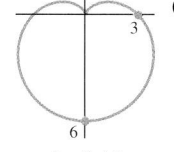 (b) 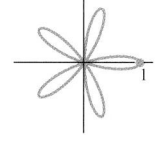 (c)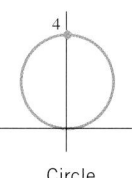

Cardioid   Five-petal rose   Circle

In Exercises 21–50, sketch the curve in polar coordinates.

21. $\theta = \dfrac{\pi}{6}$ **22.** $\theta = -\dfrac{3\pi}{4}$

23. $r = 3$ **24.** $r = 4\sin\theta$

25. $r = 6\cos\theta$ **26.** $r = 1 + \sin\theta$

27. $2r = \cos\theta$ **28.** $r - 2 = 2\cos\theta$

29. $r = 3(1 - \sin\theta)$ **30.** $r = -5 + 5\sin\theta$

31. $r = 4 - 4\cos\theta$ **32.** $r = 1 + 2\sin\theta$

33. $r = -1 - \cos\theta$ **34.** $r = 4 + 3\cos\theta$

35. $r = 2 + \sin\theta$ **36.** $r = 3 - \cos\theta$

37. $r = 3 + 4\cos\theta$ **38.** $r - 5 = 3\sin\theta$

39. $r = 5 - 2\cos\theta$ **40.** $r = -3 - 4\sin\theta$

41. $r^2 = 9\cos 2\theta$ **42.** $r^2 = \sin 2\theta$

43. $r^2 = 16\sin 2\theta$ **44.** $r = 4\theta \quad (\theta \geq 0)$

45. $r = 4\theta \quad (\theta \leq 0)$ **46.** $r = 4\theta$

47. $r = \cos 2\theta$ **48.** $r = 3\sin 2\theta$

49. $r = 9\sin 4\theta$ **50.** $r = 2\cos 3\theta$

51. For each of the curves you sketched in Exercises 21–50, check your work with a graphing utility.

In Exercises 52–55, use a graphing utility to generate the polar graph. Be sure to choose the parameter interval so that a complete graph is generated.

52. $r = \sin\dfrac{\theta}{2}$ **53.** $r = 1 + 2\cos\dfrac{\theta}{4}$

54. $r = 0.5 + \cos\dfrac{\theta}{3}$ **55.** $r = \cos\dfrac{\theta}{5}$

56. The accompanying figure shows the graph of the "butterfly curve"
$$r = e^{\cos\theta} - 2\cos 4\theta + \sin^3\dfrac{\theta}{4}$$
Generate the complete butterfly with a graphing utility, and state the parameter interval you used.

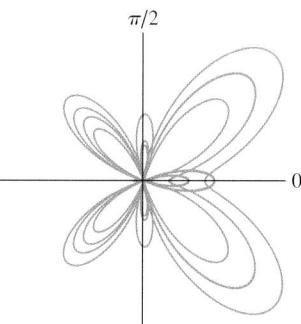

Figure Ex-56

57. Figure Ex-57 (next page) shows the Archimedean spiral $r = \theta/2$ produced with a graphing calculator.
(a) What interval of values for θ do you think was used to generate the graph?
(b) Duplicate the graph with your own graphing utility.

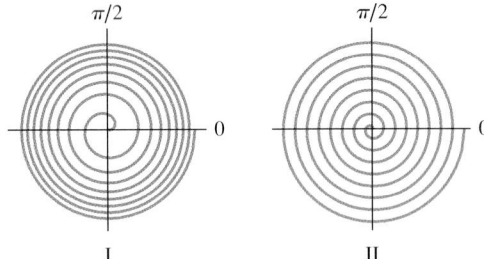

[-9, 9] × [-6, 6]
xScl = 1, yScl = 1

Figure Ex-57

58. The accompanying figure shows graphs of the Archimedean spiral $r = \theta$ and the parabolic spiral $r = \sqrt{\theta}$. Which is which? Explain your reasoning.

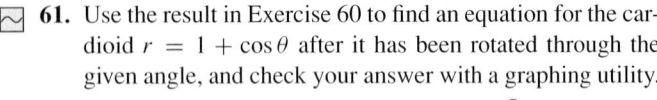

Figure Ex-58

59. (a) Show that if a varies, then the polar equation

$$r = a \sec\theta \quad (-\pi/2 < \theta < \pi/2)$$

describes a family of lines perpendicular to the polar axis.

(b) Show that if b varies, then the polar equation

$$r = b \csc\theta \quad (0 < \theta < \pi)$$

describes a family of lines parallel to the polar axis.

60. Show that if the polar graph of $r = f(\theta)$ is rotated counterclockwise around the origin through an angle α, then $r = f(\theta - \alpha)$ is an equation for the rotated curve. [*Hint:* If (r_0, θ_0) is any point on the original graph, then $(r_0, \theta_0 + \alpha)$ is a point on the rotated graph.]

61. Use the result in Exercise 60 to find an equation for the cardioid $r = 1 + \cos\theta$ after it has been rotated through the given angle, and check your answer with a graphing utility.

(a) $\dfrac{\pi}{4}$ (b) $\dfrac{\pi}{2}$ (c) π (d) $\dfrac{5\pi}{4}$

62. Use the result in Exercise 60 to find an equation for the lemniscate that results when the lemniscate in Example 8 is rotated counterclockwise through an angle of $\pi/2$.

63. Sketch the polar graph of the equation $(r - 1)(\theta - 1) = 0$.

64. (a) Show that if A and B are not both zero, then the graph of the polar equation

$$r = A \sin\theta + B \cos\theta$$

is a circle. Find its radius.

(b) Derive Formulas (4) and (5) from the formula given in part (a).

65. Find the highest point on the cardioid $r = 1 + \cos\theta$.

66. Find the leftmost point on the upper half of the cardioid $r = 1 + \cos\theta$.

67. (a) Show that in a polar coordinate system the distance d between the points (r_1, θ_1) and (r_2, θ_2) is

$$d = \sqrt{r_1^2 + r_1^2 - 2r_1 r_2 \cos(\theta_1 - \theta_2)}$$

(b) Show that if $0 \le \theta_1 < \theta_2 \le \pi$ and if r_1 and r_2 are positive, then the area A of the triangle with vertices $(0, 0)$, (r_1, θ_1), and (r_2, θ_2) is

$$A = \tfrac{1}{2} r_1 r_2 \sin(\theta_2 - \theta_1)$$

(c) Find the distance between the points whose polar coordinates are $(3, \pi/6)$ and $(2, \pi/3)$.

(d) Find the area of the triangle whose vertices in polar coordinates are $(0, 0)$, $(1, 5\pi/6)$, and $(2, \pi/3)$.

68. In the late seventeenth century the Italian astronomer Giovanni Domenico Cassini (1625–1712) introduced the family of curves

$$(x^2 + y^2 + a^2)^2 - b^4 - 4a^2 x^2 = 0 \quad (a > 0, b > 0)$$

in his studies of the relative motions of the Earth and the Sun. These curves, which are called **Cassini ovals**, have one of the three basic shapes shown in the accompanying figure.

(a) Show that if $a = b$, then the polar equation of the Cassini oval is $r^2 = 2a^2 \cos 2\theta$, which is a lemniscate.

(b) Use the formula in Exercise 67(a) to show that the lemniscate in part (a) is the curve traced by a point that moves in such a way that the product of its distances from the polar points $(a, 0)$ and (a, π) is a^2.

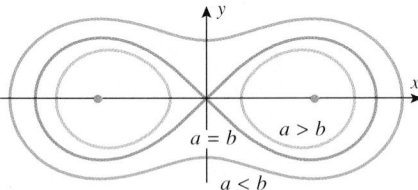

Figure Ex-68

Vertical and horizontal asymptotes of polar curves can often be detected by investigating the behavior of $x = r \cos\theta$ and $y = r \sin\theta$ as θ varies. This idea is used in Exercises 69–72.

69. Show that the **hyperbolic spiral** $r = 1/\theta$ $(\theta > 0)$ has a horizontal asymptote at $y = 1$ by showing that $y \to 1$ and $x \to +\infty$ as $\theta \to 0^+$. Confirm this result by generating the spiral with a graphing utility.

70. Show that the spiral $r = 1/\theta^2$ does not have any horizontal asymptotes.

71. (a) Show that the *kappa curve* $r = 4\tan\theta$ $(0 \le \theta \le 2\pi)$ has a vertical asymptote at $x = 4$ by showing that $x \to 4$ and $y \to +\infty$ as $\theta \to \pi/2^-$ and that $x \to 4$ and $y \to -\infty$ as $\theta \to \pi/2^+$.

    (b) Use the method in part (a) to show that the kappa curve also has a vertical asymptote at $x = -4$.

    (c) Confirm the results in parts (a) and (b) by generating the kappa curve with a graphing utility.

72. Use a graphing utility to make a conjecture about the existence of asymptotes for the *cissoid* $r = 2\sin\theta\tan\theta$, and then confirm your conjecture by calculating appropriate limits.

73. Prove that a rose with an even number of petals is traced out exactly once as θ varies over the interval $0 \le \theta < 2\pi$ and a rose with an odd number of petals is traced out exactly once as θ varies over the interval $0 \le \theta < \pi$.

12.2 TANGENT LINES AND ARC LENGTH FOR PARAMETRIC AND POLAR CURVES

In this section we will derive the formulas required to find slopes, tangent lines, and arc lengths of parametric and polar curves.

TANGENT LINES TO PARAMETRIC CURVES

We will be concerned in this section with curves that are given by parametric equations

$$x = f(t), \quad y = g(t)$$

in which $f(t)$ and $g(t)$ have continuous first derivatives with respect to t. It can be proved that if $dx/dt \ne 0$, then y is a differentiable function of x, in which case the chain rule implies that

$$\frac{dy}{dx} = \frac{dy/dt}{dx/dt} \tag{1}$$

This formula makes it possible to find dy/dx directly from the parametric equations without eliminating the parameter.

Example 1

Find the slope of the tangent line to the unit circle

$$x = \cos t, \quad y = \sin t \quad (0 \le t \le 2\pi)$$

at the point where $t = \pi/6$ (Figure 12.2.1).

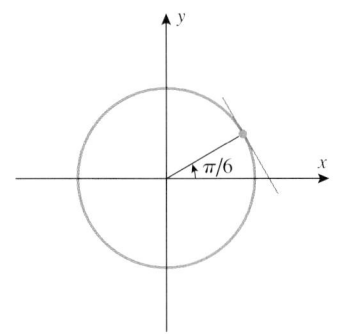

Figure 12.2.1

Solution. From (1), the slope at a general point on the circle is

$$\frac{dy}{dx} = \frac{dy/dt}{dx/dt} = \frac{\cos t}{-\sin t} = -\cot t \tag{2}$$

Thus, the slope at $t = \pi/6$ is

$$\left.\frac{dy}{dx}\right|_{t=\pi/6} = -\cot\frac{\pi}{6} = -\sqrt{3} \qquad \blacktriangleleft$$

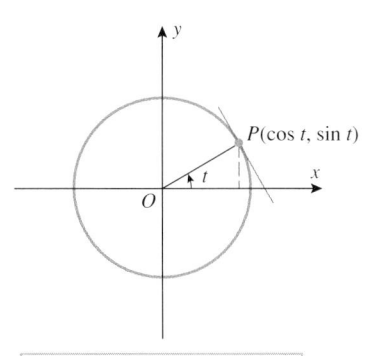

Radius *OP* has slope $m = \tan t$.

Figure 12.2.2

REMARK.   Note that Formula (2) makes sense geometrically because the radius to the point $P(\cos t, \sin t)$ has slope $m = \tan t$; hence, the tangent line at P, being perpendicular to the radius, has slope $-1/m = -1/\tan t = -\cot t$ (Figure 12.2.2).

It follows from Formula (1) that the tangent line to a parametric curve will be horizontal at those points where $dy/dt = 0$ and $dx/dt \ne 0$, since $dy/dx = 0$ at such points. Two different situations occur when $dx/dt = 0$. At points where $dx/dt = 0$ and $dy/dt \ne 0$, the right side of (1) has a nonzero numerator and a zero denominator; we will agree that the curve has ***infinite slope*** and a ***vertical tangent line*** at such points. At points where dx/dt and dy/dt are both zero, the right side of (1) becomes an indeterminate form; we call such points ***singular points***. No general statement can be made about the behavior of parametric curves at singular points; they must be analyzed case by case.

Example 2

In a disastrous first flight, an experimental paper airplane follows the trajectory

$$x = t - 3\sin t, \quad y = 4 - 3\cos t \quad (t \geq 0)$$

but crashes into a wall at time $t = 10$ (Figure 12.2.3).

(a) At what times was the airplane flying horizontally?

(b) At what times was it flying vertically?

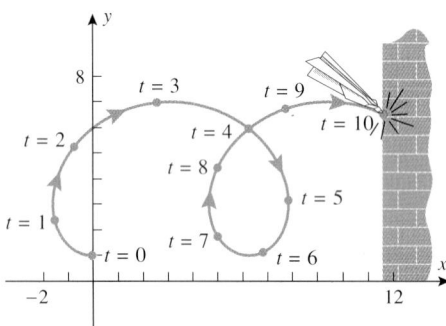

Figure 12.2.3

Solution (a). The airplane was flying horizontally at those times when $dy/dt = 0$ and $dx/dt \neq 0$. From the given trajectory we have

$$\frac{dy}{dt} = 3\sin t \quad \text{and} \quad \frac{dx}{dt} = 1 - 3\cos t \tag{3}$$

Setting $dy/dt = 0$ yields the equation $3\sin t = 0$, or, more simply, $\sin t = 0$. This equation has four solutions in the time interval $0 \leq t \leq 10$:

$$t = 0, \quad t = \pi, \quad t = 2\pi, \quad t = 3\pi$$

Since $dx/dt = 1 - 3\cos t \neq 0$ for these values of t (verify), the airplane was flying horizontally at times

$$t = 0, \quad t = \pi \approx 3.14, \quad t = 2\pi \approx 6.28, \quad \text{and} \quad t = 3\pi \approx 9.42$$

which is consistent with Figure 12.2.3.

Solution (b). The airplane was flying vertically at those times when $dx/dt = 0$ and $dy/dt \neq 0$. Setting $dx/dt = 0$ in (3) yields the equation

$$1 - 3\cos t = 0 \quad \text{or} \quad \cos t = \tfrac{1}{3}$$

This equation has three solutions in the time interval $0 \leq t \leq 10$ (Figure 12.2.4):

$$t = \cos^{-1}\tfrac{1}{3}, \quad t = 2\pi - \cos^{-1}\tfrac{1}{3}, \quad t = 2\pi + \cos^{-1}\tfrac{1}{3}$$

Since $dy/dt = 3\sin t$ is not zero at these points (why?), it follows that the airplane was flying vertically at times

$$t = \cos^{-1}\tfrac{1}{3} \approx 1.23, \quad t \approx 2\pi - 1.23 \approx 5.05, \quad t \approx 2\pi + 1.23 \approx 7.51$$

which again is consistent with Figure 12.2.3. ◄

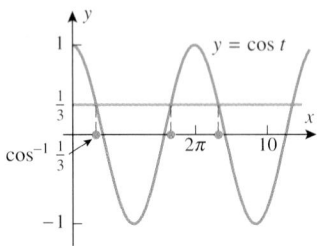

Figure 12.2.4

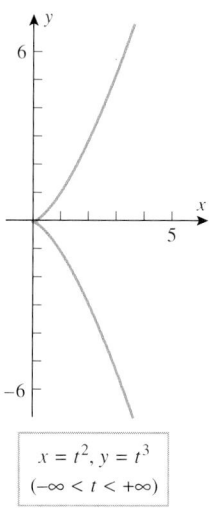

$$x = t^2, \ y = t^3$$
$$(-\infty < t < +\infty)$$

Figure 12.2.5

Example 3

The curve represented by the parametric equations

$$x = t^2, \quad y = t^3 \qquad (-\infty < t < +\infty)$$

is called a **semicubical parabola**. The parameter t can be eliminated by cubing x and squaring y, from which it follows that $y^2 = x^3$. The graph of this equation, shown in Figure 12.2.5, consists of two branches: an upper branch obtained by graphing $y = x^{3/2}$ and a lower branch obtained by graphing $y = -x^{3/2}$. The two branches meet at the origin, which corresponds to $t = 0$ in the parametric equations. This is a singular point because the derivatives $dx/dt = 2t$ and $dy/dt = 3t^2$ are both zero there. ◄

Example 4

Without eliminating the parameter, find dy/dx and d^2y/dx^2 at the points $(1, 1)$ and $(1, -1)$ on the semicubical parabola given by the parametric equations in Example 3.

Solution. From (1) we have

$$\frac{dy}{dx} = \frac{dy/dt}{dx/dt} = \frac{3t^2}{2t} = \frac{3}{2}t \quad (t \neq 0) \tag{4}$$

and from (1) applied to $y' = dy/dx$ we have

$$\frac{d^2y}{dx^2} = \frac{dy'}{dx} = \frac{dy'/dt}{dx/dt} = \frac{3/2}{2t} = \frac{3}{4t} \tag{5}$$

Since the point $(1, 1)$ on the curve corresponds to $t = 1$ in the parametric equations, it follows from (4) and (5) that

$$\left.\frac{dy}{dx}\right|_{t=1} = \frac{3}{2} \quad \text{and} \quad \left.\frac{d^2y}{dx^2}\right|_{t=1} = \frac{3}{4}$$

Similarly, the point $(1, -1)$ corresponds to $t = -1$ in the parametric equations, so applying (4) and (5) again yields

$$\left.\frac{dy}{dx}\right|_{t=-1} = -\frac{3}{2} \quad \text{and} \quad \left.\frac{d^2y}{dx^2}\right|_{t=-1} = -\frac{3}{4}$$

Note that the values we obtained for the first and second derivatives are consistent with the graph in Figure 12.2.5, since at $(1, 1)$ on the upper branch the tangent line has positive slope and the curve is concave up, and at $(1, -1)$ on the lower branch the tangent line has negative slope and the curve is concave down.

Finally, observe that we were able to apply Formulas (4) and (5) for both $t = 1$ and $t = -1$, even though the points $(1, 1)$ and $(1, -1)$ lie on different branches. In contrast, had we chosen to perform the same computations by eliminating the parameter, we would have had to obtain separate derivative formulas for $y = x^{3/2}$ and $y = -x^{3/2}$. ◄

TANGENT LINES TO POLAR CURVES

Our next objective is to find a method for obtaining slopes of tangent lines to polar curves of the form $r = f(\theta)$ in which r is a differentiable function of θ. We showed in the last section that a curve of this form can be expressed parametrically in terms of the parameter θ by substituting $f(\theta)$ for r in the equations $x = r\cos\theta$ and $y = r\sin\theta$. This yields

$$x = f(\theta)\cos\theta, \quad y = f(\theta)\sin\theta$$

from which we obtain

$$\frac{dx}{d\theta} = -f(\theta)\sin\theta + f'(\theta)\cos\theta = -r\sin\theta + \frac{dr}{d\theta}\cos\theta$$

$$\frac{dy}{d\theta} = f(\theta)\cos\theta + f'(\theta)\sin\theta = r\cos\theta + \frac{dr}{d\theta}\sin\theta \tag{6}$$

Thus, if $dx/d\theta$ and $dy/d\theta$ are continuous and if $dx/d\theta \neq 0$, then y is a differentiable function of x, and Formula (1) with θ in place of t yields

$$\frac{dy}{dx} = \frac{dy/d\theta}{dx/d\theta} = \frac{r\cos\theta + \sin\theta\dfrac{dr}{d\theta}}{-r\sin\theta + \cos\theta\dfrac{dr}{d\theta}} \tag{7}$$

Example 5

Find the slope of the tangent line to the circle $r = 4\cos\theta$ at the point where $\theta = \pi/4$.

Solution. From (7) with $r = 4\cos\theta$ we obtain (verify)

$$\frac{dy}{dx} = \frac{4\cos^2\theta - 4\sin^2\theta}{-8\sin\theta\cos\theta} = \frac{4\cos 2\theta}{-4\sin 2\theta} = -\cot 2\theta$$

Thus, at the point where $\theta = \pi/4$ the slope of the tangent line is

$$m = \left.\frac{dy}{dx}\right|_{\theta=\pi/4} = -\cot\frac{\pi}{2} = 0$$

which implies that the circle has a horizontal tangent line at the point where $\theta = \pi/4$ (Figure 12.2.6). ◀

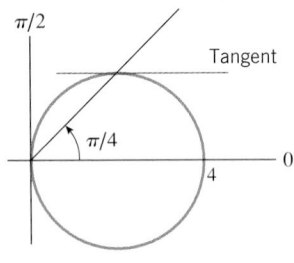

Figure 12.2.6

Example 6

Find the points on the cardioid $r = 1 - \cos\theta$ at which there is a horizontal tangent line, a vertical tangent line, or a singular point.

Solution. A horizontal tangent line will occur where $dy/d\theta = 0$ and $dx/d\theta \neq 0$, a vertical tangent line where $dy/d\theta \neq 0$ and $dx/d\theta = 0$, and a singular point where $dy/d\theta = 0$ and $dx/d\theta = 0$. We could find these derivatives from the formulas in (6). However, an alternative approach is go back to basic principles and express the cardioid parametrically by substituting $r = 1 - \cos\theta$ in the conversion formulas $x = r\cos\theta$ and $y = r\sin\theta$. This yields

$$x = (1 - \cos\theta)\cos\theta, \quad y = (1 - \cos\theta)\sin\theta \qquad (0 \leq \theta \leq 2\pi)$$

Differentiating these equations with respect to θ and then simplifying yields (verify)

$$\frac{dx}{d\theta} = \sin\theta(2\cos\theta - 1), \quad \frac{dy}{d\theta} = (1 - \cos\theta)(1 + 2\cos\theta)$$

Thus, $dx/d\theta = 0$ if $\sin\theta = 0$ or $\cos\theta = \frac{1}{2}$, and $dy/d\theta = 0$ if $\cos\theta = 1$ or $\cos\theta = -\frac{1}{2}$. We leave it for you to solve these equations and show that the solutions of $dx/d\theta = 0$ on the interval $0 \leq \theta \leq 2\pi$ are

$$\frac{dx}{d\theta} = 0: \quad \theta = 0, \; \frac{\pi}{3}, \; \pi, \; \frac{5\pi}{3}, \; 2\pi$$

and the solutions of $dy/d\theta = 0$ on the interval $0 \leq \theta \leq 2\pi$ are

$$\frac{dy}{d\theta} = 0: \quad \theta = 0, \; \frac{2\pi}{3}, \; \frac{4\pi}{3}, \; 2\pi$$

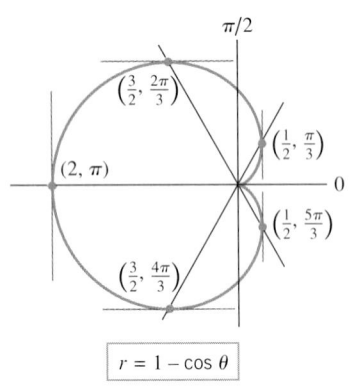

Figure 12.2.7

Thus, horizontal tangent lines occur at $\theta = 2\pi/3$ and $\theta = 4\pi/3$; vertical tangent lines occur at $\theta = \pi/3$, π, and $5\pi/3$; and singular points occur at $\theta = 0$ and $\theta = 2\pi$ (Figure 12.2.7). Note, however, that $r = 0$ at both singular points, so there is really only one singular point on the cardioid—the pole. ◀

TANGENT LINES TO POLAR CURVES AT THE ORIGIN

Formula (7) reveals some useful information about the behavior of a polar curve $r = f(\theta)$ that passes through the origin. If we assume that $r = 0$ and $dr/d\theta \neq 0$ when $\theta = \theta_0$, then it follows from Formula (7) that the slope of the tangent line to the curve at $\theta = \theta_0$ is

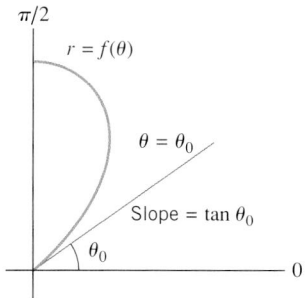

Figure 12.2.8

$$\frac{dy}{dx} = \frac{0 + \sin\theta_0 \dfrac{dr}{d\theta}}{0 + \cos\theta_0 \dfrac{dr}{d\theta}} = \frac{\sin\theta_0}{\cos\theta_0} = \tan\theta_0$$

(Figure 12.2.8). However, $\tan\theta_0$ is also the slope of the line $\theta = \theta_0$, so we can conclude that this line is tangent to the curve at the origin. Thus, we have established the following result.

12.2.1   THEOREM.   *If the polar curve $r = f(\theta)$ passes through the origin at $\theta = \theta_0$, and if $dr/d\theta \neq 0$ at $\theta = \theta_0$, then the line $\theta = \theta_0$ is tangent to the curve at the origin.*

This theorem tells us that equations of the tangent lines at the origin to the curve $r = f(\theta)$ can be obtained by solving the equation $f(\theta) = 0$. It is important to keep in mind, however, that $r = f(\theta)$ may be zero for more than one value of θ, so there may be more than one tangent line at the origin. This is illustrated in the next example.

Example 7

The three-petal rose $r = \sin 3\theta$ in Figure 12.2.9 has three tangent lines at the origin, which can be found by solving the equation

$$\sin 3\theta = 0$$

It was shown in Exercise 73 of Section 12.1 that the complete rose is traced once as θ varies over the interval $0 \leq \theta < \pi$, so we need only look for solutions in this interval. We leave it for you to confirm that these solutions are

$$\theta = 0, \quad \theta = \frac{\pi}{3}, \quad \text{and} \quad \theta = \frac{2\pi}{3}$$

Since $dr/d\theta = 3\cos 3\theta \neq 0$ for these values of θ, these three lines are tangent to the rose at the origin, which is consistent with the figure.   ◄

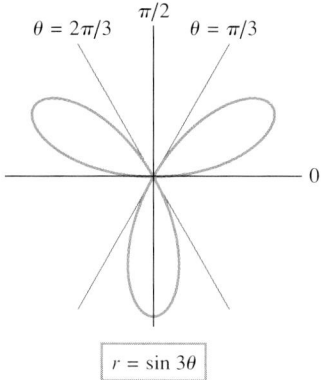

Figure 12.2.9

ARC LENGTH OF A POLAR CURVE

A formula for the arc length of a polar curve $r = f(\theta)$ can be derived by expressing the curve in parametric form and applying Formula (6) of Section 8.4 for the arc length of a parametric curve. We leave it as an exercise to show the following.

12.2.2   ARC LENGTH FORMULA FOR POLAR CURVES.   If no segment of the polar curve $r = f(\theta)$ is traced more than once as θ increases from α to β, and if $dr/d\theta$ is continuous for $\alpha \leq \theta \leq \beta$, then the arc length L from $\theta = \alpha$ to $\theta = \beta$ is

$$L = \int_\alpha^\beta \sqrt{r^2 + \left(\frac{dr}{d\theta}\right)^2}\, d\theta \qquad (8)$$

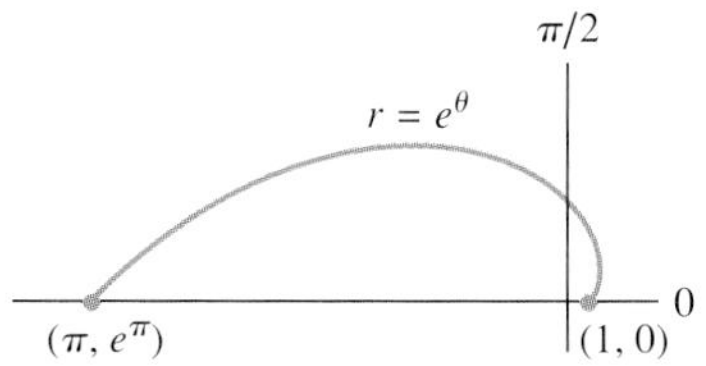

Example 8

Find the arc length of the spiral $r = e^\theta$ in Figure 12.2.10 between $\theta = 0$ and $\theta = \pi$.

Solution.

$$L = \int_\alpha^\beta \sqrt{r^2 + \left(\frac{dr}{d\theta}\right)^2}\, d\theta = \int_0^\pi \sqrt{(e^\theta)^2 + (e^\theta)^2}\, d\theta$$

$$= \int_0^\pi \sqrt{2}\, e^\theta\, d\theta = \sqrt{2}\, e^\theta \Big]_0^\pi = \sqrt{2}(e^\pi - 1) \approx 31.3 \qquad ◄$$

Figure 12.2.10

Example 9

Find the total arc length of the cardioid $r = 1 + \cos\theta$.

Solution. The cardioid is traced out once as θ varies from $\theta = 0$ to $\theta = 2\pi$. Thus,

$$L = \int_\alpha^\beta \sqrt{r^2 + \left(\frac{dr}{d\theta}\right)^2}\, d\theta = \int_0^{2\pi} \sqrt{(1+\cos\theta)^2 + (-\sin\theta)^2}\, d\theta$$

$$= \sqrt{2} \int_0^{2\pi} \sqrt{1 + \cos\theta}\, d\theta$$

$$= 2 \int_0^{2\pi} \sqrt{\cos^2 \frac{1}{2}\theta}\, d\theta \qquad \text{Identity (45) of Appendix E}$$

$$= 2 \int_0^{2\pi} \left|\cos \frac{1}{2}\theta\right| d\theta$$

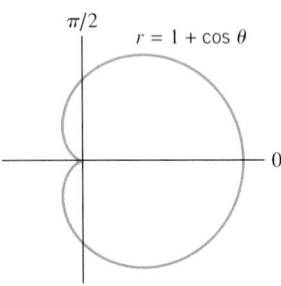

$\pi/2$

$r = 1 + \cos\theta$

0

Figure 12.2.11

Since $\cos\frac{1}{2}\theta$ changes sign at π, we must split the last integral into the sum of two integrals: the integral from 0 to π plus the integral from π to 2π. However, the integral from π to 2π is equal to the integral from 0 to π, since the cardioid is symmetric about the polar axis (Figure 12.2.11). Thus,

$$L = 2 \int_0^{2\pi} \left|\cos \frac{1}{2}\theta\right| d\theta = 4 \int_0^\pi \cos \frac{1}{2}\theta\, d\theta = 8 \sin \frac{1}{2}\theta \Big]_0^\pi = 8 \qquad \blacktriangleleft$$

EXERCISE SET 12.2 ☒ Graphing Calculator

1. (a) Find the slope of the tangent line to the parametric curve $x = t^2 + 1$, $y = t/2$ at $t = -1$ and at $t = 1$ without eliminating the parameter.
   (b) Check your answers in part (a) by eliminating the parameter and differentiating an appropriate function of x.

2. (a) Find the slope of the tangent line to the parametric curve $x = 3\cos t$, $y = 4\sin t$ at $t = \pi/4$ and at $t = 7\pi/4$ without eliminating the parameter.
   (b) Check your answers in part (a) by eliminating the parameter and differentiating an appropriate function of x.

3. For the parametric curve in Exercise 1, make a conjecture about the sign of d^2y/dx^2 at $t = -1$ and at $t = 1$, and confirm your conjecture without eliminating the parameter.

4. For the parametric curve in Exercise 2, make a conjecture about the sign of d^2y/dx^2 at $t = \pi/4$ and at $t = 7\pi/4$, and confirm your conjecture without eliminating the parameter.

In Exercises 5–10, find dy/dx and d^2y/dx^2 at the given point without eliminating the parameter.

5. $x = \sqrt{t}$, $y = 2t + 4$; $t = 1$

6. $x = \frac{1}{2}t^2$, $y = \frac{1}{3}t^3$; $t = 2$

7. $x = \sec t$, $y = \tan t$; $t = \pi/3$

8. $x = \sinh t$, $y = \cosh t$; $t = 0$

9. $x = 2\theta + \cos\theta$, $y = 1 - \sin\theta$; $\theta = \pi/3$

10. $x = \cos\phi$, $y = 3\sin\phi$; $\phi = 5\pi/6$

11. (a) Find the equation of the tangent line to the curve
   $$x = e^t, \quad y = e^{-t}$$
   at $t = 1$ without eliminating the parameter.
   (b) Check your answer in part (a) by eliminating the parameter.

12. (a) Find the equation of the tangent line to the curve
   $$x = 2t + 4, \quad y = 8t^2 - 2t + 4$$
   at $t = 1$ without eliminating the parameter.
   (b) Check your answer in part (a) by eliminating the parameter.

In Exercises 13 and 14, find all values of t at which the parametric curve has (a) a horizontal tangent line and (b) a vertical tangent line.

13. $x = 2\cos t$, $y = 4\sin t$ ($0 \le t \le 2\pi$)

14. $x = 2t^3 - 15t^2 + 24t + 7$, $y = t^2 + t + 1$

15. As shown in Figure Ex-15 (next page), the Lissajous curve
   $$x = \sin t, \quad y = \sin 2t \qquad (0 \le t \le 2\pi)$$
   crosses itself at the origin. Find equations for the two tangent lines at the origin.

16. As shown in the accompanying figure, the ***prolate cycloid***

$$x = 2 - \pi \cos t, \quad y = 2t - \pi \sin t \quad (-\pi \leq t \leq \pi)$$

crosses itself at a point on the x-axis. Find equations for the two tangent lines at that point.

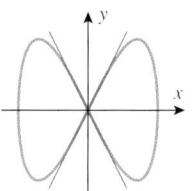

Figure Ex-15   Figure Ex-16

17. Show that the curve $x = t^3 - 4t$, $y = t^2$ intersects itself at the point $(0, 4)$, and find equations for the two tangent lines to the curve at the point of intersection.

18. Show that the curve with parametric equations

$$x = t^2 - 3t + 5, \quad y = t^3 + t^2 - 10t + 9$$

intersects itself at the point $(3, 1)$, and find equations for the two tangent lines to the curve at the point of intersection.

19. (a) Use a graphing utility to generate the graph of the parametric curve

$$x = \cos^3 t, \quad y = \sin^3 t \quad (0 \leq t \leq 2\pi)$$

and make a conjecture about the values of t at which singular points occur.

   (b) Confirm your conjecture in part (a) by calculating appropriate derivatives.

20. (a) At what values of θ would you expect the cycloid in Figure 1.7.13 to have singular points?

   (b) Confirm your answer in part (a) by calculating appropriate derivatives.

In Exercises 21–26, find the slope of the tangent line to the polar curve for the given value of θ.

21. $r = 2 \cos \theta$; $\theta = \pi/3$ **22.** $r = 1 + \sin \theta$; $\theta = \pi/4$

23. $r = 1/\theta$; $\theta = 2$ **24.** $r = a \sec 2\theta$; $\theta = \pi/6$

25. $r = \cos 3\theta$; $\theta = 3\pi/4$ **26.** $r = 4 - 3 \sin \theta$; $\theta = \pi$

In Exercises 27 and 28, calculate the slopes of the tangent lines indicated in the accompanying figures.

27. $r = 2 + 2 \sin \theta$ **28.** $r = 1 - 2 \sin \theta$

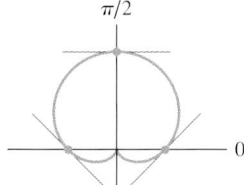
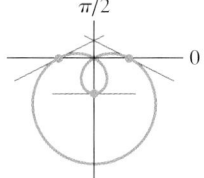

Figure Ex-27   Figure Ex-28

In Exercises 29 and 30, find polar coordinates of all points at which the polar curve has a horizontal or a vertical tangent line.

29. $r = a(1 + \cos \theta)$ **30.** $r = a \sin \theta$

In Exercises 31 and 32, use a graphing utility to make a conjecture about the number of points on the polar curve at which there is a horizontal tangent line, and confirm your conjecture by finding appropriate derivatives.

31. $r = \sin \theta \cos^2 \theta$ **32.** $r = 1 - 2 \sin \theta$

In Exercises 33–38, sketch the polar curve and find polar equations of the tangent lines to the curve at the pole.

33. $r = 2 \cos 3\theta$ **34.** $r = 4 \cos \theta$

35. $r = 4\sqrt{\cos 2\theta}$ **36.** $r = \sin 2\theta$

37. $r = 1 + 2 \cos \theta$ **38.** $r = 2\theta$

In Exercises 39–44, use Formula (8) to calculate the arc length of the polar curve.

39. The entire circle $r = a$

40. The entire circle $r = 2a \cos \theta$

41. The entire cardioid $r = a(1 - \cos \theta)$

42. $r = \sin^2(\theta/2)$ from $\theta = 0$ to $\theta = \pi$

43. $r = e^{3\theta}$ from $\theta = 0$ to $\theta = 2$

44. $r = \sin^3(\theta/3)$ from $\theta = 0$ to $\theta = \pi/2$

45. (a) What is the slope of the tangent line at time t to the trajectory of the paper airplane in Example 2?

   (b) What was the airplane's approximate angle of inclination when it crashed into the wall?

46. Suppose that a bee follows the trajectory

$$x = t - 2 \sin t, \quad y = 2 - 2 \cos t \quad (t \geq 0)$$

but lands on a wall at time $t = 10$.

   (a) At what times was the bee flying horizontally?

   (b) At what times was the bee flying vertically?

47. (a) Show that the arc length of one petal of the rose $r = \cos n\theta$ is given by

$$2 \int_0^{\pi/(2n)} \sqrt{1 + (n^2 - 1) \sin^2 n\theta} \; d\theta$$

   (b) Use the numerical integration capability of a calculating utility to approximate the arc length of one petal of the four-petal rose $r = \cos 2\theta$.

   (c) Use the numerical integration capability of a calculating utility to approximate the arc length of one petal of the n-petal rose $r = \cos n\theta$ for $n = 2, 3, 4, \ldots, 20$; then make a conjecture about the limit of these arc lengths as $n \to +\infty$.

48. (a) Sketch the spiral $r = e^{-\theta}$ $(0 \leq \theta < +\infty)$.
  (b) Find an improper integral for the total arc length of the spiral.
  (c) Show that the integral converges and find the total arc length of the spiral.

Exercises 49–54 require the formulas developed in the following discussion: If $f'(t)$ and $g'(t)$ are continuous functions and if no segment of the curve

$$x = f(t), \quad y = g(t) \qquad (a \leq t \leq b)$$

is traced more than once, then it can be shown that the area of the surface generated by revolving this curve about the x-axis is

$$S = \int_a^b 2\pi y \sqrt{\left(\frac{dx}{dt}\right)^2 + \left(\frac{dy}{dt}\right)^2} \, dt$$

and the area of the surface generated by revolving the curve about the y-axis is

$$S = \int_a^b 2\pi x \sqrt{\left(\frac{dx}{dt}\right)^2 + \left(\frac{dy}{dt}\right)^2} \, dt$$

[The derivations are similar to those used to obtain Formulas (4) and (5) in Section 8.5.]

---

49. Find the area of the surface generated by revolving $x = t^2$, $y = 2t$ $(0 \leq t \leq 4)$ about the x-axis.

50. Find the area of the surface generated by revolving the equations $x = e^t \cos t$, $y = e^t \sin t$ $(0 \leq t \leq \pi/2)$ about the x-axis.

51. Find the area of the surface generated by revolving the equations $x = \cos^2 t$, $y = \sin^2 t$ $(0 \leq t \leq \pi/2)$ about the y-axis.

52. Find the area of the surface generated by revolving $x = t$, $y = 2t^2$ $(0 \leq t \leq 1)$ about the y-axis.

53. By revolving the semicircle

$$x = r \cos t, \quad y = r \sin t \qquad (0 \leq t \leq \pi)$$

about the x-axis, show that the surface area of a sphere of radius r is $4\pi r^2$.

54. The equations

$$x = a\phi - a \sin \phi, \quad y = a - a \cos \phi \quad (0 \leq \phi \leq 2\pi)$$

represent one arch of a cycloid. Show that the surface area generated by revolving this curve about the x-axis is given by $S = 64\pi a^2/3$.

55. As illustrated in the accompanying figure, suppose that a rod with one end fixed at the pole of a polar coordinate system rotates counterclockwise at the constant rate of 1 rad/s. At time $t = 0$ a bug on the rod is 10 mm from the pole and is moving outward along the rod at the constant speed of 2 mm/s.
  (a) Find an equation of the form $r = f(\theta)$ for the path of motion of the bug, assuming that $\theta = 0$ when $t = 0$.
  (b) Find the distance the bug travels along the path in part (a) during the first 5 seconds. Round your answer to the nearest tenth of a millimeter.

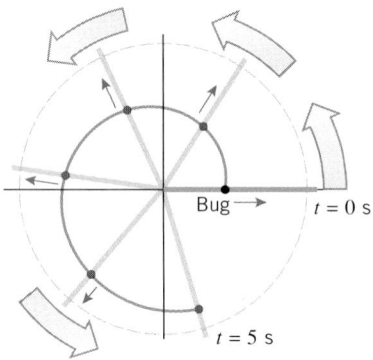

Figure Ex-55

56. Use Formula (6) of Section 8.4 to derive Formula (8).

12.3 AREA IN POLAR COORDINATES

In this section we will show how to find areas of regions that are bounded by polar curves.

AREA IN POLAR COORDINATES

12.3.1 AREA PROBLEM IN POLAR COORDINATES. Suppose that α and β are angles that satisfy the condition

$$\alpha < \beta \leq \alpha + 2\pi$$

and suppose that $f(\theta)$ is continuous for $\alpha \leq \theta \leq \beta$. Find the area of the region R enclosed by the polar curve $r = f(\theta)$ and the rays $\theta = \alpha$ and $\theta = \beta$ (Figure 12.3.1).

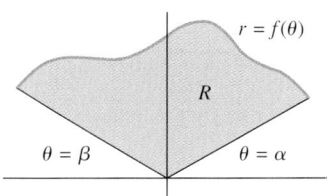

Figure 12.3.1

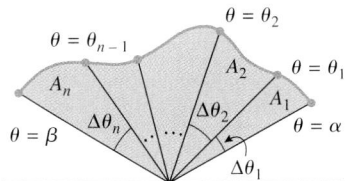

Figure 12.3.2

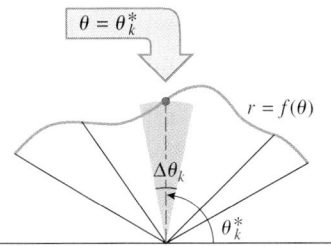

Figure 12.3.3

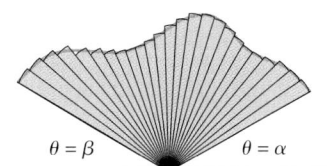

Figure 12.3.4

In rectangular coordinates we solved Area Problem 7.1.1 by dividing the region into an increasing number of vertical strips, approximating the strips by rectangles, and taking a limit. In polar coordinates rectangles are clumsy to work with, and it is better to divide the region into *wedges* by using rays

$$\theta = \theta_1, \ \theta = \theta_2, \ldots, \ \theta = \theta_{n-1}$$

such that

$$\alpha < \theta_1 < \theta_2 < \cdots < \theta_{n-1} < \beta$$

(Figure 12.3.2). As shown in that figure, the rays divide the region R into n wedges with areas $A_1, A_2, \ldots, A_n$ and central angles $\Delta\theta_1, \Delta\theta_2, \ldots, \Delta\theta_n$. The area of the entire region can be written as

$$A = A_1 + A_2 + \cdots + A_n = \sum_{k=1}^{n} A_k \tag{1}$$

If $\Delta\theta_k$ is small, and if we assume for simplicity that $f(\theta)$ is nonnegative, then we can approximate the area A_k of the kth wedge by the area of a sector with central angle $\Delta\theta_k$ and radius $f(\theta_k^*)$, where $\theta = \theta_k^*$ is any ray that lies in the kth wedge (Figure 12.3.3). Thus, from (1) and Formula (5) of Appendix E for the area of a sector, we obtain

$$A = \sum_{k=1}^{n} A_k \approx \sum_{k=1}^{n} \frac{1}{2}[f(\theta_k^*)]^2 \Delta\theta_k \tag{2}$$

If we now increase n in such a way that max $\Delta\theta_k \to 0$, then the sectors will become better and better approximations of the wedges and it is reasonable to expect that (2) will approach the exact value of the area A (Figure 12.3.4); that is,

$$A = \lim_{\max \Delta\theta_k \to 0} \sum_{k=1}^{n} \frac{1}{2}[f(\theta_k^*)]^2 \Delta\theta_k = \int_{\alpha}^{\beta} \frac{1}{2}[f(\theta)]^2 \, d\theta$$

Thus, we have the following solution of Area Problem 12.3.1.

12.3.2 AREA IN POLAR COORDINATES.   If α and β are angles that satisfy the condition

$$\alpha < \beta \leq \alpha + 2\pi$$

and if $f(\theta)$ is continuous for $\alpha \leq \theta \leq \beta$, then the area A of the region R enclosed by the polar curve $r = f(\theta)$ and the rays $\theta = \alpha$ and $\theta = \beta$ is

$$A = \int_{\alpha}^{\beta} \frac{1}{2}[f(\theta)]^2 \, d\theta = \int_{\alpha}^{\beta} \frac{1}{2}r^2 \, d\theta \tag{3}$$

The hardest part of applying (3) is determining the limits of integration. This can be done as follows:

Step 1. Sketch the region R whose area is to be determined.

Step 2. Draw an arbitrary "radial line" from the pole to the boundary curve $r = f(\theta)$.

Step 3. Ask, "Over what interval of values must θ vary in order for the radial line to sweep out the region R?"

Step 4. Your answer in Step 3 will determine the lower and upper limits of integration.

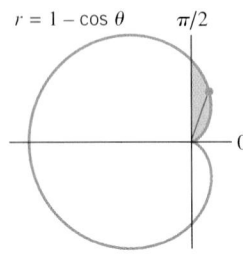

$r = 1 - \cos\theta$ $\pi/2$

The shaded region is swept
out by the radial line as θ
varies from 0 to $\pi/2$.

Figure 12.3.5

Example 1

Find the area of the region in the first quadrant within the cardioid $r = 1 - \cos\theta$.

Solution. The region and a typical radial line are shown in Figure 12.3.5. For the radial line to sweep out the region, θ must vary from 0 to $\pi/2$. Thus, from (3) with $\alpha = 0$ and $\beta = \pi/2$, we obtain

$$A = \int_0^{\pi/2} \frac{1}{2} r^2 \, d\theta = \int_0^{\pi/2} (1 - \cos\theta)^2 \, d\theta = \frac{1}{2} \int_0^{\pi/2} (1 - 2\cos\theta + \cos^2\theta) \, d\theta$$

With the help of the identity $\cos^2\theta = \frac{1}{2}(1 + \cos 2\theta)$, this can be rewritten as

$$A = \frac{1}{2} \int_0^{\pi/2} \left(\frac{3}{2} - 2\cos\theta + \frac{1}{2}\cos 2\theta \right) d\theta = \frac{1}{2} \left[\frac{3}{2}\theta - 2\sin\theta + \frac{1}{4}\sin 2\theta \right]_0^{\pi/2} = \frac{3}{8}\pi - 1 \quad \blacktriangleleft$$

Example 2

Find the entire area within the cardioid of Example 1.

Solution. For the radial line to sweep out the entire cardioid, θ must vary from 0 to 2π. Thus, from (3) with $\alpha = 0$ and $\beta = 2\pi$,

$$A = \int_0^{2\pi} \frac{1}{2} r^2 \, d\theta = \frac{1}{2} \int_0^{2\pi} (1 - \cos\theta)^2 \, d\theta$$

If we proceed as in Example 1, this reduces to

$$A = \frac{1}{2} \int_0^{2\pi} \left(\frac{3}{2} - 2\cos\theta + \frac{1}{2}\cos 2\theta \right) d\theta = \frac{3\pi}{2}$$

Alternative Solution. Since the cardioid is symmetric about the x-axis, we can calculate the portion of the area above the x-axis and double the result. In the portion of the cardioid above the x-axis, θ ranges from 0 to π, so that

$$A = 2 \int_0^{\pi} \frac{1}{2} r^2 \, d\theta = \int_0^{\pi} (1 - \cos\theta)^2 \, d\theta = \frac{3\pi}{2} \quad \blacktriangleleft$$

USING SYMMETRY

Although Formula (3) is applicable if $r = f(\theta)$ is negative, area computations can sometimes be simplified by using symmetry to restrict the limits of integration to intervals where $r \geq 0$. This is illustrated in the next example.

Example 3

Find the area of the region enclosed by the rose curve $r = \cos 2\theta$.

Solution. Referring to Figure 12.1.10 and using symmetry, the area in the first quadrant that is swept out for $0 \leq \theta \leq \pi/4$ is one-eighth of the total area inside the rose. Thus, from Formula (3)

$$A = 8 \int_0^{\pi/4} \frac{1}{2} r^2 \, d\theta = 4 \int_0^{\pi/4} \cos^2 2\theta \, d\theta$$

$$= 4 \int_0^{\pi/4} \frac{1}{2}(1 + \cos 4\theta) \, d\theta = 2 \int_0^{\pi/4} (1 + \cos 4\theta) \, d\theta$$

$$= 2\theta + \frac{1}{2}\sin 4\theta \bigg]_0^{\pi/4} = \frac{\pi}{2} \quad \blacktriangleleft$$

Sometimes the most natural way to satisfy the restriction $\alpha < \beta \leq \alpha + 2\pi$ required by Formula (3) is to use a negative value for α. For example, suppose that we are interested in finding the area of the shaded region in Figure 12.3.6a. The first step would be to determine the intersections of the cardioid $r = 4 + 4\cos\theta$ and the circle $r = 6$, since this information is needed for the limits of integration. To find the points of intersection, we can equate the two expressions for r. This yields

$$4 + 4\cos\theta = 6 \quad \text{or} \quad \cos\theta = \frac{1}{2}$$

which is satisfied by the positive angles

$$\theta = \frac{\pi}{3} \quad \text{and} \quad \theta = \frac{5\pi}{3}$$

However, there is a problem here because the radial lines to the circle and cardioid do not sweep through the shaded region shown in Figure 12.3.6 as θ varies over the interval $\pi/3 \le \theta \le 5\pi/3$. There are two ways to circumvent this problem—one is to take advantage of the symmetry by integrating over the interval $0 \le \theta \le \pi/3$ and doubling the result, and the second is to use a negative lower limit of integration and integrate over the interval $-\pi/3 \le \theta \le \pi/3$ (Figure 12.3.6c). The two methods are illustrated in the next example.

Example 4

Find the area of the region that is inside of the cardioid $r = 4 + 4\cos\theta$ and outside of the circle $r = 6$.

Solution Using a Negative Angle. The area of the region can be obtained by subtracting the areas in Figures 12.3.6d and 12.3.6e:

$$A = \int_{-\pi/3}^{\pi/3} \frac{1}{2}(4 + 4\cos\theta)^2\, d\theta - \int_{-\pi/3}^{\pi/3} \frac{1}{2}(6)^2\, d\theta \qquad \text{\small Area inside cardioid} \\ \text{\small minus area inside circle.}$$

$$= \int_{-\pi/3}^{\pi/3} \frac{1}{2}[(4 + 4\cos\theta)^2 - 36]\, d\theta = \int_{-\pi/3}^{\pi/3} (16\cos\theta + 8\cos^2\theta - 10)\, d\theta$$

$$= \Big[16\sin\theta + (4\theta + 2\sin 2\theta) - 10\theta\Big]_{-\pi/3}^{\pi/3} = 18\sqrt{3} - 4\pi$$

Solution Using Symmetry. Using symmetry, we can calculate the area above the polar axis and double it. This yields (verify)

$$A = 2\int_{0}^{\pi/3} \frac{1}{2}[(4 + 4\cos\theta)^2 - 36]\, d\theta = 2(9\sqrt{3} - 2\pi) = 18\sqrt{3} - 4\pi$$

which agrees with the preceding result. ◀

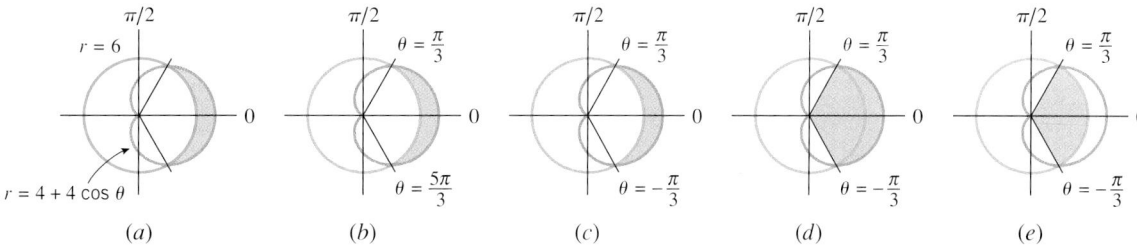

Figure 12.3.6

      (a)   (b)   (c)   (d)   (e)

INTERSECTIONS OF POLAR GRAPHS

In the last example we found the intersections of the cardioid and circle by equating their expressions for r and solving for θ. However, because a point can be represented in different ways in polar coordinates, this procedure will not always produce all of the intersections. For example, the cardioids

$$r = 1 - \cos\theta \quad \text{and} \quad r = 1 + \cos\theta \tag{4}$$

intersect at three points: the pole, the point $(1, \pi/2)$, and the point $(1, 3\pi/2)$ (Figure 12.3.7). Equating the right-hand sides of the equations in (4) yields $1 - \cos\theta = 1 + \cos\theta$

or $\cos\theta = 0$, so

$$\theta = \frac{\pi}{2} + k\pi, \quad k = 0, \pm 1, \pm 2, \ldots$$

Substituting any of these values in (4) yields $r = 1$, so that we have found only two distinct points of intersection, $(1, \pi/2)$ and $(1, 3\pi/2)$; the pole has been missed. This problem occurs because the two cardioids pass through the pole at different values of θ—the cardioid $r = 1 - \cos\theta$ passes through the pole at $\theta = 0$, and the cardioid $r = 1 + \cos\theta$ passes through the pole at $\theta = \pi$.

The situation with the cardioids is analogous to two satellites circling the Earth in intersecting orbits (Figure 12.3.8). The satellites will not collide unless they reach the same point at the same time. In general, when looking for intersections of polar curves, it is a good idea to graph the curves to determine how many intersections there should be.

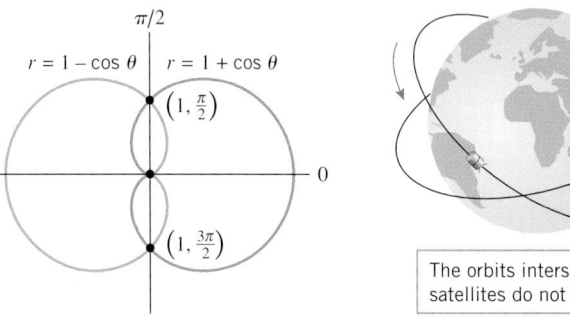

Figure 12.3.7   Figure 12.3.8

The orbits intersect, but the satellites do not collide.

EXERCISE SET 12.3 〰 Graphing Calculator [C] CAS

1. Write down, but do not evaluate, an integral for the area of each shaded region.

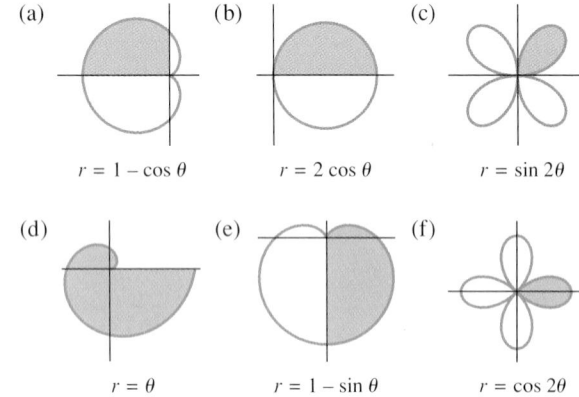

(a) $r = 1 - \cos\theta$ (b) $r = 2\cos\theta$ (c) $r = \sin 2\theta$

(d) $r = \theta$ (e) $r = 1 - \sin\theta$ (f) $r = \cos 2\theta$

2. Evaluate the integrals you obtained in Exercise 1.

3. In each part, find the area of the circle by integration.
   (a) $r = a$ (b) $r = 2a\sin\theta$ (c) $r = 2a\cos\theta$

4. (a) Show that $r = \sin\theta + \cos\theta$ is a circle.
   (b) Find the area of the circle using a geometric formula and then by integration.

In Exercises 5–10, find the area of the region described.

5. The region that is enclosed by the cardioid $r = 2 + 2\cos\theta$.

6. The region in the first quadrant within the cardioid $r = 1 + \sin\theta$.

7. The region enclosed by the rose $r = 4\cos 3\theta$.

8. The region enclosed by the rose $r = 2\sin 2\theta$.

9. The region enclosed by the inner loop of the limaçon $r = 1 + 2\cos\theta$. [*Hint:* $r \leq 0$ over the interval of integration.]

10. The region swept out by a radial line from the pole to the curve $r = 2/\theta$ as θ varies over the interval $1 \leq \theta \leq 3$.

In Exercises 11–14, find the area of the shaded region.

11. 12.

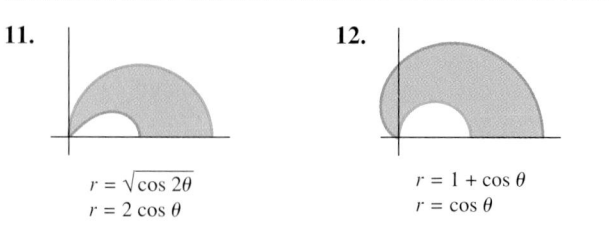

$r = \sqrt{\cos 2\theta}$ $r = 1 + \cos\theta$
$r = 2\cos\theta$ $r = \cos\theta$

13.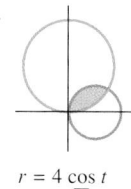

$r = 4 \cos t$
$r = 4\sqrt{3} \sin t$

14.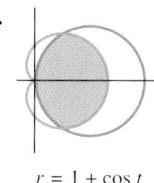

$r = 1 + \cos t$
$r = 3 \cos t$

In Exercises 15–22, find the area of the region described.

15. The region inside the circle $r = 5 \sin \theta$ and outside the limaçon $r = 2 + \sin \theta$.

16. The region outside the cardioid $r = 2 - 2 \cos \theta$ and inside the circle $r = 4$.

17. The region inside the cardioid $r = 2 + 2 \cos \theta$ and outside the circle $r = 3$.

18. The region that is common to the circles $r = 4 \cos \theta$ and $r = 4 \sin \theta$.

19. The region between the loops of the limaçon $r = \frac{1}{2} + \cos \theta$.

20. The region inside the cardioid $r = 2 + 2 \cos \theta$ and to the right of the line $r \cos \theta = \frac{3}{2}$.

21. The region inside the circle $r = 10$ and to the right of the line $r = 6 \sec \theta$.

22. The region inside the rose $r = 2a \cos 2\theta$ and outside the circle $r = a\sqrt{2}$.

23. (a) Find the error: The area that is inside the lemniscate $r^2 = a^2 \cos 2\theta$ is

$$A = \int_0^{2\pi} \frac{1}{2} r^2 \, d\theta = \int_0^{2\pi} \frac{1}{2} a^2 \cos 2\theta \, d\theta$$
$$= \frac{1}{4} a^2 \sin 2\theta \Big]_0^{2\pi} = 0$$

   (b) Find the correct area.
   (c) Find the area inside the lemniscate $r^2 = 4 \cos 2\theta$ and outside the circle $r = \sqrt{2}$.

24. Find the area inside the curve $r^2 = \sin 2\theta$.

25. A radial line is drawn from the origin to the spiral $r = a\theta$ ($a > 0$ and $\theta \geq 0$). Find the area swept out during the second revolution of the radial line that was not swept out during the first revolution.

26. (a) In the discussion associated with Exercises 49–54 of Section 12.2, formulas were given for the area of the surface of revolution that is generated by revolving a parametric curve about the x-axis or y-axis. Use those formulas to derive the following formulas for the areas of the surfaces of revolution that are generated by revolving the portion of the polar curve $r = f(\theta)$ from $\theta = \alpha$ to $\theta = \beta$ about the polar axis and about the line $\theta = \pi/2$:

$$S = \int_\alpha^\beta 2\pi r \sin \theta \sqrt{r^2 + \left(\frac{dr}{d\theta}\right)^2} \, d\theta \quad \boxed{\text{About } \theta = 0}$$

$$S = \int_\alpha^\beta 2\pi r \cos \theta \sqrt{r^2 + \left(\frac{dr}{d\theta}\right)^2} \, d\theta \quad \boxed{\text{About } \theta = \pi/2}$$

   (b) State conditions under which these formulas hold.

In Exercises 27–30, sketch the surface, and use the formulas in Exercise 26 to find the surface area.

27. The surface generated by revolving the circle $r = \cos \theta$ about the line $\theta = \pi/2$.

28. The surface generated by revolving the spiral $r = e^\theta$ ($0 \leq \theta \leq \pi/2$) about the line $\theta = \pi/2$.

29. The "apple" generated by revolving the upper half of the cardioid $r = 1 - \cos \theta$ ($0 \leq \theta \leq \pi$) about the polar axis.

30. The sphere of radius a generated by revolving the semicircle $r = a$ in the upper half-plane about the polar axis.

31. (a) Show that the Folium of Descartes $x^3 - 3xy + y^3 = 0$ can be expressed in polar coordinates as

$$r = \frac{3 \sin \theta \cos \theta}{\cos^3 \theta + \sin^3 \theta}$$

   (b) Use a CAS to show that the area inside of the loop is $\frac{3}{2}$ (Figure 4.3.2).

32. (a) What is the area that is enclosed by one petal of the rose $r = a \cos n\theta$ if n is an even integer?
   (b) What is the area that is enclosed by one petal of the rose $r = a \cos n\theta$ if n is an odd integer?
   (c) Use a CAS to show that the total area enclosed by the rose $r = a \cos n\theta$ is $\pi a^2/2$ if the number of petals is even. [*Hint:* See Exercise 73 of Section 12.1.]
   (d) Use a CAS to show that the total area enclosed by the rose $r = a \cos n\theta$ is $\pi a^2/4$ if the number of petals is odd.

33. One of the most famous problems in Greek antiquity was "squaring the circle"; that is, using a straightedge and compass to construct a square whose area is equal to that of a given circle. It was proved in the nineteenth century that no such construction is possible. However, show that the shaded areas in the accompanying figure are equal, thereby "squaring the crescent."

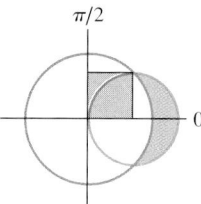

Figure Ex-33

34. Use a graphing utility to generate the polar graph of the equation $r = \cos 3\theta + 2$, and find the area that it encloses.

35. Use a graphing utility to generate the graph of the *bifolium* $r = 2 \cos \theta \sin^2 \theta$, and find the area of the upper loop.

12.4 CONIC SECTIONS IN CALCULUS

In this section we will discuss some of the basic geometric properties of parabolas, ellipses, and hyperbolas. These curves play an important role in calculus and also arise naturally in a broad range of applications in such fields as planetary motion, design of telescopes and antennas, geodetic positioning, and medicine, to name a few.

Some students may already be familiar with the material in this section, in which case it can be treated as a review. Instructors who want to spend some additional time on precalculus review may want to allocate more than one lecture on this material.

CONIC SECTIONS

Circles, ellipses, parabolas, and hyperbolas are called **conic sections** or **conics** because they can be obtained as intersections of a plane with a double-napped circular cone (Figure 12.4.1). If the plane passes through the vertex of the double-napped cone, then the intersection is a point, a pair of intersecting lines, or a single line. These are called **degenerate conic sections**.

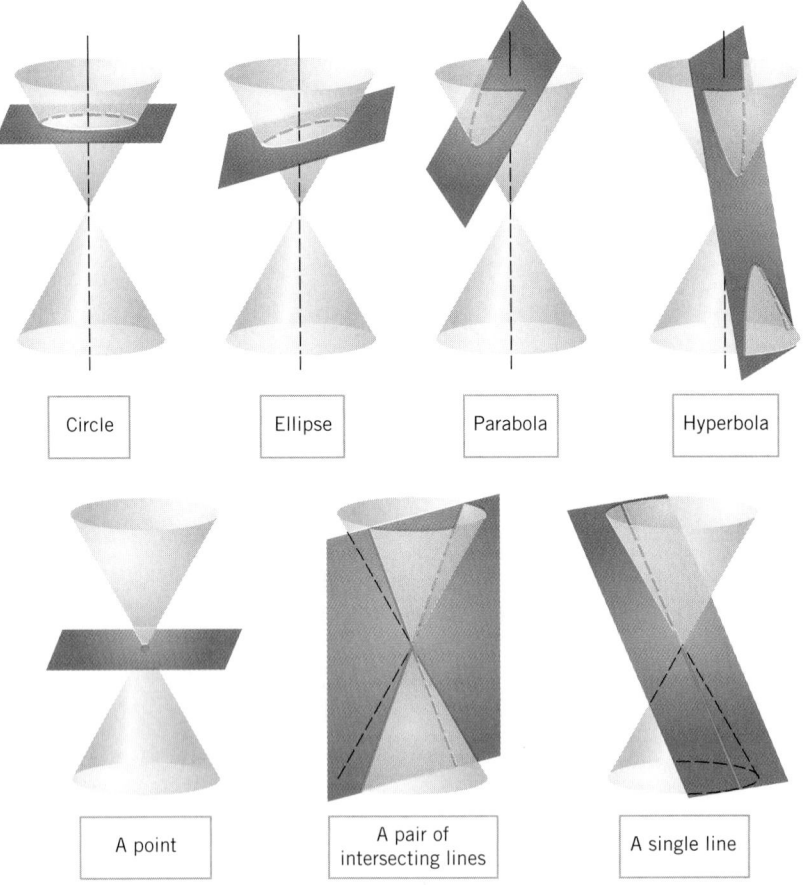

Figure 12.4.1

DEFINITIONS OF THE CONIC SECTIONS

Although we could derive properties of parabolas, ellipses, and hyperbolas by defining them as intersections with a double-napped cone, it will be better suited to calculus if we begin with equivalent definitions that are based on their geometric properties.

> **12.4.1   DEFINITION.**   A *parabola* is the set of all points in the plane that are equidistant from a fixed line and a fixed point not on the line.

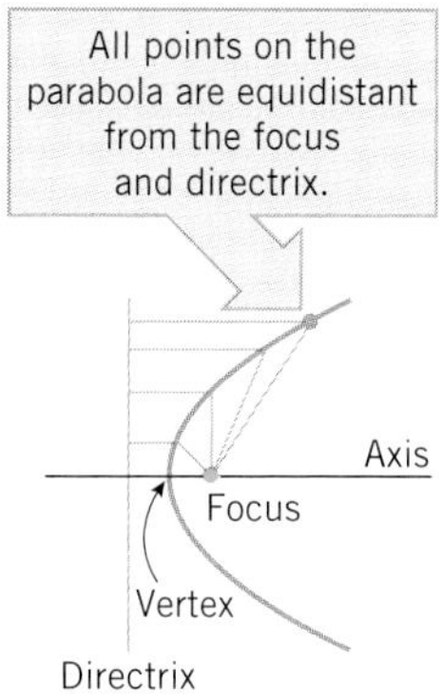

The line is called the *directrix* of the parabola, and the point is called the *focus* (Figure 12.4.2). A parabola is symmetric about the line that passes through the focus at right angles to the directrix. This line, called the *axis* or the *axis of symmetry* of the parabola, intersects the parabola at a point called the *vertex*.

> **12.4.2   DEFINITION.**   An *ellipse* is the set of all points in the plane, the sum of whose distances from two fixed points is a given positive constant that is greater than the distance between the fixed points.

The two fixed points are called the *foci* (plural of "focus") of the ellipse, and the midpoint of the line segment joining the foci is called the *center* (Figure 12.4.3a). To help visualize Definition 12.4.2, imagine that two ends of a string are tacked to the foci and a pencil traces a curve as it is held tight against the string (Figure 12.4.3b). The resulting curve will be an ellipse since the sum of the distances to the foci is a constant, namely the total length of the string. Note that if the foci coincide, the ellipse reduces to a circle. For ellipses other than circles, the line segment through the foci and across the ellipse is called the *major axis* (Figure 12.4.3c), and the line segment across the ellipse, through the center, and perpendicular to the major axis is called the *minor axis*. The endpoints of the major axis are called *vertices*.

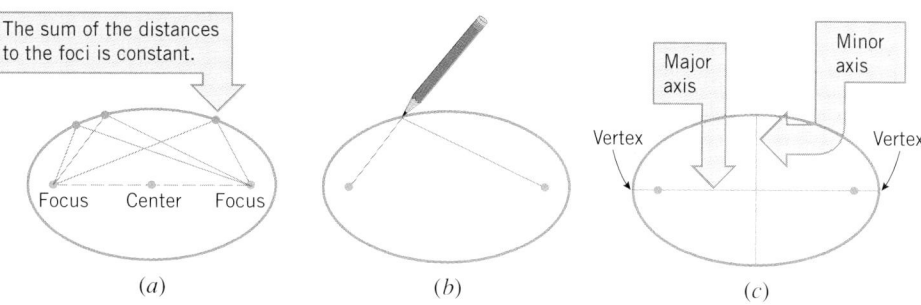

Figure 12.4.3

> **12.4.3   DEFINITION.**   A *hyperbola* is the set of all points in the plane, the difference of whose distances from two fixed distinct points is a given positive constant that is less than the distance between the fixed points.

The two fixed points are called the *foci* of the hyperbola, and the term "difference" that is used in the definition is understood to mean the distance to the farther focus minus the distance to the closer focus. As a result, the points on the hyperbola form two *branches*, each "wrapping around" the closer focus (Figure 12.4.4a). The midpoint of the line segment joining the foci is called the *center* of the hyperbola, the line through the foci is called the *focal axis*, and the line through the center that is perpendicular to the focal axis is called the *conjugate axis*. The hyperbola intersects the focal axis at two points called the *vertices*.

Associated with every hyperbola is a pair of lines, called the *asymptotes* of the hyperbola. These lines intersect at the center of the hyperbola and have the property that as a point P moves along the hyperbola away from the center, the vertical distance between P and one of the asymptotes approaches zero (Figure 12.4.4b).

All points on the parabola are equidistant from the focus and directrix.

Axis

Focus

Vertex

Directrix

Figure 12.4.2

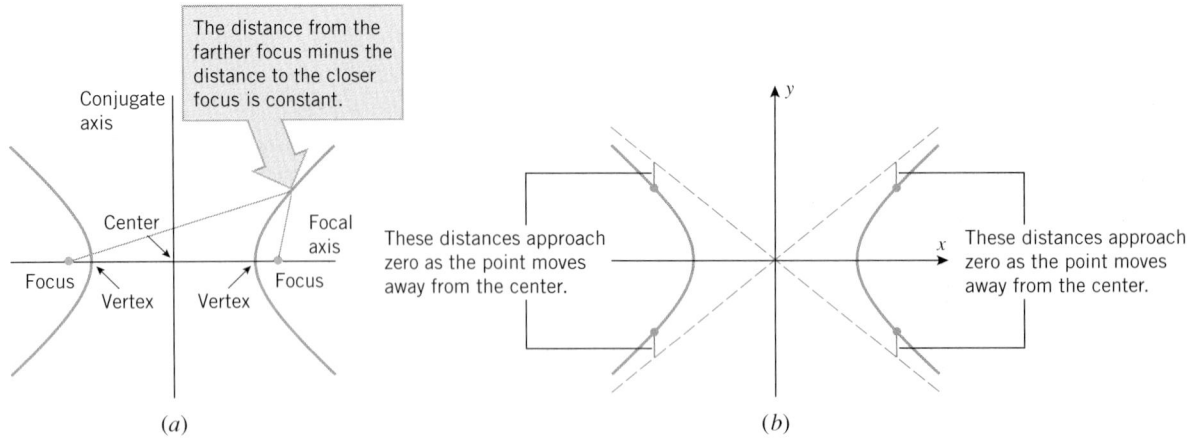

Figure 12.4.4

EQUATIONS OF PARABOLAS IN STANDARD POSITION

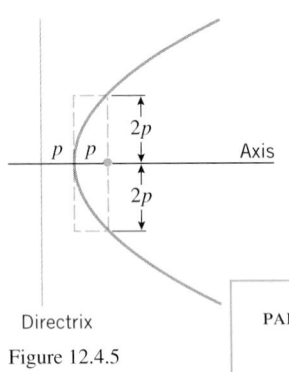

Figure 12.4.5

It is traditional in the study of parabolas to denote the distance between the focus and the vertex by p. The vertex is equidistant from the focus and the directrix, so the distance between the vertex and the directrix is also p; consequently, the distance between the focus and the directrix is $2p$ (Figure 12.4.5). As illustrated in that figure, the parabola passes through two of the corners of a box that extends from the vertex to the focus along the axis of symmetry and extends $2p$ units above and $2p$ units below the axis of symmetry.

The equation of a parabola is simplest if the vertex is the origin and the axis of symmetry is along the x-axis or y-axis. The four possible such orientations are shown in Figure 12.4.6. These are called the *standard positions* of a parabola, and the resulting equations are called the *standard equations* of a parabola.

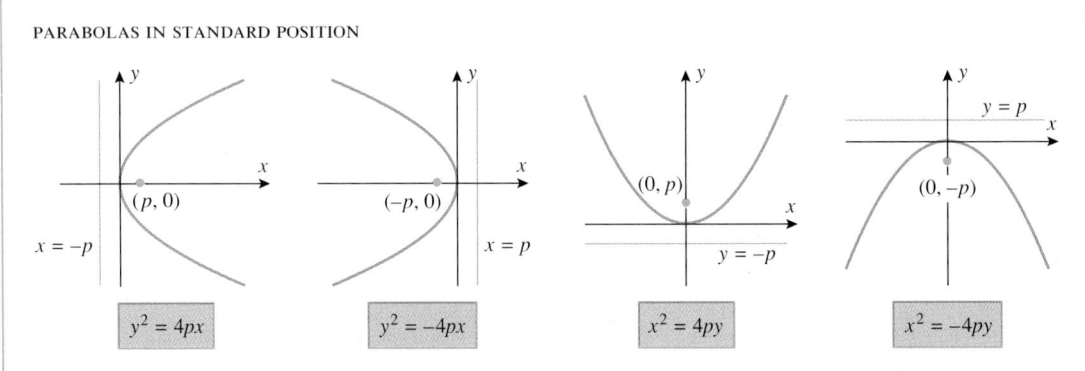

PARABOLAS IN STANDARD POSITION

Figure 12.4.6

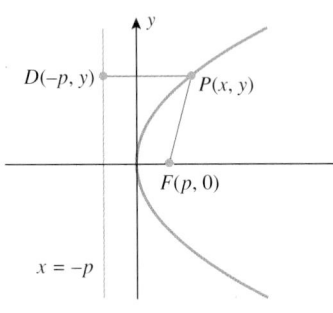

Figure 12.4.7

To illustrate how the equations in Figure 12.4.6 are obtained, we will derive the equation for the parabola with focus $(p, 0)$ and directrix $x = -p$. Let $P(x, y)$ by any point on the parabola. Since P is equidistant from the focus and directrix, the distances PF and PD in Figure 12.4.7 are equal; that is,

$$PF = PD \tag{1}$$

where $D(-p, y)$ is the foot of the perpendicular from P to the directrix. From the distance formula, the distances PF and PD are

$$PF = \sqrt{(x - p)^2 + y^2} \quad \text{and} \quad PD = \sqrt{(x + p)^2} \tag{2}$$

Substituting in (1) and squaring yields

$$(x - p)^2 + y^2 = (x + p)^2 \tag{3}$$

and after simplifying

$$y^2 = 4px \tag{4}$$

The derivations of the other equations in Figure 12.4.6 are similar.

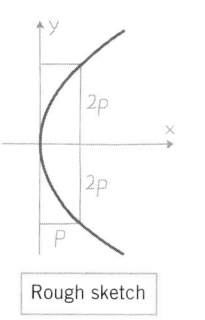

Rough sketch

Figure 12.4.8

Parabolas can be sketched from their *standard equations* using four basic steps:

- Determine whether the axis of symmetry is along the x-axis or the y-axis. Referring to Figure 12.4.6, the axis of symmetry is along the x-axis if the equation has a y^2-term, and it is along the y-axis if it has an x^2-term.
- Determine which way the parabola opens. If the axis of symmetry is along the x-axis, then the parabola opens to the right if the coefficient of x is positive, and it opens to the left if the coefficient is negative. If the axis of symmetry is along the y-axis, then the parabola opens up if the coefficient of y is positive, and it opens down if the coefficient is negative.
- Determine the value of p and draw a box extending p units from the origin along the axis of symmetry in the direction in which the parabola opens and extending $2p$ units on each side of the axis of symmetry.
- Using the box as a guide, sketch the parabola so that its vertex is at the origin and it passes through the corners of the box (Figure 12.4.8).

Example 1

Sketch the graphs of the parabolas

(a) $x^2 = 12y$ (b) $y^2 + 8x = 0$

and show the focus and directrix of each.

Solution (a). This equation involves x^2, so the axis of symmetry is along the y-axis, and the coefficient of y is positive, so the parabola opens upward. From the coefficient of y, we obtain $4p = 12$ or $p = 3$. Drawing a box extending $p = 3$ units up from the origin and $2p = 6$ units to the left and $2p = 6$ units to the right of the y-axis, then using corners of the box as a guide, yields the graph in Figure 12.4.9.

The focus is $p = 3$ units from the vertex along the axis of symmetry in the direction in which the parabola opens, so its coordinates are $(0, 3)$. The directrix is perpendicular to the axis of symmetry at a distance of $p = 3$ units from the vertex on the opposite side from the focus, so its equation is $y = -3$.

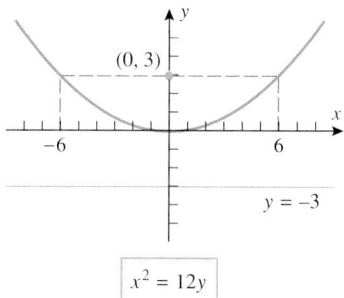

Figure 12.4.9

Solution (b). We first rewrite the equation in the standard form

$$y^2 = -8x$$

This equation involves y^2, so the axis of symmetry is along the x-axis, and the coefficient of x is negative, so the parabola opens to the left. From the coefficient of x we obtain $4p = 8$, so $p = 2$. Drawing a box extending $p = 2$ units left from the origin and $2p = 4$ units above and $2p = 4$ units below the x-axis, then using corners of the box as a guide, yields the graph in Figure 12.4.10. ◀

Example 2

Find an equation of the parabola that is symmetric about the y-axis, has its vertex at the origin, and passes through the point $(5, 2)$.

Solution. Since the parabola is symmetric about the y-axis and has its vertex at the origin, the equation is of the form

$$x^2 = 4py \quad \text{or} \quad x^2 = -4py$$

where the sign depends on whether the parabola opens up or down. But the parabola must

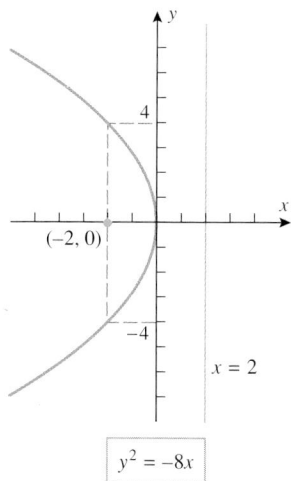

Figure 12.4.10

open up, since it passes through the point $(5, 2)$, which lies in the first quadrant. Thus, the equation is of the form

$$x^2 = 4py \tag{5}$$

Since the parabola passes through $(5, 2)$, we must have $5^2 = 4p \cdot 2$ or $4p = \frac{25}{2}$. Therefore, (5) becomes

$$x^2 = \frac{25}{2}y$$

◀

EQUATIONS OF ELLIPSES IN STANDARD POSITION

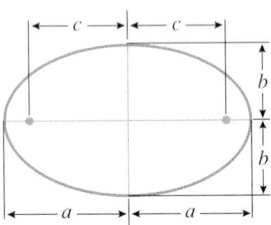

Figure 12.4.11

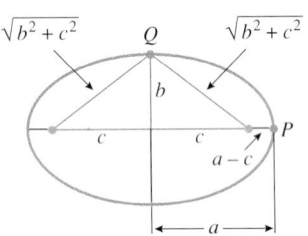

Figure 12.4.12

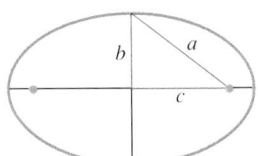

Figure 12.4.13

It is traditional in the study of ellipses to denote the length of the major axis by $2a$, the length of the minor axis by $2b$, and the distance between the foci by $2c$ (Figure 12.4.11). The number a is called the **semimajor axis** and the number b the **semiminor axis** (standard but odd terminology, since a and b are numbers, not geometric axes).

There is a basic relationship between the numbers a, b, and c that can be obtained by examining the sum of the distances to the foci from a point P at the end of the major axis and from a point Q at the end of the minor axis (Figure 12.4.12). From Definition 12.4.2, these sums must be equal, so we obtain

$$2\sqrt{b^2 + c^2} = (a - c) + (a + c)$$

from which it follows that

$$a = \sqrt{b^2 + c^2} \tag{6}$$

or, equivalently,

$$c = \sqrt{a^2 - b^2} \tag{7}$$

From (6), the distance from a focus to an end of the minor axis is a (Figure 12.4.13), which implies that for *all* points on the ellipse the sum of the distances to the foci is $2a$.

It also follows from (6) that $a \geq b$ with the equality holding only when $c = 0$. Geometrically, this means that the major axis of an ellipse is at least as large as the minor axis and that the two axes have equal length only when the foci coincide, in which case the ellipse is a circle.

The equation of an ellipse is simplest if the center of the ellipse is at the origin and the foci are on the x-axis or y-axis. The two possible such orientations are shown in Figure 12.4.14. These are called the **standard positions** of an ellipse, and the resulting equations are called the **standard equations** of an ellipse.

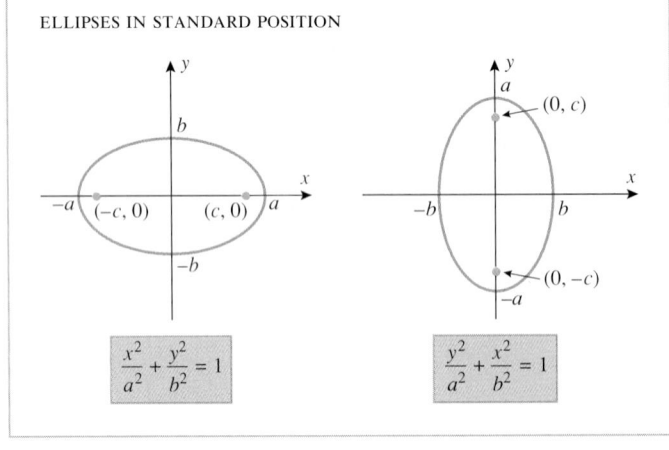

ELLIPSES IN STANDARD POSITION

$$\frac{x^2}{a^2} + \frac{y^2}{b^2} = 1$$

$$\frac{y^2}{a^2} + \frac{x^2}{b^2} = 1$$

Figure 12.4.14

To illustrate how the equations in Figure 12.4.14 are obtained, we will derive the equation for the ellipse with foci on the x-axis. Let $P(x, y)$ be any point on that ellipse. Since the sum of the distances from P to the foci is $2a$, it follows (Figure 12.4.15) that

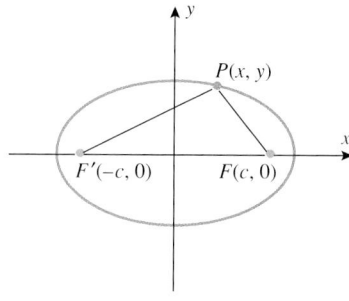

Figure 12.4.15

$$PF' + PF = 2a$$

so

$$\sqrt{(x+c)^2 + y^2} + \sqrt{(x-c)^2 + y^2} = 2a$$

Transposing the second radical to the right side of the equation and squaring yields

$$(x+c)^2 + y^2 = 4a^2 - 4a\sqrt{(x-c)^2 + y^2} + (x-c)^2 + y^2$$

and, on simplifying,

$$\sqrt{(x-c)^2 + y^2} = a - \frac{c}{a}x \qquad (8)$$

Squaring again and simplifying yields

$$\frac{x^2}{a^2} + \frac{y^2}{a^2 - c^2} = 1$$

which, by virtue of (6), can be written as

$$\frac{x^2}{a^2} + \frac{y^2}{b^2} = 1 \qquad (9)$$

Conversely, it can be shown that any point whose coordinates satisfy (9) has $2a$ as the sum of its distances from the foci, so that such a point is on the ellipse.

A TECHNIQUE FOR SKETCHING ELLIPSES

Ellipses can be sketched from their *standard equations* using three basic steps:

- Determine whether the major axis is on the x-axis or the y-axis. This can be ascertained from the sizes of the denominators in the equation. Referring to Figure 12.4.14, and keeping in mind that $a^2 > b^2$ (since $a > b$), the major axis is along the x-axis if x^2 has the larger denominator, and it is along the y-axis if y^2 has the larger denominator. If the denominators are equal, the ellipse is a circle.
- Determine the values of a and b and draw a box extending a units on each side of the center along the major axis and b units on each side of the center along the minor axis.
- Using the box as a guide, sketch the ellipse so that its center is at the origin and it touches the sides of the box where the sides intersect the coordinate axes (Figure 12.4.16).

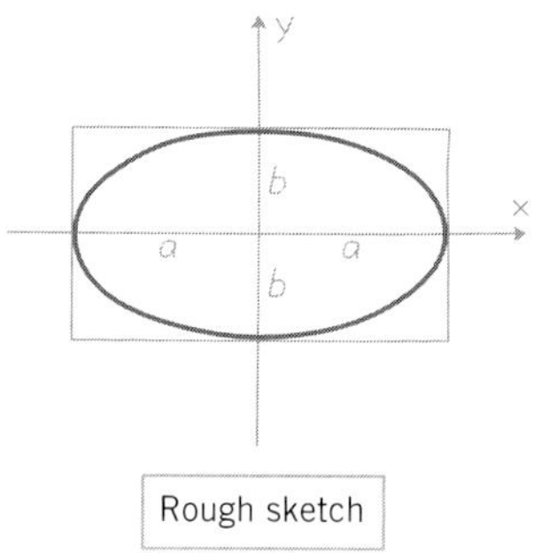

Rough sketch

Figure 12.4.16

Example 3

Sketch the graphs of the ellipses

(a) $\dfrac{x^2}{9} + \dfrac{y^2}{16} = 1$ \qquad (b) $x^2 + 2y^2 = 4$

showing the foci of each.

Solution (a). Since y^2 has the larger denominator, the major axis is along the y-axis. Moreover, since $a^2 > b^2$, we must have $a^2 = 16$ and $b^2 = 9$, so

$$a = 4 \quad \text{and} \quad b = 3$$

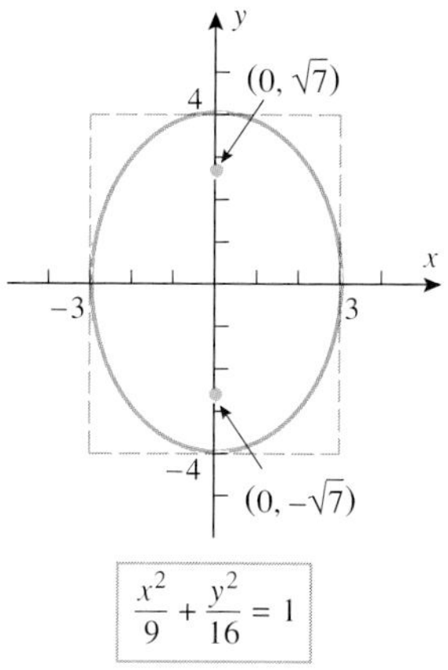

Drawing a box extending 4 units on each side of the origin along the y-axis and 3 units on each side of the origin along the x-axis as a guide yields the graph in Figure 12.4.17.

The foci lie c units on each side of the center along the major axis, where c is given by (7). From the values of a^2 and b^2 above, we obtain

$$c = \sqrt{a^2 - b^2} = \sqrt{16 - 9} = \sqrt{7} \approx 2.6$$

$$\frac{x^2}{9} + \frac{y^2}{16} = 1$$

Figure 12.4.17

Thus, the coordinates of the foci are $(0, \sqrt{7})$ and $(0, -\sqrt{7})$, since they lie on the y-axis.

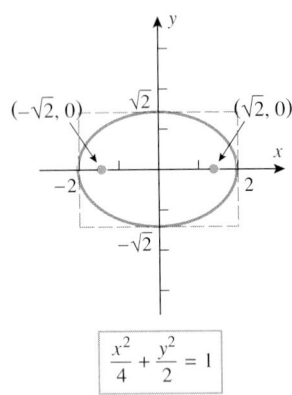

Figure 12.4.18

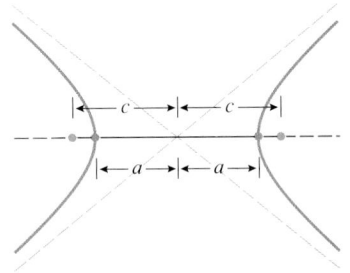

Figure 12.4.19

EQUATIONS OF HYPERBOLAS IN STANDARD POSITION

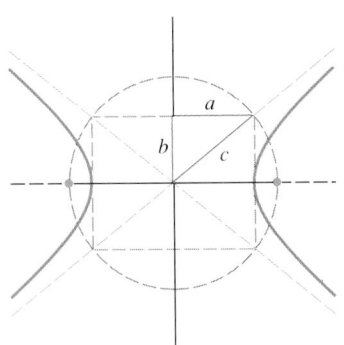

Figure 12.4.20

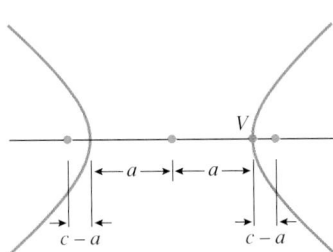

Figure 12.4.21

Solution (b). We first rewrite the equation in the standard form

$$\frac{x^2}{4} + \frac{y^2}{2} = 1$$

Since x^2 has the larger denominator, the major axis lies along the x-axis, and we have $a^2 = 4$ and $b^2 = 2$. Drawing a box extending $a = 2$ on each side of the origin along the x-axis and extending $b = \sqrt{2} \approx 1.4$ units on each side of the origin along the y-axis as a guide yields the graph in Figure 12.4.18.

From (7), we obtain

$$c = \sqrt{a^2 - b^2} = \sqrt{2} \approx 1.4$$

Thus, the coordinates of the foci are $(\sqrt{2}, 0)$ and $(-\sqrt{2}, 0)$, since they lie on the x-axis. ◄

Example 4

Find an equation for the ellipse with foci $(0, \pm 2)$ and major axis with endpoints $(0, \pm 4)$.

Solution. From Figure 12.4.14, the equation has the form

$$\frac{x^2}{b^2} + \frac{y^2}{a^2} = 1$$

and from the given information, $a = 4$ and $c = 2$. It follows from (6) that

$$b^2 = a^2 - c^2 = 16 - 4 = 12$$

so the equation of the ellipse is

$$\frac{x^2}{12} + \frac{y^2}{16} = 1$$

◄

It is traditional in the study of hyperbolas to denote the distance between the vertices by $2a$, the distance between the foci by $2c$ (Figure 12.4.19), and to define the quantity b as

$$b = \sqrt{c^2 - a^2} \tag{10}$$

This relationship, which can also be expressed as

$$c = \sqrt{a^2 + b^2} \tag{11}$$

is pictured geometrically in Figure 12.4.20. As illustrated in that figure, and as we will show later in this section, the asymptotes pass through the corners of a box extending b units on each side of the center along the conjugate axis and a units on each side of the center along the focal axis. The number a is called the *semifocal axis* of the hyperbola and the number b the *semiconjugate axis*. (As with the semimajor and semiminor axes of an ellipse, these are numbers, not geometric axes).

If V is one vertex of a hyperbola, then, as illustrated in Figure 12.4.21, the distance from V to the farther focus minus the distance from V to the closer focus is

$$[(c - a) + 2a] - (c - a) = 2a$$

Thus, for *all* points on a hyperbola, the distance to the farther focus minus the distance to the closer focus is $2a$.

The equation of a hyperbola is simplest if the center of the hyperbola is at the origin and the foci are on the x-axis or y-axis. The two possible such orientations are shown in Figure 12.4.22. These are called the *standard positions* of a hyperbola, and the resulting equations are called the *standard equations* of a hyperbola.

The derivations of these equations are similar to those already given for parabolas and ellipses, so we will leave them as exercises. However, to illustrate how the equations of the asymptotes are derived, we will derive those equations for the hyperbola

$$\frac{x^2}{a^2} - \frac{y^2}{b^2} = 1$$

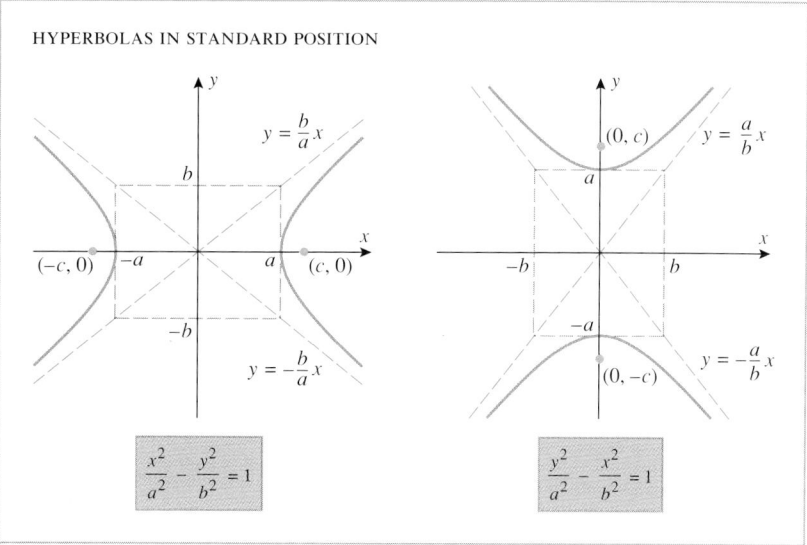

HYPERBOLAS IN STANDARD POSITION

$$\frac{x^2}{a^2} - \frac{y^2}{b^2} = 1$$

$$\frac{y^2}{a^2} - \frac{x^2}{b^2} = 1$$

Figure 12.4.22

We can rewrite this equation as

$$y^2 = \frac{b^2}{a^2}(x^2 - a^2)$$

which is equivalent to the pair of equations

$$y = \frac{b}{a}\sqrt{x^2 - a^2} \quad \text{and} \quad y = -\frac{b}{a}\sqrt{x^2 - a^2}$$

Thus, in the first quadrant, the vertical distance between the line $y = (b/a)x$ and the hyperbola can be written (Figure 12.4.23) as

$$\frac{b}{a}x - \frac{b}{a}\sqrt{x^2 - a^2}$$

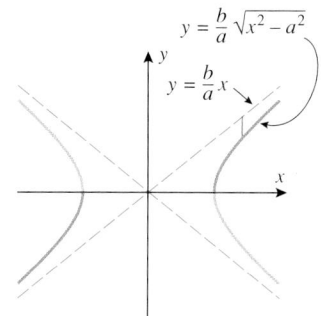

Figure 12.4.23

But this distance tends to zero as $x \to +\infty$ since

$$\lim_{x \to +\infty}\left(\frac{b}{a}x - \frac{b}{a}\sqrt{x^2 - a^2}\right) = \lim_{x \to +\infty}\frac{b}{a}(x - \sqrt{x^2 - a^2})$$

$$= \lim_{x \to +\infty}\frac{b}{a}\frac{(x - \sqrt{x^2 - a^2})(x + \sqrt{x^2 - a^2})}{x + \sqrt{x^2 - a^2}}$$

$$= \lim_{x \to +\infty}\frac{ab}{x + \sqrt{x^2 - a^2}} = 0$$

The analysis in the remaining quadrants is similar.

A QUICK WAY TO FIND ASYMPTOTES

There is a trick that can be used to avoid memorizing the equations of the asymptotes of a hyperbola. They can be obtained, when needed, by substituting 0 for the 1 on the right side of the hyperbola equation, and then solving for y in terms of x. For example, for the hyperbola

$$\frac{x^2}{a^2} - \frac{y^2}{b^2} = 1$$

we would write

$$\frac{x^2}{a^2} - \frac{y^2}{b^2} = 0 \quad \text{or} \quad y^2 = \frac{b^2}{a^2}x^2 \quad \text{or} \quad y = \pm\frac{b}{a}x$$

which are the equations for the asymptotes.

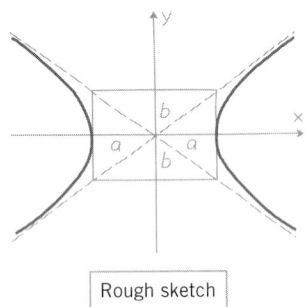

Rough sketch

Figure 12.4.24

Hyperbolas can be sketched from their *standard equations* using four basic steps:

- Determine whether the focal axis is on the x-axis or the y-axis. This can be ascertained from the location of the minus sign in the equation. Referring to Figure 12.4.22, the focal axis is along the x-axis when the minus sign precedes the y^2-term, and it is along the y-axis when the minus sign precedes the x^2-term.

- Determine the values of a and b and draw a box extending a units on either side of the center along the focal axis and b units on either side of the center along the conjugate axis. (The squares of a and b can be read directly from the equation.)

- Draw the asymptotes along the diagonals of the box.

- Using the box and the asymptotes as a guide, sketch the graph of the hyperbola (Figure 12.4.24).

Example 5

Sketch the graphs of the hyperbolas

$$\text{(a)} \quad \frac{x^2}{4} - \frac{y^2}{9} = 1 \qquad \text{(b)} \quad y^2 - x^2 = 1$$

showing their vertices, foci, and asymptotes.

Solution (a). The minus sign precedes the y^2-term, so the focal axis is along the x-axis. From the denominators in the equation we obtain

$$a^2 = 4 \quad \text{and} \quad b^2 = 9$$

Since a and b are positive, we must have $a = 2$ and $b = 3$. Recalling that the vertices lie a units on each side of the center on the focal axis, it follows that their coordinates in this case are $(2, 0)$ and $(-2, 0)$. Drawing a box extending $a = 2$ units along the x-axis on each side of the origin and $b = 3$ units on each side of the origin along the y-axis, then drawing the asymptotes along the diagonals of the box as a guide, yields the graph in Figure 12.4.25.

To obtain equations for the asymptotes, we substitute 0 for 1 in the given equation; this yields

$$\frac{x^2}{4} - \frac{y^2}{9} = 0 \quad \text{or} \quad y = \pm\frac{3}{2}x$$

The foci lie c units on each side of the center along the focal axis, where c is given by (11). From the values of a^2 and b^2 above we obtain

$$c = \sqrt{a^2 + b^2} = \sqrt{4 + 9} = \sqrt{13} \approx 3.6$$

Since the foci lie on the x-axis in this case, their coordinates are $(\sqrt{13}, 0)$ and $(-\sqrt{13}, 0)$.

Solution (b). The minus sign precedes the x^2-term, so the focal axis is along the y-axis. From the denominators in the equation we obtain $a^2 = 1$ and $b^2 = 1$, from which it follows that

$$a = 1 \quad \text{and} \quad b = 1$$

Thus, the vertices are at $(0, -1)$ and $(0, 1)$. Drawing a box extending $a = 1$ unit on either side of the origin along the y-axis and $b = 1$ unit on either side of the origin along the x-axis, then drawing the asymptotes, yields the graph in Figure 12.4.26. Since the box is actually a square, the asymptotes are perpendicular and have equations $y = \pm x$. This can also be seen by substituting 0 for 1 in the given equation, which yields $y^2 - x^2 = 0$ or $y = \pm x$. Also,

$$c = \sqrt{a^2 + b^2} = \sqrt{1 + 1} = \sqrt{2}$$

so the foci, which lie on the y-axis, are $(0, -\sqrt{2})$ and $(0, \sqrt{2})$. ◀

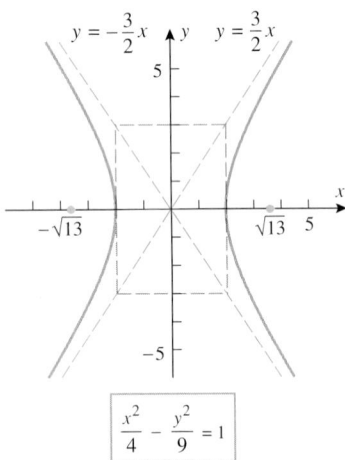

Figure 12.4.25

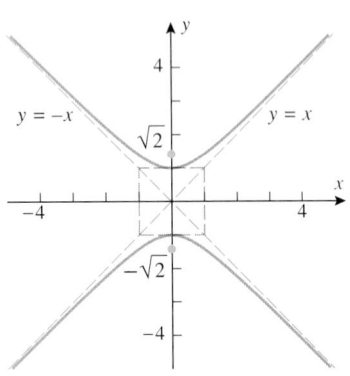

Figure 12.4.26

REMARK.   A hyperbola in which $a = b$, as in part (b) of this example, is called an ***equilateral hyperbola***. Such hyperbolas always have perpendicular asymptotes.

Example 6

Find the equation of the hyperbola with vertices $(0, \pm 8)$ and asymptotes $y = \pm \frac{4}{3} x$.

Solution.   Since the vertices are on the y-axis, the equation of the hyperbolas has the form $(y^2/a^2) - (x^2/b^2) = 1$ and the asymptotes are

$$y = \pm \frac{a}{b} x$$

From the location of the vertices we have $a = 8$, so the given equations of the asymptotes yield

$$y = \pm \frac{a}{b} x = \pm \frac{8}{b} x = \pm \frac{4}{3} x$$

from which it follows that $b = 6$. Thus, the hyperbola has the equation

$$\frac{y^2}{64} - \frac{x^2}{36} = 1 \qquad \blacktriangleleft$$

TRANSLATED CONICS

Equations of conics that are translated from their standard positions can be obtained by replacing x by $x - h$ and y by $y - k$ in their standard equations. For a parabola, this translates the vertex from the origin to the point (h, k); and for ellipses and hyperbolas, this translates the center from the origin to the point (h, k).

Parabolas with vertex (h, k) and axis parallel to x-axis

$$(y - k)^2 = 4p(x - h) \qquad \text{[Opens right]} \tag{12}$$

$$(y - k)^2 = -4p(x - h) \qquad \text{[Opens left]} \tag{13}$$

Parabolas with vertex (h, k) and axis parallel to y-axis

$$(x - h)^2 = 4p(y - k) \qquad \text{[Opens up]} \tag{14}$$

$$(x - h)^2 = -4p(y - k) \qquad \text{[Opens down]} \tag{15}$$

Ellipse with center (h, k) and major axis parallel to x-axis

$$\frac{(x - h)^2}{a^2} + \frac{(y - k)^2}{b^2} = 1 \quad [b \leq a] \tag{16}$$

Ellipse with center (h, k) and major axis parallel to y-axis

$$\frac{(x - h)^2}{b^2} + \frac{(y - k)^2}{a^2} = 1 \quad [b \leq a] \tag{17}$$

Hyperbola with center (h, k) and focal axis parallel to x-axis

$$\frac{(x - h)^2}{a^2} - \frac{(y - k)^2}{b^2} = 1 \tag{18}$$

Hyperbola with center (h, k) and focal axis parallel to y-axis

$$\frac{(y - k)^2}{a^2} - \frac{(x - h)^2}{b^2} = 1 \tag{19}$$

Example 7

Find an equation for the parabola that has its vertex at $(1, 2)$ and its focus at $(4, 2)$.

Solution. Since the focus and vertex are on a horizontal line, and since the focus is to the right of the vertex, the parabola opens to the right and its equation has the form

$$(y - k)^2 = 4p(x - h)$$

Since the vertex and focus are 3 units apart, we have $p = 3$, and since the vertex is at $(h, k) = (1, 2)$, we obtain

$$(y - 2)^2 = 12(x - 1) \qquad \blacktriangleleft$$

Sometimes the equations of translated conics occur in expanded form, in which case we are faced with the problem of identifying the graph of a *quadratic equation in x and y*:

$$Ax^2 + Cy^2 + Dx + Ey + F = 0 \qquad (20)$$

The basic procedure for determining the nature of such a graph is to complete the squares of the quadratic terms and then try to match up the resulting equation with one for the forms of a translated conic.

Example 8

Describe the graph of the equation

$$y^2 - 8x - 6y - 23 = 0$$

Solution. The equation involves quadratic terms in y but none in x, so we first take all of the y-terms to one side:

$$y^2 - 6y = 8x + 23$$

Next, we complete the square on the y-terms by adding 9 to both sides:

$$(y - 3)^2 = 8x + 32$$

Finally, we factor out the coefficient of the x-term to obtain

$$(y - 3)^2 = 8(x + 4)$$

This equation is of form (12) with $h = -4$, $k = 3$, and $p = 2$, so the graph is a parabola with vertex $(-4, 3)$ opening to the right. Since $p = 2$, the focus is 2 units to the right of the vertex, which places it at the point $(-2, 3)$; and the directrix is 2 units to the left of the vertex, which means that its equation is $x = -6$. The parabola is shown in Figure 12.4.27. $\blacktriangleleft$

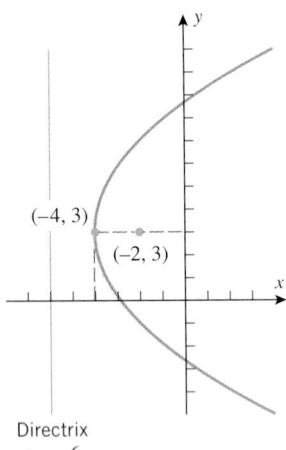

Directrix
$x = -6$

Figure 12.4.27

Example 9

Describe the graph of the equation

$$16x^2 + 9y^2 - 64x - 54y + 1 = 0$$

Solution. This equation involves quadratic terms in both x and y, so we will group the x-terms and the y-terms on one side and put the constant on the other:

$$(16x^2 - 64x) + (9y^2 - 54y) = -1$$

Next, factor out the coefficients of x^2 and y^2 and complete the squares:

$$16(x^2 - 4x + 4) + 9(y^2 - 6y + 9) = -1 + 64 + 81$$

or

$$16(x - 2)^2 + 9(y - 3)^2 = 144$$

Finally, divide through by 144 to introduce a 1 on the right side:

$$\frac{(x - 2)^2}{9} + \frac{(y - 3)^2}{16} = 1$$

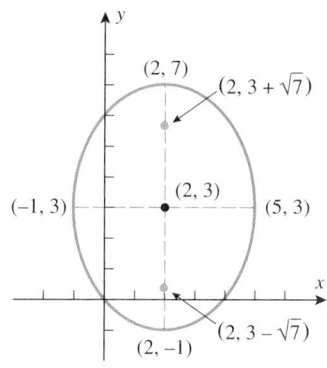

Figure 12.4.28

This is an equation of form (17), with $h = 2$, $k = 3$, $a^2 = 16$, and $b^2 = 9$. Thus, the equation is an ellipse with center $(2, 3)$ and major axis parallel to the y-axis. Since $a = 4$, the major axis extends 4 units above and 4 units below the center, so its endpoints are $(2, 7)$ and $(2, -1)$ (Figure 12.4.28). Since $b = 3$, the minor axis extends 3 units to the left and 3 units to the right of the center, so its endpoints are $(-1, 3)$ and $(5, 3)$. Since

$$c = \sqrt{a^2 - b^2} = \sqrt{16 - 9} = \sqrt{7}$$

the foci lie $\sqrt{7}$ units above and below the center, placing them at the points $(2, 3 + \sqrt{7})$ and $(2, 3 - \sqrt{7})$. ◀

Example 10

Describe the graph of the equation

$$x^2 - y^2 - 4x + 8y - 21 = 0$$

Solution. This equation involves quadratic terms in both x and y, so we will group the x-terms and the y-terms on one side and put the constant on the other:

$$(x^2 - 4x) - (y^2 - 8y) = 21$$

We leave it for you to verify by completing the squares that this equation can be written as

$$\frac{(x-2)^2}{9} - \frac{(y-4)^2}{9} = 1 \tag{21}$$

This is an equation of form (18) with $h = 2$, $k = 4$, $a^2 = 9$, and $b^2 = 9$. Thus, the equation represents a hyperbola with center $(2, 4)$ and focal axis parallel to the x-axis. Since $a = 3$, the vertices are located 3 units to the left and 3 units to the right of the center, or at the points $(-1, 4)$ and $(5, 4)$. From (11), $c = \sqrt{a^2 + b^2} = \sqrt{9 + 9} = 3\sqrt{2}$, so the foci are located $3\sqrt{2}$ units to the left and right of the center, or at the points $(2 - 3\sqrt{2}, 4)$ and $(2 + 3\sqrt{2}, 4)$.

The equations of the asymptotes may be found using the trick of substituting 0 for 1 in (21) to obtain

$$\frac{(x-2)^2}{9} - \frac{(y-4)^2}{9} = 0$$

This can be written as $y - 4 = \pm(x - 2)$, which yields the asymptotes

$$y = x + 2 \quad \text{and} \quad y = -x + 6$$

With the aid of a box extending $a = 3$ units left and right of the center and $b = 3$ units above and below the center, we obtain the sketch in Figure 12.4.29. ◀

Figure 12.4.29

ROTATED CONICS

An equation of the form

$$Ax^2 + Bxy + Cy^2 + Dx + Ey + F = 0 \tag{22}$$

is called a *second-degree equation in x and y.* The term Bxy in this equation is called the *cross-product term*. If the cross-product term is absent from the equation ($B = 0$), then the equation reduces to (20), in which case the graph is a conic section (possibly degenerate) that is either in standard position or translated from its standard position. It can be proved that if the cross-product term is present ($B \neq 0$), then the graph is a conic (possibly degenerate) that is *rotated* from its standard orientation. A discussion of rotated conics can be found in the *Student Resources.*

REFLECTION PROPERTIES OF THE CONIC SECTIONS

Parabolas, ellipses, and hyperbolas have certain reflection properties that make them extremely valuable in various applications. In the exercises we will ask you to prove the following results.

12.4.4 THEOREM (*Reflection Property of Parabolas*). *The tangent line at a point P on a parabola makes equal angles with the line through P parallel to the axis of symmetry and the line through P and the focus* (Figure 12.4.30a).

12.4.5 THEOREM (*Reflection Property of Ellipses*). *A line tangent to an ellipse at a point P makes equal angles with the lines joining P to the foci* (Figure 12.4.30*b*).

12.4.6 THEOREM (*Reflection Property of Hyperbolas*). *A line tangent to a hyperbola at a point P makes equal angles with the lines joining P to the foci* (Figure 12.4.30*c*).

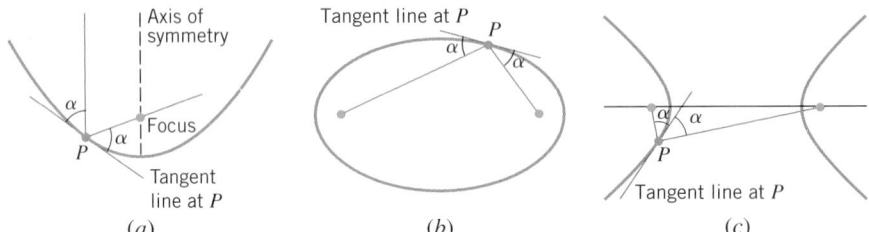

Figure 12.4.30

(*a*)   (*b*)   (*c*)

APPLICATIONS OF THE CONIC SECTIONS

Incoming signals are reflected by the parabolic antenna to the receiver at the focus.

It is a principle of physics that when light is reflected from a point P on a surface the angle between the incoming ray and the tangent line at P is equal to the angle between the outgoing ray and the tangent line at P. Therefore, if a reflecting surface has parabolic cross sections with a common focus and axis, then it follows from Theorem 12.4.4 that all light rays entering parallel to the axis will be reflected to the focus (Figure 12.4.31*a*); conversely, if a light source is located at the focus, then the reflected rays will all be parallel to the axis (Figure 12.4.31*b*). This principle is used in certain telescopes to reflect the approximately parallel rays of light from the stars and planets off of a parabolic mirror to an eyepiece at the focus; and the parabolic reflectors in flashlights and automobile headlights utilize this principle to form a parallel beam of light rays from a bulb placed at the focus. The same optical principles apply to radar signals and sound waves, which explains the parabolic shape of many antennas.

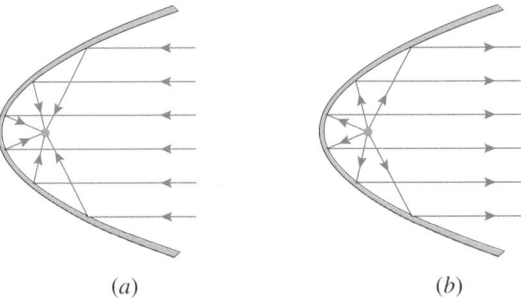

Figure 12.4.31

(*a*)   (*b*)

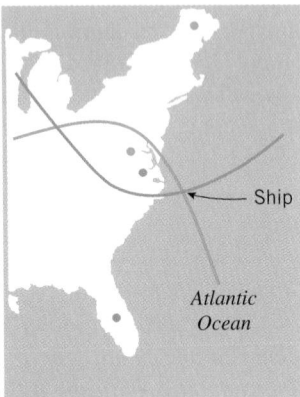

Figure 12.4.32

Visitors to various rooms in the United States Capitol Building and in St. Paul's Cathedral in Rome are often astonished by the "whispering gallery" effect in which two people at opposite ends of the room can hear one another's whispers very clearly. Such rooms have ceilings with elliptical cross sections and common foci. Thus, when the two people stand at the foci, their whispers are reflected directly to one another off of the elliptical ceiling.

Hyperbolic navigation systems, which were developed in World War II as navigational aids to ships, are based on the definition of a hyperbola. With these systems the ship receives synchronized radio signals from two widely spaced transmitters with known positions. The ship's electronic receiver measures the difference in reception times between the signals and then uses that difference to compute the difference $2a$ in its distance between the two transmitters. This information places the ship somewhere on the hyperbola whose foci are at the transmitters and whose points have $2a$ as the difference in their distances from the foci. By repeating the process with a second set of transmitters, the position of the ship can be determined as the intersection of two hyperbolas (Figure 12.4.32).

EXERCISE SET 12.4 ~ Graphing Calculator   [c] CAS

....

1. In each part, find the equation of the conic.

(a)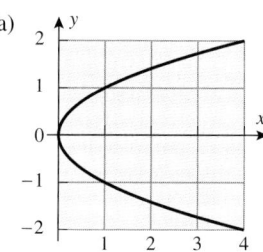

(b)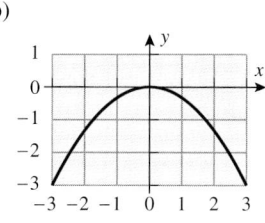

(c)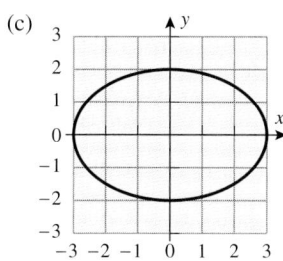

(d)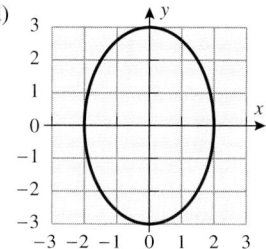

(e)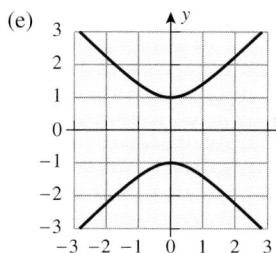

(f)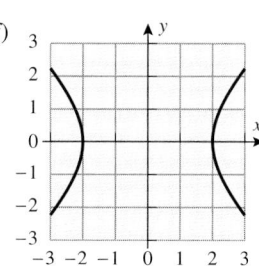

2. (a) Find the focus and directrix of the parabolas that are given in Exercise 1.

(b) Find the foci of the ellipses in Exercise 1.

(c) Find the foci and the equations of the asymptotes of the hyperbolas in Exercise 1.

In Exercises 3–8, sketch the parabola, and label the focus, vertex, and directrix.

3. (a) $y^2 = 6x$ (b) $x^2 = -9y$

4. (a) $y^2 = -10x$ (b) $x^2 = 4y$

5. (a) $(y-3)^2 = 6(x-2)$ (b) $(x+2)^2 = -(y+2)$

6. (a) $(y+1)^2 = -7(x-4)$ (b) $\left(x-\frac{1}{2}\right)^2 = 2(y-1)$

7. (a) $x^2 - 4x + 2y = 1$ (b) $x = y^2 - 4y + 2$

8. (a) $y^2 - 6y - 2x + 1 = 0$ (b) $y = 4x^2 + 8x + 5$

In Exercises 9–14, sketch the ellipse, and label the foci, the vertices, and the ends of the minor axis.

9. (a) $\dfrac{x^2}{16} + \dfrac{y^2}{9} = 1$ (b) $9x^2 + y^2 = 9$

10. (a) $\dfrac{x^2}{4} + \dfrac{y^2}{25} = 1$ (b) $4x^2 + 9y^2 = 36$

11. (a) $9(x-1)^2 + 16(y-3)^2 = 144$

(b) $3(x+2)^2 + 4(y+1)^2 = 12$

12. (a) $(x+3)^2 + 4(y-5)^2 = 16$

(b) $\frac{1}{4}x^2 + \frac{1}{9}(y+2)^2 - 1 = 0$

13. (a) $x^2 + 9y^2 + 2x - 18y + 1 = 0$

(b) $4x^2 + y^2 + 8x - 10y = -13$

14. (a) $9x^2 + 4y^2 + 18x - 24y + 9 = 0$

(b) $5x^2 + 9y^2 - 20x + 54y = -56$

In Exercises 15–20, sketch the hyperbola, and label the vertices, foci, and asymptotes.

15. (a) $\dfrac{x^2}{16} - \dfrac{y^2}{4} = 1$ (b) $9y^2 - 4x^2 = 36$

16. (a) $\dfrac{y^2}{9} - \dfrac{x^2}{25} = 1$ (b) $16x^2 - 25y^2 = 400$

17. (a) $\dfrac{(x-2)^2}{9} - \dfrac{(y-4)^2}{4} = 1$

(b) $(y+3)^2 - 9(x+2)^2 = 36$

18. (a) $\dfrac{(y+4)^2}{3} - \dfrac{(x-2)^2}{5} = 1$

(b) $16(x+1)^2 - 8(y-3)^2 = 16$

19. (a) $x^2 - 4y^2 + 2x + 8y - 7 = 0$

(b) $16x^2 - y^2 - 32x - 6y = 57$

20. (a) $4x^2 - 9y^2 + 16x + 54y - 29 = 0$

(b) $4y^2 - x^2 + 40y - 4x = -60$

In Exercises 21–26, find an equation for the parabola that satisfies the given conditions.

21. (a) Vertex $(0, 0)$; focus $(3, 0)$.

(b) Vertex $(0, 0)$; directrix $x = 7$.

22. (a) Vertex $(0, 0)$; focus $(0, -4)$.

(b) Vertex $(0, 0)$; directrix $y = \frac{1}{2}$.

23. (a) Focus $(0, -3)$; directrix $y = 3$.

(b) Vertex $(1, 1)$; directrix $y = -2$.

24. (a) Focus $(6, 0)$; directrix $x = -6$.

(b) Focus $(-1, 4)$; directrix $x = 5$.

25. Axis $y = 0$; passes through $(3, 2)$ and $(2, -3)$.

26. Vertex $(5, -3)$; axis parallel to the y-axis; passes through $(9, 5)$.

In Exercises 27–32, find an equation for the ellipse that satisfies the given conditions.

27. (a) Ends of major axis $(\pm 3, 0)$; ends of minor axis $(0, \pm 2)$.

(b) Length of major axis 26; foci $(\pm 5, 0)$.

28. (a) Ends of major axis $(0, \pm\sqrt{5})$; ends of minor axis $(\pm 1, 0)$.

(b) Length of minor axis 16; foci $(0, \pm 6)$.

29. (a) Foci $(\pm 1, 0)$; $b = \sqrt{2}$.

(b) $c = 2\sqrt{3}$; $a = 4$; center at the origin; foci on a coordinate axis (two answers).

30. (a) Foci $(\pm 3, 0)$; $a = 4$.

(b) $b = 3$; $c = 4$; center at the origin; foci on a coordinate axis (two answers).

31. (a) Ends of major axis $(\pm 6, 0)$; passes through $(2, 3)$.

(b) Foci $(1, 2)$ and $(1, 4)$; minor axis of length 2.

32. (a) Center at $(0, 0)$; major and minor axes along the coordinate axes; passes through $(3, 2)$ and $(1, 6)$.

(b) Foci $(2, 1)$ and $(2, -3)$; major axis of length 6.

In Exercises 33–38, find an equation for a hyperbola that satisfies the given conditions. (In some cases there may be more than one hyperbola.)

33. (a) Vertices $(\pm 2, 0)$; foci $(\pm 3, 0)$.

(b) Vertices $(\pm 1, 0)$; asymptotes $y = \pm 2x$.

34. (a) Vertices $(0, \pm 3)$; foci $(0, \pm 5)$.

(b) Vertices $(0, \pm 3)$; asymptotes $y = \pm x$.

35. (a) Asymptotes $y = \pm\frac{3}{2}x$; $b = 4$.

(b) Foci $(0, \pm 5)$; asymptotes $y = \pm 2x$.

36. (a) Asymptotes $y = \pm\frac{3}{4}x$; $c = 5$.

(b) Foci $(\pm 3, 0)$; asymptotes $y = \pm 2x$.

37. (a) Vertices $(2, 4)$ and $(10, 4)$; foci 10 units apart.

(b) Asymptotes $y = 2x + 1$ and $y = -2x + 3$; passes through the origin.

38. (a) Foci $(1, 8)$ and $(1, -12)$; vertices 4 units apart.

(b) Vertices $(-3, -1)$ and $(5, -1)$; $b = 4$.

39. (a) As illustrated in the accompanying figure, a parabolic arch spans a road 40 feet wide. How high is the arch if a center section of the road 20 feet wide has a minimum clearance of 12 feet?

(b) How high would the center be if the arch were the upper half of an ellipse?

40. (a) Find an equation for the parabolic arch with base b and height h, shown in the accompanying figure.

(b) Find the area under the arch.

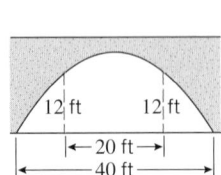

Figure Ex-39

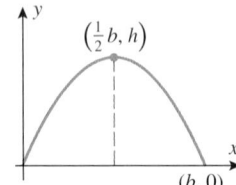

Figure Ex-40

41. Show that the vertex is the closest point on a parabola to the focus. [*Suggestion:* Introduce a convenient coordinate system and use Definition 12.4.1.]

42. As illustrated in the accompanying figure, suppose that a comet moves in a parabolic orbit with the Sun at its focus and that the line from the Sun to the comet makes an angle of $60°$ with the axis of the parabola when the comet is 40 million miles from the center of the Sun. Use the result in Exercise 41 to determine how close the comet will come to the center of the Sun.

43. For the parabolic reflector in the accompanying figure, how far from the vertex should the light source be placed to produce a beam of parallel rays?

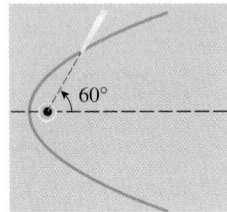

Figure Ex-42

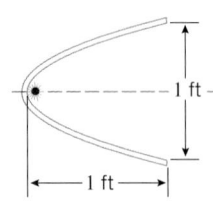

Figure Ex-43

44. In each part, find the shaded area in the figure.

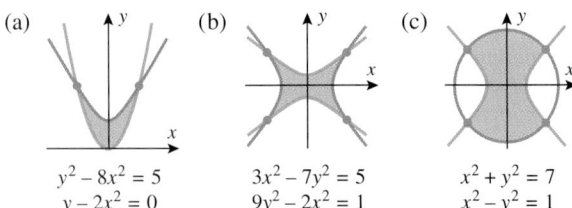

(a) $y^2 - 8x^2 = 5$
$y - 2x^2 = 0$

(b) $3x^2 - 7y^2 = 5$
$9y^2 - 2x^2 = 1$

(c) $x^2 + y^2 = 7$
$x^2 - y^2 = 1$

45. (a) The accompanying figure shows an ellipse with semimajor axis a and semiminor axis b. Express the coordinates of the points P, Q, and R in terms of t.

(b) How does the geometric interpretation of the parameter t differ between a circle

$$x = a\cos t, \quad y = a\sin t$$

and an ellipse

$$x = a\cos t, \quad y = b\sin t?$$

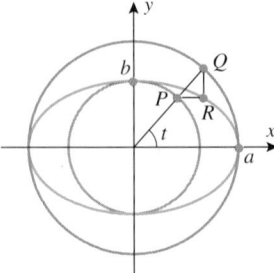

Figure Ex-45

46. (a) Show that the right and left branches of the hyperbola

$$\frac{x^2}{a^2} - \frac{y^2}{b^2} = 1$$

can be represented parametrically as

$$x = a \cosh t, \quad y = b \sinh t \quad (-\infty < t < +\infty)$$
$$x = -a \cosh t, \quad y = b \sinh t \quad (-\infty < t < +\infty)$$

(b) Use a graphing utility to generate both branches of the hyperbola $x^2 - y^2 = 1$ on the same screen.

47. (a) Show that the right and left branches of the hyperbola

$$\frac{x^2}{a^2} - \frac{y^2}{b^2} = 1$$

can be represented parametrically as

$$x = a \sec t, \quad y = b \tan t \quad (-\pi/2 < t < \pi/2)$$
$$x = -a \sec t, \quad y = b \tan t \quad (-\pi/2 < t < \pi/2)$$

(b) Use a graphing utility to generate both branches of the hyperbola $x^2 - y^2 = 1$ on the same screen.

48. Find an equation of the parabola traced by a point that moves so that its distance from $(-1, 4)$ is the same as its distance to $y = 1$.

49. Find an equation of the ellipse traced by a point that moves so that the sum of its distances to $(4, 1)$ and $(4, 5)$ is 12.

50. Find the equation of the hyperbola traced by a point that moves so that the difference between its distances to $(0, 0)$ and $(1, 1)$ is 1.

51. Suppose that the base of a solid is elliptical with a major axis of length 9 and a minor axis of length 4. Find the volume of the solid if the cross sections perpendicular to the major axis are squares (see the accompanying figure).

52. Suppose that the base of a solid is elliptical with a major axis of length 9 and a minor axis of length 4. Find the volume of the solid if the cross sections perpendicular to the minor axis are equilateral triangles (see the accompanying figure).

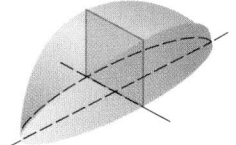

Figure Ex-51

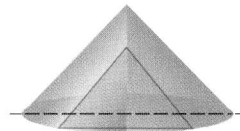

Figure Ex-52

53. Show that an ellipse with semimajor axis a and semiminor axis b has area $A = \pi ab$.

54. (a) Show that the ellipsoid that results when an ellipse with semimajor axis a and semiminor axis b is revolved about the major axis has volume $V = \frac{4}{3}\pi ab^2$.

(b) Show that the ellipsoid that results when an ellipse with semimajor axis a and semiminor axis b is revolved about the minor axis has volume $V = \frac{4}{3}\pi a^2 b$.

55. Show that the ellipsoid that results when an ellipse with semimajor axis a and semiminor axis b is revolved about the major axis has surface area

$$S = 2\pi ab \left(\frac{b}{a} + \frac{a}{c} \sin^{-1} \frac{c}{a} \right)$$

where $c = \sqrt{a^2 - b^2}$.

56. Show that the ellipsoid that results when an ellipse with semimajor axis a and semiminor axis b is revolved about the minor axis has surface area

$$S = 2\pi ab \left(\frac{a}{b} + \frac{b}{c} \ln \frac{a+c}{b} \right)$$

where $c = \sqrt{a^2 - b^2}$.

57. Suppose that you want to draw an ellipse that has given values for the lengths of the major and minor axes by using the method shown in Figure 12.4.3b. Assuming that the axes are drawn, explain how a compass can be used to locate the positions for the tacks.

58. The accompanying figure shows Kepler's method for constructing a parabola: a piece of string the length of the left edge of the drafting triangle is tacked to the vertex Q of the triangle and the other end to a fixed point F. A pencil holds the string taut against the base of the triangle as the edge opposite Q slides along a horizontal line L below F. Show that the pencil traces an arc of a parabola with focus F and directrix L.

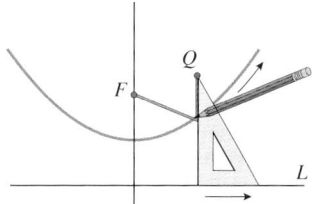

Figure Ex-58

59. The accompanying figure shows a method for constructing a hyperbola: a corner of a ruler is pinned to a fixed point F_1 and the ruler is free to rotate about that point. A piece of string whose length is less than that of the ruler is tacked to a point F_2 and to the free corner Q of the ruler on the same edge as F_1. A pencil holds the string taut against the top edge of the ruler as the ruler rotates about the point F_1. Show that the pencil traces an arc of a hyperbola with foci F_1 and F_2.

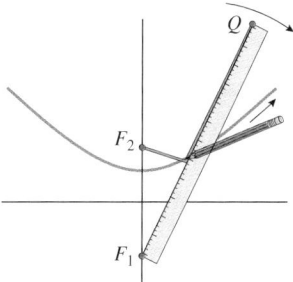

Figure Ex-59

60. Show that if a plane is not parallel to the axis of a right circular cylinder, then the intersection of the plane and cylinder is an ellipse (possibly a circle). [*Hint:* Let θ be the angle shown in Figure Ex-60 (next page), introduce coordinate axes as shown, and express x' and y' in terms of x and y.]

61. As illustrated in the accompanying figure, a carpenter needs to cut an elliptical hole in a sloped roof through which a circular vent pipe of diameter D is to be inserted vertically. The carpenter wants to draw the outline of the hole on the roof using a pencil, two tacks, and a piece of string (as in Figure 12.4.3b). The center point of the ellipse is known, and common sense suggests that its major axis must be perpendicular to the drip line of the roof. The carpenter needs to determine the length L of the string and the distance T between a tack and the center point. The architect's plans show that the pitch of the roof is p (pitch = rise over run; see the accompanying figure). Find T and L in terms of D and p. [*Note:* This exercise is based on an article by William H. Enos, which appeared in the *Mathematics Teacher*, Feb. 1991, p. 148.]

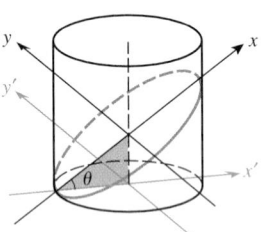

Figure Ex-60   Figure Ex-61

62. Prove: The line tangent to the parabola $x^2 = 4py$ at the point (x_0, y_0) is $x_0 x = 2p(y + y_0)$.

63. Prove: The line tangent to the ellipse

$$\frac{x^2}{a^2} + \frac{y^2}{b^2} = 1$$

at the point (x_0, y_0) has the equation

$$\frac{x x_0}{a^2} + \frac{y y_0}{b^2} = 1$$

64. Prove: The line tangent to the hyperbola

$$\frac{x^2}{a^2} - \frac{y^2}{b^2} = 1$$

at the point (x_0, y_0) has the equation

$$\frac{x x_0}{a^2} - \frac{y y_0}{b^2} = 1$$

65. Use the results in Exercises 63 and 64 to show that if an ellipse and a hyperbola have the same foci, then at each point of intersection their tangent lines are perpendicular.

66. Find two values of k such that the line $x + 2y = k$ is tangent to the ellipse $x^2 + 4y^2 = 8$. Find the points of tangency.

67. Find the coordinates of all points on the hyperbola

$$4x^2 - y^2 = 4$$

where the two lines that pass through the point and the foci are perpendicular.

68. A line tangent to the hyperbola $4x^2 - y^2 = 36$ intersects the y-axis at the point $(0, 4)$. Find the point(s) of tangency.

69. As illustrated in the accompanying figure, suppose that two observers are stationed at the points $F_1(c, 0)$ and $F_2(-c, 0)$ in an xy-coordinate system. Suppose also that the sound of an explosion in the xy-plane is heard by the F_1 observer t seconds before it is heard by the F_2 observer. Assuming that the speed of sound is a constant v, show that the explosion occurred somewhere on the hyperbola

$$\frac{x^2}{v^2 t^2 / 4} - \frac{y^2}{c^2 - (v^2 t^2 / 4)} = 1$$

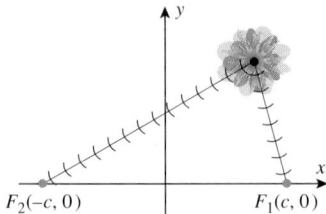

Figure Ex-69

70. As illustrated in the accompanying figure, suppose that two transmitting stations are positioned 100 km apart at points $F_1(50, 0)$ and $F_2(-50, 0)$ on a straight shoreline in an xy-coordinate system. Suppose also that a ship is traveling parallel to the shoreline but 200 km at sea. Find the coordinates of the ship if the stations transmit a pulse simultaneously, but the pulse from station F_1 is received by the ship 0.1 microsecond sooner than the pulse from station F_2. [*Hint:* Use the formula obtained in Exercise 69, assuming that the pulses travel at the speed of light (299,792,458 m/s).]

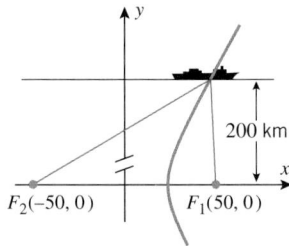

Figure Ex-70

c **71.** As illustrated in Figure Ex-71 (next page), the tank of an oil truck is 18 feet long and has elliptical cross sections that are 6 feet wide and 4 feet high.
  (a) Show that the volume V of oil in the tank (in cubic feet) when it is filled to a depth of h feet is

$$V = 27 \left[4 \sin^{-1} \frac{h - 2}{2} + (h - 2)\sqrt{4h - h^2} + 2\pi \right]$$

  (b) Use the numerical root-finding capability of a CAS to determine how many inches from the bottom of a dipstick the calibration marks should be placed to indicate when the tank is $\frac{1}{4}$, $\frac{1}{2}$, and $\frac{3}{4}$ full.

72. Consider the second-degree equation

$$Ax^2 + Cy^2 + Dx + Ey + F = 0$$

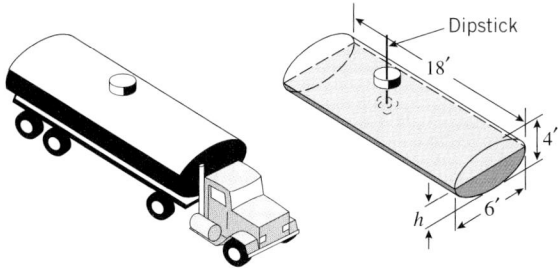

Figure Ex-71

where A and C are not both 0. Show by completing the square:

(a) If $AC > 0$, then the equation represents an ellipse, a circle, a point, or has no graph.

(b) If $AC < 0$, then the equation represents a hyperbola or a pair of intersecting lines.

(c) If $AC = 0$, then the equation represents a parabola, a pair of parallel lines, or has no graph.

73. In each part, use the result in Exercise 72 to make a statement about the graph of the equation, and then check your conclusion by completing the square and identifying the graph.

(a) $x^2 - 5y^2 - 2x - 10y - 9 = 0$

(b) $x^2 - 3y^2 - 6y - 3 = 0$

(c) $4x^2 + 8y^2 + 16x + 16y + 20 = 0$

(d) $3x^2 + y^2 + 12x + 2y + 13 = 0$

(e) $x^2 + 8x + 2y + 14 = 0$

(f) $5x^2 + 40x + 2y + 94 = 0$

74. Derive the equation $x^2 = 4py$ in Figure 12.4.6.

75. Derive the equation $(x^2/b^2) + (y^2/a^2) = 1$ given in Figure 12.4.14.

76. Derive the equation $(x^2/a^2) - (y^2/b^2) = 1$ given in Figure 12.4.22.

12.5 CONIC SECTIONS IN POLAR COORDINATES

It will be shown later in the text that if an object moves in a gravitational field that is directed toward a fixed point (such as the center of the Sun), then the path of that object must be a conic section with the fixed point at a focus. For example, planets in our solar system move along elliptical paths with the Sun at a focus, and the comets move along parabolic, elliptical, or hyperbolic paths with the Sun at a focus, depending on the conditions under which they were born. For applications of this type it is usually desirable to express the equations of the conic sections in polar coordinates with the pole at a focus. In this section we will show how to do this.

THE FOCUS–DIRECTRIX CHARACTERIZATION OF CONICS

To obtain polar equations for the conic sections we will need the following theorem.

> **12.5.1** THEOREM (*Focus–Directrix Property of Conics*).   *Suppose that a point P moves in the plane determined by a fixed point (called the **focus**) and a fixed line (called the **directrix**), where the focus does not lie on the directrix. If the point moves in such a way that its distance to the focus divided by its distance to the directrix is some constant e (called the **eccentricity**), then the curve traced by the point is a conic section. Moreover, the conic is a parabola if e = 1, an ellipse if 0 < e < 1, and a hyperbola if e > 1.*

REMARK.   It is an unfortunate historical accident that the letter e is used for the base of the natural logarithms and the eccentricity of conic sections. However, the appropriate interpretation will usually be clear from the context in which the letter is used.

We will not give a formal proof of this theorem; rather, we will use the specific cases in Figure 12.5.1 to illustrate the basic ideas. For the parabola, we will take the directrix to be $x = -p$, as usual; and for the ellipse and the hyperbola we will take the directrix to be $x = a^2/c$. We want to show in all three cases that if P is a point on the graph, F is the focus, and D is the directrix, then the ratio PF/PD is some constant e, where $e = 1$ for the parabola, $0 < e < 1$ for the ellipse, and $e > 1$ for the hyperbola. We will give the arguments for the parabola and ellipse and leave the argument for the hyperbola as an exercise.

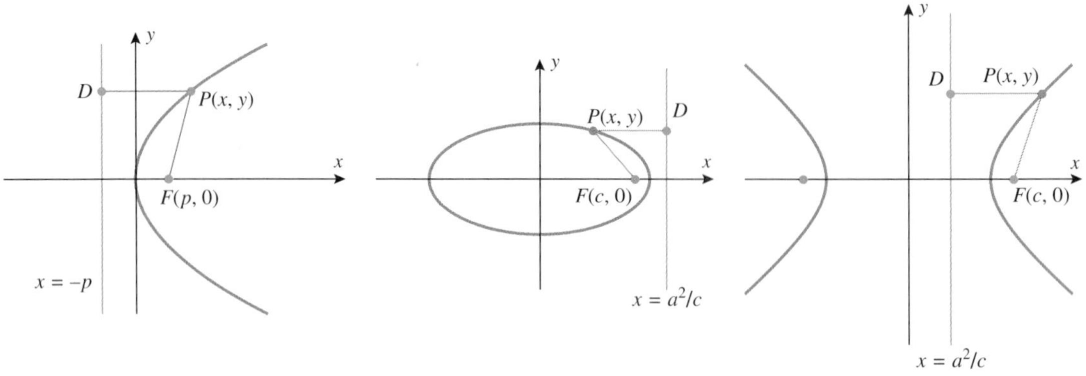

Figure 12.5.1

For the parabola, the distance PF to the focus is equal to the distance PD to the directrix, so that $PF/PD = 1$, which is what we wanted to show. For the ellipse, we rewrite Equation (8) of Section 12.4 as

$$\sqrt{(x-c)^2 + y^2} = a - \frac{c}{a}x = \frac{c}{a}\left(\frac{a^2}{c} - x\right)$$

But the expression on the left side is the distance PF, and the expression in the parentheses on the right side is the distance PD, so we have shown that

$$PF = \frac{c}{a}PD$$

Thus, PF/PD is constant, and the eccentricity is

$$e = \frac{c}{a} \qquad\qquad (1)$$

If we rule out the degenerate case where $a = 0$ or $c = 0$, then it follows from Formula (7) of Section 12.4 that $0 < c < a$, so $0 < e < 1$, which is what we wanted to show.

We will leave it as an exercise to show that the eccentricity of the hyperbola in Figure 12.5.1 is also given by Formula (1), but in this case it follows from Formula (11) of Section 12.4 that $c > a$, so $e > 1$.

ECCENTRICITY OF AN ELLIPSE AS A MEASURE OF FLATNESS

The eccentricity of an ellipse can be viewed as a measure of its flatness—as e approaches 0 the ellipses become more and more circular, and as e approaches 1 they become more and more flat (Figure 12.5.2). Table 12.5.1 shows the orbital eccentricities of various celestial objects. Note that most of the planets actually have fairly circular orbits.

Table 12.5.1

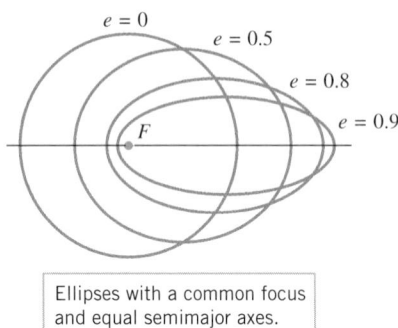

Ellipses with a common focus and equal semimajor axes.

Figure 12.5.2

CELESTIAL BODY	ECCENTRICITY
Mercury	0.206
Venus	0.007
Earth	0.017
Mars	0.093
Jupiter	0.048
Saturn	0.056
Uranus	0.046
Neptune	0.010
Pluto	0.249
Halley's comet	0.970

POLAR EQUATIONS OF CONICS

Our next objective is to derive polar equations for the conic sections from their focus–directrix characterizations. We will assume that the focus is at the pole and the directrix is either parallel or perpendicular to the polar axis. If the directrix is parallel to the polar axis, then it can be above or below the pole; and if the directrix is perpendicular to the polar axis,

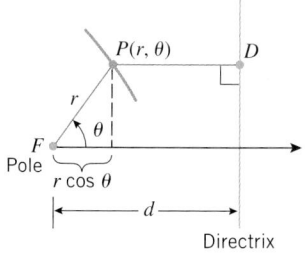

Figure 12.5.3

then it can be to the left or right of the pole. Thus, there are four cases to consider. We will derive the formulas for the case in which the directrix is perpendicular to the polar axis and to the right of the pole.

As illustrated in Figure 12.5.3, let us assume that the directrix is perpendicular to the polar axis and d units to the right of the pole, where the constant d is known. If P is a point on the conic and if the eccentricity of the conic is e, then it follows from Theorem 12.5.1 that $PF/PD = e$ or, equivalently, that

$$PF = ePD \tag{2}$$

However, it is evident from Figure 12.5.3 that $PF = r$ and $PD = d - r\cos\theta$. Thus, (2) can be written as

$$r = e(d - r\cos\theta)$$

which can be solved for r and expressed as

$$r = \frac{ed}{1 + e\cos\theta}$$

(verify). Observe that this single polar equation can represent a parabola, an ellipse, or a hyperbola, depending on the value of e. In contrast, the rectangular equations for these conics all have different forms. The derivations in the other three cases are similar.

12.5.2   THEOREM.   *If a conic section with eccentricity e is positioned in a polar coordinate system so that its focus is at the pole and the corresponding directrix is d units from the pole, then the equation of the conic has one of four possible forms, depending on its orientation:*

$$r = \frac{ed}{1 + e\cos\theta} \qquad\qquad r = \frac{ed}{1 - e\cos\theta} \tag{3-4}$$
Directrix right of pole   Directrix left of pole

$$r = \frac{ed}{1 + e\sin\theta} \qquad\qquad r = \frac{ed}{1 - e\sin\theta} \tag{5-6}$$
Directrix above pole   Directrix below pole

SKETCHING CONICS IN POLAR COORDINATES

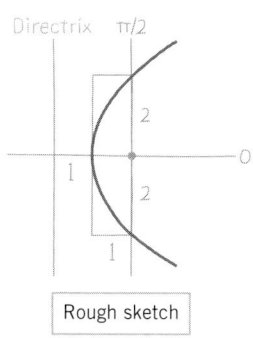

Figure 12.5.4

Precise graphs of conic sections in polar coordinates can be generated with graphing utilities. However, it is often useful to be able to make quick sketches of these graphs that show their orientation and give some sense of their dimensions. The orientation of a conic relative to the polar axis can be deduced by matching its equation with one of the four forms in Theorem 12.5.2. The key dimensions of a parabola are determined by the constant p (Figure 12.4.5) and those of ellipses and hyperbolas by the constants a, b, and c (Figures 12.4.11 and 12.4.20). Thus, we need to show how these constants can be obtained from the polar equations.

Example 1

Sketch the graph of $r = \dfrac{2}{1 - \cos\theta}$ in polar coordinates.

Solution.   The equation is an exact match to (4) with $d = 2$ and $e = 1$. Thus, the graph is a parabola with the focus at the pole and the directrix 2 units to the left of the pole. This tells us that the parabola opens to the right along the polar axis and $p = 1$. Thus, the parabola looks roughly like that sketched in Figure 12.5.4.   ◀

All of the important geometric information about an ellipse can be obtained from the values of a, b, and c in Figure 12.5.5. One way to find these values from the polar equation

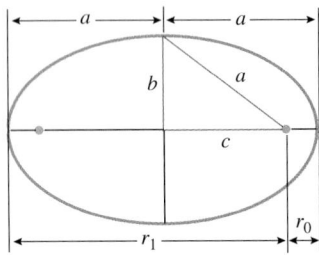

Figure 12.5.5

of an ellipse is based on finding the distances from the focus to the vertices. As shown in the figure, let r_0 be the distance from the focus to the closest vertex and r_1 the distance to the farthest vertex. Thus,

$$r_0 = a - c \quad \text{and} \quad r_1 = a + c \tag{7}$$

from which it follows that

$$a = \tfrac{1}{2}(r_1 + r_0) \tag{8}$$

and

$$c = \tfrac{1}{2}(r_1 - r_0) \tag{9}$$

Moreover, it also follows from (7) that

$$r_0 r_1 = a^2 - c^2 = b^2$$

Thus,

$$b = \sqrt{r_0 r_1} \tag{10}$$

REMARK.   In words, Formula (8) states that a is the **arithmetic average** (also called the **arithmetic mean**) of r_0 and r_1, and Formula (10) states that b is the **geometric mean** of r_0 and r_1.

Example 2

Sketch the graph of $r = \dfrac{6}{2 + \cos\theta}$ in polar coordinates.

Solution. This equation does not match any of the forms in Theorem 12.5.2 because they all require a constant term of 1 in the denominator. However, we can put the equation into one of these forms by dividing the numerator and denominator by 2 to obtain

$$r = \frac{3}{1 + \tfrac{1}{2}\cos\theta}$$

This is an exact match to (3) with $d = 6$ and $e = \tfrac{1}{2}$, so the graph is an ellipse with the directrix 6 units to the right of the pole. The distance r_0 from the focus to the closest vertex can be obtained by setting $\theta = 0$ in this equation, and the distance r_1 to the farthest vertex can be obtained by setting $\theta = \pi$. This yields

$$r_0 = \frac{3}{1 + \tfrac{1}{2}\cos 0} = \frac{3}{\tfrac{3}{2}} = 2, \quad r_1 = \frac{3}{1 + \tfrac{1}{2}\cos\pi} = \frac{3}{\tfrac{1}{2}} = 6$$

Thus, from Formulas (8), (10), and (9), respectively, we obtain

$$a = \tfrac{1}{2}(r_1 + r_0) = 4, \quad b = \sqrt{r_0 r_1} = 2\sqrt{3}, \quad c = \tfrac{1}{2}(r_1 - r_0) = 2$$

Thus, the ellipse looks roughly like that sketched in Figure 12.5.6. ◀

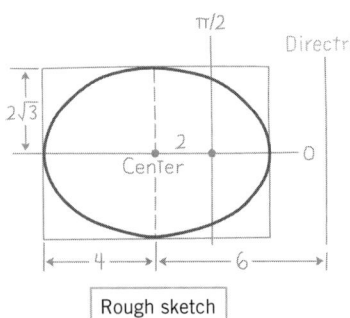

Rough sketch

Figure 12.5.6

All of the important information about a hyperbola can be obtained from the values of a, b, and c in Figure 12.5.7. As with the ellipse, one way to find these values from the polar equation of a hyperbola is based on finding the distances from the focus to the vertices. As shown in the figure, let r_0 be the distance from the focus to the closest vertex and r_1 the distance to the farthest vertex. Thus,

$$r_0 = c - a \quad \text{and} \quad r_1 = c + a \tag{11}$$

from which it follows that

$$a = \tfrac{1}{2}(r_1 - r_0) \tag{12}$$

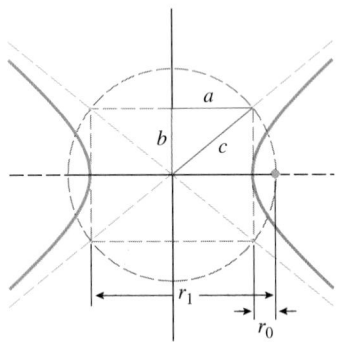

Figure 12.5.7

and

$$c = \tfrac{1}{2}(r_1 + r_0) \tag{13}$$

Moreover, it also follows from (11) that

$$r_0 r_1 = c^2 - a^2 = b^2$$

from which it follows that

$$b = \sqrt{r_0 r_1} \tag{14}$$

Example 3

Sketch the graph of $r = \dfrac{2}{1 + 2\sin\theta}$ in polar coordinates.

Solution. This equation is an exact match to (5) with $d = 1$ and $e = 2$. Thus, the graph is a hyperbola with its directrix 1 unit above the pole. However, it is not so straightforward to compute the values of r_0 and r_1, since hyperbolas in polar coordinates are generated in a strange way as θ varies from 0 to 2π. This can be seen from Figure 12.5.8a, which is the graph of the given equation in rectangular coordinates. It follows from this graph that the corresponding polar graph is generated in pieces (see Figure 12.5.8b):

- As θ varies over the interval $0 \leq \theta < 7\pi/6$, the value of r is positive and varies from 2 to $+\infty$, which generates part of the lower branch.

- As θ varies over the interval $7\pi/6 < \theta \leq 3\pi/2$, the value of r is negative and varies from $-\infty$ to -2, which generates the right part of the upper branch.

- As θ varies over the interval $3\pi/2 \leq \theta < 11\pi/6$, the value of r is negative and varies from -2 to $-\infty$, which generates the left part of the upper branch.

- As θ varies over the interval $11\pi/6 < \theta \leq 2\pi$, the value of r is positive and varies from $+\infty$ to 2, which fills in the missing piece of the lower right branch.

It is now clear that we can obtain r_0 by setting $\theta = \pi/2$ and r_1 by setting $\theta = 3\pi/2$. Keeping in mind that r_0 and r_1 are positive, this yields

$$r_0 = \frac{2}{1 + 2\sin(\pi/2)} = \frac{2}{3}, \quad r_1 = \left| \frac{2}{1 + 2\sin(3\pi/2)} \right| = \left| \frac{2}{-1} \right| = 2$$

Thus, from Formulas (12), (14), and (13), respectively, we obtain

$$a = \frac{1}{2}(r_1 - r_0) = \frac{2}{3}, \quad b = \sqrt{r_0 r_1} = \frac{2\sqrt{3}}{3}, \quad c = \frac{1}{2}(r_1 + r_0) = \frac{4}{3}$$

Thus, the hyperbola looks roughly like that sketched in Figure 12.5.8c. ◀

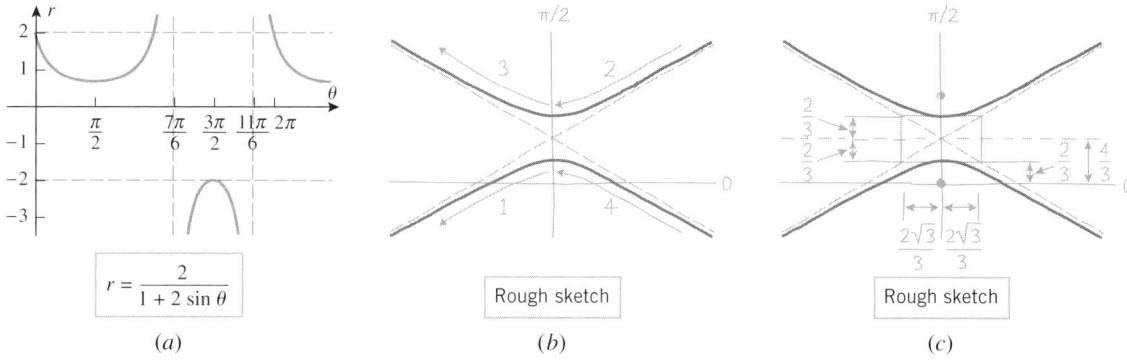

Figure 12.5.8

$$r = \frac{2}{1 + 2\sin\theta}$$

(a)   Rough sketch   (b)   Rough sketch   (c)

APPLICATIONS IN ASTRONOMY

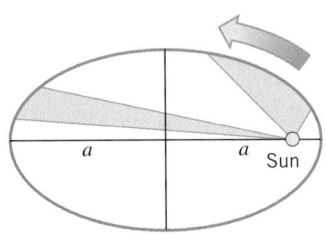

Equal areas are swept out in equal times, and the square of the period T is proportional to a^3.

Figure 12.5.9

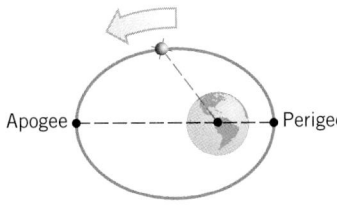

Figure 12.5.10

In 1609 Johannes Kepler[*] published a book known as *Astronomia Nova* (or sometimes *Commentaries on the Motions of Mars*) in which he succeeded in distilling thousands of years of observational astronomy into three beautiful laws of planetary motion (Figure 12.5.9).

12.5.3 KEPLER'S LAWS.

- First law (***Law of Orbits***). Each planet moves in an elliptical orbit with the Sun at a focus.

- Second law (***Law of Areas***). The radial line from the center of the Sun to the center of a planet sweeps out equal areas in equal times.

- Third law (***Law of Periods***). The square of a planet's period (the time it takes the planet to complete one orbit about the Sun) is proportional to the cube of the semimajor axis of its orbit.

Kepler's laws, although stated for planetary motion around the Sun, apply to all orbiting celestial bodies that are subjected to a *single* central gravitational force—artificial satellites subjected only to the central force of Earth's gravity and moons subjected only to the central gravitational force of a planet, for example. Later in the text we will derive Kepler's laws from basic principles, but for now we will show how they can be used in basic astronomical computations.

In an elliptical orbit, the closest point to the focus is called the ***perigee*** and the farthest point the ***apogee*** (Figure 12.5.10). The distances from the focus to the perigee and apogee are called the ***perigee distance*** and ***apogee distance***, respectively. For orbits around the Sun, it is more common to use the terms ***perihelion*** and ***aphelion***, rather than perigee and apogee, and to measure time in Earth years and distances in astronomical units (AU), where 1 AU is the semimajor axis a of the Earth's orbit (approximately 150×10^6 km or 92.9×10^6 mi). With this choice of units, the constant of proportionality in Kepler's third law is 1, since $a = 1$ AU produces a period of $T = 1$ Earth year. In this case Kepler's third law can be expressed as

$$T = a^{3/2} \tag{15}$$

Shapes of elliptical orbits are often specified by giving the eccentricity e and the semimajor axis a, so it is useful to express the polar equations of an ellipse in terms of these

---

[*] JOHANNES KEPLER (1571–1630). German astronomer and physicist, Kepler, whose work provided our contemporary view of planetary motion, led a fascinating but ill-starred life. His alcoholic father made him work in a family-owned tavern as a child, later withdrawing him from elementary school and hiring him out as a field laborer, where the boy contracted smallpox, permanently crippling his hands and impairing his eyesight. In later years, Kepler's first wife and several children died, his mother was accused of witchcraft, and being a Protestant he was often subjected to persecution by Catholic authorities. He was often impoverished, eking out a living as an astrologer and prognosticator. Looking back on his unhappy childhood, Kepler described his father as "criminally inclined" and "quarrelsome" and his mother as "garrulous" and "bad-tempered." However, it was his mother who left an indelible mark on the six-year-old Kepler by showing him the comet of 1577; and in later life he personally prepared her defense against the witchcraft charges. Kepler became acquainted with the work of Copernicus as a student at the University of Tübingen, where he received his master's degree in 1591. He continued on as a theological student, but at the urging of the university officials he abandoned his clerical studies and accepted a position as a mathematician and teacher in Graz, Austria. However, he was expelled from the city when it came under Catholic control, and in 1600 he finally moved on to Prague, where he became an assistant at the observatory of the famous Danish astronomer Tycho Brahe. Brahe was a brilliant and meticulous astronomical observer who amassed the most accurate astronomical data known at that time; and when Brahe died in 1601 Kepler inherited the treasure-trove of data. After eight years of intense labor, Kepler deciphered the underlying principles buried in the data and in 1609 published his monumental work, *Astronomia Nova*, in which he stated his first two laws of planetary motion. Commenting on his discovery of elliptical orbits, Kepler wrote, "I was almost driven to madness in considering and calculating this matter. I could not find out why the planet would rather go on an elliptical orbit (rather than a circle). Oh ridiculous me!" It ultimately remained for Isaac Newton to discover the laws of gravitation that explained the reason for elliptical orbits.

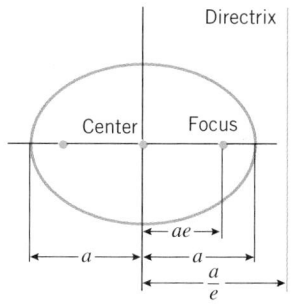

Figure 12.5.11

constants. Figure 12.5.11, which can be obtained from the ellipse in Figure 12.5.1 and the relationship $c = ea$, implies that the distance d between the focus and the directrix is

$$d = \frac{a}{e} - c = \frac{a}{e} - ea = \frac{a(1 - e^2)}{e} \tag{16}$$

from which it follows that $ed = a(1 - e^2)$. Thus, depending on the orientation of the ellipse, the formulas in Theorem 12.5.2 can be expressed in terms of a and e as

$$r = \frac{a(1 - e^2)}{1 \pm e \cos \theta} \qquad r = \frac{a(1 - e^2)}{1 \pm e \sin \theta} \tag{17–18}$$

+: Directrix right of pole   +: Directrix above pole
−: Directrix left of pole   −: Directrix below pole

Moreover, it is evident from Figure 12.5.11 that the distances from the focus to the closest and farthest vertices can be expressed in terms of a and e as

$$r_0 = a - ea = a(1 - e) \quad \text{and} \quad r_1 = a + ea = a(1 + e) \tag{19–20}$$

Example 4

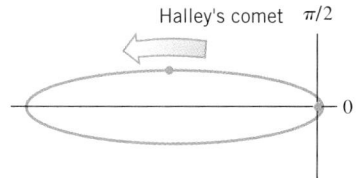

Figure 12.5.12

Halley's comet (last seen in 1986) has an eccentricity of 0.97 and a semimajor axis of $a = 18.1$ AU.

(a)   Find the equation of its orbit in the polar coordinate system shown in Figure 12.5.12.

(b)   Find the period of its orbit.

(c)   Find its perihelion and aphelion distances.

Solution (a). From (17), the polar equation of the orbit has the form

$$r = \frac{a(1 - e^2)}{1 + e \cos \theta}$$

But $a(1 - e^2) = 18.1[1 - (0.97)^2] \approx 1.07$. Thus, the equation of the orbit is

$$r = \frac{1.07}{1 + 0.97 \cos \theta}$$

Halley's comet photographed
April 21, 1910 in Peru

Solution (b). From (15), with $a = 18.1$, the period of the orbit is

$$T = (18.1)^{3/2} \approx 77 \text{ years}$$

Solution (c). Since the perihelion and aphelion distances are the distances to the closest and farthest vertices, respectively, it follows from (19) and (20) that

$$r_0 = a - ea = a(1 - e) = 18.1(1 - 0.97) \approx 0.543 \text{ AU}$$
$$r_1 = a + ea = a(1 + e) = 18.1(1 + 0.97) \approx 35.7 \text{ AU}$$

or since 1 AU $\approx 150 \times 10^6$ km, the perihelion and aphelion distances in kilometers are

$$r_0 = 18.1(1 - 0.97)(150 \times 10^6) \approx 81,500,000 \text{ km}$$
$$r_1 = 18.1(1 + 0.97)(150 \times 10^6) \approx 5,350,000,000 \text{ km} \qquad \blacktriangleleft$$

FOR THE READER.   Use the polar equation of the orbit of Halley's comet to check the values of r_0 and r_1.

Example 5

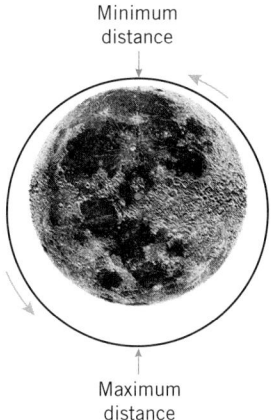

Figure 12.5.13

An Apollo lunar lander orbits the Moon in an elliptic orbit with eccentricity $e = 0.12$ and semimajor axis $a = 2015$ km. Assuming the Moon to be a sphere of radius 1740 km, find the minimum and maximum heights of the lander above the lunar surface (Figure 12.5.13).

Solution. If we let r_0 and r_1 denote the minimum and maximum distances from the center of the Moon, then the minimum and maximum distances from the surface of the Moon will be

$$d_{\min} = r_0 - 1740$$

$$d_{\max} = r_1 - 1740$$

or from Formulas (19) and (20)

$$d_{\min} = r_0 - 1740 = a(1 - e) - 1740 = 2015(0.88) - 1740 \approx 33.2 \text{ km}$$

$$d_{\max} = r_1 - 1740 = a(1 + e) - 1740 = 2015(1.12) \approx 516.8 \text{ km} \quad \blacktriangleleft$$

EXERCISE SET 12.5 ◠ Graphing Calculator ▣ CAS

For the conics in Exercises 1 and 2, find the eccentricity and the distance from the pole to the directrix, and sketch the graph in polar coordinates.

1. (a) $r = \dfrac{3}{2 - 2\cos\theta}$ (b) $r = \dfrac{3}{2 + \sin\theta}$

(c) $r = \dfrac{4}{2 + 3\cos\theta}$ (d) $r = \dfrac{5}{3 + 3\sin\theta}$

2. (a) $r = \dfrac{4}{3 - 2\cos\theta}$ (b) $r = \dfrac{3}{3 - 4\sin\theta}$

(c) $r = \dfrac{1}{3 + 3\sin\theta}$ (d) $r = \dfrac{1}{2 + 6\sin\theta}$

In Exercises 3 and 4, use Formulas (3)–(6) to name and describe the orientation of the conic, and then check your answer by generating the graph with a graphing utility.

3. (a) $r = \dfrac{8}{1 - \sin\theta}$ (b) $r = \dfrac{16}{4 + 3\sin\theta}$

(c) $r = \dfrac{4}{2 - 3\sin\theta}$ (d) $r = \dfrac{12}{4 + \cos\theta}$

4. (a) $r = \dfrac{15}{1 + \cos\theta}$ (b) $r = \dfrac{2}{3 + 3\cos\theta}$

(c) $r = \dfrac{64}{7 - 12\sin\theta}$ (d) $r = \dfrac{12}{3 - 2\cos\theta}$

In Exercises 5–8, find a polar equation for the conic that has its focus at the pole and satisfies the stated conditions. Points are in polar coordinates and directrices in rectangular coordinates for simplicity. (In some cases there may be more than one conic that satisfies the conditions.)

5. (a) Ellipse; $e = \frac{2}{3}$; directrix $x = 1$.
 (b) Parabola; directrix $x = -1$.
 (c) Hyperbola; $e = \frac{3}{2}$; directrix $y = 1$.

6. (a) Ellipse; $e = \frac{2}{3}$; directrix $y = -1$.
 (b) Parabola; directrix $y = 1$.
 (c) Hyperbola; $e = \frac{4}{3}$; directrix $x = -1$.

7. (a) Ellipse; vertices $(6, 0)$ and $(4, \pi)$.

(b) Parabola; vertex $(1, 3\pi/2)$.
(c) Hyperbola; vertices $(3, \pi/2)$ and $(-7, 3\pi/2)$.

8. (a) Ellipse; ends of major axis $(1, \pi/2)$ and $(4, 3\pi/2)$.
 (b) Parabola; vertex $(3, \pi)$.
 (c) Hyperbola; equilateral; vertex $(5, 0)$.

In Exercises 9 and 10, find the distances from the pole to the vertices, and then apply Formulas (8)–(10) to find the equation of the ellipse in rectangular coordinates.

9. (a) $r = \dfrac{6}{2 + \sin\theta}$ (b) $r = \dfrac{1}{2 - \cos\theta}$

10. (a) $r = \dfrac{6}{5 + 2\cos\theta}$ (b) $r = \dfrac{8}{4 - 3\sin\theta}$

In Exercises 11 and 12, find the distances from the pole to the vertices, and then apply Formulas (12)–(14) to find the equation of the hyperbola in rectangular coordinates.

11. (a) $r = \dfrac{2}{1 + 3\sin\theta}$ (b) $r = \dfrac{10}{6 - 9\cos\theta}$

12. (a) $r = \dfrac{4}{1 - 2\sin\theta}$ (b) $r = \dfrac{15}{2 + 8\cos\theta}$

In Exercises 13 and 14, find a polar equation for the ellipse that has its focus at the pole and satisfies the stated conditions.

13. (a) Directrix to the right of the pole; $a = 8$; $e = \frac{1}{2}$.
 (b) Directrix below the pole; $a = 4$; $e = \frac{3}{5}$.
 (c) Directrix to the left of the pole; $b = 4$; $e = \frac{3}{5}$.
 (d) Directrix above the pole; $c = 5$; $e = \frac{1}{5}$.

14. (a) Directrix above the pole; $a = 10$; $e = \frac{1}{2}$.
 (b) Directrix to the left of the pole; $a = 6$; $e = \frac{1}{5}$.
 (c) Directrix below the pole; $b = 4$; $e = \frac{3}{4}$.
 (d) Directrix to the right of the pole; $c = 10$; $e = \frac{4}{5}$.

15. (a) Show that the eccentricity of an ellipse can be expressed in terms of r_0 and r_1 as

$$e = \frac{r_1 - r_0}{r_1 + r_0}$$

(b) Show that
$$\frac{r_1}{r_0} = \frac{1+e}{1-e}$$

16. (a) Show that the eccentricity of a hyperbola can be expressed in terms of r_0 and r_1 as
$$e = \frac{r_1 + r_0}{r_1 - r_0}$$

(b) Show that
$$\frac{r_1}{r_0} = \frac{e+1}{e-1}$$

In Exercises 17–22, use the following values, where needed:

radius of the Earth = 4000 mi = 6440 km

1 year (Earth year) = 365 days (Earth days)

1 AU = 92.9×10^6 mi = 150×10^6 km

17. The planet Pluto has eccentricity $e = 0.249$ and semimajor axis $a = 39.5$ AU.
(a) Find the period T in years.
(b) Find the perihelion and aphelion distances.
(c) Choose a polar coordinate system with the center of the Sun at the pole, and find a polar equation of Pluto's orbit in that coordinate system.
(d) Make a sketch of the orbit with reasonably accurate proportions.

18. (a) Let a be the semimajor axis of a planet's orbit around the Sun, and let T be its period. Show that if T is measured in days and a in kilometers, then $T = (365 \times 10^{-9})(a/150)^{3/2}$.
(b) Use the result in part (a) to find the period of the planet Mercury in days, given that its semimajor axis is $a = 57.95 \times 10^6$ km.
(c) Choose a polar coordinate system with the Sun at the pole, and find an equation for the orbit of Mercury in that coordinate system given that the eccentricity of the orbit is $e = 0.206$.
(d) Use a graphing utility to generate the orbit of Mercury from the equation obtained in part (c).

19. The Hale–Bopp comet, discovered independently on July 23, 1995 by Alan Hale and Thomas Bopp, has an orbital eccentricity of $e = 0.9951$ and a period of 2380 years.
(a) Find its semimajor axis in astronomical units (AU).
(b) Find its perihelion and aphelion distances.
(c) Choose a polar coordinate system with the center of the Sun at the pole, and find an equation for the Hale–Bopp orbit in that coordinate system.
(d) Make a sketch of the Hale–Bopp orbit with reasonably accurate proportions.

20. Mars has a perihelion distance of 204,520,000 km and an aphelion distance of 246,280,000 km.
(a) Use these data to calculate the eccentricity, and compare your answer to the value given in Table 12.5.1.
(b) Find the period of Mars.
(c) Choose a polar coordinate system with the center of the Sun at the pole, and find an equation for the orbit of Mars in that coordinate system.
(d) Use a graphing utility to generate the orbit of Mars from the equation obtained in part (c).

21. *Vanguard 1* was launched in March 1958 into an orbit around the Earth with eccentricity $e = 0.21$ and semimajor axis 8864.5 km. Find the minimum and maximum heights of *Vanguard 1* above the surface of the Earth.

22. The planet Jupiter is believed to have a rocky core of radius 10,000 km surrounded by two layers of hydrogen—a 40,000-km-thick layer of compressed metallic-like hydrogen and a 20,000-km-thick layer of ordinary molecular hydrogen. The visible features, such as the Great Red Spot, are at the outer surface of the molecular hydrogen layer. On November 6, 1997 the spacecraft *Galileo* was placed in a Jovian orbit to study the moon Europa. The orbit had eccentricity 0.814580 and semimajor axis 3,514,918.9 km. Find *Galileo*'s minimum and maximum heights above the molecular hydrogen layer (see the accompanying figure).

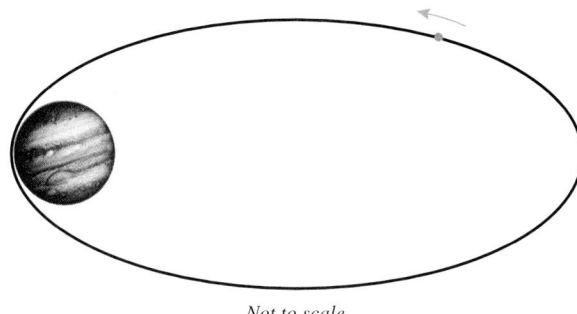

Not to scale

Figure Ex-22

23. What happens to the distance between the directrix and the center of an ellipse if the foci remain fixed and $e \to 0$?

24. (a) Show that the coordinates of the point P on the hyperbola in Figure 12.5.1 satisfy the equation
$$\sqrt{(x-c)^2 + y^2} = \frac{c}{a}x - a$$
(b) Use the result in part (a) to show that $PF/PD = c/a$.

SUPPLEMENTARY EXERCISES

1. Under what conditions does a parametric curve $x = f(t)$, $y = g(t)$ have a horizontal tangent line? A vertical tangent line? A singular point?

2. Express the point whose xy-coordinates are $(-1, 1)$ in polar coordinates with
   (a) $r > 0$, $0 \le \theta < 2\pi$ (b) $r < 0$, $0 \le \theta < 2\pi$
   (c) $r > 0$, $-\pi < \theta \le \pi$ (d) $r < 0$, $-\pi < \theta \le \pi$.

3. In each part, state the name that describes the polar curve most precisely: a rose, a line, a circle, a limaçon, a cardioid, a spiral, a lemniscate, or none of these.
   (a) $r = 3\cos\theta$ (b) $r = \cos 3\theta$
   (c) $r = \dfrac{3}{\cos\theta}$ (d) $r = 3 - \cos\theta$
   (e) $r = 1 - 3\cos\theta$ (f) $r^2 = 3\cos\theta$
   (g) $r = (3\cos\theta)^2$ (h) $r = 1 + 3\theta$

4. In each part: (i) Identify the polar graph as a parabola, an ellipse, or a hyperbola; (ii) state whether the directrix is above, below, to the left, or to the right of the pole; and (iii) find the distance from the pole to the directrix.
   (a) $r = \dfrac{1}{3 + \cos\theta}$ (b) $r = \dfrac{1}{1 - 3\cos\theta}$
   (c) $r = \dfrac{1}{3(1 + \sin\theta)}$ (d) $r = \dfrac{3}{1 - \sin\theta}$

5. The accompanying figure shows the polar graph of the equation $r = f(\theta)$. Sketch the graph of
   (a) $r = f(-\theta)$ (b) $r = f\left(\theta - \dfrac{\pi}{2}\right)$
   (c) $r = f\left(\theta + \dfrac{\pi}{2}\right)$ (d) $r = -f(\theta)$
   (e) $r = f(\theta) + 1$.

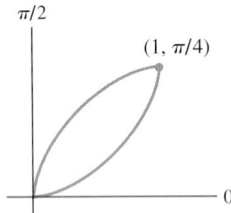

Figure Ex-5

6. Find equations for the two families of circles in the accompanying figure.

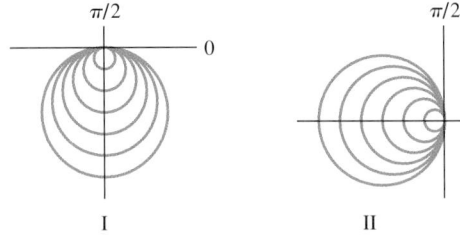

Figure Ex-6

7. In each part, identify the curve by converting the polar equation to rectangular coordinates. Assume that $a > 0$.
   (a) $r = a\sec^2\dfrac{\theta}{2}$ (b) $r^2\cos 2\theta = a^2$
   (c) $r = 4\csc\left(\theta - \dfrac{\pi}{4}\right)$ (d) $r = 4\cos\theta + 8\sin\theta$

8. Use a graphing utility to investigate how the family of polar curves $r = 1 + a\cos n\theta$ is affected by changing the values of a and n, where a is a positive real number and n is a positive integer. Write a brief paragraph to explain your conclusions.

In Exercises 9 and 10, find an equation in xy-coordinates for the conic section that satisfies the given conditions.

9. (a) Ellipse with eccentricity $e = \frac{2}{7}$ and ends of the minor axis at the points $(0, \pm 3)$.
   (b) Parabola with vertex at the origin, focus on the y-axis, and directrix passing through the point $(7, 4)$.
   (c) Hyperbola that has the same foci as the ellipse $3x^2 + 16y^2 = 48$ and asymptotes $y = \pm 2x/3$.

10. (a) Ellipse with center $(-3, 2)$, vertex $(2, 2)$, and eccentricity $e = \frac{4}{5}$.
    (b) Parabola with focus $(-2, -2)$ and vertex $(-2, 0)$.
    (c) Hyperbola with vertex $(-1, 7)$ and asymptotes $y - 5 = \pm 8(x + 1)$.

11. In each part, sketch the graph of the conic section with reasonably accurate proportions.
    (a) $x^2 - 4x + 8y + 36 = 0$
    (b) $3x^2 + 4y^2 - 30x - 8y + 67 = 0$
    (c) $4x^2 - 5y^2 - 8x - 30y - 21 = 0$

12. If you have a CAS that can graph implicit equations, use it to check your work in Exercise 11.

13. It can be shown that hanging cables form parabolic arcs rather than catenaries if they are subjected to uniformly distributed downward forces along their length. For example, if the weight of the roadway in a suspension bridge is assumed to be uniformly distributed along the supporting cables, then the cables can be modeled by parabolas.
    (a) Assuming a parabolic model, find an equation for the cable in the accompanying figure, taking the y-axis to be vertical and the origin at the low point of the cable.
    (b) Find the length of the cable between the supports.

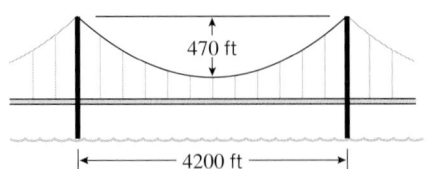

Figure Ex-13

14. A parametric curve of the form

$$x = a \cot t + b \cos t, \quad y = a + b \sin t \quad (0 < t < 2\pi)$$

is called a **conchoid of Nicomedes** (see the accompanying figure for the case $0 < a < b$).

(a) Describe how the conchoid

$$x = \cot t + 4 \cos t, \quad y = 1 + 4 \sin t$$

is generated as t varies over the interval $0 < t < 2\pi$.

(b) Find the horizontal asymptote of the conchoid given in part (a).

(c) For what values of t does the conchoid in part (a) have a horizontal tangent line? A vertical tangent line?

(d) Find a polar equation $r = f(\theta)$ for the conchoid in part (a), and then find polar equations for the tangent lines to the conchoid at the pole.

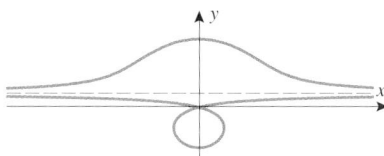

Figure Ex-14

15. Find the area of the region that is common to the circles $r = 1$, $r = 2 \cos \theta$, and $r = 2 \sin \theta$.

16. Find the area of the region that is inside the cardioid $r = a(1 + \sin \theta)$ and outside the circle $r = a \sin \theta$.

17. (a) Find the arc length of the polar curve $r = 1/\theta$ for $\pi/4 \le \theta \le \pi/2$.

(b) What can you say about the arc length of the portion of the curve that lies inside the circle $r = 1$?

18. (a) If a thread is unwound from a fixed circle while being held taut (i.e., tangent to the circle), then the end of the thread traces a curve called an **involute of a circle**. Show that if the circle is centered at the origin, has radius a, and the end of the thread is initially at the point $(a, 0)$, then the involute can be expressed parametrically as

$$x = a(\cos \theta + \theta \sin \theta), \quad y = a(\sin \theta - \theta \cos \theta)$$

where θ is the angle shown in part (a) of the accompanying figure.

(b) Assuming that the dog in part (b) of the accompanying figure unravels its leash while keeping it taut, for what values of θ in the interval $0 \le \theta \le 2\pi$ will the dog be walking North? South? East? West?

(c) Use a graphing utility to generate the curve traced by the dog, and show that it is consistent with your answer in part (b).

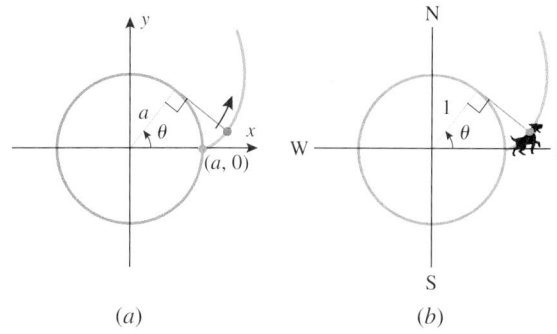

(a)   (b)

Figure Ex-18

19. Let R be the region that is above the x-axis and enclosed between the curve $b^2x^2 - a^2y^2 = a^2b^2$ and the line $x = \sqrt{a^2 + b^2}$.

(a) Sketch the solid generated by revolving R about the x-axis, and find its volume.

(b) Sketch the solid generated by revolving R about the y-axis, and find its volume.

20. (a) Sketch the curves

$$r = \frac{1}{1 + \cos \theta} \quad \text{and} \quad r = \frac{1}{1 - \cos \theta}$$

(b) Find polar coordinates of the intersections of the curves in part (a).

(c) Show that the curves are *orthogonal*, that is, their tangent lines are perpendicular at the points of intersection.

21. How is the shape of a hyperbola affected as its eccentricity approaches 1? As it approaches $+\infty$? Draw some pictures to illustrate your conclusions.

22. Use the formula obtained in part (a) of Exercise 67 of Section 12.1 to find the distance between successive tips of the three-petal rose $r = \sin 3\theta$, and check your answer using trigonometry.

23. (a) Find the minimum and maximum x-coordinates of points on the cardioid $r = 1 + \cos \theta$.

(b) Find the minimum and maximum y-coordinates of points on the cardioid in part (a).

24. (a) Show that the maximum value of the y-coordinate of points on the curve $r = 1/\sqrt{\theta}$ for θ in the interval $(0, \pi]$ occurs when $\tan \theta = 2\theta$.

(b) Use Newton's Method to solve the equation in part (a) for θ to at least four decimal-place accuracy.

(c) Use the result of part (b) to approximate the maximum value of y for $0 < \theta \le \pi$.

25. Define the width of a petal of a rose curve to be the dimension shown in the accompanying figure. Show that the width w of a petal of the four-petal rose $r = \cos 2\theta$ is $w = 2\sqrt{6}/9$. [*Hint:* Express y in terms of θ, and investigate the maximum value of y.]

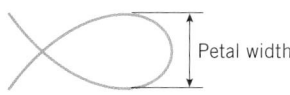

Petal width

Figure Ex-25

26. A nuclear cooling tower is to have a height of h feet and the shape of the solid that is generated by revolving the region R enclosed by the right branch of the hyperbola $1521x^2 - 225y^2 = 342{,}225$ and the lines $x = 0$, $y = -h/2$, and $y = h/2$ about the y-axis.

(a) Find the volume of the tower.

(b) Find the lateral surface area of the tower.

27. The amusement park rides illustrated in the accompanying figure consist of two connected rotating arms of length 1— an inner arm that rotates counterclockwise at 1 radian per second and an outer arm that can be programmed to rotate either clockwise at 2 radians per second (the Scrambler ride) or counterclockwise at 2 radians per second (the Calypso ride). The center of the rider cage is at the end of the outer arm.

(a) Show that in the Scrambler ride the center of the cage has parametric equations

$$x = \cos t + \cos 2t, \qquad y = \sin t - \sin 2t$$

(b) Find parametric equations for the center of the cage in the Calypso ride, and use a graphing utility to confirm that the center traces the curve shown in the accompanying figure.

(c) Do you think that a rider travels the same distance in one revolution of the Scrambler ride as in one revolution of the Calypso ride? Justify your conclusion.

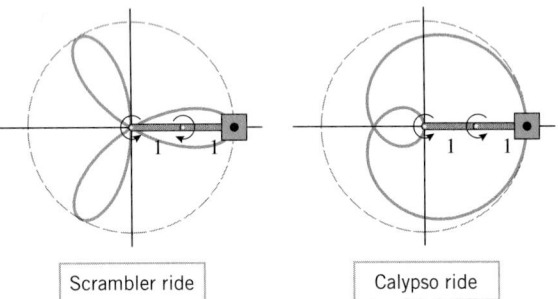

Scrambler ride   Calypso ride

Figure Ex-27

28. Use a graphing utility to explore the effect of changing the rotation rates and the arm lengths in Exercise 27.

29. Use the parametric equations $x = a\cos t$, $y = b\sin t$ to show that the circumference C of an ellipse with semimajor axis a and eccentricity e is

$$C = 4a \int_0^{\pi/2} \sqrt{1 - e^2 \sin^2 u}\, du$$

30. Use Simpson's rule or the numerical integration capability of a graphing utility to approximate the circumference of the ellipse $4x^2 + 9y^2 = 36$ from the integral obtained in Exercise 29.

31. (a) Calculate the eccentricity of the Earth's orbit, given that the ratio of the distance between the center of the Earth and the center of the Sun at perihelion to the distance between the centers at aphelion is $\frac{59}{61}$.

(b) Find the distance between the center of the Earth and the center of the Sun at perihelion, given that the average value of the perihelion and aphelion distances between the centers is 93 million miles.

(c) Use the result in Exercise 29 and Simpson's rule or the numerical integration capability of a graphing utility to approximate the distance that the Earth travels in 1 year (one revolution around the Sun).

32. It will be shown later in this text that if a projectile is launched with speed v_0 at an angle α with the horizontal and at a height y_0 above ground level, then the resulting trajectory relative to the coordinate system in the accompanying figure will have parametric equations

$$x = (v_0 \cos \alpha)t, \qquad y = y_0 + (v_0 \sin \alpha)t - \tfrac{1}{2}gt^2$$

(a) Show that the trajectory is a parabola.

(b) Find the coordinates of the vertex.

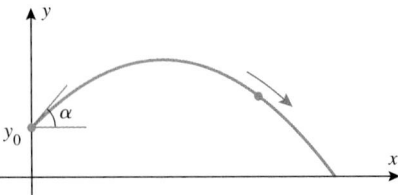

Figure Ex-32

33. Mickey Mantle is recognized as baseball's unofficial king of long home runs. On April 17, 1953 Mantle blasted a pitch by Chuck Stobbs of the hapless Washington Senators out of Griffith Stadium, just clearing the 50-ft wall at the 391-ft marker in left center. Assuming that the ball left the bat at a height of 3 ft above the ground and at an angle of $45°$, use the parametric equations in Exercise 32 with $g = 32$ ft/s^2 to find

(a) the speed of the ball as it left the bat

(b) the maximum height of the ball

(c) the distance along the ground from home plate where the ball struck the ground.

34. Recall from Section 7.9 that the Fresnel sine and cosine functions are defined as

$$S(x) = \int_0^x \sin\left(\frac{\pi t^2}{2}\right) dt \quad \text{and} \quad C(x) = \int_0^x \cos\left(\frac{\pi t^2}{2}\right) dt$$

The following parametric curve, which is used to study amplitudes of light waves in optics, is called a *clothoid* or *Cornu spiral* in honor of the French scientist Marie Alfred Cornu (1841–1902):

$$x = C(t) = \int_0^t \cos\left(\frac{\pi u^2}{2}\right) du$$
$$y = S(t) = \int_0^t \sin\left(\frac{\pi u^2}{2}\right) du \qquad (-\infty < t < +\infty)$$

(a) Use a CAS to graph the cornu spiral.

(b) Describe the behavior of the spiral as $t \to +\infty$ and as $t \to -\infty$.

(c) Find the arc length of the spiral for $-1 \le t \le 1$.

35. As illustrated in the accompanying figure, let $P(r, \theta)$ be a point on the polar curve $r = f(\theta)$, let ψ be the smallest counterclockwise angle from the extended radius OP to the tangent line at P, and let ϕ be the angle of inclination of the tangent line. Derive the formula

$$\tan \psi = \frac{r}{dr/d\theta}$$

by substituting $\tan \phi$ for dy/dx in Formula (7) of Section 12.2 and applying the trigonometric identity

$$\tan(\phi - \theta) = \frac{\tan \phi - \tan \theta}{1 + \tan \phi \tan \theta}$$

In Exercises 36 and 37, use the formula for ψ obtained in Exercise 35.

36. (a) Use the trigonometric identity

$$\tan \frac{\theta}{2} = \frac{1 - \cos \theta}{\sin \theta}$$

to show that if (r, θ) is a point on the cardioid

$$r = 1 - \cos \theta \quad (0 \le \theta < 2\pi)$$

then $\psi = \theta/2$.

(b) Sketch the cardioid and show the angle ψ at the points where the cardioid crosses the y-axis.

(c) Find the angle ψ at the points where the cardioid crosses the y-axis.

37. Show that for a logarithmic spiral $r = ae^{b\theta}$, the angle from the radial line to the tangent line is constant along the spiral (see the accompanying figure). [*Note:* For this reason, logarithmic spirals are sometimes called ***equiangular spirals***.]

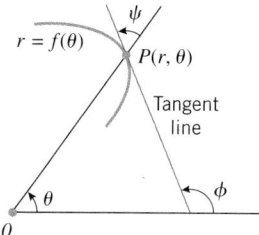

Figure Ex-35

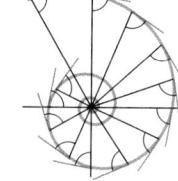

Figure Ex-37

EXPANDING THE CALCULUS HORIZON

Comet Collision

*The Earth lives in a cosmic shooting gallery of comets and asteroids. Although the probability that the Earth will be hit by a comet or asteroid in any given year is small, the consequences of such a collision are so catastrophic that the international community is now beginning to track **near Earth objects** (NEOs). Your job, as part of the international NEO tracking team, is to compute the orbits of incoming comets and asteroids, determine how close they will come to colliding with the Earth, and issue a notification if there is danger of a collision or near miss.*

At the time when the Earth is at its *aphelion* (its farthest point from the Sun), your NEO tracking team receives a notification from the NASA/Caltech Jet Propulsion Laboratory that a previously unknown comet (designation Rogue 2000) is hurtling in the direction of the Earth. You immediately transmit a request to NASA for the orbital parameters and the current positions of the Earth and Rogue 2000 and receive the following report:

ORBITAL PARAMETERS

EARTH	ROGUE 2000
Eccentricity: $e_1 = 0.017$	Eccentricity: $e_2 = 0.98$
Semimajor axis: $a_1 = 1 \text{ AU} = 1.496 \times 10^8$ km	Semimajor axis: $a_2 = 5 \text{ AU} = 7.48 \times 10^8$ km
Period: $T_1 = 1$ year	Period: $T_2 = 5\sqrt{5}$ years

INITIAL POSITION INFORMATION

The major axes of Earth and Rogue 2000 coincide.
The aphelions of Earth and Rogue 2000 are on the same side of the Sun.
Initial polar angle of Earth: $\theta = 0$ radians.
Initial polar angle of Rogue 2000: $\theta = 0.45$ radian.

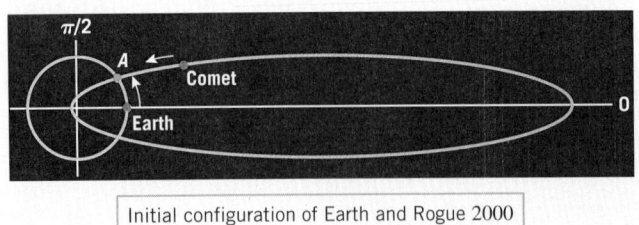

Initial configuration of Earth and Rogue 2000

Figure 1

The Calculation Strategy

Since the immediate concern is a possible collision at intersection A in Figure 1, your team works out the following plan:

Step 1. Find the polar equations for Earth and Rogue 2000.
Step 2. Find the polar coordinates of intersection A.
Step 3. Determine how long it will take the Earth to reach intersection A.
Step 4. Determine where Rogue 2000 will be when the Earth reaches intersection A.
Step 5. Determine how far Rogue 2000 will be from the Earth when the Earth is at intersection A.

Polar Equations of the Orbits

Exercise 1   Write polar equations of the form

$$r = \frac{a(1 - e^2)}{1 - e \cos \theta}$$

for the orbits of Earth and Rogue 2000 using AU units for r.

Exercise 2   Use a graphing utility to generate the two orbits on the same screen.

Intersection of the Orbits

The second step in your team's calculation plan is to find the polar coordinates of intersection A in Figure 1.

Exercise 3   For simplicity, let $k_1 = a_1(1 - e_1^2)$ and $k_2 = a_2(1 - e_2^2)$, and use the polar equations obtained in Exercise 1 to show that the angle θ at intersection A satisfies the equation

$$\cos \theta = \frac{k_1 - k_2}{k_1 e_2 - k_2 e_1}$$

Exercise 4   Use the result in Exercise 3 and the inverse cosine capability of a calculating utility to show that the angle θ at intersection A in Figure 1 is $\theta = 0.607$ radian.

Exercise 5   Use the result in Exercise 4 and either polar equation obtained in Exercise 1 to show that if r is in AU units, then the polar coordinates of intersection A are $(r, \theta) = (1.014, 0.607)$.

Time Required for Earth to Reach Intersection A

According to Kepler's second law (see 12.5.3), the radial line from the center of the Sun to the center of an object orbiting around it sweeps out equal areas in equal times. Thus, if t is the time that it takes for the radial line to sweep out an "elliptic sector" from some initial angle θ_I to some final angle θ_F (Figure 2), and if T is the period of the object (the time for one complete revolution),

then

$$\frac{t}{T} = \frac{\text{area of the "elliptic sector"}}{\text{area of the entire ellipse}} \qquad (1)$$

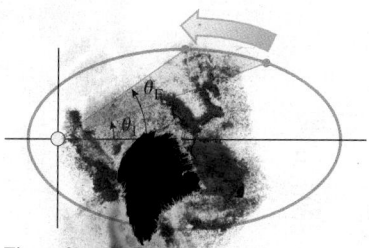

Figure 2

Exercise 6   Use Formula (1) to show that

$$t = \frac{T \displaystyle\int_{\theta_I}^{\theta_F} r^2\, d\theta}{2\pi a^2 \sqrt{1 - e^2}} \qquad (2)$$

Exercise 7   Use a calculating utility with a numerical integration capability, Formula (2), and the polar equation for the orbit of the Earth obtained in Exercise 1 to find the time t (in years) required for the Earth to move from its initial position to intersection A.

Position of Rogue 2000 When the Earth Is at Intersection A

The fourth step in your team's calculation strategy is to determine the position of Rogue 2000 when the Earth reaches intersection A.

Exercise 8   During the time that it takes for the Earth to move from its initial position to intersection A, the polar angle of Rogue 2000 will change from its initial value $\theta_I = 0.45$ radian to some final value θ_F that remains to be determined. Apply Formula (2) using the orbital data for Rogue 2000 and the time t obtained in Exercise 7 to show that θ_F satisfies the equation

$$\int_{0.45}^{\theta_F} \left[\frac{a_2(1 - e_2^2)}{1 - e_2 \cos\theta} \right]^2 d\theta = \frac{2t\pi a_2^2 \sqrt{1 - e_2^2}}{5\sqrt{5}} \qquad (3)$$

Your team is now faced with the problem of solving Equation (3) for the unknown upper limit θ_F. Some members of the team plan to use a CAS to perform the integration, some plan to use integration tables, and others plan to use hand calculation by making the substitution $u = \tan(\theta/2)$ and applying the formulas in (5) of Section 9.6.

Exercise 9

(a) Evaluate the integral in (3) using a CAS or by hand calculation.

(b) Use the root-finding capability of a calculating utility to find the polar angle of Rogue 2000 when the Earth is at intersection A.

Calculating the Critical Distance

It is the policy of your NEO tracking team to issue a notification to various governmental agencies for any asteroid or comet that will be within 4 million kilometers of the Earth at an orbital intersection. (This distance is roughly 10 times that between the Earth and the Moon.) Accordingly, the final step in your team's plan is to calculate the distance between the Earth and Rogue 2000 when the Earth is at intersection A, and then determine whether a notification should be issued.

Exercise 10   Use the polar equation of Rogue 2000 obtained in Exercise 1 and the result in Exercise 9(b) to find polar coordinates of Rogue 2000 with r in AU units when the Earth is at intersection A.

Exercise 11   Use the distance formula in Exercise 67(a) of Section 12.1 to calculate the distance between the Earth and Rogue 2000 in AU units when the Earth is at intersection A, and then use the conversion factor 1 AU $= 1.496 \times 10^8$ km to determine whether a government notification should be issued.

Note: One of the closest near misses in recent history occurred on October 30, ... the asteroid Hermes passed within 900,000 km of the Earth. More recently, on June 14 ... 68 the asteroid Icarus passed within 23,000,000 km of the Earth.

Module by Mary Ann Connors, USMA, West Point, and Howard Anton, Drexel University

Additional material for this module can be found on the World Wide Web at http://www.wiley.com/college/anton

Multivariable Calculus

13

THREE-DIMENSIONAL SPACE; VECTORS

Johann Bernoulli
Jacques Bernoulli

*I*n this chapter we will discuss rectangular coordinate systems in three dimensions, and we will study the analytic geometry of lines, planes, and other basic surfaces. The second theme of this chapter is the study of vectors. These are the mathematical objects that physicists and engineers use to study forces, displacements, and velocities of objects moving on curved paths. More generally, vectors are used to represent all physical entities that involve both a magnitude and a direction for their complete description. We will introduce various algebraic operations on vectors, and we will apply these operations to problems involving force, work, and rotational tendencies in two and three dimensions. Finally, we will discuss cylindrical and spherical coordinate systems, which are appropriate in problems that involve various kinds of symmetries and also have specific applications in navigation and celestial mechanics.

13.1 RECTANGULAR COORDINATES IN 3-SPACE; SPHERES; CYLINDRICAL SURFACES

In this section we will discuss coordinate systems in three-dimensional space and some basic facts about surfaces in three dimensions.

RECTANGULAR COORDINATE SYSTEMS

In the remainder of this text we will call three-dimensional space *3-space*, two-dimensional space (a plane) *2-space*, and one-dimensional space (a line) *1-space*. Just as points in 2-space can be placed in one-to-one correspondence with pairs of real numbers using two perpendicular coordinate lines, so points in 3-space can be placed in one-to-one correspondence with triples of real numbers by using three mutually perpendicular coordinate lines, called the *x-axis*, the *y-axis*, and the *z-axis*, positioned so that their origins coincide (Figure 13.1.1). The three coordinate axes form a three-dimensional *rectangular coordinate system* (or *Cartesian coordinate system*). The point of intersection of the coordinate axes is called the *origin* of the coordinate system.

Rectangular coordinate systems in 3-space fall into two categories: *left-handed* and *right-handed*. A right-handed system has the property that when the fingers of the right hand are cupped so that they curve from the positive *x*-axis toward the positive *y*-axis, the thumb points (roughly) in the direction of the positive *z*-axis (Figure 13.1.2*a*). Similarly for a left-handed coordinate system (Figure 13.1.2*b*). We will use only right-handed coordinate systems in this text.

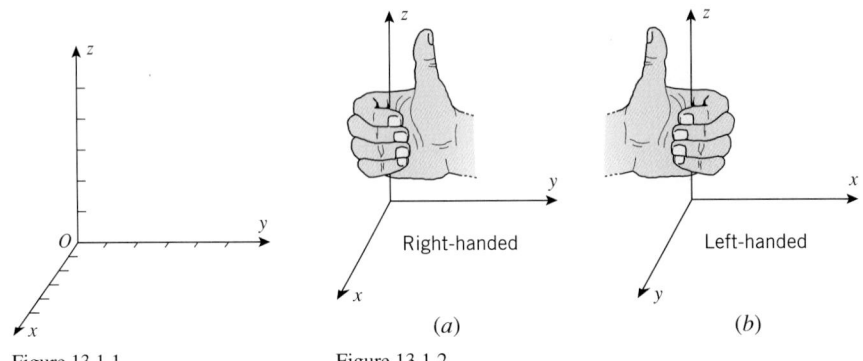

Right-handed   Left-handed

(a)   (b)

Figure 13.1.1   Figure 13.1.2

The coordinate axes, taken in pairs, determine three *coordinate planes*: the *xy-plane*, the *xz-plane*, and the *yz-plane*. To each point P in 3-space we can assign a triple of real numbers by passing three planes through P parallel to the coordinate planes and letting a, b, and c be the coordinates of the intersections of those planes with the *x*-axis, *y*-axis, and *z*-axis, respectively (Figure 13.1.3). We call a, b, and c the *x-coordinate*, *y-coordinate*, and *z-coordinate* of P, respectively, and we denote the point P by (a, b, c) or by $P(a, b, c)$. Figure 13.1.4 shows the points $(4, 5, 6)$ and $(-3, 2, -4)$.

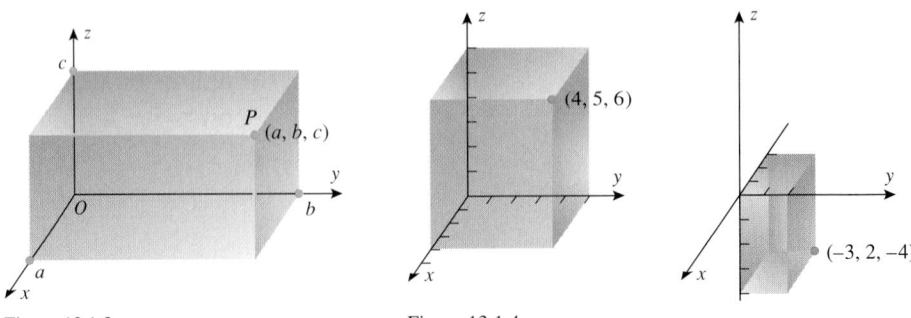

Figure 13.1.3   Figure 13.1.4

Just as the coordinate axes in a two-dimensional coordinate system divide 2-space into four quadrants, so the coordinate planes of a three-dimensional coordinate system divide 3-space into eight parts, called ***octants***. The set of points with three positive coordinates forms the ***first octant***; the remaining octants have no standard numbering.

You should be able to visualize the following facts about three-dimensional rectangular coordinate systems:

REGION	DESCRIPTION
xy-plane	Consists of all points of the form $(x, y, 0)$
xz-plane	Consists of all points of the form $(x, 0, z)$
yz-plane	Consists of all points of the form $(0, y, z)$
x-axis	Consists of all points of the form $(x, 0, 0)$
y-axis	Consists of all points of the form $(0, y, 0)$
z-axis	Consists of all points of the form $(0, 0, z)$

DISTANCE IN 3-SPACE

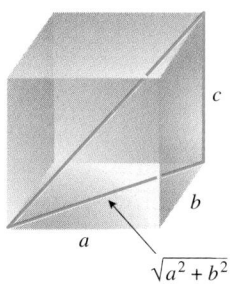

Figure 13.1.5

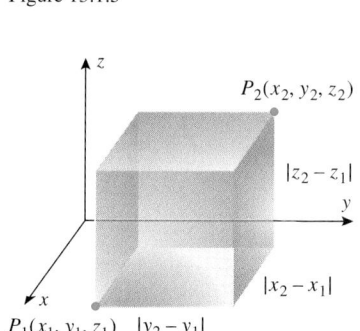

Figure 13.1.6

To derive a formula for the distance between two points in 3-space, we start by considering a box whose sides have lengths a, b, and c (Figure 13.1.5). The length d of a diagonal of the box can be obtained by applying the Theorem of Pythagoras twice: first to show that a diagonal of the base has length $\sqrt{a^2 + b^2}$, then again to show that a diagonal of the box has length

$$d = \sqrt{(\sqrt{a^2 + b^2})^2 + c^2} = \sqrt{a^2 + b^2 + c^2} \tag{1}$$

We can now obtain a formula for the distance d between two points $P_1(x_1, y_1, z_1)$ and $P_2(x_2, y_2, z_2)$ in 3-space by finding the length of the diagonal of a box that has these points as diagonal corners (Figure 13.1.6). The sides of such a box have lengths

$$|x_2 - x_1|, \quad |y_2 - y_1|, \quad \text{and} \quad |z_2 - z_1|$$

and hence from (1) the distance d between the points P_1 and P_2 is

$$d = \sqrt{(x_2 - x_1)^2 + (y_2 - y_1)^2 + (z_2 - z_1)^2} \tag{2}$$

(where we have omitted the unnecessary absolute value signs).

REMARK.   Recall that in 2-space the distance d between points $P_1(x_1, y_1)$ and $P_2(x_2, y_2)$ is

$$d = \sqrt{(x_2 - x_1)^2 + (y_2 - y_1)^2}$$

Thus, the distance formula in 3-space has the same form as the formula in 2-space, but it has a third term to account for the additional dimension. We will see that this is a common occurrence in extending formulas from 2-space to 3-space.

Example 1

Find the distance d between the points $(2, 3, -1)$ and $(4, -1, 3)$.

Solution. From Formula (2)

$$d = \sqrt{(4 - 2)^2 + (-1 - 3)^2 + (3 + 1)^2} = \sqrt{36} = 6 \qquad \blacktriangleleft$$

GRAPHS IN 3-SPACE

Recall that in an xy-coordinate system, the set of points (x, y) whose coordinates satisfy an equation in x and y is called the *graph* of the equation. Analogously, in an xyz-coordinate system, the set of points (x, y, z) whose coordinates satisfy an equation in x, y, and z is called the ***graph*** of the equation. For example, consider the equation

$$x^2 + y^2 + z^2 = 25$$

The coordinates of a point (x, y, z) satisfy this equation if and only if the distance from the origin to the point is 5 (why?). Thus, the graph of this equation is a sphere of radius 5 centered at the origin (Figure 13.1.7).

SPHERES

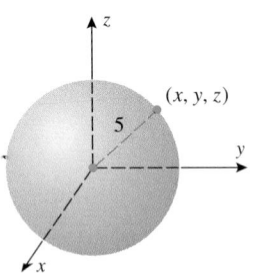

Figure 13.1.7

The sphere with center (x_0, y_0, z_0) and radius r consists of those points (x, y, z) whose coordinates satisfy

$$\sqrt{(x - x_0)^2 + (y - y_0)^2 + (z - z_0)^2} = r$$

or, equivalently,

$$(x - x_0)^2 + (y - y_0)^2 + (z - z_0)^2 = r^2 \tag{3}$$

This is called the **standard equation of the sphere** with center (x_0, y_0, z_0) and radius r. Some examples are given in the following table.

EQUATION	GRAPH
$(x - 3)^2 + (y - 2)^2 + (z - 1)^2 = 9$	Sphere with center $(3, 2, 1)$ and radius 3
$(x + 1)^2 + y^2 + (z + 4)^2 = 5$	Sphere with center $(-1, 0, -4)$ and radius $\sqrt{5}$
$x^2 + y^2 + z^2 = 1$	Sphere with center $(0, 0, 0)$ and radius 1

Recall that in 2-space the standard equation of the circle with center (x_0, y_0) and radius r is

$$(x - x_0)^2 + (y - y_0)^2 = r^2$$

Thus, the standard equation of a sphere in 3-space has the same form as the standard equation of a circle in 2-space, but with an additional term to account for the third coordinate.

If the terms in (3) are squared out and like terms are then collected, then the resulting equation has the form

$$x^2 + y^2 + z^2 + Gx + Hy + Iz + J = 0 \tag{4}$$

The following example shows how the center and radius of a sphere that is expressed in this form can be obtained by completing the squares.

Example 2

Find the center and radius of the sphere

$$x^2 + y^2 + z^2 - 2x - 4y + 8z + 17 = 0$$

Solution. We can put the equation in the form of (3) by completing the squares:

$$(x^2 - 2x) + (y^2 - 4y) + (z^2 + 8z) = -17$$
$$(x^2 - 2x + 1) + (y^2 - 4y + 4) + (z^2 + 8z + 16) = -17 + 21$$
$$(x - 1)^2 + (y - 2)^2 + (z + 4)^2 = 4$$

which is the equation of the sphere with center $(1, 2, -4)$ and radius 2.   ◀

In general, completing the squares in (4) produces an equation of the form

$$(x - x_0)^2 + (y - y_0)^2 + (z - z_0)^2 = k$$

If $k > 0$, then the graph of this equation is a sphere with center (x_0, y_0, z_0) and radius $\sqrt{k}$. If $k = 0$, then the sphere has radius zero, so the graph is the single point (x_0, y_0, z_0). If $k < 0$, the equation is not satisfied by any values of x, y, and z (why?), so it has no graph.

> **13.1.1   THEOREM.**   *An equation of the form*
> $$x^2 + y^2 + z^2 + Gx + Hy + Iz + J = 0$$
> *represents a sphere, a point, or has no graph.*

CYLINDRICAL SURFACES

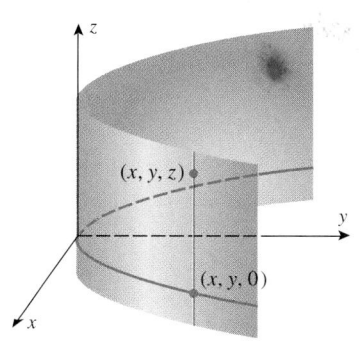

Figure 13.1.8

Although it is natural to graph equations in two variables in 2-space and equations in three variables in 3-space, it is also possible to graph equations in two variables in 3-space. For example, the graph of the equation $y = x^2$ in an xy-coordinate system is a parabola; however, there is nothing to prevent us from writing this equation as $y = x^2 + 0z$ and inquiring about its graph in an xyz-coordinate system. To obtain this graph we need only observe that the equation $y = x^2$ does not impose any restrictions on z. Thus, if we find values of x and y that satisfy this equation, then the coordinates of the point (x, y, z) will also satisfy the equation for *arbitrary* values of z. Geometrically, the point (x, y, z) lies on the vertical line through the point $(x, y, 0)$ in the xy-plane, which means that we can obtain the graph of $y = x^2$ in an xyz-coordinate system by first graphing the equation in the xy-plane and then translating that graph parallel to the z-axis to generate the entire graph (Figure 13.1.8).

The process of generating a surface by translating a plane curve parallel to some line is called ***extrusion***, and surfaces that are generated by extrusion are called ***cylindrical surfaces***. A familiar example is the surface of a right circular cylinder, which can be generated by translating a circle parallel to the axis of the cylinder. The following theorem provides basic information about graphing equations in two variables in 3-space:

> **13.1.2   THEOREM.**   *An equation that contains only two of the variables x, y, and z represents a cylindrical surface in an xyz-coordinate system. The surface can be obtained by graphing the equation in the coordinate plane of the two variables that appear in the equation and then translating that graph parallel to the axis of the missing variable.*

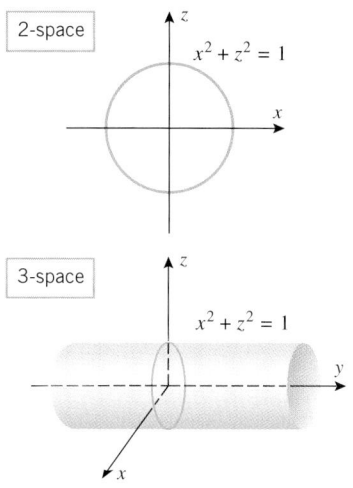

Figure 13.1.9

Example 3

Sketch the graph of $x^2 + z^2 = 1$ in 3-space.

Solution.   Since y does not appear in this equation, the graph is a cylindrical surface generated by extrusion parallel to the y-axis. In the xz-plane the graph of the equation $x^2 + z^2 = 1$ is a circle (Figure 13.1.9). Thus, in 3-space the graph is a right circular cylinder along the y-axis.   ◄

Example 4

Sketch the graph of $z = \sin y$ in 3-space.

Solution.   (See Figure 13.1.10.)   ◄

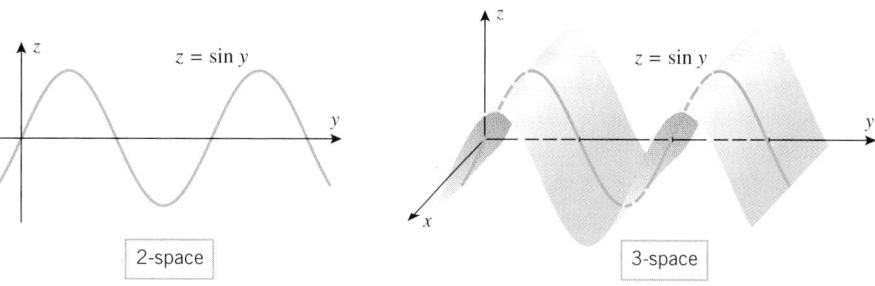

Figure 13.1.10

FOR THE READER.   Describe the graph of $x = 1$ in an xyz-coordinate system.

EXERCISE SET 13.1 ☑ Graphing Calculator

1. In each part, find the coordinates of the eight corners of the box.

(a)   (b)

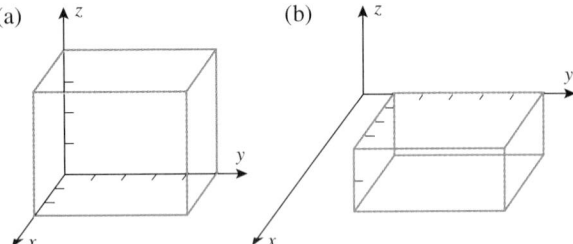

2. A cube of side 4 has its geometric center at the origin and its faces parallel to the coordinate planes. Sketch the cube and give the coordinates of the corners.

3. Suppose that a box has its faces parallel to the coordinate planes and the points $(4, 2, -2)$ and $(-6, 1, 1)$ are endpoints of a diagonal. Sketch the box and give the coordinates of the remaining six corners.

4. Suppose that a box has its faces parallel to the coordinate planes and the points (x_1, y_1, z_1) and (x_2, y_2, z_2) are endpoints of a diagonal.
   (a) Find the coordinates of the remaining six corners.
   (b) Show that the midpoint of the line segment joining (x_1, y_1, z_1) and (x_2, y_2, z_2) is
   $$\left(\tfrac{1}{2}(x_1 + x_2), \tfrac{1}{2}(y_1 + y_2), \tfrac{1}{2}(z_1 + z_2)\right)$$
   [*Suggestion:* Apply Theorem D.2 in Appendix D to three appropriate edges of the box.]

5. Find the center and radius of the sphere that has $(1, -2, 4)$ and $(3, 4, -12)$ as endpoints of a diameter. [See Exercise 4.]

6. Show that $(4, 5, 2)$, $(1, 7, 3)$, and $(2, 4, 5)$ are vertices of an equilateral triangle.

7. (a) Show that $(2, 1, 6)$, $(4, 7, 9)$, and $(8, 5, -6)$ are the vertices of a right triangle.
   (b) Which vertex is at the $90°$ angle?
   (c) Find the area of the triangle.

8. Find the distance from the point $(-5, 2, -3)$ to the
   (a) xy-plane   (b) xz-plane   (c) yz-plane
   (d) x-axis   (e) y-axis   (f) z-axis.

9. In each part, find the standard equation of the sphere that satisfies the stated conditions.
   (a) Center $(1, 0, -1)$; diameter $= 8$.
   (b) Center $(-1, 3, 2)$ and passing through the origin.
   (c) A diameter has endpoints $(-1, 2, 1)$ and $(0, 2, 3)$.

10. Find equations of two spheres that are centered at the origin and are tangent to the sphere of radius 1 centered at $(3, -2, 4)$.

11. In each part, find an equation of the sphere with center $(2, -1, -3)$ and satisfying the given condition.
    (a) Tangent to the xy-plane

(b) Tangent to the xz-plane
(c) Tangent to the yz-plane

12. (a) Find an equation of the sphere that is inscribed in the cube that is centered at the point $(-2, 1, 3)$ and has sides of length 1 that are parallel to the coordinate planes.
    (b) Find an equation of the sphere that is circumscribed about the cube in part (a).

In Exercises 13–18, describe the surface whose equation is given.

13. $x^2 + y^2 + z^2 + 10x + 4y + 2z - 19 = 0$
14. $x^2 + y^2 + z^2 - y = 0$
15. $2x^2 + 2y^2 + 2z^2 - 2x - 3y + 5z - 2 = 0$
16. $x^2 + y^2 + z^2 + 2x - 2y + 2z + 3 = 0$
17. $x^2 + y^2 + z^2 - 3x + 4y - 8z + 25 = 0$
18. $x^2 + y^2 + z^2 - 2x - 6y - 8z + 1 = 0$

19. In each part, sketch the portion of the surface that lies in the first octant.
    (a) $y - x$ (b) $y = z$ (c) $x = z$

20. In each part, sketch the graph of the equation in 3-space.
    (a) $x = 1$ (b) $y = 1$ (c) $z = 1$

21. In each part, sketch the graph of the equation in 3-space.
    (a) $x^2 + y^2 = 25$ (b) $y^2 + z^2 = 25$ (c) $x^2 + z^2 = 25$

22. In each part, sketch the graph of the equation in 3-space.
    (a) $y = x^2$ (b) $z = x^2$ (c) $y = z^2$

23. In each part, write an equation for the surface.
    (a) The plane that contains the x-axis and the point $(0, 1, 2)$.
    (b) The plane that contains the y-axis and the point $(1, 0, 2)$.
    (c) The right circular cylinder that has radius 1 and is centered on the line parallel to the z-axis that passes through the point $(1, 1, 0)$.
    (d) The right circular cylinder that has radius 1 and is centered on the line parallel to the y-axis that passes through the point $(1, 0, 1)$.

24. Find equations for the following right circular cylinders. Each cylinder has radius a and is "tangent" to two coordinate planes.

(a)   (b)   (c)

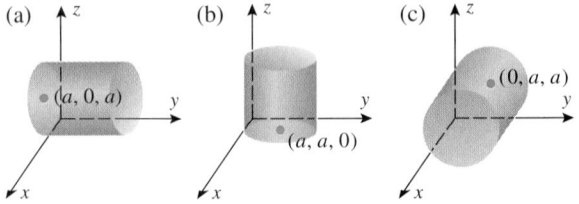

In Exercises 25–34, sketch the surface in 3-space.

25. $y = \sin x$ 26. $y = e^x$

27. $z = 1 - y^2$

28. $z = \cos x$

29. $2x + z = 3$

30. $2x + 3y = 6$

31. $4x^2 + 9z^2 = 36$

32. $z = \sqrt{3 - x}$

33. $y^2 - 4z^2 = 4$

34. $yz = 1$

35. Use a graphing utility to generate the curve $y = x^3/(1+x^2)$ in the xy-plane, and then use the graph to help sketch the surface $z = y^3/(1 + y^2)$ in 3-space.

36. Use a graphing utility to generate the curve $y = x/(1+x^4)$ in the xy-plane, and then use the graph to help sketch the surface $z = y/(1 + y^4)$ in 3-space.

37. If a bug walks on the sphere
$$x^2 + y^2 + z^2 + 2x - 2y - 4z - 3 = 0$$
how close and how far can it get from the origin?

38. Describe the set of all points in 3-space whose coordinates satisfy the inequality $x^2 + y^2 + z^2 - 2x + 8z \le 8$.

39. Describe the set of all points in 3-space whose coordinates satisfy the inequality $y^2 + z^2 + 6y - 4z > 3$.

40. The distance between a point $P(x, y, z)$ and the point $A(1, -2, 0)$ is twice the distance between P and the point $B(0, 1, 1)$. Show that the set of all such points is a sphere, and find the center and radius of the sphere.

41. As shown in the accompanying figure, a bowling ball of radius R is placed inside a box just large enough to hold it,
and it is secured for shipping by packing a Styrofoam sphere into each corner of the box. Find the radius of the largest Styrofoam sphere that can be used. [*Hint:* Take the origin of a Cartesian coordinate system at a corner of the box with the coordinate axes along the edges.]

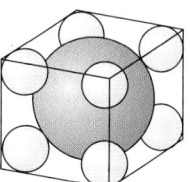

Figure Ex-41

42. Consider the equation
$$x^2 + y^2 + z^2 + Gx + Hy + Iz + J = 0$$
and let $K = G^2 + H^2 + I^2 - 4J$.
(a) Prove that the equation represents a sphere if $K > 0$, a point if $K = 0$, and has no graph if $K < 0$.
(b) In the case where $K > 0$, find the center and radius of the sphere.

43. Show that for all values of θ and ϕ, the point
$$(a \sin \phi \cos \theta, a \sin \phi \sin \theta, a \cos \phi)$$
lies on the sphere $x^2 + y^2 + z^2 = a^2$.

13.2 VECTORS

Many physical quantities such as area, length, mass, and temperature are completely described once the magnitude of the quantity is given. Such quantities are called "scalars." Other physical quantities, called "vectors," are not completely determined until both a magnitude and a direction are specified. For example, winds are usually described by giving their speed and direction, say 20 mi/h northeast. The wind speed and wind direction together form a vector quantity called the wind velocity. Other examples of vectors are force and displacement. In this section we will develop the basic mathematical properties of vectors.

VECTORS IN PHYSICS AND ENGINEERING

A particle that moves along a line can move in only two directions, so its direction of motion can be described by taking one direction to be positive and the other negative. Thus, the *displacement* or *change in position* of the point can be described by a signed real number. For example, a displacement of 3 ($= +3$) describes a position change of 3 units in the positive direction, and a displacement of -3 describes a position change of 3 units in the negative direction. However, for a particle that moves in two dimensions or three dimensions, a plus or minus sign is no longer sufficient to specify the direction of motion—other methods are required. One method is to use an arrow, called a *vector*, that points in the direction of motion and whose length represents the distance from the starting point to the ending point; this is called the *displacement vector* for the motion. For example, Figure 13.2.1*a* shows the displacement vector of a particle that moves from point A to point B along a circuitous path. Note that the length of the arrow describes the distance between the starting and ending points and not the actual distance traveled by the particle.

Arrows are not limited to describing displacements—they can be used to describe any physical quantity that involves both a magnitude and direction. Two important examples are forces and velocities. For example, the arrow in Figure 13.2.1*b* shows a force vector of 10 lb acting in a specific direction on a block, and the arrows in Figure 13.2.1*c* show the velocity vector of a boat whose motor propels it parallel to the shore at 2 mi/h and the velocity vector of a 3 mi/h wind acting at an angle of 45° with the shoreline. Intuition suggests that the two velocity vectors will combine to produce some net velocity for the boat at an angle to the shoreline. Thus, our first objective in this section is to define mathematical operations on vectors that can be used to determine the combined effect of vectors.

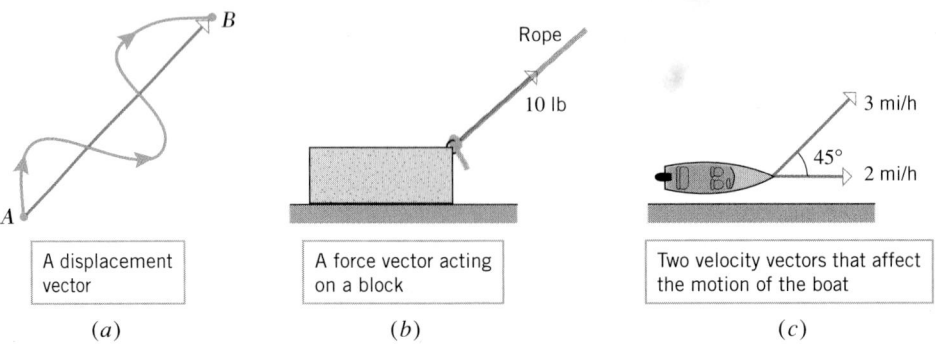

A displacement vector	A force vector acting on a block	Two velocity vectors that affect the motion of the boat
(*a*)	(*b*)	(*c*)

Figure 13.2.1

VECTORS VIEWED GEOMETRICALLY

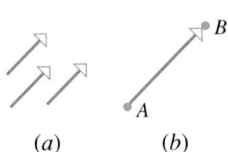

(*a*)	(*b*)

Figure 13.2.2

Vectors can be represented geometrically by arrows in 2-space or 3-space; the direction of the arrow specifies the direction of the vector and the length of the arrow describes its magnitude. The tail of the arrow is called the ***initial point*** of the vector, and the tip of the arrow the ***terminal point***. We will denote vectors with lowercase boldface type such as **a**, **k**, **v**, **w**, and **x**. When discussing vectors, we will refer to real numbers as ***scalars***. Scalars will be denoted by lowercase italic type such as a, k, v, w, and x. Two vectors, **v** and **w**, are considered to be ***equal*** (also called ***equivalent***) if they have the same length and same direction, in which case we write **v** = **w**. Geometrically, two vectors are equal if they are translations of one another; thus, the three vectors in Figure 13.2.2*a* are equal, even though they are in different positions.

Because vectors are not affected by translation, the initial point of a vector **v** can be moved to any convenient point A by making an appropriate translation. If the initial point of **v** is A and the terminal point is B, then we write $\mathbf{v} = \overrightarrow{AB}$ when we want to emphasize the initial and terminal points (Figure 13.2.2*b*). If the initial and terminal points of a vector coincide, then the vector has length zero; we call this the ***zero vector*** and denote it by **0**. The zero vector does not have a specific direction, so we will agree that it can be assigned any convenient direction in a specific problem.

There are various algebraic operations that are performed on vectors, all of whose definitions originated in physics. We begin with vector addition.

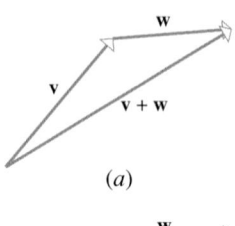

(*a*)

> **13.2.1**   DEFINITION.   If **v** and **w** are vectors, then the ***sum*** **v** + **w** is the vector from the initial point of **v** to the terminal point of **w** when the vectors are positioned so the initial point of **w** is at the terminal point of **v** (Figure 13.2.3*a*).

In Figure 13.2.3*b* we have constructed two sums, **v** + **w** (purple arrows) and **w** + **v** (green arrows). It is evident that

$$\mathbf{v} + \mathbf{w} = \mathbf{w} + \mathbf{v}$$

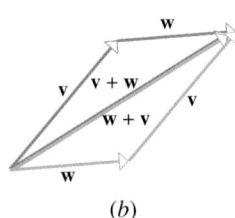

(*b*)

Figure 13.2.3

and that the sum coincides with the diagonal of the parallelogram determined by **v** and **w** when these vectors are positioned so they have the same initial point.

Since the initial and terminal points of **0** coincide, it follows that

$$\mathbf{0} + \mathbf{v} = \mathbf{v} + \mathbf{0} = \mathbf{v}$$

13.2.2 DEFINITION. If **v** is a nonzero vector and k is a nonzero real number (a scalar), then the ***scalar multiple*** $k\mathbf{v}$ is defined to be the vector whose length is $|k|$ times the length of **v** and whose direction is the same as that of **v** if $k > 0$ and opposite to that of **v** if $k < 0$. We define $k\mathbf{v} = \mathbf{0}$ if $k = 0$ or $\mathbf{v} = \mathbf{0}$.

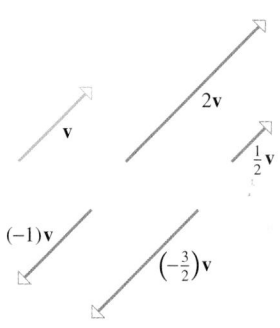

Figure 13.2.4

Figure 13.2.4 shows the geometric relationship between a vector **v** and various scalar multiples of it. Observe that if k and **v** are nonzero, then the vectors **v** and $k\mathbf{v}$ lie on the same line if their initial points coincide and lie on parallel or coincident lines if they do not. Thus, we say that **v** and $k\mathbf{v}$ are ***parallel vectors***. Observe also that the vector $(-1)\mathbf{v}$ has the same length as **v** but is oppositely directed. We call $(-1)\mathbf{v}$ the ***negative*** of **v** and denote it by $-\mathbf{v}$ (Figure 13.2.5). In particular, $-\mathbf{0} = (-1)\mathbf{0} = \mathbf{0}$.

Vector subtraction is defined in terms of addition and scalar multiplication by

$$\mathbf{v} - \mathbf{w} = \mathbf{v} + (-\mathbf{w})$$

The difference $\mathbf{v} - \mathbf{w}$ can be obtained geometrically by first constructing the vector $-\mathbf{w}$ and then adding **v** and $-\mathbf{w}$, say by the parallelogram method (Figure 13.2.6a). However, if **v** and **w** are positioned so their initial points coincide, then $\mathbf{v} - \mathbf{w}$ can be formed more directly, as shown in Figure 13.2.6b, by drawing the vector from the terminal point of **w** (the second term) to the terminal point of **v** (the first term). In the special case where $\mathbf{v} = \mathbf{w}$ the terminal points of the vectors coincide, so their difference is **0**; that is,

$$\mathbf{v} + (-\mathbf{v}) = \mathbf{v} - \mathbf{v} = \mathbf{0}$$

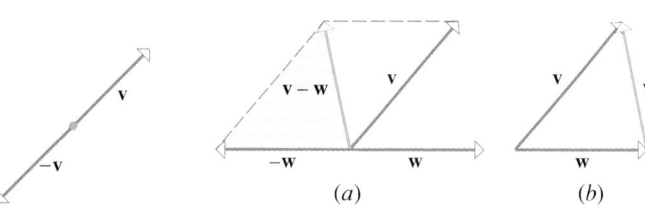

(a)   (b)

Figure 13.2.5   Figure 13.2.6

VECTORS IN COORDINATE SYSTEMS

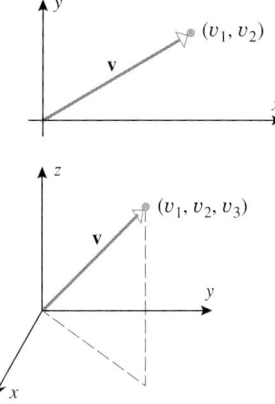

Figure 13.2.7

Problems involving vectors are often best solved by introducing a rectangular coordinate system. If a vector **v** is positioned with its initial point at the origin of a rectangular coordinate system, then its terminal point will have coordinates of the form (v_1, v_2) or (v_1, v_2, v_3), depending on whether the vector is in 2-space or 3-space (Figure 13.2.7). We call these coordinates the ***components*** of **v**, and we write

$$\mathbf{v} = \langle v_1, v_2 \rangle \quad \text{or} \quad \mathbf{v} = \langle v_1, v_2, v_3 \rangle$$

| 2-space | 3-space |

In particular, the zero vector is

$$\mathbf{0} = \langle 0, 0 \rangle \quad \text{and} \quad \mathbf{0} = \langle 0, 0, 0 \rangle$$

| 2-space | 3-space |

Components provide a simple way of identifying equivalent vectors. For example, consider the vectors $\mathbf{v} = \langle v_1, v_2 \rangle$ and $\mathbf{w} = \langle w_1, w_2 \rangle$ in 2-space. If $\mathbf{v} = \mathbf{w}$, then the vectors have the same length and same direction, and this means that their terminal points coincide when their initial points are placed at the origin. It follow that $v_1 = w_1$ and $v_2 = w_2$, so we have shown that equivalent vectors have the same components. Conversely, if $v_1 = w_1$

and $v_2 = w_2$, then the terminal points of the vectors coincide when their initial points are placed at the origin. It follows that the vectors have the same length and same direction, so we have shown that vectors with the same components are equivalent. A similar argument holds for vectors in 3-space, so we have the following result:

13.2.3 THEOREM. *Two vectors are equivalent if and only if their corresponding components are equal.*

For example,

$$\langle a, b, c \rangle = \langle 1, -4, 2 \rangle$$

if and only if $a = 1$, $b = -4$, and $c = 2$.

• •
ARITHMETIC OPERATIONS ON VECTORS

The next theorem shows how to perform arithmetic operations on vectors using components.

13.2.4 THEOREM. *If* $\mathbf{v} = \langle v_1, v_2 \rangle$ *and* $\mathbf{w} = \langle w_1, w_2 \rangle$ *are vectors in 2-space and k is any scalar, then*

$$\mathbf{v} + \mathbf{w} = \langle v_1 + w_1, v_2 + w_2 \rangle \tag{1}$$
$$\mathbf{v} - \mathbf{w} = \langle v_1 - w_1, v_2 - w_2 \rangle \tag{2}$$
$$k\mathbf{v} = \langle kv_1, kv_2 \rangle \tag{3}$$

Similarly, if $\mathbf{v} = \langle v_1, v_2, v_3 \rangle$ *and* $\mathbf{w} = \langle w_1, w_2, w_3 \rangle$ *are vectors in 3-space and k is any scalar, then*

$$\mathbf{v} + \mathbf{w} = \langle v_1 + w_1, v_2 + w_2, v_3 + w_3 \rangle \tag{4}$$
$$\mathbf{v} - \mathbf{w} = \langle v_1 - w_1, v_2 - w_2, v_3 - w_3 \rangle \tag{5}$$
$$k\mathbf{v} = \langle kv_1, kv_2, kv_3 \rangle \tag{6}$$

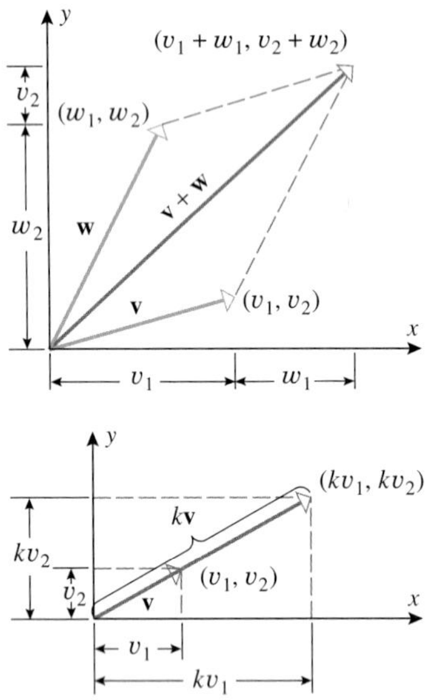

Figure 13.2.8

We will not prove this theorem. However, results (1) and (3) should be evident from Figure 13.2.8. Similar figures in 3-space can be used to motivate (4) and (6). Formulas (2) and (5) can be obtained by writing $\mathbf{v} + \mathbf{w} = \mathbf{v} + (-1)\mathbf{w}$.

Example 1

If $\mathbf{v} = \langle -2, 0, 1 \rangle$ and $\mathbf{w} = \langle 3, 5, -4 \rangle$, then

$$\mathbf{v} + \mathbf{w} = \langle -2, 0, 1 \rangle + \langle 3, 5, -4 \rangle = \langle 1, 5, -3 \rangle$$
$$3\mathbf{v} = \langle -6, 0, 3 \rangle$$
$$-\mathbf{w} = \langle -3, -5, 4 \rangle$$
$$\mathbf{w} - 2\mathbf{v} = \langle 3, 5, -4 \rangle - \langle -4, 0, 2 \rangle = \langle 7, 5, -6 \rangle \qquad \blacktriangleleft$$

• •
VECTORS WITH INITIAL POINT NOT AT THE ORIGIN

Recall that we defined the components of a vector to be the coordinates of its terminal point when its initial point is at the origin. We will now consider the problem of finding the components of a vector whose initial point is not at the origin. To be specific, suppose that $P_1(x_1, y_1)$ and $P_2(x_2, y_2)$ are points in 2-space and we are interested in finding the components of the vector $\overrightarrow{P_1P_2}$. As illustrated in Figure 13.2.9, we can write this vector as

$$\overrightarrow{P_1P_2} = \overrightarrow{OP_2} - \overrightarrow{OP_1} = \langle x_2, y_2 \rangle - \langle x_1, y_1 \rangle = \langle x_2 - x_1, y_2 - y_1 \rangle$$

Thus, we have shown that the components of the vector $\overrightarrow{P_1P_2}$ can be obtained by subtracting the coordinates of its initial point from the coordinates of its terminal point. Similar computations hold in 3-space, so we have established the following result:

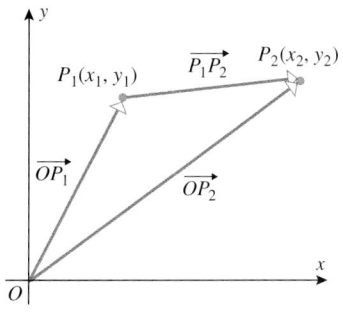

Figure 13.2.9

13.2.5 THEOREM. *If $\overrightarrow{P_1P_2}$ is a vector in 2-space with initial point $P_1(x_1, y_1)$ and terminal point $P_2(x_2, y_2)$, then*

$$\overrightarrow{P_1P_2} = \langle x_2 - x_1, y_2 - y_1 \rangle \tag{7}$$

Similarly, if $\overrightarrow{P_1P_2}$ is a vector in 3-space with initial point $P_1(x_1, y_1, z_1)$ and terminal point $P_2(x_2, y_2, z_2)$, then

$$\overrightarrow{P_1P_2} = \langle x_2 - x_1, y_2 - y_1, z_2 - z_1 \rangle \tag{8}$$

Example 2

In 2-space the vector from $P_1(1, 3)$ to $P_2(4, -2)$ is

$$\overrightarrow{P_1P_2} = \langle 4 - 1, -2 - 3 \rangle = \langle 3, -5 \rangle$$

and in 3-space the vector from $A(0, -2, 5)$ to $B(3, 4, -1)$ is

$$\overrightarrow{AB} = \langle 3 - 0, 4 - (-2), -1 - 5 \rangle = \langle 3, 6, -6 \rangle \qquad \blacktriangleleft$$

RULES OF VECTOR ARITHMETIC

The following theorem shows that many of the familiar rules of ordinary arithmetic also hold for vector arithmetic.

13.2.6 THEOREM. *For any vectors $\mathbf{u}$, $\mathbf{v}$, and $\mathbf{w}$ and any scalars k and l, the following relationships hold*:

(*a*) $\mathbf{u} + \mathbf{v} = \mathbf{v} + \mathbf{u}$

(*b*) $(\mathbf{u} + \mathbf{v}) + \mathbf{w} = \mathbf{u} + (\mathbf{v} + \mathbf{w})$

(*c*) $\mathbf{u} + \mathbf{0} = \mathbf{0} + \mathbf{u} = \mathbf{u}$

(*d*) $\mathbf{u} + (-\mathbf{u}) = \mathbf{0}$

(*e*) $k(l\mathbf{u}) = (kl)\mathbf{u}$

(*f*) $k(\mathbf{u} + \mathbf{v}) = k\mathbf{u} + k\mathbf{v}$

(*g*) $(k + l)\mathbf{u} = k\mathbf{u} + l\mathbf{u}$

(*h*) $1\mathbf{u} = \mathbf{u}$

The results in this theorem can be proved either algebraically by using components or geometrically by treating the vectors as arrows. We will prove part (*b*) both ways and leave some of the remaining proofs as exercises.

Proof (b) (Algebraic in 2-space). Let $\mathbf{u} = \langle u_1, u_2 \rangle$, $\mathbf{v} = \langle v_1, v_2 \rangle$, and $\mathbf{w} = \langle w_1, w_2 \rangle$. Then

$$
\begin{aligned}
(\mathbf{u} + \mathbf{v}) + \mathbf{w} &= (\langle u_1, u_2 \rangle + \langle v_1, v_2 \rangle) + \langle w_1, w_2 \rangle \\
&= \langle u_1 + v_1, u_2 + v_2 \rangle + \langle w_1, w_2 \rangle \\
&= \langle (u_1 + v_1) + w_1, (u_2 + v_2) + w_2 \rangle \\
&= \langle u_1 + (v_1 + w_1), u_2 + (v_2 + w_2) \rangle \\
&= \langle u_1, u_2 \rangle + \langle v_1 + w_1, v_2 + w_2 \rangle \\
&= \mathbf{u} + (\mathbf{v} + \mathbf{w})
\end{aligned}
$$

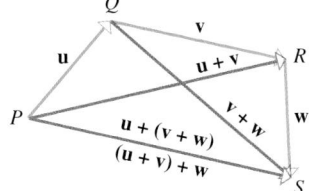

Figure 13.2.10

Proof (b) (Geometric). Let $\mathbf{u}$, $\mathbf{v}$, and $\mathbf{w}$ be represented by $\overrightarrow{PQ}$, $\overrightarrow{QR}$, and $\overrightarrow{RS}$ as shown in Figure 13.2.10. Then

$$\mathbf{v} + \mathbf{w} = \overrightarrow{QS} \quad \text{and} \quad \mathbf{u} + (\mathbf{v} + \mathbf{w}) = \overrightarrow{PS}$$

$$\mathbf{u} + \mathbf{v} = \overrightarrow{PR} \quad \text{and} \quad (\mathbf{u} + \mathbf{v}) + \mathbf{w} = \overrightarrow{PS}$$

Therefore,

$$(\mathbf{u} + \mathbf{v}) + \mathbf{w} = \mathbf{u} + (\mathbf{v} + \mathbf{w}) \qquad \blacksquare$$

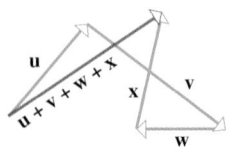

Figure 13.2.11

REMARK. It follows from part (*b*) of this theorem that the symbol $\mathbf{u} + \mathbf{v} + \mathbf{w}$ is unambiguous since the same vector results no matter how the terms are grouped. Moreover, Figure 13.2.10 shows that if the vectors $\mathbf{u}$, $\mathbf{v}$, and $\mathbf{w}$ are placed "tip to tail," then the sum $\mathbf{u} + \mathbf{v} + \mathbf{w}$ is the vector from the initial point of $\mathbf{u}$ to the terminal point of $\mathbf{w}$. This also works for four or more vectors (Figure 13.2.11).

NORM OF A VECTOR

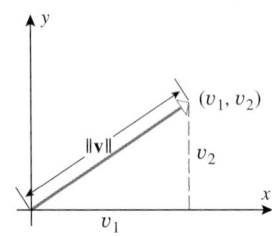

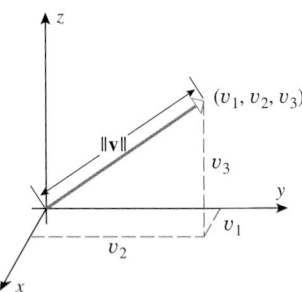

Figure 13.2.12

The distance between the initial and terminal points of a vector $\mathbf{v}$ is called the **length**, the **norm**, or the **magnitude** of $\mathbf{v}$ and is denoted by $\|\mathbf{v}\|$. This distance does not change if the vector is translated, so for purposes of calculating the norm we can assume that the vector is positioned with its initial point at the origin (Figure 13.2.12). This makes it evident that the norm of a vector $\mathbf{v} = \langle v_1, v_2 \rangle$ in 2-space is given by

$$\|\mathbf{v}\| = \sqrt{v_1^2 + v_2^2} \tag{9}$$

and the norm of a vector $\mathbf{v} = \langle v_1, v_2, v_3 \rangle$ in 3-space is given by

$$\|\mathbf{v}\| = \sqrt{v_1^2 + v_2^2 + v_3^2} \tag{10}$$

Example 3

Find the norm of $\mathbf{v} = \langle -2, 3 \rangle$ and $\mathbf{w} = \langle 2, 3, 6 \rangle$.

Solution. From (9) and (10)

$$\|\mathbf{v}\| = \sqrt{(-2)^2 + 3^2} = \sqrt{13}$$

$$\|\mathbf{w}\| = \sqrt{2^2 + 3^2 + 6^2} = \sqrt{49} = 7 \qquad \blacktriangleleft$$

Recall from Definition 13.2.2 that the length of $k\mathbf{v}$ is $|k|$ times the length of $\mathbf{v}$; that is,

$$\|k\mathbf{v}\| = |k|\|\mathbf{v}\| \tag{11}$$

Thus, for example,

$$\|3\mathbf{v}\| = |3|\|\mathbf{v}\| = 3\|\mathbf{v}\|$$

$$\|-2\mathbf{v}\| = |-2|\|\mathbf{v}\| = 2\|\mathbf{v}\|$$

$$\|-\mathbf{v}\| = |-1|\|\mathbf{v}\| = \|\mathbf{v}\|$$

This applies to vectors in 2-space and 3-space.

UNIT VECTORS

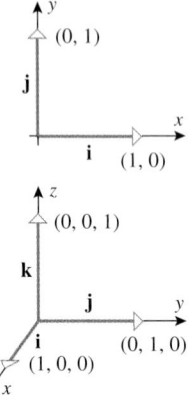

Figure 13.2.13

A vector of length 1 is called a **unit vector**. In an *xy*-coordinate system the unit vectors along the *x*- and *y*-axes are denoted by $\mathbf{i}$ and $\mathbf{j}$, respectively; and in an *xyz*-coordinate system the unit vectors along the *x*-, *y*-, and *z*-axes are denoted by $\mathbf{i}$, $\mathbf{j}$, and $\mathbf{k}$, respectively (Figure 13.2.13). Thus,

$$\mathbf{i} = \langle 1, 0 \rangle, \qquad \mathbf{j} = \langle 0, 1 \rangle \qquad \boxed{\text{In 2-space}}$$

$$\mathbf{i} = \langle 1, 0, 0 \rangle, \quad \mathbf{j} = \langle 0, 1, 0 \rangle, \quad \mathbf{k} = \langle 0, 0, 1 \rangle \qquad \boxed{\text{In 3-space}}$$

Every vector in 2-space is expressible uniquely in terms of $\mathbf{i}$ and $\mathbf{j}$, and every vector in 3-space is expressible uniquely in terms of $\mathbf{i}$, $\mathbf{j}$, and $\mathbf{k}$ as follows:

$$\mathbf{v} = \langle v_1, v_2 \rangle = \langle v_1, 0 \rangle + \langle 0, v_2 \rangle = v_1 \langle 1, 0 \rangle + v_2 \langle 0, 1 \rangle = v_1 \mathbf{i} + v_2 \mathbf{j}$$

$$\mathbf{v} = \langle v_1, v_2, v_3 \rangle = v_1 \langle 1, 0, 0 \rangle + v_2 \langle 0, 1, 0 \rangle + v_3 \langle 0, 0, 1 \rangle = v_1 \mathbf{i} + v_2 \mathbf{j} + v_3 \mathbf{k}$$

REMARK. The bracket and unit vector notations for vectors are completely interchangeable, the choice being a matter of convenience or personal preference.

Example 4

2-SPACE	3-SPACE
$\langle 2, 3 \rangle = 2\mathbf{i} + 3\mathbf{j}$	$\langle 2, -3, 4 \rangle = 2\mathbf{i} - 3\mathbf{j} + 4\mathbf{k}$
$\langle -4, 0 \rangle = -4\mathbf{i} + 0\mathbf{j} = -4\mathbf{i}$	$\langle 0, 3, 0 \rangle = 3\mathbf{j}$
$\langle 0, 0 \rangle = 0\mathbf{i} + 0\mathbf{j} = \mathbf{0}$	$\langle 0, 0, 0 \rangle = 0\mathbf{i} + 0\mathbf{j} + 0\mathbf{k} = \mathbf{0}$
$(3\mathbf{i} + 2\mathbf{j}) + (4\mathbf{i} + \mathbf{j}) = 7\mathbf{i} + 3\mathbf{j}$	$(3\mathbf{i} + 2\mathbf{j} - \mathbf{k}) - (4\mathbf{i} - \mathbf{j} + 2\mathbf{k}) = -\mathbf{i} + 3\mathbf{j} - 3\mathbf{k}$
$5(6\mathbf{i} - 2\mathbf{j}) = 30\mathbf{i} - 10\mathbf{j}$	$2(\mathbf{i} + \mathbf{j} - \mathbf{k}) + 4(\mathbf{i} - \mathbf{j}) = 6\mathbf{i} - 2\mathbf{j} - 2\mathbf{k}$
$\|2\mathbf{i} - 3\mathbf{j}\| = \sqrt{2^2 + (-3)^2} = \sqrt{13}$	$\|\mathbf{i} + 2\mathbf{j} - 3\mathbf{k}\| = \sqrt{1^2 + 2^2 + (-3)^2} = \sqrt{14}$
$\|v_1\mathbf{i} + v_2\mathbf{j}\| = \sqrt{v_1^2 + v_2^2}$	$\|\langle v_1, v_2, v_3 \rangle\| = \sqrt{v_1^2 + v_2^2 + v_3^2}$

◀

NORMALIZING A VECTOR

A common problem in applications is to find a unit vector $\mathbf{u}$ that has the same direction as some given nonzero vector $\mathbf{v}$. This can be done by multiplying $\mathbf{v}$ by the reciprocal of its length; that is,

$$\mathbf{u} = \frac{1}{\|\mathbf{v}\|}\mathbf{v} = \frac{\mathbf{v}}{\|\mathbf{v}\|}$$

is a unit vector with the same direction as $\mathbf{v}$—the direction is the same because $k = 1/\|\mathbf{v}\|$ is a positive scalar, and the length is 1 because

$$\|\mathbf{u}\| = \|k\mathbf{v}\| = |k|\|\mathbf{v}\| = k\|\mathbf{v}\| = \frac{1}{\|\mathbf{v}\|}\|\mathbf{v}\| = 1$$

The process of multiplying a vector $\mathbf{v}$ by the reciprocal of its length to obtain a unit vector with the same direction is called ***normalizing*** $\mathbf{v}$.

Example 5

Find the unit vector that has the same direction as $\mathbf{v} = 2\mathbf{i} + 2\mathbf{j} - \mathbf{k}$.

Solution. The vector $\mathbf{v}$ has length

$$\|\mathbf{v}\| = \sqrt{2^2 + 2^2 + (-1)^2} = 3$$

so the unit vector $\mathbf{u}$ in the same direction as $\mathbf{v}$ is

$$\mathbf{u} = \tfrac{1}{3}\mathbf{v} = \tfrac{2}{3}\mathbf{i} + \tfrac{2}{3}\mathbf{j} - \tfrac{1}{3}\mathbf{k}$$

◀

FOR THE READER.   Many calculating utilities can perform vector operations, and some have built-in norm and normalization operations. If your calculating utility has these capabilities, use it to check the computations in Examples 1, 3, and 5.

VECTORS DETERMINED BY LENGTH AND ANGLE

Figure 13.2.14

If $\mathbf{v}$ is a nonzero vector with its initial point at the origin of an xy-coordinate system, and if ϕ is the angle from the positive x-axis to the radial line through $\mathbf{v}$, then the x-component of $\mathbf{v}$ can be written as $\|\mathbf{v}\|\cos\phi$ and the y-component as $\|\mathbf{v}\|\sin\phi$ (Figure 13.2.14); and hence $\mathbf{v}$ can be expressed in trigonometric form as

$$\mathbf{v} = \|\mathbf{v}\|\langle \cos\phi, \sin\phi \rangle \quad \text{or} \quad \mathbf{v} = \|\mathbf{v}\|\cos\phi\,\mathbf{i} + \|\mathbf{v}\|\sin\phi\,\mathbf{j} \tag{12}$$

In the special case of a unit vector $\mathbf{u}$ this simplifies to

$$\mathbf{u} = \langle \cos\phi, \sin\phi \rangle \quad \text{or} \quad \mathbf{u} = \cos\phi\,\mathbf{i} + \sin\phi\,\mathbf{j} \tag{13}$$

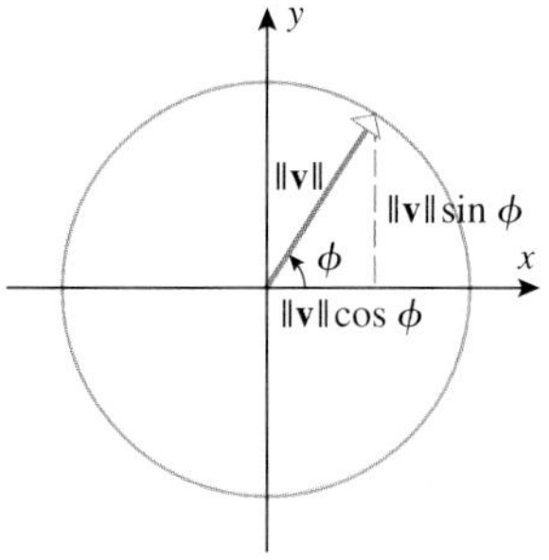

Example 6

(a)   Find the vector of length 2 that makes an angle of $\pi/4$ with the positive x-axis.

(b)   Find the angle that the vector $\mathbf{v} = -\sqrt{3}\,\mathbf{i} + \mathbf{j}$ makes with the positive x-axis.

Solution (a). From (12)

$$\mathbf{v} = 2\cos\frac{\pi}{4}\mathbf{i} + 2\sin\frac{\pi}{4}\mathbf{j} = \sqrt{2}\,\mathbf{i} + \sqrt{2}\,\mathbf{j}$$

Solution (b). We will normalize $\mathbf{v}$, then use (13) to find $\sin\phi$ and $\cos\phi$, and then use these values to find ϕ. Normalizing $\mathbf{v}$ yields

$$\frac{\mathbf{v}}{\|\mathbf{v}\|} = \frac{-\sqrt{3}\,\mathbf{i} + \mathbf{j}}{\sqrt{(-\sqrt{3})^2 + 1^2}} = -\frac{\sqrt{3}}{2}\mathbf{i} + \frac{1}{2}\mathbf{j}$$

Thus, $\cos\phi = -\sqrt{3}/2$ and $\sin\phi = \frac{1}{2}$, from which we conclude that $\phi = 5\pi/6$. ◀

VECTORS DETERMINED BY LENGTH AND A VECTOR IN THE SAME DIRECTION

It is a common problem in many applications that a direction in 2-space or 3-space is determined by some known unit vector $\mathbf{u}$, and it is of interest to find the components of a vector $\mathbf{v}$ that has the same direction as $\mathbf{u}$ and some specified length $\|\mathbf{v}\|$. This can be done by expressing $\mathbf{v}$ as

$$\mathbf{v} = \|\mathbf{v}\|\mathbf{u} \qquad \boxed{\text{v is equal to its length times a unit vector in the same direction.}}$$

and then reading off the components of $\|\mathbf{v}\|\mathbf{u}$.

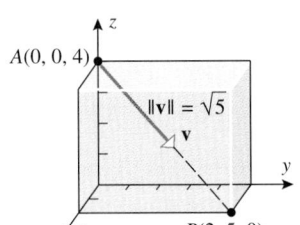

Figure 13.2.15

Example 7

Figure 13.2.15 shows a vector $\mathbf{v}$ of length $\sqrt{5}$ that extends along the line through A and B. Find the components of $\mathbf{v}$.

Solution. First we will find the components of the vector $\overrightarrow{AB}$, then we will normalize this vector to obtain a unit vector in the direction of $\mathbf{v}$, and then we will multiply this unit vector by $\|\mathbf{v}\|$ to obtain the vector $\mathbf{v}$. The computations are as follows:

$$\overrightarrow{AB} = \langle 2, 5, 0\rangle - \langle 0, 0, 4\rangle = \langle 2, 5, -4\rangle$$

$$\|\overrightarrow{AB}\| = \sqrt{2^2 + 5^2 + (-4)^2} = \sqrt{45} = 3\sqrt{5}$$

$$\frac{\overrightarrow{AB}}{\|\overrightarrow{AB}\|} = \left\langle \frac{2}{3\sqrt{5}}, \frac{5}{3\sqrt{5}}, -\frac{4}{3\sqrt{5}}\right\rangle$$

$$\mathbf{v} = \|\mathbf{v}\|\left(\frac{\overrightarrow{AB}}{\|\overrightarrow{AB}\|}\right) = \sqrt{5}\left\langle \frac{2}{3\sqrt{5}}, \frac{5}{3\sqrt{5}}, -\frac{4}{3\sqrt{5}}\right\rangle = \left\langle \frac{2}{3}, \frac{5}{3}, -\frac{4}{3}\right\rangle$$ ◀

RESULTANT OF TWO CONCURRENT FORCES

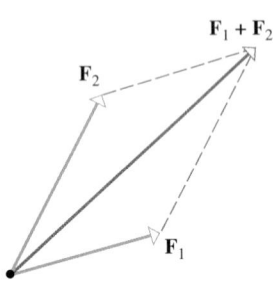

The single force $\mathbf{F}_1 + \mathbf{F}_2$ has the same effect as the two forces $\mathbf{F}_1$ and $\mathbf{F}_2$.

Figure 13.2.16

The effect that a force has on an object depends on the magnitude and direction of the force and the point at which it is applied. Thus, forces are regarded to be vector quantities and, indeed, the algebraic operations on vectors that we have defined in this section have their origin in the study of forces. For example, it is a fact of physics that if two forces $\mathbf{F}_1$ and $\mathbf{F}_2$ are applied at the same point on an object, then the two forces have the same effect on the object as the single force $\mathbf{F}_1 + \mathbf{F}_2$ applied at the point (Figure 13.2.16). Physicists and engineers call $\mathbf{F}_1 + \mathbf{F}_2$ the *resultant* of $\mathbf{F}_1$ and $\mathbf{F}_2$, and they say that the forces $\mathbf{F}_1$ and $\mathbf{F}_2$ are *concurrent* to indicate that they are applied at the same point.

In many applications, the magnitudes of two concurrent forces and the angle between them are known, and the problem is to find the magnitude and direction of the resultant. For example, referring to Figure 13.2.17, suppose that we know the magnitudes of the forces $\mathbf{F}_1$ and $\mathbf{F}_2$ and the angle ϕ between them, and we are interested in finding the magnitude of the resultant $\mathbf{F}_1 + \mathbf{F}_2$ and the angle α that the resultant makes with the force $\mathbf{F}_1$. This can be done by trigonometric methods based on the laws of sines and cosines. For this purpose, recall that the law of sines applied to the triangle in Figure 13.2.18 states that

$$\frac{a}{\sin\alpha} = \frac{b}{\sin\beta} = \frac{c}{\sin\gamma}$$

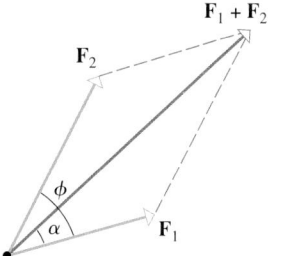

Figure 13.2.17

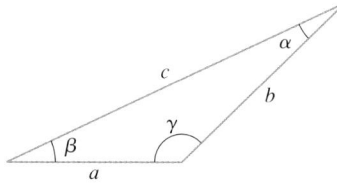

Figure 13.2.18

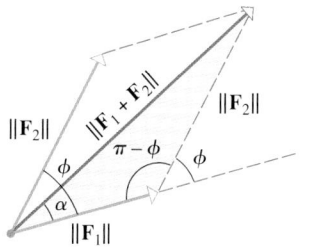

Figure 13.2.19

and the law of cosines implies that

$$c^2 = a^2 + b^2 - 2ab \cos \gamma$$

Referring to Figure 13.2.19, and using the fact that $\cos(\pi - \phi) = -\cos\phi$, it follows from the law of cosines that

$$\|\mathbf{F}_1 + \mathbf{F}_2\|^2 = \|\mathbf{F}_1\|^2 + \|\mathbf{F}_2\|^2 + 2\|\mathbf{F}_1\|\|\mathbf{F}_2\| \cos \phi \qquad (14)$$

Moreover, it follows from the law of sines that

$$\frac{\|\mathbf{F}_2\|}{\sin \alpha} = \frac{\|\mathbf{F}_1 + \mathbf{F}_2\|}{\sin(\pi - \phi)}$$

which, with the help of the identity $\sin(\pi - \phi) = \sin\phi$, can be expressed as

$$\sin \alpha = \frac{\|\mathbf{F}_2\|}{\|\mathbf{F}_1 + \mathbf{F}_2\|} \sin \phi \qquad (15)$$

Example 8

Suppose that two forces are applied to an eye bracket, as shown in Figure 13.2.20. Find the magnitude of the resultant and the angle θ that it makes with the positive x-axis.

Solution. We are given that $\|\mathbf{F}_1\| = 200$ N and $\|\mathbf{F}_2\| = 300$ N and that the angle between the vectors $\mathbf{F}_1$ and $\mathbf{F}_2$ is $\phi = 40°$. Thus, it follows from (14) that the magnitude of the resultant is

$$\|\mathbf{F}_1 + \mathbf{F}_2\| = \sqrt{\|\mathbf{F}_1\|^2 + \|\mathbf{F}_2\|^2 + 2\|\mathbf{F}_1\|\|\mathbf{F}_2\| \cos \phi}$$
$$= \sqrt{(200)^2 + (300)^2 + 2(200)(300) \cos 40°}$$
$$\approx 471 \text{ N}$$

Moreover, it follows from (15) that the angle α between $\mathbf{F}_1$ and the resultant is

$$\alpha = \sin^{-1}\left(\frac{\|\mathbf{F}_2\|}{\|\mathbf{F}_1 + \mathbf{F}_2\|} \sin \phi \right)$$
$$\approx \sin^{-1}\left(\frac{300}{471} \sin 40° \right)$$
$$\approx 24.2°$$

Thus, the angle θ that the resultant makes with the positive x-axis is

$$\theta = \alpha + 30° \approx 24.2° + 30° = 54.2°$$

(Figure 13.2.21). ◀

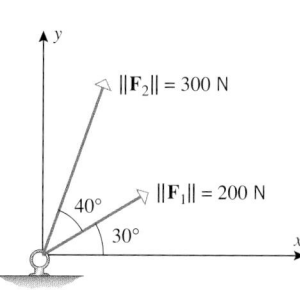

Figure 13.2.20

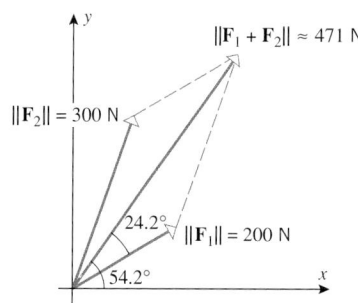

Figure 13.2.21

REMARK. The resultant of three or more concurrent forces can be found by working in pairs. For example, the resultant of three concurrent forces can be found by finding the resultant of any two of the three forces and then finding the resultant of that resultant with the third force.

EXERCISE SET 13.2

In Exercises 1–4, sketch the vectors with their initial points at the origin.

1. (a) $\langle 2, 5 \rangle$ (b) $\langle -5, -4 \rangle$ (c) $\langle 2, 0 \rangle$
   (d) $-5\mathbf{i} + 3\mathbf{j}$ (e) $3\mathbf{i} - 2\mathbf{j}$ (f) $-6\mathbf{j}$

2. (a) $\langle -3, 7 \rangle$ (b) $\langle 6, -2 \rangle$ (c) $\langle 0, -8 \rangle$
   (d) $4\mathbf{i} + 2\mathbf{j}$ (e) $-2\mathbf{i} - \mathbf{j}$ (f) $4\mathbf{i}$

3. (a) $\langle 1, -2, 2 \rangle$ (b) $\langle 2, 2, -1 \rangle$
   (c) $-\mathbf{i} + 2\mathbf{j} + 3\mathbf{k}$ (d) $2\mathbf{i} + 3\mathbf{j} - \mathbf{k}$

4. (a) $\langle -1, 3, 2 \rangle$ (b) $\langle 3, 4, 2 \rangle$
   (c) $2\mathbf{j} - \mathbf{k}$ (d) $\mathbf{i} - \mathbf{j} + 2\mathbf{k}$

In Exercises 5 and 6, find the components of the vector, and sketch an equivalent vector with its initial point at the origin.

5. (a)   (b)

6. (a)   (b)

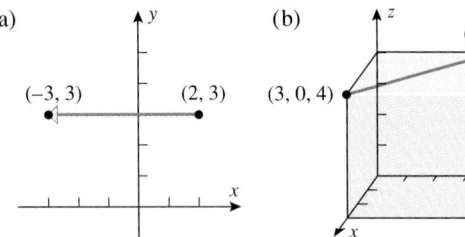

In Exercises 7 and 8, find the components of the vector $\overrightarrow{P_1P_2}$.

7. (a) $P_1(3, 5)$, $P_2(2, 8)$ (b) $P_1(7, -2)$, $P_2(0, 0)$
   (c) $P_1(5, -2, 1)$, $P_2(2, 4, 2)$

8. (a) $P_1(-6, -2)$, $P_2(-4, -1)$
   (b) $P_1(0, 0, 0)$, $P_2(-1, 6, 1)$
   (c) $P_1(4, 1, -3)$, $P_2(9, 1, -3)$

9. (a) Find the terminal point of $\mathbf{v} = 3\mathbf{i} - 2\mathbf{j}$ if the initial point is $(1, -2)$.
   (b) Find the initial point of $\mathbf{v} = \langle -3, 1, 2 \rangle$ if the terminal point is $(5, 0, -1)$.

10. (a) Find the terminal point of $\mathbf{v} = \langle 7, 6 \rangle$ if the initial point is $(2, -1)$.
   (b) Find the terminal point of $\mathbf{v} = \mathbf{i} + 2\mathbf{j} - 3\mathbf{k}$ if the initial point is $(-2, 1, 4)$.

In Exercises 11 and 12, perform the stated operations on the vectors $\mathbf{u}$, $\mathbf{v}$, and $\mathbf{w}$.

11. $\mathbf{u} = 3\mathbf{i} - \mathbf{k}$, $\mathbf{v} = \mathbf{i} - \mathbf{j} + 2\mathbf{k}$, $\mathbf{w} = 3\mathbf{j}$
   (a) $\mathbf{w} - \mathbf{v}$ (b) $6\mathbf{u} + 4\mathbf{w}$
   (c) $-\mathbf{v} - 2\mathbf{w}$ (d) $4(3\mathbf{u} + \mathbf{v})$
   (e) $-8(\mathbf{v} + \mathbf{w}) + 2\mathbf{u}$ (f) $3\mathbf{w} - (\mathbf{v} - \mathbf{w})$

12. $\mathbf{u} = \langle 2, -1, 3 \rangle$, $\mathbf{v} = \langle 4, 0, -2 \rangle$, $\mathbf{w} = \langle 1, 1, 3 \rangle$
   (a) $\mathbf{u} - \mathbf{w}$ (b) $7\mathbf{v} + 3\mathbf{w}$ (c) $-\mathbf{w} + \mathbf{v}$
   (d) $3(\mathbf{u} - 7\mathbf{v})$ (e) $-3\mathbf{v} - 8\mathbf{w}$ (f) $2\mathbf{v} - (\mathbf{u} + \mathbf{w})$

In Exercises 13 and 14, find the norm of $\mathbf{v}$.

13. (a) $\mathbf{v} = \langle 1, -1 \rangle$ (b) $\mathbf{v} = -\mathbf{i} + 7\mathbf{j}$
   (c) $\mathbf{v} = \langle -1, 2, 4 \rangle$ (d) $\mathbf{v} = -3\mathbf{i} + 2\mathbf{j} + \mathbf{k}$

14. (a) $\mathbf{v} = \langle 3, 4 \rangle$ (b) $\mathbf{v} = \sqrt{2}\mathbf{i} - \sqrt{7}\mathbf{j}$
   (c) $\mathbf{v} = \langle 0, -3, 0 \rangle$ (d) $\mathbf{v} = \mathbf{i} + \mathbf{j} + \mathbf{k}$

15. Let $\mathbf{u} = \mathbf{i} - 3\mathbf{j} + 2\mathbf{k}$, $\mathbf{v} = \mathbf{i} + \mathbf{j}$, and $\mathbf{w} = 2\mathbf{i} + 2\mathbf{j} - 4\mathbf{k}$. Find
   (a) $\|\mathbf{u} + \mathbf{v}\|$ (b) $\|\mathbf{u}\| + \|\mathbf{v}\|$
   (c) $\|-2\mathbf{u}\| + 2\|\mathbf{v}\|$ (d) $\|3\mathbf{u} - 5\mathbf{v} + \mathbf{w}\|$
   (e) $\dfrac{1}{\|\mathbf{w}\|}\mathbf{w}$ (f) $\left\| \dfrac{1}{\|\mathbf{w}\|}\mathbf{w} \right\|$.

16. Is it possible to have $\|\mathbf{u}\| + \|\mathbf{v}\| = \|\mathbf{u} + \mathbf{v}\|$ if $\mathbf{u}$ and $\mathbf{v}$ are nonzero vectors? Justify your conclusion geometrically.

In Exercises 17 and 18, find unit vectors that satisfy the stated conditions.

17. (a) Same direction as $-\mathbf{i} + 4\mathbf{j}$.
   (b) Oppositely directed to $6\mathbf{i} - 4\mathbf{j} + 2\mathbf{k}$.
   (c) Same direction as the vector from the point $A(-1, 0, 2)$ to the point $B(3, 1, 1)$.

18. (a) Oppositely directed to $3\mathbf{i} - 4\mathbf{j}$.
   (b) Same direction as $2\mathbf{i} - \mathbf{j} - 2\mathbf{k}$.
   (c) Same direction as the vector from the point $A(-3, 2)$ to the point $B(1, -1)$.

In Exercises 19 and 20, find vectors that satisfy the stated conditions.

19. (a) Oppositely directed to $\mathbf{v} = \langle 3, -4 \rangle$ and half the length of $\mathbf{v}$.
   (b) Length $\sqrt{17}$ and same direction as $\mathbf{v} = \langle 7, 0, -6 \rangle$.

20. (a) Same direction as $\mathbf{v} = -2\mathbf{i} + 3\mathbf{j}$ and three times the length of $\mathbf{v}$.
   (b) Length 2 and oppositely directed to $\mathbf{v} = -3\mathbf{i} + 4\mathbf{j} + \mathbf{k}$.

21. In each part, find the component form of the vector $\mathbf{v}$ in 2-space that has the stated length and makes the stated angle ϕ with the positive x-axis.
   (a) $\|\mathbf{v}\| = 3$; $\phi = \pi/4$ (b) $\|\mathbf{v}\| = 2$; $\phi = 90°$
   (c) $\|\mathbf{v}\| = 5$; $\phi = 120°$ (d) $\|\mathbf{v}\| = 1$; $\phi = \pi$

22. Find the component forms of $\mathbf{v} + \mathbf{w}$ and $\mathbf{v} - \mathbf{w}$ in 2-space, given that $\|\mathbf{v}\| = 1$, $\|\mathbf{w}\| = 1$, $\mathbf{v}$ makes an angle of $\pi/6$ with the positive x-axis, and $\mathbf{w}$ makes an angle of $3\pi/4$ with the positive x-axis.

In Exercises 23 and 24, find the component form of $\mathbf{v} + \mathbf{w}$, given that $\mathbf{v}$ and $\mathbf{w}$ are unit vectors.

23. 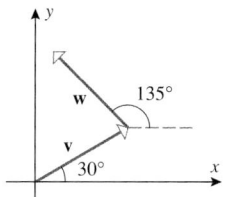   **24.**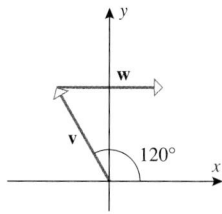

25. In each part, sketch the vector $\mathbf{u} + \mathbf{v} + \mathbf{w}$ and express it in component form.

(a) 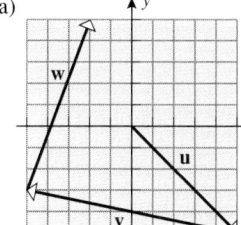   (b)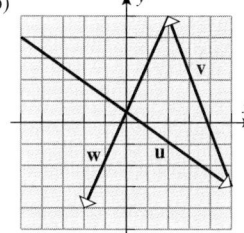

26. In each part of Exercise 25, sketch the vector $\mathbf{u} - \mathbf{v} + \mathbf{w}$ and express it in component form.

27. Let $\mathbf{u} = \langle 1, 3 \rangle$, $\mathbf{v} = \langle 2, 1 \rangle$, $\mathbf{w} = \langle 4, -1 \rangle$. Find the vector $\mathbf{x}$ that satisfies $2\mathbf{u} - \mathbf{v} + \mathbf{x} = 7\mathbf{x} + \mathbf{w}$.

28. Let $\mathbf{u} = \langle -1, 1 \rangle$, $\mathbf{v} = \langle 0, 1 \rangle$, and $\mathbf{w} = \langle 3, 4 \rangle$. Find the vector $\mathbf{x}$ that satisfies $\mathbf{u} - 2\mathbf{x} = \mathbf{x} - \mathbf{w} + 3\mathbf{v}$.

29. Find $\mathbf{u}$ and $\mathbf{v}$ if $\mathbf{u} + 2\mathbf{v} = 3\mathbf{i} - \mathbf{k}$ and $3\mathbf{u} - \mathbf{v} = \mathbf{i} + \mathbf{j} + \mathbf{k}$.

30. Find $\mathbf{u}$ and $\mathbf{v}$ if $\mathbf{u} + \mathbf{v} = \langle 2, -3 \rangle$ and $3\mathbf{u} + 2\mathbf{v} = \langle -1, 2 \rangle$.

31. Use vectors to find the lengths of the diagonals of the parallelogram that has $\mathbf{i} + \mathbf{j}$ and $\mathbf{i} - 2\mathbf{j}$ as adjacent sides.

32. Use vectors to find the fourth vertex of a parallelogram, three of whose vertices are $(0, 0)$, $(1, 3)$, and $(2, 4)$. [*Note:* There is more than one answer.]

33. (a) Given that $\|\mathbf{v}\| = 3$, find all values of k such that $\|k\mathbf{v}\| = 5$.
    (b) Given that $k = -2$ and $\|k\mathbf{v}\| = 6$, find $\|\mathbf{v}\|$.

34. What do you know about k and $\mathbf{v}$ if $\|k\mathbf{v}\| = 0$?

35. In each part, find two unit vectors in 2-space that satisfy the stated condition.
    (a) Parallel to the line $y = 3x + 2$
    (b) Parallel to the line $x + y = 4$
    (c) Perpendicular to the line $y = -5x + 1$

36. In each part, find two unit vectors in 3-space that satisfy the stated condition.
    (a) Perpendicular to the xy-plane
    (b) Perpendicular to the xz-plane
    (c) Perpendicular to the yz-plane

37. Let $\mathbf{r} = \langle x, y \rangle$ be an arbitrary vector. In each part, describe the set of all points (x, y) in 2-space that satisfy the stated condition.
    (a) $\|\mathbf{r}\| = 1$ (b) $\|\mathbf{r}\| \leq 1$ (c) $\|\mathbf{r}\| > 1$

38. Let $\mathbf{r} = \langle x, y \rangle$ and $\mathbf{r}_0 = \langle x_0, y_0 \rangle$. In each part, describe the set of all points (x, y) in 2-space that satisfy the stated condition.
    (a) $\|\mathbf{r} - \mathbf{r}_0\| = 1$ (b) $\|\mathbf{r} - \mathbf{r}_0\| \leq 1$ (c) $\|\mathbf{r} - \mathbf{r}_0\| > 1$

39. Let $\mathbf{r} = \langle x, y, z \rangle$ be an arbitrary vector. In each part, describe the set of all points (x, y, z) in 3-space that satisfy the stated condition.
    (a) $\|\mathbf{r}\| = 1$ (b) $\|\mathbf{r}\| \leq 1$ (c) $\|\mathbf{r}\| > 1$

40. Let $\mathbf{r}_1 = \langle x_1, y_1 \rangle$, $\mathbf{r}_2 = \langle x_2, y_2 \rangle$, and $\mathbf{r} = \langle x, y \rangle$. Describe the set of all points (x, y) for which $\|\mathbf{r} - \mathbf{r}_1\| + \|\mathbf{r} - \mathbf{r}_2\| = k$, assuming that $k > \|\mathbf{r}_2 - \mathbf{r}_1\|$.

In Exercises 41–46, find the magnitude of the resultant force and the angle that it makes with the positive x-axis.

41. 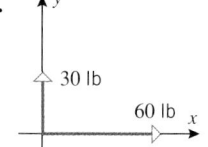   **42.**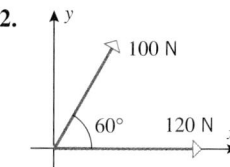

43. 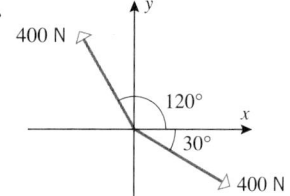   **44.**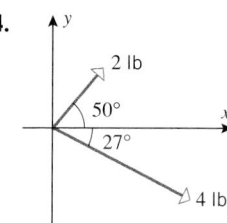

45. 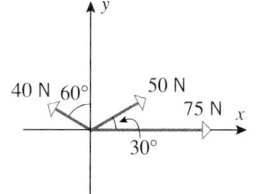   **46.**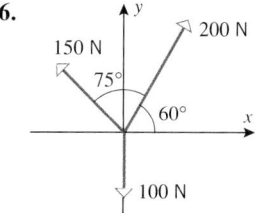

A particle is said to be in *static equilibrium* if the resultant of all forces applied to it is zero. In Exercises 47 and 48, find the force $\mathbf{F}$ that must be applied to the point to produce static equilibrium. Describe $\mathbf{F}$ by specifying its magnitude and the angle that it makes with the positive x-axis.

47. 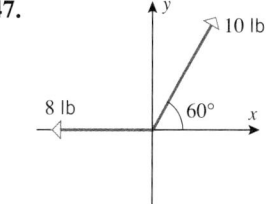   **48.**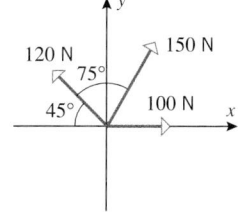

49. The accompanying figure shows a 250-lb traffic light supported by two flexible cables. The magnitudes of the forces that the cables apply to the eye ring are called the cable *tensions*. Find the tensions in the cables if the traffic light is in static equilibrium (defined above Exercise 47).

50. Find the tensions in the cables shown in the accompanying figure if the block is in static equilibrium (see Exercise 49).

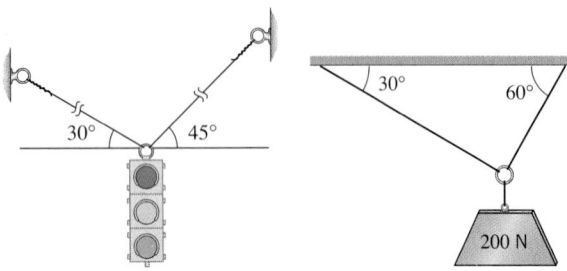

Figure Ex-49   Figure Ex-50

51. A vector **w** is said to be a *linear combination* of the vectors v_1 and v_2 if **w** can be expressed as $\mathbf{w} = c_1 \mathbf{v}_1 + c_2 \mathbf{v}_2$, where c_1 and c_2 are scalars.
(a) Find scalars c_1 and c_2 to express the vector $4\mathbf{j}$ as a linear combination of the vectors $\mathbf{v}_1 = 2\mathbf{i} - \mathbf{j}$ and $\mathbf{v}_2 = 4\mathbf{i} + 2\mathbf{j}$.
(b) Show that the vector $\langle 3, 5 \rangle$ cannot be expressed as a linear combination of the vectors $\mathbf{v}_1 = \langle 1, -3 \rangle$ and $\mathbf{v}_2 = \langle -2, 6 \rangle$.

52. A vector **w** is said to be a *linear combination* of the vectors v_1, v_2, and v_3 if **w** can be expressed as $\mathbf{w} = c_1 \mathbf{v}_1 + c_2 \mathbf{v}_2 + c_3 \mathbf{v}_3$, where c_1, c_2, and c_3 are scalars.
(a) Find scalars c_1, c_2, and c_3 to express $\langle -1, 1, 5 \rangle$ as a linear combination of the vectors $\mathbf{v}_1 = \langle 1, 0, 1 \rangle$, $\mathbf{v}_2 = \langle 3, 2, 0 \rangle$, and $\mathbf{v}_3 = \langle 0, 1, 1 \rangle$.
(b) Show that the vector $2\mathbf{i} + \mathbf{j} - \mathbf{k}$ cannot be expressed as a linear combination of the vectors $\mathbf{v}_1 = \mathbf{i} - \mathbf{j}$, $\mathbf{v}_2 = 3\mathbf{i} + \mathbf{k}$, and $\mathbf{v}_3 = 4\mathbf{i} - \mathbf{j} + \mathbf{k}$.

53. Use a theorem from plane geometry to show that if **u** and **v** are vectors in 2-space or 3-space, then

$$\|\mathbf{u} + \mathbf{v}\| \leq \|\mathbf{u}\| + \|\mathbf{v}\|$$

which is called the *triangle inequality for vectors*. Give some examples to illustrate this inequality.

54. Prove parts (a), (c), and (e) of Theorem 13.2.6 algebraically in 2-space.

55. Prove parts (d), (g), and (h) of Theorem 13.2.6 algebraically in 2-space.

56. Prove part (f) of Theorem 13.2.6 geometrically.

57. Use vectors to prove that the line segment joining the midpoints of two sides of a triangle is parallel to the third side and half as long.

58. Use vectors to prove that the midpoints of the sides of a quadrilateral are the vertices of a parallelogram.

13.3 DOT PRODUCT; PROJECTIONS

In the last section we defined three operations on vectors—addition, subtraction, and scalar multiplication. In scalar multiplication a vector is multiplied by a scalar and the result is a vector. In this section we will define a new kind of multiplication in which two vectors are multiplied to produce a scalar. This multiplication operation has many uses, some of which we will also discuss in this section.

DEFINITION OF THE DOT PRODUCT

13.3.1 DEFINITION. If $\mathbf{u} = \langle u_1, u_2 \rangle$ and $\mathbf{v} = \langle v_1, v_2 \rangle$ are vectors in 2-space, then the *dot product* of **u** and **v** is written as $\mathbf{u} \cdot \mathbf{v}$ and is defined as

$$\mathbf{u} \cdot \mathbf{v} = u_1 v_1 + u_2 v_2$$

Similarly, if $\mathbf{u} = \langle u_1, u_2, u_3 \rangle$ and $\mathbf{v} = \langle v_1, v_2, v_3 \rangle$ are vectors in 3-space, then their dot product is defined as

$$\mathbf{u} \cdot \mathbf{v} = u_1 v_1 + u_2 v_2 + u_3 v_3$$

In words, the dot product of two vectors is formed by multiplying their corresponding components and adding the products. Note that the dot product of two vectors is a scalar.

Example 1

$$\langle 3, 5 \rangle \cdot \langle -1, 2 \rangle = 3(-1) + 5(2) = 7$$
$$\langle 2, 3 \rangle \cdot \langle -3, 2 \rangle = 2(-3) + 3(2) = 0$$
$$\langle 1, -3, 4 \rangle \cdot \langle 1, 5, 2 \rangle = 1(1) + (-3)(5) + 4(2) = -6$$

Here are the same computations expressed another way:

$$(3\mathbf{i} + 5\mathbf{j}) \cdot (-\mathbf{i} + 2\mathbf{j}) = 3(-1) + 5(2) = 7$$
$$(2\mathbf{i} + 3\mathbf{j}) \cdot (-3\mathbf{i} + 2\mathbf{j}) = 2(-3) + 3(2) = 0$$
$$(\mathbf{i} - 3\mathbf{j} + 4\mathbf{k}) \cdot (\mathbf{i} + 5\mathbf{j} + 2\mathbf{k}) = 1(1) + (-3)(5) + (4)(2) = -6 \qquad \blacktriangleleft$$

FOR THE READER. Many calculating utilities have a built-in dot product operation. If your calculating utility has this capability, use it to check the computations in Example 1.

ALGEBRAIC PROPERTIES OF THE DOT PRODUCT

The following theorem provides some of the basic algebraic properties of the dot product:

> **13.3.2 THEOREM.** *If* $\mathbf{u}$, $\mathbf{v}$, *and* $\mathbf{w}$ *are vectors in 2- or 3-space and* k *is a scalar, then*
> (*a*) $\mathbf{u} \cdot \mathbf{v} = \mathbf{v} \cdot \mathbf{u}$
> (*b*) $\mathbf{u} \cdot (\mathbf{v} + \mathbf{w}) = \mathbf{u} \cdot \mathbf{v} + \mathbf{u} \cdot \mathbf{w}$
> (*c*) $k(\mathbf{u} \cdot \mathbf{v}) = (k\mathbf{u}) \cdot \mathbf{v} = \mathbf{u} \cdot (k\mathbf{v})$
> (*d*) $\mathbf{v} \cdot \mathbf{v} = \|\mathbf{v}\|^2$
> (*e*) $\mathbf{0} \cdot \mathbf{v} = 0$

We will prove parts (*c*) and (*d*) for vectors in 3-space and leave some of the others as exercises.

Proof (c). Let $\mathbf{u} = \langle u_1, u_2, u_3 \rangle$ and $\mathbf{v} = \langle v_1, v_2, v_3 \rangle$. Then

$$k(\mathbf{u} \cdot \mathbf{v}) = k(u_1 v_1 + u_2 v_2 + u_3 v_3) = (ku_1)v_1 + (ku_2)v_2 + (ku_3)v_3 = (k\mathbf{u}) \cdot \mathbf{v}$$

Similarly, $k(\mathbf{u} \cdot \mathbf{v}) = \mathbf{u} \cdot (k\mathbf{v})$.

Proof (d). $\mathbf{v} \cdot \mathbf{v} = v_1 v_1 + v_2 v_2 + v_3 v_3 = v_1^2 + v_2^2 + v_3^2 = \|\mathbf{v}\|^2$. ∎

REMARK. Pay particular attention to the two zeros that appear in part (*e*) of the last theorem—the zero on the left side is the zero vector (boldface), and the zero on the right side is the zero scalar (lightface). It is also worth noting that the result in part (*d*) can be written as

$$\|\mathbf{v}\| = \sqrt{\mathbf{v} \cdot \mathbf{v}} \qquad (1)$$

which provides a way of expressing the norm of a vector in terms of a dot product.

ANGLE BETWEEN VECTORS

Suppose that $\mathbf{u}$ and $\mathbf{v}$ are nonzero vectors in 2-space or 3-space that are positioned so their initial points coincide. We define the *angle between* $\mathbf{u}$ *and* $\mathbf{v}$ to be the angle θ determined by the vectors that satisfies the condition $0 \le \theta \le \pi$ (Figure 13.3.1). In 2-space, θ is the smallest counterclockwise angle through which one of the vectors can be rotated until it aligns with the other.

The next theorem provides a way of calculating the angle between two vectors from their components.

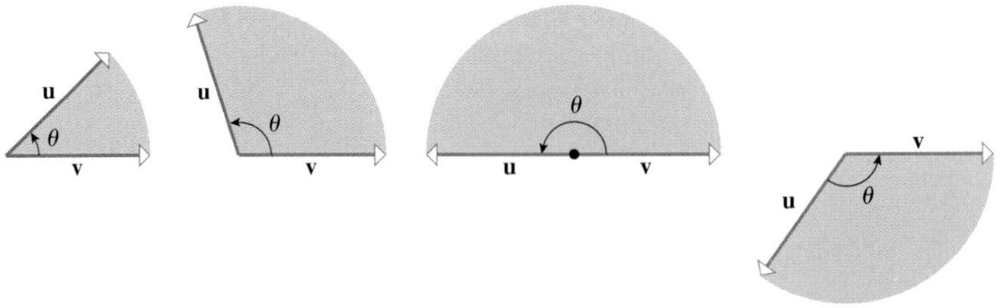

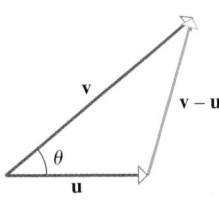

Figure 13.3.1

13.3.3 THEOREM. *If* **u** *and* **v** *are nonzero vectors in 2-space or 3-space, and if* θ *is the angle between them, then*

$$\cos \theta = \frac{\mathbf{u} \cdot \mathbf{v}}{\|\mathbf{u}\| \|\mathbf{v}\|} \tag{2}$$

Figure 13.3.2

Proof. Suppose that the vectors **u**, **v**, and **v** − **u** are positioned to form three sides of a triangle, as shown in Figure 13.3.2. It follows from the law of cosines that

$$\|\mathbf{v} - \mathbf{u}\|^2 = \|\mathbf{u}\|^2 + \|\mathbf{v}\|^2 - 2\|\mathbf{u}\|\|\mathbf{v}\| \cos \theta \tag{3}$$

Using the properties of the dot product in Theorem 13.3.2, we can rewrite the left side of this equation as

$$\|\mathbf{v} - \mathbf{u}\|^2 = (\mathbf{v} - \mathbf{u}) \cdot (\mathbf{v} - \mathbf{u})$$
$$= (\mathbf{v} - \mathbf{u}) \cdot \mathbf{v} - (\mathbf{v} - \mathbf{u}) \cdot \mathbf{u}$$
$$= \mathbf{v} \cdot \mathbf{v} - \mathbf{u} \cdot \mathbf{v} - \mathbf{v} \cdot \mathbf{u} + \mathbf{u} \cdot \mathbf{u}$$
$$= \|\mathbf{v}\|^2 - 2\mathbf{u} \cdot \mathbf{v} + \|\mathbf{u}\|^2$$

Substituting this back into (3) yields

$$\|\mathbf{v}\|^2 - 2\mathbf{u} \cdot \mathbf{v} + \|\mathbf{u}\|^2 = \|\mathbf{u}\|^2 + \|\mathbf{v}\|^2 - 2\|\mathbf{u}\|\|\mathbf{v}\| \cos \theta$$

which we can simplify and rewrite as

$$\mathbf{u} \cdot \mathbf{v} = \|\mathbf{u}\|\|\mathbf{v}\| \cos \theta$$

Finally, dividing both sides of this equation by $\|\mathbf{u}\|\|\mathbf{v}\|$ yields (2). ■

Example 2

Find the angle between the vector $\mathbf{u} = \mathbf{i} - 2\mathbf{j} + 2\mathbf{k}$ and

(a) $\mathbf{v} = -3\mathbf{i} + 6\mathbf{j} + 2\mathbf{k}$ (b) $\mathbf{w} = 2\mathbf{i} + 7\mathbf{j} + 6\mathbf{k}$ (c) $\mathbf{z} = -3\mathbf{i} + 6\mathbf{j} - 6\mathbf{k}$

Solution (a).

$$\cos \theta = \frac{\mathbf{u} \cdot \mathbf{v}}{\|\mathbf{u}\| \|\mathbf{v}\|} = \frac{-11}{(3)(7)} = -\frac{11}{21}$$

Thus,

$$\theta = \cos^{-1}\left(-\tfrac{11}{21}\right) \approx 2.12 \text{ radians} \approx 121.6°$$

Solution (b).

$$\cos \theta = \frac{\mathbf{u} \cdot \mathbf{w}}{\|\mathbf{u}\| \|\mathbf{w}\|} = \frac{0}{\|\mathbf{u}\| \|\mathbf{w}\|} = 0$$

Thus, $\theta = \pi/2$, which means that the vectors are perpendicular.

Solution (c).

$$\cos \theta = \frac{\mathbf{u} \cdot \mathbf{z}}{\|\mathbf{u}\| \|\mathbf{z}\|} = \frac{-27}{(3)(9)} = -1$$

Thus, $\theta = \pi$, which means that the vectors are oppositely directed. In retrospect, we could have seen this without computing θ, since $\mathbf{z} = -3\mathbf{u}$. ◄

**INTERPRETING THE SIGN OF THE
DOT PRODUCT**

It will often be convenient to express Formula (2) as

$$\mathbf{u} \cdot \mathbf{v} = \|\mathbf{u}\|\|\mathbf{v}\| \cos\theta \tag{4}$$

which expresses the dot product of $\mathbf{u}$ and $\mathbf{v}$ in terms of the lengths of these vectors and the angle between them. Since $\mathbf{u}$ and $\mathbf{v}$ are assumed to be nonzero vectors, this version of the formula makes it clear that the sign of $\mathbf{u} \cdot \mathbf{v}$ is the same as the sign of $\cos\theta$. Thus, we can tell from the dot product whether the angle between two vectors is acute or obtuse or whether the vectors are perpendicular (Figure 13.3.3).

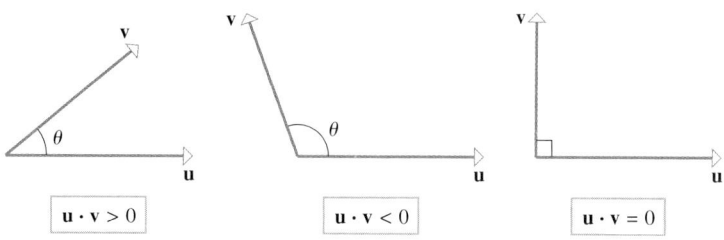

Figure 13.3.3

REMARK. The terms "perpendicular," "orthogonal," and "normal" are all commonly used to describe geometric objects that meet at right angles. For consistency, we will say that two vectors are *orthogonal*, a vector is *normal* to a plane, and two planes are *perpendicular*. Moreover, although the zero vector does not make a well-defined angle with other vectors, we will consider $\mathbf{0}$ to be orthogonal to *all* vectors. This convention allows us to say that $\mathbf{u}$ and $\mathbf{v}$ are orthogonal vectors if and only if $\mathbf{u} \cdot \mathbf{v} = 0$, and it makes Formula (4) valid if $\mathbf{u}$ or $\mathbf{v}$ (or both) is zero.

DIRECTION ANGLES

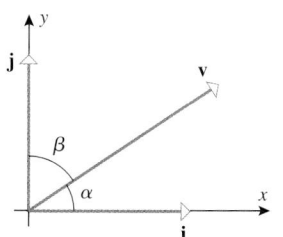

Figure 13.3.4

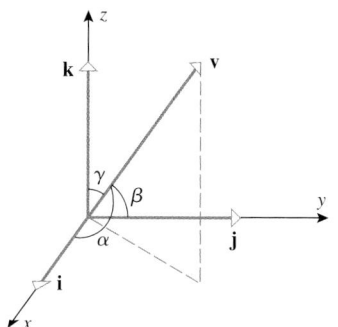

Figure 13.3.5

In an xy-coordinate system, the direction of a nonzero vector $\mathbf{v}$ is completely determined by the angles α and β between $\mathbf{v}$ and the unit vectors $\mathbf{i}$ and $\mathbf{j}$ (Figure 13.3.4), and in an xyz-coordinate system the direction is completely determined by the angles α, β, and γ between $\mathbf{v}$ and the unit vectors $\mathbf{i}$, $\mathbf{j}$, and $\mathbf{k}$ (Figure 13.3.5). In both 2-space and 3-space the angles between a nonzero vector $\mathbf{v}$ and the vectors $\mathbf{i}$, $\mathbf{j}$, and $\mathbf{k}$ are called the *direction angles* of $\mathbf{v}$, and the cosines of those angles are called the *direction cosines* of $\mathbf{v}$. Formulas for the direction cosines of a vector can be obtained from Formula (2). For example, if $\mathbf{v} = v_1\mathbf{i} + v_2\mathbf{j} + v_3\mathbf{k}$, then

$$\cos\alpha = \frac{\mathbf{v}\cdot\mathbf{i}}{\|\mathbf{v}\|\|\mathbf{i}\|} = \frac{v_1}{\|\mathbf{v}\|}, \quad \cos\beta = \frac{\mathbf{v}\cdot\mathbf{j}}{\|\mathbf{v}\|\|\mathbf{j}\|} = \frac{v_2}{\|\mathbf{v}\|}, \quad \cos\gamma = \frac{\mathbf{v}\cdot\mathbf{k}}{\|\mathbf{v}\|\|\mathbf{k}\|} = \frac{v_3}{\|\mathbf{v}\|}$$

Thus, we have the following result:

13.3.4 THEOREM. *The direction cosines of a nonzero vector* $\mathbf{v} = v_1\mathbf{i} + v_2\mathbf{j} + v_3\mathbf{k}$ *are*

$$\cos\alpha = \frac{v_1}{\|\mathbf{v}\|}, \quad \cos\beta = \frac{v_2}{\|\mathbf{v}\|}, \quad \cos\gamma = \frac{v_3}{\|\mathbf{v}\|}$$

The direction cosines of a vector $\mathbf{v} = v_1\mathbf{i} + v_2\mathbf{j} + v_3\mathbf{k}$ can be computed by normalizing $\mathbf{v}$ and reading off the components of $\mathbf{v}/\|\mathbf{v}\|$, since

$$\frac{\mathbf{v}}{\|\mathbf{v}\|} = \frac{v_1}{\|\mathbf{v}\|}\mathbf{i} + \frac{v_2}{\|\mathbf{v}\|}\mathbf{j} + \frac{v_3}{\|\mathbf{v}\|}\mathbf{k} = (\cos\alpha)\mathbf{i} + (\cos\beta)\mathbf{j} + (\cos\gamma)\mathbf{k}$$

We leave it as an exercise for you to show that the direction cosines of a vector satisfy the equation

$$\cos^2 \alpha + \cos^2 \beta + \cos^2 \gamma = 1 \tag{5}$$

Example 3

Find the direction cosines of the vector $\mathbf{v} = 2\mathbf{i} - 4\mathbf{j} + 4\mathbf{k}$, and approximate the direction angles to the nearest degree.

Solution. First we will normalize the vector $\mathbf{v}$ and then read off the components. We have $\|\mathbf{v}\| = \sqrt{4 + 16 + 16} = 6$, so that $\mathbf{v}/\|\mathbf{v}\| = \frac{1}{3}\mathbf{i} - \frac{2}{3}\mathbf{j} + \frac{2}{3}\mathbf{k}$. Thus,

$$\cos \alpha = \tfrac{1}{3}, \quad \cos \beta = -\tfrac{2}{3}, \quad \cos \gamma = \tfrac{2}{3}$$

With the help of a calculating utility we obtain

$$\alpha = \cos^{-1}\left(\tfrac{1}{3}\right) \approx 71°, \quad \beta = \cos^{-1}\left(-\tfrac{2}{3}\right) \approx 132°, \quad \gamma = \cos^{-1}\left(\tfrac{2}{3}\right) \approx 48° \quad \blacktriangleleft$$

Example 4

Find the angle between a diagonal of a cube and one of its edges.

Solution. Assume that the cube has side a, and introduce a coordinate system as shown in Figure 13.3.6. In this coordinate system the vector

$$\mathbf{d} = a\mathbf{i} + a\mathbf{j} + a\mathbf{k}$$

is a diagonal of the cube and the unit vectors $\mathbf{i}, \mathbf{j},$ and $\mathbf{k}$ run along the edges. By symmetry, the diagonal makes the same angle with each edge, so it is sufficient to find the angle between $\mathbf{d}$ and $\mathbf{i}$ (the direction angle α). Thus,

$$\cos \alpha = \frac{\mathbf{d} \cdot \mathbf{i}}{\|\mathbf{d}\| \|\mathbf{i}\|} = \frac{a}{\|\mathbf{d}\|} = \frac{a}{\sqrt{3a^2}} = \frac{1}{\sqrt{3}}$$

and hence

$$\alpha = \cos^{-1}\left(\frac{1}{\sqrt{3}}\right) \approx 0.955 \text{ radian} \approx 54.7° \quad \blacktriangleleft$$

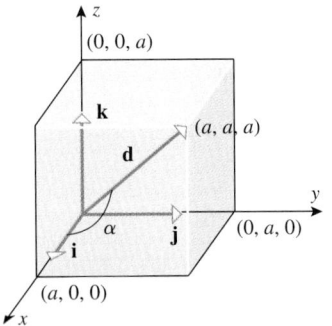

Figure 13.3.6

DECOMPOSING VECTORS INTO ORTHOGONAL COMPONENTS

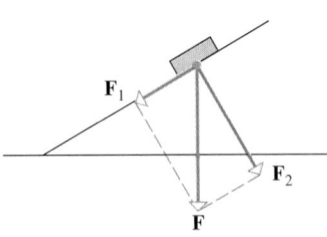

The force of gravity pulls the block against the ramp and down the ramp.

Figure 13.3.7

In many applications it is desirable to "decompose" a vector into a sum of two orthogonal vectors with convenient specified directions. For example, Figure 13.3.7 shows a block on an inclined plane. The downward force $\mathbf{F}$ that gravity exerts on the block can be decomposed into the sum

$$\mathbf{F} = \mathbf{F}_1 + \mathbf{F}_2$$

where the force $\mathbf{F}_1$ is parallel to the ramp and the force $\mathbf{F}_2$ is perpendicular to the ramp. The forces $\mathbf{F}_1$ and $\mathbf{F}_2$ are useful because $\mathbf{F}_1$ is the force that pulls the block *along* the ramp, and $\mathbf{F}_2$ is the force that the block exerts *against* the ramp.

Thus, our next objective is to develop a computational procedure for decomposing a vector into a sum of orthogonal vectors. For this purpose, suppose that $\mathbf{e}_1$ and $\mathbf{e}_2$ are two orthogonal *unit* vectors in 2-space, and suppose that we want to express a given vector $\mathbf{v}$ as a sum

$$\mathbf{v} = \mathbf{w}_1 + \mathbf{w}_2$$

where $\mathbf{w}_1$ is a scalar multiple of $\mathbf{e}_1$ and $\mathbf{w}_2$ is a scalar multiple of $\mathbf{e}_2$ (Figure 13.3.8a); that is, we want to find scalars k_1 and k_2 such that

$$\mathbf{v} = k_1 \mathbf{e}_1 + k_2 \mathbf{e}_2 \tag{6}$$

We can find k_1 by taking the dot product of $\mathbf{v}$ with $\mathbf{e}_1$. This yields

$$\mathbf{v} \cdot \mathbf{e}_1 = (k_1 \mathbf{e}_1 + k_2 \mathbf{e}_2) \cdot \mathbf{e}_1 = k_1 (\mathbf{e}_1 \cdot \mathbf{e}_1) + k_2 (\mathbf{e}_2 \cdot \mathbf{e}_1) = k_1 \|\mathbf{e}_1\|^2 + 0 = k_1$$

and, similarly,

$$\mathbf{v} \cdot \mathbf{e}_2 = (k_1\mathbf{e}_1 + k_2\mathbf{e}_2) \cdot \mathbf{e}_2 = k_1(\mathbf{e}_1 \cdot \mathbf{e}_2) + k_2(\mathbf{e}_2 \cdot \mathbf{e}_2) = 0 + k_2\|\mathbf{e}_2\|^2 = k_2$$

Substituting these expressions for k_1 and k_2 in (6) yields

$$\mathbf{v} = (\mathbf{v} \cdot \mathbf{e}_1)\mathbf{e}_1 + (\mathbf{v} \cdot \mathbf{e}_2)\mathbf{e}_2 \tag{7}$$

In this formula we call $(\mathbf{v} \cdot \mathbf{e}_1)\mathbf{e}_1$ and $(\mathbf{v} \cdot \mathbf{e}_2)\mathbf{e}_2$ the ***vector components*** of $\mathbf{v}$ along $\mathbf{e}_1$ and $\mathbf{e}_2$, respectively; and we call $\mathbf{v} \cdot \mathbf{e}_1$ and $\mathbf{v} \cdot \mathbf{e}_2$ the ***scalar components*** of $\mathbf{v}$ along $\mathbf{e}_1$ and $\mathbf{e}_2$, respectively. If θ denotes the angle between $\mathbf{v}$ and $\mathbf{e}_1$, then the scalar components of $\mathbf{v}$ can be written in trigonometric form as

$$\mathbf{v} \cdot \mathbf{e}_1 = \|\mathbf{v}\| \cos\theta \quad \text{and} \quad \mathbf{v} \cdot \mathbf{e}_2 = \|\mathbf{v}\| \sin\theta \tag{8}$$

(Figure 13.3.8*b*). Moreover, the vector components of $\mathbf{v}$ can be expressed as

$$(\mathbf{v} \cdot \mathbf{e}_1)\mathbf{e}_1 = (\|\mathbf{v}\|\cos\theta)\mathbf{e}_1 \quad \text{and} \quad (\mathbf{v} \cdot \mathbf{e}_2)\mathbf{e}_2 = (\|\mathbf{v}\|\sin\theta)\mathbf{e}_2 \tag{9}$$

and the decomposition (6) can be expressed as

$$\mathbf{v} = (\|\mathbf{v}\|\cos\theta)\mathbf{e}_1 + (\|\mathbf{v}\|\sin\theta)\mathbf{e}_2 \tag{10}$$

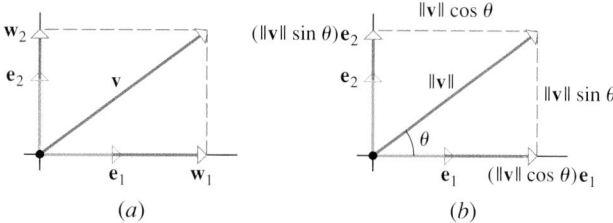

(a)   (b)

Figure 13.3.8

Example 5

A rope is attached to a 100-lb block on a ramp that is inclined at an angle of $30°$ with the ground (Figure 13.3.9*a*). How much force does the block exert against the ramp, and how much force must be applied to the rope in a direction parallel to the ramp to prevent the block from sliding down the ramp? (Assume that the ramp is smooth, that is, exerts no frictional forces.)

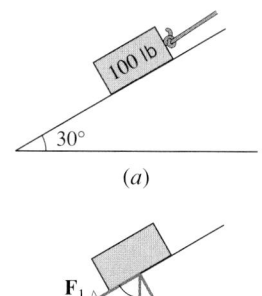

(a)

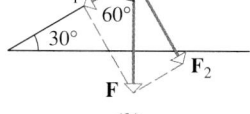

(b)

Figure 13.3.9

Solution. Let $\mathbf{F}$ denote the downward force of gravity on the block (so $\|\mathbf{F}\| = 100$ lb), and let $\mathbf{F}_1$ and $\mathbf{F}_2$ be the vector components of $\mathbf{F}$ parallel and perpendicular to the ramp (as shown in Figure 13.3.9*b*). The lengths of $\mathbf{F}_1$ and $\mathbf{F}_2$ are

$$\|\mathbf{F}_1\| = \|\mathbf{F}\| \cos 60° = 100\left(\frac{1}{2}\right) = 50 \text{ lb}$$

$$\|\mathbf{F}_2\| = \|\mathbf{F}\| \sin 60° = 100\left(\frac{\sqrt{3}}{2}\right) \approx 86.6 \text{ lb}$$

Thus, the block exerts a force of approximately 86.6 lb against the ramp, and it requires a force of 50 lb to prevent the block from sliding down the ramp. ◀

ORTHOGONAL PROJECTIONS

The vector components of $\mathbf{v}$ along $\mathbf{e}_1$ and $\mathbf{e}_2$ in (7) are also called the *orthogonal projections* of $\mathbf{v}$ on $\mathbf{e}_1$ and $\mathbf{e}_2$ and are commonly denoted by

$$\text{proj}_{\mathbf{e}_1}\mathbf{v} = (\mathbf{v} \cdot \mathbf{e}_1)\mathbf{e}_1 \quad \text{and} \quad \text{proj}_{\mathbf{e}_2}\mathbf{v} = (\mathbf{v} \cdot \mathbf{e}_2)\mathbf{e}_2$$

In general, if $\mathbf{e}$ is a unit vector, then we define the ***orthogonal projection of $\mathbf{v}$ on $\mathbf{e}$*** to be

$$\text{proj}_{\mathbf{e}}\mathbf{v} = (\mathbf{v} \cdot \mathbf{e})\mathbf{e} \tag{11}$$

The orthogonal projection of $\mathbf{v}$ on an arbitrary nonzero vector $\mathbf{b}$ can be obtained by normalizing $\mathbf{b}$ and then applying Formula (11); that is,

$$\text{proj}_\mathbf{b}\mathbf{v} = \left(\mathbf{v} \cdot \frac{\mathbf{b}}{\|\mathbf{b}\|}\right)\left(\frac{\mathbf{b}}{\|\mathbf{b}\|}\right)$$

which can be rewritten as

$$\text{proj}_\mathbf{b}\mathbf{v} = \frac{\mathbf{v} \cdot \mathbf{b}}{\|\mathbf{b}\|^2}\mathbf{b} \tag{12}$$

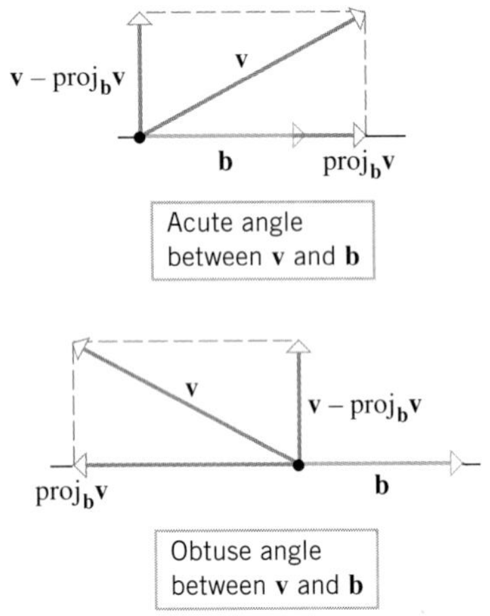

Acute angle between **v** and **b**

Obtuse angle between **v** and **b**

Figure 13.3.10

Geometrically, if $\mathbf{b}$ and $\mathbf{v}$ have a common initial point, then $\text{proj}_\mathbf{b}\mathbf{v}$ is the vector that is determined when a perpendicular is dropped from the terminal point of $\mathbf{v}$ to the line through $\mathbf{b}$ (illustrated in Figure 13.3.10 in two cases). Moreover, it is evident from Figure 13.3.10 that if we subtract $\text{proj}_\mathbf{b}\mathbf{v}$ from $\mathbf{v}$, then the resulting vector

$$\mathbf{v} - \text{proj}_\mathbf{b}\mathbf{v}$$

will be orthogonal to $\mathbf{b}$; we call this the ***vector component of v orthogonal to b***.

Example 6

Find the orthogonal projection of $\mathbf{v} = \mathbf{i} + \mathbf{j} + \mathbf{k}$ on $\mathbf{b} = 2\mathbf{i} + 2\mathbf{j}$, and then find the vector component of $\mathbf{v}$ orthogonal to $\mathbf{b}$.

Solution. We have

$$\mathbf{v} \cdot \mathbf{b} = (\mathbf{i} + \mathbf{j} + \mathbf{k}) \cdot (2\mathbf{i} + 2\mathbf{j}) = 2 + 2 + 0 = 4$$
$$\|\mathbf{b}\|^2 = 2^2 + 2^2 = 8$$

Thus, the orthogonal projection of $\mathbf{v}$ on $\mathbf{b}$ is

$$\text{proj}_\mathbf{b}\mathbf{v} = \frac{\mathbf{v} \cdot \mathbf{b}}{\|\mathbf{b}\|^2}\mathbf{b} = \frac{4}{8}(2\mathbf{i} + 2\mathbf{j}) = \mathbf{i} + \mathbf{j}$$

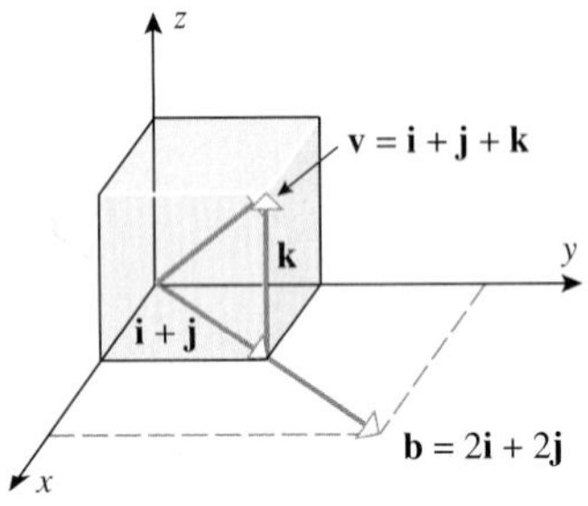

Figure 13.3.11

and the vector component of $\mathbf{v}$ orthogonal to $\mathbf{b}$ is

$$\mathbf{v} - \text{proj}_\mathbf{b}\mathbf{v} = (\mathbf{i} + \mathbf{j} + \mathbf{k}) - (\mathbf{i} + \mathbf{j}) = \mathbf{k}$$

These results are consistent with Figure 13.3.11. ◄

WORK

In Section 8.6 we discussed the work done by a constant force acting on an object that moves along a line. We defined the work W done on the object by a constant force of magnitude F acting in the direction of motion over a distance d to be

$$W = Fd = \text{force} \times \text{distance} \tag{13}$$

If we let $\mathbf{F}$ denote a force vector of magnitude $\|\mathbf{F}\| = F$ acting in the direction of motion, then we can write (13) as

$$W = \|\mathbf{F}\|d$$

Furthermore, if we assume that the object moves along a line from point P to point Q, then $d = \|\overrightarrow{PQ}\|$, so that the work can be expressed entirely in vector form as

$$W = \|\mathbf{F}\|\|\overrightarrow{PQ}\|$$

(Figure 13.3.12*a*). The vector $\overrightarrow{PQ}$ is called the ***displacement vector*** for the object. In the case where a constant force $\mathbf{F}$ is not in the direction of motion, but rather makes an angle θ with the displacement vector, then we *define* the work W done by $\mathbf{F}$ to be

$$W = (\|\mathbf{F}\|\cos\theta)\|\overrightarrow{PQ}\| = \mathbf{F} \cdot \overrightarrow{PQ} \tag{14}$$

(Figure 13.3.12*b*).

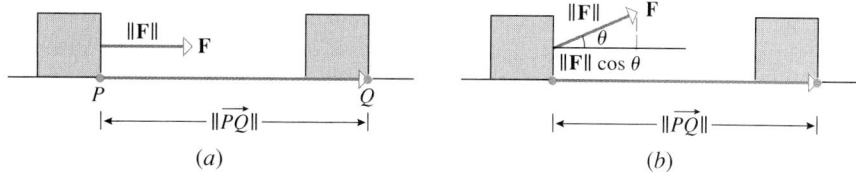

Figure 13.3.12

REMARK.   Note that in Formula (14) the quantity $\|\mathbf{F}\|\cos\theta$ is the scalar component of force along the displacement vector. Thus, in the case where $\cos\theta > 0$, a force of magnitude $\|\mathbf{F}\|$ acting at an angle θ does the same work as a force of magnitude $\|\mathbf{F}\|\cos\theta$ acting in the direction of motion.

Example 7

A wagon is pulled horizontally by exerting a constant force of 10 lb on the handle at an angle of $60°$ with the horizontal. How much work is done in moving the wagon 50 ft?

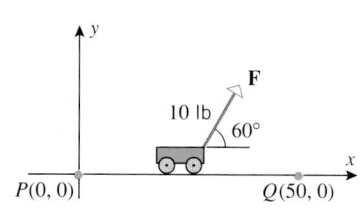

Figure 13.3.13

Solution.   Introduce an xy-coordinate system so that the wagon moves from $P(0,0)$ to $Q(50,0)$ along the x-axis (Figure 13.3.13). In this coordinate system

$$\overrightarrow{PQ} = 50\mathbf{i}$$

and

$$\mathbf{F} = (10\cos 60°)\mathbf{i} + (10\sin 60°)\mathbf{j} = 5\mathbf{i} + 5\sqrt{3}\mathbf{j}$$

so the work done is

$$W = \mathbf{F} \cdot \overrightarrow{PQ} = (5\mathbf{i} + 5\sqrt{3}\mathbf{j}) \cdot (50\mathbf{i}) = 250 \text{ (foot-pounds)} \qquad \blacktriangleleft$$

EXERCISE SET 13.3   ⌁ Graphing Calculator   ⒞ CAS

1. In each part, find the dot product of the vectors and the cosine of the angle between them.
   (a) $\mathbf{u} = \mathbf{i} + 2\mathbf{j}, \ \mathbf{v} = 6\mathbf{i} - 8\mathbf{j}$
   (b) $\mathbf{u} = \langle -7, -3 \rangle, \ \mathbf{v} = \langle 0, 1 \rangle$
   (c) $\mathbf{u} = \mathbf{i} - 3\mathbf{j} + 7\mathbf{k}, \ \mathbf{v} = 8\mathbf{i} - 2\mathbf{j} - 2\mathbf{k}$
   (d) $\mathbf{u} = \langle -3, 1, 2 \rangle, \ \mathbf{v} = \langle 4, 2, -5 \rangle$

2. In each part use the given information to find $\mathbf{u} \cdot \mathbf{v}$.
   (a) $\|\mathbf{u}\| = 1, \|\mathbf{v}\| = 2$, the angle between $\mathbf{u}$ and $\mathbf{v}$ is $\pi/6$.
   (b) $\|\mathbf{u}\| = 2, \|\mathbf{v}\| = 3$, the angle between $\mathbf{u}$ and $\mathbf{v}$ is $135°$.

3. In each part, determine whether $\mathbf{u}$ and $\mathbf{v}$ make an acute angle, an obtuse angle, or are orthogonal.
   (a) $\mathbf{u} = 7\mathbf{i} + 3\mathbf{j} + 5\mathbf{k}, \ \mathbf{v} = -8\mathbf{i} + 4\mathbf{j} + 2\mathbf{k}$
   (b) $\mathbf{u} = 6\mathbf{i} + \mathbf{j} + 3\mathbf{k}, \ \mathbf{v} = 4\mathbf{i} - 6\mathbf{k}$
   (c) $\mathbf{u} = \langle 1, 1, 1 \rangle, \ \mathbf{v} = \langle -1, 0, 0 \rangle$
   (d) $\mathbf{u} = \langle 4, 1, 6 \rangle, \ \mathbf{v} = \langle -3, 0, 2 \rangle$

4. Does the triangle in 3-space with vertices $(-1, 2, 3)$, $(2, -2, 0)$, and $(3, 1, -4)$ have an obtuse angle? Justify your answer.

5. The accompanying figure shows eight vectors that are equally spaced around a circle of radius 1. Find the dot product of $\mathbf{v}_0$ with each of the other seven vectors.

6. The accompanying figure shows six vectors that are equally spaced around a circle of radius 5. Find the dot product of $\mathbf{v}_0$ with each of the other five vectors.

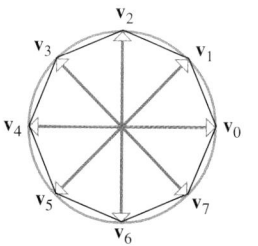

Figure Ex-5

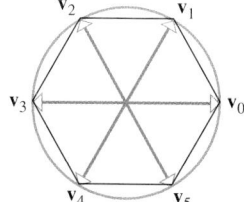

Figure Ex-6

7. (a) Use vectors to show that $A(2, -1, 1)$, $B(3, 2, -1)$, and $C(7, 0, -2)$ are vertices of a right triangle. At which vertex is the right angle?
   (b) Use vectors to find the interior angles of the triangle with vertices $(-1, 0)$, $(2, -1)$, and $(1, 4)$. Express your answers to the nearest degree.

8. Find k so that the vector from the point $A(1, -1, 3)$ to the point $B(3, 0, 5)$ is perpendicular to the vector from A to the point $P(k, k, k)$.

9. (a) Show that if $\mathbf{v} = a\mathbf{i} + b\mathbf{j}$ is a vector in 2-space, then the vectors
$$\mathbf{v}_1 = -b\mathbf{i} + a\mathbf{j} \quad \text{and} \quad \mathbf{v}_2 = b\mathbf{i} - a\mathbf{j}$$
are both orthogonal to $\mathbf{v}$.

(b) Use the result in part (a) to find two unit vectors that are orthogonal to the vector $\mathbf{v} = 3\mathbf{i} - 2\mathbf{j}$. Sketch the vectors $\mathbf{v}$, $\mathbf{v}_1$, and $\mathbf{v}_2$.

10. Find two unit vectors in 2-space that make an angle of $45°$ with $4\mathbf{i} + 3\mathbf{j}$.

11. Explain why each of the following expressions makes no sense.

(a) $\mathbf{u} \cdot (\mathbf{v} \cdot \mathbf{w})$ (b) $(\mathbf{u} \cdot \mathbf{v}) + \mathbf{w}$

(c) $\|\mathbf{u} \cdot \mathbf{v}\|$ (d) $k \cdot (\mathbf{u} + \mathbf{v})$

12. Verify parts (b) and (c) of Theorem 13.3.2 for the vectors $\mathbf{u} = 6\mathbf{i} - \mathbf{j} + 2\mathbf{k}$, $\mathbf{v} = 2\mathbf{i} + 7\mathbf{j} + 4\mathbf{k}$, $\mathbf{w} = \mathbf{i} + \mathbf{j} - 3\mathbf{k}$ and $k = -5$.

13. Let $\mathbf{u} = \langle 1, 2 \rangle$, $\mathbf{v} = \langle 4, -2 \rangle$, and $\mathbf{w} = \langle 6, 0 \rangle$. Find

(a) $\mathbf{u} \cdot (7\mathbf{v} + \mathbf{w})$ (b) $\|(\mathbf{u} \cdot \mathbf{w})\mathbf{w}\|$

(c) $\|\mathbf{u}\|(\mathbf{v} \cdot \mathbf{w})$ (d) $(\|\mathbf{u}\|\mathbf{v}) \cdot \mathbf{w}$.

14. True or False? If $\mathbf{a} \cdot \mathbf{b} = \mathbf{a} \cdot \mathbf{c}$ and if $\mathbf{a} \neq \mathbf{0}$, then $\mathbf{b} = \mathbf{c}$. Justify your conclusion.

In Exercises 15 and 16, find the direction cosines of $\mathbf{v}$, and confirm that they satisfy Equation (5). Then use the direction cosines to approximate the direction angles to the nearest degree.

---

15. (a) $\mathbf{v} = \mathbf{i} + \mathbf{j} - \mathbf{k}$ (b) $\mathbf{v} = 2\mathbf{i} - 2\mathbf{j} + \mathbf{k}$

16. (a) $\mathbf{v} = 3\mathbf{i} - 2\mathbf{j} - 6\mathbf{k}$ (b) $\mathbf{v} = 3\mathbf{i} - 4\mathbf{k}$

17. Show that the direction cosines of a vector satisfy
$$\cos^2 \alpha + \cos^2 \beta + \cos^2 \gamma = 1$$

18. Let θ and λ be the angles shown in the accompanying figure. Show that the direction cosines of $\mathbf{v}$ can be expressed as
$$\cos \alpha = \cos \lambda \cos \theta$$
$$\cos \beta = \cos \lambda \sin \theta$$
$$\cos \gamma = \sin \lambda$$

[*Hint:* Express $\mathbf{v}$ in component form and normalize.]

19. Use the result in Exercise 18 to find the direction angles of the vector shown in the accompanying figure to the nearest degree.

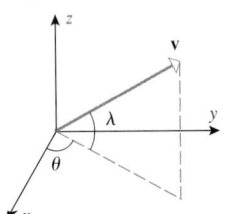

Figure Ex-18   Figure Ex-19

20. Show that two nonzero vectors $\mathbf{v}_1$ and $\mathbf{v}_2$ are orthogonal if and only if their direction cosines satisfy
$$\cos \alpha_1 \cos \alpha_2 + \cos \beta_1 \cos \beta_2 + \cos \gamma_1 \cos \gamma_2 = 0$$

21. In each part, find the vector component of $\mathbf{v}$ along $\mathbf{b}$ and the vector component of $\mathbf{v}$ orthogonal to $\mathbf{b}$. Then sketch the vectors $\mathbf{v}$, $\text{proj}_\mathbf{b}\mathbf{v}$, and $\mathbf{v} - \text{proj}_\mathbf{b}\mathbf{v}$.

(a) $\mathbf{v} = 2\mathbf{i} - \mathbf{j}$, $\mathbf{b} = 3\mathbf{i} + 4\mathbf{j}$

(b) $\mathbf{v} = \langle 4, 5 \rangle$, $\mathbf{b} = \langle 1, -2 \rangle$

(c) $\mathbf{v} = -3\mathbf{i} - 2\mathbf{j}$, $\mathbf{b} = 2\mathbf{i} + \mathbf{j}$

22. In each part, find the vector component of $\mathbf{v}$ along $\mathbf{b}$ and the vector component of $\mathbf{v}$ orthogonal to $\mathbf{b}$.

(a) $\mathbf{v} = 2\mathbf{i} - \mathbf{j} + 3\mathbf{k}$, $\mathbf{b} = \mathbf{i} + 2\mathbf{j} + 2\mathbf{k}$

(b) $\mathbf{v} = \langle 4, -1, 7 \rangle$, $\mathbf{b} = \langle 2, 3, -6 \rangle$

In Exercises 23 and 24, express the vector $\mathbf{v}$ as the sum of a vector parallel to $\mathbf{b}$ and a vector orthogonal to $\mathbf{b}$.

---

23. (a) $\mathbf{v} = 2\mathbf{i} - 4\mathbf{j}$, $\mathbf{b} = \mathbf{i} + \mathbf{j}$

(b) $\mathbf{v} = 3\mathbf{i} + \mathbf{j} - 2\mathbf{k}$, $\mathbf{b} = 2\mathbf{i} - \mathbf{k}$

24. (a) $\mathbf{v} = \langle -3, 5 \rangle$, $\mathbf{b} = \langle 1, 1 \rangle$

(b) $\mathbf{v} = \langle -2, 1, 6 \rangle$, $\mathbf{b} = \langle 0, -2, 1 \rangle$

25. If L is a line in 2-space or 3-space that passes through the points A and B, then the distance from a point P to the line L is equal to the length of the component of the vector $\overrightarrow{AP}$ that is orthogonal to the vector $\overrightarrow{AB}$ (see the accompanying figure). Use this result to find the distance from the point $P(1, 0)$ to the line through $A(2, -3)$ and $B(5, 1)$.

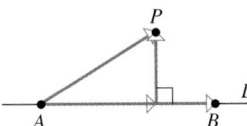

Figure Ex-25

26. Use the method of Exercise 25 to find the distance from the point $P(-3, 1, 2)$ to the line through $A(1, 1, 0)$ and $B(-2, 3, -4)$.

27. As shown in the accompanying figure, a block with a mass of 10 kg rests on a smooth (frictionless) ramp that is inclined at an angle of $45°$ with the ground. How much force does the block exert on the ramp, and how much force must be applied in the direction of $\mathbf{P}$ to prevent the block from sliding down the ramp? Take the acceleration due to gravity to be 9.8 m/s^2.

28. For the block in Exercise 27, how much force must be applied in the direction of $\mathbf{Q}$ (shown in the accompanying figure) to prevent the block from sliding down the ramp?

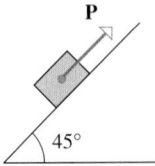 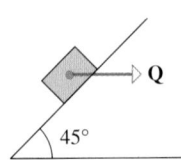

Figure Ex-27   Figure Ex-28

29. A block weighing 300 lb is suspended by cables A and B, as shown in the accompanying figure. Determine the forces that the block exerts along the cables.

30. A block weighing 100 N is suspended by cables A and B, as shown in the accompanying figure.
(a) Use a graphing utility to graph the forces that the block exerts along cables A and B as functions of the "sag" d.
(b) Does increasing the sag increase or decrease the forces on the cables?
(c) How much sag is required if the cables cannot tolerate forces in excess of 150 N?

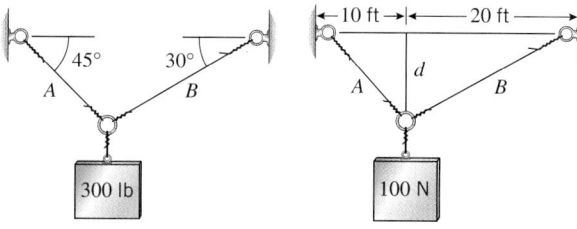

Figure Ex-29   Figure Ex-30

31. Find the work done by a force $\mathbf{F} = -3\mathbf{j}$ (pounds) applied to a point that moves on a line from $(1, 3)$ to $(4, 7)$. Assume that distance is measured in feet.

32. A boat travels 100 meters due north while the wind exerts a force of 500 newtons toward the northeast. How much work does the wind do?

33. A box is dragged along the floor by a rope that applies a force of 50 lb at an angle of $60°$ with the floor. How much work is done in moving the box 15 ft?

34. A force of $\mathbf{F} = 4\mathbf{i} - 6\mathbf{j} + \mathbf{k}$ newtons is applied to a point that moves a distance of 15 meters in the direction of the vector $\mathbf{i} + \mathbf{j} + \mathbf{k}$. How much work is done?

35. Find, to the nearest degree, the acute angle formed by two diagonals of a cube.

36. Find, to the nearest degree, the angles that a diagonal of a box with dimensions 10 cm by 15 cm by 25 cm makes with the edges of the box.

37. Let $\mathbf{u}$ and $\mathbf{v}$ be adjacent sides of a parallelogram. Use vectors to prove that the diagonals of the parallelogram are perpendicular if the sides are equal in length.

38. Let $\mathbf{u}$ and $\mathbf{v}$ be adjacent sides of a parallelogram. Use vectors to prove that the parallelogram is a rectangle if the diagonals are equal in length.

39. Prove that
$$\|\mathbf{u} + \mathbf{v}\|^2 + \|\mathbf{u} - \mathbf{v}\|^2 = 2\|\mathbf{u}\|^2 + 2\|\mathbf{v}\|^2$$
and interpret the result geometrically by translating it into a theorem about parallelograms.

40. Prove: $\mathbf{u} \cdot \mathbf{v} = \frac{1}{4}\|\mathbf{u} + \mathbf{v}\|^2 - \frac{1}{4}\|\mathbf{u} - \mathbf{v}\|^2$.

41. Show that if $\mathbf{v}_1$, $\mathbf{v}_2$, and $\mathbf{v}_3$ are mutually perpendicular nonzero vectors in 3-space, and if a vector $\mathbf{v}$ in 3-space is expressed as
$$\mathbf{v} = c_1\mathbf{v}_1 + c_2\mathbf{v}_2 + c_3\mathbf{v}_3$$
then the scalars c_1, c_2, and c_3 are given by the formulas
$$c_i = (\mathbf{v} \cdot \mathbf{v}_i)/\|\mathbf{v}_i\|^2, \quad i = 1, 2, 3$$

42. Show that the three vectors
$$\mathbf{v}_1 = 3\mathbf{i} - \mathbf{j} + 2\mathbf{k}, \mathbf{v}_2 = \mathbf{i} + \mathbf{j} - \mathbf{k}, \mathbf{v}_3 = \mathbf{i} - 5\mathbf{j} - 4\mathbf{k}$$
are mutually perpendicular, and then use the result of Exercise 41 to find scalars c_1, c_2, and c_3 so that
$$c_1\mathbf{v}_1 + c_2\mathbf{v}_2 + c_3\mathbf{v}_3 = \mathbf{i} - \mathbf{j} + \mathbf{k}$$

43. For each x in $(-\infty, +\infty)$, let $\mathbf{u}(x)$ be the vector from the origin to the point $P(x, y)$ on the curve $y = x^2 + 1$, and $\mathbf{v}(x)$ the vector from the origin to the point $Q(x, y)$ on the line $y = -x - 1$.
(a) Use a CAS to find, to the nearest degree, the minimum angle between $\mathbf{u}(x)$ and $\mathbf{v}(x)$ for x in $(-\infty, +\infty)$.
(b) Determine whether there are any real values of x for which $\mathbf{u}(x)$ and $\mathbf{v}(x)$ are orthogonal.

44. Let $\mathbf{u}$ be a unit vector in the xy-plane of an xyz-coordinate system, and let $\mathbf{v}$ be a unit vector in the yz-plane. Let θ_1 be the angle between $\mathbf{u}$ and $\mathbf{i}$, let θ_2 be the angle between $\mathbf{v}$ and $\mathbf{k}$, and let θ be the angle between $\mathbf{u}$ and $\mathbf{v}$.
(a) Show that $\cos\theta = \pm\sin\theta_1\sin\theta_2$.
(b) Find θ if θ is acute and $\theta_1 = \theta_2 = 45°$.
(c) Use a CAS to find, to the nearest degree, the maximum and minimum values of θ if θ is acute and $\theta_2 = 2\theta_1$.

45. Prove parts (b) and (e) of Theorem 13.3.2 for vectors in 3-space.

13.4 CROSS PRODUCT

In many applications of vectors in mathematics, physics, and engineering, there is a need to find a vector that is orthogonal to two given vectors. In this section we will discuss a new type of vector multiplication that can be used for this purpose.

DETERMINANTS

Some of the concepts that we will develop in this section require basic ideas about *determinants*, which are functions that assign numerical values to square arrays of numbers. For example, if a_1, a_2, b_1, and b_2 are real numbers, then we define a **2 × 2 determinant** by
$$\begin{vmatrix} a_1 & a_2 \\ b_1 & b_2 \end{vmatrix} = a_1b_2 - a_2b_1 \tag{1}$$

The purpose of the arrows is to help you remember the formula—the determinant is the product of the entries on the rightward arrow minus the product of the entries on the leftward arrow. For example,

$$\begin{vmatrix} 3 & -2 \\ 4 & 5 \end{vmatrix} = (3)(5) - (-2)(4) = 15 + 8 = 23$$

A **3 × 3 determinant** is defined in terms of 2 × 2 determinants by

$$\begin{vmatrix} a_1 & a_2 & a_3 \\ b_1 & b_2 & b_3 \\ c_1 & c_2 & c_3 \end{vmatrix} = a_1 \begin{vmatrix} b_2 & b_3 \\ c_2 & c_3 \end{vmatrix} - a_2 \begin{vmatrix} b_1 & b_3 \\ c_1 & c_3 \end{vmatrix} + a_3 \begin{vmatrix} b_1 & b_2 \\ c_1 & c_2 \end{vmatrix} \tag{2}$$

The right side of this formula is easily remembered by noting that a_1, a_2, and a_3 are the entries in the first "row" of the left side, and the 2 × 2 determinants on the right side arise by deleting the first row and an appropriate column from the left side. The pattern is as follows:

$$\begin{vmatrix} a_1 & a_2 & a_3 \\ b_1 & b_2 & b_3 \\ c_1 & c_2 & c_3 \end{vmatrix} = a_1 \begin{vmatrix} a_1 & a_2 & a_3 \\ b_1 & b_2 & b_3 \\ c_1 & c_2 & c_3 \end{vmatrix} - a_2 \begin{vmatrix} a_1 & a_2 & a_3 \\ b_1 & b_2 & b_3 \\ c_1 & c_2 & c_3 \end{vmatrix} + a_3 \begin{vmatrix} a_1 & a_2 & a_3 \\ b_1 & b_2 & b_3 \\ c_1 & c_2 & c_3 \end{vmatrix}$$

For example,

$$\begin{vmatrix} 3 & -2 & -5 \\ 1 & 4 & -4 \\ 0 & 3 & 2 \end{vmatrix} = 3 \begin{vmatrix} 4 & -4 \\ 3 & 2 \end{vmatrix} - (-2) \begin{vmatrix} 1 & -4 \\ 0 & 2 \end{vmatrix} + (-5) \begin{vmatrix} 1 & 4 \\ 0 & 3 \end{vmatrix}$$

$$= 3(20) + 2(2) - 5(3) = 49$$

There are also definitions of 4 × 4 determinants, 5 × 5 determinants, and higher, but we will not need them in this text. Properties of determinants are studied in a branch of mathematics called **linear algebra**, but we will only need the two properties stated in the following theorem:

13.4.1 THEOREM.

(a) *If two rows of a determinant are the same, then the value of the determinant is 0.*

(b) *Interchanging two rows of a determinant multiplies its value by* −1.

We will give the proofs of parts (a) and (b) for 2 × 2 determinants and leave the proofs for 3 × 3 determinants as exercises.

Proof (a).

$$\begin{vmatrix} a_1 & a_2 \\ a_1 & a_2 \end{vmatrix} = a_1 a_2 - a_2 a_1 = 0$$

Proof (b).

$$\begin{vmatrix} b_1 & b_2 \\ a_1 & a_2 \end{vmatrix} = b_1 a_2 - b_2 a_1 = -(a_1 b_2 - a_2 b_1) = - \begin{vmatrix} a_1 & a_2 \\ b_1 & b_2 \end{vmatrix} \qquad ■$$

CROSS PRODUCT

We now turn to the main concept in this section.

13.4.2 DEFINITION. If $\mathbf{u} = \langle u_1, u_2, u_3 \rangle$ and $\mathbf{v} = \langle v_1, v_2, v_3 \rangle$ are vectors in 3-space, then the **cross product** $\mathbf{u} \times \mathbf{v}$ is the vector defined by

$$\mathbf{u} \times \mathbf{v} = \begin{vmatrix} u_2 & u_3 \\ v_2 & v_3 \end{vmatrix} \mathbf{i} - \begin{vmatrix} u_1 & u_3 \\ v_1 & v_3 \end{vmatrix} \mathbf{j} + \begin{vmatrix} u_1 & u_2 \\ v_1 & v_2 \end{vmatrix} \mathbf{k} \tag{3}$$

or, equivalently,

$$\mathbf{u} \times \mathbf{v} = (u_2 v_3 - u_3 v_2) \mathbf{i} - (u_1 v_3 - u_3 v_1) \mathbf{j} + (u_1 v_2 - u_2 v_1) \mathbf{k} \tag{4}$$

Observe that the right side of Formula (3) has the same form as the right side of Formula (2), the difference being notation and the order of the factors in the three terms. Thus, we can rewrite (3) as

$$\mathbf{u} \times \mathbf{v} = \begin{vmatrix} \mathbf{i} & \mathbf{j} & \mathbf{k} \\ u_1 & u_2 & u_3 \\ v_1 & v_2 & v_3 \end{vmatrix} \tag{5}$$

However, this is just a mnemonic device and not a true determinant since the entries in a determinant are numbers, not vectors.

Example 1

Let $\mathbf{u} = \langle 1, 2, -2 \rangle$ and $\mathbf{v} = \langle 3, 0, 1 \rangle$. Find

(a) $\mathbf{u} \times \mathbf{v}$ (b) $\mathbf{v} \times \mathbf{u}$

Solution (a).

$$\mathbf{u} \times \mathbf{v} = \begin{vmatrix} \mathbf{i} & \mathbf{j} & \mathbf{k} \\ 1 & 2 & -2 \\ 3 & 0 & 1 \end{vmatrix}$$

$$= \begin{vmatrix} 2 & -2 \\ 0 & 1 \end{vmatrix} \mathbf{i} - \begin{vmatrix} 1 & -2 \\ 3 & 1 \end{vmatrix} \mathbf{j} + \begin{vmatrix} 1 & 2 \\ 3 & 0 \end{vmatrix} \mathbf{k} = 2\mathbf{i} - 7\mathbf{j} - 6\mathbf{k}$$

Solution (b). We could use the method of part (a), but it is really not necessary to perform any computations. We need only observe that reversing $\mathbf{u}$ and $\mathbf{v}$ interchanges the second and third rows in (5), which in turn interchanges the rows of the 2×2 determinants in (3). But interchanging the rows of a 2×2 determinant reverses its sign, so the net effect of reversing the factors in a cross product is to reverse the signs of the components. Thus, by inspection

$$\mathbf{v} \times \mathbf{u} = -(\mathbf{u} \times \mathbf{v}) = -2\mathbf{i} + 7\mathbf{j} + 6\mathbf{k} \qquad \blacktriangleleft$$

Example 2

Show that $\mathbf{u} \times \mathbf{u} = \mathbf{0}$ for any vector $\mathbf{u}$ in 3-space.

Solution. We could let $\mathbf{u} = u_1\mathbf{i} + u_2\mathbf{j} + u_3\mathbf{k}$ and apply the method in part (a) of Example 1 to show that

$$\mathbf{u} \times \mathbf{u} = \begin{vmatrix} \mathbf{i} & \mathbf{j} & \mathbf{k} \\ u_1 & u_2 & u_3 \\ u_1 & u_2 & u_3 \end{vmatrix} = 0$$

However, the actual computations are unnecessary. We need only observe that if the two factors in a cross product are the same, then each 2×2 determinant in (3) is zero because it has identical rows. Thus, $\mathbf{u} \times \mathbf{u} = \mathbf{0}$ by inspection. $\blacktriangleleft$

ALGEBRAIC PROPERTIES OF THE CROSS PRODUCT

Our next goal is to establish some of the basic algebraic properties of the cross product. As you read the discussion, keep in mind the essential differences between the cross product and the dot product:

- The cross product is defined only for vectors in 3-space, whereas the dot product is defined for vectors in 2-space and 3-space.

- The cross product of two vectors is a vector, whereas the dot product of two vectors is a scalar.

The main algebraic properties of the cross product are listed in the next theorem.

13.4.3   THEOREM.   *If* **u**, **v**, *and* **w** *are any vectors in 3-space and k is any scalar, then*

(a) $\mathbf{u} \times \mathbf{v} = -(\mathbf{v} \times \mathbf{u})$

(b) $\mathbf{u} \times (\mathbf{v} + \mathbf{w}) = (\mathbf{u} \times \mathbf{v}) + (\mathbf{u} \times \mathbf{w})$

(c) $(\mathbf{u} + \mathbf{v}) \times \mathbf{w} = (\mathbf{u} \times \mathbf{w}) + (\mathbf{v} \times \mathbf{w})$

(d) $k(\mathbf{u} \times \mathbf{v}) = (k\mathbf{u}) \times \mathbf{v} = \mathbf{u} \times (k\mathbf{v})$

(e) $\mathbf{u} \times \mathbf{0} = \mathbf{0} \times \mathbf{u} = \mathbf{0}$

(f) $\mathbf{u} \times \mathbf{u} = \mathbf{0}$

Parts (*a*) and (*f*) were addressed in Examples 1 and 2. The other proofs are left as exercises.

WARNING.   In ordinary multiplication and in dot products the order of the factors does not matter, but in cross products it does. Part (*a*) of the last theorem shows that reversing the order of the factors in a cross product reverses the direction of the resulting vector.

The following cross products occur so frequently that it is helpful to be familiar with them:

$$\mathbf{i} \times \mathbf{j} = \mathbf{k} \qquad \mathbf{j} \times \mathbf{k} = \mathbf{i} \qquad \mathbf{k} \times \mathbf{i} = \mathbf{j}$$
$$\mathbf{j} \times \mathbf{i} = -\mathbf{k} \qquad \mathbf{k} \times \mathbf{j} = -\mathbf{i} \qquad \mathbf{i} \times \mathbf{k} = -\mathbf{j} \tag{6}$$

These results are easy to obtain; for example,

$$\mathbf{i} \times \mathbf{j} = \begin{vmatrix} \mathbf{i} & \mathbf{j} & \mathbf{k} \\ 1 & 0 & 0 \\ 0 & 1 & 0 \end{vmatrix} = \begin{vmatrix} 0 & 0 \\ 1 & 0 \end{vmatrix} \mathbf{i} - \begin{vmatrix} 1 & 0 \\ 0 & 0 \end{vmatrix} \mathbf{j} + \begin{vmatrix} 1 & 0 \\ 0 & 1 \end{vmatrix} \mathbf{k} = \mathbf{k}$$

However, rather than computing these cross products each time you need them, you can use the diagram in Figure 13.4.1. In this diagram, the cross product of two consecutive vectors in the clockwise direction is the next vector around, and the cross product of two consecutive vectors in the counterclockwise direction is the negative of the next vector around.

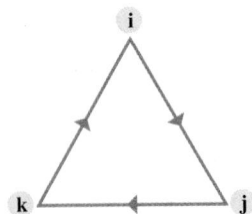

Figure 13.4.1

WARNING.   We can write a product of three real numbers as uvw because the associative law $u(vw) = (uv)w$ ensures that the same value for the product results no matter where the parentheses are inserted. However, the associative law *does not* hold for cross products. For example,

$$\mathbf{i} \times (\mathbf{j} \times \mathbf{j}) = \mathbf{i} \times \mathbf{0} = \mathbf{0} \quad \text{and} \quad (\mathbf{i} \times \mathbf{j}) \times \mathbf{j} = \mathbf{k} \times \mathbf{j} = -\mathbf{i}$$

so that $\mathbf{i} \times (\mathbf{j} \times \mathbf{j}) \neq (\mathbf{i} \times \mathbf{j}) \times \mathbf{j}$. Thus, we cannot write a cross product with three vectors as $\mathbf{u} \times \mathbf{v} \times \mathbf{w}$, since this expression is ambiguous without parentheses.

GEOMETRIC PROPERTIES OF THE CROSS PRODUCT

The following theorem shows that the cross product of two vectors is orthogonal to both factors.

13.4.4   THEOREM.   *If* **u** *and* **v** *are vectors in 3-space, then*

(a) $\mathbf{u} \cdot (\mathbf{u} \times \mathbf{v}) = 0$ (**u** × **v** *is orthogonal to* **u**)

(b) $\mathbf{v} \cdot (\mathbf{u} \times \mathbf{v}) = 0$ (**u** × **v** *is orthogonal to* **v**)

We will prove part (*a*). The proof of part (*b*) is similar.

Proof (a).   Let $\mathbf{u} = \langle u_1, u_2, u_3 \rangle$ and $\mathbf{v} = \langle v_1, v_2, v_3 \rangle$. Then from (4)

$$\mathbf{u} \times \mathbf{v} = \langle u_2 v_3 - u_3 v_2, \ u_3 v_1 - u_1 v_3, \ u_1 v_2 - u_2 v_1 \rangle \tag{7}$$

so that

$$\mathbf{u} \cdot (\mathbf{u} \times \mathbf{v}) = u_1(u_2 v_3 - u_3 v_2) + u_2(u_3 v_1 - u_1 v_3) + u_3(u_1 v_2 - u_2 v_1) = 0$$

Example 3

In Example 1 we showed that the cross product $\mathbf{u} \times \mathbf{v}$ of $\mathbf{u} = \langle 1, 2, -2 \rangle$ and $\mathbf{v} = \langle 3, 0, 1 \rangle$ is

$$\mathbf{u} \times \mathbf{v} = 2\mathbf{i} - 7\mathbf{j} - 6\mathbf{k} = \langle 2, -7, -6 \rangle$$

Theorem 13.4.4 guarantees that this vector is orthogonal to both $\mathbf{u}$ and $\mathbf{v}$; this is confirmed by the computations

$$\mathbf{u} \cdot (\mathbf{u} \times \mathbf{v}) = \langle 1, 2, -2 \rangle \cdot \langle 2, -7, -6 \rangle = (1)(2) + (2)(-7) + (-2)(-6) = 0$$
$$\mathbf{v} \cdot (\mathbf{u} \times \mathbf{v}) = \langle 3, 0, 1 \rangle \cdot \langle 2, -7, -6 \rangle = (3)(2) + (0)(-7) + (1)(-6) = 0 \qquad \blacktriangleleft$$

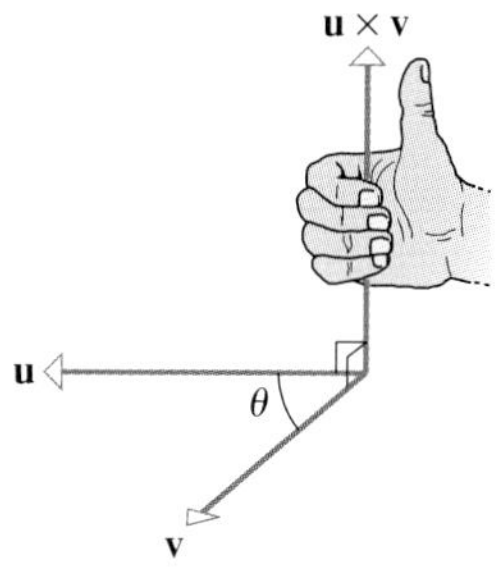

It can be proved that if $\mathbf{u}$ and $\mathbf{v}$ are nonzero and nonparallel vectors, then the direction of $\mathbf{u} \times \mathbf{v}$ relative to $\mathbf{u}$ and $\mathbf{v}$ is determined by a right-hand rule;[*] that is, if the fingers of the right hand are cupped so they curl from $\mathbf{u}$ toward $\mathbf{v}$ in the direction of rotation that takes $\mathbf{u}$ into $\mathbf{v}$ in less than $180°$, then the thumb will point (roughly) in the direction of $\mathbf{u} \times \mathbf{v}$ (Figure 13.4.2). For example, we stated in (6) that

$$\mathbf{i} \times \mathbf{j} = \mathbf{k}, \quad \mathbf{j} \times \mathbf{k} = \mathbf{i}, \quad \mathbf{k} \times \mathbf{i} = \mathbf{j}$$

all of which are consistent with the right-hand rule (verify).

The next theorem lists some more important geometric properties of the cross product.

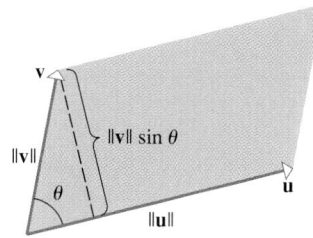

Figure 13.4.2

13.4.5   THEOREM.   *Let $\mathbf{u}$ and $\mathbf{v}$ be nonzero vectors in 3-space, and let θ be the angle between these vectors when they are positioned so their initial points coincide.*

(a) $\|\mathbf{u} \times \mathbf{v}\| = \|\mathbf{u}\|\|\mathbf{v}\|\sin\theta$

(b)   *The area A of the parallelogram that has $\mathbf{u}$ and $\mathbf{v}$ as adjacent sides is*

$$A = \|\mathbf{u} \times \mathbf{v}\| \qquad (8)$$

(c) $\mathbf{u} \times \mathbf{v} = \mathbf{0}$ *if and only if $\mathbf{u}$ and $\mathbf{v}$ are parallel vectors, that is, if and only if they are scalar multiples of one another.*

Proof (a).

$$\|\mathbf{u}\|\|\mathbf{v}\|\sin\theta = \|\mathbf{u}\|\|\mathbf{v}\|\sqrt{1 - \cos^2\theta}$$

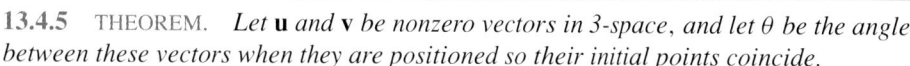

$$= \|\mathbf{u}\|\|\mathbf{v}\|\sqrt{1 - \frac{(\mathbf{u} \cdot \mathbf{v})^2}{\|\mathbf{u}\|^2\|\mathbf{v}\|^2}}$$

$$= \sqrt{\|\mathbf{u}\|^2\|\mathbf{v}\|^2 - (\mathbf{u} \cdot \mathbf{v})^2}$$

$$= \sqrt{(u_1^2 + u_2^2 + u_3^2)(v_1^2 + v_2^2 + v_3^2) - (u_1v_1 + u_2v_2 + u_3v_3)^2}$$

$$= \sqrt{(u_2v_3 - u_3v_2)^2 + (u_1v_3 - u_3v_1)^2 + (u_1v_2 - u_2v_1)^2}$$

$$= \|\mathbf{u} \times \mathbf{v}\| \qquad \boxed{\text{See Formula (4).}}$$

Proof (b).   Referring to Figure 13.4.3, the parallelogram that has $\mathbf{u}$ and $\mathbf{v}$ as adjacent sides can be viewed as having base $\|\mathbf{u}\|$ and altitude $\|\mathbf{v}\|\sin\theta$. Thus, its area A is

$$A = (\text{base})(\text{altitude}) = \|\mathbf{u}\|\|\mathbf{v}\|\sin\theta = \|\mathbf{u} \times \mathbf{v}\|$$

Proof (c).   Since $\mathbf{u}$ and $\mathbf{v}$ are assumed to be nonzero vectors, it follows from part (a) that $\mathbf{u} \times \mathbf{v} = \mathbf{0}$ if and only if $\sin\theta = 0$; this is true if and only if $\theta = 0$ or $\theta = \pi$ (since

Figure 13.4.3

---

[*]Recall that we agreed to consider only right-handed coordinate systems in this text. Had we used left-handed systems instead, a "left-hand rule" would apply here.

$0 \leq \theta \leq \pi$). Geometrically, this means that $\mathbf{u} \times \mathbf{v} = \mathbf{0}$ if and only if $\mathbf{u}$ and $\mathbf{v}$ are parallel vectors. ■

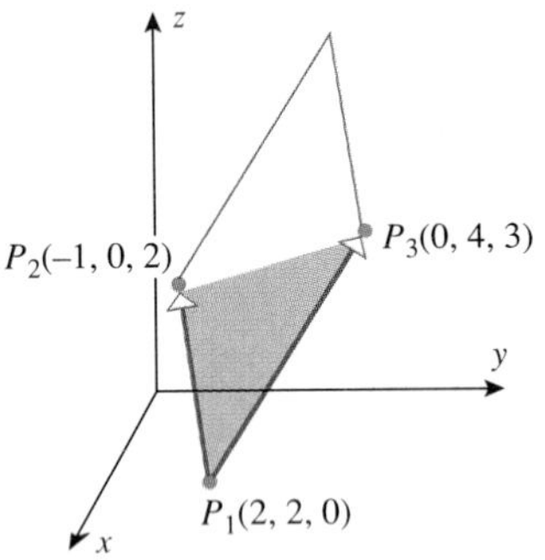

Example 4

Find the area of the triangle that is determined by the points $P_1(2, 2, 0)$, $P_2(-1, 0, 2)$, and $P_3(0, 4, 3)$.

Solution. The area A of the triangle is half the area of the parallelogram determined by the vectors $\overrightarrow{P_1P_2}$ and $\overrightarrow{P_1P_3}$ (Figure 13.4.4). But $\overrightarrow{P_1P_2} = \langle -3, -2, 2 \rangle$ and $\overrightarrow{P_1P_3} = \langle -2, 2, 3 \rangle$, so

$$\overrightarrow{P_1P_2} \times \overrightarrow{P_1P_3} = \langle -10, 5, -10 \rangle$$

(verify), and consequently

$$A = \tfrac{1}{2} \| \overrightarrow{P_1P_2} \times \overrightarrow{P_1P_3} \| = \tfrac{15}{2}$$ ◀

Figure 13.4.4

....

SCALAR TRIPLE PRODUCTS

If $\mathbf{u} = \langle u_1, u_2, u_3 \rangle$, $\mathbf{v} = \langle v_1, v_2, v_3 \rangle$, and $\mathbf{w} = \langle w_1, w_2, w_3 \rangle$ are vectors in 3-space, then the number

$$\mathbf{u} \cdot (\mathbf{v} \times \mathbf{w})$$

is called the *scalar triple product* of $\mathbf{u}$, $\mathbf{v}$, and $\mathbf{w}$. It is not necessary to compute the dot product and cross product to evaluate a scalar triple product—the value can be obtained directly from the formula

$$\mathbf{u} \cdot (\mathbf{v} \times \mathbf{w}) = \begin{vmatrix} u_1 & u_2 & u_3 \\ v_1 & v_2 & v_3 \\ w_1 & w_2 & w_3 \end{vmatrix} \qquad (9)$$

the validity of which can be seen by writing

$$\mathbf{u} \cdot (\mathbf{v} \times \mathbf{w}) = \mathbf{u} \cdot \left(\begin{vmatrix} v_2 & v_3 \\ w_2 & w_3 \end{vmatrix} \mathbf{i} - \begin{vmatrix} v_1 & v_3 \\ w_1 & w_3 \end{vmatrix} \mathbf{j} + \begin{vmatrix} v_1 & v_2 \\ w_1 & w_2 \end{vmatrix} \mathbf{k} \right)$$

$$= u_1 \begin{vmatrix} v_2 & v_3 \\ w_2 & w_3 \end{vmatrix} - u_2 \begin{vmatrix} v_1 & v_3 \\ w_1 & w_3 \end{vmatrix} + u_3 \begin{vmatrix} v_1 & v_2 \\ w_1 & w_2 \end{vmatrix}$$

$$= \begin{vmatrix} u_1 & u_2 & u_3 \\ v_1 & v_2 & v_3 \\ w_1 & w_2 & w_3 \end{vmatrix}$$

Example 5

Calculate the scalar triple product $\mathbf{u} \cdot (\mathbf{v} \times \mathbf{w})$ of the vectors

$$\mathbf{u} = 3\mathbf{i} - 2\mathbf{j} - 5\mathbf{k}, \quad \mathbf{v} = \mathbf{i} + 4\mathbf{j} - 4\mathbf{k}, \quad \mathbf{w} = 3\mathbf{j} + 2\mathbf{k}$$

Solution.

$$\mathbf{u} \cdot (\mathbf{v} \times \mathbf{w}) = \begin{vmatrix} 3 & -2 & -5 \\ 1 & 4 & -4 \\ 0 & 3 & 2 \end{vmatrix} = 49$$ ◀

FOR THE READER.   Many calculating utilities have built-in cross product and determinant operations. If your calculating utility has these capabilities, use it to check the computations in Examples 1 and 5.

....

GEOMETRIC PROPERTIES OF THE SCALAR TRIPLE PRODUCT

If $\mathbf{u}$, $\mathbf{v}$, and $\mathbf{w}$ are nonzero vectors in 3-space that are positioned so their initial points coincide, then these vectors form the adjacent sides of a parallelepiped (Figure 13.4.5). The following theorem establishes a relationship between the volume of this parallelepiped and the scalar triple product of the sides.

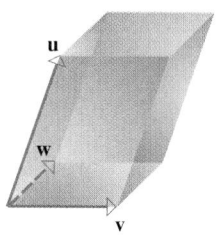

Figure 13.4.5

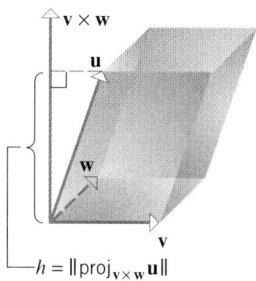

Figure 13.4.6

13.4.6 THEOREM. *Let* **u**, **v**, *and* **w** *be nonzero vectors in 3-space.*

(a) *The volume V of the parallelepiped that has* **u**, **v**, *and* **w** *as adjacent edges is*

$$V = |\mathbf{u} \cdot (\mathbf{v} \times \mathbf{w})| \tag{10}$$

(b) **u** $\cdot$ (**v** $\times$ **w**) = 0 *if and only if* **u**, **v**, *and* **w** *lie in the same plane.*

Proof (a). Referring to Figure 13.4.6, let us regard the base of the parallelepiped with **u**, **v**, and **w** as adjacent sides to be the parallelogram determined by **v** and **w**. Thus, the area of the base is $\|\mathbf{v} \times \mathbf{w}\|$, and the altitude h of the parallelepiped (shown in the figure) is the length of the orthogonal projection of **u** on the vector **v** $\times$ **w**. Therefore, from Formula (12) of Section 13.3 we have

$$h = \|\text{proj}_{\mathbf{v} \times \mathbf{w}} \mathbf{u}\| = \frac{|\mathbf{u} \cdot (\mathbf{v} \times \mathbf{w})|}{\|\mathbf{v} \times \mathbf{w}\|^2} \|\mathbf{v} \times \mathbf{w}\| = \frac{|\mathbf{u} \cdot (\mathbf{v} \times \mathbf{w})|}{\|\mathbf{v} \times \mathbf{w}\|}$$

It now follows that the volume of the parallelepiped is

$$V = (\text{area of base})(\text{height}) = \|\mathbf{v} \times \mathbf{w}\| h = |\mathbf{u} \cdot (\mathbf{v} \times \mathbf{w})|$$

Proof (b). The vectors **u**, **v**, and **w** lie in the same plane if and only if the parallelepiped with these vectors as adjacent sides has volume zero (why?). Thus, from part (a) the vectors lie in the same plane if and only if **u** $\cdot$ (**v** $\times$ **w**) = 0. ∎

REMARK. It follows from Formula (10) that

$$\mathbf{u} \cdot (\mathbf{v} \times \mathbf{w}) = \pm V$$

The + occurs when **u** makes an acute angle with **v** $\times$ **w** and the − occurs when it makes an obtuse angle.

ALGEBRAIC PROPERTIES OF THE SCALAR TRIPLE PRODUCT

We observed earlier in this section that the expression **u** $\times$ **v** $\times$ **w** must be avoided because it is ambiguous without parentheses. However, the expression **u** $\cdot$ **v** $\times$ **w** is not ambiguous—it has to mean **u** $\cdot$ (**v** $\times$ **w**) and not (**u** $\cdot$ **v**) $\times$ **w** because we cannot form the cross product of a scalar and a vector. Similarly, the expression **u** $\times$ **v** $\cdot$ **w** must mean (**u** $\times$ **v**) $\cdot$ **w** and not **u** $\times$ (**v** $\cdot$ **w**). Thus, when you see an expression of the form **u** $\cdot$ **v** $\times$ **w** or **u** $\times$ **v** $\cdot$ **w**, the cross product is formed first and the dot product second.

Since interchanging two rows of a determinant multiplies its value by −1, making two row interchanges in a determinant has no effect on its value. This being the case, it follows that

$$\mathbf{u} \cdot (\mathbf{v} \times \mathbf{w}) = \mathbf{w} \cdot (\mathbf{u} \times \mathbf{v}) = \mathbf{v} \cdot (\mathbf{w} \times \mathbf{u}) \tag{11}$$

since the 3×3 determinants that are used to compute these scalar triple products can be obtained from one another by two row interchanges (verify).

REMARK. Observe that the second expression in (11) can be obtained from the first by leaving the dot, the cross, and the parentheses fixed, moving the first two vectors to the right, and bringing the third vector to the first position. The same procedure produces the third expression from the second and the first expression from the third (verify).

Another useful formula can be obtained by rewriting the first equality in (11) as

$$\mathbf{u} \cdot (\mathbf{v} \times \mathbf{w}) = (\mathbf{u} \times \mathbf{v}) \cdot \mathbf{w}$$

and then omitting the superfluous parentheses to obtain

$$\mathbf{u} \cdot \mathbf{v} \times \mathbf{w} = \mathbf{u} \times \mathbf{v} \cdot \mathbf{w} \tag{12}$$

In words, this formula states that the dot and cross in a scalar triple product can be interchanged (provided the factors are grouped appropriately).

DOT AND CROSS PRODUCTS ARE COORDINATE INDEPENDENT

In Definitions 13.3.1 and 13.4.2 we defined the dot product and the cross product of two vectors in terms of the components of those vectors in a coordinate system. Thus, it is theoretically possible that changing the coordinate system might change $\mathbf{u} \cdot \mathbf{v}$ or $\mathbf{u} \times \mathbf{v}$, since the components of a vector depend on the coordinate system that is chosen. However, the relationships

$$\mathbf{u} \cdot \mathbf{v} = \|\mathbf{u}\| \|\mathbf{v}\| \cos\theta \tag{13}$$

$$\|\mathbf{u} \times \mathbf{v}\| = \|\mathbf{u}\| \|\mathbf{v}\| \sin\theta \tag{14}$$

that were obtained in Theorems 13.3.3 and 13.4.5 show that this is not the case. Formula (13) shows that the value of $\mathbf{u} \cdot \mathbf{v}$ depends only on the lengths of the vectors and the angle between them—not on the coordinate system. Similarly, Formula (14), in combination with the right-hand rule and Theorem 13.4.4, shows that $\mathbf{u} \times \mathbf{v}$ does not depend on the coordinate system (as long as it is right handed). These facts are important in applications because they allow us to choose any convenient coordinate system for solving a problem with full confidence that the choice will not affect computations that involve dot products or cross products.

MOMENTS AND ROTATIONAL MOTION IN 3-SPACE

Astronauts use tools that are designed to limit forces that would impart unintended rotational motion to a satellite.

Cross products play an important role in describing rotational motion in 3-space. For example, suppose that an astronaut on a satellite repair mission in space applies a force $\mathbf{F}$ at a point Q on the surface of a spherical satellite. If the force is directed along a line that passes through the center P of the satellite, then Newton's Second Law of Motion implies that the force will accelerate the satellite in the direction of $\mathbf{F}$. However, if the astronaut applies the same force at an angle θ with the vector $\overrightarrow{PQ}$, then $\mathbf{F}$ will tend to cause a rotation as well as an acceleration in the direction of $\mathbf{F}$. To see why this is so, let us resolve $\mathbf{F}$ into a sum of orthogonal components $\mathbf{F} = \mathbf{F}_1 + \mathbf{F}_2$, where $\mathbf{F}_1$ is the orthogonal projection of $\mathbf{F}$ on the vector $\overrightarrow{PQ}$ and $\mathbf{F}_2$ is the component of $\mathbf{F}$ orthogonal to $\overrightarrow{PQ}$ (Figure 13.4.7). Since the force $\mathbf{F}_1$ acts along the line through the center of the satellite, it contributes to the linear acceleration of the satellite but does not cause any rotation. However, the force $\mathbf{F}_2$ is tangent to the circle around the satellite in the plane of $\mathbf{F}$ and $\overrightarrow{PQ}$, so it causes the satellite to rotate about an axis that is perpendicular to that plane.

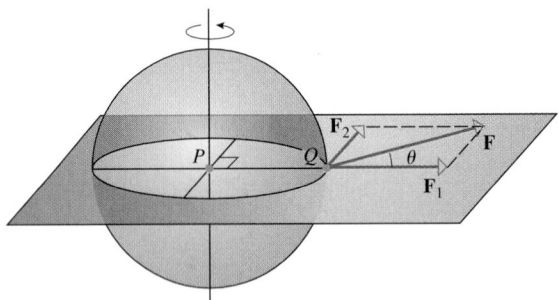

Figure 13.4.7

You know from your own experience that the "tendency" for rotation about an axis depends both on the amount of force and how far from the axis it is applied. For example, it is easier to close a door by pushing on its outer edge than applying the same force close to the hinges. Thus, the tendency of rotation of the satellite can be measured by

$$\|\overrightarrow{PQ}\| \|\mathbf{F}_2\| \qquad \boxed{\text{distance from the center} \times \text{magnitude of the force}} \tag{15}$$

However, $\|\mathbf{F}_2\| = \|\mathbf{F}\| \sin\theta$, so we can rewrite (15) as

$$\|\overrightarrow{PQ}\| \|\mathbf{F}\| \sin\theta = \|\overrightarrow{PQ} \times \mathbf{F}\|$$

This is called the ***scalar moment*** or ***torque*** of $\mathbf{F}$ about the point P. Scalar moments have units of force times distance—pound-feet or newton-meters, for example. The vector $\overrightarrow{PQ} \times \mathbf{F}$ is called the ***vector moment*** or ***torque vector*** of $\mathbf{F}$ about P.

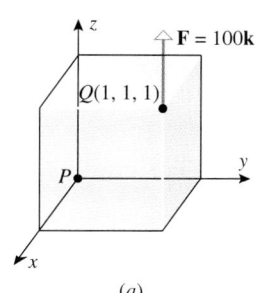

(a)

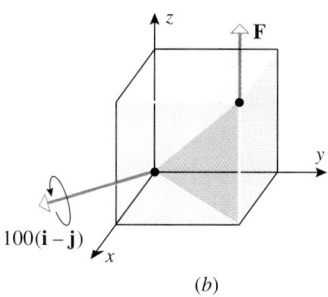

(b)

Figure 13.4.8

Recalling that the direction of $\overrightarrow{PQ} \times \mathbf{F}$ is determined by the right-hand rule, it follows that the direction of rotation about P that results by applying the force $\mathbf{F}$ at the point Q is counterclockwise looking down the axis of $\overrightarrow{PQ} \times \mathbf{F}$ (Figure 13.4.7). Thus, the vector moment $\overrightarrow{PQ} \times \mathbf{F}$ captures the essential information about the rotational effect of the force— the magnitude of the cross product provides the scalar moment of the force, and the cross product vector itself provides the axis and direction of rotation.

Example 6

Figure 13.4.8a shows a force $\mathbf{F}$ of 100 N applied in the positive z-direction at the point $Q(1, 1, 1)$ of a cube whose sides have a length of 1 m. Assuming that the cube is free to rotate about the point $P(0, 0, 0)$ (the origin), find the scalar moment of the force about P, and describe the direction of rotation.

Solution. The force vector is $\mathbf{F} = 100\mathbf{k}$, and the vector from P to Q is $\overrightarrow{PQ} = \mathbf{i} + \mathbf{j} + \mathbf{k}$, so the vector moment of $\mathbf{F}$ about P is

$$\overrightarrow{PQ} \times \mathbf{F} = \begin{vmatrix} \mathbf{i} & \mathbf{j} & \mathbf{k} \\ 1 & 1 & 1 \\ 0 & 0 & 100 \end{vmatrix} = 100\mathbf{i} - 100\mathbf{j}$$

Thus, the scalar moment of $\mathbf{F}$ about P is $\|100\mathbf{i} - 100\mathbf{j}\| = 100\sqrt{2} \approx 141$ N·m, and the direction of rotation is counterclockwise looking along the vector $100\mathbf{i} - 100\mathbf{j} = 100(\mathbf{i} - \mathbf{j})$ toward its initial point (Figure 13.4.8b). ◀

EXERCISE SET 13.4 ☐C CAS

1. (a) Use a determinant to find the cross product

   $$\mathbf{i} \times (\mathbf{i} + \mathbf{j} + \mathbf{k})$$

   (b) Check your answer in part (a) by rewriting the cross product as

   $$\mathbf{i} \times (\mathbf{i} + \mathbf{j} + \mathbf{k}) = (\mathbf{i} \times \mathbf{i}) + (\mathbf{i} \times \mathbf{j}) + (\mathbf{i} \times \mathbf{k})$$

   and evaluating each term.

2. In each part, use the two methods in Exercise 1 to find
   (a) $\mathbf{j} \times (\mathbf{i} + \mathbf{j} + \mathbf{k})$ (b) $\mathbf{k} \times (\mathbf{i} + \mathbf{j} + \mathbf{k})$.

In Exercises 3–6, find $\mathbf{u} \times \mathbf{v}$, and check that it is orthogonal to both $\mathbf{u}$ and $\mathbf{v}$.

3. $\mathbf{u} = \langle 1, 2, -3 \rangle$, $\mathbf{v} = \langle -4, 1, 2 \rangle$

4. $\mathbf{u} = 3\mathbf{i} + 2\mathbf{j} - \mathbf{k}$, $\mathbf{v} = -\mathbf{i} - 3\mathbf{j} + \mathbf{k}$

5. $\mathbf{u} = \langle 0, 1, -2 \rangle$, $\mathbf{v} = \langle 3, 0, -4 \rangle$

6. $\mathbf{u} = 4\mathbf{i} + \mathbf{k}$, $\mathbf{v} = 2\mathbf{i} - \mathbf{j}$

7. Let $\mathbf{u} = \langle 2, -1, 3 \rangle$, $\mathbf{v} = \langle 0, 1, 7 \rangle$, and $\mathbf{w} = \langle 1, 4, 5 \rangle$. Find
   (a) $\mathbf{u} \times (\mathbf{v} \times \mathbf{w})$ (b) $(\mathbf{u} \times \mathbf{v}) \times \mathbf{w}$
   (c) $(\mathbf{u} \times \mathbf{v}) \times (\mathbf{v} \times \mathbf{w})$ (d) $(\mathbf{v} \times \mathbf{w}) \times (\mathbf{u} \times \mathbf{v})$.

☐C 8. Use a CAS or a calculating utility that can compute determinants or cross products to solve Exercise 7.

9. Find the direction cosines of $\mathbf{u} \times \mathbf{v}$ for the vectors $\mathbf{u}$ and $\mathbf{v}$ in the accompanying figure.

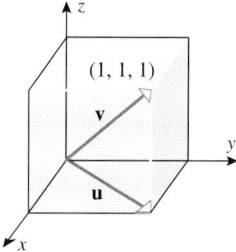

Figure Ex-9

10. Find two unit vectors that are orthogonal to both
    $$\mathbf{u} = -7\mathbf{i} + 3\mathbf{j} + \mathbf{k}, \quad \mathbf{v} = 2\mathbf{i} + 4\mathbf{k}$$

11. Find two unit vectors that are perpendicular to the plane determined by the points $A(0, -2, 1)$, $B(1, -1, -2)$, and $C(-1, 1, 0)$.

12. Find two unit vectors that are parallel to the yz-plane and are perpendicular to the vector $3\mathbf{i} - \mathbf{j} + 2\mathbf{k}$.

In Exercises 13 and 14, find the area of the parallelogram that has $\mathbf{u}$ and $\mathbf{v}$ as adjacent sides.

13. $\mathbf{u} = \mathbf{i} - \mathbf{j} + 2\mathbf{k}$, $\mathbf{v} = 3\mathbf{j} + \mathbf{k}$

14. $\mathbf{u} = 2\mathbf{i} + 3\mathbf{j}$, $\mathbf{v} = -\mathbf{i} + 2\mathbf{j} - 2\mathbf{k}$

In Exercises 15 and 16, find the area of the triangle with vertices P, Q, and R.

15. $P(1, 5, -2)$, $Q(0, 0, 0)$, $R(3, 5, 1)$

16. $P(2, 0, -3)$, $Q(1, 4, 5)$, $R(7, 2, 9)$

In Exercises 17–20, find $\mathbf{u} \cdot (\mathbf{v} \times \mathbf{w})$.

---

17. $\mathbf{u} = 2\mathbf{i} - 3\mathbf{j} + \mathbf{k}$, $\mathbf{v} = 4\mathbf{i} + \mathbf{j} - 3\mathbf{k}$, $\mathbf{w} = \mathbf{j} + 5\mathbf{k}$

18. $\mathbf{u} = \langle 1, -2, 2 \rangle$, $\mathbf{v} = \langle 0, 3, 2 \rangle$, $\mathbf{w} = \langle -4, 1, -3 \rangle$

19. $\mathbf{u} = \langle 2, 1, 0 \rangle$, $\mathbf{v} = \langle 1, -3, 1 \rangle$, $\mathbf{w} = \langle 4, 0, 1 \rangle$

20. $\mathbf{u} = \mathbf{i}$, $\mathbf{v} = \mathbf{i} + \mathbf{j}$, $\mathbf{w} = \mathbf{i} + \mathbf{j} + \mathbf{k}$

In Exercises 21 and 22, use a scalar triple product to find the volume of the parallelepiped that has $\mathbf{u}$, $\mathbf{v}$, and $\mathbf{w}$ as adjacent edges.

---

21. $\mathbf{u} = \langle 2, -6, 2 \rangle$, $\mathbf{v} = \langle 0, 4, -2 \rangle$, $\mathbf{w} = \langle 2, 2, -4 \rangle$

22. $\mathbf{u} = 3\mathbf{i} + \mathbf{j} + 2\mathbf{k}$, $\mathbf{v} = 4\mathbf{i} + 5\mathbf{j} + \mathbf{k}$, $\mathbf{w} = \mathbf{i} + 2\mathbf{j} + 4\mathbf{k}$

23. In each part, use a scalar triple product to determine whether the vectors lie in the same plane.
(a) $\mathbf{u} = \langle 1, -2, 1 \rangle$, $\mathbf{v} = \langle 3, 0, -2 \rangle$, $\mathbf{w} = \langle 5, -4, 0 \rangle$
(b) $\mathbf{u} = 5\mathbf{i} - 2\mathbf{j} + \mathbf{k}$, $\mathbf{v} = 4\mathbf{i} - \mathbf{j} + \mathbf{k}$, $\mathbf{w} = \mathbf{i} - \mathbf{j}$
(c) $\mathbf{u} = \langle 4, -8, 1 \rangle$, $\mathbf{v} = \langle 2, 1, -2 \rangle$, $\mathbf{w} = \langle 3, -4, 12 \rangle$

24. Suppose that $\mathbf{u} \cdot (\mathbf{v} \times \mathbf{w}) = 3$. Find
(a) $\mathbf{u} \cdot (\mathbf{w} \times \mathbf{v})$ (b) $(\mathbf{v} \times \mathbf{w}) \cdot \mathbf{u}$
(c) $\mathbf{w} \cdot (\mathbf{u} \times \mathbf{v})$ (d) $\mathbf{v} \cdot (\mathbf{u} \times \mathbf{w})$
(e) $(\mathbf{u} \times \mathbf{w}) \cdot \mathbf{v}$ (f) $\mathbf{v} \cdot (\mathbf{w} \times \mathbf{w})$.

25. Consider the parallelepiped with adjacent edges
$$\mathbf{u} = 3\mathbf{i} + 2\mathbf{j} + \mathbf{k}$$
$$\mathbf{v} = \mathbf{i} + \mathbf{j} + 2\mathbf{k}$$
$$\mathbf{w} = \mathbf{i} + 3\mathbf{j} + 3\mathbf{k}$$
(a) Find the volume.
(b) Find the area of the face determined by $\mathbf{u}$ and $\mathbf{w}$.
(c) Find the angle between $\mathbf{u}$ and the plane containing the face determined by $\mathbf{v}$ and $\mathbf{w}$.

26. Show that in 3-space the distance d from a point P to the line L through points A and B can be expressed as
$$d = \frac{\|\overrightarrow{AP} \times \overrightarrow{AB}\|}{\|\overrightarrow{AB}\|}$$

27. Use the result in Exercise 26 to find the distance between the point P and the line through the points A and B.
(a) $P(-3, 1, 2)$, $A(1, 1, 0)$, $B(-2, 3, -4)$
(b) $P(4, 3)$, $A(2, 1)$, $B(0, 2)$

28. It is a theorem of solid geometry that the volume of a tetrahedron is $\frac{1}{3}$(area of base) $\cdot$ (height). Use this result to prove that the volume of a tetrahedron whose edges are the vectors $\mathbf{u}$, $\mathbf{v}$, and $\mathbf{w}$ is $\frac{1}{6}|\mathbf{u} \cdot (\mathbf{v} \times \mathbf{w})|$.

29. Use the result of Exercise 28 to find the volume of the tetrahedron with vertices
$$P(-1, 2, 0), \quad Q(2, 1, -3), \quad R(1, 0, 1), \quad S(3, -2, 3)$$

30. Let θ be the angle between the vectors $\mathbf{u} = 2\mathbf{i} + 3\mathbf{j} - 6\mathbf{k}$ and $\mathbf{v} = 2\mathbf{i} + 3\mathbf{j} + 6\mathbf{k}$.

(a) Use the dot product to find $\cos\theta$.
(b) Use the cross product to find $\sin\theta$.
(c) Confirm that $\sin^2\theta + \cos^2\theta = 1$.

31. What can you say about the angle between nonzero vectors $\mathbf{u}$ and $\mathbf{v}$ if $\mathbf{u} \cdot \mathbf{v} = \|\mathbf{u} \times \mathbf{v}\|$?

32. Show that if $\mathbf{u}$ and $\mathbf{v}$ are vectors in 3-space, then
$$\|\mathbf{u} \times \mathbf{v}\|^2 = \|\mathbf{u}\|^2 \|\mathbf{v}\|^2 - (\mathbf{u} \cdot \mathbf{v})^2$$
[*Note:* This result is sometimes called **Lagrange's identity**.]

33. The accompanying figure shows a force $\mathbf{F}$ of 10 lb applied in the positive y-direction to the point $Q(1, 1, 1)$ of a cube whose sides have a length of 1 ft. In each part, find the scalar moment of $\mathbf{F}$ about the point P, and describe the direction of rotation, if any, if the cube is free to rotate about P.
(a) P is the point $(0, 0, 0)$. (b) P is the point $(1, 0, 0)$.
(c) P is the point $(1, 0, 1)$.

34. The accompanying figure shows a force $\mathbf{F}$ of 1000 N applied to the corner of a box.
(a) Find the scalar moment of $\mathbf{F}$ about the point P.
(b) Find the direction angles of the vector moment of $\mathbf{F}$ about the point P to the nearest degree.

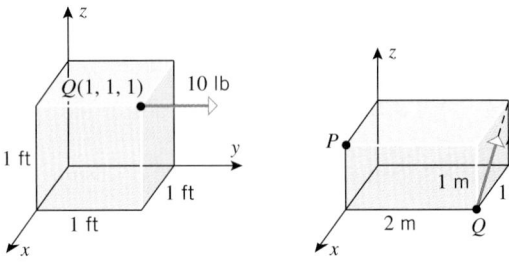

Figure Ex-33   Figure Ex-34

35. As shown in the accompanying figure, a force of 200 N is applied at an angle of $18°$ to a point near the end of a monkey wrench. Find the scalar moment of the force about the center of the bolt. [Treat this as a problem in two dimensions.]

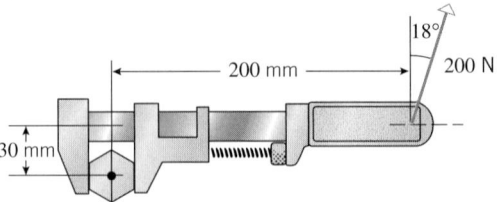

Figure Ex-35

36. Prove parts (*b*) and (*c*) of Theorem 13.4.3.

37. Prove parts (*d*) and (*e*) of Theorem 13.4.3.

38. Prove part (*b*) of Theorem 13.4.1 for 3×3 determinants. [Just give the proof for the first two rows.] Then use (*b*) to prove (*a*).

39. Expressions of the form
$$\mathbf{u} \times (\mathbf{v} \times \mathbf{w}) \quad \text{and} \quad (\mathbf{u} \times \mathbf{v}) \times \mathbf{w}$$

are called **vector triple products**. It can be proved with some effort that

$$\mathbf{u} \times (\mathbf{v} \times \mathbf{w}) = (\mathbf{u} \cdot \mathbf{w})\mathbf{v} - (\mathbf{u} \cdot \mathbf{v})\mathbf{w}$$
$$(\mathbf{u} \times \mathbf{v}) \times \mathbf{w} = (\mathbf{w} \cdot \mathbf{u})\mathbf{v} - (\mathbf{w} \cdot \mathbf{v})\mathbf{u}$$

These expressions can be summarized with the following mnemonic rule:

$$\text{vector triple product} = (\text{outer} \cdot \text{remote})\text{adjacent}$$
$$- (\text{outer} \cdot \text{adjacent})\text{remote}$$

See if you can figure out what the expressions "outer," "remote," and "adjacent" mean in this rule, and then use the rule to find the two vector triple products of the vectors

$$\mathbf{u} = \mathbf{i} + 3\mathbf{j} - \mathbf{k}, \quad \mathbf{v} = \mathbf{i} + \mathbf{j} + 2\mathbf{k}, \quad \mathbf{w} = 3\mathbf{i} - \mathbf{j} + 2\mathbf{k}$$

40. Use the result in Exercise 39 to show that
(a) $\mathbf{u} \times (\mathbf{v} \times \mathbf{w})$ lies in the same plane as $\mathbf{v}$ and $\mathbf{w}$
(b) $(\mathbf{u} \times \mathbf{v}) \times \mathbf{w}$ lies in the same plane as $\mathbf{u}$ and $\mathbf{v}$.

41. Prove: If $\mathbf{a}, \mathbf{b}, \mathbf{c},$ and $\mathbf{d}$ lie in the same plane when positioned with a common initial point, then

$$(\mathbf{a} \times \mathbf{b}) \times (\mathbf{c} \times \mathbf{d}) = \mathbf{0}$$

C **42.** Use a CAS to approximate the minimum area of a triangle if two of its vertices are $(2, -1, 0)$ and $(3, 2, 2)$ and its third vertex is on the curve $y = \ln x$ in the xy-plane.

43. If a force $\mathbf{F}$ is applied to an object at a point Q, then the line through Q parallel to $\mathbf{F}$ is called the **line of action** of the force. We defined the vector moment of $\mathbf{F}$ about a point P to be $\overrightarrow{PQ} \times \mathbf{F}$. Show that if Q' is any point on the line of action of $\mathbf{F}$, then $\overrightarrow{PQ} \times \mathbf{F} = \overrightarrow{PQ'} \times \mathbf{F}$; that is, it is not essential to use the point of application to compute the vector moment—any point on the line of action will do. [*Hint:* Write $\overrightarrow{PQ'} = \overrightarrow{PQ} + \overrightarrow{QQ'}$ and use properties of the cross product.]

13.5 PARAMETRIC EQUATIONS OF LINES

In this section we will discuss parametric equations of lines in 2-space and 3-space. In 3-space, parametric equations of lines are especially important because they generally provide the most convenient form for representing lines algebraically.

LINES DETERMINED BY A POINT AND A VECTOR

A line in 2-space or 3-space can be determined uniquely by specifying a point on the line and a nonzero vector parallel to the line (Figure 13.5.1). The following theorem gives parametric equations of the line through a point P_0 and parallel to a nonzero vector $\mathbf{v}$:

> **13.5.1 THEOREM.**
> (a) *The line in 2-space that passes through the point $P_0(x_0, y_0)$ and is parallel to the nonzero vector $\mathbf{v} = \langle a, b \rangle = a\mathbf{i} + b\mathbf{j}$ has parametric equations*
>
> $$x = x_0 + at, \quad y = y_0 + bt \tag{1}$$
>
> (b) *The line in 3-space that passes through the point $P_0(x_0, y_0, z_0)$ and is parallel to the nonzero vector $\mathbf{v} = \langle a, b, c \rangle = a\mathbf{i} + b\mathbf{j} + c\mathbf{k}$ has parametric equations*
>
> $$x = x_0 + at, \quad y = y_0 + bt, \quad z = z_0 + ct \tag{2}$$

We will prove part (b). The proof of (a) is similar.

Proof (b). If L is the line in 3-space that passes through the point $P_0(x_0, y_0, z_0)$ and is parallel to the nonzero vector $\mathbf{v} = \langle a, b, c \rangle$, then L consists precisely of those points $P(x, y, z)$ for which the vector $\overrightarrow{P_0P}$ is parallel to $\mathbf{v}$ (Figure 13.5.2). In other words, the point $P(x, y, z)$ is on L if and only if $\overrightarrow{P_0P}$ is a scalar multiple of $\mathbf{v}$, say

$$\overrightarrow{P_0P} = t\mathbf{v}$$

This equation can be written as

$$\langle x - x_0, y - y_0, z - z_0 \rangle = \langle ta, tb, tc \rangle$$

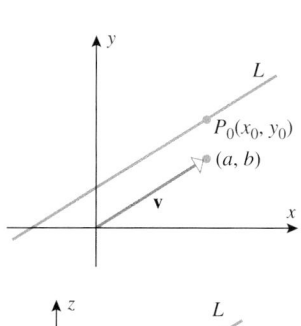

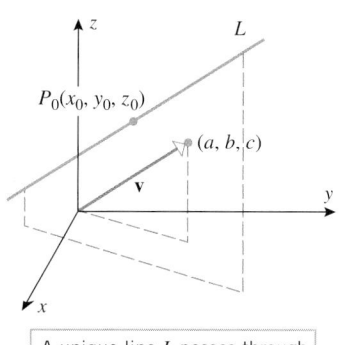

A unique line L passes through P_0 and is parallel to $\mathbf{v}$.

Figure 13.5.1

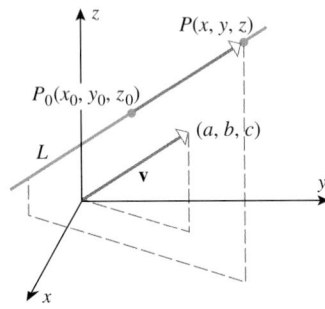

Figure 13.5.2

which implies that

$$x - x_0 = ta, \quad y - y_0 = tb, \quad z - z_0 = tc$$

from which (2) follows. ∎

REMARK. Although it is not stated explicitly, it is understood in Equations (1) and (2) that $-\infty < t < +\infty$, which reflects the fact that lines extend indefinitely.

Example 1

Find parametric equations of the line

(a)   passing through $(4, 2)$ and parallel to $\mathbf{v} = \langle -1, 5 \rangle$;

(b)   passing through $(1, 2, -3)$ and parallel to $\mathbf{v} = 4\mathbf{i} + 5\mathbf{j} - 7\mathbf{k}$;

(c)   passing through the origin in 3-space and parallel to $\mathbf{v} = \langle 1, 1, 1 \rangle$.

Solution (a). From (1) with $x_0 = 4$, $y_0 = 2$, $a = -1$, and $b = 5$ we obtain

$$x = 4 - t, \quad y = 2 + 5t$$

Solution (b). From (2) we obtain

$$x = 1 + 4t, \quad y = 2 + 5t, \quad z = -3 - 7t$$

Solution (c). From (2) with $x_0 = 0$, $y_0 = 0$, $z_0 = 0$, $a = 1$, $b = 1$, and $c = 1$ we obtain

$$x = t, \quad y = t, \quad z = t \qquad \blacktriangleleft$$

Example 2

(a)   Find parametric equations of the line L passing through the points $P_1(2, 4, -1)$ and $P_2(5, 0, 7)$.

(b)   Where does the line intersect the xy-plane?

Solution (a). The vector $\overrightarrow{P_1P_2} = \langle 3, -4, 8 \rangle$ is parallel to L and the point $P_1(2, 4, -1)$ lies on L, so it follows from (2) that L has parametric equations

$$x = 2 + 3t, \quad y = 4 - 4t, \quad z = -1 + 8t \tag{3}$$

Had we used P_2 as the point on L rather than P_1, we would have obtained the equations

$$x = 5 + 3t, \quad y = -4t, \quad z = 7 + 8t$$

Although these equations look different from those obtained using P_1, the two sets of equations are actually equivalent in that both generate L as t varies from $-\infty$ to $+\infty$.

Solution (b). It follows from (3) in part (a) that the line intersects the xy-plane at the point where $z = -1 + 8t = 0$, that is, when $t = \frac{1}{8}$. Substituting this value of t in (3) yields the point of intersection $(x, y, z) = \left(\frac{19}{8}, \frac{7}{2}, 0 \right)$. ◀

Example 3

Let L_1 and L_2 be the lines

$$L_1: x = 1 + 4t, \quad y = 5 - 4t, \quad z = -1 + 5t$$
$$L_2: x = 2 + 8t, \quad y = 4 - 3t, \quad z = 5 + t$$

(a)   Are the lines parallel?

(b)   Do the lines intersect?

Solution (a). The line L_1 is parallel to the vector $4\mathbf{i} - 4\mathbf{j} + 5\mathbf{k}$, and the line L_2 is parallel to the vector $8\mathbf{i} - 3\mathbf{j} + \mathbf{k}$. These vectors are not parallel since neither is a scalar multiple of the other. Thus, the lines are not parallel.

Solution (*b*). For L_1 and L_2 to intersect at some point (x_0, y_0, z_0) these coordinates would have to satisfy the equations of both lines. In other words, there would have to exist values t_1 and t_2 for the parameters such that

$$x_0 = 1 + 4t_1, \quad y_0 = 5 - 4t_1, \quad z_0 = -1 + 5t_1$$

and

$$x_0 = 2 + 8t_2, \quad y_0 = 4 - 3t_2, \quad z_0 = 5 + t_2$$

This leads to three conditions on t_1 and t_2,

$$
\begin{aligned}
1 + 4t_1 &= 2 + 8t_2 \\
5 - 4t_1 &= 4 - 3t_2 \\
-1 + 5t_1 &= 5 + t_2
\end{aligned}
\tag{4}
$$

Thus, the lines intersect if there are values of t_1 and t_2 that satisfy all three equations, and the lines do not intersect if there are no such values. You should be familiar with methods for solving systems of two linear equations in two unknowns; however, this is a system of three linear equations in two unknowns. To determine whether this system has a solution we will solve the first two equations for t_1 and t_2 and then check whether these values satisfy the third equation.

We will solve the first two equations by the method of elimination. We can eliminate the unknown t_1 by adding the equations. This yields the equation

$$6 = 6 + 5t_2$$

from which we obtain $t_2 = 0$. We can now find t_1 by substituting this value of t_2 in either the first or second equation. This yields $t_1 = \frac{1}{4}$. However, the values $t_1 = \frac{1}{4}$ and $t_2 = 0$ do not satisfy the third equation in (4), so the lines do not intersect. ◄

Two lines in 3-space that are not parallel and do not intersect (such as those in Example 3) are called *skew* lines. As illustrated in Figure 13.5.3, any two skew lines lie in parallel planes.

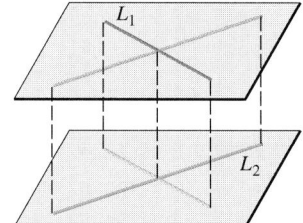

Parallel planes containing skew lines L_1 and L_2 can be determined by translating each line until it intersects the other.

Figure 13.5.3

LINE SEGMENTS

Sometimes one is not interested in an entire line, but rather some *segment* of a line. Parametric equations of a line segment can be obtained by finding parametric equations for the entire line, then restricting the parameter appropriately so that only the desired segment is generated.

Example 4

Find parametric equations for the line segment that joins the points $P_1(2, 4, -1)$ and $P_2(5, 0, 7)$.

Solution. From Example 2, the line through the points P_1 and P_2 has parametric equations $x = 2 + 3t$, $y = 4 - 4t$, $z = -1 + 8t$. With these equations, the point P_1 corresponds to $t = 0$ and P_2 to $t = 1$. Thus, the line segment that joins P_1 and P_2 is given by

$$x = 2 + 3t, \quad y = 4 - 4t, \quad z = -1 + 8t \qquad (0 \le t \le 1)$$ ◄

VECTOR EQUATIONS OF LINES

We will now show how vector notation can be used to express the parametric equations of a line more compactly. Because two vectors are equal if and only if their components are equal, (1) and (2) can be written in vector form as

$$\langle x, y \rangle = \langle x_0 + at, y_0 + bt \rangle$$
$$\langle x, y, z \rangle = \langle x_0 + at, y_0 + bt, z_0 + ct \rangle$$

or, equivalently, as

$$\langle x, y \rangle = \langle x_0, y_0 \rangle + t \langle a, b \rangle \tag{5}$$
$$\langle x, y, z \rangle = \langle x_0, y_0, z_0 \rangle + t \langle a, b, c \rangle \tag{6}$$

For the equation in 2-space we define the vectors $\mathbf{r}$, $\mathbf{r}_0$, and $\mathbf{v}$ as

$$\mathbf{r} = \langle x, y \rangle, \quad \mathbf{r}_0 = \langle x_0, y_0 \rangle, \quad \mathbf{v} = \langle a, b \rangle \tag{7}$$

and for the equation in 3-space we define them as

$$\mathbf{r} = \langle x, y, z \rangle, \quad \mathbf{r}_0 = \langle x_0, y_0, z_0 \rangle, \quad \mathbf{v} = \langle a, b, c \rangle \tag{8}$$

Substituting (7) and (8) in (5) and (6), respectively, yields the equation

$$\mathbf{r} = \mathbf{r}_0 + t\mathbf{v} \tag{9}$$

in both cases. We call this the *vector equation of a line* in 2-space or 3-space. In this equation, $\mathbf{v}$ is a nonzero vector parallel to the line, and $\mathbf{r}_0$ is a vector whose components are the coordinates of a point on the line.

We can interpret Equation (9) geometrically by positioning the vectors $\mathbf{r}_0$ and $\mathbf{v}$ with their initial points at the origin and the vector $t\mathbf{v}$ with its initial point at P_0 (Figure 13.5.4). The vector $t\mathbf{v}$ is a scalar multiple of $\mathbf{v}$ and hence is parallel to $\mathbf{v}$ and L. Moreover, since the initial point of $t\mathbf{v}$ is at the point P_0 on L, this vector actually runs along L; hence, the vector $\mathbf{r} = \mathbf{r}_0 + t\mathbf{v}$ can be interpreted as the vector from the origin to a point on L. As the parameter t varies from 0 to $+\infty$, the terminal point of $\mathbf{r}$ traces out the portion of L that extends from P_0 in the direction of $\mathbf{v}$, and as t varies from 0 to $-\infty$, the terminal point of $\mathbf{r}$ traces out the portion of L that extends from P_0 in the direction that is opposite to $\mathbf{v}$. Thus, the entire line is traced as t varies over the interval $(-\infty, +\infty)$, and it is traced in the direction of $\mathbf{v}$ as t increases.

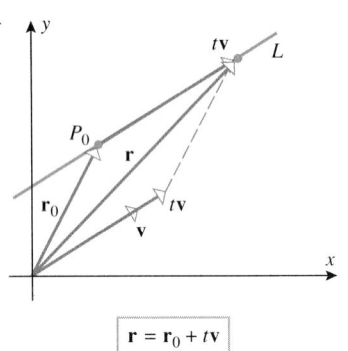

$$\mathbf{r} = \mathbf{r}_0 + t\mathbf{v}$$

Figure 13.5.4

Example 5

The equation

$$\langle x, y, z \rangle = \langle -1, 0, 2 \rangle + t \langle 1, 5, -4 \rangle$$

is of form (9) with

$$\mathbf{r}_0 = \langle -1, 0, 2 \rangle \quad \text{and} \quad \mathbf{v} = \langle 1, 5, -4 \rangle$$

Thus, the equation represents the line in 3-space that passes through the point $(-1, 0, 2)$ and is parallel to the vector $\langle 1, 5, -4 \rangle$. ◀

Example 6

Find an equation of the line in 3-space that passes through the points $P_1(2, 4, -1)$ and $P_2(5, 0, 7)$.

Solution. The vector

$$\overrightarrow{P_1 P_2} = \langle 3, -4, 8 \rangle$$

is parallel to the line, so it can be used as $\mathbf{v}$ in (9). For $\mathbf{r}_0$ we can use either the vector from the origin to P_1 or the vector from the origin to P_2. Using the former yields

$$\mathbf{r}_0 = \langle 2, 4, -1 \rangle$$

Thus, a vector equation of the line through P_1 and P_2 is

$$\langle x, y, z \rangle = \langle 2, 4, -1 \rangle + t \langle 3, -4, 8 \rangle$$

If needed, we can express the line parametrically by equating corresponding components on the two sides of this vector equation, in which case we obtain the parametric equations in Example 2 (verify). ◀

EXERCISE SET 13.5 🔲 Graphing Calculator 🅲 CAS

1. (a) Find parametric equations for the lines through the corner of the unit square shown in part (a) of the accompanying figure.
   (b) Find parametric equations for the lines through the corner of the unit cube shown in part (b) of the accompanying figure.

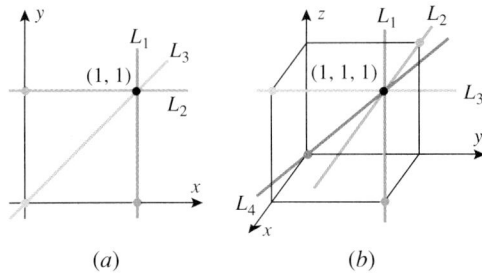

(a)   (b)

Figure Ex-1

2. (a) Find parametric equations for the line segments on the unit square in part (a) of the accompanying figure.
   (b) Find parametric equations for the line segments in the unit cube shown in part (b) of the accompanying figure.

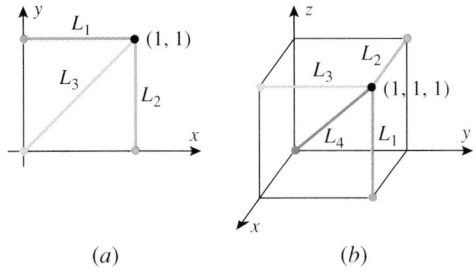

(a)   (b)

Figure Ex-2

In Exercises 3 and 4, find parametric equations for the line through P_1 and P_2 and also for the line segment joining those points.

3. (a) $P_1(3, -2)$, $P_2(5, 1)$ (b) $P_1(5, -2, 1)$, $P_2(2, 4, 2)$
4. (a) $P_1(0, 1)$, $P_2(-3, -4)$ (b) $P_1(-1, 3, 5)$, $P_2(-1, 3, 2)$

In Exercises 5 and 6, find parametric equations for the line whose vector equation is given.

5. (a) $\langle x, y \rangle = \langle 2, -3 \rangle + t \langle 1, -4 \rangle$
   (b) $x\mathbf{i} + y\mathbf{j} + z\mathbf{k} = \mathbf{k} + t(\mathbf{i} - \mathbf{j} + \mathbf{k})$
6. (a) $x\mathbf{i} + y\mathbf{j} = (3\mathbf{i} - 4\mathbf{j}) + t(2\mathbf{i} + \mathbf{j})$
   (b) $\langle x, y, z \rangle = \langle -1, 0, 2 \rangle + t \langle -1, 3, 0 \rangle$

In Exercises 7 and 8, find a point P on the line and a vector **v** parallel to the line by inspection.

7. (a) $x\mathbf{i} + y\mathbf{j} = (2\mathbf{i} - \mathbf{j}) + t(4\mathbf{i} - \mathbf{j})$
   (b) $\langle x, y, z \rangle = \langle -1, 2, 4 \rangle + t \langle 5, 7, -8 \rangle$
8. (a) $\langle x, y \rangle = \langle -1, 5 \rangle + t \langle 2, 3 \rangle$
   (b) $x\mathbf{i} + y\mathbf{j} + z\mathbf{k} = (\mathbf{i} + \mathbf{j} - 2\mathbf{k}) + t\mathbf{j}$

In Exercises 9 and 10, express the given parametric equations of a line in vector form using bracket notation and also using **i, j, k** notation.

9. (a) $x = -3 + t$, $y = 4 + 5t$
   (b) $x = 2 - t$, $y = -3 + 5t$, $z = t$
10. (a) $x = t$, $y = -2 + t$
    (b) $x = 1 + t$, $y = -7 + 3t$, $z = 4 - 5t$

In Exercises 11–18, find parametric equations of the line that satisfies the stated conditions.

11. The line through $(-5, 2)$ that is parallel to $2\mathbf{i} - 3\mathbf{j}$.
12. The line through $(0, 3)$ that is parallel to the line $x = -5 + t$, $y = 1 - 2t$.
13. The line that is tangent to the circle $x^2 + y^2 = 25$ at the point $(3, -4)$.
14. The line that is tangent to the parabola $y = x^2$ at the point $(-2, 4)$.
15. The line through $(-1, 2, 4)$ that is parallel to $3\mathbf{i} - 4\mathbf{j} + \mathbf{k}$.
16. The line through $(2, -1, 5)$ that is parallel to $\langle -1, 2, 7 \rangle$.
17. The line through $(-2, 0, 5)$ that is parallel to the line $x = 1 + 2t$, $y = 4 - t$, $z = 6 + 2t$.
18. The line through the origin that is parallel to the line $x = t$, $y = -1 + t$, $z = 2$.
19. Where does the line $x = 1 + 3t$, $y = 2 - t$ intersect
    (a) the x-axis   (b) the y-axis   (c) the parabola $y = x^2$?
20. Where does the line $\langle x, y \rangle = \langle 4t, 3t \rangle$ intersect the circle $x^2 + y^2 = 25$?

In Exercises 21 and 22, find the intersections of the lines with the xy-plane, the xz-plane, and the yz-plane.

21. $x = -2$, $y = 4 + 2t$, $z = -3 + t$
22. $x = -1 + 2t$, $y = 3 + t$, $z = 4 - t$
23. Where does the line $x = 1 + t$, $y = 3 - t$, $z = 2t$ intersect the cylinder $x^2 + y^2 = 16$?
24. Where does the line $x = 2 - t$, $y = 3t$, $z = -1 + 2t$ intersect the plane $2y + 3z = 6$?

In Exercises 25 and 26, show that the lines L_1 and L_2 intersect, and find their point of intersection.

25. $L_1: x = 2 + t$, $y = 2 + 3t$, $z = 3 + t$
    $L_2: x = 2 + t$, $y = 3 + 4t$, $z = 4 + 2t$

26. $L_1: x + 1 = 4t, \quad y - 3 = t, \quad z - 1 = 0$
$L_2: x + 13 = 12t, \quad y - 1 = 6t, \quad z - 2 = 3t$

In Exercises 27 and 28, show that the lines L_1 and L_2 are skew.

27. $L_1: x = 1 + 7t, \quad y = 3 + t, \quad z = 5 - 3t$
$L_2: x = 4 - t, \quad y = 6, \quad z = 7 + 2t$

28. $L_1: x = 2 + 8t, \quad y = 6 - 8t, \quad z = 10t$
$L_2: x = 3 + 8t, \quad y = 5 - 3t, \quad z = 6 + t$

In Exercises 29 and 30, determine whether the lines L_1 and L_2 are parallel.

29. $L_1: x = 3 - 2t, \quad y = 4 + t, \quad z = 6 - t$
$L_2: x = 5 - 4t, \quad y = -2 + 2t, \quad z = 7 - 2t$

30. $L_1: x = 5 + 3t, \quad y = 4 - 2t, \quad z = -2 + 3t$
$L_2: x = -1 + 9t, \quad y = 5 - 6t, \quad z = 3 + 8t$

In Exercises 31 and 32, determine whether the points P_1, P_2, and P_3 lie on the same line.

31. $P_1(6, 9, 7), \quad P_2(9, 2, 0), \quad P_3(0, -5, -3)$

32. $P_1(1, 0, 1), \quad P_2(3, -4, -3), \quad P_3(4, -6, -5)$

In Exercises 33 and 34, show that the lines L_1 and L_2 are the same.

33. $L_1: x = 3 - t, \quad y = 1 + 2t$
$L_2: x = -1 + 3t, \quad y = 9 - 6t$

34. $L_1: x = 1 + 3t, \quad y = -2 + t, \quad z = 2t$
$L_2: x = 4 - 6t, \quad y = -1 - 2t, \quad z = 2 - 4t$

In Exercises 35 and 36, describe the line segment represented by the vector equation.

35. $\langle x, y \rangle = \langle 1, 0 \rangle + t \langle -2, 3 \rangle \quad (0 \le t \le 2)$

36. $\langle x, y, z \rangle = \langle -2, 1, 4 \rangle + t \langle 3, 0, -1 \rangle \quad (0 \le t \le 3)$

In Exercises 37 and 38, use the method in Exercise 25 of Section 13.3 to find the distance from the point P to the line L, and then check your answer using the method in Exercise 26 of Section 13.4.

37. $P(-2, 1, 1)$
$L: x = 3 - t, \quad y = t, \quad z = 1 + 2t$

38. $P(1, 4, -3)$
$L: x = 2 + t, \quad y = -1 - t, \quad z = 3t$

In Exercises 39 and 40, show that the lines L_1 and L_2 are parallel, and find the distance between them.

39. $L_1: x = 2 - t, \quad y = 2t, \quad z = 1 + t$
$L_2: x = 1 + 2t, \quad y = 3 - 4t, \quad z = 5 - 2t$

40. $L_1: x = 2t, \quad y = 3 + 4t, \quad z = 2 - 6t$
$L_2: x = 1 + 3t, \quad y = 6t, \quad z = -9t$

41. (a) Find parametric equations for the line through the points (x_0, y_0, z_0) and (x_1, y_1, z_1).
(b) Find parametric equations for the line through the point (x_1, y_1, z_1) and parallel to the line
$$x = x_0 + at, \quad y = y_0 + bt, \quad z = z_0 + ct$$

42. Let L be the line that passes through the point (x_0, y_0, z_0) and is parallel to the vector $\mathbf{v} = \langle a, b, c \rangle$, where a, b, and c are nonzero. Show that a point (x, y, z) lies on the line L if and only if
$$\frac{x - x_0}{a} = \frac{y - y_0}{b} = \frac{z - z_0}{c}$$
These equations, which are called the **symmetric equations** of L, provide a nonparametric representation of L.

43. (a) Describe the line whose symmetric equations are
$$\frac{x - 1}{2} = \frac{y + 3}{4} = z - 5$$
[See Exercise 42.]
(b) Find parametric equations for the line in part (a).

44. Find the point on the line segment joining $P_1(1, 4, -3)$ and $P_2(1, 5, -1)$ that is $\frac{2}{3}$ of the way from P_1 to P_2.

45. Let L_1 and L_2 be the lines whose parametric equations are
$$L_1: x = 1 + 2t, \quad y = 2 - t, \quad z = 4 - 2t$$
$$L_2: x = 9 + t, \quad y = 5 + 3t, \quad z = -4 - t$$
(a) Show that L_1 and L_2 intersect at the point $(7, -1, -2)$.
(b) Find, to the nearest degree, the acute angle between L_1 and L_2 at their intersection.
(c) Find parametric equations for the line that is perpendicular to L_1 and L_2 and passes through their point of intersection.

46. Let L_1 and L_2 be the lines whose parametric equations are
$$L_1: x = 4t, \quad y = 1 - 2t, \quad z = 2 + 2t$$
$$L_2: x = 1 + t, \quad y = 1 - t, \quad z = -1 + 4t$$
(a) Show that L_1 and L_2 intersect at the point $(2, 0, 3)$.
(b) Find, to the nearest degree, the acute angle between L_1 and L_2 at their intersection.
(c) Find parametric equations for the line that is perpendicular to L_1 and L_2 and passes through their point of intersection.

In Exercises 47 and 48, find parametric equations of the line that contains the point P and intersects the line L at a right angle.

47. $P(0, 2, 1)$
$L: x = 2t, \quad y = 1 - t, \quad z = 2 + t$

48. $P(3, 1, -2)$
$L: x = -2 + 2t, \quad y = 4 + 2t, \quad z = 2 + t$

49. Two bugs are walking along lines in 3-space. At time t bug 1 is at the point (x, y, z) on the line

$$x = 4 - t, \quad y = 1 + 2t, \quad z = 2 + t$$

and at the same time t bug 2 is at the point (x, y, z) on the line

$$x = t, \quad y = 1 + t, \quad z = 1 + 2t$$

Assume that distance is in centimeters and that time is in minutes.

(a) Find the distance between the bugs at time $t = 0$.

(b) Use a graphing utility to graph the distance between the bugs as a function of time from $t = 0$ to $t = 5$.

(c) What does the graph tell you about the distance between the bugs?

(d) How close do the bugs get?

50. Suppose that the temperature T at a point (x, y, z) on the line $x = t$, $y = 1 + t$, $z = 3 - 2t$ is $T = 25x^2 yz$. Use a CAS or a calculating utility with a root-finding capability to approximate the maximum temperature on that portion of the line that extends from the xz-plane to the xy-plane.

13.6 PLANES IN 3-SPACE

In this section we will use vectors to derive equations of planes in 3-space, and then we will use these equations to solve various geometric problems.

PLANES PARALLEL TO THE COORDINATE PLANES

The graph of the equation $x = a$ in an xyz-coordinate system consists of all points of the form (a, y, z), where y and z are arbitrary. One such point is $(a, 0, 0)$, and all others are in the plane that passes through this point and is parallel to the yz-plane (Figure 13.6.1). Similarly, the graph of $y = b$ is the plane through $(0, b, 0)$ that is parallel to the xz-plane, and the graph of $z = c$ is the plane through $(0, 0, c)$ that is parallel to the xy-plane.

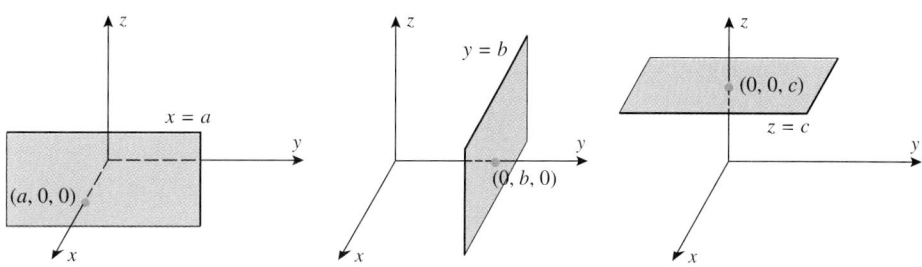

Figure 13.6.1

PLANES DETERMINED BY A POINT AND A NORMAL VECTOR

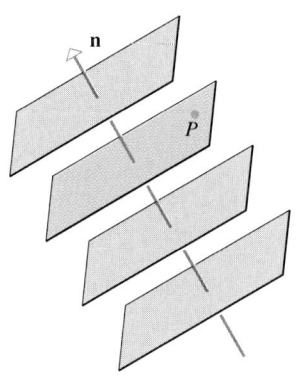

The colored plane is uniquely determined by the point P and the vector **n** perpendicular to the plane.

Figure 13.6.2

A plane in 3-space can be determined uniquely by specifying a point in the plane and a vector perpendicular to the plane (Figure 13.6.2). A vector perpendicular to a plane is called a ***normal*** to the plane.

Suppose that we want to find an equation of the plane passing through $P_0(x_0, y_0, z_0)$ and perpendicular to the vector $\mathbf{n} = \langle a, b, c \rangle$. Define the vectors $\mathbf{r}_0$ and $\mathbf{r}$ as

$$\mathbf{r}_0 = \langle x_0, y_0, z_0 \rangle \quad \text{and} \quad \mathbf{r} = \langle x, y, z \rangle$$

It should be evident from Figure 13.6.3 that the plane consists precisely of those points $P(x, y, z)$ for which the vector $\mathbf{r} - \mathbf{r}_0$ is orthogonal to $\mathbf{n}$; or, expressed as an equation,

$$\mathbf{n} \cdot (\mathbf{r} - \mathbf{r}_0) = 0 \tag{1}$$

If preferred, we can express this vector equation in terms of components as

$$\langle a, b, c \rangle \cdot \langle x - x_0, y - y_0, z - z_0 \rangle = 0 \tag{2}$$

from which we obtain

$$a(x - x_0) + b(y - y_0) + c(z - z_0) = 0 \tag{3}$$

This is called the ***point-normal form*** of the equation of a plane. Formulas (1) and (2) are vector versions of this formula.

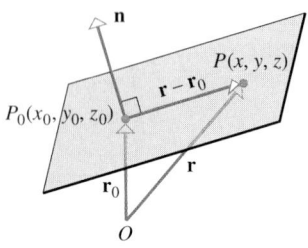

Figure 13.6.3

FOR THE READER.   What does Equation (1) represent if $\mathbf{n} = \langle a, b \rangle$, $\mathbf{r}_0 = \langle x_0, y_0 \rangle$, and $\mathbf{r} = \langle x, y \rangle$ are vectors in an xy-plane in 2-space? Draw a picture.

Example 1

Find an equation of the plane passing through the point $(3, -1, 7)$ and perpendicular to the vector $\mathbf{n} = \langle 4, 2, -5 \rangle$.

Solution.   From (3), a point-normal form of the equation is

$$4(x - 3) + 2(y + 1) - 5(z - 7) = 0 \tag{4}$$

If preferred, this equation can be written in vector form as

$$\langle 4, 2, -5 \rangle \cdot \langle x - 3, y + 1, z - 7 \rangle = 0 \qquad \blacktriangleleft$$

Observe that if we multiply out the terms in (3) and simplify, we obtain an equation of the form

$$ax + by + cz + d = 0 \tag{5}$$

For example, Equation (4) in Example 1 can be rewritten as

$$4x + 2y - 5z + 25 = 0$$

The following theorem shows that every equation of form (5) represents a plane in 3-space.

13.6.1   THEOREM.   *If $a, b, c,$ and d are constants, and $a, b,$ and c are not all zero, then the graph of the equation*

$$ax + by + cz + d = 0 \tag{6}$$

is a plane that has the vector $\mathbf{n} = \langle a, b, c \rangle$ as a normal.

Proof.   Since $a, b,$ and c are not all zero, there is at least one point (x_0, y_0, z_0) whose coordinates satisfy Equation (6). For example, if $a \neq 0$, then such a point is $(-d/a, 0, 0)$, and similarly if $b \neq 0$ or $c \neq 0$ (verify). Thus, let (x_0, y_0, z_0) be any point whose coordinates satisfy (6); that is,

$$ax_0 + by_0 + cz_0 + d = 0$$

Subtracting this equation from (6) yields

$$a(x - x_0) + b(y - y_0) + c(z - z_0) = 0$$

which is the point-normal form of a plane with normal $\mathbf{n} = \langle a, b, c \rangle$.   ▮

Equation (6) is called the ***general form*** of the equation of a plane.

Example 2

Determine whether the planes

$$3x - 4y + 5z = 0 \quad \text{and} \quad -6x + 8y - 10z - 4 = 0$$

are parallel.

Solution.   It is clear geometrically that two planes are parallel if and only if their normals are parallel vectors. A normal to the first plane is

$$\mathbf{n}_1 = \langle 3, -4, 5 \rangle$$

and a normal to the second plane is

$$\mathbf{n}_2 = \langle -6, 8, -10 \rangle$$

Since $\mathbf{n}_2$ is a scalar multiple of $\mathbf{n}_1$, the normals are parallel, and hence so are the planes.   ◀

We have seen that a unique plane is determined by a point in the plane and a nonzero vector normal to the plane. In contrast, a unique plane is not determined by a point in the plane and a nonzero vector *parallel* to the plane (Figure 13.6.4). However, a unique plane is determined by a point in the plane and two nonparallel vectors that are parallel to the plane (Figure 13.6.5). A unique plane is also determined by three noncollinear points that lie in the plane (Figure 13.6.6).

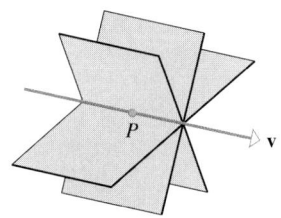

There are infinitely many planes containing P and parallel to $\mathbf{v}$.

Figure 13.6.4

Example 3

Find an equation of the plane through the points $P_1(1, 2, -1)$, $P_2(2, 3, 1)$, and $P_3(3, -1, 2)$.

Solution.   Since the points P_1, P_2, and P_3 lie in the plane, the vectors $\overrightarrow{P_1P_2} = \langle 1, 1, 2 \rangle$ and $\overrightarrow{P_1P_3} = \langle 2, -3, 3 \rangle$ are parallel to the plane. Therefore,

$$\overrightarrow{P_1P_2} \times \overrightarrow{P_1P_3} = \begin{vmatrix} \mathbf{i} & \mathbf{j} & \mathbf{k} \\ 1 & 1 & 2 \\ 2 & -3 & 3 \end{vmatrix} = 9\mathbf{i} + \mathbf{j} - 5\mathbf{k}$$

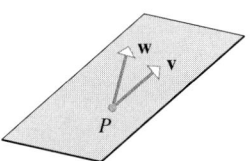

There is a unique plane through P that is parallel to both $\mathbf{v}$ and $\mathbf{w}$.

Figure 13.6.5

is normal to the plane, since it is perpendicular to both $\overrightarrow{P_1P_2}$ and $\overrightarrow{P_1P_3}$. By using this normal and the point $P_1(1, 2, -1)$ in the plane, we obtain the point-normal form

$$9(x - 1) + (y - 2) - 5(z + 1) = 0$$

which can be rewritten as

$$9x + y - 5z - 16 = 0 \qquad \blacktriangleleft$$

Example 4

Determine whether the line

$$x = 3 + 8t, \quad y = 4 + 5t, \quad z = -3 - t$$

is parallel to the plane $x - 3y + 5z = 12$.

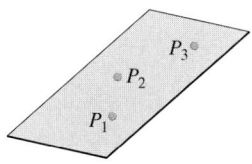

There is a unique plane through three noncollinear points.

Figure 13.6.6

Solution.   The vector $\mathbf{v} = \langle 8, 5, -1 \rangle$ is parallel to the line and the vector $\mathbf{n} = \langle 1, -3, 5 \rangle$ is normal to the plane. For the line and plane to be parallel, the vectors $\mathbf{v}$ and $\mathbf{n}$ must be orthogonal. But this is not so, since the dot product

$$\mathbf{v} \cdot \mathbf{n} = (8)(1) + (5)(-3) + (-1)(5) = -12$$

is nonzero. Thus, the line and plane are not parallel. $\blacktriangleleft$

Example 5

Find the intersection of the line and plane in Example 4.

Solution.   If we let (x_0, y_0, z_0) be the point of intersection, then the coordinates of this point satisfy both the equation of the plane and the parametric equations of the line. Thus,

$$x_0 - 3y_0 + 5z_0 = 12 \tag{7}$$

and for some value of t, say $t = t_0$,

$$x_0 = 3 + 8t_0, \quad y_0 = 4 + 5t_0, \quad z_0 = -3 - t_0 \tag{8}$$

Substituting (8) in (7) yields

$$(3 + 8t_0) - 3(4 + 5t_0) + 5(-3 - t_0) = 12$$

Solving for t_0 yields $t_0 = -3$ and on substituting this value in (8), we obtain

$$(x_0, y_0, z_0) = (-21, -11, 0) \qquad \blacktriangleleft$$

ANGLES BETWEEN PLANES

Two distinct intersecting planes determine two positive angles of intersection—an (acute) angle θ that satisfies the condition $0 \leq \theta \leq \pi/2$ and the supplement of that angle (Figure 13.6.7*a*). If $\mathbf{n}_1$ and $\mathbf{n}_2$ are normals to the planes, then depending on the directions of $\mathbf{n}_1$

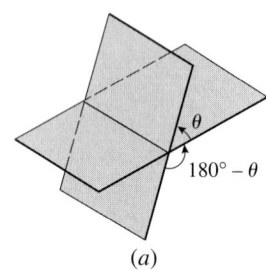

(a)

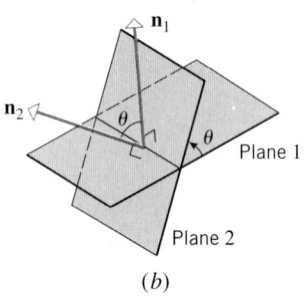

(b)

Figure 13.6.7

and $\mathbf{n}_2$, the angle θ is either the angle between $\mathbf{n}_1$ and $\mathbf{n}_2$ or the angle between $\mathbf{n}_1$ and $-\mathbf{n}_2$ (Figure 13.6.7*b*). In both cases, Theorem 13.3.3 yields the following formula for the acute angle θ between the planes:

$$\cos \theta = \frac{|\mathbf{n}_1 \cdot \mathbf{n}_2|}{\|\mathbf{n}_1\| \|\mathbf{n}_2\|} \tag{9}$$

Example 6

Find the acute angle of intersection between the two planes

$$2x - 4y + 4z = 7 \quad \text{and} \quad 6x + 2y - 3z = 2$$

Solution. The given equations yield the normals $\mathbf{n}_1 = \langle 2, -4, 4 \rangle$ and $\mathbf{n}_2 = \langle 6, 2, -3 \rangle$. Thus, Formula (9) yields

$$\cos \theta = \frac{|\mathbf{n}_1 \cdot \mathbf{n}_2|}{\|\mathbf{n}_1\| \|\mathbf{n}_2\|} = \frac{|-8|}{\sqrt{36}\sqrt{49}} = \frac{4}{21}$$

from which we obtain

$$\theta = \cos^{-1}\left(\frac{4}{21}\right) \approx 79°$$ ◀

DISTANCE PROBLEMS INVOLVING PLANES

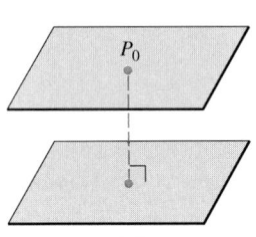

(a)

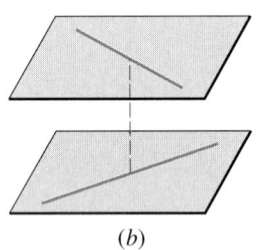

(b)

Figure 13.6.8

Next we will consider three basic "distance problems" in 3-space:

- Find the distance between a point and a plane.
- Find the distance between two parallel planes.
- Find the distance between two skew lines.

The three problems are related. If we can find the distance between a point and a plane, then we can find the distance between parallel planes by computing the distance between one of the planes and an arbitrary point P_0 in the other plane (Figure 13.6.8*a*). Moreover, we can find the distance between two skew lines by computing the distance between parallel planes containing them (Figure 13.6.8*b*).

> **13.6.2   THEOREM.**   *The distance D between a point $P_0(x_0, y_0, z_0)$ and the plane $ax + by + cz + d = 0$ is*
>
> $$D = \frac{|ax_0 + by_0 + cz_0 + d|}{\sqrt{a^2 + b^2 + c^2}} \tag{10}$$

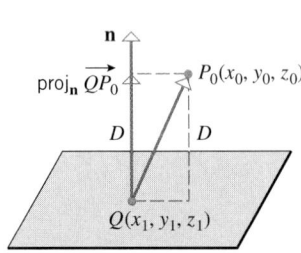

Figure 13.6.9

Proof. Let $Q(x_1, y_1, z_1)$ be any point in the plane, and position the normal $\mathbf{n} = \langle a, b, c \rangle$ so that its initial point is at Q. As illustrated in Figure 13.6.9, the distance D is equal to the length of the orthogonal projection of $\overrightarrow{QP_0}$ on $\mathbf{n}$. Thus, from (12) of Section 13.3,

$$D = \|\text{proj}_{\mathbf{n}} \overrightarrow{QP_0}\| = \left\| \frac{\overrightarrow{QP_0} \cdot \mathbf{n}}{\|\mathbf{n}\|^2} \mathbf{n} \right\| = \frac{|\overrightarrow{QP_0} \cdot \mathbf{n}|}{\|\mathbf{n}\|^2} \|\mathbf{n}\| = \frac{|\overrightarrow{QP_0} \cdot \mathbf{n}|}{\|\mathbf{n}\|}$$

But

$$\overrightarrow{QP_0} = \langle x_0 - x_1, y_0 - y_1, z_0 - z_1 \rangle$$
$$\overrightarrow{QP_0} \cdot \mathbf{n} = a(x_0 - x_1) + b(y_0 - y_1) + c(z_0 - z_1)$$
$$\|\mathbf{n}\| = \sqrt{a^2 + b^2 + c^2}$$

Thus,

$$D = \frac{|a(x_0 - x_1) + b(y_0 - y_1) + c(z_0 - z_1)|}{\sqrt{a^2 + b^2 + c^2}} \tag{11}$$

Since the point $Q(x_1, y_1, z_1)$ lies in the plane, its coordinates satisfy the equation of the plane; that is,

$$ax_1 + by_1 + cz_1 + d = 0$$

or

$$d = -ax_1 - by_1 - cz_1$$

Substituting this expression in (11) yields (10). ▮

Example 7

Find the distance D between the point $(1, -4, -3)$ and the plane

$$2x - 3y + 6z = -1$$

Solution. Formula (10) requires the plane to be rewritten in the form $ax + by + cz + d = 0$. Thus, we rewrite the equation of the given plane as

$$2x - 3y + 6z + 1 = 0$$

from which we obtain $a = 2, b = -3, c = 6$, and $d = 1$. Substituting these values and the coordinates of the given point in (10), we obtain

$$D = \frac{|(2)(1) + (-3)(-4) + 6(-3) + 1|}{\sqrt{2^2 + (-3)^2 + 6^2}} = \frac{|-3|}{7} = \frac{3}{7} \quad ◀$$

REMARK.   See Exercise 48 for an analog of Formula (10) in 2-space that can be used to compute the distance between a point to a line.

Example 8

The planes

$$x + 2y - 2z = 3 \quad \text{and} \quad 2x + 4y - 4z = 7$$

are parallel since their normals, $\langle 1, 2, -2 \rangle$ and $\langle 2, 4, -4 \rangle$, are parallel vectors. Find the distance between these planes.

Solution. To find the distance D between the planes, we can select an arbitrary point in one of the planes and compute its distance to the other plane. By setting $y = z = 0$ in the equation $x + 2y - 2z = 3$, we obtain the point $P_0(3, 0, 0)$ in this plane. From (10), the distance from P_0 to the plane $2x + 4y - 4z = 7$ is

$$D = \frac{|(2)(3) + 4(0) + (-4)(0) - 7|}{\sqrt{2^2 + 4^2 + (-4)^2}} = \frac{1}{6} \quad ◀$$

Example 9

It was shown in Example 3 of Section 13.5 that the lines

$$L_1: x = 1 + 4t, \quad y = 5 - 4t, \quad z = -1 + 5t$$
$$L_2: x = 2 + 8t, \quad y = 4 - 3t, \quad z = 5 + t$$

are skew. Find the distance between them.

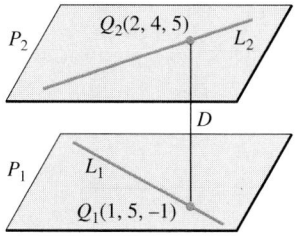

Figure 13.6.10

Solution. Let P_1 and P_2 denote parallel planes containing L_1 and L_2, respectively (Figure 13.6.10). To find the distance D between L_1 and L_2, we will calculate the distance from a point in P_1 to the plane P_2. Since L_1 lies in plane P_1, we can find a point in P_1

by finding a point on the line L_1; we can do this by substituting any convenient value of t in the parametric equations of L_1. The simplest choice is $t = 0$, which yields the point $Q_1(1, 5, -1)$.

The next step is to find an equation for the plane P_2. For this purpose, observe that the vector $\mathbf{u}_1 = \langle 4, -4, 5 \rangle$ is parallel to line L_1, and therefore also parallel to planes P_1 and P_2. Similarly, $\mathbf{u}_2 = \langle 8, -3, 1 \rangle$ is parallel to L_2 and hence parallel to P_1 and P_2. Therefore, the cross product

$$\mathbf{n} = \mathbf{u}_1 \times \mathbf{u}_2 = \begin{vmatrix} \mathbf{i} & \mathbf{j} & \mathbf{k} \\ 4 & -4 & 5 \\ 8 & -3 & 1 \end{vmatrix} = 11\mathbf{i} + 36\mathbf{j} + 20\mathbf{k}$$

is normal to both P_1 and P_2. Using this normal and the point $Q_2(2, 4, 5)$ found by setting $t = 0$ in the equations of L_2, we obtain an equation for P_2:

$$11(x - 2) + 36(y - 4) + 20(z - 5) = 0$$

or

$$11x + 36y + 20z - 266 = 0$$

The distance between $Q_1(1, 5, -1)$ and this plane is

$$D = \frac{|(11)(1) + (36)(5) + (20)(-1) - 266|}{\sqrt{11^2 + 36^2 + 20^2}} = \frac{95}{\sqrt{1817}}$$

which is also the distance between L_1 and L_2. ◀

EXERCISE SET 13.6

1. Find equations of the planes P_1, P_2, and P_3 that are parallel to the coordinate planes and pass through the corner $(3, 4, 5)$ of the box shown in the accompanying figure.

2. Find equations of the planes P_1, P_2, and P_3 that are parallel to the coordinate planes and pass through the corner (x_0, y_0, z_0) of the box shown in the accompanying figure.

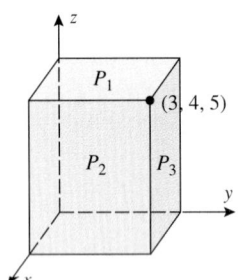

Figure Ex-1

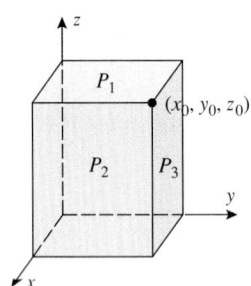

Figure Ex-2

In Exercises 3–6, find an equation of the plane that passes through the point P and has the vector $\mathbf{n}$ as a normal.

3. $P(2, 6, 1)$; $\mathbf{n} = \langle 1, 4, 2 \rangle$

4. $P(-1, -1, 2)$; $\mathbf{n} = \langle -1, 7, 6 \rangle$

5. $P(1, 0, 0)$; $\mathbf{n} = \langle 0, 0, 1 \rangle$

6. $P(0, 0, 0)$; $\mathbf{n} = \langle 2, -3, -4 \rangle$

In Exercises 7–10, find an equation of the plane indicated in the figure.

7.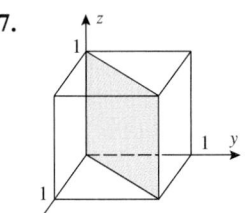

8.

9.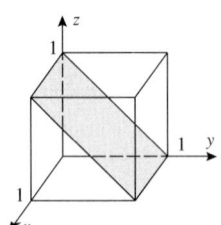

10.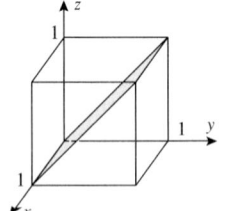

In Exercises 11 and 12, find an equation of the plane that passes through the given points.

11. $(-2, 1, 1)$, $(0, 2, 3)$, and $(1, 0, -1)$

12. $(3, 2, 1)$, $(2, 1, -1)$, and $(-1, 3, 2)$

In Exercises 13 and 14, determine whether the planes are parallel, perpendicular, or neither.

13. (a) $2x - 8y - 6z - 2 = 0$ (b) $3x - 2y + z = 1$
$\quad -x + 4y + 3z - 5 = 0$ $4x + 5y - 2z = 4$

   (c) $x - y + 3z - 2 = 0$
$\quad 2x + z = 1$

14. (a) $3x - 2y + z = 4$ (b) $y = 4x - 2z + 3$
$\quad 6x - 4y + 3z = 7$ $x = \frac{1}{4}y + \frac{1}{2}z$

   (c) $x + 4y + 7z = 3$
$\quad 5x - 3y + z = 0$

In Exercises 15 and 16, determine whether the line and plane are parallel, perpendicular, or neither.

---

15. (a) $x = 4 + 2t, \ y = -t, \ z = -1 - 4t$;
$\quad 3x + 2y + z - 7 = 0$

   (b) $x = t, \ y = 2t, \ z = 3t$;
$\quad x - y + 2z = 5$

   (c) $x = -1 + 2t, \ y = 4 + t, \ z = 1 - t$;
$\quad 4x + 2y - 2z = 7$

16. (a) $x = 3 - t, \ y = 2 + t, \ z = 1 - 3t$;
$\quad 2x + 2y - 5 = 0$

   (b) $x = 1 - 2t, \ y = t, \ z = -t$;
$\quad 6x - 3y + 3z = 1$

   (c) $x = t, \ y = 1 - t, \ z = 2 + t$;
$\quad x + y + z = 1$

In Exercises 17 and 18, determine whether the line and plane intersect; if so, find the coordinates of the intersection.

---

17. (a) $x = t, \ y = t, \ z = t$;
$\quad 3x - 2y + z - 5 = 0$

   (b) $x = 2 - t, \ y = 3 + t, \ z = t$;
$\quad 2x + y + z = 1$

18. (a) $x = 3t, \ y = 5t, \ z = -t$;
$\quad 2x - y + z + 1 = 0$

   (b) $x = 1 + t, \ y = -1 + 3t, \ z = 2 + 4t$;
$\quad x - y + 4z = 7$

In Exercises 19 and 20, find the acute angle of intersection of the planes to the nearest degree.

---

19. $x = 0$ and $2x - y + z - 4 = 0$

20. $x + 2y - 2z = 5$ and $6x - 3y + 2z = 8$

In Exercises 21–30, find an equation of the plane that satisfies the stated conditions.

---

21. The plane through the origin that is parallel to the plane $4x - 2y + 7z + 12 = 0$.

22. The plane that contains the line $x = -2 + 3t, \ y = 4 + 2t, \ z = 3 - t$ and is perpendicular to the plane $x - 2y + z = 5$.

23. The plane through the point $(-1, 4, 2)$ that contains the line of intersection of the planes $4x - y + z - 2 = 0$ and $2x + y - 2z - 3 = 0$.

24. The plane through $(-1, 4, -3)$ that is perpendicular to the line $x - 2 = t, \ y + 3 = 2t, \ z = -t$.

25. The plane through $(1, 2, -1)$ that is perpendicular to the line of intersection of the planes $2x + y + z = 2$ and $x + 2y + z = 3$.

26. The plane through the points $P_1(-2, 1, 4), \ P_2(1, 0, 3)$ that is perpendicular to the plane $4x - y + 3z = 2$.

27. The plane through $(-1, 2, -5)$ that is perpendicular to the planes $2x - y + z = 1$ and $x + y - 2z = 3$.

28. The plane that contains the point $(2, 0, 3)$ and the line $x = -1 + t, \ y = t, \ z = -4 + 2t$.

29. The plane whose points are equidistant from $(2, -1, 1)$ and $(3, 1, 5)$.

30. The plane that contains the line $x = 3t, \ y = 1 + t, \ z = 2t$ and is parallel to the intersection of the planes $2x - y + z = 0$ and $y + z + 1 = 0$.

31. Find parametric equations of the line through the point $(5, 0, -2)$ that is parallel to the planes $x - 4y + 2z = 0$ and $2x + 3y - z + 1 = 0$.

32. Do the points $(1, 0, -1), \ (0, 2, 3), \ (-2, 1, 1)$, and $(4, 2, 3)$ lie in the same plane? Justify your answer two different ways.

33. Show that the line $x = 0, \ y = t, \ z = t$
   (a) lies in the plane $6x + 4y - 4z = 0$
   (b) is parallel to and below the plane $5x - 3y + 3z = 1$
   (c) is parallel to and above the plane $6x + 2y - 2z = 3$.

34. Show that if a, b, and c are nonzero, then the plane whose intercepts with the coordinate axes are $x = a, \ y = b$, and $z = c$ is given by the equation
$$\frac{x}{a} + \frac{y}{b} + \frac{z}{c} = 1$$

35. Show that the lines
$$x = -2 + t, \quad y = 3 + 2t, \quad z = 4 - t$$
$$x = 3 - t, \quad\ y = 4 - 2t, \quad z = t$$
are parallel and find an equation of the plane they determine.

36. Show that the lines
$$L_1: x + 1 = 4t, \quad y - 3 = t, \quad\ z - 1 = 0$$
$$L_2: x + 13 = 12t, \quad y - 1 = 6t, \quad z - 2 = 3t$$
intersect and find an equation of the plane they determine.

In Exercises 37 and 38, find parametric equations of the line of intersection of the planes.

---

37. $-2x + 3y + 7z + 2 = 0$
$\quad\ x + 2y - 3z + 5 = 0$

38. $3x - 5y + 2z = 0$
$\quad z = 0$

In Exercises 39 and 40, find the distance between the point and the plane.

---

39. $(1, -2, 3); \ 2x - 2y + z = 4$

40. $(0, 1, 5);\ 3x + 6y - 2z - 5 = 0$

In Exercises 41 and 42, find the distance between the given parallel planes.

41. $-2x + y + z = 0$
$6x - 3y - 3z - 5 = 0$

42. $x + y + z = 1$
$x + y + z = -1$

In Exercises 43 and 44, find the distance between the given skew lines.

43. $x = 1 + 7t,\ y = 3 + t,\ z = 5 - 3t$
$x = 4 - t,\ y = 6,\ z = 7 + 2t$

44. $x = 3 - t,\ y = 4 + 4t,\ z = 1 + 2t$
$x = t,\ y = 3,\ z = 2t$

45. Find an equation of the sphere with center $(2, 1, -3)$ that is tangent to the plane $x - 3y + 2z = 4$.

46. Locate the point of intersection of the plane $2x + y - z = 0$ and the line through $(3, 1, 0)$ that is perpendicular to the plane.

47. Show that the line $x = -1 + t,\ y = 3 + 2t,\ z = -t$ and the plane $2x - 2y - 2z + 3 = 0$ are parallel, and find the distance between them.

48. Formulas (1), (2), (3), (5), and (10), which apply to planes in 3-space, have analogs for lines in 2-space.
   (a) Draw an analog of Figure 13.6.3 in 2-space to illustrate that the equation of the line that passes through the point $P(x_0, y_0)$ and is perpendicular to the vector $\mathbf{n} = \langle a, b \rangle$

can be expressed as

$$\mathbf{n} \cdot (\mathbf{r} - \mathbf{r}_0) = 0$$

where $\mathbf{r} = \langle x, y \rangle$ and $\mathbf{r}_0 = \langle x_0, y_0 \rangle$.
   (b) Show that the vector equation in part (a) can be expressed as

$$a(x - x_0) + b(y - y_0) = 0$$

This is called the ***point-normal form of a line***.
   (c) Using the proof of Theorem 13.6.1 as a guide, show that if a and b are not both zero, then the graph of the equation

$$ax + by + c = 0$$

is a line that has $\mathbf{n} = \langle a, b \rangle$ as a normal.
   (d) Using the proof of Theorem 13.6.2 as a guide, show that the distance D between a point $P(x_0, y_0)$ and the line $ax + by + c = 0$ is

$$D = \frac{|ax_0 + by_0 + c|}{\sqrt{a^2 + b^2}}$$

49. Use the formula in part (d) of Exercise 48 to find the distance between the point $P(-3, 5)$ and the line $y = -2x + 1$.

50. (a) Show that the distance D between parallel planes

$$ax + by + cz + d_1 = 0$$
$$ax + by + cz + d_2 = 0$$

is

$$D = \frac{|d_1 - d_2|}{\sqrt{a^2 + b^2 + c^2}}$$

   (b) Use the formula in part (a) to solve Exercise 41.

13.7 QUADRIC SURFACES

In this section we will study an important class of surfaces that are the three-dimensional analogs of the conic sections.

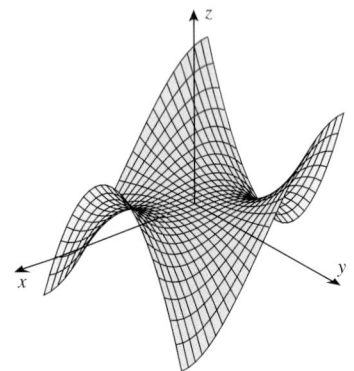

TRACES OF SURFACES

Figure 13.7.1

Although the general shape of a curve in 2-space can be obtained by plotting points, this method is not usually helpful for surfaces in 3-space because too many points are required. It is more common to build up the shape of a surface with a network of ***mesh lines***, which are curves obtained by cutting the surface with well-chosen planes. For example, Figure 13.7.1, which was generated by a CAS, shows the graph of $z = x^3 - 3xy^2$ rendered with a combination of mesh lines and colorization to produce the surface detail. This surface is called a "monkey saddle" because a monkey sitting astride the surface has a place for its two legs and tail.

The mesh line that results when a surface is cut by a plane is called the ***trace*** of the surface in the plane (Figure 13.7.2). Usually, surfaces are built up from traces in planes that are parallel to the coordinate planes, so we will begin by showing how the equations of such traces can be obtained. For this purpose, we will consider the surface

$$z = x^2 + y^2 \tag{1}$$

shown in Figure 13.7.3*a*.

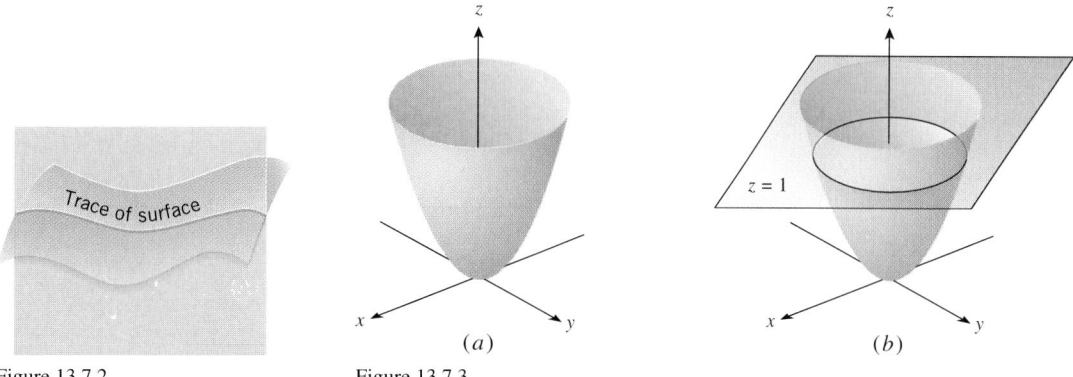

Figure 13.7.2   Figure 13.7.3

The basic procedure for finding the equation of a trace is to substitute the equation of the plane into the equation of the surface. For example, to find the trace of the surface $z = x^2 + y^2$ in the plane $z = 1$, we substitute $z = 1$ in (1), which yields

$$x^2 + y^2 = 1 \qquad (z = 1) \tag{2}$$

This is a circle of radius 1 centered at the point $(0, 0, 1)$ (Figure 13.7.3b).

REMARK. The parenthetical part of Equation (2) is a reminder that the z-coordinate of all points on the trace is $z = 1$. This needs to be stated explicitly because z does not appear in the equation $x^2 + y^2 = 1$.

Figure 13.7.4a suggests that the traces of (1) in planes that are parallel to and above the xy-plane form a family of circles that are centered on the z-axis and whose radii increase with z. To confirm this, let us consider the trace in a general plane $z = k$ that is parallel to the xy-plane. The equation of the trace is

$$x^2 + y^2 = k \qquad (z = k)$$

If $k \geq 0$, then the trace is a circle of radius $\sqrt{k}$ centered at the point $(0, 0, k)$. In particular, if $k = 0$, then the radius is zero, so the trace in the xy-plane is the single point $(0, 0, 0)$. Thus, for nonnegative values of k the traces parallel to the xy-plane form a family of circles, centered on the z-axis, whose radii start at zero and increase with k. This confirms our conjecture. If $k < 0$, then the equation $x^2 + y^2 = k$ has no graph, which means that there is no trace.

Now let us examine the traces of (1) in planes parallel to the yz-plane. Such planes have equations of the form $x = k$, so we substitute this in (1) to obtain

$$z = k^2 + y^2 \qquad (x = k)$$

which we can rewrite as

$$z - k^2 = y^2 \qquad (x = k) \tag{3}$$

For simplicity, let us start with the case where $k = 0$ (the trace in the yz-plane), in which case the trace has the equation

$$z = y^2 \qquad (x = 0)$$

You should be able to recognize that this is a parabola that has its vertex at the origin, opens in the positive z-direction, and is symmetric about the z-axis (Figure 13.7.4b shows a two-dimensional view). You should also be able to recognize that the $-k^2$ term in (3) has the effect of translating the parabola $z = y^2$ in the positive z-direction, so the new vertex falls at $(0, 0, k^2)$. Thus, the traces parallel to the yz-plane form a family of parabolas whose vertices move upward as k^2 increases. This is consistent with Figure 13.7.4c. Similarly, the traces in planes parallel to the xz-plane have equations of the form

$$z - k^2 = x^2 \qquad (y = k)$$

which again is a family of parabolas whose vertices move upward as k^2 increases (Figure 13.7.4d).

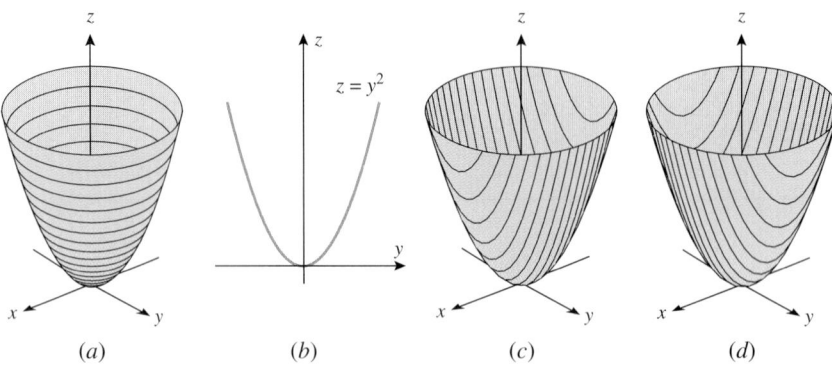

(a)   *(b)*   *(c)*   *(d)*

Figure 13.7.4

THE QUADRIC SURFACES

In the discussion of Formula (22) in Section 12.4 we noted that a second-degree equation

$$Ax^2 + Bxy + Cy^2 + Dx + Ey + F = 0$$

represents a conic section (possibly degenerate). The analog of this equation in an xyz-coordinate system is

$$Ax^2 + By^2 + Cz^2 + Dxy + Exz + Fyz + Gx + Hy + Iz + J = 0 \qquad (4)$$

which is called a **second-degree equation in x, y, and z.** The graphs of such equations are called **quadric surfaces** or sometimes **quadrics.**

The six nondegenerate types of quadric surfaces are shown in Table 13.7.1—*ellipsoids, hyperboloids of one sheet, hyperboloids of two sheets, elliptic cones, elliptic paraboloids,* and *hyperbolic paraboloids.* (The constants a, b, and c that appear in the equations in the table are assumed to be positive.) Observe that none of the quadric surfaces in the table have cross-product terms in their equations. This is because of their orientations relative to the coordinate axes. Later in this section we will discuss other possible orientations that produce equations of the quadric surfaces with no cross-product terms. In the special case where the elliptic cross sections of an elliptic cone or an elliptic paraboloid are circles, the terms *circular cone* and *circular paraboloid* are used.

TECHNIQUES FOR GRAPHING QUADRIC SURFACES

Accurate graphs of quadric surfaces are best left for graphing utilities. However, the techniques that we will now discuss can be used to generate rough sketches of these surfaces that are useful for various purposes.

A rough sketch of an ellipsoid

$$\frac{x^2}{a^2} + \frac{y^2}{b^2} + \frac{z^2}{c^2} = 1 \qquad (a > 0, b > 0, c > 0) \qquad (5)$$

can be obtained by first plotting the intersections with the coordinate axes, then sketching the elliptical traces in the coordinate planes, and then sketching the surface itself using the traces as a guide. Example 1 illustrates this technique.

Example 1

Sketch the ellipsoid

$$\frac{x^2}{4} + \frac{y^2}{16} + \frac{z^2}{9} = 1 \qquad (6)$$

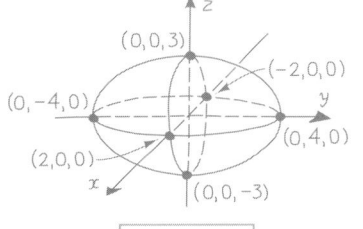

Rough sketch

Figure 13.7.5

Solution. The x-intercepts can be obtained by setting $y = 0$ and $z = 0$ in (6). This yields $x = \pm 2$. Similarly, the y-intercepts are $y = \pm 4$, and the z-intercepts are $z = \pm 3$. From these intercepts we obtain the elliptical traces and the ellipsoid sketched in Figure 13.7.5. ◀

Table 13.7.1

SURFACE	EQUATION	SURFACE	EQUATION
ELLIPSOID	$\dfrac{x^2}{a^2} + \dfrac{y^2}{b^2} + \dfrac{z^2}{c^2} = 1$ The traces in the coordinate planes are ellipses, as are the traces in those planes that are parallel to the coordinate planes and intersect the surface in more than one point.	ELLIPTIC CONE	$z^2 = \dfrac{x^2}{a^2} + \dfrac{y^2}{b^2}$ The trace in the xy-plane is a point (the origin), and the traces in planes parallel to the xy-plane are ellipses. The traces in the yz- and xz-planes are pairs of lines intersecting at the origin. The traces in planes parallel to these are hyperbolas.
HYPERBOLOID OF ONE SHEET	$\dfrac{x^2}{a^2} + \dfrac{y^2}{b^2} - \dfrac{z^2}{c^2} = 1$ The trace in the xy-plane is an ellipse, as are the traces in planes parallel to the xy-plane. The traces in the yz-plane and xz-plane are hyperbolas, as are the traces in those planes that are parallel to these and do not pass through the x- or y-intercepts. At these intercepts the traces are pairs of intersecting lines.	ELLIPTIC PARABOLOID	$z = \dfrac{x^2}{a^2} + \dfrac{y^2}{b^2}$ The trace in the xy-plane is a point (the origin), and the traces in planes parallel to and above the xy-plane are ellipses. The traces in the yz- and xz-planes are parabolas, as are the traces in planes parallel to these.
HYPERBOLOID OF TWO SHEETS	$\dfrac{z^2}{c^2} - \dfrac{x^2}{a^2} - \dfrac{y^2}{b^2} = 1$ There is no trace in the xy-plane. In planes parallel to the xy-plane that intersect the surface in more than one point the traces are ellipses. In the yz- and xz-planes, the traces are hyperbolas, as are the traces in those planes that are parallel to these and intersect the surface in more than one point.	HYPERBOLIC PARABOLOID	$z = \dfrac{y^2}{b^2} - \dfrac{x^2}{a^2}$ The trace in the xy-plane is a pair of lines intersecting at the origin. The traces in planes parallel to the xy-plane are hyperbolas. The hyperbolas above the xy-plane open in the y-direction, and those below in the x-direction. The traces in the yz- and xz-planes are parabolas, as are the traces in planes parallel to these.

A rough sketch of a hyperboloid of one sheet

$$\frac{x^2}{a^2} + \frac{y^2}{b^2} - \frac{z^2}{c^2} = 1 \qquad (a > 0, b > 0, c > 0) \tag{7}$$

can be obtained by first sketching the elliptical trace in the xy-plane, then the elliptical traces in the planes $z = \pm c$, and then the hyperbolic curves that join the endpoints of the axes of these ellipses. The next example illustrates this technique.

Example 2

Sketch the graph of the hyperboloid of one sheet

$$x^2 + y^2 - \frac{z^2}{4} = 1 \tag{8}$$

Solution. The trace in the *xy*-plane, obtained by setting $z = 0$ in (8), is

$$x^2 + y^2 = 1 \qquad (z = 0)$$

which is a circle of radius 1 centered on the *z*-axis. The traces in the planes $z = 2$ and $z = -2$, obtained by setting $z = \pm 2$ in (8), are given by

$$x^2 + y^2 = 2 \qquad (z = \pm 2)$$

which are circles of radius $\sqrt{2}$ centered on the *z*-axis. Joining these circles by the hyperbolic traces in the vertical coordinate planes yields the graph in Figure 13.7.6. ◀

A rough sketch of the hyperboloid of two sheets

$$\frac{z^2}{c^2} - \frac{x^2}{a^2} - \frac{y^2}{b^2} = 1 \qquad (a > 0, b > 0, c > 0) \tag{9}$$

can be obtained by first plotting the intersections with the *z*-axis, then sketching the elliptical traces in the planes $z = \pm 2c$, and then sketching the hyperbolic traces that connect the *z*-axis intersections and the endpoints of the axes of the ellipses. (It is not essential to use the planes $z = \pm 2c$, but these are good choices since they simplify the calculations slightly and have the right spacing for a good sketch.) The next example illustrates this technique.

Example 3

Sketch the graph of the hyperboloid of two sheets

$$z^2 - x^2 - \frac{y^2}{4} = 1 \tag{10}$$

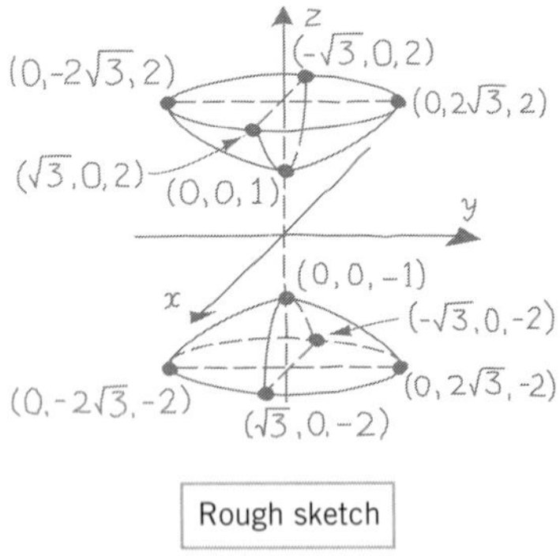

Solution. The *z*-intercepts, obtained by setting $x = 0$ and $y = 0$ in (10), are $z = \pm 1$. The traces in the planes $z = 2$ and $z = -2$, obtained by setting $z = \pm 2$ in (10), are given by

$$\frac{x^2}{3} + \frac{y^2}{12} = 1 \qquad (z = \pm 2)$$

Sketching these ellipses and the hyperbolic traces in the vertical coordinate planes yields Figure 13.7.7. ◀

A rough sketch of the elliptic cone

$$z^2 = \frac{x^2}{a^2} + \frac{y^2}{b^2} \qquad (a > 0, b > 0) \tag{11}$$

can be obtained by first sketching the elliptical traces in the planes $z = \pm 1$ and then sketching the linear traces that connect the endpoints of the axes of the ellipses. The next example illustrates this technique.

Example 4

Sketch the graph of the elliptic cone

$$z^2 = x^2 + \frac{y^2}{4} \tag{12}$$

Solution. The traces of (12) in the planes $z = \pm 1$ are given by

$$x^2 + \frac{y^2}{4} = 1 \qquad (z = \pm 1)$$

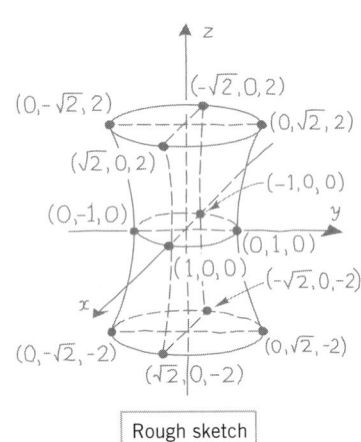

Rough sketch

Figure 13.7.6

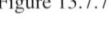

Rough sketch

Figure 13.7.7

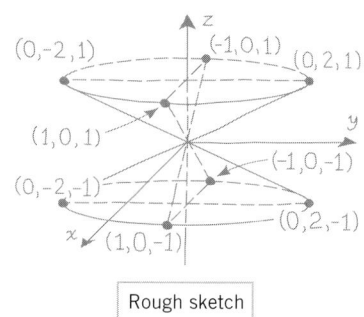

Rough sketch

Figure 13.7.8

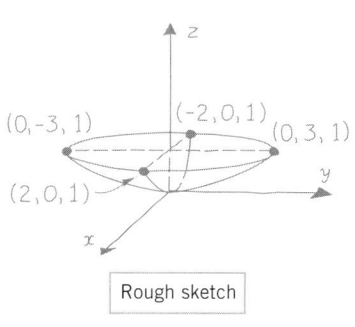

Rough sketch

Figure 13.7.9

Sketching these ellipses and the linear traces in the vertical coordinate planes yields the graph in Figure 13.7.8.

REMARK. Observe that if $a = b$ in (11), then the traces parallel to the xy-plane are circles, in which case we call the surface a *circular cone*.

A rough sketch of the elliptic paraboloid

$$z = \frac{x^2}{a^2} + \frac{y^2}{b^2} \qquad (a > 0, b > 0) \tag{13}$$

can be obtained by first sketching the elliptical trace in the plane $z = 1$ and then sketching the parabolic traces in the vertical coordinate planes to connect the origin to the ends of the axes of the ellipse. The next example illustrates this technique.

Example 5

Sketch the graph of the elliptic paraboloid

$$z = \frac{x^2}{4} + \frac{y^2}{9} \tag{14}$$

Solution. The trace of (14) in the plane $z = 1$ is

$$\frac{x^2}{4} + \frac{y^2}{9} = 1 \qquad (z = 1)$$

Sketching this ellipse and the parabolic traces in the vertical coordinate planes yields the graph in Figure 13.7.9. ◄

A rough sketch of the hyperbolic paraboloid

$$z = \frac{y^2}{b^2} - \frac{x^2}{a^2} \qquad (a > 0, b > 0) \tag{15}$$

can be obtained by first sketching the two parabolic traces that pass through the origin (one in the plane $x = 0$ and the other in the plane $y = 0$). After the parabolic traces are drawn, sketch the hyperbolic traces in the planes $z = \pm 1$ and then fill in any missing edges. The next example illustrates this technique.

Example 6

Sketch the graph of the hyperbolic paraboloid

$$z = \frac{y^2}{4} - \frac{x^2}{9} \tag{16}$$

Solution. Setting $x = 0$ in (16) yields

$$z = \frac{y^2}{4} \qquad (x = 0)$$

which is a parabola in the yz-plane with vertex at the origin and opening in the positive z-direction (since $z \geq 0$), and setting $y = 0$ yields

$$z = -\frac{x^2}{9} \qquad (y = 0)$$

which is a parabola in the xz-plane with vertex at the origin and opening in the negative z-direction.

The trace in the plane $z = 1$ is

$$\frac{y^2}{4} - \frac{x^2}{9} = 1 \qquad (z = 1)$$

which is a hyperbola that opens along a line parallel to the y-axis (verify), and the trace in the plane $z = -1$ is

$$\frac{x^2}{9} - \frac{y^2}{4} = 1 \qquad (z = -1)$$

which is a hyperbola that opens along a line parallel to the x-axis. Combining all of the above information leads to the sketch in Figure 13.7.10. ◄

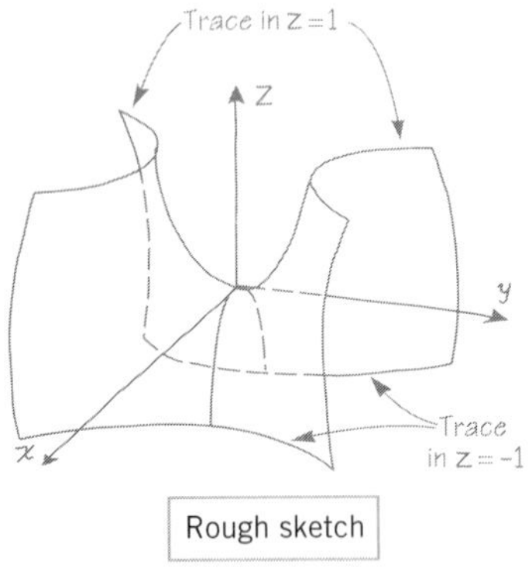

Rough sketch

Figure 13.7.10

REMARK. The hyperbolic paraboloid in Figure 13.7.10 has an interesting behavior at the origin—the trace in the xz-plane has a relative maximum at $(0, 0, 0)$, and the trace in the yz-plane has a relative minimum at $(0, 0, 0)$. Thus, a bug walking on the surface along the line of the x-axis reaches the top of a hill at the origin, while a bug walking on the surface along the line of the y-axis reaches the bottom of a valley at the origin. A point with this property is commonly called a **saddle point** or a **minimax point**.

Figure 13.7.11 shows two computer-generated views of the hyperbolic paraboloid in Example 6. The first view, which is much like our rough sketch in Figure 13.7.10, has cuts at the top and bottom that are hyperbolic traces parallel to the xy-plane. In the second view the top horizontal cut has been omitted; this helps to emphasize the parabolic traces parallel to the xz-plane.

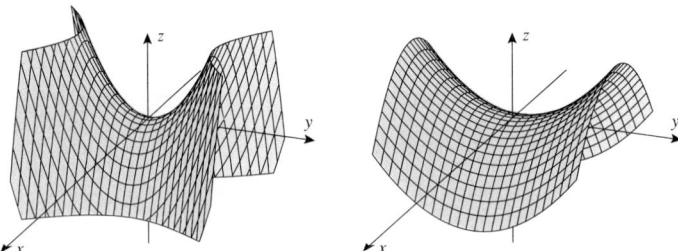

Figure 13.7.11

TRANSLATIONS OF QUADRIC SURFACES

In Section 12.4 we saw that a conic in an xy-coordinate system can be translated by substituting $x - h$ for x and $y - k$ for y in its equation. To understand why this works, think of the xy-axes as fixed, and think of the plane as a transparent sheet of plastic on which all graphs are drawn. When the coordinates of points are modified by substituting $(x - h, y - k)$ for (x, y), the geometric effect is to translate the sheet of plastic (and hence all curves) so that the point on the plastic that was initially at $(0, 0)$ is moved to the point (h, k) (see Figure 13.7.12a).

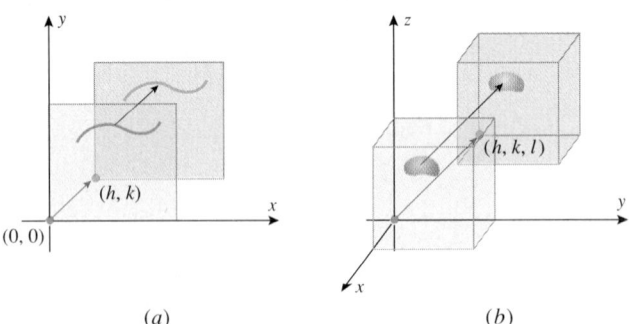

Figure 13.7.12

(a) (b)

For the analog in three dimensions, think of the xyz-axes as fixed, and think of 3-space as a transparent block of plastic in which all surfaces are embedded. When the coordinates of points are modified by substituting $(x - h, y - k, z - l)$ for (x, y, z), the geometric effect is to translate the block of plastic (and hence all surfaces) so that the point in the plastic block that was initially at $(0, 0, 0)$ is moved to the point (h, k, l) (see Figure 13.7.12b).

Example 7

Describe the surface $z = (x - 1)^2 + (y + 2)^2 + 3$.

Solution. The equation can be rewritten as

$$z - 3 = (x - 1)^2 + (y + 2)^2$$

This surface is the paraboloid that results by translating the paraboloid

$$z = x^2 + y^2$$

in Figure 13.7.3 so that the new "vertex" is at the point $(1, -2, 3)$. A rough sketch of this paraboloid is shown in Figure 13.7.13. ◄

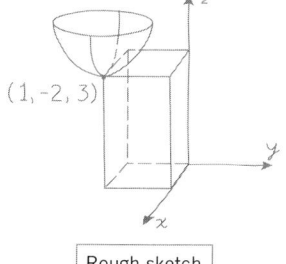

Rough sketch

Figure 13.7.13

Example 8

Describe the surface

$$4x^2 + 4y^2 + z^2 + 8y - 4z = -4$$

Solution. Completing the squares yields

$$4x^2 + 4(y + 1)^2 + (z - 2)^2 = -4 + 4 + 4$$

or

$$x^2 + (y + 1)^2 + \frac{(z - 2)^2}{4} = 1$$

Thus, the surface is the ellipsoid that results when the ellipsoid

$$x^2 + y^2 + \frac{z^2}{4} = 1$$

is translated so that the new "center" is at the point $(0, -1, 2)$. A rough sketch of this ellipsoid is shown in Figure 13.7.14. ◄

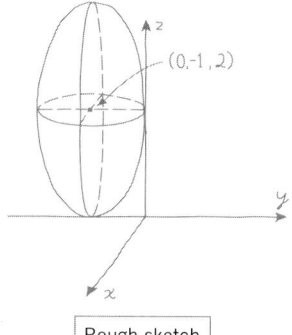

Rough sketch

Figure 13.7.14

FOR THE READER. The ellipsoid in Figure 13.7.14 was sketched with its cross section in the yz-plane tangent to the y- and z-axes. Confirm that this is correct.

REFLECTIONS OF SURFACES IN 3-SPACE

Recall that in an xy-coordinate system a point (x, y) is reflected about the x-axis if y is replaced by $-y$, and it is reflected about the y-axis if x is replaced by $-x$. In an xyz-coordinate system, a point (x, y, z) is reflected about the xy-plane if z is replaced by $-z$, it is reflected about the yz-plane if x is replaced by $-x$, and it is reflected about the xz-plane if y is replaced by $-y$ (Figure 13.7.15). It follows that *replacing a variable by its negative in the equation of a surface causes that surface to be reflected about a coordinate plane.*

Recall also that in an xy-coordinate system a point (x, y) is reflected about the line $y = x$ if x and y are interchanged. However, in an xyz-coordinate system, interchanging x and y reflects the point (x, y, z) about the plane $y = x$ (Figure 13.7.16). Similarly, interchanging

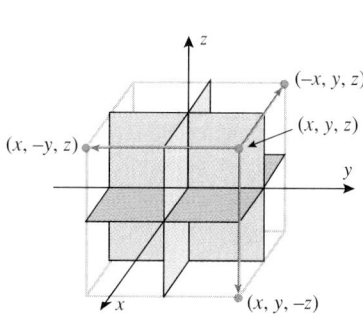

Figure 13.7.15

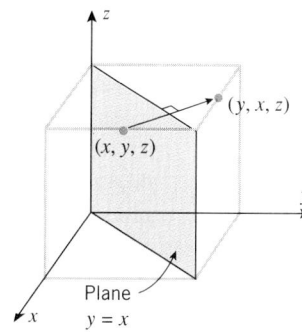

Figure 13.7.16

x and z reflects the point about the plane $x = z$, and interchanging y and z reflects it about the plane $y = z$. Thus, it follows that *interchanging two variables in the equation of a surface reflects that surface about a plane that makes a $45°$ angle with two of the coordinate planes.*

Example 9

Describe the surfaces

(a) $y^2 = x^2 + z^2$ (b) $z = -(x^2 + y^2)$

Solution (a). The graph of the equation $y^2 = x^2 + z^2$ results from interchanging y and z in the equation $z^2 = x^2 + y^2$. Thus, the graph of the equation $y^2 = x^2 + z^2$ can be obtained by reflecting the graph of $z^2 = x^2 + y^2$ about the plane $y = z$. Since the graph of $z^2 = x^2 + y^2$ is a circular cone opening along the z-axis (see Table 13.7.1), it follows that the graph of $y^2 = x^2 + z^2$ is a circular cone opening along the y-axis (Figure 13.7.17).

Solution (b). The graph of the equation $z = -(x^2 + y^2)$ can be written as $-z = x^2 + y^2$, which can be obtained by replacing z with $-z$ in the equation $z = x^2 + y^2$. Since the graph of $z = x^2 + y^2$ is a circular paraboloid opening in the positive z-direction (see Table 13.7.1), it follows that the graph of $z = -(x^2 + y^2)$ is a circular paraboloid opening in the negative z-direction (Figure 13.7.18).   ◀

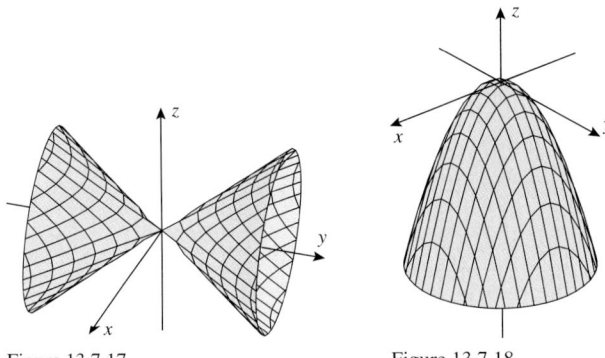

Figure 13.7.17   Figure 13.7.18

A TECHNIQUE FOR IDENTIFYING QUADRIC SURFACES

The equations of the quadric surfaces in Table 13.7.1 have certain characteristics that make it possible to identify quadric surfaces that are derived from these equations by reflections. These identifying characteristics, which are shown in Table 13.7.2, are based on writing the equation of the quadric surface so that all of the variable terms are on the left side of the equation and there is a 1 or a 0 on the right side. When there is a 1 on the right side the surface is an ellipsoid, hyperboloid of one sheet, or a hyperboloid of two sheets, and when there is a 0 on the right side it is an elliptic cone, an elliptic paraboloid, or a hyperbolic paraboloid. Within the group with a 1 on the right side, ellipsoids have no minus signs, hyperboloids of one sheet have one minus sign, and hyperboloids of two sheets have two minus signs. Within the group with a 0 on the right side, elliptic cones have no linear terms, elliptic paraboloids have one linear term and two quadratic terms with the same sign, and hyperbolic paraboloids have one linear term and two quadratic terms with opposite signs. These characteristics do not change when the surface is reflected about a coordinate plane or planes of the form $x = y$, $x = z$, or $y = z$, thereby making it possible to identify the reflected quadric surface from the form of its equation.

Example 10

Identify the surfaces

(a) $3x^2 - 4y^2 + 12z^2 + 12 = 0$ (b) $4x^2 - 4y + z^2 = 0$

Solution (a). The equation can be rewritten as

$$\frac{y^2}{3} - \frac{x^2}{4} - z^2 = 1$$

This equation has a 1 on the right side and two negative terms on the left side, so its graph is a hyperboloid of two sheets.

Solution (b). The equation has one linear term and two quadratic terms with the same sign, so its graph is an elliptic paraboloid. ◀

Table 13.7.2

EQUATION	CHARACTERISTIC	CLASSIFICATION
$\dfrac{x^2}{a^2} + \dfrac{y^2}{b^2} + \dfrac{z^2}{c^2} = 1$	No minus signs	Ellipsoid
$\dfrac{x^2}{a^2} + \dfrac{y^2}{b^2} - \dfrac{z^2}{c^2} = 1$	One minus sign	Hyperboloid of one sheet
$\dfrac{z^2}{c^2} - \dfrac{x^2}{a^2} - \dfrac{y^2}{b^2} = 1$	Two minus signs	Hyperboloid of two sheets
$z^2 - \dfrac{x^2}{a^2} - \dfrac{y^2}{b^2} = 0$	No linear terms	Elliptic cone
$z - \dfrac{x^2}{a^2} - \dfrac{y^2}{b^2} = 0$	One linear term; two quadratic terms with the same sign	Elliptic paraboloid
$z - \dfrac{y^2}{b^2} + \dfrac{x^2}{a^2} = 0$	One linear term; two quadratic terms with opposite signs	Hyperbolic paraboloid

EXERCISE SET 13.7

In Exercises 1 and 2, identify the quadric surface as an ellipsoid, hyperboloid of one sheet, hyperboloid of two sheets, elliptic cone, elliptic paraboloid, or hyperbolic paraboloid by matching the equation with one of the forms given in Table 13.7.1. State the values of a, b, and c in each case.

1. (a) $z = \dfrac{x^2}{4} + \dfrac{y^2}{9}$ (b) $z = \dfrac{y^2}{25} - x^2$
 (c) $x^2 + y^2 - z^2 = 16$ (d) $x^2 + y^2 - z^2 = 0$
 (e) $4z = x^2 + 4y^2$ (f) $z^2 - x^2 - y^2 = 1$

2. (a) $6x^2 + 3y^2 + 4z^2 = 12$ (b) $y^2 - x^2 - z = 0$
 (c) $9x^2 + y^2 - 9z^2 = 9$ (d) $4x^2 + y^2 - 4z^2 = -4$
 (e) $2z - x^2 - 4y^2 = 0$ (f) $12z^2 - 3x^2 = 4y^2$

3. Find an equation for and sketch the surface that results when the circular paraboloid $z = x^2 + y^2$ is reflected about the plane
 (a) $z = 0$ (b) $x = 0$ (c) $y = 0$

 (d) $y = x$ (e) $x = z$ (f) $y = z$.

4. Find an equation for and sketch the surface that results when the hyperboloid of one sheet $x^2 + y^2 - z^2 = 1$ is reflected about the plane
 (a) $z = 0$ (b) $x = 0$ (c) $y = 0$
 (d) $y = x$ (e) $x = z$ (f) $y = z$.

5. The given equations represent quadric surfaces whose orientations are different from those in Table 13.7.1. In each part, identify the quadric surface, and give a verbal description of its orientation (e.g., an elliptic cone opening along the z-axis or a hyperbolic paraboloid straddling the y-axis).

 (a) $\dfrac{z^2}{c^2} - \dfrac{y^2}{b^2} + \dfrac{x^2}{a^2} = 1$ (b) $\dfrac{x^2}{a^2} - \dfrac{y^2}{b^2} - \dfrac{z^2}{c^2} = 1$

 (c) $x = \dfrac{y^2}{b^2} + \dfrac{z^2}{c^2}$ (d) $x^2 = \dfrac{y^2}{b^2} + \dfrac{z^2}{c^2}$

 (e) $y = \dfrac{z^2}{c^2} - \dfrac{x^2}{a^2}$ (f) $y = -\left(\dfrac{x^2}{a^2} + \dfrac{z^2}{c^2}\right)$

6. For each of the surfaces in Exercise 5, find the equation of the surface that results if the given surface is reflected about the xz-plane and that surface is then reflected about the plane $z = 0$.

In Exercises 7 and 8, find equations of the traces in the coordinate planes, and sketch the traces in an xyz-coordinate system. [*Suggestion:* If you have trouble sketching a trace directly in three dimensions, start with a sketch in two dimensions by placing the coordinate plane in the plane of the paper; then transfer that sketch to three dimensions.]

7. (a) $\dfrac{x^2}{9} + \dfrac{y^2}{25} + \dfrac{z^2}{4} = 1$ (b) $z = x^2 + 4y^2$

    (c) $\dfrac{x^2}{9} + \dfrac{y^2}{16} - \dfrac{z^2}{4} = 1$

8. (a) $y^2 + 9z^2 = x$ (b) $4x^2 - y^2 + 4z^2 = 4$

    (c) $z^2 = x^2 + \dfrac{y^2}{4}$

In Exercises 9 and 10, the traces of the surfaces in the planes are conic sections. In each part, find an equation of the trace, and state whether it is an ellipse, a parabola, or a hyperbola.

9. (a) $4x^2 + y^2 + z^2 = 4$; $y = 1$
    (b) $4x^2 + y^2 + z^2 = 4$; $x = \frac{1}{2}$
    (c) $9x^2 - y^2 - z^2 = 16$; $x = 2$
    (d) $9x^2 - y^2 - z^2 = 16$; $z = 2$
    (e) $z = 9x^2 + 4y^2$; $y = 2$
    (f) $z = 9x^2 + 4y^2$; $z = 4$

10. (a) $9x^2 - y^2 + 4z^2 = 9$; $x = 2$
    (b) $9x^2 - y^2 + 4z^2 = 9$; $y = 4$
    (c) $x^2 + 4y^2 - 9z^2 = 0$; $y = 1$
    (d) $x^2 + 4y^2 - 9z^2 = 0$; $z = 1$
    (e) $z = x^2 - 4y^2$; $x = 1$
    (f) $z = x^2 - 4y^2$; $z = 4$

In Exercises 11–22, identify and sketch the quadric surface.

11. $x^2 + \dfrac{y^2}{4} + \dfrac{z^2}{9} = 1$ **12.** $x^2 + 4y^2 + 9z^2 = 36$

13. $\dfrac{x^2}{4} + \dfrac{y^2}{9} - \dfrac{z^2}{16} = 1$ **14.** $x^2 + y^2 - z^2 = 9$

15. $4z^2 = x^2 + 4y^2$ **16.** $9x^2 + 4y^2 - 36z^2 = 0$

17. $9z^2 - 4y^2 - 9x^2 = 36$ **18.** $y^2 - \dfrac{x^2}{4} - \dfrac{z^2}{9} = 1$

19. $z = y^2 - x^2$ **20.** $16z = y^2 - x^2$

21. $4z = x^2 + 2y^2$ **22.** $z - 3x^2 - 3y^2 = 0$

In Exercises 23–28, the given equations represent quadric surfaces whose orientations are different from those in Table 13.7.1. Identify and sketch the surface.

23. $x^2 - 3y^2 - 3z^2 = 0$ **24.** $x - y^2 - 4z^2 = 0$

25. $2y^2 - x^2 + 2z^2 = 8$ **26.** $x^2 - 3y^2 - 3z^2 = 9$

27. $z = \dfrac{x^2}{4} - \dfrac{y^2}{9}$ **28.** $4x^2 - y^2 + 4z^2 = 16$

In Exercises 29–32, sketch the surface.

29. $z = \sqrt{x^2 + y^2}$ **30.** $z = \sqrt{1 - x^2 - y^2}$

31. $z = \sqrt{x^2 + y^2 - 1}$ **32.** $z = \sqrt{1 + x^2 + y^2}$

In Exercises 33–36, identify the surface, and make a rough sketch that shows its position and orientation.

33. $z = (x + 2)^2 + (y - 3)^2 - 9$

34. $4x^2 - y^2 + 16(z - 2)^2 = 100$

35. $9x^2 + y^2 + 4z^2 - 18x + 2y + 16z = 10$

36. $z^2 = 4x^2 + y^2 + 8x - 2y + 4z$

Exercises 37 and 38 are concerned with the ellipsoid $4x^2 + 9y^2 + 18z^2 = 72$.

37. (a) Find an equation of the elliptical trace in the plane $z = \sqrt{2}$.
    (b) Find the lengths of the major and minor axes of the ellipse in part (a).
    (c) Find the coordinates of the foci of the ellipse in part (a).
    (d) Describe the orientation of the focal axis of the ellipse in part (a) relative to the coordinate axes.

38. (a) Find an equation of the elliptical trace in the plane $x = 3$.
    (b) Find the lengths of the major and minor axes of the ellipse in part (a).
    (c) Find the coordinates of the foci of the ellipse in part (a).
    (d) Describe the orientation of the focal axis of the ellipse in part (a) relative to the coordinate axes.

Exercises 39–42 refer to the hyperbolic paraboloid $z = y^2 - x^2$.

39. (a) Find an equation of the hyperbolic trace in the plane $z = 4$.
    (b) Find the vertices of the hyperbola in part (a).
    (c) Find the foci of the hyperbola in part (a).
    (d) Describe the orientation of the focal axis of the hyperbola in part (a) relative to the coordinate axes.

40. (a) Find an equation of the hyperbolic trace in the plane $z = -4$.
    (b) Find the vertices of the hyperbola in part (a).
    (c) Find the foci of the hyperbola in part (a).
    (d) Describe the orientation of the focal axis of the hyperbola in part (a) relative to the coordinate axes.

41. (a) Find an equation of the parabolic trace in the plane $x = 2$.
  (b) Find the vertex of the parabola in part (a).
  (c) Find the focus of the parabola in part (a).
  (d) Describe the orientation of the focal axis of the parabola in part (a) relative to the coordinate axes.

42. (a) Find an equation of the parabolic trace in the plane $y = 2$.
  (b) Find the vertex of the parabola in part (a).
  (c) Find the focus of the parabola in part (a).
  (d) Describe the orientation of the focal axis of the parabola in part (a) relative to the coordinate axes.

In Exercises 43 and 44, sketch the region enclosed between the surfaces and describe their curve of intersection.

43. The paraboloids $z = x^2 + y^2$ and $z = 4 - x^2 - y^2$

44. The hyperbolic paraboloid $x^2 = y^2 + z$ and the ellipsoid $x^2 = 4 - 2y^2 - 2z$

In Exercises 45 and 46, find an equation for the surface generated by revolving the curve about the axis.

45. $y = 4x^2$ $(z = 0)$ about the y-axis

46. $y = 2x$ $(z = 0)$ about the y-axis

47. Find an equation of the surface consisting of all points $P(x, y, z)$ that are equidistant from the point $(0, 0, 1)$ and the plane $z = -1$. Identify the surface.

48. Find an equation of the surface consisting of all points $P(x, y, z)$ that are twice as far from the plane $z = -1$ as from the point $(0, 0, 1)$. Identify the surface.

49. If a sphere
$$\frac{x^2}{a^2} + \frac{y^2}{a^2} + \frac{z^2}{a^2} = 1$$
of radius a is compressed in the z-direction, then the resulting surface, called an ***oblate spheroid***, has an equation of

the form
$$\frac{x^2}{a^2} + \frac{y^2}{a^2} + \frac{z^2}{c^2} = 1$$
where $c < a$. Show that the oblate spheroid has a circular trace of radius a in the xy-plane and an elliptical trace in the xz-plane with major axis of length $2a$ along the x-axis and minor axis of length $2c$ along the z-axis.

50. The Earth's rotation causes a flattening at the poles, so its shape is often modeled as an oblate spheroid rather than a sphere (see Exercise 49 for terminology). One of the models used by global positioning satellites is the ***World Geodetic System of 1984*** (WGS-84), which treats the Earth as an oblate spheroid whose equatorial radius is 6378.1370 km and whose polar radius (the distance from the Earth's center to the poles) is 6356.5231 km. Use the WGS-84 model to find an equation for the surface of the Earth relative to the coordinate system shown in the accompanying figure.

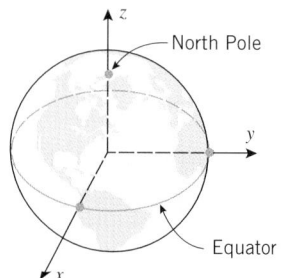

Figure Ex-50

51. Use the method of slicing to show that the volume of the ellipsoid
$$\frac{x^2}{a^2} + \frac{y^2}{b^2} + \frac{z^2}{c^2} = 1$$
is $\frac{4}{3}\pi abc$.

13.8 CYLINDRICAL AND SPHERICAL COORDINATES

In this section we will discuss two new types of coordinate systems in 3-space that are often more useful than rectangular coordinate systems for studying surfaces with symmetries. These new coordinate systems also have important applications in navigation, astronomy, and the study of rotational motion about an axis.

CYLINDRICAL AND SPHERICAL COORDINATE SYSTEMS

Three coordinates are required to establish the location of a point in 3-space. We have already done this using rectangular coordinates. However, Figure 13.8.1 shows two other possibilities: part (*a*) of the figure shows the ***rectangular coordinates*** (x, y, z) of a point P, part (*b*) shows the ***cylindrical coordinates*** (r, θ, z) of P, and part (*c*) shows the ***spherical coordinates*** (ρ, θ, ϕ) of P. In a rectangular coordinate system the coordinates can be any real numbers, but in cylindrical and spherical coordinate systems there are restrictions on the allowable values of the coordinates (as indicated in Figure 13.8.1).

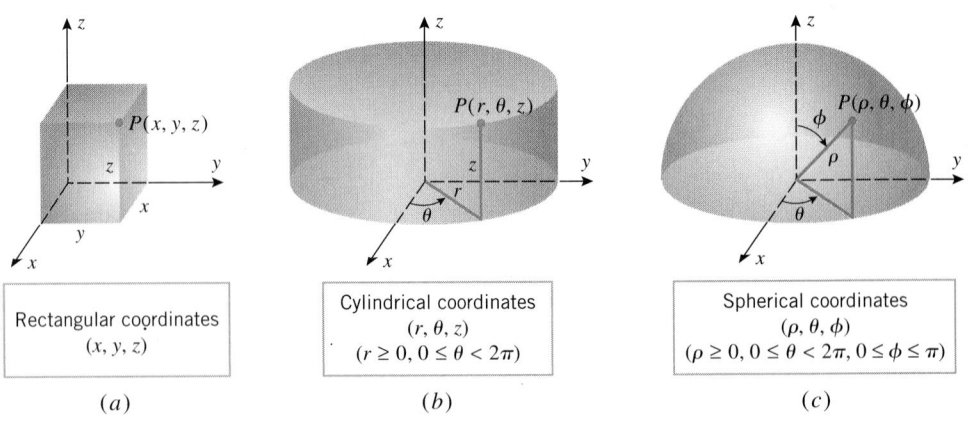

Figure 13.8.1

CONSTANT SURFACES

In rectangular coordinates the surfaces represented by equations of the form

$$x = x_0, \quad y = y_0, \quad \text{and} \quad z = z_0$$

where x_0, y_0, and z_0 are constants, are planes parallel to the yz-plane, xz-plane, and xy-plane, respectively (Figure 13.8.2a). In cylindrical coordinates the surfaces represented by equations of the form

$$r = r_0, \quad \theta = \theta_0, \quad \text{and} \quad z = z_0$$

where r_0, θ_0, and z_0 are constants, are shown in Figure 13.8.2b:

- The surface $r = r_0$ is a right circular cylinder of radius r_0 centered on the z-axis. At each point (r, θ, z) on this cylinder, r has the value r_0, but θ and z are unrestricted except for our general restriction that $0 \le \theta < 2\pi$.

- The surface $\theta = \theta_0$ is a half-plane attached along the z-axis and making an angle θ_0 with the positive x-axis. At each point (r, θ, z) on this surface, θ has the value θ_0, but r and z are unrestricted except for our general restriction that $r \ge 0$.

- The surface $z = z_0$ is a horizontal plane. At each point (r, θ, z) on this plane, z has the value z_0, but r and θ are unrestricted except for the general restrictions.

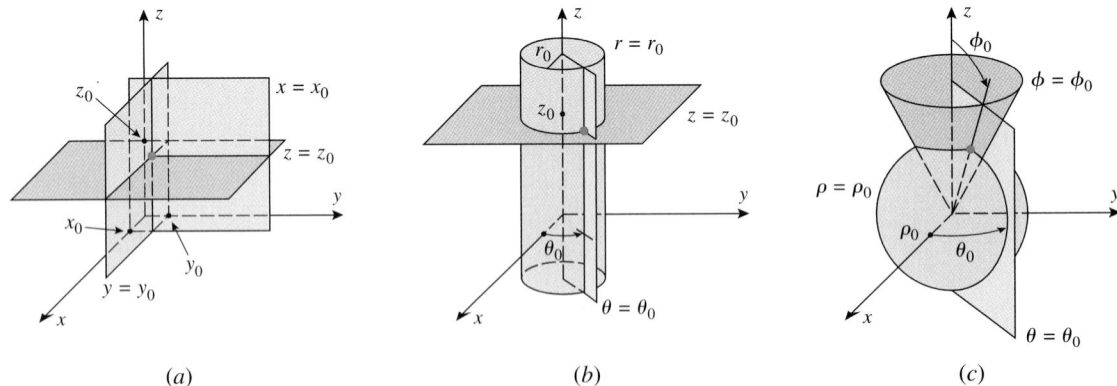

Figure 13.8.2

In spherical coordinates the surfaces represented by equations of the form

$$\rho = \rho_0, \quad \theta = \theta_0, \quad \text{and} \quad \phi = \phi_0$$

where ρ_0, θ_0, and ϕ_0 are constants, are shown in Figure 13.8.2c:

- The surface $\rho = \rho_0$ consists of all points whose distance ρ from the origin is ρ_0. Assuming ρ_0 to be nonnegative, this is a sphere of radius ρ_0 centered at the origin.

- As in cylindrical coordinates, the surface $\theta = \theta_0$ is a half-plane attached along the z-axis, making an angle of θ_0 with the positive x-axis.

- The surface $\phi = \phi_0$ consists of all points from which a line segment to the origin makes an angle of ϕ_0 with the positive z-axis. Depending on whether $0 < \phi_0 < \pi/2$ or $\pi/2 < \phi_0 < \pi$, this will be the nappe of a cone opening up or opening down. (If $\phi_0 = \pi/2$, then the cone is flat, and the surface is the xy-plane.)

CONVERTING COORDINATES

Just as we needed to convert between rectangular and polar coordinates in 2-space, so we will need to be able to convert between rectangular, cylindrical, and spherical coordinates in 3-space. Table 13.8.1 provides formulas for making these conversions.

Table 13.8.1

CONVERSION		FORMULAS	RESTRICTIONS
Cylindrical to rectangular	$(r, \theta, z) \rightarrow (x, y, z)$	$x = r\cos\theta, \quad y = r\sin\theta, \quad z = z$	
Rectangular to cylindrical	$(x, y, z) \rightarrow (r, \theta, z)$	$r = \sqrt{x^2 + y^2}, \quad \tan\theta = y/x, \quad z = z$	
Spherical to cylindrical	$(\rho, \theta, \phi) \rightarrow (r, \theta, z)$	$r = \rho\sin\phi, \quad \theta = \theta, \quad z = \rho\cos\phi$	$r \geq 0, \rho \geq 0$
Cylindrical to spherical	$(r, \theta, z) \rightarrow (\rho, \theta, \phi)$	$\rho = \sqrt{r^2 + z^2}, \quad \theta = \theta, \quad \tan\phi = r/z$	$0 \leq \theta < 2\pi$ $0 \leq \phi \leq \pi$
Spherical to rectangular	$(\rho, \theta, \phi) \rightarrow (x, y, z)$	$x = \rho\sin\phi\cos\theta, \quad y = \rho\sin\phi\sin\theta, \quad z = \rho\cos\phi$	
Rectangular to spherical	$(x, y, z) \rightarrow (\rho, \theta, \phi)$	$\rho = \sqrt{x^2 + y^2 + z^2}, \quad \tan\theta = y/x, \quad \cos\phi = z/\sqrt{x^2 + y^2 + z^2}$	

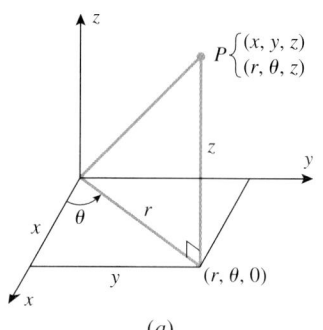

(a)

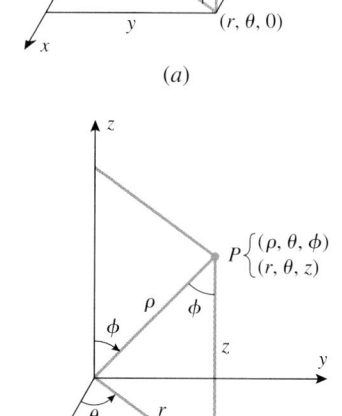

(b)

Figure 13.8.3

The diagrams in Figure 13.8.3 will help you to understand how the formulas in Table 13.8.1 are derived. For example, part (a) of the figure shows that in converting between rectangular coordinates (x, y, z) and cylindrical coordinates (r, θ, z), we can interpret (r, θ) as polar coordinates of (x, y). Thus, the polar-to-rectangular and rectangular-to-polar conversion formulas (1) and (2) of Section 12.1 provide the conversion formulas between rectangular and cylindrical coordinates in the table.

Part (b) of Figure 13.8.3 suggests that the spherical coordinates (ρ, θ, ϕ) of a point P can be converted to cylindrical coordinates (r, θ, z) by the conversion formulas

$$r = \rho\sin\phi, \quad \theta = \theta, \quad z = \rho\cos\phi \tag{1}$$

Moreover, since the cylindrical coordinates (r, θ, z) of P can be converted to rectangular coordinates (x, y, z) by the conversion formulas

$$x = r\cos\theta, \quad y = r\sin\theta, \quad z = z \tag{2}$$

we can obtain direct conversion formulas from spherical coordinates to rectangular coordinates by substituting (1) in (2). This yields

$$x = \rho\sin\phi\cos\theta, \quad y = \rho\sin\phi\sin\theta, \quad z = \rho\cos\phi \tag{3}$$

The other conversion formulas in Table 13.8.1 are left as exercises.

Example 1

(a) Find the rectangular coordinates of the point with cylindrical coordinates
$$(r, \theta, z) = (4, \pi/3, -3)$$

(b) Find the rectangular coordinates of the point with spherical coordinates
$$(\rho, \theta, \phi) = (4, \pi/3, \pi/4)$$

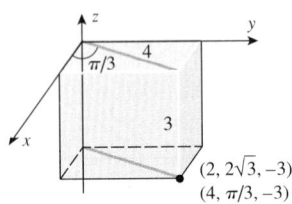

Figure 13.8.4

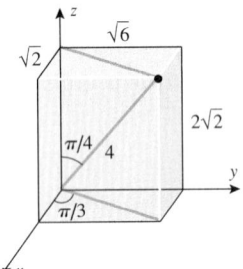

Figure 13.8.5

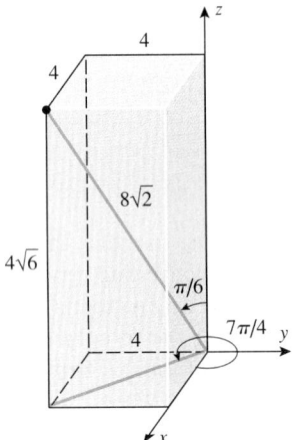

Figure 13.8.6

EQUATIONS OF SURFACES IN CYLINDRICAL AND SPHERICAL COORDINATES

Solution (a). Applying the cylindrical-to-rectangular conversion formulas in Table 13.8.1 yields

$$x = r \cos \theta = 4 \cos \frac{\pi}{3} = 2, \quad y = r \sin \theta = 4 \sin \frac{\pi}{3} = 2\sqrt{3}, \quad z = -3$$

Thus, the rectangular coordinates of the point are $(x, y, z) = (2, 2\sqrt{3}, -3)$ (Figure 13.8.4).

Solution (b). Applying the spherical-to-rectangular conversion formulas in Table 13.8.1 yields

$$x = \rho \sin \phi \cos \theta = 4 \sin \frac{\pi}{4} \cos \frac{\pi}{3} = \sqrt{2}$$

$$y = \rho \sin \phi \sin \theta = 4 \sin \frac{\pi}{4} \sin \frac{\pi}{3} = \sqrt{6}$$

$$z = \rho \cos \phi = 4 \cos \frac{\pi}{4} = 2\sqrt{2}$$

Thus, the rectangular coordinates of the point are $(x, y, z) = (\sqrt{2}, \sqrt{6}, 2\sqrt{2})$ (Figure 13.8.5). ◀

Example 2

Find the spherical coordinates of the point that has rectangular coordinates

$$(x, y, z) = (4, -4, 4\sqrt{6})$$

Solution. From the rectangular-to-spherical conversion formulas in Table 13.8.1 we obtain

$$\rho = \sqrt{x^2 + y^2 + z^2} = \sqrt{16 + 16 + 96} = \sqrt{128} = 8\sqrt{2}$$

$$\tan \theta = \frac{y}{x} = -1$$

$$\cos \phi = \frac{z}{\sqrt{x^2 + y^2 + z^2}} = \frac{4\sqrt{6}}{8\sqrt{2}} = \frac{\sqrt{3}}{2}$$

From the restriction $0 \le \theta < 2\pi$ and the computed value of $\tan \theta$, the possibilities for θ are $\theta = 3\pi/4$ and $\theta = 7\pi/4$. However, the given point has a negative y-coordinate, so we must have $\theta = 7\pi/4$. Moreover, from the restriction $0 \le \phi \le \pi$ and the computed value of $\cos \phi$, the only possibility for ϕ is $\phi = \pi/6$. Thus, the spherical coordinates of the point are $(\rho, \theta, \phi) = (8\sqrt{2}, 7\pi/4, \pi/6)$ (Figure 13.8.6). ◀

Surfaces of revolution about the z-axis of a rectangular coordinate system usually have simpler equations in cylindrical coordinates than in rectangular coordinates, and the equations of surfaces with symmetry about the origin are usually simpler in spherical coordinates than in rectangular coordinates. For example, consider the upper nappe of the circular cone whose equation in rectangular coordinates is

$$z = \sqrt{x^2 + y^2}$$

(Table 13.8.2). The corresponding equation in cylindrical coordinates can be obtained from the cylindrical-to-rectangular conversion formulas in Table 13.8.1. This yields

$$z = \sqrt{(r \cos \theta)^2 + (r \sin \theta)^2} = \sqrt{r^2} = |r| = r$$

so the equation of the cone in cylindrical coordinates is $z = r$. Going a step further, the equation of the cone in spherical coordinates can be obtained from the spherical-to-cylindrical conversion formulas from Table 13.8.1. This yields

$$\rho \cos \phi = \rho \sin \phi$$

which, if $\rho \ne 0$, can be rewritten as

$$\tan \phi = 1$$

Geometrically, this tells us that the radial line from the origin to any point on the cone makes an angle of $\pi/4$ with the z-axis.

Table 13.8.2

	CONE	CYLINDER	SPHERE	PARABOLOID	HYPERBOLOID
	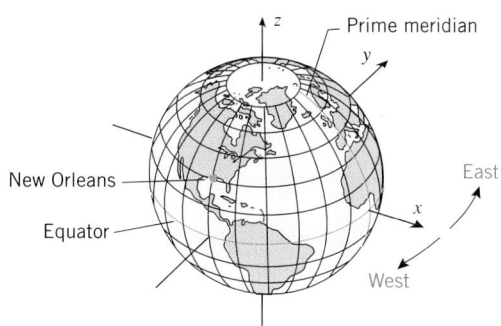				
RECTANGULAR	$z = \sqrt{x^2 + y^2}$	$x^2 + y^2 = 1$	$x^2 + y^2 + z^2 = 1$	$z = x^2 + y^2$	$x^2 + y^2 - z^2 = 1$
CYLINDRICAL	$z = r$	$r = 1$	$z^2 = 1 - r^2$	$z = r^2$	$z^2 = r^2 - 1$
SPHERICAL	$\tan \phi = 1$	$\rho = \csc \phi$	$\rho = 1$	$\rho = \cos \phi \csc^2 \phi$	$\rho^2 = -\sec 2\phi$

Example 3

Find equations of the paraboloid $z = x^2 + y^2$ in cylindrical and spherical coordinates.

Solution. The rectangular-to-cylindrical conversion formulas in Table 13.8.1 yield

$$z = r^2 \tag{4}$$

which is the equation in cylindrical coordinates. Now applying the spherical-to-cylindrical conversion formulas to (4) yields

$$\rho \cos \phi = \rho^2 \sin^2 \phi$$

which we can rewrite as

$$\rho = \cos \phi \csc^2 \phi$$

Alternatively, we could have obtained this equation directly from the equation in rectangular coordinates by applying the spherical-to-rectangular conversion formulas (verify). ◀

FOR THE READER. Confirm that the equations for the cylinder and hyperboloid in cylindrical and spherical coordinates given in Table 13.8.2 are correct.

SPHERICAL COORDINATES IN NAVIGATION

Spherical coordinates are related to longitude and latitude coordinates used in navigation. To see why this is so, let us construct a right-hand rectangular coordinate system with its origin at the center of the Earth, its positive z-axis passing through the North Pole, and its positive x-axis passing through the prime meridian (Figure 13.8.7). If we assume the Earth to be a sphere of radius $\rho = 4000$ miles, then each point on the Earth has spherical coordinates of the form $(4000, \theta, \phi)$, where ϕ and θ determine the latitude and longitude of

Figure 13.8.7

the point. It is common to specify longitudes in degrees east or west of the prime meridian and latitudes in degrees north or south of the equator. However, the next example shows that it is a simple matter to determine ϕ and θ from such data.

Example 4

The city of New Orleans is located at $90°$ west longitude and $30°$ north latitude. Find its spherical and rectangular coordinates relative to the coordinate axes of Figure 13.8.7. (Assume that distance is in miles.)

Solution. A longitude of $90°$ west corresponds to $\theta = 360° - 90° = 270°$ or $\theta = 3\pi/2$ radians; and a latitude of $30°$ north corresponds to $\phi = 90° - 30° = 60°$ or $\phi = \pi/3$ radians. Thus, the spherical coordinates (ρ, θ, ϕ) of New Orleans are $(4000, 3\pi/2, \pi/3)$.

To find the rectangular coordinates we apply the spherical-to-rectangular conversion formulas in Table 13.8.1. This yields

$$x = 4000 \sin\frac{\pi}{3} \cos\frac{3\pi}{2} = 4000\frac{\sqrt{3}}{2}(0) = 0 \text{ mi}$$

$$y = 4000 \sin\frac{\pi}{3} \sin\frac{3\pi}{2} = 4000\frac{\sqrt{3}}{2}(-1) = -2000\sqrt{3} \text{ mi}$$

$$z = 4000 \cos\frac{\pi}{3} = 4000\left(\frac{1}{2}\right) = 2000 \text{ mi} \qquad \blacktriangleleft$$

EXERCISE SET 13.8   ⌁ Graphing Calculator   [c] CAS

In Exercises 1 and 2, convert from rectangular to cylindrical coordinates.

1. (a) $(4\sqrt{3}, 4, -4)$ (b) $(-5, 5, 6)$
    (c) $(0, 2, 0)$ (d) $(4, -4\sqrt{3}, 6)$

2. (a) $(\sqrt{2}, -\sqrt{2}, 1)$ (b) $(0, 1, 1)$
    (c) $(-4, 4, -7)$ (d) $(2, -2, -2)$

In Exercises 3 and 4, convert from cylindrical to rectangular coordinates.

3. (a) $(4, \pi/6, 3)$ (b) $(8, 3\pi/4, -2)$
    (c) $(5, 0, 4)$ (d) $(7, \pi, -9)$

4. (a) $(6, 5\pi/3, 7)$ (b) $(1, \pi/2, 0)$
    (c) $(3, \pi/2, 5)$ (d) $(4, \pi/2, -1)$

In Exercises 5 and 6, convert from rectangular to spherical coordinates.

5. (a) $(1, \sqrt{3}, -2)$ (b) $(1, -1, \sqrt{2})$
    (c) $(0, 3\sqrt{3}, 3)$ (d) $(-5\sqrt{3}, 5, 0)$

6. (a) $(4, 4, 4\sqrt{6})$ (b) $(1, -\sqrt{3}, -2)$
    (c) $(2, 0, 0)$ (d) $(\sqrt{3}, 1, 2\sqrt{3})$

In Exercises 7 and 8, convert from spherical to rectangular coordinates.

7. (a) $(5, \pi/6, \pi/4)$ (b) $(7, 0, \pi/2)$
    (c) $(1, \pi, 0)$ (d) $(2, 3\pi/2, \pi/2)$

8. (a) $(1, 2\pi/3, 3\pi/4)$ (b) $(3, 7\pi/4, 5\pi/6)$
    (c) $(8, \pi/6, \pi/4)$ (d) $(4, \pi/2, \pi/3)$

In Exercises 9 and 10, convert from cylindrical to spherical coordinates.

9. (a) $(\sqrt{3}, \pi/6, 3)$ (b) $(1, \pi/4, -1)$
    (c) $(2, 3\pi/4, 0)$ (d) $(6, 1, -2\sqrt{3})$

10. (a) $(4, 5\pi/6, 4)$ (b) $(2, 0, -2)$
    (c) $(4, \pi/2, 3)$ (d) $(6, \pi, 2)$

In Exercises 11 and 12, convert from spherical to cylindrical coordinates.

11. (a) $(5, \pi/4, 2\pi/3)$ (b) $(1, 7\pi/6, \pi)$
    (c) $(3, 0, 0)$ (d) $(4, \pi/6, \pi/2)$

12. (a) $(5, \pi/2, 0)$ (b) $(6, 0, 3\pi/4)$
    (c) $(\sqrt{2}, 3\pi/4, \pi)$ (d) $(5, 2\pi/3, 5\pi/6)$

[c] **13.** Use a CAS or a programmable calculating utility to set up the conversion formulas in Table 13.8.1, and then use the CAS or calculating utility to solve the problems in Exercises 1, 3, 5, 7, 9, and 11.

[c] **14.** Use a CAS or a programmable calculating utility to set up the conversion formulas in Table 13.8.1, and then use the

CAS or calculating utility to solve the problems in Exercises 2, 4, 6, 8, 10, and 12.

In Exercises 15–22, an equation is given in cylindrical coordinates. Express the equation in rectangular coordinates and sketch the graph.

15. $r = 3$ **16.** $\theta = \pi/4$ **17.** $z = r^2$
18. $z = r\cos\theta$ **19.** $r = 4\sin\theta$ **20.** $r = 2\sec\theta$
21. $r^2 + z^2 = 1$ **22.** $r^2\cos 2\theta = z$

In Exercises 23–30, an equation is given in spherical coordinates. Express the equation in rectangular coordinates and sketch the graph.

23. $\rho = 3$ **24.** $\theta = \pi/3$

25. $\phi = \pi/4$ **26.** $\rho = 2\sec\phi$

27. $\rho = 4\cos\phi$ **28.** $\rho\sin\phi = 1$

29. $\rho\sin\phi = 2\cos\theta$ **30.** $\rho - 2\sin\phi\cos\theta = 0$

In Exercises 31–42, an equation of a surface is given in rectangular coordinates. Find an equation of the surface in (a) cylindrical coordinates and (b) spherical coordinates.

31. $z = 3$ **32.** $y = 2$

33. $z = 3x^2 + 3y^2$ **34.** $z = \sqrt{3x^2 + 3y^2}$

35. $x^2 + y^2 = 4$ **36.** $x^2 + y^2 - 6y = 0$

37. $x^2 + y^2 + z^2 = 9$ **38.** $z^2 = x^2 - y^2$

39. $2x + 3y + 4z = 1$ **40.** $x^2 + y^2 - z^2 = 1$

41. $x^2 = 16 - z^2$ **42.** $x^2 + y^2 + z^2 = 2z$

In Exercises 43–46, describe the region in 3-space that satisfies the given inequalities.

43. $r^2 \le z \le 4$

44. $0 \le r \le 2\sin\theta,\quad 0 \le z \le 3$

45. $1 \le \rho \le 3$

46. $0 \le \phi \le \pi/6,\quad 0 \le \rho \le 2$

47. St. Petersburg (Leningrad), Russia, is located at $30°$ east longitude and $60°$ north latitude. Find its spherical and rect-

angular coordinates relative to the coordinate axes of Figure 13.8.7. Take miles as the unit of distance and assume the Earth to be a sphere of radius 4000 miles.

48. (a) Show that the curve of intersection of the surfaces $z = \sin\theta$ and $r = a$ (cylindrical coordinates) is an ellipse.
   (b) Sketch the surface $z = \sin\theta$ for $0 \le \theta \le \pi/2$.

49. The accompanying figure shows a right circular cylinder of radius 10 cm spinning at 3 revolutions per minute about the z-axis. At time $t = 0$ s, a bug at the point $(0, 10, 0)$ begins walking straight up the face of the cylinder at the rate of 0.5 cm/min.
   (a) Find the cylindrical coordinates of the bug after 2 min.
   (b) Find the rectangular coordinates of the bug after 2 min.
   (c) Find the spherical coordinates of the bug after 2 min.

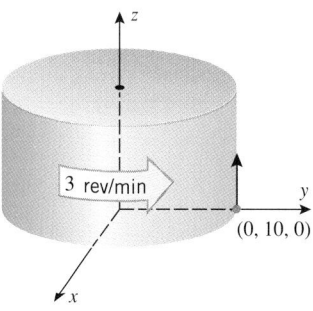

Figure Ex-49

50. Referring to Exercise 49, use a graphing utility to graph the bug's distance from the origin as a function of time.

51. A ship at sea is at point A that is $60°$ west longitude and $40°$ north latitude. The ship travels to point B that is $40°$ west longitude and $20°$ north latitude. Assuming that the Earth is a sphere with radius 6370 kilometers, find the shortest distance the ship can travel in going from A to B, given that the shortest distance between two points on a sphere is along the arc of the great circle joining the points. [*Suggestion:* Introduce an xyz-coordinate system as in Figure 13.8.7, and consider the angle between the vectors from the center of the Earth to the points A and B. If you are not familiar with the term "great circle," consult a dictionary.]

SUPPLEMENTARY EXERCISES

1. (a) What is the difference between a vector and a scalar? Give a physical example of each.
   (b) How can you determine whether or not two vectors are orthogonal?
   (c) How can you determine whether or not two vectors are parallel?
   (d) How can you determine whether or not three vectors with a common initial point lie in the same plane in 3-space?

2. (a) Sketch vectors $\mathbf{u}$ and $\mathbf{v}$ for which $\mathbf{u} + \mathbf{v}$ and $\mathbf{u} - \mathbf{v}$ are orthogonal.
   (b) How can you use vectors to determine whether four points in 3-space lie in the same plane?
   (c) If forces $\mathbf{F}_1 = \mathbf{i}$ and $\mathbf{F}_2 = \mathbf{j}$ are applied at a point in 2-space, what force would you apply at that point to cancel the combined effect of $\mathbf{F}_1$ and $\mathbf{F}_2$?
   (d) Write an equation of the sphere with center $(1, -2, 2)$ that passes through the origin.

3. (a) Draw a picture that shows the direction angles α, β, and γ of a vector.
   (b) What are the components of a unit vector in 2-space that makes an angle of $120°$ with the positive x-axis (two answers)?
   (c) How can you use vectors to determine whether a triangle with known vertices P_1, P_2, and P_3 has an obtuse angle?
   (d) True or false: The cross product of orthogonal unit vectors is a unit vector. Explain your reasoning.

4. (a) Make a table that shows all possible cross products of the vectors $\mathbf{i}$, $\mathbf{j}$, and $\mathbf{k}$.
   (b) Give a geometric interpretation of $\|\mathbf{u} \times \mathbf{v}\|$.
   (c) Give a geometric interpretation of $|\mathbf{u} \cdot (\mathbf{v} \times \mathbf{w})|$.
   (d) Write an equation of the plane that passes through the origin and is perpendicular to the line $x = t$, $y = 2t$, $z = -t$.

5. (a) List the six basic types of quadric surfaces, and describe their traces in planes parallel to the coordinate planes.
   (b) Give the coordinates of the points that result when the point (x, y, z) is reflected about the plane $y = x$, the plane $y = z$, and the plane $x = z$.
   (c) Describe the intersection of the surfaces $r = 5$ and $z = 1$ in cylindrical coordinates.
   (d) Describe the intersection of the surfaces $\phi = \pi/4$ and $\theta = 0$ in spherical coordinates.

6. In each part, find an equation of the sphere with center $(-3, 5, -4)$ and satisfying the given condition.
   (a) Tangent to the xy-plane
   (b) Tangent to the xz-plane
   (c) Tangent to the yz-plane

7. (a) Find the area of the triangle with vertices $A(1, 0, 1)$, $B(0, 2, 3)$, and $C(2, 1, 0)$.
   (b) Use the result in part (a) to find the length of the altitude from vertex C to side AB.

8. Find the largest and smallest distances between the point $P(1, 1, 1)$ and the sphere
$$x^2 + y^2 + z^2 - 2y + 6z - 6 = 0$$

9. Let $\mathbf{a} = c\mathbf{i} + \mathbf{j}$ and $\mathbf{b} = 4\mathbf{i} + 3\mathbf{j}$. Find c so that
   (a) $\mathbf{a}$ and $\mathbf{b}$ are orthogonal
   (b) the angle between $\mathbf{a}$ and $\mathbf{b}$ is $\pi/4$
   (c) the angle between $\mathbf{a}$ and $\mathbf{b}$ is $\pi/6$
   (d) $\mathbf{a}$ and $\mathbf{b}$ are parallel.

10. Given the points $P(3, 4)$, $Q(1, 1)$, and $R(5, 2)$, use vector methods to find the coordinates of the fourth vertex of the parallelogram whose adjacent sides are $\overrightarrow{PQ}$ and $\overrightarrow{QR}$.

11. Let $\mathbf{r}_0 = \langle x_0, y_0, z_0 \rangle$ and $\mathbf{r} = \langle x, y, z \rangle$. Describe the set of all points (x, y, z) for which
    (a) $\mathbf{r} \cdot \mathbf{r}_0 = 0$ (b) $(\mathbf{r} - \mathbf{r}_0) \cdot \mathbf{r}_0 = 0$.

12. What condition must the constants satisfy for the planes
$$a_1 x + b_1 y + c_1 z = d_1 \quad \text{and} \quad a_2 x + b_2 y + c_2 z = d_2$$
to be perpendicular?

13. Let $A, B, C,$ and D be four distinct points in 3-space. Explain why the line through A and B must intersect the line through C and D if $\overrightarrow{AB} \times \overrightarrow{CD} \neq \mathbf{0}$ and $\overrightarrow{AC} \cdot (\overrightarrow{AB} \times \overrightarrow{CD}) = 0$.

14. Let A, B, and C be three distinct noncollinear points in 3-space. Describe the set of all points P that satisfy the vector equation $\overrightarrow{AP} \cdot (\overrightarrow{AB} \times \overrightarrow{AC}) = 0$.

15. True or false? Explain your reasoning.
    (a) If $\mathbf{u} \cdot \mathbf{v} = 0$, then $\mathbf{u} = \mathbf{0}$ or $\mathbf{v} = \mathbf{0}$.
    (b) If $\mathbf{u} \times \mathbf{v} = \mathbf{0}$, then $\mathbf{u} = \mathbf{0}$ or $\mathbf{v} = \mathbf{0}$.
    (c) If $\mathbf{u} \cdot \mathbf{v} = 0$ and $\mathbf{u} \times \mathbf{v} = \mathbf{0}$, then $\mathbf{u} = \mathbf{0}$ or $\mathbf{v} = \mathbf{0}$.

16. In each part, use the result in Exercise 39 of Section 13.4 to prove the vector identity.
    (a) $(\mathbf{a} \times \mathbf{b}) \times (\mathbf{c} \times \mathbf{d}) = (\mathbf{a} \times \mathbf{b} \cdot \mathbf{d})\mathbf{c} - (\mathbf{a} \times \mathbf{b} \cdot \mathbf{c})\mathbf{d}$
    (b) $(\mathbf{a} \times \mathbf{b}) \times \mathbf{c} + (\mathbf{b} \times \mathbf{c}) \times \mathbf{a} + (\mathbf{c} \times \mathbf{a}) \times \mathbf{b} = \mathbf{0}$

17. Show that if $\mathbf{u}$ and $\mathbf{v}$ are unit vectors and θ is the angle between them, then $\|\mathbf{u} - \mathbf{v}\| = 2 \sin \frac{1}{2}\theta$.

18. Consider the points
$$A(1, -1, 2), B(2, -3, 0), C(-1, -2, 0), D(2, 1, -1)$$
    (a) Find the volume of the parallelepiped that has the vectors $\overrightarrow{AB}$, $\overrightarrow{AC}$, $\overrightarrow{AD}$ as adjacent edges.
    (b) Find the distance from D to the plane containing A, B, and C.

19. (a) Find parametric equations for the intersection of the planes $2x + y - z = 3$ and $x + 2y + z = 3$.
    (b) Find the acute angle between the two planes.

20. A diagonal of a box makes angles of $50°$ and $70°$ with two of its edges. Find to the nearest degree the angle that it makes with the third edge.

21. Find the vector with length 5 and direction angles $\alpha = 60°$, $\beta = 120°$, $\gamma = 135°$.

22. The accompanying figure shows a cube.
    (a) Find the angle between the vectors $\mathbf{d}$ and $\mathbf{u}$ to the nearest degree.
    (b) Make a conjecture about the angle between the vectors $\mathbf{d}$ and $\mathbf{v}$, and confirm your conjecture by computing the angle.

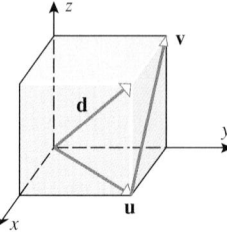

Figure Ex-22

23. In each part, identify the surface by completing the squares.
    (a) $x^2 + 4y^2 - z^2 - 6x + 8y + 4z = 0$
    (b) $x^2 + y^2 + z^2 + 6x - 4y + 12z = 0$
    (c) $x^2 + y^2 - z^2 - 2x + 4y + 5 = 0$

24. In Exercise 42 of Section 13.5 we defined the symmetric equations of a line in 3-space. Consider the lines L_1 and L_2 whose symmetric equations are

$$L_1: \frac{x-1}{2} = \frac{y+\frac{3}{2}}{1} = \frac{z+1}{2}$$

$$L_2: \frac{x-4}{-1} = \frac{y-3}{-2} = \frac{z+4}{2}$$

(a) Are L_1 and L_2 parallel? Perpendicular?
(b) Find parametric equations for L_1 and L_2.
(c) Do L_1 and L_2 intersect? If so, where?

25. In each part, express the equation in cylindrical and spherical coordinates.
(a) $x^2 + y^2 = z$ (b) $x^2 - y^2 - z^2 = 0$

26. In each part, express the equation in rectangular coordinates.
(a) $z = r^2 \cos 2\theta$ (b) $\rho^2 \sin\phi \cos\phi \cos\theta = 1$

In Exercises 27 and 28, sketch the solid in 3-space that is described in spherical coordinates by the stated inequalities.

27. (a) $0 \le \rho \le 2$ (b) $0 \le \phi \le \pi/6$
(c) $0 \le \rho \le 2$ and $0 \le \phi \le \pi/6$

28. (a) $0 \le \rho \le 5$, $0 \le \phi \le \pi/2$, and $0 \le \theta \le \pi/2$
(b) $0 \le \phi \le \pi/3$ and $0 \le \rho \le 2\sec\phi$
(c) $0 \le \rho \le 2$ and $\pi/6 \le \phi \le \pi/3$

In Exercises 29 and 30, sketch the solid in 3-space that is described in cylindrical coordinates by the stated inequalities.

29. (a) $1 \le r \le 2$ (b) $2 \le z \le 3$ (c) $\pi/6 \le \theta \le \pi/3$
(d) $1 \le r \le 2$, $2 \le z \le 3$, and $\pi/6 \le \theta \le \pi/3$

30. (a) $r^2 + z^2 \le 4$ (b) $r \le 1$ (c) $r^2 + z^2 \le 4$ and $r > 1$

31. (a) The accompanying figure shows a surface of revolution that is generated by revolving the curve $y = f(x)$ in the xy-plane about the x-axis. Show that the equation of this surface is $y^2 + z^2 = [f(x)]^2$. [Hint: Each point on the curve traces a circle as it revolves about the x-axis.]
(b) Find an equation of the surface of revolution that is generated by revolving the curve $y = e^x$ in the xy-plane about the x-axis.
(c) Show that the ellipsoid $3x^2 + 4y^2 + 4z^2 = 16$ is a surface of revolution about the x-axis by finding a curve $y = f(x)$ in the xy-plane that generates it.

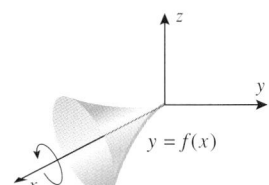

Figure Ex-31

32. In each part, use the idea in Exercise 31(a) to derive a formula for the stated surface of revolution.
(a) The surface generated by revolving the curve $x = f(y)$ in the xy-plane about the y-axis.
(b) The surface generated by revolving the curve $y = f(z)$ in the yz-plane about the z-axis.
(c) The surface generated by revolving the curve $z = f(x)$ in the xz-plane about the x-axis.

33. Sketch the surface whose equation in spherical coordinates is $\rho = a(1 - \cos\phi)$. [Hint: The surface is shaped like a familiar fruit.]

34. Assuming that force is in pounds and distance is in feet, find the work done by a constant force $\mathbf{F} = 3\mathbf{i} - 4\mathbf{j} + \mathbf{k}$ acting on a particle that moves on a straight line from $P(5, 7, 0)$ to $Q(6, 6, 6)$.

35. Assuming that force is in newtons and distance is in meters, find the work done by the resultant of the constant forces $\mathbf{F}_1 = \mathbf{i} - 3\mathbf{j} + \mathbf{k}$ and $\mathbf{F}_2 = \mathbf{i} + 2\mathbf{j} + 2\mathbf{k}$ acting on a particle that moves on a straight line from $P(-1, -2, 3)$ to $Q(0, 2, 0)$.

36. As shown in the accompanying figure, a force of 250 N is applied to a boat at an angle of $38°$ with the positive x-axis. What force $\mathbf{F}$ should be applied to the boat to produce a resultant force of 1000 N acting in the positive x-direction? State your answer by giving the magnitude of the force and its angle with the positive x-axis to the nearest degree.

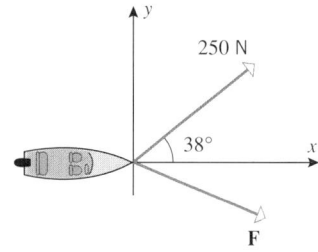

Figure Ex-36

37. Suppose that a force $\mathbf{F}$ with a magnitude of 9 lb is applied to the lever–shaft assembly shown in the accompanying figure.
(a) Express the force $\mathbf{F}$ in component form.
(b) Find the vector moment of $\mathbf{F}$ about the origin.

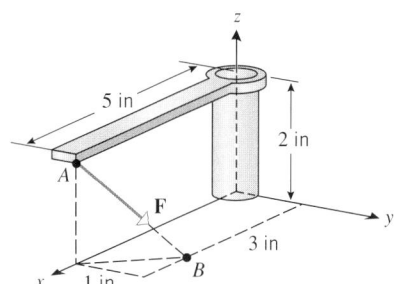

Figure Ex-37

EXPANDING THE CALCULUS HORIZON

For additional material relating to this chapter, visit the Anton Website at http://www.wiley.com/college/anton

14

VECTOR-VALUED FUNCTIONS

Karl Weierstrass

*I*n this chapter we will consider functions whose values are vectors. Such functions provide a unified way of studying parametric curves in 2-space and 3-space and are a basic tool for analyzing the motion of particles along curved paths. We will begin by developing the calculus of vector-valued functions—we will show how to differentiate and integrate such functions, and we will develop some of the basic properties of these operations. We will then apply these calculus tools to define three fundamental vectors that can be used to describe such basic characteristics of curves as curvature and twisting tendencies. Once this is done, we will develop the concept of velocity and acceleration for such motion, and we will apply these concepts to explain various physical phenomena. Finally, we will use the calculus of vector-valued functions to develop basic principles of gravitational attraction and to derive Kepler's laws of planetary motion.

14.1 INTRODUCTION TO VECTOR-VALUED FUNCTIONS

In Section 13.5 we discussed parametric equations of lines in 3-space. In this section we will discuss more general parametric curves in 3-space, and we will show how vector notation can be used to express parametric equations in 2-space and 3-space in a more compact form. This will lead us to consider a new kind of function; namely, functions that associate vectors with real numbers. Such functions have many important applications in physics and engineering.

PARAMETRIC CURVES IN 3-SPACE

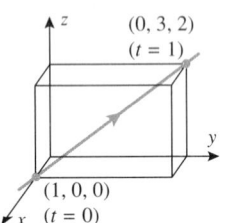

Figure 14.1.1

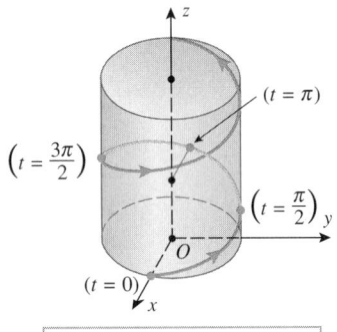

$x = a \cos t, \ y = a \sin t, \ z = ct$

Figure 14.1.2

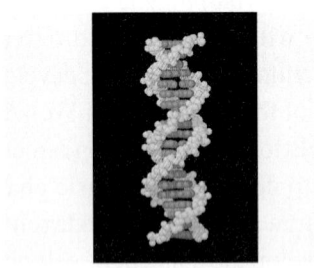

Computer representation of the twin helix DNA molecule (Deoxyribonucleic Acid). This structure contains all the inherited instructions necessary for the development of a living organism.

PARAMETRIC CURVES GENERATED WITH TECHNOLOGY

Recall from Section 1.7 that if f and g are well-behaved functions, then the pair of parametric equations

$$x = f(t), \quad y = g(t) \tag{1}$$

generates a curve in 2-space that is traced in a specific direction as the parameter t increases. We defined this direction to be the *orientation* of the curve or the *direction of increasing parameter*, and we called the curve together with its orientation the *graph* of the equations or the *parametric curve* represented by the equations. Analogously, if f, g, and h are three well-behaved functions, then the parametric equations

$$x = f(t), \quad y = g(t), \quad z = h(t) \tag{2}$$

generate a curve in 3-space that is traced in a specific direction as t increases. As in 2-space, this direction is called the **orientation** or **direction of increasing parameter**, and the curve together with its orientation is called the **graph** of the equations or the **parametric curve** represented by the equations. If no restrictions are stated explicitly or are implied by the equations, then it will be understood that t varies over the interval $(-\infty, +\infty)$.

Example 1

The parametric equations

$$x = 1 - t, \quad y = 3t, \quad z = 2t$$

represent a line in 3-space that passes through the point $(-1, 0, 0)$ and is parallel to the vector $\langle -1, 3, 2 \rangle$. Since x, y, and z increase as t increases, the line has the orientation shown in Figure 14.1.1. ◄

Example 2

Describe the parametric curve represented by the equations

$$x = a \cos t, \quad y = a \sin t, \quad z = ct$$

where a and c are positive constants.

Solution. As the parameter t increases, the value of $z = ct$ also increases, so the point (x, y, z) moves upward. However, as t increases, the point (x, y, z) also moves in a path directly over the circle

$$x = a \cos t, \quad y = a \sin t$$

in the xy-plane. The combination of these upward and circular motions produces a corkscrew-shaped curve that wraps around a right circular cylinder of radius a centered on the z-axis (Figure 14.1.2). This curve is called a **circular helix**. ◄

Except in the simplest cases, parametric curves in 3-space can be difficult to visualize and draw without the help of a graphing utility. For example, Figure 14.1.3a shows the graph of the parametric curve called a *torus knot* that was produced by a CAS. However, even this computer rendering is difficult to visualize because it is unclear whether the points of overlap are intersections or whether one portion of the curve is in front of the other. To resolve this visualization problem, some graphing utilities provide the capability of enclosing the curve within a thin tube, as in Figure 14.1.3*b*. Such graphs are called ***tube plots***.

FOR THE READER. If you have a CAS, read the documentation on graphing parametric curves in 3-space, and then use it to generate the line in Example 1 and the helix

$$x = 4\cos t, \quad y = 4\sin t, \quad z = t \quad (0 \le t \le 3\pi)$$

shown in Figure 14.1.4.

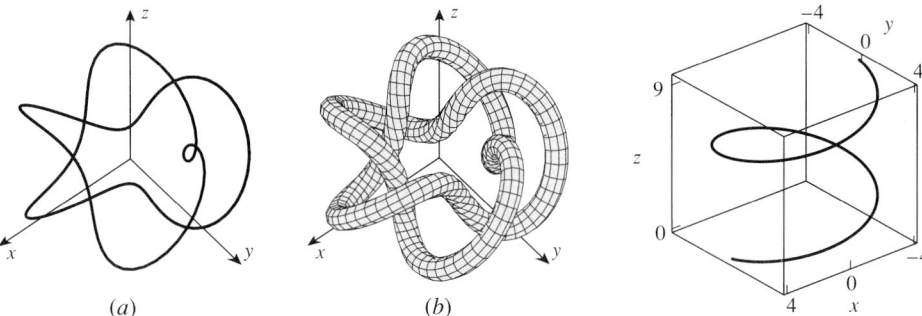

(*a*)   (*b*)

Figure 14.1.3   Figure 14.1.4

PARAMETRIC EQUATIONS FOR INTERSECTIONS OF SURFACES

Curves in 3-space often arise as intersections of surfaces. For example, Figure 14.1.5*a* shows a portion of the intersection of the cylinders $z = x^3$ and $y = x^2$. One method for finding parametric equations for the curve of intersection is to choose one of the variables as the parameter and use the two equations to express the remaining two variables in terms of that parameter. In particular, if we choose $x = t$ as the parameter and substitute this into the equations $z = x^3$ and $y = x^2$, we obtain the parametric equations

$$x = t, \quad y = t^2, \quad z = t^3 \tag{3}$$

This curve is called a ***twisted cubic***. The portion of the twisted cubic shown in Figure 14.1.5*a* corresponds to $t \ge 0$; a computer-generated graph of the twisted cubic for positive and negative values of t is shown in Figure 14.1.5*b*. Some other examples and techniques for finding intersections of surfaces are discussed in the exercises.

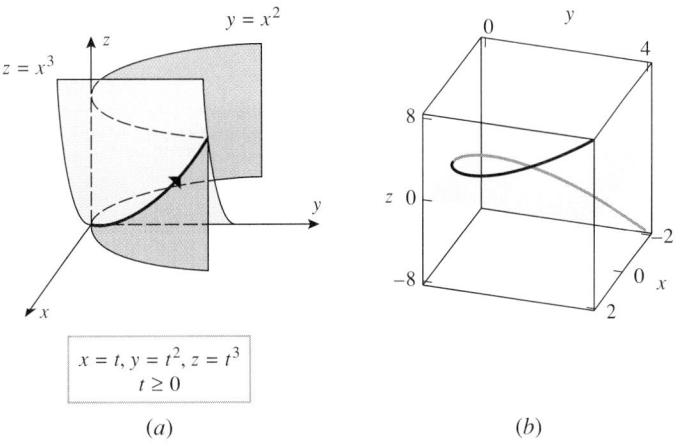

(*a*)   (*b*)

Figure 14.1.5

VECTOR-VALUED FUNCTIONS

Since two vectors are equal if and only if their corresponding components are equal, the two parametric equations in (1) can be expressed as the single vector equation

$$x\mathbf{i} + y\mathbf{j} = f(t)\mathbf{i} + g(t)\mathbf{j} \tag{4}$$

and, similarly, the three parametric equations in (2) can be expressed as

$$x\mathbf{i} + y\mathbf{j} + z\mathbf{k} = f(t)\mathbf{i} + g(t)\mathbf{j} + h(t)\mathbf{k} \tag{5}$$

It is possible to write Equations (4) and (5) even more compactly by introducing vectors $\mathbf{r}$ and $\mathbf{F}(t)$, which we define as

$$\mathbf{r} = x\mathbf{i} + y\mathbf{j} \quad \text{and} \quad \mathbf{F}(t) = f(t)\mathbf{i} + g(t)\mathbf{j} \qquad \boxed{\text{2-space}}$$
$$\mathbf{r} = x\mathbf{i} + y\mathbf{j} + z\mathbf{k} \quad \text{and} \quad \mathbf{F}(t) = f(t)\mathbf{i} + g(t)\mathbf{j} + h(t)\mathbf{k} \qquad \boxed{\text{3-space}}$$

With this notation Equations (4) and (5) both have the form

$$\mathbf{r} = \mathbf{F}(t)$$

In this equation $\mathbf{F}(t)$ is a function that associates a vector $\mathbf{r}$ with a real value of t, so we call $\mathbf{F}(t)$ a *vector-valued function of a real variable* or more simply a *vector-valued function*. As an example, the twisted cubic given parametrically in (3) can be expressed as

$$x\mathbf{i} + y\mathbf{j} + z\mathbf{k} = t\mathbf{i} + t^2\mathbf{j} + t^3\mathbf{k}$$

which has the form $\mathbf{r} = \mathbf{F}(t)$ with

$$\mathbf{r} = x\mathbf{i} + y\mathbf{j} + z\mathbf{k} \quad \text{and} \quad \mathbf{F}(t) = t\mathbf{i} + t^2\mathbf{j} + t^3\mathbf{k}$$

GRAPHS OF VECTOR-VALUED FUNCTIONS

If $\mathbf{F}(t) = f(t)\mathbf{i} + g(t)\mathbf{j} + h(t)\mathbf{k}$ is a vector-valued function in 3-space, then the real-valued functions $f(t)$, $g(t)$, and $h(t)$ are called the *component functions* or the *components* of $\mathbf{F}(t)$ (and similarly in 2-space). We define the *domain* of $\mathbf{F}(t)$ to be the set of allowable values of t. If the domain is not specified explicitly, then it will be understood that it is the set of all values of t for which *every* component is defined and yields a real value; we call this the *natural domain* of $\mathbf{F}(t)$. For example, the components of

$$\mathbf{r}(t) = \ln(t - 1)\mathbf{i} + e^t\mathbf{j} + \sqrt{t}\,\mathbf{k}$$

are

$$x(t) = \ln(t - 1), \quad y(t) = e^t, \quad z(t) = \sqrt{t}$$

and the natural domain of $\mathbf{r}(t)$ is the set of t-values such that $t > 1$.

If $\mathbf{F}(t)$ is a vector-valued function in 2-space or 3-space, then we define the *graph* of $\mathbf{F}(t)$ to be the graph of the parametric equations that correspond to the vector equation $\mathbf{r} = \mathbf{F}(t)$. For example, if

$$\mathbf{F}(t) = (1 - t)\mathbf{i} + 3t\mathbf{j} + 2t\mathbf{k} \tag{6}$$

then the equation $\mathbf{r} = \mathbf{F}(t)$ is

$$x\mathbf{i} + y\mathbf{j} + z\mathbf{k} = (1 - t)\mathbf{i} + 3t\mathbf{j} + 2t\mathbf{k}$$

and the corresponding parametric equations are

$$x = 1 - t, \quad y = 3t, \quad z = 2t$$

Thus, the graph of (6) is the line in Figure 14.1.1.

REMARK.   It is common practice to write $\mathbf{r} = \mathbf{r}(t)$ rather than $\mathbf{r} = \mathbf{F}(t)$ for a general vector-valued function and to write parametric equations $x = f(t), y = g(t), z = h(t)$ as $x = x(t), y = y(t), z = z(t)$. This dual use of letters for dependent variables and function names rarely causes confusion and has the advantage of reducing the number of letters appearing in problems.

Example 3

Describe the graph of the vector-valued function $\mathbf{r}(t) = \cos t\,\mathbf{i} + \sin t\,\mathbf{j} + t\mathbf{k}$.

Solution.   The equation $\mathbf{r} = \mathbf{r}(t)$ is

$$x\mathbf{i} + y\mathbf{j} + z\mathbf{k} = \cos t\,\mathbf{i} + \sin t\,\mathbf{j} + t\mathbf{k}$$

so the corresponding parametric equations are

$$x = \cos t, \quad y = \sin t, \quad z = t$$

Thus, as we saw in Example 2, the graph is a circular helix wrapped around a cylinder of radius 1. ◀

Up to now we have considered parametric curves to be paths traced by moving points. However, if a parametric curve is viewed as the graph of a vector-valued function, then we can also imagine the graph to be traced by the tip of a moving vector. For example, if the curve C in 3-space is the graph of

$$x\mathbf{i} + y\mathbf{j} + z\mathbf{k} = x(t)\mathbf{i} + y(t)\mathbf{j} + z(t)\mathbf{k}$$

and if we position the vector $\mathbf{r} = x\mathbf{i} + y\mathbf{j} + z\mathbf{k}$ with its initial point at the origin, then its terminal point will fall at the point (x, y, z) on the curve C (as shown in Figure 14.1.6). Thus, the terminal point of $\mathbf{r}$ will trace out the curve C as the parameter t varies. We call $\mathbf{r}$ the ***radius vector*** or the ***position vector*** for C.

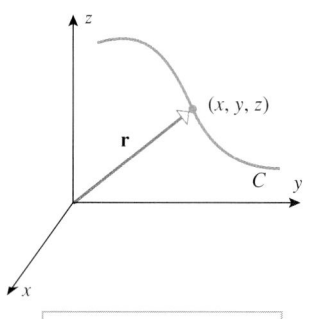

As t varies, the tip of the radius vector $\mathbf{r}$ traces out the curve C.

Figure 14.1.6

Example 4

Sketch the graph and a radius vector of

(a) $\mathbf{r}(t) = \cos t\,\mathbf{i} + \sin t\,\mathbf{j}, \quad 0 \le t \le 2\pi$

(b) $\mathbf{r}(t) = \cos t\,\mathbf{i} + \sin t\,\mathbf{j} + 2\mathbf{k}, \quad 0 \le t \le 2\pi$

Solution (a). The corresponding parametric equations are

$$x = \cos t, \quad y = \sin t \qquad (0 \le t \le 2\pi)$$

so the graph is a circle of radius 1, centered at the origin, and oriented counterclockwise. The graph and a radius vector are shown in Figure 14.1.7.

Solution (b). The corresponding parametric equations are

$$x = \cos t, \quad y = \sin t, \quad z = 2 \qquad (0 \le t \le 2\pi)$$

From the third equation, the tip of the radius vector traces a curve in the plane $z = 2$, and from the first two equations, the curve is a circle of radius 1 centered on the z-axis and traced counterclockwise looking down the z-axis. The graph and a radius vector are shown in Figure 14.1.8. ◀

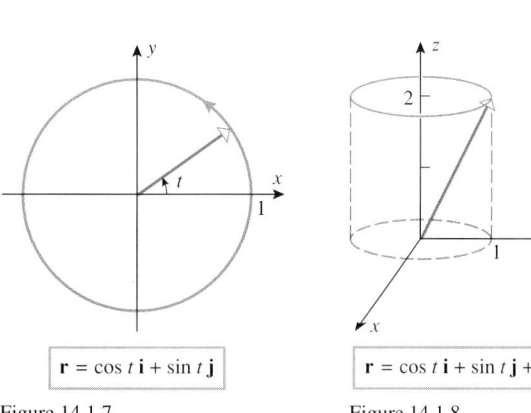

$$\mathbf{r} = \cos t\,\mathbf{i} + \sin t\,\mathbf{j}$$

Figure 14.1.7

$$\mathbf{r} = \cos t\,\mathbf{i} + \sin t\,\mathbf{j} + 2\mathbf{k}$$

Figure 14.1.8

VECTOR FORM OF A LINE SEGMENT

Recall from Formula (9) of Section 13.5 that if $\mathbf{r}_0$ is a vector in 2-space or 3-space with its initial point at the origin, then the line that passes through the terminal point of $\mathbf{r}_0$ and is parallel to the vector $\mathbf{v}$ can be expressed in vector form as

$$\mathbf{r} = \mathbf{r}_0 + t\mathbf{v}$$

In particular, if $\mathbf{r}_0$ and $\mathbf{r}_1$ are vectors in 2-space or 3-space with their initial points at the

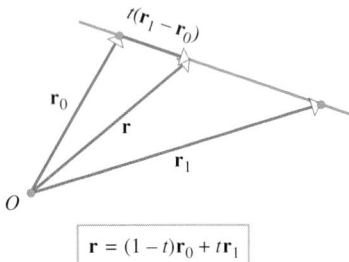

$$\mathbf{r} = (1-t)\mathbf{r}_0 + t\mathbf{r}_1$$

Figure 14.1.9

origin, then the line that passes through the terminal points of these vectors can be expressed in vector form as

$$\mathbf{r} = \mathbf{r}_0 + t(\mathbf{r}_1 - \mathbf{r}_0) \qquad \text{or} \qquad \mathbf{r} = (1-t)\mathbf{r}_0 + t\mathbf{r}_1 \qquad (7\text{–}8)$$

as indicated in Figure 14.1.9.

REMARK.   It is common to call either (7) or (8) the **two-point vector form of a line** and to say, for simplicity, that the line passes through the *points* $\mathbf{r}_0$ and $\mathbf{r}_1$ (as opposed to saying that it passes through the *terminal points* of $\mathbf{r}_0$ and $\mathbf{r}_1$).

It is understood in (7) and (8) that t varies from $-\infty$ to $+\infty$. However, if we restrict t to vary over the interval $0 \le t \le 1$, then $\mathbf{r}$ will vary from $\mathbf{r}_0$ to $\mathbf{r}_1$. Thus, for example, the equation

$$\mathbf{r} = (1-t)\mathbf{r}_0 + t\mathbf{r}_1 \qquad (0 \le t \le 1) \qquad (9)$$

represents the line segment in 2-space or 3-space that is traced from $\mathbf{r}_0$ to $\mathbf{r}_1$.

EXERCISE SET 14.1   ⌇ Graphing Calculator

In Exercises 1–4, find the domain of $\mathbf{r}(t)$ and the value of $\mathbf{r}(t_0)$.

1. $\mathbf{r}(t) = \cos t\,\mathbf{i} - 3t\mathbf{j}; \ t_0 = \pi$

2. $\mathbf{r}(t) = \langle \sqrt{3t+1}, t^2 \rangle; \ t_0 = 1$

3. $\mathbf{r}(t) = \cos \pi t\,\mathbf{i} - \ln t\,\mathbf{j} + \sqrt{t-2}\,\mathbf{k}; \ t_0 = 3$

4. $\mathbf{r}(t) = \langle 2e^{-t}, \sin^{-1} t, \ln(1-t) \rangle; \ t_0 = 0$

In Exercises 5–8, express the parametric equations as a single vector equation of the form $\mathbf{r} = x(t)\mathbf{i} + y(t)\mathbf{j}$ or $\mathbf{r} = x(t)\mathbf{i} + y(t)\mathbf{j} + z(t)\mathbf{k}$.

5. $x = 3\cos t, \ y = t + \sin t$

6. $x = t^2 + 1, \ y = e^{-2t}$

7. $x = 2t, \ y = 2\sin 3t, \ z = 5\cos 3t$

8. $x = t\sin t, \ y = \ln t, \ z = \cos^2 t$

In Exercises 9–12, find the parametric equations that correspond to the given vector equation.

9. $\mathbf{r} = 3t^2\mathbf{i} - 2\mathbf{j}$

10. $\mathbf{r} = \sin^2 t\,\mathbf{i} + (1 - \cos 2t)\mathbf{j}$

11. $\mathbf{r} = (2t-1)\mathbf{i} - 3\sqrt{t}\,\mathbf{j} + \sin 3t\,\mathbf{k}$

12. $\mathbf{r} = te^{-t}\mathbf{i} - 5t^2\mathbf{k}$

In Exercises 13–18, describe the graph of the equation.

13. $\mathbf{r} = (2 - 3t)\mathbf{i} - 4t\mathbf{j}$

14. $\mathbf{r} = 3\sin 2t\,\mathbf{i} + 3\cos 2t\,\mathbf{j}$

15. $\mathbf{r} = 2t\mathbf{i} - 3\mathbf{j} + (1 + 3t)\mathbf{k}$

16. $\mathbf{r} = 3\mathbf{i} + 2\cos t\,\mathbf{j} + 2\sin t\,\mathbf{k}$

17. $\mathbf{r} = 3\cos t\,\mathbf{i} + 2\sin t\,\mathbf{j} - \mathbf{k}$

18. $\mathbf{r} = -2\mathbf{i} + t\mathbf{j} + (t^2 - 1)\mathbf{k}$

19. (a) Find the slope of the line in 2-space that is represented by the vector equation $\mathbf{r} = (1 - 2t)\mathbf{i} - (2 - 3t)\mathbf{j}$.
(b) Find the coordinates of the point where the line

$$\mathbf{r} = (2 + t)\mathbf{i} + (1 - 2t)\mathbf{j} + 3t\mathbf{k}$$

intersects the xz-plane.

20. (a) Find the y-intercept of the line in 2-space that is represented by the vector equation $\mathbf{r} = (3 + 2t)\mathbf{i} + 5t\mathbf{j}$.
(b) Find the coordinates of the point where the line

$$\mathbf{r} = t\mathbf{i} + (1 + 2t)\mathbf{j} - 3t\mathbf{k}$$

intersects the plane $3x - y - z = 2$.

In Exercises 21 and 22, sketch the line segment represented by the vector equation.

21. (a) $\mathbf{r} = (1 - t)\mathbf{i} + t\mathbf{j}; \ 0 \le t \le 1$
(b) $\mathbf{r} = (1 - t)(\mathbf{i} + \mathbf{j}) + t(\mathbf{i} - \mathbf{j}); \ 0 \le t \le 1$

22. (a) $\mathbf{r} = (1 - t)(\mathbf{i} + \mathbf{j}) + t\mathbf{k}; \ 0 \le t \le 1$
(b) $\mathbf{r} = (1 - t)(\mathbf{i} + \mathbf{j} + \mathbf{k}) + t(\mathbf{i} + \mathbf{j}); \ 0 \le t \le 1$

In Exercises 23 and 24, write a vector equation for the line segment from P to Q.

23.

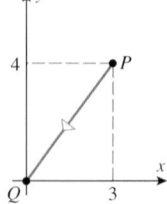

24.

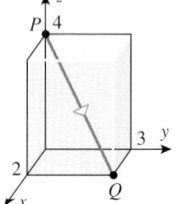

In Exercises 25–34, sketch the graph of $\mathbf{r}(t)$ and show the direction of increasing t.

25. $\mathbf{r}(t) = 2\mathbf{i} + t\mathbf{j}$

26. $\mathbf{r}(t) = \langle 3t - 4, 6t + 2 \rangle$

27. $\mathbf{r}(t) = (1 + \cos t)\mathbf{i} + (3 - \sin t)\mathbf{j}; \ 0 \leq t \leq 2\pi$

28. $\mathbf{r}(t) = \langle 2\cos t, 5\sin t \rangle; \ 0 \leq t \leq 2\pi$

29. $\mathbf{r}(t) = \cosh t\,\mathbf{i} + \sinh t\,\mathbf{j}$

30. $\mathbf{r}(t) = \sqrt{t}\,\mathbf{i} + (2t + 4)\mathbf{j}$

31. $\mathbf{r}(t) = 2\cos t\,\mathbf{i} + 2\sin t\,\mathbf{j} + t\mathbf{k}$

32. $\mathbf{r}(t) = 9\cos t\,\mathbf{i} + 4\sin t\,\mathbf{j} + t\mathbf{k}$

33. $\mathbf{r}(t) = t\mathbf{i} + t^2\mathbf{j} + 2\mathbf{k}$

34. $\mathbf{r}(t) = t\mathbf{i} + t\mathbf{j} + \sin t\,\mathbf{k}; \ 0 \leq t \leq 2\pi$

In Exercises 35 and 36, sketch the curve of intersection of the surfaces, and find parametric equations for the intersection in terms of parameter $x = t$. Check your work with a graphing utility by generating the parametric curve over the interval $-1 \leq t \leq 1$.

35. $z = x^2 + y^2, \ x - y = 0$

36. $y + x = 0, \ z = \sqrt{2 - x^2 - y^2}$

In Exercises 37 and 38, sketch the curve of intersection of the surfaces, and find a vector equation for the curve in terms of the parameter $x = t$.

37. $9x^2 + y^2 + 9z^2 = 81, \ y = x^2 \quad (z > 0)$

38. $y = x, \ x + y + z = 1$

39. Show that the graph of
$$\mathbf{r} = t\sin t\,\mathbf{i} + t\cos t\,\mathbf{j} + t^2\mathbf{k}$$
lies on the paraboloid $z = x^2 + y^2$.

40. Show that the graph of
$$\mathbf{r} = t\mathbf{i} + \frac{1+t}{t}\mathbf{j} + \frac{1-t^2}{t}\mathbf{k}, \quad t > 0$$
lies in the plane $x - y + z + 1 = 0$.

41. Show that the graph of
$$\mathbf{r} = \sin t\,\mathbf{i} + 2\cos t\,\mathbf{j} + \sqrt{3}\sin t\,\mathbf{k}$$
is a circle, and find its center and radius. [*Hint:* Show that the curve lies on both a sphere and a plane.]

42. Show that the graph of
$$\mathbf{r} = 3\cos t\,\mathbf{i} + 3\sin t\,\mathbf{j} + 3\sin t\,\mathbf{k}$$
is an ellipse, and find the lengths of the major and minor axes. [*Hint:* Show that the graph lies on both a circular cylinder and a plane and use the result in Exercise 60 of Section 12.4.]

43. For the helix $\mathbf{r} = a\cos t\,\mathbf{i} + a\sin t\,\mathbf{j} + ct\mathbf{k}$, find $c \ (c > 0)$ so that the helix will make one complete turn in a distance of 3 units measured along the z-axis.

44. How many revolutions will the circular helix
$$\mathbf{r} = a\cos t\,\mathbf{i} + a\sin t\,\mathbf{j} + 0.2t\mathbf{k}$$
make in a distance of 10 units measured along the z-axis?

45. Show that the curve $\mathbf{r} = t\cos t\,\mathbf{i} + t\sin t\,\mathbf{j} + t\mathbf{k}, \ t \geq 0$, lies on the cone $z = \sqrt{x^2 + y^2}$. Describe the curve.

46. Describe the curve $\mathbf{r} = a\cos t\,\mathbf{i} + b\sin t\,\mathbf{j} + ct\mathbf{k}$, where a, b, and c are positive constants such that $a \neq b$.

47. In each part, match the vector equation with one of the accompanying graphs, and explain your reasoning.
 (a) $\mathbf{r} = t\mathbf{i} - t\mathbf{j} + \sqrt{2 - t^2}\,\mathbf{k}$
 (b) $\mathbf{r} = \sin \pi t\,\mathbf{i} - t\mathbf{j} + t\mathbf{k}$
 (c) $\mathbf{r} = \sin t\,\mathbf{i} + \cos t\,\mathbf{j} + \sin 2t\,\mathbf{k}$
 (d) $\mathbf{r} = \frac{1}{2}t\mathbf{i} + \cos 3t\,\mathbf{j} + \sin 3t\,\mathbf{k}$

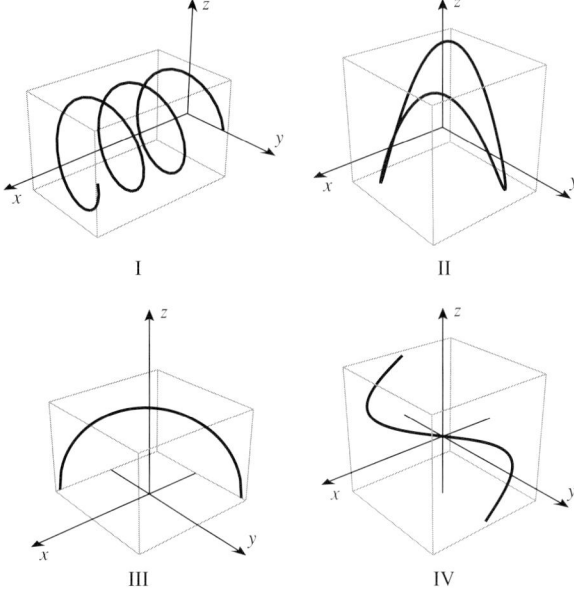

I   II

III   IV

48. Check your conclusions in Exercise 47 by generating the curves with a graphing utility. [*Note:* Your graphing utility may look at the curve from a different viewpoint. Read the documentation for your graphing utility to determine how to control the viewpoint, and see if you can generate a reasonable facsimile of the graphs shown in the figure by adjusting the viewpoint and choosing the interval of t-values appropriately.]

49. (a) Find parametric equations for the curve of intersection of the circular cylinder $x^2 + y^2 = 9$ and the parabolic cylinder $z = x^2$ in terms of a parameter t for which $x = 3\cos t$.
 (b) Use a graphing utility to generate the curve of intersection in part (a).

50. Use a graphing utility to generate the intersection of the cone $z = \sqrt{x^2 + y^2}$ and the plane $z = y + 2$. Identify the curve and explain your reasoning.

14.2 CALCULUS OF VECTOR-VALUED FUNCTIONS

In this section we will define limits, derivatives, and integrals of vector-valued functions and discuss their properties.

As shown in Table 14.2.1, limits, derivatives, and integrals of vector-valued functions can be defined by taking the limits, derivatives, and integrals of the components.

Table 14.2.1

2-SPACE $\mathbf{r}(t) = x(t)\mathbf{i} + y(t)\mathbf{j}$	3-SPACE $\mathbf{r}(t) = x(t)\mathbf{i} + y(t)\mathbf{j} + z(t)\mathbf{k}$
$\displaystyle\lim_{t\to a}\mathbf{r}(t) = (\lim_{t\to a} x(t))\mathbf{i} + (\lim_{t\to a} y(t))\mathbf{j}$	$\displaystyle\lim_{t\to a}\mathbf{r}(t) = (\lim_{t\to a} x(t))\mathbf{i} + (\lim_{t\to a} y(t))\mathbf{j} + (\lim_{t\to a} z(t))\mathbf{k}$
$\mathbf{r}'(t) = x'(t)\mathbf{i} + y'(t)\mathbf{j}$	$\mathbf{r}'(t) = x'(t)\mathbf{i} + y'(t)\mathbf{j} + z'(t)\mathbf{k}$
$\displaystyle\int \mathbf{r}(t)\,dt = \left(\int x(t)\,dt\right)\mathbf{i} + \left(\int y(t)\,dt\right)\mathbf{j}$	$\displaystyle\int \mathbf{r}(t)\,dt = \left(\int x(t)\,dt\right)\mathbf{i} + \left(\int y(t)\,dt\right)\mathbf{j} + \left(\int z(t)\,dt\right)\mathbf{k}$
$\displaystyle\int_a^b \mathbf{r}(t)\,dt = \left(\int_a^b x(t)\,dt\right)\mathbf{i} + \left(\int_a^b y(t)\,dt\right)\mathbf{j}$	$\displaystyle\int_a^b \mathbf{r}(t)\,dt = \left(\int_a^b x(t)\,dt\right)\mathbf{i} + \left(\int_a^b y(t)\,dt\right)\mathbf{j} + \left(\int_a^b z(t)\,dt\right)\mathbf{k}$

The definitions in this table assume that the coordinate system is fixed and operations on the components of $\mathbf{r}(t)$ can be performed. Thus, for a limit of $\mathbf{r}(t)$ to exist the limits of all components of $\mathbf{r}(t)$ must exist, for $\mathbf{r}(t)$ to be *differentiable* all components must be differentiable, and for $\mathbf{r}(t)$ to be *integrable* all components must be integrable. In keeping with the definition of continuity for real-valued functions, we define a vector-valued function $\mathbf{r}(t)$ to be *continuous* at a if

$$\lim_{t\to a}\mathbf{r}(t) = \mathbf{r}(a)$$

Expressed in terms of components, $\mathbf{r}(t)$ is continuous at a if and only if each component of $\mathbf{r}(t)$ is continuous at a.

All of the standard notation for derivatives continues to apply. For example, the derivative of $\mathbf{r}(t)$ can be expressed as

$$\frac{d}{dt}[\mathbf{r}(t)], \quad \frac{d\mathbf{r}}{dt}, \quad \mathbf{r}'(t), \quad \text{and} \quad \mathbf{r}'$$

Example 1

Let

$$\mathbf{r}(t) = t^2\mathbf{i} + e^t\mathbf{j} - 2\cos \pi t\,\mathbf{k}$$

Then

$$\lim_{t\to 0}\mathbf{r}(t) = (\lim_{t\to 0} t^2)\mathbf{i} + (\lim_{t\to 0} e^t)\mathbf{j} - (\lim_{t\to 0} 2\cos \pi t)\mathbf{k} = \mathbf{j} - 2\mathbf{k}$$

$$\mathbf{r}'(t) = 2t\mathbf{i} + e^t\mathbf{j} + 2\pi \sin \pi t\,\mathbf{k}$$

$$\mathbf{r}'(1) = 2\mathbf{i} + e\mathbf{j}$$

$$\int_0^1 \mathbf{r}(t)\,dt = \left.\frac{t^3}{3}\right]_0^1 \mathbf{i} + \left.e^t\right]_0^1 \mathbf{j} - \left.\frac{2}{\pi}\sin \pi t\right]_0^1 \mathbf{k} = \frac{1}{3}\mathbf{i} + (e-1)\mathbf{j} \qquad \blacktriangleleft$$

Recall that indefinite integration of a real-valued function produces a constant of integration C that is an arbitrary real number. Analogously, indefinite integration of a vector-valued function produces a constant of integration $\mathbf{C}$ that is an arbitrary vector. This is illustrated in the following example.

Example 2

$$\int (2t\mathbf{i} + 3t^2\mathbf{j})\, dt = \left(\int 2t\, dt \right)\mathbf{i} + \left(\int 3t^2\, dt \right)\mathbf{j}$$

$$= (t^2 + C_1)\mathbf{i} + (t^3 + C_2)\mathbf{j}$$

$$= (t^2\mathbf{i} + t^3\mathbf{j}) + C_1\mathbf{i} + C_2\mathbf{j} = t^2\mathbf{i} + t^3\mathbf{j} + \mathbf{C}$$

where $\mathbf{C} = C_1\mathbf{i} + C_2\mathbf{j}$ is an arbitrary vector constant of integration.   ◄

PROPERTIES OF DERIVATIVES AND INTEGRALS

Because limits, derivatives, and integrals of vector-valued functions are defined in terms of the corresponding operations on components, most of the standard theorems on limits, derivatives, and integrals of real-valued functions carry over to vector-valued functions. The following two theorems, whose proofs are left as exercises, list the standard properties of differentiation and integration of vector-valued functions.

14.2.1   THEOREM (*Rules of Differentiation*).   *Let $\mathbf{r}(t)$, $\mathbf{r}_1(t)$, and $\mathbf{r}_2(t)$ be vector-valued functions in 2-space or 3-space, and let $f(t)$ be a real-valued function, k a scalar, and $\mathbf{c}$ a constant vector (that is, a vector that does not depend on t). Then the following rules of differentiation hold:*

(a) $\dfrac{d}{dt}[\mathbf{c}] = \mathbf{0}$

(b) $\dfrac{d}{dt}[k\mathbf{r}(t)] = k\dfrac{d}{dt}[\mathbf{r}(t)]$

(c) $\dfrac{d}{dt}[\mathbf{r}_1(t) + \mathbf{r}_2(t)] = \dfrac{d}{dt}[\mathbf{r}_1(t)] + \dfrac{d}{dt}[\mathbf{r}_2(t)]$

(d) $\dfrac{d}{dt}[\mathbf{r}_1(t) - \mathbf{r}_2(t)] = \dfrac{d}{dt}[\mathbf{r}_1(t)] - \dfrac{d}{dt}[\mathbf{r}_2(t)]$

(e) $\dfrac{d}{dt}[f(t)\mathbf{r}(t)] = f(t)\dfrac{d}{dt}[\mathbf{r}(t)] + \dfrac{d}{dt}[f(t)]\mathbf{r}(t)$

14.2.2   THEOREM (*Rules of Integration*).   *Let $\mathbf{r}(t)$, $\mathbf{r}_1(t)$, and $\mathbf{r}_2(t)$ be vector-valued functions in 2-space or 3-space, and let k be a scalar. Then the following rules of integration hold:*

(a) $\displaystyle\int k\mathbf{r}(t)\, dt = k \int \mathbf{r}(t)\, dt$

(b) $\displaystyle\int [\mathbf{r}_1(t) + \mathbf{r}_2(t)]\, dt = \int \mathbf{r}_1(t)\, dt + \int \mathbf{r}_2(t)\, dt$

(c) $\displaystyle\int [\mathbf{r}_1(t) - \mathbf{r}_2(t)]\, dt = \int \mathbf{r}_1(t)\, dt - \int \mathbf{r}_2(t)\, dt$

The results in this theorem are also valid for definite integrals of vector-valued functions.

Most of the familiar integration properties have vector counterparts. For example, vector differentiation and integration are inverse operations in the sense that

$$\frac{d}{dt}\left[\int \mathbf{r}(t)\, dt \right] = \mathbf{r}(t) \qquad \text{and} \qquad \int \mathbf{r}'(t)\, dt = \mathbf{r}(t) + \mathbf{C} \tag{1–2}$$

and if $\mathbf{R}(t)$ is an antiderivative of $\mathbf{r}(t)$, that is, $\mathbf{R}'(t) = \mathbf{r}(t)$, then

$$\int_a^b \mathbf{r}(t)\, dt = \mathbf{R}(t) \Bigg]_a^b = \mathbf{R}(b) - \mathbf{R}(a) \tag{3}$$

Example 3

Evaluate the definite integral $\displaystyle\int_0^2 (2t\mathbf{i} + 3t^2\mathbf{j})\, dt$.

Solution. Integrating the components yields

$$\int_0^2 (2t\mathbf{i} + 3t^2\mathbf{j})\, dt = t^2 \Bigg]_0^2 \mathbf{i} + t^3 \Bigg]_0^2 \mathbf{j} = 4\mathbf{i} + 8\mathbf{j}$$

Alternative Solution. The function $\mathbf{R}(t) = t^2\mathbf{i} + t^3\mathbf{j}$ is an antiderivative of the integrand since $\mathbf{R}'(t) = 2t\mathbf{i} + 3t^2\mathbf{j}$. Thus, it follows from the Fundamental Theorem of Calculus that

$$\int_0^2 (2t\mathbf{i} + 3t^2\mathbf{j})\, dt = \big[\mathbf{R}(t)\big]_0^2 = \big[t^2\mathbf{i} + t^3\mathbf{j}\big]_0^2 = (4\mathbf{i} + 8\mathbf{j}) - (0\mathbf{i} + 0\mathbf{j}) = 4\mathbf{i} + 8\mathbf{j} \quad \blacktriangleleft$$

Example 4

Find $\mathbf{r}(t)$ given that $\mathbf{r}'(t) = 3\mathbf{i} + 2t\mathbf{j}$ and $\mathbf{r}(1) = 2\mathbf{i} + 5\mathbf{j}$.

Solution. Integrating $\mathbf{r}'(t)$ to obtain $\mathbf{r}(t)$ yields

$$\mathbf{r}(t) = \int \mathbf{r}'(t)\, dt = \int (3\mathbf{i} + 2t\mathbf{j})\, dt = 3t\mathbf{i} + t^2\mathbf{j} + \mathbf{C}$$

where $\mathbf{C}$ is a vector constant of integration. To find $\mathbf{C}$ we substitute $t = 1$ in this equation and use the given value of $\mathbf{r}(1)$ to obtain

$$\mathbf{r}(1) = 3\mathbf{i} + \mathbf{j} + \mathbf{C} = 2\mathbf{i} + 5\mathbf{j}$$

so that $\mathbf{C} = -\mathbf{i} + 4\mathbf{j}$. Thus,

$$\mathbf{r}(t) = 3t\mathbf{i} + t^2\mathbf{j} - \mathbf{i} + 4\mathbf{j} = (3t - 1)\mathbf{i} + (t^2 + 4)\mathbf{j} \quad \blacktriangleleft$$

GEOMETRIC INTERPRETATION OF LIMITS AND DERIVATIVES

It is desirable to have a geometric interpretation of the limit of a vector-valued function that does not require breaking the vectors into components. For this purpose, suppose that

$$\lim_{t \to a} \mathbf{r}(t) = \mathbf{L} \tag{4}$$

which means that each component of $\mathbf{r}(t)$ approaches the corresponding component of $\mathbf{L}$. However, if we position $\mathbf{r}(t)$ and $\mathbf{L}$ with their initial points at the origin, then the components of these vectors are the coordinates of their terminal points. Thus, (4) can be interpreted to mean that the terminal point of $\mathbf{r}(t)$ approaches the terminal point of $\mathbf{L}$ as t approaches a; that is, the vector $\mathbf{r}(t)$ approaches the vector $\mathbf{L}$ in both length and direction (Figure 14.2.1).

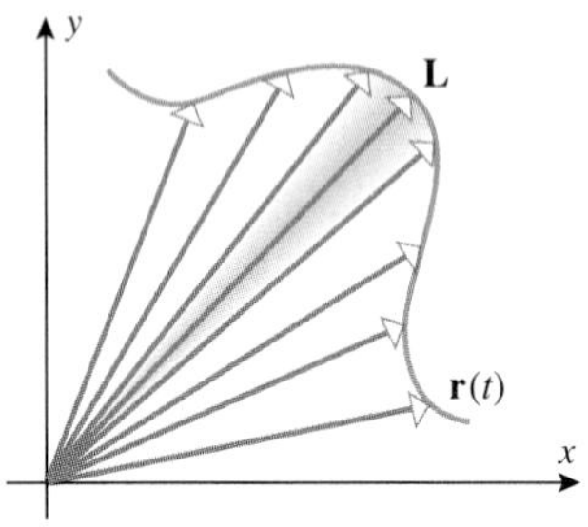

> **14.2.3** GEOMETRIC INTERPRETATION OF LIMITS. If $\mathbf{r}(t)$ is a vector-valued function in 2-space or 3-space, then
>
> $$\lim_{t \to a} \mathbf{r}(t) = \mathbf{L}$$
>
> if and only if the radius vector $\mathbf{r} = \mathbf{r}(t)$ approaches $\mathbf{L}$ in both length and direction as $t \to a$.

To obtain a geometric interpretation of the derivative of a vector-valued function $\mathbf{r}(t)$, we will first need a formula for $\mathbf{r}'(t)$ that does not involve the individual components.

$\mathbf{r}(t)$ approaches $\mathbf{L}$ in length and direction if $\lim_{t \to a} \mathbf{r}(t) = \mathbf{L}$.

Figure 14.2.1

> **14.2.4 THEOREM.** *If $\mathbf{r}(t)$ is a vector-valued function in 2-space or 3-space, then the derivative of $\mathbf{r}(t)$ can be expressed as*
>
> $$\mathbf{r}'(t) = \lim_{h \to 0} \frac{\mathbf{r}(t+h) - \mathbf{r}(t)}{h} \qquad (5)$$
>
> *provided this limit exists.*

Proof. For simplicity, we give the proof in 2-space; the proof in 3-space is identical, except for the additional component. Assume that $\mathbf{r}(t) = x(t)\mathbf{i} + y(t)\mathbf{j}$, so

$$\mathbf{r}'(t) = x'(t)\mathbf{i} + y'(t)\mathbf{j}$$

$$= \lim_{h \to 0} \frac{x(t+h) - x(t)}{h}\mathbf{i} + \lim_{h \to 0} \frac{y(t+h) - y(t)}{h}\mathbf{j}$$

$$= \lim_{h \to 0} \frac{[x(t+h)\mathbf{i} + y(t+h)\mathbf{j}] - [x(t)\mathbf{i} + y(t)\mathbf{j}]}{h}$$

$$= \lim_{h \to 0} \frac{\mathbf{r}(t+h) - \mathbf{r}(t)}{h} \qquad \blacksquare$$

With the help of this theorem we can obtain a geometric interpretation of $\mathbf{r}'(t)$. For this purpose, consider parts (*a*) and (*b*) of Figure 14.2.2. These illustrations show the graph C of $\mathbf{r}(t)$ (with its orientation) and the vectors $\mathbf{r}(t)$, $\mathbf{r}(t+h)$, and $\mathbf{r}(t+h) - \mathbf{r}(t)$ for positive h and for negative h. In both cases the vector $\mathbf{r}(t+h) - \mathbf{r}(t)$ runs along the secant line joining the terminal points of $\mathbf{r}(t+h)$ and $\mathbf{r}(t)$ but with opposite directions in the two cases. In the case where h is positive the vector $\mathbf{r}(t+h) - \mathbf{r}(t)$ points in the direction of increasing parameter, and in the case where h is negative it points in the opposite direction. However, in the case where h is negative the direction gets reversed when we multiply by $1/h$, so that in both cases the vector

$$\frac{1}{h}[\mathbf{r}(t+h) - \mathbf{r}(t)] = \frac{\mathbf{r}(t+h) - \mathbf{r}(t)}{h}$$

points in the direction of increasing parameter and runs along the secant line. As $h \to 0$, the secant line approaches the tangent line at the terminal point of $\mathbf{r}(t)$, so we conclude that the limit

$$\mathbf{r}'(t) = \lim_{h \to 0} \frac{\mathbf{r}(t+h) - \mathbf{r}(t)}{h}$$

(if it exists and is nonzero) is a vector that is tangent to the curve C at the tip of $\mathbf{r}(t)$ and points in the direction of increasing parameter (Figure 14.2.2*c*).

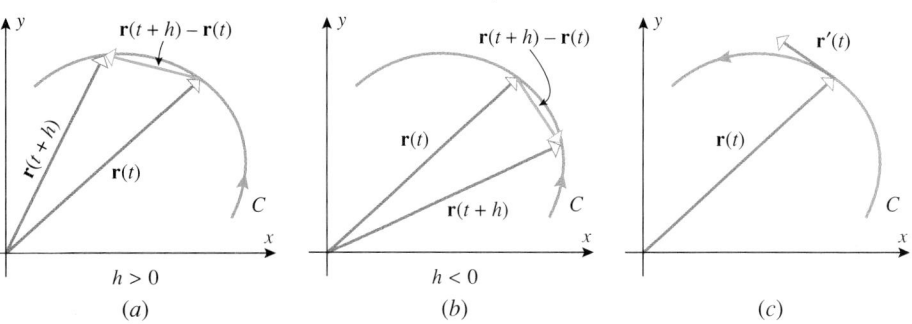

Figure 14.2.2

> **14.2.5**   GEOMETRIC INTERPRETATION OF THE DERIVATIVE.   Suppose that C is the graph of a vector-valued function $\mathbf{r}(t)$ in 2-space or 3-space and that $\mathbf{r}'(t)$ exists and is nonzero for a given value of t. If the vector $\mathbf{r}'(t)$ is positioned with its initial point at the terminal point of the radius vector $\mathbf{r}(t)$, then $\mathbf{r}'(t)$ is tangent to C and points in the direction of increasing parameter.

TANGENT LINES TO GRAPHS OF VECTOR-VALUED FUNCTIONS

Motivated by the preceding discussion, we make the following definition.

> **14.2.6**   DEFINITION.   Let P be a point on the graph of a vector-valued function $\mathbf{r}(t)$, and let $\mathbf{r}(t_0)$ be the radius vector from the origin to P (Figure 14.2.3). If $\mathbf{r}'(t_0)$ exists and $\mathbf{r}'(t_0) \neq \mathbf{0}$, then we call $\mathbf{r}'(t_0)$ the **tangent vector** to the graph of $\mathbf{r}(t)$ at $\mathbf{r}(t_0)$, and we call the line through P that is parallel to the tangent vector the **tangent line** to the graph of $\mathbf{r}(t)$ at $\mathbf{r}(t_0)$.

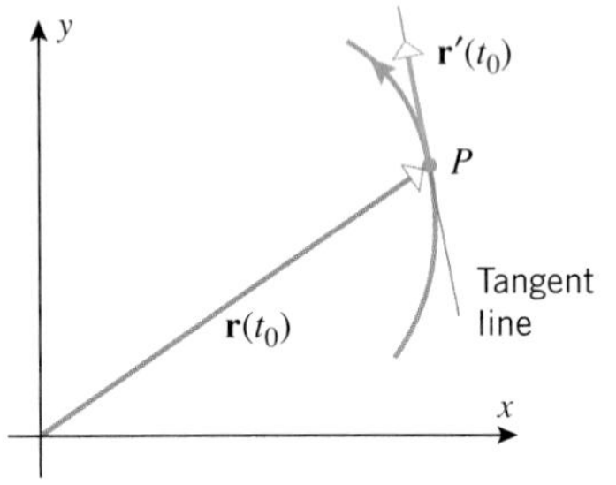

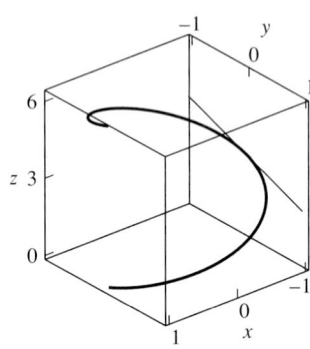

Figure 14.2.3

It follows from Formula (9) of Section 13.5 that the tangent line to the graph of $\mathbf{r}(t)$ at $\mathbf{r}(t_0)$ is given by the vector equation

$$\mathbf{r} = \mathbf{r}(t_0) + t\mathbf{r}'(t_0) \tag{6}$$

Example 5

Find parametric equations of the tangent line to the circular helix

$$x = \cos t, \quad y = \sin t, \quad z = t$$

where $t = t_0$, and use that result to find parametric equations for the tangent line at the point where $t = \pi$.

Solution.   The vector equation of the helix is

$$x\mathbf{i} + y\mathbf{j} + z\mathbf{k} = \cos t\,\mathbf{i} + \sin t\,\mathbf{j} + t\mathbf{k} \qquad \boxed{\mathbf{r} = \mathbf{r}(t)}$$

and from (6) the vector equation of the tangent line at $t = t_0$ is

$$\mathbf{r} = \mathbf{r}(t_0) + t\mathbf{r}'(t_0) \tag{7}$$

But

$$\mathbf{r} = x\mathbf{i} + y\mathbf{j} + z\mathbf{k}$$
$$\mathbf{r}(t) = \cos t\,\mathbf{i} + \sin t\,\mathbf{j} + t\mathbf{k}$$
$$\mathbf{r}'(t) = (-\sin t)\mathbf{i} + \cos t\,\mathbf{j} + \mathbf{k}$$

so that (7) yields

$$x\mathbf{i} + y\mathbf{j} + z\mathbf{k} = \cos t_0\mathbf{i} + \sin t_0\mathbf{j} + t_0\mathbf{k} + t[(-\sin t_0)\mathbf{i} + \cos t_0\mathbf{j} + \mathbf{k}]$$
$$= (\cos t_0 - t \sin t_0)\mathbf{i} + (\sin t_0 + t \cos t_0)\mathbf{j} + (t_0 + t)\mathbf{k}$$

Thus, the parametric equations of the tangent line at $t = t_0$ are

$$x = \cos t_0 - t \sin t_0, \quad y = \sin t_0 + t \cos t_0, \quad z = t_0 + t$$

In particular, the tangent line at the point where $t = \pi$ has parametric equations

$$x = -1, \quad y = -t, \quad z = \pi + t$$

The graph of the helix and this tangent line are shown in Figure 14.2.4.   ◀

Figure 14.2.4

DERIVATIVES OF DOT AND CROSS PRODUCTS

The following rules, which are derived in the exercises, provide a method for differentiating dot products in 2-space and 3-space and cross products in 3-space.

$$\frac{d}{dt}[\mathbf{r}_1(t) \cdot \mathbf{r}_2(t)] = \mathbf{r}_1(t) \cdot \frac{d\mathbf{r}_2}{dt} + \frac{d\mathbf{r}_1}{dt} \cdot \mathbf{r}_2(t) \tag{8}$$

$$\frac{d}{dt}[\mathbf{r}_1(t) \times \mathbf{r}_2(t)] = \mathbf{r}_1(t) \times \frac{d\mathbf{r}_2}{dt} + \frac{d\mathbf{r}_1}{dt} \times \mathbf{r}_2(t) \tag{9}$$

REMARK.   In (8) the order of the factors in each term on the right does not matter, but in (9) it does.

In plane geometry one learns that a tangent line to a circle is perpendicular to the radius at the point of tangency. Consequently, if a point moves along a circle in 2-space that is centered at the origin, then one would expect the radius vector and the tangent vector at any point on the circle to be orthogonal. This is the motivation for the following useful theorem, which is applicable in both 2-space and 3-space.

14.2.7   THEOREM.   *If $\mathbf{r}(t)$ is a vector-valued function in 2-space or 3-space and $\|\mathbf{r}(t)\|$ is constant for all t, then*

$$\mathbf{r}(t) \cdot \mathbf{r}'(t) = 0 \tag{10}$$

that is, $\mathbf{r}(t)$ and $\mathbf{r}'(t)$ are orthogonal vectors for all t.

Proof.   It follows from (8) with $\mathbf{r}_1(t) = \mathbf{r}_2(t) = \mathbf{r}(t)$ that

$$\frac{d}{dt}[\mathbf{r}(t) \cdot \mathbf{r}(t)] = \mathbf{r}(t) \cdot \frac{d\mathbf{r}}{dt} + \frac{d\mathbf{r}}{dt} \cdot \mathbf{r}(t)$$

or, equivalently,

$$\frac{d}{dt}[\|\mathbf{r}(t)\|^2] = 2\mathbf{r}(t) \cdot \frac{d\mathbf{r}}{dt} \tag{11}$$

But $\|\mathbf{r}(t)\|^2$ is constant, so its derivative is zero. Thus

$$2\mathbf{r}(t) \cdot \frac{d\mathbf{r}}{dt} = 0$$

from which (10) follows.   ▮

Example 6

Just as a tangent line to a circle in 2-space is perpendicular to the radius at the point of tangency, so a tangent vector to a curve on the surface of a sphere in 3-space that is centered at the origin is orthogonal to the radius vector at the point of tangency (Figure 14.2.5). To see that this is so, suppose that the graph of $\mathbf{r}(t)$ lies on the surface of a sphere of positive radius k centered at the origin. For each value of t we have $\|\mathbf{r}(t)\| = k$, so by Theorem 14.2.7

$$\mathbf{r}(t) \cdot \mathbf{r}'(t) = 0$$

and hence the radius vector $\mathbf{r}(t)$ and the tangent vector $\mathbf{r}'(t)$ are orthogonal.   ◀

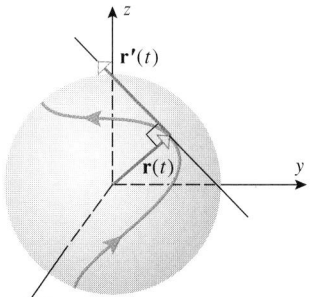

Figure 14.2.5

EXERCISE SET 14.2   ◪ Graphing Calculator

1. Sketch the circle $\mathbf{r}(t) = \cos t\,\mathbf{i} + \sin t\,\mathbf{j}$, and in each part draw the vector with its correct length.
   (a) $\mathbf{r}'(\pi/4)$ (b) $\mathbf{r}''(\pi)$ (c) $\mathbf{r}(2\pi) - \mathbf{r}(3\pi/2)$

2. Sketch the circle $\mathbf{r}(t) = \cos t\,\mathbf{i} - \sin t\,\mathbf{j}$, and in each part draw the vector with its correct length.
   (a) $\mathbf{r}'(\pi/4)$ (b) $\mathbf{r}''(\pi)$ (c) $\mathbf{r}(2\pi) - \mathbf{r}(3\pi/2)$

In Exercises 3–6, find $\mathbf{r}'(t)$.

3. $\mathbf{r}(t) = (4 + 5t)\mathbf{i} + (t - t^2)\mathbf{j}$

4. $\mathbf{r}(t) = 4\mathbf{i} - \cos t\,\mathbf{j}$

5. $\mathbf{r}(t) = \dfrac{1}{t}\mathbf{i} + \tan t\,\mathbf{j} + e^{2t}\mathbf{k}$

6. $\mathbf{r}(t) = (\tan^{-1} t)\mathbf{i} + t\cos t\,\mathbf{j} - \sqrt{t}\,\mathbf{k}$

In Exercises 7–10, find the vector $\mathbf{r}'(t_0)$; then sketch the graph of $\mathbf{r}(t)$ in 2-space and draw the tangent vector $\mathbf{r}'(t_0)$.

7. $\mathbf{r}(t) = \langle t, t^2 \rangle;\ t_0 = 2$

8. $\mathbf{r}(t) = t^3\mathbf{i} + t^2\mathbf{j};\ t_0 = 1$

9. $\mathbf{r}(t) = \sec t\,\mathbf{i} + \tan t\,\mathbf{j};\ t_0 = 0$

10. $\mathbf{r}(t) = 2\sin t\,\mathbf{i} + 3\cos t\,\mathbf{j};\ t_0 = \pi/6$

In Exercises 11 and 12, find the vector $\mathbf{r}'(t_0)$; then sketch the graph of $\mathbf{r}(t)$ in 3-space and draw the tangent vector $\mathbf{r}'(t_0)$.

11. $\mathbf{r}(t) = 2\sin t\,\mathbf{i} + \mathbf{j} + 2\cos t\,\mathbf{k};\ t_0 = \pi/2$

12. $\mathbf{r}(t) = \cos t\,\mathbf{i} + \sin t\,\mathbf{j} + t\mathbf{k};\ t_0 = \pi/4$

In Exercises 13–18, find the limit.

13. $\displaystyle\lim_{t \to 3} (t^2\mathbf{i} + 2t\mathbf{j})$

14. $\displaystyle\lim_{t \to \pi/4} \langle \cos t, \sin t \rangle$

15. $\displaystyle\lim_{t \to +\infty} \left\langle \dfrac{t^2 + 1}{3t^2 + 2}, \dfrac{1}{t} \right\rangle$

16. $\displaystyle\lim_{t \to 0^+} \left(\sqrt{t}\,\mathbf{i} + \dfrac{\sin t}{t}\mathbf{j} \right)$

17. $\displaystyle\lim_{t \to 2} (t\mathbf{i} - 3\mathbf{j} + t^2\mathbf{k})$

18. $\displaystyle\lim_{t \to 1} \left\langle \dfrac{3}{t^2}, \dfrac{\ln t}{t^2 - 1}, \sin 2t \right\rangle$

In Exercises 19 and 20, determine whether $\mathbf{r}(t)$ is continuous at $t = 0$. Explain your reasoning.

19. (a) $\mathbf{r}(t) = 3\sin t\,\mathbf{i} - 2t\mathbf{j}$

(b) $\mathbf{r}(t) = t^2\mathbf{i} + \dfrac{1}{t}\mathbf{j} + t\mathbf{k}$

20. (a) $\mathbf{r}(t) = e^t\mathbf{i} + \mathbf{j} + \csc t\,\mathbf{k}$

(b) $\mathbf{r}(t) = 5\mathbf{i} - \sqrt{3t + 1}\,\mathbf{j} + e^{2t}\mathbf{k}$

21. Let $\mathbf{r}(t) = \cos t\,\mathbf{i} + \sin t\,\mathbf{j} + \mathbf{k}$. Find

(a) $\displaystyle\lim_{t \to 0} (\mathbf{r}(t) - \mathbf{r}'(t))$

(b) $\displaystyle\lim_{t \to 0} (\mathbf{r}(t) \times \mathbf{r}'(t))$

(c) $\displaystyle\lim_{t \to 0} (\mathbf{r}(t) \cdot \mathbf{r}'(t))$.

22. Let $\mathbf{r}(t) = t\mathbf{i} + t^2\mathbf{j} + t^3\mathbf{k}$. Find

$$\lim_{t \to 1} \mathbf{r}(t) \cdot (\mathbf{r}'(t) \times \mathbf{r}''(t))$$

In Exercises 23–26, find parametric equations of the line tangent to the graph of $\mathbf{r}(t)$ at the point where $t = t_0$.

23. $\mathbf{r}(t) = t^2\mathbf{i} + (2 - \ln t)\mathbf{j};\ t_0 = 1$

24. $\mathbf{r}(t) = e^{2t}\mathbf{i} - 2\cos 3t\,\mathbf{j};\ t_0 = 0$

25. $\mathbf{r}(t) = 2\cos \pi t\,\mathbf{i} + 2\sin \pi t\,\mathbf{j} + 3t\mathbf{k};\ t_0 = \frac{1}{3}$

26. $\mathbf{r}(t) = \ln t\,\mathbf{i} + e^{-t}\mathbf{j} + t^3\mathbf{k};\ t_0 = 2$

In Exercises 27–30, find a vector equation of the line tangent to the graph of $\mathbf{r}(t)$ at the point P_0 on the curve.

27. $\mathbf{r}(t) = (2t - 1)\mathbf{i} + \sqrt{3t + 4}\,\mathbf{j};\ P_0(-1, 2)$

28. $\mathbf{r}(t) = 4\cos t\,\mathbf{i} - 3t\mathbf{j};\ P_0(2, -\pi)$

29. $\mathbf{r}(t) = t^2\mathbf{i} - \dfrac{1}{t + 1}\mathbf{j} + (4 - t^2)\mathbf{k};\ P_0(4, 1, 0)$

30. $\mathbf{r}(t) = \sin t\,\mathbf{i} + \cosh t\,\mathbf{j} + (\tan^{-1} t)\mathbf{k};\ P_0(0, 1, 0)$

In Exercises 31–36, evaluate the indefinite integral.

31. $\displaystyle\int (3\mathbf{i} + 4t\mathbf{j})\,dt$

32. $\displaystyle\int (\cos t\,\mathbf{i} + \sin t\,\mathbf{j})\,dt$

33. $\displaystyle\int (t\sin i + \mathbf{j})\,dt$

34. $\displaystyle\int \langle te^t, \ln t \rangle\,dt$

35. $\displaystyle\int \left(t^2\mathbf{i} - 2t\mathbf{j} + \dfrac{1}{t}\mathbf{k} \right)\,dt$

36. $\displaystyle\int \langle e^{-t}, e^t, 3t^2 \rangle\,dt$

In Exercises 37–42, evaluate the definite integral.

37. $\displaystyle\int_0^{\pi/3} \langle \cos 3t, -\sin 3t \rangle\,dt$

38. $\displaystyle\int_0^1 (t^2\mathbf{i} + t^3\mathbf{j})\,dt$

39. $\displaystyle\int_0^2 \|t\mathbf{i} + t^2\mathbf{j}\|\,dt$

40. $\displaystyle\int_{-3}^3 \langle (3 - t)^{3/2}, (3 + t)^{3/2}, 1 \rangle\,dt$

41. $\displaystyle\int_1^9 (t^{1/2}\mathbf{i} + t^{-1/2}\mathbf{j})\,dt$

42. $\displaystyle\int_0^1 (e^{2t}\mathbf{i} + e^{-t}\mathbf{j} + t\mathbf{k})\,dt$

In Exercises 43–46, solve the vector initial-value problem for $\mathbf{y}(t)$ by integrating and using the initial conditions to find the constants of integration.

43. $\mathbf{y}'(t) = t^2\mathbf{i} + 2t\mathbf{j},\ \mathbf{y}(0) = \mathbf{i} + \mathbf{j}$

44. $\mathbf{y}'(t) = \cos t\,\mathbf{i} + \sin t\,\mathbf{j},\ \mathbf{y}(0) = \mathbf{i} - \mathbf{j}$

45. $\mathbf{y}''(t) = \mathbf{i} + e^t\mathbf{j},\ \mathbf{y}(0) = 2\mathbf{i},\ \mathbf{y}'(0) = \mathbf{j}$

46. $\mathbf{y}''(t) = 12t^2\mathbf{i} - 2t\mathbf{j},\ \mathbf{y}(0) = 2\mathbf{i} - 4\mathbf{j},\ \mathbf{y}'(0) = \mathbf{0}$

In Exercises 47 and 48, use a graphing utility to generate the graph of $\mathbf{r}(t)$ and the graph of the tangent line at t_0 on the same screen.

47. $\mathbf{r}(t) = \sin \pi t\,\mathbf{i} + t^2\mathbf{j};\ t_0 = \frac{1}{2}$

48. $\mathbf{r}(t) = 3\sin t\,\mathbf{i} + 4\cos t\,\mathbf{j};\ t_0 = \pi/4$

In Exercises 49 and 50, let $\theta(t)$ be the angle between $\mathbf{r}(t)$ and $\mathbf{r}'(t)$. Use a graphing calculator to generate the graph of θ versus t, and make rough estimates of the t-values at which t-intercepts or relative extrema occur. What do these values tell you about the vectors $\mathbf{r}(t)$ and $\mathbf{r}'(t)$?

49. $\mathbf{r}(t) = 4\cos t\,\mathbf{i} + 3\sin t\,\mathbf{j}; \ 0 \le t \le 2\pi$

50. $\mathbf{r}(t) = t^2\mathbf{i} + t^3\mathbf{j}; \ 0 \le t \le 1$

51. (a) Find the points where the curve

$$\mathbf{r} = t\mathbf{i} + t^2\mathbf{j} - 3t\mathbf{k}$$

intersects the plane $2x - y + z = -2$.

(b) For the curve and plane in part (a), find, to the nearest degree, the acute angle that the tangent line to the curve makes with a line normal to the plane at each point of intersection.

52. Find where the tangent line to the curve

$$\mathbf{r} = e^{-2t}\mathbf{i} + \cos t\,\mathbf{j} + 3\sin t\,\mathbf{k}$$

at the point $(1, 1, 0)$ intersects the yz-plane.

In Exercises 53 and 54, show that the graphs of $\mathbf{r}_1(t)$ and $\mathbf{r}_2(t)$ intersect at the point P. Find, to the nearest degree, the acute angle between the tangent lines to the graphs of $\mathbf{r}_1(t)$ and $\mathbf{r}_2(t)$ at the point P.

53. $\mathbf{r}_1(t) = t^2\mathbf{i} + t\mathbf{j} + 3t^3\mathbf{k}$
$\mathbf{r}_2(t) = (t - 1)\mathbf{i} + \frac{1}{4}t^2\mathbf{j} + (5 - t)\mathbf{k}; \ P(1, 1, 3)$

54. $\mathbf{r}_1(t) = 2e^{-t}\mathbf{i} + \cos t\,\mathbf{j} + (t^2 + 3)\mathbf{k}$
$\mathbf{r}_2(t) = (1 - t)\mathbf{i} + t^2\mathbf{j} + (t^3 + 4)\mathbf{k}; \ P(2, 1, 3)$

In Exercises 55 and 56, calculate

$$\frac{d}{dt}[\mathbf{r}_1(t) \cdot \mathbf{r}_2(t)] \quad \text{and} \quad \frac{d}{dt}[\mathbf{r}_1(t) \times \mathbf{r}_2(t)]$$

first by differentiating the product directly and then by applying Formulas (8) and (9).

55. $\mathbf{r}_1(t) = 2t\mathbf{i} + 3t^2\mathbf{j} + t^3\mathbf{k}, \ \mathbf{r}_2(t) = t^4\mathbf{k}$

56. $\mathbf{r}_1(t) = \cos t\,\mathbf{i} + \sin t\,\mathbf{j} + t\mathbf{k}, \ \mathbf{r}_2(t) = \mathbf{i} + t\mathbf{k}$

57. Use Formula (9) to derive the differentiation formula

$$\frac{d}{dt}[\mathbf{r}(t) \times \mathbf{r}'(t)] = \mathbf{r}(t) \times \mathbf{r}''(t)$$

58. Let $\mathbf{u} = \mathbf{u}(t), \ \mathbf{v} = \mathbf{v}(t)$, and $\mathbf{w} = \mathbf{w}(t)$ be differentiable vector-valued functions. Use Formulas (8) and (9) to show that

$$\frac{d}{dt}[\mathbf{u} \cdot (\mathbf{v} \times \mathbf{w})]$$

$$= \frac{d\mathbf{u}}{dt} \cdot [\mathbf{v} \times \mathbf{w}] + \mathbf{u} \cdot \left[\frac{d\mathbf{v}}{dt} \times \mathbf{w}\right] + \mathbf{u} \cdot \left[\mathbf{v} \times \frac{d\mathbf{w}}{dt}\right]$$

59. Let $u_1, u_2, u_3, v_1, v_2, v_3, w_1, w_2$, and w_3 be differentiable functions of t. Use Exercise 58 to show that

$$\frac{d}{dt}\begin{vmatrix} u_1 & u_2 & u_3 \\ v_1 & v_2 & v_3 \\ w_1 & w_2 & w_3 \end{vmatrix}$$

$$= \begin{vmatrix} u'_1 & u'_2 & u'_3 \\ v_1 & v_2 & v_3 \\ w_1 & w_2 & w_3 \end{vmatrix} + \begin{vmatrix} u_1 & u_2 & u_3 \\ v'_1 & v'_2 & v'_3 \\ w_1 & w_2 & w_3 \end{vmatrix} + \begin{vmatrix} u_1 & u_2 & u_3 \\ v_1 & v_2 & v_3 \\ w'_1 & w'_2 & w'_3 \end{vmatrix}$$

60. Prove Theorem 14.2.1 for 2-space.

61. Derive Formulas (8) and (9) for 3-space.

62. Prove Theorem 14.2.2 for 2-space.

14.3 CHANGE OF PARAMETER; ARC LENGTH

We observed in earlier sections that a curve in 2-space or 3-space can be represented parametrically in more than one way. For example, in Section 1.7 we gave two parametric representations of a circle—one in which the circle was traced clockwise and the other in which it was traced counterclockwise. Sometimes it will be desirable to change the parameter for a parametric curve to a different parameter that is better suited for the problem at hand. In this section we will investigate issues associated with changes of parameter, and we will show that arc length plays a special role in parametric representations of curves.

SMOOTH PARAMETRIZATIONS

Graphs of vector-valued functions range from continuous and smooth to discontinuous and wildly erratic. In this text we will not be concerned with graphs of the latter type, so we will need to impose restrictions to eliminate the unwanted behavior. We will say that $\mathbf{r}(t)$ is *smoothly parametrized* or that $\mathbf{r}(t)$ is a *smooth function* of t if $\mathbf{r}'(t)$ is continuous and $\mathbf{r}'(t) \ne \mathbf{0}$ for any allowable value of t. Algebraically, smoothness implies that the components of $\mathbf{r}(t)$ have continuous derivatives that are not all zero for the same value of t, and geometrically, it implies that the tangent vector $\mathbf{r}'(t)$ varies continuously along the curve. For this reason a smoothly parametrized function is said to have a ***continuously turning tangent vector***.

Example 1

Determine whether the following vector-valued functions have continuously turning tangent vectors.

(a) $\mathbf{r}(t) = a\cos t\,\mathbf{i} + a\sin t\,\mathbf{j} + ct\,\mathbf{k}$ $(a > 0, c > 0)$

(b) $\mathbf{r}(t) = t^2\mathbf{i} + t^3\mathbf{j}$

Solution (*a*). We have

$$\mathbf{r}'(t) = -a\sin t\,\mathbf{i} + a\cos t\,\mathbf{j} + c\,\mathbf{k}$$

The components are continuous functions, and there is no value of t for which all three of them are zero (verify), so $\mathbf{r}(t)$ has a continuously turning tangent vector. The graph of $\mathbf{r}(t)$ is the circular helix in Figure 14.1.2.

Solution (*b*). We have

$$\mathbf{r}'(t) = 2t\mathbf{i} + 3t^2\mathbf{j}$$

Although the components are continuous functions, they are both equal to zero if $t = 0$, so $\mathbf{r}(t)$ does not have a continuously turning tangent vector. The graph of $\mathbf{r}(t)$, which is shown in Figure 14.3.1, is a semicubical parabola traced in the upward direction (see Example 3 of Section 12.2). Observe that for values of t slightly less than zero the angle between $\mathbf{r}'(t)$ and $\mathbf{i}$ is near π, and for values of t slightly larger than zero the angle is near 0; hence there is a sudden reversal in the direction of the tangent vector as t increases through $t = 0$. ◀

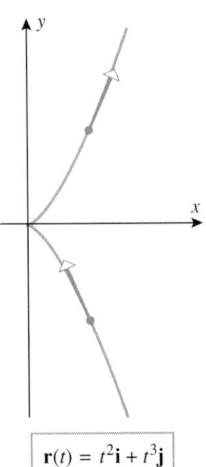

$\boxed{\mathbf{r}(t) = t^2\mathbf{i} + t^3\mathbf{j}}$

Figure 14.3.1

....
ARC LENGTH FROM THE VECTOR VIEWPOINT

.

Recall from Theorem 8.4.3 that the arc length L of a parametric curve

$$x = x(t), \quad y = y(t) \qquad (a \le t \le b) \tag{1}$$

is given by the formula

$$L = \int_a^b \sqrt{\left(\frac{dx}{dt}\right)^2 + \left(\frac{dy}{dt}\right)^2}\, dt \tag{2}$$

Analogously, the arc length L of a parametric curve

$$x = x(t), \quad y = y(t), \quad z = z(t) \qquad (a \le t \le b) \tag{3}$$

in 3-space is given by the formula

$$L = \int_a^b \sqrt{\left(\frac{dx}{dt}\right)^2 + \left(\frac{dy}{dt}\right)^2 + \left(\frac{dz}{dt}\right)^2}\, dt \tag{4}$$

Formulas (2) and (4) have vector forms that we can obtain by letting

$$\underbrace{\mathbf{r}(t) = x(t)\mathbf{i} + y(t)\mathbf{j}}_{\text{2-space}} \quad \text{or} \quad \underbrace{\mathbf{r}(t) = x(t)\mathbf{i} + y(t)\mathbf{j} + z(t)\mathbf{k}}_{\text{3-space}}$$

It follows that

$$\underbrace{\frac{d\mathbf{r}}{dt} = \frac{dx}{dt}\mathbf{i} + \frac{dy}{dt}\mathbf{j}}_{\text{2-space}} \quad \text{or} \quad \underbrace{\frac{d\mathbf{r}}{dt} = \frac{dx}{dt}\mathbf{i} + \frac{dy}{dt}\mathbf{j} + \frac{dz}{dt}\mathbf{k}}_{\text{3-space}}$$

and hence

$$\left\| \frac{d\mathbf{r}}{dt} \right\| = \sqrt{\left(\frac{dx}{dt}\right)^2 + \left(\frac{dy}{dt}\right)^2} \quad \text{or} \quad \left\| \frac{d\mathbf{r}}{dt} \right\| = \sqrt{\left(\frac{dx}{dt}\right)^2 + \left(\frac{dy}{dt}\right)^2 + \left(\frac{dz}{dt}\right)^2}$$

2-space   3-space

Substituting these expressions in (2) and (4) leads us to the following theorem.

> **14.3.1 THEOREM.** *If C is the graph in 2-space or 3-space of a smooth vector-valued function* $\mathbf{r}(t)$, *then its arc length L from* $t = a$ *to* $t = b$ *is*
>
> $$L = \int_a^b \left\| \frac{d\mathbf{r}}{dt} \right\| dt \qquad (5)$$

Example 2

Find the arc length of that portion of the circular helix

$$x = \cos t, \quad y = \sin t, \quad z = t$$

from $t = 0$ to $t = \pi$.

Solution. From (4) the arc length of the helix is

$$L = \int_0^\pi \sqrt{\left(\frac{dx}{dt}\right)^2 + \left(\frac{dy}{dt}\right)^2 + \left(\frac{dz}{dt}\right)^2} \, dt$$

$$= \int_0^\pi \sqrt{(-\sin t)^2 + (\cos t)^2 + 1} \, dt = \int_0^\pi \sqrt{2} \, dt = \sqrt{2}\,\pi \qquad \blacktriangleleft$$

ARC LENGTH AS A PARAMETER

For many purposes the best parameter to use for representing a curve in 2-space or 3-space parametrically is the length of arc measured along the curve from some fixed reference point. This can be done as follows:

Step 1. Select an arbitrary point on the curve C to serve as a *reference point*.

Step 2. Starting from the reference point, choose one direction along the curve to be the *positive direction* and the other to be the *negative direction*.

Step 3. If P is a point on the curve, let s be the "signed" arc length along C from the reference point to P, where s is positive if P is in the positive direction from the reference point, and s is negative if P is in the negative direction from the reference point. Figure 14.3.2 illustrates this idea.

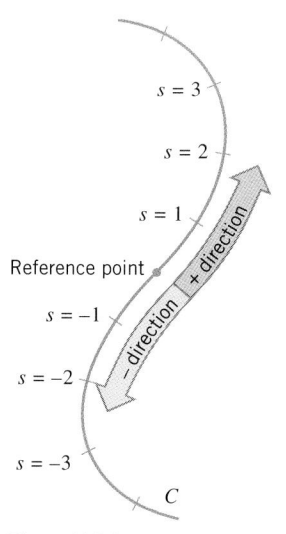

Figure 14.3.2

By this procedure, a unique point P on the curve is determined when a value for s is given. For example, $s = 2$ determines the point that is 2 units along the curve in the positive direction from the reference point, and $s = -\frac{3}{2}$ determines the point that is $\frac{3}{2}$ units along the curve in the negative direction from the reference point.

Let us now treat s as a variable. As the value of s changes, the corresponding point P moves along C and the coordinates of P become functions of s. Thus, in 2-space the coordinates of P are $(x(s), y(s))$, and in 3-space they are $(x(s), y(s), z(s))$. Therefore, in 2-space or 3-space the curve C is given by the parametric equations

$$x = x(s), \quad y = y(s) \quad \text{or} \quad x = x(s), \quad y = y(s), \quad z = z(s)$$

A parametric representation of a curve with arc length as the parameter is called an *arc length parametrization* of the curve. Note that a given curve will generally have infinitely many different arc length parametrizations, since the reference point and orientation can be chosen arbitrarily.

Example 3

Find the arc length parametrization of the circle $x^2 + y^2 = a^2$ with counterclockwise orientation and $(a, 0)$ as the reference point.

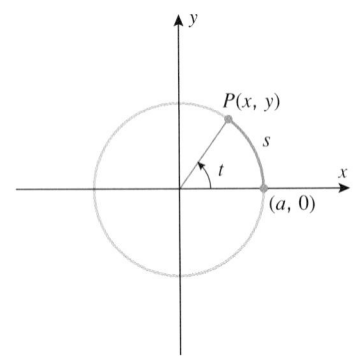

Figure 14.3.3

Solution. The circle with counterclockwise orientation can be represented by the parametric equations

$$x = a \cos t, \quad y = a \sin t \qquad (0 \le t \le 2\pi) \tag{6}$$

in which t can be interpreted as the angle in radian measure from the positive x-axis to the radius from the origin to the point $P(x, y)$ (Figure 14.3.3). If we take the positive direction for measuring the arc length to be counterclockwise, and we take $(a, 0)$ to be the reference point, then s and t are related by

$$s = at \quad \text{or} \quad t = s/a$$

Making this change of variable in (6) and noting that s increases from 0 to $2\pi a$ as t increases from 0 to 2π yields the following arc length parametrization of the circle:

$$x = a \cos(s/a), \quad y = a \sin(s/a) \qquad (0 \le s \le 2\pi a) \qquad \blacktriangleleft$$

CHANGE OF PARAMETER

In many situations the solution of a problem can be simplified by choosing the parameter in a vector-valued function or a parametric curve in the right way. The two most common parameters for curves in 2-space or 3-space are time and arc length. However, there are other useful possibilities as well. For example, in analyzing the motion of a particle in 2-space, it is often desirable to parametrize its trajectory in terms of the angle ϕ between the tangent vector and the positive x-axis (Figure 14.3.4). Thus, our next objective is to develop methods for changing the parameter in a vector-valued function or parametric curve. This will allow us to move freely between different possible parametrizations.

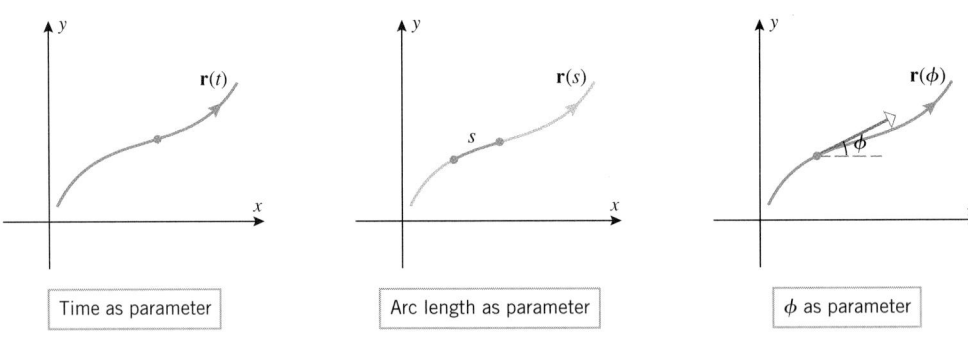

| Time as parameter | Arc length as parameter | ϕ as parameter |

Figure 14.3.4

A *change of parameter* in a vector-valued function $\mathbf{r}(t)$ is a substitution $t = g(\tau)$ that produces a new vector-valued function $\mathbf{r}(g(\tau))$ having the same graph as $\mathbf{r}(t)$, but possibly traced differently as the parameter τ increases.

Example 4

Find a change of parameter $t = g(\tau)$ for the circle

$$\mathbf{r}(t) = \cos t\, \mathbf{i} + \sin t\, \mathbf{j} \quad (0 \le t \le 2\pi)$$

such that

(a) the circle is traced counterclockwise as τ increases over the interval $[0, 1]$;

(b) the circle is traced clockwise as τ increases over the interval $[0, 1]$.

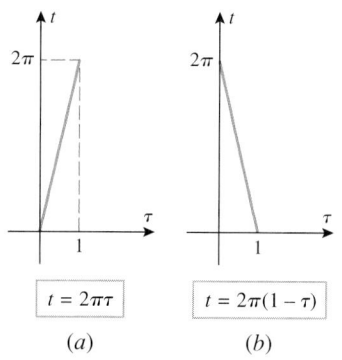

Figure 14.3.5

Solution (a). The given circle is traced counterclockwise as t increases. Thus, if we choose g to be an increasing function, then it will follow from the relationship $t = g(\tau)$ that t increases when τ increases, thereby ensuring that the circle will be traced counterclockwise as τ increases. We also want to choose g so that t increases from 0 to 2π as τ increases from 0 to 1. A simple choice of g that satisfies all of the required criteria is the linear function graphed in Figure 14.3.5*a*. The equation of this line is

$$t = g(\tau) = 2\pi\tau \tag{7}$$

which is the desired change of parameter. The resulting representation of the circle in terms of the parameter τ is

$$\mathbf{r}(g(\tau)) = \cos 2\pi\tau\,\mathbf{i} + \sin 2\pi\tau\,\mathbf{j} \quad (0 \le \tau \le 1)$$

Solution (b). To ensure that the circle is traced clockwise, we will choose g to be a decreasing function such that t decreases from 2π to 0 as τ increases from 0 to 1. A simple choice of g that achieves this is the linear function

$$t = g(\tau) = 2\pi(1 - \tau) \tag{8}$$

graphed in Figure 14.3.5*b*. The resulting representation of the circle in terms of the parameter τ is

$$\mathbf{r}(g(\tau)) = \cos(2\pi(1 - \tau))\mathbf{i} + \sin(2\pi(1 - \tau))\mathbf{j} \quad (0 \le \tau \le 1)$$

which simplifies to (verify)

$$\mathbf{r}(g(\tau)) = \cos 2\pi\tau\,\mathbf{i} - \sin 2\pi\tau\,\mathbf{j} \quad (0 \le \tau \le 1) \qquad \blacktriangleleft$$

When making a change of parameter $t = g(\tau)$ in a vector-valued function $\mathbf{r}(t)$, it will be important to ensure that the new vector-valued function $\mathbf{r}(g(\tau))$ is smooth if $\mathbf{r}(t)$ is smooth. To establish conditions under which this happens, we will need the following version of the chain rule for vector-valued functions. The proof is left as an exercise.

14.3.2 THEOREM (*Chain Rule*).   *Let $\mathbf{r}(t)$ be a vector-valued function in 2-space or 3-space that is differentiable with respect to t. If $t = g(\tau)$ is a change of parameter in which g is differentiable with respect to τ, then $\mathbf{r}(g(\tau))$ is differentiable with respect to τ and*

$$\frac{d\mathbf{r}}{d\tau} = \frac{d\mathbf{r}}{dt}\frac{dt}{d\tau} \tag{9}$$

A change of parameter $t = g(\tau)$ in which $\mathbf{r}(g(\tau))$ is smooth if $\mathbf{r}(t)$ is smooth is called a *smooth change of parameter*. It follows from (9) that $t = g(\tau)$ will be a smooth change of parameter if $dt/d\tau$ is continuous and $dt/d\tau \neq 0$ for all values of τ, since these conditions imply that $d\mathbf{r}/d\tau$ is continuous and nonzero if $d\mathbf{r}/dt$ is continuous and nonzero. Smooth changes of parameter fall into two categories—those for which $dt/d\tau > 0$ for all τ (called a *positive change of parameter*) and those for which $dt/d\tau < 0$ for all τ (called a *negative change of parameter*). A positive change of parameter preserves the orientation of a parametric curve, and a negative change of parameter reverses it.

Example 5

In Example 4 the change of parameter given by (7) is positive since $dt/d\tau = 2\pi > 0$, and the change of parameter given by (8) is negative since $dt/d\tau = -2\pi < 0$. The positive change of parameter preserved the orientation of the circle, and the negative change of parameter reversed it. $\blacktriangleleft$

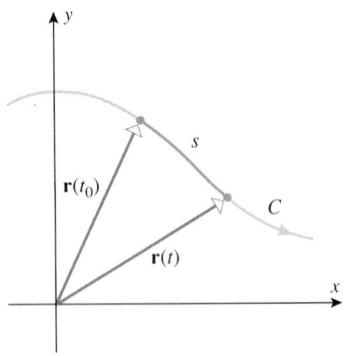

Figure 14.3.6

Next, we will consider the problem of finding an arc length parametrization of a vector-valued function that is expressed initially in terms of some other parameter t. The following theorem will provide a general method for doing this.

14.3.3   THEOREM.   *Let C be the graph of a smooth vector-valued function $\mathbf{r}(t)$ in 2-space or 3-space, and let $\mathbf{r}(t_0)$ be any point on C. Then the following formula defines a positive change of parameter from t to s, where s is an arc length parameter having $\mathbf{r}(t_0)$ as its reference point* (Figure 14.3.6):

$$s = \int_{t_0}^{t} \left\| \frac{d\mathbf{r}}{du} \right\| du \tag{10}$$

Proof.   From (5) with u as the variable of integration instead of t, the integral represents the arc length of that portion of C between $\mathbf{r}(t_0)$ and $\mathbf{r}(t)$ if $t > t_0$ and the negative of that arc length if $t < t_0$. Thus, s is the arc length parameter with $\mathbf{r}(t_0)$ as its reference point and its positive direction in the direction of increasing t.   ∎

When needed, Formula (10) can be expressed in component form as

$$s = \int_{t_0}^{t} \sqrt{\left(\frac{dx}{du}\right)^2 + \left(\frac{dy}{du}\right)^2}\, du \qquad \boxed{\text{2-space}} \tag{11}$$

$$s = \int_{t_0}^{t} \sqrt{\left(\frac{dx}{du}\right)^2 + \left(\frac{dy}{du}\right)^2 + \left(\frac{dz}{du}\right)^2}\, du \qquad \boxed{\text{3-space}} \tag{12}$$

Example 6

Find the arc length parametrization of the circular helix

$$\mathbf{r} = \cos t\, \mathbf{i} + \sin t\, \mathbf{j} + t\, \mathbf{k} \tag{13}$$

that has reference point $\mathbf{r}(0) = (1, 0, 0)$ and the same orientation as the given helix.

Solution.   Replacing t by u in $\mathbf{r}$ for integration purposes and taking $t_0 = 0$ in Formula (10), we obtain

$$\mathbf{r} = \cos u\, \mathbf{i} + \sin u\, \mathbf{j} + u\, \mathbf{k}$$

$$\frac{d\mathbf{r}}{du} = (-\sin u)\mathbf{i} + \cos u\, \mathbf{j} + \mathbf{k}$$

$$\left\| \frac{d\mathbf{r}}{du} \right\| = \sqrt{(-\sin u)^2 + \cos^2 u + 1} = \sqrt{2}$$

$$s = \int_0^t \left\| \frac{d\mathbf{r}}{du} \right\| du = \int_0^t \sqrt{2}\, du = \sqrt{2}u \Big]_0^t = \sqrt{2}t$$

Thus, $t = s/\sqrt{2}$, so (13) can be reparametrized in terms of s as

$$\mathbf{r} = \cos\left(\frac{s}{\sqrt{2}}\right)\mathbf{i} + \sin\left(\frac{s}{\sqrt{2}}\right)\mathbf{j} + \frac{s}{\sqrt{2}}\mathbf{k}$$

We are guaranteed that this reparametrization preserves the orientation of the helix since Formula (10) produces a positive change of parameter.   ◄

Example 7

A bug, starting at the reference point $(1, 0, 0)$ of the helix in Example 6, walks up the helix for a distance of 10 units. What are the bug's final coordinates?

Solution.   From Example 6, the arc length parametrization of the helix relative to the reference point $(1, 0, 0)$ is

$$\mathbf{r} = \cos\left(\frac{s}{\sqrt{2}}\right)\mathbf{i} + \sin\left(\frac{s}{\sqrt{2}}\right)\mathbf{j} + \frac{s}{\sqrt{2}}\mathbf{k}$$

or, expressed parametrically,

$$x = \cos\left(\frac{s}{\sqrt{2}}\right), \quad y = \sin\left(\frac{s}{\sqrt{2}}\right), \quad z = \frac{s}{\sqrt{2}}$$

Thus, at $s = 10$ the coordinates are

$$\left(\cos\left(\frac{10}{\sqrt{2}}\right), \sin\left(\frac{10}{\sqrt{2}}\right), \frac{10}{\sqrt{2}}\right) \approx (0.705, 0.709, 7.07)$$ ◄

Example 8

Recall from Formula (9) of Section 13.5 that the equation

$$\mathbf{r} = \mathbf{r}_0 + t\mathbf{v} \tag{14}$$

is the vector form of the line that passes through the terminal point of $\mathbf{r}_0$ and is parallel to the vector $\mathbf{v}$. Find the arc length parametrization of the line that has reference point $\mathbf{r}_0$ and the same orientation as the given line.

Solution.   Replacing t by u in $\mathbf{r}$ for integration purposes and taking $t_0 = 0$ in Formula (10), we obtain

$$\mathbf{r} = \mathbf{r}_0 + u\mathbf{v}$$

$$\frac{d\mathbf{r}}{du} = \mathbf{v} \qquad \text{Since } \mathbf{r}_0 \text{ is constant}$$

$$\left\|\frac{d\mathbf{r}}{du}\right\| = \|\mathbf{v}\|$$

$$s = \int_0^t \left\|\frac{d\mathbf{r}}{du}\right\| du = \int_0^t \|\mathbf{v}\|\, du = \|\mathbf{v}\| u \bigg]_0^t = t\|\mathbf{v}\|$$

Thus, $t = s/\|\mathbf{v}\|$, so (14) can be reparametrized in terms of s as

$$\mathbf{r} = \mathbf{r}_0 + s\left(\frac{\mathbf{v}}{\|\mathbf{v}\|}\right) \tag{15}$$

REMARK.   Comparing Formulas (14) and (15) shows that the vector equation of the line through the terminal point of $\mathbf{r}_0$ that is parallel to $\mathbf{v}$ can be reparametrized in terms of arc length with reference point $\mathbf{r}_0$ by normalizing $\mathbf{v}$ and then replacing t by s.

Example 9

Find the arc length parametrization of the line

$$x = 2t + 1, \quad y = 3t - 2$$

that has the same orientation as the given line and uses $(1, -2)$ as the reference point.

Solution.   The line passes through the point $(1, -2)$ and is parallel to the vector $\mathbf{v} = 2\mathbf{i} + 3\mathbf{j}$. To find the arc length parametrization of the line, we need only rewrite the given equations using $\mathbf{v}/\|\mathbf{v}\|$ rather than $\mathbf{v}$ to determine the direction and replace t by s. Since

$$\frac{\mathbf{v}}{\|\mathbf{v}\|} = \frac{2\mathbf{i} + 3\mathbf{j}}{\sqrt{13}} = \frac{2}{\sqrt{13}}\mathbf{i} + \frac{3}{\sqrt{13}}\mathbf{j}$$

it follows that the parametric equations for the line in terms of s are

$$x = \frac{2}{\sqrt{13}}s + 1, \quad y = \frac{3}{\sqrt{13}}s - 2 \qquad \blacktriangleleft$$

PROPERTIES OF ARC LENGTH PARAMETRIZATIONS

Because arc length parameters for a curve C are intimately related to the geometric characteristics of C, arc length parametrizations have properties that are not enjoyed by other parametrizations. For example, the following theorem shows that if a smooth curve is represented parametrically using an arc length parameter, then the tangent vectors all have length 1.

14.3.4 THEOREM.

(a) *If C is the graph of a smooth vector-valued function $\mathbf{r}(t)$ in 2-space or 3-space, where t is a general parameter, and if s is the arc length parameter for C defined by Formula (10), then for every value of t the tangent vector has length*

$$\left\|\frac{d\mathbf{r}}{dt}\right\| = \frac{ds}{dt} \tag{16}$$

(b) *If C is the graph of a smooth vector-valued function $\mathbf{r}(s)$ in 2-space or 3-space, where s is an arc length parameter, then for every value of s the tangent vector to C has length*

$$\left\|\frac{d\mathbf{r}}{ds}\right\| = 1 \tag{17}$$

(c) *If C is the graph of a smooth vector-valued function $\mathbf{r}(t)$ in 2-space or 3-space, and if*

$$\left\|\frac{d\mathbf{r}}{dt}\right\| = 1$$

for every value of t, then t is an arc length parameter that has its reference point at the point on C where $t = 0$.

Proof (a). This result follows by applying the Fundamental Theorem of Calculus (Theorem 7.6.3) to Formula (10).

Proof (b). Let $t = s$ in part (a).

Proof (c). It follows from Theorem 14.3.3 that the formula

$$s = \int_0^t \left\|\frac{d\mathbf{r}}{du}\right\| du$$

defines an arc length parameter for C with reference point $\mathbf{r}(0)$. However, $\|d\mathbf{r}/du\| = 1$ by hypothesis, so we can rewrite the formula for s as

$$s = \int_0^t du = u\Big]_0^t = t - 0 = t$$

The component forms of Formulas (16) and (17) will be of sufficient interest in later sections that we provide them here for reference:

$$\frac{ds}{dt} = \left\|\frac{d\mathbf{r}}{dt}\right\| = \sqrt{\left(\frac{dx}{dt}\right)^2 + \left(\frac{dy}{dt}\right)^2} \qquad \text{2-space} \tag{18}$$

$$\frac{ds}{dt} = \left\|\frac{d\mathbf{r}}{dt}\right\| = \sqrt{\left(\frac{dx}{dt}\right)^2 + \left(\frac{dy}{dt}\right)^2 + \left(\frac{dz}{dt}\right)^2} \qquad \text{3-space} \tag{19}$$

$$\left\|\frac{d\mathbf{r}}{ds}\right\| = \sqrt{\left(\frac{dx}{ds}\right)^2 + \left(\frac{dy}{ds}\right)^2} = 1 \qquad \text{2-space} \tag{20}$$

$$\left\|\frac{d\mathbf{r}}{ds}\right\| = \sqrt{\left(\frac{dx}{ds}\right)^2 + \left(\frac{dy}{ds}\right)^2 + \left(\frac{dz}{ds}\right)^2} = 1 \qquad \text{3-space} \tag{21}$$

REMARK.   Note that Formulas (18) and (19) do not involve t_0, and hence do not depend on where the reference point for s is chosen. This is to be expected, since changing the reference point shifts s by a constant (the arc length between the two reference points), and this constant drops out on differentiating.

EXERCISE SET 14.3

1. The accompanying figure shows the graph of the *four-cusped hypocycloid* $\mathbf{r}(t) = \cos^3 t\mathbf{i} + \sin^3 t\mathbf{j}$ $(0 \leq t \leq 2\pi)$.
 (a) Give an informal explanation of why $\mathbf{r}(t)$ is not smooth.
 (b) Confirm that $\mathbf{r}(t)$ is not smooth by examining $\mathbf{r}'(t)$.

2. The accompanying figure shows the graph of the vector-valued function $\mathbf{r}(t) = \sin t\mathbf{i} + \sin^2 t\mathbf{j}$ $(0 \leq t \leq 2\pi)$. Show that this parametric curve is not smooth, even though it has no corners. Give an informal explanation of what causes the lack of smoothness.

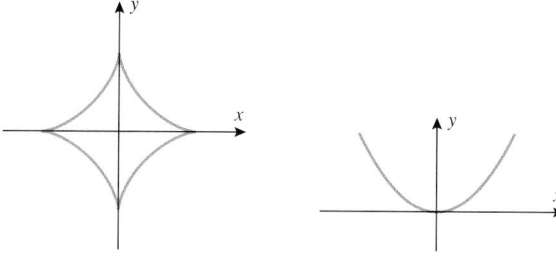

Figure Ex-1   Figure Ex-2

In Exercises 3–6, determine whether $\mathbf{r}(t)$ is a smooth function of the parameter t.

3. $\mathbf{r}(t) = t^3\mathbf{i} + (3t^2 - 2t)\mathbf{j} + t^2\mathbf{k}$

4. $\mathbf{r}(t) = \cos t^2\mathbf{i} + \sin t^2\mathbf{j} + e^{-t}\mathbf{k}$

5. $\mathbf{r}(t) = te^{-t}\mathbf{i} + (t^2 - 2t)\mathbf{j} + \cos \pi t\mathbf{k}$

6. $\mathbf{r}(t) = \sin \pi t\mathbf{i} + (2t - \ln t)\mathbf{j} + (t^2 - t)\mathbf{k}$

In Exercises 7–10, find the arc length of the parametric curve.

7. $x = \cos^3 t, y = \sin^3 t, z = 2; \ 0 \leq t \leq \pi/2$

8. $x = 3\cos t, y = 3\sin t, z = 4t; \ 0 \leq t \leq \pi$

9. $x = e^t, y = e^{-t}, z = \sqrt{2}t; \ 0 \leq t \leq 1$

10. $x = \frac{1}{2}t, y = \frac{1}{3}(1-t)^{3/2}, z = \frac{1}{3}(1+t)^{3/2}; \ -1 \leq t \leq 1$

In Exercises 11–14, find the arc length of the graph of $\mathbf{r}(t)$.

11. $\mathbf{r}(t) = t^3\mathbf{i} + t\mathbf{j} + \frac{1}{2}\sqrt{6}t^2\mathbf{k}; \ 1 \leq t \leq 3$

12. $\mathbf{r}(t) = (4 + 3t)\mathbf{i} + (2 - 2t)\mathbf{j} + (5 + t)\mathbf{k}; \ 3 \leq t \leq 4$

13. $\mathbf{r}(t) = 3\cos t\mathbf{i} + 3\sin t\mathbf{j} + t\mathbf{k}; \ 0 \leq t \leq 2\pi$

14. $\mathbf{r}(t) = t^2\mathbf{i} + (\cos t + t\sin t)\mathbf{j} + (\sin t - t\cos t)\mathbf{k}; \ 0 \leq t \leq \pi$

In Exercises 15–18, calculate $d\mathbf{r}/d\tau$ by the chain rule, and then check your result by expressing $\mathbf{r}$ in terms of τ and differentiating.

15. $\mathbf{r} = t\mathbf{i} + t^2\mathbf{j}; \ t = 4\tau + 1$

16. $\mathbf{r} = \langle 3\cos t, 3\sin t \rangle; \ t = \pi\tau$

17. $\mathbf{r} = e^t\mathbf{i} + 4e^{-t}\mathbf{j}; \ t = \tau^2$

18. $\mathbf{r} = \mathbf{i} + 3t^{3/2}\mathbf{j} + t\mathbf{k}; \ t = 1/\tau$

19. (a) Find the arc length parametrization of the line

$$x = t, \quad y = t$$

that has the same orientation as the given line and has reference point $(0, 0)$.

(b) Find the arc length parametrization of the line

$$x = t, \quad y = t, \quad z = t$$

that has the same orientation as the given line and has reference point $(0, 0, 0)$.

20. Find arc length parametrizations of the lines in Exercise 19 that have the stated reference points but are oriented opposite to the given lines.

21. (a) Find the arc length parametrization of the line

$$x = 1 + t, \quad y = 3 - 2t, \quad z = 4 + 2t$$

that has the same direction as the given line and has reference point $(1, 3, 4)$.

(b) Use the parametric equations obtained in part (a) to find the point on the line that is 25 units from the reference point in the direction of increasing parameter.

22. (a) Find the arc length parametrization of the line

$$x = -5 + 3t, \quad y = 2t, \quad z = 5 + t$$

that has the same direction as the given line and has reference point $(-5, 0, 5)$.

(b) Use the parametric equations obtained in part (a) to find the point on the line that is 10 units from the reference point in the direction of increasing parameter.

In Exercises 23–28, find an arc length parametrization of the curve that has the same orientation as the given curve and has $t = 0$ as the reference point.

23. $\mathbf{r}(t) = (3 + \cos t)\mathbf{i} + (2 + \sin t)\mathbf{j}; \ 0 \le t \le 2\pi$

24. $\mathbf{r}(t) = \cos^3 t\,\mathbf{i} + \sin^3 t\,\mathbf{j}; \ 0 \le t \le \pi/2$

25. $\mathbf{r}(t) = \frac{1}{3}t^3\mathbf{i} + \frac{1}{2}t^2\mathbf{j}; \ t \ge 0$

26. $\mathbf{r}(t) = (1 + t)^2\mathbf{i} + (1 + t)^3\mathbf{j}; \ 0 \le t \le 1$

27. $\mathbf{r}(t) = e^t\cos t\,\mathbf{i} + e^t\sin t\,\mathbf{j}; \ 0 \le t \le \pi/2$

28. $\mathbf{r}(t) = \sin e^t\mathbf{i} + \cos e^t\mathbf{j} + \sqrt{3}e^t\mathbf{k}; \ t \ge 0$

29. Show that the arc length of the circular helix $x = a\cos t$, $y = a\sin t$, $z = ct$ for $0 \le t \le t_0$ is $t_0\sqrt{a^2 + c^2}$.

30. Use the result in Exercise 29 to show the circular helix

$$\mathbf{r} = a\cos t\,\mathbf{i} + a\sin t\,\mathbf{j} + ct\mathbf{k}$$

can be expressed as

$$\mathbf{r} = \left(a\cos\frac{s}{w}\right)\mathbf{i} + \left(a\sin\frac{s}{w}\right)\mathbf{j} + \frac{cs}{w}\mathbf{k}$$

where $w = \sqrt{a^2 + c^2}$ and s is an arc length parameter with reference point at $(a, 0, 0)$.

31. Find an arc length parametrization of the cycloid

$$\begin{aligned} x &= at - a\sin t \\ y &= a - a\cos t \end{aligned} \quad (0 \le t \le 2\pi)$$

with $(0, 0)$ as the reference point.

32. Show that in cylindrical coordinates a curve given by the parametric equations $r = r(t)$, $\theta = \theta(t)$, $z = z(t)$ for $a \le t \le b$ has arc length

$$L = \int_a^b \sqrt{\left(\frac{dr}{dt}\right)^2 + r^2\left(\frac{d\theta}{dt}\right)^2 + \left(\frac{dz}{dt}\right)^2}\,dt$$

[*Hint:* Use the relationships $x = r\cos\theta$, $y = r\sin\theta$.]

33. In each part, use the formula in Exercise 32 to find the arc length of the curve.

(a) $r = e^{2t}, \theta = t, z = e^{2t}; \ 0 \le t \le \ln 2$

(b) $r = t^2, \theta = \ln t, z = \frac{1}{3}t^3; \ 1 \le t \le 2$

34. Show that in spherical coordinates a curve given by the parametric equations $\rho = \rho(t)$, $\theta = \theta(t)$, $\phi = \phi(t)$ for $a \le t \le b$ has arc length

$$L = \int_a^b \sqrt{\left(\frac{d\rho}{dt}\right)^2 + \rho^2\sin^2\phi\left(\frac{d\theta}{dt}\right)^2 + \rho^2\left(\frac{d\phi}{dt}\right)^2}\,dt$$

[*Hint:* $x = \rho\sin\phi\cos\theta$, $y = \rho\sin\phi\sin\theta$, $z = \rho\cos\phi$.]

35. In each part, use the formula in Exercise 34 to find the arc length of the curve.

(a) $\rho = e^{-t}, \theta = 2t, \phi = \pi/4; \ 0 \le t \le 2$

(b) $\rho = 2t, \theta = \ln t, \phi = \pi/6; \ 1 \le t \le 5$

36. (a) Show that $\mathbf{r}(t) = t\mathbf{i} + t^2\mathbf{j} \ (-1 \le t \le 1)$ is a smooth vector-valued function, but the change of parameter $t = \tau^3$ produces a vector-valued function that is not smooth, yet has the same graph as $\mathbf{r}(t)$.

(b) Examine how the two vector-valued functions are traced and see if you can explain what causes the problem.

37. Find a change of parameter $t = g(\tau)$ for the semicircle

$$\mathbf{r}(t) = \cos t\mathbf{i} + \sin t\mathbf{j} \quad (0 \le t \le \pi)$$

such that

(a) the semicircle is traced counterclockwise as τ varies over the interval $[0, 1]$

(b) the semicircle is traced clockwise as τ varies over the interval $[0, 1]$.

38. What change of parameter $t = g(\tau)$ would you make if you wanted to trace the graph of $\mathbf{r}(t) \ (0 \le t \le 1)$ in the opposite direction with τ varying from 0 to 1?

39. As illustrated in the accompanying figure, copper cable with a diameter of $\frac{1}{2}$ inch is to be wrapped in a circular helix around a cylinder that has a 12-inch diameter. What length of cable (measured along its center line) will make one complete turn around the cylinder in a distance of 20 inches measured along the axis of the cylinder?

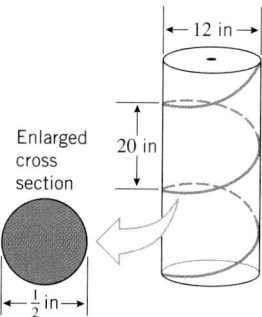

Figure Ex-39

40. Let $x = \cos t,\ y = \sin t,\ z = t^{3/2}$. Find

(a) $\|\mathbf{r}'(t)\|$ (b) $\dfrac{ds}{dt}$ (c) $\displaystyle\int_0^2 \|\mathbf{r}'(t)\|\,dt.$

41. Let $\mathbf{r}(t) = \ln t\,\mathbf{i} + 2t\,\mathbf{j} + t^2\mathbf{k}$. Find

(a) $\|\mathbf{r}'(t)\|$ (b) $\dfrac{ds}{dt}$ (c) $\displaystyle\int_1^3 \|\mathbf{r}'(t)\|\,dt.$

42. Prove: If $\mathbf{r}(t)$ is a smoothly parametrized function, then the angles between $\mathbf{r}'(t)$ and the vectors $\mathbf{i}$, $\mathbf{j}$, and $\mathbf{k}$ are continuous functions of t.

43. Prove the vector form of the chain rule for 2-space (Theorem 14.3.2) by expressing $\mathbf{r}(t)$ in terms of components.

14.4 UNIT TANGENT, NORMAL, AND BINORMAL VECTORS

In this section we will discuss some of the fundamental geometric properties of vector-valued functions. Our work here will have important applications to the study of motion along a curved path in 2-space or 3-space and to the study of the geometric properties of curves and surfaces.

UNIT TANGENT VECTORS

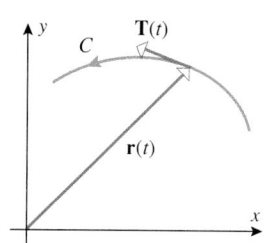

Figure 14.4.1

Recall that if C is the graph of a *smooth* vector-valued function $\mathbf{r}(t)$ in 2-space or 3-space, then the vector $\mathbf{r}'(t)$ is nonzero, tangent to C, and points in the direction of increasing parameter. Thus, by normalizing $\mathbf{r}'(t)$ we obtain a unit vector

$$\mathbf{T}(t) = \frac{\mathbf{r}'(t)}{\|\mathbf{r}'(t)\|} \tag{1}$$

that is tangent to C and points in the direction of increasing parameter. We call $\mathbf{T}(t)$ the **unit tangent vector** to C at t.

REMARK.   Unless stated otherwise, we will assume that $\mathbf{T}(t)$ is positioned with its initial point at the terminal point of $\mathbf{r}(t)$ as in Figure 14.4.1. This will ensure that $\mathbf{T}(t)$ is actually tangent to the graph of $\mathbf{r}(t)$ and not simply parallel to the tangent line.

Example 1

Find the unit tangent vector to the graph of $\mathbf{r}(t) = t^2\mathbf{i} + t^3\mathbf{j}$ at the point where $t = 2$.

Solution. Since

$$\mathbf{r}'(t) = 2t\mathbf{i} + 3t^2\mathbf{j}$$

we obtain

$$\mathbf{T}(2) = \frac{\mathbf{r}'(2)}{\|\mathbf{r}'(2)\|} = \frac{4\mathbf{i} + 12\mathbf{j}}{\sqrt{160}} = \frac{4\mathbf{i} + 12\mathbf{j}}{4\sqrt{10}} = \frac{1}{\sqrt{10}}\mathbf{i} + \frac{3}{\sqrt{10}}\mathbf{j}$$

The graph of $\mathbf{r}(t)$ and the vector $\mathbf{T}(2)$ are shown in Figure 14.4.2.   ◀

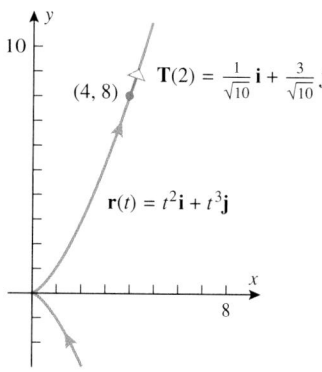

Figure 14.4.2

UNIT NORMAL VECTORS

Recall from Theorem 14.2.7 that if a vector-valued function $\mathbf{r}(t)$ has constant norm, then $\mathbf{r}(t)$ and $\mathbf{r}'(t)$ are orthogonal vectors. In particular, $\mathbf{T}(t)$ has constant norm 1, so $\mathbf{T}(t)$ and $\mathbf{T}'(t)$ are orthogonal vectors. This implies that $\mathbf{T}'(t)$ is perpendicular to the tangent line to C at t, so we say that $\mathbf{T}'(t)$ is **normal** to C at t. It follows that if $\mathbf{T}'(t) \neq \mathbf{0}$, and if we normalize $\mathbf{T}'(t)$, then we obtain a unit vector

$$\mathbf{N}(t) = \frac{\mathbf{T}'(t)}{\|\mathbf{T}'(t)\|} \tag{2}$$

that is normal to C and points in the same direction as $\mathbf{T}'(t)$. We call $\mathbf{N}(t)$ the **principal unit normal vector** to C at t or more simply the **unit normal vector**. Observe that the unit normal vector is only defined at points where $\mathbf{T}'(t) \neq \mathbf{0}$. Unless stated otherwise, we will assume that this condition is satisfied. In particular, this *excludes* straight lines.

REMARK. In 2-space there are two unit vectors that are orthogonal to $\mathbf{T}(t)$, and in 3-space there are infinitely many such vectors (Figure 14.4.3). In both cases the principal unit normal is that particular normal that points in the direction of $\mathbf{T}'(t)$. After the next example we will show that for a nonlinear parametric curve in 2-space the principal unit normal is the one that points "inward" toward the concave side of the curve.

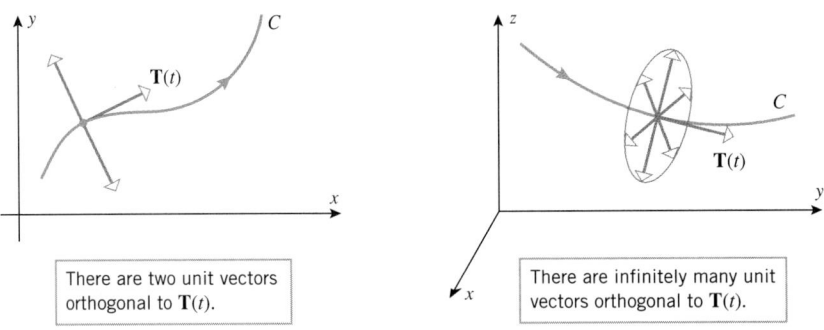

Figure 14.4.3

Example 2

Find $\mathbf{T}(t)$ and $\mathbf{N}(t)$ for the circular helix

$$x = a\cos t, \quad y = a\sin t, \quad z = ct$$

where $a > 0$.

Solution. The radius vector for the helix is

$$\mathbf{r}(t) = a\cos t\,\mathbf{i} + a\sin t\,\mathbf{j} + ct\,\mathbf{k}$$

Thus,

$$\mathbf{r}'(t) = (-a\sin t)\mathbf{i} + a\cos t\,\mathbf{j} + c\mathbf{k}$$

$$\|\mathbf{r}'(t)\| = \sqrt{(-a\sin t)^2 + (a\cos t)^2 + c^2} = \sqrt{a^2 + c^2}$$

$$\mathbf{T}(t) = \frac{\mathbf{r}'(t)}{\|\mathbf{r}'(t)\|} = -\frac{a\sin t}{\sqrt{a^2 + c^2}}\mathbf{i} + \frac{a\cos t}{\sqrt{a^2 + c^2}}\mathbf{j} + \frac{c}{\sqrt{a^2 + c^2}}\mathbf{k}$$

$$\mathbf{T}'(t) = -\frac{a\cos t}{\sqrt{a^2 + c^2}}\mathbf{i} - \frac{a\sin t}{\sqrt{a^2 + c^2}}\mathbf{j}$$

$$\|\mathbf{T}'(t)\| = \sqrt{\left(-\frac{a\cos t}{\sqrt{a^2 + c^2}}\right)^2 + \left(-\frac{a\sin t}{\sqrt{a^2 + c^2}}\right)^2} = \sqrt{\frac{a^2}{a^2 + c^2}} = \frac{a}{\sqrt{a^2 + c^2}}$$

$$\mathbf{N}(t) = \frac{\mathbf{T}'(t)}{\|\mathbf{T}'(t)\|} = (-\cos t)\mathbf{i} - (\sin t)\mathbf{j}$$

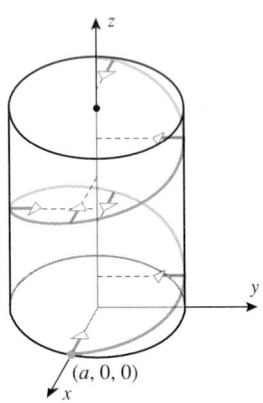

$(a, 0, 0)$

Figure 14.4.4

FOR THE READER. Because the $\mathbf{k}$ component of $\mathbf{N}(t)$ is zero, this vector lies in a horizontal plane for every value of t. Show that $\mathbf{N}(t)$ actually points directly toward the z-axis for all t (Figure 14.4.4). ◄

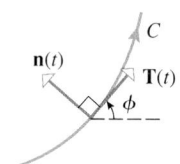

Figure 14.4.5

Our next objective is to show that for a nonlinear parametric curve C in 2-space the unit normal vector always points toward the concave side of C. For this purpose, let $\phi(t)$ be the angle from the positive x-axis to $\mathbf{T}(t)$, and let $\mathbf{n}(t)$ be the unit vector that results when $\mathbf{T}(t)$ is rotated counterclockwise through an angle of $\pi/2$ (Figure 14.4.5). Since $\mathbf{T}(t)$ and $\mathbf{n}(t)$ are unit vectors, it follows from Formula (12) of Section 13.2 that these vectors can be expressed as

$$\mathbf{T}(t) = \cos\phi(t)\mathbf{i} + \sin\phi(t)\mathbf{j} \tag{3}$$

and

$$\mathbf{n}(t) = \cos[\phi(t) + \pi/2]\mathbf{i} + \sin[\phi(t) + \pi/2]\mathbf{j} = -\sin\phi(t)\mathbf{i} + \cos\phi(t)\mathbf{j} \tag{4}$$

Observe that on intervals where $\phi(t)$ is increasing the vector $\mathbf{n}(t)$ points *toward* the concave side of C, and on intervals where $\phi(t)$ is decreasing it points *away* from the concave side (Figure 14.4.6).

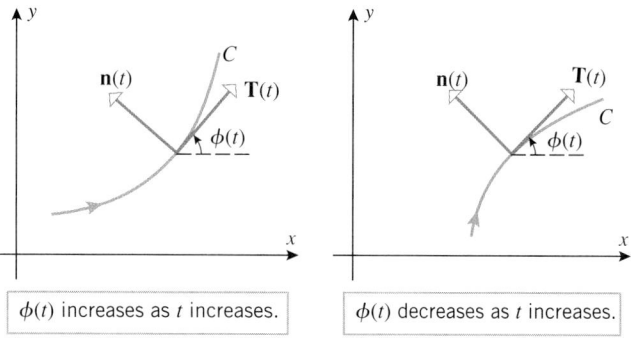

| $\phi(t)$ increases as t increases. | $\phi(t)$ decreases as t increases. |

Figure 14.4.6

Now let us differentiate $\mathbf{T}(t)$ by using Formula (3) and applying the chain rule. This yields

$$\frac{d\mathbf{T}}{dt} = \frac{d\mathbf{T}}{d\phi}\frac{d\phi}{dt} = [(-\sin\phi)\mathbf{i} + (\cos\phi)\mathbf{j}]\frac{d\phi}{dt}$$

and thus from (4)

$$\frac{d\mathbf{T}}{dt} = \mathbf{n}(t)\frac{d\phi}{dt} \tag{5}$$

But $d\phi/dt > 0$ on intervals where $\phi(t)$ is increasing and $d\phi/dt < 0$ on intervals where $\phi(t)$ is decreasing. Thus, it follows from (5) that $d\mathbf{T}/dt$ has the same direction as $\mathbf{n}(t)$ on intervals where $\phi(t)$ is increasing and the opposite direction on intervals where $\phi(t)$ is decreasing. Therefore, $\mathbf{T}'(t) = d\mathbf{T}/dt$ points "inward" toward the concave side of the curve in all cases, and hence so does $\mathbf{N}(t)$. For this reason, $\mathbf{N}(t)$ is also called the ***inward unit normal*** when applied to curves in 2-space.

COMPUTING T AND N FOR CURVES PARAMETRIZED BY ARC LENGTH

In the case where $\mathbf{r}(s)$ is parametrized by arc length, the procedures for computing the unit tangent vector $\mathbf{T}(s)$ and the unit normal vector $\mathbf{N}(s)$ are simpler than in the general case. For example, we showed in Theorem 14.3.4 that if s is an arc length parameter, then $\|\mathbf{r}'(s)\| = 1$. Thus, Formula (1) for the unit tangent vector simplifies to

$$\mathbf{T}(s) = \mathbf{r}'(s) \tag{6}$$

and consequently Formula (2) for the unit normal vector simplifies to

$$\mathbf{N}(s) = \frac{\mathbf{r}''(s)}{\|\mathbf{r}''(s)\|} \tag{7}$$

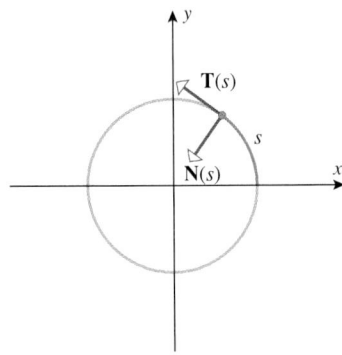

Figure 14.4.7

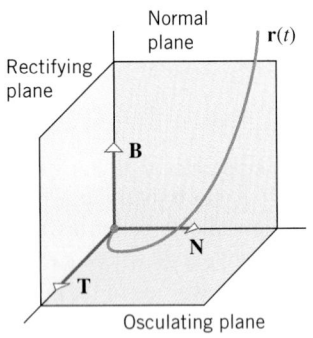

Figure 14.4.8

Example 3

The circle of radius a with counterclockwise orientation and centered at the origin can be represented by the vector-valued function

$$\mathbf{r} = a\cos t\mathbf{i} + a\sin t\mathbf{j} \quad (0 \le t \le 2\pi) \tag{8}$$

In this representation we can interpret t as the angle in radian measure from the positive x-axis to the radius vector (Figure 14.4.7). This angle subtends an arc of length $s = at$ on the circle, so we can reparametrize the circle in terms of s by substituting s/a for t in (8). This yields

$$\mathbf{r}(s) = a\cos(s/a)\mathbf{i} + a\sin(s/a)\mathbf{j} \quad (0 \le s \le 2\pi a)$$

To find $\mathbf{T}(s)$ and $\mathbf{N}(s)$ from Formulas (6) and (7), we must compute $\mathbf{r}'(s)$, $\mathbf{r}''(s)$, and $\|\mathbf{r}''(s)\|$. Doing so, we obtain

$$\mathbf{r}'(s) = -\sin(s/a)\mathbf{i} + \cos(s/a)\mathbf{j}$$

$$\mathbf{r}''(s) = -(1/a)\cos(s/a)\mathbf{i} - (1/a)\sin(s/a)\mathbf{j}$$

$$\|\mathbf{r}''(s)\| = \sqrt{(-1/a)^2\cos^2(s/a) + (-1/a)^2\sin^2(s/a)} = 1/a$$

Thus,

$$\mathbf{T}(s) = \mathbf{r}'(s) = -\sin(s/a)\mathbf{i} + \cos(s/a)\mathbf{j}$$

$$\mathbf{N}(s) = \mathbf{r}''(s)/\|\mathbf{r}''(s)\| = -\cos(s/a)\mathbf{i} - \sin(s/a)\mathbf{j}$$

so $\mathbf{N}(s)$ points toward the center of the circle for all s (Figure 14.4.8). This makes sense geometrically and is also consistent with our earlier observation that in 2-space the unit normal vector is the inward normal. ◄

....

BINORMAL VECTORS IN 3-SPACE

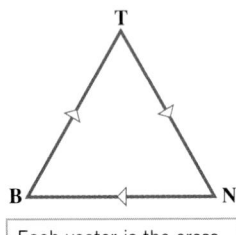

Figure 14.4.9

Each vector is the cross product of the other two taken in clockwise order.

Figure 14.4.10

If C is the graph of a vector-valued function $\mathbf{r}(t)$ in 3-space, then we define the ***binormal vector*** to C at t to be

$$\mathbf{B}(t) = \mathbf{T}(t) \times \mathbf{N}(t) \tag{9}$$

It follows from properties of the cross product that $\mathbf{B}(t)$ is orthogonal to both $\mathbf{T}(t)$ and $\mathbf{N}(t)$ and is oriented relative to $\mathbf{T}(t)$ and $\mathbf{N}(t)$ by the right-hand rule. Moreover, $\mathbf{T}(t) \times \mathbf{N}(t)$ is a unit vector since

$$\|\mathbf{T}(t) \times \mathbf{N}(t)\| = \|\mathbf{T}(t)\|\|\mathbf{N}(t)\|\sin(\pi/2) = 1$$

Thus, $\{\mathbf{T}(t), \mathbf{N}(t), \mathbf{B}(t)\}$ is a set of three mutually orthogonal unit vectors.

Just as the vectors $\mathbf{i}$, $\mathbf{j}$, and $\mathbf{k}$ determine a right-handed coordinate system in 3-space, so do the vectors $\mathbf{T}(t)$, $\mathbf{N}(t)$, and $\mathbf{B}(t)$. At each point on a smooth parametric curve C in 3-space, these vectors determine three mutually perpendicular planes that pass through the point—the **TB**-plane (called the ***rectifying plane***), the **TN**-plane (called the ***osculating plane***), and the **NB**-plane (called the ***normal plane***) (Figure 14.4.9). Moreover, one can show that a coordinate system determined by $\mathbf{T}(t)$, $\mathbf{N}(t)$, and $\mathbf{B}(t)$ is right-handed in the sense that each of these vectors is related to the other two by the right-hand rule (Figure 14.4.10):

$$\mathbf{B}(t) = \mathbf{T}(t) \times \mathbf{N}(t), \quad \mathbf{N}(t) = \mathbf{B}(t) \times \mathbf{T}(t), \quad \mathbf{T}(t) = \mathbf{N}(t) \times \mathbf{B}(t) \tag{10}$$

The coordinate system determined by $\mathbf{T}(t)$, $\mathbf{N}(t)$, and $\mathbf{B}(t)$ is called the **TNB-*frame*** or sometimes the ***Frenet frame*** in honor of the French mathematician Jean Frédéric Frenet (1816–1900) who pioneered its application to the study of space curves. Typically, the xyz-coordinate system determined by the unit vectors $\mathbf{i}$, $\mathbf{j}$, and $\mathbf{k}$ remains fixed, whereas the **TNB**-frame changes as its origin moves along the curve C (Figure 14.4.11).

Formula (9) expresses $\mathbf{B}(t)$ in terms of $\mathbf{T}(t)$ and $\mathbf{N}(t)$. Alternatively, the binormal $\mathbf{B}(t)$ can be expressed directly in terms of $\mathbf{r}(t)$ as

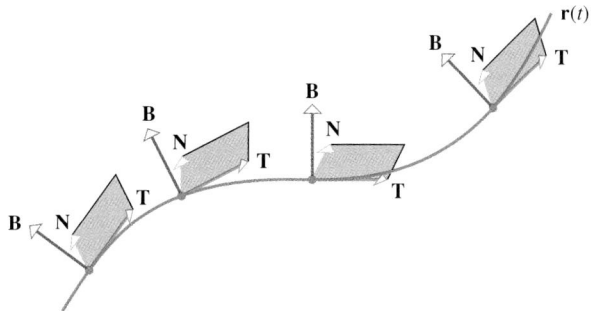

Figure 14.4.11

$$\mathbf{B}(t) = \frac{\mathbf{r}'(t) \times \mathbf{r}''(t)}{\|\mathbf{r}'(t) \times \mathbf{r}''(t)\|} \tag{11}$$

and in the case where the parameter is arc length it can be expressed in terms of $\mathbf{r}(s)$ as

$$\mathbf{B}(s) = \frac{\mathbf{r}'(s) \times \mathbf{r}''(s)}{\|\mathbf{r}''(s)\|} \tag{12}$$

We omit the proof.

EXERCISE SET 14.4

1. In each part, sketch the unit tangent and normal vectors at the points P, Q, and R, taking into account the orientation of the curve C.

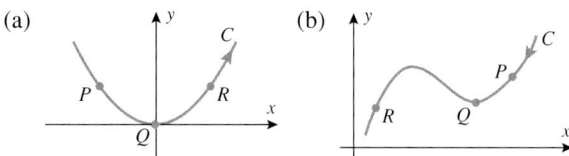

2. Make a rough sketch that shows the ellipse

    $$\mathbf{r}(t) = 3\cos t\mathbf{i} + 2\sin t\mathbf{j}$$

    for $0 \le t \le 2\pi$ and the unit tangent and normal vectors at the points $t = 0$, $t = \pi/4$, $t = \pi/2$, and $t = \pi$.

In Exercises 3–10, find $\mathbf{T}(t)$ and $\mathbf{N}(t)$ at the given point.

3. $\mathbf{r}(t) = (t^2 - 1)\mathbf{i} + t\mathbf{j}; \ t = 1$

4. $\mathbf{r}(t) = \frac{1}{2}t^2\mathbf{i} + \frac{1}{3}t^3\mathbf{j}; \ t = 1$

5. $\mathbf{r}(t) = 5\cos t\mathbf{i} + 5\sin t\mathbf{j}; \ t = \pi/3$

6. $\mathbf{r}(t) = \ln t\mathbf{i} + t\mathbf{j}; \ t = e$

7. $\mathbf{r}(t) = 4\cos t\mathbf{i} + 4\sin t\mathbf{j} + t\mathbf{k}; \ t = \pi/2$

8. $\mathbf{r}(t) = t\mathbf{i} + \frac{1}{2}t^2\mathbf{j} + \frac{1}{3}t^3\mathbf{k}; \ t = 0$

9. $x = e^t \cos t, \ y = e^t \sin t, z = e^t; \ t = 0$

10. $x = \cosh t, \ y = \sinh t, z = t; \ t = \ln 2$

11. In the remark following Example 8 of Section 14.3, we observed that a line $\mathbf{r} = \mathbf{r}_0 + t\mathbf{v}$ can be parametrized in terms of an arc length parameter s with reference point $\mathbf{r}_0$ by normalizing $\mathbf{v}$. Use this result to show that the tangent line to the graph of $\mathbf{r}(t)$ at the point t_0 can be expressed as

    $$\mathbf{r} = \mathbf{r}(t_0) + s\mathbf{T}(t_0)$$

    where s is an arc length parameter with reference point $\mathbf{r}(t_0)$.

12. Use the result in Exercise 11 to show that the tangent line to the parabola

    $$x = t, \quad y = t^2$$

    at the point $(1, 1)$ can be expressed parametrically as

    $$x = 1 + \frac{s}{\sqrt{5}}, \quad y = 1 + \frac{2s}{\sqrt{5}}$$

In Exercises 13 and 14, use the result in Exercise 11 to find parametric equations for the tangent line to the graph of $\mathbf{r}(t)$ at t_0 in terms of an arc length parameter s.

13. $\mathbf{r}(t) = \sin t\mathbf{i} + \cos t\mathbf{j} + \frac{1}{2}t^2\mathbf{k}; \ t_0 = 0$

14. $\mathbf{r}(t) = t\mathbf{i} + t\mathbf{j} + \sqrt{9 - t^2}\mathbf{k}; \ t_0 = 1$

In Exercises 15–18, use the formula $\mathbf{B}(t) = \mathbf{T}(t) \times \mathbf{N}(t)$ to find $\mathbf{B}(t)$, and then check your answer by using Formula (11) to find $\mathbf{B}(t)$ directly from $\mathbf{r}(t)$.

15. $\mathbf{r}(t) = 3\sin t\mathbf{i} + 3\cos t\mathbf{j} + 4t\mathbf{k}$

16. $\mathbf{r}(t) = e^t \sin t\mathbf{i} + e^t \cos t\mathbf{j} + 3\mathbf{k}$

17. $\mathbf{r}(t) = (\sin t - t\cos t)\mathbf{i} + (\cos t + t\sin t)\mathbf{j} + \mathbf{k}$

18. $\mathbf{r}(t) = a\cos t\mathbf{i} + a\sin t\mathbf{j} + ct\mathbf{k} \quad (a \ne 0, c \ne 0)$

In Exercises 19 and 20, find $\mathbf{T}(t)$, $\mathbf{N}(t)$, and $\mathbf{B}(t)$ for the given value of t. Then find equations for the osculating, normal, and rectifying planes at the point that corresponds to that value of t.

19. $\mathbf{r}(t) = \cos t\,\mathbf{i} + \sin t\,\mathbf{j} + \mathbf{k}; \;\; t = \pi/4$

20. $\mathbf{r}(t) = e^t\mathbf{i} + e^t\cos t\,\mathbf{j} + e^t\sin t\,\mathbf{k}; \;\; t = 0$

21. (a) Use the formula $\mathbf{N}(t) = \mathbf{B}(t) \times \mathbf{T}(t)$ and Formulas (1) and (11) to show that $\mathbf{N}(t)$ can be expressed in terms of $\mathbf{r}(t)$ as

$$\mathbf{N}(t) = \frac{\mathbf{r}'(t) \times \mathbf{r}''(t)}{\|\mathbf{r}'(t) \times \mathbf{r}''(t)\|} \times \frac{\mathbf{r}'(t)}{\|\mathbf{r}'(t)\|}$$

    (b) Use properties of cross products to show that the formula in part (a) can be expressed as

$$\mathbf{N}(t) = \frac{(\mathbf{r}'(t) \times \mathbf{r}''(t)) \times \mathbf{r}'(t)}{\|(\mathbf{r}'(t) \times \mathbf{r}''(t)) \times \mathbf{r}'(t)\|}$$

    (c) Use the result in part (b) and Exercise 39 of Section 13.4 to show that $\mathbf{N}(t)$ can be expressed directly in terms of $\mathbf{r}(t)$ as

$$\mathbf{N}(t) = \frac{\mathbf{u}(t)}{\|\mathbf{u}(t)\|}$$

where

$$\mathbf{u}(t) = \|\mathbf{r}'(t)\|^2\mathbf{r}''(t) - (\mathbf{r}'(t) \cdot \mathbf{r}''(t))\mathbf{r}'(t)$$

22. Use the result in part (b) of Exercise 21 to find the unit normal vector requested in
    (a) Exercise 3   (b) Exercise 7.

In Exercises 23 and 24, use the result in part (c) of Exercise 21 to find $\mathbf{N}(t)$.

23. $\mathbf{r}(t) = \sin t\,\mathbf{i} + \cos t\,\mathbf{j} + t\mathbf{k}$ **24.** $\mathbf{r}(t) = t\mathbf{i} + t^2\mathbf{j} + t^3\mathbf{k}$

14.5 CURVATURE

In this section we will consider the problem of obtaining a numerical measure of how sharply a curve in 2-space or 3-space bends. Our results will have applications in geometry and in the study of motion along a curved path.

DEFINITION OF CURVATURE

Suppose that C is the graph of a smooth vector-valued function in 2-space or 3-space that is parametrized in terms of arc length. Figure 14.5.1 suggests that for a curve in 2-space the "sharpness" of the bend in C is closely related to $d\mathbf{T}/ds$, which is the rate of change of the unit tangent vector $\mathbf{T}$ with respect to s. (Keep in mind that $\mathbf{T}$ has constant length, so only its direction changes.) If C is a straight line (no bend), then the direction of $\mathbf{T}$ remains constant (Figure 14.5.1a); if C bends slightly, then $\mathbf{T}$ undergoes a gradual change of direction (Figure 14.5.1b); and if C bends sharply, then $\mathbf{T}$ undergoes a rapid change of direction (Figure 14.5.1c).

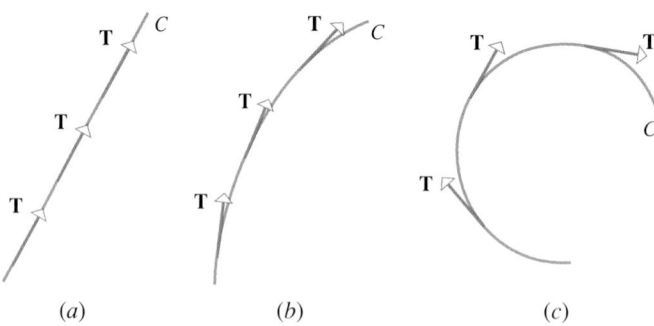

(a)   (b)   (c)

Figure 14.5.1

The situation in 3-space is more complicated because bends in a curve are not limited to a single plane—they can occur in all directions, as illustrated by the complicated tube plot in Figure 14.1.3. To describe the bending characteristics of a curve in 3-space completely, one must take into account $d\mathbf{T}/ds$, $d\mathbf{N}/ds$, and $d\mathbf{B}/ds$. A complete study of this topic would take us too far afield, so we will limit our discussion to $d\mathbf{T}/ds$, which is the most important of these derivatives in applications.

14.5.1 DEFINITION. If C is a smooth curve in 2-space or 3-space that is parametrized by arc length, then the ***curvature*** of C, denoted by $\kappa = \kappa(s)$ (κ = Greek "kappa"), is defined by

$$\kappa(s) = \left\| \frac{d\mathbf{T}}{ds} \right\| = \|\mathbf{r}''(s)\| \tag{1}$$

Observe that $\kappa(s)$ is a real-valued function of s, since it is the *length* of $d\mathbf{T}/ds$ that measures the curvature. In general, the curvature will vary from point to point along a curve; however, the following example shows that the curvature is constant for circles in 2-space, as you might expect.

Example 1

In Example 3 of Section 14.4 we showed that the circle of radius a, centered at the origin, can be parametrized in terms of arc length as

$$\mathbf{r}(s) = a \cos\left(\frac{s}{a}\right)\mathbf{i} + a \sin\left(\frac{s}{a}\right)\mathbf{j} \quad (0 \le s \le 2\pi a)$$

Thus,

$$\mathbf{r}''(s) = -\frac{1}{a}\cos\left(\frac{s}{a}\right)\mathbf{i} - \frac{1}{a}\sin\left(\frac{s}{a}\right)\mathbf{j}$$

and hence from (1)

$$\kappa(s) = \|\mathbf{r}''(s)\| = \sqrt{\left[-\frac{1}{a}\cos\left(\frac{s}{a}\right)\right]^2 + \left[-\frac{1}{a}\sin\left(\frac{s}{a}\right)\right]^2} = \frac{1}{a}$$

so the circle has constant curvature $1/a$. ◄

The next example shows that lines have zero curvature, which is consistent with the fact that they do not bend.

Example 2

Recall from the remark following Example 8 of Section 14.3 that a line in 2-space or 3-space can be parametrized in terms of arc length as

$$\mathbf{r} = \mathbf{r}_0 + \mathbf{u}s$$

where the terminal point of $\mathbf{r}_0$ is a point on the line and $\mathbf{u}$ is a unit vector parallel to the line. Thus,

$$\mathbf{r}'(s) = \frac{d\mathbf{r}}{ds} = \frac{d}{ds}[\mathbf{r}_0 + \mathbf{u}s] = \mathbf{0} + \mathbf{u} = \mathbf{u}$$

$\mathbf{r}_0$ is constant

and hence

$$\mathbf{r}''(s) = \frac{d\mathbf{r}'}{ds} = \frac{d}{ds}[\mathbf{u}] = \mathbf{0}$$

$\mathbf{u}$ is constant

Thus,

$$\kappa(s) = \|\mathbf{r}''(s)\| = 0 \qquad\qquad\qquad ◄$$

Formula (1) is only applicable if the curve is parametrized in terms of arc length. The following theorem provides two formulas for curvature in terms of a general parameter t.

14.5.2 THEOREM. *If* $\mathbf{r}(t)$ *is a smooth vector-valued function in 2-space or 3-space, then for each value of t at which* $\mathbf{T}'(t)$ *and* $\mathbf{r}''(t)$ *exist, the curvature* κ *can be expressed as*

(a) $\quad \kappa(t) = \dfrac{\|\mathbf{T}'(t)\|}{\|\mathbf{r}'(t)\|}$ $\hspace{4cm}$ (2)

(b) $\quad \kappa(t) = \dfrac{\|\mathbf{r}'(t) \times \mathbf{r}''(t)\|}{\|\mathbf{r}'(t)\|^3}$ $\hspace{3cm}$ (3)

Proof (a). It follows from Formula (1) and Formulas (16) and (17) of Section 14.3 that

$$\kappa(t) = \left\| \frac{d\mathbf{T}}{ds} \right\| = \left\| \frac{d\mathbf{T}/dt}{ds/dt} \right\| = \left\| \frac{d\mathbf{T}/dt}{\|d\mathbf{r}/dt\|} \right\| = \frac{\|\mathbf{T}'(t)\|}{\|\mathbf{r}'(t)\|}$$

Proof (b). It follows from Formula (1) of Section 14.4 that

$$\mathbf{r}'(t) = \|\mathbf{r}'(t)\|\mathbf{T}(t) \tag{4}$$

so

$$\mathbf{r}''(t) = \|\mathbf{r}'(t)\|'\mathbf{T}(t) + \|\mathbf{r}'(t)\|\mathbf{T}'(t) \tag{5}$$

But from Formula (2) of Section 14.4 and part (a) of this theorem we have

$$\mathbf{T}'(t) = \|\mathbf{T}'(t)\|\mathbf{N}(t) \quad \text{and} \quad \|\mathbf{T}'(t)\| = \kappa(t)\|\mathbf{r}'(t)\|$$

so

$$\mathbf{T}'(t) = \kappa(t)\|\mathbf{r}'(t)\|\mathbf{N}(t)$$

Substituting this into (5) yields

$$\mathbf{r}''(t) = \|\mathbf{r}'(t)\|'\mathbf{T}(t) + \kappa(t)\|\mathbf{r}'(t)\|^2\mathbf{N}(t) \tag{6}$$

Thus, from (4) and (6)

$$\mathbf{r}'(t) \times \mathbf{r}''(t) = \|\mathbf{r}'(t)\|\|\mathbf{r}'(t)\|'(\mathbf{T}(t) \times \mathbf{T}(t)) + \kappa(t)\|\mathbf{r}'(t)\|^3(\mathbf{T}(t) \times \mathbf{N}(t))$$

But the cross product of a vector with itself is zero, so this equation simplifies to

$$\mathbf{r}'(t) \times \mathbf{r}''(t) = \kappa(t)\|\mathbf{r}'(t)\|^3(\mathbf{T}(t) \times \mathbf{N}(t)) = \kappa(t)\|\mathbf{r}'(t)\|^3\mathbf{B}(t)$$

It follows from this equation and the fact that $\mathbf{B}(t)$ is a unit vector that

$$\|\mathbf{r}'(t) \times \mathbf{r}''(t)\| = \kappa(t)\|\mathbf{r}'(t)\|^3$$

Formula (3) now follows. ∎

REMARKS.   Formula (2) is useful if $\mathbf{T}(t)$ is known or is easy to obtain; however, Formula (3) will usually be easier to apply, since it involves only $\mathbf{r}(t)$ and its derivatives. We also note that cross products were defined only for vectors in 3-space, so to use Formula (3) in 2-space we must first write the 2-space function $\mathbf{r}(t) = x(t)\mathbf{i} + y(t)\mathbf{j}$ as the 3-space function $\mathbf{r}(t) = x(t)\mathbf{i} + y(t)\mathbf{j} + 0\mathbf{k}$ with a zero $\mathbf{k}$ component.

Example 3

Find $\kappa(t)$ for the circular helix

$$x = a\cos t, \quad y = a\sin t, \quad z = ct$$

where $a > 0$.

Solution. The radius vector for the helix is

$$\mathbf{r}(t) = a \cos t\,\mathbf{i} + a \sin t\,\mathbf{j} + ct\,\mathbf{k}$$

Thus,

$$\mathbf{r}'(t) = (-a \sin t)\mathbf{i} + a \cos t\,\mathbf{j} + c\mathbf{k}$$
$$\mathbf{r}''(t) = (-a \cos t)\mathbf{i} + (-a \sin t)\mathbf{j}$$

so

$$\mathbf{r}'(t) \times \mathbf{r}''(t) = \begin{vmatrix} \mathbf{i} & \mathbf{j} & \mathbf{k} \\ -a \sin t & a \cos t & c \\ -a \cos t & -a \sin t & 0 \end{vmatrix} = (ac \sin t)\mathbf{i} - (ac \cos t)\mathbf{j} + a^2\mathbf{k}$$

Therefore,

$$\|\mathbf{r}'(t)\| = \sqrt{(-a \sin t)^2 + (a \cos t)^2 + c^2} = \sqrt{a^2 + c^2}$$

and

$$\|\mathbf{r}'(t) \times \mathbf{r}''(t)\| = \sqrt{(ac \sin t)^2 + (-ac \cos t)^2 + a^4}$$
$$= \sqrt{a^2 c^2 + a^4} = a\sqrt{a^2 + c^2}$$

so

$$\kappa(t) = \frac{\|\mathbf{r}'(t) \times \mathbf{r}''(t)\|}{\|\mathbf{r}'(t)\|^3} = \frac{a\sqrt{a^2 + c^2}}{(\sqrt{a^2 + c^2})^3} = \frac{a}{a^2 + c^2}$$

Note that κ does not depend on t, which tells us that the helix has constant curvature. ◄

Example 4

The graph of the vector equation

$$\mathbf{r} = 2 \cos t\,\mathbf{i} + 3 \sin t\,\mathbf{j} \quad (0 \le t \le 2\pi)$$

is the ellipse in Figure 14.5.2. Find the curvature of the ellipse at the endpoints of the major and minor axes, and use a graphing utility to generate the graph of $\kappa(t)$.

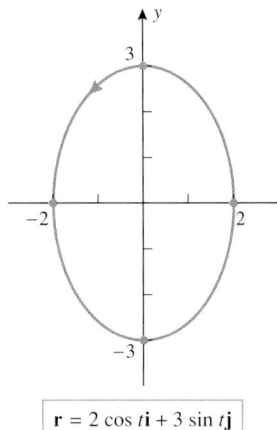

$\mathbf{r} = 2 \cos t\,\mathbf{i} + 3 \sin t\,\mathbf{j}$

Figure 14.5.2

Solution. To apply Formula (3), we must treat the ellipse as a curve in the *xy*-plane of an *xyz*-coordinate system by adding a zero **k** component and writing its equation as

$$\mathbf{r} = 2 \cos t\,\mathbf{i} + 3 \sin t\,\mathbf{j} + 0\mathbf{k}$$

It is not essential to write the zero **k** component explicitly as long as you assume it to be there when you calculate a cross product. Thus,

$$\mathbf{r}'(t) = (-2 \sin t)\mathbf{i} + 3 \cos t\,\mathbf{j}$$
$$\mathbf{r}''(t) = (-2 \cos t)\mathbf{i} + (-3 \sin t)\mathbf{j}$$

$$\mathbf{r}'(t) \times \mathbf{r}''(t) = \begin{vmatrix} \mathbf{i} & \mathbf{j} & \mathbf{k} \\ -2 \sin t & 3 \cos t & 0 \\ -2 \cos t & -3 \sin t & 0 \end{vmatrix} = [(6 \sin^2 t) + (6 \cos^2 t)]\mathbf{k} = 6\mathbf{k}$$

Therefore,

$$\|\mathbf{r}'(t)\| = \sqrt{(-2 \sin t)^2 + (3 \cos t)^2} = \sqrt{4 \sin^2 t + 9 \cos^2 t}$$
$$\|\mathbf{r}'(t) \times \mathbf{r}''(t)\| = 6$$

so

$$\kappa(t) = \frac{\|\mathbf{r}'(t) \times \mathbf{r}''(t)\|}{\|\mathbf{r}'(t)\|^3} = \frac{6}{[4 \sin^2 t + 9 \cos^2 t]^{3/2}} \qquad (7)$$

The endpoints of the minor axis are $(2, 0)$ and $(-2, 0)$, which correspond to $t = 0$ and $t = \pi$, respectively. Substituting these values in (7) yields the same curvature at both points, namely

$$\kappa = \kappa(0) = \kappa(\pi) = \frac{6}{9^{3/2}} = \frac{6}{27} = \frac{2}{9}$$

The endpoints of the major axis are $(0, 3)$ and $(0, -3)$, which correspond to $t = \pi/2$ and $t = 3\pi/2$, respectively; from (7) the curvature at these points is

$$\kappa = \kappa\left(\frac{\pi}{2}\right) = \kappa\left(\frac{3\pi}{2}\right) = \frac{6}{4^{3/2}} = \frac{3}{4}$$

Observe that the curvature is greater at the ends of the major axis than at the ends of the minor axis, as you might expect. Figure 14.5.3 shows the graph κ versus t. This graph illustrates clearly that the curvature is minimum at $t = 0$ (the right end of the minor axis), increases to a maximum at $t = \pi/2$ (the top of the major axis), decreases to a minimum again at $t = \pi$ (the left end of the minor axis), and continues cyclically in this manner. Figure 14.5.4 provides another way of picturing the curvature. ◄

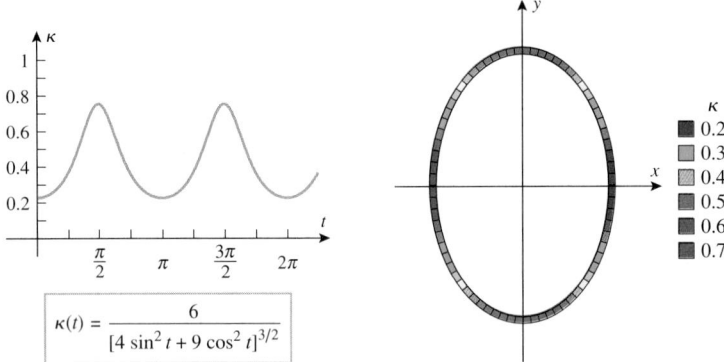

$$\kappa(t) = \frac{6}{[4\sin^2 t + 9\cos^2 t]^{3/2}}$$

Figure 14.5.3   Figure 14.5.4

RADIUS OF CURVATURE

In the last example we found the curvature at the ends of the minor axis to be $\frac{2}{9}$ and the curvature at the ends of the major axis to be $\frac{3}{4}$. To obtain a better understanding of the meaning of these numbers, recall from Example 1 that a circle of radius a has a constant curvature of $1/a$; thus, the curvature of the ellipse at the ends of the minor axis is the same as that of a circle of radius $\frac{9}{2}$, and the curvature at the ends of the major axis is the same as that of a circle of radius $\frac{4}{3}$ (Figure 14.5.5).

In general, if a curve C in 2-space has nonzero curvature κ at a point P, then the circle of radius $\rho = 1/\kappa$ sharing a common tangent with C at P, and centered on the concave side of the curve at P, is called the *circle of curvature* or *osculating circle* at P (Figure 14.5.6). The osculating circle and the curve C not only touch at P but they have equal curvatures at that point. In this sense, the osculating circle is the circle that best approximates the curve C near P. The radius ρ of the osculating circle at P is called the *radius of curvature* at P, and the center of the circle is called the *center of curvature* at P (Figure 14.5.6).

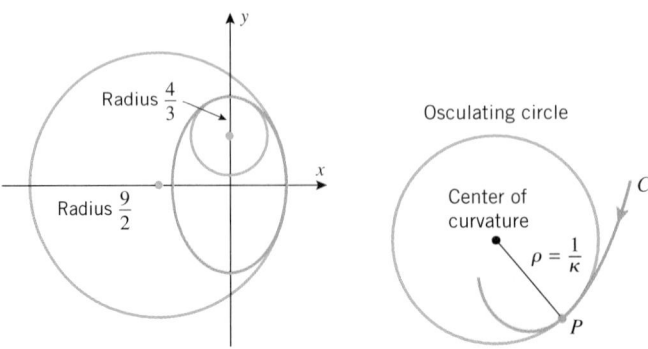

Figure 14.5.5   Figure 14.5.6

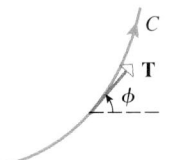

AN INTERPRETATION OF CURVATURE IN 2-SPACE

A useful geometric interpretation of curvature in 2-space can be obtained by considering the angle ϕ measured counterclockwise from the direction of the positive x-axis to the unit tangent vector $\mathbf{T}$ (Figure 14.5.7). By Formula (12) of Section 13.2, we can express $\mathbf{T}$ in terms of ϕ as

Figure 14.5.7

$$\mathbf{T}(\phi) = \cos\phi\,\mathbf{i} + \sin\phi\,\mathbf{j}$$

Thus,

$$\frac{d\mathbf{T}}{d\phi} = (-\sin\phi)\mathbf{i} + \cos\phi\,\mathbf{j}$$

$$\frac{d\mathbf{T}}{ds} = \frac{d\mathbf{T}}{d\phi}\frac{d\phi}{ds}$$

from which we obtain

$$\kappa(s) = \left\|\frac{d\mathbf{T}}{ds}\right\| = \left|\frac{d\phi}{ds}\right|\left\|\frac{d\mathbf{T}}{d\phi}\right\| = \left|\frac{d\phi}{ds}\right|\sqrt{(-\sin\phi)^2 + \cos^2\phi} = \left|\frac{d\phi}{ds}\right|$$

In summary, we have shown that

$$\kappa(s) = \left|\frac{d\phi}{ds}\right| \tag{8}$$

which tells us that curvature in 2-space can be interpreted as the magnitude of the rate of change of ϕ with respect to s—the greater the curvature, the more rapidly ϕ changes with s (Figure 14.5.8). In the case of a straight line, the angle ϕ is constant (Figure 14.5.9) and consequently $\kappa(s) = |d\phi/ds| = 0$, which is consistent with the fact that a straight line has zero curvature at every point.

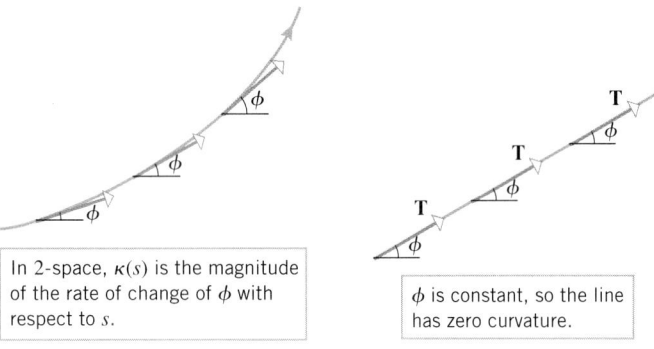

In 2-space, $\kappa(s)$ is the magnitude of the rate of change of ϕ with respect to s.

Figure 14.5.8

ϕ is constant, so the line has zero curvature.

Figure 14.5.9

FORMULA SUMMARY

We conclude this section with a summary of formulas for $\mathbf{T}$, $\mathbf{N}$, and $\mathbf{B}$. These formulas have either been derived in the text or are easily derivable from formulas we have already established.

$$\mathbf{T}(s) = \mathbf{r}'(s) \tag{9}$$

$$\mathbf{N}(s) = \frac{1}{\kappa(s)}\frac{d\mathbf{T}}{ds} = \frac{\mathbf{r}''(s)}{\|\mathbf{r}''(s)\|} = \frac{\mathbf{r}''(s)}{\kappa(s)} \tag{10}$$

$$\mathbf{B}(s) = \frac{\mathbf{r}'(s)\times\mathbf{r}''(s)}{\|\mathbf{r}''(s)\|} = \frac{\mathbf{r}'(s)\times\mathbf{r}''(s)}{\kappa(s)} \tag{11}$$

$$\mathbf{T}(t) = \frac{\mathbf{r}'(t)}{\|\mathbf{r}'(t)\|} \tag{12}$$

$$\mathbf{B}(t) = \frac{\mathbf{r}'(t) \times \mathbf{r}''(t)}{\|\mathbf{r}'(t) \times \mathbf{r}''(t)\|} \tag{13}$$

$$\mathbf{N}(t) = \mathbf{B}(t) \times \mathbf{T}(t) \tag{14}$$

EXERCISE SET 14.5   ~ Graphing Calculator   [c] CAS

In Exercises 1 and 2, use the osculating circle shown in the figure to estimate the curvature at the indicated point.

1.
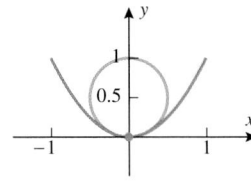

2.
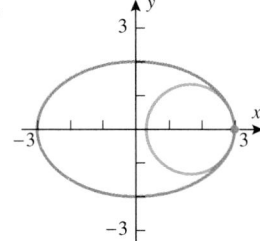

In Exercises 3–10, use Formula (3) to find $\kappa(t)$.

3. $\mathbf{r}(t) = t^2\mathbf{i} + t^3\mathbf{j}$

4. $\mathbf{r}(t) = 4\cos t\,\mathbf{i} + \sin t\,\mathbf{j}$

5. $\mathbf{r}(t) = e^{3t}\mathbf{i} + e^{-t}\mathbf{j}$

6. $x = 1 - t^3, \; y = t - t^2$

7. $\mathbf{r}(t) = 4\cos t\,\mathbf{i} + 4\sin t\,\mathbf{j} + t\mathbf{k}$

8. $\mathbf{r}(t) = t\mathbf{i} + \frac{1}{2}t^2\mathbf{j} + \frac{1}{3}t^3\mathbf{k}$

9. $x = \cosh t, \; y = \sinh t, \; z = t$

10. $\mathbf{r}(t) = \mathbf{i} + t\mathbf{j} + t^2\mathbf{k}$

In Exercises 11–14, find the curvature and the radius of curvature at the stated point.

11. $\mathbf{r}(t) = 3\cos t\,\mathbf{i} + 4\sin t\,\mathbf{j} + t\mathbf{k}; \; t = \pi/2$

12. $\mathbf{r}(t) = e^t\mathbf{i} + e^{-t}\mathbf{j} + t\mathbf{k}; \; t = 0$

13. $x = e^t\cos t, \; y = e^t\sin t, \; z = e^t; \; t = 0$

14. $x = \sin t, \; y = \cos t, \; z = \frac{1}{2}t^2; \; t = 0$

In Exercises 15 and 16, confirm that s is an arc length parameter by showing that $\|d\mathbf{r}/ds\| = 1$, and then apply Formula (1) to find $\kappa(s)$.

15. $\mathbf{r} = \sin\left(1 + \frac{s}{2}\right)\mathbf{i} + \cos\left(1 + \frac{s}{2}\right)\mathbf{j} + \sqrt{3}\left(1 + \frac{s}{2}\right)\mathbf{k}$

16. $\mathbf{r} = \left(1 - \frac{2}{3}s\right)^{3/2}\mathbf{i} + \left(\frac{2}{3}s\right)^{3/2}\mathbf{j} \quad \left(0 \le s \le \frac{3}{2}\right)$

17. (a) Use Formula (3) to show that in 2-space the curvature of a smooth parametric curve

$$x = x(t), \quad y = y(t)$$

is

$$\kappa(t) = \frac{|x'y'' - y'x''|}{(x'^2 + y'^2)^{3/2}}$$

where primes denote differentiation with respect to t.

(b) Use the result in part (a) to show that in 2-space the curvature of the plane curve given by $y = f(x)$ is

$$\kappa(x) = \frac{|d^2y/dx^2|}{[1 + (dy/dx)^2]^{3/2}}$$

[*Hint:* Express $y = f(x)$ parametrically with $x = t$ as the parameter.]

18. Use part (b) of Exercise 17 to show that the curvature of $y = f(x)$ can be expressed in terms of the angle of inclination of the tangent line as

$$\kappa(\phi) = \left| \frac{d^2y}{dx^2} \cos^3\phi \right|$$

[*Hint:* $\tan\phi = dy/dx$.]

In Exercises 19–24, use the result in Exercise 17(b) to find the curvature at the stated point.

19. $y = \sin x; \; x = \pi/2$ **20.** $y = x^3/3; \; x = 0$

21. $y = 1/x; \; x = 1$ **22.** $y = e^{-x}; \; x = 1$

23. $y = \tan x; \; x = \pi/4$ **24.** $y^2 - 4x^2 = 9; \; (2, 5)$

In Exercises 25–30, use the result in Exercise 17(a) to find the curvature at the stated point.

25. $x = t^2, \; y = t^3; \; t = \frac{1}{2}$

26. $x = 4\cos t, \; y = \sin t; \; t = \pi/2$

27. $x = e^{3t}, \; y = e^{-t}; \; t = 0$

28. $x = 1 - t^3, \; y = t - t^2; \; t = 1$

29. $x = t, \; y = 1/t; \; t = 1$

30. $x = 2\sin 2t, \; y = 3\sin t; \; t = \pi/2$

31. In each part, use the formulas in Exercise 17 to help find the radius of curvature at the stated points. Then sketch the graph together with the osculating circles at those points.

(a) $y = \cos x$ at $x = 0$ and $x = \pi$

(b) $x = 2 \cos t$, $y = \sin t$ $(0 \le t \le 2\pi)$ at $t = 0$ and $t = \pi/2$

32. Use the formula in Exercise 17(a) to find $\kappa(t)$ for the curve $x = e^{-t} \cos t$, $y = e^{-t} \sin t$. Then sketch the graph of $\kappa(t)$.

In each part of Exercises 33 and 34, the graphs of $f(x)$ and the associated curvature function $\kappa(x)$ are shown. Determine which is which, and explain your reasoning.

33. (a)   (b)

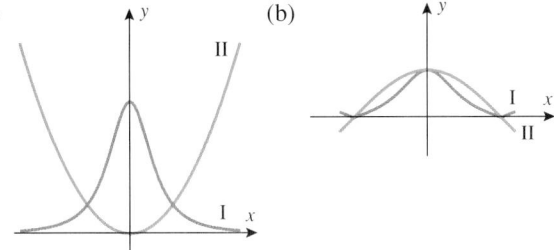

34. (a)   (b)

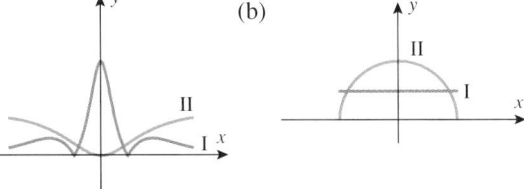

In Exercises 35 and 36, use a graphing utility to generate the graph of $y = f(x)$, and then make a conjecture about the shape of the graph of $y = \kappa(x)$. Check your conjecture by generating the graph of $y = \kappa(x)$.

 35. $f(x) = xe^{-x}$ for $0 \le x \le 5$

 36. $f(x) = x^3 - x$ for $-1 \le x \le 1$

[c] **37.** (a) If you have a CAS, read the documentation on calculating higher-order derivatives. Then use the CAS and part (b) of Exercise 17 to find $\kappa(x)$ for $f(x) = x^4 - 2x^2$.

(b) Use the CAS to generate the graphs of $f(x) = x^4 - 2x^2$ and $\kappa(x)$ on the same screen for $-2 \le x \le 2$.

(c) Find the radius of curvature at each relative extremum.

(d) Make a reasonably accurate hand-drawn sketch that shows the graph of $f(x) = x^4 - 2x^2$ and the osculating circles in their correct proportions at the relative extrema.

[c] **38.** (a) Use a CAS to graph the parametric curve $x = t \cos t$, $y = t \sin t$ for $t \ge 0$.

(b) Make a conjecture about the behavior of $\kappa(t)$ as $t \to +\infty$.

(c) Use the CAS and part (a) of Exercise 17 to find $\kappa(t)$.

(d) Check your conjecture by finding the limit of $\kappa(t)$ as $t \to +\infty$.

39. Use Formula (3) to show that for a curve in polar coordinates described by $r = f(\theta)$ the curvature is

$$\kappa(\theta) = \frac{\left| r^2 + 2 \left(\dfrac{dr}{d\theta} \right)^2 - r \dfrac{d^2r}{d\theta^2} \right|}{\left[r^2 + \left(\dfrac{dr}{d\theta} \right)^2 \right]^{3/2}}$$

[*Hint:* Let θ be the parameter and use the relationships $x = r \cos \theta$, $y = r \sin \theta$.]

40. Use the result in Exercise 39 to show that a circle has constant curvature.

In Exercises 41–44, use the formula of Exercise 39 to find the curvature at the indicated point.

41. $r = 1 + \cos \theta$; $\theta = \pi/2$ **42.** $r = e^{2\theta}$; $\theta = 1$

43. $r = \sin 3\theta$; $\theta = 0$ **44.** $r = \theta$; $\theta = 1$

45. The accompanying figure is the graph of the radius of curvature versus θ in rectangular coordinates for the cardioid $r = 1 + \cos \theta$. In words, explain what the graph tells you about the cardioid.

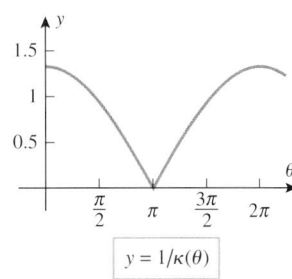

$$y = 1/\kappa(\theta)$$

Figure Ex-45

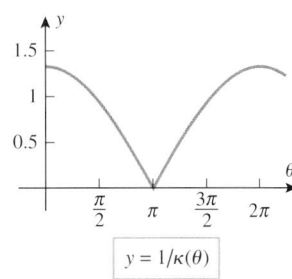 **46.** Use the formula in Exercise 39 and a graphing utility to generate the graph in Exercise 45.

47. Find the radius of curvature of the parabola $y^2 = 4px$ at $(0, 0)$.

48. At what point(s) does $y = e^x$ have maximum curvature?

49. At what point(s) does $4x^2 + 9y^2 = 36$ have minimum radius of curvature?

50. Find the value of x, $x > 0$, where $y = x^3$ has maximum curvature.

51. Find the maximum and minimum values of the radius of curvature for the curve $x = \cos t$, $y = \sin t$, $z = \cos t$.

52. Find the minimum value of the radius of curvature for the curve $x = e^t$, $y = e^{-t}$, $z = \sqrt{2}t$.

53. Use the formula in Exercise 39 to show that the curvature of the polar curve $r = e^{a\theta}$ is inversely proportional to r.

[c] **54.** Use the formula in Exercise 39 and a CAS to show that the curvature of the lemniscate $r = \sqrt{a \cos 2\theta}$ is directly proportional to r.

55. (a) Use the result in Exercise 18 to show that for the parabola $y = x^2$ the curvature $\kappa(\phi)$ at points where the tangent line has an angle of inclination of ϕ is

$$\kappa(\phi) = |2\cos^3\phi|$$

(b) Use the result in part (a) to find the radius of curvature of the parabola at the point on the parabola where the tangent line has slope 1.

(c) Make a sketch with reasonably accurate proportions that shows the osculating circles at the point on the parabola where the tangent line has slope 1.

56. The *evolute* of a smooth parametric curve C in 2-space is the curve formed from the centers of curvature of C. The accompanying figure shows the ellipse $x = 3\cos t$, $y = 2\sin t \, (0 \le t \le 2\pi)$ and its evolute graphed together.

(a) Which points on the evolute correspond to $t = 0$ and $t = \pi/2$?

(b) In what direction is the evolute traced as t increases from 0 to 2π?

(c) What does the evolute of a circle look like? Explain your reasoning.

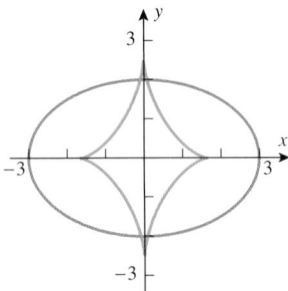

Figure Ex-56

In Exercises 57–60, we will be concerned with the problem of creating a single smooth curve by piecing together two separate smooth curves. If two smooth curves C_1 and C_2 are joined at a point P to form a curve C, then we will say that C_1 and C_2 make a *smooth transition* at P if the curvature of C is continuous at P.

57. Show that the transition at $x = 0$ from the horizontal line $y = 0$ for $x \le 0$ to the parabola $y = x^2$ for $x > 0$ is not smooth, whereas the transition to $y = x^3$ for $x > 0$ is smooth.

58. (a) Sketch the graph of the curve defined piecewise by $y = x^2$ for $x < 0$, $y = x^4$ for $x \ge 0$.

(b) Show that for the curve in part (a) the transition at $x = 0$ is not smooth.

59. The accompanying figure shows the arc of a circle of radius r with center at $(0, r)$. Find the value of a so that there is a smooth transition from the circle to the parabola $y = ax^2$ at the point where $x = 0$.

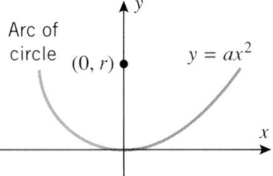

Figure Ex-59

60. Find a, b, and c so that there is a smooth transition at $x = 0$ from the curve $y = e^x$ for $x \le 0$ to the parabola $y = ax^2 + bx + c$ for $x > 0$. [*Hint:* The curvature is continuous at those points where y'' is continuous.]

In Exercises 61–64, we assume that s is an arc length parameter for a smooth vector-valued function $\mathbf{r}(s)$ in 3-space and that $d\mathbf{T}/ds$ and $d\mathbf{N}/ds$ exist at each point on the curve. This implies that $d\mathbf{B}/ds$ exists as well, since $\mathbf{B} = \mathbf{T} \times \mathbf{N}$.

61. Show that

$$\frac{d\mathbf{T}}{ds} = \kappa(s)\mathbf{N}(s)$$

and use this result to obtain the formulas in (10).

62. (a) Show that $d\mathbf{B}/ds$ is perpendicular to $\mathbf{B}(s)$.

(b) Show that $d\mathbf{B}/ds$ is perpendicular to $\mathbf{T}(s)$. [*Hint:* Use the fact that $\mathbf{B}(s)$ is perpendicular to both $\mathbf{T}(s)$ and $\mathbf{N}(s)$, and differentiate $\mathbf{B} \cdot \mathbf{T}$ with respect to s.]

(c) Use the results in parts (a) and (b) to show that $d\mathbf{B}/ds$ is a scalar multiple of $\mathbf{N}(s)$. The *negative* of this scalar is called the *torsion* of $\mathbf{r}(s)$ and is denoted by $\tau(s)$. Thus,

$$\frac{d\mathbf{B}}{ds} = -\tau(s)\mathbf{N}(s)$$

(d) Show that $\tau(s) = 0$ for all s if the graph of $\mathbf{r}(s)$ lies in a plane. [*Note:* For reasons that we cannot discuss here, the torsion is related to the "twisting" properties of the curve, and $\tau(s)$ is regarded as a numerical measure of the tendency for the curve to twist out of the osculating plane.]

63. Let κ be the curvature of C and τ the torsion (defined in Exercise 62). By differentiating $\mathbf{N} = \mathbf{B} \times \mathbf{T}$ with respect to s, show that $d\mathbf{N}/ds = -\kappa\mathbf{T} + \tau\mathbf{B}$.

64. The following derivatives, known as the *Frenet–Serret formulas*, are fundamental in the theory of curves in 3-space:

$$dT/ds = \kappa\mathbf{N} \qquad \text{[Exercise 61]}$$
$$dN/ds = -\kappa\mathbf{T} + \tau\mathbf{B} \qquad \text{[Exercise 63]}$$
$$dB/ds = -\tau\mathbf{N} \qquad \text{[Exercise 62(c)]}$$

Use the first two Frenet–Serret formulas and the fact that $\mathbf{r}'(s) = \mathbf{T}$ if $\mathbf{r} = \mathbf{r}(s)$ to show that

$$\tau = \frac{[\mathbf{r}'(s) \times \mathbf{r}''(s)] \cdot \mathbf{r}'''(s)}{\|\mathbf{r}''(s)\|^2} \quad \text{and} \quad \mathbf{B} = \frac{\mathbf{r}'(s) \times \mathbf{r}''(s)}{\|\mathbf{r}''(s)\|}$$

65. Use the results in Exercise 64 and the results in Exercise 30 of Section 14.3 to show that for the circular helix

$$\mathbf{r} = a\cos t\,\mathbf{i} + a\sin t\,\mathbf{j} + ct\,\mathbf{k}$$

with $a > 0$ the torsion and the binormal vector are

$$\tau = \frac{c}{w^2}$$

and

$$\mathbf{B} = \left(\frac{c}{w} \sin \frac{s}{w}\right)\mathbf{i} - \left(\frac{c}{w} \cos \frac{s}{w}\right)\mathbf{j} + \left(\frac{a}{w}\right)\mathbf{k}$$

where $w = \sqrt{a^2 + c^2}$ and s has reference point $(a, 0, 0)$.

66. (a) Use the chain rule and the first two Frenet–Serret formulas in Exercise 64 to show that

$$\mathbf{T}' = \kappa s'\mathbf{N} \quad \text{and} \quad \mathbf{N}' = -\kappa s'\mathbf{T} + \tau s'\mathbf{B}$$

where primes denote differentiation with respect to t.

(b) Show that Formulas (4) and (6) can be written in the form

$$\mathbf{r}'(t) = s'\mathbf{T} \quad \text{and} \quad \mathbf{r}''(t) = s''\mathbf{T} + \kappa(s')^2\mathbf{N}$$

(c) Use the results in parts (a) and (b) to show that

$$\mathbf{r}'''(t) = [s''' - \kappa^2(s')^3]\mathbf{T}$$
$$+ [3\kappa s's'' + \kappa'(s')^2]\mathbf{N} + \kappa\tau(s')^3\mathbf{B}$$

(d) Use the results in parts (b) and (c) to show that

$$\tau(t) = \frac{[\mathbf{r}'(t) \times \mathbf{r}''(t)] \cdot \mathbf{r}'''(t)}{\|\mathbf{r}'(t) \times \mathbf{r}''(t)\|^2}$$

In Exercises 67–70, use the formula in Exercise 66(d) to find the torsion $\tau = \tau(t)$.

67. The twisted cubic $\mathbf{r}(t) = 2t\mathbf{i} + t^2\mathbf{j} + \frac{1}{3}t^3\mathbf{k}$

68. The circular helix $\mathbf{r}(t) = a\cos t\,\mathbf{i} + a\sin t\,\mathbf{j} + ct\mathbf{k}$

69. $\mathbf{r}(t) = e^t\mathbf{i} + e^{-t}\mathbf{j} + \sqrt{2}t\mathbf{k}$

70. $\mathbf{r}(t) = (t - \sin t)\mathbf{i} + (1 - \cos t)\mathbf{j} + t\mathbf{k}$

14.6 MOTION ALONG A CURVE

In earlier sections we considered the motion of a particle along a line. In that situation there are only two directions in which the particle can move—the positive direction or the negative direction. Motion in 2-space or 3-space is more complicated because there are infinitely many directions in which a particle can move. In this section we will show how vectors can be used to analyze motion along curves in 2-space or 3-space.

VELOCITY, ACCELERATION, AND SPEED

Let us assume that the motion of a particle in 2-space or 3-space is described by a smooth vector-valued function $\mathbf{r}(t)$ in which the parameter t denotes time; we will call this the *position function* or *trajectory* of the particle. As the particle moves along its trajectory, its direction of motion and its speed can vary from instant to instant. Thus, before we can undertake any analysis of such motion, we must have clear answers to the following questions:

- What is the direction of motion of the particle at an instant of time?
- What is the speed of the particle at an instant of time?

We will define the direction of motion at time t to be the direction of the unit tangent vector $\mathbf{T}(t)$, and we will define the speed to be ds/dt—the instantaneous rate of change of the arc length traveled by the particle from an arbitrary reference point. Taking this a step further, we will combine the speed and the direction of motion to form the vector

$$\mathbf{v}(t) = \frac{ds}{dt}\mathbf{T}(t) \tag{1}$$

which we call the *velocity function* of the particle at time t. Thus, at each instant of time the velocity vector $\mathbf{v}(t)$ points in the direction of motion and has a magnitude that is equal to the speed of the particle (Figure 14.6.1).

Recall that for motion along a coordinate line the velocity function is the derivative of the position function. The same is true for motion along a curve, since

$$\frac{d\mathbf{r}}{dt} = \frac{d\mathbf{r}}{ds}\frac{ds}{dt} = \frac{ds}{dt}\mathbf{T}(t) = \mathbf{v}(t)$$

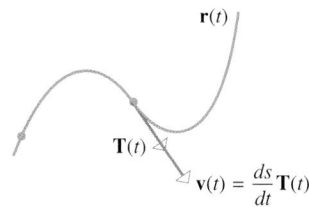

The length of the velocity vector is the speed of the particle, and the direction of the velocity vector is the direction of motion.

Figure 14.6.1

For motion along a coordinate line, the acceleration function was defined to be the derivative of the velocity function. The definition is the same for motion along a curve.

14.6.1   DEFINITION.   If $\mathbf{r}(t)$ is the position function of a particle moving along a curve in 2-space or 3-space, then the *instantaneous velocity*, *instantaneous acceleration*, and *instantaneous speed* of the particle at time t are defined by

$$\text{velocity} = \mathbf{v}(t) = \frac{d\mathbf{r}}{dt} \tag{2}$$

$$\text{acceleration} = \mathbf{a}(t) = \frac{d\mathbf{v}}{dt} = \frac{d^2\mathbf{r}}{dt^2} \tag{3}$$

$$\text{speed} = \|\mathbf{v}(t)\| = \frac{ds}{dt} \tag{4}$$

As shown in Table 14.6.1, the position, velocity, acceleration, and speed can also be expressed in component form:

Table 14.6.1

	2-SPACE	3-SPACE
POSITION	$\mathbf{r}(t) = x(t)\mathbf{i} + y(t)\mathbf{j}$	$\mathbf{r}(t) = x(t)\mathbf{i} + y(t)\mathbf{j} + z(t)\mathbf{k}$
VELOCITY	$\mathbf{v}(t) = \dfrac{dx}{dt}\mathbf{i} + \dfrac{dy}{dt}\mathbf{j}$	$\mathbf{v}(t) = \dfrac{dx}{dt}\mathbf{i} + \dfrac{dy}{dt}\mathbf{j} + \dfrac{dz}{dt}\mathbf{k}$
ACCELERATION	$\mathbf{a}(t) = \dfrac{d^2x}{dt^2}\mathbf{i} + \dfrac{d^2y}{dt^2}\mathbf{j}$	$\mathbf{a}(t) = \dfrac{d^2x}{dt^2}\mathbf{i} + \dfrac{d^2y}{dt^2}\mathbf{j} + \dfrac{d^2z}{dt^2}\mathbf{k}$
SPEED	$\|\mathbf{v}(t)\| = \sqrt{\left(\dfrac{dx}{dt}\right)^2 + \left(\dfrac{dy}{dt}\right)^2}$	$\|\mathbf{v}(t)\| = \sqrt{\left(\dfrac{dx}{dt}\right)^2 + \left(\dfrac{dy}{dt}\right)^2 + \left(\dfrac{dz}{dt}\right)^2}$

Example 1

A particle moves along a circular path in such a way that its x- and y-coordinates at time t are

$$x = 2\cos t, \quad y = 2\sin t$$

(a)   Find the instantaneous velocity and speed of the particle at time t.

(b)   Sketch the path of the particle, and show the position and velocity vectors at time $t = \pi/4$ with the velocity vector drawn so that its initial point is at the tip of the position vector.

(c)   Show that at each instant the acceleration vector is perpendicular to the velocity vector.

Solution (*a*).   At time t, the position vector is

$$\mathbf{r}(t) = 2\cos t\,\mathbf{i} + 2\sin t\,\mathbf{j}$$

so the instantaneous velocity and speed are

$$\mathbf{v}(t) = \frac{d\mathbf{r}}{dt} = -2\sin t\,\mathbf{i} + 2\cos t\,\mathbf{j}$$

$$\|\mathbf{v}(t)\| = \sqrt{(-2\sin t)^2 + (2\cos t)^2} = 2$$

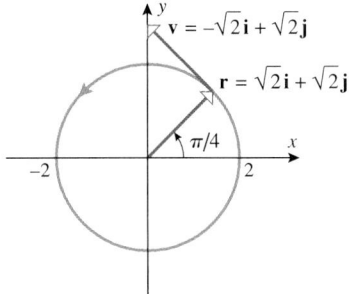

Figure 14.6.2

Solution (b). The graph of the parametric equations is a circle of radius 2 centered at the origin. At time $t = \pi/4$ the position and velocity vectors of the particles are

$$\mathbf{r}(\pi/4) = 2\cos(\pi/4)\mathbf{i} + 2\sin(\pi/4)\mathbf{j} = \sqrt{2}\mathbf{i} + \sqrt{2}\mathbf{j}$$

$$\mathbf{v}(\pi/4) = -2\sin(\pi/4)\mathbf{i} + 2\cos(\pi/4)\mathbf{j} = -\sqrt{2}\mathbf{i} + \sqrt{2}\mathbf{j}$$

These vectors and the circle are shown in Figure 14.6.2.

Solution (c). At time t, the acceleration vector is

$$\mathbf{a}(t) = \frac{d\mathbf{v}}{dt} = -2\cos t\,\mathbf{i} - 2\sin t\,\mathbf{j}$$

One way of showing that $\mathbf{v}(t)$ and $\mathbf{a}(t)$ are perpendicular is to show that their dot product is zero (try it). However, it is easier to observe that $\mathbf{a}(t)$ is the negative of $\mathbf{r}(t)$, which implies that $\mathbf{v}(t)$ and $\mathbf{a}(t)$ are perpendicular, since at each point on a circle the radius and tangent line are perpendicular. ◄

Since $\mathbf{v}(t)$ can be obtained by differentiating $\mathbf{r}(t)$, and since $\mathbf{a}(t)$ can be obtained by differentiating $\mathbf{v}(t)$, it follows that $\mathbf{r}(t)$ can be obtained by integrating $\mathbf{v}(t)$, and $\mathbf{v}(t)$ can be obtained by integrating $\mathbf{a}(t)$. However, such integrations do not produce unique functions because constants of integration occur. Typically, initial conditions are required to determine these constants.

Example 2

A particle moves through 3-space in such a way that its velocity is

$$\mathbf{v}(t) = \mathbf{i} + t\mathbf{j} + t^2\mathbf{k}$$

Find the coordinates of the particle at time $t = 1$ given that the particle is at the point $(-1, 2, 4)$ at time $t = 0$.

Solution. Integrating the velocity function to obtain the position function yields

$$\mathbf{r}(t) = \int \mathbf{v}(t)\,dt = \int (\mathbf{i} + t\mathbf{j} + t^2\mathbf{k})\,dt = t\mathbf{i} + \frac{t^2}{2}\mathbf{j} + \frac{t^3}{3}\mathbf{k} + \mathbf{C} \qquad (5)$$

where $\mathbf{C}$ is a vector constant of integration. Since the coordinates of the particle at time $t = 0$ are $(-1, 2, 4)$, the position vector at time $t = 0$ is

$$\mathbf{r}(0) = -\mathbf{i} + 2\mathbf{j} + 4\mathbf{k} \qquad (6)$$

It follows on substituting $t = 0$ in (5) and equating the result with (6) that

$$\mathbf{C} = -\mathbf{i} + 2\mathbf{j} + 4\mathbf{k}$$

Substituting this value of $\mathbf{C}$ in (5) and simplifying yields

$$\mathbf{r}(t) = (t-1)\mathbf{i} + \left(\frac{t^2}{2} + 2\right)\mathbf{j} + \left(\frac{t^3}{3} + 4\right)\mathbf{k}$$

Thus, at time $t = 1$ the position vector of the particle is

$$\mathbf{r}(1) = 0\mathbf{i} + \frac{5}{2}\mathbf{j} + \frac{13}{3}\mathbf{k}$$

so its coordinates at that instant are $\left(0, \frac{5}{2}, \frac{13}{3}\right)$. ◄

DISPLACEMENT AND DISTANCE TRAVELED

If a particle travels along a curve C in 2-space or 3-space, the *displacement* of the particle over the time interval $t_1 \leq t \leq t_2$ is commonly denoted by $\Delta\mathbf{r}$ and is defined as

$$\Delta\mathbf{r} = \mathbf{r}(t_2) - \mathbf{r}(t_1) \qquad (7)$$

(Figure 14.6.3). The displacement vector, which describes the change in position of the particle during the time interval, can be obtained by integrating the velocity function from

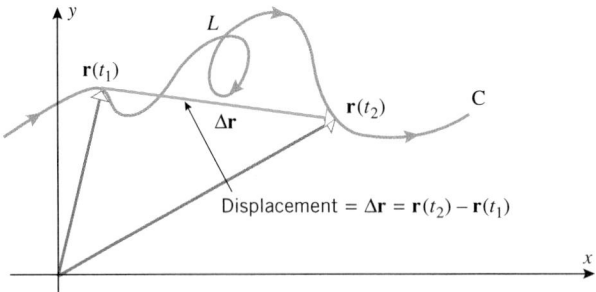

Figure 14.6.3

t_1 to t_2:

$$\Delta \mathbf{r} = \int_{t_1}^{t_2} \mathbf{v}(t)\,dt = \int_{t_1}^{t_2} \frac{d\mathbf{r}}{dt}\,dt = \mathbf{r}(t)\Big]_{t_1}^{t_2} = \mathbf{r}(t_2) - \mathbf{r}(t_1) \qquad \boxed{\text{Displacement}} \qquad (8)$$

It follows from Theorem 14.3.1 that we can find the distance s traveled by a particle over a time interval $t_1 \le t \le t_2$ by integrating the speed over that interval, since

$$s = \int_{t_1}^{t_2} \left\| \frac{d\mathbf{r}}{dt} \right\| dt = \int_{t_1}^{t_2} \| \mathbf{v}(t) \|\, dt \qquad \boxed{\text{Distance traveled}} \qquad (9)$$

Example 3

Suppose that a particle moves along a circular helix in 3-space so that its position vector at time t is

$$\mathbf{r}(t) = (4\cos \pi t)\mathbf{i} + (4\sin \pi t)\mathbf{j} + t\mathbf{k}$$

Find the distance traveled and the displacement of the particle during the time interval $1 \le t \le 5$.

Solution. We have

$$\mathbf{v}(t) = \frac{d\mathbf{r}}{dt} = (-4\pi \sin \pi t)\mathbf{i} + (4\pi \cos \pi t)\mathbf{j} + \mathbf{k}$$

$$\| \mathbf{v}(t) \| = \sqrt{(-4\pi \sin \pi t)^2 + (4\pi \cos \pi t)^2 + 1} = \sqrt{16\pi^2 + 1}$$

Thus, it follows from (9) that the distance traveled by the particle from time $t = 1$ to $t = 5$ is

$$s = \int_1^5 \sqrt{16\pi^2 + 1}\,dt = 4\sqrt{16\pi^2 + 1}$$

Moreover, it follows from (8) that the displacement over the time interval is

$$\Delta \mathbf{r} = \mathbf{r}(5) - \mathbf{r}(1)$$
$$= (4\cos 5\pi \mathbf{i} + 4\sin 5\pi \mathbf{j} + 5\mathbf{k}) - (4\cos \pi \mathbf{i} + 4\sin \pi \mathbf{j} + \mathbf{k})$$
$$= (-4\mathbf{i} + 5\mathbf{k}) - (-4\mathbf{i} + \mathbf{k}) = 4\mathbf{k}$$

which tells us that the change in the position of the particle over the time interval was 4 units straight up. ◄

NORMAL AND TANGENTIAL COMPONENTS OF ACCELERATION

You know from your experience as an automobile passenger that if a car speeds up rapidly, then your body is thrown back against the backrest of the seat. You also know that if the car rounds a turn in the road, then your body is thrown toward the outside of the curve—the greater the curvature in the road, the greater the force with which you are thrown. The explanation of these effects can be understood by resolving the velocity and acceleration components of the motion into vector components that are parallel to the unit tangent and unit normal vectors. The following theorem explains how this can be done.

> **14.6.2 THEOREM.** *If a particle moves along a smooth curve C in 2-space or 3-space, then at each point on the curve velocity and acceleration vectors can be written as*
>
> $$\mathbf{v} = \frac{ds}{dt}\mathbf{T} \tag{10}$$
>
> $$\mathbf{a} = \frac{d^2s}{dt^2}\mathbf{T} + \kappa\left(\frac{ds}{dt}\right)^2\mathbf{N} \tag{11}$$
>
> *where s is an arc length parameter for the curve, and* $\mathbf{T}$, $\mathbf{N}$, *and* κ *denote the unit tangent vector, unit normal vector, and curvature at the point* (Figure 14.6.4).

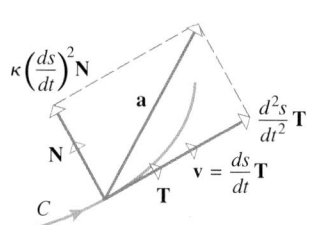

Figure 14.6.4

Proof. Formula (10) is just a restatement of (1). To obtain (11), we differentiate both sides of (10) with respect to t; this yields

$$\mathbf{a} = \frac{d}{dt}\left(\frac{ds}{dt}\mathbf{T}\right) = \frac{d^2s}{dt^2}\mathbf{T} + \frac{ds}{dt}\frac{d\mathbf{T}}{dt}$$

$$= \frac{d^2s}{dt^2}\mathbf{T} + \frac{ds}{dt}\frac{d\mathbf{T}}{ds}\frac{ds}{dt}$$

$$= \frac{d^2s}{dt^2}\mathbf{T} + \left(\frac{ds}{dt}\right)^2\frac{d\mathbf{T}}{ds}$$

$$= \frac{d^2s}{dt^2}\mathbf{T} + \left(\frac{ds}{dt}\right)^2\kappa\mathbf{N} \qquad \text{Formula (10) of Section 14.5}$$

from which (11) follows. ∎

The coefficients of $\mathbf{T}$ and $\mathbf{N}$ in (11) are commonly denoted by

$$a_T = \frac{d^2s}{dt^2} \qquad a_N = \kappa\left(\frac{ds}{dt}\right)^2 \tag{12–13}$$

in which case Formula (11) is expressed as

$$\mathbf{a} = a_T\mathbf{T} + a_N\mathbf{N} \tag{14}$$

In this formula the scalars a_T and a_N are called the ***tangential scalar component of acceleration*** and the ***normal scalar component of acceleration***, and the vectors $a_T\mathbf{T}$ and $a_N\mathbf{N}$ are called the ***tangential vector component of acceleration*** and the ***normal vector component of acceleration***.

The scalar components of acceleration explain the effect that you experience when a car speeds up rapidly or rounds a turn. The rapid increase in speed produces a large value for d^2s/dt^2, which results in a large tangential scalar component of acceleration; and by Newton's second law this produces a large tangential force on the car in the direction of motion. To understand the effect of rounding a turn, observe that the normal scalar component of acceleration has the curvature κ and the speed ds/dt as factors. Thus, sharp turns or turns taken at high speed both produce large normal forces on the car.

REMARK. Formula (14) applies to motion in both 2-space and 3-space. What is interesting is that the 3-space formula does not involve the binormal vector $\mathbf{B}$, so the acceleration vector always lies in the plane of $\mathbf{T}$ and $\mathbf{N}$ (the osculating plane), even for the most twisting paths of motion (Figure 14.6.5).

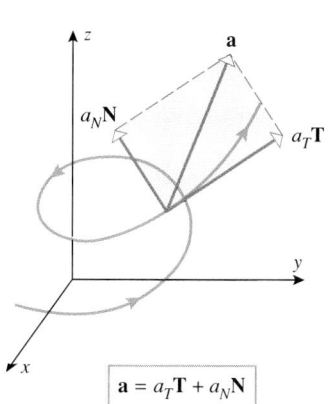

Figure 14.6.5

Although Formulas (12) and (13) provide useful insight into the behavior of particles moving along curved paths, they are not always the best formulas for computations. The following theorem provides some more useful formulas that relate a_T, a_N, and κ to the velocity $\mathbf{v}$ and acceleration $\mathbf{a}$.

14.6.3 THEOREM. *If a particle moves along a smooth curve C in 2-space or 3-space, then at each point on the curve the velocity $\mathbf{v}$ and the acceleration $\mathbf{a}$ are related to a_T, a_N, and κ by the formulas*

$$a_T = \frac{\mathbf{v} \cdot \mathbf{a}}{\|\mathbf{v}\|} \qquad a_N = \frac{\|\mathbf{v} \times \mathbf{a}\|}{\|\mathbf{v}\|} \qquad \kappa = \frac{\|\mathbf{v} \times \mathbf{a}\|}{\|\mathbf{v}\|^3} \qquad (15\text{--}17)$$

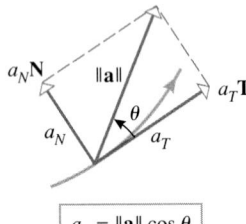

$a_T = \|\mathbf{a}\| \cos\theta$
$a_N = \|\mathbf{a}\| \sin\theta$

Figure 14.6.6

Proof. As illustrated in Figure 14.6.6, let θ be the angle between the vector $\mathbf{a}$ and the vector $a_T\mathbf{T}$. Thus,

$$a_T = \|\mathbf{a}\| \cos\theta \quad \text{and} \quad a_N = \|\mathbf{a}\| \sin\theta$$

from which we obtain

$$a_T = \|\mathbf{a}\| \cos\theta = \frac{\|\mathbf{v}\| \|\mathbf{a}\| \cos\theta}{\|\mathbf{v}\|} = \frac{\mathbf{v} \cdot \mathbf{a}}{\|\mathbf{v}\|}$$

$$a_N = \|\mathbf{a}\| \sin\theta = \frac{\|\mathbf{v}\| \|\mathbf{a}\| \sin\theta}{\|\mathbf{v}\|} = \frac{\|\mathbf{v} \times \mathbf{a}\|}{\|\mathbf{v}\|}$$

$$\kappa = \frac{a_N}{(ds/dt)^2} = \frac{a_N}{\|\mathbf{v}\|^2} = \frac{1}{\|\mathbf{v}\|^2} \frac{\|\mathbf{v} \times \mathbf{a}\|}{\|\mathbf{v}\|} = \frac{\|\mathbf{v} \times \mathbf{a}\|}{\|\mathbf{v}\|^3}$$ ∎

REMARK. Theorem 14.6.3 applies to motion in 2-space and 3-space, but for motion in 2-space you will have to add a zero $\mathbf{k}$ component to $\mathbf{v}$ and $\mathbf{a}$ to calculate the cross product. Also, recall that for nonlinear smooth curves in 2-space the unit normal vector $\mathbf{N}$ is the inward normal; that is, it points toward the concave side of the curve. Thus, the same is true for $a_N\mathbf{N}$, since a_N is a nonnegative scalar.

Example 4

Suppose that a particle moves through 3-space so that its position vector at time t is

$$\mathbf{r}(t) = t\mathbf{i} + t^2\mathbf{j} + t^3\mathbf{k}$$

(The path is the twisted cubic shown in Figure 14.1.5.)

(a) Find the scalar tangential and normal components of acceleration at time t.

(b) Find the scalar tangential and normal components of acceleration at time $t = 1$.

(c) Find the vector tangential and normal components of acceleration at time $t = 1$.

(d) Find the curvature of the path at the point where the particle is located at time $t = 1$.

Solution (a). We have

$$\mathbf{v}(t) = \mathbf{r}'(t) = \mathbf{i} + 2t\mathbf{j} + 3t^2\mathbf{k}$$

$$\mathbf{a}(t) = \mathbf{v}'(t) = 2\mathbf{j} + 6t\mathbf{k}$$

$$\|\mathbf{v}(t)\| = \sqrt{1 + 4t^2 + 9t^4}$$

$$\mathbf{v}(t) \cdot \mathbf{a}(t) = 4t + 18t^3$$

$$\mathbf{v}(t) \times \mathbf{a}(t) = \begin{vmatrix} \mathbf{i} & \mathbf{j} & \mathbf{k} \\ 1 & 2t & 3t^2 \\ 0 & 2 & 6t \end{vmatrix} = 6t^2\mathbf{i} - 6t\mathbf{j} + 2\mathbf{k}$$

Thus, from (15) and (16)

$$a_T = \frac{\mathbf{v} \cdot \mathbf{a}}{\|\mathbf{v}\|} = \frac{4t + 18t^3}{\sqrt{1 + 4t^2 + 9t^4}}$$

$$a_N = \frac{\|\mathbf{v} \times \mathbf{a}\|}{\|\mathbf{v}\|} = \frac{\sqrt{36t^4 + 36t^2 + 4}}{\sqrt{1 + 4t^2 + 9t^4}} = 2\sqrt{\frac{9t^4 + 9t^2 + 1}{9t^4 + 4t^2 + 1}}$$

Solution (b). At time $t = 1$, the components a_T and a_N in part (a) are

$$a_T = \frac{22}{\sqrt{14}} \approx 5.88 \quad \text{and} \quad a_N = 2\sqrt{\frac{19}{14}} \approx 2.33$$

Solution (c). Since $\mathbf{T}$ and $\mathbf{v}$ have the same direction, $\mathbf{T}$ can be obtained by normalizing $\mathbf{v}$, that is,

$$\mathbf{T}(t) = \frac{\mathbf{v}(t)}{\|\mathbf{v}(t)\|}$$

At time $t = 1$ we have

$$\mathbf{T}(1) = \frac{\mathbf{v}(1)}{\|\mathbf{v}(1)\|} = \frac{\mathbf{i} + 2\mathbf{j} + 3\mathbf{k}}{\|\mathbf{i} + 2\mathbf{j} + 3\mathbf{k}\|} = \frac{1}{\sqrt{14}}(\mathbf{i} + 2\mathbf{j} + 3\mathbf{k})$$

From this and part (b) we obtain the vector tangential component of acceleration:

$$a_T(1)\mathbf{T}(1) = \frac{22}{\sqrt{14}}\mathbf{T}(1) = \frac{11}{7}(\mathbf{i} + 2\mathbf{j} + 3\mathbf{k}) = \frac{11}{7}\mathbf{i} + \frac{22}{7}\mathbf{j} + \frac{33}{7}\mathbf{k}$$

To find the normal vector component of acceleration, we rewrite $\mathbf{a} = a_T\mathbf{T} + a_N\mathbf{N}$ as

$$a_N\mathbf{N} = \mathbf{a} - a_T\mathbf{T}$$

Thus, at time $t = 1$ the normal vector component of acceleration is

$$a_N(1)\mathbf{N}(1) = \mathbf{a}(1) - a_T(1)\mathbf{T}(1)$$

$$= (2\mathbf{j} + 6\mathbf{k}) - \left(\frac{11}{7}\mathbf{i} + \frac{22}{7}\mathbf{j} + \frac{33}{7}\mathbf{k}\right)$$

$$= -\frac{11}{7}\mathbf{i} - \frac{8}{7}\mathbf{j} + \frac{9}{7}\mathbf{k}$$

Solution (d). We will apply Formula (17) with $t = 1$. From part (a)

$$\|\mathbf{v}(1)\| = \sqrt{14} \quad \text{and} \quad \mathbf{v}(1) \times \mathbf{a}(1) = 6\mathbf{i} - 6\mathbf{j} + 2\mathbf{k}$$

Thus, at time $t = 1$

$$\kappa = \frac{\|\mathbf{v} \times \mathbf{a}\|}{\|\mathbf{v}\|^3} = \frac{\sqrt{76}}{(\sqrt{14})^3} = \frac{1}{14}\sqrt{\frac{38}{7}} \approx 0.17 \qquad \blacktriangleleft$$

FOR THE READER.   It follows from Figure 14.6.6 and the Theorem of Pythagoras that a_N can be expressed in terms of $\|\mathbf{v}\|$ and a_T as

$$a_N = \sqrt{\|\mathbf{a}\|^2 - a_T^2} \qquad (18)$$

Confirm that this is so in Example 4.

A MODEL OF PROJECTILE MOTION

Earlier in this text we examined various problems concerned with objects moving *vertically* in the Earth's gravitational field (see model 6.3.4, Example 4 of Section 7.7, and the subsection of Section 10.1 entitled A Model of Free-Fall Motion Retarded by Air Resistance). Now we will consider the motion of a projectile launched along a *curved* path in the Earth's gravitational field. For this purpose we will need the vector version of Newton's Second

Law of Motion (10.1.1)

$$\mathbf{F} = m\mathbf{a} \tag{19}$$

and we will need to make three modeling assumptions:

- The mass m of the object is constant.
- The only force acting on the object after it is launched is the force of the Earth's gravity. (Thus, air resistance and the gravitational effect of other planets and celestial objects are ignored.)
- The object remains sufficiently close to the Earth that we can assume the force of gravity to be constant.

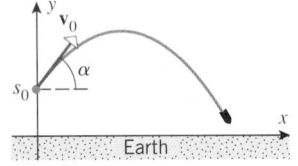

Figure 14.6.7

Let us assume that at time $t = 0$ an object of mass m is launched from a height of s_0 above the Earth with an initial velocity vector of $\mathbf{v}_0$ that makes an angle α with the horizontal. Furthermore, let us introduce an xy-coordinate system as shown in Figure 14.6.7. In this coordinate system the positive y-direction is up, the origin is at the surface of the Earth, and the initial coordinate of the object is $(0, s_0)$. Our objective is to use basic principles of physics to derive the velocity function $\mathbf{v}(t)$ and the position function $\mathbf{r}(t)$ from the acceleration function $\mathbf{a}(t)$ of the object. Our starting point is the physical observation that the downward force $\mathbf{F}$ of the Earth's gravity on an object of mass m is

$$\mathbf{F} = -mg\mathbf{j}$$

where g is the acceleration due to gravity (see 10.3.3). It follows from this fact and Newton's second law (19) that

$$m\mathbf{a} = -mg\mathbf{j}$$

or on canceling m from both sides

$$\mathbf{a} = -g\mathbf{j} \tag{20}$$

Observe that this acceleration function does not involve t and hence is constant. We can now obtain the velocity function $\mathbf{v}(t)$ by integrating this acceleration function and using the initial condition $\mathbf{v}(0) = \mathbf{v}_0$ to find the constant of integration. Integrating (20) with respect to t and keeping in mind that $-g\mathbf{j}$ is constant yields

$$\mathbf{v}(t) = \int -g\mathbf{j}\, dt = -gt\mathbf{j} + \mathbf{c}_1$$

where $\mathbf{c}_1$ is a vector constant of integration. Substituting $t = 0$ in this equation and using the initial condition $\mathbf{v}(0) = \mathbf{v}_0$ yields

$$\mathbf{v}_0 = \mathbf{c}_1$$

Thus, the velocity function of the object is

$$\mathbf{v}(t) = -gt\mathbf{j} + \mathbf{v}_0 \tag{21}$$

To obtain the position function $\mathbf{r}(t)$ of the object, we will integrate the velocity function and use the known initial position of the object to find the constant of integration. For this purpose observe that the object has coordinates $(0, s_0)$ at time $t = 0$, so the position vector at that time is

$$\mathbf{r}(0) = 0\mathbf{i} + s_0\mathbf{j} = s_0\mathbf{j} \tag{22}$$

This is the initial condition that we will need to find the constant of integration. Integrating (21) with respect to t yields

$$\mathbf{r}(t) = \int (-gt\mathbf{j} + \mathbf{v}_0)\, dt = -\tfrac{1}{2}gt^2\mathbf{j} + t\mathbf{v}_0 + \mathbf{c}_2 \tag{23}$$

where $\mathbf{c}_2$ is another vector constant of integration. Substituting $t = 0$ in (23) and using initial condition (22) yields

$$s_0 \mathbf{j} = \mathbf{c}_2$$

so that (23) can be written as

$$\mathbf{r}(t) = \left(-\tfrac{1}{2}gt^2 + s_0\right)\mathbf{j} + t\mathbf{v}_0 \tag{24}$$

This formula expresses the position function of the object in terms of its known initial position and velocity.

REMARK. Observe that the mass of the object does not enter into the final formulas for velocity and position. Physically, this means that the mass has no influence on the trajectory or the velocity of the object—these are completely determined by the initial position and velocity. This explains the famous observation of Galileo that two objects of different mass, released from the same height, will reach the ground at the same time if air resistance is neglected.

PARAMETRIC EQUATIONS OF PROJECTILE MOTION

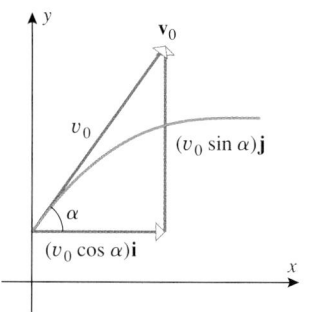

Figure 14.6.8

Formulas (21) and (24) can be used to obtain parametric equations for the position and velocity in terms of the initial speed of the object and the angle that the initial velocity vector makes with the positive x-axis. For this purpose, let $v_0 = \|\mathbf{v}_0\|$ be the initial speed, let α be the angle that the initial velocity vector $\mathbf{v}_0$ makes with the positive x-axis, let v_x and v_y be the horizontal and vertical scalar components of $\mathbf{v}(t)$ at time t, and let x and y be the horizontal and vertical components of $\mathbf{r}(t)$ at time t. As illustrated in Figure 14.6.8, the initial velocity vector can be expressed as

$$\mathbf{v}_0 = (v_0 \cos \alpha)\mathbf{i} + (v_0 \sin \alpha)\mathbf{j} \tag{25}$$

Substituting this expression in (24) and combining like components yields (verify)

$$\mathbf{r}(t) = (v_0 \cos \alpha)t\mathbf{i} + \left(s_0 + (v_0 \sin \alpha)t - \tfrac{1}{2}gt^2\right)\mathbf{j} \tag{26}$$

which is equivalent to the parametric equations

$$x = (v_0 \cos \alpha)t, \quad y = s_0 + (v_0 \sin \alpha)t - \tfrac{1}{2}gt^2 \tag{27}$$

Similarly, substituting (25) in (21) and combining like components yields

$$\mathbf{v}(t) = (v_0 \cos \alpha)\mathbf{i} + (v_0 \sin \alpha - gt)\mathbf{j}$$

which is equivalent to the parametric equations

$$v_x = v_0 \cos \alpha, \quad v_y = v_0 \sin \alpha - gt \tag{28}$$

The parameter t can be eliminated in (27) by solving the first equation for t and substituting in the second equation. We leave it for you to show that this yields

$$y = s_0 + (\tan \alpha)x - \left(\frac{g}{2v_0^2 \cos^2 \alpha}\right)x^2 \tag{29}$$

which is the equation of a parabola, since the right side is a quadratic polynomial in x. Thus, we have shown that the trajectory of the projectile is a parabolic arc.

Example 5

A shell, fired from a cannon, has a muzzle speed (the speed as it leaves the barrel) of 800 ft/s. The barrel makes an angle of $45°$ with the horizontal and, for simplicity, the barrel opening is assumed to be at ground level.

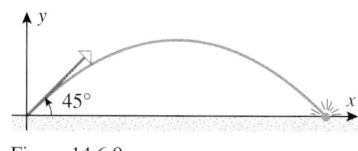

Figure 14.6.9

(a)   Find parametric equations for the shell's trajectory relative to the coordinate system in Figure 14.6.9.

(b)   How high does the shell rise?

(c)   How far does the shell travel horizontally?

(d)   What is the speed of the shell at its point of impact with the ground?

Solution (*a*).   From (27) with $v_0 = 800$ ft/s, $\alpha = 45°$, $s_0 = 0$ ft (since the shell starts at ground level), and $g = 32$ ft/s^2, we obtain the parametric equations

$$x = (800 \cos 45°)t, \quad y = (800 \sin 45°)t - 16t^2 \qquad (t \geq 0)$$

which simplify to

$$x = 400\sqrt{2}t, \quad y = 400\sqrt{2}t - 16t^2 \qquad (t \geq 0) \tag{30}$$

Solution (*b*).   The maximum height of the shell is the maximum value of y in (30), which occurs when $dy/dt = 0$, that is, when

$$400\sqrt{2} - 32t = 0 \quad \text{or} \quad t = \frac{25\sqrt{2}}{2}$$

Substituting this value of t in (30) yields

$$y = 5000 \text{ ft}$$

as the maximum height of the shell.

Solution (*c*).   The shell will hit the ground when $y = 0$. From (30), this occurs when

$$400\sqrt{2}t - 16t^2 = 0 \quad \text{or} \quad t(400\sqrt{2} - 16t) = 0$$

The solution $t = 0$ corresponds to the initial position of the shell and the solution $t = 25\sqrt{2}$ to the time of impact. Substituting the latter value in the equation for x in (30) yields

$$x = 20{,}000 \text{ ft}$$

as the horizontal distance traveled by the shell.

Solution (*d*).   From (30), the position function of the shell is

$$\mathbf{r}(t) = 400\sqrt{2}t\mathbf{i} + (400\sqrt{2}t - 16t^2)\mathbf{j}$$

so that the velocity function is

$$\mathbf{v}(t) = \mathbf{r}'(t) = 400\sqrt{2}\mathbf{i} + (400\sqrt{2} - 32t)\mathbf{j}$$

From part (c), impact occurs when $t = 25\sqrt{2}$, so the velocity vector at this point is

$$\mathbf{v}(25\sqrt{2}) = 400\sqrt{2}\mathbf{i} + [400\sqrt{2} - 32(25\sqrt{2})]\mathbf{j} = 400\sqrt{2}\mathbf{i} - 400\sqrt{2}\mathbf{j}$$

Thus, the speed at impact is

$$\|\mathbf{v}(25\sqrt{2})\| = \sqrt{(400\sqrt{2})^2 + (-400\sqrt{2})^2} = 800 \text{ ft/s} \qquad \blacktriangleleft$$

EXERCISE SET 14.6   ⬿ Graphing Calculator   ⌷ CAS

In Exercises 1–4, $\mathbf{r}(t)$ is the position vector of a particle moving in the plane. Find the velocity, acceleration, and speed at an arbitrary time t. Then sketch the path of the particle together with the velocity and acceleration vectors at the indicated time t.

1. $\mathbf{r}(t) = 3\cos t\mathbf{i} + 3\sin t\mathbf{j}$; $t = \pi/3$

2. $\mathbf{r}(t) = t\mathbf{i} + t^2\mathbf{j}$; $t = 2$

3. $\mathbf{r}(t) = e^t\mathbf{i} + e^{-t}\mathbf{j}$; $t = 0$

4. $\mathbf{r}(t) = (2 + 4t)\mathbf{i} + (1 - t)\mathbf{j}$; $t = 1$

In Exercises 5–8, find the velocity, speed, and acceleration at the given time t of a particle moving along the given curve.

5. $\mathbf{r}(t) = t\mathbf{i} + \frac{1}{2}t^2\mathbf{j} + \frac{1}{3}t^3\mathbf{k}$; $t = 1$

6. $x = 1 + 3t, \ y = 2 - 4t, \ z = 7 + t$; $t = 2$

7. $x = 2\cos t$, $y = 2\sin t$, $z = t$; $t = \pi/4$

8. $\mathbf{r}(t) = e^t \sin t\,\mathbf{i} + e^t \cos t\,\mathbf{j} + t\mathbf{k}$; $t = \pi/2$

9. As illustrated in the accompanying figure, suppose that the equations of motion of a particle moving along an elliptic path are $x = a\cos\omega t$, $y = b\sin\omega t$.
   (a) Show that the acceleration is directed toward the origin.
   (b) Show that the magnitude of the acceleration is proportional to the distance from the particle to the origin.

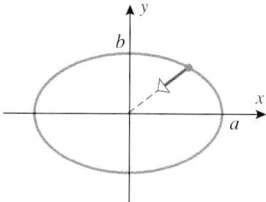

Figure Ex-9

10. Suppose that a particle vibrates in such a way that its position function is $\mathbf{r}(t) = 16\sin \pi t\,\mathbf{i} + 4\cos 2\pi t\,\mathbf{j}$, where distance is in millimeters and t is in seconds.
   (a) Find the velocity and acceleration at time $t = 1$ s.
   (b) Show that the particle moves along a parabolic curve.
   (c) Show that the particle moves back and forth along the curve.

11. Suppose that the position vector of a particle moving in the plane is $\mathbf{r} = 12\sqrt{t}\,\mathbf{i} + t^{3/2}\mathbf{j}$, $t > 0$. Find the minimum speed of the particle and its location when it has this speed.

12. Suppose that the motion of a particle is described by the position vector $\mathbf{r} = (t - t^2)\mathbf{i} - t^2\mathbf{j}$. Find the minimum speed of the particle and its location when it has this speed.

13. Suppose that the position function of a particle moving in 2-space is $\mathbf{r} = \sin 3t\,\mathbf{i} - 2\cos 3t\,\mathbf{j}$.
   (a) Use a graphing utility to graph the speed of the particle versus time from $t = 0$ to $t = 2\pi/3$.
   (b) What are the maximum and minimum speeds of the particle?
   (c) Use the graph to estimate the time at which the maximum speed first occurs.
   (d) Find the exact time at which the maximum speed first occurs.

14. Suppose that the position function of a particle moving in 3-space is $\mathbf{r} = 3\cos 2t\,\mathbf{i} + \sin 2t\,\mathbf{j} + 4t\mathbf{k}$.
   (a) Use a graphing utility to graph the speed of the particle versus time from $t = 0$ to $t = \pi$.
   (b) Use the graph to estimate the maximum and minimum speeds of the particle.
   (c) Use the graph to estimate the time at which the maximum speed first occurs.
   (d) Find the exact values of the maximum and minimum speeds and the exact time at which the maximum speed first occurs.

In Exercises 15–18, use the given information to find the position and velocity vectors of the particle.

15. $\mathbf{a}(t) = -\cos t\,\mathbf{i} - \sin t\,\mathbf{j}$; $\mathbf{v}(0) = \mathbf{i}$; $\mathbf{r}(0) = \mathbf{j}$

16. $\mathbf{a}(t) = \mathbf{i} + e^{-t}\mathbf{j}$; $\mathbf{v}(0) = 2\mathbf{i} + \mathbf{j}$; $\mathbf{r}(0) = \mathbf{i} - \mathbf{j}$

17. $\mathbf{a}(t) = \sin t\,\mathbf{i} + \cos t\,\mathbf{j} + e^t\mathbf{k}$; $\mathbf{v}(0) = \mathbf{k}$; $\mathbf{r}(0) = -\mathbf{i} + \mathbf{k}$

18. $\mathbf{a}(t) = (t + 1)^{-2}\mathbf{j} - e^{-2t}\mathbf{k}$; $\mathbf{v}(0) = 3\mathbf{i} - \mathbf{j}$; $\mathbf{r}(0) = 2\mathbf{k}$

19. What can you say about the trajectory of a particle that moves in 2-space or 3-space with zero acceleration? Justify your answer.

20. Recall from Theorem 14.2.7 that if $\mathbf{r}(t)$ is a vector-valued function in 2-space or 3-space, and if $\|\mathbf{r}(t)\|$ is constant for all t, then $\mathbf{r}(t) \cdot \mathbf{r}'(t) = 0$.
   (a) Translate this theorem into a statement about the motion of a particle in 2-space or 3-space.
   (b) Replace $\mathbf{r}(t)$ by $\mathbf{r}'(t)$ in the theorem, and translate the result into a statement about the motion of a particle in 2-space or 3-space.

21. Find, to the nearest degree, the angle between $\mathbf{v}$ and $\mathbf{a}$ for $\mathbf{r} = t^3\mathbf{i} + t^2\mathbf{j}$ when $t = 1$.

22. Show that the angle between $\mathbf{v}$ and $\mathbf{a}$ is constant for the position vector $\mathbf{r} = e^t \cos t\,\mathbf{i} + e^t \sin t\,\mathbf{j}$. Find the angle.

23. (a) Suppose that at time $t = t_0$ an electron has a position vector of $\mathbf{r} = 3.5\mathbf{i} - 1.7\mathbf{j} + \mathbf{k}$, and at a later time $t = t_1$ it has a position vector of $\mathbf{r} = 4.2\mathbf{i} + \mathbf{j} - 2.4\mathbf{k}$. What is the displacement of the electron during the time interval from t_0 to t_1?
   (b) Suppose that during a certain time interval a proton has a displacement of $\Delta\mathbf{r} = 0.7\mathbf{i} + 2.9\mathbf{j} - 1.2\mathbf{k}$ and its final position vector is known to be $\mathbf{r} = 3.6\mathbf{k}$. What was the initial position vector of the proton?

24. Suppose that the position function of a particle moving along a circle in the xy-plane is $\mathbf{r} = 5\cos 2\pi t\,\mathbf{i} + 5\sin 2\pi t\,\mathbf{j}$.
   (a) Sketch some typical displacement vectors over the time interval from $t = 0$ to $t = 1$.
   (b) What is the distance traveled by the particle during the time interval?

In Exercises 25–28, find the displacement and the distance traveled over the indicated time interval.

25. $\mathbf{r} = t^2\mathbf{i} + \tfrac{1}{3}t^3\mathbf{j}$; $1 \le t \le 3$

26. $\mathbf{r} = (1 - 3\sin t)\mathbf{i} + 3\cos t\,\mathbf{j}$; $0 \le t \le 3\pi/2$

27. $\mathbf{r} = e^t\mathbf{i} + e^{-t}\mathbf{j} + \sqrt{2}t\mathbf{k}$; $0 \le t \le \ln 3$

28. $\mathbf{r} = \cos 2t\,\mathbf{i} + (1 - \cos 2t)\mathbf{j} + \left(3 + \tfrac{1}{2}\cos 2t\right)\mathbf{k}$; $0 \le t \le \pi$

In Exercises 29 and 30, the position vectors of two particles are given. Show that the particles move along the same path but the speed of the first is constant and the speed of the second is not.

29. $\mathbf{r}_1 = 2\cos 3t\,\mathbf{i} + 2\sin 3t\,\mathbf{j}$
$\mathbf{r}_2 = 2\cos(t^2)\mathbf{i} + 2\sin(t^2)\mathbf{j}$ $(t \ge 0)$

30. $\mathbf{r}_1 = (3 + 2t)\mathbf{i} + t\mathbf{j} + (1 - t)\mathbf{k}$
$\mathbf{r}_2 = (5 - 2t^3)\mathbf{i} + (1 - t^3)\mathbf{j} + t^3\mathbf{k}$

In Exercises 31–38, the position function of a particle is given. Use Theorem 14.6.3 to find
(a) the scalar tangential and normal components of acceleration at the stated time t;
(b) the vector tangential and normal components of acceleration at the stated time t;
(c) the curvature of the path at the point where the particle is located at the stated time t.

31. $\mathbf{r} = e^{-t}\mathbf{i} + e^{t}\mathbf{j}$; $t = 0$

32. $\mathbf{r} = \cos(t^2)\mathbf{i} + \sin(t^2)\mathbf{j}$; $t = \sqrt{\pi}/2$

33. $\mathbf{r} = (t^3 - 2t)\mathbf{i} + (t^2 - 4)\mathbf{j}$; $t = 1$

34. $\mathbf{r} = e^{t}\cos t\,\mathbf{i} + e^{t}\sin t\,\mathbf{j}$; $t = \pi/4$

35. $\mathbf{r} = t\mathbf{i} + t^2\mathbf{j} + t^3\mathbf{k}$; $t = 1$

36. $\mathbf{r} = e^{t}\mathbf{i} + e^{-2t}\mathbf{j} + t\mathbf{k}$; $t = 0$

37. $\mathbf{r} = 3\sin t\,\mathbf{i} + 2\cos t\,\mathbf{j} - \sin 2t\,\mathbf{k}$; $t = \pi/2$

38. $\mathbf{r} = 2\mathbf{i} + t^3\mathbf{j} - 16\ln t\,\mathbf{k}$; $t = 1$

In Exercises 39–42, $\mathbf{v}$ and $\mathbf{a}$ are given at a certain instant of time. Find a_T, a_N, $\mathbf{T}$, and $\mathbf{N}$ at this instant.

39. $\mathbf{v} = -4\mathbf{j}$, $\mathbf{a} = 2\mathbf{i} + 3\mathbf{j}$ **40.** $\mathbf{v} = \mathbf{i} + 2\mathbf{j}$, $\mathbf{a} = 3\mathbf{i}$

41. $\mathbf{v} = 2\mathbf{i} + 2\mathbf{j} + \mathbf{k}$, $\mathbf{a} = \mathbf{i} + 2\mathbf{k}$

42. $\mathbf{v} = 3\mathbf{i} - 4\mathbf{k}$, $\mathbf{a} = \mathbf{i} - \mathbf{j} + 2\mathbf{k}$

In Exercises 43–46, the speed $\|\mathbf{v}\|$ of a particle at an arbitrary time t is given. Find the scalar tangential component of acceleration at the indicated time.

43. $\|\mathbf{v}\| = \sqrt{3t^2 + 4}$; $t = 2$

44. $\|\mathbf{v}\| = \sqrt{t^2 + e^{-3t}}$; $t = 0$

45. $\|\mathbf{v}\| = \sqrt{(4t - 1)^2 + \cos^2 \pi t}$; $t = \frac{1}{4}$

46. $\|\mathbf{v}\| = \sqrt{t^4 + 5t^2 + 3}$; $t = 1$

47. The nuclear accelerator at the Enrico Fermi Laboratory is circular with a radius of 1 km. Find the scalar normal component of acceleration of a proton moving around the accelerator with a constant speed of 2.9×10^5 km/s.

48. Suppose that a particle moves with nonzero acceleration along the curve $y = f(x)$. Use part (b) of Exercise 17 in Section 14.5 to show that the acceleration vector is tangent to the curve at each point where $f''(x) = 0$.

In Exercises 49 and 50, use the given information and Exercise 17 of Section 14.5 to find the normal scalar component of acceleration as a function of x.

49. A particle moves along the parabola $y = x^2$ with a constant speed of 3 units per second.

50. A particle moves along the curve $x = \ln y$ with a constant speed of 2 units per second.

In Exercises 51 and 52, use the given information to find the normal scalar component of acceleration at time $t = 1$.

51. $\mathbf{a}(1) = \mathbf{i} + 2\mathbf{j} - 2\mathbf{k}$; $a_T(1) = 3$

52. $\|\mathbf{a}(1)\| = 9$; $a_T(1)\mathbf{T}(1) = 2\mathbf{i} - 2\mathbf{j} + \mathbf{k}$

53. An automobile travels at a constant speed around a curve whose radius of curvature is 1000 m. What is the maximum allowable speed if the maximum acceptable value for the normal scalar component of acceleration is 1.5 m/s^2?

54. If an automobile of mass m rounds a curve, then its inward vector component of acceleration $a_N \mathbf{N}$ is caused by the frictional force $\mathbf{F}$ of the road. Thus, it follows from the vector form of Newton's second law (Equation (19)) that the frictional force and the normal scalar component of acceleration are related by the equation $\mathbf{F} = m a_N \mathbf{N}$. Thus,

$$\|\mathbf{F}\| = m\kappa \left(\frac{ds}{dt}\right)^2$$

Use this result to find the magnitude of the frictional force in newtons exerted by the road on a 500-kg go-cart driven at a speed of 10 km/h around a circular track of radius 15 m. [*Note:* 1 N $= 1$ kg·m/s^2]

55. A shell is fired from ground level with a muzzle speed of 320 ft/s and elevation angle of 60°. Find
(a) parametric equations for the shell's trajectory
(b) the maximum height reached by the shell
(c) the horizontal distance traveled by the shell
(d) the speed of the shell at impact.

56. Solve Exercise 55 assuming that the muzzle speed is 980 m/s and the elevation angle is 45°.

57. A rock is thrown downward from the top of a building, 168 ft high, at an angle of 60° with the horizontal. How far from the base of the building will the rock land if its initial speed is 80 ft/s?

58. Solve Exercise 57 assuming that the rock is thrown horizontally at a speed of 80 ft/s.

59. A shell is to be fired from ground level at an elevation angle of 30°. What should the muzzle speed be in order for the maximum height of the shell to be 2500 ft?

60. A shell, fired from ground level at an elevation angle of 45°, hits the ground 24,500 m away. Calculate the muzzle speed of the shell.

61. Find two elevation angles that will enable a shell, fired from ground level with a muzzle speed of 800 ft/s, to hit a ground-level target 10,000 ft away.

62. A ball rolls off a table 4 ft high while moving at a constant speed of 5 ft/s.
(a) How long does it take for the ball to hit the floor after it leaves the table?
(b) At what speed does the ball hit the floor?
(c) If a ball were dropped from rest at table height just as the rolling ball leaves the table, which ball would hit the ground first? Justify your answer.

63. As illustrated in the accompanying figure, a fire hose sprays water with an initial velocity of 40 ft/s at an angle of $60°$ with the horizontal.

(a) Confirm that the water will clear corner point A.

(b) Confirm that the water will hit the roof.

(c) How far from corner point A will the water hit the roof?

64. What is the minimum initial velocity that will allow the water in Exercise 63 to hit the roof?

65. As shown in the accompanying figure, water is sprayed from a hose with an initial velocity of 35 m/s at an angle of $45°$ with the horizontal.

(a) What is the radius of curvature of the stream at the point where it leaves the hose?

(b) What is the maximum height of the stream above the nozzle of the hose?

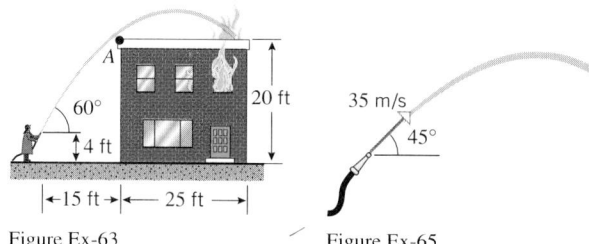

Figure Ex-63   Figure Ex-65

66. As illustrated in the accompanying figure, a train is traveling on a curved track. At a point where the train is traveling at a speed of 132 ft/s and the radius of curvature of the track is 3000 ft, the engineer hits the brakes to make the train slow down at a constant rate of 7.5 ft/s^2.

(a) Find the magnitude of the acceleration vector at the instant the engineer hits the brakes.

(b) Approximate the angle between the acceleration vector and the unit tangent vector $\mathbf{T}$ at the instant the engineer hits the brakes.

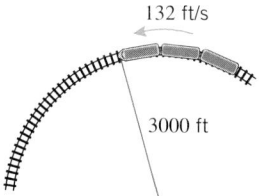

Figure Ex-66

67. A shell is fired from ground level at an elevation angle of α and a muzzle speed of v_0.

(a) Show that the maximum height reached by the shell is

$$\text{maximum height} = \frac{(v_0 \sin \alpha)^2}{2g}$$

(b) The *horizontal range* R of the shell is the horizontal distance traveled when the shell returns to ground level. Show that $R = (v_0^2 \sin 2\alpha)/g$. For what elevation angle will the range be maximum? What is the maximum range?

68. A shell is fired from ground level with an elevation angle α and a muzzle speed of v_0. Find two angles that can be used to hit a target at ground level that is a distance of three-fourths the maximum range of the shell. Express your answer to the nearest tenth of a degree. [*Hint:* See Exercise 67(b).]

69. At time $t = 0$ a baseball that is 5 ft above the ground is hit with a bat. The ball leaves the bat with a speed of 80 ft/s at an angle of $30°$ above the horizontal.

(a) How long will it take for the baseball to hit the ground? Express your answer to the nearest hundredth of a second.

(b) Use the result in part (a) to find the horizontal distance traveled by the ball. Express your answer to the nearest tenth of a foot.

70. At time $t = 0$ a projectile is fired from a height h above level ground at an elevation angle of α with a speed v. Let R be the horizontal distance to the point where the projectile hits the ground.

(a) Show that α and R must satisfy the equation

$$g(\sec^2 \alpha)R^2 - 2v^2(\tan \alpha)R - 2v^2h = 0$$

(b) If g, h, and v are constant, then the equation in part (a) defines R implicitly as a function of α. Let R_0 be the maximum value of R and α_0 the value of α when $R = R_0$. Use implicit differentiation to find $dR/d\alpha$ and show that

$$\tan \alpha_0 = \frac{v^2}{g R_0}$$

[*Hint:* Assume that $dR/d\alpha = 0$ when R is maximum.]

(c) Use the results in parts (a) and (b) to show that

$$R_0 = \frac{v}{g}\sqrt{v^2 + 2gh}$$

and

$$\alpha_0 = \tan^{-1} \frac{v}{\sqrt{v^2 + 2gh}}$$

c **71.** At time $t = 0$ a skier leaves the end of a ski jump with a speed of v_0 ft/s at an angle α with the horizontal (see the accompanying figure). The skier lands 259 ft down the incline 2.9 s later.

(a) Approximate v_0 to the nearest ft/s and α to the nearest degree.

(b) Use a CAS or a calculating utility with a numerical integration capability to approximate the distance traveled by the skier.

(Use $g = 32$ ft/s^2 as the acceleration due to gravity.)

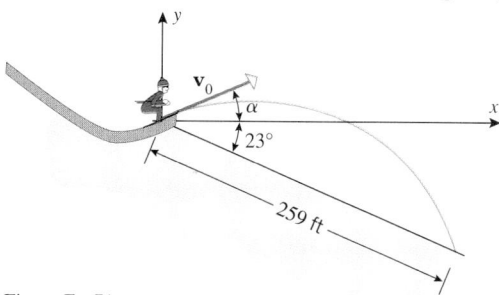

Figure Ex-71

14.7 KEPLER'S LAWS OF PLANETARY MOTION

One of the great advances in the history of astronomy occurred in the early 1600s when Johannes Kepler deduced from empirical data that all planets in our solar system move in elliptical orbits with the Sun at a focus. Subsequently, Isaac Newton showed mathematically that such planetary motion is the consequence of an inverse-square law of gravitational attraction. In this section we will use the concepts developed in the preceding sections of this chapter to derive three basic laws of planetary motion, known as **Kepler's laws**.*

KEPLER'S LAWS

In Section 12.5 we stated the following laws of planetary motion that were published by Johannes Kepler in 1609 in his book known as *Astronomia Nova*.

14.7.1 KEPLER'S LAWS.

- First law (*Law of Orbits*). Each planet moves in an elliptical orbit with the Sun at a focus.

- Second law (*Law of Areas*). Equal areas are swept out in equal times by the line from the Sun to a planet.

- Third law (*Law of Periods*). The square of a planet's period (the time it takes the planet to complete one orbit about the Sun) is proportional to the cube of the semimajor axis of its orbit.

CENTRAL FORCES

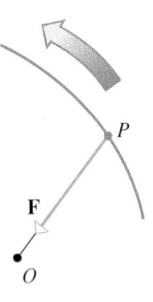

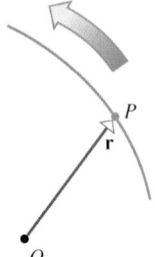

Figure 14.7.1

If a particle moves under the influence of a *single* force that is always directed toward a fixed point O, then the particle is said to be moving in a **central force field**. The force is called a **central force**, and the point O is called the **center of force**. For example, in the simplest model of planetary motion, it is assumed that the only force acting on a planet is the force of the Sun's gravity, directed toward the center of the Sun. This model, which produces Kepler's laws, ignores the forces that other celestial objects exert on the planet as well as the minor effect that the planet's gravity has on the Sun. Central force models are also used to study the motion of comets, asteroids, planetary moons, and artificial satellites. They also have important applications in electromagnetics. Our objective in this section is to develop some basic principles about central force fields and then use those results to derive Kepler's laws.

Suppose that a particle P of mass m moves in a central force field due to a force $\mathbf{F}$ that is directed toward a fixed point O, and let $\mathbf{r} = \mathbf{r}(t)$ be the position vector from O to P (Figure 14.7.1). Let $\mathbf{v} = \mathbf{v}(t)$ and $\mathbf{a} = \mathbf{a}(t)$ be the velocity and acceleration functions of the particle, and assume that $\mathbf{F}$ and $\mathbf{a}$ are related by Newton's second law ($\mathbf{F} = m\mathbf{a}$).

Our first objective is to show that the particle P moves in a plane containing the point O. For this purpose observe that $\mathbf{a}$ has the same direction as $\mathbf{F}$ by Newton's second law, and this implies that $\mathbf{a}$ and $\mathbf{r}$ are oppositely directed vectors. Thus, it follows from part (*c*) of Theorem 13.4.5 that

$$\mathbf{r} \times \mathbf{a} = \mathbf{0}$$

Since the velocity and acceleration of the particle are given by $\mathbf{v} = d\mathbf{r}/dt$ and $\mathbf{a} = d\mathbf{v}/dt$, respectively, we have

$$\frac{d}{dt}(\mathbf{r} \times \mathbf{v}) = \mathbf{r} \times \frac{d\mathbf{v}}{dt} + \frac{d\mathbf{r}}{dt} \times \mathbf{v} = (\mathbf{r} \times \mathbf{a}) + (\mathbf{v} \times \mathbf{v}) = \mathbf{0} + \mathbf{0} = \mathbf{0} \qquad (1)$$

Integrating the left and right sides of this equation with respect to t yields

$$\mathbf{r} \times \mathbf{v} = \mathbf{b} \qquad (2)$$

---

*See biography on p. 748.

where $\mathbf{b}$ is a constant (independent of t). However, $\mathbf{b}$ is orthogonal to both $\mathbf{r}$ and $\mathbf{v}$, so we can conclude that $\mathbf{r} = \mathbf{r}(t)$ and $\mathbf{v} = \mathbf{v}(t)$ lie in a fixed plane containing the point O.

REMARK. The preceding discussion shows that each planet moves in a plane through the center of the Sun. Astronomers call this plane the *ecliptic* of the planet.

NEWTON'S LAW OF UNIVERSAL GRAVITATION

Our next objective is to derive the position function of a particle moving under a central force in a polar coordinate system. For this purpose we will need the following result, known as *Newton's Law of Universal Gravitation*.

14.7.2 NEWTON'S LAW OF UNIVERSAL GRAVITATION. Every particle of matter in the Universe attracts every other particle of matter in the Universe with a force that is proportional to the product of their masses and inversely proportional to the square of the distance between them. Specifically, if a particle of mass M and a particle of mass m are at a distance r from one another, then they attract each other with equal and opposite forces, $\mathbf{F}$ and $-\mathbf{F}$, of magnitude

$$\|\mathbf{F}\| = \frac{GMm}{r^2} \tag{3}$$

where G is a constant called the *universal gravitational constant*.

To obtain a formula for the vector force $\mathbf{F}$ that mass M exerts on mass m, we will let $\mathbf{r}$ be the radius vector from mass M to mass m (Figure 14.7.2). Thus, the distance r between the masses is $\|\mathbf{r}\|$, and the force $\mathbf{F}$ can be expressed in terms of $\mathbf{r}$ as

$$\mathbf{F} = \|\mathbf{F}\| \left(-\frac{\mathbf{r}}{\|\mathbf{r}\|} \right) = \|\mathbf{F}\| \left(-\frac{\mathbf{r}}{r} \right)$$

which from (3) can be expressed as

$$\mathbf{F} = -\frac{GMm}{r^3}\mathbf{r} \tag{4}$$

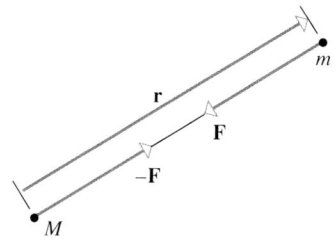

M exerts force $\mathbf{F}$ on m, and m exerts force $-\mathbf{F}$ on M.

Figure 14.7.2

One way to find a formula for the position function of the mass m is to find the acceleration function $\mathbf{a} = \mathbf{a}(t)$ and then integrate twice with respect to t to obtain $\mathbf{v} = \mathbf{v}(t)$ and then $\mathbf{r} = \mathbf{r}(t)$. We will use a slightly different approach, but we will still need to start by finding a formula for the acceleration function. To do this we use Formula (4) and Newton's second law to obtain

$$m\mathbf{a} = -\frac{GMm}{r^3}\mathbf{r}$$

from which we obtain

$$\mathbf{a} = -\frac{GM}{r^3}\mathbf{r} \tag{5}$$

REMARK. Observe that the acceleration $\mathbf{a}$ depends on the mass M but not on the mass m. Thus, for example, the acceleration of a planet is affected by the mass of the Sun but not by its own mass.

To obtain a formula for the position function of the mass m, we will need to introduce a coordinate system and make some assumptions about the initial conditions. Let us assume:

- The distance r from m to M is minimum at time $t = 0$.
- The mass m has nonzero position and velocity vectors $\mathbf{r}_0$ and $\mathbf{v}_0$ at time $t = 0$.

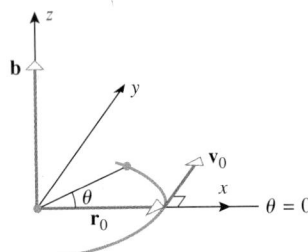

Figure 14.7.3

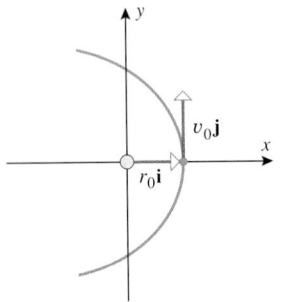

Figure 14.7.4

- A polar coordinate system is introduced with its pole at mass M and oriented so $\theta = 0$ at time $t = 0$.
- The vector $\mathbf{v}_0$ is perpendicular to the polar axis at time $t = 0$.

Moreover, to ensure that the polar angle θ increases with t, let us agree to observe this polar coordinate system looking toward the pole from the terminal point of the vector $\mathbf{b} = \mathbf{r}_0 \times \mathbf{v}_0$. We will also find it useful to superimpose an xyz-coordinate system on the polar coordinate system with the positive z-axis in the direction of $\mathbf{b}$ (Figure 14.7.3).

For computational purposes, it will be helpful to denote $\|\mathbf{r}_0\|$ by r_0 and $\|\mathbf{v}_0\|$ by v_0, in which case we can express the vectors $\mathbf{r}_0$ and $\mathbf{v}_0$ in xyz-coordinates as

$$\mathbf{r}_0 = r_0 \mathbf{i} \quad \text{and} \quad \mathbf{v}_0 = v_0 \mathbf{j}$$

and the vector $\mathbf{b}$ as

$$\mathbf{b} = \mathbf{r}_0 \times \mathbf{v}_0 = r_0 \mathbf{i} \times v_0 \mathbf{j} = r_0 v_0 \mathbf{k} \tag{6}$$

(Figure 14.7.4). It will also be useful to introduce the unit vector

$$\mathbf{u} = \cos\theta \, \mathbf{i} + \sin\theta \, \mathbf{j} \tag{7}$$

which will allow us to express the polar form of the position vector $\mathbf{r}$ as

$$\mathbf{r} = r\cos\theta \, \mathbf{i} + r\sin\theta \, \mathbf{j} = r(\cos\theta \, \mathbf{i} + \sin\theta \, \mathbf{j}) = r\mathbf{u} \tag{8}$$

and to express the acceleration vector $\mathbf{a}$ in terms of $\mathbf{u}$ by rewriting (5) as

$$\mathbf{a} = -\frac{GM}{r^2}\mathbf{u} \tag{9}$$

We are now ready to derive the position function of the mass m in polar coordinates. For this purpose, recall from (2) that the vector $\mathbf{b} = \mathbf{r} \times \mathbf{v}$ is constant, so it follows from (6) that the relationship

$$\mathbf{b} = \mathbf{r} \times \mathbf{v} = r_0 v_0 \mathbf{k} \tag{10}$$

holds for *all* values of t. Now let us examine $\mathbf{b}$ from another point of view. It follows from (8) that

$$\mathbf{v} = \frac{d\mathbf{r}}{dt} = \frac{d}{dt}(r\mathbf{u}) = r\frac{d\mathbf{u}}{dt} + \frac{dr}{dt}\mathbf{u}$$

and hence

$$\mathbf{b} = \mathbf{r} \times \mathbf{v} = (r\mathbf{u}) \times \left(r\frac{d\mathbf{u}}{dt} + \frac{dr}{dt}\mathbf{u}\right) = r^2\mathbf{u} \times \frac{d\mathbf{u}}{dt} + r\frac{dr}{dt}\mathbf{u} \times \mathbf{u} = r^2\mathbf{u} \times \frac{d\mathbf{u}}{dt} \tag{11}$$

But (7) implies that

$$\frac{d\mathbf{u}}{dt} = \frac{d\mathbf{u}}{d\theta}\frac{d\theta}{dt} = (-\sin\theta \, \mathbf{i} + \cos\theta \, \mathbf{j})\frac{d\theta}{dt}$$

so

$$\mathbf{u} \times \frac{d\mathbf{u}}{dt} = \frac{d\theta}{dt}\mathbf{k} \tag{12}$$

Substituting (12) in (11) yields

$$\mathbf{b} = r^2\frac{d\theta}{dt}\mathbf{k} \tag{13}$$

Thus, it follows from (7), (9), and (13) that

$$\mathbf{a} \times \mathbf{b} = -\frac{GM}{r^2}(\cos\theta \, \mathbf{i} + \sin\theta \, \mathbf{j}) \times \left(r^2\frac{d\theta}{dt}\mathbf{k}\right)$$

$$= GM(-\sin\theta \, \mathbf{i} + \cos\theta \, \mathbf{j})\frac{d\theta}{dt} = GM\frac{d\mathbf{u}}{dt} \tag{14}$$

From this formula and the fact that $d\mathbf{b}/dt = \mathbf{0}$ (since $\mathbf{b}$ is constant), we obtain

$$\frac{d}{dt}(\mathbf{v} \times \mathbf{b}) = \mathbf{v} \times \frac{d\mathbf{b}}{dt} + \frac{d\mathbf{v}}{dt} \times \mathbf{b} = \mathbf{a} \times \mathbf{b} = GM\frac{d\mathbf{u}}{dt}$$

Integrating both sides of this equation with respect to t yields

$$\mathbf{v} \times \mathbf{b} = GM\mathbf{u} + \mathbf{C} \qquad (15)$$

where $\mathbf{C}$ is a vector constant of integration. This constant can be obtained by evaluating both sides of the equation at $t = 0$. We leave it as an exercise to show that

$$\mathbf{C} = (r_0 v_0^2 - GM)\mathbf{i} \qquad (16)$$

from which it follows that

$$\mathbf{v} \times \mathbf{b} = GM\mathbf{u} + (r_0 v_0^2 - GM)\mathbf{i} \qquad (17)$$

We can now obtain the position function by computing the scalar triple product $\mathbf{r} \cdot (\mathbf{v} \times \mathbf{b})$ in two ways. First we use (10) and property (11) of Section 13.4 to obtain

$$\mathbf{r} \cdot (\mathbf{v} \times \mathbf{b}) = (\mathbf{r} \times \mathbf{v}) \cdot \mathbf{b} = \mathbf{b} \cdot \mathbf{b} = r_0^2 v_0^2 \qquad (18)$$

and next we use (17) to obtain

$$\mathbf{r} \cdot (\mathbf{v} \times \mathbf{b}) = \mathbf{r} \cdot (GM\mathbf{u}) + \mathbf{r} \cdot (r_0 v_0^2 - GM)\mathbf{i}$$

$$= \mathbf{r} \cdot \left(GM\frac{\mathbf{r}}{r}\right) + r\mathbf{u} \cdot (r_0 v_0^2 - GM)\mathbf{i}$$

$$= GMr + r(r_0 v_0^2 - GM)\cos\theta$$

If we now equate this to (18), we obtain

$$r_0^2 v_0^2 = GMr + r(r_0 v_0^2 - GM)\cos\theta$$

which when solved for r gives

$$r = \frac{r_0^2 v_0^2}{GM + (r_0 v_0^2 - GM)\cos\theta} = \frac{\dfrac{r_0^2 v_0^2}{GM}}{1 + \left(\dfrac{r_0 v_0^2}{GM} - 1\right)\cos\theta} \qquad (19)$$

or more simply

$$r = \frac{k}{1 + e\cos\theta} \qquad (20)$$

where

$$k = \frac{r_0^2 v_0^2}{GM} \quad \text{and} \quad e = \frac{r_0 v_0^2}{GM} - 1 \qquad (21\text{--}22)$$

We will leave it as an exercise to show that $e \geq 0$. Accepting this to be so, it follows by comparing (20) to Formula (3) of Section 12.5 that the trajectory is a conic section with eccentricity e, the focus at the pole, and $d = k/e$. Thus, depending on whether $e < 1$, $e = 1$, or $e > 1$, the trajectory will be, respectively, an ellipse, a parabola, or a hyperbola (Figure 14.7.5).

Note from Formula (22) that e depends on r_0 and v_0, so the exact form of the trajectory is determined by the mass M and the initial conditions. If the initial conditions are such that $e < 1$, then the mass m becomes trapped in an elliptical orbit; otherwise the mass m "escapes" and never returns to its initial position. Accordingly, the initial velocity that produces an eccentricity of $e = 1$ is called the ***escape speed*** and is denoted by v_{esc}. Thus, it follows from (22) that

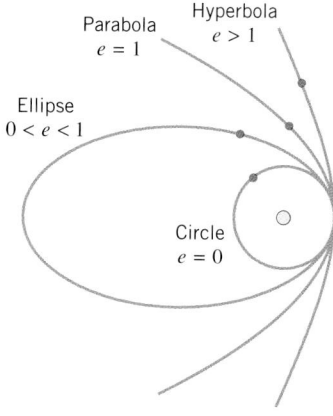

Parabola
$e = 1$

Hyperbola
$e > 1$

Ellipse
$0 < e < 1$

Circle
$e = 0$

Figure 14.7.5

$$v_{\text{esc}} = \sqrt{\frac{2GM}{r_0}} \qquad (23)$$

(verify).

It follows from our general discussion of central force fields that the planets have elliptical orbits with the Sun at the focus, which is Kepler's first law. To derive Kepler's second law, we begin by equating (10) and (13) to obtain

$$r^2 \frac{d\theta}{dt} = r_0 v_0 \tag{24}$$

To prove that the radial line from the center of the Sun to the center of a planet sweeps out equal areas in equal times, let $r = f(\theta)$ denote the polar equation of the planet, and let A denote the area swept out by the radial line as it varies from any fixed angle θ_0 to an angle θ. It follows from Theorem 12.3.2 that A can be expressed as

$$A = \int_{\theta_0}^{\theta} \frac{1}{2}[f(\phi)]^2 \, d\phi$$

where the dummy variable ϕ is introduced for the integration to reserve θ for the upper limit. It now follows from Part 2 of the Fundamental Theorem of Calculus and the chain rule that

$$\frac{dA}{dt} = \frac{dA}{d\theta}\frac{d\theta}{dt} = \frac{1}{2}[f(\theta)]^2\frac{d\theta}{dt} = \frac{1}{2}r^2\frac{d\theta}{dt}$$

Thus, it follows from (24) that

$$\frac{dA}{dt} = \frac{1}{2}r_0 v_0 \tag{25}$$

which shows that A changes at a constant rate. This implies that equal areas are swept out in equal times.

To derive Kepler's third law, we let a and b be the semimajor and semiminor axes of the elliptical orbit, and we recall that the area of this ellipse is $\pi a b$. It follows by integrating (25) that in t units of time the radial line will sweep out an area of $A = \frac{1}{2}r_0 v_0 t$. Thus, if T denotes the time required for the planet to make one revolution around the Sun (the period), then the radial line will sweep out the area of the entire ellipse during that time and hence

$$\pi a b = \tfrac{1}{2}r_0 v_0 T$$

from which we obtain

$$T^2 = \frac{4\pi^2 a^2 b^2}{r_0^2 v_0^2} \tag{26}$$

However, it follows from Formula (1) of Section 12.5 and the relationship $c^2 = a^2 - b^2$ for an ellipse that

$$e = \frac{c}{a} = \frac{\sqrt{a^2 - b^2}}{a}$$

Thus, $b^2 = a^2(1 - e^2)$ and hence (26) can be written as

$$T^2 = \frac{4\pi^2 a^4 (1 - e^2)}{r_0^2 v_0^2} \tag{27}$$

But comparing Equation (20) to Equation (17) of Section 12.5 shows that

$$k = a(1 - e^2)$$

Finally, substituting this expression and (21) in (27) yields

$$T^2 = \frac{4\pi^2 a^3}{r_0^2 v_0^2}k = \frac{4\pi^2 a^3}{r_0^2 v_0^2}\frac{r_0^2 v_0^2}{GM} = \frac{4\pi^2}{GM}a^3 \tag{28}$$

Thus, we have proved that T^2 is proportional to a^3, which is Kepler's third law. When convenient, Formula (28) can also be expressed as

$$T = \frac{2\pi}{\sqrt{GM}}a^{3/2} \tag{29}$$

....
ARTIFICIAL SATELLITES

Kepler's second and third laws and Formula (23) also apply to satellites that orbit a celestial body; we need only interpret M to be the mass of the body exerting the force and m to be the mass of the satellite. Values of GM that are required in many of the formulas in this section have been determined experimentally for various attracting bodies (Table 14.7.1).

Table 14.7.1

ATTRACTING BODY	INTERNATIONAL SYSTEM	BRITISH ENGINEERING SYSTEM
Earth	$GM = 3.99 \times 10^{14}\ \mathrm{m^3/s^2}$ $GM = 3.99 \times 10^{5}\ \mathrm{km^3/s^2}$	$GM = 1.41 \times 10^{16}\ \mathrm{ft^3/s^2}$ $GM = 1.24 \times 10^{12}\ \mathrm{mi^3/h^2}$
Sun	$GM = 1.33 \times 10^{20}\ \mathrm{m^3/s^2}$ $GM = 1.33 \times 10^{11}\ \mathrm{km^3/s^2}$	$GM = 4.69 \times 10^{21}\ \mathrm{ft^3/s^2}$ $GM = 4.13 \times 10^{17}\ \mathrm{mi^3/h^2}$
Moon	$GM = 4.90 \times 10^{12}\ \mathrm{m^3/s^2}$ $GM = 4.90 \times 10^{3}\ \mathrm{km^3/s^2}$	$GM = 1.73 \times 10^{14}\ \mathrm{ft^3/s^2}$ $GM = 1.53 \times 10^{10}\ \mathrm{mi^3/h^2}$

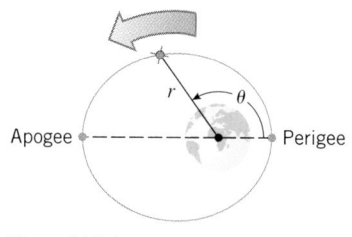

Apogee —— Perigee

Figure 14.7.6

Recall that for orbits of planets around the Sun, the point at which the distance between the center of the planet and the center of the Sun is maximum is called the *aphelion* and the point at which it is minimum the *perihelion*. For satellites around the Earth the point at which the maximum distance occurs is called the **apogee** and the point at which the minimum distance occurs is called the **perigee** (Figure 14.7.6). The actual distances between the centers at apogee and perigee are called the **apogee distance** and the **perigee distance**.

Example 1

A geosynchronous orbit for a satellite is a circular orbit about the equator of the Earth in which the satellite stays fixed over a point on the equator. Use the fact that the Earth makes one revolution about its axis every 24 hours to find the altitude in miles of a communications satellite in geosynchronous orbit. Assume the Earth to be a sphere of radius 4000 mi.

Solution. To remain fixed over a point on the equator, the satellite must have a period of $T = 24$ h. It follows from (28) or (29) and the Earth value of $GM = 1.24 \times 10^{12}\ \mathrm{mi^3/h^2}$ from Table 14.7.1 that

$$a = \sqrt[3]{\frac{GMT^2}{4\pi^2}} = \sqrt[3]{\frac{(1.24 \times 10^{12})(24)^2}{4\pi^2}} \approx 26{,}250\ \mathrm{mi}$$

and hence the altitude h of the satellite is

$$h = 26{,}252 - 4000 = 22{,}250\ \mathrm{mi}$$

◀

EXERCISE SET 14.7

....

In Exercises that require numerical values, use Table 14.7.1 and the following values, where needed:

   radius of Earth $= 4000$ mi $= 6440$ km

   radius of Moon $= 1080$ mi $= 1740$ km

   1 year (Earth year) $= 365$ days

1. Suppose that a particle is in an elliptical orbit in a central force field in which the center of force is at a focus, and let $r_{\min}$ and $r_{\max}$ denote the minimum and maximum dis-

tances from the particle to the center of force. Review the discussion of ellipses in polar coordinates in Section 12.5, and show that if the ellipse has eccentricity e and semimajor axis a, then $r_{\min} = a(1 - e)$ and $r_{\max} = a(1 + e)$.

2. (a) Use the results in Exercise 1 to show that

$$e = \frac{r_{\max} - r_{\min}}{r_{\max} + r_{\min}}$$

(b) Show that

$$r_{\max} = r_{\min}\frac{1 + e}{1 - e}$$

3. (a) Obtain the value of $\mathbf{C}$ given in Formula (16) by setting $t = 0$ in (15).
   (b) Use Formulas (7), (17), and (22) to show that
   $$\mathbf{v} \times \mathbf{b} = GM[(e + \cos\theta)\mathbf{i} + \sin\theta\,\mathbf{j}]$$
   (c) Show that $\|\mathbf{v} \times \mathbf{b}\| = \|\mathbf{v}\|\|\mathbf{b}\|$.
   (d) Use the results in parts (b) and (c) to show that the speed of a particle in an elliptical orbit is
   $$v = \frac{v_0}{1 + e}\sqrt{e^2 + 2e\cos\theta + 1}$$

4. Use the result in Exercise 3(d) to show that when a particle in an elliptical orbit with eccentricity e reaches an end of the minor axis, its speed is
   $$v = v_0\sqrt{\frac{1 - e}{1 + e}}$$

5. Use the result in Exercise 3(d) to show that for a particle in an elliptical orbit with eccentricity e, the maximum and minimum speeds are related by
   $$v_{\max} = v_{\min}\frac{1 + e}{1 - e}$$

6. Use Formula (22) and the result in part (d) of Exercise 3 to show that the speed v of a particle in a circular orbit of radius r_0 is constant and is given by
   $$v = \sqrt{\frac{GM}{r_0}}$$

7. Use the result in Exercise 6 to find the speed in km/s of a satellite in a circular orbit that is 200 km above the surface of the Earth.

8. Use the result in Exercise 6 to find the speed in mi/h of a communications satellite that is in geosynchronous orbit around the Earth. [See Example 1.]

9. Find the escape speed in km/s for a space probe in a circular orbit that is 300 km above the surface of the Earth.

10. The universal gravitational constant is approximately
    $$G = 6.67 \times 10^{-11}\ \text{m}^3/\text{kg·s}^2$$
    and the semimajor axis of the Earth's orbit is approximately
    $$a = 149.6 \times 10^6\ \text{km}$$
    Estimate the mass of the Sun in kg.

11. (a) The eccentricity of the Moon's orbit around the Earth is 0.055, and its semimajor axis is $a = 238{,}900$ mi. Find the maximum and minimum distances between the surface of the Earth and the surface of the Moon.
    (b) Find the period of the Moon's orbit in days.

12. (a) *Vanguard 1* was launched in March 1958 with perigee and apogee altitudes above the Earth of 649 km and 4340 km, respectively. Find the length of the semimajor axis of its orbit.
    (b) Use the result in part (a) of Exercise 2 to find the eccentricity of its orbit.
    (c) Find the period of *Vanguard 1* in minutes.

13. (a) Suppose that a space probe is in a circular orbit at an altitude of 180 mi above the surface of the Earth. Use the result in Exercise 6 to find its speed.
    (b) During a very short period of time, a thruster rocket on the space probe is fired to increase the speed of the probe by 600 mi/h in its direction of motion. Find the eccentricity of the resulting elliptical orbit, and use the result in part (b) of Exercise 2 to find the apogee altitude.

14. Show that the quantity e defined by Formula (22) is nonnegative. [*Hint:* The polar axis was chosen so that r is minimum when $\theta = 0$.]

SUPPLEMENTARY EXERCISES

1. In words, what is meant by the graph of a vector-valued function $\mathbf{r}(t)$?

2. Describe the graph of the vector-valued function.
   (a) $\mathbf{r} = \mathbf{r}_0 + t(\mathbf{r}_1 - \mathbf{r}_0)$
   (b) $\mathbf{r} = \mathbf{r}_0 + t(\mathbf{r}_1 - \mathbf{r}_0)$ $(0 \le t \le 1)$
   (c) $\mathbf{r} = \mathbf{r}_0 + t\mathbf{r}'(t_0)$

3. In words, describe what happens geometrically to $\mathbf{r}(t)$ if
   $$\lim_{t \to a}\mathbf{r}(t) = \mathbf{L}.$$

4. Suppose that $\mathbf{r}(t)$ is the position function of a particle moving in 2-space or 3-space. In each part, explain what the given quantity represents physically.
   (a) $\left\|\dfrac{d\mathbf{r}}{dt}\right\|$ (b) $\displaystyle\int_{t_0}^{t_1}\left\|\dfrac{d\mathbf{r}}{dt}\right\|\,dt$ (c) $\|\mathbf{r}(t)\|$

5. Suppose that $\mathbf{r}(t)$ is a smooth vector-valued function. State the definitions of $\mathbf{T}(t)$, $\mathbf{N}(t)$, and $\mathbf{B}(t)$.

6. State the definition of "curvature" and explain what it means geometrically.

7. In Supplementary Exercise 36 of Chapter 12, we defined the Cornu spiral parametrically as
   $$x = \int_0^t \cos\left(\frac{\pi u^2}{2}\right)\,du, \qquad y = \int_0^t \sin\left(\frac{\pi u^2}{2}\right)\,du$$
   This curve, which is graphed in the accompanying figure, is used in highway design to create a gradual transition from a straight road (zero curvature) to an exit ramp with positive curvature.
   (a) Express the Cornu spiral as a vector-valued function $\mathbf{r}(t)$, and then use Theorem 14.3.3 to show that $s = t$ is the arc length parameter with reference point $(0, 0)$.
   (b) Replace t by s and use Formula (1) of Section 14.5 to show that $\kappa(s) = \pi|s|$. [*Note:* If $s \ge 0$, then the curva-

ture $\kappa(s) = \pi s$ increases from 0 at a constant rate with respect to s. This makes the spiral ideal for joining a curved road to a straight road.]

(c) What happens to the curvature of the Cornu spiral as $s \to +\infty$? In words, explain why this is consistent with the graph.

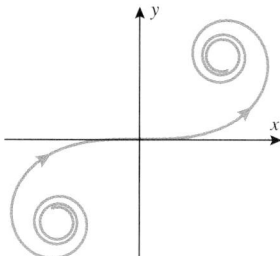

Figure Ex-7

8. (a) What does Theorem 14.2.7 tell you about the velocity vector of a particle that moves over a sphere?

   (b) What does Theorem 14.2.7 tell you about the acceleration vector of a particle that moves with constant speed?

   (c) Show that the particle with position function

$$\mathbf{r}(t) = \sqrt{1 - \tfrac{1}{4}\cos^2 t} \cos t \,\mathbf{i} + \sqrt{1 - \tfrac{1}{4}\cos^2 t}\sin t\,\mathbf{j} + \tfrac{1}{2}\cos t\,\mathbf{k}$$

moves over a sphere.

9. As illustrated in the accompanying figure, suppose that a particle moves counterclockwise around a circle of radius R centered at the origin at a constant rate of ω radians per second. This is called **uniform circular motion**. If we assume that the particle is at the point $(R, 0)$ at time $t = 0$, then its position function will be

$$\mathbf{r}(t) = R\cos\omega t\,\mathbf{i} + R\sin\omega t\,\mathbf{j}$$

(a) Show that the velocity vector $\mathbf{v}(t)$ is always tangent to the circle and that the particle has constant speed v given by

$$v = R\omega$$

(b) Show that the acceleration vector $\mathbf{a}(t)$ is always directed toward the center of the circle and has constant magnitude a given by

$$a = R\omega^2$$

(c) Show that the time T required for the particle to make one complete revolution is

$$T = \frac{2\pi}{\omega} = \frac{2\pi R}{v}$$

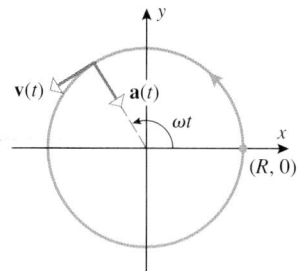

Figure Ex-9

10. If a particle of mass m has uniform circular motion (see Exercise 9), then the acceleration vector $\mathbf{a}(t)$ is called the **centripetal acceleration**. According to Newton's second law, this acceleration must be produced by some force $\mathbf{F}(t)$, called the **centripetal force**, that is related to $\mathbf{a}(t)$ by the equation $\mathbf{F}(t) = m\mathbf{a}(t)$. If this force is not present, then the particle cannot undergo uniform circular motion.

(a) Show that the direction of the centripetal force varies with time but that it has constant magnitude F given by

$$F = \frac{mv^2}{R}$$

(b) An astronaut with a mass of $m = 70$ kg orbits the Earth at an altitude of $h = 3200$ km with a constant speed of $v = 6.5$ km/s. Find her centripetal acceleration assuming that the radius of the Earth is 6440 km.

(c) What centripetal gravitational force in newtons does the Earth exert on the astronaut?

11. (a) Show that the graph of the vector-valued function $\mathbf{r}(t) = t\sin\pi t\,\mathbf{i} + t\,\mathbf{j} + t\cos\pi t\,\mathbf{k}$ lies on the surface of a cone, and sketch the cone.

   (b) Find parametric equations for the intersection of the surfaces

$$y = x^2 \quad \text{and} \quad 2x^2 + y^2 + 6z^2 = 24$$

and sketch the intersection.

12. Sketch the graph of the vector-valued function that is defined piecewise by

$$\mathbf{r}(t) = \begin{cases} 3t\mathbf{i}, & 0 \le t \le \tfrac{1}{3} \\ (2-3t)\mathbf{i} + (3t-1)\mathbf{j}, & \tfrac{1}{3} \le t \le \tfrac{2}{3} \\ 3(1-t)\mathbf{j}, & \tfrac{2}{3} \le t \le 1 \end{cases}$$

13. Suppose that the position function of a point moving in the xy-plane is

$$\mathbf{r} = x(t)\mathbf{i} + y(t)\mathbf{j}$$

This equation can be expressed in polar coordinates by making the substitution

$$x(t) = r(t)\cos\theta(t), \quad y(t) = r(t)\sin\theta(t)$$

This yields

$$\mathbf{r} = r(t)\cos\theta(t)\mathbf{i} + r(t)\sin\theta(t)\mathbf{j}$$

which can be expressed as

$$\mathbf{r} = r(t)\mathbf{e}_r(t)$$

where $\mathbf{e}_r(t) = \cos\theta(t)\mathbf{i} + \sin\theta(t)\mathbf{j}$.

(a) Show that $\mathbf{e}_r(t)$ is a unit vector that has the same direction as the radius vector $\mathbf{r}$ if $r(t) > 0$ and that $\mathbf{e}_\theta(t) = -\sin\theta(t)\mathbf{i} + \cos\theta(t)\mathbf{j}$ is the unit vector that results when $\mathbf{e}_r(t)$ is rotated counterclockwise through an angle of $\pi/2$. The vector $\mathbf{e}_r(t)$ is called the **radial unit vector**, and the vector $\mathbf{e}_\theta(t)$ is called the **transverse unit vector** (see the accompanying figure).

(b) Show that the velocity function $\mathbf{v} = \mathbf{v}(t)$ can be expressed in terms of radial and transverse components as

$$\mathbf{v} = \frac{dr}{dt}\mathbf{e}_r + r\frac{d\theta}{dt}\mathbf{e}_\theta$$

(c) Show that the acceleration function $\mathbf{a} = \mathbf{a}(t)$ can be expressed in terms of radial and transverse components as

$$\mathbf{a} = \left[\frac{d^2r}{dt^2} - r\left(\frac{d\theta}{dt}\right)^2\right]\mathbf{e}_r + \left[r\frac{d^2\theta}{dt^2} + 2\frac{dr}{dt}\frac{d\theta}{dt}\right]\mathbf{e}_\theta$$

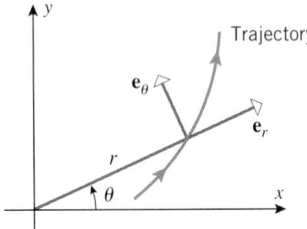

Figure Ex-13

14. As illustrated in the accompanying figure, the polar coordinates of a rocket are tracked by radar from a point that is b units from the launching pad. Show that the speed v of the rocket can be expressed in terms b, θ, and $d\theta/dt$ as

$$v = b\sec^2\theta\,\frac{d\theta}{dt}$$

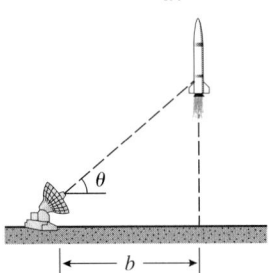

Figure Ex-14

15. Find the arc length parametrization of the line through $P(-1, 4, 3)$ and $Q(0, 2, 5)$ that has reference point P and orients the line in the direction from P to Q.

16. A player throws a ball with an initial speed of 60 ft/s at an unknown angle α with the horizontal from a point that is 4 ft above the floor. Given that the ceiling of the gymnasium is 25 ft high, determine the maximum height h at which the ball can hit a wall that is 60 ft away (see the accompanying figure).

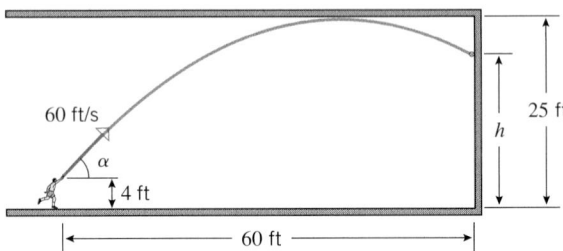

Figure Ex-16

17. Find all points on the graph of $\mathbf{r}(t) = t^3\mathbf{i} + 10t\mathbf{j} + 5t^2\mathbf{k}$ at which the tangent line is perpendicular to the tangent line at $t = 1$.

18. Solve the vector initial-value problem

$$\frac{d\mathbf{r}}{dt} = \mathbf{r}, \quad \mathbf{r}(0) = \mathbf{r}_0$$

for the unknown vector-valued function $\mathbf{r}(t)$.

19. At time $t = 0$ a particle at the origin of an xyz-coordinate system has a velocity vector of $\mathbf{v}_0 = \mathbf{i} + 2\mathbf{j} - \mathbf{k}$. The acceleration function of the particle is $\mathbf{a}(t) = 2t^2\mathbf{i} + \mathbf{j} + \cos 2t\mathbf{k}$.
(a) Find the position function of the particle.
(b) Find the speed of the particle at time $t = 1$.

20. Let $\mathbf{v} = \mathbf{v}(t)$ and $\mathbf{a} = \mathbf{a}(t)$ be the velocity and acceleration vectors for a particle moving in 2-space or 3-space. Show that the rate of change of its speed can be expressed as

$$\frac{d}{dt}(\|\mathbf{v}\|) = \frac{1}{\|\mathbf{v}\|}(\mathbf{v} \cdot \mathbf{a})$$

EXPANDING THE CALCULUS HORIZON

For additional material relating to this chapter, visit the Anton Website at http://www.wiley.com/college/anton

15

PARTIAL DERIVATIVES

Joseph Louis
Lagrange

Triceratops generated
with a CAS

*I*n this chapter we will extend many of the basic concepts of calculus to functions of two or more variables, commonly called functions of "several variables." We will begin by discussing limits and continuity for functions of two and three variables, then we will define derivatives of such functions, and then we will use these derivatives to study tangent planes, rates of change, slopes of surfaces, and maximization and minimization problems. Although many of the basic ideas that we developed for functions of one variable will carry over in a natural way, functions of several variables are intrinsically more complicated than functions of one variable, so we will need to develop new tools and new ideas to deal with such functions.

15.1 FUNCTIONS OF TWO OR MORE VARIABLES

In previous sections we studied real-valued functions of a real variable and vector-valued functions of a real variable. In this section we will consider real-valued functions of two or more real variables.

NOTATION AND TERMINOLOGY

There are many familiar formulas in which a given variable depends on two or more other variables. For example, the area A of a triangle depends on the base length b and height h by the formula $A = \frac{1}{2}bh$; the volume V of a rectangular box depends on the length l, the width w, and the height h by the formula $V = lwh$; and the arithmetic average $\bar{x}$ of n real numbers, $x_1, x_2, \ldots, x_n$, depends on those numbers by the formula $\bar{x} = (x_1 + x_2 + \cdots + x_n)/n$. Thus, we say that

A is a function of the two variables b and h;

V is a function of the three variables l, w, and h;

$\bar{x}$ is a function of the n variables $x_1, x_2, \ldots, x_n$.

The terminology and notation for functions of two or more variables is similar to that for functions of one variable. For example, the expression

$$z = f(x, y)$$

means that z is a function of x and y in the sense that a unique value of the dependent variable z is determined by specifying values for the independent variables x and y. Similarly,

$$w = f(x, y, z)$$

expresses w as a function of x, y, and z, and

$$u = f(x_1, x_2, \ldots, x_n)$$

expresses u as a function of $x_1, x_2, \ldots, x_n$.

One can think of a function f of two or more variables as a computer program that takes two or more inputs, operates on those inputs, and produces an output (Figure 15.1.1). In this section we will only be concerned with functions whose inputs and outputs are real numbers. One can also think of such functions in more geometric terms. For example, if $z = f(x, y)$, then we can view (x, y) as a point in the xy-plane and think of f as a rule that associates a unique numerical value z with the point (x, y); similarly, we can think of $w = f(x, y, z)$ as a rule that associates a unique numerical value w with a point (x, y, z) in an xyz-coordinate system (Figure 15.1.2).

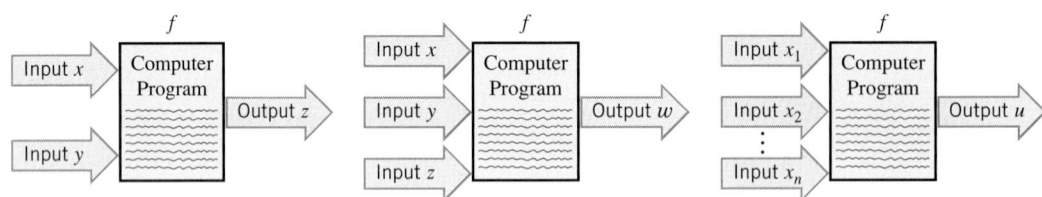

Figure 15.1.1

As with functions of one variable, the inputs of a function of two or more variables may be restricted to lie in some set D, which we call the **domain** of f. Sometimes the domain will be determined by physical restrictions on the variables. If the function is defined by a formula and if there are no physical restrictions or other restrictions stated explicitly, then it is understood that the domain consists of all points for which the formula yields a real value for the output. We call this the **natural domain** of the function. The following definitions summarize this discussion.

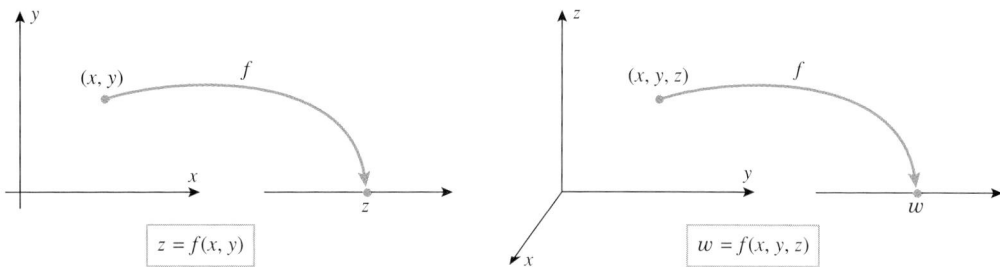

Figure 15.1.2

15.1.1 DEFINITION. A *function f of two variables*, x and y, is a rule that assigns a unique real number $f(x, y)$ to each point (x, y) in some set D in the xy-plane.

15.1.2 DEFINITION. A *function f of three variables*, x, y, and z, is a rule that assigns a unique real number $f(x, y, z)$ to each point (x, y, z) in some set D in three-dimensional space.

REMARK. In more advanced courses the notion of "n-dimensional space" for $n > 3$ is defined, and a *function f of n real variables*, $x_1, x_2, \ldots, x_n$, is regarded as a rule that assigns a unique real number $f(x_1, x_2, \ldots, x_n)$ to each "point" $(x_1, x_2, \ldots, x_n)$ in some set in n-dimensional space. However, we will not pursue that idea in this text.

Example 1

Let

$$f(x, y) = 3x^2 \sqrt{y} - 1$$

Find $f(1, 4)$, $f(0, 9)$, $f(t^2, t)$, $f(ab, 9b)$, and the natural domain of f.

Solution. By substitution

$$f(1, 4) = 3(1)^2 \sqrt{4} - 1 = 5$$
$$f(0, 9) = 3(0)^2 \sqrt{9} - 1 = -1$$
$$f(t^2, t) = 3(t^2)^2 \sqrt{t} - 1 = 3t^4 \sqrt{t} - 1$$
$$f(ab, 9b) = 3(ab)^2 \sqrt{9b} - 1 = 9a^2 b^2 \sqrt{b} - 1$$

Because of the radical $\sqrt{y}$ in the formula for f, we must have $y \geq 0$ to avoid imaginary values for $f(x, y)$. Thus, the natural domain of f consists of all points in the xy-plane that are on or above the x-axis. (See Figure 15.1.3.) ◄

Example 2

Sketch the natural domain of the function $f(x, y) = \ln(x^2 - y)$.

Solution. $\ln(x^2 - y)$ is defined only when $0 < x^2 - y$ or $y < x^2$. To sketch this region, we use the fact that the curve $y = x^2$ separates the region where $y < x^2$ from the region where $y > x^2$. To determine the region where $y < x^2$ holds, we can select an arbitrary "test point" off the boundary $y = x^2$ and determine whether $y < x^2$ or $y > x^2$ at the test point. For example, if we choose the test point $(x, y) = (0, 1)$, then $x^2 = 0$, $y = 1$, so this point lies in the region where $y > x^2$. Thus, the region where $y < x^2$ is the one that does *not* contain the test point (Figure 15.1.4). ◄

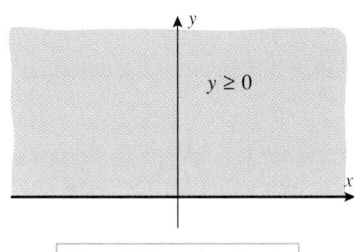

The solid boundary line is included in the domain.

Figure 15.1.3

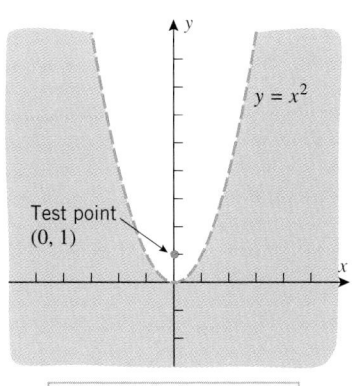

The dashed boundary does not belong to the domain.

Figure 15.1.4

Example 3

Let

$$f(x, y, z) = \sqrt{1 - x^2 - y^2 - z^2}$$

Find $f\left(0, \frac{1}{2}, -\frac{1}{2}\right)$ and the natural domain of f.

Solution. By substitution,

$$f\left(0, \tfrac{1}{2}, -\tfrac{1}{2}\right) = \sqrt{1 - (0)^2 - \left(\tfrac{1}{2}\right)^2 - \left(-\tfrac{1}{2}\right)^2} = \sqrt{\tfrac{1}{2}}$$

Because of the square root sign, we must have $0 \le 1 - x^2 - y^2 - z^2$ in order to have a real value for $f(x, y, z)$. Rewriting this inequality in the form

$$x^2 + y^2 + z^2 \le 1$$

we see that the natural domain of f consists of all points on or within the sphere

$$x^2 + y^2 + z^2 = 1 \qquad\qquad \blacktriangleleft$$

GRAPHS OF FUNCTIONS OF TWO VARIABLES

Recall that for a function f of one variable, the graph of $f(x)$ in the xy-plane was defined to be the graph of the equation $y = f(x)$. Similarly, if f is a function of two variables, we define the **graph** of $f(x, y)$ in xyz-space to be the graph of the equation $z = f(x, y)$. In general, such a graph will be a surface in 3-space.

Example 4

In each part, describe the graph of the function in an xyz-coordinate system.

(a) $f(x, y) = 1 - x - \frac{1}{2}y$ (b) $f(x, y) = \sqrt{1 - x^2 - y^2}$

(c) $f(x, y) = -\sqrt{x^2 + y^2}$

Solution (a). By definition, the graph of the given function is the graph of the equation

$$z = 1 - x - \tfrac{1}{2}y$$

which is a plane. A triangular portion of the plane can be sketched by plotting the intersections with the coordinate axes and joining them with line segments (Figure 15.1.5a).

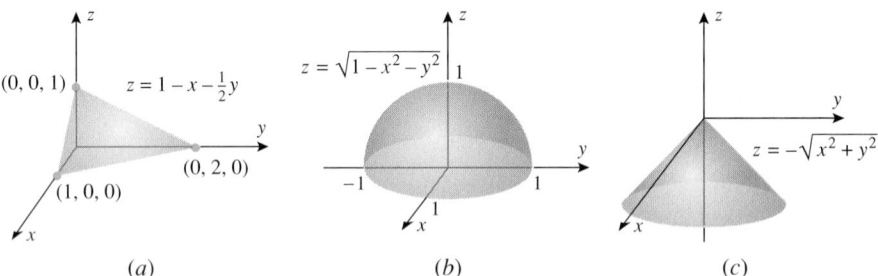

Figure 15.1.5   (a)   (b)   (c)

Solution (b). By definition, the graph of the given function is the graph of the equation

$$z = \sqrt{1 - x^2 - y^2} \qquad\qquad (1)$$

After squaring both sides, this can be rewritten as

$$x^2 + y^2 + z^2 = 1$$

which represents a sphere of radius 1, centered at the origin. Since (1) imposes the added condition that $z \ge 0$, the graph is just the upper hemisphere (Figure 15.1.5b).

Solution (c). The graph of the given function is the graph of the equation

$$z = -\sqrt{x^2 + y^2} \qquad\qquad (2)$$

After squaring, we obtain

$$z^2 = x^2 + y^2$$

which is the equation of a circular cone (see Table 13.7.1). Since (2) imposes the condition that $z \leq 0$, the graph is just the lower nappe of the cone (Figure 15.1.5c).   ◀

GRAPHS OF FUNCTIONS OF TWO VARIABLES USING TECHNOLOGY

Except in the simplest cases, graphs of functions of two variables can be difficult to visualize without the help of a graphing utility. CAS programs have extensive surface-graphing capabilities, as do many commercial computer programs specifically designed for this purpose. In addition, many newer graphing calculators incorporate surface-graphing features. Table 15.1.1 illustrates various ways that graphing technology can be used to represent the graph of a function of two variables. The table shows six typical graphical representations of the function $f(x, y) = \cos x \sin y$ over the domain $0 \leq x \leq \pi, 0 \leq y \leq 2\pi$.

Table 15.1.1

SURFACE	DESCRIPTION	SURFACE	DESCRIPTION
Wire Frame 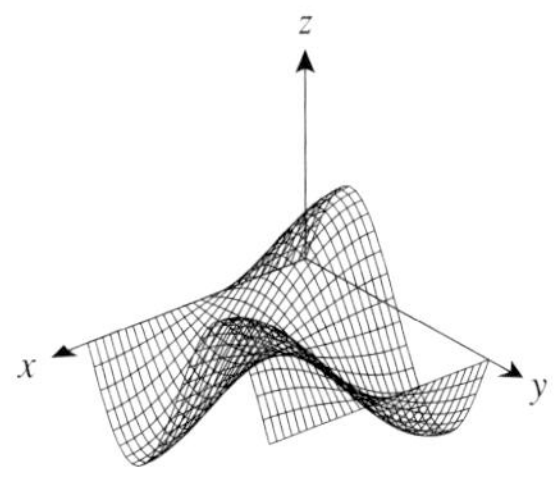	The surface is formed from mesh lines. Transparency allows the mesh in the back to be seen through the mesh in front.	Coloration by Height 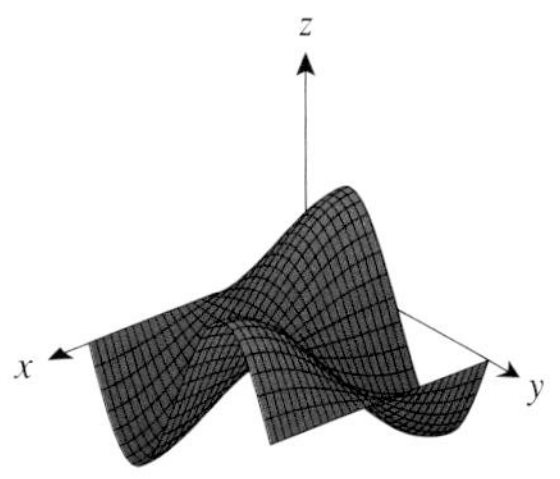	The surface is colored by height in a spectrum from blue at the lowest points to red at the highest points. This is called "temperature coloration."
Hidden Line 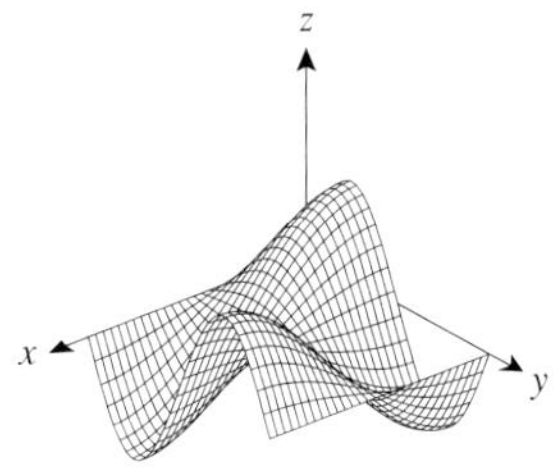	The regions enclosed by the mesh lines have an opaque white fill, and the surface is drawn from back to front so that the mesh in back becomes hidden.	Landscape Style 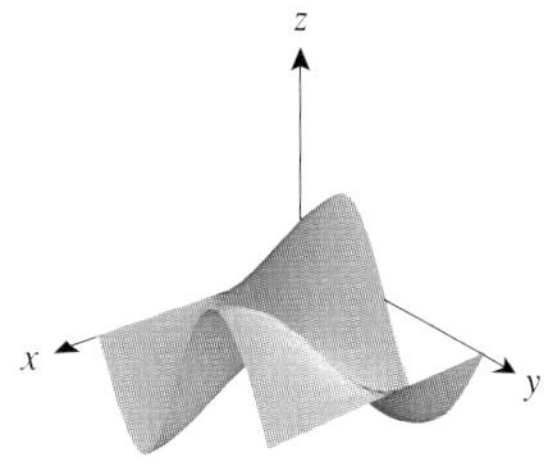	The opaque surface with no mesh lines is colorized by simulating the effect of colored lights shining on the surface from certain positions.
Lighted Surface 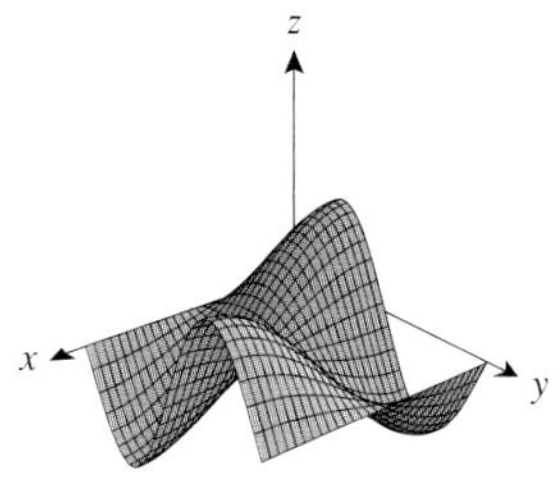	The opaque surface is colorized by simulating the effect of colored lights shining on the surface from certain positions.	Painted Faces 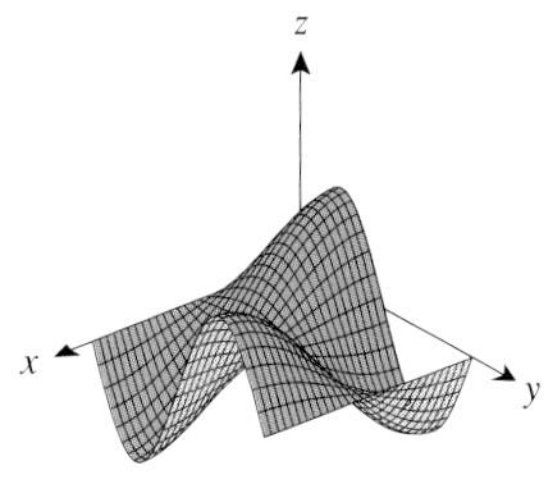	The regions enclosed by the mesh line on the "top" and "bottom" faces of the surface are painted with different solid colors.

LEVEL CURVES

We are all familiar with the topographic (or contour) maps in which a three-dimensional landscape, such as a mountain range, is represented by two-dimensional contour lines or curves of constant elevation. Consider, for example, the model hill and its contour map shown in Figure 15.1.6. The contour map is constructed by passing planes of constant elevation through the hill, projecting the resulting contours onto a flat surface, and labeling the contours with their elevations. In Figure 15.1.6, note how the two gullies appear as indentations in the contour lines and how the curves are close together on the contour map where the hill has a steep slope and become more widely spaced where the slope is gradual.

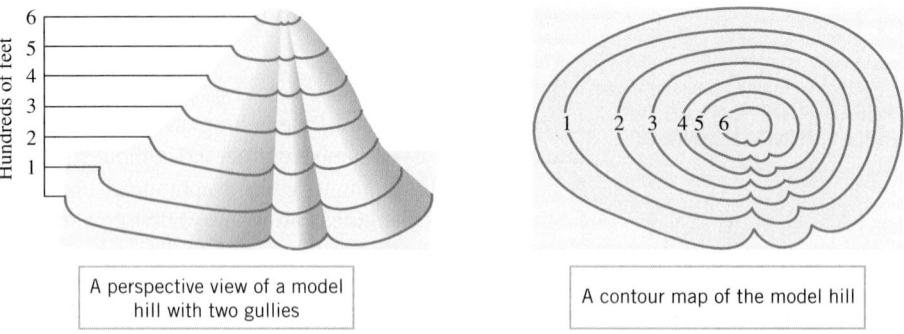

A perspective view of a model hill with two gullies

A contour map of the model hill

Figure 15.1.6

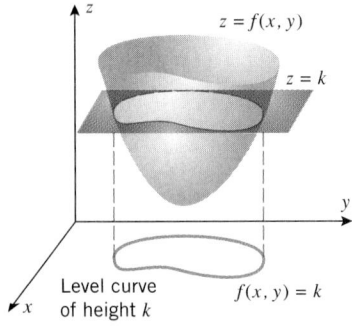

Figure 15.1.7

Contour maps are also useful for studying functions of two variables. If the surface $z = f(x, y)$ is cut by the horizontal plane $z = k$, then at all points on the intersection we have $f(x, y) = k$. The projection of this intersection onto the xy-plane is called the ***level curve of height k*** or the ***level curve with constant k*** (Figure 15.1.7). A set of level curves for $z = f(x, y)$ is called a ***contour plot*** or ***contour map*** of f.

Example 5

The graph of the function $f(x, y) = y^2 - x^2$ in xyz-space is the hyperbolic paraboloid (saddle surface) shown in Figure 15.1.8a. The level curves have equations of the form $y^2 - x^2 = k$. For $k > 0$ these curves are hyperbolas opening along lines parallel to the y-axis; for $k < 0$ they are hyperbolas opening along lines parallel to the x-axis; and for $k = 0$ the level curve consists of the intersecting lines $y + x = 0$ and $y - x = 0$ (Figure 15.1.8b). ◄

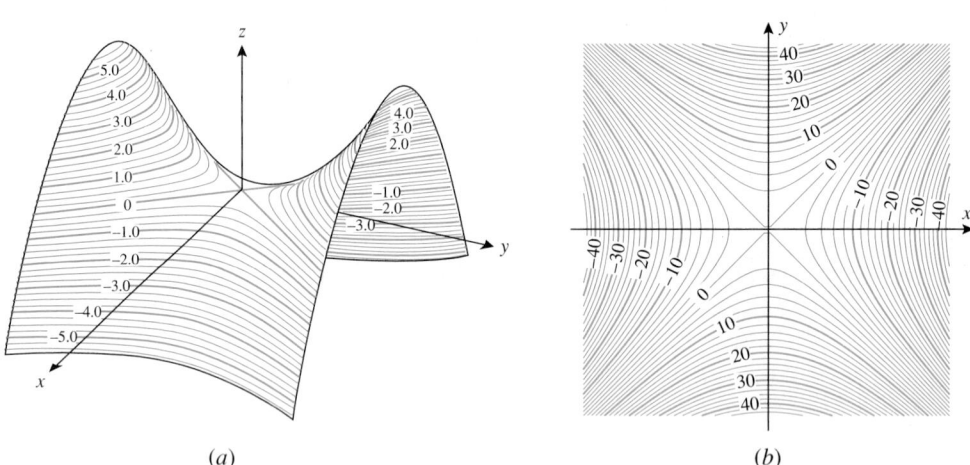

Figure 15.1.8

(a) (b)

Example 6

(a) Sketch the contour plot of $f(x, y) = 4x^2 + y^2$ using level curves of height $k = 0, 1, 2, 3, 4, 5$.

(b) Sketch the contour plot of $f(x, y) = 2 - x - y$ using level curves of height $k = -6, -4, -2, 0, 2, 4, 6$.

Solution (a). The graph of the surface $z = 4x^2 + y^2$ is the paraboloid shown in Figure 15.1.9, so we can reasonably expect the contour plot to be a family of ellipses centered at the origin. The level curve of height k has the equation $4x^2 + y^2 = k$. If $k = 0$, then the

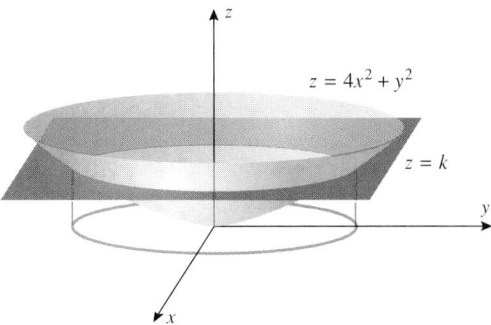

Figure 15.1.9

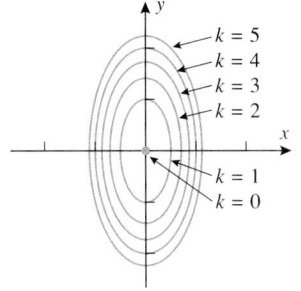

Figure 15.1.10

graph is the single point $(0, 0)$. For $k > 0$ we can rewrite the equation as

$$\frac{x^2}{k/4} + \frac{y^2}{k} = 1$$

which represents a family of ellipses with x-intercepts $\pm\sqrt{k}/2$ and y-intercepts $\pm\sqrt{k}$. The contour plot for the specified values of k is shown in Figure 15.1.10.

Solution (b). The graph of the surface $z = 2 - x - y$ is the plane shown in Figure 15.1.11, so we can reasonably expect the contour plot to be a family of parallel lines. The level curve of height k has the equation $2 - x - y = k$, which we can rewrite as

$$y = -x + (2 - k)$$

This represents a family of parallel lines of slope -1. The contour plot for the specified values of k is shown in Figure 15.1.12. ◀

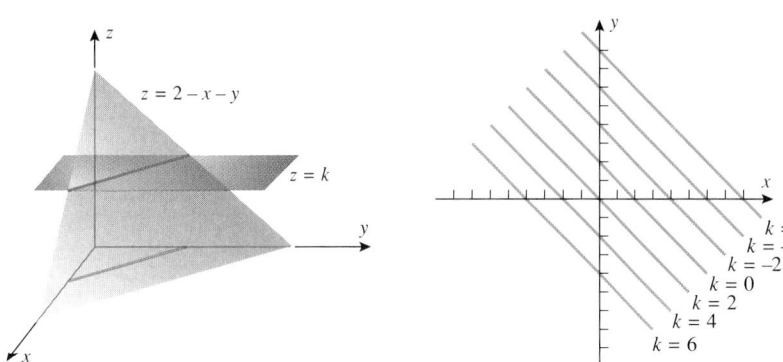

Figure 15.1.11   Figure 15.1.12

**CONTOUR PLOTS USING
TECHNOLOGY**

Except in the simplest cases, contour plots can be difficult to produce without the help of a graphing utility. Figure 15.1.13 illustrates how graphing technology can be used to display level curves. The table shows two graphical representations of the level curves of the function $f(x, y) = |\sin x \sin y|$ produced with a CAS over the domain $0 \le x \le 2\pi$, $0 \le y \le 2\pi$.

LEVEL SURFACES

Observe that the graph of $y = f(x)$ is a curve in 2-space, and the graph of $z = f(x, y)$ is a surface in 3-space, so the number of dimensions required for these graphs is one greater than the number of variables. Accordingly, there is no "direct" way to graph a function of three variables since four dimensions are required. However, if k is a constant, then the graph of the equation $f(x, y, z) = k$ will generally be a surface in 3-space (e.g., the graph of $x^2 + y^2 + z^2 = 1$ is a sphere), which we call the *level surface with constant k*. Some geometric insight into the behavior of the function f can sometimes be obtained by graphing these level surfaces for various values of k.

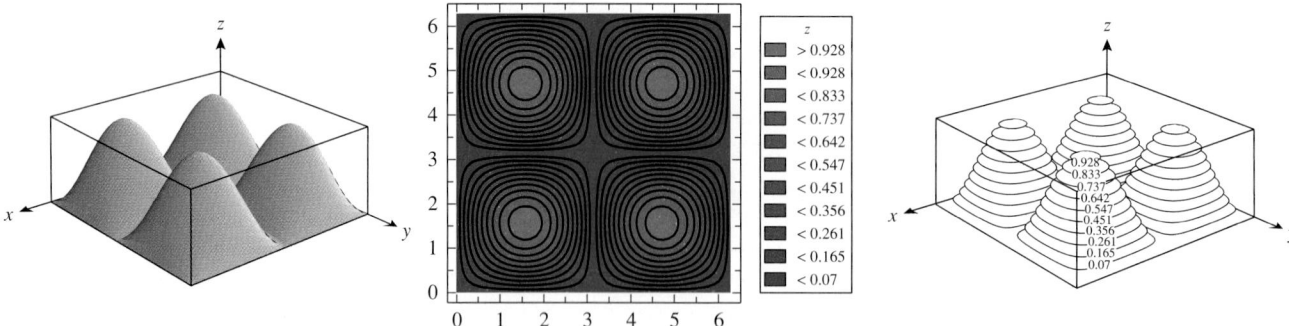

Figure 15.1.13

REMARK. The term "level surface" is standard but confusing, since a level surface need *not* be level in the sense of being horizontal; it is simply a surface on which all values of f are the same.

Example 7

Describe the level surfaces of

(a) $f(x, y, z) = x^2 + y^2 + z^2$ (b) $f(x, y, z) = z^2 - x^2 - y^2$

Solution (a). The level surfaces have equations of the form

$$x^2 + y^2 + z^2 = k$$

For $k > 0$ the graph of this equation is a sphere of radius $\sqrt{k}$, centered at the origin; for $k = 0$ the graph is the single point $(0, 0, 0)$; and for $k < 0$ there is no level surface (Figure 15.1.14).

Solution (b). The level surfaces have equations of the form

$$z^2 - x^2 - y^2 = k$$

As discussed in Section 13.7, this equation represents a cone if $k = 0$, a hyperboloid of two sheets if $k > 0$, and a hyperboloid of one sheet if $k < 0$ (Figure 15.1.15). ◄

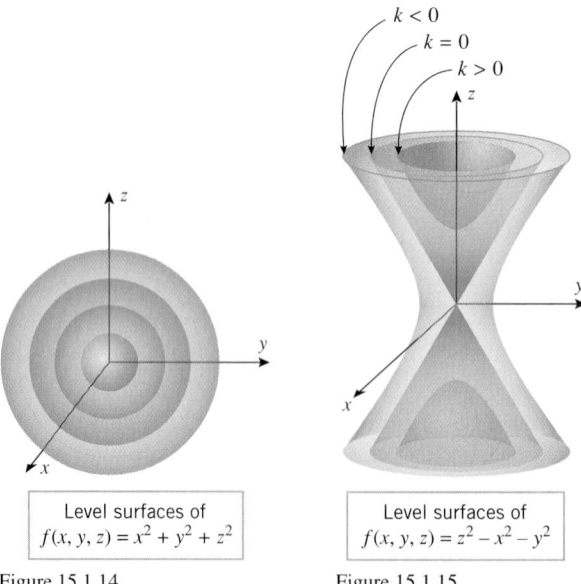

Level surfaces of
$f(x, y, z) = x^2 + y^2 + z^2$

Figure 15.1.14

Level surfaces of
$f(x, y, z) = z^2 - x^2 - y^2$

Figure 15.1.15

GRAPHING FUNCTIONS OF TWO
VARIABLES USING TECHNOLOGY

Generating surfaces with a graphing utility is more complicated than generating plane curves because there are more factors that must be taken into account. We can only touch on the ideas here, so if you want to use a graphing utility, its documentation will be your main source of information.

Graphing utilities can only show a portion of xyz-space in a viewing screen, so the first step in graphing a surface is to determine which portion of xyz-space you want to display. This region is called the *viewing window* or *viewing box*. For example, the first row of Table 15.1.2 shows the effect of graphing the paraboloid $z = x^2 + y^2$ in three different viewing windows. However, within a fixed viewing window, the appearance of the surface is also affected by the *viewpoint*, that is, the direction from which the surface is viewed, and the distance from the viewer to the surface. For example, the second row of Table 15.1.2 shows the graph of the paraboloid $z = x^2 + y^2$ from three different viewpoints using the viewing window in the first part of the figure.

Table 15.1.2

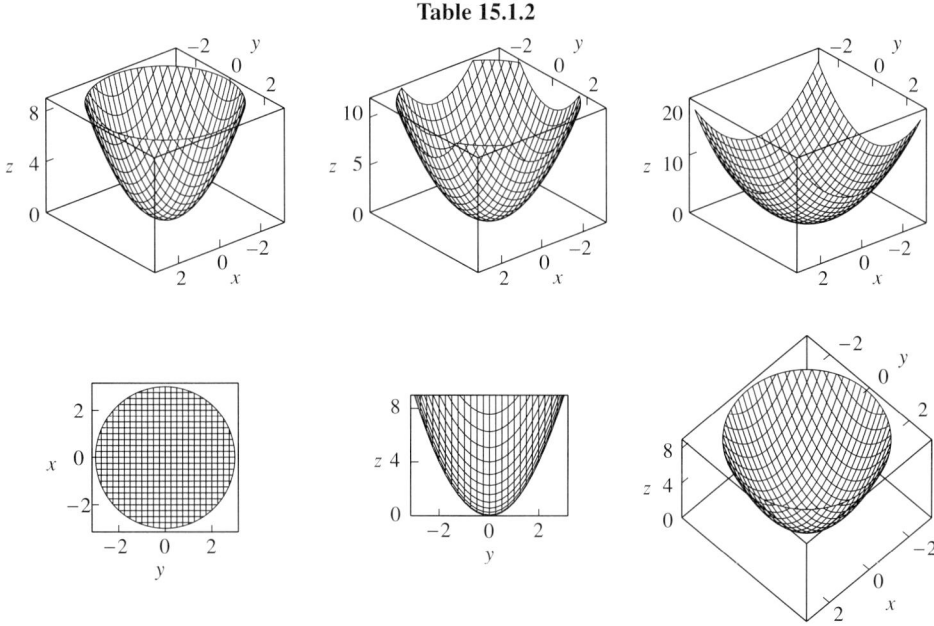

FOR THE READER. If you have a graphing utility that can generate surfaces in 3-space, read the documentation and try to duplicate Table 15.1.2.

FOR THE READER. Table 15.1.3 shows six surfaces in 3-space. Examine each surface and convince yourself that the contour plot describes its level curves. This will take a little thought because the mesh lines on the surface are traces in vertical planes, whereas the level curves are traces in horizontal planes. In these contour plots the color gradation is from dark to light as z increases. If you have a graphing utility that can generate surfaces in 3-space, try to duplicate some of these figures. You need not match the colors or generate the coordinate axes.

Table 15.1.3

SURFACE	CONTOUR PLOT	SURFACE	CONTOUR PLOT
$z = \cos y$		$z = e^x \sin y$	
$z = \sin(\sqrt{x^2 + y^2})$		$z = xye^{-\frac{1}{2}(x^2+y^2)}$	
$z = \cos(xy)$		$z = |xy|$	

EXERCISE SET 15.1 ⌁ Graphing Calculator   🅒 CAS

Exercises 1–8 are concerned with functions of two variables.

1. Let $f(x, y) = x^2 y + 1$. Find
(a) $f(2, 1)$ (b) $f(1, 2)$ (c) $f(0, 0)$
(d) $f(1, -3)$ (e) $f(3a, a)$ (f) $f(ab, a - b)$.

2. Let $f(x, y) = x + \sqrt[3]{xy}$. Find
(a) $f(t, t^2)$ (b) $f(x, x^2)$ (c) $f(2y^2, 4y)$.

3. Let $f(x, y) = xy + 3$. Find
(a) $f(x + y, x - y)$ (b) $f(xy, 3x^2y^3)$.

4. Let $g(x) = x \sin x$. Find
(a) $g(x/y)$ (b) $g(xy)$ (c) $g(x - y)$.

5. Find $F(g(x), h(y))$ if $F(x, y) = xe^{xy}$, $g(x) = x^3$, and $h(y) = 3y + 1$.

6. Find $g(u(x, y), v(x, y))$ if $g(x, y) = y \sin(x^2 y)$, $u(x, y) = x^2 y^3$, and $v(x, y) = \pi xy$.

7. Let $f(x, y) = x + 3x^2 y^2$, $x(t) = t^2$, and $y(t) = t^3$. Find
(a) $f(x(t), y(t))$ (b) $f(x(0), y(0))$
(c) $f(x(2), y(2))$.

8. Let $g(x, y) = ye^{-3x}$, $x(t) = \ln(t^2 + 1)$, and $y(t) = \sqrt{t}$. Find $g(x(t), y(t))$.

Exercises 9–12 involve functions of three variables.

9. Let $f(x, y, z) = xy^2z^3 + 3$. Find
(a) $f(2, 1, 2)$ (b) $f(-3, 2, 1)$
(c) $f(0, 0, 0)$ (d) $f(a, a, a)$
(e) $f(t, t^2, -t)$ (f) $f(a + b, a - b, b)$.

10. Let $f(x, y, z) = zxy + x$. Find
(a) $f(x + y, x - y, x^2)$ (b) $f(xy, y/x, xz)$.

11. Find $F(f(x), g(y), h(z))$ if $F(x, y, z) = ye^{xyz}$, $f(x) = x^2$, $g(y) = y + 1$, and $h(z) = z^2$.

12. Find $g(u(x, y, z), v(x, y, z), w(x, y, z))$ if $g(x, y, z) = z \sin xy$, $u(x, y, z) = x^2z^3$, $v(x, y, z) = \pi xyz$, and $w(x, y, z) = xy/z$.

Exercises 13 and 14 are concerned with functions of four or more variables.

13. (a) Let $f(x, y, z, t) = x^2y^3\sqrt{z + t}$.
Find $f(\sqrt{5}, 2, \pi, 3\pi)$.
(b) Let $f(x_1, x_2, \ldots, x_n) = \sum_{k=1}^{n} kx_k$.
Find $f(1, 1, \ldots, 1)$.

14. (a) Let $f(u, v, \lambda, \phi) = e^{u+v} \cos \lambda \tan \phi$.
Find $f(-2, 2, 0, \pi/4)$.
(b) Let $f(x_1, x_2, \ldots, x_n) = x_1^2 + x_2^2 + \cdots + x_n^2$.
Find $f(1, 2, \ldots, n)$.

In Exercises 15–18, sketch the domain of f. Use solid lines for portions of the boundary included in the domain and dashed lines for portions not included.

15. $f(x, y) = \ln(1 - x^2 - y^2)$ **16.** $f(x, y) = \sqrt{x^2 + y^2 - 4}$
17. $f(x, y) = \dfrac{1}{x - y^2}$ **18.** $f(x, y) = \ln xy$

In Exercises 19 and 20, describe the domain of f in words.

19. (a) $f(x, y) = xe^{-\sqrt{y+2}}$
(b) $f(x, y, z) = \sqrt{25 - x^2 - y^2 - z^2}$
(c) $f(x, y, z) = e^{xyz}$

20. (a) $f(x, y) = \dfrac{\sqrt{4 - x^2}}{y^2 + 3}$ (b) $f(x, y) = \ln(y - 2x)$
(c) $f(x, y, z) = \dfrac{xyz}{x + y + z}$

In Exercises 21–30, sketch the graph of f.

21. $f(x, y) = 3$ **22.** $f(x, y) = \sqrt{9 - x^2 - y^2}$
23. $f(x, y) = \sqrt{x^2 + y^2}$ **24.** $f(x, y) = x^2 + y^2$
25. $f(x, y) = x^2 - y^2$ **26.** $f(x, y) = 4 - x^2 - y^2$
27. $f(x, y) = \sqrt{x^2 + y^2 + 1}$ **28.** $f(x, y) = \sqrt{x^2 + y^2 - 1}$
29. $f(x, y) = y + 1$ **30.** $f(x, y) = x^2$

In Exercises 31–36, sketch the level curve $z = k$ for the specified values of k.

31. $z = x^2 + y^2$; $k = 0, 1, 2, 3, 4$
32. $z = y/x$; $k = -2, -1, 0, 1, 2$
33. $z = x^2 + y$; $k = -2, -1, 0, 1, 2$
34. $z = x^2 + 9y^2$; $k = 0, 1, 2, 3, 4$
35. $z = x^2 - y^2$; $k = -2, -1, 0, 1, 2$
36. $z = y \csc x$; $k = -2, -1, 0, 1, 2$

In Exercises 37–40, sketch the level surface $f(x, y, z) = k$.

37. $f(x, y, z) = 4x^2 + y^2 + 4z^2$; $k = 16$
38. $f(x, y, z) = x^2 + y^2 - z^2$; $k = 0$
39. $f(x, y, z) = z - x^2 - y^2 + 4$; $k = 7$
40. $f(x, y, z) = 4x - 2y + z$; $k = 1$

In Exercises 41–44, describe the level surfaces in words.

41. $f(x, y, z) = (x - 2)^2 + y^2 + z^2$
42. $f(x, y, z) = 3x - y + 2z$ **43.** $f(x, y, z) = x^2 + z^2$
44. $f(x, y, z) = z - x^2 - y^2$

45. Let $f(x, y) = x^2 - 2x^3 + 3xy$. Find an equation of the level curve that passes through the point
(a) $(-1, 1)$ (b) $(0, 0)$ (c) $(2, -1)$.

46. Let $f(x, y) = ye^x$. Find an equation of the level curve that passes through the point
(a) $(\ln 2, 1)$ (b) $(0, 3)$ (c) $(1, -2)$.

47. Let $f(x, y, z) = x^2 + y^2 - z$. Find an equation of the level surface that passes through the point
(a) $(1, -2, 0)$ (b) $(1, 0, 3)$ (c) $(0, 0, 0)$.

48. Let $f(x, y, z) = xyz + 3$. Find an equation of the level surface that passes through the point
(a) $(1, 0, 2)$ (b) $(-2, 4, 1)$ (c) $(0, 0, 0)$.

49. If $T(x, y)$ is the temperature at a point (x, y) on a thin metal plate in the xy-plane, then the level curves of T are called *isothermal curves*. All points on such a curve are at the same temperature. Suppose that a plate occupies the first quadrant and $T(x, y) = xy$.
(a) Sketch the isothermal curves on which $T = 1$, $T = 2$, and $T = 3$.
(b) An ant, initially at $(1, 4)$, wants to walk on the plate so that the temperature along its path remains constant. What path should the ant take and what is the temperature along that path?

50. If $V(x, y)$ is the voltage or potential at a point (x, y) in the xy-plane, then the level curves of V are called *equipotential curves*. Along such a curve, the voltage remains constant. Given that

$$V(x, y) = \frac{8}{\sqrt{16 + x^2 + y^2}}$$

sketch the equipotential curves at which $V = 2.0$, $V = 1.0$, and $V = 0.5$.

51. In each part, match the contour plot with one of the functions

$$f(x, y) = \sqrt{x^2 + y^2}, \quad f(x, y) = x^2 + y^2,$$
$$f(x, y) = 1 - x^2 - y^2$$

by inspection, and explain your reasoning. The larger the value of z, the lighter the color in the contour plot, and the contours correspond to equally spaced values of z.

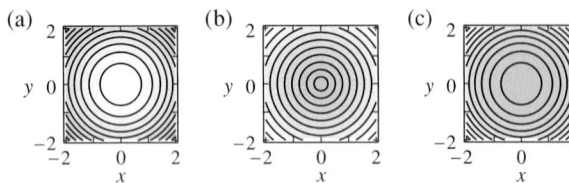

52. In each part, match the contour plot with one of the surfaces in the accompanying figure by inspection, and explain your reasoning. The larger the value of z, the lighter the color in the contour plot.

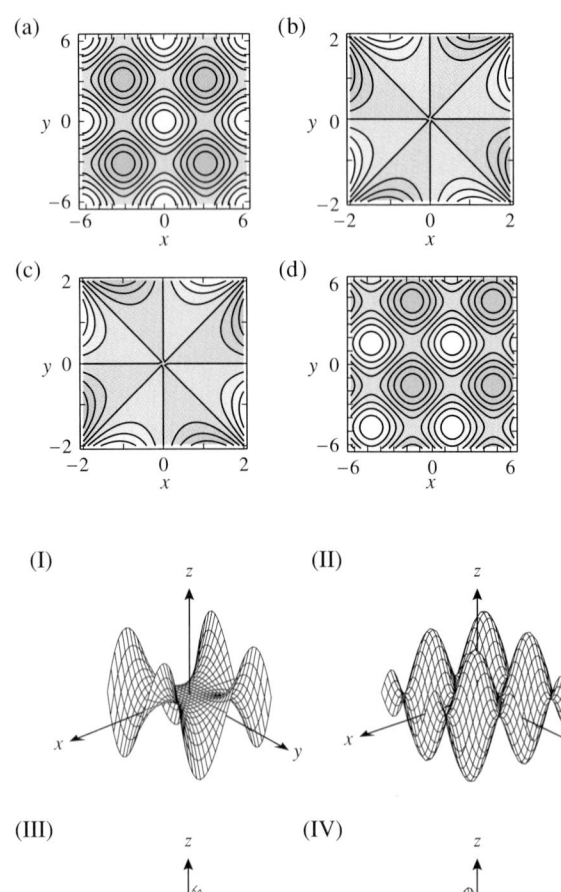

Figure Ex-52

53. In each part, the questions refer to the contour map in the accompanying figure.
   (a) Is A or B the higher point? Explain your reasoning.
   (b) Is A or B on the steeper slope? Explain your reasoning.
   (c) Starting at A and moving so that y remains constant and x increases, will the elevation begin to increase or decrease?
   (d) Starting at B and moving so that y remains constant and x increases, will the elevation begin to increase or decrease?
   (e) Starting at A and moving so that x remains constant and y decreases, will the elevation begin to increase or decrease?
   (f) Starting at B and moving so that x remains constant and y decreases, will the elevation begin to increase or decrease?

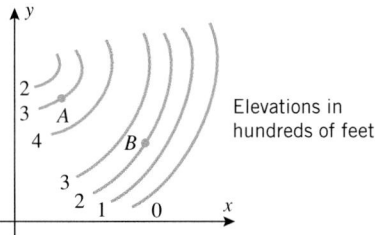

Figure Ex-53

54. A curve connecting points of equal atmospheric pressure on a weather map is called an **isobar**. On a typical weather map the isobars refer to pressure at mean sea level and are given in units of **millibars** (mb). Mathematically, isobars are level curves for the pressure function $p(x, y)$ defined at the geographic points (x, y) represented on the map. Tightly packed isobars correspond to steep slopes on the graph of the pressure function, and these are usually associated with strong winds—the steeper the slope, the greater the speed of the wind.
   (a) Referring to the accompanying weather map, is the wind speed greater in Calgary or in Chicago? Explain your reasoning.
   (b) Estimate the average rate of change in atmospheric pressure from Calgary to Chicago, given that the distance between the cities is approximately 1600 mi.

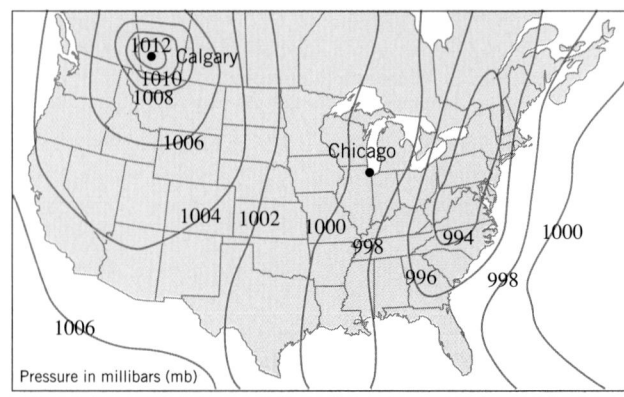

Figure Ex-54

55. Let $f(x, y) = x^2 + y^3$.
(a) Use a graphing utility to generate the level curve that passes through the point $(2, -1)$.
(b) Generate the level curve of height 1.

56. Let $f(x, y) = 2\sqrt{xy}$.
(a) Use a graphing utility to generate the level curve that passes through the point $(2, 2)$.
(b) Generate the level curve of height 8.

57. Let $f(x, y) = xe^{-(x^2+y^2)}$.
(a) Use a CAS to generate the graph of f for $-2 \le x \le 2$ and $-2 \le y \le 2$.
(b) Generate a contour plot for the surface, and confirm visually that it is consistent with the surface obtained in part (a).
(c) Read the appropriate documentation and explore the effect of generating the graph of f from various viewpoints.
(d) Read the appropriate documentation and generate the surface in different styles, as in Table 15.1.1.

58. Let $f(x, y) = \frac{1}{10}e^x \sin y$.
(a) Use a CAS to generate the graph of f for $0 \le x \le 4$ and $0 \le y \le 2\pi$.
(b) Generate a contour plot for the surface, and confirm visually that it is consistent with the surface obtained in part (a).
(c) Read the appropriate documentation and explore the effect of generating the graph of f from various viewpoints.
(d) Read the appropriate documentation and generate the surface in different styles, as in Table 15.1.1.

59. In each part, describe in words how the graph of g is related to the graph of f.
(a) $g(x, y) = f(x - 1, y)$ (b) $g(x, y) = 1 + f(x, y)$
(c) $g(x, y) = -f(x, y + 1)$

60. (a) Sketch the graph of $f(x, y) = e^{-(x^2+y^2)}$.
(b) In this part, describe in words how the graph of the function $g(x, y) = e^{-a(x^2+y^2)}$ is related to the graph of f for positive values of a.

15.2 LIMITS AND CONTINUITY

In this section we will introduce the notions of limit and continuity for functions of two or more variables. We will not go into great detail—our objective is to develop the basic concepts accurately and to obtain results needed in later sections. A more extensive study of these topics is usually given in advanced calculus.

OPEN AND CLOSED SETS

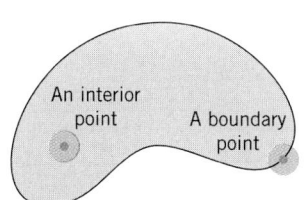

Figure 15.2.1

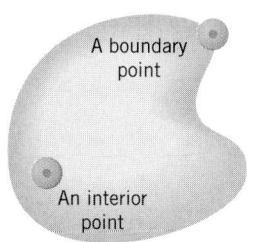

Figure 15.2.2

In our study of functions of one variable, the domains of the functions we encountered were generally intervals. For functions of two or three variables the situation is more complicated, so we will need to discuss some terminology about sets in 2-space and 3-space that will be helpful when we want to accurately describe the domain of a function of two or three variables.

If D is a set of points in 2-space, then a point (x_0, y_0) is called an ***interior point*** of D if there is *some* circular disk with positive radius, centered at (x_0, y_0), and containing only points in D (Figure 15.2.1). A point (x_0, y_0) is called a ***boundary point*** of D if *every* circular disk with positive radius and centered at (x_0, y_0) contains both points in D and points not in D (Figure 15.2.1). Similarly, if D is a set of points in 3-space, then a point (x_0, y_0, z_0) is called an ***interior point*** of D if there is *some* spherical ball with positive radius, centered at (x_0, y_0, z_0), and containing only points in D (Figure 15.2.2). A point (x_0, y_0, z_0) is called a ***boundary point*** of D if *every* spherical ball with positive radius and centered at (x_0, y_0, z_0) contains both points in D and points not in D (Figure 15.2.2).

For a set D in either 2-space or 3-space, the set of all boundary points of D is called the ***boundary*** of D and the set of all interior points of D is called the ***interior*** of D.

Recall that an open interval (a, b) on a coordinate line contains *neither* of its endpoints and a closed interval $[a, b]$ contains *both* of its endpoints. Analogously, a set D in 2-space or 3-space is called ***open*** if it contains *none* of its boundary points and ***closed*** if it contains *all* of its boundary points. The set D of all points in 2-space has no boundary points; it is regarded as both open and closed. Similarly, the set D of all points in 3-space is both open and closed.

Example 1

Let D be the set containing points in the xy-plane that are inside or on the circle of radius 1 centered at the origin. The set D, its interior I, and its boundary B can be expressed in set notation as

$$D = \{(x, y) : x^2 + y^2 \le 1\}, \quad I = \{(x, y) : x^2 + y^2 < 1\}, \quad B = \{(x, y) : x^2 + y^2 = 1\}$$

respectively (Figure 15.2.3). The set D is closed and the set I is open.   ◄

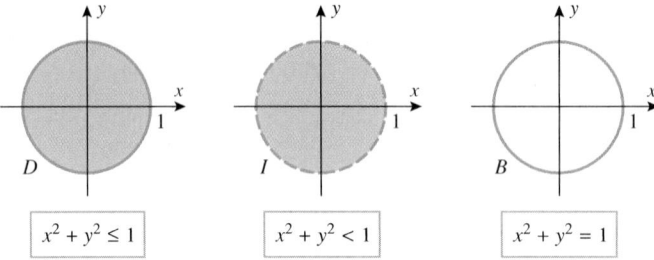

Figure 15.2.3

BOUNDED SETS

Just as we distinguished between finite intervals and infinite intervals on the real line, so we will want to distinguish between regions of "finite extent" and regions of "infinite extent" in 2-space and 3-space. A set of points in 2-space is called **bounded** if the entire set can be contained within some rectangle, and is called **unbounded** if there is no rectangle that contains all the points of the set. Similarly, a set of points in 3-space is **bounded** if the entire set can be contained within some box, and is unbounded otherwise (Figure 15.2.4).

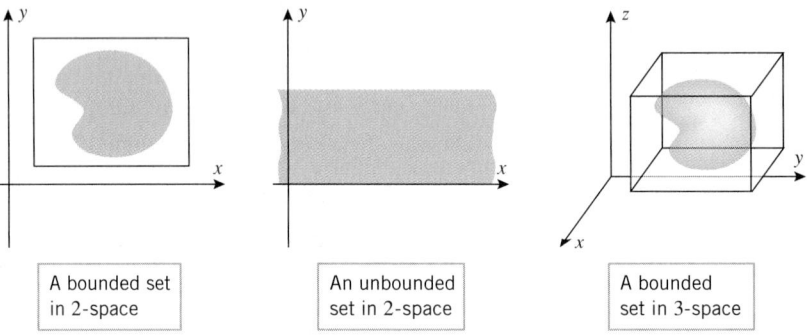

A bounded set in 2-space

An unbounded set in 2-space

A bounded set in 3-space

Figure 15.2.4

LIMITS ALONG CURVES

For a function of one variable there are two one-sided limits at a point x_0, namely

$$\lim_{x \to x_0^+} f(x) \quad \text{and} \quad \lim_{x \to x_0^-} f(x)$$

reflecting the fact that there are only two directions from which x can approach x_0, the right or the left. For functions of two or three variables the situation is more complicated because there are infinitely many different curves along which one point can approach another (Figure 15.2.5). Our first objective in this section is to define the limit of $f(x, y)$ as (x, y) approaches a point (x_0, y_0) along a curve C (and similarly for functions of three variables).

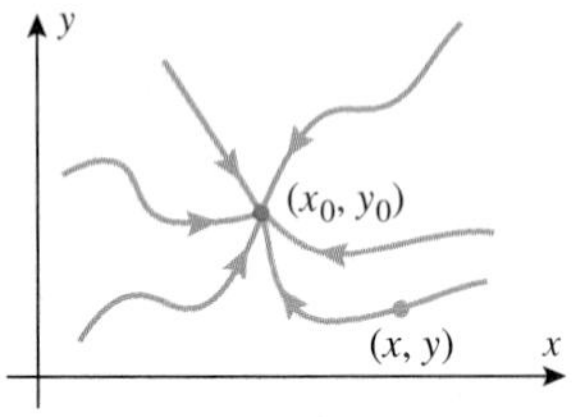

If C is a smooth parametric curve in 2-space or 3-space that is represented by the equations

$$x = x(t), \quad y = y(t) \qquad \text{or} \qquad x = x(t), \quad y = y(t), \quad z = z(t)$$

and if $x_0 = x(t_0)$, $y_0 = y(t_0)$, and $z_0 = z(t_0)$, then the limits

$$\lim_{\substack{(x, y) \to (x_0, y_0) \\ \text{(along } C)}} f(x, y) \quad \text{and} \quad \lim_{\substack{(x, y, z) \to (x_0, y_0, z_0) \\ \text{(along } C)}} f(x, y, z)$$

Figure 15.2.5

are defined by

$$\lim_{\substack{(x,\,y)\to(x_0,\,y_0)\\ (\text{along } C)}} f(x, y) = \lim_{t\to t_0} f(x(t), y(t)) \tag{1}$$

$$\lim_{\substack{(x,\,y,\,z)\to(x_0,\,y_0,\,z_0)\\ (\text{along } C)}} f(x, y, z) = \lim_{t\to t_0} f(x(t), y(t), z(t)) \tag{2}$$

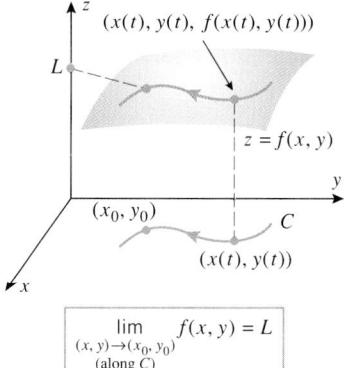

$$\lim_{\substack{(x,\,y)\to(x_0,\,y_0)\\ (\text{along } C)}} f(x, y) = L$$

Figure 15.2.6

Simply stated, limits along parametric curves are obtained by substituting the parametric equations into the formula for the function f and computing the appropriate limit of the resulting function of one variable. A geometric interpretation of the limit along a curve for a function of two variables is shown in Figure 15.2.6: As the point $(x(t), y(t))$ moves along the curve C in the xy-plane toward (x_0, y_0), the point $(x(t), y(t), f(x(t), y(t)))$ moves directly above it along the graph of $z = f(x, y)$ with $f(x(t), y(t))$ approaching the limiting value L. In the figure we followed a common practice of omitting the zero z-coordinate for points in the xy-plane.

REMARK.   In both (1) and (2), the limit of the function of t has to be treated as a one-sided limit if (x_0, y_0) or (x_0, y_0, z_0) is an endpoint of C.

Example 2

Figure 15.2.7a shows a computer-generated graph of the function

$$f(x, y) = -\frac{xy}{x^2 + y^2}$$

The graph reveals that the surface has a ridge above the line $y = -x$, which is to be expected since $f(x, y)$ has a constant value of $\frac{1}{2}$ for $y = -x$, except at $(0, 0)$ where f is undefined (verify). Moreover, the graph suggests that the limit of $f(x, y)$ as $(x, y) \to (0, 0)$ along a line through the origin varies with the direction of the line. Find this limit along

(a) the x-axis   (b) the y-axis   (c) the line $y = x$

(d) the line $y = -x$ (e) the parabola $y = x^2$

Solution (a).   The x-axis has parametric equations $x = t$, $y = 0$, with $(0, 0)$ corresponding to $t = 0$, so

$$\lim_{\substack{(x,\,y)\to(0,\,0)\\ (\text{along } y=0)}} f(x, y) = \lim_{t\to 0} f(t, 0) = \lim_{t\to 0}\left(-\frac{0}{t^2}\right) = \lim_{t\to 0} 0 = 0$$

which is consistent with Figure 15.2.7b.

Solution (b).   The y-axis has parametric equations $x = 0$, $y = t$, with $(0, 0)$ corresponding to $t = 0$, so

$$\lim_{\substack{(x,\,y)\to(0,\,0)\\ (\text{along } x=0)}} f(x, y) = \lim_{t\to 0} f(0, t) = \lim_{t\to 0}\left(-\frac{0}{t^2}\right) = \lim_{t\to 0} 0 = 0$$

which is consistent with Figure 15.2.7b.

Solution (c).   The line $y = x$ has parametric equations $x = t$, $y = t$, with $(0, 0)$ corresponding to $t = 0$, so

$$\lim_{\substack{(x,\,y)\to(0,\,0)\\ (\text{along } y=x)}} f(x, y) = \lim_{t\to 0} f(t, t) = \lim_{t\to 0}\left(-\frac{t^2}{2t^2}\right) = \lim_{t\to 0}\left(-\frac{1}{2}\right) = -\frac{1}{2}$$

which is consistent with Figure 15.2.7b.

Solution (d). The line $y = -x$ has parametric equations $x = t$, $y = -t$, with $(0, 0)$ corresponding to $t = 0$, so

$$\lim_{\substack{(x,y) \to (0,0) \\ (\text{along } y = -x)}} f(x, y) = \lim_{t \to 0} f(t, -t) = \lim_{t \to 0} \frac{t^2}{2t^2} = \lim_{t \to 0} \frac{1}{2} = \frac{1}{2}$$

which is consistent with Figure 15.2.7*b*.

Solution (e). The parabola $y = x^2$ has parametric equations $x = t$, $y = t^2$, with $(0, 0)$ corresponding to $t = 0$, so

$$\lim_{\substack{(x,y) \to (0,0) \\ (\text{along } y = x^2)}} f(x, y) = \lim_{t \to 0} f(t, t^2) = \lim_{t \to 0} \left(-\frac{t^3}{t^2 + t^4} \right) = \lim_{t \to 0} \left(-\frac{t}{1 + t^2} \right) = 0$$

This is consistent with Figure 15.2.7*c*, which shows the parametric curve

$$x = t, \quad y = t^2, \quad z = -\frac{t}{1 + t^2}$$

superimposed on the surface. ◀

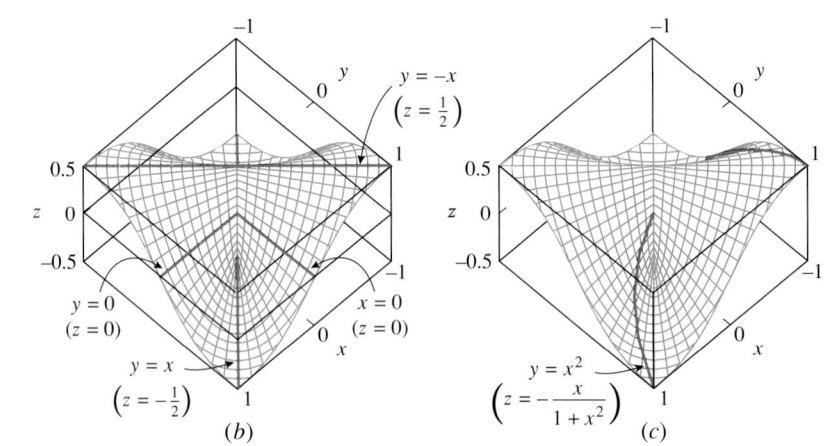

Figure 15.2.7

GENERAL LIMITS OF FUNCTIONS OF TWO VARIABLES

Although limits along specific curves are useful for many purposes, they do not always tell the complete story about the limiting behavior of a function; what is required is a limit concept that accounts for the behavior of the function in an *entire vicinity* of a point, not just along smooth curves passing through the point. As illustrated in Figure 15.2.8, we will want the statement

$$\lim_{(x,y) \to (x_0, y_0)} f(x, y) = L$$

to mean that the value of $f(x, y)$ can be made as close as we like to L (say within ϵ units of L) by restricting (x, y) to lie within (but not at the center of) some sufficiently small circle centered at (x_0, y_0) (say a circle of radius δ). This idea is conveyed by Definition 15.2.1.

15.2.1   DEFINITION.   Let f be a function of two variables. We will write

$$\lim_{(x,y) \to (x_0, y_0)} f(x, y) = L \tag{3}$$

if given any number $\epsilon > 0$, we can find a number $\delta > 0$ such that $f(x, y)$ satisfies

$$|f(x, y) - L| < \epsilon$$

whenever (x, y) lies in the domain of f and the distance between (x, y) and (x_0, y_0) satisfies

$$0 < \sqrt{(x - x_0)^2 + (y - y_0)^2} < \delta$$

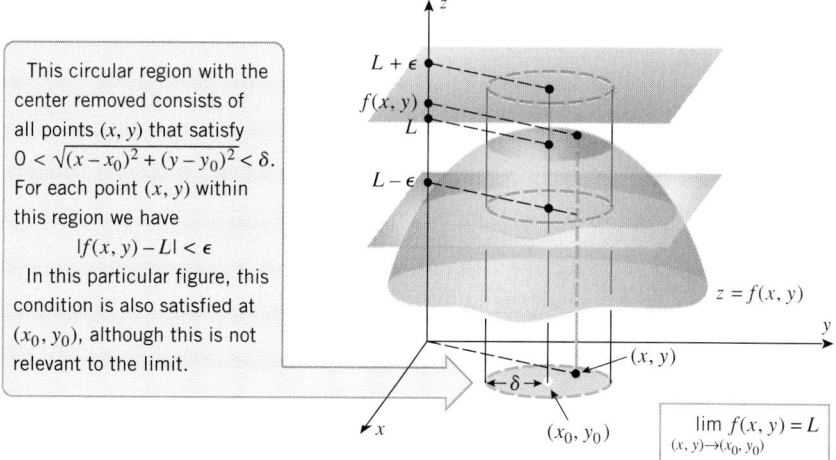

This circular region with the center removed consists of all points (x, y) that satisfy $0 < \sqrt{(x-x_0)^2 + (y-y_0)^2} < \delta$. For each point (x, y) within this region we have
$$|f(x, y) - L| < \epsilon$$
In this particular figure, this condition is also satisfied at (x_0, y_0), although this is not relevant to the limit.

$$\lim_{(x, y) \to (x_0, y_0)} f(x, y) = L$$

Figure 15.2.8

When convenient, (3) can also be written as

$$f(x, y) \to L \quad \text{as} \quad (x, y) \to (x_0, y_0)$$

PROPERTIES OF LIMITS

We note without proof that the standard properties of limits hold for limits along curves and for general limits of functions of two variables, so that computations involving such limits can be performed in the usual way.

Example 3

$$\lim_{(x,y) \to (1,4)} [5x^3 y^2 - 9] = \lim_{(x,y) \to (1,4)} [5x^3 y^2] - \lim_{(x,y) \to (1,4)} 9$$

$$= 5 \left[\lim_{(x,y) \to (1,4)} x \right]^3 \left[\lim_{(x,y) \to (1,4)} y \right]^2 - 9$$

$$= 5(1)^3 (4)^2 - 9 = 71 \qquad \blacktriangleleft$$

RELATIONSHIPS BETWEEN GENERAL LIMITS AND LIMITS ALONG SMOOTH CURVES

The following theorem, which we state without proof, establishes an important relationship between general limits and limits along smooth curves.

15.2.2 THEOREM.

(a) If $f(x, y) \to L$ as $(x, y) \to (x_0, y_0)$, then $f(x, y) \to L$ as $(x, y) \to (x_0, y_0)$ along any smooth curve that lies in the domain of f.

(b) If the limit of $f(x, y)$ fails to exist as $(x, y) \to (x_0, y_0)$ along some smooth curve in the domain of f, or if $f(x, y)$ has different limits as $(x, y) \to (x_0, y_0)$ along two different smooth curves in the domain of f, then the limit of $f(x, y)$ does not exist as $(x, y) \to (x_0, y_0)$.

Example 4

The limit

$$\lim_{(x,y) \to (0,0)} -\frac{xy}{x^2 + y^2}$$

does not exist because in Example 2 we found two different smooth curves along which this limit had different values. Specifically,

$$\lim_{\substack{(x, y) \to (0, 0) \\ (\text{along } x = 0)}} -\frac{xy}{x^2 + y^2} = 0 \quad \text{and} \quad \lim_{\substack{(x, y) \to (0, 0) \\ (\text{along } y = x)}} -\frac{xy}{x^2 + y^2} = -\frac{1}{2} \qquad \blacktriangleleft$$

REMARK.   One cannot prove that $f(x, y) \to L$ as $(x, y) \to (x_0, y_0)$ by showing that $f(x, y) \to L$ as $(x, y) \to (x_0, y_0)$ along a specific curve or even an entire family of curves. The problem is that there may be some curve outside of the family for which the limit does not exist or has a limit that is different from L (see Exercise 38, for example).

CONTINUITY

Stated informally, a function of one variable is continuous if its graph is an unbroken curve without jumps or holes. To extend this idea to functions of two variables, imagine that the graph of $z = f(x, y)$ is molded from a thin sheet of clay that has been hollowed or pinched into peaks and valleys. We will regard f as being continuous if the clay surface has no tears or holes. The functions graphed in Figure 15.2.9 fail to be continuous because of their behavior at $(0, 0)$.

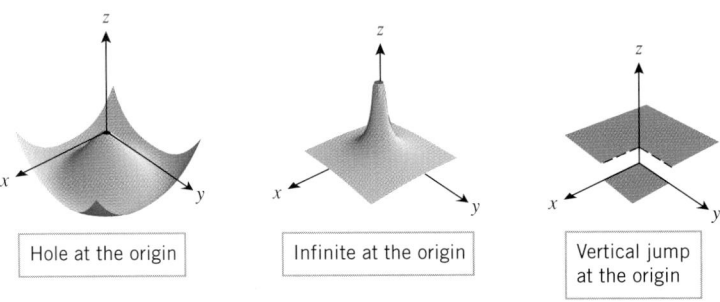

| Hole at the origin | Infinite at the origin | Vertical jump at the origin |

Figure 15.2.9

The precise definition of continuity at a point for functions of two variables is similar to that for functions of one variable—we require the limit of the function and the value of the function to be the same at the point.

15.2.3   DEFINITION.   A function $f(x)$ is said to be **continuous at (x_0, y_0)** if

$$\lim_{(x,y) \to (x_0,y_0)} f(x, y) = f(x_0, y_0)$$

Moreover, if f is continuous at each point of a region R in the xy-plane, then we say that **f is continuous on R**; and if f is continuous at every point in the xy-plane, then we say that f is **continuous everywhere**. In addition, we will say that f is a **continuous function** if it is continuous at each point of its domain.

The following theorem, which we state without proof, will help us to identify continuous functions of two variables.

15.2.4   THEOREM.
(a)   *If g and h are continuous functions of one variable, then $f(x, y) = g(x)h(y)$ is a continuous function of x and y.*
(b)   *If g is a continuous function of one variable and h is a continuous function of two variables, then their composition $f(x, y) = g(h(x, y))$ is a continuous function of x and y.*

Example 5

Use Theorem 15.2.4 to show that $f(x, y) = 3x^2 y^5$ and $f(x, y) = \sin(3x^2 y^5)$ are continuous functions.

Solution.   The function $f(x, y) = 3x^2 y^5$ is continuous because it is the product of the continuous functions $g(x) = 3x^2$ and $h(y) = y^5$, and the function $f(x, y) = \sin(3x^2 y^5)$ is continuous because it is the composition of the continuous function $\sin x$ and the continuous function $3x^2 y^5$. ◄

Theorem 15.2.4 is one of a whole class of theorems about continuity of functions in two or more variables. The content of these theorems can be summarized informally with three basic principles:

- A composition of continuous functions is continuous.
- A sum, difference, or product of continuous functions is continuous.
- A quotient of continuous functions is continuous, except where the denominator is zero.

By using these principles and Theorem 15.2.4, you should be able to confirm that the following functions are all continuous:

$$x e^{xy} + y^{2/3}, \quad \cosh(xy^3) - |xy|, \quad \frac{xy}{1 + x^2 + y^2}$$

Example 6

Evaluate $\lim\limits_{(x,y) \to (-1,2)} \dfrac{xy}{x^2 + y^2}$.

Solution. Since $f(x, y) = xy/(x^2 + y^2)$ is continuous at $(-1, 2)$ (why?), it follows from the definition of continuity for functions of two variables that

$$\lim_{(x,y) \to (-1,2)} \frac{xy}{x^2 + y^2} = \frac{(-1)(2)}{(-1)^2 + (2)^2} = -\frac{2}{5}$$ ◄

Example 7

Since the function

$$f(x, y) = \frac{x^3 y^2}{1 - xy}$$

is a quotient of continuous functions, it is continuous except where $1 - xy = 0$. Thus, $f(x, y)$ is continuous everywhere except on the hyperbola $xy = 1$. ◄

LIMITS AT POINTS OF DISCONTINUITY

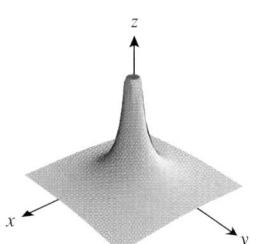

Figure 15.2.10

Sometimes it is easy to recognize when a limit does not exist. For example, it is evident that

$$\lim_{(x,y) \to (0,0)} \frac{1}{x^2 + y^2} = +\infty$$

which implies that the values of the function approach $+\infty$ as $(x, y) \to (0, 0)$ along any smooth curve (Figure 15.2.10). However, it is not evident whether the limit

$$\lim_{(x,y) \to (0,0)} (x^2 + y^2) \ln(x^2 + y^2)$$

exists because it is an indeterminate form of type $0 \cdot \infty$. Although L'Hôpital's rule cannot be applied directly, the following example illustrates a method for finding this limit by converting to polar coordinates.

Example 8

Find $\lim\limits_{(x,y) \to (0,0)} (x^2 + y^2) \ln(x^2 + y^2)$.

Solution. Let (r, θ) be polar coordinates of the point (x, y) with $r \geq 0$. Then we have

$$x = r \cos \theta, \quad y = r \sin \theta, \quad r^2 = x^2 + y^2$$

Moreover, since $r \geq 0$ we have $r = \sqrt{x^2 + y^2}$, so that $r \to 0^+$ if and only if $(x, y) \to (0, 0)$.

Thus, we can rewrite the given limit as

$$\lim_{(x,y) \to (0,0)} (x^2 + y^2) \ln(x^2 + y^2) = \lim_{r \to 0^+} r^2 \ln r^2$$

$$= \lim_{r \to 0^+} \frac{2 \ln r}{1/r^2} \quad \boxed{\text{This converts the limit to an indeterminate form of type } \infty/\infty.}$$

$$= \lim_{r \to 0^+} \frac{2/r}{-2/r^3} \quad \boxed{\text{L'Hôpital's rule}}$$

$$= \lim_{r \to 0^+} (-r^2) = 0 \quad \blacktriangleleft$$

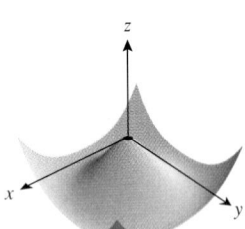

Figure 15.2.11

REMARK.   Recall that for a function f of one variable, a hole occurs in the graph of f at x_0 if $f(x_0)$ is undefined but $f(x)$ has a limit as $x \to x_0$ (Figure 2.4.2a, for example). Similarly, a hole will occur in the graph of $f(x, y)$ at (x_0, y_0) if $f(x_0, y_0)$ is undefined but $f(x, y)$ has a limit as $(x, y) \to (x_0, y_0)$. In particular, it follows from the last example that the graph of $f(x, y) = (x^2 + y^2) \ln(x^2 + y^2)$ has a hole at $(0, 0)$ (Figure 15.2.11).

EXTENSIONS TO THREE VARIABLES

All of the results in this section can be extended to functions of three or more variables. For example, the distance between the points (x, y, z) and (x_0, y_0, z_0) in 3-space is

$$\sqrt{(x - x_0)^2 + (y - y_0)^2 + (z - z_0)^2}$$

so the natural extension of Definition 15.2.1 to 3-space is as follows:

15.2.5   DEFINITION.   Let f be a function of three variables. We will write

$$\lim_{(x,y,z) \to (x_0,y_0,z_0)} f(x, y, z) = L \tag{4}$$

if given any number $\epsilon > 0$, we can find a number $\delta > 0$ such that $f(x, y, z)$ satisfies

$$|f(x, y, z) - L| < \epsilon$$

whenever (x, y, z) lies in the domain of f and the distance between (x, y, z) and (x_0, y_0, z_0) satisfies

$$0 < \sqrt{(x - x_0)^2 + (y - y_0)^2 + (z - z_0)^2} < \delta$$

As with functions of one and two variables, we define a function $f(x, y, z)$ of three variables to be continuous at a point (x_0, y_0, z_0) if the limit of the function and the value of the function are the same at this point; that is,

$$\lim_{(x,y,z) \to (x_0,y_0,z_0)} f(x, y, z) = f(x_0, y_0, z_0)$$

Although we will omit the details, the properties of limits and continuity that we discussed for functions of two variables carry over to functions of three variables.

EXERCISE SET 15.2

In Exercises 1–8, sketch the region where the function f is continuous.

1. $f(x, y) = y \ln(1 + x)$ **2.** $f(x, y) = \sqrt{x - y}$

3. $f(x, y) = \dfrac{x^2 y}{\sqrt{25 - x^2 - y^2}}$ **4.** $f(x, y) = \ln(2x - y + 1)$

5. $f(x, y) = \cos\left(\dfrac{xy}{1 + x^2 + y^2}\right)$

6. $f(x, y) = e^{1-xy}$ **7.** $f(x, y) = \sin^{-1}(xy)$

8. $f(x, y) = \tan^{-1}(y - x)$

In Exercises 9–12, describe the region on which the function f is continuous.

9. $f(x, y, z) = 3x^2 e^{yz} \cos(xyz)$

10. $f(x, y, z) = \ln(4 - x^2 - y^2 - z^2)$

11. $f(x, y, z) = \dfrac{y + 1}{x^2 + z^2 - 1}$

12. $f(x, y, z) = \sin\sqrt{x^2 + y^2 + 3z^2}$

In Exercises 13–18, use limit laws and continuity properties to evaluate the limit.

13. $\displaystyle\lim_{(x,y)\to(1,3)} (4xy^2 - x)$

14. $\displaystyle\lim_{(x,y)\to(1/2,\pi)} (xy^2 \sin xy)$

15. $\displaystyle\lim_{(x,y)\to(-1,2)} \frac{xy^3}{x+y}$

16. $\displaystyle\lim_{(x,y)\to(1,-3)} e^{2x-y^2}$

17. $\displaystyle\lim_{(x,y)\to(0,0)} \ln(1 + x^2 y^3)$

18. $\displaystyle\lim_{(x,y)\to(4,-2)} x\sqrt[3]{y^3 + 2x}$

In Exercises 19 and 20, show that the limit does not exist by considering the limits as $(x, y) \to (0, 0)$ along the coordinate axes.

19. (a) $\displaystyle\lim_{(x,y)\to(0,0)} \frac{3}{x^2 + 2y^2}$ (b) $\displaystyle\lim_{(x,y)\to(0,0)} \frac{x+y}{x+y^2}$

20. (a) $\displaystyle\lim_{(x,y)\to(0,0)} \frac{x-y}{x^2 + y^2}$ (b) $\displaystyle\lim_{(x,y)\to(0,0)} \frac{\cos xy}{x+y}$

In Exercises 21–24, evaluate the limit by making the substitution $z = x^2 + y^2$ and observing that $z \to 0^+$ as $(x, y) \to (0, 0)$.

21. $\displaystyle\lim_{(x,y)\to(0,0)} \frac{\sin(x^2 + y^2)}{x^2 + y^2}$

22. $\displaystyle\lim_{(x,y)\to(0,0)} \frac{1 - \cos(x^2 + y^2)}{x^2 + y^2}$

23. $\displaystyle\lim_{(x,y)\to(0,0)} e^{-1/(x^2+y^2)}$

24. $\displaystyle\lim_{(x,y)\to(0,0)} \frac{e^{-1/\sqrt{x^2+y^2}}}{\sqrt{x^2 + y^2}}$

In Exercises 25–32, determine whether the limit exists. If so, find its value.

25. $\displaystyle\lim_{(x,y)\to(0,0)} \frac{x^4 - y^4}{x^2 + y^2}$

26. $\displaystyle\lim_{(x,y)\to(0,0)} \frac{x^4 - 16y^4}{x^2 + 4y^2}$

27. $\displaystyle\lim_{(x,y)\to(0,0)} \frac{xy}{3x^2 + 2y^2}$

28. $\displaystyle\lim_{(x,y)\to(0,0)} \frac{1 - x^2 - y^2}{x^2 + y^2}$

29. $\displaystyle\lim_{(x,y,z)\to(2,-1,2)} \frac{xz^2}{\sqrt{x^2 + y^2 + z^2}}$

30. $\displaystyle\lim_{(x,y,z)\to(2,0,-1)} \ln(2x + y - z)$

31. $\displaystyle\lim_{(x,y,z)\to(0,0,0)} \frac{\sin(x^2 + y^2 + z^2)}{\sqrt{x^2 + y^2 + z^2}}$

32. $\displaystyle\lim_{(x,y,z)\to(0,0,0)} \frac{\sin\sqrt{x^2 + y^2 + z^2}}{x^2 + y^2 + z^2}$

In Exercises 33 and 34, evaluate the limit, if it exists, by converting to polar coordinates, as in Example 8.

33. $\displaystyle\lim_{(x,y)\to(0,0)} y\ln(x^2 + y^2)$

34. $\displaystyle\lim_{(x,y)\to(0,0)} \frac{x^2 y^2}{\sqrt{x^2 + y^2}}$

In Exercises 35 and 36, evaluate the limit, if it exists, by converting to spherical coordinates; that is, let $x = \rho \sin\phi \cos\theta$, $y = \rho \sin\phi \sin\theta$, $z = \rho \cos\phi$ and observe that $\rho \to 0^+$ as $(x, y, z) \to (0, 0, 0)$, since $\rho = \sqrt{x^2 + y^2 + z^2}$.

35. $\displaystyle\lim_{(x,y,z)\to(0,0,0)} \frac{e^{\sqrt{x^2+y^2+z^2}}}{\sqrt{x^2 + y^2 + z^2}}$

36. $\displaystyle\lim_{(x,y,z)\to(0,0,0)} \tan^{-1}\left[\frac{1}{x^2 + y^2 + z^2}\right]$

37. The accompanying figure shows a portion of the graph of
$$f(x, y) = \frac{x^2 y}{x^4 + y^2}$$

(a) Based on the graph in the figure, does $f(x, y)$ have a limit as $(x, y) \to (0, 0)$? Explain your reasoning.

(b) Show that $f(x, y) \to 0$ as $(x, y) \to (0, 0)$ along any line $y = mx$. Does this imply that $f(x, y) \to 0$ as $(x, y) \to (0, 0)$? Explain.

(c) Show that $f(x, y) \to \frac{1}{2}$ as $(x, y) \to (0, 0)$ along the parabola $y = x^2$, and confirm visually that this is consistent with the graph of $f(x, y)$.

(d) Based on parts (b) and (c), does $f(x, y)$ have a limit as $(x, y) \to (0, 0)$? Is this consistent with your answer to part (a)?

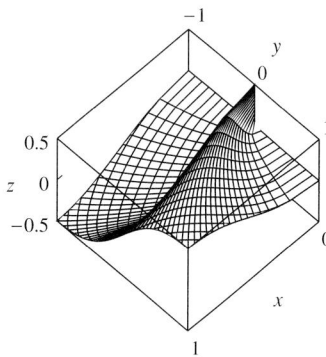

Figure Ex-37

38. (a) Show that the value of $\dfrac{x^3 y}{2x^6 + y^2}$ approaches 0 as $(x, y) \to (0, 0)$ along any straight line $y = mx$, or along any parabola $y = kx^2$.

(b) Show that $\displaystyle\lim_{(x,y)\to(0,0)} \frac{x^3 y}{2x^6 + y^2}$ does not exist by letting $(x, y) \to (0, 0)$ along the curve $y = x^3$.

39. (a) Show that the value of $\dfrac{xyz}{x^2 + y^4 + z^4}$ approaches 0 as $(x, y, z) \to (0, 0, 0)$ along any line $x = at$, $y = bt$, $z = ct$.

(b) Show that the limit $\displaystyle\lim_{(x,y,z)\to(0,0,0)} \frac{xyz}{x^2 + y^4 + z^4}$ does not exist by letting $(x, y, z) \to (0, 0, 0)$ along the curve $x = t^2$, $y = t$, $z = t$.

40. Find
$$\lim_{(x,y)\to(0,1)} \tan^{-1}\left[\frac{x^2 + 1}{x^2 + (y-1)^2}\right]$$

41. Find
$$\lim_{(x,y)\to(0,1)} \tan^{-1}\left[\frac{x^2 - 1}{x^2 + (y-1)^2}\right]$$

42. Let $f(x, y) = \begin{cases} \dfrac{\sin(x^2 + y^2)}{x^2 + y^2}, & (x, y) \neq (0, 0) \\ 1, & (x, y) = (0, 0). \end{cases}$

Show that f is continuous at $(0, 0)$.

43. Let $f(x, y) = \dfrac{x^2}{x^2 + y^2}$. Is it possible to define $f(0, 0)$ so that f will be continuous at $(0, 0)$?

44. Let $f(x, y) = xy \ln(x^2 + y^2)$. Is it possible to define $f(0, 0)$ so that f will be continuous at $(0, 0)$?

15.3 PARTIAL DERIVATIVES

If $z = f(x, y)$, then one can inquire how the value of z changes if x is held fixed and y is allowed to vary or if y is held fixed and x is allowed to vary. For example, the ideal gas law in physics states that under appropriate conditions the pressure exerted by a gas is a function of the volume of the gas and its temperature. Thus, a physicist studying gases might be interested in the rate of change of the pressure if the volume is held fixed and the temperature is allowed to vary or if the temperature is held fixed and the volume is allowed to vary. In this section we will develop the mathematical tools for studying rates of change that involve two or more independent variables.

PARTIAL DERIVATIVES OF FUNCTIONS OF TWO VARIABLES

Recall that if $y = f(x)$, then the rate of change of y with respect to x is given by the derivative of f with respect to x, which can be expressed using Formula (10) of Section 3.2 as

$$f'(x) = \lim_{\Delta x \to 0} \frac{f(x + \Delta x) - f(x)}{\Delta x}$$

This suggests the following definition.

15.3.1 DEFINITION. If $z = f(x, y)$, then the ***partial derivative of f with respect to x*** (also called the ***partial derivative of z with respect to x***) is the derivative with respect to x of the function that results when y is held fixed and x is allowed to vary. This partial derivative is denoted by $f_x(x, y)$ and can be expressed as the limit

$$f_x(x, y) = \lim_{\Delta x \to 0} \frac{f(x + \Delta x, y) - f(x, y)}{\Delta x} \tag{1}$$

Similarly, the ***partial derivative of f with respect to y*** (also called the ***partial derivative of z with respect to y***) is the derivative with respect to y of the function that results when x is held fixed and y is allowed to vary. This partial derivative is denoted by $f_y(x, y)$ and can be expressed as the limit

$$f_y(x, y) = \lim_{\Delta y \to 0} \frac{f(x, y + \Delta y) - f(x, y)}{\Delta y} \tag{2}$$

Example 1

Find the partial derivatives of $f(x, y) = 2x^3 y^2 + 2y + 4x$.

Solution. Treating y as a constant and differentiating with respect to x, we obtain

$$f_x(x, y) = 6x^2 y^2 + 4$$

Treating x as a constant and differentiating with respect to y, we obtain

$$f_y(x, y) = 4x^3 y + 2$$

◀

FOR THE READER.   If you have a CAS, read the relevant documentation on calculating partial derivatives, and then use the CAS to perform the computations in Example 1.

PARTIAL DERIVATIVES VIEWED AS RATES OF CHANGE AND SLOPES

Recall that if $y = f(x)$, then the value of $f'(x_0)$ can be interpreted either as the rate of change of y with respect to x at the point x_0 or as the slope of the tangent line to the graph of f at the point x_0. Partial derivatives have analogous interpretations. To see that this is so, suppose that C_1 is the intersection of the surface $z = f(x, y)$ with the plane $y = y_0$ and that C_2 is its intersection with the plane $x = x_0$ (Figure 15.3.1). Thus, $f_x(x, y_0)$ can be interpreted as the rate of change of z with respect to x along the curve C_1, and $f_y(x_0, y)$ can be interpreted as the rate of change of z with respect to y along the curve C_2. In particular, $f_x(x_0, y_0)$ is the rate of change of z with respect to x along the curve C_1 at the point (x_0, y_0), and $f_y(x_0, y_0)$ is the rate of change of z with respect to y along the curve C_2 at the point (x_0, y_0). Geometrically, $f_x(x_0, y_0)$ can be viewed as the slope of the tangent line to the curve C_1 at the point (x_0, y_0), and $f_y(x_0, y_0)$ can be viewed as the slope of the tangent line to curve C_2 at the point (x_0, y_0) (Figure 15.3.1). We will call $f_x(x_0, y_0)$ the **slope of the surface in the x-direction** at (x_0, y_0), and $f_y(x_0, y_0)$ the **slope of the surface in the y-direction** at (x_0, y_0).

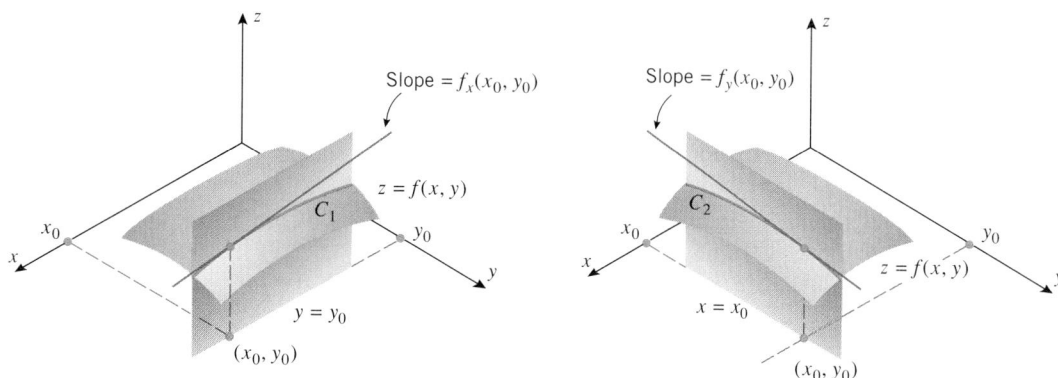

Figure 15.3.1

Example 2

Let $f(x, y) = x^2 y + 5y^3$.

(a)   Find the slope of the surface $z = f(x, y)$ in the x-direction at the point $(1, -2)$.

(b)   Find the slope of the surface $z = f(x, y)$ in the y-direction at the point $(1, -2)$.

Solution (a). Differentiating f with respect to x with y held fixed yields

$$f_x(x, y) = 2xy$$

Thus, the slope in the x-direction is $f_x(1, -2) = -4$; that is, z is decreasing at the rate of 4 units per unit increase in x.

Solution (b). Differentiating f with respect to y with x held fixed yields

$$f_y(x, y) = x^2 + 15y^2$$

Thus, the slope in the y-direction is $f_y(1, -2) = 61$; that is, z is increasing at the rate of 61 units per unit increase in y.   ◀

Example 3

Figure 15.3.2 shows the graph of the function

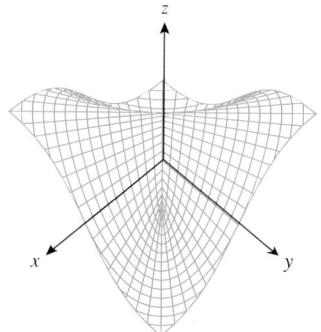

Figure 15.3.2

$$f(x, y) = \begin{cases} -\dfrac{xy}{x^2 + y^2}, & \text{if } (x, y) \neq (0, 0) \\ 0, & \text{if } (x, y) = (0, 0) \end{cases} \tag{3}$$

This is similar to the function considered in Example 2 of Section 15.2, except that here we have assigned f a value at $(0, 0)$. Except at this point, the partial derivatives of f are

$$f_x(x, y) = -\frac{(x^2 + y^2)y - xy(2x)}{(x^2 + y^2)^2} = \frac{x^2 y - y^3}{(x^2 + y^2)^2} \tag{4}$$

$$f_y(x, y) = -\frac{(x^2 + y^2)x - xy(2y)}{(x^2 + y^2)^2} = \frac{xy^2 - x^3}{(x^2 + y^2)^2} \tag{5}$$

Figure 15.3.2 suggests that at each point on the x-axis [except possibly $(0, 0)$] the surface has slope 0 in the x-direction and at each point on the y-axis [except possibly $(0, 0)$] the surface has slope 0 in the y-direction. This can be confirmed by evaluating f_x at a typical point $(x, 0)$ on the x-axis and evaluating f_y at a typical point $(0, y)$ on the y-axis. Setting $y = 0$ in (4) and $x = 0$ in (5) yields

$$f_x(x, 0) = 0 \quad \text{and} \quad f_y(0, y) = 0$$

which confirms our conjecture.

It is not evident from Formula (3) whether f has partial derivatives at $(0, 0)$ and if so, what the values of those derivatives are. To answer that question we will have to use the definitions of the partial derivatives (Definition 15.3.1). Applying Formulas (1) and (2) to (3) we obtain

$$f_x(0, 0) = \lim_{\Delta x \to 0} \frac{f(\Delta x, 0) - f(0, 0)}{\Delta x} = \lim_{\Delta x \to 0} \frac{0 - 0}{\Delta x} = 0$$

$$f_y(0, 0) = \lim_{\Delta y \to 0} \frac{f(0, \Delta y) - f(0, 0)}{\Delta y} = \lim_{\Delta y \to 0} \frac{0 - 0}{\Delta y} = 0$$

This shows that f has partial derivatives at $(0, 0)$ and the values of both partial derivatives are 0 at that point. ◀

PARTIAL DERIVATIVE NOTATION

If $z = f(x, y)$, then the partial derivatives f_x and f_y are also denoted by the symbols[*]

$$\frac{\partial f}{\partial x}, \quad \frac{\partial z}{\partial x} \quad \text{and} \quad \frac{\partial f}{\partial y}, \quad \frac{\partial z}{\partial y}$$

Some typical notations for the partial derivatives of $z = f(x, y)$ at a point (x_0, y_0) are

$$\frac{\partial f}{\partial x}\bigg|_{x=x_0, y=y_0}, \quad \frac{\partial z}{\partial x}\bigg|_{(x_0, y_0)}, \quad \frac{\partial f}{\partial x}\bigg|_{(x_0, y_0)}, \quad \frac{\partial f}{\partial x}(x_0, y_0), \quad \frac{\partial z}{\partial x}(x_0, y_0)$$

Example 4

Find $\partial z / \partial x$ and $\partial z / \partial y$ if $z = x^4 \sin(xy^3)$.

Solution.

$$\frac{\partial z}{\partial x} = \frac{\partial}{\partial x}[x^4 \sin(xy^3)] = x^4 \frac{\partial}{\partial x}[\sin(xy^3)] + \sin(xy^3) \cdot \frac{\partial}{\partial x}(x^4)$$

$$= x^4 \cos(xy^3) \cdot y^3 + \sin(xy^3) \cdot 4x^3 = x^4 y^3 \cos(xy^3) + 4x^3 \sin(xy^3)$$

$$\frac{\partial z}{\partial y} = \frac{\partial}{\partial y}[x^4 \sin(xy^3)] = x^4 \frac{\partial}{\partial y}[\sin(xy^3)] + \sin(xy^3) \cdot \frac{\partial}{\partial y}(x^4)$$

$$= x^4 \cos(xy^3) \cdot 3xy^2 + \sin(xy^3) \cdot 0 = 3x^5 y^2 \cos(xy^3) \quad ◀$$

IMPLICIT PARTIAL DIFFERENTIATION

Example 5

Find the slope of the sphere $x^2 + y^2 + z^2 = 1$ in the y-direction at the points $\left(\frac{2}{3}, \frac{1}{3}, \frac{2}{3}\right)$ and $\left(\frac{2}{3}, \frac{1}{3}, -\frac{2}{3}\right)$ (Figure 15.3.3).

---

[*]The symbol ∂ is called a partial derivative sign. It is derived from the Cyrillic alphabet.

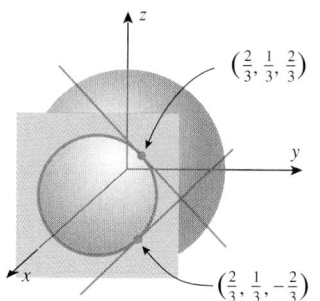

$\left(\frac{2}{3}, \frac{1}{3}, \frac{2}{3}\right)$

$\left(\frac{2}{3}, \frac{1}{3}, -\frac{2}{3}\right)$

Figure 15.3.3

Solution. The point $\left(\frac{2}{3}, \frac{1}{3}, \frac{2}{3}\right)$ lies on the upper hemisphere $z = \sqrt{1 - x^2 - y^2}$, and the point $\left(\frac{2}{3}, \frac{1}{3}, -\frac{2}{3}\right)$ lies on the lower hemisphere $z = -\sqrt{1 - x^2 - y^2}$. We could find the slopes by differentiating each expression for z separately with respect to y and then evaluating the derivatives at $x = \frac{2}{3}$ and $y = \frac{1}{3}$. However, it is more efficient to differentiate the given equation

$$x^2 + y^2 + z^2 = 1$$

implicitly with respect to y, since this will give us both slopes with one differentiation. To perform the implicit differentiation, we view z as a function of x and y and differentiate both sides with respect to y, taking x to be fixed. The computations are as follows:

$$\frac{\partial}{\partial y}[x^2 + y^2 + z^2] = \frac{\partial}{\partial y}[1]$$

$$0 + 2y + 2z\frac{\partial z}{\partial y} = 0$$

$$\frac{\partial z}{\partial y} = -\frac{y}{z}$$

Substituting the y- and z-coordinates of the points $\left(\frac{2}{3}, \frac{1}{3}, \frac{2}{3}\right)$ and $\left(\frac{2}{3}, \frac{1}{3}, -\frac{2}{3}\right)$ in this expression, we find that the slope at the point $\left(\frac{2}{3}, \frac{1}{3}, \frac{2}{3}\right)$ is $-\frac{1}{2}$ and the slope at $\left(\frac{2}{3}, \frac{1}{3}, -\frac{2}{3}\right)$ is $\frac{1}{2}$. ◀

FOR THE READER. Check the results obtained in Example 5 by differentiating the functions $z = \sqrt{1 - x^2 - y^2}$ and $z = -\sqrt{1 - x^2 - y^2}$ directly.

Example 6

Suppose that $D = \sqrt{x^2 + y^2}$ is the length of the diagonal of a rectangle whose sides have lengths x and y that are allowed to vary. Find a formula for the rate of change of D with respect to x if x varies with y held constant, and use this formula to find the rate of change of D with respect to x at the point where $x = 3$ and $y = 4$.

Solution. The instantaneous rate of change of D with respect to x with y held constant is

$$\frac{\partial D}{\partial x} = \frac{1}{2}(x^2 + y^2)^{-1/2}(2x) = \frac{x}{\sqrt{x^2 + y^2}}$$

from which it follows that

$$\frac{\partial D}{\partial x}\bigg|_{x=3, y=4} = \frac{3}{\sqrt{3^2 + 4^2}} = \frac{3}{5}$$

Thus, D is increasing at a rate of $\frac{3}{5}$ unit per unit increase in x at the point $(3, 4)$. ◀

HIGHER-ORDER PARTIAL DERIVATIVES

Since the partial derivatives $\partial f/\partial x$ and $\partial f/\partial y$ are functions of x and y, these functions may themselves have partial derivatives. This gives rise to four possible *second-order* partial derivatives of f, which are defined by

$$\frac{\partial^2 f}{\partial x^2} = \frac{\partial}{\partial x}\left(\frac{\partial f}{\partial x}\right) = f_{xx} \qquad \frac{\partial^2 f}{\partial y^2} = \frac{\partial}{\partial y}\left(\frac{\partial f}{\partial y}\right) = f_{yy}$$

Differentiate twice with respect to x.

Differentiate twice with respect to y.

$$\frac{\partial^2 f}{\partial y \partial x} = \frac{\partial}{\partial y}\left(\frac{\partial f}{\partial x}\right) = f_{xy} \qquad \frac{\partial^2 f}{\partial x \partial y} = \frac{\partial}{\partial x}\left(\frac{\partial f}{\partial y}\right) = f_{yx}$$

Differentiate first with respect to x and then with respect to y.

Differentiate first with respect to y and then with respect to x.

The last two cases are called the ***mixed second-order partial derivatives*** or the ***mixed second partials***. Also, the derivatives $\partial f/\partial x$ and $\partial f/\partial y$ are often called the ***first-order partial derivatives*** when it is necessary to distinguish them from higher-order partial derivatives.

WARNING. Observe that the two notations for the mixed second partials have opposite conventions for the order of differentiation. In the "∂" notation the derivatives are taken right to left and in the "subscript" notation they are taken left to right. However, the conventions are logical if you insert parentheses:

$$\frac{\partial^2 f}{\partial y \partial x} = \frac{\partial}{\partial y}\left(\frac{\partial f}{\partial x}\right)$$

Right to left.
Differentiate inside
the parentheses first.

$$f_{xy} = (f_x)_y$$

Left to right.
Differentiate inside
the parentheses first.

Example 7

Find the second-order partial derivatives of $f(x, y) = x^2 y^3 + x^4 y$.

Solution. We have

$$\frac{\partial f}{\partial x} = 2xy^3 + 4x^3 y \quad \text{and} \quad \frac{\partial f}{\partial y} = 3x^2 y^2 + x^4$$

so that

$$\frac{\partial^2 f}{\partial x^2} = \frac{\partial}{\partial x}\left(\frac{\partial f}{\partial x}\right) = \frac{\partial}{\partial x}(2xy^3 + 4x^3 y) = 2y^3 + 12x^2 y$$

$$\frac{\partial^2 f}{\partial y^2} = \frac{\partial}{\partial y}\left(\frac{\partial f}{\partial y}\right) = \frac{\partial}{\partial y}(3x^2 y^2 + x^4) = 6x^2 y$$

$$\frac{\partial^2 f}{\partial x \partial y} = \frac{\partial}{\partial x}\left(\frac{\partial f}{\partial y}\right) = \frac{\partial}{\partial x}(3x^2 y^2 + x^4) = 6xy^2 + 4x^3$$

$$\frac{\partial^2 f}{\partial y \partial x} = \frac{\partial}{\partial y}\left(\frac{\partial f}{\partial x}\right) = \frac{\partial}{\partial y}(2xy^3 + 4x^3 y) = 6xy^2 + 4x^3 \qquad \blacktriangleleft$$

REMARK. Observe that the mixed second partials in this example are equal. In the next section we will state precise conditions under which this occurs, and we will see that most of the standard functions have this property.

Third-order, fourth-order, and higher-order partial derivatives can be obtained by successive differentiation. Some possibilities are

$$\frac{\partial^3 f}{\partial x^3} = \frac{\partial}{\partial x}\left(\frac{\partial^2 f}{\partial x^2}\right) = f_{xxx} \qquad\qquad \frac{\partial^4 f}{\partial y^4} = \frac{\partial}{\partial y}\left(\frac{\partial^3 f}{\partial y^3}\right) = f_{yyyy}$$

$$\frac{\partial^3 f}{\partial y^2 \partial x} = \frac{\partial}{\partial y}\left(\frac{\partial^2 f}{\partial y \partial x}\right) = f_{xyy} \qquad\qquad \frac{\partial^4 f}{\partial y^2 \partial x^2} = \frac{\partial}{\partial y}\left(\frac{\partial^3 f}{\partial y \partial x^2}\right) = f_{xxyy}$$

Example 8

Let $f(x, y) = y^2 e^x + y$. Find f_{xyy}.

Solution.

$$f_{xyy} = \frac{\partial^3 f}{\partial y^2 \partial x} = \frac{\partial^2}{\partial y^2}\left(\frac{\partial f}{\partial x}\right) = \frac{\partial^2}{\partial y^2}(y^2 e^x) = \frac{\partial}{\partial y}(2ye^x) = 2e^x \qquad \blacktriangleleft$$

THE WAVE EQUATION

Consider a string of length L that is stretched taut between the points $x = 0$ and $x = L$ on an x-axis, and suppose that the string is set into vibratory motion by "plucking" it at time $t = 0$ (Figure 15.3.4a). The displacement of a point on the string depends both on its coordinate x and the elapsed time t, and hence is described by a function $u(x, t)$ of two variables. For a fixed value t, the function $u(x, t)$ depends on x alone, and the graph of u versus x describes the shape of the string—think of it as a "snapshot" of the string at time t (Figure 15.3.4b). It follows that at a fixed time t, the partial derivative $\partial u / \partial x$ represents the slope of the string at the point x, and the sign of the second partial derivative $\partial^2 u / \partial x^2$ tells us whether the string is concave up or concave down at the point x (Figure 15.3.4c).

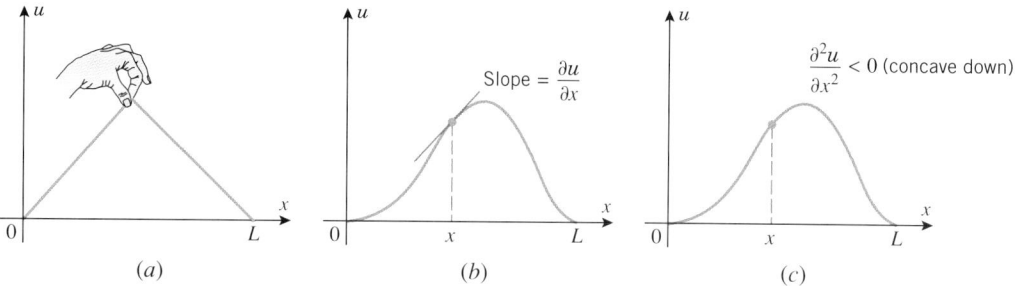

$$(a) \qquad\qquad (b) \qquad\qquad (c)$$

Figure 15.3.4

The vibration of a plucked string is governed by the wave equation.

For a fixed value of x, the function $u(x, t)$ depends on t alone, and the graph of u versus t is the position versus time curve of the point on the string with coordinate x. Thus, for a fixed value of x, the partial derivative $\partial u / \partial t$ is the velocity of the point with coordinate x, and $\partial^2 u / \partial t^2$ is the acceleration of that point.

It can be proved that under appropriate conditions the function $u(x, t)$ satisfies an equation of the form

$$\frac{\partial^2 u}{\partial t^2} = c^2 \frac{\partial^2 u}{\partial x^2} \tag{6}$$

where c is a positive constant that depends on the physical characteristics of the string. This equation, which is called the ***one-dimensional wave equation***, involves partial derivatives of the unknown function $u(x, t)$ and hence is classified as a ***partial differential equation***. Techniques for solving partial differential equations are studied in advanced courses and will not be discussed in this text.

Example 9

Show that the function $u(x, t) = \sin(x - ct)$ is a solution of Equation (6).

Solution. We have

$$\frac{\partial u}{\partial x} = \cos(x - ct), \qquad \frac{\partial^2 u}{\partial x^2} = -\sin(x - ct)$$

$$\frac{\partial u}{\partial t} = -c \cos(x - ct), \qquad \frac{\partial^2 u}{\partial t^2} = -c^2 \sin(x - ct)$$

Thus, $u(x, t)$ satisfies (6). ◄

PARTIAL DERIVATIVES OF FUNCTIONS WITH MORE THAN TWO VARIABLES

For a function $f(x, y, z)$ of three variables, there are three ***partial derivatives***:

$$f_x(x, y, z), \quad f_y(x, y, z), \quad f_z(x, y, z)$$

The partial derivative f_x is calculated by holding y and z constant and differentiating with respect to x. For f_y the variables x and z are held constant, and for f_z the variables x and y are held constant. If a dependent variable

$$w = f(x, y, z)$$

is used, then the three partial derivatives of f can be denoted by

$$\frac{\partial w}{\partial x}, \quad \frac{\partial w}{\partial y}, \quad \text{and} \quad \frac{\partial w}{\partial z}$$

Example 10

If $f(x, y, z) = x^3 y^2 z^4 + 2xy + z$, then

$$f_x(x, y, z) = 3x^2 y^2 z^4 + 2y$$
$$f_y(x, y, z) = 2x^3 y z^4 + 2x$$
$$f_z(x, y, z) = 4x^3 y^2 z^3 + 1$$
$$f_z(-1, 1, 2) = 4(-1)^3 (1)^2 (2)^3 + 1 = -31$$ ◀

Example 11

If $f(\rho, \theta, \phi) = \rho^2 \cos \phi \sin \theta$, then

$$f_\rho(\rho, \theta, \phi) = 2\rho \cos \phi \sin \theta$$
$$f_{\rho\phi}(\rho, \theta, \phi) = -2\rho \sin \phi \sin \theta$$
$$f_{\rho\phi\theta}(\rho, \theta, \phi) = -2\rho \sin \phi \cos \theta$$ ◀

In general, if $f(v_1, v_2, \ldots, v_n)$ is a function of n variables, there are n partial derivatives of f, each of which is obtained by holding $n-1$ of the variables fixed and differentiating the function f with respect to the remaining variable. If $w = f(v_1, v_2, \ldots, v_n)$, then these partial derivatives are denoted by

$$\frac{\partial w}{\partial v_1}, \frac{\partial w}{\partial v_2}, \ldots, \frac{\partial w}{\partial v_n}$$

where $\partial w / \partial v_i$ is obtained by holding all variables except v_i fixed and differentiating with respect to v_i.

Example 12

Find

$$\frac{\partial}{\partial x_i}[\sqrt{x_1^2 + x_2^2 + \cdots + x_n^2}]$$

for $i = 1, 2, \ldots, n$.

Solution.   For each $i = 1, 2, \ldots, n$ we obtain

$$\frac{\partial}{\partial x_i}[\sqrt{x_1^2 + x_2^2 + \cdots + x_n^2}] = \frac{1}{2\sqrt{x_1^2 + x_2^2 + \cdots + x_n^2}} \cdot \frac{\partial}{\partial x_i}[x_1^2 + x_2^2 + \cdots + x_n^2]$$

$$= \frac{1}{2\sqrt{x_1^2 + x_2^2 + \cdots + x_n^2}}[2x_i] \qquad \boxed{\text{All terms except } x_i^2 \text{ are constant.}}$$

$$= \frac{x_i}{\sqrt{x_1^2 + x_2^2 + \cdots + x_n^2}}$$ ◀

EXERCISE SET 15.3   ◨ Graphing Calculator   [c] CAS
. . . .

1. Let $f(x, y) = 3x^3 y^2$. Find
(a) $f_x(x, y)$
(b) $f_y(x, y)$
(c) $f_x(1, y)$
(d) $f_x(x, 1)$
(e) $f_y(1, y)$
(f) $f_y(x, 1)$
(g) $f_x(1, 2)$
(h) $f_y(1, 2)$.

2. Let $z = e^{2x} \sin y$. Find
(a) $\partial z / \partial x$
(b) $\partial z / \partial y$
(c) $\partial z / \partial x|_{(0, y)}$
(d) $\partial z / \partial x|_{(x, 0)}$
(e) $\partial z / \partial y|_{(0, y)}$
(f) $\partial z / \partial y|_{(x, 0)}$
(g) $\partial z / \partial x|_{(\ln 2, 0)}$
(h) $\partial z / \partial y|_{(\ln 2, 0)}$.

3. Let $z = \sqrt{x}\,\cos y$. Find
   (a) $\partial^2 z/\partial x^2$
   (b) $\partial^2 z/\partial y^2$
   (c) $\partial^2 z/\partial x \partial y$
   (d) $\partial^2 z/\partial y \partial x$.

4. Let $f(x, y) = 4x^2 - 2y + 7x^4 y^5$. Find
   (a) f_{xx}
   (b) f_{yy}
   (c) f_{xy}
   (d) f_{yx}.

5. Let $f(x, y) = \sqrt{3x + 2y}$.
   (a) Find the slope of the surface $z = f(x, y)$ in the x-direction at the point $(4, 2)$.
   (b) Find the slope of the surface $z = f(x, y)$ in the y-direction at the point $(4, 2)$.

6. Let $f(x, y) = xe^{-y} + 5y$.
   (a) Find the slope of the surface $z = f(x, y)$ in the x-direction at the point $(3, 0)$.
   (b) Find the slope of the surface $z = f(x, y)$ in the y-direction at the point $(3, 0)$.

7. Let $z = \sin(y^2 - 4x)$.
   (a) Find the rate of change of z with respect to x at the point $(2, 1)$ with y held fixed.
   (b) Find the rate of change of z with respect to y at the point $(2, 1)$ with x held fixed.

8. Let $z = (x + y)^{-1}$.
   (a) Find the rate of change of z with respect to x at the point $(-2, 4)$ with y held fixed.
   (b) Find the rate of change of z with respect to y at the point $(-2, 4)$ with x held fixed.

9. Use the information in the accompanying figure to find the values of the first-order partial derivatives of f at the point $(1, 2)$.

10. What can you say about the signs of $\partial z/\partial x$, $\partial^2 z/\partial x^2$, $\partial z/\partial y$, and $\partial^2 z/\partial y^2$ at the point P in the accompanying figure? Explain your reasoning.

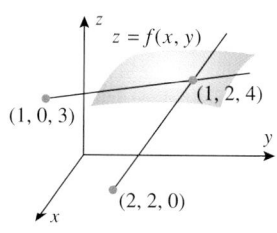

Figure Ex-9

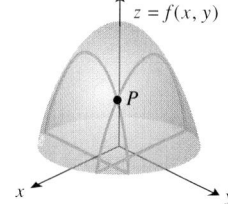

Figure Ex-10

In Exercises 11–16, find $\partial z/\partial x$ and $\partial z/\partial y$.

11. $z = 4e^{x^2 y^3}$

12. $z = \cos(x^5 y^4)$

13. $z = x^3 \ln(1 + xy^{-3/5})$

14. $z = e^{xy} \sin 4y^2$

15. $z = \dfrac{xy}{x^2 + y^2}$

16. $z = \dfrac{x^2 y^3}{\sqrt{x + y}}$

In Exercises 17–22, find $f_x(x, y)$ and $f_y(x, y)$.

17. $f(x, y) = \sqrt{3x^5 y - 7x^3 y}$

18. $f(x, y) = \dfrac{x + y}{x - y}$

19. $f(x, y) = y^{-3/2} \tan^{-1}(x/y)$

20. $f(x, y) = x^3 e^{-y} + y^3 \sec \sqrt{x}$

21. $f(x, y) = (y^2 \tan x)^{-4/3}$

22. $f(x, y) = \cosh(\sqrt{x}) \sinh^2(xy^2)$

In Exercises 23–26, evaluate the indicated partial derivatives.

23. $f(x, y) = 9 - x^2 - 7y^3$; $f_x(3, 1)$, $f_y(3, 1)$

24. $f(x, y) = x^2 y e^{xy}$; $\partial f/\partial x(1, 1)$, $\partial f/\partial y(1, 1)$

25. $z = \sqrt{x^2 + 4y^2}$; $\partial z/\partial x(1, 2)$, $\partial z/\partial y(1, 2)$

26. $w = x^2 \cos xy$; $\partial w/\partial x\left(\frac{1}{2}, \pi\right)$, $\partial w/\partial y\left(\frac{1}{2}, \pi\right)$

In Exercises 27–32, confirm that the mixed second-order partial derivatives of f are the same.

27. $f(x, y) = 4x^2 - 8xy^4 + 7y^5 - 3$

28. $f(x, y) = \sqrt{x^2 + y^2}$

29. $f(x, y) = e^x \cos y$

30. $f(x, y) = e^{x - y^2}$

31. $f(x, y) = \ln(4x - 5y)$

32. $f(x, y) = (x^2 - y^2)/(x^2 + y^2)$

c 33. Use a CAS to check the answers to the problems that you solved in Exercises 11–22.

c 34. Use a CAS to check the calculations in the problems that you solved in Exercises 27–32.

35. (a) By differentiating implicitly, find the slope of the hyperboloid $x^2 + y^2 - z^2 = 1$ in the x-direction at the points $(3, 4, 2\sqrt{6})$ and $(3, 4, -2\sqrt{6})$.
    (b) Check the results in part (a) by solving for z and differentiating the resulting functions directly.

36. (a) By differentiating implicitly, find the slope of the hyperboloid $x^2 + y^2 - z^2 = 1$ in the y-direction at the points $(3, 4, 2\sqrt{6})$ and $(3, 4, -2\sqrt{6})$.
    (b) Check the results in part (a) by solving for z and differentiating the resulting functions directly.

In Exercises 37–40, calculate $\partial z/\partial x$ and $\partial z/\partial y$ using implicit differentiation. Leave your answers in terms of x, y, and z.

37. $(x^2 + y^2 + z^2)^{3/2} = 1$

38. $\ln(2x^2 + y - z^3) = x$

39. $x^2 + z \sin xyz = 0$

40. $e^{xy} \sinh z - z^2 x + 1 = 0$

41. The accompanying figure shows the graphs of an unspecified function $f(x, y)$ and its partial derivatives $f_x(x, y)$ and $f_y(x, y)$. Determine which is which, and explain your reasoning.

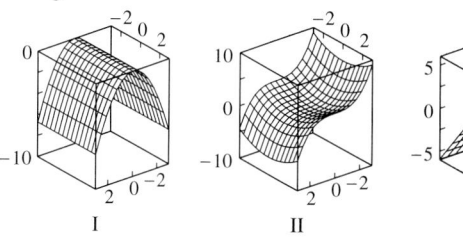

I   II   III

Figure Ex-41

42. The accompanying figure shows a contour plot for an un-specified function $f(x, y)$. Make a conjecture about the signs of the partial derivatives $f_x(x_0, y_0)$ and $f_y(x_0, y_0)$, and explain your reasoning.

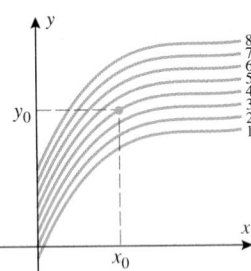

Figure Ex-42

43. Given $f(x, y) = x^3 y^5 - 2x^2 y + x$, find
(a) f_{xxy}
(b) f_{yxy}
(c) f_{yyy}.

44. Given $z = (2x - y)^5$, find
(a) $\dfrac{\partial^3 z}{\partial y \partial x \partial y}$
(b) $\dfrac{\partial^3 z}{\partial x^2 \partial y}$
(c) $\dfrac{\partial^4 z}{\partial x^2 \partial y^2}$.

45. Given $f(x, y) = y^3 e^{-5x}$, find
(a) $f_{xyy}(0, 1)$
(b) $f_{xxx}(0, 1)$
(c) $f_{yyxx}(0, 1)$.

46. Given $w = e^y \cos x$, find
(a) $\dfrac{\partial^3 w}{\partial y^2 \partial x}\bigg|_{(\pi/4, 0)}$
(b) $\dfrac{\partial^3 w}{\partial x^2 \partial y}\bigg|_{(\pi/4, 0)}$

47. Express the following derivatives in "∂" notation.
(a) f_{xxx}
(b) f_{xyy}
(c) f_{yyxx}
(d) f_{xyyy}

48. Express the derivatives in "subscript" notation.
(a) $\dfrac{\partial^3 f}{\partial y^2 \partial x}$
(b) $\dfrac{\partial^4 f}{\partial x^4}$
(c) $\dfrac{\partial^4 f}{\partial y^2 \partial x^2}$
(d) $\dfrac{\partial^5 f}{\partial x^2 \partial y^3}$

49. Let $f(x, y, z) = x^2 y^4 z^3 + xy + z^2 + 1$. Find
(a) $f_x(x, y, z)$
(b) $f_y(x, y, z)$
(c) $f_z(x, y, z)$
(d) $f_x(1, y, z)$
(e) $f_y(1, 2, z)$
(f) $f_z(1, 2, 3)$.

50. Let $w = x^2 y \cos z$. Find
(a) $\partial w / \partial x (x, y, z)$
(b) $\partial w / \partial y (x, y, z)$
(c) $\partial w / \partial z (x, y, z)$
(d) $\partial w / \partial x (2, y, z)$
(e) $\partial w / \partial y (2, 1, z)$
(f) $\partial w / \partial z (2, 1, 0)$.

In Exercises 51–54, find f_x, f_y, and f_z.

51. $f(x, y, z) = z \ln(x^2 y \cos z)$

52. $f(x, y, z) = y^{-3/2} \sec\left(\dfrac{xz}{y}\right)$

53. $f(x, y, z) = \tan^{-1}\left(\dfrac{1}{xy^2 z^3}\right)$

54. $f(x, y, z) = \cosh(\sqrt{z}\,) \sinh^2(x^2 yz)$

In Exercises 55–58, find $\partial w / \partial x$, $\partial w / \partial y$, and $\partial w / \partial z$.

55. $w = y e^z \sin xz$

56. $w = \dfrac{x^2 - y^2}{y^2 + z^2}$

57. $w = \sqrt{x^2 + y^2 + z^2}$

58. $w = y^3 e^{2x + 3z}$

59. Let $f(x, y, z) = y^2 e^{xz}$. Find
(a) $\partial f / \partial x|_{(1,1,1)}$
(b) $\partial f / \partial y|_{(1,1,1)}$
(c) $\partial f / \partial z|_{(1,1,1)}$.

60. Let $w = \sqrt{x^2 + 4y^2 - z^2}$. Find
(a) $\partial w / \partial x|_{(2,1,-1)}$
(b) $\partial w / \partial y|_{(2,1,-1)}$
(c) $\partial w / \partial z|_{(2,1,-1)}$.

[c] **61.** Use a CAS to check the answers to the problems you solved in Exercises 51–58.

[~] **62.** Let $f(x, y) = e^x \sin y$. Use a graphing utility to graph the functions $f_x(0, y)$ and $f_y(x, 0)$.

In Exercises 63–66, find $\partial w / \partial x$, $\partial w / \partial y$, and $\partial w / \partial z$ using implicit differentiation. Leave your answers in terms of x, y, z, and w.

63. $(x^2 + y^2 + z^2 + w^2)^{3/2} = 4$

64. $\ln(2x^2 + y - z^3 + 3w) = z$

65. $w^2 + w \sin xyz = 1$

66. $e^{xy} \sinh w - z^2 w + 1 = 0$

67. Let $f(x, y, z) = x^3 y^5 z^7 + xy^2 + y^3 z$. Find
(a) f_{xy}
(b) f_{yz}
(c) f_{xz}
(d) f_{zz}
(e) f_{zyy}
(f) f_{xxy}
(g) f_{zyx}
(h) f_{xxyz}.

68. Let $w = (4x - 3y + 2z)^5$. Find
(a) $\dfrac{\partial^2 w}{\partial x \partial z}$
(b) $\dfrac{\partial^3 w}{\partial x \partial y \partial z}$
(c) $\dfrac{\partial^4 w}{\partial z^2 \partial y \partial x}$.

In Exercises 69 and 70, find f_x and f_y.

69. $f(x, y) = \displaystyle\int_y^x e^{t^2}\, dt$

70. $f(x, y) = \displaystyle\int_1^{xy} e^{t^2}\, dt$

In Exercises 71 and 72, find $\partial w / \partial x_i$ for $i = 1, 2, \ldots, n$.

71. $w = \cos(x_1 + 2x_2 + \cdots + nx_n)$

72. $w = \left(\displaystyle\sum_{k=1}^n x_k\right)^{1/n}$

73. Show that the function satisfies *Laplace's equation*
$$\frac{\partial^2 z}{\partial x^2} + \frac{\partial^2 z}{\partial y^2} = 0$$
(a) $z = x^2 - y^2 + 2xy$
(b) $z = e^x \sin y + e^y \cos x$
(c) $z = \ln(x^2 + y^2) + 2 \tan^{-1}(y/x)$

74. Show that the function satisfies the *heat equation*
$$\frac{\partial z}{\partial t} = c^2 \frac{\partial^2 z}{\partial x^2} \quad (c > 0, \text{ constant})$$
(a) $z = e^{-t} \sin(x/c)$
(b) $z = e^{-t} \cos(x/c)$

75. Show that the function $u(x, t) = \sin c\omega t \sin \omega x$ satisfies the wave equation [Equation (6)] for all real values of ω.

76. In each part, show that $u(x, y)$ and $v(x, y)$ satisfy the **Cauchy–Riemann equations**

$$\frac{\partial u}{\partial x} = \frac{\partial v}{\partial y} \quad \text{and} \quad \frac{\partial u}{\partial y} = -\frac{\partial v}{\partial x}$$

(a) $u = x^2 - y^2$, $\quad\quad v = 2xy$
(b) $u = e^x \cos y$, $\quad\quad v = e^x \sin y$
(c) $u = \ln(x^2 + y^2)$, $\quad v = 2 \tan^{-1}(y/x)$

77. Show that if $u(x, y)$ and $v(x, y)$ each have equal mixed second partials, and if u and v satisfy the Cauchy–Riemann equations (Exercise 76), then u, v, and $u + v$ satisfy Laplace's equation (Exercise 73).

78. A point moves along the intersection of the elliptic paraboloid $z = x^2 + 3y^2$ and the plane $x = 2$. At what rate is z changing with y when the point is at $(2, 1, 7)$?

79. A point moves along the intersection of the plane $y = 3$ and the surface $z = \sqrt{29 - x^2 - y^2}$. At what rate is z changing with respect to x when the point is at $(4, 3, 2)$?

80. Find the slope of the tangent line at $(-1, 1, 5)$ to the curve of intersection of the surface $z = x^2 + 4y^2$ and
(a) the plane $x = -1$ $\quad\quad$ (b) the plane $y = 1$.

81. The volume V of a right circular cylinder is given by the formula $V = \pi r^2 h$, where r is the radius and h is the height.
(a) Find a formula for the instantaneous rate of change of V with respect to r if r changes and h remains constant.
(b) Find a formula for the instantaneous rate of change of V with respect to h if h changes and r remains constant.
(c) Suppose that h has a constant value of 4 in, but r varies. Find the rate of change of V with respect to r at the point where $r = 6$ in.
(d) Suppose that r has a constant value of 8 in, but h varies. Find the instantaneous rate of change of V with respect to h at the point where $h = 10$ in.

82. The volume V of a right circular cone is given by

$$V = \frac{\pi}{24} d^2 \sqrt{4s^2 - d^2}$$

where s is the slant height and d is the diameter of the base.
(a) Find a formula for the instantaneous rate of change of V with respect to s if d remains constant.
(b) Find a formula for the instantaneous rate of change of V with respect to d if s remains constant.
(c) Suppose that d has a constant value of 16 cm, but s varies. Find the rate of change of V with respect to s when $s = 10$ cm.
(d) Suppose that s has a constant value of 10 cm, but d varies. Find the rate of change of V with respect to d when $d = 16$ cm.

83. According to the ideal gas law, the pressure, temperature, and volume of a gas are related by $P = kT/V$, where k is a constant of proportionality. Suppose that V is measured in cubic inches (in^3), T is measured in kelvins (K), and that for a certain gas the constant of proportionality is $k = 10$ in·lb/K.
(a) Find the instantaneous rate of change of pressure with respect to temperature if the temperature is 80 K and the volume remains fixed at 50 in^3.
(b) Find the instantaneous rate of change of volume with respect to pressure if the volume is 50 in^3 and the temperature remains fixed at 80 K.

84. Find parametric equations for the tangent line at $(1, 3, 3)$ to the curve of intersection of the surface $z = x^2 y$ and
(a) the plane $x = 1$ $\quad\quad$ (b) the plane $y = 3$.

85. Suppose that $\sin(x + z) + \sin(x - y) = 1$. Use implicit differentiation to find $\partial z/\partial x$, $\partial z/\partial y$, and $\partial^2 z/\partial x \partial y$ in terms of x, y, and z.

86. The volume of a right circular cone of radius r and height h is $V = \frac{1}{3}\pi r^2 h$. Show that if the height remains constant while the radius changes, then the volume satisfies

$$\frac{\partial V}{\partial r} = \frac{2V}{r}$$

87. The temperature at a point (x, y) on a metal plate in the xy-plane is $T(x, y) = x^3 + 2y^2 + x$ degrees. Assume that distance is measured in centimeters and find the rate at which temperature changes with distance if we start at the point $(1, 2)$ and move
(a) to the right and parallel to the x-axis
(b) upward and parallel to the y-axis.

88. When two resistors having resistances R_1 ohms and R_2 ohms are connected in parallel, their combined resistance R in ohms is $R = R_1 R_2/(R_1 + R_2)$. Show that

$$\frac{\partial^2 R}{\partial R_1^2} \frac{\partial^2 R}{\partial R_2^2} = \frac{4R^2}{(R_1 + R_2)^4}$$

89. Let $f(x, y) = 2x^2 - 3xy + y^2$. Use Definition 15.3.1 to find $f_x(2, -1)$ and $f_y(2, -1)$. Then check your work by calculating the derivative in the usual way.

90. Let $f(x, y) = (x^2 + y^2)^{2/3}$. Show that

$$f_x(x, y) = \begin{cases} \dfrac{4x}{3(x^2 + y^2)^{1/3}}, & (x, y) \neq (0, 0) \\[2mm] 0, & (x, y) = (0, 0) \end{cases}$$

[This problem, due to Don Cohen, appeared in *Mathematics and Computer Education*, Vol. 25, No. 2, 1991, p. 179.]

91. Let $f(x, y) = (x^3 + y^3)^{1/3}$.
(a) Show that $f_y(0, 0) = 1$.
(b) At what points, if any, does $f_y(x, y)$ fail to exist?

15.4 DIFFERENTIABILITY AND CHAIN RULES

In this section we will extend the notion of differentiability to functions of two variables and derive versions of the chain rule for these functions. We have restricted the discussion in this section to functions of two variables because some of the results we will discuss have geometric interpretations that only apply to such functions. Later, we will extend the concepts developed here to functions of three or more variables.

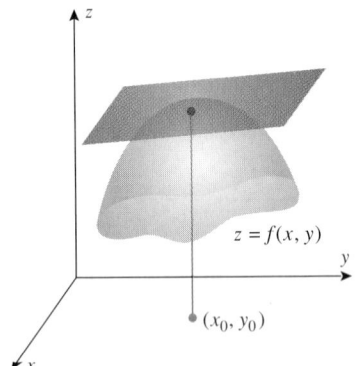

DIFFERENTIABILITY OF FUNCTIONS OF TWO VARIABLES

Recall that a function f of one variable is called differentiable at x_0 if it has a derivative at x_0, that is, if the limit

$$f'(x_0) = \lim_{\Delta x \to 0} \frac{f(x_0 + \Delta x) - f(x_0)}{\Delta x} \tag{1}$$

exists. A function f that is differentiable at a point x_0 has two important properties:

- $f(x)$ is continuous at x_0.
- The graph of $y = f(x)$ has a nonvertical tangent line at x_0.

Our primary objective in this section is to extend the notion of differentiability to functions of two variables in such a way that the natural analogs of these two properties hold. More precisely, when $f(x, y)$ is differentiable at (x_0, y_0), we will want it to be the case that

- $f(x, y)$ is continuous at (x_0, y_0);
- the surface $z = f(x, y)$ has a nonvertical tangent plane at (x_0, y_0) (Figure 15.4.1).

It would not be unreasonable to conjecture that a function f of two variables should be called differentiable at (x_0, y_0) if the two partial derivatives $f_x(x_0, y_0)$ and $f_y(x_0, y_0)$ exist at (x_0, y_0). Unfortunately, this condition is not strong enough to meet our objectives, since there are functions that have partial derivatives at a point but are not continuous at that point. For example, consider the function

$$f(x, y) = \begin{cases} -1 & \text{if } x > 0 \text{ and } y > 0 \\ 0 & \text{otherwise} \end{cases}$$

Figure 15.4.1

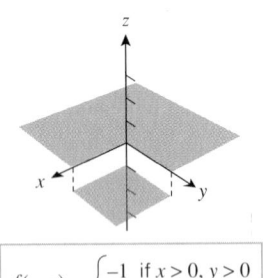

$f(x, y) = \begin{cases} -1 & \text{if } x > 0, \, y > 0 \\ 0 & \text{otherwise} \end{cases}$

Figure 15.4.2

whose graph is shown in Figure 15.4.2. This function is discontinuous at $(0, 0)$ but has partial derivatives at $(0, 0)$; these derivatives are $f_x(0, 0) = 0$ and $f_y(0, 0) = 0$ (why?).

To motivate an appropriate definition of differentiability for functions of two variables, it will be helpful to reexamine the definition of differentiability for a function f of one variable. To say that f is differentiable at x_0 means that there exists a number $f'(x_0)$, which we call the derivative of f at x_0, such that

$$f'(x_0) = \lim_{\Delta x \to 0} \frac{f(x_0 + \Delta x) - f(x_0)}{\Delta x} \tag{2}$$

For convenience, let us express the numerator as

$$\Delta f = f(x_0 + \Delta x) - f(x_0)$$

which allows us to rewrite (2) more compactly as

$$f'(x_0) = \lim_{\Delta x \to 0} \frac{\Delta f}{\Delta x}$$

or alternatively as

$$\lim_{\Delta x \to 0} \left[\frac{\Delta f}{\Delta x} - f'(x_0) \right] = 0 \tag{3}$$

Now, let us define ϵ [or more accurately, $\epsilon(\Delta x)$] to be the error in approximating $f'(x_0)$ by $\Delta f / \Delta x$; that is,

$$\epsilon = \frac{\Delta f}{\Delta x} - f'(x_0)$$

Thus, we can rewrite (3) as

$$\lim_{\Delta x \to 0} \epsilon = 0 \tag{4}$$

which suggests the following alternative definition of differentiability for functions of one variable.

15.4.1 DEFINITION. A function f of one variable is said to be ***differentiable*** at x_0 if there exists a number $f'(x_0)$ such that Δf can be written in the form

$$\Delta f = f'(x_0)\Delta x + \epsilon \Delta x \tag{5}$$

where ϵ is a function of Δx such that $\epsilon \to 0$ as $\Delta x \to 0$, and $\epsilon = 0$ if $\Delta x = 0$.

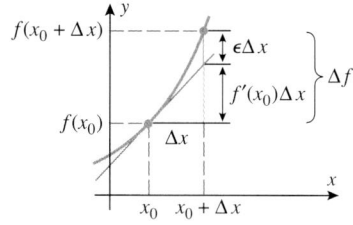

Figure 15.4.3

Although this definition of differentiability is more complicated than that given earlier in the text, it provides the basis for extending the notion of differentiability to functions of two or more variables. A geometric interpretation of the terms appearing in (5) is shown in Figure 15.4.3. The term Δf represents the change in height that results when a point moves along the graph of f as the x-coordinate changes from x_0 to $x_0 + \Delta x$; the term $f'(x_0)\Delta x$ represents the change in height that results when a point moves along the tangent line at $(x_0, f(x_0))$ as the x-coordinate changes from x_0 to $x_0 + \Delta x$; finally, the term $\epsilon \Delta x$ represents the difference between Δf and $f'(x_0)\Delta x$.

REMARK. It is evident from Figure 15.4.3 that $\epsilon \Delta x \to 0$ as $\Delta x \to 0$. However, (5) actually makes the stronger statement that $\epsilon \to 0$ as $\Delta x \to 0$. This is not at all evident from Figure 15.4.3, but it follows from (4). It is this property that distinguishes the tangent line from all other lines through the point $(x_0, f(x_0))$.

If $z = f(x, y)$, then the change in the value of $f(x, y)$ as (x, y) moves from some initial position (x_0, y_0) to some new position $(x_0 + \Delta x, y_0 + \Delta y)$ is called the ***increment in f*** or the ***increment in z*** and is denoted by Δf or Δz. Thus,

$$\Delta f = f(x_0 + \Delta x, y_0 + \Delta y) - f(x_0, y_0) \tag{6}$$

(See Figure 15.4.4.)

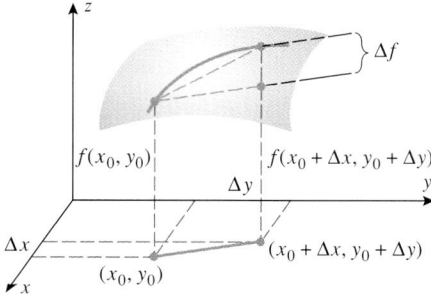

Figure 15.4.4

Motivated by Definition 15.4.1, we now make the following definition of differentiability for functions of two variables.

15.4.2 DEFINITION. A function f of two variables is said to be ***differentiable*** at (x_0, y_0) if $f_x(x_0, y_0)$ and $f_y(x_0, y_0)$ exist and Δf can be written in the form

$$\Delta f = f_x(x_0, y_0)\Delta x + f_y(x_0, y_0)\Delta y + \epsilon_1 \Delta x + \epsilon_2 \Delta y \tag{7}$$

where ϵ_1 and ϵ_2 are functions of Δx and Δy such that $\epsilon_1 \to 0$ and $\epsilon_2 \to 0$ as $(\Delta x, \Delta y) \to (0, 0)$ and $\epsilon_1 = \epsilon_2 = 0$ if $(\Delta x, \Delta y) = (0, 0)$.

If a function f of two variables is differentiable at each point of a region R in the xy-plane, then we say that f is **differentiable on R**; and if f is differentiable at every point in the xy-plane, then we say that f is **differentiable everywhere**. Moreover, we will say that f is a **differentiable function** if it is differentiable at each point of its domain.

SUFFICIENT CONDITIONS FOR DIFFERENTIABILITY

The term "differentiable" has different implications for a function of two variables than for a function of one variable. A function of one variable is differentiable at a point if it has a derivative at that point. However, it is not necessarily true that a function of two variables is differentiable at a point if both of its first-order partial derivatives exist at that point. (We will give an example to illustrate this later in this section.) The following theorem, which is proved in advanced calculus courses, provides sufficient conditions for a function of two variables to be differentiable at a point.

> **15.4.3** THEOREM. *If f has first-order partial derivatives at each point in some circular region centered at (x_0, y_0), and if these partial derivatives are continuous at (x_0, y_0), then f is differentiable at (x_0, y_0).*

Example 1

Show that $f(x, y) = x^3 y^4$ is a differentiable function.

Solution. The partial derivatives $f_x = 3x^2 y^4$ and $f_y = 4x^3 y^3$ are defined and continuous everywhere in the xy-plane. Thus, the hypotheses of Theorem 15.4.3 are satisfied at each point (x_0, y_0) in the xy-plane, so $f(x, y) = x^3 y^4$ is everywhere differentiable. ◄

RELATIONSHIP BETWEEN DIFFERENTIABILITY AND CONTINUITY

Earlier, we set two goals for our definition of differentiability: we wanted a function that is differentiable at (x_0, y_0) to be continuous at (x_0, y_0), and we wanted its graph to have a nonvertical tangent plane at (x_0, y_0). The next theorem shows that the continuity criterion is met; the existence of a nonvertical tangent plane will be demonstrated in the next section.

> **15.4.4** THEOREM. *If f is differentiable at (x_0, y_0), then f is continuous at (x_0, y_0).*

Proof. We must prove that
$$\lim_{(x,y) \to (x_0, y_0)} f(x, y) = f(x_0, y_0)$$
which, on letting $x = x_0 + \Delta x$ and $y = y_0 + \Delta y$, is equivalent to
$$\lim_{(\Delta x, \Delta y) \to (0,0)} f(x_0 + \Delta x, y_0 + \Delta y) = f(x_0, y_0)$$
which from (6) can be rewritten as
$$\lim_{(\Delta x, \Delta y) \to (0,0)} \Delta f = 0$$
But f is assumed to be differentiable at (x_0, y_0), so it follows from (7) that
$$\Delta f = f_x(x_0, y_0)\Delta x + f_y(x_0, y_0)\Delta y + \epsilon_1 \Delta x + \epsilon_2 \Delta y$$
where $\epsilon_1 \to 0$, $\epsilon_2 \to 0$ as $(\Delta x, \Delta y) \to (0, 0)$. Thus,
$$\lim_{(\Delta x, \Delta y) \to (0,0)} \Delta f = \lim_{(\Delta x, \Delta y) \to (0,0)} [f_x(x_0, y_0)\Delta x + f_y(x_0, y_0)\Delta y + \epsilon_1 \Delta x + \epsilon_2 \Delta y] = 0$$
which completes the proof. ∎

By combining Theorems 15.4.3 and 15.4.4, we obtain sufficient conditions for continuity in terms of partial derivatives.

15.4.5 COROLLARY. *If f has first-order partial derivatives at each point of some circular region centered at (x_0, y_0), and if these partial derivatives are continuous at (x_0, y_0), then f is continuous at (x_0, y_0).*

REMARK. We stated earlier in this section that for a function of two variables the existence of the two first-order partial derivatives at a point does not imply that the function is differentiable at that point. For example, the function graphed in Figure 15.4.2 has first-order partial derivatives at $(0, 0)$, but it is not differentiable at $(0, 0)$ because it is not continuous at that point.

EQUALITY OF MIXED PARTIALS

The following theorem, which we state without proof, shows that with appropriate continuity restrictions the mixed second-order partial derivatives of a function of two variables are equal.

15.4.6 THEOREM. *Let f be a function of two variables. If f_x, f_y, f_{xy}, and f_{yx} are continuous on an open set, then $f_{xy} = f_{yx}$ at each point of the set.*

Example 2

Let $f(x, y) = 2e^{xy} \sin y$. It should be evident from the form of this function and from Theorem 15.2.4 that f and all its partial derivatives are continuous everywhere. Thus, Theorem 15.4.6 guarantees that $f_{xy} = f_{yx}$ everywhere. This is confirmed by the following computations:

$$f_x(x, y) = 2ye^{xy} \sin y = (2y \sin y)e^{xy}$$
$$f_{xy}(x, y) = (2y \sin y)(xe^{xy}) + e^{xy}(2y \cos y + 2 \sin y)$$
$$= 2e^{xy}(xy \sin y + y \cos y + \sin y)$$
$$f_y(x, y) = 2e^{xy} \cos y + 2xe^{xy} \sin y$$
$$f_{yx}(x, y) = 2ye^{xy} \cos y + 2xye^{xy} \sin y + 2e^{xy} \sin y$$
$$= 2e^{xy}(xy \sin y + y \cos y + \sin y)$$

Thus, $f_{xy}(x, y) = f_{yx}(x, y)$ for all (x, y). ◄

In general, the order of differentiation in an nth-order partial derivative can be changed without affecting the final result whenever the function and all its partial derivatives of order n or less are continuous. For example, if f and its partial derivatives of the first, second, and third orders are continuous on an open set, then at each point of that set,

$$f_{xyy} = f_{yxy} = f_{yyx}$$

or in "∂" notation,

$$\frac{\partial^3 f}{\partial y^2 \partial x} = \frac{\partial^3 f}{\partial y \partial x \partial y} = \frac{\partial^3 f}{\partial x \partial y^2}$$

CHAIN RULES

If y is a differentiable function of x and x is a differentiable function of t, then the chain rule for functions of one variable states that

$$\frac{dy}{dt} = \frac{dy}{dx}\frac{dx}{dt}$$

We will now derive versions of the chain rule for functions of two variables.

Assume that z is a function of x and y, say

$$z = f(x, y) \tag{8}$$

and suppose that x and y, in turn, are functions of a single variable t, say

$$x = x(t), \quad y = y(t)$$

On substituting these functions of t in (8), we obtain the relationship

$$z = f(x(t), y(t))$$

which expresses z as a function of the single variable t. Thus, we can ask for the derivative dz/dt, and we can inquire about its relationship to the derivatives $\partial z/\partial x$, $\partial z/\partial y$, dx/dt, and dy/dt.

15.4.7 THEOREM (*Chain Rule*). *If $x = x(t)$ and $y = y(t)$ are differentiable at t, and if $z = f(x, y)$ is differentiable at the point $(x(t), y(t))$, then $z = f(x(t), y(t))$ is differentiable at t, and*

$$\frac{dz}{dt} = \frac{\partial z}{\partial x}\frac{dx}{dt} + \frac{\partial z}{\partial y}\frac{dy}{dt} \tag{9}$$

Proof. From the derivative definition for functions of one variable,

$$\frac{dz}{dt} = \lim_{\Delta t \to 0} \frac{\Delta z}{\Delta t} \tag{10}$$

Since $z = f(x, y)$ is differentiable at the point $(x, y) = (x(t), y(t))$, we can express Δz in the form

$$\Delta z = \frac{\partial z}{\partial x}\Delta x + \frac{\partial z}{\partial y}\Delta y + \epsilon_1 \Delta x + \epsilon_2 \Delta y \tag{11}$$

where the partial derivatives are evaluated at the point $(x(t), y(t))$ and $\epsilon_1 \to 0$, $\epsilon_2 \to 0$ as $(\Delta x, \Delta y) \to (0, 0)$. Thus, from (10) and (11),

$$\frac{dz}{dt} = \lim_{\Delta t \to 0} \frac{\Delta z}{\Delta t} = \lim_{\Delta t \to 0}\left[\frac{\partial z}{\partial x}\frac{\Delta x}{\Delta t} + \frac{\partial z}{\partial y}\frac{\Delta y}{\Delta t} + \epsilon_1\frac{\Delta x}{\Delta t} + \epsilon_2\frac{\Delta y}{\Delta t}\right] \tag{12}$$

But

$$\lim_{\Delta t \to 0}\frac{\Delta x}{\Delta t} = \frac{dx}{dt} \quad \text{and} \quad \lim_{\Delta t \to 0}\frac{\Delta y}{\Delta t} = \frac{dy}{dt}$$

Therefore, if we can show that $\epsilon_1 \to 0$, $\epsilon_2 \to 0$ as $\Delta t \to 0$, then the proof will be complete, since (12) will reduce to (9). But $\Delta x \to 0$ and $\Delta y \to 0$ as $\Delta t \to 0$, since

$$\lim_{\Delta t \to 0}\Delta x = \lim_{\Delta t \to 0}\frac{\Delta x}{\Delta t}\Delta t = \frac{dx}{dt}\cdot 0 = 0$$

and similarly for Δy. Thus, as Δt tends to zero, $(\Delta x, \Delta y) \to (0, 0)$, which implies that $\epsilon_1 \to 0$, $\epsilon_2 \to 0$. ∎

Formula (9) can be represented schematically by a "tree diagram" that is constructed as follows (Figure 15.4.5). Starting with z at the top of the tree and moving downward, join each variable by lines (or branches) to those variables on which it depends *directly*. Thus, z is joined to x and y and these in turn are joined to t. Next, label each branch with a derivative whose "numerator" contains the variable at the top end of that branch and whose "denominator" contains the variable at the bottom end of that branch. This completes the "tree." To find the formula for dz/dt, follow the two paths through the tree that start with z and end with t. Each such path corresponds to a term in Formula (9).

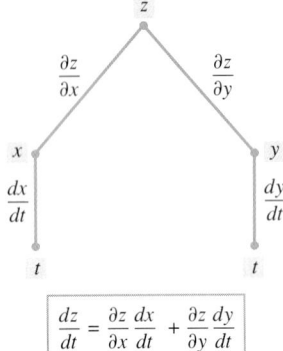

$$\frac{dz}{dt} = \frac{\partial z}{\partial x}\frac{dx}{dt} + \frac{\partial z}{\partial y}\frac{dy}{dt}$$

Figure 15.4.5

Example 3

Suppose that

$$z = x^2 y, \quad x = t^2, \quad y = t^3$$

Use the chain rule to find dz/dt, and check the result by expressing z as a function of t and differentiating directly.

Solution. By the chain rule

$$\frac{dz}{dt} = \frac{\partial z}{\partial x}\frac{dx}{dt} + \frac{\partial z}{\partial y}\frac{dy}{dt} = (2xy)(2t) + (x^2)(3t^2)$$

$$= (2t^5)(2t) + (t^4)(3t^2) = 7t^6$$

Alternatively, we can express z directly as a function of t,

$$z = x^2 y = (t^2)^2 (t^3) = t^7$$

and then differentiate to obtain $dz/dt = 7t^6$. However, this procedure may not always be convenient. ◄

Example 4

Suppose that

$$z = \sqrt{xy + y}, \quad x = \cos\theta, \quad y = \sin\theta$$

Use the chain rule to find $dz/d\theta$ when $\theta = \pi/2$.

Solution. From the chain rule with θ in place of t,

$$\frac{dz}{d\theta} = \frac{\partial z}{\partial x}\frac{dx}{d\theta} + \frac{\partial z}{\partial y}\frac{dy}{d\theta}$$

we obtain

$$\frac{dz}{d\theta} = \frac{1}{2}(xy + y)^{-1/2}(y)(-\sin\theta) + \frac{1}{2}(xy + y)^{-1/2}(x + 1)(\cos\theta)$$

When $\theta = \pi/2$, we have

$$x = \cos\frac{\pi}{2} = 0, \quad y = \sin\frac{\pi}{2} = 1$$

Substituting $x = 0$, $y = 1$, $\theta = \pi/2$ in the formula for $dz/d\theta$ yields

$$\left.\frac{dz}{d\theta}\right|_{\theta=\pi/2} = \frac{1}{2}(1)(1)(-1) + \frac{1}{2}(1)(1)(0) = -\frac{1}{2} \qquad ◄$$

REMARK.   There are many variations in derivative notations, each of which gives the chain rule a different look. If $z = f(x, y)$, where x and y are functions of t, then some possibilities are

$$\frac{dz}{dt} = f_x\frac{dx}{dt} + f_y\frac{dy}{dt}$$

$$\frac{df}{dt} = \frac{\partial f}{\partial x}\frac{dx}{dt} + \frac{\partial f}{\partial y}\frac{dy}{dt}$$

$$\frac{df}{dt} = f_x x'(t) + f_y y'(t)$$

In the special case where $z = F(x, y)$ and y is a differentiable function of x, chain-rule formula (9) yields

$$\frac{dz}{dx} = \frac{\partial F}{\partial x}\frac{dx}{dx} + \frac{\partial F}{\partial y}\frac{dy}{dx} = \frac{\partial F}{\partial x} + \frac{\partial F}{\partial y}\frac{dy}{dx} \qquad (13)$$

This result can be used to find derivatives of functions that are defined implicitly. Suppose that the equation

$$F(x, y) = 0 \qquad (14)$$

defines y implicitly as a differentiable function of x, and we are interested in finding dy/dx. Differentiating both sides of (14) with respect to x and applying (13) yields

$$\frac{\partial F}{\partial x} + \frac{\partial F}{\partial y}\frac{dy}{dx} = 0$$

Thus, if $\partial F/\partial y \neq 0$, we obtain

$$\frac{dy}{dx} = -\frac{\partial F/\partial x}{\partial F/\partial y}$$

In summary, we have the following result.

15.4.8 THEOREM. *If the equation $F(x, y) = 0$ defines y implicitly as a differentiable function of x, and if $\partial F/\partial y \neq 0$, then*

$$\frac{dy}{dx} = -\frac{\partial F/\partial x}{\partial F/\partial y} \tag{15}$$

Example 5

Given that

$$x^3 + y^2 x - 3 = 0$$

find dy/dx using (15), and check the result using implicit differentiation.

Solution. By (15) with $F(x, y) = x^3 + y^2 x - 3$,

$$\frac{dy}{dx} = -\frac{\partial F/\partial x}{\partial F/\partial y} = -\frac{3x^2 + y^2}{2yx}$$

Alternatively, differentiating the given equation implicitly yields

$$3x^2 + y^2 + x\left(2y\frac{dy}{dx}\right) - 0 = 0 \quad \text{or} \quad \frac{dy}{dx} = -\frac{3x^2 + y^2}{2yx}$$

which agrees with the result obtained by (15). ◄

In Theorem 15.4.7 the variables x and y are each functions of a single variable t. We now consider the case where x and y are each functions of two variables. Let

$$z = f(x, y) \tag{16}$$

and suppose that x and y are functions of u and v, say

$$x = x(u, v), \quad y = y(u, v)$$

On substituting these functions of u and v into (16), we obtain the relationship

$$z = f(x(u, v), y(u, v))$$

which expresses z as a function of the two variables u and v. Thus, we can ask for the partial derivatives $\partial z/\partial u$ and $\partial z/\partial v$; and we can inquire about the relationship between these derivatives and the derivatives $\partial z/\partial x$, $\partial z/\partial y$, $\partial x/\partial u$, $\partial x/\partial v$, $\partial y/\partial u$, and $\partial y/\partial v$.

15.4.9 THEOREM (Chain Rule). *If $x = x(u, v)$ and $y = y(u, v)$ have first-order partial derivatives at the point (u, v), and if $z = f(x, y)$ is differentiable at the point $(x(u, v), y(u, v))$, then $z = f(x(u, v), y(u, v))$ has first-order partial derivatives at (u, v) given by*

$$\frac{\partial z}{\partial u} = \frac{\partial z}{\partial x}\frac{\partial x}{\partial u} + \frac{\partial z}{\partial y}\frac{\partial y}{\partial u} \quad \text{and} \quad \frac{\partial z}{\partial v} = \frac{\partial z}{\partial x}\frac{\partial x}{\partial v} + \frac{\partial z}{\partial y}\frac{\partial y}{\partial v}$$

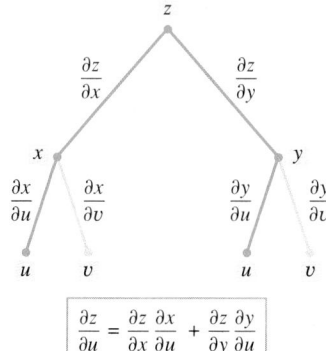

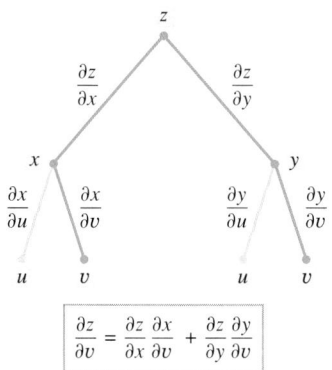

Figure 15.4.6

Proof. If v is held fixed, then $x = x(u, v)$ and $y = y(u, v)$ become functions of u alone. Thus, we are back to the case of Theorem 15.4.7. If we apply that theorem with u in place of t, and if we use ∂ rather than d to indicate that the variable v is fixed, we obtain

$$\frac{\partial z}{\partial u} = \frac{\partial z}{\partial x}\frac{\partial x}{\partial u} + \frac{\partial z}{\partial y}\frac{\partial y}{\partial u}$$

The formula for $\partial z/\partial v$ is derived similarly. ∎

Figure 15.4.6 shows tree diagrams for the formulas in Theorem 15.4.9. The formula for $\partial z/\partial u$ can be obtained by tracing all paths through the tree that start with z and end with u, and the formula for $\partial z/\partial v$ can be obtained by tracing all paths through the tree that start with z and end with v.

Example 6

Given that

$$z = e^{xy}, \quad x = 2u + v, \quad y = u/v$$

find $\partial z/\partial u$ and $\partial z/\partial v$ using the chain rule.

Solution.

$$\frac{\partial z}{\partial u} = \frac{\partial z}{\partial x}\frac{\partial x}{\partial u} + \frac{\partial z}{\partial y}\frac{\partial y}{\partial u} = (ye^{xy})(2) + (xe^{xy})\left(\frac{1}{v}\right) = \left[2y + \frac{x}{v}\right]e^{xy}$$

$$= \left[\frac{2u}{v} + \frac{2u+v}{v}\right]e^{(2u+v)(u/v)} = \left[\frac{4u}{v} + 1\right]e^{(2u+v)(u/v)}$$

$$\frac{\partial z}{\partial v} = \frac{\partial z}{\partial x}\frac{\partial x}{\partial v} + \frac{\partial z}{\partial y}\frac{\partial y}{\partial v} = (ye^{xy})(1) + (xe^{xy})\left(-\frac{u}{v^2}\right)$$

$$= \left[y - x\left(\frac{u}{v^2}\right)\right]e^{xy} = \left[\frac{u}{v} - (2u+v)\left(\frac{u}{v^2}\right)\right]e^{(2u+v)(u/v)}$$

$$= -\frac{2u^2}{v^2}e^{(2u+v)(u/v)} \quad \blacktriangleleft$$

RELATED RATES IN TWO VARIABLES

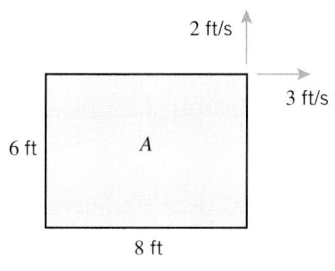

Figure 15.4.7

Example 7

At what rate is the area of a rectangle changing if its length is 8 ft and increasing at 3 ft/s while its width is 6 ft and increasing at 2 ft/s (Figure 15.4.7)?

Solution. Let

x = length of the rectangle in feet

y = width of the rectangle in feet

A = area of the rectangle in square feet

t = time in seconds

We are given that

$$\frac{dx}{dt} = 3 \quad \text{and} \quad \frac{dy}{dt} = 2 \tag{17}$$

at the instant when

$$x = 8, \quad y = 6 \tag{18}$$

We want to find dA/dt at that instant. For this purpose we use the area formula $A = xy$ to obtain

$$\frac{dA}{dt} = \frac{\partial A}{\partial x}\frac{dx}{dt} + \frac{\partial A}{\partial y}\frac{dy}{dt} = y\frac{dx}{dt} + x\frac{dy}{dt}$$

Substituting (17) and (18) in this equation yields

$$\frac{dA}{dt} = 6(3) + 8(2) = 34$$

Thus, the area is increasing at a rate of 34 ft^2/s at the given instant. ◀

EXERCISE SET 15.4

In Exercises 1–6, use an appropriate form of the chain rule to find dz/dt.

1. $z = 3x^2y^3$; $x = t^4$, $y = t^2$

2. $z = \ln(2x^2 + y)$; $x = \sqrt{t}$, $y = t^{2/3}$

3. $z = 3\cos x - \sin xy$; $x = 1/t$, $y = 3t$

4. $z = \sqrt{1 + x - 2xy^4}$; $x = \ln t$, $y = t$

5. $z = e^{1-xy}$; $x = t^{1/3}$, $y = t^3$

6. $z = \cosh^2 xy$; $x = t/2$, $y = e^t$

In Exercises 7–12, use appropriate forms of the chain rule to find $\partial z/\partial u$ and $\partial z/\partial v$.

7. $z = 8x^2y - 2x + 3y$; $x = uv$, $y = u - v$

8. $z = x^2 - y\tan x$; $x = u/v$, $y = u^2v^2$

9. $z = x/y$; $x = 2\cos u$, $y = 3\sin v$

10. $z = 3x - 2y$; $x = u + v\ln u$, $y = u^2 - v\ln v$

11. $z = e^{x^2y}$; $x = \sqrt{uv}$, $y = 1/v$

12. $z = \cos x \sin y$; $x = u - v$, $y = u^2 + v^2$

13. Let $T = x^2y - xy^3 + 2$; $x = r\cos\theta$, $y = r\sin\theta$. Use a chain rule to find $\partial T/\partial r$ and $\partial T/\partial\theta$.

14. Let $R = e^{2s-t^2}$; $s = 3\phi$, $t = \phi^{1/2}$. Use a chain rule to find $dR/d\phi$.

15. Let $t = u/v$; $u = x^2 - y^2$, $v = 4xy^3$. Use a chain rule to find $\partial t/\partial x$ and $\partial t/\partial y$.

16. Let $w = rs/(r^2 + s^2)$; $r = uv$, $s = u - 2v$. Use a chain rule to find $\partial w/\partial u$ and $\partial w/\partial v$.

17. Use a chain rule to find the value of $\left.\dfrac{dw}{ds}\right|_{s=1/4}$ if $w = r^2 - r\tan\theta$; $r = \sqrt{s}$, $\theta = \pi s$.

18. Use a chain rule to find the value of
$$\left.\frac{\partial f}{\partial u}\right|_{u=1,v=-2} \quad \text{and} \quad \left.\frac{\partial f}{\partial v}\right|_{u=1,v=-2}$$
if $f(x, y) = x^2y^2 - x + 2y$; $x = \sqrt{u}$, $y = uv^3$.

19. Use a chain rule to find the value of
$$\left.\frac{\partial z}{\partial r}\right|_{r=2,\theta=\pi/6} \quad \text{and} \quad \left.\frac{\partial z}{\partial\theta}\right|_{r=2,\theta=\pi/6}$$
if $z = xye^{x/y}$; $x = r\cos\theta$, $y = r\sin\theta$.

20. Use a chain rule to find $\left.\dfrac{dz}{dt}\right|_{t=3}$ if $z = x^2y$; $x = t^2$, $y = t + 7$.

In Exercises 21–24, use (15) to find dy/dx and check your result using implicit differentiation.

21. $x^2y^3 + \cos y = 0$

22. $x^3 - 3xy^2 + y^3 = 5$

23. $e^{xy} + ye^y = 1$

24. $x - \sqrt{xy} + 3y = 4$

25. Two straight roads intersect at right angles. Car A, moving on one of the roads, approaches the intersection at 25 mi/h and car B, moving on the other road, approaches the intersection at 30 mi/h. At what rate is the distance between the cars changing when A is 0.3 mile from the intersection and B is 0.4 mile from the intersection?

26. Use the ideal gas law $P = kT/V$ with V in cubic inches (in^3), T in kelvins (K), and $k = 10$ in·lb/K to find the rate at which the temperature of a gas is changing when the volume is 200 in^3 and increasing at the rate of 4 in^3/s, while the pressure is 5 lb/in^2 and decreasing at the rate of 1 lb/in^2/s.

27. Two sides of a triangle have lengths $a = 4$ cm and $b = 3$ cm, but are increasing at the rate of 1 cm/s. If the area of the triangle remains constant, at what rate is the angle θ between a and b changing when $\theta = \pi/6$?

28. Two sides of a triangle have lengths $a = 5$ cm and $b = 10$ cm, and the included angle is $\theta = \pi/3$. If a is increasing at a rate of 2 cm/s, b is increasing at a rate of 1 cm/s, and θ remains constant, at what rate is the third side changing? Is it increasing or decreasing? [*Hint:* Use the law of cosines.]

29. Suppose that the portion of a tree that is usable for lumber is a right circular cylinder. If the usable height of a tree increases 2 ft per year and the usable diameter increases 3 in per year, how fast is the volume of usable lumber increasing when the usable height of the tree is 20 ft and the usable diameter is 30 in?

30. Suppose that a particle moving along a metal plate in the xy-plane has velocity $\mathbf{v} = \mathbf{i} - 4\mathbf{j}$ (cm/s) at the point $(3, 2)$. Given that the temperature of the plate at points in the xy-plane is $T(x, y) = y^2\ln x$, $x \geq 1$, in degrees Celsius, find dT/dt at the point $(3, 2)$.

In Exercises 31 and 32, describe the largest open set on which the hypotheses of Theorem 15.4.6 are satisfied, and confirm that f_{xy} and f_{yx} are equal on that set.

31. (a) $f(x, y) = 4x^3y + 3x^2y$ (b) $f(x, y) = x^3/y$

32. (a) $f(x, y) = \sqrt{x^2 + y^2 - 1}$

(b) $f(x, y) = \sin(x^2 + y^3)$

33. Let f be a function of two variables with continuous third- and fourth-order partial derivatives.

(a) How many of the third-order partial derivatives can be distinct?

(b) How many of the fourth order?

34. Let $f(x, y) = e^{xy^2}$.

(a) In words, explain why f_{xyx}, f_{xxy}, and f_{yxx} are equal.

(b) Calculate the derivatives in part (a), and confirm their equality.

> A function $f(x, y)$ is said to be ***homogeneous of degree n*** if $f(tx, ty) = t^n f(x, y)$ for $t > 0$. This terminology is needed in Exercises 35 and 36.

35. In each part, show that the function is homogeneous, and find its degree.

(a) $f(x, y) = 3x^2 + y^2$ (b) $f(x, y) = \sqrt{x^2 + y^2}$

(c) $f(x, y) = x^2 y - 2y^3$ (d) $f(x, y) = \dfrac{5}{(x^2 + 2y^2)^2}$

36. (a) Show that if $f(x, y)$ is a homogeneous function of degree n, then

$$x \frac{\partial f}{\partial x} + y \frac{\partial f}{\partial y} = nf$$

[*Hint:* Let $u = tx$ and $v = ty$ in $f(tx, ty)$, and differentiate both sides of $f(u, v) = t^n f(x, y)$ with respect to t.]

(b) Confirm that the functions in Exercise 35 satisfy the equation in part (a).

37. (a) Suppose that $z = f(u)$ and $u = g(x, y)$. Draw a tree diagram, and use it to construct chain rules that express $\partial z/\partial x$ and $\partial z/\partial y$ in terms of dz/du, $\partial u/\partial x$, and $\partial u/\partial y$.

(b) Show that

$$\frac{\partial^2 z}{\partial x^2} = \frac{dz}{du} \frac{\partial^2 u}{\partial x^2} + \frac{d^2 z}{du^2} \left(\frac{\partial u}{\partial x} \right)^2$$

$$\frac{\partial^2 z}{\partial y^2} = \frac{dz}{du} \frac{\partial^2 u}{\partial y^2} + \frac{d^2 z}{du^2} \left(\frac{\partial u}{\partial y} \right)^2$$

$$\frac{\partial^2 z}{\partial y \partial x} = \frac{dz}{du} \frac{\partial^2 u}{\partial y \partial x} + \frac{d^2 z}{du^2} \frac{\partial u}{\partial x} \frac{\partial u}{\partial y}$$

38. (a) Let $z = f(x^2 - y^2)$. Use the result in Exercise 37(a) to show that

$$y \frac{\partial z}{\partial x} + x \frac{\partial z}{\partial y} = 0$$

(b) Let $z = f(xy)$. Use the result in Exercise 37(a) to show that

$$x \frac{\partial z}{\partial x} - y \frac{\partial z}{\partial y} = 0$$

(c) Confirm the result in part (a) in the case where $z = \sin(x^2 - y^2)$.

(d) Confirm the result in part (b) in the case where $z = e^{xy}$.

39. Suppose that the equation $z = f(x, y)$ is expressed in the polar form $z = g(r, \theta)$ by making the substitution $x = r \cos \theta$ and $y = r \sin \theta$.

(a) View r and θ as functions of x and y and use implicit differentiation to show that

$$\frac{\partial r}{\partial x} = \cos \theta \quad \text{and} \quad \frac{\partial \theta}{\partial x} = -\frac{\sin \theta}{r}$$

(b) View r and θ as functions of x and y and use implicit differentiation to show that

$$\frac{\partial r}{\partial y} = \sin \theta \quad \text{and} \quad \frac{\partial \theta}{\partial y} = \frac{\cos \theta}{r}$$

(c) Use the results in parts (a) and (b) to show that

$$\frac{\partial z}{\partial x} = \frac{\partial z}{\partial r} \cos \theta - \frac{1}{r} \frac{\partial z}{\partial \theta} \sin \theta$$

$$\frac{\partial z}{\partial y} = \frac{\partial z}{\partial r} \sin \theta + \frac{1}{r} \frac{\partial z}{\partial \theta} \cos \theta$$

(d) Use the result in part (c) to show that

$$\left(\frac{\partial z}{\partial x} \right)^2 + \left(\frac{\partial z}{\partial y} \right)^2 = \left(\frac{\partial z}{\partial r} \right)^2 + \frac{1}{r^2} \left(\frac{\partial z}{\partial \theta} \right)^2$$

(e) Use the result in part (c) to show that if $z = f(x, y)$ satisfies Laplace's equation

$$\frac{\partial^2 z}{\partial x^2} + \frac{\partial^2 z}{\partial y^2} = 0$$

then $z = g(r, \theta)$ satisfies the equation

$$\frac{\partial^2 z}{\partial r^2} + \frac{1}{r^2} \frac{\partial^2 z}{\partial \theta^2} + \frac{1}{r} \frac{\partial z}{\partial r} = 0$$

and conversely. The latter equation is called the ***polar form of Laplace's equation***.

40. Show that the function

$$z = \tan^{-1} \frac{2xy}{x^2 - y^2}$$

satisfies Laplace's equation; then make the substitution $x = r \cos \theta$, $y = r \sin \theta$, and show that the resulting function of r and θ satisfies the polar form of Laplace's equation given in part (e) of Exercise 39.

41. (a) Show that if $u(x, y)$ and $v(x, y)$ satisfy the Cauchy–Riemann equations (Exercise 76, Section 15.3), and if $x = r \cos \theta$ and $y = r \sin \theta$, then

$$\frac{\partial u}{\partial r} = \frac{1}{r} \frac{\partial v}{\partial \theta} \quad \text{and} \quad \frac{\partial v}{\partial r} = -\frac{1}{r} \frac{\partial u}{\partial \theta}$$

This is called the ***polar form of the Cauchy–Riemann equations***.

(b) Show that the functions

$$u = \ln(x^2 + y^2), \quad v = 2 \tan^{-1}(y/x)$$

satisfy the Cauchy–Riemann equations; then make the substitution $x = r \cos \theta$, $y = r \sin \theta$, and show that the resulting functions of r and θ satisfy the polar form of the Cauchy–Riemann equations.

42. Let $z = f(x - y, y - x)$. Show that $\partial z/\partial x + \partial z/\partial y = 0$.

43. Recall from Formula (6) of Section 15.3 that under appropriate conditions a plucked string satisfies the wave equation

$$\frac{\partial^2 u}{\partial t^2} = c^2 \frac{\partial^2 u}{\partial x^2}$$

where c is a positive constant.
(a) Show that a function of the form $u(x, t) = f(x + ct)$ satisfies the wave equation.
(b) Show that a function of the form $u(x, t) = g(x - ct)$ satisfies the wave equation.
(c) Show that a function of the form

$$u(x, t) = f(x + ct) + g(x - ct)$$

satisfies the wave equation.
(d) It can be proved that every solution of the wave equation is expressible in the form stated in part (c). Confirm that $u(x, t) = \sin t \sin x$ satisfies the wave equation in which $c = 1$, and then use appropriate trigonometric identities to express this function in the form $f(x + t) + g(x - t)$.

In Exercises 44–47, use Definition 15.4.2 to establish the differentiability of the given function. [*Remark:* ϵ_1 and ϵ_2 are not unique.]

---

44. $f(x, y) = xy$ **45.** $f(x, y) = x^2 + y^2$

46. $f(x, y) = x^2 y$ **47.** $f(x, y) = 3x + y^2$

48. Let $f(x, y) = \sqrt{x^2 + y^2}$
(a) Show that f is continuous at $(0, 0)$.
(b) Use Definition 15.3.1 to show that $f_x(0, 0)$ does not exist, and hence that f is not differentiable at $(0, 0)$.

49. Let

$$f(x, y) = \begin{cases} 5 - 3x - 2y, & x \geq 0 \text{ or } y \geq 0 \\ 0, & x < 0 \text{ and } y < 0 \end{cases}$$

Show that $f_x(0, 0)$ and $f_y(0, 0)$ exist, but f is not continuous at $(0, 0)$.

50. Let

$$f(x, y) = \begin{cases} \dfrac{xy}{x^2 + y^2}, & (x, y) \neq (0, 0) \\ 0, & (x, y) = (0, 0) \end{cases}$$

(a) Use Definition 15.3.1 to show that $f_x(0, 0)$ and $f_y(0, 0)$ exist.
(b) Show that f is not continuous at $(0, 0)$. [*Hint:* Examine the limit of $f(x, y)$ as $(x, y) \to (0, 0)$ along $y = 0$ and along $y = x$.]

51. Let

$$f(x, y) = \begin{cases} \dfrac{xy(x^2 - y^2)}{x^2 + y^2}, & (x, y) \neq (0, 0) \\ 0, & (x, y) = (0, 0) \end{cases}$$

(a) Use Definition 15.3.1 to show that $f_x(0, 0) = 0$ and $f_y(0, 0) = 0$.
(b) Show that $f_x(0, y) = -y$ for all y and $f_y(x, 0) = x$ for all x.
(c) Use Definition 15.3.1 to show that $f_{xy}(0, 0) = -1$ and $f_{yx}(0, 0) = 1$.
(d) Does this violate Theorem 15.4.6? Explain.

52. Prove: If f, f_x, and f_y are continuous on a circular region containing $A(x_0, y_0)$ and $B(x_1, y_1)$, then there is a point (x^*, y^*) on the line segment joining A and B such that

$$f(x_1, y_1) - f(x_0, y_0)$$
$$= f_x(x^*, y^*)(x_1 - x_0) + f_y(x^*, y^*)(y_1 - y_0)$$

This result is the two-dimensional version of the Mean-Value Theorem. [*Hint:* Express the line segment joining A and B in parametric form and use the Mean-Value Theorem for functions of one variable.]

53. Prove: If $f_x(x, y) = 0$ and $f_y(x, y) = 0$ throughout a circular region, then $f(x, y)$ is constant on that region. [*Hint:* Use the result of Exercise 52.]

15.5 TANGENT PLANES; TOTAL DIFFERENTIALS FOR FUNCTIONS OF TWO VARIABLES

In this section we will discuss tangent planes to surfaces in three-dimensional space. We will be concerned with three main questions: What is a tangent plane? When do tangent planes exist? How do we find equations of tangent planes? Once we have answered these questions, we will use our results on tangent planes to extend the concept of a differential to functions of two variables.

TANGENT PLANES

Recall that if C is a smooth parametric curve in 3-space, then the tangent line to C at a point P_0 is the line through P_0 along the unit tangent vector to C at P_0 (Figure 15.5.1). The concept of a *tangent plane* builds on this definition. If $P_0(x_0, y_0, z_0)$ is a point on a surface S, and if the tangent lines at P_0 to all smooth curves on the surface that pass through P_0 lie in a common plane, then we will regard that plane to be the ***tangent plane*** to the surface

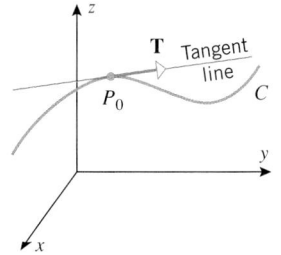

Figure 15.5.1

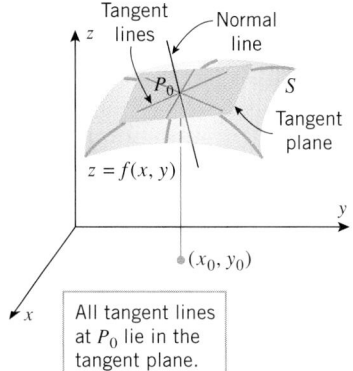

All tangent lines at P_0 lie in the tangent plane.

Figure 15.5.2

at P_0, and we will call the line through P_0 that is perpendicular to the tangent plane the **normal line** to the surface at P_0 (Figure 15.5.2).

The following theorem states conditions that ensure the existence of a tangent plane and gives an equation for that plane.

15.5.1 THEOREM. *Let $P_0(x_0, y_0, z_0)$ be any point on the surface $z = f(x, y)$. If $f(x, y)$ is differentiable at (x_0, y_0), then the surface has a tangent plane at P_0, and this plane has the equation*

$$f_x(x_0, y_0)(x - x_0) + f_y(x_0, y_0)(y - y_0) - (z - z_0) = 0 \qquad (1)$$

Proof. To prove the existence of a tangent plane at the point $P_0(x_0, y_0, z_0)$, we must show that all smooth curves on the surface $z = f(x, y)$ that pass through P_0 have tangent lines at P_0 that lie in a common plane. We will do this by showing that the unit tangent vectors to these curves at P_0 are all orthogonal to the vector

$$\mathbf{n} = f_x(x_0, y_0)\mathbf{i} + f_y(x_0, y_0)\mathbf{j} - \mathbf{k} \qquad (2)$$

This will show that the tangent lines at $P_0(x_0, y_0, z_0)$ lie in a common plane that has $\mathbf{n}$ as a normal and that (1) is the point-normal equation of the plane, thereby completing the proof.

Assume that C is any smooth curve that lies on the surface $z = f(x, y)$ and passes through $P_0(x_0, y_0, z_0)$, and suppose that C can be expressed parametrically as

$$x = x(s), \quad y = y(s), \quad z = z(s)$$

where s is an arc length parameter with reference point (x_0, y_0, z_0). Because C lies on the surface $z = f(x, y)$, we must have

$$z(s) = f(x(s), y(s))$$

for all values of s. Since f is differentiable at the point (x_0, y_0, z_0), and since this point corresponds to $s = 0$, it follows that $z(s)$ is differentiable at $s = 0$ and its derivative at that point is given by the chain rule

$$\frac{dz}{ds} = \frac{\partial f}{\partial x}\frac{dx}{ds} + \frac{\partial f}{\partial y}\frac{dy}{ds}$$

or, equivalently,

$$\frac{\partial f}{\partial x}\frac{dx}{ds} + \frac{\partial f}{\partial y}\frac{dy}{ds} - \frac{dz}{ds} = 0$$

This equation can be written in vector form as

$$\left(\frac{\partial f}{\partial x}\mathbf{i} + \frac{\partial f}{\partial y}\mathbf{j} - \mathbf{k}\right) \cdot \left(\frac{dx}{ds}\mathbf{i} + \frac{dy}{ds}\mathbf{j} + \frac{dz}{ds}\mathbf{k}\right) = 0$$

By Formula (6) of Section 14.4, the second factor in this dot product is the unit tangent vector $\mathbf{T}(s)$. Thus, at $s = 0$ this equation can be written as

$$(f_x(x_0, y_0)\mathbf{i} + f_y(x_0, y_0)\mathbf{j} - \mathbf{k}) \cdot \mathbf{T}(0) = 0$$

which shows that all smooth curves on the surface that pass through the point $P_0(x_0, y_0, z_0)$ have unit tangent vectors at that point that are orthogonal to the vector

$$\mathbf{n} = f_x(x_0, y_0)\mathbf{i} + f_y(x_0, y_0)\mathbf{j} - \mathbf{k} \qquad \blacksquare$$

Since the normal line to the surface at the point $P_0(x_0, y_0, z_0)$ is parallel to the vector $\mathbf{n}$ in (2), it follows that this line can be expressed parametrically as

$$x = x_0 + f_x(x_0, y_0)t, \quad y = y_0 + f_y(x_0, y_0)t, \quad z = z_0 - t \qquad (3)$$

Example 1

Find an equation for the tangent plane and parametric equations for the normal line to the surface $z = x^2 y$ at the point $(2, 1, 4)$.

Solution. Since $f(x, y) = x^2 y$, it follows that

$$f_x(x, y) = 2xy \quad \text{and} \quad f_y(x, y) = x^2$$

so that with $x = 2$, $y = 1$,

$$f_x(2, 1) = 4 \quad \text{and} \quad f_y(2, 1) = 4$$

Thus, a vector normal to the surface at $(2, 1, 4)$ is

$$\mathbf{n} = f_x(2, 1)\mathbf{i} + f_y(2, 1)\mathbf{j} - \mathbf{k} = 4\mathbf{i} + 4\mathbf{j} - \mathbf{k}$$

Therefore, the tangent plane has the equation

$$4(x - 2) + 4(y - 1) - (z - 4) = 0 \quad \text{or} \quad 4x + 4y - z = 8$$

and the normal line has equations

$$x = 2 + 4t, \quad y = 1 + 4t, \quad z = 4 - t \qquad \blacktriangleleft$$

THE GEOMETRIC SIGNIFICANCE OF DIFFERENTIABILITY

In the last section we set two goals for the definition of differentiability of a function $f(x, y)$ of two variables at a point (x_0, y_0)—we wanted f to be continuous at (x_0, y_0) and we wanted the surface $z = f(x, y)$ to have a nonvertical tangent plane at (x_0, y_0). Both of these goals have now been achieved: we showed in Theorem 15.4.4 that differentiability implies continuity, and now Theorem 15.5.1 shows that differentiability implies the existence of a nonvertical tangent plane. [The tangent plane is nonvertical because the third component of its normal vector $\mathbf{n} = f_x(x, y)\mathbf{i} + f_y(x, y)\mathbf{j} - \mathbf{k}$ is nonzero.]

TOTAL DIFFERENTIALS

Recall that if $y = f(x)$ is a function of one variable, then the differential

$$dy = f'(x_0)\, dx$$

represents the change in y along the *tangent line* at (x_0, y_0) produced by a change dx in x and

$$\Delta y = f(x_0 + \Delta x) - f(x_0)$$

represents the change in y along the *curve* $y = f(x)$ produced by a change Δx in x. Analogously, if $z = f(x, y)$ is a function of two variables, we will define dz to be the change in z along the *tangent plane* at (x_0, y_0, z_0) to the surface $z = f(x, y)$ produced by changes dx and dy in x and y, respectively. This is in contrast to

$$\Delta z = f(x_0 + \Delta x, y_0 + \Delta y) - f(x_0, y_0) \tag{4}$$

which represents the change in z *along the surface* produced by changes Δx and Δy in x and y. A comparison of dz and Δz is shown in Figure 15.5.3 in the case where $dx = \Delta x$ and $dy = \Delta y$.

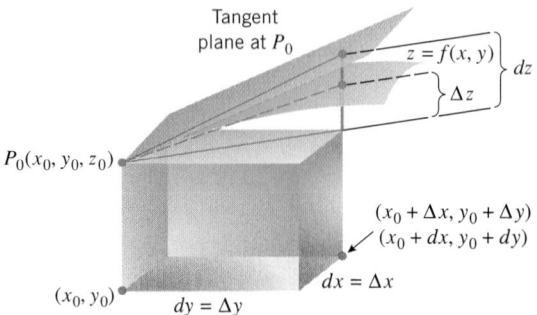

Figure 15.5.3

To derive a formula for dz, let $P_0(x_0, y_0, z_0)$ be a fixed point on the surface $z = f(x, y)$. If we assume f to be differentiable at (x_0, y_0), then the surface has a tangent plane at P_0, given by the equation

$$f_x(x_0, y_0)(x - x_0) + f_y(x_0, y_0)(y - y_0) - (z - z_0) = 0$$

or

$$z = z_0 + f_x(x_0, y_0)(x - x_0) + f_y(x_0, y_0)(y - y_0) \tag{5}$$

It follows from (5) that the tangent plane has height z_0 when $x = x_0$, $y = y_0$, and it has height

$$z_0 + f_x(x_0, y_0)\, dx + f_y(x_0, y_0)\, dy \tag{6}$$

when $x = x_0 + dx$, $y = y_0 + dy$. Thus, the change dz in the height of the tangent plane as (x, y) varies from (x_0, y_0) to $(x_0 + dx, y_0 + dy)$ is obtained by subtracting z_0 from expression (6). This yields

$$dz = f_x(x_0, y_0)\, dx + f_y(x_0, y_0)\, dy \tag{7}$$

Often, we will omit the subscripts on x_0 and y_0 in Formulas (4) and (7) and write these equations as

$$\Delta z = f(x + \Delta x, y + \Delta y) - f(x, y) \tag{8}$$

$$dz = f_x(x, y)\, dx + f_y(x, y)\, dy \tag{9}$$

We call dz the **total differential of** z or the **total differential of** f. In these formulas, Δx, Δy, dx, and dy are usually viewed as variables and x and y as fixed numbers. When it is desirable not to use a dependent variable, we will write df for the total differential of f.

LOCAL LINEAR APPROXIMATION

If $z = f(x, y)$ is differentiable at the point (x_0, y_0), then it follows from Definition 15.4.2 that

$$\Delta f = f_x(x_0, y_0)\Delta x + f_y(x_0, y_0)\Delta y + \epsilon_1 \Delta x + \epsilon_2 \Delta y$$

where $\epsilon_1 \to 0$ and $\epsilon_2 \to 0$ as $(\Delta x, \Delta y) \to (0, 0)$; and from (4) with Δf in place of Δz we can rewrite this as

$$f(x_0 + \Delta x, y_0 + \Delta y) = f(x_0, y_0) + f_x(x_0, y_0)\Delta x + f_y(x_0, y_0)\Delta y + \epsilon_1 \Delta x + \epsilon_2 \Delta y$$

Thus, if $\Delta x \approx 0$ and $\Delta y \approx 0$, the terms involving ϵ_1 and ϵ_2 will be small, and we can approximate $f(x_0 + \Delta x, y_0 + \Delta y)$ as

$$f(x_0 + \Delta x, y_0 + \Delta y) \approx f(x_0, y_0) + f_x(x_0, y_0)\Delta x + f_y(x_0, y_0)\Delta y \tag{10}$$

This formula, which is the two-variable analog of Formula (7) in Section 3.6, is called the **local linear approximation of** f **at** (x_0, y_0). When convenient, we can replace Δx and Δy in this formula by $\Delta x = x - x_0$ and $\Delta y = y - y_0$ and rewrite the formula as

$$f(x, y) \approx f(x_0, y_0) + f_x(x_0, y_0)(x - x_0) + f_y(x_0, y_0)(y - y_0) \tag{11}$$

which is the two-variable analog of Formula (6) in Section 3.6.

REMARK.   It follows from Formula (5) with $z_0 = f(x_0, y_0)$ that the right side of Formula (11) is the height of the tangent plane at (x_0, y_0) above the point (x, y). Thus, the local linear approximation approximates the value of $f(x, y)$ at a point (x, y) near (x_0, y_0) by the height of the tangent plane above the point (x, y). This is analogous to approximating the value of $f(x)$ at a point x near x_0 by the height of the tangent line above the point x (Figure 3.6.5).

Example 2

(a) Find the local linear approximation of $f(x, y) = \sqrt{x^2 + y^2}$ at a point (x_0, y_0).

(b) Use the local linear approximation that you found in part (a) to approximate
$$f(3.04, 3.98) = \sqrt{(3.04)^2 + (3.98)^2}.$$

Solution (a). We have
$$f_x(x, y) = \frac{x}{\sqrt{x^2 + y^2}} \quad \text{and} \quad f_y(x, y) = \frac{y}{\sqrt{x^2 + y^2}}$$

Thus, it follows from (11) that the local linear approximation of f at (x_0, y_0) is

$$\sqrt{x^2 + y^2} \approx \sqrt{x_0^2 + y_0^2} + \frac{x_0}{\sqrt{x_0^2 + y_0^2}}(x - x_0) + \frac{y_0}{\sqrt{x_0^2 + y_0^2}}(y - y_0) \qquad (12)$$

Solution (b). Applying Formula (12) with $x_0 = 3$, $y_0 = 4$, $x = 3.04$, and $y = 3.98$ yields

$$\sqrt{(3.04)^2 + (3.98)^2} \approx 5 + \tfrac{3}{5}(0.04) + \tfrac{4}{5}(-0.02) = 5.008$$

We leave it for you to confirm with a calculator that $\sqrt{(3.04)^2 + (3.98)^2} \approx 5.00819$ to five decimal places. ◀

APPROXIMATIONS USING TOTAL DIFFERENTIALS

If $z = f(x, y)$ is differentiable at the point (x, y), then it follows from Definition 15.4.2 that the increment Δz can be written as

$$\Delta z = f_x(x, y)\Delta x + f_y(x, y)\Delta y + \epsilon_1 \Delta x + \epsilon_2 \Delta y \qquad (13)$$

where $\epsilon_1 \to 0$ and $\epsilon_2 \to 0$ as $(\Delta x, \Delta y) \to (0, 0)$. In the case where $dx = \Delta x$ and $dy = \Delta y$, it follows from (13) that this formula can be rewritten as

$$\Delta z = dz + \epsilon_1 \Delta x + \epsilon_2 \Delta y$$

Thus, when $\Delta x = dx$ and $\Delta y = dy$ are small, we can approximate Δz by

$$\Delta z \approx dz \qquad (14)$$

Geometrically, this approximation tells us that the change in z along the surface and the change in z along the tangent plane are approximately equal when $\Delta x = dx$ and $\Delta y = dy$ are small (see Figure 15.5.3). This is the extension to functions of two variables of Formula (12) of Section 3.6.

Example 3

Let $z = 4x^3 y^2$. Find dz.

Solution. Since $f(x, y) = 4x^3 y^2$,
$$f_x(x, y) = 12x^2 y^2 \quad \text{and} \quad f_y(x, y) = 8x^3 y$$

so

$$dz = 12x^2 y^2\, dx + 8x^3 y\, dy \qquad ◀$$

Example 4

The radius of a right circular cylinder is measured with an error of at most 2%, and the height is measured with an error of at most 4%. Approximate the maximum possible percentage error in the volume V calculated from these measurements.

Solution. Let r, h, and V be the true radius, height, and volume of the cylinder, and let $\Delta r, \Delta h$, and ΔV be the errors in these quantities. We are given that

$$\left| \frac{\Delta r}{r} \right| \leq 0.02 \quad \text{and} \quad \left| \frac{\Delta h}{h} \right| \leq 0.04$$

We want to find the maximum possible value of $|\Delta V / V|$. Since the volume of the cylinder is $V = \pi r^2 h$, it follows from (9) that

$$dV = \frac{\partial V}{\partial r}\, dr + \frac{\partial V}{\partial h}\, dh = 2\pi r h\, dr + \pi r^2\, dh$$

If we choose $dr = \Delta r$ and $dh = \Delta h$, then we can use the approximations

$$\Delta V \approx dV \quad \text{and} \quad \frac{\Delta V}{V} \approx \frac{dV}{V}$$

But

$$\frac{dV}{V} = \frac{2\pi r h\, dr + \pi r^2\, dh}{\pi r^2 h} = 2\frac{dr}{r} + \frac{dh}{h}$$

so by the triangle inequality (1.2.2)

$$\left|\frac{dV}{V}\right| = \left|2\frac{dr}{r} + \frac{dh}{h}\right| \le 2\left|\frac{dr}{r}\right| + \left|\frac{dh}{h}\right| \le 2(0.02) + (0.04) = 0.08$$

Thus, the maximum percentage error in V is approximately 8%. ◀

EXERCISE SET 15.5

In Exercises 1–8, find an equation for the tangent plane and parametric equations for the normal line to the surface at the point P.

1. $z = 4x^3 y^2 + 2y$; $P(1, -2, 12)$

2. $z = \frac{1}{2}x^7 y^{-2}$; $P(2, 4, 4)$

3. $z = xe^{-y}$; $P(1, 0, 1)$

4. $z = \ln \sqrt{x^2 + y^2}$; $P(-1, 0, 0)$

5. $z = e^{3y} \sin 3x$; $P(\pi/6, 0, 1)$

6. $z = x^{1/2} + y^{1/2}$; $P(4, 9, 5)$

7. $x^2 + y^2 + z^2 = 25$; $P(-3, 0, 4)$

8. $x^2 y - 4z^2 = -7$; $P(-3, 1, -2)$

9. Find all points on the surface at which the tangent plane is horizontal.
(a) $z = x^3 y^2$
(b) $z = x^2 - xy + y^2 - 2x + 4y$

10. Find a point on the surface $z = 3x^2 - y^2$ at which the tangent plane is parallel to the plane $6x + 4y - z = 5$.

11. Find a point on the surface $z = 8 - 3x^2 - 2y^2$ at which the tangent plane is perpendicular to the line $x = 2 - 3t$, $y = 7 + 8t$, $z = 5 - t$.

12. Show that the surfaces

$$z = \sqrt{x^2 + y^2} \quad \text{and} \quad z = \frac{1}{10}(x^2 + y^2) + \frac{5}{2}$$

intersect at $(3, 4, 5)$ and have a common tangent plane at that point.

13. Let $f(x, y) = x^2 y$. Find df and Δf at the point $(1, 3)$ with $\Delta x = dx = 0.1$ and $\Delta y = dy = 0.2$.

14. Let $z = 3x^2 - 2y$. Find dz and Δz at the point $(-2, 4)$ with $\Delta x = dx = 0.02$ and $\Delta y = dy = -0.03$.

15. Let $z = x/y$. Find the increment in z as (x, y) varies from $(-1, 2)$ to $(3, 1)$.

16. Let $g(u, v) = 2uv - v^3$. Find the increment in g as (u, v) varies from $(0, 1)$ to $(4, -2)$.

In Exercises 17 and 18, find formulas for dz and Δz at a general point (x, y).

17. $z = x^3 y^2$

18. $z = e^{xy}$

In Exercises 19–22, find dz.

19. $z = 7x - 2y$

20. $z = 5x^2 y^5 - 2x + 4y + 7$

21. $z = \tan^{-1} xy$

22. $z = \sec^2(x - 3y)$

23. In each part, confirm that the stated formula is the local linear approximation at $(0, 0)$.
(a) $e^x \sin y \approx y$
(b) $\dfrac{2x + 1}{y + 1} \approx 1 + 2x - y$

24. Show that if $\alpha \neq 1$ and $\beta \neq 1$, then the local linear approximation of the function $f(x, y) = x^\alpha y^\beta$ at $(1, 1)$ is
$$x^\alpha y^\beta \approx 1 + \alpha(x - 1) + \beta(y - 1)$$

25. Suppose that $T(x, y)$ is the Fahrenheit temperature at a point (x, y) on a metal plate. Given that $T(1, 3) = 93^\circ$F, $T_x(1, 3) = 2^\circ$F/cm, and $T_y(1, 3) = -1^\circ$F/cm, use a local linear approximation to estimate the temperature at the point $T(0.98, 3.02)$.

26. Suppose that $p(x, y)$ denotes the atmospheric pressure at a point (x, y). Given that $p(100, 98) = 1008$ mb (millibars), $p_x(100, 98) = -2$ mb/km, and $p_y(100, 98) = 1$ mb/km, use a local linear approximation to estimate the atmospheric pressure at the point $(104, 103)$.

In Exercises 27 and 28, use an appropriate local linear approximation to estimate the value of the given quantity, and then check your answer using a calculating utility.

27. $\dfrac{1}{\sqrt{(3.92)^2 + (3.01)^2}}$

28. $(1.05)^{0.5}(0.97)^{0.3}$

In Exercises 29–32, use a total differential to approximate the change in $f(x, y)$ as (x, y) varies from P to Q.

29. $f(x, y) = x^2 + 2xy - 4x$; $P(1, 2)$, $Q(1.01, 2.04)$

30. $f(x, y) = x^{1/3}y^{1/2}$; $P(8, 9)$, $Q(7.78, 9.03)$

31. $f(x, y) = \dfrac{x + y}{xy}$; $P(-1, -2)$, $Q(-1.02, -2.04)$

32. $f(x, y) = \ln \sqrt{1 + xy}$; $P(0, 2)$, $Q(-0.09, 1.98)$

33. One leg of a right triangle increases from 3 cm to 3.2 cm, while the other leg decreases from 4 cm to 3.96 cm. Use a total differential to approximate the change in the length of the hypotenuse.

34. The volume V of a right circular cone of radius r and height h is given by $V = \frac{1}{3}\pi r^2 h$. Suppose that the height decreases from 20 in to 19.95 in, while the radius increases from 4 in to 4.05 in. Use a total differential to approximate the change in volume.

35. The length and width of a rectangle are measured with errors of at most 3% and 5%, respectively. Use differentials to approximate the maximum percentage error in the calculated area.

36. The radius and height of a right circular cone are measured with errors of at most 1% and 4%, respectively. Use differentials to approximate the maximum percentage error in the calculated volume.

37. The length and width of a rectangle are measured with errors of at most $r\%$, where r is small. Use differentials to approximate the maximum percentage error in the calculated length of the diagonal.

38. The legs of a right triangle are measured to be 3 cm and 4 cm, with a maximum error of 0.05 cm in each measurement. Use differentials to approximate the maximum possible error in the calculated value of (a) the hypotenuse and (b) the area of the triangle.

39. The total resistance R of two resistances R_1 and R_2, connected in parallel, is

$$R = \frac{R_1 R_2}{R_1 + R_2}$$

Suppose that R_1 and R_2 are measured to be 200 ohms and 400 ohms, respectively, with a maximum error of 2% in each. Use differentials to approximate the maximum percentage error in the calculated value of R.

40. According to the ideal gas law, the pressure, temperature, and volume of a confined gas are related by $P = kT/V$,

where k is a constant. Use differentials to approximate the percentage change in pressure if the temperature of a gas is increased 3% and the volume is increased 5%.

41. An angle θ of a right triangle is calculated by the formula

$$\theta = \sin^{-1}\frac{a}{c}$$

where a is the length of the side opposite to θ and c is the length of the hypotenuse. Suppose that the measurements $a = 3$ inches and $c = 5$ inches each have a maximum possible error of 0.01 inch. Use differentials to approximate the maximum possible error in the calculated value of θ.

42. A cylindrical can that is open at one end has an inside radius of 2 cm and an inside height of 5 cm. Use differentials to approximate the volume of metal in the can if it is 0.01 cm thick. [*Hint:* The volume of metal is the difference, ΔV, in the volumes of two cylinders.]

43. The period T of a simple pendulum with small oscillations is calculated from the formula $T = 2\pi\sqrt{L/g}$, where L is the length of the pendulum and g is the acceleration due to gravity. Suppose that values of L and g have errors of at most 0.5% and 0.1%, respectively. Use differentials to approximate the maximum percentage error in the calculated value of T.

44. The angle of elevation from a point on the ground to the top of a building is measured as $60°$ with a maximum possible error of $0.2°$. Suppose that the distance from the point to the building is measured as 100 ft with a maximum possible error of 2 in. Use differentials to approximate the maximum possible error in the calculated height of the building.

45. Suppose that x and y have errors of at most $r\%$ and $s\%$, respectively. For each of the following formulas in x and y, use differentials to approximate the maximum possible error in the calculated result.
(a) xy (b) x/y (c) $x^2 y^3$ (d) $x^3\sqrt{y}$

46. Show that the volume of the solid bounded by the coordinate planes and the plane tangent to the portion of the surface $xyz = k$, $k > 0$, in the first octant does not depend on the point of tangency.

47. (a) Find all points of intersection of the line $x = -1 + t$, $y = 2 + t$, $z = 2t + 7$ and the surface $z = x^2 + y^2$.
(b) At each point of intersection, find the cosine of the acute angle between the given line and the line normal to the surface.

48. Show that if f is differentiable and $z = xf(x/y)$, then all tangent planes to the graph of this equation pass through the origin.

49. Show that the equation of the plane that is tangent to the ellipsoid

$$\frac{x^2}{a^2} + \frac{y^2}{b^2} + \frac{z^2}{c^2} = 1$$

at (x_0, y_0, z_0) can be written in the form

$$\frac{x_0 x}{a^2} + \frac{y_0 y}{b^2} + \frac{z_0 z}{c^2} = 1$$

50. Show that the equation of the plane that is tangent to the paraboloid

$$z = \frac{x^2}{a^2} + \frac{y^2}{b^2}$$

at (x_0, y_0, z_0) can be written in the form

$$z + z_0 = \frac{2x_0 x}{a^2} + \frac{2y_0 y}{b^2}$$

51. Prove: If the surfaces $z = f(x, y)$ and $z = g(x, y)$ intersect at $P(x_0, y_0, z_0)$, and if f and g are differentiable at (x_0, y_0), then the normal lines at P are perpendicular if and only if

$$f_x(x_0, y_0)g_x(x_0, y_0) + f_y(x_0, y_0)g_y(x_0, y_0) = -1$$

52. Use the result in Exercise 51 to show that the normal lines to the cones $z = \sqrt{x^2 + y^2}$ and $z = -\sqrt{x^2 + y^2}$ are perpendicular to the normal lines to the sphere $x^2 + y^2 + z^2 = a^2$ at every point of intersection (see the accompanying figure).

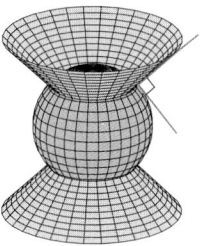

Figure Ex-52

15.6 DIRECTIONAL DERIVATIVES AND GRADIENTS FOR FUNCTIONS OF TWO VARIABLES

The partial derivatives $f_x(x, y)$ and $f_y(x, y)$ represent the rates of change of $f(x, y)$ in directions parallel to the x- and y-axes. In this section we will investigate rates of change of $f(x, y)$ in other directions.

DIRECTIONAL DERIVATIVES

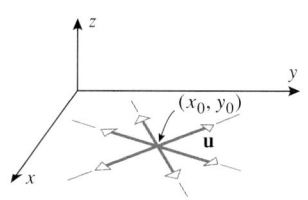

Figure 15.6.1

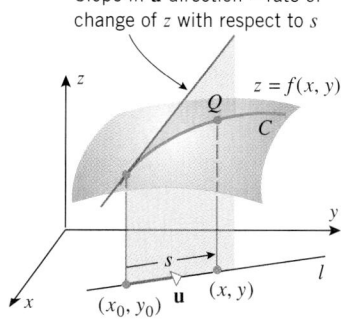

Slope in **u** direction = rate of change of z with respect to s

Figure 15.6.2

Our first objective is to determine how to find the slope of a surface $z = f(x, y)$ at a point (x_0, y_0) in an *arbitrary* specified direction. Since there are infinitely many directions in which a point can move in the xy-plane, we need some method for describing a specific direction starting at (x_0, y_0). One way to do this is to use a unit vector

$$\mathbf{u} = u_1\mathbf{i} + u_2\mathbf{j}$$

that has its initial point at (x_0, y_0) and points in the desired direction (Figure 15.6.1). This vector determines a line l in the xy-plane that can be expressed parametrically as

$$x = x_0 + su_1, \quad y = y_0 + su_2 \tag{1}$$

where s is the arc length parameter that has its reference point at (x_0, y_0) and has positive values in the direction of **u**. For $s = 0$, the point (x, y) is at the reference point (x_0, y_0) and it moves along the line l in the direction of **u** as s increases. Concurrently, a companion point $Q(x, y, f(x, y))$ on the vertical line through (x, y) moves along the surface $z = f(x, y)$, tracing out a curve C (Figure 15.6.2). We will call the rate of change of z with respect to s along C at $s = 0$ the *slope of the surface in the direction of* **u** at (x_0, y_0). If f is differentiable at (x_0, y_0), then this slope can be calculated by finding the value of dz/ds at $s = 0$. To do this, we start with the chain rule

$$\frac{dz}{ds} = f_x(x, y)\frac{dx}{ds} + f_y(x, y)\frac{dy}{ds}$$

or from (1)

$$\frac{dz}{ds} = f_x(x, y)u_1 + f_y(x, y)u_2$$

At $s = 0$, the point (x, y) is (x_0, y_0), so

$$\left.\frac{dz}{ds}\right|_{s=0} = f_x(x_0, y_0)u_1 + f_y(x_0, y_0)u_2 \tag{2}$$

which is a formula for the slope of the surface in the direction of **u** at (x_0, y_0).

If we now keep the vector $\mathbf{u} = u_1\mathbf{i} + u_2\mathbf{j}$ in Formula (2) fixed but replace the specific point (x_0, y_0) by a general point (x, y), then the resulting expression

$$f_x(x, y)u_1 + f_y(x, y)u_2$$

is a function of x and y whose value at the point (x, y) is the slope of the surface in the direction of $\mathbf{u}$ at that point. This "slope-producing function" is sufficiently important that it has some terminology and notation associated with it.

15.6.1 DEFINITION. If f is a differentiable function of x and y, and if $\mathbf{u} = u_1\mathbf{i} + u_2\mathbf{j}$ is a unit vector, then the *directional derivative of f in the direction of* $\mathbf{u}$ is denoted by $D_{\mathbf{u}}f$ and is defined by

$$D_{\mathbf{u}}f(x, y) = f_x(x, y)u_1 + f_y(x, y)u_2 \qquad (3)$$

Note that the value of $D_{\mathbf{u}}f(x, y)$ depends on the point (x, y) and the direction of $\mathbf{u}$. Thus, at a fixed point the slope of the surface may vary with the direction (Figure 15.6.3).

Recall from Formula (13) of Section 13.2 that a unit vector $\mathbf{u}$ can be expressed as

$$\mathbf{u} = \cos\phi\,\mathbf{i} + \sin\phi\,\mathbf{j}$$

where ϕ is the angle from the positive x-axis to $\mathbf{u}$. Thus, Formula (3) can also be expressed as

$$D_{\mathbf{u}}f(x, y) = f_x(x, y)\cos\phi + f_y(x, y)\sin\phi \qquad (4)$$

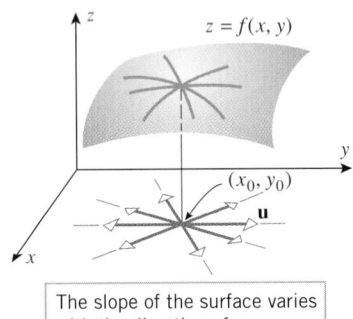

The slope of the surface varies with the direction of **u**.

Figure 15.6.3

Example 1

Find the directional derivative of $f(x, y) = 3x^2y$ at the point $(1, 2)$ in the direction of the vector $\mathbf{a} = 3\mathbf{i} + 4\mathbf{j}$.

Solution. The vector $\mathbf{a}$ is not a unit vector, so we must normalize it to apply Formula (3). This yields

$$\mathbf{u} = \frac{\mathbf{a}}{\|\mathbf{a}\|} = \frac{1}{\sqrt{25}}(3\mathbf{i} + 4\mathbf{j}) = \frac{3}{5}\mathbf{i} + \frac{4}{5}\mathbf{j}$$

from which we obtain $u_1 = \frac{3}{5}$ and $u_2 = \frac{4}{5}$. Since the partial derivatives of f are

$$f_x(x, y) = 6xy, \qquad f_y(x, y) = 3x^2$$

it follows from (3) that the directional derivative of f in the direction of $\mathbf{u}$ is

$$D_{\mathbf{u}}f(x, y) = (6xy)\left(\tfrac{3}{5}\right) + (3x^2)\left(\tfrac{4}{5}\right) = \tfrac{18}{5}xy + \tfrac{12}{5}x^2$$

Thus,

$$D_{\mathbf{u}}f(1, 2) = \tfrac{36}{5} + \tfrac{12}{5} = \tfrac{48}{5} \qquad \blacktriangleleft$$

Example 2

Find the directional derivative of $f(x, y) = e^{xy}$ at $(-2, 0)$ in the direction of the unit vector that makes an angle of $\pi/3$ with the positive x-axis.

Solution. The partial derivatives of f are

$$f_x(x, y) = ye^{xy}, \qquad f_y(x, y) = xe^{xy}$$
$$f_x(-2, 0) = 0, \qquad f_y(-2, 0) = -2$$

The unit vector $\mathbf{u}$ that makes an angle of $\pi/3$ with the positive x-axis is

$$\mathbf{u} = \cos(\pi/3)\mathbf{i} + \sin(\pi/3)\mathbf{j}$$

Thus, from (4)

$$D_{\mathbf{u}} f(-2, 0) = f_x(-2, 0) \cos(\pi/3) + f_y(-2, 0) \sin(\pi/3)$$
$$= 0(1/2) + (-2)(\sqrt{3}/2) = -\sqrt{3}$$ ◀

THE RELATIONSHIP BETWEEN DIRECTIONAL DERIVATIVES AND PARTIAL DERIVATIVES

It follows from Formula (4) that the directional derivative of a function f in the direction of the unit vector $\mathbf{i}$ is

$$D_{\mathbf{i}} f(x, y) = f_x(x, y) \cos 0 + f_y(x, y) \sin 0 = f_x(x, y)$$

and the directional derivative of f in the direction of the unit vector $\mathbf{j}$ is

$$D_{\mathbf{j}} f(x, y) = f_x(x, y) \cos(\pi/2) + f_y(x, y) \sin(\pi/2) = f_y(x, y)$$

Thus, the partial derivative of f with respect to x is the directional derivative in the positive x-direction, and the partial derivative of f with respect to y is the directional derivative in the positive y-direction.

Also note that Formula (3) expresses $D_{\mathbf{u}} f(x, y)$ in terms of $f_x(x, y)$ and $f_y(x, y)$ and the components of $\mathbf{u}$, so the values of *all* directional derivatives at a point are completely determined by values of the directional derivatives in the directions of $\mathbf{i}$ and $\mathbf{j}$.

THE EFFECT OF REVERSING DIRECTION

Since reversing the direction of a vector $\mathbf{u}$ reverses the signs of its components, it follows from Formula (3) that

$$D_{-\mathbf{u}} f(x, y) = -D_{\mathbf{u}} f(x, y) \tag{5}$$

Thus, for example,

$$D_{-\mathbf{i}} f(x, y) = -D_{\mathbf{i}} f(x, y) = -f_x(x, y) \quad \text{and} \quad D_{-\mathbf{j}} f(x, y) = -D_{\mathbf{j}} f(x, y) = -f_y(x, y)$$

Formula (5) makes sense intuitively if you view directional derivatives as slopes and consider the effect of walking in opposite directions on the side of a hill. At each point on the side of the hill the "uphill grade" and the "downhill grade" are the same, but in one direction you are ascending and in the opposite direction you are descending.

THE GRADIENT

The directional derivative

$$D_{\mathbf{u}} f(x, y) = f_x(x, y) u_1 + f_y(x, y) u_2$$

can be expressed in the form of a dot product as

$$D_{\mathbf{u}} f(x, y) = (f_x(x, y)\mathbf{i} + f_y(x, y)\mathbf{j}) \cdot (u_1\mathbf{i} + u_2\mathbf{j}) \tag{6}$$

The second vector in the dot product is $\mathbf{u}$. However, the first vector is new; it is called the *gradient of f* and is denoted by the symbol ∇f or $\nabla f(x, y)$.[*] The notation ∇z is also used if $z = f(x, y)$.

15.6.2   DEFINITION.   If f is a function of x and y, then the ***gradient of f*** is defined by

$$\nabla f(x, y) = f_x(x, y)\mathbf{i} + f_y(x, y)\mathbf{j} \tag{7}$$

With this notation Formula (6) can be expressed as

$$D_{\mathbf{u}} f(x, y) = \nabla f(x, y) \cdot \mathbf{u} \tag{8}$$

which states that the slope of the surface $z = f(x, y)$ at the point (x, y) in the direction of $\mathbf{u}$ is the dot product of the gradient with $\mathbf{u}$ (Figure 15.6.4).

REMARK.   It is important to keep in mind that ∇f is not the product of ∇ and f. The symbol ∇ does not have a value in and of itself; rather, you should think of it as an operator

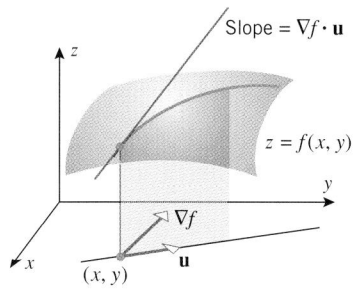

Figure 15.6.4

---
[*] The symbol ∇ (read "del") is an inverted delta. In older books this symbol is sometimes called a "nabla" because of its similarity in form to an ancient Hebrew ten-stringed harp of that name.

that acts on the function f to produce the gradient ∇f in the same sense that d/dx is an operator that acts on a function f to produce the derivative f'.

Example 3

Find the gradient of $f(x, y) = 3x^2 y$ at the point $(1, 2)$ and use it to calculate the directional derivative of f at $(1, 2)$ in the direction of the vector $\mathbf{a} = 3\mathbf{i} + 4\mathbf{j}$.

Solution. From (7)

$$\nabla f(x, y) = f_x(x, y)\mathbf{i} + f_y(x, y)\mathbf{j} = 6xy\mathbf{i} + 3x^2\mathbf{j}$$

so the gradient of f at $(1, 2)$ is

$$\nabla f(1, 2) = 12\mathbf{i} + 3\mathbf{j}$$

The unit vector in the direction of $\mathbf{a}$ is

$$\mathbf{u} = \frac{\mathbf{a}}{\|\mathbf{a}\|} = \frac{1}{5}(3\mathbf{i} + 4\mathbf{j}) = \frac{3}{5}\mathbf{i} + \frac{4}{5}\mathbf{j}$$

Thus, from (8)

$$D_{\mathbf{u}} f(1, 2) = \nabla f(1, 2) \cdot \mathbf{u} = (12\mathbf{i} + 3\mathbf{j}) \cdot \left(\tfrac{3}{5}\mathbf{i} + \tfrac{4}{5}\mathbf{j}\right) = \tfrac{48}{5}$$

which agrees with the result obtained in Example 1.   ◀

PROPERTIES OF THE GRADIENT

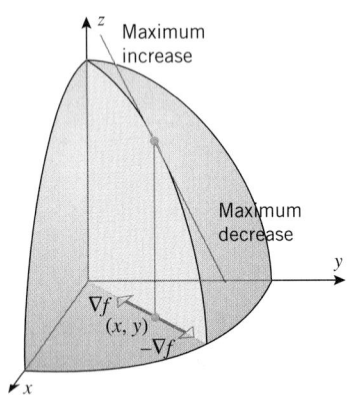

Figure 15.6.5

The gradient is not merely a notational device to simplify the formula for the directional derivative; we will see that the length and direction of the gradient ∇f provide important information about the function f and the surface $z = f(x, y)$. For example, suppose that $\nabla f(x, y) \neq \mathbf{0}$, and let us use Formula (4) of Section 13.3 to rewrite (8) as

$$D_{\mathbf{u}} f(x, y) = \nabla f(x, y) \cdot \mathbf{u} = \|\nabla f(x, y)\|\|\mathbf{u}\|\cos\theta = \|\nabla f(x, y)\|\cos\theta \qquad (9)$$

where θ is the angle between $\nabla f(x, y)$ and $\mathbf{u}$. This equation tells us that the maximum value of $D_{\mathbf{u}} f(x, y)$ is $\|\nabla f(x, y)\|$, and this maximum occurs when $\theta = 0$, that is, when $\mathbf{u}$ is in the direction of $\nabla f(x, y)$. Geometrically, this means that *the surface $z = f(x, y)$ has its maximum slope at a point (x, y) in the direction of the gradient, and the maximum slope is* $\|\nabla f(x, y)\|$ (Figure 15.6.5). Similarly, (9) tells us that the minimum value of $D_{\mathbf{u}} f(x, y)$ is $-\|\nabla f(x, y)\|$, and this minimum occurs when $\theta = \pi$, that is, when $\mathbf{u}$ is oppositely directed to $\nabla f(x, y)$. Geometrically, this means that *the surface $z = f(x, y)$ has its minimum slope at a point (x, y) in the direction that is opposite to the gradient, and the minimum slope is* $-\|\nabla f(x, y)\|$ (Figure 15.6.5).

Finally, in the case where $\nabla f(x, y) = \mathbf{0}$, it follows from (9) that $D_{\mathbf{u}} f(x, y) = 0$ in all directions at the point (x, y). This typically occurs where the surface $z = f(x, y)$ has a "relative maximum," a "relative minimum," or a saddle point.

15.6.3   THEOREM.   *Let f be a function of two variables that is differentiable at (x, y).*

(a) *If $\nabla f(x, y) = \mathbf{0}$, then all directional derivatives of f at (x, y) are zero.*

(b) *If $\nabla f(x, y) \neq \mathbf{0}$, then among all possible directional derivatives of f at (x, y), the derivative in the direction of $\nabla f(x, y)$ has the largest value. The value of that directional derivative is $\|\nabla f(x, y)\|$.*

(c) *If $\nabla f(x, y) \neq \mathbf{0}$, then among all possible directional derivatives of f at (x, y), the derivative in the direction opposite to that of $\nabla f(x, y)$ has the smallest value. The value of that directional derivative is $-\|\nabla f(x, y)\|$.*

Example 4

Let $f(x, y) = x^2 e^y$. Find the maximum value of a directional derivative at $(-2, 0)$, and find the unit vector in the direction in which the maximum value occurs.

Solution. Since

$$\nabla f(x, y) = f_x(x, y)\mathbf{i} + f_y(x, y)\mathbf{j} = 2xe^y\mathbf{i} + x^2e^y\mathbf{j}$$

the gradient of f at $(-2, 0)$ is

$$\nabla f(-2, 0) = -4\mathbf{i} + 4\mathbf{j}$$

By Theorem 15.6.3, the maximum value of the directional derivative is

$$\|\nabla f(-2, 0)\| = \sqrt{(-4)^2 + 4^2} = \sqrt{32} = 4\sqrt{2}$$

This maximum occurs in the direction of $\nabla f(-2, 0)$. The unit vector in this direction is

$$\mathbf{u} = \frac{\nabla f(-2, 0)}{\|\nabla f(-2, 0)\|} = \frac{1}{4\sqrt{2}}(-4\mathbf{i} + 4\mathbf{j}) = -\frac{1}{\sqrt{2}}\mathbf{i} + \frac{1}{\sqrt{2}}\mathbf{j} \qquad ◄$$

GRADIENTS ARE NORMAL TO LEVEL CURVES

We have seen that the gradient points in the direction in which a surface $z = f(x, y)$ has its maximum slope. We will now consider how the direction of maximum slope can be determined from a contour map of the function. To do this we will need to investigate the geometric relationship between the gradient and the level curves of a function.

Suppose that $\nabla f(x, y)$ is the gradient at a point on the level curve $f(x, y) = c$, and assume that this level curve can be smoothly parametrized as

$$x = x(s), \quad y = y(s) \qquad (10)$$

where s is an arc length parameter.[*] Our objective is to show that at each point on the level curve the gradient is orthogonal to the unit tangent; that is, the gradient is *normal* to the level curve. For this purpose, recall from Formula (6) of Section 14.4 that the unit tangent vector to (10) at s is

$$\mathbf{T}(s) = \left(\frac{dx}{ds}\right)\mathbf{i} + \left(\frac{dy}{ds}\right)\mathbf{j}$$

If we now differentiate both sides of the equation $f(x, y) = c$ with respect to s using the chain rule, we obtain

$$\frac{\partial f}{\partial x}\frac{dx}{ds} + \frac{\partial f}{\partial y}\frac{dy}{ds} = 0$$

which we can rewrite as

$$\left(\frac{\partial f}{\partial x}\mathbf{i} + \frac{\partial f}{\partial y}\mathbf{j}\right) \cdot \left(\frac{dx}{ds}\mathbf{i} + \frac{dy}{ds}\mathbf{j}\right) = 0$$

or alternatively as

$$\nabla f(x, y) \cdot \mathbf{T}(s) = 0$$

Thus, the gradient is orthogonal to the unit tangent vector, which is what we wanted to show.

15.6.4   THEOREM.   *If f is differentiable at (x_0, y_0) and if $\nabla f(x_0, y_0) \neq \mathbf{0}$, then $\nabla f(x_0, y_0)$ is normal to the level curve of f through (x_0, y_0).*

REMARK.   If (x_0, y_0) is a point on the level curve $f(x, y) = c$, then the slope of the surface $z = f(x, y)$ at that point in the direction of $\mathbf{u}$ is

$$D_\mathbf{u}f(x_0, y_0) = \nabla f(x_0, y_0) \cdot \mathbf{u}$$

---

[*] It is proved in advanced courses that if the gradient of a differentiable function f is nonzero at a point, then there is a unique level curve through the point, and this level curve can be smoothly parametrized in terms of arc length.

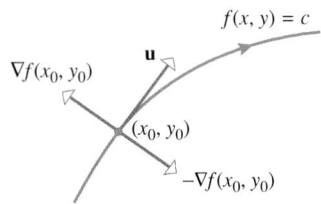

$f(x, y) = c$

$\nabla f(x_0, y_0)$

(x_0, y_0)

$-\nabla f(x_0, y_0)$

Figure 15.6.6

If $\mathbf{u}$ is tangent to the level curve at (x_0, y_0), then $f(x, y)$ is neither increasing nor decreasing in that direction, so $D_\mathbf{u} f(x_0, y_0) = 0$. Thus, $\nabla f(x_0, y_0)$, $-\nabla f(x_0, y_0)$, and the tangent vector $\mathbf{u}$ mark the directions of maximum slope, minimum slope, and zero slope at a point (x_0, y_0) on a level curve (Figure 15.6.6). Good skiers use these facts intuitively to control their speed by zigzagging down ski slopes—they ski across the slope with their skis tangential to a level curve to stop their downhill motion, and they point their skis down the slope and normal to the level curve to obtain the most rapid descent.

Example 5

Sketch the level curve for the function $f(x, y) = x^2 + y^2$ through the point $(3, 4)$, and draw the gradient vector at this point.

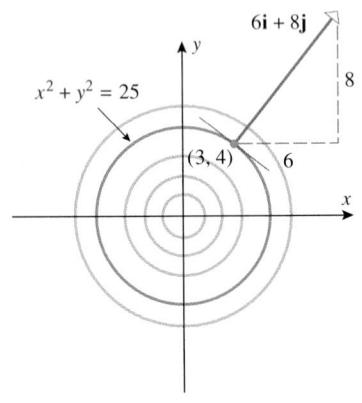

$6\mathbf{i} + 8\mathbf{j}$

$x^2 + y^2 = 25$

$(3, 4)$

Figure 15.6.7

Solution. Since $f(3, 4) = 25$, the level curve through the point $(3, 4)$ has the equation $f(x, y) = 25$, which is the circle

$$x^2 + y^2 = 25$$

Since

$$\nabla f(x, y) = f_x(x, y)\mathbf{i} + f_y(x, y)\mathbf{j} = 2x\mathbf{i} + 2y\mathbf{j}$$

the gradient vector at $(3, 4)$ is

$$\nabla f(3, 4) = 6\mathbf{i} + 8\mathbf{j}$$

(Figure 15.6.7). Note that the gradient vector is perpendicular to the circle at $(3, 4)$, as guaranteed by Theorem 15.6.4.   ◀

AN APPLICATION OF GRADIENTS

There are numerous applications in which the motion of an object must be controlled so that it moves toward a heat source. For example, in medical applications the operation of certain diagnostic equipment is designed to locate heat sources generated by tumors or infections, and in military applications the trajectories of heat-seeking missiles are controlled to seek and destroy enemy aircraft. The following example illustrates how gradients are used to solve such problems.

Example 6

A heat-seeking particle is located at the point $(2, 3)$ on a flat metal plate whose temperature at a point (x, y) is

$$T(x, y) = 10 - 8x^2 - 2y^2$$

Find an equation for the trajectory of the particle if it moves continuously in the direction of maximum temperature increase.

Solution. Assume that the trajectory is represented parametrically by the equations

$$x = x(t), \quad y = y(t)$$

where the particle is at the point $(2, 3)$ at time $t = 0$. Because the particle moves in the direction of maximum temperature increase, its direction of motion at time t is in the direction of the gradient of $T(x, y)$, and hence its velocity vector $\mathbf{v}(t)$ at time t points in the direction of the gradient. Thus, there is a scalar k that depends on t such that

$$\mathbf{v}(t) = k\nabla T(x, y)$$

from which we obtain

$$\frac{dx}{dt}\mathbf{i} + \frac{dy}{dt}\mathbf{j} = k(-16x\mathbf{i} - 4y\mathbf{j})$$

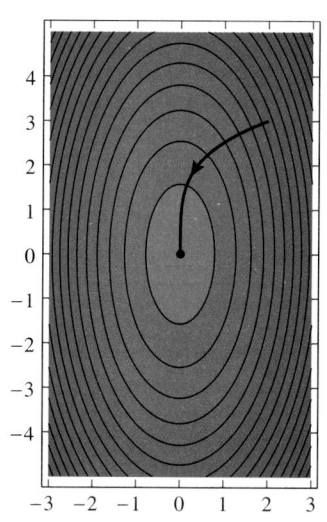

Figure 15.6.8

Equating components yields

$$\frac{dx}{dt} = -16kx, \quad \frac{dy}{dt} = -4ky$$

and dividing to eliminate k yields

$$\frac{dy}{dx} = \frac{-4ky}{-16kx} = \frac{y}{4x}$$

Thus, we can obtain the trajectory by solving the initial-value problem

$$\frac{dy}{dx} - \frac{y}{4x} = 0, \quad y(2) = 3$$

The differential equation is a separable first-order linear equation and hence can be solved by separating the variables or by the method of integrating factors discussed in Section 10.1. We leave it for you to show that the solution of the initial-value problem is

$$y = \frac{3}{\sqrt[4]{2}} x^{1/4}$$

The graph of the trajectory and a contour plot of the temperature function are shown in Figure 15.6.8. ◀

EXERCISE SET 15.6 ⌇ Graphing Calculator   🇨 CAS

1. The accompanying figure shows some level curves of an unspecified function $f(x, y)$. Which of the three vectors shown in the figure is most likely to be ∇f? Explain.

2. The accompanying figure shows some level curves of an unspecified function $f(x, y)$. Of the gradients at P and Q, which probably has the greater length? Explain.

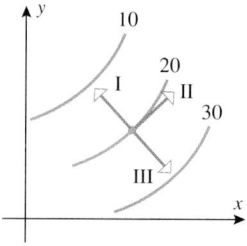

Figure Ex-1

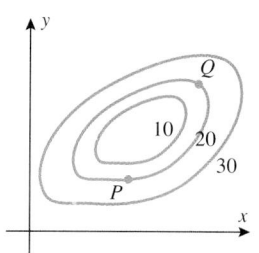

Figure Ex-2

In Exercises 3–6, find ∇z.

3. $z = 4x - 8y$

4. $z = e^{-3y} \cos 4x$

5. $z = \ln \sqrt{x^2 + y^2}$

6. $z = e^{-5x} \sec x^2 y$

In Exercises 7–10, find the gradient of f at the indicated point.

7. $f(x, y) = (x^2 + xy)^3$; $(-1, -1)$

8. $f(x, y) = (x^2 + y^2)^{-1/2}$; $(3, 4)$

9. $f(x, y) = y \ln(x + y)$; $(-3, 4)$

10. $f(x, y) = y^2 \tan^3 x$; $(\pi/4, -3)$

In Exercises 11–14, find $D_{\mathbf{u}} f$ at P.

11. $f(x, y) = (1 + xy)^{3/2}$; $P(3, 1)$; $\mathbf{u} = \frac{1}{\sqrt{2}}\mathbf{i} + \frac{1}{\sqrt{2}}\mathbf{j}$

12. $f(x, y) = e^{2xy}$; $P(4, 0)$; $\mathbf{u} = -\frac{3}{5}\mathbf{i} + \frac{4}{5}\mathbf{j}$

13. $f(x, y) = \ln(1 + x^2 + y)$; $P(0, 0)$; $\mathbf{u} = -\frac{1}{\sqrt{10}}\mathbf{i} - \frac{3}{\sqrt{10}}\mathbf{j}$

14. $f(x, y) = \frac{cx + dy}{x - y}$; $P(3, 4)$; $\mathbf{u} = \frac{4}{5}\mathbf{i} + \frac{3}{5}\mathbf{j}$

In Exercises 15–20, find the directional derivative of f at P in the direction of $\mathbf{a}$.

15. $f(x, y) = 4x^3 y^2$; $P(2, 1)$; $\mathbf{a} = 4\mathbf{i} - 3\mathbf{j}$

16. $f(x, y) = x^2 - 3xy + 4y^3$; $P(-2, 0)$; $\mathbf{a} = \mathbf{i} + 2\mathbf{j}$

17. $f(x, y) = y^2 \ln x$; $P(1, 4)$; $\mathbf{a} = -3\mathbf{i} + 3\mathbf{j}$

18. $f(x, y) = e^x \cos y$; $P(0, \pi/4)$; $\mathbf{a} = 5\mathbf{i} - 2\mathbf{j}$

19. $f(x, y) = \tan^{-1}(y/x)$; $P(-2, 2)$; $\mathbf{a} = -\mathbf{i} - \mathbf{j}$

20. $f(x, y) = xe^y - ye^x$; $P(0, 0)$; $\mathbf{a} = 5\mathbf{i} - 2\mathbf{j}$

In Exercises 21–24, find the directional derivative of f at P in the direction of a vector making the counterclockwise angle θ with the positive x-axis.

21. $f(x, y) = \sqrt{xy}$; $P(1, 4)$; $\theta = \pi/3$

22. $f(x, y) = \frac{x - y}{x + y}$; $P(-1, -2)$; $\theta = \pi/2$

23. $f(x, y) = \tan(2x + y)$; $P(\pi/6, \pi/3)$; $\theta = 7\pi/4$

24. $f(x, y) = \sinh x \cosh y$; $P(0, 0)$; $\theta = \pi$

In Exercises 25–28, sketch the level curve of $f(x, y)$ that passes through P and draw the gradient vector at P.

---

25. $f(x, y) = 4x - 2y + 3$; $P(1, 2)$

26. $f(x, y) = y/x^2$; $P(-2, 2)$

27. $f(x, y) = x^2 + 4y^2$; $P(-2, 0)$

28. $f(x, y) = x^2 - y^2$; $P(2, -1)$

In Exercises 29–32, find a unit vector in the direction in which f increases most rapidly at P; and find the rate of change of f at P in that direction.

---

29. $f(x, y) = 4x^3y^2$; $P(-1, 1)$

30. $f(x, y) = 3x - \ln y$; $P(2, 4)$

31. $f(x, y) = \sqrt{x^2 + y^2}$; $P(4, -3)$

32. $f(x, y) = \dfrac{x}{x + y}$; $P(0, 2)$

In Exercises 33–36, find a unit vector in the direction in which f decreases most rapidly at P; and find the rate of change of f at P in that direction.

---

33. $f(x, y) = 20 - x^2 - y^2$; $P(-1, -3)$

34. $f(x, y) = e^{xy}$; $P(2, 3)$

35. $f(x, y) = \cos(3x - y)$; $P(\pi/6, \pi/4)$

36. $f(x, y) = \sqrt{\dfrac{x - y}{x + y}}$; $P(3, 1)$

37. Find the directional derivative of $f(x, y) = \dfrac{x}{x + y}$ at $P(1, 0)$ in the direction of $Q(-1, -1)$.

38. Find the directional derivative of $f(x, y) = e^{-x} \sec y$ at $P(0, \pi/4)$ in the direction of the origin.

39. Find the directional derivative of $f(x, y) = \sqrt{xy}e^y$ at $P(1, 1)$ in the direction of the negative y-axis.

40. Let $f(x, y) = \dfrac{y}{x + y}$. Find a unit vector **u** for which $D_{\mathbf{u}} f(2, 3) = 0$.

41. Find a unit vector **u** that is normal at $P(1, -2)$ to the level curve of $f(x, y) = 4x^2y$ through P.

42. Find a unit vector **u** that is normal at $P(2, -3)$ to the level curve of $f(x, y) = 3x^2y - xy$ through P.

43. Suppose that $D_{\mathbf{u}} f(1, 2) = -5$ and $D_{\mathbf{v}} f(1, 2) = 10$, where $\mathbf{u} = \frac{3}{5}\mathbf{i} - \frac{4}{5}\mathbf{j}$ and $\mathbf{v} = \frac{4}{5}\mathbf{i} + \frac{3}{5}\mathbf{j}$. Find
(a) $f_x(1, 2)$ (b) $f_y(1, 2)$
(c) the directional derivative of f at $(1, 2)$ in the direction of the origin.

44. Given that $f_x(-5, 1) = -3$ and $f_y(-5, 1) = 2$, find the directional derivative of f at $P(-5, 1)$ in the direction of the vector from P to $Q(-4, 3)$.

45. Given that $\nabla f(4, -5) = 2\mathbf{i} - \mathbf{j}$, find the directional derivative of the function f at the point $(4, -5)$ in the direction of $\mathbf{a} = 5\mathbf{i} + 2\mathbf{j}$.

46. Given that $\nabla f(x_0, y_0) = \mathbf{i} - 2\mathbf{j}$ and $D_{\mathbf{u}} f(x_0, y_0) = -2$, find **u** (two answers).

47. The accompanying figure shows some level curves of an unspecified function $f(x, y)$.
(a) Use the available information to approximate the length of the vector $\nabla f(1, 2)$, and sketch the approximation. Explain how you approximated the length and determined the direction of the vector.
(b) Sketch an approximation of the vector $-\nabla f(4, 4)$.

48. (a) The accompanying figure shows a topographic map of a hill and a point P at the bottom of the hill. Suppose that you want to climb from the point P toward the top of the hill in such a way that you are always ascending in the direction of steepest slope. Sketch the projection of your path on the contour map. This is called the **path of steepest ascent**. Explain how you determined the path.
(b) Suppose that when you are at the top you want to climb down the hill in such a way that you are always descending in the direction of steepest slope. Sketch the projection of your path on the contour map. This is called the **path of steepest descent**. Explain how you determined the path.

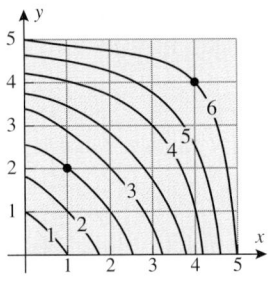

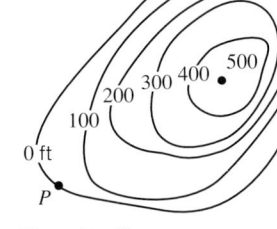

Figure Ex-47   Figure Ex-48

49. Let $z = 3x^2 - y^2$. Find all points at which $\|\nabla z\| = 6$.

50. Given that $z = 3x + y^2$, find $\nabla \|\nabla z\|$ at the point $(5, 2)$.

51. A particle moves along a path C given by the equations $x = t$ and $y = -t^2$. If $z = x^2 + y^2$, find dz/ds along C at the instant when the particle is at the point $(2, -4)$.

52. The temperature at a point (x, y) on a metal plate in the xy-plane is $T(x, y) = \dfrac{xy}{1 + x^2 + y^2}$ degrees Celsius.
(a) Find the rate of change of temperature at $(1, 1)$ in the direction of $\mathbf{a} = 2\mathbf{i} - \mathbf{j}$.
(b) An ant at $(1, 1)$ wants to walk in the direction in which the temperature drops most rapidly. Find a unit vector in that direction.

53. If the electric potential at a point (x, y) in the xy-plane is $V(x, y)$, then the ***electric intensity vector*** at the point (x, y) is $\mathbf{E} = -\nabla V(x, y)$. Suppose that $V(x, y) = e^{-2x} \cos 2y$.
  (a) Find the electric intensity vector at $(\pi/4, 0)$.
  (b) Show that at each point in the plane, the electric potential decreases most rapidly in the direction of the vector $\mathbf{E}$.

54. On a certain mountain, the elevation z in miles above a point (x, y) in an xy-plane at sea level is $z = 2000 - 2x^2 - 4y^2$ ft. The positive x-axis points east, and the positive y-axis north. A climber is at the point $(-20, 5, 1100)$.
  (a) If the climber uses a compass reading to walk due west, will she begin to ascend or descend?
  (b) If the climber uses a compass reading to walk northeast, will she ascend or descend? At what rate?
  (c) In what compass direction should the climber begin walking to travel a level path (two answers)?

55. Let $r = \sqrt{x^2 + y^2}$.
  (a) Show that $\nabla r = \dfrac{\mathbf{r}}{r}$, where $\mathbf{r} = x\mathbf{i} + y\mathbf{j}$.
  (b) Show that $\nabla f(r) = f'(r) \nabla r = \dfrac{f'(r)}{r} \mathbf{r}$.

56. Use the formula in part (b) of Exercise 55 to find
  (a) $\nabla f(r)$ if $f(r) = re^{-3r}$
  (b) $f(r)$ if $\nabla f(r) = 3r^2 \mathbf{r}$ and $f(2) = 1$.

57. Let $\mathbf{u}_r$ be a unit vector whose counterclockwise angle from the positive x-axis is θ, and let $\mathbf{u}_\theta$ be a unit vector $90°$ counterclockwise from $\mathbf{u}_r$. Show that if $z = f(x, y)$, $x = r \cos \theta$, and $y = r \sin \theta$, then

$$\nabla z = \frac{\partial z}{\partial r} \mathbf{u}_r + \frac{1}{r} \frac{\partial z}{\partial \theta} \mathbf{u}_\theta$$

[*Hint:* Use part (c) of Exercise 39, Section 15.4.]

58. Prove: If f and g are differentiable, then
  (a) $\nabla(f + g) = \nabla f + \nabla g$
  (b) $\nabla(cf) = c\nabla f$ (c constant)
  (c) $\nabla(fg) = f\nabla g + g\nabla f$
  (d) $\nabla \left(\dfrac{f}{g} \right) = \dfrac{g\nabla f - f\nabla g}{g^2}$
  (e) $\nabla(f^n) = nf^{n-1}\nabla f$.

In Exercises 59 and 60, a heat-seeking particle is located at the point P on a flat metal plate whose temperature at a point (x, y) is $T(x, y)$. Find parametric equations for the trajectory of the particle if it moves continuously in the direction of maximum temperature increase.

59. $T(x, y) = 5 - 4x^2 - y^2$; $P(1, 4)$

60. $T(x, y) = 100 - x^2 - 2y^2$; $P(5, 3)$

61. Use a graphing utility to generate the trajectory of the particle together with some representative level curves of the temperature function in Exercise 59.

62. Use a graphing utility to generate the trajectory of the particle together with some representative level curves of the temperature function in Exercise 60.

63. (a) Use a CAS to graph $f(x, y) = (x^2 + 3y^2)e^{-(x^2+y^2)}$.
  (b) At how many points do you think it is true that $D_{\mathbf{u}} f(x, y) = 0$ for all unit vectors $\mathbf{u}$?
  (c) Use a CAS to find ∇f.
  (d) Use a CAS to solve the equation $\nabla f(x, y) = 0$ for x and y.
  (e) Use the result in part (d) together with Theorem 15.6.3(a) to check your conjecture in part (b).

64. Prove: If $x = x(t)$ and $y = y(t)$ are differentiable at t, and if $z = f(x, y)$ is differentiable at the point $(x(t), y(t))$, then

$$\frac{dz}{dt} = \nabla z \cdot \mathbf{r}'(t)$$

where $\mathbf{r}(t) = x(t)\mathbf{i} + y(t)\mathbf{j}$.

65. Prove: If f, f_x, and f_y are continuous on a circular region, and if $\nabla f(x, y) = \mathbf{0}$ throughout the region, then $f(x, y)$ is constant on the region. [*Hint:* See Exercise 52, Section 15.4.]

66. Prove: If the function f is differentiable at the point (x, y) and if $D_{\mathbf{u}} f(x, y) = 0$ in two nonparallel directions, then $D_{\mathbf{u}} f(x, y) = 0$ in all directions.

15.7 DIFFERENTIABILITY, DIRECTIONAL DERIVATIVES, AND GRADIENTS FOR FUNCTIONS OF THREE OR MORE VARIABLES

In this section we will extend most of the results obtained in the last two sections to functions of three or more variables.

DIFFERENTIABILITY

The definition of differentiability for functions of three variables and the basic theorems about differentiability are direct generalizations of the corresponding results for functions of two variables. (See Definition 15.4.2, Theorem 15.4.3, Theorem 15.4.4, and Theorem 15.4.7.)

15.7.1 DEFINITION. A function f of three variables is said to be ***differentiable*** at the point (x_0, y_0, z_0) if the partial derivatives $f_x(x_0, y_0, z_0)$, $f_y(x_0, y_0, z_0)$, and $f_z(x_0, y_0, z_0)$ exist and

$$\Delta f = f(x_0 + \Delta x, y_0 + \Delta y, z_0 + \Delta z) - f(x_0, y_0, z_0)$$

can be written in the form

$$\Delta f = f_x(x_0, y_0, z_0)\Delta x + f_y(x_0, y_0, z_0)\Delta y + f_z(x_0, y_0, z_0)\Delta z + \epsilon_1 \Delta x + \epsilon_2 \Delta y + \epsilon_3 \Delta z$$

where $\epsilon_1, \epsilon_2,$ and ϵ_3 are functions of $\Delta x, \Delta y,$ and Δz such that $\epsilon_1 \to 0, \epsilon_2 \to 0,$ and $\epsilon_3 \to 0$ as $(\Delta x, \Delta y, \Delta z) \to (0, 0, 0)$, and $\epsilon_1 = \epsilon_2 = \epsilon_3 = 0$ if $(\Delta x, \Delta y, \Delta z) = (0, 0, 0)$.

15.7.2 THEOREM. *If f has first-order partial derivatives at each point of some spherical region centered at (x_0, y_0, z_0), and if these partial derivatives are continuous at (x_0, y_0, z_0), then f is differentiable at (x_0, y_0, z_0).*

15.7.3 THEOREM. *If f is differentiable at the point (x_0, y_0, z_0), then f is continuous at that point.*

15.7.4 THEOREM (*Chain Rule*). *If $x = x(t)$, $y = y(t)$, and $z = z(t)$ are differentiable at the point t and $w = f(x, y, z)$ is differentiable at the point $(x(t), y(t), z(t))$, then $w = f(x(t), y(t), z(t))$ is differentiable at t, and*

$$\frac{dw}{dt} = \frac{\partial w}{\partial x}\frac{dx}{dt} + \frac{\partial w}{\partial y}\frac{dy}{dt} + \frac{\partial w}{\partial z}\frac{dz}{dt} \tag{1}$$

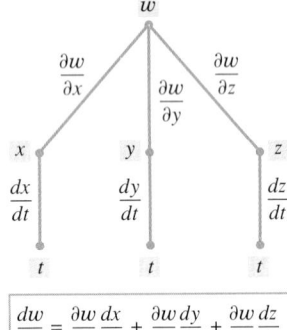

Figure 15.7.1

A tree diagram for this version of the chain rule is shown in Figure 15.7.1.

REMARK. The most significant difference between working with functions of two variables and functions of three variables is geometric. For a function of two variables the equation $z = f(x, y)$ can be graphed as a surface in three-dimensional space. However, for a function of three variables, there is no direct way to graph $w = f(x, y, z)$, since "four dimensions" would be required (one dimension for each variable). This is not devastating, however; it simply means that we must rely more heavily on the analytic formulas than on the geometry.

DIRECTIONAL DERIVATIVES AND GRADIENTS

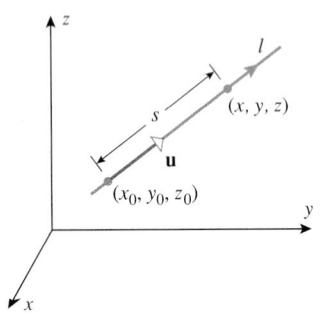

Figure 15.7.2

Recall that the directional derivative $D_{\mathbf{u}} f(x_0, y_0)$ can be interpreted either as the slope of the surface $z = f(x, y)$ at the point (x_0, y_0) in the direction of $\mathbf{u}$ or as the rate of change of z with respect to s, where s is the arc length measured from (x_0, y_0) in the direction of $\mathbf{u}$ along a line. For a function of three variables, we will use the latter interpretation as the basis for the definition of the directional derivative of $w = f(x, y, z)$ at the point (x_0, y_0, z_0) in the direction of a unit vector $\mathbf{u}$ in 3-space; that is, we will interpret $D_{\mathbf{u}}(x_0, y_0, z_0)$ to be the rate of change of w with respect to the arc length s measured from (x_0, y_0, z_0) in the direction of $\mathbf{u}$ along a line (Figure 15.7.2).

We leave it as an exercise for you to obtain the following formal definition by imitating the motivational computations that led to Formula (3) in Section 15.6.

15.7.5 DEFINITION. If f is differentiable at (x, y, z), and if $\mathbf{u} = u_1\mathbf{i} + u_2\mathbf{j} + u_3\mathbf{k}$ is a unit vector, then the ***directional derivative*** of f at (x, y, z) in the direction of $\mathbf{u}$ is defined by

$$D_{\mathbf{u}} f(x, y, z) = f_x(x, y, z)u_1 + f_y(x, y, z)u_2 + f_z(x, y, z)u_3 \tag{2}$$

Formula (2) can be expressed in vector form as

$$D_{\mathbf{u}} f(x, y, z) = (f_x(x, y, z)\mathbf{i} + f_y(x, y, z)\mathbf{j} + f_z(x, y, z)\mathbf{k}) \cdot (u_1\mathbf{i} + u_2\mathbf{j} + u_3\mathbf{k})$$

Thus, we define the gradient of f as

$$\nabla f(x, y, z) = f_x(x, y, z)\mathbf{i} + f_y(x, y, z)\mathbf{j} + f_z(x, y, z)\mathbf{k} \tag{3}$$

which enables us to express (2) as

$$D_{\mathbf{u}} f(x, y, z) = \nabla f(x, y, z) \cdot \mathbf{u} \tag{4}$$

FOR THE READER.   What is the generalization of Formula (4) of Section 15.6 to functions of three variables?

The following theorem is the three-variable analog of Theorem 15.6.3.

15.7.6 THEOREM. *Let f be a function of three variables that is differentiable at (x, y, z).*

(a) *If $\nabla f(x, y, z) = \mathbf{0}$, then all directional derivatives of f at (x, y, z) are zero.*

(b) *If $\nabla f(x, y, z) \neq \mathbf{0}$, then among all possible directional derivatives of f at (x, y, z), the derivative in the direction of $\nabla f(x, y, z)$ has the largest value. The value of that directional derivative is $\|\nabla f(x, y, z)\|$.*

(c) *If $\nabla f(x, y, z) \neq \mathbf{0}$, then among all possible directional derivatives of f at (x, y, z), the derivative in the direction opposite to that of $\nabla f(x, y, z)$ has the smallest value. The value of that directional derivative is $-\|\nabla f(x, y, z)\|$.*

Example 1

Find the directional derivative of $f(x, y, z) = x^2 y - yz^3 + z$ at the point $P(1, -2, 0)$ in the direction of the vector $\mathbf{a} = 2\mathbf{i} + \mathbf{j} - 2\mathbf{k}$, and find the maximum rate of increase of f at P.

Solution. Since

$$f_x(x, y, z) = 2xy, \quad f_y(x, y, z) = x^2 - z^3, \quad f_z(x, y, z) = -3yz^2 + 1$$

it follows that

$$\nabla f(x, y, z) = 2xy\mathbf{i} + (x^2 - z^3)\mathbf{j} + (-3yz^2 + 1)\mathbf{k}$$

$$\nabla f(1, -2, 0) = -4\mathbf{i} + \mathbf{j} + \mathbf{k}$$

A unit vector in the direction of $\mathbf{a}$ is

$$\mathbf{u} = \frac{\mathbf{a}}{\|\mathbf{a}\|} = \frac{1}{\sqrt{9}}(2\mathbf{i} + \mathbf{j} - 2\mathbf{k}) = \frac{2}{3}\mathbf{i} + \frac{1}{3}\mathbf{j} - \frac{2}{3}\mathbf{k}$$

Therefore,

$$D_{\mathbf{u}} f(1, -2, 0) = \nabla f(1, -2, 0) \cdot \mathbf{u} = (-4)\left(\tfrac{2}{3}\right) + (1)\left(\tfrac{1}{3}\right) + (1)\left(-\tfrac{2}{3}\right) = -3$$

The maximum rate of increase of f at P is

$$\|\nabla f(1, -2, 0)\| = \sqrt{(-4)^2 + (1)^2 + (1)^2} = 3\sqrt{2} \qquad \blacktriangleleft$$

GRADIENTS ARE NORMAL TO LEVEL SURFACES

We saw in Section 15.6 that for a function of two variables the gradient at a point is normal to the level curve through that point. For functions of three variables the gradient at a point is normal to the level *surface* through that point. Although we will not give a formal proof of this result, it is not hard to motivate. For this purpose, suppose that $f(x, y, z) = c$ is a level surface through the point (x_0, y_0, z_0) and that $\mathbf{u}$ is any unit vector in the tangent plane

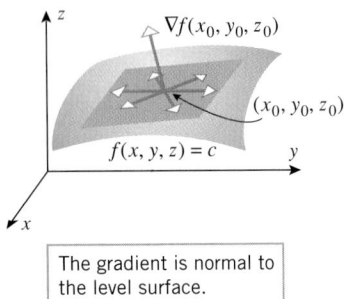

The gradient is normal to the level surface.

Figure 15.7.3

to the level surface at this point (Figure 15.7.3). Because the function f is constant on the level surface, it is reasonable to expect that the directional derivative $D_{\mathbf{u}} f(x_0, y_0, z_0)$ will be zero. Thus, for all unit vectors in the tangent plane at (x_0, y_0, z_0) we have

$$D_{\mathbf{u}} f(x_0, y_0, z_0) = \nabla f(x_0, y_0, z_0) \cdot \mathbf{u} = 0$$

Geometrically, this means that $\nabla f(x_0, y_0, z_0)$ is orthogonal to every unit vector $\mathbf{u}$ in the tangent plane. This implies that $\nabla f(x_0, y_0, z_0)$ is normal to the tangent plane itself and hence is normal to the level surface.

15.7.7 THEOREM. *If f is differentiable at (x_0, y_0, z_0) and if $\nabla f(x_0, y_0, z_0) \neq \mathbf{0}$, then $\nabla f(x_0, y_0, z_0)$ is normal to the level surface of $f(x, y, z)$ through (x_0, y_0, z_0).*

USING GRADIENTS TO FIND TANGENT PLANES

In Section 15.5 we showed how to find the tangent plane to a surface of the form $z = f(x, y)$ (see Theorem 15.5.1). We will now consider the problem of finding tangent planes to surfaces that are represented implicitly by equations of the form $F(x, y, z) = c$. Such an equation represents a level surface of the function $F(x, y, z)$, so the vector $\nabla F(x_0, y_0, z_0)$ will be normal to the surface, and hence normal to the tangent plane, at any point where $\nabla F(x_0, y_0, z_0) \neq \mathbf{0}$. Thus, where this condition is satisfied it follows from Formula (3) of Section 13.6 that the tangent plane at (x_0, y_0, z_0) can be expressed in point-normal form as

$$F_x(x_0, y_0, z_0)(x - x_0) + F_y(x_0, y_0, z_0)(y - y_0) + F_z(x_0, y_0, z_0)(z - z_0) = 0 \qquad (5)$$

Alternatively, we can express this equation in vector form by letting $\mathbf{r} = x\mathbf{i} + y\mathbf{j} + z\mathbf{k}$ and $\mathbf{r}_0 = x_0\mathbf{i} + y_0\mathbf{j} + z_0\mathbf{k}$, in which case it becomes

$$\nabla F(x_0, y_0, z_0) \cdot (\mathbf{r} - \mathbf{r}_0) = 0 \qquad (6)$$

Example 2

Find an equation of the tangent plane to the ellipsoid $x^2 + 4y^2 + z^2 = 18$ at the point $(1, 2, 1)$, and determine the acute angle that this plane makes with the xy-plane.

Solution. The ellipsoid is a level surface of the function $F(x, y, z) = x^2 + 4y^2 + z^2$, so we begin by finding the gradient of this function at the point $(1, 2, 1)$. The computations are

$$\nabla F(x, y, z) = \frac{\partial F}{\partial x}\mathbf{i} + \frac{\partial F}{\partial y}\mathbf{j} + \frac{\partial F}{\partial z}\mathbf{k} = 2x\mathbf{i} + 8y\mathbf{j} + 2z\mathbf{k}$$

$$\nabla F(1, 2, 1) = 2\mathbf{i} + 16\mathbf{j} + 2\mathbf{k}$$

Thus,

$$F_x(1, 2, 1) = 2, \quad F_y(1, 2, 1) = 16, \quad F_z(1, 2, 1) = 2$$

and hence from (5) the equation of the tangent plane is

$$2(x - 1) + 16(y - 2) + 2(z - 1) = 0 \quad \text{or} \quad x + 8y + z = 18$$

To find the acute angle θ between the tangent plane and the xy-plane, we will apply Formula (9) of Section 13.6 with $\mathbf{n}_1 = \nabla F(1, 2, 1) = 2\mathbf{i} + 16\mathbf{j} + 2\mathbf{k}$ and $\mathbf{n}_2 = \mathbf{k}$. This yields

$$\cos \theta = \frac{|\nabla F(1, 2, 1) \cdot \mathbf{k}|}{\|\nabla F(1, 2, 1)\|\|\mathbf{k}\|} = \frac{2}{(2\sqrt{66})(1)} = \frac{1}{\sqrt{66}}$$

Thus,

$$\theta = \cos^{-1}\left(\frac{1}{\sqrt{66}}\right) \approx 83°$$

(Figure 15.7.4). ◄

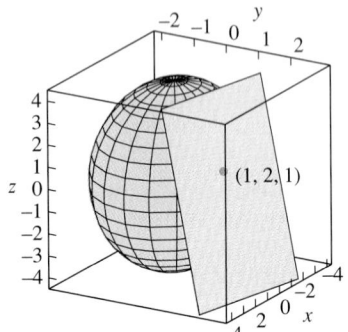

Figure 15.7.4

USING GRADIENTS TO FIND TANGENT LINES TO INTERSECTIONS OF SURFACES

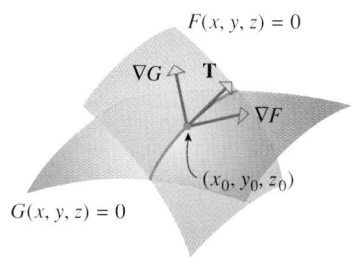

Figure 15.7.5

In general, the intersection of two surfaces $F(x, y, z) = 0$ and $G(x, y, z) = 0$ will be a curve in 3-space. If (x_0, y_0, z_0) is a point on this curve, then $\nabla F(x_0, y_0, z_0)$ will be normal to the surface $F(x, y, z) = 0$ at (x_0, y_0, z_0) and $\nabla G(x_0, y_0, z_0)$ will be normal to the surface $G(x, y, z) = 0$ at (x_0, y_0, z_0). Thus, if the curve of intersection can be smoothly parametrized, then its unit tangent vector **T** at (x_0, y_0, z_0) will be orthogonal to both $\nabla F(x_0, y_0, z_0)$ and $\nabla G(x_0, y_0, z_0)$ (Figure 15.7.5). Consequently, if

$$\nabla F(x_0, y_0, z_0) \times \nabla G(x_0, y_0, z_0) \neq \mathbf{0}$$

then this cross product will be parallel to **T** and hence will be tangent to the curve of intersection. This tangent vector can be used to determine the direction of the tangent line to the curve of intersection at the point (x_0, y_0, z_0).

Example 3

Find parametric equations of the tangent line to the curve of intersection of the paraboloid $z = x^2 + y^2$ and the ellipsoid $3x^2 + 2y^2 + z^2 = 9$ at the point $(1, 1, 2)$ (Figure 15.7.6).

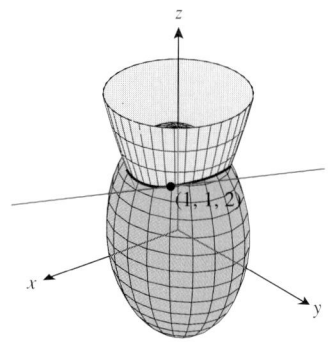

Figure 15.7.6

Solution.   We begin by rewriting the equations of the surfaces as

$$x^2 + y^2 - z = 0 \quad \text{and} \quad 3x^2 + 2y^2 + z^2 - 9 = 0$$

and we take

$$F(x, y, z) = x^2 + y^2 - z \quad \text{and} \quad G(x, y, z) = 3x^2 + 2y^2 + z^2 - 9$$

We will need the gradients of these functions at the point $(1, 1, 2)$. The computations are

$$\nabla F(x, y, z) = 2x\mathbf{i} + 2y\mathbf{j} - \mathbf{k}$$
$$\nabla G(x, y, z) = 6x\mathbf{i} + 4y\mathbf{j} + 2z\mathbf{k}$$
$$\nabla F(1, 1, 2) = 2\mathbf{i} + 2\mathbf{j} - \mathbf{k}$$
$$\nabla G(1, 1, 2) = 6\mathbf{i} + 4\mathbf{j} + 4\mathbf{k}$$

Thus, a tangent vector at $(1, 1, 2)$ to the curve of intersection is

$$\nabla F(1, 1, 2) \times \nabla G(1, 1, 2) = \begin{vmatrix} \mathbf{i} & \mathbf{j} & \mathbf{k} \\ 2 & 2 & -1 \\ 6 & 4 & 4 \end{vmatrix} = 12\mathbf{i} - 14\mathbf{j} - 4\mathbf{k}$$

Since any scalar multiple of this vector will do just as well, we can multiply by $\frac{1}{2}$ to reduce the size of the coefficients and use the vector of $6\mathbf{i} - 7\mathbf{j} - 2\mathbf{k}$ to determine the direction of the tangent line. This vector and the point $(1, 1, 2)$ yield the parametric equations

$$x = 1 + 6t, \quad y = 1 - 7t, \quad z = 2 - 2t \qquad \blacktriangleleft$$

TOTAL DIFFERENTIALS

The following formulas are natural extensions of Formulas (8) and (9) of Section 15.5 to functions $w = f(x, y, z)$ of three variables:

$$\Delta w = f(x + \Delta x, y + \Delta y, z + \Delta z) - f(x, y, z) \tag{7}$$

$$dw = f_x(x, y, z)\, dx + f_y(x, y, z)\, dy + f_z(x, y, z)\, dz \tag{8}$$

We call Δw (also written Δf) the **increment in w** or the **increment in f**, and we call dw (also written df) the **total differential of w** or the **total differential of f**.

Although increments and total differentials have geometric interpretations for functions of two variables (see Figure 15.5.3), there are no corresponding geometric interpretations for functions of three variables—these quantities must be viewed algebraically. For example, you should think of the increment Δw as the change in the value of $w = f(x, y, z)$ when x, y, and z are changed by the amounts Δx, Δy, and Δz, respectively.

If $w = f(x, y, z)$ is differentiable at the point (x, y, z), then it follows from Definition 15.7.1 that the increment in w can be written as

$$\Delta w = f_x(x, y, z)\Delta x + f_y(x, y, z)\Delta y + f_z(x, y, z)\Delta z + \epsilon_1\Delta x + \epsilon_2\Delta y + \epsilon_3\Delta z$$

where $\epsilon_1 \to 0$, $\epsilon_2 \to 0$, $\epsilon_3 \to 0$ as $(\Delta x, \Delta y, \Delta z) \to (0, 0, 0)$. If $dx = \Delta x$, $dy = \Delta y$, and $dz = \Delta z$, then it follows from (8) that this formula can be rewritten as

$$\Delta w = dw + \epsilon_1\Delta x + \epsilon_2\Delta y + \epsilon_3\Delta z$$

Thus, when $dx = \Delta x$, $dy = \Delta y$, and $dz = \Delta z$ are small, we can approximate Δw by

$$\Delta w \approx dw$$

This is the extension to functions of three variables of Formula (12) in Section 3.6 and Formula (14) of Section 15.5.

Example 4

The length, width, and height of a rectangular box are each measured with an error of at most 5%. Estimate the maximum percentage error that results if these quantities are used to calculate the diagonal of the box.

Solution. Let x, y, z, and D be the true length, width, height, and diagonal of the box, respectively; and let Δx, Δy, Δz, and ΔD be the errors in these quantities. We are given that

$$|\Delta x/x| \leq 0.05, \quad |\Delta y/y| \leq 0.05, \quad |\Delta z/z| \leq 0.05$$

We want to estimate $|\Delta D/D|$. Since the diagonal D is related to the length, width, and height by

$$D = \sqrt{x^2 + y^2 + z^2}$$

it follows that

$$dD = \frac{\partial D}{\partial x}\,dx + \frac{\partial D}{\partial y}\,dy + \frac{\partial D}{\partial z}\,dz$$

$$= \frac{x}{\sqrt{x^2 + y^2 + z^2}}\,dx + \frac{y}{\sqrt{x^2 + y^2 + z^2}}\,dy + \frac{z}{\sqrt{x^2 + y^2 + z^2}}\,dz$$

If we choose $\Delta x = dx$, $\Delta y = dy$, and $\Delta z = dz$, then we can use the approximation $\Delta D/D \approx dD/D$. But,

$$\frac{dD}{D} = \frac{x}{x^2 + y^2 + z^2}\,dx + \frac{y}{x^2 + y^2 + z^2}\,dy + \frac{z}{x^2 + y^2 + z^2}\,dz$$

or

$$\frac{dD}{D} = \frac{x^2}{x^2 + y^2 + z^2}\frac{dx}{x} + \frac{y^2}{x^2 + y^2 + z^2}\frac{dy}{y} + \frac{z^2}{x^2 + y^2 + z^2}\frac{dz}{z}$$

Thus,

$$\left|\frac{dD}{D}\right| = \left|\frac{x^2}{x^2 + y^2 + z^2}\frac{dx}{x} + \frac{y^2}{x^2 + y^2 + z^2}\frac{dy}{y} + \frac{z^2}{x^2 + y^2 + z^2}\frac{dz}{z}\right|$$

$$\leq \left|\frac{x^2}{x^2 + y^2 + z^2}\frac{dx}{x}\right| + \left|\frac{y^2}{x^2 + y^2 + z^2}\frac{dy}{y}\right| + \left|\frac{z^2}{x^2 + y^2 + z^2}\frac{dz}{z}\right|$$

$$\leq \frac{x^2}{x^2 + y^2 + z^2}(0.05) + \frac{y^2}{x^2 + y^2 + z^2}(0.05) + \frac{z^2}{x^2 + y^2 + z^2}(0.05)$$

$$= 0.05$$

Therefore, we estimate the maximum percentage error in D to be 5%. ◀

EXTENSIONS TO FUNCTIONS OF *n* VARIABLES

Most of the definitions and theorems we have stated for functions of two and three variables can be extended to functions of four or more variables. Recall that if

$$w = f(v_1, v_2, \ldots, v_n)$$

is a function of n variables, then there are n partial derivatives

$$\frac{\partial w}{\partial v_1}, \frac{\partial w}{\partial v_2}, \ldots, \frac{\partial w}{\partial v_n}$$

each of which is calculated by holding $n-1$ of the variables fixed and differentiating with respect to the remaining variable.

TOTAL DIFFERENTIALS

If $w = f(v_1, v_2, \ldots, v_n)$, then we define the ***increment*** Δw to be

$$\Delta w = f(v_1 + \Delta v_1, v_2 + \Delta v_2, \ldots, v_n + \Delta v_n) - f(v_1, v_2, \ldots, v_n) \tag{9}$$

and we define the ***total differential*** to be

$$dw = \frac{\partial w}{\partial v_1}dv_1 + \frac{\partial w}{\partial v_2}dv_2 + \cdots + \frac{\partial w}{\partial v_n}dv_n \tag{10}$$

where $\Delta v_1, \Delta v_2, \ldots, \Delta v_n$ and $dv_1, dv_2, \ldots, dv_n$ are variables representing changes in the values of $v_1, v_2, \ldots, v_n$. These are the natural extensions of Formulas (7) and (8).

CHAIN RULES

If $v_1, v_2, \ldots, v_n$ are functions of a single variable t, then $w = f(v_1, v_2, \ldots, v_n)$ is a function of t, and the chain rule for dw/dt is

$$\frac{dw}{dt} = \frac{\partial w}{\partial v_1}\frac{dv_1}{dt} + \frac{\partial w}{\partial v_2}\frac{dv_2}{dt} + \cdots + \frac{\partial w}{\partial v_n}\frac{dv_n}{dt} \tag{11}$$

This is a natural extension of Formula (9) in Theorem 15.4.7 and Formula (1) in Theorem 15.7.4. Observe that (11) is the formula that results if we formally divide both sides of (10) by dt.

There are infinitely many variations of the chain rule, depending on the number of variables and the choice of independent and dependent variables. A good working procedure is to use tree diagrams to derive new versions of the chain rule as needed.

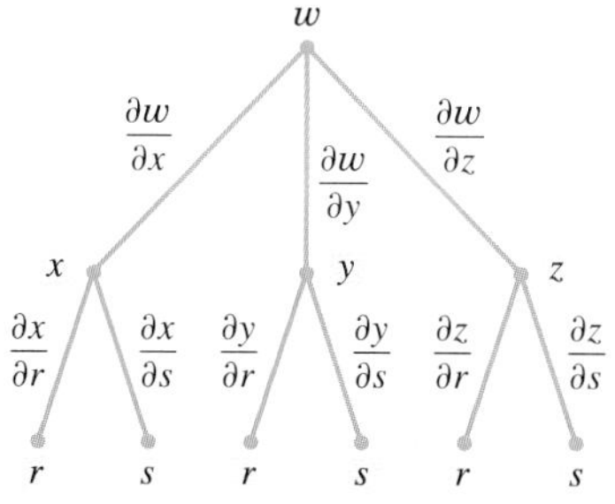

$$\frac{\partial w}{\partial r} = \frac{\partial w}{\partial x}\frac{\partial x}{\partial r} + \frac{\partial w}{\partial y}\frac{\partial y}{\partial r} + \frac{\partial w}{\partial z}\frac{\partial z}{\partial r}$$

$$\frac{\partial w}{\partial s} = \frac{\partial w}{\partial x}\frac{\partial x}{\partial s} + \frac{\partial w}{\partial y}\frac{\partial y}{\partial s} + \frac{\partial w}{\partial z}\frac{\partial z}{\partial s}$$

Figure 15.7.7

Example 5

Suppose that

$$w = e^{xyz}, \quad x = 3r + s, \quad y = 3r - s, \quad z = r^2 s$$

Use appropriate forms of the chain rule to find $\partial w/\partial r$ and $\partial w/\partial s$.

Solution. From the tree diagram and corresponding formulas in Figure 15.7.7 we obtain

$$\frac{\partial w}{\partial r} = yze^{xyz}(3) + xze^{xyz}(3) + xye^{xyz}(2rs) = e^{xyz}(3yz + 3xz + 2xyrs)$$

and

$$\frac{\partial w}{\partial s} = yze^{xyz}(1) + xze^{xyz}(-1) + xye^{xyz}(r^2) = e^{xyz}(yz - xz + xyr^2)$$

If desired, we can express $\partial w/\partial r$ and $\partial w/\partial s$ in terms of r and s alone by replacing x, y, and z by their expressions in terms of r and s. ◀

Example 6

Suppose that $w = x^2 + y^2 - z^2$ and

$$x = \rho\sin\phi\cos\theta, \quad y = \rho\sin\phi\sin\theta, \quad z = \rho\cos\phi$$

Use appropriate forms of the chain rule to find $\partial w/\partial\rho$ and $\partial w/\partial\theta$.

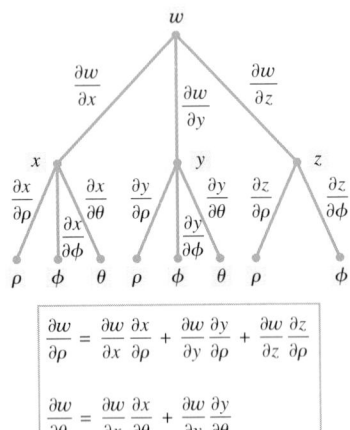

$$\frac{\partial w}{\partial \rho} = \frac{\partial w}{\partial x}\frac{\partial x}{\partial \rho} + \frac{\partial w}{\partial y}\frac{\partial y}{\partial \rho} + \frac{\partial w}{\partial z}\frac{\partial z}{\partial \rho}$$

$$\frac{\partial w}{\partial \theta} = \frac{\partial w}{\partial x}\frac{\partial x}{\partial \theta} + \frac{\partial w}{\partial y}\frac{\partial y}{\partial \theta}$$

Figure 15.7.8

Solution. From the tree diagram and corresponding formulas in Figure 15.7.8 we obtain

$$\frac{\partial w}{\partial \rho} = 2x \sin \phi \cos \theta + 2y \sin \phi \sin \theta - 2z \cos \phi$$

$$= 2\rho \sin^2 \phi \cos^2 \theta + 2\rho \sin^2 \phi \sin^2 \theta - 2\rho \cos^2 \phi$$

$$= 2\rho \sin^2 \phi (\cos^2 \theta + \sin^2 \theta) - 2\rho \cos^2 \phi$$

$$= 2\rho(\sin^2 \phi - \cos^2 \phi)$$

$$= -2\rho \cos 2\phi$$

$$\frac{\partial w}{\partial \theta} = (2x)(-\rho \sin \phi \sin \theta) + (2y)\rho \sin \phi \cos \theta$$

$$= -2\rho^2 \sin^2 \phi \sin \theta \cos \theta + 2\rho^2 \sin^2 \phi \sin \theta \cos \theta$$

$$= 0$$

This result is explained by the fact that w does not vary with θ. You can see this directly by expressing the variables x, y, and z in terms of r, ϕ, and θ in the formula for w. (Verify that $w = -\rho^2 \cos 2\phi$.) ◀

It is possible for some of the variables in a function $w = f(v_1, v_2, \ldots, v_n)$ to be functions of the remaining variables. Tree diagrams are especially helpful in such situations.

Example 7

Suppose that

$$w = xy + yz, \quad y = \sin x, \quad z = e^x$$

Use an appropriate form of the chain rule to find dw/dx.

Solution. From the tree diagram and corresponding formulas in Figure 15.7.9 we obtain

$$\frac{dw}{dx} = y + (x + z) \cos x + ye^x$$

$$= \sin x + (x + e^x) \cos x + e^x \sin x$$

This result can also be obtained by first expressing w explicitly in terms of x as

$$w = x \sin x + e^x \sin x$$

and then differentiating with respect to x; however, such direct substitution is not always possible. ◀

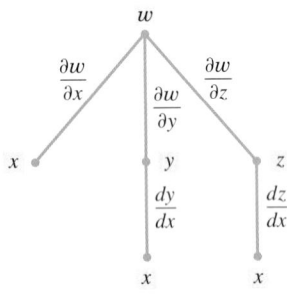

$$\frac{dw}{dx} = \frac{\partial w}{\partial x} + \frac{\partial w}{\partial y}\frac{dy}{dx} + \frac{\partial w}{\partial z}\frac{dz}{dx}$$

Figure 15.7.9

WARNING.   The symbol ∂z, unlike the differential dz, has no meaning of its own. For example, if we were to "cancel" partial symbols in the chain-rule formula

$$\frac{\partial z}{\partial u} = \frac{\partial z}{\partial x}\frac{\partial x}{\partial u} + \frac{\partial z}{\partial y}\frac{\partial y}{\partial u}$$

we would obtain

$$\frac{\partial z}{\partial u} = \frac{\partial z}{\partial u} + \frac{\partial z}{\partial u}$$

which is false in cases where $\partial z/\partial u \neq 0$.

In each of the expressions

$$z = \sin xy, \quad z = \frac{xy}{1 + xy}, \quad z = e^{xy}$$

the independent variables occur only in the combination xy, so the substitution $t = xy$ reduces the expression to a function of one variable:

$$z = \sin t, \quad z = \frac{t}{1 + t}, \quad z = e^t$$

Conversely, if we begin with a function of one variable $z = f(t)$ and substitute $t = xy$, we obtain a function $z = f(xy)$ in which the variables appear only in the combination xy. Functions whose variables occur in fixed combinations arise frequently in applications.

Example 8

Show that if $\partial z/\partial x$ and $\partial z/\partial y$ exist, then a function of the form $z = f(xy)$ satisfies the equation

$$x\frac{\partial z}{\partial x} - y\frac{\partial z}{\partial y} = 0$$

Solution. Let $t = xy$, so that $z = f(t)$. From the tree diagram in Figure 15.7.10 we obtain the formulas

$$\frac{\partial z}{\partial x} = \frac{dz}{dt}\frac{\partial t}{\partial x} = y\frac{dz}{dt} \quad \text{and} \quad \frac{\partial z}{\partial y} = \frac{dz}{dt}\frac{\partial t}{\partial y} = x\frac{dz}{dt}$$

from which it follows that

$$x\frac{\partial z}{\partial x} - y\frac{\partial z}{\partial y} = xy\frac{dz}{dt} - yx\frac{dz}{dt} = 0$$ ◀

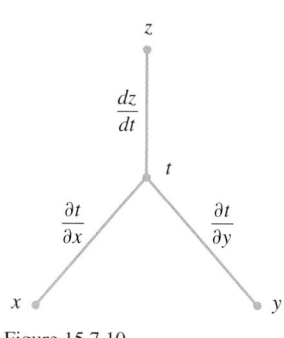

Figure 15.7.10

EXERCISE SET 15.7 C CAS

In Exercises 1–4, use an appropriate form of the chain rule to find dw/dt.

1. $w = 5x^2y^3z^4$; $x = t^2$, $y = t^3$, $z = t^5$

2. $w = \ln(3x^2 - 2y + 4z^3)$; $x = t^{1/2}$, $y = t^{2/3}$, $z = t^{-2}$

3. $w = 5\cos xy - \sin xz$; $x = 1/t$, $y = t$, $z = t^3$

4. $w = \sqrt{1 + x - 2yz^4x}$; $x = \ln t$, $y = t$, $z = 4t$

5. Suppose that

$$w = x^3y^2z^4; \quad x = t^2, \ y = t + 2, \ z = 2t^4$$

Find the rate of change of w with respect to t at $t = 1$ by using the chain rule, and then check your work by expressing w as a function of t and differentiating.

6. Suppose that

$$w = x\sin yz^2; \quad x = \cos t, \ y = t^2, \ z = e^t$$

Find the rate of change of w with respect to t at $t = 0$ by using the chain rule, and then check your work by expressing w as a function of t and differentiating.

In Exercises 7–10, find the gradient of f at P, and then use the gradient to calculate $D_{\mathbf{u}}f$ at P.

7. $f(x, y, z) = 4x^5y^2z^3$; $P(2, -1, 1)$; $\mathbf{u} = \frac{1}{3}\mathbf{i} + \frac{2}{3}\mathbf{j} - \frac{2}{3}\mathbf{k}$

8. $f(x, y, z) = ye^{xz} + z^2$; $P(0, 2, 3)$; $\mathbf{u} = \frac{2}{7}\mathbf{i} - \frac{3}{7}\mathbf{j} + \frac{6}{7}\mathbf{k}$

9. $f(x, y, z) = \ln(x^2 + 2y^2 + 3z^2)$; $P(-1, 2, 4)$; $\mathbf{u} = -\frac{3}{13}\mathbf{i} - \frac{4}{13}\mathbf{j} - \frac{12}{13}\mathbf{k}$

10. $f(x, y, z) = \sin xyz$; $P\left(\frac{1}{2}, \frac{1}{3}, \pi\right)$; $\mathbf{u} = \frac{1}{\sqrt{3}}\mathbf{i} - \frac{1}{\sqrt{3}}\mathbf{j} + \frac{1}{\sqrt{3}}\mathbf{k}$

In Exercises 11–14, find the directional derivative of f at P in the direction of $\mathbf{a}$.

11. $f(x, y, z) = x^3z - yx^2 + z^2$; $P(2, -1, 1)$; $\mathbf{a} = 3\mathbf{i} - \mathbf{j} + 2\mathbf{k}$

12. $f(x, y, z) = y - \sqrt{x^2 + z^2}$; $P(-3, 1, 4)$; $\mathbf{a} = 2\mathbf{i} - 2\mathbf{j} - \mathbf{k}$

13. $f(x, y, z) = \dfrac{z - x}{z + y}$; $P(1, 0, -3)$; $\mathbf{a} = -6\mathbf{i} + 3\mathbf{j} - 2\mathbf{k}$

14. $f(x, y, z) = e^{x+y+3z}$; $P(-2, 2, -1)$; $\mathbf{a} = 20\mathbf{i} - 4\mathbf{j} + 5\mathbf{k}$

In Exercises 15–18, find a unit vector in the direction in which f increases most rapidly at P, and find the rate of increase of f in that direction.

15. $f(x, y, z) = x^3z^2 + y^3z + z - 1$; $P(1, 1, -1)$

16. $f(x, y, z) = \sqrt{x - 3y + 4z}$; $P(0, -3, 0)$

17. $f(x, y, z) = \dfrac{x}{z} + \dfrac{z}{y^2}$; $P(1, 2, -2)$

18. $f(x, y, z) = \tan^{-1}\left(\dfrac{x}{y+z}\right)$; $P(4, 2, 2)$

In Exercises 19 and 20, find a unit vector in the direction in which f decreases most rapidly at P, and find the rate of change of f in that direction.

19. $f(x, y, z) = \dfrac{x + z}{z - y}$; $P(5, 7, 6)$

20. $f(x, y, z) = 4e^{xy}\cos z$; $P(0, 1, \pi/4)$

21. Find the directional derivative of

$$f(x, y, z) = \frac{y}{x + z}$$

at $P(2, 1, -1)$ in the direction from P to $Q(-1, 2, 0)$.

22. Find the directional derivative of the function

$$f(x, y, z) = x^3y^2z^5 - 2xz + yz + 3x$$

at $P(-1, -2, 1)$ in the direction of the negative z-axis.

23. Given that the directional derivative of $f(x, y, z)$ at the point $(3, -2, 1)$ in the direction of $\mathbf{a} = 2\mathbf{i} - \mathbf{j} - 2\mathbf{k}$ is -5 and that $\|\nabla f(3, -2, 1)\| = 5$, find $\nabla f(3, -2, 1)$.

24. The temperature (in degrees Celsius) at a point (x, y, z) in a metal solid is

$$T(x, y, z) = \frac{xyz}{1 + x^2 + y^2 + z^2}$$

(a) Find the rate of change of temperature with respect to distance at $(1, 1, 1)$ in the direction of the origin.

(b) Find the direction in which the temperature rises most rapidly at the point $(1, 1, 1)$. (Express your answer as a unit vector.)

(c) Find the rate at which the temperature rises moving from $(1, 1, 1)$ in the direction obtained in part (b).

25. Consider the ellipsoid $x^2 + y^2 + 4z^2 = 12$.

(a) Use the method of Example 2 to find an equation of the tangent plane to the ellipsoid at the point $(2, 2, 1)$.

(b) Find parametric equations of the line that is normal to the ellipsoid at the point $(2, 2, 1)$.

(c) Find the acute angle that the tangent plane at the point $(2, 2, 1)$ makes with the xy-plane.

26. Consider the surface $xz - yz^3 + yz^2 = 2$.

(a) Use the method of Example 2 to find an equation of the tangent plane to the surface at the point $(2, -1, 1)$.

(b) Find parametric equations of the line that is normal to the surface at the point $(2, -1, 1)$.

(c) Find the acute angle that the tangent plane at the point $(2, -1, 1)$ makes with the xy-plane.

In Exercises 27 and 28, find two unit vectors that are normal to the given surface at the point P.

27. $\sqrt{\dfrac{z + x}{y - 1}} = z^2$; $P(3, 5, 1)$

28. $\sin xz - 4 \cos yz = 4$; $P(\pi, \pi, 1)$

29. Show that every line that is normal to the sphere

$$x^2 + y^2 + z^2 = 1$$

passes through the origin.

30. Find all points on the ellipsoid $2x^2 + 3y^2 + 4z^2 = 9$ at which the tangent plane is parallel to the plane $x - 2y + 3z = 5$.

31. Find all points on the surface $x^2 + y^2 - z^2 = 1$ at which the normal line is parallel to the line through $P(1, -2, 1)$ and $Q(4, 0, -1)$.

32. Show that the ellipsoid $2x^2 + 3y^2 + z^2 = 9$ and the sphere

$$x^2 + y^2 + z^2 - 6x - 8y - 8z + 24 = 0$$

have a common tangent plane at the point $(1, 1, 2)$.

33. Find parametric equations for the tangent line to the curve of intersection of the paraboloid $z = x^2 + y^2$ and the ellipsoid $x^2 + 4y^2 + z^2 = 9$ at the point $(1, -1, 2)$.

34. Find parametric equations for the tangent line to the curve of intersection of the cone $z = \sqrt{x^2 + y^2}$ and the plane $x + 2y + 2z = 20$ at the point $(4, 3, 5)$.

35. Find parametric equations for the tangent line to the curve of intersection of the cylinders $x^2 + z^2 = 25$ and $y^2 + z^2 = 25$ at the point $(3, -3, 4)$.

c **36.** The accompanying figure shows the intersection of the surfaces $z = 8 - x^2 - y^2$ and $4x + 2y - z = 0$.

(a) Find parametric equations for the tangent line to the curve of intersection at the point $(0, 2, 4)$.

(b) Use a CAS to generate a reasonable facsimile of the figure. You need not generate the colors, but try to obtain a similar viewpoint.

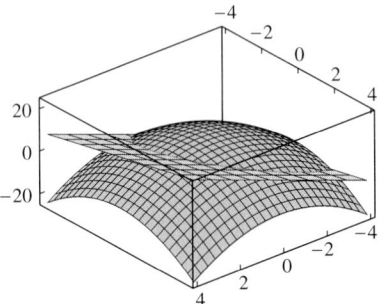

Figure Ex-36

In Exercises 37 and 38, find formulas for dw and Δw at a general point (x, y, z).

37. $w = x^3 y^2 z$ **38.** $w = e^{xyz}$

In Exercises 39–42, find dw.

39. $w = 8x - 3y + 4z$

40. $w = 4x^2 y^3 z^7 - 3xy + z + 5$

41. $w = \tan^{-1}(xyz)$ **42.** $w = \sqrt{x} + \sqrt{y} + \sqrt{z}$

43. Use a total differential to approximate the change in $f(x, y, z) = 2xy^2 z^3$ as (x, y, z) varies from $P(1, -1, 2)$ to $Q(0.99, -1.02, 2.02)$.

44. Use a total differential to approximate the change in $f(x, y, z) = xyz/(x + y + z)$ as (x, y, z) varies from $P(-1, -2, 4)$ to $Q(-1.04, -1.98, 3.97)$.

45. The length, width, and height of a rectangular box are measured to be 3 cm, 4 cm, and 5 cm, respectively, with a maximum error of 0.05 cm in each measurement. Use differentials to approximate the maximum error in the calculated volume.

46. The total resistance R of three resistances R_1, R_2, and R_3, connected in parallel, is given by

$$\frac{1}{R} = \frac{1}{R_1} + \frac{1}{R_2} + \frac{1}{R_3}$$

Suppose that R_1, R_2, and R_3 are measured to be 100 ohms, 200 ohms, and 500 ohms, respectively, with a maximum error of 10% in each. Use differentials to approximate the maximum percentage error in the calculated value of R.

47. The area of a triangle is to be computed from the formula $A = \frac{1}{2}ab\sin\theta$, where a and b are the lengths of two sides and θ is the included angle. Suppose that a, b, and θ are measured to be 40 ft, 50 ft, and $30°$, respectively. Use differentials to approximate the maximum error in the calculated value of A if the maximum errors in a, b, and θ are $\frac{1}{2}$ ft, $\frac{1}{4}$ ft, and $2°$, respectively.

48. The length, width, and height of a rectangular box are measured with errors of at most $r\%$ (where r is small). Use differentials to approximate the maximum percentage error in the computed value of the volume.

In Exercises 49–52, find the indicated partial derivatives.

49. $f(v, w, x, y) = 4v^2w^3x^4y^5$;
$\partial f/\partial v$, $\partial f/\partial w$, $\partial f/\partial x$, $\partial f/\partial y$

50. $w = r\cos st + e^u \sin ur$;
$\partial w/\partial r$, $\partial w/\partial s$, $\partial w/\partial t$, $\partial w/\partial u$

51. $f(v_1, v_2, v_3, v_4) = \dfrac{v_1^2 - v_2^2}{v_3^2 + v_4^2}$;
$\partial f/\partial v_1$, $\partial f/\partial v_2$, $\partial f/\partial v_3$, $\partial f/\partial v_4$

52. $V = xe^{2x-y} + we^{zw} + yw$;
$\partial V/\partial x$, $\partial V/\partial y$, $\partial V/\partial z$, $\partial V/\partial w$

53. Let $u(w, x, y, z) = xe^{yw}\sin^2 z$. Find

(a) $\dfrac{\partial u}{\partial x}(0, 0, 1, \pi)$ (b) $\dfrac{\partial u}{\partial y}(0, 0, 1, \pi)$

(c) $\dfrac{\partial u}{\partial w}(0, 0, 1, \pi)$ (d) $\dfrac{\partial u}{\partial z}(0, 0, 1, \pi)$

(e) $\dfrac{\partial^4 u}{\partial x\partial y\partial w\partial z}$ (f) $\dfrac{\partial^4 u}{\partial w\partial z\partial y^2}$.

54. Let $f(v, w, x, y) = 2v^{1/2}w^4x^{1/2}y^{2/3}$. Find $f_v(1, -2, 4, 8)$, $f_w(1, -2, 4, 8)$, $f_x(1, -2, 4, 8)$, and $f_y(1, -2, 4, 8)$.

In Exercises 55–58, use appropriate forms of the chain rule to find the derivatives.

55. Let $z = \ln(x^2 + 1)$, where $x = r\cos\theta$. Find $\partial z/\partial r$ and $\partial z/\partial\theta$.

56. Let $u = rs^2\ln t$, $r = x^2$, $s = 4y + 1$, $t = xy^3$. Find $\partial u/\partial x$ and $\partial u/\partial y$.

57. Let $w = 4x^2 + 4y^2 + z^2$, $x = \rho\sin\phi\cos\theta$, $y = \rho\sin\phi\sin\theta$, $z = \rho\cos\phi$. Find $\partial w/\partial\rho$, $\partial w/\partial\phi$, and $\partial w/\partial\theta$.

58. Let $w = 3xy^2z^3$, $y = 3x^2 + 2$, $z = \sqrt{x - 1}$. Find dw/dx.

59. The length, width, and height of a rectangular box are increasing at rates of 1 in/s, 2 in/s, and 3 in/s, respectively.
(a) At what rate is the volume increasing when the length is 2 in, the width is 3 in, and the height is 6 in?
(b) At what rate is the length of the diagonal increasing at that instant?

60. The area A of a triangle is given by $A = \frac{1}{2}ab\sin\theta$, where a and b are the lengths of two sides and θ is the angle between these sides. Suppose that $a = 5$, $b = 10$, and $\theta = \pi/3$.
(a) Find the rate at which A changes with respect to a if b and θ are held constant.

(b) Find the rate at which A changes with respect to θ if a and b are held constant.
(c) Find the rate at which b changes with respect to a if A and θ are held constant.

61. Let f be a differentiable function of one variable, and let $z = f(x + 2y)$. Show that
$$2\frac{\partial z}{\partial x} - \frac{\partial z}{\partial y} = 0$$

62. Let f be a differentiable function of one variable, and let $z = f(x^2 + y^2)$. Show that
$$y\frac{\partial z}{\partial x} - x\frac{\partial z}{\partial y} = 0$$

63. Let f be a differentiable function of one variable, and let $w = f(\rho)$, where $\rho = (x^2 + y^2 + z^2)^{1/2}$. Show that
$$\left(\frac{\partial w}{\partial x}\right)^2 + \left(\frac{\partial w}{\partial y}\right)^2 + \left(\frac{\partial w}{\partial z}\right)^2 = \left(\frac{dw}{d\rho}\right)^2$$

64. Let f be a differentiable function of three variables and suppose that $w = f(x - y, y - z, z - x)$. Show that
$$\frac{\partial w}{\partial x} + \frac{\partial w}{\partial y} + \frac{\partial w}{\partial z} = 0$$

65. Let f be a differentiable function of three variables, and let $w = f(x, y, z)$, $x = \rho\sin\phi\cos\theta$, $y = \rho\sin\phi\sin\theta$, and $z = \rho\cos\phi$. Express $\partial w/\partial\rho$, $\partial w/\partial\phi$, and $\partial w/\partial\theta$ in terms of $\partial w/\partial x$, $\partial w/\partial y$, and $\partial w/\partial z$.

66. Assume that $F(x, y, z) = 0$ defines z implicitly as a function of x and y. Show that if $\partial F/\partial z \neq 0$, then
$$\frac{\partial z}{\partial x} = -\frac{\partial F/\partial x}{\partial F/\partial z} \quad \text{and} \quad \frac{\partial z}{\partial y} = -\frac{\partial F/\partial y}{\partial F/\partial z}$$

In Exercises 67–70, find $\partial z/\partial x$ and $\partial z/\partial y$ by implicit differentiation, and confirm that the results obtained agree with those produced by the formulas in Exercise 66.

67. $x^2 - 3yz^2 + xyz - 2 = 0$
68. $\ln(1 + z) + xy^2 + z = 1$
69. $ye^x - 5\sin 3z = 3z$
70. $e^{xy}\cos yz - e^{yz}\sin xz + 2 = 0$
71. Given that the equations $u = u(x, y, z)$, $v = v(x, y, z)$, and $w = w(x, y, z)$ are all differentiable, show that
$$\nabla f(u, v, w) = \frac{\partial f}{\partial u}\nabla u + \frac{\partial f}{\partial v}\nabla v + \frac{\partial f}{\partial w}\nabla w$$

72. Let $w = f(x, y, z)$, where $z = g(x, y)$. Taking x and y as the independent variables, express each of the following in terms of $\partial f/\partial x$, $\partial f/\partial y$, $\partial f/\partial z$, $\partial z/\partial x$, and $\partial z/\partial y$.
(a) $\partial w/\partial x$ (b) $\partial w/\partial y$

73. Let $w = \ln(e^r + e^s + e^t + e^u)$. Show that
$$w_{rstu} = -6e^{r+s+t+u-4w}$$
[*Hint:* Take advantage of the relationship $e^w = e^r + e^s + e^t + e^u$.]

74. Suppose that w is a differentiable function of x_1, x_2, and x_3, and

$$x_1 = a_1 y_1 + b_1 y_2$$
$$x_2 = a_2 y_1 + b_2 y_2$$
$$x_3 = a_3 y_1 + b_3 y_2$$

where the a's and b's are constants. Express $\partial w/\partial y_1$ and $\partial w/\partial y_2$ in terms of $\partial w/\partial x_1$, $\partial w/\partial x_2$, and $\partial w/\partial x_3$.

75. (a) Let w be a differentiable function of x_1, x_2, x_3, and x_4, and let each x_i be a function of t. Find a chain-rule formula for dw/dt.

(b) Let w be a differentiable function of x_1, x_2, x_3, and x_4, and let each x_i be a differentiable function of v_1, v_2, and v_3. Find chain-rule formulas for $\partial w/\partial v_1$, $\partial w/\partial v_2$, and $\partial w/\partial v_3$.

76. Let $w = (x_1^2 + x_2^2 + \cdots + x_n^2)^k$, where $n > 2$. For what values of k does

$$\frac{\partial^2 w}{\partial x_1^2} + \frac{\partial^2 w}{\partial x_2^2} + \cdots + \frac{\partial^2 w}{\partial x_n^2} = 0$$

hold?

77. We showed in Exercise 24 of Section 7.9 that

$$\frac{d}{dx} \int_{h(x)}^{g(x)} f(t)\,dt = f(g(x))g'(x) - f(h(x))h'(x)$$

Derive this same result by letting $u = g(x)$ and $v = h(x)$

and then differentiating the function

$$F(u, v) = \int_v^u f(t)\,dt$$

with respect to x.

78. Two surfaces $f(x, y, z) = 0$ and $g(x, y, z) = 0$ are said to be **orthogonal** at a point P of intersection if ∇f and ∇g are nonzero at P and the normal lines to the surfaces are perpendicular at P. Show that if $\nabla f(x_0, y_0, z_0) \neq \mathbf{0}$ and $\nabla g(x_0, y_0, z_0) \neq \mathbf{0}$, then the surfaces $f(x, y, z) = 0$ and $g(x, y, z) = 0$ are orthogonal at the point (x_0, y_0, z_0) if and only if

$$f_x g_x + f_y g_y + f_z g_z = 0$$

at this point. [*Note:* This is a more general version of the result in Exercise 51 of Section 15.5.]

79. Use the result of Exercise 78 to show that the sphere $x^2 + y^2 + z^2 = a^2$ and the cone $z^2 = x^2 + y^2$ are orthogonal at every point of intersection (see the accompanying figure).

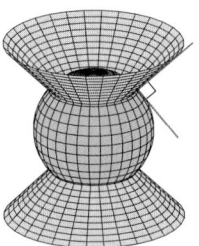

Figure Ex-79

15.8 MAXIMA AND MINIMA OF FUNCTIONS OF TWO VARIABLES

Earlier in this text we learned how to find maximum and minimum values of a function of one variable. In this section we will develop similar techniques for functions of two variables.

EXTREMA

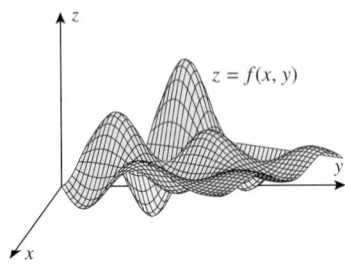

Figure 15.8.1

If we imagine the graph of a function f of two variables to be a mountain range (Figure 15.8.1), then the mountaintops, which are the high points in their immediate vicinity, are called *relative maxima* of f, and the valley bottoms, which are the low points in their immediate vicinity, are called *relative minima* of f.

Just as a geologist might be interested in finding the highest mountain and deepest valley in an entire mountain range, so a mathematician might be interested in finding the largest and smallest values of $f(x, y)$ over the *entire* domain of f. These are called the *absolute maximum* and *absolute minimum values* of f. The following definitions make these informal ideas precise.

15.8.1 DEFINITION. A function f of two variables is said to have a **relative maximum** at a point (x_0, y_0) if there is a circle centered at (x_0, y_0) such that $f(x_0, y_0) \geq f(x, y)$ for all points (x, y) in the domain of f that lie inside the circle, and f is said to have an **absolute maximum** at (x_0, y_0) if $f(x_0, y_0) \geq f(x, y)$ for all points (x, y) in the domain of f.

15.8.2 DEFINITION. A function f of two variables is said to have a ***relative minimum*** at a point (x_0, y_0) if there is a circle centered at (x_0, y_0) such that $f(x_0, y_0) \leq f(x, y)$ for all points (x, y) in the domain of f that lie inside the circle, and f is said to have an ***absolute minimum*** at (x_0, y_0) if $f(x_0, y_0) \leq f(x, y)$ for all points (x, y) in the domain of f.

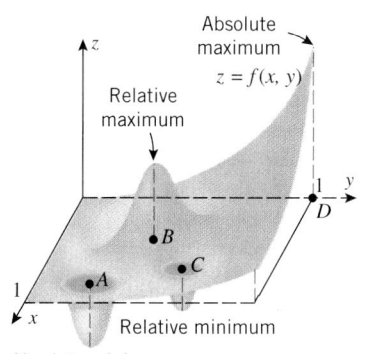

Figure 15.8.2

If f has a relative maximum or a relative minimum at (x_0, y_0), then we say that f has a ***relative extremum*** at (x_0, y_0), and if f has an absolute maximum or absolute minimum at (x_0, y_0), then we say that f has an ***absolute extremum*** at (x_0, y_0).

Figure 15.8.2 shows the graph of a function f whose domain is the closed square region in the xy-plane whose points satisfy the inequalities $0 \leq x \leq 1, 0 \leq y \leq 1$. The function f has relative minima at the points A and C and a relative maximum at B. There is an absolute minimum at A and an absolute (and relative) maximum at D.

For functions of two variables we will be concerned with two important questions:

- Are there any relative or absolute extrema?
- If so, where are they located?

THE EXTREME-VALUE THEOREM

For functions of one variable that are continuous on a closed interval, the Extreme-Value Theorem (Theorem 6.1.3) answered the existence question for absolute extrema. The following theorem, which we state without proof, is the corresponding result for functions of two variables.

15.8.3 THEOREM (***Extreme-Value Theorem***). *If $f(x, y)$ is continuous on a closed and bounded set R, then f has both an absolute maximum and an absolute minimum on R.*

Example 1

The square region R whose points satisfy the inequalities

$$0 \leq x \leq 1 \quad \text{and} \quad 0 \leq y \leq 1$$

is a closed and bounded set in the xy-plane. The function f whose graph is shown in Figure 15.8.2 is continuous on R; thus, it is guaranteed to have an absolute maximum and minimum on R by the last theorem. These occur at points D and A that are shown in the figure. ◄

REMARK. If any of the conditions in the Extreme-Value Theorem fail to hold, then there is no guarantee that an absolute maximum or absolute minimum exists on the region R. Thus, a discontinuous function on a closed and bounded set need not have any absolute extrema, and a continuous function on a set that is not closed and bounded also need not have any absolute extrema.

FINDING RELATIVE EXTREMA

Recall that if a function g of one variable has a relative extremum at a point x_0 where g is differentiable, then $g'(x_0) = 0$. To obtain the analog of this result for functions of two variables, suppose that $f(x, y)$ has a relative maximum at a point (x_0, y_0) and that the partial derivatives of f exist at (x_0, y_0). It seems plausible geometrically that the traces of the surface $z = f(x, y)$ on the planes $x = x_0$ and $y = y_0$ have horizontal tangent lines at (x_0, y_0) (Figure 15.8.3), so

$$f_x(x_0, y_0) = 0 \quad \text{and} \quad f_y(x_0, y_0) = 0$$

The same conclusion holds if f has a relative minimum at (x_0, y_0), all of which suggests the following result, which we state without formal proof.

15.8.4   THEOREM.   *If f has a relative extremum at a point (x_0, y_0), and if the first-order partial derivatives of f exist at this point, then*

$$f_x(x_0, y_0) = 0 \quad and \quad f_y(x_0, y_0) = 0$$

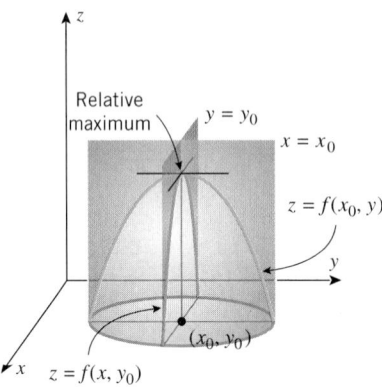

Figure 15.8.3

Recall that the *critical points* of a function f of one variable are those values of x at which $f'(x) = 0$ or f is not differentiable. The following definition is the analog for functions of two variables.

15.8.5   DEFINITION.   A point (x_0, y_0) is called a ***critical point*** of the function $f(x, y)$ if $f_x(x_0, y_0) = 0$ and $f_y(x_0, y_0) = 0$ or if one or both partial derivatives does not exist at (x_0, y_0).

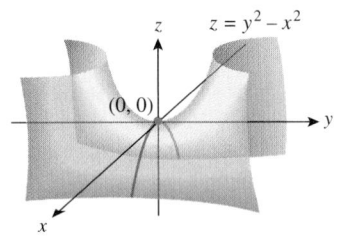

The function $f(x, y) = y^2 - x^2$ has neither a relative maximum nor a relative minimum at the critical point $(0, 0)$.

Figure 15.8.4

It follows from this definition and Theorem 15.8.4 that if f has first-order partial derivatives, then the relative extrema occur at critical points, just as for a function of one variable. However, recall that for a function of one variable a relative extremum need not occur at *every* critical point. For example, the function might have an inflection point with a horizontal tangent line at the critical point (see Figure 5.2.4). Similarly, a function of two variables need not have a relative extremum at every critical point. For example, consider the function

$$f(x, y) = y^2 - x^2$$

This function, whose graph is the hyperbolic paraboloid shown in Figure 15.8.4, has a critical point at $(0, 0)$, since

$$f_x(x, y) = -2x \quad and \quad f_y(x, y) = 2y$$

from which it follows that

$$f_x(0, 0) = 0 \quad and \quad f_y(0, 0) = 0$$

However, the function f has neither a relative maximum nor a relative minimum at $(0, 0)$. For obvious reasons, the point $(0, 0)$ is called a *saddle point* of f. In general, we will say that a surface $z = f(x, y)$ has a ***saddle point*** at (x_0, y_0) if there are two distinct vertical planes through this point such that the trace of the surface in one of the planes has a relative maximum at (x_0, y_0) and the trace in the other has a relative minimum at (x_0, y_0).

Example 2

The three functions graphed in Figure 15.8.5 all have critical points at $(0, 0)$. For the paraboloids, the partial derivatives at the origin are zero. You can check this algebraically

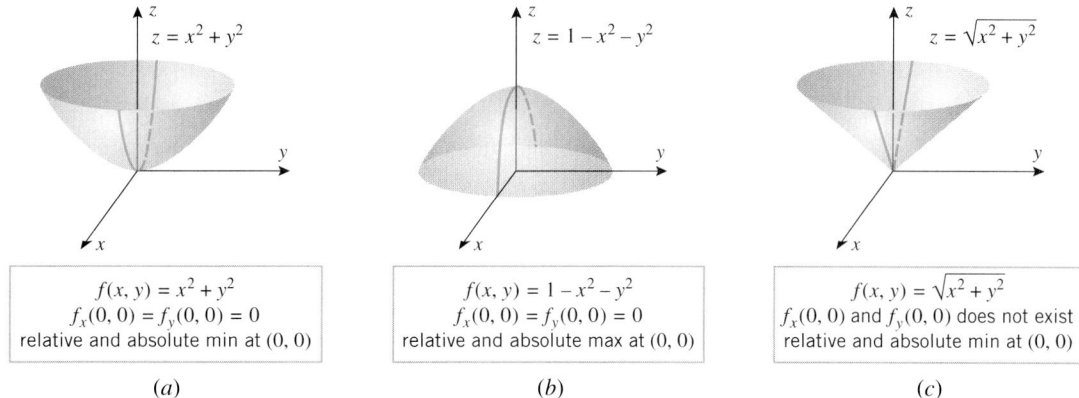

$f(x, y) = x^2 + y^2$
$f_x(0, 0) = f_y(0, 0) = 0$
relative and absolute min at $(0, 0)$

$f(x, y) = 1 - x^2 - y^2$
$f_x(0, 0) = f_y(0, 0) = 0$
relative and absolute max at $(0, 0)$

$f(x, y) = \sqrt{x^2 + y^2}$
$f_x(0, 0)$ and $f_y(0, 0)$ does not exist
relative and absolute min at $(0, 0)$

Figure 15.8.5 (a) (b) (c)

by evaluating the partial derivatives at $(0, 0)$, but you can see it geometrically by observing that the traces in the xz-plane and yz-plane have horizontal tangent lines at $(0, 0)$. For the cone neither partial derivative exists at the origin because the traces in the xz-plane and the yz-plane have corners there. The paraboloid in part (a) and the cone in part (c) have a relative minimum and absolute minimum at the origin, and the paraboloid in part (b) has a relative maximum and an absolute maximum at the origin.   ◀

THE SECOND PARTIALS TEST

For functions of one variable the second derivative test (Theorem 5.2.4) was used to determine the behavior of a function at a critical point. The following theorem, which is usually proved in advanced calculus, is the analog of that theorem for functions of two variables.

> **15.8.6**   THEOREM (**The Second Partials Test**).   *Let f be a function of two variables with continuous second-order partial derivatives in some circle centered at a critical point (x_0, y_0), and let*
>
> $$D = f_{xx}(x_0, y_0) f_{yy}(x_0, y_0) - f_{xy}^2(x_0, y_0)$$
>
> *(a)   If $D > 0$ and $f_{xx}(x_0, y_0) > 0$, then f has a relative minimum at (x_0, y_0).*
> *(b)   If $D > 0$ and $f_{xx}(x_0, y_0) < 0$, then f has a relative maximum at (x_0, y_0).*
> *(c)   If $D < 0$, then f has a saddle point at (x_0, y_0).*
> *(d)   If $D = 0$, then no conclusion can be drawn.*

Example 3

Locate all relative extrema and saddle points of

$$f(x, y) = 3x^2 - 2xy + y^2 - 8y$$

Solution. Since $f_x(x, y) = 6x - 2y$ and $f_y(x, y) = -2x + 2y - 8$, the critical points of f satisfy the equations

$$6x - 2y = 0$$
$$-2x + 2y - 8 = 0$$

Solving these for x and y yields $x = 2$, $y = 6$ (verify), so $(2, 6)$ is the only critical point. To apply Theorem 15.8.6 we need the second-order partial derivatives

$$f_{xx}(x, y) = 6, \quad f_{yy}(x, y) = 2, \quad f_{xy}(x, y) = -2$$

At the point $(2, 6)$ we have

$$D = f_{xx}(2, 6) f_{yy}(2, 6) - f_{xy}^2(2, 6) = (6)(2) - (-2)^2 = 8 > 0$$

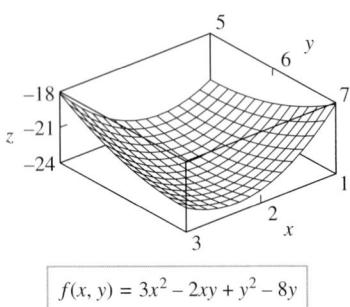

$$f(x, y) = 3x^2 - 2xy + y^2 - 8y$$

Figure 15.8.6

and

$$f_{xx}(2, 6) = 6 > 0$$

so f has a relative minimum at $(2, 6)$ by part (a) of the second partials test. Figure 15.8.6 shows a graph of f in the vicinity of the relative minimum. ◀

Example 4

Locate all relative extrema and saddle points of

$$f(x, y) = 4xy - x^4 - y^4$$

Solution. Since

$$f_x(x, y) = 4y - 4x^3$$
$$f_y(x, y) = 4x - 4y^3 \tag{1}$$

the critical points of f have coordinates satisfying the equations

$$4y - 4x^3 = 0 \quad \text{or} \quad y = x^3$$
$$4x - 4y^3 = 0 \qquad\qquad x = y^3 \tag{2}$$

Substituting the top equation in the bottom yields $x = (x^3)^3$ or $x^9 - x = 0$ or $x(x^8 - 1) = 0$, which has solutions $x = 0, x = 1, x = -1$. Substituting these values in the top equation of (2), we obtain the corresponding y-values $y = 0, y = 1, y = -1$. Thus, the critical points of f are $(0, 0), (1, 1)$, and $(-1, -1)$.

From (1),

$$f_{xx}(x, y) = -12x^2, \quad f_{yy}(x, y) = -12y^2, \quad f_{xy}(x, y) = 4$$

which yields the following table:

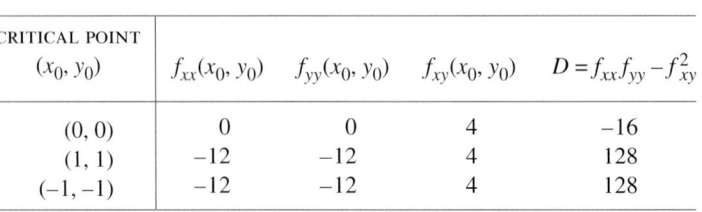

CRITICAL POINT (x_0, y_0)	$f_{xx}(x_0, y_0)$	$f_{yy}(x_0, y_0)$	$f_{xy}(x_0, y_0)$	$D = f_{xx}f_{yy} - f_{xy}^2$
$(0, 0)$	0	0	4	-16
$(1, 1)$	-12	-12	4	128
$(-1, -1)$	-12	-12	4	128

At the points $(1, 1)$ and $(-1, -1)$, we have $D > 0$ and $f_{xx} < 0$, so relative maxima occur at these critical points. At $(0, 0)$ there is a saddle point since $D < 0$. The surface and a contour plot are shown in Figure 15.8.7. ◀

FOR THE READER.   The "figure eight" pattern at $(0, 0)$ in the contour plot for the surface in Figure 15.8.7 is typical for level curves that pass through a saddle point. If a bug starts at the point $(0, 0, 0)$ on the surface, in how many directions can it walk and remain in the xy-plane?

The following theorem, which is the analog for functions of two variables of Theorem 6.1.4, will lead to an important method for finding absolute extrema.

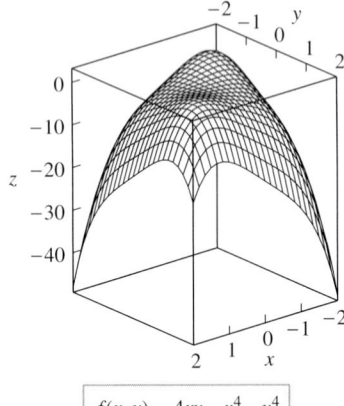

$$f(x, y) = 4xy - x^4 - y^4$$

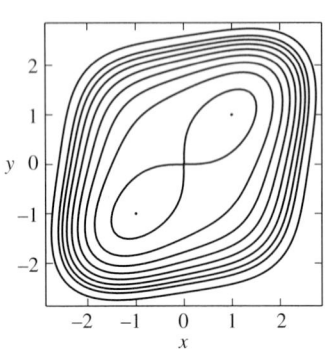

Figure 15.8.7

15.8.7   THEOREM.   *If a function f of two variables has an absolute extremum (either an absolute maximum or an absolute minimum) at an interior point of its domain, then this extremum occurs at a critical point.*

Proof. If f has an absolute maximum at the point (x_0, y_0) in the interior of the domain of f, then f has a relative maximum at (x_0, y_0). If both partial derivatives exist at (x_0, y_0),

then

$$f_x(x_0, y_0) = 0 \quad \text{and} \quad f_y(x_0, y_0) = 0$$

by Theorem 15.8.4, so (x_0, y_0) is a critical point of f. If either partial derivative does not exist, then again (x_0, y_0) is a critical point, so (x_0, y_0) is a critical point in all cases. The proof for an absolute minimum is similar. ∎

FINDING ABSOLUTE EXTREMA ON CLOSED AND BOUNDED SETS

If $f(x, y)$ is continuous on a closed and bounded set R, then the Extreme-Value Theorem (Theorem 15.8.3) guarantees the existence of an absolute maximum and an absolute minimum of f on R. These absolute extrema can occur either on the boundary of R or in the interior of R, but if an absolute extremum occurs in the interior, then it occurs at a critical point by Theorem 15.8.7. Thus, we are led to the following procedure for finding absolute extrema:

> **How to Find the Absolute Extrema of a Continuous Function f of Two Variables on a Closed and Bounded Set R**
>
> **Step 1.** Find the critical points of f that lie in the interior of R.
>
> **Step 2.** Find all boundary points at which the absolute extrema can occur.
>
> **Step 3.** Evaluate $f(x, y)$ at the points obtained in the preceding steps. The largest of these values is the absolute maximum and the smallest the absolute minimum.

Example 5

Find the absolute maximum and minimum values of

$$f(x, y) = 3xy - 6x - 3y + 7 \tag{3}$$

on the closed triangular region R with vertices $(0, 0)$, $(3, 0)$, and $(0, 5)$.

Solution. The region R is shown in Figure 15.8.8. We have

$$\frac{\partial f}{\partial x} = 3y - 6 \quad \text{and} \quad \frac{\partial f}{\partial y} = 3x - 3$$

so all critical points occur where

$$3y - 6 = 0 \quad \text{and} \quad 3x - 3 = 0$$

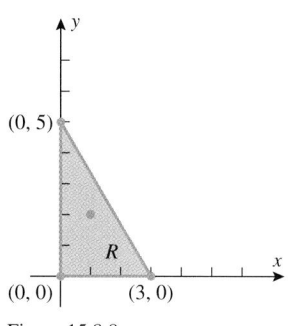

Figure 15.8.8

Solving these equations yields $x = 1$ and $y = 2$, so $(1, 2)$ is the only critical point. As shown in Figure 15.8.8, this critical point is in the interior of R.

Next, we want to determine the location of the points on the boundary of R at which the absolute extrema might occur. The boundary of R consists of three line segments, each of which we will treat separately:

The line segment between $(0, 0)$ *and* $(3, 0)$: On this line segment we have $y = 0$, so (3) simplifies to a function of the single variable x,

$$u(x) = f(x, 0) = -6x + 7, \quad 0 \le x \le 3$$

This function has no critical points because $u'(x) = -6$ is nonzero for all x. Thus the extreme values of $u(x)$ occur at the endpoints $x = 0$ and $x = 3$, which correspond to the points $(0, 0)$ and $(3, 0)$ of R.

The line segment between $(0, 0)$ *and* $(0, 5)$: On this line segment we have $x = 0$, so (3) simplifies to a function of the single variable y,

$$v(y) = f(0, y) = -3y + 7, \quad 0 \le y \le 5$$

This function has no critical points because $v'(y) = -3$ is nonzero for all y. Thus, the extreme values of $v(y)$ occur at the endpoints $y = 0$ and $y = 5$, which correspond to the points $(0, 0)$ and $(0, 5)$ of R.

The line segment between (3, 0) *and* (0, 5): In the xy-plane, an equation for this line segment is

$$y = -\tfrac{5}{3}x + 5, \quad 0 \le x \le 3 \tag{4}$$

so (3) simplifies to a function of the single variable x,

$$w(x) = f\left(x, -\tfrac{5}{3}x + 5\right) = 3x\left(-\tfrac{5}{3}x + 5\right) - 6x - 3\left(-\tfrac{5}{3}x + 5\right) + 7$$
$$= -5x^2 + 14x - 8, \quad 0 \le x \le 3$$

Since $w'(x) = -10x + 14$, the equation $w'(x) = 0$ yields $x = \tfrac{7}{5}$ as the only critical point of w. Thus, the extreme values of w occur either at the critical point $x = \tfrac{7}{5}$ or at the endpoints $x = 0$ and $x = 3$. The endpoints correspond to the points (0, 5) and (3, 0) of R, and from (4) the critical point corresponds to $\left(\tfrac{7}{5}, \tfrac{8}{3}\right)$.

Finally, Table 15.8.1 lists the values of $f(x, y)$ at the interior critical point and at the points on the boundary where an absolute extremum can occur. From the table we conclude that the absolute maximum value of f is $f(0, 0) = 7$ and the absolute minimum value is $f(3, 0) = -11$. ◀

Table 15.8.1

(x, y)	$(0, 0)$	$(3, 0)$	$(0, 5)$	$\left(\tfrac{7}{5}, \tfrac{8}{3}\right)$	$(1, 2)$
$f(x, y)$	7	−11	−8	$\tfrac{9}{5}$	1

Example 6

Determine the dimensions of a rectangular box, open at the top, having a volume of 32 ft^3, and requiring the least amount of material for its construction.

Solution. Let

$$x = \text{length of the box (in feet)}$$
$$y = \text{width of the box (in feet)}$$
$$z = \text{height of the box (in feet)}$$
$$S = \text{surface area of the box (in square feet)}$$

We may reasonably assume that the box with least surface area requires the least amount of material, so our objective is to minimize the surface area

$$S = xy + 2xz + 2yz \tag{5}$$

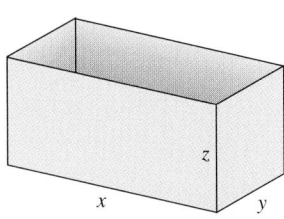

Two sides each have area xz.
Two sides each have area yz.
The base has area xy.

Figure 15.8.9

(Figure 15.8.9) subject to the volume requirement

$$xyz = 32 \tag{6}$$

From (6) we obtain $z = 32/xy$, so (5) can be rewritten as

$$S = xy + \frac{64}{y} + \frac{64}{x} \tag{7}$$

which expresses S as a function of two variables. The dimensions x and y in this formula must be positive, but otherwise have no limitation, so our problem reduces to finding the absolute minimum value of S over the region for which $x > 0$ and $y > 0$ (Figure 15.8.10). Because this region is not bounded, we have no mathematical guarantee at this stage that an absolute minimum exists. However, if it does, then it occurs at a critical point of S, so we will begin by finding the critical points. Differentiating (7) we obtain

$$\frac{\partial S}{\partial x} = y - \frac{64}{x^2}, \quad \frac{\partial S}{\partial y} = x - \frac{64}{y^2} \tag{8}$$

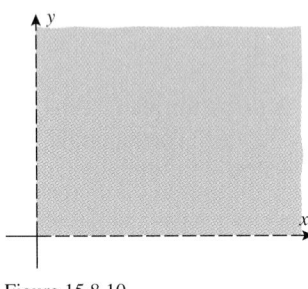

Figure 15.8.10

so the coordinates of the critical points of S satisfy

$$y - \frac{64}{x^2} = 0, \quad x - \frac{64}{y^2} = 0$$

Solving the first equation for y yields

$$y = \frac{64}{x^2} \qquad (9)$$

and substituting this expression in the second equation yields

$$x - \frac{64}{(64/x^2)^2} = 0$$

which can be rewritten as

$$x\left(1 - \frac{x^3}{64}\right) = 0$$

The solutions of this equation are $x = 0$ and $x = 4$. Since we require $x > 0$, the only solution of significance is $x = 4$. Substituting this value in (9) yields $y = 4$. To see that we have located a relative minimum, we use the second partials test. From (8),

$$\frac{\partial^2 S}{\partial x^2} = \frac{128}{x^3}, \quad \frac{\partial^2 S}{\partial y^2} = \frac{128}{y^3}, \quad \frac{\partial^2 S}{\partial y \partial x} = 1$$

Thus, when $x = 4$ and $y = 4$, we have

$$\frac{\partial^2 S}{\partial x^2} = 2, \quad \frac{\partial^2 S}{\partial y^2} = 2, \quad \frac{\partial^2 S}{\partial y \partial x} = 1$$

and

$$D = \frac{\partial^2 S}{\partial x^2}\frac{\partial^2 S}{\partial y^2} - \left(\frac{\partial^2 S}{\partial y \partial x}\right)^2 = (2)(2) - (1)^2 = 3$$

Since $\partial^2 S/\partial x^2 > 0$ and $D > 0$, it follows from the second partials test that a relative minimum occurs when $x = y = 4$. Substituting these values in (6) yields $z = 2$, so the box using least material has a height of 2 ft and a square base whose edges are 4 ft long. ◀

REMARK. Strictly speaking, the solution in the last example is incomplete since we have not shown that an *absolute minimum* for S occurs when $x = y = 4$ and $z = 2$, only a relative minimum. The problem of showing that a relative extremum is also an absolute extremum can be difficult for functions of two or more variables and will not be considered in this text. However, in applied problems we can sometimes use physical considerations to deduce that an absolute extremum has been found. Another possibility is to use graphical evidence. For example, the graph of Equation (7) in Figure 15.8.11 strongly suggests that the relative minimum at $x = 4$ and $y = 4$ is also an absolute minimum.

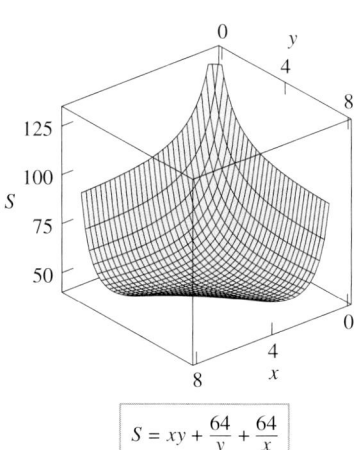

$$S = xy + \frac{64}{y} + \frac{64}{x}$$

Figure 15.8.11

EXERCISE SET 15.8 ∿ Graphing Calculator [c] CAS

In Exercises 1 and 2, locate all absolute maxima and minima, if any, by inspection. Then check your answers using calculus.

1. (a) $f(x, y) = (x - 2)^2 + (y + 1)^2$
(b) $f(x, y) = 1 - x^2 - y^2$ (c) $f(x, y) = x + 2y - 5$

2. (a) $f(x, y) = 1 - (x + 1)^2 - (y - 5)^2$
(b) $f(x, y) = e^{xy}$ (c) $f(x, y) = x^2 - y^2$

In Exercises 3 and 4, complete the squares and locate all absolute maxima and minima, if any, by inspection. Then check your answers using calculus.

3. $f(x, y) = 13 - 6x + x^2 + 4y + y^2$

4. $f(x, y) = 1 - 2x - x^2 + 4y - 2y^2$

In Exercises 5–8, the contour plots show all significant features of the function. Make a conjecture about the number and the location of all relative extrema and saddle points, and then use calculus to check your conjecture.

5.

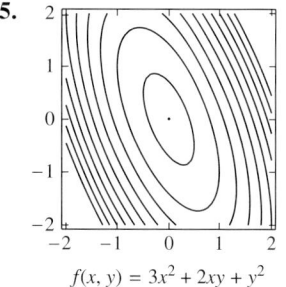

$f(x, y) = 3x^2 + 2xy + y^2$

6.

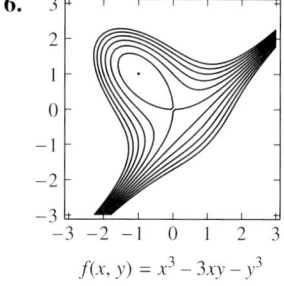

$f(x, y) = x^3 - 3xy - y^3$

7.

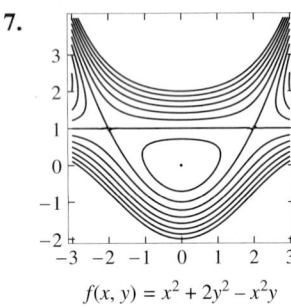

$f(x, y) = x^2 + 2y^2 - x^2y$

8.

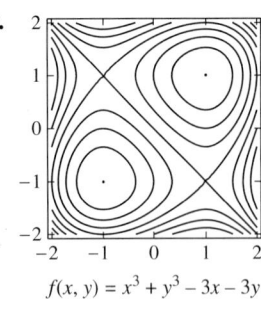

$f(x, y) = x^3 + y^3 - 3x - 3y$

In Exercises 9–20, locate all relative maxima, relative minima, and saddle points, if any.

9. $f(x, y) = y^2 + xy + 3y + 2x + 3$

10. $f(x, y) = x^2 + xy - 2y - 2x + 1$

11. $f(x, y) = x^2 + xy + y^2 - 3x$

12. $f(x, y) = xy - x^3 - y^2$　　**13.** $f(x, y) = x^2 + y^2 + \dfrac{2}{xy}$

14. $f(x, y) = xe^y$　　**15.** $f(x, y) = x^2 + y - e^y$

16. $f(x, y) = xy + \dfrac{2}{x} + \dfrac{4}{y}$　　**17.** $f(x, y) = e^x \sin y$

18. $f(x, y) = y \sin x$　　**19.** $f(x, y) = e^{-(x^2+y^2+2x)}$

20. $f(x, y) = xy + \dfrac{a^3}{x} + \dfrac{b^3}{y}$　　$(a \neq 0, b \neq 0)$

C **21.** Use a CAS to generate a contour plot of

$$f(x, y) = 2x^2 - 4xy + y^4 + 2$$

for $-2 \leq x \leq 2$ and $-2 \leq y \leq 2$, and use the plot to approximate the location of all relative extrema and saddle points in the region. Check your answer using calculus, and identify the relative extrema as relative maxima or minima.

C **22.** Use a CAS to generate a contour plot of

$$f(x, y) = 2y^2x - yx^2 + 4xy$$

for $-5 \leq x \leq 5$ and $-5 \leq y \leq 5$, and use the plot to approximate the location of all relative extrema and saddle points in the region. Check your answer using calculus, and identify the relative extrema as relative maxima or minima.

23. (a) Show that the second partials test provides no information about the critical points of $f(x, y) = x^4 + y^4$.
(b) Classify all critical points of f as relative maxima, relative minima, or saddle points.

24. (a) Show that the second partials test provides no information about the critical points of $f(x, y) = x^4 - y^4$.
(b) Classify all critical points of f as relative maxima, relative minima, or saddle points.

25. Recall from Theorem 6.1.5 that if a continuous function of one variable has exactly one relative extremum on an interval, then that relative extremum is an absolute extremum on the interval. This exercise shows that this result does not extend to functions of two variables.

(a) Show that $f(x, y) = 3xe^y - x^3 - e^{3y}$ has only one critical point and that a relative maximum occurs there. (See the accompanying figure.)
(b) Show that f does not have an absolute maximum.

[This exercise is based on the article "The Only Critical Point in Town Test" by Ira Rosenholtz and Lowell Smylie, *Mathematics Magazine*, Vol. 58, No. 3, May 1985, pp. 149–150.]

26. If f is a continuous function of one variable with two relative maxima on an interval, then there must be a relative minimum between the relative maxima. (Convince yourself of this by drawing some pictures.) The purpose of this exercise is to show that this result does not extend to functions of two variables. Show that $f(x, y) = 4x^2e^y - 2x^4 - e^{4y}$ has two relative maxima but no other critical points (see the accompanying figure).

[This exercise is based on the problem "Two Mountains Without a Valley" proposed and solved by Ira Rosenholtz, *Mathematics Magazine*, Vol. 60, No. 1, February 1987, p. 48.]

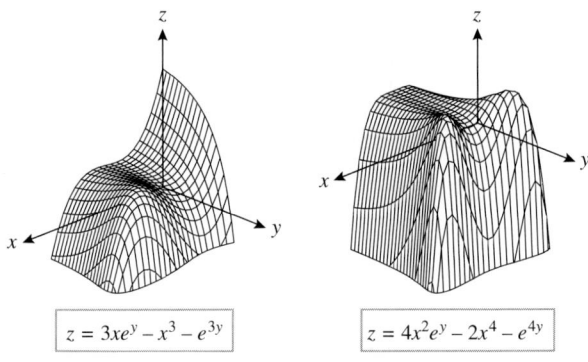

$z = 3xe^y - x^3 - e^{3y}$　　$z = 4x^2e^y - 2x^4 - e^{4y}$

Figure Ex-25　　Figure Ex-26

In Exercises 27–32, find the absolute extrema of the given function on the indicated closed and bounded set R.

27. $f(x, y) = xy - x - 3y$; R is the triangular region with vertices $(0, 0)$, $(0, 4)$, and $(5, 0)$.

28. $f(x, y) = xy - 2x$; R is the triangular region with vertices $(0, 0)$, $(0, 4)$, and $(4, 0)$.

29. $f(x, y) = x^2 - 3y^2 - 2x + 6y$; R is the square region with vertices $(0, 0)$, $(0, 2)$, $(2, 2)$, and $(2, 0)$.

30. $f(x, y) = xe^y - x^2 - e^y$; R is the rectangular region with vertices $(0, 0)$, $(0, 1)$, $(2, 1)$, and $(2, 0)$.

31. $f(x, y) = x^2 + 2y^2 - x$; R is the circular region $x^2 + y^2 \leq 4$.

32. $f(x, y) = xy^2$; R is the region that satisfies the inequalities $x \geq 0$, $y \geq 0$, and $x^2 + y^2 \leq 1$.

33. Find three positive numbers whose sum is 48 and such that their product is as large as possible.

34. Find three positive numbers whose sum is 27 and such that the sum of their squares is as small as possible.

35. Find all points on the portion of the plane $x + y + z = 5$ in the first octant at which $f(x, y, z) = xy^2z^2$ has a maximum value.

36. Find the points on the surface $x^2 - yz = 5$ that are closest to the origin.

37. Find the dimensions of the rectangular box of maximum volume that can be inscribed in a sphere of radius a.

38. Find the maximum volume of a rectangular box with three faces in the coordinate planes and a vertex in the first octant on the plane $x + y + z = 1$.

39. A closed rectangular box with a volume of 16 ft^3 is made from two kinds of materials. The top and bottom are made of material costing 10¢ per square foot and the sides from material costing 5¢ per square foot. Find the dimensions of the box so that the cost of materials is minimized.

40. A manufacturer makes two models of an item, standard and deluxe. It costs $40 to manufacture the standard model and $60 for the deluxe. A market research firm estimates that if the standard model is priced at x dollars and the deluxe at y dollars, then the manufacturer will sell $500(y - x)$ of the standard items and $45,000 + 500(x - 2y)$ of the deluxe each year. How should the items be priced to maximize the profit?

41. Consider the function

$$f(x, y) = 4x^2 - 3y^2 + 2xy$$

over the unit square $0 \le x \le 1, 0 \le y \le 1$.
(a) Find the maximum and minimum values of f on each edge of the square.
(b) Find the maximum and minimum values of f on each diagonal of the square.
(c) Find the maximum and minimum values of f on the entire square.

42. Show that among all parallelograms with perimeter l, a square with sides of length $l/4$ has maximum area. [*Hint:* The area of a parallelogram is given by the formula $A = ab \sin \alpha$, where a and b are the lengths of two adjacent sides and α is the angle between them.]

43. Determine the dimensions of a rectangular box, open at the top, having volume V, and requiring the least amount of material for its construction.

44. A length of sheet metal 27 inches wide is to be made into a water trough by bending up two sides as shown in the accompanying figure. Find x and ϕ so that the trapezoid-shaped cross section has a maximum area.

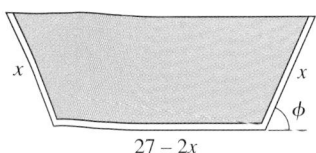

$27 - 2x$

Figure Ex-44

A common problem in experimental work is ... ematical relationship $y = f(x)$ between t... y by "fitting" a curve to points in the pl... to experimentally determined values of x and y,

$$(x_1, y_1), (x_2, y_2), \ldots, (x_n, y_n)$$

The curve $y = f(x)$ is called a ***mathematical model*** of the data. The general form of the function f is commonly determined by some underlying physical principle, but sometimes it is just determined by the pattern of the data. We are concerned with fitting a straight line $y = mx + b$ to data. Usually, the data will not lie on a line (possibly due to experimental error or variations in experimental conditions), so the problem is to find a line that fits the data "best" according to some criterion. One criterion for selecting the line of best fit is to choose m and b to minimize the function

$$g(m, b) = \sum_{i=1}^{n} (mx_i + b - y_i)^2$$

This is called the ***method of least squares***, and the resulting line is called the ***regression line*** or the ***least-squares line of best fit***. Geometrically, $|mx_i + b - y_i|$ is the vertical distance between the data point (x_i, y_i) and the line $y = mx + b$.

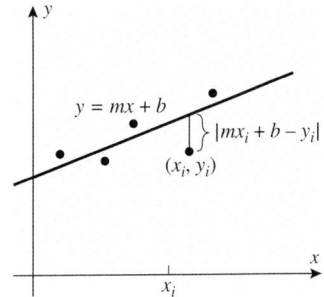

These vertical distances are called the ***residuals*** of the data points, so the effect of minimizing $g(m, b)$ is to minimize the sum of the squares of the residuals. In Exercises 45 and 46, we will derive a formula for the regression line. More on this topic can be found in the module entitled "Functions from Data" at the end of Chapter 5.

45. The purpose of this exercise is to find the values of m and b that produce the regression line.
(a) To minimize $g(m, b)$, we start by finding values of m and b such that $\partial g / \partial m = 0$ and $\partial g / \partial b = 0$. Show that these equations are satisfied if m and b satisfy the conditions

$$\left(\sum_{i=1}^{n} x_i^2 \right) m + \left(\sum_{i=1}^{n} x_i \right) b = \sum_{i=1}^{n} x_i y_i$$

$$\left(\sum_{i=1}^{n} x_i \right) m + nb = \sum_{i=1}^{n} y_i$$

(b) The equations in part (a) are two linear equations in the two unknowns m and b. Solve these equations to obtain

$$m = \frac{n \sum\limits_{i=1}^{n} x_i y_i - \sum\limits_{i=1}^{n} x_i \sum\limits_{i=1}^{n} y_i}{n \sum\limits_{i=1}^{n} x_i^2 - \left(\sum\limits_{i=1}^{n} x_i \right)^2}$$

$$b = \frac{1}{n} \left(\sum\limits_{i=1}^{n} y_i - m \sum\limits_{i=1}^{n} x_i \right)$$

[*Note:* We have shown that g has a critical point at these values of m and b. In the next exercise we will show that g has an absolute minimum at this critical point. Accepting this to be so, we have shown that the line $y = mx+b$ is the regression line for these values of m and b.]

46. The purpose of this exercise is to show that $g(m, b)$ has an absolute minimum for the values of m and b obtained in Exercise 45.

(a) It was shown in Exercise 58 of Section 6.2 that the value of $\sum_{i=1}^{n}(x_i - \bar{x})^2$ is minimized by taking $\bar{x}$ to be the arithmetic average

$$\bar{x} = \frac{1}{n} \sum_{i=1}^{n} x_i$$

Use this fact to show that

$$n \left(\sum_{i=1}^{n} x_i^2 \right) - \left(\sum_{i=1}^{n} x_i \right)^2 \geq 0$$

[*Note:* $\sum_{i=1}^{n}(x_i - \bar{x})^2 > 0$ if the x_i's are not all the same.]

(b) Find the partial derivatives $g_{mm}(m, b)$, $g_{bb}(m, b)$, and $g_{mb}(m, b)$, and then apply the second partials test to show that g has a relative minimum at the critical point obtained in Exercise 45.

(c) Show that the graph of the equation $z = g(m, b)$ is a quadric surface. [*Hint:* See Formula (4) of Section 13.7.]

(d) It can be proved that the graph of $z = g(m, b)$ is an elliptic paraboloid. Accepting this to be so, show that this paraboloid opens in the positive z-direction, and explain how this shows that g has an absolute minimum at the critical point obtained in Exercise 45.

In Exercises 47–50, use the formulas obtained in Exercise 45 to find and draw the regression line. If you have a calculating utility that can calculate regression lines, use it to check your work.

47.

48.

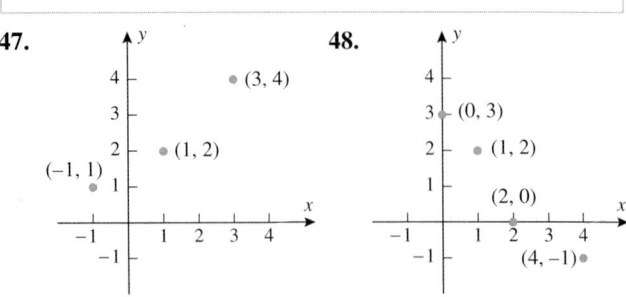

49.

x	1	2	3	4
y	1.5	1.6	2.1	3.0

50.

x	1	2	3	4	5
y	4.2	3.5	3.0	2.4	2.0

51. The following table shows the life expectancy by year of birth of females in the United States:

YEAR OF BIRTH	1930	1940	1950	1960	1970	1980	1990
LIFE EXPECTANCY	61.6	65.2	71.1	73.1	74.7	77.5	78.8

(a) Take $t = 0$ to be the year 1930, and let y be the life expectancy for birth year t. Use the regression capability of a calculating utility to find the regression line of y as a function of t.

(b) Use a graphing utility to make a graph that shows the data points and the regression line.

(c) Use the regression line to make a conjecture about the life expectancy of females born in the year 2000.

52. A company manager wants to establish a relationship between the sales of a certain product and the price. The company research department provides the following data:

PRICE (x) IN DOLLARS	$35.00	$40.00	$45.00	$48.00	$50.00
DAILY SALES VOLUME (y) IN UNITS	80	75	68	66	63

(a) Use a calculating utility to find the regression line of y as a function of x.

(b) Use a graphing utility to make a graph that shows the data points and the regression line.

(c) Use the regression line to make a conjecture about the number of units that would be sold at a price of $60.00.

53. If a gas is cooled with its volume held constant, then it follows from the *ideal gas law* in physics that its pressure drops proportionally to the drop in temperature. The temperature that, in theory, corresponds to a pressure of zero is called *absolute zero*. Suppose that an experiment produces the following data for pressure P versus temperature T with the volume held constant:

P (KILOPASCALS)	134	142	155	160	171	184
T (°CELSIUS)	0	20	40	60	80	100

(a) Use a calculating utility to find the regression line of P as a function of T.

(b) Use a graphing utility to make a graph that shows the data points and the regression line.

(c) Use the regression line to estimate the value of absolute zero in °Celsius.

54. Find:

(a) a continuous function $f(x, y)$ that is defined on the entire xy-plane and has no absolute extrema on the xy-plane;

(b) a function $f(x, y)$ that is defined everywhere on the rectangle $0 \leq x \leq 1, 0 \leq y \leq 1$ and has no absolute extrema on the rectangle.

55. Show that if f has a relative maximum at (x_0, y_0), then $G(x) = f(x, y_0)$ has a relative maximum at $x = x_0$ and $H(y) = f(x_0, y)$ has a relative maximum at $y = y_0$.

15.9 LAGRANGE MULTIPLIERS

In this section we will study a powerful new method for maximizing or minimizing a function subject to constraints on the variables. This method will help us to solve certain optimization problems that are difficult or impossible to solve using the methods studied in the last section.

EXTREMUM PROBLEMS WITH CONSTRAINTS

In Example 6 of the last section, we solved the problem of minimizing

$$S = xy + 2xz + 2yz \tag{1}$$

subject to the constraint

$$xyz - 32 = 0 \tag{2}$$

This is a special case of the following general problem:

15.9.1 Three-Variable Extremum Problem with One Constraint
Maximize or minimize the function $f(x, y, z)$ subject to the constraint $g(x, y, z) = 0$.

We will also be interested in the following two-variable version of this problem:

15.9.2 Two-Variable Extremum Problem with One Constraint
Maximize or minimize the function $f(x, y)$ subject to the constraint $g(x, y) = 0$.

LAGRANGE MULTIPLIERS

One way to attack problems of these types is to solve the constraint equation for one of the variables in terms of the others and substitute the result into f. This produces a new function of one or two variables that incorporates the constraint and can be maximized or minimized by applying standard methods. For example, to solve the problem in Example 6 of the last section we substituted (2) into (1) to obtain

$$S = xy + \frac{64}{y} + \frac{64}{x}$$

which we then minimized by finding the critical points and applying the second partials test. However, this approach hinges on our ability to solve the constraint equation for one of the variables in terms of the others. If this cannot be done, then other methods must be used. One such method, called the *method of Lagrange*[*] *multipliers*, will be discussed in this section.

To motivate the method of Lagrange multipliers, suppose that we are trying to maximize a function $f(x, y)$ subject to the constraint $g(x, y) = 0$. Geometrically, this means that we are looking for a point (x_0, y_0) on the graph of the constraint curve at which $f(x, y)$ is as large as possible. To help locate such a point, let us construct a contour plot of $f(x, y)$ in the same coordinate system as the graph of $g(x, y) = 0$. For example, Figure 15.9.1a shows some typical level curves of $f(x, y) = c$, which we have labeled $c = 100, 200, 300, 400,$ and 500 for purposes of illustration. In this figure, each point of intersection of $g(x, y) = 0$ with a level curve is a candidate for a solution, since these points lie on the constraint curve.

---

[*] See biography on page 968.

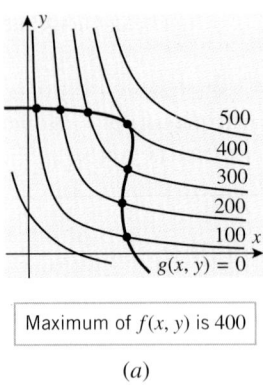

Maximum of $f(x, y)$ is 400

(a)

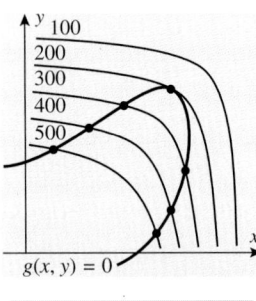

Minimum of $f(x, y)$ is 200

(b)

Figure 15.9.1

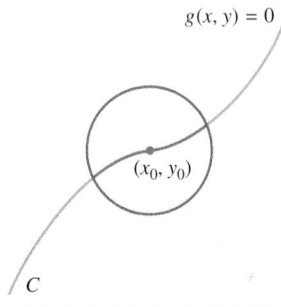

A constrained relative maximum occurs at (x_0, y_0) if $f(x_0, y_0) \geq f(x, y)$ on some segment of C that extends on both sides of (x_0, y_0).

Figure 15.9.2

Among the seven such intersections shown in the figure, the maximum value of $f(x, y)$ occurs at the intersection where $f(x, y)$ has a value of 400, which is the point where the constraint curve and the level curve just touch. Observe that at this point the level curve and the constraint curve have a common normal line. This suggests that the maximum of $f(x, y)$, if it exists, occurs at a point (x_0, y_0) on the constraint curve at which the gradient vectors ∇f and ∇g are scalar multiples of one another; that is,

$$\nabla f(x_0, y_0) = \lambda \nabla g(x_0, y_0) \tag{3}$$

for some scalar λ. The same condition holds at points on the constraint curve where $f(x, y)$ has a minimum. For example, if the level curves are as shown in Figure 15.9.1b, then the minimum value of $f(x, y)$ occurs where the constraint curve just touches a level curve. Thus, to find the maximum or minimum of $f(x, y)$ subject to the constraint $g(x, y) = 0$, we look for points at which (3) holds—this is the method of Lagrange multipliers.

Our next objective in this section is to make the preceding intuitive argument more precise. For this purpose it will help to begin with some terminology about the problem of maximizing or minimizing a function $f(x, y)$ subject to a constraint $g(x, y) = 0$. As with other kinds of maximization and minimization problems, we need to distinguish between relative and absolute extrema. We will say that f has a ***constrained absolute maximum (minimum)*** at (x_0, y_0) if $f(x_0, y_0)$ is the largest (smallest) value of f on the constraint curve, and we will say that f has a ***constrained relative maximum (minimum)*** at (x_0, y_0) if $f(x_0, y_0)$ is the largest (smallest) value of f on some segment of the constraint curve that extends on both sides of the point (x_0, y_0) (Figure 15.9.2).

Let us assume that a constrained relative maximum or minimum occurs at the point (x_0, y_0) and for simplicity, let us further assume that the equation $g(x, y) = 0$ can be smoothly parametrized as

$$x = x(s), \quad y = y(s)$$

where s is an arc length parameter with reference point (x_0, y_0) at $s = 0$. Thus, the quantity

$$z = f(x(s), y(s))$$

has a relative maximum or minimum at $s = 0$, and this implies that $dz/ds = 0$ at that point. From the chain rule, this equation can be expressed as

$$\frac{dz}{ds} = \frac{\partial f}{\partial x}\frac{dx}{ds} + \frac{\partial f}{\partial y}\frac{dy}{ds} = \left(\frac{\partial f}{\partial x}\mathbf{i} + \frac{\partial f}{\partial y}\mathbf{j}\right) \cdot \left(\frac{dx}{ds}\mathbf{i} + \frac{dy}{ds}\mathbf{j}\right) = 0$$

where the derivatives are all evaluated at $s = 0$. However, the first factor in the dot product is the gradient of f, and the second factor is the unit tangent vector to the constraint curve.

---

*JOSEPH LOUIS LAGRANGE (1736–1813). French–Italian mathematician and astronomer. Lagrange, the son of a public official, was born in Turin, Italy. (Baptismal records list his name as Giuseppe Lodovico Lagrangia.) Although his father wanted him to be a lawyer, Lagrange was attracted to mathematics and astronomy after reading a memoir by the astronomer Halley. At age 16 he began to study mathematics on his own and by age 19 was appointed to a professorship at the Royal Artillery School in Turin. The following year Lagrange sent Euler solutions to some famous problems using new methods that eventually blossomed into a branch of mathematics called calculus of variations. These methods and Lagrange's applications of them to problems in celestial mechanics were so monumental that by age 25 he was regarded by many of his contemporaries as the greatest living mathematician.

In 1776, on the recommendations of Euler, he was chosen to succeed Euler as the director of the Berlin Academy. During his stay in Berlin, Lagrange distinguished himself not only in celestial mechanics, but also in algebraic equations and the theory of numbers. After twenty years in Berlin, he moved to Paris at the invitation of Louis XVI. He was given apartments in the Louvre and treated with great honor, even during the revolution.

Napoleon was a great admirer of Lagrange and showered him with honors—count, senator, and Legion of Honor. The years Lagrange spent in Paris were devoted primarily to didactic treatises summarizing his mathematical conceptions. One of Lagrange's most famous works is a memoir, *Mécanique Analytique*, in which he reduced the theory of mechanics to a few general formulas from which all other necessary equations could be derived.

It is an interesting historical fact that Lagrange's father speculated unsuccessfully in several financial ventures, so his family was forced to live quite modestly. Lagrange himself stated that if his family had money, he would not have made mathematics his vocation. In spite of his fame, Lagrange was always a shy and modest man. On his death, he was buried with honor in the Pantheon.

Since the point (x_0, y_0) corresponds to $s = 0$, it follows from this equation that

$$\nabla f(x_0, y_0) \cdot \mathbf{T}(0) = 0$$

which implies that the gradient is normal to the constraint curve at a constrained relative extremum. However, the constraint curve $g(x, y) = 0$ is a level curve for the function $g(x, y)$, so $\nabla g(x_0, y_0)$ is also normal to this curve at (x_0, y_0). Thus, we have shown that $\nabla f(x_0, y_0)$ and $\nabla g(x_0, y_0)$ are parallel vectors at any point on the constraint curve where a constrained relative extremum occurs. Stated another way, at a constrained relative extremum there is some scalar λ such that

$$\nabla f(x_0, y_0) = \lambda \nabla g(x_0, y_0) \tag{4}$$

This scalar is called a **Lagrange multiplier**. Thus, the **method of Lagrange multipliers** for finding constrained relative extrema is to look for points on the constraint curve $g(x, y) = 0$ at which Equation (4) is satisfied for some scalar λ.

15.9.3 THEOREM (*Constrained-Extremum Principle for Two Variables and One Constraint*). *Let f and g be functions of two variables with continuous first partial derivatives on some open set containing the constraint curve $g(x, y) = 0$, and assume that $\nabla g \neq \mathbf{0}$ at any point on this curve. If f has a constrained relative extremum, then this extremum occurs at a point (x_0, y_0) on the constraint curve at which the gradient vectors $\nabla f(x_0, y_0)$ and $\nabla g(x_0, y_0)$ are parallel; that is, there is some number λ such that*

$$\nabla f(x_0, y_0) = \lambda \nabla g(x_0, y_0)$$

Example 1

At what point or points on the circle $x^2 + y^2 = 1$ does $f(x, y) = xy$ have an absolute maximum, and what is that maximum?

Solution. Since the circle $x^2 + y^2 = 1$ is a closed and bounded set, and since $f(x, y) = xy$ is a continuous function, it follows from the Extreme-Value Theorem (Theorem 15.8.3) that f has an absolute maximum and an absolute minimum on the circle. To find these extrema, we will use Lagrange multipliers to find the constrained relative extrema, and then we will evaluate f at those relative extrema to find the absolute extrema.

We want to maximize $f(x, y) = xy$ subject to the constraint

$$g(x, y) = x^2 + y^2 - 1 = 0 \tag{5}$$

First we will look for constrained *relative* extrema. For this purpose we will need the gradients

$$\nabla f = y\mathbf{i} + x\mathbf{j} \quad \text{and} \quad \nabla g = 2x\mathbf{i} + 2y\mathbf{j}$$

From the formula for ∇g we see that $\nabla g = \mathbf{0}$ if and only if $x = 0$ and $y = 0$, so $\nabla g \neq \mathbf{0}$ at any point on the circle $x^2 + y^2 = 1$. Thus, at a constrained relative extremum we must have

$$\nabla f = \lambda \nabla g \quad \text{or} \quad y\mathbf{i} + x\mathbf{j} = \lambda(2x\mathbf{i} + 2y\mathbf{j})$$

which is equivalent to the pair of equations

$$y = 2x\lambda \quad \text{and} \quad x = 2y\lambda$$

It follows from these equations that if $x = 0$, then $y = 0$, and if $y = 0$, then $x = 0$. In either case we have $x^2 + y^2 = 0$, so the contraint equation $x^2 + y^2 = 1$ is not satisfied. Thus, we can assume that x and y are nonzero, and we can rewrite the equations as

$$\lambda = \frac{y}{2x} \quad \text{and} \quad \lambda = \frac{x}{2y}$$

from which we obtain

$$\frac{y}{2x} = \frac{x}{2y}$$

or

$$y^2 = x^2 \qquad\qquad (6)$$

Substituting this in (5) yields

$$2x^2 - 1 = 0$$

from which we obtain $x = \pm 1/\sqrt{2}$. Each of these values, when substituted in Equation (6), produces y-values of $y = \pm 1/\sqrt{2}$. Thus, constrained relative extrema occur at the points $(1/\sqrt{2}, 1/\sqrt{2})$, $(1/\sqrt{2}, -1/\sqrt{2})$, $(-1/\sqrt{2}, 1/\sqrt{2})$, and $(-1/\sqrt{2}, -1/\sqrt{2})$. The values of xy at these points are as follows:

(x, y)	$(1/\sqrt{2}, 1/\sqrt{2})$	$(1/\sqrt{2}, -1/\sqrt{2})$	$(-1/\sqrt{2}, 1/\sqrt{2})$	$(-1/\sqrt{2}, -1/\sqrt{2})$
xy	1/2	-1/2	-1/2	1/2

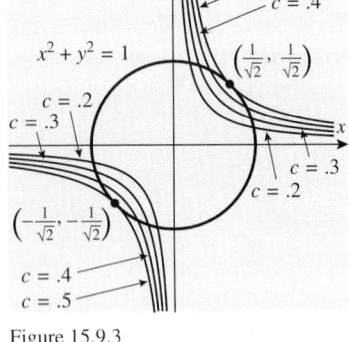

Figure 15.9.3

Thus, the function $f(x, y) = xy$ has an absolute maximum of $\frac{1}{2}$ occurring at the two points $(1/\sqrt{2}, 1/\sqrt{2})$ and $(-1/\sqrt{2}, -1/\sqrt{2})$. Although it was not asked for, we can also see that f has an absolute minimum of $-\frac{1}{2}$ occurring at the points $(1/\sqrt{2}, -1/\sqrt{2})$ and $(-1/\sqrt{2}, 1/\sqrt{2})$. Figure 15.9.3 shows some level curves $xy = c$ and the constraint curve in the vicinity of the maxima. A similar figure for the minima can be obtained using negative values of c for the level curves $xy = c$. ◀

REMARK.   If c is a constant, then the functions $g(x, y)$ and $g(x, y) - c$ have the same gradient since the constant c drops out when we differentiate. Consequently, it is *not* essential to rewrite a constraint of the form $g(x, y) = c$ as $g(x, y) - c = 0$ in order to apply the constrained-extremum principle. Thus, in the last example, we could have kept the constraint in the form $x^2 + y^2 = 1$ and then taken $g(x, y) = x^2 + y^2$ rather than $g(x, y) = x^2 + y^2 - 1$.

Example 2

Use the method of Lagrange multipliers to find the dimensions of a rectangle with perimeter p and maximum area.

Solution. Let

$$x = \text{length of the rectangle}$$
$$y = \text{width of the rectangle}$$
$$A = \text{area of the rectangle}$$

We want to maximize $A = xy$ subject to the perimeter constraint

$$2x + 2y = p \qquad\qquad (7)$$

If we let $f(x, y) = xy$ and $g(x, y) = 2x + 2y$, then we have

$$\nabla f = y\mathbf{i} + x\mathbf{j} \quad \text{and} \quad \nabla g = 2\mathbf{i} + 2\mathbf{j}$$

Noting that $\nabla g \neq \mathbf{0}$, it follows from (4) that

$$y\mathbf{i} + x\mathbf{j} = \lambda(2\mathbf{i} + 2\mathbf{j})$$

at a constrained relative maximum. This is equivalent to the two equations

$$y = 2\lambda \quad \text{and} \quad x = 2\lambda$$

Eliminating λ from these equations we obtain $x = y$, which shows that the rectangle is actually a square. Using this condition and constraint (7), we obtain $x = p/4$, $y = p/4$. ◀

**THREE VARIABLES AND ONE
CONSTRAINT**

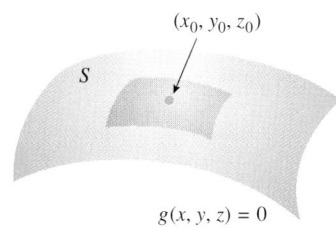

(x_0, y_0, z_0)

S

$g(x, y, z) = 0$

A constrained relative maximum occurs
at (x_0, y_0, z_0) if $f(x_0, y_0, z_0) \geq f(x, y, z)$
on some patch of S that contains
(x_0, y_0, z_0) in its interior.

Figure 15.9.4

The method of Lagrange multipliers can also be used to maximize or minimize a function
of three variables $f(x, y, z)$ subject to a constraint $g(x, y, z) = 0$. As a rule, the graph
of $g(x, y, z) = 0$ will be some surface S in 3-space. Thus, from a geometric viewpoint,
the problem is to maximize or minimize $f(x, y, z)$ as (x, y, z) varies over the surface S
(Figure 15.9.4). As usual, we distinguish between relative and absolute extrema. We will
say that f has a ***constrained absolute maximum (minimum)*** at (x_0, y_0, z_0) if $f(x_0, y_0, z_0)$
is the largest (smallest) value of $f(x, y, z)$ on S, and we will say that f has a ***constrained
relative maximum (minimum)*** at (x_0, y_0, z_0) if $f(x_0, y_0, z_0)$ is the largest (smallest) value
of $f(x, y, z)$ on some patch of the surface S that contains (x_0, y_0, z_0) in its interior.

The following theorem, which we state without proof, is the three-variable analog of
Theorem 15.9.3.

> **15.9.4** THEOREM (***Constrained-Extremum Principle for Three Variables and One Constraint***). *Let
> f and g be functions of three variables with continuous first partial derivatives on some
> open set containing the constraint surface $g(x, y, z) = 0$, and assume that $\nabla g \neq \mathbf{0}$ at
> any point on this surface. If f has a constrained relative extremum, then this extremum
> occurs at a point (x_0, y_0, z_0) on the constraint surface at which the gradient vectors
> $\nabla f(x_0, y_0, z_0)$ and $\nabla g(x_0, y_0, z_0)$ are parallel; that is, there is some number λ such that*
>
> $$\nabla f(x_0, y_0, z_0) = \lambda \nabla g(x_0, y_0, z_0)$$

Example 3

Find the points on the sphere $x^2 + y^2 + z^2 = 36$ that are closest to and farthest from the
point $(1, 2, 2)$.

Solution. To avoid radicals, we will find points on the sphere that minimize and maximize
the *square* of the distance to $(1, 2, 2)$. Thus, we want to find the relative extrema of

$$f(x, y, z) = (x - 1)^2 + (y - 2)^2 + (z - 2)^2$$

subject to the constraint

$$x^2 + y^2 + z^2 = 36 \tag{8}$$

If we let $g(x, y, z) = x^2 + y^2 + z^2$, then $\nabla g = 2x\mathbf{i} + 2y\mathbf{j} + 2z\mathbf{k}$. Thus, $\nabla g = \mathbf{0}$ if and only
if $x = y = z = 0$. It follows that $\nabla g \neq \mathbf{0}$ at any point of the constraint curve (8), and hence
the constrained relative extrema must occur at points where

$$\nabla f(x, y, z) = \lambda \nabla g(x, y, z)$$

That is,

$$2(x - 1)\mathbf{i} + 2(y - 2)\mathbf{j} + 2(z - 2)\mathbf{k} = \lambda(2x\mathbf{i} + 2y\mathbf{j} + 2z\mathbf{k})$$

which leads to the equations

$$2(x - 1) = 2x\lambda, \quad 2(y - 2) = 2y\lambda, \quad 2(z - 2) = 2z\lambda \tag{9}$$

We may assume that x, y, and z are nonzero since $x = 0$ does not satisfy the first equation,
$y = 0$ does not satisfy the second, and $z = 0$ does not satisfy the third. Thus, we can rewrite
(9) as

$$\frac{x - 1}{x} = \lambda, \quad \frac{y - 2}{y} = \lambda, \quad \frac{z - 2}{z} = \lambda$$

The first two equations imply that

$$\frac{x - 1}{x} = \frac{y - 2}{y}$$

from which it follows that

$$y = 2x \qquad (10)$$

Similarly, the first and third equations imply that

$$z = 2x \qquad (11)$$

Substituting (10) and (11) in the constraint equation (8), we obtain

$$9x^2 = 36 \quad \text{or} \quad x = \pm 2$$

Substituting these values in (10) and (11) yields two points

$$(2, 4, 4) \quad \text{and} \quad (-2, -4, -4)$$

Since $f(2, 4, 4) = 9$ and $f(-2, -4, -4) = 81$, it follows that $(2, 4, 4)$ is the point on the sphere closest to $(1, 2, 2)$, and $(-2, -4, -4)$ is the point that is farthest (Figure 15.9.5).   ◄

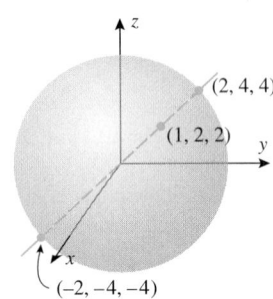

(2, 4, 4)

(1, 2, 2)

(−2, −4, −4)

Figure 15.9.5

Next we will use Lagrange multipliers to solve the problem of Example 6 in the last section.

Example 4

Use Lagrange multipliers to determine the dimensions of a rectangular box, open at the top, having a volume of 32 ft³, and requiring the least amount of material for its construction.

Solution. With the notation introduced in Example 6 of the last section, the problem is to minimize the surface area

$$S = xy + 2xz + 2yz$$

subject to the volume constraint

$$xyz = 32 \qquad (12)$$

If we let $f(x, y, z) = xy + 2xz + 2yz$ and $g(x, y, z) = xyz$, then

$$\nabla f = (y + 2z)\mathbf{i} + (x + 2z)\mathbf{j} + (2x + 2y)\mathbf{k} \quad \text{and} \quad \nabla g = yz\mathbf{i} + xz\mathbf{j} + xy\mathbf{k}$$

It follows that $\nabla g \neq \mathbf{0}$ at any point on the surface $xyz = 32$, since x, y, and z are all nonzero on this surface. Thus, at a constrained relative extremum we must have $\nabla f = \lambda \nabla g$, that is,

$$(y + 2z)\mathbf{i} + (x + 2z)\mathbf{j} + (2x + 2y)\mathbf{k} = \lambda(yz\mathbf{i} + xz\mathbf{j} + xy\mathbf{k})$$

This condition yields the three equations

$$y + 2z = \lambda yz, \quad x + 2z = \lambda xz, \quad 2x + 2y = \lambda xy$$

Because x, y, and z are nonzero, these equations can be rewritten as

$$\frac{1}{z} + \frac{2}{y} = \lambda, \quad \frac{1}{z} + \frac{2}{x} = \lambda, \quad \frac{2}{y} + \frac{2}{x} = \lambda$$

From the first two equations,

$$y = x \qquad (13)$$

and from the first and third equations,

$$z = \tfrac{1}{2}x \qquad (14)$$

Substituting (13) and (14) in the volume constraint (12) yields

$$\tfrac{1}{2}x^3 = 32$$

This equation, together with (13) and (14), yields

$$x = 4, \quad y = 4, \quad z = 2$$

which agrees with the result that was obtained in Example 6 of the last section.   ◄

There are variations in the method of Lagrange multipliers that can be used to solve problems with two or more constraints. However, we will not discuss that topic here.

EXERCISE SET 15.9 ~ Graphing Calculator   [c] CAS

1. The accompanying figure shows graphs of the line $x + y = 4$ and the level curves of height $c = 2, 4, 6,$ and 8 for the function $f(x, y) = xy$.
   (a) Use the figure to find the maximum value of the function $f(x, y) = xy$ subject to the constraint $x + y = 4$, and explain your reasoning.
   (b) How can you tell from the figure that you have not obtained the minimum value of f subject to the constraint?
   (c) Use Lagrange multipliers to check your work.

2. The accompanying figure shows the graphs of the line $3x + 4y = 25$ and the level curves of height $c = 9, 16, 25, 36,$ and 49 for the function $f(x, y) = x^2 + y^2$.
   (a) Use the accompanying figure to find the minimum value of the function $f(x, y) = x^2 + y^2$ subject to the constraint $3x + 4y = 25$, and explain your reasoning.
   (b) How can you tell from the accompanying figure that you have not obtained the maximum value of f subject to the constraint?
   (c) Use Lagrange multipliers to check your work.

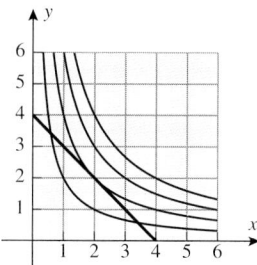

Figure Ex-1   Figure Ex-2

In Exercises 3–10, use Lagrange multipliers to find the maximum and minimum values of f subject to the given constraint. Also, find the points at which these extreme values occur.

3. $f(x, y) = xy$; $4x^2 + 8y^2 = 16$

4. $f(x, y) = x^2 - y$; $x^2 + y^2 = 25$

5. $f(x, y) = 4x^3 + y^2$; $2x^2 + y^2 = 1$

6. $f(x, y) = x - 3y - 1$; $x^2 + 3y^2 = 16$

7. $f(x, y, z) = 2x + y - 2z$; $x^2 + y^2 + z^2 = 4$

8. $f(x, y, z) = 3x + 6y + 2z$; $2x^2 + 4y^2 + z^2 = 70$

9. $f(x, y, z) = xyz$; $x^2 + y^2 + z^2 = 1$

10. $f(x, y, z) = x^4 + y^4 + z^4$; $x + y + z = 1$

In Exercises 11–18, solve using Lagrange multipliers.

11. Find the point on the line $2x - 4y = 3$ that is closest to the origin.

12. Find the point on the line $y = 2x + 3$ that is closest to $(4, 2)$.

13. Find the point on the plane $x + 2y + z = 1$ that is closest to the origin.

14. Find the point on the plane $4x + 3y + z = 2$ that is closest to $(1, -1, 1)$.

15. Find the points on the circle $x^2 + y^2 = 45$ that are closest to and farthest from $(1, 2)$.

16. Find the points on the surface $xy - z^2 = 1$ that are closest to the origin.

17. Find a vector in 3-space whose length is 5 and whose components have the largest possible sum.

18. Suppose that the temperature at a point (x, y) on a metal plate is $T(x, y) = 4x^2 - 4xy + y^2$. An ant, walking on the plate, traverses a circle of radius 5 centered at the origin. What are the highest and lowest temperatures encountered by the ant?

~ 19. (a) Use a graphing utility to graph the circle $x^2 + y^2 = 25$ and two distinct level curves of $f(x, y) = x^2 - y$ that just touch the circle.
   (b) Use the results you obtained in part (a) to approximate the maximum and minimum values of f subject to the constraint $x^2 + y^2 = 25$.
   (c) Check your approximations in part (b) using Lagrange multipliers.

[c] 20. (a) If you have a CAS that can generate implicit curves, use it to graph the circle $(x - 4)^2 + (y - 4)^2 = 4$ and two level curves of the function $f(x, y) = x^3 + y^3 - 3xy$ that just touch the circle.
   (b) Use the result you obtained in part (a) to approximate the minimum value of f subject to the constraint $(x - 4)^2 + (y - 4)^2 = 4$.
   (c) Confirm graphically that you have found a minimum and not a maximum.
   (d) Check your approximation using Lagrange multipliers and solving the required equations numerically.

In Exercises 21–28, use Lagrange multipliers to solve the indicated problems from Section 15.8.

21. Exercise 34   22. Exercise 35

23. Exercise 36   24. Exercise 37

25. Exercise 39   **26.** Exercise 41

27. Exercise 42   **28.** Exercise 43

[c] **29.** Let α, β, and γ be the angles of a triangle.
(a) Use Lagrange multipliers to find the maximum value of $f(\alpha, \beta, \gamma) = \cos\alpha \cos\beta \cos\gamma$, and determine the angles for which the maximum occurs.
(b) Express $f(\alpha, \beta, \gamma)$ as a function of α and β alone, and use a CAS to graph this function of two variables. Confirm that the result obtained in part (a) is consistent with the graph.

30. The accompanying figure shows the intersection of the elliptic paraboloid $z = x^2 + 4y^2$ and the right circular cylinder $x^2 + y^2 = 1$. Use Lagrange multipliers to find the highest and lowest points on the curve of intersection.

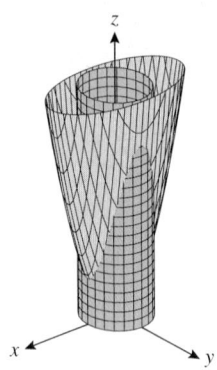

Figure Ex-30

SUPPLEMENTARY EXERCISES

1. (a) A company manufactures two types of computer monitors: standard and high resolution. Suppose that $P(x, y)$ is the profit that results from producing and selling x standard monitors and y high-resolution monitors. What do the two partial derivatives $\partial P/\partial x$ and $\partial P/\partial y$ represent?
(b) Suppose that the temperature at time t at a point (x, y) on the surface of a lake is $T(x, y, t)$. What do the partial derivatives $\partial T/\partial x$, $\partial T/\partial y$, and $\partial T/\partial t$ represent?

2. Let $z = f(x, y)$.
(a) Express $\partial z/\partial x$ and $\partial z/\partial y$ as limits.
(b) In words, what do the derivatives $f_x(x_0, y_0)$ and $f_y(x_0, y_0)$ tell you about the surface $z = f(x, y)$?
(c) In words, what do the derivatives $\partial z/\partial x(x_0, y_0)$ and $\partial z/\partial y(x_0, y_0)$ tell you about the rates of change of z with respect to x and y?
(d) In words, what does the derivative $D_{\mathbf{u}}f(x_0, y_0)$ tell you about the surface $z = f(x, y)$?

3. Show that the level curves of the cone $z = \sqrt{x^2 + y^2}$ and the paraboloid $z = x^2 + y^2$ are circles, and make a sketch that illustrates the difference between the contour plots of the two functions.

4. (a) How are the directional derivative and the gradient of a function related?
(b) Under what conditions is the directional derivative of a differentiable function 0?
(c) In what direction does the directional derivative of a differentiable function have its maximum value? Its minimum value?

5. (a) In words, describe the level surfaces of the function $f(x, y, z) = a^2x^2 + a^2y^2 + z^2$, where $a > 0$.
(b) Find a function $f(x, y, z)$ whose level surfaces form a family of circular paraboloids that open in the positive z-direction.

6. What do Δf and df represent, and how are they related?

7. Let $f(x, y) = e^x \ln y$. Find
(a) $f(\ln y, e^x)$ (b) $f(r + s, rs)$.

8. Sketch the domain of f using solid lines for portions of the boundary included in the domain and dashed lines for portions not included.
(a) $f(x, y) = \ln(xy - 1)$ (b) $f(x, y) = (\sin^{-1} x)/e^y$

In Exercises 9–12, verify the assertion.

9. If $w = \tan(x^2 + y^2) + x\sqrt{y}$, then $w_{xy} = w_{yx}$.

10. If $w = \ln(3x - 3y) + \cos(x + y)$, then $\partial^2 w/\partial x^2 = \partial^2 w/\partial y^2$.

11. If $F(x, y, z) = 2z^3 - 3(x^2 + y^2)z$, then $F_{xx} + F_{yy} + F_{zz} = 0$.

12. If $f(x, y, z) = xyz + x^2 + \ln(y/z)$, then $f_{xyzx} = f_{zxxy}$.

13. The pressure in N/m^2 of a gas in a cylinder is given by $P = 10T/V$ with T in kelvins (K) and V in m^3.
(a) If T is increasing at a rate of 3 K/min with V held fixed at 2.5 m^3, find the rate at which the pressure is changing when $T = 50$ K.
(b) If T is held fixed at 50 K while V is decreasing at the rate of 3 m^3/min, find the rate at which the pressure is changing when $V = 2.5$ m^3.

14. Find the slope of the tangent line at the point $(1, -2, -3)$ on the curve of intersection of the surface $z = 5 - 4x^2 - y^2$ with
(a) the plane $x = 1$ (b) the plane $y = -2$.

In Exercises 15 and 16, (a) find the limit of the function $f(x, y)$ as $(x, y) \to (0, 0)$ if it exists, and (b) determine whether f is continuous at $(0, 0)$.

15. $f(x, y) = \dfrac{x^4 - x + y - x^3y}{x - y}$

16. $f(x, y) = \begin{cases} \dfrac{x^4 - y^4}{x^2 + y^2} & \text{if } (x, y) \neq (0, 0) \\ 0 & \text{if } (x, y) = (0, 0) \end{cases}$

17. At the point $(1, 2)$, the directional derivative $D_{\mathbf{u}} f$ is $2\sqrt{2}$ toward $P_1(2, 3)$ and -3 toward $P_2(1, 0)$. Find $D_{\mathbf{u}} f(1, 2)$ toward the origin.

18. Find equations for the tangent plane and normal line to the given surface at P_0.
 (a) $z = x^2 e^{2y}$; $P_0(1, \ln 2, 4)$
 (b) $x^2 y^3 z^4 + xyz = 2$; $P_0(2, 1, -1)$

19. Find all points P_0 on the surface $z = 2 - xy$ at which the normal line passes through the origin.

20. Show that for all tangent planes to the surface

$$x^{2/3} + y^{2/3} + z^{2/3} = 1$$

the sum of the squares of the x-, y-, and z-intercepts is 1.

21. Find all points on the paraboloid $z = 9x^2 + 4y^2$ at which the normal line is parallel to the line through the points $P(4, -2, 5)$ and $Q(-2, -6, 4)$.

22. If $w = x^2 y - 2xy + y^2 x$, find the increment Δw and the differential dw if (x, y) varies from $(1, 0)$ to $(1.1, -0.1)$.

23. Use differentials to estimate the change in the volume $V = \frac{1}{3} x^2 h$ of a pyramid with a square base when its height h is increased from 2 to 2.2 m, while its base dimension x is decreased from 1 to 0.9 m. Compare this to ΔV.

In Exercises 24–26, locate all relative minima, relative maxima, and saddle points.

---

24. $f(x, y) = x^2 + 3xy + 3y^2 - 6x + 3y$

25. $f(x, y) = x^2 y - 6y^2 - 3x^2$

26. $f(x, y) = x^3 - 3xy + \frac{1}{2} y^2$

In economics, a ***production model*** is a mathematical relationship between the output of a company or a country and the labor and capital equipment required to produce that output. Much of the pioneering work in the field of production models occurred in the 1920s when Paul Douglas of the University of Chicago and his collaborator Charles Cobb proposed that the output P can be expressed in terms of the labor L and the capital equipment K by an equation of the form

$$P = cL^{\alpha} K^{\beta}$$

where c is a constant of proportionality and α and β are constants such that $0 < \alpha < 1$ and $0 < \beta < 1$. This is called the ***Cobb–Douglas production model***. Typically, P, L, and K are all expressed in terms of their equivalent monetary values. Exercises 27–29 explore properties of this model.

---

27. (a) Consider the Cobb–Douglas production model given by the formula $P = L^{0.75} K^{0.25}$. Sketch the level curves $P(L, K) = 1$, $P(L, K) = 2$, and $P(L, K) = 3$ in an LK-coordinate system (L horizontal and K vertical). Your sketch need not be accurate numerically, but it

should show the general shape of the curves and their relative positions.
 (b) Use a graphing utility to make a more extensive contour plot of the model.

28. (a) Find $\partial P / \partial L$ and $\partial P / \partial K$ for the Cobb–Douglas production model $P = cL^{\alpha} K^{\beta}$.
 (b) The derivative $\partial P / \partial L$ is called the ***marginal productivity of labor***, and the derivative $\partial P / \partial K$ is called the ***marginal productivity of capital***. Explain what these quantities mean in practical terms.
 (c) Show that if $\beta = 1 - \alpha$, then P satisfies the partial differential equation

$$K \frac{\partial P}{\partial K} + L \frac{\partial P}{\partial L} = P$$

29. Consider the Cobb–Douglas production model

$$P = 1000 L^{0.6} K^{0.4}$$

 (a) Find the maximum output value of P if labor costs $\$50.00$ per unit, capital costs $\$100.00$ per unit, and the total cost of labor and capital is set at $\$200,000$.
 (b) How should the $\$200,000$ be allocated between labor and capital to achieve the maximum?

Solve Exercises 30 and 31 two ways:
(a) Use the constraint to eliminate a variable.
(b) Use Lagrange multipliers.

---

30. Find all relative extrema of $x^2 y^2$ subject to the constraint $4x^2 + y^2 = 8$.

31. Find the dimensions of the rectangular box of maximum volume that can be inscribed in the ellipsoid

$$(x/a)^2 + (y/b)^2 + (z/c)^2 = 1$$

32. In each part, use Theorem 15.4.8 to find dy/dx.
 (a) $3x^2 - 5xy + \tan xy = 0$
 (b) $x \ln y + \sin(x - y) = \pi$

33. Given that $F(x, y) = 0$, use Theorem 15.4.8 to express $d^2 y/dx^2$ in terms of partial derivatives of F.

34. As illustrated in the accompanying figure, suppose that a current I branches into currents I_1, I_2, and I_3 through resistors R_1, R_2, and R_3 in such a way that the total energy to the three resistors is a minimum. Find the ratios $I_1 : I_2 : I_3$ if the energy delivered to R_i is $I_i^2 R_i (i = 1, 2, 3)$ and $I_1 + I_2 + I_3 = I$.

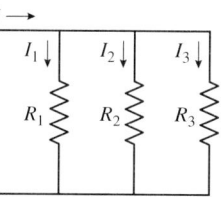

Figure Ex-34

35. Suppose the equations of motion of a particle are $x = t - 1$, $y = 4e^{-t}$, $z = 2 - \sqrt{t}$, where $t > 0$. Find, to the nearest tenth of a degree, the acute angle between the velocity vector and the normal line to the surface $(x^2/4) + y^2 + z^2 = 1$

at the points where the particle collides with the surface. Use a calculating utility with a root-finding capability where needed.

[c] **36.** Let

$$F(x) = \int_c^d f(x, y)\,dy, \quad a \le x \le b$$

It can be shown that if $f(x, y)$ and $\partial f/\partial x$ are continuous for $a \le x \le b$ and $c \le y \le d$, then

$$F'(x) = \int_c^d \frac{\partial f}{\partial x}\,dy$$

(a) Use this result to find $F'(x)$ if

$$F(x) = \int_0^1 \sin(xe^y)\,dy$$

(b) Use a CAS and the result in part (a) to find the maximum value of $F(x)$ for $0 \le x \le 2$. Express your answer to six decimal places.

37. Angle A of triangle ABC is increasing at a rate of $\pi/60$ rad/s, side AB is increasing at a rate of 2 cm/s, and side AC is increasing at a rate of 4 cm/s. At what rate is the length of BC changing when angle A is $\pi/3$ rad, $AB = 20$ cm, and $AC = 10$ cm? Is the length of BC increasing or decreasing? [*Hint:* Use the law of cosines.]

38. Let $z = f(x, y)$, where $x = g(t)$ and $y = h(t)$.
(a) Show that

$$\frac{d}{dt}\left(\frac{\partial z}{\partial x}\right) = \frac{\partial^2 z}{\partial x^2}\frac{dx}{dt} + \frac{\partial^2 z}{\partial y \partial x}\frac{dy}{dt}$$

and

$$\frac{d}{dt}\left(\frac{\partial z}{\partial y}\right) = \frac{\partial^2 z}{\partial x \partial y}\frac{dx}{dt} + \frac{\partial^2 z}{\partial y^2}\frac{dy}{dt}$$

(b) Use the formulas in part (a) to help find a formula for d^2z/dt^2.

EXPANDING THE CALCULUS HORIZON

For additional material relating to this chapter, visit the Anton Website at http://www.wiley.com/college/anton

16

MULTIPLE INTEGRALS

Pierre-Simon
de Laplace

Alexander Calder, "Lobster Trap and Fish Tail,"
1939. Hanging mobile: painted steel wire and
sheet aluminum, about 6" high x 9'6" diameter.
The Museum of Modern Art, New York.
Commissioned by the Advisory Committee
for the stairwell of the Museum. Photograph
©1998, The Museum of Modern Art, New York.

Calder's work reflects his intuitive sense for
finding the perfect aesthetic and physical
balance of complex objects.

*I*n this chapter we will extend the concept of a definite integral to functions of two and three variables. Whereas functions of one variable are usually integrated over intervals, functions of two variables are usually integrated over regions in 2-space and functions of three variables over regions in 3-space. Calculating such integrals will require some new techniques that will be a central focus in this chapter. Once we have developed the basic methods for integrating functions of two and three variables, we will show how such integrals can be used to calculate surface areas and volumes of solids; and we will also show how they can be used to find masses and centers of gravity of flat plates and three-dimensional solids. In addition to our study of integration, we will generalize the concept of a parametric curve in 2-space to a parametric surface in 3-space. This will allow us to work with a wider variety of surfaces than previously possible and will provide a powerful tool for generating surfaces using computers and other graphing utilities.

16.1 DOUBLE INTEGRALS

The notion of a definite integral can be extended to functions of two or more variables. In this section we will discuss the double integral, which is the extension to functions of two variables.

Recall that the definite integral of a function of one variable

$$\int_a^b f(x)\,dx = \lim_{n \to +\infty} \sum_{k=1}^n f(x_k^*)\Delta x_k \tag{1}$$

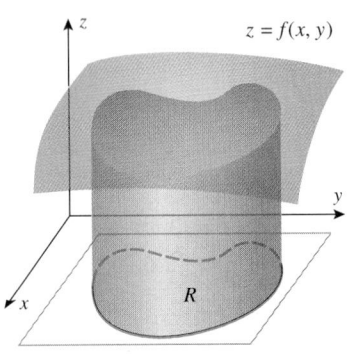

Figure 16.1.1

arose from the problem of finding areas under curves. Integrals of functions of two variables arise from the problem of finding volumes under surfaces:

> **16.1.1 THE VOLUME PROBLEM.** Given a function f of two variables that is continuous and nonnegative on a region R in the xy-plane, find the volume of the solid enclosed between the surface $z = f(x, y)$ and the region R (Figure 16.1.1).

Later, we will place more restrictions on the region R, but for now we will just assume that the entire region can be enclosed within some suitably large rectangle with sides parallel to the coordinate axes. This ensures that R does not extend indefinitely in any direction.

The procedure for finding the volume V of the solid in Figure 16.1.1 will be similar to the limiting process used for finding areas, except that now the approximating elements will be rectangular parallelepipeds rather than rectangles. We proceed as follows:

- Using lines parallel to the coordinate axes, divide the rectangle enclosing the region R into subrectangles, and exclude from consideration all those subrectangles that contain any points outside of R. This leaves only rectangles that are subsets of R (Figure 16.1.2). Assume that there are n such rectangles, and denote the area of the kth such rectangle by ΔA_k.

- Choose any arbitrary point in each subrectangle, and denote the point in the kth subrectangle by (x_k^*, y_k^*). As shown in Figure 16.1.3, the product $f(x_k^*, y_k^*)\Delta A_k$ is the volume of a rectangular parallelepiped with base area ΔA_k and height $f(x_k^*, y_k^*)$, so the sum

$$\sum_{k=1}^n f(x_k^*, y_k^*)\Delta A_k$$

can be viewed as an approximation to the volume V of the entire solid.

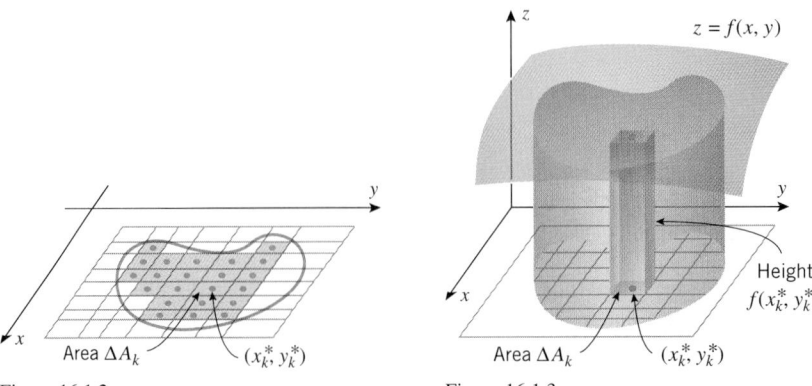

Figure 16.1.2

Figure 16.1.3

- There are two sources of error in the approximation: first, the parallelepipeds have flat tops, whereas the surface $z = f(x, y)$ may be curved; second, the rectangles that form the bases of the parallelepipeds do not completely cover the region R. However, if we repeat the above process with more and more subdivisions in such a way that the lengths and widths of the base rectangles approach zero, then it is plausible that the errors of both types approach zero, and the exact volume of the solid is

$$V = \lim_{n \to +\infty} \sum_{k=1}^{n} f(x_k^*, y_k^*) \Delta A_k \qquad (2)$$

This suggests the following definition.

16.1.2 DEFINITION (*Volume Under a Surface*). If f is a function of two variables that is continuous and nonnegative on a region R in the xy-plane, then the volume of the solid enclosed between the surface $z = f(x, y)$ and the region R is defined by

$$V = \lim_{n \to +\infty} \sum_{k=1}^{n} f(x_k^*, y_k^*) \Delta A_k \qquad (3)$$

REMARK.   Although this definition is satisfactory for our present purposes, there are various issues that would have to be resolved before it could be regarded as a rigorous mathematical definition. For example, we would have to prove that the limit actually exists and that its value does not depend on how the points $(x_1^*, y_1^*), (x_2^*, y_2^*), \ldots, (x_n^*, y_n^*)$ are chosen. It can be proved that this is true if f is continuous on the region R and this region is not too "complicated." The details are beyond the scope of this text.

DEFINITION OF A DOUBLE INTEGRAL

It is assumed in Definition 16.1.2 that f is nonnegative on the region R. If f is continuous on R and has both positive and negative values, then the limit

$$\lim_{n \to +\infty} \sum_{k=1}^{n} f(x_k^*, y_k^*) \Delta A_k \qquad (4)$$

no longer represents the volume between R and the surface $z = f(x, y)$; rather, it represents a *difference* of volumes—the volume between R and the portion of the surface that is above the xy-plane minus the volume between R and the portion of the surface below the xy-plane. We call this the **net signed volume** between the region R and the surface $z = f(x, y)$.

The limit in (4) is sufficiently important that there is some notation and terminology associated with it—the sums in (4) are called **Riemann sums**, and the limit of the Riemann sums is denoted by

$$\iint_R f(x, y)\, dA = \lim_{n \to +\infty} \sum_{k=1}^{n} f(x_k^*, y_k^*) \Delta A_k \qquad (5)$$

which is called the **double integral** of $f(x, y)$ over R.

If f is continuous and nonnegative on the region R, then volume formula (3) can be expressed as

$$V = \iint_R f(x, y)\, dA \qquad (6)$$

If f has both positive and negative values on R, then a positive value for the double integral of f over R means that there is more volume above R than below, a negative value for the double integral means that there is more volume below than above, and a value of zero means that the volume above is the same as the volume below.

To distinguish between double integrals of functions of two variables and definite integrals of functions of one variable, we will refer to the latter as *single integrals*. Because double integrals, like single integrals, are defined as limits, they inherit many of the properties of limits. The following results, which we state without proof, are analogs of those in Theorem 7.5.4.

$$\iint\limits_{R} cf(x, y)\, dA = c \iint\limits_{R} f(x, y)\, dA \quad (c \text{ a constant}) \tag{7}$$

$$\iint\limits_{R} [f(x, y) + g(x, y)]\, dA = \iint\limits_{R} f(x, y)\, dA + \iint\limits_{R} g(x, y)\, dA \tag{8}$$

$$\iint\limits_{R} [f(x, y) - g(x, y)]\, dA = \iint\limits_{R} f(x, y)\, dA - \iint\limits_{R} g(x, y)\, dA \tag{9}$$

It is evident intuitively that if $f(x, y)$ is nonnegative on a region R, then subdividing R into two regions R_1 and R_2 has the effect of subdividing the solid between R and $z = f(x, y)$ into two solids, the sum of whose volumes is the volume of the entire solid (Figure 16.1.4). This suggests the following result, which holds even if f has negative values:

$$\iint\limits_{R} f(x, y)\, dA = \iint\limits_{R_1} f(x, y)\, dA + \iint\limits_{R_2} f(x, y)\, dA \tag{10}$$

The proof of this result will be omitted.

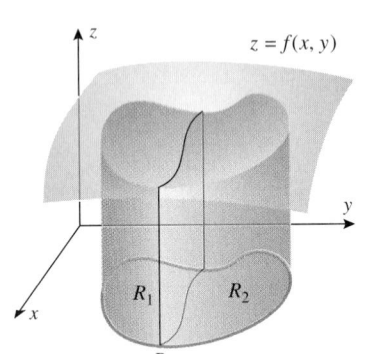

The volume of the entire solid is the sum of the volumes of the solids above R_1 and R_2.

Figure 16.1.4

Except in the simplest cases, it is impractical to obtain the value of a double integral from the limit in (5). However, we will now show how to evaluate double integrals by calculating two successive single integrals. For the rest of this section, we will limit our discussion to the case where R is a rectangle; in the next section we will consider double integrals over more complicated regions.

The partial derivatives of a function $f(x, y)$ are calculated by holding one of the variables fixed and differentiating with respect to the other variable. Let us consider the reverse of this process, *partial integration*. The symbols

$$\int_a^b f(x, y)\, dx \quad \text{and} \quad \int_c^d f(x, y)\, dy$$

denote *partial definite integrals*; the first integral, called the *partial definite integral with respect to x*, is evaluated by holding y fixed and integrating with respect to x, and the second integral, called the *partial definite integral with respect to y*, is evaluated by holding x fixed and integrating with respect to y. As the following example shows, the partial definite integral with respect to x is a function of y, and the partial definite integral with respect to y is a function of x.

Example 1

$$\int_0^1 xy^2\, dx = y^2 \int_0^1 x\, dx = \left.\frac{y^2 x^2}{2}\right]_{x=0}^{1} = \frac{y^2}{2}$$

$$\int_0^1 xy^2\,dy = x\int_0^1 y^2\,dy = \frac{xy^3}{3}\Bigg]_{y=0}^1 = \frac{x}{3}$$ ◀

A partial definite integral with respect to x is a function of y and hence can be integrated with respect to y; similarly, a partial definite integral with respect to y can be integrated with respect to x. This two-stage integration process is called *iterated* (or *repeated*) *integration*. We introduce the following notation:

$$\int_c^d \int_a^b f(x,y)\,dx\,dy = \int_c^d \left[\int_a^b f(x,y)\,dx\right] dy \qquad (11)$$

$$\int_a^b \int_c^d f(x,y)\,dy\,dx = \int_a^b \left[\int_c^d f(x,y)\,dy\right] dx \qquad (12)$$

These integrals are called *iterated integrals*.

Example 2

Evaluate

(a) $\displaystyle\int_0^3 \int_1^2 (1+8xy)\,dy\,dx$ (b) $\displaystyle\int_1^2 \int_0^3 (1+8xy)\,dx\,dy$

Solution (a).

$$\int_0^3 \int_1^2 (1+8xy)\,dy\,dx = \int_0^3 \left[\int_1^2 (1+8xy)\,dy\right] dx$$

$$= \int_0^3 \left[y + 4xy^2\right]_{y=1}^2 dx$$

$$= \int_0^3 [(2+16x)-(1+4x)]\,dx$$

$$= \int_0^3 (1+12x)\,dx = (x+6x^2)\big]_0^3 = 57$$

Solution (b).

$$\int_1^2 \int_0^3 (1+8xy)\,dx\,dy = \int_1^2 \left[\int_0^3 (1+8xy)\,dx\right] dy$$

$$= \int_1^2 \left[x + 4x^2 y\right]_{x=0}^3 dy$$

$$= \int_1^2 (3+36y)\,dy = (3y+18y^2)\big]_1^2 = 57$$ ◀

The following theorem shows that it is no accident that the two iterated integrals in the last example have the same value.

16.1.3   THEOREM.   *Let R be the rectangle defined by the inequalities*

$a \le x \le b, \quad c \le y \le d$

If $f(x,y)$ is continuous on this rectangle, then

$$\iint\limits_R f(x,y)\,dA = \int_c^d \int_a^b f(x,y)\,dx\,dy = \int_a^b \int_c^d f(x,y)\,dy\,dx$$

This important theorem allows us to evaluate a double integral over a rectangle by converting it to an iterated integral. This can be done in two ways, both of which produce the value of the double integral. We will not formally prove this result; however, we will give a geometric motivation of the result for the case where $f(x, y)$ is nonnegative on R. In this case the double integral can be interpreted as the volume of the solid S bounded above by the surface $z = f(x, y)$ and below by the region R, so it suffices to show that the two iterated integrals also represent this volume.

For a fixed value of y, the function $f(x, y)$ is a function of x, and hence the integral

$$A(y) = \int_a^b f(x, y)\, dx$$

represents the area under the graph of this function of x. This area, shown in yellow in Figure 16.1.5, is the cross-sectional area at y of the solid S bounded above by $z = f(x, y)$ and below by the region R. Thus, by the method of slicing discussed in Section 8.2, the volume V of the solid S is

$$V = \int_c^d A(y)\, dy = \int_c^d \left[\int_a^b f(x, y)\, dx \right] dy = \int_c^d \int_a^b f(x, y)\, dx\, dy \qquad (13)$$

Similarly, the integral

$$A(x) = \int_c^d f(x, y)\, dy$$

represents the area of the cross section of S at x (Figure 16.1.6), and the method of slicing again yields

$$V = \int_a^b A(x)\, dx = \int_a^b \left[\int_c^d f(x, y)\, dy \right] dx = \int_a^b \int_c^d f(x, y)\, dy\, dx \qquad (14)$$

This establishes the result in Theorem 16.1.3 for the case where $f(x, y)$ is continuous and nonnegative on R.

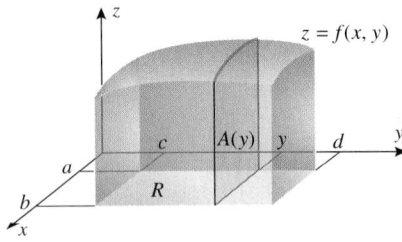

Figure 16.1.5

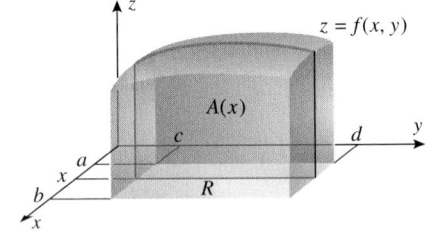

Figure 16.1.6

Example 3

Evaluate the double integral

$$\iint_R y^2 x\, dA$$

over the rectangle $R = \{(x, y) : -3 \le x \le 2, 0 \le y \le 1\}$.

Solution. In view of Theorem 16.1.3, the value of the double integral may be obtained from either of the iterated integrals

$$\int_{-3}^2 \int_0^1 y^2 x\, dy\, dx \quad \text{or} \quad \int_0^1 \int_{-3}^2 y^2 x\, dx\, dy \qquad (15)$$

Using the first of these, we obtain

$$\iint\limits_{R} y^2 x \, dA = \int_{-3}^{2} \int_{0}^{1} y^2 x \, dy \, dx = \int_{-3}^{2} \left[\frac{1}{3} y^3 x \right]_{y=0}^{1} dx$$

$$= \int_{-3}^{2} \frac{1}{3} x \, dx = \frac{x^2}{6} \bigg]_{-3}^{2} = -\frac{5}{6}$$

You can check this result by evaluating the second integral in (15). ◄

REMARK. We will often express the rectangle $\{(x, y) : a \le x \le b, 0 \le y \le d\}$ as $[a, b] \times [c, d]$ for simplicity.

Example 4

Use a double integral to find the volume of the solid that is bounded above by the plane $z = 4 - x - y$ and below by the rectangle $R = [0, 1] \times [0, 2]$ (Figure 16.1.7).

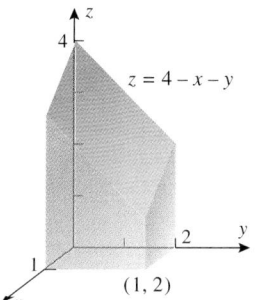

Figure 16.1.7

Solution.

$$V = \iint\limits_{R} (4 - x - y) \, dA = \int_{0}^{2} \int_{0}^{1} (4 - x - y) \, dx \, dy$$

$$= \int_{0}^{2} \left[4x - \frac{x^2}{2} - xy \right]_{x=0}^{1} dy = \int_{0}^{2} \left(\frac{7}{2} - y \right) dy$$

$$= \left[\frac{7}{2} y - \frac{y^2}{2} \right]_{0}^{2} = 5$$

The volume can also be obtained by first integrating with respect to y and then with respect to x. ◄

FOR THE READER. Most computer algebra systems have a built-in capability for computing iterated double integrals. If you have a CAS, read the relevant documentation and use the CAS to check Examples 3 and 4.

EXERCISE SET 16.1   [C] CAS
· ·

In Exercises 1–12, evaluate the iterated integrals.

1. $\int_{0}^{1} \int_{0}^{2} (x + 3) \, dy \, dx$

2. $\int_{1}^{3} \int_{-1}^{1} (2x - 4y) \, dy \, dx$

3. $\int_{2}^{4} \int_{0}^{1} x^2 y \, dx \, dy$

4. $\int_{-2}^{0} \int_{-1}^{2} (x^2 + y^2) \, dx \, dy$

5. $\int_{0}^{\ln 3} \int_{0}^{\ln 2} e^{x+y} \, dy \, dx$

6. $\int_{0}^{2} \int_{0}^{1} y \sin x \, dy \, dx$

7. $\int_{-1}^{0} \int_{2}^{5} dx \, dy$

8. $\int_{4}^{6} \int_{-3}^{7} dy \, dx$

9. $\int_{0}^{1} \int_{0}^{1} \frac{x}{(xy + 1)^2} \, dy \, dx$

10. $\int_{\pi/2}^{\pi} \int_{1}^{2} x \cos xy \, dy \, dx$

11. $\int_{0}^{\ln 2} \int_{0}^{1} xy e^{y^2 x} \, dy \, dx$

12. $\int_{3}^{4} \int_{1}^{2} \frac{1}{(x + y)^2} \, dy \, dx$

In Exercises 13–16, evaluate the double integral over the rectangular region R.

13. $\iint\limits_{R} 4xy^3 \, dA; \ R = \{(x, y) : -1 \le x \le 1, -2 \le y \le 2\}$

14. $\iint\limits_{R} \frac{xy}{\sqrt{x^2 + y^2 + 1}} \, dA;$
$R = \{(x, y) : 0 \le x \le 1, 0 \le y \le 1\}$

15. $\iint\limits_{R} x\sqrt{1 - x^2} \, dA; \ R = \{(x, y) : 0 \le x \le 1, 2 \le y \le 3\}$

16. $\iint\limits_{R} (x \sin y - y \sin x) \, dA;$
$R = \{(x, y) : 0 \le x \le \pi/2, 0 \le y \le \pi/3\}$

[C] **17.** Use a CAS to check the answers to the problems you solved in Exercises 1–16.

c **18.** Use a CAS to show that the volume V under the surface $z = xy^3 \sin xy$ over the rectangle shown in the accompanying figure is $V = 3/\pi$.

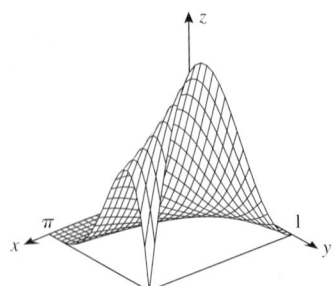

Figure Ex-18

In Exercises 19–22, use a double integral to find the volume.

19. The volume under the plane $z = 2x + y$ and over the rectangle $R = \{(x, y) : 3 \le x \le 5, 1 \le y \le 2\}$.

20. The volume under the surface $z = 3x^3 + 3x^2 y$ and over the rectangle $R = \{(x, y) : 1 \le x \le 3, 0 \le y \le 2\}$.

21. The volume of the solid enclosed by the surface $z = x^2$ and the planes $x = 0$, $x = 2$, $y = 3$, $y = 0$, and $z = 0$.

22. The volume in the first octant bounded by the coordinate planes, the plane $y = 4$, and the plane $(x/3) + (z/5) = 1$.

In Exercises 23 and 24, the iterated integral represents the volume of a solid. Make a sketch of the solid. (You do *not* have to find the volume.)

23. (a) $\displaystyle\int_0^5 \int_1^2 4\,dx\,dy$ (b) $\displaystyle\int_0^3 \int_0^4 \sqrt{25 - x^2 - y^2}\,dy\,dx$

24. (a) $\displaystyle\int_0^1 \int_0^1 (2 - x - y)\,dy\,dx$ (b) $\displaystyle\int_{-2}^2 \int_{-2}^2 (x^2 + y^2)\,dx\,dy$

25. Evaluate the integral by choosing a convenient order of integration:
$$\iint\limits_R x\cos(xy)\cos^2 \pi x\,dA;\ R = \left[0, \tfrac{1}{2}\right] \times [0, \pi]$$

26. (a) Sketch the solid in the first octant that is enclosed by the planes $x = 0$, $z = 0$, $x = 5$, $z - y = 0$, and $z = -2y + 6$.
(b) Find the volume of the solid by breaking it into two parts.

The ***average value*** or ***mean value*** of a continuous function $f(x, y)$ over a rectangle $R = [a, b] \times [c, d]$ is defined as
$$f_{ave} = \frac{1}{A(R)} \iint\limits_R f(x, y)\,dA$$
where $A(R) = (b - a)(d - c)$ is the area of the rectangle R (compare to Definition 7.7.5). Use this definition in Exercises 27–30.

27. Find the average value of $f(x, y) = y \sin xy$ over the rectangle $[0, 1] \times [0, \pi/2]$.

28. Find the average value of $f(x, y) = x(x^2 + y)^{1/2}$ over the interval $[0, 1] \times [0, 3]$.

29. Suppose that the temperature in degrees Celsius at a point (x, y) on a flat metal plate is $T(x, y) = 10 - 8x^2 - 2y^2$, where x and y are in meters. Find the average temperature of the rectangular portion of the plate for which $0 \le x \le 1$ and $0 \le y \le 2$.

30. Show that if $f(x, y)$ is constant on the rectangle $R = [a, b] \times [c, d]$, say $f(x, y) = k$, then $f_{ave} = k$ over R.

Most computer algebra systems have commands for approximating double integrals numerically. For Exercises 31 and 32, read the relevant documentation and use a CAS to find a numerical approximation of the double integral.

31. $\displaystyle\int_0^2 \int_0^1 \sin\sqrt{x^3 + y^3}\,dx\,dy$

32. $\displaystyle\int_{-1}^1 \int_{-1}^1 e^{-(x^2 + y^2)}\,dx\,dy$

33. In this exercise, suppose that $f(x, y) = g(x)h(y)$ and $R = \{(x, y) : a \le x \le b, c \le y \le d\}$. Show that
$$\iint\limits_R f(x, y)\,dA = \left[\int_a^b g(x)\,dx\right]\left[\int_c^d h(y)\,dy\right]$$

34. Use the result in Exercise 33 to evaluate the integral
$$\int_0^{\ln 2} \int_{-1}^1 \sqrt{e^y + 1}\,\tan x\,dx\,dy$$
by inspection. Explain your reasoning.

c **35.** Use a CAS to evaluate the iterated integrals
$$\int_0^1 \int_0^1 \frac{y - x}{(x + y)^3}\,dx\,dy \quad \text{and} \quad \int_0^1 \int_0^1 \frac{y - x}{(x + y)^3}\,dy\,dx$$
Does this violate Theorem 16.1.3? Explain.

36. (a) Let $f(x, y) = x - 2y$, and as shown in the accompanying figure, let the rectangle $R = [0, 2] \times [0, 2]$ be subdivided into the 16 subrectangles. Take (x_k^*, y_k^*) to be the center of the kth rectangle, and approximate the double integral of f over R by the resulting Riemann sum.
(b) Compare the result in part (a) to the exact value of the integral.

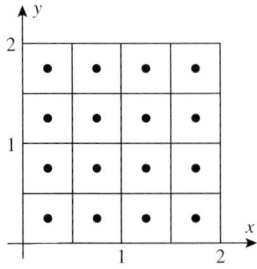

Figure Ex-36

16.2 DOUBLE INTEGRALS OVER NONRECTANGULAR REGIONS

In this section we will show how to evaluate double integrals over regions other than rectangles.

ITERATED INTEGRALS WITH NONCONSTANT LIMITS OF INTEGRATION

Later in this section we will see that double integrals over nonrectangular regions can often be evaluated as iterated integrals of the following types:

$$\int_a^b \int_{g_1(x)}^{g_2(x)} f(x, y)\, dy\, dx = \int_a^b \left[\int_{g_1(x)}^{g_2(x)} f(x, y)\, dy \right] dx \tag{1}$$

$$\int_c^d \int_{h_1(y)}^{h_2(y)} f(x, y)\, dx\, dy = \int_c^d \left[\int_{h_1(y)}^{h_2(y)} f(x, y)\, dx \right] dy \tag{2}$$

We begin with an example that illustrates how to evaluate such integrals.

Example 1

Evaluate

(a) $\displaystyle \int_0^2 \int_{x^2}^x y^2 x\, dy\, dx$ (b) $\displaystyle \int_0^\pi \int_0^{\cos y} x \sin y\, dx\, dy$

Solution (a).

$$\int_0^2 \int_{x^2}^x y^2 x\, dy\, dx = \int_0^2 \left[\int_{x^2}^x y^2 x\, dy \right] dx = \int_0^2 \left[\frac{y^3 x}{3} \right]_{y=x^2}^x dx$$

$$= \int_0^2 \left(\frac{x^4}{3} - \frac{x^7}{3} \right) dx = \left[\frac{x^5}{15} - \frac{x^8}{24} \right]_0^2$$

$$= \frac{32}{15} - \frac{256}{24} = -\frac{128}{15}$$

Solution (b).

$$\int_0^\pi \int_0^{\cos y} x \sin y\, dx\, dy = \int_0^\pi \left[\int_0^{\cos y} x \sin y\, dx \right] dy = \int_0^\pi \left[\frac{x^2}{2} \sin y \right]_{x=0}^{\cos y} dy$$

$$= \int_0^\pi \frac{1}{2} \cos^2 y \sin y\, dy = \left[-\frac{1}{6} \cos^3 y \right]_0^\pi = \frac{1}{3} \quad \blacktriangleleft$$

DOUBLE INTEGRALS OVER NONRECTANGULAR REGIONS

Plane regions can be extremely complex, and the theory of double integrals over very general regions is a topic for advanced courses in mathematics. We will limit our study of double integrals to two basic types of regions, which we will call *type I* and *type II*; they are defined as follows:

16.2.1 DEFINITION.

(a) A *type I region* is bounded on the left and right by vertical lines $x = a$ and $x = b$ and is bounded below and above by continuous curves $y = g_1(x)$ and $y = g_2(x)$, where $g_1(x) \le g_2(x)$ for $a \le x \le b$ (Figure 16.2.1a).

(b) A *type II region* is bounded below and above by horizontal lines $y = c$ and $y = d$ and is bounded on the left and right by continuous curves $x = h_1(y)$ and $x = h_2(y)$ satisfying $h_1(y) \le h_2(y)$ for $c \le y \le d$ (Figure 16.2.1b).

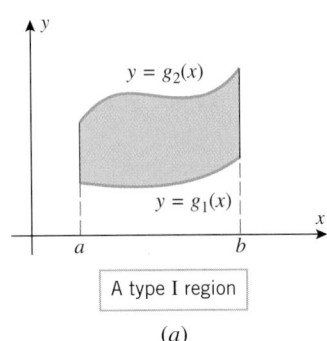

A type I region

(a)

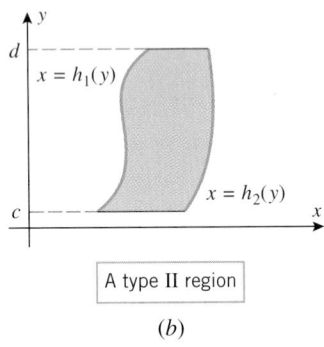

A type II region

(b)

Figure 16.2.1

The following theorem will enable us to evaluate double integrals over type I and type II regions using iterated integrals.

16.2.2 THEOREM.

(a) *If R is a type I region on which $f(x, y)$ is continuous, then*

$$\iint\limits_{R} f(x, y)\, dA = \int_{a}^{b} \int_{g_1(x)}^{g_2(x)} f(x, y)\, dy\, dx \tag{3}$$

(b) *If R is a type II region on which $f(x, y)$ is continuous, then*

$$\iint\limits_{R} f(x, y)\, dA = \int_{c}^{d} \int_{h_1(y)}^{h_2(y)} f(x, y)\, dx\, dy \tag{4}$$

We will not prove this theorem, but for the case where $f(x, y)$ is nonnegative on the region R, it can be made plausible by a geometric argument that is similar to that given for Theorem 16.1.3. Since $f(x, y)$ is nonnegative, the double integral can be interpreted as the volume of the solid S that is bounded above by the surface $z = f(x, y)$ and below by the region R, so it suffices to show that the iterated integrals also represent this volume. Consider the iterated integral in (3), for example. For a fixed value of x, the function $f(x, y)$ is a function of y, and hence the integral

$$A(x) = \int_{g_1(x)}^{g_2(x)} f(x, y)\, dy$$

represents the area under the graph of this function of y between the points $y = g_1(x)$ and $y = g_2(x)$. This area, shown in yellow in Figure 16.2.2, is the cross-sectional area at x of the solid S, and hence by the method of slicing, the volume V of the solid S is

$$V = \int_{a}^{b} \int_{g_1(x)}^{g_2(x)} f(x, y)\, dy\, dx$$

which shows that in (3) the iterated integral is equal to the double integral. Similarly for (4).

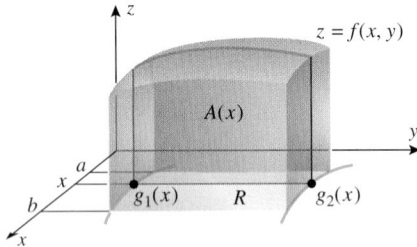

Figure 16.2.2

SETTING UP LIMITS OF INTEGRATION FOR EVALUATING DOUBLE INTEGRALS

To apply Theorem 16.2.2, it is helpful to start with a two-dimensional sketch of the region R. [It is not necessary to graph $f(x, y)$.] For a type I region, the limits of integration in Formula (3) can be obtained as follows:

Step 1. Since x is held fixed for the first integration, we draw a vertical line through the region R at an arbitrary fixed point x (Figure 16.2.3). This line crosses the boundary of R twice. The lower point of intersection is on the curve $y = g_1(x)$ and the higher point is on the curve $y = g_2(x)$. These two intersections determine the lower and upper y-limits of integration in Formula (3).

Step 2. Imagine moving the line drawn in Step 1 first to the left and then to the right (Figure 16.2.3). The leftmost position where the line intersects the region R is $x = a$ and the rightmost position where the line intersects the region R is $x = b$. This yields the limits for the x-integration in Formula (3).

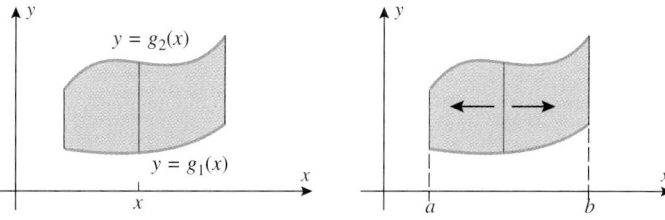

Figure 16.2.3

Example 2

Evaluate

$$\iint_R xy \, dA$$

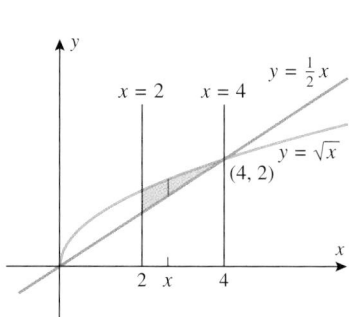

Figure 16.2.4

over the region R enclosed between $y = \frac{1}{2}x$, $y = \sqrt{x}$, $x = 2$, and $x = 4$.

Solution. We view R as a type I region. The region R and a vertical line corresponding to a fixed x are shown in Figure 16.2.4. This line meets the region R at the lower boundary $y = \frac{1}{2}x$ and the upper boundary $y = \sqrt{x}$. These are the y-limits of integration. Moving this line first left and then right yields the x-limits of integration, $x = 2$ and $x = 4$. Thus,

$$\iint_R xy \, dA = \int_2^4 \int_{x/2}^{\sqrt{x}} xy \, dy \, dx = \int_2^4 \left[\frac{xy^2}{2} \right]_{y=x/2}^{\sqrt{x}} dx = \int_2^4 \left(\frac{x^2}{2} - \frac{x^3}{8} \right) dx$$

$$= \left[\frac{x^3}{6} - \frac{x^4}{32} \right]_2^4 = \left(\frac{64}{6} - \frac{256}{32} \right) - \left(\frac{8}{6} - \frac{16}{32} \right) = \frac{11}{6} \quad \blacktriangleleft$$

If R is a type II region, then the limits of integration in Formula (4) can be obtained as follows:

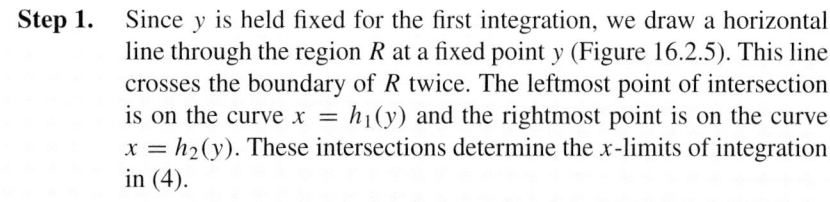

Step 1. Since y is held fixed for the first integration, we draw a horizontal line through the region R at a fixed point y (Figure 16.2.5). This line crosses the boundary of R twice. The leftmost point of intersection is on the curve $x = h_1(y)$ and the rightmost point is on the curve $x = h_2(y)$. These intersections determine the x-limits of integration in (4).

Step 2. Imagine moving the line drawn in Step 1 first down and then up (Figure 16.2.5). The lowest position where the line intersects the region R is $y = c$, and the highest position where the line intersects the region R is $y = d$. This yields the y-limits of integration in (4).

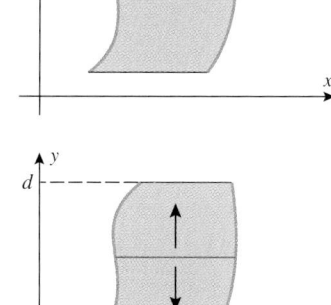

Figure 16.2.5

Example 3

Evaluate

$$\iint_R (2x - y^2) \, dA$$

over the triangular region R enclosed between the lines $y = -x + 1$, $y = x + 1$, and $y = 3$.

Solution. We view R as a type II region. The region R and a horizontal line corresponding to a fixed y are shown in Figure 16.2.6. This line meets the region R at its left-hand boundary $x = 1 - y$ and its right-hand boundary $x = y - 1$. These are the x-limits of integration. Moving this line first down and then up yields the y-limits, $y = 1$ and $y = 3$. Thus,

$$\iint_R (2x - y^2)\, dA = \int_1^3 \int_{1-y}^{y-1} (2x - y^2)\, dx\, dy = \int_1^3 \left[x^2 - y^2 x \right]_{x=1-y}^{y-1} dy$$

$$= \int_1^3 [(1 - 2y + 2y^2 - y^3) - (1 - 2y + y^3)]\, dy$$

$$= \int_1^3 (2y^2 - 2y^3)\, dy = \left[\frac{2y^3}{3} - \frac{y^4}{2} \right]_1^3 = -\frac{68}{3} \qquad \blacktriangleleft$$

REMARK.   To integrate over a type II region, the left- and right-hand boundaries must be expressed in the form $x = h_1(y)$ and $x = h_2(y)$. This is why we rewrote the boundary equations $y = -x + 1$ and $y = x + 1$ as $x = 1 - y$ and $x = y - 1$ in the last example.

In Example 3 we could have treated R as a type I region, but with an added complication: Viewed as a type I region, the upper boundary of R is the line $y = 3$ (Figure 16.2.7) and the lower boundary consists of two parts, the line $y = -x + 1$ to the left of the origin and the line $y = x + 1$ to the right of the origin. To carry out the integration it is necessary to decompose the region R into two parts, R_1 and R_2, as shown in Figure 16.2.7, and write

$$\iint_R (2x - y^2)\, dA = \iint_{R_1} (2x - y^2)\, dA + \iint_{R_2} (2x - y^2)\, dA$$

$$= \int_{-2}^0 \int_{-x+1}^3 (2x - y^2)\, dy\, dx + \int_0^2 \int_{x+1}^3 (2x - y^2)\, dy\, dx$$

This will yield the same result that was obtained in Example 3.

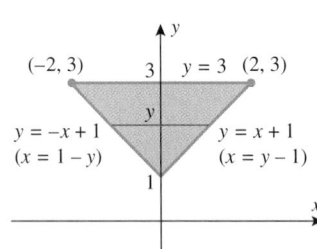

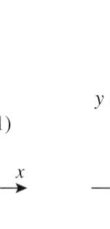

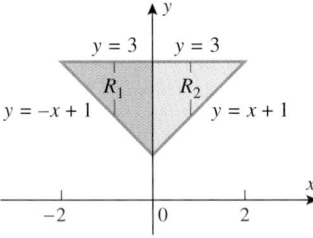

Figure 16.2.6   Figure 16.2.7

Example 4

Use a double integral to find the volume of the tetrahedron bounded by the coordinate planes and the plane $z = 4 - 4x - 2y$.

Solution. The tetrahedron in question is bounded above by the plane

$$z = 4 - 4x - 2y \qquad (5)$$

and below by the triangular region R shown in Figure 16.2.8. Thus, the volume is given by

$$V = \iint_R (4 - 4x - 2y)\, dA$$

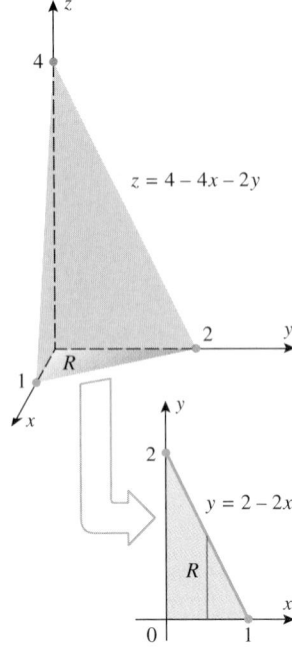

Figure 16.2.8

The region R is bounded by the x-axis, the y-axis, and the line $y = 2 - 2x$ [set $z = 0$ in (5)], so that treating R as a type I region yields

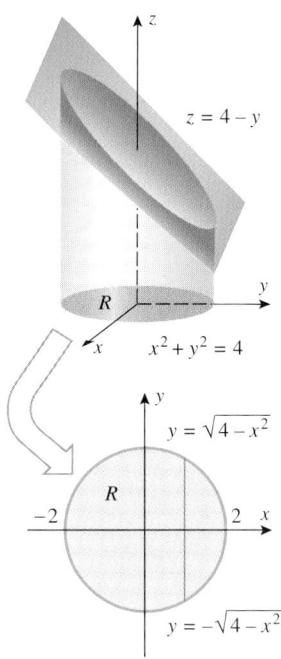

Figure 16.2.9

$$V = \iint\limits_{R} (4 - 4x - 2y)\, dA = \int_0^1 \int_0^{2-2x} (4 - 4x - 2y)\, dy\, dx$$

$$= \int_0^1 \left[4y - 4xy - y^2 \right]_{y=0}^{2-2x} dx = \int_0^1 (4 - 8x + 4x^2)\, dx = \frac{4}{3}$$

Example 5

Find the volume of the solid bounded by the cylinder $x^2 + y^2 = 4$ and the planes $y + z = 4$ and $z = 0$.

Solution. The solid shown in Figure 16.2.9 is bounded above by the plane $z = 4 - y$ and below by the region R within the circle $x^2 + y^2 = 4$. The volume is given by

$$V = \iint\limits_{R} (4 - y)\, dA$$

Treating R as a type I region we obtain

$$V = \int_{-2}^{2} \int_{-\sqrt{4-x^2}}^{\sqrt{4-x^2}} (4 - y)\, dy\, dx = \int_{-2}^{2} \left[4y - \frac{1}{2} y^2 \right]_{y=-\sqrt{4-x^2}}^{\sqrt{4-x^2}} dx$$

$$= \int_{-2}^{2} 8\sqrt{4 - x^2}\, dx = 8(2\pi) = 16\pi \qquad \boxed{\text{See Formula (3) of Section 9.4.}}$$

- -

REVERSING THE ORDER OF INTEGRATION

Sometimes the evaluation of an iterated integral can be simplified by reversing the order of integration. The next example illustrates how this is done.

Example 6

Since there is no elementary antiderivative of e^{x^2}, the integral

$$\int_0^2 \int_{y/2}^1 e^{x^2}\, dx\, dy$$

cannot be evaluated by performing the x-integration first. Evaluate this integral by expressing it as an equivalent iterated integral with the order of integration reversed.

Solution. For the inside integration, y is fixed and x varies from the line $x = y/2$ to the line $x = 1$ (Figure 16.2.10). For the outside integration, y varies from 0 to 2, so the given iterated integral is equal to a double integral over the triangular region R in Figure 16.2.10.

To reverse the order of integration, we treat R as a type I region, which enables us to write the given integral as

$$\int_0^2 \int_{y/2}^1 e^{x^2}\, dx\, dy = \iint\limits_{R} e^{x^2}\, dA = \int_0^1 \int_0^{2x} e^{x^2}\, dy\, dx = \int_0^1 \left[e^{x^2} y \right]_{y=0}^{2x} dx$$

$$= \int_0^1 2x e^{x^2}\, dx = e^{x^2} \Big]_0^1 = e - 1$$

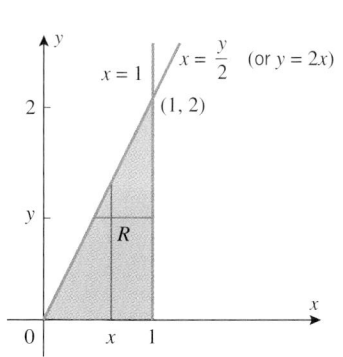

Figure 16.2.10

- -

AREA CALCULATED AS A DOUBLE INTEGRAL

Although double integrals arose in the context of calculating volumes, they can also be used to calculate areas. To see why this is so, recall that a *right cylinder* is a solid that is generated when a plane region is translated along a line that is perpendicular to the region. In Formula (2) of Section 8.2 we stated that the volume V of a right cylinder with cross-sectional area A and height h is

$$V = A \cdot h \tag{6}$$

Now suppose that we are interested in finding the area A of a region R in the xy-plane. If we translate the region R upward 1 unit, then the resulting solid will be a right cylinder that

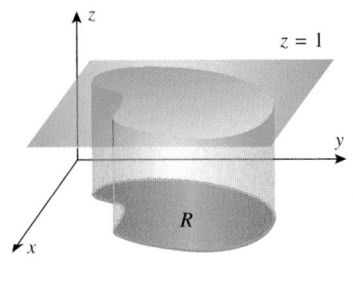

Cylinder with base R and height 1

Figure 16.2.11

has cross-sectional area A, base R, and the plane $z = 1$ as its top (Figure 16.2.11). Thus, it follows from (6) that

$$\iint_R 1 \, dA = (\text{area of } R) \cdot 1$$

which we can rewrite as

$$\text{area of } R = \iint_R 1 \, dA = \iint_R dA \tag{7}$$

REMARK. Formula (7) is sometimes confusing because it equates an area and a volume; the formula is intended to equate only the *numerical values* of the area and volume and not the units, which must, of course, be different.

Example 7

Use a double integral to find the area of the region R enclosed between the parabola $y = \frac{1}{2}x^2$ and the line $y = 2x$.

Solution. The region R may be treated equally well as type I (Figure 16.2.12a) or type II (Figure 16.2.12b). Treating R as type I yields

$$\text{area of } R = \iint_R dA = \int_0^4 \int_{x^2/2}^{2x} dy \, dx = \int_0^4 \left[y\right]_{y=x^2/2}^{2x} dx$$

$$= \int_0^4 \left(2x - \frac{1}{2}x^2\right) dx = \left[x^2 - \frac{x^3}{6}\right]_0^4 = \frac{16}{3}$$

Treating R as type II yields

$$\text{area of } R = \iint_R dA = \int_0^8 \int_{y/2}^{\sqrt{2y}} dx \, dy = \int_0^8 \left[x\right]_{x=y/2}^{\sqrt{2y}} dy$$

$$= \int_0^8 \left(\sqrt{2y} - \frac{1}{2}y\right) dy = \left[\frac{2\sqrt{2}}{3}y^{3/2} - \frac{y^2}{4}\right]_0^8 = \frac{16}{3} \qquad \blacktriangleleft$$

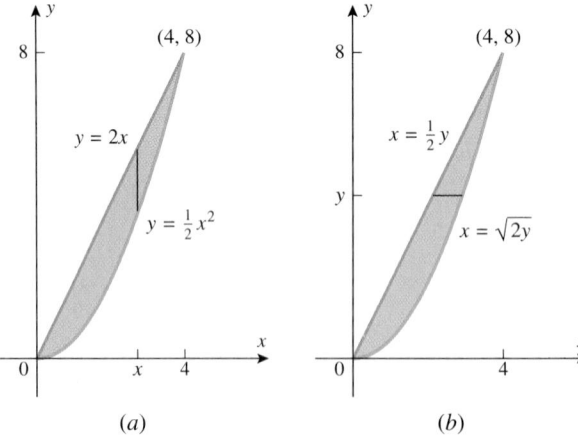

Figure 16.2.12

EXERCISE SET 16.2 ~ Graphing Calculator   c CAS

In Exercises 1–10, evaluate the iterated integral.

1. $\displaystyle\int_0^1 \int_{x^2}^x xy^2 \, dy \, dx$ **2.** $\displaystyle\int_1^{3/2} \int_y^{3-y} y \, dx \, dy$

3. $\displaystyle\int_0^3 \int_0^{\sqrt{9-y^2}} y \, dx \, dy$ **4.** $\displaystyle\int_{1/4}^1 \int_{x^2}^x \sqrt{\frac{x}{y}} \, dy \, dx$

5. $\displaystyle\int_{\sqrt{\pi}}^{\sqrt{2\pi}} \int_0^{x^3} \sin \frac{y}{x} \, dy \, dx$ **6.** $\displaystyle\int_{-1}^1 \int_{-x^2}^{x^2} (x^2 - y) \, dy \, dx$

7. $\displaystyle\int_{\pi/2}^\pi \int_0^{x^2} \frac{1}{x} \cos \frac{y}{x} \, dy \, dx$ **8.** $\displaystyle\int_0^1 \int_0^x e^{x^2} \, dy \, dx$

9. $\displaystyle\int_0^1 \int_0^x y\sqrt{x^2 - y^2} \, dy \, dx$ **10.** $\displaystyle\int_1^2 \int_0^{y^2} e^{x/y^2} \, dx \, dy$

11. In each part, find $\displaystyle\iint_R xy \, dA$ over the shaded region R.

(a)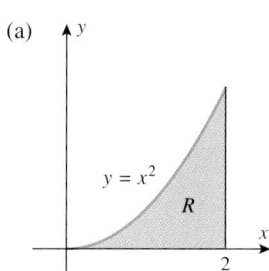

$y = x^2$

R

(b)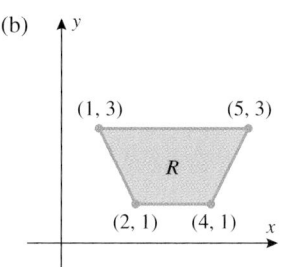

$(1, 3)$ $(5, 3)$

R

$(2, 1)$ $(4, 1)$

12. In each part, find $\displaystyle\iint_R (x + y) \, dA$ over the shaded region R.

(a)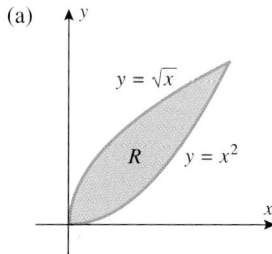

$y = \sqrt{x}$

R $y = x^2$

(b)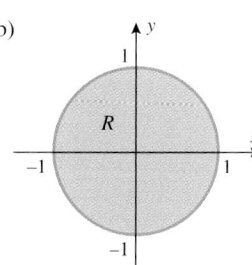

R

In Exercises 13–22, evaluate the double integral.

13. $\displaystyle\iint_R x^2 \, dA$; R is the region bounded by $y = 16/x$, $y = x$, and $x = 8$.

14. $\displaystyle\iint_R xy^2 \, dA$; R is the region enclosed by $y = 1$, $y = 2$, $x = 0$, and $y = x$.

15. $\displaystyle\iint_R x(1 + y^2)^{-1/2} \, dA$; R is the region in the first quadrant enclosed by $y = x^2$, $y = 4$, and $x = 0$.

16. $\displaystyle\iint_R x \cos y \, dA$; R is the triangular region bounded by the lines $y = x$, $y = 0$, and $x = \pi$.

17. $\displaystyle\iint_R (3x - 2y) \, dA$; R is the region enclosed by the circle $x^2 + y^2 = 1$.

18. $\displaystyle\iint_R y \, dA$; R is the region in the first quadrant enclosed between the circle $x^2 + y^2 = 25$ and the line $x + y = 5$.

19. $\displaystyle\iint_R xy \, dA$; R is the region enclosed by $y = \sqrt{x}$, $y = 6 - x$, and $y = 0$.

20. $\displaystyle\iint_R x \, dA$; R is the region enclosed by $y = \sin^{-1} x$, $x = 1/\sqrt{2}$, and $y = 0$.

21. $\displaystyle\iint_R (x - 1) \, dA$; R is the region in the first quadrant enclosed between $y = x$ and $y = x^3$.

22. $\displaystyle\iint_R x^2 \, dA$; R is the region in the first quadrant enclosed by $xy = 1$, $y = x$, and $y = 2x$.

c **23.** Use a CAS to check the answers to the problems you solved in Exercises 1–22.

~ **24.** (a) By hand or with the help of a graphing utility, make a sketch of the region R enclosed between the curves $y = 4x^3 - x^4$ and $y = 3 - 4x + 4x^2$.
     (b) Find the intersections of the curves in part (a).
     (c) Find $\displaystyle\iint_R x \, dA$.

In Exercises 25–28, use double integration to find the area of the plane region enclosed by the given curves.

25. $y = \sin x$ and $y = \cos x$, for $0 \le x \le \pi/4$.

26. $y^2 = -x$ and $3y - x = 4$.

27. $y^2 = 9 - x$ and $y^2 = 9 - 9x$.

28. $y = \cosh x$, $y = \sinh x$, $x = 0$, and $x = 1$.

In Exercises 29 and 30, use double integration to find the volume of the solid.

29.

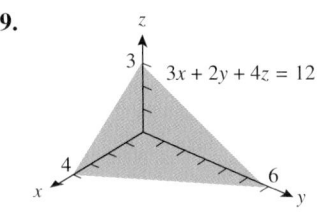

$3x + 2y + 4z = 12$

30.

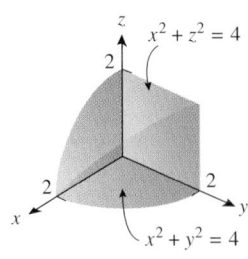

$x^2 + z^2 = 4$

$x^2 + y^2 = 4$

In Exercises 31–38, use double integration to find the volume of each solid.

31. The solid bounded by the cylinder $x^2 + y^2 = 9$ and the planes $z = 0$ and $z = 3 - x$.

32. The solid in the first octant bounded above by the paraboloid $z = x^2 + 3y^2$, below by the plane $z = 0$, and laterally by $y = x^2$ and $y = x$.

33. The solid bounded above by the paraboloid $z = 9x^2 + y^2$, below by the plane $z = 0$, and laterally by the planes $x = 0$, $y = 0$, $x = 3$, and $y = 2$.

34. The solid enclosed by $y^2 = x$, $z = 0$, and $x + z = 1$.

35. The wedge cut from the cylinder $4x^2 + y^2 = 9$ by the planes $z = 0$ and $z = y + 3$.

36. The solid in the first octant bounded above by $z = 9 - x^2$, below by $z = 0$, and laterally by $y^2 = 3x$.

37. The solid that is common to the cylinders $x^2 + y^2 = 25$ and $x^2 + z^2 = 25$.

38. The solid bounded above by the paraboloid $z = x^2 + y^2$, bounded laterally by the circular cylinder $x^2 + (y-1)^2 = 1$, and bounded below by the xy-plane.

In Exercises 39 and 40, use a double integral and a CAS to find the volume of the solid.

C **39.** The solid bounded above by the paraboloid $z = 1 - x^2 - y^2$ and below by the xy-plane.

C **40.** The solid in the first octant that is bounded by the paraboloid $z = x^2 + y^2$, the cylinder $x^2 + y^2 = 4$ and the coordinate planes.

In Exercises 41–46, express the integral as an equivalent integral with the order of integration reversed.

41. $\int_0^2 \int_0^{\sqrt{x}} f(x, y)\, dy\, dx$

42. $\int_0^4 \int_{2y}^8 f(x, y)\, dx\, dy$

43. $\int_0^2 \int_1^{e^y} f(x, y)\, dx\, dy$

44. $\int_1^e \int_0^{\ln x} f(x, y)\, dy\, dx$

45. $\int_0^1 \int_{\sin^{-1} y}^{\pi/2} f(x, y)\, dx\, dy$

46. $\int_0^1 \int_{y^2}^{\sqrt{y}} f(x, y)\, dx\, dy$

In Exercises 47–50, evaluate the integral by first reversing the order of integration.

47. $\int_0^1 \int_{4x}^4 e^{-y^2}\, dy\, dx$

48. $\int_0^2 \int_{y/2}^1 \cos(x^2)\, dx\, dy$

49. $\int_0^4 \int_{\sqrt{y}}^2 e^{x^3}\, dx\, dy$

50. $\int_1^3 \int_0^{\ln x} x\, dy\, dx$

51. Evaluate $\iint\limits_R \sin(y^3)\, dA$, where R is the region bounded by $y = \sqrt{x}$, $y = 2$, and $x = 0$. [*Hint:* Choose the order of integration carefully.]

52. Evaluate $\iint\limits_R x\, dA$, where R is the region bounded by $x = \ln y$, $x = 0$, and $y = e$. [*Hint:* Choose the order of integration carefully.]

C **53.** Try to evaluate the integral with a CAS using the stated order of integration, and then by reversing the order of integration.

(a) $\int_0^4 \int_{\sqrt{x}}^2 \sin \pi y^3\, dy\, dx$

(b) $\int_0^1 \int_{\sin^{-1} y}^{\pi/2} \sec^2(\cos x)\, dx\, dy$

54. Use the appropriate Wallis formula (see Exercise Set 9.3) to find the volume of the solid enclosed between the circular paraboloid $z = x^2 + y^2$, the right circular cylinder $x^2 + y^2 = 4$, and the xy-plane (see the accompanying figure for cut view).

55. Evaluate $\iint\limits_R xy^2\, dA$ over the region R shown in the accompanying figure.

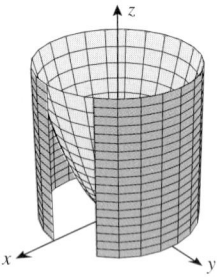

Figure Ex-54

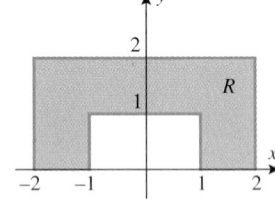

Figure Ex-55

56. Give a geometric argument to show that
$$\int_0^1 \int_0^{\sqrt{1-y^2}} \sqrt{1 - x^2 - y^2}\, dx\, dy = \frac{\pi}{6}$$

The *average value* or *mean value* of a continuous function $f(x, y)$ over a region R in the xy-plane is defined as
$$f_{\text{ave}} = \frac{1}{A(R)} \iint\limits_R f(x, y)\, dA$$
where $A(R)$ is the area of the region R (compare to the definition preceding Exercise 27 of Section 16.1). Use this definition in Exercises 57 and 58.

57. Find the average value of $1/(1 + x^2)$ over the triangular region with vertices $(0, 0)$, $(1, 1)$, and $(0, 1)$.

58. Find the average value of $f(x, y) = x^2 - xy$ over the region enclosed by $y = x$ and $y = 3x - x^2$.

c **59.** Use a CAS to approximate the intersections of the curves $y = \sin x$ and $y = x/2$, and then approximate the volume of the solid in the first octant that is below the surface $z = \sqrt{1 + x + y}$ and above the region in the xy-plane that is enclosed by the curves.

16.3 DOUBLE INTEGRALS IN POLAR COORDINATES

In this section we will study double integrals in which the integrand and the region of integration are expressed in polar coordinates. Such integrals are important for two reasons: first, they arise naturally in many applications, and second, many double integrals in rectangular coordinates can be evaluated more easily if they are converted to polar coordinates.

SIMPLE POLAR REGIONS

Some double integrals are easier to evaluate if the region of integration is expressed in polar coordinates. This is usually true if the region is bounded by a cardioid, a rose curve, a spiral, or, more generally, by any curve whose equation is simpler in polar coordinates than in rectangular coordinates. Moreover, double integrals whose integrands involve $x^2 + y^2$ also tend to be easier to evaluate in polar coordinates because this sum simplifies to r^2 when the conversion formulas $x = r \cos \theta$ and $y = r \sin \theta$ are applied.

Figure 16.3.1*a* shows a region R in a polar coordinate system that is enclosed between two rays, $\theta = \alpha$ and $\theta = \beta$, and two polar curves, $r = r_1(\theta)$ and $r = r_2(\theta)$. If, as shown in that figure, the functions $r_1(\theta)$ and $r_2(\theta)$ are continuous and their graphs do not cross, then the region R is called a *simple polar region*. If $r_1(\theta)$ is identically zero, then the boundary $r = r_1(\theta)$ reduces to a point (the origin), and the region has the general shape shown in Figure 16.3.1*b*. If, in addition, $\beta = \alpha + 2\pi$, then the rays coincide, and the region has the general shape shown in Figure 16.3.1*c*. The following definition expresses these geometric ideas algebraically.

> **16.3.1**   DEFINITION.   A *simple polar region* in a polar coordinate system is a region that is enclosed between two rays, $\theta = \alpha$ and $\theta = \beta$, and two continuous polar curves, $r = r_1(\theta)$ and $r = r_2(\theta)$, where the equations of the rays and the polar curves satisfy the following conditions:
>
> (i) $\alpha \leq \beta$ (ii) $\beta - \alpha \leq 2\pi$ (iii) $0 \leq r_1(\theta) \leq r_2(\theta)$

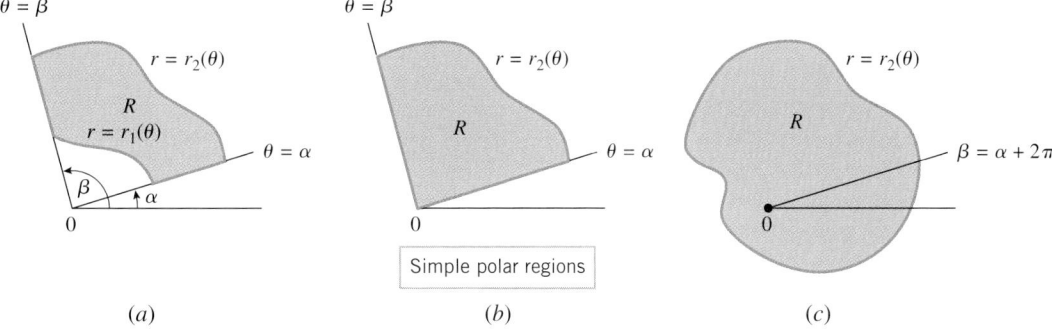

(a)   (b)   (c)

Simple polar regions

Figure 16.3.1

REMARK.   Conditions (i) and (ii) together imply that the ray $\theta = \beta$ can be obtained by rotating the ray $\theta = \alpha$ counterclockwise through an angle that is at most 2π radians. This is consistent with Figure 16.3.1. Condition (iii) implies that the boundary curves $r = r_1(\theta)$ and $r = r_2(\theta)$ can touch but cannot actually cross over one another (why?). Thus, in keeping with Figure 16.3.1, it is appropriate to describe $r = r_1(\theta)$ as the **inner boundary** of the region and $r = r_2(\theta)$ as the **outer boundary**.

DOUBLE INTEGRALS IN POLAR COORDINATES

Next, we will consider the polar version of Problem 16.1.1.

> **16.3.2**   THE VOLUME PROBLEM IN POLAR COORDINATES.   Given a function $f(r, \theta)$ that is continuous and nonnegative on a simple polar region R, find the volume of the solid that is enclosed between the region R and the surface whose equation in cylindrical coordinates is $z = f(r, \theta)$ (Figure 16.3.2).

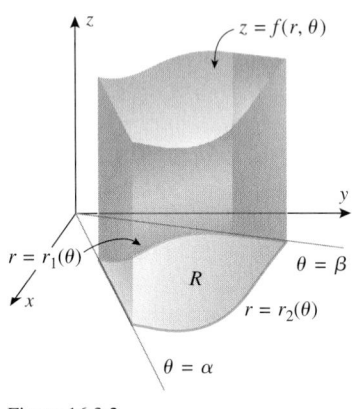

Figure 16.3.2

To motivate a formula for the volume V of the solid in Figure 16.3.2, we will use a limit process similar to that used to obtain Formula (3) of Section 16.1, except that here we will use circular arcs and rays to subdivide the region R into blocks, called **polar rectangles**. As shown in Figure 16.3.3, we will exclude from consideration all polar rectangles that contain any points outside of R, leaving only polar rectangles that are subsets of R. Assume that there are n such polar rectangles, and denote the area of the kth polar rectangle by ΔA_k. Let (r_k^*, θ_k^*) be any point in this polar rectangle. As shown in Figure 16.3.4, the product $f(r_k^*, \theta_k^*)\Delta A_k$ is the volume of a solid with base area ΔA_k and height $f(r_k^*, \theta_k^*)$, so the sum

$$\sum_{k=1}^{n} f(r_k^*, \theta_k^*)\Delta A_k$$

can be viewed as an approximation to the volume V of the entire solid.

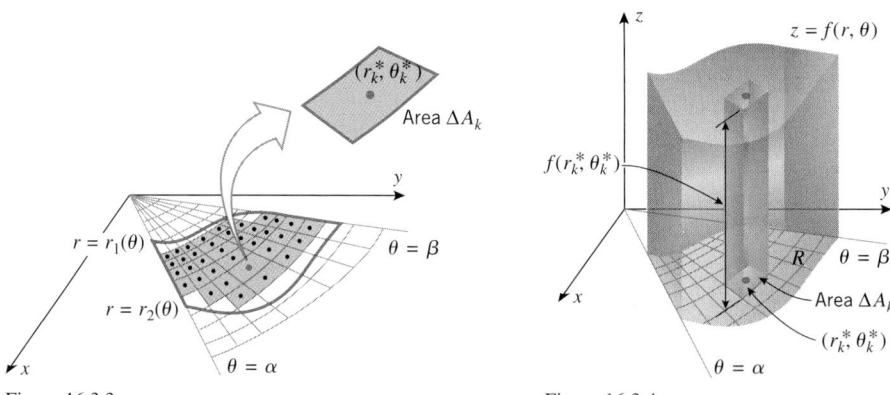

Figure 16.3.3   Figure 16.3.4

If we now increase the number of subdivisions in such a way that the dimensions of the polar rectangles approach zero, then it seems plausible that the errors in the approximations approach zero, and the exact volume of the solid is

$$V = \lim_{n \to +\infty} \sum_{k=1}^{n} f(r_k^*, \theta_k^*)\Delta A_k \tag{1}$$

If $f(r, \theta)$ is continuous on R and has both positive and negative values, then the limit

$$\lim_{n \to +\infty} \sum_{k=1}^{n} f(r_k^*, \theta_k^*)\Delta A_k \tag{2}$$

represents the net signed volume between the region R and the surface $z = f(r, \theta)$ (as with double integrals in rectangular coordinates). The sums in (2) are called **polar Riemann sums**, and the limit of the polar Riemann sums is denoted by

$$\iint\limits_{R} f(r, \theta)\, dA = \lim_{n \to +\infty} \sum_{k=1}^{n} f(r_k^*, \theta_k^*)\Delta A_k \tag{3}$$

which is called the ***polar double integral*** of $f(r, \theta)$ over R. If $f(r, \theta)$ is continuous and nonnegative on R, then the volume formula (1) can be expressed as

$$V = \iint\limits_{R} f(r, \theta)\, dA \tag{4}$$

REMARK.   Polar double integrals are also called ***double integrals in polar coordinates*** to distinguish them from double integrals over regions in the xy-plane, which are called ***double integrals in rectangular coordinates***. Because double integrals in polar coordinates are defined as limits, they have the usual integral properties, such as those stated in Formulas (7), (8), and (9) of Section 16.1.

EVALUATING POLAR DOUBLE INTEGRALS

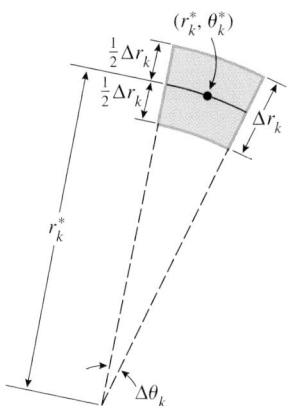

Figure 16.3.5

In Sections 16.1 and 16.2 we evaluated double integrals in rectangular coordinates by expressing them as iterated integrals. Polar double integrals are evaluated the same way. To motivate the formula that expresses a double polar integral as an iterated integral, we will assume that $f(r, \theta)$ is nonnegative so that we can interpret (3) as a volume. However, the results that we will obtain will also be applicable if f has negative values. To begin, let us choose the arbitrary point (r_k^*, θ_k^*) in (3) to be at the "center" of the kth polar rectangle as shown in Figure 16.3.5. Suppose also that this polar rectangle has a central angle $\Delta\theta_k$ and a "radial thickness" Δr_k. Thus, the inner radius of this polar rectangle is $r_k^* - \frac{1}{2}\Delta r_k$ and the outer radius is $r_k^* + \frac{1}{2}\Delta r_k$. Treating the area ΔA_k of this polar rectangle as the difference in area of two sectors, we obtain

$$\Delta A_k = \tfrac{1}{2}\left(r_k^* + \tfrac{1}{2}\Delta r_k\right)^2 \Delta\theta_k - \tfrac{1}{2}\left(r_k^* - \tfrac{1}{2}\Delta r_k\right)^2 \Delta\theta_k$$

which simplifies to

$$\Delta A_k = r_k^* \Delta r_k \Delta\theta_k \tag{5}$$

Thus, from (3) and (4)

$$V = \iint\limits_{R} f(r, \theta)\, dA = \lim_{n \to +\infty} \sum_{k=1}^{n} f(r_k^*, \theta_k^*)r_k^*\Delta r_k \Delta\theta_k$$

which suggests that the volume V can be expressed as the iterated integral

$$V = \iint\limits_{R} f(r, \theta)\, dA = \int_{\alpha}^{\beta}\int_{r_1(\theta)}^{r_2(\theta)} f(r, \theta)r\, dr\, d\theta \tag{6}$$

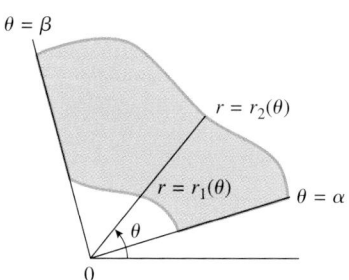

Figure 16.3.6

in which the limits of integration are chosen to cover the region R; that is, with θ fixed between α and β, the value of r varies from $r_1(\theta)$ to $r_2(\theta)$ (Figure 16.3.6).

Although we assumed $f(r, \theta)$ to be nonnegative in deriving Formula (6), it can be proved that the relationship between the polar double integral and the iterated integral in this formula also holds if f has negative values. Accepting this to be so, we obtain the following theorem, which we state without formal proof.

16.3.3   THEOREM.   *If R is a simple polar region whose boundaries are the rays $\theta = \alpha$ and $\theta = \beta$ and the curves $r = r_1(\theta)$ and $r = r_2(\theta)$ shown in Figure 16.3.6, and if $f(r, \theta)$ is continuous on R, then*

$$\iint\limits_{R} f(r, \theta)\, dA = \int_{\alpha}^{\beta}\int_{r_1(\theta)}^{r_2(\theta)} f(r, \theta)r\, dr\, d\theta \tag{7}$$

To apply this theorem, you will need to be able to find the rays and the curves that form the boundary of the region R, since these determine the limits of integration in the iterated integral. This can be done as follows:

Step 1. Since θ is held fixed for the first integration, draw a radial line from the origin through the region R at a fixed angle θ (Figure 16.3.7a). This line crosses the boundary of R at most twice. The innermost point of intersection is on the inner boundary curve $r = r_1(\theta)$ and the outermost point is on the outer boundary curve $r = r_2(\theta)$. These intersections determine the r-limits of integration in (7).

Step 2. Imagine rotating a ray along the polar x-axis one revolution counterclockwise about the origin. The smallest angle at which this ray intersects the region R is $\theta = \alpha$ and the largest angle is $\theta = \beta$ (Figure 16.3.7b). This determines the θ-limits of integration.

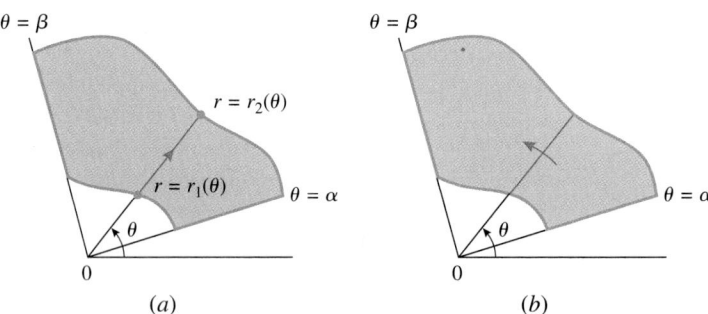

Figure 16.3.7

Example 1

Evaluate

$$\iint_R \sin\theta \, dA$$

where R is the region in the first quadrant that is outside the circle $r = 2$ and inside the cardioid $r = 2(1 + \cos\theta)$.

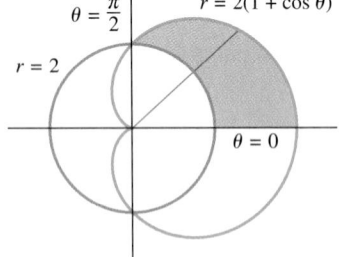

Figure 16.3.8

Solution. The region R is sketched in Figure 16.3.8. Following the two steps outlined above we obtain

$$\iint_R \sin\theta \, dA = \int_0^{\pi/2} \int_2^{2(1+\cos\theta)} (\sin\theta) r \, dr \, d\theta$$

$$= \int_0^{\pi/2} \frac{1}{2} r^2 \sin\theta \Big]_{r=2}^{2(1+\cos\theta)} d\theta$$

$$= 2 \int_0^{\pi/2} [(1 + \cos\theta)^2 \sin\theta - \sin\theta] \, d\theta$$

$$= 2 \left[-\frac{1}{3}(1 + \cos\theta)^3 + \cos\theta \right]_0^{\pi/2}$$

$$= 2 \left[-\frac{1}{3} - \left(-\frac{5}{3} \right) \right] = \frac{8}{3} \qquad \blacktriangleleft$$

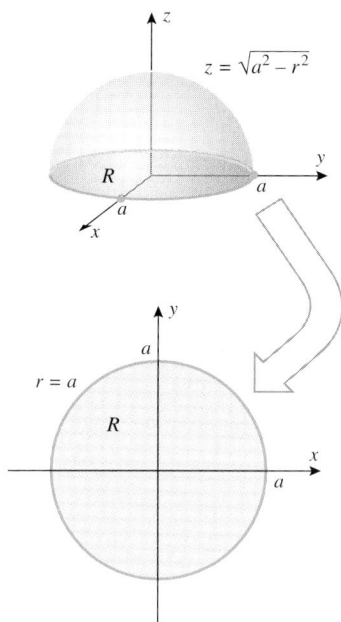

Example 2

The sphere of radius a centered at the origin is expressed in rectangular coordinates as $x^2 + y^2 + z^2 = a^2$, and hence its equation in cylindrical coordinates is $r^2 + z^2 = a^2$. Use this equation and a polar double integral to find the volume of the sphere.

Solution. In cylindrical coordinates the upper hemisphere is given by the equation

$$z = \sqrt{a^2 - r^2}$$

so the volume enclosed by the entire sphere is

$$V = 2 \iint_R \sqrt{a^2 - r^2} \, dA$$

where R is the circular region shown in Figure 16.3.9. Thus,

$$V = 2 \iint_R \sqrt{a^2 - r^2} \, dA = \int_0^{2\pi} \int_0^a \sqrt{a^2 - r^2}\,(2r) \, dr \, d\theta$$

$$= \int_0^{2\pi} \left[-\frac{2}{3}(a^2 - r^2)^{3/2} \right]_{r=0}^a d\theta = \int_0^{2\pi} \frac{2}{3}a^3 \, d\theta$$

$$= \left[\frac{2}{3}a^3 \theta \right]_0^{2\pi} = \frac{4}{3}\pi a^3 \qquad \blacktriangleleft$$

Figure 16.3.9

FINDING AREAS USING POLAR DOUBLE INTEGRALS

Recall from Formula (7) of Section 16.2 that the area of a region R in the xy-plane can be expressed as

$$\text{area of } R = \iint_R 1 \, dA = \iint_R dA \tag{8}$$

The argument used to derive this result can also be used to show that the formula applies to polar double integrals over regions in polar coordinates.

Example 3

Use a polar double integral to find the area enclosed by the three-petaled rose $r = \sin 3\theta$.

Solution. The rose is sketched in Figure 16.3.10. We will use Formula (8) to calculate the area of the petal R in the first quadrant and multiply by three.

$$A = 3 \iint_R dA = 3 \int_0^{\pi/3} \int_0^{\sin 3\theta} r \, dr \, d\theta$$

$$= \frac{3}{2} \int_0^{\pi/3} \sin^2 3\theta \, d\theta = \frac{3}{4} \int_0^{\pi/3} (1 - \cos 6\theta) \, d\theta$$

$$= \frac{3}{4} \left[\theta - \frac{\sin 6\theta}{6} \right]_0^{\pi/3} = \frac{1}{4}\pi \qquad \blacktriangleleft$$

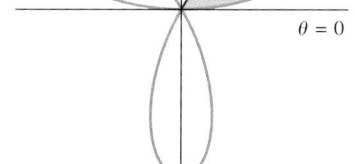

Figure 16.3.10

CONVERTING DOUBLE INTEGRALS FROM RECTANGULAR TO POLAR COORDINATES

Sometimes a double integral that is difficult to evaluate in rectangular coordinates can be evaluated more easily in polar coordinates by making the substitution $x = r \cos\theta$, $y = r \sin\theta$ and expressing the region of integration in polar form; that is, we rewrite the double integral in rectangular coordinates as

$$\iint_R f(x, y) \, dA = \iint_R f(r \cos\theta, r \sin\theta) \, dA = \iint_{\substack{\text{appropriate} \\ \text{limits}}} f(r \cos\theta, r \sin\theta) r \, dr \, d\theta \tag{9}$$

Example 4

Use polar coordinates to evaluate $\displaystyle\int_{-1}^{1}\int_{0}^{\sqrt{1-x^2}} (x^2 + y^2)^{3/2}\, dy\, dx$.

Solution. In this problem we are starting with an iterated integral in rectangular coordinates rather than a double integral, so before we can make the conversion to polar coordinates we will have to identify the region of integration. To do this, we observe that for fixed x the y-integration runs from $y = 0$ to $y = \sqrt{1 - x^2}$, which tells us that the lower boundary of the region is the x-axis and the upper boundary is a semicircle of radius 1 centered at the origin. From the x-integration we see that x varies from -1 to 1, so we conclude that the region of integration is as shown in Figure 16.3.11. In polar coordinates, this is the region swept out as r varies between 0 and 1 and θ varies between 0 and π. Thus,

$$\int_{-1}^{1}\int_{0}^{\sqrt{1-x^2}} (x^2 + y^2)^{3/2}\, dy\, dx = \iint_{R} (x^2 + y^2)^{3/2}\, dA$$

$$= \int_{0}^{\pi}\int_{0}^{1} (r^3)\, r\, dr\, d\theta = \int_{0}^{\pi} \frac{1}{5}\, d\theta = \frac{\pi}{5}$$ ◀

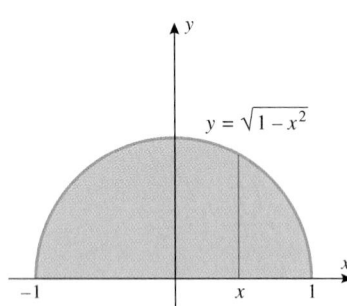

$y = \sqrt{1 - x^2}$

Figure 16.3.11

REMARK. The conversion to polar coordinates worked so nicely in this example because the substitution $x = r\cos\theta$, $y = r\sin\theta$ collapsed the sum $x^2 + y^2$ into the single term r^2, thereby simplifying the integrand. Whenever you see an expression involving $x^2 + y^2$ in the integrand, you should consider the possibility of converting to polar coordinates.

EXERCISE SET 16.3   C CAS

In Exercises 1–6, evaluate the iterated integral.

1. $\displaystyle\int_{0}^{\pi/2}\int_{0}^{\sin\theta} r\cos\theta\, dr\, d\theta$ **2.** $\displaystyle\int_{0}^{\pi}\int_{0}^{1+\cos\theta} r\, dr\, d\theta$

3. $\displaystyle\int_{0}^{\pi/2}\int_{0}^{a\sin\theta} r^2\, dr\, d\theta$ **4.** $\displaystyle\int_{0}^{\pi/6}\int_{0}^{\cos 3\theta} r\, dr\, d\theta$

5. $\displaystyle\int_{0}^{\pi}\int_{0}^{1-\sin\theta} r^2\cos\theta\, dr\, d\theta$ **6.** $\displaystyle\int_{0}^{\pi/2}\int_{0}^{\cos\theta} r^3\, dr\, d\theta$

In Exercises 7–12, use a double integral in polar coordinates to find the area of the region described.

7. The region enclosed by the cardioid $r = 1 - \cos\theta$.

8. The region enclosed by the rose $r = \sin 2\theta$.

9. The region in the first quadrant bounded by $r = 1$ and $r = \sin 2\theta$, with $\pi/4 \le \theta \le \pi/2$.

10. The region inside the circle $x^2 + y^2 = 4$ and to the right of the line $x = 1$.

11. The region inside the circle $r = 4\sin\theta$ and outside the circle $r = 2$.

12. The region inside the circle $r = 1$ and outside the cardioid $r = 1 + \cos\theta$.

In Exercises 13–18, use a double integral in polar coordinates to find the volume of the solid that is described.

13.

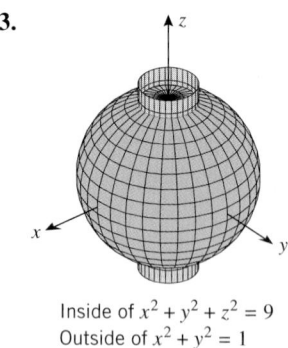

Inside of $x^2 + y^2 + z^2 = 9$
Outside of $x^2 + y^2 = 1$

14.

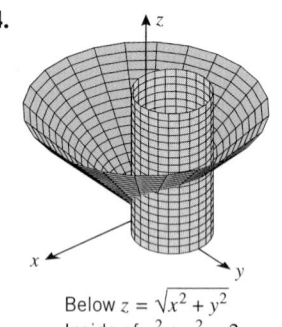

Below $z = \sqrt{x^2 + y^2}$
Inside of $x^2 + y^2 = 2y$
Above $z = 0$

15.

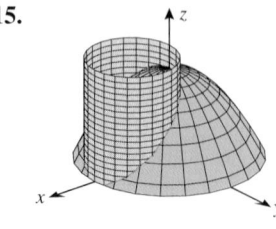

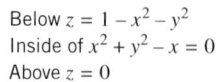

Below $z = 1 - x^2 - y^2$
Inside of $x^2 + y^2 - x = 0$
Above $z = 0$

16.

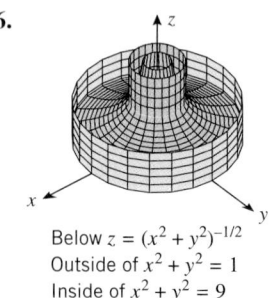

Below $z = (x^2 + y^2)^{-1/2}$
Outside of $x^2 + y^2 = 1$
Inside of $x^2 + y^2 = 9$
Above $z = 0$

17. The solid in the first octant bounded above by the surface $z = r \sin\theta$, below by the xy-plane, and laterally by the plane $x = 0$ and the surface $r = 3\sin\theta$.

18. The solid inside of the surface $r^2 + z^2 = 4$ and outside of the surface $r = 2\cos\theta$.

In Exercises 19–22, use polar coordinates to evaluate the double integral.

19. $\displaystyle\iint\limits_R e^{-(x^2+y^2)}\,dA$, where R is the region enclosed by the circle $x^2 + y^2 = 1$.

20. $\displaystyle\iint\limits_R \sqrt{9 - x^2 - y^2}\,dA$, where R is the region in the first quadrant within the circle $x^2 + y^2 = 9$.

21. $\displaystyle\iint\limits_R \frac{1}{1 + x^2 + y^2}\,dA$, where R is the sector in the first quadrant bounded by $y = 0$, $y = x$, and $x^2 + y^2 = 4$.

22. $\displaystyle\iint\limits_R 2y\,dA$, where R is the region in the first quadrant bounded above by the circle $(x - 1)^2 + y^2 = 1$ and below by the line $y = x$.

In Exercises 23–30, evaluate the iterated integral by converting to polar coordinates.

23. $\displaystyle\int_0^1 \int_0^{\sqrt{1-x^2}} (x^2 + y^2)\,dy\,dx$

24. $\displaystyle\int_{-2}^2 \int_{-\sqrt{4-y^2}}^{\sqrt{4-y^2}} e^{-(x^2+y^2)}\,dx\,dy$

25. $\displaystyle\int_0^2 \int_0^{\sqrt{2x-x^2}} \sqrt{x^2 + y^2}\,dy\,dx$

26. $\displaystyle\int_0^1 \int_0^{\sqrt{1-y^2}} \cos(x^2 + y^2)\,dx\,dy$

27. $\displaystyle\int_0^a \int_0^{\sqrt{a^2-x^2}} \frac{dy\,dx}{(1 + x^2 + y^2)^{3/2}}\quad (a > 0)$

28. $\displaystyle\int_0^1 \int_y^{\sqrt{y}} \sqrt{x^2 + y^2}\,dx\,dy$

29. $\displaystyle\int_0^{\sqrt{2}} \int_y^{\sqrt{4-y^2}} \frac{1}{\sqrt{1 + x^2 + y^2}}\,dx\,dy$

30. $\displaystyle\int_0^4 \int_3^{\sqrt{25-x^2}} dy\,dx$

31. Use a double integral in polar coordinates to find the volume of a cylinder of radius a and height h.

32. (a) Use a double integral in polar coordinates to find the volume of the oblate spheroid
$$\frac{x^2}{a^2} + \frac{y^2}{a^2} + \frac{z^2}{c^2} = 1 \quad (0 < c < a)$$

(b) Use the result in part (a) and the World Geodetic System of 1984 (WGS-84) discussed in Exercise 50 of Section 13.7 to find the volume of the Earth in cubic meters.

33. Use polar coordinates to find the volume of the solid that is inside of the ellipsoid
$$\frac{x^2}{a^2} + \frac{y^2}{a^2} + \frac{z^2}{c^2} = 1$$
above the xy-plane, and inside of the cylinder
$$x^2 + y^2 - ay = 0.$$

34. Find the area of the region enclosed by the lemniscate $r^2 = 2a^2 \cos 2\theta$.

35. Find the area in the first quadrant that is inside of the circle $r = 4\sin\theta$ and outside of the lemniscate $r^2 = 8\cos 2\theta$.

36. Show that the shaded area in the accompanying figure is $a^2\phi - \frac{1}{2}a^2 \sin 2\phi$.

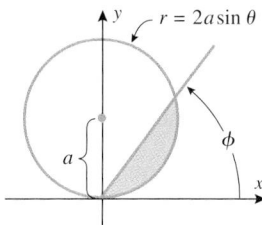

Figure Ex-36

37. The integral $\displaystyle\int_0^{+\infty} e^{-x^2}\,dx$, which arises in probability theory, can be evaluated using the following method. Let the value of the integral be I. Thus,
$$I = \int_0^{+\infty} e^{-x^2}\,dx = \int_0^{+\infty} e^{-y^2}\,dy$$
since the letter used for the variable of integration in a definite integral does not matter.

(a) Give a reasonable argument to show that
$$I^2 = \int_0^{+\infty} \int_0^{+\infty} e^{-(x^2+y^2)}\,dx\,dy$$

(b) Evaluate the iterated integral in part (a) by converting to polar coordinates.

(c) Use the result in part (b) to show that $I = \sqrt{\pi}/2$.

[c] **38.** (a) Use the numerical integration capability of a CAS to approximate the value of the double integral
$$\int_{-1}^1 \int_0^{\sqrt{1-x^2}} e^{-(x^2+y^2)^2}\,dy\,dx$$

(b) Compare the approximation obtained in part (a) to the approximation that results if the integral is first converted to polar coordinates.

39. Suppose that a geyser, centered at the origin of a polar coordinate system, sprays water in a circular pattern in such a way that the depth D of water that reaches a point at a distance of r feet from the origin in 1 hour is $D = ke^{-r}$. Find the total volume of water that the geyser sprays inside a circle of radius R centered at the origin.

40. Evaluate $\displaystyle\iint\limits_{R} x^2 \, dA$ over the region R shown in the accompanying figure.

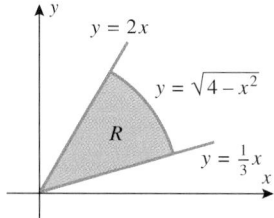

Figure Ex-40

16.4 PARAMETRIC SURFACES; SURFACE AREA

In previous sections we considered parametric curves in 2-space and 3-space. In this section we will discuss parametric surfaces in 3-space. As we will see, parametric representations of surfaces are not only important in computer graphics but also allow us to study more general kinds of surfaces than those encountered so far. In Section 8.5 we showed how to find the surface area of a surface of revolution. Our work on parametric surfaces will enable us to derive area formulas for more general kinds of surfaces.

PARAMETRIC REPRESENTATION OF SURFACES

We have seen that curves in 3-space can be represented by three equations involving one parameter, say

$$x = x(t), \quad y = y(t), \quad z = z(t)$$

Surfaces in 3-space can be represented parametrically by three equations involving two parameters, say

$$x = x(u, v), \quad y = y(u, v), \quad z = z(u, v) \tag{1}$$

To visualize why such equations represent a surface, think of (u, v) as a point that varies over some region in a uv-plane. If u is held constant, then v is the only varying parameter in (1), and hence these equations represent a curve in 3-space. We call this a ***constant u-curve*** (Figure 16.4.1). Similarly, if v is held constant, then u is the only varying parameter in (1), so again these equations represent a curve in 3-space. We call this a ***constant v-curve***. By varying the constants we generate a family of u-curves and a family of v-curves that together form a surface.

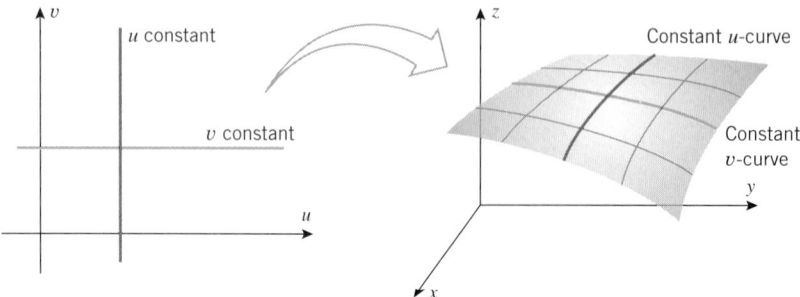

Figure 16.4.1

Example 1

Consider the paraboloid $z = 4 - x^2 - y^2$. One way to parametrize this surface is to take $x = u$ and $y = v$ as the parameters, in which case the surface is represented by the parametric equations

$$x = u, \quad y = v, \quad z = 4 - u^2 - v^2 \tag{2}$$

Figure 16.4.2*a* shows a computer-generated graph of this surface. The constant *u*-curves correspond to constant *x*-values and hence appear on the surface as traces parallel to the *yz*-plane. Similarly, the constant *v*-curves correspond to constant *y*-values and hence appear on the surface as traces parallel to the *xz*-plane. ◀

Example 2

The paraboloid $z = 4 - x^2 - y^2$ that was considered in Example 1 can also be parametrized by first expressing the equation in cylindrical coordinates. For this purpose, we make the substitution $x = r\cos\theta$, $y = r\sin\theta$, which yields $z = 4 - r^2$. Thus, the paraboloid can be represented parametrically in terms of *r* and θ as

$$x = r\cos\theta, \quad y = r\sin\theta, \quad z = 4 - r^2 \tag{3}$$

Figure 16.4.2*b* shows a computer-generated graph of this surface for $0 \le r \le 2$ and $0 \le \theta \le 2\pi$. The constant *r*-curves correspond to constant *z*-values and hence appear on the surface as traces parallel to the *xy*-plane. The constant θ-curves appear on the surface as traces from vertical planes through the origin at varying angles with the *x*-axis. Parts (*c*) and (*d*) of Figure 16.4.2 show the effect of restrictions on the parameters *r* and θ. ◀

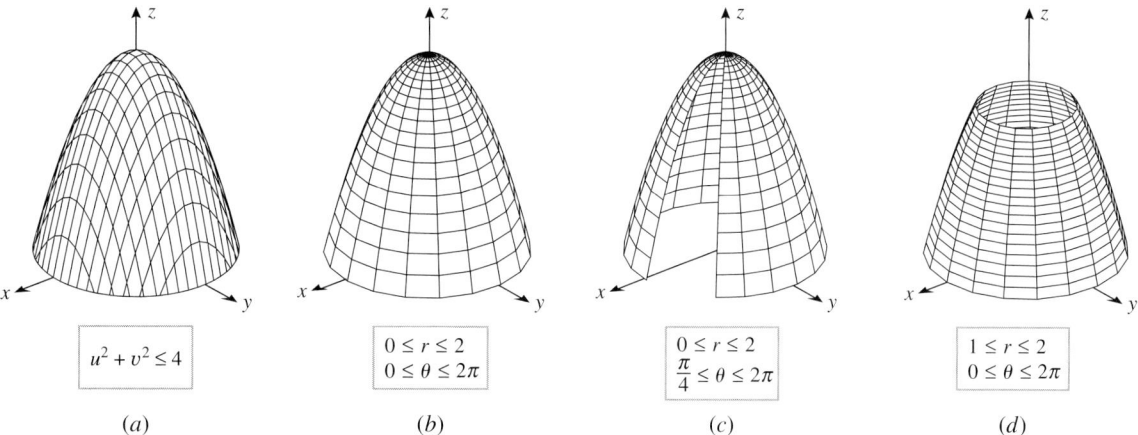

(*a*)   (*b*)   (*c*)   (*d*)

Figure 16.4.2

FOR THE READER. If you have a graphing utility that can generate parametric surfaces, read the relevant documentation and then try to make reasonable duplicates of the surfaces in Figure 16.4.2.

Example 3

One way to generate the sphere $x^2 + y^2 + z^2 = 1$ with a graphing utility is to graph the upper and lower hemispheres

$$z = \sqrt{1 - x^2 - y^2} \quad \text{and} \quad z = -\sqrt{1 - x^2 - y^2}$$

on the same screen. However, this usually produces a fragmented sphere (Figure 16.4.3*a*) because roundoff error sporadically produces negative values inside the radical when $1 - x^2 - y^2$ is near zero. A better graph can be generated by first expressing the sphere in spherical coordinates as $\rho = 1$ and then using the spherical-to-rectangular conversion formulas in Table 13.8.1 to obtain the parametric equations

$$x = \sin\phi\cos\theta, \quad y = \sin\phi\sin\theta, \quad z = \cos\phi$$

with parameters θ and ϕ. Figure 16.4.3*b* shows the graph of this parametric surface for

$0 \leq \theta \leq 2\pi$ and $0 \leq \phi \leq \pi$. In the language of cartographers, the constant ϕ-curves are the ***lines of latitude*** and the constant θ-curves are the ***lines of longitude***. ◄

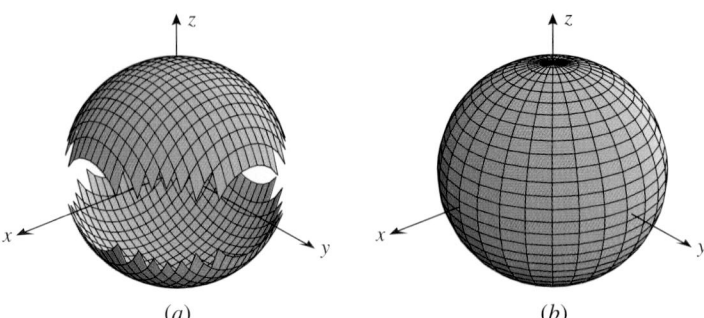

(a) (b)

Figure 16.4.3

Example 4

Find parametric equations for the portion of the right circular cylinder $x^2 + z^2 = 9$ for which $0 \leq y \leq 5$ in terms of the parameters u and v shown in Figure 16.4.4a. The parameter u is the y-coordinate of a point $P(x, y, z)$ on the surface, and v is the angle shown in the figure.

Solution. The radius of the cylinder is 3, so it is evident from the figure that $y = u$, $x = 3\cos v$, and $z = 3\sin v$. Thus, the surface can be represented parametrically as

$$x = 3\cos v, \quad y = u, \quad z = 3\sin v$$

To obtain the portion of the surface from $y = 0$ to $y = 5$, we let the parameter u vary over the interval $0 \leq u \leq 5$, and to ensure that the entire lateral surface is covered, we let the parameter v vary over the interval $0 \leq v \leq 2\pi$. Figure 16.4.4b shows a computer-generated graph of the surface in which u and v vary over these intervals. Constant u-curves appear as circular traces parallel to the xz-plane, and constant v-curves appear as lines parallel to the y-axis. ◄

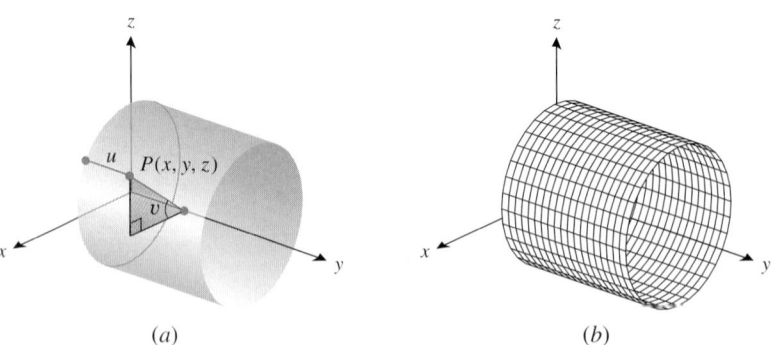

(a) (b)

Figure 16.4.4

**REPRESENTING SURFACES OF
REVOLUTION PARAMETRICALLY**

The basic idea of Example 4 can be adapted to obtain parametric equations for surfaces of revolution. For example, suppose that we want to find parametric equations for the surface generated by revolving the plane curve $y = f(x)$ about the x-axis. Figure 16.4.5 suggests that the surface can be represented parametrically as

$$x = u, \quad y = f(u)\cos v, \quad z = f(u)\sin v \tag{4}$$

where v is the angle shown. In the exercises we will discuss analogous formulas for surfaces of revolution about other axes.

Example 5

Find parametric equations for the surface generated by revolving the curve $y = 1/x$ about the x-axis.

Solution. From (4) this surface can be represented parametrically as

$$x = u, \quad y = \frac{1}{u}\cos v, \quad z = \frac{1}{u}\sin v$$

Figure 16.4.6 shows a computer-generated graph of the surface for $0.7 \le u \le 5$ and $0 \le v \le 2\pi$. This surface is a portion of Gabriel's horn, which was discussed in Exercise 49 of Section 9.8. ◄

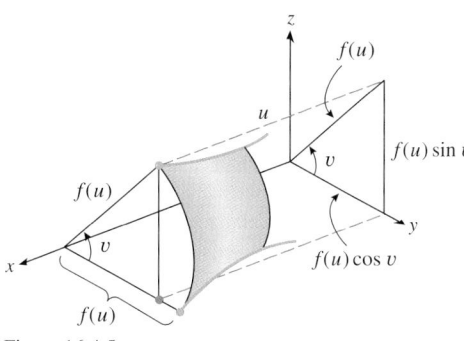

Figure 16.4.5   Figure 16.4.6

VECTOR-VALUED FUNCTIONS OF TWO VARIABLES

Recall that the parametric equations

$$x = x(t), \quad y = y(t), \quad z = z(t)$$

can be expressed in vector form as

$$\mathbf{r} = x(t)\mathbf{i} + y(t)\mathbf{j} + z(t)\mathbf{k}$$

where $\mathbf{r} = x\mathbf{i} + y\mathbf{j} + z\mathbf{k}$ is the radius vector and $\mathbf{r}(t) = x(t)\mathbf{i} + y(t)\mathbf{j} + z(t)\mathbf{k}$ is a vector-valued function of one variable. Similarly, the parametric equations

$$x = x(u, v), \quad y = y(u, v), \quad z = z(u, v)$$

can be expressed in vector form as

$$\mathbf{r} = x(u, v)\mathbf{i} + y(u, v)\mathbf{j} + z(u, v)\mathbf{k}$$

where the vector $\mathbf{r} = x\mathbf{i} + y\mathbf{j} + z\mathbf{k}$ is called the ***radius vector*** and where the function $\mathbf{r}(u, v) = x(u, v)\mathbf{i} + y(u, v)\mathbf{j} + z(u, v)\mathbf{k}$ is a ***vector-valued function of two variables***. We define the ***graph*** of $\mathbf{r}(u, v)$ to be the graph of the corresponding parametric equations. Geometrically, we can view $\mathbf{r}$ as a vector from the origin to a point (x, y, z) that moves over the surface $\mathbf{r} = \mathbf{r}(u, v)$ as u and v vary (Figure 16.4.7). As with vector-valued functions of one variable, we say that $\mathbf{r}(u, v)$ is ***continuous*** if each component is continuous.

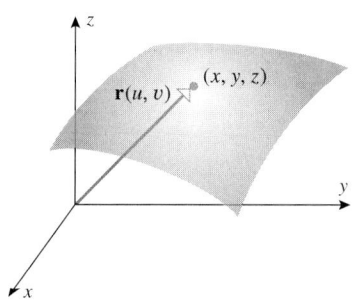

Figure 16.4.7

Example 6

The paraboloid in Example 1 was expressed parametrically as

$$x = u, \quad y = v, \quad z = 4 - u^2 - v^2$$

These equations can be expressed in vector form as

$$\mathbf{r} = u\mathbf{i} + v\mathbf{j} + (4 - u^2 - v^2)\mathbf{k} \qquad\qquad ◄$$

PARTIAL DERIVATIVES OF VECTOR-VALUED FUNCTIONS

Partial derivatives of vector-valued functions of two variables are obtained by taking partial derivatives of the components. For example, if

$$\mathbf{r}(u, v) = x(u, v)\mathbf{i} + y(u, v)\mathbf{j} + z(u, v)\mathbf{k}$$

then

$$\frac{\partial \mathbf{r}}{\partial u} = \frac{\partial x}{\partial u}\mathbf{i} + \frac{\partial y}{\partial u}\mathbf{j} + \frac{\partial z}{\partial u}\mathbf{k}$$

$$\frac{\partial \mathbf{r}}{\partial v} = \frac{\partial x}{\partial v}\mathbf{i} + \frac{\partial y}{\partial v}\mathbf{j} + \frac{\partial z}{\partial v}\mathbf{k}$$

These derivatives can also be written as $\mathbf{r}_u$ and $\mathbf{r}_v$ or $\mathbf{r}_u(u, v)$ and $\mathbf{r}_v(u, v)$ and can be expressed as the limits

$$\frac{\partial \mathbf{r}}{\partial u} = \lim_{h \to 0} \frac{\mathbf{r}(u + h, v) - \mathbf{r}(u, v)}{h} \tag{5}$$

$$\frac{\partial \mathbf{r}}{\partial v} = \lim_{k \to 0} \frac{\mathbf{r}(u, v + k) - \mathbf{r}(u, v)}{k} \tag{6}$$

Example 7

For the vector-valued function in Example 6, we have

$$\frac{\partial \mathbf{r}}{\partial u} = \frac{\partial}{\partial u}[u\mathbf{i} + v\mathbf{j} + (4 - u^2 - v^2)\mathbf{k}] = \mathbf{i} - 2u\mathbf{k}$$

$$\frac{\partial \mathbf{r}}{\partial v} = \frac{\partial}{\partial v}[u\mathbf{i} + v\mathbf{j} + (4 - u^2 - v^2)\mathbf{k}] = \mathbf{j} - 2v\mathbf{k} \qquad \blacktriangleleft$$

TANGENT PLANES TO PARAMETRIC SURFACES

Our next objective is to show how to find tangent planes to parametric surfaces. Recall from Section 15.5 that a surface has a tangent plane at a point if all smooth curves on the surface that pass through the point have tangent lines and those tangent lines lie in a common plane (the tangent plane). Moreover, we showed that if $z = f(x, y)$, then the graph of f has a tangent plane at a point if f is differentiable at that point. It is beyond the scope of this text to obtain precise conditions under which a parametric surface has a tangent plane at a point, so we will simply assume the existence of tangent planes at points of interest and focus on finding their equations.

Suppose that the parametric surface σ is the graph of the vector-valued function $\mathbf{r}(u, v)$ and that we are interested in the tangent plane at the point (x_0, y_0, z_0) on the surface that corresponds to the parameter values $u = u_0$ and $v = v_0$; that is,

$$\mathbf{r}(u_0, v_0) = x_0\mathbf{i} + y_0\mathbf{j} + z_0\mathbf{k}$$

If $v = v_0$ is kept fixed and u is allowed to vary, then $\mathbf{r}(u, v_0)$ is a vector-valued function of one variable whose graph is the constant v-curve through the point (u_0, v_0); similarly, if $u = u_0$ is kept fixed and v is allowed to vary, then $\mathbf{r}(u_0, v)$ is a vector-valued function of one variable whose graph is the constant u-curve through the point (u_0, v_0). Moreover, it follows from 14.2.5 that if $\partial \mathbf{r}/\partial u \neq \mathbf{0}$ at (u_0, v_0), then this vector is tangent to the constant v-curve through (u_0, v_0); and if $\partial \mathbf{r}/\partial v \neq \mathbf{0}$ at (u_0, v_0), then this vector is tangent to the constant u-curve through (u_0, v_0) (Figure 16.4.8). Thus, if $\partial \mathbf{r}/\partial u \times \partial \mathbf{r}/\partial v \neq \mathbf{0}$ at (u_0, v_0), then the vector

$$\frac{\partial \mathbf{r}}{\partial u} \times \frac{\partial \mathbf{r}}{\partial v} = \begin{vmatrix} \mathbf{i} & \mathbf{j} & \mathbf{k} \\ \dfrac{\partial x}{\partial u} & \dfrac{\partial y}{\partial u} & \dfrac{\partial z}{\partial u} \\ \dfrac{\partial x}{\partial v} & \dfrac{\partial y}{\partial v} & \dfrac{\partial z}{\partial v} \end{vmatrix} \tag{7}$$

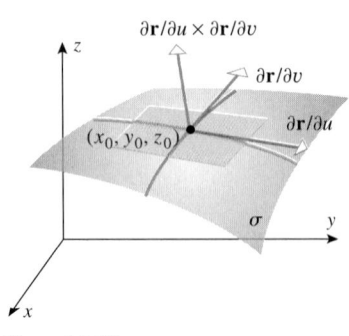

Figure 16.4.8

is orthogonal to both tangent vectors at the point (u_0, v_0) and hence is normal to the tangent plane and the surface at this point (Figure 16.4.8). Accordingly, we make the following definition.

16.4.1 DEFINITION. If a parametric surface σ is the graph of $\mathbf{r} = \mathbf{r}(u, v)$, and if $\partial \mathbf{r}/\partial u \times \partial \mathbf{r}/\partial v \neq \mathbf{0}$ at a point on the surface, then the ***principal unit normal vector*** to the surface at that point is denoted by $\mathbf{n}$ or $\mathbf{n}(u, v)$ and is defined as

$$\mathbf{n} = \frac{\dfrac{\partial \mathbf{r}}{\partial u} \times \dfrac{\partial \mathbf{r}}{\partial v}}{\left\| \dfrac{\partial \mathbf{r}}{\partial u} \times \dfrac{\partial \mathbf{r}}{\partial v} \right\|} \tag{8}$$

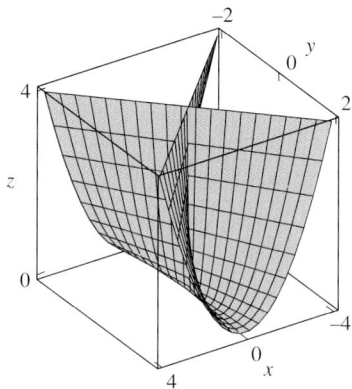

Figure 16.4.9

Example 8

Find an equation of the tangent plane to the parametric surface

$$x = uv, \quad y = u, \quad z = v^2$$

at the point where $u = 2$ and $v = -1$. This surface, called *Whitney's umbrella*, is an example of a self-intersecting parametric surface (Figure 16.4.9).

Solution. We start by writing the equations in the vector form

$$\mathbf{r} = uv\mathbf{i} + u\mathbf{j} + v^2\mathbf{k}$$

The partial derivatives of $\mathbf{r}$ are

$$\frac{\partial \mathbf{r}}{\partial u}(u, v) = v\mathbf{i} + \mathbf{j}$$

$$\frac{\partial \mathbf{r}}{\partial v}(u, v) = u\mathbf{i} + 2v\mathbf{k}$$

and at $u = 2$ and $v = -1$ these partial derivatives are

$$\frac{\partial \mathbf{r}}{\partial u}(2, -1) = -\mathbf{i} + \mathbf{j}$$

$$\frac{\partial \mathbf{r}}{\partial v}(2, -1) = 2\mathbf{i} - 2\mathbf{k}$$

Thus, from (7) and (8) a normal to the surface at this point is

$$\frac{\partial \mathbf{r}}{\partial u}(2, -1) \times \frac{\partial \mathbf{r}}{\partial v}(2, -1) = \begin{vmatrix} \mathbf{i} & \mathbf{j} & \mathbf{k} \\ -1 & 1 & 0 \\ 2 & 0 & -2 \end{vmatrix} = -2\mathbf{i} - 2\mathbf{j} - 2\mathbf{k}$$

Since any normal will suffice to find the tangent plane, it makes sense to multiply this vector by $-\frac{1}{2}$ and use the simpler normal $\mathbf{i} + \mathbf{j} + \mathbf{k}$. It follows from the given parametric equations that the point on the surface corresponding to $u = 2$ and $v = -1$ is $(-2, 2, 1)$, so the tangent plane at this point can be expressed in point-normal form as

$$(x + 2) + (y - 2) + (z - 1) = 0 \quad \text{or} \quad x + y + z = 1 \qquad \blacktriangleleft$$

FOR THE READER. Convince yourself that the result obtained in this example is consistent with Figure 16.4.9.

Example 9

The sphere $x^2 + y^2 + z^2 = a^2$ can be expressed in spherical coordinates as $\rho = a$, and the spherical-to-rectangular conversion formulas in Table 13.8.1 can then be used to express the sphere as the graph of the vector-valued function

$$\mathbf{r}(\phi, \theta) = a \sin \phi \cos \theta \mathbf{i} + a \sin \phi \sin \theta \mathbf{j} + a \cos \phi \mathbf{k}$$

where $0 \leq \phi \leq \pi$ and $0 \leq \theta \leq 2\pi$ (verify). Use this function to show that at each point on the sphere the tangent plane is perpendicular to the radius vector.

Solution. We will show that at each point of the sphere the unit normal vector **n** is a scalar multiple of **r** (and hence is parallel to **r**). But

$$\frac{\partial \mathbf{r}}{\partial \phi} \times \frac{\partial \mathbf{r}}{\partial \theta} = \begin{vmatrix} \mathbf{i} & \mathbf{j} & \mathbf{k} \\ \dfrac{\partial x}{\partial \phi} & \dfrac{\partial y}{\partial \phi} & \dfrac{\partial z}{\partial \phi} \\ \dfrac{\partial x}{\partial \theta} & \dfrac{\partial y}{\partial \theta} & \dfrac{\partial z}{\partial \theta} \end{vmatrix} = \begin{vmatrix} \mathbf{i} & \mathbf{j} & \mathbf{k} \\ a\cos\phi\cos\theta & a\cos\phi\sin\theta & -a\sin\phi \\ -a\sin\phi\sin\theta & a\sin\phi\cos\theta & 0 \end{vmatrix}$$

$$= a^2 \sin^2\phi\cos\theta\,\mathbf{i} + a^2\sin^2\phi\sin\theta\,\mathbf{j} + a^2\sin\phi\cos\phi\,\mathbf{k}$$

and hence

$$\left\| \frac{\partial \mathbf{r}}{\partial \phi} \times \frac{\partial \mathbf{r}}{\partial \theta} \right\| = \sqrt{a^4\sin^4\phi\cos^2\theta + a^4\sin^4\phi\sin^2\theta + a^4\sin^2\phi\cos^2\phi}$$

$$= \sqrt{a^4\sin^4\phi + a^4\sin^2\phi\cos^2\phi}$$

$$= a^2\sqrt{\sin^2\phi} = a^2|\sin\phi| = a^2\sin\phi$$

Thus, it follows from (8) that

$$\mathbf{n} = \sin\phi\cos\theta\,\mathbf{i} + \sin\phi\sin\theta\,\mathbf{j} + \cos\phi\,\mathbf{k} = \frac{1}{a}\mathbf{r} \qquad \blacktriangleleft$$

....

SURFACE AREA OF PARAMETRIC SURFACES

In Section 8.5 we obtained formulas for the surface area of a surface of revolution [see Formulas (4) and (5) and the discussion preceding Exercise 20 in that section]. We will now obtain a formula for the surface area S of a parametric surface σ and from that formula we will then derive a formula for the surface area of surfaces of the form $z = f(x, y)$.

Let σ be a parametric surface whose vector equation is

$$\mathbf{r} = x(u, v)\mathbf{i} + y(u, v)\mathbf{j} + z(u, v)\mathbf{k}$$

We will say that σ is a ***smooth parametric surface*** on a region R of the uv-plane if $\partial\mathbf{r}/\partial u$ and $\partial\mathbf{r}/\partial v$ are continuous on R and $\partial\mathbf{r}/\partial u \times \partial\mathbf{r}/\partial v \neq \mathbf{0}$ on R. Geometrically, this means that σ has a principal unit normal vector (and hence a tangent plane) for all (u, v) in R and $\mathbf{n} = \mathbf{n}(u, v)$ is a continuous function on R. Thus, on a smooth parametric surface the unit normal vector **n** varies continuously and has no abrupt changes in direction. We will derive a surface area formula for smooth surfaces that have no self-intersections.

We will begin by subdividing R into rectangular regions by lines parallel to the u- and v-axes and discarding any nonrectangular portions that contain points of the boundary. Assume that there are n rectangles, and let R_k denote the kth rectangle. Let (u_k, v_k) be the lower left corner of R_k, and assume that R_k has area $\Delta A_k = \Delta u_k \Delta v_k$, where Δu_k and Δv_k are the dimensions of R_k (Figure 16.4.10a). The image of R_k will be some *curvilinear patch* σ_k on the surface σ that has a corner at $\mathbf{r}(u_k, v_k)$; denote the area of this patch by ΔS_k (Figure 16.4.10b).

As suggested by Figure 16.4.10c, the two edges of the patch that meet at $\mathbf{r}(u_k, v_k)$ can be approximated by the "secant" vectors

$$\mathbf{r}(u_k + \Delta u_k, v_k) - \mathbf{r}(u_k, v_k)$$

$$\mathbf{r}(u_k, v_k + \Delta v_k) - \mathbf{r}(u_k, v_k)$$

and hence the area of σ_k can be approximated by the area of the parallelogram determined by these vectors. However, it follows from Formulas (5) and (6) that if Δu_k and Δv_k are small, then these secant vectors can in turn be approximated by the tangent vectors

$$\frac{\partial \mathbf{r}}{\partial u}\Delta u_k \quad \text{and} \quad \frac{\partial \mathbf{r}}{\partial v}\Delta v_k$$

where the partial derivatives are evaluated at (u_k, v_k). Thus, the area of the patch σ_k can be approximated by the area of the parallelogram determined by these vectors (Figure 16.4.10d);

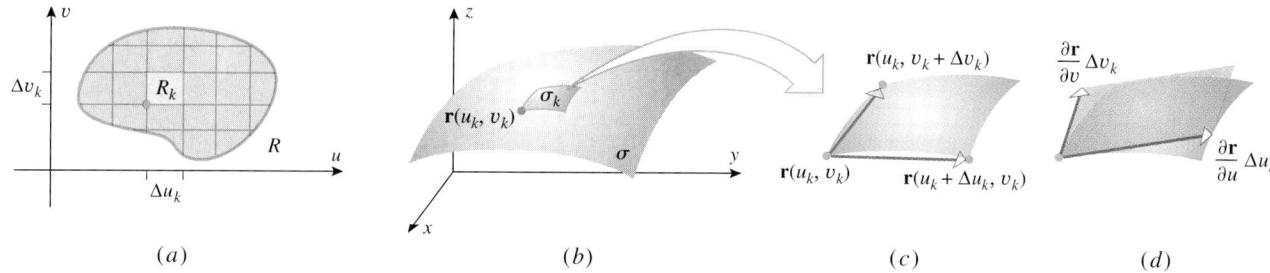

(a) (b) (c) (d)

Figure 16.4.10

that is,

$$\Delta S_k \approx \left\| \frac{\partial \mathbf{r}}{\partial u} \Delta u_k \times \frac{\partial \mathbf{r}}{\partial v} \Delta v_k \right\| = \left\| \frac{\partial \mathbf{r}}{\partial u} \times \frac{\partial \mathbf{r}}{\partial v} \right\| \Delta u_k \Delta v_k = \left\| \frac{\partial \mathbf{r}}{\partial u} \times \frac{\partial \mathbf{r}}{\partial v} \right\| \Delta A_k \qquad (9)$$

It follows that the surface area S of the entire surface σ can be approximated as

$$S \approx \sum_{k=1}^{n} \left\| \frac{\partial \mathbf{r}}{\partial u} \times \frac{\partial \mathbf{r}}{\partial v} \right\| \Delta A_k$$

Thus, if we assume that the errors in the approximations approach zero as n increases in such a way that the dimensions of the rectangles approach zero, then it is plausible that the exact value of S is

$$S = \lim_{n \to +\infty} \sum_{k=1}^{n} \left\| \frac{\partial \mathbf{r}}{\partial u} \times \frac{\partial \mathbf{r}}{\partial v} \right\| \Delta A_k$$

or, equivalently,

$$S = \iint_R \left\| \frac{\partial \mathbf{r}}{\partial u} \times \frac{\partial \mathbf{r}}{\partial v} \right\| \, dA \qquad (10)$$

Example 10

It follows from (4) that the parametric equations

$$x = u, \quad y = u \cos v, \quad z = u \sin v$$

represent the cone that results when the line $y = x$ in the xy-plane is revolved about the x-axis. Use Formula (10) to find the surface area of that portion of the cone for which $0 \le u \le 2$ and $0 \le v \le 2\pi$ (Figure 16.4.11).

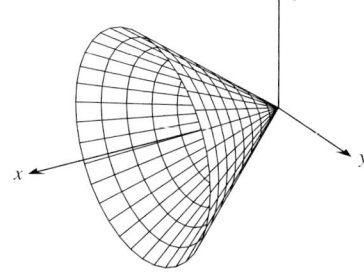

Figure 16.4.11

Solution. The surface can be expressed in vector form as

$$\mathbf{r} = u\mathbf{i} + u \cos v\mathbf{j} + u \sin v\mathbf{k} \quad (0 \le u \le 2, \ 0 \le v \le 2\pi)$$

Thus,

$$\frac{\partial \mathbf{r}}{\partial u} = \mathbf{i} + \cos v\mathbf{j} + \sin v\mathbf{k}$$

$$\frac{\partial \mathbf{r}}{\partial v} = -u \sin v\mathbf{j} + u \cos v\mathbf{k}$$

$$\frac{\partial \mathbf{r}}{\partial u} \times \frac{\partial \mathbf{r}}{\partial v} = \begin{vmatrix} \mathbf{i} & \mathbf{j} & \mathbf{k} \\ 1 & \cos v & \sin v \\ 0 & -u \sin v & u \cos v \end{vmatrix} = u\mathbf{i} - u \cos v\mathbf{j} - u \sin v\mathbf{k}$$

$$\left\| \frac{\partial \mathbf{r}}{\partial u} \times \frac{\partial \mathbf{r}}{\partial v} \right\| = \sqrt{u^2 + (-u \cos v)^2 + (-u \sin v)^2} = |u|\sqrt{2} = u\sqrt{2}$$

Thus, from (10)

$$S = \iint_R \left\| \frac{\partial \mathbf{r}}{\partial u} \times \frac{\partial \mathbf{r}}{\partial v} \right\| dA = \int_0^{2\pi} \int_0^2 \sqrt{2}u \, du \, dv = 2\sqrt{2} \int_0^{2\pi} dv = 4\pi\sqrt{2}$$ ◀

SURFACE AREA OF SURFACES OF THE FORM $z = f(x, y)$

In the case where σ is a surface of the form $z = f(x, y)$, we can take $x = u$ and $y = v$ as parameters and express the surface parametrically as

$$x = u, \quad y = v, \quad z = f(u, v)$$

or in vector form as

$$\mathbf{r} = u\mathbf{i} + v\mathbf{j} + f(u, v)\mathbf{k}$$

Thus,

$$\frac{\partial \mathbf{r}}{\partial u} = \mathbf{i} + \frac{\partial f}{\partial u}\mathbf{k} = \mathbf{i} + \frac{\partial z}{\partial x}\mathbf{k}$$

$$\frac{\partial \mathbf{r}}{\partial v} = \mathbf{j} + \frac{\partial f}{\partial v}\mathbf{k} = \mathbf{j} + \frac{\partial z}{\partial y}\mathbf{k}$$

$$\frac{\partial \mathbf{r}}{\partial u} \times \frac{\partial \mathbf{r}}{\partial v} = \begin{vmatrix} \mathbf{i} & \mathbf{j} & \mathbf{k} \\ 1 & 0 & \dfrac{\partial z}{\partial x} \\ 0 & 1 & \dfrac{\partial z}{\partial y} \end{vmatrix} = -\frac{\partial z}{\partial x}\mathbf{i} - \frac{\partial z}{\partial y}\mathbf{j} + \mathbf{k}$$

$$\left\| \frac{\partial \mathbf{r}}{\partial u} \times \frac{\partial \mathbf{r}}{\partial v} \right\| = \sqrt{\left(\frac{\partial z}{\partial x}\right)^2 + \left(\frac{\partial z}{\partial y}\right)^2 + 1}$$

Thus, it follows from (10) that

$$S = \iint_R \sqrt{\left(\frac{\partial z}{\partial x}\right)^2 + \left(\frac{\partial z}{\partial y}\right)^2 + 1} \, dA \tag{11}$$

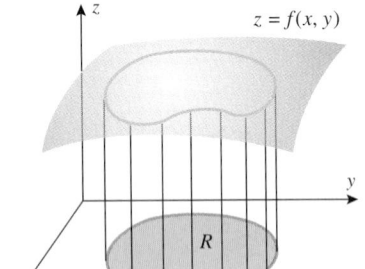

$z = f(x, y)$

Figure 16.4.12

REMARK.   In this formula the region R lies in the xy-plane because the parameters are x and y. Geometrically, this region is the projection on the xy-plane of that portion of the surface $z = f(x, y)$ whose area is being determined by the formula (Figure 16.4.12).

Example 11

Find the surface area of that portion of the surface $z = \sqrt{4 - x^2}$ that lies above the rectangle R in the xy-plane whose coordinates satisfy $0 \le x \le 1$ and $0 \le y \le 4$.

Solution.   As shown in Figure 16.4.13, the surface is a portion of the cylinder $x^2 + z^2 = 4$. It follows from (11) that the surface area is

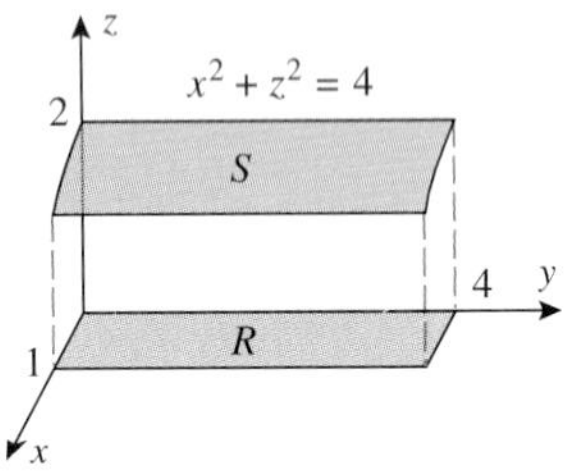

$$S = \iint_R \sqrt{\left(\frac{\partial z}{\partial x}\right)^2 + \left(\frac{\partial z}{\partial y}\right)^2 + 1} \, dA$$

$$= \iint_R \sqrt{\left(-\frac{x}{\sqrt{4 - x^2}}\right)^2 + 0 + 1} \, dA = \int_0^4 \int_0^1 \frac{2}{\sqrt{4 - x^2}} \, dx \, dy$$

$$= 2 \int_0^4 \left[\sin^{-1}\left(\frac{1}{2}x\right) \right]_{x=0}^1 dy = 2 \int_0^4 \frac{\pi}{6} \, dy = \frac{4}{3}\pi$$ ◀

$x^2 + z^2 = 4$

S

R

Figure 16.4.13

Formula (21)
of Section 9.1

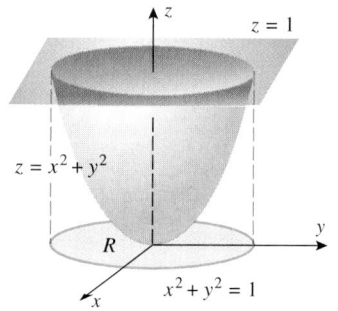

Figure 16.4.14

Example 12

Find the surface area of the portion of the paraboloid $z = x^2 + y^2$ below the plane $z = 1$.

Solution. The surface is shown in Figure 16.4.14. The trace of the paraboloid $z = x^2 + y^2$ in the plane $z = 1$ projects onto the circle $x^2 + y^2 = 1$ in the xy-plane, and the portion of the paraboloid that lies below the plane $z = 1$ projects onto the region R that is enclosed by this circle. Thus, it follows from (11) that the surface area is

$$S = \iint_R \sqrt{4x^2 + 4y^2 + 1}\, dA$$

The expression $4x^2 + 4y^2 + 1 = 4(x^2 + y^2) + 1$ in the integrand suggests that we evaluate the integral in polar coordinates. In accordance with Formula (9) of Section 16.3, we substitute $x = r\cos\theta$ and $y = r\sin\theta$ in the integrand, replace dA by $r\, dr\, d\theta$, and find the limits of integration by expressing the region R in polar coordinates. This yields

$$S = \int_0^{2\pi} \int_0^1 \sqrt{4r^2 + 1}\, r\, dr\, d\theta = \int_0^{2\pi} \left[\frac{1}{12}(4r^2 + 1)^{3/2} \right]_{r=0}^1 d\theta$$

$$= \int_0^{2\pi} \frac{1}{12}(5\sqrt{5} - 1)\, d\theta = \frac{1}{6}\pi(5\sqrt{5} - 1) \qquad \blacktriangleleft$$

EXERCISE SET 16.4   ⌇ Graphing Calculator   [C] CAS

In Exercises 1 and 2, sketch the parametric surface.

1. (a) $x = u$, $y = v$, $z = \sqrt{u^2 + v^2}$
   (b) $x = u$, $y = \sqrt{u^2 + v^2}$, $z = v$
   (c) $x = \sqrt{u^2 + v^2}$, $y = u$, $z = v$

2. (a) $x = u$, $y = v$, $z = u^2 + v^2$
   (b) $x = u$, $y = u^2 + v^2$, $z = v$
   (c) $x = u^2 + v^2$, $y = u$, $z = v$

In Exercises 3 and 4, find a parametric representation of the surface in terms of the parameters $u = x$ and $v = y$.

3. (a) $2z - 3x + 4y = 5$
   (b) $z = x^2$

4. (a) $z + zx^2 - y = 0$
   (b) $y^2 - 3z = 5$

5. (a) Find parametric equations for the portion of the cylinder $x^2 + y^2 = 5$ that extends between the planes $z = 0$ and $z = 1$.
   (b) Find parametric equations for the portion of the cylinder $x^2 + z^2 = 4$ that extends between the planes $y = 1$ and $y = 3$.

6. (a) Find parametric equations for the portion of the plane $x + y = 1$ that extends between the planes $z = -1$ and $z = 1$.
   (b) Find parametric equations for the portion of the plane $y - 2z = 5$ that extends between the planes $x = 0$ and $x = 3$.

7. Find parametric equations for the surface generated by revolving the curve $y = \sin x$ about the x-axis.

8. Find parametric equations for the surface generated by revolving the curve $y - e^x = 0$ about the x-axis.

In Exercises 9–14, find a parametric representation of the surface in terms of the parameters r and θ, where (r, θ, z) are the cylindrical coordinates of a point on the surface.

9. $z = \dfrac{1}{1 + x^2 + y^2}$ **10.** $z = e^{-(x^2 + y^2)}$

11. $z = 2xy$ **12.** $z = x^2 - y^2$

13. The portion of the sphere $x^2 + y^2 + z^2 = 9$ on or above the plane $z = 2$.

14. The portion of the cone $z = \sqrt{x^2 + y^2}$ on or below the plane $z = 3$.

15. Find a parametric representation of the cone
$$z = \sqrt{3x^2 + 3y^2}$$
in terms of parameters ρ and θ, where (ρ, θ, ϕ) are spherical coordinates of a point on the surface.

16. Find a parametric representation of the cylinder $x^2 + y^2 = 9$ in terms of parameters θ and ϕ, where (ρ, θ, ϕ) are spherical coordinates of a point on the surface.

In Exercises 17–22, eliminate the parameters to obtain an equation in rectangular coordinates, and describe the surface.

17. $x = 2u + v$, $y = u - v$, $z = 3v$ for $-\infty < u < +\infty$ and $-\infty < v < +\infty$.

18. $x = u \cos v$, $y = u^2$, $z = u \sin v$ for $0 \le u \le 2$ and $0 \le v < 2\pi$.

19. $x = 3 \sin u$, $y = 2 \cos u$, $z = 2v$ for $0 \le u < 2\pi$ and $1 \le v \le 2$.

20. $x = \sqrt{u} \cos v$, $y = \sqrt{u} \sin v$, $z = u$ for $0 \le u \le 4$ and $0 \le v < 2\pi$.

21. $\mathbf{r}(u, v) = 3u \cos v\mathbf{i} + 4u \sin v\mathbf{j} + u\mathbf{k}$ for $0 \le u \le 1$ and $0 \le v < 2\pi$.

22. $\mathbf{r}(u, v) = \sin u \cos v\mathbf{i} + 2 \sin u \sin v\mathbf{j} + 3 \cos u\mathbf{k}$ for $0 \le u \le \pi$ and $0 \le v < 2\pi$.

23. The accompanying figure shows the graphs of two parametric representations of the cone $z = \sqrt{x^2 + y^2}$ for $0 \le z \le 2$.
 (a) Find parametric equations that produce reasonable facsimiles of these surfaces.
 (b) Use a graphing utility to check your answer to part (a).

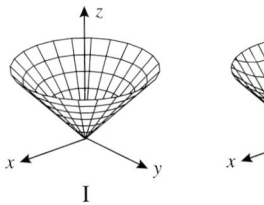

I   II

Figure Ex-23

24. The accompanying figure shows the graphs of two parametric representations of the paraboloid $z = x^2 + y^2$ for $0 \le z \le 2$.
 (a) Find parametric equations that produce reasonable facsimiles of these surfaces.
 (b) Use a graphing utility to check your answer to part (a).

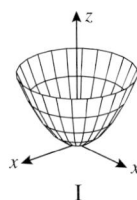

I   II

Figure Ex-24

25. In each part, the figure shows a portion of the parametric surface $x = 3 \cos v$, $y = u$, $z = 3 \sin v$. Find restrictions on u and v that produce the surface, and check your answer with a graphing utility.

(a)   (b)

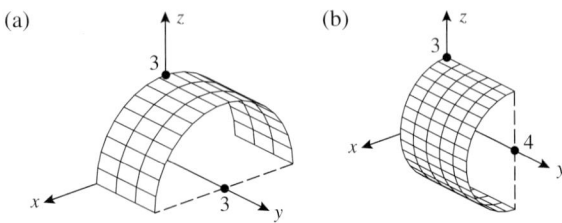

26. In each part, the figure shows a portion of the parametric surface $x = 3 \cos v$, $y = 3 \sin v$, $z = u$. Find restrictions

on u and v that produce the surface, and check your answer with a graphing utility.

(a)   (b)

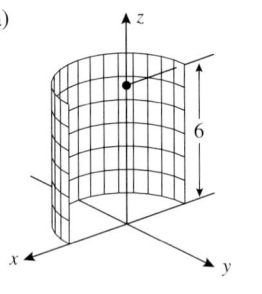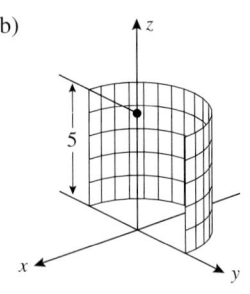

27. In each part, the figure shows a hemisphere that is a portion of the sphere $x = \sin \phi \cos \theta$, $y = \sin \phi \sin \theta$, $z = \cos \phi$. Find restrictions on ϕ and θ that produce the hemisphere, and check your answer with a graphing utility.

(a)   (b)

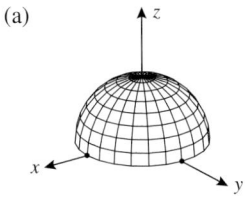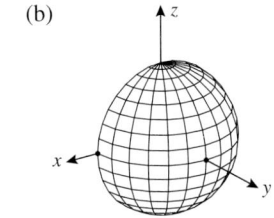

28. Each figure shows a portion of the sphere $x = \sin \phi \cos \theta$, $y = \sin \phi \sin \theta$, $z = \cos \phi$. Find restrictions on ϕ and θ that produce the surface, and check your answer with a graphing utility.

(a)   (b)

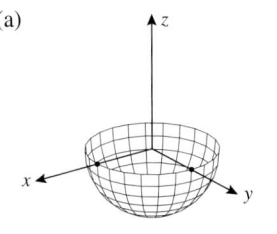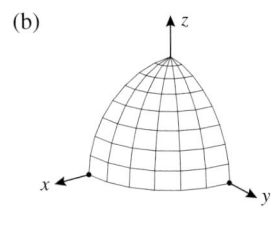

In Exercises 29–34, find an equation of the tangent plane to the parametric surface at the stated point.

29. $x = u$, $y = v$, $z = u^2 + v^2$; $(1, 2, 5)$

30. $x = u^2$, $y = v^2$, $z = u + v$; $(1, 4, 3)$

31. $x = 3v \sin u$, $y = 2v \cos u$, $z = u^2$; $(0, 2, 0)$

32. $\mathbf{r} = uv\mathbf{i} + (u - v)\mathbf{j} + (u + v)\mathbf{k}$; $u = 1$, $v = 2$

33. $\mathbf{r} = u \cos v\mathbf{i} + u \sin v\mathbf{j} + v\mathbf{k}$; $u = 1/2$, $v = \pi/4$

34. $\mathbf{r} = uv\mathbf{i} + ue^v\mathbf{j} + ve^u\mathbf{k}$; $u = \ln 2$, $v = 0$

In Exercises 35–46, find the area of the given surface.

35. The portion of the cylinder $y^2 + z^2 = 9$ that is above the rectangle $R = \{(x, y) : 0 \le x \le 2, -3 \le y \le 3\}$.

36. The portion of the plane $2x + 2y + z = 8$ in the first octant.

37. The portion of the cone $z^2 = 4x^2 + 4y^2$ that is above the region in the first quadrant bounded by the line $y = x$ and the parabola $y = x^2$.

38. The portion of the cone $z = \sqrt{x^2 + y^2}$ that lies inside the cylinder $x^2 + y^2 = 2x$.

39. The portion of the paraboloid $z = 1 - x^2 - y^2$ that is above the xy-plane.

40. The portion of the surface $z = 2x + y^2$ that is above the triangular region with vertices $(0, 0)$, $(0, 1)$, and $(1, 1)$.

41. The portion of the paraboloid

$$\mathbf{r}(u, v) = u \cos v \mathbf{i} + u \sin v \mathbf{j} + u^2 \mathbf{k}$$

for which $1 \leq u \leq 2, 0 \leq v \leq 2\pi$.

42. The portion of the cone

$$\mathbf{r}(u, v) = u \cos v \mathbf{i} + u \sin v \mathbf{j} + u \mathbf{k}$$

for which $0 \leq u \leq 2v, 0 \leq v \leq \pi/2$.

43. The portion of the surface $z = xy$ that is above the sector in the first quadrant bounded by the lines $y = x/\sqrt{3}$, $y = 0$, and the circle $x^2 + y^2 = 9$.

44. The portion of the paraboloid $2z = x^2 + y^2$ that is inside the cylinder $x^2 + y^2 = 8$.

45. The portion of the sphere $x^2 + y^2 + z^2 = 16$ between the planes $z = 1$ and $z = 2$.

46. The portion of the sphere $x^2 + y^2 + z^2 = 8$ that is inside of the cone $z = \sqrt{x^2 + y^2}$.

47. Use parametric equations to derive the formula for the surface area of a sphere of radius a.

48. Use parametric equations to derive the formula for the lateral surface area of a right circular cylinder of radius r and height h.

49. The portion of the surface

$$z = \frac{h}{a}\sqrt{x^2 + y^2} \quad (a, h > 0)$$

between the xy-plane and the plane $z = h$ is a right circular cone of height h and radius a. Use a double integral to show that the lateral surface area of this cone is $S = \pi a \sqrt{a^2 + h^2}$.

50. The accompanying figure shows the ***torus*** that is generated by revolving the circle

$$(x - a)^2 + z^2 = b^2 \quad (0 < b < a)$$

in the xz-plane about the z-axis.
(a) Show that this torus can be expressed parametrically as

$$x = (a + b \cos v) \cos u$$
$$y = (a + b \cos v) \sin u$$
$$z = b \sin v$$

where u and v are the parameters shown in the figure and $0 \leq u \leq 2\pi, 0 \leq v \leq 2\pi$.
(b) Use a graphing utility to generate a torus.

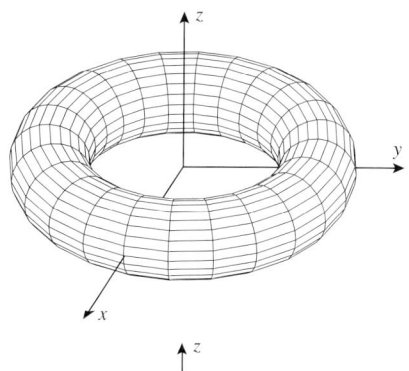

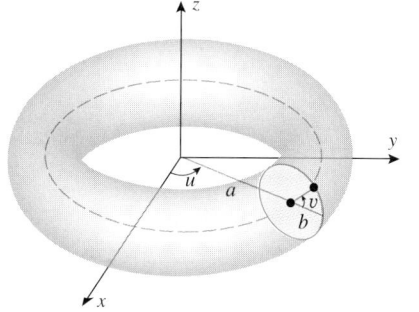

Figure Ex-50

51. Find the surface area of the torus in Exercise 50(a).

52. Use a CAS to graph the ***helicoid***

$$x = u \cos v, \quad y = u \sin v, \quad z = v$$

for $0 \leq u \leq 5$ and $0 \leq v \leq 4\pi$ (see the accompanying figure), and then use the numerical double integration operation of the CAS to approximate the surface area.

53. Use a CAS to graph the ***pseudosphere***

$$x = \cos u \sin v$$
$$y = \sin u \sin v$$
$$z = \cos v + \ln \left(\tan \frac{v}{2} \right)$$

for $0 \leq u \leq 2\pi, 0 < v < \pi$ (see the accompanying figure), and then use the numerical double integration operation of the CAS to approximate the surface area between the planes $z = -1$ and $z = 1$.

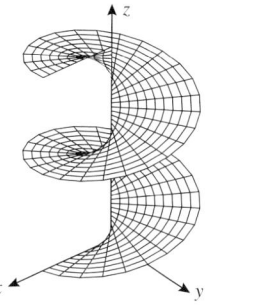

Figure Ex-52

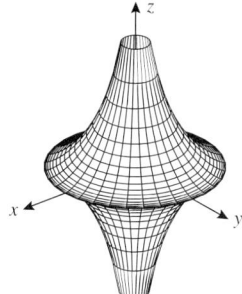

Figure Ex-53

54. (a) Find parametric equations for the surface of revolution that is generated by revolving the curve $z = f(x)$ in the xz-plane about the z-axis.

(b) Use the result obtained in part (a) to find parametric equations for the surface of revolution that is generated by revolving the curve $z = 1/x^2$ in the xz-plane about the z-axis.

(c) Use a graphing utility to check your work by graphing the parametric surface.

In Exercises 55–57, the parametric equations represent a quadric surface for positive values of a, b, and c. Identify the type of surface by eliminating the parameters u and v. Check your conclusion by choosing specific values for the constants and generating the surface with a graphing utility.

55. $x = a \cos u \cos v, \ y = b \sin u \cos v, \ z = c \sin v$

56. $x = a \cos u \cosh v, \ y = b \sin u \cosh v, \ z = c \sinh v$

57. $x = a \sinh v, \ y = b \sinh u \cosh v, \ z = c \cosh u \cosh v$

16.5 TRIPLE INTEGRALS

In the preceding sections we defined and discussed properties of double integrals for functions of two variables. In this section we will define triple integrals for functions of three variables.

DEFINITION OF A TRIPLE INTEGRAL

Volume $= \Delta V_k$

(x_k^*, y_k^*, z_k^*)

Figure 16.5.1

A single integral of a function $f(x)$ is defined over a finite closed interval on the x-axis, and a double integral of a function $f(x, y)$ is defined over a finite closed region R in the xy-plane. Our first goal in this section is to define what is meant by a *triple integral* of $f(x, y, z)$ over a closed solid region G in an xyz-coordinate system. To ensure that G does not extend indefinitely in some direction, we will assume that it can be enclosed in a suitably large box whose sides are parallel to the coordinate planes (Figure 16.5.1). In this case we say that G is a *finite solid*.

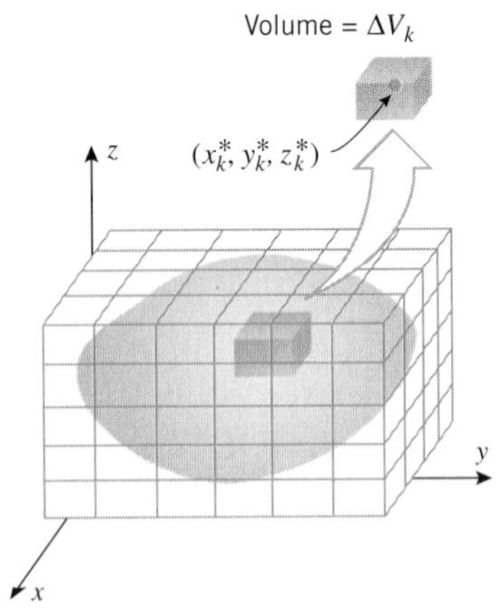

To define the triple integral of $f(x, y, z)$ over G, we first divide the box into n "subboxes" by planes parallel to the coordinate planes. We then discard those subboxes that contain any points outside of G and choose an arbitrary point in each of the remaining subboxes. As shown in Figure 16.5.1, we denote the volume of the kth remaining subbox by ΔV_k and the point selected in the kth subbox by (x_k^*, y_k^*, z_k^*). Next, we form the product

$$f(x_k^*, y_k^*, z_k^*) \Delta V_k$$

for each subbox, then add the products for all of the subboxes to obtain the *Riemann sum*

$$\sum_{k=1}^{n} f(x_k^*, y_k^*, z_k^*) \Delta V_k$$

Finally, we repeat this process with more and more subdivisions in such a way that the length, width, and height of each subbox approach zero, and n approaches $+\infty$. The limit

$$\iiint_G f(x, y, z) \, dV = \lim_{n \to +\infty} \sum_{k=1}^{n} f(x_k^*, y_k^*, z_k^*) \Delta V_k \tag{1}$$

is called the *triple integral* of $f(x, y, z)$ over the region G. Conditions under which the triple integral exists are studied in advanced calculus. However, for our purposes it suffices to say that existence is ensured when f is continuous on G and the region G is not too "complicated."

PROPERTIES OF TRIPLE INTEGRALS

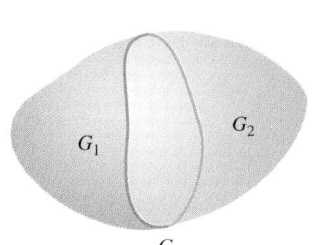

Figure 16.5.2

Triple integrals enjoy many properties of single and double integrals:

$$\iiint_G cf(x, y, z)\, dV = c \iiint_G f(x, y, z)\, dV \quad (c \text{ a constant})$$

$$\iiint_G [f(x, y, z) + g(x, y, z)]\, dV = \iiint_G f(x, y, z)\, dV + \iiint_G g(x, y, z)\, dV$$

$$\iiint_G [f(x, y, z) - g(x, y, z)]\, dV = \iiint_G f(x, y, z)\, dV - \iiint_G g(x, y, z)\, dV$$

Moreover, if the region G is subdivided into two subregions G_1 and G_2 (Figure 16.5.2), then

$$\iiint_G f(x, y, z)\, dV = \iiint_{G_1} f(x, y, z)\, dV + \iiint_{G_2} f(x, y, z)\, dV$$

We omit the proofs.

EVALUATING TRIPLE INTEGRALS OVER RECTANGULAR BOXES

Just as a double integral can be evaluated by two successive single integrations, so a triple integral can be evaluated by three successive integrations. The following theorem, which we state without proof, is the analog of Theorem 16.1.3.

16.5.1 THEOREM. *Let G be the rectangular box defined by the inequalities*

$$a \le x \le b, \quad c \le y \le d, \quad k \le z \le l$$

If f is continuous on the region G, then

$$\iiint_G f(x, y, z)\, dV = \int_a^b \int_c^d \int_k^l f(x, y, z)\, dz\, dy\, dx \tag{2}$$

Moreover, the iterated integral on the right can be replaced with any of the five other iterated integrals that result by altering the order of integration.

Example 1

Evaluate the triple integral

$$\iiint_G 12xy^2z^3\, dV$$

over the rectangular box G defined by the inequalities $-1 \le x \le 2, 0 \le y \le 3, 0 \le z \le 2$.

Solution. Of the six possible iterated integrals we might use, we will choose the one in (2). Thus, we will first integrate with respect to z, holding x and y fixed, then with respect to y, holding x fixed, and finally with respect to x.

$$\iiint_G 12xy^2z^3\, dV = \int_{-1}^2 \int_0^3 \int_0^2 12xy^2z^3\, dz\, dy\, dx$$

$$= \int_{-1}^2 \int_0^3 \left[3xy^2z^4\right]_{z=0}^2 dy\, dx = \int_{-1}^2 \int_0^3 48xy^2\, dy\, dx$$

$$= \int_{-1}^2 \left[16xy^3\right]_{y=0}^3 dx = \int_{-1}^2 432x\, dx$$

$$= 216x^2\Big]_{-1}^2 = 648 \quad \blacktriangleleft$$

Next we will consider how triple integrals can be evaluated over solids that are not rectangular boxes. For the moment we will limit our discussion to solids of the type shown in Figure 16.5.3. Specifically, we will assume that the solid G is bounded above by a surface $z = g_2(x, y)$ and below by a surface $z = g_1(x, y)$ and that the projection of the solid on the xy-plane is a type I or type II region R (see Definition 16.2.1). In addition, we will assume that $g_1(x, y)$ and $g_2(x, y)$ are continuous on R and that $g_1(x, y) \leq g_2(x, y)$ on R. Geometrically, this means that the surfaces may touch but cannot cross. We call a solid of this type a **simple xy-solid**.

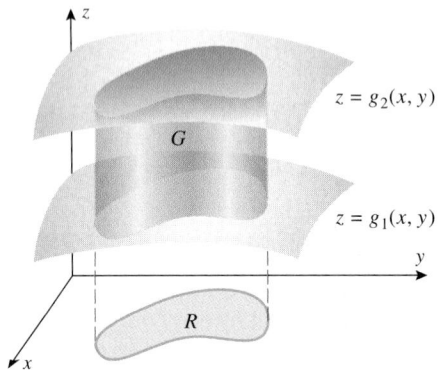

Figure 16.5.3

The following theorem, which we state without proof, will enable us to evaluate triple integrals over simple xy-solids.

16.5.2 THEOREM. *Let G be a simple xy-solid with upper surface $z = g_2(x, y)$ and lower surface $z = g_1(x, y)$, and let R be the projection of G on the xy-plane. If $f(x, y, z)$ is continuous on G, then*

$$\iiint\limits_{G} f(x, y, z)\, dV = \iint\limits_{R} \left[\int_{g_1(x,y)}^{g_2(x,y)} f(x, y, z)\, dz \right] dA \tag{3}$$

In (3), the first integration is with respect to z, after which a function of x and y remains. This function of x and y is then integrated over the region R in the xy-plane. To apply (3), it is helpful to begin with a three-dimensional sketch of the solid G. The limits of integration can be obtained from the sketch as follows:

Step 1. Find an equation $z = g_2(x, y)$ for the upper surface and an equation $z = g_1(x, y)$ for the lower surface of G. The functions $g_1(x, y)$ and $g_2(x, y)$ determine the lower and upper z-limits of integration.

Step 2. Make a two-dimensional sketch of the projection R of the solid on the xy-plane. From this sketch determine the limits of integration for the double integral over R in (3).

Example 2

Let G be the wedge in the first octant cut from the cylindrical solid $y^2 + z^2 \leq 1$ by the planes $y = x$ and $x = 0$. Evaluate

$$\iiint\limits_{G} z\, dV$$

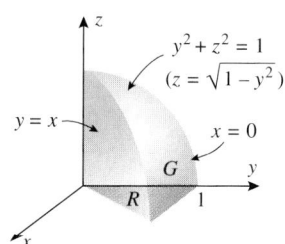

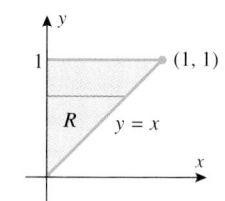

Figure 16.5.4

Solution. The solid G and its projection R on the xy-plane are shown in Figure 16.5.4. The upper surface of the solid is formed by the cylinder and the lower surface by the xy-plane. Since the portion of the cylinder $y^2 + z^2 = 1$ that lies above the xy-plane has the equation $z = \sqrt{1 - y^2}$, and the xy-plane has the equation $z = 0$, it follows from (3) that

$$\iiint_G z\, dV = \iint_R \left[\int_0^{\sqrt{1-y^2}} z\, dz \right] dA \tag{4}$$

For the double integral over R, the x- and y-integrations can be performed in either order, since R is both a type I and type II region. We will integrate with respect to x first. With this choice, (4) yields

$$\iiint_G z\, dV = \int_0^1 \int_0^y \int_0^{\sqrt{1-y^2}} z\, dz\, dx\, dy = \int_0^1 \int_0^y \frac{1}{2} z^2 \Big]_{z=0}^{\sqrt{1-y^2}} dx\, dy$$

$$= \int_0^1 \int_0^y \frac{1}{2}(1 - y^2)\, dx\, dy = \frac{1}{2} \int_0^1 (1 - y^2)x \Big]_{x=0}^y dy$$

$$= \frac{1}{2} \int_0^1 (y - y^3)\, dy = \frac{1}{2}\left[\frac{1}{2} y^2 - \frac{1}{4} y^4 \right]_0^1 = \frac{1}{8} \quad \blacktriangleleft$$

FOR THE READER. Most computer algebra systems have a built-in capability for computing iterated triple integrals. If you have a CAS, read the relevant documentation and use the CAS to check Examples 1 and 2.

VOLUME CALCULATED AS A TRIPLE INTEGRAL

Triple integrals have many physical interpretations, some of which we will consider in the next section. However, in the special case where $f(x, y, z) = 1$, Formula (1) yields

$$\iiint_G dV = \lim_{n \to +\infty} \sum_{k=1}^n \Delta V_k$$

which Figure 16.5.1 suggests is the volume of G; that is,

$$\text{volume of } G = \iiint_G dV \tag{5}$$

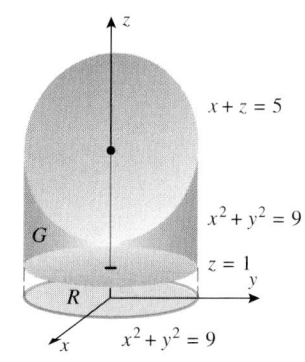

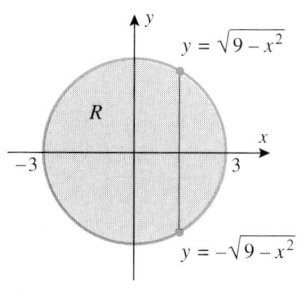

Figure 16.5.5

Example 3

Use a triple integral to find the volume of the solid within the cylinder $x^2 + y^2 = 9$ and between the planes $z = 1$ and $x + z = 5$.

Solution. The solid G and its projection R on the xy-plane are shown in Figure 16.5.5. The lower surface of the solid is the plane $z = 1$ and the upper surface is the plane $x + z = 5$ or, equivalently, $z = 5 - x$. Thus, from (3) and (5)

$$\text{volume of } G = \iiint_G dV = \iint_R \left[\int_1^{5-x} dz \right] dA \tag{6}$$

For the double integral over R, we will integrate with respect to y first. Thus, (6) yields

$$\text{volume of } G = \int_{-3}^{3} \int_{-\sqrt{9-x^2}}^{\sqrt{9-x^2}} \int_{1}^{5-x} dz\,dy\,dx = \int_{-3}^{3} \int_{-\sqrt{9-x^2}}^{\sqrt{9-x^2}} z \Bigg]_{z=1}^{5-x} dy\,dx$$

$$= \int_{-3}^{3} \int_{-\sqrt{9-x^2}}^{\sqrt{9-x^2}} (4-x)\,dy\,dx = \int_{-3}^{3} (8-2x)\sqrt{9-x^2}\,dx$$

$$= 8\int_{-3}^{3} \sqrt{9-x^2}\,dx - \int_{-3}^{3} 2x\sqrt{9-x^2}\,dx \qquad \begin{array}{l}\text{For the first integral,}\\ \text{see Formula (3) of}\\ \text{Section 9.4.}\end{array}$$

$$= 8\left(\frac{9}{2}\pi\right) - \int_{-3}^{3} 2x\sqrt{9-x^2}\,dx \qquad \begin{array}{l}\text{The second integral is 0 because}\\ \text{the integrand is an odd function.}\\ \text{See Exercise 41 of Section 7.6.}\end{array}$$

$$= 8\left(\frac{9}{2}\pi\right) - 0 = 36\pi \qquad \blacktriangleleft$$

Example 4

Find the volume of the solid enclosed between the paraboloids

$$z = 5x^2 + 5y^2 \quad\text{and}\quad z = 6 - 7x^2 - y^2$$

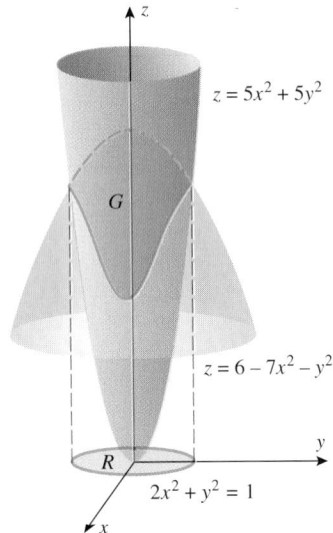

$z = 5x^2 + 5y^2$

G

$z = 6 - 7x^2 - y^2$

R

$2x^2 + y^2 = 1$

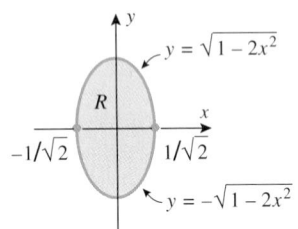

$y = \sqrt{1 - 2x^2}$

R

$-1/\sqrt{2}$ $1/\sqrt{2}$

$y = -\sqrt{1 - 2x^2}$

Figure 16.5.6

Solution. The solid G and its projection R on the xy-plane are shown in Figure 16.5.6. The projection R is obtained by solving the given equations simultaneously to determine where the paraboloids intersect. We obtain

$$5x^2 + 5y^2 = 6 - 7x^2 - y^2$$

or

$$2x^2 + y^2 = 1 \tag{7}$$

which tells us that the paraboloids intersect in a curve on the elliptic cylinder given by (7).

The projection of this intersection on the xy-plane is an ellipse with this same equation. Therefore,

$$\text{volume of } G = \iiint\limits_{G} dV = \iint\limits_{R} \left[\int_{5x^2+5y^2}^{6-7x^2-y^2} dz \right] dA$$

$$= \int_{-1/\sqrt{2}}^{1/\sqrt{2}} \int_{-\sqrt{1-2x^2}}^{\sqrt{1-2x^2}} \int_{5x^2+5y^2}^{6-7x^2-y^2} dz\,dy\,dx$$

$$= \int_{-1/\sqrt{2}}^{1/\sqrt{2}} \int_{-\sqrt{1-2x^2}}^{\sqrt{1-2x^2}} (6 - 12x^2 - 6y^2)\,dy\,dx$$

$$= \int_{-1/\sqrt{2}}^{1/\sqrt{2}} \left[6(1-2x^2)y - 2y^3 \right]_{y=-\sqrt{1-2x^2}}^{\sqrt{1-2x^2}} dx$$

$$= 8\int_{-1/\sqrt{2}}^{1/\sqrt{2}} (1-2x^2)^{3/2}\,dx = \frac{8}{\sqrt{2}} \int_{-\pi/2}^{\pi/2} \cos^4\theta\,d\theta = \frac{3\pi}{\sqrt{2}} \qquad \blacktriangleleft$$

Let $x = \dfrac{1}{\sqrt{2}}\sin\theta$.

Use the Wallis cosine formula in Exercise 66 of Section 9.3.

....

INTEGRATION IN OTHER ORDERS

In Formula (3) for integrating over a simple xy-solid, the z-integration was performed first. However, there are situations in which it is preferable to integrate in a different order. For example, Figure 16.5.7a shows a **simple xz-solid**, and Figure 16.5.7b shows a **simple yz-solid**. For a simple xz-solid it is usually best to integrate with respect to y first, and for a simple yz-solid it is usually best to integrate with respect to x first:

$$\iiint\limits_{\substack{G \\ \text{simple } xz\text{-solid}}} f(x, y, z)\, dV = \iint\limits_{R} \left[\int_{g_1(x,z)}^{g_2(x,z)} f(x, y, z)\, dy \right] dA \tag{8}$$

$$\iiint\limits_{\substack{G \\ \text{simple } yz\text{-solid}}} f(x, y, z)\, dV = \iint\limits_{R} \left[\int_{g_1(y,z)}^{g_2(y,z)} f(x, y, z)\, dx \right] dA \tag{9}$$

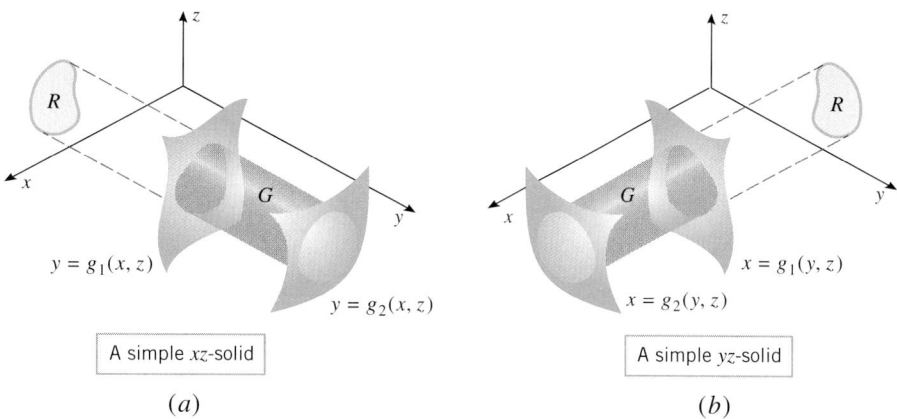

Figure 16.5.7

Sometimes a solid G can be viewed as a simple xy-solid, a simple xz-solid, and a simple yz-solid, in which case the order of integration can be chosen to simplify the computations.

Example 5

In Example 2, we evaluated

$$\iiint\limits_{G} z\, dV$$

over the wedge in Figure 16.5.4 by integrating first with respect to z. Evaluate this integral by integrating first with respect to x.

Solution. The solid is bounded in the back by the plane $x = 0$ and in the front by the plane $x = y$, so

$$\iiint\limits_{G} z\, dV = \iint\limits_{R} \left[\int_0^y z\, dx \right] dA$$

where R is the projection of G on the yz-plane (Figure 16.5.8). The integration over R can be performed first with respect to z and then y or vice versa. Performing the z-integration first yields

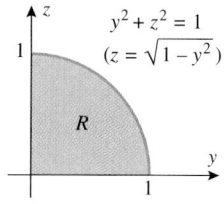

Figure 16.5.8

$$\iiint\limits_{G} z\, dV = \int_0^1 \int_0^{\sqrt{1-y^2}} \int_0^y z\, dx\, dz\, dy = \int_0^1 \int_0^{\sqrt{1-y^2}} zx \Big]_{x=0}^{y} dz\, dy$$

$$= \int_0^1 \int_0^{\sqrt{1-y^2}} zy\, dz\, dy = \int_0^1 \frac{1}{2}z^2 y \Big]_{z=0}^{\sqrt{1-y^2}} dy = \int_0^1 \frac{1}{2}(1 - y^2)y\, dy = \frac{1}{8}$$

which agrees with the result in Example 2. ◄

In Exercises 1–8, evaluate the iterated integral.

1. $\displaystyle\int_{-1}^{1}\int_{0}^{2}\int_{0}^{1}(x^2+y^2+z^2)\,dx\,dy\,dz$

2. $\displaystyle\int_{1/3}^{1/2}\int_{0}^{\pi}\int_{0}^{1}zx\sin xy\,dz\,dy\,dx$

3. $\displaystyle\int_{0}^{2}\int_{-1}^{y^2}\int_{-1}^{z}yz\,dx\,dz\,dy$

4. $\displaystyle\int_{0}^{\pi/4}\int_{0}^{1}\int_{0}^{x^2}x\cos y\,dz\,dx\,dy$

5. $\displaystyle\int_{0}^{3}\int_{0}^{\sqrt{9-z^2}}\int_{0}^{x}xy\,dy\,dx\,dz$

6. $\displaystyle\int_{1}^{3}\int_{x}^{x^2}\int_{0}^{\ln z}xe^y\,dy\,dz\,dx$

7. $\displaystyle\int_{0}^{2}\int_{0}^{\sqrt{4-x^2}}\int_{-5+x^2+y^2}^{3-x^2-y^2}x\,dz\,dy\,dx$

8. $\displaystyle\int_{1}^{2}\int_{z}^{2}\int_{0}^{\sqrt{3}y}\frac{y}{x^2+y^2}\,dx\,dy\,dz$

In Exercises 9–12, evaluate the triple integral.

9. $\displaystyle\iiint_{G}xy\sin yz\,dV$, where G is the rectangular box defined by the inequalities $0\le x\le\pi,\,0\le y\le1,\,0\le z\le\pi/6$.

10. $\displaystyle\iiint_{G}y\,dV$, where G is the solid enclosed by the plane $z=y$, the xy-plane, and the parabolic cylinder $y=1-x^2$.

11. $\displaystyle\iiint_{G}xyz\,dV$, where G is the solid in the first octant that is bounded by the parabolic cylinder $z=2-x^2$ and the planes $z=0$, $y=x$, and $y=0$.

12. $\displaystyle\iiint_{G}\cos(z/y)\,dV$, where G is the solid defined by the inequalities $\pi/6\le y\le\pi/2,\,y\le x\le\pi/2,\,0\le z\le xy$.

[C] **13.** Use a CAS to check the answers to the problems you solved in Exercises 1–12.

[C] **14.** Use the numerical triple integral operation of a CAS to approximate
$$\iiint_{G}e^{-x^2-y^2-z^2}\,dV$$
where G is the spherical region $x^2+y^2+z^2\le1$.

In Exercises 15–18, use a triple integral to find the volume of the solid.

15. The solid in the first octant bounded by the coordinate planes and the plane $3x+6y+4z=12$.

16. The solid bounded by the surface $z=\sqrt{y}$ and the planes $x+y=1$, $x=0$, and $z=0$.

17. The solid bounded by the surface $y=x^2$ and the planes $y+z=4$ and $z=0$.

18. The wedge in the first octant that is cut from the cylinder $y^2+z^2\le1$ by the planes $y=x$ and $x=0$.

In Exercises 19–22, set up (but do not evaluate) an iterated triple integral for the volume of the solid enclosed between the given surfaces.

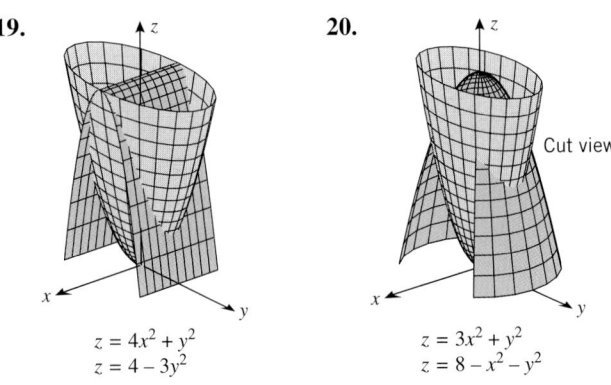

19.
$z=4x^2+y^2$
$z=4-3y^2$

20. Cut view
$z=3x^2+y^2$
$z=8-x^2-y^2$

21. The elliptic cylinder $x^2+9y^2=9$ and the planes $z=0$ and $z=x+3$.

22. The cylinders $x^2+y^2=1$ and $x^2+z^2=1$.

In Exercises 23 and 24, sketch the solid whose volume is given by the integral.

23. (a) $\displaystyle\int_{-1}^{1}\int_{-\sqrt{1-x^2}}^{\sqrt{1-x^2}}\int_{0}^{y+1}dz\,dy\,dx$

(b) $\displaystyle\int_{0}^{9}\int_{0}^{y/3}\int_{0}^{\sqrt{y^2-9x^2}}dz\,dx\,dy$

(c) $\displaystyle\int_{0}^{1}\int_{0}^{\sqrt{1-x^2}}\int_{0}^{2}dy\,dz\,dx$

24. (a) $\displaystyle\int_{0}^{3}\int_{x^2}^{9}\int_{0}^{2}dz\,dy\,dx$

(b) $\displaystyle\int_{0}^{2}\int_{0}^{2-y}\int_{0}^{2-x-y}dz\,dx\,dy$

(c) $\displaystyle\int_{-2}^{2}\int_{0}^{4-y^2}\int_{0}^{2}dx\,dz\,dy$

The **average value** or **mean value** of a continuous function $f(x, y, z)$ over a solid G is defined as

$$f_{\text{ave}} = \frac{1}{V(G)} \iiint\limits_{G} f(x, y, z)\, dV$$

where $V(G)$ is the volume of the solid (compare to the definition preceding Exercise 57 of Section 16.2). Use this definition in Exercises 25 and 26.

25. Find the average value of $f(x, y, z) = x + y + z$ over the tetrahedron shown in the accompanying figure.

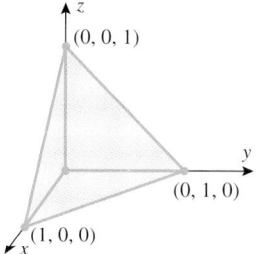

Figure Ex-25

26. Find the average value of $f(x, y, z) = xyz$ over the spherical region $x^2 + y^2 + z^2 \leq 1$.

27. Let G be the tetrahedron in the first octant bounded by the coordinate planes and the plane

$$\frac{x}{a} + \frac{y}{b} + \frac{z}{c} = 1 \quad (a > 0, b > 0, c > 0)$$

(a) List six different iterated integrals that represent the volume of G.

(b) Evaluate any one of the six to show that the volume of G is $\frac{1}{6}abc$.

28. In parts (a)–(c), express the integral as an equivalent integral in which the z-integration is performed first, the y-integration second, and the x-integration last.

(a) $\displaystyle\int_0^3 \int_0^{\sqrt{9-z^2}} \int_0^{\sqrt{9-y^2-z^2}} f(x, y, z)\, dx\, dy\, dz$

(b) $\displaystyle\int_0^4 \int_0^2 \int_0^{x/2} f(x, y, z)\, dy\, dz\, dx$

(c) $\displaystyle\int_0^4 \int_0^{4-y} \int_0^{\sqrt{z}} f(x, y, z)\, dx\, dz\, dy$

29. Use a triple integral to derive the formula for the volume of the ellipsoid

$$\frac{x^2}{a^2} + \frac{y^2}{b^2} + \frac{z^2}{c^2} = 1$$

30. Let G be the rectangular box defined by the inequalities $a \leq x \leq b, c \leq y \leq d, k \leq z \leq l$. Show that

$$\iiint\limits_{G} f(x)g(y)h(z)\, dV$$

$$= \left[\int_a^b f(x)\, dx\right]\left[\int_c^d g(y)\, dy\right]\left[\int_k^l h(z)\, dz\right]$$

31. Use the result of Exercise 30 to evaluate

(a) $\displaystyle\iiint\limits_{G} xy^2 \sin z\, dV$, where G is the set of points satisfying $-1 \leq x \leq 1, 0 \leq y \leq 1, 0 \leq z \leq \pi/2$.

(b) $\displaystyle\iiint\limits_{G} e^{2x+y-z}\, dV$, where G is the set of points satisfying $0 \leq x \leq 1, 0 \leq y \leq \ln 3, 0 \leq z \leq \ln 2$.

C **32.** (a) Find the region G over which the triple integral

$$\iiint\limits_{G} (1 - x^2 - y^2 - z^2)\, dV$$

has its maximum value.

(b) Use the numerical triple integral operation of a CAS to approximate the maximum value.

(c) Find the exact maximum value.

16.6 CENTROID, CENTER OF GRAVITY, THEOREM OF PAPPUS

*Suppose that a rigid physical body is acted on by a gravitational field. Because the body is composed of many particles, each of which is affected by gravity, the action of a constant gravitational field on the body consists of a large number of forces distributed over the entire body. However, these individual forces can be replaced by a single force acting at a point called the **center of gravity** of the body. In this section we will show how double and triple integrals can be used to locate centers of gravity.*

DENSITY OF A LAMINA

Let us consider an idealized flat object that is thin enough to be viewed as a two-dimensional plane region (Figure 16.6.1). Such an object is called a **lamina**. A lamina is called **homogeneous** if its composition is uniform throughout and **inhomogeneous** otherwise. The **density** of a *homogeneous* lamina is defined to be its mass per unit area. Thus, the density δ of a homogeneous lamina of mass M and area A is given by $\delta = M/A$.

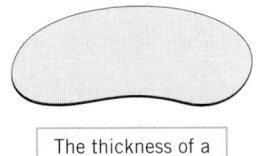

The thickness of a lamina is negligible.

Figure 16.6.1

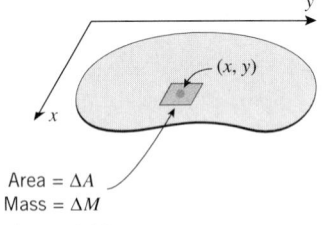

Area = ΔA
Mass = ΔM

Figure 16.6.2

For an inhomogeneous lamina the composition may vary from point to point, and hence an appropriate definition of "density" must reflect this. To motivate such a definition, suppose that the lamina is placed in an xy-plane. The density at a point (x, y) can be specified by a function $\delta(x, y)$, called the ***density function***, which can be interpreted as follows. Construct a small rectangle centered at (x, y) and let ΔM and ΔA be the mass and area of the portion of the lamina enclosed by this rectangle (Figure 16.6.2). If the ratio $\Delta M / \Delta A$ approaches a limiting value as the dimensions (and hence the area) of the rectangle approach zero, then this limit is considered to be the density of the lamina at (x, y). Symbolically,

$$\delta(x, y) = \lim_{\Delta A \to 0} \frac{\Delta M}{\Delta A} \tag{1}$$

From this relationship we obtain the approximation

$$\Delta M \approx \delta(x, y) \Delta A \tag{2}$$

which relates the mass and area of a small rectangular portion of the lamina centered at (x, y). It is assumed that as the dimensions of the rectangle tend to zero, the error in this approximation also tends to zero.

The following result shows how to find the mass of a lamina from its density function.

MASS OF A LAMINA

> **16.6.1**   MASS OF A LAMINA.   If a lamina with a continuous density function $\delta(x, y)$ occupies a region R in the xy-plane, then its total mass M is given by
>
> $$M = \iint\limits_{R} \delta(x, y)\, dA \tag{3}$$

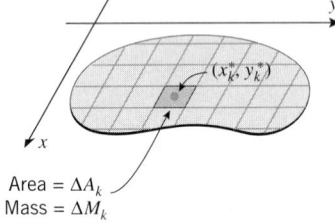

Area = ΔA_k
Mass = ΔM_k

Figure 16.6.3

This formula can be motivated by a familiar limiting process that can be outlined as follows: Imagine the lamina to be subdivided into rectangular pieces using lines parallel to the coordinate axes and excluding from consideration any nonrectangular parts at the boundary (Figure 16.6.3). Assume that there are n such rectangular pieces, and suppose that the kth piece has area ΔA_k. If we let (x_k^*, y_k^*) denote the center of the kth piece, then from Formula (2), the mass ΔM_k of this piece can be approximated by

$$\Delta M_k \approx \delta(x_k^*, y_k^*) \Delta A_k \tag{4}$$

and hence the mass M of the entire lamina can be approximated by

$$M \approx \sum_{k=1}^{n} \delta(x_k^*, y_k^*) \Delta A_k$$

If we now increase n in such a way that the dimensions of the rectangles tend to zero, then it is plausible that the errors in our approximations will approach zero, so

$$M = \lim_{n \to +\infty} \sum_{k=1}^{n} \delta(x_k^*, y_k^*) \Delta A_k = \iint\limits_{R} \delta(x, y)\, dA$$

Example 1

A triangular lamina with vertices $(0, 0)$, $(0, 1)$, and $(1, 0)$ has density function $\delta(x, y) = xy$. Find its total mass.

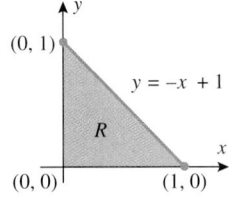

$(0, 1)$

$y = -x + 1$

R

$(0, 0)$ $(1, 0)$

Figure 16.6.4

Solution.   Referring to (3) and Figure 16.6.4, the mass M of the lamina is

$$M = \iint\limits_{R} \delta(x, y)\, dA = \iint\limits_{R} xy\, dA = \int_0^1 \int_0^{-x+1} xy\, dy\, dx$$

$$= \int_0^1 \left[\frac{1}{2} xy^2 \right]_{y=0}^{-x+1} dx = \int_0^1 \left[\frac{1}{2} x^3 - x^2 + \frac{1}{2} x \right] dx = \frac{1}{24} \text{ (unit of mass)} \quad \blacktriangleleft$$

CENTER OF GRAVITY OF A LAMINA

Assuming that the force of gravity is constant and acts downward, consider the following problem.

> **16.6.2**   PROBLEM.   Suppose that a lamina with a continuous density function $\delta(x, y)$ occupies a region R in a horizontal xy-plane. Find the coordinates $(\bar{x}, \bar{y})$ of the center of gravity.

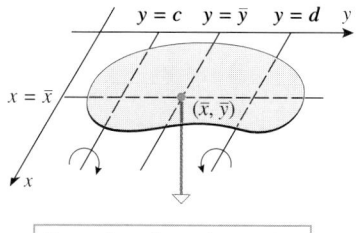

Force of gravity acting on the center of gravity of the lamina

Figure 16.6.5

To motivate the solution, consider what happens if we try to balance the lamina on a knife-edge parallel to the x-axis. Suppose the lamina in Figure 16.6.5 is placed on a knife-edge along a line $y = c$ that does not pass through the center of gravity. Because the lamina behaves as if its entire mass is concentrated at the center of gravity $(\bar{x}, \bar{y})$, the lamina will be rotationally unstable and the force of gravity will cause a rotation about $y = c$. Similarly, the lamina will undergo a rotation if placed on a knife-edge along $y = d$. However, if the knife-edge runs along the line $y = \bar{y}$ through the center of gravity, the lamina will be in perfect balance. Similarly, the lamina will be in perfect balance on a knife-edge along the line $x = \bar{x}$ through the center of gravity. This suggests that the center of gravity of a lamina can be determined as the intersection of two lines of balance, one parallel to the x-axis and the other parallel to the y-axis. In order to find these lines of balance, we will need some preliminary results about rotations.

Children on a seesaw learn by experience that a lighter child can balance a heavier one by sitting farther from the fulcrum or pivot point. This is because the tendency for an object to produce rotation is proportional not only to its mass but also to the distance between the object and the fulcrum. To make this more precise, consider an x-axis, which we view as a weightless beam. If a point-mass m is located on the axis at a point x, then the tendency for that mass to produce a rotation of the beam about a point a on the axis is measured by the following quantity, called the ***moment of m about x = a***:

$$\begin{bmatrix} \text{moment of } m \\ \text{about } a \end{bmatrix} = m(x - a)$$

The number $x - a$ is called the ***lever arm***. Depending on whether the mass is to the right or left of a, the lever arm is either the distance between x and a or the negative of this distance (Figure 16.6.6). Positive lever arms result in positive moments and clockwise rotations, while negative lever arms result in negative moments and counterclockwise rotations.

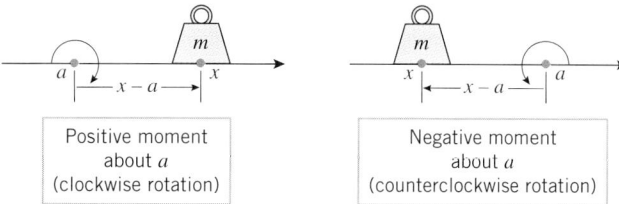

Positive moment about a (clockwise rotation)

Negative moment about a (counterclockwise rotation)

Figure 16.6.6

Suppose that masses $m_1, m_2, \ldots, m_n$ are located at points $x_1, x_2, \ldots, x_n$ on a coordinate axis and a fulcrum is positioned at the point a (Figure 16.6.7). Depending on whether the sum of the moments about a,

$$\sum_{k=1}^{n} m_k(x_k - a) = m_1(x_1 - a) + m_2(x_2 - a) + \cdots + m_n(x_n - a)$$

is positive, negative, or zero, a weightless beam along the axis will rotate clockwise about a, rotate counterclockwise about a, or balance perfectly. In the last case, the system of masses is is said to be in ***equilibrium***.

Figure 16.6.7

Fulcrum

The preceding ideas can be extended to masses distributed in two-dimensional space. If we imagine the xy-plane to be a weightless sheet supporting a point-mass m located at a point (x, y), then the tendency for the mass to produce a rotation of the sheet about the line $x = a$ is $m(x - a)$, called the ***moment of m about $x = a$***, and the tendency for the mass to produce a rotation about the line $y = c$ is $m(y - c)$, called the ***moment of m about $y = c$*** (Figure 16.6.8). In summary,

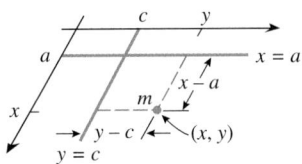

Figure 16.6.8

$$\begin{bmatrix} \text{moment of } m \\ \text{about the} \\ \text{line } x = a \end{bmatrix} = m(x - a) \quad \text{and} \quad \begin{bmatrix} \text{moment of } m \\ \text{about the} \\ \text{line } y = c \end{bmatrix} = m(y - c) \tag{5–6}$$

If a number of masses are distributed throughout the xy-plane, then the plane (viewed as a weightless sheet) will balance on a knife-edge along the line $x = a$ if the sum of the moments about the line is zero. Similarly for the line $y = c$.

We are now ready to solve Problem 16.6.2. We imagine the lamina to be subdivided into rectangular pieces using lines parallel to the coordinate axes and excluding from consideration any nonrectangular pieces at the boundary (Figure 16.6.3). We assume that there are n such rectangular pieces and that the kth piece has area ΔA_k and mass ΔM_k. We will let (x_k^*, y_k^*) be the center of the kth piece, and we will assume that the entire mass of the kth piece is concentrated at its center. From (4), the mass of the kth piece can be approximated by

$$\Delta M_k \approx \delta(x_k^*, y_k^*) \Delta A_k$$

Since the lamina balances on the lines $x = \bar{x}$ and $y = \bar{y}$, the sum of the moments of the rectangular pieces about those lines should be close to zero; that is,

$$\sum_{k=1}^{n} (x_k^* - \bar{x}) \Delta M_k = \sum_{k=1}^{n} (x_k^* - \bar{x}) \delta(x_k^*, y_k^*) \Delta A_k \approx 0$$

$$\sum_{k=1}^{n} (y_k^* - \bar{y}) \Delta M_k = \sum_{k=1}^{n} (y_k^* - \bar{y}) \delta(x_k^*, y_k^*) \Delta A_k \approx 0$$

If we now increase n in such a way that the dimensions of the rectangles tend to zero, then it is plausible that the errors in our approximations will approach zero, so that

$$\lim_{n \to +\infty} \sum_{k=1}^{n} (x_k^* - \bar{x}) \delta(x_k^*, y_k^*) \Delta A_k = 0$$

$$\lim_{n \to +\infty} \sum_{k=1}^{n} (y_k^* - \bar{y}) \delta(x_k^*, y_k^*) \Delta A_k = 0$$

from which we obtain

$$\iint\limits_{R} (x - \bar{x}) \delta(x, y) \, dA = 0$$

$$\iint\limits_{R} (y - \bar{y}) \delta(x, y) \, dA = 0$$

Since $\bar{x}$ and $\bar{y}$ are constant, these equations can be rewritten as

$$\iint\limits_{R} x \delta(x, y) \, dA = \bar{x} \iint\limits_{R} \delta(x, y) \, dA$$

$$\iint\limits_{R} y \delta(x, y) \, dA = \bar{y} \iint\limits_{R} \delta(x, y) \, dA$$

from which we obtain the following formulas for the center of gravity of the lamina:

Center of Gravity $(\bar{x}, \bar{y})$ ***of a Lamina***

$$\bar{x} = \frac{\displaystyle\iint_R x\delta(x, y)\, dA}{\displaystyle\iint_R \delta(x, y)\, dA}, \qquad \bar{y} = \frac{\displaystyle\iint_R y\delta(x, y)\, dA}{\displaystyle\iint_R \delta(x, y)\, dA} \tag{7-8}$$

Observe that in both formulas the denominator is the mass M of the lamina [see (3)]. The numerator in the formula for $\bar{x}$ is denoted by M_y and is called the ***first moment of the lamina about the y-axis***; the numerator of the formula for $\bar{y}$ is denoted by M_x and is called the ***first moment of the lamina about the x-axis***. Thus, Formulas (7) and (8) can be expressed as

$$\bar{x} = \frac{M_y}{M} = \frac{1}{\text{mass of } R} \iint_R x\delta(x, y)\, dA \tag{9}$$

$$\bar{y} = \frac{M_x}{M} = \frac{1}{\text{mass of } R} \iint_R y\delta(x, y)\, dA \tag{10}$$

Example 2

Find the center of gravity of the triangular lamina with vertices $(0, 0)$, $(0, 1)$, and $(1, 0)$ and density function $\delta(x, y) = xy$.

Solution. The lamina is shown in Figure 16.6.4. In Example 1 we found the mass of the lamina to be

$$M = \iint_R \delta(x, y)\, dA = \iint_R xy\, dA = \frac{1}{24}$$

The moment of the lamina about the y-axis is

$$M_y = \iint_R x\delta(x, y)\, dA = \iint_R x^2 y\, dA = \int_0^1 \int_0^{-x+1} x^2 y\, dy\, dx$$

$$= \int_0^1 \left[\frac{1}{2}x^2 y^2\right]_{y=0}^{-x+1} dx = \int_0^1 \left(\frac{1}{2}x^4 - x^3 + \frac{1}{2}x^2\right) dx = \frac{1}{60}$$

and the moment about the x-axis is

$$M_x = \iint_R y\delta(x, y)\, dA = \iint_R xy^2\, dA = \int_0^1 \int_0^{-x+1} xy^2\, dy\, dx$$

$$= \int_0^1 \left[\frac{1}{3}xy^3\right]_{y=0}^{-x+1} dx = \int_0^1 \left(-\frac{1}{3}x^4 + x^3 - x^2 + \frac{1}{3}x\right) dx = \frac{1}{60}$$

From (9) and (10),

$$\bar{x} = \frac{M_y}{M} = \frac{1/60}{1/24} = \frac{2}{5}, \qquad \bar{y} = \frac{M_x}{M} = \frac{1/60}{1/24} = \frac{2}{5}$$

so the center of gravity is $\left(\frac{2}{5}, \frac{2}{5}\right)$. ◀

CENTROIDS

In the special case of a *homogeneous* lamina, the center of gravity is called the ***centroid of the lamina*** or sometimes the ***centroid of the region R***. Because the density function δ is constant for a homogeneous lamina, the factor δ may be moved through the integral signs in (7) and (8) and canceled. Thus, the centroid $(\bar{x}, \bar{y})$ is a geometric property of the region

R and is given by the following formulas:

Centroid of a Region R

$$\bar{x} = \frac{\displaystyle\iint_R x \, dA}{\displaystyle\iint_R dA} = \frac{1}{\text{area of } R} \iint_R x \, dA \qquad (11)$$

$$\bar{y} = \frac{\displaystyle\iint_R y \, dA}{\displaystyle\iint_R dA} = \frac{1}{\text{area of } R} \iint_R y \, dA \qquad (12)$$

Example 3

Find the centroid of the semicircular region in Figure 16.6.9.

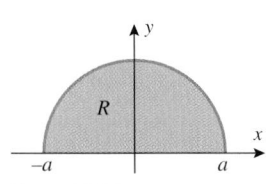

Figure 16.6.9

Solution. By symmetry, $\bar{x} = 0$ since the y-axis is obviously a line of balance. From (12),

$$\bar{y} = \frac{1}{\text{area of } R} \iint_R y \, dA = \frac{1}{\frac{1}{2}\pi a^2} \iint_R y \, dA$$

$$= \frac{1}{\frac{1}{2}\pi a^2} \int_0^\pi \int_0^a (r \sin\theta) r \, dr \, d\theta \qquad \boxed{\begin{array}{l}\text{Evaluating in}\\ \text{polar coordinates}\end{array}}$$

$$= \frac{1}{\frac{1}{2}\pi a^2} \int_0^\pi \left[\frac{1}{3} r^3 \sin\theta\right]_{r=0}^a d\theta$$

$$= \frac{1}{\frac{1}{2}\pi a^2} \left(\frac{1}{3} a^3\right) \int_0^\pi \sin\theta \, d\theta = \frac{1}{\frac{1}{2}\pi a^2} \left(\frac{2}{3} a^3\right) = \frac{4a}{3\pi}$$

so the centroid is $\left(0, \dfrac{4a}{3\pi}\right)$. ◀

CENTER OF GRAVITY AND CENTROID OF A SOLID

For a three-dimensional solid G, the formulas for moments, center of gravity, and centroid are similar to those for laminas. If G is *homogeneous*, then its *density* is defined to be its mass per unit volume. Thus, if G is a homogeneous solid of mass M and volume V, then its density δ is given by $\delta = M/V$. If G is inhomogeneous and is in an *xyz*-coordinate system, then its density at a general point (x, y, z) is specified by a *density function* $\delta(x, y, z)$ whose value at a point can be viewed as a limit:

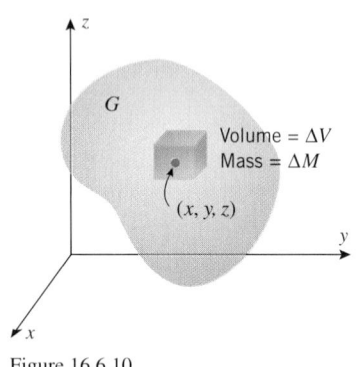

Figure 16.6.10

$$\delta(x, y, z) = \lim_{\Delta V \to 0} \frac{\Delta M}{\Delta V}$$

where ΔM and ΔV represent the mass and volume of a rectangular parallelepiped, centered at (x, y, z), whose dimensions tend to zero (Figure 16.6.10).

Using the discussion of laminas as a model, you should be able to show that the mass M of a solid with a continuous density function $\delta(x, y, z)$ is

$$M = \text{mass of } G = \iiint_G \delta(x, y, z) \, dV \qquad (13)$$

The formulas for center of gravity and centroid are

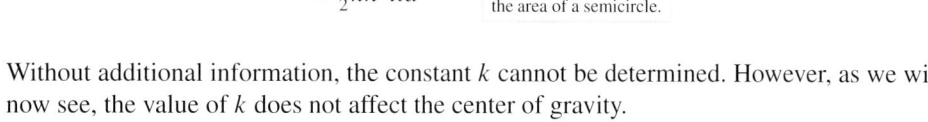

Center of Gravity $(\bar{x}, \bar{y}, \bar{z})$ of a Solid G

$$\bar{x} = \frac{1}{M} \iiint\limits_{G} x\delta(x, y, z)\, dV$$

$$\bar{y} = \frac{1}{M} \iiint\limits_{G} y\delta(x, y, z)\, dV$$

$$\bar{z} = \frac{1}{M} \iiint\limits_{G} z\delta(x, y, z)\, dV$$

Centroid $(\bar{x}, \bar{y}, \bar{z})$ of a Solid G

$$\bar{x} = \frac{1}{V} \iiint\limits_{G} x\, dV$$

$$\bar{y} = \frac{1}{V} \iiint\limits_{G} y\, dV$$ (14–15)

$$\bar{z} = \frac{1}{V} \iiint\limits_{G} z\, dV$$

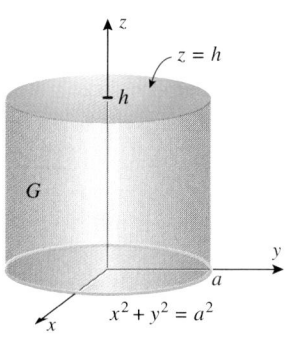

Figure 16.6.11

Example 4

Find the mass and the center of gravity of a cylindrical solid of height h and radius a (Figure 16.6.11), assuming that the density at each point is proportional to the distance between the point and the base of the solid.

Solution.   Since the density is proportional to the distance z from the base, the density function has the form $\delta(x, y, z) = kz$, where k is some (unknown) positive constant of proportionality. From (13) the mass of the solid is

$$M = \iiint\limits_{G} \delta(x, y, z)\, dV = \int_{-a}^{a} \int_{-\sqrt{a^2-x^2}}^{\sqrt{a^2-x^2}} \int_{0}^{h} kz\, dz\, dy\, dx$$

$$= k \int_{-a}^{a} \int_{-\sqrt{a^2-x^2}}^{\sqrt{a^2-x^2}} \frac{1}{2}h^2\, dy\, dx$$

$$= kh^2 \int_{-a}^{a} \sqrt{a^2 - x^2}\, dx$$

$$= \tfrac{1}{2}kh^2\pi a^2 \qquad \boxed{\text{Interpret the integral as the area of a semicircle.}}$$

Without additional information, the constant k cannot be determined. However, as we will now see, the value of k does not affect the center of gravity.

From (14),

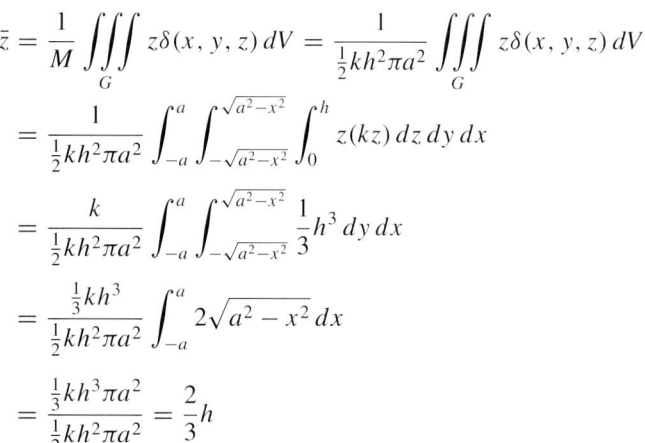

$$\bar{z} = \frac{1}{M} \iiint\limits_{G} z\delta(x, y, z)\, dV = \frac{1}{\frac{1}{2}kh^2\pi a^2} \iiint\limits_{G} z\delta(x, y, z)\, dV$$

$$= \frac{1}{\frac{1}{2}kh^2\pi a^2} \int_{-a}^{a} \int_{-\sqrt{a^2-x^2}}^{\sqrt{a^2-x^2}} \int_{0}^{h} z(kz)\, dz\, dy\, dx$$

$$= \frac{k}{\frac{1}{2}kh^2\pi a^2} \int_{-a}^{a} \int_{-\sqrt{a^2-x^2}}^{\sqrt{a^2-x^2}} \frac{1}{3}h^3\, dy\, dx$$

$$= \frac{\frac{1}{3}kh^3}{\frac{1}{2}kh^2\pi a^2} \int_{-a}^{a} 2\sqrt{a^2 - x^2}\, dx$$

$$= \frac{\frac{1}{3}kh^3\pi a^2}{\frac{1}{2}kh^2\pi a^2} = \frac{2}{3}h$$

Similar calculations using (14) will yield $\bar{x} = \bar{y} = 0$. However, this is evident by inspection, since it follows from the symmetry of the solid and the form of its density function that the center of gravity is on the z-axis. Thus, the center of gravity is $\left(0, 0, \frac{2}{3}h\right)$. ◀

....

THEOREM OF PAPPUS

The following theorem, due to the Greek mathematician Pappus,[*] gives an important relationship between the centroid of a plane region R and the volume of the solid generated when the region is revolved about a line.

16.6.3 THEOREM. *If R is a bounded plane region and L is a line that lies in the plane of R but is entirely on one side of R, then the volume of the solid formed by revolving R about L is given by*

$$volume = (area\ of\ R) \cdot \binom{distance\ traveled}{by\ the\ centroid}$$

Proof. Introduce an xy-coordinate system so that L is along the y-axis and the region R is in the first quadrant (Figure 16.6.12). Let R be partitioned into subregions in the usual way and let R_k be a typical rectangle interior to R. If (x_k^*, y_k^*) is the center of R_k, and if the area of R_k is $\Delta A_k = \Delta x_k \Delta y_k$, then from Formula (1) of Section 8.3 the volume generated by R_k as it revolves about L is

$$2\pi x_k^* \Delta x_k \Delta y_k = 2\pi x_k^* \Delta A_k$$

Therefore, the total volume of the solid is approximately

$$V \approx \sum_{k=1}^{n} 2\pi x_k^* \Delta A_k$$

from which it follows that the exact volume is

$$V = \iint\limits_{R} 2\pi x\, dA = 2\pi \iint\limits_{R} x\, dA$$

Thus, it follows from (11) that

$$V = 2\pi \cdot \bar{x} \cdot [area\ of\ R]$$

This completes the proof since $2\pi\bar{x}$ is the distance traveled by the centroid when R is revolved about the y-axis. ∎

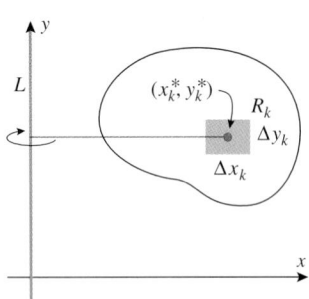

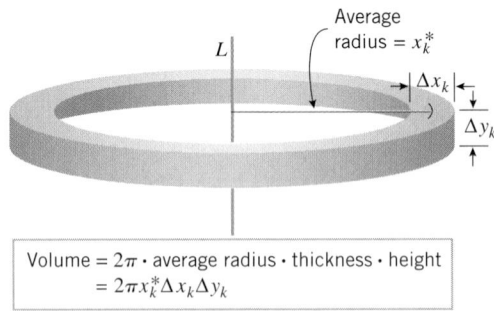

Volume = 2π · average radius · thickness · height
= $2\pi x_k^* \Delta x_k \Delta y_k$

Figure 16.6.12

---

[*] PAPPUS OF ALEXANDRIA (4th century A.D.). Greek mathematician. Pappus lived during the early Christian era when mathematical activity was in a period of decline. His main contributions to mathematics appeared in a series of eight books called *The Collection* (written about 340 A.D.). This work, which survives only partially, contained some original results but was devoted mostly to statements, refinements, and proofs of results by earlier mathematicians. Pappus' Theorem, stated without proof in Book VII of *The Collection*, was probably known and proved in earlier times. This result is sometimes called Guldin's Theorem in recognition of the Swiss mathematician, Paul Guldin (1577–1643), who rediscovered it independently.

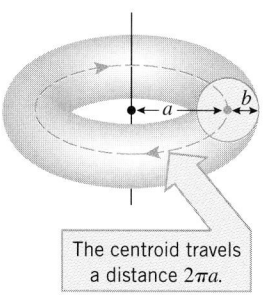

The centroid travels
a distance $2\pi a$.

Figure 16.6.13

Example 5

Use Pappus' Theorem to find the volume V of the torus generated by revolving a circular region of radius b about a line at a distance a (greater than b) from the center of the circle (Figure 16.6.13).

Solution.   By symmetry, the centroid of a circular region is its center. Thus, the distance traveled by the centroid is $2\pi a$. Since the area of a circle of radius b is πb^2, it follows from Pappus' Theorem that the volume of the torus is

$$V = (2\pi a)(\pi b^2) = 2\pi^2 ab^2$$ ◀

EXERCISE SET 16.6   ⃝ Graphing Calculator   🄲 CAS

1. Where should the fulcrum be placed so that the beam in the accompanying figure is in equilibrium?

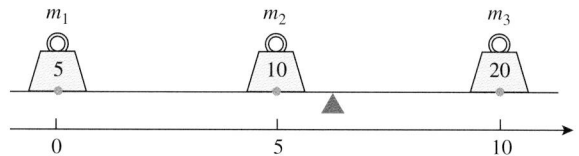

Figure Ex-1

2. Given that the beam in the accompanying figure is in equilibrium, what is the mass m?

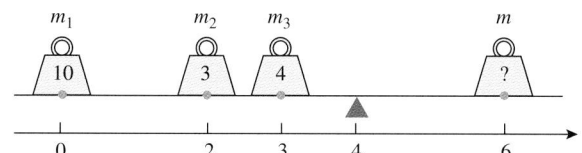

Figure Ex-2

For the regions in Exercises 3 and 4, make a conjecture about the coordinates of the centroid, and confirm your conjecture by integrating.

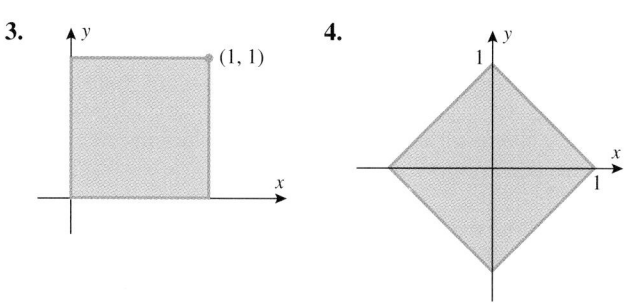

In Exercises 5–10, find the centroid of the region.

5.
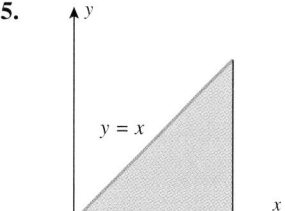

6.
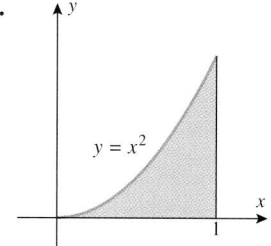

7.
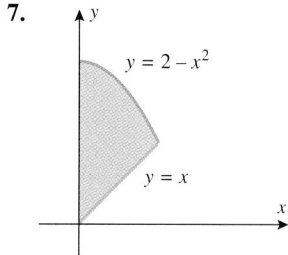

8.
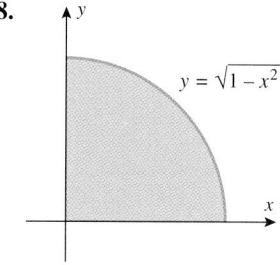

9. The region above the x-axis and between the circles $x^2 + y^2 = a^2$ and $x^2 + y^2 = b^2$ $(a < b)$.

10. The region enclosed between the y-axis and the right half of the circle $x^2 + y^2 = a^2$.

In Exercises 11–14, find the mass and center of gravity of the lamina.

11. A lamina with density $\delta(x, y) = x + y$ is bounded by the x-axis, the line $x = 1$, and the curve $y = \sqrt{x}$.

12. A lamina with density $\delta(x, y) = y$ is bounded by $y = \sin x$, $y = 0$, $x = 0$, and $x = \pi$.

13. A lamina with density $\delta(x, y) = xy$ is in the first quadrant and is bounded by the circle $x^2 + y^2 = a^2$ and the coordinate axes.

14. A lamina with density $\delta(x, y) = x^2 + y^2$ is bounded by the x-axis and the upper half of the circle $x^2 + y^2 = 1$.

For the solids in Exercises 15 and 16, make a conjecture about the coordinates of the centroid, and confirm your conjecture by integrating.

15.   **16.**

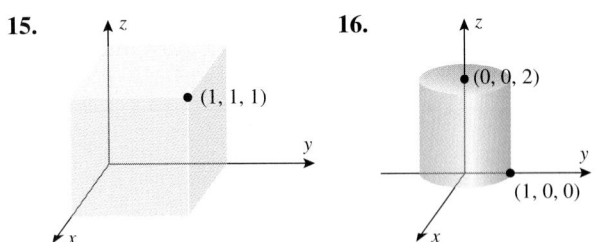

In Exercises 17–22, find the centroid of the solid.

17. The tetrahedron in the first octant enclosed by the coordinate planes and the plane $x + y + z = 1$.

18. The solid bounded by the parabolic cylinder $z = 1 - y^2$ and the planes $x + z = 1$, $x = 0$, and $z = 0$.

19. The solid bounded by the surface $z = y^2$ and the planes $x = 0$, $x = 1$, and $z = 1$.

20. The solid in the first octant bounded by the surface $z = xy$ and the planes $z = 0$, $x = 2$, and $y = 2$.

21. The solid in the first octant that is bounded by the sphere $x^2 + y^2 + z^2 = a^2$ and the coordinate planes.

22. The solid enclosed by the xy-plane and the hemisphere $z = \sqrt{a^2 - x^2 - y^2}$.

In Exercises 23–26, find the mass and center of gravity of the solid.

23. The cube that has density $\delta(x, y, z) = a - x$ and is defined by the inequalities $0 \le x \le a$, $0 \le y \le a$, and $0 \le z \le a$.

24. The cylindrical solid that has density $\delta(x, y, z) = h - z$ and is enclosed by $x^2 + y^2 = a^2$, $z = 0$, and $z = h$.

25. The solid that has density $\delta(x, y, z) = yz$ and is enclosed by $z = 1 - y^2$ (for $y \ge 0$), $z = 0$, $x = -1$, and $x = 1$.

26. The solid that has density $\delta(x, y, z) = xz$ and is enclosed by $y = 9 - x^2$ (for $x \ge 0$), $x = 0$, $y = 0$, $z = 0$, and $z = 1$.

27. Find the center of gravity of the square lamina with vertices $(0, 0)$, $(1, 0)$, $(0, 1)$, and $(1, 1)$ if
   (a) the density is proportional to the square of the distance from the origin
   (b) the density is proportional to the distance from the y-axis.

28. Find the center of gravity of the cube that is determined by the inequalities $0 \le x \le 1$, $0 \le y \le 1$, $0 \le z \le 1$ if
   (a) the density is proportional to the square of the distance to the origin

   (b) the density is proportional to the sum of the distances to the faces that lie in the coordinate planes.

c 29. Use the numerical triple integral capability of a CAS to approximate the location of the centroid of the solid that is bounded above by the surface $z = 1/(1 + x^2 + y^2)$, below by the xy-plane, and laterally by the plane $y = 0$ and the surface $y = \sin x$ for $0 \le x \le \pi$ (see the accompanying figure).

30. The accompanying figure shows the solid that is bounded above by the surface $z = 1/(x^2 + y^2 + 1)$, below by the xy-plane, and laterally by the surface $x^2 + y^2 = a^2$.
   (a) By symmetry, the centroid of the solid lies on the z-axis. Make a conjecture about the behavior of the z-coordinate of the centroid as $a \to 0^+$ and as $a \to +\infty$.
   (b) Find the z-coordinate of the centroid, and check your conjecture by calculating the appropriate limits.
   (c) Use a graphing utility to plot the z-coordinate of the centroid versus a, and use the graph to estimate the value of a for which the centroid is $(0, 0, 0.25)$.

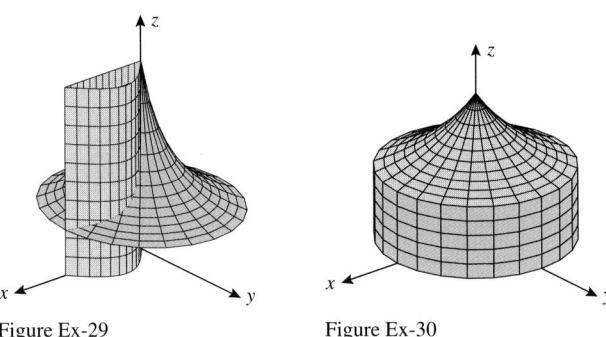

Figure Ex-29   Figure Ex-30

31. Show that in polar coordinates the formulas for the centroid $(\bar{x}, \bar{y})$ of a region R are

$$\bar{x} = \frac{1}{\text{area of } R} \iint\limits_{R} r^2 \cos\theta \, dr \, d\theta$$

$$\bar{y} = \frac{1}{\text{area of } R} \iint\limits_{R} r^2 \sin\theta \, dr \, d\theta$$

32. Use the result of Exercise 31 to find the centroid $(\bar{x}, \bar{y})$ of the region enclosed by the cardioid $r = a(1 + \sin\theta)$.

33. Use the result of Exercise 31 to find the centroid $(\bar{x}, \bar{y})$ of the petal of the rose $r = \sin 2\theta$ in the first quadrant.

34. Let R be the rectangle bounded by the lines $x = 0$, $x = 3$, $y = 0$, and $y = 2$. By inspection, find the centroid of R and use it to evaluate

$$\iint\limits_{R} x \, dA \quad \text{and} \quad \iint\limits_{R} y \, dA$$

35. Use the Theorem of Pappus and the fact that the volume of a sphere of radius a is $V = \frac{4}{3}\pi a^3$ to show that the centroid of the lamina that is bounded by the x-axis and the semicircle $y = \sqrt{a^2 - x^2}$ is $(0, 4a/(3\pi))$. (This problem was solved directly in Example 3.)

36. Use the Theorem of Pappus and the result of Exercise 35 to find the volume of the solid generated when the region bounded by the x-axis and the semicircle $y = \sqrt{a^2 - x^2}$ is revolved about
(a) the line $y = -a$ 　　　(b) the line $y = x - a$.

37. Use the Theorem of Pappus and the fact that the area of an ellipse with semiaxes a and b is πab to find the volume of the elliptical torus generated by revolving the ellipse
$$\frac{(x - k)^2}{a^2} + \frac{y^2}{b^2} = 1$$
about the y-axis. Assume that $k > a$.

38. Use the Theorem of Pappus to find the volume of the solid that is generated when the region enclosed by $y = x^2$ and $y = 8 - x^2$ is revolved about the x-axis.

39. Use the Theorem of Pappus to find the centroid of the triangular region with vertices $(0, 0)$, $(a, 0)$, and $(0, b)$, where $a > 0$ and $b > 0$. [Hint: Revolve the region about the x-axis to obtain $\bar{y}$ and about the y-axis to obtain $\bar{x}$.]

The tendency of a lamina to resist a change in rotational motion about an axis is measured by its ***moment of inertia*** about that axis. If the lamina occupies a region R of the xy-plane, and if its density function $\delta(x, y)$ is continuous on R, then the moments of inertia about the x-axis, the y-axis, and the z-axis are denoted by I_x, I_y, and I_z, respectively, and are defined by

$$I_x = \iint\limits_{R} y^2\, \delta(x, y)\, dA, \quad I_y = \iint\limits_{R} x^2\, \delta(x, y)\, dA,$$

$$I_z = \iint\limits_{R} (x^2 + y^2)\, \delta(x, y)\, dA$$

These definitions will be used in Exercises 40 and 41.

40. Consider the rectangular lamina that occupies the region described by the inequalities $0 \le x \le a$ and $0 \le y \le b$. Assuming that the lamina has constant density δ, show that
$$I_x = \frac{\delta ab^3}{3}, \quad I_y = \frac{\delta a^3 b}{3}, \quad I_z = \frac{\delta ab(a^2 + b^2)}{3}$$

41. Consider the circular lamina that occupies the region described by the inequalities $0 \le x^2 + y^2 \le a^2$. Assuming that the lamina has constant density δ, show that
$$I_x = I_y = \frac{\delta \pi a^4}{4}, \quad I_z = \frac{\delta \pi a^4}{2}$$

16.7 TRIPLE INTEGRALS IN CYLINDRICAL AND SPHERICAL COORDINATES

Earlier we saw that some double integrals are easier to evaluate in polar coordinates than in rectangular coordinates. Similarly, some triple integrals are easier to evaluate in cylindrical or spherical coordinates than in rectangular coordinates. In this section we will study triple integrals in these coordinate systems.

TRIPLE INTEGRALS IN CYLINDRICAL COORDINATES

Recall that in rectangular coordinates the triple integral of a continuous function f over a solid region G is defined as

$$\iiint\limits_{G} f(x, y, z)\, dV = \lim_{n \to +\infty} \sum_{k=1}^{n} f(x_k^*, y_k^*, z_k^*)\Delta V_k$$

where ΔV_k denotes the volume of a rectangular parallelepiped interior to G and (x_k^*, y_k^*, z_k^*) is a point in this parallelepiped (see Figure 16.5.1). Triple integrals in cylindrical and spherical coordinates are defined similarly, except that the region G is divided not into rectangular parallelepipeds but into regions more appropriate to these coordinate systems.

In cylindrical coordinates, the simplest equations are of the form

$$r = \text{constant}, \quad \theta = \text{constant}, \quad z = \text{constant}$$

As indicated in Figure 13.8.2b, the first equation represents a right circular cylinder centered on the z-axis, the second a vertical half-plane hinged on the z-axis, and the third a horizontal plane. These surfaces can be paired up to determine solids called ***cylindrical wedges*** or

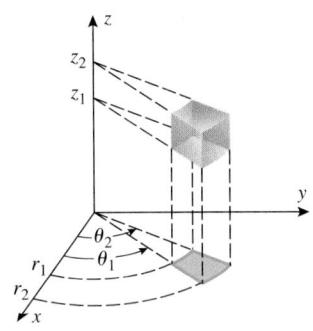

Figure 16.7.1

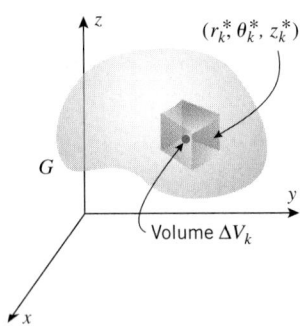

Figure 16.7.2

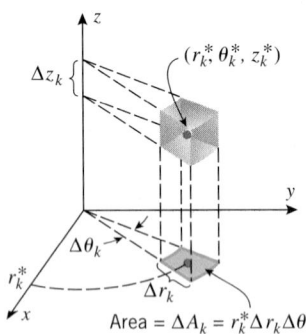

Figure 16.7.3

cylindrical elements of volume. To be precise, a cylindrical wedge is a solid enclosed between six surfaces of the following form:

two cylinders	$r = r_1,$	$r = r_2$	$(r_1 < r_2)$
two half-planes	$\theta = \theta_1,$	$\theta = \theta_2$	$(\theta_1 < \theta_2)$
two planes	$z = z_1,$	$z = z_2$	$(z_1 < z_2)$

(Figure 16.7.1). The dimensions $\theta_2 - \theta_1$, $r_2 - r_1$, and $z_2 - z_1$ are called the *central angle*, *thickness*, and *height* of the wedge.

To define the triple integral over G of a function $f(r, \theta, z)$ in cylindrical coordinates we proceed as follows:

- Subdivide G into pieces by a three-dimensional grid consisting of concentric circular cylinders centered on the z-axis, half-planes hinged on the z-axis, and horizontal planes. Exclude from consideration all pieces that contain any points outside of G, thereby leaving only cylindrical wedges that are subsets of G.

- Assume that there are n such cylindrical wedges, and denote the volume of the kth cylindrical wedge by ΔV_k. As indicated in Figure 16.7.2, let $(r_k^*, \theta_k^*, z_k^*)$ be any point in the kth cylindrical wedge.

- Repeat this process with more and more subdivisions so that as n increases, the height, thickness, and central angle of the cylindrical wedges approach zero. Define

$$\iiint\limits_{G} f(r, \theta, z)\, dV = \lim_{n \to +\infty} \sum_{k=1}^{n} f(r_k^*, \theta_k^*, z_k^*)\Delta V_k \tag{1}$$

For computational purposes, it will be helpful to express (1) as an iterated integral. Toward this end we note that the volume ΔV_k of the kth cylindrical wedge can be expressed as

$$\Delta V_k = [\text{area of base}] \cdot [\text{height}] \tag{2}$$

If we denote the thickness, central angle, and height of this wedge by Δr_k, $\Delta \theta_k$, and Δz_k, and if we choose the arbitrary point $(r_k^*, \theta_k^*, z_k^*)$ to lie above the "center" of the base (Figures 16.3.5 and 16.7.3), then it follows from (5) of Section 16.3 that the base has area $\Delta A_k = r_k^* \Delta r_k \Delta \theta_k$. Thus, (2) can be written as

$$\Delta V_k = r_k^* \Delta r_k \Delta \theta_k \Delta z_k = r_k^* \Delta z_k \Delta r_k \Delta \theta_k$$

Substituting this expression in (1) yields

$$\iiint\limits_{G} f(r, \theta, z)\, dV = \lim_{n \to +\infty} \sum_{k=1}^{n} f(r_k^*, \theta_k^*, z_k^*) r_k^* \Delta z_k \Delta r_k \Delta \theta_k$$

which suggests that a triple integral in cylindrical coordinates can be evaluated as an iterated integral of the form

$$\iiint\limits_{G} f(r, \theta, z)\, dV = \iiint\limits_{\substack{\text{appropriate} \\ \text{limits}}} f(r, \theta, z) r\, dz\, dr\, d\theta \tag{3}$$

REMARK.   Note the extra factor of r that appears in the integrand on converting from the triple integral to the iterated integral. In this formula the integration with respect to z is done first, then with respect to r, and then with respect to θ, but any order of integration is allowable.

The following theorem, which we state without proof, makes the preceding ideas more precise.

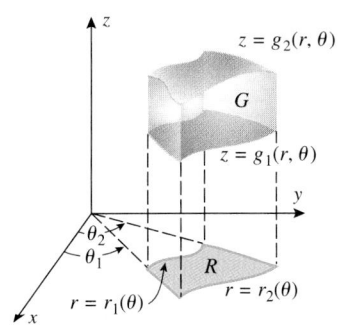

Figure 16.7.4

16.7.1 THEOREM. *Let G be a simple xy-solid whose upper surface has the equation* $z = g_2(r, \theta)$ *and whose lower surface has the equation* $z = g_1(r, \theta)$ *in cylindrical coordinates. If R is the projection of the solid on the xy-plane, and if* $f(r, \theta, z)$ *is continuous on G, then*

$$\iiint\limits_{G} f(r, \theta, z)\, dV = \iint\limits_{R} \left[\int_{g_1(r,\theta)}^{g_2(r,\theta)} f(r, \theta, z)\, dz \right] dA \tag{4}$$

where the double integral over R is evaluated in polar coordinates. In particular, if the projection R is as shown in Figure 16.7.4, then (4) can be written as

$$\iiint\limits_{G} f(r, \theta, z)\, dV = \int_{\theta_1}^{\theta_2} \int_{r_1(\theta)}^{r_2(\theta)} \int_{g_1(r,\theta)}^{g_2(r,\theta)} f(r, \theta, z)\, r\, dz\, dr\, d\theta \tag{5}$$

The type of solid to which Formula (5) applies is illustrated in Figure 16.7.4. To apply (4) and (5) it is best to begin with a three-dimensional sketch of the solid G, from which the limits of integration can be obtained as follows:

Step 1. Identify the upper surface $z = g_2(r, \theta)$ and the lower surface $z = g_1(r, \theta)$ of the solid. The functions $g_1(r, \theta)$ and $g_2(r, \theta)$ determine the z-limits of integration. (If the upper and lower surfaces are given in rectangular coordinates, convert them to cylindrical coordinates.)

Step 2. Make a two-dimensional sketch of the projection R of the solid on the xy-plane. From this sketch the r- and θ-limits of integration may be obtained exactly as with double integrals in polar coordinates.

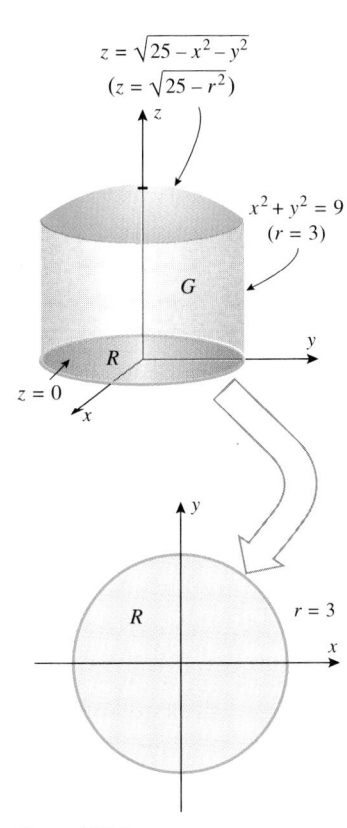

Figure 16.7.5

Example 1

Use triple integration in cylindrical coordinates to find the volume and the centroid of the solid G that is bounded above by the hemisphere $z = \sqrt{25 - x^2 - y^2}$, below by the xy-plane, and laterally by the cylinder $x^2 + y^2 = 9$.

Solution. The solid G and its projection R on the xy-plane are shown in Figure 16.7.5. In cylindrical coordinates, the upper surface of G is the hemisphere $z = \sqrt{25 - r^2}$ and the lower surface is the plane $z = 0$. Thus, from (4), the volume of G is

$$V = \iiint\limits_{G} dV = \iint\limits_{R} \left[\int_0^{\sqrt{25-r^2}} dz \right] dA$$

For the double integral over R, we use polar coordinates:

$$V = \int_0^{2\pi} \int_0^3 \int_0^{\sqrt{25-r^2}} r\, dz\, dr\, d\theta = \int_0^{2\pi} \int_0^3 \left[rz \right]_{z=0}^{\sqrt{25-r^2}} dr\, d\theta$$

$$= \int_0^{2\pi} \int_0^3 r\sqrt{25 - r^2}\, dr\, d\theta = \int_0^{2\pi} \left[-\frac{1}{3}(25 - r^2)^{3/2} \right]_{r=0}^{3} d\theta$$

$$= \int_0^{2\pi} \frac{61}{3}\, d\theta = \frac{122}{3}\pi \qquad \boxed{\begin{array}{l} u = 25 - r^2 \\ du = -2r\, dr \end{array}}$$

From this result and (15) of Section 16.6,

$$\bar{z} = \frac{1}{V} \iiint\limits_{G} z \, dV = \frac{3}{122\pi} \iiint\limits_{G} z \, dV = \frac{3}{122\pi} \iint\limits_{R} \left[\int_{0}^{\sqrt{25-r^2}} z \, dz \right] dA$$

$$= \frac{3}{122\pi} \int_{0}^{2\pi} \int_{0}^{3} \int_{0}^{\sqrt{25-r^2}} zr \, dz \, dr \, d\theta = \frac{3}{122\pi} \int_{0}^{2\pi} \int_{0}^{3} \left[\frac{1}{2}rz^2 \right]_{z=0}^{\sqrt{25-r^2}} dr \, d\theta$$

$$= \frac{3}{244\pi} \int_{0}^{2\pi} \int_{0}^{3} (25r - r^3) \, dr \, d\theta = \frac{3}{244\pi} \int_{0}^{2\pi} \frac{369}{4} \, d\theta = \frac{1107}{488}$$

By symmetry, the centroid $(\bar{x}, \bar{y}, \bar{z})$ of G lies on the z-axis, so $\bar{x} = \bar{y} = 0$. Thus, the centroid is at the point $(0, 0, 1107/488)$. ◀

··

CONVERTING TRIPLE INTEGRALS FROM RECTANGULAR TO CYLINDRICAL COORDINATES

Sometimes a triple integral that is difficult to integrate in rectangular coordinates can be evaluated more easily by making the substitution $x = r\cos\theta$, $y = r\sin\theta$, $z = z$ to convert it to an integral in cylindrical coordinates. Under such a substitution, a rectangular triple integral can be expressed as an iterated integral in cylindrical coordinates as

$$\iiint\limits_{G} f(x, y, z) \, dV = \iiint\limits_{\substack{\text{appropriate} \\ \text{limits}}} f(r\cos\theta, r\sin\theta, z) r \, dz \, dr \, d\theta \tag{6}$$

REMARK.   In (6), the order of integration is first with respect to z, then r, and then θ. However, the order of integration can be changed, provided the limits of integration are adjusted accordingly.

Example 2

Use cylindrical coordinates to evaluate

$$\int_{-3}^{3} \int_{-\sqrt{9-x^2}}^{\sqrt{9-x^2}} \int_{0}^{9-x^2-y^2} x^2 \, dz \, dy \, dx$$

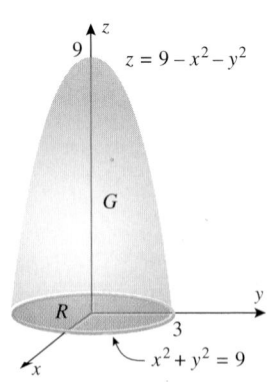

Figure 16.7.6

Solution. In problems of this type, it is helpful to sketch the region of integration G and its projection R on the xy-plane. From the z-limits of integration, the upper surface of G is the paraboloid $z = 9 - x^2 - y^2$ and the lower surface is the xy-plane $z = 0$. From the x- and y-limits of integration, the projection R is the region in the xy-plane enclosed by the circle $x^2 + y^2 = 9$ (Figure 16.7.6). Thus,

$$\int_{-3}^{3} \int_{-\sqrt{9-x^2}}^{\sqrt{9-x^2}} \int_{0}^{9-x^2-y^2} x^2 \, dz \, dy \, dx = \iiint\limits_{G} x^2 \, dV$$

$$= \iint\limits_{R} \left[\int_{0}^{9-r^2} r^2 \cos^2\theta \, dz \right] dA = \int_{0}^{2\pi} \int_{0}^{3} \int_{0}^{9-r^2} (r^2 \cos^2\theta) r \, dz \, dr \, d\theta$$

$$= \int_{0}^{2\pi} \int_{0}^{3} \int_{0}^{9-r^2} r^3 \cos^2\theta \, dz \, dr \, d\theta = \int_{0}^{2\pi} \int_{0}^{3} \left[zr^3 \cos^2\theta \right]_{z=0}^{9-r^2} dr \, d\theta$$

$$= \int_{0}^{2\pi} \int_{0}^{3} (9r^3 - r^5) \cos^2\theta \, dr \, d\theta = \int_{0}^{2\pi} \left[\left(\frac{9r^4}{4} - \frac{r^6}{6} \right) \cos^2\theta \right]_{r=0}^{3} d\theta$$

$$= \frac{243}{4} \int_{0}^{2\pi} \cos^2\theta \, d\theta = \frac{243}{4} \int_{0}^{2\pi} \frac{1}{2}(1 + \cos 2\theta) \, d\theta = \frac{243\pi}{4} \qquad ◀$$

TRIPLE INTEGRALS IN SPHERICAL COORDINATES

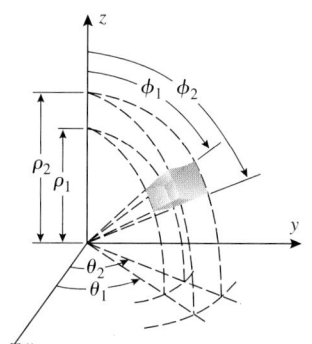

Figure 16.7.7

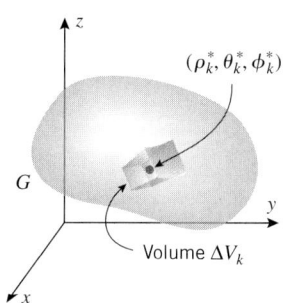

Figure 16.7.8

In spherical coordinates, the simplest equations are of the form

$$\rho = \text{constant}, \quad \theta = \text{constant}, \quad \phi = \text{constant}$$

As indicated in Figure 13.8.2c, the first equation represents a sphere centered at the origin, the second a half-plane hinged on the z-axis, and the third a right circular cone with its vertex at the origin and its line of symmetry along the z-axis. By a *spherical wedge* or *spherical element of volume* we mean a solid enclosed between six surfaces of the following form:

two spheres	$\rho = \rho_1, \quad \rho = \rho_2$	$(\rho_1 < \rho_2)$
two half-planes	$\theta = \theta_1, \quad \theta = \theta_2$	$(\theta_1 < \theta_2)$
nappes of two right circular cones	$\phi = \phi_1, \quad \phi = \phi_2$	$(\phi_1 < \phi_2)$

(Figure 16.7.7). We will refer to the numbers $\rho_2 - \rho_1$, $\theta_2 - \theta_1$, and $\phi_2 - \phi_1$ as the *dimensions* of a spherical wedge.

If G is a solid region in three-dimensional space, then the triple integral over G of a continuous function $f(\rho, \theta, \phi)$ in spherical coordinates is similar in definition to the triple integral in cylindrical coordinates, except that the solid G is partitioned into *spherical wedges* by a three-dimensional grid consisting of spheres centered at the origin, half-planes hinged on the z-axis, and nappes of right circular cones with vertices at the origin and lines of symmetry along the z-axis (Figure 16.7.8).

The defining equation of a triple integral in spherical coordinates is

$$\iiint\limits_{G} f(\rho, \theta, \phi)\, dV = \lim_{n \to +\infty} \sum_{k=1}^{n} f(\rho_k^*, \theta_k^*, \phi_k^*)\Delta V_k \tag{7}$$

where ΔV_k is the volume of the kth spherical wedge that is interior to G, $(\rho_k^*, \theta_k^*, \phi_k^*)$ is an arbitrary point in this wedge, and n increases in such a way that the dimensions of each interior spherical wedge tend to zero.

For computational purposes, it will be desirable to express (7) as an iterated integral. In the exercises we will help you to show that if the point $(\rho_k^*, \theta_k^*, \phi_k^*)$ is suitably chosen, then the volume ΔV_k in (7) can be written as

$$\Delta V_k = \rho_k^{*2} \sin \phi_k^* \Delta \rho_k \Delta \phi_k \Delta \theta_k \tag{8}$$

where $\Delta \rho_k$, $\Delta \phi_k$, and $\Delta \theta_k$ are the dimensions of the wedge (Exercise 38). Substituting this in (7) we obtain

$$\iiint\limits_{G} f(\rho, \theta, \phi)\, dV = \lim_{n \to +\infty} \sum_{k=1}^{n} f(\rho_k^*, \theta_k^*, \phi_k^*)\rho_k^{*2} \sin \phi_k^* \Delta \rho_k \Delta \phi_k \Delta \theta_k$$

which suggests that a triple integral in spherical coordinates can be evaluated as an iterated integral of the form

$$\iiint\limits_{G} f(\rho, \theta, \phi)\, dV = \iiint\limits_{\substack{\text{appropriate} \\ \text{limits}}} f(\rho, \theta, \phi)\rho^2 \sin \phi\, d\rho\, d\phi\, d\theta \tag{9}$$

REMARK. Note the extra factor of $\rho^2 \sin \phi$ that appears in the integrand of the iterated integral. This is analogous to the extra factor of r that appeared when we integrated in cylindrical coordinates.

The analog of Theorem 16.7.1 for triple integrals in spherical coordinates is tedious to state, so instead we will give some examples that illustrate techniques for obtaining the limits of integration. In all of our examples we will use the same order of integration—first with respect to ρ, then ϕ, and then θ. Once you have mastered the basic ideas, there should be no trouble using other orders of integration.

Suppose that we want to integrate $f(\rho, \theta, \phi)$ over the spherical solid G enclosed by the sphere $\rho = \rho_0$. The basic idea is to choose the limits of integration so that every point of the solid is accounted for in the integration process. Figure 16.7.9 illustrates one way of doing this. Holding θ and ϕ fixed for the first integration, we let ρ vary from 0 to ρ_0. This covers a radial line from the origin to the surface of the sphere. Next, keeping θ fixed, we let ϕ vary from 0 to π so that the radial line sweeps out a fan-shaped region. Finally, we let θ vary from 0 to 2π so that the fan-shaped region makes a complete revolution, thereby sweeping out the entire sphere. Thus, the triple integral of $f(\rho, \theta, \phi)$ over the spherical solid G may be evaluated by writing

$$\iiint_{G} f(\rho, \theta, \phi)\, dV = \int_{0}^{2\pi} \int_{0}^{\pi} \int_{0}^{\rho_0} f(\rho, \theta, \phi)\rho^2 \sin\phi\, d\rho\, d\phi\, d\theta$$

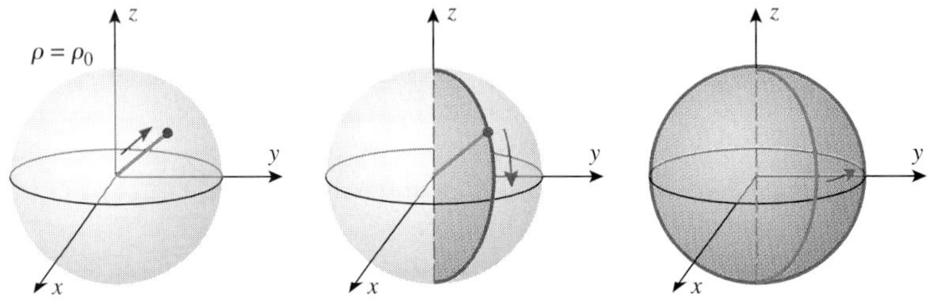

$\rho = \rho_0$

| ρ varies from 0 to ρ_0 with θ and ϕ fixed. | ϕ varies from 0 to π with θ fixed. | θ varies from 0 to 2π. |

Figure 16.7.9

Table 16.7.1 suggests how the limits of integration in spherical coordinates can be obtained for some other common solids.

Example 3

Use spherical coordinates to find the volume and the centroid of the solid G bounded above by the sphere $x^2 + y^2 + z^2 = 16$ and below by the cone $z = \sqrt{x^2 + y^2}$.

Solution. The solid G is sketched in Figure 16.7.10.

In spherical coordinates, the equation of the sphere $x^2 + y^2 + z^2 = 16$ is $\rho = 4$ and the equation of the cone $z = \sqrt{x^2 + y^2}$ is

$$\rho \cos\phi = \sqrt{\rho^2 \sin^2\phi \cos^2\theta + \rho^2 \sin^2\phi \sin^2\theta}$$

which simplifies to

$$\rho \cos\phi = \rho \sin\phi$$

or, on dividing both sides by $\rho \cos\phi$,

$$\tan\phi = 1$$

Thus $\phi = \pi/4$, so the volume of G is

$$V = \iiint_{G} dV = \int_{0}^{2\pi} \int_{0}^{\pi/4} \int_{0}^{4} \rho^2 \sin\phi\, d\rho\, d\phi\, d\theta$$

$$= \int_{0}^{2\pi} \int_{0}^{\pi/4} \left[\frac{\rho^3}{3} \sin\phi\right]_{\rho=0}^{4} d\phi\, d\theta$$

$$= \int_{0}^{2\pi} \int_{0}^{\pi/4} \frac{64}{3} \sin\phi\, d\phi\, d\theta$$

$$= \frac{64}{3} \int_{0}^{2\pi} \left[-\cos\phi\right]_{\phi=0}^{\pi/4} d\theta = \frac{64}{3} \int_{0}^{2\pi} \left(1 - \frac{\sqrt{2}}{2}\right) d\theta$$

$$= \frac{64\pi}{3}(2 - \sqrt{2})$$

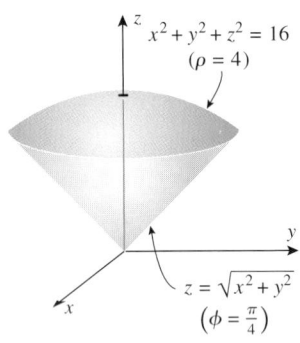

Figure 16.7.10

Table 16.7.1

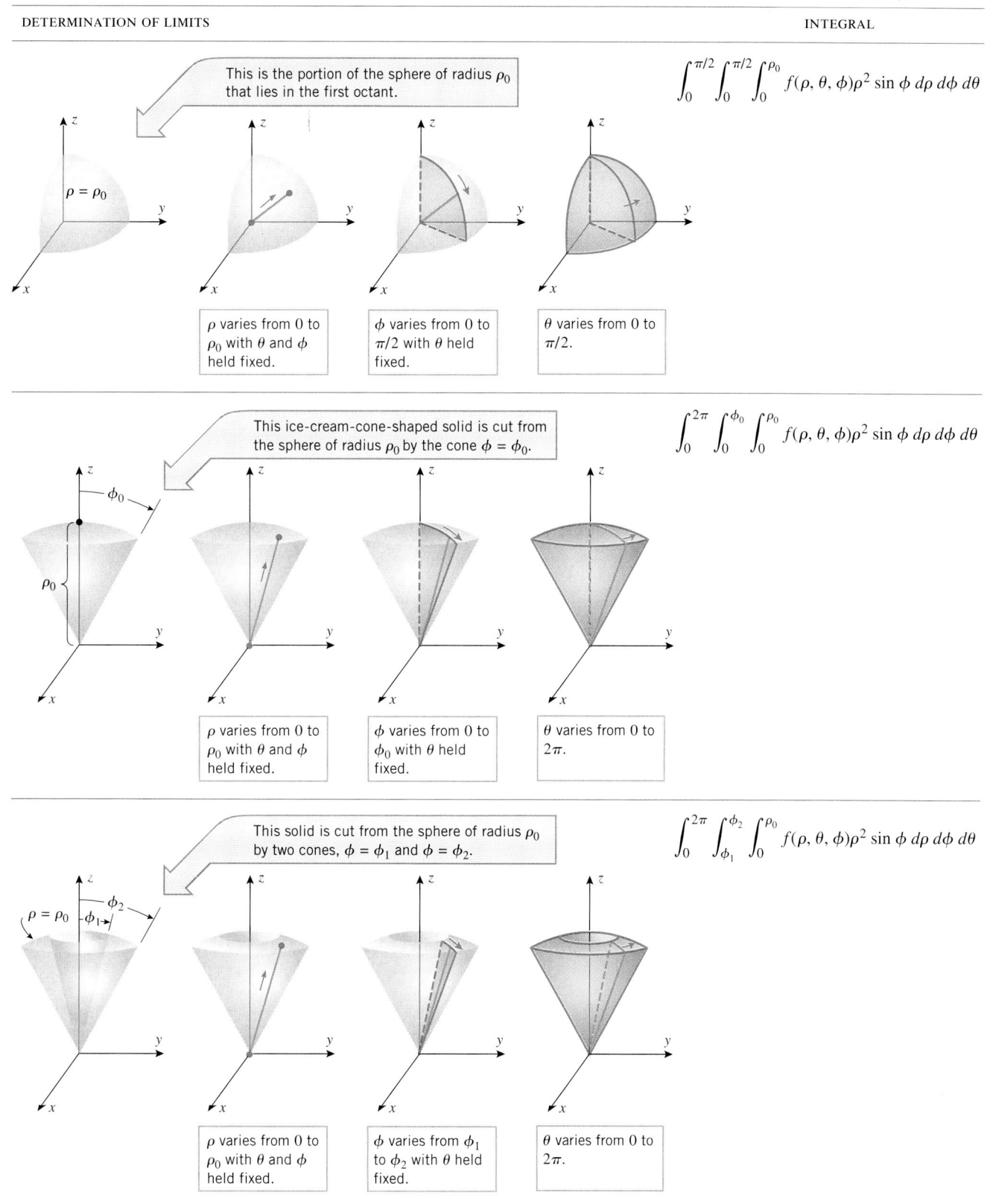

This is the portion of the sphere of radius ρ_0 that lies in the first octant.

$$\int_0^{\pi/2}\int_0^{\pi/2}\int_0^{\rho_0} f(\rho, \theta, \phi)\rho^2 \sin\phi \, d\rho \, d\phi \, d\theta$$

$\rho = \rho_0$

| ρ varies from 0 to ρ_0 with θ and ϕ held fixed. | ϕ varies from 0 to $\pi/2$ with θ held fixed. | θ varies from 0 to $\pi/2$. |

This ice-cream-cone-shaped solid is cut from the sphere of radius ρ_0 by the cone $\phi = \phi_0$.

$$\int_0^{2\pi}\int_0^{\phi_0}\int_0^{\rho_0} f(\rho, \theta, \phi)\rho^2 \sin\phi \, d\rho \, d\phi \, d\theta$$

ϕ_0

ρ_0

| ρ varies from 0 to ρ_0 with θ and ϕ held fixed. | ϕ varies from 0 to ϕ_0 with θ held fixed. | θ varies from 0 to 2π. |

This solid is cut from the sphere of radius ρ_0 by two cones, $\phi = \phi_1$ and $\phi = \phi_2$.

$$\int_0^{2\pi}\int_{\phi_1}^{\phi_2}\int_0^{\rho_0} f(\rho, \theta, \phi)\rho^2 \sin\phi \, d\rho \, d\phi \, d\theta$$

$\rho = \rho_0$ ϕ_1 ϕ_2

| ρ varies from 0 to ρ_0 with θ and ϕ held fixed. | ϕ varies from ϕ_1 to ϕ_2 with θ held fixed. | θ varies from 0 to 2π. |

Table 16.7.1 (*continued*)

DETERMINATION OF LIMITS	INTEGRAL

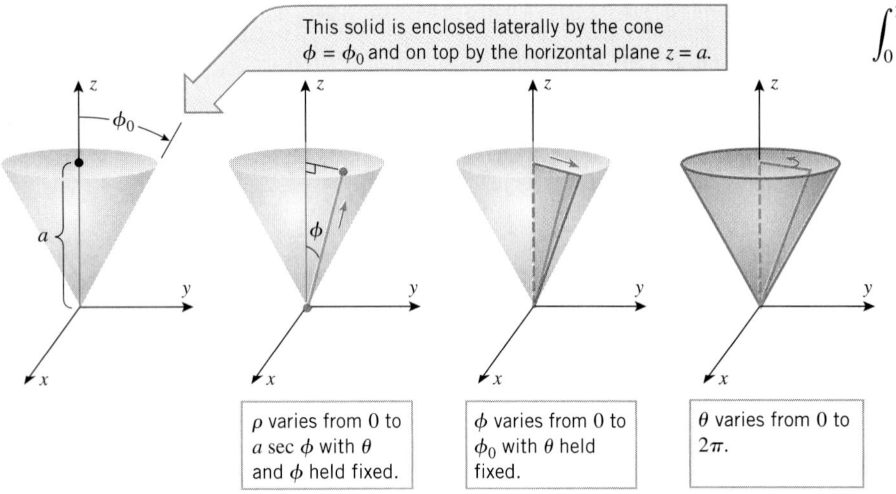

This solid is enclosed laterally by the cone $\phi = \phi_0$ and on top by the horizontal plane $z = a$.

$$\int_0^{2\pi} \int_0^{\phi_0} \int_0^{a \sec \phi} f(\rho, \theta, \phi)\rho^2 \sin \phi \, d\rho \, d\phi \, d\theta$$

ρ varies from 0 to $a \sec \phi$ with θ and ϕ held fixed.

ϕ varies from 0 to ϕ_0 with θ held fixed.

θ varies from 0 to 2π.

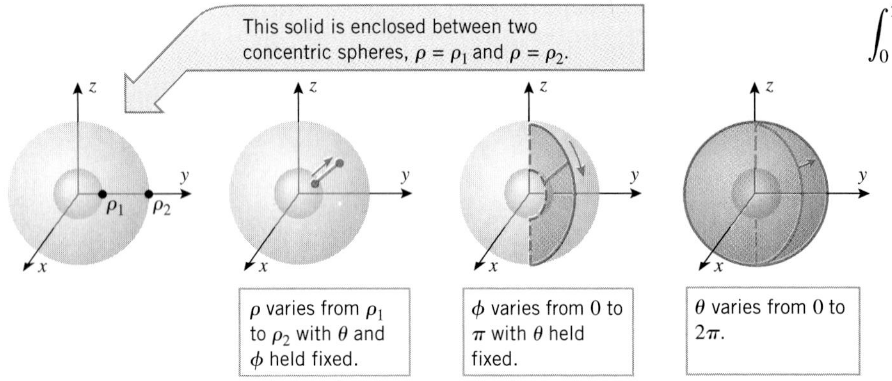

This solid is enclosed between two concentric spheres, $\rho = \rho_1$ and $\rho = \rho_2$.

$$\int_0^{2\pi} \int_0^{\pi} \int_{\rho_1}^{\rho_2} f(\rho, \theta, \phi)\rho^2 \sin \phi \, d\rho \, d\phi \, d\theta$$

ρ varies from ρ_1 to ρ_2 with θ and ϕ held fixed.

ϕ varies from 0 to π with θ held fixed.

θ varies from 0 to 2π.

By symmetry, the centroid $(\bar{x}, \bar{y}, \bar{z})$ is on the z-axis, so $\bar{x} = \bar{y} = 0$. From (15) of Section 16.6 and the volume calculated above,

$$\bar{z} = \frac{1}{V} \iiint\limits_G z \, dV = \frac{1}{V} \int_0^{2\pi} \int_0^{\pi/4} \int_0^4 (\rho \cos \phi)\rho^2 \sin \phi \, d\rho \, d\phi \, d\theta$$

$$= \frac{1}{V} \int_0^{2\pi} \int_0^{\pi/4} \left[\frac{\rho^4}{4} \cos \phi \sin \phi\right]_{\rho=0}^4 d\phi \, d\theta$$

$$= \frac{64}{V} \int_0^{2\pi} \int_0^{\pi/4} \sin \phi \cos \phi \, d\phi \, d\theta = \frac{64}{V} \int_0^{2\pi} \left[\frac{1}{2} \sin^2 \phi\right]_{\phi=0}^{\pi/4} d\theta$$

$$= \frac{16}{V} \int_0^{2\pi} d\theta = \frac{32\pi}{V} = \frac{3}{2(2 - \sqrt{2})}$$

With the help of a calculator, $\bar{z} \approx 2.56$ (to two decimal places), so the approximate location of the centroid in the xyz-coordinate system is $(0, 0, 2.56)$. ◄

CONVERTING TRIPLE INTEGRALS FROM RECTANGULAR TO SPHERICAL COORDINATES

Referring to Table 13.8.1, triple integrals can be converted from rectangular coordinates to spherical coordinates by making the substitution $x = \rho \sin \phi \cos \theta$, $y = \rho \sin \phi \sin \theta$, $z = \rho \cos \phi$. The two integrals are related by the equation

$$\iiint\limits_{G} f(x, y, z)\, dV = \iiint\limits_{\substack{\text{appropriate} \\ \text{limits}}} f(\rho \sin \phi \cos \theta,\, \rho \sin \phi \sin \theta,\, \rho \cos \phi) \rho^2 \sin \phi \, d\rho \, d\phi \, d\theta \quad (10)$$

Example 4

Use spherical coordinates to evaluate

$$\int_{-2}^{2} \int_{-\sqrt{4-x^2}}^{\sqrt{4-x^2}} \int_{0}^{\sqrt{4-x^2-y^2}} z^2 \sqrt{x^2 + y^2 + z^2}\, dz \, dy \, dx$$

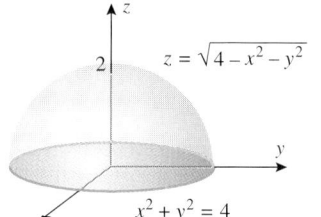

$z = \sqrt{4 - x^2 - y^2}$

$x^2 + y^2 = 4$

Figure 16.7.11

Solution. In problems like this, it is helpful to begin (when possible) with a sketch of the region G of integration. From the z-limits of integration, the upper surface of G is the hemisphere $z = \sqrt{4 - x^2 - y^2}$ and the lower surface is the xy-plane $z = 0$. From the x- and y-limits of integration, the projection of the solid G on the xy-plane is the region enclosed by the circle $x^2 + y^2 = 4$. From this information we obtain the sketch of G in Figure 16.7.11. Thus,

$$\int_{-2}^{2} \int_{-\sqrt{4-x^2}}^{\sqrt{4-x^2}} \int_{0}^{\sqrt{4-x^2-y^2}} z^2 \sqrt{x^2 + y^2 + z^2}\, dz \, dy \, dx$$

$$= \iiint\limits_{G} z^2 \sqrt{x^2 + y^2 + z^2}\, dV$$

$$= \int_{0}^{2\pi} \int_{0}^{\pi/2} \int_{0}^{2} \rho^5 \cos^2 \phi \sin \phi \, d\rho \, d\phi \, d\theta$$

$$= \int_{0}^{2\pi} \int_{0}^{\pi/2} \frac{32}{3} \cos^2 \phi \sin \phi \, d\phi \, d\theta$$

$$= \frac{32}{3} \int_{0}^{2\pi} \left[-\frac{1}{3} \cos^3 \phi \right]_{\phi=0}^{\pi/2} d\theta = \frac{32}{9} \int_{0}^{2\pi} d\theta = \frac{64}{9}\pi \quad \blacktriangleleft$$

EXERCISE SET 16.7   [C] CAS

In Exercises 1–4, evaluate the iterated integral.

1. $\displaystyle \int_{0}^{2\pi} \int_{0}^{1} \int_{0}^{\sqrt{1-r^2}} zr \, dz \, dr \, d\theta$

2. $\displaystyle \int_{0}^{\pi/2} \int_{0}^{\cos\theta} \int_{0}^{r^2} r \sin\theta \, dz \, dr \, d\theta$

3. $\displaystyle \int_{0}^{\pi/2} \int_{0}^{\pi/2} \int_{0}^{1} \rho^3 \sin\phi \cos\phi \, d\rho \, d\phi \, d\theta$

4. $\displaystyle \int_{0}^{2\pi} \int_{0}^{\pi/4} \int_{0}^{a \sec\phi} \rho^2 \sin\phi \, d\rho \, d\phi \, d\theta \quad (a > 0)$

In Exercises 5–8, use cylindrical coordinates to find the volume of the solid.

5. The solid enclosed by the paraboloid $z = x^2 + y^2$ and the plane $z = 9$.

6. The solid that is bounded above and below by the sphere $x^2 + y^2 + z^2 = 9$ and inside the cylinder $x^2 + y^2 = 4$.

7. The solid that is inside the surface $r^2 + z^2 = 20$ and below the surface $z = r^2$.

8. The solid enclosed between the cone $z = (hr)/a$ and the plane $z = h$.

In Exercises 9–12, use spherical coordinates to find the volume of the solid.

9. The solid bounded above by the sphere $\rho = 4$ and below by the cone $\phi = \pi/3$.

10. The solid within the cone $\phi = \pi/4$ and between the spheres $\rho = 1$ and $\rho = 2$.

11. The solid enclosed by the sphere $x^2 + y^2 + z^2 = 4a^2$ and the planes $z = 0$ and $z = a$.

12. The solid within the sphere $x^2 + y^2 + z^2 = 9$, outside the cone $z = \sqrt{x^2 + y^2}$, and above the xy-plane.

In Exercises 13–16, use cylindrical or spherical coordinates to evaluate the integral.

13. $\displaystyle\int_0^a \int_0^{\sqrt{a^2-x^2}} \int_0^{a^2-x^2-y^2} x^2\, dz\, dy\, dx \quad (a>0)$

14. $\displaystyle\int_{-1}^1 \int_0^{\sqrt{1-x^2}} \int_0^{\sqrt{1-x^2-y^2}} e^{-(x^2+y^2+z^2)^{3/2}}\, dz\, dy\, dx$

15. $\displaystyle\int_0^2 \int_0^{\sqrt{4-y^2}} \int_{\sqrt{x^2+y^2}}^{\sqrt{8-x^2-y^2}} z^2\, dz\, dx\, dy$

16. $\displaystyle\int_{-3}^3 \int_{-\sqrt{9-y^2}}^{\sqrt{9-y^2}} \int_{-\sqrt{9-x^2-y^2}}^{\sqrt{9-x^2-y^2}} \sqrt{x^2+y^2+z^2}\, dz\, dx\, dy$

C **17.** Use a CAS to check your answers to the problems you solved in Exercises 1–16.

C **18.** Use a CAS to evaluate

$$\int_0^{\pi/2} \int_0^{\pi/4} \int_0^{\cos\theta} \rho^{17} \cos\phi \cos^{19}\theta\, d\rho\, d\phi\, d\theta$$

19. Find the volume of $x^2+y^2+z^2=a^2$ using
 (a) cylindrical coordinates
 (b) spherical coordinates.

20. Let G be the solid in the first octant bounded by the sphere $x^2+y^2+z^2=4$ and the coordinate planes. Evaluate

$$\iiint_G xyz\, dV$$

 (a) using rectangular coordinates
 (b) using cylindrical coordinates
 (c) using spherical coordinates.

In Exercises 21 and 22, use cylindrical coordinates.

21. Find the mass of the solid with density $\delta(x,y,z)=3-z$ that is bounded by the cone $z=\sqrt{x^2+y^2}$ and the plane $z=3$.

22. Find the mass of a right circular cylinder of radius a and height h if the density is proportional to the distance from the base. (Let k be the constant of proportionality.)

In Exercises 23 and 24, use spherical coordinates.

23. Find the mass of a spherical solid of radius a if the density is proportional to the distance from the center. (Let k be the constant of proportionality.)

24. Find the mass of the solid enclosed between the spheres $x^2+y^2+z^2=1$ and $x^2+y^2+z^2=4$ if the density is $\delta(x,y,z)=(x^2+y^2+z^2)^{-1/2}$.

In Exercises 25 and 26, use cylindrical coordinates to find the centroid of the solid.

25. The solid that is bounded above by the sphere

$$x^2+y^2+z^2=2$$

and below by the paraboloid $z=x^2+y^2$.

26. The solid that is bounded by the cone $z=\sqrt{x^2+y^2}$ and the plane $z=2$.

In Exercises 27 and 28, use spherical coordinates to find the centroid of the solid.

27. The solid in the first octant bounded by the coordinate planes and the sphere $x^2+y^2+z^2=a^2$.

28. The solid bounded above by the sphere $\rho=4$ and below by the cone $\phi=\pi/3$.

In Exercises 29 and 30, use the Wallis formulas in Exercises 64 and 66 of Section 9.3.

29. Find the centroid of the solid bounded above by the paraboloid $z=x^2+y^2$, below by the plane $z=0$, and laterally by the cylinder $(x-1)^2+y^2=1$.

30. Find the mass of the solid in the first octant bounded above by the paraboloid $z=4-x^2-y^2$, below by the plane $z=0$, and laterally by the cylinder $x^2+y^2=2x$ and the plane $y=0$, assuming the density to be $\delta(x,y,z)=z$.

In Exercises 31–36, solve the problem using either cylindrical or spherical coordinates (whichever seems appropriate).

31. Find the volume of the solid in the first octant bounded by the sphere $\rho=2$, the coordinate planes, and the cones $\phi=\pi/6$ and $\phi=\pi/3$.

32. Find the mass of the solid that is enclosed by the sphere $x^2+y^2+z^2=1$ and lies within the cone $z=\sqrt{x^2+y^2}$ if the density is $\delta(x,y,z)=\sqrt{x^2+y^2+z^2}$.

33. Find the center of gravity of the solid bounded by the paraboloid $z=1-x^2-y^2$ and the xy-plane, assuming the density to be $\delta(x,y,z)=x^2+y^2+z^2$.

34. Find the center of gravity of the solid that is bounded by the cylinder $x^2+y^2=1$, the cone $z=\sqrt{x^2+y^2}$, and the xy-plane if the density is $\delta(x,y,z)=z$.

35. Find the center of gravity of the solid hemisphere bounded by $z=\sqrt{a^2-x^2-y^2}$ and $z=0$ if the density is proportional to the distance from the origin.

36. Find the centroid of the solid that is enclosed by the hemispheres $y=\sqrt{9-x^2-z^2}$, $y=\sqrt{4-x^2-z^2}$, and the plane $y=0$.

37. Suppose that the density at a point in a gaseous spherical star is modeled by the formula

$$\delta=\delta_0 e^{-(\rho/R)^3}$$

where δ_0 is a positive constant, R is the radius of the star, and ρ is the distance from the point to the star's center. Find the mass of the star.

38. In this exercise we will obtain a formula for the volume of the spherical wedge in Figure 16.7.7.

(a) Use a triple integral in cylindrical coordinates to show that the volume of the solid bounded above by a sphere $\rho = \rho_0$, below by a cone $\phi = \phi_0$, and on the sides by $\theta = \theta_1$ and $\theta = \theta_2$ ($\theta_1 < \theta_2$) is

$$V = \tfrac{1}{3}\rho_0^3(1 - \cos\phi_0)(\theta_2 - \theta_1)$$

[*Hint:* In cylindrical coordinates, the sphere has the equation $r^2 + z^2 = \rho_0^2$ and the cone has the equation $z = r\cot\phi_0$. For simplicity, consider only the case $0 < \phi_0 < \pi/2$.]

(b) Subtract appropriate volumes and use the result in part (a) to deduce that the volume ΔV of the spherical wedge is

$$\Delta V = \frac{\rho_2^3 - \rho_1^3}{3}(\cos\phi_1 - \cos\phi_2)(\theta_2 - \theta_1)$$

(c) Apply the Mean-Value Theorem to the functions $\cos\phi$ and ρ^3 to deduce that the formula in part (b) can be written as

$$\Delta V = \rho^{*2}\sin\phi^* \, \Delta\rho \, \Delta\phi \, \Delta\theta$$

where ρ^* is between ρ_1 and ρ_2, ϕ^* is between ϕ_1 and ϕ_2, and $\Delta\rho = \rho_2 - \rho_1$, $\Delta\phi = \phi_2 - \phi_1$, $\Delta\theta = \theta_2 - \theta_1$.

The tendency of a solid to resist a change in rotational motion about an axis is measured by its **moment of inertia** about that axis. If the solid occupies a region G in an xyz-coordinate system, and if its density function $\delta(x, y, z)$ is continuous on G, then the moments of inertia about the x-axis, the y-axis, and the z-axis are denoted by I_x, I_y, and I_z, respectively, and are defined by

$$I_x = \iiint\limits_G (y^2 + z^2)\,\delta(x, y, z)\,dV$$

$$I_y = \iiint\limits_G (x^2 + z^2)\,\delta(x, y, z)\,dV$$

$$I_z = \iiint\limits_G (x^2 + y^2)\,\delta(x, y, z)\,dV$$

(compare with the discussion preceding Exercises 40 and 41 of Section 16.6). In Exercises 39–42, find the indicated moment of inertia of the solid, assuming that it has constant density δ.

39. I_z for the solid cylinder $x^2 + y^2 \leq a^2$, $0 \leq z \leq h$.

40. I_y for the solid cylinder $x^2 + y^2 \leq a^2$, $0 \leq z \leq h$.

41. I_z for the hollow cylinder $a_1^2 \leq x^2 + y^2 \leq a_2^2$, $0 \leq z \leq h$.

42. I_z for the solid sphere $x^2 + y^2 + z^2 \leq a^2$.

16.8 CHANGE OF VARIABLES IN MULTIPLE INTEGRALS; JACOBIANS

In this section we will discuss a general method for evaluating double and triple integrals by substitution. Most of the results in this section are very difficult to prove, so our approach will be informal and motivational. Our goal is to provide a geometric understanding of the basic principles and an exposure to computational techniques.

CHANGE OF VARIABLE IN A SINGLE INTEGRAL

To motivate techniques for evaluating double and triple integrals by substitution, it will be helpful to consider the effect of a substitution $x = g(u)$ on a single integral over an interval $[a, b]$. If g is differentiable and either increasing or decreasing, then g is one-to-one and

$$\int_a^b f(x)\,dx = \int_{g^{-1}(a)}^{g^{-1}(b)} f(g(u))g'(u)\,du$$

In this relationship $f(x)$ and dx are expressed in terms of u, and the u-limits of integration result from solving the equations

$$a = g(u) \quad \text{and} \quad b = g(u)$$

In the case where g is decreasing we have $g^{-1}(b) < g^{-1}(a)$, which is contrary to our usual convention of writing definite integrals with the larger limit of integration at the top. We can remedy this by reversing the limits of integration and writing

$$\int_a^b f(x)\,dx = -\int_{g^{-1}(b)}^{g^{-1}(a)} f(g(u))g'(u)\,du = \int_{g^{-1}(b)}^{g^{-1}(a)} f(g(u))|g'(u)|\,du$$

where the absolute value results from the fact that $g'(u)$ is negative. Thus, regardless of whether g is increasing or decreasing we can write

$$\int_a^b f(x)\,dx = \int_\alpha^\beta f(g(u))|g'(u)|\,du \tag{1}$$

where α and β are the u-limits of integration and $\alpha < \beta$.

The expression $g'(u)$ that appears in (1) is called the *Jacobian* of the change of variable $x = g(u)$ in honor of C. G. J. Jacobi,[*] who made the first serious study of change of variables in multiple integrals in the mid 1800s. Formula (1) reveals three effects of the change of variable $x = g(u)$:

- The new integrand becomes $f(g(u))$ times the absolute value of the Jacobian.
- dx becomes du.
- The x-interval of integration is transformed into a u-interval of integration.

Our goal in this section is to show that analogous results hold for changing variables in double and triple integrals.

TRANSFORMATIONS OF THE PLANE

In earlier sections we considered parametric equations of three kinds:

$x = x(t), \quad y = y(t)$ | A curve in the plane |

$x = x(t), \quad y = y(t), \quad z = z(t)$ | A curve in 3-space |

$x = x(u, v), \quad y = y(u, v), \quad z = z(u, v)$ | A surface in 3-space |

Now, we will consider parametric equations of the form

$$x = x(u, v), \quad y = y(u, v) \tag{2}$$

Parametric equations of this type associate points in the xy-plane with points in the uv-plane. These equations can be written in vector form as

$$\mathbf{r} = \mathbf{r}(u, v) = x(u, v)\mathbf{i} + y(u, v)\mathbf{j}$$

where $\mathbf{r} = x\mathbf{i} + y\mathbf{j}$ is a position vector in the xy-plane and $\mathbf{r}(u, v)$ is a vector-valued function of the variables u and v.

It will also be useful in this section to think of the parametric equations in (2) in terms of inputs and outputs. If we think of the pair of numbers (u, v) as an input, then the two

---

[*]CARL GUSTAV JACOB JACOBI (1804–1851). German mathematician, Jacobi, the son of a banker, grew up in a background of wealth and culture and showed brilliance in mathematics early. He resisted studying mathematics by rote, preferring instead to learn general principles from the works of the masters, Euler and Lagrange. He entered the University of Berlin at age 16 as a student of mathematics and classical studies. However, he soon realized that he could not do both and turned fully to mathematics with a blazing intensity that he would maintain throughout his life. He received his Ph.D. in 1825 and was able to secure a position as a lecturer at the University of Berlin by giving up Judaism and becoming a Christian. However, his promotion opportunities remained limited and he moved on to the University of Königsberg. Jacobi was born to teach—he had a dynamic personality and delivered his lectures with a clarity and enthusiasm that frequently left his audience spellbound. In spite of extensive teaching commitments, he was able to publish volumes of revolutionary mathematical research that eventually made him the leading European mathematician after Gauss. His main body of research was in the area of elliptic functions, a branch of mathematics with important applications in astronomy and physics as well as in other fields of mathematics. Because of his family wealth, Jacobi was not dependent on his teaching salary in his early years. However, his comfortable world eventually collapsed. In 1840 his family went bankrupt and he was personally wiped out financially. In 1842 he had a nervous breakdown from overwork. In 1843 he became seriously ill with diabetes and moved to Berlin with the help of a government grant to defray his medical expenses. In 1848 he made a stupid political remark that caused the government to withdraw the grant, eventually resulting in the loss of his home. His health continued to decline and in 1851 he finally succumbed to successive bouts of influenza and smallpox. In spite of all his problems, Jacobi was a tireless worker to the end. When a friend expressed concern about the effect of the hard work on his health, Jacobi replied, "Certainly, I have sometimes endangered my health by overwork, but what of it? Only cabbages have no nerves, no worries. And what do they get out of their perfect well-being?"

equations, in combination, produce a unique output (x, y), and hence define a function T that associates points in the xy-plane with points in the uv-plane. This function is described by the formula

$$T(u, v) = (x(u, v), y(u, v))$$

We call T a ***transformation*** from the uv-plane to the xy-plane and (x, y) the ***image*** of (u, v) under the transformation T. We also say that T ***maps*** (u, v) into (x, y). The set R of all images in the xy-plane of a set S in the uv-plane is called the ***image of S under T***. If distinct points in the uv-plane have distinct images in the xy-plane, then T is said to be ***one-to-one***. In this case the equations in (2) define u and v as functions of x and y, say

$$u = u(x, y), \quad v = v(x, y)$$

These equations, which can often be obtained by solving (2) for u and v in terms of x and y, define a transformation from the xy-plane to the uv-plane that maps the image of (u, v) under T back into (u, v). This transformation is denoted by T^{-1} and is called the ***inverse of T*** (Figure 16.8.1).

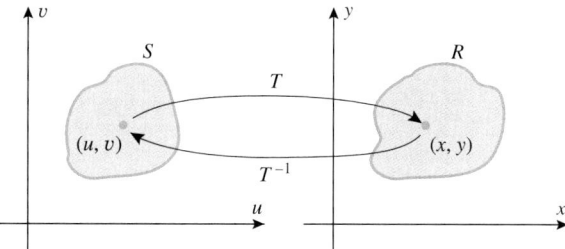

Figure 16.8.1

One way to visualize the geometric effect of a transformation T is to determine the images in the xy-plane of the vertical and horizontal lines in the uv-plane. Sets of points in the xy-plane that are images of horizontal lines (v constant) are called ***constant v-curves***, and sets of points that are images of vertical lines (u constant) are called ***constant u-curves*** (Figure 16.8.2).

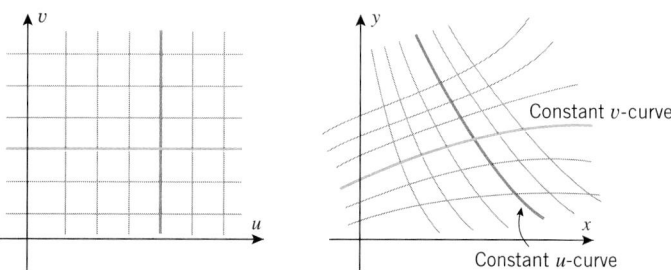

Figure 16.8.2

Example 1

Let T be the transformation from the uv-plane to the xy-plane defined by the equations

$$x = \tfrac{1}{4}(u + v), \quad y = \tfrac{1}{2}(u - v) \tag{3}$$

(a)   Find $T(1, 3)$.

(b)   Sketch the constant v-curves corresponding to $v = -2, -1, 0, 1, 2$.

(c)   Sketch the constant u-curves corresponding to $u = -2, -1, 0, 1, 2$.

(d)   Sketch the image under T of the square region in the uv-plane bounded by the lines $u = -2, u = 2, v = -2$, and $v = 2$.

Solution (a). Substituting $u = 1$ and $v = 3$ in (3) yields $T(1, 3) = (1, -1)$.

Solutions (b and c). In these parts it will be convenient to express the transformation equations with u and v as functions of x and y. We leave it for you to show that

$$u = 2x + y, \quad v = 2x - y$$

Thus, the constant v-curves corresponding to $v = -2, -1, 0, 1$, and 2 are

$$2x - y = -2, \quad 2x - y = -1, \quad 2x - y = 0, \quad 2x - y = 1, \quad 2x - y = 2$$

and the constant u-curves corresponding to $u = -2, -1, 0, 1$, and 2 are

$$2x + y = -2, \quad 2x + y = -1, \quad 2x + y = 0, \quad 2x + y = 1, \quad 2x + y = 2$$

In Figure 16.8.3 the constant v-curves are shown in green and the constant u-curves in red.

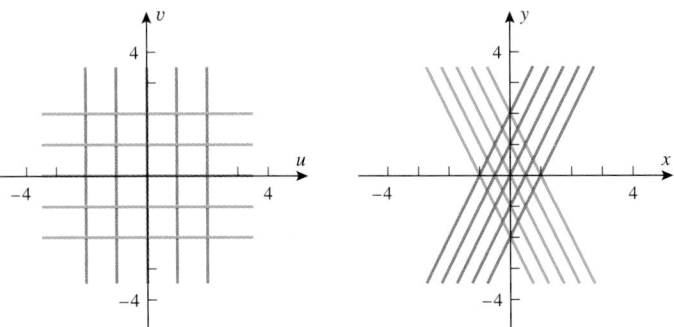

Figure 16.8.3

Solution (d). The image of a region can often be found by finding the image of its boundary. In this case the images of the boundary lines $u = -2$, $u = 2$, $v = -2$, and $v = 2$ enclose the diamond-shaped region in the xy-plane shown in Figure 16.8.4.   ◀

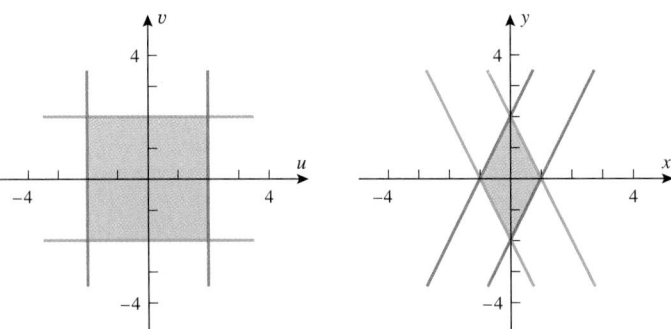

Figure 16.8.4

JACOBIANS IN TWO VARIABLES

To derive the change-of-variable formula for double integrals, we will need to understand the relationship between the area of a *small* rectangular region in the uv-plane and the area of its image in the xy-plane under a transformation T given by the equations

$$x = x(u, v), \quad y = y(u, v)$$

For this purpose, suppose that Δu and Δv are positive, and consider a rectangular region S in the uv-plane enclosed by the lines

$$u = u_0, \quad u = u_0 + \Delta u, \quad v = v_0, \quad v = v_0 + \Delta v$$

If the functions $x(u, v)$ and $y(u, v)$ are continuous, and if Δu and Δv are not too large, then the image of S in the xy-plane will be a region R that looks like a slightly distorted parallelogram (Figure 16.8.5). The sides of R are the constant u-curves and v-curves that correspond to the sides of S.

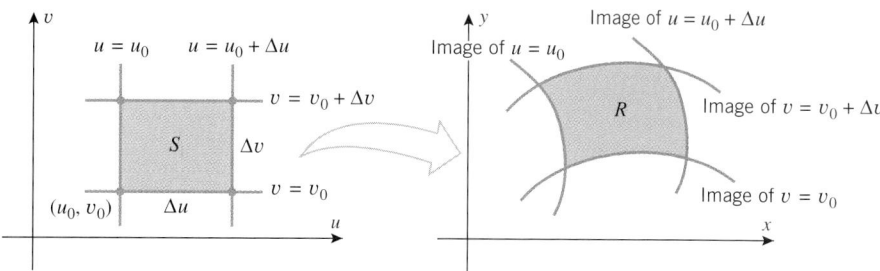

Figure 16.8.5

If we let

$$\mathbf{r} = \mathbf{r}(u, v) = x(u, v)\mathbf{i} + y(u, v)\mathbf{j}$$

be the position vector of the point in the xy-plane that corresponds to the point (u, v) in the uv-plane, then the constant v-curve corresponding to $v = v_0$ and the constant u-curve corresponding to $u = u_0$ can be represented in vector form as

$$\mathbf{r}(u, v_0) = x(u, v_0)\mathbf{i} + y(u, v_0)\mathbf{j} \qquad \text{Constant } v\text{-curve}$$

$$\mathbf{r}(u_0, v) = x(u_0, v)\mathbf{i} + y(u_0, v)\mathbf{j} \qquad \text{Constant } u\text{-curve}$$

Since we are assuming Δu and Δv to be small, the region R can be approximated by a parallelogram determined by the "secant vectors"

$$\mathbf{a} = \mathbf{r}(u_0 + \Delta u, v_0) - \mathbf{r}(u_0, v_0) \tag{4}$$

$$\mathbf{b} = \mathbf{r}(u_0, v_0 + \Delta v) - \mathbf{r}(u_0, v_0) \tag{5}$$

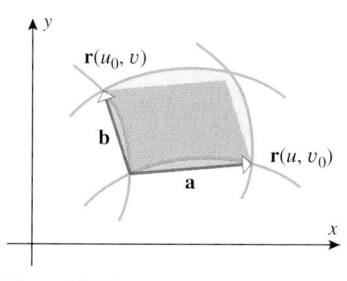

Figure 16.8.6

shown in Figure 16.8.6. A more useful approximation of R can be obtained by using Formulas (5) and (6) of Section 16.4 to approximate these secant vectors by tangent vectors as follows:

$$\mathbf{a} = \frac{\mathbf{r}(u_0 + \Delta u, v_0) - \mathbf{r}(u_0, v_0)}{\Delta u} \Delta u$$

$$\approx \frac{\partial \mathbf{r}}{\partial u} \Delta u = \left(\frac{\partial x}{\partial u}\mathbf{i} + \frac{\partial y}{\partial u}\mathbf{j} \right) \Delta u$$

$$\mathbf{b} = \frac{\mathbf{r}(u_0, v_0 + \Delta v) - \mathbf{r}(u_0, v_0)}{\Delta v} \Delta v$$

$$\approx \frac{\partial \mathbf{r}}{\partial v} \Delta v = \left(\frac{\partial x}{\partial v}\mathbf{i} + \frac{\partial y}{\partial v}\mathbf{j} \right) \Delta v$$

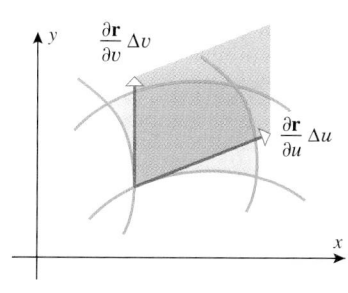

Figure 16.8.7

where the partial derivatives are evaluated at (u_0, v_0) (Figure 16.8.7). Hence, it follows that the area of the region R, which we will denote by ΔA, can be approximated by the area of the parallelogram determined by these vectors. Thus, from Formula (8) of Section 13.4 we have

$$\Delta A \approx \left\| \frac{\partial \mathbf{r}}{\partial u} \Delta u \times \frac{\partial \mathbf{r}}{\partial v} \Delta v \right\| = \left\| \frac{\partial \mathbf{r}}{\partial u} \times \frac{\partial \mathbf{r}}{\partial v} \right\| \Delta u \, \Delta v \tag{6}$$

where the derivatives are evaluated at (u_0, v_0). Computing the cross product, we obtain

$$\frac{\partial \mathbf{r}}{\partial u} \times \frac{\partial \mathbf{r}}{\partial v} = \begin{vmatrix} \mathbf{i} & \mathbf{j} & \mathbf{k} \\ \dfrac{\partial x}{\partial u} & \dfrac{\partial y}{\partial u} & 0 \\ \dfrac{\partial x}{\partial v} & \dfrac{\partial y}{\partial v} & 0 \end{vmatrix} = \begin{vmatrix} \dfrac{\partial x}{\partial u} & \dfrac{\partial y}{\partial u} \\ \dfrac{\partial x}{\partial v} & \dfrac{\partial y}{\partial v} \end{vmatrix} \mathbf{k} = \begin{vmatrix} \dfrac{\partial x}{\partial u} & \dfrac{\partial x}{\partial v} \\ \dfrac{\partial y}{\partial u} & \dfrac{\partial y}{\partial v} \end{vmatrix} \mathbf{k} \tag{7}$$

The determinant in (7) is sufficiently important that it has its own terminology and notation.

16.8.1 DEFINITION. If T is the transformation from the uv-plane to the xy-plane defined by the equations $x = x(u, v)$, $y = y(u, v)$, then the **_Jacobian of T_** is denoted by $J(u, v)$ or by $\partial(x, y)/\partial(u, v)$ and is defined by

$$J(u, v) = \frac{\partial(x, y)}{\partial(u, v)} = \begin{vmatrix} \dfrac{\partial x}{\partial u} & \dfrac{\partial x}{\partial v} \\[2mm] \dfrac{\partial y}{\partial u} & \dfrac{\partial y}{\partial v} \end{vmatrix} = \frac{\partial x}{\partial u}\frac{\partial y}{\partial v} - \frac{\partial y}{\partial u}\frac{\partial x}{\partial v}$$

Using the notation in this definition, it follows from (6) and (7) that

$$\Delta A \approx \left\| \frac{\partial(x, y)}{\partial(u, v)}\mathbf{k} \right\| \Delta u \, \Delta v$$

or, since $\mathbf{k}$ is a unit vector,

$$\Delta A \approx \left| \frac{\partial(x, y)}{\partial(u, v)} \right| \Delta u \, \Delta v \tag{8}$$

At the point (u_0, v_0) this important formula relates the areas of the regions R and S in Figure 16.8.5: it tells us that *for small values of Δu and Δv, the area of R is approximately the absolute value of the Jacobian times the area of S.* Moreover, it is proved in advanced calculus courses that the relative error in the approximation approaches zero as $\Delta u \to 0$ and $\Delta v \to 0$.

CHANGE OF VARIABLES IN DOUBLE INTEGRALS

Our next objective is to provide a geometric motivation for the following result.

16.8.2 CHANGE-OF-VARIABLE FORMULA FOR DOUBLE INTEGRALS. If the transformation $x = x(u, v)$, $y = y(u, v)$ maps the region S in the uv-plane into the region R in the xy-plane, and if the Jacobian $\partial(x, y)/\partial(u, v)$ is nonzero and does not change sign on S, then with appropriate restrictions on the transformation and the regions it follows that

$$\iint\limits_{R} f(x, y)\, dA_{xy} = \iint\limits_{S} f(x(u, v), y(u, v)) \left| \frac{\partial(x, y)}{\partial(u, v)} \right| dA_{uv} \tag{9}$$

where we have attached subscripts to the dA's to help identify the associated variables.

REMARK. A precise statement of conditions under which Formula (9) holds would take us beyond the scope of this course. Suffice it to say that the formula holds if T is a one-to-one transformation, $f(x, y)$ is continuous on R, the partial derivatives of $x(u, v)$ and $y(u, v)$ exist and are continuous on S, and the regions R and S are not too complicated.

To motivate Formula (9), we proceed as follows:

- Subdivide the region S in the uv-plane into pieces by lines parallel to the coordinate axes, and exclude from consideration any pieces that contain points outside of S. This leaves only rectangular regions that are subsets of S. Assume that there are n such regions and denote the kth such region by S_k. Assume that S_k has dimensions Δu_k by Δv_k and, as shown in Figure 16.8.8a, let (u_k^*, v_k^*) be its "lower left corner."

- As shown in Figure 16.8.8b, the transformation T defined by the equations $x = x(u, v)$, $y = y(u, v)$ maps S_k into a curvilinear parallelogram R_k in the xy-plane and maps the point (u_k^*, v_k^*) into the point $(x_k^*, y_k^*) = (x(u_k^*, v_k^*), y(u_k^*, v_k^*))$ in R_k. Denote the area of R_k by ΔA_k.

- In rectangular coordinates the double integral of $f(x, y)$ over a region R is defined as a limit of Riemann sums in which R is subdivided into *rectangular* subregions. It is proved in advanced calculus courses that under appropriate conditions subdivisions into *curvilinear* parallelograms can be used instead. Accepting this to be so, we can approximate the double integral of $f(x, y)$ over R as

$$\iint_R f(x, y) \, dA_{xy} \approx \sum_{k=1}^{n} f(x_k^*, y_k^*) \, \Delta A_k$$

$$\approx \sum_{k=1}^{n} f(x(u_k^*, v_k^*), y(u_k^*, v_k^*)) \left| \frac{\partial(x, y)}{\partial(u, v)} \right| \Delta u_k \, \Delta v_k$$

where the Jacobian is evaluated at (u_k^*, v_k^*). But the last expression is a Riemann sum for the integral

$$\iint_S f(x(u, v), y(u, v)) \left| \frac{\partial(x, y)}{\partial(u, v)} \right| dA_{uv}$$

so Formula (9) follows if we assume that the errors in the approximations approach zero as $n \to +\infty$.

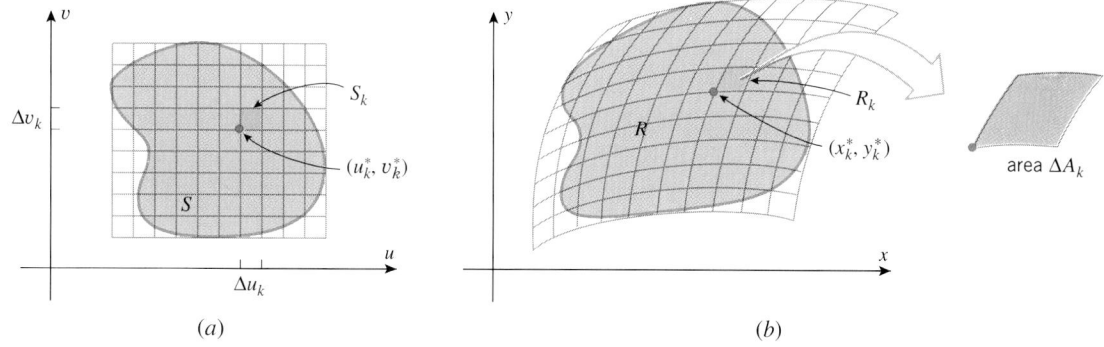

Figure 16.8.8

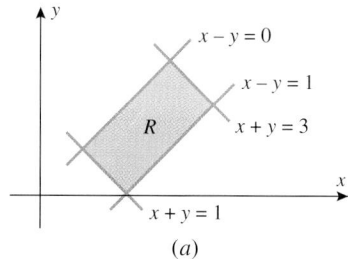

(a)

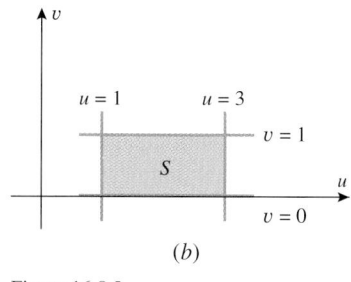

(b)

Figure 16.8.9

Example 2

Evaluate

$$\iint_R \frac{x - y}{x + y} \, dA$$

where R is the region enclosed by the lines $x - y = 0$, $x - y = 1$, $x + y = 1$, and $x + y = 3$ (Figure 16.8.9a).

Solution. This integral would be tedious to evaluate directly because the region R is oriented in such a way that we would have to subdivide it and integrate over each part separately. However, the occurrence of the expressions $x - y$ and $x + y$ in the equations of the boundary suggests that the transformation

$$u = x + y, \quad v = x - y \tag{10}$$

would be helpful, since with this transformation the boundary lines

$$x + y = 1, \quad x + y = 3, \quad x - y = 0, \quad x - y = 1$$

are constant u-curves and constant v-curves corresponding to the lines

$$u = 1, \quad u = 3, \quad v = 0, \quad v = 1$$

in the uv-plane. These lines enclose the rectangular region S shown in Figure 16.8.9b. To find the Jacobian $\partial(x, y)/\partial(u, v)$ of this transformation, we first solve (10) for x and y in terms of u and v. This yields

$$x = \tfrac{1}{2}(u + v), \qquad y = \tfrac{1}{2}(u - v)$$

from which we obtain

$$\frac{\partial(x, y)}{\partial(u, v)} = \begin{vmatrix} \dfrac{\partial x}{\partial u} & \dfrac{\partial x}{\partial v} \\[2mm] \dfrac{\partial y}{\partial u} & \dfrac{\partial y}{\partial v} \end{vmatrix} = \begin{vmatrix} \tfrac{1}{2} & \tfrac{1}{2} \\[1mm] \tfrac{1}{2} & -\tfrac{1}{2} \end{vmatrix} = -\tfrac{1}{4} - \tfrac{1}{4} = -\tfrac{1}{2}$$

Thus, from Formula (9), but with the notation dA rather than dA_{xy},

$$\iint\limits_{R} \frac{x - y}{x + y}\, dA = \iint\limits_{S} \frac{v}{u} \left| \frac{\partial(x, y)}{\partial(u, v)} \right| dA_{uv}$$

$$= \iint\limits_{S} \frac{v}{u} \left| -\frac{1}{2} \right| dA_{uv} = \frac{1}{2} \int_{0}^{1} \int_{1}^{3} \frac{v}{u}\, du\, dv$$

$$= \frac{1}{2} \int_{0}^{1} v \ln |u| \Bigg]_{u=1}^{3} dv$$

$$= \frac{1}{2} \ln 3 \int_{0}^{1} v\, dv = \frac{1}{4} \ln 3 \qquad \blacktriangleleft$$

REMARK.   In retrospect, the underlying idea illustrated in this example is to find a one-to-one transformation that maps a rectangle S in the uv-plane into the region R of integration, and then use that transformation as a substitution in the integral to produce an equivalent integral over S.

Example 3

Evaluate

$$\iint\limits_{R} e^{xy}\, dA$$

where R is the region enclosed by the lines $y = \tfrac{1}{2}x$ and $y = x$ and the hyperbolas $y = 1/x$ and $y = 2/x$ (Figure 16.8.10a).

Solution. As in the last example, we look for a transformation in which the boundary curves in the xy-plane become constant v-curves and constant u-curves. For this purpose we rewrite the four boundary curves as

$$\frac{y}{x} = \frac{1}{2}, \quad \frac{y}{x} = 1, \quad xy = 1, \quad xy = 2$$

which suggests the transformation

$$u = \frac{y}{x}, \quad v = xy \tag{11}$$

With this transformation the boundary curves in the xy-plane are constant u-curves and constant v-curves corresponding to the lines

$$u = \tfrac{1}{2}, \quad u = 1, \quad v = 1, \quad v = 2$$

in the uv-plane. These lines enclose the region S shown in Figure 16.8.10b. To find the Jacobian $\partial(x, y)/\partial(u, v)$ of this transformation, we first solve (11) for x and y in terms of u and v. This yields

$$x = \sqrt{v/u}, \quad y = \sqrt{uv}$$

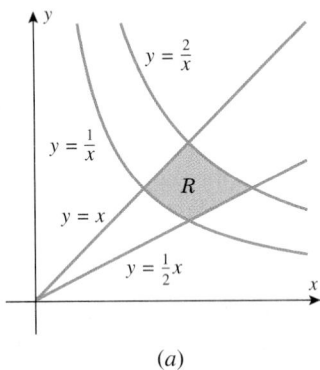

(a)

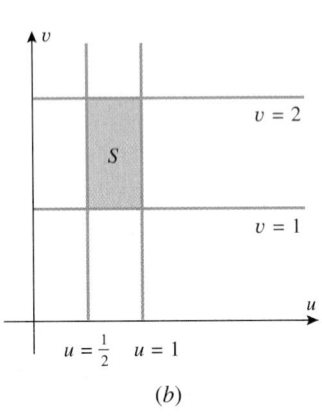

(b)

Figure 16.8.10

from which we obtain

$$\frac{\partial(x, y)}{\partial(u, v)} = \begin{vmatrix} \dfrac{\partial x}{\partial u} & \dfrac{\partial x}{\partial v} \\[2mm] \dfrac{\partial y}{\partial u} & \dfrac{\partial y}{\partial v} \end{vmatrix} = \begin{vmatrix} -\dfrac{1}{2u}\sqrt{\dfrac{v}{u}} & \dfrac{1}{2\sqrt{uv}} \\[3mm] \dfrac{1}{2}\sqrt{\dfrac{v}{u}} & \dfrac{1}{2}\sqrt{\dfrac{u}{v}} \end{vmatrix} = -\frac{1}{4u} - \frac{1}{4u} = -\frac{1}{2u}$$

Thus, from Formula (9), but with the notation dA rather than dA_{xy},

$$\iint\limits_{R} e^{xy}\,dA = \iint\limits_{S} e^{v}\left|-\frac{1}{2u}\right|\,dA_{uv} = \frac{1}{2}\iint\limits_{S}\frac{1}{u}e^{v}\,dA_{uv}$$

$$= \frac{1}{2}\int_{1}^{2}\int_{1/2}^{1}\frac{1}{u}e^{v}\,du\,dv = \frac{1}{2}\int_{1}^{2}e^{v}\ln|u|\Big]_{u=1/2}^{1}\,dv$$

$$= \frac{1}{2}\ln 2\int_{1}^{2}e^{v}\,dv = \frac{1}{2}(e^{2}-e)\ln 2 \qquad \blacktriangleleft$$

CHANGE OF VARIABLES IN TRIPLE INTEGRALS

Equations of the form

$$x = x(u, v, w), \quad y = y(u, v, w), \quad z = z(u, v, w) \tag{12}$$

define a **transformation** T from uvw-space to xyz-space. Just as a transformation $x = x(u, v)$, $y = y(u, v)$ in two variables maps small rectangles in the uv-plane into curvilinear parallelograms in the xy-plane, so (12) maps small rectangular parallelepipeds in uvw-space into curvilinear parallelepipeds in xyz-space (Figure 16.8.11). The definition of the Jacobian of (12) is similar to Definition 16.8.1.

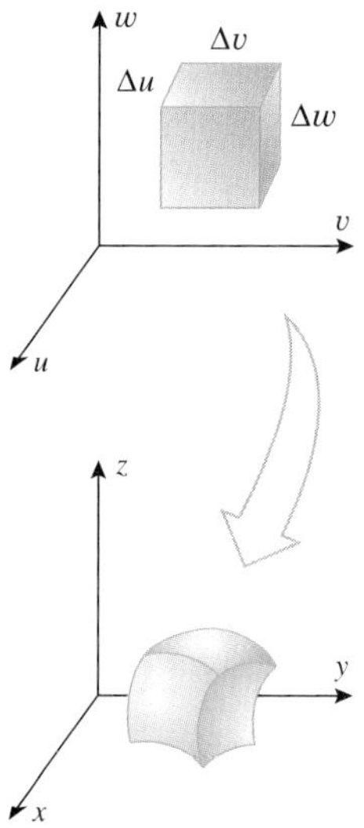

Figure 16.8.11

16.8.3 DEFINITION. If T is the transformation from uvw-space to xyz-space defined by the equations $x = x(u, v, w)$, $y = y(u, v, w)$, $z = z(u, v, w)$, then the **Jacobian of T** is denoted by $J(u, v, w)$ or $\partial(x, y, z)/\partial(u, v, w)$ and is defined by

$$J(u, v, w) = \frac{\partial(x, y, z)}{\partial(u, v, w)} = \begin{vmatrix} \dfrac{\partial x}{\partial u} & \dfrac{\partial x}{\partial v} & \dfrac{\partial x}{\partial w} \\[3mm] \dfrac{\partial y}{\partial u} & \dfrac{\partial y}{\partial v} & \dfrac{\partial y}{\partial w} \\[3mm] \dfrac{\partial z}{\partial u} & \dfrac{\partial z}{\partial v} & \dfrac{\partial z}{\partial w} \end{vmatrix}$$

For small values of Δu, Δv, and Δw, the volume ΔV of the curvilinear parallelepiped in Figure 16.8.11 is related to the volume $\Delta u\,\Delta v\,\Delta w$ of the rectangular parallelepiped by

$$\Delta V \approx \left|\frac{\partial(x, y, z)}{\partial(u, v, w)}\right|\Delta u\,\Delta v\,\Delta w \tag{13}$$

which is the analog of Formula (8). Using this relationship and an argument similar to the one that led to Formula (9), we can obtain the following result.

16.8.4 CHANGE-OF-VARIABLE FORMULA FOR TRIPLE INTEGRALS. If the transformation $x = x(u, v, w)$, $y = y(u, v, w)$, $z = z(u, v, w)$ maps the region S in uvw-space into the region R in xyz-space, and if the Jacobian $\partial(x, y, z)/\partial(u, v, w)$ is nonzero and does not change sign on S, then with appropriate restrictions on the transformation and the regions it follows that

$$\iiint\limits_{R} f(x, y, z)\,dV_{xyz} = \iiint\limits_{S} f(x(u, v, w), y(u, v, w), z(u, v, w))\left|\frac{\partial(x, y, z)}{\partial(u, v, w)}\right|\,dV_{uvw} \tag{14}$$

Example 4

Find the volume of the region G enclosed by the ellipsoid

$$\frac{x^2}{a^2} + \frac{y^2}{b^2} + \frac{z^2}{c^2} = 1$$

Solution. The volume V is given by the triple integral

$$V = \iiint\limits_{G} dV$$

To evaluate this integral, we make the change of variables

$$x = au, \quad y = bv, \quad z = cw \tag{15}$$

which maps the region S is uvw-space enclosed by a sphere of radius 1 into the region G in xyz-space. This can be seen from (15) by noting that

$$\frac{x^2}{a^2} + \frac{y^2}{b^2} + \frac{z^2}{c^2} = 1 \quad \text{becomes} \quad u^2 + v^2 + w^2 = 1$$

The Jacobian of (15) is

$$\frac{\partial(x, y, z)}{\partial(u, v, w)} = \begin{vmatrix} \dfrac{\partial x}{\partial u} & \dfrac{\partial x}{\partial v} & \dfrac{\partial x}{\partial w} \\[2mm] \dfrac{\partial y}{\partial u} & \dfrac{\partial y}{\partial v} & \dfrac{\partial y}{\partial w} \\[2mm] \dfrac{\partial z}{\partial u} & \dfrac{\partial z}{\partial v} & \dfrac{\partial z}{\partial w} \end{vmatrix} = \begin{vmatrix} a & 0 & 0 \\ 0 & b & 0 \\ 0 & 0 & c \end{vmatrix} = abc$$

Thus, from Formula (14), but with the notation dV rather than dV_{xyz},

$$V = \iiint\limits_{G} dV = \iiint\limits_{S} \left| \frac{\partial(x, y, z)}{\partial(u, v, w)} \right| dV_{uvw} = abc \iiint\limits_{S} dV_{uvw}$$

The last integral is the volume enclosed by a sphere of radius 1, which we know to be $\frac{4}{3}\pi$. Thus, the volume enclosed by the ellipsoid is $V = \frac{4}{3}\pi abc$.   ◀

Jacobians also arise in converting triple integrals in rectangular coordinates to iterated integrals in cylindrical and spherical coordinates. For example, we will ask you to show in Exercise 42 that the Jacobian of the transformation

$$x = r\cos\theta, \quad y = r\sin\theta, \quad z = z$$

is

$$\frac{\partial(x, y, z)}{\partial(r, \theta, z)} = r$$

and the Jacobian of the transformation

$$x = \rho\sin\phi\cos\theta, \quad y = \rho\sin\phi\sin\theta, \quad z = \rho\cos\phi$$

is

$$\frac{\partial(x, y, z)}{\partial(\rho, \phi, \theta)} = \rho^2 \sin\phi$$

Thus, Formulas (6) and (10) of Section 16.7 can be expressed in terms of Jacobians as

$$\iiint\limits_{G} f(x, y, z)\, dV = \iiint\limits_{\substack{\text{appropriate} \\ \text{limits}}} f(r\cos\theta, r\sin\theta, z) \frac{\partial(x, y, z)}{\partial(r, \theta, z)} dz\, dr\, d\theta \tag{16}$$

$$\iiint\limits_{G} f(x, y, z)\,dV = \iiint\limits_{\substack{\text{appropriate} \\ \text{limits}}} f(\rho\sin\phi\cos\theta,\ \rho\sin\phi\sin\theta,\ \rho\cos\phi)\,\frac{\partial(x, y, z)}{\partial(\rho, \phi, \theta)}\,d\rho\,d\phi\,d\theta$$

(17)

REMARK. The absolute value signs are omitted in these formulas because the Jacobians are nonnegative (see the restrictions in Table 13.8.1).

EXERCISE SET 16.8 C CAS

In Exercises 1–4, find the Jacobian $\partial(x, y)/\partial(u, v)$.

1. $x = u + 4v,\ y = 3u - 5v$

2. $x = u + 2v^2,\ y = 2u^2 - v$

3. $x = \sin u + \cos v,\ y = -\cos u + \sin v$

4. $x = \dfrac{2u}{u^2 + v^2},\ y = -\dfrac{2v}{u^2 + v^2}$

In Exercises 5–8, solve for x and y in terms of u and v, and then find the Jacobian $\partial(x, y)/\partial(u, v)$.

5. $u = 2x - 5y,\ v = x + 2y$

6. $u = e^x,\ v = ye^{-x}$

7. $u = x^2 - y^2,\ v = x^2 + y^2 \quad (x > 0, y > 0)$

8. $u = xy,\ v = xy^3 \quad (x > 0, y > 0)$

In Exercises 9–12, find the Jacobian $\partial(x, y, z)/\partial(u, v, w)$.

9. $x = 3u + v,\ y = u - 2w,\ z = v + w$

10. $x = u - uv,\ y = uv - uvw,\ z = uvw$

11. $u = xy,\ v = y,\ w = x + z$

12. $u = x + y + z,\ v = x + y - z,\ w = x - y + z$

In Exercises 13–16, sketch the image in the xy-plane of the set S under the given transformation.

13.

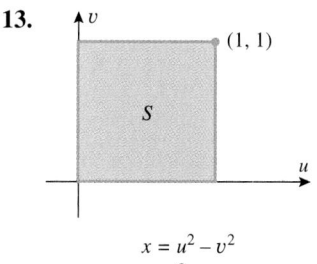

$x = u^2 - v^2$
$y = 2uv$

14.

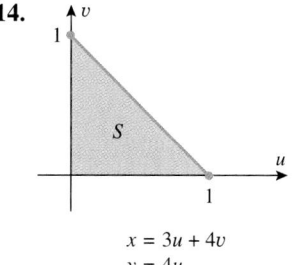

$x = 3u + 4v$
$y = 4u$

15.

$x = 2u$
$y = 3v$

16.

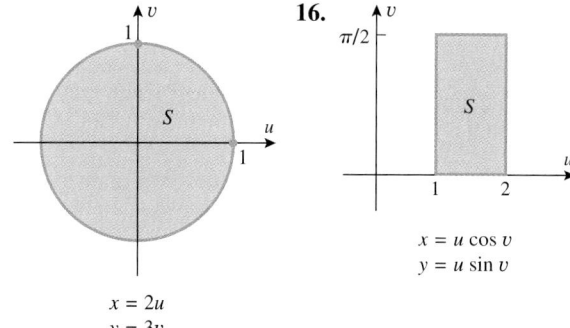

$x = u\cos v$
$y = u\sin v$

17. Use the transformation $u = x - 2y,\ v = 2x + y$ to find

$$\iint\limits_{R} \frac{x - 2y}{2x + y}\,dA$$

where R is the rectangular region enclosed by the lines $x - 2y = 1,\ x - 2y = 4,\ 2x + y = 1,\ 2x + y = 3$.

18. Use the transformation $u = x + y,\ v = x - y$ to find

$$\iint\limits_{R} (x - y)e^{x^2 - y^2}\,dA$$

over the rectangular region R enclosed by the lines $x + y = 0,\ x + y = 1,\ x - y = 1,\ x - y = 4$.

19. Use the transformation $u = \tfrac{1}{2}(x + y),\ v = \tfrac{1}{2}(x - y)$ to find

$$\iint\limits_{R} \sin\tfrac{1}{2}(x + y)\cos\tfrac{1}{2}(x - y)\,dA$$

over the triangular region R with vertices $(0, 0),\ (2, 0),\ (1, 1)$.

20. Use the transformation $u = y/x,\ v = xy$ to find

$$\iint\limits_{R} xy^3\,dA$$

over the region R in the first quadrant enclosed by $y = x,\ y = 3x,\ xy = 1,\ xy = 4$.

The transformation $x = au$, $y = bv$ ($a > 0, b > 0$) can be rewritten as $x/a = u$, $y/b = v$, and hence it maps the circular region

$$u^2 + v^2 \leq 1$$

into the elliptical region

$$\frac{x^2}{a^2} + \frac{y^2}{b^2} \leq 1$$

In Exercises 21–24, perform the integration by transforming the elliptical region of integration into a circular region of integration and then evaluating the transformed integral in polar coordinates.

21. $\displaystyle\iint\limits_{R} \sqrt{16x^2 + 9y^2}\, dA$, where R is the region enclosed by the ellipse $(x^2/9) + (y^2/16) = 1$.

22. $\displaystyle\iint\limits_{R} e^{-(x^2+4y^2)}\, dA$, where R is the region enclosed by the ellipse $(x^2/4) + y^2 = 1$.

23. $\displaystyle\iint\limits_{R} \sin(4x^2 + 9y^2)\, dA$, where R is the region in the first quadrant enclosed by the ellipse $4x^2 + 9y^2 = 1$ and the coordinate axes.

24. Show that the area of the ellipse

$$\frac{x^2}{a^2} + \frac{y^2}{b^2} = 1$$

is πab.

If a, b, and c are positive constants, then the transformation $x = au$, $y = bv$, $z = cw$ can be rewritten as $x/a = u$, $y/b = v$, $z/c = w$, and hence it maps the spherical region

$$u^2 + v^2 + w^2 \leq 1$$

into the ellipsoidal region

$$\frac{x^2}{a^2} + \frac{y^2}{b^2} + \frac{z^2}{c^2} \leq 1$$

In Exercises 25 and 26, perform the integration by transforming the ellipsoidal region of integration into a spherical region of integration and then evaluating the transformed integral in spherical coordinates.

25. $\displaystyle\iiint\limits_{G} x^2\, dV$, where G is the region enclosed by the ellipsoid $9x^2 + 4y^2 + z^2 = 36$.

26. Find the moment of inertia about the x-axis of the solid ellipsoid bounded by

$$\frac{x^2}{a^2} + \frac{y^2}{b^2} + \frac{z^2}{c^2} = 1$$

given that $\delta(x, y, z) = 1$. [See the definition preceding Exercise 39 of Section 16.7.]

In Exercises 27–30, evaluate the integral by making an appropriate change of variables.

27. $\displaystyle\iint\limits_{R} \frac{y - 4x}{y + 4x}\, dA$, where R is the region enclosed by the lines $y = 4x$, $y = 4x + 2$, $y = 2 - 4x$, $y = 5 - 4x$.

28. $\displaystyle\iint\limits_{R} (x^2 - y^2)\, dA$, where R is the rectangular region enclosed by the lines $y = -x$, $y = 1 - x$, $y = x$, $y = x + 2$.

29. $\displaystyle\iint\limits_{R} \frac{\sin(x - y)}{\cos(x + y)}\, dA$, where R is the triangular region enclosed by the lines $y = 0$, $y = x$, $x + y = \pi/4$.

30. $\displaystyle\iint\limits_{R} e^{(y-x)/(y+x)}\, dA$, where R is the region in the first quadrant enclosed by the trapezoid with vertices $(0, 1)$, $(1, 0)$, $(0, 4)$, $(4, 0)$.

31. Use an appropriate change of variables to find the area of the region in the first quadrant enclosed by the curves $y = x$, $y = 2x$, $x = y^2$, $x = 4y^2$.

32. Use an appropriate change of variables to find the volume of the solid bounded above by the plane $x + y + z = 9$, below by the xy-plane, and laterally by the elliptic cylinder $4x^2 + 9y^2 = 36$. [*Hint:* Express the volume as a double integral in xy-coordinates, then use polar coordinates to evaluate the transformed integral.]

33. Use the transformation $u = x$, $v = z - y$, $w = xy$ to find

$$\iiint\limits_{G} (z - y)^2 xy\, dV$$

where G is the region enclosed by the surfaces $x = 1$, $x = 3$, $z = y$, $z = y + 1$, $xy = 2$, $xy = 4$.

34. Use the transformation $u = xy$, $v = yz$, $w = xz$ to find the volume of the region in the first octant that is enclosed by the hyperbolic cylinders $xy = 1$, $xy = 2$, $yz = 1$, $yz = 3$, $xz = 1$, $xz = 4$.

35. (a) Verify that

$$\begin{vmatrix} a_1 & b_1 \\ c_1 & d_1 \end{vmatrix} \begin{vmatrix} a_2 & b_2 \\ c_2 & d_2 \end{vmatrix} = \begin{vmatrix} a_1 a_2 + b_1 c_2 & a_1 b_2 + b_1 d_2 \\ c_1 a_2 + d_1 c_2 & c_1 b_2 + d_1 d_2 \end{vmatrix}$$

(b) If $x = x(u, v)$, $y = y(u, v)$ is a one-to-one transformation, then $u = u(x, y)$, $v = v(x, y)$. Assuming the necessary differentiability, use the result in part (a) and the chain rule to show that

$$\frac{\partial(x, y)}{\partial(u, v)} \cdot \frac{\partial(u, v)}{\partial(x, y)} = 1$$

36. In each part, confirm that the formula obtained in part (b) of Exercise 35 holds for the given transformation.
(a) $x = u - uv$, $y = uv$
(b) $x = uv$, $y = v^2$ ($v > 0$)
(c) $x = \frac{1}{2}(u^2 + v^2)$, $y = \frac{1}{2}(u^2 - v^2)$ ($u > 0, v > 0$)

The formula obtained in part (b) of Exercise 35 is useful in integration problems where it is inconvenient or impossible to solve the transformation equations $u = f(x, y)$, $y = g(x, y)$ explicitly for x and y in terms of u and v. In Exercises 37–39, use the relationship

$$\frac{\partial(x, y)}{\partial(u, v)} = 1 \bigg/ \frac{\partial(u, v)}{\partial(x, y)}$$

to avoid solving for x and y in terms of u and v.

37. Use the transformation $u = xy$, $v = xy^4$ to find

$$\iint_R \sin(xy)\, dA$$

where R is the region enclosed by the curves $xy = \pi$, $xy = 2\pi$, $xy^4 = 1$, $xy^4 = 2$.

38. Use the transformation $u = x^2 - y^2$, $v = x^2 + y^2$ to find

$$\iint_R xy\, dA$$

where R is the region in the first quadrant that is enclosed by the hyperbolas $x^2 - y^2 = 1$, $x^2 - y^2 = 4$ and the circles $x^2 + y^2 = 9$, $x^2 + y^2 = 16$.

39. Use the transformation $u = xy$, $v = x^2 - y^2$ to find

$$\iint_R (x^4 - y^4)e^{xy}\, dA$$

where R is the region in the first quadrant enclosed by the hyperbolas $xy = 1$, $xy = 3$, $x^2 - y^2 = 3$, $x^2 - y^2 = 4$.

40. The three-variable analog of the formula derived in part (b) of Exercise 35 is

$$\frac{\partial(x, y, z)}{\partial(u, v, w)} \cdot \frac{\partial(u, v, w)}{\partial(x, y, z)} = 1$$

Use this result to show that the volume V of the oblique parallelepiped that is bounded by the planes $x + y + 2z = \pm 3$, $x - 2y + z = \pm 2$, $4x + y + z = \pm 6$ is $V = 16$.

41. (a) Show that if R is the triangular region with vertices $(0, 0)$, $(1, 0)$, and $(0, 1)$, then

$$\iint_R f(x + y)\, dA = \int_0^1 u f(u)\, du$$

(b) Use the result in part (a) to evaluate the integral

$$\iint_R e^{x+y}\, dA$$

42. (a) Consider the transformation $x = r\cos\theta$, $y = r\sin\theta$ from rectangular to polar coordinates, where $r \geq 0$. Show that

$$\left| \frac{\partial(x, y)}{\partial(r, \theta)} \right| = r$$

(b) Consider the transformation $x = \rho\sin\phi\cos\theta$, $y = \rho\sin\phi\sin\theta$, $z = \rho\cos\phi$ from spherical to rectangular coordinates, where $0 \leq \phi \leq \pi$. Show that

$$\left| \frac{\partial(x, y, z)}{\partial(\rho, \theta, \phi)} \right| = \rho^2 \sin\phi$$

SUPPLEMENTARY EXERCISES

1. The double integral over a region R in the xy-plane is defined as

$$\iint_R f(x, y)\, dA = \lim_{n \to +\infty} \sum_{k=1}^{n} f(x_k^*, y_k^*)\, \Delta A_k$$

Describe the procedure on which this definition is based.

2. The triple integral over a solid G in an xyz-coordinate system is defined as

$$\iiint_G f(x, y, z)\, dV = \lim_{n \to +\infty} \sum_{k=1}^{n} f(x_k^*, y_k^*, z_k^*)\, \Delta V_k$$

Describe the procedure on which this definition is based.

3. (a) Express the area of a region R in the xy-plane as a double integral.

(b) Express the volume of a region G in an xyz-coordinate system as a triple integral.

(c) Express the area of the portion of the surface $z = f(x, y)$ that lies above the region R in the xy-plane as a double integral.

4. (a) Write down parametric equations for a sphere of radius a centered at the origin.

(b) Write down parametric equations for the right circular cylinder of radius a and height h that is centered on the z-axis, has its base in the xy-plane, and extends in the positive z-direction.

5. (a) In physical terms, what is meant by the center of gravity of a lamina?

(b) What is meant by the centroid of a lamina?

(c) Write down formulas for the coordinates of the center of gravity of a lamina in the xy-plane.

(d) Write down formulas for the coordinates of the centroid of a lamina in the xy-plane.

6. Suppose that you have a double integral over a region R in the xy-plane and you want to transform that integral into an equivalent double integral over a region S in the uv-plane. Describe the procedure you would use.

7. Let R be the region in the accompanying figure. Fill in the missing limits of integration in the iterated integral

$$\int_{\square}^{\square} \int_{\square}^{\square} f(x, y)\, dx\, dy$$

over R.

8. Let R be the region shown in the accompanying figure. Fill in the missing limits of integration in the sum of the iterated integrals

$$\int_0^2 \int_{\square}^{\square} f(x, y)\, dy\, dx + \int_2^3 \int_{\square}^{\square} f(x, y)\, dy\, dx$$

over R.

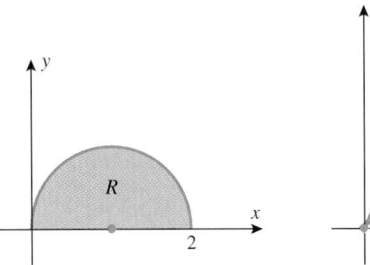

Figure Ex-7

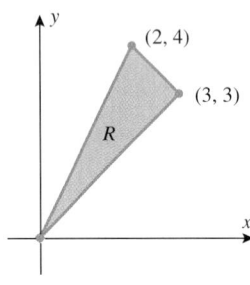

Figure Ex-8

9. (a) Find constants a, b, c, and d such that the transformation $x = au + bv$, $y = cu + dv$ maps the region S in the accompanying figure into the region R.

   (b) Find the area of the parallelogram R by integrating over the region S, and check your answer using a formula from geometry.

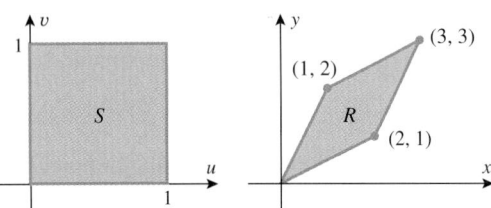

Figure Ex-9

10. Give a geometric argument to show that

$$0 < \int_0^\pi \int_0^\pi \sin\sqrt{xy}\, dy\, dx < \pi^2$$

In Exercises 11 and 12, evaluate the iterated integral.

11. $\displaystyle\int_{1/2}^1 \int_0^{2x} \cos(\pi x^2)\, dy\, dx$

12. $\displaystyle\int_0^2 \int_{-y}^{2y} x e^{y^3}\, dx\, dy$

In Exercises 13 and 14, express the iterated integral as an equivalent iterated integral with the order of integration reversed.

13. $\displaystyle\int_0^2 \int_0^{x/2} e^x e^y\, dy\, dx$

14. $\displaystyle\int_0^\pi \int_y^\pi \frac{\sin x}{x}\, dx\, dy$

In Exercises 15 and 16, sketch the region whose area is represented by the iterated integral.

15. $\displaystyle\int_0^{\pi/2} \int_{\tan(x/2)}^{\sin x} dy\, dx$

16. $\displaystyle\int_{\pi/6}^{\pi/2} \int_a^{a(1+\cos\theta)} r\, dr\, d\theta \quad (a > 0)$

In Exercises 17 and 18, evaluate the double integral.

17. $\displaystyle\iint_R x^2 \sin y^2\, dA$; R is the region that is bounded by $y = x^3$, $y = -x^3$, and $y = 8$.

18. $\displaystyle\iint_R (4 - x^2 - y^2)\, dA$; R is the sector in the first quadrant bounded by the circle $x^2 + y^2 = 4$ and the coordinate axes.

19. Convert to rectangular coordinates and evaluate:

$$\int_0^{\pi/2} \int_0^{2a\sin\theta} r \sin 2\theta\, dr\, d\theta$$

20. Convert to polar coordinates and evaluate:

$$\int_0^{\sqrt{2}} \int_x^{\sqrt{4-x^2}} 4xy\, dy\, dx$$

21. Convert to cylindrical coordinates and evaluate:

$$\int_{-2}^2 \int_{-\sqrt{4-x^2}}^{\sqrt{4-x^2}} \int_{(x^2+y^2)^2}^{16} x^2\, dz\, dy\, dx$$

22. Convert to spherical coordinates and evaluate:

$$\int_0^1 \int_0^{\sqrt{1-x^2}} \int_0^{\sqrt{1-x^2-y^2}} \frac{1}{1 + x^2 + y^2 + z^2}\, dz\, dy\, dx$$

23. Let G be the region bounded above by the sphere $\rho = a$ and below by the cone $\phi = \pi/3$. Express

$$\iiint_G (x^2 + y^2)\, dV$$

as an iterated integral in
   (a) spherical coordinates   (b) cylindrical coordinates
   (c) rectangular coordinates.

24. Let $G = \{(x, y, z) : x^2 + y^2 \le z \le 4x\}$. Express the volume of G as an iterated integral in
   (a) rectangular coordinates   (b) cylindrical coordinates.

In Exercises 25 and 26, find the area of the region using a double integral.

25. The region bounded by $y = 2x^3$, $2x + y = 4$, and the x-axis.

26. The region enclosed by the rose $r = \cos 3\theta$.

In Exercises 27 and 28, find the volume of the solid using a triple integral.

27. The solid bounded below by the cone $\phi = \pi/6$ and above by the plane $z = a$.

28. The solid enclosed between the surfaces $x = y^2 + z^2$ and $x = 1 - y^2$.

29. Find the surface area of the portion of the hyperbolic paraboloid

$$\mathbf{r}(u, v) = (u + v)\mathbf{i} + (u - v)\mathbf{j} + uv\mathbf{k}$$

for which $u^2 + v^2 \leq 4$.

30. Find the surface area of the portion of the spiral ramp

$$\mathbf{r}(u, v) = u \cos v\mathbf{i} + u \sin v\mathbf{j} + v\mathbf{k}$$

for which $0 \leq u \leq 2, 0 \leq v \leq 3u$.

In Exercises 31 and 32, find the equation of the tangent plane to the surface at the specified point.

31. $\mathbf{r} = u\mathbf{i} + v\mathbf{j} + (u^2 + v^2)\mathbf{k};\; u = 1, v = 2$

32. $x = u \cosh v, y = u \sinh v, z = u^2;\; (-3, 0, 9)$

In Exercises 33 and 34, find the centroid of the region.

33. The region bounded by $y^2 = 4x$ and $y^2 = 8(x - 2)$.

34. The upper half of the ellipse $(x/a)^2 + (y/b)^2 = 1$.

In Exercises 35 and 36, find the centroid of the solid.

35. The solid bounded above by the cone with vertex $(0, 0, h)$, bounded below by the xy-plane, and with base $x^2 + y^2 \leq a^2$ in the xy-plane.

36. The solid bounded by $y = x^2$, $z = 0$, and $y + z = 4$.

37. Show that

$$\int_0^{+\infty} \int_0^{+\infty} \frac{1}{(1 + x^2 + y^2)^2} \, dx \, dy = \frac{\pi}{4}$$

[*Hint:* See Exercise 37 of Section 16.3.]

38. It can be proved that if a bounded plane region slides along a helix in such a way that the region is always orthogonal to the helix (i.e., orthogonal to the unit tangent vector to the helix), then the volume swept out by the region is equal to the area of the region times the distance traveled by its centroid. Use this result to find the volume of the "tube" in

the accompanying figure that is swept out by sliding a circle of radius $\frac{1}{2}$ along the helix

$$x = \cos t, \quad y = \sin t, \quad z = \frac{t}{4} \quad (0 \leq t \leq 4\pi)$$

in such a way that the circle is always centered on the helix and lies in the plane perpendicular to the helix.

C **39.** The accompanying figure shows the graph of an ***astroidal sphere***

$$x^{2/3} + y^{2/3} + z^{2/3} = a$$

(a) Show that this surface can be represented parametrically as

$$x = a(\cos u \cos v)^3$$
$$y = a(\sin u \cos v)^3 \quad (0 \leq u \leq \pi,\; 0 \leq v \leq 2\pi)$$
$$z = a(\sin v)^3$$

(b) Use a CAS to approximate the surface area in the case where $a = 1$.

(c) Use a CAS to approximate the volume in the case where $a = 1$.

(d) Show that the exact volume is $4\pi/35$.
   [*Hint:* Let $x = t \cos^3 u$ and $y = t \sin^3 u$, where $0 \leq t \leq 1$ and $0 \leq u \leq \pi/2$.]

Figure Ex-38   Figure Ex-39

40. Find the average distance from a point inside a sphere of radius a to the center. [See the definition preceding Exercise 25 of Section 16.5.]

C **41.** (a) Describe the surface that is represented by the parametric equations

$$x = a \sin \phi \cos \theta$$
$$y = b \sin \phi \sin \theta \quad (0 \leq \phi \leq \pi,\; 0 \leq \theta \leq 2\pi)$$
$$z = c \cos \phi$$

where $a > 0, b > 0$, and $c > 0$.

(b) Use a CAS to approximate the area of the surface for $a = 2, b = 3, c = 4$.

EXPANDING THE CALCULUS HORIZON

For additional material relating to this chapter, visit the Anton Website at http://www.wiley.com/college/anton

17

TOPICS
IN VECTOR
CALCULUS

Carl Friedrich
Gauss

$\mathscr{Y}$ou have reached the final chapter in this text, and in a sense you have come full circle back to the roots of calculus. The main theme of this chapter is the concept of a *flow*, and the body of mathematics that we will study here is concerned with analyzing flows of various types—the flow of a fluid or the flow of electricity, for example. Indeed, the early writings of Isaac Newton on calculus are replete with such nouns as "fluxion" and "fluent," which are rooted in the Latin *fluens* (to flow). We will begin this chapter by introducing the concept of a *vector field*, which is the mathematical description of a flow. In subsequent sections, we will introduce two new kinds of integrals that are used in a variety of applications to analyze properties of vector fields and flows. Finally, we conclude with three major theorems, *Green's Theorem*, the *Divergence Theorem*, and *Stokes' Theorem*. These theorems provide a deep insight into the nature of flows and are the basis for many of the most important principles in physics and engineering.

17.1 VECTOR FIELDS

In this section we will consider functions that associate vectors with points in 2-space or 3-space. We will see that such functions play an important role in the study of fluid flow, gravitational force fields, electromagnetic force fields, and a wide range of other applied problems.

VECTOR FIELDS

Figure 17.1.1

To motivate the mathematical ideas in this section, consider a *unit* point mass located at any point in the Universe. According to Newton's Universal Law of Gravitation, the Earth exerts an attractive force on the mass that is directed toward the center of the Earth and has a magnitude that is inversely proportional to the square of the distance from the mass to the Earth's center (Figure 17.1.1). This association of force vectors with points in space is called the Earth's *gravitational field*. A similar idea arises in fluid flow. Imagine a stream in which the water flows horizontally at every level, and consider the layer of water at a specific depth. At each point of the layer, the water has a certain velocity, which we can represent by a vector at that point (Figure 17.1.2). This association of velocity vectors with points in the two-dimensional layer is called the *velocity field* at that layer. These ideas are captured in the following definition.

> **17.1.1** DEFINITION. A ***vector field*** is a function that associates a unique vector $\mathbf{F}(P)$ with each point P in a region of 2-space or 3-space.

Observe that in this definition there is no reference to a coordinate system. However, for computational purposes it is usually desirable to introduce a coordinate system so that vectors can be assigned components. Specifically, if $\mathbf{F}(P)$ is a vector field in an xy-coordinate system, then the point P will have some coordinates (x, y) and the associated vector will have components that are functions of x and y. Thus, the vector field $\mathbf{F}(P)$ can be expressed as

$$\mathbf{F}(x, y) = f(x, y)\mathbf{i} + g(x, y)\mathbf{j}$$

Similarly, in 3-space with an xyz-coordinate system, a vector field $\mathbf{F}(P)$ can be expressed as

$$\mathbf{F}(x, y, z) = f(x, y, z)\mathbf{i} + g(x, y, z)\mathbf{j} + h(x, y, z)\mathbf{k}$$

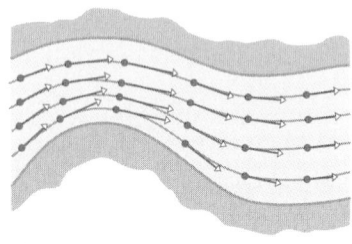

Figure 17.1.2

GRAPHICAL REPRESENTATIONS OF VECTOR FIELDS

A vector field in 2-space can be pictured geometrically by drawing representative field vectors $\mathbf{F}(x, y)$ at some well-chosen points in the xy-plane. But, just as it is usually not possible to describe a plane curve completely by plotting finitely many points, so it is usually not possible to describe a vector field completely by drawing finitely many vectors. Nevertheless, such graphical representations can provide useful information about the general behavior of the field if the vectors are chosen appropriately. However, graphical representations of vector fields require a substantial amount of computation, so they are usually created using computers. Figure 17.1.3 shows four computer-generated vector fields. The vector field in part (*a*) might describe the velocity of the current in a stream at various depths. At the bottom of the stream the velocity is zero, but the speed of the current increases as the depth decreases. Points at the same depth have the same speed. The vector field in part (*b*) might describe the velocity at points on a rotating wheel. At the center of the wheel the velocity is zero, but the speed increases with the distance from the center. Points at the same distance from the center have the same speed. The vector field in part (*c*) might describe the repulsive force of an electrical charge—the closer to the charge, the greater the force of repulsion. Part (*d*) shows a vector field in 3-space. Such pictures tend to be cluttered and hence are of lesser value than graphical representations of vector fields in 2-space. Note also that the vectors in parts (*b*) and (*c*) are not to scale—their lengths have been compressed for clarity. We will follow this procedure throughout this chapter.

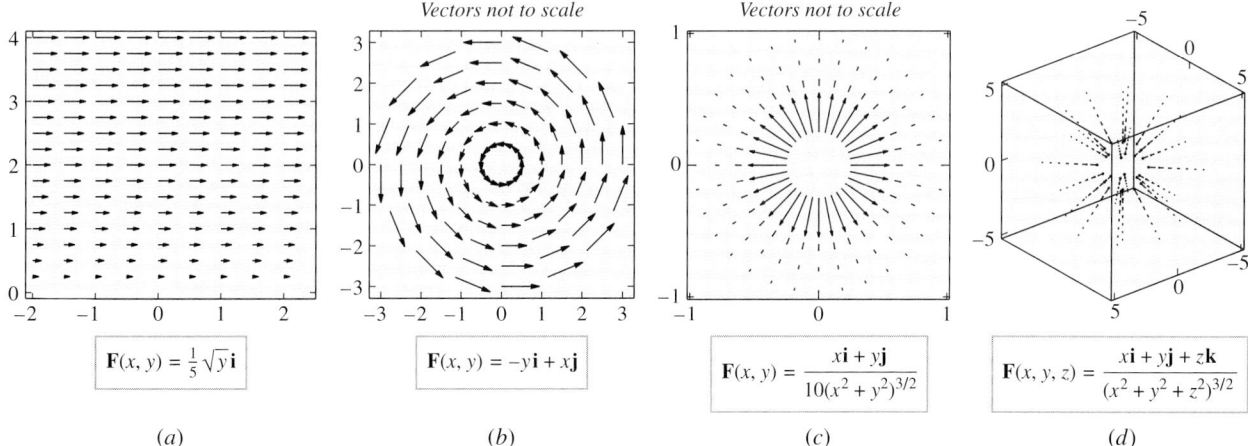

$$\mathbf{F}(x, y) = \tfrac{1}{5}\sqrt{y}\,\mathbf{i}$$

$$\mathbf{F}(x, y) = -y\mathbf{i} + x\mathbf{j}$$

$$\mathbf{F}(x, y) = \frac{x\mathbf{i} + y\mathbf{j}}{10(x^2 + y^2)^{3/2}}$$

$$\mathbf{F}(x, y, z) = \frac{x\mathbf{i} + y\mathbf{j} + z\mathbf{k}}{(x^2 + y^2 + z^2)^{3/2}}$$

(a)   (b)   (c)   (d)

Figure 17.1.3

FOR THE READER.   If you have a graphing utility that can generate vector fields, read the relevant documentation and try to make reasonable duplicates of parts (a) and (b) of Figure 17.1.3.

A COMPACT NOTATION FOR VECTOR FIELDS

Sometimes it is helpful to denote the vector fields $\mathbf{F}(x, y)$ and $\mathbf{F}(x, y, z)$ entirely in vector notation by identifying (x, y) with the radius vector $\mathbf{r} = x\mathbf{i} + y\mathbf{j}$ and (x, y, z) with the radius vector $\mathbf{r} = x\mathbf{i} + y\mathbf{j} + z\mathbf{k}$. With this notation a vector field in either 2-space or 3-space can be written as $\mathbf{F}(\mathbf{r})$. When no confusion is likely to arise, we will sometimes omit the $\mathbf{r}$ altogether and denote the vector field as $\mathbf{F}$.

INVERSE-SQUARE FIELDS

According to Newton's Universal Law of Gravitation, objects with masses m and M attract each other with a force $\mathbf{F}$ of magnitude

$$\|\mathbf{F}\| = \frac{GmM}{r^2} \tag{1}$$

where r is the distance between the objects (treated as point masses) and G is a constant. If we assume that the object of mass M is located at the origin of an xyz-coordinate system and $\mathbf{r}$ is the radius vector to the object of mass m, then $r = \|\mathbf{r}\|$, and the force $\mathbf{F}(\mathbf{r})$ exerted by the object of mass M on the object of mass m is in the direction of the unit vector $-\mathbf{r}/\|\mathbf{r}\|$. Thus, from (1)

$$\mathbf{F}(\mathbf{r}) = -\frac{GmM}{\|\mathbf{r}\|^2}\frac{\mathbf{r}}{\|\mathbf{r}\|} = -\frac{GmM}{\|\mathbf{r}\|^3}\mathbf{r} \tag{2}$$

If m and M are constant, and we let $c = -GmM$, then this formula can be expressed as

$$\mathbf{F}(\mathbf{r}) = \frac{c}{\|\mathbf{r}\|^3}\mathbf{r}$$

Vector fields of this form arise in electromagnetic as well as gravitational problems. Such fields are so important that they have their own terminology.

17.1.2   DEFINITION.   If $\mathbf{r}$ is a radius vector in 2-space or 3-space, and if c is a constant, then a vector field of the form

$$\mathbf{F}(\mathbf{r}) = \frac{c}{\|\mathbf{r}\|^3}\mathbf{r} \tag{3}$$

is called an ***inverse-square field***.

Observe that if $c > 0$ in (3), then $\mathbf{F}(\mathbf{r})$ has the same direction as $\mathbf{r}$, so each vector in the field is directed away from the origin; and if $c < 0$, then $\mathbf{F}(\mathbf{r})$ is oppositely directed to $\mathbf{r}$, so each vector in the field is directed toward the origin. In either case the magnitude of $\mathbf{F}(\mathbf{r})$ is inversely proportional to the square of the distance from the terminal point of $\mathbf{r}$ to the origin, since

$$\|\mathbf{F}(\mathbf{r})\| = \frac{|c|}{\|\mathbf{r}\|^3}\|\mathbf{r}\| = \frac{|c|}{\|\mathbf{r}\|^2}$$

We leave it for you to verify that in 2-space Formula (3) can be written in component form as

$$\mathbf{F}(x, y) = \frac{c}{(x^2 + y^2)^{3/2}}(x\mathbf{i} + y\mathbf{j}) \tag{4}$$

and in 3-space as

$$\mathbf{F}(x, y, z) = \frac{c}{(x^2 + y^2 + z^2)^{3/2}}(x\mathbf{i} + y\mathbf{j} + z\mathbf{k}) \tag{5}$$

[see parts (c) and (d) of Figure 17.1.3].

Example 1

Coulomb's law states that *the electrostatic force exerted by one charged particle on another is directly proportional to the product of the charges and inversely proportional to the square of the distance between them.* This has the same form as Newton's Universal Law of Gravitation, so the electrostatic force field exerted by a charged particle is an inverse-square field. Specifically, if a particle of charge Q is at the origin of a coordinate system, and if $\mathbf{r}$ is the radius vector to a particle of charge q, then the force $\mathbf{F}(\mathbf{r})$ that the particle of charge Q exerts on the particle of charge q is of the form

$$\mathbf{F}(\mathbf{r}) = \frac{qQ}{4\pi\epsilon_0\|\mathbf{r}\|^3}\mathbf{r}$$

where ϵ_0 is a positive constant (called the **permittivity constant**). This formula is of form (3) with $c = qQ/4\pi\epsilon_0$. ◄

GRADIENT FIELDS

An important class of vector fields arises from the process of finding gradients. Recall that if ϕ is a function of three variables, then the gradient of ϕ is defined as

$$\nabla\phi = \frac{\partial\phi}{\partial x}\mathbf{i} + \frac{\partial\phi}{\partial y}\mathbf{j} + \frac{\partial\phi}{\partial z}\mathbf{k}$$

This formula defines a vector field in 3-space called the **gradient field of** ϕ. Similarly, the gradient of a function of two variables defines a gradient field in 2-space. At each point in a gradient field where the gradient is nonzero, the vector points in the direction in which the rate of increase of ϕ is maximum.

Example 2

Sketch the gradient field of $\phi(x, y) = x + y$.

Solution. The gradient of ϕ is

$$\nabla\phi = \frac{\partial\phi}{\partial x}\mathbf{i} + \frac{\partial\phi}{\partial y}\mathbf{j} = \mathbf{i} + \mathbf{j}$$

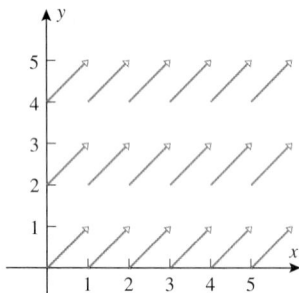

Figure 17.1.4

which is the same at each point. A portion of the vector field is sketched in Figure 17.1.4. ◄

CONSERVATIVE FIELDS AND POTENTIAL FUNCTIONS

If $\mathbf{F}(\mathbf{r})$ is an arbitrary vector field in 2-space or 3-space, we can ask whether it is the gradient field of some function ϕ, and if so, how we can find ϕ. This is an important problem in various applications, and we will study it in more detail later. However, there is some terminology for such fields that we will introduce now.

> **17.1.3 DEFINITION.** A vector field **F** in 2-space or 3-space is said to be ***conservative*** in a region if it is the gradient field for some function ϕ in that region. The function ϕ is called a ***potential function*** for **F** in the region.

Example 3

Inverse-square fields are conservative in any region that does not contain the origin. For example, in the two-dimensional case the function

$$\phi(x, y) = -\frac{c}{(x^2 + y^2)^{1/2}} \tag{6}$$

is a potential function for (4) in any region not containing the origin, since

$$\nabla \phi(x, y) = \frac{\partial \phi}{\partial x}\mathbf{i} + \frac{\partial \phi}{\partial y}\mathbf{j}$$

$$= \frac{cx}{(x^2 + y^2)^{3/2}}\mathbf{i} + \frac{cy}{(x^2 + y^2)^{3/2}}\mathbf{j}$$

$$= \frac{c}{(x^2 + y^2)^{3/2}}(x\mathbf{i} + y\mathbf{j})$$

$$= \mathbf{F}(x, y)$$

In a later section we will discuss methods for finding potential functions for conservative vector fields. ◄

DIVERGENCE AND CURL

We will now define two important operations on vector fields in 3-space—the *divergence* and the *curl* of the field. These names originate in the study of fluid flow, in which case the divergence relates to the way in which fluid flows toward or away from a point and the curl relates to the rotational properties of the fluid at a point. We will investigate the physical interpretations of these operations in more detail later, but for now we will focus only on their computation.

> **17.1.4 DEFINITION.** If $\mathbf{F}(x, y, z) = f(x, y, z)\mathbf{i} + g(x, y, z)\mathbf{j} + h(x, y, z)\mathbf{k}$, then we define the ***divergence of*** **F**, written div **F**, by
>
> $$\text{div } \mathbf{F} = \frac{\partial f}{\partial x} + \frac{\partial g}{\partial y} + \frac{\partial h}{\partial z} \tag{7}$$

> **17.1.5 DEFINITION.** If $\mathbf{F}(x, y, z) = f(x, y, z)\mathbf{i} + g(x, y, z)\mathbf{j} + h(x, y, z)\mathbf{k}$, then we define the ***curl of*** **F**, written curl **F**, by
>
> $$\text{curl } \mathbf{F} = \left(\frac{\partial h}{\partial y} - \frac{\partial g}{\partial z}\right)\mathbf{i} + \left(\frac{\partial f}{\partial z} - \frac{\partial h}{\partial x}\right)\mathbf{j} + \left(\frac{\partial g}{\partial x} - \frac{\partial f}{\partial y}\right)\mathbf{k} \tag{8}$$

REMARK. Observe that div **F** and curl **F** depend on the point at which they are computed, and hence are more properly written as div $\mathbf{F}(x, y, z)$ and curl $\mathbf{F}(x, y, z)$. However, even though these functions are expressed in terms of x, y, and z, it can be proved that their values at a fixed point depend on the point but not on the coordinate system selected. This is important in applications, since it allows physicists and engineers to compute the curl and divergence in any convenient coordinate system.

Before proceeding to some examples, we note that div **F** has scalar values, whereas curl **F** has vector values (i.e., curl **F** is itself a vector field). Moreover, for computational purposes it is useful to note that the formula for the curl can be expressed in the determinant form

$$
\text{curl } \mathbf{F} = \begin{vmatrix} \mathbf{i} & \mathbf{j} & \mathbf{k} \\ \dfrac{\partial}{\partial x} & \dfrac{\partial}{\partial y} & \dfrac{\partial}{\partial z} \\ f & g & h \end{vmatrix}
\tag{9}
$$

You should verify that Formula (8) results if the determinant is computed by interpreting a "product" such as $(\partial/\partial x)(g)$ to mean $\partial g/\partial x$. Keep in mind, however, that (9) is just a mnemonic device and not a true determinant, since the entries in a determinant must be numbers, not vectors and partial derivative symbols.

Example 4

Find the divergence and the curl of the vector field

$$\mathbf{F}(x, y, z) = x^2 y \mathbf{i} + 2y^3 z \mathbf{j} + 3z \mathbf{k}$$

Solution. From (7)

$$
\text{div } \mathbf{F} = \frac{\partial}{\partial x}(x^2 y) + \frac{\partial}{\partial y}(2y^3 z) + \frac{\partial}{\partial z}(3z)
$$

$$
= 2xy + 6y^2 z + 3
$$

and from (9)

$$
\text{curl } \mathbf{F} = \begin{vmatrix} \mathbf{i} & \mathbf{j} & \mathbf{k} \\ \dfrac{\partial}{\partial x} & \dfrac{\partial}{\partial y} & \dfrac{\partial}{\partial z} \\ x^2 y & 2y^3 z & 3z \end{vmatrix}
$$

$$
= \left[\frac{\partial}{\partial y}(3z) - \frac{\partial}{\partial z}(2y^3 z) \right] \mathbf{i} + \left[\frac{\partial}{\partial z}(x^2 y) - \frac{\partial}{\partial x}(3z) \right] \mathbf{j}
$$

$$
+ \left[\frac{\partial}{\partial x}(2y^3 z) - \frac{\partial}{\partial y}(x^2 y) \right] \mathbf{k}
$$

$$
= -2y^3 \mathbf{i} - x^2 \mathbf{k} \qquad \blacktriangleleft
$$

FOR THE READER.   Most computer algebra systems can compute gradient fields, divergence, and curl. If you have a CAS with these capabilities, read the relevant documentation and use your CAS to check the computations in Examples 2 and 4.

Example 5

Show that the divergence of the inverse-square field

$$\mathbf{F}(x, y, z) = \frac{c}{(x^2 + y^2 + z^2)^{3/2}}(x\mathbf{i} + y\mathbf{j} + z\mathbf{k})$$

is zero.

Solution. The computations can be simplified by letting $r = (x^2 + y^2 + z^2)^{1/2}$, in which case $\mathbf{F}$ can be expressed as

$$\mathbf{F}(x, y, z) = \frac{cx\mathbf{i} + cy\mathbf{j} + cz\mathbf{k}}{r^3} = \frac{cx}{r^3}\mathbf{i} + \frac{cy}{r^3}\mathbf{j} + \frac{cz}{r^3}\mathbf{k}$$

We leave it for you to show that

$$\frac{\partial r}{\partial x} = \frac{x}{r}, \quad \frac{\partial r}{\partial y} = \frac{y}{r}, \quad \frac{\partial r}{\partial z} = \frac{z}{r}$$

Thus

$$\text{div } \mathbf{F} = c\left[\frac{\partial}{\partial x}\left(\frac{x}{r^3}\right) + \frac{\partial}{\partial y}\left(\frac{y}{r^3}\right) + \frac{\partial}{\partial z}\left(\frac{z}{r^3}\right) \right] \tag{10}$$

But

$$\frac{\partial}{\partial x}\left(\frac{x}{r^3}\right) = \frac{r^3 - x(3r^2)(x/r)}{(r^3)^2} = \frac{1}{r^3} - \frac{3x^2}{r^5}$$

$$\frac{\partial}{\partial y}\left(\frac{y}{r^3}\right) = \frac{1}{r^3} - \frac{3y^2}{r^5}$$

$$\frac{\partial}{\partial z}\left(\frac{z}{r^3}\right) = \frac{1}{r^3} - \frac{3z^2}{r^5}$$

Substituting these expressions in (10) yields

$$\text{div } \mathbf{F} = c\left[\frac{3}{r^3} - \frac{3x^2 + 3y^2 + 3z^2}{r^5}\right] = c\left[\frac{3}{r^3} - \frac{3r^2}{r^5}\right] = 0 \qquad \blacktriangleleft$$

THE ∇ OPERATOR

Thus far, the symbol ∇ that appears in the gradient expression $\nabla\phi$ has not been given a meaning of its own. However, it is often convenient to view ∇ as an operator

$$\nabla = \frac{\partial}{\partial x}\mathbf{i} + \frac{\partial}{\partial y}\mathbf{j} + \frac{\partial}{\partial z}\mathbf{k} \tag{11}$$

which when applied to $\phi(x, y, z)$ produces the gradient

$$\nabla\phi = \frac{\partial\phi}{\partial x}\mathbf{i} + \frac{\partial\phi}{\partial y}\mathbf{j} + \frac{\partial\phi}{\partial z}\mathbf{k}$$

We call (11) the ***del operator***. This is analogous to the derivative operator d/dx, which when applied to $f(x)$ produces the derivative $f'(x)$.

The del operator allows us to express the divergence of a vector field

$$\mathbf{F} = f(x, y, z)\mathbf{i} + g(x, y, z)\mathbf{j} + h(x, y, z)\mathbf{k}$$

in dot product notation as

$$\text{div } \mathbf{F} = \nabla \cdot \mathbf{F} = \frac{\partial f}{\partial x} + \frac{\partial g}{\partial y} + \frac{\partial h}{\partial z} \tag{12}$$

and the curl of this field in cross-product notation as

$$\text{curl } \mathbf{F} = \nabla \times \mathbf{F} = \begin{vmatrix} \mathbf{i} & \mathbf{j} & \mathbf{k} \\ \dfrac{\partial}{\partial x} & \dfrac{\partial}{\partial y} & \dfrac{\partial}{\partial z} \\ f & g & h \end{vmatrix} \tag{13}$$

THE LAPLACIAN ∇^2

The operator that results by taking the dot product of the del operator with itself is denoted by ∇^2 and is called the ***Laplacian**** operator. This operator has the form

$$\nabla^2 = \nabla \cdot \nabla = \frac{\partial^2}{\partial x^2} + \frac{\partial^2}{\partial y^2} + \frac{\partial^2}{\partial z^2} \tag{14}$$

When applied to $\phi(x, y, z)$ the Laplacian operator produces the function

$$\nabla^2\phi = \frac{\partial^2\phi}{\partial x^2} + \frac{\partial^2\phi}{\partial y^2} + \frac{\partial^2\phi}{\partial z^2}$$

Note that $\nabla^2\phi$ can also be expressed as div $(\nabla\phi)$. The equation $\nabla^2\phi = 0$ or, equivalently,

$$\frac{\partial^2\phi}{\partial x^2} + \frac{\partial^2\phi}{\partial y^2} + \frac{\partial^2\phi}{\partial z^2} = 0$$

is known as ***Laplace's equation***. This partial differential equation plays an important role in a wide variety of applications, resulting from the fact that it is satisfied by the potential function for the inverse-square field.

---

*See biography on page 1062.

EXERCISE SET 17.1 ⌐∿ Graphing Calculator [c] CAS

In Exercises 1 and 2, match the vector field $\mathbf{F}(x, y)$ with one of the plots, and explain your reasoning.

1. (a) $\mathbf{F}(x, y) = x\mathbf{i}$ (b) $\mathbf{F}(x, y) = \sin x\mathbf{i} + \mathbf{j}$

2. (a) $\mathbf{F}(x, y) = \mathbf{i} + \mathbf{j}$

  (b) $\mathbf{F}(x, y) = \dfrac{x}{\sqrt{x^2 + y^2}}\mathbf{i} + \dfrac{y}{\sqrt{x^2 + y^2}}\mathbf{j}$

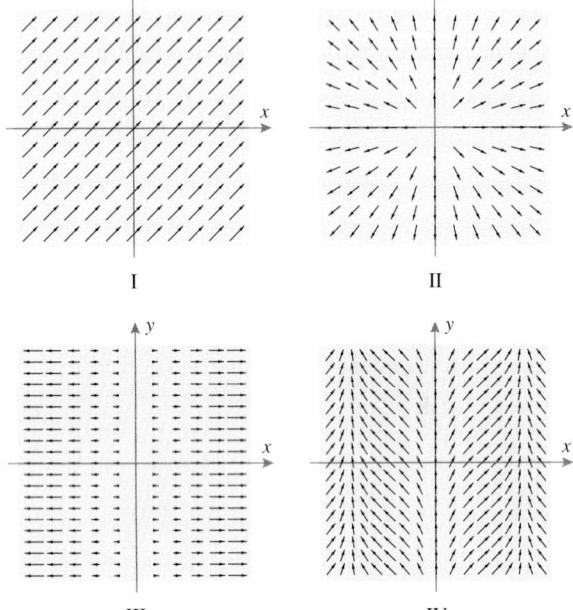

I

II

III

IV

In Exercises 3 and 4, determine whether the statement about the vector field $\mathbf{F}(x, y)$ is true or false. If false, explain why.

3. $\mathbf{F}(x, y) = x^2\mathbf{i} - y\mathbf{j}$.
  (a) As (x, y) gets closer to the origin, the lengths of the vectors decrease.
  (b) If (x, y) is on the positive y-axis, then the vector points in the negative y-direction.
  (c) If (x, y) is in the first quadrant, then the vector points down and to the right.

4. $\mathbf{F}(x, y) = \dfrac{x}{\sqrt{x^2 + y^2}}\mathbf{i} - \dfrac{y}{\sqrt{x^2 + y^2}}\mathbf{j}$.
  (a) As (x, y) moves away from the origin, the lengths of the vectors decrease.
  (b) If (x, y) is a point on the positive x-axis, then the vector points up.
  (c) If (x, y) is a point on the positive y-axis, the vector points to the right.

In Exercises 5–8, sketch the vector field by drawing some representative nonintersecting vectors. The vectors need not be drawn to scale, but they should be in reasonably correct proportion relative to each other.

5. $\mathbf{F}(x, y) = 2\mathbf{i} - \mathbf{j}$ **6.** $\mathbf{F}(x, y) = y\mathbf{j}, \quad y > 0$

7. $\mathbf{F}(x, y) = y\mathbf{i} - x\mathbf{j}$. [*Note:* Each vector in the field is perpendicular to the position vector $\mathbf{r} = x\mathbf{i} + y\mathbf{j}$.]

8. $\mathbf{F}(x, y) = \dfrac{x\mathbf{i} + y\mathbf{j}}{\sqrt{x^2 + y^2}}$. [*Note:* Each vector in the field is a unit vector in the same direction as the position vector $\mathbf{r} = x\mathbf{i} + y\mathbf{j}$.]

---

*PIERRE-SIMON DE LAPLACE (1749–1827). French mathematician and physicist. Laplace is sometimes referred to as the French Isaac Newton because of his work in celestial mechanics. In a five-volume treatise entitled *Traité de Mécanique Céleste*, he solved extremely difficult problems involving gravitational interactions between the planets. In particular, he was able to show that our solar system is stable and not prone to catastrophic collapse as a result of these interactions. This was an issue of major concern at the time because Jupiter's orbit appeared to be shrinking and Saturn's expanding; Laplace showed that these were expected periodic anomalies. In addition to his work in celestial mechanics, he founded modern probability theory, showed with Lavoisier that respiration is a form of combustion, and developed methods that fostered many new branches of pure mathematics.

Laplace was born to moderately successful parents in Normandy, his father being a farmer and cider merchant. He matriculated in the theology program at the University of Caen at age 16 but left for Paris at age 18 with a letter of introduction to the influential mathematician d'Alembert, who eventually helped him undertake a career in mathematics. Laplace was a prolific writer, and after his election to the Academy of Sciences in 1773, the secretary wrote that the Academy had never received so many important research papers by so young a person in such a short time. Laplace had little interest in pure mathematics—he regarded mathematics merely as a tool for solving applied problems. In his impatience with mathematical detail, he frequently omitted complicated arguments with the statement, "It is easy to show that...." He admitted, however, that as time passed he often had trouble reconstructing the omitted details himself!

At the height of his fame, Laplace served on many government committees and held the posts of Minister of the Interior and chancellor of the Senate. He barely escaped imprisonment and execution during the period of the Revolution, probably because he was able to convince each opposing party that he sided with them. Napoleon described him as a great mathematician but a poor administrator who "sought subtleties everywhere, had only doubtful ideas, and ... carried the spirit of the infinitely small into administration." In spite of his genius, Laplace was both egotistic and insecure, attempting to ensure his place in history by conveniently failing to credit mathematicians whose work he used—an unnecessary pettiness since his own work was so brilliant. However, on the positive side he was supportive of young mathematicians, often treating them as his own children. Laplace ranks as one of the most influential mathematicians in history.*

In Exercises 9 and 10, use a graphing utility to generate a plot of the vector field.

9. $\mathbf{F}(x, y) = \mathbf{i} + \cos y\,\mathbf{j}$ **10.** $\mathbf{F}(x, y) = y\mathbf{i} - x\mathbf{j}$

In Exercises 11 and 12, confirm that ϕ is a potential function for $\mathbf{F(r)}$ on some region, and state the region.

11. (a) $\phi(x, y) = \tan^{-1} xy$
$$\mathbf{F}(x, y) = \frac{y}{1 + x^2y^2}\mathbf{i} + \frac{x}{1 + x^2y^2}\mathbf{j}$$
   (b) $\phi(x, y, z) = x^2 - 3y^2 + 4z^2$
   $\mathbf{F}(x, y, z) = 2x\mathbf{i} - 6y\mathbf{j} + 8z\mathbf{k}$

12. (a) $\phi(x, y) = 2y^2 + 3x^2y - xy^3$
   $\mathbf{F}(x, y) = (6xy - y^3)\mathbf{i} + (4y + 3x^2 - 3xy^2)\mathbf{j}$
   (b) $\phi(x, y, z) = x\sin z + y\sin x + z\sin y$
   $\mathbf{F}(x, y, z) = (\sin z + y\cos x)\mathbf{i} + (\sin x + z\cos y)\mathbf{j}$
   $\qquad\qquad\qquad\qquad + (x\cos z + \sin y)\mathbf{k}$

In Exercises 13–18, find div $\mathbf{F}$ and curl $\mathbf{F}$.

13. $\mathbf{F}(x, y, z) = x^2\mathbf{i} - 2\mathbf{j} + yz\mathbf{k}$

14. $\mathbf{F}(x, y, z) = xz^3\mathbf{i} + 2y^4x^2\mathbf{j} + 5z^2y\mathbf{k}$

15. $\mathbf{F}(x, y, z) = 7y^3z^2\mathbf{i} - 8x^2z^5\mathbf{j} - 3xy^4\mathbf{k}$

16. $\mathbf{F}(x, y, z) = e^{xy}\mathbf{i} - \cos y\,\mathbf{j} + \sin^2 z\,\mathbf{k}$

17. $\mathbf{F}(x, y, z) = \dfrac{1}{\sqrt{x^2 + y^2 + z^2}}(x\mathbf{i} + y\mathbf{j} + z\mathbf{k})$

18. $\mathbf{F}(x, y, z) = \ln x\mathbf{i} + e^{xyz}\mathbf{j} + \tan^{-1}(z/x)\mathbf{k}$

In Exercises 19 and 20, find $\nabla \cdot (\mathbf{F} \times \mathbf{G})$.

19. $\mathbf{F}(x, y, z) = 2x\mathbf{i} + \mathbf{j} + 4y\mathbf{k}$
   $\mathbf{G}(x, y, z) = x\mathbf{i} + y\mathbf{j} - z\mathbf{k}$

20. $\mathbf{F}(x, y, z) = yz\mathbf{i} + xz\mathbf{j} + xy\mathbf{k}$
   $\mathbf{G}(x, y, z) = xy\mathbf{j} + xyz\mathbf{k}$

In Exercises 21 and 22, find $\nabla \cdot (\nabla \times \mathbf{F})$.

21. $\mathbf{F}(x, y, z) = \sin x\mathbf{i} + \cos(x - y)\mathbf{j} + z\mathbf{k}$
22. $\mathbf{F}(x, y, z) = e^{xz}\mathbf{i} + 3xe^y\mathbf{j} - e^{yz}\mathbf{k}$

In Exercises 23 and 24, find $\nabla \times (\nabla \times \mathbf{F})$.

23. $\mathbf{F}(x, y, z) = xy\mathbf{j} + xyz\mathbf{k}$
24. $\mathbf{F}(x, y, z) = y^2x\mathbf{i} - 3yz\mathbf{j} + xy\mathbf{k}$

In Exercises 25–32, let k be a constant, and let $\mathbf{F} = \mathbf{F}(x, y, z)$, $\mathbf{G} = \mathbf{G}(x, y, z)$, and $\phi = \phi(x, y, z)$. Prove the following identities, assuming that all derivatives involved exist and are continuous.

25. $\mathrm{div}(k\mathbf{F}) = k\,\mathrm{div}\,\mathbf{F}$ **26.** $\mathrm{curl}(k\mathbf{F}) = k\,\mathrm{curl}\,\mathbf{F}$
27. $\mathrm{div}(\mathbf{F} + \mathbf{G}) = \mathrm{div}\,\mathbf{F} + \mathrm{div}\,\mathbf{G}$
28. $\mathrm{curl}(\mathbf{F} + \mathbf{G}) = \mathrm{curl}\,\mathbf{F} + \mathrm{curl}\,\mathbf{G}$
29. $\mathrm{div}(\phi\mathbf{F}) = \phi\,\mathrm{div}\,\mathbf{F} + \nabla\phi \cdot \mathbf{F}$

30. $\mathrm{curl}(\phi\mathbf{F}) = \phi\,\mathrm{curl}\,\mathbf{F} + \nabla\phi \times \mathbf{F}$
31. $\mathrm{div}(\mathrm{curl}\,\mathbf{F}) = 0$ **32.** $\mathrm{curl}(\nabla\phi) = \mathbf{0}$
33. Rewrite the identities in Exercises 25, 27, 29, and 31 in an equivalent form using the notation $\nabla \cdot$ for divergence and $\nabla \times$ for curl.
34. Rewrite the identities in Exercises 26, 28, 30, and 32 in an equivalent form using the notation $\nabla \cdot$ for divergence and $\nabla \times$ for curl.

35. Use a CAS to check the calculations in Exercises 19, 21, and 23.
36. Use a CAS to check the calculations in Exercises 20, 22, and 24.

In Exercises 37 and 38, verify that the radius vector $\mathbf{r} = x\mathbf{i} + y\mathbf{j} + z\mathbf{k}$ has the stated property.

37. (a) $\mathrm{curl}\,\mathbf{r} = \mathbf{0}$ (b) $\nabla\|\mathbf{r}\| = \dfrac{\mathbf{r}}{\|\mathbf{r}\|}$

38. (a) $\mathrm{div}\,\mathbf{r} = 3$ (b) $\nabla\dfrac{1}{\|\mathbf{r}\|} = -\dfrac{\mathbf{r}}{\|\mathbf{r}\|^3}$

In Exercises 39 and 40, let $\mathbf{r} = x\mathbf{i} + y\mathbf{j} + z\mathbf{k}$, let $r = \|\mathbf{r}\|$, let f be a differentiable function of one variable, and let $\mathbf{F(r)} = f(r)\mathbf{r}$.

39. (a) Use the chain rule and Exercise 37(b) to show that
$$\nabla f(r) = \frac{f'(r)}{r}\mathbf{r}$$
   (b) Use the result in part (a) and Exercises 29 and 38(a) to show that
$$\mathrm{div}\,\mathbf{F} = 3f(r) + rf'(r)$$

40. (a) Use part (a) of Exercise 39, Exercise 30, and Exercise 37(a) to show that
$$\mathrm{curl}\,\mathbf{F} = \mathbf{0}$$
   (b) Use the result in part (a) of Exercise 39 and Exercises 29 and 38(a) to show that
$$\nabla^2 f(r) = 2\frac{f'(r)}{r} + f''(r)$$

41. Use the result in Exercise 39(b) to show that the divergence of the inverse-square field $\mathbf{F} = \mathbf{r}/\|\mathbf{r}\|^3$ is zero.

42. Use the result of Exercise 39(b) to show that if $\mathbf{F}$ is a vector field of the form $\mathbf{F} = f(\|\mathbf{r}\|)\mathbf{r}$ and if div $\mathbf{F} = 0$, then $\mathbf{F}$ is an inverse-square field. [*Suggestion:* Let $r = \|\mathbf{r}\|$ and multiply $3f(r) + rf'(r) = 0$ through by r^2. Then write the result as a derivative of a product.]

43. A curve C is called a *flow line* of a vector field $\mathbf{F}$ if $\mathbf{F}$ is a tangent vector to C at each point along C (see Figure Ex-43 on page 1064).
   (a) Let C be a flow line for $\mathbf{F}(x, y) = -y\mathbf{i} + x\mathbf{j}$, and let (x, y) be a point on C for which $y \neq 0$. Show that the flow lines satisfy the differential equation
$$\frac{dy}{dx} = -\frac{x}{y}$$

(b) Solve the differential equation in part (a) by separation of variables, and show that the flow lines are concentric circles centered at the origin.

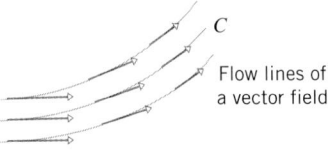

C

Flow lines of
a vector field

Figure Ex-43

In Exercises 44–46, find a differential equation satisfied by the flow lines of **F** (see Exercise 43), and solve it to find equations for the flow lines of **F**. Sketch some typical flow lines and tangent vectors.

44. $\mathbf{F}(x, y) = \mathbf{i} + x\mathbf{j}$ **45.** $\mathbf{F}(x, y) = x\mathbf{i} + \mathbf{j}, \quad x > 0$

46. $\mathbf{F}(x, y) = x\mathbf{i} - y\mathbf{j}, \quad x > 0 \text{ and } y > 0$

17.2 LINE INTEGRALS

In earlier chapters we considered three kinds of integrals in rectangular coordinates: single integrals over intervals, double integrals over two-dimensional regions, and triple integrals over three-dimensional regions. In this section we will discuss integrals along curves in two- or three-dimensional space.

LINE INTEGRALS

Integrals along curves arise in a variety of problems. One such problem can be stated as follows:

17.2.1 AN AREA PROBLEM. Let C be a smooth curve that extends between two points in the xy-plane, and let $f(x, y)$ be continuous and nonnegative on C. Find the area of the "sheet" that is swept out by the vertical line segment that extends upward from the point (x, y) to a height of $f(x, y)$ and moves along C from one endpoint to the other (Figure 17.2.1).

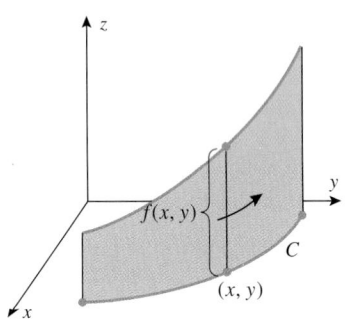

Figure 17.2.1

We use the following limit process to find the area of the sheet:

- Divide C into n arcs by inserting a succession of distinct points $P_1, P_2, \ldots, P_{n-1}$ between the initial and terminal points of C in the direction of increasing parameter. As illustrated on the left side of Figure 17.2.2, these points divide the surface into n strips. If we denote the area of the kth strip by ΔA_k, then the total area A of the sheet can be expressed as

$$A = \Delta A_1 + \Delta A_2 + \cdots + \Delta A_n = \sum_{k=1}^{n} \Delta A_k$$

- The next step is to approximate the area ΔA_k of the kth strip, assuming that this strip is narrow. For this purpose, let Δs_k be the length of the arc along C at the base of the

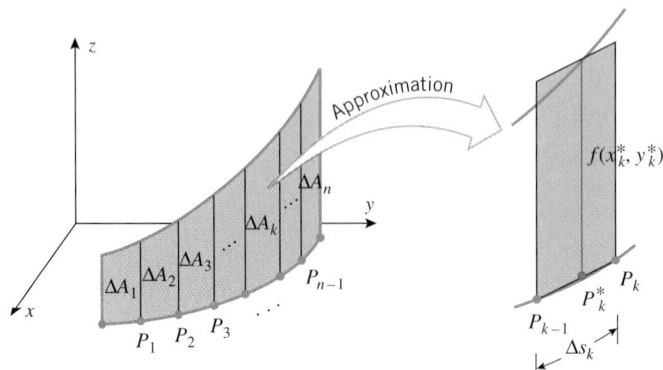

Figure 17.2.2

kth strip, and choose an arbitrary point $P_k^*(x_k^*, y_k^*)$ on this arc. Since the strip is narrow and f is continuous, the value of f will not vary much along the kth arc, so we can assume that f has a constant value of $f(x_k^*, y_k^*)$ on this arc. Thus, the area ΔA_k of the kth strip can be closely approximated by the area of a rectangle with base Δs_k and height $f(x_k^*, y_k^*)$, as shown in the right part of Figure 17.2.2; that is,

$$\Delta A_k \approx f(x_k^*, y_k^*)\Delta s_k$$

from which it follows that

$$A \approx \sum_{k=1}^{n} f(x_k^*, y_k^*)\Delta s_k$$

• If we now increase n so that the length of each arc approaches zero, then it is plausible that the error in this approximation approaches zero, and the exact surface area is

$$A = \lim_{n \to +\infty} \sum_{k=1}^{n} f(x_k^*, y_k^*)\Delta s_k \tag{1}$$

In deriving Formula (1) we assumed that f is continuous and nonnegative on the curve C. If f is continuous on C and has both positive and negative values, then the limit

$$\lim_{n \to +\infty} \sum_{k=1}^{n} f(x_k^*, y_k^*)\Delta s_k$$

does not represent the area of the surface over C; rather, it represents a *difference* of areas— the area between the curve C and the graph of $f(x, y)$ above the xy-plane minus the area between C and the graph of $f(x, y)$ below the xy-plane. We call this the **net signed area** between the curve C and the graph of $f(x, y)$. Also, we call the limit in (1) the **line integral of f with respect to s along C** and denote it by

$$\int_C f(x, y)\, ds = \lim_{n \to +\infty} \sum_{k=1}^{n} f(x_k^*, y_k^*)\Delta s_k \tag{2}$$

With this notation, the area of the surface in Figure 17.2.1 can be expressed as

$$A = \int_C f(x, y)\, ds \tag{3}$$

REMARK.   In Section 8.1 we observed that the area of a region in the xy-plane under a curve or between two curves over an interval $[a, b]$ is obtained by integrating the length of a vertical cross section of the region from a to b (see the remark preceding Example 1 in Section 8.1). Similarly, Formula (3) states that the area of a sheet along a curve C is obtained by integrating the length of a vertical cross section of the sheet along the curve C.

EVALUATING LINE INTEGRALS

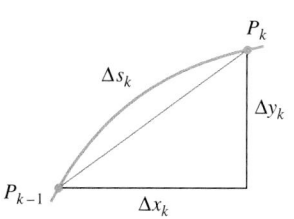

Figure 17.2.3

Except in simple cases, it will not be feasible to evaluate a line integral directly from (2). However, we will now show that it is possible to express a line integral as an ordinary definite integral, so that no special methods of evaluation are required. To see how this can be done, suppose that the curve C is represented by the parametric equations

$$x = x(t), \quad y = y(t) \qquad (a \le t \le b)$$

Moreover, suppose that the points P_{k-1} and P_k in Figure 17.2.3 correspond to parameter values of t_{k-1} and t_k, respectively, and that $P_k^*(x_k^*, y_k^*)$ corresponds to the parameter value t_k^*. If we let $\Delta t_k = t_k - t_{k-1}$, then we can approximate Δs_k as

$$\Delta s_k \approx \sqrt{(\Delta x_k)^2 + (\Delta y_k)^2} = \sqrt{\left(\frac{\Delta x_k}{\Delta t_k}\right)^2 + \left(\frac{\Delta y_k}{\Delta t_k}\right)^2}\, \Delta t_k \tag{4}$$

from which it follows that (2) can be expressed as

$$\int_C f(x, y)\, ds = \lim_{n \to +\infty} \sum_{k=1}^{n} f(x(t_k^*), y(t_k^*)) \sqrt{\left(\frac{\Delta x_k}{\Delta t_k}\right)^2 + \left(\frac{\Delta y_k}{\Delta t_k}\right)^2}\, \Delta t_k$$

which suggests that

$$\int_C f(x, y)\, ds = \int_a^b f(x(t), y(t)) \sqrt{\left(\frac{dx}{dt}\right)^2 + \left(\frac{dy}{dt}\right)^2}\, dt \qquad (5)$$

In words, this formula states that *a line integral can be evaluated by expressing the integrand in terms of the parameter, multiplying the integrand by an appropriate "radical," and then integrating from the initial value of the parameter to the final value of the parameter.*

In the special case where t is an arc length parameter, say $t = s$, it follows from Formula (20) of Section 14.3 that the radical in (5) reduces to 1, so the integration formula simplifies to

$$\int_C f(x, y)\, ds = \int_a^b f(x(s), y(s))\, ds \qquad (6)$$

Example 1

Evaluate the line integral $\int_C (1 + xy^2)\, ds$ from $(0, 0)$ to $(1, 2)$ along the line segment C that is represented by the parametric equations $x = t$, $y = 2t$ $(0 \le t \le 1)$.

Solution. It follows from Formula (5) that

$$\int_C (1 + xy^2)\, ds = \int_0^1 (1 + (t)(4t^2)) \sqrt{\left(\frac{dx}{dt}\right)^2 + \left(\frac{dy}{dt}\right)^2}\, dt$$

$$= \int_0^1 (1 + 4t^3)\sqrt{5}\, dt$$

$$= \sqrt{5}\left[t + t^4\right]_0^1 = 2\sqrt{5} \qquad \blacktriangleleft$$

Example 2

Find the area of the surface extending upward from the circle $x^2 + y^2 = 1$ in the xy-plane to the parabolic cylinder $z = 1 - x^2$ (Figure 17.2.4).

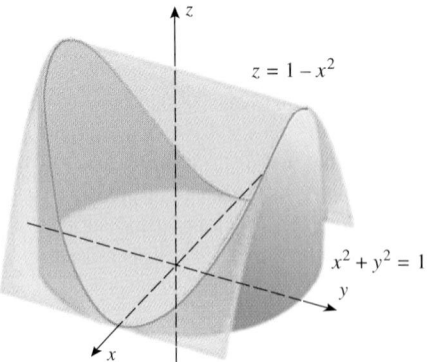

Figure 17.2.4

Solution. The area A of the surface can be expressed as the line integral

$$A = \int_C (1 - x^2)\, ds \qquad (7)$$

where C is the circle $x^2 + y^2 = 1$. This circle can be parametrized in terms of arc length as

$$x = \cos s, \quad y = \sin s \quad (0 \le s \le 2\pi)$$

Thus, it follows from (6) and (7) that

$$A = \int_C (1 - x^2)\, ds = \int_0^{2\pi} (1 - \cos^2 s)\, ds$$

$$= \int_0^{2\pi} \sin^2 s\, ds = \frac{1}{2} \int_0^{2\pi} (1 - \cos 2s)\, ds = \pi \quad \blacktriangleleft$$

REMARK.   We will show later in this section that we would have obtained the same value for (7) had we used any other smooth parametrization of the circle $x^2 + y^2 = 1$ in the xy-plane.

LINE INTEGRALS IN 3-SPACE

If C is a smooth curve that extends between two points in an xyz-coordinate system in 3-space, and if $f(x, y, z)$ is continuous on C, then the *line integral of f with respect to s along C* is defined as

$$\int_C f(x, y, z)\, ds = \lim_{n \to +\infty} \sum_{k=1}^{n} f(x_k^*, y_k^*, z_k^*)\Delta s_k \tag{8}$$

where the sum on the right side is obtained by subdividing the curve C into n arcs, choosing an arbitrary point (x_k^*, y_k^*, z_k^*) in the kth arc, multiplying $f(x_k^*, y_k^*, z_k^*)$ by the length Δs_k of the kth arc, and summing over all n arcs. If the curve C is represented by the parametric equations

$$x = x(t), \quad y = y(t), \quad z = z(t) \quad (a \le t \le b)$$

then (8) can be evaluated from the formula

$$\int_C f(x, y, z)\, ds = \int_a^b f(x(t), y(t), z(t))\sqrt{\left(\frac{dx}{dt}\right)^2 + \left(\frac{dy}{dt}\right)^2 + \left(\frac{dz}{dt}\right)^2}\, dt \tag{9}$$

and if t is an arc length parameter, say $t = s$, then it follows from Formula (21) of Section 14.3 that the radical in (9) reduces to 1, so the integration formula simplifies to

$$\int_C f(x, y, z)\, ds = \int_a^b f(x(s), y(s), z(s))\, ds \tag{10}$$

REMARK.   Observe that Formulas (9) and (10) have the same form as (5) and (6) but with an additional z-component. In general, line integrals along curves in 3-space do not have a simple area interpretation, so there is no analog of Formula (3). However, we will see later in this section that line integrals along curves in 3-space have other important interpretations.

Example 3

Evaluate the line integral $\int_C (xy + z^3)\, ds$ from $(1, 0, 0)$ to $(-1, 0, \pi)$ along the helix C that is represented by the parametric equations

$$x = \cos t, \quad y = \sin t, \quad z = t \quad (0 \le t \le \pi)$$

(Figure 17.2.5).

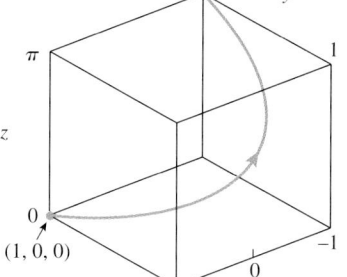

Figure 17.2.5

Solution. From (9)

$$\int_C (xy + z^3)\, ds = \int_0^\pi (\cos t \sin t + t^3)\sqrt{\left(\frac{dx}{dt}\right)^2 + \left(\frac{dy}{dt}\right)^2 + \left(\frac{dz}{dt}\right)^2}\, dt$$

$$= \int_0^\pi (\cos t \sin t + t^3)\sqrt{(-\sin t)^2 + (\cos t)^2 + 1}\, dt$$

$$= \sqrt{2} \int_0^\pi (\cos t \sin t + t^3)\, dt$$

$$= \sqrt{2}\left[\frac{\sin^2 t}{2} + \frac{t^4}{4}\right]_0^\pi = \frac{\sqrt{2}\pi^4}{4}$$

◀

MASS OF A WIRE AS A LINE INTEGRAL

We will now show how a line integral can be used to calculate the mass of a thin wire. For this purpose consider an idealized thin wire in 2-space or 3-space that is bent in the shape of a curve C. If the composition of the wire is uniform so that its mass is distributed uniformly, then the wire is said to be *homogeneous*, and we define the *linear mass density* of the wire to be the total mass divided by the total length. For example, a homogeneous wire with a mass of 2 g and a length of 8 cm would have a linear mass density of $\frac{2}{8} = 0.25$ g/cm. However, if the mass of the wire is not uniformly distributed, then the linear mass density is not a useful measure, since it does not account for variations in mass concentration. In this case we describe the mass concentration at a point by a *mass density function* δ, which we view as a limit; that is,

$$\delta = \lim_{\Delta s \to 0} \frac{\Delta M}{\Delta s} \tag{11}$$

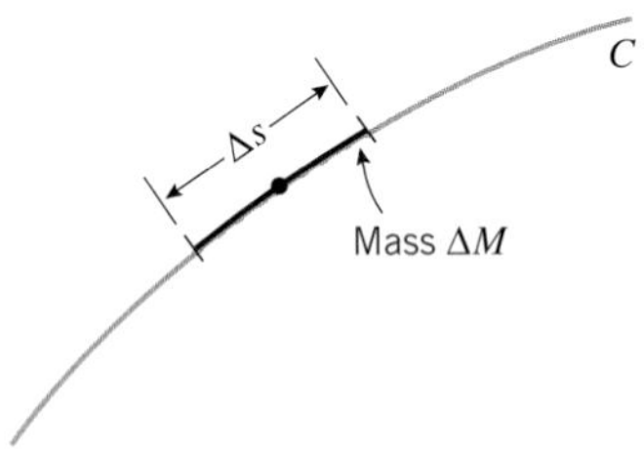

where ΔM and Δs denote the mass and length of a small section of wire centered at the point (Figure 17.2.6). Observe that $\Delta M/\Delta s$ is the linear mass density of the small section of wire, so that the mass density function at a point can be viewed informally as *the limit of the linear mass densities of small wire sections centered at the point*.

To translate this informal idea into a useful formula, suppose that $\delta = \delta(x, y)$ is the density function for a thin smooth wire in C in 2-space. Assume that the wire is subdivided into n small sections; let (x_k^*, y_k^*) be the center of the kth section, let ΔM_k be the mass of the kth section, and let Δs_k be the length of the kth section. Since we are assuming that the sections are small, it follows from (11) that the mass of the kth section can be approximated as

$$\Delta M_k \approx \delta(x_k^*, y_k^*)\Delta s_k$$

and hence the mass M of the entire wire can be approximated as

$$M = \sum_{k=1}^n \Delta M_k \approx \sum_{k=1}^n \delta(x_k^*, y_k^*)\Delta s_k \tag{12}$$

If we now increase n in such a way that the lengths of the sections approach zero, then it is plausible that the error in (12) will approach zero, and the exact value of M will be given by the line integral

$$M = \int_C \delta(x, y)\, ds \tag{13}$$

Similarly, the mass M of a wire C in 3-space with density function $\delta(x, y, z)$ is given by

$$M = \int_C \delta(x, y, z)\, ds \tag{14}$$

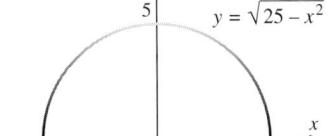

Figure 17.2.6

Example 4

Suppose that a semicircular wire has the equation $y = \sqrt{25 - x^2}$ and that its mass density is $\delta(x, y) = 15 - y$ (Figure 17.2.7). Physically, this means the wire has a maximum density

Figure 17.2.7

of 15 units at the base ($y = 0$) and that the density of the wire decreases linearly with respect to y to a value of 10 units at the top ($y = 5$). Find the mass of the wire.

Solution.   The mass M of the wire can be expressed as the line integral

$$M = \int_C \delta(x, y)\, ds = \int_C (15 - y)\, ds \tag{15}$$

along the semicircle C. To evaluate this integral we will express C parametrically as

$$x = 5 \cos t, \quad y = 5 \sin t \qquad (0 \le t \le \pi)$$

Thus, it follows from (5) and (15) that

$$
\begin{aligned}
M = \int_C (15 - y)\, ds &= \int_0^\pi (15 - 5 \sin t)\sqrt{\left(\frac{dx}{dt}\right)^2 + \left(\frac{dy}{dt}\right)^2}\, dt \\
&= \int_0^\pi (15 - 5 \sin t)\sqrt{(-5 \sin t)^2 + (5 \cos t)^2}\, dt \\
&= 5 \int_0^\pi (15 - 5 \sin t)\, dt \\
&= 5 \left[15t + 5 \cos t\right]_0^\pi \\
&\approx 75\pi - 50 \approx 185.6 \text{ units of mass} \qquad\blacktriangleleft
\end{aligned}
$$

ARC LENGTH AS A LINE INTEGRAL

In the special cases where $f(x, y)$ and $f(x, y, z)$ are 1, Formulas (5) and (9) become

$$\int_C ds = \int_a^b \sqrt{\left(\frac{dx}{dt}\right)^2 + \left(\frac{dy}{dt}\right)^2}\, dt$$

$$\int_C ds = \int_a^b \sqrt{\left(\frac{dx}{dt}\right)^2 + \left(\frac{dy}{dt}\right)^2 + \left(\frac{dz}{dt}\right)^2}\, dt$$

However, it follows from Formulas (2) and (4) of Section 14.3 that these integrals represent the arc length of C. Thus, we have established the following result.

> **17.2.2**   THEOREM.   *If C is a smooth parametric curve in 2-space or 3-space, then its arc length L can be expressed as*
>
> $$L = \int_C ds \tag{16}$$

REMARK.   This result adds nothing new computationally, since Formula (16) is just a reformulation of the arc length formulas in Section 14.3. However, the relationship between line integrals and arc length is important to know.

LINE INTEGRALS WITH RESPECT TO x, y, AND z

There are other important types of line integrals that result by replacing Δs_k in definitions (2) and (8) by $\Delta x_k = x_k - x_{k-1}$, $\Delta y_k = y_k - y_{k-1}$, or $\Delta z_k = z_k - z_{k-1}$, where (x_k, y_k, z_k) and $(x_{k-1}, y_{k-1}, z_{k-1})$ are the coordinates of the points P_k and P_{k-1} in Figure 17.2.2. For example, in 2-space we define

$$\int_C f(x, y)\, dx = \lim_{n \to +\infty} \sum_{k=1}^n f(x_k^*, y_k^*)\Delta x_k \tag{17}$$

$$\int_C f(x, y)\, dy = \lim_{n \to +\infty} \sum_{k=1}^n f(x_k^*, y_k^*)\Delta y_k \tag{18}$$

and in 3-space we define

$$\int_C f(x, y, z)\, dx = \lim_{n \to +\infty} \sum_{k=1}^{n} f(x_k^*, y_k^*, z_k^*) \Delta x_k \qquad (19)$$

$$\int_C f(x, y, z)\, dy = \lim_{n \to +\infty} \sum_{k=1}^{n} f(x_k^*, y_k^*, z_k^*) \Delta y_k \qquad (20)$$

$$\int_C f(x, y, z)\, dz = \lim_{n \to +\infty} \sum_{k=1}^{n} f(x_k^*, y_k^*, z_k^*) \Delta z_k \qquad (21)$$

We will call these *line integrals with respect to x, y, and z* (as appropriate) in contrast to (2) and (8), which are line integrals with respect to s (also called *line integrals with respect to arc length*).

The basic procedure for evaluating these line integrals is to find parametric equations for C, say

$$x = x(t), \quad y = y(t), \quad z = z(t) \qquad (a \le t \le b)$$

and then express the integrand in terms of t. For example,

$$\int_C f(x, y)\, dx = \int_a^b \left[f(x(t), y(t)) \frac{dx}{dt} \right] dt = \int_a^b f(x(t), y(t)) x'(t)\, dt$$

We omit the formal proof.

For reference, we list the relevant formulas.

$$\int_C f(x, y)\, dx = \int_a^b f(x(t), y(t)) x'(t)\, dt \qquad (22)$$

$$\int_C f(x, y)\, dy = \int_a^b f(x(t), y(t)) y'(t)\, dt \qquad (23)$$

$$\int_C f(x, y, z)\, dx = \int_a^b f(x(t), y(t), z(t)) x'(t)\, dt \qquad (24)$$

$$\int_C f(x, y, z)\, dy = \int_a^b f(x(t), y(t), z(t)) y'(t)\, dt \qquad (25)$$

$$\int_C f(x, y, z)\, dz = \int_a^b f(x(t), y(t), z(t)) z'(t)\, dt \qquad (26)$$

Frequently, the line integrals with respect to x and y occur in combination, in which case we dispense with one of the integral signs and write

$$\int_C f(x, y)\, dx + g(x, y)\, dy = \int_C f(x, y)\, dx + \int_C g(x, y)\, dy \qquad (27)$$

and similarly,

$$\int_C f(x, y, z)\, dx + g(x, y, z)\, dy + h(x, y, z)\, dz$$
$$= \int_C f(x, y, z)\, dx + \int_C g(x, y, z)\, dy + \int_C h(x, y, z)\, dz \qquad (28)$$

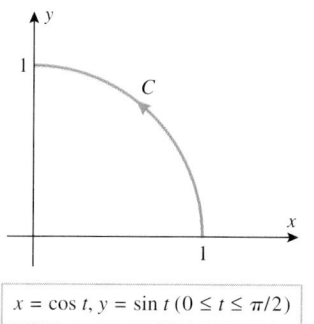

$x = \cos t, \, y = \sin t \, (0 \le t \le \pi/2)$

Figure 17.2.8

Example 5

Evaluate

$$\int_C 2xy \, dx + (x^2 + y^2) \, dy$$

along the circular arc C given by $x = \cos t$, $y = \sin t$ ($0 \le t \le \pi/2$) (Figure 17.2.8).

Solution. From (22) and (23)

$$\int_C 2xy \, dx = \int_0^{\pi/2} (2 \cos t \sin t) \left[\frac{d}{dt} (\cos t) \right] dt$$

$$= -2 \int_0^{\pi/2} \sin^2 t \cos t \, dt = -\frac{2}{3} \sin^3 t \bigg]_0^{\pi/2} = -\frac{2}{3}$$

$$\int_C (x^2 + y^2) \, dy = \int_0^{\pi/2} (\cos^2 t + \sin^2 t) \left[\frac{d}{dt} (\sin t) \right] dt$$

$$= \int_0^{\pi/2} \cos t \, dt = \sin t \bigg]_0^{\pi/2} = 1$$

Thus, from (27)

$$\int_C 2xy \, dx + (x^2 + y^2) \, dy = \int_C 2xy \, dx + \int_C (x^2 + y^2) \, dy$$

$$= -\frac{2}{3} + 1 = \frac{1}{3}$$ ◄

Example 6

(a) Show that $\int_C f(x, y) \, dx = 0$ along any line segment parallel to the y-axis.

(b) Show that $\int_C f(x, y) \, dy = 0$ along any line segment parallel to the x-axis.

Solution. A line segment parallel to the y-axis can be represented parametrically by equations of the form $x = k$, $y = t$, where k is a constant. Thus, $x'(t) = 0$ in (22). Similarly, a line segment parallel to the x-axis can be represented parametrically by equations of the form $x = t$, $y = k$, where k is a constant. Thus, $y'(t) = 0$ in (23). ◄

┊ FOR THE READER.   What is the analog of Example 6 in 3-space?

LINE INTEGRALS ALONG PIECEWISE SMOOTH CURVES

Thus far, we have only considered line integrals along smooth curves. However, the notion of a line integral can be extended to curves formed from finitely many smooth curves $C_1, C_2, \ldots, C_n$ joined end to end. Such a curve is called *piecewise smooth* (Figure 17.2.9). We define a line integral along a piecewise smooth curve C to be the sum of the integrals along the sections:

$$\int_C = \int_{C_1} + \int_{C_2} + \cdots + \int_{C_n}$$

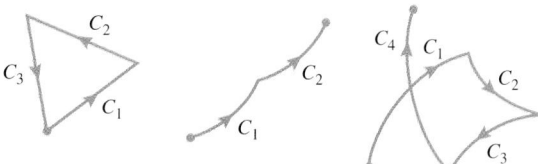

Figure 17.2.9

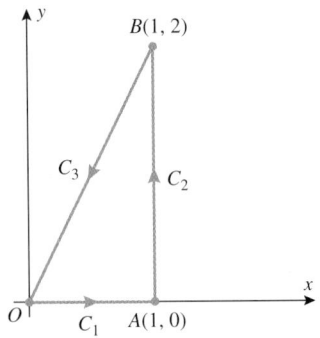

Figure 17.2.10

Example 7

Evaluate

$$\int_C x^2 y \, dx + x \, dy$$

in a counterclockwise direction around the triangular path shown in Figure 17.2.10.

Solution. We will integrate over C_1, C_2, and C_3 separately and add the results. For each of the three integrals we must find parametric equations that trace the path of integration in the correct direction. For this purpose recall from Formula (9) of Section 14.1 that the graph of the vector-valued function

$$\mathbf{r}(t) = (1 - t)\mathbf{r}_0 + t\mathbf{r}_1 \quad (0 \le t \le 1)$$

is the line segment joining $\mathbf{r}_0$ and $\mathbf{r}_1$, oriented in the direction from $\mathbf{r}_0$ to $\mathbf{r}_1$. Thus, the line segments C_1, C_2, and C_3 can be represented in vector notation as

$$C_1: \mathbf{r}(t) = (1 - t)\langle 0, 0\rangle + t\langle 1, 0\rangle = \langle t, 0\rangle$$
$$C_2: \mathbf{r}(t) = (1 - t)\langle 1, 0\rangle + t\langle 1, 2\rangle = \langle 1, 2t\rangle$$
$$C_3: \mathbf{r}(t) = (1 - t)\langle 1, 2\rangle + t\langle 0, 0\rangle = \langle 1 - t, 2 - 2t\rangle$$

where t varies from 0 to 1 in each case. From these equations and Example 6 we obtain

$$\int_{C_1} x^2 y \, dx + x \, dy = \int_{C_1} x^2 y \, dx = \int_0^1 (t^2)(0) \frac{d}{dt}[t] \, dt = 0$$

$$\int_{C_2} x^2 y \, dx + x \, dy = \int_{C_2} x \, dy = \int_0^1 (1) \frac{d}{dt}[2t] \, dt = 2$$

$$\int_{C_3} x^2 y \, dx + x \, dy = \int_0^1 (1 - t)^2 (2 - 2t) \frac{d}{dt}[1 - t] \, dt + \int_0^1 (1 - t) \frac{d}{dt}[2 - 2t] \, dt$$

$$= 2 \int_0^1 (t - 1)^3 \, dt + 2 \int_0^1 (t - 1) \, dt = -\tfrac{1}{2} - 1 = -\tfrac{3}{2}$$

Thus,

$$\int_C x^2 y \, dx + x \, dy = 0 + 2 + \left(-\tfrac{3}{2}\right) = \tfrac{1}{2} \quad \blacktriangleleft$$

CHANGE OF PARAMETER IN LINE INTEGRALS

Since the parametric equations of a curve are used to evaluate line integrals along that curve, it seems possible that two different parametrizations of a curve C might produce different values for the same line integral along C. The following theorem, which we state without proof, shows that this is not the case.

> **17.2.3** THEOREM (*Independence of Parametrization*). *The value of a line integral along a curve C does not depend on the parametrization of C in the sense that any two parametrizations of C with the same orientation produce the same value for the line integral.*

REMARK.   This is an extremely important theorem because it allows us to choose any convenient parametrization for the path of integration without concern that the choice will affect the value of the integral. Indeed, we have tacitly used this result in all of the examples in this section where we chose the parametric equations for C.

REVERSING THE DIRECTION OF INTEGRATION

Suppose that C is a parametric curve that begins at point A and ends at point B when traced in the direction of increasing parameter. If the curve C is reparametrized so that it is traced from B to A as the parameter increases, then we denote the reparametrized curve by $-C$. Thus, C and $-C$ consist of the same points but have opposite orientations (Figure 17.2.11).

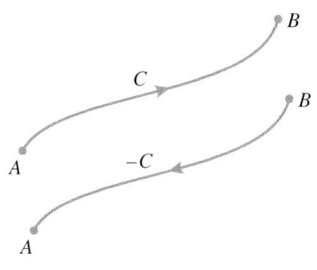

Figure 17.2.11

When the orientation of C is reversed, the signs of Δx_k, Δy_k, and Δz_k in (17) to (21) are reversed, so the effect is to reverse the signs of the line integrals with respect to x, y, and z. However, reversing the orientation of C has no effect on a line integral with respect to s because the quantity Δs_k in (2) and (8) denotes an arc length, which is positive regardless of the orientation. Thus, we have the following result, which we state without formal proof.

17.2.4 THEOREM (*Reversal of Orientation*). *If C is a smooth parametric curve, then a smooth change of parameter that reverses the orientation of C changes the sign of a line integral along C with respect to x, y, or z, but leaves the value of a line integral along C with respect to arc length unchanged.*

It follows from this theorem that

$$\int_{-C} f(x, y)\, dx + g(x, y)\, dy = -\int_{C} f(x, y)\, dx + g(x, y)\, dy \tag{29}$$

$$\int_{-C} f(x, y)\, ds = \int_{C} f(x, y)\, ds \tag{30}$$

and similarly for line integrals in 3-space.

WORK AS A LINE INTEGRAL

In Section 8.6 we first defined the work W performed by a force of constant magnitude acting on an object in the direction of motion (Definition 8.6.1), and later in that section we extended the definition to allow for a force of variable magnitude acting in the direction of motion (Definition 8.6.3). In Section 13.3 we took the concept of work a step further by defining the work W performed by a constant force $\mathbf{F}$ acting at a fixed angle to the displacement vector $\overrightarrow{PQ}$ to be

$$W = \mathbf{F} \cdot \overrightarrow{PQ} \tag{31}$$

[Formula (14) of Section 13.3]. Our next goal is to define a more general concept of work—the work performed by a variable force acting on a particle that moves along a curved path in 2-space or 3-space.

In many applications variable forces arise from force fields (gravitational fields, electromagnetic fields, and so forth), so we will consider the problem of work in that context. More precisely, let us assume that a particle moves along a smooth parametric curve C through a continuous force field $\mathbf{F}(x, y)$ in 2-space or $\mathbf{F}(x, y, z)$ in 3-space. We will call the work done by $\mathbf{F}$ the ***work performed by the force field***. To motivate an appropriate definition for the work performed by the force field, we will use a limit process, and since the procedure is the same in 2-space and 3-space, we will discuss it for 3-space only. The idea is as follows:

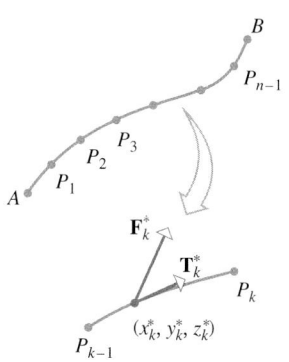

Figure 17.2.12

- Assume that the particle moves along C from a point A to a point B as the parameter increases, and divide C into n arcs by inserting a succession of distinct points $P_1, P_2, \ldots, P_{n-1}$ between A and B in the direction of increasing parameter. Denote the length of the kth arc by Δs_k. Let (x_k^*, y_k^*, z_k^*) be any point on the kth arc, and let $\mathbf{T}_k^* = \mathbf{T}(x_k^*, y_k^*, z_k^*)$ be the unit tangent vector and $\mathbf{F}_k^* = \mathbf{F}(x_k^*, y_k^*, z_k^*)$ the force vector at this point (Figure 17.2.12).

- If the kth arc is small, then the force will not vary much, so we can assume that the force has a constant value of $\mathbf{F}_k^*$ on this arc. Moreover, the direction of motion will not vary much over the small arc, so we can assume that the particle moves in the direction of $\mathbf{T}_k^*$ for a distance of Δs_k; that is, the particle has a linear displacement $\Delta s_k \mathbf{T}_k^*$. Thus, it follows from (31) that the work ΔW_k performed by the vector field along the kth arc can be approximated as

$$\Delta W_k \approx \mathbf{F}_k^* \cdot (\Delta s_k \mathbf{T}_k^*) = (\mathbf{F}_k^* \cdot \mathbf{T}_k^*)\Delta s_k$$

and the total work W performed by the vector field as the particle moves along C from A to B can be approximated as

$$W \approx \sum_{k=1}^{n}(\mathbf{F}_k^* \cdot \mathbf{T}_k^*)\Delta s_k$$

- If we now increase n so that the length of each arc approaches zero, then it is plausible that the error in the approximations approaches zero, and the exact work performed by the vector field is

$$W = \lim_{n \to +\infty} \sum_{k=1}^{n}(\mathbf{F}_k^* \cdot \mathbf{T}_k^*)\Delta s_k = \int_C \mathbf{F}(x, y, z) \cdot \mathbf{T}(x, y, z)\, ds$$

Thus, we are led to the following definition:

17.2.5   DEFINITION.   If $\mathbf{F}$ is a continuous vector field and C is a smooth parametric curve in 2-space or 3-space with unit tangent vector $\mathbf{T}$, then the **work performed by the vector field** on a particle that moves along C in the direction of increasing parameter is

$$W = \int_C \mathbf{F} \cdot \mathbf{T}\, ds \tag{32}$$

REMARK.   In words, this definition states that *the work performed by a vector field on a particle moving along a parametric curve C is obtained by integrating the scalar tangential component of force along C.*

A METHOD FOR CALCULATING WORK

Although Formula (32) can be used to calculate work, it is not usually the best choice. A more useful formula can be obtained by using Formula (6) of Section 14.4 to express $\mathbf{T}$ as

$$\mathbf{T} = \frac{d\mathbf{r}}{ds}$$

This suggests that (32) can be expressed as

$$W = \int_C \mathbf{F} \cdot d\mathbf{r} \tag{33}$$

in which $d\mathbf{r}$ is interpreted as

$$d\mathbf{r} = dx\,\mathbf{i} + dy\,\mathbf{j} \qquad \text{or} \qquad d\mathbf{r} = dx\,\mathbf{i} + dy\,\mathbf{j} + dz\,\mathbf{k} \tag{34}$$

depending on whether C is in 2-space or 3-space.

Example 8

Find the work done by the force field

$$\mathbf{F}(x, y) = x^3 y\,\mathbf{i} + (x - y)\,\mathbf{j}$$

on a particle that moves along the parabola $y = x^2$ from $(-2, 4)$ to $(1, 1)$ (see Figure 17.2.13).

Solution.   If we use $x = t$ as the parameter, the path C of the particle can be expressed parametrically as

$$x = t, \quad y = t^2 \qquad (-2 \le t \le 1)$$

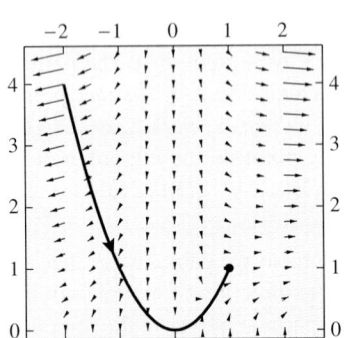

Figure 17.2.13

or in vector notation as

$$\mathbf{r}(t) = t\mathbf{i} + t^2\mathbf{j} \quad (-2 \le t \le 1)$$

Thus, from (33) the work W done by $\mathbf{F}$ is

$$W = \int_C \mathbf{F} \cdot d\mathbf{r} = \int_C (x^3 y\mathbf{i} + (x - y)\mathbf{j}) \cdot (dx\mathbf{i} + dy\mathbf{j})$$

$$= \int_C x^3 y \, dx + (x - y) \, dy = \int_{-2}^{1} (t^5 + (t - t^2)(2t)) \, dt$$

$$= \frac{1}{6}t^6 + \frac{2}{3}t^3 - \frac{1}{2}t^4 \bigg]_{-2}^{1} = 3$$

where the units for W depend on the units chosen for force and distance.   ◀

REMARK.   In light of Theorem 17.2.4, you might expect that reversing the orientation of C in Formula (32) would have no effect on the work W performed by the vector field. However, reversing the orientation of C reverses the orientation of $\mathbf{T}$ in the integrand and hence reverses the sign of the integral; that is,

$$\int_{-C} \mathbf{F} \cdot \mathbf{T} \, ds = -\int_C \mathbf{F} \cdot \mathbf{T} \, ds \qquad (35)$$

$$\int_{-C} \mathbf{F} \cdot d\mathbf{r} = -\int_C \mathbf{F} \cdot d\mathbf{r} \qquad (36)$$

Thus, in Example 8 the work performed on a particle that moves along the given parabola from $(1, 1)$ to $(-2, 4)$ is -3, and the work performed on a particle that moves along the parabola from $(-2, 4)$ to $(1, 1)$ and then back along the parabola to $(-2, 4)$ is zero.

WORK EXPRESSED IN SCALAR FORM

We conclude this section by noting that it is sometimes useful to express Formula (33) in scalar form. For example, if $\mathbf{F} = \mathbf{F}(x, y) = f(x, y)\mathbf{i} + g(x, y)\mathbf{j}$ is a vector field in 2-space, then

$$\mathbf{F} \cdot d\mathbf{r} = f(x, y) \, dx + g(x, y) \, dy$$

so (33) can be expressed as

$$W = \int_C f(x, y) \, dx + g(x, y) \, dy \qquad (37)$$

and similarly in 3-space as

$$W = \int_C f(x, y, z) \, dx + g(x, y, z) \, dy + h(x, y, z) \, dz \qquad (38)$$

EXERCISE SET 17.2 ☐c CAS

1. Let C be the line segment from $(0, 0)$ to $(0, 1)$. In each part, evaluate the line integral along C by inspection, and explain your reasoning.

(a) $\int_C ds$

(b) $\int_C \sin xy \, dy$

2. Let C be the line segment from $(0, 2)$ to $(0, 4)$. In each part, evaluate the line integral along C by inspection, and explain your reasoning.

(a) $\int_C ds$

(b) $\int_C e^{xy} \, dx$

3. Let C be the curve represented by the equations

$$x = 2t, \quad y = 3t^2 \quad (0 \le t \le 1)$$

In each part, evaluate the line integral along C.

(a) $\displaystyle\int_C (x - y)\,ds$ (b) $\displaystyle\int_C (x - y)\,dx$

(c) $\displaystyle\int_C (x - y)\,dy$

4. Let C be the curve represented by the equations

$$x = t, \quad y = 3t^2, \quad z = 6t^3 \quad (0 \le t \le 1)$$

In each part, evaluate the line integral along C.

(a) $\displaystyle\int_C xyz^2\,ds$ (b) $\displaystyle\int_C xyz^2\,dx$

(c) $\displaystyle\int_C xyz^2\,dy$ (d) $\displaystyle\int_C xyz^2\,dz$

5. In each part, evaluate the integral

$$\int_C (3x + 2y)\,dx + (2x - y)\,dy$$

along the stated curve.

(a) The line segment from $(0, 0)$ to $(1, 1)$.
(b) The parabolic arc $y = x^2$ from $(0, 0)$ to $(1, 1)$.
(c) The curve $y = \sin(\pi x/2)$ from $(0, 0)$ to $(1, 1)$.
(d) The curve $x = y^3$ from $(0, 0)$ to $(1, 1)$.

6. In each part, evaluate the integral

$$\int y\,dx + z\,dy - x\,dz$$

along the stated curve.

(a) The line segment from $(0, 0, 0)$ to $(1, 1, 1)$.
(b) The twisted cubic $x = t, y = t^2, z = t^3$ from $(0, 0, 0)$ to $(1, 1, 1)$.
(c) The helix $x = \cos \pi t, y = \sin \pi t, z = t$ from $(1, 0, 0)$ to $(-1, 0, 1)$.

In Exercises 7–10, evaluate the line integral with respect to s along the parametric curve C.

7. $\displaystyle\int_C \frac{1}{1 + x}\,ds$

$C: x = t, \ y = \frac{2}{3}t^{3/2} \quad (0 \le t \le 3)$

8. $\displaystyle\int_C \frac{x}{1 + y^2}\,ds$

$C: x = 1 + 2t, \ y = t \quad (0 \le t \le 1)$

9. $\displaystyle\int_C 3x^2 yz\,ds$

$C: x = t, \ y = t^2, \ z = \frac{2}{3}t^3 \quad (0 \le t \le 1)$

10. $\displaystyle\int_C \frac{e^{-z}}{x^2 + y^2}\,ds$

$C: x = 2\cos t, \ y = 2\sin t, \ z = t \quad (0 \le t \le 2\pi)$

In Exercises 11–18, evaluate the line integral along the parametric curve C.

11. $\displaystyle\int_C (x + 2y)\,dx + (x - y)\,dy$

$C: x = 2\cos t, \ y = 4\sin t \quad (0 \le t \le \pi/4)$

12. $\displaystyle\int_C (x^2 - y^2)\,dx + x\,dy$

$C: x = t^{2/3}, \ y = t \quad (-1 \le t \le 1)$

13. $\displaystyle\int_C -y\,dx + x\,dy$

$C: y^2 = 3x$ from $(3, 3)$ to $(0, 0)$

14. $\displaystyle\int_C (y - x)\,dx + x^2 y\,dy$

$C: y^2 = x^3$ from $(1, -1)$ to $(1, 1)$

15. $\displaystyle\int_C (x^2 + y^2)\,dx - x\,dy$

$C: x^2 + y^2 = 1$, counterclockwise from $(1, 0)$ to $(0, 1)$

16. $\displaystyle\int_C (y - x)\,dx + xy\,dy$

$C:$ the line segment from $(3, 4)$ to $(2, 1)$

17. $\displaystyle\int_C yz\,dx - xz\,dy + xy\,dz$

$C: x = e^t, \ y = e^{3t}, \ z = e^{-t} \quad (0 \le t \le 1)$

18. $\displaystyle\int_C x^2\,dx + xy\,dy + z^2\,dz$

$C: x = \sin t, \ y = \cos t, \ z = t^2 \quad (0 \le t \le \pi/2)$

C **19.** Use a CAS to check the answers to the problems that you solved in Exercises 7–18 by evaluating the integral with respect to t that you obtained.

C **20.** In each part, use a CAS to evaluate the line integral along the parametric curve C.

(a) $\displaystyle\int_C x^7 y^3\,ds$

$C: x = \cos^3 t, \ y = \sin^3 t \quad (0 \le t \le \pi/2)$

(b) $\displaystyle\int_C x^5 z\,dx + 7y\,dy + y^2 z\,dz$

$C: x = t, \ y = t^2, \ z = \ln t \quad (1 \le t \le e)$

In Exercises 21 and 22, evaluate $\int_C y\,dx - x\,dy$ along the curve C shown in the figure.

21. (a) 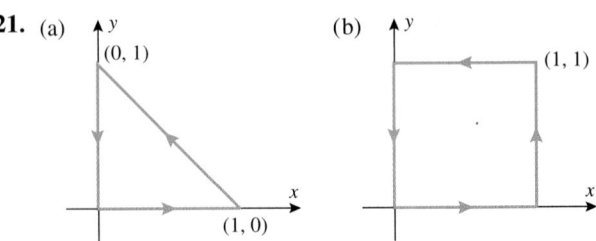 (b)

22. (a) 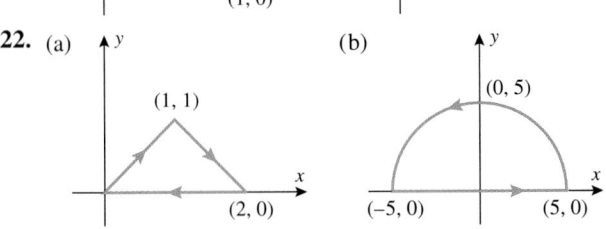 (b)

In Exercises 23 and 24, evaluate $\int_C x^2 z\, dx - yx^2\, dy + 3\, dz$ along the curve C shown in the figure.

23.

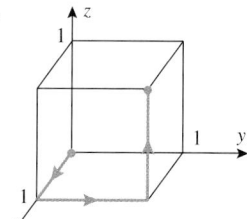

24.

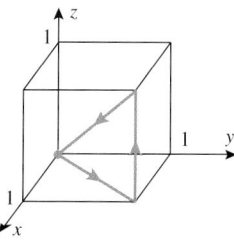

In Exercises 25–28, evaluate $\int_C \mathbf{F} \cdot d\mathbf{r}$ along the curve C.

25. $\mathbf{F}(x, y) = x^2\mathbf{i} + xy\mathbf{j}$
$C: \mathbf{r}(t) = 2\cos t\,\mathbf{i} + 2\sin t\,\mathbf{j}$ $(0 \le t \le \pi)$

26. $\mathbf{F}(x, y) = x^2 y\mathbf{i} + 4\mathbf{j}$
$C: \mathbf{r}(t) = e^t\mathbf{i} + e^{-t}\mathbf{j}$ $(0 \le t \le 1)$

27. $\mathbf{F}(x, y) = (x^2 + y^2)^{-3/2}(x\mathbf{i} + y\mathbf{j})$
$C: \mathbf{r}(t) = e^t \sin t\,\mathbf{i} + e^t \cos t\,\mathbf{j}$ $(0 \le t \le 1)$

28. $\mathbf{F}(x, y, z) = z\mathbf{i} + x\mathbf{j} + y\mathbf{k}$
$C: \mathbf{r}(t) = \sin t\,\mathbf{i} + 3\sin t\,\mathbf{j} + \sin^2 t\,\mathbf{k}$ $(0 \le t \le \pi/2)$

29. Find the mass of a thin wire shaped in the form of the circular arc $y = \sqrt{9 - x^2}$ $(0 \le x \le 3)$ if the density function is $\delta(x, y) = x\sqrt{y}$.

30. Find the mass of a thin wire shaped in the form of the curve $x = e^t \cos t$, $y = e^t \sin t$ $(0 \le t \le 1)$ if the density function δ is proportional to the distance from the origin.

31. Find the mass of a thin wire shaped in the form of the helix $x = 3\cos t$, $y = 3\sin t$, $z = 4t$ $(0 \le t \le \pi/2)$ if the density function is $\delta = kx/(1 + y^2)$ $(k > 0)$.

32. Find the mass of a thin wire shaped in the form of the curve $x = 2t$, $y = \ln t$, $z = 4\sqrt{t}$ $(1 \le t \le 4)$ if the density function is proportional to the distance above the xy-plane.

In Exercises 33–36, find the work done by the force field $\mathbf{F}$ on a particle that moves along the curve C.

33. $\mathbf{F}(x, y) = xy\mathbf{i} + x^2\mathbf{j}$
$C: x = y^2$ from $(0, 0)$ to $(1, 1)$

34. $\mathbf{F}(x, y) = (x^2 + xy)\mathbf{i} + (y - x^2 y)\mathbf{j}$
$C: x = t$, $y = 1/t$ $(1 \le t \le 3)$

35. $\mathbf{F}(x, y, z) = xy\mathbf{i} + yz\mathbf{j} + xz\mathbf{k}$
$C: \mathbf{r}(t) = t\mathbf{i} + t^2\mathbf{j} + t^3\mathbf{k}$ $(0 \le t \le 1)$

36. $\mathbf{F}(x, y, z) = (x + y)\mathbf{i} + xy\mathbf{j} - z^2\mathbf{k}$
$C:$ along line segments from $(0, 0, 0)$ to $(1, 3, 1)$ to $(2, -1, 4)$

In Exercises 37 and 38, find $\int_C \mathbf{F} \cdot d\mathbf{r}$ by inspection for the force field $\mathbf{F}(x, y) = \mathbf{i} + \mathbf{j}$ and the curve C shown in the figure. Explain your reasoning. [For clarity, the vectors in the force field are shown at less than true scale.]

37.

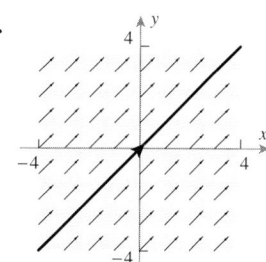

38.

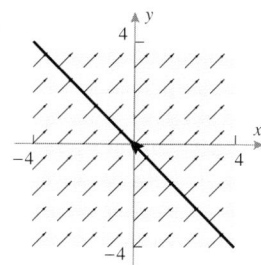

In Exercises 39 and 40, find the work done by the force field

$$\mathbf{F}(x, y) = \frac{1}{x^2 + y^2}\mathbf{i} + \frac{4}{x^2 + y^2}\mathbf{j}$$

on a particle that moves along the curve C shown in the figure.

39.   **40.**

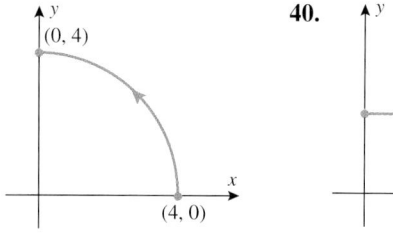

In Exercises 41 and 42, use a line integral to find the area of the surface.

41. The surface that extends upward from the parabola $y = x^2$ $(0 \le x \le 2)$ in the xy-plane to the plane $z = 3x$.

42. The surface that extends upward from the semicircle $y = \sqrt{4 - x^2}$ in the xy-plane to the surface $z = x^2 y$.

43. As illustrated in the accompanying figure, a sinusoidal cut is made in the top of a cylindrical tin can. Suppose that the base is modeled by the parametric equations $x = \cos t$, $y = \sin t$, $z = 0$ $(0 \le t \le 2\pi)$, and the height of the cut as a function of t is $z = 2 + 0.5\sin 3t$.

(a) Use a geometric argument to find the lateral surface area of the cut can.

(b) Write down a line integral for the surface area.

(c) Use the line integral to calculate the surface area.

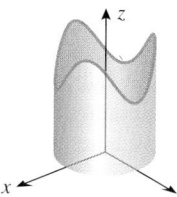

Figure Ex-43

44. Evaluate the integral $\displaystyle\int_{-C} \frac{x\,dy - y\,dx}{x^2 + y^2}$, where C is the circle $x^2 + y^2 = a^2$ traversed counterclockwise.

45. Suppose that a particle moves through the force field $\mathbf{F}(x, y) = xy\mathbf{i} + (x - y)\mathbf{j}$ from the point $(0, 0)$ to the point

$(1, 0)$ along the curve $x = t$, $y = \lambda t (1 - t)$. For what value of λ will the work done by the force field be 1?

46. A farmer weighing 150 lb carries a sack of grain weighing 20 lb up a circular helical staircase around a silo of radius 25 ft. As the farmer climbs, grain leaks from the sack at a rate of 1 lb per 10 ft of ascent. How much work is performed by the farmer in climbing through a vertical distance of 60 ft in exactly four revolutions? [*Hint:* Find a vector field that represents the force exerted by the farmer in lifting his own weight plus the weight of the sack upward at each point along his path.]

17.3 INDEPENDENCE OF PATH; CONSERVATIVE VECTOR FIELDS

In this section we will study properties of vector fields that relate to the work they perform on particles moving along various curves. In particular, we will show that for certain kinds of vector fields the work that the field performs on a particle moving along a curve depends only on the endpoints of the curve and not on the curve itself. Such vector fields are of special importance in physics and engineering.

WORK INTEGRALS

We saw in the last section that if $\mathbf{F}$ is a vector field in 2-space or 3-space, then the work performed by the field on a particle moving along a parametric curve C from an initial point A to a final point B is given by the integral

$$\int_C \mathbf{F} \cdot \mathbf{T} \, ds \quad \text{or, equivalently,} \quad \int_C \mathbf{F} \cdot d\mathbf{r}$$

Accordingly, we call an integral of this type a *work integral*. At the end of the last section we noted that a work integral can be expressed in scalar form as

$$\int_C \mathbf{F} \cdot d\mathbf{r} = \int_C f(x, y) \, dx + g(x, y) \, dy \qquad \boxed{\text{2-space}} \tag{1}$$

$$\int_C \mathbf{F} \cdot d\mathbf{r} = \int_C f(x, y, z) \, dx + g(x, y, z) \, dy + h(x, y, z) \, dz \qquad \boxed{\text{3-space}} \tag{2}$$

where f, g, and h are the component functions of $\mathbf{F}$.

INDEPENDENCE OF PATH

The parametric curve C in a work integral is called the *path of integration*. One of the important problems in applications is to determine how the path of integration affects the work performed by a vector field on a particle that moves from a fixed point P to a fixed point Q. We will show shortly that if the vector field $\mathbf{F}$ is conservative (i.e., is the gradient of some potential function ϕ), then the work that the field performs on a particle that moves from P to Q does not depend on the particular path C that the particle follows. This is illustrated in the following example.

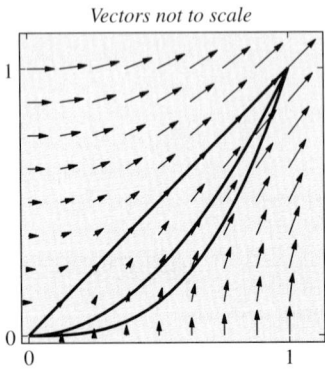
Vectors not to scale

Figure 17.3.1

Example 1

The vector field $\mathbf{F}(x, y) = y\mathbf{i} + x\mathbf{j}$ is conservative since it is the gradient of $\phi(x, y) = xy$ (verify). Thus, the preceding discussion suggests that the work performed by the field on a particle that moves from the point $(0, 0)$ to the point $(1, 1)$ should be the same along different paths. Confirm that the value of the work integral

$$\int_C \mathbf{F} \cdot d\mathbf{r}$$

is the same along the following paths (Figure 17.3.1):

(a)   The line segment $y = x$ from $(0, 0)$ to $(1, 1)$.

(b)   The parabola $y = x^2$ from $(0, 0)$ to $(1, 1)$.

(c)   The cubic $y = x^3$ from $(0, 0)$ to $(1, 1)$.

Solution (a). With $x = t$ as the parameter, the path of integration is given by

$$x = t, \quad y = t \qquad (0 \le t \le 1)$$

Thus,

$$\int_C \mathbf{F} \cdot d\mathbf{r} = \int_C (y\mathbf{i} + x\mathbf{j}) \cdot (dx\mathbf{i} + dy\mathbf{j}) = \int_C y\,dx + x\,dy$$

$$= \int_0^1 2t\,dt = 1$$

Solution (b). With $x = t$ as the parameter, the path of integration is given by

$$x = t, \quad y = t^2 \qquad (0 \le t \le 1)$$

Thus,

$$\int_C \mathbf{F} \cdot d\mathbf{r} = \int_C y\,dx + x\,dy = \int_0^1 3t^2\,dt = 1$$

Solution (c). With $x = t$ as the parameter, the path of integration is given by

$$x = t, \quad y = t^3 \qquad (0 \le t \le 1)$$

Thus,

$$\int_C \mathbf{F} \cdot d\mathbf{r} = \int_C y\,dx + x\,dy = \int_0^1 4t^3\,dt = 1 \qquad \blacktriangleleft$$

THE FUNDAMENTAL THEOREM OF WORK INTEGRALS

Recall from the Fundamental Theorem of Calculus (Theorem 7.6.1) that if F is an antiderivative of f, then

$$\int_a^b f(x)\,dx = F(b) - F(a)$$

The following result is the analog of that theorem for work integrals in 2-space.

17.3.1 THEOREM (*The Fundamental Theorem of Work Integrals*). *Suppose that*

$$\mathbf{F}(x, y) = f(x, y)\mathbf{i} + g(x, y)\mathbf{j}$$

is a conservative vector field in some open region D containing the points (x_0, y_0) and (x_1, y_1) and that $f(x, y)$ and $g(x, y)$ are continuous in this region. If

$$\mathbf{F}(x, y) = \nabla\phi(x, y)$$

and if C is any piecewise smooth parametric curve that starts at (x_0, y_0), ends at (x_1, y_1), and lies in the region D, then

$$\int_C \mathbf{F}(x, y) \cdot d\mathbf{r} = \phi(x_1, y_1) - \phi(x_0, y_0) \tag{3}$$

or, equivalently,

$$\int_C \nabla\phi \cdot d\mathbf{r} = \phi(x_1, y_1) - \phi(x_0, y_0) \tag{4}$$

Proof. We will give the proof for a smooth curve C. The proof for a piecewise smooth curve, which is left as an exercise, can be obtained by applying the theorem to each individual smooth piece and adding the results. Suppose that C is given parametrically by $x = x(t)$,

$y = y(t)$ $(a \le t \le b)$, so that the initial and final points of the curve are

$$(x_0, y_0) = (x(a), y(a)) \quad \text{and} \quad (x_1, y_1) = (x(b), y(b))$$

Since $\mathbf{F}(x, y) = \nabla \phi$, it follows that

$$\mathbf{F}(x, y) = \frac{\partial \phi}{\partial x}\mathbf{i} + \frac{\partial \phi}{\partial y}\mathbf{j}$$

so

$$\int_C \mathbf{F}(x, y) \cdot d\mathbf{r} = \int_C \frac{\partial \phi}{\partial x}\, dx + \frac{\partial \phi}{\partial y}\, dy = \int_a^b \left[\frac{\partial \phi}{\partial x}\frac{dx}{dt} + \frac{\partial \phi}{\partial y}\frac{dy}{dt} \right] dt$$

$$= \int_a^b \frac{d}{dt}[\phi(x(t), y(t))]\, dt = \phi(x(t), y(t)) \Big]_{t=a}^{b}$$

$$= \phi(x(b), y(b)) - \phi(x(a), y(a))$$

$$= \phi(x_1, y_1) - \phi(x_0, y_0)$$
■

Stated informally, this theorem shows that *the value of a work integral along a piecewise smooth path in a conservative vector field is **independent of the path***; that is, the value of the integral depends on the endpoints and not on the actual path C. Accordingly, for work integrals along paths in conservative vector fields, it is common to express (3) and (4) as

$$\int_{(x_0, y_0)}^{(x_1, y_1)} \mathbf{F} \cdot d\mathbf{r} = \int_{(x_0, y_0)}^{(x_1, y_1)} \nabla \phi \cdot d\mathbf{r} = \phi(x_1, y_1) - \phi(x_0, y_0) \tag{5}$$

Example 2

(a)   Confirm that the vector field $\mathbf{F}(x, y) = y\mathbf{i} + x\mathbf{j}$ in Example 1 is conservative by showing that $\mathbf{F}(x, y)$ is the gradient of $\phi(x, y) = xy$.

(b)   Use the Fundamental Theorem of Work Integrals to evaluate $\int_{(0,0)}^{(1,1)} \mathbf{F} \cdot d\mathbf{r}$.

Solution (a).

$$\nabla \phi = \frac{\partial \phi}{\partial x}\mathbf{i} + \frac{\partial \phi}{\partial y}\mathbf{j} = y\mathbf{i} + x\mathbf{j}$$

Solution (b). From (5) we obtain

$$\int_{(0,0)}^{(1,1)} \mathbf{F} \cdot d\mathbf{r} = \phi(1, 1) - \phi(0, 0) = 1 - 0 = 1$$

which agrees with the results obtained in Example 1 by integrating from $(0, 0)$ to $(1, 1)$ along specific paths.   ◄

REMARK.   You can visualize the result in this example geometrically from the picture of the vector field shown in Figure 17.3.1 and the relationship

$$\int_C \mathbf{F} \cdot d\mathbf{r} = \int_C \mathbf{F} \cdot \mathbf{T}\, ds$$

We see from this that the more closely the unit tangent vector $\mathbf{T}$ to C aligns with $\mathbf{F}$ along C, the greater the integrand and hence the greater the value of the integral. However, the length of the curve C also affects the value of the integral. Thus, in comparing the three curves in Figure 17.3.1, we see that the alignment of $\mathbf{T}$ with $\mathbf{F}$ is best for the line, but the line has the shortest length. The alignments are not as good for $y = x^2$ and $y = x^3$, but they have greater lengths to compensate. Thus, it seems plausible that the integrals have the same value.

WORK INTEGRALS ALONG CLOSED PATHS

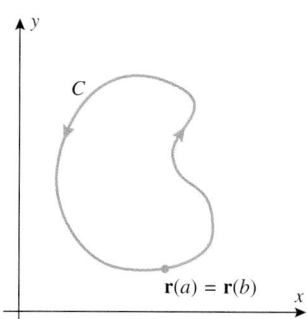

Figure 17.3.2

Parametric curves that begin and end at the same point play an important role in the study of vector fields, so there is some special terminology associated with them. A parametric curve C that is represented by the vector-valued function $\mathbf{r}(t)$ for $a \le t \le b$ is said to be *closed* if the initial point $\mathbf{r}(a)$ and the terminal point $\mathbf{r}(b)$ coincide; that is, $\mathbf{r}(a) = \mathbf{r}(b)$ (Figure 17.3.2).

It follows from (5) that if a particle moving in a conservative vector field traverses a closed path C that begins and ends at (x_0, y_0), then the work performed by the field is zero. This is because the point (x_1, y_1) in (5) is the same as (x_0, y_0) and hence

$$\int_C \mathbf{F} \cdot d\mathbf{r} = \phi(x_1, y_1) - \phi(x_0, y_0) = 0$$

Our next objective is to show that the converse of this result is also true. That is, we want to show that under appropriate conditions a vector field in which the work is zero along *all* closed paths must be conservative. For this to be true we will need to require that the domain D of the vector field be **connected**, by which we mean that any two points in D can be joined by some piecewise smooth curve that lies entirely in D. Stated informally, D is connected if it does not consist of two or more separate pieces (Figure 17.3.3).

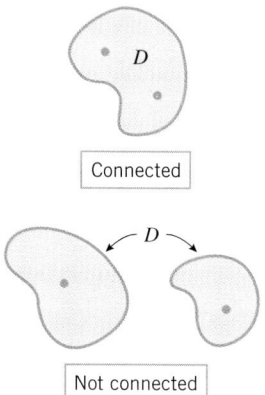

Connected

Not connected

Figure 17.3.3

17.3.2 THEOREM. *If $f(x, y)$ and $g(x, y)$ are continuous on some open connected region D, then the following statements are equivalent (all true or all false):*

(a) $\mathbf{F}(x, y) = f(x, y)\mathbf{i} + g(x, y)\mathbf{j}$ *is a conservative vector field on the region D.*

(b) $\displaystyle\int_C \mathbf{F} \cdot d\mathbf{r} = 0$ *for every piecewise smooth closed curve C in D.*

(c) $\displaystyle\int_C \mathbf{F} \cdot d\mathbf{r}$ *is independent of the path from any point P in D to any point Q in D for every piecewise smooth curve C in D.*

This theorem can be established by proving three implications: $(a) \Rightarrow (b)$, $(b) \Rightarrow (c)$, and $(c) \Rightarrow (a)$. Since we showed above that $(a) \Rightarrow (b)$, we need only prove the last two implications. We will prove $(c) \Rightarrow (a)$ and leave the other implication as an exercise.

Proof. $(c) \Rightarrow (a)$. We are assuming that $\int_C \mathbf{F} \cdot d\mathbf{r}$ is independent of the path for every piecewise smooth curve C in the region, and we want to show that there is a function $\phi = \phi(x, y)$ such that $\nabla\phi = \mathbf{F}(x, y)$ at each point of the region; that is,

$$\frac{\partial\phi}{\partial x} = f(x, y) \quad \text{and} \quad \frac{\partial\phi}{\partial y} = g(x, y) \tag{6}$$

Now choose a fixed point (a, b) in D, let (x, y) be any point in D, and define

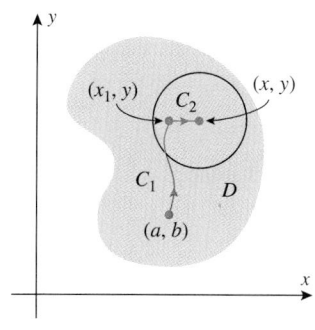

Figure 17.3.4

$$\phi(x, y) = \int_{(a,b)}^{(x,y)} \mathbf{F} \cdot d\mathbf{r} \tag{7}$$

This is an unambiguous definition because we have assumed that the integral is independent of the path. We will show that $\nabla\phi = \mathbf{F}$. Since D is open, we can find a circular disk centered at (x, y) whose points lie entirely in D. As shown in Figure 17.3.4, choose any point (x_1, y) in this disk that lies on the same horizontal line as (x, y) but that is different from (x, y). Because the integral in (7) is independent of path, we can evaluate it by first integrating from (a, b) to (x_1, y) along an arbitrary piecewise smooth curve C_1 in D, and then continuing along the horizontal line segment C_2 from (x_1, y) to (x, y). This yields

$$\phi(x, y) = \int_{C_1} \mathbf{F} \cdot d\mathbf{r} + \int_{C_2} \mathbf{F} \cdot d\mathbf{r} = \int_{(a,b)}^{(x_1,y)} \mathbf{F} \cdot d\mathbf{r} + \int_{C_2} \mathbf{F} \cdot d\mathbf{r}$$

Since the first term does not depend on x, its partial derivative with respect to x is zero and hence

$$\frac{\partial \phi}{\partial x} = \frac{\partial}{\partial x} \int_{C_2} \mathbf{F} \cdot d\mathbf{r} = \frac{\partial}{\partial x} \int_{C_2} f(x, y)\, dx + g(x, y)\, dy$$

However, the line integral with respect to y is zero along the horizontal line segment C_2, so this equation simplifies to

$$\frac{\partial \phi}{\partial x} = \frac{\partial}{\partial x} \int_{C_2} f(x, y)\, dx \tag{8}$$

To evaluate the integral in this expression, we treat y as a constant and express the line C_2 parametrically as

$$x = t, \quad y = y \qquad (x_1 \leq t \leq x)$$

At the risk of confusion, but to avoid complicating the notation, we have used x both as the dependent variable in the parametric equations and as the endpoint of the line segment. With the latter interpretation of x, it follows that (8) can be expressed as

$$\frac{\partial \phi}{\partial x} = \frac{\partial}{\partial x} \int_{x_1}^{x} f(t, y)\, dt$$

Now we apply Part 2 of the Fundamental Theorem of Calculus (Theorem 7.6.3), treating y as constant. This yields

$$\frac{\partial \phi}{\partial x} = f(x, y)$$

which proves the first part of (6). The proof that $\partial \phi / \partial y = g(x, y)$ can be obtained in a similar manner by joining (x, y) to a point (x, y_1) with a vertical line segment (Exercise 33). ∎

A TEST FOR CONSERVATIVE VECTOR FIELDS

Although Theorem 17.3.2 is an important characterization of conservative vector fields, it is not an effective computational tool because it is usually not possible to evaluate the work integral over all possible piecewise smooth curves in D, as required in parts (b) and (c). To develop a method for determining whether a vector field is conservative, we will need to introduce some new concepts about parametric curves and connected sets. We will say that a parametric curve is *simple* if it does not intersect itself between its endpoints. A simple parametric curve may or may not be closed (Figure 17.3.5). In addition, we will say that a connected set D in 2-space is *simply connected* if no simple closed curve in D encloses points that are not in D. Stated informally, a connected set D is simply connected if it has no holes; a connected set with one or more holes is said to be *multiply connected* (Figure 17.3.6).

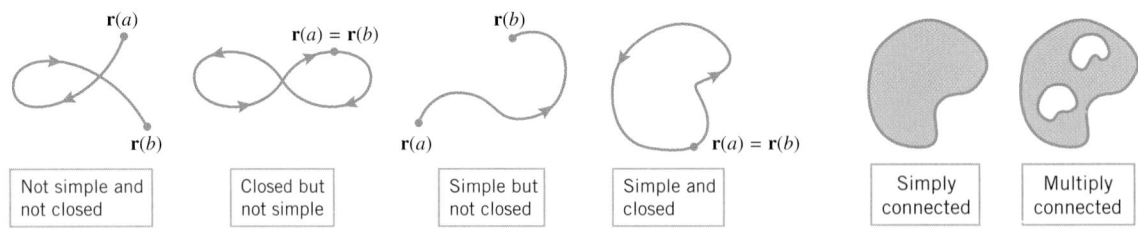

| Not simple and not closed | Closed but not simple | Simple but not closed | Simple and closed | Simply connected | Multiply connected |

Figure 17.3.5   Figure 17.3.6

The following theorem is the primary tool for determining whether a vector field in 2-space is conservative.

17.3.3   THEOREM (*Conservative Field Test*).   *If $f(x, y)$ and $g(x, y)$ are continuous and have continuous first partial derivatives on some open simply connected region D, then $\mathbf{F}(x, y) = f(x, y)\mathbf{i} + g(x, y)\mathbf{j}$ is a conservative vector field on D if and only if*

$$\frac{\partial f}{\partial y} = \frac{\partial g}{\partial x} \tag{9}$$

at each point in D.

A complete proof of this theorem requires results from advanced calculus and will be omitted. However, it is not hard to see why (9) must hold if $\mathbf{F}$ is conservative. For this purpose suppose that $\mathbf{F} = \nabla\phi$, in which case we can express the functions f and g as

$$\frac{\partial\phi}{\partial x} = f \quad \text{and} \quad \frac{\partial\phi}{\partial y} = g \tag{10}$$

Thus,

$$\frac{\partial f}{\partial y} = \frac{\partial}{\partial y}\left(\frac{\partial\phi}{\partial x}\right) = \frac{\partial^2\phi}{\partial y\,\partial x} \quad \text{and} \quad \frac{\partial g}{\partial x} = \frac{\partial}{\partial x}\left(\frac{\partial\phi}{\partial y}\right) = \frac{\partial^2\phi}{\partial x\,\partial y}$$

But the mixed partial derivatives in these equations are equal (Theorem 15.4.6), so (9) follows.

WARNING. In (9), the $\mathbf{i}$-component of $\mathbf{F}$ is differentiated with respect to y and the $\mathbf{j}$-component with respect to x. It is easy to get this backwards by mistake.

Example 3

Use Theorem 17.3.3 to determine whether the vector field $\mathbf{F}(x, y) = (y + x)\mathbf{i} + (y - x)\mathbf{j}$ is conservative on some simply connected open set.

Solution. Let $f(x, y) = y + x$ and $g(x, y) = y - x$. Then

$$\frac{\partial f}{\partial y} = 1 \quad \text{and} \quad \frac{\partial g}{\partial x} = -1$$

Thus, there are no points in the xy-plane at which condition (9) holds, and hence $\mathbf{F}$ is not conservative on any simply connected open set. ◀

REMARK. Since the vector field $\mathbf{F}$ in this example is not conservative, it follows from Theorem 17.3.3 that there must exist piecewise smooth closed curves in every simply connected open set in the xy-plane on which

$$\int_C \mathbf{F} \cdot d\mathbf{r} = \int_C \mathbf{F} \cdot \mathbf{T}\,ds \neq 0$$

One such curve is the circle shown in Figure 17.3.7. The figure suggests that $\mathbf{F} \cdot \mathbf{T} < 0$ at each point of C (why?), so $\int_C \mathbf{F} \cdot \mathbf{T}\,ds < 0$.

Once it is established that a vector field is conservative, a potential function for the field can be obtained by first integrating either of the equations in (10). This is illustrated in the following example.

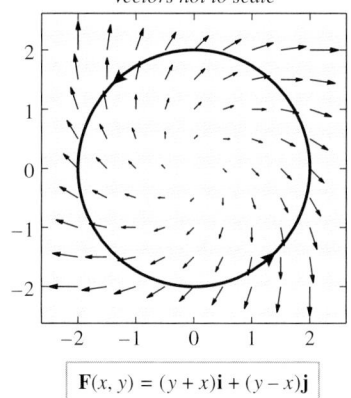

Vectors not to scale

$\mathbf{F}(x, y) = (y + x)\mathbf{i} + (y - x)\mathbf{j}$

Figure 17.3.7

Example 4

Let $\mathbf{F}(x, y) = 2xy^3\mathbf{i} + (1 + 3x^2y^2)\mathbf{j}$.

(a) Show that $\mathbf{F}$ is a conservative vector field on the entire xy-plane.

(b) Find ϕ by first integrating $\partial\phi/\partial x$.

(c) Find ϕ by first integrating $\partial\phi/\partial y$.

Solution (a). Since $f(x, y) = 2xy^3$ and $g(x, y) = 1 + 3x^2y^2$, we have

$$\frac{\partial f}{\partial y} = 6xy^2 = \frac{\partial g}{\partial x}$$

so (9) holds for all (x, y).

Solution (b). Since the field $\mathbf{F}$ is conservative, there is a potential function ϕ such that

$$\frac{\partial\phi}{\partial x} = 2xy^3 \quad \text{and} \quad \frac{\partial\phi}{\partial y} = 1 + 3x^2y^2 \tag{11}$$

Integrating the first of these equations with respect to x (and treating y as a constant) yields

$$\phi = \int 2xy^3 \, dx = x^2y^3 + k(y) \tag{12}$$

where $k(y)$ represents the "constant" of integration. We are justified in treating the constant of integration as a function of y, since y is held constant in the integration process. To find $k(y)$ we differentiate (12) with respect to y and use the second equation in (11) to obtain

$$\frac{\partial \phi}{\partial y} = 3x^2y^2 + k'(y) = 1 + 3x^2y^2$$

from which it follows that $k'(y) = 1$. Thus,

$$k(y) = \int k'(y) \, dy = \int 1 \, dy = y + K$$

where K is a (numerical) constant of integration. Substituting in (12) we obtain

$$\phi = x^2y^3 + y + K$$

The appearance of the arbitrary constant K tells us that ϕ is not unique. As a check on the computations, you may want to verify that $\nabla \phi = \mathbf{F}$.

Solution (c). Integrating the second equation in (11) with respect to y (and treating x as a constant) yields

$$\phi = \int (1 + 3x^2y^2) \, dy = y + x^2y^3 + k(x) \tag{13}$$

where $k(x)$ is the "constant" of integration. Differentiating (13) with respect to x and using the first equation in (11) yields

$$\frac{\partial \phi}{\partial x} = 2xy^3 + k'(x) = 2xy^3$$

from which it follows that $k'(x) = 0$ and consequently that $k(x) = K$, where K is a numerical constant of integration. Substituting this in (13) yields

$$\phi = y + x^2y^3 + K$$

which agrees with the solution in part (b). ◀

Example 5

Use the potential function obtained in Example 4 to evaluate the integral

$$\int_{(1,4)}^{(3,1)} 2xy^3 \, dx + (1 + 3x^2y^2) \, dy$$

Solution. The integrand can be expressed as $\mathbf{F} \cdot d\mathbf{r}$, where $\mathbf{F}$ is the vector field in Example 4. Thus, using Formula (3) and the potential function $\phi = y + x^2y^3 + K$ for $\mathbf{F}$, we obtain

$$\int_{(1,4)}^{(3,1)} 2xy^3 \, dx + (1 + 3x^2y^2) \, dy = \int_{(1,4)}^{(3,1)} \mathbf{F} \cdot d\mathbf{r} = \phi(3, 1) - \phi(1, 4)$$

$$= (10 + K) - (68 + K) = -58 \qquad ◀$$

REMARK. Note that the constant K drops out. In future integration problems we will omit K from the computations.

Example 6

Let $\mathbf{F}(x, y) = e^y \mathbf{i} + xe^y \mathbf{j}$.

(a) Verify that the vector field $\mathbf{F}$ is conservative on the entire xy-plane.

(b) Find the work done by the field on a particle that moves from $(1, 0)$ to $(-1, 0)$ along the semicircular path C shown in Figure 17.3.8.

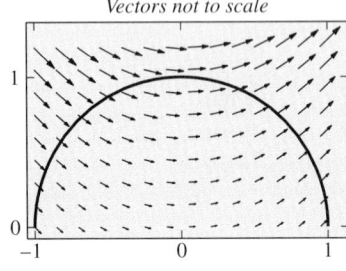

Vectors not to scale

Figure 17.3.8

Solution (*a*). For the given field we have $f(x, y) = e^y$ and $g(x, y) = xe^y$. Thus,

$$\frac{\partial}{\partial y}(e^y) = e^y = \frac{\partial}{\partial x}(xe^y)$$

so (9) holds for all (x, y) and hence **F** is conservative on the entire xy-plane.

Solution (*b*). From Formula (33) of Section 17.2, the work done by the field is

$$W = \int_C \mathbf{F} \cdot d\mathbf{r} = \int_C e^y\, dx + xe^y\, dy \tag{14}$$

However, the calculations involved in integrating along C are tedious, so it is preferable to apply Theorem 17.3.1, taking advantage of the fact that the field is conservative and the integral is independent of path. Thus, we write (14) as

$$W = \int_{(1,0)}^{(-1,0)} e^y\, dx + xe^y\, dy = \phi(-1, 0) - \phi(1, 0) \tag{15}$$

As illustrated in Example 4, we can find ϕ by integrating either of the equations

$$\frac{\partial \phi}{\partial x} = e^y \quad \text{and} \quad \frac{\partial \phi}{\partial y} = xe^y \tag{16}$$

We will integrate the first. We obtain

$$\phi = \int e^y\, dx = xe^y + k(y) \tag{17}$$

Differentiating this equation with respect to y and using the second equation in (16) yields

$$\frac{\partial \phi}{\partial y} = xe^y + k'(y) = xe^y$$

from which it follows that $k'(y) = 0$ or $k(y) = K$. Thus, from (17)

$$\phi = xe^y + K$$

and hence from (15)

$$W = \phi(-1, 0) - \phi(1, 0) = (-1)e^0 - 1e^0 = -2 \qquad \blacktriangleleft$$

CONSERVATIVE VECTOR FIELDS IN 3-SPACE

All of the results in this section have analogs in 3-space: Theorems 17.3.1 and 17.3.2 can be extended to vector fields in 3-space simply by adding a third variable and modifying the hypotheses appropriately. For example, in 3-space, Formula (3) becomes

$$\int_C \mathbf{F}(x, y, z) \cdot d\mathbf{r} = \phi(x_1, y_1, z_1) - \phi(x_0, y_0, z_0) \tag{18}$$

Theorem 17.3.3 can also be extended to vector fields in 3-space. We leave it for the exercises to show that if $\mathbf{F}(x, y, z) = f(x, y, z)\mathbf{i} + g(x, y, z)\mathbf{j} + h(x, y, z)\mathbf{k}$ is a conservative field, then

$$\frac{\partial f}{\partial y} = \frac{\partial g}{\partial x}, \quad \frac{\partial f}{\partial z} = \frac{\partial h}{\partial x}, \quad \frac{\partial g}{\partial z} = \frac{\partial h}{\partial y} \tag{19}$$

that is, curl $\mathbf{F} = \mathbf{0}$. Conversely, a vector field satisfying these conditions on a suitably restricted region is conservative on that region if f, g, and h are continuous and have continuous first partial derivatives in the region. Some problems involving Formulas (18) and (19) are given in the supplementary exercises at the end of this chapter.

CONSERVATION OF ENERGY

If $\mathbf{F}(x, y, z)$ is a conservative force field with a potential function $\phi(x, y, z)$, then we call $V(x, y, z) = -\phi(x, y, z)$ the **potential energy** of the field at the point (x, y, z). Thus, it follows from the 3-space version of Theorem 17.3.1 that the work W done by $\mathbf{F}$ on a particle that moves along any path C from a point (x_0, y_0, z_0) to a point (x_1, y_1, z_1) is related to the potential energy by the equation

$$W = \int_C \mathbf{F} \cdot d\mathbf{r} = \phi(x_1, y_1, z_1) - \phi(x_0, y_0, z_0) = -[V(x_1, y_1, z_1) - V(x_0, y_0, z_0)] \quad (20)$$

That is, the work done by the field is the negative of the change in potential energy. In particular, it follows from the 3-space analog of Theorem 17.3.2 that if a particle traverses a piecewise smooth closed path in a conservative vector field, then the work done by the field is zero, and there is no change in potential energy. To take this a step further, suppose that a particle of mass m moves along any piecewise smooth curve (not necessarily closed) in a conservative vector field, starting at (x_0, y_0, z_0) with velocity v_i and ending at (x_1, y_1, z_1) with velocity v_f. If we let V_i denote the potential energy at the starting point and V_f the potential energy at the final point, then it follows from the Work–Energy Theorem (Theorem 8.6.4) that

$$\tfrac{1}{2}mv_f^2 - \tfrac{1}{2}mv_i^2 = -[V_f - V_i]$$

which we can rewrite as

$$\tfrac{1}{2}mv_f^2 + V_f = \tfrac{1}{2}mv_i^2 + V_i$$

This equation states that the total energy of the particle (kinetic energy + potential energy) does not change as the particle moves along a path in a conservative vector field. This result, called the **conservation of energy principle**, explains the origin of the term "conservative vector field."

EXERCISE SET 17.3 ⬚C CAS

In Exercises 1–6, determine whether $\mathbf{F}$ is a conservative vector field. If so, find a potential function for it.

1. $\mathbf{F}(x, y) = x\mathbf{i} + y\mathbf{j}$

2. $\mathbf{F}(x, y) = 3y^2\mathbf{i} + 6xy\mathbf{j}$

3. $\mathbf{F}(x, y) = x^2y\mathbf{i} + 5xy^2\mathbf{j}$

4. $\mathbf{F}(x, y) = e^x \cos y\mathbf{i} - e^x \sin y\mathbf{j}$

5. $\mathbf{F}(x, y) = (\cos y + y \cos x)\mathbf{i} + (\sin x - x \sin y)\mathbf{j}$

6. $\mathbf{F}(x, y) = x \ln y\mathbf{i} + y \ln x\mathbf{j}$

7. (a) Show that the line integral $\int_C y^2\, dx + 2xy\, dy$ is independent of the path.

  (b) Evaluate the integral in part (a) along the line segment from $(-1, 2)$ to $(1, 3)$.

  (c) Evaluate the integral $\int_{(-1,2)}^{(1,3)} y^2\, dx + 2xy\, dy$ using Theorem 17.3.1, and confirm that the value is the same as that obtained in part (b).

8. (a) Show that the line integral $\int_C y \sin x\, dx - \cos x\, dy$ is independent of the path.

  (b) Evaluate the integral in part (a) along the line segment from $(0, 1)$ to $(\pi, -1)$.

  (c) Evaluate the integral $\int_{(0,1)}^{(\pi,-1)} y \sin x\, dx - \cos x\, dy$ using Theorem 17.3.1, and confirm that the value is the same as that obtained in part (b).

In Exercises 9–14, show that the integral is independent of the path, and use Theorem 17.3.1 to find its value.

9. $\displaystyle\int_{(1,2)}^{(4,0)} 3y\, dx + 3x\, dy$

10. $\displaystyle\int_{(0,0)}^{(1,\pi/2)} e^x \sin y\, dx + e^x \cos y\, dy$

11. $\displaystyle\int_{(0,0)}^{(3,2)} 2xe^y\, dx + x^2e^y\, dy$

12. $\displaystyle\int_{(-1,2)}^{(0,1)} (3x - y + 1)\, dx - (x + 4y + 2)\, dy$

13. $\displaystyle\int_{(2,-2)}^{(-1,0)} 2xy^3\, dx + 3y^2x^2\, dy$

14. $\displaystyle\int_{(1,1)}^{(3,3)} \left(e^x \ln y - \dfrac{e^y}{x}\right) dx + \left(\dfrac{e^x}{y} - e^y \ln x\right) dy$, where x and y are positive.

In Exercises 15–18, confirm that the force field $\mathbf{F}$ is conservative in some open connected region containing the points P and Q, and then find the work done by the force field on a particle moving along an arbitrary smooth curve in the region from P to Q.

15. $\mathbf{F}(x, y) = xy^2\mathbf{i} + x^2y\mathbf{j}$; $P(1, 1)$, $Q(0, 0)$

16. $\mathbf{F}(x, y) = 2xy^3\mathbf{i} + 3x^2y^2\mathbf{j}$; $P(-3, 0)$, $Q(4, 1)$

17. $\mathbf{F}(x, y) = ye^{xy}\mathbf{i} + xe^{xy}\mathbf{j}$; $P(-1, 1)$, $Q(2, 0)$

18. $\mathbf{F}(x, y) = e^{-y} \cos x\mathbf{i} - e^{-y} \sin x\mathbf{j}$; $P(\pi/2, 1)$, $Q(-\pi/2, 0)$

In Exercises 19 and 20, find the exact value of $\int_C \mathbf{F} \cdot d\mathbf{r}$ using any method.

19. $\mathbf{F}(x, y) = (e^y + ye^x)\mathbf{i} + (xe^y + e^x)\mathbf{j}$
$C : \mathbf{r}(t) = \sin(\pi t/2)\mathbf{i} + \ln t\,\mathbf{j} \quad (1 \le t \le 2)$

20. $\mathbf{F}(x, y) = 2xy\mathbf{i} + (x^2 + \cos y)\mathbf{j}$
$C: \mathbf{r}(t) = t\mathbf{i} + t\cos(t/3)\mathbf{j} \quad (0 \le t \le \pi)$

C **21.** Use the numerical integration capability of a CAS or other calculating utility to approximate the value of the integral in Exercise 19 by direct integration. Confirm that the numerical approximation is consistent with the exact value.

C **22.** Use the numerical integration capability of a CAS or other calculating utility to approximate the value of the integral in Exercise 20 by direct integration. Confirm that the numerical approximation is consistent with the exact value.

In Exercises 23 and 24, is the vector field conservative? Explain your reasoning.

23.

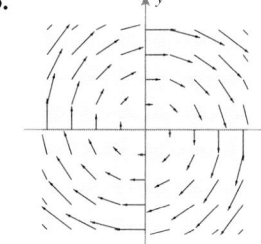

24.

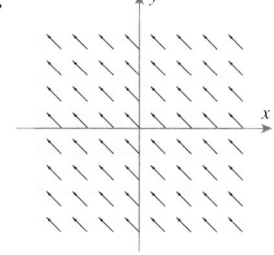

25. Prove: If $\mathbf{F}(x, y, z) = f(x, y, z)\mathbf{i} + g(x, y, z)\mathbf{j} + h(x, y, z)\mathbf{k}$ is a conservative field and f, g, and h are continuous and have continuous first partial derivatives in a region, then

$$\frac{\partial f}{\partial y} = \frac{\partial g}{\partial x}, \quad \frac{\partial f}{\partial z} = \frac{\partial h}{\partial x}, \quad \frac{\partial g}{\partial z} = \frac{\partial h}{\partial y}$$

in the region.

26. Use the result in Exercise 25 to show that the integral

$$\int_C yz\,dx + xz\,dy + yx^2\,dz$$

is not independent of the path.

27. Find a nonzero function h for which

$$\mathbf{F}(x, y) = h(x)[x \sin y + y \cos y]\mathbf{i}$$
$$+ h(x)[x \cos y - y \sin y]\mathbf{j}$$

is conservative.

28. (a) In Example 3 of Section 17.1 we showed that

$$\phi(x, y) = -\frac{c}{(x^2 + y^2)^{1/2}}$$

is a potential function for the two-dimensional inverse-square field

$$\mathbf{F}(x, y) = \frac{c}{(x^2 + y^2)^{3/2}}(x\mathbf{i} + y\mathbf{j})$$

but we did not explain how the potential function $\phi(x, y)$ was obtained. Use Theorem 17.3.3 to show that the two-dimensional inverse-square field is conservative everywhere except at the origin, and then use the method of Example 4 to derive the formula for $\phi(x, y)$.

(b) Use an appropriate generalization of the method of Example 4 to derive the potential function

$$\phi(x, y, z) = -\frac{c}{(x^2 + y^2 + z^2)^{1/2}}$$

for the three-dimensional inverse-square field given by Formula (5) of Section 17.1.

In Exercises 29 and 30, use the result in Exercise 28(b).

29. In each part, find the work done by the three-dimensional inverse-square field

$$\mathbf{F}(\mathbf{r}) = \frac{1}{\|\mathbf{r}\|^3}\mathbf{r}$$

on a particle that moves along the curve C.
(a) C is the line segment from $P(1, 1, 2)$ to $Q(3, 2, 1)$.
(b) C is the curve $\mathbf{r}(t) = (2t^2 + 1)\mathbf{i} + (t^3 + 1)\mathbf{j} + (2 - \sqrt{t})\mathbf{k}$, where $0 \le t \le 1$.
(c) C is the circle of radius 1 centered at $(2, 0, 0)$ traversed counterclockwise.

30. Let $\mathbf{F}(x, y) = \dfrac{y}{x^2 + y^2}\mathbf{i} - \dfrac{x}{x^2 + y^2}\mathbf{j}$.
(a) Show that

$$\int_{C_1} \mathbf{F} \cdot d\mathbf{r} \ne \int_{C_2} \mathbf{F} \cdot d\mathbf{r}$$

if C_1 and C_2 are the semicircular paths from $(1, 0)$ to $(-1, 0)$ given by

$C_1: x = \cos t, \quad y = \sin t \qquad (0 \le t \le \pi)$
$C_2: x = \cos t, \quad y = -\sin t \qquad (0 \le t \le \pi)$

(b) Show that the components of $\mathbf{F}$ satisfy Formula (9).
(c) Do the results in parts (a) and (b) violate Theorem 17.3.3? Explain.

31. Prove Theorem 17.3.1 if C is a piecewise smooth curve composed of smooth curves $C_1, C_2, \ldots, C_n$.

32. Prove that (b) implies (c) in Theorem 17.3.2. [*Hint:* Consider any two piecewise smooth oriented curves C_1 and C_2 in the region from a point P to a point Q, and integrate around the closed curve consisting of C_1 and $-C_2$.]

33. Complete the proof of Theorem 17.3.2 by showing that $\partial \phi / \partial y = g(x, y)$, where $\phi(x, y)$ is the function in (7).

17.4 GREEN'S THEOREM

In this section we will discuss a remarkable and beautiful theorem that expresses the double integral over a plane region in terms of a line integral around its boundary.

17.4.1 THEOREM (*Green's* [*] *Theorem*). *Let R be a simply connected plane region whose boundary is a simple, closed, piecewise smooth curve C oriented counterclockwise. If* $f(x, y)$ *and* $g(x, y)$ *are continuous and have continuous first partial derivatives on some open set containing R, then*

$$\int_C f(x, y)\, dx + g(x, y)\, dy = \iint_R \left(\frac{\partial g}{\partial x} - \frac{\partial f}{\partial y} \right) dA \tag{1}$$

Proof. For simplicity, we will prove the theorem for regions that are simultaneously type I and type II (see Definition 16.2.1). Such a region is shown in Figure 17.4.1. The crux of the proof is to show that

$$\int_C f(x, y)\, dx = -\iint_R \frac{\partial f}{\partial y}\, dA \quad \text{and} \quad \int_C g(x, y)\, dy = \iint_R \frac{\partial g}{\partial x}\, dA \tag{2-3}$$

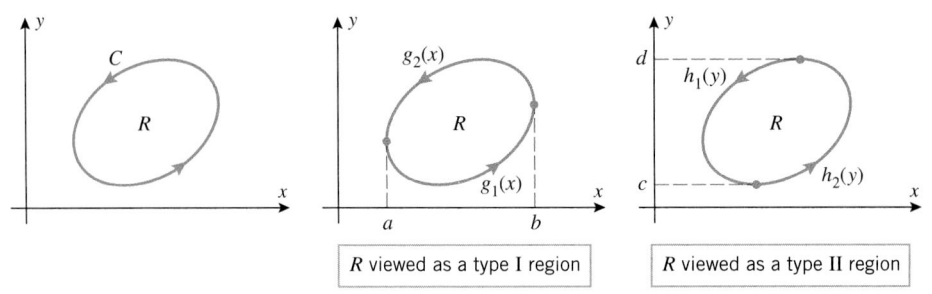

R viewed as a type I region

R viewed as a type II region

Figure 17.4.1

To prove (2), view R as a type I region and let C_1 and C_2 be the lower and upper boundary curves, oriented as in Figure 17.4.2. Then

$$\int_C f(x, y)\, dx = \int_{C_1} f(x, y)\, dx + \int_{C_2} f(x, y)\, dx$$

or, equivalently,

$$\int_C f(x, y)\, dx = \int_{C_1} f(x, y)\, dx - \int_{-C_2} f(x, y)\, dx \tag{4}$$

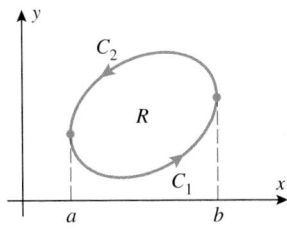

Figure 17.4.2

---

[*] GEORGE GREEN (1793–1841). English mathematician and physicist. Green left school at an early age to work in his father's bakery and consequently had little early formal education. When his father opened a mill, the boy used the top room as a study in which he taught himself physics and mathematics from library books. In 1828 Green published his most important work, *An Essay on the Application of Mathematical Analysis to the Theories of Electricity and Magnetism*. Although Green's Theorem appeared in that paper, the result went virtually unnoticed because of the small pressrun and local distribution. Following the death of his father in 1829, Green was urged by friends to seek a college education. In 1833, after four years of self-study to close the gaps in his elementary education, Green was admitted to Caius College, Cambridge. He graduated four years later, but with a disappointing performance on his final examinations—possibly because he was more interested in his own research. After a succession of works on light and sound, he was named to be Perse Fellow at Caius College. Two years later he died. In 1845, four years after his death, his paper of 1828 was published and the theories developed therein by this obscure, self-taught baker's son helped pave the way to the modern theories of electricity and magnetism.

(This step will help simplify our calculations since C_1 and $-C_2$ are then both oriented left to right.) The curves C_1 and $-C_2$ can be expressed parametrically as

$$C_1: x = t, \quad y = g_1(t) \quad (a \leq t \leq b)$$
$$-C_2: x = t, \quad y = g_2(t) \quad (a \leq t \leq b)$$

Thus, we can rewrite (4) as

$$\int_C f(x, y)\, dx = \int_a^b f(t, g_1(t))x'(t)\, dt - \int_a^b f(t, g_2(t))x'(t)\, dt$$

$$= \int_a^b f(t, g_1(t))\, dt - \int_a^b f(t, g_2(t))\, dt$$

$$= -\int_a^b [f(t, g_2(t)) - f(t, g_1(t))]\, dt$$

$$= -\int_a^b \left[f(t, y) \right]_{y=g_1(t)}^{y=g_2(t)} dt = -\int_a^b \left[\int_{g_1(t)}^{g_2(t)} \frac{\partial f}{\partial y}\, dy \right] dt$$

$$= -\int_a^b \int_{g_1(x)}^{g_2(x)} \frac{\partial f}{\partial y}\, dy\, dx = -\iint_R \frac{\partial f}{\partial y}\, dA$$

Since $x = t$

The proof of (3) is obtained similarly by treating R as a type II region. We omit the details. ▮

Example 1

Use Green's Theorem to evaluate

$$\int_C x^2 y\, dx + x\, dy$$

along the triangular path shown in Figure 17.4.3.

Solution. Since $f(x, y) = x^2 y$ and $g(x, y) = x$, it follows from (1) that

$$\int_C x^2 y\, dx + x\, dy = \iint_R \left[\frac{\partial}{\partial x}(x) - \frac{\partial}{\partial y}(x^2 y) \right] dA = \int_0^1 \int_0^{2x} (1 - x^2)\, dy\, dx$$

$$= \int_0^1 (2x - 2x^3)\, dx = \left[x^2 - \frac{x^4}{2} \right]_0^1 = \frac{1}{2}$$

This agrees with the result obtained in Example 7 of Section 17.2, where we evaluated the line integral directly. Note how much simpler this solution is. ◄

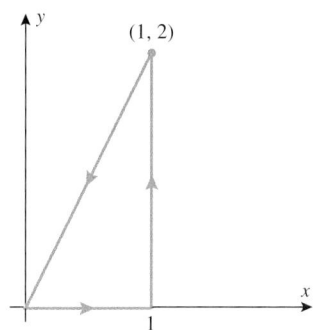

Figure 17.4.3

A NOTATION FOR LINE INTEGRALS AROUND SIMPLE CLOSED CURVES

It is common practice to denote a line integral around a simple closed curve by an integral sign with a superimposed circle. With this notation Formula (1) would be written as

$$\oint_C f(x, y)\, dx + g(x, y)\, dy = \iint_R \left(\frac{\partial g}{\partial x} - \frac{\partial f}{\partial y} \right) dA$$

Sometimes a direction arrow is added to the circle to indicate whether the integration is clockwise or counterclockwise. Thus, if we wanted to emphasize the counterclockwise direction of integration required by Theorem 17.4.1, we could express (1) as

$$\oint_C f(x, y)\, dx + g(x, y)\, dy = \iint_R \left(\frac{\partial g}{\partial x} - \frac{\partial f}{\partial y} \right) dA \tag{5}$$

FINDING WORK USING GREEN'S THEOREM

It follows from Formula (37) of Section 17.2 that the integral on the left side of (5) is the work performed by the vector field $\mathbf{F}(x, y) = f(x, y)\mathbf{i} + g(x, y)\mathbf{j}$ on a particle moving counterclockwise around the simple closed curve C. In the case where this vector field is conservative, it follows from Theorem 17.3.2 that the integrand in the double integral on the right side of (5) is zero, so the work performed by the field is zero, as expected. For vector fields that are not conservative, it is often more efficient to calculate the work around simple closed curves by using Green's Theorem than by parametrizing the curve.

Example 2

Find the work done by the force field $\mathbf{F}(x, y) = (e^x - y^3)\mathbf{i} + (\cos y + x^3)\mathbf{j}$ on a particle that travels once around the unit circle $x^2 + y^2 = 1$ in the counterclockwise direction (Figure 17.4.4).

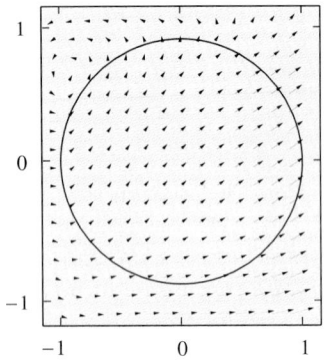

Figure 17.4.4

Solution. The work W performed by the field is

$$W = \oint_C \mathbf{F} \cdot d\mathbf{r} = \oint_C (e^x - y^3)\,dx + (\cos y + x^3)\,dy$$

$$= \iint_R \left[\frac{\partial}{\partial x}(\cos y + x^3) - \frac{\partial}{\partial y}(e^x - y^3) \right] dA \qquad \boxed{\text{Green's Theorem}}$$

$$= \iint_R (3x^2 + 3y^2)\,dA = 3 \iint_R (x^2 + y^2)\,dA$$

$$= 3 \int_0^{2\pi} \int_0^1 (r^2)r\,dr\,d\theta = \frac{3}{4} \int_0^{2\pi} d\theta = \frac{3\pi}{2} \qquad \blacktriangleleft$$

$\boxed{\text{We converted to polar coordinates.}}$

FINDING AREAS USING GREEN'S THEOREM

Green's Theorem leads to some useful new formulas for the area A of a region R that satisfies the conditions of the theorem. Two such formulas can be obtained as follows:

$$A = \iint_R dA = \oint_C x\,dy \quad \text{and} \quad A = \iint_R dA = \oint_C (-y)\,dx$$

$\boxed{\text{Set } f(x, y) = 0 \text{ and } g(x, y) = x \text{ in (1).}}$ $\boxed{\text{Set } f(x, y) = -y \text{ and } g(x, y) = 0 \text{ in (1).}}$

A third formula can be obtained by adding these two equations together. Thus, we have the following three formulas that express the area A of a region R in terms of line integrals around the boundary:

$$A = \oint_C x\,dy = -\oint_C y\,dx = \frac{1}{2}\oint_C -y\,dx + x\,dy \qquad (6)$$

REMARK. Although the third formula in (6) looks more complicated than the other two, it often leads to simpler integrations; but each has advantages in certain situations.

Example 3

Use a line integral to find the area enclosed by the ellipse

$$\frac{x^2}{a^2} + \frac{y^2}{b^2} = 1$$

Solution. The ellipse, with counterclockwise orientation, can be represented parametrically by

$$x = a\cos t, \quad y = b\sin t \qquad (0 \le t \le 2\pi)$$

If we denote this curve by C, then from the third formula in (6) the area A enclosed by the ellipse is

$$A = \frac{1}{2} \oint_C -y\,dx + x\,dy$$

$$= \frac{1}{2} \int_0^{2\pi} [(-b\sin t)(-a\sin t) + (a\cos t)(b\cos t)]\,dt$$

$$= \frac{1}{2}\,ab \int_0^{2\pi} (\sin^2 t + \cos^2 t)\,dt = \frac{1}{2}\,ab \int_0^{2\pi} dt = \pi ab \qquad \blacktriangleleft$$

GREEN'S THEOREM FOR MULTIPLY CONNECTED REGIONS

Recall that a plane region is said to be simply connected if it has no holes and is said to be multiply connected if it has one or more holes (see Figure 17.3.6). At the beginning of this section we stated Green's Theorem for a counterclockwise integration around the boundary of a simply connected region R (Theorem 17.4.1). Our next goal is to extend this theorem to multiply connected regions. To make this extension we will need to assume that *the region lies on the left when any portion of the boundary is traversed in the direction of its orientation.* This implies that the outer boundary curve of the region is oriented counterclockwise and the boundary curves that enclose holes have clockwise orientation (Figure 17.4.5a). If all portions of the boundary of a multiply connected region R are oriented in this way, then we say that the boundary of R has *positive orientation*.

We will now derive a version of Green's Theorem that applies to multiply connected regions with positively oriented boundaries. For simplicity, we will consider a multiply connected region R with one hole, and we will assume that $f(x, y)$ and $g(x, y)$ have continuous first partial derivatives on some open set containing R. As shown in Figure 17.4.5b, let us divide R into two regions R' and R'' by introducing two "cuts" in R. The cuts are shown as line segments, but any piecewise smooth curves will suffice. If we assume that f and g satisfy the hypotheses of Green's Theorem on R (and hence on R' and R''), then we can apply this theorem to both R' and R'' to obtain

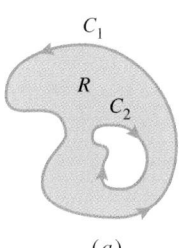

(a)

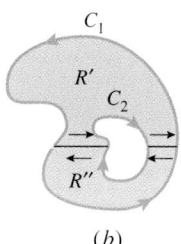

(b)

Figure 17.4.5

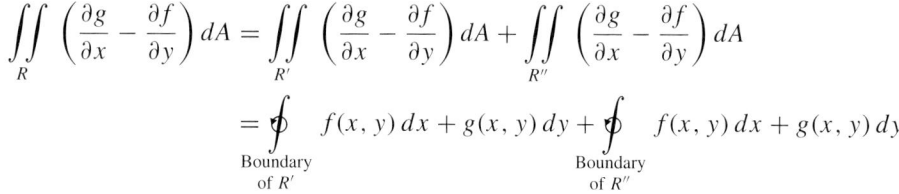

$$\iint\limits_{R} \left(\frac{\partial g}{\partial x} - \frac{\partial f}{\partial y} \right) dA = \iint\limits_{R'} \left(\frac{\partial g}{\partial x} - \frac{\partial f}{\partial y} \right) dA + \iint\limits_{R''} \left(\frac{\partial g}{\partial x} - \frac{\partial f}{\partial y} \right) dA$$

$$= \underset{\substack{\text{Boundary} \\ \text{of } R'}}{\oint} f(x, y)\,dx + g(x, y)\,dy + \underset{\substack{\text{Boundary} \\ \text{of } R''}}{\oint} f(x, y)\,dx + g(x, y)\,dy$$

However, the two line integrals are taken in opposite directions along the cuts, and hence cancel there, leaving only the contributions along C_1 and C_2. Thus,

$$\iint\limits_{R} \left(\frac{\partial g}{\partial x} - \frac{\partial f}{\partial y} \right) dA = \oint_{C_1} f(x, y)\,dx + g(x, y)\,dy + \oint_{C_2} f(x, y)\,dx + g(x, y)\,dy \qquad (7)$$

which is an extension of Green's Theorem to a multiply connected region with one hole. Observe that the integral around the outer boundary is taken counterclockwise and the integral around the hole is taken clockwise. More generally, if R is a multiply connected region with n holes, then the analog of (7) involves a sum of $n + 1$ integrals, one taken counterclockwise around the outer boundary of R and the rest taken clockwise around the holes.

Example 4

Evaluate the integral

$$\oint_C \frac{-y\,dx + x\,dy}{x^2 + y^2}$$

if C is a piecewise smooth simple closed curve oriented counterclockwise such that (a) C does not enclose the origin and (b) C encloses the origin.

Solution (*a*). Let

$$f(x, y) = -\frac{y}{x^2 + y^2}, \quad g(x, y) = \frac{x}{x^2 + y^2} \tag{8}$$

so that

$$\frac{\partial g}{\partial x} = \frac{y^2 - x^2}{(x^2 + y^2)^2} = \frac{\partial f}{\partial y}$$

if x and y are not both zero. Thus, if C does not enclose the origin, we have

$$\frac{\partial g}{\partial x} - \frac{\partial f}{\partial y} = 0 \tag{9}$$

on the simply connected region enclosed by C, and hence the given integral is zero by Green's Theorem.

Solution (*b*). Unlike the situation in part (a), we cannot apply Green's Theorem directly because the functions $f(x, y)$ and $g(x, y)$ in (8) are discontinuous at the origin. Our problems are further compounded by the fact that we do not have a specific curve C that we can parametrize to evaluate the integral. Our strategy circumventing these problems will be to replace C with a specific curve that produces the same value for the integral and then use that curve for the evaluation. To obtain such a curve, we will apply Green's Theorem for multiply connected regions to a region that does not contain the origin. For this purpose we construct a circle C_a with *clockwise* orientation, centered at the origin, and with sufficiently small radius a that it lies inside the region enclosed by C (Figure 17.4.6). This creates a multiply connected region R whose boundary curves C and C_a have the orientations required by Formula (7) and such that within R the functions $f(x, y)$ and $g(x, y)$ in (8) satisfy the hypotheses of Green's Theorem (the origin is outside of R). Thus, it follows from (7) and (9) that

$$\oint_C \frac{-y\,dx + x\,dy}{x^2 + y^2} + \oint_{C_a} \frac{-y\,dx + x\,dy}{x^2 + y^2} = \iint_R 0\,dA = 0$$

It follows from this equation that

$$\oint_C \frac{-y\,dx + x\,dy}{x^2 + y^2} = -\oint_{C_a} \frac{-y\,dx + x\,dy}{x^2 + y^2}$$

which we can rewrite as

$$\oint_C \frac{-y\,dx + x\,dy}{x^2 + y^2} = \oint_{-C_a} \frac{-y\,dx + x\,dy}{x^2 + y^2}$$

> Reversing the orientation of C_a reverses the sign of the integral.

But C_a has clockwise orientation, so $-C_a$ has counterclockwise orientation. Thus, we have shown that the original integral can be evaluated by integrating clockwise around a circle of radius a that is centered at the origin and lies within the region enclosed by C. Such a circle can be expressed parametrically as $x = a\cos t$, $y = a\sin t$ $(0 \le t \le 2\pi)$; and hence

$$\oint_C \frac{-y\,dx + x\,dy}{x^2 + y^2} = \int_0^{2\pi} \frac{(-a\sin t)(-a\sin t)\,dt + (a\cos t)(a\cos t)\,dt}{(a\cos t)^2 + (a\sin t)^2}$$

$$= \int_0^{2\pi} 1\,dt = 2\pi \qquad \blacktriangleleft$$

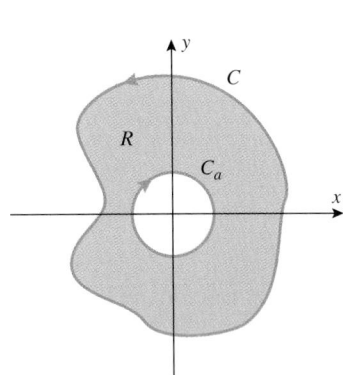

Figure 17.4.6

EXERCISE SET 17.4 ⊆ CAS

In Exercises 1 and 2, evaluate the line integral using Green's Theorem and check the answer by evaluating it directly.

1. $\oint_C y^2\,dx + x^2\,dy$, where C is the square with vertices $(0, 0)$, $(1, 0)$, $(1, 1)$, and $(0, 1)$ oriented counterclockwise.

2. $\oint_C y\,dx + x\,dy$, where C is the unit circle oriented counterclockwise.

In Exercises 3–13, use Green's Theorem to evaluate the integral. In each exercise, assume that the curve C is oriented counterclockwise.

3. $\oint_C 3xy\,dx + 2xy\,dy$, where C is the rectangle bounded by $x = -2$, $x = 4$, $y = 1$, and $y = 2$.

4. $\oint_C (x^2 - y^2)\,dx + x\,dy$, where C is the circle $x^2 + y^2 = 9$.

5. $\oint_C x \cos y\,dx - y \sin x\,dy$, where C is the square with vertices $(0, 0)$, $(\pi/2, 0)$, $(\pi/2, \pi/2)$, and $(0, \pi/2)$.

6. $\oint_C y \tan^2 x\,dx + \tan x\,dy$, where C is the circle $x^2 + (y + 1)^2 = 1$.

7. $\oint_C (x^2 - y)\,dx + x\,dy$, where C is the circle $x^2 + y^2 = 4$.

8. $\oint_C (e^x + y^2)\,dx + (e^y + x^2)\,dy$, where C is the boundary of the region between $y = x^2$ and $y = x$.

9. $\oint_C \ln(1 + y)\,dx - \dfrac{xy}{1 + y}\,dy$, where C is the triangle with vertices $(0, 0)$, $(2, 0)$, and $(0, 4)$.

10. $\oint_C x^2 y\,dx - y^2 x\,dy$, where C is the boundary of the region in the first quadrant, enclosed between the coordinate axes and the circle $x^2 + y^2 = 16$.

11. $\oint_C \tan^{-1} y\,dx - \dfrac{y^2 x}{1 + y^2}\,dy$, where C is the square with vertices $(0, 0)$, $(1, 0)$, $(1, 1)$, and $(0, 1)$.

12. $\oint_C \cos x \sin y\,dx + \sin x \cos y\,dy$, where C is the triangle with vertices $(0, 0)$, $(3, 3)$, and $(0, 3)$.

13. $\oint_C x^2 y\,dx + (y + xy^2)\,dy$, where C is the boundary of the region enclosed by $y = x^2$ and $x = y^2$.

C **14.** For the problems that you solved in Exercises 1–13, use a CAS to check Green's Theorem by evaluating the line integral and the double integral.

C **15.** Use a CAS to check Green's Theorem by evaluating both integrals in the equation

$$\oint_C e^y\,dx + ye^x\,dy = \iint_R \left[\frac{\partial}{\partial x}(ye^x) - \frac{\partial}{\partial y}(e^y) \right]dA$$

where
(a) C is the circle $x^2 + y^2 = 1$
(b) C is the boundary of the region enclosed by $y = x^2$ and $x = y^2$.

16. In Example 3, we used Green's Theorem to obtain the area of an ellipse. Obtain this area using the first and then the second formula in (6).

17. Use a line integral to find the area of the region enclosed by the astroid

$$x = \cos^3 \phi, \quad y = a \sin^3 \phi \qquad (0 \leq \phi \leq 2\pi)$$

[See Exercise 18 of Section 8.4.]

18. Use a line integral to find the area of the triangle with vertices $(0, 0)$, $(a, 0)$, and $(0, b)$, where $a > 0$ and $b > 0$.

19. Use the formula

$$A = \frac{1}{2} \oint_C -y\,dx + x\,dy$$

to find the area of the region swept out by the line from the origin to the ellipse $x = a \cos t$, $y = b \sin t$ if t varies from $t = 0$ to $t = t_0$ ($0 \leq t_0 \leq 2\pi$).

20. Use the formula

$$A = \frac{1}{2} \oint_C -y\,dx + x\,dy$$

to find the area of the region swept out by the line from the origin to the hyperbola $x = a \cosh t$, $y = b \sinh t$ if t varies from $t = 0$ to $t = t_0$ ($t_0 \geq 0$).

In Exercises 21 and 22, use Green's Theorem to find the work done by the force field $\mathbf{F}$ on a particle that moves along the stated path.

21. $\mathbf{F}(x, y) = xy\mathbf{i} + \left(\frac{1}{2}x^2 + xy\right)\mathbf{j}$; the particle starts at $(5, 0)$, traverses the upper semicircle $x^2 + y^2 = 25$, and returns to its starting point along the x-axis.

22. $\mathbf{F}(x, y) = \sqrt{y}\,\mathbf{i} + \sqrt{x}\,\mathbf{j}$; the particle moves counterclockwise one time around the closed curve given by the equations $y = 0$, $x = 2$, and $y = x^3/4$.

23. Evaluate $\oint_C y\,dx - x\,dy$, where C is the cardioid

$$r = a(1 + \cos \theta) \quad (0 \leq \theta \leq 2\pi)$$

24. Let R be a plane region with area A whose boundary is a piecewise smooth simple closed curve C. Use Green's Theorem to prove that the centroid $(\bar{x}, \bar{y})$ of R is given by

$$\bar{x} = \frac{1}{2A} \oint_C x^2\,dy, \quad \bar{y} = -\frac{1}{2A} \oint_C y^2\,dx$$

In Exercises 25–28, use the result in Exercise 24 to find the centroid of the region.

25.

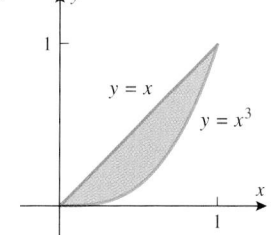

26.

27.

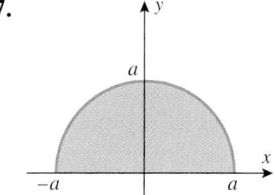

28.

29. Find a simple closed curve C with counterclockwise orientation that maximizes the value of

$$\oint_C \tfrac{1}{3} y^3 \, dx + \left(x - \tfrac{1}{3} x^3\right) dy$$

and explain your reasoning.

30. (a) Let C be the line segment from a point (a, b) to a point (c, d). Show that

$$\int_C -y \, dx + x \, dy = ad - bc$$

(b) Use the result in part (a) to show that the area A of a triangle with successive vertices (x_1, y_1), (x_2, y_2), and (x_3, y_3) going counterclockwise is

$$A = \tfrac{1}{2}[(x_1 y_2 - x_2 y_1) \\ + (x_2 y_3 - x_3 y_2) + (x_3 y_1 - x_1 y_3)]$$

(c) Find a formula for the area of a polygon with successive vertices (x_1, y_1), (x_2, y_2), ..., (x_n, y_n) going counterclockwise.

(d) Use the result in part (c) to find the area of a quadrilateral with vertices $(0, 0)$, $(3, 4)$, $(-2, 2)$, $(-1, 0)$.

> In Exercises 31 and 32, evaluate the integral $\int_C \mathbf{F} \cdot d\mathbf{r}$, where C is the boundary of the region R and C is oriented so that the region is on the left when the boundary is traversed in the direction of its orientation.

31. $\mathbf{F}(x, y) = (x^2 + y)\mathbf{i} + (4x - \cos y)\mathbf{j}$; C is the boundary of the region R that is inside the square with vertices $(0, 0)$, $(5, 0)$, $(5, 5)$, $(0, 5)$ but is outside the rectangle with vertices $(1, 1)$, $(3, 1)$, $(3, 2)$, $(1, 2)$.

32. $\mathbf{F}(x, y) = (e^{-x} + 3y)\mathbf{i} + x\mathbf{j}$; C is the boundary of the region R between the circles $x^2 + y^2 = 16$ and $x^2 - 2x + y^2 = 3$.

17.5 SURFACE INTEGRALS

In previous sections we considered four kinds of integrals—integrals over intervals, double integrals over two-dimensional regions, triple integrals over three-dimensional solids, and line integrals along curves in two- or three-dimensional space. In this section we will discuss integrals over surfaces in three-dimensional space. Such integrals occur in problems involving fluid and heat flow, electricity, magnetism, mass, and center of gravity.

....
DEFINITION OF A SURFACE INTEGRAL

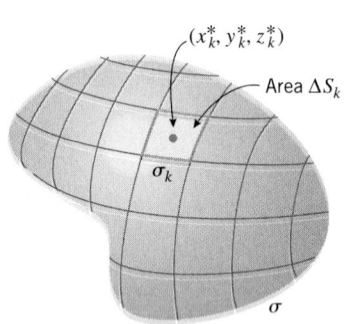

Figure 17.5.1

Recall that if C is a smooth parametric curve in 3-space, and $f(x, y, z)$ is continuous on C, then the line integral of f along C with respect to arc length is defined by subdividing C into n arcs and defining the line integral as the limit

$$\int_C f(x, y, z) \, ds = \lim_{n \to +\infty} \sum_{k=1}^n f(x_k^*, y_k^*, z_k^*) \Delta s_k$$

where (x_k^*, y_k^*, z_k^*) is a point on the kth arc and Δs_k is the length of the kth arc. We will define *surface integrals* in an analogous manner.

Let σ be a surface in 3-space with finite surface area, and let $f(x, y, z)$ be a continuous function defined on σ. As shown in Figure 17.5.1, subdivide σ into patches, $\sigma_1, \sigma_2, \ldots, \sigma_n$ with areas $\Delta S_1, \Delta S_2, \ldots, \Delta S_n$, and form the sum

$$\sum_{k=1}^n f(x_k^*, y_k^*, z_k^*) \Delta S_k \tag{1}$$

where (x_k^*, y_k^*, z_k^*) is an arbitrary point on σ_k. Now repeat the subdivision process, dividing σ into more and more patches in such a way that the maximum dimension of each patch approaches zero as $n \to +\infty$. If (1) approaches a limit that does not depend on the way the subdivisions are made or how the points (x_k^*, y_k^*, z_k^*) are chosen, then this limit is called the *surface integral* of $f(x, y, z)$ over σ and is denoted by

$$\iint_\sigma f(x, y, z) \, dS = \lim_{n \to +\infty} \sum_{k=1}^n f(x_k^*, y_k^*, z_k^*) \Delta S_k \tag{2}$$

....
EVALUATING SURFACE INTEGRALS

There are various procedures for evaluating surface integrals that depend on how the surface σ is represented. The following theorem provides a method for evaluating a surface integral when σ is represented parametrically.

> **17.5.1   THEOREM.**   *Let σ be a smooth parametric surface whose vector equation is*
> $$\mathbf{r} = x(u, v)\mathbf{i} + y(u, v)\mathbf{j} + z(u, v)\mathbf{k}$$
> *where (u, v) varies over a region R in the uv-plane. If $f(x, y, z)$ is continuous on σ, then*
> $$\iint_{\sigma} f(x, y, z)\, dS = \iint_{R} f(x(u, v), y(u, v), z(u, v)) \left\| \frac{\partial \mathbf{r}}{\partial u} \times \frac{\partial \mathbf{r}}{\partial v} \right\| dA \qquad (3)$$

To motivate this result, suppose that the parameter domain R is subdivided as in Figure 16.4.10, and suppose that the point (x_k^*, y_k^*, z_k^*) in (2) corresponds to parameter values of u_k^* and v_k^*. If we use Formula (9) of Section 16.4 to approximate ΔS_k, and if we assume that the errors in the approximations approach zero as $n \to +\infty$, then it follows from (2) that

$$\iint_{\sigma} f(x, y, z)\, dS = \lim_{n \to +\infty} \sum_{k=1}^{n} f(x(u_k^*, v_k^*), y(u_k^*, v_k^*), z(u_k^*, v_k^*)) \left\| \frac{\partial \mathbf{r}}{\partial u} \times \frac{\partial \mathbf{r}}{\partial v} \right\| \Delta A_k$$

which suggests Formula (3).

We will discuss various applications and interpretations of surface integrals later in this section and in subsequent sections, but for now we will focus on techniques for evaluating such integrals.

Example 1

Evaluate the surface integral $\iint_{\sigma} x^2\, dS$ over the sphere $x^2 + y^2 + z^2 = 1$.

Solution.   As in Example 9 of Section 16.4 (with $a = 1$), the sphere is the graph of the vector-valued function

$$\mathbf{r}(\phi, \theta) = \sin\phi \cos\theta\, \mathbf{i} + \sin\phi \sin\theta\, \mathbf{j} + \cos\phi\, \mathbf{k} \quad (0 \le \phi \le \pi, \ 0 \le \theta \le 2\pi) \qquad (4)$$

and

$$\left\| \frac{\partial \mathbf{r}}{\partial \phi} \times \frac{\partial \mathbf{r}}{\partial \theta} \right\| = \sin\phi$$

From the **i**-component of $\mathbf{r}$, the integrand in the surface integral can be expressed in terms of ϕ and θ as $x^2 = \sin^2\phi \cos^2\theta$. Thus, it follows from (3) with ϕ and θ in place of u and v and R as the rectangular region in the $\phi\theta$-plane determined by the inequalities in (4) that

$$\iint_{\sigma} x^2\, dS = \iint_{R} (\sin^2\phi \cos^2\theta) \left\| \frac{\partial \mathbf{r}}{\partial \phi} \times \frac{\partial \mathbf{r}}{\partial \theta} \right\| dA$$

$$= \int_{0}^{2\pi} \int_{0}^{\pi} \sin^3\phi \cos^2\theta\, d\phi\, d\theta$$

$$= \int_{0}^{2\pi} \left[\int_{0}^{\pi} \sin^3\phi\, d\phi \right] \cos^2\theta\, d\theta$$

$$= \int_{0}^{2\pi} \left[\frac{1}{3}\cos^3\phi - \cos\phi \right]_{0}^{\pi} \cos^2\theta\, d\theta \qquad \boxed{\text{Formula (11),}\ \text{Section 9.3}}$$

$$= \frac{4}{3} \int_{0}^{2\pi} \cos^2\theta\, d\theta$$

$$= \frac{4}{3} \left[\frac{1}{2}\theta + \frac{1}{4}\sin 2\theta \right]_{0}^{2\pi} = \frac{4\pi}{3} \qquad \boxed{\text{Formula (8),}\ \text{Section 9.3}} \qquad \blacktriangleleft$$

SURFACE INTEGRALS OVER
z = g(x, y), y = g(x, z), AND
x = g(y, z)

In the case where σ is a surface of the form $z = g(x, y)$, we can take $x = u$ and $y = v$ as parameters and express the equation of the surface as

$$\mathbf{r} = u\mathbf{i} + v\mathbf{j} + g(u, v)\mathbf{k}$$

in which case we obtain

$$\left\| \frac{\partial \mathbf{r}}{\partial u} \times \frac{\partial \mathbf{r}}{\partial v} \right\| = \sqrt{\left(\frac{\partial z}{\partial x}\right)^2 + \left(\frac{\partial z}{\partial y}\right)^2 + 1}$$

[see the derivation of Formula (11) in Section 16.4]. Thus, it follows from (3) that

$$\iint\limits_{\sigma} f(x, y, z)\, dS = \iint\limits_{R} f(x, y, g(x, y)) \sqrt{\left(\frac{\partial z}{\partial x}\right)^2 + \left(\frac{\partial z}{\partial y}\right)^2 + 1}\, dA$$

Note that in this formula the region R lies in the xy-plane because the parameters are x and y. Geometrically, this region is the projection of σ on the xy-plane. The following theorem summarizes this result and gives analogous formulas for surface integrals over surfaces of the form $y = g(x, z)$ and $x = g(y, z)$.

17.5.2 THEOREM.

(a) *Let σ be a surface with equation $z = g(x, y)$ and let R be its projection on the xy-plane. If g has continuous first partial derivatives on R and $f(x, y, z)$ is continuous on σ, then*

$$\iint\limits_{\sigma} f(x, y, z)\, dS = \iint\limits_{R} f(x, y, g(x, y)) \sqrt{\left(\frac{\partial z}{\partial x}\right)^2 + \left(\frac{\partial z}{\partial y}\right)^2 + 1}\, dA \qquad (5)$$

(b) *Let σ be a surface with equation $y = g(x, z)$ and let R be its projection on the xz-plane. If g has continuous first partial derivatives on R and $f(x, y, z)$ is continuous on σ, then*

$$\iint\limits_{\sigma} f(x, y, z)\, dS = \iint\limits_{R} f(x, g(x, z), z) \sqrt{\left(\frac{\partial y}{\partial x}\right)^2 + \left(\frac{\partial y}{\partial z}\right)^2 + 1}\, dA \qquad (6)$$

(c) *Let σ be a surface with equation $x = g(y, z)$ and let R be its projection on the yz-plane. If g has continuous first partial derivatives on R and $f(x, y, z)$ is continuous on σ, then*

$$\iint\limits_{\sigma} f(x, y, z)\, dS = \iint\limits_{R} f(g(y, z), y, z) \sqrt{\left(\frac{\partial x}{\partial y}\right)^2 + \left(\frac{\partial x}{\partial z}\right)^2 + 1}\, dA \qquad (7)$$

Example 2

Evaluate the surface integral

$$\iint\limits_{\sigma} xz\, dS$$

where σ is the part of the plane $x + y + z = 1$ that lies in the first octant.

Solution. The equation of the plane can be written as

$$z = 1 - x - y$$

which is of the form $z = g(x, y)$. Consequently, we can apply Formula (5) with $z = g(x, y) = 1 - x - y$ and $f(x, y, z) = xz$. We have

$$\frac{\partial z}{\partial x} = -1 \quad \text{and} \quad \frac{\partial z}{\partial y} = -1$$

Figure 17.5.2

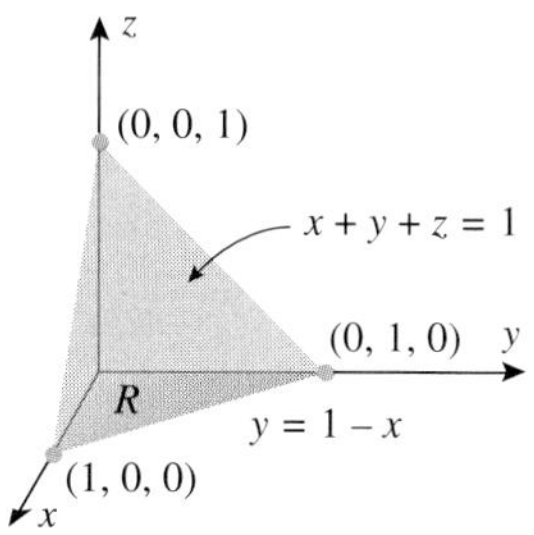

so (5) becomes

$$\iint_{\sigma} xz\, dS = \iint_{R} x(1 - x - y)\sqrt{(-1)^2 + (-1)^2 + 1}\, dA \tag{8}$$

where R is the projection of σ on the xy-plane (Figure 17.5.2). Rewriting the double integral in (8) as an iterated integral yields

$$\begin{aligned}
\iint_{\sigma} xz\, dS &= \sqrt{3}\int_0^1 \int_0^{1-x} (x - x^2 - xy)\, dy\, dx \\
&= \sqrt{3}\int_0^1 \left[xy - x^2y - \frac{xy^2}{2}\right]_{y=0}^{1-x} dx \\
&= \sqrt{3}\int_0^1 \left(\frac{x}{2} - x^2 + \frac{x^3}{2}\right) dx \\
&= \sqrt{3}\left[\frac{x^2}{4} - \frac{x^3}{3} + \frac{x^4}{8}\right]_0^1 = \frac{\sqrt{3}}{24}
\end{aligned}$$

◀

Example 3

Evaluate the surface integral

$$\iint_{\sigma} y^2z^2\, dS$$

where σ is the part of the cone $z = \sqrt{x^2 + y^2}$ that lies between the planes $z = 1$ and $z = 2$ (Figure 17.5.3).

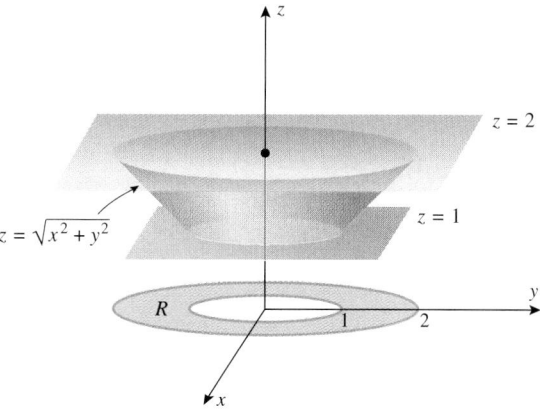

Figure 17.5.3

Solution. We will apply Formula (5) with

$$z = g(x, y) = \sqrt{x^2 + y^2} \quad \text{and} \quad f(x, y, z) = y^2z^2$$

Thus,

$$\frac{\partial z}{\partial x} = \frac{x}{\sqrt{x^2 + y^2}} \quad \text{and} \quad \frac{\partial z}{\partial y} = \frac{y}{\sqrt{x^2 + y^2}}$$

so

$$\sqrt{\left(\frac{\partial z}{\partial x}\right)^2 + \left(\frac{\partial z}{\partial y}\right)^2 + 1} = \sqrt{2}$$

(verify), and (5) yields

$$\iint_{\sigma} y^2 z^2 \, dS = \iint_R y^2 \left(\sqrt{x^2 + y^2}\right)^2 \sqrt{2} \, dA = \sqrt{2} \iint_R y^2 (x^2 + y^2) \, dA$$

where R is the annulus enclosed between $x^2 + y^2 = 1$ and $x^2 + y^2 = 4$ (Figure 17.5.3). Using polar coordinates to evaluate this double integral over the annulus R yields

$$\iint_{\sigma} y^2 z^2 \, dS = \sqrt{2} \int_0^{2\pi} \int_1^2 (r \sin\theta)^2 (r^2) r \, dr \, d\theta$$

$$= \sqrt{2} \int_0^{2\pi} \int_1^2 r^5 \sin^2\theta \, dr \, d\theta$$

$$= \sqrt{2} \int_0^{2\pi} \frac{r^6}{6} \sin^2\theta \Big]_{r=1}^2 \, d\theta = \frac{21}{\sqrt{2}} \int_0^{2\pi} \sin^2\theta \, d\theta$$

$$= \frac{21}{\sqrt{2}} \left[\frac{1}{2}\theta - \frac{1}{4}\sin 2\theta\right]_0^{2\pi} = \frac{21\pi}{\sqrt{2}} \quad \boxed{\text{Formula (7), Section 9.3}} \quad \blacktriangleleft$$

MASS OF A CURVED LAMINA AS A SURFACE INTEGRAL

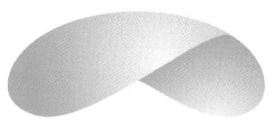

The thickness of a curved lamina is negligible.

Figure 17.5.4

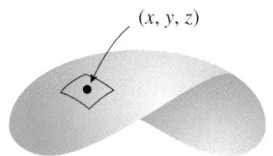

Figure 17.5.5

In Section 16.6 we defined a *lamina* to be an idealized flat object that is thin enough to be viewed as a plane region. Analogously, a ***curved lamina*** is an idealized object that is thin enough to be viewed as a surface in 3-space. A curved lamina may look like a bent plate, as in Figure 17.5.4, or it may enclose a region in 3-space, like the shell of an egg. If the composition of a curved lamina is uniform so that its mass is distributed uniformly, then it is said to be ***homogeneous***, and we define its ***mass density*** to be the total mass divided by the total surface area. However, if the mass of the lamina is not uniformly distributed, then this is not a useful measure, since it does not account for the variations in mass concentration. In this case we describe the mass concentration at a point by a ***mass density function*** δ, which we view as a limit; that is,

$$\delta = \lim_{\Delta S \to 0} \frac{\Delta M}{\Delta S} \tag{9}$$

where ΔM and ΔS denote the mass and surface area of a small section of lamina containing the point (Figure 17.5.5).

To translate this informal idea into a useful formula, suppose that $\delta = \delta(x, y, z)$ is the density function of a smooth curved lamina σ. Assume that the lamina is subdivided into n small sections; let (x_k^*, y_k^*, z_k^*) be a point in the kth section, let ΔM_k be the mass of the kth section, and let ΔS_k be the surface area of the kth section. Since we are assuming that the sections are small, it follows from (9) that the mass of the kth section can be approximated as

$$\Delta M_k \approx \delta(x_k^*, y_k^*, z_k^*) \Delta S_k$$

and hence the mass M of the entire lamina can be approximated as

$$M = \sum_{k=1}^n \Delta M_k \approx \sum_{k=1}^n \delta(x_k^*, y_k^*, z_k^*) \Delta S_k \tag{10}$$

If we now increase n in such a way that the dimensions of the sections approach zero, then it is plausible that the error in (10) will approach zero, and the exact value of M will be given by the surface integral

$$M = \iint_{\sigma} \delta(x, y, z) \, dS \tag{11}$$

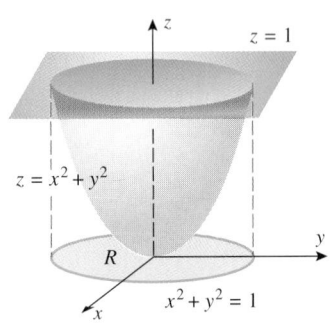

Figure 17.5.6

Example 4

Suppose that a curved lamina σ with constant density $\delta(x, y, z) = \delta_0$ is the portion of the paraboloid $z = x^2 + y^2$ below the plane $z = 1$ (Figure 17.5.6). Find the mass of the lamina.

Solution.   Since $z = g(x, y) = x^2 + y^2$, it follows that

$$\frac{\partial z}{\partial x} = 2x \quad \text{and} \quad \frac{\partial z}{\partial y} = 2y$$

Substituting these expressions and $\delta(x, y, z) = \delta(x, y, g(x, y)) = \delta_0$ into (11) yields

$$M = \iint_\sigma \delta_0 \, dS = \iint_R \delta_0 \sqrt{(2x)^2 + (2y)^2 + 1} \, dA = \delta_0 \iint_R \sqrt{4x^2 + 4y^2 + 1} \, dA \quad (12)$$

where R is the circular region enclosed by $x^2 + y^2 = 1$. To evaluate (12) we use polar coordinates:

$$M = \delta_0 \int_0^{2\pi} \int_0^1 \sqrt{4r^2 + 1} \, r \, dr \, d\theta = \frac{\delta_0}{12} \int_0^{2\pi} (4r^2 + 1)^{3/2} \Big]_{r=0}^1 d\theta$$

$$= \frac{\delta_0}{12} \int_0^{2\pi} (5^{3/2} - 1) \, d\theta = \frac{\pi \delta_0}{6} (5\sqrt{5} - 1) \qquad \blacktriangleleft$$

SURFACE AREA AS A SURFACE INTEGRAL

In the special case where $f(x, y, z)$ is 1, Formula (3) becomes

$$\iint_\sigma dS = \iint_R \left\| \frac{\partial \mathbf{r}}{\partial u} \times \frac{\partial \mathbf{r}}{\partial v} \right\| dA$$

However, it follows from Formula (10) of Section 16.4 that this integral represents the surface area of σ. Thus, we have established the following result.

17.5.3   THEOREM.   *If σ is a smooth parametric surface in 3-space, then its surface area S can be expressed as*

$$S = \iint_\sigma dS \qquad (13)$$

REMARK.   This result adds nothing new computationally, since Formula (13) is just a reformulation of Formula (10) in Section 16.4. However, the relationship between surface integrals and surface area is important to understand.

EXERCISE SET 17.5   [C] CAS

In Exercises 1–10, evaluate the surface integral

$$\iint_\sigma f(x, y, z) \, dS$$

1. $f(x, y, z) = z^2$; σ is the portion of the cone $z = \sqrt{x^2 + y^2}$ between the planes $z = 1$ and $z = 2$.

2. $f(x, y, z) = xy$; σ is the portion of the plane $x + y + z = 1$ lying in the first octant.

3. $f(x, y, z) = x^2 y$; σ is the portion of the cylinder $x^2 + z^2 = 1$ between the planes $y = 0$, $y = 1$, and above the xy-plane.

4. $f(x, y, z) = (x^2 + y^2)z$; σ is the portion of the sphere $x^2 + y^2 + z^2 = 4$ above the plane $z = 1$.

5. $f(x, y, z) = x - y - z$; σ is the portion of the plane $x + y = 1$ in the first octant between $z = 0$ and $z = 1$.

6. $f(x, y, z) = x + y$; σ is the portion of the plane $z = 6 - 2x - 3y$ in the first octant.

7. $f(x, y, z) = x + y + z$; σ is the surface of the cube defined by the inequalities $0 \le x \le 1$, $0 \le y \le 1$, $0 \le z \le 1$. [*Hint:* Integrate over each face separately.]

8. $f(x, y, z) = z + 1$; σ is the upper hemisphere $z = \sqrt{1 - x^2 - y^2}$.

9. $f(x, y, z) = \sqrt{x^2 + y^2 + z^2}$; σ is the portion of the cone $z = \sqrt{x^2 + y^2}$ below the plane $z = 1$.

10. $f(x, y, z) = x^2 + y^2$; σ is the surface of the sphere $x^2 + y^2 + z^2 = a^2$.

In Exercises 11 and 12, set up, but do not evaluate, an iterated integral equal to the given surface integral by projecting σ on (a) the xy-plane, (b) the yz-plane, and (c) the xz-plane.

11. $\iint\limits_{\sigma} xyz\, dS$, where σ is the portion of the plane $2x + 3y + 4z = 12$ in the first octant.

12. $\iint\limits_{\sigma} xz\, dS$, where σ is the portion of the sphere $x^2 + y^2 + z^2 = a^2$ in the first octant.

c 13. Use a CAS to confirm that the three integrals you obtained in Exercise 11 are equal, and find the exact value of the surface integral.

c 14. Try to confirm with a CAS that the three integrals you obtained in Exercise 12 are equal. If you did not succeed, what was the difficulty?

In Exercises 15 and 16, set up, but do not evaluate, two different iterated integrals equal to the given integral.

15. $\iint\limits_{\sigma} xyz\, dS$, where σ is the portion of the surface $y^2 = x$ between the planes $z = 0$, $z = 4$, $y = 1$, and $y = 2$.

16. $\iint\limits_{\sigma} x^2y\, dS$, where σ is the portion of the cylinder $y^2 + z^2 = a^2$ in the first octant between the planes $x = 0$, $x = 9$, $z = y$, and $z = 2y$.

c 17. Use a CAS to confirm that the two integrals you obtained in Exercise 15 are equal, and find the exact value of the surface integral.

c 18. Use a CAS to find the value of the surface integral

$$\iint\limits_{\sigma} x^2yz\, dS$$

over the portion of the elliptic paraboloid $z = 5 - 3x^2 - 2y^2$ that lies above the xy-plane.

In Exercises 19 and 20, find the mass of the lamina with constant density δ_0.

19. The lamina that is the portion of the circular cylinder $x^2 + z^2 = 4$ that lies directly above the rectangle $R = \{(x, y) : 0 \le x \le 1, 0 \le y \le 4\}$ in the xy-plane.

20. The lamina that is the portion of the paraboloid $2z = x^2 + y^2$ inside the cylinder $x^2 + y^2 = 8$.

21. Find the mass of the lamina that is the portion of the surface $y^2 = 4 - z$ between the planes $x = 0$, $x = 3$, $y = 0$, and $y = 3$ if the density is $\delta(x, y, z) = y$.

22. Find the mass of the lamina that is the portion of the cone $z = \sqrt{x^2 + y^2}$ between $z = 1$ and $z = 4$ if the density is $\delta(x, y, z) = x^2z$.

23. If a curved lamina has constant density δ_0, what relationship must exist between its mass and surface area? Explain your reasoning.

24. Show that if the density of the lamina $x^2 + y^2 + z^2 = a^2$ at each point is equal to the distance between that point and the xy-plane, then the mass of the lamina is $2\pi a^3$.

The centroid of a surface σ is defined by

$$\bar{x} = \frac{\iint\limits_{\sigma} x\, dS}{\text{area of } \sigma}, \quad \bar{y} = \frac{\iint\limits_{\sigma} y\, dS}{\text{area of } \sigma}, \quad \bar{z} = \frac{\iint\limits_{\sigma} z\, dS}{\text{area of } \sigma}$$

In Exercises 25 and 26, find the centroid of the surface.

25. The portion of the paraboloid $z = \frac{1}{2}(x^2 + y^2)$ below the plane $z = 4$.

26. The portion of the sphere $x^2 + y^2 + z^2 = 4$ above the plane $z = 1$.

In Exercises 27–30, evaluate the integral $\iint_{\sigma} f(x, y, z)\, dS$ over the surface σ represented by the vector-valued function $\mathbf{r}(u, v)$.

27. $f(x, y, z) = xyz$; $\mathbf{r}(u, v) = u\cos v\mathbf{i} + u\sin v\mathbf{j} + 3u\mathbf{k}$
$(1 \le u \le 2, 0 \le v \le \pi/2)$

28. $f(x, y, z) = \dfrac{x^2 + z^2}{y}$; $\mathbf{r}(u, v) = 2\cos v\mathbf{i} + u\mathbf{j} + 2\sin v\mathbf{k}$
$(1 \le u \le 3, 0 \le v \le 2\pi)$

29. $f(x, y, z) = \dfrac{1}{\sqrt{1 + 4x^2 + 4y^2}}$;
$\mathbf{r}(u, v) = u\cos v\mathbf{i} + u\sin v\mathbf{j} + u^2\mathbf{k}$
$(0 \le u \le \sin v, 0 \le v \le \pi)$

30. $f(x, y, z) = e^{-z}$;
$\mathbf{r}(u, v) = 2\sin u\cos v\mathbf{i} + 2\sin u\sin v\mathbf{j} + 2\cos u\mathbf{k}$
$(0 \le u \le \pi/2, 0 \le v \le 2\pi)$

c 31. Use a CAS to approximate the mass of the curved lamina $z = e^{-x^2-y^2}$ that lies above the region in the xy-plane enclosed by $x^2 + y^2 = 9$ given that the density function is $\delta(x, y, z) = \sqrt{x^2 + y^2}$.

c 32. The surface σ shown in the accompanying figure, called a **Möbius strip**, is represented by the parametric equations
$$x = (5 + u\cos(v/2))\cos v$$
$$y = (5 + u\cos(v/2))\sin v \quad (-1 \le u \le 1, 0 \le v \le 2\pi)$$
$$z = u\sin(v/2)$$

(a) Use a CAS to generate a reasonable facsimile of this surface.

(b) Use a CAS to approximate the location of the centroid of σ (see the definition preceding Exercise 25).

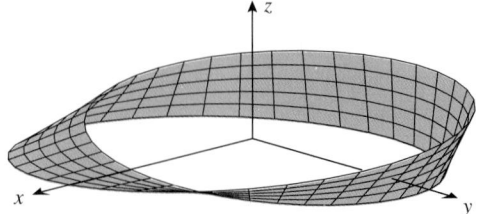

Figure Ex-32

17.6 APPLICATIONS OF SURFACE INTEGRALS; FLUX

In this section we will discuss applications of surface integrals in vector fields associated with fluid flow and electrostatic forces. However, the ideas that we will develop will be general in nature and applicable to other kinds of vector fields as well.

FLOW FIELDS

We will be concerned in this section with vector fields in 3-space that involve some type of "flow"—the flow of a fluid or the flow of charged particles in an electrostatic field, for example. In the case of fluid flow, the vector field $\mathbf{F}(x, y, z)$ represents the velocity of a fluid particle at the point (x, y, z), and the fluid particles flow along "streamlines" that are tangential to the velocity vectors (Figure 17.6.1a). In the case of an electrostatic field, $\mathbf{F}(x, y, z)$ is the force that the field exerts on a small unit of positive charge at the point (x, y, z), and such charges accelerate along "electric lines" that are tangential to the force vectors (Figures 17.6.1b and 17.6.1c).

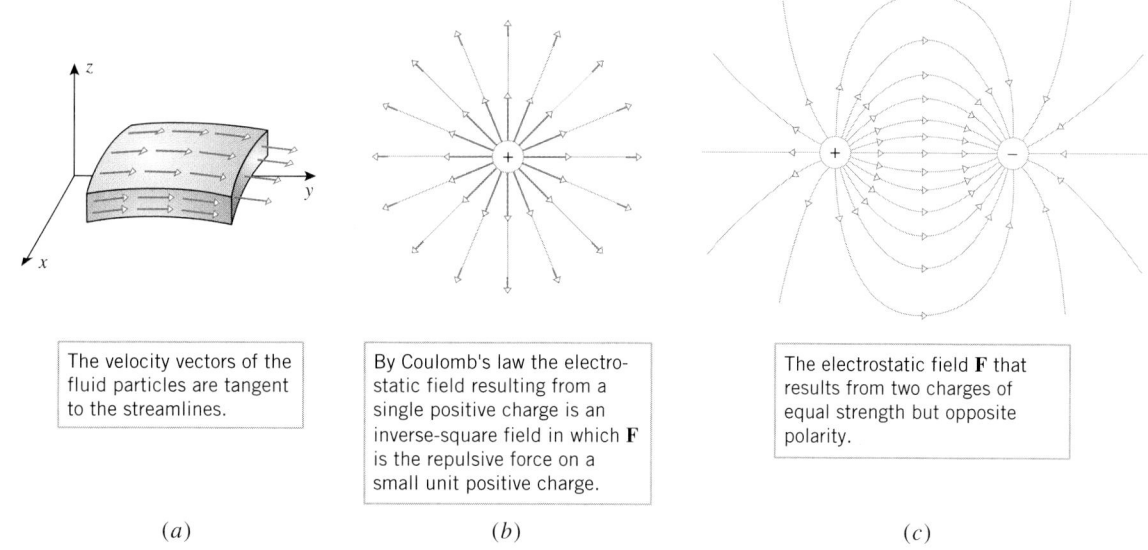

The velocity vectors of the fluid particles are tangent to the streamlines.

By Coulomb's law the electrostatic field resulting from a single positive charge is an inverse-square field in which $\mathbf{F}$ is the repulsive force on a small unit positive charge.

The electrostatic field $\mathbf{F}$ that results from two charges of equal strength but opposite polarity.

(*a*)   (*b*)   (*c*)

Figure 17.6.1

ORIENTED SURFACES

Our main goal in this section is to study flows of vector fields through permeable surfaces placed in the field. For this purpose we will need to consider some basic ideas about surfaces. Most surfaces that we encounter in applications have two sides—a sphere has an inside and an outside, and an infinite horizontal plane has a top side and a bottom side, for example. However, there exist mathematical surfaces with only one side. For example, Figure 17.6.2a shows the construction of a surface called a ***Möbius strip*** [in honor of the German mathematician August Möbius (1790–1868)]. The Möbius strip has only one side in the sense that a bug can traverse the *entire* surface without crossing an edge (Figure 17.6.2b). In contrast, a sphere is two-sided in the sense that a bug walking on the sphere can traverse the inside surface or the outside surface but cannot traverse both without somehow passing

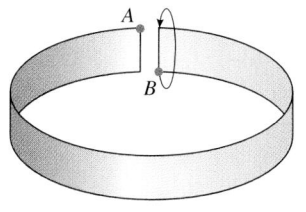

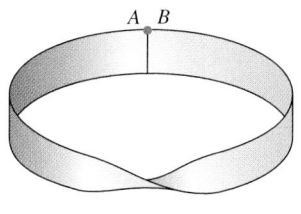

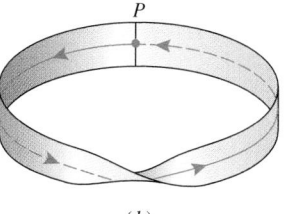

If an ant starts at P with its back facing you and makes one circuit around the strip, then its back will face away from you when it returns to P. Thus, the Möbius strip has only one side.

(*a*)   (*b*)

Figure 17.6.2

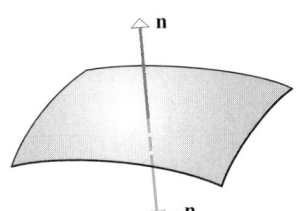

Figure 17.6.3

through the sphere. A two-sided surface is said to be ***orientable***, and a one-sided surface is said to be ***nonorientable***. In the rest of this text we will only be concerned with orientable surfaces.

In applications, it is important to have some way of distinguishing between the two sides of an orientable surface. For this purpose let us suppose that σ is an orientable surface that has a unit normal vector **n** at each point. As illustrated in Figure 17.6.3, the vectors **n** and $-\mathbf{n}$ point to opposite sides of the surface and hence serve to distinguish between the two sides. It can be proved that if σ is a smooth orientable surface, then it is always possible to choose the direction of **n** at each point so that $\mathbf{n} = \mathbf{n}(x, y, z)$ varies continuously over the surface. These unit vectors are then said to form an ***orientation*** of the surface. It can also be proved that a smooth orientable surface has only two possible orientations. For example, the surface in Figure 17.6.4 is oriented up by the purple vectors and down by the green vectors. However, we cannot create a third orientation by mixing the two since this produces points on the surface at which there is an abrupt change in direction (across the black curve in the figure, for example).

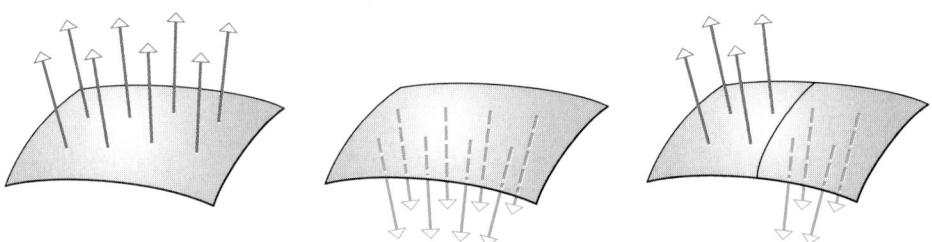

Figure 17.6.4

ORIENTATION OF A SMOOTH PARAMETRIC SURFACE

When a surface is expressed parametrically, the parametric equations create a natural orientation of the surface. To see why this is so, recall from Section 16.4 that if a smooth parametric surface σ is given by the vector equation

$$\mathbf{r} = x(u, v)\mathbf{i} + y(u, v)\mathbf{j} + z(u, v)\mathbf{k}$$

then the unit normal

$$\mathbf{n} = \mathbf{n}(u, v) = \dfrac{\dfrac{\partial \mathbf{r}}{\partial u} \times \dfrac{\partial \mathbf{r}}{\partial v}}{\left\| \dfrac{\partial \mathbf{r}}{\partial u} \times \dfrac{\partial \mathbf{r}}{\partial v} \right\|} \qquad (1)$$

is a continuous vector-valued function of u and v. Thus, Formula (1) defines an orientation of the surface; we call this the ***positive orientation*** of the parametric surface and we say that **n** points in the ***positive direction*** from the surface. The orientation determined by $-\mathbf{n}$ is called the ***negative orientation*** of the surface and we say that $-\mathbf{n}$ points in the ***negative direction*** from the surface. For example, consider the sphere that is represented parametrically by the vector equation

$$\mathbf{r}(\phi, \theta) = a \sin \phi \cos \theta\, \mathbf{i} + a \sin \phi \sin \theta\, \mathbf{j} + a \cos \phi\, \mathbf{k} \qquad (0 \le \phi \le \pi, \ \ 0 \le \theta \le 2\pi)$$

We showed in Example 9 of Section 16.4 that

$$\mathbf{n} = \frac{1}{a}\mathbf{r}$$

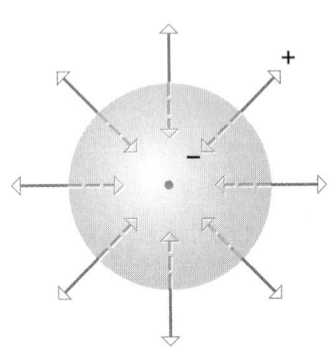

Figure 17.6.5

This vector points in the same direction as the radius vector **r** (outward from the center). Thus, for the given parametrization, the positive orientation of the sphere is *outward* and the negative orientation is *inward* (Figure 17.6.5).

FOR THE READER.   See if you can find a parametrization of the sphere in which the positive direction is inward.

FLUX

In physics, the term *fluid* is used to describe both liquids and gases. Liquids are usually regarded to be ***incompressible***, meaning that the liquid has a uniform density (mass per unit volume) that cannot be altered by compressive forces. Gases are regarded to be ***compressible***, meaning that the density may vary from point to point and can be altered by compressive forces. In this text we will be concerned primarily with incompressible fluids. Moreover, we will assume that the velocity of the fluid at a fixed point does not vary with time. Fluid flows with this property are said to be in a ***steady state***.

Our next goal in this section is to define a fundamental concept of physics known as *flux* (from the Latin word *fluxus*, meaning "flow"). This concept is applicable in any vector field, but we will motivate it in the context of steady-state flow of an incompressible fluid. We consider the following problem:

> **17.6.1** PROBLEM.   Suppose that an oriented surface σ is immersed in an incompressible, steady-state fluid flow, and assume further that the surface is permeable so that the fluid can flow through it freely in either direction. Find the net volume of fluid Φ that passes through the surface per unit of time, where the net volume is interpreted to mean the volume that passes through the surface in the positive direction minus the volume that passes through the surface in the negative direction.

To solve this problem, suppose that the velocity of the fluid at a point (x, y, z) on the surface σ is given by

$$\mathbf{F}(x, y, z) = f(x, y, z)\mathbf{i} + g(x, y, z)\mathbf{j} + h(x, y, z)\mathbf{k}$$

Let $\mathbf{n}$ be the unit normal toward the positive side of σ at the point (x, y, z), and let $\mathbf{T}$ be a unit vector that is orthogonal to $\mathbf{n}$ and lies in the plane of $\mathbf{F}$ and $\mathbf{n}$. As illustrated in Figure 17.6.6, the velocity vector $\mathbf{F}$ can be resolved into two orthogonal components— a component $(\mathbf{F} \cdot \mathbf{T})\mathbf{T}$ along the "face" of the surface σ and a component $(\mathbf{F} \cdot \mathbf{n})\mathbf{n}$ that is perpendicular to σ. The component of velocity along the face of the surface does not contribute to the flow through σ and hence can be ignored in our computations. Moreover, observe that the sign of $\mathbf{F} \cdot \mathbf{n}$ determines the direction of flow—a positive value means the flow is in the direction of $\mathbf{n}$ and a negative value means that it is opposite to $\mathbf{n}$.

To solve Problem 17.6.1, we subdivide σ into n patches $\sigma_1, \sigma_2, \ldots, \sigma_n$ with areas

$$\Delta S_1, \Delta S_2, \ldots, \Delta S_n$$

If the patches are small and the flow is not too erratic, it is reasonable to assume that the velocity does not vary much on each patch. Thus, if (x_k^*, y_k^*, z_k^*) is any point in the kth patch, we can assume that $\mathbf{F}(x, y, z)$ is constant and equal to $\mathbf{F}(x_k^*, y_k^*, z_k^*)$ throughout the patch and that the component of velocity across the surface σ_k is

$$\mathbf{F}(x_k^*, y_k^*, z_k^*) \cdot \mathbf{n}(x_k^*, y_k^*, z_k^*) \tag{2}$$

(Figure 17.6.7). Thus, we can interpret

$$\mathbf{F}(x_k^*, y_k^*, z_k^*) \cdot \mathbf{n}(x_k^*, y_k^*, z_k^*)\Delta S_k$$

as the approximate volume of fluid crossing the patch σ_k in the direction of $\mathbf{n}$ per unit of time (Figure 17.6.8). For example, if the component of velocity in the direction of $\mathbf{n}$ is $\mathbf{F}(x_k^*, y_k^*, z_k^*) \cdot \mathbf{n} = 25$ cm/s, and the area of the patch is $\Delta S_k = 2$ cm^2, then the volume of fluid ΔV_k crossing the patch in the direction of $\mathbf{n}$ per unit of time is approximately

$$\Delta V_k \approx \mathbf{F}(x_k^*, y_k^*, z_k^*) \cdot \mathbf{n}(x_k^*, y_k^*, z_k^*)\Delta S_k = 25 \text{ cm/s} \cdot 2 \text{ cm}^2 = 50 \text{ cm}^3/\text{s}$$

In the case where the velocity component $\mathbf{F}(x_k^*, y_k^*, z_k^*) \cdot \mathbf{n}(x_k^*, y_k^*, z_k^*)$ is negative, the flow is in the direction opposite to $\mathbf{n}$, so that $-\Delta V_k$ is the approximate volume of fluid crossing the patch σ_k in the direction opposite to $\mathbf{n}$ per unit time. Thus, the sum

$$\sum_{k=1}^{n} \mathbf{F}(x_k^*, y_k^*, z_k^*) \cdot \mathbf{n}(x_k^*, y_k^*, z_k^*)\Delta S_k$$

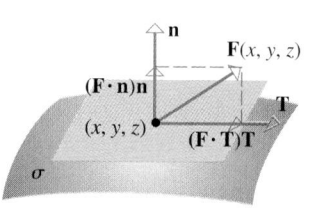

Figure 17.6.6

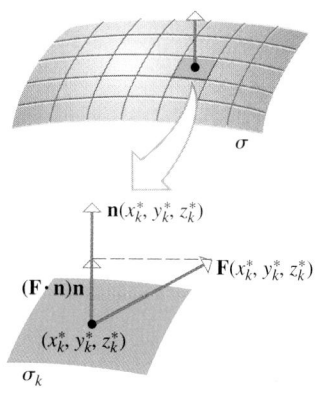

Figure 17.6.7

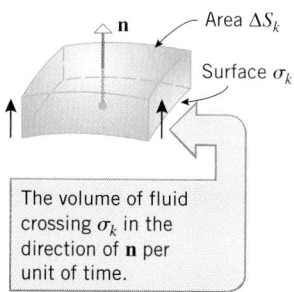

The volume of fluid crossing σ_k in the direction of $\mathbf{n}$ per unit of time.

Figure 17.6.8

measures the approximate net volume of fluid that crosses the surface σ in the direction of its orientation $\mathbf{n}$ per unit of time.

If we now increase n in such a way that the maximum dimension of each patch approaches zero, then it is plausible that the errors in the approximations approach zero, and the limit

$$\Phi = \lim_{n \to +\infty} \sum_{k=1}^{n} \mathbf{F}(x_k^*, y_k^*, z_k^*) \cdot \mathbf{n}(x_k^*, y_k^*, z_k^*)\Delta S_k \tag{3}$$

represents the exact net volume of fluid that crosses the surface σ in the direction of its orientation $\mathbf{n}$ per unit of time. The quantity Φ defined by Equation (3) is called the **flux of F across** σ. The flux can also be expressed as the surface integral

$$\Phi = \iint_{\sigma} \mathbf{F}(x, y, z) \cdot \mathbf{n}(x, y, z)\, dS \tag{4}$$

A positive flux means that in one unit of time a greater volume of fluid passes through σ in the positive direction than in the negative direction, a negative flux means that a greater volume passes through the surface in the negative direction than in the positive direction, and a zero flux means that the same volume passes through the surface in each direction. Integrals of form (4) arise in other contexts as well and are called **flux integrals**.

REMARK.   If the fluid has mass density δ, then $\Phi\delta$ (volume $\times$ density) represents the net mass of fluid that passes through σ per unit of time.

EVALUATING FLUX INTEGRALS

An effective formula for evaluating flux integrals can be obtained by applying Theorem 17.5.1 and using Formula (1) for $\mathbf{n}$. This yields

$$\iint_{\sigma} \mathbf{F} \cdot \mathbf{n}\, dS = \iint_{R} \mathbf{F} \cdot \mathbf{n} \left\| \frac{\partial \mathbf{r}}{\partial u} \times \frac{\partial \mathbf{r}}{\partial v} \right\| dA$$

$$= \iint_{R} \mathbf{F} \cdot \frac{\dfrac{\partial \mathbf{r}}{\partial u} \times \dfrac{\partial \mathbf{r}}{\partial v}}{\left\| \dfrac{\partial \mathbf{r}}{\partial u} \times \dfrac{\partial \mathbf{r}}{\partial v} \right\|} \left\| \frac{\partial \mathbf{r}}{\partial u} \times \frac{\partial \mathbf{r}}{\partial v} \right\| dA$$

$$= \iint_{R} \mathbf{F} \cdot \left(\frac{\partial \mathbf{r}}{\partial u} \times \frac{\partial \mathbf{r}}{\partial v} \right) dA$$

In summary, we have the following result.

17.6.2   THEOREM.   *Let σ be a smooth parametric surface represented by the vector equation $\mathbf{r} = \mathbf{r}(u, v)$ in which (u, v) varies over a region R in the uv-plane. If the component functions of the vector field $\mathbf{F}$ are continuous on σ, and if $\mathbf{n}$ determines the positive orientation of σ, then*

$$\Phi = \iint_{\sigma} \mathbf{F} \cdot \mathbf{n}\, dS = \iint_{R} \mathbf{F} \cdot \left(\frac{\partial \mathbf{r}}{\partial u} \times \frac{\partial \mathbf{r}}{\partial v} \right) dA \tag{5}$$

where it is understood that the integrand on the right side of the equation is expressed in terms of u and v.

Example 1

Find the flux of the vector field $\mathbf{F}(x, y, z) = z\mathbf{k}$ across the sphere $x^2 + y^2 + z^2 = a^2$ oriented outward.

Solution. The sphere with outward positive orientation can be represented by the vector-valued function

$$\mathbf{r}(\phi, \theta) = a \sin \phi \cos \theta \mathbf{i} + a \sin \phi \sin \theta \mathbf{j} + a \cos \phi \mathbf{k} \quad (0 \le \phi \le \pi, \ 0 \le \theta \le 2\pi)$$

From this formula we obtain (see Example 9 of Section 16.4 for the computations)

$$\frac{\partial \mathbf{r}}{\partial \phi} \times \frac{\partial \mathbf{r}}{\partial \theta} = a^2 \sin^2 \phi \cos \theta \mathbf{i} + a^2 \sin^2 \phi \sin \theta \mathbf{j} + a^2 \sin \phi \cos \phi \mathbf{k}$$

Moreover, for points on the sphere we have $\mathbf{F} = z\mathbf{k} = a \cos \phi \mathbf{k}$; hence,

$$\mathbf{F} \cdot \left(\frac{\partial \mathbf{r}}{\partial \phi} \times \frac{\partial \mathbf{r}}{\partial \theta} \right) = a^3 \sin \phi \cos^2 \phi$$

Thus, it follows from (5) with the parameters u and v replaced by ϕ and θ that

$$\Phi = \iint_\sigma \mathbf{F} \cdot \mathbf{n} \, dS$$

$$= \iint_R \mathbf{F} \cdot \left(\frac{\partial \mathbf{r}}{\partial \phi} \times \frac{\partial \mathbf{r}}{\partial \theta} \right) dA$$

$$= \int_0^{2\pi} \int_0^\pi a^3 \sin \phi \cos^2 \phi \, d\phi \, d\theta$$

$$= a^3 \int_0^{2\pi} \left[-\frac{\cos^3 \phi}{3} \right]_0^\pi d\theta$$

$$= \frac{2a^3}{3} \int_0^{2\pi} d\theta = \frac{4\pi a^3}{3} \qquad \blacktriangleleft$$

REMARK. Although the computations in this example give a correct result, they are technically flawed in that the parametric representation used for the sphere is not smooth at $\phi = 0$ or $\phi = \pi$ (see Example 9 of Section 16.4). However, this difficulty can be circumvented by cutting holes with a small radius in the sphere around the z-axis (to avoid the problem areas), performing the required computations on the cut surface, and then taking the limit as the radius approaches zero. It can be shown that this leads to the same result that we obtained in our formal computations. In general, no problems occur when Formula (5) is applied directly to spheres that are parametrized as in this example.

REMARK. Reversing the orientation of the surface σ in (5) reverses the sign $\mathbf{n}$, hence the sign of $\mathbf{F} \cdot \mathbf{n}$, and hence reverses the sign of Φ. This can also be seen physically by interpreting the flux integral as the volume of fluid per unit time that crosses σ in the positive direction minus the volume per unit time that crosses in the negative direction—reversing the orientation of σ changes the sign of the difference. Thus, in Example 1 an inward orientation of the sphere would produce a flux of $-4\pi a^3/3$.

ORIENTATION OF NONPARAMETRIC SURFACES

Nonparametric surfaces of the form $z = g(x, y)$, $y = g(z, x)$, and $x = g(y, z)$ can be expressed parametrically using the independent variables as parameters. More precisely, these surfaces can be represented by the vector equations

$$\mathbf{r} = u\mathbf{i} + v\mathbf{j} + g(u, v)\mathbf{k}, \quad \mathbf{r} = v\mathbf{i} + g(u, v)\mathbf{j} + u\mathbf{k}, \quad \mathbf{r} = g(u, v)\mathbf{i} + u\mathbf{j} + v\mathbf{k} \quad (6\text{–}8)$$

$$\underbrace{\qquad}_{z = g(x, y)} \qquad \underbrace{\qquad}_{y = g(z, x)} \qquad \underbrace{\qquad}_{x = g(y, z)}$$

These representations impose positive and negative orientations on the surfaces in accordance with Formula (1). We leave it as an exercise to calculate $\mathbf{n}$ and $-\mathbf{n}$ in each case and to show that the positive and negative orientations are as shown in Table 17.6.1.

Table 17.6.1

$z = g(x, y)$	$y = g(z, x)$	$x = g(y, z)$
$\mathbf{n} = \dfrac{-\dfrac{\partial z}{\partial x}\,\mathbf{i} - \dfrac{\partial z}{\partial y}\,\mathbf{j} + \mathbf{k}}{\sqrt{\left(\dfrac{\partial z}{\partial x}\right)^2 + \left(\dfrac{\partial z}{\partial y}\right)^2 + 1}}$ Positive **k**-component Positive orientation	$\mathbf{n} = \dfrac{-\dfrac{\partial y}{\partial x}\,\mathbf{i} + \mathbf{j} - \dfrac{\partial y}{\partial z}\,\mathbf{k}}{\sqrt{\left(\dfrac{\partial y}{\partial x}\right)^2 + \left(\dfrac{\partial y}{\partial z}\right)^2 + 1}}$ Positive **j**-component Positive orientation	$\mathbf{n} = \dfrac{\mathbf{i} - \dfrac{\partial x}{\partial y}\,\mathbf{j} - \dfrac{\partial x}{\partial z}\,\mathbf{k}}{\sqrt{\left(\dfrac{\partial x}{\partial y}\right)^2 + \left(\dfrac{\partial x}{\partial z}\right)^2 + 1}}$ Positive **i**-component Positive orientation
$-\mathbf{n} = \dfrac{\dfrac{\partial z}{\partial x}\,\mathbf{i} + \dfrac{\partial z}{\partial y}\,\mathbf{j} - \mathbf{k}}{\sqrt{\left(\dfrac{\partial z}{\partial x}\right)^2 + \left(\dfrac{\partial z}{\partial y}\right)^2 + 1}}$ Negative **k**-component Negative orientation	$-\mathbf{n} = \dfrac{\dfrac{\partial y}{\partial x}\,\mathbf{i} - \mathbf{j} + \dfrac{\partial y}{\partial z}\,\mathbf{k}}{\sqrt{\left(\dfrac{\partial y}{\partial x}\right)^2 + \left(\dfrac{\partial y}{\partial z}\right)^2 + 1}}$ Negative **j**-component Negative orientation	$-\mathbf{n} = \dfrac{-\mathbf{i} + \dfrac{\partial x}{\partial y}\,\mathbf{j} + \dfrac{\partial x}{\partial z}\,\mathbf{k}}{\sqrt{\left(\dfrac{\partial x}{\partial y}\right)^2 + \left(\dfrac{\partial x}{\partial z}\right)^2 + 1}}$ Negative **i**-component Negative orientation

The results in Table 17.6.1 can also be obtained using gradients. To see how this can be done, rewrite the equations of the surfaces as

$$z - g(x, y) = 0, \quad y - g(z, x) = 0, \quad x - g(y, z) = 0$$

Each of these equations has the form $G(x, y, z) = 0$ and hence can be viewed as a level surface of a function $G(x, y, z)$. Since the gradient of G is normal to the level surface, it follows that the unit normal $\mathbf{n}$ is either $\nabla G / \|\nabla G\|$ or $-\nabla G / \|\nabla G\|$. However, if $G(x, y, z) = z - g(x, y)$, then ∇G has a **k**-component of 1; if $G(x, y, z) = y - g(z, x)$, then ∇G has a **j**-component of 1; and if $G(x, y, z) = x - g(y, z)$, then ∇G has an **i**-component of 1. Thus, it is evident from Table 17.6.1 that in all three cases we have

$$\mathbf{n} = \frac{\nabla G}{\|\nabla G\|} \tag{9}$$

Moreover, we leave it as an exercise to show that if the surfaces $z = g(x, y)$, $y = g(z, x)$, and $x = g(y, z)$ are expressed in vector forms (6), (7), and (8), then

$$\nabla G = \frac{\partial \mathbf{r}}{\partial u} \times \frac{\partial \mathbf{r}}{\partial v} \tag{10}$$

[compare (1) and (9)]. Thus, we are led to the following version of Theorem 17.6.2 for nonparametric surfaces.

17.6.3   THEOREM.   *Let σ be a smooth surface of the form $z = g(x, y)$, $y = g(z, x)$, or $x = g(y, z)$, and suppose that the component functions of the vector field $\mathbf{F}$ are continuous on σ. Suppose also that the equation for σ is rewritten as $G(x, y, z) = 0$ by taking g to the left side of the equation, and let R be the projection of σ on the coordinate plane determined by the independent variables of g. If σ has positive orientation, then*

$$\Phi = \iint_{\sigma} \mathbf{F} \cdot \mathbf{n}\, dS = \iint_{R} \mathbf{F} \cdot \nabla G\, dA \tag{11}$$

Formula (11) can either be used directly for computations or to derive some more specific formulas for each of the three surface types. For example, if $z = g(x, y)$, then we have $G(x, y, z) = z - g(x, y)$, so

$$\nabla G = -\frac{\partial g}{\partial x}\mathbf{i} - \frac{\partial g}{\partial y}\mathbf{j} + \mathbf{k} = -\frac{\partial z}{\partial x}\mathbf{i} - \frac{\partial z}{\partial y}\mathbf{j} + \mathbf{k}$$

Substituting this expression for ∇G in (11) and taking R to be the projection of the surface $z = g(x, y)$ on the xy-plane yields

$$\iint\limits_{\sigma} \mathbf{F} \cdot \mathbf{n}\, dS = \iint\limits_{R} \mathbf{F} \cdot \left(-\frac{\partial z}{\partial x}\mathbf{i} - \frac{\partial z}{\partial y}\mathbf{j} + \mathbf{k} \right) dA \qquad \begin{array}{l}\sigma \text{ of the form } z = f(x, y)\\ \text{and oriented up}\end{array} \qquad (12)$$

$$\iint\limits_{\sigma} \mathbf{F} \cdot \mathbf{n}\, dS = \iint\limits_{R} \mathbf{F} \cdot \left(\frac{\partial z}{\partial x}\mathbf{i} + \frac{\partial z}{\partial y}\mathbf{j} - \mathbf{k} \right) dA \qquad \begin{array}{l}\sigma \text{ of the form } z = f(x, y)\\ \text{and oriented down}\end{array} \qquad (13)$$

The derivation of the corresponding formulas when $y = g(z, x)$ and $x = g(y, z)$ are left as exercises.

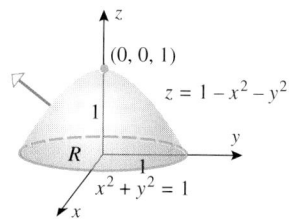

Figure 17.6.9

Example 2

Let σ be the portion of the surface $z = 1 - x^2 - y^2$ that lies above the xy-plane, and suppose that σ is oriented up, as shown in Figure 17.6.9. Find the flux of the vector field $\mathbf{F}(x, y, z) = x\mathbf{i} + y\mathbf{j} + z\mathbf{k}$ across σ.

Solution. From (12) the flux Φ is given by

$$\Phi = \iint\limits_{\sigma} \mathbf{F} \cdot \mathbf{n}\, dS = \iint\limits_{R} \mathbf{F} \cdot \left(-\frac{\partial z}{\partial x}\mathbf{i} - \frac{\partial z}{\partial y}\mathbf{j} + \mathbf{k} \right) dA$$

$$= \iint\limits_{R} (x\mathbf{i} + y\mathbf{j} + z\mathbf{k}) \cdot (2x\mathbf{i} + 2y\mathbf{j} + \mathbf{k})\, dA$$

$$= \iint\limits_{R} (x^2 + y^2 + 1)\, dA \qquad \begin{array}{l}\text{Since } z = 1 - x^2 - y^2\\ \text{on the surface}\end{array}$$

$$= \int_0^{2\pi} \int_0^1 (r^2 + 1)r\, dr\, d\theta \qquad \begin{array}{l}\text{Using polar coordinates}\\ \text{to evaluate the integral}\end{array}$$

$$= \int_0^{2\pi} \left(\frac{3}{4} \right) d\theta = \frac{3\pi}{2} \qquad \blacktriangleleft$$

EXERCISE SET 17.6

· ·

1. Suppose that the surface σ of unit cube in the accompanying figure has an outward orientation. In each part, determine whether the flux of the vector field $\mathbf{F}(x, y, z) = z\mathbf{j}$ across the specified face is positive, negative, or zero.

   (a) The face $x = 1$ (b) The face $x = 0$

   (c) The face $y = 1$ (d) The face $y = 0$

   (e) The face $z = 1$ (f) The face $z = 0$

2. Answer the questions posed in Exercise 1 for the vector field $\mathbf{F}(x, y, z) = x\mathbf{i} - z\mathbf{k}$.

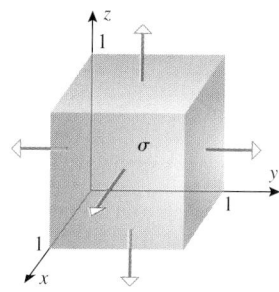

Figure Ex-1

3. Answer the questions posed in Exercise 1 for the vector field $\mathbf{F}(x, y, z) = x\mathbf{i} + y\mathbf{j} + z\mathbf{k}$.

4. What is the flux of the constant vector field $\mathbf{F}(x, y, z) = \mathbf{i}$ across the entire surface σ in Figure Ex-1? Explain your reasoning.

5. Let σ be the cylindrical surface that is represented by the vector-valued function $\mathbf{r}(u, v) = \cos v\mathbf{i} + \sin v\mathbf{j} + u\mathbf{k}$ with $0 \le u \le 1$ and $0 \le v \le 2\pi$.
   (a) Find the unit normal $\mathbf{n} = \mathbf{n}(u, v)$ that defines the positive orientation of σ.
   (b) Is the positive orientation inward or outward? Justify your answer.

6. Let σ be the conical surface that is represented by the parametric equations $x = r \cos\theta$, $y = r \sin\theta$, $z = r$ with $0 \le r \le 1$ and $0 \le \theta \le 2\pi$.
   (a) Find the unit normal $\mathbf{n} = \mathbf{n}(r, \theta)$ that defines the positive orientation of σ.
   (b) Is the positive orientation inward or outward? Justify your answer.

In Exercises 7–12, find the flux of the vector field $\mathbf{F}$ across σ.

7. $\mathbf{F}(x, y, z) = x\mathbf{i} + y\mathbf{j} + 2z\mathbf{k}$; σ is the portion of the surface $z = 1 - x^2 - y^2$ above the xy-plane, oriented by upward normals.

8. $\mathbf{F}(x, y, z) = (x + y)\mathbf{i} + (y + z)\mathbf{j} + (z + x)\mathbf{k}$; σ is the portion of the plane $x + y + z = 1$ in the first octant, oriented by unit normals with positive components.

9. $\mathbf{F}(x, y, z) = x\mathbf{i} + y\mathbf{j} + 2z\mathbf{k}$; σ is the portion of the cone $z^2 = x^2 + y^2$ between the planes $z = 1$ and $z = 2$, oriented by upward unit normals.

10. $\mathbf{F}(x, y, z) = y\mathbf{j} + k$; σ is the portion of the paraboloid $z = x^2 + y^2$ below the plane $z = 4$, oriented by downward unit normals.

11. $\mathbf{F}(x, y, z) = x\mathbf{k}$; the surface σ is the portion of the paraboloid $z = x^2 + y^2$ below the plane $z = y$, oriented by downward unit normals.

12. $\mathbf{F}(x, y, z) = x^2\mathbf{i} + yx\mathbf{j} + zx\mathbf{k}$; σ is the portion of the plane $6x + 3y + 2z = 6$ in the first octant, oriented by unit normals with positive components.

In Exercises 13–16, find the flux of the vector field $\mathbf{F}$ across σ in the direction of positive orientation.

13. $\mathbf{F}(x, y, z) = x\mathbf{i} + y\mathbf{j} + \mathbf{k}$; σ is the portion of the paraboloid
$$\mathbf{r}(u, v) = u \cos v\mathbf{i} + u \sin v\mathbf{j} + (1 - u^2)\mathbf{k}$$
with $1 \le u \le 2$, $0 \le v \le 2\pi$.

14. $\mathbf{F}(x, y, z) = e^{-y}\mathbf{i} - y\mathbf{j} + x \sin z\mathbf{k}$; σ is the portion of the elliptic cylinder
$$\mathbf{r}(u, v) = 2 \cos v\mathbf{i} + \sin v\mathbf{j} + u\mathbf{k}$$
with $0 \le u \le 5$, $0 \le v \le 2\pi$.

15. $\mathbf{F}(x, y, z) = \sqrt{x^2 + y^2}\,\mathbf{k}$; σ is the portion of the cone
$$\mathbf{r}(u, v) = u \cos v\mathbf{i} + u \sin v\mathbf{j} + 2u\mathbf{k}$$
with $0 \le u \le \sin v$, $0 \le v \le \pi$.

16. $\mathbf{F}(x, y, z) = x\mathbf{i} + y\mathbf{j} + z\mathbf{k}$; σ is the portion of the sphere
$$\mathbf{r}(u, v) = 2 \sin u \cos v\mathbf{i} + 2 \sin u \sin v\mathbf{j} + 2 \cos u\mathbf{k}$$
with $0 \le u \le \pi/3$, $0 \le v \le 2\pi$.

17. Let σ be the surface of the cube bounded by the planes $x = \pm 1$, $y = \pm 1$, $z = \pm 1$, oriented by outward unit normals. In each part, find the flux of $\mathbf{F}$ across σ.
   (a) $\mathbf{F}(x, y, z) = x\mathbf{i}$
   (b) $\mathbf{F}(x, y, z) = x\mathbf{i} + y\mathbf{j} + z\mathbf{k}$
   (c) $\mathbf{F}(x, y, z) = x^2\mathbf{i} + y^2\mathbf{j} + z^2\mathbf{k}$

18. Let σ be the closed surface consisting of the portion of the paraboloid $z = x^2 + y^2$ for which $0 \le z \le 1$ and capped by the disk $x^2 + y^2 \le 1$ in the plane $z = 1$. Find the flux of the vector field $\mathbf{F}(x, y, z) = z\mathbf{j} - y\mathbf{k}$ in the outward direction across σ.

In Exercises 19 and 20, find the flux of $\mathbf{F}$ across σ by expressing σ parametrically.

19. $\mathbf{F}(x, y, z) = \mathbf{i} + \mathbf{j} + \mathbf{k}$; the surface σ is the portion of the cone $z = \sqrt{x^2 + y^2}$ below the plane $z = 1$, oriented by downward unit normals.

20. $\mathbf{F}(x, y, z) = x\mathbf{i} + y\mathbf{j} + z\mathbf{k}$; σ is the portion of the cylinder $z^2 = 1 - x^2$ between the planes $y = 1$ and $y = -2$, oriented by outward unit normals.

21. Let $\mathbf{F}(x, y, z) = 2x\mathbf{i} - 3y\mathbf{j} + z\mathbf{k}$ be the velocity vector (in m/s) of a fluid particle at the point (x, y, z) in a steady-state fluid flow.
   (a) Find the net volume of fluid that passes in the upward direction through the portion of the plane $x + y + z = 1$ in the first octant in 1 s.
   (b) Assuming that the fluid has a mass density of 806 kg/m³, find the net mass of fluid that passes in the upward direction through the surface in part (a) in 1 s.

22. Let x, y, and z be measured in meters, and suppose that $\mathbf{F}(x, y, z) = -y\mathbf{i} + z\mathbf{j} + 3x\mathbf{k}$ is the velocity vector (in m/s) of a fluid particle at the point (x, y, z) in a steady-state incompressible fluid flow.
   (a) Find the net volume of fluid that passes in the upward direction through the hemisphere $z = \sqrt{9 - x^2 - y^2}$ in 1 s.
   (b) Assuming that the fluid has a mass density of 1060 kg/m³, find the net mass of fluid that passes in the upward direction through the surface in part (a) in 1 s.

23. (a) Derive the analogs of Formulas (12) and (13) for surfaces of the form $x = g(y, z)$.
   (b) Let σ be the portion of the paraboloid $x = y^2 + z^2$ for $x \le 1$ and $z \ge 0$ oriented by unit normals with negative

x-components. Use the result in part (a) to find the flux of

$$\mathbf{F}(x, y, z) = y\mathbf{i} - z\mathbf{j} + 8\mathbf{k}$$

across σ.

24. (a) Derive the analogs of Formulas (12) and (13) for surfaces of the form $y = g(z, x)$.

(b) Let σ be the portion of the paraboloid $y = z^2 + x^2$ for $y \leq 1$ and $z \geq 0$ oriented by unit normals with positive y-components. Use the result in part (a) to find the flux of

$$\mathbf{F}(x, y, z) = x\mathbf{i} + y\mathbf{j} + z\mathbf{k}$$

across σ.

25. Let $\mathbf{F} = \|\mathbf{r}\|^k \mathbf{r}$, where $\mathbf{r} = x\mathbf{i} + y\mathbf{j} + z\mathbf{k}$ and k is a constant. (Note that if $k = -3$, this is an inverse-square field.) Let σ be the sphere of radius a centered at the origin and oriented by the outward normal $\mathbf{n} = \mathbf{r}/\|\mathbf{r}\| = \mathbf{r}/a$.

(a) Find the flux of $\mathbf{F}$ across σ without performing any integrations. [*Hint:* The surface area of a sphere of radius a is $4\pi a^2$.]

(b) For what value of k is the flux independent of the radius of the sphere?

26. Let

$$\mathbf{F}(x, y, z) = a^2 x\mathbf{i} + (y/a)\mathbf{j} + az^2\mathbf{k}$$

and let σ be the sphere of radius 1, centered at the origin and oriented outward. Approximate all values of a such that the flux of $\mathbf{F}$ across σ is 10.

17.7 THE DIVERGENCE THEOREM

In this section we will be concerned with flux across surfaces, such as spheres, that "enclose" a region of space. We will show that the flux across such surfaces can be expressed in terms of the divergence of the vector field, and we will use this result to give a physical interpretation of the concept of divergence.

ORIENTATION OF PIECEWISE SMOOTH CLOSED SURFACES

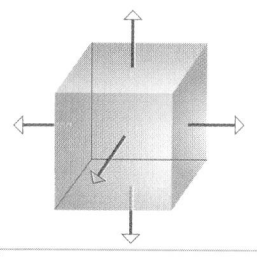

Box with outward orientation

Figure 17.7.1

THE DIVERGENCE THEOREM

In the last section we studied flux across general surfaces. Here we will be concerned exclusively with surfaces that are boundaries of finite solids—the surface of a solid sphere, the surface of a solid box, or the surface of a solid cylinder, for example. Such surfaces are said to be **closed**. A closed surface may or may not be smooth, but most of the surfaces that arise in applications are generally **piecewise smooth**; that is, they consist of finitely many smooth surfaces joined together at the edges (a box, for example). We will limit our discussion to piecewise smooth surfaces that can be assigned an **inward orientation** (toward the interior of the solid) and an **outward orientation** (away from the interior). It is very difficult to make this concept mathematically precise, but the basic idea is that each piece of the surface is orientable, and oriented pieces fit together in such a way that the entire surface can be assigned an orientation (Figure 17.7.1).

In Section 17.1 we defined the divergence of a vector field

$$\mathbf{F}(x, y, z) = f(x, y, z)\mathbf{i} + g(x, y, z)\mathbf{j} + h(x, y, z)\mathbf{k}$$

as

$$\text{div } \mathbf{F} = \frac{\partial f}{\partial x} + \frac{\partial g}{\partial y} + \frac{\partial h}{\partial z}$$

but we did not attempt to give a physical explanation of its meaning at that time. The following result, known as the **Divergence Theorem** or **Gauss's** [*] **Theorem**, will provide us with a physical interpretation of divergence in the context of fluid flow.

---

[*]See biography on page 1110.

17.7.1 THEOREM (*The Divergence Theorem*). *Let G be a solid whose surface σ is oriented outward. If*

$$\mathbf{F}(x, y, z) = f(x, y, z)\mathbf{i} + g(x, y, z)\mathbf{j} + h(x, y, z)\mathbf{k}$$

where f, g, and h have continuous first partial derivatives on some open set containing G, then

$$\iint\limits_{\sigma} \mathbf{F} \cdot \mathbf{n}\, dS = \iiint\limits_{G} \operatorname{div} \mathbf{F}\, dV \tag{1}$$

---

*CARL FRIEDRICH GAUSS (1777–1855). German mathematician and scientist. Sometimes called the "prince of mathematicians," Gauss ranks with Newton and Archimedes as one of the three greatest mathematicians who ever lived. His father, a laborer, was an uncouth but honest man who would have liked Gauss to take up a trade such as gardening or bricklaying; but the boy's genius for mathematics was not to be denied. In the entire history of mathematics there may never have been a child so precocious as Gauss—by his own account he worked out the rudiments of arithmetic before he could talk. One day, before he was even three years old, his genius became apparent to his parents in a very dramatic way. His father was preparing the weekly payroll for the laborers under his charge while the boy watched quietly from a corner. At the end of the long and tedious calculation, Gauss informed his father that there was an error in the result and stated the answer, which he had worked out in his head. To the astonishment of his parents, a check of the computations showed Gauss to be correct!

For his elementary education Gauss was enrolled in a squalid school run by a man named Büttner whose main teaching technique was thrashing. Büttner was in the habit of assigning long addition problems which, unknown to his students, were arithmetic progressions that he could sum up using formulas. On the first day that Gauss entered the arithmetic class, the students were asked to sum the numbers from 1 to 100. But no sooner had Büttner stated the problem than Gauss turned over his slate and exclaimed in his peasant dialect, "Ligget se'." (Here it lies.) For nearly an hour Büttner glared at Gauss, who sat with folded hands while his classmates toiled away. When Büttner examined the slates at the end of the period, Gauss's slate contained a single number, 5050—the only correct solution in the class. To his credit, Büttner recognized the genius of Gauss and with the help of his assistant, John Bartels, had him brought to the attention of Karl Wilhelm Ferdinand, Duke of Brunswick. The shy and awkward boy, who was then fourteen, so captivated the Duke that he subsidized him through preparatory school, college, and the early part of his career.

From 1795 to 1798 Gauss studied mathematics at the University of Göttingen, receiving his degree in absentia from the University of Helmstadt. For his dissertation, he gave the first complete proof of the fundamental theorem of algebra, which states that every polynomial equation has as many solutions as its degree. At age 19 he solved a problem that baffled Euclid, inscribing a regular polygon of 17 sides in a circle using straightedge and compass; and in 1801, at age 24, he published his first masterpiece, Disquisitiones Arithmeticae, considered by many to be one of the most brilliant achievements in mathematics. In that book Gauss systematized the study of number theory (properties of the integers) and formulated the basic concepts that form the foundation of that subject.

In the same year that the Disquisitiones was published, Gauss again applied his phenomenal computational skills in a dramatic way. The astronomer Giuseppi Piazzi had observed the asteroid Ceres for $\frac{1}{40}$ of its orbit, but lost it in the Sun. Using only three observations and the "method of least squares" that he had developed in 1795, Gauss computed the orbit with such accuracy that astronomers had no trouble relocating it the following year. This achievement brought him instant recognition as the premier mathematician in Europe, and in 1807 he was made Professor of Astronomy and head of the astronomical observatory at Göttingen.

In the years that followed, Gauss revolutionized mathematics by bringing to it standards of precision and rigor undreamed of by his predecessors. He had a passion for perfection that drove him to polish and rework his papers rather than publish less finished work in greater numbers—his favorite saying was "Pauca, sed matura" (Few, but ripe). As a result, many of his important discoveries were squirreled away in diaries that remained unpublished until years after his death.

Among his myriad achievements, Gauss discovered the Gaussian or "bell-shaped" error curve fundamental in probability, gave the first geometric interpretation of complex numbers and established their fundamental role in mathematics, developed methods of characterizing surfaces intrinsically by means of the curves that they contain, developed the theory of conformal (angle-preserving) maps, and discovered non-Euclidean geometry 30 years before the ideas were published by others. In physics he made major contributions to the theory of lenses and capillary action, and with Wilhelm Weber he did fundamental work in electromagnetism. Gauss invented the heliotrope, bifilar magnetometer, and an electrotelegraph.

Gauss was deeply religious and aristocratic in demeanor. He mastered foreign languages with ease, read extensively, and enjoyed mineralogy and botany as hobbies. He disliked teaching and was usually cool and discouraging to other mathematicians, possibly because he had already anticipated their work. It has been said that if Gauss had published all of his discoveries, the current state of mathematics would be advanced by 50 years. He was without a doubt the greatest mathematician of the modern era.*

The proof of this theorem for a general solid G is too difficult to present here. However, we can give a proof for the special case where G is simultaneously a simple xy-solid, a simple yz-solid, and a simple zx-solid (see Figure 16.5.3 and the related discussion for terminology).

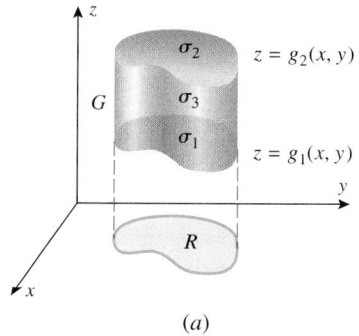

(a)

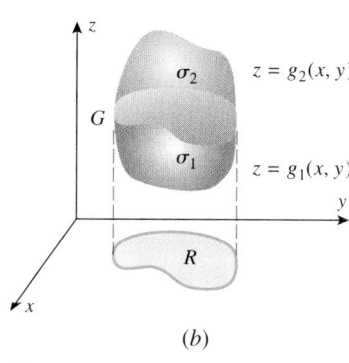

(b)

Figure 17.7.2

Proof. Suppose that G has upper surface $z = g_2(x, y)$, lower surface $z = g_1(x, y)$, and projection R on the xy-plane. Let σ_1 denote the lower surface, σ_2 the upper surface, and σ_3 the lateral surface (Figure 17.7.2a). If the upper surface and lower surface meet as in Figure 17.7.2b, then there is no lateral surface σ_3. Our proof will allow for both cases shown in those figures.

Formula (1) can be expressed as

$$\iint_\sigma [f(x, y, z)\mathbf{i} + g(x, y, z)\mathbf{j} + h(x, y, z)\mathbf{k}] \cdot \mathbf{n} \, dS = \iiint_G \left(\frac{\partial f}{\partial x} + \frac{\partial g}{\partial y} + \frac{\partial h}{\partial z} \right) dV$$

so it suffices to prove the three equalities

$$\iint_\sigma f(x, y, z)\mathbf{i} \cdot \mathbf{n} \, dS = \iiint_G \frac{\partial f}{\partial x} \, dV \tag{2a}$$

$$\iint_\sigma g(x, y, z)\mathbf{j} \cdot \mathbf{n} \, dS = \iiint_G \frac{\partial g}{\partial y} \, dV \tag{2b}$$

$$\iint_\sigma h(x, y, z)\mathbf{k} \cdot \mathbf{n} \, dS = \iiint_G \frac{\partial h}{\partial z} \, dV \tag{2c}$$

Since the proofs of all three equalities are similar, we will prove only the third.

It follows from Theorem 16.5.2 that

$$\iiint_G \frac{\partial h}{\partial z} \, dV = \iint_R \left[\int_{g_1(x,y)}^{g_2(x,y)} \frac{\partial h}{\partial z} dz \right] dA = \iint_R \left[h(x, y, z) \right]_{z=g_1(x,y)}^{g_2(x,y)} dA$$

so

$$\iiint_G \frac{\partial h}{\partial z} \, dV = \iint_R [h(x, y, g_2(x, y)) - h(x, y, g_1(x, y))] \, dA \tag{3}$$

Next, we will evaluate the surface integral in (2c) by integrating over each surface of G separately. If there is a lateral surface σ_3, then at each point of this surface $\mathbf{n} \cdot \mathbf{k} = 0$ since $\mathbf{n}$ is horizontal and $\mathbf{k}$ is vertical. Thus,

$$\iint_{\sigma_3} h(x, y, z)\mathbf{k} \cdot \mathbf{n} \, dS = 0$$

Therefore, regardless of whether G has a lateral surface, we can write

$$\iint_\sigma h(x, y, z)\mathbf{k} \cdot \mathbf{n} \, dS = \iint_{\sigma_1} h(x, y, z)\mathbf{k} \cdot \mathbf{n} \, dS + \iint_{\sigma_2} h(x, y, z)\mathbf{k} \cdot \mathbf{n} \, dS \tag{4}$$

On the upper surface σ_2, the outer normal is an upward normal, and on the lower surface σ_1, the outer normal is a downward normal. Thus, Formulas (12) and (13) of Section 17.6 imply that

$$\iint_{\sigma_2} h(x, y, z)\mathbf{k} \cdot \mathbf{n} \, dS = \iint_R h(x, y, g_2(x, y))\mathbf{k} \cdot \left(-\frac{\partial z}{\partial x}\mathbf{i} - \frac{\partial z}{\partial y}\mathbf{j} + \mathbf{k} \right) dA$$

$$= \iint_R h(x, y, g_2(x, y)) \, dA \tag{5}$$

and

$$\iint\limits_{\sigma_1} h(x, y, z)\mathbf{k} \cdot \mathbf{n}\, dS = \iint\limits_{R} h(x, y, g_1(x, y))\mathbf{k} \cdot \left(\frac{\partial z}{\partial x}\mathbf{i} + \frac{\partial z}{\partial y}\mathbf{j} - \mathbf{k} \right) dA$$

$$= -\iint\limits_{R} h(x, y, g_1(x, y))\, dA \qquad (6)$$

Substituting (5) and (6) into (4) and combining the terms into a single integral yields

$$\iint\limits_{\sigma} h(x, y, z)\mathbf{k} \cdot \mathbf{n}\, dS = \iint\limits_{R} [h(x, y, g_2(x, y)) - h(x, y, g_1(x, y))]\, dA \qquad (7)$$

Equation (2c) now follows from (3) and (7). ∎

REMARK.   In words, the Divergence Theorem states that *the flux of a vector field across a closed surface with outward orientation is equal to the triple integral of the divergence over the region enclosed by the surface.* This is sometimes called the **outward flux** across the surface.

USING THE DIVERGENCE THEOREM TO FIND FLUX

Sometimes it is easier to find the flux across a closed surface by using the Divergence Theorem than by evaluating the flux integral directly. This is illustrated in the following example.

Example 1

Use the Divergence Theorem to find the outward flux of the vector field $\mathbf{F}(x, y, z) = z\mathbf{k}$ across the sphere $x^2 + y^2 + z^2 = a^2$.

Solution. Let σ denote the outward-oriented spherical surface and G the region that it encloses. The divergence of the vector field is

$$\operatorname{div} \mathbf{F} = \frac{\partial z}{\partial z} = 1$$

so from (1) the flux across σ is

$$\Phi = \iint\limits_{\sigma} \mathbf{F} \cdot \mathbf{n}\, dS = \iiint\limits_{G} dV = \text{volume of } G = \frac{4\pi a^3}{3}$$

Note how much simpler this calculation is than that in Example 1 of Section 17.6. ◀

The Divergence Theorem is usually the method of choice for finding the flux across closed piecewise smooth surfaces with multiple sections, since it eliminates the need for a separate integral evaluation over each section. This is illustrated in the next three examples.

Example 2

Use the Divergence Theorem to find the outward flux of the vector field

$$\mathbf{F}(x, y, z) = 2x\mathbf{i} + 3y\mathbf{j} + z^2\mathbf{k}$$

across the unit cube in Figure 17.7.3.

Solution. Let σ denote the outward-oriented surface of the cube and G the region that it encloses. The divergence of the vector field is

$$\operatorname{div} \mathbf{F} = \frac{\partial}{\partial x}(2x) + \frac{\partial}{\partial y}(3y) + \frac{\partial}{\partial z}(z^2) = 5 + 2z$$

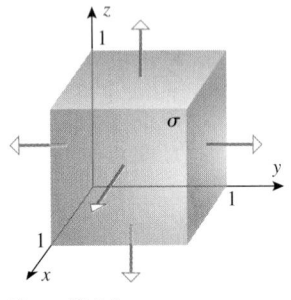

Figure 17.7.3

so from (1) the flux across σ is

$$\Phi = \iint_\sigma \mathbf{F} \cdot \mathbf{n}\, dS = \iiint_G (5 + 2z)\, dV = \int_0^1 \int_0^1 \int_0^1 (5 + 2z)\, dz\, dy\, dx$$

$$= \int_0^1 \int_0^1 \left[5z + z^2\right]_{z=0}^1 dy\, dx = \int_0^1 \int_0^1 6\, dy\, dx = 6 \qquad \blacktriangleleft$$

Example 3

Use the Divergence Theorem to find the outward flux of the vector field

$$\mathbf{F}(x, y, z) = x^3\mathbf{i} + y^3\mathbf{j} + z^2\mathbf{k}$$

across the surface of the region that is enclosed by the circular cylinder $x^2 + y^2 = 9$ and the planes $z = 0$ and $z = 2$ (Figure 17.7.4).

Solution. Let σ denote the outward-oriented surface and G the region that it encloses. The divergence of the vector field is

$$\operatorname{div} \mathbf{F} = \frac{\partial}{\partial x}(x^3) + \frac{\partial}{\partial y}(y^3) + \frac{\partial}{\partial z}(z^2) = 3x^2 + 3y^2 + 2z$$

so from (1) the flux across σ is

$$\Phi = \iint_\sigma \mathbf{F} \cdot \mathbf{n}\, dS = \iiint_G (3x^2 + 3y^2 + 2z)\, dV$$

$$= \int_0^{2\pi} \int_0^3 \int_0^2 (3r^2 + 2z) r\, dz\, dr\, d\theta \qquad \text{Using cylindrical coordinates}$$

$$= \int_0^{2\pi} \int_0^3 \left[3r^3 z + z^2 r\right]_{z=0}^2 dr\, d\theta$$

$$= \int_0^{2\pi} \int_0^3 (6r^3 + 4r)\, dr\, d\theta$$

$$= \int_0^{2\pi} \left[\frac{3r^4}{2} + 2r^2\right]_0^3 d\theta$$

$$= \int_0^{2\pi} \frac{279}{2} d\theta = 279\pi \qquad \blacktriangleleft$$

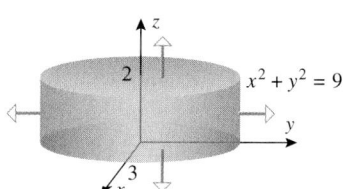

Figure 17.7.4

Example 4

Use the Divergence Theorem to find the outward flux of the vector field

$$\mathbf{F}(x, y, z) = x^3\mathbf{i} + y^3\mathbf{j} + z^3\mathbf{k}$$

across the surface of the region that is enclosed by the hemisphere $z = \sqrt{a^2 - x^2 - y^2}$ and the plane $z = 0$ (Figure 17.7.5).

Solution. Let σ denote the outward-oriented surface and G the region that it encloses. The divergence of the vector field is

$$\operatorname{div} \mathbf{F} = \frac{\partial}{\partial x}(x^3) + \frac{\partial}{\partial y}(y^3) + \frac{\partial}{\partial z}(z^3) = 3x^2 + 3y^2 + 3z^2$$

so from (1) the flux across σ is

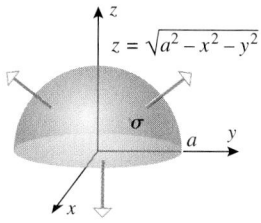

Figure 17.7.5

$$\Phi = \iint\limits_{\sigma} \mathbf{F} \cdot \mathbf{n}\, dS = \iiint\limits_{G} (3x^2 + 3y^2 + 3z^2)\, dV$$

$$= \int_0^{2\pi} \int_0^{\pi/2} \int_0^{a} (3\rho^2)\rho^2 \sin\phi\, d\rho\, d\phi\, d\theta \qquad \text{Using spherical coordinates}$$

$$= 3\int_0^{2\pi} \int_0^{\pi/2} \int_0^{a} \rho^4 \sin\phi\, d\rho\, d\phi\, d\theta$$

$$= 3\int_0^{2\pi} \int_0^{\pi/2} \left[\frac{\rho^5}{5}\sin\phi\right]_{\rho=0}^{a} d\phi\, d\theta$$

$$= \frac{3a^5}{5}\int_0^{2\pi} \int_0^{\pi/2} \sin\phi\, d\phi\, d\theta$$

$$= \frac{3a^5}{5}\int_0^{2\pi} \left[-\cos\phi\right]_0^{\pi/2} d\theta$$

$$= \frac{3a^5}{5}\int_0^{2\pi} d\theta = \frac{6\pi a^5}{5} \qquad \blacktriangleleft$$

DIVERGENCE VIEWED AS FLUX DENSITY

The Divergence Theorem provides a way of interpreting the divergence of a vector field **F**. Suppose that G is a *small* spherical region centered at the point P_0 and that its surface, denoted by $\sigma(G)$, is oriented outward. Denote the volume of the region by vol(G) and the flux of **F** across $\sigma(G)$ by $\Phi(G)$. If div **F** is continuous on G, then across the small region G the value of div **F** will not vary much from its value div $\mathbf{F}(P_0)$ at the center, and we can reasonably approximate div **F** by the constant div $\mathbf{F}(P_0)$ on G. Thus, the Divergence Theorem implies that the flux $\Phi(G)$ of **F** across $\sigma(G)$ can be approximated as

$$\Phi(G) = \iint\limits_{\sigma(G)} \mathbf{F} \cdot \mathbf{n}\, dS = \iiint\limits_{G} \text{div } \mathbf{F}\, dV \approx \text{div } \mathbf{F}(P_0) \iiint\limits_{G} dV = \text{div } \mathbf{F}(P_0)\, \text{vol}(G)$$

from which we obtain the approximation

$$\text{div } \mathbf{F}(P_0) \approx \frac{\Phi(G)}{\text{vol}(G)} \qquad (8)$$

The expression on the right side of (8) is called the ***outward flux density of* F *across* G**. If we now let the radius of the sphere approach zero [so that vol(G) approaches zero], then it is plausible that the error in this approximation will approach zero, and the divergence of **F** at the point P_0 will be given exactly by

$$\text{div } \mathbf{F}(P_0) = \lim_{\text{vol}(G) \to 0} \frac{\Phi(G)}{\text{vol}(G)}$$

which we can express as

$$\text{div } \mathbf{F}(P_0) = \lim_{\text{vol}(G) \to 0} \frac{1}{\text{vol}(G)} \iint\limits_{\sigma(G)} \mathbf{F} \cdot \mathbf{n}\, dS \qquad (9)$$

This limit, which is called the ***outward flux density of* F *at* P_0**, tells us that *in a steady-state fluid flow*, div **F** *can be interpreted as the limiting flux per unit volume at a point*. Moreover, it follows from (8) that for a small spherical region G centered at a point P_0 in the flow, the outward flux across the surface of G can be approximated as

$$\Phi(G) \approx (\text{div } \mathbf{F}(P_0))(\text{Vol}(G)) \qquad (10)$$

REMARK. Formula (9) is sometimes taken as the definition of divergence. This is a useful alternative to Definition 17.1.4 because it does not require a coordinate system.

SOURCES AND SINKS

If P_0 is a point in an incompressible fluid at which div $\mathbf{F}(P_0) > 0$, then it follows from (8) that $\Phi(G) > 0$ for a sufficiently small sphere G centered at P_0. Thus, there is a greater volume of fluid going out through the surface of G than coming in. But this can only happen if there is some point *inside* the sphere at which fluid is entering the flow from an external source (say by condensation, melting of a solid, or a chemical reaction); otherwise the net outward flow through the surface would result in a decrease in density within the sphere, contradicting the incompressibility assumption. Similarly, if div $\mathbf{F}(P_0) < 0$, there would have to be a point *inside* the sphere at which fluid is leaving the flow (say by evaporation); otherwise the net inward flow through the surface would result in an increase in density within the sphere. In an incompressible fluid, points at which div $\mathbf{F}(P_0) > 0$ are called *sources* and points at which div $\mathbf{F}(P_0) < 0$ are called *sinks*. Fluid enters the flow at a source and drains out at a sink. In an incompressible fluid without sources or sinks we must have

$$\text{div } \mathbf{F}(P) = 0$$

at every point P. In hydrodynamics this is called the ***continuity equation for incompressible fluids*** and is sometimes taken as the defining characteristic of an incompressible fluid.

GAUSS'S LAW FOR INVERSE-SQUARE FIELDS

The Divergence Theorem applied to inverse-square fields (see Definition 17.1.2) produces a result called ***Gauss's Law for Inverse-Square Fields***. This result is the basis for many important principles in physics.

17.7.2 GAUSS'S LAW FOR INVERSE-SQUARE FIELDS. If

$$\mathbf{F}(\mathbf{r}) = \frac{c}{\|\mathbf{r}\|^3}\mathbf{r}$$

is an inverse-square field in 3-space, and if σ is a closed orientable surface that surrounds the origin, then the outward flux of $\mathbf{F}$ across σ is

$$\Phi = \iint\limits_{\sigma} \mathbf{F} \cdot \mathbf{n}\, dS = 4\pi c \tag{11}$$

Recall from Formula (5) of Section 17.1 that $\mathbf{F}$ can be expressed in component form as

$$\mathbf{F}(x, y, z) = \frac{c}{(x^2 + y^2 + z^2)^{3/2}}(x\mathbf{i} + y\mathbf{j} + z\mathbf{k}) \tag{12}$$

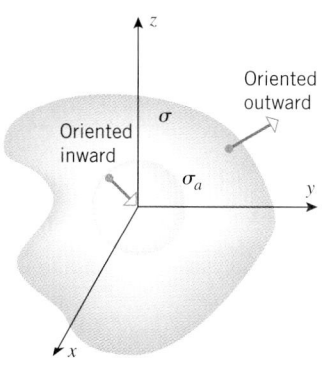

Figure 17.7.6

Since the components of $\mathbf{F}$ are not continuous at the origin, we cannot apply the Divergence Theorem across the solid enclosed by σ. However, we can circumvent this difficulty by constructing a sphere of radius a centered at the origin, where the radius is sufficiently small that the sphere lies entirely within the region enclosed by σ (Figure 17.7.6). We will denote the surface of this sphere by σ_a. The solid G enclosed between σ_a and σ is an example of a three-dimensional solid with an internal "cavity." Just as we were able to extend Green's Theorem to multiply connected regions in the plane (regions with holes), so it is possible to extend the Divergence Theorem to solids in 3-space with internal cavities, provided the surface integral in the theorem is taken over the *entire* boundary with the outside boundary of the solid oriented outward and the boundaries of the cavities oriented inward. Thus, if $\mathbf{F}$ is the inverse-square field in (12), and if σ_a is oriented inward, then the Divergence Theorem yields

$$\iiint\limits_{G} \text{div } \mathbf{F}\, dV = \iint\limits_{\sigma} \mathbf{F} \cdot \mathbf{n}\, dS + \iint\limits_{\sigma_a} \mathbf{F} \cdot \mathbf{n}\, dS \tag{13}$$

But we showed in Example 5 of Section 17.1 that div $\mathbf{F} = 0$, so (13) yields

$$\iint\limits_{\sigma} \mathbf{F} \cdot \mathbf{n}\, dS = -\iint\limits_{\sigma_a} \mathbf{F} \cdot \mathbf{n}\, dS \tag{14}$$

We can evaluate the surface integral over σ_a by expressing the integrand in terms of components; however, it is easier to leave it in vector form. At each point on the sphere the unit normal **n** points inward along a radius from the origin, and hence $\mathbf{n} = -\mathbf{r}/\|\mathbf{r}\|$. Thus, (14) yields

$$\iint\limits_{\sigma} \mathbf{F} \cdot \mathbf{n}\, dS = -\iint\limits_{\sigma_a} \frac{c}{\|\mathbf{r}\|^3}\mathbf{r} \cdot \left(-\frac{\mathbf{r}}{\|\mathbf{r}\|}\right) dS$$

$$= \iint\limits_{\sigma_a} \frac{c}{\|\mathbf{r}\|^4}(\mathbf{r} \cdot \mathbf{r})\, dS$$

$$= \iint\limits_{\sigma_a} \frac{c}{\|\mathbf{r}\|^2}\, dS$$

$$= \frac{c}{a^2}\iint\limits_{\sigma_a} dS \qquad \boxed{\|\mathbf{r}\| = a \text{ on } \sigma_a}$$

$$= \frac{c}{a^2}(4\pi a^2) \qquad \boxed{\begin{array}{l}\text{The integral is the surface}\\\text{area of the sphere.}\end{array}}$$

$$= 4\pi c$$

which establishs (11).

It follows from Example 1 of Section 17.1 with $q = 1$ that a single charged particle of charge Q located at the origin creates an inverse-square field

$$\mathbf{F}(\mathbf{r}) = \frac{Q}{4\pi\epsilon_0\|\mathbf{r}\|^3}\mathbf{r}$$

in which $\mathbf{F}(\mathbf{r})$ is the electrical force exerted by Q on a unit positive charge ($q = 1$) located at the point with position vector $\mathbf{r}$. In this case Gauss's law (17.7.2) states that the outward flux Φ across any closed orientable surface σ that surrounds Q is

$$\Phi = \iint\limits_{\sigma} \mathbf{F} \cdot \mathbf{n}\, dS = 4\pi\left(\frac{Q}{4\pi\epsilon_0}\right) = \frac{Q}{\epsilon_0}$$

This result, which is called *Gauss's Law for Electric Fields*, can be extended to more than one charge. It is one of the fundamental laws in electricity and magnetism.

EXERCISE SET 17.7 ☐C CAS

In Exercises 1–4, verify Formula (1) in the Divergence Theorem by evaluating the surface integral and the triple integral.

1. $\mathbf{F}(x, y, z) = x\mathbf{i} + y\mathbf{j} + z\mathbf{k}$; σ is the surface of the cube bounded by the planes $x = 0$, $x = 1$, $y = 0$, $y = 1$, $z = 0$, $z = 1$.

2. $\mathbf{F}(x, y, z) = x\mathbf{i} + y\mathbf{j} + z\mathbf{k}$; σ is the spherical surface $x^2 + y^2 + z^2 = 1$.

3. $\mathbf{F}(x, y, z) = 2x\mathbf{i} - yz\mathbf{j} + z^2\mathbf{k}$; the surface σ is the paraboloid $z = x^2 + y^2$ capped by the disk $x^2 + y^2 \le 1$ in the plane $z = 1$.

4. $\mathbf{F}(x, y, z) = xy\mathbf{i} + yz\mathbf{j} + xz\mathbf{k}$; σ is the surface of the cube bounded by the planes $x = 0$, $x = 2$, $y = 0$, $y = 2$, $z = 0$, $z = 2$.

In Exercises 5–15, use the Divergence Theorem to find the flux of **F** across the surface σ with outward orientation.

5. $\mathbf{F}(x, y, z) = (x^2 + y)\mathbf{i} + z^2\mathbf{j} + (e^y - z)\mathbf{k}$; σ is the surface of the rectangular solid bounded by the coordinate planes and the planes $x = 3$, $y = 1$, and $z = 2$.

6. $\mathbf{F}(x, y, z) = z^3\mathbf{i} - x^3\mathbf{j} + y^3\mathbf{k}$, where σ is the sphere $x^2 + y^2 + z^2 = a^2$.

7. $\mathbf{F}(x, y, z) = (x - z)\mathbf{i} + (y - x)\mathbf{j} + (z - y)\mathbf{k}$; σ is the surface of the cylindrical solid bounded by $x^2 + y^2 = a^2$, $z = 0$, and $z = 1$.

8. $\mathbf{F}(x, y, z) = x\mathbf{i} + y\mathbf{j} + z\mathbf{k}$; σ is the surface of the solid bounded by the paraboloid $z = 1 - x^2 - y^2$ and the xy-plane.

9. $\mathbf{F}(x, y, z) = x^3\mathbf{i} + y^3\mathbf{j} + z^3\mathbf{k}$; σ is the surface of the cylindrical solid bounded by $x^2 + y^2 = 4$, $z = 0$, and $z = 3$.

10. $\mathbf{F}(x, y, z) = (x^2 + y)\mathbf{i} + xy\mathbf{j} - (2xz + y)\mathbf{k}$; σ is the surface of the tetrahedron in the first octant bounded by $x + y + z = 1$ and the coordinate planes.

11. $\mathbf{F}(x, y, z) = (x^3 - e^y)\mathbf{i} + (y^3 + \sin z)\mathbf{j} + (z^3 - xy)\mathbf{k}$, where σ is the surface of the solid bounded by $z = \sqrt{4 - x^2 - y^2}$ and the xy-plane. [*Hint:* Use spherical coordinates.]

12. $\mathbf{F}(x, y, z) = 2xz\mathbf{i} + yz\mathbf{j} + z^2\mathbf{k}$, where σ is the surface of the hemispherical solid bounded above by $z = \sqrt{a^2 - x^2 - y^2}$ and below by the xy-plane.

13. $\mathbf{F}(x, y, z) = x^2\mathbf{i} + y^2\mathbf{j} + z^2\mathbf{k}$; σ is the surface of the conical solid bounded by $z = \sqrt{x^2 + y^2}$ and $z = 1$.

14. $\mathbf{F}(x, y, z) = x^2y\mathbf{i} - xy^2\mathbf{j} + (z + 2)\mathbf{k}$; σ is the surface of the solid bounded above by the plane $z = 2x$ and below by the paraboloid $z = x^2 + y^2$.

15. $\mathbf{F}(x, y, z) = x^3\mathbf{i} + x^2y\mathbf{j} + xy\mathbf{k}$; σ is the surface of the solid bounded by $z = 4 - x^2$, $y + z = 5$, $z = 0$, and $y = 0$.

16. Let $\mathbf{F}(x, y, z) = a\mathbf{i} + b\mathbf{j} + c\mathbf{k}$ be a constant vector field and let σ be the surface a solid G. Use the Divergence Theorem to show that the flux of $\mathbf{F}$ across σ is zero. Give an informal physical explanation of this result.

17. Prove that if $\mathbf{r} = x\mathbf{i} + y\mathbf{j} + z\mathbf{k}$ and σ is the surface of a solid G oriented by outward unit normals, then

$$\text{vol}(G) = \frac{1}{3} \iint_\sigma \mathbf{r} \cdot \mathbf{n} \, dS$$

where $\text{vol}(G)$ is the volume of G.

18. Use the result in Exercise 17 to find the outward flux of the vector field $\mathbf{F}(x, y, z) = x\mathbf{i} + y\mathbf{j} + z\mathbf{k}$ across the surface σ of the cylindrical solid bounded by $x^2 + 4x + y^2 = 5$, $z = -1$, and $z = 4$.

In Exercises 19–24, prove the identity, assuming that $\mathbf{F}$, σ, and G satisfy the hypotheses of the Divergence Theorem and that all necessary differentiability requirements for the functions $f(x, y, z)$ and $g(x, y, z)$ are met.

19. $\displaystyle\iint_\sigma \text{curl } \mathbf{F} \cdot \mathbf{n} \, dS = 0$ [*Hint:* See Exercise 31, Section 17.1.]

20. $\displaystyle\iint_\sigma \nabla f \cdot \mathbf{n} \, dS = \iiint_G \nabla^2 f \, dV$

$$\left(\nabla^2 f = \frac{\partial^2 f}{\partial x^2} + \frac{\partial^2 f}{\partial y^2} + \frac{\partial^2 f}{\partial z^2} \right)$$

21. $\displaystyle\iint_\sigma (f\nabla g) \cdot \mathbf{n} \, dS = \iiint_G (f\nabla^2 g + \nabla f \cdot \nabla g) \, dV$

22. $\displaystyle\iint_\sigma (f\nabla g - g\nabla f) \cdot \mathbf{n} \, dS = \iiint_G (f\nabla^2 g - g\nabla^2 f) \, dV$
[*Hint:* Interchange f and g in 21.]

23. $\displaystyle\iint_\sigma f \cdot \mathbf{n} \, dS = \iiint_G \nabla f \, dV$

24. Show that if $\mathbf{F}(\mathbf{r}) = c\dfrac{\mathbf{r}}{\|\mathbf{r}\|^3}$ is an inverse-square field, then $\text{div } \mathbf{r} = 0$, except at $\mathbf{r} = \mathbf{0}$.

25. In each part, the figure shows a horizontal layer of the vector field of a fluid flow in which the flow is parallel to the xy-plane at every point and is identical in each layer (i.e., is independent of z). For each flow, what can you say about the sign of the divergence at the origin? Explain your reasoning.

(a)   (b)

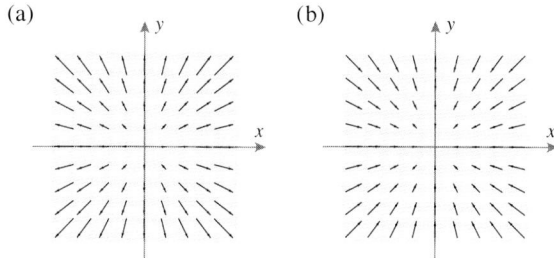

26. Find a vector field $\mathbf{F}(x, y, z)$ that has
(a) positive divergence everywhere
(b) negative divergence everywhere.

In Exercises 27–30, determine whether the vector field $\mathbf{F}(x, y, z)$ is free of sources and sinks. If it is not, locate them.

27. $\mathbf{F}(x, y, z) = (y + z)\mathbf{i} - xz^3\mathbf{j} + (x^2 \sin y)\mathbf{k}$

28. $\mathbf{F}(x, y, z) = xy\mathbf{i} - xy\mathbf{j} + y^2\mathbf{k}$

29. $\mathbf{F}(x, y, z) = x^3\mathbf{i} + y^3\mathbf{j} + z^3\mathbf{k}$

30. $\mathbf{F}(x, y, z) = (x^3 - x)\mathbf{i} + (y^3 - y)\mathbf{j} + (z^3 - z)\mathbf{k}$

$\boxed{\text{c}}$ **31.** Let σ be the surface of the solid G that is enclosed by the paraboloid $z = 1 - x^2 - y^2$ and the plane $z = 0$. Use a CAS to verify Formula (1) in the Divergence Theorem for the vector field

$$\mathbf{F} = (x^2y - z^2)\mathbf{i} + (y^3 - x)\mathbf{j} + (2x + 3z - 1)\mathbf{k}$$

by evaluating the surface integral and the triple integral.

17.8 STOKES' THEOREM

In this section we will discuss a generalization of Green's Theorem to three dimensions that has important applications in the study of vector fields, particularly in the analysis of rotational motion of fluids. This theorem will also provide us with a physical interpretation of the curl of a vector field.

RELATIVE ORIENTATION OF CURVES AND SURFACES

We will be concerned in this section with oriented surfaces in 3-space that are bounded by simple closed parametric curves (Figure 17.8.1a). If σ is an oriented surface bounded by a simple closed parametric curve C, then there are two possible relationships between the orientations of σ and C, which can be described as follows. Imagine a person walking along the curve C with his or her head in the direction of the orientation of σ. The person is said to be walking in the **positive direction** of C relative to the orientation of σ if the surface is on the person's left (Figure 17.8.1b), and the person is said to be walking in the **negative direction** of C relative to the orientation of σ if the surface is on the person's right (Figure 17.8.1c). The positive direction of C establishes a right-hand relationship between the orientations of σ and C in the sense that if the fingers of the right hand are cupped in the positive direction of C, then the thumb points (roughly) in the direction of the orientation of σ.

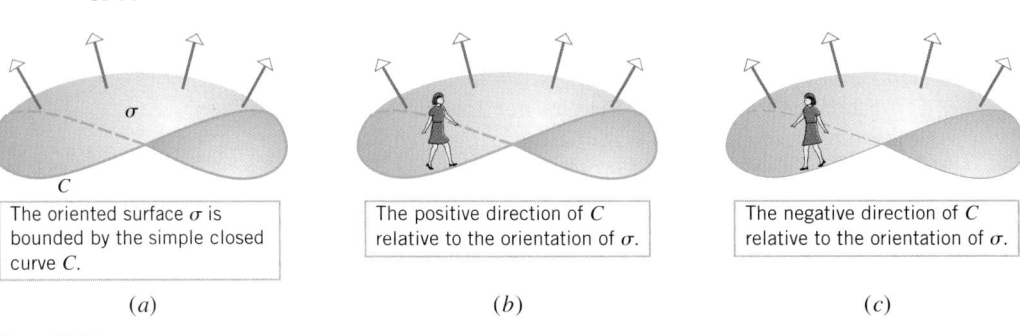

The oriented surface σ is bounded by the simple closed curve C.	The positive direction of C relative to the orientation of σ.	The negative direction of C relative to the orientation of σ.
(a)	(b)	(c)

Figure 17.8.1

STOKES' THEOREM

In Section 17.1 we defined the curl of a vector field

$$\mathbf{F}(x, y, z) = f(x, y, z)\mathbf{i} + g(x, y, z)\mathbf{j} + h(x, y, z)\mathbf{k}$$

as

$$\text{curl } \mathbf{F} = \left(\frac{\partial h}{\partial y} - \frac{\partial g}{\partial z}\right)\mathbf{i} + \left(\frac{\partial f}{\partial z} - \frac{\partial h}{\partial x}\right)\mathbf{j} + \left(\frac{\partial g}{\partial x} - \frac{\partial f}{\partial y}\right)\mathbf{k} = \begin{vmatrix} \mathbf{i} & \mathbf{j} & \mathbf{k} \\ \dfrac{\partial}{\partial x} & \dfrac{\partial}{\partial y} & \dfrac{\partial}{\partial z} \\ f & g & h \end{vmatrix} \quad (1)$$

but we did not attempt to give a physical explanation of its meaning at that time. The following result, known as **Stokes'* Theorem**, will provide us with a physical interpretation of the curl in the context of fluid flow.

---

* GEORGE GABRIEL STOKES (1819–1903). Irish mathematician and physicist. Born in Skreen, Ireland, Stokes came from a family deeply rooted in the Church of Ireland. His father was a rector, his mother the daughter of a rector, and three of his brothers took holy orders. He received his early education from his father and a local parish clerk. In 1837, he entered Pembroke College and after graduating with top honors accepted a fellowship at the college. In 1847 he was appointed Lucasian professor of mathematics at Cambridge, a position once held by Isaac Newton, but one that had lost its esteem through the years. By virtue of his accomplishments, Stokes ultimately restored the position to the eminence it once held. Unfortunately, the position paid very little and Stokes was forced to teach at the Government School of Mines during the 1850s to supplement his income.

Stokes was one of several outstanding nineteenth century scientists who helped turn the physical sciences in a more empirical direction. He systematically studied hydrodynamics, elasticity of solids, behavior of waves in elastic solids, and diffraction of light. For Stokes, mathematics was a tool for his physical studies. He wrote classic papers on the motion of viscous fluids that laid the foundation for modern hydrodynamics; he elaborated on the wave theory of light; and he wrote papers on gravitational variation that established him as a founder of the modern science of geodesy.

Stokes was honored in his later years with degrees, medals, and memberships in foreign societies. He was knighted in 1889. Throughout his life, Stokes gave generously of his time to learned societies and readily assisted those who sought his help in solving problems. He was deeply religious and vitally concerned with the relationship between science and religion.

17.8.1 THEOREM (*Stokes' Theorem*). *Let σ be a piecewise smooth oriented surface that is bounded by a simple, closed, piecewise smooth curve C with positive orientation. If the components of the vector field*

$$\mathbf{F}(x, y, z) = f(x, y, z)\mathbf{i} + g(x, y, z)\mathbf{j} + h(x, y, z)\mathbf{k}$$

*are continuous and have continuous first partial derivatives on some open set containing σ, and if **T** is the unit tangent vector to C, then*

$$\oint_C \mathbf{F} \cdot \mathbf{T}\, ds = \iint_\sigma (\text{curl } \mathbf{F}) \cdot \mathbf{n}\, dS \tag{2}$$

The proof of this theorem is beyond the scope of this text, so we will focus on its applications.

REMARK. Recall from Formula (32) of Section 17.2 that the integral on the left side of (2) represents the work performed by the vector field **F** on a particle that traverses the curve C. Thus, loosely phrased, Stokes' Theorem states that *the work performed by a vector field on a particle that traverses a simple, closed, piecewise smooth curve C in the positive direction can be obtained by integrating the normal component of the curl over an oriented surface σ bounded by C.*

USING STOKES' THEOREM TO CALCULATE WORK

For computational purposes it is usually preferable to use Formula (33) of Section 17.2 to rewrite the formula in Stokes' Theorem as

$$\oint_C \mathbf{F} \cdot d\mathbf{r} = \iint_\sigma (\text{curl } \mathbf{F}) \cdot \mathbf{n}\, dS \tag{3}$$

Stokes' Theorem is usually the method of choice for calculating work around piecewise smooth curves with multiple sections, since it eliminates the need for a separate integral evaluation over each section. This is illustrated in the following example.

Example 1

Find the work performed by the vector field

$$\mathbf{F}(x, y, z) = x^2\mathbf{i} + 4xy^3\mathbf{j} + y^2x\mathbf{k}$$

on a particle that traverses the rectangle C in the plane $z = y$ shown in Figure 17.8.2.

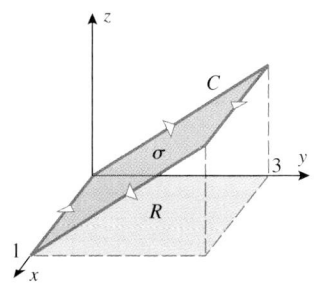

Figure 17.8.2

Solution. The work performed by the field is

$$W = \oint_C \mathbf{F} \cdot d\mathbf{r}$$

However, to evaluate this integral directly would require four separate integrations, one over each side of the rectangle. Instead, we will use Formula (3) to express the work as the surface integral

$$W = \iint_\sigma (\text{curl } \mathbf{F}) \cdot \mathbf{n}\, dS$$

in which the plane surface $σ$ enclosed by C is assigned a *downward* orientation to make the orientation of C positive, as required by Stokes' Theorem.

Since the surface σ has equation $z = y$ and

$$\text{curl } \mathbf{F} = \begin{vmatrix} \mathbf{i} & \mathbf{j} & \mathbf{k} \\ \dfrac{\partial}{\partial x} & \dfrac{\partial}{\partial y} & \dfrac{\partial}{\partial z} \\ x^2 & 4xy^3 & xy^2 \end{vmatrix} = 2xy\mathbf{i} - y^2\mathbf{j} + 4y^3\mathbf{k}$$

it follows from Formula (13) of Section 17.6 with curl $\mathbf{F}$ replacing $\mathbf{F}$ that

$$W = \iint_\sigma (\text{curl } \mathbf{F}) \cdot \mathbf{n}\, dS = \iint_R (\text{curl } \mathbf{F}) \cdot \left(\frac{\partial z}{\partial x}\mathbf{i} + \frac{\partial z}{\partial y}\mathbf{j} - \mathbf{k} \right) dA$$

$$= \iint_R (2xy\mathbf{i} - y^2\mathbf{j} + 4y^3\mathbf{k}) \cdot (0\mathbf{i} + \mathbf{j} - \mathbf{k})\, dA$$

$$= \int_0^1 \int_0^3 (-y^2 - 4y^3)\, dy\, dx$$

$$= -\int_0^1 \left[\frac{y^3}{3} + y^4 \right]_{y=0}^3 dx$$

$$= -\int_0^1 90\, dx = -90 \quad \blacktriangleleft$$

Example 2

Verify Stokes' Theorem for the vector field $\mathbf{F}(x, y, z) = 2z\mathbf{i} + 3x\mathbf{j} + 5y\mathbf{k}$, taking σ to be the portion of the paraboloid $z = 4 - x^2 - y^2$ for which $z \geq 0$ with upward orientation, and C to be the positively oriented circle $x^2 + y^2 = 4$ that forms the boundary of σ in the xy-plane (Figure 17.8.3).

Solution. We will verify Formula (3). Since σ is oriented up, the positive orientation of C is counterclockwise looking down the positive z-axis. Thus, C can be represented parametrically (with positive orientation) by

$$x = 2\cos t, \quad y = 2\sin t, \quad z = 0 \quad (0 \leq t \leq 2\pi) \tag{4}$$

Therefore,

$$\oint_C \mathbf{F} \cdot d\mathbf{r} = \oint_C 2z\, dx + 3x\, dy + 5y\, dz$$

$$= \int_0^{2\pi} [0 + (6\cos t)(2\cos t) + 0]\, dt$$

$$= \int_0^{2\pi} 12\cos^2 t\, dt = 12 \left[\frac{1}{2}t + \frac{1}{4}\sin 2t \right]_0^{2\pi} = 12\pi$$

To evaluate the right side of (3), we start by finding curl $\mathbf{F}$. We obtain

$$\text{curl } \mathbf{F} = \begin{vmatrix} \mathbf{i} & \mathbf{j} & \mathbf{k} \\ \dfrac{\partial}{\partial x} & \dfrac{\partial}{\partial y} & \dfrac{\partial}{\partial z} \\ 2z & 3x & 5y \end{vmatrix} = 5\mathbf{i} + 2\mathbf{j} + 3\mathbf{k}$$

Since σ is oriented up and is expressed in the form $z = g(x, y) = 4 - x^2 - y^2$, it follows from Formula (12) of Section 17.6 with curl $\mathbf{F}$ replacing $\mathbf{F}$ that

$$\iint_\sigma (\text{curl } \mathbf{F}) \cdot \mathbf{n}\, dS = \iint_R (\text{curl } \mathbf{F}) \cdot \left(-\frac{\partial z}{\partial x}\mathbf{i} - \frac{\partial z}{\partial y}\mathbf{j} + \mathbf{k} \right) dA$$

$$= \iint_R (5\mathbf{i} + 2\mathbf{j} + 3\mathbf{k}) \cdot (2x\mathbf{i} + 2y\mathbf{j} + \mathbf{k})\, dA$$

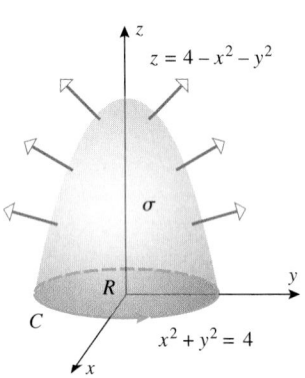

Figure 17.8.3

$$= \iint\limits_{R} (10x + 4y + 3)\, dA$$

$$= \int_{0}^{2\pi} \int_{0}^{2} (10r\cos\theta + 4r\sin\theta + 3)r\, dr\, d\theta$$

$$= \int_{0}^{2\pi} \left[\frac{10r^3}{3}\cos\theta + \frac{4r^3}{3}\sin\theta + \frac{3r^2}{2} \right]_{r=0}^{2} d\theta$$

$$= \int_{0}^{2\pi} \left(\frac{80}{3}\cos\theta + \frac{32}{3}\sin\theta + 6 \right) d\theta$$

$$= \left[\frac{80}{3}\sin\theta - \frac{32}{3}\cos\theta + 6\theta \right]_{0}^{2\pi} = 12\pi$$

As guaranteed by Stokes' Theorem, the value of this surface integral is the same as the value of the line integral obtained above. Note, however, that the line integral was simpler to evaluate and hence would be the method of choice in this case. ◄

REMARK. Observe that in Formula (3) the only relationship required between σ and C is that C be the boundary of σ and that C be positively oriented relative to the orientation of σ. Thus, if σ_1 and σ_2 are *different* oriented surfaces but have the *same* positively oriented boundary curve C, then it follows from (3) that

$$\iint\limits_{\sigma_1} \text{curl } \mathbf{F} \cdot \mathbf{n}\, dS = \iint\limits_{\sigma_2} \text{curl } \mathbf{F} \cdot \mathbf{n}\, dS$$

For example, the parabolic surface in Example 2 has the same positively oriented boundary C as the disk R in Figure 17.8.3 with upper orientation. Thus, the value of the surface integral in that example would not change if σ is replaced by R (or by any other oriented surface that has the positively oriented circle C as its boundary). This can be useful in computations because it is sometimes possible to circumvent a difficult integration by changing the surface of integration.

RELATIONSHIP BETWEEN GREEN'S THEOREM AND STOKES' THEOREM

It is sometimes convenient to regard a vector field

$$\mathbf{F}(x, y) = f(x, y)\mathbf{i} + g(x, y)\mathbf{j}$$

in 2-space as a vector field in 3-space by expressing it as

$$\mathbf{F}(x, y) = f(x, y)\mathbf{i} + g(x, y)\mathbf{j} + 0\mathbf{k} \tag{5}$$

If R is a region in the xy-plane enclosed by a simple, closed, piecewise smooth curve C, then we can treat R as a *flat* surface, and we can treat a surface integral over R as an ordinary double integral over R. Thus, if we orient R up and C counterclockwise looking down the positive z-axis, then Formula (3) applied to (5) yields

$$\oint_{C} \mathbf{F} \cdot d\mathbf{r} = \iint\limits_{R} \text{curl } \mathbf{F} \cdot \mathbf{k}\, dA \tag{6}$$

But

$$\text{curl } \mathbf{F} = \begin{vmatrix} \mathbf{i} & \mathbf{j} & \mathbf{k} \\ \dfrac{\partial}{\partial x} & \dfrac{\partial}{\partial y} & \dfrac{\partial}{\partial z} \\ f & g & 0 \end{vmatrix} = -\frac{\partial g}{\partial z}\mathbf{i} + \frac{\partial f}{\partial z}\mathbf{j} + \left(\frac{\partial g}{\partial x} - \frac{\partial f}{\partial y} \right)\mathbf{k} = \left(\frac{\partial g}{\partial x} - \frac{\partial f}{\partial y} \right)\mathbf{k}$$

since $\partial g / \partial z = \partial f / \partial z = 0$. Substituting this expression in (6) and expressing the integrals in terms of components yields

$$\oint_{C} f\, dx + g\, dy = \iint\limits_{R} \left(\frac{\partial g}{\partial x} - \frac{\partial f}{\partial y} \right) dA$$

which is Green's Theorem [Formula (1) of Section 17.4]. Thus, we have shown that Green's Theorem can be viewed as a special case of Stokes' Theorem.

Stokes' Theorem provides a way of interpreting the curl of a vector field $\mathbf{F}$ in the context of fluid flow. For this purpose let σ_a be a small oriented disk of radius a centered at a point P_0 in an incompressible steady-state fluid flow, and let $\mathbf{n}$ be a unit normal vector at the center of the disk that points in the direction of orientation. Let us assume that the flow of liquid past the disk causes it to spin around the axis through $\mathbf{n}$, and let us try to find the direction of $\mathbf{n}$ that will produce the maximum rotation rate in the positive direction of the boundary curve C_a (Figure 17.8.4). For convenience, we will denote the area of the disk σ_a by $A(\sigma_a)$; that is, $A(\sigma_a) = \pi a^2$.

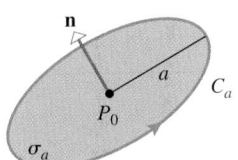

Figure 17.8.4

If the direction of $\mathbf{n}$ is fixed, then at each point of C_a the only component of $\mathbf{F}$ that contributes to the rotation of the disk about $\mathbf{n}$ is the component $\mathbf{F} \cdot \mathbf{T}$ tangent to C_a (Figure 17.8.5). Thus, for a fixed $\mathbf{n}$ the integral

$$\oint_{C_a} \mathbf{F} \cdot \mathbf{T} \, ds \qquad (7)$$

can be viewed as a measure of the tendency for the fluid to flow in the positive direction around C_a. Accordingly, (7) is called the ***circulation of $\mathbf{F}$ around C_a***. For example, in the extreme case where the flow is normal to the circle at each point, the circulation around C_a is zero, since $\mathbf{F} \cdot \mathbf{T} = 0$ at each point. The more closely that $\mathbf{F}$ aligns with $\mathbf{T}$ along the circle, the larger the value of $\mathbf{F} \cdot \mathbf{T}$ and the larger the value of the circulation.

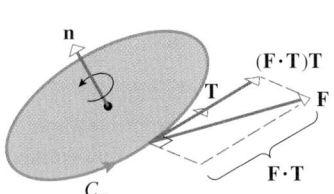

Figure 17.8.5

To see the relationship between circulation and curl, suppose that curl $\mathbf{F}$ is continuous on σ_a, so that when σ_a is small the value of curl $\mathbf{F}$ at any point of σ_a will not vary much from the value of curl $\mathbf{F}(P_0)$ at the center. Thus, for a small disk σ_a we can reasonably assume that curl $\mathbf{F}$ has a constant value of curl $\mathbf{F}(P_0)$ on σ_a. Moreover, because the surface σ_a is flat, the unit normal vectors that orient σ_a are all equal. Thus, the vector quantity $\mathbf{n}$ in Formula (3) can be treated as a constant, and we can write

$$\oint_{C_a} \mathbf{F} \cdot \mathbf{T} \, ds = \iint_{\sigma_a} (\text{curl } \mathbf{F}) \cdot \mathbf{n} \, dS \approx \text{curl } \mathbf{F}(P_0) \cdot \mathbf{n} \iint_{\sigma_a} dS$$

where the line integral is taken in the positive direction of C_a. But the double integral in this equation represents the surface area of σ_a, so

$$\oint_{C_a} \mathbf{F} \cdot \mathbf{T} \, ds \approx [\text{curl } \mathbf{F}(P_0) \cdot \mathbf{n}] A(\sigma_a)$$

from which we obtain

$$\text{curl } \mathbf{F}(P_0) \cdot \mathbf{n} \approx \frac{1}{A(\sigma_a)} \oint_{C_a} \mathbf{F} \cdot \mathbf{T} \, ds \qquad (8)$$

The quantity on the right side of (8) is called the ***circulation density of $\mathbf{F}$ around C_a***. If we now let the radius a of the disk approach zero (with $\mathbf{n}$ fixed), then it is plausible that the error in this approximation will approach zero and the exact value of curl $\mathbf{F}(P_0) \cdot \mathbf{n}$ will be given by

$$\text{curl } \mathbf{F}(P_0) \cdot \mathbf{n} = \lim_{a \to 0} \frac{1}{A(\sigma_a)} \oint_{C_a} \mathbf{F} \cdot \mathbf{T} \, ds \qquad (9)$$

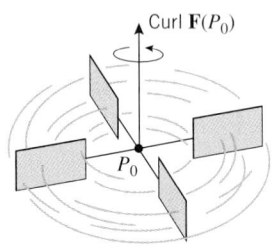

Figure 17.8.6

We call curl $\mathbf{F}(P_0) \cdot \mathbf{n}$ the ***circulation density of $\mathbf{F}$ at P_0 in the direction of $\mathbf{n}$***. This quantity has its maximum value when $\mathbf{n}$ is in the same direction as curl $\mathbf{F}(P_0)$; this tells us that *at each point in a steady-state fluid flow the maximum circulation density occurs in the direction of the curl*. Physically, this means that if a small paddle wheel is immersed in the fluid so that the pivot point is at P_0, then the paddles will turn most rapidly when the spindle is aligned with curl $\mathbf{F}(P_0)$ (Figure 17.8.6). If curl $\mathbf{F} = \mathbf{0}$ at each point of a region, then $\mathbf{F}$ is said to be ***irrotational*** in that region, since no circulation occurs about any point of the region.

REMARK. Formula (9) is sometimes taken as a definition of curl. This is a useful alternative to Definition 17.1.5 because it does not require a coordinate system.

EXERCISE SET 17.8 ⬚c CAS

The figures in Exercises 1 and 2 show a horizontal layer of the vector field of a fluid flow in which the flow is parallel to the xy-plane at every point and is identical in each layer (i.e., is independent of z). For each flow, state whether you believe that the curl is nonzero at the origin, and explain your reasoning. If you believe that it is nonzero, then state whether it points in the positive or negative z-direction.

1. (a)   (b)

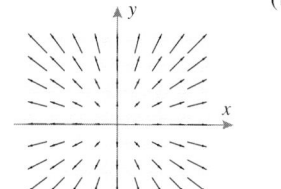 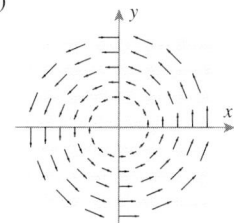

2. (a)   (b)

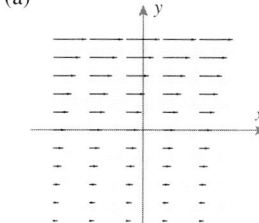 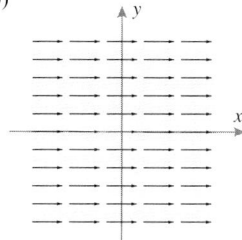

In Exercises 3–6, verify Formula (2) in Stokes' Theorem by evaluating the line integral and the double integral. Assume that the surface has an upward orientation.

3. $\mathbf{F}(x, y, z) = (x - y)\mathbf{i} + (y - z)\mathbf{j} + (z - x)\mathbf{k}$; σ is the portion of the plane $x + y + z = 1$ in the first octant.

4. $\mathbf{F}(x, y, z) = x^2\mathbf{i} + y^2\mathbf{j} + z^2\mathbf{k}$; σ is the portion of the cone $z = \sqrt{x^2 + y^2}$ below the plane $z = 1$.

5. $\mathbf{F}(x, y, z) = x\mathbf{i} + y\mathbf{j} + z\mathbf{k}$; σ is the upper hemisphere $z = \sqrt{a^2 - x^2 - y^2}$.

6. $\mathbf{F}(x, y, z) = (z - y)\mathbf{i} + (z + x)\mathbf{j} - (x + y)\mathbf{k}$; σ is the portion of the paraboloid $z = 9 - x^2 - y^2$ above the xy-plane.

In Exercises 7–14, use Stokes' Theorem to evaluate the integral $\oint_C \mathbf{F} \cdot d\mathbf{r}$.

7. $\mathbf{F}(x, y, z) = z^2\mathbf{i} + 2x\mathbf{j} - y^3\mathbf{k}$; C is the circle $x^2 + y^2 = 1$ in the xy-plane with counterclockwise orientation looking down the positive z-axis.

8. $\mathbf{F}(x, y, z) = xz\mathbf{i} + 3x^2y^2\mathbf{j} + yx\mathbf{k}$; C is the rectangle in the plane $z = y$ shown in Figure 17.8.2.

9. $\mathbf{F}(x, y, z) = 3z\mathbf{i} + 4x\mathbf{j} + 2y\mathbf{k}$; C is the boundary of the paraboloid shown in Figure 17.8.3.

10. $\mathbf{F}(x, y, z) = -3y^2\mathbf{i} + 4z\mathbf{j} + 6x\mathbf{k}$; C is the triangle in the plane $z = \frac{1}{2}y$ with vertices $(2, 0, 0)$, $(0, 2, 1)$, and $(0, 0, 0)$ with a counterclockwise orientation looking down the positive z-axis.

11. $\mathbf{F}(x, y, z) = xy\mathbf{i} + x^2\mathbf{j} + z^2\mathbf{k}$; C is the intersection of the paraboloid $z = x^2 + y^2$ and the plane $z = y$ with a counterclockwise orientation looking down the positive z-axis.

12. $\mathbf{F}(x, y, z) = xy\mathbf{i} + yz\mathbf{j} + zx\mathbf{k}$; C is the triangle in the plane $x + y + z = 1$ with vertices $(1, 0, 0)$, $(0, 1, 0)$, and $(0, 0, 1)$ with a counterclockwise orientation looking from the first octant toward the origin.

13. $\mathbf{F}(x, y, z) = (x - y)\mathbf{i} + (y - z)\mathbf{j} + (z - x)\mathbf{k}$; C is the circle $x^2 + y^2 = a^2$ in the xy-plane with counterclockwise orientation looking down the positive z-axis.

14. $\mathbf{F}(x, y, z) = (z + \sin x)\mathbf{i} + (x + y^2)\mathbf{j} + (y + e^z)\mathbf{k}$; C is the intersection of the sphere $x^2 + y^2 + z^2 = 1$ and the cone $z = \sqrt{x^2 + y^2}$ with counterclockwise orientation looking down the positive z-axis.

15. Consider the vector field given by the formula

$$\mathbf{F}(x, y, z) = (x - z)\mathbf{i} + (y - x)\mathbf{j} + (z - xy)\mathbf{k}$$

(a) Use Stokes' Theorem to find the circulation around the triangle with vertices $A(1, 0, 0)$, $B(0, 2, 0)$, and $C(0, 0, 1)$ oriented counterclockwise looking from the origin toward the first octant.

(b) Find the circulation density of $\mathbf{F}$ at the origin in the direction of $\mathbf{k}$.

(c) Find the unit vector $\mathbf{n}$ such that the circulation density of $\mathbf{F}$ at the origin is maximum in the direction of $\mathbf{n}$.

16. (a) Show that if $\mathbf{F}$ is a vector field whose components have continuous second-order partial derivatives, then $\text{div}(\text{curl }\mathbf{F}) = 0$.

(b) Use the result in part (a) to show that if the surface σ of a solid G has outward orientation, $\mathbf{n}$ is the outward unit normal to σ, and the components of $\mathbf{F}$ have continuous first partial derivatives on and within σ, then

$$\iint_\sigma (\text{curl }\mathbf{F} \cdot \mathbf{n}) \, dS = 0$$

(c) The vector field curl($\mathbf{F}$) is called the ***curl field*** of $\mathbf{F}$. In words, interpret the formula in part (b) as a statement about the flux of the curl field.

17. In 1831 the physicist Michael Faraday discovered that an electric current can be produced by varying the magnetic flux through a conducting loop. His experiments showed that the electromotive force **E** is related to the magnetic induction **B** by the equation

$$\oint_C \mathbf{E} \cdot d\mathbf{r} = -\iint_\sigma \frac{\partial \mathbf{B}}{\partial t} \cdot \mathbf{n} \, dS$$

Use this result to make a conjecture about the relationship between curl **E** and **B**, and explain your reasoning.

C **18.** Let σ be the portion of the paraboloid $z = 1 - x^2 - y^2$ for which $z \geq 0$, and let C be the circle $x^2 + y^2 = 1$ that forms the boundary of σ in the xy-plane. Assuming that σ is oriented up, use a CAS to verify Formula (2) in Stokes' Theorem for the vector field

$$\mathbf{F} = (x^2 y - z^2)\mathbf{i} + (y^3 - x)\mathbf{j} + (2x + 3z - 1)\mathbf{k}$$

by evaluating the line integral and the surface integral.

SUPPLEMENTARY EXERCISES

1. In words, what is a vector field? Give some physical examples of vector fields.

2. (a) Give a physical example of an inverse-square field $\mathbf{F}(\mathbf{r})$ in 3-space.
  (b) Write a formula for a general inverse-square field $\mathbf{F}(\mathbf{r})$ in terms of the radius vector $\mathbf{r}$.
  (c) Write a formula for a general inverse-square field $\mathbf{F}(x, y, z)$ in 3-space using rectangular coordinates.

3. Assume that C is the parametric curve $x = x(t)$, $y = y(t)$, where t varies from a to b. In each part, express the line integral as a definite integral with variable of integration t.
  (a) $\displaystyle\int_C f(x, y)\, dx + g(x, y)\, dy$ (b) $\displaystyle\int_C f(x, y)\, ds$

4. (a) Express the mass M of a thin wire in 3-space as a line integral.
  (b) Express the length of a curve as a line integral.
  (c) Express the area of a surface as a surface integral.
  (d) Express the area of a plane region as a line integral.

5. In each part, give a physical interpretation of the integral.
  (a) $\displaystyle\int_C \mathbf{F} \cdot \mathbf{T}\, ds$ (b) $\displaystyle\iint_\sigma \mathbf{F} \cdot \mathbf{n}\, dS$

6. State some alternative notations for $\displaystyle\int_C \mathbf{F} \cdot \mathbf{T}\, ds$.

7. (a) State the Fundamental Theorem of Work Integrals, including all required hypotheses.
  (b) State Green's Theorem, including all of the required hypotheses.

8. What conditions must C, D, and $\mathbf{F}$ satisfy to be assured that

$$\int_C \mathbf{F} \cdot d\mathbf{r} = 0$$

around every piecewise smooth curve C in the region D in 2-space?

9. How can you tell whether the vector field

$$\mathbf{F}(x, y) = f(x, y)\mathbf{i} + g(x, y)\mathbf{j}$$

is conservative on a simply connected open region D?

10. Make a sketch of a vector field that is not conservative, and give an argument in support of your answer.

11. Assume that σ is the parametric surface

$$\mathbf{r} = x(u, v)\mathbf{i} + y(u, v)\mathbf{j} + z(u, v)\mathbf{k}$$

where (u, v) varies over a region R. Express the surface integral

$$\iint_\sigma f(x, y, z)\, dS$$

as a double integral with variables of integration u and v.

12. State the Divergence Theorem and Stokes' Theorem, including all required hypotheses.

13. In our discussion of hyperbolic functions in Section 8.8, we stated without proof that the parameter t in the parametric equations

$$x = \cosh t, \quad y = \sinh t$$

of a hyperbola can be interpreted as twice the shaded area in the accompanying figure (see the discussion relating to Figure 8.8.3). Use Green's Theorem to prove this.

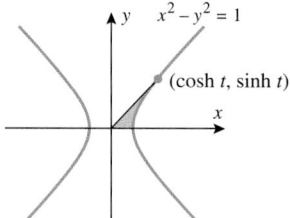

Figure Ex-13

14. As discussed in Example 1 of Section 17.1, *Coulomb's law* states that the electrostatic force $\mathbf{F}(\mathbf{r})$ that a particle of charge Q exerts on a particle of charge q is given by the formula

$$\mathbf{F}(\mathbf{r}) = \frac{qQ}{4\pi\epsilon_0 \|\mathbf{r}\|^3}\mathbf{r}$$

where $\mathbf{r}$ is the radius vector from Q to q and ϵ_0 is the permittivity constant.
  (a) Express the vector field $\mathbf{F}(\mathbf{r})$ in coordinate form $\mathbf{F}(x, y, z)$ with Q at the origin.
  (b) Find the work performed by the vector field $\mathbf{F}$ on a charge q that moves along a straight line from $(3, 0, 0)$ to $(3, 1, 5)$.

15. As discussed in Section 17.1, it follows from *Newton's Universal Law of Gravitation* that the gravitational force $\mathbf{F}(\mathbf{r})$ exerted by an object of mass M on an object of mass m is given by the formula

$$\mathbf{F}(\mathbf{r}) = -\frac{GmM}{\|\mathbf{r}\|^3}\mathbf{r}$$

where $\mathbf{r}$ is the radius vector from M to m and G is the universal gravitational constant.

(a) Show that the work W done by the gravitational field $\mathbf{F}(\mathbf{r})$ when the mass m moves from a distance of r_1 from M to a distance of r_2 from M is

$$W = GmM\left(\frac{1}{r_2} - \frac{1}{r_1}\right)$$

(b) The value of the constant GM for the Earth is approximately $3.99 \times 10^5 \text{ km}^3/\text{s}^2$. Find the work done by the Earth's gravitational field on a satellite with a mass of 1000 kg that moves from a perigee of 600 km above the surface of the Earth to an apogee of 800 km above the surface of the Earth, assuming the Earth to be a sphere of radius 6370 km.

16. Use Green's Theorem to show that the coordinates of the centroid (x, y) of a region with area A that is bounded by a simple closed piecewise smooth curve C can be expressed as

$$\bar{x} = \frac{1}{2A}\int_C x^2\,dy, \quad \bar{y} = -\frac{1}{2A}\int_C y^2\,dx$$

where C has counterclockwise orientation.

In Exercises 17–20, use the result in Exercise 16 to confirm that the centroid of the region is as shown in the figure.

17.
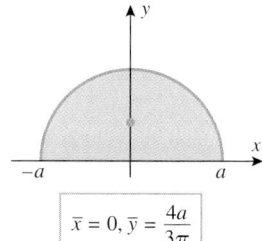
$$\bar{x} = 0, \bar{y} = \frac{4a}{3\pi}$$

18.
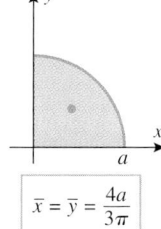
$$\bar{x} = \bar{y} = \frac{4a}{3\pi}$$

19.
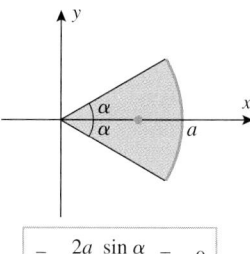
$$\bar{x} = \frac{2a}{3}\frac{\sin\alpha}{\alpha}, \bar{y} = 0$$

20.
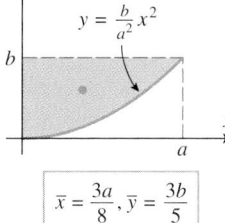
$$y = \frac{b}{a^2}x^2$$
$$\bar{x} = \frac{3a}{8}, \bar{y} = \frac{3b}{5}$$

21. (a) Use Green's theorem to prove that

$$\int f(x)\,dx + g(y)\,dy = 0$$

if f and g are differentiable functions and C is a simple, closed, piecewise smooth curve.

(b) What does this tell you about the vector field $\mathbf{F}(x, y) = f(x)\mathbf{i} + g(y)\mathbf{j}$?

22. The purpose of this exercise is to establish the role of the curl in describing the rotation of a rigid body. As illustrated in the accompanying figure, consider a rigid body rotating about the z-axis of an xyz-coordinate system at a constant *angular speed* of ω rad/s. Let P be a point on the body, and let $\mathbf{r}$ be the position vector of P. Thus, the velocity of P is $\mathbf{v} = d\mathbf{r}/dt$, where $\mathbf{v}$ is tangent to the circle of rotation of P. Let θ and ϕ be the angles shown in the figure, and define the *angular velocity* of the point P to be $\boldsymbol{\omega} = \omega\mathbf{k}$.

(a) Show that $\mathbf{v} = \boldsymbol{\omega} \times \mathbf{r}$.

(b) Show that $\mathbf{v} = -\omega y\mathbf{i} + \omega x\mathbf{j}$.

(c) Show that curl $\mathbf{v} = 2\boldsymbol{\omega}$.

(d) Is the velocity field $\mathbf{v}$ conservative? Justify your answer.

23. Do you think that the surface in the accompanying figure is orientable? Explain your reasoning.

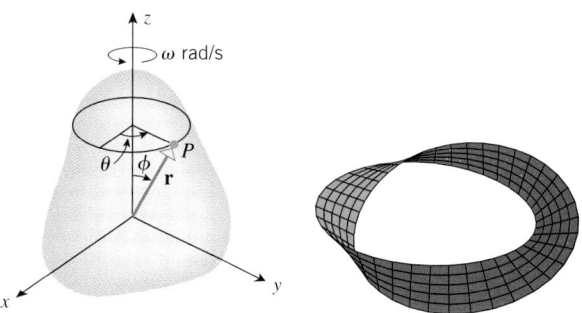

Figure Ex-22   Figure Ex-23

24. Let G be a solid with the surface σ oriented by outward unit normals, suppose that ϕ has continuous first and second partial derivatives in some open set containing G, and let $D_{\mathbf{n}}\phi$ be the directional derivative of ϕ. Show that

$$\iint_\sigma D_{\mathbf{n}}\phi\,dS = \iiint_G \left[\frac{\partial^2\phi}{\partial x^2} + \frac{\partial^2\phi}{\partial y^2} + \frac{\partial^2\phi}{\partial z^2}\right]dV$$

25. Let σ be the sphere $x^2 + y^2 + z^2 = 1$, let $\mathbf{n}$ be an inward unit normal, and let $D_{\mathbf{n}}f$ be the directional derivative of $f(x, y, z) = x^2 + y^2 + z^2$. Use the result in Exercise 24 to evaluate the surface integral

$$\iint_\sigma D_{\mathbf{n}}f\,dS$$

26. Let $\mathbf{F}(x, y) = (ye^{xy} - 1)\mathbf{i} + xe^{xy}\mathbf{j}$.

(a) Show that $\mathbf{F}$ is a conservative vector field.

(b) Find a potential function for $\mathbf{F}$.

(c) Find the work performed by the vector field on a particle that moves along the sawtooth curve represented by the parametric equations

$$x = t + \sin^{-1}(\sin t)$$
$$y = (2/\pi)\sin^{-1}(\sin t)$$
$$(0 \le t \le 8\pi)$$

(see Figure Ex-26 on page 1126).

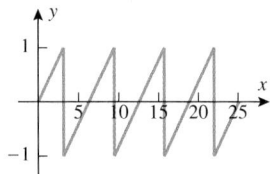

Figure Ex-26

27. Let $\mathbf{F}(x, y) = y\mathbf{i} - 2x\mathbf{j}$.

(a) Find a nonzero function $h(x)$ such that $h(x)\mathbf{F}(x, y)$ is a conservative vector field.

(b) Find a nonzero function $g(y)$ such that $g(y)\mathbf{F}(x, y)$ is a conservative vector field.

28. Let $\mathbf{F}(x, y, z) = f(x, y, z)\mathbf{i} + g(x, y, z)\mathbf{j} + h(x, y, z)\mathbf{k}$ and suppose that f, g, and h are continuous and have continuous first partial derivatives in a region. It was shown in Exercise 25 of Section 17.3 that if $\mathbf{F}$ is conservative in the region, then

$$\frac{\partial f}{\partial y} = \frac{\partial g}{\partial x}, \quad \frac{\partial f}{\partial z} = \frac{\partial h}{\partial x}, \quad \frac{\partial g}{\partial z} = \frac{\partial h}{\partial y}$$

there. Use this result and Stokes' Theorem to help show that $\mathbf{F}$ is conservative in an open spherical region if and only if curl $\mathbf{F} = \mathbf{0}$ in that region.

In Exercises 29 and 30, use the result in Exercise 28 to determine whether $\mathbf{F}$ is conservative. If so, find a potential function for it.

29. (a) $\mathbf{F}(x, y, z) = z^2\mathbf{i} + e^{-y}\mathbf{j} + 2xz\mathbf{k}$

(b) $\mathbf{F}(x, y, z) = xy\mathbf{i} + x^2\mathbf{j} + \sin z\mathbf{k}$

30. (a) $\mathbf{F}(x, y, z) = \sin x\mathbf{i} + z\mathbf{j} + y\mathbf{k}$

(b) $\mathbf{F}(x, y, z) = z\mathbf{i} + 2yz\mathbf{j} + y^2\mathbf{k}$

EXPANDING THE CALCULUS HORIZON

Hurricane Modeling

Each year population centers throughout the world are ravaged by hurricanes, and it is the mission of the National Hurricane Center to minimize the damage and loss of life by issuing warnings and forecasts of hurricanes developing in the Caribbean, Atlantic, Gulf of Mexico, and Eastern Pacific regions. Your assignment as a trainee at the Center is to construct a simple mathematical model of a hurricane using basic principles of fluid flow and properties of vector fields.

Modeling Assumptions

You have been notified of a developing hurricane in the Bahamas (designated hurricane *Isaac*) and have been asked to construct a model of its velocity field. Because hurricanes are complicated three-dimensional fluid flows, you will have to make many simplifying assumptions about the structure of a hurricane and the properties of the fluid flow. Accordingly, you decide to model the moisture in Isaac as an ***ideal fluid***, meaning that it is ***incompressible*** and its ***viscosity*** can be ignored. An incompressible fluid is one in which the density of the fluid is the same at all points and cannot be altered by compressive forces. Experience has shown that water can be regarded as incompressible but water vapor cannot. However, incompressibility is a reasonable assumption for a basic hurricane model because a hurricane is not restricted to a closed container that would produce compressive forces.

All fluids have a certain amount of viscosity, which is a resistance to flow—oil and molasses have a high viscosity, whereas water has almost none at subsonic speeds. Thus, it is reasonable to ignore viscosity in a basic model. Next, you decide to assume that the flow is in a ***steady state***, meaning that the velocity of the fluid at any point does not vary with time. This is reasonable over very short time periods for hurricanes that move and change slowly. Finally, although hurricanes are three-dimensional flows, you decide to model a two-dimensional horizontal cross section, so you make the simplifying assumption that the fluid in the cross section flows horizontally.

   The photograph of Isaac shown at the beginning of this module reveals a typical pattern of a Caribbean hurricane—a counterclockwise swirl of fluid around the *eye* through which the fluid exits the flow in the form of rain. The lower pressure in the eye causes an inward-rushing air mass, and circular winds around the eye contribute to the swirling effect.

   Your first objective is to find an explicit formula for Isaac's velocity field $\mathbf{F}(x, y)$, so you begin by introducing a rectangular coordinate system with its origin at the eye and its y-axis pointing north. Moreover, based on the hurricane picture and your knowledge of meteorological theory, you decide to build up the velocity field for Isaac from the velocity fields of simpler flows—a counterclockwise "vortex flow" $\mathbf{F}_1(x, y)$ in which fluid flows counterclockwise in concentric circles around the eye and a "sink flow" $\mathbf{F}_2(x, y)$ in which the fluid flows in straight lines toward a sink at the eye. Once you find explicit formulas for $\mathbf{F}_1(x, y)$ and $\mathbf{F}_2(x, y)$, your plan is to use the *superposition principle* from fluid dynamics to express the velocity field for Isaac as $\mathbf{F}(x, y) = \mathbf{F}_1(x, y) + \mathbf{F}_2(x, y)$.

Modeling a Vortex Flow

A *counterclockwise vortex flow* of an ideal fluid around the origin has four defining characteristics (Figure 1*a* on the following page):

- The velocity vector at a point (x, y) is tangent to the circle that is centered at the origin and passes through the point (x, y).

- The direction of the velocity vector at a point (x, y) indicates a counterclockwise motion.

- The speed of the fluid is constant on circles centered at the origin.

- The speed of the fluid along a circle is inversely proportional to the radius of the circle (and hence the speed approaches $+\infty$ as the radius of the circle approaches 0).

In fluid dynamics, the *strength* k of a vortex flow is defined to be 2π times the speed of the fluid along the unit circle. If the strength of a vortex flow is known, then the speed of the fluid along any other circle can be found from the fact that speed is inversely proportional to the radius of the circle. Thus, your first objective is to find a formula for a vortex flow $\mathbf{F}_1(x, y)$ with a specified strength k.

· · · · · · · · · · ·
Exercise 1   Show that

$$\mathbf{F}_1(x, y) = -\frac{k}{2\pi(x^2 + y^2)}(y\mathbf{i} - x\mathbf{j})$$

is a model for the velocity field of a counterclockwise vortex flow around the origin of strength k by confirming that

(a) $\mathbf{F}_1(x, y)$ has the four properties required of a counterclockwise vortex flow around the origin;

(b) k is 2π times the speed of the fluid along the unit circle.

· · · · · · · · · · ·
Exercise 2   Use a graphing utility that can generate vector fields to generate a vortex flow of strength 2π.

Modeling a Sink Flow

A *uniform sink flow* of an ideal fluid toward the origin has four defining characteristics (Figure 1*b*):

- The velocity vector at every point (x, y) is directed toward the origin.

- The speed of the fluid is the same at all points on a circle centered at the origin.

- The speed of the fluid at a point is inversely proportional to its distance from the origin (from which it follows that the speed approaches $+\infty$ as the distance from the origin approaches 0).

- There is a sink at the origin at which fluid leaves the flow.

As with a vortex flow, the *strength* q of a uniform sink flow is defined to be 2π times the speed of the fluid at points on the unit circle. If the strength of a sink flow is known, then the speed of the fluid at any point in the flow can be found using the fact that the speed is inversely proportional to the distance from the origin. Thus, your next objective is to find a formula for a uniform sink flow $\mathbf{F}_2(x, y)$ with a specified strength q.

Exercise 3   Show that

$$\mathbf{F}_2(x, y) = -\frac{q}{2\pi(x^2 + y^2)}(x\mathbf{i} + y\mathbf{j})$$

is a model for the velocity field of a uniform sink flow toward the origin of strength q by confirming the following facts:

(a) $\mathbf{F}_2(x, y)$ has the four properties required of a uniform sink flow toward the origin.
    [A reasonable physical argument to confirm the existence of the sink will suffice.]

(b) q is 2π times the speed of the fluid at points on the unit circle.

Exercise 4   Use a graphing utility that can generate vector fields to generate a uniform sink flow of strength 2π.

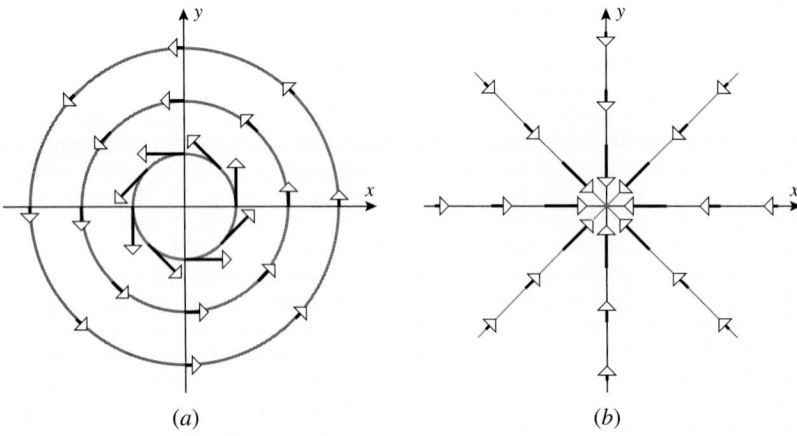

(a)   (b)

Figure 1

A Basic Hurricane Model

It now follows from Exercises 1 and 3 that the vector field $\mathbf{F}(x, y)$ for a hurricane model that combines a vortex flow around the origin of strength k and a uniform sink flow toward the origin of strength q is

$$\mathbf{F}(x, y) = -\frac{1}{2\pi(x^2 + y^2)}[(qx + ky)\mathbf{i} + (qy - kx)\mathbf{j}] \qquad (1)$$

Exercise 5

(a) Figure 2 shows a vector field for a hurricane with vortex strength $k = 2\pi$ and sink strength $q = 2\pi$. Use a graphing utility that can generate vector fields to produce a reasonable facsimile of this figure.

(b) Make a conjecture about the effect of increasing k and keeping q fixed, and check your conjecture using a graphing utility.

(c) Make a conjecture about the effect of increasing q and keeping k fixed, and check your conjecture using a graphing utility.

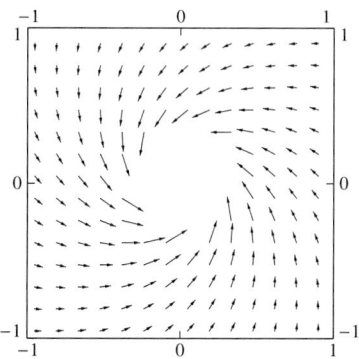

Figure 2

Modeling Hurricane Isaac

You are now ready to apply Formula (1) to obtain a model of the vector field $\mathbf{F}(x, y)$ of hurricane Isaac. You need some observational data to determine the constants k and q, so you call the Technical Support Branch of the Center for the latest information on hurricane Isaac. They report that 20 km from the eye the wind velocity has a component of 15 km/h toward the eye and a counterclockwise tangential component of 45 km/h.

············
Exercise 6

(a) Find the strengths k and q of the vortex and sink for hurricane Isaac.

(b) Find the vector field $\mathbf{F}(x, y)$ for hurricane Isaac.

(c) Estimate the size of hurricane Isaac by finding a radius beyond which the wind speed is less than 5 km/h.

Streamlines for the Basic Hurricane Model

The paths followed by the fluid particles in a fluid flow are called the *streamlines* of the flow. Thus, the vectors $\mathbf{F}(x, y)$ in the velocity field of a fluid flow are tangent to the streamlines. If the streamlines can be represented as the level curves of some function $\psi(x, y)$, then the function ψ is called a *stream function* for the flow. Since $\nabla \psi$ is normal to the level curves $\psi(x, y) = c$, it follows that $\nabla \psi$ is normal to the streamlines; and this in turn implies that

$$\nabla \psi \cdot \mathbf{F} = 0 \tag{2}$$

Your plan is to use this equation to find the stream function and then the streamlines of the basic hurricane model.

Since the vortex and sink flows that produce the basic hurricane model have a central symmetry, intuition suggests that polar coordinates may lead to simpler equations for the streamlines than rectangular coordinates. Thus, you decide to express the velocity vector $\mathbf{F}$ at a point (r, θ) in terms of the orthogonal unit vectors

$$\mathbf{u}_r = \cos \theta \mathbf{i} + \sin \theta \mathbf{j} \quad \text{and} \quad \mathbf{u}_\theta = -\sin \theta \mathbf{i} + \cos \theta \mathbf{j}$$

The vector $\mathbf{u}_r$, called the *radial unit vector*, points away from the origin, and the vector $\mathbf{u}_\theta$, called the *transverse unit vector*, is obtained by rotating $\mathbf{u}_r$ counterclockwise $90°$ (Figure 3).

············
Exercise 7   Show that the vector field for the basic hurricane model given in (1) can be expressed in terms of $\mathbf{u}_r$ and $\mathbf{u}_\theta$ as

$$\mathbf{F} = -\frac{1}{2\pi r}(q\mathbf{u}_r - k\mathbf{u}_\theta)$$

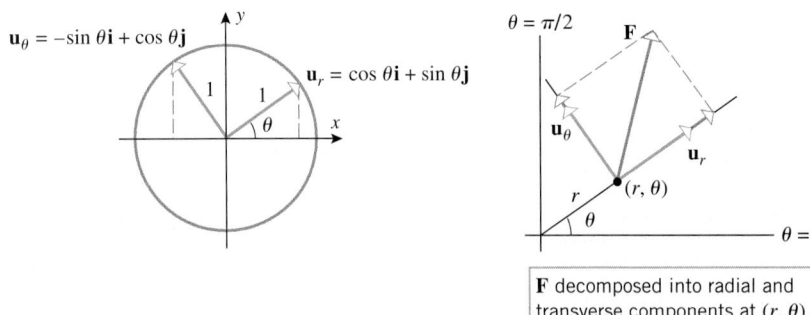

Figure 3

It follows from Exercise 57 of Section 15.6 that the gradient of the stream function can be expressed in terms of $\mathbf{u}_r$ and $\mathbf{u}_\theta$ as

$$\nabla \psi = \frac{\partial \psi}{\partial r}\mathbf{u}_r + \frac{1}{r}\frac{\partial \psi}{\partial \theta}\mathbf{u}_\theta$$

Exercise 8 Confirm that for the basic hurricane model the orthogonality condition in (2) is satisfied if

$$\frac{\partial \psi}{\partial r} = \frac{k}{r} \quad \text{and} \quad \frac{\partial \psi}{\partial \theta} = q$$

Exercise 9 By integrating the equations in Exercise 8, show that

$$\psi = k \ln r + q\theta$$

is a stream function for the basic hurricane model.

Exercise 10 Show that the streamlines for the basic hurricane model are logarithmic spirals of the form

$$r = Ke^{-q\theta/k} \quad (K > 0)$$

Exercise 11 Use a graphing utility to generate some typical streamlines for the basic hurricane model with vortex strength 2π and sink strength 2π.

Streamlines for Hurricane Isaac

Exercise 12 In Exercise 6 you found the strengths k and q of the vortex and sink for hurricane Isaac. Use that information to find a formula for the family of streamlines for Isaac; and then use a graphing utility to graph the streamline that passes through the point that is 20 km from the eye in the direction that is $45°$ NE from the eye.

Module by: Josef S. Torok, Rochester Institute of Technology
Howard Anton, Drexel University

Additional material for this module can be found on the World Wide Web at http://www.wiley.com/college/anton

APPENDIX A

Real Numbers, Intervals, and Inequalities

REAL NUMBERS

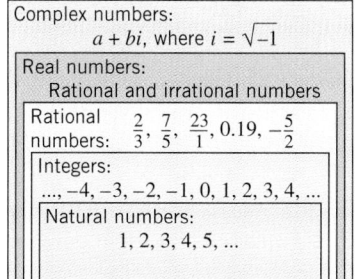

Figure A.1

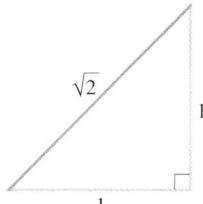

Figure A.2

COMPLEX NUMBERS

Figure A.1 describes the various categories of numbers that we will encounter in this text. The simplest numbers are the *natural numbers*

$$1, \quad 2, \quad 3, \quad 4, \quad 5, \dots$$

These are a subset of the *integers*

$$\dots, \quad -4, \quad -3, \quad -2, \quad -1, \quad 0, \quad 1, \quad 2, \quad 3, \quad 4, \dots$$

and these in turn are a subset of the *rational numbers*, which are the numbers formed by taking ratios of integers (avoiding division by 0). Some examples are

$$\tfrac{2}{3}, \quad \tfrac{7}{5}, \quad 23 = \tfrac{23}{1}, \quad 0.19 = \tfrac{19}{100}, \quad -\tfrac{5}{2} = \tfrac{-5}{2} = \tfrac{5}{-2}$$

The early Greeks believed that every measurable quantity had to be a rational number. However, this idea was overturned in the fifth century B.C. by Hippasus of Metapontum[*] who demonstrated the existence of *irrational numbers*, that is, numbers that cannot be expressed as the ratio of two integers. Using geometric methods, he showed that the length of the hypotenuse of the triangle in Figure A.2 could not be expressed as a ratio of integers, thereby proving that $\sqrt{2}$ is an irrational number. Some other examples of irrational numbers are

$$\sqrt{3}, \quad \sqrt{5}, \quad 1 + \sqrt{2}, \quad \sqrt[3]{7}, \quad \pi, \quad \cos 19°$$

The rational and irrational numbers together comprise what is called the *real number system*, and both the rational and irrational numbers are called *real numbers*.

Because the square of a real number cannot be negative, the equation

$$x^2 = -1$$

has no solutions in the real number system. In the eighteenth century mathematicians remedied this problem by inventing a new number, which they denoted by

$$i = \sqrt{-1}$$

and which they defined to have the property $i^2 = -1$. This, in turn, led to the development

---

[*] HIPPASUS OF METAPONTUM (circa 500 B.C.). A Greek Pythagorean philosopher. According to legend, Hippasus made his discovery at sea and was thrown overboard by fanatic Pythagoreans because his result contradicted their doctrine. The discovery of Hippasus is one of the most fundamental in the entire history of science.

of the **complex numbers**, which are numbers of the form

$$a + bi$$

where a and b are real numbers. Some examples are

$2 + 3i$	$3 - 4i$	$6i$	$\frac{2}{3}$
$|a = 2, b = 3|$	$|a = 3, b = -4|$	$|a = 0, b = 6|$	$|a = \frac{2}{3}, b = 0|$

Observe that every real number a is also a complex number because it can be written as

$$a = a + 0i$$

Thus, the real numbers are a subset of the complex numbers. Those complex numbers that are not real numbers are called **imaginary numbers**. Although we will be concerned primarily with real numbers in this text, imaginary numbers will arise in the course of solving equations. For example, the solutions of the quadratic equation

$$ax^2 + bx + c = 0$$

which are given by the **quadratic formula**

$$x = \frac{-b \pm \sqrt{b^2 - 4ac}}{2a}$$

are imaginary if the quantity $b^2 - 4ac$ is negative.

DIVISION BY ZERO

Division by zero is not allowed in numerical computations because it leads to mathematical inconsistencies. For example, if $1/0$ were assigned some numerical value, say $1/0 = p$, then it would follow that $0 \cdot p = 1$, which is incorrect.

DECIMAL REPRESENTATION OF REAL NUMBERS

Rational and irrational numbers can be distinguished by their decimal representations. Rational numbers have decimals that are **repeating**, by which we mean that at some point in the decimal some fixed block of numbers begins to repeat indefinitely. For example,

$$\frac{4}{3} = 1.333\ldots, \quad \frac{3}{11} = .272727\ldots, \quad \frac{1}{2} = .50000\ldots, \quad \frac{5}{7} = .714285714285714285\ldots$$

$$\underset{\text{3 repeats}}{} \qquad \underset{\text{27 repeats}}{} \qquad \underset{\text{0 repeats}}{} \qquad \underset{\text{714285 repeats}}{}$$

Decimals in which zero repeats from some point on are called **terminating decimals**. For brevity, it is usual to omit the repetitive zeros in terminating decimals and for other repeating decimals to write the repeating digits only once but with a bar over them to indicate the repetition. For example,

$$\frac{1}{2} = .5, \quad \frac{12}{4} = 3, \quad \frac{8}{25} = .32, \quad \frac{4}{3} = 1.\overline{3}, \quad \frac{3}{11} = .\overline{27}, \quad \frac{5}{7} = .\overline{714285}$$

Irrational numbers have nonrepeating decimals, so we can be certain that the decimals

$$\sqrt{2} = 1.414213562373095\ldots \quad \text{and} \quad \pi = 3.141592653589793\ldots$$

do not repeat from some point on. Moreover, if we stop the decimal expansion of an irrational number at some point, we get only an approximation to the number, never an exact value. For example, even if we compute π to 1000 decimal places, as in Figure A.3, we still have only an approximation.

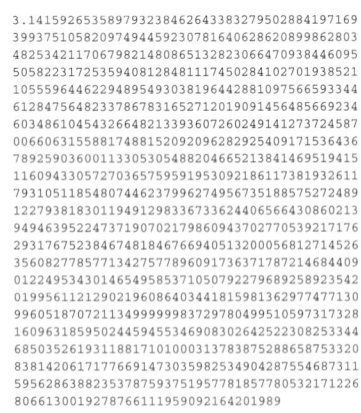

```
3.1415926535897932384626433832795028841971 69
399375105820974944592307816406286208998 62803
482534211706798214808651328230664709384446095
505822317253594081284811174502841027019 38521
105559644622948954930381964428810975665 93344
612847564823378678316527120190914564856 69234
603486104543266482133936072602491412737 24587
006606315588174881520920962829254091715 36436
789259036001133053054882046652138414695 19415
116094330572703657595919530921861173819 32611
793105118548074462379962749567351885752 72489
122793818301194912983367336244065664308 60213
949463952247371907021798609437027705392 17176
293176752384674818467669405132000568127 14526
356082778577134275778960917363717872146 84409
012249534301465495853710507922796892589 23542
019956112129021960864034418159813629774 77130
996051870721134999999837297804995105597 317328
160963185950244594553469083026425225308 253344
685035261931188171010003137838752886587 53320
838142061717669147303598253490428755468 7311
595628638823537875937519577818577805321 71226
806613001927876611195909216420 1989
```

Figure A.3

REMARK.   Beginning mathematics students are sometimes taught to approximate π by $\frac{22}{7}$. Keep in mind, however, that this is only an approximation, since

$$\frac{22}{7} = 3.\overline{142857}$$

is a rational number whose decimal representation begins to differ from π in the third decimal place.

COORDINATE LINES

Figure A.4

In 1637 René Descartes[*] published a philosophical work called *Discourse on the Method of Rightly Conducting the Reason*. In the back of that book was an appendix that the British philosopher John Stuart Mill described as "the greatest single step ever made in the progress of the exact sciences." In that appendix René Descartes linked together algebra and geometry, thereby creating a new subject called **analytic geometry**; it gave a way of describing algebraic formulas by geometric curves and, conversely, geometric curves by algebraic formulas.

The key step in analytic geometry is to establish a correspondence between real numbers and points on a line. To do this, choose any point on the line as a reference point, and call it the **origin**; and then arbitrarily choose one of the two directions along the line to be the **positive direction**, and let the other be the **negative direction**. It is usual to mark the positive direction with an arrowhead, as in Figure A.4, and to take the positive direction to the right when the line is horizontal. Next, choose a convenient unit of measure, and represent each positive number r by the point that is r units from the origin in the positive direction, each negative number $-r$ by the point that is r units from the origin in the negative direction from the origin, and 0 by the origin itself (Figure A.5). The number associated with a point P is called the **coordinate** of P, and the line is called a **coordinate line**, a **real number line**, or a **real line**.

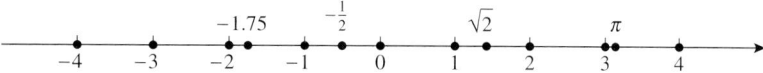

Figure A.5

INEQUALITY NOTATION

The real numbers can be ordered by size as follows: If $b - a$ is positive, then we write either $a < b$ (read "a is less than b") or $b > a$ (read "b is greater than a"). We write $a \leq b$ to mean $a < b$ or $a = b$, and we write $a < b < c$ to mean that $a < b$ and $b < c$. As one traverses a coordinate line in the positive direction, the real numbers increase in size, so on a horizontal coordinate line the inequality $a < b$ implies that a is to left of b, and the inequalities $a < b < c$ imply that a is to the left of c, and b lies between a and c. The meaning of such symbols as

$$a \leq b < c, \quad a \leq b \leq c, \quad \text{and} \quad a < b < c < d$$

should be clear. For example, you should be able to confirm that all of the following are true statements:

$$3 < 8, \quad -7 < 1.5, \quad -12 \leq -\pi, \quad 5 \leq 5, \quad 0 \leq 2 \leq 4,$$
$$8 \geq 3, \quad 1.5 > -7, \quad -\pi > -12, \quad 5 \geq 5, \quad 3 > 0 > -1 > -3$$

REVIEW OF SETS

In the following discussion we will be concerned with certain sets of real numbers, so it will be helpful to review the basic ideas about sets. Recall that a **set** is a collection of objects, called **elements** or **members** of the set. In this text we will be concerned primarily with sets whose members are numbers or points that lie on a line, a plane, or in three-dimensional

[*] RENÉ DESCARTES (1596–1650). Descartes, a French aristocrat, was the son of a government official. He graduated from the University of Poitiers with a law degree at age 20. After a brief probe into the pleasures of Paris he became a military engineer, first for the Dutch Prince of Nassau and then for the German Duke of Bavaria. It was during his service as a soldier that Descartes began to pursue mathematics seriously and develop his analytic geometry. After the wars, he returned to Paris where he stalked the city as an eccentric, wearing a sword in his belt and a plumed hat. He lived in leisure, seldom arose before 11 A.M., and dabbled in the study of human physiology, philosophy, glaciers, meteors, and rainbows. He eventually moved to Holland, where he published his *Discourse on the Method*, and finally to Sweden where he died while serving as tutor to Queen Christina. Descartes is regarded as a genius of the first magnitude. In addition to major contributions in mathematics and philosophy, he is considered, along with William Harvey, to be a founder of modern physiology.

space. We will denote sets by capital letters and elements by lowercase letters. To indicate that a is a member of the set A we will write $a \in A$ (read "a belongs to A"), and to indicate that a is not a member of the set A we will write $a \notin A$ (read "a does not belong to A"). For example, if A is the set of positive integers, then $5 \in A$, but $-5 \notin A$. Sometimes sets arise that have no members (e.g., the set of odd integers that are divisible by 2). A set with no members is called an ***empty set*** or a ***null set*** and is denoted by the symbol $\varnothing$.

Some sets can be described by listing their members between braces. The order in which the members are listed does not matter, so, for example, the set A of positive integers that are less than 6 can be expressed as

$$A = \{1, 2, 3, 4, 5\} \quad \text{or} \quad A = \{2, 3, 1, 5, 4\}$$

We can also write A in *set-builder notation* as

$$A = \{x : x \text{ is an integer and } 0 < x < 6\}$$

which is read "A is the set of all x such that x is an integer and $0 < x < 6$." In general, to express a set S in set-builder notation we write $S = \{x : \underline{\hspace{1cm}}\}$ in which the line is replaced by a property that uniquely defines the set S.

INTERVALS

In calculus we will be concerned with sets of real numbers, called ***intervals***, that correspond to line segments on a coordinate line. For example, if $a < b$, then the ***open interval*** from a to b, denoted by (a, b), is the line segment extending from a to b, *excluding* the endpoints; and the ***closed interval*** from a to b, denoted by $[a, b]$, is the line segment extending from a to b, *including* the endpoints (Figure A.6). These sets can be expressed in set-builder notation as

$$(a, b) = \{x : a < x < b\} \qquad \boxed{\text{The open interval from } a \text{ to } b}$$

$$[a, b] = \{x : a \le x \le b\} \qquad \boxed{\text{The closed interval from } a \text{ to } b}$$

The open interval (a, b)

The closed interval $[a, b]$

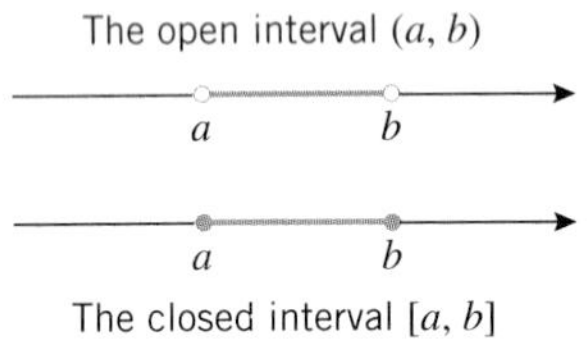

Figure A.6

REMARK.   Observe that in this notation and in the corresponding Figure A.6, parentheses and open dots mark endpoints that are excluded from the interval, whereas brackets and closed dots mark endpoints that are included in the interval. Observe also, that in set-builder notation for the intervals, it is understood that x is a real number, even though it is not stated explicitly.

As shown in Table 1, an interval can include one endpoint and not the other; such intervals are called ***half-open*** (or sometimes ***half-closed***). Moreover, the table also shows that it is possible for an interval to extend indefinitely in one or both directions. To indicate that an interval extends indefinitely in the positive direction we write $+\infty$ (read "positive infinity") in place of a right endpoint, and to indicate that an interval extends indefinitely in the negative direction we write $-\infty$ (read "negative infinity") in place of a left endpoint. Intervals that extend between two real numbers are called ***finite intervals***, whereas intervals that extend indefinitely in one or both directions are called ***infinite intervals***.

REMARK.   By convention, infinite intervals of the form $[a, +\infty)$ or $(-\infty, b]$ are considered to be closed because they contain their endpoint, and intervals of the form $(a, +\infty)$ and $(-\infty, b)$ are considered to be open because they do not include their endpoint. The interval $(-\infty, +\infty)$, which is the set of all real numbers, has no endpoints and can be regarded as either open or closed, as convenient. This set is often denoted by the special symbol $\mathbb{R}$. To distinguish verbally between the open interval $(0, +\infty) = \{x : x > 0\}$ and the closed interval $[0, +\infty) = \{x : x \ge 0\}$, we will call x ***positive*** if $x > 0$ and ***nonnegative*** if $x \ge 0$. Thus, a positive number must be nonnegative, but a nonnegative number need not be positive, since it might possibly be 0.

Table 1

INTERVAL NOTATION	SET NOTATION	GEOMETRIC PICTURE	CLASSIFICATION
(a, b)	$\{x : a < x < b\}$		Finite; open
$[a, b]$	$\{x : a \le x \le b\}$		Finite; closed
$[a, b)$	$\{x : a \le x < b\}$		Finite; half-open
$(a, b]$	$\{x : a < x \le b\}$		Finite; half-open
$(-\infty, b]$	$\{x : x \le b\}$		Infinite; closed
$(-\infty, b)$	$\{x : x < b\}$		Infinite; open
$[a, +\infty)$	$\{x : x \ge a\}$		Infinite; closed
$(a, +\infty)$	$\{x : x > a\}$		Infinite; open
$(-\infty, +\infty)$	$\mathbb{R}$		Infinite; open and closed

UNIONS AND INTERSECTIONS OF INTERVALS

If A and B are sets, then the **union** of A and B (denoted by $A \cup B$) is the set whose members belong to A or B (or both), and the **intersection** of A and B (denoted by $A \cap B$) is the set whose members belong to both A and B. For example,

$$\{x : 0 < x < 5\} \cup \{x : 1 < x < 7\} = \{x : 0 < x < 7\}$$

$$\{x : x < 1\} \cap \{x : x \ge 0\} = \{x : 0 \le x < 1\}$$

$$\{x : x < 0\} \cap \{x : x > 0\} = \varnothing$$

or in interval notation,

$$(0, 5) \cup (1, 7) = (0, 7)$$

$$(-\infty, 1) \cap [0, +\infty) = [0, 1)$$

$$(-\infty, 0) \cap (0, +\infty) = \varnothing$$

ALGEBRAIC PROPERTIES OF INEQUALITIES

The following algebraic properties of inequalities will be used frequently in this text. We omit the proofs.

A.1   THEOREM (*Properties of Inequalities*).   *Let a, b, c, and d be real numbers.*
(a)   *If $a < b$ and $b < c$, then $a < c$.*
(b)   *If $a < b$, then $a + c < b + c$ and $a - c < b - c$.*
(c)   *If $a < b$, then $ac < bc$ when c is positive and $ac > bc$ when c is negative.*
(d)   *If $a < b$ and $c < d$, then $a + c < b + d$.*
(e)   *If a and b are both positive or both negative and $a < b$, then $1/a > 1/b$.*

If we call the direction of an inequality its *sense*, then these properties can be paraphrased as follows:

(b)   *The sense of an inequality is unchanged if the same number is added to or subtracted from both sides.*

(c)   *The sense of an inequality is unchanged if both sides are multiplied by the same positive number, but the sense is reversed if both sides are multiplied by the same negative number.*

(d)   *Inequalities with the same sense can be added.*

(e)   *If both sides of an inequality have the same sign, then the sense of the inequality is reversed by taking the reciprocal of each side.*

REMARK.   These properties remain true if the symbols $<$ and $>$ are replaced by $\leq$ and $\geq$ in Theorem A.1.

Example 1

STARTING INEQUALITY	OPERATION	RESULTING INEQUALITY
$-2 < 6$	Add 7 to both sides.	$5 < 13$
$-2 < 6$	Subtract 8 from both sides.	$-10 < -2$
$-2 < 6$	Multiply both sides by 3.	$-6 < 18$
$-2 < 6$	Multiply both sides by -3.	$6 > -18$
$3 < 7$	Multiply both sides by 4.	$12 < 28$
$3 < 7$	Multiply both sides by -4.	$-12 > -28$
$3 < 7$	Take reciprocals of both sides.	$\frac{1}{3} > \frac{1}{7}$
$-8 < -6$	Take reciprocals of both sides.	$-\frac{1}{8} > -\frac{1}{6}$
$4 < 5, -7 < 8$	Add corresponding sides.	$-3 < 13$

◀

SOLVING INEQUALITIES

A *solution* of an inequality in an unknown x is a value for x that makes the inequality a true statement. For example, $x = 1$ is a solution of the inequality $x < 5$, but $x = 7$ is not. The set of all solutions of an inequality is called its *solution set*. It can be shown that if one does not multiply both sides of an inequality by zero or an expression involving an unknown, then the operations in Theorem A.1 will not change the solution set of the inequality. The process of finding the solution set of an inequality is called *solving* the inequality.

Example 2

Solve $3 + 7x \leq 2x - 9$.

Solution.   We will use the operations of Theorem A.1 to isolate x on one side of the inequality.

$$3 + 7x \leq 2x - 9 \quad \boxed{\text{Given.}}$$

$$7x \leq 2x - 12 \quad \boxed{\text{We subtracted 3 from both sides.}}$$

$$5x \leq -12 \quad \boxed{\text{We subtracted } 2x \text{ from both sides.}}$$

$$x \leq -\tfrac{12}{5} \quad \boxed{\text{We multiplied both sides by } \tfrac{1}{5}.}$$

Because we have not multiplied by any expressions involving the unknown x, the last inequality has the same solution set as the first. Thus, the solution set is the interval $\left(-\infty, -\frac{12}{5}\right]$ shown in Figure A.7.   ◀

Figure A.7

Example 3

Solve $7 \leq 2 - 5x < 9$.

Solution.   The given inequality is actually a combination of the two inequalities

$$7 \leq 2 - 5x \quad \text{and} \quad 2 - 5x < 9$$

We could solve the two inequalities separately, then determine the values of x that satisfy both by taking the intersection of the two solution sets. However, it is possible to work with the combined inequalities in this problem:

$$7 \le 2 - 5x < 9 \quad \boxed{\text{Given.}}$$

$$5 \le -5x < 7 \quad \boxed{\text{We subtracted 2 from each member.}}$$

$$-1 \ge x > -\frac{7}{5} \quad \boxed{\begin{array}{l}\text{We multiplied by } -\frac{1}{5} \text{ and reversed} \\ \text{the sense of the inequalities.}\end{array}}$$

$$-\frac{7}{5} < x \le -1 \quad \boxed{\begin{array}{l}\text{For clarity, we rewrote the inequalities} \\ \text{with the smaller number on the left.}\end{array}}$$

Thus, the solution set is the interval $\left(-\frac{7}{5}, -1\right]$ shown in Figure A.8.   ◀

Figure A.8

Example 4

Solve $x^2 - 3x > 10$.

Solution. By subtracting 10 from both sides, the inequality can be rewritten as

$$x^2 - 3x - 10 > 0$$

Factoring the left side yields

$$(x + 2)(x - 5) > 0$$

The values of x for which $x + 2 = 0$ or $x - 5 = 0$ are $x = -2$ and $x = 5$. These points divide the coordinate line into three open intervals,

$$(-\infty, -2), \quad (-2, 5), \quad (5, +\infty)$$

on each of which the product $(x + 2)(x - 5)$ has constant sign. To determine those signs we will choose an *arbitrary* point in each interval at which we will determine the sign; these are called ***test points***. As shown in Figure A.9, we will use -3, 0, and 6 as our test points. The results can be organized as follows:

INTERVAL	TEST POINT	SIGN OF $(x + 2)(x - 5)$ AT THE TEST POINT
$(-\infty, -2)$	-3	$(-)(-) = +$
$(-2, 5)$	0	$(+)(-) = -$
$(5, +\infty)$	6	$(+)(+) = +$

The pattern of signs in the intervals is shown on the number line in the middle of Figure A.9. We deduce that the solution set is $(-\infty, -2) \cup (5, +\infty)$, which is shown at the bottom of Figure A.9.   ◀

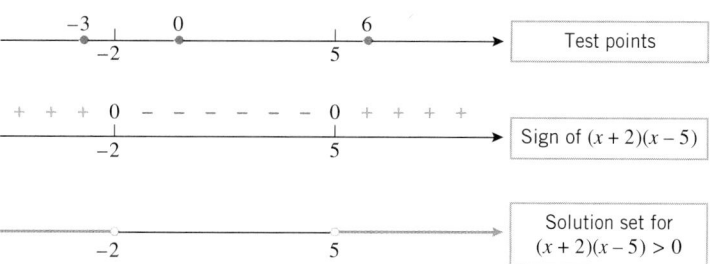

Figure A.9

Example 5

Solve $\dfrac{2x - 5}{x - 2} < 1$.

Solution. We could start by multiplying both sides by $x - 2$ to eliminate the fraction. However, this would require us to consider the cases $x - 2 > 0$ and $x - 2 < 0$ separately because the sense of the inequality would be reversed in the second case, but not the first. The following approach is simpler:

$$\frac{2x-5}{x-2} < 1 \qquad \boxed{\text{Given.}}$$

$$\frac{2x-5}{x-2} - 1 < 0 \qquad \boxed{\begin{array}{l}\text{We subtracted 1 from both sides}\\\text{to obtain a 0 on the right.}\end{array}}$$

$$\frac{(2x-5)-(x-2)}{x-2} < 0 \qquad \boxed{\text{We combined terms.}}$$

$$\frac{x-3}{x-2} < 0 \qquad \boxed{\text{We simplified.}}$$

The quantity $x - 3$ is zero if $x = 3$, and the quantity $x - 2$ is zero if $x = 2$. These points divide the coordinate line into three open intervals,

$$(-\infty, 2), \quad (2, 3), \quad (3, +\infty)$$

on each of which the quotient $(x - 3)/(x - 2)$ has constant sign. Using 0, 2.5, and 4 as test points (Figure A.10), we obtain the following results:

INTERVAL	TEST POINT	SIGN OF $(x - 3)(x - 2)$ AT THE TEST POINT
$(-\infty, 2)$	0	$(-)/(-) = +$
$(2, 3)$	2.5	$(-)/(+) = -$
$(3, +\infty)$	4	$(+)/(+) = +$

The signs of the quotient are shown in the middle of Figure A.10. From the figure we see that the solution set consists of all real values of x such that $2 < x < 3$. This is the interval $(2, 3)$ shown at the bottom of Figure A.10. ◀

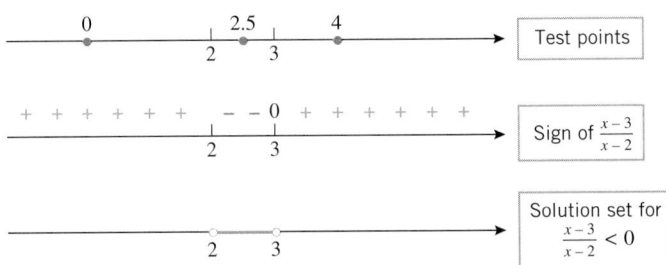

Figure A.10

EXERCISE SET A

1. Among the terms *integer*, *rational*, and *irrational*, which ones apply to the given number?
(a) $-\frac{3}{4}$ (b) 0 (c) $\frac{24}{8}$
(d) 0.25 (e) $-\sqrt{16}$ (f) $2^{1/2}$
(g) $0.020202\ldots$ (h) $7.000\ldots$

2. Which of the terms *integer*, *rational*, and *irrational* apply to the given number?
(a) $0.31311311131111\ldots$ (b) $0.729999\ldots$
(c) $0.376237623762\ldots$ (d) $17\frac{4}{5}$

3. The repeating decimal $0.137137137\ldots$ can be expressed as a ratio of integers by writing

$$x = 0.137137137\ldots$$

$$1000x = 137.137137137\ldots$$

and subtracting to obtain $999x = 137$ or $x = \frac{137}{999}$. Use this idea, where needed, to express the following decimals as ratios of integers.
(a) $0.123123123\ldots$ (b) $12.7777\ldots$
(c) $38.07818181\ldots$ (d) $0.4296000\ldots$

4. Show that the repeating decimal 0.99999... represents the number 1. Since 1.000... is also a decimal representation of 1, this problem shows that a real number can have two different decimal representations. [*Hint:* Use the technique of Exercise 3.]

5. The Rhind Papyrus, which is a fragment of Egyptian mathematical writing from about 1650 B.C., is one of the oldest known examples of written mathematics. It is stated in the papyrus that the area A of a circle is related to its diameter D by

$$A = \left(\tfrac{8}{9}D\right)^2$$

(a) What approximation to π were the Egyptians using?

(b) Use a calculating utility to determine if this approximation is better or worse than the approximation of $\frac{22}{7}$.

6. The following are all famous approximations to π:

$$\frac{333}{106} \qquad \text{Adrian Athoniszoon, c. 1583}$$

$$\frac{355}{113} \qquad \text{Tsu Chung-Chi and others}$$

$$\frac{63}{25}\left(\frac{17 + 15\sqrt{5}}{7 + 15\sqrt{5}}\right) \qquad \text{Ramanujan}$$

$$\frac{22}{7} \qquad \text{Archimedes}$$

$$\frac{223}{71} \qquad \text{Archimedes}$$

(a) Use a calculating utility to order these approximations according to size.

(b) Which of these approximations is closest to but larger than π?

(c) Which of these approximations is closest to but smaller than π?

(d) Which of these approximations is most accurate?

7. In each line of the table in the accompanying figure, check the blocks, if any, that describe a valid relationship between the real numbers a and b. The first line is already completed as an illustration.

a	b	$a < b$	$a \le b$	$a > b$	$a \ge b$	$a = b$
1	6	✓	✓			
6	1					
-3	5					
5	-3					
-4	-4					
0.25	$\frac{1}{3}$					
$-\frac{1}{4}$	$-\frac{3}{4}$					

Figure Ex-7

8. In each line of the table in the accompanying figure, check the blocks, if any, that describe a valid relationship between the real numbers a, b, and c.

a	b	c	$a < b < c$	$a \le b \le c$	$a < b \le c$	$a \le b < c$
-1	0	2				
2	4	-3				
$\frac{1}{2}$	$\frac{1}{2}$	$\frac{3}{4}$				
-5	-5	-5				
0.75	1.25	1.25				

Figure Ex-8

9. Which of the following are always correct if $a \le b$?

(a) $a - 3 \le b - 3$ (b) $-a \le -b$

(c) $3 - a \le 3 - b$ (d) $6a \le 6b$

(e) $a^2 \le ab$ (f) $a^3 \le a^2 b$

10. Which of the following are always correct if $a \le b$ and $c \le d$?

(a) $a + 2c \le b + 2d$ (b) $a - 2c \le b - 2d$

(c) $a - 2c \ge b - 2d$

11. For what values of a are the following inequalities valid?

(a) $a \le a$ (b) $a < a$

12. If $a \le b$ and $b \le a$, what can you say about a and b?

13. (a) If $a < b$ is true, does it follow that $a \le b$ must also be true?

(b) If $a \le b$ is true, does it follow that $a < b$ must also be true?

14. In each part, list the elements in the set.

(a) $\{x : x^2 - 5x = 0\}$

(b) $\{x : x$ is an integer satisfying $-2 < x < 3\}$

15. In each part, express the set in the notation $\{x : \underline{\hspace{1cm}}\}$.

(a) $\{1, 3, 5, 7, 9, \ldots\}$

(b) the set of even integers

(c) the set of irrational numbers

(d) $\{7, 8, 9, 10\}$

16. Let $A = \{1, 2, 3\}$. Which of the following sets are equal to A?

(a) $\{0, 1, 2, 3\}$ (b) $\{3, 2, 1\}$

(c) $\{x : (x - 3)(x^2 - 3x + 2) = 0\}$

17. In the accompanying figure, let

$$S = \text{the set of points inside the square}$$
$$T = \text{the set of points inside the triangle}$$
$$C = \text{the set of points inside the circle}$$

and let a, b, and c be the points shown. Answer the following as true or false.

(a) $T \subset C$

(b) $T \subset S$

(c) $a \notin T$

(d) $a \notin S$

(e) $b \in T$ and $b \in C$

(f) $a \in C$ or $a \in T$

(g) $c \in T$ and $c \notin C$

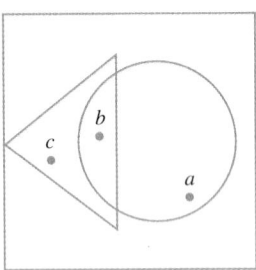

Figure Ex-17

18. List all subsets of
   (a) $\{a_1, a_2, a_3\}$ (b) $\varnothing$.

19. In each part, sketch on a coordinate line all values of x that satisfy the stated condition.
   (a) $x \leq 4$ (b) $x \geq -3$ (c) $-1 \leq x \leq 7$
   (d) $x^2 = 9$ (e) $x^2 \leq 9$ (f) $x^2 \geq 9$

20. In parts (a)–(d), sketch on a coordinate line all values of x, if any, that satisfy the stated conditions.
   (a) $x > 4$ and $x \leq 8$
   (b) $x \leq 2$ or $x \geq 5$
   (c) $x > -2$ and $x \geq 3$
   (d) $x \leq 5$ and $x > 7$

21. Express in interval notation.
   (a) $\{x : x^2 \leq 4\}$ (b) $\{x : x^2 > 4\}$

22. In each part, sketch the set on a coordinate line.
   (a) $[-3, 2] \cup [1, 4]$ (b) $[4, 6] \cup [8, 11]$
   (c) $(-4, 0) \cup (-5, 1)$ (d) $[2, 4) \cup (4, 7)$
   (e) $(-2, 4) \cap (0, 5]$ (f) $[1, 2.3) \cup (1.4, \sqrt{2})$
   (g) $(-\infty, -1) \cup (-3, +\infty)$ (h) $(-\infty, 5) \cap [0, +\infty)$

In Exercises 23–44, solve the inequality and sketch the solution on a coordinate line.

23. $3x - 2 < 8$

24. $\frac{1}{5}x + 6 \geq 14$

25. $4 + 5x \leq 3x - 7$

26. $2x - 1 > 11x + 9$

27. $3 \leq 4 - 2x < 7$

28. $-2 \geq 3 - 8x \geq -11$

29. $\dfrac{x}{x - 3} < 4$

30. $\dfrac{x}{8 - x} \geq -2$

31. $\dfrac{3x + 1}{x - 2} < 1$

32. $\dfrac{\frac{1}{2}x - 3}{4 + x} > 1$

33. $\dfrac{4}{2 - x} \leq 1$

34. $\dfrac{3}{x - 5} \leq 2$

35. $x^2 > 9$

36. $x^2 \leq 5$

37. $(x - 4)(x + 2) > 0$

38. $(x - 3)(x + 4) < 0$

39. $x^2 - 9x + 20 \leq 0$

40. $2 - 3x + x^2 \geq 0$

41. $\dfrac{2}{x} < \dfrac{3}{x - 4}$

42. $\dfrac{1}{x + 1} \geq \dfrac{3}{x - 2}$

43. $x^3 - x^2 - x - 2 > 0$

44. $x^3 - 3x + 2 \leq 0$

In Exercises 45 and 46, find all values of x for which the given expression yields a real number.

45. $\sqrt{x^2 + x - 6}$

46. $\sqrt{\dfrac{x + 2}{x - 1}}$

47. Fahrenheit and Celsius temperatures are related by the formula $C = \frac{5}{9}(F - 32)$. If the temperature in degrees Celsius ranges over the interval $25 \leq C \leq 40$ on a certain day, what is the temperature range in degrees Fahrenheit that day?

48. Every integer is either even or odd. The even integers are those that are divisible by 2, so n is even if and only if $n = 2k$ for some integer k. Each odd integer is one unit larger than an even integer, so n is odd if and only if $n = 2k + 1$ for some integer k. Show:
   (a) If n is even, then so is n^2
   (b) If n is odd, then so is n^2.

49. Prove the following results about sums of rational and irrational numbers:
   (a) rational + rational = rational
   (b) rational + irrational = irrational.

50. Prove the following results about products of rational and irrational numbers:
   (a) rational $\cdot$ rational = rational
   (b) rational $\cdot$ irrational = irrational   (provided the rational factor is nonzero).

51. Show that the sum or product of two irrational numbers can be rational or irrational.

52. Classify the following as rational or irrational and justify your conclusion.
   (a) $3 + \pi$ (b) $\frac{3}{4}\sqrt{2}$
   (c) $\sqrt{8}\sqrt{2}$ (d) $\sqrt{\pi}$
   (See Exercises 49 and 50.)

53. Prove: The average of two rational numbers is a rational number, but the average of two irrational numbers can be rational or irrational.

54. Can a rational number satisfy $10^x = 3$?

55. Solve: $8x^3 - 4x^2 - 2x + 1 < 0$.

56. Solve: $12x^3 - 20x^2 \geq -11x + 2$.

57. Prove: If a, b, c, and d are positive numbers such that $a < b$ and $c < d$, then $ac < bd$. (This result gives conditions under which inequalities can be "multiplied together.")

58. Is the number represented by the decimal

   $0.101001000100001000001\ldots$

   rational or irrational? Explain your reasoning.

APPENDIX B

Absolute Value

B.1 DEFINITION. The *absolute value* or *magnitude* of a real number a is denoted by $|a|$ and is defined by

$$|a| = \begin{cases} a & \text{if } a \geq 0 \\ -a & \text{if } a < 0 \end{cases}$$

Example 1

$$|5| = 5 \qquad \left|-\tfrac{4}{7}\right| = -\left(-\tfrac{4}{7}\right) = \tfrac{4}{7} \qquad |0| = 0 \qquad \blacktriangleleft$$

Since $5 > 0$ Since $-\tfrac{4}{7} < 0$ Since $0 \geq 0$

Note that the effect of taking the absolute value of a number is to strip away the minus sign if the number is negative and to leave the number unchanged if it is nonnegative.

Example 2

Solve $|x - 3| = 4$.

Solution. Depending on whether $x - 3$ is positive or negative, the equation $|x - 3| = 4$ can be written as

$$x - 3 = 4 \quad \text{or} \quad x - 3 = -4$$

Solving these two equations gives $x = 7$ and $x = -1$. $\blacktriangleleft$

Example 3

Solve $|3x - 2| = |5x + 4|$.

Solution. Because two numbers with the same absolute value are either equal or differ in sign, the given equation will be satisfied if either

$$3x - 2 = 5x + 4 \quad \text{or} \quad 3x - 2 = -(5x + 4)$$

Solving the first equation yields $x = -3$ and solving the second yields $x = -\tfrac{1}{4}$; thus, the given equation has the solutions $x = -3$ and $x = -\tfrac{1}{4}$. $\blacktriangleleft$

RELATIONSHIP BETWEEN SQUARE ROOTS AND ABSOLUTE VALUES

Recall from algebra that a number is called a *square root* of a if its square is a. Recall also that every positive real number has two square roots, one positive and one negative; the positive square root is denoted by $\sqrt{a}$ and the negative square root by $-\sqrt{a}$. For example, the positive square root of 9 is $\sqrt{9} = 3$, and the negative square root of 9 is $-\sqrt{9} = -3$.

REMARK. Readers who may have been taught to write $\sqrt{9} = \pm 3$ should stop doing so, since it is incorrect.

It is a common error to write $\sqrt{a^2} = a$. Although this equality is correct when a is nonnegative, it is false for negative a. For example, if $a = -4$, then

$$\sqrt{a^2} = \sqrt{(-4)^2} = \sqrt{16} = 4 \neq a$$

A result that is correct for all a is given in the following theorem.

B.2 THEOREM. *For any real number a,*

$$\sqrt{a^2} = |a|$$

Proof. Since $a^2 = (+a)^2 = (-a)^2$, the numbers $+a$ and $-a$ are square roots of a^2. If $a \geq 0$, then $+a$ is the nonnegative square root of a^2, and if $a < 0$, then $-a$ is the nonnegative square root of a^2. Since $\sqrt{a^2}$ denotes the nonnegative square root of a^2, it follows that

$$\sqrt{a^2} = +a \quad \text{if} \quad a \geq 0$$
$$\sqrt{a^2} = -a \quad \text{if} \quad a < 0$$

That is, $\sqrt{a^2} = |a|$. ∎

PROPERTIES OF ABSOLUTE VALUE

B.3 THEOREM. *If a and b are real numbers, then*

(a) $|-a| = |a|$ A number and its negative have the same absolute value.

(b) $|ab| = |a||b|$ The absolute value of a product is the product of the absolute values.

(c) $|a/b| = |a|/|b|$ The absolute value of a ratio is the ratio of the absolute values.

We will prove parts (a) and (b) only.

Proof (a). From Theorem B.2,

$$|-a| = \sqrt{(-a)^2} = \sqrt{a^2} = |a|$$

Proof (b). From Theorem B.2 and a basic property of square roots,

$$|ab| = \sqrt{(ab)^2} = \sqrt{a^2 b^2} = \sqrt{a^2}\sqrt{b^2} = |a||b|$$ ∎

REMARK. In part (c) of Theorem B.3 we did not explicitly state that $b \neq 0$, but this must be so since division by zero is not allowed. Whenever divisions occur in this text, it will be assumed that the denominator is not zero, even if we do not mention it explicitly.

The result in part (b) of Theorem B.3 can be extended to three or more factors. More precisely, for any n real numbers, $a_1, a_2, \ldots, a_n$, it follows that

$$|a_1 a_2 \cdots a_n| = |a_1||a_2| \cdots |a_n| \tag{1}$$

In the special case where $a_1, a_2, \ldots, a_n$ have the same value, a, it follows from (1) that

$$|a^n| = |a|^n \tag{2}$$

GEOMETRIC INTERPRETATION OF ABSOLUTE VALUE

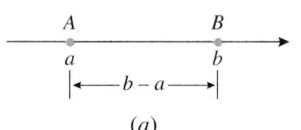

(a)

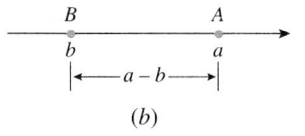

(b)

Figure B.1

The notion of absolute value arises naturally in distance problems. For example, suppose that A and B are points on a coordinate line that have coordinates a and b, respectively. Depending on the relative positions of the points, the distance d between them will be $b - a$ or $a - b$ (Figure B.1). In either case, the distance can be written as $d = |b - a|$, so we have the following result.

> **B.4**   THEOREM (**Distance Formula**).   *If A and B are points on a coordinate line with coordinates a and b, respectively, then the distance d between A and B is $d = |b - a|$.*

This theorem provides useful geometric interpretations of some common mathematical expressions:

EXPRESSION	GEOMETRIC INTERPRETATION ON A COORDINATE LINE
$|x - a|$	The distance between x and a
$|x + a|$	The distance between x and $-a$ (since $|x + a| = |x - (-a)|$)
$|x|$	The distance between x and the origin (since $|x| = |x - 0|$)

INEQUALITIES WITH ABSOLUTE VALUES

Inequalities of the form $|x - a| < k$ and $|x - a| > k$ arise so often that we have summarized the key facts about them in Table 1.

Table 1

INEQUALITY ($k > 0$)	GEOMETRIC INTERPRETATION	FIGURE	ALTERNATIVE FORMS OF THE INEQUALITY
$|x - a| < k$	x is within k units of a.	⊢k units→⊢k units→⊣ $a-k$ a x $a+k$	$-k < x - a < k$ $a - k < x < a + k$
$|x - a| > k$	x is more than k units away from a.	⊢k units→⊢k units→⊣ $a-k$ a $a+k$ x	$x - a < -k$ or $x - a > k$ $x < a - k$ or $x > a + k$

REMARK.   The statements in this table remain true if $<$ is replaced by $\leq$ and $>$ by $\geq$, and if the open dots are replaced by closed dots in the illustrations.

Example 4

Solve

(a) $|x - 3| < 4$ (b) $|x + 4| \geq 2$ (c) $\dfrac{1}{|2x - 3|} > 5$

Solution (a). The inequality $|x - 3| < 4$ can be rewritten as

$$-4 < x - 3 < 4$$

Adding 3 throughout yields

$$-1 < x < 7$$

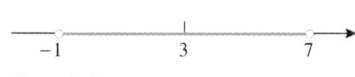

Figure B.2

which can be written in interval notation as $(-1, 7)$. Observe that this solution set consists of all x that are within 4 units of 3 on a number line (Figure B.2), which is consistent with Table 1.

Solution (b). The inequality $|x + 4| \geq 2$ will be satisfied if

$$x + 4 \leq -2 \quad \text{or} \quad x + 4 \geq 2$$

Solving for x in the two cases yields

$$x \leq -6 \quad \text{or} \quad x \geq -2$$

which can be expressed in interval notation as

$$(-\infty, -6] \cup [-2, +\infty)$$

Observe that the solution set consists of all x that are at least 2 units away from -4 on a number line (Figure B.3), which is consistent with Table 1 and the remark that follows it.

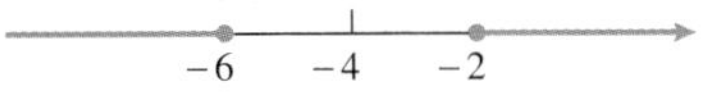

Figure B.3

Solution (c). Observe first that $x = \frac{3}{2}$ results in a division by zero, so this value of x cannot be in the solution set. Putting this aside for the moment, we will begin by taking reciprocals on both sides and reversing the sense of the inequality in accordance with Theorem A.1(d) of Appendix A; then we will use Theorem B.3 to rewrite the inequality $1/|2x - 3| > 5$ in a more familiar form:

$$|2x - 3| < \tfrac{1}{5}$$

$$|2||x - \tfrac{3}{2}| < \tfrac{1}{5} \qquad \boxed{\text{Theorem B.3}(b)}$$

$$|x - \tfrac{3}{2}| < \tfrac{1}{10} \qquad \boxed{\text{We multiplied both sides by } 1/|2| = 1/2.}$$

$$-\tfrac{1}{10} < x - \tfrac{3}{2} < \tfrac{1}{10} \qquad \boxed{\text{Table 1}}$$

$$\tfrac{7}{5} < x < \tfrac{8}{5} \qquad \boxed{\text{We added } 3/2 \text{ throughout.}}$$

As noted earlier, we must eliminate $x = \frac{3}{2}$ to avoid a division by zero, so the solution set is

$$\tfrac{7}{5} < x < \tfrac{3}{2} \quad \text{or} \quad \tfrac{3}{2} < x < \tfrac{8}{5}$$

Figure B.4

which can be expressed in interval notation as $\left(\frac{7}{5}, \frac{3}{2}\right) \cup \left(\frac{3}{2}, \frac{8}{5}\right)$. (See Figure B.4.)   ◀

AN INEQUALITY FROM CALCULUS

One of the most important inequalities in calculus is

$$0 < |x - a| < \delta \tag{3}$$

where δ (Greek "delta") is a positive real number. This is equivalent to the two inequalities

$$0 < |x - a| \quad \text{and} \quad |x - a| < \delta$$

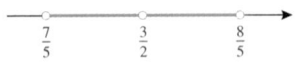

the first of which is satisfied by all x except $x = a$, and the second of which is satisfied by all x that are within δ units of a on a coordinate line. Combining these two restrictions, we conclude that the solution set of (3) consists of all x in the interval $(a - \delta, a + \delta)$ except $x = a$ (Figure B.5). Stated another way, the solution set of (3) is

$$(a - \delta, a) \cup (a, a + \delta) \tag{4}$$

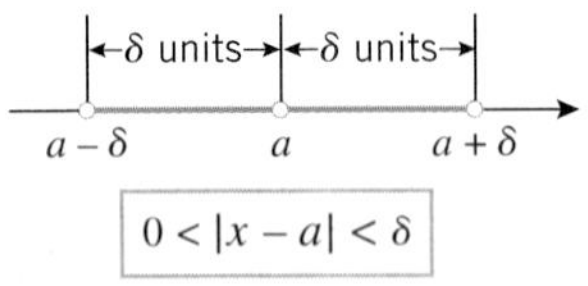

Figure B.5

THE TRIANGLE INEQUALITY

It is *not* generally true that $|a + b| = |a| + |b|$. For example, if $a = 1$ and $b = -1$, then $|a + b| = 0$, whereas $|a| + |b| = 2$. It is true, however, that *the absolute value of a sum is always less than or equal to the sum of the absolute values*. This is the content of the following useful theorem, called the ***triangle inequality***.

B.5   THEOREM (*Triangle Inequality*).   *If a and b are any real numbers, then*

$$|a + b| \leq |a| + |b| \tag{5}$$

Proof.   Observe first that a satisfies the inequality

$$-|a| \leq a \leq |a|$$

because either $a = |a|$ or $a = -|a|$, depending on the sign of a. The corresponding inequality for b is

$$-|b| \le b \le |b|$$

Adding the two inequalities we obtain

$$-(|a| + |b|) \le a + b \le (|a| + |b|) \qquad (6)$$

Let us now consider the cases $a + b \ge 0$ and $a + b < 0$ separately. In the first case, $a + b = |a + b|$, so the right-hand inequality in (6) yields the triangle inequality (5). In the second case, $a + b = -|a + b|$, so the left-hand inequality in (6) can be written as

$$-(|a| + |b|) \le -|a + b|$$

which yields the triangle inequality (5) on multiplying by -1. ■

REMARK.   The name "triangle inequality" arises from a geometric interpretation of the inequality that can be made when a and b are complex numbers. A more detailed explanation is outside the scope of this text.

EXERCISE SET B

1. Compute $|x|$ if
   (a) $x = 7$
   (b) $x = -\sqrt{2}$
   (c) $x = k^2$
   (d) $x = -k^2$.

2. Rewrite $\sqrt{(x-6)^2}$ without using a square root or absolute value sign.

In Exercises 3–10, find all values of x for which the given statement is true.

3. $|x - 3| = 3 - x$
4. $|x + 2| = x + 2$
5. $|x^2 + 9| = x^2 + 9$
6. $|x^2 + 5x| = x^2 + 5x$
7. $|3x^2 + 2x| = x|3x + 2|$
8. $|6 - 2x| = 2|x - 3|$
9. $\sqrt{(x + 5)^2} = x + 5$
10. $\sqrt{(3x - 2)^2} = 2 - 3x$
11. Verify $\sqrt{a^2} = |a|$ for $a = 7$ and $a = -7$.
12. Verify the inequalities $-|a| \le a \le |a|$ for $a = 2$ and for $a = -5$.
13. Let A and B be points with coordinates a and b. In each part find the distance between A and B.
    (a) $a = 9$, $b = 7$
    (b) $a = 2$, $b = 3$
    (c) $a = -8$, $b = 6$
    (d) $a = \sqrt{2}$, $b = -3$
    (e) $a = -11$, $b = -4$
    (f) $a = 0$, $b = -5$
14. Is the equality $\sqrt{a^4} = a^2$ valid for all values of a? Explain.
15. Let A and B be points with coordinates a and b. In each part, use the given information to find b.
    (a) $a = -3$, B is to the left of A, and $|b - a| = 6$.
    (b) $a = -2$, B is to the right of A, and $|b - a| = 9$.
    (c) $a = 5$, $|b - a| = 7$, and $b > 0$.
16. Let E and F be points with coordinates e and f. In each part, determine whether E is to the left or to the right of F on a coordinate line.
    (a) $f - e = 4$
    (b) $e - f = 4$
    (c) $f - e = -6$
    (d) $e - f = -7$

In Exercises 17–24, solve for x.

17. $|6x - 2| = 7$
18. $|3 + 2x| = 11$
19. $|6x - 7| = |3 + 2x|$
20. $|4x + 5| = |8x - 3|$
21. $|9x| - 11 = x$
22. $2x - 7 = |x + 1|$
23. $\left|\dfrac{x + 5}{2 - x}\right| = 6$
24. $\left|\dfrac{x - 3}{x + 4}\right| = 5$

In Exercises 25–36, solve for x and express the solution in terms of intervals.

25. $|x + 6| < 3$
26. $|7 - x| \le 5$
27. $|2x - 3| \le 6$
28. $|3x + 1| < 4$
29. $|x + 2| > 1$
30. $|\frac{1}{2}x - 1| \ge 2$
31. $|5 - 2x| \ge 4$
32. $|7x + 1| > 3$
33. $\dfrac{1}{|x - 1|} < 2$
34. $\dfrac{1}{|3x + 1|} \ge 5$
35. $\dfrac{3}{|2x - 1|} \ge 4$
36. $\dfrac{2}{|x + 3|} < 1$
37. For which values of x is $\sqrt{\left(x^2 - 5x + 6\right)^2} = x^2 - 5x + 6$?
38. Solve $3 \le |x - 2| \le 7$ for x.
39. Solve $|x - 3|^2 - 4|x - 3| = 12$ for x. [*Hint:* Begin by letting $u = |x - 3|$.]
40. Verify the triangle inequality $|a + b| \le |a| + |b|$ (Theorem B.5) for
    (a) $a = 3$, $b = 4$
    (b) $a = -2$, $b = 6$
    (c) $a = -7$, $b = -8$
    (d) $a = -4$, $b = 4$.
41. Prove: $|a - b| \le |a| + |b|$.
42. Prove: $|a| - |b| \le |a - b|$.
43. Prove: $\big| |a| - |b| \big| \le |a - b|$. [*Hint:* Use Exercise 42.]

APPENDIX C

Coordinate Planes and Lines

Just as points on a coordinate line can be associated with real numbers, so points in a plane can be associated with pairs of real numbers by introducing a *rectangular coordinate system* (also called a *Cartesian coordinate system*). A rectangular coordinate system consists of two perpendicular coordinate lines, called *coordinate axes*, that intersect at their origins. Usually, but not always, one axis is horizontal with its positive direction to the right, and the other is vertical with its positive direction up. The intersection of the axes is called the *origin* of the coordinate system.

It is common to call the horizontal axis the *x-axis* and the vertical axis the *y-axis*, in which case the plane and the axes together are referred to as the *xy-plane* (Figure C.1). Although labeling the axes with the letters x and y is common, other letters may be more appropriate in specific applications. Figure C.2 shows a uv-plane and a ts-plane—the first letter in the name of the plane always refers to the horizontal axis and the second to the vertical axis.

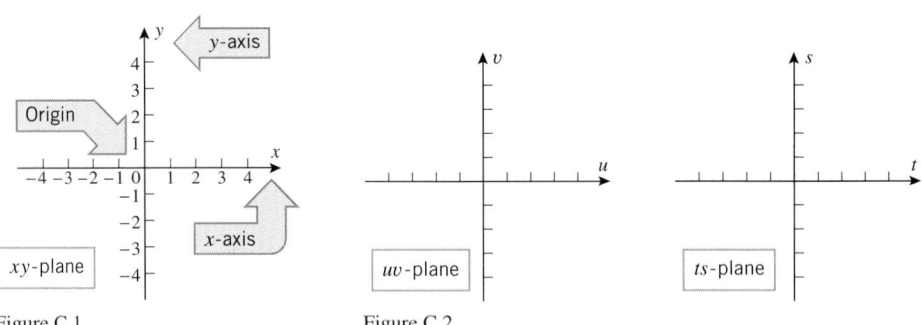

Figure C.1   Figure C.2

Every point P in a coordinate plane can be associated with a unique ordered pair of real numbers by drawing two lines through P, one perpendicular to the x-axis and the other perpendicular to the y-axis (Figure C.3). If the first line intersects the x-axis at the point with coordinate a and the second line intersects the y-axis at the point with coordinate b, then we associate the ordered pair of real numbers (a, b) with the point P. The number a is called the *x-coordinate* or *abscissa* of P and the number b is called the *y-coordinate* or *ordinate* of P. We will say that P has *coordinates* (a, b) and write $P(a, b)$ when we want to emphasize that the coordinates of P are (a, b). We can also reverse the above procedure and find the point P associated with the coordinates (a, b) by locating the intersection of the dashed

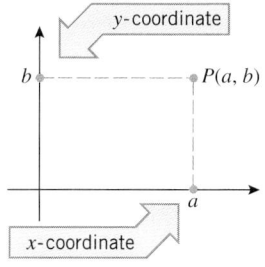

Figure C.3

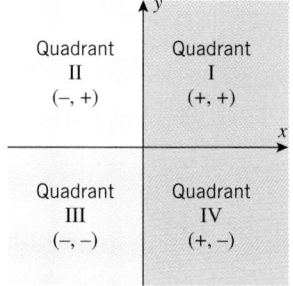

Figure C.4

lines in Figure C.3. Because of this one-to-one correspondence between coordinates and points, we will sometimes blur the distinction between points and ordered pairs of numbers by talking about the *point* (a, b).

REMARK.   Recall that the symbol (a, b) also denotes the open interval between a and b; the appropriate interpretation will usually be clear from the context.

In a rectangular coordinate system the coordinate axes divide the plane into four regions called **quadrants**. These are numbered counterclockwise with roman numerals as shown in Figure C.4. As indicated in that figure, it is easy to determine the quadrant in which a given point lies from the signs of its coordinates: a point with two positive coordinates $(+, +)$ lies in Quadrant I, a point with a negative x-coordinate and a positive y-coordinate $(-, +)$ lies in Quadrant II, and so forth. Points with a zero x-coordinate lie on the y-axis and points with a zero y-coordinate lie on the x-axis.

To **plot** a point $P(a, b)$ means to locate the point with coordinates (a, b) in a coordinate plane. For example, in Figure C.5 we have plotted the points

$$P(2, 5), \quad Q(-4, 3), \quad R(-5, -2), \quad \text{and} \quad S(4, -3)$$

Observe how the signs of the coordinates identify the quadrants in which the points lie.

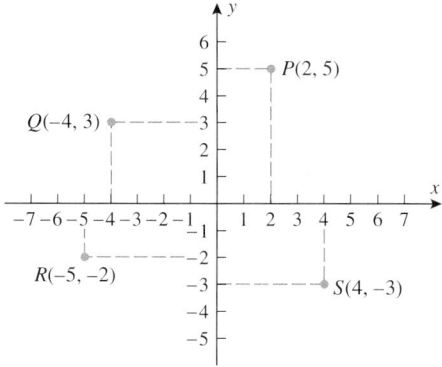

Figure C.5

GRAPHS

The correspondence between points in a plane and ordered pairs of real numbers makes it possible to visualize algebraic equations as geometric curves, and, conversely, to represent geometric curves by algebraic equations. To understand how this is done, suppose that we have an xy-coordinate system and an equation involving two variables x and y, say

$$6x - 4y = 10, \quad y = \sqrt{x}, \quad x = y^3 + 1, \quad \text{or} \quad x^2 + y^2 = 1$$

We define a **solution** of such an equation to be any ordered pair of real numbers (a, b) whose coordinates satisfy the equation when we substitute $x = a$ and $y = b$. For example, the ordered pair $(3, 2)$ is a solution of the equation $6x - 4y = 10$, since the equation is satisfied by $x = 3$ and $y = 2$ (verify). However, the ordered pair $(2, 0)$ is not a solution of this equation, since the equation is not satisfied by $x = 2$ and $y = 0$ (verify).

The following definition makes the association between equations in x and y and curves in the xy-plane.

> **C.1**   DEFINITION.   The set of all solutions of an equation in x and y is called the **solution set** of the equation, and the set of all points in the xy-plane whose coordinates are members of the solution set is called the **graph** of the equation.

One of the main themes in calculus is to identify the exact shape of a graph. Point plotting is one approach to obtaining a graph, but this method has limitations, as discussed in the following example.

Example 1

Sketch the graph of $y = x^2$.

Solution. The solution set of the equation has infinitely many members, since we can substitute an arbitrary value for x into the right side of $y = x^2$ and compute the associated y to obtain a point (x, y) in the solution set. The fact that the solution set has infinitely many members means that we cannot obtain the *entire* graph of $y = x^2$ by point plotting. However, we can obtain an *approximation* to the graph by plotting some sample members of the solution set and connecting them with a smooth curve, as in Figure C.6. The problem with this method is that we cannot be sure how the graph behaves *between* the plotted points. For example, the curves in Figure C.7 also pass through the plotted points and hence are legitimate candidates for the graph in the absence of additional information. Moreover, even if we use a graphing calculator or a computer program to generate the graph, as in Figure C.8, we have the same problem because graphing technology uses point-plotting algorithms to generate graphs. Indeed, in Section 1.3 of the text we see examples where graphing technology can be fooled into producing grossly inaccurate graphs.   ◀

x	$y = x^2$	(x, y)
0	0	(0, 0)
1	1	(1, 1)
2	4	(2, 4)
3	9	(3, 9)
−1	1	(−1, 1)
−2	4	(−2, 4)
−3	9	(−3, 9)

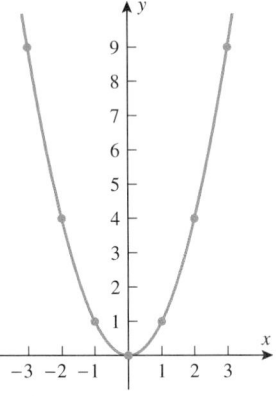

Figure C.6

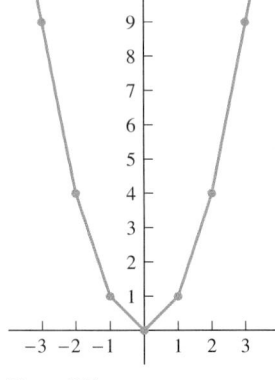

Figure C.7

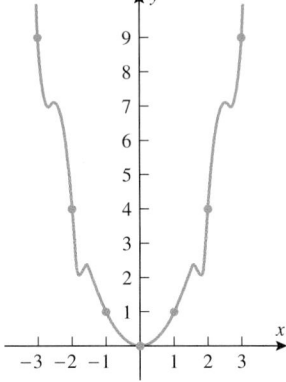

[−4, 4] × [0, 10]
xScl = 1, yScl = 2

Figure C.8

In spite of its limitations, point plotting by hand or with the help of graphing technology can be useful, so here are two more examples.

Example 2

Sketch the graph of $y = \sqrt{x}$.

Solution. If $x < 0$, then $\sqrt{x}$ is an imaginary number. Thus, we can only plot points for which $x \geq 0$, since points in the xy-plane have real coordinates. Figure C.9 shows the graph obtained by point plotting and a graph obtained with a graphing calculator.   ◀

Example 3

Sketch the graph of $y^2 - 2y - x = 0$.

Solution. To calculate coordinates of points on the graph of an equation in x and y, it is desirable to have y expressed in terms of x or of x in terms of y. In this case it is easier to

x	$y = \sqrt{x}$	(x, y)
0	0	$(0, 0)$
1	1	$(1, 1)$
2	$\sqrt{2}$	$(2, \sqrt{2}) \approx (2, 1.4)$
3	$\sqrt{3}$	$(3, \sqrt{3}) \approx (3, 1.7)$
4	2	$(4, 2)$

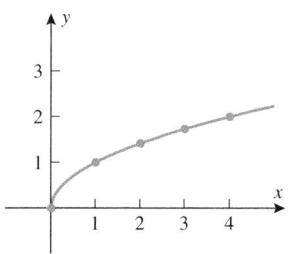

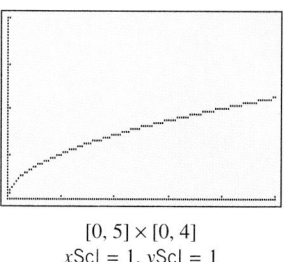

$[0, 5] \times [0, 4]$
$x\text{Scl} = 1, \, y\text{Scl} = 1$

Figure C.9

express x in terms of y, so we rewrite the equation as

$$x = y^2 - 2y$$

Members of the solution set can be obtained from this equation by substituting arbitrary values for y in the right side and computing the associated values of x (Figure C.10). ◀

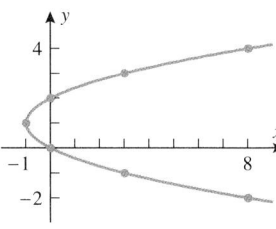

y	$x = y^2 - 2y$	(x, y)
-2	8	$(8, -2)$
-1	3	$(3, -1)$
0	0	$(0, 0)$
1	-1	$(-1, 1)$
2	0	$(0, 2)$
3	3	$(3, 3)$
4	8	$(8, 4)$

Figure C.10

REMARK.   Most graphing calculators and computer graphing programs require that y be expressed in terms of x to generate a graph in the xy-plane. In Section 1.7 we discuss a method for circumventing this restriction.

Example 4

Sketch the graph of $y = 1/x$.

Solution.   Because $1/x$ is undefined at $x = 0$, we can only plot points for which $x \neq 0$. This forces a break, called a *discontinuity*, in the graph at $x = 0$ (Figure C.11). ◀

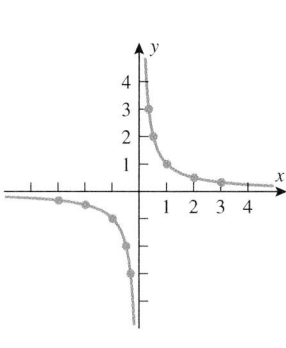

x	$y = 1/x$	(x, y)
$\frac{1}{3}$	3	$\left(\frac{1}{3}, 3\right)$
$\frac{1}{2}$	2	$\left(\frac{1}{2}, 2\right)$
1	1	$(1, 1)$
2	$\frac{1}{2}$	$\left(2, \frac{1}{2}\right)$
3	$\frac{1}{3}$	$\left(3, \frac{1}{3}\right)$
$-\frac{1}{3}$	-3	$\left(-\frac{1}{3}, -3\right)$
$-\frac{1}{2}$	-2	$\left(-\frac{1}{2}, -2\right)$
-1	-1	$(-1, -1)$
-2	$-\frac{1}{2}$	$\left(-2, -\frac{1}{2}\right)$
-3	$-\frac{1}{3}$	$\left(-3, -\frac{1}{3}\right)$

Figure C.11

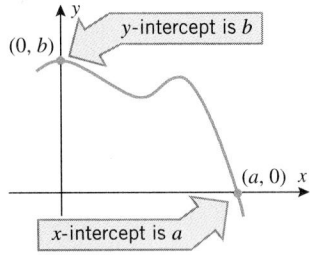

Figure C.12

Points where a graph intersects the coordinate axes are of special interest in many problems. As illustrated in Figure C.12, intersections of a graph with the x-axis have the form $(a, 0)$ and intersections with the y-axis have the form $(0, b)$. The number a is called an ***x-intercept*** of the graph and the number b a ***y-intercept***.

Example 5

Find all intercepts of

   (a) $3x + 2y = 6$ (b) $x = y^2 - 2y$ (c) $y = 1/x$

Solution (a). To find the x-intercepts we set $y = 0$ and solve for x:

$$3x = 6 \quad \text{or} \quad x = 2$$

To find the y-intercepts we set $x = 0$ and solve for y:

$$2y = 6 \quad \text{or} \quad y = 3$$

As we will see later, the graph of $3x + 2y = 6$ is the line shown in Figure C.13.

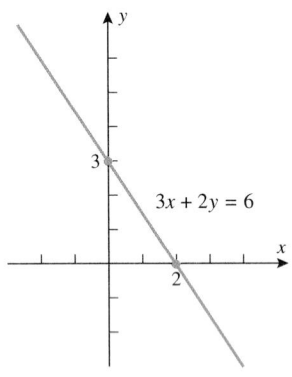

Figure C.13

Solution (b). To find the x-intercepts, set $y = 0$ and solve for x:

$$x = 0$$

Thus, $x = 0$ is the only x-intercept. To find the y-intercepts, set $x = 0$ and solve for y:

$$y^2 - 2y = 0$$
$$y(y - 2) = 0$$

So the y-intercepts are $y = 0$ and $y = 2$. The graph is shown in Figure C.10.

Solution (c). To find the x-intercepts, set $y = 0$:

$$\frac{1}{x} = 0$$

This equation has no solutions (why?), so there are no x-intercepts. To find y-intercepts we would set $x = 0$ and solve for y. But, substituting $x = 0$ leads to a division by zero, which is not allowed, so there are no y-intercepts either. The graph of the equation is shown in Figure C.11.   ◀

To obtain equations of lines we will first need to discuss the concept of *slope*, which is a numerical measure of the "steepness" of a line.

Consider a particle moving left to right along a *nonvertical* line from a point $P_1(x_1, y_1)$ to a point $P_2(x_2, y_2)$. As shown in Figure C.14, the particle moves $y_2 - y_1$ units in the y-direction as it travels $x_2 - x_1$ units in the positive x-direction. The vertical change $y_2 - y_1$ is called the ***rise***, and the horizontal change $x_2 - x_1$ the ***run***. The ratio of the rise over the run can be used to measure the steepness of the line, which leads us to the following definition.

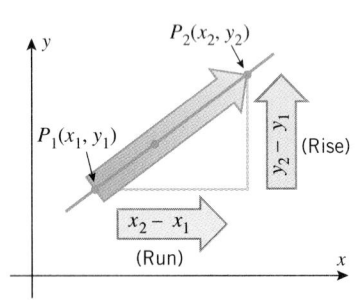

Figure C.14

> **C.2   DEFINITION.**   If $P_1(x_1, y_1)$ and $P_2(x_2, y_2)$ are points on a nonvertical line, then the ***slope*** m of the line is defined by
>
> $$m = \frac{\text{rise}}{\text{run}} = \frac{y_2 - y_1}{x_2 - x_1} \tag{1}$$

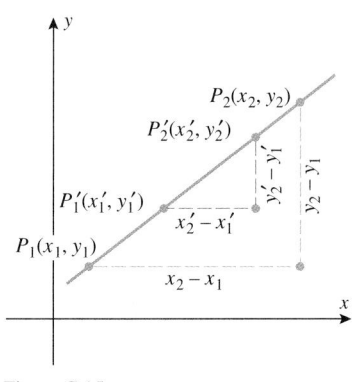

Figure C.15

REMARK.   Observe that this definition does not apply to vertical lines. For such lines we have $x_2 = x_1$ (a zero run), which means that the formula for m involves a division by zero. For this reason, the slope of a vertical line is **undefined**, which is sometimes described informally by stating that a vertical line has **infinite slope**.

When calculating the slope of a nonvertical line from Formula (1), it does not matter which two points on the line you use for the calculation, as long as they are distinct. This can be proved using Figure C.15 and similar triangles to show that

$$ m = \frac{y_2 - y_1}{x_2 - x_1} = \frac{y_2' - y_1'}{x_2' - x_1'} $$

Moreover, once you choose two points to use for the calculation, it does not matter which one you call P_1 and which one you call P_2 because reversing the points reverses the sign of both the numerator and denominator of (1) and hence has no effect on the ratio.

Example 6

In each part find the slope of the line through

(a)   the points $(6, 2)$ and $(9, 8)$

(b)   the points $(2, 9)$ and $(4, 3)$

(c)   the points $(-2, 7)$ and $(5, 7)$.

Solution.

(a) $m = \dfrac{8 - 2}{9 - 6} = \dfrac{6}{3} = 2$ (b) $m = \dfrac{3 - 9}{4 - 2} = \dfrac{-6}{2} = -3$ (c) $m = \dfrac{7 - 7}{5 - (-2)} = 0$ ◀

Example 7

Figure C.16 shows the three lines determined by the points in Example 6 and explains the significance of their slopes.   ◀

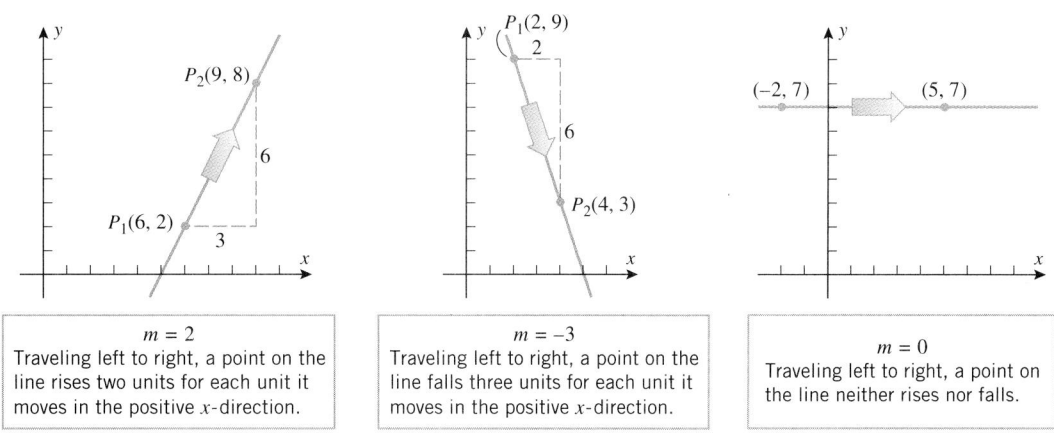

Figure C.16

As illustrated in this example, the slope of a line can be positive, negative, or zero. A positive slope means that the line is inclined upward to the right, a negative slope means that the line is inclined downward to the right, and a zero slope means that the line is horizontal.

An undefined slope means that the line is vertical. Figure C.17 shows various lines through the origin with their slopes.

The following theorem shows how slopes can be used to tell whether two lines are parallel or perpendicular.

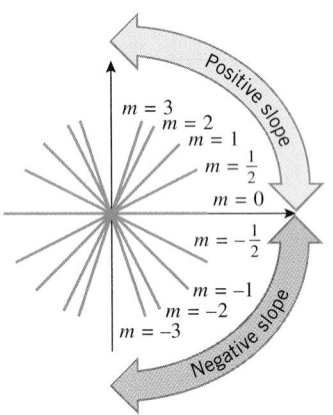

Figure C.17

C.3   THEOREM.

(a)   *Two nonvertical lines with slopes m_1 and m_2 are parallel if and only if they have the same slope, that is,*

$$m_1 = m_2$$

(b)   *Two nonvertical lines with slopes m_1 and m_2 are perpendicular if and only if the product of their slopes is -1, that is,*

$$m_1 m_2 = -1$$

This relationship can also be expressed as $m_1 = -1/m_2$ or $m_2 = -1/m_1$, which states that nonvertical lines are perpendicular if and only if their slopes are negative reciprocals of one another.

A complete proof of this theorem is a little tedious, but it is not hard to motivate the results informally. Let us start with part (*a*).

Suppose that L_1 and L_2 are nonvertical parallel lines with slopes m_1 and m_2, respectively. If the lines are parallel to the x-axis, then $m_1 = m_2 = 0$, and we are done. If they are not parallel to the x-axis, then both lines intersect the x-axis; and for simplicity assume that they are oriented as in Figure C.18*a*. On each line choose the point whose run relative to the point of intersection with the x-axis is 1. On line L_1 the corresponding rise will be m_1 and on L_2 it will be m_2. However, because the lines are parallel, the shaded triangles in the figure must be congruent (verify), so $m_1 = m_2$. Conversely, the condition $m_1 = m_2$ can be used to show that the shaded triangles are congruent, from which it follows that the lines make the same angle with the x-axis and hence are parallel (verify).

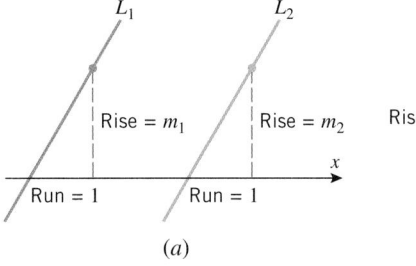
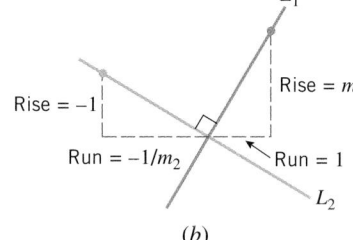

Figure C.18

Now suppose that L_1 and L_2 are nonvertical perpendicular lines with slopes m_1 and m_2, respectively; and for simplicity assume that they are oriented as in Figure C.18*b*. On line L_1 choose the point whose run relative to the point of intersection of the lines is 1, in which case the corresponding rise will be m_1; and on line L_2 choose the point whose rise relative to the point of intersection is -1, in which case the corresponding run will be $-1/m_2$. Because the lines are perpendicular, the shaded triangles in the figure must be congruent (verify), and hence the ratios of corresponding sides of the triangles must be equal. Taking into account that for line L_2 the vertical side of the triangle has length 1 and the horizontal side has length $-1/m_2$ (since m_2 is negative), the congruence of the triangles implies that

$m_1/1 = (-1/m_2)/1$ or $m_1m_2 = -1$. Conversely, the condition $m_1 = -1/m_2$ can be used to show that the shaded triangles are congruent, from which it can be deduced that the lines are perpendicular (verify).

Example 8

Use slopes to show that the points $A(1, 3)$, $B(3, 7)$, and $C(7, 5)$ are vertices of a right triangle.

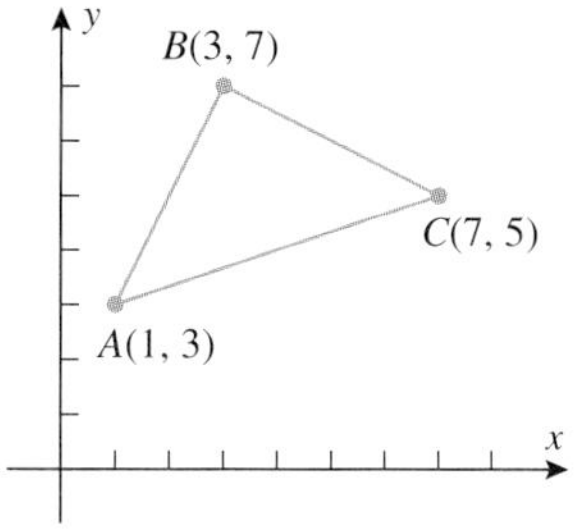

Solution. We will show that the line through A and B is perpendicular to the line through B and C. The slopes of these lines are

$$m_1 = \frac{7-3}{3-1} = 2 \qquad \text{and} \qquad m_2 = \frac{5-7}{7-3} = -\frac{1}{2}$$

Slope of the line through A and B ⎴ Slope of the line through B and C

Since $m_1m_2 = -1$, the line through A and B is perpendicular to the line through B and C; thus, ABC is a right triangle (Figure C.19). ◀

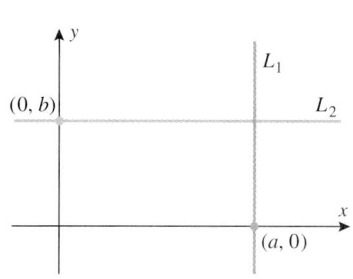

Figure C.19

LINES PARALLEL TO THE COORDINATE AXES

We now turn to the problem of finding equations of lines that satisfy specified conditions. The simplest cases are lines parallel to the coordinate axes. A line parallel to the y-axis intersects the x-axis at some point $(a, 0)$. This line consists precisely of those points whose x-coordinate is equal to a (Figure C.20). Similarly, a line parallel to the x-axis intersects the y-axis at some point $(0, b)$. This line consists precisely of those points whose y-coordinate is equal to b (Figure C.20). Thus, we have the following theorem.

> **C.4**   THEOREM.   *The vertical line through $(a, 0)$ and the horizontal line through $(0, b)$ are represented, respectively, by the equations*
>
> $$x = a \qquad and \qquad y = b$$

Every point on L_1 has an x-coordinate of a and every point on L_2 has a y-coordinate of b.

Figure C.20

Example 9

The graph of $x = -5$ is the vertical line through $(-5, 0)$, and the graph of $y = 7$ is the horizontal line through $(0, 7)$ (Figure C.21). ◀

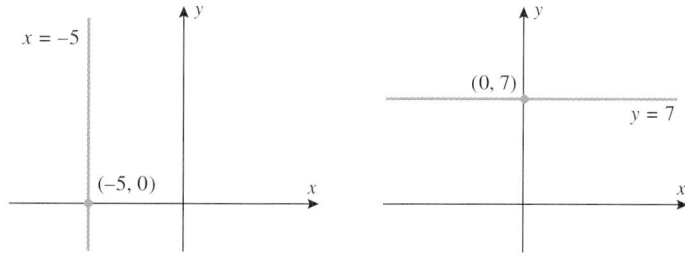

Figure C.21

LINES DETERMINED BY POINT AND SLOPE

There are infinitely many lines that pass through any given point in the plane. However, if we specify the slope of the line in addition to a point on it, then the point and the slope together determine a unique line (Figure C.22).

Let us now consider how to find an equation of a nonvertical line L that passes through a point $P_1(x_1, y_1)$ and has slope m. If $P(x, y)$ is any point on L, different from P_1, then the

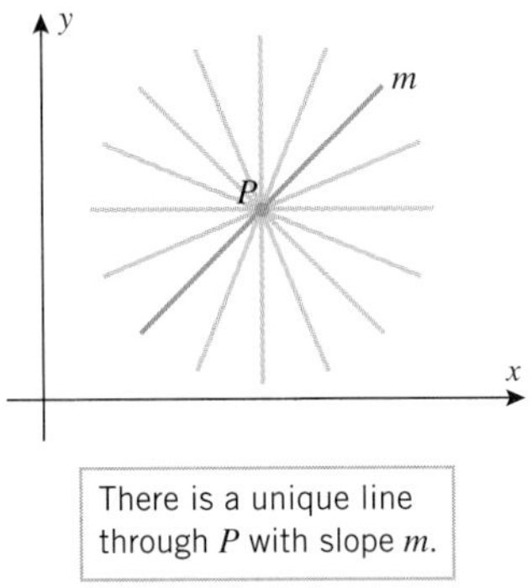

There is a unique line through P with slope m.

Figure C.22

slope m can be obtained from the points $P(x, y)$ and $P_1(x_1, y_1)$; this gives

$$m = \frac{y - y_1}{x - x_1}$$

which can be rewritten as

$$y - y_1 = m(x - x_1) \tag{2}$$

With the possible exception of (x_1, y_1), we have shown that every point on L satisfies (2). But $x = x_1$, $y = y_1$ satisfies (2), so that all points on L satisfy (2). We leave it as an exercise to show that every point satisfying (2) lies on L.

In summary, we have the following theorem.

C.5   THEOREM.   *The line passing through $P_1(x_1, y_1)$ and having slope m is given by the equation*

$$y - y_1 = m(x - x_1) \tag{3}$$

*This is called the **point-slope form** of the line.*

Example 10

Find the point-slope form of the line through $(4, -3)$ with slope 5.

Solution.   Substituting the values $x_1 = 4$, $y_1 = -3$, and $m = 5$ in (3) yields the point-slope form $y + 3 = 5(x - 4)$.   ◀

LINES DETERMINED BY SLOPE AND y-INTERCEPT

A nonvertical line crosses the y-axis at some point $(0, b)$. If we use this point in the point-slope form of its equation, we obtain

$$y - b = m(x - 0)$$

which we can rewrite as $y = mx + b$. To summarize:

C.6   THEOREM.   *The line with y-intercept b and slope m is given by the equation*

$$y = mx + b \tag{4}$$

*This is called the **slope-intercept form** of the line.*

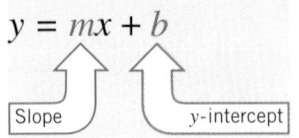

Figure C.23

REMARK.   Note that y is alone on one side of Equation (4). When the equation of a line is written in this way the slope of the line and its y-intercept can be determined by inspection of the equation—the slope is the coefficient of x and the y-intercept is the constant term (Figure C.23).

Example 11

EQUATION	SLOPE	y-INTERCEPT
$y = 3x + 7$	$m = 3$	$b = 7$
$y = -x + \frac{1}{2}$	$m = -1$	$b = \frac{1}{2}$
$y = x$	$m = 1$	$b = 0$
$y = \sqrt{2}x - 8$	$m = \sqrt{2}$	$b = -8$
$y = 2$	$m = 0$	$b = 2$

◀

Example 12

Find the slope-intercept form of the equation of the line that satisfies the stated conditions:

(a) slope is -9; crosses the y-axis at $(0, -4)$

(b) slope is 1; passes through the origin

(c) passes through $(5, -1)$; perpendicular to $y = 3x + 4$

(d) passes through $(3, 4)$ and $(2, -5)$.

Solution (a). From the given conditions we have $m = -9$ and $b = -4$, so (4) yields $y = -9x - 4$.

Solution (b). From the given conditions $m = 1$ and the line passes through $(0, 0)$, so $b = 0$. Thus, it follows from (4) that $y = x + 0$ or $y = x$.

Solution (c). The given line has slope 3, so the line to be determined will have slope $m = -\frac{1}{3}$. Substituting this slope and the given point in the point-slope form (3) and then simplifying yields

$$y - (-1) = -\tfrac{1}{3}(x - 5)$$
$$y = -\tfrac{1}{3}x + \tfrac{2}{3}$$

Solution (d). We will first find the point-slope form, then solve for y in terms of x to obtain the slope-intercept form. From the given points the slope of the line is

$$m = \frac{-5 - 4}{2 - 3} = 9$$

We can use either of the given points for (x_1, y_1) in (3). We will use $(3, 4)$. This yields the point-slope form

$$y - 4 = 9(x - 3)$$

Solving for y in terms of x yields the slope-intercept form

$$y = 9x - 23$$

We leave it for the reader to show that the same equation results if $(2, -5)$ rather than $(3, 4)$ is used for (x_1, y_1) in (3). ◄

THE GENERAL EQUATION OF A LINE

An equation that is expressible in the form

$$Ax + By + C = 0 \tag{5}$$

where A, B, and C are constants and A and B are not both zero, is called a ***first-degree equation*** in x and y. For example,

$$4x + 6y - 5 = 0$$

is a first-degree equation in x and y since it has form (5) with

$$A = 4, \quad B = 6, \quad C = -5$$

In fact, all the equations of lines studied in this section are first-degree equations in x and y.

   The following theorem states that the first-degree equations in x and y are precisely the equations whose graphs in the xy-plane are straight lines.

> **C.7** THEOREM. *Every first-degree equation in x and y has a straight line as its graph and, conversely, every straight line can be represented by a first-degree equation in x and y.*

Because of this theorem, (5) is sometimes called the **general equation** of a line or a **linear equation** in x and y.

Example 13

Graph the equation $3x - 4y + 12 = 0$.

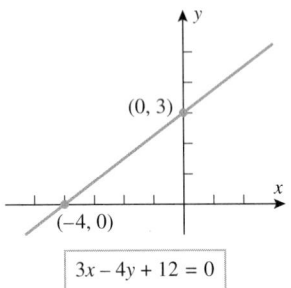

(0, 3)

(−4, 0)

$3x - 4y + 12 = 0$

Figure C.24

Solution. Since this is a linear equation in x and y, its graph is a straight line. Thus, to sketch the graph we need only plot any two points on the graph and draw the line through them. It is particularly convenient to plot the points where the line crosses the coordinate axes. These points are $(0, 3)$ and $(-4, 0)$ (verify), so the graph is the line in Figure C.24. ◄

Example 14

Find the slope of the line in Example 13.

Solution. Solving the equation for y yields

$$y = \tfrac{3}{4}x + 3$$

which is the slope-intercept form of the line. Thus, the slope is $m = \tfrac{3}{4}$. ◄

EXERCISE SET C

1. Draw the rectangle, three of whose vertices are $(6, 1)$, $(-4, 1)$, and $(6, 7)$, and find the coordinates of the fourth vertex.

2. Draw the triangle whose vertices are $(-3, 2)$, $(5, 2)$, and $(4, 3)$, and find its area.

In Exercises 3 and 4, draw a rectangular coordinate system and sketch the set of points whose coordinates (x, y) satisfy the given conditions.

3. (a) $x = 2$ (b) $y = -3$ (c) $x \geq 0$
      (d) $y = x$ (e) $y \geq x$ (f) $|x| \geq 1$

4. (a) $x = 0$ (b) $y = 0$
      (c) $y < 0$ (d) $x \geq 1$ and $y \leq 2$
      (e) $x = 3$ (f) $|x| = 5$

In Exercises 5–12, sketch the graph of the equation. (A calculating utility will be helpful in some of these problems.)

5. $y = 4 - x^2$ **6.** $y = 1 + x^2$

7. $y = \sqrt{x - 4}$ **8.** $y = -\sqrt{x + 1}$

9. $x^2 - x + y = 0$ **10.** $x = y^3 - y^2$

11. $x^2 y = 2$ **12.** $xy = -1$

13. Find the slope of the line through
      (a) $(-1, 2)$ and $(3, 4)$ (b) $(5, 3)$ and $(7, 1)$
      (c) $(4, \sqrt{2})$ and $(-3, \sqrt{2})$ (d) $(-2, -6)$ and $(-2, 12)$.

14. Find the slopes of the sides of the triangle with vertices $(-1, 2)$, $(6, 5)$, and $(2, 7)$.

15. Use slopes to determine whether the given points lie on the same line.
      (a) $(1, 1)$, $(-2, -5)$, and $(0, -1)$
      (b) $(-2, 4)$, $(0, 2)$, and $(1, 5)$

16. Draw the line through $(4, 2)$ with slope
      (a) $m = 3$ (b) $m = -2$ (c) $m = -\tfrac{3}{4}$.

17. Draw the line through $(-1, -2)$ with slope
      (a) $m = \tfrac{3}{5}$ (b) $m = -1$ (c) $m = \sqrt{2}$.

18. An equilateral triangle has one vertex at the origin, another on the x-axis, and the third in the first quadrant. Find the slopes of its sides.

19. List the lines in the accompanying figure in the order of increasing slope.

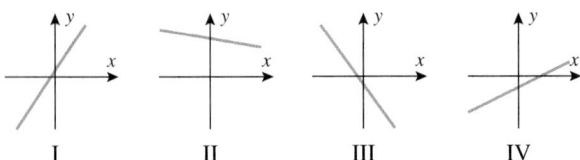

I   II   III   IV

20. List the lines in the accompanying figure in the order of increasing slope.

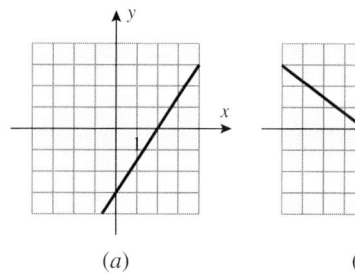

I   II   III   IV

21. A particle, initially at $(1, 2)$, moves along a line of slope $m = 3$ to a new position (x, y).
(a) Find y if $x = 5$.   (b) Find x if $y = -2$.

22. A particle, initially at $(7, 5)$, moves along a line of slope $m = -2$ to a new position (x, y).
(a) Find y if $x = 9$.   (b) Find x if $y = 12$.

23. Let the point $(3, k)$ lie on the line of slope $m = 5$ through $(-2, 4)$; find k.

24. Given that the point $(k, 4)$ is on the line through $(1, 5)$ and $(2, -3)$, find k.

25. Find x if the slope of the line through $(1, 2)$ and $(x, 0)$ is the negative of the slope of the line through $(4, 5)$ and $(x, 0)$.

26. Find x and y if the line through $(0, 0)$ and (x, y) has slope $\frac{1}{2}$, and the line through (x, y) and $(7, 5)$ has slope 2.

27. Use slopes to show that $(3, -1)$, $(6, 4)$, $(-3, 2)$, and $(-6, -3)$ are vertices of a parallelogram.

28. Use slopes to show that $(3, 1)$, $(6, 3)$, and $(2, 9)$ are vertices of a right triangle.

29. Graph the equations
(a) $2x + 5y = 15$ (b) $x = 3$
(c) $y = -2$ (d) $y = 2x - 7$.

30. Graph the equations
(a) $\dfrac{x}{3} - \dfrac{y}{4} = 1$ (b) $x = -8$
(c) $y = 0$ (d) $x = 3y + 2$.

31. Graph the equations
(a) $y = 2x - 1$ (b) $y = 3$
(c) $y = -2x$.

32. Graph the equations
(a) $y = 2 - 3x$ (b) $y = \frac{1}{4}x$
(c) $y = -\sqrt{3}$.

33. Find the slope and y-intercept of
(a) $y = 3x + 2$ (b) $y = 3 - \frac{1}{4}x$
(c) $3x + 5y = 8$ (d) $y = 1$
(e) $\dfrac{x}{a} + \dfrac{y}{b} = 1$.

34. Find the slope and y-intercept of
(a) $y = -4x + 2$ (b) $x = 3y + 2$
(c) $\dfrac{x}{2} + \dfrac{y}{3} = 1$ (d) $y - 3 = 0$
(e) $a_0 x + a_1 y = 0$ $(a_1 \neq 0)$.

In Exercises 35 and 36, use the graph to find the equation of the line in slope-intercept form.

35.

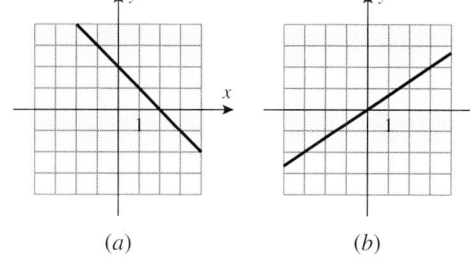

(a) (b)

Figure Ex-35

36.

(a) (b)

Figure Ex-36

In Exercises 37–48, find the slope-intercept form of the line satisfying the given conditions.

37. Slope $= -2$, y-intercept $= 4$.

38. $m = 5$, $b = -3$.

39. The line is parallel to $y = 4x - 2$ and its y-intercept is 7.

40. The line is parallel to $3x + 2y = 5$ and passes through $(-1, 2)$.

41. The line is perpendicular to $y = 5x + 9$ and its y-intercept is 6.

42. The line is perpendicular to $x - 4y = 7$ and passes through $(3, -4)$.

43. The line passes through $(2, 4)$ and $(1, -7)$.

44. The line passes through $(-3, 6)$ and $(-2, 1)$.

45. The y-intercept is 2 and the x-intercept is -4.

46. The y-intercept is b and the x-intercept is a.

47. The line is perpendicular to the y-axis and passes through $(-4, 1)$.

48. The line is parallel to $y = -5$ and passes through $(-1, -8)$.

49. In each part, classify the lines as parallel, perpendicular, or neither.
(a) $y = 4x - 7$ and $y = 4x + 9$
(b) $y = 2x - 3$ and $y = 7 - \frac{1}{2}x$
(c) $5x - 3y + 6 = 0$ and $10x - 6y + 7 = 0$
(d) $Ax + By + C = 0$ and $Bx - Ay + D = 0$
(e) $y - 2 = 4(x - 3)$ and $y - 7 = \frac{1}{4}(x - 3)$

50. In each part, classify the lines as parallel, perpendicular, or neither.
(a) $y = -5x + 1$ and $y = 3 - 5x$

(b) $y - 1 = 2(x - 3)$ and $y - 4 = -\frac{1}{2}(x + 7)$

(c) $4x + 5y + 7 = 0$ and $5x - 4y + 9 = 0$

(d) $Ax + By + C = 0$ and $Ax + By + D = 0$

(e) $y = \frac{1}{2}x$ and $x = \frac{1}{2}y$

51. For what value of k will the line $3x + ky = 4$
   (a) have slope 2
   (b) have y-intercept 5
   (c) pass through the point $(-2, 4)$
   (d) be parallel to the line $2x - 5y = 1$
   (e) be perpendicular to the line $4x + 3y = 2$?

52. Sketch the graph of $y^2 = 3x$ and explain how this graph is related to the graphs of $y = \sqrt{3x}$ and $y = -\sqrt{3x}$.

53. Sketch the graph of $(x - y)(x + y) = 0$ and explain how it is related to the graphs of $x - y = 0$ and $x + y = 0$.

54. Graph $F = \frac{9}{5}C + 32$ in a CF-coordinate system.

55. Graph $u = 3v^2$ in a uv-coordinate system.

56. Graph $Y = 4X + 5$ in a YX-coordinate system.

57. A point moves in the xy-plane in such a way that at any time t its coordinates are given by $x = 5t + 2$ and $y = t - 3$. By expressing y in terms of x, show that the point moves along a straight line.

58. A point moves in the xy-plane in such a way that at any time t its coordinates are given by $x = 1 + 3t^2$ and $y = 2 - t^2$. By expressing y in terms of x, show that the point moves along a straight-line path and specify the values of x for which the equation is valid.

59. Find the area of the triangle formed by the coordinate axes and the line through $(1, 4)$ and $(2, 1)$.

60. Draw the graph of $4x^2 - 9y^2 = 0$.

61. In each part, name an appropriate coordinate system for graphing the equation [e.g., an $\alpha\beta$-coordinate system in part (a)], and state whether the graph of the equation is a line in that coordinate system.
   (a) $3\alpha - 2\beta = 5$
   (b) $A = 2000(1 + 0.06t)$
   (c) $A = \pi r^2$
   (d) $E = mc^2$ (c constant)
   (e) $V = C(1 - rt)$ (r and C constant)
   (f) $V = \frac{1}{3}\pi r^2 h$ (r constant)
   (g) $V = \frac{1}{3}\pi r^2 h$ (h constant)

APPENDIX D

Distance, Circles, and Quadratic Equations

Suppose that we are interested in finding the distance d between two points $P_1(x_1, y_1)$ and $P_2(x_2, y_2)$ in the xy-plane. If, as in Figure D.1, we form a right triangle with P_1 and P_2 as vertices, then it follows from Theorem B.4 in Appendix B that the sides of that triangle have lengths $|x_2 - x_1|$ and $|y_2 - y_1|$. Thus, it follows from the Theorem of Pythagoras that

$$d = \sqrt{|x_2 - x_1|^2 + |y_2 - y_1|^2} = \sqrt{(x_2 - x_1)^2 + (y_2 - y_1)^2}$$

and hence we have the following result.

D.1 THEOREM. *The distance d between two points $P_1(x_1, y_1)$ and $P_2(x_2, y_2)$ in a coordinate plane is given by*

$$d = \sqrt{(x_2 - x_1)^2 + (y_2 - y_1)^2} \tag{1}$$

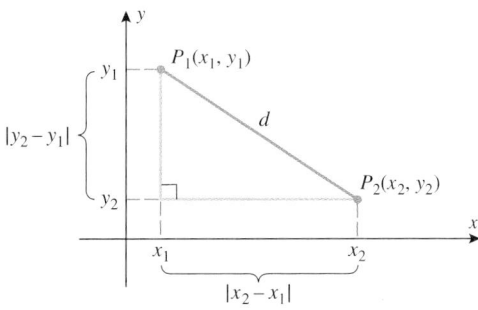

Figure D.1

REMARK. To apply Formula (1) the scales on the coordinate axes must be the same; otherwise, we would not have been able to use the Theorem of Pythagoras in the derivation. Moreover, when using Formula (1) it does not matter which point is labeled P_1 and which one is labeled P_2, since reversing the points changes the signs of $x_2 - x_1$ and $y_2 - y_1$; this has no effect on the value of d because these quantities are squared in the formula. When it is important to emphasize the points, the distance between P_1 and P_2 is denoted by $d(P_1, P_2)$ or $d(P_2, P_1)$.

Example 1

Find the distance between the points $(-2, 3)$ and $(1, 7)$.

Solution. If we let (x_1, y_1) be $(-2, 3)$ and let (x_2, y_2) be $(1, 7)$, then (1) yields

$$d = \sqrt{[1 - (-2)]^2 + [7 - 3]^2} = \sqrt{3^2 + 4^2} = \sqrt{25} = 5 \quad \blacktriangleleft$$

Example 2

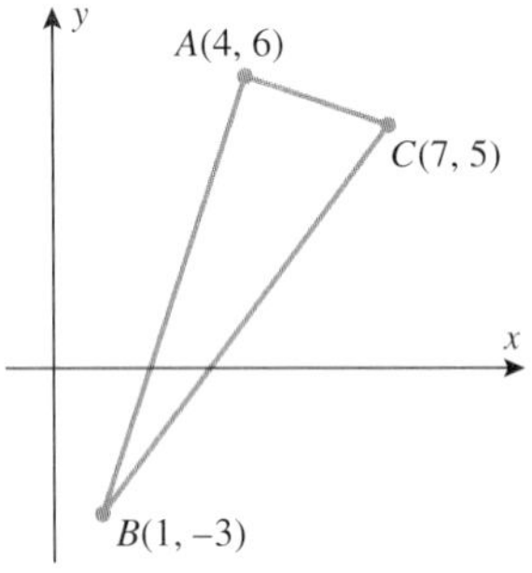

It can be shown that the converse of the Theorem of Pythagoras is true; that is, if the sides of a triangle satisfy the relationship $a^2 + b^2 = c^2$, then the triangle must be a right triangle. Use this result to show that the points $A(4, 6)$, $B(1, -3)$, and $C(7, 5)$ are vertices of a right triangle.

Solution. The points and the triangle are shown in Figure D.2. From (1), the lengths of the sides of the triangles are

$$d(A, B) = \sqrt{(1 - 4)^2 + (-3 - 6)^2} = \sqrt{9 + 81} = \sqrt{90}$$

$$d(A, C) = \sqrt{(7 - 4)^2 + (5 - 6)^2} = \sqrt{9 + 1} = \sqrt{10}$$

$$d(B, C) = \sqrt{(7 - 1)^2 + [5 - (-3)]^2} = \sqrt{36 + 64} = \sqrt{100} = 10$$

Since

$$[d(A, B)]^2 + [d(A, C)]^2 = [d(B, C)]^2$$

it follows that $\triangle ABC$ is a right triangle with hypotenuse BC. $\blacktriangleleft$

Figure D.2

THE MIDPOINT FORMULA

It is often necessary to find the coordinates of the midpoint of a line segment joining two points in the plane. To derive the midpoint formula, we will start with two points on a coordinate line. If we assume that the points have coordinates a and b and that $a \leq b$, then, as shown in Figure D.3, the distance between a and b is $b - a$, and the coordinate of the midpoint between a and b is

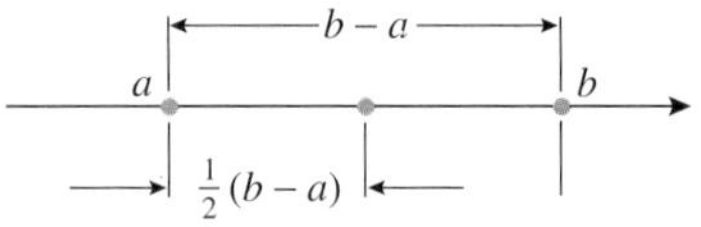

$$a + \tfrac{1}{2}(b - a) = \tfrac{1}{2}a + \tfrac{1}{2}b = \tfrac{1}{2}(a + b)$$

which is the arithmetic average of a and b. Had the points been labeled with $b \leq a$, the same formula would have resulted (verify). Therefore, *the midpoint of two points on a coordinate line is the arithmetic average of their coordinates, regardless of their relative positions.*

If we now let $P_1(x_1, y_1)$ and $P_2(x_2, y_2)$ be any two points in the plane and $M(x, y)$ the midpoint of the line segment joining them (Figure D.4), then it can be shown using similar triangles that x is the midpoint of x_1 and x_2 on the x-axis and y is the midpoint of y_1 and y_2 on the y-axis, so

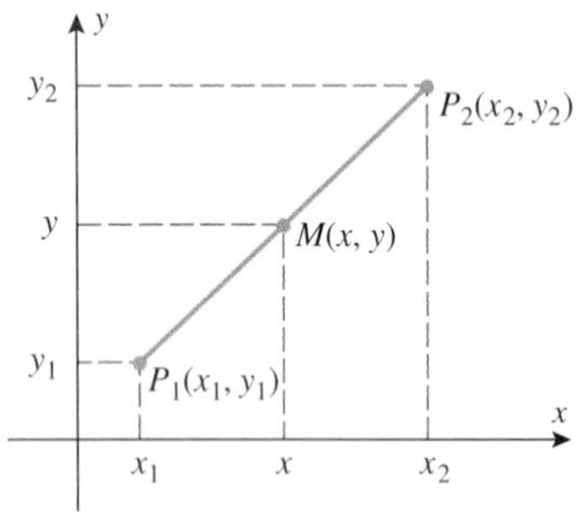

$$x = \tfrac{1}{2}(x_1 + x_2) \quad \text{and} \quad y = \tfrac{1}{2}(y_1 + y_2)$$

Thus, we have the following result.

Figure D.3

Figure D.4

> **D.2**   THEOREM (*The Midpoint Formula*).   *The midpoint of the line segment joining two points (x_1, y_1) and (x_2, y_2) in a coordinate plane is*
>
> $$\left(\tfrac{1}{2}(x_1 + x_2), \tfrac{1}{2}(y_1 + y_2)\right) \tag{2}$$

Example 3

Find the midpoint of the line segment joining $(3, -4)$ and $(7, 2)$.

Solution. From (2) the midpoint is

$$\left(\tfrac{1}{2}(3 + 7), \tfrac{1}{2}(-4 + 2)\right) = (5, -1) \quad \blacktriangleleft$$

CIRCLES

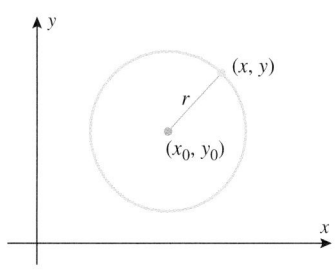

Figure D.5

If (x_0, y_0) is a fixed point in the plane, then the circle of radius r centered at (x_0, y_0) is the set of all points in the plane whose distance from (x_0, y_0) is r (Figure D.5). Thus, a point (x, y) will lie on this circle if and only if

$$\sqrt{(x - x_0)^2 + (y - y_0)^2} = r$$

or equivalently,

$$(x - x_0)^2 + (y - y_0)^2 = r^2 \qquad (3)$$

This is called the *standard form of the equation of a circle*.

Example 4

Find an equation for the circle of radius 4 centered at $(-5, 3)$.

Solution. From (3) with $x_0 = -5$, $y_0 = 3$, and $r = 4$ we obtain

$$(x + 5)^2 + (y - 3)^2 = 16$$

If desired, this equation can be written in an expanded form by squaring the terms and then simplifying:

$$(x^2 + 10x + 25) + (y^2 - 6y + 9) - 16 = 0$$
$$x^2 + y^2 + 10x - 6y + 18 = 0 \qquad \blacktriangleleft$$

Example 5

Find an equation for the circle with center $(1, -2)$ that passes through $(4, 2)$.

Solution. The radius r of the circle is the distance between $(4, 2)$ and $(1, -2)$, so

$$r = \sqrt{(1 - 4)^2 + (-2 - 2)^2} = 5$$

We now know the center and radius, so we can use (3) to obtain the equation

$$(x - 1)^2 + (y + 2)^2 = 25 \quad \text{or} \quad x^2 + y^2 - 2x + 4y - 20 = 0 \qquad \blacktriangleleft$$

FINDING THE CENTER AND RADIUS OF A CIRCLE

When you encounter an equation of form (3), you will know immediately that its graph is a circle; its center and radius can then be found from the constants that appear in the equation:

$$\underbrace{(x - x_0)^2}_{\text{x-coordinate of the center is x_0}} + \underbrace{(y - y_0)^2}_{\text{y-coordinate of the center is y_0}} = \underbrace{r^2}_{\text{radius squared}}$$

Example 6

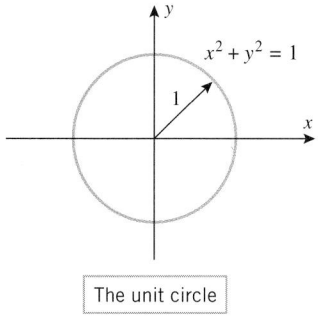

The unit circle

Figure D.6

EQUATION OF A CIRCLE	CENTER (x_0, y_0)	RADIUS r
$(x - 2)^2 + (y - 5)^2 = 9$	$(2, 5)$	3
$(x + 7)^2 + (y + 1)^2 = 16$	$(-7, -1)$	4
$x^2 + y^2 = 25$	$(0, 0)$	5
$(x - 4)^2 + y^2 = 5$	$(4, 0)$	$\sqrt{5}$

$\blacktriangleleft$

The circle $x^2 + y^2 = 1$, which is centered at the origin and has radius 1, is of special importance; it is called the *unit circle* (Figure D.6).

OTHER FORMS FOR THE EQUATION OF A CIRCLE

An alternative version of Equation (3) can be obtained by squaring the terms and simplifying. This yields an equation of the form

$$x^2 + y^2 + dx + ey + f = 0 \qquad (4)$$

where d, e, and f are constants. (See the final equations in Examples 4 and 5.)

Still another version of the equation of a circle can be obtained by multiplying both sides of (4) by a nonzero constant A. This yields an equation of the form

$$Ax^2 + Ay^2 + Dx + Ey + F = 0 \tag{5}$$

where A, D, E, and F are constants and $A \neq 0$.

If the equation of a circle is given by (4) or (5), then the center and radius can be found by first rewriting the equation in standard form, then reading off the center and radius from that equation. The following example shows how to do this using the technique of **completing the square**. However, in preparation for the example, recall that completing the square is a method for rewriting an expression of the form

$$x^2 + bx$$

as a difference of two squares. The procedure is to take half the coefficient of x, square it, and then add and subtract that result from the original expression to obtain

$$x^2 + bx = x^2 + bx + (b/2)^2 - (b/2)^2 = [x + (b/2)]^2 - (b/2)^2$$

Example 7

Find the center and radius of the circle with equation

(a) $x^2 + y^2 - 8x + 2y + 8 = 0$ (b) $2x^2 + 2y^2 + 24x - 81 = 0$

Solution (a). First, group the x-terms, group the y-terms, and take the constant to the right side:

$$(x^2 - 8x) + (y^2 + 2y) = -8$$

Next we want to add the appropriate constant within each set of parentheses to complete the square, and subtract the same constant outside the parentheses to maintain equality. The appropriate constant is obtained by taking half the coefficient of the first-degree term and squaring it. This yields

$$(x^2 - 8x + 16) - 16 + (y^2 + 2y + 1) - 1 = -8$$

from which we obtain

$$(x - 4)^2 + (y + 1)^2 = -8 + 16 + 1 \quad \text{or} \quad (x - 4)^2 + (y + 1)^2 = 9$$

Thus from (3), the circle has center $(4, -1)$ and radius 3.

Solution (b). The given equation is of form (5). We will first divide through by 2 (the coefficient of the squared terms) to reduce the equation to form (4). Then we will proceed as in part (a) of this example. The computations are as follows:

$$x^2 + y^2 + 12x - \tfrac{81}{2} = 0 \quad \boxed{\text{We divided through by 2.}}$$
$$(x^2 + 12x) + y^2 = \tfrac{81}{2}$$
$$(x^2 + 12x + 36) + y^2 = \tfrac{81}{2} + 36 \quad \boxed{\text{We completed the square.}}$$
$$(x + 6)^2 + y^2 = \tfrac{153}{2}$$

From (3), the circle has center $(-6, 0)$ and radius $\sqrt{\tfrac{153}{2}}$. ◀

DEGENERATE CASES OF A CIRCLE

There is no guarantee that an equation of form (5) represents a circle. For example, suppose that we divide both sides of (5) by A, then complete the squares to obtain

$$(x - x_0)^2 + (y - y_0)^2 = k$$

Depending on the value of k, the following situations occur:

- $(k > 0)$ The graph is a circle with center (x_0, y_0) and radius $\sqrt{k}$.
- $(k = 0)$ The only solution of the equation is $x = x_0$, $y = y_0$, so the graph is the single point (x_0, y_0).
- $(k < 0)$ The equation has no real solutions and consequently no graph.

Example 8

Describe the graphs of

(a) $(x - 1)^2 + (y + 4)^2 = -9$ (b) $(x - 1)^2 + (y + 4)^2 = 0$

Solution (a). There are no real values of x and y that will make the left side of the equation negative. Thus, the solution set of the equation is empty, and the equation has no graph.

Solution (b). The only values of x and y that will make the left side of the equation 0 are $x = 1$, $y = -4$. Thus, the graph of the equation is the single point $(1, -4)$. ◀

The following theorem summarizes our observations.

D.3 THEOREM. *An equation of the form*

$$Ax^2 + Ay^2 + Dx + Ey + F = 0 \tag{6}$$

where $A \neq 0$, represents a circle, or a point, or else has no graph.

REMARK. The last two cases in Theorem D.3 are called ***degenerate cases***. In spite of the fact that these degenerate cases can occur, (6) is often called the ***general equation of a circle***.

THE GRAPH of $y = ax^2 + bx + c$

An equation of the form

$$y = ax^2 + bx + c \quad (a \neq 0) \tag{7}$$

is called a ***quadratic equation in x***. Depending on whether a is positive or negative, the graph, which is called a ***parabola***, has one of the two forms shown in Figure D.7. In both cases the parabola is symmetric about a vertical line parallel to the y-axis. This line of symmetry cuts the parabola at a point called the ***vertex***. The vertex is the low point on the curve if $a > 0$ and the high point if $a < 0$.

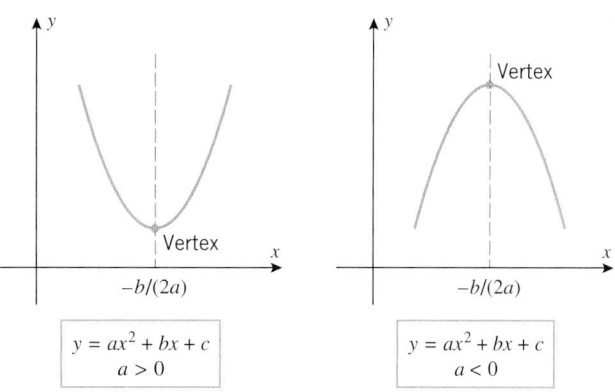

Figure D.7

x	$y = x^2 - 2x - 2$
-1	1
0	-2
1	-3
2	-2
3	1

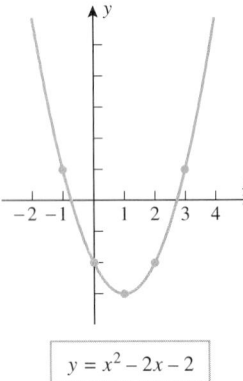

$y = x^2 - 2x - 2$

Figure D.8

In the exercises (Exercise 78) we will help the reader show that the x-coordinate of the vertex is given by the formula

$$x = -\frac{b}{2a} \tag{8}$$

With the aid of this formula, a reasonably accurate graph of a quadratic equation in x can be obtained by plotting the vertex and two points on each side of it.

Example 9

Sketch the graph of

   (a) $y = x^2 - 2x - 2$ (b) $y = -x^2 + 4x - 5$

Solution (a). The equation is of form (7) with $a = 1$, $b = -2$, and $c = -2$, so by (8) the x-coordinate of the vertex is

$$x = -\frac{b}{2a} = 1$$

Using this value and two additional values on each side, we obtain Figure D.8.

Solution (b). The equation is of form (7) with $a = -1$, $b = 4$, and $c = -5$, so by (8) the x-coordinate of the vertex is

$$x = -\frac{b}{2a} = 2$$

Using this value and two additional values on each side, we obtain the table and graph in Figure D.9.   ◄

x	$y = -x^2 + 4x - 5$
0	-5
1	-2
2	-1
3	-2
4	-5

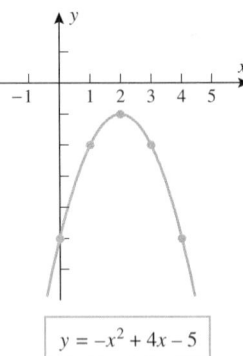

$y = -x^2 + 4x - 5$

Figure D.9

Quite often the intercepts of a parabola $y = ax^2 + bx + c$ are important to know. The y-intercept, $y = c$, results immediately by setting $x = 0$. However, in order to obtain the x-intercepts, if any, we must set $y = 0$ and then solve the resulting quadratic equation $ax^2 + bx + c = 0$.

Example 10

Solve the inequality

$$x^2 - 2x - 2 > 0$$

Solution. Because the left side of the inequality does not have readily discernible factors, the test-point method illustrated in Example 4 of Appendix A is not convenient to use. Instead, we will give a graphical solution. The given inequality is satisfied for those values of x where the graph of $y = x^2 - 2x - 2$ is above the x-axis. From Figure D.8 those are the values of x to the left of the smaller intercept or to the right of the larger intercept. To find these intercepts we set $y = 0$ to obtain

$$x^2 - 2x - 2 = 0$$

Solving by the quadratic formula gives

$$x = \frac{-b \pm \sqrt{b^2 - 4ac}}{2a} = \frac{2 \pm \sqrt{12}}{2} = 1 \pm \sqrt{3}$$

Thus, the x-intercepts are

$$x = 1 + \sqrt{3} \approx 2.7 \quad \text{and} \quad x = 1 - \sqrt{3} \approx -0.7$$

and the solution set of the inequality is

$$(-\infty, 1 - \sqrt{3}) \cup (1 + \sqrt{3}, +\infty)$$ ◄

REMARK.   Note that the decimal approximations of the intercepts calculated in the preceding example agree with the graph in Figure D.8. Observe, however, that we used the exact values of the intercepts to express the solution. The choice of exact versus approximate values is often a matter of judgment that depends on the purpose for which the values are to be used. Numerical approximations often provide a sense of size that exact values do not, but they can introduce severe errors if not used with care.

Example 11

From Figure D.9 we see that the parabola $y = -x^2 + 4x - 5$ has no x-intercepts. This can also be seen algebraically by solving for the x-intercepts. Setting $y = 0$ and solving the resulting equation

$$-x^2 + 4x - 5 = 0$$

by the quadratic formula yields

$$y = \frac{-4 \pm \sqrt{16 - 20}}{-2} = 2 \pm i$$

Because the solutions are complex numbers, there are no (real) x-intercepts. ◄

Example 12

A ball is thrown straight up from the surface of the Earth at time $t = 0$ s with an initial velocity of 24.5 m/s. If air resistance is ignored, it can be shown that the distance s (in meters) of the ball above the ground after t seconds is given by

$$s = 24.5t - 4.9t^2 \tag{9}$$

(a)   Graph s versus t, making the t-axis horizontal and the s-axis vertical.

(b)   How high does the ball rise above the ground?

Solution (a). Equation (9) is of form (7) with $a = -4.9$, $b = 24.5$, and $c = 0$, so by (8) the t-coordinate of the vertex is

$$t = -\frac{b}{2a} = -\frac{24.5}{2(-4.9)} = 2.5 \text{ s}$$

and consequently the s-coordinate of the vertex is

$$s = 24.5(2.5) - 4.9(2.5)^2 = 30.625 \text{ m}$$

The factored form of (9) is

$$s = 4.9t(5 - t)$$

so the graph has t-intercepts $t = 0$ and $t = 5$. From the vertex and the intercepts we obtain the graph shown in Figure D.10.

Solution (b). From the s-coordinate of the vertex we deduce that the ball rises 30.625 m above the ground. ◄

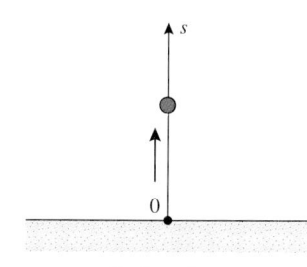

Earth surface

Figure D.10

THE GRAPH of $x = ay^2 + by + c$

If x and y are interchanged in (7), the resulting equation,

$$x = ay^2 + by + c$$

is called a ***quadratic equation in y***. The graph of such an equation is a parabola with its line of symmetry parallel to the x-axis and its vertex at the point with y-coordinate $y = -b/(2a)$ (Figure D.11). Some problems relating to such equations appear in the exercises.

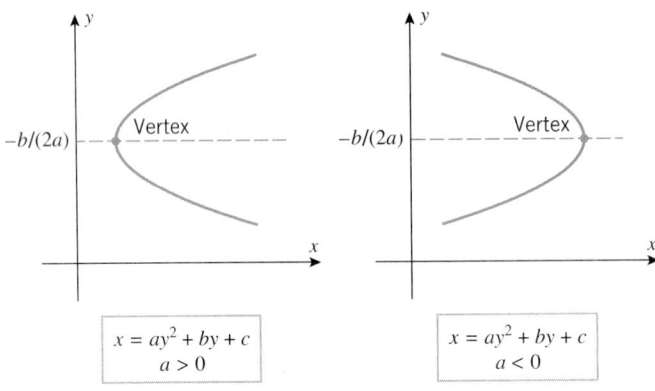

Figure D.11

Exercise Set D

1. Where in this section did we use the fact that the same scale was used on both coordinate axes?

> In Exercises 2–5, find
> (a) the distance between A and B
> (b) the midpoint of the line segment joining A and B.

2. $A(2, 5)$, $B(-1, 1)$ 3. $A(7, 1)$, $B(1, 9)$

4. $A(2, 0)$, $B(-3, 6)$ 5. $A(-2, -6)$, $B(-7, -4)$

> In Exercises 6–10, use the distance formula to solve the given problem.

6. Prove that $(1, 1)$, $(-2, -8)$, and $(4, 10)$ lie on a straight line.

7. Prove that the triangle with vertices $(5, -2)$, $(6, 5)$, $(2, 2)$ is isosceles.

8. Prove that $(1, 3)$, $(4, 2)$, and $(-2, -6)$ are vertices of a right triangle and then specify the vertex at which the right angle occurs.

9. Prove that $(0, -2)$, $(-4, 8)$, and $(3, 1)$ lie on a circle with center $(-2, 3)$.

10. Prove that for all values of t the point $(t, 2t - 6)$ is equidistant from $(0, 4)$ and $(8, 0)$.

11. Find k, given that $(2, k)$ is equidistant from $(3, 7)$ and $(9, 1)$.

12. Find x and y if $(4, -5)$ is the midpoint of the line segment joining $(-3, 2)$ and (x, y).

> In Exercises 13 and 14, find an equation of the given line.

13. The line is the perpendicular bisector of the line segment joining $(2, 8)$ and $(-4, 6)$.

14. The line is the perpendicular bisector of the line segment joining $(5, -1)$ and $(4, 8)$.

15. Find the point on the line $4x - 2y + 3 = 0$ that is equidistant from $(3, 3)$ and $(7, -3)$. [*Hint:* First find an equation of the line that is the perpendicular bisector of the line segment joining $(3, 3)$ and $(7, -3)$.]

16. Find the distance from the point $(3, -2)$ to the line
    (a) $y = 4$ (b) $x = -1$.

17. Find the distance from $(2, 1)$ to the line $4x - 3y + 10 = 0$. [*Hint:* Find the foot of the perpendicular dropped from the point to the line.]

18. Find the distance from $(8, 4)$ to the line $5x + 12y - 36 = 0$. [*Hint:* See the hint in Exercise 17.]

19. Use the method described in Exercise 17 to prove that the distance d from (x_0, y_0) to the line $Ax + By + C = 0$ is

$$d = \frac{|Ax_0 + By_0 + C|}{\sqrt{A^2 + B^2}}$$

20. Use the formula in Exercise 19 to solve Exercise 17.

21. Use the formula in Exercise 19 to solve Exercise 18.

22. Prove: For any triangle, the perpendicular bisectors of the sides meet at a point. [*Hint:* Position the triangle with one vertex on the y-axis and the opposite side on the x-axis, so that the vertices are $(0, a)$, $(b, 0)$, and $(c, 0)$.]

In Exercises 23 and 24, find the center and radius of each circle.

23. (a) $x^2 + y^2 = 25$
(b) $(x - 1)^2 + (y - 4)^2 = 16$
(c) $(x + 1)^2 + (y + 3)^2 = 5$
(d) $x^2 + (y + 2)^2 = 1$

24. (a) $x^2 + y^2 = 9$
(b) $(x - 3)^2 + (y - 5)^2 = 36$
(c) $(x + 4)^2 + (y + 1)^2 = 8$
(d) $(x + 1)^2 + y^2 = 1$

In Exercises 25–32, find the standard equation of the circle satisfying the given conditions.

25. Center $(3, -2)$; radius $= 4$.

26. Center $(1, 0)$; diameter $= \sqrt{8}$.

27. Center $(-4, 8)$; circle is tangent to the x-axis.

28. Center $(5, 8)$; circle is tangent to the y-axis.

29. Center $(-3, -4)$; circle passes through the origin.

30. Center $(4, -5)$; circle passes through $(1, 3)$.

31. A diameter has endpoints $(2, 0)$ and $(0, 2)$.

32. A diameter has endpoints $(6, 1)$ and $(-2, 3)$.

In Exercises 33–44, determine whether the equation represents a circle, a point, or no graph. If the equation represents a circle, find the center and radius.

33. $x^2 + y^2 - 2x - 4y - 11 = 0$

34. $x^2 + y^2 + 8x + 8 = 0$

35. $2x^2 + 2y^2 + 4x - 4y = 0$

36. $6x^2 + 6y^2 - 6x + 6y = 3$

37. $x^2 + y^2 + 2x + 2y + 2 = 0$

38. $x^2 + y^2 - 4x - 6y + 13 = 0$

39. $9x^2 + 9y^2 = 1$

40. $(x^2/4) + (y^2/4) = 1$

41. $x^2 + y^2 + 10y + 26 = 0$

42. $x^2 + y^2 - 10x - 2y + 29 = 0$

43. $16x^2 + 16y^2 + 40x + 16y - 7 = 0$

44. $4x^2 + 4y^2 - 16x - 24y = 9$

45. Find an equation of
(a) the bottom half of the circle $x^2 + y^2 = 16$
(b) the top half of the circle $x^2 + y^2 + 2x - 4y + 1 = 0$.

46. Find an equation of
(a) the right half of the circle $x^2 + y^2 = 9$
(b) the left half of the circle $x^2 + y^2 - 4x + 3 = 0$.

47. Graph
(a) $y = \sqrt{25 - x^2}$
(b) $y = \sqrt{5 + 4x - x^2}$.

48. Graph
(a) $x = -\sqrt{4 - y^2}$
(b) $x = 3 + \sqrt{4 - y^2}$.

49. Find an equation of the line that is tangent to the circle

$$x^2 + y^2 = 25$$

at the point $(3, 4)$ on the circle.

50. Find an equation of the line that is tangent to the circle at the point P on the circle
(a) $x^2 + y^2 + 2x = 9$; $P(2, -1)$
(b) $x^2 + y^2 - 6x + 4y = 13$; $P(4, 3)$.

51. For the circle $x^2 + y^2 = 20$ and the point $P(-1, 2)$:
(a) Is P inside, outside, or on the circle?
(b) Find the largest and smallest distances between P and points on the circle.

52. Follow the directions of Exercise 51 for the circle

$$x^2 + y^2 - 2y - 4 = 0$$

and the point $P\left(3, \frac{5}{2}\right)$.

53. Referring to the accompanying figure, find the coordinates of the points T and T', where the lines L and L' are tangent to the circle of radius 1 with center at the origin.

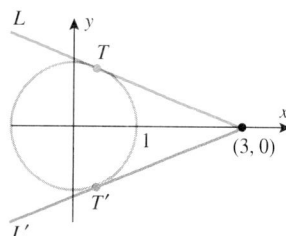

Figure Ex-53

54. A point (x, y) moves so that its distance to $(2, 0)$ is $\sqrt{2}$ times its distance to $(0, 1)$.
(a) Show that the point moves along a circle.
(b) Find the center and radius.

55. A point (x, y) moves so that the sum of the squares of its distances from $(4, 1)$ and $(2, -5)$ is 45.
(a) Show that the point moves along a circle.
(b) Find the center and radius.

56. Find all values of c for which the system of equations

$$\begin{cases} x^2 - y^2 = 0 \\ (x - c)^2 + y^2 = 1 \end{cases}$$

has 0, 1, 2, 3, or 4 solutions. [*Hint:* Sketch a graph.]

In Exercises 57–70, graph the parabola and label the coordinates of the vertex and the intersections with the coordinate axes.

57. $y = x^2 + 2$

58. $y = x^2 - 3$

59. $y = x^2 + 2x - 3$

60. $y = x^2 - 3x - 4$

61. $y = -x^2 + 4x + 5$

62. $y = -x^2 + x$

63. $y = (x - 2)^2$

64. $y = (3 + x)^2$

65. $x^2 - 2x + y = 0$

66. $x^2 + 8x + 8y = 0$

67. $y = 3x^2 - 2x + 1$

68. $y = x^2 + x + 2$

69. $x = -y^2 + 2y + 2$

70. $x = y^2 - 4y + 5$

71. Find an equation of
 (a) the right half of the parabola $y = 3 - x^2$
 (b) the left half of the parabola $y = x^2 - 2x$.

72. Find an equation of
 (a) the upper half of the parabola $x = y^2 - 5$
 (b) the lower half of the parabola $x = y^2 - y - 2$.

73. Graph
 (a) $y = \sqrt{x + 5}$
 (b) $x = -\sqrt{4 - y}$.

74. Graph
 (a) $y = 1 + \sqrt{4 - x}$
 (b) $x = 3 + \sqrt{y}$.

75. If a ball is thrown straight up with an initial velocity of 32 ft/s, then after t seconds the distance s above its starting height, in feet, is given by $s = 32t - 16t^2$.
 (a) Graph this equation in a ts-coordinate system (t-axis horizontal).
 (b) At what time t will the ball be at its highest point, and how high will it rise?

76. A rectangular field is to be enclosed with 500 ft of fencing along three sides and by a straight stream on the fourth side. Let x be the length of each side perpendicular to the stream, and let y be the length of the side parallel to the stream.
 (a) Express y in terms of x.
 (b) Express the area A of the field in terms of x.
 (c) What is the largest area that can be enclosed?

77. A rectangular plot of land is to be enclosed using two kinds of fencing. Two opposite sides will have heavy-duty fencing costing \$3/ft, while the other two sides will have standard fencing costing \$2/ft. A total of \$600 is available for the fencing. Let x be the length of each side with the heavy-duty fencing, and let y be the length of each side with the standard fencing.
 (a) Express y in terms of x.
 (b) Find a formula for the area A of the rectangular plot in terms of x.
 (c) What is the largest area that can be enclosed?

78. (a) By completing the square, show that the quadratic equation $y = ax^2 + bx + c$ can be rewritten as
$$y = a\left(x + \frac{b}{2a}\right)^2 + \left(c - \frac{b^2}{4a}\right)$$
if $a \neq 0$.
 (b) Use the result in part (a) to show that the graph of the quadratic equation $y = ax^2 + bx + c$ has its high point at $x = -b/(2a)$ if $a < 0$ and its low point there if $a > 0$.

In Exercises 79 and 80, solve the given inequality.

79. (a) $2x^2 + 5x - 1 < 0$
 (b) $x^2 - 2x + 3 > 0$

80. (a) $x^2 + x - 1 > 0$
 (b) $x^2 - 4x + 6 < 0$

81. At time $t = 0$ a ball is thrown straight up from a height of 5 ft above the ground. After t seconds its distance s, in feet, above the ground is given by $s = 5 + 40t - 16t^2$.
 (a) Find the maximum height of the ball above the ground.
 (b) Find, to the nearest tenth of a second, the time when the ball strikes the ground.
 (c) Find, to the nearest tenth of a second, how long the ball will be more than 12 ft above the ground.

82. Find all values of x at which points on the parabola $y = x^2$ lie below the line $y = x + 3$.

APPENDIX E

Trigonometry Review

TRIGONOMETRIC FUNCTIONS AND IDENTITIES

ANGLES

Angles in the plane can be generated by rotating a ray about its endpoint. The starting position of the ray is called the *initial side* of the angle, the final position is called the *terminal side* of the angle, and the point at which the initial and terminal sides meet is called the *vertex* of the angle. We allow for the possibility that the ray may make more than one complete revolution. Angles are considered to be *positive* if generated counterclockwise and *negative* if generated clockwise (Figure E.1).

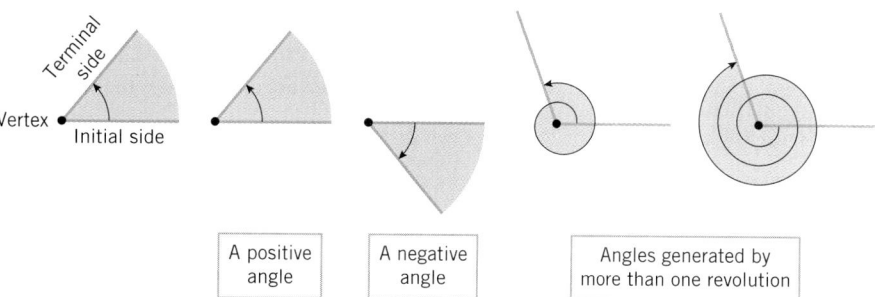

Figure E.1

There are two standard measurement systems for describing the size of an angle: *degree measure* and *radian measure*. In degree measure, one degree (written 1°) is the measure of an angle generated by 1/360 of one revolution. Thus, there are 360° in an angle of one revolution, 180° in an angle of one-half revolution, 90° in an angle of one-quarter revolution (a *right angle*), and so forth. Degrees are divided into sixty equal parts, called *minutes*, and minutes are divided into sixty equal parts, called *seconds*. Thus, one minute (written 1′) is 1/60 of a degree, and one second (written 1″) is 1/60 of a minute. Smaller subdivisions of a degree are expressed as fractions of a second.

In radian measure, angles are measured by the length of the arc that the angle subtends on a circle of radius 1 when the vertex is at the center. One unit of arc on a circle of radius 1 is called one *radian* (written 1 radian or 1 rad) (Figure E.2), and hence the entire circumference of a circle of radius 1 is 2π radians. It follows that an angle of 360° subtends an arc of 2π radians, an angle of 180° subtends an arc of π radians, an angle of 90° subtends an arc of $\pi/2$ radians, and so forth. Figure E.3 and Table 1 show the relationship between degree measure and radian measure for some important positive angles.

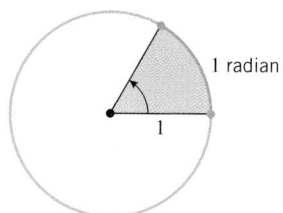

Figure E.2

REMARK.   Observe that in Table 1, angles in degrees are designated by the degree symbol, but angles in radians have no units specified. This is standard practice—when no units are specified for an angle, it is understood that the units are radians.

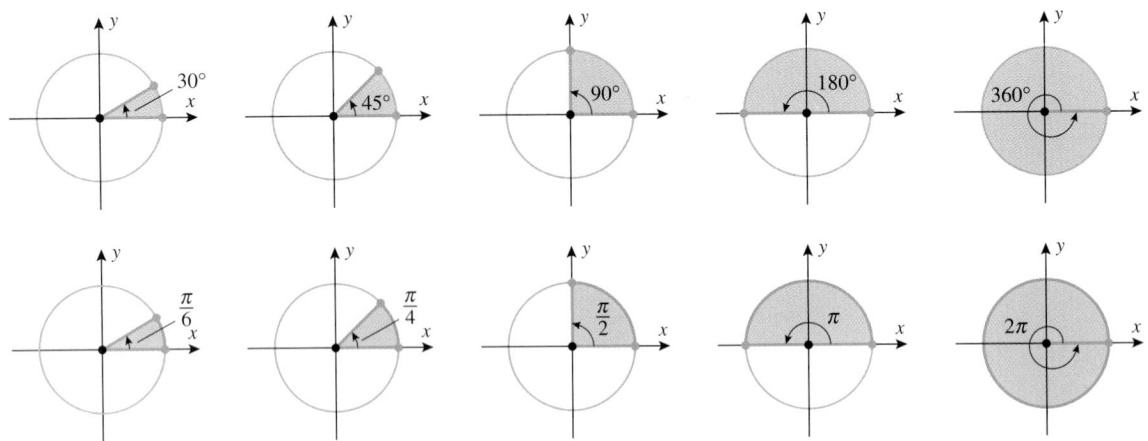

Figure E.3

Table 1

DEGREES	30°	45°	60°	90°	120°	135°	150°	180°	270°	360°
RADIANS	$\dfrac{\pi}{6}$	$\dfrac{\pi}{4}$	$\dfrac{\pi}{3}$	$\dfrac{\pi}{2}$	$\dfrac{2\pi}{3}$	$\dfrac{3\pi}{4}$	$\dfrac{5\pi}{6}$	π	$\dfrac{3\pi}{2}$	2π

From the fact that π radians corresponds to $180°$, we obtain the following formulas, which are useful for converting from degrees to radians and conversely.

$$1° = \frac{\pi}{180}\text{rad} \approx 0.01745 \text{ rad} \tag{1}$$

$$1 \text{ rad} = \left(\frac{180}{\pi}\right)^{\circ} \approx 57°\,17'\,44.8'' \tag{2}$$

Example 1

(a) Express $146°$ in radians.   (b) Express 3 radians in degrees.

Solution (a). From (1), degrees can be converted to radians by multiplying by a conversion factor of $\pi/180$. Thus,

$$146° = \left(\frac{\pi}{180} \cdot 146\right) \text{rad} = \frac{73\pi}{90} \text{ rad} \approx 2.5482 \text{ rad}$$

Solution (b). From (2), radians can be converted to degrees by multiplying by a conversion factor of $180/\pi$. Thus,

$$3 \text{ rad} = \left(3 \cdot \frac{180}{\pi}\right)^{\circ} = \left(\frac{540}{\pi}\right)^{\circ} \approx 171.9° \qquad \blacktriangleleft$$

RELATIONSHIPS BETWEEN ARC LENGTH, ANGLE, RADIUS, AND AREA

There is a theorem from plane geometry which states that for two concentric circles, the ratio of the arc lengths subtended by a central angle is equal to the ratio of the corresponding radii (Figure E.4). In particular, if s is the arc length subtended on a circle of radius r by a central angle of θ radians, then by comparison with the arc length subtended by that angle on a circle of radius 1 we obtain

$$\frac{s}{\theta} = \frac{r}{1}$$

from which we obtain the following relationships between the central angle θ, the radius r, and the subtended arc length s when θ is in radians (Figure E.5):

$$\theta = s/r \qquad \text{and} \qquad s = r\theta \tag{3-4}$$

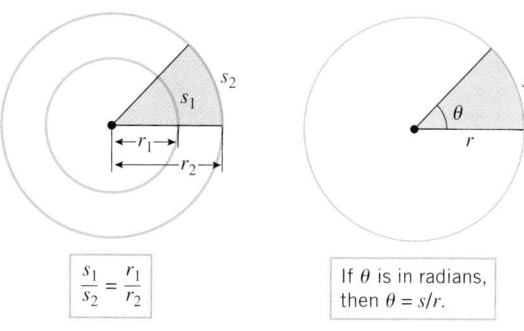

$$\frac{s_1}{s_2} = \frac{r_1}{r_2}$$

Figure E.4

If θ is in radians, then $\theta = s/r$.

Figure E.5

The shaded region in Figure E.5 is called a *sector*. It is a theorem from plane geometry that the ratio of the area A of this sector to the area of the entire circle is the same as the ratio of the central angle of the sector to the central angle of the entire circle; thus, if the angles are in radians, we have

$$\frac{A}{\pi r^2} = \frac{\theta}{2\pi}$$

Solving for A yields the following formula for the area of a sector in terms of the radius r and the angle θ in radians:

$$A = \tfrac{1}{2}r^2\theta \tag{5}$$

TRIGONOMETRIC FUNCTIONS FOR RIGHT TRIANGLES

The *sine*, *cosine*, *tangent*, *cosecant*, *secant*, and *cotangent* of a positive acute angle θ can be defined as ratios of the sides of a right triangle. Using the notation from Figure E.6, these definitions take the following form:

$$\sin\theta = \frac{\text{side opposite }\theta}{\text{hypotenuse}} = \frac{y}{r}, \qquad \csc\theta = \frac{\text{hypotenuse}}{\text{side opposite }\theta} = \frac{r}{y}$$

$$\cos\theta = \frac{\text{side adjacent to }\theta}{\text{hypotenuse}} = \frac{x}{r}, \qquad \sec\theta = \frac{\text{hypotenuse}}{\text{side adjacent to }\theta} = \frac{r}{x} \tag{6}$$

$$\tan\theta = \frac{\text{side opposite }\theta}{\text{side adjacent to }\theta} = \frac{y}{x}, \qquad \cot\theta = \frac{\text{side adjacent to }\theta}{\text{side opposite }\theta} = \frac{x}{y}$$

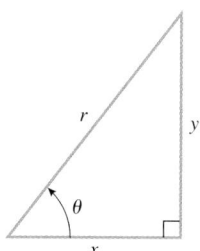

Figure E.6

We will call sin, cos, tan, csc, sec, and cot the *trigonometric functions*. Because similar triangles have proportional sides, the values of the trigonometric functions depend only on the size of θ and not on the particular right triangle used to compute the ratios. Moreover, in these definitions it does not matter whether θ is measured in degrees or radians.

Example 2

Recall from geometry that the two legs of a $45°-45°-90°$ triangle are of equal size and that the hypotenuse of a $30°-60°-90°$ triangle is twice the shorter leg, where the shorter leg is opposite the $30°$ angle. These facts and the Theorem of Pythagoras yield Figure E.7. From that figure we obtain the results in Table 2.

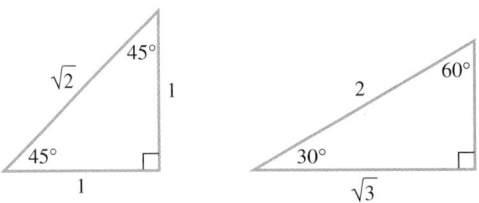

Figure E.7

Table 2

$\sin 45° = 1/\sqrt{2},$	$\cos 45° = 1/\sqrt{2},$	$\tan 45° = 1$
$\csc 45° = \sqrt{2},$	$\sec 45° = \sqrt{2},$	$\cot 45° = 1$
$\sin 30° = 1/2,$	$\cos 30° = \sqrt{3}/2,$	$\tan 30° = 1/\sqrt{3}$
$\csc 30° = 2,$	$\sec 30° = 2/\sqrt{3},$	$\cot 30° = \sqrt{3}$
$\sin 60° = \sqrt{3}/2,$	$\cos 60° = 1/2,$	$\tan 60° = \sqrt{3}$
$\csc 60° = 2/\sqrt{3},$	$\sec 60° = 2,$	$\cot 60° = 1/\sqrt{3}$

◄

ANGLES IN RECTANGULAR COORDINATE SYSTEMS

Because the angles of a right triangle are between $0°$ and $90°$, the formulas in (6) are not directly applicable to negative angles or to angles greater than $90°$. To extend the trigonometric functions to include these cases, it will be convenient to consider angles in rectangular coordinate systems. An angle is said to be in ***standard position*** in an xy-coordinate system if its vertex is at the origin and its initial side is on the positive x-axis (Figure E.8).

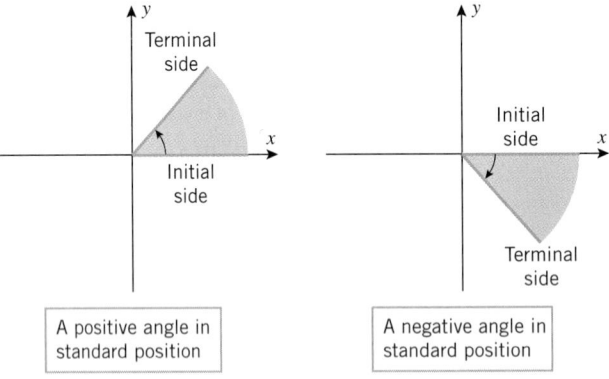

Figure E.8

To define the trigonometric functions of an angle θ in standard position, construct a circle of radius r, centered at the origin, and let $P(x, y)$ be the intersection of the terminal side of θ with this circle (Figure E.9). We make the following definition.

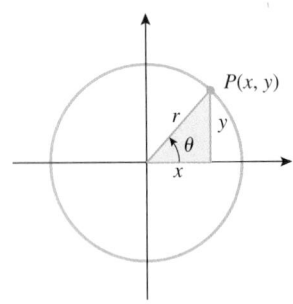

Figure E.9

E.1 DEFINITION.

$$\sin \theta = \frac{y}{r}, \quad \cos \theta = \frac{x}{r}, \quad \tan \theta = \frac{y}{x}$$

$$\csc \theta = \frac{r}{y}, \quad \sec \theta = \frac{r}{x}, \quad \cot \theta = \frac{x}{y}$$

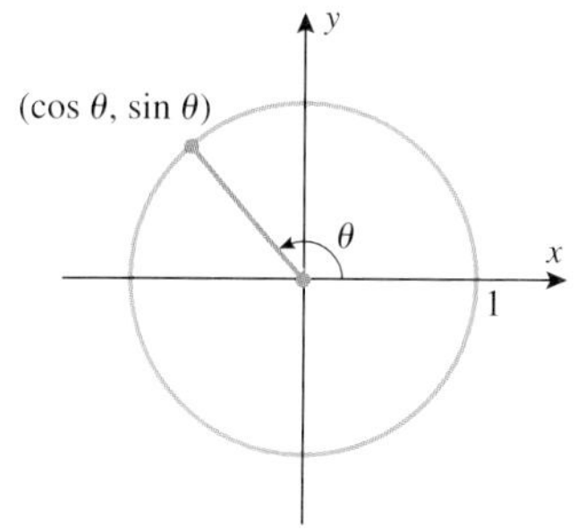

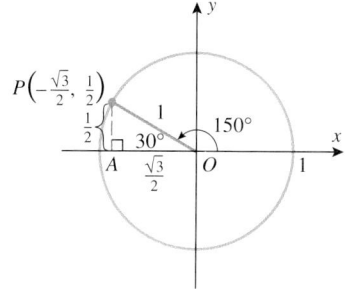

Figure E.10

Note that the formulas in this definition agree with those in (6), so there is no conflict with the earlier definition of the trigonometric functions for triangles. However, this definition applies to all angles (except for cases where a zero denominator occurs).

In the special case where $r = 1$, we have $\sin\theta = y$ and $\cos\theta = x$, so the terminal side of the angle θ intersects the unit circle at the point $(\cos\theta, \sin\theta)$ (Figure E.10). It follows from Definition E.1 that the remaining trigonometric functions of θ are expressible as (verify)

$$\tan\theta = \frac{\sin\theta}{\cos\theta}, \quad \cot\theta = \frac{\cos\theta}{\sin\theta} = \frac{1}{\tan\theta}, \quad \sec\theta = \frac{1}{\cos\theta}, \quad \csc\theta = \frac{1}{\sin\theta} \quad (7\text{--}10)$$

These observations suggest the following procedure for evaluating the trigonometric functions of common angles:

- Construct the angle θ in standard position in an xy-coordinate system.
- Find the coordinates of the intersection of the terminal side of the angle and the unit circle; the x- and y-coordinates of this intersection are the values of $\cos\theta$ and $\sin\theta$, respectively.
- Use Formulas (7) through (10) to find the values of the remaining trigonometric functions from the values of $\cos\theta$ and $\sin\theta$.

Example 3

Evaluate the trigonometric functions of $\theta = 150°$.

Solution. Construct a unit circle and place the angle $\theta = 150°$ in standard position (Figure E.11). Since $\angle AOP$ is $30°$ and $\triangle OAP$ is a $30°\text{--}60°\text{--}90°$ triangle, the leg AP has length $\frac{1}{2}$ (half the hypotenuse) and the leg OA has length $\sqrt{3}/2$ by the Theorem of Pythagoras. Thus, the coordinates of P are $(-\sqrt{3}/2, 1/2)$, from which we obtain

$$\sin 150° = \frac{1}{2}, \quad \cos 150° = -\frac{\sqrt{3}}{2}, \quad \tan 150° = \frac{\sin 150°}{\cos 150°} = \frac{1/2}{-\sqrt{3}/2} = -\frac{1}{\sqrt{3}}$$

$$\csc 150° = \frac{1}{\sin 150°} = 2, \quad \sec 150° = \frac{1}{\cos 150°} = -\frac{2}{\sqrt{3}}$$

$$\cot 150° = \frac{1}{\tan 150°} = -\sqrt{3} \quad\blacktriangleleft$$

Figure E.11

Example 4

Evaluate the trigonometric functions of $\theta = 5\pi/6$.

Solution. Since $5\pi/6 = 150°$, this problem is equivalent to that of Example 3. From that example we obtain

$$\sin\frac{5\pi}{6} = \frac{1}{2}, \quad \cos\frac{5\pi}{6} = -\frac{\sqrt{3}}{2}, \quad \tan\frac{5\pi}{6} = -\frac{1}{\sqrt{3}}$$

$$\csc\frac{5\pi}{6} = 2, \quad \sec\frac{5\pi}{6} = -\frac{2}{\sqrt{3}}, \quad \cot\frac{5\pi}{6} = -\sqrt{3} \quad\blacktriangleleft$$

Example 5

Evaluate the trigonometric functions of $\theta = -\pi/2$.

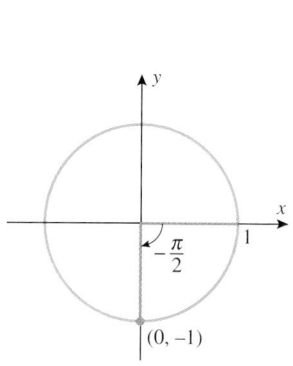

Figure E.12

Solution. As shown in Figure E.12, the terminal side of $\theta = -\pi/2$ intersects the unit circle at the point $(0, -1)$, so

$$\sin(-\pi/2) = -1, \quad \cos(-\pi/2) = 0$$

and from Formulas (7) through (10),

$$\tan(-\pi/2) = \frac{\sin(-\pi/2)}{\cos(-\pi/2)} = \frac{-1}{0} \quad \text{(undefined)}$$

$$\cot(-\pi/2) = \frac{\cos(-\pi/2)}{\sin(-\pi/2)} = \frac{0}{-1} = 0$$

$$\sec(-\pi/2) = \frac{1}{\cos(-\pi/2)} = \frac{1}{0} \quad \text{(undefined)}$$

$$\csc(-\pi/2) = \frac{1}{\sin(-\pi/2)} = \frac{1}{-1} = -1 \qquad \blacktriangleleft$$

The reader should be able to obtain all of the results in Table 3 by the methods illustrated in the last three examples. The dashes indicate quantities that are undefined.

Table 3

	$\theta = 0$ (0°)	$\pi/6$ (30°)	$\pi/4$ (45°)	$\pi/3$ (60°)	$\pi/2$ (90°)	$2\pi/3$ (120°)	$3\pi/4$ (135°)	$5\pi/6$ (150°)	π (180°)	$3\pi/2$ (270°)	2π (360°)
$\sin\theta$	0	1/2	$1/\sqrt{2}$	$\sqrt{3}/2$	1	$\sqrt{3}/2$	$1/\sqrt{2}$	1/2	0	−1	0
$\cos\theta$	1	$\sqrt{3}/2$	$1/\sqrt{2}$	1/2	0	−1/2	$-1/\sqrt{2}$	$-\sqrt{3}/2$	−1	0	1
$\tan\theta$	0	$1/\sqrt{3}$	1	$\sqrt{3}$	—	$-\sqrt{3}$	−1	$-1/\sqrt{3}$	0	—	0
$\csc\theta$	—	2	$\sqrt{2}$	$2/\sqrt{3}$	1	$2/\sqrt{3}$	$\sqrt{2}$	2	—	−1	—
$\sec\theta$	1	$2/\sqrt{3}$	$\sqrt{2}$	2	—	−2	$-\sqrt{2}$	$-2/\sqrt{3}$	−1	—	1
$\cot\theta$	—	$\sqrt{3}$	1	$1/\sqrt{3}$	0	$-1/\sqrt{3}$	−1	$-\sqrt{3}$	—	0	—

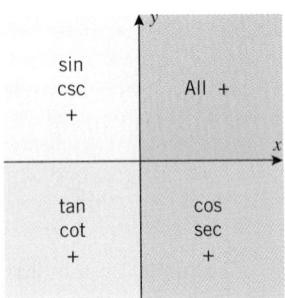

Figure E.13

TRIGONOMETRIC IDENTITIES

REMARK.   It is only in special cases that exact values for trigonometric functions can be obtained; usually, a calculating utility or a computer program will be required.

The signs of the trigonometric functions of an angle are determined by the quadrant in which the terminal side of the angle falls. For example, if the terminal side falls in the first quadrant, then x and y are positive in Definition E.1, so all of the trigonometric functions have positive values. If the terminal side falls in the second quadrant, then x is negative and y is positive, so sin and csc are positive, but all other trigonometric functions are negative. The diagram in Figure E.13 shows which trigonometric functions are positive in the various quadrants. The reader will find it instructive to check that the results in Table 3 are consistent with Figure E.13.

A *trigonometric identity* is an equation involving trigonometric functions that is true for all angles for which both sides of the equation are defined. One of the most important identities in trigonometry can be derived by applying the Theorem of Pythagoras to the triangle in Figure E.9 to obtain

$$x^2 + y^2 = r^2$$

Dividing both sides by r^2 and using the definitions of $\sin\theta$ and $\cos\theta$ (Definition E.1), we obtain the following fundamental result:

$$\sin^2\theta + \cos^2\theta = 1 \tag{11}$$

The following identities can be obtained from (11) by dividing through by $\cos^2\theta$ and $\sin^2\theta$, respectively, then applying Formulas (7) through (10):

$$\tan^2 \theta + 1 = \sec^2 \theta \tag{12}$$

$$1 + \cot^2 \theta = \csc^2 \theta \tag{13}$$

If (x, y) is a point on the unit circle, then the points $(-x, y)$, $(-x, -y)$, and $(x, -y)$ also lie on the unit circle (why?), and the four points form corners of a rectangle with sides parallel to the coordinate axes (Figure E.14a). The x- and y-coordinates of each corner represent the cosine and sine of an angle in standard position whose terminal side passes through the corner; hence we obtain the identities in parts (b), (c), and (d) of Figure E.14 for sine and cosine. Dividing those identities leads to identities for the tangent. In summary:

$$\sin(\pi - \theta) = \sin\theta, \qquad \sin(\pi + \theta) = -\sin\theta, \qquad \sin(-\theta) = -\sin\theta \tag{14–16}$$
$$\cos(\pi - \theta) = -\cos\theta, \qquad \cos(\pi + \theta) = -\cos\theta, \qquad \cos(-\theta) = \cos\theta \tag{17–19}$$
$$\tan(\pi - \theta) = -\tan\theta, \qquad \tan(\pi + \theta) = \tan\theta, \qquad \tan(-\theta) = -\tan\theta \tag{20–22}$$

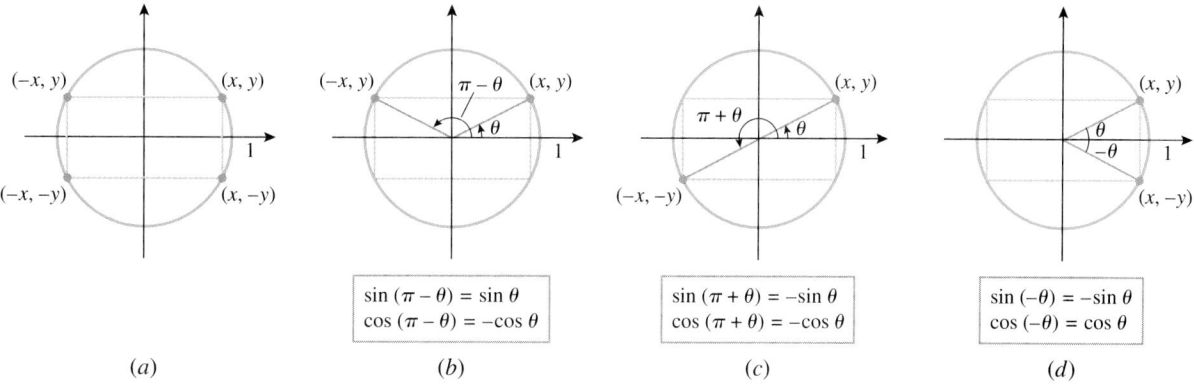

(a) (b) (c) (d)

Figure E.14

Two angles in standard position that have the same terminal side must have the same values for their trigonometric functions since their terminal sides intersect the unit circle at the same point. In particular, two angles whose radian measures differ by a multiple of 2π have the same terminal side and hence have the same values for their trigonometric functions. This yields the identities

$$\sin\theta = \sin(\theta + 2\pi) = \sin(\theta - 2\pi) \tag{23}$$
$$\cos\theta = \cos(\theta + 2\pi) = \cos(\theta - 2\pi) \tag{24}$$

and more generally,

$$\sin\theta = \sin(\theta \pm 2n\pi), \quad n = 0, 1, 2, \ldots \tag{25}$$
$$\cos\theta = \cos(\theta \pm 2n\pi), \quad n = 0, 1, 2, \ldots \tag{26}$$

Identities (20) through (22) imply that

$$\tan\theta = \tan(\theta + \pi) \qquad \text{and} \qquad \tan\theta = \tan(\theta - \pi) \tag{27–28}$$

Identity (27) is just (21) with the terms in the sum reversed, and identity (28) follows from (20) and (22) (verify). These two identities state that adding or subtracting π from an angle does not affect the value of the tangent of the angle. It follows that the same is true for any

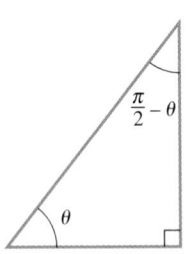

Figure E.15

multiple of π; thus,

$$\tan \theta = \tan(\theta \pm n\pi), \quad n = 0, 1, 2, \ldots \tag{29}$$

Figure E.15 shows complementary angles θ and $(\pi/2) - \theta$ of a right triangle. It follows from (6) that

$$\sin \theta = \frac{\text{side opposite } \theta}{\text{hypotenuse}} = \frac{\text{side adjacent to } (\pi/2) - \theta}{\text{hypotenuse}} = \cos \left(\frac{\pi}{2} - \theta \right)$$

$$\cos \theta = \frac{\text{side adjacent to } \theta}{\text{hypotenuse}} = \frac{\text{side opposite } (\pi/2) - \theta}{\text{hypotenuse}} = \sin \left(\frac{\pi}{2} - \theta \right)$$

which yields the identities

$$\sin \left(\frac{\pi}{2} - \theta \right) = \cos \theta, \quad \cos \left(\frac{\pi}{2} - \theta \right) = \sin \theta, \quad \tan \left(\frac{\pi}{2} - \theta \right) = \cot \theta \tag{30–32}$$

where the third identity results from dividing the first two. These identities are also valid for angles that are not acute and for negative angles as well.

THE LAW OF COSINES

The next theorem, called the *law of cosines*, generalizes the Theorem of Pythagoras. This result is important in its own right and is also the starting point for some important trigonometric identities.

E.2 THEOREM (*Law of Cosines*). *If the sides of a triangle have lengths a, b, and c, and if θ is the angle between the sides with lengths a and b, then*

$$c^2 = a^2 + b^2 - 2ab \cos \theta$$

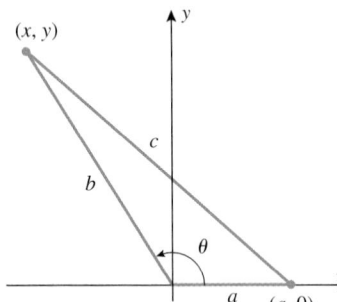

Figure E.16

Proof. Introduce a coordinate system so that θ is in standard position and the side of length a falls along the positive x-axis. As shown in Figure E.16, the side of length a extends from the origin to $(a, 0)$ and the side of length b extends from the origin to some point (x, y). From the definition of $\sin \theta$ and $\cos \theta$ we have $\sin \theta = y/b$ and $\cos \theta = x/b$, so

$$y = b \sin \theta, \quad x = b \cos \theta \tag{33}$$

From the distance formula in Theorem D.1 of Appendix D, we obtain

$$c^2 = (x - a)^2 + (y - 0)^2$$

so that, from (33),

$$c^2 = (b \cos \theta - a)^2 + b^2 \sin^2 \theta$$

$$= a^2 + b^2 (\cos^2 \theta + \sin^2 \theta) - 2ab \cos \theta$$

$$= a^2 + b^2 - 2ab \cos \theta$$

which completes the proof. ■

We will now show how the law of cosines can be used to obtain the following identities, called the *addition formulas* for sine and cosine:

$$\sin(\alpha + \beta) = \sin \alpha \cos \beta + \cos \alpha \sin \beta \tag{34}$$

$$\cos(\alpha + \beta) = \cos \alpha \cos \beta - \sin \alpha \sin \beta \tag{35}$$

$$\sin(\alpha - \beta) = \sin \alpha \cos \beta - \cos \alpha \sin \beta \tag{36}$$

$$\cos(\alpha - \beta) = \cos \alpha \cos \beta + \sin \alpha \sin \beta \tag{37}$$

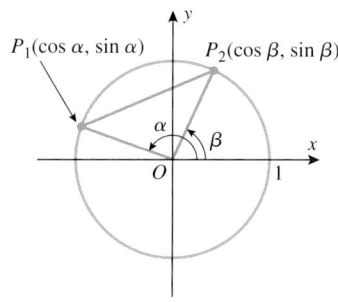

$P_1(\cos\alpha, \sin\alpha)$ $P_2(\cos\beta, \sin\beta)$

Figure E.17

We will derive (37) first. In our derivation we will assume that $0 \le \beta < \alpha < 2\pi$ (Figure E.17). As shown in the figure, the terminal sides of α and β intersect the unit circle at the points $P_1(\cos\alpha, \sin\alpha)$ and $P_2(\cos\beta, \sin\beta)$. If we denote the lengths of the sides of triangle OP_1P_2 by OP_1, P_1P_2, and OP_2, then $OP_1 = OP_2 = 1$ and, from the distance formula in Theorem D.1 of Appendix D,

$$(P_1P_2)^2 = (\cos\beta - \cos\alpha)^2 + (\sin\beta - \sin\alpha)^2$$
$$= (\sin^2\alpha + \cos^2\alpha) + (\sin^2\beta + \cos^2\beta) - 2(\cos\alpha\cos\beta + \sin\alpha\sin\beta)$$
$$= 2 - 2(\cos\alpha\cos\beta + \sin\alpha\sin\beta)$$

But angle $P_2OP_1 = \alpha - \beta$, so that the law of cosines yields

$$(P_1P_2)^2 = (OP_1)^2 + (OP_2)^2 - 2(OP_1)(OP_2)\cos(\alpha - \beta)$$
$$= 2 - 2\cos(\alpha - \beta)$$

Equating the two expressions for $(P_1P_2)^2$ and simplifying, we obtain

$$\cos(\alpha - \beta) = \cos\alpha\cos\beta + \sin\alpha\sin\beta$$

which completes the derivation of (37).

We can use (31) and (37) to derive (36) as follows:

$$\sin(\alpha - \beta) = \cos\left[\frac{\pi}{2} - (\alpha - \beta)\right] = \cos\left[\left(\frac{\pi}{2} - \alpha\right) - (-\beta)\right]$$
$$= \cos\left(\frac{\pi}{2} - \alpha\right)\cos(-\beta) + \sin\left(\frac{\pi}{2} - \alpha\right)\sin(-\beta)$$
$$= \cos\left(\frac{\pi}{2} - \alpha\right)\cos\beta - \sin\left(\frac{\pi}{2} - \alpha\right)\sin\beta$$
$$= \sin\alpha\cos\beta - \cos\alpha\sin\beta$$

Identities (34) and (35) can be obtained from (36) and (37) by substituting $-\beta$ for β and using the identities

$$\sin(-\beta) = -\sin\beta, \quad \cos(-\beta) = \cos\beta$$

We leave it for the reader to derive the identities

$$\tan(\alpha + \beta) = \frac{\tan\alpha + \tan\beta}{1 - \tan\alpha\tan\beta} \qquad \tan(\alpha - \beta) = \frac{\tan\alpha - \tan\beta}{1 + \tan\alpha\tan\beta} \qquad (38\text{--}39)$$

Identity (38) can be obtained by dividing (34) by (35) and then simplifying. Identity (39) can be obtained from (38) by substituting $-\beta$ for β and simplifying.

In the special case where $\alpha = \beta$, identities (34), (35), and (38) yield the ***double-angle formulas***

$$\sin 2\alpha = 2\sin\alpha\cos\alpha \tag{40}$$

$$\cos 2\alpha = \cos^2\alpha - \sin^2\alpha \tag{41}$$

$$\tan 2\alpha = \frac{2\tan\alpha}{1 - \tan^2\alpha} \tag{42}$$

By using the identity $\sin^2\alpha + \cos^2\alpha = 1$, (41) can be rewritten in the alternative forms

$$\cos 2\alpha = 2\cos^2\alpha - 1 \qquad \text{and} \qquad \cos 2\alpha = 1 - 2\sin^2\alpha \qquad (43\text{--}44)$$

If we replace α by $\alpha/2$ in (43) and (44) and use some algebra, we obtain the ***half-angle formulas***

$$\cos^2\frac{\alpha}{2} = \frac{1 + \cos\alpha}{2} \qquad \text{and} \qquad \sin^2\frac{\alpha}{2} = \frac{1 - \cos\alpha}{2} \qquad (45\text{--}46)$$

We leave it for the exercises to derive the following ***product-to-sum formulas*** from (34) through (37):

$$\sin\alpha \cos\beta = \frac{1}{2}[\sin(\alpha - \beta) + \sin(\alpha + \beta)] \tag{47}$$

$$\sin\alpha \sin\beta = \frac{1}{2}[\cos(\alpha - \beta) - \cos(\alpha + \beta)] \tag{48}$$

$$\cos\alpha \cos\beta = \frac{1}{2}[\cos(\alpha - \beta) + \cos(\alpha + \beta)] \tag{49}$$

We also leave it for the exercises to derive the following ***sum-to-product formulas***:

$$\sin\alpha + \sin\beta = 2\sin\frac{\alpha + \beta}{2}\cos\frac{\alpha - \beta}{2} \tag{50}$$

$$\sin\alpha - \sin\beta = 2\cos\frac{\alpha + \beta}{2}\sin\frac{\alpha - \beta}{2} \tag{51}$$

$$\cos\alpha + \cos\beta = 2\cos\frac{\alpha + \beta}{2}\cos\frac{\alpha - \beta}{2} \tag{52}$$

$$\cos\alpha - \cos\beta = -2\sin\frac{\alpha + \beta}{2}\sin\frac{\alpha - \beta}{2} \tag{53}$$

FINDING AN ANGLE FROM THE VALUE OF ITS TRIGONOMETRIC FUNCTIONS

There are numerous situations in which it is necessary to find an unknown angle from a known value of one of its trigonometric functions. The following example illustrates a method for doing this.

Example 6

Find θ if $\sin\theta = \frac{1}{2}$.

Solution. We begin by looking for positive angles that satisfy the equation. Because $\sin\theta$ is positive, the angle θ must terminate in the first or second quadrant. If it terminates in the first quadrant, then the hypotenuse of $\triangle OAP$ in Figure E.18a is double the leg AP, so

$$\theta = 30° = \frac{\pi}{6} \text{ radians}$$

If θ terminates in the second quadrant (Figure E.18b), then the hypotenuse of $\triangle OAP$ is double the leg AP, so $\angle AOP = 30°$, which implies that

$$\theta = 180° - 30° = 150° = \frac{5\pi}{6} \text{ radians}$$

Now that we have found these two solutions, all other solutions are obtained by adding or subtracting multiples of $360°$ (2π radians) to them. Thus, the entire set of solutions is given by the formulas

$$\theta = 30° \pm n \cdot 360°, \quad n = 0, 1, 2, \ldots$$

and

$$\theta = 150° \pm n \cdot 360°, \quad n = 0, 1, 2, \ldots$$

or in radian measure,

$$\theta = \frac{\pi}{6} \pm n \cdot 2\pi, \quad n = 0, 1, 2, \ldots$$

and

$$\theta = \frac{5\pi}{6} \pm n \cdot 2\pi, \quad n = 0, 1, 2, \ldots \quad \blacktriangleleft$$

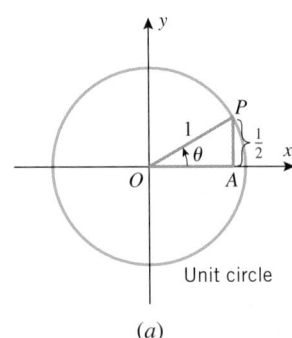

Unit circle

(a)

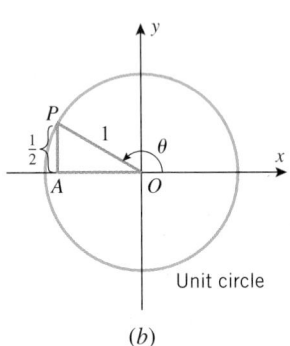

Unit circle

(b)

Figure E.18

EXERCISE SET E

In Exercises 1 and 2, express the angles in radians.

1. (a) $75°$ (b) $390°$ (c) $20°$ (d) $138°$

2. (a) $420°$ (b) $15°$ (c) $225°$ (d) $165°$

In Exercises 3 and 4, express the angles in degrees.

3. (a) $\pi/15$ (b) 1.5 (c) $8\pi/5$ (d) 3π

4. (a) $\pi/10$ (b) 2 (c) $2\pi/5$ (d) $7\pi/6$

In Exercises 5 and 6, find the exact values of all six trigonometric functions of θ.

5. (a)   (b)   (c)

   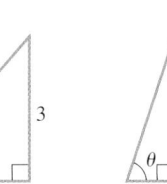

6. (a)   (b)   (c)

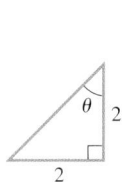   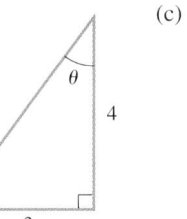   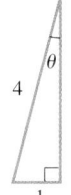

In Exercises 7–12, the angle θ is an acute angle of a right triangle. Solve the problems by drawing an appropriate right triangle. Do *not* use a calculator.

7. Find $\sin\theta$ and $\cos\theta$ given that $\tan\theta = 3$.

8. Find $\sin\theta$ and $\tan\theta$ given that $\cos\theta = \frac{2}{3}$.

9. Find $\tan\theta$ and $\csc\theta$ given that $\sec\theta = \frac{5}{2}$.

10. Find $\cot\theta$ and $\sec\theta$ given that $\csc\theta = 4$.

11. Find the length of the side adjacent to θ given that the hypotenuse has length 6 and $\cos\theta = 0.3$.

12. Find the length of the hypotenuse given that the side opposite θ has length 2.4 and $\sin\theta = 0.8$.

In Exercises 13 and 14, the value of an angle θ is given. Find the values of all six trigonometric functions of θ without using a calculator.

13. (a) $225°$ (b) $-210°$ (c) $5\pi/3$ (d) $-3\pi/2$

14. (a) $330°$ (b) $-120°$ (c) $9\pi/4$ (d) -3π

In Exercises 15 and 16, use the information to find the exact values of the remaining five trigonometric functions of θ.

15. (a) $\cos\theta = \frac{3}{5}$, $0 < \theta < \pi/2$
   (b) $\cos\theta = \frac{3}{5}$, $-\pi/2 < \theta < 0$
   (c) $\tan\theta = -1/\sqrt{3}$, $\pi/2 < \theta < \pi$
   (d) $\tan\theta = -1/\sqrt{3}$, $-\pi/2 < \theta < 0$
   (e) $\csc\theta = \sqrt{2}$, $0 < \theta < \pi/2$
   (f) $\csc\theta = \sqrt{2}$, $\pi/2 < \theta < \pi$

16. (a) $\sin\theta = \frac{1}{4}$, $0 < \theta < \pi/2$
   (b) $\sin\theta = \frac{1}{4}$, $\pi/2 < \theta < \pi$
   (c) $\cot\theta = \frac{1}{3}$, $0 < \theta < \pi/2$
   (d) $\cot\theta = \frac{1}{3}$, $\pi < \theta < 3\pi/2$
   (e) $\sec\theta = -\frac{5}{2}$, $\pi/2 < \theta < \pi$
   (f) $\sec\theta = -\frac{5}{2}$, $\pi < \theta < 3\pi/2$

In Exercises 17 and 18, use a calculating utility to find x to four decimal places.

17. (a)   (b)

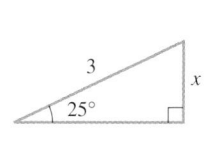   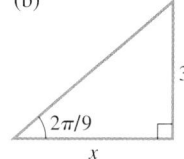

18. (a)   (b)

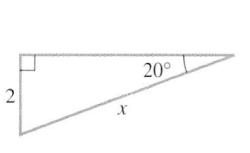   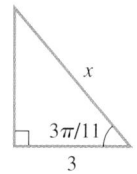

19. In each part, let θ be an acute angle of a right triangle. Express the remaining five trigonometric functions in terms of a.
   (a) $\sin\theta = a/3$ (b) $\tan\theta = a/5$ (c) $\sec\theta = a$

In Exercises 20–27, find all values of θ (in radians) that satisfy the given equation. Do not use a calculator.

20. (a) $\cos\theta = -1/\sqrt{2}$ (b) $\sin\theta = -1/\sqrt{2}$

21. (a) $\tan\theta = -1$ (b) $\cos\theta = \frac{1}{2}$

22. (a) $\sin\theta = -\frac{1}{2}$ (b) $\tan\theta = \sqrt{3}$

23. (a) $\tan\theta = 1/\sqrt{3}$ (b) $\sin\theta = -\sqrt{3}/2$

24. (a) $\sin\theta = -1$ (b) $\cos\theta = -1$

25. (a) $\cot\theta = -1$ (b) $\cot\theta = \sqrt{3}$

26. (a) $\sec\theta = -2$ (b) $\csc\theta = -2$

27. (a) $\csc\theta = 2/\sqrt{3}$ (b) $\sec\theta = 2/\sqrt{3}$

In Exercises 28 and 29, find the values of all six trigonometric functions of θ.

28.

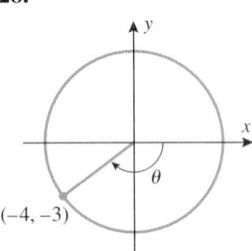

29.

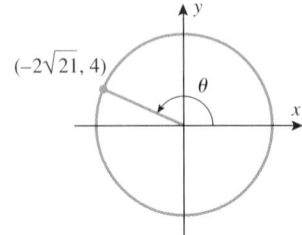

30. Find all values of θ (in radians) such that
(a) $\sin\theta = 1$ (b) $\cos\theta = 1$ (c) $\tan\theta = 1$
(d) $\csc\theta = 1$ (e) $\sec\theta = 1$ (f) $\cot\theta = 1$.

31. Find all values of θ (in radians) such that
(a) $\sin\theta = 0$ (b) $\cos\theta = 0$ (c) $\tan\theta = 0$
(d) $\csc\theta$ is undefined   (e) $\sec\theta$ is undefined
(f) $\cot\theta$ is undefined.

32. How could you use a ruler and protractor to approximate $\sin 17°$ and $\cos 17°$?

33. Find the length of the circular arc on a circle of radius 4 cm subtended by an angle of
(a) $\pi/6$ (b) $150°$.

34. Find the radius of a circular sector that has an angle of $\pi/3$ and a circular arc length of 7 units.

35. A point P moving counterclockwise on a circle of radius 5 cm traverses an arc length of 2 cm. What is the angle swept out by a radius from the center to P?

36. Find a formula for the area A of a circular sector in terms of its radius r and arc length s.

37. As shown in the accompanying figure, a right circular cone is made from a circular piece of paper of radius R by cutting out a sector of angle θ radians and gluing the cut edges of the remaining piece together. Find
(a) the radius r of the base of the cone in terms of R and θ
(b) the height h of the cone in terms of R and θ.

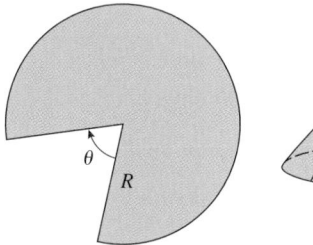
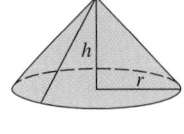

Figure Ex-37

38. As shown in the accompanying figure, let r and L be the radius of the base and the slant height of a right circular cone. Show that the lateral surface area, S, of the cone is $S = \pi r L$. [*Hint:* As shown in the figure in Exercise 37, the lateral surface of the cone becomes a circular sector when cut along a line from the vertex to the base and flattened.]

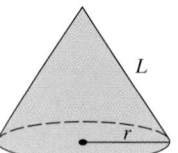

Figure Ex-38

39. Two sides of a triangle have lengths of 3 cm and 7 cm and meet at an angle of $60°$. Find the area of the triangle.

40. Let ABC be a triangle whose angles at A and B are $30°$ and $45°$. If the side opposite the angle B has length 9, find the lengths of the remaining sides and the size of the angle C.

41. A 10-foot ladder leans against a house and makes an angle of $67°$ with level ground. How far is the top of the ladder above the ground? Express your answer to the nearest tenth of a foot.

42. From a point 120 feet on level ground from a building, the angle of elevation to the top of the building is $76°$. Find the height of the building. Express your answer to the nearest foot.

43. An observer on level ground is at a distance d from a building. The angles of elevation to the bottoms of the windows on the second and third floors are α and β, respectively. Find the distance h between the bottoms of the windows in terms of α, β, and d.

44. From a point on level ground, the angle of elevation to the top of a tower is α. From a point that is d units closer to the tower, the angle of elevation is β. Find the height h of the tower in terms of α, β, and d.

In Exercises 45 and 46, do *not* use a calculator.

45. If $\cos\theta = \frac{2}{3}$ and $0 < \theta < \pi/2$, find
(a) $\sin 2\theta$ (b) $\cos 2\theta$.

46. If $\tan\alpha = \frac{3}{4}$ and $\tan\beta = 2$, where $0 < \alpha < \pi/2$ and $0 < \beta < \pi/2$, find
(a) $\sin(\alpha - \beta)$ (b) $\cos(\alpha + \beta)$.

47. Express $\sin 3\theta$ and $\cos 3\theta$ in terms of $\sin\theta$ and $\cos\theta$.

In Exercises 48–58, derive the given identities.

48. $\dfrac{\cos\theta\sec\theta}{1 + \tan^2\theta} = \cos^2\theta$

49. $\dfrac{\cos\theta\tan\theta + \sin\theta}{\tan\theta} = 2\cos\theta$

50. $2\csc 2\theta = \sec\theta\csc\theta$ **51.** $\tan\theta + \cot\theta = 2\csc 2\theta$

52. $\dfrac{\sin 2\theta}{\sin \theta} - \dfrac{\cos 2\theta}{\cos \theta} = \sec \theta$

53. $\dfrac{\sin \theta + \cos 2\theta - 1}{\cos \theta - \sin 2\theta} = \tan \theta$

54. $\sin 3\theta + \sin \theta = 2 \sin 2\theta \cos \theta$

55. $\sin 3\theta - \sin \theta = 2 \cos 2\theta \sin \theta$

56. $\tan \dfrac{\theta}{2} = \dfrac{1 - \cos \theta}{\sin \theta}$ **57.** $\tan \dfrac{\theta}{2} = \dfrac{\sin \theta}{1 + \cos \theta}$

58. $\cos \left(\dfrac{\pi}{3} + \theta\right) + \cos \left(\dfrac{\pi}{3} - \theta\right) = \cos \theta$

Exercises 59 and 60 refer to an arbitrary triangle ABC in which the side of length a is opposite angle A, the side of length b is opposite angle B, and the side of length c is opposite angle C.

---

59. Prove: The area of a triangle ABC can be written as

$$\text{area} = \tfrac{1}{2} bc \sin A$$

Find two other similar formulas for the area.

60. Prove the *law of sines*: In any triangle, the ratios of the sides to the sines of the opposite angles are equal; that is,

$$\dfrac{a}{\sin A} = \dfrac{b}{\sin B} = \dfrac{c}{\sin C}$$

61. Use identities (34) through (37) to express each of the following in terms of $\sin \theta$ or $\cos \theta$.

(a) $\sin \left(\dfrac{\pi}{2} + \theta\right)$ (b) $\cos \left(\dfrac{\pi}{2} + \theta\right)$

(c) $\sin \left(\dfrac{3\pi}{2} - \theta\right)$ (d) $\cos \left(\dfrac{3\pi}{2} + \theta\right)$

62. Derive identities (38) and (39).

63. Derive identity
(a) (47)   (b) (48)   (c) (49).

64. If $A = \alpha + \beta$ and $B = \alpha - \beta$, then $\alpha = \tfrac{1}{2}(A + B)$ and $\beta = \tfrac{1}{2}(A - B)$ (verify). Use this result and identities (47) through (49) to derive identity
(a) (50)   (b) (52)   (c) (53).

65. Substitute $-\beta$ for β in identity (50) to derive identity (51).

66. (a) Express $3 \sin \alpha + 5 \cos \alpha$ in the form

$$C \sin(\alpha + \phi)$$

(b) Show that a sum of the form

$$A \sin \alpha + B \cos \alpha$$

can be rewritten in the form $C \sin(\alpha + \phi)$.

67. Show that the length of the diagonal of the parallelogram in the accompanying figure is

$$d = \sqrt{a^2 + b^2 + 2ab \cos \theta}$$

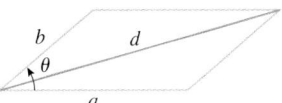

Figure Ex-67

APPENDIX F

Solving Polynomial Equations

In the subsection of Section 5.3 entitled A Brief Review of Polynomials, we reviewed some of the basic ideas and terminology concerning polynomials. We will assume in this appendix that you have read that material, and we will also assume that you know how to divide polynomials using long division and synthetic division. If you need to review those techniques, refer to an algebra book.

THE REMAINDER THEOREM

When two positive integers are divided, the numerator can be expressed as the quotient plus the remainder over the divisor, where the remainder is less than the divisor. For example,

$$\tfrac{17}{5} = 3 + \tfrac{2}{5}$$

If we multiply this equation through by 5, we obtain

$$17 = 5 \cdot 3 + 2$$

which states that the *numerator is the divisor times the quotient plus the remainder*.

The following theorem, which we state without proof, is an analogous result for division of polynomials.

F.1 THEOREM. *If $p(x)$ and $s(x)$ are polynomials, and if $s(x)$ is not the zero polynomial, then $p(x)$ can be expressed as*

$$p(x) = s(x)q(x) + r(x)$$

where $q(x)$ and $r(x)$ are the quotient and remainder that result when $p(x)$ is divided by $s(x)$, and the degree of $r(x)$ is less than the degree of $s(x)$.

In the special case where $p(x)$ is divided by a first-degree polynomial of the form $x - c$, the degree of the remainder must be 0, since it is less than the degree of $x - c$, which is 1. This implies that the remainder is a constant, say r. Thus, Theorem F.1 implies that

$$p(x) = (x - c)q(x) + r$$

and this in turn implies that $p(c) = r$. In summary, we have the following theorem.

F.2 THEOREM (*Remainder Theorem*). *If a polynomial $p(x)$ is divided by $x - c$, then the remainder is $p(c)$.*

Example 1

According to the Remainder Theorem, the remainder on dividing

$$p(x) = 2x^3 + 3x^2 - 4x - 3$$

by $x + 4$ should be

$$p(-4) = 2(-4)^3 + 3(-4)^2 - 4(-4) - 3 = -67$$

Show that this is so.

Solution. By long division

$$
\begin{array}{r}
2x^2 - 5x + 16 \\
x + 4\overline{\smash{)}2x^3 + 3x^2 - 4x - 3} \\
\underline{2x^3 + 8x^2} \\
-5x^2 - 4x \\
\underline{-5x^2 - 20x} \\
16x - 3 \\
\underline{16x + 64} \\
-67
\end{array}
$$

which shows that the remainder is -67.

Alternative Solution. Because we are dividing by an expression of the form $x - c$ (where $c = -4$), we can use synthetic division rather than long division. The computations are

$$
\begin{array}{r|rrrr}
-4 & 2 & 3 & -4 & -3 \\
   &   & -8 & 20 & -64 \\
\hline
   & 2 & -5 & 16 & -67
\end{array}
$$

which again shows that the remainder is -67.   ◀

THE FACTOR THEOREM

To *factor* a polynomial $p(x)$ is to write it as a product of lower-degree polynomials, called *factors* of $p(x)$. For $s(x)$ to be a factor of $p(x)$ there must be no remainder when $p(x)$ is divided by $s(x)$. For example, if $p(x)$ can be factored as

$$p(x) = s(x)q(x) \tag{1}$$

then

$$\frac{p(x)}{s(x)} = q(x) \tag{2}$$

so dividing $p(x)$ by $s(x)$ produces a quotient $q(x)$ with no remainder. Conversely, (2) implies (1), so $s(x)$ is a factor of $p(x)$ if there is no remainder when $p(x)$ is divided by $s(x)$.

In the special case where $x - c$ is a factor of $p(x)$, the polynomial $p(x)$ can be expressed as

$$p(x) = (x - c)q(x)$$

which implies that $p(c) = 0$. Conversely, if $p(c) = 0$, then the Remainder Theorem implies that $x - c$ is a factor of $p(x)$, since the remainder is 0 when $p(x)$ is divided by $x - c$. These results are summarized in the following theorem.

> **F.3** THEOREM (*Factor Theorem*).   *A polynomial $p(x)$ has a factor $x - c$ if and only if $p(c) = 0$.*

It follows from this theorem that the statements below say the same thing in different ways:

- $x - c$ is a factor of $p(x)$.
- $p(c) = 0$.
- c is a zero of $p(x)$.
- c is a root of the equation $p(x) = 0$.
- c is a solution of the equation $p(x) = 0$.
- c is an x-intercept of $y = p(x)$.

Example 2

Confirm that $x - 1$ is a factor of

$$p(x) = x^3 - 3x^2 - 13x + 15$$

by dividing $x - 1$ into $p(x)$ and checking that the remainder is zero.

Solution. By long division

$$
\require{enclose}
\begin{array}{r}
x^2 - 2x - 15 \\
x - 1 \enclose{longdiv}{x^3 - 3x^2 - 13x + 15} \\
\underline{x^3 - x^2} \\
-2x^2 - 13x \\
\underline{-2x^2 + 2x} \\
-15x + 15 \\
\underline{-15x + 15} \\
0
\end{array}
$$

which shows that the remainder is zero.

Alternative Solution. Because we are dividing by an expression of the form $x - c$, we can use synthetic division rather than long division. The computations are

$$
\begin{array}{r|rrrr}
1\underline{|} & 1 & -3 & -13 & 15 \\
  &   & 1 & -2 & -15 \\
\hline
  & 1 & -2 & -15 & 0
\end{array}
$$

which again confirms that the remainder is zero. ◀

USING ONE FACTOR TO FIND OTHER FACTORS

If $x - c$ is a factor of $p(x)$, and if $q(x) = p(x)/(x - c)$, then

$$p(x) = (x - c)q(x) \tag{3}$$

so that additional linear factors of $p(x)$ can be obtained by factoring the quotient $q(x)$.

Example 3

Factor

$$p(x) = x^3 - 3x^2 - 13x + 15 \tag{4}$$

completely into linear factors.

Solution. We showed in Example 2 that $x - 1$ is a factor of $p(x)$ and we also showed that $p(x)/(x - 1) = x^2 - 2x - 15$. Thus,

$$x^3 - 3x^2 - 13x + 15 = (x - 1)(x^2 - 2x - 15)$$

Factoring $x^2 - 2x - 15$ by inspection yields

$$x^3 - 3x^2 - 13x + 15 = (x - 1)(x - 5)(x + 3)$$

which is the complete linear factorization of $p(x)$. ◀

METHODS FOR FINDING ROOTS

A general quadratic equation $ax^2 + bx + c = 0$ can be solved by using the quadratic formula to express the solutions of the equation in terms of the coefficients. Versions of this formula were known since Babylonian times, and by the seventeenth century formulas had been obtained for solving general cubic and quartic equations. However, attempts to find formulas for the solutions of general fifth-degree equations and higher proved fruitless. The reason for this became clear in 1829 when the French mathematician Evariste Galois (1811–1832) proved that it is impossible to express the solutions of a general fifth-degree equation or higher in terms of its coefficients using algebraic operations.

Today, we have powerful computer programs for finding the zeros of specific polynomials. For example, it takes only seconds for a computer algebra system, such as *Mathematica*, *Maple*, or *Derive*, to show that the zeros of the polynomial

$$p(x) = 10x^4 - 23x^3 - 10x^2 + 29x + 6 \tag{5}$$

are

$$x = -1, \quad x = -\tfrac{1}{5}, \quad x = \tfrac{3}{2}, \quad \text{and} \quad x = 2 \tag{6}$$

The algorithms that these programs use to find the integer and rational zeros of a polynomial, if any, are based on the following theorem, which is proved in advanced algebra courses.

F.4　THEOREM.　*Suppose that*

$$p(x) = c_n x^n + c_{n-1} x^{n-1} + \cdots + c_1 x + c_0$$

is a polynomial with integer coefficients.

(a)　If r is an integer zero of $p(x)$, then r must be a divisor of the constant term c_0.

(b)　If $r = a/b$ is a rational zero of $p(x)$ in which all common factors of a and b have been canceled, then a must be a divisor of the constant term c_0, and b must be a divisor of the leading coefficient c_n.

For example, in (5) the constant term is 6 (which has divisors $\pm 1, \pm 2, \pm 3,$ and ± 6) and the leading coefficient is 10 (which has divisors $\pm 1, \pm 2, \pm 5,$ and ± 10). Thus, the only possible integer zeros of $p(x)$ are

$$\pm 1, \quad \pm 2, \quad \pm 3, \quad \pm 6$$

and the only possible noninteger rational zeros are

$$\pm \tfrac{1}{2}, \quad \pm \tfrac{1}{5}, \quad \pm \tfrac{1}{10}, \quad \pm \tfrac{2}{5}, \quad \pm \tfrac{3}{2}, \quad \pm \tfrac{3}{5}, \quad \pm \tfrac{3}{10}, \quad \pm \tfrac{6}{5}$$

Using a computer, it is a simple matter to evaluate $p(x)$ at each of the numbers in these lists to show that its only zeros are the numbers in (6).

Example 4

Solve the equation $x^3 + 3x^2 - 7x - 21 = 0$.

Solution.　The solutions of the equation are the zeros of the polynomial

$$p(x) = x^3 + 3x^2 - 7x - 21$$

We will look for integer zeros first. All such zeros must divide the constant term, so the only possibilities are $\pm 1, \pm 3, \pm 7,$ and ± 21. Substituting these values into $p(x)$ (or using the method of Exercise 6) shows that $x = -3$ is an integer zero. This tells us that $x + 3$ is a factor of $p(x)$ and that $p(x)$ can be written as

$$x^3 + 3x^2 - 7x - 21 = (x + 3)q(x)$$

where $q(x)$ is the quotient that results when $x^3 + 3x^2 - 7x - 21$ is divided by $x + 3$. We leave it for you to perform the division and show that $q(x) = x^2 - 7$; hence,

$$x^3 + 3x^2 - 7x - 21 = (x + 3)(x^2 - 7) = (x + 1)(x + \sqrt{7})(x - \sqrt{7})$$

which tells us that the solutions of the given equation are $x = 3$, $x = \sqrt{7} \approx 2.65$, and $x = -\sqrt{7} \approx -2.65$. ◄

EXERCISE SET F [c] CAS

In Exercises 1 and 2, find the quotient $q(x)$ and the remainder $r(x)$ that result when $p(x)$ is divided by $s(x)$.

1. (a) $p(x) = x^4 + 3x^3 - 5x + 10$; $s(x) = x^2 - x + 2$
(b) $p(x) = 6x^4 + 10x^2 + 5$; $s(x) = 3x^2 - 1$
(c) $p(x) = x^5 + x^3 + 1$; $s(x) = x^2 + x$

2. (a) $p(x) = 2x^4 - 3x^3 + 5x^2 + 2x + 7$; $s(x) = x^2 - x + 1$
(b) $p(x) = 2x^5 + 5x^4 - 4x^3 + 8x^2 + 1$; $s(x) = 2x^2 - x + 1$
(c) $p(x) = 5x^6 + 4x^2 + 5$; $s(x) = x^3 + 1$

In Exercises 3 and 4, use synthetic division to find the quotient $q(x)$ and the remainder r that result when $p(x)$ is divided by $s(x)$.

3. (a) $p(x) = 3x^3 - 4x - 1$; $s(x) = x - 2$
(b) $p(x) = x^4 - 5x^2 + 4$; $s(x) = x + 5$
(c) $p(x) = x^5 - 1$; $s(x) = x - 1$

4. (a) $p(x) = 2x^3 - x^2 - 2x + 1$; $s(x) = x - 1$
(b) $p(x) = 2x^4 + 3x^3 - 17x^2 - 27x - 9$; $s(x) = x + 4$
(c) $p(x) = x^7 + 1$; $s(x) = x - 1$

5. Let $p(x) = 2x^4 + x^3 - 3x^2 + x - 4$. Use synthetic division and the Remainder Theorem to find $p(0)$, $p(1)$, $p(-3)$, and $p(7)$.

6. Let $p(x)$ be the polynomial in Example 4. Use synthetic division and the Remainder Theorem to evaluate $p(x)$ at $x = \pm 1, \pm 3, \pm 7$, and ± 21.

7. Let $p(x) = x^3 + 4x^2 + x - 6$. Find a polynomial $q(x)$ and a constant r such that
(a) $p(x) = (x - 2)q(x) + r$
(b) $p(x) = (x + 1)q(x) + r$.

8. Let $p(x) = x^5 - 1$. Find a polynomial $q(x)$ and a constant r such that
(a) $p(x) = (x + 1)q(x) + r$
(b) $p(x) = (x - 1)q(x) + r$.

9. In each part, make a list of all possible candidates for the rational zeros of $p(x)$.
(a) $p(x) = x^7 + 3x^3 - x + 24$
(b) $p(x) = 3x^4 - 2x^2 + 7x - 10$
(c) $p(x) = x^{35} - 17$

10. Find all integer zeros of

$$p(x) = x^6 + 5x^5 - 16x^4 - 15x^3 - 12x^2 - 38x - 21$$

In Exercises 11–15, factor the polynomials completely.

11. $p(x) = x^3 - 2x^2 - x + 2$
12. $p(x) = 3x^3 + x^2 - 12x - 4$
13. $p(x) = x^4 + 10x^3 + 36x^2 + 54x + 27$
14. $p(x) = 2x^4 + x^3 - 19x^2 + 9$
15. $p(x) = x^5 + 4x^4 - 4x^3 - 34x^2 - 45x - 18$

[c] **16.** For each of the factorizations that you obtained in Exercises 11–15, check your answer using a CAS.

In Exercises 17–21, find all real solutions of the equations.

17. $x^3 + 3x^2 + 4x + 12 = 0$
18. $2x^3 - 5x^2 - 10x + 3 = 0$
19. $3x^4 + 14x^3 + 14x^2 - 8x - 8 = 0$
20. $2x^4 - x^3 - 14x^2 - 5x + 6 = 0$
21. $x^5 - 2x^4 - 6x^3 + 5x^2 + 8x + 12 = 0$

[c] **22.** For each of the equations you solved in Exercises 17–21, check your answer using a CAS.

23. Find all values of k for which $x - 1$ is a factor of the polynomial $p(x) = k^2 x^3 - 7kx + 10$.

24. Is $x + 3$ a factor of $x^7 + 2187$? Justify your answer.

[c] **25.** A 3-cm-thick slice is cut from a cube, leaving a volume of 196 cm³. Use a CAS to find the length of a side of the original cube.

26. (a) Show that there is no positive rational number that exceeds its cube by 1.
(b) Does there exist a real number that exceeds its cube by 1? Justify your answer.

27. Use the Factor Theorem to show each of the following.
(a) $x - y$ is a factor of $x^n - y^n$ for all positive integer values of n.
(b) $x + y$ is a factor of $x^n - y^n$ for all positive even integer values of n.
(c) $x + y$ is a factor of $x^n + y^n$ for all positive odd integer values of n.

APPENDIX G

Selected Proofs

An extensive excursion into proofs of limit theorems would be too time consuming to undertake, so we have selected a few proofs of results from Section 2.2 that illustrate some of the basic ideas.

G.1 THEOREM. *Let k be a constant, and suppose that $\lim\limits_{x \to a} f(x) = L_1$ and that $\lim\limits_{x \to a} g(x) = L_2$. Then*

(a) $\lim\limits_{x \to a} k = k$

(b) $\lim\limits_{x \to a} [f(x) + g(x)] = \lim\limits_{x \to a} f(x) + \lim\limits_{x \to a} g(x) = L_1 + L_2$

(c) $\lim\limits_{x \to a} [f(x)g(x)] = \lim\limits_{x \to a} f(x) \lim\limits_{x \to a} g(x) = L_1 L_2$

Proof (a). We will apply Definition 2.3.3 with $f(x) = k$ and $L = k$. Thus, given $\epsilon > 0$, we must find a number $\delta > 0$ such that

$$|k - k| < \epsilon \quad \text{if} \quad 0 < |x - a| < \delta$$

or equivalently,

$$0 < \epsilon \quad \text{if} \quad 0 < |x - a| < \delta$$

But the condition on the left side of this statement is *always* true, no matter how δ is chosen. Thus, any positive value for δ will suffice.

Proof (b). We must show that given $\epsilon > 0$ we can find a number $\delta > 0$ such that

$$|(f(x) + g(x)) - (L_1 + L_2)| < \epsilon \quad \text{if} \quad 0 < |x - a| < \delta \tag{1}$$

However, from the limits of f and g in the hypothesis of the theorem we can find numbers δ_1 and δ_2 such that

$$|f(x) - L_1| < \epsilon/2 \quad \text{if} \quad 0 < |x - a| < \delta_1$$

$$|g(x) - L_2| < \epsilon/2 \quad \text{if} \quad 0 < |x - a| < \delta_2$$

Moreover, the inequalities on the left sides of these statements *both* hold if we replace δ_1 and δ_2 by any positive number δ that is less than both δ_1 and δ_2. Thus, for any such δ it follows that

$$|f(x) - L_1| + |g(x) - L_2| < \epsilon \quad \text{if} \quad 0 < |x - a| < \delta \tag{2}$$

However, it follows from the triangle inequality [Theorem 1.2.2(d)] that

$$|(f(x) + g(x)) - (L_1 + L_2)| = |(f(x) - L_1) + (g(x) - L_2)|$$
$$\leq |f(x) - L_1| + |g(x) - L_2|$$

so that (1) follows from (2).

Proof (c). We must show that given $\epsilon > 0$ we can find a number $\delta > 0$ such that

$$|f(x)g(x) - L_1L_2| < \epsilon \quad \text{if} \quad 0 < |x - a| < \delta \tag{3}$$

To find δ it will be helpful to express (3) in a different form. If we rewrite $f(x)$ and $g(x)$ as

$$f(x) = L_1 + (f(x) - L_1) \quad \text{and} \quad g(x) = L_2 + (g(x) - L_2)$$

then the inequality on the left side of (3) can be expressed as (verify)

$$|L_1(g(x) - L_2) + L_2(f(x) - L_1) + (f(x) - L_1)(g(x) - L_2)| < \epsilon \tag{4}$$

Since

$$\lim_{x \to a} f(x) = L_1 \quad \text{and} \quad \lim_{x \to a} g(x) = L_2$$

we can find positive numbers $\delta_1, \delta_2, \delta_3,$ and δ_4 such that

$$
\begin{aligned}
|f(x) - L_1| &< \sqrt{\epsilon/3} & &\text{if} \quad 0 < |x - a| < \delta_1 \\
|f(x) - L_1| &< \frac{\epsilon}{3(1 + |L_2|)} & &\text{if} \quad 0 < |x - a| < \delta_2 \\
|g(x) - L_2| &< \sqrt{\epsilon/3} & &\text{if} \quad 0 < |x - a| < \delta_3 \\
|g(x) - L_2| &< \frac{\epsilon}{3(1 + |L_1|)} & &\text{if} \quad 0 < |x - a| < \delta_4
\end{aligned}
\tag{5}
$$

Moreover, the inequalities on the left sides of these four statements *all* hold if we replace $\delta_1, \delta_2, \delta_3,$ and δ_4 by any number δ that is smaller than $\delta_1, \delta_2, \delta_3,$ and δ_4. Thus, for any such δ it follows with the help of the triangle inequality that

$$
\begin{aligned}
|L_1(g(x) &- L_2) + L_2(f(x) - L_1) + (f(x) - L_1)(g(x) - L_2)| \\
&\leq |L_1(g(x) - L_2)| + |L_2(f(x) - L_1)| + |(f(x) - L_1)(g(x) - L_2)| \\
&= |L_1||g(x) - L_2| + |L_2||f(x) - L_1| + |f(x) - L_1||g(x) - L_2| \\
&< |L_1|\frac{\epsilon}{3(1 + |L_1|)} + |L_2|\frac{\epsilon}{3(1 + |L_2|)} + \sqrt{\epsilon/3}\sqrt{\epsilon/3} \quad \boxed{\text{From (5)}} \\
&= \frac{\epsilon}{3}\frac{|L_1|}{1 + |L_1|} + \frac{\epsilon}{3}\frac{|L_2|}{1 + |L_2|} + \frac{\epsilon}{3} \\
&< \frac{\epsilon}{3} + \frac{\epsilon}{3} + \frac{\epsilon}{3} = \epsilon \quad \boxed{\text{Since } \frac{|L_1|}{1 + |L_1|} < 1 \text{ and } \frac{|L_2|}{1 + |L_2|} < 1}
\end{aligned}
$$

which shows that (4) holds for the δ selected. ■

REMARK.   Do not be alarmed if the proof of part (c) seems difficult; it takes some experience with proofs of this type to develop a feel for choosing the right δ. Your initial goal should be to understand the ideas and the computations.

PROOF OF A BASIC CONTINUITY PROPERTY

Next, we will prove Theorem 2.4.5 for two-sided limits.

> **G.2**   THEOREM (*Theorem 2.4.5*).   *If* $\lim_{x \to c} g(x) = L$ *and if the function f is continuous at L, then* $\lim_{x \to c} f(g(x)) = f(L)$; *that is,* $\lim_{x \to c} f(g(x)) = f(\lim_{x \to c} g(x))$.

Proof. We must show that given $\epsilon > 0$, we can find a number $\delta > 0$ such that

$$|f(g(x)) - f(L)| < \epsilon \quad \text{if} \quad 0 < |x - c| < \delta \tag{6}$$

Since f is continuous at L, we have

$$\lim_{u \to L} f(u) = f(L)$$

and hence we can find a number $\delta_1 > 0$ such that

$$|f(u) - f(L)| < \epsilon \quad \text{if} \quad |u - L| < \delta_1$$

In particular, if $u = g(x)$, then

$$|f(g(x)) - f(L)| < \epsilon \quad \text{if} \quad |g(x) - L| < \delta_1 \tag{7}$$

But $\lim\limits_{x \to c} g(x) = L$, and hence there is a number $\delta > 0$ such that

$$|g(x) - L| < \delta_1 \quad \text{if} \quad 0 < |x - c| < \delta \tag{8}$$

Thus, if x satisfies the condition on the right side of statement (8), then it follows that $g(x)$ satisfies the condition on the right side of statement (7), and this implies that the condition on the left side of statement (6) is satisfied, completing the proof. ▮

PROOF OF THE CHAIN RULE

Next, we will prove the chain rule (Theorem 3.5.2), but first we need a preliminary result.

G.3   THEOREM.   *If f is differentiable at x and if $y = f(x)$, then*

$$\Delta y = f'(x)\Delta x + \epsilon \Delta x$$

where $\epsilon \to 0$ as $\Delta x \to 0$ and $\epsilon = 0$ if $\Delta x = 0$.

Proof.   Define

$$\epsilon = \begin{cases} \dfrac{f(x + \Delta x) - f(x)}{\Delta x} - f'(x) & \text{if } \Delta x \neq 0 \\ \\ 0 & \text{if } \Delta x = 0 \end{cases} \tag{9}$$

If $\Delta x \neq 0$, it follows from (9) that

$$\epsilon \Delta x = [f(x + \Delta x) - f(x)] - f'(x)\Delta x \tag{10}$$

But

$$\Delta y = f(x + \Delta x) - f(x) \tag{11}$$

so (10) can be written as

$$\epsilon \Delta x = \Delta y - f'(x)\Delta x$$

or

$$\Delta y = f'(x)\Delta x + \epsilon \Delta x \tag{12}$$

If $\Delta x = 0$, then (12) still holds (why?), so (12) is valid for all values of Δx. It remains to show that $\epsilon \to 0$ as $\Delta x \to 0$. But this follows from the assumption that f is differentiable at x, since

$$\lim_{\Delta x \to 0} \epsilon = \lim_{\Delta x \to 0} \left[\frac{f(x + \Delta x) - f(x)}{\Delta x} - f'(x) \right] = f'(x) - f'(x) = 0 \qquad ▮$$

We are now ready to prove the chain rule.

G.4   THEOREM (*Theorem 3.5.2*).   *If g is differentiable at the point x and f is differentiable at the point $g(x)$, then the composition $f \circ g$ is differentiable at the point x. Moreover, if $y = f(g(x))$ and $u = g(x)$, then*

$$\frac{dy}{dx} = \frac{dy}{du} \cdot \frac{du}{dx}$$

Proof. Since g is differentiable at x and $u = g(x)$, it follows from Theorem G.3 that

$$\Delta u = g'(x)\Delta x + \epsilon_1 \Delta x \tag{13}$$

where $\epsilon_1 \to 0$ as $\Delta x \to 0$. And since $y = f(u)$ is differentiable at $u = g(x)$, it follows from Theorem G.3 that

$$\Delta y = f'(u)\Delta u + \epsilon_2 \Delta u \tag{14}$$

where $\epsilon_2 \to 0$ as $\Delta u \to 0$.

Factoring out the Δu in (14) and then substituting (13) yields

$$\Delta y = [f'(u) + \epsilon_2][g'(x)\Delta x + \epsilon_1 \Delta x]$$

or

$$\Delta y = [f'(u) + \epsilon_2][g'(x) + \epsilon_1]\Delta x$$

or if $\Delta x \neq 0$,

$$\frac{\Delta y}{\Delta x} = [f'(u) + \epsilon_2][g'(x) + \epsilon_1] \tag{15}$$

But (13) implies that $\Delta u \to 0$ as $\Delta x \to 0$, and hence $\epsilon_1 \to 0$ and $\epsilon_2 \to 0$ as $\Delta x \to 0$. Thus, from (15)

$$\lim_{\Delta x \to 0}\frac{\Delta y}{\Delta x} = f'(u)g'(x)$$

or

$$\frac{dy}{dx} = f'(u)g'(x) = \frac{dy}{du} \cdot \frac{du}{dx}$$ ∎

PROOF THAT RELATIVE EXTREMA OCCUR AT CRITICAL POINTS

In this subsection we will prove Theorem 5.2.2, which states that the relative extrema of a function occur at critical points.

> **G.5** THEOREM (*Theorem 5.2.2*).   *If a function f has any relative extrema, then they occur either at points where $f'(x) = 0$ or at points where f is not differentiable.*

Proof. There are two possibilities—either f is differentiable at a point x_0 or it is not. If it is not, then x_0 is a critical point for f and we are done. If f is differentiable at x_0, then we must show that $f'(x_0) = 0$. We will do this by showing that $f'(x_0) \geq 0$ and $f'(x_0) \leq 0$, from which it follows that $f'(x_0) = 0$. From the definition of a derivative we have

$$f'(x_0) = \lim_{h \to 0}\frac{f(x_0 + h) - f(x_0)}{h}$$

so that

$$f'(x_0) = \lim_{h \to 0^+}\frac{f(x_0 + h) - f(x_0)}{h} \tag{16}$$

and

$$f'(x_0) = \lim_{h \to 0^-}\frac{f(x_0 + h) - f(x_0)}{h} \tag{17}$$

Because f has a relative maximum at x_0, there is an open interval (a, b) containing x_0 in which $f(x) \leq f(x_0)$ for all x in (a, b).

Assume that h is sufficiently small so that $x_0 + h$ lies in the interval (a, b). Thus,

$$f(x_0 + h) \leq f(x_0) \quad \text{or equivalently,} \quad f(x_0 + h) - f(x_0) \leq 0$$

Thus, if h is negative,

$$\frac{f(x_0 + h) - f(x_0)}{h} \geq 0 \tag{18}$$

and if h is positive,

$$\frac{f(x_0 + h) - f(x_0)}{h} \leq 0 \tag{19}$$

But an expression that never assumes negative values cannot approach a negative limit and an expression that never assumes positive values cannot approach a positive limit, so that

$$f'(x_0) = \lim_{h \to 0^-} \frac{f(x_0 + h) - f(x_0)}{h} \geq 0 \qquad \boxed{\text{From (17) and (18)}}$$

and

$$f'(x_0) = \lim_{h \to 0^+} \frac{f(x_0 + h) - f(x_0)}{h} \leq 0 \qquad \boxed{\text{From (16) and (19)}}$$

Since $f'(x_0) \geq 0$ and $f'(x_0) \leq 0$, it must be that $f'(x_0) = 0$. ∎

PROOF OF THE LIMIT COMPARISON TEST

> **G.6** THEOREM (*Theorem 11.6.4*). *Let $\sum a_k$ and $\sum b_k$ be a series with positive terms and suppose that*
> $$\rho = \lim_{k \to +\infty} \frac{a_k}{b_k}$$
> *If ρ is finite and $\rho > 0$, then the series both converge or both diverge.*

Proof. We need only show that $\sum b_k$ converges when $\sum a_k$ converges and that $\sum b_k$ diverges when $\sum a_k$ diverges, since the remaining cases are logical implications of these (why?). The idea of the proof is to apply the comparison test to $\sum a_k$ and suitable multiples of $\sum b_k$. For this purpose let ϵ be any positive number. Since

$$\rho = \lim_{k \to +\infty} \frac{a_k}{b_k}$$

it follows that eventually the terms in the sequence $\{a_k/b_k\}$ must be within ϵ units of ρ; that is, there is a positive integer K such that for $k \geq K$ we have

$$\rho - \epsilon < \frac{a_k}{b_k} < \rho + \epsilon$$

In particular, if we take $\epsilon = \rho/2$, then for $k \geq K$ we have

$$\frac{1}{2}\rho < \frac{a_k}{b_k} < \frac{3}{2}\rho \quad \text{or} \quad \frac{1}{2}\rho b_k < a_k < \frac{3}{2}\rho b_k$$

Thus, by the comparison test we can conclude that

$$\sum_{k=K}^{\infty} \frac{1}{2}\rho b_k \quad \text{converges if} \quad \sum_{k=K}^{\infty} a_k \quad \text{converges} \tag{20}$$

$$\sum_{k=K}^{\infty} \frac{3}{2}\rho b_k \quad \text{diverges if} \quad \sum_{k=K}^{\infty} a_k \quad \text{diverges} \tag{21}$$

But the convergence or divergence of a series is not affected by deleting finitely many terms or by multiplying the general term by a nonzero constant, so (20) and (21) imply that

$$\sum_{k=1}^{\infty} b_k \quad \text{converges if} \quad \sum_{k=1}^{\infty} a_k \quad \text{converges}$$

$$\sum_{k=1}^{\infty} b_k \quad \text{diverges if} \quad \sum_{k=1}^{\infty} a_k \quad \text{diverges} \qquad ∎$$

PROOF OF THE RATIO TEST

G.7 THEOREM (*Theorem 11.6.5*). *Let $\sum u_k$ be a series with positive terms and suppose that*

$$\rho = \lim_{k \to +\infty} \frac{u_{k+1}}{u_k}$$

(a) *If $\rho < 1$, the series converges.*

(b) *If $\rho > 1$ or $\rho = +\infty$, the series diverges.*

(c) *If $\rho = 1$, the series may converge or diverge, so that another test must be tried.*

Proof (a). The number ρ must be nonnegative since it is the limit of u_{k+1}/u_k, which is positive for all k. In this part of the proof we assume that $\rho < 1$, so that $0 \le \rho < 1$.

We will prove convergence by showing that the terms of the given series are eventually less than the terms of a convergent geometric series. For this purpose, choose any real number r such that $0 < \rho < r < 1$. Since the limit of u_{k+1}/u_k is ρ, and $\rho < r$, the terms of the sequence $\{u_{k+1}/u_k\}$ must eventually be less than r. Thus, there is a positive integer K such that for $k \ge K$ we have

$$\frac{u_{k+1}}{u_k} < r \quad \text{or} \quad u_{k+1} < r u_k$$

This yields the inequalities

$$u_{K+1} < r u_K$$
$$u_{K+2} < r u_{K+1} < r^2 u_K$$
$$u_{K+3} < r u_{K+2} < r^3 u_K \qquad (22)$$
$$u_{K+4} < r u_{K+3} < r^4 u_K$$
$$\vdots$$

But $0 < r < 1$, so

$$r u_K + r^2 u_K + r^3 u_K + \cdots$$

is a convergent geometric series. From the inequalities in (22) and the comparison test it follows that

$$u_{K+1} + u_{K+2} + u_{K+3} + \cdots$$

must also be a convergent series. Thus, $u_1 + u_2 + u_3 + \cdots + u_k + \cdots$ converges by Theorem 11.4.3(c).

Proof (b). In this part we will prove divergence by showing that the limit of the general term is not zero. Since the limit of u_{k+1}/u_k is ρ and $\rho > 1$, the terms in the sequence $\{u_{k+1}/u_k\}$ must eventually be greater than 1. Thus, there is a positive integer K such that for $k \ge K$ we have

$$\frac{u_{k+1}}{u_k} > 1 \quad \text{or} \quad u_{k+1} > u_k$$

This yields the inequalities

$$u_{K+1} > u_K$$
$$u_{K+2} > u_{K+1} > u_K$$
$$u_{K+3} > u_{K+2} > u_K \qquad (23)$$
$$u_{K+4} > u_{K+3} > u_K$$
$$\vdots$$

Since $u_K > 0$, it follows from the inequalities in (23) that $\lim_{k \to +\infty} u_k \ne 0$, and thus the series

$u_1 + u_2 + \cdots + u_k + \cdots$ diverges by part (*a*) of Theorem 11.4.1. The proof in the case where $\rho = +\infty$ is omitted.

Proof (*c*). The divergent harmonic series and the convergent *p*-series with $p = 2$ both have $\rho = 1$ (verify), so the ratio test does not distinguish between convergence and divergence when $\rho = 1$. ∎

PROOF OF THE REMAINDER ESTIMATION THEOREM

G.8 THEOREM (*Theorem 11.9.3*).   *If the function* f *can be differentiated* $n + 1$ *times on an interval I containing the point* x_0, *and if* $|f^{(n+1)}(x)| \leq M$ *for all x in I, then*

$$|R_n(x)| \leq \frac{M}{(n+1)!}|x - x_0|^{n+1}$$

for all x in I.

Proof.   We are assuming that f can be differentiated $n + 1$ times on an interval I containing the point x_0 and that

$$|f^{(n+1)}(x)| \leq M \tag{24}$$

for all x in I. We want to show that

$$|R_n(x)| \leq \frac{M}{(n+1)!}|x - x_0|^{n+1} \tag{25}$$

for all x in I, where

$$R_n(x) = f(x) - \sum_{k=0}^{n} \frac{f^{(k)}(x_0)}{k!}(x - x_0)^k \tag{26}$$

In our proof we will need the following two properties of $R_n(x)$:

$$R_n(x_0) = R_n'(x_0) = \cdots = R_n^{(n)}(x_0) = 0 \tag{27}$$

$$R_n^{(n+1)}(x) = f^{(n+1)}(x) \quad \text{for all } x \text{ in } I \tag{28}$$

These properties can be obtained by analyzing what happens if the expression for $R_n(x)$ in Formula (26) is differentiated j times and x_0 is then substituted in that derivative. If $j < n$, then the jth derivative of the summation in Formula (26) consists of a constant term $f^{(j)}(x_0)$ plus terms involving powers of $x - x_0$ (verify). Thus, $R_n^{(j)}(x_0) = 0$ for $j < n$, which proves all but the last equation in (27). For the last equation, observe that the nth derivative of the summation in (26) is the constant $f^{(n)}(x_0)$, so $R_n^{(n)}(x_0) = 0$. Formula (28) follows from the observation that the $(n + 1)$-st derivative of the summation in (26) is zero (why?).

Now to the main part of the proof. For simplicity we will give the proof for the case where $x \geq x_0$ and leave the case where $x < x_0$ for the reader. It follows from (24) and (28) that $|R_n^{(n+1)}(x)| \leq M$, and hence

$$-M \leq R_n^{(n+1)}(x) \leq M$$

Thus,

$$\int_{x_0}^{x} -M \, dt \leq \int_{x_0}^{x} R_n^{(n+1)}(t) \, dt \leq \int_{x_0}^{x} M \, dt \tag{29}$$

However, it follows from (27) that $R_n^{(n)}(x_0) = 0$, so

$$\int_{x_0}^{x} R_n^{(n+1)}(t) \, dt = R_n^{(n)}(t) \Big]_{x_0}^{x} = R_n^{(n)}(x)$$

Thus, performing the integrations in (29) we obtain the inequalities

$$-M(x - x_0) \leq R_n^{(n)}(x) \leq M(x - x_0)$$

Now we will integrate again. Replacing x by t in these inequalities, integrating from x_0 to x, and using $R_n^{(n-1)}(x_0) = 0$ yields

$$-\frac{M}{2}(x-x_0)^2 \le R_n^{(n-1)}(x) \le \frac{M}{2}(x-x_0)^2$$

If we keep repeating this process, then after n integrations we will obtain

$$-\frac{M}{(n+1)!}(x-x_0)^{n+1} \le R_n(x) \le \frac{M}{(n+1)!}(x-x_0)^{n+1}$$

which we can rewrite as

$$|R_n(x)| \le \frac{M}{(n+1)!}(x-x_0)^{n+1}$$

This completes the proof of (25), since the absolute value signs can be omitted in that formula when $x \ge x_0$ (which is the case we are considering). ∎

ANSWERS TO ODD-NUMBERED EXERCISES

··

▶ Exercise Set for Introduction **(Page 12)**

1. **(a)** $\frac{41}{333}$ **(b)** $\frac{115}{9}$ **(c)** $\frac{20943}{550}$ **(d)** $\frac{537}{1250}$

3. **(a)** $\frac{223}{71} < \frac{333}{106} < \frac{63}{25}\left(\frac{17+15\sqrt{5}}{7+15\sqrt{5}}\right) < \frac{355}{113} < \frac{22}{7}$ **(b)** $\frac{63}{25}\left(\frac{17+15\sqrt{5}}{7+15\sqrt{5}}\right)$ **(c)** $\frac{333}{106}$ **(d)** $\frac{63}{25}\left(\frac{17+15\sqrt{5}}{7+15\sqrt{5}}\right)$ 5. 3.1416 (Machin); 3.0418

7. **(a)** $\frac{7}{11} = 0.636363\ldots = \frac{6}{10} + \frac{3}{100} + \frac{6}{1000} + \frac{3}{10000} + \frac{6}{100000} + \frac{3}{1000000} + \cdots$

   (b) $\frac{8}{33} = 0.242424\ldots = \frac{2}{10} + \frac{4}{100} + \frac{2}{1000} + \frac{4}{10000} + \frac{2}{100000} + \frac{4}{1000000} + \cdots$

   (c) $\frac{5}{12} = 0.416666\ldots = \frac{4}{10} + \frac{1}{100} + \frac{6}{1000} + \frac{6}{10000} + \frac{6}{100000} + \frac{6}{1000000} + \cdots$ 9. **(a)** 2.6458 **(b)** 7.0711

▶ Exercise Set 1.1 **(Page 22)**

1. **(a)** 1943 **(b)** 1960; 4200 **(c)** no, you need the year's population **(d)** war, marketing

   (e) news of health risk, social pressure, antismoking campaigns, increased taxation

3. **(a)** $-2.9, -2.0, 2.35, 2.9$ **(b)** none **(c)** 0 **(d)** $-1.75 \le x \le 2.15$ **(e)** $y_{\max} = 2.8$ at $x = -2.6$; $y_{\min} = -2.2$ at $x = 1.2$

5. **(a)** 2, 4 **(b)** none **(c)** $x \le 2; 4 \le x$ **(d)** $y_{\min} = -1$; no maximum

7. **(a)** no; war, pestilence, flood, earthquakes **(b)** decreases for 8 hours, takes a jump upward, and repeats

9. **(a)** $L = x + 2000/x$ **(b)** $x > 0$; x must be smaller than the width of the building, which was not given.

   (c) 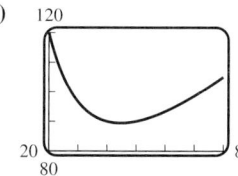   **(d)** 89.44   11. **(a)** $r \approx 3.4, h \approx 13.8$

   (b) taller

   (c) $r \approx 3.1, h \approx 16.0, C \approx 4.76$

▶ Exercise Set 1.2 **(Page 33)**

1. **(a)** $-2; 10; 10; 25; 4; 27t^2 - 2$ **(b)** $0; 4; -4; 6; 2\sqrt{2}$; $f(3t) = 1/3t$ for $t > 1$ and $f(3t) = 6t$ for $t \le 1$

3. **(a)** $x \ne 3$ **(b)** $x \le -\sqrt{3}, x \ge \sqrt{3}$ **(c)** $(-\infty, +\infty)$ **(d)** $x \ne 0$ **(e)** $x \ne \left(2n + \frac{1}{2}\right)\pi, n = 0, \pm 1, \pm 2, \ldots$

5. **(a)** $x \le 3$ **(b)** $-2 \le x \le 2$ **(c)** $x \ge 0$ **(d)** all x **(e)** all x 7. **(a)** yes **(b)** yes **(c)** no **(d)** no

9. $h = L(1 - \cos\theta)$ 11.   13.
   (a) $f(x) = \begin{cases} 2x + 1, & x < 0 \\ 4x + 1, & x \ge 0 \end{cases}$ 15. **(a)** $V = (8 - 2x)(15 - 2x)x$

   (b) $-\infty < x < \infty, -\infty < V < \infty$

   (b) $g(x) = \begin{cases} 1 - 2x, & x < 0 \\ 1, & 0 \le x < 1 \\ 2x - 1, & x \ge 1 \end{cases}$ **(c)** $0 < x < 4$

17. **(i)** $x = 1, -2$ **(ii)** $g(x) = x + 1$, all x 19. **(a)** $25°F$ **(b)** $2°F$ **(c)** $-15°F$ 21. $5°F$ 23. $D(t) = 1000 - 20t$

▶ Exercise Set 1.3 **(Page 45)**

1. (e)   3. (b), (c)   5. $[-3, 3] \times [0, 5]$ 9. $[-5, 14] \times [-60, 40]$ 11. $[-0.1, 0.1] \times [-3, 3]$

13. $[-1000, 1050] \times [-1500000, 10000]$ 15. $[-2, 2] \times [-20, 20]$ 19. (a) $f(x) = \sqrt{16 - x^2}$ 35. $4.6455, -0.6455$

(b) $f(x) = -\sqrt{16 - x^2}$

(e) no

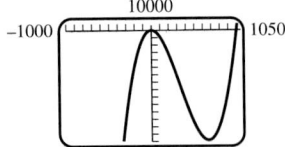

▶ Exercise Set 1.4 **(Page 57)**

1. (a) 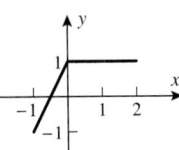   (b) 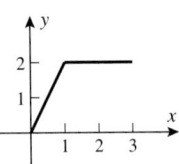   (c) 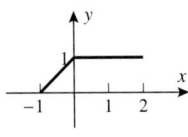   (d)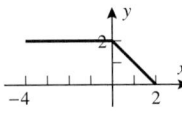

3. (a) 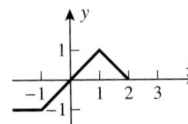   (b) 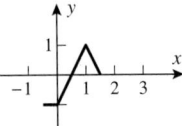   (c) 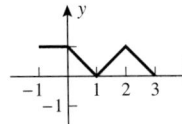   (d)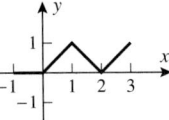

5. 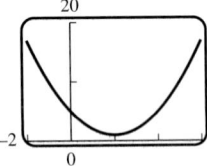   7. 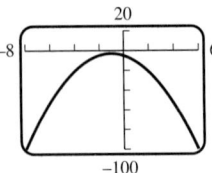   9. 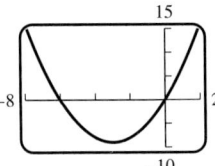   11.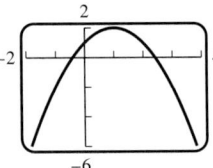

13. 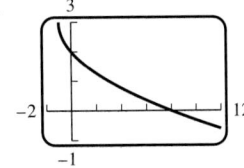   15. 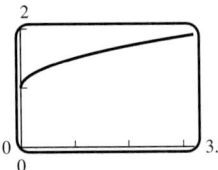   17. 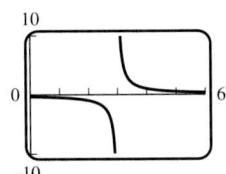   19.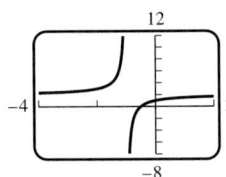

21. 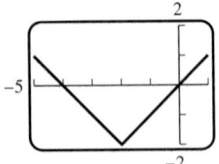   23. 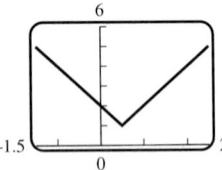   25. 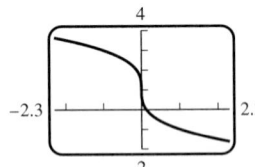   27.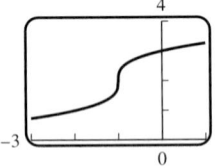

29. (a) 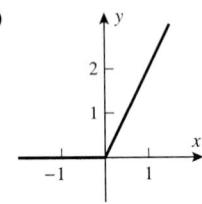   (b) $y = \begin{cases} 0, & x \leq 0 \\ 2x, & x > 0 \end{cases}$ 31. $x^2 + 2x + 1$, all x;   33. $3\sqrt{x - 1}, x \geq 1$;   35. (a) 3

$2x - x^2 - 1$, all x; $\sqrt{x - 1}, x \geq 1$;   (b) 9

$2x^3 + 2x$, all x; $2x - 2, x \geq 1$;   (c) 2

$2x/(x^2 + 1)$, all x $2, x > 1$ (d) 2

37. (a) $t^4 + 1$ (b) $t^2 + 4t + 5$ (c) $x^2 + 4x + 5$ (d) $\dfrac{1}{x^2} + 1$ (e) $x^2 + 2xh + h^2 + 1$ (f) $x^2 + 1$ (g) $x + 1$ (h) $9x^2 + 1$

39. $2x^2 - 2x + 1$, all x; $4x^2 + 2x$, all x 41. $1 - x, x \leq 1$; $\sqrt{1 - x^2}, |x| \leq 1$ 43. $\dfrac{1}{1 - 2x}, x \neq \dfrac{1}{2}, 1; -\dfrac{1}{2x} - \dfrac{1}{2}, x \neq 0, 1$

45. $x^{-6} + 1$ 47. (a) $g(x) = \sqrt{x}, h(x) = x + 2$ (b) $g(x) = |x|, h(x) = x^2 - 3x + 5$

49. (a) $g(x) = x^2, h(x) = \sin x$ **(b)** $g(x) = 3/x, h(x) = 5 + \cos x$

51. (a) $f(x) = x^3, g(x) = 1 + \sin x, h(x) = x^2$ **(b)** $f(x) = \sqrt{x}, g(x) = 1 - x, h(x) = \sqrt[3]{x}$

53. 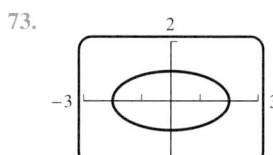 **55.** 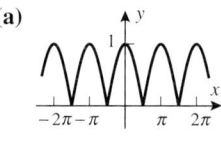 **57.** ± 2 **59.** $6x + 3h$ **61.** $-\dfrac{1}{x(x+h)}$ **63. (a)** origin
(b) x-axis
(c) y-axis
(d) none

65. (a)

x	-3	-2	-1	0	1	2	3
$f(x)$	1	-5	-1	0	-1	-5	1

(b)

x	-3	-2	-1	0	1	2	3
$f(x)$	1	5	-1	0	1	-5	-1

67. (a) even **(b)** odd **(c)** odd **(d)** neither **69. (a)** even **(b)** odd **(c)** even **(d)** neither **(e)** odd **(f)** even

71. (a) y-axis **(b)** origin **(c)** x-axis, y-axis, origin

73. 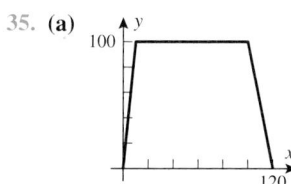 **77. (a)** 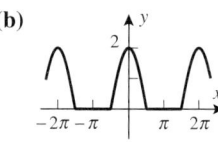 **(b)** 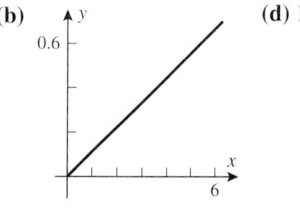 **79.** yes; $f(x) = x^k, g(x) = x^n$

▶ Exercise Set 1.5 **(Page 71)**

1. (a) $-\frac{3}{2}, -\frac{1}{18}, \frac{2}{3}$ **(b)** yes **3.** III $<$ II $<$ IV $<$ I

5. (a) The slopes are equal; the points lie on the same line.

 (b) The slopes $-1, 3, \frac{1}{3}$ are not equal; the points do not lie on a line. **7. (a)** 14 **(b)** $-\frac{1}{3}$

9. $\frac{13}{7}$ **11. (a)** $153°$ **(b)** $45°$ **(c)** $117°$ **(d)** $89°$ **13. (a)** $60°$ **(b)** $117°$ **15.** $y = -2x + 4$ **17.** $y = 4x + 7$

19. $y = -\frac{1}{5}x + 6$ **21.** $y = 11x - 18$ **23. (a)** parallel **(b)** perpendicular **(c)** parallel **(d)** perpendicular **(e)** neither

25. (a) $y = \frac{3}{2}x - 3$ **(b)** $y = -\frac{3}{4}x$ **27. (a)** $\frac{9}{10}$ ft/s **(b)** -4 **(c)** -2.2 **(d)** $\frac{80}{9}$ s

29. (a) $-\frac{4}{3}$ ft/s^2 **(b)** $v = -\frac{4}{3}t + \frac{13}{3}$ **(c)** $v = \frac{13}{3}$ ft/s **31. (b)** $-\frac{9}{10}$ cm/s **(c)** $\frac{9}{10}$ cm/s

33. (a) 0 mi/h **(b)** 48 mi/h **(c)** 240 mi

35. (a)

```
     y
100 ┤
    │
    │
    │              x
    └──────────
         120
```

(b) $v = \begin{cases} 10t & \text{if} & 0 \le t \le 10 \\ 100 & \text{if} & 10 \le t \le 100 \\ 600 - 5t & \text{if} & 100 \le t \le 120 \end{cases}$

37. (a) $y = x/9$ **(c)** 26.11 in

(b)
```
    y
0.6 ┤        /
    │      /
    │    /
    │  /         x
    └──────────
          6
```
(d) 135 lb

39. $y = 1.2x + 2$ **41. (a)** $T_C = \frac{5}{9}(T_F - 32)$ **(b)** $\frac{5}{9}$ **(c)** $-40°$ (F or C) **(d)** $37°$ **43. (a)** $p = 0.098h + 1$ **(b)** 10.20 m

45. (a) $r = -0.0125t + 0.8$ **(b)** 64 days

47. (a) $C_1 = 2x, C_2 = 25 + (x/4)$ **(b)** $x = 15$ **49. (a)** $H \approx 181$

(c) The Universe would be even older.

```
    C
60 ┤
    │        C₁
    │      ╱
    │   ──────  C₂
    │ ╱
    └──────────  x
          30
```

▶ **Exercise Set 1.6 (Page 89)**

1. (a) $y = 3x + b$ (c)
   (b) $y = 3x + 6$

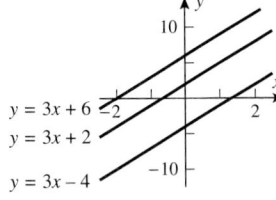

$y = 3x + 6$
$y = 3x + 2$
$y = 3x - 4$

3. (a) $y = mx + 2$ (c)
   (b) $y = -x + 2$

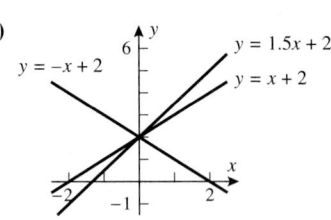

$y = -x + 2$
$y = 1.5x + 2$
$y = x + 2$

5. (a) slope: -1 (b) y-intercept: $y = -1$

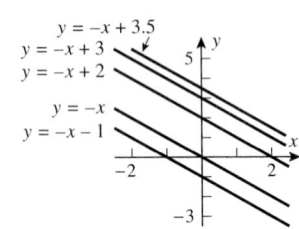

$y = -x + 3.5$
$y = -x + 3$
$y = -x + 2$
$y = -x$
$y = -x - 1$

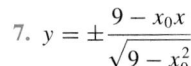

$y = 2.5x - 1$
$y = 2x - 1$
$y = -1.5x - 1$
$y = -x - 1$
$y = -1$

(c) pass through $(-4, 2)$ (d) x-intercept: $x = 1$

7. $y = \pm \dfrac{9 - x_0 x}{\sqrt{9 - x_0^2}}$ 9.

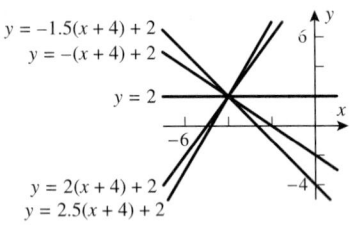

$y = -1.5(x + 4) + 2$
$y = -(x + 4) + 2$
$y = 2$
$y = 2(x + 4) + 2$
$y = 2.5(x + 4) + 2$

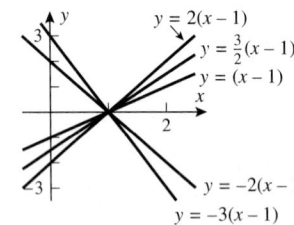

$y = 2(x - 1)$
$y = \frac{3}{2}(x - 1)$
$y = (x - 1)$
$y = -2(x - 1)$
$y = -3(x - 1)$

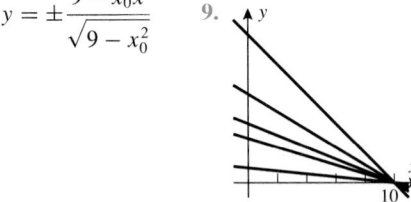

11. (a) VI   (b) IV   (c) III   (d) V   (e) I   (f) II

13. (a)

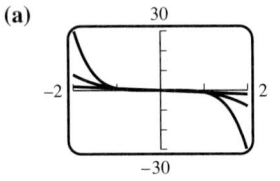

(b)

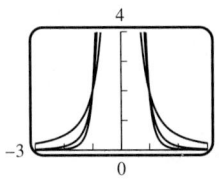

(c)

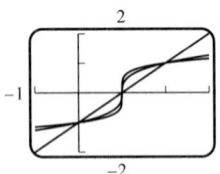

15. (a)

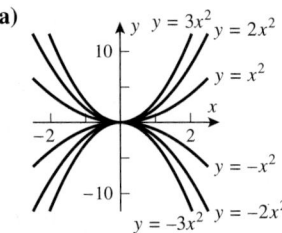

$y = 3x^2$ $y = 2x^2$
$y = x^2$
$y = -x^2$
$y = -3x^2$ $y = -2x^2$

(b)

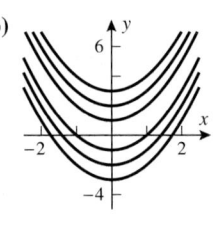

(c)

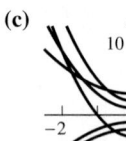

17. (a)

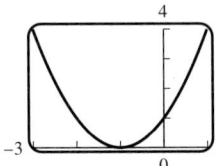

(b)

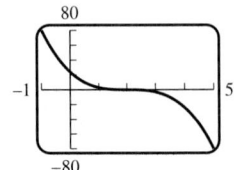

(c)

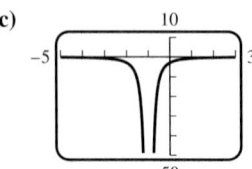

(d)

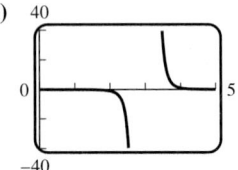

19. (a) 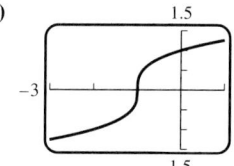   **(b)** 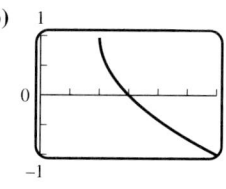   **(c)** 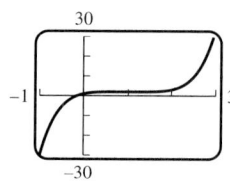   **(d)**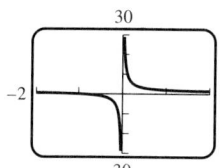

21. 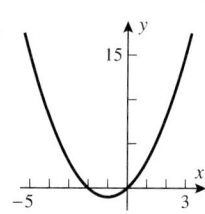   **23.** $t = 0.445\sqrt{d}$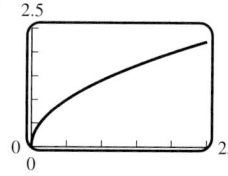

25. (a) newton-meters (N·m)   **27. (a)** $k = 0.000045$ N·m^2 **29. (a)** II; $y = 1, x = -1, 2$

(b) 20 N·m   **(b)** 0.000005 N   **(b)** I; $y = 0, x = -2, 3$

(c)

V (L)	0.25	0.5	1.0	1.5	2.0
P (N/m^2)	80×10^3	40×10^3	20×10^3	13.3×10^3	10×10^3

(c) 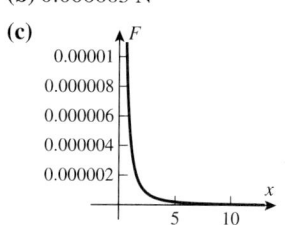   **(c)** IV; $y = 2$

(d) III; $y = 0, x = -2$

(d)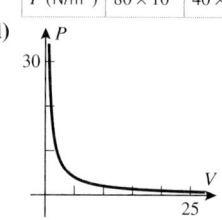

(d) The force becomes infinite; the force tends to zero.

31. Order the six trigonometric functions as sin, cos, tan, cot, sec, csc: **(a)** pos, pos, pos, pos, pos, pos

(b) neg, zero, undefined, zero, undefined, neg   **(c)** pos, neg, neg, neg, neg, pos   **(d)** neg, pos, neg, neg, pos, neg

(e) neg, neg, pos, pos, neg, neg   **(f)** neg, pos, neg, neg, pos, neg

33. (a) use $\sin(\pi - x) = \sin x$; 0.588   **(b)** use $\cos(-x) = \cos x$; 0.924   **(c)** use $\sin(2\pi + x) = \sin x$; 0.588

(d) use $\cos(\pi - x) = -\cos x$; -0.924 **(e)** use $\sin 2x = 2\sin x\sqrt{1 - \sin^2 x}$; 0.951.   **(f)** use $\cos^2 x = 1 - \sin^2 x$; 0.654

35. (a) $-a$ **(b)** b **(c)** $-c$ **(d)** $\pm\sqrt{1 - a^2}$ **(e)** $-b$ **(f)** $-a$ **(g)** $\pm 2b\sqrt{1 - b^2}$ **(h)** $2b^2 - 1$.

(i) $1/b$ **(j)** $-1/a$ **(k)** $1/c$ **(l)** $(1 - b)/2$ **37.** 80,936 km

39. The second quarter revolves twice (720°) about its own center.   **41. (a)** $y = 3\sin(x/2)$ **(b)** $y = 4\cos 2x$ **(c)** $y = -5\sin 4x$

43. (a) $y = \sin[x + (\pi/2)]$ **(b)** $y = 3 + 3\sin(2x/9)$ **(c)** $y = 1 + 2\sin\left[2\left(x - \dfrac{\pi}{4}\right)\right]$

45. (a) $3, \pi/2, 0$ 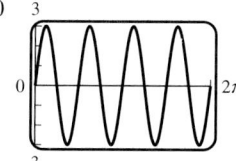   **(b)** $2, 2, 0$ 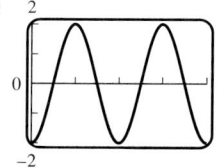   **(c)** $1, 4\pi, 0$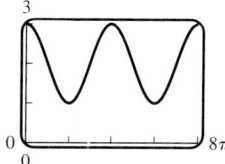

47. (b) $A = \sqrt{A_1^2 + A_2^2}, \theta = \tan^{-1}(A_2/A_1)$ **(c)** $x = \dfrac{5\sqrt{13}}{2}\sin\left(2\pi t + \tan^{-1}\dfrac{1}{2\sqrt{3}}\right)$

▶ **Exercise Set 1.7 (Page 100)**

1. (a) 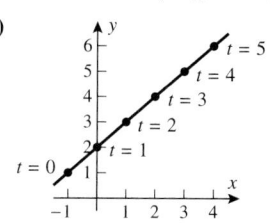   **(c)**

t	0	1	2	3	4	5
x	-1	0	1	2	3	4
y	1	2	3	4	5	6

3.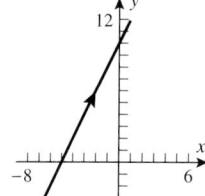

5. 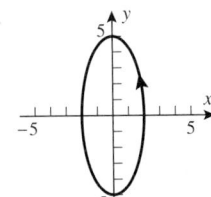   **7.** 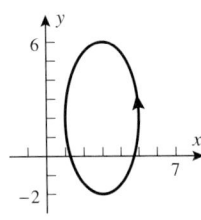   **9.** 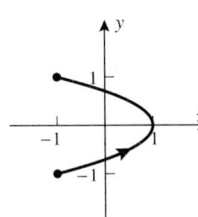   **11.**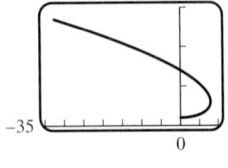

13. $x = 5\cos t,\ y = -5\sin t,\ \ 0 \le t \le 2\pi$ **15.** $x = 2,\ y = t$ **17.** $x = t^2,\ y = t,\ \ -1 \le t \le 1$

19. **(a)** IV   **21.** **(a)**    **(b)**

t	0	1	2	3	4	5
x	0	5.5	8	4.5	−8	−32.5
y	1	1.5	3	5.5	9	13.5

(b) II

(c) V

(d) VI

(e) III

(f) I

(c) $t = 0, 2\sqrt{3}$

(d) $0 < t < 2\sqrt{2}$

(e) 2

23. **(a)** 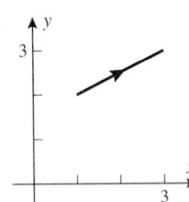   **(b)**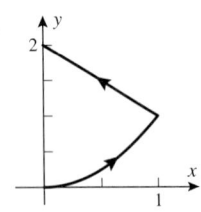

25. **(a)** $\dfrac{x - x_0}{x_1 - x_0} = \dfrac{y - y_0}{y_1 - y_0}$

(c) $x = 1 + t,\ y = -2 + 6t$

(d) $x = 2 - t,\ y = 4 - 6t$

27. **(b)** $\frac{1}{2}$ **(c)** $\frac{3}{4}$

31. **(b)** 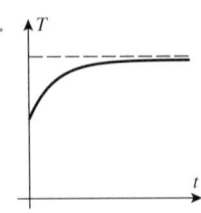   **33.**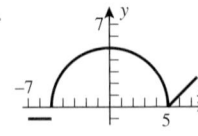

35. **(a)** $x = 4\cos t,$
$y = 3\sin t$

(b) $x = -1 + 4\cos t,$
$y = 2 + 3\sin t$

37. **(a)** $x = 400\sqrt{2}\,t,$
$y = 400\sqrt{2}\,t - 4.9t^2$

(b) 16,326.53 m

(c) 65,306.12 m

39. **(a)** ellipses with fixed center, varying axes of symmetry

    (b) (assume $a \ne 0, b \ne 0$) ellipses with varying center, fixed axes of symmetry

    (c) circles of radius 1 with centers on line $y = x - 1$

41. **(a)**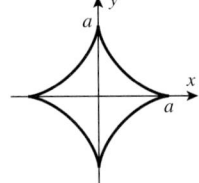

▶ **Chapter 1 Supplementary Exercises (Page 103)**

1. 1940–1945   **3.** 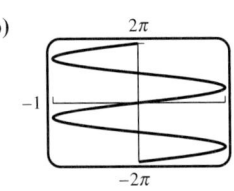   **5.** $C = 5x^2 + (64/x)$ **7.**

9. **(a)** $V = (6 - 2x)(5 - x)x$

    (b) $0 < x < 3$

    (c) 3.57 ft × 3.79 ft × 1.21 ft

11. no solution   **13.** $1/(2 - x^2)$ **15.**

x	−4	−3	−2	−1	0	1	2	3	4
$f(x)$	0	−1	2	1	3	−2	−3	4	−4
$g(x)$	3	2	1	−3	−1	−4	4	−2	0
$(f \circ g)(x)$	4	−3	−2	−1	1	0	−4	2	3
$(g \circ f)(x)$	−1	−3	4	−4	−2	1	2	0	3

17. **(a)** odd

    (b) even

    (c) neither

    (d) even

19. **(b)** 295.72 ft   **21.** C: (2.0944, 1.9132); D: (4.1888, 1.2284); B: (−2.0944, −1.9132); A: (−4.1888, −1.2284)

23. **(a)** circles of radius 1 centered on the parabola $y = x^2$ **(b)** parabolas that open up with vertices on the line $y = x/2$

25. 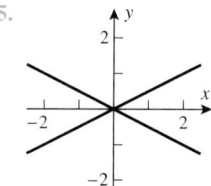 **27.** 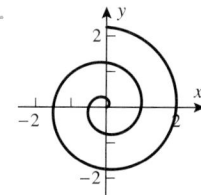 **29.** $d = \sqrt{(x-1)^2 + (\sqrt{x}-2)^2}$; **31.** 0.48 ft
$x = 1.358094$

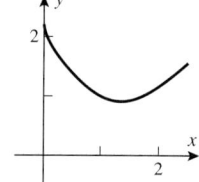

33. (a) 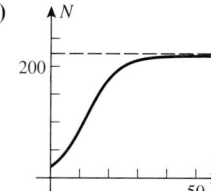 **(b)** about 10 years **35. (a)** 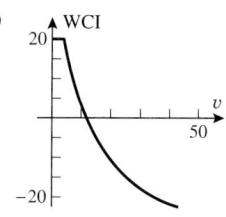 **(b)** $3°F$, $-11°F$, $-18°F$, $-22°F$
(c) 220 sheep **(c)** $v = 35, 19, 12, 7$ mi/h

37. $-0.7245 \le x \le 1.2207$; $-1.0551 \le y \le 1.4902$

39. (a) 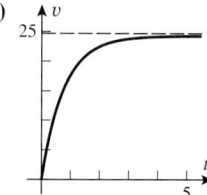 **(c)** For large t the velocity approaches c.
(d) No, but it comes arbitrarily close.
(e) 3.013 s

▶ Exercise Set 2.1 **(Page 124)**

1. (a) -1 **(b)** 3 **(c)** does not exist **(d)** 1 **(e)** -1 **(f)** 3 **3. (a)** 1 **(b)** 1 **(c)** 1 **(d)** 1 **(e)** $-\infty$ **(f)** $+\infty$

5. (a) 0 **(b)** 0 **(c)** 0 **(d)** 3 **(e)** $+\infty$ **(f)** $+\infty$ **7. (a)** $-\infty$ **(b)** $+\infty$ **(c)** does not exist **(d)** undefined **(e)** 2 **(f)** 0

9. (a) $-\infty$ **(b)** $-\infty$ **(c)** $-\infty$ **(d)** 1 **(e)** 1 **(f)** 2 **11. (a)** 0 **(b)** 0 **(c)** 0 **(d)** 0 **(e)** does not exist **(f)** does not exist

13. for all $x_0 \ne -4$ **15. (a)** At $x = 3$ the one-sided limits fail to exist.

(b) At $x = -2$ the two-sided limit exists but is not equal to $F(-2)$. **(c)** At $x = 3$ the limit fails to exist.

17. (a) $\frac{1}{3}$ **(b)** $+\infty$ **(c)** $-\infty$ **19. (a)** 3 **(b)** does not exist **21. (a)** $y = 2$ **(b)** $y = 20.086$ **(c)** no horizontal asymptote

23. (a) $\lim\limits_{x \to 0^+} \dfrac{\sin x}{x}$ **(b)** $\lim\limits_{x \to 0^+} \dfrac{x-1}{x+1}$ **(c)** $\lim\limits_{x \to 0^-} (1+2x)^{1/x}$ **25. (a)** $f(x) = \begin{cases} 1 + (1/x), & x < 0 \\ -1 + (1/x), & x \ge 0 \end{cases}$ **(b)** yes; $f(x) = (\sin x)/x$

29. (a) catastrophic subtraction when the x-interval is small (the size depending on the calculating utility). **(c)** no.

▶ Exercise Set 2.2 **(Page 137)**

1. (a) -6 **(b)** 13 **(c)** -8 **(d)** 16 **(e)** 2 **(f)** $-\frac{1}{2}$ **(g)** The denominator tends to zero but the numerator does not.

(h) The denominator tends to zero but the numerator does not. **3. (a)** 7 **(b)** -3 **(c)** π **(d)** -6 **(e)** 36 **(f)** $-\infty$ **5.** 0

7. 8 **9.** 4 **11.** $-\frac{4}{5}$ **13.** $\frac{3}{2}$ **15.** 0 **17.** 0 **19.** $-\sqrt{5}$ **21.** $1/\sqrt{6}$ **23.** $\sqrt{3}$ **25.** $+\infty$ **27.** does not exist

29. $-\infty$ **31.** $+\infty$ **33.** does not exist **35.** $+\infty$ **37.** $-\infty$ **39.** $-\frac{1}{7}$ **41.** 6 **43.** $+\infty$ **45.** $+\infty$ **47.** $-\infty$

49. (a) 2 **51. (a)** 3 **(b)** 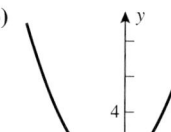 **53. (a)** Theorem 2.2.2(a) does not apply. **55.** $\frac{1}{4}$ **57.** 0
(b) 2
(c) 2
(b) $\lim\limits_{x \to 0^+} \left(\dfrac{1}{x} - \dfrac{1}{x^2} \right) = \lim\limits_{x \to 0^+} \left(\dfrac{x-1}{x^2} \right) = -\infty$

59. $a/2$ **61.** $\lim\limits_{x \to +\infty} p(x) = (-1)^n \infty$ and $\lim\limits_{x \to -\infty} p(x) = +\infty$

63. For $m > n$, the limits are both zero; for $m = n$, the limits are equal to the leading coefficient of p; for $n > m$, the limits are $\pm\infty$.

65. The left and/or right limits could be $\pm\infty$; or the limit could exist and equal any preassigned real number.

▶ **Exercise Set 2.3 (Page 145)**

1. **(a)** $|x| < 0.1$ **(b)** $|x - 3| < 0.0025$ **(c)** $|x - 4| < 0.000125$ 3. **(a)** $x_1 = 3.8025, x_2 = 4.2025$ **(b)** $\delta = 0.1975$

5. $\delta = 0.05$ 7. $\delta = \frac{1}{700}$ 9. $\delta = 0.05$ 11. $\delta = \frac{1}{9000}$ 13. $\delta = 1$ 15. $\delta = \frac{1}{3}\epsilon$ 17. $\delta = \frac{1}{2}\epsilon$ 19. $\delta = \epsilon$

21. $\delta = \min\left(1, \frac{1}{6}\epsilon\right)$ 23. $\delta = \min\left(\frac{1}{6}, \frac{\epsilon}{18}\right)$ 25. $\delta = 2\epsilon$ 27. $\delta = \epsilon$ 29. **(a)** $\sqrt{10}$ **(b)** 99 **(c)** -10 **(d)** -101

31. **(a)** $-\sqrt{\dfrac{1 - \epsilon}{\epsilon}}; \sqrt{\dfrac{1 - \epsilon}{\epsilon}}$ **(b)** $\sqrt{\dfrac{1 - \epsilon}{\epsilon}}$ **(c)** $-\sqrt{\dfrac{1 - \epsilon}{\epsilon}}$ 33. 10   35. 999   37. -202 39. -57.5 41. $\dfrac{1}{\sqrt{\epsilon}}$

43. $-2 - \dfrac{1}{\epsilon}$ 45. $\dfrac{1}{\epsilon} - 1$ 47. $-\dfrac{5}{2} - \dfrac{11}{2\epsilon}$ 49. **(a)** $|x| < \frac{1}{10}$ **(b)** $|x - 1| < \frac{1}{1000}$ **(c)** $|x - 3| < \frac{1}{10\sqrt{10}}$ **(d)** $|x| < \frac{1}{10}$

51. $\delta = 1/\sqrt{M}$ 53. $\delta = 1/M$ 55. $\delta = 1/(-M)^{1/4}$ 57. $\delta = \epsilon$ 59. $\delta = \epsilon^2$ 61. $\delta = \epsilon$

63. **(a)** $\delta = -1/M$ 65. **(a)** $N = M - 1$ 67. $\delta = \min\left(2, \frac{1}{8}\epsilon\right)$ 69. $\delta = 0.0442$

(b) $\delta = 1/M$ **(b)** $N = M - 1$

▶ **Exercise Set 2.4 (Page 156)**

1. **(a)** not continuous, $x = 2$ **(b)** not continuous, $x = 2$ **(c)** not continuous, $x = 2$ **(d)** continuous **(e)** continuous **(f)** continuous

3. **(a)** not continuous, $x = 1, 3$ **(b)** continuous **(c)** not continuous, $x = 1$ **(d)** continuous **(e)** not continuous, $x = 3$

(f) continuous   5. **(a)** 3 **(b)** 3

7. **(a)**   **(b)**   **(c)**   **(d)**

9. **(a)**

(b) One second could cost you one dollar.   11. none   13. none

15. f is not defined at $x = \pm 4$.   17. f is not defined at $x = \pm 3$.   19. none   21. none   23. **(a)** $k = 5$ **(b)** $k = \frac{4}{3}$

25. **(a)**   **(b)**   27. **(a)** $x = 0$, not removable

(b) $x = -3$, removable

(c) $x = 2$, removable;

$x = -2$, not removable

29. **(a)** $x = \frac{1}{2}$, not removable;   **(b)** $(2x - 1)(x + 3)$

at $x = -3$, removable

33. **(a)** $f(x) = k$ for $x \neq c$, $f(c) = 0$; $g(x) = l$ for $x \neq c$, $g(c) = 0$. If $k = -l$, then $f + g$ is continuous; otherwise it is not.

　　(b) $f(x) = k$ for $x \neq c$, $f(c) = 1$; $g(x) = l \neq 0$ for $x \neq c$, $g(c) = 1$. If $kl = 1$, then fg is continuous; otherwise it is not.

37. $f(x) = 1$ for $0 \leq x < 1$, $f(x) = -1$ for $1 \leq x \leq 2$　　43. $x = -1.25, x = 0.75$

45. $x = -1.605, x = 1.375$　　47. $x = 2.24$　　49. $x = 4.847$ cm

▶ Exercise Set 2.5 **(Page 163)**

1. none　　3. $x = n\pi, n = 0, \pm1, \pm2, \ldots$　　5. $x = n\pi, n = 0, \pm1, \pm2, \ldots$　　7. none

9. $2n\pi + (\pi/6), 2n\pi + (5\pi/6), n = 0, \pm1, \pm2, \ldots$

11. **(a)** $\sin x, x^3 + 7x + 1$　**(b)** $|x|, \sin x$　**(c)** $x^3, \cos x, x + 1$　**(d)** $\sqrt{x}, 3 + x, \sin x, 2x$　**(e)** $\sin x, \sin x$　**(f)** $x^5 - 2x^3 + 1, \cos x$

13. 1　15. $-\sqrt{3}/2$　17. 3　19. -1　21. 0　23. $\frac{7}{3}$　25. 1　27. 2　29. 0　31. $-\frac{25}{49}$　33. does not exist

35. 3　37. $k = \frac{1}{2}$　39. **(a)** 1　**(b)** 0　**(c)** 1　41. $-\pi$　43. $-x \leq x\cos(50\pi/x) \leq x$

45. $\lim\limits_{x \to 0} f(x) = 1$ by the Squeezing Theorem.　　47. $g(x) = -\dfrac{1}{x}, h(x) = \dfrac{1}{x}$;

$\lim\limits_{x \to +\infty} \dfrac{\sin x}{x} = 0$ by the Squeezing Theorem.

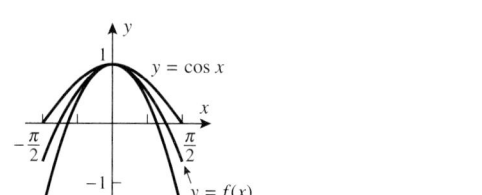

51. **(a)** 0.17365　53. **(a)** 0.08749　55. **(b)**　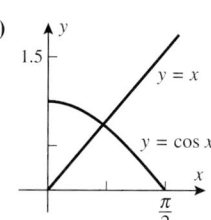　**(c)** 0.739　57. **(a)** symmetry about the equatorial plane

　　(b) 0.17453　　　**(b)** 0.08727

▶ Chapter 2 Supplementary Exercises **(Page 165)**

1. **(a)** 1　**(b)** does not exist　**(c)** does not exist　**(d)** 1　**(e)** 3　**(f)** 0　**(g)** 0　**(h)** 2　**(i)** $\frac{1}{2}$

5. **(a)** $\frac{1}{4}$　**(b)** 4　7. **(a)** 0.405　17. **(a)** 1.449　**(b)** $x = 0, \pm1.896$

19. **(a)** $\sqrt{5}$, does not exist, $\sqrt{10}, \sqrt{10}$, does not exist, $+\infty$, does not exist　**(b)** 5, 10, 0, 0, 10, $-\infty$, $+\infty$

21. a/b　23. does not exist　25. 0　27. $3 - k$　31. 2.71828　33. 0.54030　35. 0.49996　37. 0.07747

39. **(b)**　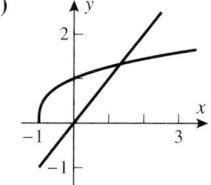　**(d)** 1, 1.26, 1.31, 1.322, 1.324, 1.3246, 1.3247　41. $x = \sqrt[5]{x + 2}$; 1.267168

▶ Exercise Set 3.1 **(Page 175)**

1. **(a)** $\frac{7}{2}$　**(d)**　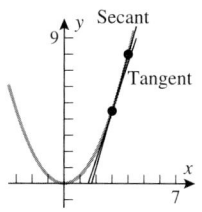　3. **(a)** $-\frac{1}{6}$　**(d)**　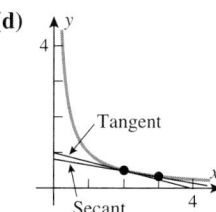　5. **(a)** $2x_0$　7. **(a)** $1/(2\sqrt{x_0})$

(b) 3　　　**(b)** $-\frac{1}{4}$　　　**(b)** 4　**(b)** $\frac{1}{2}$

(c) x_0　　　**(c)** $-1/x_0^2$

9. (a) 4 m/s   **11. (a)** t_0 **(b)** 0   **(c)** speeding up   **(d)** slowing down   **13.** straight line with slope equal to the velocity

15. (a) $72°$F at about 4:30 P.M.   **17. (a)** first year   **(d)** Growth rate (cm/year)   **19. (a)** 320,000 ft

(b) $4°$F/h   **(b)** 6 cm/year   **(b)** 8000 ft/s

(c) $-7°$F/h at about 9 P.M.   **(c)** 10 cm/year at about age 14   **(c)** 45 ft/s

(d) 24,000 ft/s

21. (a) 720 ft/min   **(b)** 192 ft/min

▶ Exercise Set 3.2 **(Page 186)**

1. $2, 0, -2, -1$ **3. (b)** 3   **5.**   **7.** $y = 5x - 16$ **9.** $6x; y = 18x - 27$ **11.** $3x^2; y = 0$

(c) 3   **13.** $\dfrac{1}{2\sqrt{x+1}}; y = \frac{1}{6}x + \frac{5}{3}$ **15.** $-1/x^2$

17. $2ax$ **19.** $-1/2x^{3/2}$ **21.** $8t + 1$

23. (a) D   **25. (a)**   **(b)**   **(c)**   **27. (a)** $x^2, 3$ **29.** 8

(b) F   **(b)** $\sqrt{x}, 1$

(c) B

(d) C

(e) A

(f) E

31. $y = -2x + 1$ 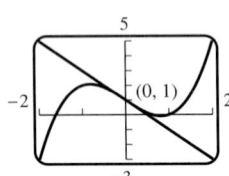   **33. (b)**

h	0.5	0.1	0.01	0.001	0.0001	0.00001
$[f(1+h)-f(1)]/h$	1.6569	1.4355	1.3911	1.3868	1.3863	1.3863

35. (a) dollars per foot   **(b)** the price per additional foot   **(c)** positive   **(d)** $1000

37. (a) $F \approx 200$ lb, $dF/d\theta \approx 60$ lb/rad   **(b)** $\mu \approx 0.3$ **39. (a)** $T \approx 120°$F, $dT/dt \approx -4.5°$F/min   **(b)** $k = -0.1$

45. $f(1) = 0, f'(1) = 5$

▶ Exercise Set 3.3 **(Page 197)**

1. $28x^6$ **3.** $24x^7 + 2$ **5.** 0   **7.** $-\frac{1}{3}(7x^6 + 2)$ **9.** $3ax^2 + 2bx + c$ **11.** $24x^{-9} + (1/\sqrt{x})$ **13.** $-3x^{-4} - 7x^{-8}$

15. $18x^2 - \frac{3}{2}x + 12$ **17.** $-15x^{-2} - 14x^{-3} + 48x^{-4} + 32x^{-5}$ **19.** $12x(3x^2 + 1)$ **21.** $-\frac{5}{4}$ **23.** $\dfrac{3}{(2t+1)^2}$ **25.** $\frac{7}{16}$

27. -29 **29.** $32t$ **31.** $3\pi r^2$ **33. (a)** $4\pi r^2$ **(b)** 100π **35. (a)** $-\frac{37}{4}$ **(b)** $-\frac{23}{16}$

37. (a) 10   **(b)** 19   **(c)** 9   **(d)** -1 **39.** $y = 5x + 17$ **41. (a)** $42x - 10$ **(b)** 24   **(c)** $2/x^3$ **(d)** $700x^3 - 96x$

43. (a) $-210x^{-8} + 60x^2$ **45. (a)** 0   **49.** $F'(x) = xf''(x) + 2f'(x)$ **51.** $\left(1, \frac{5}{6}\right)\left(2, \frac{2}{3}\right)$ 1.5   **53.** 0

(b) $-6x^{-4}$ **(b)** 112

(c) $6a$ **(c)** 360

55. 0   **57.** $y = 3x^2 - x - 2$ **59.** $x = \frac{1}{2}$ **61.** $2 \pm \sqrt{3}$ **63.** $-2x_0$ **67.** $-\dfrac{2GmM}{r^3}$ **69.** $f'(x) > 0$ for all $x \neq 0$

73. (a) $2(1 + x^{-1})(x^{-3} + 7) + (2x + 1)(-x^{-2})(x^{-3} + 7) + (2x + 1)(1 + x^{-1})(-3x^{-4})$ **(b)** $3(7x^6 + 2)(x^7 + 2x - 3)^2$

75. not differentiable at $x = 1$ **77.** $a = 6, b = -3$ **79. (a)** $x = \frac{2}{3}$ **(b)** $x = \pm 2$

83. (a) $n(n-1)(n-2)\cdots 1$ **(b)** 0   **(c)** $a_n n(n-1)(n-2)\cdots 1$

85. (b) f and all its derivatives up to $f^{(n-1)}(x)$ are continuous on (a, b).

▶ Exercise Set 3.4 **(Page 202)**

1. $-2\sin x - 3\cos x$ 3. $\dfrac{x\cos x - \sin x}{x^2}$ 5. $x^3\cos x + (3x^2+5)\sin x$ 7. $\sec x\tan x - \sqrt{2}\sec^2 x$ 9. $\sec^3 x + \sec x\tan^2 x$

11. $-\csc^3 x - \csc x\cot^2 x$ 13. $-\dfrac{\csc x}{1+\csc x}$ 15. 0 17. $\dfrac{1}{(1+x\tan x)^2}$ 19. $-x\cos x - 2\sin x$ 21. $-x\sin x + 5\cos x$

23. $-4\sin x\cos x$ 27. **(a)** $y = x$ **(b)** $y = 2x - (\pi/2) + 1$ **(c)** $y = 2x + (\pi/2) - 1$

29. **(a)** $x = \pm\pi/2, \pm3\pi/2$ **(b)** $x = -3\pi/2, \pi/2$ **(c)** no horizontal tangent line   **(d)** $x = \pm2\pi, \pm\pi, 0$ 31. 0.087 ft/degree

33. 1.75 m/degree   35. **(a)** $-\cos x$ **(b)** $\cos x$

37. **(a)** all x **(b)** all x **(c)** $x \neq (\pi/2) + n\pi, n = 0, \pm1, \pm2, \ldots$

(d) $x \neq n\pi, n = 0, \pm1, \pm2, \ldots$ **(e)** $x \neq (\pi/2) + n\pi, n = 0, \pm1, \pm2, \ldots$ **(f)** $x \neq n\pi, n = 0, \pm1, \pm2, \ldots$

(g) $x \neq (2n+1)\pi, n = 0, \pm1, \pm2, \ldots$ **(h)** $x \neq n\pi/2, n = 0, \pm1, \pm2, \ldots$ **(i)** all x 39. $3, 7, 11, \ldots$ 41. $\sec^2 y$

▶ Exercise Set 3.5 **(Page 208)**

1. $37(x^3+2x)^{36}(3x^2+2)$ 3. $-2\left(x^3 - \dfrac{7}{x}\right)^{-3}\left(3x^2 + \dfrac{7}{x^2}\right)$ 5. $\dfrac{24(1-3x)}{(3x^2-2x+1)^4}$ 7. $\dfrac{3}{4\sqrt{x}\sqrt{4+3\sqrt{x}}}$ 9. $3x^2\cos(x^3)$

11. $8x\sec^2(4x^2)$ 13. $-20\cos^4 x\sin x$ 15. $-\dfrac{2}{x^3}\cos\left(\dfrac{1}{x^2}\right)$ 17. $28x^6\sec^2(x^7)\tan(x^7)$ 19. $-\dfrac{5\sin(5x)}{2\sqrt{\cos(5x)}}$

21. $-3\left[x + \csc(x^3+3)\right]^{-4}\left[1 - 3x^2\csc(x^3+3)\cot(x^3+3)\right]$ 23. $\dfrac{x(10-3x^2)}{\sqrt{5-x^2}}$ 25. $10x^3\sin 5x\cos 5x + 3x^2\sin^2 5x$

27. $-x^3\sec\left(\dfrac{1}{x}\right)\tan\left(\dfrac{1}{x}\right) + 5x^4\sec\left(\dfrac{1}{x}\right)$ 29. $\sin(\cos x)\sin x$ 31. $-6\cos^2(\sin 2x)\sin(\sin 2x)\cos 2x$

33. $12(5x+8)^{13}(x^3+7x)^{11}(3x^2+7) + 65(x^3+7x)^{12}(5x+8)^{12}$ 35. $\dfrac{33(x-5)^2}{(2x+1)^4}$ 37. $-\dfrac{2(2x+3)^2(52x^2+96x+3)}{(4x^2-1)^9}$

39. $5\left[x\sin 2x + \tan^4(x^7)\right]^4\left[2x\cos 2x + \sin 2x + 28x^6\tan^3(x^7)\sec^2(x^7)\right]$ 41. $-25x\cos(5x) - 10\sin(5x) - 2\cos(2x)$

43. $4(1-x)^{-3}$ 45. $y = -x$ 47. $y = -1$ 49. $3\cot^2\theta\csc^2\theta$ 51. $\pi(b-a)\sin 2\pi\omega$

53. **(a)**   **(c)** $\dfrac{4-2x^2}{\sqrt{4-x^2}}$ **(d)** $y - \sqrt{3} = \dfrac{2}{\sqrt{3}}(x-1)$

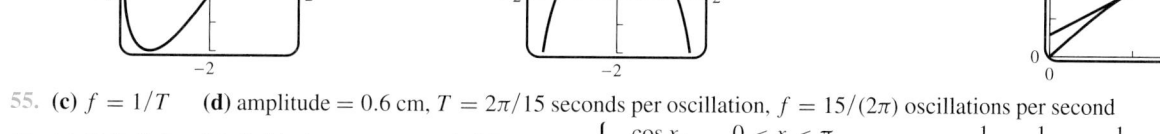

55. **(c)** $f = 1/T$ **(d)** amplitude $= 0.6$ cm, $T = 2\pi/15$ seconds per oscillation, $f = 15/(2\pi)$ oscillations per second

57. **(a)** 10 lb/in², -2 lb/in²/mi   **(b)** -0.6 lb/in²/s   59. $\begin{cases} \cos x, & 0 < x < \pi \\ -\cos x, & -\pi < x < 0 \end{cases}$ 61. **(a)** $-\dfrac{1}{x}\cos\dfrac{1}{x} + \sin\dfrac{1}{x}$

63. **(a)** 21   **(b)** -36 65. 6   67. $1/2x$ 69. $\frac{2}{3}x$ 73. $f'(g(h(x)))g'(h(x))h'(x)$

▶ Exercise Set 3.6 **(Page 217)**

1. **(a)** $4, 5$ **(b)**   3. **(a)** $0.5, 1$ **(b)**   5. $3x^2\,dx, 3x^2\,\Delta x + 3x(\Delta x)^2 + (\Delta x)^3$

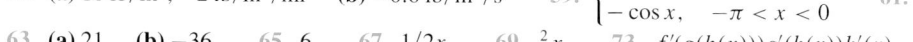

7. $(2x-2)\,dx, 2x\,\Delta x + (\Delta x)^2 - 2\,\Delta x$ 9. **(a)** $(12x^2 - 14x)\,dx$ **(b)** $(-x\sin x + \cos x)\,dx$

11. **(a)** $\dfrac{2-3x}{2\sqrt{1-x}}\,dx$ **(b)** $-17(1+x)^{-18}\,dx$ 13. **(a)** $f(x) \approx 1 + 3(x-1)$ **(b)** $f(1+\Delta x) \approx 1 + 3\,\Delta x$ **(c)** 1.06   15. **(a)** $1 + \frac{1}{2}x, 0.95, 1.05$ **(b)**

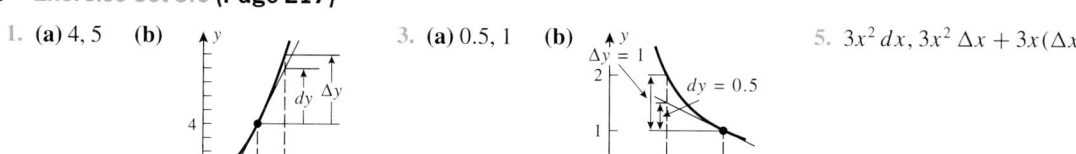

25. (a) 0.0174533   **(b)** $x_0 = 45°$ **(c)** 0.694765   **27.** 83.16   **29.** 8.0625   **31.** 8.9944   **33.** 0.1   **35.** 0.8573

37. $|x| < 0.1692$ **39.** $|x| < 0.6316$ **41.** 0.0225   **43.** 0.0048   **45. (a)** $\pm 2\ \text{ft}^2$

(b) side: $\pm 1\%$; area: $\pm 2\%$

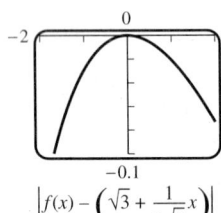

$$\left| f(x) - \left(\sqrt{3} + \frac{1}{2\sqrt{3}} x \right) \right|$$

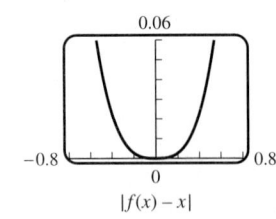

$$|f(x) - x|$$

47. (a) opposite: ± 0.151 in; adjacent: ± 0.087 in   **(b)** opposite: $\pm 3.0\%$; adjacent: $\pm 1.0\%$ **49.** $\pm 10\%$ **51.** $\pm 0.017\ \text{cm}^2$

53. $\pm 6\%$ **55.** $\pm 0.5\%$ **57.** 0.236 cm^3 **59. (a)** $\alpha = 1.5 \times 10^{-5}/°\text{C}$ **(b)** 180.1 cm long

---

▶ **Chapter 3 Supplementary Exercises (Page 219)**

5. $x = -\frac{7}{2}, 2, -\frac{1}{2}$ **7.** $x = 1, -\frac{1}{15}$ **9. (a)** $x = -2, -1, 1, 3$ **(b)** $(-\infty, -2), (-1, 1), (3, +\infty)$ **(c)** $(-2, -1), (1, 3)$ **(d)** 4

11. $y = -16x, y = -145x/4$ **13.** $x = n\pi \pm (\pi/4), n = 0, \pm 1, \pm 2, \ldots$

17. (a) $-0.5, 1, 0.5$ **(b)** $\pi/4, 1, \pi/2$ **(c)** $3, -1, 0$ **19. (a)** 2000 gal/min   **(b)** 2500 gal/min

21. (a) between 139.48 m and 144.55 m   **(b)** $|d\phi| \le 0.98°$ **23. (a)** 3.6   **(b)** -0.777778 **25.** 2.772589

27. 58.75 ft/s   **29.** ± 0.535428

▶ **Exercise Set 4.1 (Page 233)**

1. (a) yes   **3. (a)** yes   **5. (a)** yes   **(d)** yes   **7. (a)** no   **9. (b)** $[-2, 2], [-8, 8]$ **(c)**

   (b) no   **(b)** no   **(b)** yes   **(e)** no   **(b)** no

   (c) yes   **(c)** no   **(f)** no   **(c)** yes

   (d) no

11. (a) no   **(b)** yes   **(c)** yes   **13.** $x^{1/5}$ **15.** $\frac{1}{7}(x+6)$ **17.** $\sqrt[3]{(x+5)/3}$ **19.** $(x^3 + 1)/2$ **21.** $-\sqrt{3/x}$

23. $\begin{cases} (5/2) - x, & x > 1/2 \\ 1/x, & 0 < x \le 1/2 \end{cases}$ **25.** $x^{1/4} - 2$ for $x \ge 16$ **27.** $\frac{1}{2}(3 - x^2)$ for $x \le 0$ **29.** $\frac{1}{10}(1 + \sqrt{1 - 20x})$ for $x \le -4$

31. (a) $y = (6.214 \times 10^{-4})x$ **33. (b)**   **(c)** No, because $f(g(x)) = x$ for $x > 1$

   (b) $x = \dfrac{10^4}{6.214} y$ but the domain of g is $x \ge 0$.

   (c) how many meters in y miles

35. (b) symmetric about the line $y = x$ **37. (b)** $1 - (\sqrt{3}/3)$ **39.** 10

41.   **43.**   **47.**   **49.** $\frac{88}{7}$

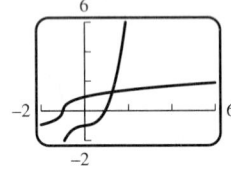

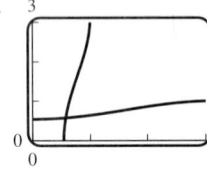

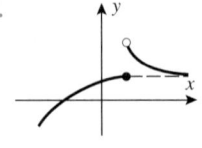

▶ **Exercise Set 4.2 (Page 243)**

1. (a) -4 **(b)** 4   **(c)** $\frac{1}{4}$ **3. (a)** 2.9690   **(b)** 0.0341   **5. (a)** 4   **(b)** -5 **(c)** 1   **(d)** $\frac{1}{2}$ **7. (a)** 1.3655   **(b)** -0.3011

9. (a) $2r + \dfrac{s}{2} + \dfrac{t}{2}$ **(b)** $s - 3r - t$ **11. (a)** $1 + \log x + \frac{1}{2}\log(x - 3)$ **(b)** $2\ln|x| + 3\ln\sin x - \frac{1}{2}\ln(x^2 + 1)$ **13.** $\log\frac{256}{3}$

15. $\ln \dfrac{\sqrt[3]{x}(x+1)^2}{\cos x}$ **17.** 0.01   **19.** e^2 **21.** 4   **23.** 10^5 **25.** $\sqrt{3/2}$ **27.** $-\dfrac{\ln 3}{2\ln 5}$ **29.** $\frac{1}{3}\ln\frac{7}{2}$ **31.** -2

33. $0, -\ln 2$ **35. (a)**    **(b)**    **37.** $2.8777, -0.3174$ **39.**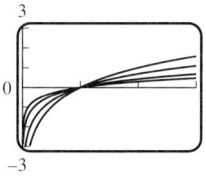

41. $x = 3.6541, y = 1.2958$ **43. (a)** no   **(d)** 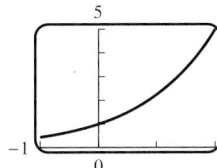   **45.** $\log\frac{1}{2} < 0$, so $3\log\frac{1}{2} < 2\log\frac{1}{2}$

(b) $y = 2^{x/4}$

(c) $y = 2^{-x}$

47. 201 days   **49. (a)** 7.4, basic   **(b)** 4.2, acidic   **(c)** 6.4, acidic   **(d)** 5.9, acidic

51. (a) 140 dB, damage   **(b)** 120 dB, damage   **(c)** 80 dB, no damage   **(d)** 75 dB, no damage

53. ≈ 200 **55. (a)** $\approx 5 \times 10^{16}$ J   **(b)** ≈ 0.67 **57.** e^{-2}

▶ Exercise Set 4.3 **(Page 253)**

1. $\frac{2}{3}(2x-5)^{-2/3}$ **3.** $\dfrac{9}{2(x+2)^2}\left[\dfrac{x-1}{x+2}\right]^{1/2}$ **5.** $\frac{1}{3}x^2(5x^2+1)^{-5/3}(25x^2+9)$ **7.** $-\dfrac{15[\sin(3/x)]^{3/2}\cos(3/x)}{2x^2}$

9. (a) $\dfrac{2-3x^2-y}{x}$ **(b)** $-\dfrac{1}{x^2}-2x$ **11.** $-\dfrac{x}{y}$ **13.** $\dfrac{1-2xy-3y^3}{x^2+9xy^2}$ **15.** $-\dfrac{y^2}{x^2}$ **17.** $\dfrac{1-2xy^2\cos(x^2y^2)}{2x^2y\cos(x^2y^2)}$

19. $\dfrac{1-3y^2\tan^2(xy^2+y)\sec^2(xy^2+y)}{3(2xy+1)\tan^2(xy^2+y)\sec^2(xy^2+y)}$ **21.** $-\dfrac{21}{16y^3}$ **23.** $\dfrac{2y}{x^2}$ **25.** $\dfrac{\sin y}{(1+\cos y)^3}$ **27.** $-1, +1$ **29.** -0.1312

31. $-\frac{9}{13}$ **35. (a)** 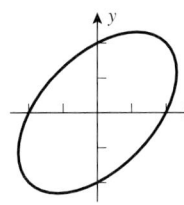   **(b)** ± 1.1547 **37.** $\dfrac{2t^3+3a^2}{2a^3-6at}$ **39.** $-\dfrac{b^2\lambda}{a^2\omega}$ **41.** $a = \frac{1}{4}, b = \frac{5}{4}$

(c) $x = \pm\dfrac{2}{\sqrt{3}}$

43. $y = (\sqrt{3}/3)x, y = -(\sqrt{3}/3)x$ **45.** $-\dfrac{2y^3+3t^2y}{(6ty^2+t^3)\cos t}$ **47.** $\dfrac{dy}{dt} = \dfrac{3\cos 3x - y^2}{2xy}\dfrac{dx}{dt}$ **49.** $-1, \frac{2}{3}$

53. $\dfrac{1}{15y^2+1}$ **55.** $\dfrac{1}{10y^4+3y^2}$

▶ Exercise Set 4.4 **(Page 260)**

1. $\dfrac{1}{x}$ **3.** $\dfrac{2\ln x}{x}$ **5.** $\dfrac{\sec^2 x}{\tan x}$ **7.** $\dfrac{1-x^2}{x(1+x^2)}$ **9.** $\dfrac{3x^2-14x}{x^3-7x^2-3}$ **11.** $\dfrac{1}{2x\sqrt{\ln x}}$ **13.** $-\dfrac{1}{x}\sin(\ln x)$

15. $3x^2\log_2(3-2x) - \dfrac{2x^3}{(\ln 2)(3-2x)}$ **17.** $\dfrac{2x(1+\log x) - x/(\ln 10)}{(1+\log x)^2}$ **19.** $7e^{7x}$ **21.** $x^2 e^x(x+3)$ **23.** $\dfrac{4}{(e^x+e^{-x})^2}$

25. $(x\sec^2 x + \tan x)e^{x\tan x}$ **27.** $(1-3e^{3x})e^{x-e^{3x}}$ **29.** $\dfrac{x-1}{e^x-x}$ **31.** $-\dfrac{y}{x(y+1)}$ **33.** $-\tan x + \dfrac{3x}{4-3x^2}$

35. $x\sqrt[3]{1+x^2}\left[\dfrac{1}{x} + \dfrac{2x}{3(1+x^2)}\right]$ **37.** $\dfrac{(x^2-8)^{1/3}\sqrt{x^3+1}}{x^6-7x+5}\left[\dfrac{2x}{3(x^2-8)} + \dfrac{3x^2}{2(x^3+1)} - \dfrac{6x^5-7}{x^6-7x+5}\right]$ **39.** $2^x \ln 2$

41. $\pi^{\sin x}(\ln \pi)\cos x$ **43.** $(x^3-2x)^{\ln x}\left[\dfrac{3x^2-2}{x^3-2x}\ln x + \dfrac{1}{x}\ln(x^3-2x)\right]$ **45.** $(\ln x)^{\tan x}\left[\dfrac{\tan x}{x\ln x} + (\sec^2 x)\ln(\ln x)\right]$

49. (a) $k^n e^{kx}$ **(b)** $(-1)^n k^n e^{-kx}$ **51.** $-\dfrac{1}{\sqrt{2\pi}\sigma^3}(x-\mu)\exp\left[-\dfrac{1}{2}\left(\dfrac{x-\mu}{\sigma}\right)^2\right]$ **53. (a)** $-\dfrac{1}{x(\ln x)^2}$ **(b)** $-\dfrac{\ln 2}{x(\ln x)^2}$

55. $-\dfrac{qk_0}{2T^2}\exp\left[-\dfrac{q(T-T_0)}{2T_0T}\right]$ **57.** ex^{e-1} **59. (a)** 1   **61. (b)**    **(d)** must take the value zero in between

     (b) ln 10   **(e)** $x = 2$

63. (b) 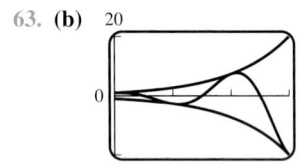   **65. (a)** 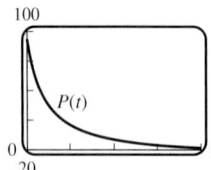   **(b)** The population tends to 19.   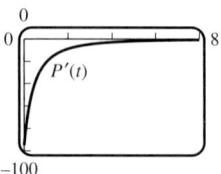

     (c) The *rate* tends to zero.

▶ Exercise Set 4.5 **(Page 267)**

1. (a) $-\pi/2$ **(b)** π **(c)** $-\pi/4$ **(d)** 0   **3.** $1/2, -\sqrt{3}, -1/\sqrt{3}, 2, -2/\sqrt{3}$ **5.** $\frac{4}{5}, \frac{3}{5}, \frac{3}{4}, \frac{5}{3}, \frac{5}{4}$ **7. (a)** $\pi/7$ **(b)** 0   **(c)** $2\pi/7$

   (d) $201\pi - 630$ **9. (a)** $0 \le x \le \pi$ **(b)** $-1 \le x \le 1$ **(c)** $-\pi/2 < x < \pi/2$ **(d)** $-\infty < x < +\infty$ **11.** $\frac{24}{25}$

13. (a) $\dfrac{1}{\sqrt{1+x^2}}$ **(c)** $\dfrac{\sqrt{x^2-1}}{x}$ **15. (a)** 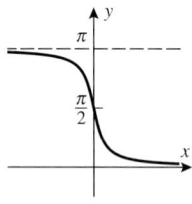   **(b)** domain of $\cot^{-1}x$ is $(-\infty, +\infty)$,

   (b) $\dfrac{1}{x}$ **(d)** $\sqrt{x^2-1}$ range is $(0, \pi)$;

     domain of $\csc^{-1}x$ is $(-\infty, -1] \cup [1, +\infty)$,

     range is $[-\pi/2, 0) \cup (0, \pi/2]$.

17. (a) $55.0°$ **(b)** $33.6°$ **(c)** $25.8°$ **19. (a)** $x = 3.6964$ rad   **(b)** $\theta = -76.7°$

21. (a) $\dfrac{1}{\sqrt{9-x^2}}$ **(b)** $-\dfrac{2}{\sqrt{1-(2x+1)^2}}$ **23. (a)** $\dfrac{7}{x\sqrt{x^{14}-1}}$ **(b)** $-\dfrac{1}{\sqrt{e^{2x}-1}}$ **25. (a)** $-\dfrac{1}{|x|\sqrt{x^2-1}}$ **(b)** $\begin{cases} 1, & \sin x > 0 \\ -1, & \sin x < 0 \end{cases}$

27. (a) $\dfrac{e^x}{x\sqrt{x^2-1}} + e^x\sec^{-1}x$ **(b)** $\dfrac{3x^2(\sin^{-1}x)^2}{\sqrt{1-x^2}} + 2x(\sin^{-1}x)^3$ **29.** $\dfrac{(3x^2 + \tan^{-1}y)(1+y^2)}{(1+y^2)e^y - x}$

31. (a) 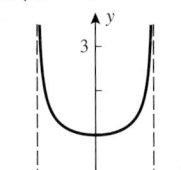   **33. (b)** 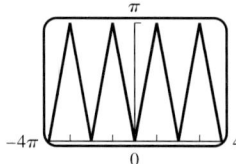   **35. (a)** 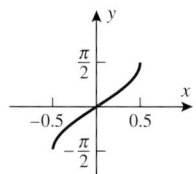   **(b)**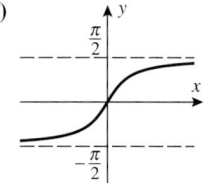

37. (b) $23°$ **39.** $32°$ or $58°$; $32°$ **41.** $29°$

▶ Exercise Set 4.6 **(Page 274)**

1. (b) $A = x^2$ **(c)** $\dfrac{dA}{dt} = 2x\dfrac{dx}{dt}$ **(d)** 12 ft²/min   **3. (a)** $\dfrac{dV}{dt} = \pi\left(r^2\dfrac{dh}{dt} + 2rh\dfrac{dr}{dt}\right)$ **(b)** -20π in³/s; decreasing

5. (a) $\dfrac{d\theta}{dt} = \dfrac{\cos^2\theta}{x^2}\left(x\dfrac{dy}{dt} - y\dfrac{dx}{dt}\right)$ **(b)** $-\frac{5}{16}$ rad/s; decreasing   **7.** $\dfrac{4\pi}{15}$ in²/min   **9.** $\dfrac{1}{\sqrt{\pi}}$ mi/h   **11.** 4860π cm³/min

13. $\frac{5}{6}$ ft/s   **15.** $\dfrac{125}{\sqrt{61}}$ ft/s   **17.** 704 ft/s   **19. (a)** 500 mi, 1716 mi   **(b)** 1354 mi; 27.7 mi/min   **21.** $\dfrac{9}{20\pi}$ ft/min

23. 125π ft³/min   **25.** 250 mi/h   **27.** $\dfrac{36\sqrt{69}}{25}$ ft/min   **29.** $\dfrac{8\pi}{5}$ km/s   **31.** $600\sqrt{7}$ mi/h   **33. (a)** $-\dfrac{60}{7}$ units per second

   (b) falling   **35.** -4 units per second   **37.** e^2 **39.** 4.5 cm/s; away   **43.** $\dfrac{20}{9\pi}$ cm/s

---

▶ Exercise Set 4.7 **(Page 284)**

1. (a) $\frac{2}{3}$ **(b)** $\frac{2}{3}$ **3.** 1   **5.** 1   **7.** 1   **9.** -1 **11.** 0   **13.** $-\infty$ **15.** 0   **17.** 2   **19.** 0

21. π **23.** $-\frac{5}{3}$ **25.** e^{-3} **27.** e^2 **29.** $e^{2/\pi}$ **31.** 0   **33.** $\frac{1}{2}$ **35.** $+\infty$ **39. (b)** 2

41. 0 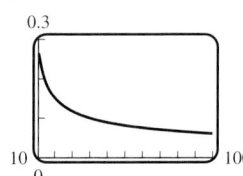   **43.** e^3 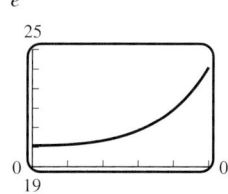   **45.** no horizontal asymptote 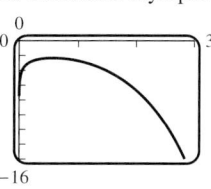   **47.** $y = 1$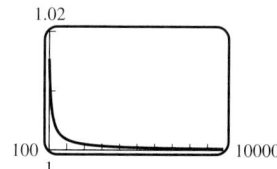

49. (a) 0   **(b)** $+\infty$ **(c)** 0   **(d)** $-\infty$ **(f)** $+\infty$ **(g)** $-\infty$ **51.** 1   **53.** does not exist

55. Vt/L **57. (c)** $1024\left(\sqrt[1024]{0.3} - 1\right) = -1.20327;\ 1024\left(\sqrt[1024]{2} - 1\right) = 0.69338$

59. $k = -1, l = \pm 2\sqrt{2}$ **61.** does not exist

▶ **Chapter 4 Supplementary Exercises (Page 286)**

3. (a) $0/0, \infty/\infty$ **(b)** no   **5. (a)** $\frac{1}{2}(x + 1)^{1/3}$ **(b)** none   **(c)** $\frac{1}{2}\ln(x - 1)$ **(d)** $\frac{x + 2}{x - 1}$

7. (a) $y + 1 = 2(x - 1)$ **(b)** $y = 1$ **9.** $15x + 2$ **11. (a)** $+\infty$ **(b)** $\frac{1}{2}$ **(c)** $\ln a$

15. $-b - \dfrac{a}{\sqrt{2}}$ cm/s   **17. (a)**    **(b)** $y = e^{-x/2} \sin 2x$ intersects $y = e^{-x/2}$ at $x = \pi/4$
and $y = -e^{-x/2}$ at $x = -\pi/4, 3\pi/4$

19. (a) $\dfrac{1}{x}(1 + x)^{(1/x)-1} - \dfrac{(1 + x)^{1/x}}{x^2}\ln(1 + x)$ **(b)** $e^x\left[x^{e^x - 1} + x^{e^x}\ln x\right]$ **(c)** $3x^2$ **(d)** $\dfrac{abe^{-x}}{(1 + be^{-x})^2}$

(e) $\dfrac{6x^{4/3}y^{1/3} - 3x^{1/3}y - 2y^{4/3}}{2x^{4/3} + 3xy^{1/3}}$ **(f)** $\dfrac{5x + 3}{6x(x + 1)} - \cot x - \tan x$ **21. (b)** $x = 3.654$ **23.** $e^{1/e}$

▶ **Exercise Set 5.1 (Page 296)**

1. (a) $f' > 0, f'' > 0$ **(b)** $f' > 0, f'' < 0$ **(c)** $f' < 0, f'' > 0$ **(d)** $f' < 0, f'' < 0$ **3.** A: $dy/dx < 0,\ d^2y/dx^2 > 0$
B: $dy/dx > 0,\ d^2y/dx^2 < 0$
C: $dy/dx < 0,\ d^2y/dx^2 < 0$

5. $x = -1, 0, 1, 2$ **7. (a)** $[4, 6]$ **(b)** $[1, 4], [6, 7]$ **(c)** $(1, 2), (3, 5)$ **(d)** $(2, 3), (5, 7)$ **(e)** $x = 2, 3, 5$

9. (a) $\left[\frac{5}{2}, +\infty\right)$ **(b)** $\left(-\infty, \frac{5}{2}\right]$ **(c)** $(-\infty, +\infty)$ **(d)** none   **(e)** none

11. (a) $(-\infty, +\infty)$ **(b)** none   **(c)** $(-2, +\infty)$ **(d)** $(-\infty, -2)$ **(e)** -2

13. (a) $[1, +\infty)$ **(b)** $(-\infty, 1]$ **(c)** $(-\infty, 0), \left(\frac{2}{3}, +\infty\right)$ **(d)** $\left(0, \frac{2}{3}\right)$ **(e)** $0, \frac{2}{3}$

15. (a) $[0, +\infty)$ **(b)** $(-\infty, 0]$ **(c)** $(-\sqrt{2/3}, \sqrt{2/3})$ **(d)** $(-\infty, -\sqrt{2/3}), (\sqrt{2/3}, +\infty)$ **(e)** $-\sqrt{2/3}, \sqrt{2/3}$

17. (a) $(-\infty, +\infty)$ **(b)** none   **(c)** $(-\infty, -2)$ **(d)** $(-2, +\infty)$ **(e)** -2

19. (a) $[-1, +\infty)$ **(b)** $(-\infty, -1]$ **(c)** $(-\infty, 0), (2, +\infty)$ **(d)** $(0, 2)$ **(e)** $0, 2$

21. (a) $(-\infty, 0]$ **(b)** $[0, +\infty)$ **(c)** $(-\infty, -1), (1, +\infty)$ **(d)** $(-1, 1)$ **(e)** $-1, 1$

23. (a) $[0, +\infty)$ **(b)** $(-\infty, 0]$ **(c)** $(-1, 1)$ **(d)** $(-\infty, -1), (1, +\infty)$ **(e)** $-1, 1$

25. (a) $[\pi, 2\pi]$
(b) $[0, \pi]$
(c) $(\pi/2, 3\pi/2)$
(d) $(0, \pi/2), (3\pi/2, 2\pi)$
(e) $\pi/2, 3\pi/2$

27. (a) $(-\pi/2, \pi/2)$
(b) none
(c) $(0, \pi/2)$
(d) $(-\pi/2, 0)$
(e) 0

29. (a) $[0, \pi/4]$, $[3\pi/4, \pi]$
(b) $[\pi/4, 3\pi/4]$
(c) $(\pi/2, \pi)$
(d) $(0, \pi/2)$
(e) $\pi/2$

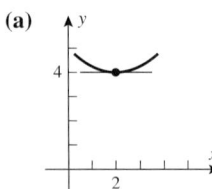

31. (a)

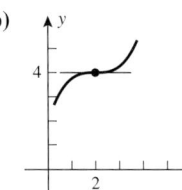

(b)

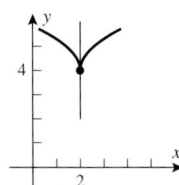

(c)

33. (a) $(a, 0)$
(b) none

35.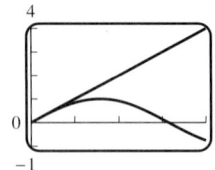

37. $x \geq \sin x$

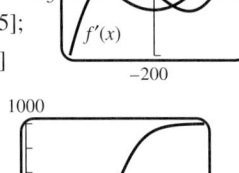

39. points of inflection at $x = -2, 2$;
concave up on $(-5, -2)$, $(2, 5)$;
concave down on $(-2, 2)$;
increasing on $[-3.5829, 0.2513]$ and $[3.3316, 5]$;
decreasing on $[-5, -3.5829]$, $[0.2513, 3.3316]$

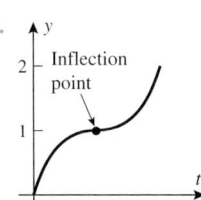

41. $-0.175, 0$ **43.** $-2.45, 0.65, 2.75$

47. (a) true   **49. (c)** 1   **53.** the eighth day

55.

Exercise Set 5.2 (Page 304)

1. (a)   **(b)**   **(c)**   **(d)**

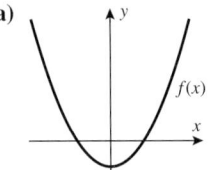 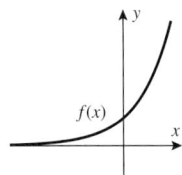 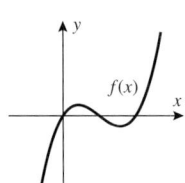 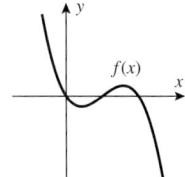

5. (b) nothing   **(c)** f has a relative minimum at $x = 1$, g has no relative extremum at $x = 1$.

7. (a) $x = -3, 1$ (stationary points)   **(b)** $x = 0, \pm\sqrt{3}$ (stationary points)   **9. (a)** $x = \pm\sqrt{2}$ (stationary points)   **(b)** no critical points

11. (a) $x = -1$ (stationary point)   **(b)** $n\pi/3$, $n = 0, \pm1, \pm2, \ldots$ (stationary points)   **13. (a)** $x = 2$ **(b)** $x = 0$ **(c)** $x = 1, 3$

15. (a) $x = 0$, relative max; $x = \pm\sqrt{5}$, relative min   **(b)** $x = 0$, relative min   **17.** relative max of 5 at $x = -2$

19. relative min of 0 at $x = \pi$, relative max of 1 at $x = \pi/2, 3\pi/2$ **21.** no relative extrema

23. relative min of 0 at $x = 1$, relative max of $\frac{4}{27}$ at $x = \frac{1}{3}$ **25.** relative min of 0 at $x = 0$, relative max of 1 at $x = 1, -1$

27. relative min of 0 at $x = 0$ **29.** relative min of 0 at $x = 0$ **31.** relative min of 0 at $x = 0$

33. relative min of 0 at $x = 2, -2$, relative max of 4 at $x = 0$

35. relative min of 0 at $x = \pi/2, \pi, 3\pi/2$;
relative max of 1 at $x = \pi/4$,
$3\pi/4, 5\pi/4, 7\pi/4$

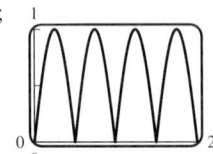

37. relative min of 0 at $x = \pi/2, 3\pi/2$;
relative max of 1 at $x = \pi$

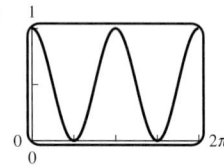

39. relative min of $-1/e$ at $x = 1/e$

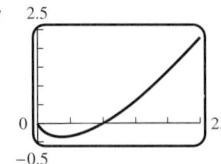

41. relative min of 0 at $x = 0$;
relative max of $1/e^2$ at $x = 1$

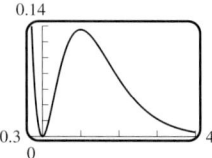

43. relative minima at $x = -3.58, 3.33$; relative max at $x = 0.25$

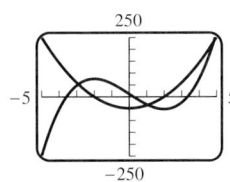

45. relative max at $x = 0.255$

47. relative min at $x = -1.20$; relative max at $x = 1.80$

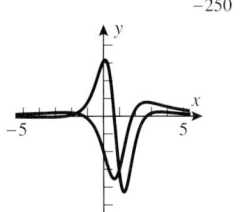

49. **(a)** 54

(b) 9

51. **(b)**

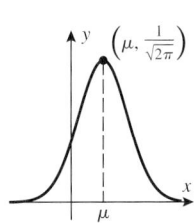

53. $f(x) = -2x^3 + 3x^2$

55. **(a)**

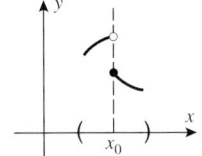

$f(x_0)$ is not an extreme value

(b)

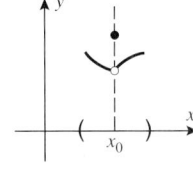

$f(x_0)$ is a relative maximum

(c)

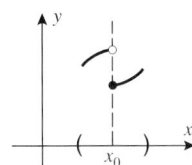

$f(x_0)$ is a relative minimum

▶ **Exercise Set 5.3 (Page 319)**

1.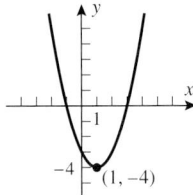

3.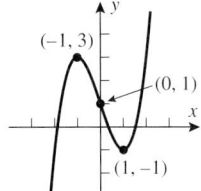

5.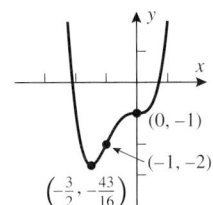

7.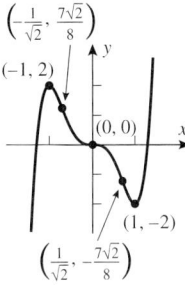

9.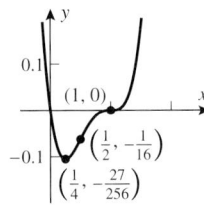

11.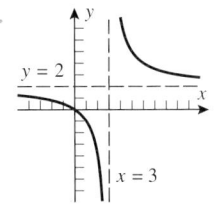

13.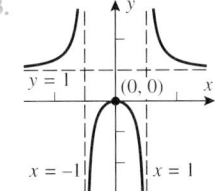

15.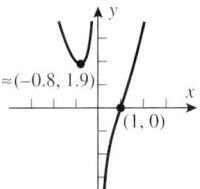

17.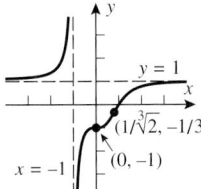

19.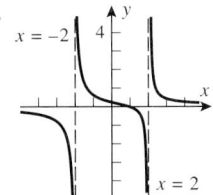

21.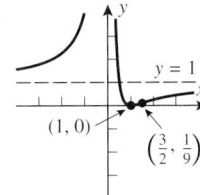

23. **(a)** VI

(b) I

(c) III

(d) V

(e) IV

(f) II

25.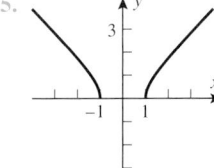

27.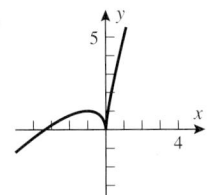

29.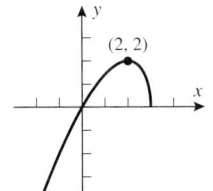

31.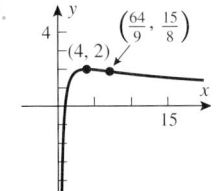

33.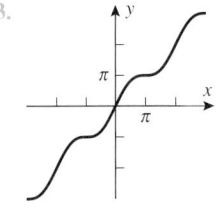

35.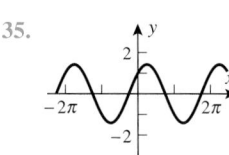

37.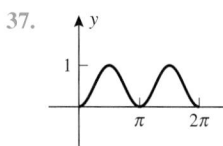

39. (a) $+\infty$, 0

(b)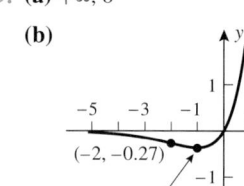

41. (a) 0, $+\infty$

(b)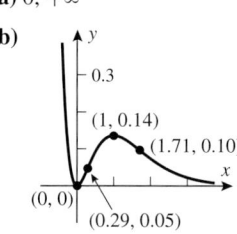

43. (a) $+\infty$, $-\infty$

(b)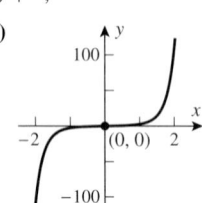

45. (a) 0; $+\infty$

(b)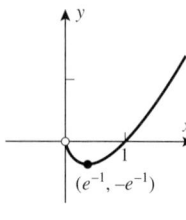

47. (a) $-\infty$; 0

(b)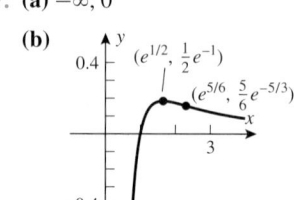

49. (a) 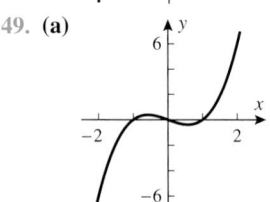 **(b)** 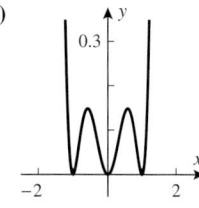 **(c)** 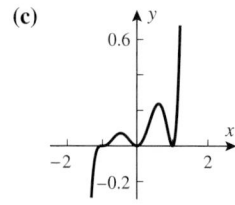 **(d)**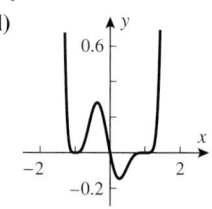

51. (a) 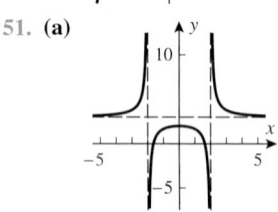 **(b)** 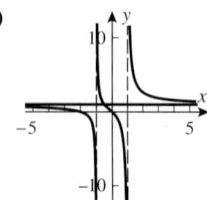 **(c)** 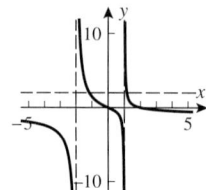 **(d)**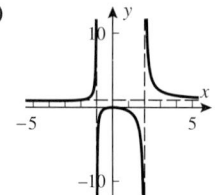

53. (a)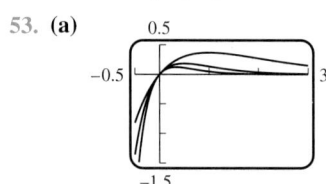

55. (a) The limit does not exist. **(c)**

(b)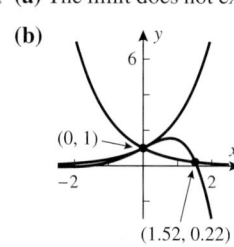

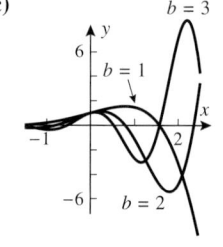

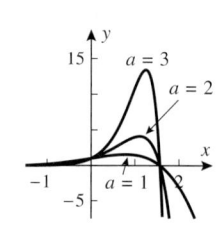

57. 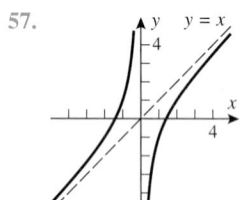 **59.** 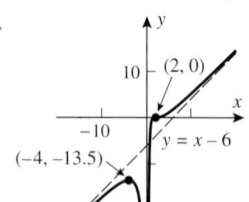 **61.** 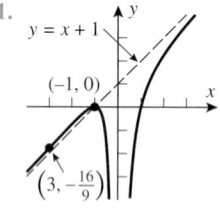 **63.**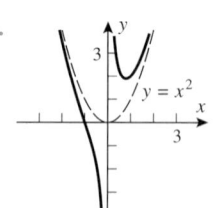

65. 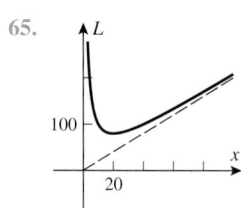 **67.**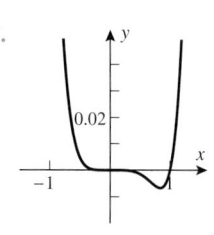

69. (a) $\dfrac{kL^2A}{(1+A)^2}$

(c) $t = \dfrac{1}{Lk}\ln A$

▶ Chapter 5 Supplementary Exercises **(Page 321)**

7. **(a)** relative max at $x = 1$, relative min at $x = 7$, neither at $x = 0$

   (b) relative max at $x = \pi/2, 3\pi/2$;

   relative min at $x = 7\pi/6, 11\pi/6$

   (c) relative max at $x = 5$

9. $\lim\limits_{x \to -\infty} f(x) = +\infty$, $\lim\limits_{x \to +\infty} f(x) = +\infty$;

   relative min at $x = 0$;

   points of inflection at $x = \frac{1}{2}, 1$;

   no asymptotes

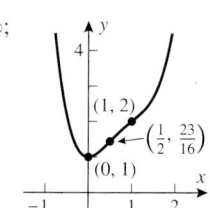

11. $\lim\limits_{x \to \pm\infty} f(x)$ does not exist;

    critical point at $x = 0$;

    relative min at $x = 0$;

    point of inflection when $1 + 4x^2 \tan(x^2 + 1) = 0$;

    vertical asymptotes at $x = \pm\sqrt{\pi(n + \frac{1}{2}) - 1}, n = 0, 1, 2, \ldots$

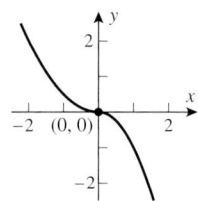

13. critical points at $x = -5, 0$;

    relative max at $x = -5$, relative min at $x = 0$;

    points of inflection at $x = -7.26, -1.44, 1.20$;

    horizontal asymptote $y = 1$ as $x \to \pm\infty$

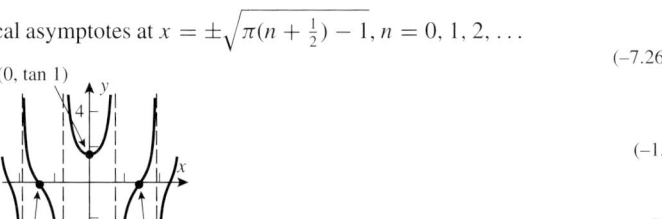

15. $\lim\limits_{x \to -\infty} f(x) = +\infty$, $\lim\limits_{x \to +\infty} f(x) = -\infty$;

    critical point at $x = 0$;

    no extrema;

    inflection point at $x = 0$ (f changes concavity);

    no asymptotes

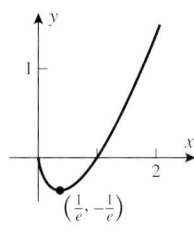

17. $\lim\limits_{x \to +\infty} f(x) = +\infty$, $\lim\limits_{x \to 0^+} f(x) = 0$, $\lim\limits_{x \to 0^+} f'(x) = -\infty$;

    critical point at $x = 1/e$;

    relative min at $x = 1/e$;

    no points of inflection;

    no asymptotes

19. critical point at $x = e^{1/2}$;

    relative max at $x = e^{1/2}$;

    point of inflection at $x = e^{5/6}$;

    horizontal asymptote $y = 0$ as $x \to +\infty$

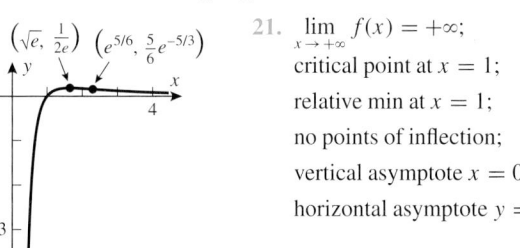

21. $\lim\limits_{x \to +\infty} f(x) = +\infty$;

    critical point at $x = 1$;

    relative min at $x = 1$;

    no points of inflection;

    vertical asymptote $x = 0$;

    horizontal asymptote $y = 0$ for $x \to -\infty$

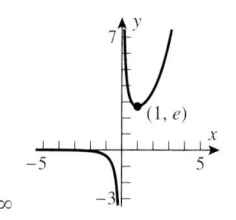

23. critical points at $x = 0, 2$;

    relative min at $x = 0$, relative max at $x = 2$;

    points of inflection at $x = 2 \pm \sqrt{2}$;

    horizontal asymptote $y = 0$ as $x \to +\infty$;

    $\lim\limits_{x \to -\infty} f(x) = +\infty$

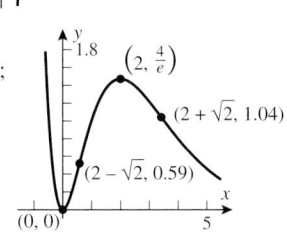

25. **(a)**
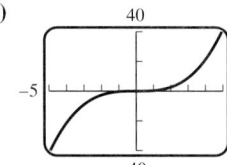

(b) relative max at $x = -\frac{1}{20}$, relative min at $x = \frac{1}{20}$

(c) The finer details can be seen when graphing over a much smaller x-window.

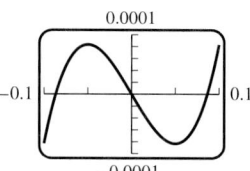

27. **(a)**
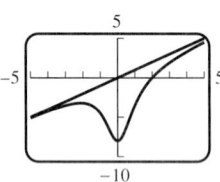

29. $y = 2x$, $y = 3$

31. $f(x) = \dfrac{x^2 + x - 7}{3x^2 + x - 1}$, $x \neq \frac{1}{2}$

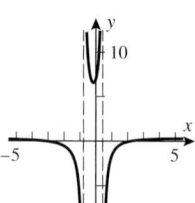

33. **(a)** $\sin x = 1$, $\sin x = -1$

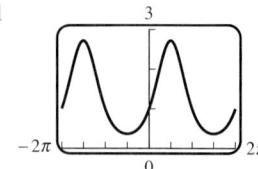

(b) relative maxima at $x = 2n\pi + \pi/2$, $y = e$; relative minima at $x = 2n\pi - \pi/2$, $y = 1/e$, $n = 0, \pm 1, \pm 2, \ldots$

(c) when $\sin x = \dfrac{-1 + \sqrt{5}}{2}$

35. **(a)** relative min -0.232466 at $x = 0.450184$ **(b)** relative max 0 at $x = 0$; relative min -0.107587 at $x = \pm 0.674841$

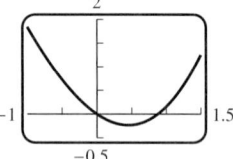

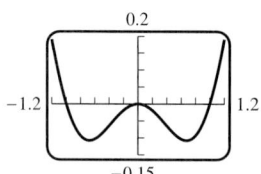

(c) relative max 0.876839 at $x = 0.886352$; relative min -0.355977 at $x = -1.244155$

37. **(a)**
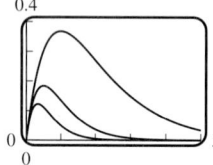

(b) $y = 0$ at $x = 0$; $\lim\limits_{x \to +\infty} y = 0$

(c) relative max at $x = 1/a$; inflection point at $x = 2/a$

(d) The maximum and the inflection point move toward the origin.

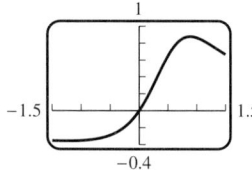

▶ **Exercise Set 6.1 (Page 337)**

1. relative maxima at $x = 2, 6$; absolute max at $x = 6$; relative and absolute min at $x = 4$

3. **(a)**
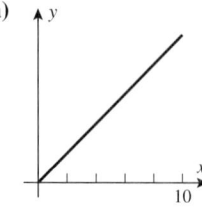

(b)
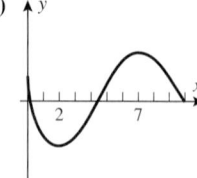

(c)
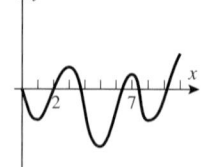

5. maximum value 1 at $x = 0, 1$; minimum value 0 at $x = \frac{1}{2}$ **7.** maximum value 27 at $x = 4$, minimum value -1 at $x = 0$

9. maximum value $3/\sqrt{5}$ at $x = 1$, minimum value $-3/\sqrt{5}$ at $x = -1$

11. maximum value $1 - (\pi/4)$ at $x = -\pi/4$, minimum value $(\pi/4) - 1$ at $x = \pi/4$

13. maximum value 17 at $x = -5$, minimum value 1 at $x = -3$ **15.** minimum value $f\left(\frac{3}{2}\right) = -\frac{13}{4}$, no maximum

17. maximum value $f(1) = 1$, no minimum   **19.** no maximum or minimum   **21.** maximum value $f(-2) = -4$, no minimum

23. minimum value 0 for $x = \pm 1$, no maximum   **25.** maximum value 48 at $x = 8$, minimum value 0 at $x = 0, 20$

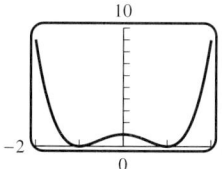

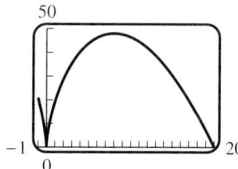

27. no maximum or minimum   **29.** maximum value 2 at $x = 0$, minimum value $\sqrt{3}$ at $x = \pi/6$

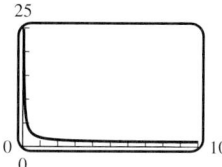

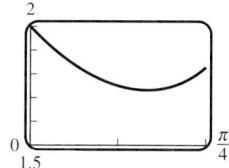

31. maximum value $\frac{27}{8}e^{-3}$ at $x = \frac{3}{2}$,   **33.** maximum value $\sin(1) \approx 0.84147$,
minimum value $64/e^8$ at $x = 4$ minimum value $-\sin(1) \approx -0.84147$

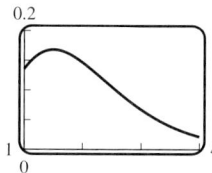

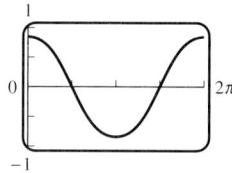

35. maximum value 2, minimum value $-\frac{1}{4}$

37. maximum value $3\sqrt{3}/2$ at $x = (\pi/6) + n\pi$, minimum value $-3\sqrt{3}/2$ at $x = (5\pi/6) + n\pi$, $n = 0, \pm 1, \pm 2, \ldots$

41. $f'(1) = 2$ **45.** $\left(\frac{1}{2}, -\frac{1}{4}\right)$ is closest, $(-1, -1)$ is farthest   **47.** maximum $y = 4$ at $t = \pi, 3\pi$; minimum $y = 0$ at $t = 0, 2\pi$

▶ **Exercise Set 6.2 (Page 348)**

1. 5, 5   **3. (a)** 1   **(b)** $\frac{1}{2}$ **5.** 500 ft × 750 ft   **7.** 5 in × $\frac{12}{5}$ in   **9.** $10\sqrt{2}$ in × $10\sqrt{2}$ in

11. 80 ft ($1 fencing), 40 ft ($2 fencing)   **15. (a)** maximum $N = 161{,}788$, minimum $N = 125{,}000$ **(b)** 40   **17.** 2 in square

19. $\frac{200}{27}$ ft^3 **21.** base 10 cm square, height 20 cm   **23.** ends $\sqrt[3]{3V/4}$ units square, height $\frac{4}{3}\sqrt[3]{3V/4}$

25. height $= 2\sqrt{(5 - \sqrt{5})/10}\,R$, radius $= \sqrt{(5 + \sqrt{5})/10}\,R$ **29.** height $=$ radius $= \sqrt[3]{500/\pi}$ **31.** $L/12$ by $L/12$ by $L/12$

33. height $= L/\sqrt{3}$, radius $= \sqrt{2/3}\,L$ **35.** radius $= \sqrt[6]{450/\pi^2}$, height $= \dfrac{30}{\pi}\sqrt[3]{\pi^2/450}$ cm   **37.** height $= 4R$, radius $= \sqrt{2}R$

39. $\pi/3$ **41.** $5\sqrt{5}$ ft   **43. (a)** 7000   **(b)** yes   **45.** 13,722 lb   **47.** $1/\sqrt{5}$ **51.** $(-\sqrt{2}, 1), (\sqrt{2}, 1)$ **53.** $\left(\sqrt{2}, \frac{1}{2}\right)$

55. $\left(-1/\sqrt{3}, \frac{3}{4}\right)$ **57.** $4(1 + 2^{2/3})^{3/2}$ ft   **59.** 30 cm from the weaker source   **63. (c)** $\frac{1}{4}$ mile downstream from the house

▶ **Exercise Set 6.3 (Page 359)**

1. (a) positive, negative, slowing down   **3. (a)** left   **5.**
(b) positive, positive, speeding up   **(b)** negative
(c) negative, positive, slowing down   **(c)** speeding up
     (d) slowing down

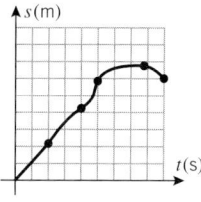

7.

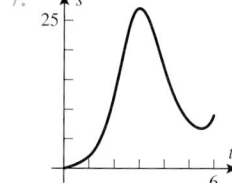
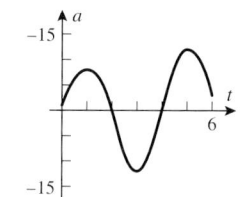

9. (a) 6.7 ft/s^2 **11. (a)** $v(t) = 3t^2 - 12t$, $a(t) = 6t - 12$
(b) $t = 0$ s   **(b)** $s(1) = -5$ ft, $v(1) = -9$ ft/s,
     $|v(1)| = 9$ ft/s, $a(1) = -6$ ft/s^2
(c) 0, 4
(d) speeding up for $0 < t < 2$ and $4 < t$,
     slowing down for $2 < t < 4$
(e) 39 ft

13. **(a)** $v(t) = -(3\pi/2)\sin(\pi t/2), a(t) = -(3\pi^2/4)\cos(\pi t/2)$

(b) $s(1) = 0$ ft, $v(1) = -3\pi/2$ ft/s, $|v(1)| = 3\pi/2$ ft/s, $a(1) = 0$ ft/s^2 **(c)** 0, 2, 4

(d) speeding up for $0 < t < 1, 2 < t < 3$, and $4 < t < 5$; slowing down for $1 < t < 2$ and $3 < t < 4$ **(e)** 15 ft

15. **(a)** $\sqrt{5}$ **(b)** $\sqrt{5}/10$ **(c)** speeding up for $\sqrt{5} < t < \sqrt{15}$, slowing down for $0 < t < \sqrt{5}$ and $\sqrt{15} < t$

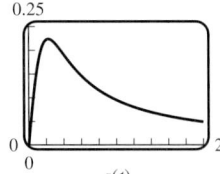

17.

Constant speed
$t = 0$
s
2

19.

(Stopped) $t = 4$
Slowing down
Speeding up
$t = 0$
$t = 3$
$t = 2$
(Stopped)
s
0
16 18 20
Slowing down

21.

Slowing down
(Stopped permanently)
$t = 3\pi/2$
$t = 2\pi$
$t = \pi$
$t = \pi/2$
$t = 0$
s
-1
0
1
Speeding up

23. **(a)** 12 ft/s   **(b)** $t = 2.2, s = -24.2$ **25.** **(a)** 11.025 m   **(b)** -14.7 m/s   **(c)** 1.2245 s   **(d)** $t = 4.5175$ s

27. **(a)** 6.12 s   **29.** 113.42 ft/s   **31.** 29.39 m   **33.** **(b)** 113.42 ft/s   **35.** **(a)**   1.5   **(b)** $\sqrt{2}$

(b) 183.67 m

(c) 6.12 s

(d) 60 m/s

0
0
5

37. **(b)** $\frac{2}{3}$ unit   **(c)** $0 \le t < 1$ and $t > 2$ **39.** **(a)** -1.25 ft/s/ft   **(b)** -2500 ft/s^2

▶ **Exercise Set 6.4 (Page 366)**

1. 1.414213562   **3.** 1.817120593   **5.** -1.671699882 **7.** 1.224439550

9. -1.452626879 **11.** 1.895494267   **13.** 4.493409458

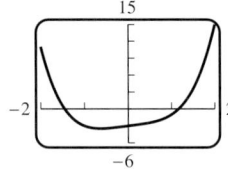
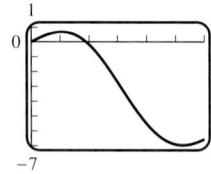
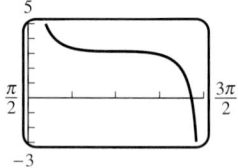

15. -1.165373043 **17.** $-0.474626618, 1.395336994$ **19.** **(b)** 3.162277660   **21.** -4.098859132

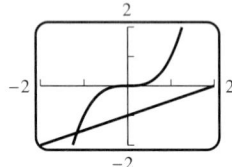
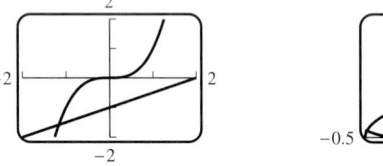

23. $(0.589754512, 0.347810385)$ **25.** **(b)** 171°   **27.** $-1.220744085, 0.724491959$ **29.** 5.3362%

▶ **Exercise Set 6.5 (Page 372)**

1. $[0, 4], c = 3$ **3.** $c = 3$ **5.** $c = \pi$ **7.** $c = 1$ **9.** $c = 1.54$ **11.** 1   **13.** $\frac{5}{4}$ **15.** $-\sqrt{5}$

17. (a) $[-2, 1]$ **(b)** $c = -1.29$ **(c)** -1.2885843 **37.** $f(x) = x^3 - 4x + 5$ **39.**

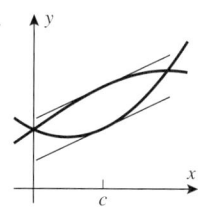

▶ **Chapter 6 Supplementary Exercises (Page 374)**

3. (a) true   **(b)** false   **5.** true

7. (a) $M = -\frac{1}{2}$ at $x = -2$; $m = -1$ at $x = -1$ **(b)** $M = \frac{27}{256}$ at $x = \frac{3}{4}$; $m = -2$ at $x = -1$

(c) $m \approx -1.9356$ at $x = \frac{12}{7}$; $M = 9$ at $x = 3$ **(d)** $m = e^2/4$ at $x = 2$ **9.** 2.3561945

11. (a) yes, $c = 0$ **(b)** no   **(c)** yes, $c = \sqrt{\pi/2}$ **13.** $r = 2P/(8 + 3\pi)$ ft   **15. (a)** yes   **(b)** yes

17. (a) 0.3501 s, 0.0820 s   **19.**   **(b)** minimum: $(-2.111985, -0.355116)$;
maximum: $(0.372591, 2.012931)$

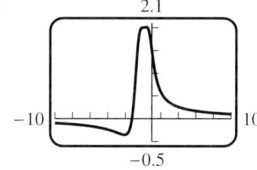

21. (a) $v = -2\dfrac{t(t^4 + 2t^2 - 1)}{(t^4 + 1)^2}$, $a = 2\dfrac{3t^8 + 10t^6 - 12t^4 - 6t^2 + 1}{(t^4 + 1)^3}$ **(c)** $t = 0.64$, $s = 1.2$ **(d)** $0 \le t \le 0.64$ s

(e) speeding up when $0 \le t < 0.36$ and $0.64 < t < 1.1$, otherwise slowing down

(f) maximum speed $= 1.05$ m/s when $t = 1.10$ s   **23.** 149.988×10^6 km

▶ **Exercise Set 7.1 (Page 382)**

1. $A = 1/2$ **3.** $A = 16$ **5.** $\frac{1}{2}$ **7.** 16 **9.** $e - 1$

n	1	2	3	4	5
A_n	1.0000	0.7500	0.6666	0.6250	0.6000

n	1	2	3	4	5
A_n	28.0000	22.0000	20.0000	19.0000	18.4000

n	6	7	8	9	10
A_n	0.5833	0.5714	0.5625	0.5556	0.5500

n	6	7	8	9	10
A_n	18.0000	17.7143	17.5000	17.3333	17.2000

▶ **Exercise Set 7.2 (Page 389)**

1. (a) $\displaystyle \int \frac{x}{\sqrt{1 + x^2}}\, dx = \sqrt{1 + x^2} + C$ **(b)** $\displaystyle \int (x + 1)e^x\, dx = xe^x + C$

3. $\dfrac{d}{dx}\left[\sqrt{x^3 + 5}\right] = \dfrac{3x^2}{2\sqrt{x^3 + 5}}$, so $\displaystyle \int \frac{3x^2}{2\sqrt{x^3 + 5}}\, dx = \sqrt{x^3 + 5} + C$.

5. $\dfrac{d}{dx}\left[\sin(2\sqrt{x})\right] = \dfrac{\cos(2\sqrt{x})}{\sqrt{x}}$, so $\displaystyle \int \frac{\cos(2\sqrt{x})}{\sqrt{x}}\, dx = \sin(2\sqrt{x}) + C$.

7. (a) $(x^9/9) + C$ **(b)** $\frac{7}{12}x^{12/7} + C$ **(c)** $\frac{2}{9}x^{9/2} + C$ **9. (a)** $-\frac{1}{4}x^{-2} + C$ **(b)** $(u^4/4) - u^2 + 7u + C$

11. $-\frac{1}{2}x^{-2} + \frac{2}{3}x^{3/2} - \frac{12}{5}x^{5/4} + \frac{1}{3}x^3 + C$ **13.** $(x^2/2) + (x^5/5) + C$ **15.** $3x^{4/3} - \frac{12}{7}x^{7/3} + \frac{3}{10}x^{10/3} + C$ **17.** $\dfrac{x^2}{2} - \dfrac{2}{x} + \dfrac{1}{3x^3} + C$

19. $2\ln x + 3e^x + C$ **21.** $-4\cos x + 2\sin x + C$ **23.** $\tan x + \sec x + C$ **25.** $\ln\theta - 2e^\theta + \cot\theta + C$ **27.** $\sec x + C$

29. $\theta - \cos\theta + C$ **31.** $\tan x - \sec x + C$ **33. (a)**   **(b)** $f(x) = (x^2/2) + 5$ **35.**

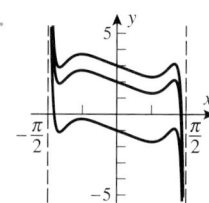

37. $f(x) = \cos x + 1$ 39. **(a)** $y(x) = \frac{3}{4}x^{4/3} + \frac{5}{4}$ **(b)** $y(t) = \ln|t| + 5$ **(c)** $y(x) = \frac{2}{3}x^{3/2} + 2x^{1/2} - \frac{8}{3}$

41. $f(x) = \frac{4}{15}x^{5/2} + C_1 x + C_2$ 43. $y = x^2 + x - 6$ 45. $y = x^3 - 6x + 7$ 47. **(b)** $F(0) - G(0) = \frac{8}{3}$ 49. $\tan x - x + C$

51. **(a)** $\frac{1}{2}(x - \sin x) + C$ **(b)** $\frac{1}{2}(x + \sin x) + C$ 53. $v = \dfrac{1087}{\sqrt{273}}T^{1/2}$ ft/s

---

▶ Exercise Set 7.3 **(Page 395)**

1. **(a)** $\dfrac{(x^2 + 1)^{24}}{24} + C$ **(b)** $-\dfrac{\cos^4 x}{4} + C$ **(c)** $-2\cos\sqrt{x} + C$ **(d)** $\frac{3}{4}\sqrt{4x^2 + 5} + C$ **(e)** $\frac{1}{3}\ln(x^3 - 4) + C$

3. **(a)** $-\frac{1}{2}\cot^2 x + C$ **(b)** $\frac{1}{10}(1 + \sin t)^{10} + C$ **(c)** $\ln|\ln x| + C$ **(d)** $-\frac{1}{5}e^{-5x} + C$ **(e)** $-\frac{1}{3}\ln|(1 + \cos 3\theta)| + C$

5. $\frac{1}{2}e^{2x} + C$ 7. $-\dfrac{(2 - x^2)^4}{8} + C$ 9. $\frac{1}{8}\sin 8x + C$ 11. $\frac{1}{4}\sec 4x + C$ 13. $\frac{1}{21}(7t^2 + 12)^{3/2} + C$ 15. $\frac{2}{3}\sqrt{x^3 + 1} + C$

17. $-\frac{1}{16}(4x^2 + 1)^{-2} + C$ 19. $e^{\sin x} + C$ 21. $-\frac{1}{6}e^{-2x^3} + C$ 23. $\frac{1}{5}\cos(5/x) + C$ 25. $\frac{1}{3}\tan(x^3) + C$ 27. $-e^{-x} + C$

29. $\frac{1}{18}\sin^6 3t + C$ 31. $-\frac{1}{6}(2 - \sin 4\theta)^{3/2} + C$ 33. $\frac{1}{6}\sec^3 2x + C$ 35. $2e^{\sqrt{y}} + C$ 39. $\dfrac{1}{b(n+1)}\sin^{n+1}(a + bx) + C$

41. $\frac{2}{5}(x - 3)^{5/2} + 2(x - 3)^{3/2} + C$ 43. $\frac{1}{3}(\tan 3\theta - 3\theta) + C$ 45. $t + \ln|t| + C$ 47. $\displaystyle\int [\ln(e^x) + \ln(e^{-x})]\,dx = C$

49. **(a)** with $u = \sin x$, $\frac{1}{2}\sin^2 x + C_1$; with $u = \cos x$, $-\frac{1}{2}\cos^2 x + C_2$ **(b)** because they differ by a constant

51. $y(x) = \frac{2}{9}(3x + 1)^{3/2} + \frac{29}{9}$ 53. $f(x) = \frac{2}{9}(3x + 1)^{3/2} + \frac{7}{9}$ 55. 100,416

▶ Exercise Set 7.4 **(Page 402)**

1. **(a)** 36   **(b)** 55   **(c)** 40   **(d)** 6   **(e)** 11   **(f)** 0   3. $\displaystyle\sum_{k=1}^{10} k$ 5. $\displaystyle\sum_{k=1}^{49} k(k + 1)$ 7. $\displaystyle\sum_{k=1}^{10} 2k$ 9. $\displaystyle\sum_{k=1}^{6} (-1)^{k+1}(2k - 1)$

11. $\displaystyle\sum_{k=1}^{5} (-1)^k \frac{1}{k}$ 13. **(a)** $\displaystyle\sum_{1}^{50} 2k$ **(b)** $\displaystyle\sum_{1}^{50}(2k - 1)$ 15. 5050   17. 2870   19. 1728   21. 214,365   23. $2n^2 - n$

25. $\frac{3}{2}(n+1)$ 27. $\frac{1}{4}(n-1)^2$ 31. **(a)** $\displaystyle\sum_{k=0}^{19} 3^{k+1} = \frac{3}{2}(3^{20}-1)$ **(b)** $\displaystyle\sum_{k=0}^{25} 2^{k+5} = 2^{31} - 2^5$ **(c)** $\displaystyle\sum_{k=0}^{100}(-1)\left(\frac{-1}{2}\right)^k = -\frac{2}{3}\left(1 + \frac{1}{2^{101}}\right)$

33. $\dfrac{n+1}{2n}$; $\frac{1}{2}$ 35. $\dfrac{5(n+1)}{2n}$; $\frac{5}{2}$ 37. **(a)** $\displaystyle\sum_{j=0}^{5} 2^j$ **(b)** $\displaystyle\sum_{j=1}^{6} 2^{j-1}$ **(c)** $\displaystyle\sum_{j=2}^{7} 2^{j-2}$ 39. **(a)** $\displaystyle\sum_{k=1}^{18} \sin\left(\frac{\pi}{k}\right)$ **(b)** $\displaystyle\sum_{k=0}^{6} e^k = \dfrac{e^7 - 1}{e - 1}$

43. $3^{17} - 3^4$ 45. $-\frac{399}{400}$ 47. **(b)** $\frac{1}{2}$ 49. Both identities are valid.   55. 18,755

▶ Exercise Set 7.5 **(Page 414)**

1. **(a)** 8   **(b)** greater than A **(c)** less than A **(d)** equal to A

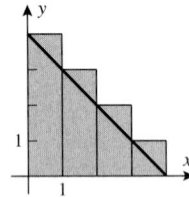 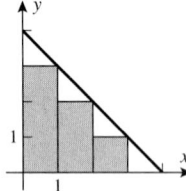 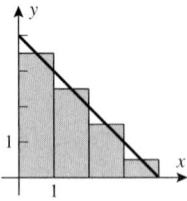

3. 35, 60, 46.25   5. $\dfrac{(1 + \sqrt{2})\pi}{4} \approx 1.896$, $\dfrac{(1 + \sqrt{2})\pi}{4} \approx 1.896$, $\dfrac{\pi\sqrt{2}\cos(\pi/8)}{2} \approx 2.052$

7. left endpoints: $A \approx (2 + 3 + 2 + 1)(1) = 8$; right endpoints: $A \approx (3 + 2 + 1 + 2)(1) = 8$

9. 0.718771403, 0.668771403, 0.692835360   11. 0.919403170, 1.07648280, 1.001028825

13. 0.351220577, 0.420535296, 0.386502483

15. **(a)** 0.693097198, 0.666154270, 1.000164512, 5.336963538, 0.386327689, 1.718167282

   (b) 0.693134682, 0.666538346, 1.000041125, 5.334644416, 0.386302694, 1.718253191

   (c) 0.693144056, 0.666634573, 1.000010281, 5.333803776, 0.3862964444, 1.718274669

17. (a) $A = \frac{9}{2}$ (b) $-A = -\frac{3}{2}$ (c) $-A_1 + A_2 = \frac{15}{2}$ (d) $-A_1 + A_2 = 0$

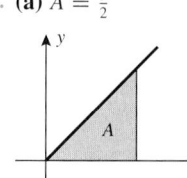 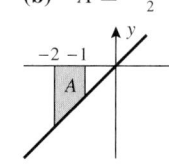 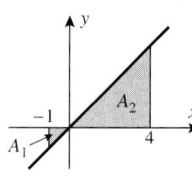 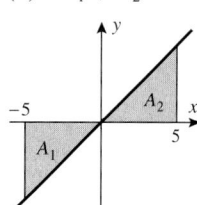

19. (a) $A = 10$ (b) $A_1 - A_2 = 0$ by symmetry   (c) $A_1 + A_2 = \frac{13}{2}$ (d) $\pi/2$ **21.** (a) 0.8
    (b) -2.6
    (c) -1.8
    (d) -0.3

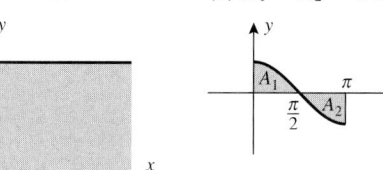 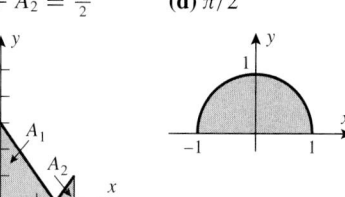

23. -1 **25.** 3   **27.** (a) $(1 + \pi)/2$ (b) -4 **29.** (a) negative   (b) positive   **31.** $25\pi/2$

33. (a) $\displaystyle\int_{-3}^{3} 4x(1 - 3x)\,dx$ (b) $\displaystyle\int_{0}^{1} e^x\,dx$ **35.** $\frac{5}{2}$

37. (a) $\displaystyle\lim_{\max \Delta x_k \to 0} \sum_{k=1}^{n} 2x_k^* \Delta x_k;\ a = 1, b = 2$ (b) $\displaystyle\lim_{\max \Delta x_k \to 0} \sum_{k=1}^{n} \frac{x_k^*}{x_k^* + 1} \Delta x_k;\ a = 0, b = 1$

39. (d) $\frac{3}{2}$ **41.** $\frac{1}{3}$ **43.** 320   **45.** (a) yes   (b) yes   (c) no   (d) yes

▶ Exercise Set 7.6 **(Page 425)**

1. (a) $\displaystyle\int_{0}^{2} (2 - x)\,dx = 2$ (b) $\displaystyle\int_{-1}^{1} 2\,dx = 4$ (c) $\displaystyle\int_{1}^{3} (x + 1)\,dx = 6$ **3.** $\frac{65}{4}$ **5.** $\frac{52}{3}$ **7.** $e^3 - e$ **9.** 48   **11.** $\frac{2}{3}$

13. $\frac{844}{5}$ **15.** 0   **17.** $\sqrt{2}$ **19.** $5e^3 - 10$ **21.** $-\frac{55}{3}$ **23.** $\frac{\pi^2}{9} + 2\sqrt{3}$ **27.** (a) $\frac{5}{2}$ (b) $2 - \frac{\sqrt{2}}{2}$ **29.** $-\frac{11}{6}$

31. $0.665867079;\ \frac{2}{3}$ **33.** $1.098242635;\ \ln 3 \approx 1.098612289$ **35.** 12   **37.** $\frac{9}{2}$

39. $A_1 = \frac{23}{6}, A_2 = \frac{343}{6}, A_3 = \frac{243}{6}, A = \frac{203}{2}$

41. (a) The integral is zero.   **43.** (a) $x^3 + 1$
    (c) $\displaystyle\int_{-a}^{a} f(x)\,dx = 2\int_{0}^{a} f(x)\,dx$

45. (a) $\sin\sqrt{x}$ (b) e^{x^2} **47.** $-\dfrac{x}{\cos x}$ **49.** (a) 0   (b) $\sqrt{13}$ (c) $6/\sqrt{13}$

51. (a) $x = 3$ (b) increasing on $[3, +\infty)$, decreasing on $(-\infty, 3]$ (c) concave up on $(-1, 7)$, concave down on $(-\infty, -1)$ and $(7, +\infty)$

53. (a) $(0, +\infty)$ (b) $x = 1$ **55.** (a) $x^* = 4$ (b) $x^* = e - 1$ **57.** $3\sqrt{2} \le \displaystyle\int_{0}^{3} \sqrt{x^3 + 2}\,dx \le 3\sqrt{29}$

▶ Exercise Set 7.7 **(Page 437)**

1. (a) the increase in height in inches, during the first 10 years
    (b) the change in the radius in cm, during the time interval $t = 1$ to $t = 2$ seconds
    (c) the change in the speed of sound in ft/s, during an increase in temperature from $t = 32°$F to $t = 100°$F
    (d) the displacement of the particle in cm, during the time interval $t = t_1$ to $t = t_2$ seconds

3. (a) displacement $= -\frac{1}{2}$; distance $= \frac{1}{2}$ (b) displacement $= 5$; distance $= \frac{5}{2}$

5. (a) 31.3 m/s   (b) 55.15 m/s   **7.** (a) $\frac{1}{4}t^4 - \frac{2}{3}t^3 + t + 1$ (b) $-\cos 2t - t - 2$ **9.** (a) $t^2 - 3t + 7$ (b) $-\cos t + t - (\pi/2)$

11. (a) displacement $= 1$; distance $= 1$ (b) displacement $= -1$; distance $= 3$

13. (a) displacement $= \frac{9}{4}$; distance $= \frac{11}{4}$ (b) displacement $= e^3 - 7$; distance $= e^3 - 9 + 4\ln 2$

15. displacement $= -6$; distance $= \frac{13}{2}$ 17. displacement $= \frac{204}{25}$; distance $= \frac{204}{25}$

19. **(a)** $s = 2/\pi$, $v = 1$, $|v| = 1$, $a = 0$ **(b)** $s = \frac{1}{2}$, $v = -\frac{3}{2}$, $|v| = \frac{3}{2}$, $a = -3$ 21. $\frac{22}{3}$ 23. $(1/e) + e - 2$

25. **(a)** 150   **(b)** 150   **(c)** 50   27. **(a)** The displacement is always positive.

$s(t)$

$v(t)$

$a(t)$

29. **(a)** $a(t) = \begin{cases} 0, & t < 4 \\ -10, & t > 4 \end{cases}$ **(b)** $v(t) = \begin{cases} 25, & t < 4 \\ 65 - 10t, & t > 4 \end{cases}$ **(c)** $x(t) = \begin{cases} 25t, & t < 4 \\ 65t - 5t^2 - 80, & t > 4 \end{cases}$ **(d)** $x(6.5) = 131.25$

so $x(8) = 120$, $x(12) = -20$.

31. **(a)** $-\frac{22}{15}$ ft/s^2 **(b)** $\frac{1}{7200}$ km/s^2 33. **(a)** $-\frac{121}{5}$ ft/s^2 **(b)** $\frac{70}{33}$ s   **(c)** $\frac{60}{11}$ s

35. 280 m   37. 100 s; 10,000 ft   39. **(a)** -48 ft/s   **(b)** 196 ft   **(c)** 112 ft/s

41. **(a)** 1 s   **(b)** $\frac{1}{2}$ s   43. **(a)** $(5 + 5\sqrt{33})/8$ s   **(b)** $20\sqrt{33}$ ft/s

45. **(a)** 5 s   **(b)** 272.5 m   **(c)** 10 s   **(d)** -49 m/s   **(e)** 12.46 s   **(f)** 73.1 m/s   47. 4.04 m/s   49. 6   51. $2/\pi$ 53. $\dfrac{1}{e-1}$

55. **(a)** $\frac{4}{3}$ **(c)**   57. **(a)** $\frac{263}{4}$ 59. 1404π lb   61. **(a)** 120 gal   63. **(b)** no
    (b) $2/\sqrt{3}$ **(b)** 31   **(b)** 420 gal
                                                (c) 2076.36 gal

▶ Exercise Set 7.8 **(Page 444)**

1. **(a)** $\displaystyle\int_1^3 u^7\,du$ **(b)** $-\dfrac{1}{2}\displaystyle\int_7^4 u^{1/2}\,du$ **(c)** $\dfrac{1}{\pi}\displaystyle\int_{-\pi}^{\pi} \sin u\,du$ **(d)** $\displaystyle\int_{-3}^0 (u+5)u^{20}\,du$ 3. $\frac{121}{5}$ 5. 10   7. $\frac{1192}{15}$

9. $8 - (4\sqrt{2})$ 11. $\ln\frac{21}{13}$ 13. $\frac{25}{12}\pi$ 15. $\pi/8$ 17. $2/\pi$ 19. $\frac{1}{24}$ 21. $\dfrac{1 - e^{-8}}{8}$ 23. $\frac{2}{3}$ 25. $\frac{2}{3}(\sqrt{10} - 2\sqrt{2})$

27. $2(\sqrt{7} - \sqrt{3})$ 29. 0   31. 0   33. $(\sqrt{3} - 1)/3$ 35. $\frac{106}{405}$ 37. $\ln 2$ 41. **(a)** $\frac{5}{3}$ **(b)** $\frac{5}{3}$ **(c)** $-\frac{1}{2}$

45. 48,233,525,650   47. **(a)** 328.69 ft   **(b)** yes   49. **(b)** 169.7 V   51. **(b)** $\frac{3}{2}$ **(c)** $\pi/4$ 53. $2/\pi$

▶ Exercise Set 7.9 **(Page 451)**

1. **(a)**   **(b)**   **(c)**   3. **(a)** 7
                                                (b) -5
                                                (c) -3
                                                (d) 6

5. 1.603210678; magnitude of error is < 0.0063 7. **(a)** x^{-1}, $x > 0$ **(b)** x^2, $x \neq 0$ **(c)** $-x^2$, $-\infty < x < +\infty$

(d) $-x$, $-\infty < x < +\infty$ **(e)** x^3, $x > 0$ **(f)** $\ln x + x$, $x > 0$ **(g)** $x - \sqrt[3]{x}$, $-\infty < x < +\infty$ **(h)** $\dfrac{e^x}{x}$, $x > 0$

9. **(a)** $e^{\pi \ln 3}$ **(b)** $e^{\sqrt{2}\ln 2}$ 11. **(a)** e^2 **(b)** e^2 13. $x^2 - x$ 15. **(a)** $3/x$ **(b)** 1   17. **(a)** 0   **(b)** $\frac{1}{3}$ **(c)** 0

19. **(a)** $2x^3\sqrt{1 + x^2}$ **(b)** $-\dfrac{2}{3}(x^2 + 1)^{3/2} + \dfrac{2}{5}(x^2 + 1)^{5/2} - \dfrac{4\sqrt{2}}{15}$ 21. **(a)** $-\sin x^2$ **(b)** $-\tan^2 x$

23. $-3\dfrac{3x-1}{9x^2+1}+2x\dfrac{x^2-1}{x^4+1}$ 25. **(a)** $3x^2\sin^2(x^3)-2x\sin^2(x^2)$ **(b)** $\dfrac{2}{1-x^2}$

27. **(a)** $F(0)=0,\ F(3)=0,\ F(5)=6,\ F(7)=6,\ F(10)=3$ **(d)**   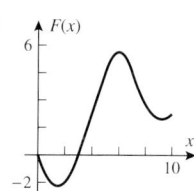   29. $F(x)=\begin{cases}(1-x^2)/2, & x<0\\(1+x^2)/2, & x\ge 0\end{cases}$

    (b) increasing on $\left[\frac{3}{2},6\right]$ and $\left[\frac{37}{4},10\right]$, decreasing on $\left[0,\frac{3}{2}\right]$ and $\left[6,\frac{37}{4}\right]$

    (c) maximum $\frac{15}{2}$ at $x=6$, minimum $-\frac{9}{4}$ at $x=\frac{3}{2}$

31. $y(x)=\frac{5}{4}+\frac{3}{4}x^{4/3}$ 33. $y(x)=\tan x+\cos x-(\sqrt{2}/2)$ 35. $P(x)=P_0+\displaystyle\int_0^x r(t)\,dt$ individuals

37. I is the derivative of II.   39. **(a)** $t=3$ **(f)**

                            (b) $t=1$

                            (c) $t=5$

                            (d) $t=3$

                            (e) F is concave up on $\left(0,\frac{1}{2}\right)$ and $(2,4)$,

                                 concave down on $\left(\frac{1}{2},2\right)$ and $(4,5)$.

41. **(a)** relative maxima at $x=\pm\sqrt{4k+1},\ k=0,1,\dots$;   43. $f(x)=3e^{3x},\ a=\frac{1}{3}\ln 2$ 45. 0.06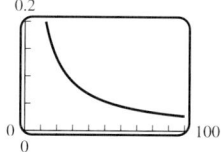

        relative minima at $x=\pm\sqrt{4k-1},\ k=1,2,\dots$

    (b) $x=\pm\sqrt{2k},\ k=1,2,\dots,$ and at $x=0$

▶ Chapter 7 Supplementary Exercises **(Page 454)**

5. $s(t)=\frac{1}{2}at^2+v_0t+s_0,\ v(t)=a(t)+v_0$

7. **(a)** $\frac{3}{4}$ **(b)** $-\frac{3}{2}$ **(c)** $-\frac{35}{4}$ **(d)** -2 **(e)** not enough information   **(f)** not enough information

9. **(a)** $2+(\pi/2)$ **(b)** $\dfrac{1}{3}(10^{3/2}-1)-\dfrac{9\pi}{4}$ **(c)** $\pi/8$ 11. $35\pi/128$ 15. **(d)** $n\ge 1000$

21. $(x^{2/3}+1)^{3/2}+C$ 23. **(a)** $\displaystyle\int_1^x\dfrac{1}{1+t^2}\,dt$ **(b)** $\displaystyle\int_{\tan[(\pi/4)-2]}^x\dfrac{1}{1+t^2}\,dt$

27. **(a)** $F(x)$ is 0 if $x=1$, positive if $x>1$, and negative if $x<1$.

    (b) $F(x)$ is 0 if $x=-1$, positive if $-1<x\le 2$, and negative if $-2\le x<-1$.

29. **(a)** 37,773.06 kW   **(c)**   **(d)** 2285.32 kW/h   31. **(a)** no   **(d)** 141.5 ft

    (b) 2200.32 kW/h   **(b)** $25<t<40$ **(e)** no

                                                           (c) 3.54 ft/s   **(f)** no

33. $\frac{1}{3}\sqrt{5+2\sin 3x}+C$ 35. $-\dfrac{1}{3a^2x^3+3ab}+C$ 37. C 39. $\ln 2$ 41. $\frac{3}{8}+\frac{1}{2}\left(\sin 1-\sin\frac{1}{4}\right)$ 43. 1.007514

45. **(a)** $k=2.073948$ 47. **(a)**   **(b)** 0.7651976866   49. The integral is better.

    (b) $k=1.837992$ **(c)** $x=2.404826$

▶ Exercise Set 8.1 **(Page 467)**

1. $9/2$ 3. 1   5. **(a)** $32/3$ **(b)** $32/3$ 7. $49/192$ 9. $1/2$ 11. $\sqrt{2}$ 13. $1/2$ 15. 24

17. $37/12$ 19. $4\sqrt{2}$ 21. $1/2$ 23. $9152/105$ 25. $9/\sqrt[3]{4}$ 27. **(a)** $4/3$ **(b)** $m=2-\sqrt[3]{4}$ 31. 1.180898334

33. **(a)** 1800 ft   **(b)** $\frac{3}{2}T^2-\frac{1}{60}T^3$ ft   35. $a^2/6$

▶ **Exercise Set 8.2 (Page 473)**

1. 8π 3. $13\pi/6$ 5. $32\pi/5$ 7. $(1 - \sqrt{2}/2)\pi$ 9. $256\pi/3$ 11. 4π 13. $2048\pi/15$ 15. $3\pi/5$ 17. 8π

19. 2π 21. $72\pi/5$ 23. $4\pi ab^2/3$ 25. π 27. $648\pi/5$ 29. $\pi/2$ 31. $40{,}000\pi$ ft^3 33. $1/30$

35. **(a)** $2\pi/3$ **(b)** $16/3$ **(c)** $4\sqrt{3}/3$ 37. 0.710172176 41. **(b)** left ≈ 11.157; right ≈ 11.771; $V \approx$ average $= 11.464$ cm^3

43. $V = \begin{cases} 3\pi h^2, & 0 \le h < 2 \\ \frac{1}{3}\pi(12h^2 - h^3 - 4), & 2 \le h \le 4 \end{cases}$ 45. $r^3 \tan\theta$ 47. $16r^3/3$

▶ **Exercise Set 8.3 (Page 479)**

1. $15\pi/2$ 3. $\pi/3$ 5. $2\pi/5$ 7. 4π 9. $20\pi/3$ 11. $\pi(e^3 - e)$ 13. $\pi/2$ 15. $\pi/5$ 17. $2\pi^2$

19. **(a)** $7\pi/30$ **(b)** easier   21. $9\pi/14$ 23. $\pi r^2 h/3$ 25. $V = \dfrac{4\pi}{3}[r^3 - (r^2 - a^2)^{3/2}]$ 27. $b = 1$

▶ **Exercise Set 8.4 (Page 483)**

1. $L = \sqrt{5}$ 3. $(85\sqrt{85} - 8)/243$ 5. $\frac{1}{27}(80\sqrt{10} - 13\sqrt{13})$ 7. $(e^3 - e^{-3})/2$ 9. $(2\sqrt{2} - 1)/3$ 11. π

13. $\sqrt{2}(e^{\pi/2} - 1)$ 15. $\ln(1 + \sqrt{2})$ 19. **(a)**   **(b)** dy/dx does not exist at $x = 0$.   **(c)** $L = (13\sqrt{13} + 80\sqrt{10} - 16)/27$

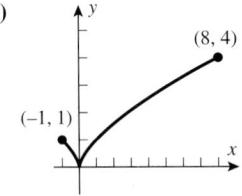

21. 4.645975301 23. 3.820197788 27. **(b)** 9.69 **(c)** 5.16 cm   29. $k = 1.83$

▶ **Exercise Set 8.5 (Page 488)**

1. $35\pi\sqrt{2}$ 3. 8π 5. $40\pi\sqrt{82}$ 7. 24π 9. $16\pi/9$ 11. $16{,}911\pi/1024$ 13. 22.94 15. 7.05

21. $\dfrac{2\sqrt{2}\pi}{5}(2e^\pi + 1)$ 23. $\dfrac{\pi}{24}(17\sqrt{17} - 1)$ 29. **(b)** for $f(x)$ constant on $[a, b]$

▶ **Exercise Set 8.6 (Page 494)**

1. **(a)** 210 ft·lb   **(b)** $5/6$ ft·lb   3. 100 ft·lb   5. 160 J   7. 20 lb/ft   9. $900\pi\rho$ ft·lb   11. $261{,}600$ J

13. **(a)** $926{,}640$ ft·lb   **(b)** hp of motor $= 0.468$ 15. $75{,}000$ ft·lb

17. **(a)** $2{,}400{,}000{,}000/x^2$ lb   **(b)** $(9.6 \times 10^{10})/(x + 4000)^2$ lb   **(c)** 2.5344×10^{10} ft·lb   19. $v_f = 100$ m/s

21. **(a)** decrease of 4.5×10^{14} J   **(b)** ≈ 0.107 **(c)** ≈ 8.24 bombs

▶ **Exercise Set 8.7 (Page 499)**

1. **(a)** $F = 31{,}200$ lb; $P = 312$ lb/ft^2 **(b)** $F = 2{,}452{,}500$ N; $P = 98.1$ kPa   3. 499.2 lb   5. 8.175×10^5 N

7. $1{,}098{,}720$ N   9. yes   11. $\rho a^3/\sqrt{2}$ lb   13. $14{,}976\sqrt{17}$ lb   15. **(b)** $80\rho_0$ lb/min

▶ **Exercise Set 8.8 (Page 508)**

1. **(a)** 10.0179 **(b)** 3.7622 3. **(a)** $\frac{4}{3}$
   (c) $15/17 \approx 0.8824$ **(d)** -1.4436 **(b)** $\frac{5}{4}$
   (e) 1.7627 **(f)** 0.9730 **(c)** $\frac{312}{313}$
   (d) $-\frac{63}{16}$

5.

	$\sinh x_0$	$\cosh x_0$	$\tanh x_0$	$\coth x_0$	$\operatorname{sech} x_0$	$\operatorname{csch} x_0$
(a)	2	$\sqrt{5}$	$2/\sqrt{5}$	$\sqrt{5}/2$	$1/\sqrt{5}$	$1/2$
(b)	$3/4$	$5/4$	$3/5$	$5/3$	$4/5$	$4/3$
(c)	$4/3$	$5/3$	$4/5$	$5/4$	$3/5$	$3/4$

9. $4\cosh(4x-8)$ **11.** $-\dfrac{1}{x}\operatorname{csch}^2(\ln x)$ **13.** $\dfrac{1}{x^2}\operatorname{csch}\left(\dfrac{1}{x}\right)\coth\left(\dfrac{1}{x}\right)$ **15.** $\dfrac{2+5\cosh(5x)\sinh(5x)}{\sqrt{4x+\cosh^2(5x)}}$

17. $x^{5/2}\tanh(\sqrt{x}\,)\operatorname{sech}^2(\sqrt{x}\,)+3x^2\tanh^2(\sqrt{x}\,)$ **19.** $\dfrac{1}{\sqrt{9+x^2}}$ **21.** $\dfrac{1}{(\cosh^{-1}x)\sqrt{x^2-1}}$ **23.** $\dfrac{-(\tanh^{-1}x)^{-2}}{1-x^2}$

25. $\dfrac{\sinh x}{|\sinh x|}=\begin{cases}1, & x>0\\ -1, & x<0\end{cases}$ **27.** $-\dfrac{e^x}{2x\sqrt{1-x}}+e^x\operatorname{sech}^{-1}x$ **31.** $\tfrac{1}{7}\sinh^7 x+C$ **33.** $\tfrac{2}{3}(\tanh x)^{3/2}+C$ **35.** $\ln(\cosh x)+C$

37. $37/375$ **39.** $\tfrac{1}{3}\sinh^{-1}3x+C$ **41.** $-\operatorname{sech}^{-1}(e^x)+C$ **43.** $-\operatorname{csch}^{-1}|2x|+C$ **45.** $\tfrac{1}{2}\ln 3$ **49.** $16/9$ **51.** 5π

53. $\tfrac{3}{4}$ **61.** $|u|<1:\tanh^{-1}u+C;\ |u|>1:\tanh^{-1}(1/u)+C$ **63.** **(a)** $+\infty$ **(b)** $-\infty$ **(c)** 1 **(d)** -1 **(e)** $+\infty$ **(f)** $+\infty$

71. 405.9 ft

▶ **Chapter 8 Supplementary Exercises (Page 510)**

7. (a) $\displaystyle\int_a^b (f(x)-g(x))\,dx+\int_b^c (g(x)-f(x))\,dx+\int_c^d (f(x)-g(x))\,dx$ **(b)** $\tfrac{11}{4}$ **9.** $9a/8$

13. Set $a=68.7672,\ b=0.0100333,$
$c=693.8597,\ d=299.2239.$

(a)

650
−300   300
0

(b) 1480.2798 ft
(c) 283.6249 ft
(d) 82°

15. (a) $\sinh^{-1}(x/2)+C$
(b) $\cosh^{-1}(x/3)+C$

(c) 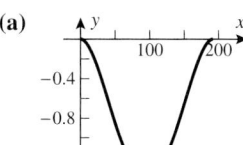 or $\dfrac{1}{2\sqrt{2}}\ln\left|\dfrac{\sqrt{2}+x}{\sqrt{2}-x}\right|+C$

(d) $\dfrac{1}{\sqrt{5}}\sinh^{-1}\left(\dfrac{\sqrt{5}x}{4}\right)+C$

17. (a) $W=\tfrac{1}{16}$ J   **19. (a)** $F=\displaystyle\int_0^1 \rho x3\,dx$ N
(b) 5 m

(b) $F=\displaystyle\int_1^4 \rho(1+x)2x\,dx$ lb/ft²

(c) $\displaystyle\int_{-10}^{0} 9810|y|2\sqrt{\tfrac{125}{8}(y+10)}\,dy$ N

21. (a)

y
100   200   x
−0.4
−0.8
−1.2
−1.6

(b) 1.42 in
(c) The length of the centerline is 192.026 in.

23. $k\approx 0.724611$ **25. (a)** $\displaystyle\int_a^{a+2}\dfrac{x}{\sqrt{1+x^3}}\,dx$ **(b)** $a\approx 0.683772$; maximum work $=1.347655$ J

▶ **Exercise Set 9.1 (Page 515)**

1. $-\tfrac{1}{8}(3-2x)^4+C$ **3.** $\tfrac{1}{2}\tan(x^2)+C$ **5.** $-\tfrac{1}{3}\ln(2+\cos 3x)+C$ **7.** $\cosh e^x+C$ **9.** $-e^{\cot x}+C$

11. $-\tfrac{1}{42}\cos^6 7x+C$ **13.** $\ln(e^x+\sqrt{e^{2x}+4}\,)+C$ **15.** $2e^{\sqrt{x-2}}+C$ **17.** $2\sinh\sqrt{x}+C$ **19.** $-\dfrac{2}{\ln 3}3^{-\sqrt{x}}+C$

21. $\dfrac{1}{2}\coth\dfrac{2}{x}+C$ **23.** $-\dfrac{1}{4}\ln\left|\dfrac{2+e^{-x}}{2-e^{-x}}\right|+C$ **25.** $\sin^{-1}e^x+C$ **27.** $\tfrac{1}{2}\sin(x^2)+C$ **29.** $-\dfrac{1}{\ln 16}4^{-x^2}+C$

▶ **Exercise Set 9.2 (Page 521)**

1. $-xe^{-x}-e^{-x}+C$ **3.** $x^2e^x-2xe^x+2e^x+C$ **5.** $-\tfrac{1}{2}x\cos 2x+\tfrac{1}{4}\sin 2x+C$ **7.** $x^2\sin x+2x\cos x-2\sin x+C$

9. $\tfrac{2}{3}x^{3/2}\ln x-\tfrac{4}{9}x^{3/2}+C$ **11.** $x(\ln x)^2-2x\ln x+2x+C$ **13.** $x\ln(2x+3)-x+\tfrac{3}{2}\ln(2x+3)+C$

15. $x\sin^{-1}x+\sqrt{1-x^2}+C$ **17.** $x\tan^{-1}(2x)-\tfrac{1}{4}\ln(1+4x^2)+C$ **19.** $\tfrac{1}{2}e^x(\sin x-\cos x)+C$

21. $\dfrac{e^{ax}}{a^2+b^2}(a\sin bx-b\cos bx)+C$ **23.** $(x/2)[\sin(\ln x)-\cos(\ln x)]+C$ **25.** $x\tan x+\ln|\cos x|+C$

27. $\tfrac{1}{2}x^2e^{x^2}-\tfrac{1}{2}e^{x^2}+C$ **29.** $(1-6e^{-5})/25$ **31.** $(2e^3+1)/9$ **33.** $5\ln 5-4$ **35.** $\dfrac{5\pi}{6}-\sqrt{3}+1$ **37.** $-\pi/8$

39. $\dfrac{1}{3}\left(2\sqrt{3}\pi-\dfrac{\pi}{2}-2+\ln 2\right)$ **41. (a)** $2(\sqrt{x}-1)e^{\sqrt{x}}+C$ **(b)** $2\sqrt{x}\sin\sqrt{x}+2\cos\sqrt{x}+C$ **43. (a)** $A=1$ **(b)** $V=\pi(e-2)$

45. $V=2\pi^2$ **47.** distance $=-37e^{-5}+2$ **49. (a)** $-\tfrac{1}{3}\sin^2 x\cos x-\tfrac{2}{3}\cos x+C$ **(b)** $\dfrac{3\pi}{32}-\dfrac{1}{4}$

53. (a) $\tfrac{1}{3}\tan^3 x-\tan x+x+C$ **(b)** $\tfrac{1}{3}\sec^2 x\tan x+\tfrac{2}{3}\tan x+C$ **(c)** $x^3e^x-3x^2e^x+6xe^x-6e^x+C$

▶ Exercise Set 9.3 **(Page 529)**

1. $-\frac{1}{6}\cos^6 x + C$ 3. $\frac{1}{2a}\sin^2 ax + C, \quad a \neq 0$ 5. $\frac{1}{2}\theta - \frac{1}{20}\sin 10\theta + C$ 7. $\sin\theta - \frac{2}{3}\sin^3\theta + \frac{1}{5}\sin^5\theta + C$

9. $\frac{1}{6}\sin^3 2t - \frac{1}{10}\sin^5 2t + C$ 11. $\frac{1}{8}x - \frac{1}{32}\sin 4x + C$ 13. $-\frac{1}{6}\cos 3x + \frac{1}{2}\cos x + C$ 15. $-\frac{1}{3}\cos(3x/2) - \cos(x/2) + C$

17. $(5\sqrt{2})/12$ 19. 0 21. $\frac{1}{24}$ 23. $\frac{1}{3}\tan(3x+1) + C$ 25. $\frac{1}{2}\ln|\cos(e^{-2x})| + C$ 27. $\frac{1}{2}\ln|\sec 2x + \tan 2x| + C$

29. $\frac{1}{3}\tan^3 x + C$ 31. $\frac{1}{16}\tan^4 4x + \frac{1}{24}\tan^6 4x + C$ 33. $\frac{1}{7}\sec^7 x - \frac{1}{5}\sec^5 x + C$

35. $\frac{1}{4}\sec^3 x\tan x - \frac{5}{8}\sec x\tan x + \frac{3}{8}\ln|\sec x + \tan x| + C$ 37. $\frac{1}{6}\sec^3 2t + C$ 39. $\tan x + \frac{1}{3}\tan^3 x + C$

41. $\frac{1}{3}\tan^3 x - \tan x + x + C$ 43. $\frac{2}{3}\tan^{3/2}x + \frac{2}{7}\tan^{7/2}x + C$ 45. $\frac{\sqrt{3}}{2} - \frac{\pi}{6}$ 47. $-\frac{1}{2} + \ln 2$

49. $-\frac{1}{5}\csc^5 x + \frac{1}{3}\csc^3 x + C$ 51. $-\frac{1}{2}\csc^2 x - \ln|\sin x| + C$ 55. $L = \ln(\sqrt{2}+1)$ 57. $V = \pi/2$

63. $-\dfrac{1}{\sqrt{a^2+b^2}}\ln\left|\dfrac{\sqrt{a^2+b^2} + a\cos x - b\sin x}{a\sin x + b\cos x}\right| + C$ 65. (a) $\frac{2}{3}$ (b) $3\pi/16$ (c) $\frac{8}{15}$ (d) $5\pi/32$

▶ Exercise Set 9.4 **(Page 535)**

1. $2\sin^{-1}(x/2) + \frac{1}{2}x\sqrt{4-x^2} + C$ 3. $\frac{9}{2}\sin^{-1}(x/3) - \frac{1}{2}x\sqrt{9-x^2} + C$ 5. $\frac{1}{16}\tan^{-1}(x/2) + \dfrac{x}{8(4+x^2)} + C$

7. $\sqrt{x^2-9} - 3\sec^{-1}(x/3) + C$ 9. $-2\sqrt{2-x^2} + \frac{1}{3}(2-x^2)^{3/2} + C$ 11. $\dfrac{\sqrt{4x^2-9}}{9x} + C$ 13. $\dfrac{x}{\sqrt{1-x^2}} + C$

15. $\ln|x + \sqrt{x^2-1}| + C$ 17. $-(x/\sqrt{9x^2-1}) + C$ 19. $\frac{1}{2}\sin^{-1}(e^x) + \frac{1}{2}e^x\sqrt{1-e^{2x}} + C$ 21. $\frac{2048}{15}$ 23. $(\sqrt{3} - \sqrt{2})/2$

25. $\dfrac{10\sqrt{3}+18}{243}$ 27. $\frac{1}{2}\ln(x^2+4) + C$ 29. $L = \sqrt{5} - \sqrt{2} + \ln\dfrac{2+2\sqrt{2}}{1+\sqrt{5}}$ 31. $S = \dfrac{\pi}{32}[18\sqrt{5} - \ln(2+\sqrt{5})]$

33. (a) $\sinh^{-1}(x/3) + C$ (b) $\ln\left(\dfrac{\sqrt{x^2+9}}{3} + \dfrac{x}{3}\right) + C$ (c) $\frac{1}{2}x\sqrt{x^2-1} - \frac{1}{2}\cosh^{-1}x + C$ 35. $\frac{1}{3}\tan^{-1}\left(\dfrac{x-2}{3}\right) + C$

37. $\sin^{-1}\left(\dfrac{x-1}{3}\right) + C$ 39. $\ln(x-3+\sqrt{(x-3)^2+1}) + C$ 41. $2\sin^{-1}\left(\dfrac{x+1}{2}\right) + \frac{1}{2}(x+1)\sqrt{3-2x-x^2} + C$

43. $\dfrac{1}{\sqrt{10}}\tan^{-1}\sqrt{\frac{2}{5}}(x+1) + C$ 45. $\pi/6$ 49. $u = \sin^2 x, \ \dfrac{1}{2}\int\sqrt{1-u^2}\,du = \dfrac{1}{4}[\sin^2 x\sqrt{1-\sin^4 x} + \sin^{-1}(\sin^2 x)] + C$

▶ Exercise Set 9.5 **(Page 542)**

1. $\dfrac{A}{(x-2)} + \dfrac{B}{(x+5)}$ 3. $\dfrac{A}{x} + \dfrac{B}{x^2} + \dfrac{C}{x-1}$ 5. $\dfrac{A}{x} + \dfrac{B}{x^2} + \dfrac{C}{x^3} + \dfrac{Dx+E}{x^2+1}$ 7. $\dfrac{Ax+B}{x^2+5} + \dfrac{Cx+D}{(x^2+5)^2}$

9. $\dfrac{1}{5}\ln\left|\dfrac{x-1}{x+4}\right| + C$ 11. $\frac{5}{2}\ln|2x-1| + 3\ln|x+4| + C$ 13. $\ln\left|\dfrac{x(x+3)^2}{x-3}\right| + C$ 15. $\frac{1}{2}x^2 - 2x + 6\ln|x+2| + C$

17. $3x + 12\ln|x-2| - \dfrac{2}{x-2} + C$ 19. $\frac{1}{3}x^3 + x + \ln\left|\dfrac{(x+1)(x-1)^2}{x}\right| + C$ 21. $3\ln|x| - \ln|x-1| - \dfrac{5}{x-1} + C$

23. $\ln\dfrac{(x-3)^2}{|x+1|} + \dfrac{1}{x-3} + C$ 25. $\ln|x+2| + \dfrac{4}{x+2} - \dfrac{2}{(x+2)^2} + C$ 27. $-\frac{7}{34}\ln|4x-1| + \frac{6}{17}\ln(x^2+1) + \frac{3}{17}\tan^{-1}x + C$

29. $3\tan^{-1}x + \frac{1}{2}\ln(x^2+3) + C$ 31. $\frac{1}{2}x^2 - 3x + \frac{1}{2}\ln(x^2+1) + C$ 33. $\dfrac{1}{6}\ln\left(\dfrac{1-\sin\theta}{5+\sin\theta}\right) + C$ 35. $V = \pi\left(\frac{19}{5} - \frac{9}{4}\ln 5\right)$

37. $\dfrac{1}{\sqrt{2}}\tan^{-1}\left(\dfrac{x+1}{\sqrt{2}}\right) + \dfrac{1}{x^2+2x+3} + C$ 39. $\frac{1}{8}\ln|x-1| - \frac{1}{5}\ln|x-2| + \frac{1}{12}\ln|x-3| - \frac{1}{120}\ln|x+3| + C$

▶ Exercise Set 9.6 **(Page 551)**

1. Formula (60): $\frac{3}{16}[4x + \ln|-1+4x|] + C$ 3. Formula (65): $\dfrac{1}{5}\ln\left|\dfrac{x}{5+2x}\right| + C$ 5. Formula (102): $\frac{1}{5}(x+1)(-3+2x)^{3/2} + C$

7. Formula (108): $\dfrac{1}{2}\ln\left|\dfrac{\sqrt{4-3x}-2}{\sqrt{4-3x}+2}\right| + C$ 9. Formula (69): $\dfrac{1}{2\sqrt{5}}\ln\left|\dfrac{x+\sqrt{5}}{x-\sqrt{5}}\right| + C$

11. Formula (73): $\dfrac{x}{2}\sqrt{x^2-3} - \dfrac{3}{2}\ln|x+\sqrt{x^2-3}| + C$ 13. Formula (95): $\dfrac{x}{2}\sqrt{x^2+4} - 2\ln(x+\sqrt{x^2+4}) + C$

15. Formula (74): $\dfrac{x}{2}\sqrt{9-x^2} + \dfrac{9}{2}\sin^{-1}\dfrac{x}{3} + C$ 17. Formula (79): $\sqrt{3-x^2} - \sqrt{3}\ln\left|\dfrac{\sqrt{3}+\sqrt{9-x^2}}{x}\right| + C$

19. Formula (38): $-\dfrac{1}{10}\sin(5x) + \dfrac{1}{2}\sin x + C$ 21. Formula (50): $\dfrac{x^4}{16}[4\ln x - 1] + C$

23. Formula (42): $\dfrac{e^{-2x}}{13}[-2\sin(3x) - 3\cos(3x)] + C$ 25. Formula (62): $\dfrac{1}{2}\displaystyle\int \dfrac{u\,du}{(4-3u)^2} = \dfrac{1}{18}\left[\dfrac{4}{4-3e^{2x}} + \ln|4-3e^{2x}|\right] + C$

27. Formula (68): $\dfrac{2}{3}\displaystyle\int \dfrac{du}{u^2+4} = \dfrac{1}{3}\tan^{-1}\dfrac{3\sqrt{x}}{2} + C$ 29. Formula (76): $\dfrac{1}{3}\displaystyle\int \dfrac{du}{\sqrt{u^2-4}} = \dfrac{1}{3}\ln|3x + \sqrt{9x^2-4}| + C$

31. Formula (81): $\dfrac{1}{54}\displaystyle\int \dfrac{u^2\,du}{\sqrt{5-u^2}} = -\dfrac{x^2}{36}\sqrt{5-9x^4} + \dfrac{5}{108}\sin^{-1}\dfrac{3x^2}{\sqrt{5}} + C$

33. Formula (26): $\displaystyle\int \sin^2 u\,du = \dfrac{1}{2}\ln x + \dfrac{1}{4}\sin(2\ln x) + C$ 35. Formula (51): $\dfrac{1}{4}\displaystyle\int ue^u\,du = \dfrac{1}{4}(-2x-1)e^{-2x} + C$

37. $u = \cos 3x$, Formula (67): $-\displaystyle\int \dfrac{du}{u(u+1)^2} = -\dfrac{1}{3}\left[\dfrac{1}{1+\cos 3x} + \ln\left|\dfrac{\cos 3x}{1+\cos 3x}\right|\right] + C$

39. $u = 4x^2$, Formula (70): $\dfrac{1}{8}\displaystyle\int \dfrac{du}{u^2-1} = \dfrac{1}{16}\ln\left|\dfrac{4x^2-1}{4x^2+1}\right| + C$

41. $u = 2e^x$, Formula (74): $\dfrac{1}{2}\displaystyle\int \sqrt{3-u^2}\,du = \dfrac{1}{2}e^x\sqrt{3-4e^{2x}} + \dfrac{3}{4}\sin^{-1}\left(\dfrac{2e^x}{\sqrt{3}}\right) + C$

43. $u = 3x$, Formula (112): $\dfrac{1}{3}\displaystyle\int \sqrt{\tfrac{5}{3}u - u^2}\,du = \dfrac{18x-5}{36}\sqrt{5x-9x^2} + \dfrac{25}{216}\sin^{-1}\left(\dfrac{18x-5}{5}\right) + C$

45. $u = 3x$, Formula (44): $\dfrac{1}{9}\displaystyle\int u\sin u\,du = \dfrac{1}{9}(\sin 3x - 3x\cos 3x) + C$

47. $u = -\sqrt{x}$, Formula (51): $2\displaystyle\int ue^u\,du = -2(\sqrt{x}+1)e^{-\sqrt{x}} + C$

49. $x^2 + 4x - 5 = (x+2)^2 - 9$; $u = x+2$, Formula (70): $\displaystyle\int \dfrac{du}{u^2-9} = \dfrac{1}{6}\ln\left|\dfrac{x-1}{x+5}\right| + C$

51. $x^2 - 4x - 5 = (x-2)^2 - 9$, $u = x-2$, Formula (77): $\displaystyle\int \dfrac{u+2}{\sqrt{9-u^2}}\,du = -\sqrt{5+4x-x^2} + 2\sin^{-1}\left(\dfrac{x-2}{3}\right) + C$

53. $u = \sqrt{x-2}$, $\dfrac{2}{5}(x-2)^{5/2} + \dfrac{4}{3}(x-2)^{3/2} + C$ 55. $u = \sqrt{x^3+1}$, $\dfrac{2}{3}\displaystyle\int u^2(u^2-1)\,du = \dfrac{2}{15}(x^3+1)^{5/2} - \dfrac{2}{9}(x^3+1)^{3/2} + C$

57. $u = x^{1/6}$, $\displaystyle\int \dfrac{6u^5}{u^3+u^2}\,du = 2x^{1/2} - 3x^{1/3} + 6x^{1/6} - 6\ln(x^{1/6}+1) + C$ 59. $u = x^{1/4}$, $4\displaystyle\int \dfrac{1}{u(1-u)}\,du = 4\ln\dfrac{x^{1/4}}{|1-x^{1/4}|} + C$

61. $u = x^{1/6}$, $6\displaystyle\int \dfrac{u^3}{u-1}\,du = 2x^{1/2} + 3x^{1/3} + 6x^{1/6} + 6\ln|x^{1/6}-1| + C$

63. $u = \sqrt{1+x^2}$, $\displaystyle\int (u^2-1)\,du = \dfrac{1}{3}(1+x^2)^{3/2} - (1+x^2)^{1/2} + C$ 65. $u = \sqrt{x}$, $2\displaystyle\int u\sin u\,du = 2\sin\sqrt{x} - 2\sqrt{x}\cos\sqrt{x} + C$

67. $\displaystyle\int \dfrac{1}{1 + \dfrac{2u}{1+u^2} + \dfrac{1-u^2}{1+u^2}}\,\dfrac{2}{1+u^2}\,du = \displaystyle\int \dfrac{1}{u+1}\,du = \ln|\tan(x/2)+1| + C$

69. $\displaystyle\int \dfrac{d\theta}{1-\cos\theta} = \displaystyle\int \dfrac{1}{u^2}\,du = -\dfrac{1}{u} + C = -\cot(\theta/2) + C$ 71. $2\displaystyle\int \dfrac{1-u^2}{(3u^2+1)(u^2+1)}\,du$, $\dfrac{4}{\sqrt{3}}\tan^{-1}[\sqrt{3}\tan(x/2)] - x + C$

73. $x \approx 3.523188312$ 75. $A \approx 17.59119023$ 77. $A \approx 0.054930614$ 79. $V \approx 3.586419094$ 81. $V \approx 5.031899801$

83. $L \approx 8.409316783$ 85. $S \approx 14.42359945$ 87. **(a)** $s(t) = 2 + \displaystyle\int_0^t 20\cos^6 u\sin^3 u\,du = -\dfrac{20}{9}\sin^2 t\cos^7 t - \dfrac{40}{63}\cos^7 t + \dfrac{166}{63}$

91. $\dfrac{2}{\sqrt{3}}\tan^{-1}\left(\dfrac{2\tanh(x/2)+1}{\sqrt{3}} + C\right)$ **(b)**

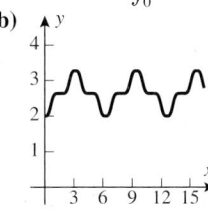

▶ Exercise Set 9.7 **(Page 562)**

1. exact value $= 14/3 \approx 4.666666667$
   (a) 4.667600663, $|E_M| \approx 0.000933996$
   (b) 4.664795679, $|E_T| \approx 0.001870988$
   (c) 4.666651630, $|E_S| \approx 0.000015037$

3. exact value $= 2$
   (a) 2.008248408, $|E_M| \approx 0.008248408$
   (b) 1.983523538, $|E_T| \approx 0.016476462$
   (c) 2.000109517, $|E_S| \approx 0.000109517$

5. exact value $= e^{-1} - e^{-3} \approx 0.318092373$
   (a) 0.317562837, $|E_M| \approx 0.000529536$
   (b) 0.319151975, $|E_T| \approx 0.001059602$
   (c) 0.318095187, $|E_S| \approx 0.000002814$

7. (a) $|E_M| \le \dfrac{27}{2400}(1/4) = 0.002812500$ (b) $|E_T| \le \dfrac{27}{1200}(1/4) = 0.005625000$ (c) $|E_S| \le \dfrac{243}{180 \times 10^4}(15/16) \approx 0.000126563$

9. (a) $|E_M| \le \dfrac{\pi^3}{2400}(1) \approx 0.012919282$ (b) $|E_T| \le \dfrac{\pi^3}{1200}(1) \approx 0.025838564$ (c) $|E_S| \le \dfrac{\pi^5}{180 \times 10^4}(1) \approx 0.000170011$

11. (a) $|E_M| \le \dfrac{8}{2400}(e^{-1}) \approx 0.001226265$ (b) $|E_T| \le \dfrac{8}{1200}(e^{-1}) \approx 0.002452530$ (c) $|E_S| \le \dfrac{32}{180 \times 10^4}(e^{-1}) \approx 0.000006540$

13. (a) $n = 24$ (b) $n = 34$ (c) $n = 8$ 15. (a) $n = 36$ (b) $n = 51$ (c) $n = 8$ 17. (a) $n = 351$ (b) $n = 496$ (c) $n = 16$

19. 0.746824948, 0.746824133 21. 2.129861595, 2.129861293 23. 0.805376152, 0.804776489

25. (a) 3.142425985, $|E_M| \approx 0.000833331$ (b) 3.139925989, $|E_T| \approx 0.001666665$ (c) 3.141592614, $|E_S| \approx 0.000000040$

27. $S_{14} = 0.693147984$, $|E_S| \approx 0.000000803 = 8.03 \times 10^{-7}$ 29. $n = 116$ 33. $L \approx 3.820187623$ 35. 1604 ft

37. 37.9 mi   39. 9.3 L   43. (a) $\max |f''(x)| \approx 3.844880$ (b) $n = 18$ (c) 0.904741

45. (a) $\max |f^{(4)}(x)| \approx 42.551816$ (b) $n = 8$ (c) 0.904524

▶ Exercise Set 9.8 **(Page 571)**

1. (a) improper; infinite discontinuity at $x = 3$ (b) not improper   (c) improper; infinite discontinuity at $x = 0$ (d) improper; infinite
   interval of integration   (e) improper; infinite interval of integration and infinite discontinuity at $x = 1$ (f) not improper

3. 1 5. $\ln \frac{5}{3}$ 7. $\frac{1}{2}$ 9. $-\frac{1}{4}$ 11. $\frac{1}{3}$ 13. divergent   15. 0 17. divergent   19. divergent   21. $\pi/2$

23. 1 25. divergent   27. $\frac{9}{2}$ 29. divergent   31. 2 33. 2 37. $\frac{1}{2}$

39. (a) 2.726585 (b) 2.804364 (c) 0.219384 (d) 0.504067 41. -1 43. $\frac{1}{9}$

45. (a) $V = \pi/2$ (b) $S = \pi[\sqrt{2} + \ln(1 + \sqrt{2})]$ 47. (b) $1/e$ (c) It is convergent.   51. $\dfrac{8\sqrt{2}}{5}$

53. $\dfrac{2\pi N I}{kr}\left(1 - \dfrac{a}{\sqrt{r^2 + a^2}}\right)$ 55. (b) 2.4×10^7 mi·lb   57. (a) $\dfrac{1}{s^2}$ (b) $\dfrac{2}{s^3}$ (c) $\dfrac{e^{-3s}}{s}$ 61. (a) 1.047 65. 1.809

▶ Chapter 9 Supplementary Exercises **(Page 574)**

1. (a) parts   (b) substitution   (c) reduction formula   (d) substitution   (e) substitution   (f) substitution   (g) parts
   (h) substitution   (i) $u = 4 - x^2$ 5. (a) 40 (b) 57 (c) 113 (d) 108 (e) 52 (f) 71

7. (a) $-\frac{1}{8}\sin^3 2x \cos 2x - \frac{3}{16}\sin 2x \cos 2x + \frac{3}{8}x + C$
   (b) $\frac{1}{8}\cos^3(x^2)\sin(x^2) + \frac{3}{16}\cos(x^2)\sin(x^2) + \frac{3}{16}x^2 + C$

9. $2\sin^{-1}(\sqrt{x/2}) + C$
   $-2\sin^{-1}(\sqrt{2-x}/\sqrt{2}) + C$
   $\sin^{-1}(x-1) + C$

11.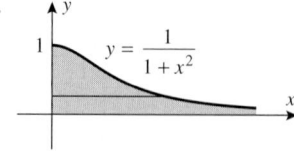

13. $V = 2\pi$ 15. $-\frac{2}{3}\cos^{3/2}\theta + C$ 17. $\frac{1}{6}\tan^3(x^2) + C$ 19. $\dfrac{x}{3\sqrt{3+x^2}} + C$

21. $\sqrt{x^2 + 2x + 2} + 2\ln(\sqrt{x^2 + 2x + 2} + x + 1) + C$ 23. $-\frac{1}{6}\ln|x-1| + \frac{1}{15}\ln|x+2| + \frac{1}{10}\ln|x-3| + C$ 25. $4 - \pi$

27. $\ln \dfrac{\sqrt{e^x + 1} - 1}{\sqrt{e^x + 1} + 1} + C$ 29. $\dfrac{1}{2(a^2 + 1)}$ 31. $\frac{1}{4}\sin^{-1}(x^4) + C$ 33. $\dfrac{\sqrt{2}}{3}[(x+2)^{3/2} - (x-2)^{3/2}] + C$

35. (a) $(x+4)(x-5)(x^2+1)^2$; $\dfrac{A}{x+4} + \dfrac{B}{x-5} + \dfrac{Cx+D}{x^2+1} + \dfrac{Ex+F}{(x^2+1)^2}$
   (b) $-\dfrac{3}{x+4} + \dfrac{2}{x-5} - \dfrac{x-2}{x^2+1} - \dfrac{3}{(x^2+1)^2}$ (c) $-3\ln|x+4| + 2\ln|x-5| + 2\tan^{-1}x - \dfrac{1}{2}\ln(x^2+1) - \dfrac{3}{2}\left(\dfrac{x}{x^2+1} + \tan^{-1}x\right)$

▶ Exercise Set 10.1 **(Page 589)**

3. **(a)** first order   **(b)** second order   7. **(a)** $y = Ce^{-3x}$ **(b)** $y = Ce^{2t}$ 9. $y = Cx$ 11. $y = Ce^{-\sqrt{1+x^2}} - 1$

13. $\ln|y| + y^2/2 = e^x + C$ and $y = 0$ 15. $y = \ln(\sec x + C)$ 17. $y = \dfrac{1}{1 - C(\csc x - \cot x)}$ and $y = 0$ 19. $y = e^{-2x} + Ce^{-3x}$

21. $y = e^{-x}\sin(e^x) + Ce^{-x}$ 23. $y = \dfrac{C}{\sqrt{x^2+1}}$ 25. **(a)** $y = \dfrac{x}{2} + \dfrac{3}{2x}$ **(b)** $y = \dfrac{x}{2} - \dfrac{5}{2x}$

27. $y = -1 + 4e^{x^2/2}$ 29. $3y^2 + 6\sin y = 8x^3 + 3\pi^2 - 8$ 31. $y^2 - 2y = t^2 + t + 3$

33. **(a)**

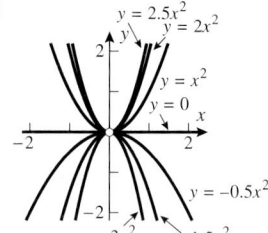

(b) $x = 2y^2$ 35. $y = \dfrac{C}{\sqrt{x^2+4}}$

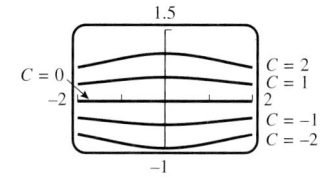

37. $x^3 + y^3 - 3y = C$

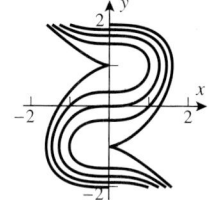

41. $x^2 + 2e^{-y} = 6$ 43. **(a)** $200 - 175e^{-t/25}$ oz   **(b)** 136 oz   45. 25 lb   49. **(a)** $I(t) = \frac{6}{5}(1 - e^{-5t/2})$ A   **(b)** It tends to $\frac{6}{5}$ A.

51. **(a)** $v = c\ln\dfrac{m_0}{m_0 - kt} - gt$ **(b)** 3044 m/s   53. **(a)** $h \approx (2 - 0.003979t)^2$ **(b)** 8.4 min

55. $v = \dfrac{50}{2t+1}$ cm/s, $x = 25\ln(2t+1)$ cm   57. $\dfrac{dy}{dx} = -\sin x + e^{-x^2}$, $y(0) = 1$

▶ Exercise Set 10.2 **(Page 597)**

1. 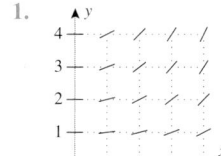   3. 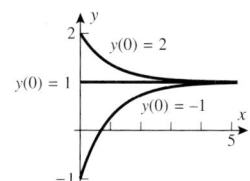   5. 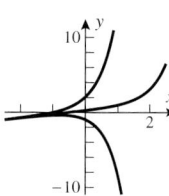   9. **(a)** IV
    (b) VI
    (c) V
    (d) II
    (e) I
    (f) III

11. **(a)**

n	0	1	2	3	4	5
x_n	0	0.2	0.4	0.6	0.8	1.0
y_n	1	1.20	1.48	1.86	2.35	2.98

(b) $y = -(x+1) + 2e^x$

x_n	0	0.2	0.4	0.6	0.8	1.0
$y(x_n)$	1	1.24	1.58	2.04	2.65	3.44
absolute error	0	0.04	0.10	0.19	0.30	0.46
percentage error	0	3	7	9	11	13

(c)

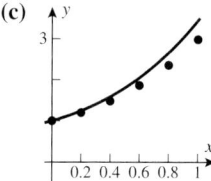

13.

n	0	1	2	3	4	5	6	7	8
x_n	0	0.5	1	1.5	2	2.5	3	3.5	4
y_n	1	1.50	2.11	2.84	3.68	4.64	5.72	6.91	8.23

15.

n	0	1	2	3	4
t_n	0	0.5	1	1.5	2
y_n	1	1.42	1.92	2.39	2.73

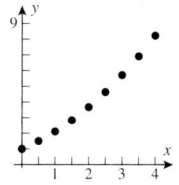

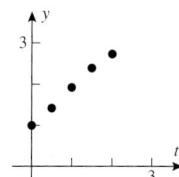

17. $y_5 \approx 1$

n	0	1	2	3	4	5
t_n	0	0.2	0.4	0.6	0.8	1.0
y_n	1.00	1.06	0.90	0.74	0.80	1.00

19. **(b)** $y(1/2) = \sqrt{3}/2$

▶ Exercise Set 10.3 **(Page 609)**

1. **(a)** $\dfrac{dy}{dt} = ky^2$, $y(0) = y_0 (k > 0)$ **(b)** $\dfrac{dy}{dt} = -ky^2$, $y(0) = y_0 (k > 0)$ 3. **(a)** $\dfrac{ds}{dt} = \dfrac{1}{2}s$ **(b)** $\dfrac{d^2s}{dt^2} = 2\dfrac{ds}{dt}$

5. (a) $\dfrac{dy}{dt} = 0.01y,\ y_0 = 10{,}000$ **(b)** $y = 10{,}000e^{t/100}$ **(c)** 69.31 h **(d)** 150.41 h

7. (a) $\dfrac{dy}{dt} = -ky,\ k \approx 0.1810$ **(b)** $y = 5.0 \times 10^7 e^{-0.181t}$ **(c)** 219,297 atoms **(d)** 12.72 days **9.** 196 days **11.** 3.30 days

13. (a) $y \approx 2e^{0.1386t}$ **(b)** $y = 5e^{0.015t}$ **(c)** $y \approx 0.5995e^{0.5117t}$ **(d)** $y \approx 0.8706e^{0.1386t}$ **17. (b)** 70 years **(c)** 20 years **(d)** 7%

21. $y_0 \approx 2,\ L \approx 8,\ k \approx 0.5493$ **23. (a)** $y_0 = 5$ **(b)** $L = 12$ **(c)** $k = 1$ **(d)** $t = 0.3365$ **(e)** $\dfrac{dy}{dt} = \dfrac{1}{12}y(12 - y),\ y(0) = 5$

25. (a) $L = 10$ **(b)** $k = 10$ **(c)** $y = 5$

27. Assume that $y(t)$ students have had the flu t days after semester break. Then $y(0) = 20,\ y(5) = 35$.

(a) $\dfrac{dy}{dt} = ky(1000 - y),\ y_0 = 20$

(b) $y = \dfrac{1000}{1 + 49e^{-1000kt}};\ k = 0.000115$

(c)

t	0	1	2	3	4	5	6	7	8	9	10	11	12	13	14
$y(t)$	20	22	25	28	31	35	39	44	49	54	61	67	75	83	93

(d)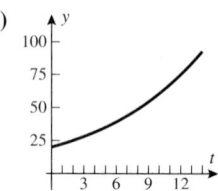

29. (a) $\dfrac{dT}{dt} = -k(T - 21),\ T(0) = 95$
$T = 21 + 74e^{-kt}$
(b) 6.22 min

33. (a) $y = 0.3\cos(t/2)$
(b) $T = 4\pi$ s, $f = 1/(4\pi)$ Hz
(c)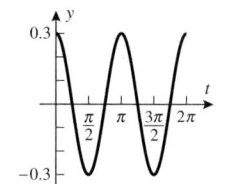
(d) $t = \pi$ s
(e) $t = 2\pi$ s

35. (a) $y = -0.12\cos 14t$
(b) $T = \pi/7$ s, $f = 7/\pi$ Hz
(c)

(d) $t = \pi/28$ s
(e) $t = \pi/14$ s

37. (a) Maximum speed occurs when $y = 0$.
(b) Minimum speed occurs when $y = \pm y_0$.

39. $mx''(t) + kx(t) = 0,\ x(0) = x_0,\ x'(0) = 0$ **41. (c)** $y = 4e^{t\ln 2}$ **(d)** $y = 4e^{-t\ln 2}$

▶ **Chapter 10 Supplementary Exercises (Page 612)**

5. (a) linear **(b)** both **(c)** separable **(d)** neither **7.** $y = L/2$ **9.** $r = 4 - t$ m

11. (a) $P = 4(1 - e^{-t/12{,}000})$ percent **(b)** 35.95 min **13.** $y^{-4} + 4\ln(x/y) = 1$ **15.** $y = \dfrac{1}{3 - 2\tan 2x}$

17. (a) $y = \left(-\dfrac{3}{10}x - \dfrac{3}{50}\right)\cos 3x + \left(-\dfrac{1}{10}x + \dfrac{2}{25}\right)\sin 3x + \dfrac{53}{50}e^x$ **19. (a)** no **21. (b)** $y = C_1 e^x + C_2 e^{-x}$
(b) $100\left|e^{ktr/(100+r)} - 1\right|$ percent **(c)** $y = e^x$
(c)

23. (a) 7.77 years **(b)** $\dfrac{dy}{dt} = k\left(1 - \dfrac{y}{95}\right)y,\ y(0) = 19$ **27. (a)** \$1491.82 **(b)** \$4493.29 **(c)** 8.7 years

---

▶ **Exercise Set 11.1 (Page 624)**

1. (a) $\dfrac{1}{3^{n-1}}$ **(b)** $\dfrac{(-1)^{n-1}}{3^{n-1}}$ **(c)** $\dfrac{2n-1}{2n}$ **(d)** $\dfrac{n^2}{\pi^{1/(n+1)}}$ **3. (a)** $2, 0, 2, 0$ **(b)** $1, -1, 1, -1$ **(c)** $2(1 + (-1)^n);\ 2 + 2\cos n\pi$

5. $\dfrac{1}{3}, \dfrac{2}{4}, \dfrac{3}{5}, \dfrac{4}{6}, \dfrac{5}{7}$; converges, $\displaystyle\lim_{n \to +\infty} \dfrac{n}{n+2} = 1$ **7.** $2, 2, 2, 2, 2$; converges, $\displaystyle\lim_{n \to +\infty} 2 = 2$

9. $\dfrac{\ln 1}{1}, \dfrac{\ln 2}{2}, \dfrac{\ln 3}{3}, \dfrac{\ln 4}{4}, \dfrac{\ln 5}{5}$; converges, $\displaystyle\lim_{n \to +\infty} \dfrac{\ln n}{n} = 0$ **11.** $0, 2, 0, 2, 0$; diverges

13. $-1, \frac{16}{9}, -\frac{54}{28}, \frac{128}{65}, -\frac{250}{126}$; diverges   15. $\frac{6}{2}, \frac{12}{8}, \frac{20}{18}, \frac{30}{32}, \frac{42}{50}$; converges, $\lim\limits_{n \to +\infty} \frac{1}{2}\left(1+\frac{1}{n}\right)\left(1+\frac{2}{n}\right) = \frac{1}{2}$

17. $\cos 3, \cos\frac{3}{2}, \cos 1, \cos\frac{3}{4}, \cos\frac{3}{5}$; converges, $\lim\limits_{n \to +\infty} \cos(3/n) = 1$

19. $e^{-1}, 4e^{-2}, 9e^{-3}, 16e^{-4}, 25e^{-5}$; converges, $\lim\limits_{n \to +\infty} n^2 e^{-n} = 0$ 21. $2, \left(\frac{5}{3}\right)^2, \left(\frac{6}{4}\right)^3, \left(\frac{7}{5}\right)^4, \left(\frac{8}{6}\right)^5$; converges, $\lim\limits_{n \to +\infty}\left[\frac{n+3}{n+1}\right]^n = e^2$

23. $\left\{\frac{2n-1}{2n}\right\}_{n=1}^{+\infty}$; converges, $\lim\limits_{n \to +\infty} \frac{2n-1}{2n} = 1$ 25. $\left\{\frac{1}{3^n}\right\}_{n=1}^{+\infty}$; converges, $\lim\limits_{n \to +\infty} \frac{1}{3^n} = 0$

27. $\left\{\frac{1}{n} - \frac{1}{n+1}\right\}_{n=1}^{+\infty}$; converges, $\lim\limits_{n \to +\infty}\left(\frac{1}{n} - \frac{1}{n+1}\right) = 0$ 29. $\{\sqrt{n+1} - \sqrt{n+2}\}_{n=1}^{+\infty}$; converges, $\lim\limits_{n \to +\infty}(\sqrt{n+1} - \sqrt{n+2}) = 0$

33. (a) $1, 2, 1, 4, 1, 6$ (b) $a_n = \begin{cases} n, & n \text{ odd} \\ 1/2^n, & n \text{ even} \end{cases}$ (c) $a_n = \begin{cases} 1/n, & n \text{ odd} \\ 1/(n+1), & n \text{ even} \end{cases}$ (d) (a) diverges; (b) diverges; (c) $\lim\limits_{n \to +\infty} a_n = 0$

37. (a) $1, \frac{3}{4}, \frac{2}{3}, \frac{5}{8}$ 41. (a) $(0.5)^{2^n}$ 43. (a)   45. converges to 0   47. (a) $N = 3$
   (d) $-1 \le a_0 \le 1$ (b) $N = 11$
                                                                                       (c) $N = 1001$

(b) $\lim\limits_{n \to +\infty}(2^n + 3^n)^{1/n} = 3$

▶ **Exercise Set 11.2 (Page 631)**

1. strictly decreasing   3. strictly increasing   5. strictly decreasing   7. strictly increasing   9. strictly decreasing

11. strictly increasing   13. strictly increasing   15. strictly decreasing   17. strictly decreasing

19. eventually strictly increasing   21. eventually strictly decreasing   23. eventually strictly increasing

25. (a) Yes; the limit lies in the interval $[1, 2]$.   (b) No, but if so, then the limit is ≤ 2.   27. $\sqrt{2}, \sqrt{2 + \sqrt{2}}, \sqrt{2 + \sqrt{2 + \sqrt{2}}}$

▶ **Exercise Set 11.3 (Page 638)**

1. (a) $2, \frac{12}{5}, \frac{62}{25}, \frac{312}{125}, \frac{5}{2}\left(1 - \left(\frac{1}{5}\right)^n\right)$, $\lim\limits_{n \to +\infty} s_n = \frac{5}{2}$, converges   (b) $\frac{1}{4}, \frac{3}{4}, \frac{7}{4}, \frac{15}{4}, -\frac{1}{4}(1 - 2^n)$, $\lim\limits_{n \to +\infty} s_n = +\infty$, diverges

   (c) $\frac{1}{6}, \frac{1}{4}, \frac{3}{10}, \frac{1}{3}, \frac{1}{2} - \frac{1}{n+2}$, $\lim\limits_{n \to +\infty} s_n = \frac{1}{2}$, converges   3. $\frac{4}{7}$ 5. 6 7. $\lim\limits_{n \to +\infty} s_n = \frac{1}{3}$

9. $\lim\limits_{n \to +\infty} s_n = \frac{1}{6}$ 11. diverges   13. $\frac{448}{3}$ 15. $\frac{4}{9}$ 17. $\frac{532}{99}$ 19. $\frac{869}{1111}$ 23. 70 m

25. (a) $s_n = -\ln(n+1)$, $\lim\limits_{n \to +\infty} s_n = -\infty$, diverges   (b) $s_n = \sum\limits_{k=2}^{n+1}\left[\ln\frac{k-1}{k} - \ln\frac{k}{k+1}\right]$, $\lim\limits_{n \to +\infty} s_n = -\ln 2$

27. (a) converges for $|x| < 1$; $S = \frac{x}{1 + x^2}$ (b) converges for $|x| > 2$; $S = \frac{1}{x^2 - 2x}$ (c) converges for $x > 0$; $S = \frac{1}{e^x - 1}$

33. $a_n = \frac{1}{2^{n-1}}a_1 + \frac{1}{2^{n-1}} + \frac{1}{2^{n-2}} + \cdots + \frac{1}{2}$, $\lim\limits_{n \to +\infty} a_n = 1$ 35. The series converges only if $-1 < x < 1$.

39. (b) $A = 1, B = -2$ (c) $s_n = 2 - \frac{2^{n+1}}{3^{n+1} - 2^{n+1}}$, $\lim\limits_{n \to +\infty} s_n = \lim\limits_{n \to +\infty}\left[2 - \frac{(2/3)^{n+1}}{1 - (2/3)^{n+1}}\right] = 2$

▶ **Exercise Set 11.4 (Page 645)**

1. (a) $\frac{4}{3}$ (b) $-\frac{3}{4}$ 3. (a) $p = 3$, converges   (b) $p = \frac{1}{2}$, diverges   (c) $p = 1$, diverges   (d) $p = \frac{2}{3}$, diverges

5. (a) diverges   (b) diverges   (c) diverges   (d) no information   7. (a) diverges   (b) converges   9. diverges

11. diverges   13. diverges   15. diverges   17. diverges   19. converges   21. diverges   23. converges

25. converges for $p > 1$ 27. (a) $\left(\frac{\pi^2}{2}\right) - \left(\frac{\pi^4}{90}\right)$ (b) $\left(\frac{\pi^2}{6}\right) - \left(\frac{5}{4}\right)$ (c) $\pi^4/90$ 29. (a) diverge   (b) diverges   (c) converges

31. (c) $\frac{1}{11} < \frac{1}{6}\pi^2 - s_{10} < \frac{1}{10}$ 33. (b) $n = 5$ (c) $S \approx 1.203$ 35. (b) $13 < s_{1,000,000} < 15$ (d) $n > 2.69 \times 10^{43}$ 37. converges

---

▶ **Exercise Set 11.5 (Page 655)**

1. (a) $1 - x + \frac{1}{2}x^2, 1 - x$ (b) $1 - \frac{1}{2}x^2, 1$ (c) $1 - \frac{1}{2}(x - \pi/2)^2, 1$ (d) $1 + \frac{1}{2}(x - 1) - \frac{1}{8}(x - 1)^2, 1 + \frac{1}{2}(x - 1)$

3. **(a)** $1 + \frac{1}{2}(x-1) - \frac{1}{8}(x-1)^2$ **(b)** 1.04875   **5.** 1.80397443

7. $p_0(x) = 1$, $p_1(x) = 1 - x$, $p_2(x) = 1 - x + \frac{1}{2}x^2$,

$p_3(x) = 1 - x + \frac{1}{2}x^2 - \frac{1}{3!}x^3$, $p_4(x) = 1 - x + \frac{1}{2}x^2 - \frac{1}{3!}x^3 + \frac{1}{4!}x^4$; $\sum\limits_{k=0}^{\infty} \frac{(-1)^k}{k!}x^k$

9. $p_0(x) = 1$, $p_1(x) = 1$, $p_2(x) = 1 - \frac{\pi^2}{2!}x^2$; $p_3(x) = 1 - \frac{\pi^2}{2!}x^2$, $p_4(x) = 1 - \frac{\pi^2}{2!}x^2 + \frac{\pi^4}{4!}x^4$; $\sum\limits_{k=0}^{\infty} \frac{(-1)^k \pi^{2k}}{(2k)!}x^{2k}$

11. $p_0(x) = 0$, $p_1(x) = x$, $p_2(x) = x - \frac{1}{2}x^2$, $p_3(x) = x - \frac{1}{2}x^2 + \frac{1}{3}x^3$, $p_4(x) = x - \frac{1}{2}x^2 + \frac{1}{3}x^3 - \frac{1}{4}x^4$; $\sum\limits_{k=1}^{\infty} \frac{(-1)^{k+1}}{k}x^k$

13. $p_0(x) = 1$, $p_1(x) = 1$, $p_2(x) = 1 + \frac{x^2}{2}$, $p_3(x) = 1 + \frac{x^2}{2}$, $p_4(x) = 1 + \frac{x^2}{2} + \frac{x^4}{4!}$; $\sum\limits_{k=0}^{\infty} \frac{1}{(2k)!}x^{2k}$

15. $p_0(x) = 0$, $p_1(x) = 0$, $p_2(x) = x^2$, $p_3(x) = x^2$, $p_4(x) = x^2 - \frac{1}{6}x^4$; $\sum\limits_{k=0}^{\infty} \frac{(-1)^k}{(2k+1)!}x^{2k+2}$ **17.** **(a)** $1 + 2x - x^2 + x^3$

(b) $c_0 + c_1 x + c_2 x^2 + \cdots + c_n x^n$

19. $p_0(x) = e$, $p_1(x) = e + e(x-1)$, $p_2(x) = e + e(x-1) + \frac{e}{2}(x-1)^2$, $p_3(x) = e + e(x-1) + \frac{e}{2}(x-1)^2 + \frac{e}{3!}(x-1)^3$,

$p_4(x) = e + e(x-1) + \frac{e}{2}(x-1)^2 + \frac{e}{3!}(x-1)^3 + \frac{e}{4!}(x-1)^4$; $\sum\limits_{k=0}^{\infty} \frac{e}{k!}(x-1)^k$

21. $p_0(x) = -1$; $p_1(x) = -1 - (x+1)$; $p_2(x) = -1 - (x+1) - (x+1)^2$; $p_3(x) = -1 - (x+1) - (x+1)^2 - (x+1)^3$;

$p_4(x) = -1 - (x+1) - (x+1)^2 - (x+1)^3 - (x+1)^4$; $\sum\limits_{k=0}^{\infty}(-1)(x+1)^k$

23. $p_0(x) = p_1(x) = 1$, $p_2(x) = p_3(x) = 1 - \frac{\pi^2}{2}\left(x - \frac{1}{2}\right)^2$,

$p_4(x) = 1 - \frac{\pi^2}{2}\left(x - \frac{1}{2}\right)^2 + \frac{\pi^4}{4!}\left(x - \frac{1}{2}\right)^4$; $\sum\limits_{k=0}^{\infty} \frac{(-1)^k \pi^{2k}}{(2k)!}\left(x - \frac{1}{2}\right)^{2k}$

25. $p_0(x) = 0$, $p_1(x) = (x-1)$; $p_2(x) = (x-1) - \frac{1}{2}(x-1)^2$; $p_3(x) = (x-1) - \frac{1}{2}(x-1)^2 + \frac{1}{3}(x-1)^3$,

$p_4(x) = (x-1) - \frac{1}{2}(x-1)^2 + \frac{1}{3}(x-1)^3 - \frac{1}{4}(x-1)^4$; $\sum\limits_{k=1}^{\infty} \frac{(-1)^{k-1}}{k}(x-1)^k$

27. **(a)** $1 + 2(x-1) - (x-1)^2 + (x-1)^3$ **(b)** $c_0 + c_1(x - x_0) + c_2(x - x_0)^2 + \cdots + c_n(x - x_0)^n$

29. $p_0(x) = 1$, $p_1(x) = 1 - 2x$,

$p_2(x) = 1 - 2x + 2x^2$, $p_3(x) = 1 - 2x + 2x^2 - \frac{4}{3}x^3$

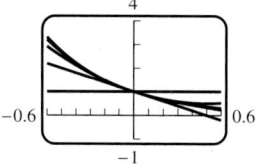

31. $p_0(x) = -1$, $p_2(x) = -1 + \frac{1}{2}(x - \pi)^2$,

$p_4(x) = -1 + \frac{1}{2}(x - \pi)^2 - \frac{1}{24}(x - \pi)^4$,

$p_6(x) = -1 + \frac{1}{2}(x - \pi)^2 - \frac{1}{24}(x - \pi)^4 + \frac{1}{720}(x - \pi)^6$

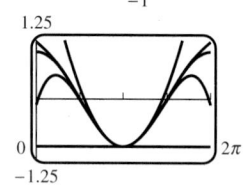

33. IV

37. **(a)**

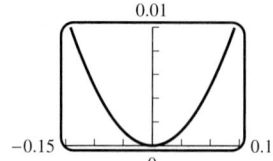

(b)

x	−1.000	−0.750	−0.500	−0.250	0.000	0.250	0.500	0.750	1.000
$f(x)$	0.431	0.506	0.619	0.781	1.000	1.281	1.615	1.977	2.320
$p_1(x)$	0.000	0.250	0.500	0.750	1.000	1.250	1.500	1.750	2.000
$p_2(x)$	0.500	0.531	0.625	0.781	1.000	1.281	1.625	2.031	2.500

(c) $|e^{\sin x} - (1 + x)| < 0.01$ for $-0.14 < x < 0.14$ **(d)** $\left| e^{\sin x} - \left(1 + x + \frac{x^2}{2}\right)\right| < 0.01$ for $-0.50 < x < 0.50$

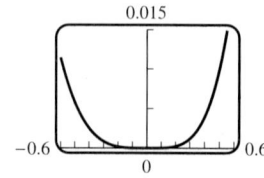

▶ Exercise Set 11.6 **(Page 661)**

1. **(a)** converges   **(b)** diverges   3. **(a)** converges   **(b)** converges   5. converges   7. converges   9. diverges

11. converges   13. inconclusive   15. diverges   17. diverges   19. converges   21. converges   23. converges

25. converges   27. converges   29. diverges   31. converges   33. diverges   35. converges   37. converges

39. diverges   41. converges   43. converges   45. $u_k = \dfrac{k!}{1 \cdot 3 \cdot 5 \cdots (2k-1)}$, $\rho = \lim\limits_{k \to +\infty} \dfrac{k+1}{2k+1} = \dfrac{1}{2}$; converges

47. converges   51. **(a)** converges   **(b)** diverges

▶ Exercise Set 11.7 **(Page 669)**

3. diverges   5. converges   7. converges absolutely   9. diverges   11. converges absolutely   13. conditionally convergent

15. divergent   17. conditionally convergent   19. conditionally convergent   21. divergent   23. conditionally convergent

25. absolutely convergent   27. conditionally convergent   29. absolutely convergent   31. |error| < 0.125   33. |error| < 0.1

35. $n = 9999$ 37. $n = 39{,}999$ 39. |error| < 0.00074; $s_{10} \approx 0.4995$; $S = 0.5$ 41. 0.84   43. 0.41

45. **(c)** $n = 50$ 51. **(a)** $124.58 < d < 124.77$ **(b)** $1243 < s < 1424$

▶ Exercise Set 11.8 **(Page 675)**

1. $-1 < x < 1$, $\dfrac{1}{1+x}$ 3. $1 < x < 3$, $\dfrac{1}{3-x}$ 5. **(a)** $-2 < x < 2$ **(b)** $f(0) = 1$; $f(1) = \frac{2}{3}$ 7. $R = 1, [-1, 1)$

9. $R = +\infty, (-\infty, +\infty)$ 11. $R = \frac{1}{5}, [-\frac{1}{5}, \frac{1}{5}]$ 13. $R = 1, [-1, 1]$ 15. $R = 1, (-1, 1]$ 17. $R = +\infty, (-\infty, +\infty)$

19. $R = +\infty, (-\infty, +\infty)$ 21. $R = 1, [-1, 1]$ 23. $R = 1, (-2, 0]$ 25. $R = \frac{4}{3}, \left(-\frac{19}{3}, -\frac{11}{3}\right)$ 27. $R = 1, [-2, 0]$

29. $R = +\infty, (-\infty, +\infty)$ 31. $(-\infty, +\infty)$ 33. 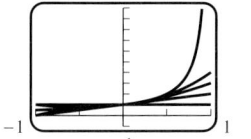   37. **(a)** radius = R
   (b) radius = R
   (c) radius $\geq \min(R_1, R_2)$

▶ Exercise Set 11.9 **(Page 684)**

1. 0.069756   3. 1.64872   5. 0.995004   7. 0.99619   9. 0.5208   13. **(a)** $\displaystyle\sum_{k=1}^{\infty} 2\frac{(1/9)^{2k-1}}{2k-1}$ **(b)** 0.2231

15. **(a)** 0.4635, 0.3218   17. **(a)** $(-0.569, 0.569)$ 19. **(a)** $|R_5(x)| < 9 \times 10^{-8}$
   (b) 3.1412
   (c) no

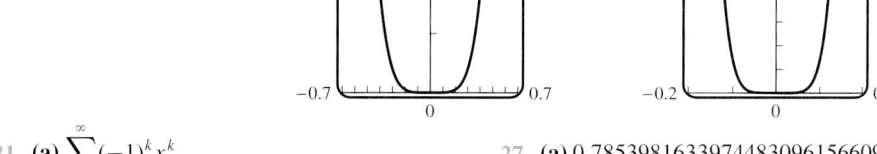

21. **(a)** $\displaystyle\sum_{k=0}^{\infty} (-1)^k x^k$ 27. **(a)** 0.785398163397448309615 6609
   (b) $1 + \dfrac{x}{3} + \displaystyle\sum_{k=2}^{\infty} (-1)^{k-1} \dfrac{2 \cdot 5 \cdots (3k-4)}{3^k k!} x^k$
   (c) $\displaystyle\sum_{k=0}^{\infty} (-1)^k \dfrac{(k+2)(k+1)}{2} x^k$

   (b)

n	s_n
0	0.3183098 78 . . .
1	0.3183098 861837906 067 . . .
2	0.3183098 861837906 7153776 695 . . .
3	0.3183098 861837906 7153776 752674502 34 . . .
$1/\pi$	0.3183098 861837906 7153776 752674502 87 . . .

▶ Exercise Set 11.10 (Page 693)

1. (a) $1 - x + x^2 - \cdots + (-1)^k x^k + \cdots$; $R = 1$

(b) $1 + x^2 + x^4 + \cdots + x^{2k} + \cdots$; $R = 1$

(c) $1 + 2x + 4x^2 + \cdots + 2^k x^k + \cdots$; $R = \frac{1}{2}$

(d) $\dfrac{1}{2} + \dfrac{1}{2^2}x + \dfrac{1}{2^3}x^2 + \cdots + \dfrac{1}{2^{k+1}}x^k + \cdots$; $R = 2$

3. (a) $(2+x)^{-1/2} = \dfrac{1}{2^{1/2}} - \dfrac{1}{2^{5/2}}x + \dfrac{1\cdot 3}{2^{9/2}\cdot 2!}x^2 - \dfrac{1\cdot 3\cdot 5}{2^{13/2}\cdot 3!}x^3 + \cdots$

(b) $(1 - x^2)^{-2} = 1 + 2x^2 + 3x^4 + 4x^6 + \cdots$

5. (a) $2x - \dfrac{2^3}{3!}x^3 + \dfrac{2^5}{5!}x^5 - \dfrac{2^7}{7!}x^7 + \cdots$; $R = +\infty$

(b) $1 - 2x + 2x^2 - \dfrac{4}{3}x^3 + \cdots$; $R = +\infty$

(c) $1 + x^2 + \dfrac{1}{2!}x^4 + \dfrac{1}{3!}x^6 + \cdots$; $R = +\infty$

(d) $x^2 - \dfrac{\pi^2}{2}x^4 + \dfrac{\pi^4}{4!}x^6 - \dfrac{\pi^6}{6!}x^8 + \cdots$; $R = +\infty$

7. (a) $x^2 - 3x^3 + 9x^4 - 27x^5 + \cdots$; $R = \frac{1}{3}$

(b) $2x^2 + \dfrac{2^3}{3!}x^4 + \dfrac{2^5}{5!}x^6 + \dfrac{2^7}{7!}x^8 + \cdots$; $R = +\infty$

(c) $x - \dfrac{3}{2}x^3 + \dfrac{3}{8}x^5 + \dfrac{1}{16}x^7 + \cdots$; $R = 1$

9. (a) $x^2 - \dfrac{2^3}{4!}x^4 + \dfrac{2^5}{6!}x^6 - \dfrac{2^7}{8!}x^8 + \cdots$ **(b)** $12x^3 - 6x^6 + 4x^9 - 3x^{12} + \cdots$

11. (a) $1 - (x-1) + (x-1)^2 - \cdots + (-1)^k(x-1)^k + \cdots$ **(b)** $(0, 2)$

13. (a) $x + x^2 + \dfrac{x^3}{3} - \dfrac{x^5}{30} + \cdots$ **(b)** $x - \dfrac{x^3}{24} + \dfrac{x^4}{24} - \dfrac{71}{1920}x^5 + \cdots$

15. (a) $1 + \frac{1}{2}x^2 + \frac{5}{24}x^4 + \frac{61}{720}x^6 + \cdots$ **(b)** $x - x^2 + \frac{1}{3}x^3 - \frac{1}{30}x^5 + \cdots$ **19.** $2 - 4x + 2x^2 - 4x^3 + 2x^4 + \cdots$

25. (a) $\displaystyle\sum_{k=0}^{\infty} x^{2k+1}$ **(b)** $f^{(5)}(0) = 5$, $f^{(6)}(0) = 0$ **(c)** $f^{(n)}(0) = n!c_n = \begin{cases} n! & \text{if } n \text{ odd} \\ 0 & \text{if } n \text{ even} \end{cases}$ **27. (a)** 1 **(b)** $-\frac{1}{3}$ **29.** 0.3103

31. 0.200 **35. (a)** $x - \frac{1}{6}x^3 + \frac{3}{40}x^5 - \frac{5}{112}x^7 + \cdots$ **(b)** $x + \displaystyle\sum_{k=1}^{\infty}(-1)^k \dfrac{1\cdot 3\cdot 5\cdots(2k-1)}{2^k k!(2k+1)}x^{2k+1}$ **(c)** $R = 1$

37. (a) $y(t) = y_0 \displaystyle\sum_{k=0}^{\infty} \dfrac{(-1)^k(0.000121)^k t^k}{k!}$ **(c)** $0.9998790073\,y_0$ **39. (a)** $T \approx 2.00709$ **(b)** $T \approx 2.008044621$ **(c)** 2.008045644

41. (a) $F = mg\left(1 - \dfrac{2h}{R} + \dfrac{3h^2}{R^2} - \dfrac{4h^3}{R^3} + \cdots\right)$ **(d)** about 0.27% less

▶ Chapter 11 Supplementary Exercises (Page 696)

9. (a) true **(b)** sometimes false **(c)** sometimes false **(d)** true **(e)** sometimes false **(f)** sometimes false **(g)** false

(h) sometimes false **(i)** true **(j)** true **(k)** sometimes false **(l)** sometimes false **11. (a)** converges **(b)** converges **(c)** diverges

13. (a) converges **(b)** diverges **(c)** converges **15.** $\dfrac{1}{4\cdot 5^{99}}$

17. (a) $p_0(x) = 1$, $p_1(x) = 1 - 7x$, $p_2(x) = 1 - 7x + 5x^2$, $p_3(x) = 1 - 7x + 5x^2 - 4x^3$, $p_4(x) = 1 - 7x + 5x^2 - 4x^3$

21. (a) converges **(b)** diverges **23. (a)** $u_{100} = \dfrac{1}{9900}$ **(b)** 0 **(c)** 2 **25. (a)** $e^2 - 1$ **(b)** 0 **(c)** $\cos e$ **(d)** $\frac{1}{3}$

27. 22.07% **31. (a)** $x + \frac{1}{2}x^2 + \frac{3}{14}x^3 + \frac{3}{35}x^4$; $R = 3$ **(b)** $-x^3 + \frac{2}{3}x^5 - \frac{2}{5}x^7 + \frac{8}{35}x^9$; $R = \sqrt{2}$

▶ Exercise Set 12.1 (Page 710)

1.

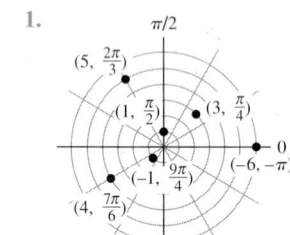

3. (a) $(3\sqrt{3}, 3)$

(b) $(-7/2, 7\sqrt{3}/2)$

(c) $(3\sqrt{3}, 3)$

(d) $(0, 0)$

(e) $(-7\sqrt{3}/2, 7/2)$

(f) $(-5, 0)$

5. (a) both $(5, \pi)$

(b) $(4, 11\pi/6)$, $(4, -\pi/6)$

(c) $(2, 3\pi/2)$, $(2, -\pi/2)$

(d) $(8\sqrt{2}, 5\pi/4)$, $(8\sqrt{2}, -3\pi/4)$

(d) both $(6, 2\pi/3)$

(d) both $(\sqrt{2}, \pi/4)$

7. (a) $(5, 0.6435)$

(b) $(\sqrt{29}, 5.0929)$

(c) $(1.2716, 0.6658)$

9. **(a)** circle
   (b) line
   (c) circle
   (d) line

11. **(a)** $r \cos \theta = 7$
    (b) $r = 3$
    (c) $r = 6 \sin \theta$
    (d) $r^2 \sin 2\theta = 9/2$

13.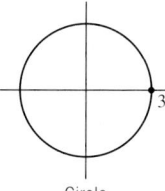

15.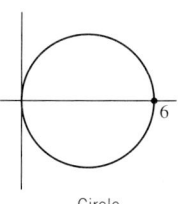

17. **(a)** $r = 5$
    (b) $r = 6 \cos \theta$
    (c) $r = 1 - \cos \theta$

19. **(a)** $r = 3 \sin 2\theta$ **(b)** $r = 3 + 2 \sin \theta$ **(c)** $r^2 = 9 \cos 2\theta$

21.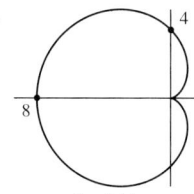
Line

23.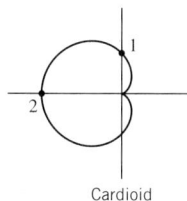
Circle

25.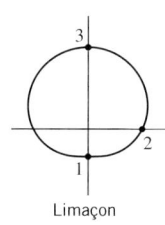
Circle

27.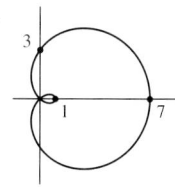
Circle

29.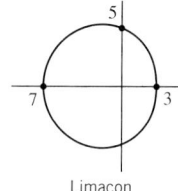
Cardioid

31.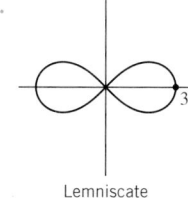
Cardioid

33.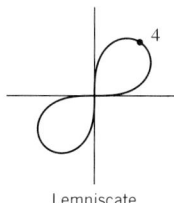
Cardioid

35.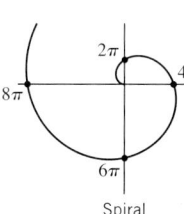
Limaçon

37.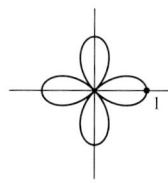
Limaçon

39.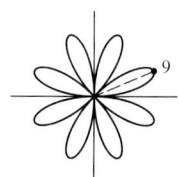
Limaçon

41.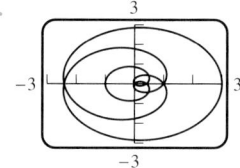
Lemniscate

43.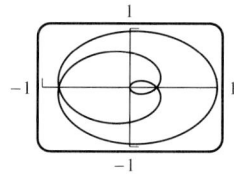
Lemniscate

45.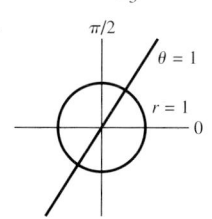
Spiral

47.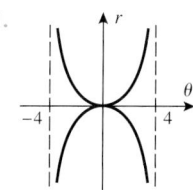
Four-petal rose

49.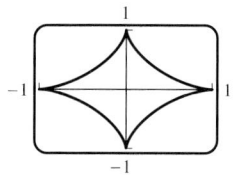
Eight-petal rose

53.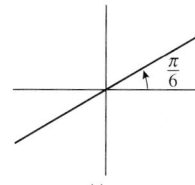

55.

57. $-4\pi < \theta < 4\pi$

61. **(a)** $r = 1 + \dfrac{\sqrt{2}}{2}(\cos \theta + \sin \theta)$
    (b) $r = 1 + \sin \theta$
    (c) $r = 1 - \cos \theta$
    (d) $r = 1 - \dfrac{\sqrt{2}}{2}(\cos \theta + \sin \theta)$

63.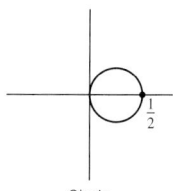
$\theta = 1$
$r = 1$

65. $(3/2, \pi/3)$

67. **(c)** $\sqrt{13 - 6\sqrt{3}} \approx 1.615$ **(d)** $A = 1$

71.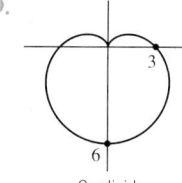

▶ Exercise Set 12.2 **(Page 718)**

1. **(a)** $-1/4$, $1/4$
3. positive when $t = -1$, negative when $t = 1$
5. $4, 4$
7. $2/\sqrt{3}$, $-1/(3\sqrt{3})$
9. $\dfrac{-1}{4 - \sqrt{3}}$, $\dfrac{8}{(4 - \sqrt{3})^3}$

11. **(a)** $y = -e^{-2}x + 2e^{-1}$
13. **(a)** $t = \pi/2 + n\pi$ for $n = 0, \pm 1, \cdots$
    (b) $t = n\pi$ for $n = 0, \pm 1, \cdots$
15. $y = -2x$, $y = 2x$
19.

21. $1/\sqrt{3}$ **23.** $\dfrac{\tan 2 - 2}{2\tan 2 + 1}$ **25.** -2 **27.** $1, 0, -1$ **29.** horizontal: $(3a/2, \pi/3), (0, \pi), (3a/2, 5\pi/3)$;
vertical: $(2a, 0), (a/2, 2\pi/3), (a/2, 4\pi/3)$

31. $(0, 0), (\sqrt{2}/4, \pi/4), (\sqrt{2}/4, 3\pi/4)$ **33.**

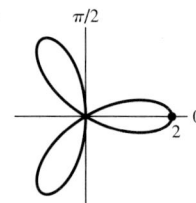

$\theta_0 = \pi/6, \pi/2, 5\pi/6$

35.

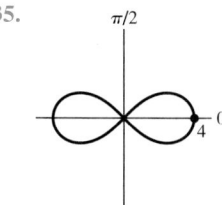

$\theta_0 = \pm\pi/4$

37.

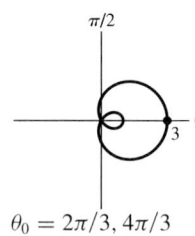

$\theta_0 = 2\pi/3, 4\pi/3$

39. $L = 2\pi a$ **41.** $L = 8a$ **43.** $L = \sqrt{10}(e^6 - 1)/3$ **45. (a)** $\dfrac{dy}{dx} = \dfrac{3\sin t}{1 - 3\cos t}$
(b) $\theta = -0.4344$

47. (b) ≈ 2.42 **(c)**

n	2	3	4	5	6	7	8	9	10	11
L	2.42211	2.22748	2.14461	2.10100	2.07501	2.05816	2.04656	2.03821	2.03199	2.02721

n	12	13	14	15	16	17	18	19	20
L	2.02346	2.02046	2.01802	2.01600	2.01431	2.01288	2.01167	2.01062	2.00971

49. $S = \dfrac{8\pi}{3}(17\sqrt{17} - 1)$

51. $S = \sqrt{2}\pi$ **55. (a)** $r = 2\theta + 10$
(b) 75.7 mm

▶ **Exercise Set 12.3. (Page 724)**

1. (a) $\displaystyle\int_{\pi/2}^{\pi} \dfrac{1}{2}(1 - \cos\theta)^2\, d\theta$ **(d)** $\displaystyle\int_0^{2\pi} \dfrac{1}{2}\theta^2\, d\theta$ **3. (a)** πa^2 **5.** 6π **7.** 4π **9.** $\pi - 3\sqrt{3}/2$ **11.** $\pi/2 - \frac{1}{4}$
(b) $\displaystyle\int_0^{\pi/2} \dfrac{1}{2}4\cos^2\theta\, d\theta$ **(e)** $\displaystyle\int_{-\pi/2}^{\pi/2} \dfrac{1}{2}(1 - \sin\theta)^2\, d\theta$ **(b)** πa^2
(c) $\displaystyle\int_0^{\pi/2} \dfrac{1}{2}\sin^2 2\theta\, d\theta$ **(f)** $\displaystyle\int_{\pi}^{\pi/3} \dfrac{1}{2}(1 + \cos 3\theta)^2\, d\theta$ **(c)** πa^2

13. $10\pi/3 - 4\sqrt{3}$ **15.** $8\pi/3 + \sqrt{3}$ **17.** $9\sqrt{3}/2 - \pi$ **19.** $(\pi + 3\sqrt{3})/4$ **21.** $100\cos^{-1}(3/5) - 48$ **23. (b)** a^2 **(c)** $2\sqrt{3} - \dfrac{2\pi}{3}$
25. $8\pi^3 a^2$ **27.** π^2 **29.** $32\pi/5$ **35.** $\pi/16$

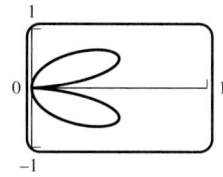

▶ **Exercise Set 12.4 (Page 739)**

1. (a) $x = y^2$ **(b)** $-3y = x^2$ **(c)** $\dfrac{x^2}{9} + \dfrac{y^2}{4} = 1$ **(d)** $\dfrac{x^2}{4} + \dfrac{y^2}{9} = 1$ **(e)** $y^2 - x^2 = 1$ **(f)** $\dfrac{x^2}{4} - \dfrac{y^2}{4} = 1$

3. (a)

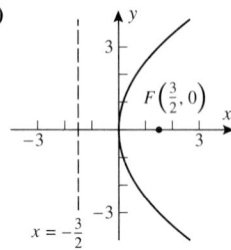

(b)

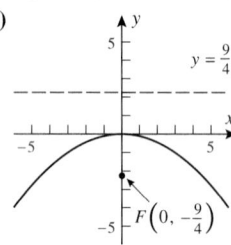

5. (a)

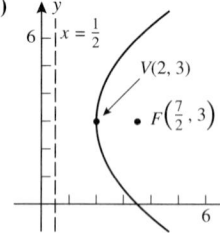

(b)

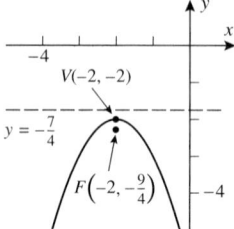

7. (a)

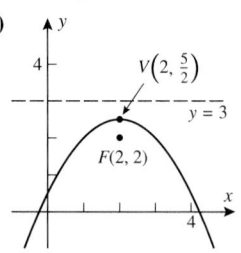

(b)

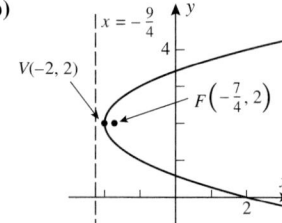

9. (a)

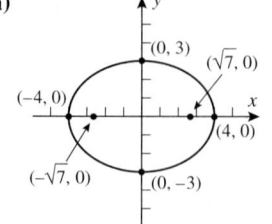

(b)

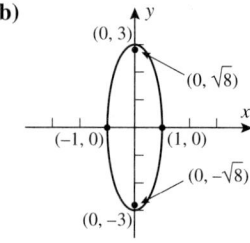

11. (a) 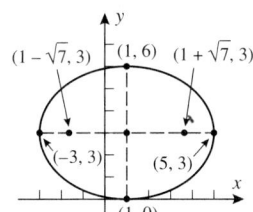 **(b)** 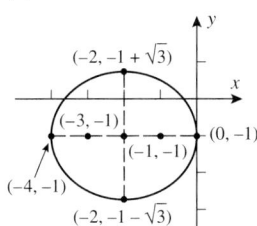   **13. (a)** 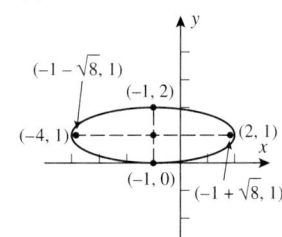 **(b)**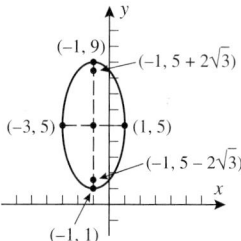

15. (a) 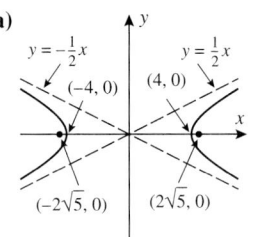 **(b)** 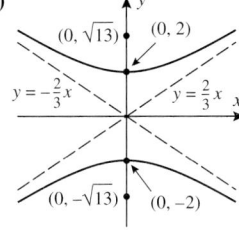   **17. (a)** 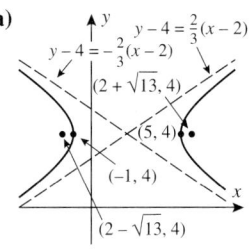 **(b)**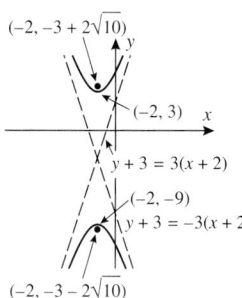

19. (a) 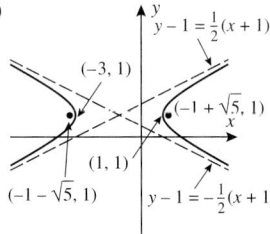 **(b)** 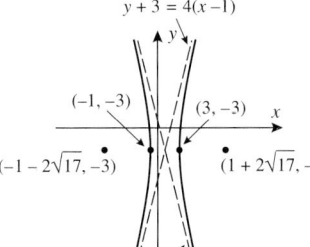   **21. (a)** $y^2 = 12x$ **23. (a)** $x^2 = -12y$
(b) $y^2 = -28x$ **(b)** $(x-1)^2 = 12(y-1)$

25. $y^2 = -5\left(x - \frac{19}{5}\right)$ **27. (a)** $\frac{1}{9}x^2 + \frac{1}{4}y^2 = 1$ **(b)** $\frac{1}{169}x^2 + \frac{1}{144}y^2 = 1$ **29. (a)** $\frac{1}{3}x^2 + \frac{1}{2}y^2 = 1$ **(b)** $\frac{1}{4}x^2 + \frac{1}{16}y^2 = 1$

31. (a) $\frac{1}{36}x^2 + \frac{8}{81}y^2 = 1$ **(b)** $(x-1)^2 + \frac{1}{2}(y-3)^2 = 1$ **33. (a)** $\frac{1}{4}x^2 - \frac{1}{5}y^2 = 1$ **(b)** $x^2 - \frac{1}{4}y^2 = 1$

35. (a) $\frac{9}{64}x^2 - \frac{1}{16}y^2 = 1$, $\frac{1}{36}y^2 - \frac{1}{16}x^2 = 1$ **(b)** $\frac{1}{20}y^2 - \frac{1}{5}x^2 = 1$

37. (a) $\frac{1}{16}(x-6)^2 - \frac{1}{9}(y-4)^2 = 1$ **39. (a)** 16 ft **(b)** $8\sqrt{3}$ ft   **43.** $\frac{1}{16}$ ft   **45. (a)** $P : (b\cos t, b\sin t)$;
(b) $\frac{1}{3}(y-2)^2 - \frac{4}{3}\left(x - \frac{1}{2}\right)^2 = 1$ $Q : (a\cos t, a\sin t)$;
    $R : (a\cos t, b\sin t)$

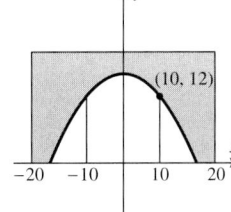

49. $\frac{1}{32}(x-4)^2 + \frac{1}{36}(y-3)^2 = 1$ **51.** 96   **61.** $L = D\sqrt{1 + p^2},\ T = \frac{1}{2}pD$ **67.** $\left(\pm\frac{3}{\sqrt{5}}, \frac{4}{\sqrt{5}}\right), \left(\pm\frac{3}{\sqrt{5}}, -\frac{4}{\sqrt{5}}\right)$

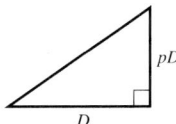

71. (b) 14.30465, 24, 33.69535 in

73. (a) $(x-1)^2 - 5(y+1)^2 = 5$, hyperbola **(b)** $x^2 - 3(y+1)^2 = 0$, $x = \pm\sqrt{3}(y+1)$, two lines
(c) $4(x+2)^2 + 8(y+1)^2 = 4$, ellipse **(d)** $3(x+2)^2 + (y+1)^2 = 0$, the point $(-2, -1)$ (degenerate case)
(e) $(x+4)^2 + 2y = 2$, parabola **(f)** $5(x+4)^2 + 2y = -14$, parabola

▶ **Exercise Set 12.5 (Page 750)**

1. **(a)** $e = 1, d = \frac{3}{2}$ **(b)** $e = \frac{1}{2}, d = 3$ **(c)** $e = \frac{3}{2}, d = \frac{4}{3}$ **(d)** $e = 1, d = \frac{5}{3}$

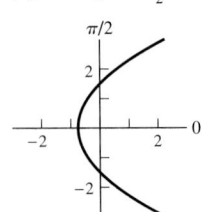 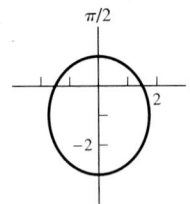 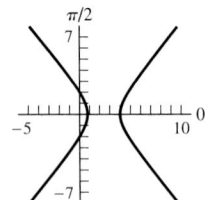 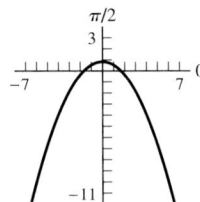

3. **(a)** parabola, opens up   **(b)** ellipse, directrix above the pole   **(c)** hyperbola, directrix below the pole   **(d)** ellipse, directrix to the right of the pole

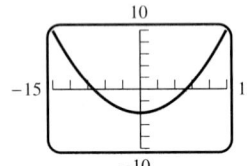 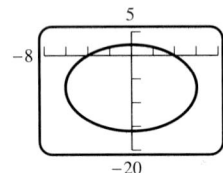 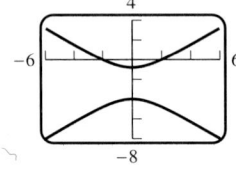 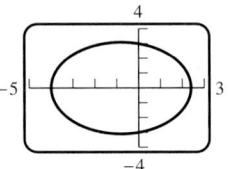

5. **(a)** $r = \dfrac{2}{3 + 2\cos\theta}$ 7. **(a)** $r = \dfrac{24}{5 - 5\cos\theta}$ 9. **(a)** $d = 6, \frac{1}{12}x^2 + \frac{1}{16}(y+2)^2 = 1$

   (b) $r = \dfrac{1}{1 - \cos\theta}$ **(b)** $r = \dfrac{2}{1 - \sin\theta}$ **(b)** $d = 1, \frac{9}{4}\left(x - \frac{1}{3}\right)^2 + 3y^2 = 1$

   (c) $r = \dfrac{3}{2 + 3\sin\theta}$ **(c)** $r = \dfrac{21}{2 + 5\sin\theta}$

11. **(a)** $d = \frac{2}{3}, -2x^2 + 16\left(y - \frac{3}{4}\right)^2 = 1$ **(b)** $d = \frac{10}{9}, \frac{9}{16}(x+2)^2 - \frac{9}{20}y^2 = 1$

13. **(a)** $r = \dfrac{12}{2 + \cos\theta}$ **(b)** $r = \dfrac{64}{25 - 15\sin\theta}$ **(c)** $r = \dfrac{16}{5 - 3\cos\theta}$ **(d)** $r = \dfrac{120}{5 + \sin\theta}$

17. **(a)** $T \approx 248$ yr   **(d)**

   (b) $r_0 \approx 4{,}449{,}675{,}000$ km

     $r_1 \approx 7{,}400{,}325{,}000$ km

   (c) $r \approx \dfrac{37.05}{1 + 0.249\cos\theta}$ AU

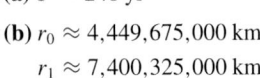

19. **(a)** $a \approx 178.26$ AU   **(d)**

   (b) $r_0 \approx 0.8735$ AU, $r_1 \approx 355.64$ AU

   (c) $r \approx \dfrac{1.74}{1 + 0.9951\cos\theta}$ AU

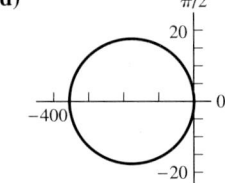

21. 563 km, 4286 km

▶ **Chapter 12 Supplementary Exercises (Page 752)**

3. **(a)** circle   **(b)** rose   **(c)** line   **(d)** limaçon   **(e)** limaçon   **(f)** none   **(g)** none   **(h)** spiral

5. **(a)**   **(b)**   **(c)**   **(d)**   **(e)**

 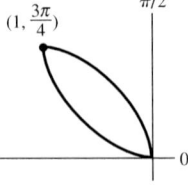 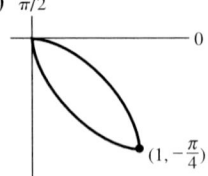 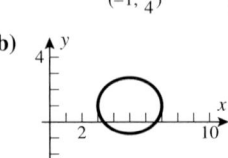 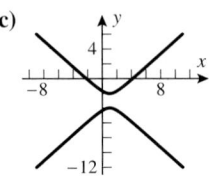

7. **(a)** parabola   9. **(a)** $\frac{5}{49}x^2 + \frac{1}{9}y^2 = 1$ 11. **(a)**   **(b)**   **(c)**

   (b) hyperbola   **(b)** $x^2 = -16y$

   (c) line

   (d) circle   **(c)** $\dfrac{x^2}{9} - \dfrac{y^2}{4} = 1$

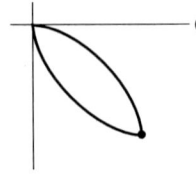

13. **(a)** $y = \dfrac{470}{2100^2}x^2$ **(b)** $L \approx 4336.3$ ft   15. $A = \dfrac{5\pi}{12} - \dfrac{\sqrt{3}}{2}$

17. (a) $L = \int_{\pi/4}^{\pi/2} \frac{1}{\theta^2} \sqrt{1 + \theta^2} \, d\theta \approx 0.9457$ (b) The arc length is infinite.

19. (a) $V = \dfrac{\pi b^2}{3a^2}(b^2 - 2a^2)\sqrt{a^2 + b^2} + \dfrac{2}{3}ab^2\pi$ (b) $V = \dfrac{2b^4}{3a}\pi$ 23. (a) $-1/4, 2$ 27. (b) $x = \cos t + \cos 2t,$
(b) $-3\sqrt{3}/4, 3\sqrt{3}/4$ $y = \sin t + \sin 2t$

(c) yes

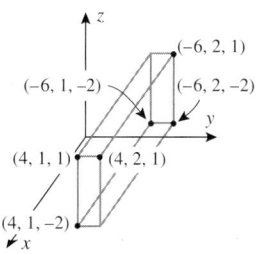

31. (a) $1/60$ (b) $91{,}450{,}000$ mi   (c) $584{,}295{,}652.5$ mi   33. (a) 119.3 ft/s   (b) 114.2 ft   (c) 447.9 ft

▶ Exercise Set 13.1 **(Page 764)**

1. (a) $(0, 0, 0), (3, 0, 0), (3, 5, 0), (0, 5, 0), (0, 0, 4), (3, 0, 4), (3, 5, 4), (0, 5, 4)$
   (b) $(0, 1, 0), (4, 1, 0), (4, 6, 0), (0, 6, 0), (0, 1, -2), (4, 1, -2), (4, 6, -2), (0, 6, -2)$

3. $(4, 2, 1), (4, 1, 1), (4, 1, -2), (-6, 2, 1), (-6, 2, -2), (-6, 1, -2)$

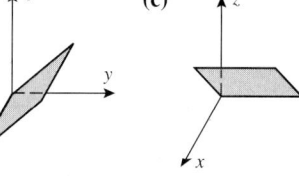

5. radius $\sqrt{74}$, center $(2, 1, -4)$ 7. (b) $(2, 1, 6)$ (c) area 49

9. (a) $(x - 1)^2 + y^2 + (z + 1)^2 = 16$ (b) $(x + 1)^2 + (y - 3)^2 + (z - 2)^2 = 14$ (c) $\left(x + \frac{1}{2}\right)^2 + (y - 2)^2 + (z - 2)^2 = \frac{5}{4}$

11. $(x - 2)^2 + (y + 1)^2 + (z + 3)^2 = r^2$;   (a) $r^2 = 9$ (b) $r^2 = 1$ (c) $r^2 = 4$ 13. sphere, center $(-5, -2, -1)$, radius 7

15. sphere; center $\left(\frac{1}{2}, \frac{3}{4}, -\frac{5}{4}\right)$, radius $\dfrac{3\sqrt{6}}{4}$ 17. no graph   19. (a)   (b)   (c)

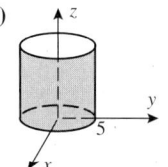 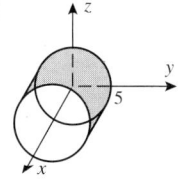 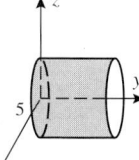

21. (a)   (b)   (c)   23. (a) $-2y + z = 0$ 25.
(b) $-2x + z = 0$
(c) $(x - 1)^2 + (y - 1)^2 = 1$
(d) $(x - 1)^2 + (z - 1)^2 = 1$

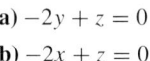 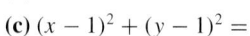 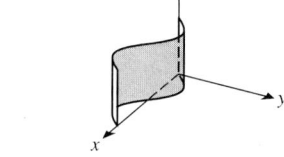

27.   29.   31.   33.

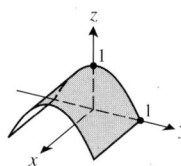 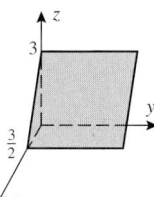 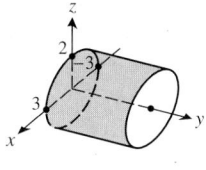 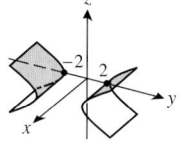

35.   37. largest distance $3 + \sqrt{6}$, smallest $3 - \sqrt{6}$

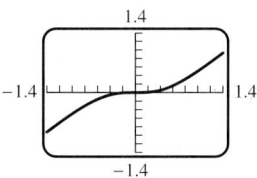 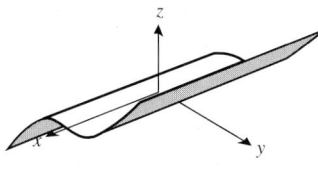

39. all points outside the circular cylinder $(y + 3)^2 + (z - 2)^2 = 16$ 41. $r = (2 - \sqrt{3})R$

▶ **Exercise Set 13.2 (Page 774)**

1. **(a–c)** 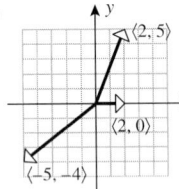 **(d–f)** 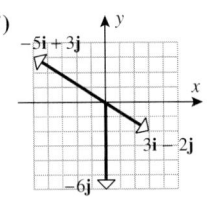 3. **(a, b)** 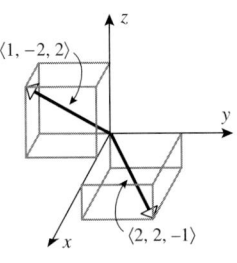 **(c, d)**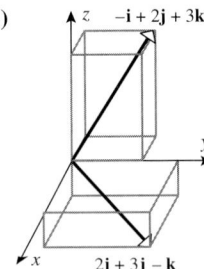

5. **(a)** $\langle 3, -4 \rangle$ **(b)** $\langle -2, -3, 4 \rangle$ 7. **(a)** $\langle -1, 3 \rangle$ 9. **(a)** $(4, -4)$

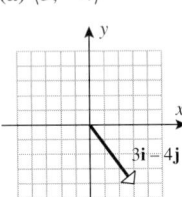 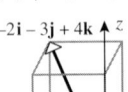 **(b)** $\langle -7, 2 \rangle$ **(b)** $(8, -1, -3)$

(c) $\langle -3, 6, 1 \rangle$

11. **(a)** $-\mathbf{i} + 4\mathbf{j} - 2\mathbf{k}$ **(b)** $18\mathbf{i} + 12\mathbf{j} - 6\mathbf{k}$ **(c)** $-\mathbf{i} - 5\mathbf{j} - 2\mathbf{k}$ **(d)** $40\mathbf{i} - 4\mathbf{j} - 4\mathbf{k}$ **(e)** $-2\mathbf{i} - 16\mathbf{j} - 18\mathbf{k}$ **(f)** $-\mathbf{i} + 13\mathbf{j} - 2\mathbf{k}$

13. **(a)** $\sqrt{2}$ **(b)** $5\sqrt{2}$ **(c)** $\sqrt{21}$ **(d)** $\sqrt{14}$

15. **(a)** $2\sqrt{3}$ **(b)** $\sqrt{14} + \sqrt{2}$ **(c)** $2\sqrt{14} + 2\sqrt{2}$ **(d)** $2\sqrt{37}$ **(e)** $(1/\sqrt{6})\mathbf{i} + (1/\sqrt{6})\mathbf{j} - (2/\sqrt{6})\mathbf{k}$ **(f)** 1

17. **(a)** $(-1/\sqrt{17})\mathbf{i} + (4/\sqrt{17})\mathbf{j}$ **(b)** $(-3\mathbf{i} + 2\mathbf{j} - \mathbf{k})/\sqrt{14}$ **(c)** $(4\mathbf{i} + \mathbf{j} - \mathbf{k})/(3\sqrt{2})$ 19. **(a)** $\langle -\frac{3}{2}, 2 \rangle$ **(b)** $\dfrac{1}{\sqrt{5}} \langle 7, 0, -6 \rangle$

21. **(a)** $\langle 3\sqrt{2}/2, 3\sqrt{2}/2 \rangle$ **(b)** $\langle 0, 2 \rangle$ **(c)** $\langle -5/2, 5\sqrt{3}/2 \rangle$ **(d)** $\langle -1, 0 \rangle$

23. $\langle (\sqrt{3} - \sqrt{2})/2, (1 + \sqrt{2})/2 \rangle$ 25. **(a)** $\langle -2, 5 \rangle$ **(b)** $\langle 3, -8 \rangle$ 27. $\langle -\frac{2}{3}, 1 \rangle$

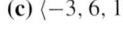

29. $\mathbf{u} = \frac{5}{7}\mathbf{i} + \frac{2}{7}\mathbf{j} + \frac{1}{7}\mathbf{k}, \mathbf{v} = \frac{8}{7}\mathbf{i} - \frac{1}{7}\mathbf{j} - \frac{4}{7}\mathbf{k}$ 31. $\sqrt{5}, 3$ 33. **(a)** $\pm\frac{5}{3}$ **(b)** 3

35. **(a)** $\langle 1/\sqrt{10}, 3/\sqrt{10} \rangle, \langle -1/\sqrt{10}, -3/\sqrt{10} \rangle$ **(b)** $\langle 1/\sqrt{2}, -1/\sqrt{2} \rangle, \langle -1/\sqrt{2}, 1/\sqrt{2} \rangle$ **(c)** $\pm\dfrac{1}{\sqrt{26}} \langle 5, 1 \rangle$

37. **(a)** the circle of radius 1 about the origin **(b)** the solid disk of radius 1 about the origin

(c) all points outside the solid disk of radius 1 about the origin

39. **(a)** the (hollow) sphere of radius 1 about the origin **(b)** the solid ball of radius 1 about the origin

(c) all points outside the solid ball of radius 1 about the origin

41. magnitude $= 30\sqrt{5}$ lb, $\theta \approx 26.57°$ 43. magnitude ≈ 207.06 N, $\theta = 45°$ 45. magnitude ≈ 94.995 N, $\theta \approx 28.28°$

47. magnitude ≈ 9.165 lb, angle $\approx -70.890°$ 49. ≈ 183.02 lb, 224.13 lb 51. **(a)** $c_1 = 2, c_2 = -1$ **(b)** no solution

▶ **Exercise Set 13.3 (Page 783)**

1. **(a)** $-10; \cos\theta = -1/\sqrt{5}$ **(b)** $-3; \cos\theta = -3/\sqrt{58}$ **(c)** $0; \cos\theta = 0$ **(d)** $-20; \cos\theta = -20/(3\sqrt{70})$

3. **(a)** obtuse **(b)** acute **(c)** obtuse **(d)** orthogonal 5. $\sqrt{2}/2, 0, -\sqrt{2}/2, -1, -\sqrt{2}/2, 0, \sqrt{2}/2$

7. **(a)** vertex B 9. **(b)** $\mathbf{u}_1 = \dfrac{2}{\sqrt{13}}\mathbf{i} + \dfrac{3}{\sqrt{13}}\mathbf{j}, \mathbf{u}_2 = -\mathbf{u}_1$ 13. **(a)** 6 15. **(a)** $\alpha = \beta \approx 55°, \gamma \approx 125°$

(b) $82°, 60°, 38°$ **(b)** 36 **(b)** $\alpha \approx 48°, \beta \approx 132°, \gamma \approx 71°$

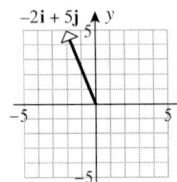

(c) $24\sqrt{5}$

(d) $24\sqrt{5}$

19. $64°, 41°, 60°$ **21. (a)** $\text{proj}_\mathbf{b}\mathbf{v} = \langle \frac{6}{25}, \frac{8}{25} \rangle$,   **(b)** $\text{proj}_\mathbf{b}\mathbf{v} = \langle -\frac{6}{5}, \frac{12}{5} \rangle$,   **(c)** $\text{proj}_\mathbf{b}\mathbf{v} = \langle -\frac{16}{5}, -\frac{8}{5} \rangle$,

$\mathbf{v} - \text{proj}_\mathbf{b}\mathbf{v} = \langle \frac{44}{25}, -\frac{33}{25} \rangle$ $\mathbf{v} - \text{proj}_\mathbf{b}\mathbf{v} = \langle \frac{26}{5}, \frac{13}{5} \rangle$ $\mathbf{v} - \text{proj}_\mathbf{b}\mathbf{v} = \langle \frac{1}{5}, -\frac{2}{5} \rangle$

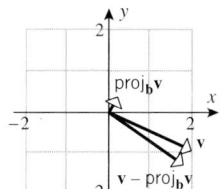 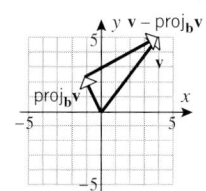 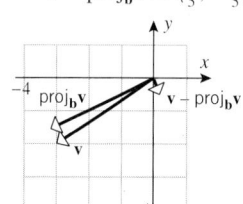

23. (a) $\mathbf{v} = \langle -1, -1 \rangle + \langle 3, -3 \rangle$ **(b)** $\mathbf{v} = \langle \frac{16}{5}, 0, -\frac{8}{5} \rangle + \langle -\frac{1}{5}, 1, -\frac{2}{5} \rangle$ **25.** $\frac{13}{5}$

27. force $49\sqrt{2}$ N against the ramp, force $49\sqrt{2}$ N to prevent the block from sliding down the ramp

29. $450\sqrt{2}+150\sqrt{6}$ lb, $300+300\sqrt{3}$ lb   **31.** $W = -12$ ft·lb   **33.** $W = 375$ ft·lb   **35.** $71°$ **43. (a)** $40°$ **(b)** $x \approx -0.682328$

▶ Exercise Set 13.4 **(Page 793)**

1. (a) $-\mathbf{j} + \mathbf{k}$ **(b)** $\mathbf{k} - \mathbf{j}$ **3.** $\langle 7, 10, 9 \rangle$ **5.** $\langle -4, -6, -3 \rangle$

7. (a) $\langle -20, -67, -9 \rangle$ **(b)** $\langle -78, 52, -26 \rangle$ **(c)** $\langle 0, -56, -392 \rangle$ **(d)** $\langle 0, 56, 392 \rangle$ **9.** $\frac{1}{\sqrt{2}}, -\frac{1}{\sqrt{2}}, 0$ **11.** $\pm \frac{1}{\sqrt{6}} \langle 2, 1, 1 \rangle$

13. $\sqrt{59}$ **15.** $\sqrt{374}/2$ **17.** 80 **19.** -3 **21.** 16 **23. (a)** yes   **(b)** yes   **(c)** no

25. (a) 9 **(b)** $\sqrt{122}$ **(c)** $\sin^{-1}\left(\frac{9}{14}\right)$ **27. (a)** $2\sqrt{141/29}$ **(b)** $6/\sqrt{5}$ **29.** $\frac{2}{3}$ **31.** $\theta = \pi/4$

33. (a) $10\sqrt{2}$ lb·ft, direction of rotation about P is counterclockwise looking along $\overrightarrow{PQ} \times \mathbf{F} = -10\mathbf{i} + 10\mathbf{k}$ toward its initial point

(b) 10 lb·ft, direction of rotation about P is counterclockwise looking along $-10\mathbf{i}$ toward its initial point   **(c)** 0 lb·ft, no rotation about P

35. ≈ 36.19 N·m   **39.** $-8\mathbf{i} - 20\mathbf{j} + 2\mathbf{k}, -8\mathbf{i} - 8\mathbf{k}$

▶ Exercise Set 13.5 **(Page 799)**

1. (a) $L_1: x = 1, y = t$ **(b)** $L_1: x = 1, y = 1, z = t$ **3. (a)** $x = 3 + 2t, y = -2 + 3t$; line segment: $0 \le t \le 1$

$L_2: x = t, y = 1$ $L_2: x = t, y = 1, z = 1$ **(b)** $x = 5 - 3t, y = -2 + 6t, z = 1 + t$; line segment: $0 \le t \le 1$

$L_3: x = t, y = t$ $L_3: x = 1, y = t, z = 1$

$L_4: x = t, y = t, z = t$

5. (a) $x = 2 + t, y = -3 - 4t$ **(b)** $x = t, y = -t, z = 1 + t$ **7. (a)** $P(2, -1), \mathbf{v} = 4\mathbf{i} - \mathbf{j}$ **(b)** $P(-1, 2, 4), \mathbf{v} = 5\mathbf{i} + 7\mathbf{j} - 8\mathbf{k}$

9. (a) $\langle -3, 4 \rangle + t \langle 1, 5 \rangle$; $-3\mathbf{i} + 4\mathbf{j} + t(\mathbf{i} + 5\mathbf{j})$ **(b)** $\langle 2, -3, 0 \rangle + t \langle -1, 5, 1 \rangle$; $2\mathbf{i} - 3\mathbf{j} + t(-\mathbf{i} + 5\mathbf{j} + \mathbf{k})$

11. $x = -5 + 2t, y = 2 - 3t$ **13.** $x = 3 + 4t, y = -4 + 3t$ **15.** $x = -1 + 3t, y = 2 - 4t, z = 4 + t$

17. $x = -2 + 2t, y = -t, z = 5 + 2t$ **19. (a)** $x = 7$ **(b)** $y = \frac{7}{3}$ **(c)** $x = \frac{-1 \pm \sqrt{85}}{6}, y = \frac{43 \mp \sqrt{85}}{18}$

21. (a) $(-2, 10, 0)$ **(b)** $(-2, 0, -5)$ **(c)** The line does not intersect the yz-plane.   **23.** $(0, 4, -2), (4, 0, 6)$

29. The lines are parallel.   **31.** The points do not lie on the same line.   **35.** the line segment joining the points $(1, 0)$ and $(-3, 6)$

37. $2\sqrt{5}$ **39.** distance $= \sqrt{35/6}$

41. (a) $x = x_0 + (x_1 - x_0)t, y = y_0 + (y_1 - y_0)t, z = z_0 + (z_1 - z_0)t$ **(b)** $x = x_1 + at, y = y_1 + bt, z = z_1 + ct$

43. (b) $\langle x, y, z \rangle = \langle 1 + 2t, -3 + 4t, 5 + t \rangle$ **45. (b)** $84°$ **(c)** $x = 7 + t, y = -1, z = -2 + t$ **47.** $x = t, y = 2 + t, z = 1 - t$

49. (a) $\sqrt{17}$ cm   **(b)** 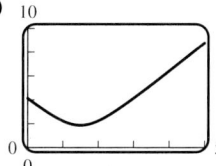   **(d)** $\sqrt{14}/2$ cm

---

▶ Exercise Set 13.6 **(Page 806)**

1. $x = 3, y = 4, z = 5$ **3.** $x + 4y + 2z = 28$ **5.** $z = 0$ **7.** $x - y = 0$ **9.** $y + z = 1$ **11.** $2y - z = 1$

13. (a) parallel   **(b)** perpendicular   **(c)** neither   **15. (a)** parallel   **(b)** neither   **(c)** perpendicular

17. (a) point of intersection is $\left(\frac{5}{2}, \frac{5}{2}, \frac{5}{2}\right)$ **(b)** no intersection   **19.** $35°$ **21.** $4x - 2y + 7z = 0$ **23.** $4x - 13y + 21z = -14$

25. $x + y - 3z = 6$ **27.** $x + 5y + 3z = -6$ **29.** $x + 2y + 4z = \frac{29}{2}$ **31.** $x = 5 - 2t,\ y = 5t,\ z = -2 + 11t$

35. $7x + y + 9z = 25$ **37.** $x = -\frac{11}{7} - 23t,\ y = -\frac{12}{7} + t,\ z = -7t$ **39.** $\frac{5}{3}$ **41.** $5/\sqrt{54}$ **43.** $25/\sqrt{126}$

45. $(x - 2)^2 + (y - 1)^2 + (z + 3)^2 = \frac{121}{14}$ **47.** $5/\sqrt{12}$ **49.** $2/\sqrt{5}$

▶ **Exercise Set 13.7 (Page 817)**

1. (a) elliptic paraboloid, $a = 2, b = 3$ **(b)** hyperbolic paraboloid, $a = 1, b = 5$ **(c)** hyperboloid of one sheet, $a = b = c = 4$

(d) circular cone, $a = b = 1$ **(e)** elliptic paraboloid, $a = 2, b = 1$ **(f)** hyperboloid of two sheets, $a = b = c = 1$

3. (a) $-z = x^2 + y^2$, circular paraboloid opening down the negative z-axis   **(b)** $z = x^2 + y^2$, circular paraboloid, no change

(c) $z = x^2 + y^2$, circular paraboloid, no change

(d) $z = x^2 + y^2$, circular paraboloid, no change

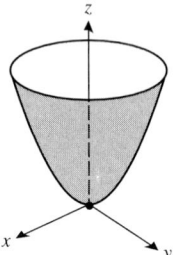

(e) $x = y^2 + z^2$, circular paraboloid opening along the positive x-axis   **(f)** $y = x^2 + z^2$, circular paraboloid opening along the positive y-axis

5. (a) hyperboloid of one sheet, axis is y-axis

(b) hyperboloid of two sheets separated by yz-plane

(c) elliptic paraboloid opening along the positive x-axis

(d) elliptic cone with x-axis as axis

(e) hyperbolic paraboloid straddling x- and z-axes

(f) paraboloid opening along the negative y-axis

7. (a) $x = 0 : \dfrac{y^2}{25} + \dfrac{z^2}{4} = 1;$

$y = 0 : \dfrac{x^2}{9} + \dfrac{z^2}{4} = 1;$

$z = 0 : \dfrac{x^2}{9} + \dfrac{y^2}{25} = 1$

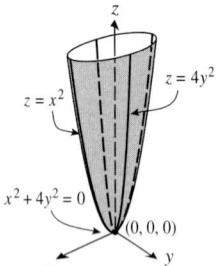
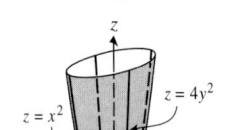
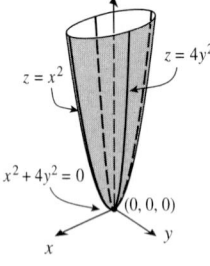

(b) $x = 0 : z = 4y^2;$

$y = 0 : z = x^2;$

$z = 0 : x = y = 0$

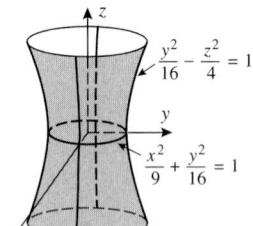

(c) $x = 0 : \dfrac{y^2}{16} - \dfrac{z^2}{4} = 1;$

$y = 0 : \dfrac{x^2}{9} - \dfrac{z^2}{4} = 1;$

$z = 0 : \dfrac{x^2}{9} + \dfrac{y^2}{16} = 1$

9. (a) $4x^2 + z^2 = 3$; ellipse

(b) $y^2 + z^2 = 3$; circle

(c) $y^2 + z^2 = 20$; circle

(d) $9x^2 - y^2 = 20$; hyperbola

(e) $z = 9x^2 + 16$; parabola

(f) $9x^2 + 4y^2 = 4$; ellipse

11.

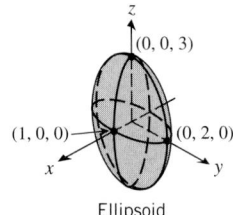

Ellipsoid

13.

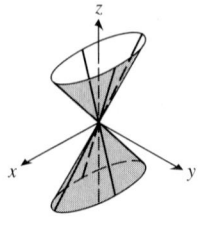

Hyperboloid of one sheet

15.

Elliptic cone

17.

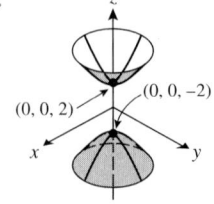

Hyperboloid of two sheets

19.

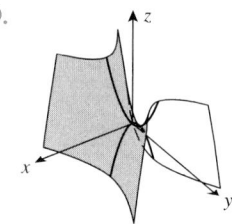

Hyperbolic paraboloid

21.

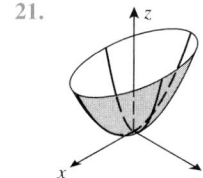

Elliptic paraboloid

23.

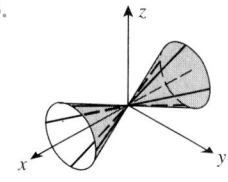

Circular cone

25.

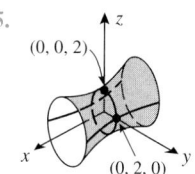

$(0, 0, 2)$ $(0, 2, 0)$

Hyperboloid
of one sheet

27.

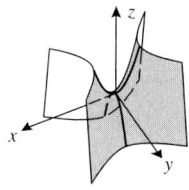

Hyperbolic
paraboloid

29.

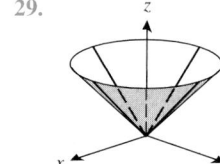

31.

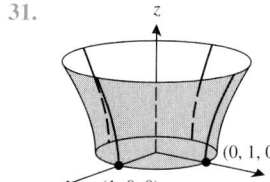

$(1, 0, 0)$ $(0, 1, 0)$

33.

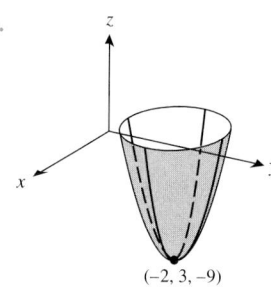

$(-2, 3, -9)$

Circular paraboloid

35.

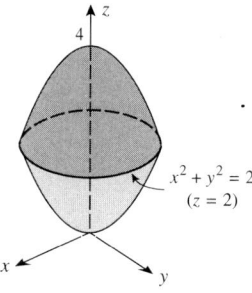

$(1, -1, -2)$

Ellipsoid

37. (a) $\dfrac{x^2}{9} + \dfrac{y^2}{4} = 1$
(b) $6, 4$
(c) $(\pm\sqrt{5}, 0, \sqrt{2})$
(d) The focal axis is parallel to the x-axis.

39. (a) $\dfrac{y^2}{4} - \dfrac{x^2}{4} = 1$
(b) $(0, \pm 2, 4)$
(c) $(0, \pm 2\sqrt{2}, 4)$
(d) The focal axis is parallel to the y-axis.

41. (a) $z + 4 = y^2$
(b) $(2, 0, -4)$
(c) $\left(2, 0, -\tfrac{15}{4}\right)$
(d) The focal axis is parallel to the z-axis.

43. circle of radius $\sqrt{2}$ in the plane $z = 2$, centered at $(0, 0, 2)$

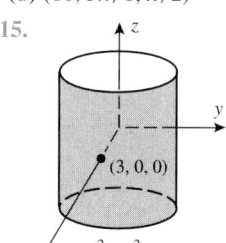

$x^2 + y^2 = 2$
$(z = 2)$

45. $y = 4(x^2 + z^2)$ **47.** $z = (x^2 + y^2)/4$ (circular paraboloid)

▶ **Exercise Set 13.8 (Page 824)**

1. (a) $(8, \pi/6, -4)$
(b) $(5\sqrt{2}, 3\pi/4, 6)$
(c) $(2, \pi/2, 0)$
(d) $(8, 5\pi/3, 6)$

9. (a) $(2\sqrt{3}, \pi/6, \pi/6)$
(b) $(\sqrt{2}, \pi/4, 3\pi/4)$
(c) $(2, 3\pi/4, \pi/2)$
(d) $(4\sqrt{3}, 1, 2\pi/3)$

3. (a) $(2\sqrt{3}, 2, 3)$
(b) $(-4\sqrt{2}, 4\sqrt{2}, -2)$
(c) $(5, 0, 4)$
(d) $(-7, 0, -9)$

11. (a) $(5\sqrt{3}/2, \pi/4, -5/2)$
(b) $(0, 7\pi/6, -1)$
(c) $(0, 0, 3)$
(d) $(4, \pi/6, 0)$

5. (a) $(2\sqrt{2}, \pi/3, 3\pi/4)$
(b) $(2, 7\pi/4, \pi/4)$
(c) $(6, \pi/2, \pi/3)$
(d) $(10, 5\pi/6, \pi/2)$

15.

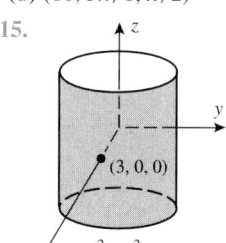

$(3, 0, 0)$

$x^2 + y^2 = 9$

7. (a) $(5\sqrt{6}/4, 5\sqrt{2}/4, 5\sqrt{2}/2)$
(b) $(7, 0, 0)$
(c) $(0, 0, 1)$
(d) $(0, -2, 0)$

17.

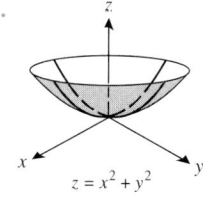

$z = x^2 + y^2$

19.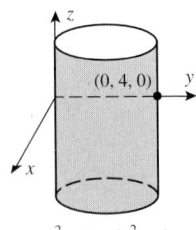
$x^2 + (y-2)^2 = 4$

21.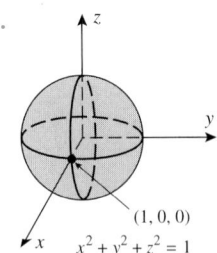
$x^2 + y^2 + z^2 = 1$

23.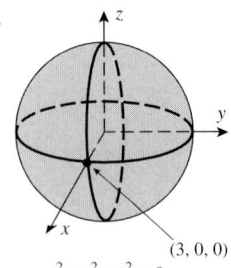
$x^2 + y^2 + z^2 = 9$

25.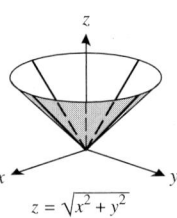

$z = \sqrt{x^2 + y^2}$

27.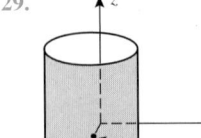
$x^2 + y^2 + (z-2)^2 = 4$

29. 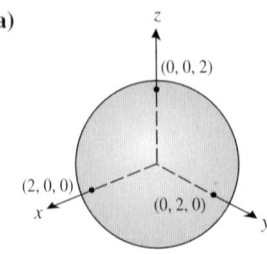

$(x-1)^2 + y^2 = 1$

31. (a) $z = 3$
(b) $\rho = 3 \sec \phi$

33. (a) $z = 3r^2$
(b) $\rho = \frac{1}{3} \csc \phi \cot \phi$

35. (a) $r = 2$
(b) $\rho = 2 \csc \phi$

37. (a) $r^2 + z^2 = 9$ **(b)** $\rho = 3$ **39. (a)** $2r \cos \theta + 3r \sin \theta + 4z = 1$ **(b)** $2\rho \sin \phi \cos \theta + 3\rho \sin \phi \sin \theta + 4\rho \cos \phi = 1$

41. (a) $r^2 \cos^2 \theta = 16 - z^2$ **(b)** $\rho^2 (1 - \sin^2 \phi \sin^2 \theta) = 16$

43. all points on or above the paraboloid $z = x^2 + y^2$ that are also on or below the plane $z = 4$

45. all points on or between concentric spheres of radii 1 and 3   **47.** spherical $(4000, \pi/6, \pi/6)$, rectangular $(1000\sqrt{3}, 1000, 2000\sqrt{3})$

49. (a) $(10, \pi/2, 1)$ **(b)** $(0, 10, 1)$ **(c)** $(\sqrt{101}, \pi/2, \tan^{-1} 10)$ **51.** ≈ 2927 km

---

▶ **Chapter 13 Supplementary Exercises (Page 825)**

3. (b) $-1/2, \pm\sqrt{3}/2$ **(d)** true

5. (b) $(y, x, z), (x, z, y), (z, y, x)$ **(c)** circle of radius 5 in plane $z = 1$ with center at $(0, 0, 1)$ (rectangular coordinates)
(d) the two half-lines $z = \pm x$ ($x \geq 0$) in the xz-plane

7. (a) $\sqrt{26}/2$ **(b)** $\sqrt{26}/3$ **9. (a)** $-\frac{3}{4}$ **(b)** $\frac{1}{7}$ **(c)** $(48 \pm 25\sqrt{3})/11$ **(d)** $c = \frac{4}{3}$

11. (a) the plane through the origin and perpendicular to $\mathbf{r}_0$ **(b)** the plane through the tip of $\mathbf{r}_0$ and perpendicular to $\mathbf{r}_0$

15. (a) false   **(b)** false   **(c)** true   **19. (a)** $x = t, y = 2 - t, z = -1 + t$ **(b)** $\pi/3$ **21.** $\langle 5/2, -5/2, -5\sqrt{2}/2 \rangle$

23. (a) hyperboloid of one sheet   **(b)** sphere   **(c)** circular cone

25. (a) $r^2 = z$; $\rho = \cot \phi \csc \phi$ **(b)** $z^2 = r^2 \cos 2\theta$; $\cos 2\theta = \cot^2 \phi$

27. (a)

(b)

(c)
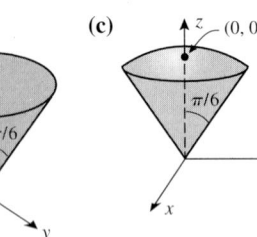

29. (a)
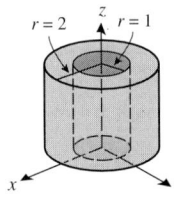

(b)
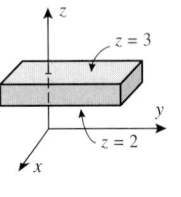

(c)
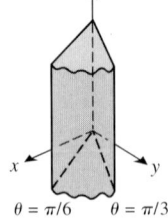

(d)
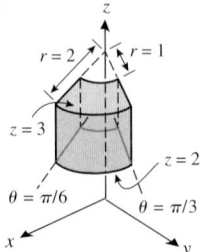

31. (b) $y^2 + z^2 = e^{2x}$ **33.**   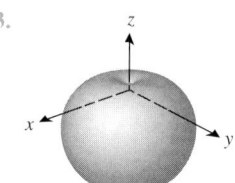   **35.** -11 N·m   **37. (a)** $\mathbf{F} = -6\mathbf{i} + 3\mathbf{j} - 6\mathbf{k}$

(b) $-6\mathbf{i} + 18\mathbf{j} + 15\mathbf{k}$

▶ Exercise Set 14.1 **(Page 834)**

1. $(-\infty, +\infty)$; $\mathbf{r}(\pi) = -\mathbf{i} - 3\pi\mathbf{j}$ 3. $[2, +\infty)$; $\mathbf{r}(3) = -\mathbf{i} - \ln 3\mathbf{j} + \mathbf{k}$ 5. $\mathbf{r} = 3\cos t\,\mathbf{i} + (t + \sin t)\mathbf{j}$

7. $\mathbf{r} = 2t\mathbf{i} + 2\sin 3t\,\mathbf{j} + 5\cos 3t\,\mathbf{k}$ 9. $x = 3t^2$, $y = -2$, $z = 0$ 11. $x = 2t - 1$, $y = -3\sqrt{t}$, $z = \sin 3t$

13. the line in 2-space through the point $(2, 0)$ and parallel to the vector $-3\mathbf{i} - 4\mathbf{j}$

15. the line in 3-space through the point $(0, -3, 1)$ and parallel to the vector $2\mathbf{i} + 3\mathbf{k}$

17. an ellipse in the plane $z = -1$, center at $(0, 0, -1)$, major axis of length 6 parallel to x-axis, minor axis of length 4 parallel to y-axis

19. **(a)** slope $-\frac{3}{2}$ 21. **(a)** 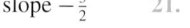 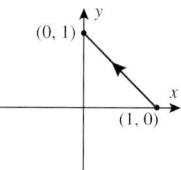   **(b)** 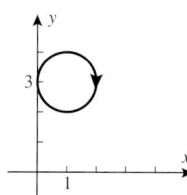   23. $\mathbf{r} = (1 - t)(3\mathbf{i} + 4\mathbf{j})$, $0 \le t \le 1$

(b) $\left(\frac{5}{2}, 0, \frac{3}{2}\right)$

25. 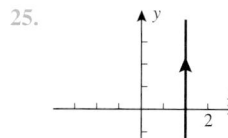   27.    29. 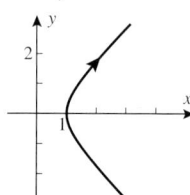   31.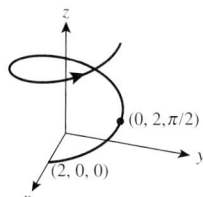

33. 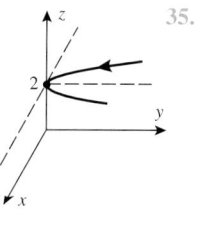   35. $x = t$, $y = t$, $z = 2t^2$ 37. $\mathbf{r} = t\mathbf{i} + t^2\mathbf{j} \pm \frac{1}{3}\sqrt{81 - 9t^2 - t^4}\,\mathbf{k}$ 43. $c = 3/(2\pi)$

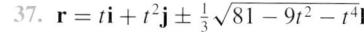 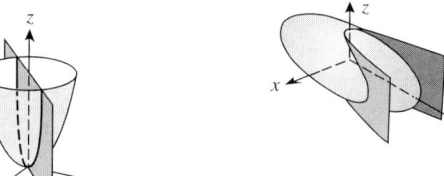

47. **(a)** III, since the curve is a subset of the plane $y = -x$

(b) IV, since only x is periodic in t and y, z increase without bound

(c) II, since all three components are periodic in t

(d) I, since the projection onto the yz-plane is a circle and the curve increases without bound in the x-direction

49. **(a)** $x = 3\cos t$, $y = 3\sin t$, $z = 9\cos^2 t$

(b)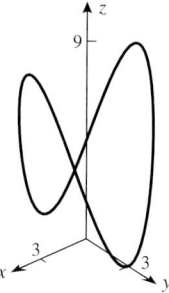

▶ **Exercise Set 14.2 (Page 841)**

1.

3. $\mathbf{r}'(t) = 5\mathbf{i} + (1 - 2t)\mathbf{j}$ 5. $\mathbf{r}'(t) = -\dfrac{1}{t^2}\mathbf{i} + \sec^2 t\,\mathbf{j} + 2e^{2t}\mathbf{k}$ 7. $\mathbf{r}'(2) = \langle 1, 4 \rangle$

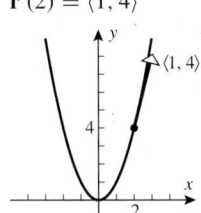

9. $\mathbf{r}'(0) = \mathbf{j}$ 11. $\mathbf{r}'(\pi/2) = -2\mathbf{k}$ 13. $9\mathbf{i} + 6\mathbf{j}$ 15. $\langle \frac{1}{3}, 0 \rangle$ 17. $2\mathbf{i} - 3\mathbf{j} + 4\mathbf{k}$

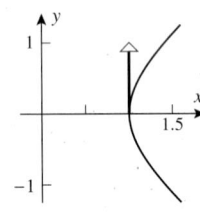

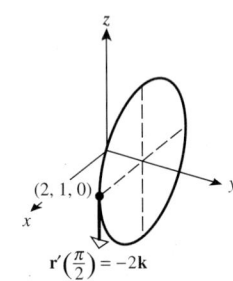

19. **(a)** continuous   **(b)** not continuous   21. **(a)** $\mathbf{i} - \mathbf{j} + \mathbf{k}$ **(b)** $-\mathbf{i} + \mathbf{k}$ **(c)** 0 23. $x = 1 + 2t, y = 2 - t, z = 0$

25. $x = 1 - \sqrt{3}\,\pi t, y = \sqrt{3} + \pi t, z = 1 + 3t$ 27. $\mathbf{r} = (-\mathbf{i} + 2\mathbf{j}) + t\left(2\mathbf{i} + \frac{3}{4}\mathbf{j}\right)$ 29. $\mathbf{r} = (4\mathbf{i} + \mathbf{j}) + t(-4\mathbf{i} + \mathbf{j} + 4\mathbf{k})$

31. $3t\mathbf{i} + 2t^2\mathbf{j} + \mathbf{C}$ 33. $(-t\cos t + \sin t)\mathbf{i} + t\mathbf{j} + \mathbf{C}$ 35. $(t^3/3)\mathbf{i} - t^2\mathbf{j} + \ln|t|\mathbf{k} + \mathbf{C}$ 37. $\langle 0, -\frac{2}{3} \rangle$ 39. $(5\sqrt{5} - 1)/3$

41. $\frac{52}{3}\mathbf{i} + 4\mathbf{j}$ 43. $\mathbf{y}(t) = \left(\frac{1}{3}t^3 + 1\right)\mathbf{i} + (t^2 + 1)\mathbf{j}$ 45. $\mathbf{y}(t) = \left(\frac{1}{2}t^2 + 2\right)\mathbf{i} + (e^t - 1)\mathbf{j}$ 47.

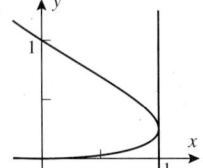

49. intercepts at $t = 0, \pi, 2\pi$;   51. **(a)** $(-2, 4, 6)$ and $(1, 1, -3)$ 53. $\theta \approx 68°$ 55. $7t^6$; $18t^5\mathbf{i} - 10t^4\mathbf{j}$
    extrema at $t = \pi/4, 3\pi/4, 5\pi/4, 7\pi/4$;   **(b)** $\theta \approx 76°, 71°$
    $\mathbf{r}$ and $\mathbf{r}'$ are never perpendicular

▶ **Exercise Set 14.3 (Page 851)**

3. smooth   5. not smooth, $\mathbf{r}'(1) = \mathbf{0}$ 7. $L = \frac{3}{2}$ 9. $L = e - e^{-1}$ 11. $L = 28$ 13. $L = 2\pi\sqrt{10}$

15. $\mathbf{r}'(\tau) = 4\mathbf{i} + 8(4\tau + 1)\mathbf{j}$ 17. $\mathbf{r}'(\tau) = 2\tau e^{\tau^2}\mathbf{i} - 8\tau e^{-\tau^2}\mathbf{j}$ 19. **(a)** $x = \dfrac{s}{\sqrt{2}}, y = \dfrac{s}{\sqrt{2}}$ **(b)** $x = y = z = \dfrac{s}{\sqrt{3}}$

21. **(a)** $x = 1 + \dfrac{s}{3}, y = 3 - \dfrac{2s}{3}, z = 4 + \dfrac{2s}{3}$ **(b)** $\langle \frac{28}{3}, -\frac{41}{3}, \frac{62}{3} \rangle$ 23. $x = 3 + \cos s, y = 2 + \sin s, 0 \le s \le 2\pi$

25. $x = \frac{1}{3}[(3s + 1)^{2/3} - 1]^{3/2}, y = \frac{1}{2}[(3s + 1)^{2/3} - 1], s \ge 0$

27. $x = \left(\dfrac{s}{\sqrt{2}} + 1\right)\cos\left[\ln\left(\dfrac{s}{\sqrt{2}} + 1\right)\right], y = \left(\dfrac{s}{\sqrt{2}} + 1\right)\sin\left[\ln\left(\dfrac{s}{\sqrt{2}} + 1\right)\right], 0 \le s \le \sqrt{2}(e^{\pi/2} - 1)$

31. $x = 2a\cos^{-1}\left(1 - \dfrac{s}{4a}\right) - 2a\left[1 - \left(1 - \dfrac{s}{4a}\right)^2\right]^{1/2}\left[2\left(1 - \dfrac{s}{4a}\right)^2 - 1\right], y = \dfrac{s(8a - s)}{8a}, 0 \le s \le 8a$

33. **(a)** $\frac{9}{2}$ **(b)** $9 - 2\sqrt{6}$ 35. **(a)** $\sqrt{3}(1 - e^{-2})$ **(b)** $4\sqrt{5}$ 37. **(a)** $g(\tau) = \pi\tau$ **(b)** $g(\tau) = \pi(1 - \tau)$

39. $2\pi\sqrt{6.25^2 + (10/\pi)^2} \approx 44$ in   41. **(a)** $2t + (1/t)$ **(b)** $2t + (1/t)$ **(c)** $8 + \ln 3$

▶ Exercise Set 14.4 **(Page 857)**

1. (a)   **(b)**

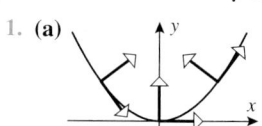

3. $\mathbf{T}(1) = \dfrac{2}{\sqrt{5}}\mathbf{i} + \dfrac{1}{\sqrt{5}}\mathbf{j}$, $\mathbf{N}(1) = \dfrac{1}{\sqrt{5}}\mathbf{i} - \dfrac{2}{\sqrt{5}}\mathbf{j}$

5. $\mathbf{T}\!\left(\dfrac{\pi}{3}\right) = -\dfrac{\sqrt{3}}{2}\mathbf{i} + \dfrac{1}{2}\mathbf{j}$, $\mathbf{N}\!\left(\dfrac{\pi}{3}\right) = -\dfrac{1}{2}\mathbf{i} - \dfrac{\sqrt{3}}{2}\mathbf{j}$ **7.** $\mathbf{T}\!\left(\dfrac{\pi}{2}\right) = -\dfrac{4}{\sqrt{17}}\mathbf{i} + \dfrac{1}{\sqrt{17}}\mathbf{k}$, $\mathbf{N}\!\left(\dfrac{\pi}{2}\right) = -\mathbf{j}$

9. $\mathbf{T}(0) = \dfrac{1}{\sqrt{3}}\mathbf{i} + \dfrac{1}{\sqrt{3}}\mathbf{j} + \dfrac{1}{\sqrt{3}}\mathbf{k}$, $\mathbf{N}(0) = -\dfrac{1}{\sqrt{2}}\mathbf{i} + \dfrac{1}{\sqrt{2}}\mathbf{j}$ **13.** $x = s$, $y = 1$ **15.** $\mathbf{B} = \frac{4}{5}\cos t\,\mathbf{i} - \frac{4}{5}\sin t\,\mathbf{j} - \frac{3}{5}\mathbf{k}$ **17.** $\mathbf{B} = -\mathbf{k}$

19. $\mathbf{T}\!\left(\dfrac{\pi}{4}\right) = \dfrac{\sqrt{2}}{2}(-\mathbf{i} + \mathbf{j})$, $\mathbf{N}\!\left(\dfrac{\pi}{4}\right) = -\dfrac{\sqrt{2}}{2}(\mathbf{i} + \mathbf{j})$, $\mathbf{B}\!\left(\dfrac{\pi}{4}\right) = \mathbf{k}$ rectifying: $x + y = \sqrt{2}$; osculating: $z = 1$; normal: $-x + y = 0$

23. $\mathbf{N} = -\sin t\,\mathbf{i} - \cos t\,\mathbf{j}$

▶ Exercise Set 14.5 **(Page 864)**

1. $\kappa \approx 2$ **3.** $\dfrac{6}{t(4 + 9t^2)^{3/2}}$ **5.** $\dfrac{12e^{2t}}{\left(9e^{6t} + e^{-2t}\right)^{3/2}}$ **7.** $\frac{4}{17}$ **9.** $\dfrac{1}{2\cosh^2 t}$ **11.** $\kappa = \frac{2}{5}$, $\rho = \frac{5}{2}$ **13.** $\kappa = \dfrac{\sqrt{2}}{3}$, $\rho = \dfrac{3\sqrt{2}}{2}$

15. $\kappa = \frac{1}{4}$ **19.** 1 **21.** $\dfrac{1}{\sqrt{2}}$ **23.** $\dfrac{4}{5\sqrt{5}}$ **25.** $\frac{96}{125}$ **27.** $\dfrac{6}{5\sqrt{10}}$ **29.** $\dfrac{1}{\sqrt{2}}$

31. (a)   **(b)**   **33. (a)** I is the curvature of II.

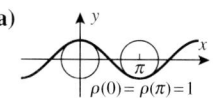   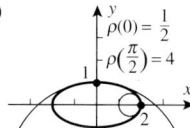

(b) I is the curvature of II.

35. (a)   **(b)**

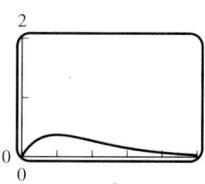   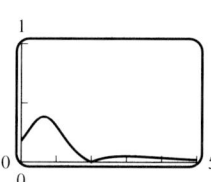

37. (a) $\kappa = \dfrac{|12x^2 - 4|}{[1 + (4x^3 - 4x)^2]^{3/2}}$ **(b)**

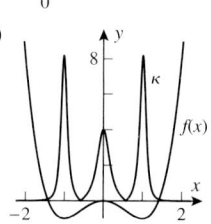

(c) $\rho = \frac{1}{4}$ for $x = 0$ and $\rho = \frac{1}{8}$ when $x = \pm 1$

41. $\dfrac{3}{2\sqrt{2}}$ **43.** $\frac{2}{3}$ **47.** $\rho = 2|p|$ **49.** $(3, 0), (-3, 0)$ **51.** $\rho_{\min} = 1/\sqrt{2}$; $\rho_{\max} = 2$

55. (b) $\rho = \sqrt{2}$ **(c)**

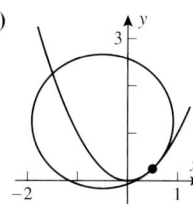

59. $a = \dfrac{1}{2r}$ **67.** $\tau = \dfrac{2}{(t^2 + 2)^2}$ **69.** $\tau = -\dfrac{\sqrt{2}}{(e^t + e^{-t})^2}$

▶ **Exercise Set 14.6 (Page 876)**

1. $\mathbf{v}(t) = -3\sin t\mathbf{i} + 3\cos t\mathbf{j}$
$\mathbf{a}(t) = -3\cos t\mathbf{i} - 3\sin t\mathbf{j}$
$\|\mathbf{v}(t)\| = 3$

3. $\mathbf{v}(t) = e^t\mathbf{i} - e^{-t}\mathbf{j}$
$\mathbf{a}(t) = e^t\mathbf{i} + e^{-t}\mathbf{j}$
$\|\mathbf{v}(t)\| = \sqrt{e^{2t} + e^{-2t}}$

5. $\mathbf{v} = \mathbf{i} + \mathbf{j} + \mathbf{k}$, $\|\mathbf{v}\| = \sqrt{3}$, $\mathbf{a} = \mathbf{j} + 2\mathbf{k}$

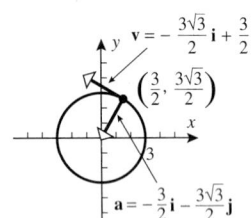

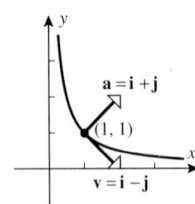

7. $\mathbf{v} = -\sqrt{2}\mathbf{i} + \sqrt{2}\mathbf{j} + \mathbf{k}$, $\|\mathbf{v}\| = \sqrt{5}$, $\mathbf{a} = -\sqrt{2}\mathbf{i} - \sqrt{2}\mathbf{j}$ 11. minimum speed $3\sqrt{2}$ when $\mathbf{r} = 24\mathbf{i} + 8\mathbf{j}$

13. (a)

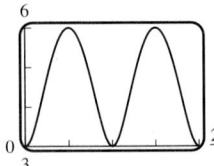

(b) maximum speed $= 6$, minimum speed $= 3$

(d) The maximum speed first occurs when $t = \pi/6$.

15. $\mathbf{v}(t) = (1 - \sin t)\mathbf{i} + (\cos t - 1)\mathbf{j}$; $\mathbf{r}(t) = (t + \cos t - 1)\mathbf{i} + (\sin t - t + 1)\mathbf{j}$

17. $\mathbf{v}(t) = (1 - \cos t)\mathbf{i} + \sin t\mathbf{j} + e^t\mathbf{k}$; $\mathbf{r}(t) = (t - \sin t - 1)\mathbf{i} + (1 - \cos t)\mathbf{j} + e^t\mathbf{k}$

19. The motion is along a straight line and has constant speed.   21. $15°$ 23. (a) $0.7\mathbf{i} + 2.7\mathbf{j} - 3.4\mathbf{k}$ (b) $\mathbf{r}_0 = -0.7\mathbf{i} - 2.9\mathbf{j} + 4.8\mathbf{k}$

25. $\Delta\mathbf{r} = 8\mathbf{i} + \frac{26}{3}\mathbf{j}$, $s = (13\sqrt{13} - 5\sqrt{5})/3$ 27. $\Delta\mathbf{r} = 2\mathbf{i} - \frac{2}{3}\mathbf{j} + \sqrt{2}\ln 3\mathbf{k}$; $s = \frac{8}{3}$

31. (a) $a_T = 0$, $a_N = \sqrt{2}$ (b) $a_T\mathbf{T} = \mathbf{0}$, $a_N\mathbf{N} = \mathbf{i} + \mathbf{j}$ (c) $1/\sqrt{2}$

33. (a) $a_T = 2\sqrt{5}$, $a_N = 2\sqrt{5}$ (b) $a_T\mathbf{T} = 2\mathbf{i} + 4\mathbf{j}$, $a_N\mathbf{N} = 4\mathbf{i} - 2\mathbf{j}$ (c) $2/\sqrt{5}$

35. (a) $a_T = 22/\sqrt{14}$, $a_N = \sqrt{38/7}$ (b) $a_T\mathbf{T} = \frac{11}{7}\mathbf{i} + \frac{22}{7}\mathbf{j} + \frac{33}{7}\mathbf{k}$, $a_N\mathbf{N} = -\frac{11}{7}\mathbf{i} - \frac{8}{7}\mathbf{j} + \frac{9}{7}\mathbf{k}$ (c) $\dfrac{\sqrt{19}}{7\sqrt{14}}$

37. (a) $a_T = 0$, $a_N = 3$ (b) $a_T\mathbf{T} = \mathbf{0}$, $a_N\mathbf{N} = -3\mathbf{i}$ (c) $\frac{3}{8}$ 39. $a_T = -3$, $a_N = 2$, $\mathbf{T} = -\mathbf{j}$, $\mathbf{N} = \mathbf{i}$

41. $a_T = \frac{4}{3}$, $a_N = \sqrt{29}/3$, $\mathbf{T} = \frac{1}{3}(2\mathbf{i} + 2\mathbf{j} + \mathbf{k})$, $\mathbf{N} = (\mathbf{i} - 8\mathbf{j} + 14\mathbf{k})/(3\sqrt{29})$

43. $\frac{3}{2}$ 45. $-\pi/\sqrt{2}$ 47. $a_N = 8.41 \times 10^{10}$ km/s^2 49. $a_N = 18/(1 + 4x^2)^{3/2}$ 51. $a_N = 0$ 53. ≈ 38.73 m/s

55. (a) $x = 160t$, $y = 160\sqrt{3}t - 16t^2$ (b) 1200 ft   (c) $1600\sqrt{3}$ ft   (d) 320 ft/s   57. $40\sqrt{3}$ ft   59. 800 ft/s

61. $15°$ or $75°$ 63. (c) 15 ft   65. (a) ≈ 0.00566 m   (b) $\frac{125}{4}$ m   67. (b) R is maximum when $\alpha = 45°$, maximum value v_0^2/g

69. (a) 2.62 s   (b) 181.5 ft   71. (a) $v_0 \approx 83$ ft/s, $\alpha \approx 8°$ (b) 268.76 ft

▶ **Exercise Set 14.7 (Page 885)**

7. 7.75 km/s   9. 10.88 km/s   11. (a) minimum distance $= 220{,}680$ mi, maximum distance $= 246{,}960$ mi   (b) 27.5 days

13. (a) $17{,}224$ mi/h   (b) $e \approx 0.07094$, apogee altitude $= 818$ mi

▶ **Chapter 14 Supplementary Exercises (Page 886)**

7. (a) $\mathbf{r}(t) = \displaystyle\int_0^t \cos\left(\frac{\pi u^2}{2}\right) du\,\mathbf{i} + \int_0^t \sin\left(\frac{\pi u^2}{2}\right) du\,\mathbf{j}$ (c) $\kappa(s) \to +\infty$

11. (a)

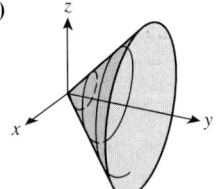

(b) $x = t$, $y = t^2$, $z = \pm\sqrt{4 - (t^2/3) - (t^4/6)}$

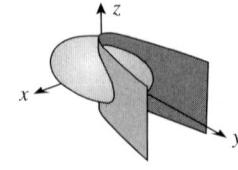

15. $\mathbf{r}(s) = \dfrac{s-3}{3}\mathbf{i} + \dfrac{12-2s}{3}\mathbf{j} + \dfrac{9+2s}{3}\mathbf{k}$

17. $(-1000, -100, 500)$, $\left(-\dfrac{1000}{729}, -\dfrac{100}{9}, \dfrac{500}{81}\right)$

19. (a) $\mathbf{r}(t) = \left(\tfrac{1}{6}t^4 + t\right)\mathbf{i} + \left(\tfrac{1}{2}t^2 + 2t\right)\mathbf{j} - \left(\tfrac{1}{4}\cos 2t + t - \tfrac{1}{4}\right)\mathbf{k}$ **(b)** 3475

▶ **Exercise Set 15.1 (Page 898)**

1. (a) 5   **(b)** 3   **(c)** 1   **(d)** -2 **(e)** $9a^3 + 1$ **(f)** $a^3b^2 - a^2b^3 + 1$ **3. (a)** $x^2 - y^2 + 3$ **(b)** $3x^3y^4 + 3$ **5.** $x^3e^{x^3(3y+1)}$

7. (a) $t^2 + 3t^{10}$ **(b)** 0   **(c)** 3076   **9. (a)** 19   **(b)** -9 **(c)** 3   **(d)** $a^6 + 3$ **(e)** $-t^8 + 3$ **(f)** $(a+b)(a-b)^2b^3 + 3$

11. $(y+1)e^{x^2(y+1)z^2}$ **13. (a)** $80\sqrt{\pi}$ **15.** 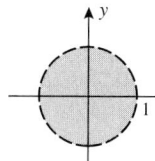   **17.**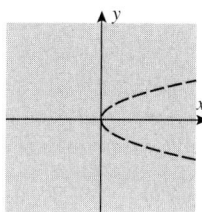

   (b) $n(n+1)/2$

19. (a) all points above or on the line $y = -2$ **(b)** all points on or within the sphere $x^2 + y^2 + z^2 = 25$ **(c)** all points in 3-space

21. 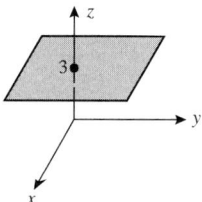   **23.** 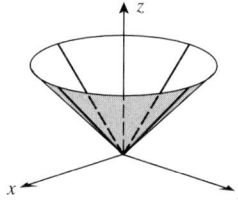   **25.** 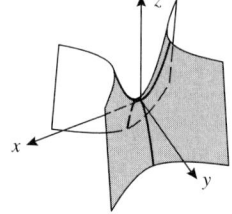   **27.**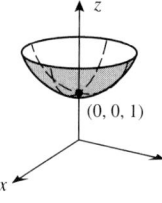

29. 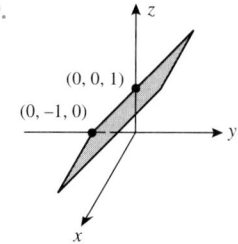   **31.** 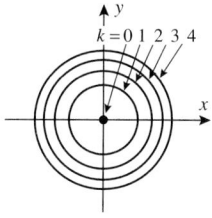   **33.** 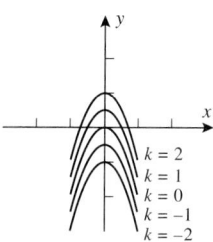   **35.**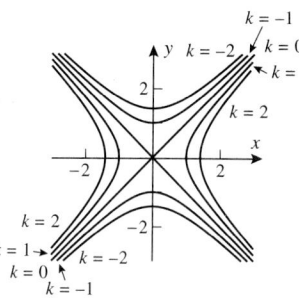

37. 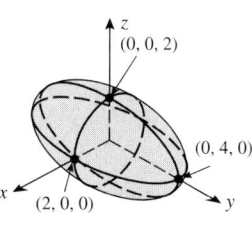   **39.** 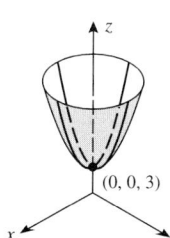   **41.** concentric spheres, common center at $(2, 0, 0)$

43. concentric cylinders, common axis the y-axis   **45. (a)** $x^2 - 2x^3 + 3xy = 0$ **(b)** $x^2 - 2x^3 + 3xy = 0$ **(c)** $x^2 - 2x^3 + 3xy = -18$

47. (a) $x^2 + y^2 - z = 5$ **(b)** $x^2 + y^2 - z = -2$ **(c)** $x^2 + y^2 - z = 0$

49. (a) 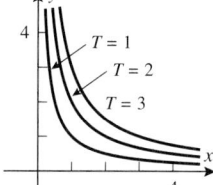   **(b)** the path $xy = 4$ **51. (a)** $1 - x^2 - y^2$ **53. (a)** A

   (b) $\sqrt{x^2 + y^2}$ **(b)** B

   (c) $x^2 + y^2$ **(c)** increase

   (d) decrease

   (e) increase

   (f) decrease

55. (a) 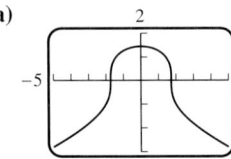 **(b)** 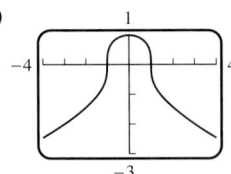 **57. (a)** 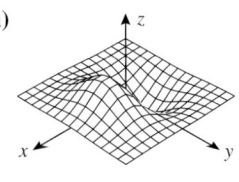 **(b)**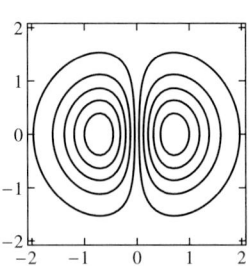

59. (a) The graph of g is the graph of f shifted one unit in the positive x-direction.

    (b) The graph of g is the graph of f shifted one unit up the z-axis.

    (c) The graph of g is the graph of f shifted one unit down the y-axis and then inverted with respected to the plane $z = 0$.

▶ **Exercise Set 15.2 (Page 908)**

1. 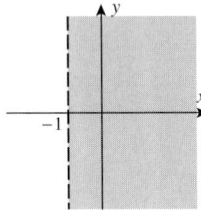   **3.** 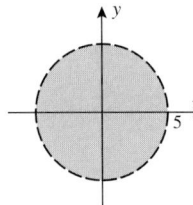   **5.** 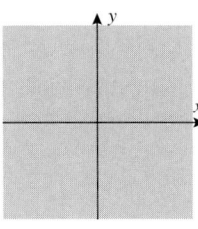   **7.**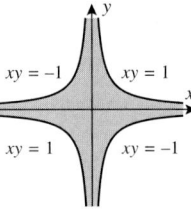

9. all of 3-space   **11.** all points not on the cylinder $x^2 + z^2 = 1$ **13.** 35   **15.** -8 **17.** 0   **21.** 1   **23.** 0   **25.** 0

27. limit does not exist   **29.** $\frac{8}{3}$ **31.** 0   **33.** 0   **35.** limit does not exist   **37. (a)** no   **(d)** no; yes   **41.** $-\pi/2$ **43.** no

▶ **Exercise Set 15.3 (Page 916)**

1. (a) $9x^2 y^2$ **(b)** $6x^3 y$ **(c)** $9y^2$ **(d)** $9x^2$ **(e)** $6y$ **(f)** $6x^3$ **(g)** 36   **(h)** 12

3. (a) $-\dfrac{1}{4x^{3/2}}\cos y$ **(b)** $-\sqrt{x}\cos y$ **(c)** $-\dfrac{\sin y}{2\sqrt{x}}$ **(d)** $-\dfrac{\sin y}{2\sqrt{x}}$ **5. (a)** $\frac{3}{8}$ **(b)** $\frac{1}{4}$ **7. (a)** $-4\cos 7$ **(b)** $2\cos 7$

9. $\partial z/\partial x = -4;\ \partial z/\partial y = \frac{1}{2}$ **11.** $8xy^3 e^{x^2 y^3},\ 12x^2 y^2 e^{x^2 y^3}$ **13.** $x^3/(y^{3/5} + x) + 3x^2\ln(1 + xy^{-3/5}),\ -\frac{3}{5}x^4/(y^{8/5} + xy)$

15. $-\dfrac{y(x^2 - y^2)}{(x^2 + y^2)^2},\ \dfrac{x(x^2 - y^2)}{(x^2 + y^2)^2}$ **17.** $f_x(x, y) = (3/2)x^2 y(5x^2 - 7)(3x^5 y - 7x^3 y)^{-1/2}$ **19.** $\dfrac{y^{-1/2}}{y^2 + x^2},\ -\dfrac{xy^{-3/2}}{y^2 + x^2} - \dfrac{3}{2}y^{-5/2}\tan^{-1}\left(\dfrac{x}{y}\right)$

    $f_y(x, y) = (1/2)x^3(3x^2 - 7)(3x^5 y - 7x^3 y)^{-1/2}$

21. $-\frac{4}{3}y^2\sec^2 x(y^2\tan x)^{-7/3},\ -\frac{8}{3}y\tan x(y^2\tan x)^{-7/3}$ **23.** $-6, -21$ **25.** $1/\sqrt{17}, 8/\sqrt{17}$

27. $f_{xy} = f_{yx} = -32y^3$ **29.** $f_{xy} = f_{yx} = -e^x\sin y$ **31.** $f_{xy} = f_{yx} = 20/(4x - 5y)^2$ **35. (a)** $\pm\sqrt{6}/4$ **37.** $-x/z, -y/z$

39. $-\dfrac{2x + yz^2\cos xyz}{xyz\cos xyz + \sin xyz};\ -\dfrac{xz^2\cos xyz}{xyz\cos xyz + \sin xyz}$

41. $z = f(x, y)$ has II as its graph, f_x has I as its graph and f_y has III as its graph.

43. (a) $30xy^4 - 4$ **(b)** $60x^2 y^3$ **(c)** $60x^3 y^2$ **45. (a)** -30 **(b)** -125 **(c)** 150

47. (a) $\dfrac{\partial^3 f}{\partial x^3}$ **(b)** $\dfrac{\partial^3 f}{\partial y^2\partial x}$ **(c)** $\dfrac{\partial^4 f}{\partial x^2\partial y^2}$ **(d)** $\dfrac{\partial^4 f}{\partial y^3\partial x}$

49. (a) $2xy^4 z^3 + y$ **(b)** $4x^2 y^3 z^3 + x$ **(c)** $3x^2 y^4 z^2 + 2z$ **(d)** $2y^4 z^3 + y$ **(e)** $32z^3 + 1$ **(f)** 438

51. $2z/x, z/y, \ln(x^2 y\cos z) - z\tan z$ **53.** $-y^2 z^3/(1 + x^2 y^4 z^6), -2xyz^3/(1 + x^2 y^4 z^6), -3xy^2 z^2/(1 + x^2 y^4 z^6)$

55. $yze^z\cos xz, e^z\sin xz, ye^z(\sin xz + x\cos xz)$ **57.** $x/\sqrt{x^2 + y^2 + z^2}, y/\sqrt{x^2 + y^2 + z^2}, z/\sqrt{x^2 + y^2 + z^2}$

59. (a) e **(b)** $2e$ **(c)** e **63.** $-x/w, -y/w, -z/w$ **65.** $-\dfrac{yzw\cos xyz}{2w + \sin xyz}, -\dfrac{xzw\cos xyz}{2w + \sin xyz}, -\dfrac{xyw\cos xyz}{2w + \sin xyz}$

67. (a) $15x^2 y^4 z^7 + 2y$ **(b)** $35x^3 y^4 z^6 + 3y^2$ **(c)** $21x^2 y^5 z^6$ **(d)** $42x^3 y^5 z^5$ **(e)** $140x^3 y^3 z^6 + 6y$

    (f) $30xy^4 z^7$ **(g)** $105x^2 y^4 z^6$ **(h)** $210xy^4 z^6$

69. $e^{x^2}, -e^{y^2}$ **71.** $-i\sin(x_1 + 2x_2 + \cdots + nx_n)$ **79.** -2 **81. (a)** $\partial V/\partial r = 2\pi rh$ **(b)** $\partial V/\partial h = \pi r^2$ **(c)** 48π **(d)** 64π

83. (a) $\dfrac{1}{5}\dfrac{\text{lb}}{\text{in}^2\cdot\text{K}}$ **(b)** $-\dfrac{25}{8}\dfrac{\text{in}^5}{\text{lb}^3}$ **85.** $-1 - \dfrac{\cos(x - y)}{\cos(x + z)};\ \dfrac{\cos(x - y)}{\cos(x + z)};\ -\dfrac{\cos^2(x + z)\sin(x - y) + \cos^2(x - y)\sin(x + z)}{\cos^3(x + z)}$

87. (a) 4   **(b)** 8   **89.** $11, -8$ **91. (b)** does not exist where $y = -x, x \neq 0$

▶ Exercise Set 15.4 **(Page 928)**

1. $42t^{13}$ 3. $3t^{-2}\sin(1/t)$ 5. $-\frac{10}{3}t^{7/3}e^{1-t^{10/3}}$ 7. $24u^2v^2 - 16uv^3 - 2v + 3$, $16u^3v - 24u^2v^2 - 2u - 3$

9. $-\dfrac{2\sin u}{3\sin v}$, $-\dfrac{2\cos u\cos v}{3\sin^2 v}$ 11. $e^u, 0$

13. $3r^2\sin\theta\cos^2\theta - 4r^3\sin^3\theta\cos\theta$, $-2r^3\sin^2\theta\cos\theta + r^4\sin^4\theta + r^3\cos^3\theta - 3r^4\sin^2\theta\cos^2\theta$

15. $\dfrac{x^2 + y^2}{4x^2y^3}$, $\dfrac{y^2 - 3x^2}{4xy^4}$ 17. $-\pi$ 19. $\sqrt{3}e^{\sqrt{3}}$, $(2 - 4\sqrt{3})e^{\sqrt{3}}$ 21. $-\dfrac{2xy^3}{3x^2y^2 - \sin y}$ 23. $-\dfrac{ye^{xy}}{xe^{xy} + ye^y + e^y}$

25. -39 mi/h   27. $-\frac{7}{36}\sqrt{3}$ rad/s   29. $16{,}200\pi$ in^3/year   31. **(a)** xy-plane **(b)** $y \neq 0$

33. **(a)** 4 **(b)** 5   35. **(a)** 2 **(b)** 1 **(c)** 3 **(d)** -4 37. **(a)** $\dfrac{\partial z}{\partial x} = \dfrac{dz}{du}\dfrac{\partial u}{\partial x}$, $\dfrac{\partial z}{\partial y} = \dfrac{dz}{du}\dfrac{\partial u}{\partial y}$

43. **(d)** $\frac{1}{2}(-\cos(x + t) + \cos(x - t))$

▶ Exercise Set 15.5 **(Page 935)**

1. tangent plane: $48x - 14y - z = 64$; normal line: $x = 1 + 48t$, $y = -2 - 14t$, $z = 12 - t$

3. tangent plane: $x - y - z = 0$; normal line: $x = 1 + t$, $y = -t$, $z = 1 - t$

5. tangent plane: $3y - z = -1$; normal line: $x = \pi/6$, $y = 3t$, $z = 1 - t$

7. tangent plane: $3x - 4z = -25$; normal line: $x = -3 + (3t/4)$, $y = 0$, $z = 4 - t$

9. **(a)** all points on the x-axis or y-axis **(b)** $(0, -2, -4)$

11. $\left(\frac{1}{2}, -2, -\frac{3}{4}\right)$ 13. $df = 0.8$, $\Delta f = 0.872$ 15. $\frac{7}{2}$

17. $dz = 3x^2y^2\,dx + 2x^3y\,dy$, $\Delta z = (x + \Delta x)^3(y + \Delta y)^2 - x^3y^2$ 19. $dz = 7\,dx - 2\,dy$

21. $dz = \dfrac{y}{1 + x^2y^2}\,dx + \dfrac{x}{1 + x^2y^2}\,dy$ 25. $\approx 92.94°$ 27. 0.20232; actual value ≈ 0.202334 29. 0.10 31. 0.03

33. 0.088 cm   35. 8% 37. $r\%$ 39. 2% 41. 0.004 radians   43. 0.3%

45. **(a)** $(r + s)\%$ **(b)** $(r + s)\%$ **(c)** $(2r + 3s)\%$ **(d)** $\left(3r + \dfrac{s}{2}\right)\%$

47. **(a)** $(-2, 1, 5)$, $(0, 3, 9)$

(b) At $(-2, 1, 5)$ the cosine of the acute angle is $4/(3\sqrt{14})$; at $(0, 3, 9)$ the cosine of the acute angle is $4/\sqrt{222}$.

▶ Exercise Set 15.6 **(Page 943)**

1. III   3. $4\mathbf{i} - 8\mathbf{j}$ 5. $\dfrac{x}{x^2 + y^2}\mathbf{i} + \dfrac{y}{x^2 + y^2}\mathbf{j}$ 7. $-36\mathbf{i} - 12\mathbf{j}$ 9. $4\mathbf{i} + 4\mathbf{j}$ 11. $6\sqrt{2}$ 13. $-3/\sqrt{10}$ 15. 0

17. $-8\sqrt{2}$ 19. $\sqrt{2}/4$ 21. $1/2 + \sqrt{3}/8$ 23. $2\sqrt{2}$ 25.   27.

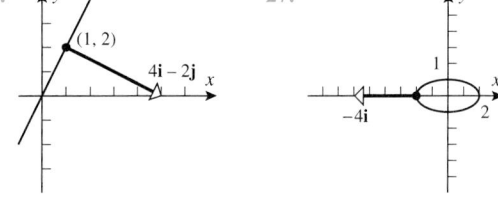

29. $\mathbf{u} = (3\mathbf{i} - 2\mathbf{j})/\sqrt{13}$, $\|\nabla f(-1, 1)\| = 4\sqrt{13}$ 31. $\mathbf{u} = (4\mathbf{i} - 3\mathbf{j})/5$, $\|\nabla f(4, -3)\| = 1$

33. $\mathbf{u} = -(\mathbf{i} + 3\mathbf{j})/\sqrt{10}$, $-\|\nabla f(-1, -3)\| = -2\sqrt{10}$ 35. $\mathbf{u} = (3\mathbf{i} - \mathbf{j})/\sqrt{10}$, $-\|\nabla f(\pi/6, \pi/4)\| = -\sqrt{5}$ 37. $1/\sqrt{5}$

39. $-3e/2$ 41. $\pm(-4\mathbf{i} + \mathbf{j})/\sqrt{17}$ 43. **(a)** 5 **(b)** 10 **(c)** $-5\sqrt{5}$ 45. $8/\sqrt{29}$

47. **(a)** $\approx 1/\sqrt{2}$ **(b)**   49. all points on the ellipse $9x^2 + y^2 = 9$ 51. $36/\sqrt{17}$ 53. **(a)** $2e^{-\pi/2}\mathbf{i}$

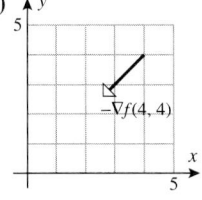

59. $x(t) = e^{-8t}$, $y(t) = 4e^{-2t}$ **61.**

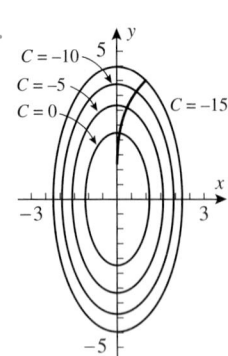

$C = -10$
$C = -5$
$C = 0$
$C = -15$

63. (a)

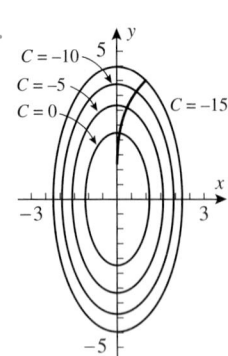

(c) $\nabla f = [2x - 2x(x^2 + 3y^2)]e^{-(x^2+y^2)}\mathbf{i} + [6y - 2y(x^2 + 3y^2)]e^{-(x^2+y^2)}\mathbf{j}$
(d) $x = y = 0$ or $x = 0$, $y = \pm 1$ or $x = \pm 1$, $y = 0$

▶ **Exercise Set 15.7 (Page 953)**

1. $165t^{32}$ **3.** $-2t \cos(t^2)$ **5.** 3264 **7.** $\nabla f(2, -1, 1) = 320\mathbf{i} - 256\mathbf{j} + 384\mathbf{k}$

9. $\nabla f(x, y, z) = \nabla f(-1, 2, 4) = -\frac{2}{57}\mathbf{i} + \frac{8}{57}\mathbf{j} + \frac{24}{57}\mathbf{k}$, $D_{\mathbf{u}}f = -\frac{314}{741}$ **11.** $72/\sqrt{14}$ **13.** $-\frac{8}{63}$

15. $\mathbf{u} = (\mathbf{i} - \mathbf{j})/\sqrt{2}$, $\|\nabla f(1, 1, -1)\| = 3\sqrt{2}$ **17.** $\mathbf{u} = (-\mathbf{i} + \mathbf{j})/\sqrt{2}$, $\|\nabla f(1, 2, -2)\| = 1/\sqrt{2}$

19. $\mathbf{u} = (\mathbf{i} - 11\mathbf{j} + 12\mathbf{k})/\sqrt{266}$, $-\|\nabla f(5, 7, 6)\| = -\sqrt{266}$ **21.** $3/\sqrt{11}$ **23.** $-\frac{10}{3}\mathbf{i} + \frac{5}{3}\mathbf{j} + \frac{10}{3}\mathbf{k}$

25. (a) $x + y + 2z = 6$ **(b)** $x(t) = 2 + t$, $y(t) = 2 + t$, $z(t) = 1 + 2t$ **(c)** $\approx 35.26°$ **27.** $\pm\frac{1}{363}(\mathbf{i} - \mathbf{j} - 19\mathbf{k})$

31. $\left(1, \frac{2}{3}, \frac{2}{3}\right)$, $\left(-1, -\frac{2}{3}, -\frac{2}{3}\right)$ **33.** $x(t) = 1 + 8t$, $y(t) = -1 + 5t$, $z(t) = 2 + 6t$

35. $x(t) = 3 + 4t$, $y(t) = -3 - 4t$, $z(t) = 4 - 3t$

37. $dw = 3x^2 y^2 z \, dx + 2x^3 yz \, dy + x^3 y^2 \, dz$, $\Delta w = (x + \Delta x)^3 (y + \Delta y)^2 (z + \Delta z) - x^3 y^2 z$

39. $8 \, dx - 3 \, dy + 4 \, dz$ **41.** $\dfrac{yz}{1 + x^2 y^2 z^2} \, dx + \dfrac{xz}{1 + x^2 y^2 z^2} \, dy + \dfrac{xy}{1 + x^2 y^2 z^2} \, dz$ **43.** 0.96 **45.** 2.35 cm^3 **47.** 39 ft^2

49. $\partial f/\partial v = 8vw^3 x^4 y^5$, $\partial f/\partial w = 12v^2 w^2 x^4 y^5$, $\partial f/\partial x = 16v^2 w^3 x^3 y^5$, $\partial f/\partial y = 20v^2 w^3 x^4 y^4$

51. $\partial f/\partial v_1 = 2v_1/(v_3^2 + v_4^2)$, $\partial f/\partial v_2 = -2v_2/(v_3^2 + v_4^2)$, $\partial f/\partial v_3 = -2v_3(v_1^2 - v_2^2)/(v_3^2 + v_4^2)^2$, $\partial f/\partial v_4 = -2v_4(v_1^2 - v_2^2)/(v_3^2 + v_4^2)^2$

53. (a) 0 **(b)** 0 **(c)** 0 **(d)** 0 **(e)** $2(yw + 1)e^{yw} \sin z \cos z$ **(f)** $2xw(yw + 2)e^{yw} \sin z \cos z$

55. $2r \cos^2 \theta/(r^2 \cos^2 \theta + 1)$, $-2r^2 \sin \theta \cos \theta/(r^2 \cos^2 \theta + 1)$ **57.** $2\rho(4 \sin^2 \phi + \cos^2 \phi)$, $6\rho^2 \sin \phi \cos \phi$, 0

59. (a) $60 \text{ in}^3/\text{s}$ **(b)** $\frac{26}{7} \text{ in/s}$ **65.** $\dfrac{\partial w}{\partial \rho} = \sin \phi \cos \theta \dfrac{\partial w}{\partial x} + \sin \phi \sin \theta \dfrac{\partial w}{\partial y} + \cos \phi \dfrac{\partial w}{\partial z}$

$\dfrac{\partial w}{\partial \phi} = \rho \cos \phi \cos \theta \dfrac{\partial w}{\partial x} + \rho \cos \phi \sin \theta \dfrac{\partial w}{\partial y} - \rho \sin \phi \dfrac{\partial w}{\partial z}$

$\dfrac{\partial w}{\partial \theta} = -\rho \sin \phi \sin \theta \dfrac{\partial w}{\partial x} + \rho \sin \phi \cos \theta \dfrac{\partial w}{\partial y}$

67. $\dfrac{\partial z}{\partial x} = \dfrac{2x + yz}{6yz - xy}$, $\dfrac{\partial z}{\partial y} = \dfrac{xz - 3z^2}{6yz - xy}$ **69.** $\dfrac{\partial z}{\partial x} = \dfrac{ye^x}{15 \cos 3z + 3}$, $\dfrac{\partial z}{\partial y} = \dfrac{e^x}{15 \cos 3z + 3}$

75. (a) $dw/dt = \displaystyle\sum_{i=1}^{4}(\partial w/\partial x_i)(dx_i/dt)$ **(b)** $\partial w/\partial v_j = \displaystyle\sum_{i=1}^{4}(\partial w/\partial x_i)(\partial x_i/\partial v_j)$ for $j = 1, 2, 3$

▶ **Exercise Set 15.8 (Page 963)**

1. (a) minimum at $(2, -1)$, no maxima   **(b)** maximum at $(0, 0)$, no minima   **(c)** no maxima or minima

3. minimum at $(3, -2)$, no maxima   **5.** relative minimum at $(0, 0)$

7. relative minimum at $(0, 0)$; saddle points at $(\pm 2, 1)$ **9.** saddle point at $(1, -2)$

11. relative minimum at $(2, -1)$

13. relative minima at $(-1, -1)$ and $(1, 1)$ **15.** saddle point at $(0, 0)$

17. no critical points   19. relative maximum at $(-1, 0)$ 21. saddle point at $(0, 0)$;

relative minima at $(1, 1)$ and $(-1, -1)$

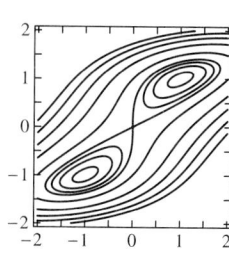

23. **(b)** relative minimum at $(0, 0)$ 27. absolute maximum 0, absolute minimum -12

29. absolute maximum 3, absolute minimum -1 31. absolute maximum $\frac{33}{4}$, absolute minimum $-\frac{1}{4}$ 33. 16, 16, 16

35. maximum at $(1, 2, 2)$ 37. $2a/\sqrt{3}, 2a/\sqrt{3}, 2a/\sqrt{3}$ 39. length and width 2 ft, height 4 ft

41. **(a)** $x = 0$: minimum -3, maximum 0;   **(b)** $y = x$: minimum 0, maximum 3; $y = 1 - x$: maximum 4, minimum -3

$x = 1$: minimum 3, maximum 13/3;   **(c)** minimum -3, maximum 13/3

$y = 0$: minimum 0, maximum 4;

$y = 1$: minimum -3, maximum 3

43. length and width $\sqrt[3]{2V}$, height $\sqrt[3]{2V}/2$ 47. $y = \frac{3}{4}x + \frac{19}{12}$ 49. $y = 0.5x + 0.8$

51. **(a)** $y = \dfrac{8843}{140} + \dfrac{57}{200}t$ **(b)** 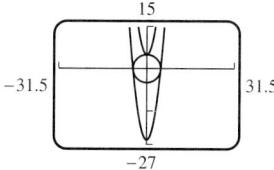   **(c)** $y = \dfrac{2909}{35}$

53. **(a)** $P = \dfrac{2798}{21} + \dfrac{171}{350}T$ **(b)** 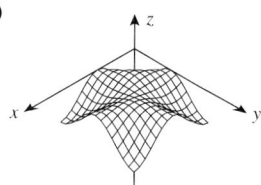   **(c)** $T \approx -272.7096°\text{C}$

▶ Exercise Set 15.9 **(Page 973)**

1. **(a)** 4   3. maximum $\sqrt{2}$ at $(-\sqrt{2}, -1)$ and $(\sqrt{2}, 1)$, minimum $-\sqrt{2}$ at $(-\sqrt{2}, 1)$ and $(\sqrt{2}, -1)$

5. maximum $\sqrt{2}$ at $(1/\sqrt{2}, 0)$, minimum $-\sqrt{2}$ at $(-1/\sqrt{2}, 0)$ 7. maximum 6 at $\left(\frac{4}{3}, \frac{2}{3}, -\frac{4}{3}\right)$, minimum -6 at $\left(-\frac{4}{3}, -\frac{2}{3}, \frac{4}{3}\right)$

9. maximum is $1/(3\sqrt{3})$ at $(1/\sqrt{3}, 1/\sqrt{3}, 1/\sqrt{3})$, $(1/\sqrt{3}, -1/\sqrt{3}, -1/\sqrt{3})$, $(-1/\sqrt{3}, 1/\sqrt{3}, -1/\sqrt{3})$, and $(-1/\sqrt{3}, -1/\sqrt{3}, 1/\sqrt{3})$;

minimum is $-1/(3\sqrt{3})$ at $(1/\sqrt{3}, 1/\sqrt{3}, -1/\sqrt{3})$, $(1/\sqrt{3}, -1/\sqrt{3}, 1/\sqrt{3})$, $(-1/\sqrt{3}, 1/\sqrt{3}, 1/\sqrt{3})$, and $(-1/\sqrt{3}, -1/\sqrt{3}, -1/\sqrt{3})$

11. $\left(\frac{3}{10}, -\frac{3}{5}\right)$ 13. $\left(\frac{1}{6}, \frac{1}{3}, \frac{1}{6}\right)$ 15. $(3, 6)$ is closest and $(-3, -6)$ is farthest

17. $5(\mathbf{i} + \mathbf{j} + \mathbf{k})/\sqrt{3}$ 19. **(a)**   **(c)** maximum $\frac{101}{4}$, minimum -5 21. 9, 9, 9

23. $(\pm\sqrt{5}, 0, 0)$ 25. length and width 2 ft, height 4 ft

29. **(a)** $\alpha = \beta = \gamma = \pi/3$, maximum $(3\sqrt{3})/8$ **(b)**

▶ Chapter 15 Supplementary Exercises **(Page 974)**

3.

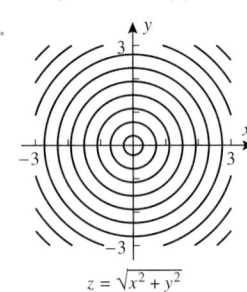

$z = \sqrt{x^2 + y^2}$

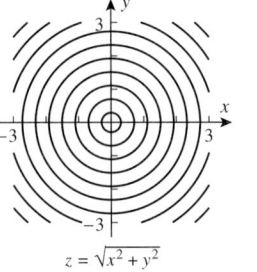

$z = x^2 + y^2$

5. **(b)** $f(x, y, z) = z - x^2 - y^2$ 7. **(a)** xy **(b)** $e^{r+s} \ln rs$

13. **(a)** 12 N/(m²·min) **(b)** 240 N/(m²·min)   15. limit $= -1$, continuous   17. $-7/\sqrt{5}$ 19. $(0, 0, 2), (1, 1, 1), (-1, -1, 1)$

21. $\left(-\frac{1}{3}, -\frac{1}{2}, 2\right)$ 23. $dV = -0.06667$ m³; $\Delta V = 0.07267$ m³   25. saddle points at $(\pm 6, 3)$; maximum at $(0, 0)$

27. **(a)**

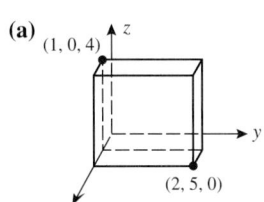

(b)

29. **(a)** 102,033,960.1   **(b)** $L = 120,000, K = 80,000$

31. $\dfrac{2a}{\sqrt{3}} \times \dfrac{2b}{\sqrt{3}} \times \dfrac{2c}{\sqrt{3}}$

33. $-\dfrac{F_{xx}F_y - 2F_x F_{xy}}{F_y^2}$ 35. $67.7°, 61.1°$ 37. increasing

▶ Exercise Set 16.1 **(Page 983)**

1. 7   3. 2   5. 2   7. 3   9. $1 - \ln 2$ 11. $\dfrac{1 - \ln 2}{2}$ 13. 0   15. $\frac{1}{3}$ 19. 19   21. 8

23. **(a)** $(1, 0, 4)$, $(2, 5, 0)$ **(b)** $(0, 0, 5)$, $(3, 4, 0)$ 25. $\dfrac{1}{3\pi}$ 27. $1 - \dfrac{2}{\pi}$ 29. $\dfrac{14}{3}°$ 31. 0.6211310829

35. first integral equals $\frac{1}{2}$, second equals $-\frac{1}{2}$; no

▶ Exercise Set 16.2 **(Page 991)**

1. $\frac{1}{40}$ 3. 9   5. $\dfrac{\pi}{2}$ 7. 1   9. $\frac{1}{12}$ 11. **(a)** $\frac{16}{3}$ **(b)** 38   13. 576   15. $\dfrac{\sqrt{17} - 1}{2}$ 17. 0   19. $\frac{50}{3}$

21. $-\frac{1}{2}$ 25. $\sqrt{2} - 1$ 27. 32   29. 12   31. 27π 33. 170   35. $\dfrac{27\pi}{2}$ 37. $\frac{2000}{3}$ 39. $\dfrac{\pi}{2}$

41. $\displaystyle\int_0^{\sqrt{2}} \int_{y^2}^2 f(x, y)\, dx\, dy$ 43. $\displaystyle\int_1^{e^2} \int_{\ln x}^2 f(x, y)\, dy\, dx$ 45. $\displaystyle\int_0^{\pi/2} \int_0^{\sin x} f(x, y)\, dy\, dx$ 47. $\dfrac{1 - e^{-16}}{8}$ 49. $\dfrac{e^8 - 1}{3}$

51. $\dfrac{1 - \cos 8}{3}$ 53. **(a)** 0   **(b)** $\tan 1$ 55. 0   57. $\dfrac{\pi}{2} - \ln 2$ 59. 0.676089

▶ Exercise Set 16.3 **(Page 998)**

1. $\frac{1}{6}$ 3. $\frac{2}{9}a^3$ 5. 0   7. $\dfrac{3\pi}{2}$ 9. $\dfrac{\pi}{16}$ 11. $\dfrac{4\pi}{3} + 2\sqrt{3}$ 13. $\dfrac{64\sqrt{2}}{3}\pi$ 15. $\dfrac{5\pi}{32}$ 17. $\dfrac{27\pi}{16}$

19. $(1 - e^{-1})\pi$ 21. $\frac{\pi}{8}\ln 5$ 23. $\frac{\pi}{8}$ 25. $\frac{16}{9}$ 27. $\frac{\pi}{2}\left(1 - \frac{1}{\sqrt{1+a^2}}\right)$ 29. $\frac{\pi}{4}(\sqrt{5}-1)$ 31. $\frac{4}{3}a^3$

33. $\frac{(3\pi - 4)a^2 c}{9}$ 35. $\frac{4\pi}{3} + 2\sqrt{3} - 2$ 37. **(b)** $\frac{\pi}{4}$ 39. $2\pi k[1 - (R+1)e^{-R}]$

▶ **Exercise Set 16.4 (Page 1009)**

1. **(a)** 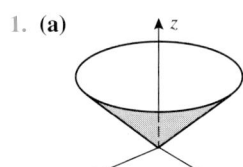   **(b)** 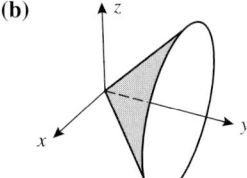   **(c)** 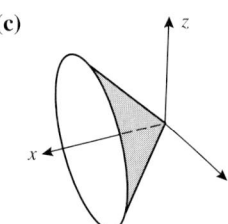   3. **(a)** $x = u, y = v, z = \frac{5}{2} + \frac{3}{2}u - 2v$
   (b) $x = u, y = v, z = u^2$

5. **(a)** $x = 5\cos u, y = 5\sin u, z = v; 0 \le u \le 2\pi, 0 \le v \le 1$ **(b)** $x = 2\cos u, y = v, z = 2\sin u; 0 \le u \le 2\pi, 1 \le v \le 3$

7. $x = u, y = \sin u \cos v, z = \sin u \sin v$ 9. $x = r\cos\theta, y = r\sin\theta, z = \frac{1}{1+r^2}$

11. $x = r\cos\theta, y = r\sin\theta, z = 2r^2\cos\theta\sin\theta$ 13. $x = r\cos\theta, y = r\sin\theta, z = \sqrt{9 - r^2}; r \le \sqrt{5}$

15. $x = \frac{1}{2}\rho\cos\theta, y = \frac{1}{2}\rho\sin\theta, z = \frac{\sqrt{3}}{2}\rho$ 17. $z = x - 2y$; a plane

19. $(x/3)^2 + (y/2)^2 = 1; 2 \le z \le 4$; part of an elliptic cylinder   21. $(x/3)^2 + (y/4)^2 = z^2; 0 \le z \le 1$; part of an elliptic cone

23. **(a)** $x = r\cos\theta, y = r\sin\theta, z = r; x = u, y = v, z = \sqrt{u^2 + v^2}; 0 \le z \le 2$

25. **(a)** $0 \le u \le 3, 0 \le v \le \pi$ **(b)** $0 \le u \le 4, -\pi/2 \le v \le \pi/2$ 27. **(a)** $0 \le \phi \le \pi/2, 0 \le \theta \le 2\pi$ **(b)** $0 \le \phi \le \pi, 0 \le \theta \le \pi$

29. $2x + 4y - z = 5$ 31. $z = 0$ 33. $x - y + \frac{\sqrt{2}}{2}z = \frac{\pi\sqrt{2}}{8}$ 35. 6π 37. $\frac{\sqrt{5}}{6}$ 39. $\frac{(5\sqrt{5}-1)\pi}{6}$

41. $\frac{(17\sqrt{17} - 5\sqrt{5})\pi}{6}$ 43. $\frac{(10\sqrt{10}-1)\pi}{18}$ 45. 8π 47. $4\pi a^2$ 49. $\pi a\sqrt{a^2 + h^2}$ 51. $4\pi^2 ab$ 53. 9.099

55. ellipsoid   57. hyperboloid of two sheets

▶ **Exercise Set 16.5 (Page 1018)**

1. 8 3. 7 5. $\frac{81}{5}$ 7. $\frac{128}{15}$ 9. $\pi(\pi - 3)/2$ 11. $\frac{1}{6}$ 15. 4 17. $\frac{256}{15}$ 19. $4\int_0^1 \int_0^{\sqrt{1-x^2}} \int_{4x^2+y^2}^{4-3y^2} dz\, dy\, dx$

21. $2\int_{-3}^3 \int_0^{\frac{1}{3}\sqrt{9-x^2}} \int_0^{x+3} dz\, dy\, dx$ 23. **(a)** **(b)** **(c)**

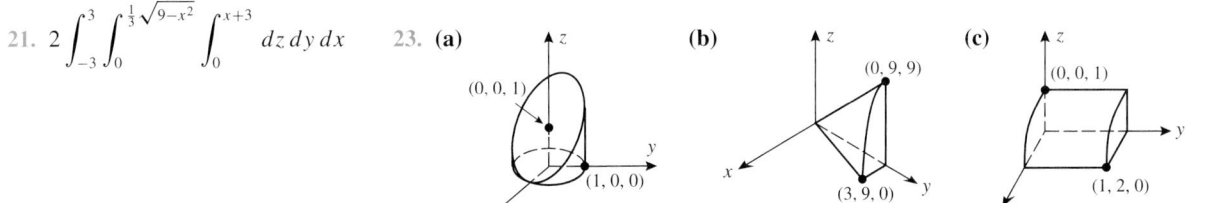

25. $\frac{3}{4}$ 27. **(a)** $\int_0^a \int_0^{b(1-x/a)} \int_0^{c(1-x/a-y/b)} dz\, dy\, dx, \int_0^b \int_0^{a(1-y/b)} \int_0^{c(1-x/a-y/b)} dz\, dx\, dy,$

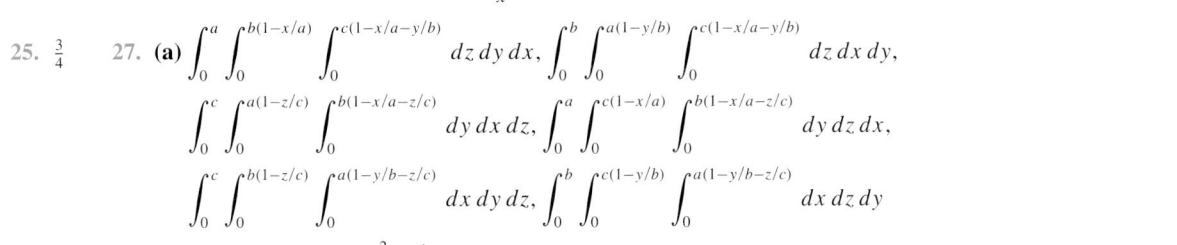

$\int_0^c \int_0^{a(1-z/c)} \int_0^{b(1-x/a-z/c)} dy\, dx\, dz, \int_0^a \int_0^{c(1-x/a)} \int_0^{b(1-x/a-z/c)} dy\, dz\, dx,$

$\int_0^c \int_0^{b(1-z/c)} \int_0^{a(1-y/b-z/c)} dx\, dy\, dz, \int_0^b \int_0^{c(1-y/b)} \int_0^{a(1-y/b-z/c)} dx\, dz\, dy$

29. $V = \frac{4}{3}\pi abc$ 31. **(a)** 0 **(b)** $\frac{e^2 - 1}{2}$

▶ **Exercise Set 16.6 (Page 1027)**

1. The fulcrum should be placed $\frac{50}{7}$ ft to the right of m_1.   3. $\left(\frac{1}{2}, \frac{1}{2}\right)$ 5. $\left(\frac{2}{3}, \frac{1}{3}\right)$ 7. $\left(\frac{5}{14}, \frac{38}{35}\right)$ 9. $\left(0, \frac{4(b^3 - a^3)}{3\pi(b^2 - a^2)}\right)$

11. $M = \frac{13}{20}$, center of gravity $\left(\frac{190}{273}, \frac{6}{13}\right)$ 13. $M = a^4/8$, center of gravity $(8a/15, 8a/15)$ 15. $\left(\frac{1}{2}, \frac{1}{2}, \frac{1}{2}\right)$ 17. $\left(\frac{1}{4}, \frac{1}{4}, \frac{1}{4}\right)$

19. $\left(\frac{1}{2}, 0, \frac{3}{5}\right)$ 21. $(3a/8, 3a/8, 3a/8)$ 23. $M = a^4/2$, center of gravity $(a/3, a/2, a/2)$ 25. $M = \frac{1}{6}$, center of gravity $\left(0, \frac{16}{35}, \frac{1}{2}\right)$

27. **(a)** $\left(\frac{5}{8}, \frac{5}{8}\right)$ **(b)** $\left(\frac{2}{3}, \frac{1}{2}\right)$ 29. $(1.177406, 0.353554, 0.231557)$ 33. $\left(\frac{128}{105\pi}, \frac{128}{105\pi}\right)$ 37. $2\pi^2 abk$ 39. $(a/3, b/3)$

▶ **Exercise Set 16.7 (Page 1037)**

1. $\frac{\pi}{4}$ 3. $\frac{\pi}{16}$ 5. $\frac{81\pi}{2}$ 7. $\frac{8(10\sqrt{5} - 19)\pi}{3}$ 9. $\frac{64\pi}{3}$ 11. $\frac{11\pi a^3}{3}$ 13. $\frac{\pi a^6}{48}$ 15. $\frac{32(2\sqrt{2} - 1)\pi}{15}$

19. $\frac{4\pi a^3}{3}$ 21. $\frac{27\pi}{4}$ 23. $\pi k a^4$ 25. $\left(0, 0, \frac{7}{16\sqrt{2} - 14}\right)$ 27. $(3a/8, 3a/8, 3a/8)$ 29. $\left(\frac{4}{3}, 0, \frac{10}{9}\right)$

31. $\frac{2(\sqrt{3} - 1)\pi}{3}$ 33. $\left(0, 0, \frac{11}{30}\right)$ 35. $(0, 0, 2a/5)$ 37. $\frac{4}{3}\pi(1 - e^{-1})\delta_0 R^3$ 39. $\frac{1}{2}\delta\pi a^4 h$ 41. $\frac{1}{2}\delta\pi(a_2^4 - a_1^4)h$

▶ **Exercise Set 16.8 (Page 1049)**

1. -17 3. $\cos(u - v)$ 5. $x = \frac{2}{9}u + \frac{5}{9}v, y = -\frac{1}{9}u + \frac{2}{9}v; \frac{1}{9}$ 7. $x = \frac{\sqrt{u + v}}{\sqrt{2}}, y = \frac{\sqrt{v - u}}{\sqrt{2}}; \frac{1}{4\sqrt{v^2 - u^2}}$ 9. 5

11. $\frac{1}{v}$ 13.   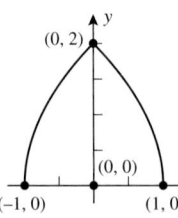   15.   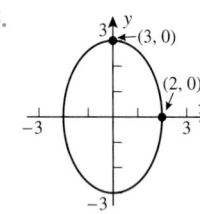   17. $\frac{3}{2}\ln 3$ 19. $1 - \frac{1}{2}\sin 2$ 21. 96π 23. $\frac{\pi}{24}(1 - \cos 1)$

25. $\frac{192}{5}\pi$ 27. $\frac{1}{4}\ln\frac{5}{2}$ 29. $\frac{1}{2}\left[\ln(\sqrt{2} + 1) - \frac{\pi}{4}\right]$ 31. $\frac{35}{256}$ 33. $2\ln 3$ 37. $-\frac{2}{3}\ln 2$ 39. $\frac{7}{4}(e^3 - e)$ 41. **(b)** 1

▶ **Chapter 16 Supplementary Exercises (Page 1051)**

3. **(a)** $\iint\limits_R dA$ **(b)** $\iiint\limits_G dV$ **(c)** $\iint\limits_R \sqrt{\left(\frac{\partial z}{\partial x}\right)^2 + \left(\frac{\partial z}{\partial y}\right)^2} \, dA$ 7. $\int_0^1 \int_{1-\sqrt{1-y^2}}^{1+\sqrt{1-y^2}} f(x, y) \, dx \, dy$ 9. **(a)** $a = 2, b = 1, c = 1, d = 2$
(b) 3

11. $-\frac{1}{\sqrt{2\pi}}$ 13. $\int_0^1 \int_{2y}^2 e^x e^y \, dx \, dy$ 15.   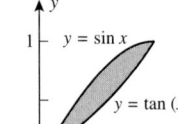   17. $\frac{1}{3}(1 - \cos 64)$ 19. a^2 21. 32π

23. **(a)** $\int_0^{2\pi} \int_0^{\pi/3} \int_0^a \rho^4 \sin^3\phi \, d\rho \, d\phi \, d\theta$ **(b)** $\int_0^{2\pi} \int_0^{\sqrt{3}a/2} \int_{r/\sqrt{3}}^{\sqrt{a^2 - r^2}} r^3 \, dz \, dr \, d\theta$ **(c)** $\int_{-\sqrt{3}a/2}^{\sqrt{3}a/2} \int_{-\sqrt{(3a^2/4)-x^2}}^{\sqrt{(3a^2/4)-x^2}} \int_{\sqrt{x^2+y^2}/\sqrt{3}}^{\sqrt{a^2-x^2-y^2}} (x^2 + y^2) \, dz \, dy \, dx$

25. $\frac{3}{2}$ 27. $\frac{\pi a^3}{9}$ 29. $\frac{8\pi}{3}(3\sqrt{3} - 1)$ 31. $2x + 4y - z = 5$ 33. $\left(\frac{8}{5}, 0\right)$ 35. $(0, 0, h/4)$

39. **(b)** ≈ 4.451 **(c)** ≈ 0.3590 41. **(a)** ellipsoid   **(b)** ≈ 111.5457699

▶ **Exercise Set 17.1 (Page 1062)**

1. **(a)** III   3. **(a)** true   5.   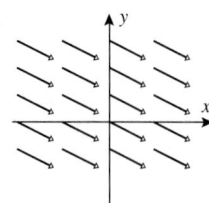   7.   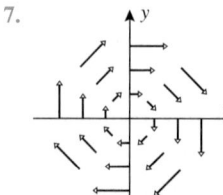   9.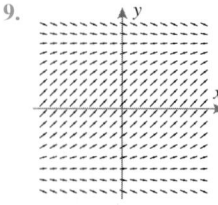
   (b) IV   **(b)** true
                      (c) true

11. **(a)** all x, y **(b)** all x, y 13. div $\mathbf{F} = 2x + y$, curl $\mathbf{F} = z\mathbf{i}$

15. div $\mathbf{F} = 0$, curl $\mathbf{F} = (40x^2z^4 - 12xy^3)\mathbf{i} + (14y^3z + 3y^4)\mathbf{j} - (16xz^5 + 21y^2z^2)\mathbf{k}$ 17. div $\mathbf{F} = \dfrac{2}{\sqrt{x^2 + y^2 + z^2}}$, curl $\mathbf{F} = \mathbf{0}$

19. $4x$ 21. 0 23. $(1 + y)\mathbf{i} + x\mathbf{j}$ 33. $\nabla \cdot (k\mathbf{F}) = k\nabla \cdot \mathbf{F}$, $\nabla \cdot (\mathbf{F} + \mathbf{G}) = \nabla \cdot \mathbf{F} + \nabla \cdot \mathbf{G}$,   43. **(b)** $x^2 + y^2 = K$
$\nabla \cdot (\phi\mathbf{F}) = \phi\nabla \cdot \mathbf{F} + \nabla\phi \cdot \mathbf{F}$, $\nabla \cdot (\nabla \times \mathbf{F}) = 0$

45. $\dfrac{dy}{dx} = \dfrac{1}{x}$, $y = \ln x + K$

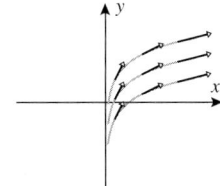

▶ Exercise Set 17.2 **(Page 1075)**

1. **(a)** 1   **(b)** 0   3. **(a)** $-\frac{11}{108}\sqrt{10} - \frac{1}{36}\ln(\sqrt{10} - 3) - \frac{4}{27}$ **(b)** 0   **(c)** $-\frac{1}{2}$ 5. **(a)** 3   **(b)** 3   **(c)** 3   **(d)** 3   7. 2

9. $\frac{13}{20}$ 11. $1 - \pi$ 13. 3   15. $-1 - (\pi/4)$ 17. $1 - e^3$ 21. **(a)** -1 **(b)** -2 23. $\frac{5}{2}$ 25. 0   27. $1 - e^{-1}$

29. $6\sqrt{3}$ 31. $5k\tan^{-1} 3$ 33. $\frac{3}{5}$ 35. $\frac{27}{28}$ 37. 16   39. $\frac{3}{4}$ 41. $\dfrac{17\sqrt{17} - 1}{4}$ 43. **(b)** $S = \displaystyle\int_C z(t)\, dt$

(c) 4π 45. $\lambda = -12$

▶ Exercise Set 17.3 **(Page 1086)**

1. conservative, $\phi = \dfrac{x^2}{2} + \dfrac{y^2}{2} + K$ 3. not conservative   5. conservative, $\phi = x\cos y + y\sin x + K$ 7. **(b)** 13   9. -6

11. $9e^2$ 13. 32   15. $W = -\frac{1}{2}$ 17. $W = 1 - e^{-1}$ 19. $\ln 2 - 1$ 21. ≈ -0.307 23. no   27. $h(x) = Ce^x$

29. **(a)** $W = -\dfrac{1}{\sqrt{14}} + \dfrac{1}{\sqrt{6}}$ **(b)** $W = -\dfrac{1}{\sqrt{14}} + \dfrac{1}{\sqrt{6}}$ **(c)** $W = 0$

▶ Exercise Set 17.4 **(Page 1092)**

1. 0   3. 0   5. 0   7. 8π 9. -4 11. -1 13. 0   15. **(a)** ≈ -3.550999378 **(b)** ≈ -0.269616482

17. $\dfrac{3\pi a^2}{8}$ 19. $\frac{1}{2}abt_0$ 21. $\frac{250}{3}$ 23. $-3\pi a^2$ 25. $\left(\frac{8}{15}, \frac{8}{21}\right)$ 27. $\left(0, \dfrac{4a}{3\pi}\right)$ 29. the circle $x^2 + y^2 = 1$
31. 69

▶ Exercise Set 17.5 **(Page 1099)**

1. $\dfrac{15}{2}\pi\sqrt{2}$ 3. $\dfrac{\pi}{4}$ 5. $-\dfrac{\sqrt{2}}{2}$ 7. 9   9. $\dfrac{4\pi}{3}$ 11. **(a)** $\dfrac{\sqrt{29}}{16}\displaystyle\int_0^6 \int_0^{(12-2x)/3} xy(12 - 2x - 3y)\, dy\, dx$

(b) $\dfrac{\sqrt{29}}{4}\displaystyle\int_0^3 \int_0^{(12-4z)/3} yz(12 - 3y - 4z)\, dy\, dz$ **(c)** $\dfrac{\sqrt{29}}{9}\displaystyle\int_0^3 \int_0^{6-2z} xz(12 - 2x - 4z)\, dx\, dz$ 13. $\dfrac{18\sqrt{29}}{5}$

15. $\displaystyle\int_0^4 \int_1^2 y^3z\sqrt{4y^2 + 1}\, dy\, dz$; $\dfrac{1}{2}\displaystyle\int_0^4 \int_1^4 xz\sqrt{1 + 4x}\, dx\, dz$ 17. $\dfrac{391\sqrt{17}}{15} - \dfrac{5\sqrt{5}}{3}$ 19. $\frac{4}{3}\pi\delta_0$ 21. $\frac{1}{4}(37\sqrt{37} - 1)$

23. $M = \delta_0 S$ 25. $(0, 0, 149/65)$ 27. $\dfrac{93}{\sqrt{10}}$ 29. $\dfrac{\pi}{4}$ 31. 57.895751

▶ Exercise Set 17.6 **(Page 1107)**

1. **(a)** zero   **(b)** zero   **(c)** positive   **(d)** negative   **(e)** zero   **(f)** zero   3. **(a)** positive   **(b)** zero   **(c)** positive   **(d)** zero

(e) positive   **(f)** zero   5. **(a)** $\mathbf{n} = -\cos v\mathbf{i} - \sin v\mathbf{j}$ **(b)** inward   7. 2π 9. $\dfrac{14\pi}{3}$ 11. 0   13. 18π

15. $\frac{4}{9}$ 17. **(a)** 8   **(b)** 24   **(c)** 0   19. $-\pi$ 21. **(b)** $\frac{32}{3}$ 23. **(a)** $4\pi a^{k+3}$ **(b)** $k = -3$

▶ Exercise Set 17.7 **(Page 1116)**

1. 3   **3.** $\dfrac{4\pi}{3}$ **5.** 12   **7.** $3\pi a^2$ **9.** 180π **11.** $\dfrac{192\pi}{5}$ **13.** $\dfrac{\pi}{2}$ **15.** $\dfrac{4608}{35}$ **25. (a)** ≥ 0 **(b)** ≤ 0

27. no sources or sinks   **29.** sources at all points except the origin, no sinks   **31.** $\dfrac{7\pi}{4}$

▶ Exercise Set 17.8 **(Page 1123)**

1. (a) The curl is zero.   **(b)** The curl is nonzero and points in the positive z-direction.   **3.** $\frac{3}{2}$ **5.** 0   **7.** 2π **9.** 16π

11. 0   **13.** πa^2 **15. (a)** $\frac{3}{2}$ **(b)** -1 **(c)** $-\dfrac{1}{\sqrt{2}}\mathbf{j} - \dfrac{1}{\sqrt{2}}\mathbf{k}$ **17.** $\text{curl }\mathbf{E} = -\dfrac{\partial \mathbf{B}}{\partial t}$

▶ Chapter 17 Supplementary Exercises **(Page 1124)**

3. (a) $\displaystyle\int_a^b \left[f(x(t), y(t))\frac{dx}{dt} + g(x(t), y(t))\frac{dy}{dt} \right] dt$ **(b)** $\displaystyle\int_a^b f(x(t), y(t))\sqrt{x'(t)^2 + y'(t)^2}\, dt$

11. $\displaystyle\iint_R f(x(u, v), y(u, v), z(u, v))\|\mathbf{r}_u \times \mathbf{r}_v\|\, du\, dv$ **15. (b)** $\approx -1.597 \times 10^9$ J   **21. (b)** The field is conservative.

23. yes   **25.** -8π **27. (a)** $h(x) = Cx^{-3/2}$ **(b)** $g(y) = C/y^3$ **29. (a)** conservative, $\phi(x, y, z) = xz^2 - e^{-y}$

(b) not conservative

▶ Exercise Set A **(Page A8)**

1. (a) rational   **(e)** integer, rational   **3. (a)** $\frac{41}{333}$ **5. (a)** $\frac{256}{81}$ **7.**
(b) integer, rational   **(f)** irrational   **(b)** $\frac{115}{9}$ **(b)** worse
(c) integer, rational   **(g)** rational   **(c)** $\frac{20943}{550}$
(d) rational   **(h)** integer, rational   **(d)** $\frac{537}{1250}$

Line	2	3	4	5	6	7
Blocks	3, 4	1, 2	3, 4	2, 4, 5	1, 2	3, 4

9. (a), (d), (f)   **11. (a)** all values   **(b)** none   **13. (a)** yes   **(b)** no

15. (a) $\{x : x$ is a positive odd integer$\}$ **(b)** $\{x : x$ is an even integer$\}$ **(c)** $\{x : x$ is irrational$\}$ **(d)** $\{x : x$ is an integer and $7 \leq x \leq 10\}$

17. (a) false   **(b)** true   **(c)** true   **(d)** false   **(e)** true   **(f)** true   **(g)** true

19. (a)   **(b)**   **(c)**

(d)   **(e)**   **(f)**

21. (a) $[-2, 2]$ **(b)** $(-\infty, -2) \cup (2, +\infty)$ **23.** $\left(-\infty, \frac{10}{3}\right)$ **25.** $\left(-\infty, -\frac{11}{2}\right]$

27. $\left(-\frac{3}{2}, \frac{1}{2}\right]$ **29.** $(-\infty, 3) \cup (4, +\infty)$ **31.** $\left(-\frac{3}{2}, 2\right)$

33. $(-\infty, -2] \cup (2, +\infty)$ **35.** $(-\infty, -3) \cup (3, +\infty)$

37. $(-\infty, -2) \cup (4, +\infty)$ **39.** $[4, 5]$ **41.** $(-8, 0) \cup (4, +\infty)$

43. $(2, +\infty)$ **45.** $(-\infty, -3) \cup [2, +\infty)$ **47.** $77 \leq F \leq 104$ **55.** $\left(-\infty, -\frac{1}{2}\right)$

▶ Exercise Set B **(Page A15)**

1. (a) 7   **(b)** $\sqrt{2}$ **(c)** k^2 **(d)** k^2 **3.** $x \leq 3$ **5.** all real x **7.** $x \geq 0$ or $x = -\frac{2}{3}$ **9.** $x \geq -5$

13. (a) 2   **(b)** 1   **(c)** 14   **(d)** $3 + \sqrt{2}$ **(e)** 7   **(f)** 5   **15. (a)** -9 **(b)** 7   **(c)** 12   **17.** $-\frac{5}{6}, \frac{3}{2}$ **19.** $\frac{1}{2}, \frac{5}{2}$

21. $-\frac{11}{10}, \frac{11}{8}$ **23.** $1, \frac{17}{5}$ **25.** $(-9, -3)$ **27.** $\left[-\frac{3}{2}, \frac{9}{2}\right]$ **29.** $(-\infty, -3) \cup (-1, +\infty)$ **31.** $\left(-\infty, \frac{1}{2}\right] \cup \left[\frac{9}{2}, +\infty\right)$

33. $\left(-\infty, \frac{1}{2}\right) \cup \left(\frac{3}{2}, +\infty\right)$ **35.** $\left[\frac{1}{8}, \frac{1}{2}\right) \cup \left(\frac{1}{2}, \frac{7}{8}\right]$ **37.** $x \in (-\infty, 2] \cup [3, +\infty)$ **39.** $-3, 9$

► Exercise Set 26C **(Page A26)**

1. $(-4, 7)$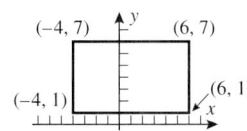

3. **(a)** 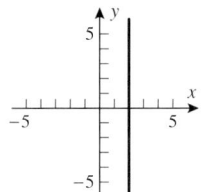   **(b)** 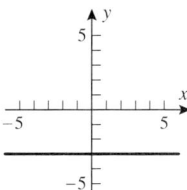   **(c)**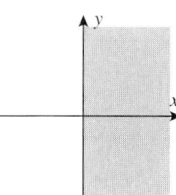

(d) 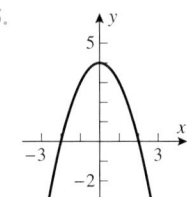   **(e)** 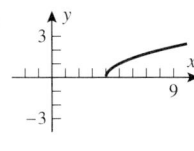   **(f)**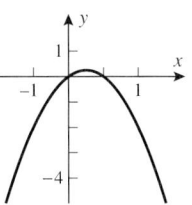

5. 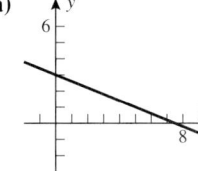   7. 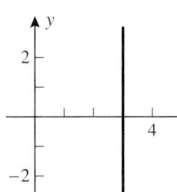   9. 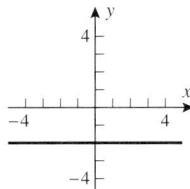   11. 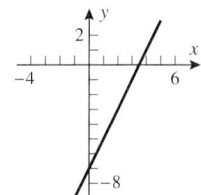   13. **(a)** $\frac{1}{2}$
   (b) -1
   (c) 0
   (d) not defined

15. **(a)** yes   17.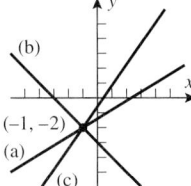

(b) no

19. III $<$ II $<$ IV $<$ I   21. **(a)** 14 23. 29 25. $\frac{13}{7}$

(b) $-\frac{1}{3}$

29. **(a)** 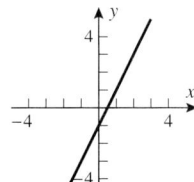   **(b)** 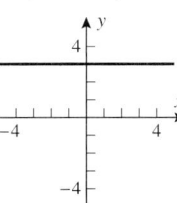   **(c)** 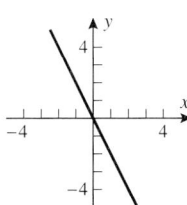   **(d)**

31. **(a)**   **(b)**   **(c)**   33.

	(a)	(b)	(c)	(d)	(e)
Slope	3	−1/4	−3/5	0	−b/a
y-intercept	2	3	8/5	1	b

35. **(a)** $y = \frac{3}{2}x - 3$ **(b)** $y = -\frac{3}{4}x$ 37. $y = -2x + 4$ 39. $y = 4x + 7$ 41. $y = -\frac{1}{5}x + 6$ 43. $y = 11x - 18$

45. $y = \frac{1}{2}x + 2$ 47. $y = 1$ 49. **(a)** parallel   **(b)** perpendicular   **(c)** parallel   **(d)** perpendicular   **(e)** neither

51. **(a)** $-\frac{3}{2}$ 53. the union of the graphs of $x - y = 0$ and $x + y = 0$ 55.   59. $\frac{49}{6}$ 61. **(a)** yes
   (b) $\frac{4}{5}$
   (c) $\frac{5}{2}$
   (d) $-\frac{15}{2}$
   (e) -4
   (b) yes
   (c) no
   (d) yes
   (e) yes
   (f) yes
   (g) no

▶ **Exercise Set D (Page A36)**

1. in the proof of Theorem D.1   3. **(a)** 10   **(b)** (4, 5)   5. **(a)** $\sqrt{29}$ **(b)** $\left(-\frac{9}{2}, -5\right)$ 11. 0   13. $y = -3x + 4$

15. $\left(-\frac{29}{8}, -\frac{23}{4}\right)$ 17. 3   21. 4   23. **(a)** (0, 0); 5   **(b)** (1, 4); 4   **(c)** (−1, −3); $\sqrt{5}$ **(d)** (0, −2); 1

25. $(x - 3)^2 + (y + 2)^2 = 16$ 27. $(x + 4)^2 + (y - 8)^2 = 64$ 29. $(x + 3)^2 + (y + 4)^2 = 25$ 31. $(x - 1)^2 + (y - 1)^2 = 2$

33. circle; center (1, 2), radius 4   35. circle; center (−1, 1), radius $\sqrt{2}$ 37. the point (−1, −1)   39. circle; center (0, 0), radius $\frac{1}{3}$

41. no graph   43. circle; center $\left(-\frac{5}{4}, -\frac{1}{2}\right)$, radius $\frac{3}{2}$ 45. **(a)** $y = -\sqrt{16 - x^2}$ **(b)** $y = 2 + \sqrt{3 - 2x - x^2}$

47. **(a)**   **(b)**   49. $y = -\frac{3}{4}x + \frac{25}{4}$ 51. **(a)** inside   53. $(1/3, \pm\sqrt{8}/3)$
(b) largest $3\sqrt{5}$, smallest $\sqrt{5}$

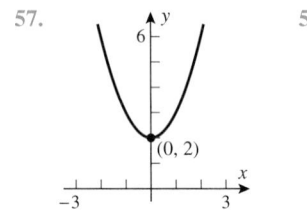
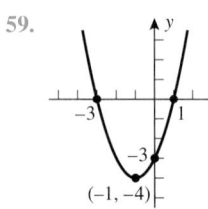

55. **(a)** equation: $2x^2 + 2y^2 - 12x + 8y + 1 = 0$ **(b)** center (3, −2), radius $5/\sqrt{2}$

57.   59.   61.   63.

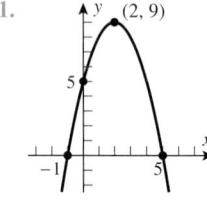
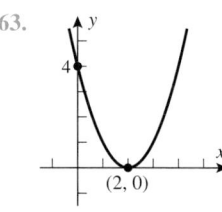

65.   67.   69.   71. **(a)** $x = \sqrt{3 - y}$
(b) $x = 1 - \sqrt{y + 1}$

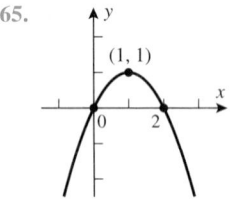
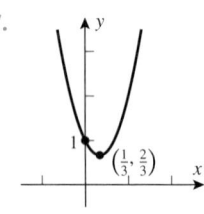
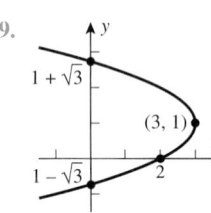

73. **(a)**   **(b)**   75. **(a)**   77. **(a)** $y = 150 - \frac{3}{2}x$
(b) $A = 150x - \frac{3}{2}x^2$
(c) 3750 ft²

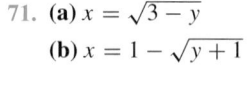
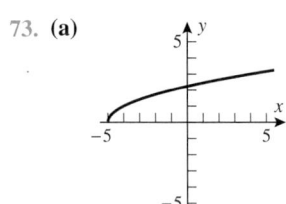
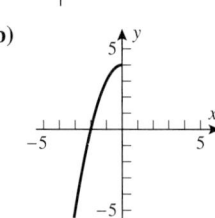
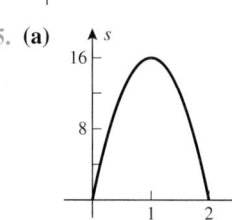

79. **(a)** $(-5 - \sqrt{33})/4 < x < (-5 + \sqrt{33})/4$ **(b)** $-\infty < x < +\infty$ 81. **(a)** 30 ft   **(b)** 2.6 s   **(c)** 2.1 s

▶ **Exercise Set E (Page A49)**

1. **(a)** $\frac{5}{12}\pi$ 3. **(a)** 12°   5.   7. $\sin\theta = 3/\sqrt{10}, \cos\theta = 1/\sqrt{10}$
(b) $\frac{13}{6}\pi$ **(b)** $(270/\pi)°$
(c) $\frac{1}{9}\pi$ **(c)** 288°
(d) $\frac{23}{30}\pi$ **(d)** 540°

	$\sin\theta$	$\cos\theta$	$\tan\theta$	$\csc\theta$	$\sec\theta$	$\cot\theta$
(a)	$\sqrt{21}/5$	2/5	$\sqrt{21}/2$	$5/\sqrt{21}$	5/2	$2/\sqrt{21}$
(b)	3/4	$\sqrt{7}/4$	$3/\sqrt{7}$	4/3	$4/\sqrt{7}$	$\sqrt{7}/3$
(c)	$3/\sqrt{10}$	$1/\sqrt{10}$	3	$\sqrt{10}/3$	$\sqrt{10}$	1/3

9. $\tan\theta = \sqrt{21}/2, \csc\theta = 5/\sqrt{21}$ 11. 1.8   13.

	θ	$\sin\theta$	$\cos\theta$	$\tan\theta$	$\csc\theta$	$\sec\theta$	$\cot\theta$
(a)	225°	$-1/\sqrt{2}$	$-1/\sqrt{2}$	1	$-\sqrt{2}$	$-\sqrt{2}$	1
(b)	−210°	1/2	$-\sqrt{3}/2$	$-1/\sqrt{3}$	2	$-2/\sqrt{3}$	$-\sqrt{3}$
(c)	$5\pi/3$	$-\sqrt{3}/2$	1/2	$-\sqrt{3}$	$-2/\sqrt{3}$	2	$-1/\sqrt{3}$
(d)	$-3\pi/2$	1	0	—	1	—	0

15.

	$\sin\theta$	$\cos\theta$	$\tan\theta$	$\csc\theta$	$\sec\theta$	$\cot\theta$
(a)	4/5	3/5	4/3	5/4	5/3	3/4
(b)	$-4/5$	3/5	$-4/3$	$-5/4$	5/3	$-3/4$
(c)	1/2	$-\sqrt{3}/2$	$-1/\sqrt{3}$	2	$-2/\sqrt{3}$	$-\sqrt{3}$
(d)	$-1/2$	$\sqrt{3}/2$	$-1/\sqrt{3}$	-2	$2/\sqrt{3}$	$-\sqrt{3}$
(e)	$1/\sqrt{2}$	$1/\sqrt{2}$	1	$\sqrt{2}$	$\sqrt{2}$	1
(f)	$1/\sqrt{2}$	$-1/\sqrt{2}$	-1	$\sqrt{2}$	$-\sqrt{2}$	-1

17. **(a)** 1.2679   **(b)** 3.5753

19.

	$\sin\theta$	$\cos\theta$	$\tan\theta$	$\csc\theta$	$\sec\theta$	$\cot\theta$
(a)	$a/3$	$\sqrt{9-a^2}/3$	$a/\sqrt{9-a^2}$	$3/a$	$3/\sqrt{9-a^2}$	$\sqrt{9-a^2}/a$
(b)	$a/\sqrt{a^2+25}$	$5/\sqrt{a^2+25}$	$a/5$	$\sqrt{a^2+25}/a$	$\sqrt{a^2+25}/5$	$5/a$
(c)	$\sqrt{a^2-1}/a$	$1/a$	$\sqrt{a^2-1}$	$a/\sqrt{a^2-1}$	a	$1/\sqrt{a^2-1}$

21. **(a)** $3\pi/4 \pm n\pi, n = 0, 1, 2, \ldots$ **(b)** $\pi/3 \pm 2n\pi$ and $5\pi/3 \pm 2n\pi, n = 0, 1, 2, \ldots$

23. **(a)** $\pi/6 \pm n\pi, n = 0, 1, 2, \ldots$ **(b)** $4\pi/3 \pm 2n\pi$ and $5\pi/3 \pm 2n\pi, n = 0, 1, 2, \ldots$

25. **(a)** $3\pi/4 \pm n\pi, n = 0, 1, 2, \ldots$ **(b)** $\pi/6 \pm n\pi, n = 0, 1, 2, \ldots$

27. **(a)** $\pi/3 \pm 2n\pi$ and $2\pi/3 \pm 2n\pi, n = 0, 1, 2, \ldots$ **(b)** $\pi/6 \pm 2n\pi$ and $11\pi/6 \pm 2n\pi, n = 0, 1, 2, \ldots$

29. $\sin\theta = 2/5, \cos\theta = -\sqrt{21}/5, \tan\theta = -2/\sqrt{21}, \csc\theta = 5/2, \sec\theta = -5/\sqrt{21}, \cot\theta = -\sqrt{21}/2$

31. **(a)** $\theta = \pm n\pi, n = 0, 1, 2, \ldots$ **(b)** $\theta = \pi/2 \pm n\pi, n = 0, 1, 2, \ldots$ **(c)** $\theta = \pm n\pi, n = 0, 1, 2, \ldots$

   (d) $\theta = \pm n\pi, n = 0, 1, 2, \ldots$ **(e)** $\theta = \pi/2 \pm n\pi, n = 0, 1, 2, \ldots$ **(f)** $\theta = \pm n\pi, n = 0, 1, 2, \ldots$

33. **(a)** $2\pi/3$ cm   **(b)** $10\pi/3$ cm   35. $\frac{2}{5}$ 37. **(a)** $\dfrac{2\pi - \theta}{2\pi}R$ **(b)** $\dfrac{\sqrt{4\pi\theta - \theta^2}}{2\pi}R$ 39. $\frac{21}{4}\sqrt{3}$ 41. 9.2 ft

43. $h = d(\tan\beta - \tan\alpha)$ 45. **(a)** $4\sqrt{5}/9$ **(b)** $-\frac{1}{9}$ 47. $\sin 3\theta = 3\sin\theta\cos^2\theta - \sin^3\theta, \cos 3\theta = \cos^3\theta - 3\sin^2\theta\cos\theta$

61. **(a)** $\cos\theta$ **(b)** $-\sin\theta$ **(c)** $-\cos\theta$ **(d)** $\sin\theta$

▶ Exercise Set F **(Page A56)**

1. **(a)** $x^2 + 4x + 2, -11x + 6$ 3. **(a)** $3x^2 + 6x + 8, 15$ 5.   7. **(a)** $x^2 + 6x + 13, 20$

   (b) $2x^2 + 4, 9$ **(b)** $x^3 - 5x^2 + 20x - 100, 504$ **(b)** $x^2 + 3x - 2, -4$

   (c) $x^3 - x^2 + 2x - 2, 2x + 1$ **(c)** $x^4 + x^3 + x^2 + x + 1, 0$

5.

x	0	1	-3	7
$p(x)$	-4	-3	101	5001

9. **(a)** $\pm 1, \pm 2, \pm 3, \pm 4, \pm 6, \pm 8, \pm 12, \pm 24$ **(b)** $\pm 1, \pm 2, \pm 5, \pm 10, \pm\frac{1}{3}, \pm\frac{2}{3}, \pm\frac{5}{3}, \pm\frac{10}{3}$ **(c)** $\pm 1, \pm 17$ 11. $(x+1)(x-1)(x-2)$

13. $(x+3)^3(x+1)$ 15. $(x+3)(x+2)(x+1)^2(x-3)$ 17. -3 19. $-2, -\frac{2}{3}$ 21. $-2, 2, 3$ 23. $2, 5$ 25. 7 cm

PHOTO CREDITS

Introduction
Page 1 (top): Granger Collection. Page 1 (bottom): Ed Honowitz/Tony Stone Images/New York, Inc. Page 2 (top): Courtesy Chris Brislawn, Los Alamos National Laboratory. Page 2 (bottom): Courtesy Dr. Adrian Maudsley, UCSF/NASA. Page 3 (top): Image generated from simulations carried out by the Aerospace Engineering and Mechanics Group at the Army HPC Research Center, Minneapolis, MN. Page 3 (center): Courtesy NOAA. Page 3 (just above bottom): Scott Camazine/Photo Researchers. Page 3 (bottom): Courtesy Applied Chaos Laboratory, Georgia Institute of Technology. Pages 10–11: Corbis-Bettmann.

Chapter 1
Page 15 (signature): Courtesy Smithsonian Institution. Page 15 (portrait): Corbis-Bettmann. Page 15 (bottom): Stephen Johnson/Tony Stone Images/New York, Inc. Page 17 (left): from *Visual Display of Quantitative Information*, ©1983, Edward R. Tufte. Published by Graphics Press, Cheshire, CT. Page 17 (right): from U.S. Department of Health and Human Services. Page 100: Corbis-Bettmann.

Chapter 2
Page 111 (signature): Courtesy Smithsonian Institution. Page 111 (portrait): Corbis-Bettmann. Page 111 (bottom): Barry Blackman/Tony Stone Images/New York, Inc.

Chapter 3
Page 169 (signature): Corbis-Bettmann. Page 169 (portrait): Courtesy the David Eugene Smith Collection, Columbia University. Page 169 (bottom): J. W. Burkey/Tony Stone Images/New York, Inc.

Chapter 4
Page 225 (signature): Courtesy New York Public Library. Page 225 (portrait): Corbis-Bettmann. Page 225 (bottom): Ken Fisher/Tony Stone Images/New York, Inc. Page 242: Bob Gruen/Star File.

Chapter 5
Page 289 (signature): Courtesy of the New York Public Library. Page 289 (portrait): Corbis-Bettmann. Page 289 (bottom): David Chambers/Tony Stone Images/New York, Inc.

Chapter 6
Page 329 (signature): Courtesy The British Library. Page 329 (portrait): Corbis-Bettmann. Page 329 (bottom): Donald Johnston/Tony Stone Images/New York, Inc. Page 359: UPI/Corbis-Bettmann.

Chapter 7
Page 377 (signature): Courtesy New York Public Library. Page 377 (portrait): Granger Collection. Page 377 (bottom): Cyber image/Tony Stone Images/New York, Inc. Page 379: Courtesy New York Public Library. Page 383: Reproduced from C. I. Gerhardt's *Briefwechsel von G.W. Leibniz mit Mathematikern* (1899). Page 457: J. Yulsman/The Image Bank.

Chapter 8
Page 461 (signature): Courtesy Smithsonian Institution. Page 461 (portrait): Corbis-Bettmann. Page 461 (bottom): Jane Sterrett/The Image Bank. Page 490: Stephen Sutton/Duomo Photography, Inc. Page 502: Glen Allison/Tony Stone Images/New York, Inc.

Chapter 9
Page 513 (signature): Courtesy Smithsonian Institution. Page 513 (portrait): Corbis-Bettmann. Page 513 (bottom): Reproduced from C. I. Gerhardt's *Briefwechsel von G.W. Leibniz mit Mathematikern* (1899). Page 528: Courtesy Catherine Watson & David Young/NASA Langley Research Center, Hampton, VA.

Chapter 10
Page 579 (signature): Courtesy Smithsonian Institution. Page 579 (portrait): Corbis-Bettmann. Page 579 (bottom): Alan Hicks/Tony Stone Images/New York, Inc. Page 604: Patrick Mesner/Gamma Liaison.

Chapter 11
Page 615 (signature): Courtesy The British Library. Page 615 (portrait): Corbis-Bettmann. Page 615 (bottom): Nick Dolding/Tony Stone Images/New York, Inc. Page 638: Courtesy Lilly Library, Indiana University.

Chapter 12
Page 699 (signature): Granger Collection. Page 699 (portrait): Archive Photos. Page 699 (bottom): Paul Morrell/Tony Stone Images/New York, Inc. Page 709 (left): Thomas Taylor/Photo Researchers. Page 709 (right): Rex Ziak/AllStock/Picture Network International, Ltd. Page 738: John Mead/Science Photo Library/Photo Researchers. Page 749 (top): Science Photo Library/Photo Researchers. Page 749 (bottom): ©Lick Observatory/Astronomical Society of the Pacific. Page 751: Courtesy NASA.

Chapter 13
Page 759 (Johann Bernoulli signature): Courtesy Smithsonian Institution. Page 759 (Jacques Bernoulli signature): Courtesy Universitäts-Bibliothek Basel. Page 759 (portraits): Corbis-Bettman. Page 759 (bottom): Rick Etkin/Tony Stone Images/New York, Inc. Page 792: World Perspectives/Tony Stone Images/New York, Inc.

Chapter 14
Page 829 (signature): Courtesy Smithsonian Institution. Page 829 (portrait): Courtesy AIP Neils Bohr Library. Page 829 (bottom): Jerome Prevost/Tony Stone Images/New York, Inc. Page 830: Ken Eward/Biografx/Photo Researchers.

Chapter 15
Page 889 (signature): Courtesy Smithsonian Institution. Page 889 (portrait): Corbis-Bettmann. Page 889 (bottom): Image created by Terry Robb, courtesy Wolfram Research, Inc. Page 915: Leverett Bradley/Tony Stone Images/New York, Inc.

Chapter 16
Page 977 (signature): Courtesy Smithsonian Institution. Page 977 (portrait): Corbis-Bettmann. Page 977 (bottom): Alexander Calder, *Lobster Trap and Fish Tail*, 1939. Hanging mobile: painted steel wire and sheet aluminum, 260 × 290 cm, The Museum of Modern Art, New York. Commissioned by the Advisory Committee for the stairwell of the Museum. Photograph ©1998 The Museum of Modern Art, New York.

Chapter 17
Page 1055 (signature): Courtesy The British Library. Page 1055 (portrait) and page 1118: Corbis-Bettmann. Page 1055 (bottom): Pal Hermansen/Tony Stone Images/New York, Inc. Page 1126: Kevin Kelley/Tony Stone Images/New York, Inc.

INDEX

Abel, Niels Henrik, 363
Abscissa, A16–A17
Absolute convergence, 665–666
  ratio test for, 667
Absolute divergence, 665
Absolute error, 154, 596
Absolute extrema, 330–337, 957
Absolute maxima and minima (extrema), 330–337,
  956–957, 963
  constrained, 968
Absolute value(s), A11–A15
Absolute value function, 27–28
Absolute zero, 966
Acceleration, 66–67, 354–355, 867–869
  centripetal, 887
  constant, 66
  gravity, due to, 164, 358–359, 430
  interpreting the sign of, 355
  normal and tangential components, 870–873
  tangential scalar component of, 871
  tangential vector component of, 871
  uniform, 428–430
Addition formulas, for sine and cosine, A46–A47
Agnesi, Maria, 289
Air resistance, free-fall motion retarded by, 587–589
Algebraic function(s), 82, 514–515
Algebraic properties
  of cross product, 787–788
  of dot product, 777
  of logarithms, 239
Algorithms, 4
Alternating current, 445
Alternating series, 662–665
  harmonic, 663
Alternating series test, 662–663, 668
Amplitude, 86, 209, 445
Angle(s), A39–A41
  elevation, 458
  finding, from trigonometric functions, A48
  of inclination, 63
  polar, 700
  in standard position, A42
  between vectors, 777
Angular frequency, 209
Angular speed, 1125
Angular velocity, 1125
Annual compounding, 367, 614
Annuities, 367
Antiderivative(s), 382–383, 838
  properties of, 385
  see also Indefinite integral(s)
Antiderivative method (for finding areas), 379,
  380–382
Antidifferentiation, 381, see also Integration
Antiphon, 6
Aphelion, 748, 755, 885
Apogee, 275, 748
  distance, 748, 885

Approximation(s)
  of area under a curve, 405–407
  left, right, and midpoint, 406–407
  local quadratic, 647–649
  midpoint, 554–556
  principle, 155
  Riemann sum, 553–554
  using Taylor series, 680–683
  trapezoidal, 554–556
Archimedean spiral, 705, 709, 711
Archimedes, 1, 4, 6, 378
Archimedes' principle, 105
Arc length, 480–482, 570–571, 843–853, A39–A40
  of curves defined parametrically, 482–483
  as a line integral, 1069
  parametrization, 845–846
    finding, 848–850
    properties of, 850–851
  of a polar curve, 717–718
  signed, 83
  using a CAS, 483
Area(s)
  antiderivative method for finding, 380–382
  under a curve, 405–407
  defining, 378–379, 404–407
  finding, using Green's Theorem, 1090
  formula, 463
  and limits, 113
  an overview of, 6–7, 378–382
  net signed, 407–409
  in polar coordinates, 720–724
  problem, 112, 378, 462
  rectangle method for finding, 379–380
  of surface of revolution, 485–488
  between two curves, 462–466
Arithmetic mean (or average), 435, 746
Arithmetic operations
  on functions, 47–48
  on vectors, 768
Arrow Paradox, 8
Artificial satellites, 885
Aspect ratio distortion, 42
Astroid, 484
Astroidal sphere, 1053
Astronomia Nova, 880
Astronomy, 748–750, 755–758
Asymptotes
  curvilinear, 501
  of a hyperbola, 727, 733
  oblique, 320
  vertical vs. horizontal, 82
Autonomous differential equations, 594
Average, arithmetic, 435, 746
Average rate of change, 173–174
Average speed, 9
Average (mean) value, 421, 435–436, 427–440, 984,
  992, 1019
  of a continuous function, 435–436

Average velocity, 65, 170–173, 436–437
Axis (axes), 469
  coordinate, A16
  of an ellipse, 727, 730
  of a hyperbola, 727
  of a parabola, 727
  polar, 700
  of revolution, 470

Badel, Hannskarl, 511
Ballistics, 457
Bar graphs, 16
Barrow, Isaac, 10, 11, 329
Bell, Alexander Graham, 242
Bernoulli, 19, 99–100, 278, 638, 759
Bernoulli's principle, 99
Bessel, Friedrich Wilhelm, 675
Bessel equation, 696
Bessel function(s), 413, 457, 675
Bifolium, 725
Binomial coefficient, 685
Binomial series, 683–684
Binormal vectors in 3-space, 856–857
Bipartite cubic(s), 46, 254
Bjerknes, Vilhelm, 3
Bolzano, Bernhard, 183
Boundary point, 901
Bounded functions, 413
Bounded set, 902
Boyle's law, 80
Brachistochrone problem, 99
Brahe, Tycho, 748
Branches, of a hyperbola, 727
Breakpoints, 28
British Engineering system (BE), 490
Butterfly curve, 711

Cap, spherical, 474
Carbon dating, 603–604
Cardioids, 704, 708
Carrying capacity, 600
Cartesian coordinate system, 760, A16
CAS, *see* Computer algebra system(s)
Cassini ovals, 712
Catastrophic subtraction, 127
Catenary, 502
Cauchy–Riemann equations, 919
  polar form of, 929
Cavalieri, Bonaventura, 475
Cavalieri's principle, 475
Center
  of curvature, 862
  of an ellipse, 727
  of a hyperbola, 727
Center of gravity, 1019
  and centroid of a solid, 1024–1025
  of a lamina, 1021–1023

Central angle (of a wedge), 1030
Central force, 880–881
Central force field, 880
Centripetal acceleration, 887
Centripetal force, 887
Centroid(s), 1023–1024
    of a lamina, 1023
    of a region R, 1024
    of a solid, 1024–1025
CGS system, 490
Chain rule(s), 204–208, 923–927, 951–953
    proof of, A59–A60
    rates of change using, 270–274
    for vector-valued functions, 847
Change in position, 432, 765
Change of parameter, 843–853, 846–847
Chaos theory, 3, 110
Chaotic ventricular fibrillation, 3
Circle(s)
    as a conic section, 726
    of curvature, 862
    degenerate cases of, A32–A33
    equations of, A31–A32
    families of, 707
    involute of a, 753
    sectors of, A41
    segment of, 367
    unit, 83
Circular cone, 810, 813
Circular helix, 830
Circular paraboloid, 810
Circulation, 1122–1123
Cissoids, 713
Closed form, of sum, 401–402
Closed intervals, A4
    absolute extrema on finite, 331–333
    continuity on, 153–154
Closed parametric curve, 1081
Closed set, 901–902
Closed surfaces, 1109
Clothoid, 754
Cobb–Douglas production model, 975
Cobweb diagram, 109
Coefficients, 81, 307
    correlation, 325
    of friction, 188, 209
    leading, 307
    of linear expansion, 218
    of sliding friction, 209
    of volume expansion, 218
Cofunctions, 203
Comets, 749, 751, 755–758
The Commercial and Political Atlas, 16
Common logarithms, 237
Comparison test, 656–658, 668
Complete elliptic integral of the first kind, 576, 693
Complete linear factorization (of p), 308
Completeness Axiom, 630–631
Completing the square, A32
Complex numbers, A1–A2
Components, 832, 836
    vector, 767
Composition(s), 49–52
    continuity of, 152–153
    derivatives of, 204–205
Composition of functions, 49–52
Compounding, 614
Compression, 41–42
    with viewing windows, 41–42
Computer algebra system(s) (CAS), 4–5, 35–45
    differentiating using, 207–208
    integration using, 395, 549–551

limitations of, 550–551
    using to evaluate unfamiliar integrals, 514
Concave down, 292
Concave up, 292
Concavity
    of functions, 292–293
    and graphing rational functions, 312–314
Conchoids of Nicomedes, 103, 753
Concurrent forces, 772
Conditional convergence, 666–667
Conic sections, 726–727
    applications of, 738
    degenerate, 726
    focus–directrix characterization of, 743–744
    in polar coordinates, 743–750
    reflection properties of, 737–738
    rotated, 737
    translated, 735–737
Conjugate axis, of a hyperbola, 727
Connected sets, 1081, 1082
Conservation of Energy Principle, 1085–1086
Conservative vector field(s), 1058–1059
    path independence and, 1080
    test for, 1082–1085
    in 3-space, 1085
Constant(s)
    derivative of, 189–190
    derivative of function times, 190–191
    function, 75
    of integration, 383, 419
    spring, 492, 606
Constant acceleration, 428, 430
Constant Difference Theorem, 371
Constant force, work done by a, 489–491
Constant functions, 75–76, 290, 291
Constant of proportionality, 67, 79–80, 90
Constant surfaces, 820–821
Constant u-curve, 1000, 1041
Constant v-curve, 1000, 1041
Constant velocity, 66, 67
Constrained absolute maximum/minimum, 968
Constrained-extremum principle, 969
Constraints, extremum problems with, 967
Continued fraction, 108
Continuity, 119–120, 148–159, 901–910
    in applications, 150
    on a closed interval, 153
    of compositions, 152–153
    differentiability vs., 183–184
    equation, for incompressible fluids, 1115
    of inverse functions, 232
    from left and right, 153
    at a point, 149
    relationship between differentiability and, 183–184, 922
    of trigonometric functions, 159–160
Continuous compounding, 614
Continuous data, 21
Continuous function(s)
    average value of a, 435–436
    definite integral of a, 407–409
    properties of, 151, A58–A59
Continuously turning tangent vector, 844
Contour plot (or map), 576–578, 893–894, 898
    using technology, 895
Control angles, *see* Robotics
Convergence, 629, 634
    absolute, 665–667
    conditional, 666–667
    interval of, 672–673
    intuitive view of, 629
    of sequences, 619–622, 629–630

of series, 634
    radius/interval of, for power series, 671–673
    set, 671
    of Taylor series, 676–678
Convergence tests, 640–645, 656–660
    comparison test, 656–658
    integral test, 643–645
    limit comparison test, 658–659
    ratio test, 659–660
    root test, 660
    summary of, 667–668
Converting coordinates, 821–822
Cooling
    modeling, 327–328
    Newton's Law of, 611
Coordinate(s), A3, A16–A17
    polar, *see* Polar coordinates
    radial, 700
Coordinate axes, A16
Coordinate line(s), A3
Coordinate planes, 760
Coordinate system(s)
    Cartesian, 760
    cylindrical, 759
    left-handed rectangular, 760
    rectangular, 760–762, A16
    right-handed rectangular, 760
    spherical, 759
Cornu, Marie Alfred, 754
Cornu spiral, 754
Correlation coefficient, 325–326
Cosecant, A41
    hyperbolic, 501
Cosine(s), 84–85, A41
    formulas, Wallis, 530
    hyperbolic, 501
    integrating powers of, 522–524, 527–528
    integrating products of sines and, 524–525
    law of, 269, 773, A46–A48
Cost function, 346
    marginal, 347
Cotangent, A41
    hyperbolic, 501
Coulomb's law, 91, 495, 1058, 1101
Counterclockwise vortex flow, 1127
CRC Standard Mathematical Tables and Formulae, 543
Critical point(s), 300, 958, A60–A61
Cross product(s), 785–795, 1061
    algebraic properties of, 787–788
    derivatives of, 841
    differentiating, 841
    geometric properties of, 788–789
Cross-product term, 737
Cross section, 468–469
Cubic polynomials, 81, 307
Curl, 1059
    viewed as circulation, 1122
Curvature, 292, 858–867
    center of, 862
    circle of, 862
    definition of, 858–859
    formulas for, 860–862
    interpretation of, in 2-space, 863
    radius of, 862
Curve(s)
    area between two, 462–468
    area under, 405–407
    continuous, 119
    discontinuous, 119
    integral, 387, 581

length of, 480–483, 844–845
level, 893–895
limits along, 902–904
Lissajous, 97–98
logistic growth, 318–319
orientation of, 830, 1118
orthogonal, 254
parametric, 94, 713–715
piecewise smooth, 1071
polar, 707–710, 715–718
position versus time, 65–66, 352–353, 355–358
smooth, 480
velocity versus time, 67, 433–435
Curved lamina, 1098
Curve fitting, 20–21, 324
Curvilinear asymptotes, 501
Cusps, 314
Cycle, 445
Cycloid(s), 98–100
prolate, 719
Cylinders, right, 469
Cylindrical coordinates, 819–824
Cylindrical elements of volume, 1030
Cylindrical shells, 475–479
Cylindrical surfaces, 763–764
Cylindrical wedges, 1029

da Pisa, Leonardo "Fibonacci," 108, 625
Data
discrete vs. continuous, 21
finding functions from, 324–328
paired, 16
De Analysi, 379
Decay
constants, 601–602
exponential, 601–604
radioactive, 603
Decibels (dB), 242
Decimal(s)
repeating vs. nonrepeating, 12, A2
terminating, 639, A2
Decomposing vectors, 780–781
Decomposition, partial fraction, 537
Decreasing functions, 290–291
Decreasing sequences, 626
Decreasing without bound, 120
Definite integral(s), 404–416
and area under a curve, 404–407
evaluating, by substitution, 441–447
integration by parts of, 519–520
partial, with respect to x and y, 980
properties of, 411–413
relationship between indefinite and, 419–420
Riemann integral, 409–410
Degenerate conic sections, 726, A32–A33
Degree (of a polynomial), 81
Degree measure, A39
Del operator, 1061
Density
of fluid(s), 497
function, 1024
weight, 497
Dependent variable, 24
Derivative(s), 169, 177–216, 384, 836–837
antiderivative vs., 383
applications of the, 329–376
of compositions, 204–205
of a constant, 189–190
definition of, 179
and differentiability, 181–182
of dot and cross products, 841

dual interpretations of, 179
at endpoints of an interval, 186
generalized formulas for, 205–207
higher, 196–197
and increments, 210–211
of inverse functions, 252–253
of irrational powers of x, 258
from the left and right, 186
notations for, 184–186
order of the, 196
and power rule for integer exponents, 195–196
of a product, 192–193
of a quotient, 193–194
of rational powers of x, 250–251
of a reciprocal, 195
of sums/differences, 191–192
and tangent line, 177–178
of x to a power, 190
Derive, 36
de Sarasa, Alfons A., 446
Descartes, René, 15, 352, 368, 496, A3
Determinants, 785–786
Differentiability, 181–182, 184, 920–930
continuity vs., 183–184
everywhere, 181
of functions defined implicitly, 250
of functions of two variables, 920–922
geometric significance of, 932
on an interval, 186
of inverse functions, 232–233
relationship between differentiability and
continuity, 183–184, 922
from the right and left, 186
sufficient conditions for, 922
at x_0, 181
Differentiable everywhere, 181
Differential calculus, 112, 377
Differential equation(s)
autonomous, 594
definition of, 388
direction/slope field for, 592–593
first-order, 582–589
initial-value problems and, 581–582
integral curve and, 581
integration from viewpoint of, 388
modeling with, 598–609
order of, 580
solutions of, 580–581
Differential form, 212
Differential(s), 210–216, 384
Differentiation, 184
implicit, 246–253
as inverse process of integration, 424–425
logarithmic, 257–258
of power series, 686–689
techniques of, 189–197
using a CAS, 207–208
Diffraction of light waves, 450
Diophantus, 352, 353
Directional derivatives, 937–939
and gradients, 946
Direction angles, 779–780
Direction cosines, 779
Direction fields, 388–389, 592–594
and Euler's Method, 595
Direction of increasing parameter, 94, 830
Directrix, 727, 743
Discontinuity(-ies), 78, 119, 149, 150, A19
infinite, 565
jump, 119–120
limits at points of, 907–908
removable, 120, 157

Discrete data, 21
Discrete vs. continuous data, 21
Disease, spread of, 599
Disks, method of, 471
Displacement, 25, 29–32, 66, 869–870
finding, from velocity versus time curve, 433–434
in rectilinear motion, 432
Displacement vector, 765, 766, 782
Distance, 433–435
between parallel planes, 804–805
between point and plane, 804
between skew lines, 804–805
traveled in rectilinear motion, 433
Distance formula, 761, A13, A29–A30
Divergence, 634, 1059
absolute, 665
of improper integral, 566–569
of a series, 634
of a sequence, 619–620
Theorem, 1109–1112
Divergence Theorem, 1055, 1109–1114
Gauss's law and, 1115
Divergence test, 641, 668
Division by zero, A2
Domain(s), 25, 29–32, 890, 1081
in applied problems, 32–33
effect of algebraic operations on, 30–31
and range of inverse functions, 227–228
of inverse functions, 227–228
natural, 29–30
restricting, to make functions invertible, 231–232
Dot product, 776–778
derivative of, 841
differentiating, 841
and directional derivative, 939
operator and, 1061
Double-angle formulas, A47
Double integral(s), 978–984, 979
area calculated as a, 989–991
change of variables in, 1044–1047
converting, from rectangular to polar coordinates,
997–998
definition of, 979–980
over nonrectangular regions, 985–993
polar, 995–997
in polar coordinates, 993–1000
properties of, 980
in rectangular coordinates, 995
setting up limits of integration for evaluating,
985–989
Doubling time, 602–603
Downward curvature, 292, *see also* Concave down
Drag force, 588
Dummy variables, 420–421
Dynamical systems, 106–110
Dynes (dyn), 490

e, 237–238, 448
Earth
and Mercator projections, 528
spherical coordinates on, 823–824
Earth weight, 492, 607, 695
Eccentricity, 743
Ecliptic, 881
Ecology, 108–109
Economics, 346–347
basic principle of, 347
Einstein, Albert, 61, 410, 698
Electric intensity vector, 945
Electric lines, 1101
Electrostatic field, 1101

Elementary functions, 448
Elements, of a set, A3–A4
Elevation angle, 458
Ellipse(s)
    area of, 1090
    as a conic section, 726
    definition of, 727
    eccentricity of, 744
    equations of, in standard position, 730–731
    reflection property of, 738
    semiaxes of, 97
    sketching, 731–732
    translated, 735
Ellipsoids, 810, 811
Elliptic cones, 810, 811
Elliptic integral of the first kind, complete, 576
Elliptic paraboloids, 810, 811
Empty set, A4
End behavior, 122
End effector, *see* Robotics
Endpaper Integral Table, 543–548
Endpoints, derivatives at, 186
Energy, 489
    conservation of, 1085–1086
    kinetic, 489, 493–494, 698
    potential, 1085
Enrico Fermi Laboratory, 878
Equality of mixed partials, 923
Equal vectors, 766, 768
Equations
    graphs of, 18
    solving, involving exponentials and logarithms, 240–241
    see also Parametric equations
Equiangular spirals, 755
Equilateral hyperbola, 78, 735
Equilibrium, 1021
    position, and spring vibration, 606
Equipotential curves, 899
Erg, 490
Error(s), 154
    absolute, 596, 695
    percentage, 596
    roundoff, 680
    truncation, 680
Error estimates, 559–562
Error function erf(x), 449
Error propagation, 215–216
Errors of omission, 301
Escape speed, 883
Euclid, 9, 1110
Eudoxus, 6
Euler, Leonhard, 19, 111, 237, 363, 968, 1040
Euler's Method, 594–596
    accuracy of, 596
Even function, 53
Eventually held properties, of sequences, 628–629
Existence theorems, 331
Explicit functions, 246
Exponential decay model, 601, 603
Exponential function(s), 236–237
    approximating, with Maclaurin series, 680–681
    derivatives of, 258–259
    family of, 236–237
    logarithmic functions vs., 239
    natural, 238
    solving equations involving logarithms and, 240–241
Exponential growth, 242–243
Exponential growth model, 601, 603
Exponential models, 109, 326, 600–601

Exponentials
    review of, 514
    solving equations involving logarithms and, 240–241
Exponents, irrational, 235–236
Extrema, 956–957
    absolute, *see* Absolute extrema
    finding, on closed and bounded sets, 961–963
    problems with constraints, 967
    relative, 299–305
Extreme-Value Theorem, 331, 339, 368, 413, 422, 957, 961
Extrusion, 763

Factorial, 622
Factoring, A53–A54
Factor Theorem, A53–A54
False gaps, 43–44
False line segments, 44
Families
    of cardioids and limaçons, 708
    of circles, 707
    of functions, 75–89
    of lines and rays, 706
    of rose curves, 707
Fermat, Pierre de, 225, 352–353, 496
Fermat's last theorem, 353
Fermat's principle, 351
Fibonacci sequence, 108, 625
Field(s)
    conservative, 1058–1059
    curl, 1059
    direction, 592–594
    divergence, 1059
    flow, 1056, 1101
    gradient, 1058
    inverse-square, 1057–1058
Fingerprint compression, 2
Finite closed intervals, 331
    finding absolute extrema on, 331–333
    problems involving, 339–343
Finite intervals, A4
Finite solid, 1012
First-degree (linear) equations, 61–62, A25
First derivative test, 299–305
    proof of, 303
First octant, 761
First-order differential equations, 582–589
    free-fall motion and, 587–589
    geometry in, 586
    mixing problems and, 586–587, 593–594
    separable, 582–583
First-order initial-value problems, 581–582
First-order models, 693
First-order partial derivatives, 914
Flow field, 1056, 1101
Flow line, 1063
    of a vector field, 1064
Fluent, 1055
Fluid(s), 496–499
    definition of, 496
    density, 497
    dynamics, 496
    force, 499
        on vertical surface, 498–499
    homogeneous, 497
    pressure, fluid, 496–499
    statics, 496
Fluid flow, 1056, 1059
    steady-state, 1103
Flux, 1103–1107
    defined, 1103
    integrals, 1104–1105

    of **F** across σ, 1104
    using Divergence Theorem to find, 1112–1114
Flux density
    divergence viewed as, 1114
    outward, **F** across G, 1114
Flux integrals, 1104–1105
Fluxion, 1055
Focal axis, of a hyperbola, 727
Focus (foci)
    of an ellipse, 727
    of a hyperbola, 727
    of a parabola, 727
Focus–directrix property, 743–744
Folium of Descartes, 247, 725
Foot-pounds (ft·lb), 490
Force
    constant, 489–491
    fluid, 499
    restoring, 606
    variable, 491–492
Force field
    work performed by, 1073
Force vector, 766
Forward kinematic equations, *see* Robotics
Four-cusped hypocycloid, 60, 484, 851
Four-petal rose curve, 703
Fractals, 110, 184
Free-fall motion, 358–359, 430–431, 587–589
Frenet, Jean-Frédéric, 856
Frenet frame (or **TNB**-frame), 856
Frenet–Serret formulas, 866
Frequency, 86, 209, 445
    angular, 209
Fresnel, Augustin, 450
Fresnel sine and cosine functions, 450
Friction, coefficient of sliding, 209
Function(s), 15, 18–19, 890–892
    absolute extremum of, 330
    absolute value, 27–28
    algebraic, 82
    antiderivative of a, 382
    arithmetic operations on, 47–48
    bounded, 413
    composition of, 49–52
    concavity of, 292–293
    constant, 75–76, 290, 291
    continuous, 119, 149, 906–907, *see also* Continuous function(s)
    critical points of, 300
    from data, 324–328
    decomposing, 51–52
    defined explicitly and implicitly, 246–247
    defined piecewise, 28–29
    definite integral of a, 978
    density, 1024
    derivative of a, 383
    differentiable, 922
    domain of, 25, 29–32
    elementary, 448
    end behavior of, 122
    error, 449
    even and odd, 53–54
    expressed as a composition, 51–52
    expressed parametrically, 96
    families of, 75–93
    Fresnel sine/cosine, 450
    Gamma, 575
    and graphical analysis, 16–22
    graphs of, 25–26, 35–45, 892–893
        using technology, 893, 897–898
    homogeneous, of degree n, 929

hyperbolic, 500–508 *see also* Hyperbolic functions
increasing/decreasing, 290–292
inflection points of, 293–296, 302–303
integrals, defined by, 448–450
integrals with, as limits of integration, 450–451
interpolating, 324
inverse, *see* Inverse function(s)
invertible, 229
linear, *see* Linear functions
logarithmic, 238–240
natural domain of, 29–30, 891
natural exponential, 238
new from old, 47–60
one-to-one, 229
piecewise defined, 28–29
polynomial, 80–81
position, 171–172, 352
potential, 1058–1059
power, 77–79, 236
power series, defined by, 674–675
proper rational, 537
properties of, 24–35
range of, 25, 31–32
rational, 81–82
reflections of, 56–57
relative maxima/minima of, 299–300
scale and units with, 33
second derivative test with, 302, 303
smooth, 480
stretching/compressing, 48–49
sums of, 49
symmetry in, 52–53
translations of, 54–56
trigonometric, 82–84, *see also* Inverse trigonometric function(s); Trigonometric function(s)
vertical line test with, 26–27
in viewing windows, *see* Viewing window(s)
Fundamental Theorem of Calculus, 377, 416–427, 442, 570, 838, 850, 1079, 1082
and dummy variables, 420–421
and Mean-Value Theorem for Integrals, 421–423
part two, 423–424, 450
and relationship between definite and indefinite integrals, 419–420
Fundamental Theorem of Work Integrals, 1079, 1080

Gabriel's Horn, 572, 1003
Galileo, 100, 875
Galois, Evariste, A55
Gamma function, 575
Gauss, Karl Friedrich, 363, 378, 410, 1110
Gauss's Law in Electric Fields, 1116
Gauss's Law for Inverse-Square Fields, 1115
Gauss's Theorem, 1109
General equation
of a circle, A33
of a line, 61–62, A25–A26
Generalized derivative formulas, 205–207
General solution, of differential equations, 580–581
General term, 616, 641
Geometric mean, 746
Geometric properties of cross products, 788–790
Geometric series, 635–637
Geometric sums, 403
Geosynchronous orbit, 885
Golden ratio, 108
Gradient(s)
and directional derivatives, 946–947
of f (∇f), 939
normal to level curves, 941–942
normal to level surfaces, 947–948
properties of, 940–941

using to find tangent lines to intersections of surfaces, 949
using to find tangent planes, 948
Gradient fields, 1058
Graph(s), 761, 830, A17–A19
analysis of functions and their, 289–328
bar, 16
complete, 306
of conics, in polar coordinates, 702, 745–747
curve fitting with, 20–21
definition of, A17
determining shapes of, 44–45
discontinuity in, A19
of equations, 18, 247
of functions, 25–26
defined by integrals, 450
of two variables, 892–893
of integral curves, 387
intersections of polar, 723–724
of inverse functions, 230–231, 233
of parametric equations, 93
of polynomials, 309–311
as problem-solving tools, 21–22
procedure for analyzing, 306–307
properties of, 306
of a quadratic equation in x and y, 736
of rational functions, 311–314
reflections of, 56–57
scale and units with, 33
as scatter plots, 16
of sequences, 618–619
stretches and compressions of, 48–49
in three-dimensional space, 761–762
translation of, 54–55
of vector-valued functions, 832–833
using technology, 893, 897–898
of vector fields, 1056–1057
Graphing utilities, 35–36, 44–45
approximating roots by zooming with, 155–156
errors of omission with, 44
generating parametric curves with, 96
generating polar curves with, 709–710
graphing inverse functions with, 233
slopes of tangent lines estimated by zooming with, 178–179
using with functions of two variables, 897–898
viewing window, 897, *see also* Viewing window(s)
Gravitational field (of Earth), 1056
Gravity, acceleration due to, 164, 358–359, 430
Greatest integer function, 60
Greatest lower bound, 630
Green, George, 1088
Green's Theorem, 1055, 1088
finding areas using, 1090–1091
finding work using, 1090
for multiply connected regions, 1091–1092
relationship between Stokes' and, 1121–1122
Gregory, James, 682
Gregory of St. Vincent, 446
Growth
exponential/logarithmic, 242–243, 601–603
population, 598–600
relative rate of, 601–602
Growth constant, 601
interpreting, 601–602
Growth rate, 188
Guldin, Paul, 1026
Guldin's Theorem, 1026

Hale–Bopp comet, 751
Half-angle formulas, A47

Half-life, 602–603
Half-open/closed interval, A4
Halley, Edmond, 11
Halley's comet, 749
Harmonic motion, simple, 607
Harmonic series, 638
alternating, 663
Harvey, William, A3
Heat equation, 918
Helicoid, 1011
Higher derivatives, 196–197
Higher-order partial derivatives, 913–914
Hippasus of Metapontum, A1
Homogeneity, 1068, 1098
Homogeneous fluids, 497
Hooke, Robert, 68, 492
Hooke's law, 68, 492, 606
Horizontal asymptote, 82, 123
and limits at infinity, 122–123
Horizontal line test, 229–230
Horizontal range, 459, 879
Hours of daylight, 268
Hubble, Edwin P., 74
Hubble's constant, 75
Hubble's law, 74–75
Human cannonball, 457–460
Humason, Milton L., 75
Hurricanes, modeling, 1126–1130
Huygens, Christian, 11
Hydrodynamics, 1115
Hyperbola(s)
as a conic section, 726
definition of, 727
equations of, in standard position, 732–733
equilateral, 78, 735
finding asymptotes of, 733
reflection property of, 738
sketching, 734–735
translated, 735
unit, 503
Hyperbolic cosecant, 501
Hyperbolic cosine, 501
Hyperbolic cotangent, 501
Hyperbolic functions, 500–508
applications of, 502
derivative and integral formulas for, 503–505
derivatives and integrals involving inverse, 507–508
graphs of the, 501–502
and hanging cables and other applications, 502
identities for, 502–503
integrals of, 515
inverses of, 505
logarithmic forms of inverse, 505–507
review of, 514
Hyperbolic navigation systems, 738
Hyperbolic paraboloid, 810, 811, 894
Hyperbolic secant, 501
Hyperbolic sine, 501
Hyperbolic spiral, 709, 712
Hyperbolic substitutions, 535
Hyperbolic tangent, 501
Hyperboloid of one sheet, 810, 811
Hyperboloid of two sheets, 810, 811
Hyperharmonic series, 643–644
Hypocycloid, 60, 102, 484

Ideal fluid, 1126
Ideal gas law, 966
Ill-posed optimization problems, 345
Image of S under T, 1041
Image of x under f, 24

Imaginary numbers, A2
Implicit differentiation, 246–253
Implicit functions, 246
Implicit partial differentiation, 912–913
Improper integrals, 564–571
Improper rational functions, integrating, 542
Incompressible fluids, 1103, 1126
    continuity equation for, 1115
    sources and sinks of, 1115
Increasing functions, 290–292
Increasing sequences, 626
Increasing without bound, 120
Increment(s), 63, 210–211, 595
Increment in f, 921
Increment in z, 921
Indefinite integral(s), 383
    properties of the, 385
    relationship between definite and, 419–420
    and trigonometric substitution, 531
Independence of parametrization, 1072
Independence of path, 1080
Independent variable, 24
Indeterminate forms, 160, 277–284
Index, for a sequence, 617–618
Index of summation, 397
    changing the, 398
Induction, 61
Inequalities, A3, A14
    with absolute values, A13–A14
    algebraic properties of, A5–A6
    solving, A6–A8
    triangle inequality, A14–A15
Infinite discontinuities, 565
Infinite intervals, A4
    absolute extrema on, 333–334
    integrals over, 565–568
Infinite limits, 120–122, 144–145
    and vertical asymptotes, 120–122
Infinite processes, 4–5
Infinite sequence, 4, 616
Infinite series, 5–6, 615, 632–638
    absolute convergence of, 665–667
    algebraic properties of, 642–643
    alternating series, 662–665
    binomial series, 683
    comparison test for, 656–658
    conditional convergence of, 666–667
    convergence/divergence of, 634
    convergence tests for, 640–645, 656–660, 667–668
    divergence test for, 641
    general term of, 616
    geometric series, 635–637
    harmonic series, 638
    integral test for, 643–645
    limit comparison test for, 658–659
    Maclaurin series, 652–655, 678, 680–684, 689–693
    nth partial sum of, 634
    power series, 670–675
    p-series, 643–644
    ratio test for, 659–660
    root test for, 660
    roundoff error with, 680
    sequences, see Sequences
    sums of, 632–635
    Taylor series, 652–655, 676–678, 689–693
    terms of, 616, 632
    truncation error with, 680
Infinite slope, 714
    of a line, A21
Infinity, limits at, 122–124

Inflection points, 293–295, 302
    in applications, 295–296
    significance of, 302–303
Inhibited growth model, 600
Initial condition(s), 388, 428–429, 581
Initial point, 766
Initial-value problems, 388, 449, 581–582
Instantaneous acceleration, 868
Instantaneous rate of change, 174
Instantaneous speed, 9, 353–354, 868
Instantaneous velocity, 171–173, 353–354, 868
    finding position and, by integration, 427–428
    and free-fall motion, 358
    geometric interpretation of, 173
    and limits, 114
    problem, 112
Integrability, 410
    conditions for, 413
    Riemann, 410
Integrable components, 836
Integral(s)
    of algebraic functions, 515
    involving $ax^2 + bx + c$, 534
    change of variable in a single, 1039–1040
    complete elliptic, of the first kind, 693
    curves, 387, 581, see also Integral curve(s)
    definite, see Definite integral(s)
    double, 978–984
    evaluating and graphing functions defined by, 450
    evaluating with a CAS, 514
    with functions as limits of integration, 450–451
    functions defined by, 448–449
    improper, see Improper integrals
    indefinite, see Indefinite integral(s)
    iterated, 981, 985
    line, 1064–1075
    link between natural logarithms and, 446–447
    Mean-Value Theorem for, 421–423
    partial definite, with respect to x and y, 980
    relationship between definite and indefinite, 419–420
    Riemann, 410
    single, 980
    surface, 1094–1099
    tables, 514
    triple, 1012–1019
Integral calculus, 112, 377
Integral curve(s), 387, 581, 583, 592–593
Integral sign ($\int$), 383
Integral tables, 543–549
Integral test, 643, 668
    proof of, 644–645
Integrand, 383, 408
Integrating factors, 584–585
Integration
    and area problem, 378–382
    constant of, 383, 517
    and differential equations, 388
    and direction fields, 388–389
    formulas, 384–385, 514–515
    integrals with functions as limits of, 450–451
    as inverse process of differentiation, 424–425
    iterated (or repeated), 981, 985
    limits of, 450–451, 987
    lower and upper limits of, 408
    methods for approaching problems, 514
    notation for, 383
    numerical, see Numerical integration
    an overview of, 514–515
    partial, 980
    by partial fractions, 536–542

by parts, 516–521
    for definite integrals, 519–520
    with reduction formulas, 520–521
path of, 1078
of power series, 689
of powers of sine and cosine, 522–524
of powers of tangent and secant, 525–526
of products of sines and cosines, 524–525
of products of tangents and secants, 526–527
of rates of change, 431–432
reversing the direction of, 1072–1073
reversing the order of, 989
setting up limits of, 986–989
by substitution, 391–395
using a CAS, 395, 549–551
from viewpoint of differential equations, 388
Intensity (of sound), 242
Intercepts, A20
Interest, 614
Interior point, 901
Intermediate-Value Theorem, 154, 486
    approximating roots using, 154
International System of Units (SI), 490
Interpolating function, 324
Intersection, of sets, A5
Interval of convergence, 671–673
Intervals, A4
    finite closed, 331
    unions and intersections of, A5
Intrinsic error, 559
Inverse function(s), 226–233
    continuity of, 232
    definition of, 226
    derivatives (differentiability) of, 232–253
    domain and range of, 227–228
    existence of, 229–230
    finding, 228–229
    graphs of, 230–231, 233
        using a graphing utility, 233
Inverse hyperbolic functions, 505–508
    derivatives/integrals involving, 507–508
    logarithmic forms of, 505–507
Inverse proportionality, 79, 90
Inverse-square fields, 1057–1058
    divergence and, 1060
    and electrostatic fields, 1101
    Gauss's law for, 1115
Inverse trigonometric function(s), 261–267
    definition of, 262
    derivatives (differentiability) of, 265–267
    evaluating, 263–264
    graphs of, 262–263
    identities for, 264–265
    and integration by parts, 519–520
Invertible functions, 229
Involute of a circle, 753
Inward orientation, 1109
Inward unit normal vectors in 2-space, 855
Irrational exponents, 235–236
Irrational numbers, 12, A1
Irreducible quadratic factors, 307
Isobar, 900
Isothermal curves, 899
Iterated function sequences, 106–108
Iterated integrals, 981
    with nonconstant limits of integration, 985
Iterated (or repeated) integration, 981
Iterated quadratic systems, 110
Iteration, 106–110

Jacobi, C. G. J., 1040
Jacobian(s), 1040–1047

Joule (J), 490, 494
Jump discontinuity, 119–120

Kappa curve, 713
Kepler, Johannes, 699, 748, 880
Kepler's laws, 376, 748
    first law (Law of Orbits), 880, 884
    law of planetary motion, 880–886
    second law (Law of Areas), 880, 884
    third law (Law of Periods), 327, 880, 884
Kinetic energy, 489, 493–494, 698
Kline, Morris, 171
Kowa, Seki, 7
Kramer, Gerhard, 528

Lagrange, Joseph Louis, 889, 968
Lagrange multipliers, 967–970
Lagrange's identity, 794
Lamé's special quartic, 254
Lamina
    center of gravity of a, 1021–1023
    centroid of, 1023
    density of a, 1019–1020
    first moment of a, about x- and y-axes, 1023
    homogeneous/inhomogenous, 1019
    mass of a, 1020, 1098–1099
    and moment of inertia of, 1029
Laplace, Pierre-Simon de, 1062
Laplace's equation, 918, 1061
    polar form of, 929
Laplace transform, 573
Laplacian operator (∇^2), 1061
Law of Areas, 748, 880
Law of cosines, 269, 773, A46–A48
Law of Large Numbers, 99
Law of Orbits, 748, 880
Law of Periods, 748, 880
Law of sines, 773, A51
Leading coefficient (term), 307
Least-squares line of best fit, 325, 965
Least squares principle, 351
Least upper bound, 630
Left endpoint approximation, 406, 554
Left-handed rectangular coordinate systems, 760
Left-hand rule, 789
Leibniz, Gottfried Wilhelm, 4, 11–12, 18, 99, 100, 211, 278, 377, 379, 380, 383, 682
Leibniz notation, 211
Lemniscates, 706
Length
    arc, 480–481, 482–483
    of plane curves, finding, 480–483
    of a vector, 770
Leonardo da Pisa, 625
Leontief, Wassily, 3
Level curves, 893–895
Level surfaces, 895–896
Lever arm, 1021
L'Hôpital, Guillaume François Antoine de, 278
L'Hôpital's rule, 277–284, 403, 567, 621, 622, 907, 908
Libby, W. F., 604
Limaçons, 708
Limit(s), 112–145, 836–837, 901–910
    along curves, 902–904
    areas and, 113
    and continuity of trigonometric functions, 159–164
    basic, 127–130
    computational techniques, 127–138
    definition of, 115, 138–141
    of functions defined piecewise, 136

of functions of two variables, 904–905
    infinite, 120–122, 144–145
        and vertical asymptotes, 120–122
    at infinity, 122–124
    instantaneous velocity and, 114
    involving radicals, 135–136
    nonexistent, 118, 123–124
    obtaining, by squeezing, 160–162
    one-sided, 117–118
    of piecewise defined functions, 136
    pitfalls of guessing at, 116–117
    at points of discontinuity, 907–908
    of polynomials as $x \to a$, 130–131
    properties of, 128–132, 905
    radicals involving, 135–136
    of rational functions, 132–135
        as $x \to a$, 132–134
    as $x \to a$, 132–134
    of Riemann sums, 979
    of a sequence, 619–922
    of summation, 397–398
    two-sided, 118–119
    as $x \to +\infty$, 142–144
Limit comparison test, 658–659, 668
    proof of, A61
Limit(s) of integration, 408
    and the area between two curves, 463
    integrals with functions as, 450–451
Limit theorems, proofs of basic, A57–A58
Line(s)
    angle of inclination, 63
    coordinate/real number/real, A3
    families of, 75–77, 706
    general equation of, 61–62, A25–A26
    interpretations of slope, 62–63
    latitude/longitude, 1002
    normal, 931
    parallel, A22–A23
    parametric equations of, 795–797
    perpendicular, A22–A23
    point-slope form of, A23–A24
    segments, 797
    slope-intercept form of, A24–A25
    slopes in applied problems, 63–64
    vector equations of, 797–798
Linear algebra, 541, 786
Linear combination of vectors, 776
Linear differential equations, first-order, 583–586
Linear equation, 61–62, A26
Linear factor rule, 537
Linear factors, partial fractions with, 537–539
Linear mathematical models, 64–65, 325–326
    from graphical data, 68–69
    from numerical data, 69–70
Linear mass density, 1068
Linear models, 61–75, 325
    from direct proportion, 67–68
    from graphical data, 68–69
    least-squares line of best fit in, 325
    from numerical data, 69–70
    regression line in, 325
Linear polynomials, 81, 307
Line graphs, 16
Line integral(s), 1064–1075
    arc length as, 1069
    change of parameter in, 1072
    evaluating, 1065–1067
    mass of a wire as a, 1068–1069
    notation, around simple closed curves, 1089
    with respect to arc length, 1070
    with respect to x, y, and z, 1069–1071

    in 3-space, 1067–1068
    work as a, 1073–1074
Line of action, 795
Line segment(s), 797
    vector form of, 833
Liquids, 496
Lissajous, Jules Antoine, 97
Lissajous curves, 97–98
Lituus spiral, 709
Local linear approximation, 210–216
    error in, 214
    and error propagation, 215–216
Local quadratic approximations, 647–649
Logarithm(s), 237–238
    algebraic properties of, 239–240
    change of base formula for, 241
    common, 237
    natural, 237, 446–447, see also Natural logarithms
    solving equations involving exponentials and, 240–241
Logarithmic differentiation, 257–258
Logarithmic function(s), 238–240
    comparison of exponential and, 239
    derivatives of, 255–257
    derivatives of exponential and, 255–261
    growth, 242–243
    from the integral point of view, 446–454
Logarithmic growth, 242–243
Logarithmic growth curves, 318–319
Logarithmic models, 326
Logarithmic scales in science and engineering, 242
Logarithmic spiral, 709
Logistic differential equation, 600
    typical solutions of, 605
Logistic growth curves, 318–319
Logistic models, 109, 600, 604–606
Lower bound
    greatest, 630
    of a sequence, 629
    of a set of numbers, 630

Machin, John, 12, 686
Machin's formula, 686
Maclaurin, Colin, 579, 650
Maclaurin polynomials, 649–651, 689–693
    sigma notation for, 652
Maclaurin series, 652–655
    approximating exponential functions with, 680–681
    approximating logarithms with, 681–682
    finding, by multiplication/division, 692–693
    table, 684
MAGLEV trains, 495
Magnetic resonance imaging (MRI), 3
Magnitude
    of a real number, 27, A11
    of a vector, 770
Major axis, of an ellipse, 727
Manufacturing cost, 346
Maple, 36
Marginal analysis, 347
Marginal cost, 347
Marginal productivity of labor and capital, 975
Marginal profit, 347
Marginal revenue, 347
Mass of a lamina, 1020, 1098–1099
Mass density, 497, 1098
Mass density function δ, 1068, 1098
Mass of a wire, 1068
Mathematica, 36
Mathematical modeling, 61

Mathematical model(s), 61–75, 324, 965
inductive and deductive, 61
linear, 64–65, 325–326
nonlinear, 326–327
Maximum, absolute, 330
Mean
arithmetic, 435, 746
geometric, 746
Mean (average) value, 435–436, 984, 992, 1919
Mean-Value Theorem, 368–371, 417, 427, 436, 481, 486
proof of the, 370–371
velocity interpretation of the, 369
Mean-Value Theorem for Integrals, 421–423, 435
Mécanique Analytique, 968
Mechanic's rule, 4, 13, 367
Members, of a set, A3–A4
Mercator (or Gerhard Kramer), 528–529
Mercator projection, 528–529
Mesh lines, 808
Mesh size, 409
Method of cylindrical shells, 475–479
variations of the, 477–479
Method of disks, 471, 479
Method of exhaustion, 6, 378
Method of integrating factors, 584–585
Method of Lagrange multiplier, 967–970
Method of least squares, 965
Method of u-substitution, 391–394, 441–443
Method of washers, 472, 478
Midpoint approximation, 406–407, 554–556
Midpoint error, 559
Midpoint formula, A30
Millibars, 900
Minimax point, 814
Minimum, absolute, 330
Minor axis, of an ellipse, 727
Minutes, A39
Mixed second-order partial derivatives, 914
Mixing problems, 586–587
differential equations and, 586–587, 593–594
Möbius, August, 1101
Möbius strip, 1100, 1101
Modeling with differential equations, 598–609
carbon dating, 603–604
disease, spread of, 599
doubling time, 602–603
exponential growth/decay, 600–602
half-life, 602–603
logistic models, 604–606
pharmacology, 599
population growth, 598–599
inhibited, 599–600
radioactive decay, 603
vibrations of springs, 606–609
Model of projectile motion, 873–875
Moment of inertia, 1029, 1039
Moment, 1021, 1022
Monkey saddle, 808
Monotone sequences, 626–630
and the Completeness Axiom, 630–631
convergence of, 629–630
definition of, 626
eventually, 628–629
testing for, 626–628
Motion
along a curve, 867–880
displacement in rectilinear, 432
free-fall, 358, 430–431, 587–589
model of, 430–431
initial conditions for, 428

rectilinear, 352–362, *see also* Rectilinear motion
simple harmonic, 607
in 3-space, 792–793
uniformly accelerated, 428–430
work done by constant force applied in direction of, 489–491
work done by variable force applied in direction of, 491–492
Multiplicity, 308–311
geometric implications of, 309–311
roots of even/odd, 310
Multiply connected regions, Green's Theorem and, 1091–1092
Muzzle velocity, 457

Natural domain, 29–30, 832, 890
of functions, 891
Natural exponential function, 238
Natural logarithms, 237, 446–448
approximating, numerically, 447
approximating with Maclaurin series, 681–682
continuity and differentiability of, 448
link between integrals and, 446–447
Natural numbers, A1
Navigation, spherical coordinates in, 823–824
Navigation systems, hyperbolic, 738
NB-plane (or normal plane), 856
n-dimensional space, 891
Near Earth objects (NEOs), 755
Negative change of parameter, 847
Negative direction, A3
Negative displacement, 432
Negative of **v**, 767
Negative orientation, 1102, 1106
Negative/positive direction, 1102, 1118
Net signed area, 407, 434, 1065
Net signed volume, 979
Newton, Isaac, 4, 10–11, 19, 20, 61, 99, 100, 169, 211, 363, 378, 379, 380, 446, 650, 651, 748, 880, 1055, 1062
Newtonian kinetic energy, 698
Newton-meters (N·m), 490, 792
Newtons (N), 490, 497
Newton's Law of Cooling, 189, 259, 328, 611
Newton's Law of Universal Gravitation, 881–883
Newton's Method, 363–368
Newton's Second Law of Motion, 588, 606, 607, 792, 871, 873, 880
Newton's Universal Law of Gravitation, 1056–1058
Nonexistent limits, 118, 123–124
Nonlinear mathematical models, 326–327
Nonnegative, A4
Nonorientable surface, 1102
Nonparametric surfaces, orientation of, 1105–1107
Nonrectangular regions, and double integrals, 985–993
Nonrepeating decimals, 12
Normal components of acceleration, 870–837
Normal plane, 856
Normal scalar component of acceleration, 871
Normal vector, 801
Norm of a vector, 770
nth Maclaurin polynomial for f, 650
nth partial sum, 634, 637
nth remainder, 676–677
estimating, 677–678
Nuclear magnetic resonance (NMR), 3
Null set, A4
Numbers
complex, A1–A2
imaginary, A2
irrational, 12, A1

natural, A1
rational, 12, A1
real, A1, A2
Numerical analysis, 6, 559, 680
Numerical integration, 553–562
Riemann sum approximations, 553–554
and Simpson's rule, 557–562
trapezoidal approximation, 554–556
Numerical methods, and finding arc length, 483
Numerical values, 990

Oblate spheroid, 819
Oblique asymptote, 320
Octants, 761
Odd function, 53
Ohm's law, 165
One-dimensional space, 760
One-dimensional wave equation, 915
One-sided limits, 117–118
One-to-one functions, 229
Open form, of a sum, 401
Open intervals, A4
absolute extrema on, 334–335
Open set, 901–902
Operator (∇), 1061
Optimization problems, 339–352
Order, of differential equations, 580
Ordinate, A16–A17
Oresme, Nicole, 638, 640
Orientation, 1091–1092
of a curve, 94–95, 830
of nonparametric surfaces, 1105–1107
of piecewise smooth closed surfaces, 1109
positive/negative, 1091, 1102
of a surface, 1102
relative, of curves and surfaces, 1118
Reversal of orientation, 1073
Oriented surfaces, 1101–1102
Origin, 700, A3
symmetry about, 52–53
Origin of coordinate system, 760
Orthogonal components, 780
Orthogonal curves, 254
Orthogonal projections, 781–782
Orthogonal surfaces, 956
Orthogonal trajectories, 254
Oscillation, limit failing to exist because of, 124
Osculating circle, 862
Osculating plane, 856
Outside functions, 51, 52
Outward flux, 1112, 1114
Outward orientation, 1109
Overhead, 346

Paired data, 16
Pappus of Alexandria, 1026
Pappus' Theorem, 1026–1027
Parabola(s), 809, A33–A34
as a conic section, 726
definition of, 727
equations of, in standard position, 728–729
reflection property of, 737
semicubical, 715
sketching, 729–730
translated, 735
Parabolic spiral, 709
Paradox(es), 8–9
Paradox of Achilles and the Tortoise, 8–9
Parallelepiped, 790
Parallelogram method, 767

Parallel vectors, 767
Parameter(s), 76
  arc length as, 845–846
  change of, 843–853, 846–847
    in line integrals, 1072
  direction of increasing, 830
  of parametric equations, 93
Parametric curve(s), 94, 830
  absolute extrema and, 336–337
  arc length of, 482–483, 844
  changing parameter in, 846
  generated with technology, 96, 830–831
  limits along, 903
  scaling, 97
  simple, 1082
  tangent lines to, 713–715
  translation of, 97
  vector fields and, 1081
Parametric equations, 93–103
  for intersections of surfaces, 831
  line integrals and, 1072
  ordinary functions expressed as, 96
  orientation in, 94–95
  of projectile motion, 875–876
  of a tangent line, 840
Parametric equations of lines, 795–801
  lines determined by point and vector, 795–797
  line segments, 797
  for projectile motion, 875
  vector form of, 797–798
Parametric surface(s), 1000–1012
  orientation, 1102
  representation of, 1000–1002
  smooth, 1006
  surface area of, 1006–1008
  tangent planes to, 1004–1006
Partial definite integral, with respect to x and y, 980
Partial derivative(s), 889–976
  directional derivatives and relationship between, 939
  implicit, 912–913
  mixed, 914
  notation (∂), 912
  of vector-valued functions, 1003–1004
  viewed as rates of change and slopes, 911–912
Partial differential equation, 915
Partial fraction decomposition, 537
  integrating improper rational functions and, 542
  linear factors and, 537–539
  quadratic factors and, 539–541
Partial fractions, 536–542
  definition of, 537
  with improper rational functions, 542
  inappropriate applications of, 542
  with linear factors, 537–539
  with quadratic factors, 539–541
Partial integration, 980
Partial sums, sequence of, 634
Partition (of an interval), 409
Parts, integration by, 516–521
  for definite integrals, 519–520
  derivation of formula for, 516–519
  and derivation of reduction formulas, 520–521
Pascal, Blaise, 496
Pascal (Pa), 496, 497
Pascal's principle, 497–498
Path of integration, 1078
Path of steepest ascent and descent, 944
Peak voltage, 445
Pendulum, simple, 576
Percentage error, 216, 596

Perigee, 275, 748
Perigee distance, 748, 885
Perihelion, 748, 885
Period, 86, 209, 218, 445
  of a pendulum, 576
Permittivity constant, 1058
Pharmacology, 599
Phase shift, 87
Philosophiae Naturalis Principia Mathematica
    (Isaac Newton), 11
pH scale, 242
Pi (π), 6, 12–13, A2
  approximating, 682
Piazzi, Giuseppi, 1110
Piecewise defined functions, 28–29
  limits of, 136
Piecewise smooth curve, 1071, 1109
Pixels, 43
Planes in 3-space, 801–808
  angles between, 803–804
  determined by a point and normal vector, 801–803
  distance problems involving, 804–806
  parallel to coordinate planes, 801
Planetary motion, 327
  laws of, 748, 880, *see also* Kepler's laws
Planets, eccentricity of, 744
Playfair, William, 16–17
Playfair's Graph of 1786, 17
Point-normal form, 801, 808
  of a line, 808
Polar angle, 700
Polar axis, 700
Polar coordinates, 700–710
  arc length, 717–718
  area in, 720–724
  cardioids/limaçons, families of, 708
  circles, families of, 707
  conic sections in, 743–750
  converting double integrals from rectangular to, 997–998
  double integrals in, 993–1000
  with graphing utilities, 709–710
  graphs in, 702–703
  lines/rays through the pole, families of, 706
  rose curves, families of, 707–708
  sketching conics in, 745–747
  spirals, families of, 708–709
  surface integrals and, 1098
  symmetry tests with, 703–706
  tangent lines, 713–717
  volume problem in, 994
Polar coordinate systems, 700
Polar double integrals
  evaluating, 995–997
  finding areas using, 997
Polar equations, of conics, 744–745
Polar form of Cauchy–Riemann equations, 929
Polar form of Laplace's equation, 929
Polar graphs, intersections of, 723–724
Polar rectangles, 994
Polar Riemann sums, 994
Pole, 700
Polygonal path, 481
Polynomial of degree n, 307
Polynomials, 80–82, 307–309
  coefficients of the, 307
  continuity of, 150–151
  degree of, 307
  graphing, 309–311
  limits of, 130–132
  Maclaurin, 649–651

  review of, 307–308
  solving (for roots), A52–A56
  Taylor, 651–652
Population growth, 598–599
  inhibited, 599–600
Position, 427–428
  change in, 432
  finding velocity and, by integration, 427–428
Position function, 171–172, 352, 867
Position vector, 833
Position versus time curve, 65–66, 352–353
  analyzing, 355–358
  and the Mean-Value Theorem, 369
Positive change of parameter, 847
Positive direction, A3
Positive displacement, 432
Positive/negative direction, 1102
Positive orientation, 1091, 1106, 1102
Potential energy, 1085
Potential function, 1059
Pounds (lb), 490, 497
Pound-feet (lb·ft), 792
Power function models, 326
Power functions, 77–79, 236
Power rule, 190
  for integer exponents, 195–196
Power series, 670–675
  convergence of, 671–673
  differentiating, 686–688
  functions defined by, 674–675
  integrating, 688–689
  representations of, 689–690
  in x, 670–671
  in $x - x_0$, 673–674
Pressure, 496–497
  definition of, 496
  fluid, 497–498
Prime notation, differential equation and, 580
Principal unit normal vector, 854, 1005
Production model, 975
Product rule, 192
Products of functions, derivatives of, 192–193
Product-to-sum formulas, A48
Profit, marginal, 347
Profit function, 346
Projectile motion, 457–460
  model of, 873–875
  parametric equations of, 875–876
Prolate cycloid, 719
Proper rational functions, 537
Properties
  of arc length parametrizations, 850–851
  of derivatives and integrals, 837–838
  of inequalities, algebraic, A5
p-series, 643–644
Pseudosphere, 1011
Ptolemy, 353
Pythagorean Theorem, *see* Theorem of Pythagoras

Quadrants, A17
Quadratic approximations, local, 647–649
Quadratic equations
  in x, A33
  in x and y, graph of, 736
  in y, A35–A36
Quadratic factor rule, 539
Quadratic factors, partial fractions with, 539–541
Quadratic formula, 363, A2
Quadratic polynomials, 81, 307

Quadrics, 810
Quadric surfaces, 808–819
    techniques for graphing, 810–814
    techniques for identifying, 816–817
    translations of, 814–815
Quarterly compounding, 614
Quartic polynomials, 81, 307
Quintic polynomials, 81, 307
Quotients, derivatives of, 193–194
Quotient rule, 193

Radial coordinate, 700
Radial unit vector, 887, 1129
Radian(s), 85, A39
    as a dimensionless unit, 85
    measure, A39
Radicals, and trigonometric substitution, 530–534
Radioactive decay, 603
Radius of convergence, 671–674
Radius of curvature, 862
Radius vector, 833, 1003
Railroad design, 576–578
Ramanujan, Srinivasa, 686
Ramanujan's formula, 686
Range, 25, 31–32
    in applied problems, 32–33
    and domain in applied problems, 32–33
    horizontal, 459
    of inverse functions, 227–228
Rates of change, 170–177, 431–432
    in applications, 174–175
    average and instantaneous, 173–174
    integrating, 431–432
    using the chain rule, 270–274
    of y with respect to x, 64
Ratio, in geometric series, 635–637
Rational functions, 81–82
    continuity of, 151–152
    graphing, 311–314
    integrating improper, 542
    limits of, 132–135
    proper, 537
    of $\sin x$ and $\cos x$, 548
Rational numbers, 12, A1
    repeating decimals and, A2
Ratio test, 659–660, 667, 668
    proof of, A62–A63
Rays, families of, through the pole, 706
Real number line, A3
Real numbers, 630, A1
    decimal representation of, A2
    sum of infinitely many, 632
Real number system, A1
Real-valued function, 836, 837
    of real variables, 24
Reciprocal rule, 195
Reciprocals, derivatives of, 195
Rectangle method (for finding areas), 379–380
Rectangular coordinates, 819, 820
    converting double integrals from polar coordinates to, 997–998
    converting triple integrals from cylindrical coordinates to, 1032
    double integrals in, 995
    graphs in, 702
    polar coordinates vs., 701
Rectangular coordinate systems, 470, 760, A16
    angles in, A42–A44
    and distance in three-space, 761
    and graphs in three-space, 761–762
    left-handed, 760

right-handed, 760
    in 3-space, 760–761
Rectifying plane, 856
Rectilinear motion, 65–66, 170, 352–362, 427–440
    and acceleration, 354–355
    definition of, 352
    displacement in, 432
    distance traveled in, 433
    and free-fall motion, 358–359, 430–431
    and instantaneous speed, 353–354
    and instantaneous velocity, 353–354
    position function for finding, 427–428
    and the position versus time curve, 355–356
    and rates of change, integrating, 431–432
    uniform, 66, 170–171
    and uniformly accelerated motion, 428–430
    and velocity versus time curve, 433–435
Recursion formulas, 107, 623
Recursively defined sequences, 623
Reduction formulas, 520–521, 527–528, 546
    integral tables, matches requiring, 546
    integrating powers of sine and cosine using, 522–524
    integrating powers of tangent and secant using, 525–526
    integrating products of sines and cosines using, 524–525
    integrating products of tangents and secants using, 526–527
Reflection properties, of conic sections, 737–738
Reflections, 56–57
    of graphs, 56–57
    of surfaces in 3-space, 815–816
Regression line, 325, 965
Related rates, 270–274
    in two variables, 927–928
Relative decay rate, 602
Relative error, 216
Relative extrema, 299–305, 957, A60–A61
    and critical points, 300
    finding, 957–959
    first derivative test for, 300–302
    occurrence of, at critical points, A60–A61
    second derivative test for, 302
Relative growth rate, 601–602
Relative maximum, 299, 956
    constrained, 968
Relative minimum, 299, 957
    constrained, 968
Relativistic kinetic energy, 698
Relativity theory, 698
Remainder Estimation Theorem, 677–678
    proof of, A63–A64
Remainder Theorem, A52–A53
Removable discontinuity, 120, 157
Repeating decimals, 12, A2
Representations, power series, 674–675
Residuals, 325, 965
Resolution, 43
Restoring force, 606
Resultant of two concurrent forces, 772–773
Revenue, marginal, 347
Revenue function, 346
Reversal of orientation, 1073
Revolution
    solids of, 470–471
    surface of, 485–488
Rhind Papyrus, 12, A9
Richter scale, 242, 245
Riemann, Georg Friedrich Bernhard, 410, 461
Riemann integrability, 410

Riemann integral, 409, 410
Riemann sums, 410, 461, 462, 553–554, 979, 1012, 1045
    arc length and, 481
    calculating work with, 493
    fluid force and, 498
    limits of, 979
    polar, 994
    surface area and, 485–486
    volume and, 469
Riemann zeta function, 698
Right cylinders, 469, 989
Right endpoint approximation, 554
Right-handed rectangular coordinate system, 760
Right-hand rule, 789
Right triangles, trigonometric functions for, A41–A42
Rise, 63, A20
Rms (root-mean-square), 445
Robotics, 221–224
    control angles, 221
    end effector, 221
    forward kinematic equations, 221
Rolle, Michel, 368
Rolle's Theorem, 368–369
Root-mean-square (rms), 445
Roots, 26, 308
    methods for finding, A55–A56
Root test, 660, 668
Rose curves, 703, 707–708
Rotated conic sections, 737
Rotational motion, 792
Roundoff error, 127, 178, 559, 680
Rule of 70 (seventy), 610
Rules
    of differentiation, 837
    of integration, 837
    of vector arithmetic, 769–770
Run, 63, A20

Saarinan, Eero, 511
Saddle point, 814, 958
Sampling error, 43
Sarasa, Alfons A. de, 446
Satellites
    artificial, 885
    geosynchronous orbit for, 885
Scalar components, 781
Scalar moment, 792–793
Scalar multiple, 767
Scalars, 765, 766
    work expressed in form of, 1075
Scalar triple product, 790
Scale factors, 37
Scale marks, 37–38, see also Tick marks
Scale variables, 37
Scaling, of parametric curves, 97
Scatter plots, 16
Secant(s), A41
    alternative method for integrating powers of, 527–528
    hyperbolic, 501
    integrating powers of, 525–528
    integrating products of tangents and, 526–527
Secant line(s), 112–113, 839
Second-degree equations
    in x and y, 737
    in x, y, and z, 810
Second derivative, 196
Second derivative test, 302
    proof of, 303
Second Fundamental Theorem of Calculus, 884

Second-order partial derivatives, 913
Second partials test, 959–961
Seconds, A39
Sectors, A41
Seed value, 106
Segment, of a circle, 367
Semiannual compounding, 614
Semiaxes, of an ellipse, 97, 730
Semicubical parabola, 715
Semimajor axis, of an ellipse, 730
Semiminor axis, of an ellipse, 730
Separable differential equations, first-order, 582–583
Separating variables, 582–583
Sequence of partial sums, 634
Sequence(s), 616–623
   convergence of, 619
   general term of a, 616
   graphs of, 618–619
   increasing/decreasing, 626
   index for, 617–618
   limits of, 619–622
   lower bound of, 629
   monotone, 626–630
   of partial sums, 634
   properties of eventual, 628–629
   recursively defined, 623
   Squeezing Theorem for, 622–623
   strictly increasing/decreasing, 626
   upper bound of, 629
Series
   absolute convergence/divergence of a, 665
   alternating, 662–663
   binomial, 683
   conditional convergence/divergence of a , 667
   geometric, 635–637
   harmonic, 638
   hyperharmonic, 643–644
   infinite, 5–6, 615, 632–638, *see also* Infinite series
   Maclaurin series, 652–655, 678, 680–684, 689–693
   power series, 670–675
   p-series, 643–644
   sums of alternating, 663–665
   sums of infinite, defined, 634
   Taylor series, 652–655
   terms of infinite, 632
Set-builder notation, A4
Sets, A3–A4
   bounded, 902
   closed, 901–902
   open, 901–902
   unbounded, 902
Shells, cylindrical, 475–479
Shroud of Turin, 604
Sigma (summation) notation (Σ), 397–404
   index of summation, changing, 398–399
   properties of, 399–400
   for Taylor/Maclaurin polynomials, 652
Signed arc length, 83
Simple harmonic model, 607
Simple harmonic motion, 209, 607
Simple parametric curve, 1082
Simple pendulum, 576
Simple polar regions, 993–994
Simple root, 308, 310
Simple xy-solid, 1014, 1016
Simple zy-solid, 1016
Simpson, Thomas, 557
Simpson error estimate, 559
Simpson's rule, 557–562

Sine(s), 84–85, A41
   integrating powers of, 527–528
   function, 450
   hyperbolic, 501
   integrating powers of, 522–524, 527–528
   integrating products of cosines and, 524–525
Single integrals, 980
Singular points, 714
Sinks, of incompressible fluid, 1115
Skew lines, 797
Slicing
   finding area by, 468–470
   and the method of washers, 472
   and solids of revolution, 471
   volumes by, 468–470
Slope(s), 62–64, A20–A21
   field(s), 389, 592–594
   infinite, 62, 714, A21
   interpretations of, 62–63
   and partial derivatives, 911–912
   of a surface, 911, 937
   of a tangent line, 170, 177–179
      by zooming, 178–179
   undefined, A21
   zero, 291
Slope fields, 388–389, 592–594
Slope-intercept form, A24
Slope-producing function, 179
Smooth change of parameter, 847
Smooth curve(s), 480
   line integrals along piecewise, 1071–1072
Smooth function, 480, 843
Smooth parametric surface, 1006
Smooth parametrizations, 843–844
Snell, Willebrord van Roijen, 353
Snell's law, 353
Solids of revolution, 470–471
Solutions, of differential equations, 580–581
Solution set, A6, A17
Sound intensity (level), 242
   synthesis, 2
Source, of fluid, 1115
Space exploration, 3–4
Speed, 9, 65, 170, 353, 867–869
   instantaneous, 353–355
   terminal, 589
   versus time curve, 355
Sphere(s), 762–763
   zone of, 488
Spherical cap, 474
Spherical coordinates, 819, 820
   in navigation, 823–824
Spherical coordinate systems, 759
Spherical element of volume, 1033
Spherical wedge, 1033
Spirals
   Archimedean, 705, 709
   equiangular, 755
   families of, 708–709
Spring constant, 492, 606
Springs, vibrations of, 606–609
Spring stiffness, 492
Square roots, and absolute values, A11–A12
Squeezing Theorem, 160–162, 277, 622–623, 678
Standard equations
   of a circle, A31
   of an ellipse, 730–731
   of a hyperbola, 732–733
   of a parabola, 728–729
   of a sphere, 762

Standard position(s)
   angles in, A42
   of an ellipse, 730–731
   of a hyperbola, 732–733
   of a parabola, 728–729
Static equilibrium, 775
Stationary points, 300
Steady-state flow, 1126, 1103
Step size, 595
Stokes, George Gabriel, 1118–1119
Stokes' Theorem, 1055, 1118–1119
   relationship between Green's Theorem and, 1121–1122
   using to calculate work, 1119–1121
Stream function, 1129
Streamlines, 1101, 1129
Strictly increasing/decreasing sequences, 626
Strictly monotone sequences, 626
Substitution(s)
   in definite integrals, 441–443
   evaluating definite integrals by, 441–446
   hyperbolic, 535
   integral tables, using, 544–549
   integration by, 392
   trigonometric, 530–535
   two methods for making, in definite integrals, 441–443
   u-substitution method, 391–394, 441–443
Summation
   formulas, 400–402
   index of, 397–399
   lower and upper limits of, 397–398
   notation (Σ), 397–404, *see also* Sigma notation
Sum(s):
   of functions, 49
      derivatives of, 191–192
   of infinite series, defined, 634
   in open/closed form, 401–402
   partial, 634
   telescoping, 401, 403
Sum-to-product formulas, A48
Superconductors, 165
Superposition principle, 1127
Surface(s)
   graphing of, 893–895
   oriented, 1101–1102
   relative orientation of, 1118
Surface area, 485–488, 1000–1012
   application of improper integrals to, 570–571
   of parametric surfaces, 1006–1008
   as a surface integral, 1099
Surface integral(s), 1094–1099
   applications of, 1101–1107
   definition of, 1094
   evaluation of, 1094–1095
   mass of a curved lamina as a, 1098–1099
   surface area as a, 1099
Surface of revolution, 485
   area of a, 485–488
   representing parametrically, 1002–1003
Symmetric equations, 800
Symmetry, 52–53
   in rectangular coordinates, 52–53
   in polar coordinates, 703–706, 722–723

Tables of integrals, 543–549, *see also* Integral tables
   using for evaluating unfamiliar integrals, 514
Tabular data, 16
Tangent(s), 84–85, A41
   hyperbolic, 501
   integrating powers of, 525–528
   integrating products of secants and, 526–527

Tangential vector component of acceleration, 871
Tangent line(s), 7–8, 112–113, 170–177, 840
    approximation, 556
    to a circle, 841
    defined precisely, 177–178
    to graphs of vector-valued functions, 840–841
    and limits, 112–113
    to parametric curves, 713–715
    parametric equations of, 840
    to polar curves, 715–717
    problem, 112
    slope of, 170, 177–179, 291
    vertical, 714
Tangent planes, 930–932
    to parametric surfaces, 1004–1006
    using gradients to find, 948
Tangent vector, 840
    continuously turning, 844
Tautochrone problem, 99
Taylor, Brook, 615, 651
Taylor polynomials, 651–652
    sigma notation for, 652
Taylor series, 652–655
    approximating trigonometric functions with,
        678–680
    convergence of, 676–678
    modeling physical laws with, 693
    power series as, 689–693
    method for finding, 690–692
Taylor's formula with remainder, 677
TB-plane (or rectifying plane), 856
Telescoping sums, 401, 403
Term(s)
    cross-product, 737
    of an infinite series, 632
    of a sequence, 616
Terminal point(s), 766, 834
Terminal side, of an angle, 700–701
Terminal speed, 589
Terminal velocity, 106, 589
Terminating decimal(s), 639, A2
Test(s)
    alternating series, 662–663, 668
    comparison, 656–658, 668
    conservative vector field, 1082
    divergence, 641, 668
    integral, 643, 668
    limit comparison, 658–659, 668
    ratio, 659–660, 667, 668
        for absolute convergence, 667, 668
    root, 660, 668
    symmetry, 703–706
Test points, 891, A7
Theorem of Pappus, 1026–1027
Theorem of Pythagoras, 22, 264, 271, 342, 761, 873,
        A29–A30, A44
Thin lens equation, 276
Tick marks, 37–38
TNB-frame (or Frenet frame), 856
TN-plane (or osculating plane), 856
Torque, 792
Torricelli's law, 591
Torsion, 866
Torus knot, 830
Total differentials, 949, 951
    approximations using, 934–935, 950
    increments and, 932–933
    of w, 949
    of z and f, 933
Trace of a surface, 808, 809
Tractrix, 510

Trajectory, 93, 867
Transformation (T), of the plane, 1040–1042
Transforms, 573
Translated conic sections, 735–737
Translation(s), 54–56
    of graphs, 54–55
    of parametric curves, 97
Transverse unit vector, 887, 1129
Trapezoidal approximation, 554–557
Trapezoidal error, 559
Tree diagram, 924
Triangle inequality, 27, A14–A15
    for vectors, 776
Trigonometric function(s), 82–88, A41
    approximating, with Taylor series, 678–680
    continuity of, 159–164
    derivatives of, 200–201
    families of, 86–88
    integrals of, 514
    review of, 514
    of right triangles, A41–A42
    see also Inverse trigonometric functions
Trigonometric identities, A44–A46
Trigonometric integrals, 522–529
    and Mercator projection, 528–529
    powers of sine and cosine, 522–524, 527–528
    powers of tangent and secant, 525–528
    products of sine and cosine, 524–525
    products of tangent and secant, 526–527
Trigonometric substitution(s), 530–535
Triple integrals, 1012–1019
    change of variables in, 1047–1049
    converting, from rectangular to cylindrical
        coordinates, 1032
    converting, from rectangular to spherical
        coordinates, 1036–1037
    in cylindrical coordinates, 1029–1032
    definition of, 1012
    properties of, 1013
    in spherical coordinates, 1033–1036
    volume calculated as a, 1015–1016
Triple point, 74
Triple scalar products, 790
    algebraic properties of, 791–792
    geometric properties of, 790–791
Truncation error, 559, 680
Tube plots, 831
Twisted cubic, 831
Two-point vector form of a line, 834
Two-sided limits, 117–119
Type I/type II region, 985

Unbounded set, 902
Uniform circular motion, 887
Uniformly accelerated motion, 428–430
Uniform rectilinear motion, 66, 170–171
Uniform sink flow, 1127
Uninhibited growth model, 599
Union, of sets, A5
Unit circle, 83, A31
Unit hyperbola, 503
Unit normal vectors, 853–854
Unit tangent vectors, 853
Unit vector, 770–771
Universal gravitational constant, 881
Universal Law of Gravitation, 20
Universe, age of, 74–75
Upper bound
    least, 630
    of a sequence, 629
    of a set of numbers, 630

Upper limit of integration, 408
Upper limit of summation, 397–398
Upward curvature, 292, see also Concave up
u-substitution
    method of, 391–394, 441–443, 514

Value(s)
    absolute, A11–A15
    average (mean), 421, 435–436, 984, 992, 1019
        of a continuous function, 435–436
    of f at x, 24
Variable(s), 1
    change of
        in a single integral, 1039–1040
        in double integrals, 1044–1047
        in triple integrals, 1047–1049
    dummy, 420–421
    independent vs. dependent, 24
    and integrals, 978
    separation of, 582–583
Variable force, work done by a, 491–492
Variable velocity, 171
Vector(s), 759–827, 765–776
    addition, 766
    arithmetic operations on, 768
    components, 781
    in coordinate systems, 767–768
    decomposing, 780–781
    determined by length and angle, 771–772
    determined by length and direction, 772
    determined by length and vector in same direction,
        772
    displacement, 765, 766, 869
    equal, 766
    equations of lines, 797–798
    force, 766
    with initial point not at origin, 768–769
    magnitude, 770
    moment, 792
    norm of, 770
    normalizing, 771
    orthogonal, 841
    position, 833
    principal unit normal vector, 1005
    radius, 833, 1003
    rules of arithmetic, 769–770
    subtraction, 767
    tangent, 840
    triple products, 795
    unit, 770–771
    velocity, 766
    zero, 766
Vector calculus, 1055–1130
Vector field(s), 1056–1061
    conservative, 1078–1086
    divergence and curl of, 1060
    flow lines of, 1064
    flows of, 1101–1107
    graphical representation of, 1056
    test for conservative, 1082–1085
    work performed by, 1074
Vector form of a line segment, 833–834
Vector-valued function(s)
    arc length and, 843–853
    calculus of, 836–843
    chain rule for, 847
    change of parameter, 843–853, 846
    continuous, 836
    and continuously turning tangent vectors, 844
    cross product, 785–795, 786–787, 792
    curvature, 858–867

derivatives of, 836–837
geometric interpretation of the limit of, 838
graphs of, 832
indefinite integration of, 836
integrals of, 836–837
introduction to, 830–835
limits of, 836–837
line integrals and, 1072
motion along a curve, 867–880
partial derivatives of, 1003–1004
of a real variable, 832
tangent lines to graphs of, 840–841
of two variables, 1003
Velocity, 65, 170–172, 353, 427–428, 867–869
and acceleration, 355
average, 65, 172–173, 436–437
constant, 66, 67, 171
drag force and, 588
function, 867
instantaneous, 114, 171–173, 353–354, 427–428
muzzle, 457
terminal, 106, 589
variable, 170–171
versus time curve, 67
Velocity vectors of fluid particles, 1101
Velocity versus time curve, 67, 429, 433–435
analyzing, 433–435
finding displacement from, 434
finding distance traveled from, 434
Verhulst, P. F., 600
Vertex (vertices)
of an angle, A39
of an ellipse, 727
of a hyperbola, 727
of a parabola, 727, A33–A34
Vertical asymptotes, 82, 122
and infinite limits, 120–122

Vertical line test, 26–27
Vertical tangency, 182, 314
Vertical tangent line, 314–315, 714
Vibrations, of springs, 606–609
Viewing window(s) (or rectangle), 36–45, 897–898
Viscosity, 1126
Voltage, peak, 445
Volume(s), 978–979
calculated as a triple integral, 1015–1016
by cylindrical shells, 475–479
by disks, 471–473
formula, 469, 470
net signed, 979
by slicing, 468–470
of a sphere, 378
by washers, 471–473

Wallis formulas, 530
Washers, method of, 472
Wave equation, 915
Wavelets, 2
Weather prediction, 3
Wedges
cylindrical, 1029
and polar coordinates, 721
Weierstrass, Karl, 184, 548, 829
Weight, 492, 607, 695
Weight density (ρ), 497
Whitney's umbrella, 1005
Wiles, Andrew, 353
Windchill index, 28
Wind velocity, 765
Wolszczan, Alexander, 3–4
Work, 489–494, 782–783
done by a constant force, 489–491
done by a variable force, 491–492
energy vs., 489
expressed in scalar form, 1075

finding, using Green's Theorem, 1090
integrals, 1078
as a line integral, 1073–1074
method for calculating, 1074–1075
performed by vector field, 1073–1074
using Stokes' Theorem, 1119–1121
Work–Energy Theorem, 493–494, 1086
Work integrals, 1078
along closed paths, 1081–1082
Fundamental Theorem of, 1079–1080
World Geodetic System of 1984, 819

x, power series in, 670–671
x-intercepts, 26, A20, A24–A25
x-interval, 36
x-limits of integration, 987
$x - x_0$, power series in, 673–674
xy-plane, 760, 761, A16
xyz-coordinate system, 763
xz-plane, 760, 761

y-axis, 760, 761, A16
Yenri (circle principle), 7
y-intercept, A20
y-interval, 36
y-limits of integration, 987
yz-plane, 760, 761

Zeno, 8–9
Zero(s), 26, 308
division by, A2
of p, 308
scalar, 777
Zone, of a sphere, 488
Zoom factors, 40–41
Zooming, 40–41, 155–156, 178–179

RATIONAL·FUNCTIONS CONTAINING POWERS OF $a + bu$ IN THE DENOMINATOR

60. $\displaystyle\int \frac{u\,du}{a+bu} = \frac{1}{b^2}[bu - a\ln|a+bu|] + C$

61. $\displaystyle\int \frac{u^2\,du}{a+bu} = \frac{1}{b^3}\left[\frac{1}{2}(a+bu)^2 - 2a(a+bu) + a^2\ln|a+bu|\right] + C$

62. $\displaystyle\int \frac{u\,du}{(a+bu)^2} = \frac{1}{b^2}\left[\frac{a}{a+bu} + \ln|a+bu|\right] + C$

63. $\displaystyle\int \frac{u^2\,du}{(a+bu)^2} = \frac{1}{b^3}\left[bu - \frac{a^2}{a+bu} - 2a\ln|a+bu|\right] + C$

64. $\displaystyle\int \frac{u\,du}{(a+bu)^3} = \frac{1}{b^2}\left[\frac{a}{2(a+bu)^2} - \frac{1}{a+bu}\right] + C$

65. $\displaystyle\int \frac{du}{u(a+bu)} = \frac{1}{a}\ln\left|\frac{u}{a+bu}\right| + C$

66. $\displaystyle\int \frac{du}{u^2(a+bu)} = -\frac{1}{au} + \frac{b}{a^2}\ln\left|\frac{a+bu}{u}\right| + C$

67. $\displaystyle\int \frac{du}{u(a+bu)^2} = \frac{1}{a(a+bu)} + \frac{1}{a^2}\ln\left|\frac{u}{a+bu}\right| + C$

RATIONAL FUNCTIONS CONTAINING $a^2 \pm u^2$ IN THE DENOMINATOR ($a > 0$)

68. $\displaystyle\int \frac{du}{a^2+u^2} = \frac{1}{a}\tan^{-1}\frac{u}{a} + C$

69. $\displaystyle\int \frac{du}{a^2-u^2} = \frac{1}{2a}\ln\left|\frac{u+a}{u-a}\right| + C$

70. $\displaystyle\int \frac{du}{u^2-a^2} = \frac{1}{2a}\ln\left|\frac{u-a}{u+a}\right| + C$

71. $\displaystyle\int \frac{bu+c}{a^2+u^2}\,du = \frac{b}{2}\ln(a^2+u^2) + \frac{c}{a}\tan^{-1}\frac{u}{a} + C$

INTEGRALS OF $\sqrt{a^2 + u^2}$, $\sqrt{a^2 - u^2}$, $\sqrt{u^2 - a^2}$ AND THEIR RECIPROCALS ($a > 0$)

72. $\displaystyle\int \sqrt{u^2+a^2}\,du = \frac{u}{2}\sqrt{u^2+a^2} + \frac{a^2}{2}\ln(u+\sqrt{u^2+a^2}) + C$

73. $\displaystyle\int \sqrt{u^2-a^2}\,du = \frac{u}{2}\sqrt{u^2-a^2} - \frac{a^2}{2}\ln|u+\sqrt{u^2-a^2}| + C$

74. $\displaystyle\int \sqrt{a^2-u^2}\,du = \frac{u}{2}\sqrt{a^2-u^2} + \frac{a^2}{2}\sin^{-1}\frac{u}{a} + C$

75. $\displaystyle\int \frac{du}{\sqrt{u^2+a^2}} = \ln(u+\sqrt{u^2+a^2}) + C$

76. $\displaystyle\int \frac{du}{\sqrt{u^2-a^2}} = \ln|u+\sqrt{u^2-a^2}| + C$

77. $\displaystyle\int \frac{du}{\sqrt{a^2-u^2}} = \sin^{-1}\frac{u}{a} + C$

POWERS OF u MULTIPLYING OR DIVIDING $\sqrt{a^2 - u^2}$ OR ITS RECIPROCAL

78. $\displaystyle\int u^2\sqrt{a^2-u^2}\,du = \frac{u}{8}(2u^2-a^2)\sqrt{a^2-u^2} + \frac{a^4}{8}\sin^{-1}\frac{u}{a} + C$

79. $\displaystyle\int \frac{\sqrt{a^2-u^2}\,du}{u} = \sqrt{a^2-u^2} - a\ln\left|\frac{a+\sqrt{a^2-u^2}}{u}\right| + C$

80. $\displaystyle\int \frac{\sqrt{a^2-u^2}\,du}{u^2} = -\frac{\sqrt{a^2-u^2}}{u} - \sin^{-1}\frac{u}{a} + C$

81. $\displaystyle\int \frac{u^2\,du}{\sqrt{a^2-u^2}} = -\frac{u}{2}\sqrt{a^2-u^2} + \frac{a^2}{2}\sin^{-1}\frac{u}{a} + C$

82. $\displaystyle\int \frac{du}{u\sqrt{a^2-u^2}} = -\frac{1}{a}\ln\left|\frac{a+\sqrt{a^2-u^2}}{u}\right| + C$

83. $\displaystyle\int \frac{du}{u^2\sqrt{a^2-u^2}} = -\frac{\sqrt{a^2-u^2}}{a^2 u} + C$

POWERS OF u MULTIPLYING OR DIVIDING $\sqrt{u^2 \pm a^2}$ OR THEIR RECIPROCALS

84. $\displaystyle\int u\sqrt{u^2+a^2}\,du = \frac{1}{3}(u^2+a^2)^{3/2} + C$

85. $\displaystyle\int u\sqrt{u^2-a^2}\,du = \frac{1}{3}(u^2-a^2)^{3/2} + C$

86. $\displaystyle\int \frac{du}{u\sqrt{u^2+a^2}} = -\frac{1}{a}\ln\left|\frac{a+\sqrt{u^2+a^2}}{u}\right| + C$

87. $\displaystyle\int \frac{du}{u\sqrt{u^2-a^2}} = \frac{1}{a}\sec^{-1}\left|\frac{u}{a}\right| + C$

88. $\displaystyle\int \frac{\sqrt{u^2-a^2}\,du}{u} = \sqrt{u^2-a^2} - a\sec^{-1}\left|\frac{u}{a}\right| + C$

89. $\displaystyle\int \frac{\sqrt{u^2+a^2}\,du}{u} = \sqrt{u^2+a^2} - a\ln\left|\frac{a+\sqrt{u^2+a^2}}{u}\right| + C$

90. $\displaystyle\int \frac{du}{u^2\sqrt{u^2 \pm a^2}} = \mp\frac{\sqrt{u^2 \pm a^2}}{a^2 u} + C$

91. $\displaystyle\int u^2\sqrt{u^2+a^2}\,du = \frac{u}{8}(2u^2+a^2)\sqrt{u^2+a^2} - \frac{a^4}{8}\ln(u+\sqrt{u^2+a^2}) + C$

92. $\displaystyle\int u^2\sqrt{u^2-a^2}\,du = \frac{u}{8}(2u^2-a^2)\sqrt{u^2-a^2} - \frac{a^4}{8}\ln|u+\sqrt{u^2-a^2}| + C$

93. $\displaystyle\int \frac{\sqrt{u^2+a^2}}{u^2}\,du = -\frac{\sqrt{u^2+a^2}}{u} + \ln(u+\sqrt{u^2+a^2}) + C$

94. $\displaystyle\int \frac{\sqrt{u^2-a^2}}{u^2}\,du = -\frac{\sqrt{u^2-a^2}}{u} + \ln|u+\sqrt{u^2-a^2}| + C$

95. $\displaystyle\int \frac{u^2}{\sqrt{u^2+a^2}}\,du = \frac{u}{2}\sqrt{u^2+a^2} - \frac{a^2}{2}\ln(u+\sqrt{u^2+a^2}) + C$

96. $\displaystyle\int \frac{u^2}{\sqrt{u^2-a^2}}\,du = \frac{u}{2}\sqrt{u^2-a^2} + \frac{a^2}{2}\ln|u+\sqrt{u^2-a^2}| + C$

INTEGRALS CONTAINING $(a^2 + u^2)^{3/2}$, $(a^2 - u^2)^{3/2}$, $(u^2 - a^2)^{3/2}$ ($a > 0$)

97. $\displaystyle\int \frac{du}{(a^2-u^2)^{3/2}} = \frac{u}{a^2\sqrt{a^2-u^2}} + C$

98. $\displaystyle\int \frac{du}{(u^2 \pm a^2)^{3/2}} = \pm\frac{u}{a^2\sqrt{u^2 \pm a^2}} + C$

99. $\displaystyle\int (a^2-u^2)^{3/2}\,du = -\frac{u}{8}(2u^2-5a^2)\sqrt{a^2-u^2} + \frac{3a^4}{8}\sin^{-1}\frac{u}{a} + C$

100. $\displaystyle\int (u^2+a^2)^{3/2}\,du = \frac{u}{8}(2u^2+5a^2)\sqrt{u^2+a^2} + \frac{3a^4}{8}\ln(u+\sqrt{u^2+a^2}) + C$

101. $\displaystyle\int (u^2-a^2)^{3/2}\,du = \frac{u}{8}(2u^2-5a^2)\sqrt{u^2-a^2} + \frac{3a^4}{8}\ln|u+\sqrt{u^2-a^2}| + C$